Comprehensive Organic Functional Group Transformations

Comprehensive Organic Functional Group Transformations

Editors-in-Chief
Alan R. Katritzky, FRS
University of Florida, Gainesville, FL, USA
Otto Meth-Cohn
University of Sunderland, UK
Charles W. Rees, FRS
Imperial College of Science, Technology and Medicine, London, UK

Volume 6
SYNTHESIS: CARBON WITH THREE OR FOUR ATTACHED HETEROATOMS

Volume Editor
Thomas L. Gilchrist
University of Liverpool, UK

PERGAMON

UK	Elsevier Science Ltd., The Boulevard, Langford Lane, Kidlington, Oxford OX5 1GB, UK
USA	Elsevier Science Inc., 660 White Plains Road, Tarrytown, New York 10591-5153, USA
JAPAN	Elsevier Science Japan, Tsunashima Building Annex, 3-20-12 Yushima, Bunkyo-ku, Tokyo 113, Japan

First edition 1995

Library of Congress Cataloging in Publication Data
Comprehensive organic functional group transformations / editors-in-chief, Alan R. Katritzky, Otto Meth-Cohn, Charles W. Rees. — 1st ed.
p. cm.
Includes indexes.
Contents: v. 1. Synthesis: carbon with no attached heteroatoms / edited by Stanley M. Roberts — v. 2. Synthesis: carbon with one heteroatom attached by a single bond / edited by Steven V. Ley — v. 3. Synthesis: carbon with one heteroatom attached by a multiple bond / edited by Gerald Pattenden — v. 4. Synthesis: carbon with two heteroatoms, each attached by a single bond / edited by Gordon W. Kirby — v. 5. Synthesis: carbon with two attached heteroatoms with at least one carbon-to-heteroatom multiple link / edited by Christopher J. Moody — v. 6. Synthesis: carbon with three or four attached heteroatoms / edited by Thomas L. Gilchrist — v. 7. Indexes.
1. Organic compounds—Synthesis. I. Katritzky, Alan R. II. Meth-Cohn, Otto. III. Rees, Charles W.
QD262.C534 1995
547'.2—dc20 95–31088

British Library Cataloguing in Publication Data
A catalogue record for this book is available from the British Library.

ISBN 0–08–040604–1 (set : alk. paper)
ISBN 0–08–042704–9 (Volume 6)

∞™ The paper used in this publication meets the minimum requirements of the American National Standard for Information Sciences—Permanence of Paper for Printed Library Materials, ANSI Z39.48–1984.

Printed and bound in Great Britain by Cambridge University Press, Cambridge, UK.

Contents

Preface

Some years ago the three of us met in a London club reviewing an ongoing publishing venture in Organic Synthesis. The conversation drifted to a consideration of volumes on the synthesis of key functional groups. No doubt the good wine helped since we actually broached the idea of a work on the synthesis of *all* functional groups. Would it be useful? Definitely. Would it be feasible? How would it be organized? Where do you start? We recognized that functionality was based on the coordination and heteroatom attachment of a carbon atom. But putting together a complete framework seemed particularly daunting. Two of us became very interested in the fascinating bouquet of the Muscat de Beaumes de Venise.

At our next dinner together Alan announced that he had solved the problems posed last time—problems that Charles and I hoped he had forgotten! He brought out a remarkable matrix analysis of *all* functional groups, analysed rigorously and logically. Even unknown functions were covered. Although we were all very impressed, the practicalities of the idea still seemed daunting. Those who know Alan's terrier instincts will appreciate that he would not give up such a challenge so easily. Our twice yearly club get-togethers, occasionally with friends from Pergamon, refined our thinking. Alan's cosmic vision was tempered by Charles's intuitive realism and fully supported by the publishers.

Another major problem remained: how to reduce our thinking into a practical handbook for authors—a dismaying task for three busy chemists. We settled on a seven-volume work and the indomitable ARK produced a rough breakdown to fit such a format. Putting flesh on these bones became feasible during a fortuitous three-month break between jobs by myself, and the largest handbook ever assembled by Pergamon (120 pages) was written and page allocations agreed—even for little or unknown functional groups. Sample chapters were commissioned and finally proved very encouraging, despite our first chosen topic uncovering virtually no known examples!

Contracts were defined and agreed, volume editors approached, and potential authors considered during a pleasant preconference stay in Grasmere. Following the sale of Pergamon to Elsevier Science Ltd there was a lull in the project but soon *Comprehensive Organic Functional Group Transformations* was back on track, and everyone adhered to a very businesslike timetable.

OTTO METH-COHN
Sunderland

CHARLES W. REES
London

ALAN R. KATRITZKY
Florida

Contributors to Volume 6

Professor J. Barluenga
Departamento de Quimica Organometalica, Universidad de Oviedo, 33071 Oviedo, Spain

Dr D. Carmichael
Laboratoire Hétéroéléments et Coordination, CNRS URA 1499, DCPH-Ecole Polytechnique, F-91128 Palaiseau Cedex, France

Professor R. D. Chambers
Department of Chemistry, University of Durham, Science Laboratories, South Road, Durham DH1 3LE, UK

Dr I. A. Cliffe
Wyeth Research (UK) Ltd., Huntercombe Lane South, Taplow, Maidenhead, Berkshire SL6 0PH, UK

Dr L. J. Drennan
Chemistry Department, University College Dublin, Belfield, Dublin 4, Republic of Ireland

Dr H. Eckert
Organische-Chemisches Institut, Technische Universität München, Lichtenbergstrasse 4, D-8046 Garching, Germany

Dr T. L. Gilchrist
Department of Chemistry, University of Liverpool, PO Box 147, Liverpool L69 3BX, UK

Mr A. H. Gouliaev
Kemisk Institut, Aarhus Universitet, Langelandsgade 140, DK-8000 Århus C, Denmark

Professor F. S. Guziec, Jr.
Department of Chemistry and Biochemistry, New Mexico State University, Las Cruces, NM 88003-0001, USA

Dr L. J. Guziec
Department of Chemistry and Biochemistry, New Mexico State University, Las Cruces, NM 88003-0001, USA

Dr G. E. Gymer
Pfizer Central Research, Sandwich, Kent CT13 9NJ, UK

Professor A. F. Hegarty
Chemistry Department, University College Dublin, Belfield, Dublin 4, Republic of Ireland

Dr J. Hutchinson
BNFL Fluorochemicals Ltd., South Road, Durham DH1 3LE, UK

Dr G. B. Jones
Department of Chemistry, Clemson University, Clemson, SC 29634-1905, USA

Professor E. Klienpeter
Department of Chemistry, University of Potsdam, Am Neuen Palais 10, D-14415 Potsdam, Germany

Dr L. Lamandé
Laboratoire de Synthese, Structure et Réactivité de Molécules Phosphorées, Université Paul Sabatier, 118 Route de Narbonne, F-31062 Toulouse Cedex, France

Dr P. D. Lickiss
Department of Chemistry, Imperial College of Science, Technology and Medicine, London SW7 2AY, UK

Dr A. Marinetti
Laboratoire Hétéroéléments et Coordination, CNRS URA 1499, DCPH-Ecole Polytechnique, F-91128 Palaiseau Cedex, France

Dr J. E. Matthews
Department of Chemistry, Clemson University, Clemson, SC 29634-1905, USA

Dr G. Mitchell
ZENECA Agrochemicals, Jealott's Hill Research Station, Bracknell, Berkshire RG12 6EY, UK

Dr S. Narayanaswami
Pfizer Central Research, Sandwich, Kent CT13 9NJ, UK

Dr D. P. J. Pearson
ZENECA Agrochemicals, Jealott's Hill Research Station, Bracknell, Berkshire RG12 6EY, UK

Professor W. Petz
Gmelin-Institut der Max-Planck-Gesellschaft, Varrentrappstrasse 40-42, D-6000 Frankfurt, Germany

Professor K. Pihlaja
Department of Chemistry, University of Turku, SF-20500 Turku, Finland

Dr M. J. Rice
ZENECA Agrochemicals, Jealott's Hill Research Station, Bracknell, Berkshire RG12 6EY, UK

Professor V. Romanenko
Institute of Organic Chemistry, Academy of Sciences of the Ukraine, Murmanskaya 5, 253660 Kiev 94, Ukraine

Dr E. Rubio
Departamento de Quimica Organometalica, Universidad de Oviedo, 33071 Oviedo, Spain

Professor M. Sanchez
Laboratoire de Synthese, Structure et Réactivité de Molécules Phosphorées, Université Paul Sabatier, 118 Route de Narbonne, F-31062 Toulouse Cedex, France

Professor P. Savignac
Laboratoire Hétéroéléments et Coordination, CNRS URA 1499, DCPH-Ecole Polytechnique, F-91128 Palaiseau Cedex, France

Professor A. Senning
Department of Applied Chemistry, Technical University of Denmark, Building 376, DK-2800 Lyngby, Denmark

Dr J.-M. Sotiropoulos
Laboratoire Chimie Theorique et Physicochemie Moleculaire, Université de Pau et Pays de l'Adour. F-6400 Pau Cedex, France

Dr M. Tomas
Departamento de Quimica Organometalica, Universidad de Oviedo, 33071 Oviedo, Spain

Professor A. Vargolis
Laboratory of Organic Chemistry, University of Thessaloniki, 540 06 Thessaloniki, Greece

Dr F. Weller
Fachbereich Chemie der Universität Marburg, Hans-Meerwienstrasse, W-3550 Marburg, Germany

Contents of All Volumes

Volume 6 Synthesis: Carbon with Three or Four Attached Heteroatoms

Volume 7 Indexes

Introduction

OBJECTIVES, SCOPE, AND COVERAGE

Comprehensive Organic Functional Group Transformations (COFGT) aims to present the vast subject of organic synthesis in terms of the introduction and interconversion of functional groups. All organic structures can be considered as skeletal frameworks of carbon atoms to which functional groups are attached[a]; it is the latter which are mainly responsible for chemical reactivity and which are highlighted in COFGT. All known functional groups fit a logical and comprehensive pattern and this forms the basis for the detailed list of contents. The format of the present work was designed with the intention to cover systematically all the possible arrangements of atoms around a carbon, including those which are quite unfamiliar. The work also considers the possibility of as yet unknown functional groups which may be constructed in the future and prove to be important; thus COFGT also indicates what is not known and so points the way to new research areas.

The philosophy of the present work has been to rationalize this enormous subject within as logical and formal a framework as possible, in a scholarly and critical fashion. COFGT is designed to provide the first point of entry to the literature for synthetic organic chemists, together with an unrivalled source for anyone interested in less common, obscure, or unknown functional groups.

All functional groups are viewed as being carbon based (even if the group contains no carbon). Thus, a nitro compound is considered from the standpoint of the immediately attached carbon atom, whether di- (*sp*), tri- (sp^2), or tetracoordinated(sp^3). The work is organized on the basis of formation or rupture of bonds to a carbon atom and it is the nature of the carbon atom left after the transformation that determines the classification of the overall sequence. Several key criteria have been used to organize the work and to minimize overlap. These are, in order of priority:

1. the number of attached heteroatoms;
2. the coordination of the carbon atom involved in the functional group;
3. the nature of the immediately attached heteroatom(s); and
4. the Latest Placement Principle.

These four key principles have been used to determine the content of each volume, and to develop the detailed chapter breakdown within each volume.

Thus, according to the number of attached heteroatoms:

Volume 1 deals with synthetic reactions which result in the alteration of bonding at carbon atoms which are left with *no* attached heteroatoms.

Volume 2 deals with syntheses which result in carbon atoms attached to *one* heteroatom by a single bond.

Volume 3 deals with syntheses which result in carbon atoms attached to *one* heteroatom by a double or by a triple bond.

Volume 4 deals with syntheses which result in carbon atoms attached to *two* heteroatoms, each by a single bond.

Volume 5 deals with syntheses which result in carbon atoms attached to *two* heteroatoms by one single and one double bond, or by two double bonds, or by one single and one triple bond.

Volume 6 deals with syntheses which result in carbon atoms attached to *three* or *four* heteroatoms.

Volume 7 comprises the author and subject indexes.

Certain key principles apply to all the volumes because all functional groups are viewed as carbon based (e.g. a nitro group is either alkyl-, vinyl-, aryl-, or alkynyl-); these are:

(a) Volumes are subdivided according to the *coordination* of the carbon atom which is the product of the reaction, i.e., tetra- coming before tri- before di- before monocoordinated carbon functions.

---

[a] The major exception to this lies in heterocyclic compounds, where the cyclic heteroatoms are more logically considered as part of the framework. The subject of heterocycles has been treated elsewhere in the companion work *Comprehensive Heterocyclic Chemistry* published in 1984 with a second edition to be published in 1996.

In Volumes 1 and 6, reactions producing four-coordinated carbon are considered first, followed by three- and then two-coordinated carbon. The other volumes contain a more limited range of coordination types (Volumes 2 and 4 only four-coordinated, Volumes 3 and 5 only two- or three-coordinated). Each type of coordination is allocated a separate section in each volume.

(b) Attached *heteroatoms* are discussed in the following order of priority:

Halogens—F, Cl, Br, I
Chalcogens—O, S, Se, Te
Nitrogen—N
Other group 15 elements—P, As, Sb, Bi
Metalloids—B, Si, Ge
Main group metals—Sn, Pb, Al, Ga, In, Tl, Be, Mg, Ca, Sr, Ba, Li, Na, K, Rb, Cs
Transition metals—Cu, Ag, Au, Zn, Cd, Hg, Ti, Zr, Hf, Cr, Mo, W, Mn, Fe, Co, Ni, Pd, Pt, and others.

Higher coordination of heteroatoms is treated after lower. Thus, in sections dealing with iodo compounds, monocoordinate (e.g. iodides) are discussed before dicoordinate (e.g. iodoxyls) and tricoordinate functions.

(c) The *Latest Placement Principle* (or Last Position Principle) is used to avoid undue overlap in the work. Thus, the carbon attached to the heteroatom is discussed at the last possible position in the above prioritizing of heteroatoms. Examples of its application are noted later. On this basis, for example, when both C—C and C—H bonds are formed the reaction will appear in the latest chapter (i.e. Chapters 1.04–1.10 rather than in the earlier Chapters 1.01–1.03), and when both C—C and C═C bonds are formed, this will be found in the later Chapter 1.17. The Latest Placement Principle is particularly important in determining where to find electrocyclic reactions in Volume 1. If C—H, ═C—C, and C═C bonds are all formed in a reaction, then the latest appropriate chapter will deal with the reaction. Only if a change in heterofunction occurs is the reaction left to a later volume.

Exceptions to the above principles are rare. However the reactions of heteroarenes are mentioned along with those of arenes. If on reduction no change in the heterofunction occurs (e.g. in going from thiophene to tetrahydrothiophene or from pyridine to 2,3,4,5-tetrahydropyridine) the reaction is found in Volume 1. However, when the function changes (e.g. pyridine to piperidine), the conversion is considered in Volume 2. Conversion of methyl phenyl sulfone into methyl cyclohexyl sulfone appears in Volume 2, whereas the formation of cyclohexyl methyl ketone is treated in Volume 1, since the coordination of the carbon atom to the heteroatoms is changed in the first but not in the second hydrogenation.

Some further exceptions to the rigorous ordering of the work have been made for the purpose of easy reference. Thus, in Volume 1, a special chapter on ions, radicals, and carbenes is added: this chapter is limited to the treatment of species capable of more than a transitory existence. Throughout the work aspects of the Latest Placement Principle are occasionally ignored for reasons of clarity. Thus metal ligands that are incidental to the chemistry under discussion are not considered when prioritizing. Also, references to aromatic substituents, some of which involve a heteroatom (e.g. pyridyl, thienyl, etc.) but are incidental to the chemistry being described, are not viewed as changing the priority.

Within each section, we have endeavored to explain the influence of important secondary effects such as inclusion in a ring, degree of strain, degree of substitution, various types of activation, influence of stereochemistry, and so on, on the transformation under consideration. General synthetic methods are treated before specific methods.

Transient intermediates, as such, do not fall within the scope of this work. Although there is clearly no sharp division, we have attempted to restrict coverage of radicals, etc., to more stable, longer lived species. It is the aim of this work to consider all organic functional groups provided that the molecules which incorporate them, though they may be unstable, can have a finite lifetime and chemistry. The whole work deals with the generation and transformation of functional groups, *not* of molecules such as CO_2, COS, CS_2, ClCN, etc. Such simple carbon derivatives are not treated unless a further carbon is attached (e.g. RN═C═O).

VOLUME 1 SYNTHESIS: CARBON WITH NO ATTACHED HETEROATOMS

Volume 1 deals solely with the formation of nonheteroatom functional groups and as such is different in style to the remaining volumes.

In addition to the general principles, Volume 1 is further organized as follows:

1. By the type of bond formed (i.e. C—H before C—C).
2. By the type of reaction involved (i.e. substitution, then addition, then rearrangement). With C=C bond formation the order is addition, elimination, condensation, then electrocyclic and other methods. One rearrangement chapter only is devoted to each of the Parts I and II.
3. In Parts II and III the treatment of formation of ions, radicals, and carbenes is added at the end of the section dealing solely with those species with a significant rather than a transient lifetime.

In Volume 1, the heteroatom sequence is a secondary feature since only remote heteroatom functions are involved in the products: but the standard order pertains in reactants that contain heteroatoms (see, e.g. Chapters 1.01 and 1.02).

All the major structural influences that are treated throughout this work apply equally (or perhaps more importantly) in Volume 1. Thus the effects of conjugation, remote substituents, rings, stereochemistry, strain, kinetic or thermodynamic factors, solvation, primary, secondary and tertiary nature, etc., are mentioned whenever relevant.

VOLUME 2 SYNTHESIS: CARBON WITH ONE HETEROATOM ATTACHED BY A SINGLE BOND

Volume 2 is arranged in three parts: I, II and III, dealing respectively with sp^3, sp^2, and *sp* carbon linked to the heteroatom. In each chapter we have endeavored to explain important effects due to such features as the primary, secondary, tertiary nature, ring effects, strain activation, effect of beta, gamma, and more remote functionality, stereochemical effects, and so on. Methods that are common to a larger group are dealt with at their first appearance and suitably cross-referenced.

Volumes 2–6 all deal with the synthesis of functions involving at least one heteroatom. To avoid major overlap we have applied the Latest Placement Principle; that is, the chemistry is discussed at the last possible position based on the prioritization of the carbon attached to the heteroatom. Thus the compound CH_3ONH_2 is treated under "Alkyl Chalcogenides" in the subsection "Functions Based on the RON-Unit" (i.e. 2.02.6). However, CH_3ONHCH_3 appears under "Alkyl Nitrogen Compounds" (2.06.2.3) since the Latest Placement Principle prevails. Also, dialkyl ethers appear in Part I of Volume 2 (Functions Linked by a Single Bond to an sp^3 Carbon Atom), while alkyl aryl ethers appear in Part II of Volume 2 (Functions Linked by a Single Bond to an sp^2 Carbon Atom). Exceptions to the rule are:

(a) When a fully unsaturated heterocyclic substituent (e.g. thienyl, pyridyl, etc.) is used as an example of an aryl group, the ring heteroatom(s) is (are) not taken into account (e.g. 2-methoxypyridine should strictly appear in Volume 6, but is covered in Volume 2 along with 3- and 4-methoxypyridine).

(b) Carbon-based metal ligands that are incidental to the synthesis under discussion (e.g. carbonyls, cyclopentadienyls, etc.) are not taken into consideration.

VOLUME 3 SYNTHESIS: CARBON WITH ONE HETEROATOM ATTACHED BY A MULTIPLE BOND

Volume 3 follows the logical development indicated in Volume 2. Thus, according to the Last Placement Principle, the imines, RCH=N—R, appear in Volume 3 rather than in Volume 2 (where functions singly bonded to carbon are treated). Furthermore, acetophenone, $PhCOCH_3$, is treated under α,β-unsaturated ketones (3.05) rather than saturated ketones (3.04). Chloronitroacrylonitriles would appear under the section "α,β-Vinylic Nitriles with Nitrogen-based Substituents" (3.19.2.7), not under the related earlier section dealing with halo-substituents (3.19.2.3).

VOLUME 4 SYNTHESIS: CARBON WITH TWO HETEROATOMS, EACH ATTACHED BY A SINGLE BOND

Volume 4 is in three parts. Part I deals with tetracoördinated carbon bearing two heteroatoms, Part II with tricoordinated carbon bearing two heteroatoms, and Part III (a brief chapter) with stabilized radicals, ions, and the like bearing two heteroatoms. The material is arranged according to the Latest Placement Principle: thus, the synthesis of $CHBr_2CHI_2$ would appear in the section dealing with diiodo, not dibromo functions (i.e. in 4.01.5, not 4.01.4), and the synthesis of CF_3CH-BrCl is discussed in Volume 6 (carbons bearing three heteroatoms), rather than in Volume 4.

VOLUME 5 SYNTHESIS: CARBON WITH TWO ATTACHED HETEROATOMS WITH AT LEAST ONE CARBON-TO-HETEROATOM MULTIPLE BOND

Volume 5 is in three parts. Part I deals with functions with one doubly bonded and one singly bonded heteroatom, Part II with functions containing two doubly bonded heteroatoms and Part III with one triply bonded and one singly bonded heteroatom. Part I constitutes the bulk of Volume 5.

The arrangement of the chemistry in each part follows the same logical sequence. The multiply bonded heteroatom is focused on first and then the other heteroatom in a secondary classification, both following the priority rules already described. Each section excludes the coverage of the previous sections. Thus, all carbonyl derivatives will appear in Chapters 5.01–5.10 but not in Chapters 5.11, *et seq.*

According to the Latest Placement Principle structure RC(O)OC(S)R is discussed in the chapter dealing with carbons bearing a doubly bonded sulfur and singly bonded oxygen (5.12.3), *not* in that dealing with doubly and singly bonded oxygen (5.04.1). Another effect of the Latest Placement Principle is that the amides RCONMePh are discussed under *N*-arylalkanoamides (5.06.2.4), rather than *N*-alkylalkanoamides (5.06.2.2). Again, exceptions are made to the latest placement rules for: (a) hetaryl rings used as examples of aryl substituents which are not viewed as functional groups. Thus, 2-methylimidazole is not considered as an example of an amidine function and 2-methoxypyridine is not an example of a doubly bonded nitrogen, singly bonded oxygen function; (b) metal ligands that are incidental to the organic chemistry under discussion are not viewed as functions in priority considerations.

VOLUME 6 SYNTHESIS: CARBON WITH THREE OR FOUR ATTACHED HETEROATOMS

Volume 6 is in four parts. Part I deals with tetracoordinate carbons bearing three heteroatoms. Part II covers tetracoordinate compounds bearing four heteroatoms, i.e. substituted methanes, and Part III deals with tricoordinate systems bearing three heteroatoms, i.e. where one heteroatom is attached by a double bond. Part IV is brief and deals with stabilized radicals and ions. Not surprisingly, the coverage of Volume 6 is very large—and also shows that many gaps in the development of organic chemistry still exist.

The organization within the three sections not only follows the same broad logic developed in the previous volumes, but also has a structure unique to the multiheteroatom volume. According to the Latest Placement Principle $CF_3C(NR_2)_3$ appears in the section dealing with carbons bearing three nitrogens (6.05.1.1), not that dealing with carbons bearing three halogens (6.01.2), while $(CF_3CH_2O)_2CO$ appears in Part III, not in Part I.

In the chapter dealing with iminocarbonyl functions in Part III, the substituents on nitrogen are discussed in each appropriate subsection in the order outlined above. Thus, the RN= group would be first considered with R = H, then alkyl, alkenyl, aryl and hetaryl, alkynyl and then heteroatom substituents in the usual order.

In each relevant section, we have endeavored to explain the influence of important secondary effects on the synthesis such as structure (primary, secondary, etc.), ring effects, strain, activation, stereochemistry, remote substituent effects, etc.

The arrangement of the chemistry in each of Parts I–III follows a similar pattern. Thus, each section commences with functions containing at least one halogen. This section deals with all combinations of halogen with other heteroatoms in the described order. The next section deals with functions containing at least one chalcogen in combination with any other heteroatoms except halogens. Subsequent sections each exclude the previous title heteroatom functions.

VOLUME 7 INDEXES

Subject Indexes are included in each of Volumes 1–6 and Cumulative Subject and Author Indexes appear in Volume 7. Most entries in the Subject Index consist of two or three lines: the first line is the entry itself (e.g. Lactones) and the second line is descriptive of that entry (e.g. reduction); in many cases more detail is given (e.g. with 9-BBN).

REFERENCES

The references are handled by the system previously used successfully in *Comprehensive Heterocyclic Chemistry*. In this system reference numbers appear neither in the text, nor as footnotes, nor at the end of chapters. Instead, each time a reference is cited in the text there appears (in parentheses) a two-letter code assigned to the journal being cited, which is preceded by the year (tens and units only for twentieth-century references) and followed by the page number. For example: "It was shown ⟨80TL1327⟩ that . . .". In this phrase, "80" refers to 1980, "TL" to *Tetrahedron Letters*, and "1327" to the page number. For those journals which are published in parts, or which have more than one volume number per year, the appropriate part of the volume is indicated, e.g. as in ⟨73JCS(P2)1594⟩ or ⟨78JOM(162)611⟩, where the first example refers to *J. Chem. Soc., Perkin Trans 2*, 1973, page 1594, and the second to *J. Organomet. Chem.*, 1978, volume 162, page 611.

This reference system is adopted because it is far more useful to the reader than the conventional "superscript number" system. It enables readers to go directly to the literature reference cited, without first having to consult the bibliography at the end of each chapter.

References to the last century quote the year in full. Books have a prefix "B-" and if they are commonly quoted (e.g. *Organic Reactions*) they will have a code. Otherwise, as with uncommon journals, they are given a miscellaneous code (MI) and numbered arbitrarily *abb1*, *abb2*, etc., where *abb* refers to the volume and chapter number and *1*, *2*, etc., are assigned sequentially. Patents are assigned appropriate three-letter codes.

The references are given in full at the end of each volume. They include *Chemical Abstract* references when these are likely to help; in particular, they are given for all patents, and for less accessible sources such as journals in languages other than English, French, or German, company reports, obscure books, and theses.

6.01 Trihalides

RICHARD D. CHAMBERS
University of Durham, UK

and

JOHN HUTCHINSON
BNFL Fluorochemicals Ltd., Durham, UK

6.01.1 GENERAL METHODS

There are several general methods which can lead to trihalomethyl compounds where the halogen atoms can be the same or different. It will be seen that some of these methods lead to compounds in which there are two such trihalomethyl groups. For simplicity, the general methods are discussed first.

6.01.1.1 The Addition of Halogens and Interhalogen Compounds to Fluoroalkenes

The addition of halogens and interhalogen compounds to fluoroalkenes affords a valuable method for the preparation of compounds containing the group —$CHal_3$ where the halogen atoms may be the same or different. The addition of halogens is conveniently carried out under ultraviolet irradiation (Equations (1) and (2)) ⟨41JA3476, 68JOC1016⟩, but the addition of iodine to tetrafluoroethene occurs only at elevated temperatures (Equation (3)) ⟨49JCS2948, 49JOC747, 53JCS1548, 63USP3076041⟩. Generally the yields of the adducts are high, but when 1,1-difluoroethene is treated with iodine a mixture of compounds is obtained, the most abundant of which is CF_3CH_2I (Equation (4)) ⟨58JOC322⟩.

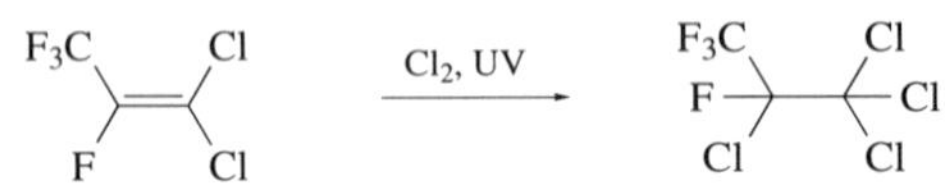

(1)

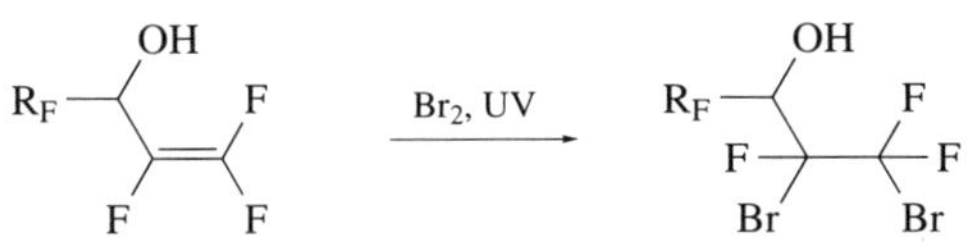

(2)

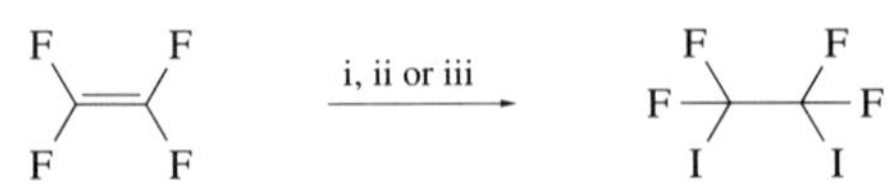

(3)

i, I_2, Et_2O, 60 °C, 15 h; ii, I_2, 150 °C, 24 h; iii, I_2, KI, H_2O, 100 °C, 5 h

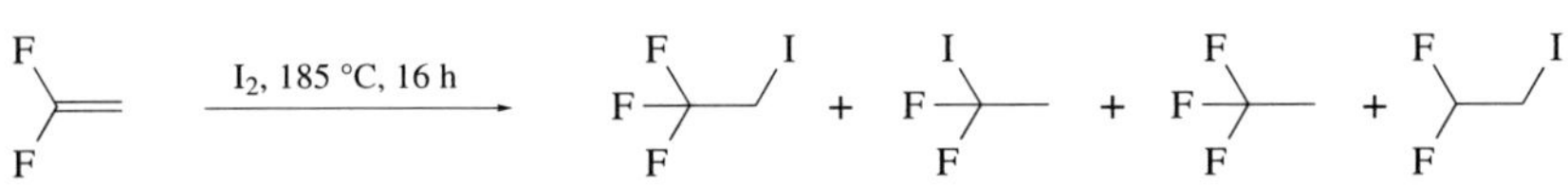

(4)

The addition of the interhalogen compounds ICl and IBr to fluoroalkenes also can be used to give compounds having the group —$CHal_3$. With asymmetric alkenes, the addition is regioselective, but the precise composition of the product is influenced by the reaction conditions and the presence of catalysts. The reactions shown in Equations (5)–(9) were carried out by several workers and thus under different sets of conditions ⟨52JCS4423, 54JCS923, 61JCS3779, 61JA2495, 62JOC1482, 64JOC252⟩; the yields and, where more than one isomer was obtained, isomer ratios were correspondingly different.

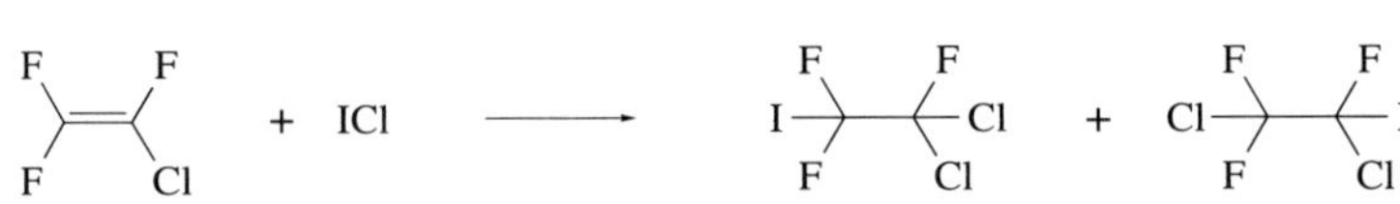

(5)

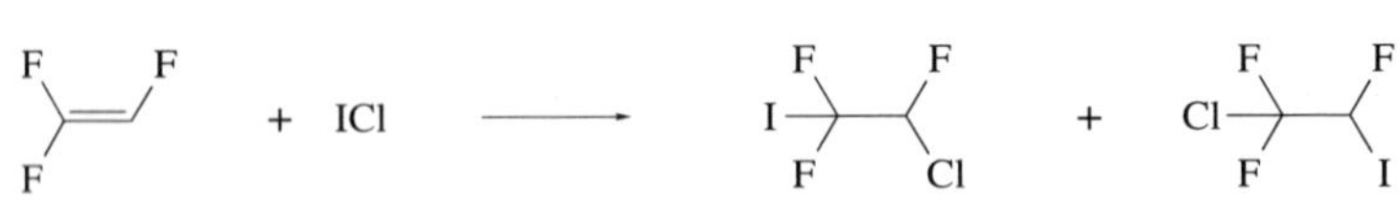

(6)

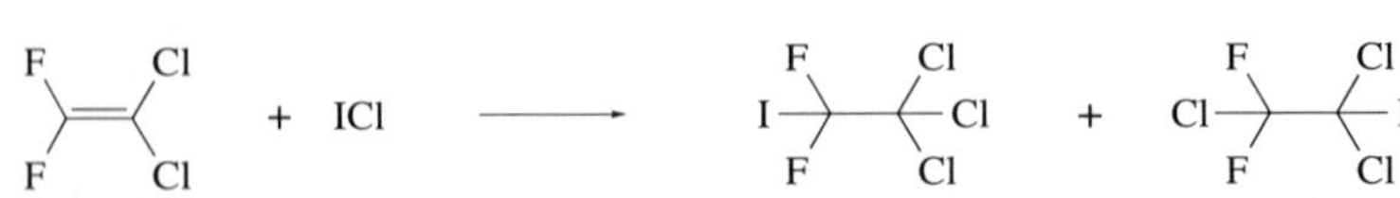

(7)

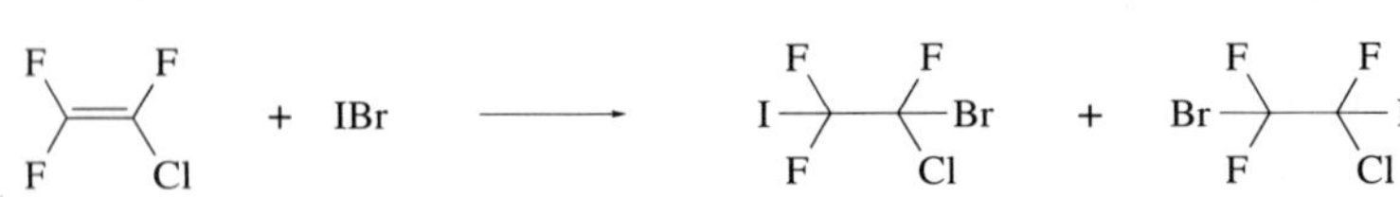

(8)

(9)

The elements bromine and fluorine, or iodine and fluorine, can be added to fluoroalkenes by treating the alkene with a mixture of bromine trifluoride and bromine, or iodine pentafluoride and iodine (Equations (10)–(22)). Here, 'IF' refers to a stoichiometric mixture of iodine pentafluoride and iodine ($IF_5 + 2I_2 \equiv 5$'IF'), and 'BrF' to a stoichiometric mixture of bromine trifluoride and bromine ($BrF_3 + Br_2 \equiv 3$'BrF'). While the addition of 'BrF' to fluoroalkenes is vigorous and needs to be moderated, the corresponding reactions of 'IF' are carried out at elevated temperatures ⟨61JCS3779⟩, and catalysts may be added ⟨61JA2383, 84JAP5951225⟩. The addition of 'IF' to tetrafluoroethene is particularly important because the product, pentafluoroiodoethane, is a starting material in one of the commercial routes to fluorochemical surfactants and surface treatments ⟨B-79MI 601-01⟩. Additions to hexafluoropropene and 1,1-difluoroethene give single isomers, but additions to chlorotrifluoroethene and 1,1-dichlorodifluoroethene yield mixtures of isomers whose composition is related to the reaction conditions. The reaction of 'IF' to tetrachloroethene results in fluorination only (Equation (18)). For the addition of 'IF' to internal alkenes with the formula $CF_3(CF_2)_nCF{=}C(CF_3)_2$, special procedures are necessary (Equation (22)). Either potassium fluoride must be added to the 'IF' mixture, or the alkene may first be treated with AgF or AgF/KF in a polar aprotic solvent to make the silver salt, which in turn is treated with iodine to give the desired product ⟨87JFC(37)223⟩. These products are tertiary perfluoroiodoalkanes and are highly toxic, so special care must be taken during their preparation.

(10)

(11)

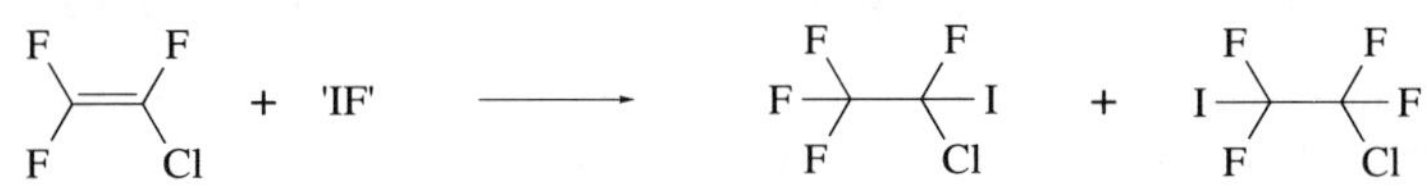
(12)

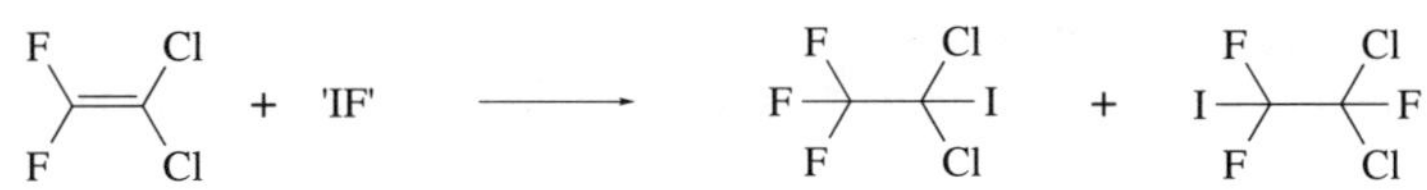
(13)

(14)

(15)

(16)

(17)

(18)

(19)

(20)

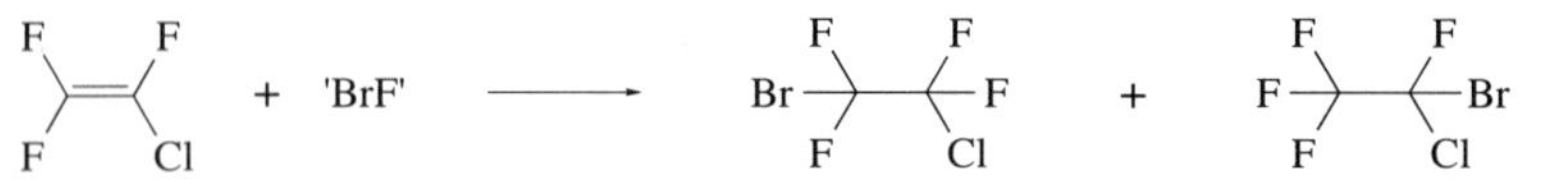

(21)

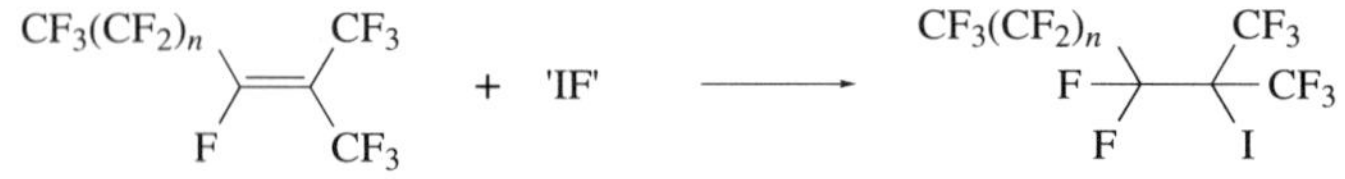

(22)

6.01.1.2 The Addition of Haloalkanes to Haloalkenes

The addition of haloalkanes to haloalkenes can be used to generate alkanes which bear a trihalomethyl group (Equation (23)). If the alkene is perhalogenated, then it can be seen that the adduct will have two trihalomethyl groups. These addition reactions can be initiated in four main ways, that is, by Lewis acids, by free radical initiators, by salts and complexes of certain transition metals, and electrochemically.

(23)

6.01.1.2.1 Lewis acid-catalysed addition of trihalomethyl cations to haloalkenes—the Prins reaction

The aluminum chloride-catalysed addition of chloroalkanes to chloroalkenes has been known since the beginning of this century, and is known as the Prins reaction ⟨14JPR415, 35RTC249, 35RTC307⟩. More recently, the reaction has been extended to include the addition of chlorofluoromethanes to fluoroalkenes and chlorofluoroalkenes. While most of the activity in this area occurred up to the early 1970s, and was reviewed by Paleta ⟨77FCR39⟩ and by Paleta and Posta ⟨72CLY937⟩, interest has been renewed recently because the adducts may be intermediates in the manufacture of potential replacements for the higher-boiling chlorofluorocarbons (e.g., ⟨91MIP601-01, 91EUP421322, 92EUP473105, 92JAP04164039⟩). Most commonly, the halomethanes used in these reactions are CCl_4, $CHCl_3$, $CFCl_3$ and $CFHCl_2$, and these have been added to the alkenes $CF_2{=}CF_2$, $CF_2{=}CFCl$, $CF_2{=}CCl_2$, $CFCl{=}CCl_2$, $CCl_2{=}CFCl$, C_2Cl_4, $CHCl{=}CHCl$, $CF_2{=}CFH$, $CF_2{=}CH_2$ and $CF_2{=}CHCl$ (see Table 1). The addition of halomethanes to asymmetric perhalogenated fluoroalkenes proceeds nonregiospecifically, but corresponding additions to the hydrogen-containing alkenes $CFCl{=}CFH$, $CF_2{=}CFH$, $CF_2{=}CH_2$ and $CF_2{=}CHCl$ are highly regioselective, with the electrophilic haloalkyl group becoming attached to the carbon which bears the hydrogen. This suggests that the preferred isomer is that in which most of the fluorine atoms on the precursor alkene are attached to the carbon where positive charge develops in the initial stages of the reaction. In reactions with fluorotrichloromethane, both C—F and C—Cl bonds are cleaved, although cleavage of the C—F bond is preferred due to the greater Al—F bond strength. In its reaction with the alkenes $CFCl{=}CHF$ and $CF_2{=}CHCl$ the C—F bond is cleaved exclusively. Some examples of these reactions are given in Table 1, where it can be seen how the approach can be used to give a variety of propanes with various trihalomethyl groups attached to carbon atoms. It is probable that the primary products in each of these reactions undergo rearrangement or further reaction, and therefore the composition and nature of the observed products are dependent upon reaction conditions.

6.01.1.2.2 Additions initiated by free radicals, heat or radiation

When subjected to heat, light, or free radical initiators such as peroxides or azo compounds, haloalkanes add to alkenes and haloalkenes (Scheme 1).

Whether a simple adduct or a mixture of high molecular-weight telomers is obtained is determined by the relative rates of the propagation, transfer and termination reactions. Thus, compounds with the weakest C—X bond, that is, iodides, give lower molecular-weight products, as do alkenes which are difficult to polymerize under the conditions of the reaction. But if the alkene is easily polymerized, for example tetrafluoroethene, or if the telogen is not particularly reactive, high molecular-weight

Table 1 Lewis acid-catalysed addition of haloalkanes to haloalkenes.

Reactants		*Conditions*	*Yield* (%)	*Products*	*Composition* (%)	*Ref.*
$CFCl_3$	CF_2CF_2	~25°C/3 h	70	$CF_3CF_2CCl_3$, $CF_2ClCF_2CFCl_2$	83, 17	71CCC1867
$CHFCl_2$	CF_2CF_2	15°C/3.5 h	58	$CF_3CF_2CHCl_2$, CF_2ClCF_2CHFCl	59, 41	71CCC1867
$CFCl_3$	CF_2CCl_2	22°C/17 h	42	$CF_3CCl_2CCl_3$, $CFCl_2CF_2CCl_3$, $CF_2ClCCl_2CFCl_2$	45, 48, 7	66CCC3584
$CHFCl_2$	CF_2CCl_2	7°C/12 h	78	$CFCl_2CF_2CHCl_2$, $CF_3CCl_2CHCl_2$, CCl_3CF_2CHFCl, CF_2ClCCl_2CHFCl, $CF_2ClCFClCHCl_2$	81, 19	67CCC3888
$CFCl_3$	CF_2CHF	0–14°C/7 h	65	$CF_3CHFCCl_3$, $CF_2ClCHFCFCl_2$	70, 30	74CCC1330
$CHFCl_2$	CF_2CHF	0°C/7 h	83	$CF_3CHFCHCl_2$, $CF_2ClCHFCHFCl$	58, 42	74CCC1330

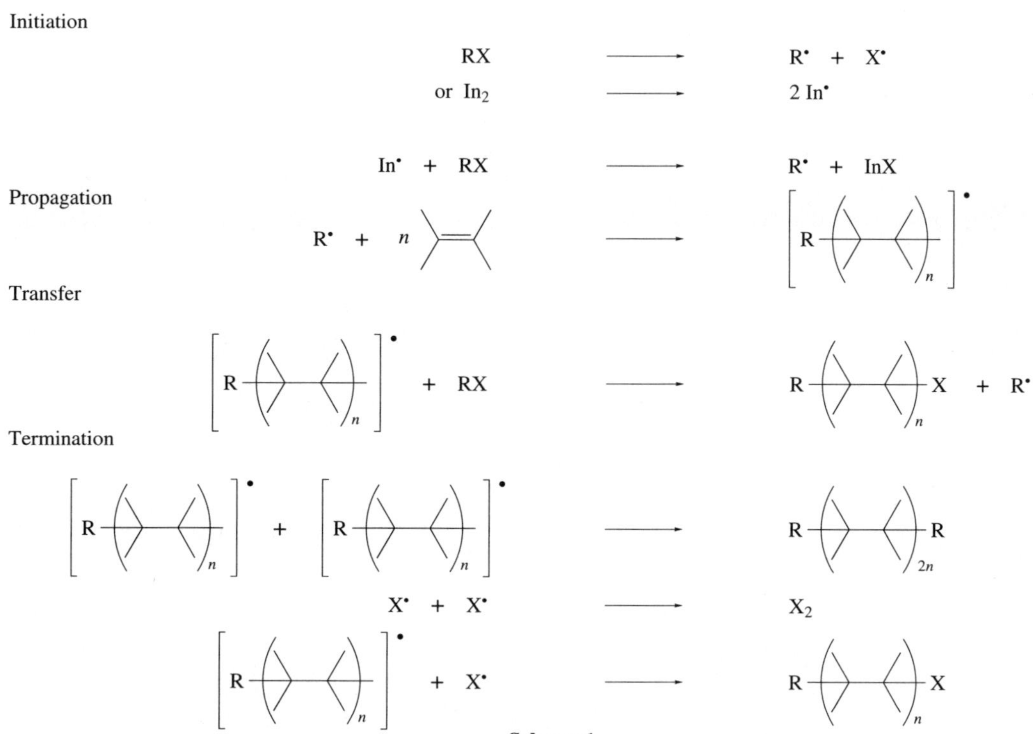

Scheme 1

material is obtained. With easily polymerized alkenes, the telomerization can be controlled by using a large excess of the telogen (preferably an iodide or diiodide) and recycling the lower telomers. This addition of a haloalkane is often regioselective, and it is observed that the incoming radical generally attacks the least sterically hindered carbon. However, there is still debate on the factors which control the products of these reactions and the relative stability of the potential intermediate radicals, and polar effects (note that trihalomethyl radicals are electrophilic) as well as steric effects all play their part.

This approach is used in the manufacture of C_8 or C_9 perfluoroalkyl iodides, which are intermediates in the preparation of perfluorocarbon surfactants and surface treatments ⟨B-79MI 601-01⟩. Illustrative examples of these addition reactions are given in Equations (24)–(35) (NB. Higher telomers are also produced in these reactions) ⟨47JA1100, 90JFC(47)261, 50JA2213, 53JOC328, 53JCS922, 53JCS3761, 52JCS3490, 64JOC1198, 58JA851, 87IZV808, 55JA768, 85T4503⟩ (references refer to the respective Equations (24)–(35)), but many more are recorded elsewhere ⟨63OR(13)91, B-74MI 601-01, B-76MI 601-01⟩. It is noteworthy that, unlike other polyhaloalkanes, trichloromethane adds to alkenes by hydrogen transfer (Equation (26)).

(24)

(25)

(26)

(27)

(28)

(29)

(30)

(31)

(32)

(33)

(34)

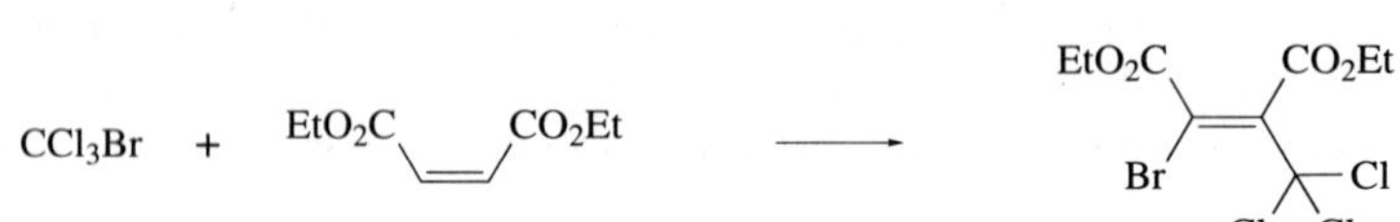
(35)

6.01.1.2.3 Additions catalysed by salts and complexes of transition metals

In 1956 ⟨56CI(M)371⟩, unexpected results were obtained when the thermal additions of CCl_4 and CCl_3H to acrylonitrile were carried out using a steel autoclave. More of the 1 : 1 adduct was obtained than expected, and trichloromethane gave the adduct $CHCl_2CH_2CHClCN$ rather than

$CCl_3CH_2CH_2CN$, which is normally obtained in free radical additions to alkenes (*vide supra*). Later ⟨63JCS1887⟩, it was established that copper(II) and iron(III) both catalysed the addition of tetrachloromethane to a variety of vinylic monomers, and that 1 : 1 adducts were frequently the sole products, even with easily polymerized alkenes. This discovery led to extensive work in the area, and copper(II) has now been used to catalyse the addition of CCl_4, CCl_3Br, CCl_2Br_2, CF_3CCl_3 and other chloro compounds to a large variety of alkenes and dienes. It is worth noting that the addition of CF_3CCl_3 followed by hydrogenation of the adduct can be used to introduce a CF_3CH_2 group. While most investigations in this area have used copper- and iron-based catalysts, other metal salts such as samarium diiodide ⟨90JCS(P1)2031⟩ and vanadium dichloride ⟨90SL217⟩ have been used more recently. As well as alkenes which contain only hydrogen, halogenated alkenes and those which contain other groups have also been used. Generally the 1 : 1 adduct is the only product, but higher telomers are produced when the alkene is halogenated. However, this method of initation always gives lower molecular-weight material than the corresponding peroxide initiated reaction. Amines are often added to form complexes and 'increase the solubility' of the copper, and consequently to increase the reaction rate. The following is not a comprehensive list of references but will serve to lead the reader into the area: ⟨B-74MI 601-01, 75ACR165, 76T2295, 77TL4315, 80CCC3488, 80CCC3502, 85PAC1827, 87JPS(A)3025, 89JFC(45)215, 90JFC(47)95, 91MI 601-01, 92CCC1291, 92JMOC51, 93JFC(61)133⟩.

As well as metal oxides, salts and their complexes with amines, other transition metal complexes such as carbonyls of iron ⟨67JCS(C)1150, 80JOC3957⟩ and cobalt ⟨70JOC2982⟩, carbonyl complexes of molybdenum ⟨82JCS(D)2281, 90JOM(397)51⟩, iron ⟨90JOM(386)229⟩ and chromium ⟨84JOM(260)C75⟩, and phosphine complexes of ruthenium ⟨73TL5147, 75TL899, 78JOC1734, 85JOM(280)397, 87JCS(P1)1515, 92JOM51⟩, rhodium ⟨81AG(E)475⟩, rhenium ⟨87JCS(P1)1515⟩ and palladium ⟨81CL1169, 85T393⟩ have all been found to be efficient catalysts in this reaction. When the chiral $Ru_3Cl_4(diop)_3$ (diop, 2,3-*O*-isopropylidene-2,3-dihydroxy-1,4-bis(diphenylphosphino)butane) was the catalyst, chiral adducts were obtained ⟨87BCJ3687⟩.

Clearly the metal-catalysed additions of haloalkanes to haloalkenes proceed by a different mechanism to those promoted by free radical initiators such as peroxides. Any mechanism needs to be able to explain the 'abnormal' addition of trichloromethane, the suppression of the propagation step (1 : 1 adducts are the main products), and the fact that the addition of tetrachloromethane to cyclohexene is highly stereoselective (with a ruthenium catalyst) compared with the situation which obtains when conventional radical initiators are used. It is unlikely that the details of the mechanism will be the same for all the catalysts that have been identified, but it is generally believed that it is not a simple redox process but one in which (e.g., for CCl_4) a trichloromethyl radical is formed on and bound to the catalyst prior to the addition to the alkene which is itself coordinated to the complex.

6.01.1.2.4 *Additions induced electrochemically*

The free radical chain addition of various polyhalomethanes to alkenes has been initiated by electrochemically *in situ* generated manganese salts used in a catalytic amount associated with an equimolecular amount of a manganese(III) oxidizable compound such as methyl cyano- or acetoacetate. In particular, tetrabromomethane, bromotrichloromethane, dibromodifluoromethane and *n*-heptadecafluorooctyl iodide have been added to a variety of alkenes in high yield by this means (Equation (36)) ⟨92TL213⟩.

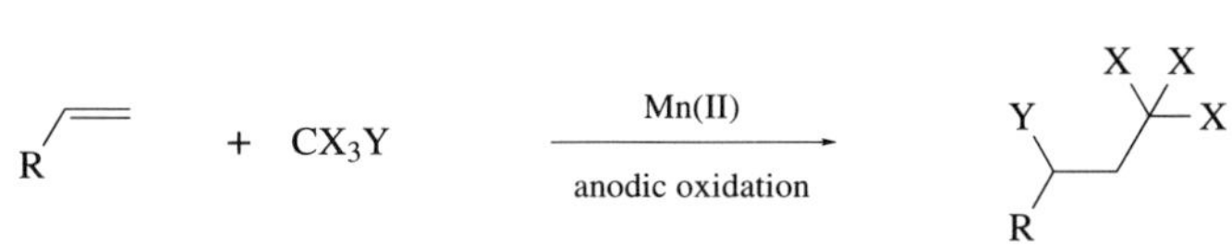

(36)

$CX_3 = CBr_3$, Y = Br; $CX_3 = CCl_3$, Y = Br; $CX_3 = CBrF_2$, Y = Br; $CX_3 = C_7F_{15}$, Y = I

6.01.1.3 The Preparation of C_2 Chlorofluorohydrocarbons

The manufacture of saturated compounds having the general formula $C_2H_xCl_yF_z$ represents an important part of the fluorochemical industry. These compounds ($x = O$) (chlorofluorocarbons—'CFC's) have been used as refrigerants, solvents and foam-blowing agents. However, because they have been implicated in the destruction of ozone in the earth's upper atmosphere, they are being replaced, under the terms of the Montreal Protocol, by hydrofluorocarbons ($y = O$) ('HFC's) and, for an interim period, by hydrochlorofluorocarbons ('HCFC's). The manufacture of these compounds is largely brought about by treating a chlorocarbon with hydrogen fluoride and sometimes chlorine. Other reactions in the process may include isomerization and hydrogenation. The fluorinations are carried out in either the liquid phase using an antimony(V) catalyst (Swarts' catalyst) or in the gas phase using a chromia-based catalyst. The main routes to the C_2 chlorofluorocarbons and those compounds which are replacing them are outlined in Scheme 2. 1,1,1,2-Tetrafluoroethane (R134a) is to be the main 'high-temperature' refrigerant (e.g., mobile air conditioning, and domestic and industrial refrigeration) while 1,1,1-trifluoroethane (R143a) and 1,1,1,2,2-pentafluoroethane (R125) will be used in compositions for 'low-temperature' refrigeration and static air conditioning (e.g., ⟨93MI 601-01⟩). As well as having the uses outlined above, some of these compounds are intermediates in the manufacture of products such as the anaesthetic halothane ($CF_3CHClBr$), and monomers such as 1,1-difluoroethene and chlorotrifluoroethene.

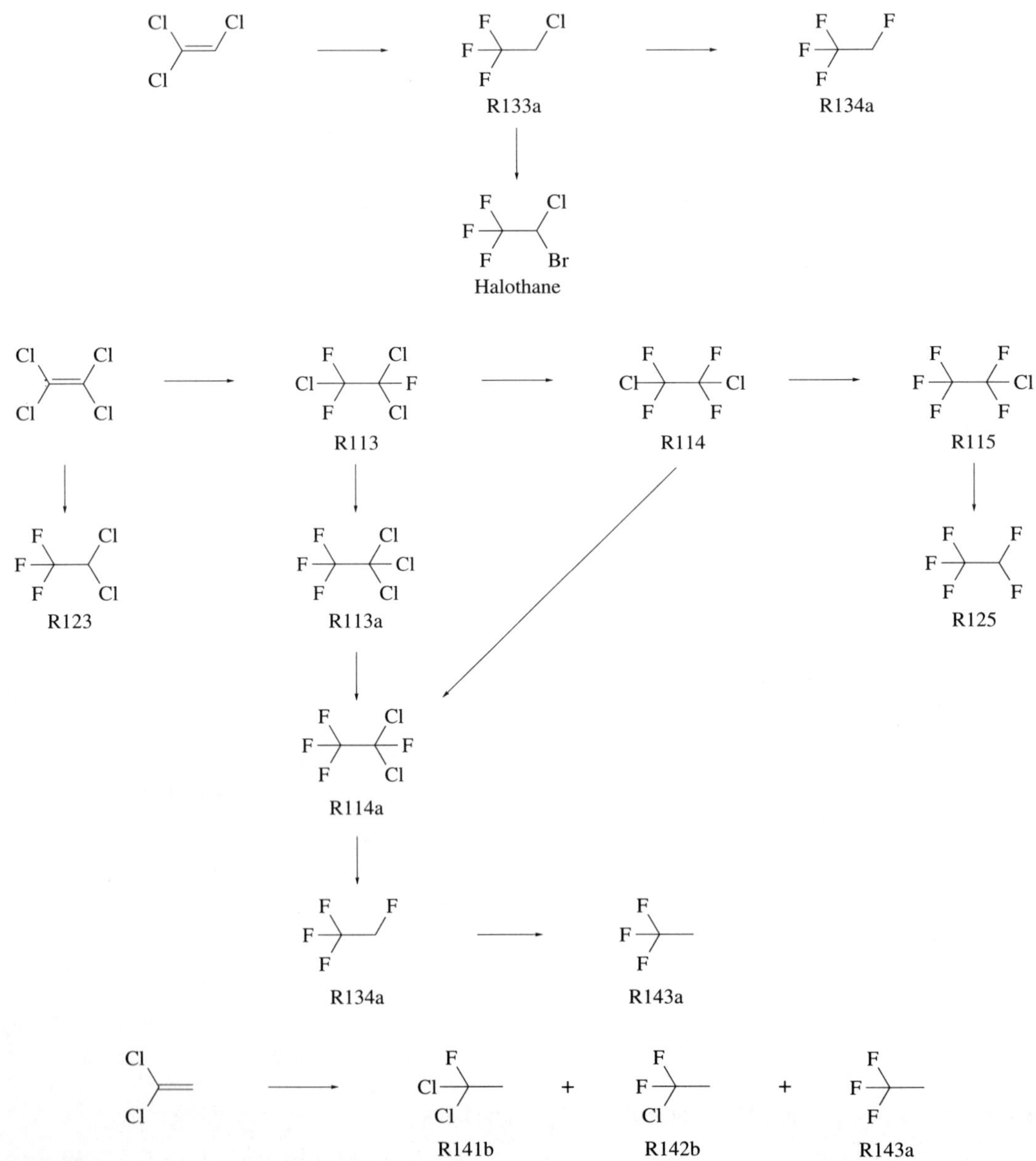

Scheme 2

6.01.2 TRIFLUOROMETHYL DERIVATIVES—RCF_3

6.01.2.1 General

There is a very considerable literature relating to the synthesis of organic compounds containing the trifluoromethyl group, and this stems principally from the importance of such compounds in the plant protection industry ⟨91MI 601-02⟩ and, to a lesser extent, the pharmaceutical industry ⟨B-91MI 601-03, B-82MI 601-01⟩. Amongst other effects, the introduction of the trifluoromethyl group increases the lipophilicity of bioactive molecules, enhancing transport across cell walls, and can reduce metabolism of a drug. A major review ⟨92T6555⟩ is focused on the synthesis of trifluoromethyl derivatives, and others ⟨91T3207, 92T189, 94TA937⟩ include discussions of relevant methodology. In considering this area, it is appropriate to draw attention to the fact that processes which involve halogen exchange for fluorine, using anhydrous hydrogen fluoride (AHF) as the source of fluorine, will be the most economically favourable for large-scale manufacture by industry, which is remarkably adept at handling AHF. Such procedures are not, however, particularly easy on the laboratory scale, and other approaches described here will frequently be preferred. Furthermore, the sensitivity of the molecule under consideration may well dictate the route that is most appropriate.

The considerations described above relate to compounds containing the trifluoromethyl group as the only halocarbon unit, but, of course, there is a very wide range of applications for more highly fluorinated systems, including their use as refrigerants and for other volatile inert fluid applications, membranes, polymers and anaesthetics.

6.01.2.2 Aryl Derivatives

6.01.2.2.1 Conversion of groups attached to an aromatic ring into the trifluoromethyl group

Only the most reactive of organic halides will react with anhydrous hydrogen fluoride without catalysis (see Section 6.01.1.3), but benzotrichloride is converted to the trifluoride (Equation (37)) ⟨52MI 601-01⟩, and analogous systems, including heterocyclic compounds, may be obtained in this way (Equation (38)) ⟨81AHC(28)1⟩. More sophisticated techniques have been used in processes to chlorofluorinate side chains in a continuous process (Equation (39)) ⟨80GEP3008081⟩. Alternatively, antimony trifluoride may be used in classical fluorination processes (Equation (40) ⟨B-76MI 601-02⟩, where the product is obtained by simply distilling from the fluorinating agent, but the trichloroethyl moiety is much less efficient for conversion to the trifluoroethyl group (Equation (41)) ⟨88JOC3637⟩. A clever 'one-pot' methodology is available for acid-induced trichloromethylation accompanied by exchange of chlorine for fluorine (Equation (42)) ⟨81JFC(18)281⟩.

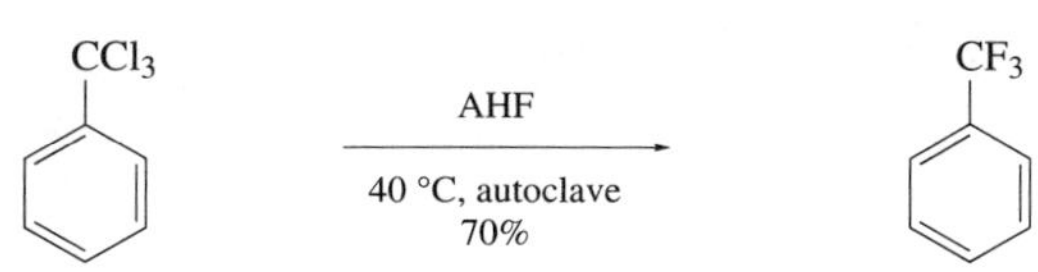

(37)

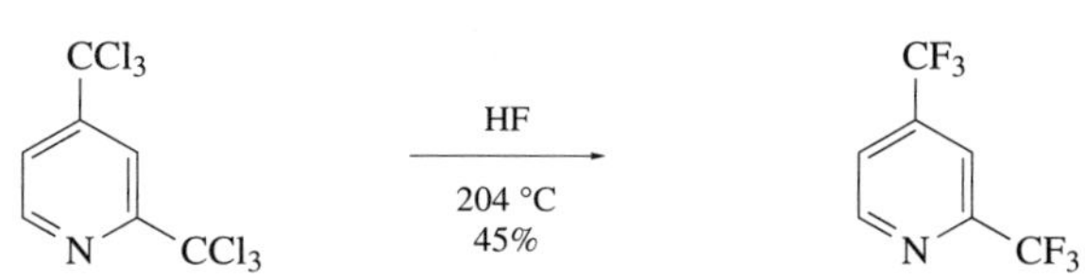

(38)

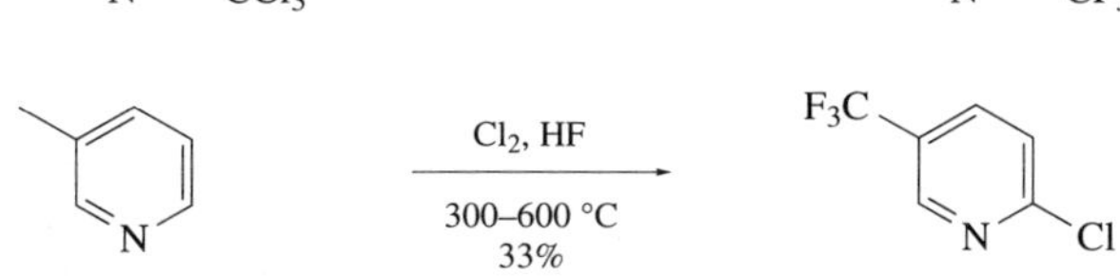

(39)

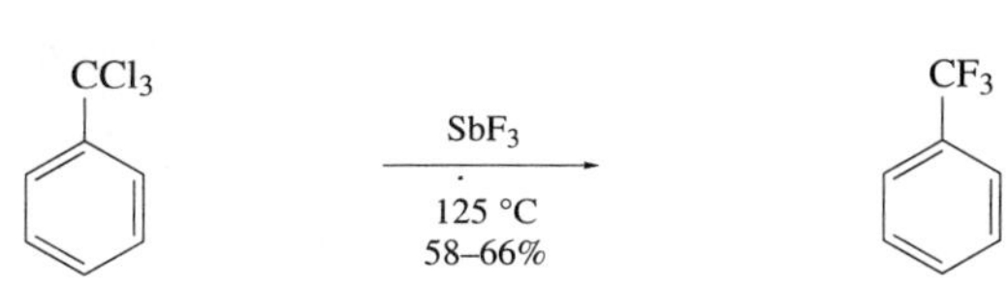

(40)

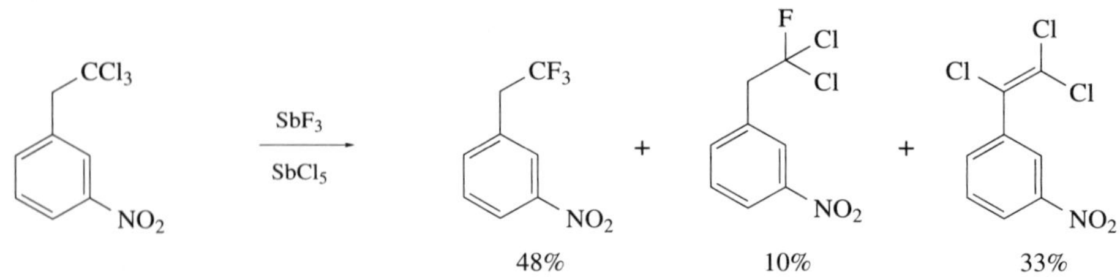

(41)

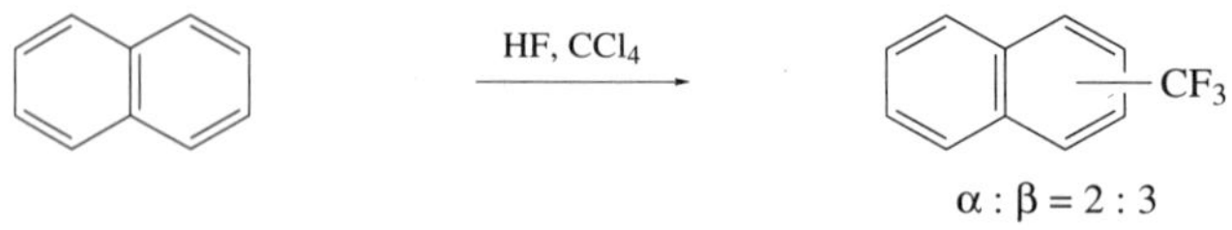

(42)

One of the most versatile methods available for introducing the trifluoromethyl group involves reactions of benzoic acid derivatives and corresponding heterocyclic compounds with sulfur tetrafluoride (Equations (43), (44) and (45)) ⟨85OR(34)319, 81AHC(28)1, 87JOM(325)13, 93JFC(60)233⟩, or the more convenient but less reactive diethylaminosulfur trifluoride (DAST) ⟨75USP3914265, 76USP3976691⟩ and related systems (Equation (46)) ⟨88OR(35)513⟩.

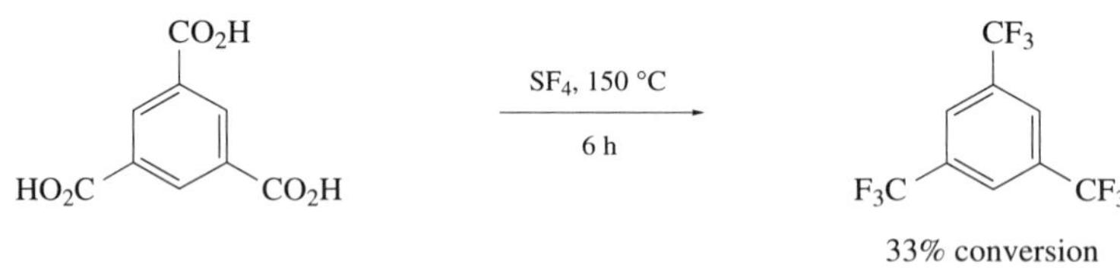

(43)

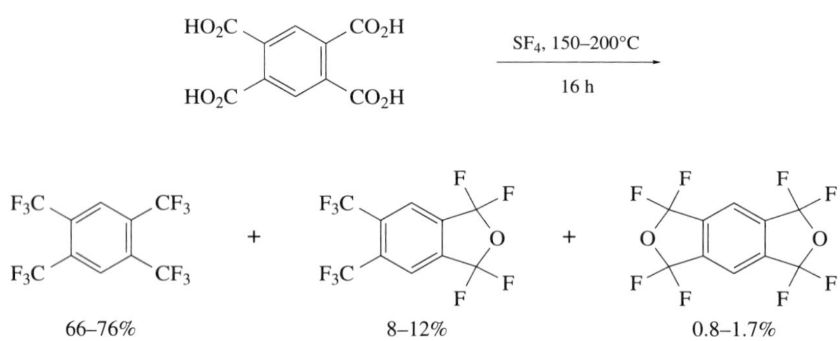

(44)

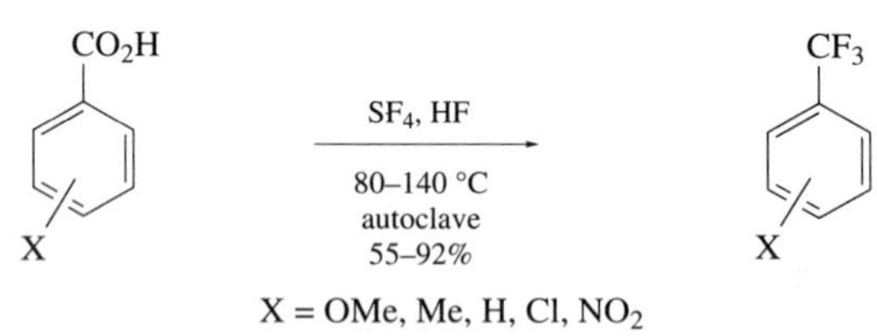

(45)

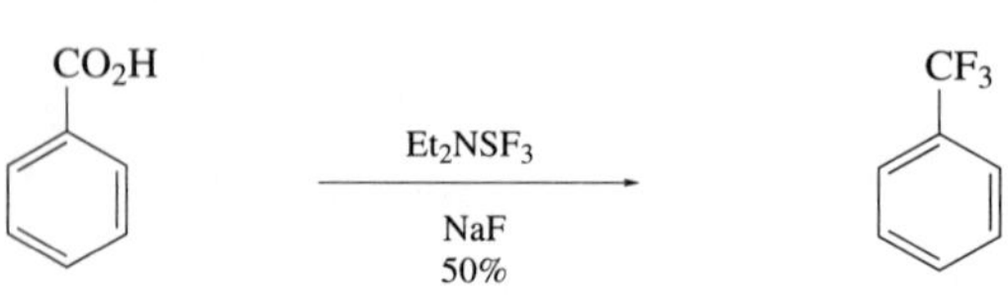

(46)

Thio derivatives (e.g., *ortho*-thioesters) may also be used for conversion to the trifluoromethyl moiety, using NBS, or its equivalent, followed by pyridine–hydrogen fluoride (Equation (47)) ⟨86TL4861⟩. Dithiocarboxylate esters are converted in a similar process (Equation (48)) ⟨92CL827⟩, and dithiocarboxylate derivatives are also converted in an interesting process that involves xenon difluoride (Equation (49)) ⟨90TL3357⟩. A very unusual conversion of a nitrile to the trifluoromethyl derivative occurs, using carbonyl fluoride, accompanied by other products (Equation (50)) ⟨78IJ129⟩.

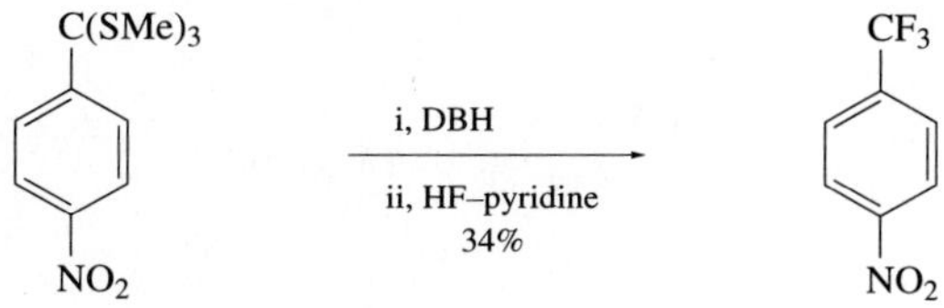 (47)

DBH = 1,3-dibromo-5,5-dimethylhydantoin

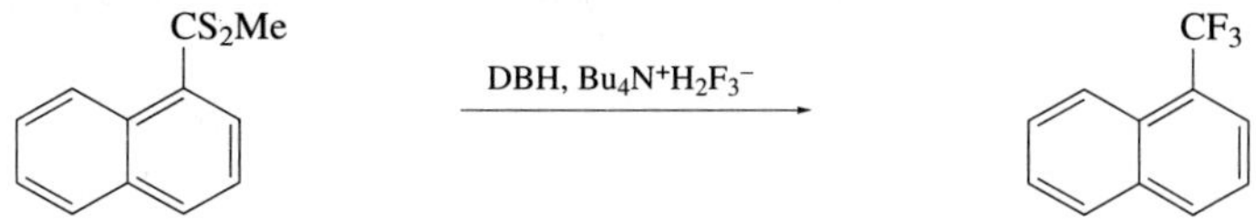 (48)

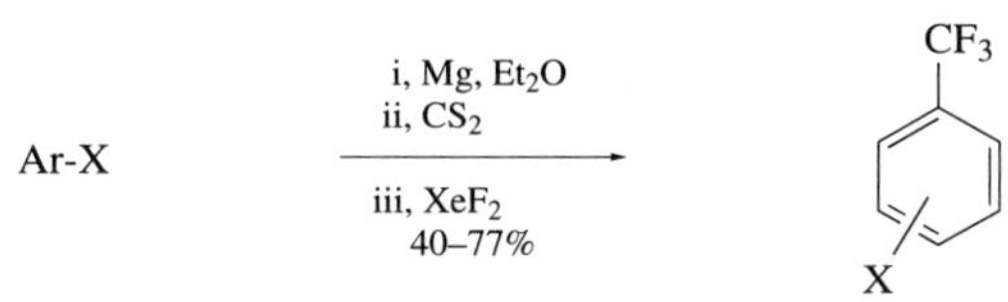 (49)

Ar = Ph, *p*-MeC_6H_4, *m*-$CF_3C_6H_4$, α-naphthyl
X = halogen (e.g., Br)

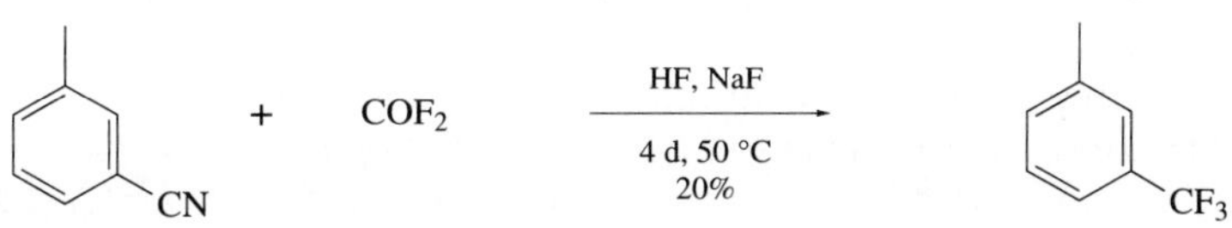 (50)

6.01.2.2.2 Substitution by trifluoromethyl radicals

Primary sources of trifluoromethyl radicals include trifluoroethanoic acid and trifluoromethanesulfonic acid, but iodo- or bromotrifluoromethanes are important sources that have been used extensively. Various methods have been applied to generate trifluoromethyl radicals from these sources ⟨92T6585, 91TL7525⟩, and these radicals are, of course, electrophilic in character. Consequently, a wide range of radical aromatic substitutions occur, and these probably proceed more effectively with the more electron-rich aromatic compounds, but, in general, they are not high-yielding processes (Equations (51)–(58)) ⟨81JFC(17)345, 87CC1701⟩.

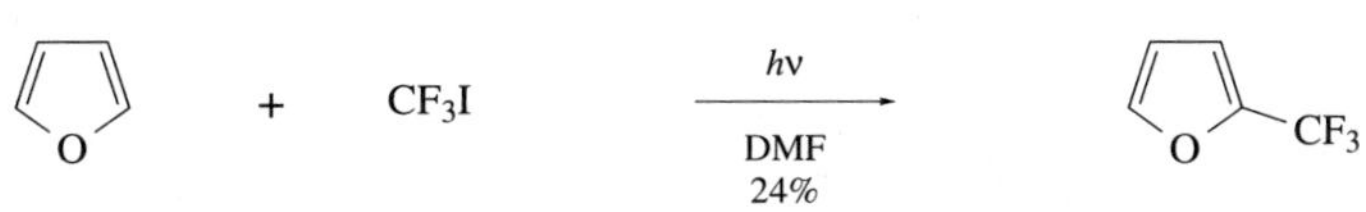 (51)

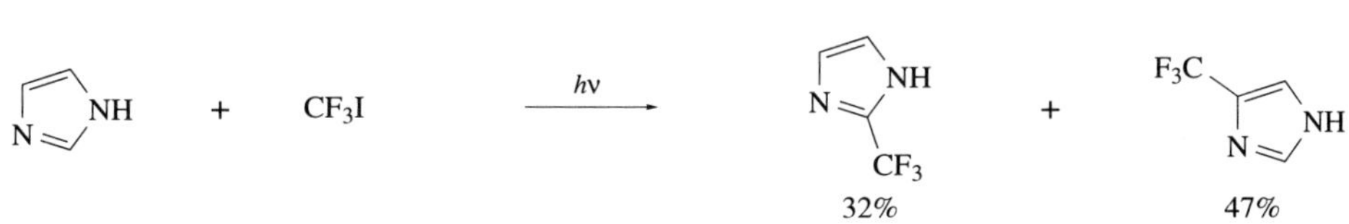 (52)

 (53)

61%, $o:m:p = 40:30:22$

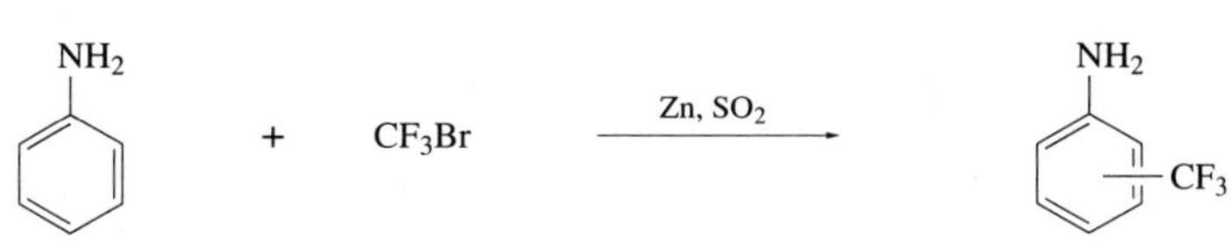 (54)

56%, $o:p = 1.8:1$

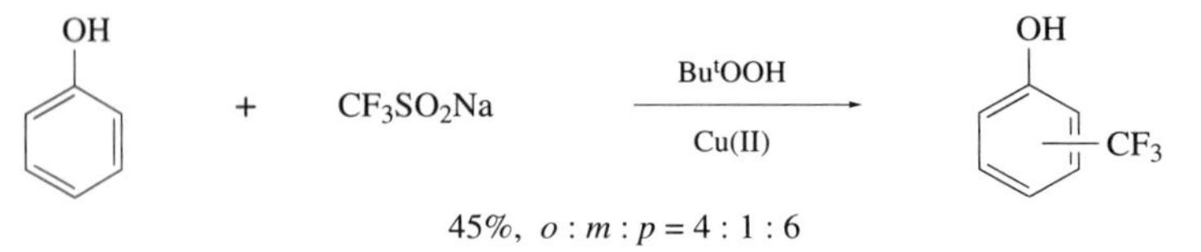

(55)

$$C_6H_6 + (CF_3CO)_2O \xrightarrow[54\%]{70\,^\circ C} C_6H_5CF_3 \qquad (56)$$

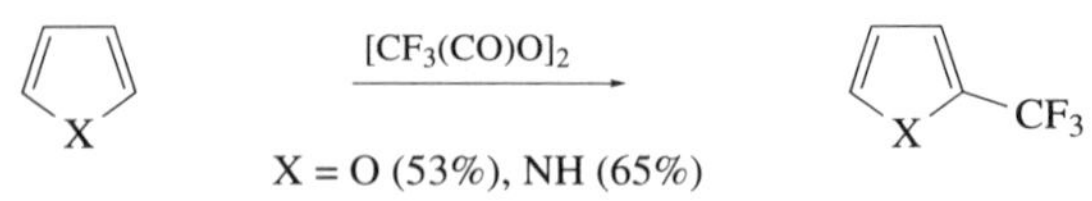

(57)

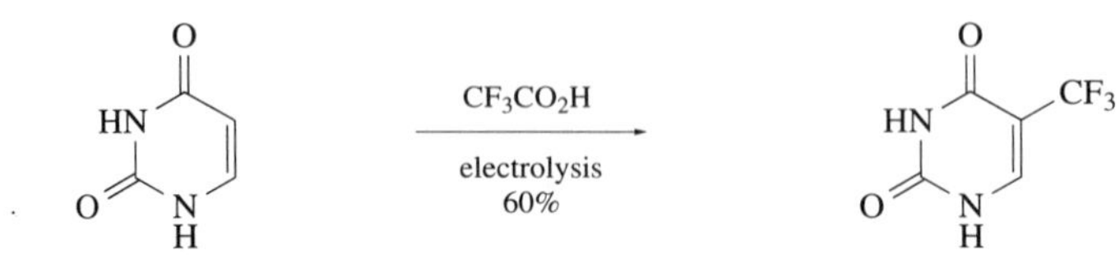

(58)

Bis(trifluoroacetyl) peroxide decomposes thermally and can be used to trifluoromethylate electron-rich aromatic compounds (Equations (59) and (60)) ⟨90JFC(46)423⟩. An atmosphere of ammonia has been used to promote the reaction of iodotrifluoromethane with pyridine (Equation (61)) ⟨79JAP79283⟩. Remarkably, some cross-coupling between hexafluorobenzene and bromotrifluoromethane occurs over copper chromite (Equation (62)) ⟨93JFC(61)1⟩, and, in other high-temperature chemistry, polytetrafluoroethylene is used as a source of difluorocarbene, which, under the conditions used, is able to insert into carbon–fluorine bonds (Equation (63)) ⟨69IZV196, 74JCS(P1)108⟩.

(59)

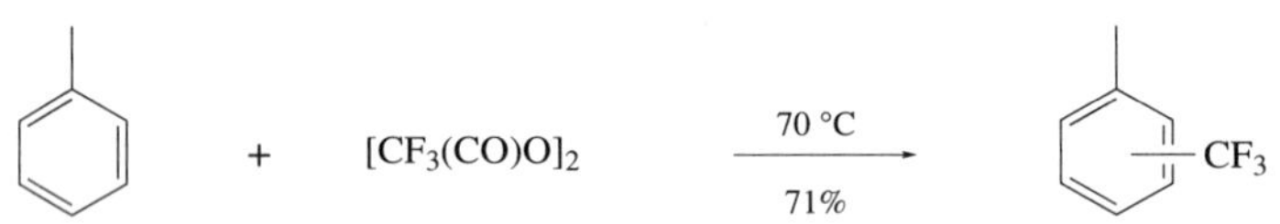

(60)

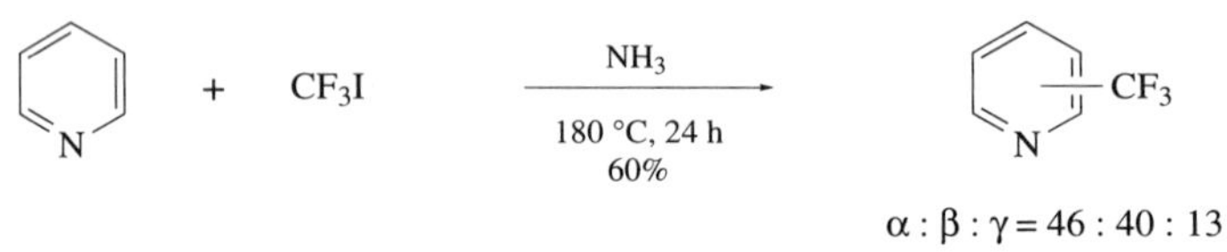

(61)

$$C_6F_6 + CF_3Br \xrightarrow[600\,^\circ C,\ \text{flow};\ 40\%]{\text{copper chromite}} C_6F_5CF_3 \qquad (62)$$

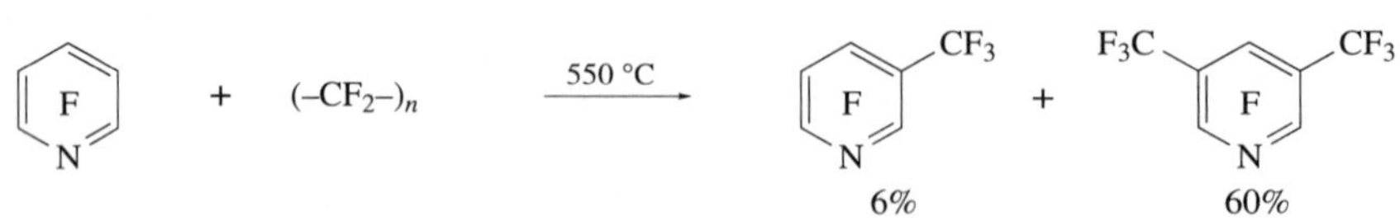

(63)

6.01.2.2.3 *Substitution of hydrogen by the trifluoromethyl group acting as an electrophile*

Nucleophilic attack at a saturated carbon contained in a highly halogenated system is, in general, very difficult, but Umemoto and co-workers have demonstrated that with superior leaving groups it is possible to effect what is, formally, nucleophilic attack on the trifluoromethyl group (Equation (64)). Very powerful perfluoroalkylating agents are available through so-called 'FITS' reagents ((perfluoroalkyl)phenyliodonium iodides) ⟨84TL81, 86JFC(31)37⟩. So far, the corresponding reagents are not available for the introduction of the trifluoromethyl group, although this may be a practical difficulty, rather than a fundamental difference in reactivity between trifluoromethyl and other perfluoroalkyl systems. However, a less powerful but effective series of salts of tellurium, selenium and sulfur heterocycles has been developed which will transfer the trifluoromethyl moiety ⟨93JA2156⟩, and it is concluded that the process involves nucleophilic attack on the trifluoromethyl group. The overall process is illustrated in Scheme 3, where the reactivity of the salts (**1**) varies in the series X = Te < Se < S, while electron-withdrawing substituents in the aromatic ring also increase reactivity (**2**). Examples of trifluoromethyl transfer from the salts (**1**) are shown (Equations (65)–(67)). The trifluoromethylation of hydroquinone (Equation (66)) effectively rules out a free radical process because hydroquinone is known to be an efficient radical scavenger.

$$[Ar-H] + F_3C-L \longrightarrow [ArHCF_3]^+ + L^- \longrightarrow ArCF_3 + H^+ + L^- \quad (64)$$

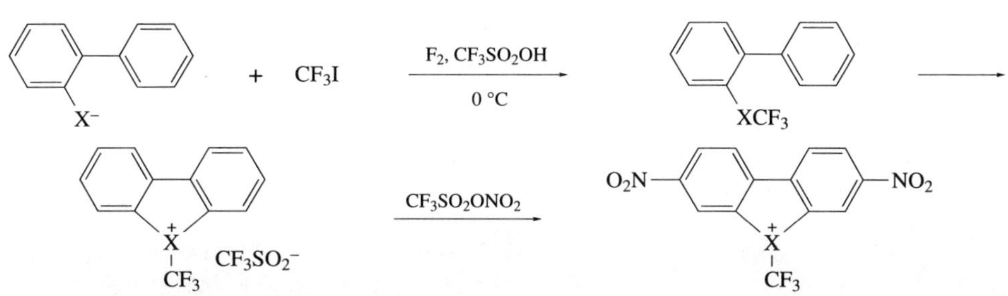

(**1**) (**2**)

X = S, Se, Te

Scheme 3

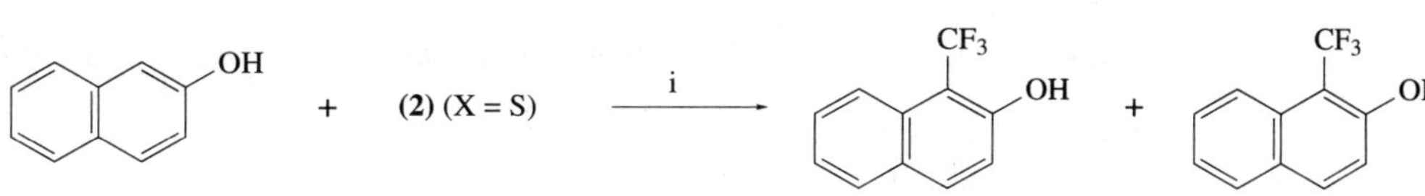

(65)

52% 6%

i, DMF (+ 4-dimethylaminopyridine)

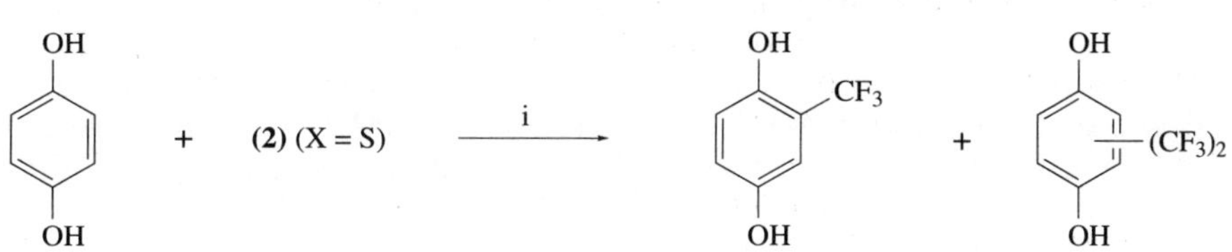

(66)

61% 11%

i, DMF (+ pyridine)

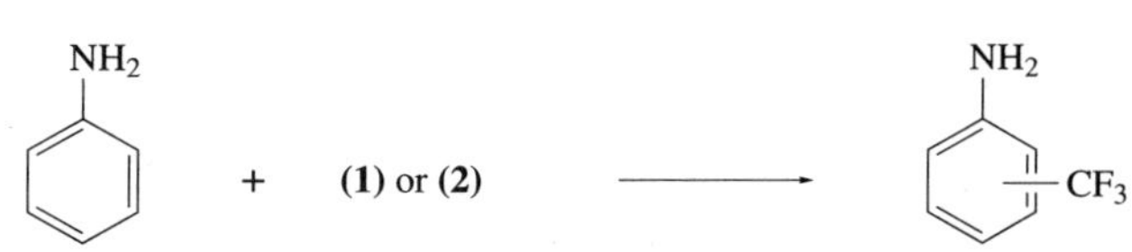

(67)

Reactant	*Conditions*	*Products*
(**1**) (X = S)	80 °C, 1 h	*o*-CF_3 (31%) + *p*-CF_3 (15%)
(**2**) (X = S)	RT, 0.5 h	*o*-CF_3 (54%) + *p*-CF_3 (20%)

6.01.2.2.4 Substitution of halogens by the trifluoromethyl group acting as a nucleophile

Trifluoromethyllithium and -magnesium derivatives are unstable and, therefore, a methodology for the transfer of a trifluoromethyl anion to electrophilic centres via silanes has been developed (Equation (68)) ⟨B-92MI 601-01⟩. A convenient synthesis of trimethyltrifluoromethylsilane has been reported (Equation (69)) ⟨80ZOB1897, 84TL2195, 93RHA232, 92MI 601-02⟩, and a process using tetrakis(dimethylamino)ethene has also been described (Equation (70)) ⟨89JFC(42)429⟩. The key step involves displacement of the trifluoromethyl group from silicon, using tetrabutylammonium fluoride (which inevitably contains amine). Understandably, reaction only occurs with the more activated aromatic systems (Equations (71) and (72)) ⟨B-92MI 601-03⟩.

$$CF_3^- + Ar-L \longrightarrow [ArLCF_3]^- \longrightarrow ArCF_3 + L^- \quad (68)$$

$$\text{TMS-Cl} + CF_3Br + (Et_3N)_3P \xrightarrow{\text{PhCN}} \text{TMS-CF}_3 \quad (69)$$

(70)

(71)

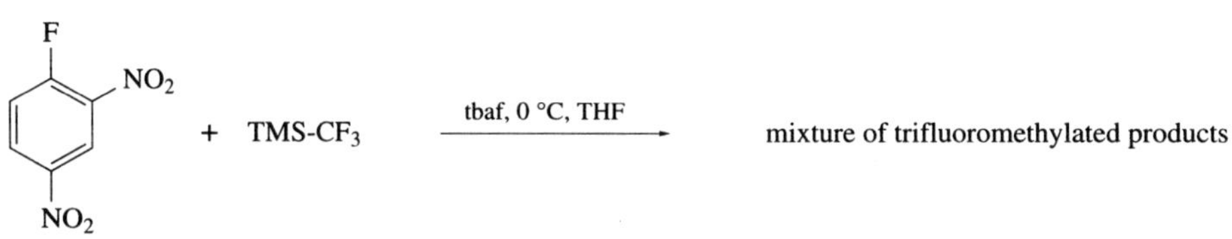

(72)

6.01.2.2.5 Substitution of halogens by the trifluoromethyl group using derivatives of metals

As indicated above, lithium and magnesium derivatives are unstable and, therefore, are not viable reagents for the introduction of the trifluoromethyl group. Metals that form weaker bonds to fluorine generally give more stable trifluoromethyl derivatives, and consequently copper, cadmium and zinc compounds are particularly useful in synthesis ⟨92T189, 92T6555⟩. Therefore, access to trifluoromethyl derivatives of these metals is of some importance. Easily accessible sources for the effective use of the trifluoromethyl moiety include CF_3CO_2H, CF_3SO_2OH, CF_3I (usually made from trifluoroacetic acid by Hunsdiecker processes), CF_3Br and CF_2Br_2.

Trifluoromethylation of aryl halides occurs using CF_3I or CF_3Br in the presence of copper powder in aprotic solvents at elevated temperatures (Equation (73)) ⟨80JCS(P1)2755⟩, but direct reaction of sodium trifluoroacetate with CuI in *N*-methyl-2-pyrrolidone (NMP) is also possible ⟨81CL1719⟩, and it has been concluded that the reaction proceeds via an intermediate like $[CF_3CuI]^{\bullet -}$ rather than $[CF_3CuI]^{\bullet}$ (Equations (74)–(76)) ⟨88JCS(P1)921⟩. Trifluoromethanesulfonyl chloride can be used in an analogous way (Equation (77)) ⟨89JFC(45)86⟩.

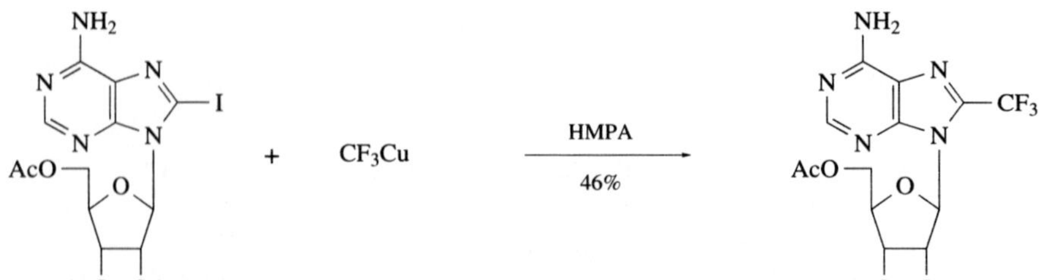

(73)

HMPA = hexamethylphosphoramide

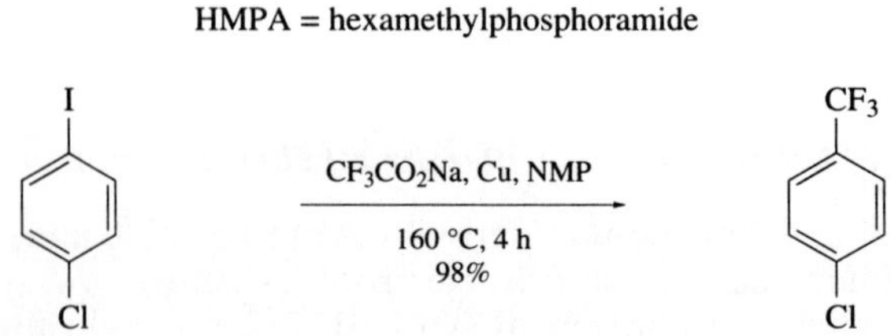

(74)

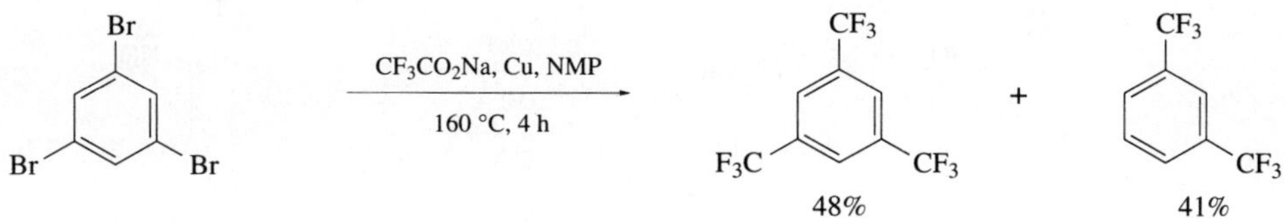

(75)

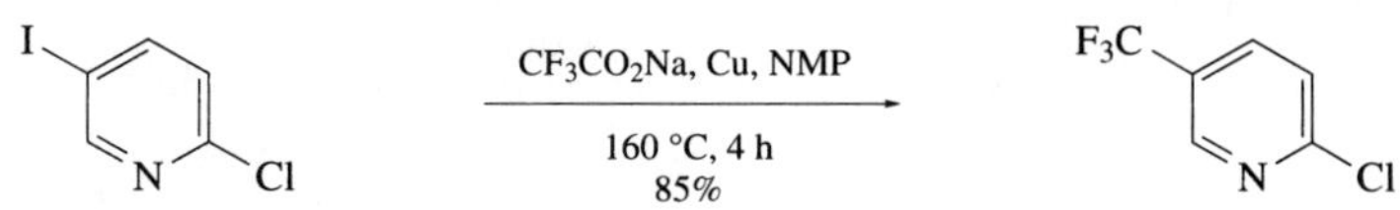

(76)

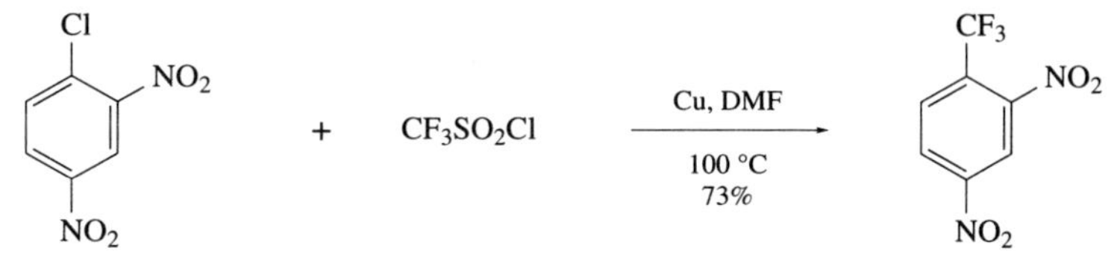

(77)

A fascinating route to trifluoromethylcadmium and -zinc has been developed by Burton and co-workers ⟨85JA5014⟩ in which reaction of dihalodifluoromethanes with cadmium or zinc powders in DMF gives stable solutions of the corresponding trifluoromethyl derivatives (Scheme 4), which then react further. Copper reacts in an analogous way but a mixture of homologues is obtained unless fluoride ion is added, or other procedures used to prevent the difluorocarbene generated reacting further with CF_3MX (Scheme 4). Cadmium derivatives can also be converted to copper reagents ⟨86JA832⟩, and these can be stabilized against formation of higher homologues by adding HMPA ⟨89JA8502⟩ (Equations (78) and (79)). Trifluoromethylation in dimethylacetamide (DMA) is also efficient (Equation (80)) ⟨88CC638⟩. Transfer of the trifluoromethyl group to amines, using dibromodifluoromethane, has also been reported ⟨91JFC(52)229⟩ (Equation (81)). Methyl fluorosulfonyldifluoroacetate in the presence of copper(I) iodide functions in a similar way ⟨89JCS(P1)2385⟩, and the corresponding iodide has also been used (Equations (82) and (83)). The source of the starting materials in the latter cases is tetrafluoroethene, via formation and ring opening of a β-sulfone (Scheme 5) ⟨60JA6181⟩; this is an attractively direct but, unfortunately, potentially hazardous process for laboratory application.

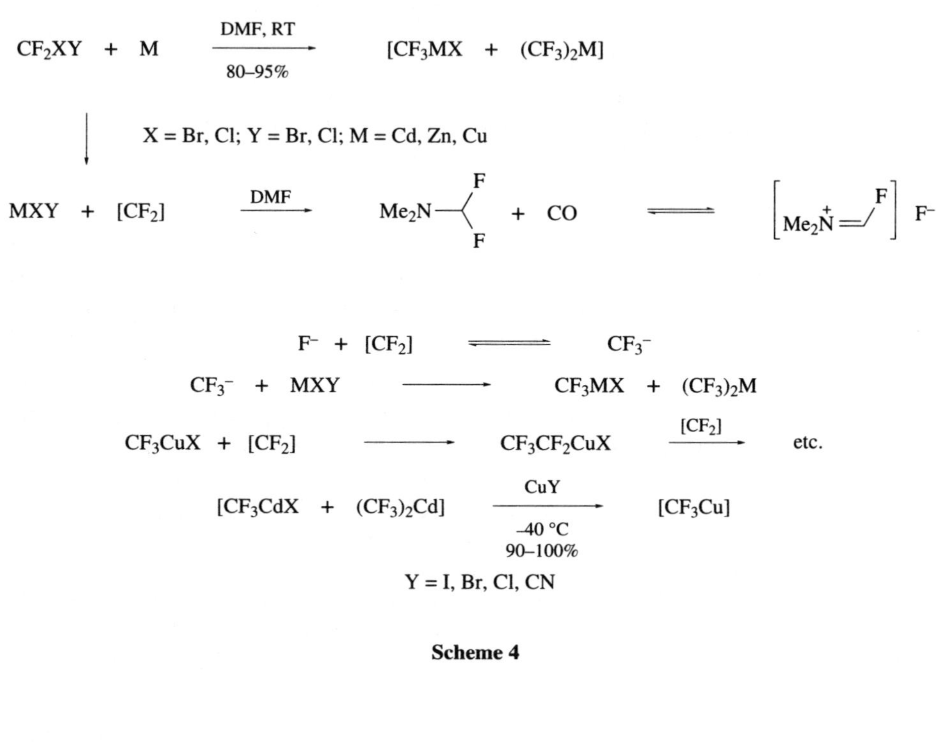

Scheme 4

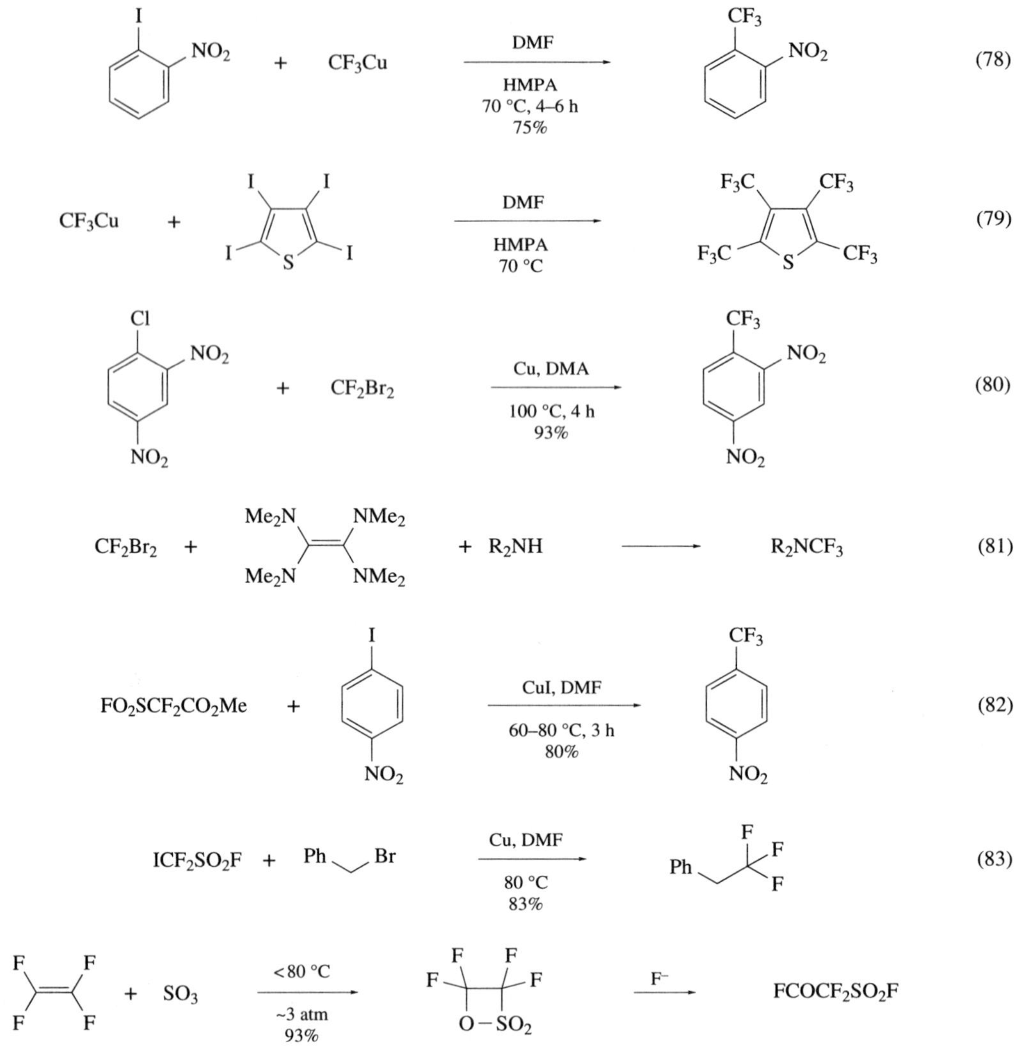

Scheme 5

6.01.2.3 Derivatives of Alkanes, Alkenes, Alkynes and Other Saturated Compounds

6.01.2.3.1 Halogen exchange

These reactions ⟨63AFC(3)181, B-89MI 601-01, 41JA3478, 47JA1820⟩ involve, essentially, the nucleophilic displacement of other halogens by fluorine using a metal fluoride ⟨B-73MI 601-01, B-76MI 601-03⟩, frequently in combination with hydrogen fluoride. An important part of the function of the metal fluoride is to act as a Lewis acid, to assist the removal of the halogen as a halide ion. Some of the reactions could involve carbocations (Scheme 6).

As the number of fluorine atoms attached to the chlorine-bearing carbon increases, it becomes progressively more difficult to effect further fluorination (Equation (84) and (Scheme 7)). Nevertheless, the inhalation anaesthetic $CF_3CHBrCl$ may be obtained by an exchange process (Scheme 8) ⟨57BRP767779⟩. For a carbocation-type process, it is understandable that the introduction of unsaturated sites makes such exchange reactions much easier (Equations (85) ⟨41JA3478⟩ and (86) ⟨47JA1820⟩). Fluorination of hexachlorobutadiene provides an accessible route to hexafluoro-2-butyne (Equations (86) and (87)) ⟨49JA298⟩. Functional derivatives are also accessible by these procedures, for example halo-ethers and -thioethers (Equation (88)) ⟨52JA3594⟩ as well as hexafluoroacetone, which is made commercially by this route (Equation (89)) ⟨64FRP1369784⟩. Antimony fluorides and various other fluorides have been used to promote exchange with hydrogen fluoride,

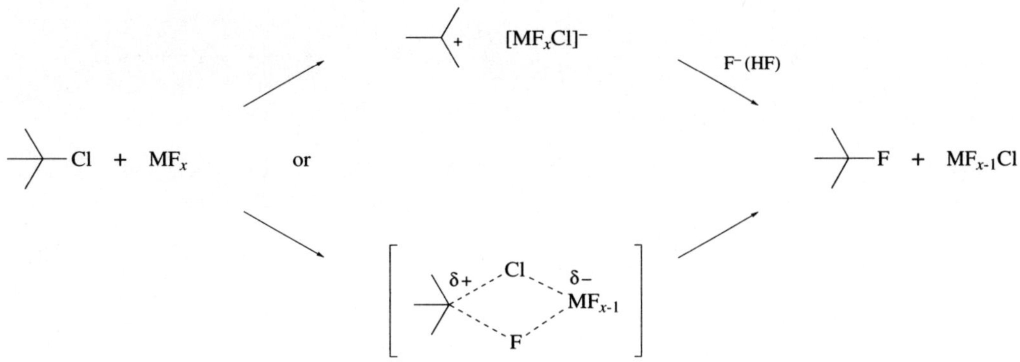

Scheme 6

and very efficient catalysts have been developed by industry, including chromia catalysts for use in vapour phase processes; these are particularly important in the refrigerant industry (see Section 6.01.1.3).

$$CCl_4 \xrightarrow[110\ ^\circ C,\ 3\ atm]{HF,\ SbCl_3,\ Cl_2} \underset{9\%}{CCl_3F} + \underset{90\%}{CCl_2F_2} + \underset{0.5\%}{CClF_3} \qquad (84)$$

$$CCl_3CCl_3 \longrightarrow CCl_3CCl_2F \longrightarrow CCl_2FCCl_2F \longrightarrow CCl_2FCClF_2 \longrightarrow CClF_2CClF_2 \longrightarrow CF_3CClF_2$$

Scheme 7

$$CCl_3Me \xrightarrow{HF,\ SbF_5} CF_3Me \xrightarrow{Cl_2} CF_3CH_2Cl \xrightarrow[425\text{–}475\ ^\circ C]{Br_2} CF_3CHBrCl$$

Scheme 8

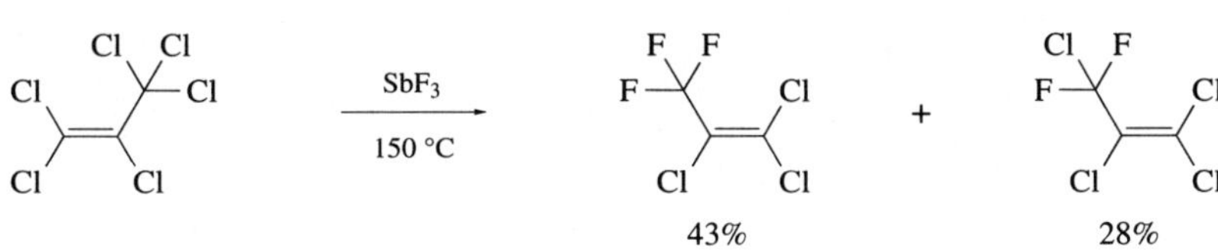

(85)

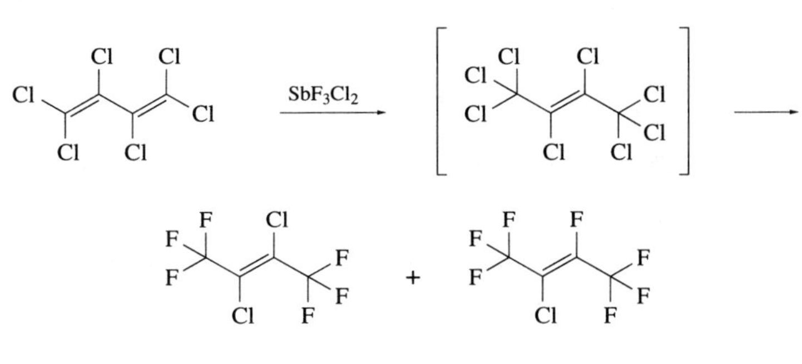

(86)

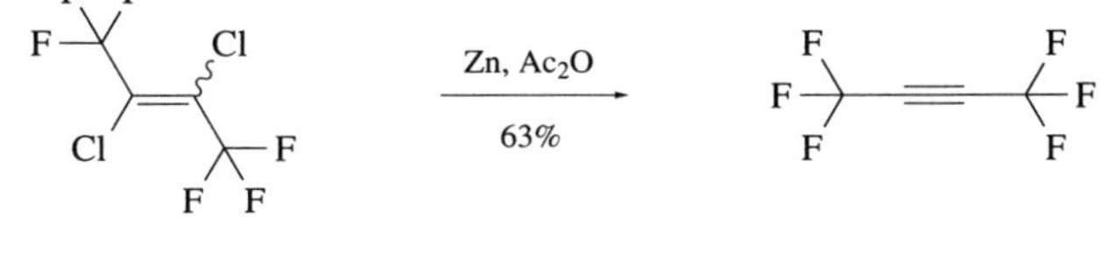

(87)

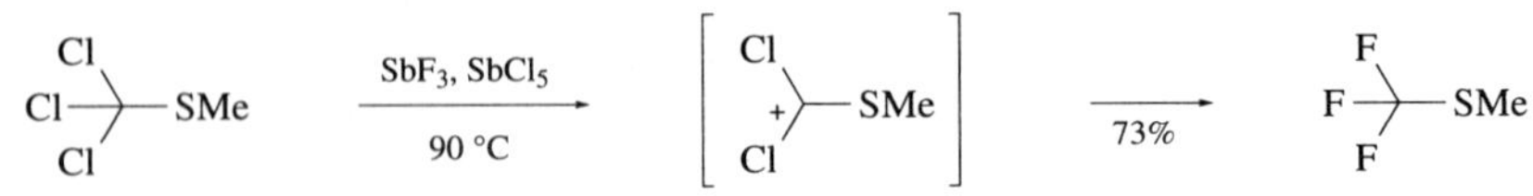
(88)

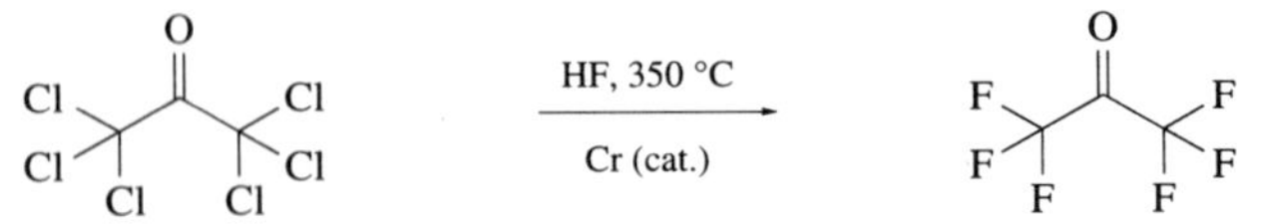
(89)

Alkali metal fluorides, although easily used, have limited application for saturated sites (Equation (90)) ⟨63JOC112⟩, but such systems are extremely effective for attack at unsaturated sites (Equation (91)), where even lithium fluoride, normally the least reactive of the alkali metal fluorides, can give some fluorination (Equation (92)) ⟨59JA2078, 60JA3091⟩. These reactions probably involve nucleophilic attack with allylic rearrangement, and other metal fluorides will also promote this process (Equation (93)) ⟨60BRP823519⟩. Terminal fluorinated alkenes are rearranged to the thermodynamically more stable isomers by the fluoride ion (Equation (94)) ⟨B-73MI 601-02⟩. Furthermore, hexafluorobutadiene is converted into hexafluoro-2-butyne by fluoride ion (Equation (95)) ⟨61JA1767⟩ and, more surprising, hexafluorocyclobutene also gives hexafluoro-2-butyne when passed over hot caesium fluoride or potassium fluoride (Equation (96)) ⟨79CC964⟩. An even more remarkable rearrangement induced by hot metal fluorides involves conversion of perfluorobicyclobutylidene to a product containing three trifluoromethyl groups (Equation (97)).

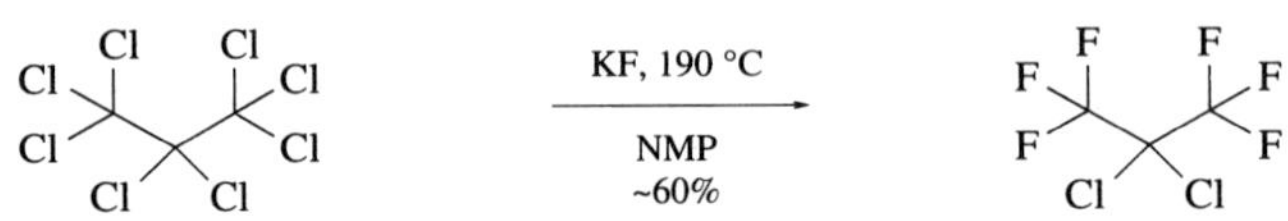
(90)

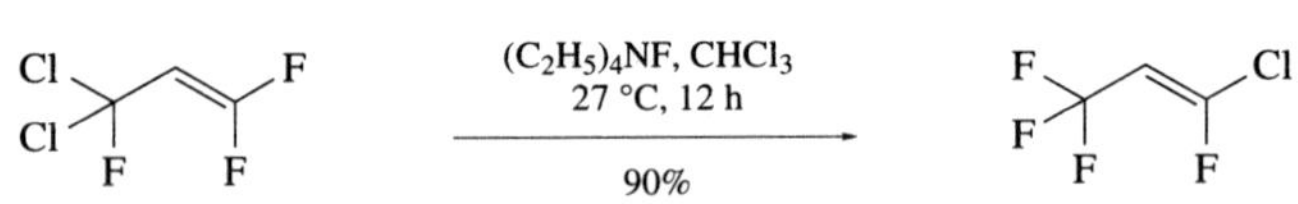
(91)

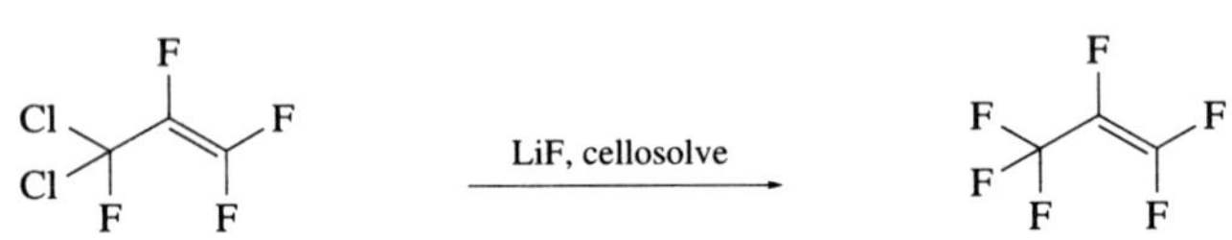
(92)

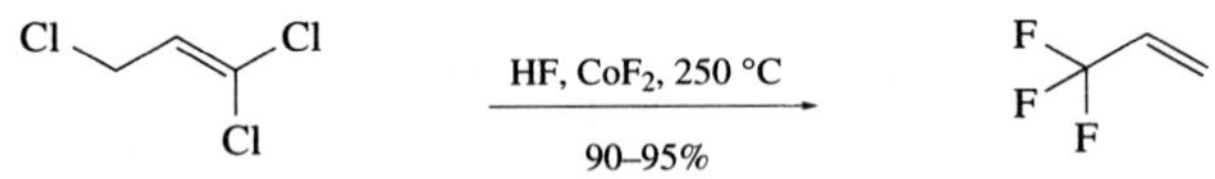
(93)

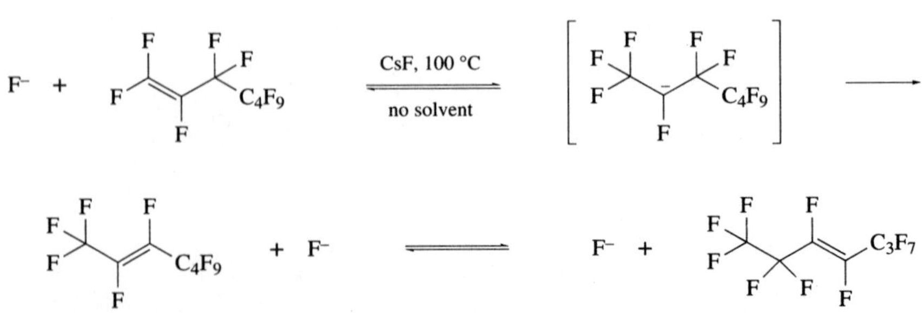
(94)

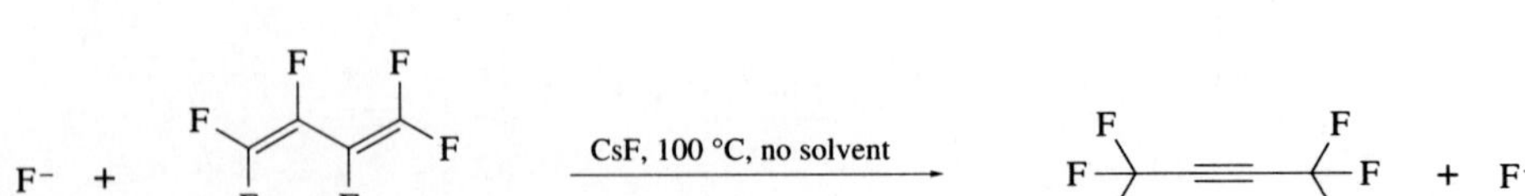
(95)

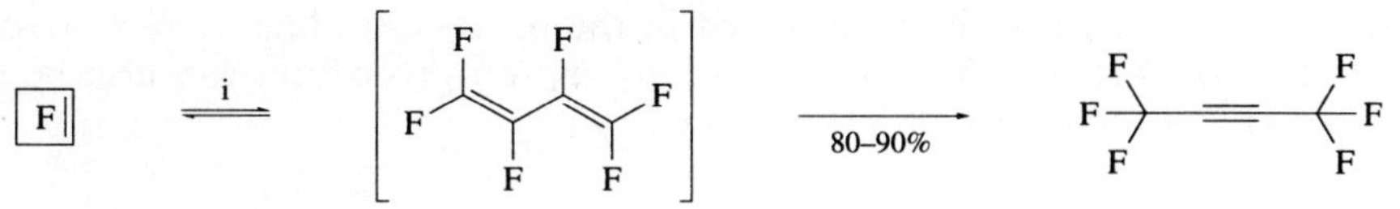

(96)

i, CsF or KF, 510–590 °C, flow system in N_2

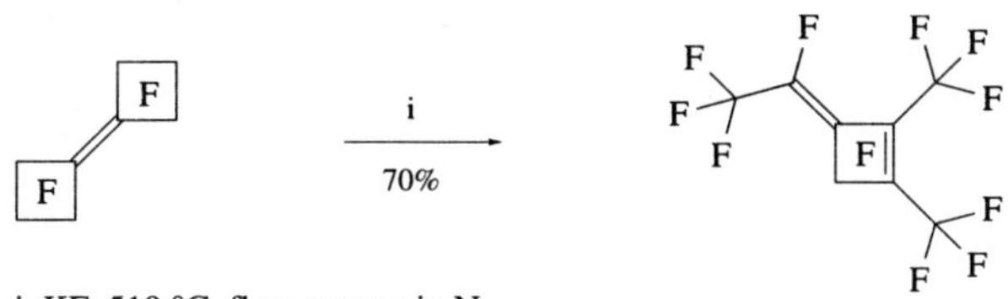

(97)

i, KF, 510 °C, flow system in N_2

6.01.2.3.2 Conversions of other groups to the trifluoromethyl group

Use of sulfur tetrafluoride ⟨60JA543⟩ to convert the carboxyl group to the trifluoromethyl group proceeds well, except where formation of anhydrides is in competition (Equations (98) and (99)) ⟨78PJC71⟩ and in some cases other functional groups present also react preferentially (Scheme 9) ⟨B-89MI 601-02⟩. Treatment of fluorinated enols with DAST gives trifluoromethyl derivatives with a high degree of regio- and stereoselectivity (Equation (100)) ⟨91TL5963⟩. An interesting *in situ* electrophilic formylation and subsequent reaction with sulfur tetrafluoride has been reported (Equation (101)) ⟨82ZOR228⟩.

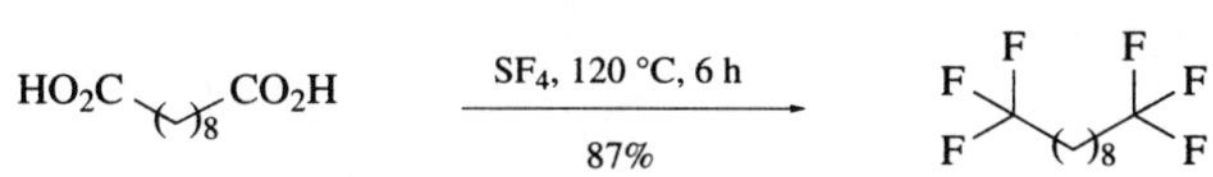

(98)

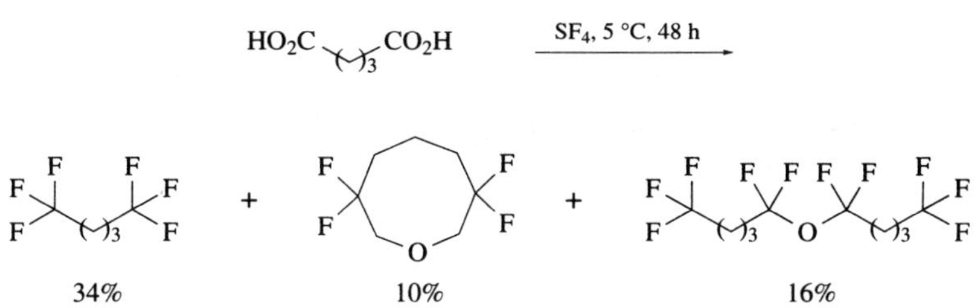

(99)

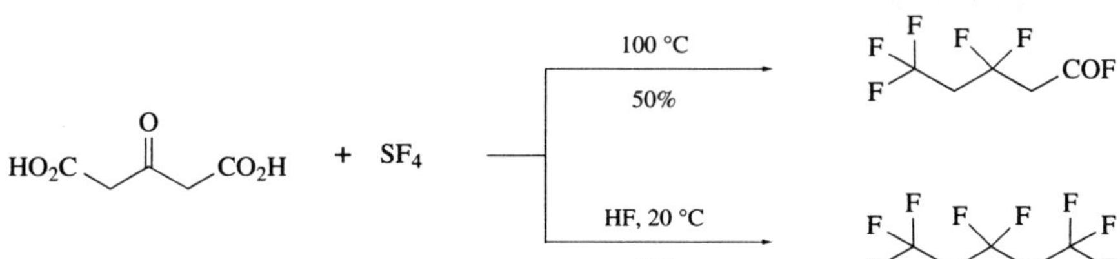

Scheme 9

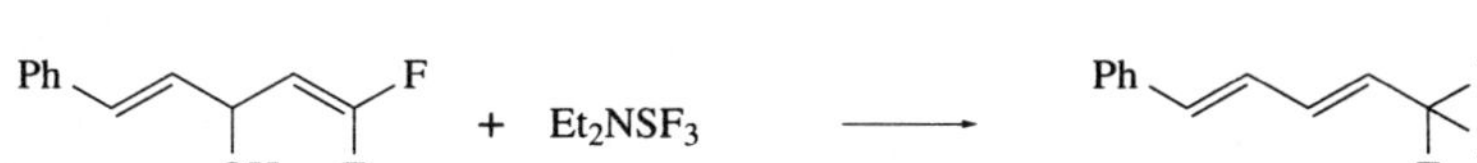

(100)

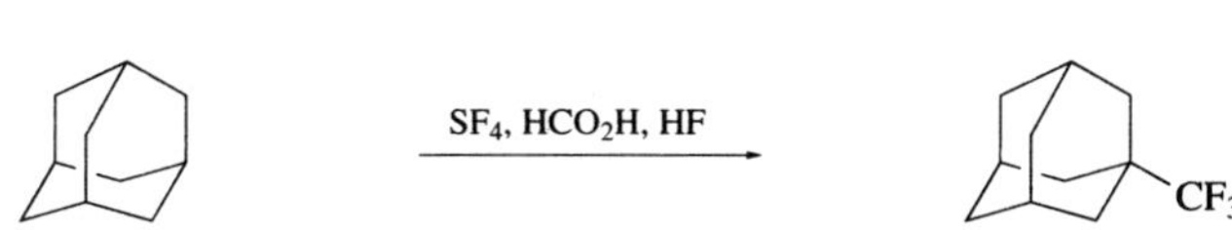

(101)

Although halogen exchange to form the trifluoromethyl group proceeds well (see above), procedures to convert the methyl group to the trifluoromethyl moiety directly are limited. High-valency metal fluorides, for example cobalt trifluoride, will carry out this conversion at high temperatures, but the rest of the molecule is also fluorinated in the process, leading to very useful inert fluids (Equation (102)) ⟨60AFC(I)166⟩. Various direct fluorination procedures can also be used for such transformations (Equation (103)) ⟨B-73MI 601-03⟩.

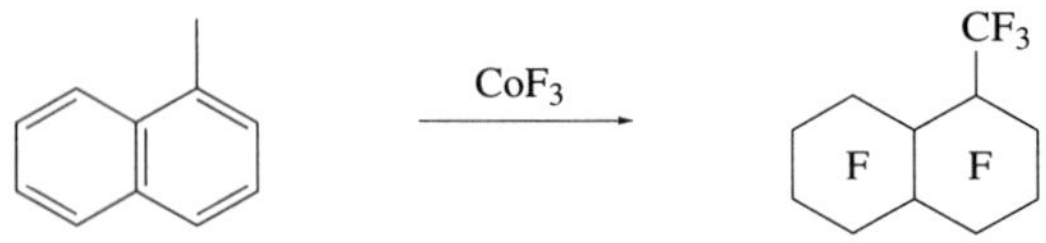

(102)

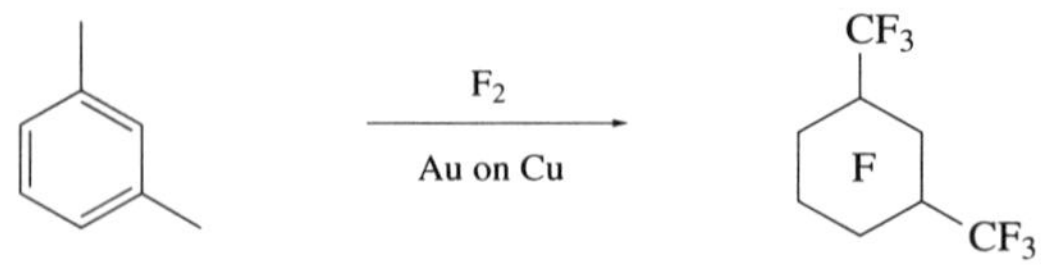

(103)

Electrochemical fluorination (ECF) is a procedure for converting C—H bonds to C—F bonds at the anode of an electrochemical cell during the electrolysis of hydrogen fluoride ⟨67FCR77⟩ under conditions that do not generate elemental fluorine. The process is operated on an industrial scale and is particularly effective for polar systems, for example the synthesis of trifluoromethanesulfonic acid (Scheme 10) and other functions (Equation (104) and Scheme 11). An interesting conversion of the amino group to the trifluoromethyl group involves conversion to an azo derivative, using nitrosotrifluoromethane, followed by photolytic elimination of nitrogen (Scheme 12) ⟨77AG(E)854⟩.

$$MeSO_2F \xrightarrow[96\%]{ECF} CF_3SO_2F \xrightarrow{H_2O} CF_3SO_2OH$$

Scheme 10

$$CS_2 \xrightarrow[90\%]{ECF} CF_3SF_5 \qquad (104)$$

$$MeCOF \xrightarrow{ECF} CF_3COF \xrightarrow{H_2O} CF_3CO_2H$$

Scheme 11

$$C_8H_{17}NH_2 + CF_3NO \longrightarrow C_8H_{17}N{=}NCF_3 \xrightarrow[69\%]{h\nu} C_8H_{17}CF_3$$

Scheme 12

6.01.2.3.3 Transfer of trifluoromethyl groups as radicals

One of the most effective and efficient procedures for transferring trifluoromethyl radicals involves the use of iodo- or bromotrifluoromethane for addition to an unsaturated unit (see Section 6.01.1.2). These are generally radical chain reactions, and they proceed with very high efficiency when a relatively nucleophilic centre is involved, and a range of catalytic procedures have been developed, as illustrated in Table 2.

Electrolysis of trifluoroacetic acid is a very direct and attractive route for generation of trifluoromethyl radicals. Mixed Kolbé processes lead to trifluoromethyl derivatives by combination of radicals (Equation (105)) ⟨75CJC529⟩, although the process is not efficient. Similarly, addition of trifluoromethyl radicals, generated this way, to alkenes and derivatives leads to mixtures of products where dimerization of the radical arising from addition of the trifluoromethyl group can be a significant pathway (Equation (106)) ⟨74CC323, 78JCS(P1)202⟩. Products that arise from cyclization of the intermediate radical (Equation (107)) ⟨91T549⟩ and from di-addition have been observed (Equations (108) ⟨79CJC2617⟩ and (109) ⟨89TL109⟩). Enzymes have been used to promote radical additions

Table 2 Addition of halofluoromethanes to alkenes.

Reactants	*Catalyst/conditions*	*Product*	*Yield* (%)	*Ref.*
$CH_2=CHSiMe_3 + CF_3I$	$Ru_3(CO)_{12}$	$CF_3CH_2CHISiMe_3$	89	84TL303
$CH_3CO(CH_2)_8CH=CH_2 + CF_3I$	Et_3B	$CH_3OCO(CH_2)_8CHICH_2CF_3$	86	89TL3159
$CH_2=CHC_6H_{13} + CF_3Br$	NaSePh	$CF_3CH_2CH(SePh)C_6H_{13}$ + CF_3SePh + PhSeSePh	26, 34, 31	91TL375, 91TL7425
$CH_2=CHR + CF_3I$	$Na_2S_2O_4$ $NaHCO_3$	CF_3CH_2CHIR		B-94MI 601-01
$CH_2=CHR$ (R = Ph, CO_2Et, alkyl) + CF_3SO_2Cl	$Ru(PPh_3)_3Cl_2$	CF_3CH_2CHClR		89CC1559

(Equation (110)) ⟨88JOC2350⟩, and it is probable that these are reacting as electron transfer agents, in the manner that various other transition metal systems behave. Reactions of plasma-generated radicals can be very efficient (Scheme 13) ⟨79CC128⟩.

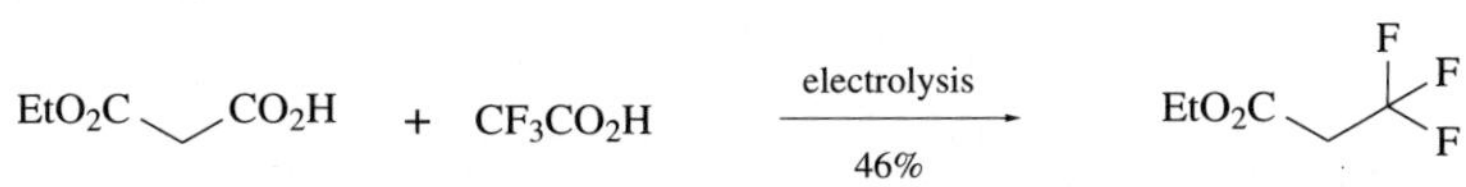

(105)

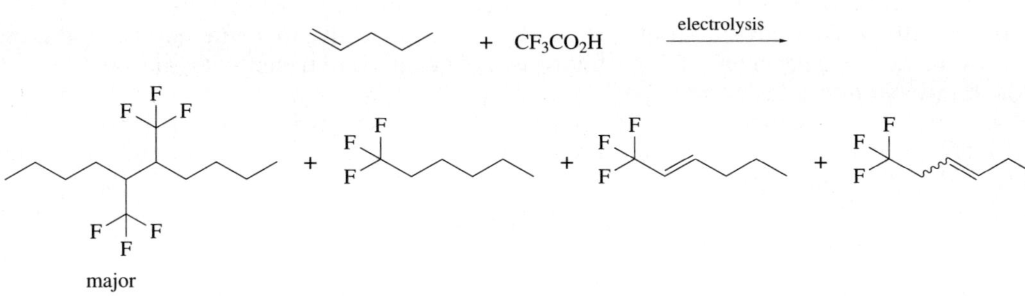

(106)

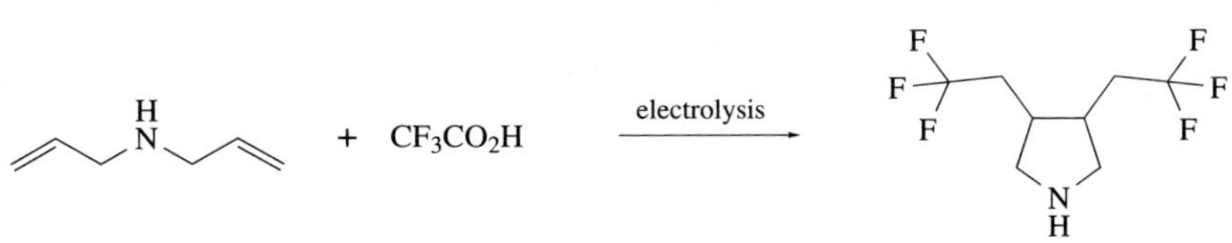

(107)

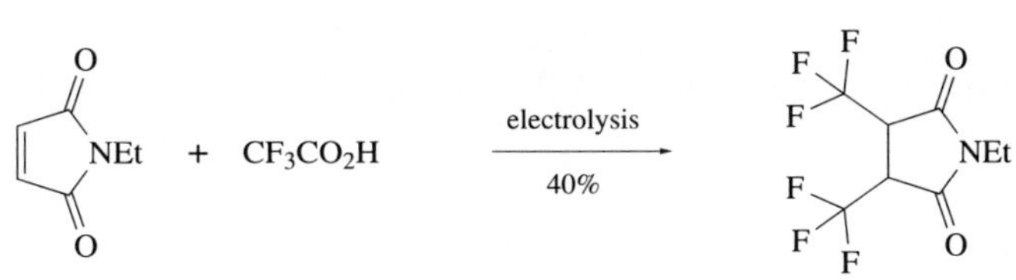

(108)

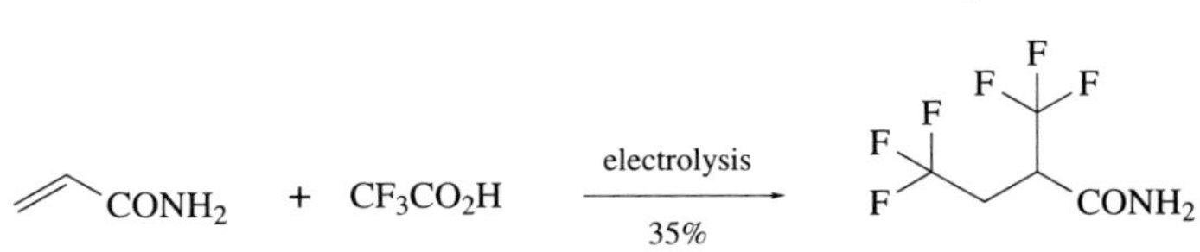

(109)

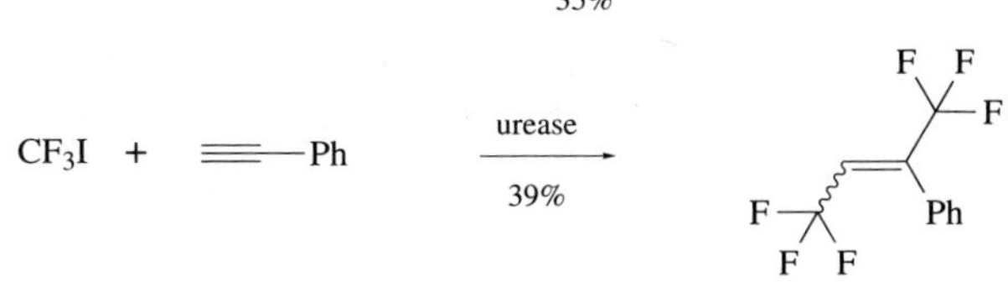

(110)

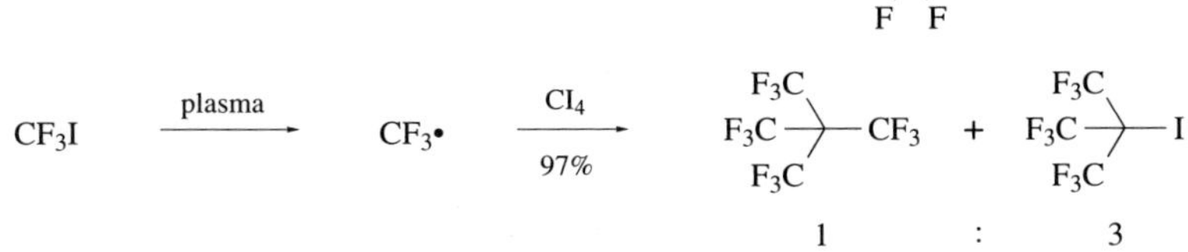

Scheme 13

6.01.2.3.4 Reactions involving trifluoromethyl derivatives of metals and metalloids

Electrophilic trifluoromethylating agents (see Section 6.01.2.2.3) can be applied to transferring the trifluoromethyl group to an alkyne (Equation (111)) ⟨90TL3579⟩, and acidic sites derived from alkyl pyridines may be trifluoromethylated using triflic anhydride, together with formation of corresponding triflates (Equation (112)) ⟨83JOC1776⟩.

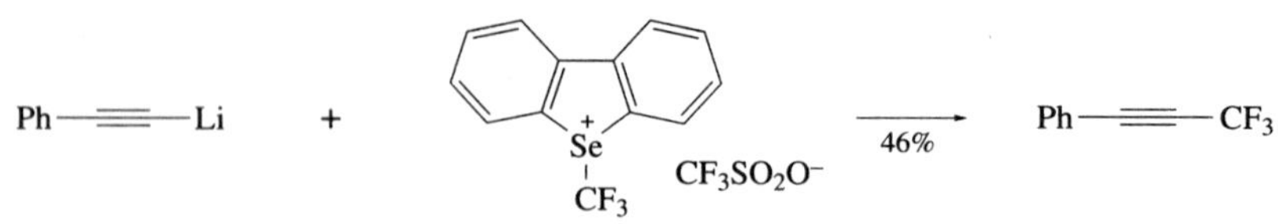

(111)

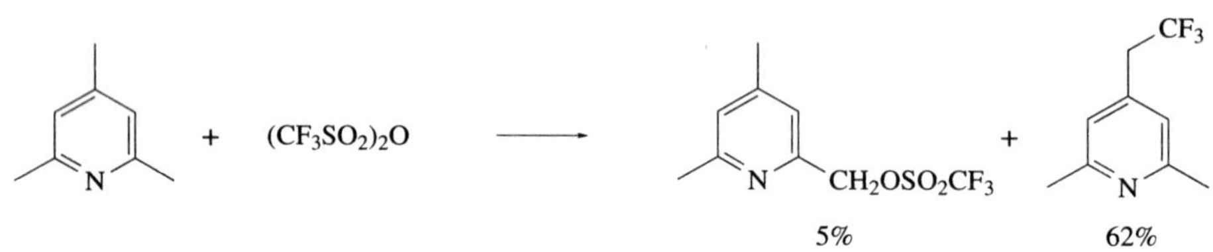

(112)

Nucleophilic trifluoromethylation has been carried out directly with a system that is highly susceptible to nucleophilic attack (Equation (113)) ⟨90IZV491⟩, although the nature of the intermediate is not clear. Displacement of the trifluoromethyl group from trimethyltrifluoromethylsilane, using the fluoride ion (see Section 6.01.2.2.4), is applicable to a range of systems (Scheme 14) ⟨B-92MI 601-01⟩, and, used in conjunction with copper(I) salts, trimethyltrifluoromethylsilane functions as a trifluoromethylcopper reagent which will trifluoromethylate aryl, benzyl and allyl halides (Equation (114)) ⟨91TL91⟩.

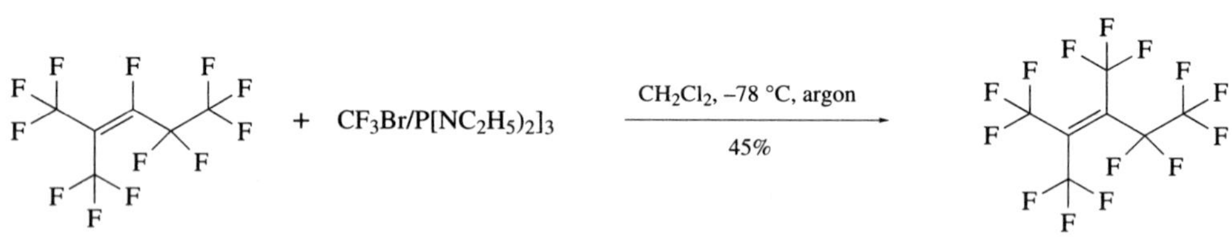

(113)

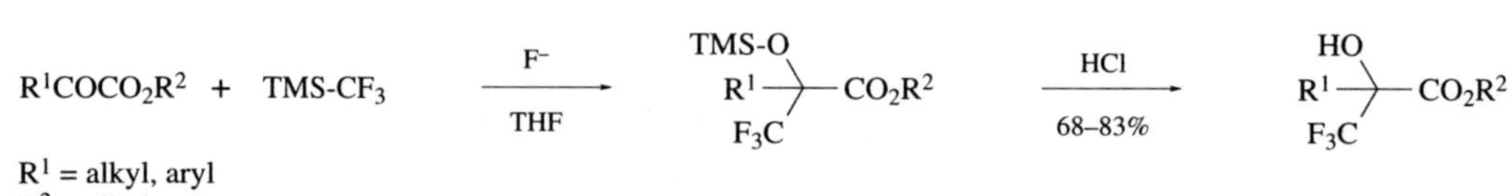

Scheme 14

$$R^1X + CF_3SiR^2_3 \xrightarrow[\text{DMF, 60–80 °C; 23–94\%}]{\text{KF, Cu(I)}} R^1CF_3 \qquad (114)$$

Trifluoromethylcopper reagents, made by other routes, react with various substrates (Equations (115) ⟨92CC53⟩, (116) ⟨79TL4071⟩ and (117) ⟨82CL1453⟩), and zinc reagents are also useful ⟨92T189⟩ in a variety of contexts.

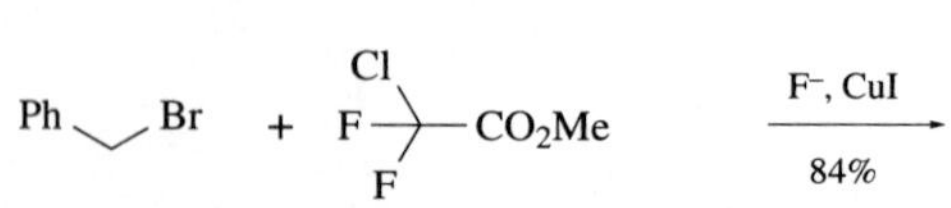

(115)

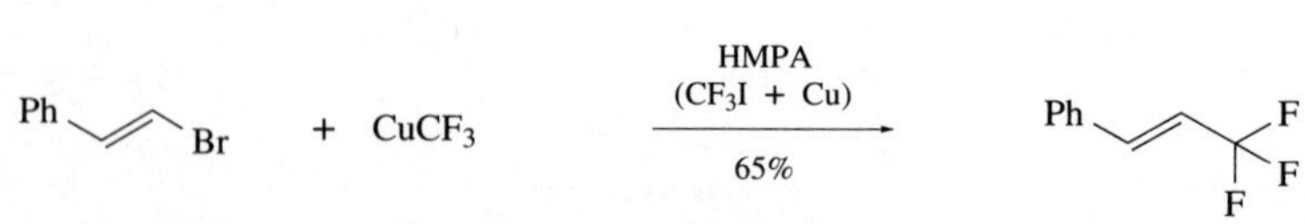

(116)

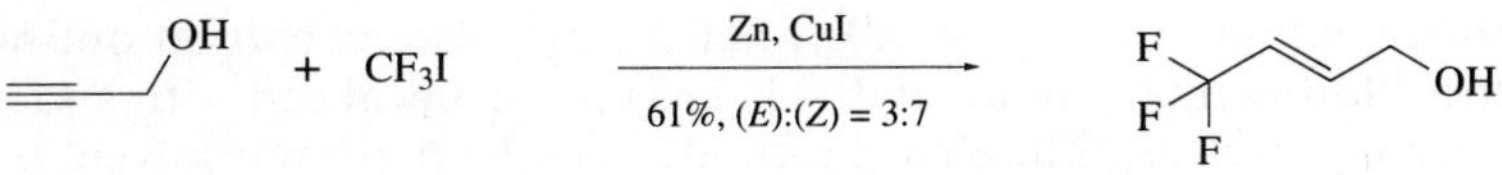

(117)

6.01.3 TRICHLOROMETHYL DERIVATIVES—$RCCl_3$

6.01.3.1 Trichloromethyl Groups Attached to an Aliphatic Centre

Trichloromethyl compounds can be prepared either by converting an existing group into a trichloromethyl group or by introducing a trichloromethyl group on to a carbon atom in an existing molecule (see also Section 6.01.1).

6.01.3.1.1 Conversion of groups attached to an aliphatic centre into the trichloromethyl group

As with chlorofluorocarbons, the manufacture of many chlorinated solvents is being phased out. Of these, only 1,1,1-trichloroethane is relevant to this chapter. Several routes are available, and the choice, which makes use of permutations of chlorination, hydrochlorination and dehydrochlorination, is outlined in Scheme 15. Hydrochlorination is usually effected by Lewis acid catalysis (e.g., $FeCl_3$), and dehydrochlorination by thermal cracking or by the action of base. Chlorination is catalysed by free radical sources in the liquid or gas phase. The route chosen depends in part on how much hydrogen chloride the operator is prepared to make.

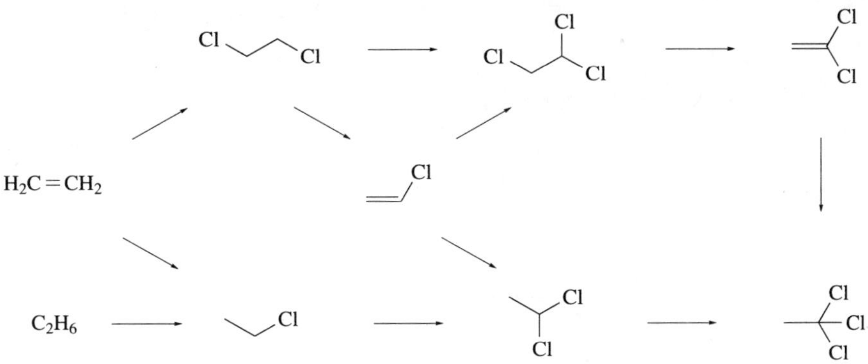

Scheme 15

6.01.3.1.2 Transfer of the trichloromethyl group to an aliphatic centre

(i) Preparation of trichloromethylalkanes

Trichloromethyl anions are an unstable species and decompose to give dichlorocarbenes. However, the problem of their instability can be minimized by running reactions at low temperature and by generating the trichloromethyl anion in the presence of the substrate. Trichloromethyllithium, which can be made at about $-100\,^{\circ}C$ from trichloromethane and butyllithium in THF, reacts successfully with iodoalkanes at this same temperature to give a series of trichloromethyl alkanes (Equation (118)) ⟨90JOC1281⟩.

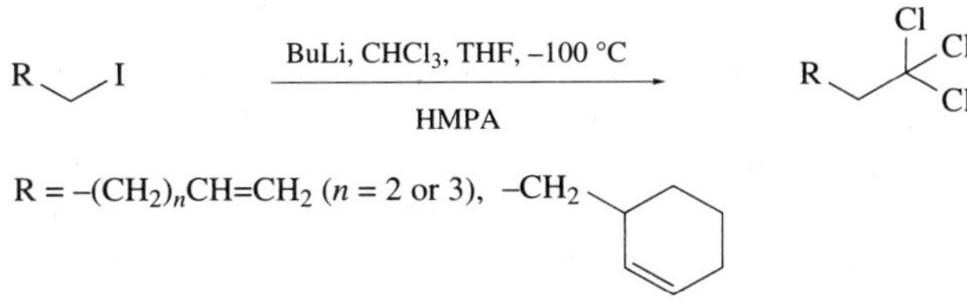

(118)

The trichloromethylation of activated alkyl halides can also be carried out by a cross-coupling reaction with tetrachloromethane in an undivided electrochemical cell with a sacrificial anode. The preferred anode is zinc, with a stainless steel cathode, and the preferred solvent is a mixture of THF and tetramethyl urea (Scheme 16) ⟨88TL1699⟩.

Anode $\quad M \longrightarrow M^{n+} + ne^-$

Cathode $\quad CCl_4 + 2e^- \longrightarrow CCl_3^- + Cl^-$

$CCl_3^- + RX \longrightarrow CCl_3R + X^-$

RX = $PhCH_2Br$, $MeCHClOCO_2Et$, $BrCH_2CO_2Me$, $Me(CH_2)_9Br$

Scheme 16

Trichloromethylated compounds can also be prepared by the action of trichloromethyl anions generated by the action of a base on trichloromethane. Thus, by using catalytic amounts of electrochemically generated base, a relatively stable anion is produced which reacts with α,β-unsaturated esters and nitriles to give the corresponding β-trichloromethyl adducts in good yield. The base is made by the electroreduction of 2-pyrrolidone in DMF using tetraalkylammonium salts as supporting electrolytes (Scheme 17) ⟨90TL7181⟩.

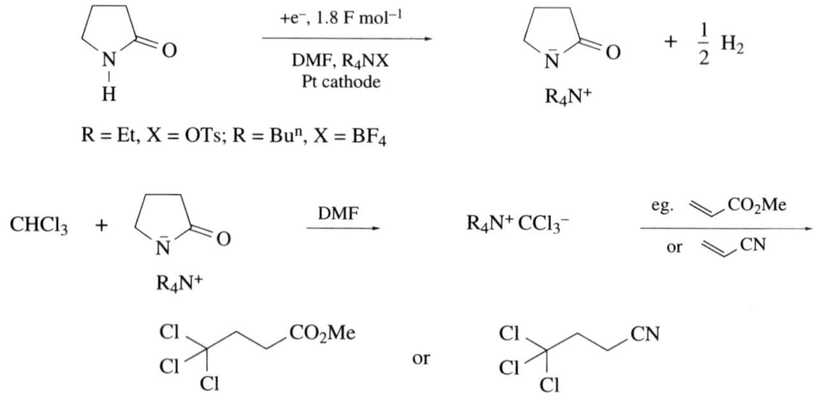

Scheme 17

β-(Trichloromethyl)nitroalkanes can be prepared in good yield by the reaction between trichloromethyl anions, generated from trichloromethyltrimethylsilane and caesium fluoride (see also 6.01.1.2), and nitroalkenes (Equation (119)) ⟨91SC2189⟩.

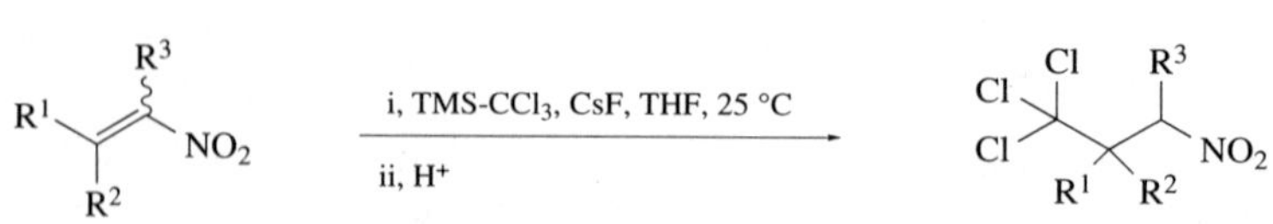

(119)

R^1	R^2	R^3
Buⁿ	Me	H
Ph	H	H
Me	H	Et
Et	H	H
Me	Me	Me

Closely related to the trichloromethyl radical additions outlined in Section 6.01.1.2.3 is the reaction of trichloromethylsufonyl chloride with 1-alkenes in the presence of ruthenium complexes such as dichlorotris(triphenylphosphine)ruthenium(II). In this reaction, the elements of CCl_3 and chlorine are added to the double bond in high yield under mild conditions with the extrusion of sulfur dioxide ⟨83SUL131⟩. When the chiral $Ru_3Cl_4(diop)_3$ is the catalyst, chiral adducts are obtained ⟨87BCJ3687⟩.

(ii) Preparation of α-trichloromethyl carbinols

α-Trichloromethyl carbinols are valuable intermediates for making compounds in which the trichloromethyl group is preserved and also for making those in which the trichloromethyl group has been transformed into other functions (e.g., ⟨71S131, 82TL1609, 88M1427, 89S466, 91RCR1318, 92TL3435⟩). Essentially, their preparation comprises the formation of a trichloromethyl anion followed by its attack on an aldehyde or ketone (Scheme 18).

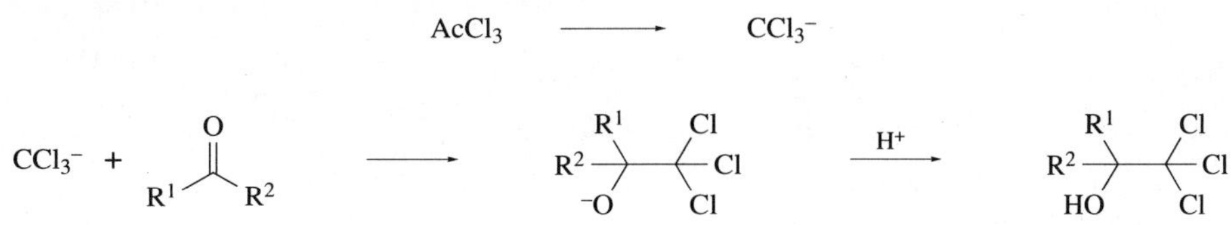

Scheme 18

Carbonyl compounds have been treated successfully with trichloromethyllithium, but because the reactions have to be carried out at −100 °C, only the more reactive carbonyls can be used ⟨64AG(E)513, 67JOC1217, 88M1427, 90JOC1281⟩. Reactions can be carried out at higher temperatures (−78 °C) if lithium dicyclohexylamide is used to generate the trichloromethyllithium from trichloromethane in the presence of the carbonyl (Equations (120)–(123)). Equation (123) gives equally good results with tribromomethane in place of trichloromethane ⟨74JA3010⟩.

$$Ph_2CO \xrightarrow[\text{ii}]{\text{i}} Ph_2COHCCl_3 \qquad (120)$$

i, Bu^nLi, $CHCl_3$, −110 °C, THF/Et_2O, petroleum ether; ii, H^+

$$(CF_3)_2CO \xrightarrow[\text{ii, } H^+]{\text{i, } Bu^nLi,\ CHCl_3,\ -100\ °C,\ THF} (CF_3)_2COHCCl_3 \qquad (121)$$

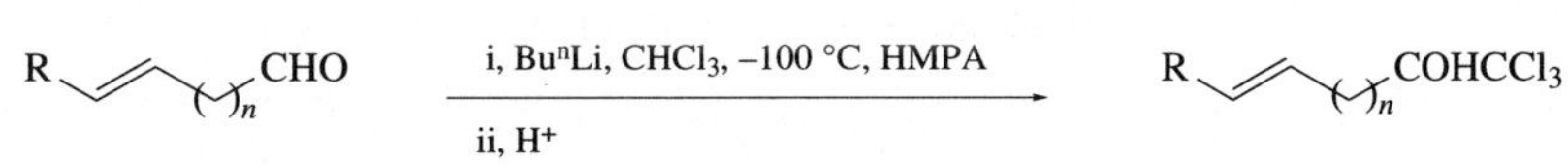

(122)

R	n
H	2
H	3
Me	2

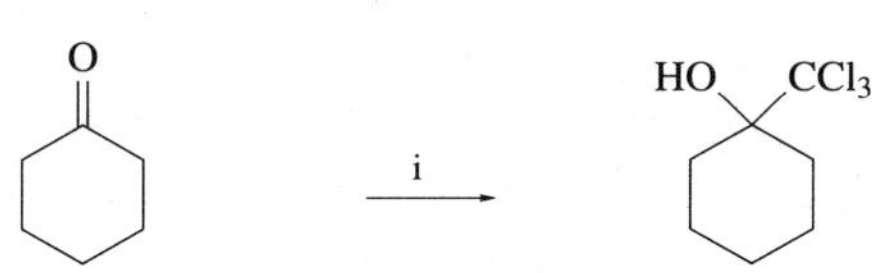

(123)

i, Lithium dicyclohexylamide, $CHCl_3$, −78 °C, THF, hexane

Trichloromethane and a base have been used extensively to generate the trichloromethyl anion. The base may be sodium or potassium hydroxide ⟨44JCS74, 50JA5012, 83M813, 87JOC944⟩, an alkali metal alkoxide ⟨58CB2664, 64JOC1148⟩, or potassium fluoride supported on alumina ⟨86TL3845⟩. The amount of base, its nature and also the choice of solvent are particularly important when the carbonyl compound is inclined to undergo an aldol condensation or the Cannizzaro reaction. Reactions are generally carried out below 0 °C to help suppress dichlorocarbene formation. Phase transfer catalysts have been used in these reactions, but they were only really successful with higher aldehydes and ketones ⟨80JOC5214, 79TL1473⟩.

The thermal decomposition of sodium trichloroacetate in a solvent yields dichlorocarbene ⟨59PCS229⟩. However, the intermediate trichloromethyl anion can be trapped by aldehydes to yield trichloromethyl carbinols. Under similar conditions, anhydrides react to give trichloromethyl lactols or their tautomeric keto acids (Equation (124)) ⟨67JOC2166⟩. Trichloromethyl carbinols are also obtained in high yield when, in the presence of an aldehyde, trichloroacetic acid decomposes in DMSO ⟨83CC283, 84JCS(P2)1247, 89JCS(P2)251⟩ or HMPA ⟨90S327⟩, or a 1 : 1 mixture of the trichloroacetic acid and its sodium salt decomposes in DMF ⟨92TL3435⟩. These reactions are successful with aliphatic as well as aromatic aldehydes. They are less successful with ketones unless the ketone is (partially) fluorinated and therefore more electrophilic, in which case the yields are high ⟨92BAU370⟩.

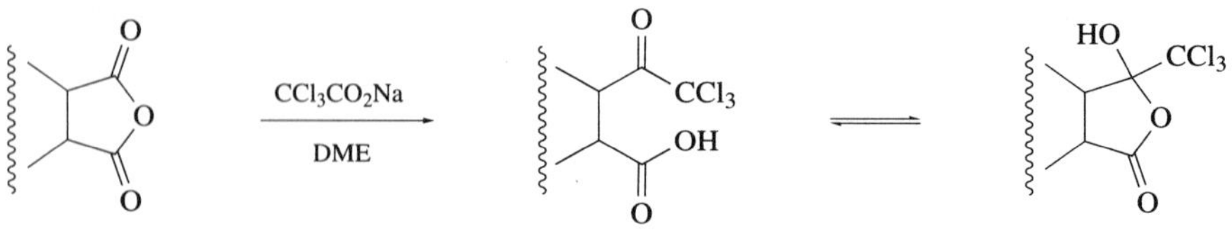

(124)

Trichloromethyl compounds of silicon and tin have been used as sources of trichloromethyl anions in reactions with carbonyl compounds. For example, the reaction of trichloromethyltrimethylsilane and trimethylsilyl trichloroacetate in the presence of the fluoride ion with aromatic and aliphatic aldehydes ⟨85JA4085, 87SC1047, 88T4135⟩, trimethylsilyl trichloroacetate in the presence of potassium carbonate and 18-crown-6 with aromatic aldehydes and ketones ⟨85TL1175⟩, and tributyl(trichloromethyl)stannane with aliphatic and aromatic aldehydes ⟨75JOM(102)423⟩ all give the corresponding trichloromethylcarbinols in good yield (Equations (125)–(128)). Equation (128) proceeds with equal success to give tribromomethylcarbinols when tribromomethyltributyltin is used instead of the trichloromethyl tin compound.

(125)

R = Ph, n-$C_{10}H_{21}$, PhCH(Me); tas = tris(diethylamino)sulfonium

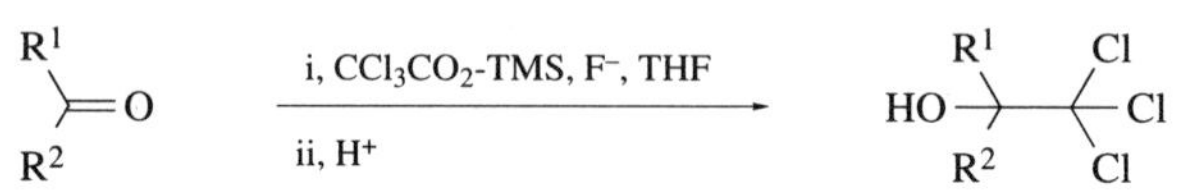

(126)

F = KF or tas-F

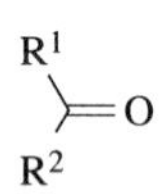

= cyclohexanone, benzaldehyde, pivaldehyde, 2-cyclohexenone, crotonaldehyde

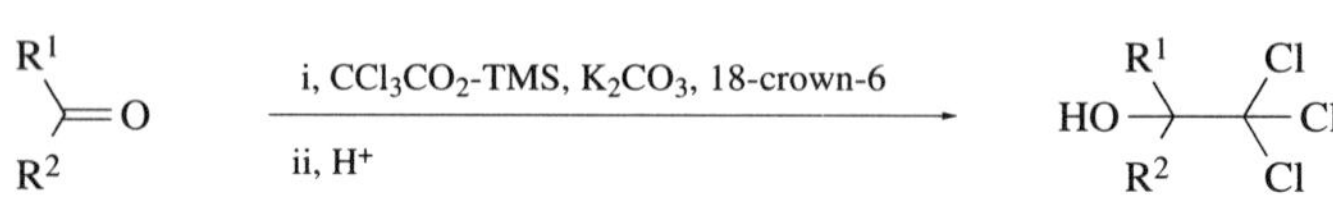

(127)

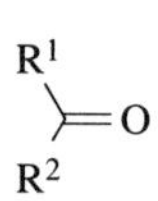

= e.g. benzaldehyde, 3,5-dichlorobenzaldehyde, 4-hydroxybenzaldehyde, cyclobutanone, cyclopentanone, cyclohexanone

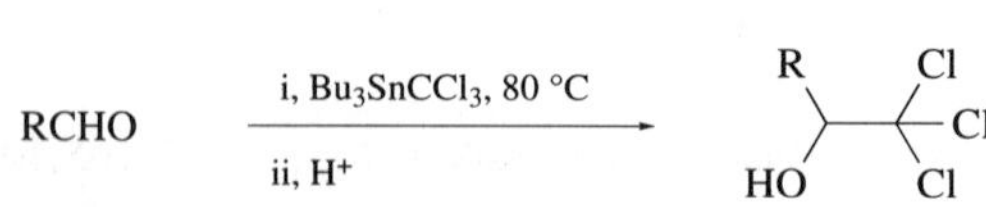

(128)

R = Me, Et, Pr^i, Ph, PhCH = CH

The reductive addition of tetrachloromethane to aldehydes using a lead/aluminum bimetal redox system has been reported to give high yields of trihalocarbinols. While a variety of aliphatic and aromatic aldehydes give good yields, ketones give less satisfactory results. The use of bromotrichloromethane instead of tetrachloromethane gives exclusively the same product (Scheme 19) ⟨89JOC444⟩.

A variety of aldehydes have been treated with trichloromethyl anions generated by the cathodic reduction of tetrachloromethane in trichloromethane. Only relatively small amounts of tetrachloromethane are used in this reaction, since it is the trichloromethane which supplies the proton to the intermediate anion and thereby generates another trichloromethyl anion—steps (ii) and (iii) in Scheme 20. The use of vinyl acetate instead of an aldehyde yields the acetate $MeCH(OAc)CCl_3$ ⟨81TL871, 82TL1609, 82TL4801⟩. These same workers oxidized trichloromethyl carbinols with chromic oxide to give the corresponding trichloromethyl ketones ⟨82TL1609⟩.

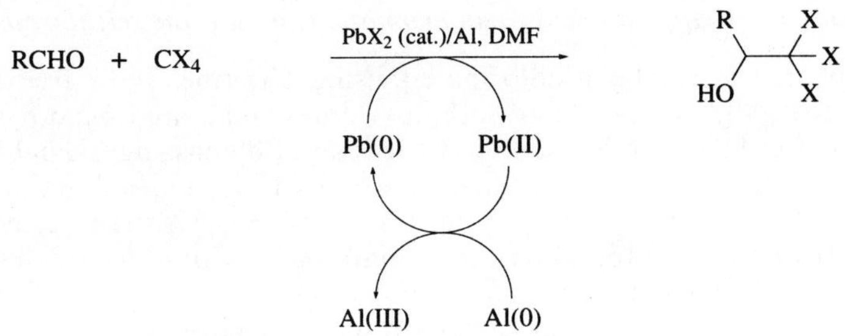

X = Cl or Br; R = alkyl, aryl

Scheme 19

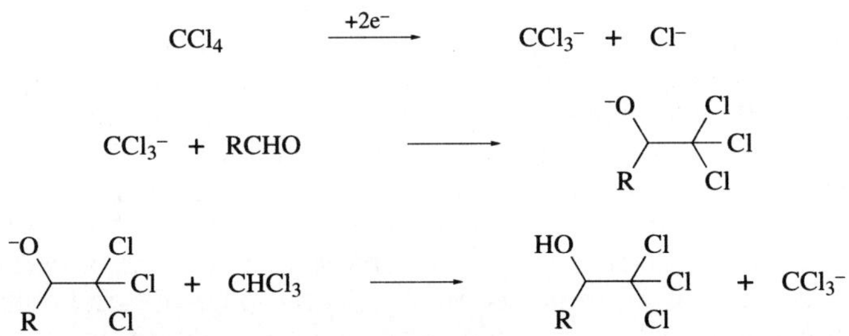

(i)

(ii)

(iii)

Scheme 20

The other approach to making trichloromethyl carbinols is to start with trichloroacetaldehyde (chloral). This compound will condense with aromatic hydrocarbons with Friedel–Craft catalysts (e.g., H_2SO_4, HCl, $AlCl_3$, BF_3 and $ZnCl_2$) ⟨23CB979, 42JA2515, 66CJC575, 68CJC2233, 80CJC485⟩ or under the influence of basic catalysts (e.g., potassium hydroxide, potassium carbonate and pyridine) ⟨56CB2578, 25JPR125, 52CB901⟩ to yield trichloromethyl carbinols. Certain Grignard reagents will react with chloral to give trichloromethyl carbinols, although reduction of the chloral to trichloroethanol can be a problem (Equation (129)) ⟨51M1008, 56BSF1441⟩. Finally, silanes, R-TMS, in which the group R is electron-withdrawing, have been shown to react with chloral in the presence of a Lewis acid to give the corresponding trichloromethyl carbinols (Scheme 21) ⟨75JOM(93)43⟩.

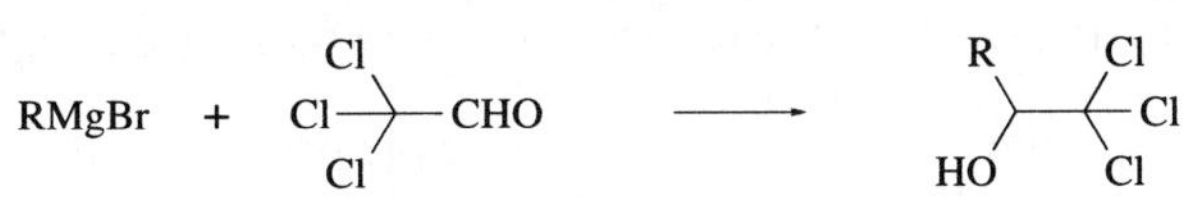

(129)

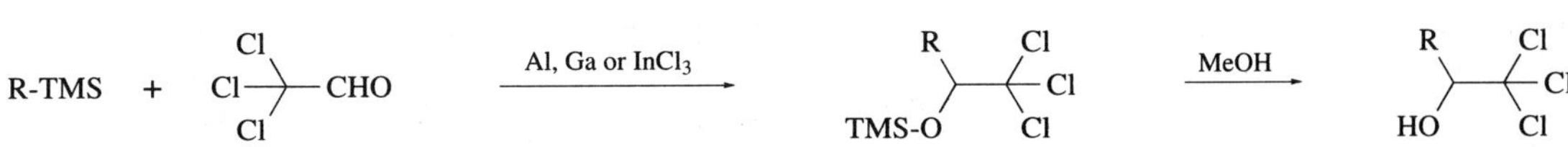

R = allyl, aryl, vinyl, ethynyl, propargyl

Scheme 21

6.01.3.2 Trichloromethyl Groups Attached to an Aromatic Ring

The high reactivity of trichloromethylarenes makes them valuable intermediates in the synthesis of acids and their derivatives, heterocyclic compounds and the trifluoromethyl group (which we have seen is of great interest to the pharmaceutical and agrochemical industries). Similarly to trichloromethyl aliphatic compounds, trichloromethylarenes may be made by converting a group which is already attached to a ring into a trichloromethyl group or by transferring a trichloromethyl group on to a ring.

6.01.3.2.1 Conversion of groups attached to an aromatic ring into the trichloromethyl group

Methyl groups in arenes can be trichlorinated using chlorine, by a free radical mechanism (Equation (130)) ⟨62HOU(5/3)735⟩. 2- and 4-Methylpyridines and 2- and 4-methylquinolines are also readily trichlorinated, but here reactions are best carried out by passing the halogen into a solution of the substrate in glacial acetic acid, optionally acetic anhydride, and sodium or potassium acetate, where the reaction is believed to proceed by an ionic process, with the acetate acting as a base ⟨23JCS2882, 39JCS781, 51JCS1145⟩. 3-Methylpyridine cannot be trichlorinated by this method, but when its vapour and chlorine are passed over a chromia catalyst, pretreated with hydrogen fluoride, 3-trichloromethylpyridine is produced ⟨82EUP65358⟩ in reasonable yield.

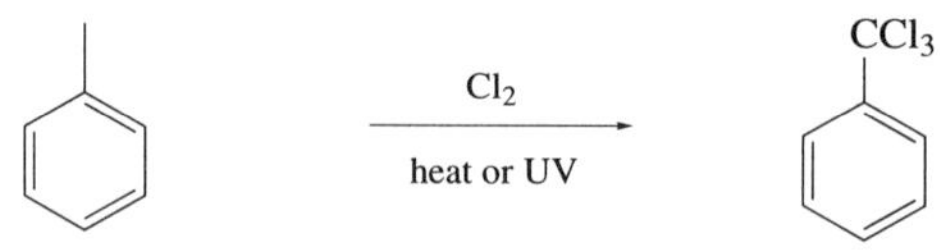

(130)

While these methods are fairly general, there are situations where other methods are necessary; for example, in the preparation of methylbenzotrichlorides where chlorination of dimethyl compounds leads to chlorination in both side chains, and in the preparation of compounds such as 4-nitrobenzotrichloride where the electron-withdrawing group inhibits chlorination. Thus, carboxylic acid groups in aromatic and heteroaromatic rings are converted into trichloromethyl groups by treatment with phosphorus pentachloride in thionyl chloride ⟨77JHC881, 78JHC893⟩, or with phosphorus pentachloride alone ⟨89JCS(P1)283⟩. This conversion can also be accomplished by treating acids with chlorine in a mixture of phosphorus trichloride and phenylphosphonic dichloride ⟨83USP4419514⟩ or phenyl dichlorophosphorane and phenylphosphonic dichloride ⟨89USP4833250⟩. Substituted benzal chlorides, which can be prepared from the corresponding aldehydes ⟨78JOC4367⟩, can be chlorinated further to trichloromethyl compounds by treatment with isobutyl hypochlorite in aqueous sodium hydroxide, optionally in the presence of an alcohol, and a phase transfer agent ⟨78USP4098831⟩. Substituted benzyl and benzal chlorides are also converted into the corresponding benzotrichlorides when they are treated with tetrachloromethane or hexachloroethane and aqueous sodium hydroxide in the presence of a phase transfer catalyst (PTC) by a process which probably involves nucleophilic attack on chlorine (Equation (131)) ⟨86S224⟩.

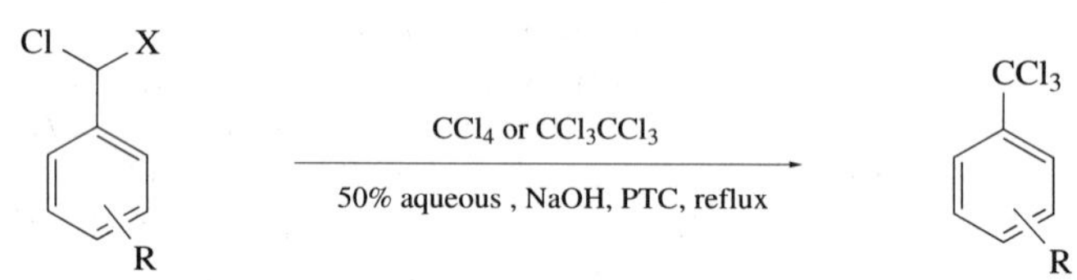

(131)

X = H; R = 2-NO_2, 2-NO_2, 3-Me
X = Cl; R = 2-NO_2, 2-NO_2, 3-Me, 2-Me, 3-Me, 4-Me, 2-Me, 2-Cl, 4-Et

A similar range of compounds is made by treating appropriate dibenzyl sulphides with chlorine or sulfuryl chloride (Equation (132)) ⟨82S951⟩; or by chlorinating with sulfuryl chloride the product obtained from treating an aldehyde with a dithiol in the presence of acid (Scheme 22) ⟨86USP4575565⟩. Dichlorine monoxide can be used to chlorinate methyl groups in methylnitrobenzenes and other methylbenzenes which have electron-withdrawing groups on the aromatic ring (Equation (133)) ⟨B-82MI 601-02⟩. While 4-nitrotoluene and highly deactivated compounds such as 2-chloro-4-nitro-, 3,4-dinitro- and 3,5-dinitrotoluene are converted into the corresponding benzotrichlorides in very high yield, 2-nitrotoluene is only dichlorinated by dichlorine monoxide to give the benzal chloride. (NB. This compound could be chlorinated further by methods outlined above.) In the case of the other substituted toluenes with deactivating groups in the 4 position, chlorination takes place exclusively in the methyl group with the exception of *p*-toluenesulfonic acid, which undergoes chlorination on the ring.

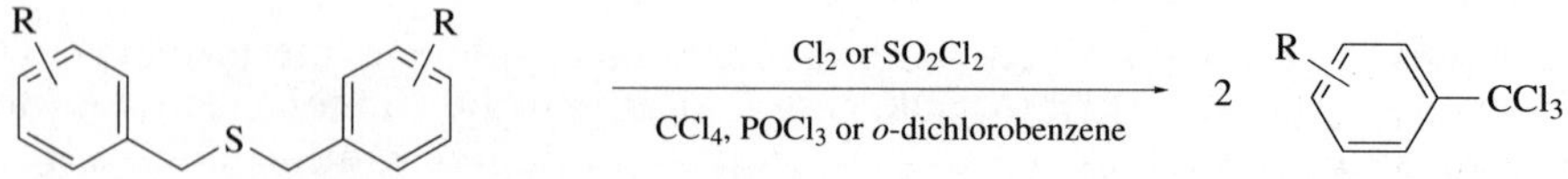

(132)

R = 2-Me[a], 3-Me[b,c], 4-Me[a], 2-Br[a], 4-NO_2[d,e], 2,3-CH=CH–CH=CH[a]
aCl_2/CCl_4, [b]Cl_2, [c]SO_2Cl_2/CCl_4, [d]Cl_2/$POCl_3$, [e]Cl_2/*o*-dichlorobenzene

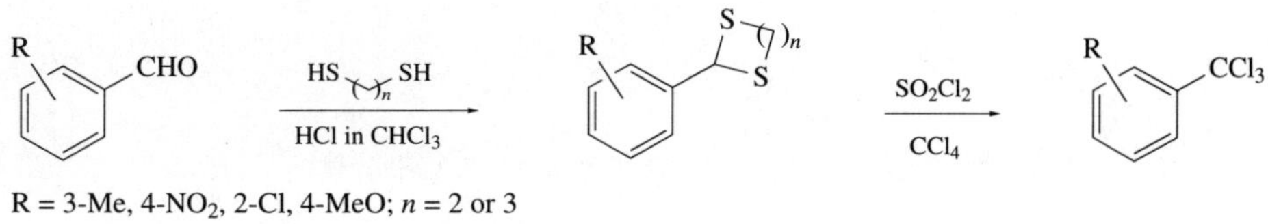

Scheme 22

(133)

6.01.3.2.2 *Transfer of the trichloromethyl group to an aromatic ring*

Aromatic compounds can be trichloromethylated by reaction with tetrachloromethane in the presence of excess Lewis acid, which is most commonly aluminum trichloride (e.g., ⟨87JPR1131⟩). Whilst the trichloromethyl compounds can be the main product, diaryldichloromethanes are also produced in significant amounts (Equation (134)), although it is claimed that the addition of mordenite can reduce dimer formation ⟨92JAP04149143⟩. The reaction is fairly general, but there are complications when polymethyl aromatics are trichloromethylated. Thus, while *m*-xylene gives the expected 2,4-dimethyltrichloromethylbenzene, a Jacobson-type rearrangement occurs in the reaction with *p*-xylene to give 2,4-dimethyltrichloromethylbenzene as the main product (*o*-xylene gives no trichloromethyl compounds but yields the dichloromethane, $(3,4\text{-}Me_2C_6H_3)_2CCl_2$) ⟨89CS81⟩. Similarly, the trichloromethylation of mesitylene gives the expected product ⟨61JA4460⟩, but during the trichloromethylation of pseudocumene ⟨89CS81⟩ both the expected 2,4,5-trimethyltrichloromethylbenzene and about 15% of the rearranged 2,4,6 isomer are produced (Scheme 23). Although mesitylene itself gives only the expected trichloromethylmesitylene, 1,3,5-trimethylhalobenzenes give significant amounts of the rearranged product (Scheme 24; in this scheme the products were isolated as the corresponding methyl benzoates) ⟨70JOC3637⟩. In the case of the tetramethylbenzenes (Scheme 25), durene gives almost exclusively the rearranged product, 2,3,4,5-tetramethyltrichloromethylbenzene (a small amount of disproportionation took place under the conditions used), whereas isodurene gives predominantly 2,3,4,6-tetramethyltrichloromethylbenzene, some of the 2,3,4,5 isomer and a trace of the 2,3,5,6 isomer ⟨61JA4460⟩. It was suggested that rearranged products in these reactions could arise by the trichloromethyl group attacking a ring carbon that is already substituted followed by a 1,2-methyl migration and proton loss, but rearrangements induced by acid, $HCl/AlCl_3$, are also possible.

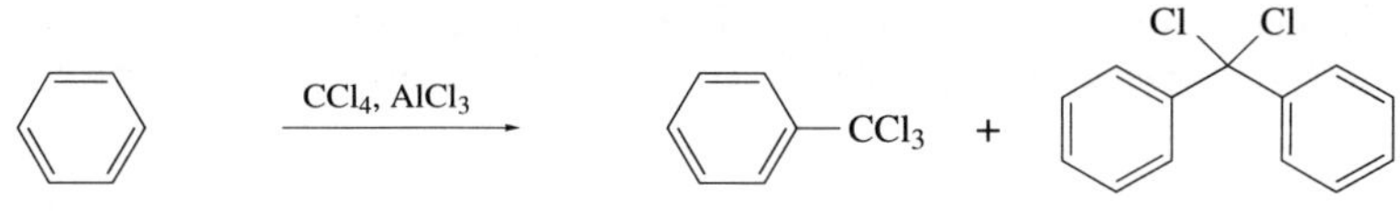

(134)

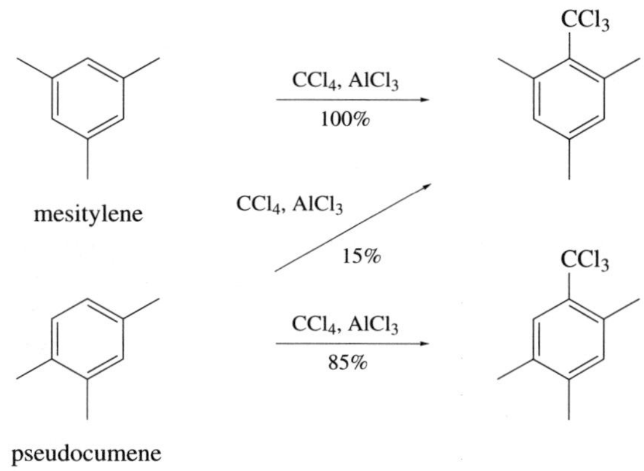

Scheme 23

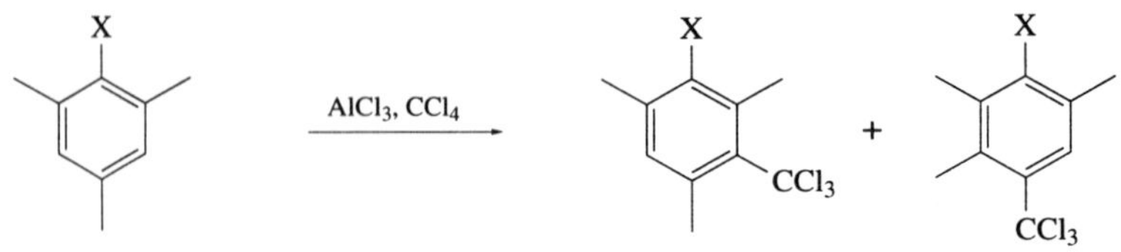

X			
H	100	:	0
F	70	:	30
Cl	88	:	12
Br	93	:	7

Scheme 24

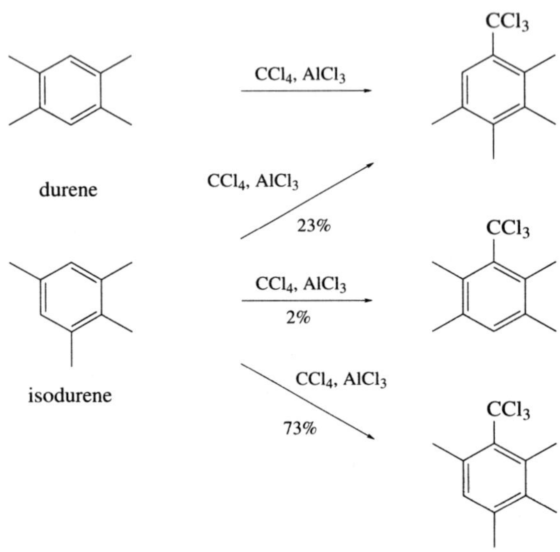

Scheme 25

Highly chlorinated trichloromethyl and trifluoromethyl aromatics are quite difficult to prepare, but Castaner *et al.* ⟨91JOC103⟩ found that treatment of polychlorinated benzenes with aluminum trichloride and fluorotrichloromethane gives the trifluoromethyl compound. Treatment of this with more aluminum trichloride in carbon disulfide yields the corresponding trichloromethyl compound. This reaction is reversed by treating the trichloromethyl compound with aluminum trichloride and fluorotrichloromethane. Bis(trichloromethyl) and bis(trifluoromethyl) compounds were also made by these methods (Scheme 26).

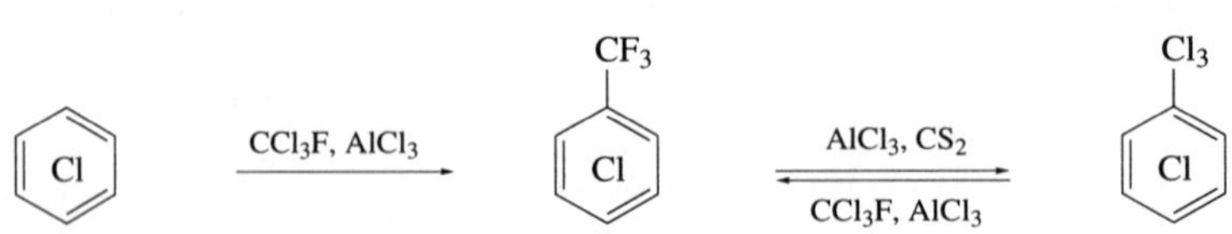

Scheme 26

Trichloromethyl anions can be used to prepare trichloromethylated aromatics ⟨90TL6831, 91JOM(419)357⟩. In a one-pot reaction sequence, a cyclopentadieneyliron-complexed arene is added to a solution made from potassium *t*-butoxide and chloroform in THF at $-78\,^{\circ}C$. The mixture is allowed to warm to room temperature, and the intermediate is demetallated *in situ* with iodine to give the trichloromethylarene in good yield (Scheme 27).

Aromatic compounds that are susceptible to nucleophilic attack, such as 1,3,5-trinitrobenzene, can be trichloromethylated by reaction with trichloromethyl anions generated by the decomposition of trichloroacetic acid in DMSO. The first step is the formation of the Meisenheimer complex, which is then oxidized to give the trichloromethylated product (Scheme 28) ⟨83CC140, 84JCS(P2)1239⟩.

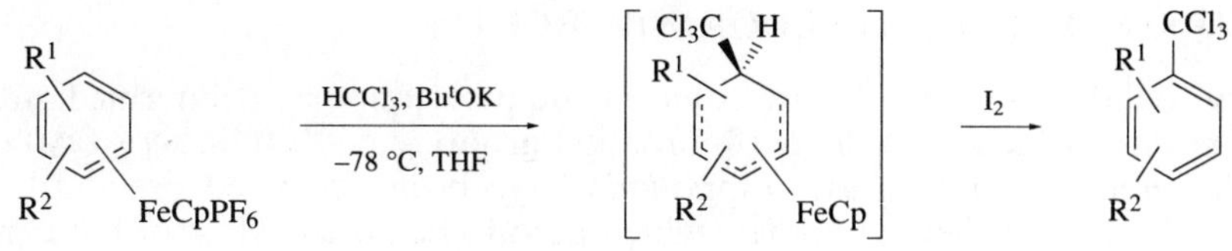

R¹	*R²*
Me	2-Me
Me	3-Me
Me	4-Me
H	*p*-TolSO$_2$
NO$_2$	4-Me
H	Bu^t
Cl	4-Cl
indane	
tetralin	
benzosuberane	

Scheme 27

Similarly, hexachloroacetone can be used as the source of trichloromethyl anions. The decomposition of hexachloroacetone in DMSO is slower than that of trichloroacetic acid, but it is accelerated by the addition of water. It is believed that it is actually the hexachloroacetone hydrate that decomposes into the trichloromethyl anion, and trichloroacetic acid which in turn decomposes to generate a second trichloromethyl anion ⟨86JCS(P2)471⟩

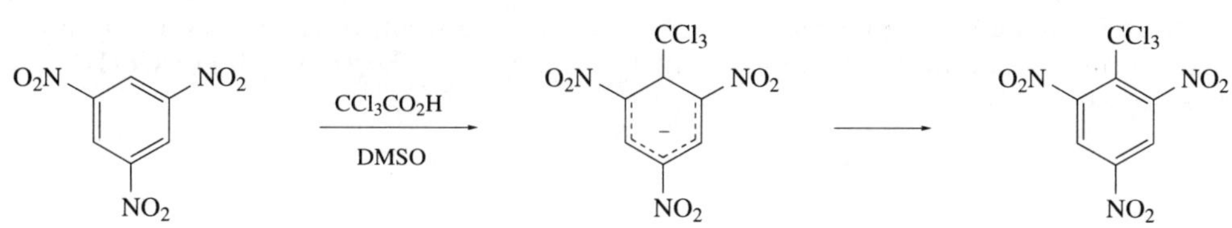

Scheme 28

6.01.4 TRIBROMOMETHYL DERIVATIVES—RBr$_3$

Considerably less work has been carried out on making compounds bearing a tribromomethyl group than on those bearing a trichloromethyl group. With few exceptions, the essentials of the methods for making tribromomethyl compounds are the same as those for making the trichloromethyl analogues.

Thus, tribromomethyl carbinols are produced when cyclohexanone is treated with tribromomethyl lithium at $-78\,^{\circ}$C in THF ⟨74JA3010⟩, or when various 3-pyridyl ketones are reacted at $-100\,^{\circ}$C ⟨88M1427⟩.

Corresponding to the decarboxylation of trichloroacetic acid and its sodium salt to give the trichloromethyl anion is the decarboxylation of the analogous tribromo compounds. The main difference between the two systems is that the tribromo compounds are less stable. Thus, tribromomethyl carbinols are obtained when sodium tribromoacetate decomposes in the presence of aldehydes. In a similar reaction, anhydrides react to give tribromolactols or their tautomeric keto acids ⟨67JOC2166⟩. Tribromomethyl carbinols are also obtained when, in the presence of aldehydes, tribromoacetic acid decarboxylates in DMSO ⟨83CC283, 84JCS(P2)1247⟩, or a mixture of the acid and its sodium salt decarboxylates in DMF ⟨92TL3435⟩.

Again corresponding to the trichloromethyl case, the reaction between tributyl(tribromomethyl)stannane and an aldehyde gives a tribromomethyl carbinol. The main difference between the two reactions is that tribromomethylation is exothermic at room temperature whereas trichloromethylation requires some heating ⟨75JOM(102)423⟩.

Finally, tribromomethyl carbinols can be made by the reductive addition of tetrabromomethane to aldehydes using a lead/aluminum bimetal system ⟨89JOC444⟩.

6.01.5 MIXED SYSTEMS WITH FLUORINE—RCF_2Hal

Perfluoroalkyl iodides—$C_nF_{2n+1}I$—are some of the most important intermediates in the preparation of compounds which bear the perfluoroalkyl group (e.g., ⟨84JFC(25)69, 88MI 601-01⟩). Some methods for the preparation of these compounds have been described previously (see Sections 6.01.1.1 and 6.01.1.2.2), but there are others which need to be noted. Most of these methods take a perfluorocarboxylic acid as their starting point. For example, the thermal decomposition of the acid in the presence of iodine ⟨53USP2647933⟩, the decomposition of perfluoroacyl chlorides in the presence of potassium iodide at 200°C ⟨58JOC2016⟩, the decomposition of perfluoroacyl anhydrides in the presence of iodine at 350–400°C ⟨57BRP757893⟩ and the thermal decomposition of dry metal salts in the presence of iodine all give perfluoroalkyl iodides. Sodium, potassium, barium, mercury and lead salts have all been used in this last reaction, but it is silver salts which give the best yields ⟨51JCS584, 52JA848, 52JA849⟩. This is known as the Hunsdiecker reaction, and it can also be used to obtain 'dihalides' from the corresponding diacid salts, although lactone formation can be a problem ⟨52JA848, 52JA1974⟩. More recently, it has been shown that treatment of sodium and potassium salts of perfluorocarboxylic acids with iodine in refluxing DMF gives perfluoroalkyl iodides in high yield ⟨67JOC833⟩.

One compound in this class which has attracted a lot of interest of late is perfluorooctyl bromide (PFOB) because of its potential use as a contrast agent for diagnostic imaging. The telomerization reaction between bromopentafluoroethane and tetrafluoroethylene does not afford a convenient route to PFOB because the propagation step is so much faster than chain transfer, and a mixture of high telomers is obtained. Most of the practical routes for the manufacture of PFOB start from perfluorooctyl iodide, which of course is manufactured on a substantial scale, and replace the iodine by bromine. For example, iodides can be treated with bromine in the vapour phase at 300–350°C ⟨91EUP450584, 92EUP515258⟩, or with ultraviolet irradiation for 10 h ⟨84MI 601-01⟩. Also, PFOB can be made by refluxing perfluorooctyl iodide with tetraalkylphosphonium bromide ⟨92GEP4025227⟩, or with sodium bromide ⟨91GEP3937567⟩ in DMF (a certain amount of reduction to the monohydro compound occurs here), and by heating with metal (Cs, Mn(II), Fe(II) Fe(III) or Co(II)) bromides to 300°C in an autoclave ⟨91EUP461563⟩.

6.01.6 MIXED HALOFORMS—$CHXY_2$ AND CHXYZ

Most methods for the preparation of mixed haloforms take the simple haloforms and replace one or two of the original halogens by others. Trichloromethane (chloroform) is fluorinated on an industrial scale in the liquid phase using Swarts' catalyst or in the vapour phase using chromia-based catalysts with hydrogen fluoride as the fluorinating agent (see also Section 6.01.1.3). The compound of greatest interest to the manufacturer is chlorodifluoromethane, since it is used in the manufacture of tetrafluoroethene, but dichlorofluoromethane can also be made by this method. Similarly, tribromomethane (bromoform) can be fluorinated using antimony trifluoride and bromine ⟨65BRP1014252⟩, or a fluoride-type ion exchange resin ⟨62NKZ936⟩, to give dibromofluoromethane (Equation (135)).

$$CHBr_3 \xrightarrow{\text{i, ii}} CHBr_2F \qquad (135)$$

i, SbF_3, Br_2, reflux; ii, amberlite IRA-400 (F type), 80 °C

Chloroform, bromoform and iodoform can all be treated with an alkali metal halide and a base in the presence of a phase transfer agent to give mixed haloforms (e.g., Equations (136)–(140)) ⟨89LA187, 87TL2769⟩, but satisfactory yields are not always obtained. A single iodine in iodoform can be replaced by bromine simply by stirring a mixture of iodoform and bromine in tetrachloromethane for some 15 h (Equation (141)) ⟨82CB3894⟩.

$$CHCl_3 \xrightarrow{\text{NaBr, NaOH, PTC}} CHBrCl_2 \qquad (136)$$

$$CHCl_3 \xrightarrow{\text{NaI, NaOH, PTC}} CHCl_2I + CHClI_2 \qquad (137)$$

$$CHBr_3 \xrightarrow{\text{NaCl, NaOH, PTC}} CHBr_2Cl \quad (138)$$

$$CHBr_3 \xrightarrow{\text{NaI, NaOH, PTC}} CHBr_2I \quad (139)$$

$$CHI_3 \xrightarrow{\text{NaCl, NaOH, PTC}} CHClI_2 + CHCl_2I \quad (140)$$

$$CHI_3 \xrightarrow{Br_2,\ CCl_4} CHBrI_2 \quad (141)$$

Dibromochloromethane reacts with sodium iodide and sodium methoxide in methanol to give bromochloroiodomethane (Equation (142)) ⟨58JA4282⟩, and with mercuric fluoride to give bromochlorofluoromethane (Equation (143)) ⟨56JA479, 85JA6993⟩. Bromochlorofluoromethane having high optical purity has been made by pyrolysing the optically pure strychnine salt of bromochlorofluoroacetic acid ⟨89JA8510⟩.

$$CHBr_2Cl \xrightarrow{\text{NaI, NaOH, MeONa}} CHBrClI \quad (142)$$

$$CHBr_2Cl \xrightarrow{HgF_2} CHBrClF \quad (143)$$

The thermal decarboxylation of silver salts of halogenated acetic acids in the presence of a halogen (the Hunsdiecker reaction—see Section 6.01.5) has been used to make the haloforms $CHCl_2F$, $CHClF_2$, $CHClBrF$, $CHClFI$, $CHBr_2F$, $CHBrFI$, $CHFI_2$, $CHBrF_2$ and CHF_2I (e.g., Equations (144) and (145)) ⟨52JCS4259⟩.

$$CHF_2CO_2Ag + I_2 \longrightarrow CHF_2I \quad (144)$$

$$CHClFCO_2Ag + Br_2 \longrightarrow CHClBrF \quad (145)$$

6.02
Functions Containing Halogens and Any Other Elements

GRAHAM B. JONES and JUDE E. MATTHEWS
Clemson University, SC, USA

6.02.1 FUNCTIONS CONTAINING HALOGEN AND A CHALCOGEN

A general review on *ortho*-ester derivatives covering the literature up to 1984 has been published ⟨85HOU(E5)⟩ and provides a broad overview of several categories of haloalkyl systems relevant to this chapter.

6.02.1.1 Halogen and Oxygen Derivatives ($R^1CHal_2OR^2$ and $R^1CHal(OR^2)_2$)

6.02.1.1.1 Tetracoordinated carbon atoms with two identical halogen and one oxygen function attached ($R^1CHal_2OR^2$)

(i) From ethers

Direct halogenation of ethers has proved a versatile technique for the synthesis of a variety of perhalogenated compounds of this class. Novel perfluorinated ethers have been obtained by direct fluorination of low molecular weight ethers with product ratios a function of the flow rate of fluorine and carrier gas (usually helium) (Scheme 1) ⟨73JOC3617⟩. The method has also been applied successfully to the preparation of perfluorinated polymers, although in these cases the mixtures required heating up to 110°C for efficient conversion ⟨77CC259, 81JCS(P1)1321⟩. Electrochemical fluorination has also been investigated, but it was found that, under the optimum conditions used, primary alcohols were converted into cyclic perfluorinated ethers in only moderate yield (Equation (1)) ⟨76BCJ1888⟩. The electrochemical fluorination of ethylene glycol ethers has also been reported, yielding complex mixtures of linear and cyclic perfluorinated compounds ⟨74ZOR2031⟩.

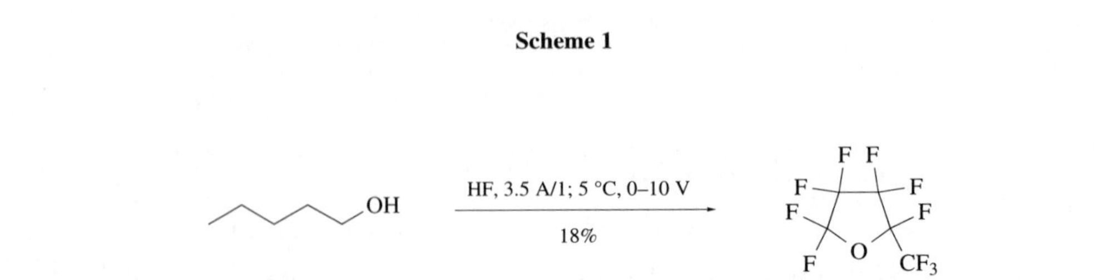

Scheme 1

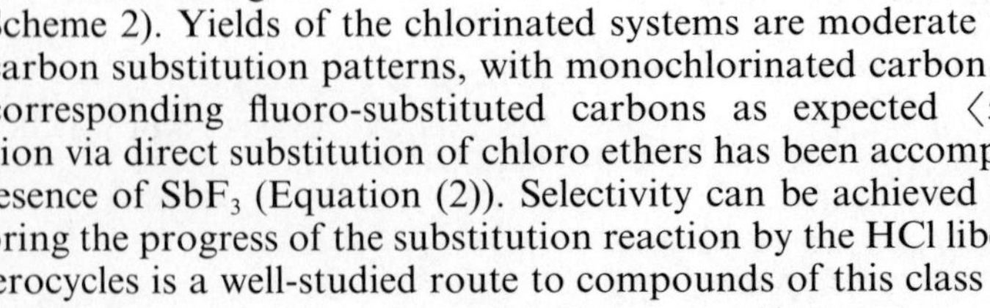

(1)

Photochlorination of mixed halogenated ethers has been used in the synthesis of a number of anaesthetic agents (Scheme 2). Yields of the chlorinated systems are moderate to good, reactivity being a function of carbon substitution patterns, with monochlorinated carbon atoms being more reactive than the corresponding fluoro-substituted carbons as expected ⟨52JA2292, 71JMC517, 71JMC593⟩. Fluorination via direct substitution of chloro ethers has been accomplished using either HF or HF in the presence of SbF_3 (Equation (2)). Selectivity can be achieved by controlling HF introduction, monitoring the progress of the substitution reaction by the HCl liberated ⟨71JMC593⟩. Halogenation of heterocycles is a well-studied route to compounds of this class with trisubstituted carbon atoms. Electrochemical fluorination of a range of oxanes gave the corresponding perfluorinated systems in moderate yield (Equation (3)) ⟨79JFC(13)519, 79JFC(15)353⟩.

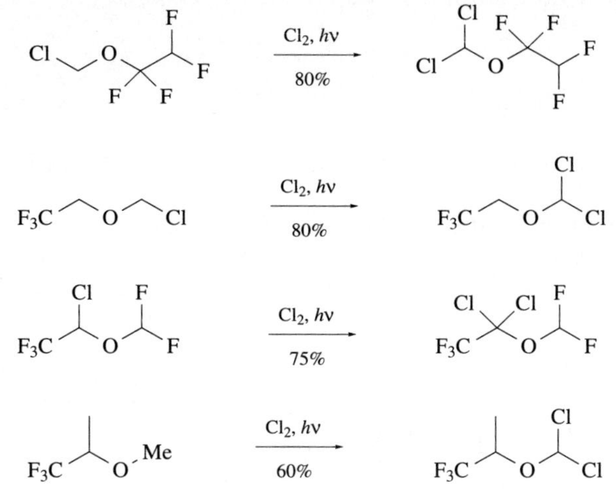

Scheme 2

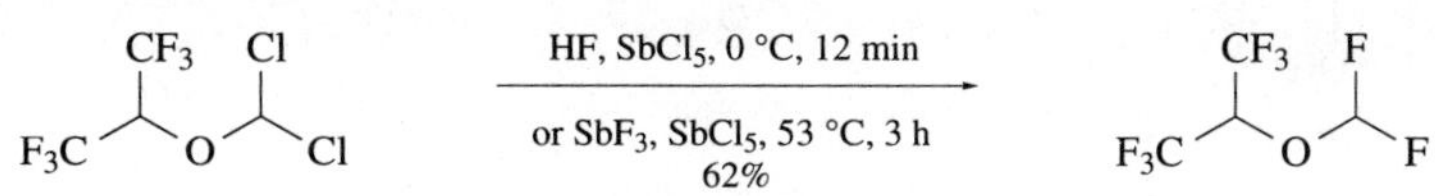

(2)

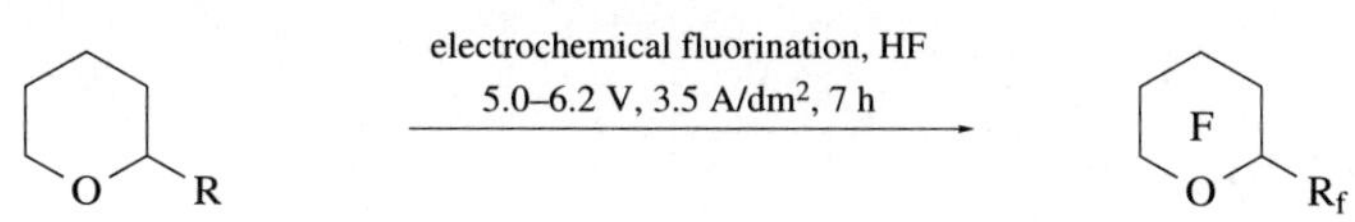

(3)

R = Me, 28%; Et, 26%; Pr^n, 30%; Pr^i, 23%; Bu^n, 30%; Am^n, 23%

Direct fluorination of dioxane using elemental fluorine is reported to give the perfluorinated system in moderate yield (Scheme 3) ⟨75JOC3271⟩, whereas treatment with a mixture of sulfur tetrafluoride and hydrogen fluoride gave the 2,2,3-trifluorodioxane in high yield (Scheme 3) ⟨83JFC(22)105⟩. The same product is also reported using the potassium salt of cobalt tetrafluoride ⟨71T4533⟩. Finally, fluorination of 4-methylmorpholine using cobalt trifluoride is reported to give the difluoromethyl derivative shown in moderate yield (Scheme 3) ⟨78T197⟩.

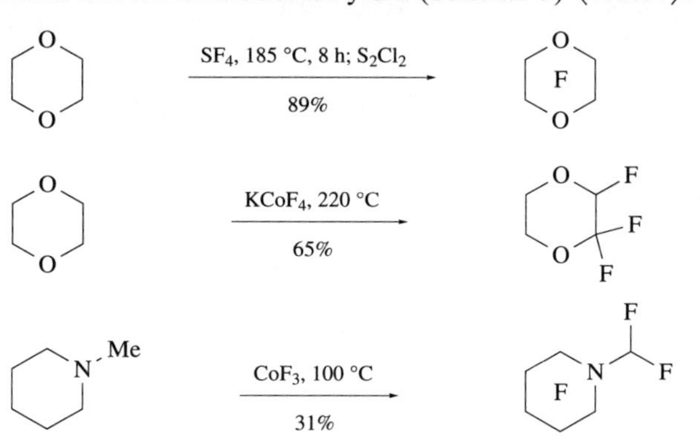

Scheme 3

(ii) From mixed trihalomethanes

Chlorodifluoromethane has been used extensively as a source of difluoromethylene, which can be intercepted by alkoxide nucleophiles to give alkoxydifluoromethane adducts. The conditions used are similar to those employed in the Reimer–Tiemann reaction, and give good yields of product difluoro ethers (Equation (4)) ⟨58JA3002, 60JOC2009⟩. A phase transfer method has also been used to generate difluorocarbene from chlorodifluoromethane ⟨88JFC(41)247⟩.

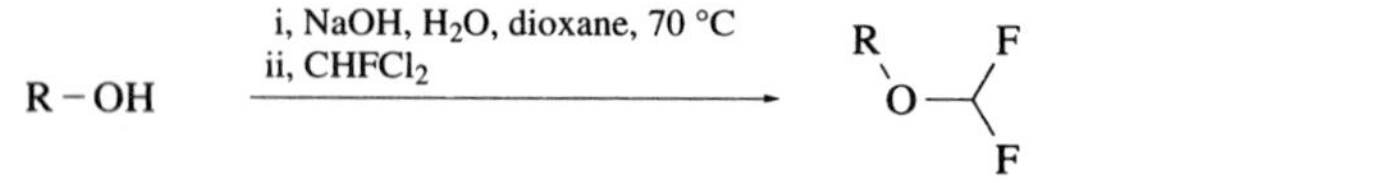

(4)

R = Ph, 65%; 4-MeC_6H_4, 66%; 4-$MeOC_6H_4$, 53%; 2,4-$Me_2C_6H_3$, 56%; 2,4-$Cl_2C_6H_3$, 44%; 2-naphthyl, 66%; Pr^i, 60%

The range of functionality tolerated by the reaction includes substituted primary and secondary alkyl halides, which provides access to heavily functionalized systems (Scheme 4) ⟨76CB2351⟩. Trihalomethylmercury compounds have also been used to deliver the dihalomethyl moiety to alcohols and carboxyl groups ⟨64JA2961, 74JOM(67)341⟩. Phenyl(bromodichloromethyl)mercury thus adds to carboxylic acids to yield the dichloromethyl esters shown (Scheme 5), and with butanol to give the dichloro ether shown in high yield, producing phenylmercuric bromide as by-product in both cases. Difluorocarbene has also been generated from $FSO_2CF_2CO_2H$, and it inserts into alcohols in the presence of CuI ⟨89JFC(44)433⟩.

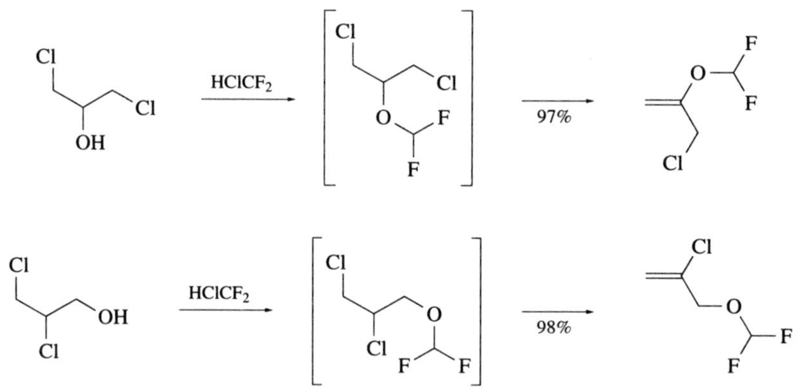

Scheme 4

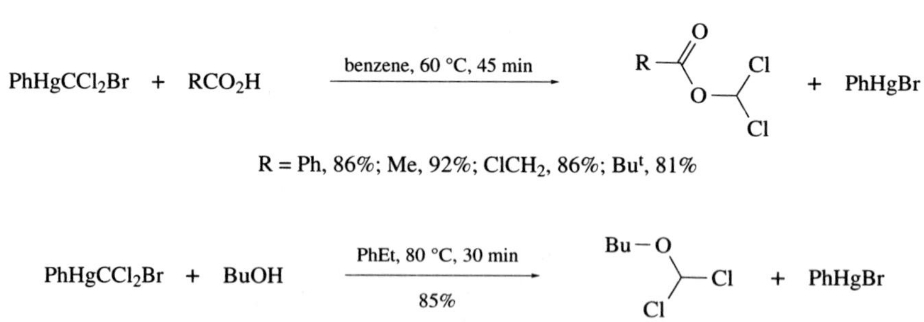

Scheme 5

(iii) From dihalo ethers

Treatment of dichloro ethers and bromochloro ethers with antimony trifluoride results in halogen substitution to yield the corresponding difluoro ether (Scheme 6) ⟨63AFC(3)181, 70ZOR144, 72JMC604, 72JMC606⟩. Alternatively, anhydrous hydrogen fluoride can be used in the presence of antimony pentafluoride giving high yields of the difluoro adducts (Equation (5)) ⟨72JMC604⟩. Treatment of cyclic perfluorinated ethers with aluminum trichloride is reported to give the corresponding α,α,α′-trichloro ether in good yield (Equation (6)) ⟨78JFC(12)359⟩.

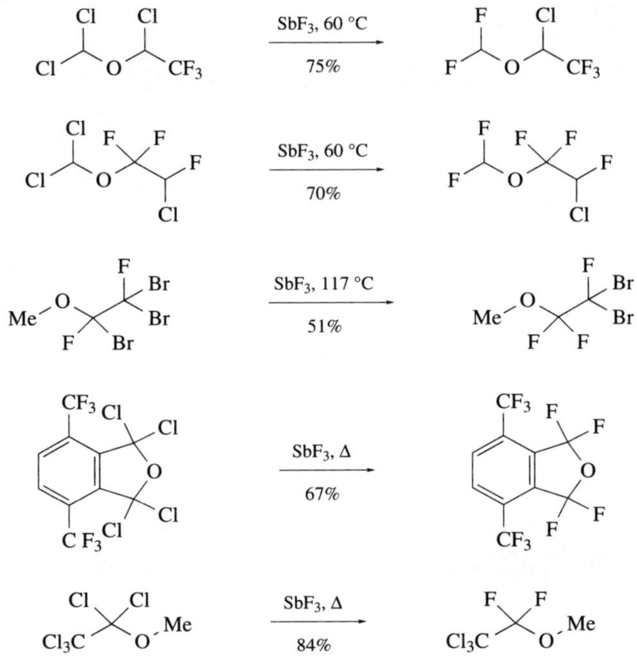

Scheme 6

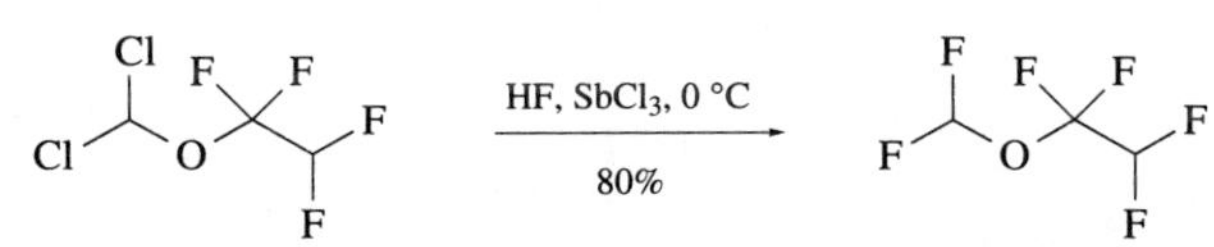

(5)

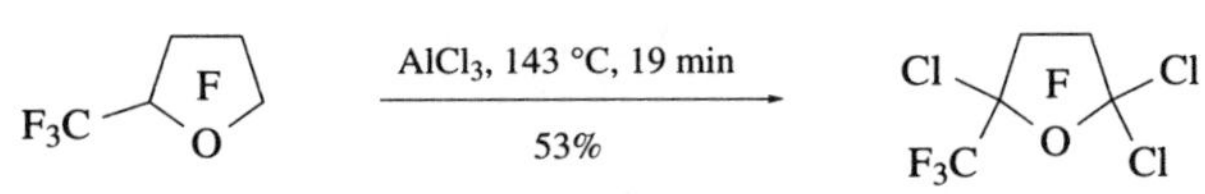

(6)

(iv) From carbonyl derivatives

Electrochemical fluorination of acyl chlorides and esters has been used to prepare compounds of this class. The process gives moderate yields of cyclic perfluorinated ethers, but often results in formation of mixtures (Scheme 7) ⟨78JFC(12)1, 83JFC(23)123⟩.

Sulfur tetrafluoride has been shown to be effective for the conversion of the carbonyl group to a difluoromethylene group. Thus, treatment of a variety of anhydrides, esters, and formate esters with sulfur tetrafluoride yields the corresponding difluoromethylene systems, often in high yield (Scheme 8) ⟨60JA543, 64JOC1, 70ZOR144, 75ZOR1672, 76ZOR1287, 76ZOR1798, 83JFC(22)207⟩. Exposure of aryl-*ortho*-diacid systems to the same conditions results in ring formation in addition to partial perfluorination (Scheme 9) ⟨70ZOR2498⟩.

Phosphorus pentachloride has been shown to react with a variety of formate esters to yield the corresponding α,α-dichloro ethers (Equation (7)) ⟨63CB1387, 67OS(47)47, 71RTC556⟩. The analogous addition reactions are also applicable to oxalates and other carboxylic esters, which constitute high-yielding routes to dichloro ethers, trichloro ethers, and dichloroacetates (Scheme 10) ⟨51JA5168, 56M323, 75S527⟩. α-Chlorination of α-chloro-α-alkoxyacyl chlorides has also been employed, giving the corresponding dichloroalkoxyacyl chloride in good yield (Scheme 10) ⟨59CB3170⟩.

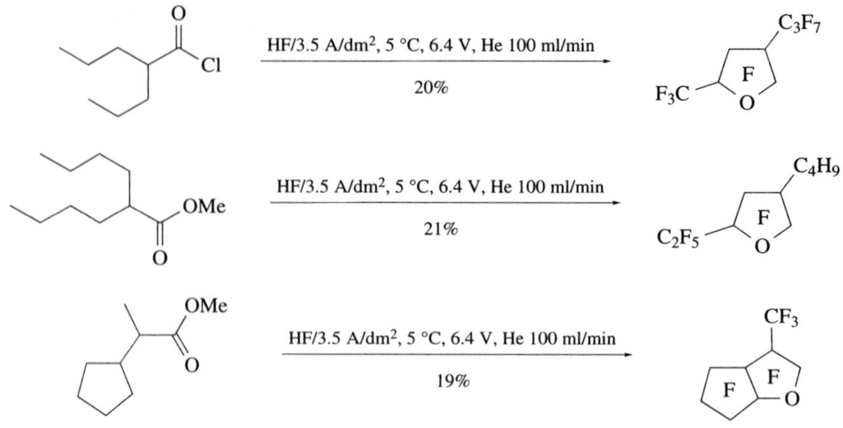

Scheme 7

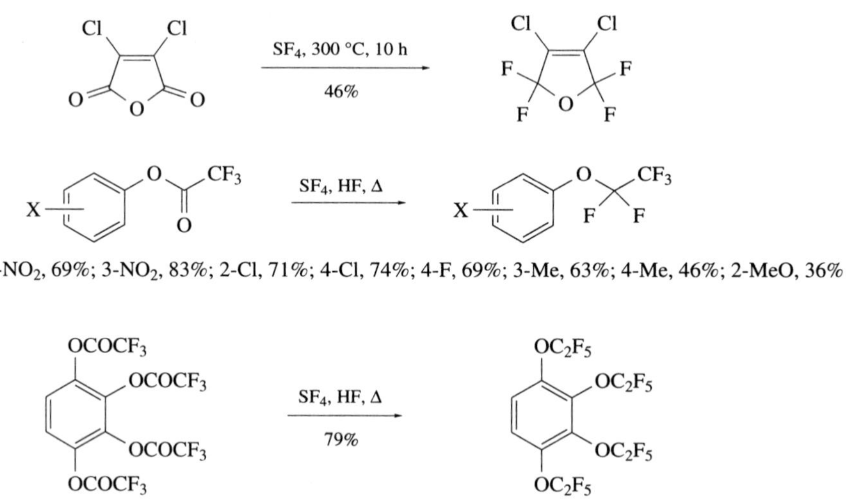

X = H, 61%; 4-NO_2, 69%; 3-NO_2, 83%; 2-Cl, 71%; 4-Cl, 74%; 4-F, 69%; 3-Me, 63%; 4-Me, 46%; 2-MeO, 36%

Scheme 8

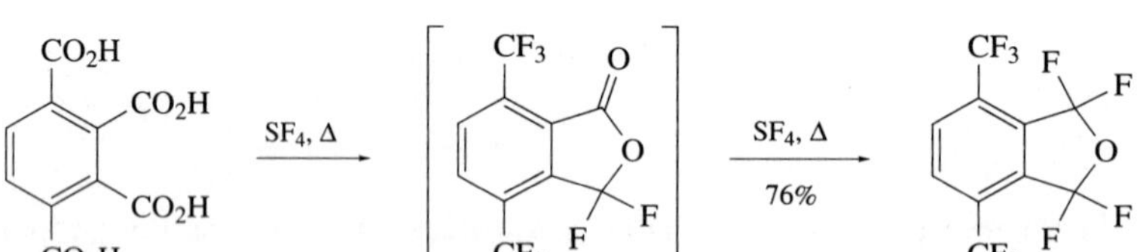

Scheme 9

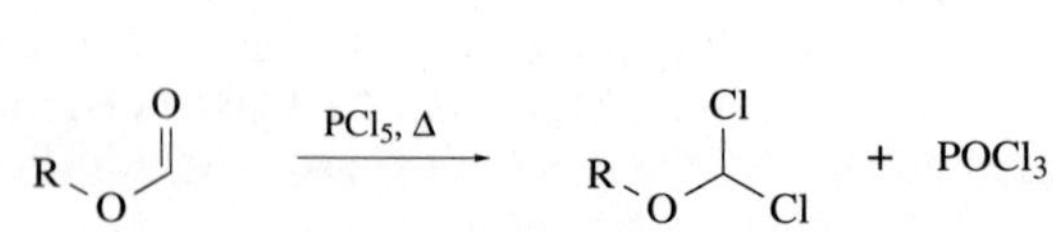

(7)

R = Me, 84%; Bu^t, 68%; Ph, 85%

Another versatile route is offered using α,α-dichloromethyl methyl ether as a transfer agent, in the presence of mercuric chloride. A variety of aryl formate esters have been converted to the corresponding dichloromethyl ethers using this very high-yielding method (Equation (8)) ⟨71RTC556⟩.

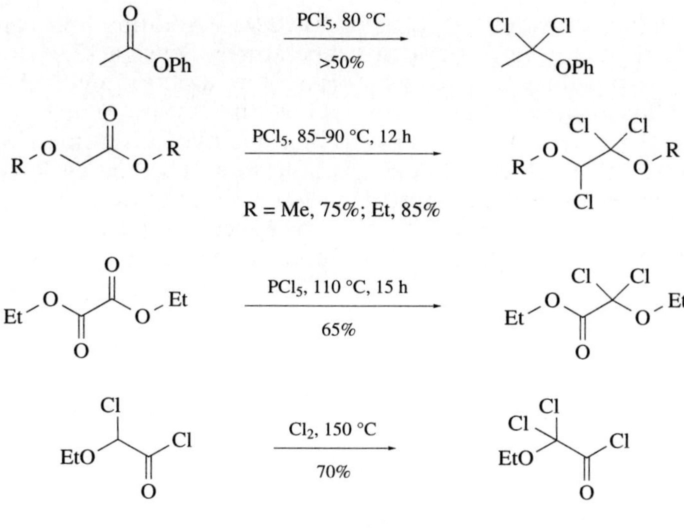

Scheme 10

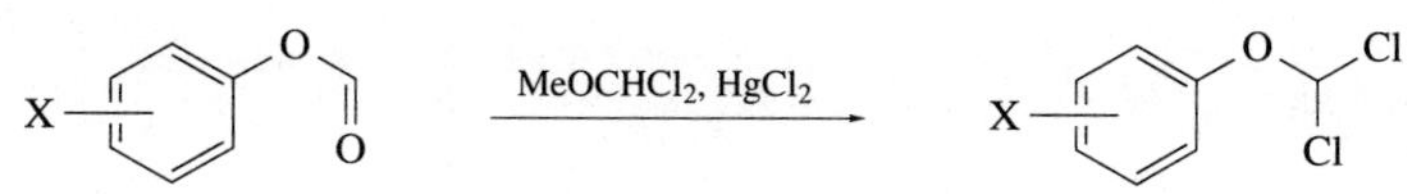

(8)

X = H, 85%; 2-Me, 94%; 3-Me, 90%; 4-Me, 90%; 3-Cl, 90%; 4-NO_2, 86%; 3-HCl_2CO, 90%

(v) From haloalkoxyalkenes

Chlorination of chloroalkoxyalkenes has been used successfully to give the corresponding dichloroalkoxyalkanes in good yield (Scheme 11). Typical conditions involve exposure of the alkene to a stream of chlorine gas or dry hydrogen chloride (generated *in situ*) at −5°C followed by distillation of the product ⟨57RTC969, 58CB806⟩.

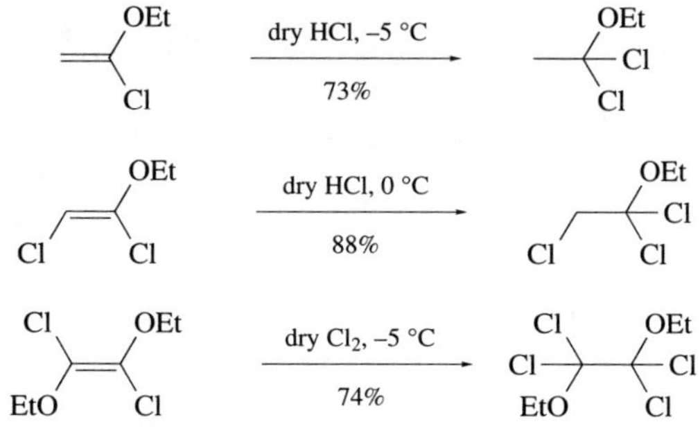

Scheme 11

(vi) From haloalkenes

The addition of primary alcohols to halogenated alkenes proceeds rapidly, giving dihaloalkoxyalkanes in high yield. The reactions are often catalyzed by trace amounts of the corresponding metal alkoxide; this is often required for addition of less reactive secondary and tertiary alcohols, which usually give mixtures of products (Scheme 12) ⟨60JOC993, 65AFC(4)50, 74BSF2072⟩. Electron-withdrawing substituents on the alkene also accelerate the reaction; thus, methyl trifluoroacrylate reacts faster than methyl acrylate ⟨73CCC66, 86ZOR1834⟩. Addition of oximes has also been accomplished, merely requiring more elevated temperatures (Scheme 12) ⟨73CCC66⟩. Addition of alkyl and trihaloalkyl hypohalites to haloalkenes is another useful route to this class. A variety of primary and tertiary alkyl hypohalites add cleanly at low temperature (Scheme 12) to give high yields of the adducts ⟨70JOC3730, 75JA13, 75JFC(5)25⟩. Bis(trifluoromethyl)trioxide has also been employed for this type of addition reaction, which is applicable both to acyclic and cyclic alkenes (Scheme 12) ⟨74JOC1298⟩. Perfluorinated alkenes are reported to react with sulfur trioxide to give cyclic perfluoro-2-β-sulfones in excellent yield ⟨86ZOR1842⟩. Perfluoroalkenes also react with halogen monofluorosulfates to yield both 2-haloalkylfluoroalkyl fluorosulfates and vicinal bis(halosulfonyloxy) fluoroalkanes ⟨85IZV659⟩.

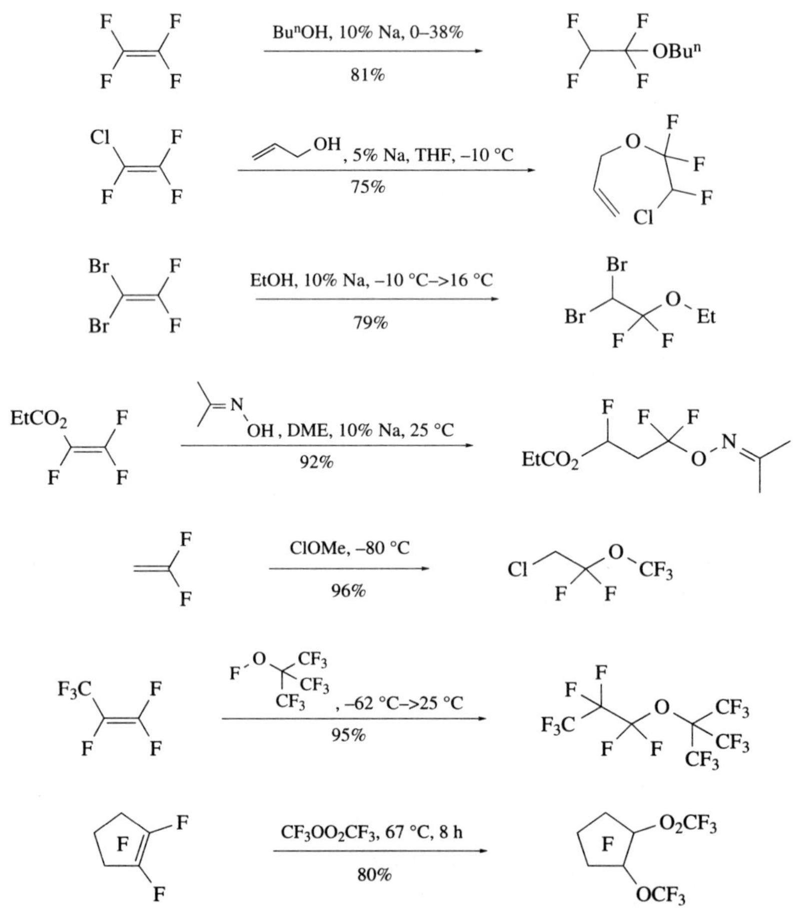

Scheme 12

Epoxidation of fluoroalkenes is a convenient route to α,α-diflurorooxiranes. High yields are attainable at low temperature using moderate to high strength hydrogen peroxide (Equation (9)) ⟨66JOC2312, 70JOC2054, 72MI 602-01, 73ZOR2013⟩.

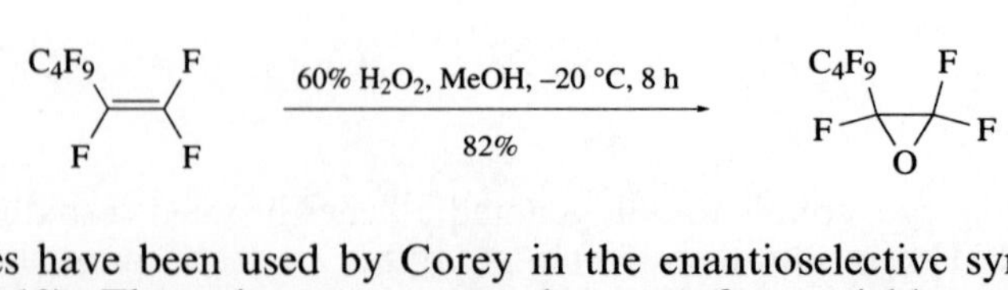

(9)

α,α-Dichlorooxiranes have been used by Corey in the enantioselective synthesis of azido and amino acids (Scheme 13). The oxiranes, generated *in situ* from trichloromethyl carbinols, are immediately ring-opened by the S_N2 attack of aqueous azide, subsequent oxidation giving α-azido acids in high yield ⟨92JA1906⟩.

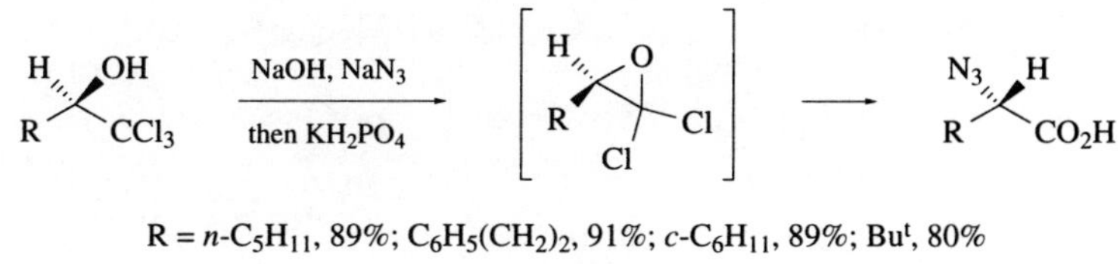

Scheme 13

6.02.1.1.2 *Tetracoordinated carbon atoms bearing mixed halogens and one oxygen function ($R^1CHal_2OR^2$)*

Selective monofluorination of dichloro ethers has been used to prepare chlorofluoro ethers in moderate to good yield (Scheme 14) ⟨71JMC593, 72JMC604, 75JFC(6)37⟩. Another useful strategy is addition of trifluoromethyl hypochlorite to mixed 1,1-dihaloalkenes giving the dihalo ethers in excellent yield (Scheme 14) ⟨70JOC3730, 72JMC606⟩. The ester $BrCF(OPh)CO_2Et$ has been prepared by bromination of ethylfluoro(phenoxy)acetate ⟨86JOC955⟩.

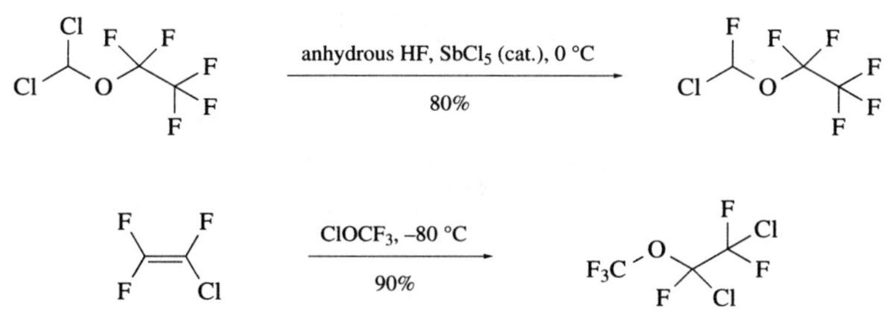

Scheme 14

6.02.1.1.3 *Tetracoordinated carbon atoms bearing one halogen and two oxygen functions ($R^1CHal(OR)^2_2$)*

Numerous examples of compounds of this class exist, as the corresponding carbenium ion halide salt, a consequence of stabilizing influence of the alkoxy functions ⟨72CRV357, 82S1, 85HOU(E5)⟩. The following examples described are clear-cut cases where the chemistry is best described by the covalent species.

(i) From 1,3-dioxolanes

Direct formation of 2-fluoro-2-trifluoromethyl-1,3-dioxolanes has been accomplished by treating aryl *ortho*-bis(trifluoroacetic) esters with a mixture of sulfur tetrafluoride and hydrogen fluoride ⟨76ZOR1287⟩. Formation of 2-chloro-1,3-dioxolanes has been achieved using a variety of protocols ⟨67LA(707)35⟩. Chlorination of derived *ortho*-esters with phosphorus pentachloride at elevated temperatures is a high-yielding process, as is direct chlorination of unsubstituted 1,3-dioxolanes (Scheme 15) ⟨61CB544, 66CB2625, 67LA(707)35, 68LA(720)146⟩. A related procedure used substitutive chlorination of a 2-acetoxy-1,3-dioxolane using dichloromethyl ether ⟨69TL5003⟩. Alternatively, displacement of chloride from 2,2-dichloro-1,3-dioxolanes has been used ⟨66CB2625⟩.

(ii) From ortho-*formates*

Decomposition of vinyl *ortho*-formates is a versatile route to dialkoxymethyl halides. Treatment of dialkoxy vinyl *ortho*-formates with either hydrogen chloride or hydrogen bromide results in

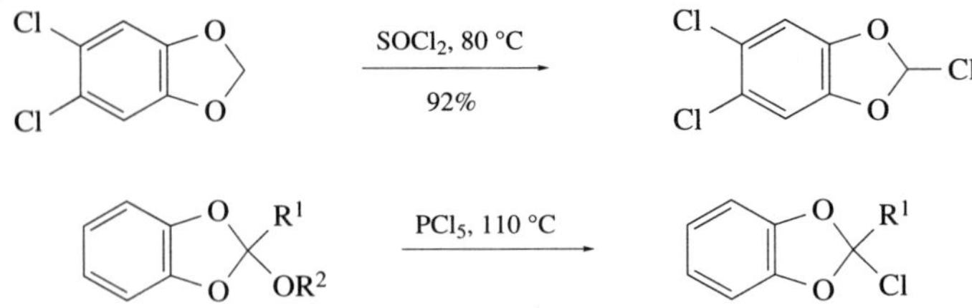

R^1 = Ph, R^2 = Me, 88%
R^1 = CO_2Et, R^2 = Et, 53%
R^1 = H, R^2 = Et, 87%

Scheme 15

liberation of the dialkoxyhalomethane (Scheme 16) ⟨71RTC1123⟩. Alternatively, substitution of triaryl *ortho*-formates with acetyl chloride has been reported to give good yields of the chlorodiaryloxymethane ⟨62CB1859⟩.

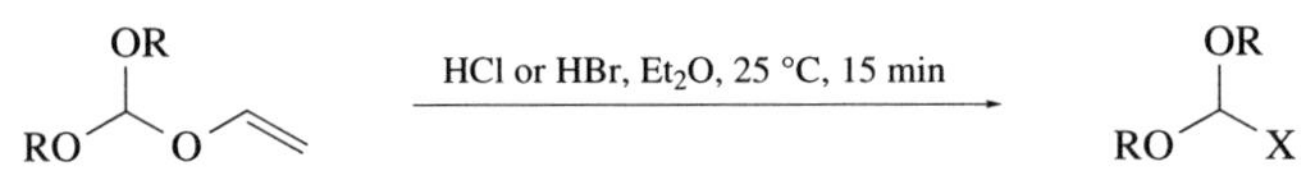

R = CH=CH_2, X = Cl, 50%
R = Ph, X = Cl, 60%
R = Ph, X = Br, 50%
R = *p*-Tol, X = Cl, 65%
R = C(Me)=CH_2, X = Cl, 90%

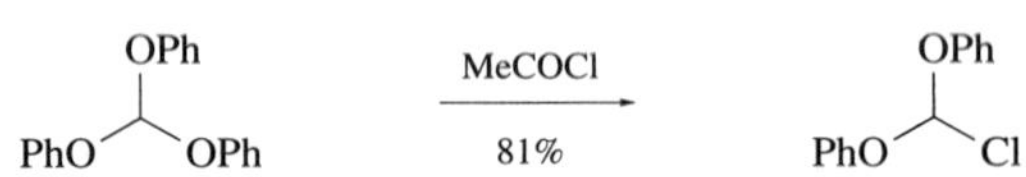

Scheme 16

6.02.1.2 Halogen and Sulfur Derivatives ($R^1CHal_2SR^2$ and $R^1CHal(SR^2)_2$)

6.02.1.2.1 Tetracoordinated carbon atoms with two identical halogen and one sulfur function attached ($R^1CHal_2SR^2$)

(i) From sulfides

α-Chlorination of sulfides is a popular route to these compounds, and a variety of different protocols have proved effective including photochemical chlorination and use of thionyl chloride (Scheme 17) ⟨52JA3594, 53LA(581)133, 58LA(616)1, 59LA(621)8, 65JOC4011⟩. The ester $MeSCH_2CO_2Me$ has been converted into the corresponding α,α-dichloro ester using thionyl chloride and into the dibromo ester by reaction with bromine ⟨81AG(E)585⟩.

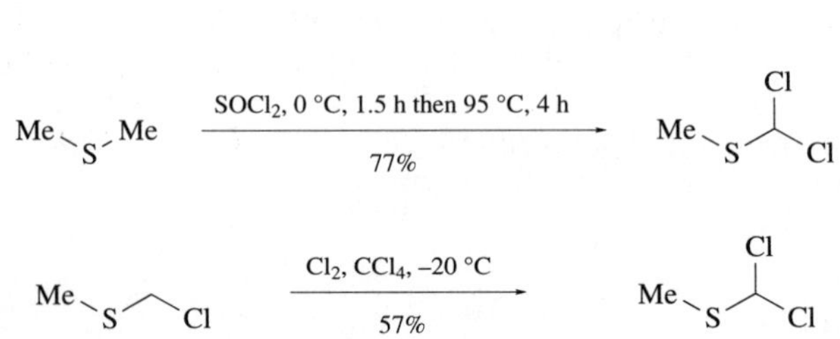

Scheme 17

(ii) From iodofluoroalkanes

The photochemical reaction of fluoroiodo alkanes with both methyl sulfide and dimethyl disulfide has been investigated extensively, and the radical addition process was found to give good yields of the corresponding difluoro sulfides (Scheme 18), the only drawback being the sluggishness of the reaction ⟨72JCS(P1)155, 72JCS(P1)159, 72JCS(P1)1506, 72JCS(P1)2438⟩. Metallation of terminal iodofluoroalkanes with methyllithium at −78 °C, followed by addition of sulfur dioxide, results in formation of a lithium difluoroalkylsulfinate by attack of the intermediate organolithium species ⟨89S463⟩.

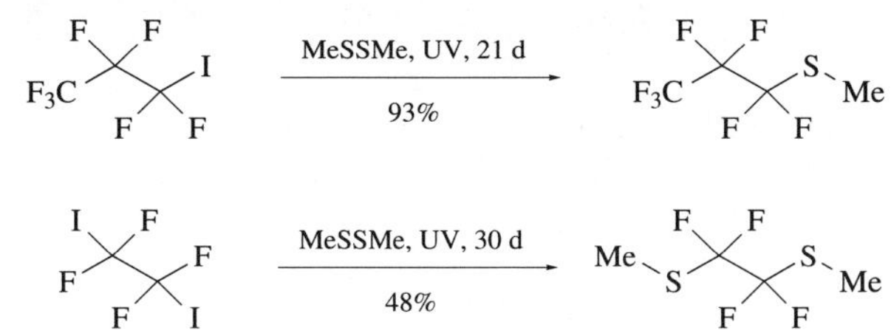

Scheme 18

(iii) From α,α-dichloroalkyl sulfides

Dichloroalkyl sulfides react smoothly with antimony trifluoride in the presence of catalytic quantities of antimony pentafluoride to yield the corresponding difluoro sulfides (Scheme 19) ⟨52JA3594, 65JOC4011⟩. The substitutions are regioselective for the carbons alpha to the sulfur center as shown by control reactions.

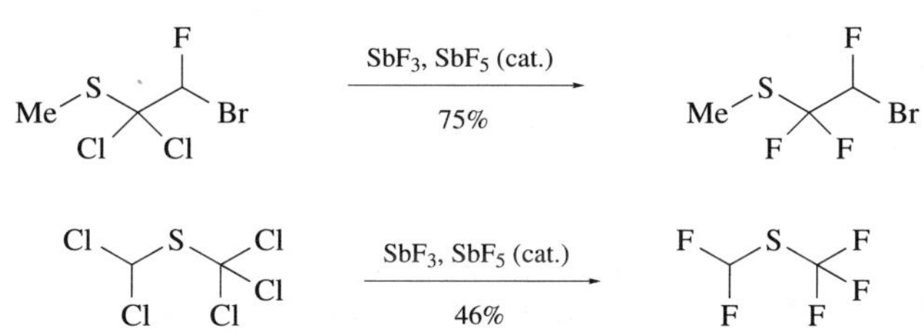

Scheme 19

(iv) From chlorodifluoromethane

Substitution of the chloro group of this haloform with thiols has been achieved using a variety of conditions, typically involving addition of a sodium thiolate followed by thermolysis (Scheme 20) ⟨57JA5493, 79JOC1708, 85S497⟩. Some debate exists concerning the mechanism of the substitution, which appears to have both a carbene and S_N2 type character. A phase-transfer method, which undoubtedly involves difluorocarbene, has been described for preparation of sulfides $ArSCHF_2$ from aromatic thiols ⟨88JFC(41)247⟩.

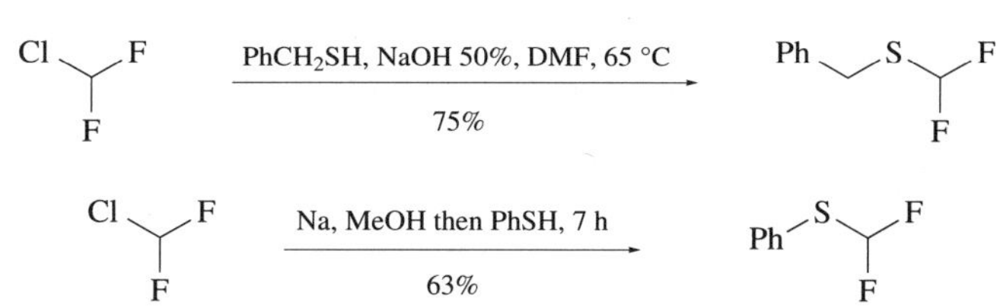

Scheme 20

(v) From thioformates

Thioformates react cleanly with phosphorus pentachloride in carbon tetrachloride to yield the corresponding α,α-dichloromethyl sulfides in high yield (Scheme 21). An alternative but slightly less efficient route to such compounds is via halomethylene transfer from dichloromethyl methyl ether, catalyzed by mercuric chloride ⟨72RTC349⟩.

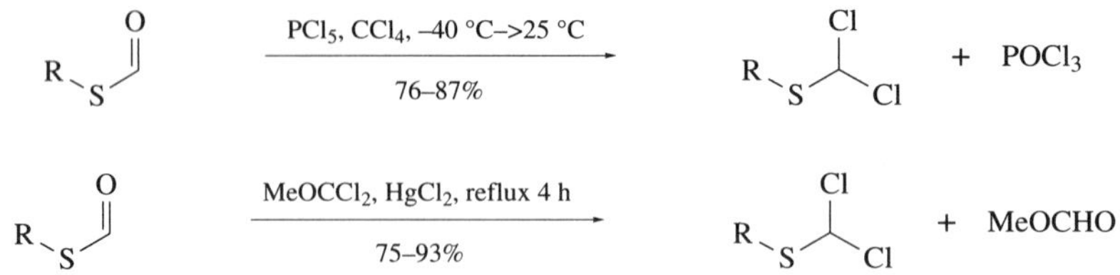

Scheme 21

(vi) From alkenes

(a) Vinyl sulfides. Chlorination of 1-chlorovinyl sulfides has been used to generate bis(alkylthio)tetrachloroethanes (Scheme 22). In this example the alkene is readily available via thiolate-induced substitution and elimination from the corresponding hexachloroethane ⟨58CB806⟩.

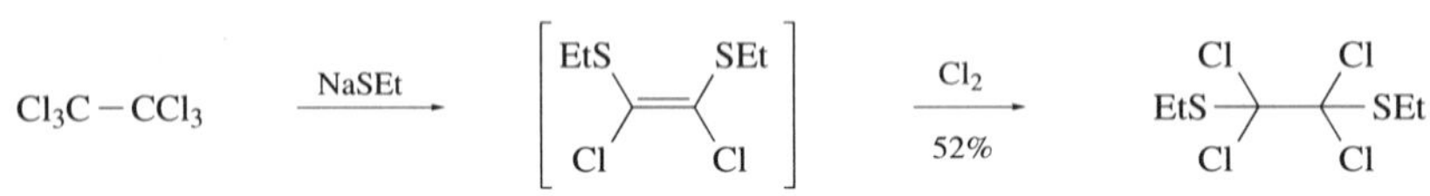

Scheme 22

(b) 1,1-Dihaloalkenes. Addition of thiols and disulfides to halogenated alkenes proceeds rapidly when they are subjected to either ultraviolet or x-ray radiation to give the corresponding addition products (Scheme 23). As expected, addition of thiolates proceeds without photolytic conditions, but the reaction rates are heavily dependent on the nature of the haloalkene substitutents. Mixed chloro- and bromofluoroalkenes are far more reactive than tetrafluoroethylene, which requires forcing conditions ⟨60JA5116, 61JA840, 65JOC4011, 66BCJ2191, 72JCS(P1)34, 74JFC(4)107, 76JCS(P1)1178⟩. Free-radical addition of trifluoromethanesulfonyl chloride to haloalkenes under UV irradiation has been employed and gives moderate yields of the trifluoromethylsulfide adducts (Scheme 23) ⟨62JA3148⟩. Thermolysis of tetrafluoroethylene with elemental sulfur is reported to produce five-, and six-membered cyclic sulfides bearing α,α-difluoromethyl groups in yields ranging from 10% to 60% ⟨62JOC3995⟩. Addition of thiols to difluoroalkenes catalyzed by Triton B is also reported to produce α,α-difluoroalkyl sulfides in high yield ⟨50JA3642⟩. Activated haloalkenes, including pentafluoropropene-2-sulfonyl fluoride, undergo rapid addition of alkyl hydrosulfides, yielding 1-alkylthio-2*H*-pentafluoropropane-2-sulfonyl fluorides (Scheme 23) ⟨86ZOR1834⟩.

(vii) From carbon disulfide

Treatment of carbon disulfide with elemental fluorine at −120°C is reported to yield octafluoroethyl methyl sulfide in 60% yield ⟨77IC2974⟩.

(viii) From α-fluoroalkyl sulfoxides

The reaction of the sulfoxide of 4-$MeOC_6H_4SOCH_2F$ with (diethylamino)sulfur trifluoride gives the difluorosulfide 4-$MeOC_6H_4SCHF_2$ (68%) by a Pummerer-type reaction ⟨90JOC4757⟩. The sulfide

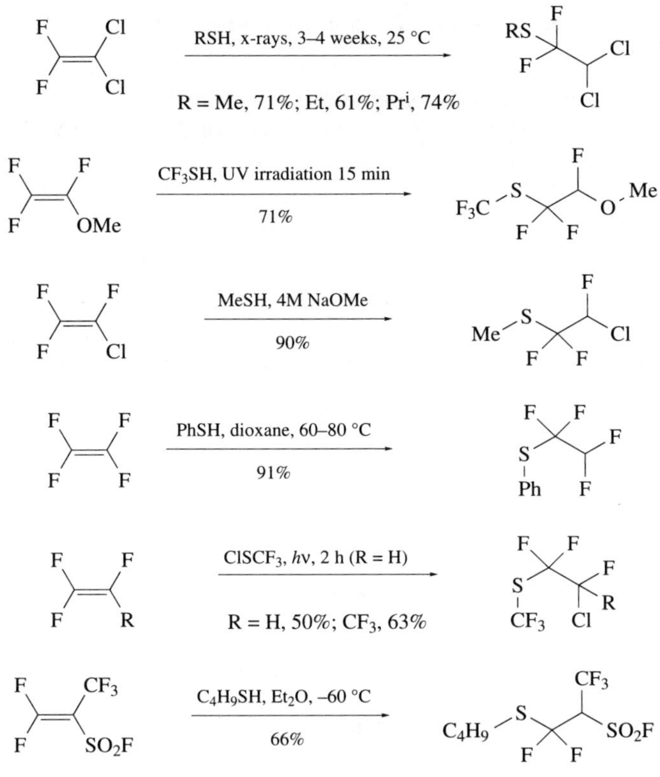

Scheme 23

can be oxidized to the corresponding sulfoxide and sulfone. The reaction is, however, unsuccessful with $PhSOCH_2F$.

6.02.1.2.2 *Tetracoordinated carbon atoms bearing mixed halogens and one sulfur function ($R^1CHal_2SR^2$)*

Thiolate additions to mixed chlorofluoroalkenes under both radical and ionic conditions have been employed to synthesize mixed dihalo sulfides of this class (see Section 6.02.1.2.1). Typical conditions parallel those for identically substituted haloalkenes, and product yields are comparable (Scheme 24) ⟨59LA(621)8, 62JA3148, 66BCJ2191, 74JFC(4)107, 76JCS(P1)1178⟩. Alternatively, treatment of dichloro sulfides with antimony trifluoride results in monosubstitution of chlorine, giving the mixed (chlorofluoro)sulfide in good yield (Scheme 24) ⟨59LA(621)8⟩. The esters $BrCF(SR)CO_2Et$ (R = Et and Ph) have been prepared by NBS bromination of the corresponding (ethylthio)fluoroacetate or fluoro(phenylthio)acetate ⟨86JOC955⟩.

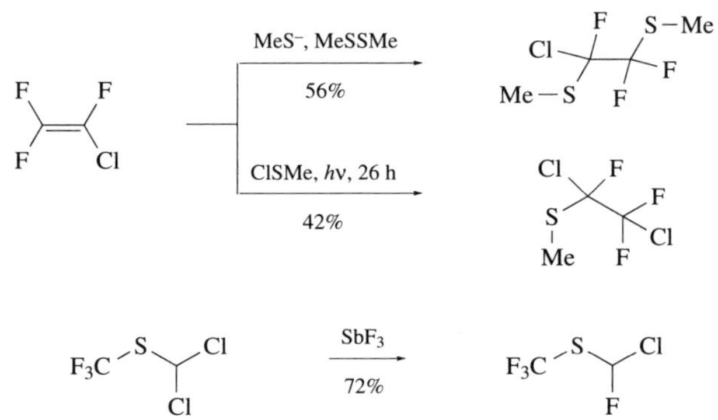

Scheme 24

6.02.1.2.3 Tetracoordinated carbon atoms bearing one halogen and two sulfur functions ($R^1CHal(SR^2)_2$)

As in the case of many compounds described in Section 6.02.1.1.3, much of the chemistry of compounds of this class is represented by the corresponding halothiolium salts ⟨66AHC(7)39, 80AHC(27)151⟩. Cases illustrated below are confined to systems where the chemistry is best represented by the covalently bound species.

(i) From thioacetals

Chlorination of 1,3-dithianes has been used to generate versatile 2-chloro-1,3-dithianes which are useful for coupling reactions of this masked carbonyl. Typical conditions involve either sulfuryl chloride or *N*-chlorosuccinimide, and recovered yields are high (Scheme 25) ⟨76BCJ553, 77TL885, 79JOC1847⟩.

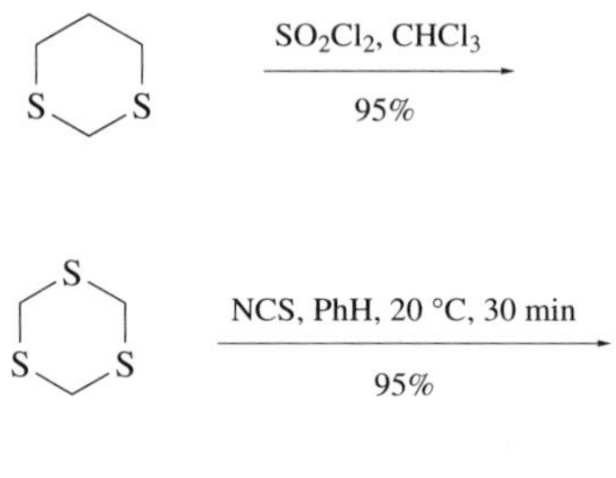

Scheme 25

(ii) From thioformates and ortho-*thioesters*

Ortho-thioesters can be monochlorinated successfully using either phosphorus chlorides or acetyl chloride as donors, giving moderate yields at elevated temperatures (Scheme 26) ⟨61LA(648)21, 77JPR17⟩. An alternative procedure involves treatment of thioesters with a dichloromethyl aryl ether in the presence of mercuric chloride (Scheme 26). This procedure, however, is less efficient in that it requires 2 equivalents of thioester ⟨72RTC349⟩.

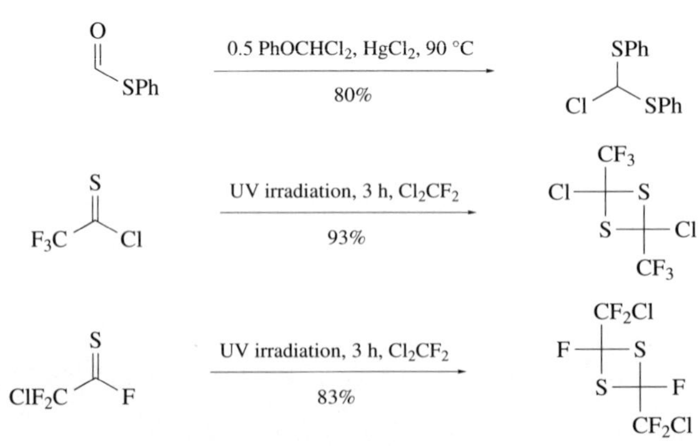

Scheme 26

(iii) Dimerization of thiocarbonyl compounds

A variety of substituted thiocarbonyl compounds are reported to undergo rapid dimerization on irradiation with UV light ⟨65JOC1375⟩. The corresponding 1,3-dithietanes are formed in high yields, and are easily purified by distillation (Scheme 26).

6.02.1.3 Halogen and Se or Te Derivatives ($R^1CHal_2SeR^2$ or $R^1CHal_2TeR^2$)

6.02.1.3.1 Tetracoordinated carbon atoms bearing two halogens and one selenium function ($R^1CHal_2SeR^2$)

A variety of examples of compounds from this class have been reported, and their preparation often involves transmetallation reactions of metal selenyl dihalomethyl derivatives. Bis(pentafluoroethyl) monoselenide has been prepared by reaction of selenium powder with the corresponding pentafluoroethyl silver salt ⟨63JCS1268, 73JCS(D)2314⟩. Thermolysis of selenium metal with tetrafluoroethylene in the presence of catalytic amounts of iodine, however, results in the formation of a mixture of octafluoroselenolane and octafluoro-1,4-diselenane (Scheme 27) ⟨62JOC3584⟩. Tetrafluoroethylene also reacts with tetraselenium bis(hexafluoroarsenate(V)) over extended periods of time to yield bis(perfluoroethyl) diselenide in good yield (Scheme 27) ⟨77CJC3136⟩. Similar reactions have been perfomed using $Se_8(AsF_6)_2$, and $Se_8(Sb_2F_{11})_2$, which both react with tetrafluoroethylene to give the bis(perfluoroethyl) diselenide in nearly quantitative yield ⟨73JCS(D)2314⟩. Reaction of selenium metal with trifluoroiodomethane results in the formation of bis(trifluoromethyl) mono- and diselenide ⟨58JCS2939⟩. Similarly, reaction of silver heptafluorobutyrate with selenium metal results in the formation of heptafluoropropyl mono- and diselenide ⟨59ZOB942⟩. Reaction of this diselenide with mercury results in the formation of bis(heptafluoropropylseleno)mercury, which is a versatile alkylseleno transfer agent (Scheme 27) ⟨63JCS1268⟩. Treatment of this, and related mercurials, with iodine results in near quantitative conversion back to the bis(heptafluoropropyl) diselenide ⟨65JCS7511⟩, whereas combination with 2 equivalents of iodosilane results in good conversion to the corresponding silyl selenide (Scheme 27) ⟨62JCS2290⟩. Diselenides also react with manganese pentacarbonyl hydride to yield the corresponding organoselenium manganese complexes ⟨65JCS7515⟩. It has been reported that chlorination of 2,5-dibromoselenophene results in the formation of 2,5-dibromoselenophene-2,3,4,5-tetrachloride, although no details of conditions or yield were provided ⟨36BCJ157⟩. Diphenyl diselenide reacts under dissolving metal reduction conditions to form an intermediate sodium selenide. This species can be intercepted with perfluoroalkyl iodides under photolytic conditions to yield perfluoroalkyl selenides in moderate yield (Scheme 27) ⟨77ZOR2008⟩. Bis(perfluoroethyl) monoselenide reacts with iodine monochloride to form difluoro(perfluoroethyl)selenium in quantitative yield ⟨76JFC(7)261⟩. Selenocarbonyl difluoride is reported to undergo addition with amines to yield an unstable amino difluoromethyl adduct ⟨90JOM(386)321⟩. Trifluoromethyl selenocarbonyl fluoride ($Se{=}C(F)CF_3$), produced by thermal 1,2 elimination of trimethyltin fluoride from trimethylstannylpentafluoroethylselenane, is also reported to undergo similar addition reactions ⟨90JOM(386)321⟩. Difluoromethyl selenoethers $ArSeCHF_2$ can be prepared from chlorodifluoromethane and $ArSe^-$ ⟨85S497⟩.

6.02.1.3.2 Tetracoordinated carbon atoms bearing two halogen and one tellurium function ($R^1CHal_2TeR^2$)

Difluoromethyl compounds of tellurium were unknown until the mid-1990s. A general protocol has been reported in which dimethyl telluride is reacted with either a bis(trifluoromethyl) cadmium or trifluoromethylzinc bromide in the presence of acetonitrile and boron trifluoride (Scheme 28) ⟨94AG(E)323⟩. The route is reported to be general for all chalcogens, thus in a similar manner, methyl (difluoromethyl) selenide is also formed; however, isolated yields were not reported in either case. Perfluoroalkyl tellurides, however, have been prepared by a variety of means ⟨75JCS(D)488⟩. One method involves reaction of perfluoroalkyl radicals on a tellurium mirror, giving low yields of perfluoroalkyl ditelluride ⟨63AJC722⟩. A more efficient method for the preparation of bisfluoroethyl monotelluride is to react tetrafluoroethylene with $Te_4(AsF_6)_2$ in a Monel reactor (Scheme 28). This method is general, and manipulation of conditions allows formation of varying amounts of the

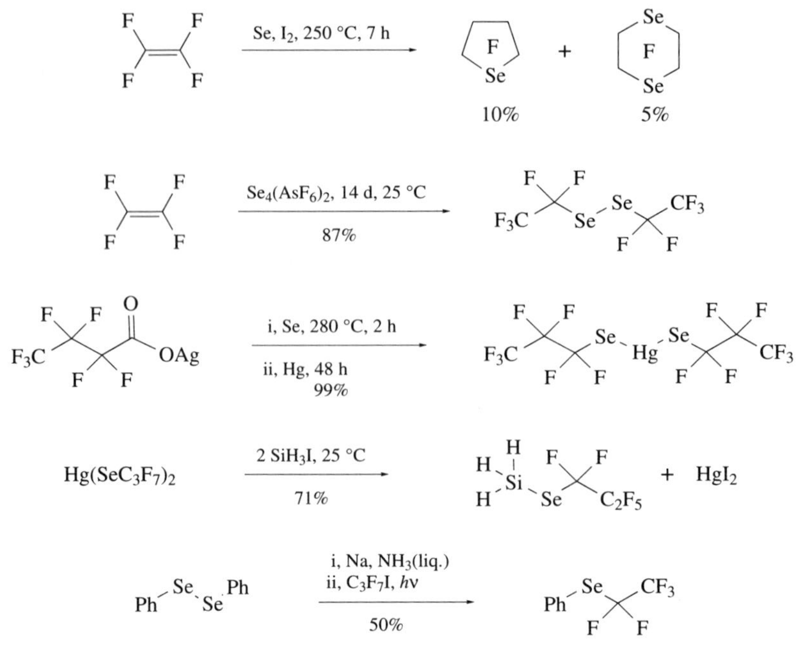

Scheme 27

corresponding ditellurides ⟨75JCS(D)488⟩. Perfluoroalkyl tellurides have also been prepared from lithium aryl tellurides (obtained by reduction of diaryl ditellurides with $LiAlH_4$) under photochemical conditions (Scheme 28) ⟨79ZOR1561⟩.

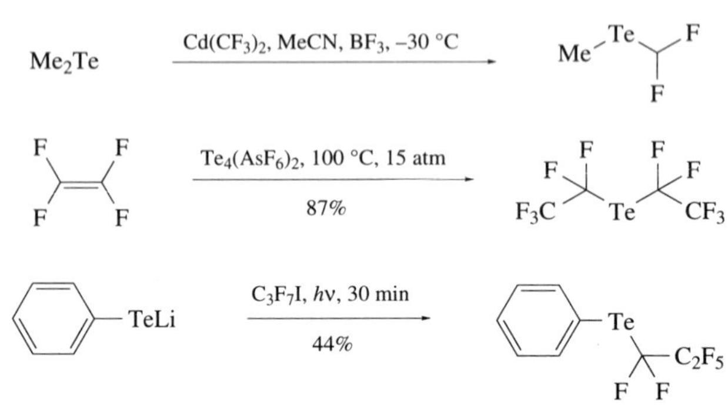

Scheme 28

6.02.1.4 Halogen and Mixed Chalcogen Derivatives ($R^1CHal(OR^2)(SR^3)$)

Chemistry of these systems is chiefly described by the corresponding oxathiolium salts, thus excluding their detailed coverage in this section. The reader is encouraged to consult references describing such systems ⟨74JHC943⟩.

6.02.2 FUNCTIONS CONTAINING HALOGEN AND A GROUP 15 ELEMENT AND POSSIBLY A CHALCOGEN

6.02.2.1 Halogen and Nitrogen Derivatives: General Considerations

Compounds of this general class are somewhat unstable, with the major contributing isomeric forms being the corresponding haloiminium salts, for example, $RCHal_2NR^1R^2$, which behaves as

RC=N$^+$R^{1}R^2 Hal$^-$ ⟨60AG836⟩. The degree to which such quaternary salts are formed depends on the overall basicity of the amino (and chalcogen) groups; thus, for most compounds of the classes RCHal$_2$NR^{1}R^2, RCHalNR^{1}R^{2}OR^3 and RCHalNR^{1}R^{2}SR^3 both species must usually be considered whereas, for compounds of class RCHal(NR^{1}R^2)$_2$, the overriding chemistry is that of the halide salt, the only exceptions being in cases where (NR^{1}R^2)$_2$ = (NO$_2$)$_2$, where the electron-withdrawing capacity of the nitro groups diminishes the tendency to ionize. Several review works discuss this dynamic equilibrium and should be consulted further ⟨B-76MI 602-01, B-79MI 602-01, B-79MI 602-02, B-79MI 602-03, B-79MI 602-04, B-79MI 602-05, 85HOU(E5)⟩.

6.02.2.1.1 *Tetracoordinate carbon atoms bearing two halogens and one nitrogen function*

Numerous examples exist of compounds in this category. In many cases, however, the chemistry of the product is often described by the corresponding haloiminium salt ⟨79S241⟩. All examples described here reflect cases where product equilibria lie in favour of the covalent species.

(i) From imines

Addition of halocarbenes to imines has been successfully used to form 2,2-dihaloaziridines bearing a variety of substituents (Scheme 29) ⟨59CI(L)1216, 62JOC3686, 71JOC3627, 73ZOR2346, 74JOC158, 76JOC3794, 77ZOR1857, 78JOC1346, 79H(12)637⟩. Methods for formation of the requisite dihalocarbene vary as does the efficiency of the addition reaction (Scheme 29). Addition of ClF to imines takes place at room temperature, to give *N*-chlorofluoroalkanes in high yield (Scheme 29) ⟨75IC1223⟩.

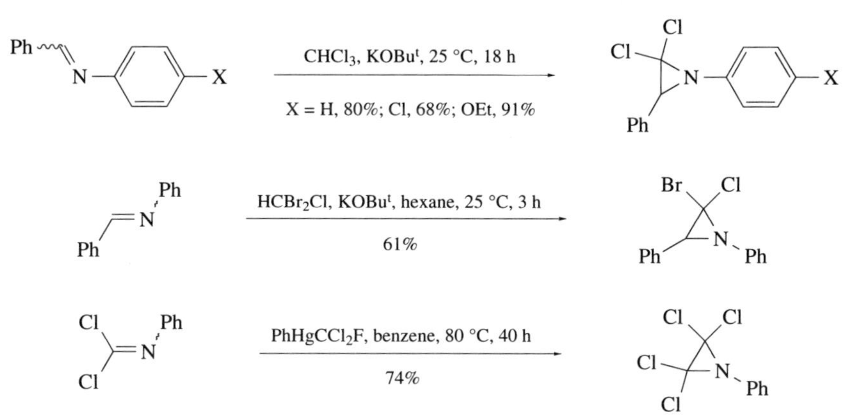

Scheme 29

(ii) From amines

A variety of halogenation procedures have been applied to tertiary amines, yielding the corresponding dihaloaminoalkanes with good conversion. Yields obtained for difluoroaminoalkanes using electrochemical fluorination methods are usually modest, halogen substitution using sulfur tetrafluoride being a far superior method (Equation (10)). Perchlorination is also an effective route, giving high yields of α,α-dichloroalkylamines (Scheme 30) ⟨59CCC4048, 69LA(730)140, 77JFC(9)279, 80JFC(17)65, 81AG(E)647, 81ZAAC(474)7⟩.

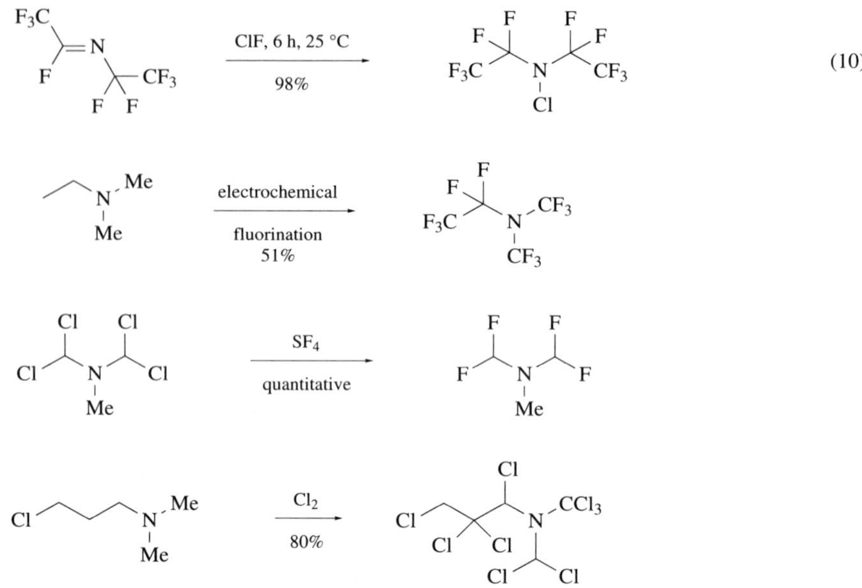

Scheme 30

(iii) From amides and thioamides

Treatment of formamides with chlorinating agents such as thionyl chloride and phosgene often results in formation of dichloroaminoalkanes which exist in equilibrium with the derived chloroiminium salts (Scheme 31) ⟨59HCA1653, 64CB1361, 79MI 602-02⟩.

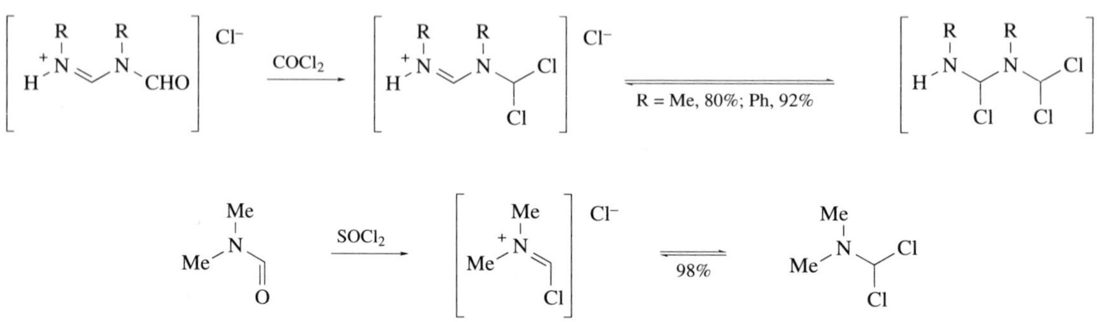

Scheme 31

Many such systems of this class have been prepared where electron-withdrawing groups limit the possibility of salt formation. A wide range of halogenation protocols have been successfully used in these instances, with a generally high level of conversion to product. These methods are useful not only for amides but also for thioamides (Scheme 32) ⟨60JA543, 62JA4275, 74JA925, 75IC1223, 76ZOR2213, 78CB921, 78JOC1727, 79AG(E)615, 81JFC(18)93, 82PJC1369⟩. Treatment of DMF with carbonyl difluoride is also reported to give a good (82%) yield of difluoro(dimethylamino)methane ⟨62JA4275⟩.

(iv) From acyl halides

Treatment of acyl fluorides with tetrafluorohydrazine under photolytic conditions results in formation of the corresponding *N*,*N*-difluoroaminoalkane (Equation (11)). One drawback of this radical addition protocol is the requirement that quartz reaction vessels be used ⟨68JOC3675⟩.

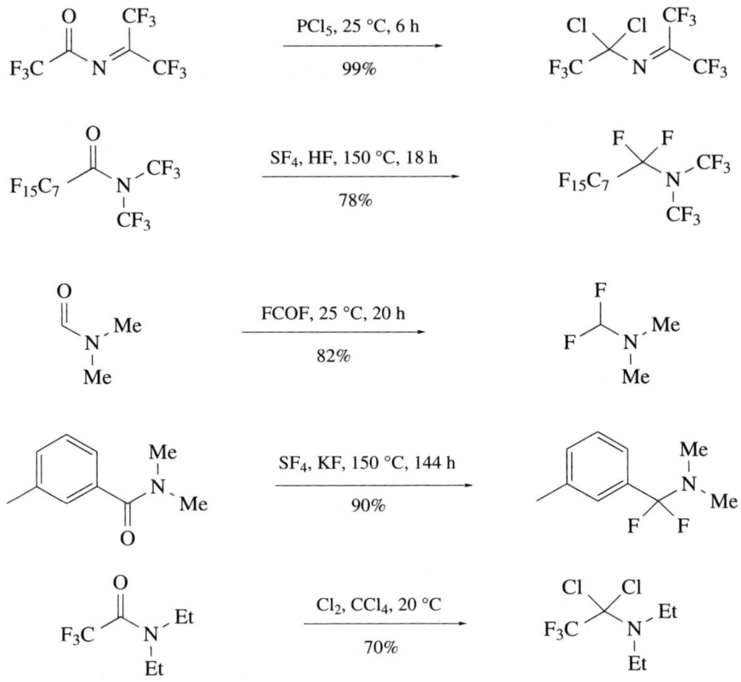

Scheme 32

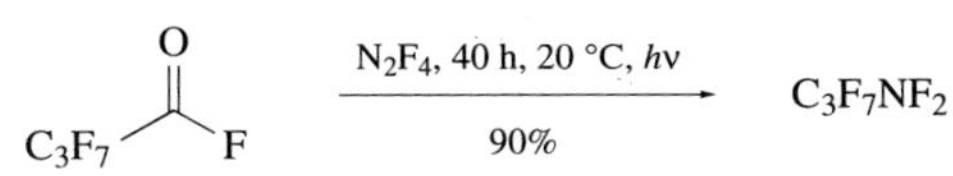

(11)

(v) From nitriles

Chlorofluorination of nitriles is a mild and convenient method for preparation of *N,N*-dihaloaminoalkanes. The reactions have only been reported for nitriles bearing electron-withdrawing groups, but yields are good to excellent (Scheme 33). The intermediate haloimine can be exposed to elemental fluorine to give *N*-fluorochloroamines or it can be reacted sequentially with ClF to yield *N,N*-dichloroamines ⟨66IC488, 79JA7640, 81IC1⟩. Addition of hydrogen fluoride to a mixture of HCN and carbonyl fluoride, catalyzed by caesium fluoride, is reported to yield difluoromethylcarbamoyl fluoride, CHF_2NHCOF (70%) ⟨62JA4275⟩. Treatment of perfluorinated cycloalkyl nitriles with silver fluoride at elevated temperature is reported to yield the corresponding perfluorinated cycloalkylmethylazo compounds ⟨84JCS(P1)455⟩.

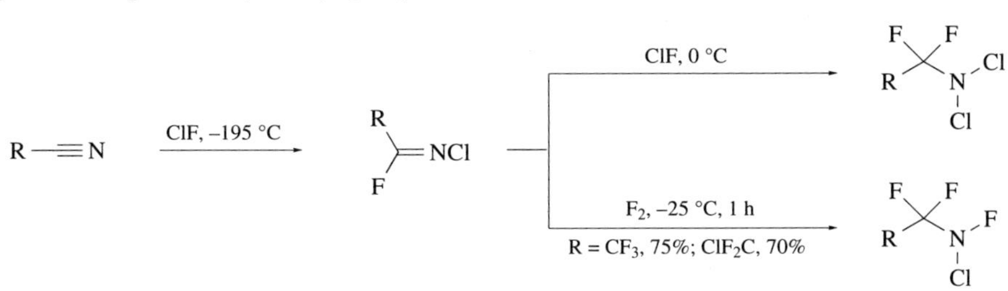

Scheme 33

(vi) From haloiminium salts

Treatment of haloiminium salts with hydrogen halides provides a facile route to α-dihaloamines (Equation (12)). The reactions are typically conducted at or below 0 °C in chlorinated solvents, and yields of addition products, which are in equilibrium with their corresponding hydrohalide salts, are good to excellent ⟨61CCC3059, 62CCC2886, 63CCC2047⟩.

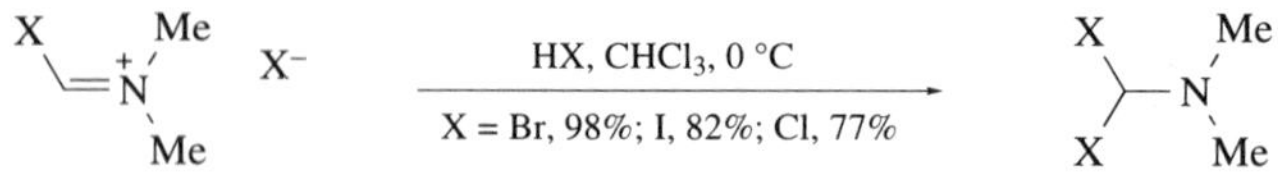

(12)

(vii) From haloalkenes and haloalkanes

Addition of diethylamine to chlorotrifluoroethene results in the formation of the corresponding diethylaminoalkane in moderate yield (Equation (13)) ⟨81AG(E)647⟩. The product, 2-chloro-1,1,2-trifluorotriethylamine (CTT), is a fluorinating agent ⟨59JGU2125⟩. A similar addition reaction has been applied to related secondary amines ⟨81JFC(18)93⟩. *N*-Iodobis(ditrifluoromethyl)amine is reported to add to hexafluoropropene under photochemical conditions to give $(F_3C)_2NCF_2CFICF_3$ in low yield ⟨71JCS(C)3833⟩. [3 + 2] Cycloaddition of benzyl azide to hexafluoropropene gives an intermediate triazoline in 85% yield, which undergoes thermal decomposition at 250 °C to give the corresponding 1-benzyl-2,3,3-trifluoro-2-(trifluoromethyl)aziridine (74%) ⟨66JOC789⟩. Cycloaddition of perfluorobutadiene to trifluoronitrosomethane is reported to give perfluoro-2-methyl-3,6-dihydro-1,2-oxazine in 60% yield ⟨65JCS6149, 66ZOB728, 67JCS(C)2263⟩. The azido ester $BrCF(N_3)CO_2Et$ has been prepared (47%) by reaction of ethyl dibromofluoroacetate with sodium azide ⟨86JOC955⟩.

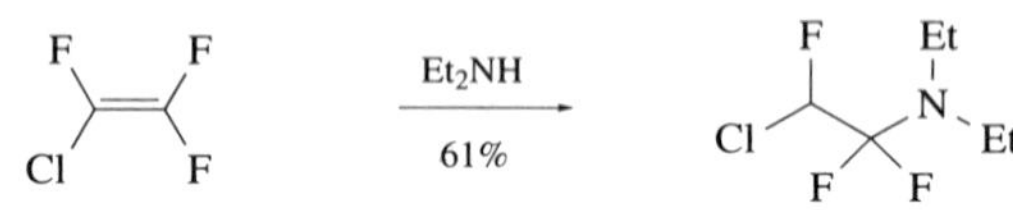

CTT

(13)

(viii) Routes specific to dihalonitroalkanes

(a) From nitromethanes. Difluoronitromethane reacts with formaldehyde under basic conditions to give 1,1-difluoro-1-nitroethanol in 88% yield (Scheme 34) ⟨67ZOB152⟩. Reactivity of its stabilized anion extends to Michael-type processes, but only moderate yields of addition products are obtained even for addition to such activated systems as acrylonitrile and ethyl acrylate ⟨79IZV1911⟩. Dibromonitromethane is reportedly formed when nitromethane is exposed to *t*-butyl hypobromite ⟨76JOC1285⟩, and 1,1-dibromoethane has been prepared from nitroethane by sequential reaction with butyllithium and bromine ⟨89ZOR2490⟩.

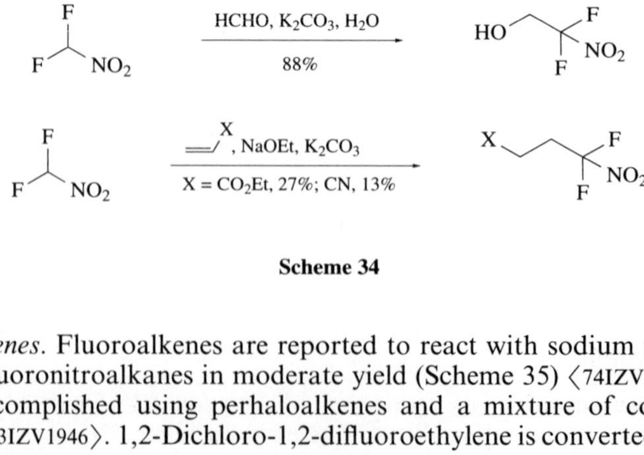

Scheme 34

(b) From haloalkenes. Fluoroalkenes are reported to react with sodium nitrite in polar aprotic solvents to furnish fluoronitroalkanes in moderate yield (Scheme 35) ⟨74IZV2144⟩. Similar addition reactions can be accomplished using perhaloalkenes and a mixture of concentrated nitric and hydrofluoric acids ⟨63IZV1946⟩. 1,2-Dichloro-1,2-difluoroethylene is converted by nitration followed

by reaction with fuming nitric acid into chlorofluoronitrosomethane, which can then be oxidized to chlorofluoronitromethane ⟨90MI 602-01⟩.

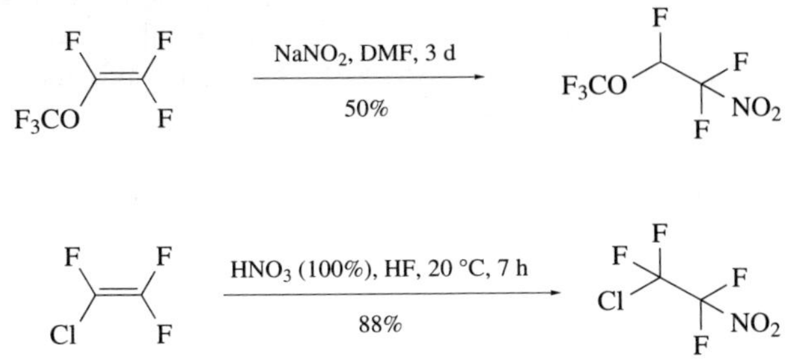

Scheme 35

(c) From nitro alcohols. 2-Fluoro-2-nitro-1-butanol reacts with aqueous sodium hypobromite to give 1-bromo-1-fluoro-1-nitropropane (Equation (14)). The reaction is believed to proceed via a transient α-fluoronitronate salt ⟨70JOC846⟩.

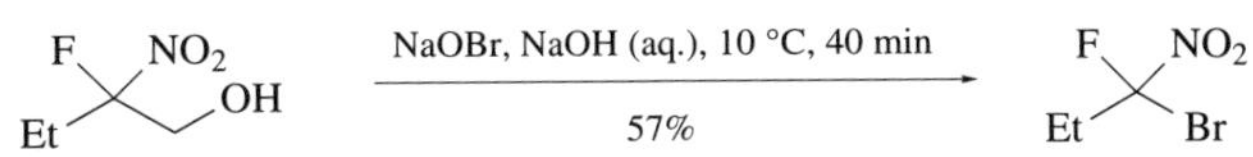

(14)

6.02.2.1.2 Tetracoordinate carbon atoms bearing one halogen and two nitrogen functions

In all cases in which the nitrogen functions are amino groups, attempts to prepare compounds of this class results in the formation of the corresponding haloiminium salts. Examples are widespread, covering simple amines ⟨65CB1078, 73LA40, 75LA195⟩, cyclic amines ⟨61CB2594, 69RTC289⟩ and hydrazines ⟨71JOC1155⟩.

One class of compounds within this category which retains stability, however, are the halodiazirines (Scheme 36). Typical conditions involve exposure of an acetamidine to bromide saturated DMSO with sodium chlorite: the so-called 'Graham conditions' ⟨65JA4396, 67JA182, 68JOC1847, 81JA6164, 87TL5801, 89ACR15⟩ or treating *N*,*N*-dihaloamidines with base (Scheme 36) ⟨81JOC5048⟩.

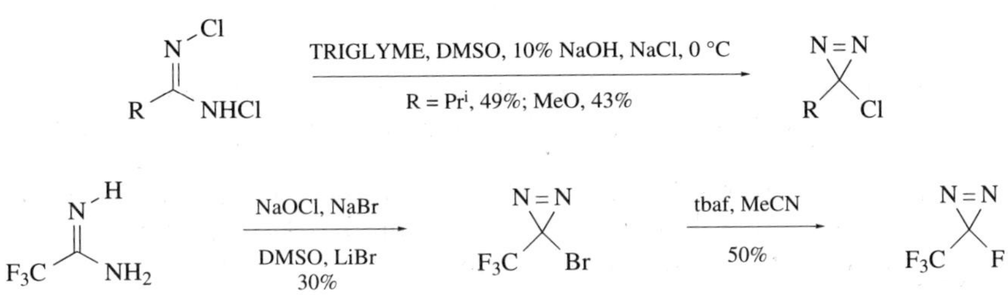

Scheme 36

(i) Tetracoordinate carbon atoms bearing one halogen and two nitro groups ($RCHal(NO_2)_2$)

Nearly all compounds in this class have been prepared from the salts of dinitroalkane anions, reacting with a halogenating agent. They are grouped according to the nature of the halogenation procedure employed.

(a) Using elemental halogens. Typically, conditions involve formation of the potassium salt of the dinitroalkane, followed by exposure to the halogen source *in situ*. The salts are either prepared by treating the dinitroalkane with the required potassium base or, in the case of dinitromethane, via a Ter Meer reaction with chloronitromethane, itself available by chlorination of nitromethane. The most commonly used procedure is to pass a fluorine–nitrogen mixture into the aqueous solution of the salt at or close to 0 °C (Scheme 37) ⟨68IZV431, 68IZV2307, 69IZV1331, 73IZV1424, 78JOC3485⟩; xenon difluoride is also effective as a fluorinating agent ⟨87ZOR1657⟩. Similarly high yields have also

been obtained using the sodium or ammonium salts of dinitroalkanes ⟨68IZV912⟩ and of 2,2-dinitropropane-1,3-diol in which one of the hydroxymethyl groups is cleaved (Scheme 37) ⟨68JOC3080⟩. Chlorination of the sodium salts are also reported to give excellent yields of the derived chlorodinitroalkane (Scheme 37) ⟨69IZV2617, 87IZV306⟩, and bromodinitromethane has been prepared by reaction of the sodium salt of dinitromethane with bromine. ICl has been used to form the corresponding iododinitroalkanes ⟨77IZV2058⟩.

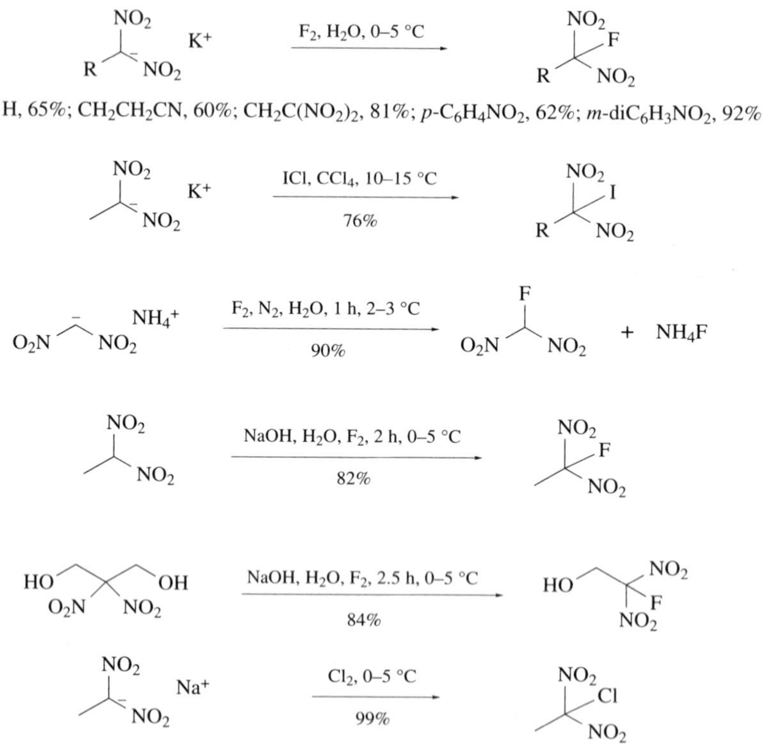

Scheme 37

(b) Using perchloryl fluoride. Most work in this category has been conducted using potassium salts of the dinitroalkanes, providing a generally high yielding approach to the desired fluorodinitroalkanes (Equation (15)) ⟨68JOC3073, 70IZV387, 71IZV1487⟩; similar results (albeit with inferior yields) are obtained using sodium salts ⟨72JOC152⟩. For systems where the corresponding salts of dinitroalkanes are unstable, use of hexamethylphosphoramide (HMPA) and perchloryl fluoride is reported to allow efficient fluorination within 2 hours (Scheme 38) ⟨69IZV1188⟩. Interestingly, exposure of $ClOSO_2F$ to cyano-substituted dinitroalkane salts results in formation of the derived *N*-chloroimidates (Scheme 38) ⟨76IZV489⟩.

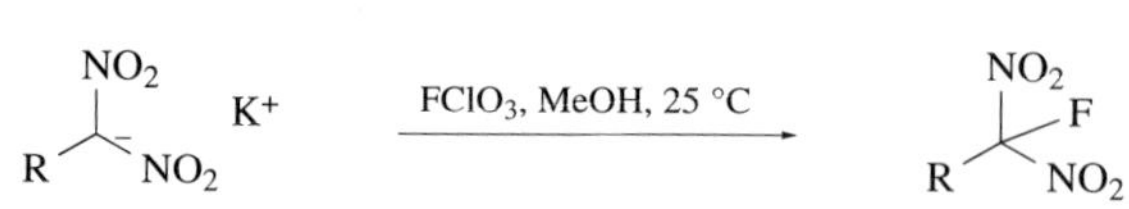

(15)

R = Ph, 95%; $CH_2CH_2CO_2Me$, 88%; $CH_2NHCOPh$, 88%; $CH_2CH_2NHCOMe$, 93%; $CH_2CH{=}CH_2$, 69%

(c) Using hydrogen halides. Treatment of sulfonium dinitroylides with hydrogen halides results in rapid formation of the corresponding halodinitroalkane (Equation (16)). The cleavage method is applicable to a range of electrophilic halogen sources including elemental halogens and hypohalites, with a dialkyl sulfide being the chief by-product ⟨77IZV139⟩.

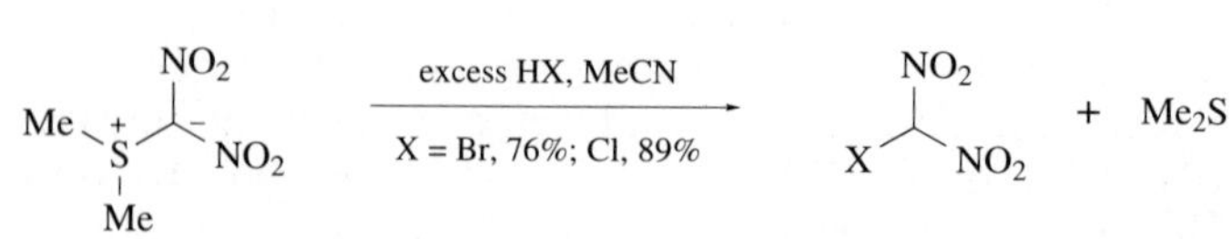

(16)

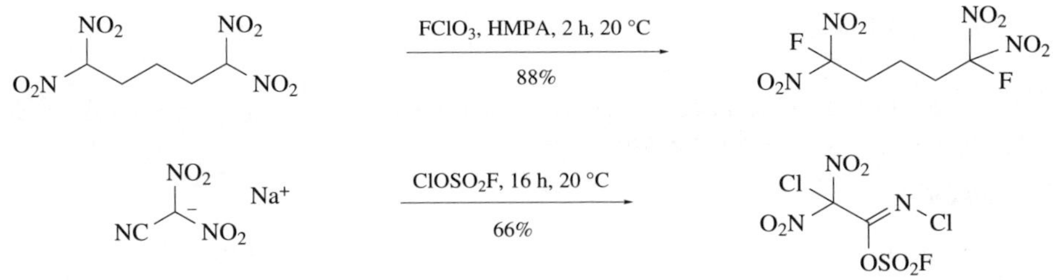

Scheme 38

(d) Functional group interconversion of dinitrohaloalkanes. The fluorodinitromethanide anion undergoes a variety of additional transformations relevant for this section. Most importantly, under basic conditions, 1, 2 additions to aldehydes are possible to yield secondary alcohols ⟨70JOC3188⟩, as are 1, 4 additions to conjugated alkenes ⟨79S311⟩, giving high yields of addition products in both cases. Related examples of compounds incorporating this functionality where the dinitrohalo group is left intact include acid-catalyzed dimerization of dinitrofluoroalcohols ⟨69JOC45⟩, hydrolysis of fluorodinitroacetonitrile to yield fluorodinitroacetamide ⟨68JOC1257⟩ and dehydroxymethylation of 2,2-dinitro-2-fluoroethanol ⟨69JOC45⟩.

(ii) Tetracoordinate carbon atoms bearing one halogen, one nitro group and one other nitrogen function

An example of this class is the ester $BrC(N_3)(NO_2)CO_2Et$, which was obtained in low yield from ethyl dibromonitroacetate by reaction with sodium azide ⟨86JOC955⟩.

6.02.2.1.3 *Tetracoordinate carbon atoms bearing one halogen, one nitrogen and one oxygen function*

Based on the previous discussions concerning the stability of *ortho*-ester halides attached to basic functionality, salt formation is predominant. Probably the best studied example of this class is the Vilsmeier reagent and related systems, formed when formamides are exposed to chlorinating agents (Scheme 39) ⟨58CCC452, 59HCA1659⟩. The equilibrium lies heavily in favour of salt formation, explaining in this case, and related examples ⟨69TL2161, 71CPB2629⟩, why members of this class have received scant attention. Other examples can be found describing intermediate formation of this class of compound, which quickly undergoes rearrangement to give the halide salt ⟨68JOC1084, 70CPB784, 72TL4217, 73TL4511, B-79MI 602-03, 80IZV2356⟩, and in some cases to give enamines ⟨65JOC4303, 68T4217⟩.

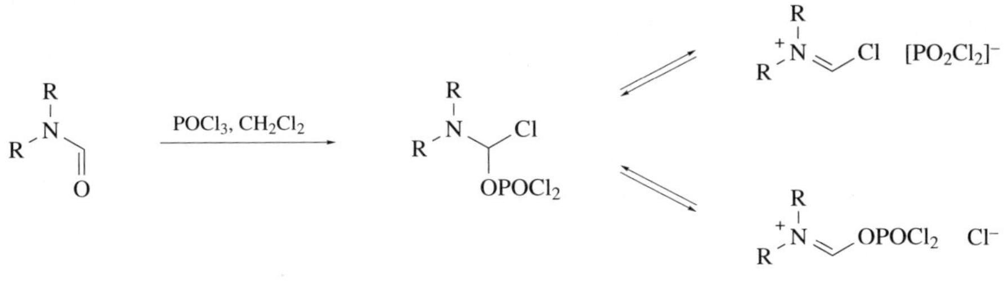

Scheme 39

6.02.2.1.4 *Tetracoordinate carbon atoms bearing one halogen, one nitrogen and one sulfur function*

Compounds of this class in which the nitrogen function is an amino group are nearly always found in the form of the corresponding iminium halide salt ⟨B-79MI 602-05⟩. The exceptions are nitrohalosulfides. 1-Chloro-1-chlorosulfenylnitroalkanes have been reported ⟨85HOU(E5)⟩ and the esters $PhS(NO_2)CXCO_2Et$ (X = Br, and F) have been prepared by halogenation of the ester PhS-$(NO_2)CHCO_2Et$ ⟨86JOC955⟩.

6.02.2.2 Halogen and Phosphorus Derivatives

6.02.2.2.1 Tetracoordinate carbon atoms bearing two halogen and one phosphorus function and one halogen and two phosphorus functions

(i) From haloalkenes

Addition of phosphine to fluoro and chlorofluoroalkenes is a general route to compounds of this class. Generally, the addition reactions are conducted at elevated temperatures, often in sealed tubes. Phenylphosphine has also been employed, with only slight loss of efficiency resulting (Scheme 40) ⟨59JA4801, 65AFC(4)50⟩. If the reactions are conducted under photolytic conditions, however, yields are often substantially improved, with 84% for the addition to tetrafluoroethylene reported ⟨63JCS1083⟩.

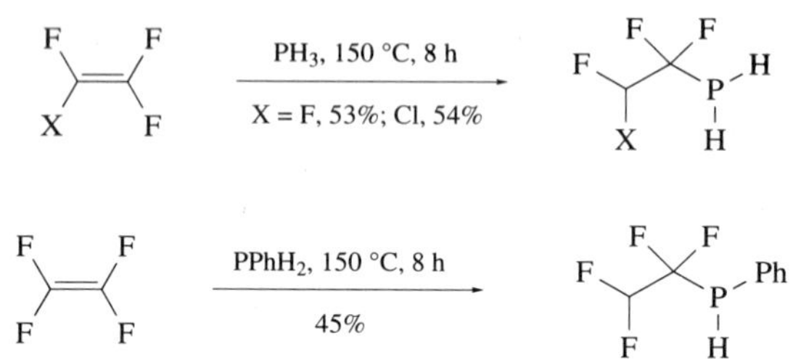

Scheme 40

(ii) From haloalkanes

Thermolysis of carbon tetrafluoride with red phosphorus in the presence of iodine results in the formation of a cyclic bis(phosphine) in moderate yield (Scheme 41) ⟨62JOC3584⟩. 2,2,2-Trichloroacetic amidines and imidates react with trialkyl phosphites to yield 2,2-dichloro-2-(dialkoxyphosphinyl)acetamidines and acetimidates, respectively (Scheme 41) in good to high levels of conversion ⟨79ZOB1025⟩. If the reactions are prolonged, a second substitution occurs giving a monohalobis(dialkylphosphinyl) derivative. Addition of chlorodifluoromethane to sodium diethylphosphonate results in the formation of diethyl difluoromethylphosphonate ⟨59JGU1115⟩. The mechanism of this reaction appears to involve capture of difluorocarbene rather than an S_N2 type process. Prompted by research on purine nucleoside phosphorylase inhibitors, it has been shown that diethylphosphinyldifluoromethyllithium condenses with *ortho*-α-bromoxylene to give the corresponding difluorophosphonate in 43% yield (Scheme 41) ⟨92BMC407⟩. Interestingly, other metal dialkylphosphinyldifluoroalkane derivatives proved vastly inferior in this reaction. Diisopropylphosphinyldifluoromethyllithium was shown to give good yields of the 1,2-addition product to *o*-tolualdehyde at low temperature (Scheme 41); the benzylic alcohol was subsequently converted into the trifluoro derivative using diethylaminosulfur trifluoride ⟨92BMC407⟩. Diethyl (lithiodifluoromethyl)phosphonate has also been used to introduce the α,α-difluorophosphonate moiety into phosphate isosteres ⟨92TL1839, 94TL3227⟩. Though alkyl halides are often poor substrates in such coupling reactions, Martin has reported conditions for addition to aldehydes, where the resulting product carbinols are deoxygenated using a Barton-type protocol ⟨92TL1839⟩. The process is versatile, tolerant of latent functionality, and complements existing methods for the introduction of this important moiety ⟨81JFC(18)197, 82TL2323, 89JOC613, 89TL7013, 91TL1019⟩. Dialkyl (iodofluoromethyl)phosphonates, available from dialkyl (bromodifluoromethyl)phosphonates, are reported to undergo reductive dimerization to yield tetrafluoroethyl derivatives using cadmium powder, refluxing in methylene chloride (Scheme 41) ⟨94JOC2393⟩. The corresponding dialkyl (bromodifluoromethyl)phosphonates fail to undergo this coupling using cadmium, but this was finally effected using a nickel derivative. The nickel approach also facilitated dimerization of a dialkyl (iodotetrafluoroethyl)phosphonate, giving the coupled product in 48% yield ⟨94JOC2393⟩. The addition of diethyl trichloromethylphosphonate to alkenes, catalyzed by copper chloride, has been reported ⟨94TL3537⟩. Ethyl acrylate thus gives the corresponding dichlorophosphonate in moderate yield (Scheme 41).

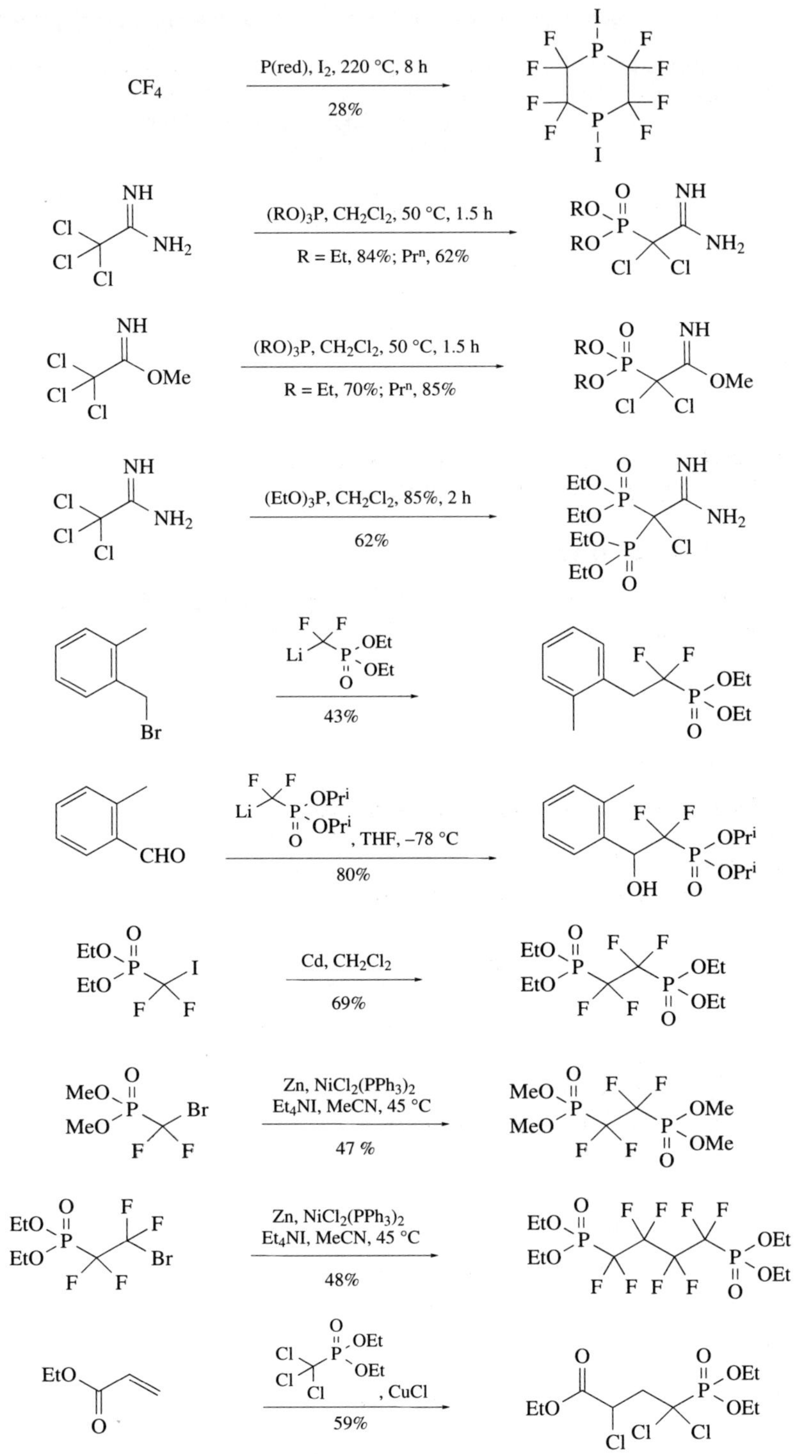
CF4
P(red), I2, 220 °C, 8 h
28%
(RO)3P, CH2Cl2, 50 °C, 1.5 h
R = Et, 84%; Prn, 62%
(RO)3P, CH2Cl2, 50 °C, 1.5 h
R = Et, 70%; Prn, 85%
(EtO)3P, CH2Cl2, 85%, 2 h
62%
43%
, THF, –78 °C
80%
Cd, CH2Cl2
69%
Zn, NiCl2(PPh3)2
Et4NI, MeCN, 45 °C
47 %
Zn, NiCl2(PPh3)2
Et4NI, MeCN, 45 °C
48%
, CuCl
59%

Scheme 41

6.02.2.3 Halogen and Arsenic Derivatives

6.02.2.3.1 Tetracoordinate carbon atoms bearing two halogens and one arsenic function

Dimethylarsine is reported to add to hexafluoropropene to give the corresponding dimethylarsenic derivative (Scheme 42) ⟨65AFC(4)50⟩. Unfortunately, no details either of yield or of conditions for this process were reported. Another report describes the reaction of tetramethyldiarsine ('cacodyl') with hexafluoropropene to produce a fluorodiarsine in 18% yield (Scheme 42) ⟨64CJC1123⟩.

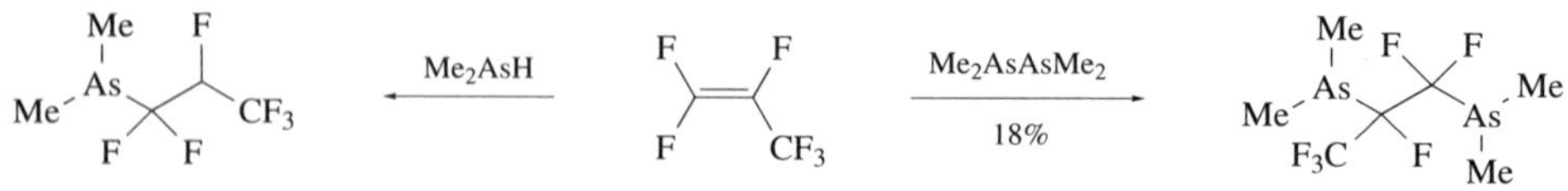

Scheme 42

6.02.3 FUNCTIONS CONTAINING HALOGEN AND A METALLOID AND POSSIBLY A CHALCOGEN AND/OR GROUP 15 ELEMENT

6.02.3.1 Halogen and Silicon Derivatives

A variety of α,α-dihaloalkylsilanes are known but few general methods for their preparation exist ⟨64CB1673⟩. The following examples cover specific cases where conditions have been examined and optimized.

(i) Dichlorocarbene additions

Addition of dichlorocarbene (generated from phenyl(bromodichloromethyl)mercury) to a variety of silanes has been reported, giving good yields of a dichloromethyl silanes (Scheme 43). Competition experiments have been carried out in order to determine the kinetics and the precise mechanism of this electrophilic addition process ⟨67JA1538, 68JA2944, 93T8487⟩.

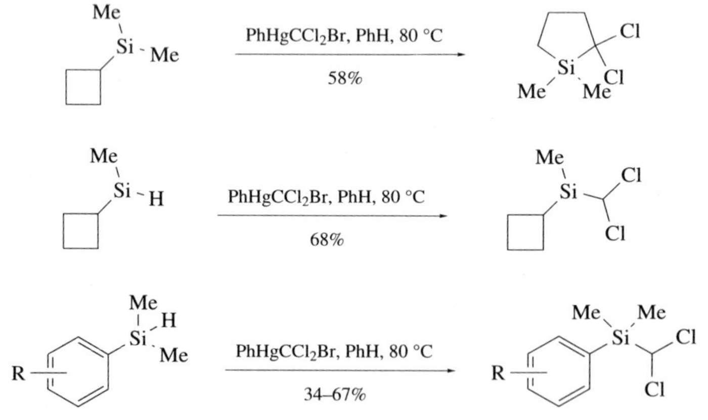

Scheme 43

(ii) Other additions to silanes

Low temperature metal halogen exchange of bis(trimethylsilyl)dichloromethane followed by protonation has been used to prepare bis(trimethylsilyl)chloromethane (Scheme 44) ⟨71JOM(29)389, 90T3859⟩. Bromination of alkynylsilanes has also been reported, the major isolated addition product being the tetrabromosilane in nearly quantitative yield (Scheme 44) ⟨67IZV693⟩. An alternative approach to dihalosilanes is cleavage of a bis(trichloromethylsilyl)dichloromethane using hydrochloric acid (Scheme 44), with the α,α-dichloromethylsilane being formed in 87% yield ⟨64CB1115⟩. Another report describes alkyl Grignard addition to trichlorosilyldichloromethane, which yields using methylmagnesium bromide, trimethylsilyldichloromethane (61%) ⟨66CB793⟩. Photochemical addition of silyl radicals to perfluorinated alkenes has also been employed, UV irradiation of silane in the presence of tetrafluoroethylene giving a 61% yield of 1,1,2,2 tetrafluoroethylsilane in 18 h ⟨65JCS2101⟩. Similarly, photochemical addition of the chlorosilyl radical derived from methyldichlorosilane to tetrafluoroethylene results in a 98% yield of the α,α-difluorosilane adduct ⟨68ZOB2813⟩. Lithiation of 1,6-dibromodecafluorohexane, followed by addition of trimethylsilyl chloride, results in formation of 1,6-bis(trimethylsilyl)decafluorohexane in 66% yield ⟨69JOM(16)33⟩.

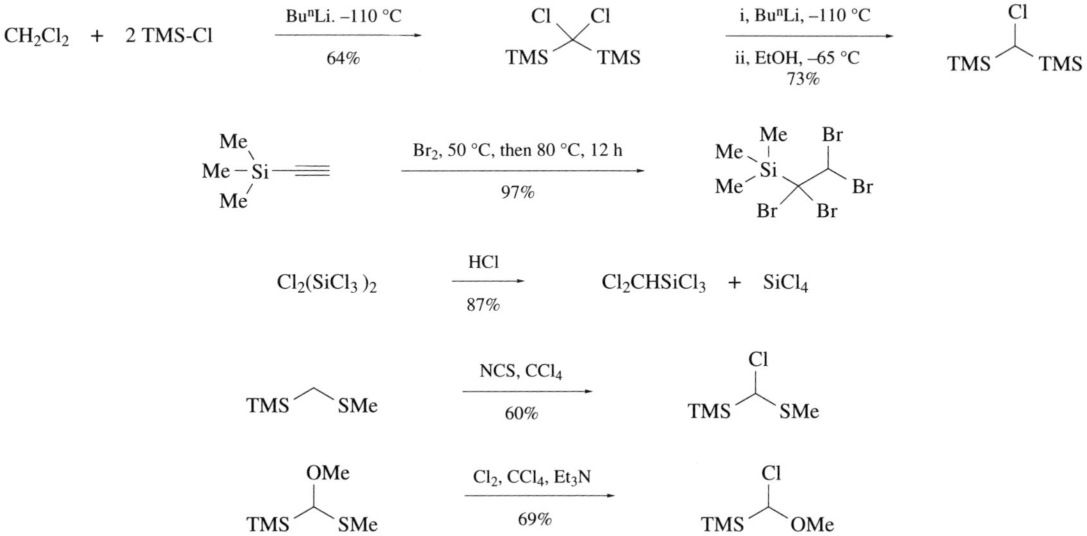

Scheme 44

Preparation of systems containing a metalloid, halogen and chalcogen have also been reported. Chlorination of (methylthiomethyl)trimethylsilane has been accomplished, giving [chloro-(methylthio)methyl]trimethylsilane in 60% yield (Scheme 44) ⟨89T637⟩. Additionally, [methoxy(methylthio)methyl]trimethylsilane (itself prepared from [chloro(methylthio)methyl] trimethylsilane by addition of sodium methoxide) has been converted to [chloro(methoxy) methyl]trimethylsilane (Scheme 44) in 69% yield, using traditional chlorination conditions ⟨89T637⟩.

6.02.3.2 Halogen and Boron Derivatives

Compounds of this class are extremely unstable, which accounts in part for their general lack of utility in synthesis. The general mechanism for the decomposition of α-bromoalkylboranes (formed from trialkylboranes and bromine) appears to involve HBr-induced cleavage which may proceed via a radical pathway, rather than by simple rupture of the C—B bond ⟨70JA7212⟩. Treatment of α,α-dihaloalkylboranes with nucleophiles such as water also induces rearrangement, providing a route, after oxidation, to secondary alcohols ⟨71JA2796⟩. An α,α-dichloroalkylborane was prepared by Seyferth by dichlorocarbene insertion into a trialkylborane (Scheme 45); however, spontaneous rearrangement occurred with subsequent elimination from the chloroborane species ⟨66JA1834⟩. It is reported that the stability of the intermediate α,α-dichloroalkylborane is a function of the borane R group, and that triarylboranes (R = Ph) yield the more stable adducts.

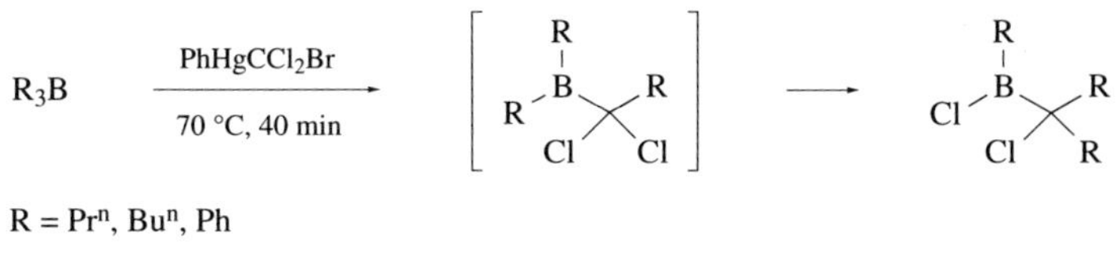

Scheme 45

6.02.3.3 Halogen and Germanium Derivatives

Trichlorogermane has been shown to undergo addition to chlorinated alkenes, and this constitutes an excellent route to α,α-dichloroalkylgermanes (Scheme 46) ⟨60DOK(131)98⟩. Photochemical chlorination of alkylgermanium compounds has also been reported, giving moderate yields of the α,α-dichloro species (Scheme 46) ⟨67ZOB1040⟩. Bis(perfluoroalkyl)mercury compounds have been shown to react with germanium tetrahalides to produce perfluorogermanium alkyls in good yield (Scheme 46) ⟨78JA1722⟩. The trihalo(perfluoroalkyl)germanium compounds produced are versatile synthetic intermediates, since they react with dialkylcadmium reagents to give alkyl (perfluoroalkyl)germaniums ⟨78JA1722⟩. Bis(triethylgermyl)mercury undergoes exchange with bis(haloalkyl)mercury compounds, to yield mixed haloalkyl (triethylgermyl)mercurials ⟨72JOM(34)299⟩. These mixed mercurials react with caesium fluoride to produce haloalkyl germaniums in good yield (Scheme 46) ⟨72JOM(34)299⟩. Bis(trialkylgermyl)mercury reacts with bis(haloalkyl)mercury compounds to produce a variety of mixed haloalkyl germanium species ⟨72IZV85⟩. The intermediate is presumably a mixed (triethylgermyl)mercury species, and the isolated yields are good to excellent ⟨72IZV85⟩.

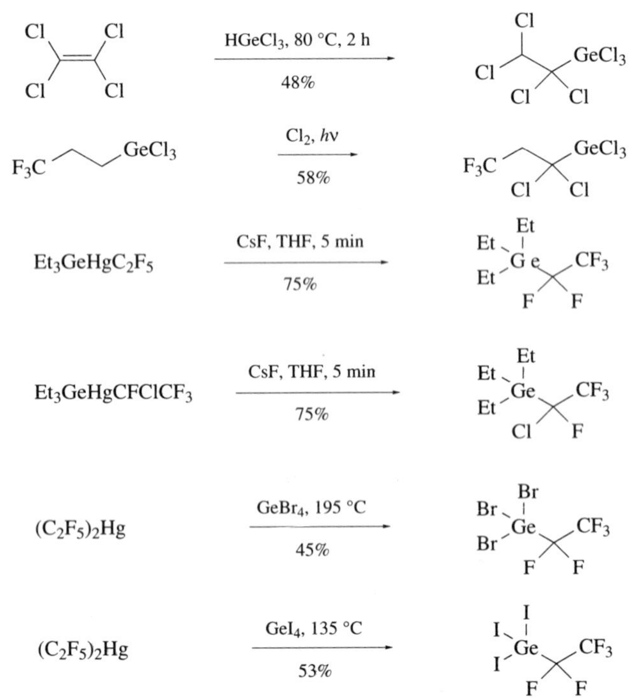

Scheme 46

6.02.4 FUNCTIONS CONTAINING HALOGEN AND A METAL AND POSSIBLY A GROUP 15 ELEMENT, A CHALCOGEN OR A METALLOID

6.02.4.1 Halogen and Lithium Derivatives

Under the general heading of 'lithium carbenoids', haloalkyllithium compounds have enjoyed widespread use due to their unusual reactivity and their ability to deliver the versatile halomethylene functional group to suitable acceptors. A critical aspect of these species is the role of solvent used in their preparation, THF being the solvent of choice, owing to its ability to stabilize the carbenoid species. However, the increased viscosity of THF at the low temperatures typically employed to generate carbenoids sometimes poses problems, and often ether or petroleum ethers are used as co-solvents. Typical conditions involve treating a dilute solution of either a dihaloalkane or a trihaloalkene at −100 °C with an alkyllithium reagent in THF (Scheme 47) ⟨65JA4147, 65TL969, 93T8487⟩. Since the carbenoids are typically thermally unstable (decomposing completely above −65 °C), *in situ* reaction with a suitable electrophile is usually carried out, otherwise intramolecular carbenoid insertions may take place. Examples of lithiodifluoroalkanes are common, available either by metallation of difluoroiodoalkanes at low temperature ⟨85TL5243⟩ or by metallation of difluoroalkanes with lithium amide bases at elevated temperature ⟨77JOC565⟩. Terminal lithioperfluoroalkanes often undergo spontaneous elimination of LiF to yield perfluorinated alkenes ⟨85TL5243⟩. The alkoxycarbenoid $ClCH_2CH_2OCHClLi$ has been generated and used to cyclopropanate cyclohexene ⟨69OS(49)86, 72JA125⟩.

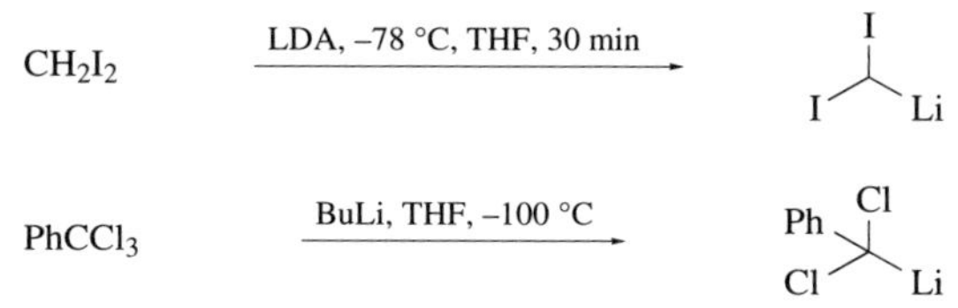

Scheme 47

6.02.4.2 Halogen and Magnesium Derivatives

Transmetallation reactions between Grignard reagents and perhalogenated systems is a versatile route to compounds of this class. Thus phenylmagnesium bromide reacts with 1-iodoheptafluoropropane at low temperature to produce heptafluoropropylmagnesium bromide in good yield (Scheme 48) ⟨57JA3351⟩. Such Grignard reagents are useful for addition of perhaloalkyl functionality to carbonyl compounds ⟨88BCJ3321⟩. Other examples of this transmetallation protocol demonstrate its synthetic utility (Scheme 48) including formation of bis(bromomagnesium) compounds ⟨75JFC(5)475⟩.

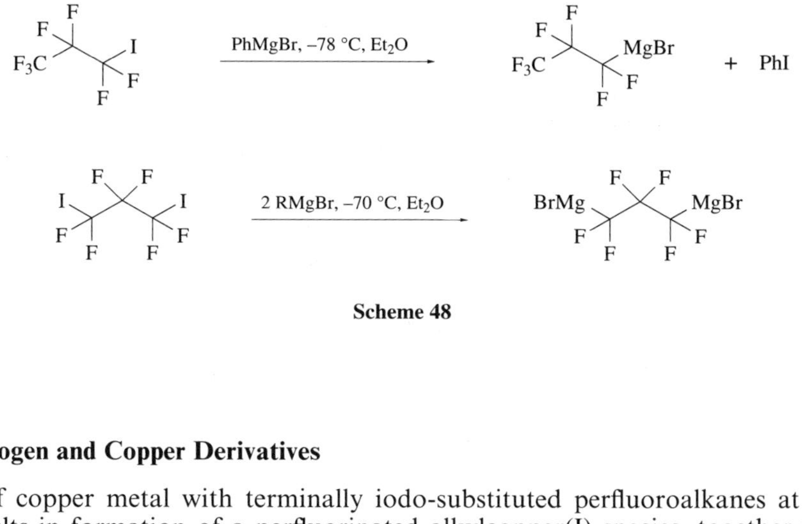

Scheme 48

6.02.4.3 Halogen and Copper Derivatives

Reaction of copper metal with terminally iodo-substituted perfluoroalkanes at elevated temperatures results in formation of a perfluorinated alkylcopper(I) species, together with copper(I) iodide (Equation (17)) ⟨87JFC(37)171⟩. Chain polymerization of these species has been reported on thermolysis in DMF ⟨88DIS(B)2973⟩.

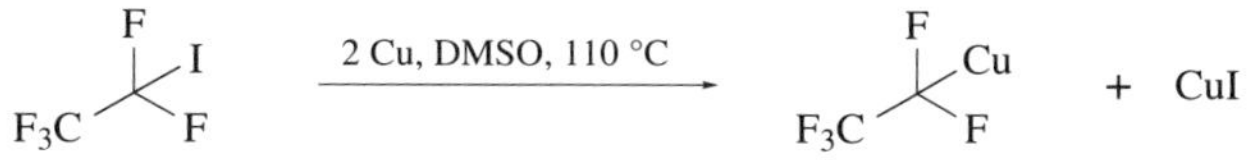

(17)

6.02.4.4 Halogen and Silver Derivatives

Addition of silver trifluoroacetate to chlorotrifluoroethene in the presence of metal fluorides results in the formation of an organosilver compound bearing an α,α-dihalo alkyl group (Equation

(18)) ⟨73JOM(57)423⟩. Caesium fluoride is reported to promote this reaction more efficiently than other alkali metals.

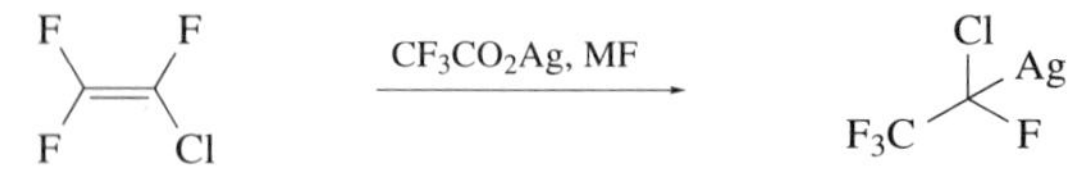

(18)

6.02.4.5 Halogen and Zinc Derivatives

Zinc metal reacts with alkyl iodides, chlorides and bromides to give the corresponding halozinc reagents (Scheme 49) ⟨53JCS3607, 57JA4159, 84JFC(26)435, 88JOM(339)17⟩. Yields of organozinc reagent are heavily dependent on both concentration and temperature of reaction and, where ethers are not employed as solvents, solvated products are formed, for example, with DMF.

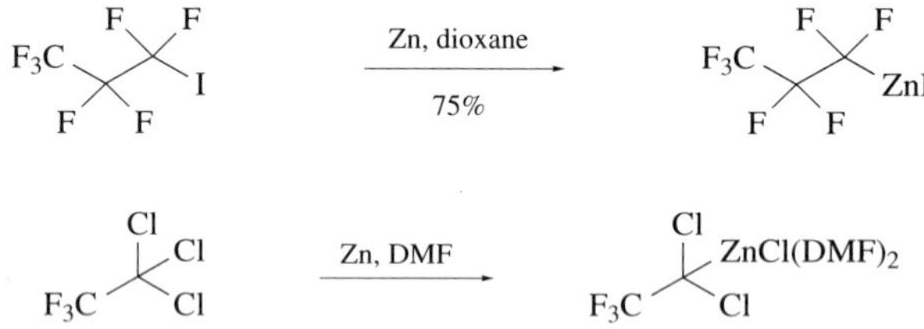

Scheme 49

6.02.4.6 Halogen and Cadmium Derivatives

Cadmium metal will insert between the carbon–bromine or carbon–iodine bond in mixed perhaloalkanes to yield a variety of organocadmium reagents (Scheme 50) ⟨88JFC(39)425, 88JFC(41)185⟩. Alternatively, transmetallation from dialkylcadmium reagents to dihaloiodoalkanes can be employed to form perhalodialkylcadmium reagents of this class.

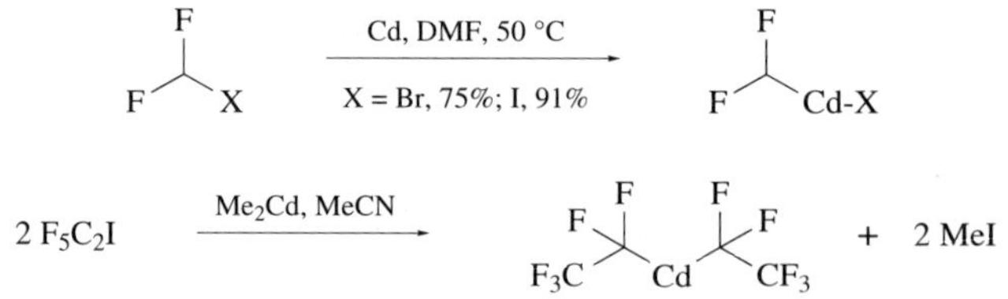

Scheme 50

6.02.4.7 Halogen and Mercury Derivatives

Mercuric fluoride will add to haloalkenes quite readily at elevated temperatures to yield perhalodialkylmercury compounds (Scheme 51) ⟨60JOC105, 61JCS3825⟩. Strict regiocontrol in the addition reactions to mixed haloalkenes is usually observed. It has been reported that perfluoropropen-2-ol, the *meta*-stable enol form of pentafluoroacetone, reacts with mercury trifluoroacetate to yield trifluoroacetoxymercuricperfluoroacetone (Scheme 51) ⟨75DOK(221)1331, 76ZOR1377⟩. Care must be taken when handling this species, as it is reported to undergo rapid hydration to form the corresponding hydrate.

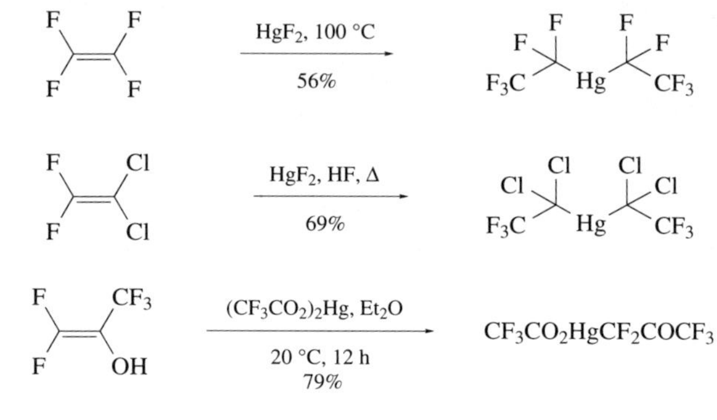

Scheme 51

6.02.4.8 Halogen and Tin Derivatives

Reaction of dialkyltin hydrides with tetrafluoroethylene is reported to give the 1:2 adducts (Scheme 52) ⟨65AFC(4)50⟩. Another route to these compounds is via insertion to iodoperfluoroalkanes using either tin(II) chloride or tin(II) fluoride (Scheme 52). The resulting organotin compounds retain the halogen atoms originally attached in this rapid reaction ⟨81CL1337⟩. Reaction of trimethyltin hydride with trialkyl(trifluoromethyl)stannanes results in reduction to the corresponding difluoroalkyltin species (Scheme 52) ⟨70IC1682⟩. Bis(perfluoroalkyl)cadmium compounds are reported to undergo transmetallation with $Me_3SnOCOCF_3$ to yield the corresponding perfluoro organotin derivatives (Scheme 52) ⟨85JFC(27)309⟩. Tetrakis(trifluoromethyl)germanium is reported to react with trimethyltin hydride to produce, in addition to tris(trifluoromethyl)germane, trimethyl(difluoromethyl)tin in low yield ⟨94JOM(465)153⟩.

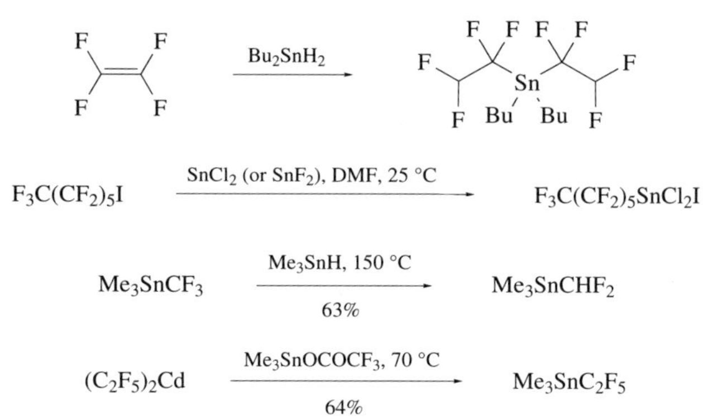

Scheme 52

6.02.4.9 Halogen and Lead Derivatives

Addition of tetramethyllead to iodopentafluoroethane at elevated temperature results in formation of pentafluoroethyltrimethyllead in moderate yield (Scheme 53) ⟨60JA6228⟩. In an analogous manner to the corresponding organotin compounds, bisperfluoroalkyl cadmium species undergo transmetallation with $Me_3PbOCOCF_3$ to yield perfluoro organolead derivatives (Scheme 53) ⟨85JFC(27)309⟩.

$$CF_3CF_2I \xrightarrow[28\%]{Me_4Pb,\ 150\ °C} Me_3PbCF_2CF_3$$

$$(C_2F_5)_2Cd \xrightarrow[91\%]{Me_3PbOCOCF_3,\ 70\ °C} Me_3PbC_2F_5$$

Scheme 53

6.03
Functions Containing Three Chalcogens (and No Halogens)

GLYNN MITCHELL
ZENECA Agrochemicals, Bracknell, UK

6.03.1 INTRODUCTION

The scope of this chapter is intended to cover the synthesis of compounds bearing the C(X)(Y)Z group in which each of X, Y and Z are independently linked through O, S, Se and Te atoms. However, at the time of writing no such compounds containing Te had been reported in the literature, so the discussion is limited to O, S and Se containing compounds.

6.03.2 FUNCTIONS BEARING THREE OXYGEN ATOMS

Carboxylic *ortho*-esters contain a carbon atom bearing three alkoxy or aryloxy groups. Closely related to these are the corresponding acyloxy esters (bearing at least one RCOO group), peroxy esters (bearing at least one ROO group) and aminoxy esters (bearing at least one R^1R^2NO group). *Ortho*-esters which contain one or more hydroxy groups are generally unstable and decompose to carboxylic esters and alcohols, though a few examples are known in which *ortho*-hydrogen esters are stabilised by electronic and/or ring strain effects 〈26BSB412, 65T2059, 80JOC2096〉. Methods for the preparation of *ortho*-esters and related compounds have been reviewed by Post 〈B-43MI 603-01〉, DeWolfe 〈B-70MI 603-01, 74S153〉, Sandler and Karo 〈B-86MI 603-01〉 and Simchen 〈85HOU(E5)3〉. Other reviews concentrate on the reactions of *ortho*-esters 〈B-69MI 603-01, 86UK1803〉, and on their use in polymerisation reactions 〈85MI 603-01〉.

6.03.2.1 Methods for the Preparation of Carboxylic *Ortho*-esters and Related Compounds

6.03.2.1.1 **Ortho-*esters from 1,1,1-trihaloalkanes, α,α-dihalo ethers and α-haloacetals***

Williamson and Kay reported the first synthesis of triethyl and tripentyl *ortho*-formates by reaction of the appropriate sodium alkoxide with chloroform (Equation (1)) 〈1854LA(92)346, 1854PRS135〉. This procedure has since been used to prepare a number of simple trialkyl *ortho*-formates 〈1879CB115, 32JA2964, 33JA3851, 64RTC119〉. The reaction is carried out either by adding chloroform to the alkoxide, or by adding sodium metal to a mixture of the alcohol and chloroform; a typical procedure is described in *Organic Syntheses* for the preparation of triethyl *ortho*-formate 〈51OSC(1)258〉. Yields are generally low (30–45%), irrespective of the method used.

(1)

The use of fluorodichloromethane or difluorochloromethane in place of chloroform has been reported to give higher yields of *ortho*-formates ⟨57JA5493, 60JA1398⟩. Carbon tetrachloride and dichlorodifluoromethane have also been shown to give *ortho*-formates on reaction with sodium alkoxides ⟨21JCS1222, 58USP2853531⟩, these reactions proceeding via the corresponding haloforms which are generated *in situ* under the alkaline conditions.

Triaryl *ortho*-formates are not obtained in appreciable amounts from the reaction of activated phenols with chloroform since the preferred pathway is that of the Reimer–Tiemann reaction; less activated phenols do give *ortho*-formates, but yields are generally poor ⟨1882CB2685, 24JA2090, 55JA6644⟩.

The Williamson synthesis has been extended to the preparation of certain higher *ortho*-esters; trialkyl *ortho*-benzoates (**1**) ⟨42JA2525, 50JA1661⟩ and heterocyclic analogues, for example (**2**) ⟨80LA1216⟩, and trialkyl and triaryl *ortho*-esters of perhalogenated-acrylates (**3**) ⟨67CB2946, 71ZOR2161⟩ and -crotonates (**4**) ⟨60IZV231⟩ have all been prepared from the appropriate trichloromethyl compounds. However, the general utility of the method for the synthesis of *ortho*-esters from trihalomethyl compounds bearing α-hydrogen atoms is limited, since these substrates may also undergo competing elimination reactions under the strongly basic reaction conditions ⟨B-70MI 603-01, 74S153⟩.

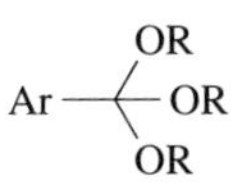

(**1**)

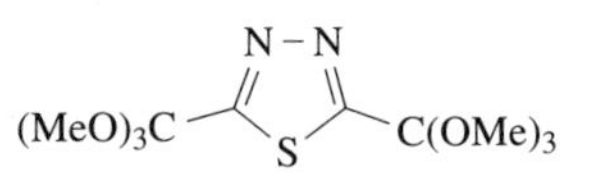

(**2**)

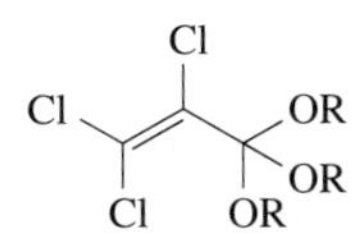

(**3**)

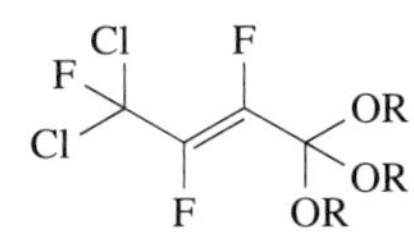

(**4**)

A variation of the Williamson synthesis was reported by Hill *et al.* ⟨65JOC411⟩, who obtained *ortho*-formates (**5**, R = H) and *ortho*-benzoates (**5**, R = Ph) derived from fluorinated alcohols, which do not undergo the base promoted reaction, by the iron(III) chloride catalysed reaction of the alcohol with chloroform and α,α,α-trichlorotoluene respectively (Equation (2)).

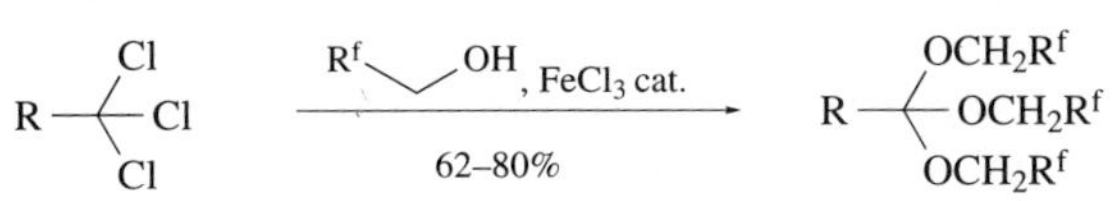

(**5**) (2)

Nucleophilic displacement of the halogen atoms of α,α-dihalo ethers and α-haloacetals with alkoxides also gives *ortho*-esters. In these cases, mixed *ortho*-esters (**6**), in which R^2 and R^3 are different, may be obtained (Scheme 1) ⟨35CB2151, 58JPR60, 61CB538⟩. The cyclic diacyloxy ester (**7**) was obtained from α,α-dichlorophthalide and 3-hydroxybutanoic acid (Equation (3)) ⟨54LA(587)226⟩.

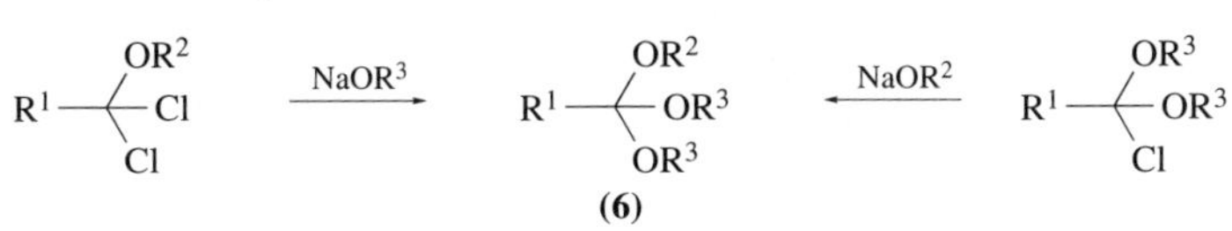

(**6**)

Scheme 1

α,α-Dihalo ethers and α-haloacetals may form as intermediates in the reactions between activated 1,1-dihaloalkenes and alkoxides which yield *ortho*-esters; these are discussed in Section 6.03.2.1.5.

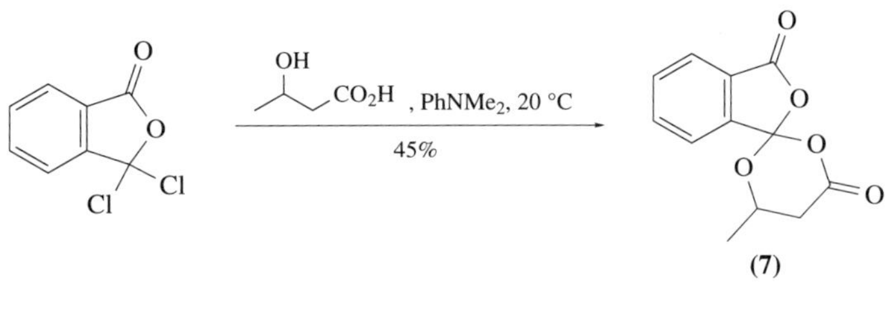

(**7**) (3)

6.03.2.1.2 **Ortho-***esters from imidate ester salts*

Imidate ester salts are transformed into *ortho*-esters by reaction with alcohols (Equation (4)). The reaction is normally carried out using imidate ester hydrochloride salts, which are prepared by the hydrogen chloride promoted addition of alcohols to nitriles, and the two-step sequence (nitrile to *ortho*-ester) is named after Pinner, who first described the preparation of a series of trialkyl *ortho*-formates using this procedure ⟨1883CB352, 1883CB1643⟩. The method is widely applicable to the synthesis of *ortho*-formic and higher aliphatic *ortho*-esters; yields are generally in the 60–80% range, but are lower (30% or less) for aliphatic *ortho*-esters bearing two substituents α- and/or β- to the ester function ⟨42JA1825, 51JA2233⟩. *Ortho*-benzoate esters may be prepared from benzimidate ester hydrochlorides, but reported yields are low ⟨74S153⟩. The procedure is normally used for the preparation of trialkyl *ortho*-esters (**8**) in which R^2 and R^3 are the same, but is also suitable for the synthesis of mixed *ortho*-esters in which R^2 and R^3 are different ⟨1883CB352, 1883CB1643, 33MI 603-01⟩.

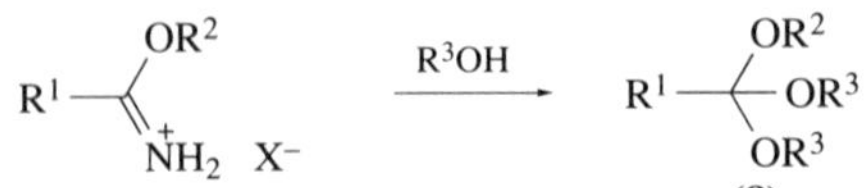

(**8**) (4)

The simplest procedure for the alcoholysis of an imidate ester is to allow a solution of the imidate ester hydrochloride in an excess of the alcohol to stand at room temperature ⟨28JA516, 35JA2480⟩. However, reaction times can be long, and yields are often low. Improved procedures include heating the imidate ester salt with an excess of the alcohol in diethyl ether ⟨42JA1825, 46JA1922⟩, or petroleum ether ⟨53JA3987, 58JA3915⟩. It is important to maintain strictly anhydrous conditions during these alcoholysis reactions, since the *ortho*-ester products are extremely prone to acid-catalysed hydrolysis. High temperatures also need to be avoided, as imidate ester hydrochlorides decompose to alkyl chlorides and the corresponding carboxamide at elevated temperatures ⟨51JA2233⟩.

Ortho-esters may be prepared from nitriles without isolation of the intermediate imidate ester salt. Erickson synthesised a number of trialkyl *ortho*-formates in moderate yields (up to 56%) by a one-step process which involved the introduction of hydrogen chloride into a large excess of both hydrogen cyanide and the alcohol ⟨55JOC1573⟩. Later workers found that better yields were obtained if the acidity of the reaction medium was reduced to pH 2–6 after initial formation of the formimidate ester hydrochloride ⟨76JAP(K)76108012, 80JAP(K)8083727, 84MIP521372⟩.

Other methods for the preparation of *ortho*-esters which proceed via imidate ester salts have been described. A variation of the Pinner method involves the alcoholysis of *O*-methyl imidate salts (**9**; Equation (5)), which are prepared from amides and dimethyl sulfate ⟨80LA1677⟩. Ethanolysis of imidoyl chloride salts (**10**, R = Et, Ph) produced triethyl *ortho*-propionate and triethyl *ortho*-benzoate in yields of 62% and 45% respectively, via the corresponding *O*-ethylimidate ester salts ⟨63CB2671⟩. Reaction of formamide with benzoyl chloride or ethyl chloroformate in the presence of alcohols gave trialkyl *ortho*-formates in yields of 40–50% (Scheme 2) ⟨68LA(716)207⟩, and mixed *ortho*-esters, (**11**) for example, were similarly obtained by treatment of formamide with benzoyl chloride in the presence of a mixture of a simple alcohol and a diol ⟨81MI 603-01⟩.

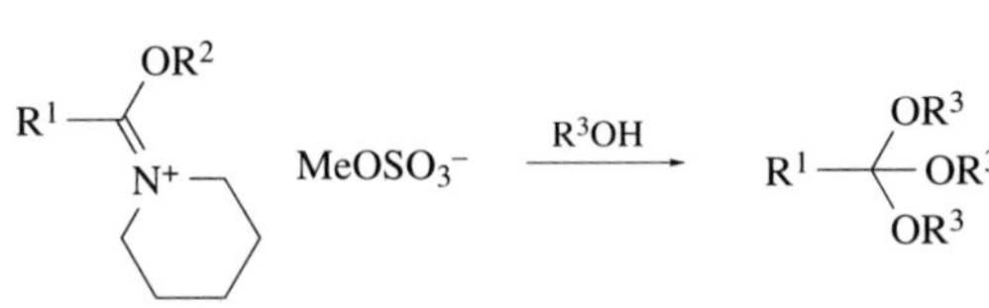

(**9**) (5)

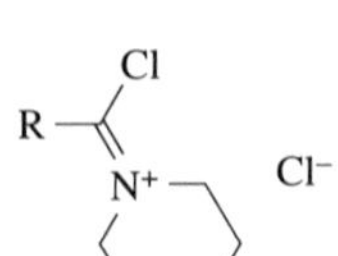

(**10**)

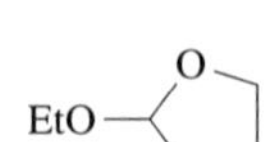

(**11**)

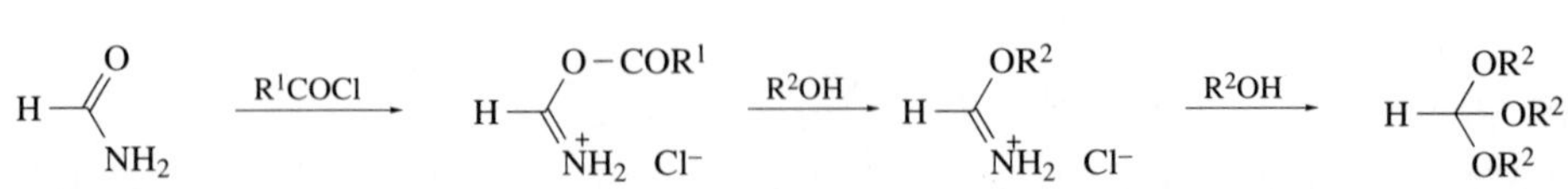

Scheme 2

6.03.2.1.3 **Ortho-*esters from carboxylic acids and derivatives***

(i) From carboxylic acids and esters

Tricyclic *ortho*-esters are obtained from the reaction of carboxylic acids with 2-alkyl-2-hydroxymethyl-1,3-dihydroxypropanes (Equation (6)) ⟨57MI 603-01, 65NEP6412636⟩. The method works best for carboxylic acids bearing electron withdrawing substituents, and the reactions are forced to completion by azeotropic removal of water using benzene or xylene. *Ortho*-esters derived from diols or simple alcohols are not generally available by this method.

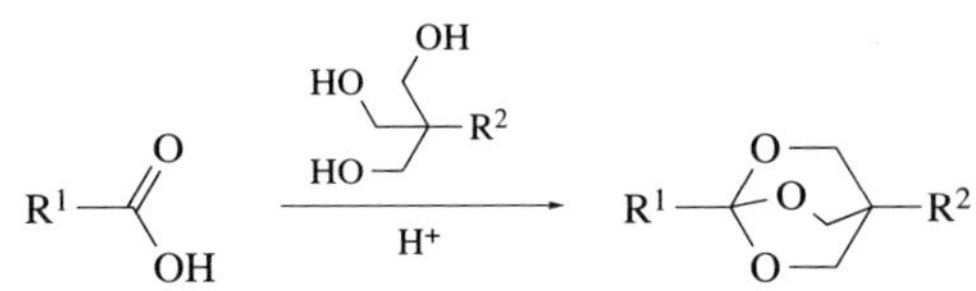

(6)

Aliphatic carboxylic esters of 3-methyl-3-hydroxymethyloxetane have been shown to rearrange smoothly to bridged tricyclic *ortho*-esters in high yield on exposure to boron trifluoride etherate at temperatures of 0 °C or less (Scheme 3) ⟨83TL5571, 90JCS(P1)375⟩. The reaction is postulated to proceed via cyclisation of the zwitterionic species (**12**).

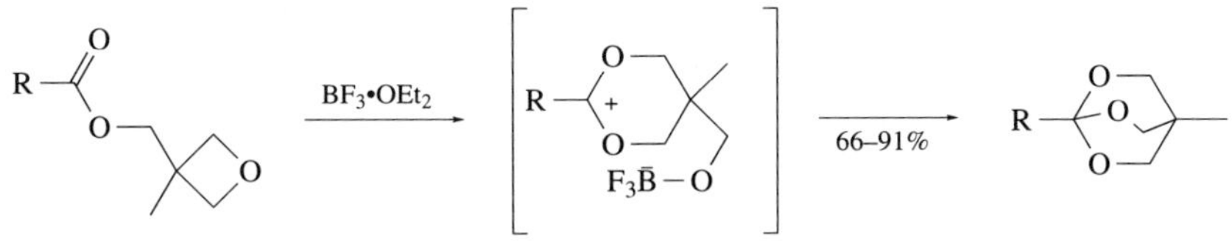

(12)
Scheme 3

Formate esters react with epoxides in the presence of a catalytic amount of boron trifluoride to give cyclic *ortho*-formate esters (Equation (7)) ⟨55AG374⟩. Acetate esters do not react with epoxides under these conditions.

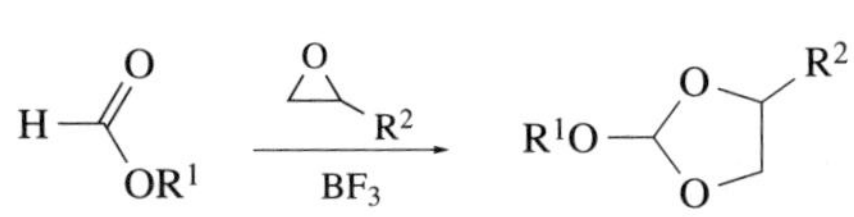

(7)

Ethanolysis of the acetoxy *p*-toluenesulfonate (**13**) under basic conditions produced the cyclic *ortho*-ester (**14**) ⟨43JA613⟩. This product is formed by addition of ethoxide to the intermediate dioxocarbenium ion (**15**), which is itself formed by neighbouring group displacement of toluenesulfonate by the acetoxy group (Scheme 4). The *ortho*-ester (**16**) was formed similarly (Equation (8)) ⟨63PCS374⟩. Acyloxy *ortho*-esters were formed by a similar mechanism when acetoxyacyl chlorides were treated with alcohols in the presence of triethylamine (Equation (9)) ⟨73JA4016⟩; the use of alkyl hydroperoxides in place of alcohols produced the analogous peroxy acyloxy *ortho*-esters in high yields (70–100%) ⟨67AG(E)949⟩. This method of *ortho*-ester synthesis has received considerable attention in the field of carbohydrate chemistry for the conversion of 1-halo-2-acyloxy carbohydrate derivatives (both furanose and pyranose forms) into the corresponding ortho-esters. Triethylamine ⟨90LA499⟩, 2,6-lutidine ⟨65CJC1918⟩, 2,4,6-collidine ⟨56CB314⟩ and silver(I) salts ⟨71CAR(19)139⟩ have been used to catalyse the reactions of 1-halo-2-acyloxy carbohydrate derivatives with alcohols, and amide acetals ⟨80MI 603-01⟩ and alkoxystannanes ⟨76CAR(51)C13⟩ have been used as alternative sources of the alkoxy group. The *exo*-form of the *ortho*-ester product normally predominates, and is often formed exclusively. A representative example is illustrated in Equation (10) ⟨90LA499⟩.

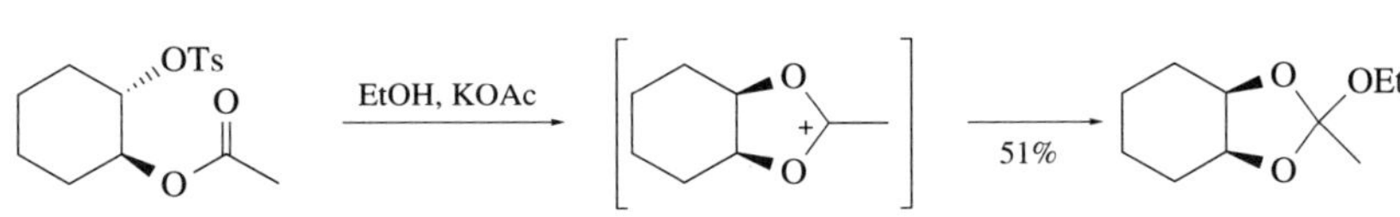

(13) **(15)** **(14)**
Scheme 4

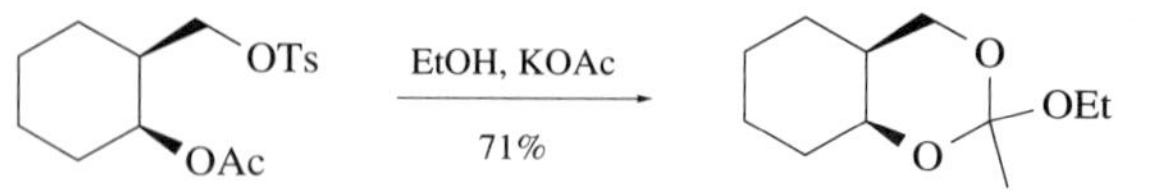

(16)

(8)

(9)

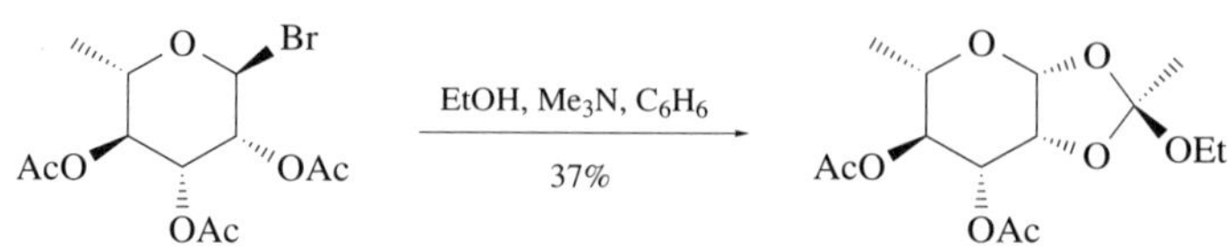

(10)

Aliphatic *O*-trimethylsilyl *ortho*-esters (**18**, R = alkyl) were obtained from hydroxy esters (**17**) by deprotonation using a strong base (*n*-butyllithium or lithium diisopropylamide) followed by reaction with chlorotrimethylsilane (Equation (11)) ⟨91SL160⟩. This method is not suitable for the preparation of *ortho*-esters from α-branched carboxylic esters.

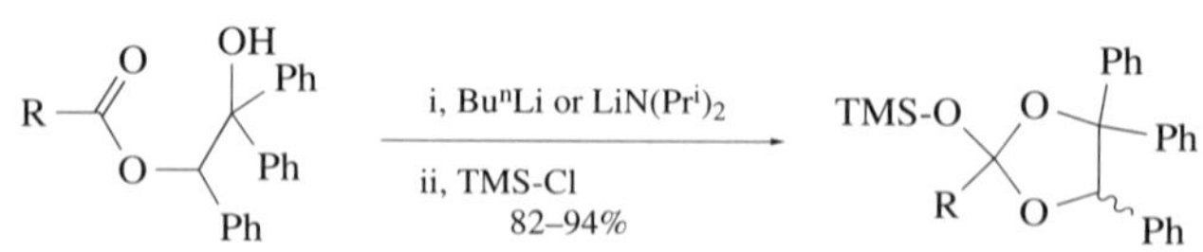

(17) (18)

(11)

(ii) From lactones

Several examples of the direct formation of *ortho*-esters from lactones and diols have been reported. Le Mahieu and Kierstead ⟨70TL5111⟩ described the acid catalysed preparation of the steroid *ortho*-esters (**19**, R = H, Br) from the corresponding lactones and 1,2-dihydroxyethane (Equation (12)), and Tamaru *et al.* ⟨80BCJ3687⟩ prepared the carbohydrate *ortho*-esters (**20**), (**21**) and (**22**) similarly from 2,3,4,6-tetra-*O*-benzylgluconolactone and the appropriate diol in yields of 77%, 73% and 43% respectively. A variation of this method, which is reported to give higher yields of *ortho*-esters, involves the trimethylsilyl trifluoromethanesulfonate catalysed reaction between lactones and bis-*O*-trimethylsilyl-1,2-diols. Yoshimura *et al.* ⟨81CL375⟩ prepared the *ortho*-esters (**20**), (**21**) and (**22**) from 2,3,4,6-tetra-*O*-benzylgluconolactone and the corresponding bis-*O*-trimethylsilyl-1,2-diols in yields of 91%, 89%, and 64% respectively. The *ortho*-ester (**23**) was obtained in 84% yield from 3,4-dihydrocoumarin and 1,2-bis-*O*-trimethylsilyloxyethane ⟨89AJC1235⟩.

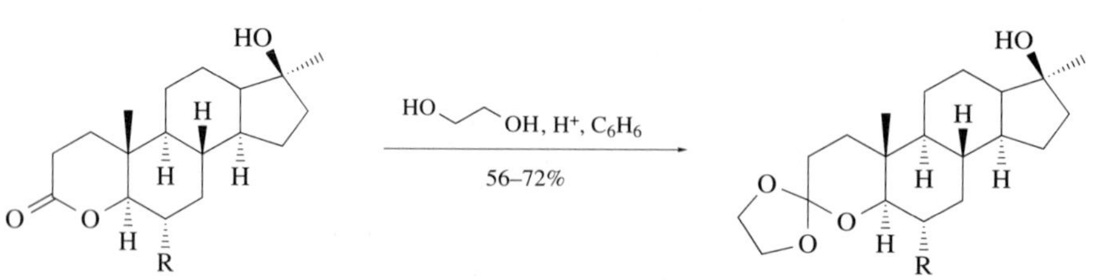

(19)

(12)

Lactones undergo Lewis acid catalysed reaction with epoxides in a similar manner to formate esters to give spirocyclic *ortho*-esters (Equation (13)). Lewis acids commonly used are boron trifluoride, tin(IV) chloride and antimony pentachloride. The reaction is general, and has been used to prepare a number of *ortho*-esters. Thus, butyrolactone reacts with 1,2-epoxyethane, 1,2-epoxypropane, 1, 2-epoxy-3-phenoxypropane, 1, 2-epoxy-3-(4-methoxyphenoxy)propane, 1, 2-epoxy-

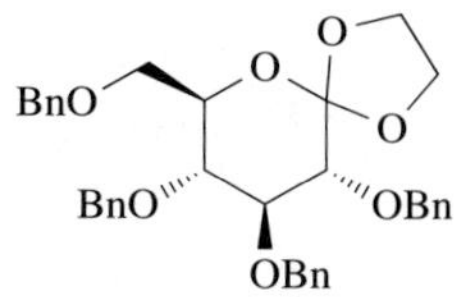

(20)

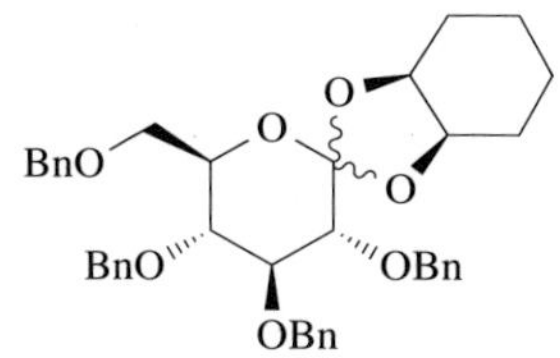

(21)

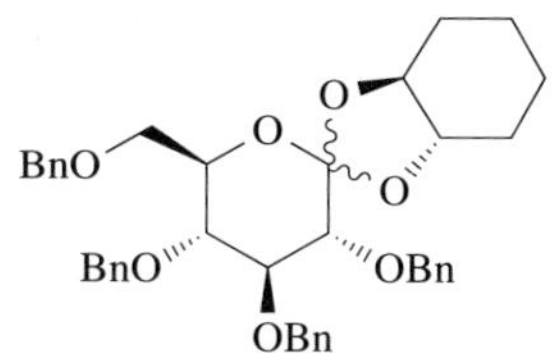

(22)

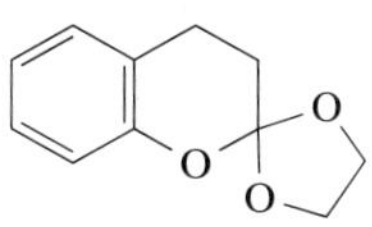

(23)

1-phenylethane and 1,2-epoxycyclohexane to give the *ortho*-esters (**24**, R = H, Me, CH_2OPh, $CH_2OC_6H_4OMe$-*p*, Ph) and (**25**) respectively ⟨59LA(623)183, 60GEP1084733, 82MI 603-01⟩.

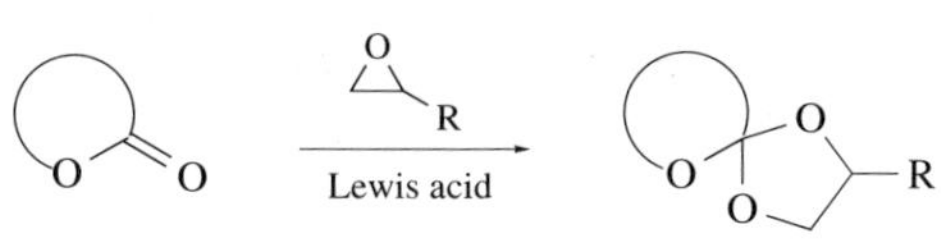

(13)

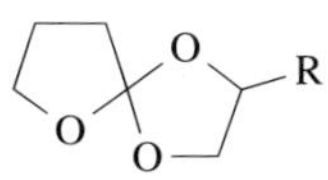

(24)

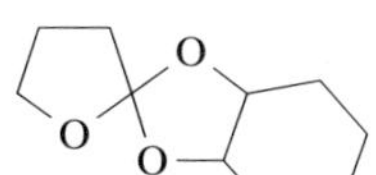

(25)

(iii) From dithiocarboxylic esters

Dithiocarboxylic esters react with dialkoxystannanes at temperatures of 60–95 °C to afford *ortho*-esters (Equation (14)) ⟨76CL891⟩. The method has been used to prepare aliphatic *ortho*-esters and substituted *ortho*-benzoate esters, and reported yields (generally 60% or greater) are favourable compared to some other methods.

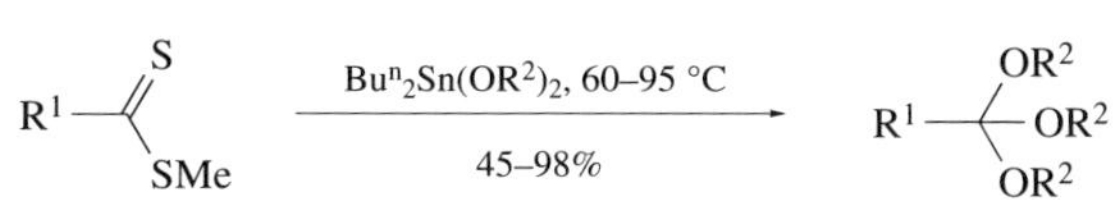

(14)

6.03.2.1.4 Ortho-*esters from dioxocarbenium salts*

Meerwein *et al.* ⟨56CB2060⟩ demonstrated that the *O*-alkyllactonium tetrafluoroborate salts (**26**, R = Me, Et) react with alkoxides to form the cyclic *ortho*-esters (**27**, R = Me, Et) in high yields (Equation (15)). *Ortho*-esters derived from the *O*-ethyl tetrafluoroborate salts of coumarin and 3,4-dihydrocoumarin have been prepared similarly ⟨65HOU(6)361⟩. Alkoxides have also been added to dioxolenium tetrafluoroborates (**28**, R^1 = Me, Ph) and to 2-phenylbenzo-1,3-dioxolium tetrafluoroborate to give the cyclic *ortho*-esters (**29**; Equation (16)) ⟨60LA(632)38⟩ and (**30**; Equation (17)) ⟨65AG(E)873⟩ respectively. Several further examples of the addition of alcohols to dioxolenium and dioxenium salts have been described ⟨67CC13, 67JOC2630, 67T4181⟩.

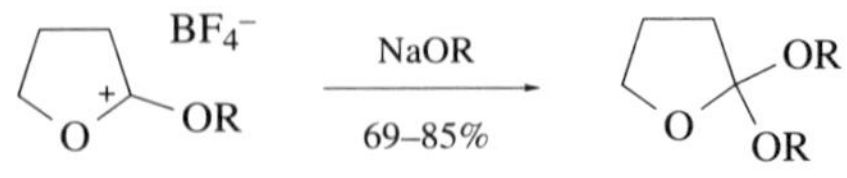

(26) (27) (15)

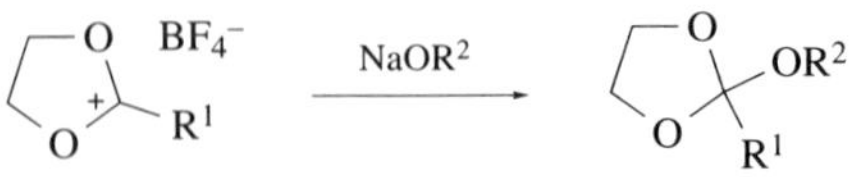

(28) (29) (16)

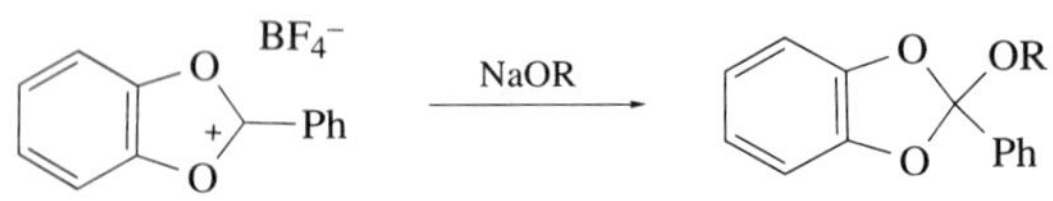

(30) (17)

6.03.2.1.5 Ortho-*esters from 1,1-dialkoxyalkenes, 1,1-dihaloalkenes and 1-alkoxyalkynes*

Alcohols and phenols add to 1,1-dialkoxyalkenes (ketene acetals) to give high yields of *ortho*-esters (Equation (18)) ⟨36JA529⟩. The reaction is catalysed by acids ⟨58JA1247⟩, though *ortho*-esters are also formed under neutral and basic conditions ⟨37JA2266, 42JA1963⟩. This method is particularly useful for the preparation of hindered *ortho*-esters, for example trimethyl *ortho*-diphenylacetate ⟨51JA3807⟩, trimethyl *ortho*-dimethoxyacetate ⟨66CB1892⟩ and triphenyl *ortho*-acetate ⟨45JA650⟩, which are difficult to prepare by other methods. The procedure is also suitable for the synthesis of mixed *ortho*-esters (**31**) in which R^3 and R^4 are different ⟨36JA529, 42JA2525, 64JOC2773⟩; use of benzaldoximes in place of alcohols or phenols gives rise to aminoxy *ortho*-ester derivatives (**31**, $R^4 = NCHAr$) ⟨61JOC2202⟩.

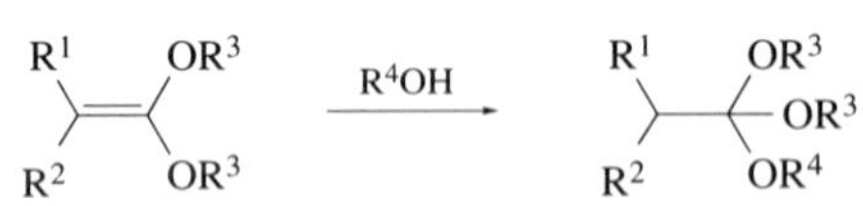

(31) (18)

Activated 1,1-dihaloalkenes undergo reaction with sodium alkoxides, giving rise to *ortho*-ester products. Thus, *ortho*-ester (**32**, R = Me; X = H) was isolated in 79% yield from the reaction between 1,1-dichloro-2-nitroethene with sodium methoxide (Equation (19)) ⟨78BSB693⟩, (**32**, R = Et; X = Cl) was obtained in 34% yield from 1,1,2-trichloro-2-nitroethene and sodium ethoxide ⟨78ZOR2229⟩, and (**33**) was prepared from perfluoropropene and sodium methoxide (Equation (20)) ⟨73USP3745220⟩. It is not known whether these reactions proceed by an addition–elimination mechanism via a 1,1-dialkoxyalkene intermediate, or whether they proceed by an addition–substitution mechanism via α,α-dihaloether and α-haloacetal intermediates.

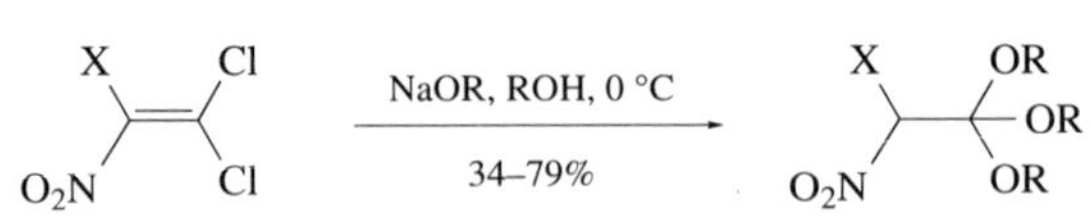

(32) (19)

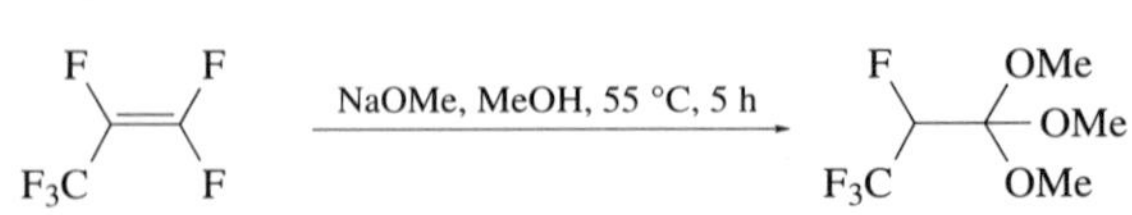

(33) (20)

In a reaction related to the addition of alcohols to 1,1-dialkoxyalkenes, mixed *ortho*-acetate esters have been prepared by the Lewis acid catalysed addition of alcohols and phenols to 1-ethoxyethyne (Equation (21)) ⟨75RTC209⟩. Zinc chloride is the preferred catalyst for the reaction with alcohols, whereas mercury(II) acetate is preferred for phenols. Triethyl *ortho*-acetate has also been obtained by treatment of 1-ethoxyethyne with ethanolic sodium ethoxide ⟨53MI 603-01⟩. Addition of oximes to 1-ethoxyethyne afforded the mixed *ortho*-ester derivatives (**34**, $R = NCR^1R^2$) ⟨60RTC888⟩.

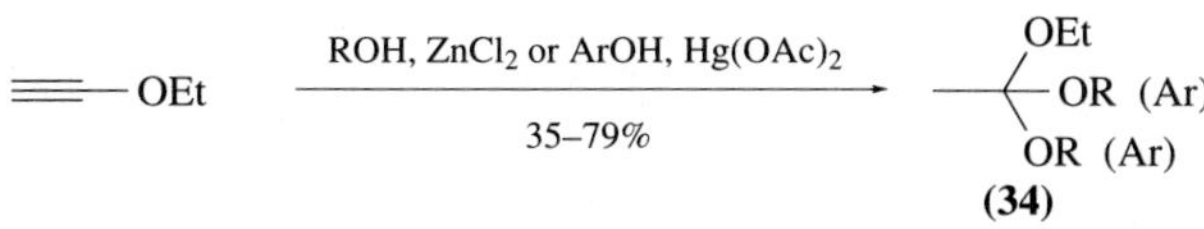

(34) (21)

Tris(β-alkoxyalkyl) *ortho*-acetates have been obtained by treatment of 1,1-dichloroethene with the appropriate sodium alkoxide ⟨64JOC2773⟩. It has been proposed that these reactions proceed through a series of elimination and addition steps via a 1-alkoxyalkyne intermediate (Scheme 5).

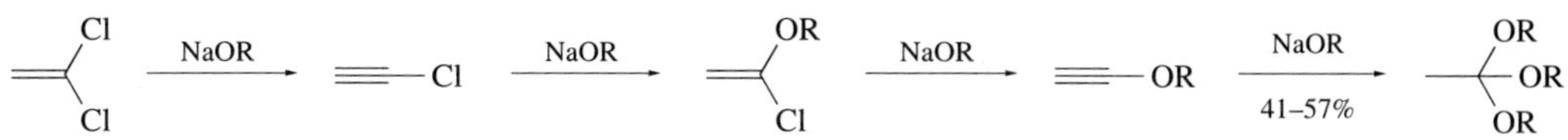

Scheme 5

1,1-Dialkoxyalkenes undergo zinc chloride promoted reactions with epoxides to give cyclic *ortho*-esters (Equation (22)) ⟨79TL2925⟩. The epoxide generally undergoes attack at the least hindered position, unless attack at the more substituted position is favoured on electronic grounds.

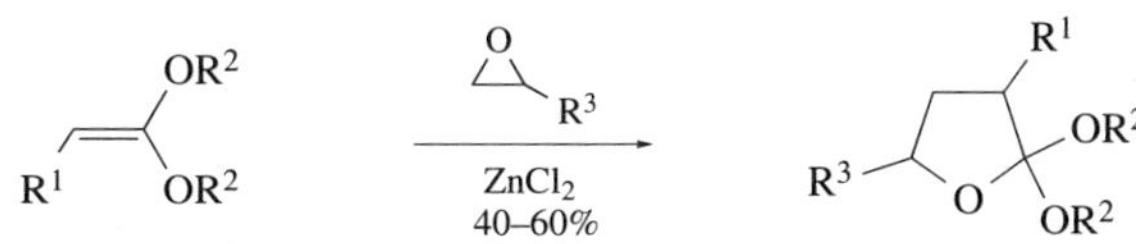

(22)

Triethyl *ortho*-cinnamate esters have been prepared by addition of ethanol to diethoxyethenylidinetriphenylphosphorane (**35**), followed by a Wittig reaction of the resulting phosphorane with aromatic aldehydes ⟨86S397⟩ (Scheme 6). A related method for the preparation of a series of unsaturated cyclic *ortho*-esters (**36**) involves the addition of an enolised 1,2-diketone to phosphorane (**35**), followed by intramolecular Wittig cyclisation of the intermediate ketophosphorane (Scheme 7) ⟨83CB1463, 83CB3482, 88T5095⟩.

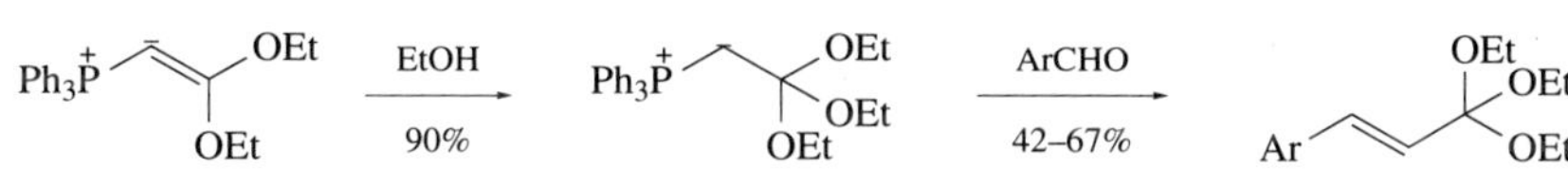

(35)

Scheme 6

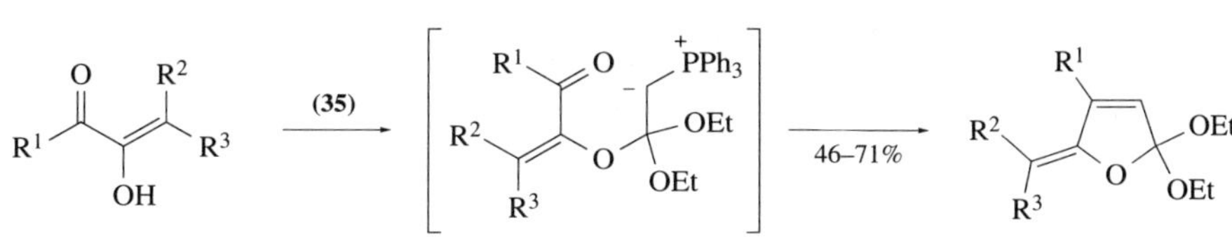

(36)

Scheme 7

The preparation of cyclic *ortho*-esters by thermal [4 + 2] cycloaddition reactions of 1,1-dialkoxyalkenes and α,β-unsaturated carbonyl compounds was first described by McElvain *et al.* (Equation (23)) ⟨54JA5736⟩. Many further examples of *ortho*-esters (**37**) have since been prepared by reaction of 1,1-dimethoxyethene, 1,1-dimethoxy-1-propene, 1,1-dimethoxy-2-methyl-1-propene, 1,1,2-trimethoxyethene, 1,1,2,2-tetramethoxyethene and 2-chloro-1,1-dimethoxyethene with a range of substituted α,β-unsaturated carbonyl compounds ⟨55JA5601, 71CC383, 72CC863, 81RTC13⟩. 1,1-Dialkoxyalkenes also undergo [2 + 2] cycloaddition reactions with electron deficient carbonyl compounds, such as acyl cyanides and aldehydes, giving rise to the cyclic *ortho*-esters (**38**) (Equation (24)) ⟨76JCS(P1)1048, 77JOC3128⟩. It has been shown that [2 + 2] cycloadducts are the products of the low temperature reaction between 1,1-dialkoxyalkenes and α,β-unsaturated carbonyl compounds, and that this process is reversible at higher temperature, when formation of the [4 + 2] cycloadducts is favoured on thermodynamic grounds ⟨81RTC13⟩.

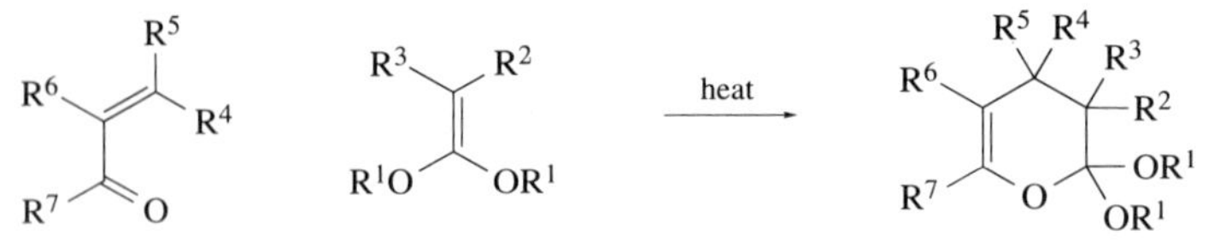

(37) (23)

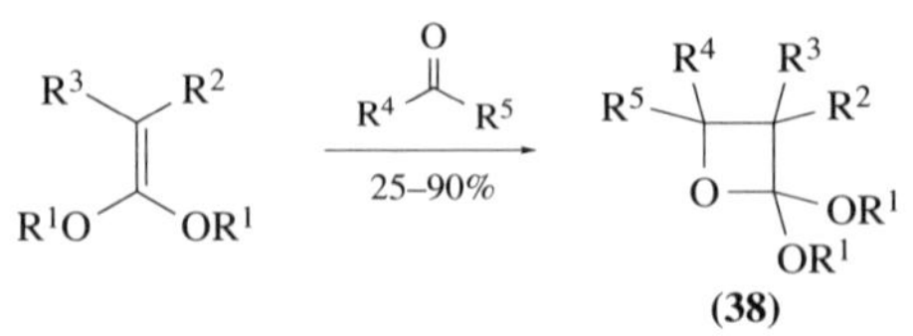

(38) (24)

A number of five-membered ring *ortho*-esters have been prepared by [3 + 2] cycloaddition reactions of 1,3-dipoles to 1,1-dialkoxyalkenes. Some examples are given in Equations (25) ⟨86JMC553⟩, (26) ⟨59G1511⟩ and (27) ⟨66G375⟩.

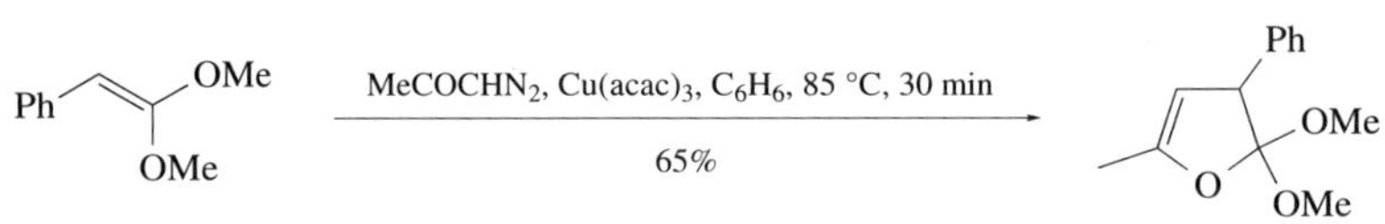

(25)

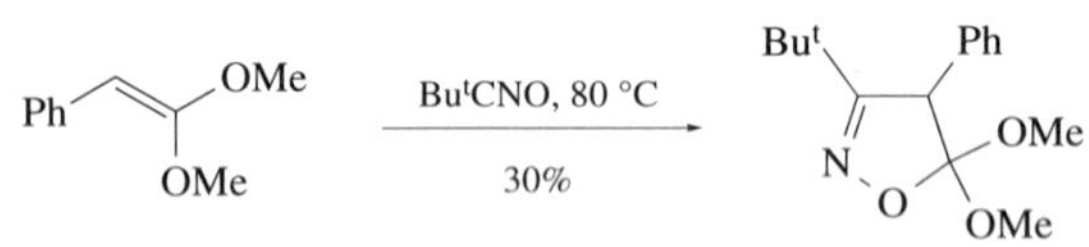

(26)

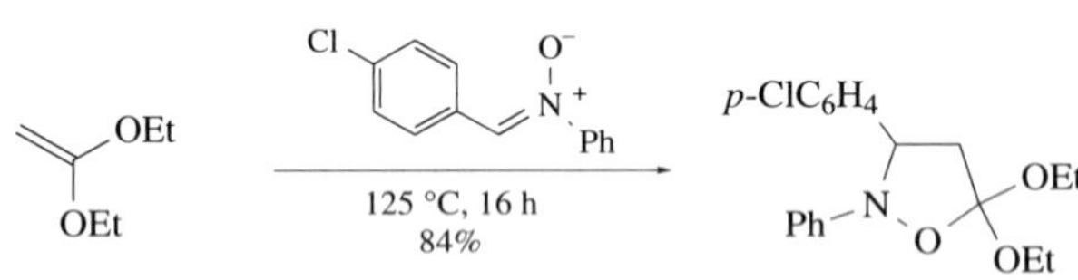

(27)

6.03.2.1.6 Ortho-*esters from 1,1-dialkoxycyclopropanes and related compounds*

Alcohols add to 1,1-dialkoxycyclopropanes which bear an electron withdrawing group (EWG) (carboxylic ester or sulfone) at the 2-position to give *ortho*-esters (Equation (28)) ⟨73JOC1361, 85S762⟩. Mixed *ortho*-esters (**39**), in which R^1 and R^4 are different, have been prepared by this method, which is analogous to the addition of alcohols to 1,1-dialkoxyalkenes. Examples of *ortho*-esters prepared using this procedure are listed in Table 1.

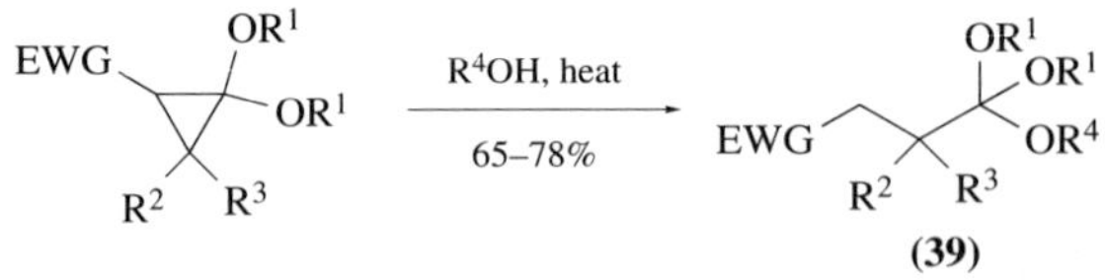

(39) (28)

Ring opening of 1,1-dialkoxy-2,2-dichlorocyclopropanes with potassium *t*-butoxide is accompanied by elimination to produce alkynyl *ortho*-esters (Equation (29)) ⟨59JA2579⟩. Addition of alcohols to 1,1-dimethoxycyclopropene produces the unsaturated *ortho*-esters (**40**, R = Me, Et, Pr^n, Pr^i, Bu^t; Equation (30)) ⟨77JOC679⟩.

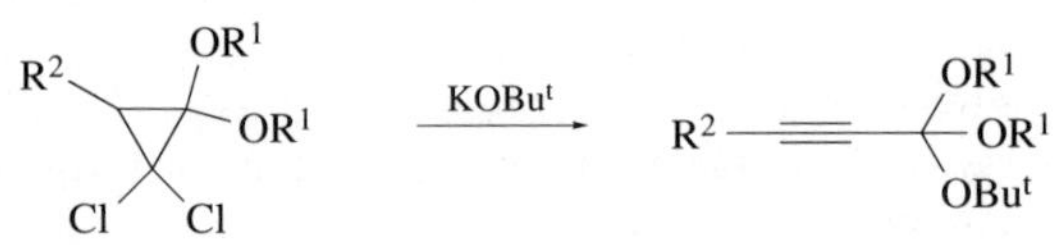

(29)

Table 1 *Ortho*-esters (**39**) prepared from 1,1-dialkoxypropanes.

R^1	R^2	R^3	R^4	*EWG*	*Yield* (%)	*Ref.*
Me	H	H	Me	CO_2Et	73	85S762
Et	H	H	Et	CO_2Et	70	85S762
Me	Me	H	Me	CO_2Et	72	85S762
Et	Me	H	Et	CO_2Et	76	85S762
Me	Et	H	Me	CO_2Et	78	85S762
Me	Ph	H	Me	CO_2Et	67	85S762
Me	Me	Me	Me	CO_2Et	66	85S762
Me	Me	Me	Me	SO_2Ph	65	73JOC1361
Et	Me	Me	Et	SO_2Ph	71	73JOC1361
Et	Me	Me	Me	SO_2Ph		73JOC1361

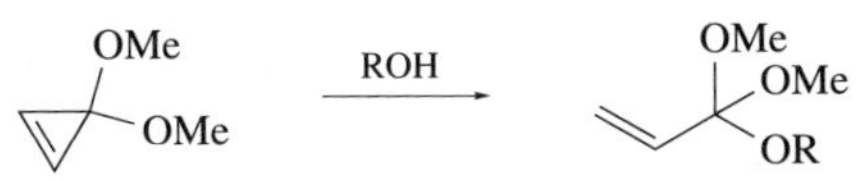

(40) (30)

Alcohols also add to 2,2-dialkoxyaziridines and 2,2-dialkoxy-2*H*-azirines to give α-amino-and α-imino-*ortho*-esters (**41**; Equation (31)) ⟨69TL2223, 73JMC978⟩ and (**42**; Equation (32)) ⟨67JA5724⟩ respectively. These compounds are difficult to prepare by other methods.

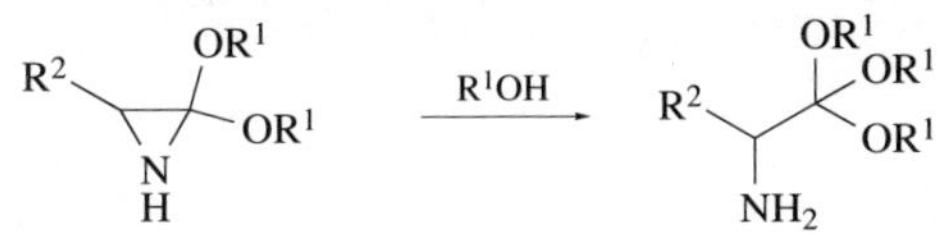

(41) (31)

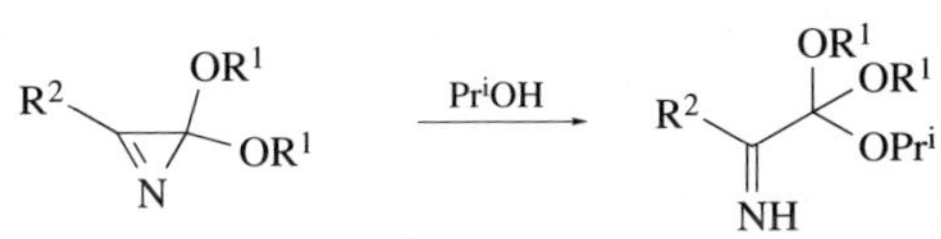

(42) (32)

6.03.2.1.7 Ortho-*esters from acetals*

Some acetals undergo electrochemical oxidation in the presence of methanol to give *ortho*-esters in reasonable yields (Equation (33)). The reaction has been shown to be successful for 2-alkyl-1,3-dioxolanes ⟨78S283⟩, 2-alkyl- and 2-aryl-1,3-benzodioxoles ⟨85S31⟩ and dimethyl acetals of benzaldehydes ⟨86T553⟩. Dimethyl acetals of simple aliphatic aldehydes gave lower yields of *ortho*-esters, and the reaction failed completely for the corresponding diethyl acetals ⟨78S283⟩. Table 2 lists some representative examples of *ortho*-esters (**43**) prepared by this method.

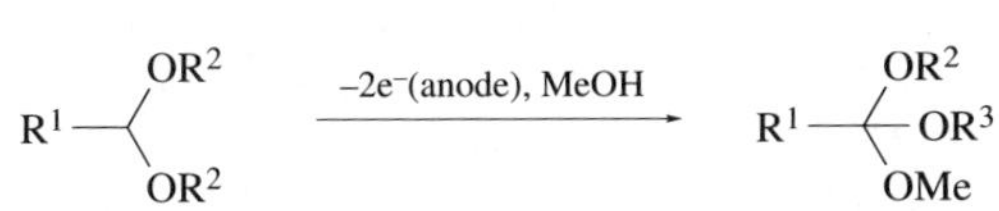

(43) (33)

6.03.2.1.8 Ortho-*esters from* ortho-*carbonate esters and trialkoxyacetonitriles*

Trialkyl *ortho*-benzoates have been reported to be formed in up to 77% yield by treatment of the appropriate tetraalkyl *ortho*-carbonate ester with phenylmagnesium bromide (Equation (34)) ⟨05CB561, B-43MI 603-01⟩, and similar syntheses of triethyl *ortho*-phenylpropiolate ⟨11BSF1308⟩ and triethyl *ortho*-3-butynoate ⟨70AG(E)456⟩ have been described. However, the reaction is not always successful, and other workers have isolated only ketals and ethers from reactions of *ortho*-carbonate esters with Grignard reagents ⟨42JA1825, 49MI 603-01⟩. No systematic survey of this method has been

Table 2 Examples of *ortho*-esters (**43**) prepared by electrochemical oxidation of acetals.

R^1	R^2	R^3	*Yield* (%)	*Ref.*
H	—$(CH_2)_2$—		44	78S283
H	—C_6H_4—		49	85S31
Me	—$(CH_2)_2$—		64	78S283
Me	—C_6H_4—		52	85S31
Pr^i	—$(CH_2)_2$—		54	78S283
Pr^i	—C_6H_4—		61	85S31
Bu^t	—$(CH_2)_2$—		21	78S283
Bu^t	—C_6H_4—		69	85S31
Ph	—C_6H_4—		50	85S31
Ph	Me	Me	90	86T553
4-ClC_6H_4	—C_6H_4—		48	85S31
4-ClC_6H_4	Me	Me	84	86T553
4-$MeOC_6H_4$	—C_6H_4—		44	85S31
4-$MeOC_6H_4$	Me	Me	84	86T553

carried out, though DeWolfe ⟨74S153⟩ has suggested that the Grignard reagent needs to bear an electron withdrawing organic moiety for the reaction to work successfully.

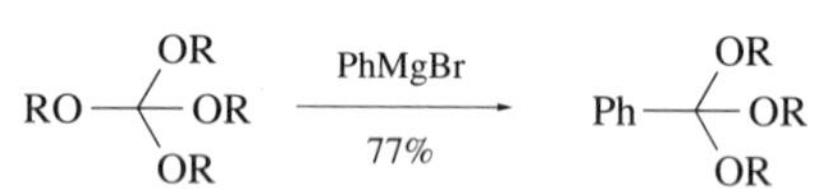

(34)

More recently, it was reported that trialkyl *ortho*-benzoate esters are obtained from the reactions of Grignard reagents with trialkoxyacetonitriles (Equation (35)) ⟨81S380⟩. Trimethyl *ortho*-benzoate, trimethyl 2-methyl-*ortho*-benzoate and triethyl 2-6-difluoro-*ortho*-benzoate were prepared in yields of 84%, 64% and 20% respectively. Triethyl *ortho*-phenylpropiolate was prepared in 64% yield from phenylethynyl magnesium bromide and triethoxyacetonitrile using this method ⟨81S380⟩.

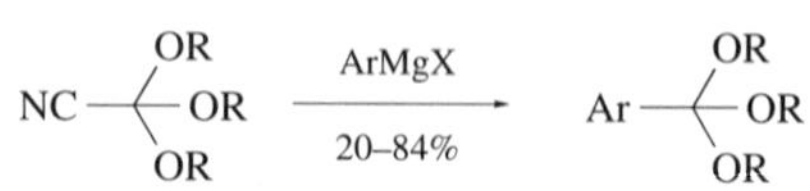

(35)

6.03.2.2 Preparation of Carboxylic *Ortho*-Esters from Other *Ortho*-Esters

6.03.2.2.1 Trans-*esterification reactions*

Ortho-formate esters undergo exchange reactions on treatment with alcohols (Scheme 8). The reaction, which is acid catalysed, can be forced to completion by distilling out R^1OH (preferably methanol or ethanol) as it is formed, and a number of trialkyl *ortho*-formates derived from higher alcohols, including functionalised alcohols such as 2-chloroethanol and 1-chloro-2-propanol, have been prepared by this method ⟨52JA554, 55JA3801, 70CB639⟩. The reaction is subject to steric constraints, proceeding readily with primary alcohols and less readily with secondary alcohols. *Trans*-esterification with tertiary alcohols does not proceed to completion; treatment of triethyl *ortho*-formate with *t*-butanol gave a mixture of *t*-butyldiethyl and di-*t*-butylethyl *ortho*-formates ⟨63PIA(A)45⟩. Relatively little has been published on the *trans*-esterification reactions of higher *ortho*-esters, though work that has been reported suggests that these proceed in much the same way as for *ortho*-formate esters ⟨50DOK(70)231, 56ACS1006, 70CB639⟩.

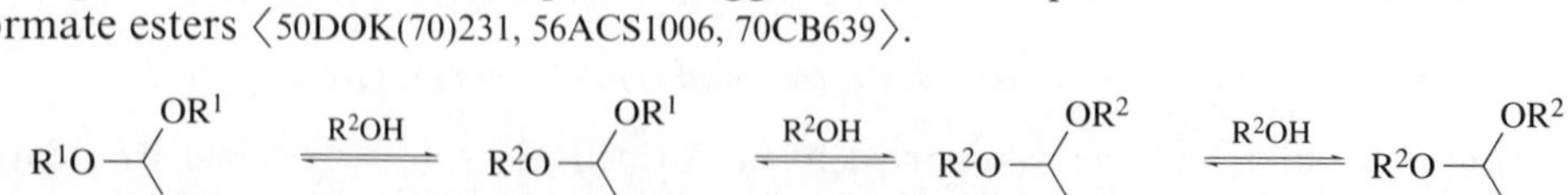

Scheme 8

The preparation of mixed *ortho*-esters by exchange reactions of trialkyl *ortho*-esters with simple alcohols is not usually feasible since equilibration reactions lead to all possible *ortho*-ester products

⟨33JA3851⟩. Among the few reported exceptions to this general trend are the reactions of *ortho*-esters with phenol; treatment of triethyl *ortho*-formate and triethyl *ortho*-acetate with one equivalent of phenol produced good yields of the diethylphenyl *ortho*-esters (**44**, R = H, Me), and reaction of triethyl *ortho*-formate with two equivalents of phenol gave a reasonable yield of the ethyldiphenyl *ortho*-ester (**45**; Scheme 9) ⟨56ACS1006, 70CB643⟩. A recently reported *trans*-esterification route to mixed *ortho*-esters, which is potentially general, involves the magnesium chloride catalysed exchange reaction with alcohols (Equation (36)) ⟨88TL2023⟩. Mixed *ortho*-esters have also been obtained from the *O*-acyl derivative (**46**) ⟨69RTC897⟩, which was itself prepared by treatment of triethyl *ortho*-formate with one equivalent of acetic formic anhydride (Scheme 10) ⟨66RTC793⟩. In contrast to their reactions with alcohols, *ortho*-esters react cleanly with peroxides to form monoperoxy *ortho*-ester derivatives in high yield ⟨58CB1942⟩.

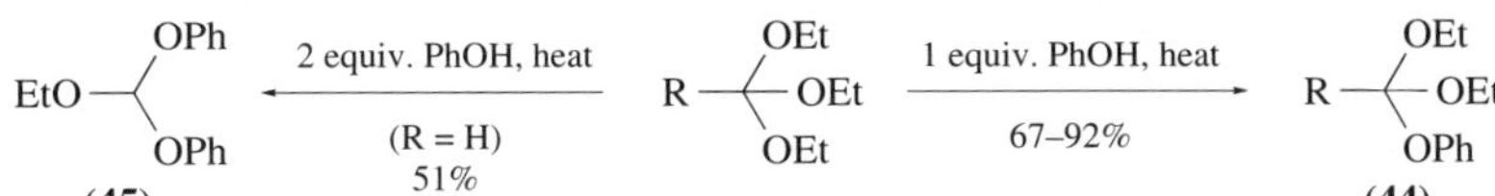

(**45**) (**44**)

Scheme 9

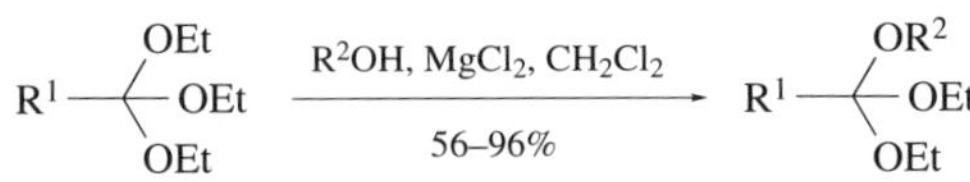

(36)

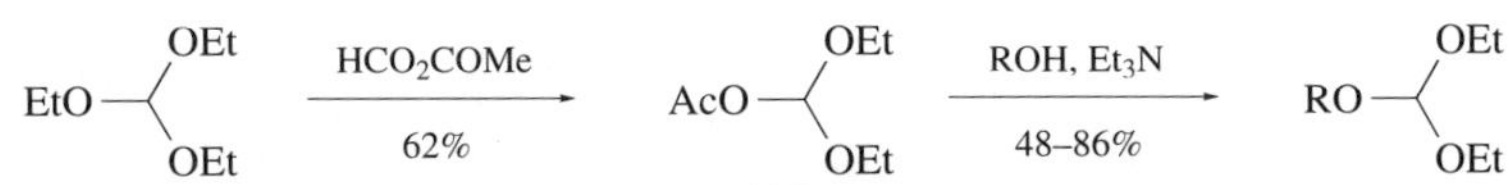

(**46**)

Scheme 10

Trans-esterification of simple trialkyl *ortho*-esters with aliphatic 1,2- and 1,3-diols proceeds cleanly to give mixed cyclic *ortho*-esters in high yields (Equation (37)). The reaction is catalysed by acids such as sulfuric acid ⟨58CB650⟩, *p*-toluenesulfonic acid ⟨70JA584⟩ and benzoic acid ⟨73JA278⟩. *Trans*-esterification reactions of trialkyl *ortho*-benzoates with benzene-1,2-diol proceed in the absence of a catalyst ⟨65AG(E)873⟩. Treatment of trialkyl *ortho*-esters with *β*-hydroxycarboxylic acids yields cyclic acyloxy *ortho*-esters (**47**) ⟨69TL1047, 87HCA1320⟩. In contrast to the equilibration reactions between simple trialkyl *ortho*-esters and alcohols, the cyclic *ortho*-ester (**11**) reacts moderately cleanly with alcohols to give the exchanged products (**48**) in yields of 41–58% (Equation (38)) ⟨80JCS(P1)756⟩.

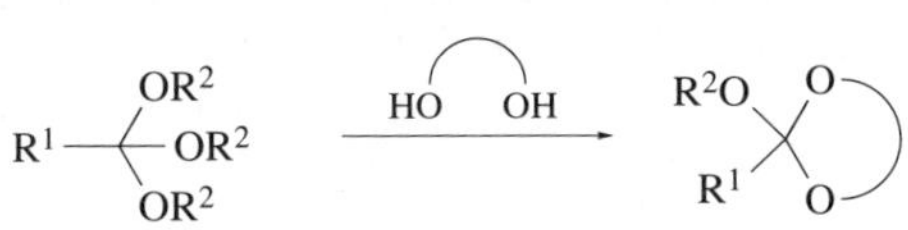

(37)

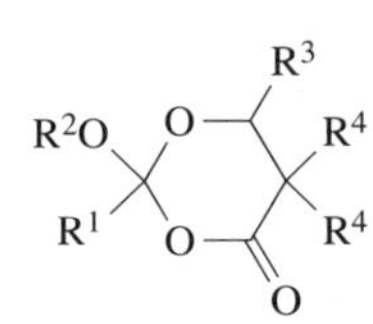

(**47**)

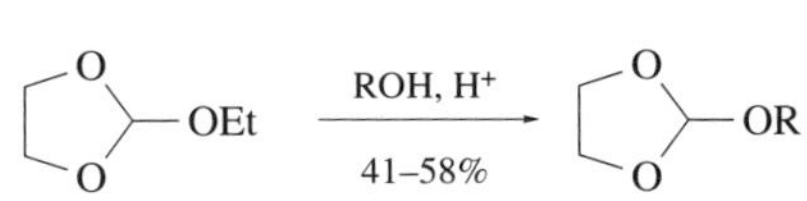

(38)

(**11**) (**48**)

Trialkyl *ortho*-esters react with conformationally flexible triols such as glycerol, 1,3-dihydroxy-2-hydroxymethyl-2-methylpropane and 1,2,4-trihydroxybutane to give good yields of the corresponding tricyclic *ortho*-esters (**49**), (**50**) and (**51**) respectively ⟨80MM252, 80CC207, 88JA4144⟩, and exchange with all *cis*-1,3,5-trihydroxycyclohexane gives high yields of polycyclic *ortho*-esters (**52**)

⟨53CB790, 54CB205⟩. *Trans*-esterification with conformationally restricted triols normally gives products arising from exchange of only two of the alkoxy groups of the *ortho*-ester. For example, exchange with 2,3,5-trihydroxy nucleoside analogues generally leads to exclusive formation of the *ortho*-ester derived from the 2,3-dihydroxy unit of the triol (Equation (39)) ⟨64CI(L)581, 67CCC3064, 85S408⟩. The *trans*-esterification method has been used for the preparation of many cyclic, bicyclic and tricyclic *ortho*-esters from dihydroxy- and trihydroxy-carbohydrate, nucleoside and steroid derivatives ⟨B-70MI 603-01, 77USP4021459, 88EUP260979⟩.

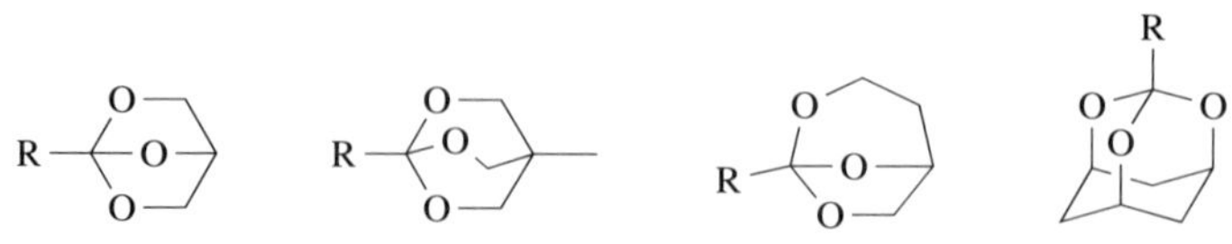

(49) **(50)** **(51)** **(52)**

(39)

Ortho-esters have also been prepared by the alcoholysis of amide acetals ⟨81JOC886⟩ and trithio-*ortho*-esters ⟨48JA2268, 80JOC740⟩.

6.03.2.2.2 Modification of R^1 and R^2 of $R^1C(OR^2)_3$

Ortho-esters bearing α-hydrogen atoms react with one equivalent of bromine to give α-bromo *ortho*-esters (**53**) in up to 80% yield ⟨37JA1273, 42JA2525, 46JA1922⟩; use of two equivalents of bromine gives α,α-dibromo *ortho*-esters ⟨42JA1963⟩. Dehydrobromination of α-bromo *ortho*-esters, (e.g., (**54**)), to unsaturated *ortho*-esters, (e.g., (**55**)), has been achieved, but yields are low (19–32%) ⟨73S207, 85JAP(K)60208983⟩. Reaction of triethyl *ortho*-acetate with trifluoromethylselenenyl chloride gave triethyl α-(trifluoromethylselenenyl)*ortho*-acetate, which was converted to the bis(trifluoromethylselenenyl) derivative by deprotonation with sodium hydride, followed by treatment with a further equivalent of trifluoromethylselenenyl chloride (Scheme 11) ⟨92CB571⟩. Trialkyl *ortho*-acetates have also been shown to react with perhalogenated aldehydes and ketones, producing the adducts (**56**, $R^1 = H$, CF_3, $CClF_2$; $R^2 = Me$, Et; X = F, Cl) in yields of 85–94% (Equation (40)) ⟨77ZOR943⟩.

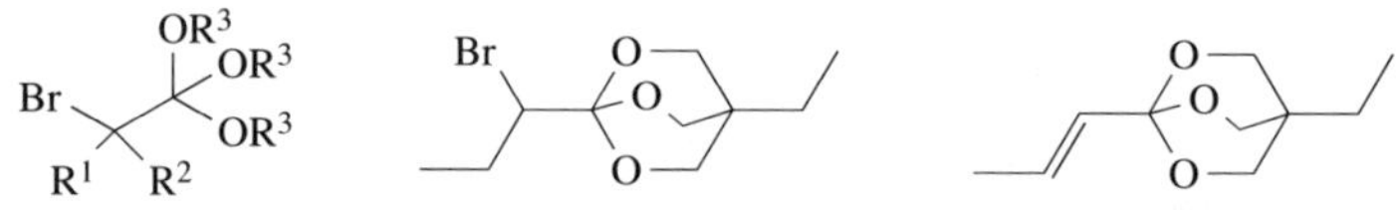

(53) **(54)** **(55)**

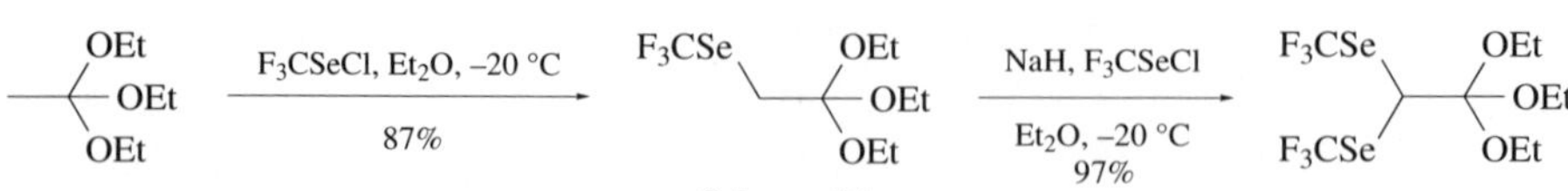

Scheme 11

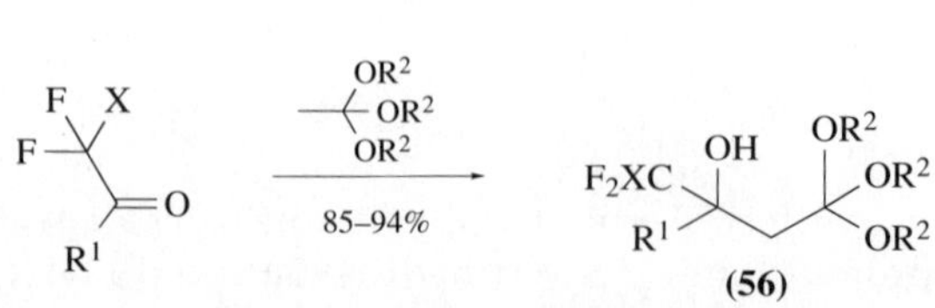

(40)

Dehydrohalogenation of tris(β-chloroalkyl) *ortho*-esters yielded trialkenyl *ortho*-esters (Equation

(41)) ⟨70CB639⟩. Ozonolysis of tris(2-propenyl) *ortho*-formate gave tri-*O*-acetyl *ortho*-formate (Equation (42)) ⟨70CB639⟩.

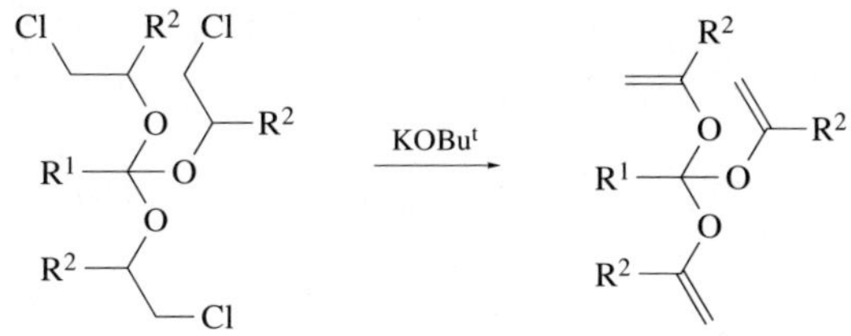

(41)

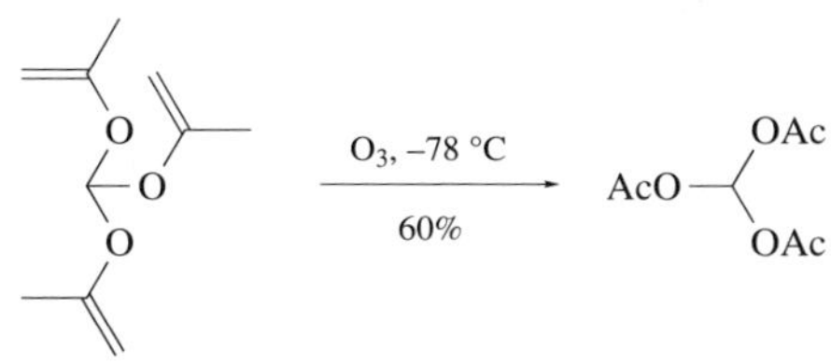

(42)

6.03.3 FUNCTIONS BEARING THREE SULFUR ATOMS

Trithio-*ortho*-esters contain a carbon atom bearing three alkanethio or arenethio groups. Related to these are the analogous acylthio esters (containing at least one RCOS group), hydrogen esters (containing at least one HS group) and oxidised derivatives bearing sulfoxide, sulfone and sulfonyl groups. Methods for the preparation of trithio-*ortho*-esters and related compounds have been reviewed by Post ⟨B-43MI 603-01⟩, DeWolfe ⟨B-70MI 603-01⟩ and Simchen ⟨85HOU(E5)3⟩.

6.03.3.1 Methods for the Preparation of Trithio-*ortho*-esters and Related Compounds

*6.03.3.1.1 Trithio-*ortho*-esters from $RC(X^1)(X^2)X^3$*

(i) From 1,1,1-trihaloalkanes and α,α-dihalo ethers

Gabriel reported the first synthesis of triethyl trithio-*ortho*-formate and triphenyl trithio-*ortho*-formate by reaction of the appropriate thiolate salt with chloroform (Equation (43)) ⟨1877CB185⟩. This method has since been used to prepare a range of trialkyl and triaryl trithio-*ortho*-formate esters ⟨1878CB2265, 33RTC437, 64CR(259)4051⟩; yields from reactions with arenethiols are usually high (70% or greater), but are lower from reactions with aliphatic thiols. Bromoform and chlorodifluoromethane also react with sodium arenethiolate salts to give high yields of triaryl trithio-*ortho*-esters ⟨56JA479, 60JA6118⟩. Carbon tetrachloride reacts with alkanethiolates to give trialkyl trithio-*ortho*-esters, the reactions proceeding via formation of chloroform under the alkaline reaction conditions ⟨33RTC437⟩. Tris(trifluoromethyl) trithio-*ortho*-formate (**57**) was prepared by reaction of iodoform with $Hg(SCF_3)_2$ (Equation (44)) ⟨67JOC2063⟩.

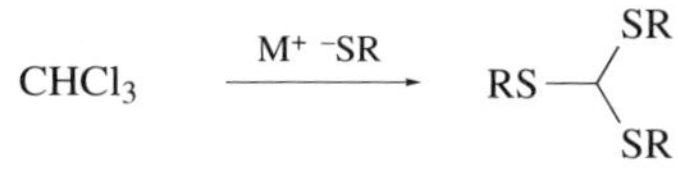

(43)

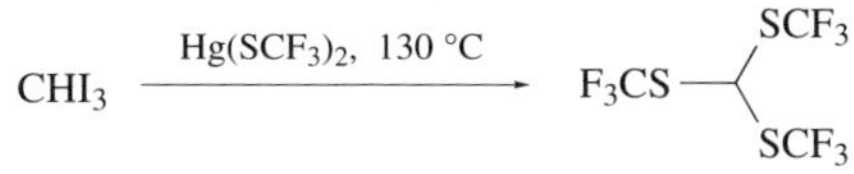

(57)

(44)

Relatively few reports of the preparation of higher trithio-*ortho*-esters from 1,1,1-trihaloalkanes have appeared. 2-Trichloromethylbenzimidazole was converted to the trithio-*ortho*-esters (**58**, R = Me, *p*-ClC_6H_4) by reaction with the corresponding thiol ⟨67JCS(C)30⟩, and the activated trifluoromethyl group of the thiazole (**59**) underwent reaction with methanethiol to give the trithio-

ortho-ester (**60**) ⟨91JHC1013⟩. A series of α,α,α-trihalotoluenes have been converted to tris (trifluoromethyl) trithio-*ortho*-benzoate esters (**61**) by treatment with $AgSCF_3$ ⟨69ZOB1755⟩.

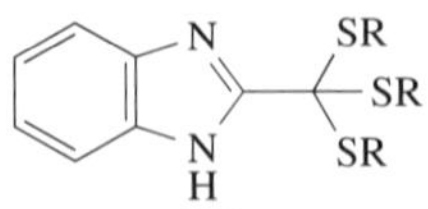

(58)

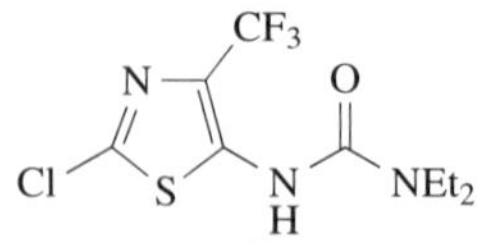

(59)

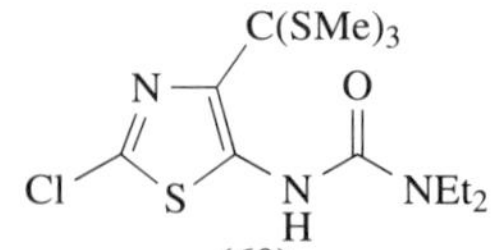

(60)

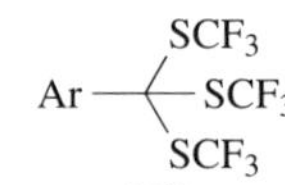

(61)

Trithio-*ortho*-formate esters may also be prepared from dichloromethyl alkyl ethers; treatment with lead(II) alkanethiolates gives trialkyl trithio-*ortho*-formates ⟨61CB538⟩, whereas reaction with arenethiols in the presence of zinc produces triaryl trithio-*ortho*-formates ⟨71ZOR1887⟩ (Scheme 12).

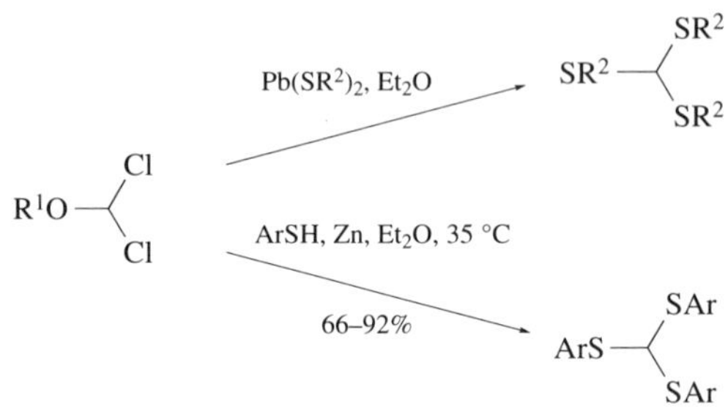

Scheme 12

(ii) From α-halodithioacetals and α-acyloxy- and α-alkoxydithioacetals

Treatment of chlorobis(methanethio)methane with methanethiol produced trimethyl trithio-*ortho*-formate, though reaction with higher alkanethiols was accompanied by disproportionation, giving rise to mixtures of trithio-*ortho*-esters ⟨61LA(648)21⟩. Cyclic α-chlorodithioacetals (**62**, X = S, CH_2) reacted with benzenethiol to give high yields of the trithio-*ortho*-esters (**63**, X = S, CH_2; Equation (45)) ⟨76BCJ553⟩. Reaction of the benzoyloxytrithiane (**64**) with a range of alkane- and arene-thiols produced the trithio-*ortho*-esters (**65**, R = alkyl, aryl; Equation (46)) ⟨74BCJ2457, 75BCJ2496⟩. The 2-alkoxybenzodithiole (**66**) reacted similarly with ethanethiol in the presence of acetic acid to give 2-ethylthio-1,3-benzodithiole (Equation (47)) ⟨75S436⟩.

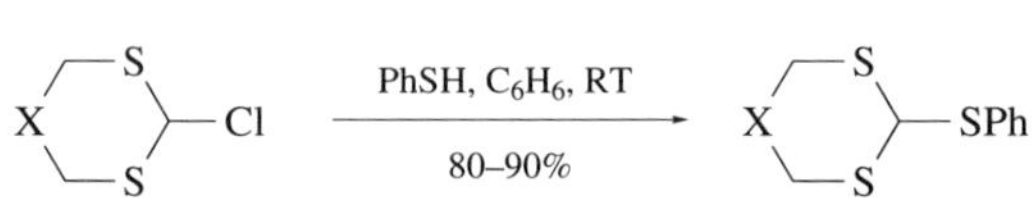

(62) **(63)** (45)

(64) **(65)** (46)

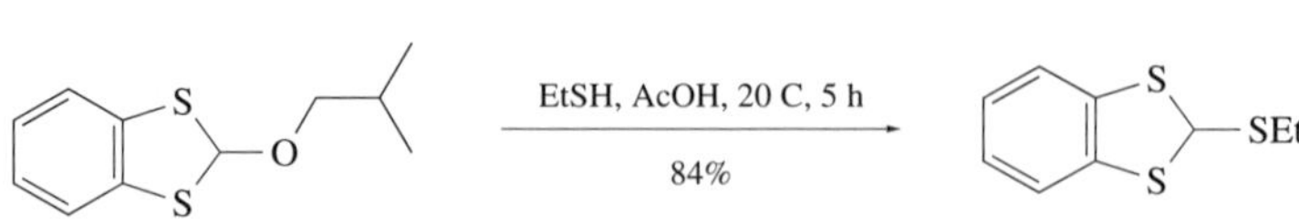

(66) (47)

(iii) From carboxylic ortho*-esters and related compounds*

Carboxylic *ortho*-formate esters undergo exchange reactions with alkanethiols and arenethiols to give trithio-*ortho*-esters (Equation (48)). The reaction may be catalysed using hydrogen chloride ⟨67AG(E)442⟩, *p*-toluenesulfonic acid ⟨55JA509⟩, zinc chloride ⟨64T417⟩, boron trifluoride etherate ⟨72CB3280⟩ and montmorillonite KSF ⟨89SC31⟩, though it will often proceed in the absence of a catalyst ⟨60IZV1901⟩. Exchange of trialkyl *ortho*-formates with ethanedithiol and propanedithiol gave (**67**) and (**68**) respectively ⟨26JCS2263, 67AG(E)442⟩, and exchange of triethyl *ortho*-formate with 2-methyl-2-thiohydroxymethylpropane-1,3-dithiol produced the tricyclic trithio-*ortho*-ester (**69**, R = H) ⟨55JA509⟩. Little work has been reported on the exchange of higher *ortho*-esters with thiols, though triethyl *ortho*-acetate does undergo reaction with 2-methyl-2-thiohydroxymethylpropane-1,3-dithiol to give the tricyclic trithio-*ortho*-ester (**69**, R = Me) ⟨55JA509⟩.

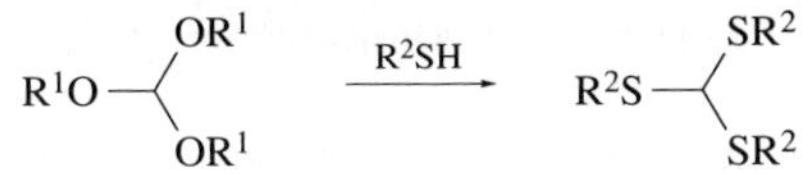
(48)

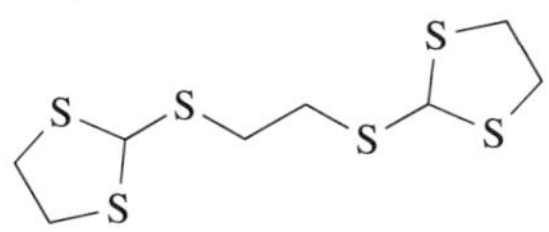
(67)

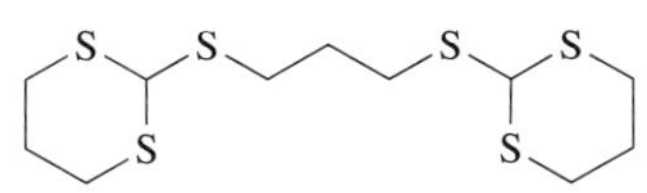
(68)

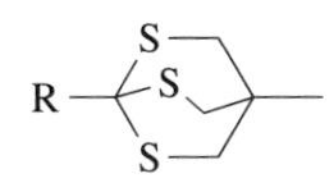
(69)

*6.03.3.1.2 Trithio-***ortho***-esters from thioimidate ester salts*

Thioimidate ester hydrochloride salts react with alkanethiols and arenethiols to yield the trithio-*ortho*-esters (**70**) ⟨53JA1668, 62JOC2858⟩. Although this method is analogous to the widely used Pinner synthesis of *ortho*-esters, relatively few reports of the preparation of trithio-*ortho*-esters using this procedure have appeared. Some trithio-*ortho*-esters prepared by this method are listed in Table 3.

(70)
(49)

Table 3 Trithio-*ortho*-esters (**70**) prepared from thioimidate ester hydrochloride salts.

R[1]	*R*[3]	*Ref.*
H	Ph	53JA1668
C_2F_5	Me	62JOC2858
C_3F_7	Me	62JOC2858
C_3F_7	Et	62JOC2858

Other methods for the preparation of trithio-*ortho*-esters which proceed via thioimidate intermediates have been described. Formamide undergoes an acid catalysed reaction with thiols to give trithio-*ortho*-esters (Scheme 13) ⟨07CB1740, 07LA(353)131⟩, and the formanilide (**71**) reacts with phosphorus oxychloride in the presence of thiols to give trithio-*ortho*-esters in yields of 44–69% (Scheme 14) ⟨60IZV1828⟩. Acetamide derivatives do not react under similar conditions ⟨59RTC354⟩.

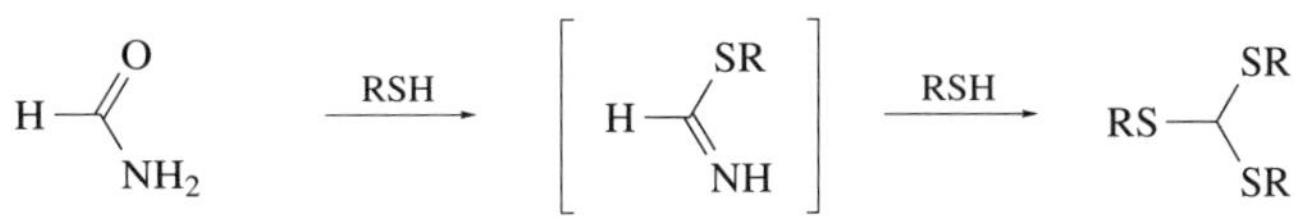
Scheme 13

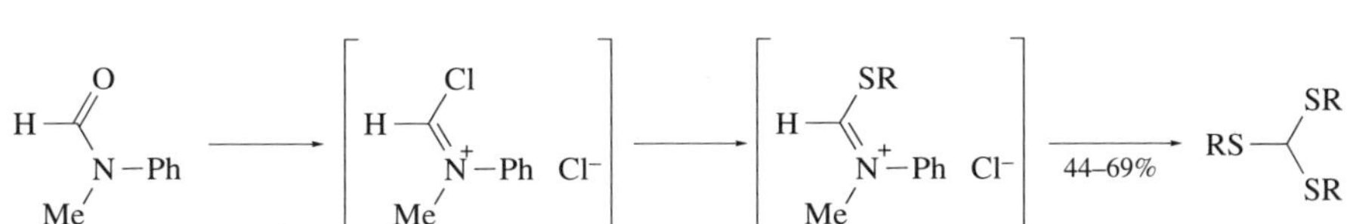
(71)

Scheme 14

6.03.3.1.3 *Trithio-***ortho***-esters from carboxylic acids and derivatives*

(i) From acyl chlorides and anhydrides

A general method for the preparation of trithio-*ortho*-esters involves the treatment of acyl chlorides with thiols in the presence of zinc chloride or aluminium trichloride (Equation (50)). This procedure has been used to prepare a range of trithio-*ortho*-esters derived from aliphatic ⟨59RTC354⟩ and aromatic ⟨53JA1668, 80JOC740, 86TL4861⟩ acyl chlorides; yields are usually greater than 40%. A limitation of the method is that acyl chlorides containing one or more α-hydrogen atoms give poor yields of trithio-*ortho*-esters due to competing elimination reactions, though this problem can be overcome to some extent by using an excess of the thiol reagent ⟨59RTC354⟩. Reaction of acetic anhydride with 3-methylbenzenethiol in the presence of boron trifluoride has been reported to give tris(3-methylphenyl) trithio-*ortho*-acetate in 30% yield ⟨53JA1668⟩.

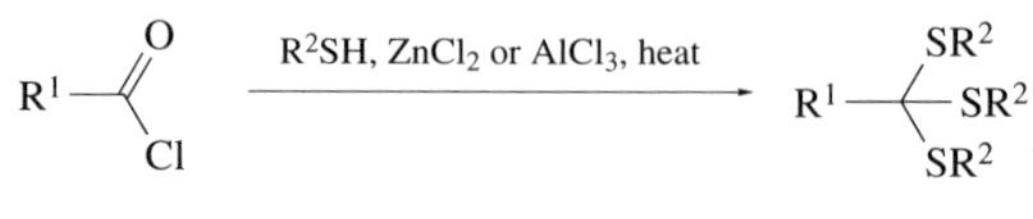

(50)

(ii) From carboxylic esters and thioesters

In 1907, Holmberg ⟨07CB1740, 07LA(353)131⟩ discovered that formate esters undergo an acid promoted reaction with alkanethiols and arenethiols to give trithio-*ortho*-esters (Equation (51)). A range of trialkyl and triaryl trithio-*ortho*-formates have been prepared using this method ⟨07CB1740, 07LA(353)131, 48JCS687, 62RTC1009⟩. These reactions simply involve saturating a mixture of the ester and thiol with hydrogen chloride, and allowing the mixture to stand; yields are normally greater than 50%. Triaryl *ortho*-formates have also been obtained from the reactions of arenethiomagnesium bromides with ethyl formate ⟨62ZOB745⟩.

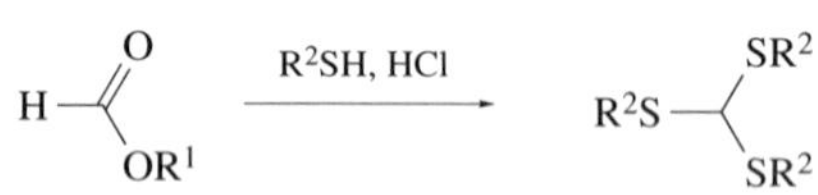

(51)

The Holmberg method is not suitable for the preparation of trithio-*ortho*-esters from higher carboxylic esters. In a modification of the reaction, triethyl trithio-*ortho*-acetate has been obtained in low yield from the zinc chloride catalysed reaction of ethanethiol with ethyl acetate ⟨59RTC354⟩. More recently, the synthesis of trialkyl trithio-*ortho*-benzoate esters by the aluminum trichloride catalysed reaction of methyl benzoates with alkanethiotrimethylsilanes has been reported (Equation (52)) ⟨80JOC740, 86TL4861⟩.

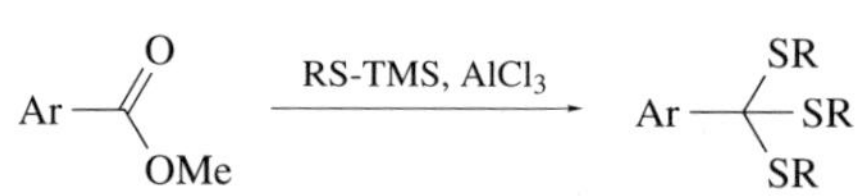

(52)

Thiocarboxylic esters (**72**, $R^1 = H$, alkyl, aryl) undergo acid or Lewis acid catalysed reactions with thiols to give trithio-*ortho*-esters ⟨45USP2389153, 53JA1668, 75JOC963⟩. Mixed trithio-*ortho*-esters (**73**), in which R^2 and R^3 are different, may be prepared by this method (Equation (53)) ⟨45USP2389153, 53JA1668⟩.

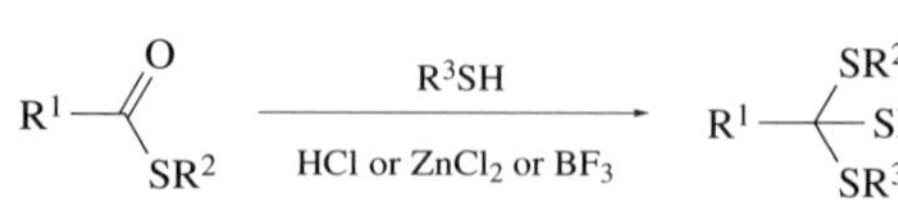

(72) **(73)** (53)

(iii) From carboxylic acids

Formic acid reacts with alkanethiols and arenethiols to give trithio-*ortho*-formate esters ⟨11CB3235, 12CB2942⟩. Higher carboxylic acids do not generally undergo this reaction; exceptions are the reactions of trifluoroacetic acid with ethane-1,2-dithiol and propane-1,3-dithiol, which give the cyclic bistrithio-*ortho*-esters (**74**) and (**75**) respectively ⟨67CC1089, 68JOC2504⟩. Reaction of formic acid with hydrogen sulfide and acetic anhydride in the presence of zinc chloride afforded tris (acetylthio)methane (**76**; Equation (54)) in 40% yield ⟨76CS122⟩. This compound may also be prepared by treatment of formic acid with thioacetic acid and acetic anhydride in the presence of zinc chloride ⟨76CS122⟩.

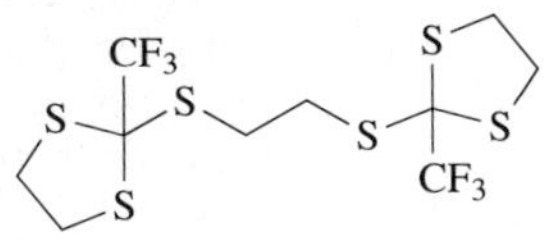

(74)

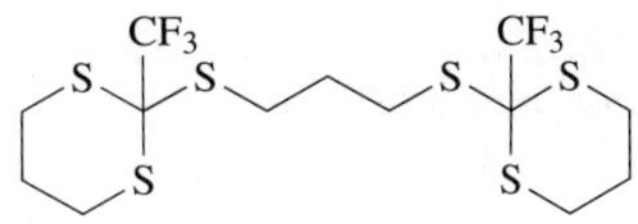

(75)

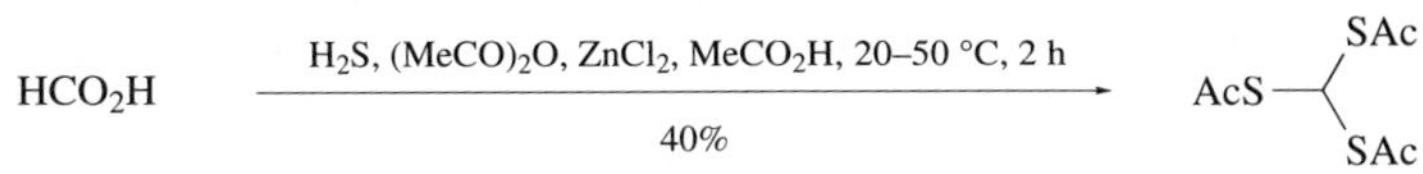

(76) (54)

6.03.3.1.4 *Trithio-*ortho*-esters from dithio- and trithiocarbenium salts*

Trithio-*ortho*-esters and related compounds are obtained from the addition of thiols and other sulfur nucleophiles to dithiocarbenium salts. Reaction of the dithiolium tetrafluoroborate salt (**77**) with sodium benzenethiolate produced the 2-phenylthio-1,3-dithiole (**78**; Equation (55)), and reaction with *N*,*N*-dialkyldithiocarbamate salts gave 2-dithiocarbamoyldithioles (**79**) ⟨69CPB1931⟩. Treatment of the dithiolium perchlorate salt (**80**) with one equivalent of potassium ethyl xanthate yielded the analogous ethoxythiocarbonythio derivative (**81**), whereas the use of an excess of potassium ethyl xanthate produced the sulfide (**82**), the reaction being thought to proceed via further reaction of (**81**) with the xanthate (Scheme 15) ⟨77JOC1543⟩.

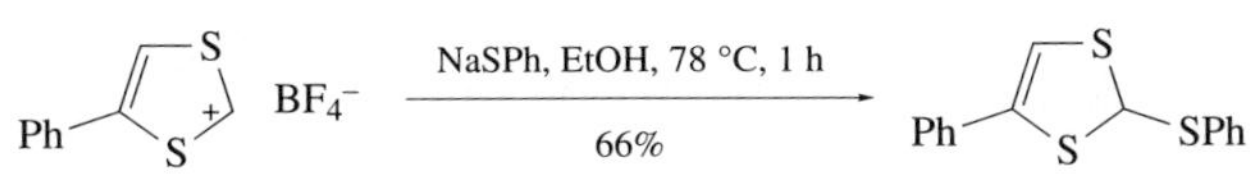

(77) **(78)** (55)

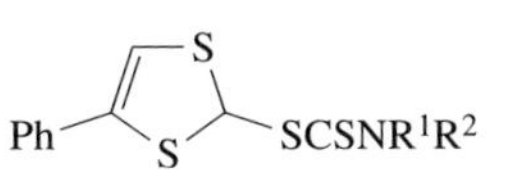

(79)

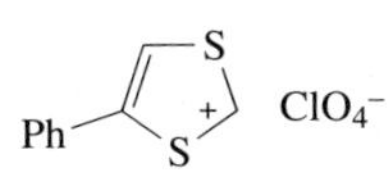

(80)

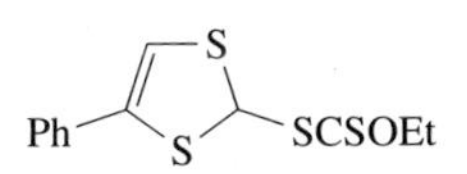

(81)

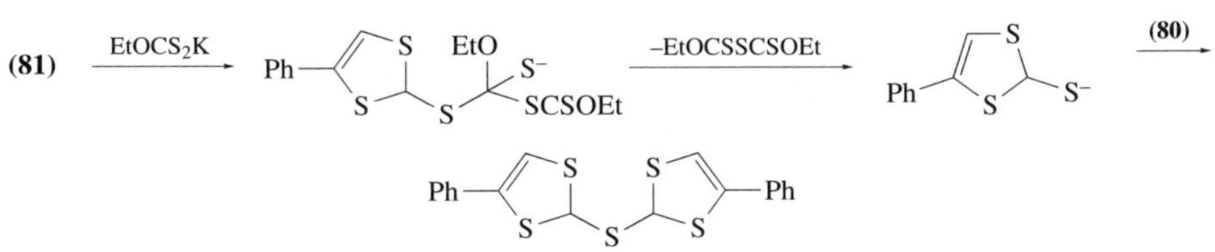

(82)

Scheme 15

2-Methanethio-1,3-dithiolium salts may be reduced to 2-methanethio-1,3-dithioles using sodium borohydride. The dithiole (**84**) was prepared in 78% yield from the salt (**83**) by this method (Equation (56)) ⟨92JOC1696⟩.

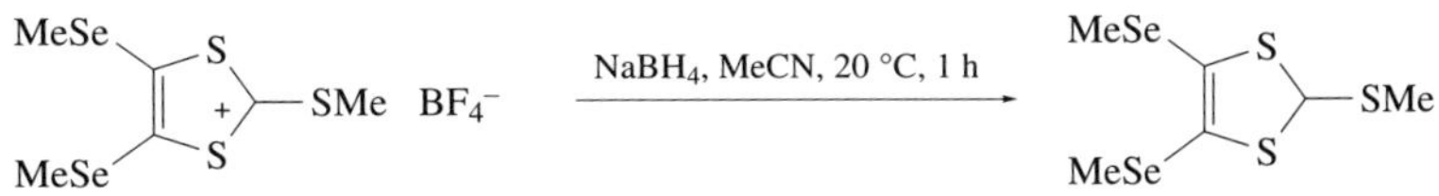

(83) **(84)** (56)

6.03.3.1.5 *Trithio-*ortho*-esters from dithioacetals*

Dithioacetals may be converted into trithio-*ortho*-esters by deprotonation with a strong base, followed by reaction of the metallated species (**85**) with a disulfide (Scheme 16). The reaction was first described by Frohling and Arens ⟨62RTC1009⟩, who prepared triethyl trithio-*ortho*-benzoate in 76% yield from benzaldehyde diethyl dithioacetal using sodium amide and diethyl disulfide (Equation (57)). A procedure for the preparation of trithio-*ortho*-ester (**86**) from the cyclic dithioacetal (**87**) using *n*-butyl lithium and dimethyl disulfide is given in *Organic Syntheses* ⟨77OS(56)8⟩, and this method has also been used for the preparation of a range of benzoic, cinnamic and aliphatic trithio-*ortho*-esters ⟨72JOC2757⟩. However, the reaction may not be general for the preparation of aliphatic trithio-*ortho*-esters—trithioacetates and trithiopropionates could not be prepared from the corresponding dithioacetals using sodium amide and dialkyl disulfides, presumably because of competing elimination reactions ⟨62RTC1009⟩. The preparation of trithio-*ortho*-formate esters by this method appears to be dependent on the nature of the substituents on sulfur, and on the base used. Seebach *et al.* prepared the mixed trithio-*ortho*-formate ester (**88**) in 95% yield from bis(benzenethio)methane using *n*-butyllithium and dimethyl disulfide (Equation (58)) ⟨72CB3280⟩, whereas Frohling and Arens obtained only tetraethyl tetrathio-*ortho*-carbonate from the reaction of the bis(ethanethio)methane with sodium amide and diethyl disulfide (Scheme 17) ⟨62RTC1009⟩. Doubly ^{13}C-labelled trimethyl trithio-*ortho*-formate has been prepared by a related method, in which the lithiated dithioacetal (**89**) is treated sequentially with elemental sulfur followed by ^{13}C-labelled methyl iodide (Scheme 18) ⟨84HCA1070⟩.

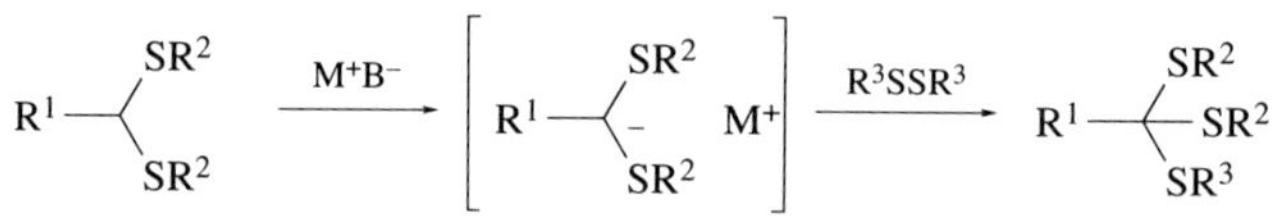

(85)

Scheme 16

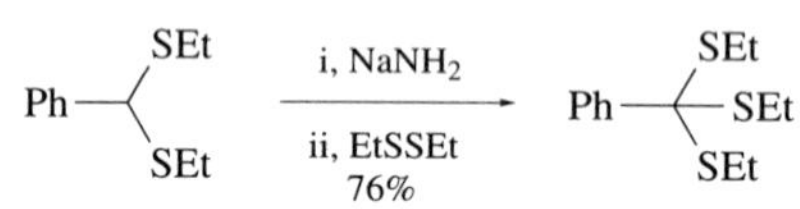

(57)

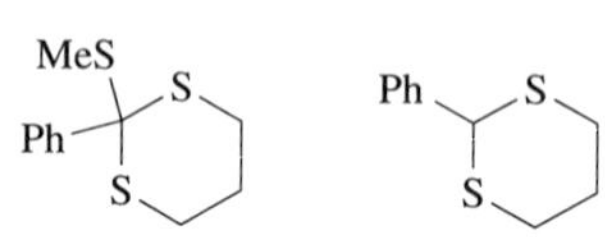

(86) **(87)**

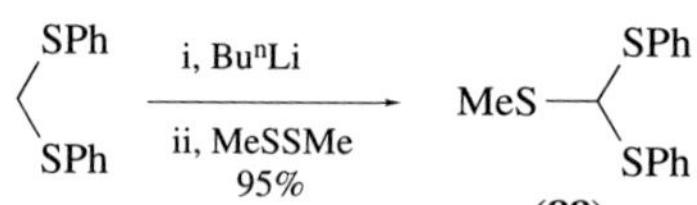

(88)

(58)

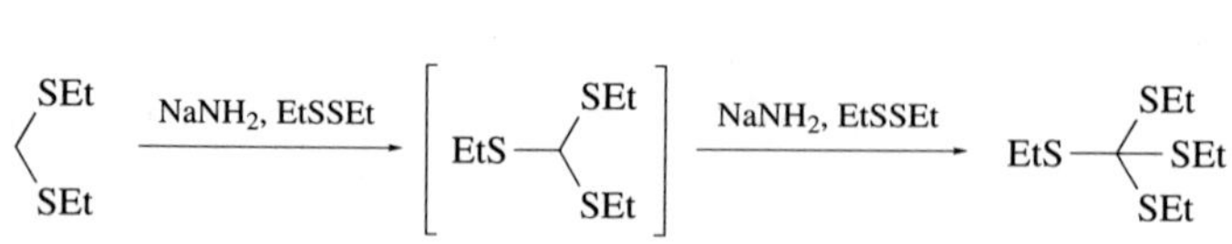

Scheme 17

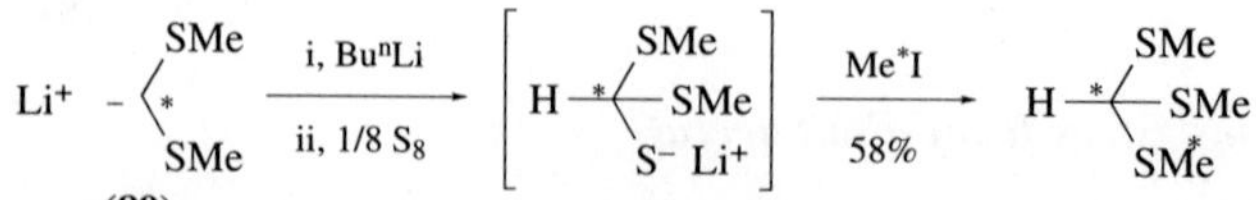

(89)

Scheme 18

Trithio-*ortho*-formates have been prepared by reaction of *N-p*-toluenesulfonylsulfilimines (**90**) derived from dithioacetals of formaldehyde with alkanethiols and arenethiols under alkaline conditions (Equation (59)) ⟨76S551⟩. Mixed trithio-*ortho*-esters are available by this method.

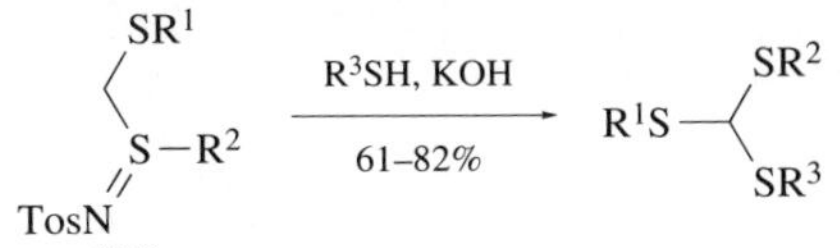

(**90**) (59)

6.03.3.1.6 *Trithio*-ortho-*formates from trithiocarbonates*

Addition of phenyllithium to diphenyl trithiocarbonate at −78 °C yields the triphenylthiomethylide anion (**91**; Equation (60)). This anion, when prepared by a different method, has been quenched with deuterated water to give the deuterated triphenyl trithio-*ortho*-formate (**92**) ⟨72CB487⟩. Similar addition of methyllithium to the dithiolone (**93**), followed by quenching with acetic acid, yielded the 2-methanethio-1,3-dithiole (**94**; Scheme 19) ⟨87TL4153⟩. The use of lithiated trithio-*ortho*-formates for the preparation of higher trithio-*ortho*-esters is discussed in Section 6.03.3.2.1.

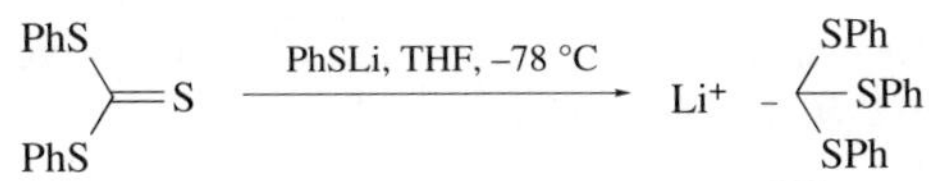

(**91**) (60)

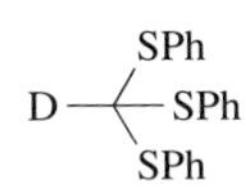

(**92**)

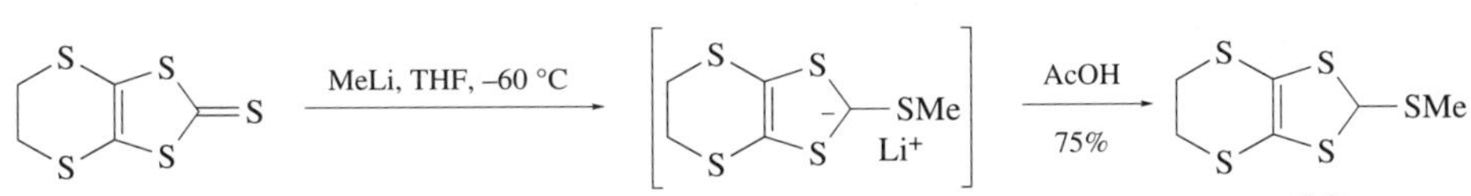

(**93**) (**94**)

Scheme 19

Diphenyl trithiocarbonate undergoes cycloaddition reactions with diaryldiazomethanes to produce the cyclic trithio-*ortho*-ester derivatives (**95**; Equation (61)) ⟨65CB3303⟩.

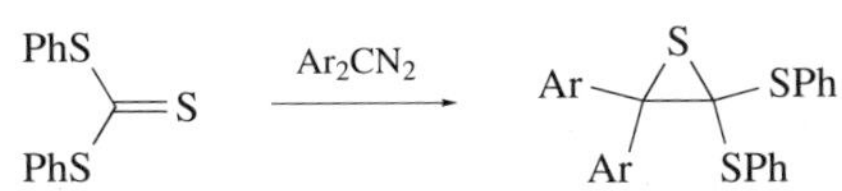

(**95**) (61)

6.03.3.1.7 *Trithio*-ortho-*formates from tetrathio*-ortho-*carbonates*

Seebach ⟨67AG(E)442⟩ reported the preparation of the trithio-*ortho*-formate (**97**) by treatment of the tetrathio-*ortho*-carbonate (**96**) with *n*-butyllithium at −78 °C, followed by quenching the reaction with water (Scheme 20). The reaction proceeds via the lithiated trithio-*ortho*-formate ester (**98**). Similar treatment of tetraphenyl tetrathio-*ortho*-carbonate with *n*-butyllithium gave the triphenylthiomethylide anion (**91**), which has been used to prepare the deuterated trithio-*ortho*-formate (**92**) ⟨72CB487⟩. The use of lithiated trithio-*ortho*-formates for the preparation of higher trithio-*ortho*-esters is discussed in Section 6.03.3.2.1.

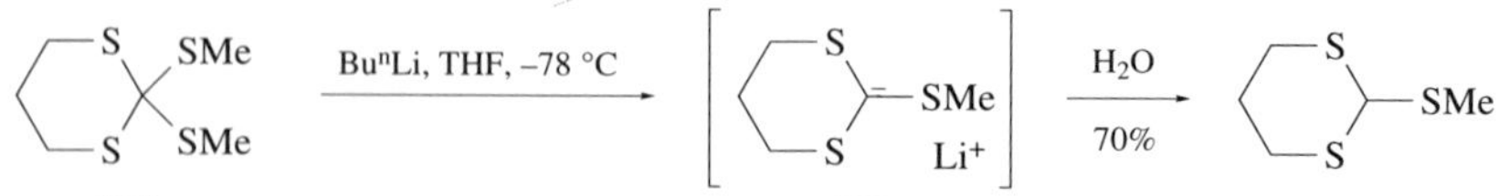

Scheme 20

6.03.3.2 Preparation of Trithio-*ortho*-esters from other Trithio-*ortho*-esters

*6.03.3.2.1 Higher trithio-*ortho*-esters from trithioformate esters*

In 1962, Hine *et al.* ⟨62JA1751⟩ reported the generation of the potassium trithio-*ortho*-formate salt (**99**) by deprotonation of trimethyl trithio-*ortho*-formate with potassium amide, and the reaction of this with methyl iodide to produce trimethyl trithio-*ortho*-acetate in 75% yield (Scheme 21). Seebach *et al.* ⟨67AG(E)442, 72CB487, 72CB3280⟩ later described the generation of several lithiated trithio-*ortho*-formate esters from the appropriate trithio-*ortho*-formate with *n*-butyllithium at −78 °C, and the subsequent reactions of these with a range of electrophiles to yield higher trithio-*ortho*-esters. This has since developed into a general method for the homologation of trithio-*ortho*-formates. Thus, lithiated trithio-*ortho*-formates react with alkyl halides to give aliphatic trithio-*ortho*-esters ⟨67AG(E)442, 72CB487, 72CB3280, 76CL43, 82TL3407⟩, with aldehydes and ketones to give α-hydroxy-trithio-*ortho*-esters ⟨67AG(E)442, 72CB487, 76CL43, 81TL4009⟩, with epoxides to give β-hydroxy trithio-*ortho*-esters ⟨67AG(E)442, 72CB487⟩, with cyclic α,β-unsaturated ketones to give γ-ketotrithio-*ortho*-esters ⟨75CC216, 89TL5481⟩, with carbon dioxide to give trithiomono-*ortho*-oxalates ⟨67M1043⟩, with carbon disulfide to give trithiomono-*ortho*-dithiooxalates ⟨88T2063⟩, with chloroformate esters to give trithiomono-*ortho*-oxalate esters ⟨67AG(E)442, 72CB487⟩ and with iodine to give coupled hexa-thiodi-*ortho*-oxalates ⟨72CB3892⟩ (Scheme 22). An example of the acylation of a lithiated trithio-*ortho*-formate by reaction with a lactone has been reported (Equation (62)), but this reaction is not general ⟨91CJC415⟩. Since lithiated trithio-*ortho*-esters may also be obtained from trithiocarbonates (Section 6.03.3.1.6) and tetrathio-*ortho*-carbonates (Section 6.03.3.1.7), these compounds should also be regarded as precursors of higher trithio-*ortho*-esters.

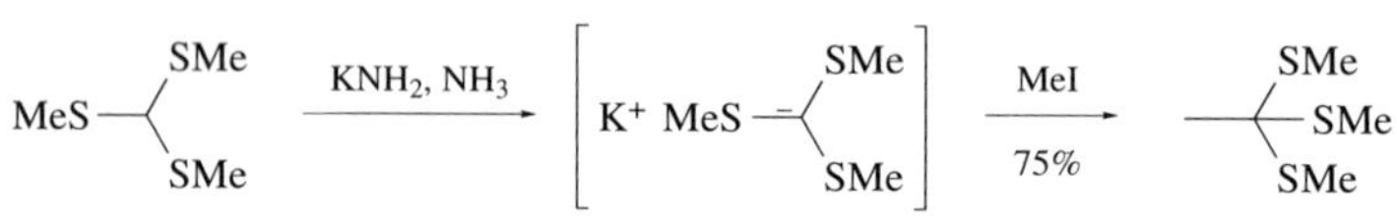

Scheme 21

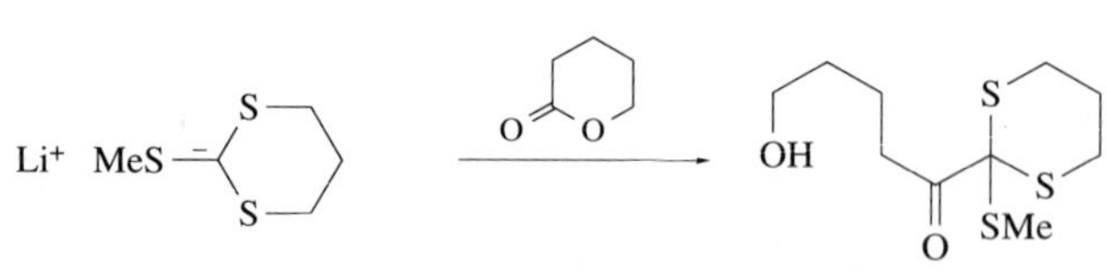

(62)

Triethyl trithio-*ortho*-heptanoate has been prepared by a radical homologation procedure, involving treatment of triethyl trithio-*ortho*-formate with 1-hexene in the presence of *t*-butyl hydroperoxide (Equation (63)) ⟨88MI 603-01⟩.

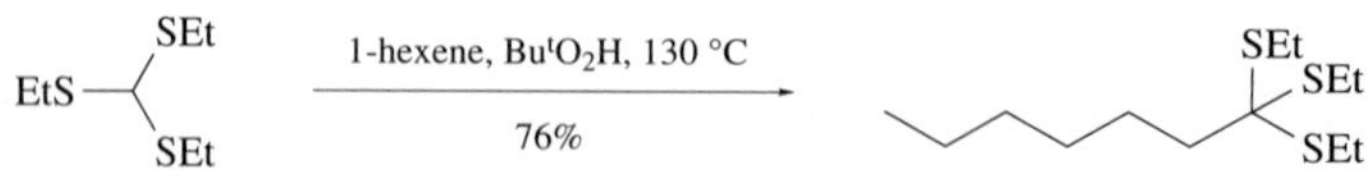

(63)

*6.03.3.2.2 Trans-esterification of trithio-*ortho*-esters*

In contrast to the chemistry of *ortho*-esters, relatively little work has been published on the preparation of trithio-*ortho*-esters by *trans*-esterification reactions of other trithio-*ortho*-esters with

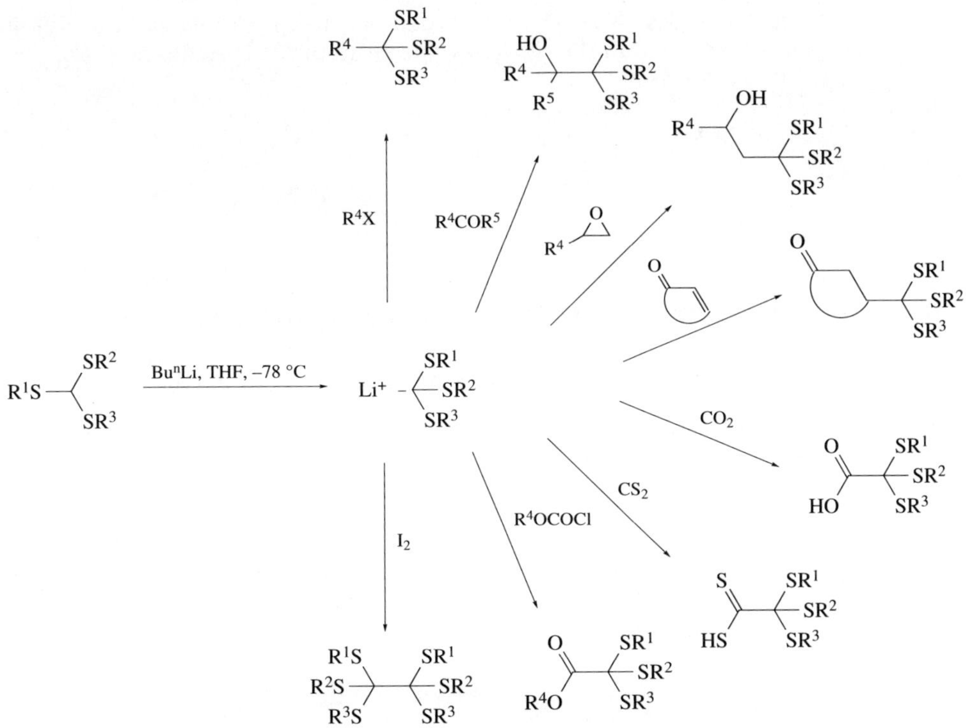

Scheme 22

thiols. Mixed trithio-*ortho*-esters undergo rapid acid catalysed disproportionation reactions to yield mixtures of trithio-*ortho*-esters (Scheme 23) ⟨61LA(648)21⟩. The dithiolane (**100**) and dithiane (**101**) have been prepared by heating triethyl trithio-*ortho*-formate with the appropriate dithiol in the presence of zinc chloride ⟨88URP1421744⟩.

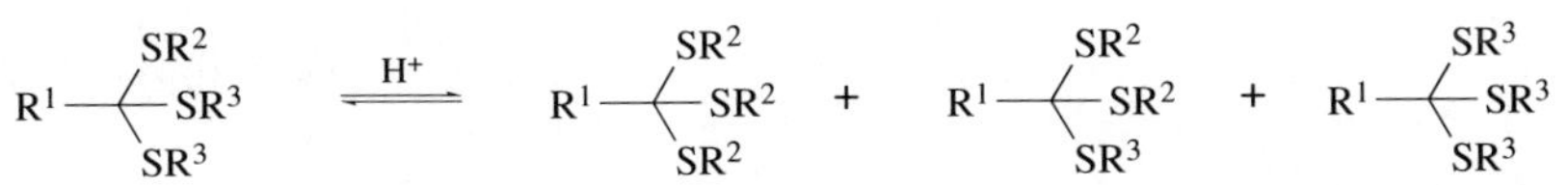

Scheme 23

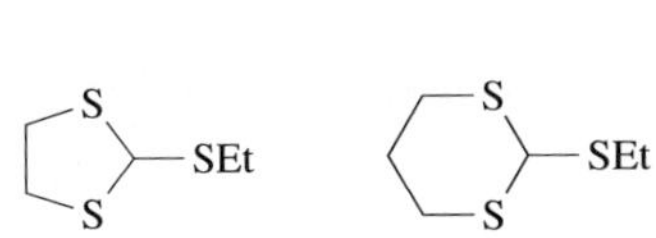

(100) **(101)**

6.03.3.2.3 *Modification of R^1 and R^2 of $R^1C(SR^2)_3$*

Treatment of triaryl trithio-*ortho*-acetates with deuterated trifluoroacetic acid in deuterochloroform results in a rapid proton exchange reaction to give the deuterated trithio-*ortho*-acetates (**102**; Equation (64)) ⟨75JOC963⟩. However, attempts to homologate triaryl trithio-*ortho*-acetates by reaction with electrophilic acylating agents have not been successful ⟨75JOC963⟩.

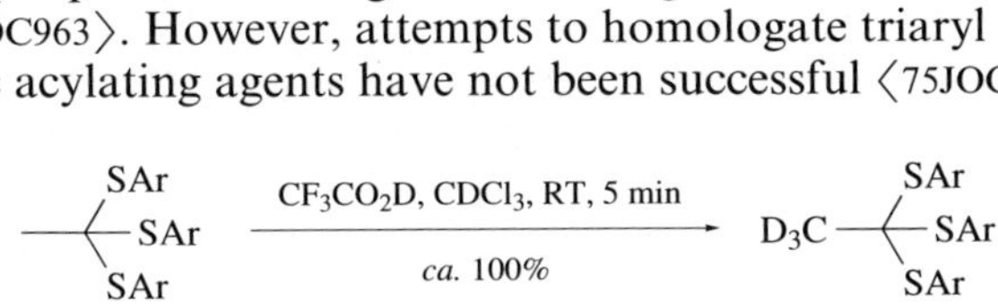

(64)

(102)

Deacetylation of tris(acetylthio)methane using methanolic hydrogen chloride produced methanetrithiol (**103**; Equation (65)) in 50% yield ⟨76CS122⟩. This product is stable at −30°C, although it polymerises slowly at 20 °C. The hydrolysis proceeds via di- and monoacetyl trithiomethane intermediates, which can be detected in the ^{1}H NMR spectrum of the reaction mixture.

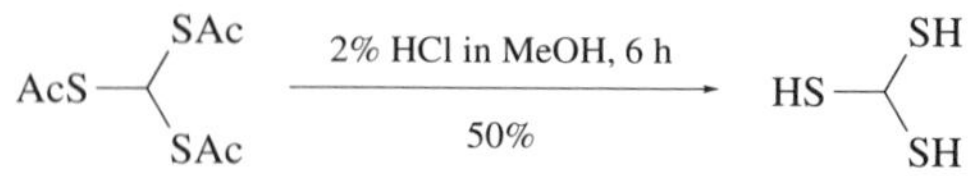

(103) (65)

6.03.3.3 Methods for the Preparation of Oxidised Derivatives of Trithio-*ortho*-esters

*6.03.3.3.1 Oxidised trithio-*ortho*-esters containing at least one sulfoxide group*

Triaryl trithio-*ortho*-formate esters are oxidised to the triaryl sulfoxide derivatives in high yields on treatment with peroxyacetic acid (Equation (66)) ⟨64CR(259)4051⟩. The sulfoxide (**104**) has been prepared by oxidation of the corresponding sulfide with peroxybenzoic acid (Equation (67)) ⟨48RTC884⟩.

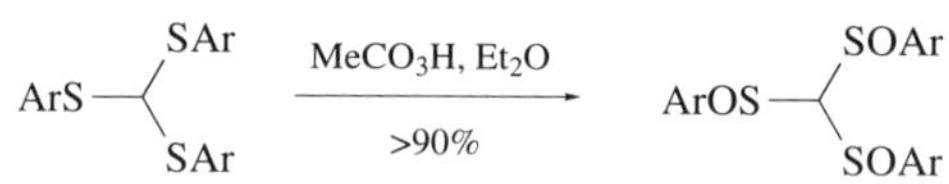

(66)

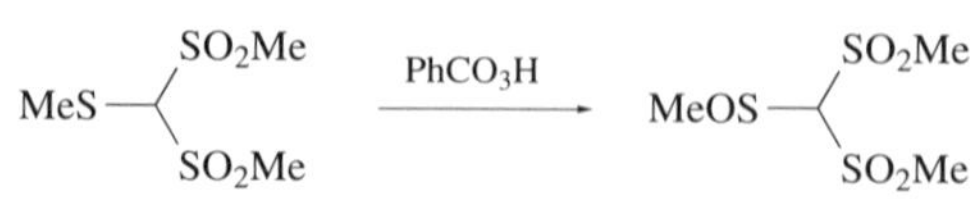

(104) (67)

*6.03.3.3.2 Oxidised trithio-*ortho*-esters containing at least one sulfone group*

Treatment of the sulfone (**105**) with two equivalents of *N*-ethanethiophthalimide in the presence of potassium *t*-butoxide produced the sulfonyldithio-*ortho*-ester derivative (**106**) in high yield (Equation (68)) ⟨82CC1183⟩. Bis(sulfonyl)methanes (**107**) react with alkyl thioalkyl sulfones and alkanesulfonyl or arenesulfonyl chlorides in the presence of sodium ethoxide to give disulfonylthiomethanes (**108**) ⟨33JCS306⟩ and trisulfonylmethanes (**109**) ⟨41CB1667⟩ respectively (Scheme 24). These oxidised trithio-*ortho*-formate derivatives produce the analogous trithio-*ortho*-acetate derivatives on treatment with methyl iodide in aqueous alkali (Equation (69)) ⟨1892CB347, 1892CB361⟩.

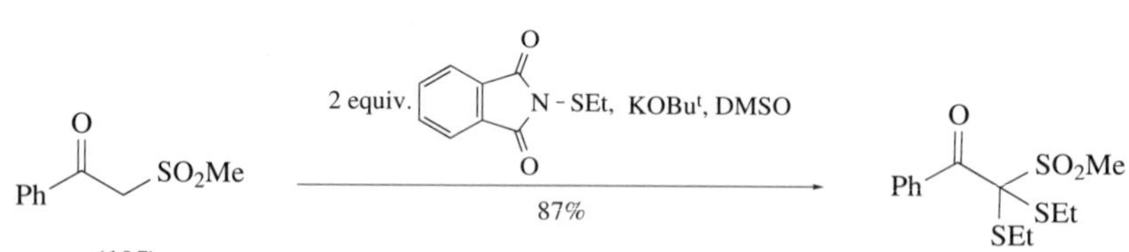

(105) (106) (68)

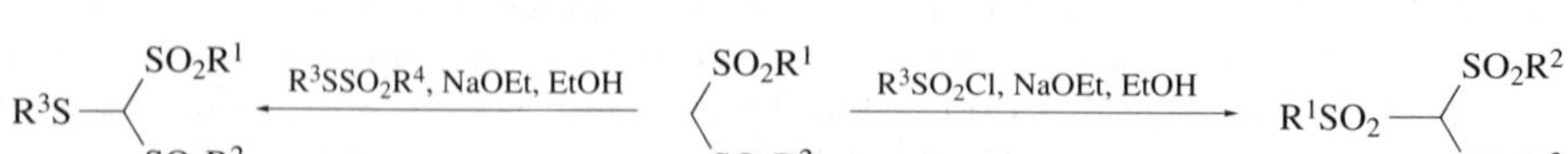

(108) (107) (109)

Scheme 24

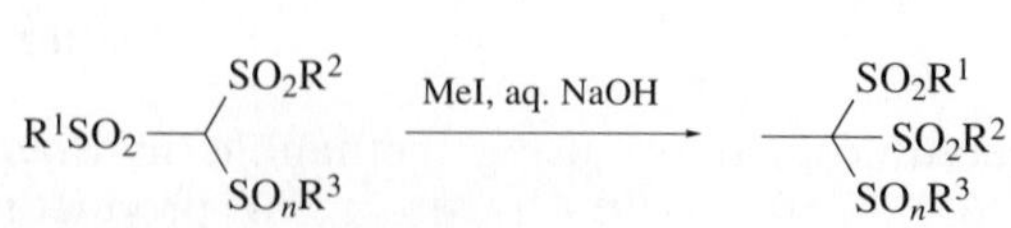

(69)

Trisulfones (**110**) are the products normally obtained from the oxidation of trialkyl and triaryl trithio-*ortho*-formates using acidic permanganate ⟨1892CB347, 33RTC437⟩, though Laves ⟨1890CB1414⟩ has reported that controlled oxidation of triphenyl trithio-*ortho*-formate using this reagent produces the disulfone (**111**). Yields from these permanganate oxidations are often low due to the competing formation of disulfonylmethanes and sulfonic acids ⟨07CB1740, 33RTC437⟩. Trialkyl trisulfones have been obtained from the oxidation of trialkyl trithio-*ortho*-formates with peroxyacetic and monoperoxyphthalic acids ⟨46RTC53, 53LA(581)133, 67USP3333007⟩. The tricyclic trisulfones (**112**, R = H, Me) and triaryl sulfones (**113**) were prepared by oxidation of the corresponding trithio-*ortho*-esters using hydrogen peroxide in acetic acid ⟨55JA509, 64CR(259)4051⟩. Oxidation of disulfones (**114**, R = H, Me) with alkaline permanganate or hydrogen peroxide produced the corresponding trisulfones in high yields (Equation (70)) ⟨1892CB347, 31JCS2637, 46RTC53⟩.

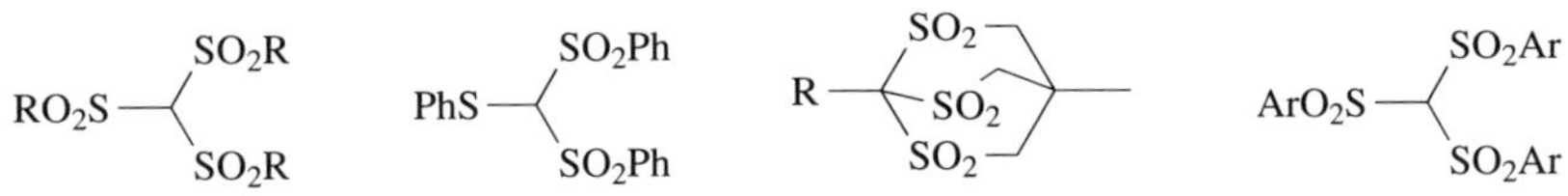

(110) **(111)** **(112)** **(113)**

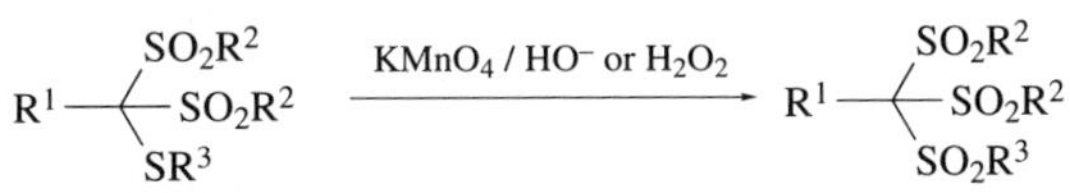

(114) (70)

The disulfonylthiomethane (**116**) was obtained in 94% yield from the sodium ethanethiolate reduction of the disulfonyldithiomethane (**115**; Equation (71)) ⟨82CC1183⟩.

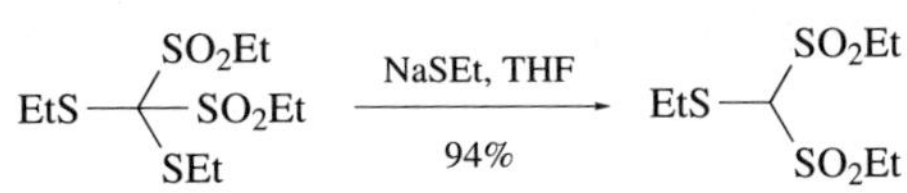

(115) **(116)** (71)

6.03.3.3.3 *Oxidised trithio*-ortho-*esters containing at least one sulfonate group*

Methanetrisulfonic acid (**117**) has been prepared by various methods, including sulfonation of methanedisulfonic acid and oxidation of thiol methanetrisulfonic acid; these have been reviewed by Backer ⟨30RTC1107⟩ and summarised by DeWolfe ⟨B-70MI 603-01⟩. The trimethyl and triethyl esters (**119**, R = Me, Et) were obtained by treatment of the silver(I) salt (**118**) with methyl iodide and ethyl iodide respectively (Equation (72)) ⟨49AK(1)231, 50ACS397⟩.

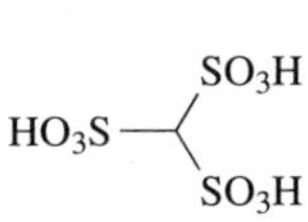

(117)

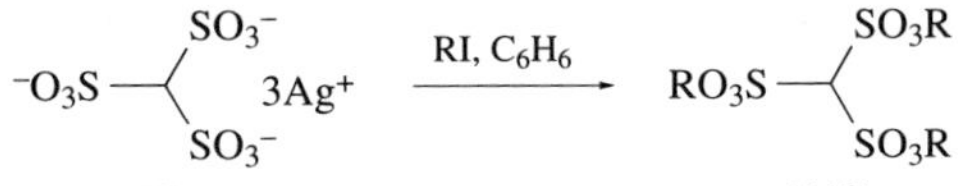

(118) **(119)** (72)

6.03.4 FUNCTIONS BEARING THREE SELENIUM ATOMS

Triseleno-*ortho*-esters contain a carbon atom bearing three alkaneseleno or areneseleno groups. Methods for the preparation of triseleno-*ortho*-esters have been reviewed briefly by Krief and Hevesi ⟨84MI 603-01⟩.

6.03.4.1 Methods for the Preparation of Triseleno-*ortho*-esters

*6.03.4.1.1 Triseleno-*ortho*-esters from $RC(X^1)(X^2)X^3$*

(i) From 1,1,1-trihaloalkanes

Triseleno-*ortho*-formates (**120**, R = Me, Ph) have been obtained from the treatment of bromoform with the appropriate sodium selenolate (Equation (73)) ⟨84HCA1070, 84S439⟩. The sodium selenolates were prepared by reduction of diselenides *in situ*; sodium methaneselenolate was prepared by the sodium/liquid ammonia reduction of dimethyl diselenide ⟨84HCA1070⟩, and sodium benzeneselenolate was prepared by reduction of diphenyl diselenide using hydrazine ⟨84S439⟩. Seebach ⟨84HCA1070⟩ prepared ^{13}C-labelled trimethyl triseleno-*ortho*-formate from ^{13}C-labelled bromoform using this method.

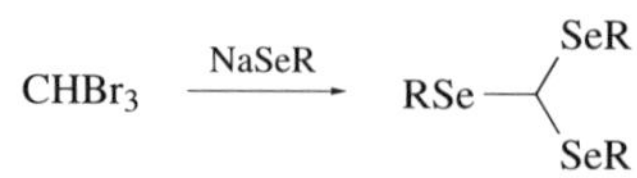

(120) (73)

(ii) From α,α-dihalo ethers

A range of alkane- and areneselenolates were shown to undergo a zinc promoted reaction with dichloromethyl methyl ether to give trialkyl and triaryl triseleno-*ortho*-formate esters in yields of 40–46% (Equation (74)) ⟨71ZOR473⟩.

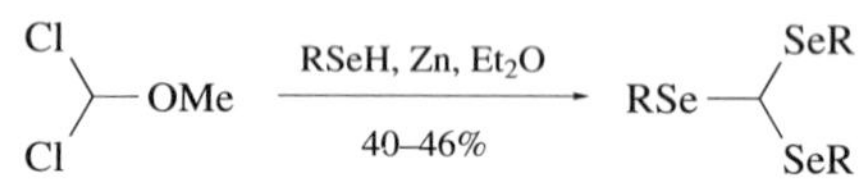

(74)

(iii) From ortho-*esters*

Trans-esterification of *ortho*-formate esters with methane- or benzeneselenol in the presence of boron trifluoride etherate produced the triseleno-*ortho*-formates (**121**, R^2 = Me) and (**121**, R^2 = Ph) in yields of 81% and 64% respectively (Equation (75)) ⟨84JA3784, 72CB511⟩. Trimethyl *ortho*-acetate and methaneselenol gave trimethyl triseleno-*ortho*-acetate in 66% yield under these conditions ⟨79JOM(177I)⟩. *Ortho*-esters have also been shown to undergo exchange reactions with boron triselenides to give triseleno-*ortho*-esters (Equation (76)), though these reactions are subject to steric limitations. Thus, boron tris(benzeneselenide) and triethyl *ortho*-formate gave triphenyl triseleno-*ortho*-formate in 76% yield, boron tris(methaneselenide) and trimethyl *ortho*-acetate produced trimethyl triseleno-*ortho*-acetate in 39% yield, whereas the reaction between boron tris-(benzeneselenide) and trimethyl *ortho*-acetate proceeded only very slowly, affording little of the triseleno-*ortho*-ester product ⟨79JOC1883, 79JOC4279⟩.

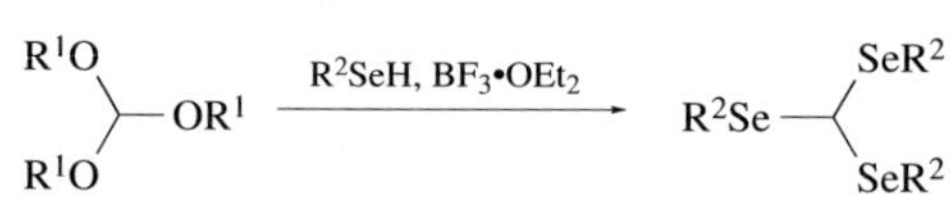

(121) (75)

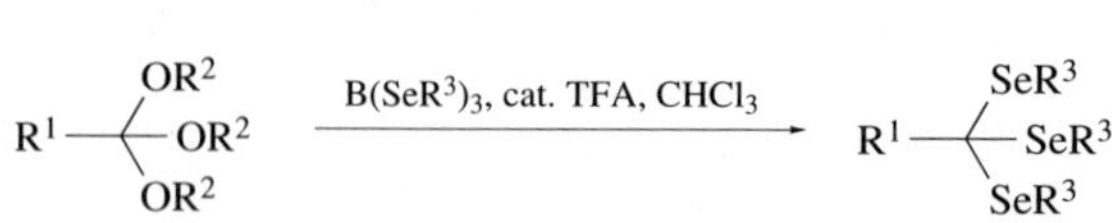

(76)

*6.03.4.1.2 Triseleno-*ortho*-esters from carboxylic acids*

^{13}C-Labelled triphenyl triseleno-*ortho*-formate has been prepared in 48% yield by the acid catalysed reaction of ^{13}C-labelled formic acid with benzeneselenol (Equation (77)) ⟨84HCA1070⟩.

The preparation of triseleno-*ortho*-esters from other carboxylic acid derivatives has not been reported.

$$HC^{*}O_2H \xrightarrow[\text{then RT, 30 d}\atop 48\%]{\text{PhSeH, HCl, 0 °C, 1 h}} HC^{*}(SePh)_3 \qquad (77)$$

6.03.4.1.3 *Triseleno-*ortho*-esters from triselenocarbenium salts*

In a manner analogous to the reduction of trithiocarbenium salts discussed in Section 6.03.3.1.4, 2-alkaneseleno-1,3-diselenoles may be prepared by sodium borohydride reduction of the corresponding triselenocarbenium salts. 2-Methaneseleno-1,3-diselenole (**122**, R = Me) was prepared from the corresponding triselenocarbenium iodide in a yield of greater than 80% using this method (Equation (78)) ⟨75TL1259⟩, and the 2-ethaneseleno derivative (**122**, R = Et) was similarly prepared from the analogous tetrafluoroborate salt ⟨86ZC138⟩.

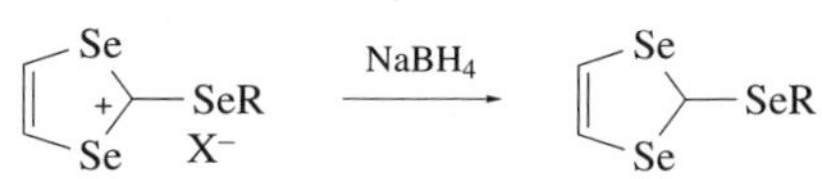

(122) (78)

6.03.4.1.4 *Triseleno-*ortho*-esters from diselenoacetals and related compounds*

A few derivatives of triseleno-*ortho*-esters have been prepared by the direct electrophilic selenation of diselenoacetals and related compounds. Reaction of α,α-bis(benzeneseleno)acetaldehyde with morpholinobenzeneselenamide produced the triseleno-*ortho*-ester (**123**; Equation (79)) ⟨82TL1557, 85BSF1219⟩, and reaction of the selenated enamine (**124**) with the same reagent gave the addition product (**125**) ⟨85BSF1219⟩. Treatment of acetone with red selenium under basic conditions produced the cyclic selenium salt (**126**) ⟨92CC1539⟩. This reaction proceeds through the diselenoacetal-like intermediates (**127**) and (**128**), and involves three electrophilic α-selenation steps (Scheme 25); it is therefore a counterpart of the haloform reaction.

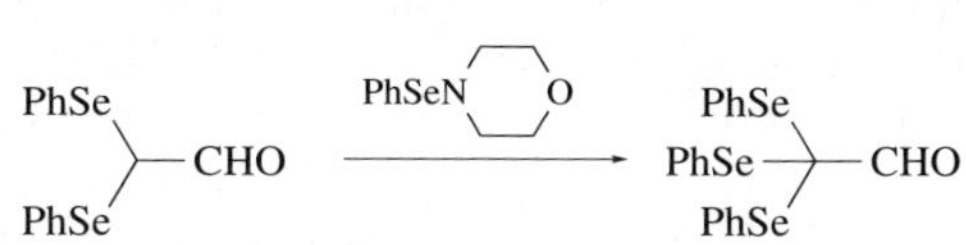

(123) (79)

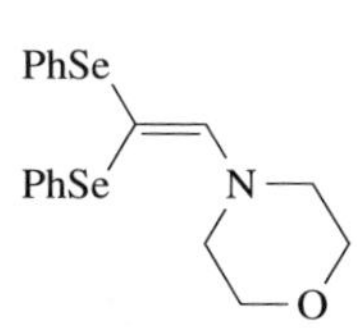

(124)

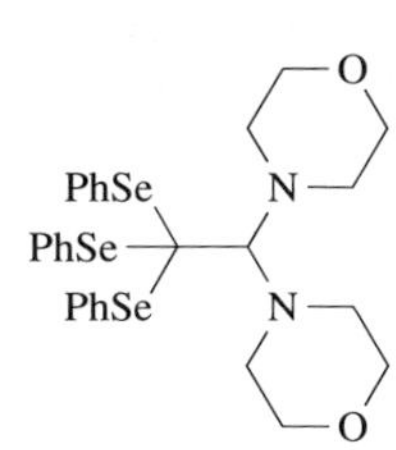

(125)

6.03.4.1.5 *Triseleno-*ortho*-formates from tetraseleno-*ortho*-carbonates*

Seebach ⟨69AG(E)450⟩ showed that the lithium triselenomethylide (**129**) is generated by treatment of tetraphenyl tetraseleno-*ortho*-carbonate with *n*-butyllithium at −78 °C (Equation (80)), and that this reacts with deuterated water to give the deuterated triseleno-*ortho*-formate (**130**). The use of lithiated triseleno-*ortho*-formates, for example (**129**), for the preparation of higher triseleno-*ortho*-esters is discussed in the next section.

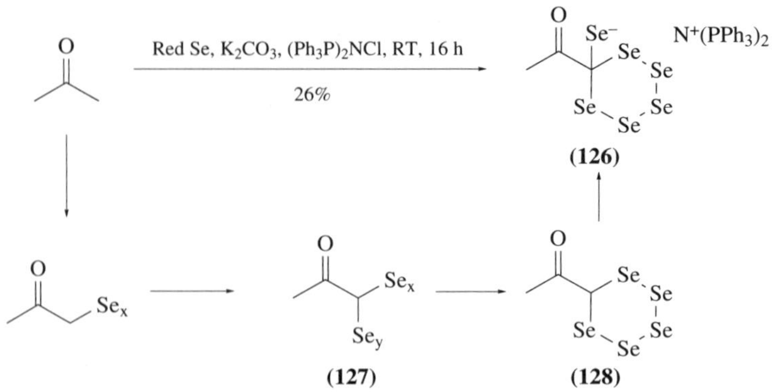

Scheme 25

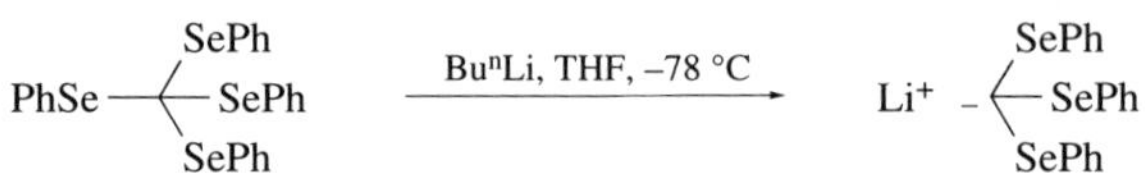

(80)

(129)

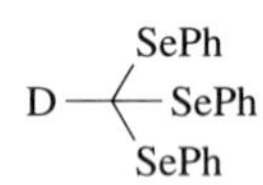

(130)

6.03.4.1.6 *Higher triseleno-*ortho*-esters from triseleno-*ortho*-formate esters*

Metallated triseleno-*ortho*-formates (**131**) are generated on treatment of triseleno-*ortho*-formate esters with strong bases; the lithiated triseleno-*ortho*-formate (**131**, R = Ph) is obtained from triphenyl triseleno-*ortho*-formate and lithium di-*s*-butylamide at $-78\,^\circ C$ ⟨69AG(E)450⟩, and (**131**, R = Me) is generated from trimethyl triseleno-*ortho*-formate and *n*-butyllithium at $-78\,^\circ C$ ⟨79JOM(1771)⟩. These lithiated triseleno-*ortho*-formates react with a range of electrophiles to yield higher triseleno-*ortho*-esters. Thus, lithiated triseleno-*ortho*-formates react with: alkyl halides to give aliphatic triseleno-*ortho*-esters ⟨69AG(E)450, 79JOM(177)1, 82TL3407⟩; aldehydes to give α-hydroxytriseleno-*ortho*-esters ⟨69AG(E)450, 81TL4009⟩, and epoxides to give β-hydroxytriseleno-*ortho*-esters ⟨78TL3971⟩ (Scheme 26). Tris(trifluoromethyl) triseleno-*ortho*-formate yielded the corresponding triseleno-*ortho*-acetate (**132**) on metallation with potassium di-*i*-propylamide, followed by treatment with methyl iodide (Equation (81)) ⟨84T4963⟩. Since metallated triseleno-*ortho*-formates may also be obtained from tetraseleno-*ortho*-carbonates (Section 6.03.4.1.5), these should also be regarded as precursors of higher triseleno-*ortho*-esters.

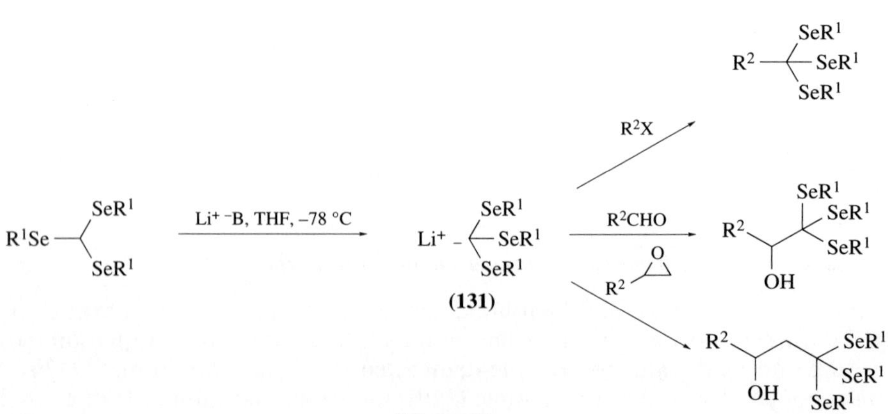

Scheme 26

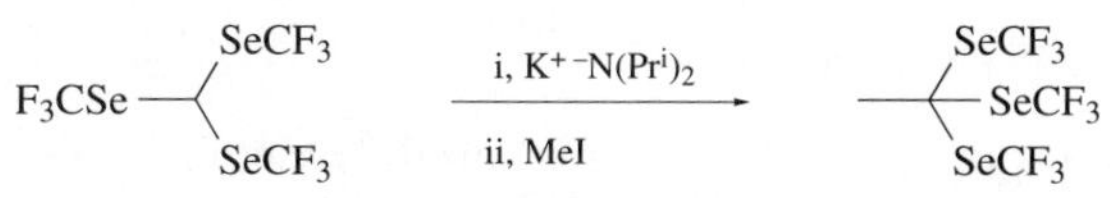

(132)

(81)

6.03.5 MIXED CHALCOGEN FUNCTIONS INCLUDING OXYGEN

Mono- and dithio-*ortho*-esters are derivatives of carboxylic *ortho*-esters in which one and two of the oxygen substituents respectively are replaced with sulfur substituents. Similarly, mono- and diseleno-*ortho*-esters are mixed oxygen and selenium functions containing one and two selenium substituents respectively. Monoselenomonothio-*ortho*-esters bear one oxygen, one sulfur and one selenium substituent. Oxidised derivatives of several mixed functions containing sulfur are known. Methods for the preparation of mixed oxygen and sulfur functions have been reviewed by Simchen ⟨85HOU(E5)3⟩.

6.03.5.1 Methods for the Preparation of Functions $R^1C(OR^2)(OR^3)SR^4$

6.03.5.1.1 From $R^1C(X^1)(X^2)X^3$

Triphenyl monothio-*ortho*-formate has been prepared in 42% yield by reaction of chlorodiphenoxymethane with benzenethiol in the presence of pyridine (Equation (82)) ⟨62CB1859⟩.

(82)

Monothio-*ortho*-esters have been prepared from the reactions of *ortho*-ester derivatives with thiols. *Ortho*-esters react with one equivalent of *n*-butanethiol in the presence of aluminum trichloride to give monothio-*ortho*-esters (Equation (83)); yields are reasonable (65–80%) for monothio-*ortho*-formates and -acetates, but are lower (ca. 35%) for thio-*ortho*-benzoates ⟨68CR(C)1506⟩. This reaction will also proceed in the absence of a catalyst; monothio-*ortho*-formates are obtained in high yields ⟨69BSF325⟩, though yields are low for monothio-*ortho*-acetates and -*ortho*-benzoates, which are obtained as mixtures with dithio-*ortho*-esters ⟨70MI 603-02⟩. Treatment of the *O*-acyl *ortho*-formate ester (**46**) with a range of thiols in the presence of triethylamine produced the monothio-*ortho*-esters (**133**) in yields of 45–75% (Equation (84)) ⟨69RTC897⟩. The acid catalysed *trans*-esterification of *ortho*-esters with 2-hydroxybenzenethiol produced the cyclic monothio-*ortho*-esters (**134**) in good yields (70–77%) (Equation (85)) ⟨75S660⟩. Cyclic monothio-*ortho*-ester derivatives have also been obtained from the exchange reactions of *ortho*-esters with 2-hydroxyethanethiol ⟨77MI 603-01⟩ and 2-thiohydroxyacetic acid ⟨78IZV468⟩.

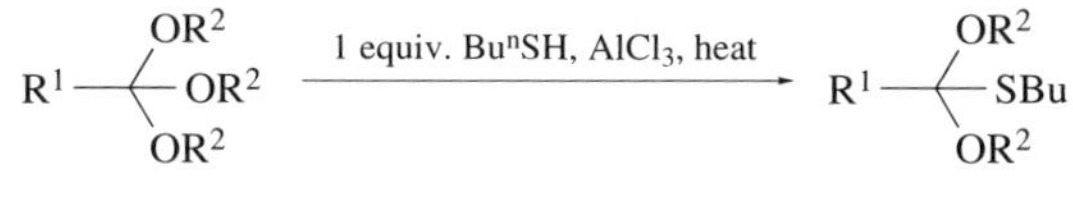

(83)

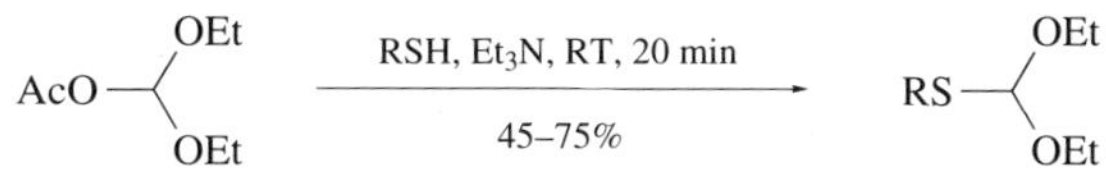

(46) **(133)**

(84)

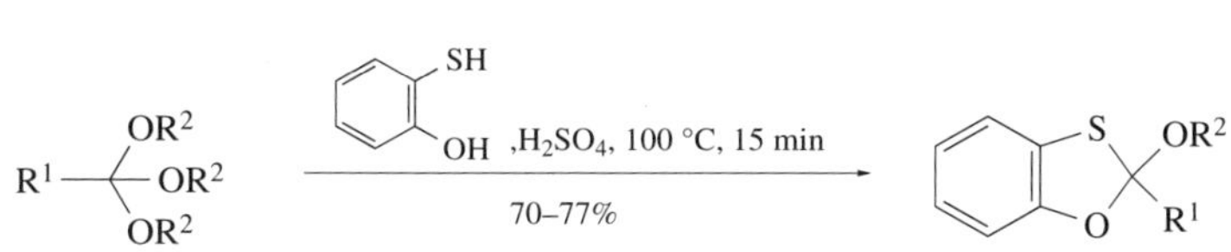

(134)

(85)

6.03.5.1.2 From monothiocarboxylic esters

Monothio-*ortho*-benzoate esters have been prepared in yields of 66% or more by reaction of a monothiobenzoate ester with a sodium alkoxide, followed by alkylation of the thiolate intermediate with an alkyl iodide (Scheme 27) ⟨89S684⟩. However, this procedure is not suitable for the preparation of monothio-*ortho*-esters from simple aliphatic monothio esters, which may undergo deprotonation under the reaction conditions ⟨89S684⟩. Treatment of the hydroxy monothioesters (**135**, n = 0, 1) with sodium hydride gave the cyclic thiolate intermediates (**136**, n = 0, 1), which were methylated *in situ* to yield monothio-*ortho*-esters (**137**, n = 0, 1) in overall yields of 35–78% (Scheme 28) ⟨86JA6683⟩. Unlike the reaction of simple monothio-*ortho*-esters with alkoxides, this procedure is applicable to the preparation of cyclic monothio-*ortho*-acetate esters as well as monothio-*ortho*-benzoates.

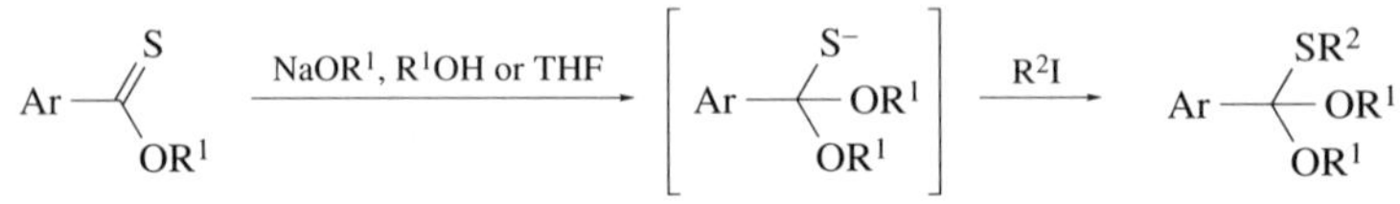

Scheme 27

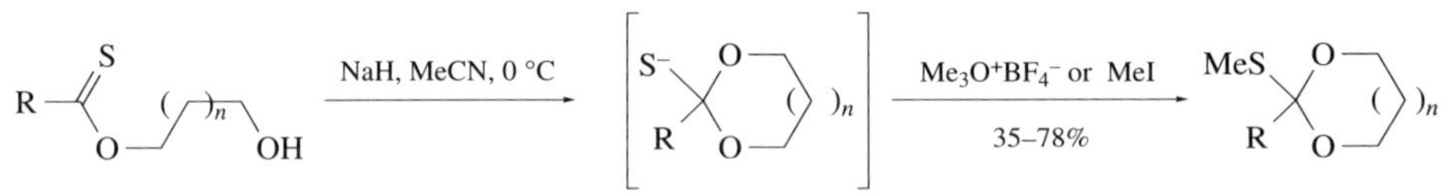

(135) **(136)** **(137)**

Scheme 28

6.03.5.1.3 From dioxo- and oxothiocarbenium salts

Monothio-*ortho*-esters are obtained from the treatment of dioxocarbenium salts with thiols. This reaction is analogous to the preparation of *ortho*-esters from dioxocarbenium salts and alcohols described in Section 6.03.2.1.4. Reaction of the dimethyl dioxocarbenium salt (**138**) with methanethiol in the presence of triethylamine produced the monothio-*ortho*-ester (**139**; Equation (86)) ⟨87JOC2657⟩, and the cyclic dioxocarbenium tetrafluoroborate salts (**140**, R = Me, Ph; n = 0, 1) gave monothio-*ortho*-esters (**141**, R = Me, Ph; n = 0, 1) on treatment with lithium methanethiolate (Equation (87)) ⟨86JA6683⟩. Alternatively, the salts (**140**) could be converted to monothio-*ortho*-esters (**141**) by treatment with sodium sulfide, followed by methylation of the thiolate intermediate with trimethyloxonium tetrafluoroborate or methyl iodide ⟨86JA6683⟩. The stable monothio-*ortho*-hydrogen ester (**143**) was obtained by treatment of the bicyclic dioxocarbenium tetrafluoroborate (**142**) with sodium hydrosulfide (Equation (88)) ⟨80JA7579⟩.

(138) **(139)** (86)

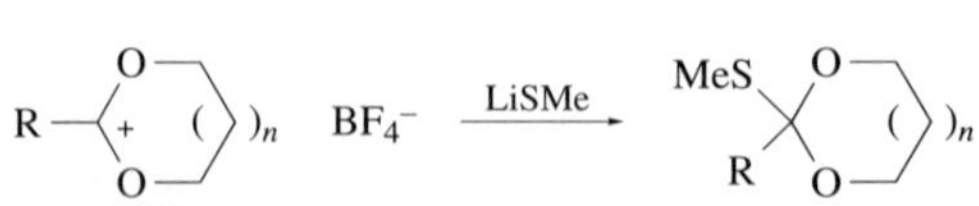

(140) **(141)** (87)

(142) **(143)** (88)

Monothio-*ortho*-esters may also be prepared from oxothiocarbenium salts by addition of alkoxides. Reaction of 2-ethoxythiolenium tetrafluoroborate with sodium ethoxide gave 2,2-diethoxythiolane (Equation (89)) ⟨56CB2060⟩, and the 2-alkyl- and 2-aryl-2-alkoxy-1,3-benzoxathioles (**145**) were prepared by treatment of the oxathiolium salts (**144**) with alcohols in the presence of sodium hydrogen carbonate (Equation (90)) ⟨79S223⟩.

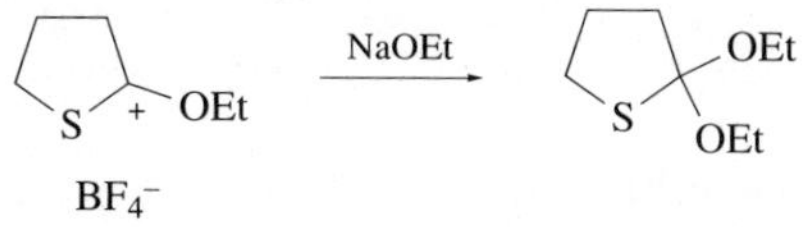

(89)

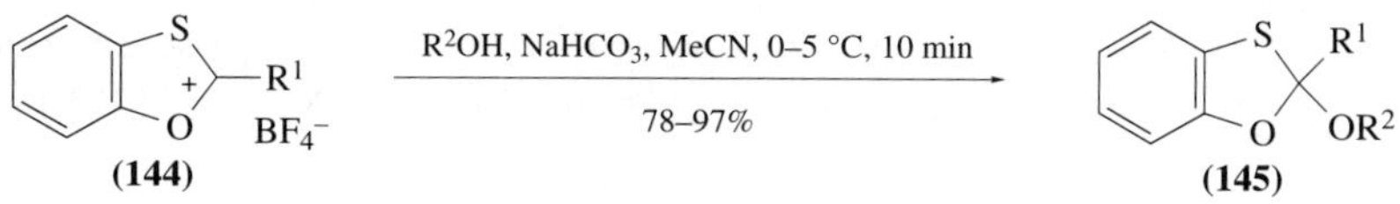

(90)

6.03.5.1.4 *From monothioacetals and related compounds*

The cyclic monothio-*ortho*-formate (**147**) has been prepared in 34% yield by methanolysis of the *N*-*p*-toluenesulfonylsulfilimine (**146**; Equation (91)) ⟨85JOC657⟩. The alkaline reaction conditions help prevent the product from undergoing disproportionation reactions. This procedure is analogous to the preparation of trithio-*ortho*-esters discussed in Section 6.03.3.1.5.

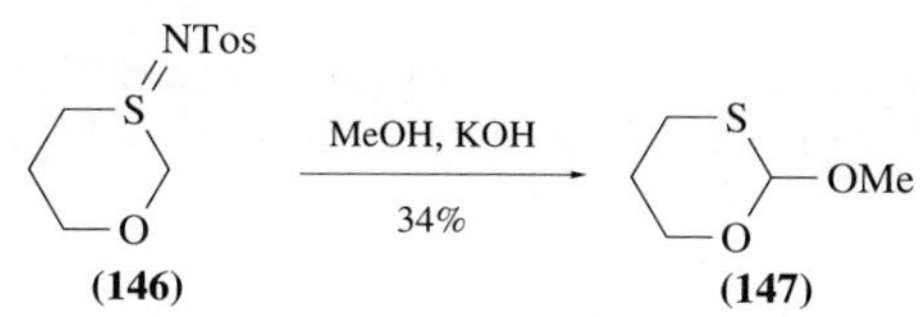

(91)

6.03.5.1.5 *From 2-alkoxythiophenes*

Terrier *et al.* demonstrated that the highly electron deficient 2,4-dinitro-5-methoxythiophene undergoes addition of methanol on treatment with potassium methoxide to produce the stable salt (**148**), which was isolated as a purple solid in virtually quantitative yield (Equation (92)) ⟨75JOC2911⟩.

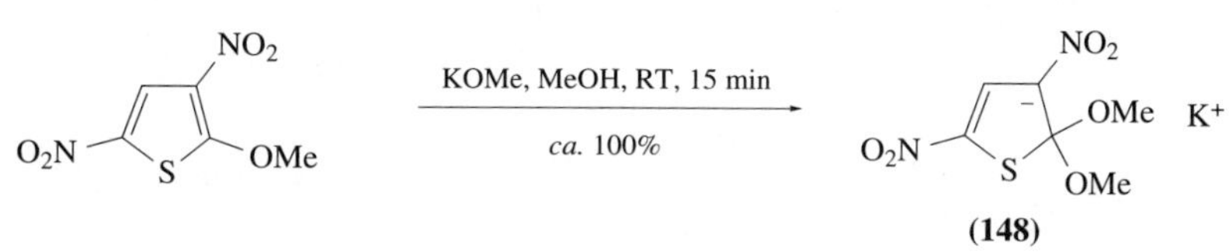

(92)

6.03.5.2 Methods for the Preparation of Functions $R^1C(OR^2)(SR^3)SR^4$

6.03.5.2.1 *From $R^1C(X^1)(X^2)X^3$*

Cyclic dithio-*ortho*-esters have been obtained from the reactions of α,α-dichloro ethers with dithiolates. Treatment of dichloromethyl methyl ether with disodium ethane-1,2-dithiolate (generated from ethanedithiol and sodium) gave 2-methoxy-1,3-dithiolane in 68% yield, and dilithium propane-1,3-dithiolate (generated from propane-1,3-dithiol and *n*-butyllithium) reacted with dichloromethyl methyl ether to produce 2-methoxy-1,3-dithiane in 45% yield (Scheme 29) ⟨72HCA75⟩.

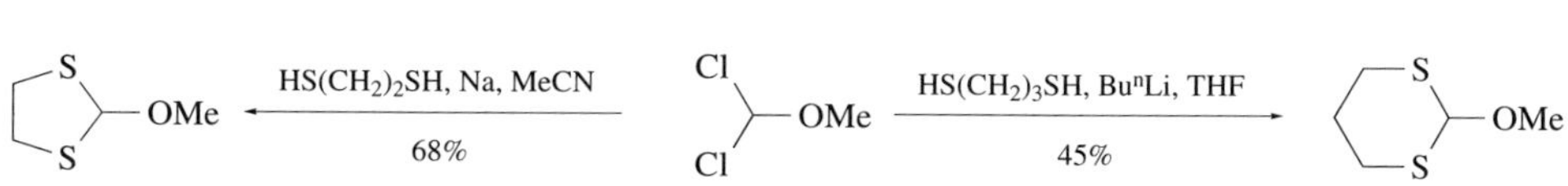

Scheme 29

Dithio-*ortho*-esters have also been obtained from exchange reactions of *ortho*-esters with *n*-butanethiol ⟨69BSF325⟩, though these are normally isolated as side-products from the preparation of monothio-*ortho*-esters by this method (see Section 6.03.5.1.1) ⟨70MI 603-02⟩.

6.03.5.2.2 From dithiocarbenium salts

Okuyama *et al.* ⟨87JOC2657⟩ prepared the dithio-*ortho*-ester (**150**) by addition of sodium methoxide to the dithiocarbenium perchlorate (**149**; Equation (93)). This procedure is analogous to the preparation of trithio-*ortho*-esters by the addition of thiols to dithiocarbenium salts discussed in Section 6.03.3.1.4.

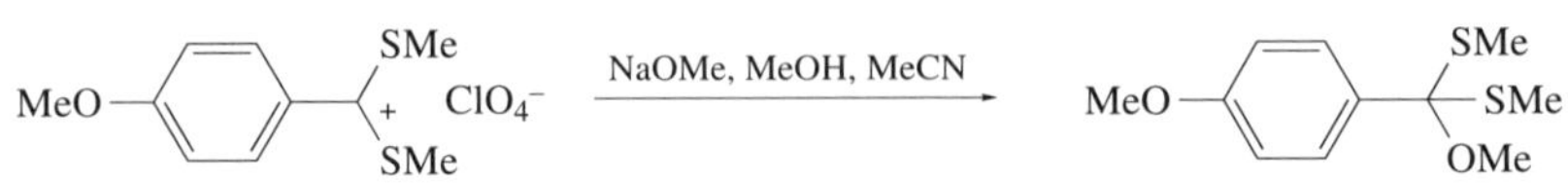

(149) (150) (93)

6.03.5.2.3 From mono- and dithioacetals and related compounds

Dithio-*ortho*-ester derivatives may be prepared by the reaction of metallated monothioacetals with disulfides. Deprotonation of 1,3-oxathiane by treatment with *n*-butyllithium at −78 °C, followed by reaction with dimethyl disulfide produced 2-methanethio-1,3-oxathiane in 82% yield (Scheme 30) ⟨81TL2005, 85JOC657⟩. In a similar manner, lithiation of a series of sulfonyl oxiranes (**151**) followed by treatment with diphenyl disulfide gave the oxidised dithio-*ortho*-ester derivatives (**152**) in yields of 45–90% (Scheme 31) ⟨91JCS(P1)3091⟩.

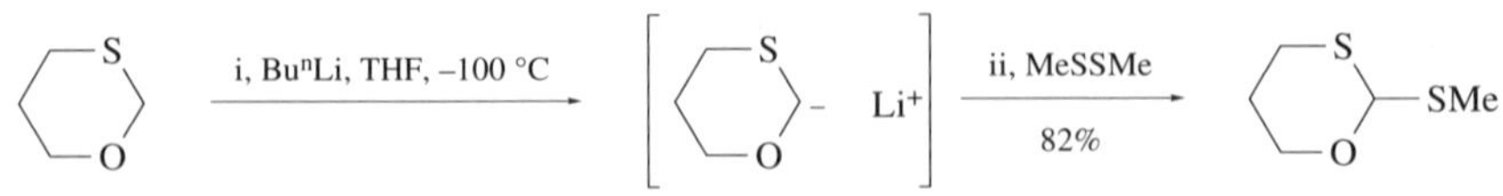

Scheme 30

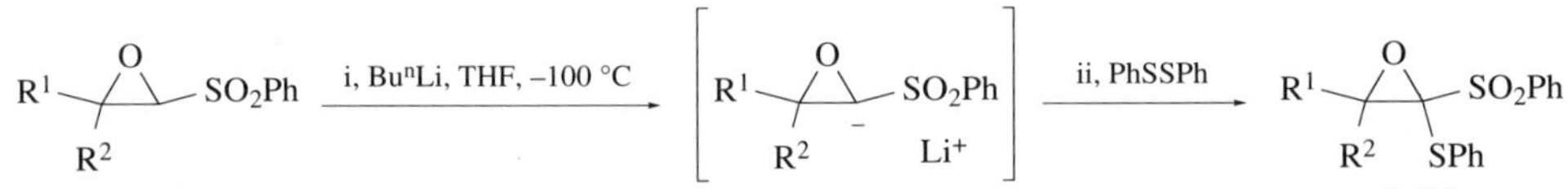

(151) (152)

Scheme 31

Lithiation of the unsaturated dithioacetal (**153**) followed by reaction with benzophenone produced the bicyclic dithio-*ortho*-ester (**154**) in 98% yield (Scheme 32) ⟨69S17⟩. This reaction proceeds via cyclisation of the intermediate ketene dithioacetal (**155**). Selenocyclisation of the related hydroxy ketenedithioacetal (**156**) gave the spirobicyclic dithio-*ortho*-ester (**157**; Equation (94)) ⟨80JOC2236⟩.

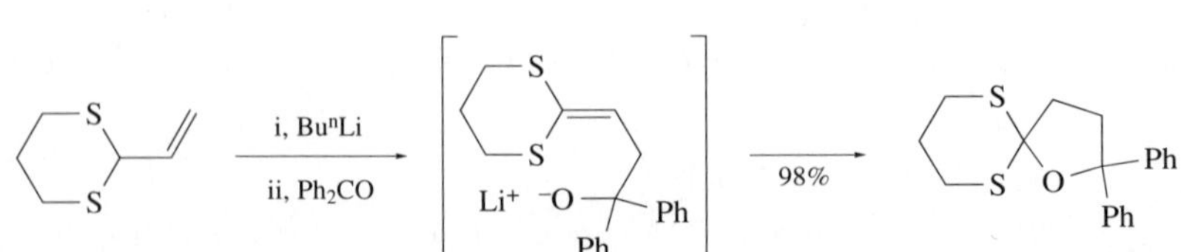

(153) (155) (154)

Scheme 32

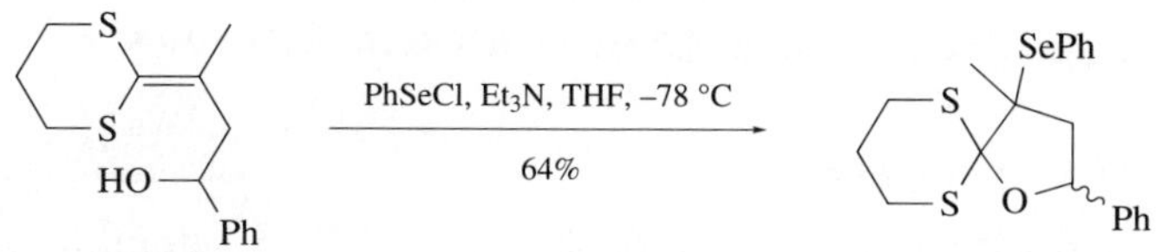

(156) (157) (94)

Some dithioacetals undergo direct oxygenation on treatment with organic peroxides. Reaction of trithiane with *t*-butyl peroxybenzoate in the presence of copper(I) chloride gave the benzoyloxytrithiane (**64**) and dibenzoyloxytrithiane (**158**; Equation (95)); the yield of (**64**) was at an optimum (70–80% based on recovered trithiane) when 0.7 equivalents of *t*-butyl peroxybenzoate were used, and (**158**) was obtained as the major product when 3.0 equivalents of *t*-butyl peroxybenzoate were used ⟨74BCJ1496⟩. 2-Benzoyloxydithiane has been generated by an analogous method, though this product was too unstable to isolate and characterise ⟨74BCJ1496⟩.

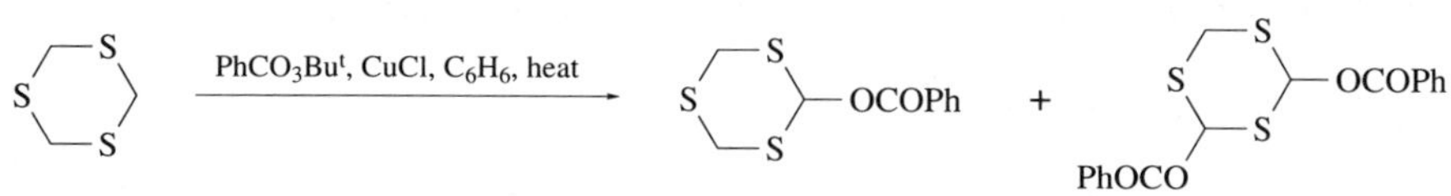

(64) (158) (95)

Cyclic dithio-*ortho*-formates have been prepared by reaction of *N*-*p*-toluenesulfonylsulfilimines (**90**) derived from dithioacetals of formaldehyde with alcohols under alkaline conditions (Equation (96)) ⟨76S551⟩.

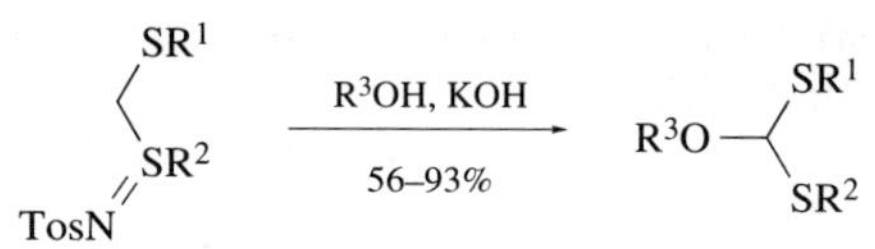

(90) (96)

6.03.5.2.4 *Miscellaneous methods*

Generation of benzyne in the presence of carbon disulfide and methanol gave 2-methoxy-1,3-benzodithiole in 78% yield (Scheme 33) ⟨74CC166⟩. The reaction proceeds via addition of methanol to the carbene (**159**), which is formed by the cycloaddition of carbon disulfide to benzyne. A modification of this reaction involves the generation of benzyne by diazotisation of anthranilic acid using an alkyl nitrite; in this case, the alkoxy group of the alkyl nitrite is the source of the 2-substituent of the product benzodithiole (Equation (97)) ⟨75S38⟩.

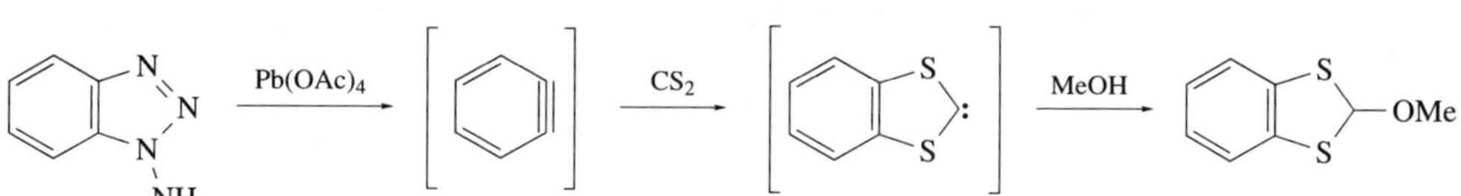

(159)

Scheme 33

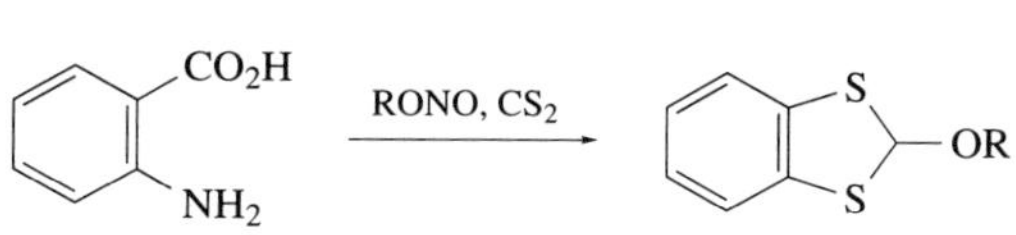

(97)

A series of cyclic oxidised derivatives of dithio-*ortho*-esters (**161**) has been prepared by the lithium *t*-butyl peroxide oxidation of alkenes (**160**; (Equation (98)) ⟨91JCS(P1)3091⟩.

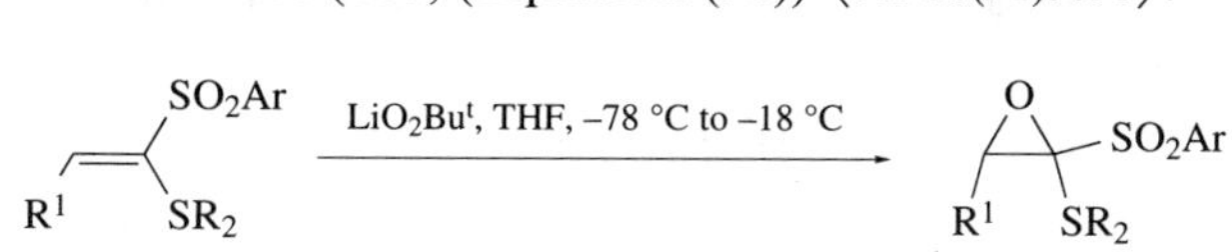

(160) (161) (98)

6.03.5.3 Methods for the Preparation of Functions $R^1C(OR^2)(OR^3)SeR^4$

6.03.5.3.1 *From 2-alkoxyselenophenes*

In an analogous reaction to that described in Section 6.03.5.1.5, addition of potassium methoxide to 2,4-dinitro-5-methoxyselenophene produced the stable salt (**162**), which was isolated as a purple solid in virtually quantitative yield (Equation (99)) ⟨75JOC2911⟩.

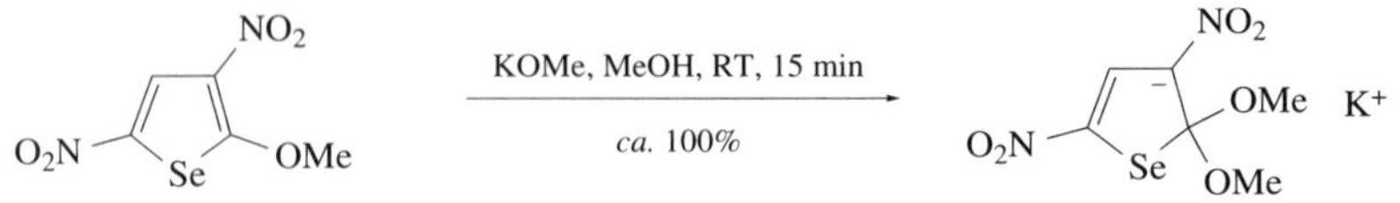

(**162**) (99)

6.03.5.4 Methods for the Preparation of Functions $R^1C(OR^2)(SeR^3)SeR^4$

6.03.5.4.1 *From $R^1C(X^1)(X^2)X^3$*

Dihalomethyl methyl ethers react with selenolates to give diseleno-*ortho*-esters. Treatment of dichloromethyl methyl ether with sodium benzeneselenolate (generated by the sodium borohydride reduction of diphenyl diselenide) gave the diseleno-*ortho*-formate ester (**163**) in 78% yield (Equation (100)) ⟨79JA6638⟩. Other dichloromethyl alkyl and aryl ethers gave diseleno-*ortho*-formates (**164**, R = alkyl, aryl) in yields of 42–76% on treatment with sodium areneselenolates (generated from the selenophenol and metallic sodium) (Equation (101)) ⟨86ZOR107⟩. Dichloromethyl ethers also react with areneselenomagnesium halides to give diseleno-*ortho*-formates, though yields are usually lower than those obtained from reaction with the corresponding sodium selenolates ⟨86ZOR107⟩. Reaction of dibromomethyl methyl ether with dilithium benzene-1,2-diselenolate gave 2-methoxy-1,3-diselenole in 38% yield (Scheme 34) ⟨88HCA1242⟩.

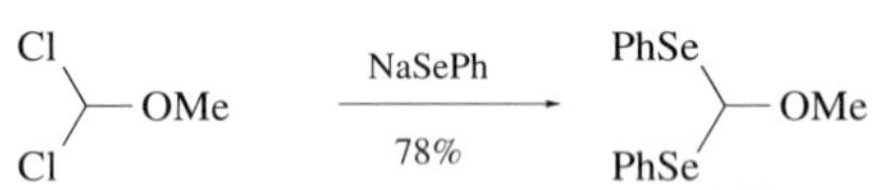

(**163**) (100)

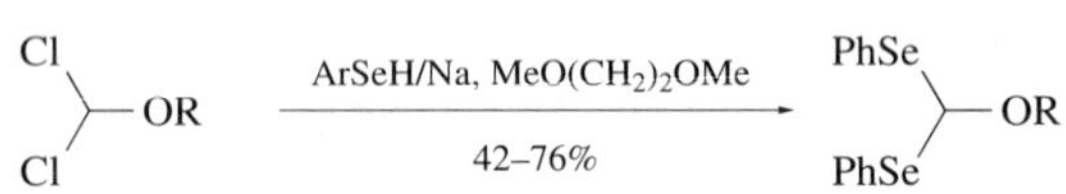

(**164**) (101)

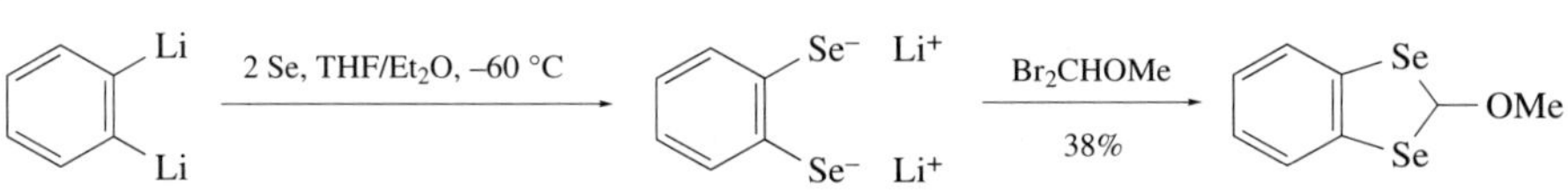

Scheme 34

The cyclic diseleno-*ortho*-esters (**166**, R = H, Me) have been generated from the boron trifluoride etherate catalysed exchange reactions between triethyl *ortho*-formate and the 1,3-diselenolates (**165**, R = H, Me; (Equation (102)). However, these products were not isolated and characterised, but were used *in situ* for the preparation of diselenothio-*ortho*-esters (see Section 6.03.6.2.1) ⟨89H(28)389⟩.

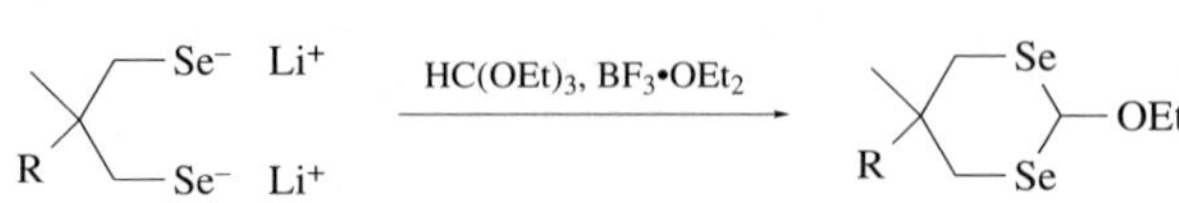

(**165**) (**166**) (102)

6.03.5.5 Methods for the Preparation of Functions $R^1C(OR^2)(SR^3)SeR^4$

6.03.5.5.1 *From monothioacetals and related compounds*

Monoselenomonothio-*ortho*-ester derivatives have been prepared from the reactions of lithiated monothioacetals with diselenides. Lithiation of 1,3-oxathiane with *n*-butyllithium followed by reaction with dimethyl diselenide produced 2-methaneseleno-1,3-oxathiane in 50% yield (Scheme 35) ⟨81TL2005, 85JOC657⟩. In a similar manner, lithiation of the sulfonyl oxiranes (**151**) followed by treatment with diphenyl diselenide gave the oxidised selenothio-*ortho*-ester derivatives (**167**) in yields of 25–92% (Scheme 36) ⟨91JCS(P1)3091⟩.

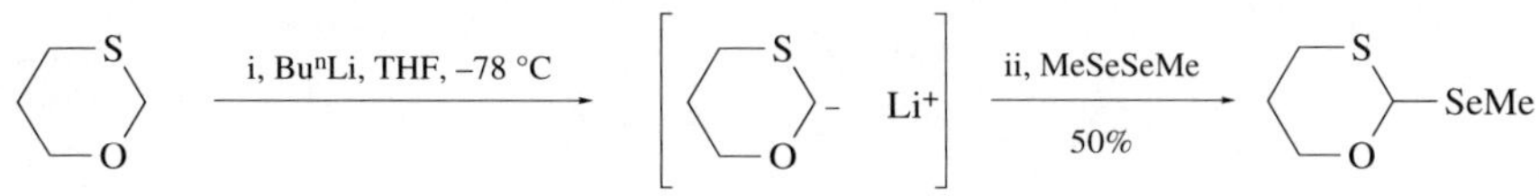

Scheme 35

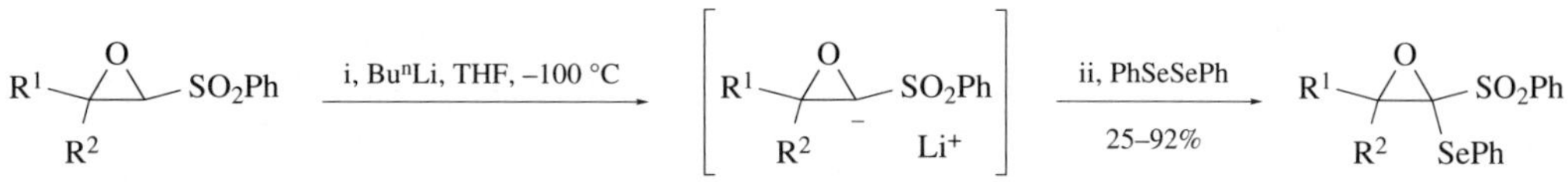

(151) **(167)**

Scheme 36

6.03.6 MIXED SULFUR AND SELENIUM FUNCTIONS

Monoselenodithio- and diselenomonothio-*ortho*-esters are mixed sulfur and selenium functions which bear one and two selenium substituents respectively.

6.03.6.1 Methods for the Preparation of Functions $R^1C(SR^2)(SR^3)SeR^4$

6.03.6.1.1 From $R^1C(X^1)(X^2)X^3$

2-Chloro-1,3-dithianes react with areneselenolates to give 2-areneseleno-1,3-dithianes. Thus, dithianes (**169**, $R^1 = H$; $R^2 = H$, Me, OMe, F, Cl, NMe_2, CF_3) were prepared from 2-chloro-1,3-dithiane and the appropriate sodium areneselenolates (prepared by reduction of the corresponding diaryl diselenide using either sodium borohydride or metallic sodium and ultrasound) ⟨88JOC3766⟩. Similar reaction of the 2-chloro-4,6-dimethyl-1,3-dithiane (**168**, R^1 = Me) with sodium (*p*-methoxybenzene)selenolate gave the selenodithio-*ortho*-ester (**169**, R^1 = Me; R^2 = OMe), which was obtained as a ca. 5 : 1 mixture of diastereoisomers (Equation (103)) ⟨88JOC5668⟩.

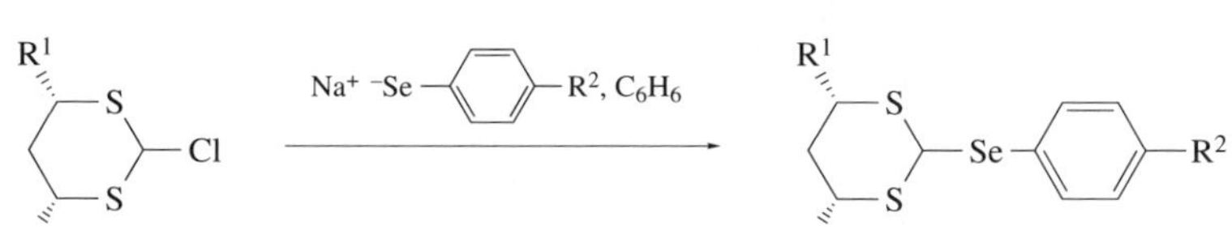

(168) **(169)** (103)

6.03.6.1.2 From selenodithiocarbenium salts

2-Methanethio-1,3-selenothiolium salts are reduced to 2-methanethio-1,3-selenothioles by sodium borohydride. Reduction of the fluorosulfonate salt (**170**) gave the 2-methanethio-1,3-selenothiole (**171**) in 70% yield (Equation (104)) ⟨80H(14)271⟩, and similar treatment of the iodide salts (**172**, R = H, Me) produced the analogous selenothioles (**173**, R = H, Me) ⟨75TL1259⟩.

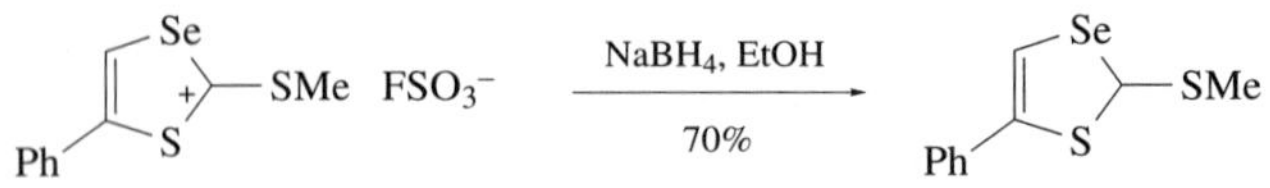

(170) (171) (104)

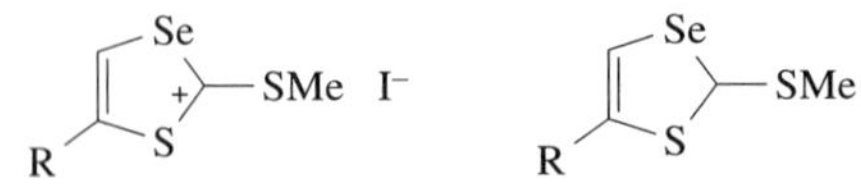

(172) (173)

6.03.6.1.3 From dithioacetals

Selenodithio-*ortho*-esters may be prepared by the selenation of lithiated dithioacetals. Deprotonation of 1,3-dithiane using *n*-butyllithium followed by reaction with *p*-nitrobenzeneselenyl cyanide produced the 2-seleno-1,3-dithiane (**174**) in 23% yield (Scheme 37) ⟨88JOC3766⟩. However, treatment of lithiated 1,3-dithiane with the diaryl diselenides (**175**, R = OMe, CF_3) gave only very low yields of the selenated compounds (**176**, R = OMe, CF_3) ⟨86CJC732⟩.

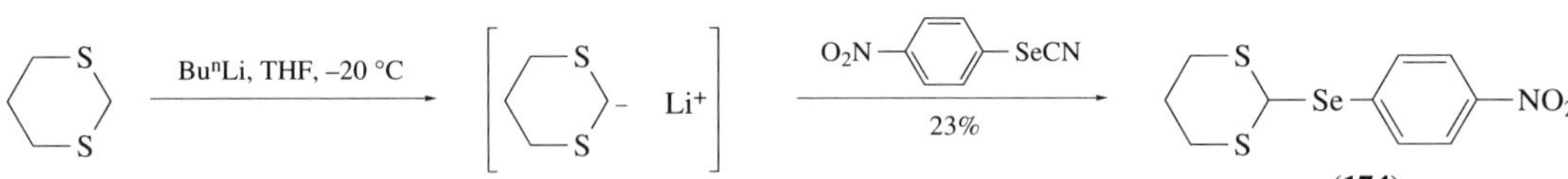

(174)

Scheme 37

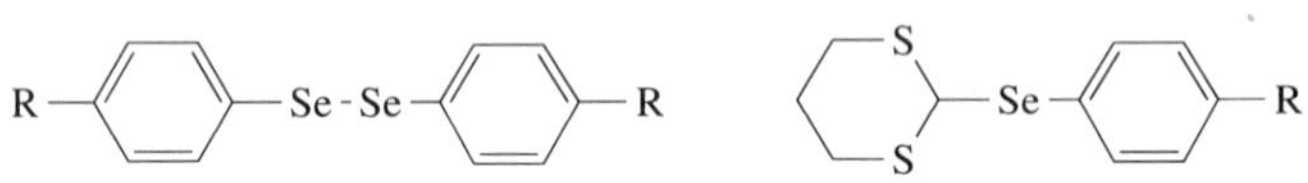

(175) (176)

6.03.6.2 Methods for the Preparation of Functions $R^1C(SR^2)(SeR^3)SeR^4$

*6.03.6.2.1 From diseleno-***ortho***-esters*

Pinto *et al.* ⟨89H(28)389⟩ reported the preparation of the diselenothio-*ortho*-esters (**177**, R = H, Me) from the corresponding *O*-ethyl diseleno-*ortho*-esters by treatment with benzenethiol. The starting diseleno-*ortho*-esters were generated by the boron trifluoride etherate catalysed reaction between dilithium diselenides and triethyl *ortho*-formate (see Section 6.03.5.4.1), and were used *in situ*.

(177) (105)

6.04
Functions Containing a Chalcogen and Any Other Heteroatoms Other Than a Halogen

MARTIN J. RICE
ZENECA Agrochemicals, Bracknell, UK

6.04.1 FUNCTIONS CONTAINING CHALCOGEN AND A GROUP 15 ELEMENT

The compounds in this class have been the subject of several reviews ⟨69ZC201, B-70MI 604-01, 71S16, 77UK685, B-79MI 604-01, 79T1675, 85HOU(E5)3⟩.

6.04.1.1 Functions Bearing Chalcogen and Nitrogen

Functional groups bearing chalcogen and nitrogen are available by modification of precursors which contain *sp*^{3}- or *sp*^{2}-carbon atoms, via cycloadditions or by other miscellaneous methods. This general order is adhered to throughout this section.

6.04.1.1.1 Functions bearing oxygen and nitrogen

(i) Functions bearing two oxygen and one nitrogen substituent

The functional group containing two ether type oxygens and one amine type nitrogen singly bonded to a carbon is most commonly referred to as an amide acetal. In the majority of cases, all the heteroatoms are substituted further. Exceptions are the cyclols which have one unsubstituted oxygen atom.

(a) From sp^3*-carbon compounds.* Amide acetals can be accessed by transacetalation of other amide acetals ⟨61LA(641)1, 63AG296, 64CCC645, 65HCA1746, 68CB41⟩ with alcohols. Typically, the starting material is *N,N*-dimethylformamide dimethyl acetal (**1**). The uncatalysed reaction is driven to completion by the removal of generated methanol by distillation. The use of sterically hindered alcohols can, however, lower the reaction yields significantly. Cyclic amide acetals (**2**) are available from diols (Scheme 1) ⟨61LA(641)1, 64CCC645, 68CB41⟩. Transamidation can be achieved with high boiling amines such as morpholine and piperidine ⟨61LA(641)1, 66USP3239519⟩. Simultaneous transacetalation and transamidation has also been reported ⟨64CCC645⟩.

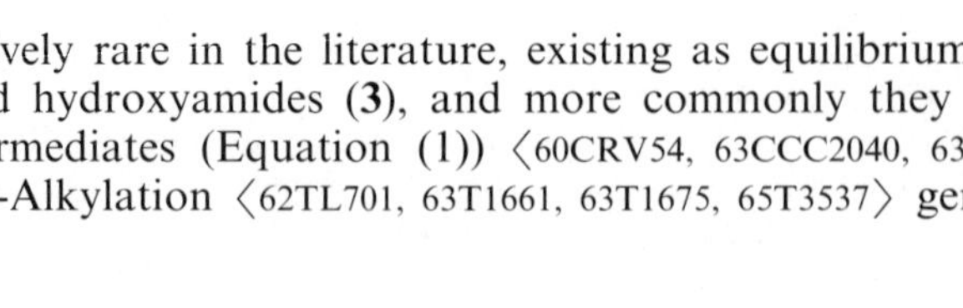

Scheme 1

Cyclols (**4**) are relatively rare in the literature, existing as equilibrium mixtures with the corresponding ring-opened hydroxyamides (**3**), and more commonly they have been proposed as transient reaction intermediates (Equation (1)) ⟨60CRV54, 63CCC2040, 63TL439, 64JOC2769, 64TL47, 66UKZ204, 68TL2713⟩. *O*-Alkylation ⟨62TL701, 63T1661, 63T1675, 65T3537⟩ generates fully substituted amide acetals.

(1)

(3) **(4)**

Symmetrical amide acetals (**5**) are obtained from *ortho*-esters with a variety of nitrogen nucleophiles such as sulfonamides (Equation (2)) ⟨61JOC4315, 65AJC1967, 80JOC4038, 82JCS(P1)2871, 88JHC907⟩, ureas ⟨53JA671, 63HCA2306⟩, imidazoles ⟨80JOC4038⟩ and difluoroamine ⟨67JA716⟩.

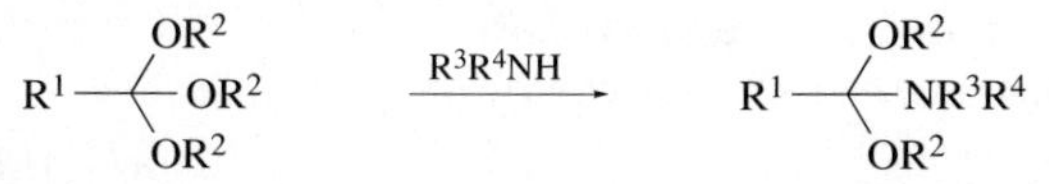

(5) (2)

Under Lewis acid catalysis, isocyanates will react to give *N*-carboxy compounds (**6**) (Equation (3)) ⟨62AG(E)592, 66GEP1156780⟩. Cyanide can be displaced from (α-alkoxy-α-alkylamino)acetonitriles (**7**) ⟨72CB1340⟩ or, in addition to one of the alkylamino groups, from α,α-bis(dialkylamino) acetonitriles (**8**) by alkoxides ⟨49LA(562)229, 72CB1340⟩. Both halogens in α,α-difluorotrialkylamines (**9**) have been replaced by alcohols ⟨63USP3092637, 64USP3121084, 65USP3214412, 83TL1035⟩. Chlorodiaryloxymethanes have been used to generate diaryloxymethyltriethylammonium chlorides (**10**) with triethylamine ⟨58JPR(7)70⟩. Such reactive compounds give amide acetals on further reaction with amines ⟨72S32, 72S418, 73S423, 75S272⟩.

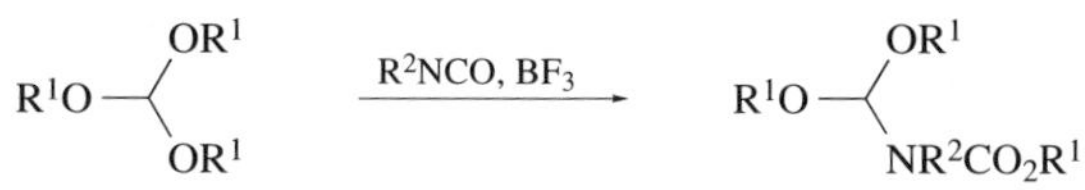

(6) (3)

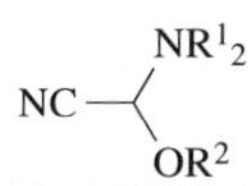

(7)

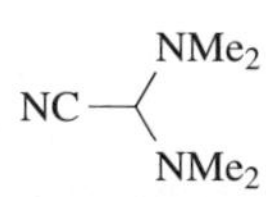

(8)

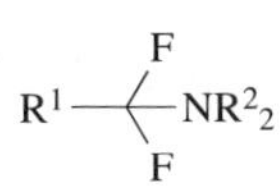

(9)

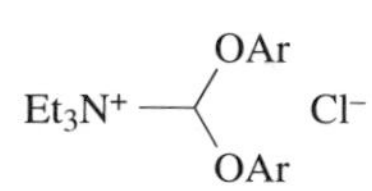

(10)

(b) From sp²*-carbon compounds.* The halogen in *N*,*N*-dialkyliminium chlorides (**11**) can be displaced on treatment with alkoxides to yield amide acetals (**12**) (Equation (4)) ⟨B-70MI 604-01, 72TL4217, B-79MI 604-01⟩. In a similar manner, alkoxide can be displaced from alkoxyiminium salts (**13**) ⟨61LA(641)1, 68PAC519, B-70MI 604-01, B-79MI 604-01⟩ and thiolate from alkylthioiminium salts (**14**) ⟨67BCJ2641⟩. Lactam acetals are also accessible by this methodology ⟨63USP3092637, B-79MI 604-01⟩. In addition, pyridine has been demonstrated to react with the oxonium salt (**15**) to give the pyridinium species (**16**) ⟨71S312⟩. The adduct (**17**) is formed by capture of the *N*-acyloxazolium salt with methoxide ⟨75JCS(P1)1302⟩.

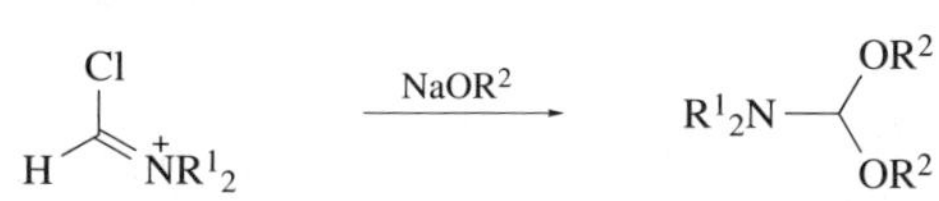

(11) (12) (4)

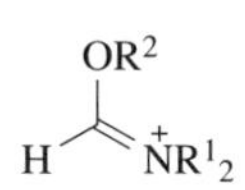

(13)

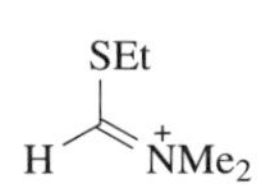

(14)

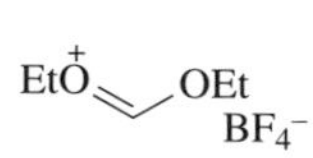

(15)

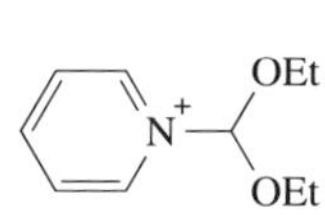

(16)

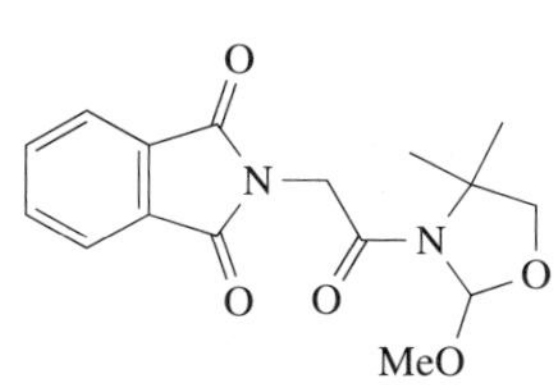

(17)

Formamide acetals are produced by the reaction of *N*,*N*,*N′*,*N′*-tetrasubstituted formamidium salts with sodium alkoxides ⟨60AG956, 62JOC3664, 63GEP1146892, 64CCC645⟩. Tetrakis(dimethylamino)ethene (**18**) is converted by methanol into a mixture of amide acetals (**19**) and (**20**) ⟨72T1965⟩. Hexafluoropropene (**21**) combines with diethanolamine (**22**) to generate an azadioxabicyclo[3.3.0]octane

(23) ⟨72TL3957⟩. Similar systems, such as **(25)**, are obtained by the reaction of the sodium salts of dialkanolamines such as **(24)** with nitriles (Scheme 2) ⟨73AG(E)996⟩.

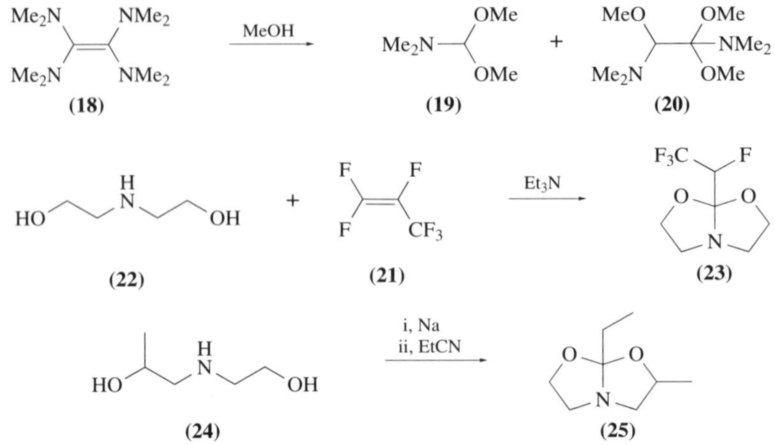

Scheme 2

(c) Cycloaddition methods. The unsubstituted ketene acetal **(26)** reacts with isocyanates to give barbituric acid acetals **(27)** ⟨59MI 604-01, 62AG(E)331, 66CB3892⟩ while the cyclic analogue **(28)** generates the β-lactam **(29)** ⟨88TL2327⟩. Similarly, tetramethoxyethene gives the azetidin-2-one **(30)** (Scheme 3) ⟨59MI 604-01, 64AG(E)380⟩. Ketene acetals will also react with nitrosobenzene, with dialkyl azodicarboxylates ⟨71CB873⟩ and with ketenimines ⟨76IZV858⟩ to give four-membered ring products **(31)**, **(32)** and **(33)**.

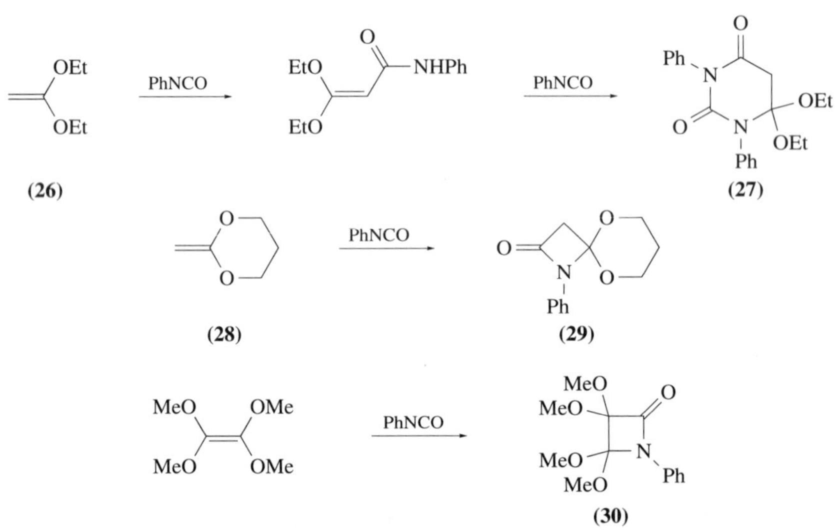

Scheme 3

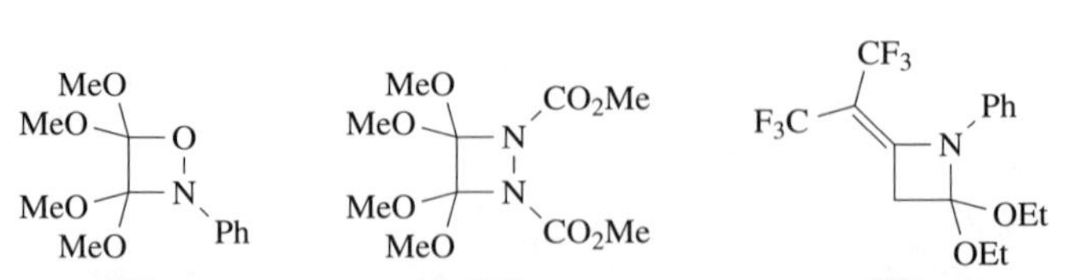

(31) **(32)** **(33)**

Dichloroketene acetals such as **(34)** react with sulfamyl chlorides to give four-membered ring products **(35)** (Equation (5)) ⟨67JA2502, 72JA6135⟩. α,β-Unsaturated ketones combine with ynamines ⟨70CR(C)468⟩ and with 1-dialkylamino-1-alkylthioethenes ⟨67BCJ2641⟩ to give bicyclic amide acetals (Scheme 4) ⟨68TL497⟩.

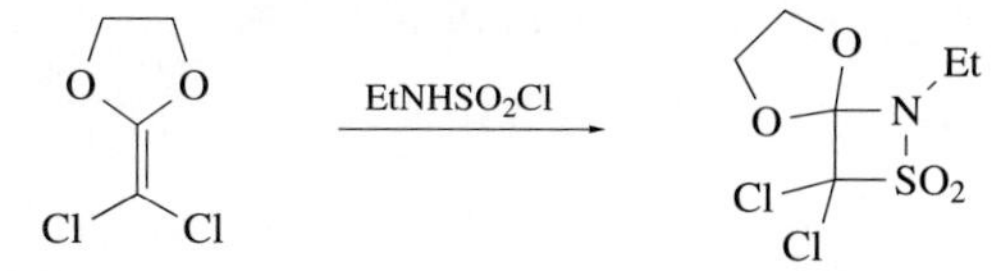

(5)

(34) **(35)**

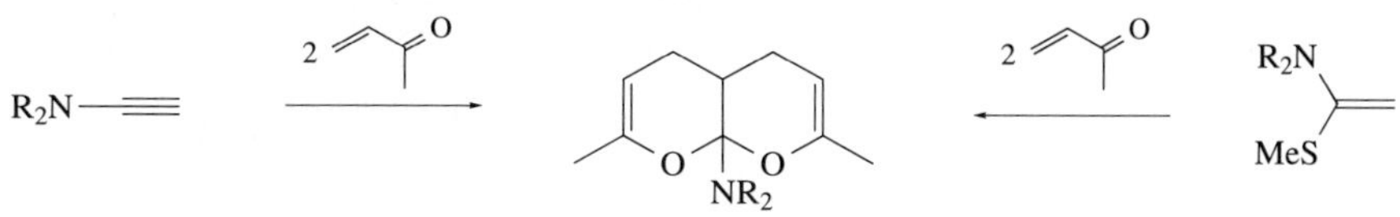

Scheme 4

An unsymmetrical bicyclic amide acetal (**37**) has been obtained by treatment of the dihydrooxazole (**36**) with benzonitrile oxide ⟨71AG(E)810⟩. Dialkoxydihydro-1,2,3-triazoles such as (**38**) are produced by the reaction of azides with some ketene acetals ⟨63G942, 68G681, 72JHC1087, 76JHC205⟩. Cyclopropanone acetals such as (**39**) undergo ring expansion with aryl isocyanates to give γ-lactam acetals (**40**) (Scheme 5) ⟨87JCR(S)362⟩.

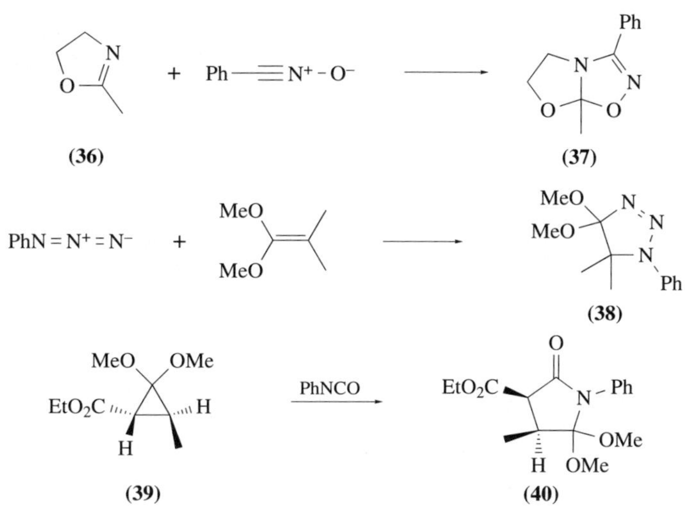

Scheme 5

(d) Miscellaneous reactions yielding amide acetals. The imidate (**41**) has been oxidized by mcpba to the corresponding oxaziridine ⟨71TL4519, 73TL1807⟩. *t*-Butyl hypochlorite has been used to oxidise a 4-oxa-β-lactam (**42**) to the amide acetal ⟨82H(18)201⟩. Benzoylhydrazones (**43**) react with lead tetraacetate to give 2,5-dihydro-1,3,4-oxadiazoles ⟨67TL3501, 68CB3851⟩. Reductive conditions are exemplified by the reaction of *N,N*-dimethylformamide and *N,N*-dimethylacetamide with 1,3-dibromoketones and a zinc/copper couple to give amide acetals (**44**) (Scheme 6) ⟨72JA3201⟩.

The insertion of ethoxycarbonylnitrene, generated by photolysis of ethyl azidoformate, into acetals ⟨67T45⟩ and ketene acetals ⟨72JHC1087⟩ leads to *N*-carboxy amide acetals such as (**45**). Dichlorocarbene reacts with dialkylamines and sodium alkoxides to give the amide acetals $HC(OR^1)NR^2_2$ resulting from insertion and double halogen displacement ⟨64GEP1161285, 69LA(725)15, 69RTC289⟩ while the diazo compound (**46**) on photolysis undergoes cyclisation to the dihydrooxazole (**47**) ⟨74JHC529⟩. Epoxides add to dihydrooxazoles (**48**) and 5,6-dihydro-4*H*-1,3-oxazines to give bicyclic amide acetals such as (**49**) ⟨66AG(E)875, 66AG(E)894, 66LA(698)174⟩. The bicyclic amide acetal (**50**) was obtained by the addition of diketene to the C=N bond of a chiral oxazoline (Scheme 7) ⟨91TL597⟩.

N-Methylpyrrole undergoes anodic oxidation in methanol to generate the symmetrical product (**51**) ⟨66JOC4054⟩. Under basic conditions, the diazene oxide (**52**) reacts with methanol to give the azo compound (**53**) ⟨78JOC1459⟩. Ozonides (**54**) are formed by the reaction of ozone with indoles

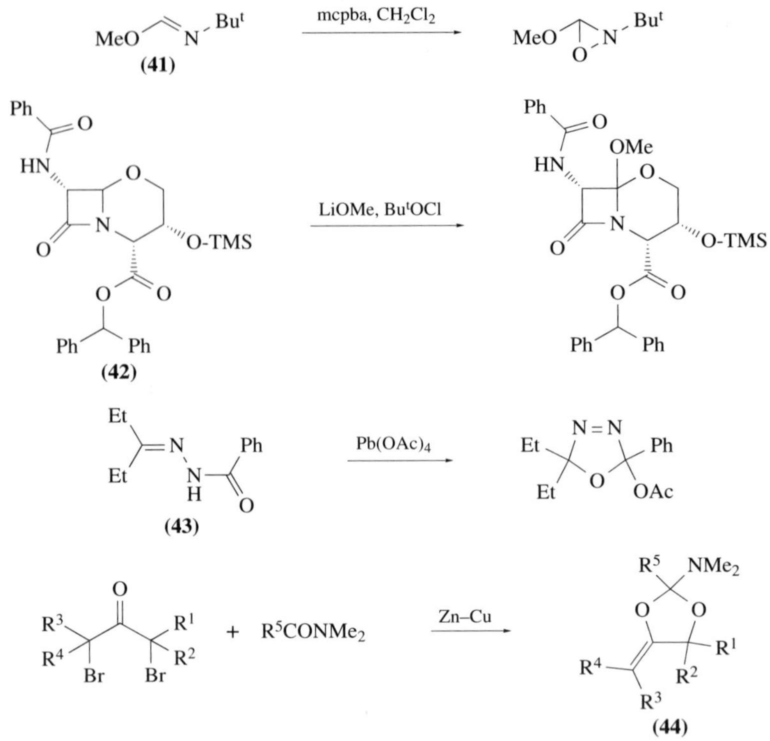

Scheme 6

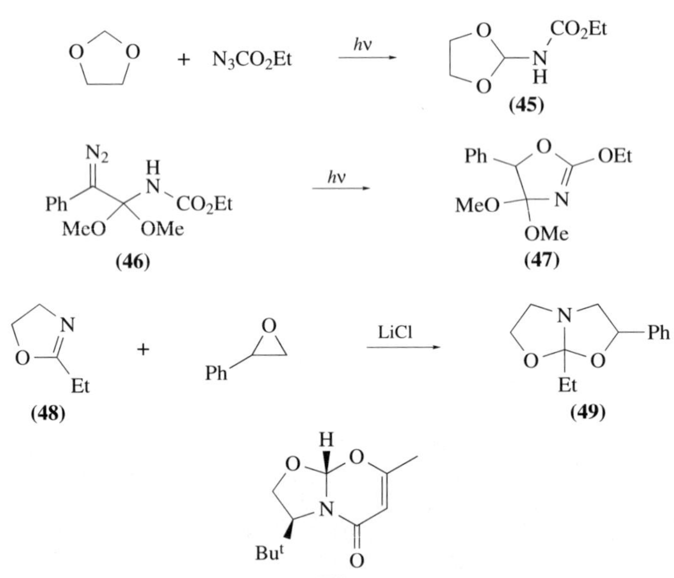

Scheme 7

⟨52JA3855⟩. Compounds of this type have also been obtained by cyclisation of hydroperoxides ⟨52CB949⟩. The reaction of some cyclic *ortho*-esters with TMS azide leads to the formation of displacement products such as **(55)** (Scheme 8) ⟨79TL513⟩.

(ii) Functions bearing one oxygen and two nitrogen substituents

Functions bearing two nitrogens and one oxygen are known commonly as ester aminals. As with the amide acetals, they are synthetically accessible from sp^3- and sp^2-carbon compounds, by cyclisations, by cycloadditions and by other miscellaneous methods.

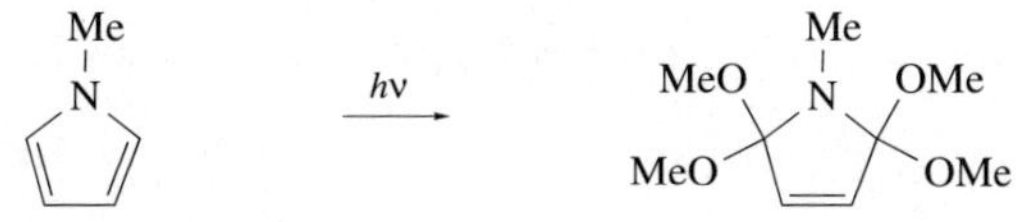

(51)

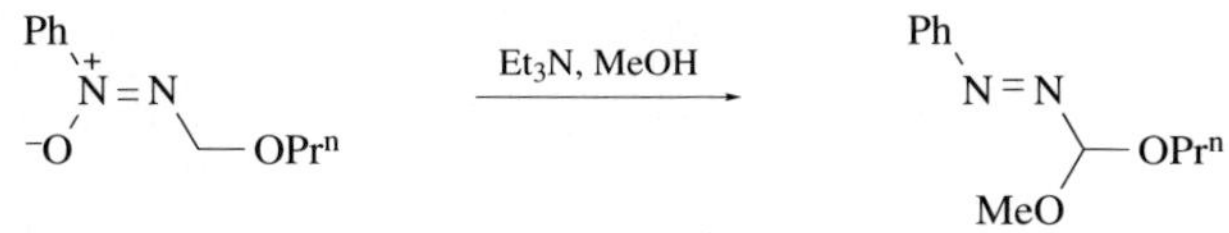

(52) (53)

Scheme 8

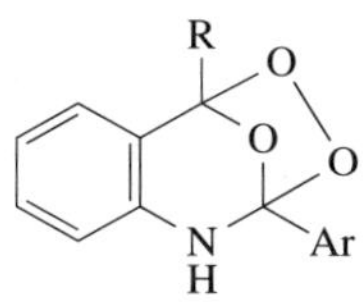

(54)

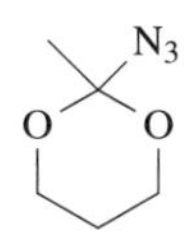

(55)

(a) From sp^3*-carbon compounds.* Some examples of the formation of ester aminals by displacement reactions are shown in Scheme 9. Treatment of bis(dimethylamino)acetonitrile with sodium ethoxide yields the ester aminal **(56)** ⟨72CB1340⟩. The base induced cyclisation of 1-chloro-5,5,5-trinitropentan-2-ol **(57)** occurs with the elimination of nitrite ⟨65AG427⟩. All three halogens are displaced in the reaction of chloroform with aziridine and sodium methoxide ⟨69LA(725)15⟩. Reaction of triethyl *ortho*-acetate with the 2-aminobenzenecarboxamide **(58)** leads to the 2-ethoxyquinazolinone **(59)** ⟨90JHC1953⟩.

(56)

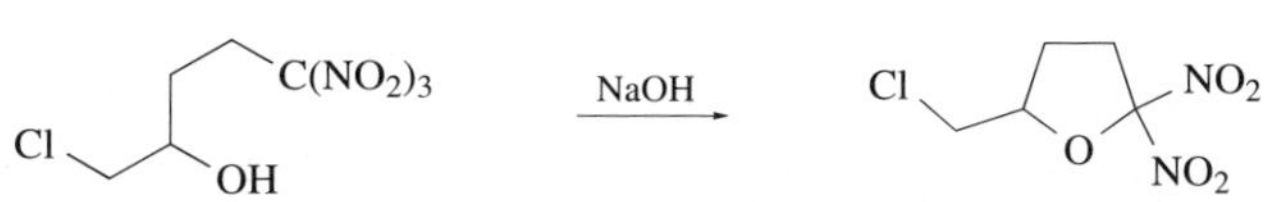

(57)

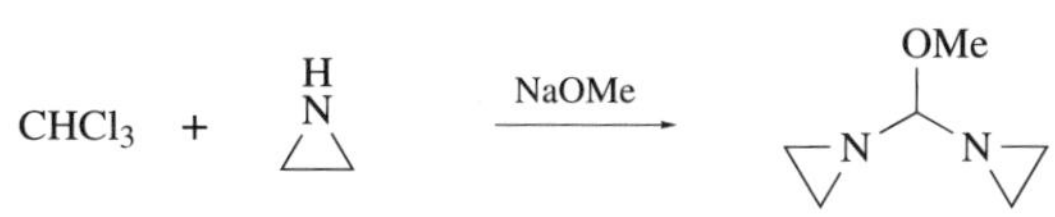

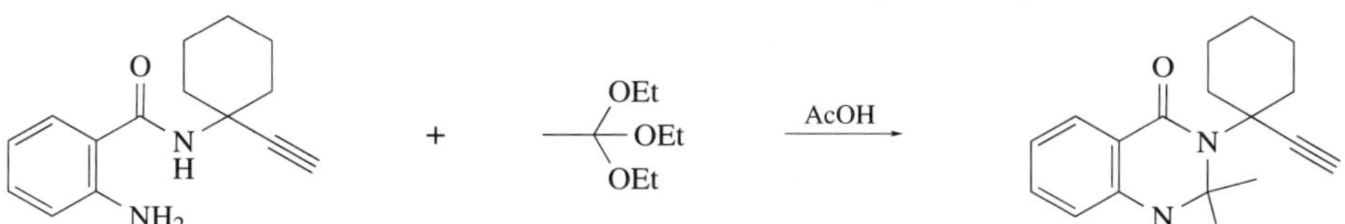

(58) (59)

Scheme 9

(b) From sp^2*-carbon compounds.* Electron-deficient heterocycles undergo a range of related additions of water and alcohols ⟨84CHEC(3)91⟩. Some examples of these reactions are shown in Scheme 10. 'Covalent hydration' is exemplified by the addition of the elements of water to the tetrazoloquinazoline **(60)** ⟨65JOC829, 67KGS1096⟩. An intramolecular addition occurs when the dinitropyridine **(61)** is treated with base ⟨68TL659⟩. Diethylamine reacts with 5-phenyl-1,2,4-dioxazolin-3-one **(62)** ⟨72JPR145⟩. An analogous reaction occurs when 1,3,4-oxadiazolium salts are treated

with secondary amines leading to tertiary products; primary amines lead to ring-opened products ⟨71JCS(C)409⟩.

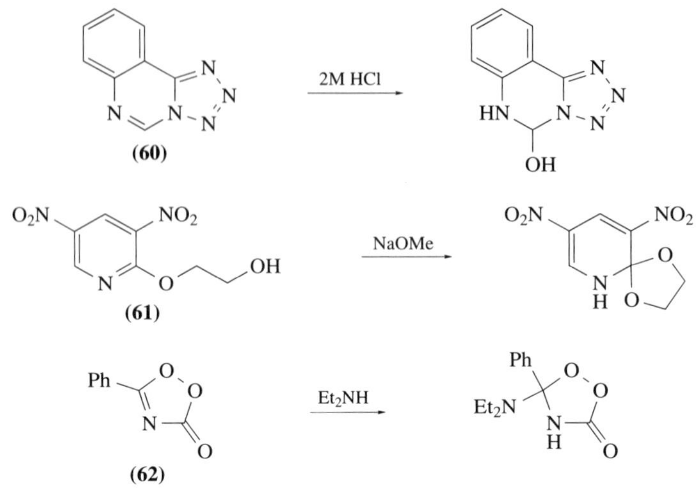

Scheme 10

In a reaction analogous to that with the oxazoline (**48**), epoxides have been demonstrated to react with dihydroimidazoles and tetrahydropyrimidines to give ester aminals ⟨70LA(742)128⟩.

(c) Cycloaddition methods. Reaction of *p*-nitrophenyl isocyanate with *O*-alkyllactims such as (**63**) leads to 4-alkoxy-1,3-diazetin-2-ones (Equation (6)) ⟨73TL1219⟩.

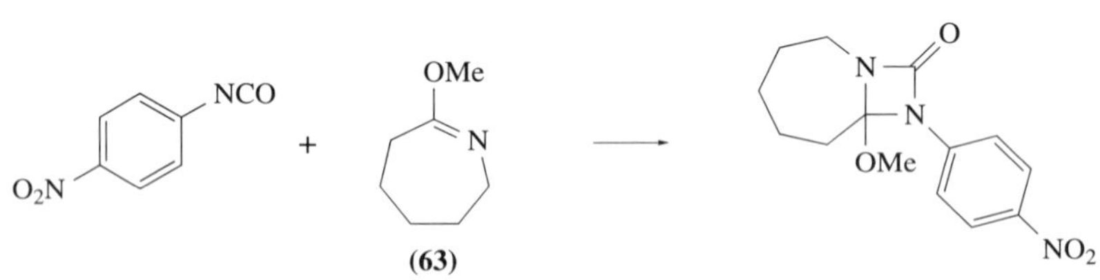

(6)

(d) Miscellaneous reactions yielding ester aminals. Addition of α-tetralone to the bisurea (**64**) in ethanolic hydrogen chloride affords a 2-hydroxyhexahydroquinazoline (**65**) ⟨66IZV98⟩. A 3-(dimethylamino)oxaziridine (**66**) is formed by photocyclisation of the nitrone (**67**) ⟨71JA4075⟩. The reaction of hydrazoic acid with ethoxypropyne gives bisazide (**68**) (Scheme 11) ⟨57RTC949⟩.

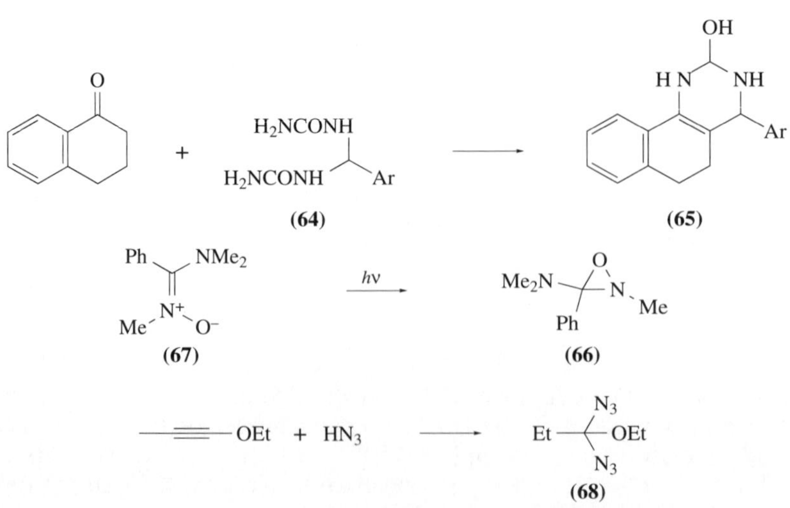

Scheme 11

6.04.1.1.2 Functions bearing sulfur and nitrogen

(i) Functions bearing one sulfur, one nitrogen and one oxygen substituent

Functional groups bearing oxygen, sulfur and nitrogen are nonsymmetrical, and this reduces the number of synthetic routes available for their preparation compared to their symmetrical analogues.

(a) From sp^3*-carbon compounds.* Two examples of formation of this functional group by displacement are shown in Scheme 12. Triethyl *ortho*-formate reacts with 2-thiolbenzamide to give a dihydrobenzothiazine (**69**) ⟨69T5995⟩. Displacement of aniline from the 1,3,4-thiadiazoline (**70**) by ethanol produces the 5-ethoxy analogue (**71**) ⟨76ACS(B)837⟩.

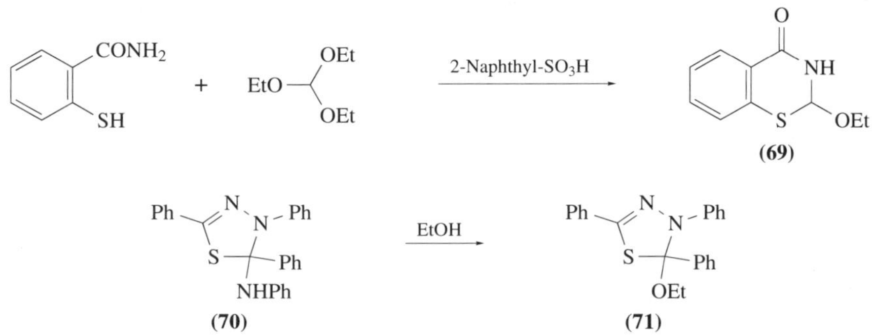

Scheme 12

(b) From sp^2*-carbon compounds.* The *N*-alkylbenzothiazolium cation (**72**) reacts with the 2-hydroxybenzaldehyde (**73**) to give the spiro compound (**74**) ⟨72HCA1782⟩; there are other examples of intramolecular cyclisation of benzothiazolium cations with phenolic ⟨69T5995⟩ and ketonic functions ⟨90IJC(B)180⟩ in the presence of base, and related reactions of thiazoles ⟨84H(22)281⟩. The dimethylformamide monothioacetal (**75**) is prepared by *O*-methylation of *N,N*-dimethylformamide with dimethyl sulfate, and subsequent interception of the iminium cation (**76**) with sodium ethanethiolate ⟨71BSF3354⟩. An alternative approach, the interception of a thioimidate with an alcohol, has also been reported ⟨84H(22)13⟩. Reaction of 2-(2′-thiolphenyl)dihydroimidazole (**77**) with phthalolyl chloride gave the spiro thioacetal (**78**). In this reaction it may be the cyclic tautomer (**79**) of phthaloyl chloride which is the reactive species (Scheme 13) ⟨88LA599⟩.

(c) Cycloaddition methods. Benzonitrile oxide can undergo 1,3-dipolar cycloadditions to C=S or to C=N bonds which lead to amide thioacetals. Two examples of such reactions, leading to the oxathiazole (**80**) ⟨79CB1873⟩ and to the fused azetidine (**81**) ⟨90TL123, 91JHC481⟩ are shown in Scheme 14. The reaction of ketene–sulfur dioxide adduct with the 2-aryldihydrooxazoles (**82**) gives the cycloadduct (**83**) ⟨69JHC729⟩. The dihydroindole (**85**) was obtained from the reaction of dichloroketene with the indole (**84**) (Scheme 15) ⟨89JOC1782⟩.

(d) Miscellaneous reactions yielding amide monothioacetals. 4-Thio-β-lactam (**86**; R = H) undergoes oxidation with lead tetraacetate ⟨74JCS(P1)22⟩ or with *t*-butyl perbenzoate in benzene ⟨79CC485⟩ to give *O*-acylated derivatives (**86**; R = OCOMe or OCOPh). Phenacyl bromide reacts with phenylpropiolic thioanilide (**87**) in the presence of acetone to produce a 1,3-oxathiole (**88**) (Scheme 16) ⟨75LA958⟩.

(ii) Functions bearing two sulfur and one nitrogen substituent

There are many parallels that can be drawn between the preparation of thioamide acetals and their oxa analogues. It is likely that many of the principles behind the methods for preparation of the latter could be applied to the former.

(a) From sp^3*-carbon compounds.* Reaction of amide acetals with thiols provides a route to both acyclic and cyclic thioamide acetals (**89**) and (**90**) (Scheme 17) ⟨69BSF332, 70BSF2013⟩. Replacement of the alkoxy function of the dithio-*ortho*-ester (**91**) has been demonstrated to give the amide dithioacetal (**92**) ⟨80PJC145⟩. α-Chlorothioacetals such as (**93**) are converted into amide thioacetals (e.g. (**94**)) by treatment with amines ⟨61LA(641)21⟩.

(b) From sp^2*-carbon compounds.* The reaction of alkoxyiminium cations such as (**76**) with thiolate salts in excess leads to amide thioacetals $HC(SR^1)_2NR^2_2$ ⟨65IZV2179, 73CB3725⟩. Chloroiminium salts react with thiols in a similar way ⟨60MI 604-01⟩ and (alkylthio)iminium cations (**95**) have also been

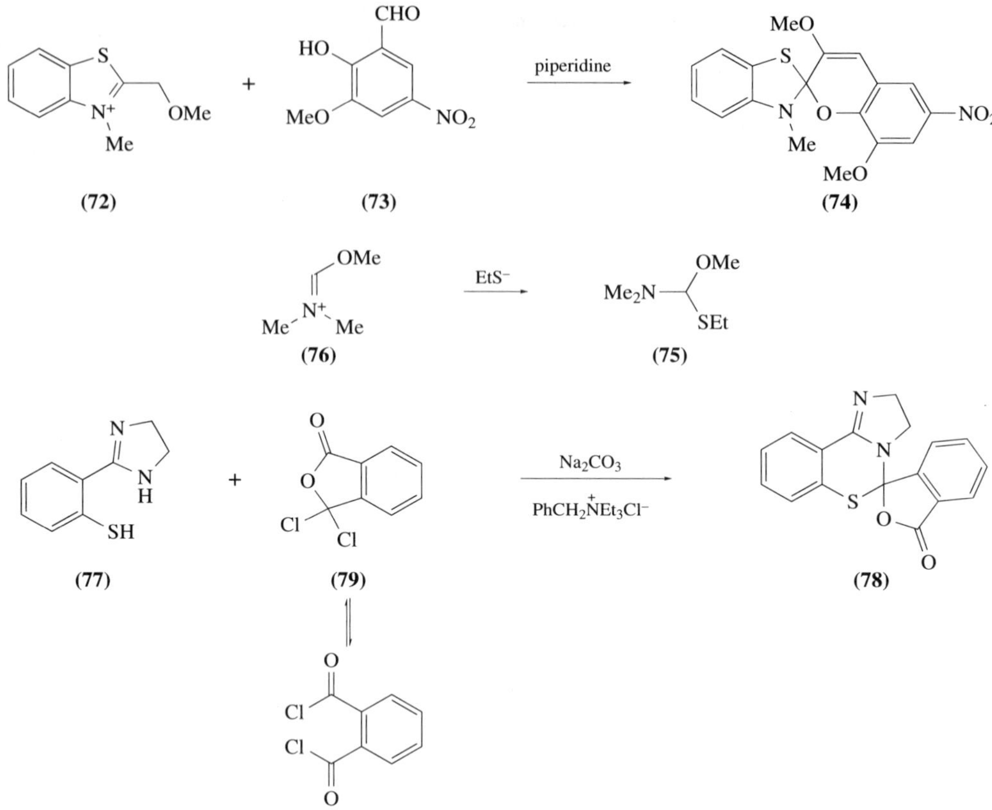

(72) (73) (74) (76) (75) (77) (79) (78)

Scheme 13

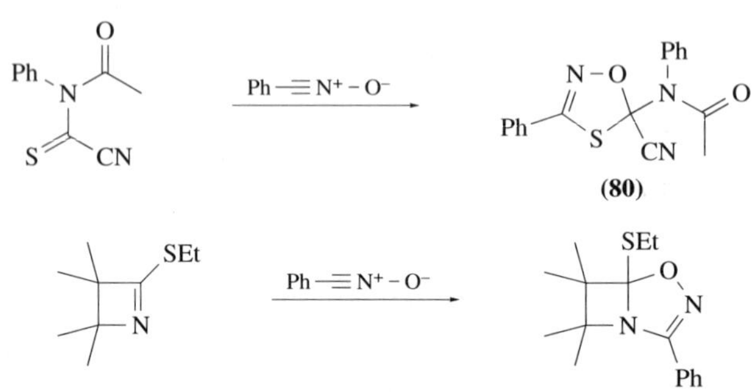

(80) (81)

Scheme 14

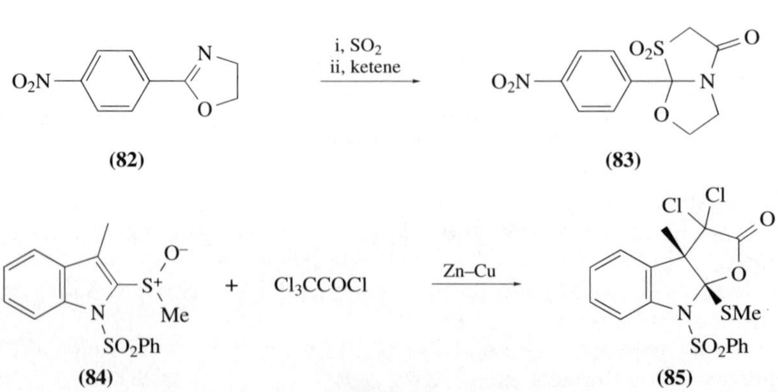

(82) (83) (84) (85)

Scheme 15

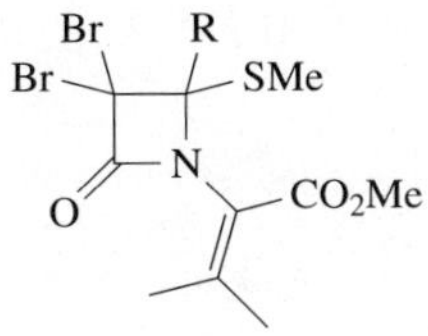

(86)

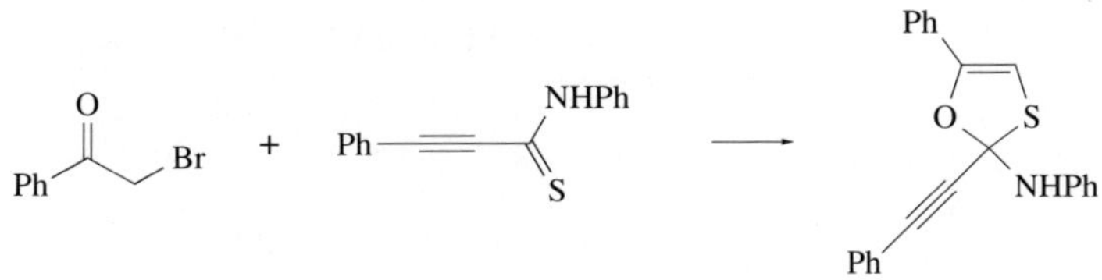

(87) **(88)**

Scheme 16

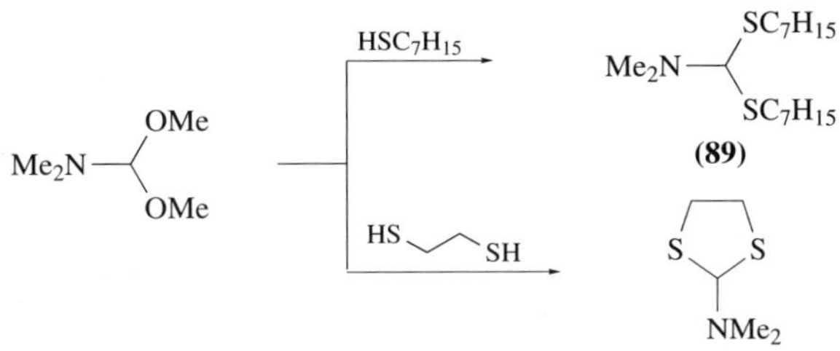

Scheme 17

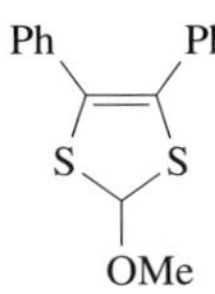

(91)

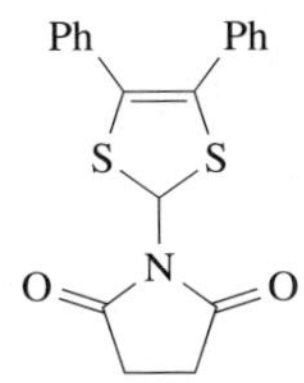

(92)

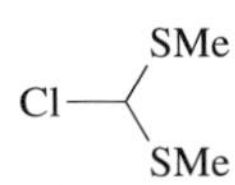

(93)

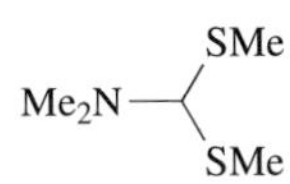

(94)

intercepted with thiols to give thioamide acetals ⟨65BCJ2107, 66BCJ2005, 83CI(L)205⟩. 1,3-Thiazolium cations (**96**) undergo reaction with azide anions and with amines to give the addition products (**97**; $X = N_3$ or NR_2) ⟨65JCS32, 76BCJ3567, 76JPR127, 80JOC2024, 81JCS(P1)618⟩. Reduction of the C═N bond of the iminium salt (**98**) with sodium borohydride yields the corresponding amide thioacetal ⟨69CPB1924⟩. A more unusual example is that provided by the reaction of the tris(alkylthio)borane derivatives (**99**) with amides (Equation (7)) ⟨66IZV364, 79ZN(B)999⟩.

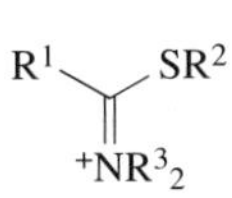

(95)

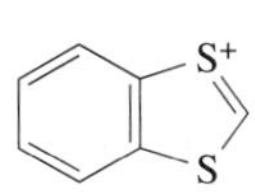

(96)

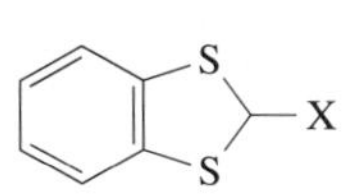

(97)

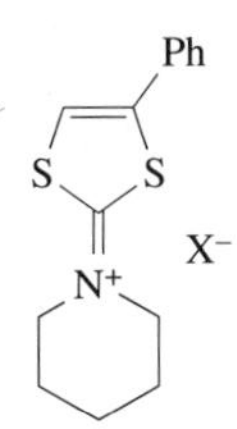

(98)

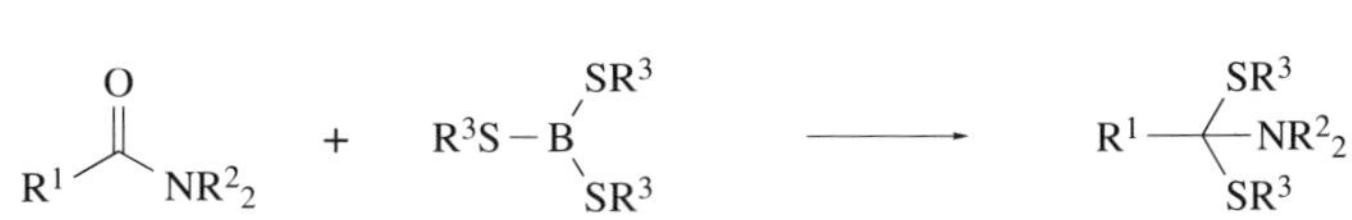

(99) (7)

(c) Cycloaddition methods. The [2 + 2] cycloaddition of ketenes to iminodithiocarbonate esters has been used as a route to β-lactams which incorporate the amide dithioacetal function. Examples with cyclic ⟨57CB2460, 73JHC791⟩ and acyclic ⟨76JOC1112, 87S990⟩ iminodithiocarbonates are shown in Scheme 18. Dihydrothiazoles also react with the ketene–sulfur dioxide adduct (in a manner

analogous to that shown for the oxazoline (**82**)) to give products which contain this functionality ⟨69JHC729⟩.

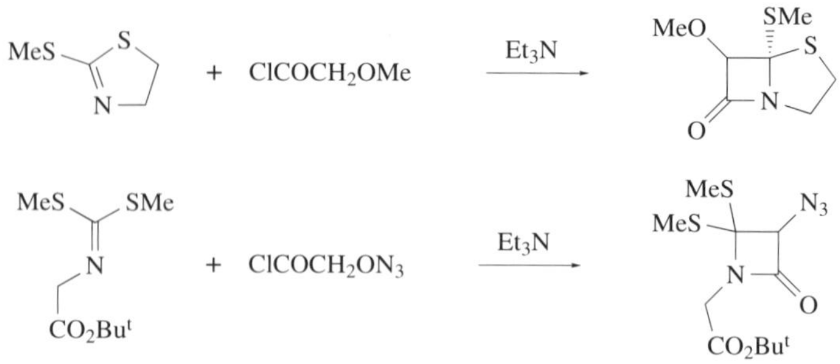

Scheme 18

(d) Miscellaneous reactions yielding amide thioacetals. Photo-rearrangement of *N*-(ethoxyacetyl)tetrahydro-1,3-thiazine-2-thione gives the thiazolidinone (**100**) (Equation (8)) ⟨86TL1335⟩. There are several routes to compounds containing methanesulfonyl functions. The azo compound (**101**; R = H) has been obtained by the reaction of bis(methanesulfonyl)methane with benzenediazonium chloride ⟨51RTC733⟩ and from its analogue (**101**; R = SMe) by reaction with piperidine ⟨51RTC892⟩. Reaction of diazomethane with the oxime (**102**) gave the *N*-methoxyaziridine (**103**) ⟨50RTC1223⟩.

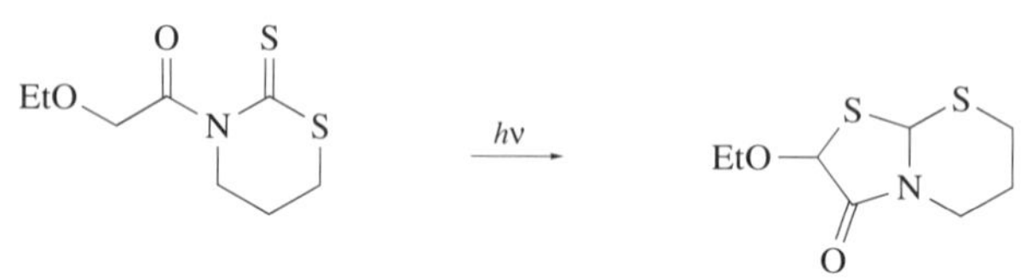

(**100**) (8)

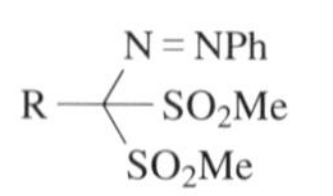

(**101**)

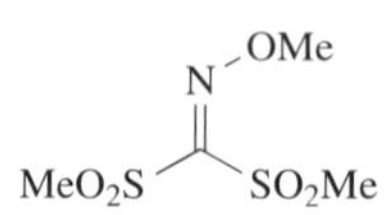

(**102**)

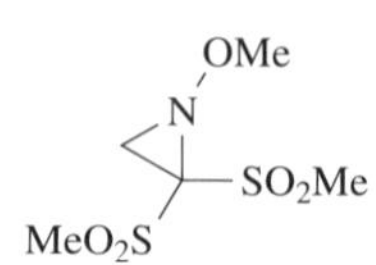

(**103**)

(iii) Functions bearing one sulfur and two nitrogen substituents

(a) From sp^2*-carbon compounds.* Addition to alkylthioiminium salts (**104**) by sodium azide and subsequent cyclisation gives thiatriazoles (**105**) ⟨74JCS(P1)540⟩. Amines will add in a similar manner to the C=N bond of thiazolium cations such as (**106**) ⟨71CPB2222, 87JOC2015⟩ and to 3-methylthio-1,2,4-dithiazolium cations (**107**) ⟨74JCS(P1)540⟩. Phenylmagnesium bromide adds to both the C=C and the C=N bonds of 5-arylidine-2-phenyliminothiazolidin-4-ones (**108**) ⟨64T25⟩. Treatment of hydrazonyl halides with thioamides leads to 2,2-disubstituted dihydro-1,3,4-thiadiazoles (**109**) ⟨76ACS(B)837⟩.

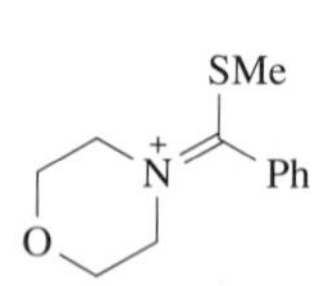

(**104**)

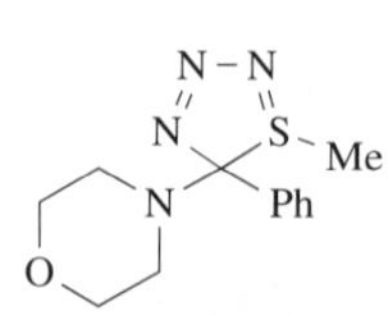

(**105**)

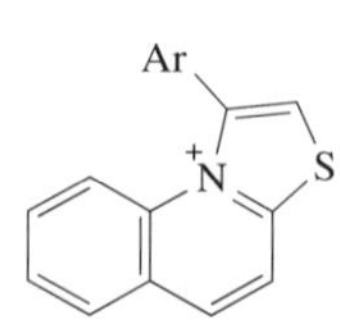

(**106**)

(b) Cycloaddition methods. Cycloaddition of cyclopentadiene to thiocarbonylbis(1,2,4-triazole) (**110**) gives the adduct (**111**) ⟨80JOC3713⟩. Diphenylethyne reacts under photochemical conditions to the C=S bond of (**112**) to give both four- (**113**) and six- (**114**) membered ring adducts (Scheme 19) ⟨76TL3563, 80LA873⟩ depending on the wavelength employed. Acid chloride derived ketenes undergo

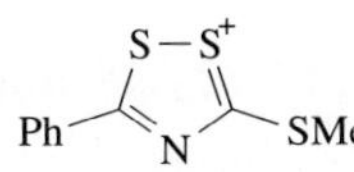

(107)

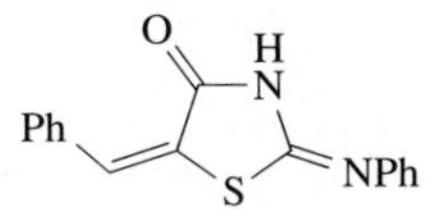

(108)

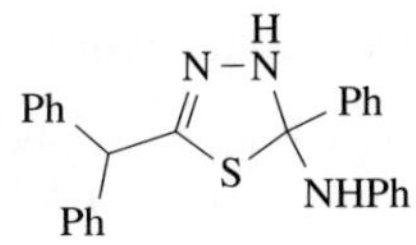

(109)

[2 + 2] cycloaddition with tetrahydro-2-alkylthiopyrimidines to yield thio substituted β-lactams such as (**115**) ⟨84TL1849⟩. Trichloroacetyl isocyanate reacts with mesoionic thiazoles to give cycloadducts such as (**116**) ⟨76ACS(B)837, 76JOC813⟩. Amide thioacetals react with aryl isocyanates to furnish the imidazolinediones (**117**) ⟨73CB3725, B-79MI 604-02⟩; a similar reaction occurs with *N*-alkyl-1,3,4-thiadiazolium salts ⟨88LA605⟩.

(c) Photoreduction. The C═S bond of 1,3-diphenyl-2-thioparabonate (**118**) undergoes photoreduction in ethanol ⟨69BCJ2323⟩.

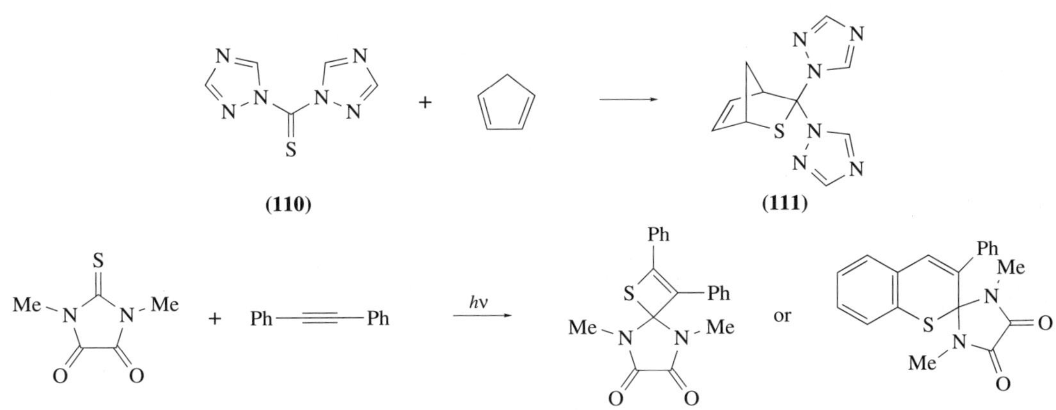

(110) **(111)** **(112)** **(113)** **(114)**

Scheme 19

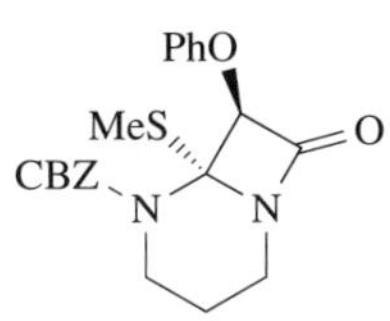

(115)

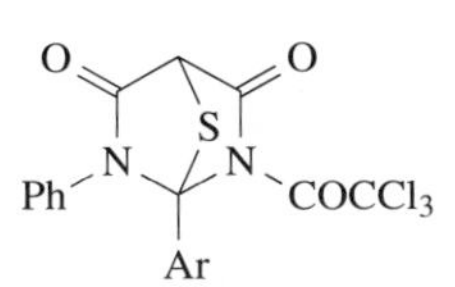

(116)

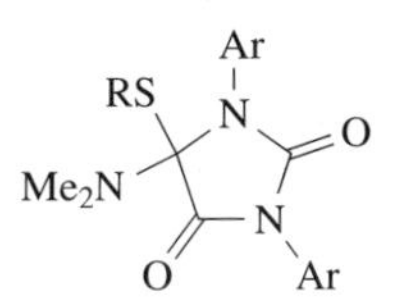

(117)

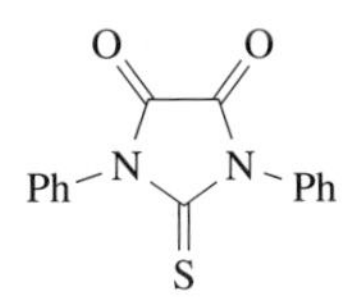

(118)

6.04.1.1.3 Functions bearing selenium and nitrogen

(i) Functions bearing one selenium, one oxygen and one nitrogen substituent

The 2-methoxydihydroselenazole (**119**; Ar = 4-PhC_6H_4) has been obtained from the corresponding *N*-methylselenazolium cation by reaction with sodium methoxide ⟨81ZOB1842⟩. Similarly, treatment of a *N*-alkylbenzoselenazolium cation (**120**) with a 2-hydroxybenzaldehyde derivative generates a 2-spiro derivative (**121**) ⟨71BSF556, 91CL1873, 93CL13⟩.

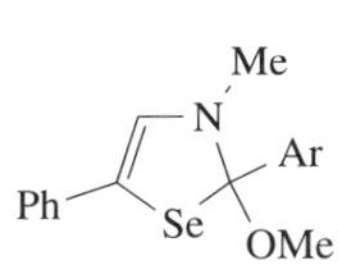

(119)

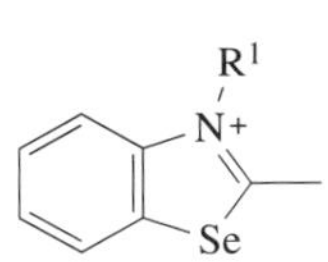

(120)

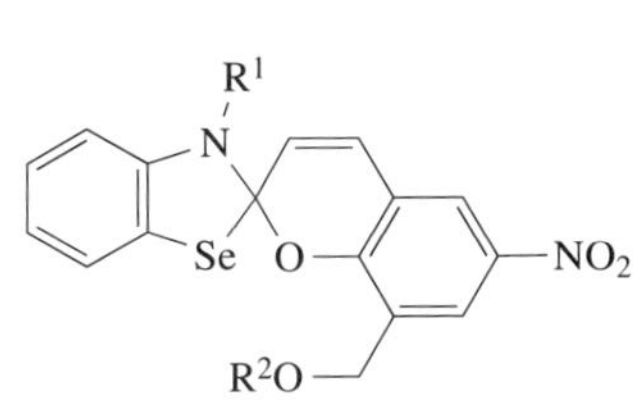

(121)

(ii) Functions bearing one selenium, one sulfur and one nitrogen substituent

Selenium substituted β-lactams such as **(122)** have been obtained by cycloaddition of ketenes to 2-alkylselenothiazolines ⟨86JAP61158986, 86JOC4737, 87NKK1447⟩.

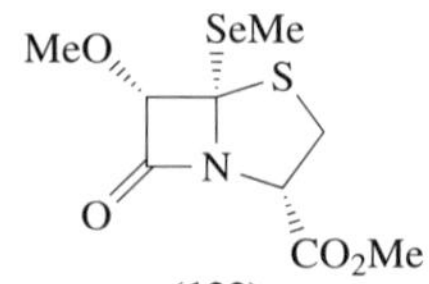

(122)

(iii) Functions bearing one selenium and two nitrogen substituents

Two examples of such compounds are provided by the selenide **(123)** which was obtained from 1,1-dinitroethane and benzeneselenyl chloride ⟨82IZV161⟩ and by the 1,3,4-selenadiazole **(124)**, which is formed by addition of a hydrazonyl chloride to the C=Se bond of a selenamide ⟨90ZOR1129⟩.

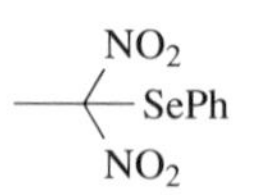

(123)

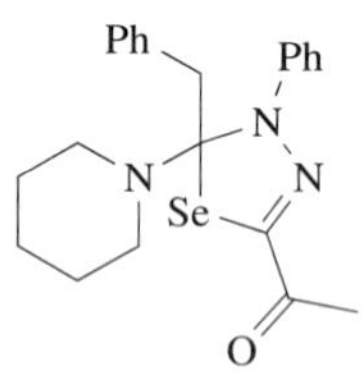

(124)

6.04.1.2 Functions Bearing Chalcogen and P, As, Sb or Bi

6.04.1.2.1 Functions bearing oxygen and P, As, Sb or Bi

(i) Functions bearing two oxygen and one phosphorus substituent

Ortho-esters and amide acetals may be converted into phosphorus containing functions by treatment with a phosphorus(III) nucleophile ⟨86MI 604-01⟩. A 2-trimethylammonium-1-3-dioxane **(125)** reacts with isopropyl diphenylphosphinate to give the phosphine oxide **(126)** (Equation (9)) ⟨88JOC3609⟩. Similarly, triethyl phosphite combines with 2-trimethylammonium-1,3-dioxolane affording the phosphonate ester ⟨74JPR913⟩. Chlorodiphenylphosphine and dichloromethylphosphine react with *ortho*-formate esters to give the phosphine oxides **(127)** and phosphonate esters **(128)**, respectively ⟨83TL1303, 89T3787⟩. Related reactions with hypophosphorous acid ⟨80AJC287⟩ and with diethyl phosphinate ⟨77TL2987⟩ have been reported. Triethyl phosphite, in combination with phosphorus trichloride and zinc chloride, reacts with triethyl *ortho*-acetate to form the phosphonate diester acetals **(129)** ⟨73S547⟩. The adduct **(130)** formed from phosphorus trichloride and acetaldehyde reacts with triethyl *ortho*-formate to give the mixed phosphonate diester acetal **(131)** ⟨90CC1133⟩.

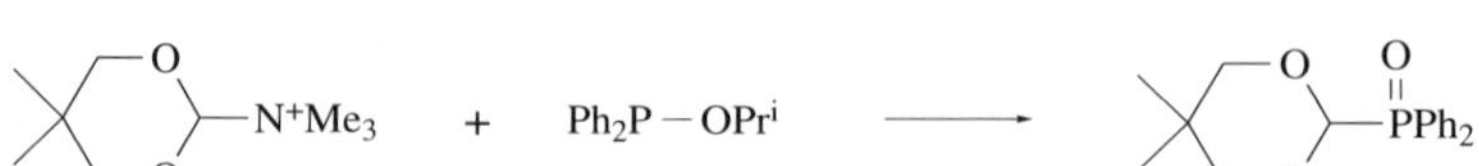

(125) **(126)** (9)

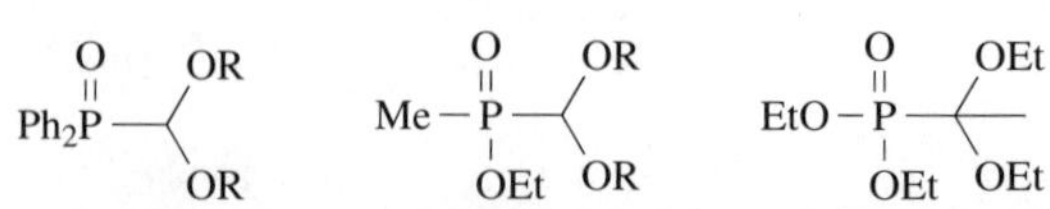

(127) **(128)** **(129)**

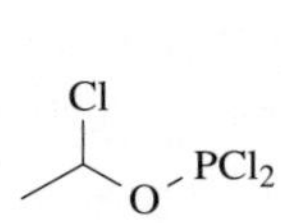

(130)

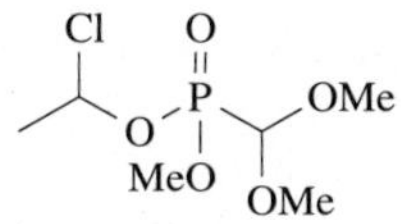

(131)

Dialkyl phosphinates react with the 3-acetylcoumarin to give the hemiacetal (**132**) ⟨91PS(57)211⟩. Hydration of α-phosphono-α-oxo esters (**133**) proceeds in high yield to give the hydrates (**134**; R = H). Similarly, alcohols yield the corresponding hemiketals (**134**; R = alkyl) (Scheme 20). The yield is higher for methanol than for *t*-amyl alcohol as branching leads to steric congestion at the quaternary centre ⟨89CC246⟩. 2,2-Dimethylpropylidynylphosphine (**135**) undergoes a double 1,3-dipolar cycloaddition with 4-chlorobenzonitrile oxide to yield the bicyclic product (**136**) ⟨84CC1634⟩.

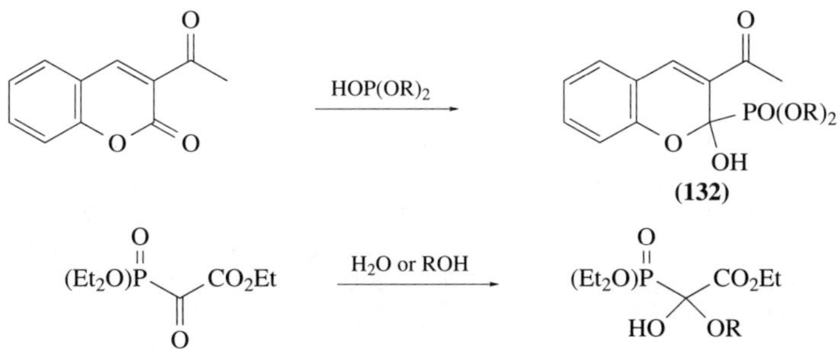

Scheme 20

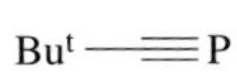

(135)

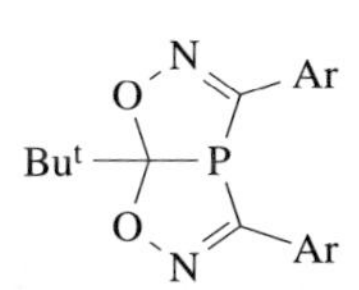

(136) (Ar = 4-ClC$_6$H$_4$)

Functions which already contain two oxygens and a phosphorus can be elaborated further at the phosphorus atom by standard methods. Phosphinite acetals (**137**) undergo an Arbuzov reaction with alkyl iodides (Equation (10)) ⟨88PS(35)329⟩, and the trimethylsilyl group of the phosphinites (**138**) is lost in the Arbuzov reaction with alkyl chloroformates which gives the phosphinates (**139**) ⟨86ZOB2427⟩. Examples of aminoalkylation of phosphorus are also known ⟨89JCS(P1)1319⟩.

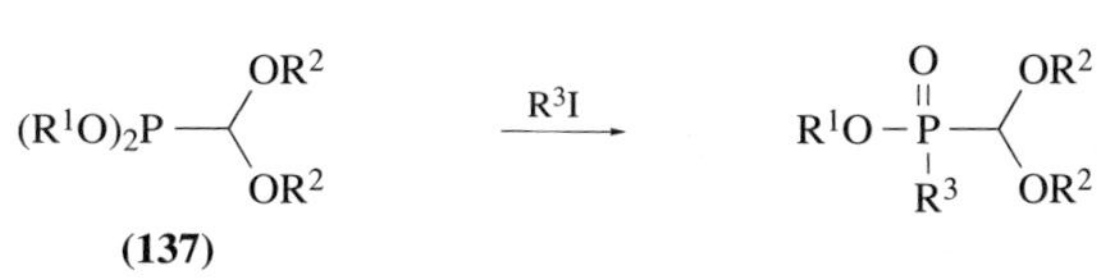

(10)

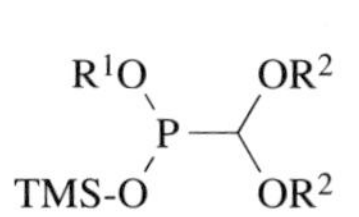

(138)

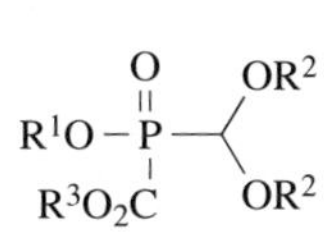

(139)

(ii) Functions bearing one oxygen, one phosphorus and one nitrogen substituent

Electrochemical oxidation of the α-acylamino phosphonate (**140**) in methanol with a carbon anode gives the methoxylated compound (**141**) (Equation (11)) ⟨89T1691⟩. In an unexpected reaction, the phosphonate (**142**) is generated from the treatment of the tosylate (**143**) with sodium diethylphosphinite ⟨93HCA2407⟩.

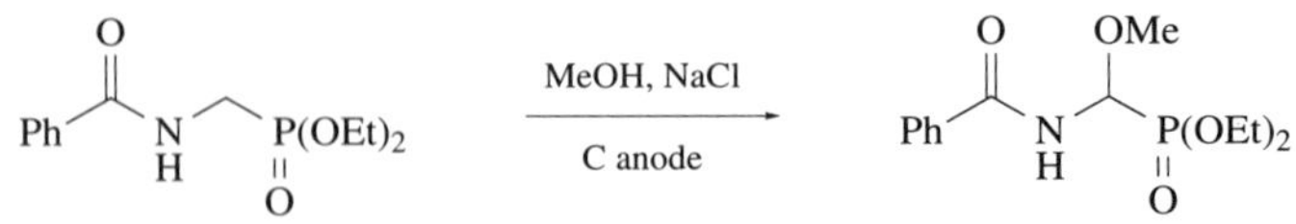

(11)

(**140**) (**141**)

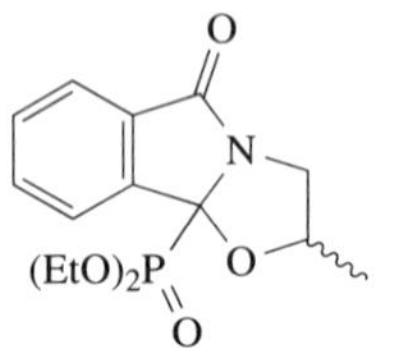

(**142**)

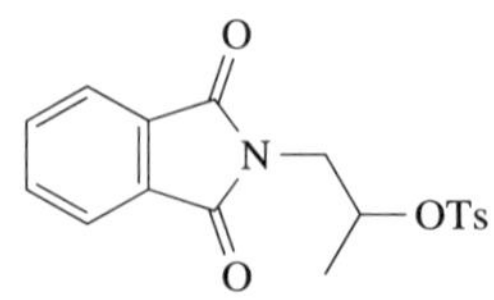

(**143**)

(iii) Functions bearing one oxygen and two phosphorus substituents

Double addition of phosphinic acids and phosphites to activated carboxylic acid derivatives such as acid chlorides and anhydrides leads to α,α-diphosphino and α,α-diphosphono alcohols, respectively ⟨56JA4450, 67JCS(C)1547, 71JOC3843, 73ZAAC(399)1, 77TL1065, 78CC285, 79S81⟩. An example of the reaction sequence is shown in Scheme 21. Carboxylic acids combine with phosphorus trichloride bisphosphonic acids ⟨72MI 604-01, 82MI 604-01⟩; thus, acetic acid gives $MeC(OH)[PO(OH)_2]_2$. Reaction of chlorodiphenylphosphine with acetic acid and then with acetic anhydride affords the bisphosphine oxide (**144**; R = H) via the intermediate ketone (**145**) (Scheme 22). The acetylated analogue (**144**; R = Ac) is obtained after prolonged reaction times ⟨77JCS(P1)1898⟩. Products analogous to (**144**; R = H) have also been obtained by hydrolysis of acylphosphine oxides ⟨78ZN(B)849⟩.

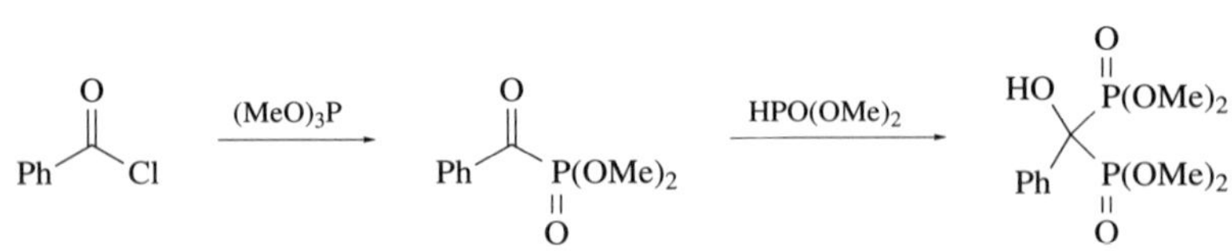

Scheme 21

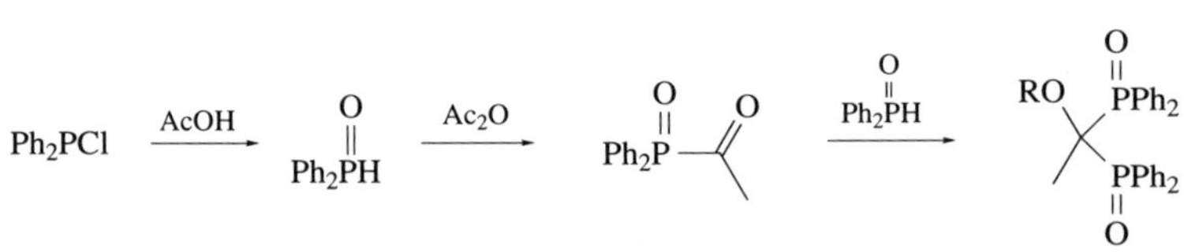

(**145**) (**144**)

Scheme 22

The diazophosphonate (**146**) reacts with cyclohexenol in the presence of rhodium acetate to give the ether (**147**) (Equation (12)) ⟨92JOC178⟩. Photochemical oxidation of α-diazophosphonates in methanol leads to the formation of products of an analogous type ⟨82JOC1284⟩. Cycloaddition of the silylnitronate (**148**) to the vinylbis(phosphonate) (**149**) leads to the formation of the dihydroisoxazole (**150**) (Equation (13)) ⟨92PS(69)75⟩. Oxygen functions can also be introduced by reaction with hydrogen peroxide, as shown in Scheme 23 ⟨89JOC4272, 93JOC4159⟩.

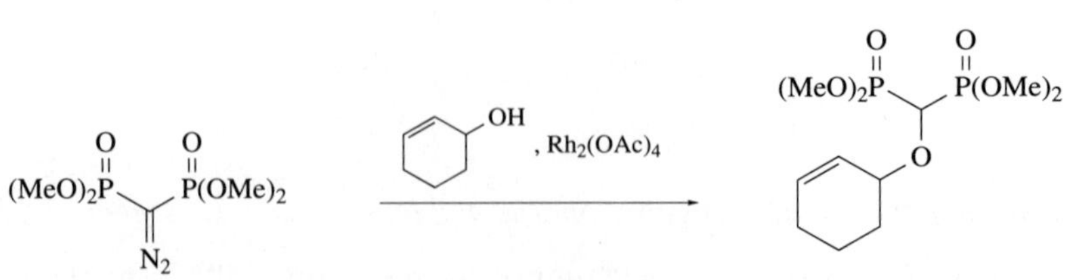

(12)

(**146**) (**147**)

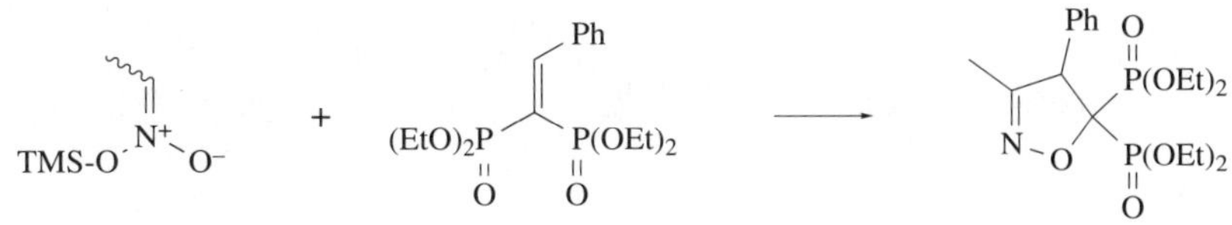

(13)

(148) **(149)** **(150)**

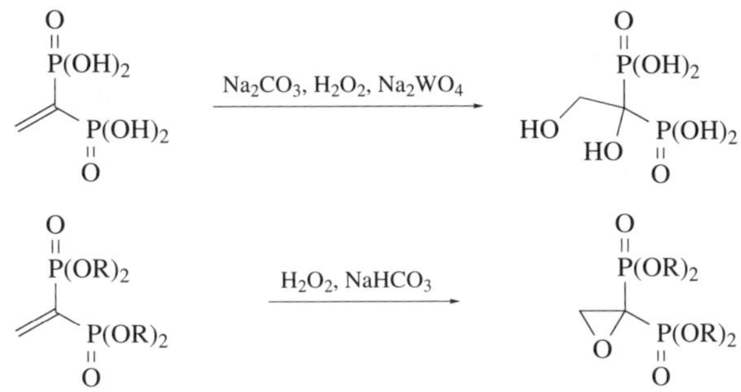

Scheme 23

(iv) Functions bearing two oxygen and one arsenic substituent

The cycloaddition of the benzoxarsole (**151**) to tetrachloro-*o*-benzoquinone gives a product (**152**) bearing this array of functional groups (Equation (14)) ⟨83TL5481⟩.

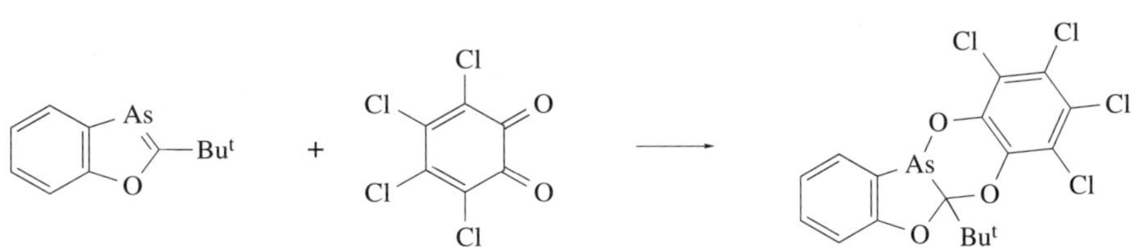

(14)

(151) **(152)**

6.04.1.2.2 Functions bearing sulfur and P, As, Sb or Bi

(i) Functions bearing one sulfur, one oxygen and one phosphorus substituent

Removal of an acidic proton from α-oxygen substituted phosphonic acid derivatives gives the opportunity for reaction with electrophilic sulfenylating reagents. Diethyl α-methoxy phosphonate (**153**) when treated sequentially with a mixture of *n*-butyllithium and potassium *t*-butoxide, elemental sulfur and iodomethane gives the methylthio product (**154**) ⟨76TL477⟩. The use of *n*-butyllithium alone gives compounds resulting from transesterification ⟨79JOC2967⟩. Similarly, the α-silyloxyphosphonate (**155**) reacts with LDA and diphenyl disulfide affording the trisubstituted product (**156**) ⟨78TL363⟩. Activation of the α-carbon atom in α-(alkylthio)phosphonates can be achieved by halogenation. Treatment of the phosphonate (**157**) with *N*-chlorosuccinimide gives the α-chloro derivative (**158**) which, in the presence of methanol, is converted into the α-methoxy compound (**159**) (Scheme 24) ⟨85TL3479⟩. The same substitution can be achieved using electrochemical oxidation in methanol ⟨87S44⟩, and a similar halogenation–substitution sequence takes place with sulfoxides analogous to (**157**) ⟨78JOC2518⟩.

2-Lithio-1,3-oxathiane reacts with dimethyl thiophosphonyl chloride to give the substitution product (**160**) ⟨85JOC662, 89JPO349⟩. The ability of a sulfoxide to activate a neighbouring carbon atom is employed in the Pummerer rearrangement of α-sulfinyl phosphonates such as (**161**) in the presence of acid anhydrides ⟨77S181, 78JOC2518, 86BCJ3293⟩. In a reaction analogous to that of the diazo compound (**146**) the ether (**162**) is formed from the rhodium catalysed solvolysis of the α-diazophosphonate (**163**) in 2-propanol (Scheme 25) ⟨92SL975⟩. Oxygen substitution can also be achieved by photolysis ⟨82JOC1284⟩.

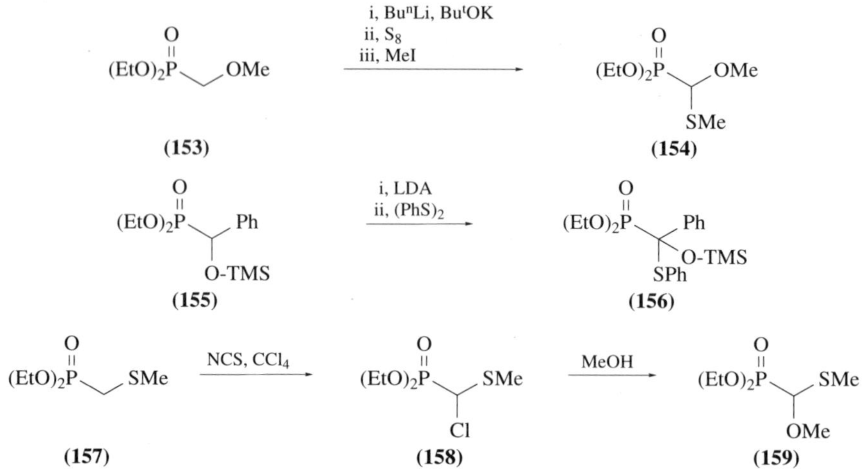

Scheme 24

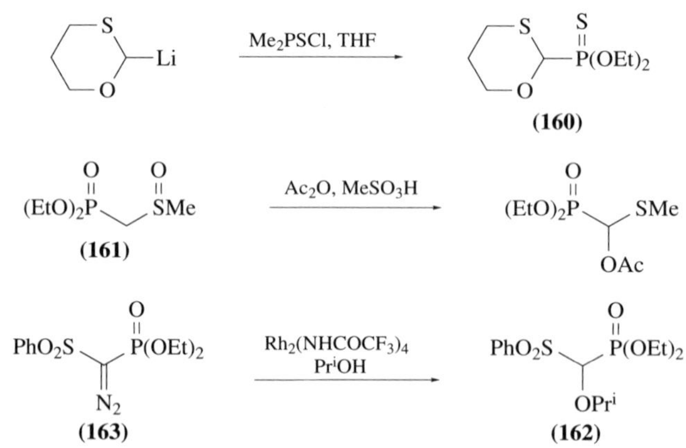

Scheme 25

(ii) Functions bearing two sulfur and one phosphorus substituent

1,3-Dithianes and 1,3-dithiolanes substituted at the 2-position with leaving groups can react with triphenylphosphine ⟨77TL885, 81S53⟩, to give phosphonium salts such as (**164**) and (**165**). Similarly, trialkyl phosphites undergo Arbuzov reactions forming phosphonate diesters such as (**166**) (Equation (15)) ⟨77TL885, 78JPR255, 80S127, 86MI 604-02⟩. The alternative approach, the reaction of 2-lithio-1,3-dithiane with an electrophilic phosphorus species, chlorodiphenylphosphine ⟨86T1963, 91JOC5919⟩ or diethyl chlorophosphonate ⟨78S309⟩, has also been reported. Phosphorus substituents can also be introduced by nucleophilic addition, as with the benzo-1,3-dithiolium salt (**167**) ⟨76TL3695, 90CC470, 91JCS(P1)157⟩. Reaction of the dithiocarbonate (**168**) with trialkyl phosphites generates the phosphonates (**169**) ⟨75JOC2577⟩.

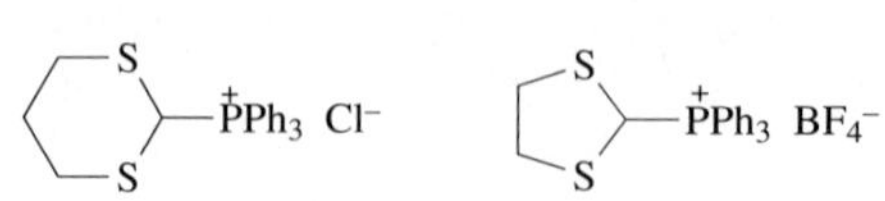

(**164**) (**165**)

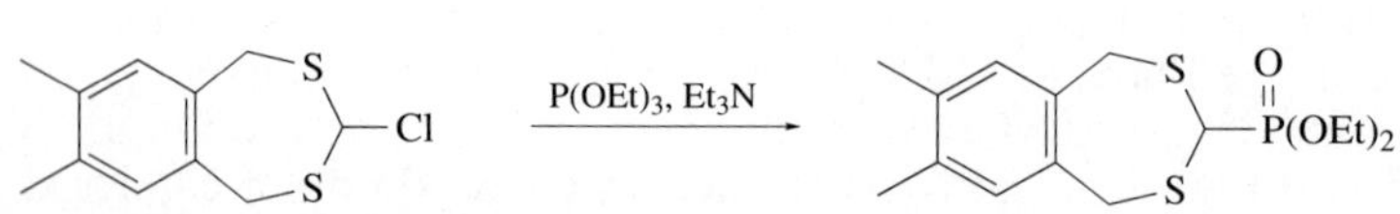

(**166**) (15)

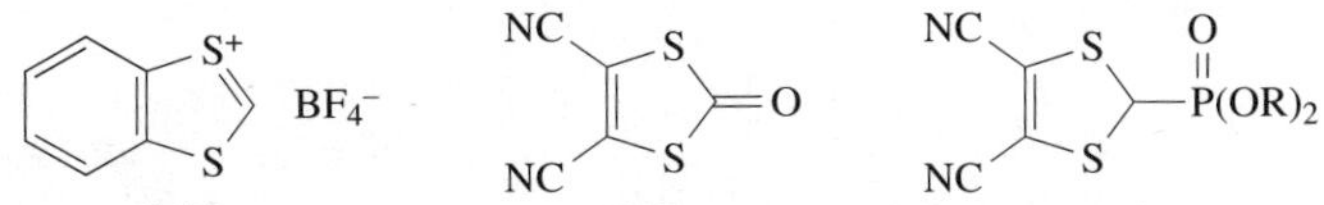

(167) **(168)** **(169)**

Sulfur functionality can be introduced into phosphorus-containing compounds by nucleophilic displacement of chloride by thiols, in a manner analogous to that in which **(158)** is formed from **(157)** ⟨88SC1611⟩. More commonly, sulfur is introduced as an electrophile. Sulfenylation of phosphorus stabilised carbanions is a useful general method for the formation of compounds containing two sulfurs and phosphorus. A further sulfur substituent can be introduced into a species containing one sulfur and one phosphorus substituent, as in sulfenylation of the ylide $Ph_3P{=}CHSPh$ ⟨90TL4359⟩ and of the phosphonate Ph_2POCH_2SPh ⟨77JCS(P1)2263, 80S127⟩. Alternatively, both sulfur substituents can be introduced by sequential sulfenylation; two examples are shown in Scheme 26 ⟨88S562, 92SC1359⟩, and other examples are known ⟨85JCS(P1)2585⟩.

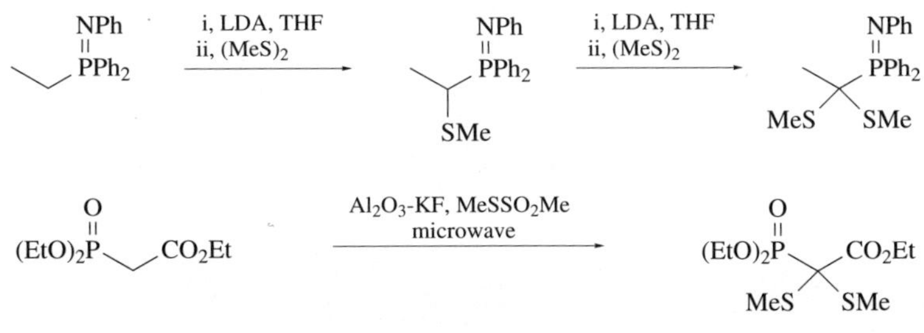

Scheme 26

S-Methyl (diisopropoxyphosphonyl)dithioformate **(170)** reacts with carbon nucleophiles on sulfur to give *S*-alkylated products such as **(171)** (Equation (16)) ⟨87JCS(P1)181, 89TL3415⟩. A similar attack on sulfur occurs with thiols ⟨92JOC4507⟩.

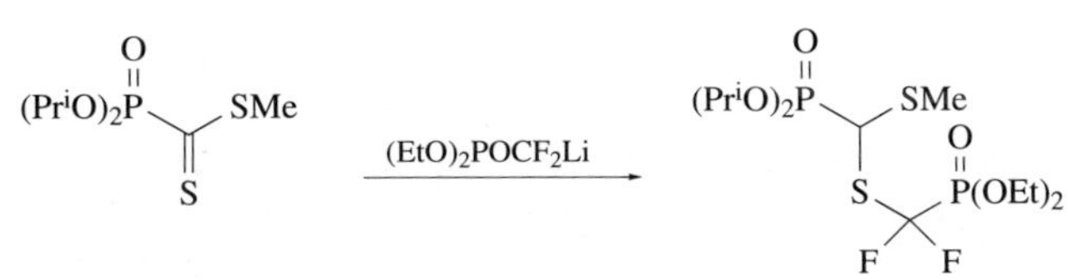

(170) **(171)** (16)

Trialkylphosphines add to carbon disulfide to give zwitterionic adducts **(172)**. These will undergo cycloaddition with acetylenedicarboxylic esters leading to the ylides **(173)** ⟨75CC960, 76BCJ1996, 79JOC930, 91JOC1816⟩. Benzyl thiol reacts with the bisisothiocyanate **(174)** to give the symmetrical product **(175)** arising from double cyclisation ⟨92HAC115⟩. Similar compounds are obtained by the treatment of compound **(176)** with acid chlorides with aluminum trichloride catalysis. The product **(177)** is a result of a carbon for phosphorus replacement. Oxidation with hydrogen peroxide gives the phosphine oxide **(178)** (Scheme 27) ⟨90JCS(P1)19⟩.

Preformed diethyl α,α-bis(methylthio)phosphonate undergoes lithiation and subsequent 1,2-addition on to α,β-unsaturated aldehydes and ketones to give the alcohols **(179)** (Equation (17)) ⟨92T8697⟩. The phosphine sulfide **(180)** can be methylated with methyl triflate to give the *S*-methylphosphonium salt, which on reaction with HMPA is reduced to the corresponding phosphine ⟨91TL7329⟩.

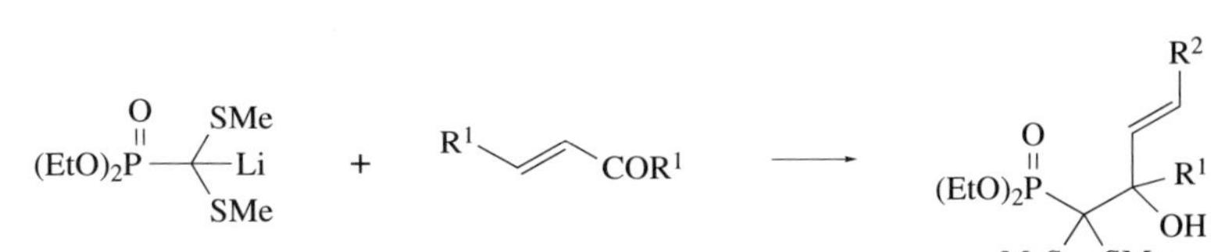

(179) (17)

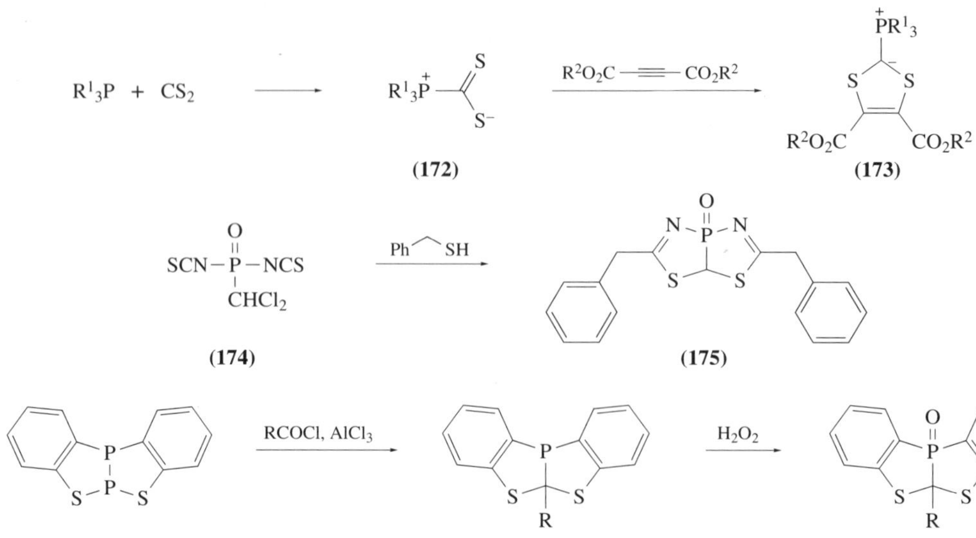

(172) (173) (174) (175) (176) (177) (178)

Scheme 27

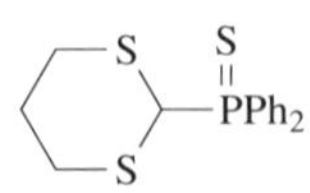

(180)

(iii) Functions bearing one sulfur, one phosphorus and one nitrogen substituent

This functionality can be produced by reaction of suitably substituted carbanions with electrophilic sulfur reagents. Compounds (**181**) ⟨92SC2381⟩ and (**182**) ⟨85RTC177⟩ are examples of compounds formed in this way, by sulfenylation α- to phosphorus. The same type of sulfenylation can be achieved starting with a haloalkane and elemental sulfur ⟨82TL913⟩. A different method is the addition of a nucleophilic phosphorus reagent to a thioimidate, as in the conversion of the thioimidate (**183**) into (**184**) with sodium diethyl phosphite ⟨78JCR(S)41⟩. The same product can be formed by addition of methylmagnesium iodide to the C=N bond of compound (**185**) (Scheme 28) ⟨75S785⟩.

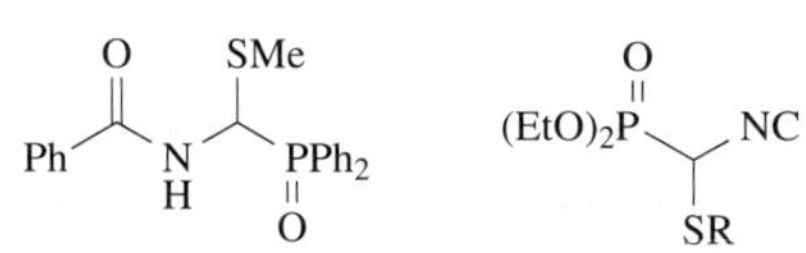

(181) **(182)**

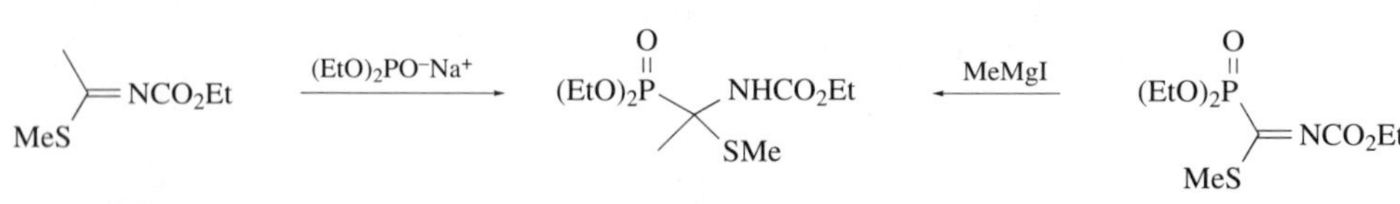

(183) **(184)** **(185)**

Scheme 28

Diazomethane undergoes cycloaddition with the vinyl sulfone (**186**) yielding initially the dihydropyrazole (**187**) (Equation (18)) ⟨85CL1099⟩.

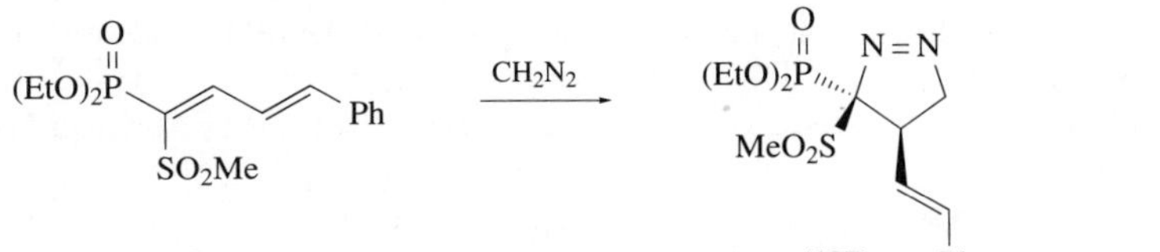

(186) (187)

(18)

(iv) Functions bearing one sulfur and two phosphorus substituents

The sulfenylation of carbanions α- to phosphorus is used to produce this type of compound ⟨80S127, 90H(30)855⟩; an example is the bis(phosphonate) **(188)** ⟨80S127⟩. Trialkyl phosphites react with thiophosgene to give compounds containing this functionality (Scheme 29) ⟨88PS(40)1⟩. A related reaction has also been reported ⟨90TL1151⟩.

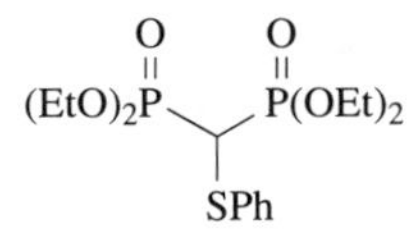

(188)

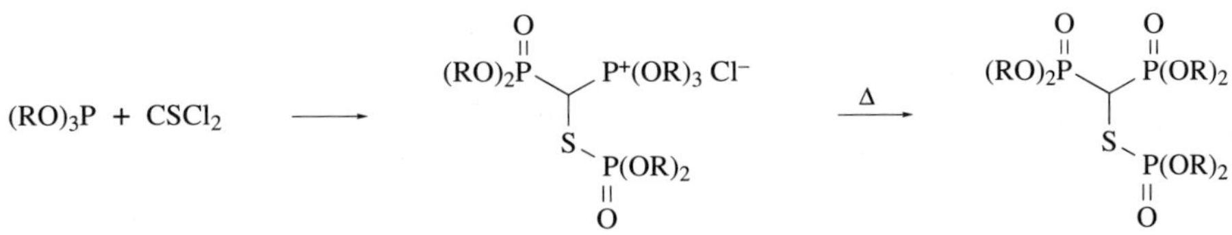

Scheme 29

6.04.1.2.3 Functions bearing selenium and P, As, Sb or Bi

(i) Functions bearing one selenium, one sulfur and one phosphorus substituent

Selenylation of the anion of the α-sulfinylphosphonate **(189)** can be accomplished with phenylselenyl bromide (Equation (19)) ⟨92TA1515⟩. Similar selenylations can be achieved using selenium and iodomethane ⟨81TL3097⟩. Phenylselenyl chloride also adds to the phosponium ylide $Ph_3P{=}CHSPh$ ⟨83CB1955⟩. Alternatively, a compound with this functionality is formed by addition of trimethyl phosphite to the heterocyclic cation **(190)** ⟨90CC470, 91JCS(P1)157⟩.

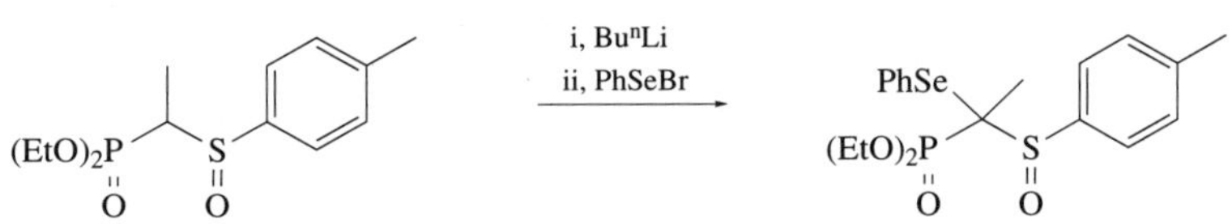

(189)

(19)

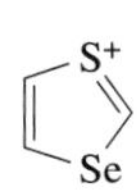

(190)

(ii) Functions bearing two selenium and one phosphorus substituent

Reaction of diethyl methylphosphonate with LDA and then with phenylselenyl bromide gives the diselenated product ⟨92TL5375⟩. In a similar fashion, double selenation of the ylide $Ph_3P{=}CHMe$ has been observed ⟨79CB355⟩. Trialkyl phosphites add to 1,3-diselenonium cations in a manner analogous to that with the cation (**190**) ⟨88HCA1242, 92TA1515⟩; a similar reaction is known with tributylphosphine ⟨84TL4227⟩. The α-diazophosphonate (**192**) undergoes insertion into the diseleno heterocycle (**191**) in the presence of boron trifluoride etherate (Equation (20)) ⟨91TL4189⟩.

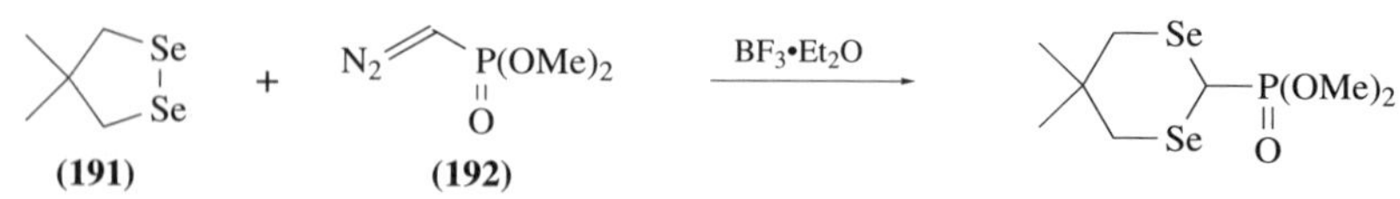

(**191**) (**192**) (20)

6.04.2 FUNCTIONS CONTAINING CHALCOGEN AND A METALLOID AND POSSIBLY A GROUP 15 ELEMENT

6.04.2.1 Functions Bearing Chalcogen and Boron

No compounds in this category have been noted. It is possible, however, that the anion which can be formed from α-phenylthioalkylboronic esters (**193**) ⟨78JA1325⟩ could react with a range of sulfur, selenium and phosphorus electrophiles.

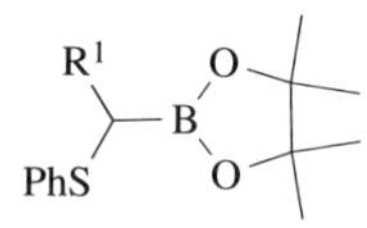

(**193**)

6.04.2.2 Functions Bearing Chalcogen and Silicon

6.04.2.2.1 Functions bearing oxygen and silicon

(i) Functions bearing two oxygen and one silicon substituent

Silyl ketals such as (**194**) are produced from the reaction of alkyl benzoates with TMS-Cl, magnesium and HMPA ⟨71JOM(26)183, 79JOC420⟩. Silylated hemiketals are also accessible from the TMS-Cl, magnesium and HMPA reagent combination ⟨71JOM(26)183⟩. Acetals with a sufficiently acidic α-hydrogen atom can be deprotonated and silylated using TMS-Cl ⟨91T3171⟩. An alternative approach is to form acetals from silyl ketones such as (**195**) (Scheme 30) ⟨69CJC4347, 80TL1357, 90JA1962⟩.

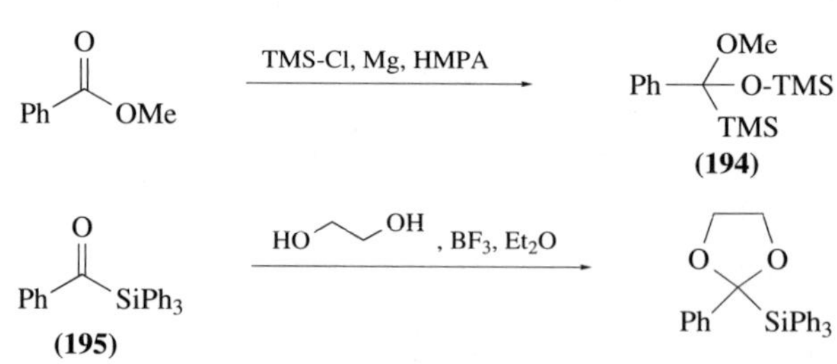

Scheme 30

Oxidation of compounds containing a vinylsilane moiety provides another route into functions bearing two oxygens and a silicon. Treatment of a highly strained 1-trimethylsilylcyclopropene derivative (**196**) with excess mcpba leads to the formation of the epoxide (**197**) ⟨86TL5143⟩. Oxidation of the compound (**198**) containing two furan rings with singlet oxygen selectively gives the product (**199**) resulting from cycloaddition to the silylated ring ⟨90TL7201⟩. Two other methods starting from vinylsilanes are the rearrangement of the trimethylsilylallene (**200**) to the dihydrofuran (**201**) ⟨82TL309⟩ and the addition of dichlorocarbene to 2-trimethylsilyldihydrofuran (**202**) and subsequent reaction with methanol ⟨84JOM(265)237⟩. 1-Ethoxy-1-trimethylsilylallene also undergoes cycloaddition with enones at the substituted double bond to give products containing this functionality (Scheme 31) ⟨91CB1425⟩.

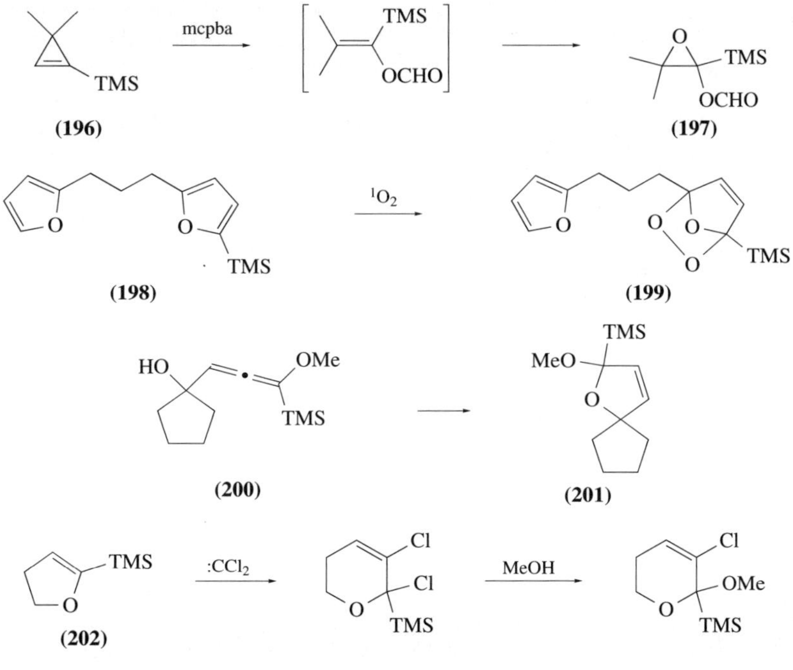

Scheme 31

The disilene (**203**) and methylfuran-2-carboxylate give a [2 + 2] cycloaddition product (**204**). This reaction is particular to the methyl ester, higher homologues failing to react (Equation (21)) ⟨91OM3466⟩.

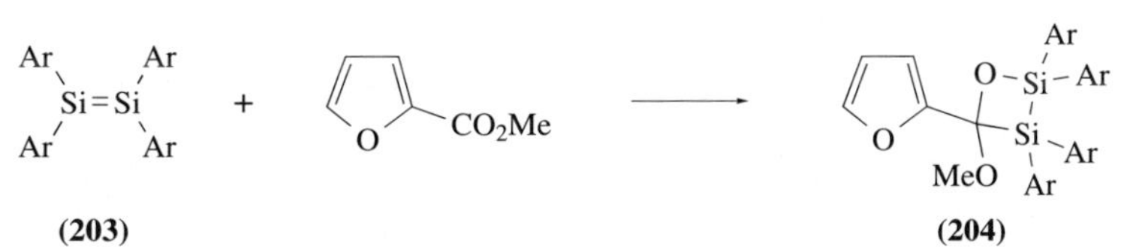

(21)

(ii) Functions bearing one oxygen, one phosphorus and one silicon substituent

Deprotonation of the α-(trimethylsilyloxy)phosphonate ester (**205**) with LDA and quenching with TMS-Cl gives the silylated adduct (**206**) ⟨81CC969⟩. Quenching of the lithiated intermediate generated from the α-substituted-α-trimethylsilyloxy phosphonate (**207**) affords some of the alcohol (**208**) by Wittig rearrangement in addition to recovered starting material ⟨79TL4475, 82BCJ224⟩. The α-diazophosphonamide (**209**) reacts with aldehydes to yield epoxides (**210**) (Scheme 32) ⟨89AG617, 91NJC393⟩.

(iii) Functions bearing one oxygen and two silicon substituents

Two silicons can be introduced into a variety of carboxylic acid derivatives by reaction with TMS-Cl and Group 1 or Group 2 metals (Equation (22)). The addition of HMPA and the use of coordinating solvents often leads to improved yields. Thus, primary amides react in the presence of lithium ⟨79JOM(177)137⟩; carboxylic acids ⟨76TL1591⟩, esters ⟨78BCJ2391⟩ and silyl ketones ⟨76TL1591⟩ utilise sodium while silyl ketones ⟨89JOC5613⟩ and acid chlorides ⟨72JOM(39)C49⟩ are converted with the addition of magnesium. Other reactions of this type which have been described include the addition of triphenylsilyllithium to ethyl formate ⟨68CJC2119⟩ and the interaction of methyl benzoate with aluminum chloride and tris(trimethylsilyl)aluminum ⟨81AG(E)581⟩.

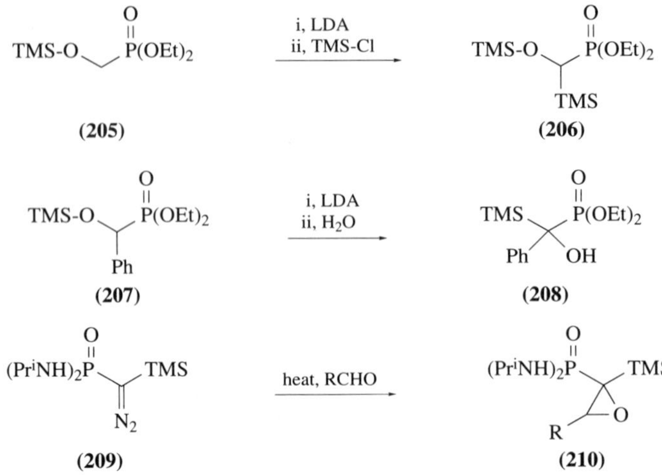

Scheme 32

 (22)

Triphenylsilylsilyloxirane can be deprotonated with *n*-butyl lithium at the substituted carbon atom. Quenching of the intermediate with TMS-Cl affords the bis(silyl)oxirane (**211**). 1,1-Bis(trimethylsilyl)ethene undergoes 1,3-dipolar cycloaddition with benzonitrile oxide giving the dihydroisoxazole (**212**) ⟨83JOC3189⟩.

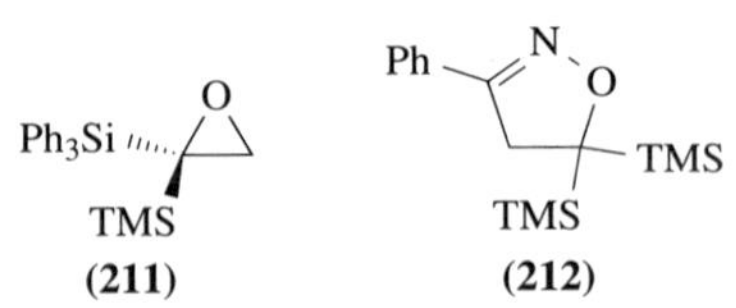

6.04.2.2.2 *Functions bearing sulfur and silicon*

(i) Functions bearing one sulfur, one oxygen and one silicon substituent

The acidity of the proton located on a methylene or methyne unit between a sulfur and an oxygen facilitates deprotonation and subsequent quenching with activated silyl compounds. Prop-1-enyloxathiane (**213**), on treatment with *s*-butyllithium and TMS-Cl gives the α-silylated product (**214**) (Equation (23)) ⟨92T2501⟩. The acidity of the proton is increased when the sulfur atom is present at its highest oxidation state, i.e., as a sulfone, when *n*-butyllithium will suffice to cause reaction ⟨79TL3375, 88CC645, 90TL1877, 91JCS(P1)897⟩.

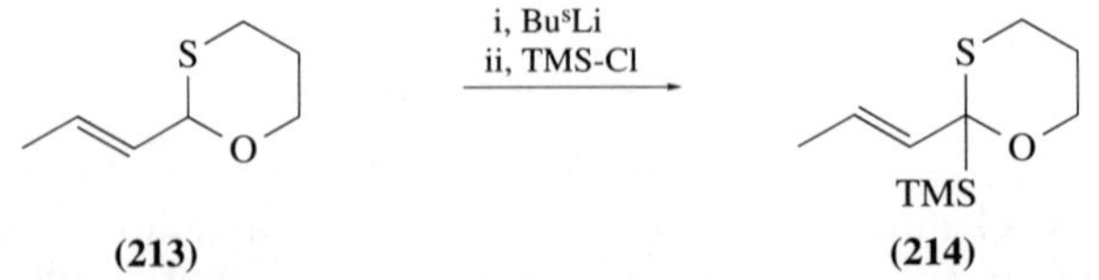 (23)

Once formed, these silylated compounds can undergo further deprotonation and alkylation. Thus, 2-trimethylsilyl-1,3-oxathiane (**215**) is methylated after treatment with *s*-butyllithium (Equation (24)) ⟨85JOC662⟩. Other alkylations of this type are known ⟨85TL2675, 87JAP6261987, 87TL2147, 89JOC5003, 92JAP4149153⟩.

(215)

(24)

Cycloadditions offer a further route into this trisubstituted functionality. Phenyl trimethylsilyl thioketone, which can be generated by the rearrangement of 2-phenyl-2-silylthiirane, ⟨67JA431⟩ will react with nitrile oxides ⟨81CC822, 90H(31)47⟩, α-nitrosostyrene ⟨85TL2131, 87JCS(P1)2647⟩ and silyl enones ⟨91TL2971⟩ to afford cyclic products (Scheme 33).

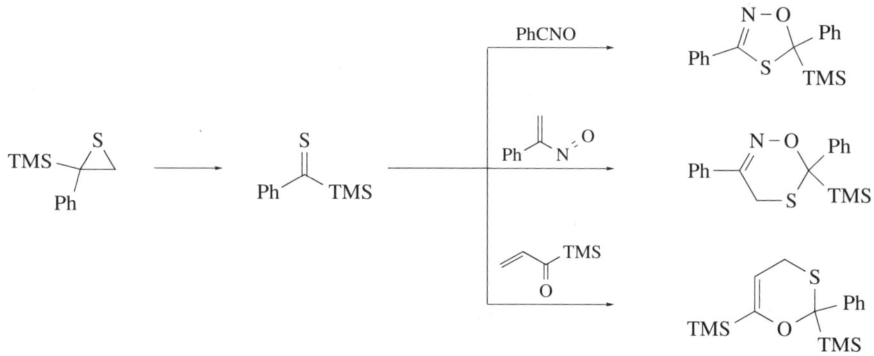

Scheme 33

(ii) Functions bearing two sulfur and one silicon substituent

Bis(alkylthio)methanes such as (**216**), on treatment with base and on subsequent addition of a silylating agent offer facile access to both tertiary (**217**) ⟨75ZAAC(418)208, 90CL1411, 90JOM(388)47, 91TL6801⟩ and quaternary ⟨77JOC1667, 85S698, 86HCA44, 86JCS(P1)195, 88ZOB1066⟩ bis(alkylthio)trialkylsilylmethanes (Equation (25)). The preferred reagent for deprotonation is *n*-butyllithium, and trialkylsilyl chlorides are used most often to quench the lithiated intermediate. These conditions are general enough to allow silylation at hindered positions ⟨86HCA44⟩. The presence of TMEDA can lead to higher yields ⟨88ZOB1066⟩. Dichlorodimethylsilane has been demonstrated to undergo double substitution of chloride by 2-lithio-2-methyl-1,3-dithiane ⟨67JA434⟩.

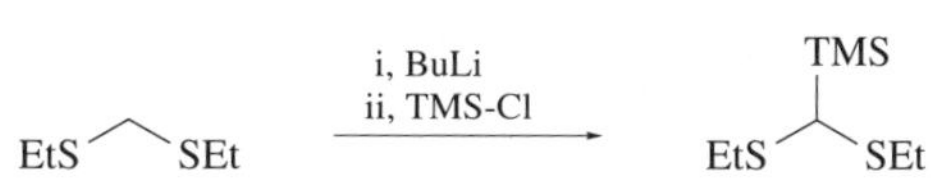

(216) **(217)**

(25)

With 2-allyl substituted 1,3-dithianes, the lithiated intermediate conceivably could be silylated at either the α- or the γ-position of the allyl substitutent. It is found that silylation usually occurs mainly at the α-position, even when this leads to the more sterically congested of the two possible products (Scheme 34) ⟨79TL1827, 80JCS(P1)2678, 86HCA1378, 87JOC855⟩. In the case of the dithiane derivatives (**218**), deprotonation by LDA and silylation takes place α- to the ring for R = Me but at both the α- and the γ-positions for R = Ph ⟨88JCS(P1)3367⟩.

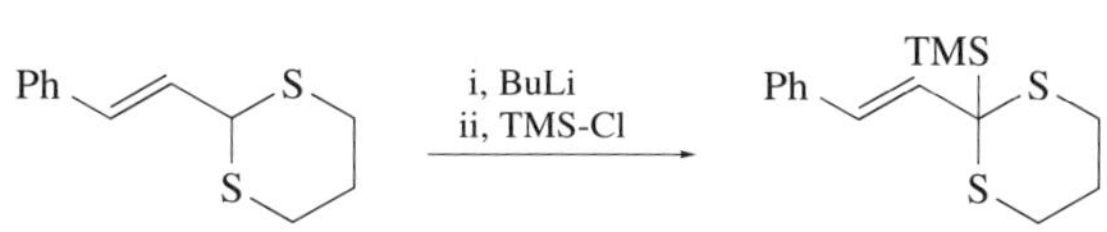

Scheme 34

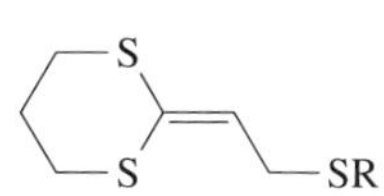

(218)

Products which contain sulfur in a higher oxidation state can be accessed by the use of starting materials at the required level of oxidation. Thus, the sulfone (**219**) reacts with trialkylsilyl perfluorobutanesulfonates and two equivalents of *n*-butyllithium to give the products (**220**) (Equation (26)) ⟨93CB537⟩. Alternatively, oxidation can be performed on the silylated compound ⟨76JOC3975⟩.

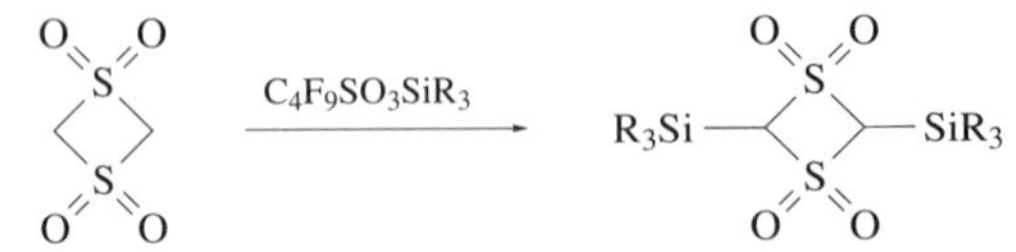

(**219**) (**220**) (26)

There are many examples of the lithiation of bis(alkylthio)trimethylsilylmethanes and reaction with electrophiles to generate quaternary products (Equation (27)). Alkylations have been performed with iodoalkanes ⟨80JA6161, 91TL2821⟩ with the optional addition of HMPA ⟨84HCA1734⟩. The sulfone (**220**; R = Me) is methylated with methyl perfluorobutanesulfonate after deprotonation with two equivalents of *n*-butyllithium ⟨91CB1805⟩. Acetylation of 2-trimethylsilyl-1,3-dithiane has been accomplished by the use of acetyl chloride ⟨73JCS(P1)2272⟩. Michael addition of lithiated bis(phenylthio)- ⟨89JOC1290⟩ and bis(methylthio)- ⟨77CB841⟩ trimethylsilylmethanes occurs with cyclopentenone; the bis(methylthio) intermediate also adds 1,4- to *t*-butyl cinnamate ⟨85TL3031⟩. 2-Lithio-2-trimethylsilyl-1,3-dithiane undergoes 1,2-addition to cyclohex-2-enone but addition of HMPA to the reaction mixture promotes 1,4-addition ⟨79CC100⟩.

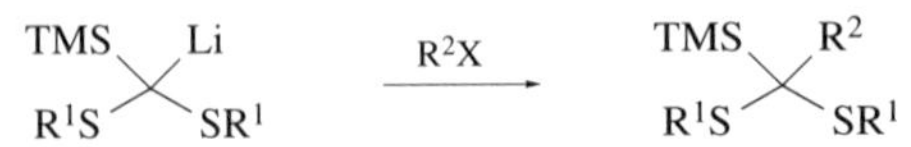

(27)

(iii) *Functions bearing one sulfur, one nitrogen and one silicon substituent*

The examples of compounds falling into this class in the literature are formed exclusively by cycloaddition. The thioketone (**221**) reacts with diazomethane to give the dihydro-1,3,4-thiadiazole (**222**) ⟨91MI 604-01⟩. A dihydro-1,3,4-thiadiazole (**223**) is obtained from the treatment of thioketone (**224**) with the hydrazonyl chloride (**225**) in the presence of triethylamine ⟨90H(31)47⟩. The trimethylsilylallene (**226**) undergoes a [2 + 2] cycloaddition with chlorosulfonyl isocyanate leading to the *β*-lactam (**227**) (Scheme 35) ⟨85TL5001⟩.

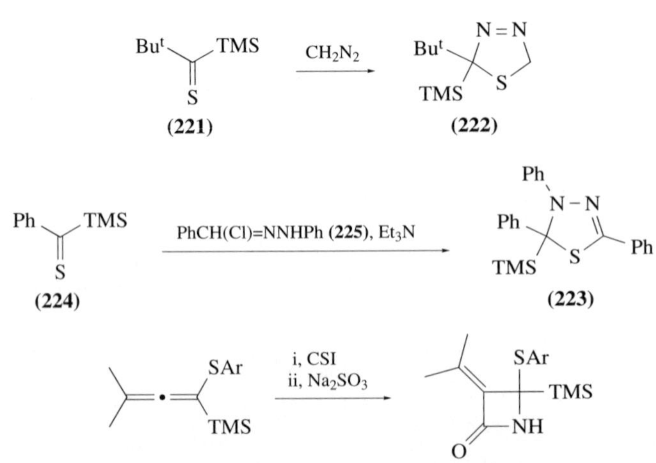

(**226**) (Ar = 4-ClC$_6$H$_4$) (**227**)

Scheme 35

(iv) Functions bearing one sulfur, one phosphorus and one silicon substituent

Pairwise combinations of sulfur, phosphorus and silicon attached to the same carbon atom will all activate the carbon atom towards anion formation or radical halogenation. The introduction of the remaining element as an electrophile or nucleophile, as appropriate, will allow the formation of the complete functional group. Thus, deprotonation of diethyl (methylthio)methylphosphonate or diethyl (trimethylsilyl)methylphosphonate with *n*-butyllithium and quenching with TMS-Cl or with dimethyl disulfide, respectively, leads to the same trisubstituted product (Scheme 36). The same compound can be prepared by the radical bromination of trimethylsilyl(methylthio)methane and subsequent Arbuzov reaction with triethyl phosphite. Of these reactions, the silylation confers the best yield ⟨89S101⟩. 1,3-Dipolar cycloaddition of nitrile sulfides to the silylated phosphaalkene (**228**) and subsequent elimination of a nitrile leads to the formation of the three-membered ring species (**229**) (Scheme 37) ⟨89TL4501⟩. An alternative route to this ring system has been described ⟨87TL6121⟩.

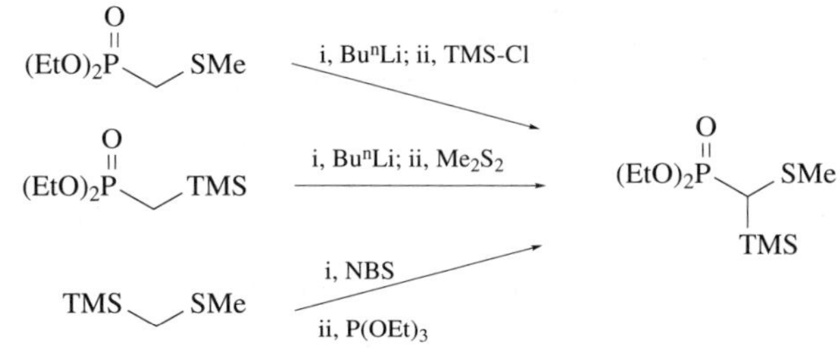

Scheme 36

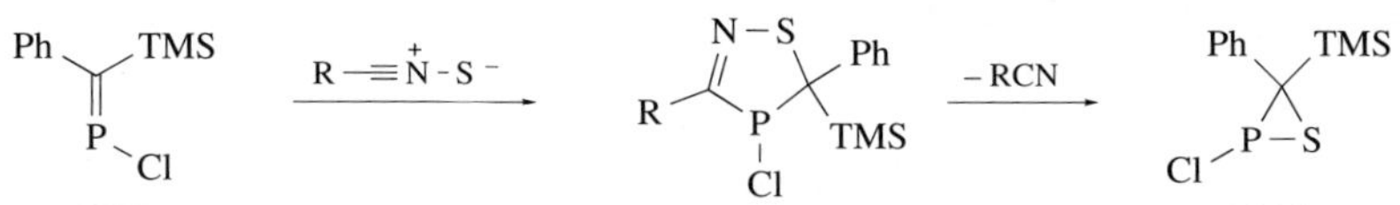

(228) **(229)**

Scheme 37

(v) Functions bearing one sulfur and two silicon substituents

Treatment of phenylthio(trimethylsilyl)methane sequentially with *n*-butyllithium and TMS-Cl gives phenylthiobis(trimethylsilyl)methane. This can be alkylated by the addition of further *n*-butyllithium and an alkyl halide ⟨91T615⟩ or it can be oxidised to the sulfone with peroxybenzoic acid ⟨84JOC5087⟩ (Scheme 38). Examples of double silylation by TMS-Cl of a carbanion adjacent to a sulfone function are also known ⟨89LA975⟩. More bulky silylating agents such as triethylsilyl chloride give selective monosilylation ⟨89TL2873⟩.

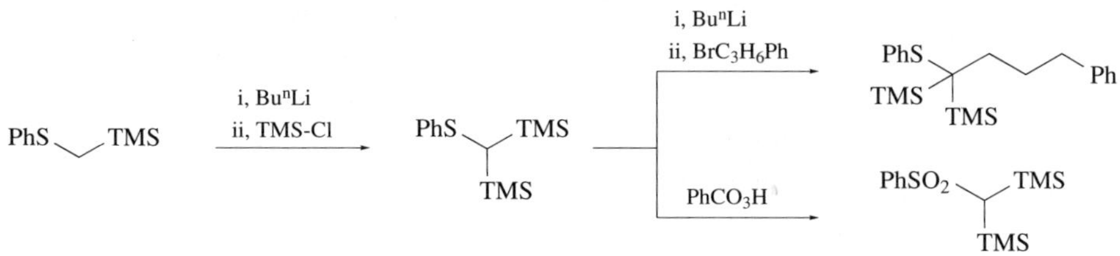

Scheme 38

The cycloaddition of the thiocarbonyl ylide (**230**) to dipolarophiles such as methyl acrylate (Equation 28) ⟨87CPB1734⟩ and acrylonitrile ⟨86H(24)1571⟩ gives bis(trimethylsilyl) substituted tetrahydrothiophenes. The alkene (**231**) undergoes thermal isomerisation to the bis(trimethylsilyl) thiirane (**232**) (Equation (29)) ⟨87CL2177⟩.

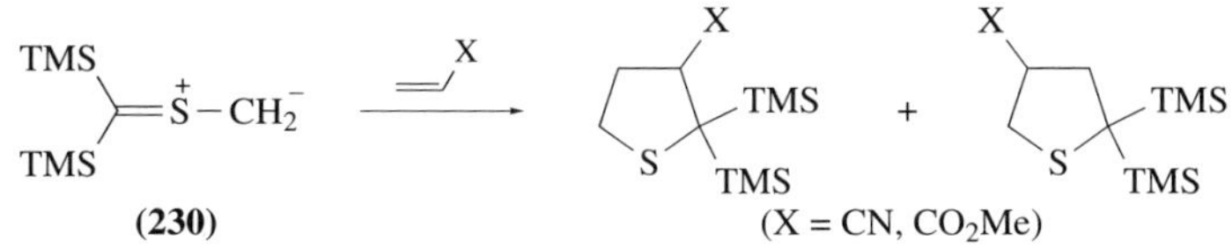

(28)

(230) (X = CN, CO_2Me)

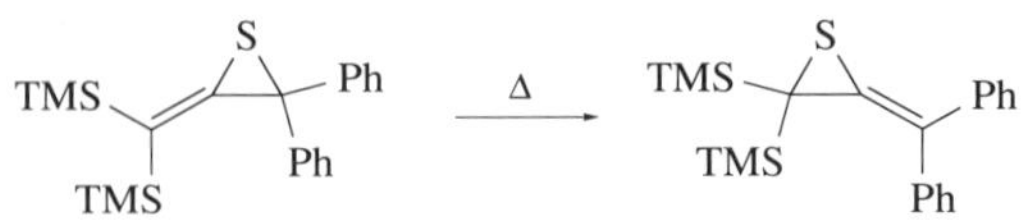

(231) (232)

(29)

6.04.2.2.3 *Functions bearing selenium and silicon*

(i) Functions bearing one selenium, one sulfur and one silicon substituent

Deprotonation of methyl phenyl sulfone with LDA and reaction of the carbanion with TBDMS-Cl gives the silylated product. Addition of a further equivalent of LDA and conversion to the cuprate can be followed by reaction with selenocyanogen to give the selenocyanate (**233**) ⟨87TL5121, 88JA8671⟩. 1-Benzenesulfonyl-1-trimethylsilylalkenes such as (**234**) undergo Michael addition by alkyllithium reagents and the resultant lithiated species can be quenched with phenylselenyl chloride (Scheme 39) ⟨81TL4287, 84TL2021⟩.

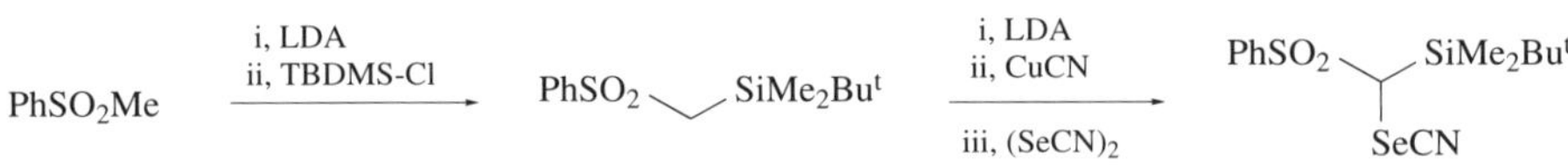

(233)

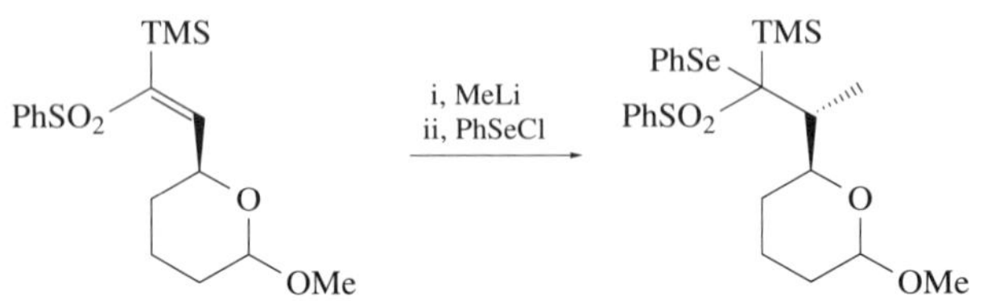

(234)

Scheme 39

(ii) Functions bearing one selenium and two silicon substituents

Reaction of α-trimethylsilyl-α-phenylselenyltoluene (**235**) with LDA and TMS-Cl gives the product (**236**) bearing two trimethylsilyl groups (Equation (30)) ⟨77JOC1773⟩.

(235) (236)

(30)

6.04.2.3 Functions Bearing Chalcogen and Germanium

6.04.2.3.1 *Functions bearing oxygen and germanium*

(i) Functions bearing two oxygen and one germanium substituent

Only one example of this functional group has been reported. Reaction of the ketal (**237**) with trimethylgermyl chloride gives the substitution product (**238**) (Equation (31)) ⟨83CC239, 83T3073⟩.

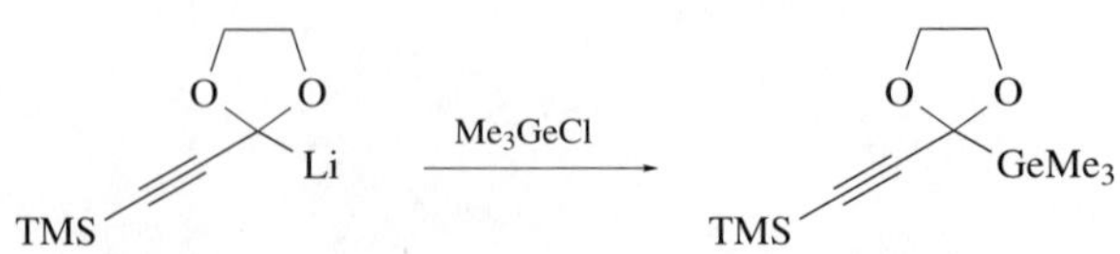

(237) (238)

(31)

(ii) Functions bearing one oxygen, one nitrogen and one germanium substituent

One class of compound has been noted with this grouping. Treatment of dimethylamides with triethylgermyllithium gives the *O*-lithiated intermediates (**239**) resulting from addition to the carbonyl group. On workup, these intermediates are converted into the ketones (**240**) (Scheme 40) ⟨83JOM(248)51, 88JOM(341)293, 88JOM(348)317⟩.

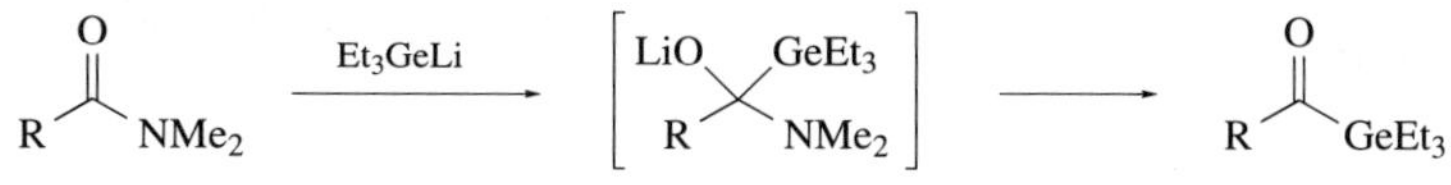

(**239**) (**240**)

Scheme 40

(iii) Functions bearing one oxygen, one phosphorus and one germanium substituent

Addition of dimethyl(trimethylgermyl)phosphine to phenyl trimethylsilyl ketone generates the *O*-silylylated product (**242**). This is believed to be obtained via the *O*-germyl isomer (**241**) (Scheme 41) ⟨77JOM(141)35⟩.

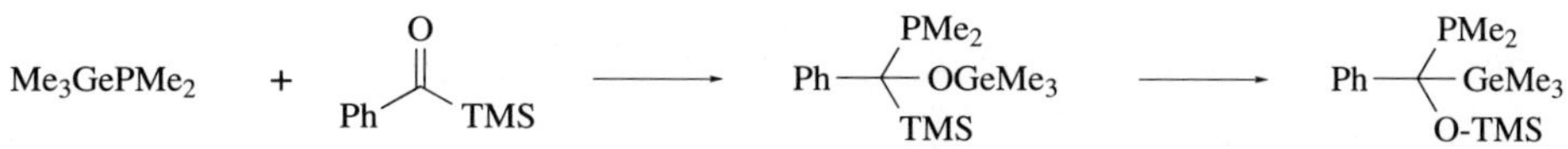

(**241**) (**242**)

Scheme 41

(iv) Functions bearing one oxygen, one germanium and one silicon substituent

Treatment of phenyl triethylgermyl ketone (**243**) with triethylsilyllithium yields the alcohol (**244**) ⟨88MI 604-01⟩. *O*-Alkylated functions such as (**245**) are accessible from the reaction of bromotrialkylgermanes with lithiated trimethylsilylmethyl methyl ether. These compounds can be further alkylated at the central carbon atom (Scheme 42) ⟨92SL843⟩.

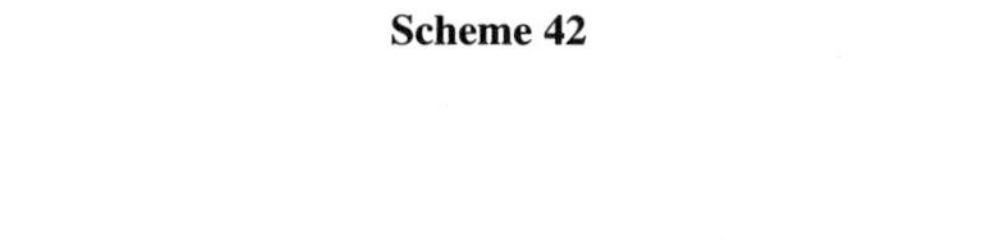

(**245**)

Scheme 42

(v) Functions bearing one oxygen and two germanium substituents

In a reaction analogous to that shown with triethylsilyllithium, the ketone (**243**) reacts with triethylgermyllithium to give bis(triethylgermyl) carbinol (**246**) ⟨87IZV2872, 88MI 604-01⟩. Ethyl formate reacts with triphenylgermyllithium to give a mixture of bis(triphenylgermyl)methanol (**247**) and its formate ester (**248**). The latter can be converted to the former by treatment with lithium aluminum hydride ⟨68CJC2119⟩.

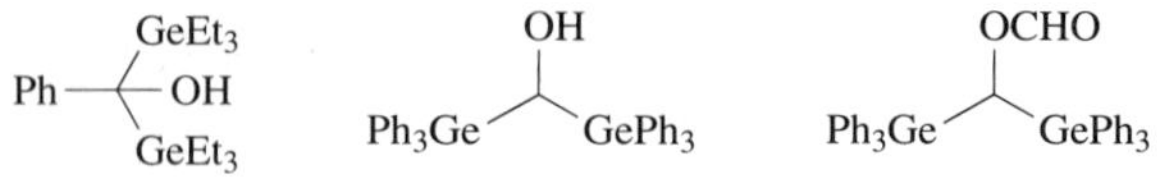

(246) (247) (248)

6.04.2.3.2 Functions bearing sulfur and germanium

(i) Functions bearing one sulfur, one oxygen and one germanium substituent

1,3-Oxathiane, when reacted with *s*-butyllithium, reacts with trimethylgermyl chloride to give the 2-substitution product (**249**) (Equation (32)) ⟨81TL2005, 85JOC657, 85JOC662⟩.

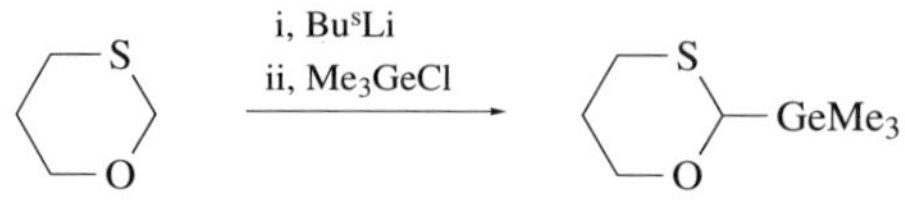

(249) (32)

(ii) Functions bearing two sulfur and one germanium substituent

The reported preparations of these compounds are all the results of the formation of a carbanion between two sulfur atoms and quenching with an electrophilic germanium species. 2-Lithio-1,3-dithiane has been reacted with triethylgermyl chloride ⟨68CJC2119⟩, triethylgermyl bromide ⟨67JA431⟩ and triphenylgermyl bromide ⟨67JA431⟩ to give products (**250**). Quaternary systems are also accessible starting from 2-alkylated dithianes ⟨67JA431⟩. Both 1,3,5-trithiane and 1,3,5,7-tetrathiocane can be substituted in the same manner ⟨75ZAAC(418)208, 76CB1239⟩. The difunctional 1-gerana-1,1-dichloro-3-cyclopentene (**251**) reacts with two equivalents of 2-lithio-1,3-dithiane affording the quaternary germanium species (**252**) ⟨84JAP59193897⟩. The bis(sulfone) (**219**) undergoes substitution at both methylene groups by trimethylgermyl perfluorobutanesulfonate ⟨93CB537⟩.

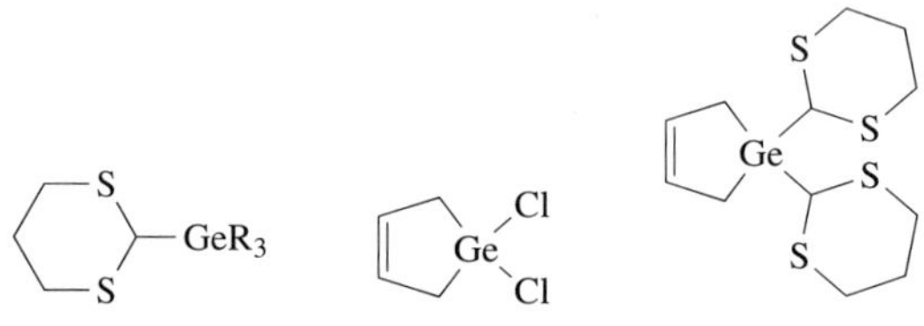

(250) (251) (252)

(iii) Functions bearing one sulfur and two germanium substituents

The only example in the literature of compounds containing this grouping (**253**) is provided by the reaction of methyl alkanesulfonates $R^1CH_2SO_3Me$ with trialkylgermyl halides and sodium hexamethyldisilazide ⟨84ZOB1842⟩; the reaction gives a mixture of products including (**253**).

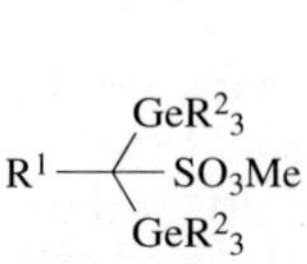

(253)

6.04.3 FUNCTIONS CONTAINING CHALCOGEN AND A METAL AND POSSIBLY A GROUP 15 ELEMENT OR A METALLOID

6.04.3.1 Functions Bearing Oxygen and a Metal

6.04.3.1.1 Functions bearing two oxygens and a metal

The 1,3-dioxane (**254**) reacts with *n*-butyllithium to generate the lithiated species (**255**) ⟨83T3073⟩. Further reaction with trimethyltin chloride affords the trimethylstannyl derivative (**256**) (Scheme 43) ⟨83CC239, 83T3073⟩.

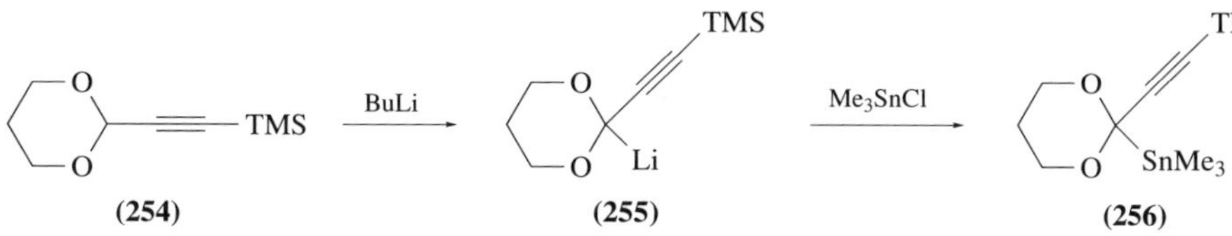

Scheme 43

6.04.3.1.2 Functions bearing oxygen, silicon and a metal

Trimethylsilylmethyl methyl ether is lithiated at the methylene group by *s*-butyllithium ⟨89TL219⟩. The proton at the 2-position of 2-trimethylsilyloxiranes can be removed with retention of configuration by *n*-butyl- or *s*-butyllithium ⟨88JOM(341)293, 89JOC4042⟩. Further reaction with trimethyltin bromide gives the trimethylstannyl derivatives (**257**) (Scheme 44) ⟨89JOC4042⟩.

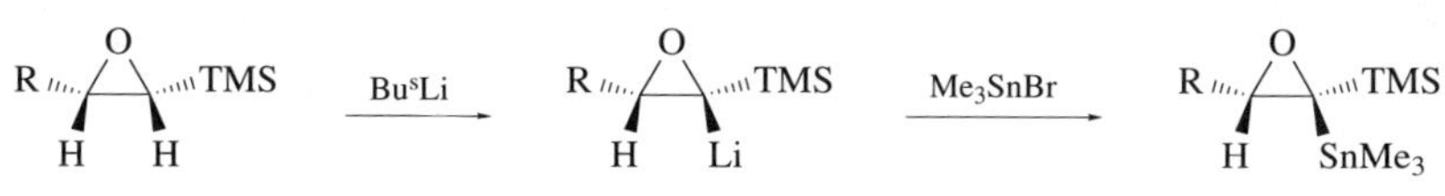

Scheme 44

6.04.3.1.3 Functions bearing oxygen and two metals

Reaction of 2-(tributylstannyl)oxirane with *n*-butyllithium results in the formation of the 2-lithio-2-(trimethylstannyl)oxirane (**258**) (Equation (33)) ⟨88JOM(341)293⟩.

(33)

6.04.3.2 Functions Bearing Sulfur and a Metal

6.04.3.2.1 Functions bearing sulfur, oxygen and a metal

Functions bearing oxygen, sulfur and lithium are generated as intermediates in reactions which involve their subsequent reaction with electrophiles; examples are provided by the reactions of 1,3-oxathianes (**213**) and (**215**) discussed in Section 6.04.2.2.2. Sulfur can be present in any of its common oxidation states. Lithiated compounds can also act as precursors to other metalled derivatives. Treatment of 2-lithio-1,3-oxathiane with trimethyllead acetate results in the formation of compound (**259**) by metal exchange ⟨81TL2005, 85JOC657⟩. Another approach to compounds with these functional groups is exemplified by the reaction of tributylstannyllithium with the thionolactone (**260**) and subsequent alkylation with iodomethane (Scheme 45) ⟨87JA2504, 90JA3696⟩.

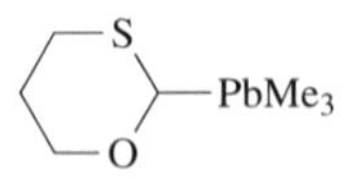

(259)

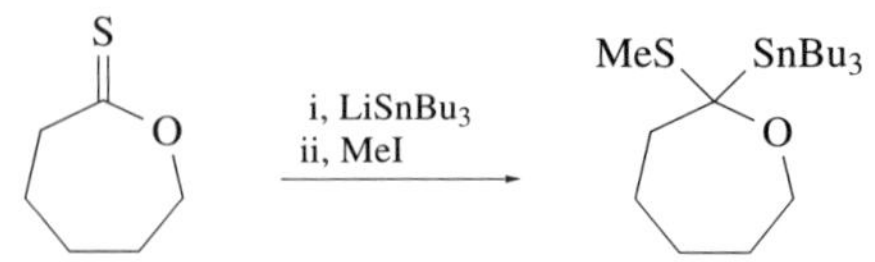

(260)

Scheme 45

6.04.3.2.2 *Functions bearing two sulfurs and a metal*

As with functions bearing oxygen, sulfur and lithium, lithiated derivatives are often used as intermediates, generated *in situ*, in the derivatisation at the carbon atom between two sulfur atoms. A range of strong bases such as the butyllithiums, LDA, lithium hexamethyldisilazide (LHMDS) and lithium phenylamide can be used in aprotic solvents at low temperatures to lithiate such compounds ⟨78JOC4235, 78MI 604-01, 81JOC1512, 82JOC1145, 86S633, 90MI 604-01⟩. In the lithiation of the 1,3-dithiane derived from of 2,4-hexadienal, reaction occurs exclusively at the 2-position ⟨92MI 604-01⟩. Lithiated dithianes (**262**) are also available from the reaction of the dithioketene acetal (**261**) with alkyllithium reagents ⟨69TL173⟩. Treatment of compounds bearing three phenylthio residues attached to the same carbon atom with *s*-butyllithium leads to replacement of one of the heteroatoms with lithium ⟨84JOC605, 85JOC3266⟩. Exchange of trimethyltin for lithium has also been used to produce a compound with this functionality (Equation (34)) ⟨88BCJ2147⟩.

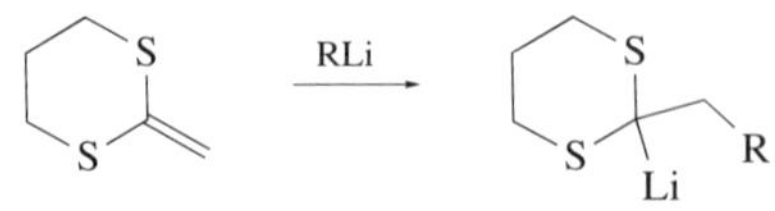

(34)

(261) **(262)**

Reaction of a dithioester with ethylmagnesium iodide results in addition to the C=S bond (Scheme 46) ⟨79JA4732⟩.

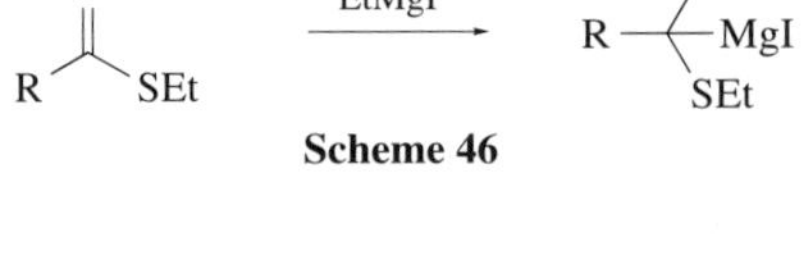

Scheme 46

A variety of metallated 1,3-dithianes and related species can be prepared from the 2-lithio intermediate by metal exchange. Thus, the tris(isopropoxy)titanium derivative (**263**) can be formed by reaction with tris(isopropoxy)titanium chloride ⟨81HCA357⟩. In a similar manner, trialkyltin or triaryltin derivatives can be obtained from 2-lithio-1,3-dithianes and -1,3,5-trithianes by reaction with the appropriate trialkyltin chlorides or triaryltin chlorides ⟨75ZAAC(418)208, 79PS(7)203, 86JOM(303)189⟩. Alternative routes to functions with two sulfur and one trialkyltin substituent are exemplified by the alkylation of 2-lithio-2-tributylstannyl-1,3-dithiane ⟨89TL15, 90JA4552, 92CJC2335⟩

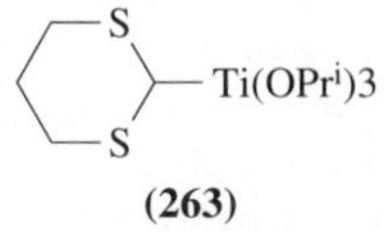

(263)

and by the conjugate addition of the intermediate (**264**) to cyclohexenone with the option of alkylation of the generated lithium enolate (Equation (35)) ⟨75AG37, 77CB841⟩.

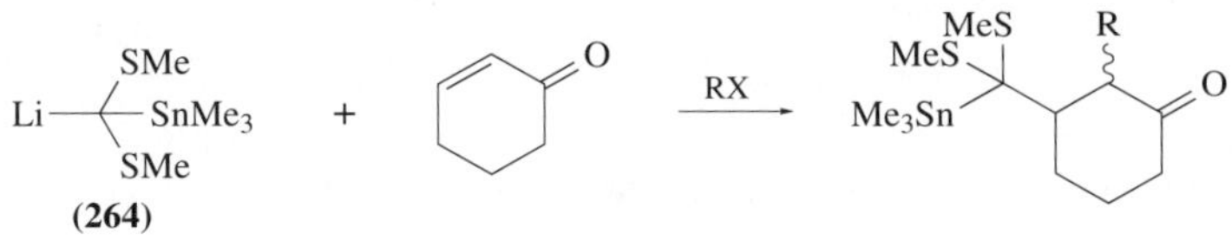

(**264**)

(35)

6.04.3.2.3 *Functions bearing sulfur, boron and a metal*

α-(Phenylthio)alkylboronate esters such as (**265**) have been demonstrated to undergo lithiation with LDA (Equation (36)) ⟨78JA1325⟩.

(**265**)

(36)

6.04.3.2.4 *Functions bearing sulfur, silicon and a metal*

α-(Trimethylsilyl)methyl sulfides and sulfones are readily metallated at the methylene group by butyllithium ⟨86JCS(P1)195⟩. Subsequent exchange reactions have been used to produce trialkyltin derivatives such as (**266**) ⟨86JCS(P1)195⟩ and copper derivatives ⟨86JOC3983⟩. An alternative route to lithiated species is the addition of alkyllithium reagents to vinylsilanes; thus, the intermediate (**267**) is produced from $H_2C{=}C(SPh)$-TMS and methyllithium ⟨86JCS(P1)195⟩. Multinuclear NMR studies have demonstrated the conformational stability of the species (**268**) in ether–THF mixtures due to ion pairing. Addition of greater than three equivalents of HMPA causes a degree of racemisation to occur ⟨93AG(E)1469⟩.

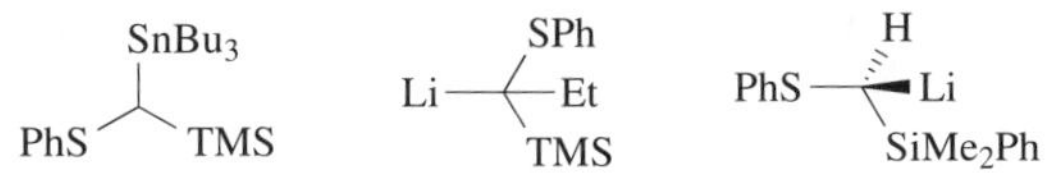

(**266**) (**267**) (**268**)

6.04.3.2.5 *Functions bearing sulfur and two metals*

Two tributyltin functions can be introduced on to a methylene group α- to a sulfone group by lithiation and reaction with tributyltin chloride ⟨92CC870⟩.

6.04.3.3 Functions Bearing Selenium and a Metal

6.04.3.3.1 *Functions bearing selenium, sulfur and a metal*

LDA will lithiate (phenylthio)(phenylseleno)methane ⟨90JA5609⟩.

6.04.3.3.2 *Functions bearing two seleniums and a metal*

Bis(phenylseleno)methane can be lithiated with the hindered base LITMP and HMPA as a cosolvent ⟨79JOM(177)1⟩. As an alternative, reaction of tris(methylseleno)methane with *n*-butyllithium results in replacement of one methylseleno group by lithium ⟨79TL3375⟩.

6.04.3.3.3 Functions bearing selenium, phosphorus and a metal

Treatment of the α-(phenylseleno)methylphosphonate $(EtO)_2POCH_2SePh$ with *n*-butyllithium results in lithiation at the methylene group ⟨87S169⟩.

6.04.3.3.4 Functions bearing selenium, silicon and a metal

Lithium derivatives can be generated by hydrogen–lithium exchange, as in the generation of the species (**269**) from the corresponding allyl(phenylseleno)silane and lithium diethylamide ⟨75JOC2570⟩ or by exchange of a phenylseleno function for lithium, as in the generation of the intermediate (**270**) from (**271**) and *n*-butyllithium (Scheme 47) ⟨78JOM(149)C10⟩.

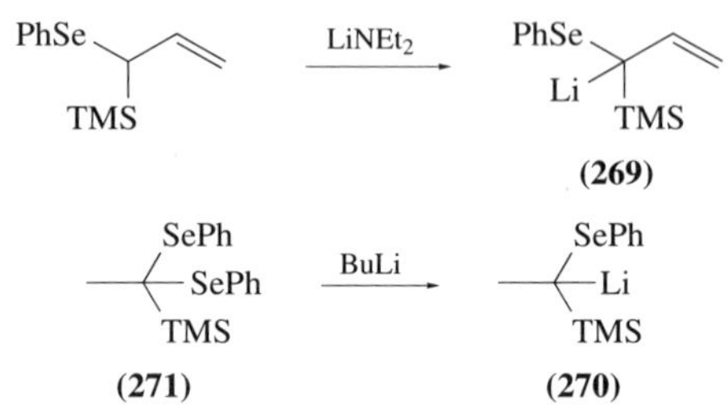

Scheme 47

6.05
Functions Containing at Least One Group 15 Element (and No Halogen or Chalcogen)

DAVID P. J. PEARSON
ZENECA Agrochemicals, Bracknell, UK

6.05.1 FUNCTIONS CONTAINING THREE GROUP 15 ELEMENTS

6.05.1.1 Functions Bearing Three Nitrogen Atoms

6.05.1.1.1 1,1,1-Triaminoalkanes

(i) Introduction

Methods for the formation of compounds of general structure (**1**), commonly known as *ortho*-amides, have been reviewed, most recently by Simchen, who has written two accounts ⟨B-74MI 605-02⟩

and ⟨85HOU(E5)3⟩, but also by deWolfe ⟨B-70MI 605-03⟩ and Kantlehner ⟨B-79MI 605-01⟩. Each of these reviews is part of a more general treatment of *ortho*-acid derivatives but gives extensive references to their preparation.

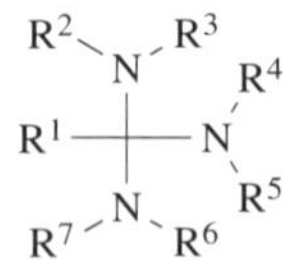

(1)

As expected, compounds of this type are moderate to strong bases (unless the nitrogen atoms are acylated) and dissociate to a significant degree in solution as evidenced by the conductivity measurements of Bredereck ⟨69MI 605-01⟩ (quoted in ⟨B-79MI 605-01⟩). This gives rise to their propensity for elimination, displacement and disproportionation reactions, as shown in Scheme 1 ⟨68CB51⟩, and explains why examples from carboxylic acids higher than formic are relatively rare, why the amines involved are generally fully substituted and why the most common structural type has three identically substituted nitrogen atoms.

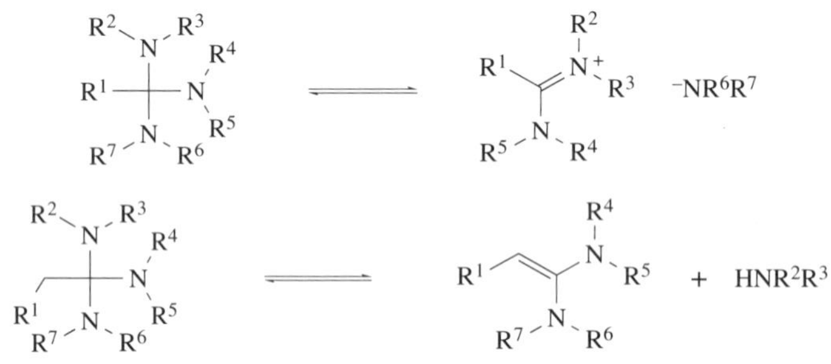

Scheme 1

This functional group occurs in some natural products, e.g., oxaline (**2**) ⟨76T2625, 80CPB2987⟩ and dibromoagelaspongin (**3**) ⟨89T3487⟩.

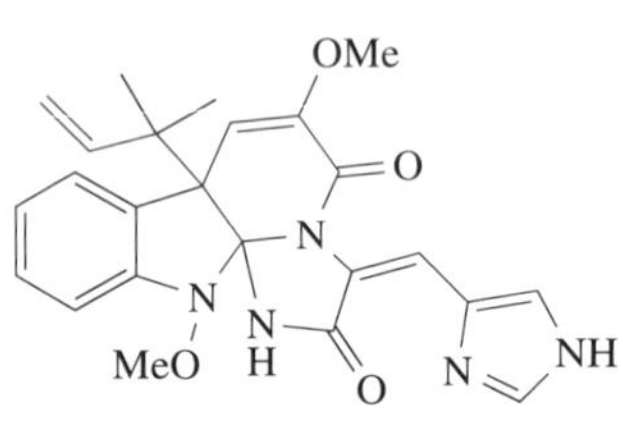

(2)

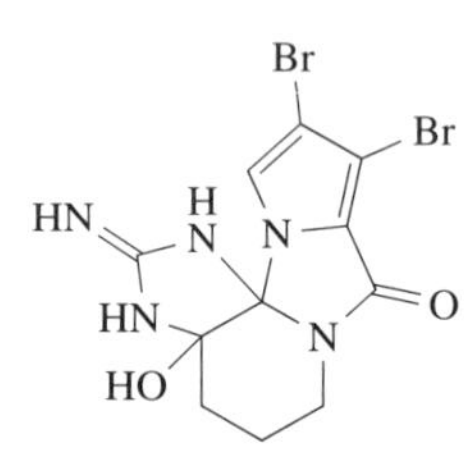

(3)

(ii) Preparation

(a) From amidinium salts. Formamidinium salts are reported by a number of authors to undergo reaction with secondary amines or their metal salts (Scheme 2). For example, the sodium salts of *N*-methylanilines react with a salt (**4**) to give the *ortho*-amides (**5**) in modest yield ⟨62JOC3664⟩. Lithium dialkylamides have been similarly used ⟨66AG(E)132⟩ and this method was applied to the preparation of unsymmetrical examples such as (**6**) ⟨68CB1885⟩. Potassium *t*-butoxide has been employed as the base in a similar synthesis ⟨68CB3058⟩; in this instance the reaction may occur by the initial addition of the alkoxide moiety. The unusual displacement of cyanide in Structure (**7**) is presumed to occur via the intermediacy of an amidinium salt (**8**) ⟨79S342⟩.

(b) From other amide salts. The Vilsmeier reagent (**9**) reacts with the potassium salts of saccharin and phthalimide to give the unsymmetrically substituted adducts (**10**) ⟨61CB3109⟩. Bredereck *et al.* report the alkylation of *N*,*N*-diethylformamide (**11**) and its subsequent reaction with diethylamine to give (**12**) in low yield ⟨68CB3058⟩. In previous studies by the same group triformamidomethane (**14**) was constructed from formamide via similar intermediates (**13**) (Scheme 3) and a comparison

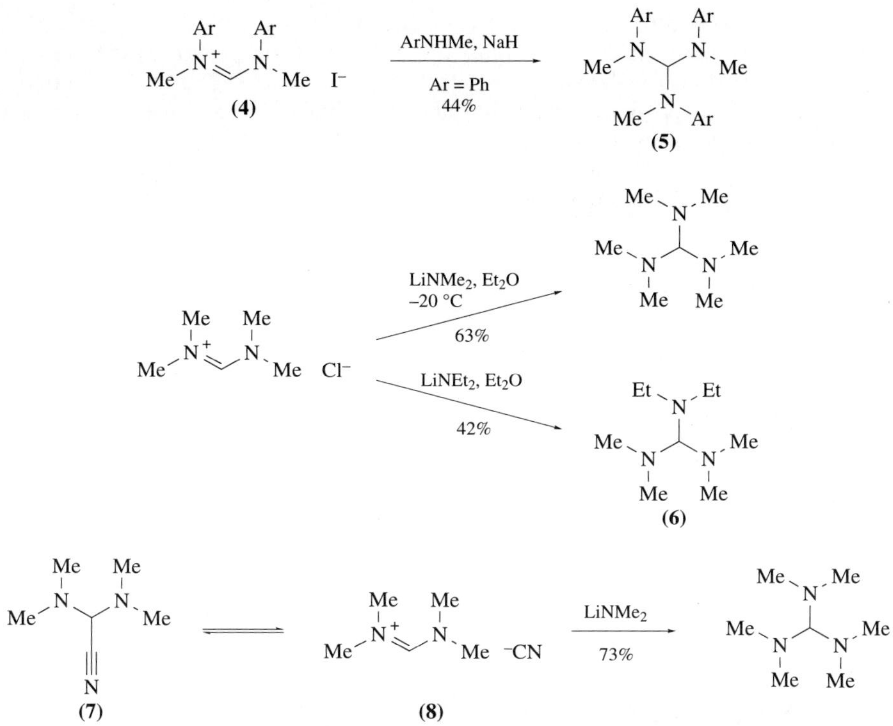

Scheme 2

was made of various alkylating and acylating agents ⟨59CB329⟩. When dialkyl sulfates were used the yield increased with the bulk of the alkyl group, perhaps reflecting the ease of the subsequent displacement step.

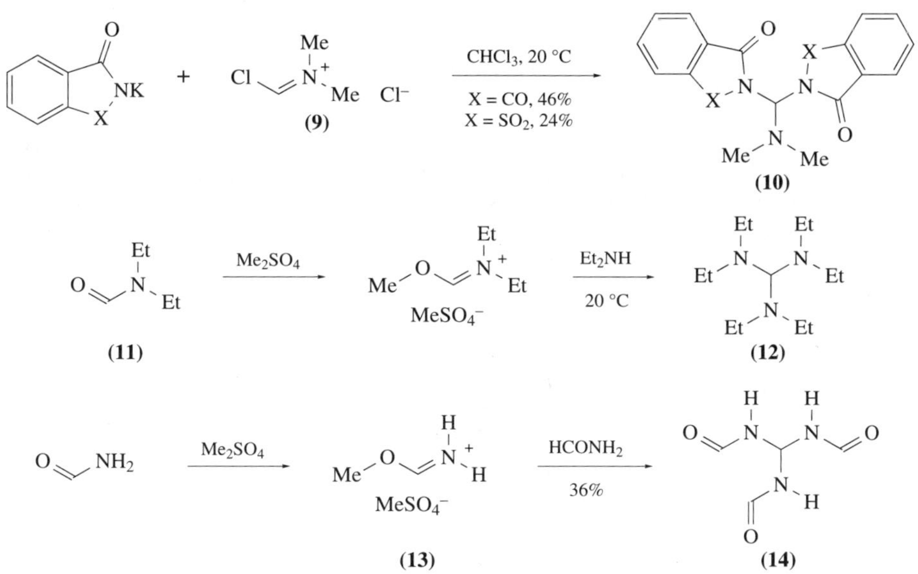

Scheme 3

(c) From guanidines and their salts. Weingarten ⟨68T2767⟩ described the displacement of a single dimethylamino group from the tetramine (**15**) by heating it with phenylacetylene. In the early 1990s, Kantlehner *et al.* used the sodium salt of alkyne to achieve a similar reaction with hexamethylguanidinium chloride: the monoadduct (**16**) giving rise to a proportion of the interesting 2:1 adduct (**17**) on distillation ⟨90CZ176⟩. This latter material could be more conveniently prepared

directly by bubbling alkyne through a mixture of the other reagents. The reaction of the same guanidinium salt with phenyllithium gave a reasonable yield of the triamine (**18**) ⟨71JOC2885⟩.

Kantlehner *et al.* have also published a route to triaminomethanes (**19**) by mixed hydride reduction of guanidinium salts ⟨83S905⟩; in this case the yield steadily declines with the increasing size of substituents on nitrogen, perhaps reflecting the stability of the products.

The alkyltitanium complex (**20**) has been found to insert into tetramethylguanidine (**21**) to give the cyclic triamine complex (**22**) in very high yield (Scheme 4) ⟨88BCJ171⟩.

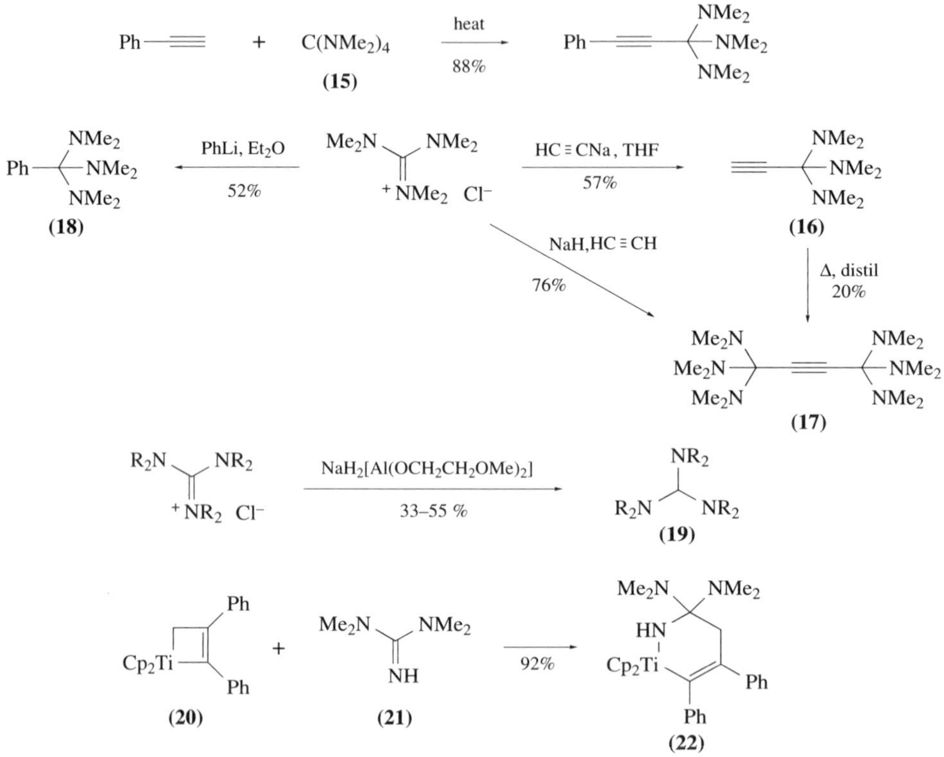

Scheme 4

(d) From ortho-*acid derivatives*. The first reported synthesis of this type came from Bredereck *et al.* who prepared triformamidomethane (**14**) by treating triethyl *ortho*-formate with formamide in the presence of an acid catalyst (Scheme 5) ⟨59CB329⟩. Follow-up studies were made in which the nature of the amide was varied and, by controlling the stoichiometry of the reaction together with the use of toluene as solvent, the yields were significantly improved ⟨60CB1398, 63CB1505⟩. The formyl groups in (**14**) could be exchanged for other acyl derivatives to give (**23**) by reaction with an appropriate acid anhydride ⟨60CB1398⟩. Urethane has been used in this context with high yields of the tricarbamate (**24**) reported ⟨67ZOR1749⟩.

Triethyl *ortho*-formate will also react with *N*-methylaniline to give the product (**5**) (Scheme 6) in moderate yield ⟨62JOC3664⟩. Triethoxyacetonitrile (**25**) reacts with pyrrolidine among other amines to give the unusual nitrile (**26**) ⟨79GEP2824028⟩.

Heating diaminoalkoxymethanes of type (**27**) gives rise to disproportionation to amide acetals (**28**) and triaminomethanes (**19**) (Scheme 7) ⟨68CB51⟩. Dithioacetals such as (**30**) react with succinimide (**29**) to give the substitution products (**31**) ⟨65IZV2179⟩; phthalimide behaves similarly.

Weisman *et al.* synthesised the interesting tricyclic examples (**34**) by reaction of formamide or acetamide acetal (**32**) with macrocyclic triamines (**33**) ⟨81TL4365⟩. They found that acetals such as (**32**) were preferable to simple *ortho*-esters. Triazaadamantane (**36**) was made by condensation of triethyl orthoformate with cyclohexanetriamine (**35**) (Scheme 8) ⟨73CB2523⟩.

(e) From alkyl halides. Clemens et al. prepared triamines (**5**) by treating the sodium salts of *N*-alkylanilines with trihalomethanes ⟨62JOC3664⟩; for this conversion difluorochloromethane was better than either chloroform or dichlorofluoromethane. This reaction may well proceed through a carbene intermediate and in support of this the pyrolysis of sodium trichloroacetate in the presence of such anilines gave (**5**) in modest yields (Scheme 9). Similar chemistry has been successful using a

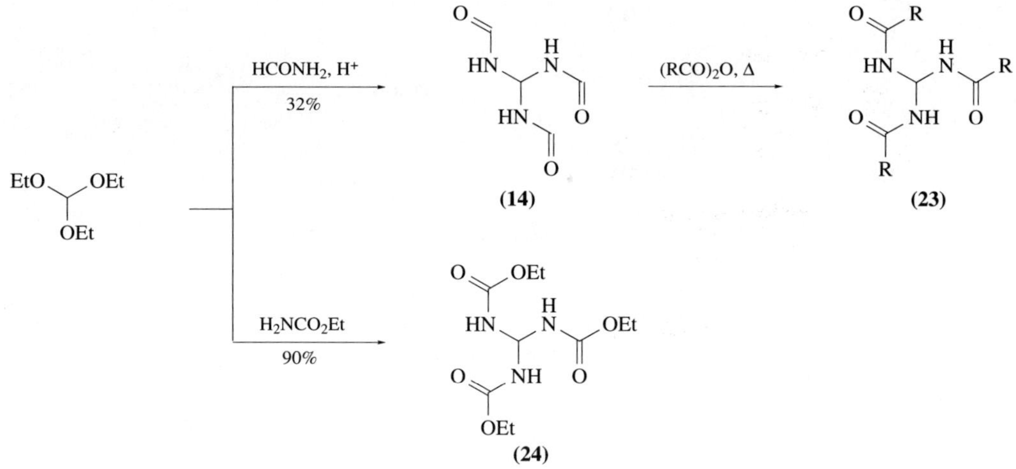

Scheme 5

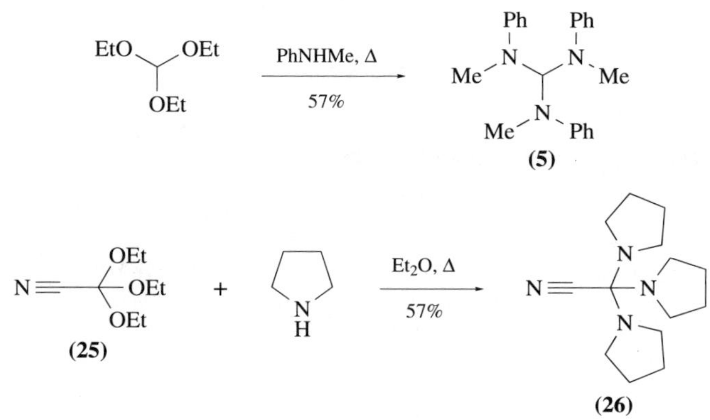

Scheme 6

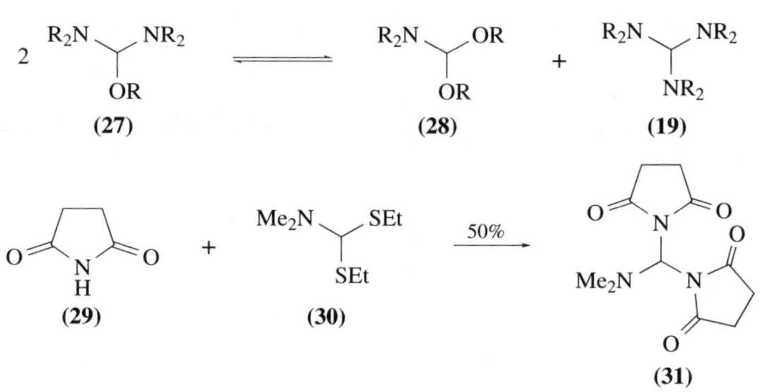

Scheme 7

pyrazole as the amine component ⟨76S798, 79LA1456⟩ although 1,2,4-triazoles ⟨69JCS(C)2251⟩ and indazoles ⟨79LA1456⟩ gave very poor yields.

(f) From titanium salts. Tetrakis(dimethylamino)titanium (**37**) reacts with DMF in ether to give a high yield of triaminomethane (**38**), whereas the corresponding acetamide derivative gave the diaminoethylene (**39**) ⟨66JA850, 66JOC2874⟩. Similarly, oxamide (**40**) gave the same product (**38**) although no yield was reported (Scheme 10) ⟨68JOC1246⟩.

(g) Addition to polyaminoethenes. Tetraaminoethenes (**41**) react with lactams and imides to give the cyclic triamines (**42**) ⟨75CB215⟩; this type of reaction is covered in a review of similar chemistry ⟨72AG(E)964⟩. Exomethylene triazinediones (**43**) have been shown to combine with some substituted

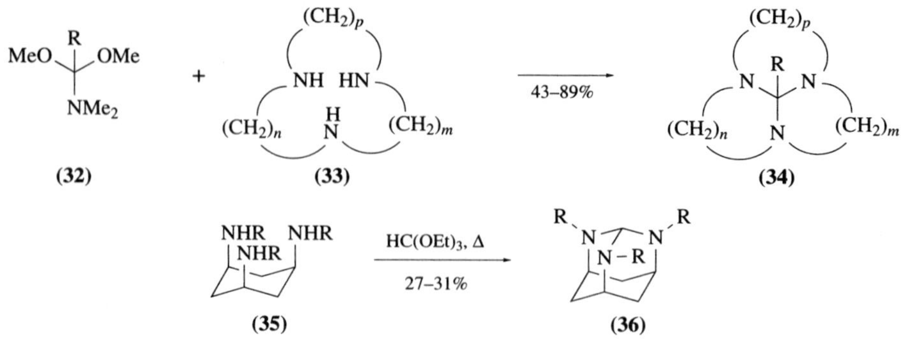

Scheme 8

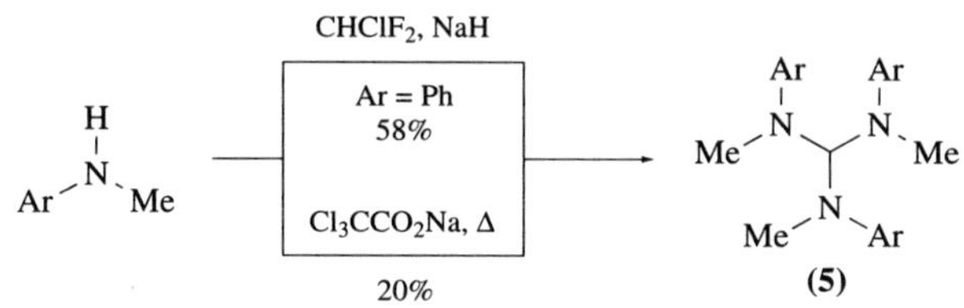

Scheme 9

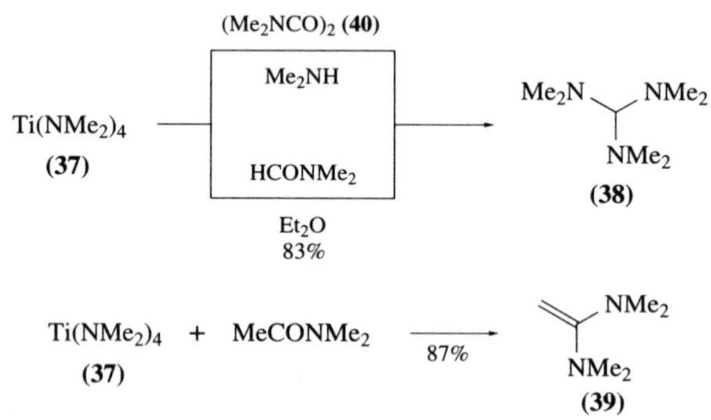

Scheme 10

hydrazines by reaction at the ring carbon giving products such as Structure (**44**) (Scheme 11) ⟨75CR(C)563⟩.

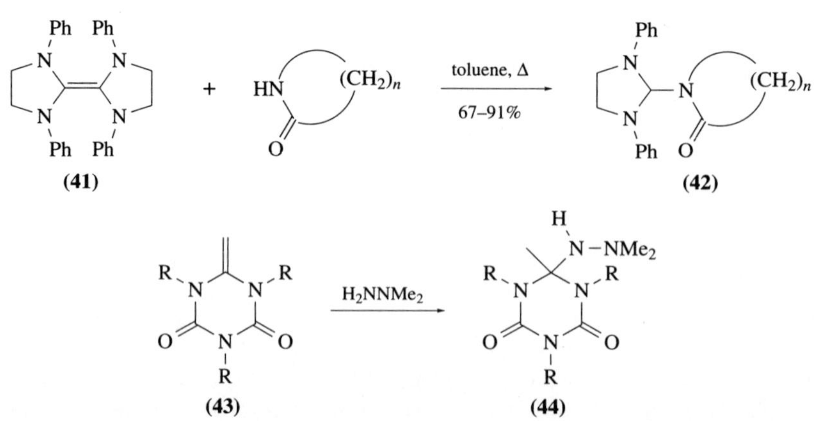

Scheme 11

(*h*) *Heterocyclisation reactions.* In an attempt to prepare an oxazoline (**46**) by pyrolysis of hydroxyethylamide (**45**), Slusarczuk and Joullié obtained the tricyclic compound (**47**) in modest yield ⟨70CC469⟩. The bis(amide) (**48**) is reported to extrude one molecule of benzimidazole on sublimation giving triamine (**49**) ⟨77JCS(P1)1162⟩, and in a similar transformation, the acylation of diamide (**50**) with chloroacetyl chloride gave the tetracycle (**51**) in high yield (Scheme 12) ⟨81JOC1571⟩.

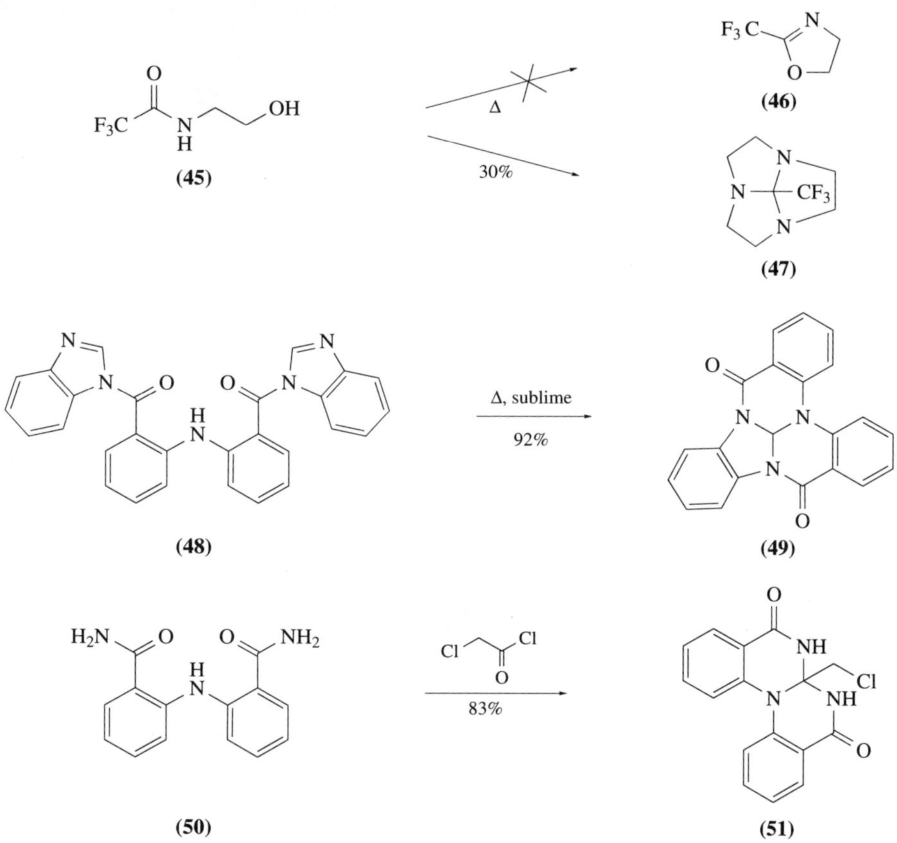

Scheme 12

(i) Radical methods. The dimerisation of triarylimidazoles by treatment with potassium ferricyanide under appropriate conditions gave the photochromic dimers (**52**) among other products (Scheme 13) ⟨66JA3825, 71JOC2262⟩. Treatment of acetonitrile with the *N*-bromosuccinimide complex (**53**) gave the trisubstituted acetonitrile (**54**) ⟨88ACS(B)666⟩; other aminoacetonitrile derivatives (**55**) and (**56**) gave better yields of similar products.

(j) From other triamines. As part of a wider study, Katritzky *et al.* have trapped the anion of tris(benzotriazolyl)methane (**57**) with a variety of electrophiles such as alkyl halides and alkyl chloroformates (Scheme 14) ⟨90S666⟩.

(k) From isocyanates. Working independently, two research groups have found that DMF reacts with phenyl isocyanate to give the bicyclic triamine (**58**) ⟨68JOC3928, 68JOC3931⟩. According to the latter report this product was also obtained from amidine (**59**). Since then a number of similar reactions have been described: among other similar substrates, lactams ⟨73JOC2614⟩, amidines ⟨68CB3002, 69CB931⟩, amidinium salts ⟨71LA(751)145, 80HCA1958, 82CB1721⟩, tetraaminoethylenes ⟨71LA(748)1, 74CB1931⟩ and 1,1-diaminoethylenes ⟨77CR(C)321⟩ have all been utilised in place of DMF (Scheme 15). Thiocyanates also reacted similarly ⟨75CB1142⟩. The cyclic selenourea (**60**) has been used as a source of the carbene (**61**) (which can also be formulated as an amidinium ylide) ⟨83CB2068⟩; this reacts with phenyl isocyanate to afford the product (**62**). Reduction of isocyanates *in situ* has been used to provide the source of the spiro carbon atom in (**63**) ⟨85CB3959⟩.

6.05.1.1.2 1,1,1-Trinitroalkanes

(i) Introduction

Much work has been done in this area of chemistry but the special advantages offered by these materials in the field of propellants and explosives represent a significant hazard in the hands of the

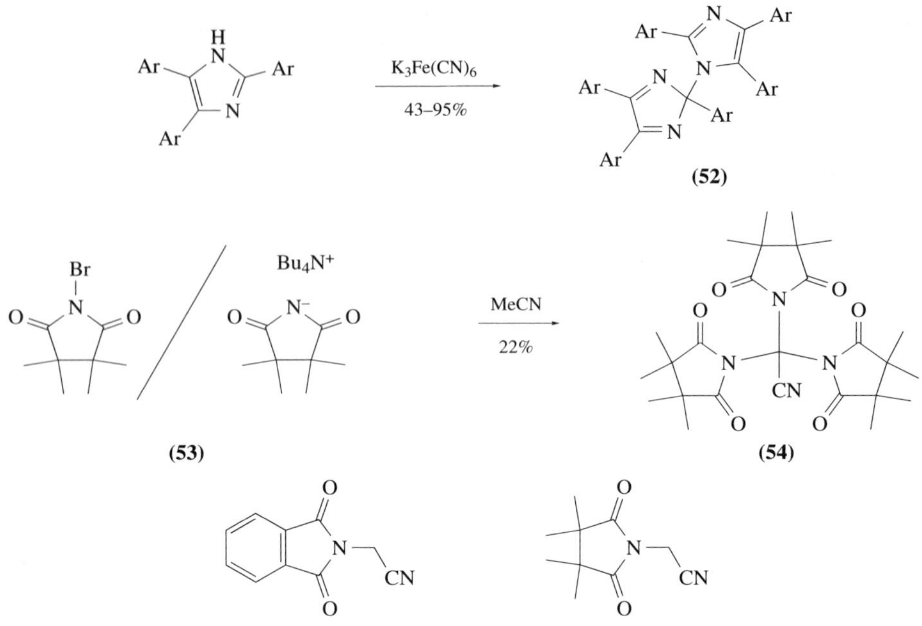

(52) (53) (54) (55) (56)

Scheme 13

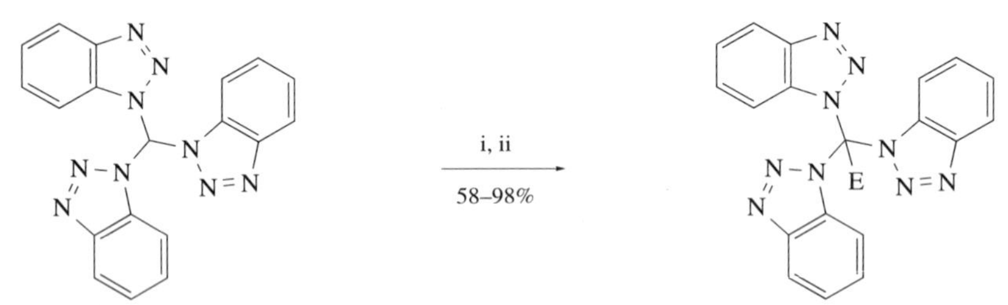

(57)

i, BuLi, THF; ii, EX
E = $PhCH_2$, CO_2Et, Bu^n *inter alia*

Scheme 14

unwary. Several reviews have been published, the most recent being in the Patai series ⟨B-70MI 605-02⟩; others cover more general aspects of nitroalkane chemistry as well as their synthesis ⟨59UK484, 63T(S)155, 63T(S)177, 64CRV19, 66UK1740⟩.

The presence of three strongly electron-withdrawing nitro groups attached to a single carbon atom has a profound influence on the chemistry of such compounds and even though not all the nitro groups are coplanar, they are still prone to elimination and substitution reactions as well as rendering β-heteroatoms virtually non-basic.

(ii) Preparation

There are two fundamental approaches to the synthesis of such compounds: the introduction of an intact trinitromethyl group (most commonly as a nucleophile but also as a radical) and the stepwise introduction of the required nitro groups.

(a) Addition of an intact trinitromethyl group. The trinitromethyl anion adds readily to Michael acceptors as shown in Equation (1). Such reactions are typically carried out in protic solvents and in the absence of added base which can give rise to complicating by-products. Kaplan and Kamlet report that for addition to acrylates the optimum pH is around 3.5 ⟨62JOC780⟩. Addition occurs to unsaturated ketones ⟨68JOC1247⟩, aldehydes ⟨68JCED437⟩, phosphonates ⟨78MI 605-01⟩, oximes

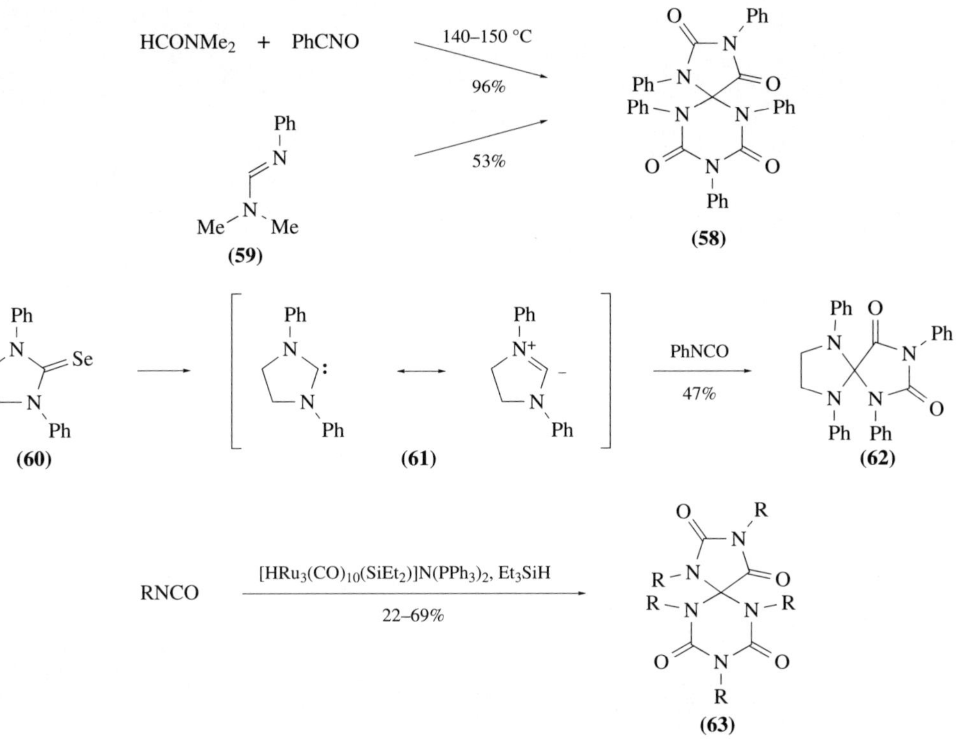

Scheme 15

⟨80MI 605-02⟩, lactones ⟨86JHC81⟩ and to acrylonitrile ⟨71ZOR30⟩. However, according to Kaplan, addition to symmetrically disubstituted Michael acceptors does not occur ⟨B-70MI 605-02⟩.

(1)

EWG = CO_2R, CHO, COR, $PO(OR)_2$, CHNOH, CN

A detailed study of the alkylation of the silver salt of trinitromethane has been made ⟨63T(S)177⟩, and although this method worked well for simple alkyl, allyl and benzyl groups it failed for other electrophiles such as haloacetates. One particular complication is the occurrence of competing *O*-alkylation ⟨68IZV447, 77IZV136⟩. Nevertheless, variations on this method, including the use of other salts, have allowed reactions to occur with diiodomethane ⟨71ZOR1304⟩, 1-bromoadamantane ⟨76KFZ35⟩, α-chloroamines ⟨81IZV2146⟩, ferrocene derivatives ⟨86ZOR1393⟩, triazolyl diazonium salts ⟨72KGS713⟩ and α-chloroethers ⟨90SC3303⟩ as illustrated in Scheme 16. At the beginning of the 1990s, a study of the phase transfer catalysed methylation of the potassium salt of trinitromethane was published ⟨90IZV1816⟩ in which the addition of a crown ether or polythylene glycols was shown to be advantageous. This salt has also been used to substitute the pyridinium salt **(64)** in modest yields ⟨83IZV2655, 90TL7379⟩.

Iodotrinitromethane and tetranitromethane are methylated in dipolar aprotic solvents (see Scheme 17) either via intermediate solvent alkylated species or perhaps through preliminary homolytic dissociation ⟨70ZOR189, 74ZOR2003⟩.

Trinitromethane is methylated by diazomethane ⟨70IZV948⟩. Diazo compounds such as **(65)** insert into bromo- or iodotrinitromethane to give trinitro derivatives **(66)** ⟨71ZOR1126, 77ZOR1559⟩. Tri- and tetranitromethane react with other nitroalkanes to give trinitromethyl derivatives such as **(67)** ⟨73IZV122⟩ and **(68)** ⟨67USP3316311⟩, and, in the presence of a silylating agent, with **(69)** to give ethers of **(70)** (Scheme 18) ⟨89MI 605-01⟩.

Shechter and Cates, Jr. found that trinitromethane reacted with enol ethers **(71)** at the oxygen-bearing carbon to give **(72)** ⟨61JOC51⟩; this reaction is also reported for enol acetates in alcoholic solvents giving products of type **(73)** (Scheme 19) ⟨69IZV2566⟩.

Trinitromethane reacts with formaldehyde to give alcohol **(74)** ⟨50JA5329, 60JOC2069⟩. In the

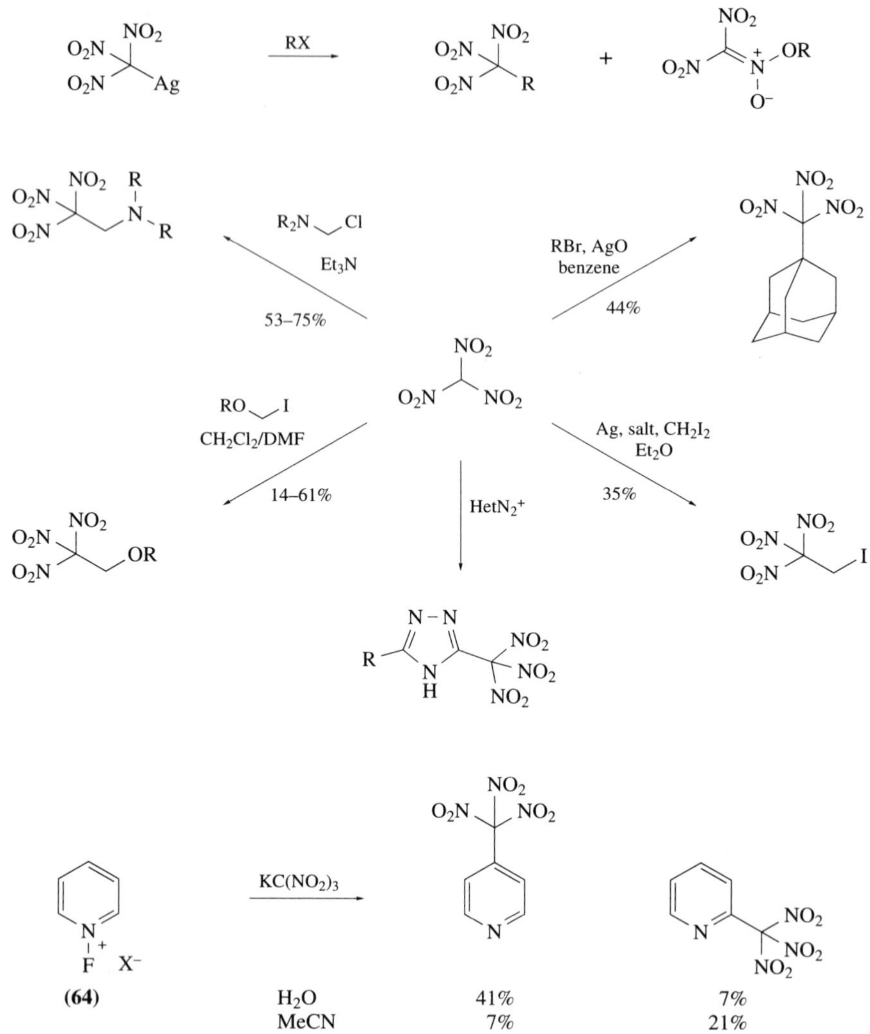

Scheme 16

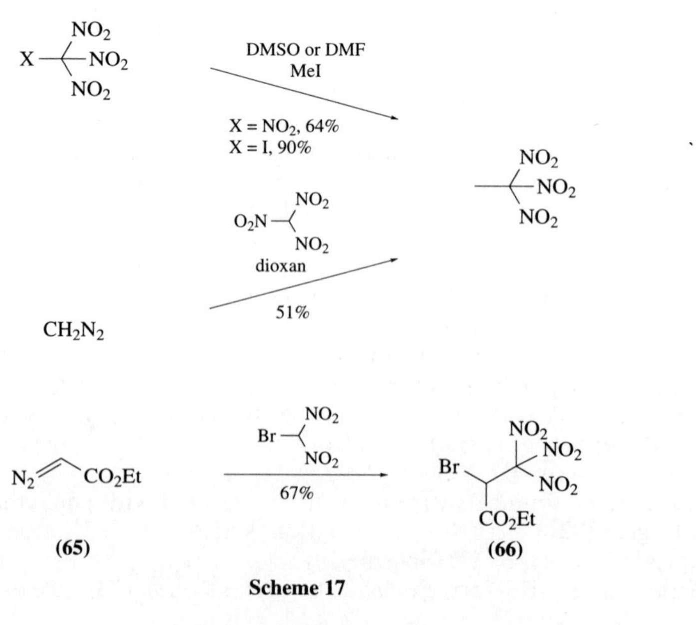

Scheme 17

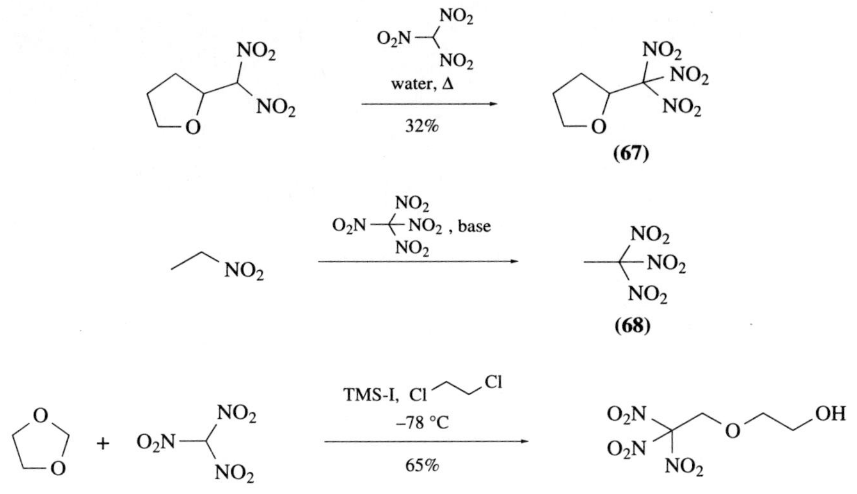

(67)

(68)

(69)

(70)

Scheme 18

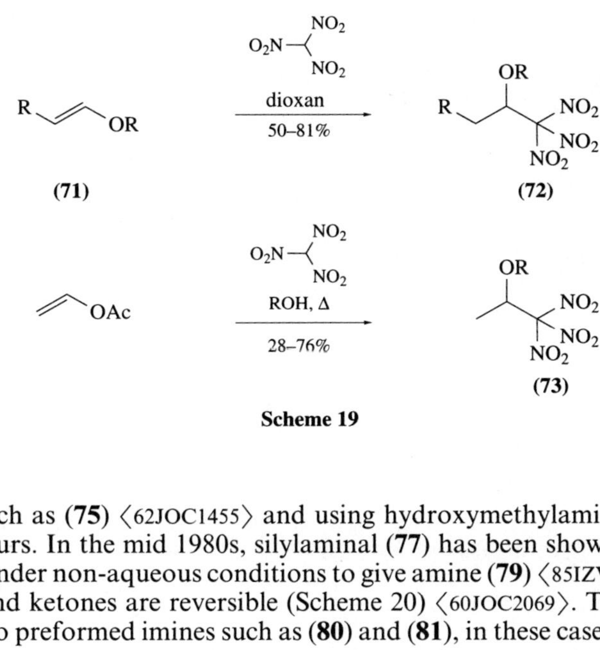

(71)

(72)

(73)

Scheme 19

presence of amines such as (**75**) ⟨62JOC1455⟩ and using hydroxymethylamides (**76**) ⟨61JOC391⟩, the Mannich reaction occurs. In the mid 1980s, silylaminal (**77**) has been shown to react with silylated trinitromethane (**78**) under non-aqueous conditions to give amine (**79**) ⟨85IZV2635⟩. Similar additions to higher aldehydes and ketones are reversible (Scheme 20) ⟨60JOC2069⟩. Trinitromethane has also been reported to add to preformed imines such as (**80**) and (**81**), in these cases in halocarbon solvents ⟨68IZV2379, 69IZV2309⟩.

Polynitromethanes have been added to various substrates via a radical mechanism to provide trinitromethane derivatives. Tetranitromethane is reported to form charge transfer complexes with electron-rich aromatics such as anthracene; subsequent photolysis can give rise to trinitroarenes (**82**) (Equation (2)) ⟨85JOC5245⟩. Other aromatic systems can lead to products such as (**83**) ⟨86RTC278⟩ and (**84**) ⟨80MI 605-01⟩. Products such as those of direct nitration may also be formed and the exact course of the reaction can be difficult to predict ⟨86RTC278⟩.

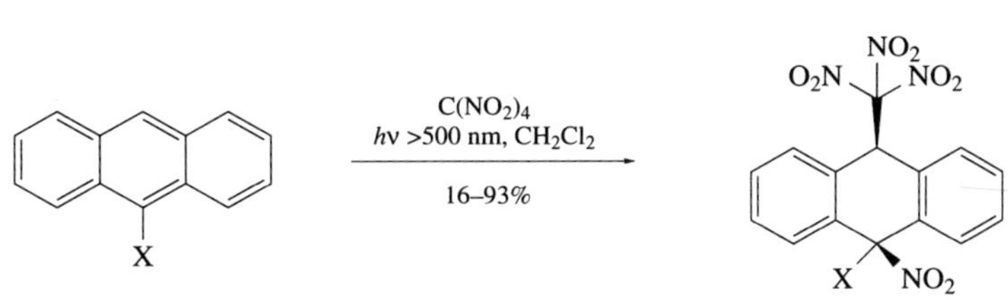

(82)

(2)

Tetranitromethane will add across alkenes such as (**85**) ⟨69ZOR2246⟩ and (**86**) ⟨86RTC286⟩ and in

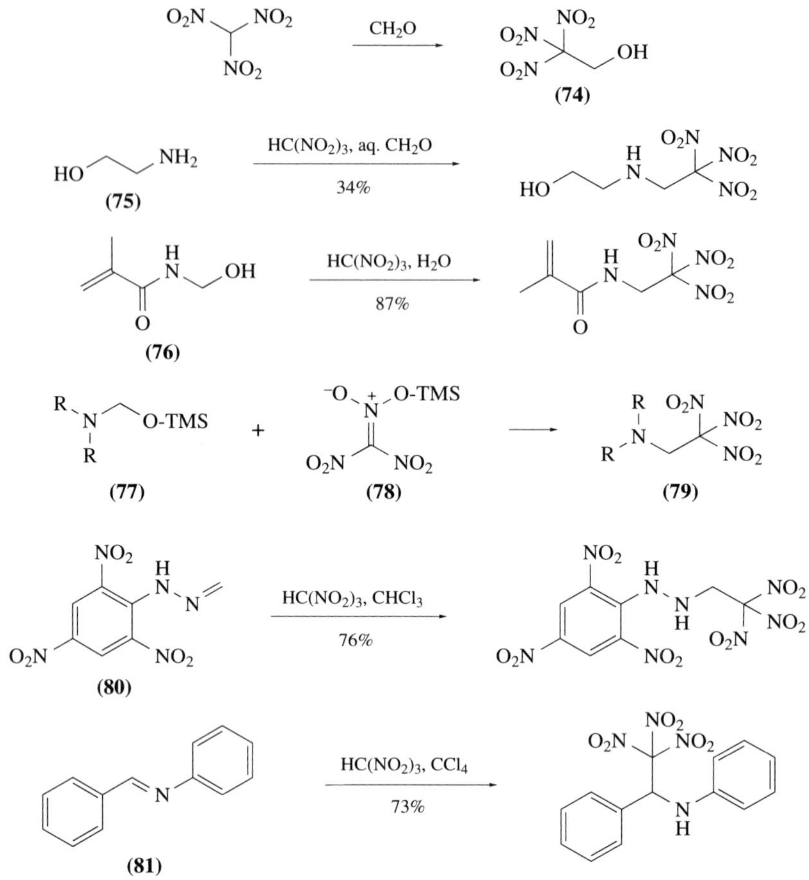

Scheme 20

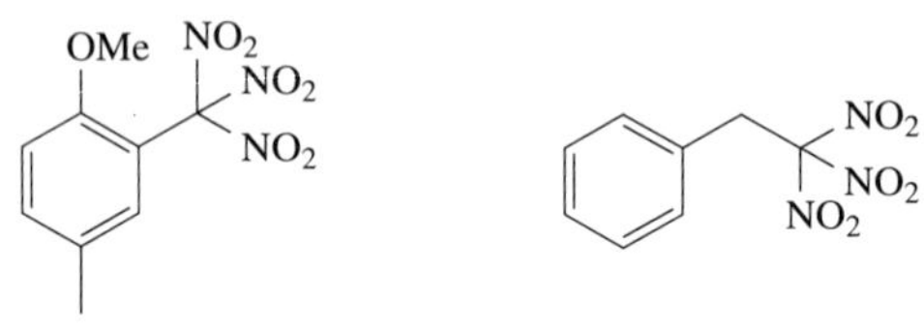

(83) **(84)**

such cases light is not always required (Scheme 21). Insertion into a C—H bond of THF afforded **(67)** ⟨73IZV1149⟩; a similar reaction occured with iodotrinitromethane (Scheme 22) ⟨72IZV490⟩.

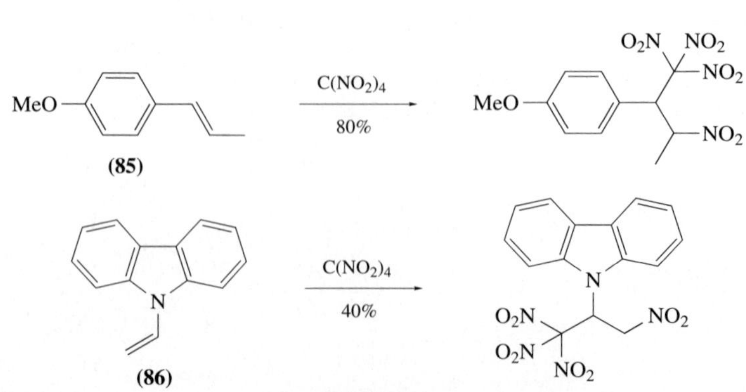

Scheme 21

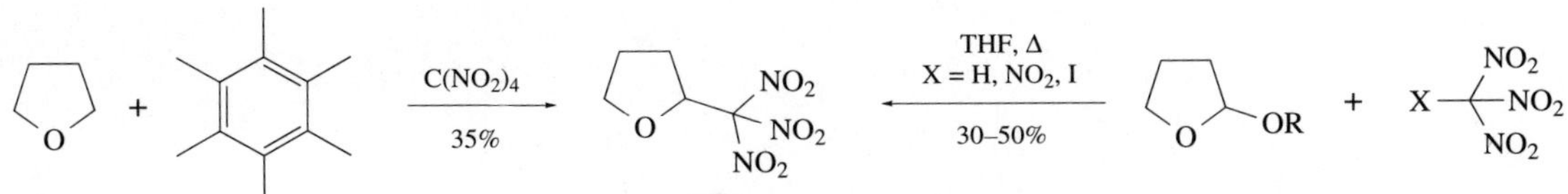

(67)

Scheme 22

Electrolysis of salts of trinitromethane as shown in Scheme 23 leads to radicals which react with cyclic ethers ⟨72IZV2603⟩ or with toluene ⟨72IZV1717⟩ and other aromatic systems ⟨82IZV2637⟩.

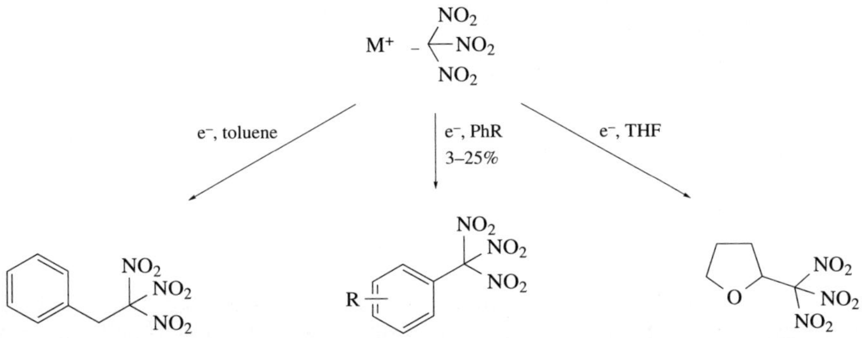

(67)

Scheme 23

The mercury salt (**87**) has been shown to add across the double bonds of allyl alcohol ⟨62IZV272⟩, methyl acrylate ⟨69IZV1845⟩ and vinyltrimethylsilane (Scheme 24) ⟨73RZC1243⟩.

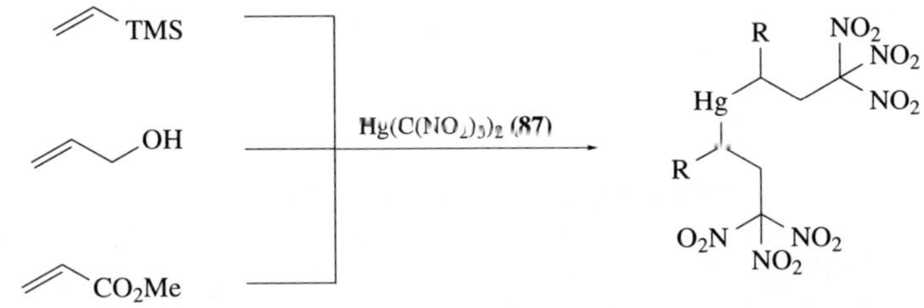

Scheme 24

Trinitroacetonitrile undergoes cycloaddition reactions with TMS azide, diazo compounds and nitrile oxides to give trinitromethyl-substituted tetrazoles (**88**) ⟨81JHC1477⟩, 1,2,3-triazoles (**89**) ⟨87ZOR2624, 88ZOR644⟩ and 1,2,4-oxadiazoles (**90**) ⟨86ZOR2618⟩ respectively.

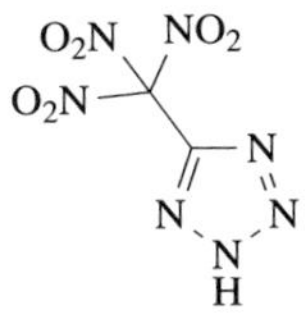

(88)

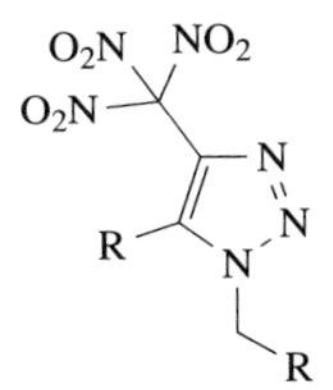

(89)

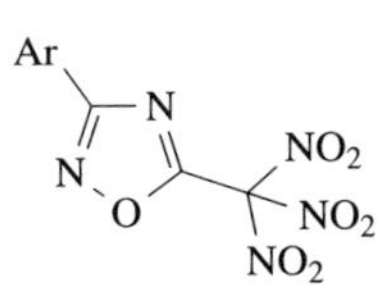

(90)

There have been many reports of the cycloaddition of trinitromethane derivatives with alkenes (Equation (3)) to give products of type (**91**). Tetranitromethane has been shown to react directly with cyclohexene ⟨67ACS(C)1392⟩, ethylene ⟨66ZOR1902⟩, vinyl ethers ⟨69ZOR220⟩ and butadiene ⟨69ZOR1313⟩. Bromo- and iodotrinitromethane have taken part in similar reactions ⟨67ZOB1163, 68IZV621, 73ZOR269⟩. Trinitromethane can be *O*-silylated ⟨73ZOR896⟩ and the resulting adduct reacts with styrene to afford (**92**). *O*-methylation with diazomethane followed by reaction with butadiene yields (**93**) (Scheme 25) ⟨68ZOR231⟩.

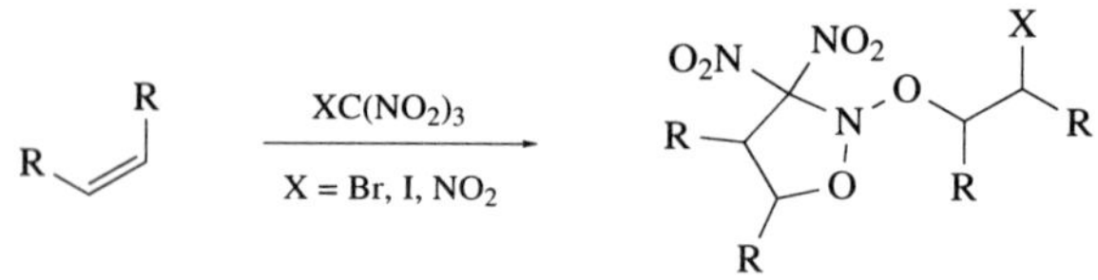

(91) (3)

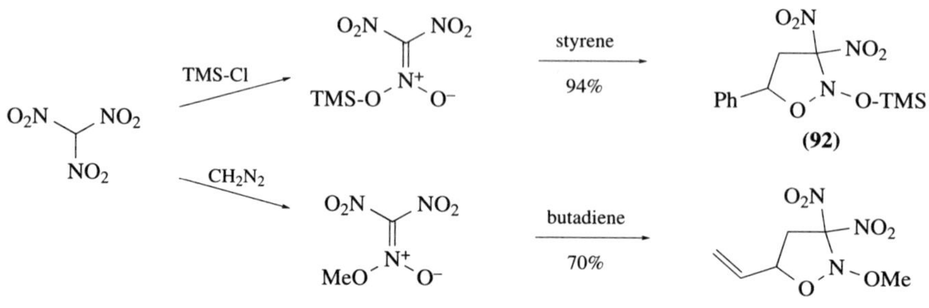

(92)

(93)

Scheme 25

(b) Successive nitration reactions. The trinitromethyl group has been assembled by successive nitration from various starting materials. For example, trinitromethane itself can be made on a commercial scale by treatment of acetylene with concentrated nitric acid in 74% yield ⟨63T(S)155⟩. Cyanoacetic acid has been converted to trinitroacetonitrile in good yield using a mixture of nitric acid and sulfur trioxide ⟨62T79⟩. Nitrile oxides (**94**) react with dinitrogen tetroxide to give aryl derivatives (**95**) (Equation (4)) ⟨90IZV1620, 90IZV1623⟩. Oximes have been converted to either di- ⟨88BCJ2927⟩ or trinitromethanes ⟨60IZV1783, 70KGS590⟩ using nitrogen dioxide.

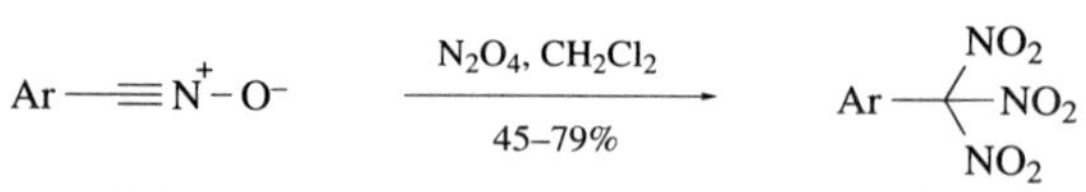

(94) (95) (4)

Dinitromethanes may be prepared as precursors to their trinitro counterparts; Kaplan and Shechter used silver nitrate/sodium nitrite on the silver nitronates (**96**) ⟨61JA3535⟩. Less conveniently, chloronitromethanes can be nitrated ⟨64CRV19⟩ and tetranitromethane treatment of nitroalkanes has also been used ⟨61USP2991315, 67USP3316311⟩. Other reagents have been used for the further nitration of dinitro analogues, e.g., nitryl fluoride (**97**) ⟨71IZV1594⟩. Trifluoroacetic anhydride (TFAA) and nitric acid react with dinitroalkenes such as (**98**) but less successfully with mononitro analogues such as (**99**) (Scheme 26) ⟨92JOC3026⟩.

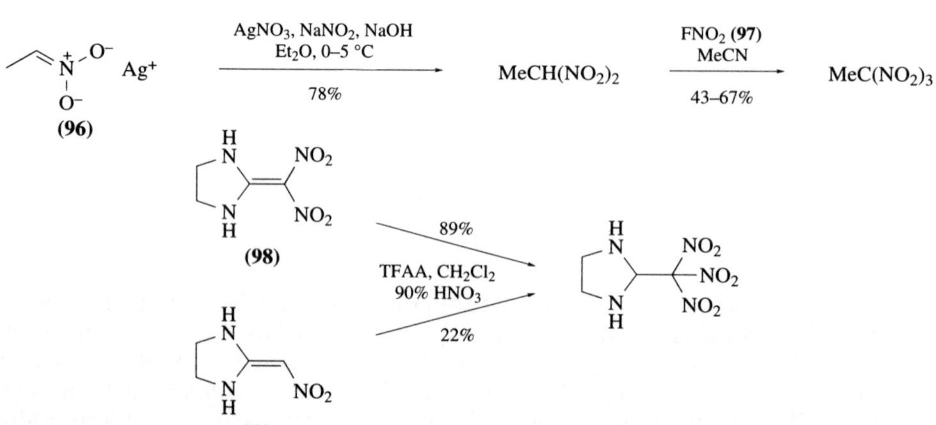

(96)

(98)

(99)

Scheme 26

6.05.1.1.3 *Mixed nitro and other functions*

One nitro group in trinitroethane has been displaced by azide in DMF as shown in Scheme 27 ⟨84BRP2123829⟩. The same product was obtained by electrolysis of the sodium salt of dinitroethane in dilute sodium azide ⟨75USP3883377⟩ and other examples have been similarly prepared ⟨90MI 605-01⟩.

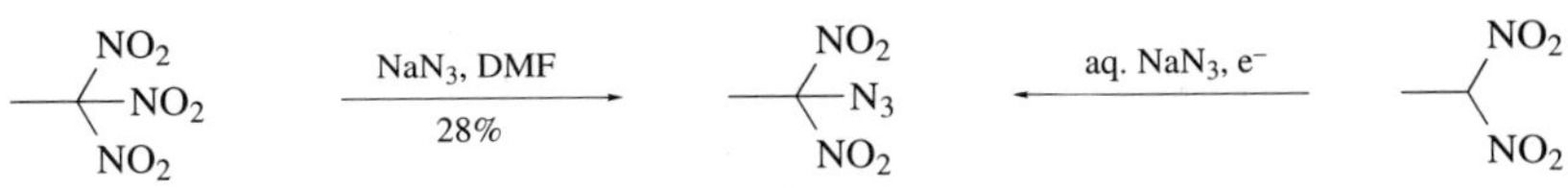

Scheme 27

6.05.1.2 Functions Bearing Three Phosphorus Atoms

6.05.1.2.1 *Introduction*

Functional groups of this type fall into two main classes: those with tri- or pentavalent phosphorus which are predominantly used as ligands in organometallic chemistry (e.g., ⟨92OM26⟩), and tris (phosphonato)methanes which have been used as surfactants (e.g., ⟨77USP4020091⟩).

6.05.1.2.2 *Functions bearing tri- or pentavalent phosphorus*

(i) From functions bearing two phosphorus groups

The method first reported by Issleib and Abicht for the reaction of methylenebisphosphines with a chlorophosphine has been subsequently much utilised to prepare ligands for organometallic complexes (Scheme 28) ⟨70JPR456⟩. The original workers described the reaction of lithium salts of bisphosphine (**100**) and its corresponding bisphosphine oxide (**101**) with diphenylphosphinyl chloride to form tris(phosphino)methanes in good yields. Karsch *et al.* extended the work to include alkyl substituents, preparing methylphosphines (**102**) ⟨79AG(E)484, 79ZN(B)1171⟩. Grim and Walton report phosphine sulfides (**103**) made by the same basic method ⟨80PS(9)123⟩; the paper contains no experimental details even though a report from the same group mentions the preparation of asymmetric tris(phosphino)methane (**104**) ⟨86IC2699⟩. In this case, the reaction was carried out on transition metal carbonyl complexes of the carbanion of bis(phosphino)methane (**105**) to improve the chemoselectivity ⟨82CC286⟩.

(ii) By further reaction of tris(phosphino)methanes

Tris(phosphino)methanes have been modified at phosphorus to produce a number of oxides, sulfides and selenides in various combinations. The first report was by Issleib and Abicht ⟨70JPR(312)456⟩ who converted tris(diphenylphosphino)methane (**106**) into its trisulfide (**107**) by heating with sulfur. This method ⟨82CB818⟩ failed to produce the alkyl analogue (**108**) which could, however, be prepared from a bis(thiophosphinoyl)methane by the standard method (Scheme 29).

Grim *et al.* report the modification of a variety of tris(phosphino)methanes: the mild oxidation of phosphine (**106**) to phosphine oxide (**109**) avoids the cleavage of one of the carbon–phosphorus bonds which occurs under harsher conditions ⟨86IC2699⟩. This method was then applied to a number of phosphine sulfides and selenides. Red selenium was also used to convert phosphine sulfide (**110**) into (**111**) although an earlier paper mentions that carbon–phosphorus bond cleavage occurred in similar reactions ⟨80PS(9)123⟩. Grim also reports that salts of phosphines react more readily with selenium because of a reduction in steric crowding ⟨87PS(38)79⟩. The functionalisation of phosphorus in metal carbonyl complexes by oxidation using either hydrogen peroxide or sulfur has been claimed although no details are given ⟨82CC286⟩. The phosphine imine (**112**) was prepared in excellent yield by reaction of phosphine (**110**) with *p*-tolyl azide (Scheme 30) ⟨91CC979⟩.

Karsch *et al.* describe the methylation of the lithium salt of tris(dimethylphosphino)methane

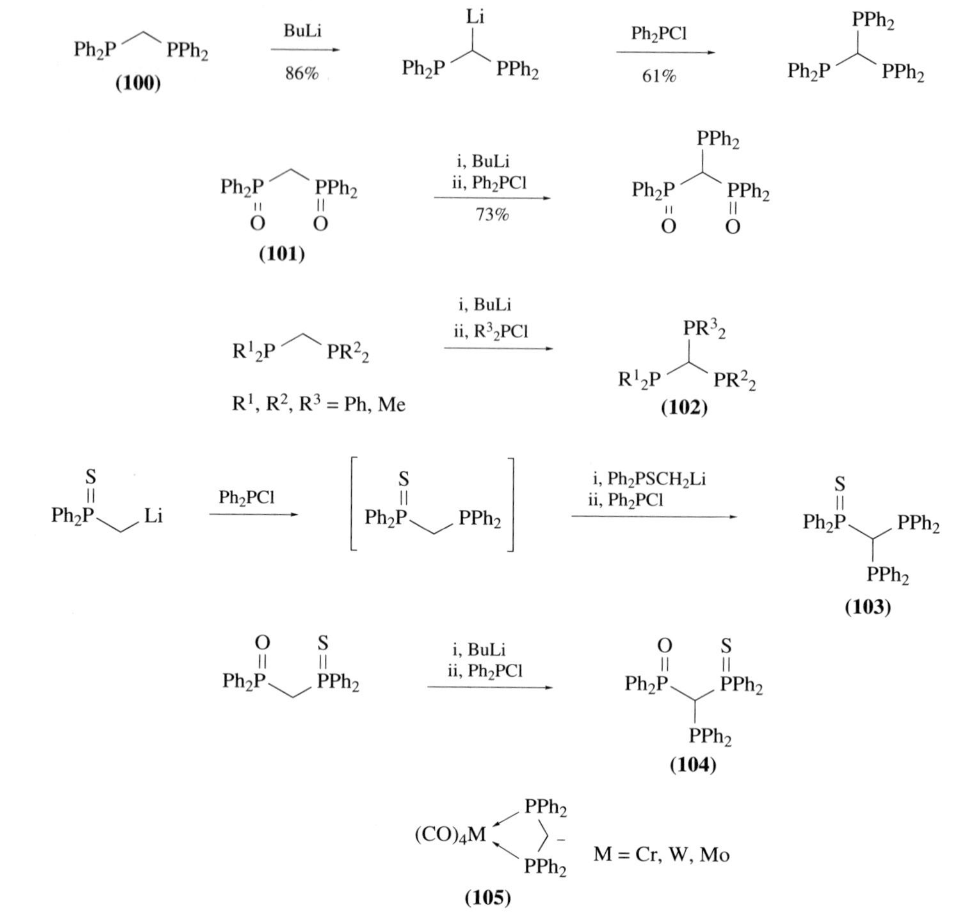

Scheme 28

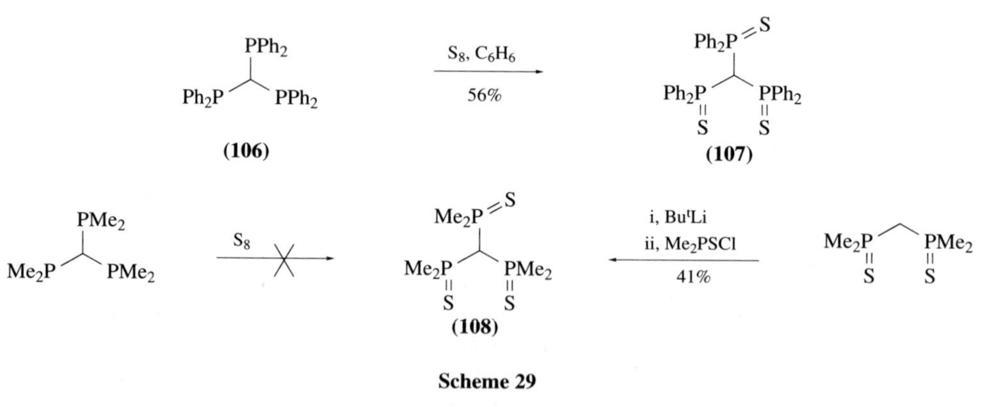

Scheme 29

(**113**): using low temperatures and ether solvents alkylation occurs at carbon to give (**114**) whereas in pentane with TMEDA alkylation takes place on phosphorus to afford the ylide (**115**) (Scheme 31) ⟨84ZN(B)1518⟩.

(iii) From phosphaalkynes

Phosphaalkynes, their oligomers and derived metal complexes have been used to prepare complex polycyclic analogues of tris(phosphino)methanes. This chemistry has been reviewed ⟨88AG(E)1484,

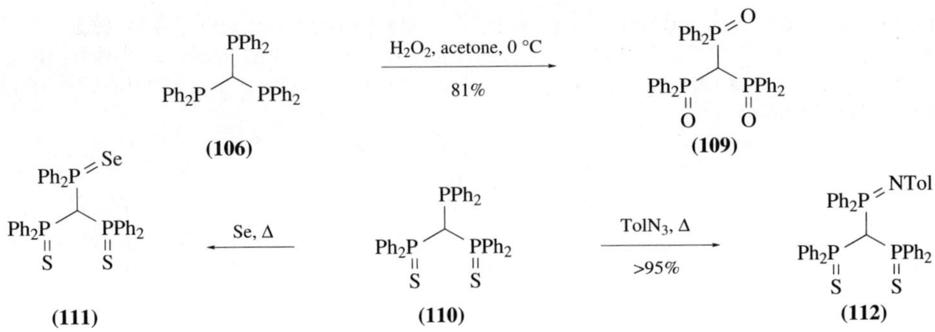

Scheme 30

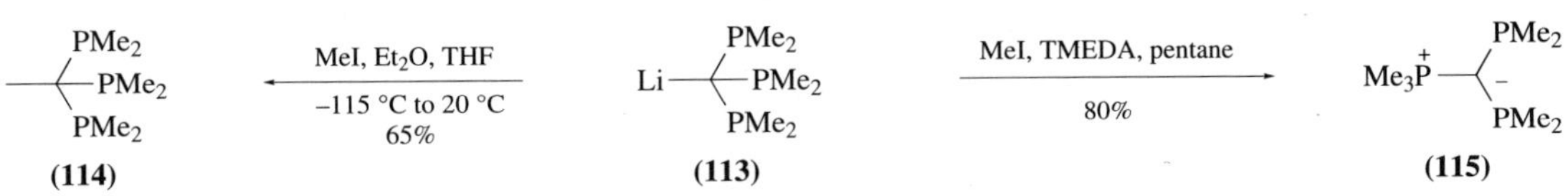

Scheme 31

90CRV191⟩. Wettling *et al.* report that the prolonged pyrolysis of **(116)** gave the tetraphosphacubane derivative **(117)**, albeit in very low yield ⟨89AG(E)1013⟩. Prior reaction of the alkyne to form a zirconium complex **(118)** greatly improves the conversion ⟨92AG(E)758⟩. The phosphacubane **(117)** has been functionalised at a single phosphorus vertex using a number of reagents ⟨92AG(E)879, 93JOC4105⟩. The zirconium complex **(118)** has been treated with phosphorus trichloride to form the bicyclic compound **(119)** (Scheme 32) ⟨91AG(E)207⟩.

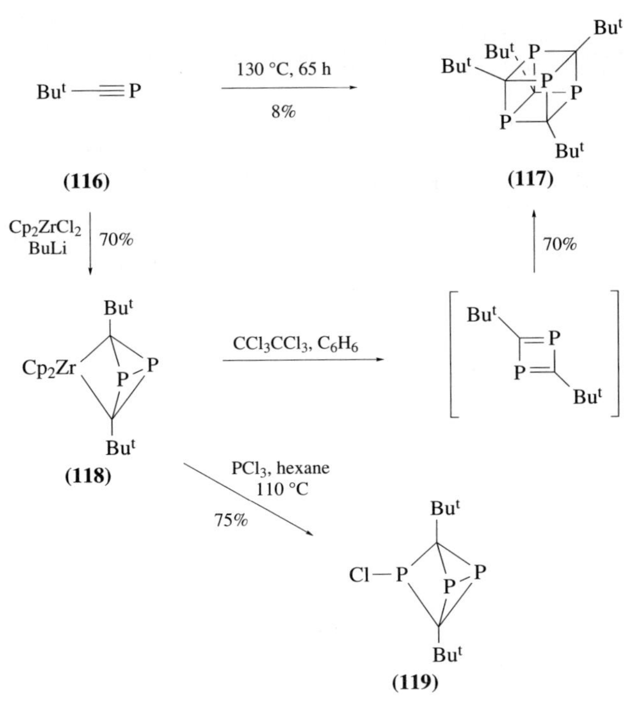

Scheme 32

Aluminum chloride catalysed tetramerisation of the phosphaalkyne (**116**) yielded polycyclic derivatives such as (**120**) ⟨92AG(E)1055⟩. Tungsten ⟨91CC1305⟩, manganese ⟨93AG(E)1424⟩, iron ⟨89AG929⟩ and vanadium complexes ⟨87AG(E)908⟩ have been converted to related polycyclic systems, for example (**121**) (Scheme 33).

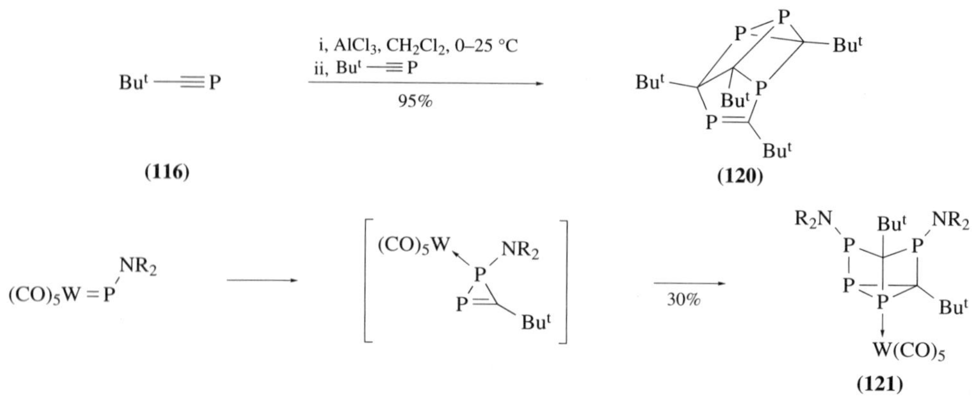

Scheme 33

Phosphaalkyne (**116**) reacts with phosphorus ⟨88AG(E)389⟩ and arsenic ⟨88AG(E)709⟩ heterocycles (**122**). In the former case a 2 : 1 cycloaddition product is obtained whereas in the latter the reaction leads to a 3 : 1 cycloaddition product minus the elements of benzonitrile (Scheme 34).

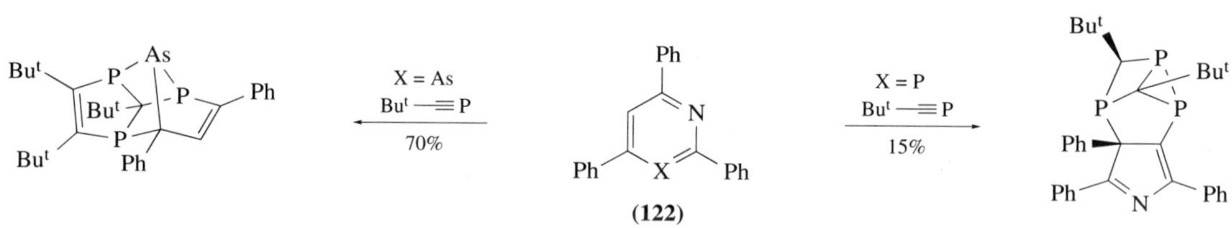

Scheme 34

(iv) From phosphorus analogues of cyclopentadiene

Di- and triphosphacyclopentadiene anions oligomerise under the influence of transition metals to give complex polycyclic products such as (**123**) (Equation (5)) ⟨89CC1046, 92JOM(433)C11, 93CC311⟩.

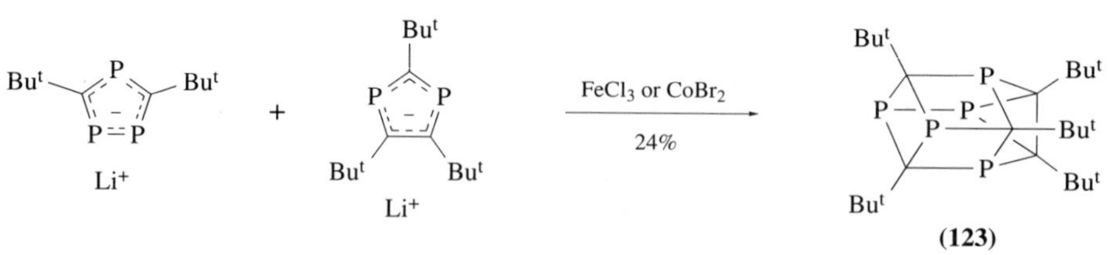

(5)

6.05.1.2.3 Functions bearing heptavalent phosphorus

The first report of this type of function utilised the reaction of triethyl phosphite with benzoyl peroxide in chloroform to form the tris(phosphonate) (**124**) ⟨63JCS1527⟩; the same product was obtained from the benzoylphosphonate (**125**). The authors report that neither carbon tetrachloride nor diethyl ether were satisfactory solvents for this reaction. Several patents claiming similar compounds as detergents appeared subsequent to this report ⟨69USP3471406, 75USP3892676, 77USP4020091, 84USP4440646⟩. More recently, generation of the quinonoid **(126)** and subsequent reaction with diethyl phosphite in the presence of a base has been used to prepare the tris(phosphonate) **(127)** in good yield (Scheme 35) ⟨90PS(47)7⟩. Gross *et al.* have developed this basic method further to produce compounds bearing differently substituted phosphorus functions, e.g., **(128)** ⟨91PS(62)35⟩.

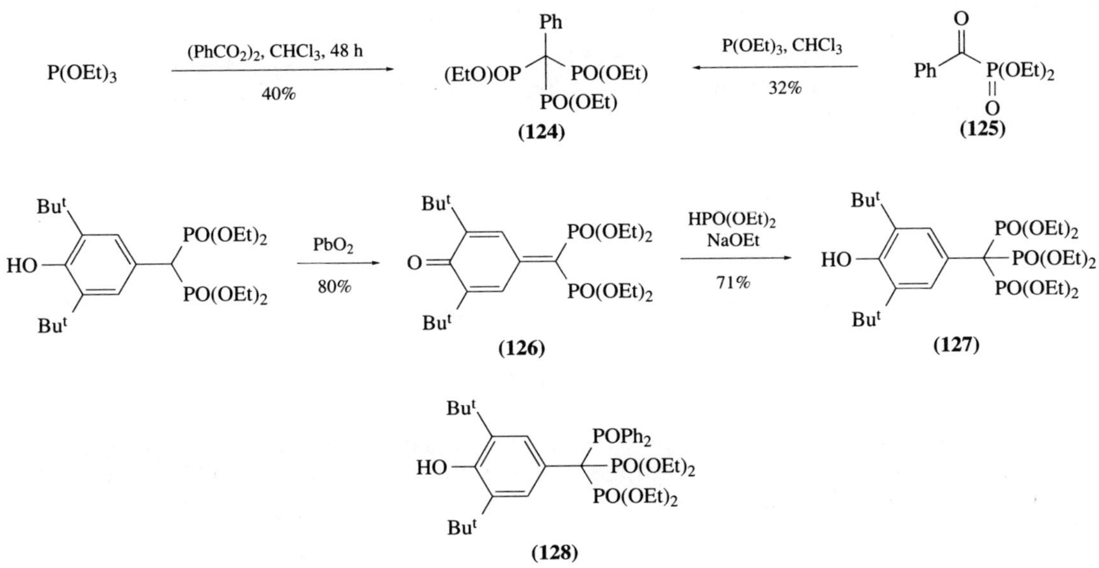

Scheme 35

6.05.1.3 Functions Bearing Three Arsenic Atoms

This author could locate only two reports referring to this functional group. Tris(diphenylarsino)methane **(130)** was prepared by reaction of the lithium salt of bis(arsine) **(129)** with chlorodiphenylarsine ⟨82OM1114⟩. The product was used in a study of its complexation properties. In the second report ⟨93AG(E)103⟩ tris(trimethylsilyl)phosphine was desilylated with butyllithium and the resulting lithium salt **(131)** treated with the arsaalkene **(132)** in the presence of cobalt(II) chloride to give tetraarsacubane **(133)** in moderate yield. The reaction probably proceeds via an intermediate arsaalkyne (Scheme 36).

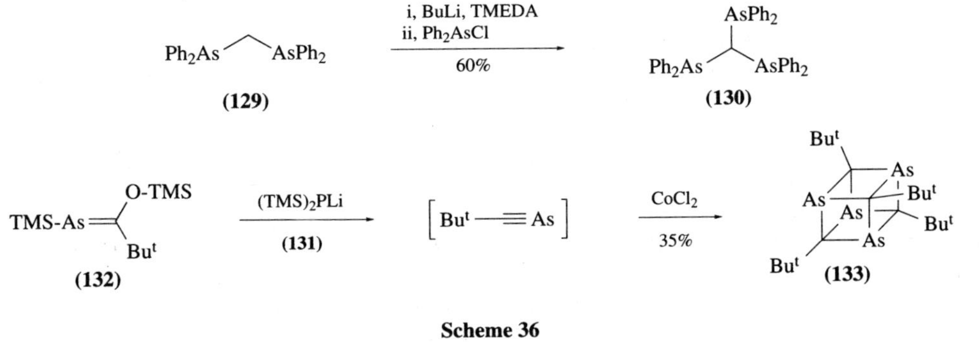

Scheme 36

6.05.1.4 Functions Bearing Three Antimony or Bismuth Atoms

This author was unable to locate any references to the functional groups of either of these types in the literature.

6.05.1.5 Functions Bearing Nitrogen and Other Group 15 Elements

6.05.1.5.1 Functions bearing two nitrogen atoms and one phosphorus atom

Gross and Costisella prepared compound (**135**) from the amidinium salt (**134**) by reaction with the sodium salt of diethyl phosphite ⟨69JPR925⟩; Structure (**135**) was also made by treating uronium salts (**136**) with diethyl phosphite (Scheme 37) ⟨88ZOB2167⟩.

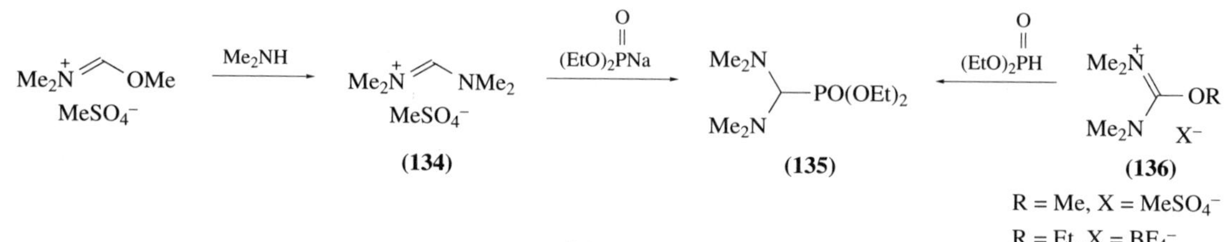

Scheme 37

The iminium salt (**138**), which was prepared from phosphonate (**137**) ⟨71LA(750)44⟩, reacted readily with a variety of amines to give the phosphonates (**139**) or with arylsulfonamides to give (**140**). Similarly, ammonia was added across the imine bond of phosphonates (**141**) (Scheme 38) ⟨91PS(63)95⟩.

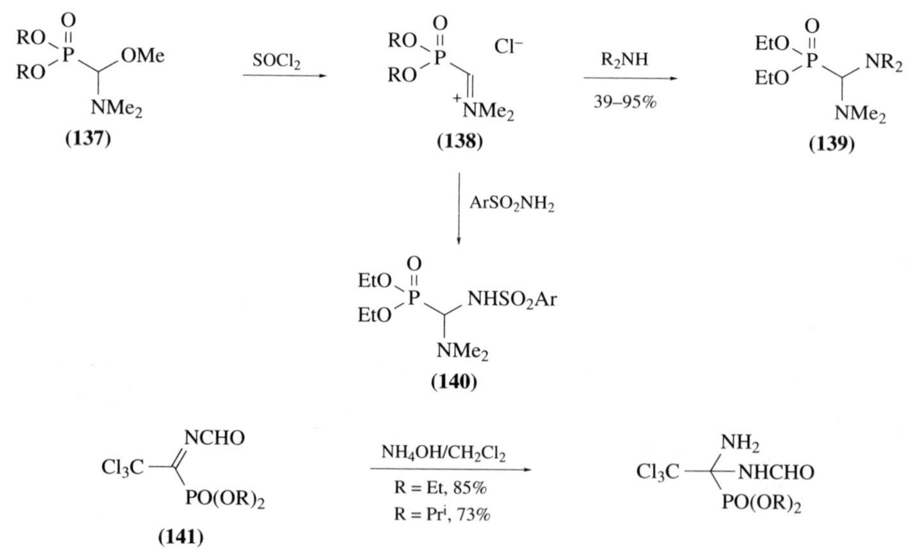

Scheme 38

The cyclic orthoamide (**142**) was treated first with diethyl phosphite then with benzylamine to yield the bis(amino)methylphosphonate (**143**) ⟨90PS(47)261⟩. The reaction of the tetraaminoalkene (**144**) with two equivalents of a dialkyl phosphite gave good yields of the imidazolidines (**145**) as illustrated in Scheme 39 ⟨78LA16⟩.

Heating a mixture of pyrazolone (**146**), diethanolamine and phosphite (**147**) gave the heterocycle (**148**) ⟨78MIP87868⟩. Other cyclic examples (**150**) were prepared by the cycloaddition of an azide to aminophosphenes (**149**) but these derivatives were unstable above −60 °C ⟨89NJC891⟩. The reaction of *N*-phenylbenzohydrazonoyl chloride to 2,5-dimethyl-1,2,3-diazaphosphole (**151**) occurs by a multistage cycloaddition process involving two equivalents of the derived nitrile imine and the elimination of benzonitrile to give the bicyclic example (**152**), albeit in low yield (Scheme 40) ⟨83CB549⟩.

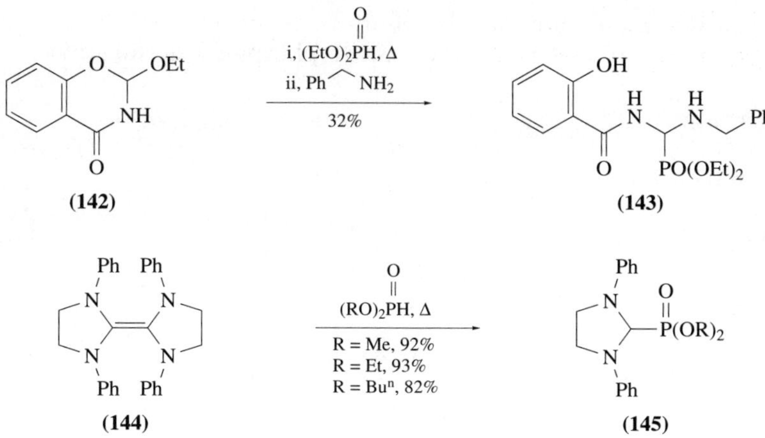

(142) (143) (144) (145)

Scheme 39

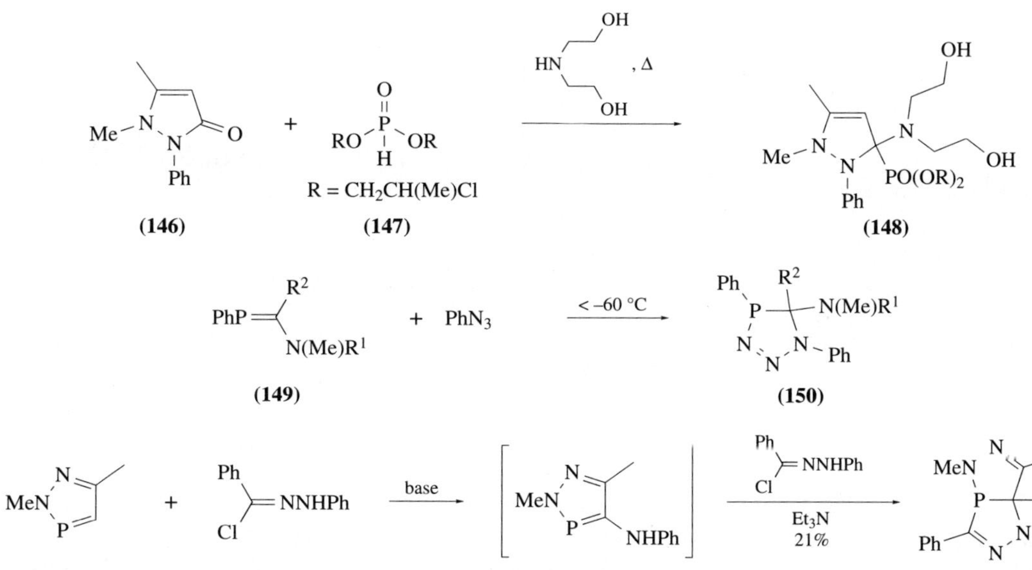

(146) (147) (148) (149) (150) (151) (152)

Scheme 40

6.05.1.5.2 *Functions bearing two phosphorus atoms and one nitrogen atom*

(i) Introduction

This type of functional group has been most widely prepared in the pharmaceutical industry, where such compounds are patented as agents for treating abnormal calcium and phosphate metabolism such as the bone disorder Paget's disease, (e.g., ⟨89EUP298553⟩).

(ii) Synthesis from ortho-*esters and similar derivatives*

Gross *et al.* have published details of the preparation of aminobis(phosphonato)methanes **(153)** and the corresponding bis(phosphine) oxides **(154)** by a method which involves heating *ortho*-amides either with dialkyl phosphites or with phosphine oxides (Scheme 41) ⟨68AG(E)391, 68AG(E)463, 69JPR577⟩. Thus, the cyclic *ortho*-amide **(142)** was treated first with diethyl phosphite and then diethyl trimethylsilylphosphite to give bis(phosphonato)methane **(155)** in moderate yield ⟨90PS(47)261⟩. The aminopyridine derivative **(157)** was prepared by the reaction of triethyl *ortho*-formate with 3-methyl-2-aminopyridine and two equivalents of phosphinite **(156)** ⟨81PS(11)311⟩.

Similar chemistry using a mixture of phosphorus species resulted in the analogues **(158)** ⟨90PS(51/52)23⟩, which were also made from the diethoxyphosphinyl acetal **(159)**.

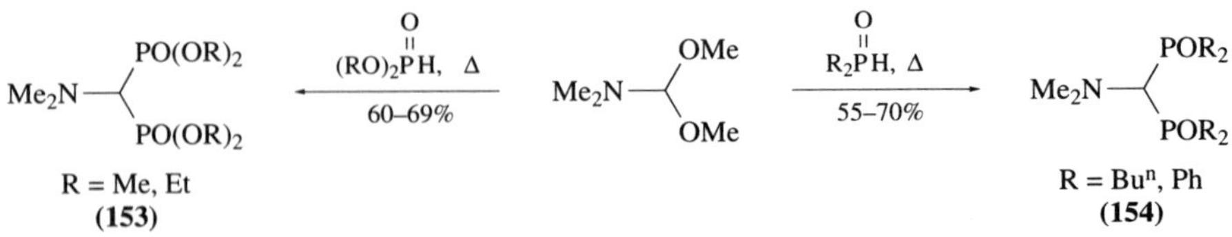

Scheme 41

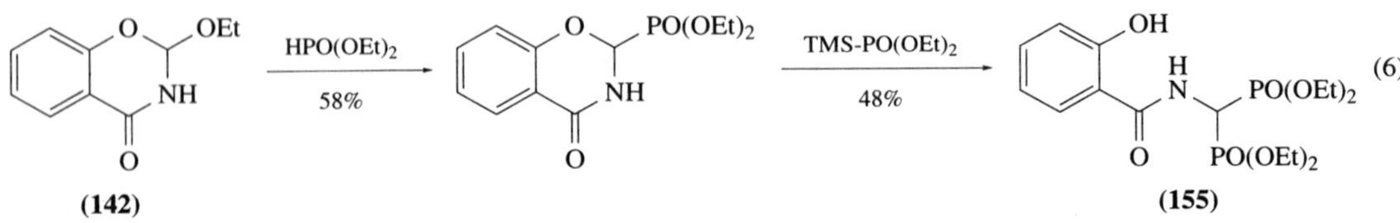

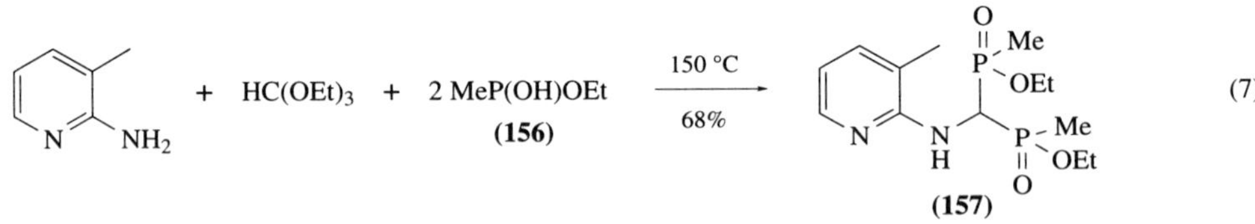

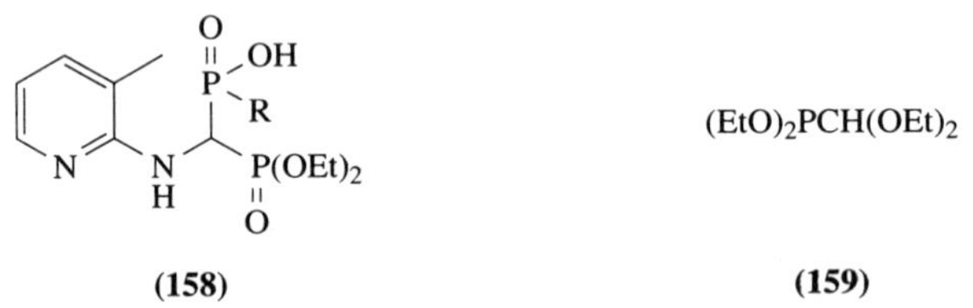

(iii) Synthesis from amides

Amides have been treated with hydrochloric and phosphorous acids to give variable yields of aminobis(phosphonates) **(160)** ⟨72ZAAC(389)119⟩; the reaction of formamides with phosphorus trichloride or tribromide followed by acid hydrolysis resulted in similar compounds ⟨75BCJ1030⟩. Using a similar procedure, chloroacetamide **(161)** was transformed into the corresponding derivative **(162)** by a mixture of phosphorus trichloride and phosphorous acid ⟨79ZAAC(457)214⟩ and the same method was used to prepare **(163)** ⟨85LA555⟩. The unusual phosphorous oxide (P_4O_6) has also been applied to the synthesis of these functions (Scheme 42) ⟨90PS(51/52)153⟩. Wet DMF when treated with two equivalents of phosphorus trichloride and then more water, gave the dimethylamino derivative **(164)** in good yield, as shown in Equation (8) ⟨91PS(56)117⟩.

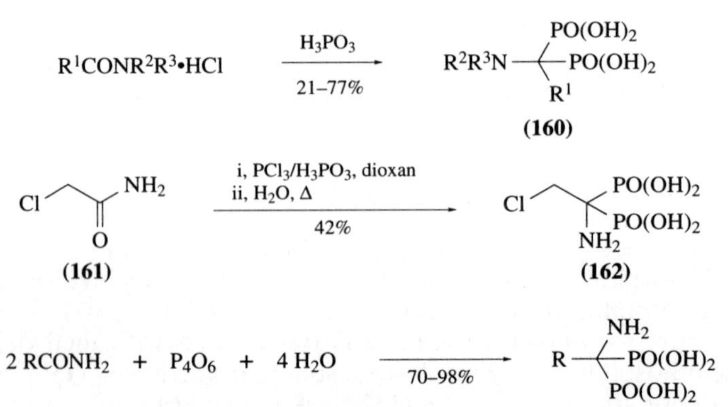

Scheme 42

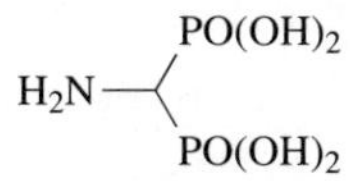

(163)

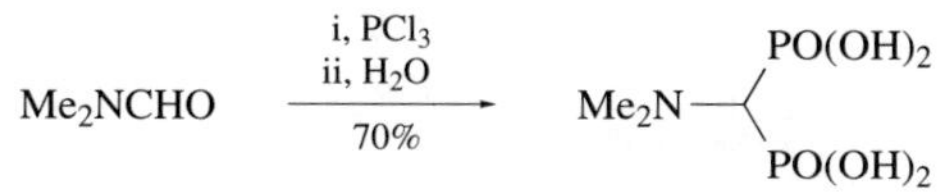 (8)

(164)

(iv) *Synthesis from nitriles*

The earliest report of the preparation of this type of functional group utilised the reaction of nitriles with phosphorus trihalides and water ⟨57GEP102355⟩. This method was subsequently used by Worms and Blum to prepare aryl derivatives (**165**) ⟨79ZAAC(457)209⟩. In addition, these workers used the reaction of cyanocyclohexane with phosphorous acid to obtain the cyclohexyl analogue (**166**). This transformation can be carried out in the presence of a ketone: Blum and Hemman reported that ketone (**168**) was made from the nitrile (**167**) and a mixture of phosphorus tribromide and phosphoric acid (Scheme 43) ⟨88GEP3611522, 88ZN(B)75⟩.

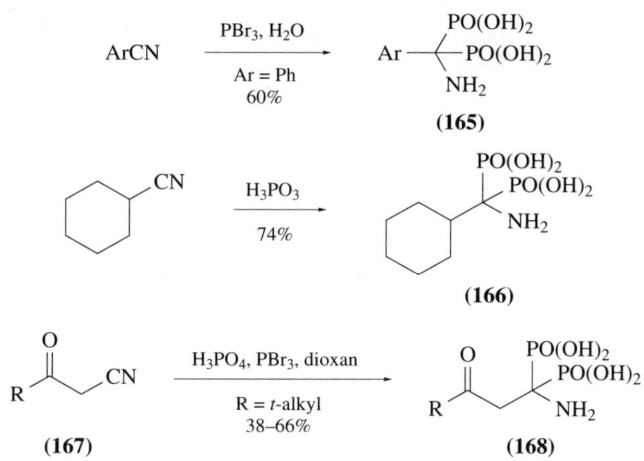

Scheme 43

(v) *Synthesis from imidoyl compounds*

The reaction of imidate (**169**) with sodium diethyl phosphite in benzene gave tetraethyl bis(phosphonato)benzylamine (**170**) ⟨59LA(623)103⟩. In a similar way, diethyl phosphite was added to imidoylphosphonate (**171**) to give aminobis(phosphonato)methane (**172**) (Scheme 44) ⟨71LA(750)44⟩. The treatment of imidoyl halides with phosphorous acid gave rise to the parent acids (**173**) ⟨72ZAAC(389)119⟩ and similar functions (**174**) and (**175**) were obtained using triethyl phosphite ⟨81PS(11)311, 82SC415⟩ and diethyl phosphinite ⟨81PS(11)311⟩, respectively.

(vi) *Synthesis from other bisphosphorus species*

In attempts to prepare aminotris(phosphonato)methanes (**177**) by addition of diethylphosphite to the imine bond in (**176**) Gross and Costisella actually obtained the product of reduction (**178**) via an unexpected carbon to nitrogen phosphorus migration (Scheme 45) ⟨72JPR969⟩. This method has since been used to prepare similar derivatives (**179**) ⟨74PS(4)241, 76JPR272⟩.

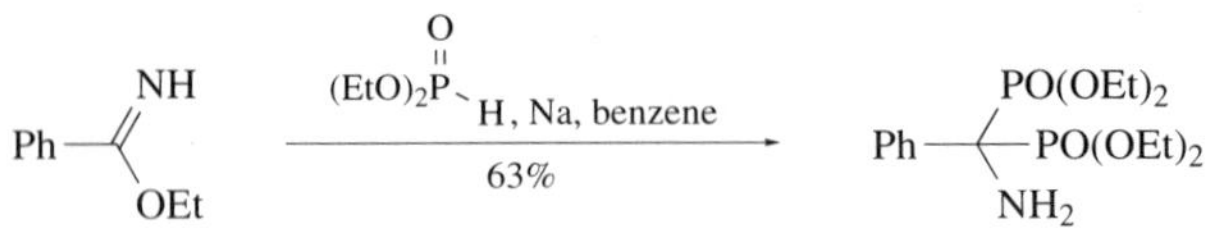

(169) (170)

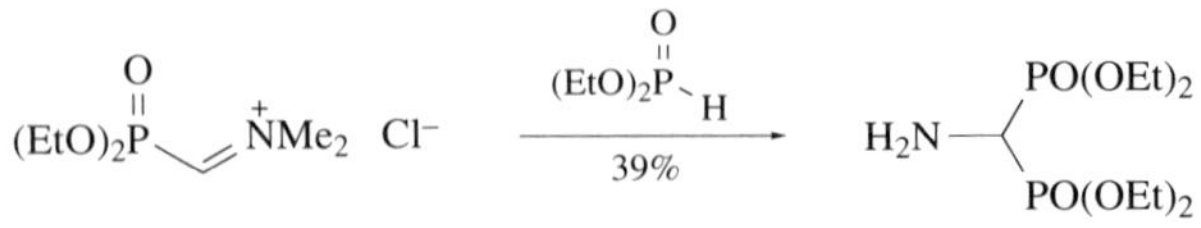

(171) (172)

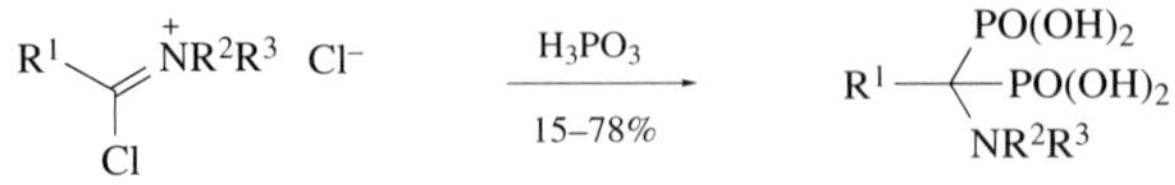

(173)

Scheme 44

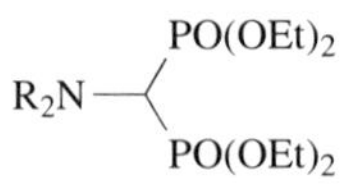

(174)

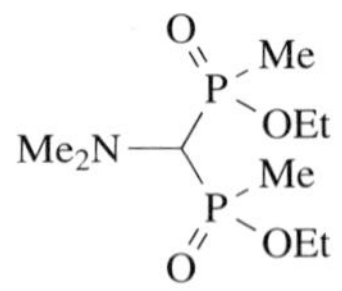

(175)

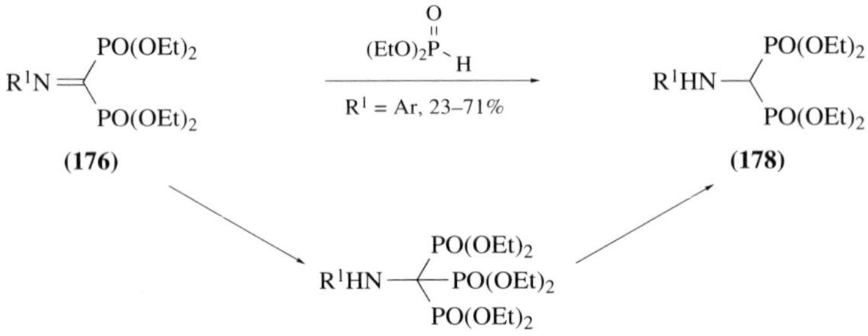

(176) (178) (177)

Scheme 45

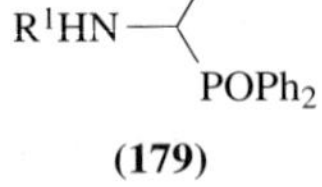

(179)

(vii) Miscellaneous syntheses

The reaction of aryl isocyanides (**180**) with dialkyl phosphites and base is reported to give the anilines (**181**) ⟨75ZOB1450⟩. Ferric chloride catalysed displacement of chloride from isocyanate (**182**) by fluorophosphite yields the unusual isocyanate (**183**) ⟨77ZOB321⟩. Dehydration of the aminoalcohol (**184**) by heating with hexamethyldisilazane gave rise to a mixture of bis(phosphonato)pyrrolidine (**185**) and the product of further loss of phosphite (Scheme 46) ⟨90PS(54)197⟩. Diphosphacyclobutanes (**187**) have been obtained by dimerisation of the phosphorous analogues (**186**) of amidines ⟨80ZAAC(462)130, 82ZAAC(485)23, 83PS(18)27, 91PS(61)361⟩.

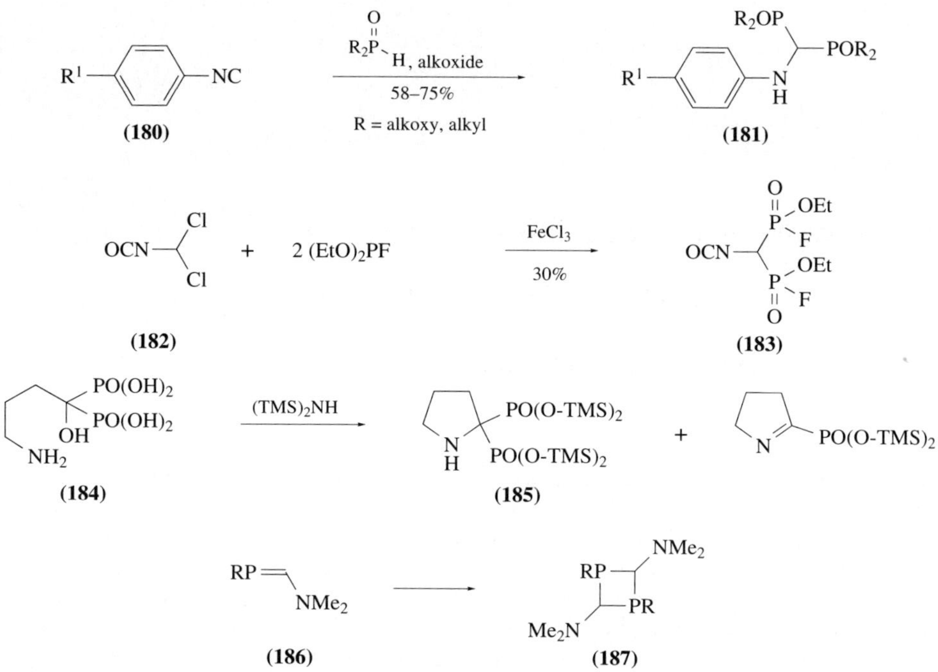

Scheme 46

6.05.1.5.3 *Functions bearing nitrogen and other group 15 elements*

Kellner *et al.* described the reaction of diphenylarsane with tetramethylamidinium chloride in acetonitrile to give bis(dimethylamino)methyldiphenylarsane (**188**) ⟨78JOM(149)167⟩. This reaction proceeds in poorer yield when the sodium salt of the arsane is used in THF. Imidoyl chloride (**190**) forms the cyclic derivative (**191**) by reaction with aminoarsane (**189**) (Scheme 47) ⟨78ZC452⟩. Treatment of arsenadiazole (**192**) with diphenylnitrile imine gives the bicyclic analogue (**193**) ⟨86TL2957⟩.

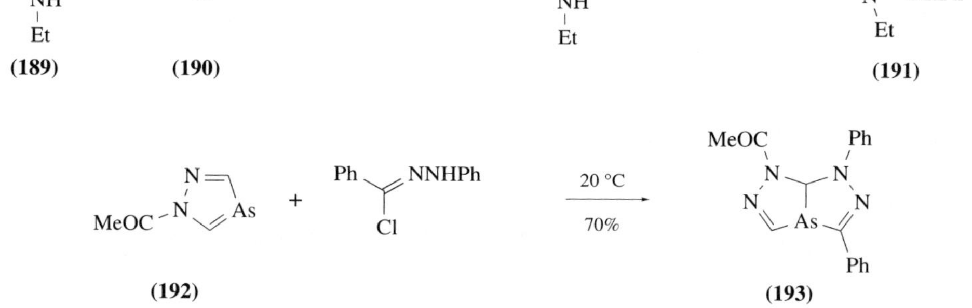

Scheme 47

The bis(nitro)methylstibane (**194**) is formed when the silver salt of 1,1-dinitroethane is treated with diphenylstibyl chloride, as shown in Equation (9) ⟨85IZV439⟩.

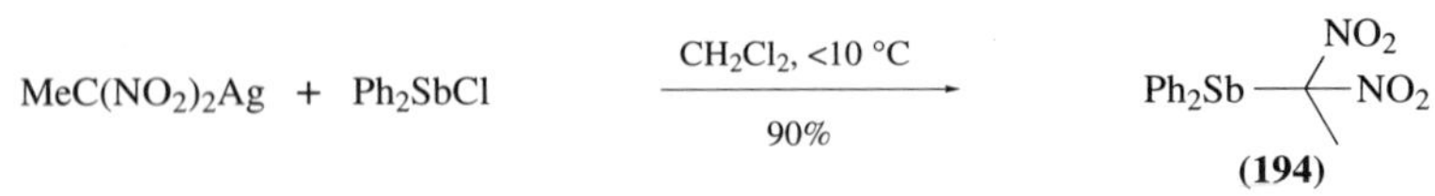

(194) (9)

6.05.2 FUNCTIONS CONTAINING GROUP 15 ELEMENTS WITH METALLOIDS

6.05.2.1 Functions Bearing Group 15 Elements and Silicon

6.05.2.1.1 Functions bearing nitrogen and silicon

(i) From silyl halides

This general method involves deprotonation of the carbon atom attached to a nitrogen function followed by silylation. Thus, methyl isocyanide has been successively deprotonated and silylated to afford bis(trimethylsilyl)methyl isocyanide (**195**) ⟨70JOM(25)385⟩, the second silylation going in much poorer yield than the first. The product (**195**) was converted to the corresponding isothiocyanate using sulfur ⟨82BCJ1163⟩. Katritzky *et al.* prepared bis(trimethylsilyl)methylbenzotriazole (**196**) by the same method and studied its further alkylation (Scheme 48) ⟨88RTC641⟩. More recently, Snieckus and co-workers showed ⟨89TL5841⟩ that whereas *N*,*N*-dimethylbenzamide could be readily silylated using lithium tetramethylpiperidide to yield (**197**), the corresponding diethylamide or piperidide gave *O*-silyl products.

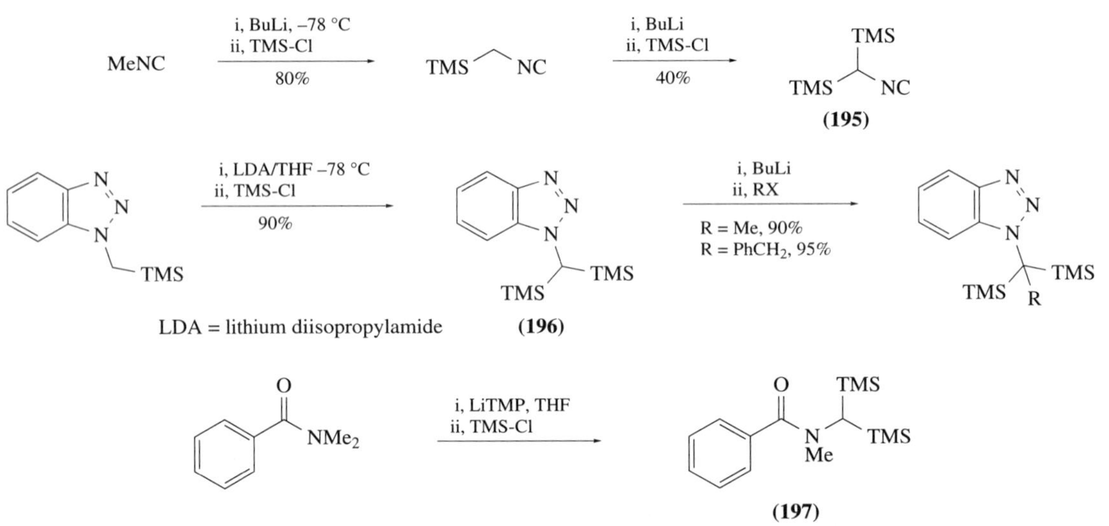

Scheme 48

(ii) By reductive silylation

Amides have been treated with silyl halides in the presence of metals to give bis(trimethylsilyl)aminomethanes. Thus, *N*,*N*-dimethylbenzamide, when added to TMS-Cl and magnesium in hexamethylphosphoramide (HMPA), gave (**198**) in modest yield ⟨71JOM(32)79⟩. The authors report that the reaction fails using *N*-methylbenzamide. Ekouya *et al.* studied the same reaction using formamides and lithium metal in THF and found the addition of triethylamine to be advantageous ⟨79JOM(177)137⟩. In this case, the product (**199**) was formed in only trace amounts in the absence of base (Scheme 49). Picard *et al.* made bis(trimethylsilyl)aminomethane (**200**) in good yield by treatment of TMS-CN with TMS-Cl and lithium in HMPA ⟨91JOM(419)C1⟩.

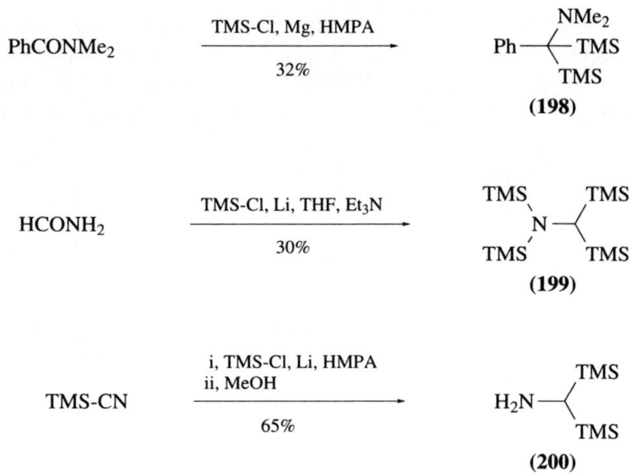

Scheme 49

(iii) Miscellaneous methods

Bis(trimethylsilyl)chloromethane has been reported to react with sodium azide in HMPA; subsequent reduction using LAH leads to the amine (**200**) (Equation (10)) (⟨91JOM(419)C1, 92TL3903⟩).

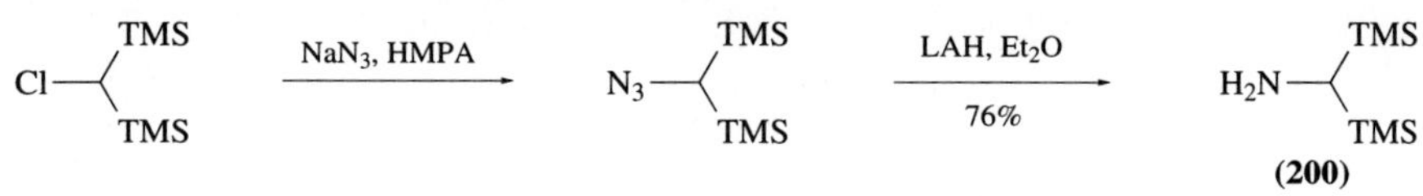

(10)

Silaneimine (**201**), when treated with dimethylethylamine at elevated temperatures, inserts into two of the C—H bonds of a methyl group of the amine resulting in the formation of the silaneamine (**202**) (Scheme 50) ⟨87CB1357⟩.

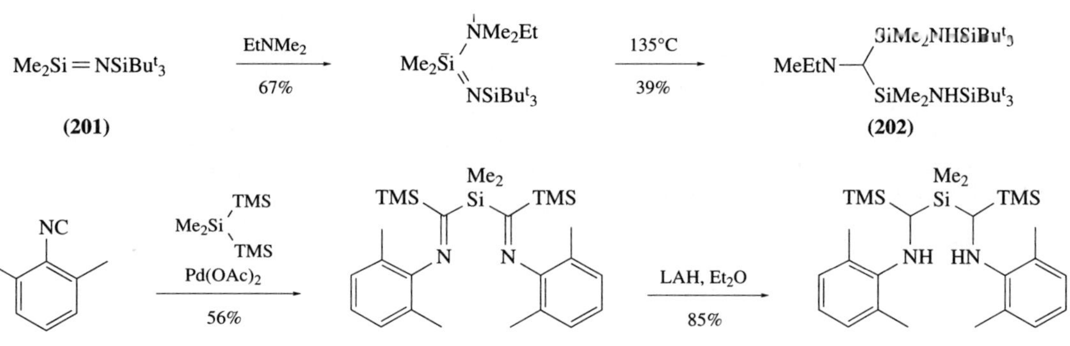

Scheme 50

The aryl isocyanide (**203**) reacts with octamethyltrisilane in the presence of a palladium catalyst to give the bis(silylimine) (**204**) which when reduced with LAH results in the formation of the corresponding amine (**205**) in good yield ⟨88JA3692⟩.

6.05.2.1.2 Functions bearing phosphorus and silicon

(i) From silyl halides

In a manner exactly analogous to the nitrogen series, the quenching of a phosphorus stabilised carbanion by a silylating agent has been widely used to prepare this functional group. Phosphine sulfide (**207**) was prepared by reaction of the lithium salt (**206**) with excess TMS-Cl ⟨74ZAAC(404)204⟩. Phosphines (**208**) have been further silylated in good yields ⟨81ZAAC(475)18, 84CB2063⟩ and phosphonates (**209**) have been similarly treated (Scheme 51) ⟨87JOM(323)135⟩. Trimethylphosphine has been doubly deprotonated using *t*-butyllithium and successively silylated to give (**210**) ⟨88ZN(B)1416⟩.

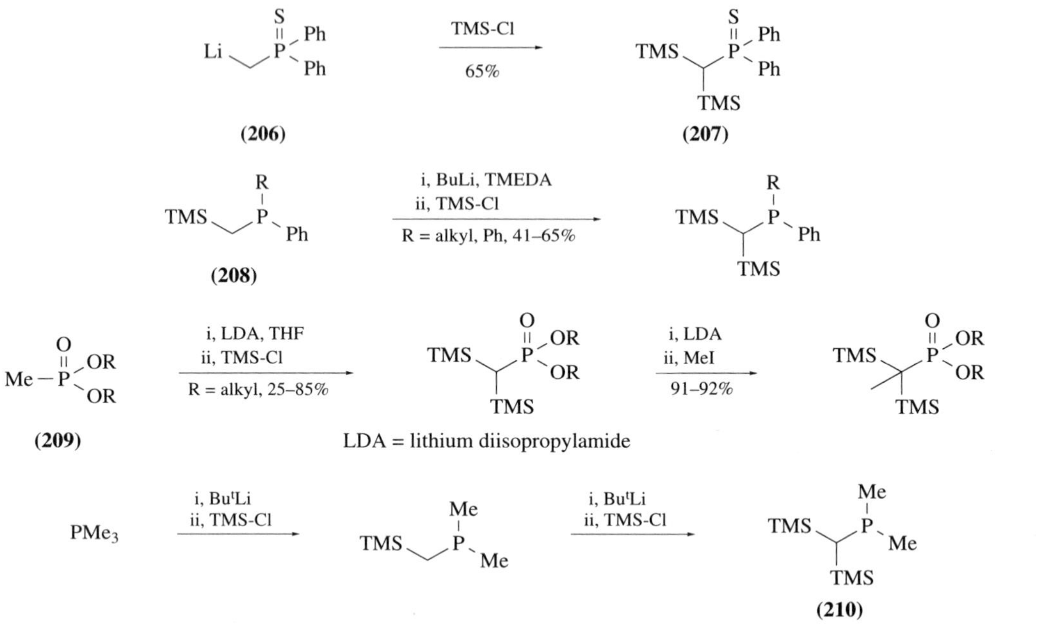

Scheme 51

Bis(phosphino)methanes (**211**) have been silylated using butyllithium as a base to afford bis(phosphino)trimethylsilylmethanes (**212**) (Equation (11)) ⟨79CB648, 88ZN(B)1416⟩.

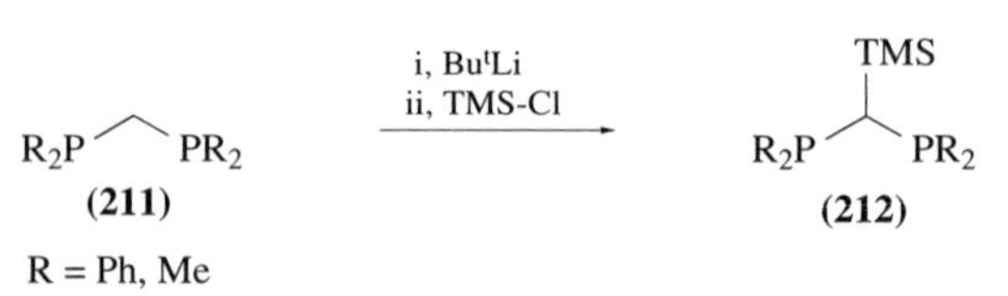

(11)

Phosphorus double bonded systems such as (**213**) and (**214**) react with nucleophiles at phosphorus (Scheme 52) ⟨83PS(18)43⟩. Subsequent silylation in the case of (**213**) gave (**215**) whereas protonation at carbon in the case of (**214**) yielded (**216**).

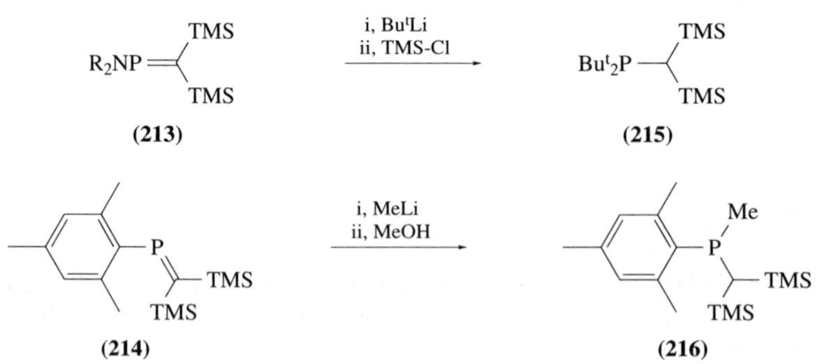

Scheme 52

(ii) From phosphorus halides

Sequential replacement of halogens in phosphorus halides using silyl stabilised carbanions is another common method for the synthesis of this group. Phosphorus trichloride when treated with the lithium salt of bis(trimethylsilyl)methane was converted to either the dichloro compound (**217**) or its analogous monochloride (**218**), depending on the proportions of the reagents ⟨80JCS(D)2428⟩. Dialkyl- ⟨91AG(E)709⟩ and diarylchlorophosphines ⟨83JCS(D)905⟩ produce similar products and Grignard reagents have also been used with phosphorus trichloride ⟨90TL3859⟩ and aryldichlorophosphine (**219**) ⟨85OM339⟩ (Scheme 53). Phosphinylidene chloride (**220**) was converted to the unsaturated derivative (**221**) in high yield ⟨82OM1720⟩.

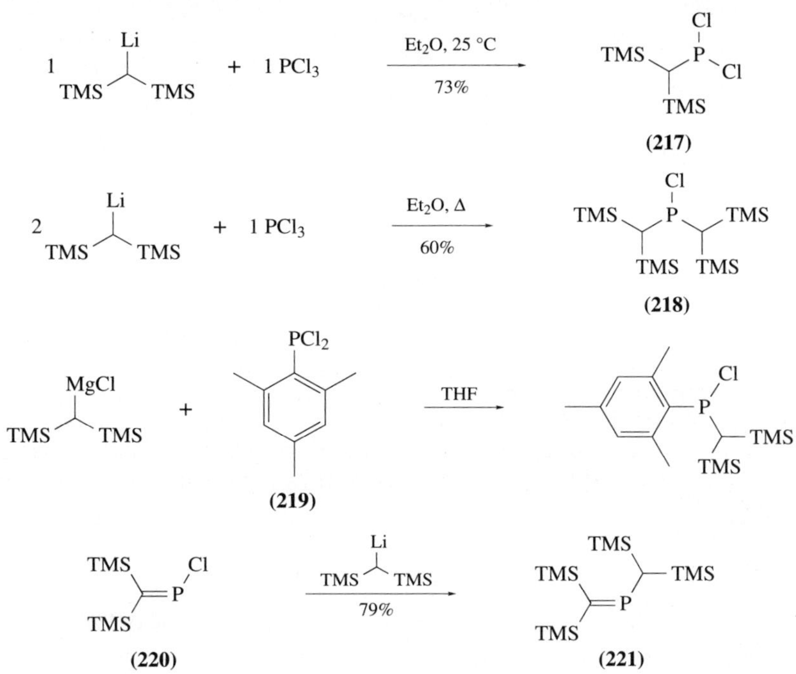

Scheme 53

The reaction of tris(trimethylsilyl)methyllithium with phosphorus trichloride produced the dichloride (**222**) ⟨89JOM(366)39⟩, which, after replacement of the halides using sodium methoxide, gave dimethoxyphosphine (**223**) which readily desilylated in methanol to give bis(trimethylsilyl)methane (**224**). This desilylation was previously noted in diphenylphosphine (**225**), which although stable under basic conditions was much less so at lower pH (Scheme 54) ⟨83JCS(D)905⟩.

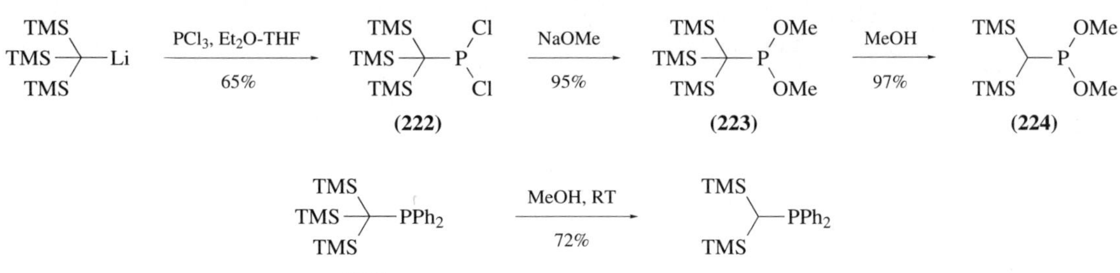

Scheme 54

(iii) Miscellaneous methods

Phosphinylidene derivatives have been reduced under a variety of conditions. For example, the phosphine (**226**) was produced effectively by the transfer of hydride from a phosphinyl methyl group ⟨84ZAAC(511)108⟩. In another case, phenylphosphinylidene disilane (**227**), when treated with an equimolar mixture of phenylphosphine and its sodium salt ⟨87CB1707⟩, gave the product (**228**) in high yield and the chloro analogue (**229**) has been reduced with concomitant disproportionation using LAH (Scheme 55) ⟨87CC1092⟩.

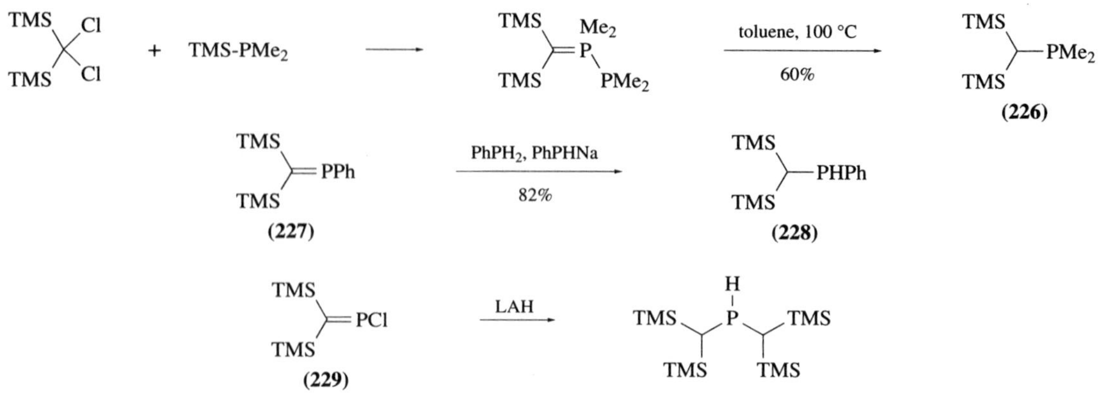

Scheme 55

The displacement of chloride ion from phosphinylidene chloride (**230**) with acetylide anions gives rise to the diphosphacyclobutane (**231**) after dimerisation as a mixture of stereoisomers ⟨84CB2693⟩. Interestingly, the aminophosphinylidene compound (**232**) dimerises with rearrangement to the unusual small ring compound (**233**) (Scheme 56) ⟨90CB1245⟩. Phosphinylidene compounds also undergo cycloadditions. For example, 2,3-dimethylbutadiene gave the phosphorus heterocycle (**234**) ⟨85CB814⟩ and diphenyldiazomethane the three-membered epiphosphine (**235**) ⟨86CB1977⟩.

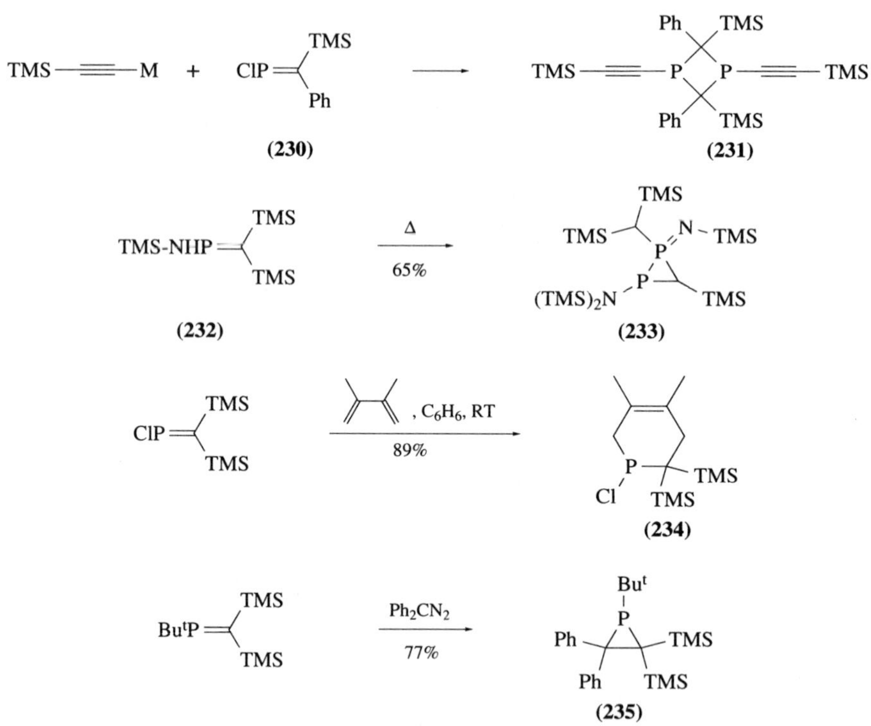

Scheme 56

Pentamethylcyclopentadiene groups in the diphosphene (**236**) have been displaced by the lithium salt of bis(trimethylsilyl)methane to form the new diphosphene (**237**) as shown in Equation (12) ⟨86AG(E)919⟩.

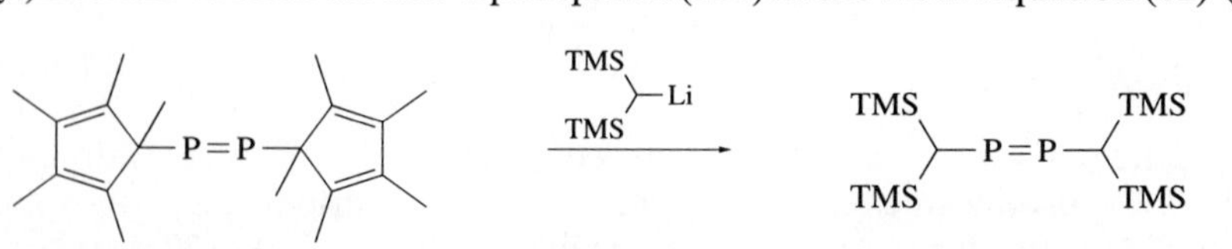

(12)

(**236**) (**237**)

(iv) Modification at phosphorus

There have been a number of reports in which previously formed functions have been modified at phosphorus. Chlorophosphines such as (**238**) have been reduced using, for example, sodium metal or LAH ⟨82OM1720, 85OM339⟩. Alternatively, the halogen was replaced by a methyl group (Scheme 57) ⟨84CC1669⟩. The phosphine oxide (**240**) has been produced from its parent compound (**239**) by a mild oxidation which avoids decomposition ⟨84JOC5087⟩, and the reaction of bis(phosphino)methane (**241**) with sulfur gives the analogous thio compound (**242**) ⟨92CB1333⟩.

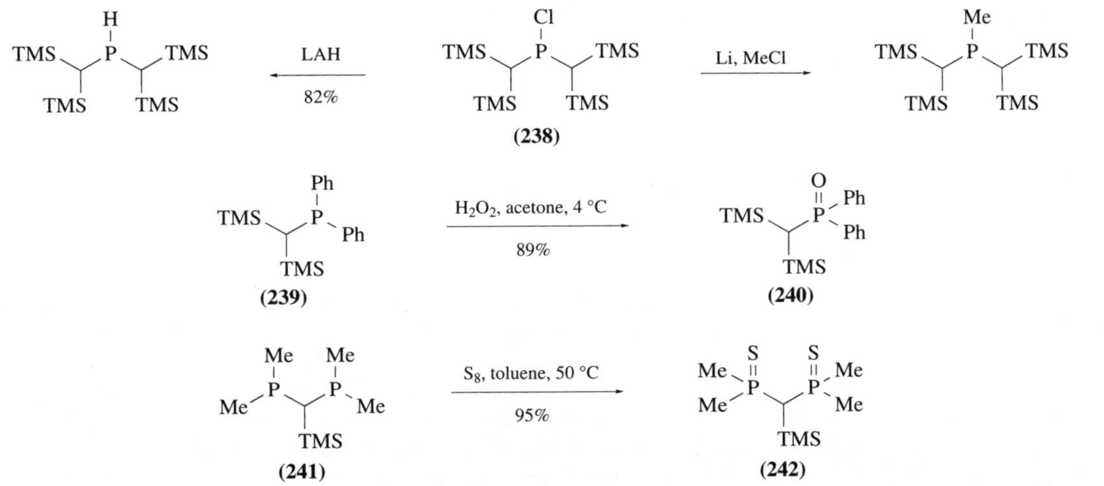

Scheme 57

The bis(trimethylsilyl)methyl group has, by virtue of its steric bulk, been used as a stabilising group in the preparation of bisphosphenes (for a review see ⟨84POL389⟩) e.g., the reaction of dichloride (**217**) with hindered phosphine (**243**) in the presence of 1,5-diazabicyclo [5.4.0] undec-5-ene (dbu) gives bisphosphene (**244**) ⟨83CC528⟩. The analogue (**237**), prepared by treating difluoride (**245**) with lithiophosphine (**246**), had a half-life of one week at 20°C ⟨87PS(31)81⟩. The same stabilising effect has been utilised in producing phosphinyl radicals (**247**) ⟨80JCS(D)2428⟩ which were subsequently shown to add to 2,3-dimethylbutadiene to give the corresponding bis(phosphines) (**248**) (Scheme 58) ⟨84TL3519⟩.

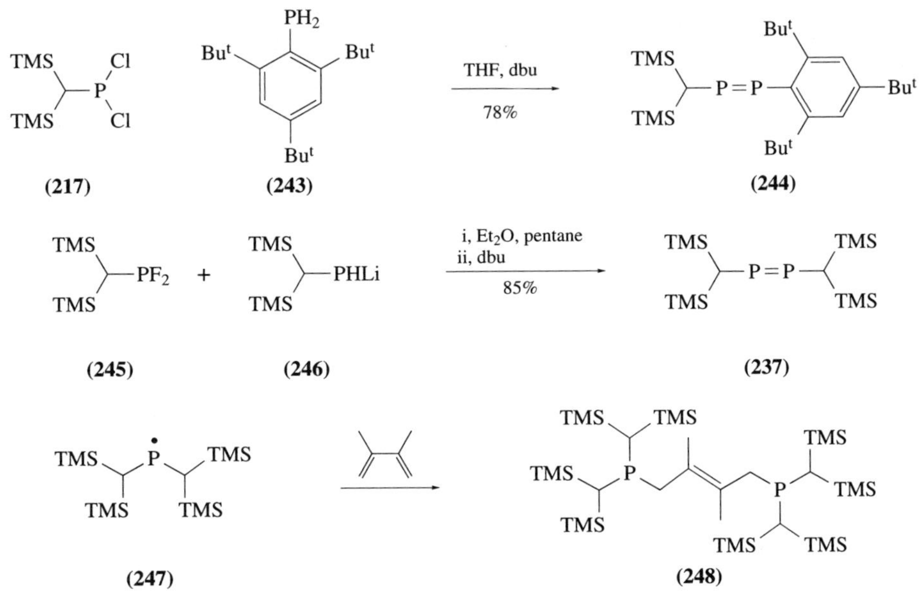

Scheme 58

6.05.2.1.3 Functions bearing nitrogen, phosphorus and silicon

The cycloadditions of *t*-butyl- ⟨81AG(E)131⟩ and trimethylsilyldiazomethane ⟨83CC1171⟩ with phosphinylidenemethylsilane (**249**) gave the corresponding heterocyclic examples (**250**) shown in Equation (13).

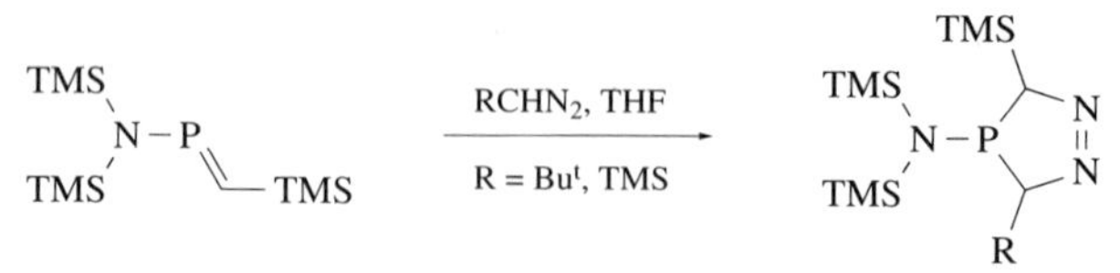

(**249**) (**250**) (13)

6.05.2.1.4 Functions bearing arsenic, antimony or bismuth

(i) From trihalides

The displacement of one or more halides from the appropriate trihalides has proved the method of choice for the preparation of these functions. Gynane *et al.* ⟨80JCS(D)2428⟩ produced arsenic and antimony derivatives with either one (**251**) or two bis(trimethylsilyl)methyl groups (**252**) by the reaction of appropriate trichlorides with the correct proportions of bis(trimethylsilyl)methyllithium. Bismuth derivatives with three such substituents (**253**) were similarly prepared ⟨83IC3421⟩ and Grignard reagents have also been used for the synthesis of stibanes (**254**) ⟨83POL291, 84IC2582⟩. More recently, the pyridyl derivatives (**255**) were produced in a similar way ⟨91CC1560⟩ (Scheme 59).

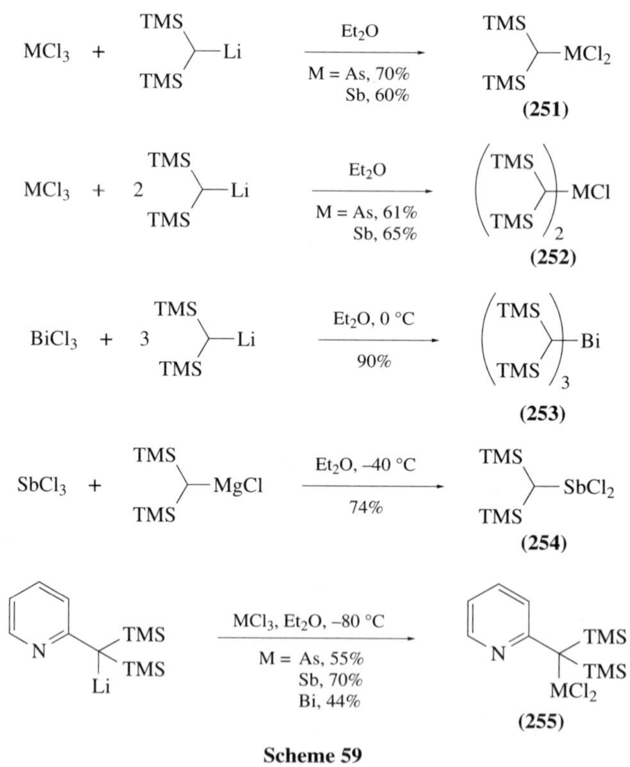

Scheme 59

(ii) By modification of the group 15 element

As in the case of the diphosphenes, bis(trimethylsilyl)methyl groups have been used to stabilise heteroatomic double bonds. Diarsenes (**258**) were prepared ⟨83JA5506⟩ by reaction of aryl arsane

(**256**) with dichloroarsane (**257**) (Equation (14)), whereas mixed functions (**259**) were made by reacting the corresponding phosphine with arsenic and antimony dichlorides ⟨83CC881⟩.

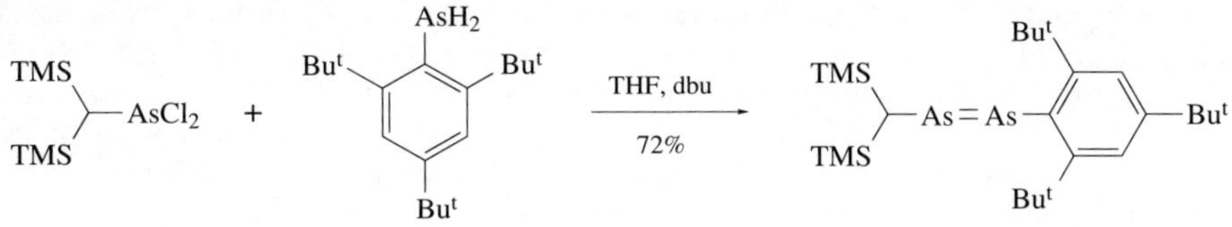

(14)

(**256**) (**257**) (**258**)

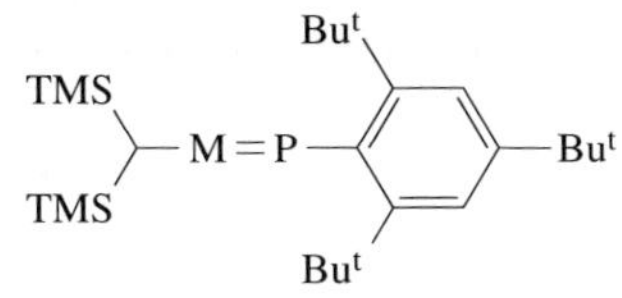

(**259**) M = As, Sb

The chloroarsane (**260**) was metallated on treatment with lithium. Subsequent protonation or methylation produced the arsanes (**261**) whereas reaction with stannous chloride gave diarsane (**262**) ⟨87JOM(320)C27⟩. Treatment of dichlorostibane (**254**) with magnesium in THF resulted in the formation of the unusual cyclotetrastibane (**263**) (Scheme 60) ⟨92OM145⟩.

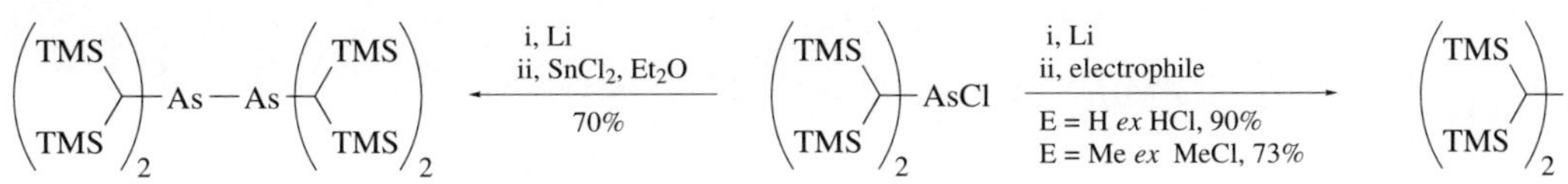

(**262**) (**260**) (**261**)

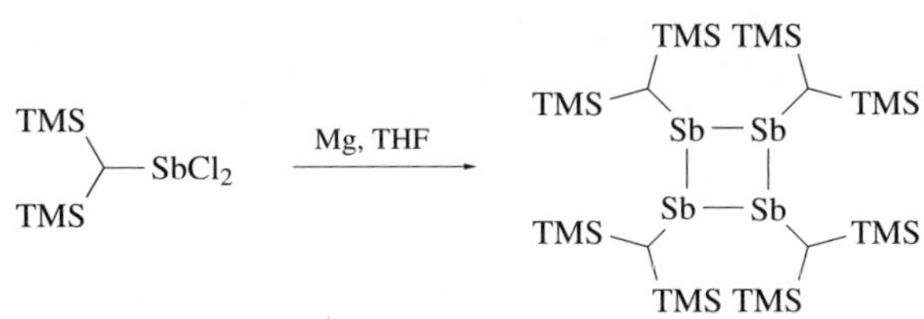

(**254**) (**263**)

Scheme 60

6.05.2.2 Functions Bearing Group 15 Elements and Boron

A search of the literature revealed many references to carborane chemistry and readers are referred to specialist texts for further information on this complex subject ⟨B-70MI 605-01, 82COMC-I411, 82COMCI-I543, 92CRV209⟩.

In the sole report of a simple example uncovered by this author, Ioffe *et al.*, disclosed the synthesis of (**264**) by treatment of dinitromethane salts with chlorodiethylborane (Equation (15)) ⟨74MI 605-01⟩.

$$Et_2BCl + (NO_2)_2CH_2 \xrightarrow{\text{base}} Et_2BCH(NO_2)_2 \quad (15)$$

(**264**)

6.05.2.3 Functions Bearing Group 15 Elements and Germanium

Karsch *et al.* ⟨86JOM(312)C1⟩ made germanium compounds (**265**) and analogous lead derivatives by the reaction of the lithium salt of bis(diphenylphosphinyl)methane with germanium(II) chloride (Equation (16)).

$$(Ph_2P)_2CHLi \xrightarrow[93\%]{GeCl_2,\ -40\ °C} Ge[CH(PPh_2)_2]_2 \quad (16)$$

(**265**)

6.05.3 FUNCTIONS CONTAINING GROUP 15 AND METAL ELEMENTS, AND POSSIBLY A METALLOID

Many such compounds have been prepared without isolation or characterisation as precursors to the functional groups mentioned earlier in this chapter and hence relevant references can be found above. The subsequent discussion will refer only to those areas previously uncovered or where the chemistry is significant enough to be considered separately. Work has been done in the area of organometallic cluster compounds but this falls outside the scope of this chapter.

Many alkali metal salts of dinitromethane derivatives have been made ⟨57JA4708, 61JOC4734⟩. It has been shown by x-ray diffraction that the rubidium salt of cyanodinitromethane exists in a form in which the metal atom is closely associated with one of the oxygens ⟨72ACS1874⟩. It seems highly likely that alkali metal salts of other systems, where delocalisation is possible, will behave similarly. The reaction of dinitromethane with triethylthallium gave a compound of Structure (**266**) in which the metal is reported to coordinate to both carbon and oxygen atoms (Equation (17)) ⟨71IZV1505⟩.

$$CH_2(NO_2)_2 \;+\; TlEt_3 \xrightarrow[95\%]{Et_2O} \underset{\textbf{(266)}}{Et_2TlCH(NO_2)_2} \qquad (17)$$

The deprotonation of bispyrrole (**267**) has been achieved using methyllithium and although formal resonance structures are difficult to write for the lithium salt, NMR studies indicate extensive delocalisation of charge ⟨80HCA1190⟩. Katritzky *et al.* have studied the use of benzotriazole as a stabilising group for anionic centres; among recent publications is one concerning the generation and use of organolithium species (**268**) ⟨91JOC2143⟩.

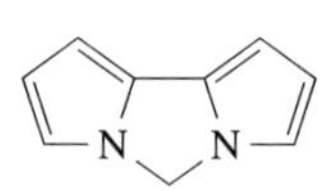
(267)

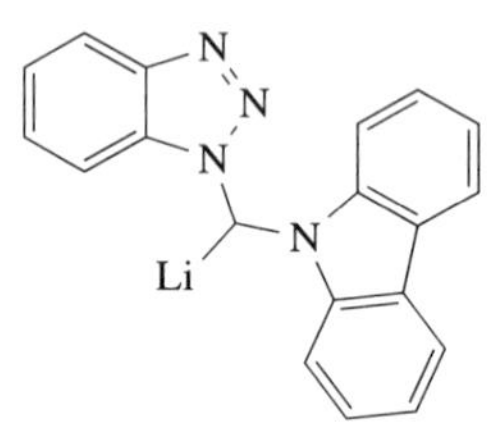
(268)

Kauffmann and his group have studied the chemistry of the deprotonation of carbon atoms attached to heavy metals and metalloids, and a review has appeared ⟨82AG(E)410⟩. In some cases the loss of one of the metal substituents is observed. The preparation of lithium salts of bis(diphenylstibanyl)methane and bis(diphenylarsanyl)methane and the subsequent deuteration of the products (**269**) has also been described (Equation (18)) ⟨79TL501⟩.

$$\underset{\textbf{(269)}}{(Ph_2M)_2CH_2} \xrightarrow{c\text{-}Hex_2NLi,\ HMPA} (Ph_2M)_2CHLi \xrightarrow[\substack{M = Sb,\ 68\% \\ M = As,\ 63\%}]{D_2O} (Ph_2M)CHD \qquad (18)$$

6.06
Functions Containing at Least One Metalloid (Si, Ge, or B) and No Halogen, Chalcogen or Group 15 Element; Also Functions Containing Three Metals

VADIM D. ROMANENKO
Institute of Organic Chemistry, Kiev, Ukraine

MICHEL SANCHEZ
Université Paul Sabatier, Toulouse, France

and

JEAN-MARC SOTIROPOULOS
Université de Pau et Pays de l'Adour, Pau, France

6.06.1 INTRODUCTION

This chapter deals with compounds of the general formula $RC(EX_m)_n(ML_k)_{3-n}$ (E = Si, Ge, or B; M = metal; R = H or organyl; X = any substituent; $n = 0$–3) having three heteroatoms (E, M) bonded to the same carbon. Most of the organoelement species of this type contain classical two-center, two-electron "single" bonds and formally arise through the replacement of hydrogen atoms with the heteroatoms mentioned. It is the purpose of this chapter to discuss the preparation and some of the chemical transformations of these compounds. A discussion of carboranes, metal clusters, and other electron-deficient molecules which cannot be portrayed in the same manner and exhibit quite different reactivities, is not included. For the same reason, π-metal complexes containing a $C(EX_m)_n(ML_k)_{3-n}$ fragment are also ignored. However, the organometallic compounds of the type $RC(EX_m)_nM_{3-n}$ (M = alkali metal), often better described as heteroatom-stabilized carbanions, are considered.

As will become evident, the methods of preparation of the above functions are wide and varied. Some of these methods are discussed in monographs by Colvin ⟨B-81MI 606-01⟩, Weber ⟨B-83MI 606-01⟩, Pereyre *et al.* ⟨B-87MI 606-01⟩, and Negishi ⟨B-80MI 606-01⟩. The latter also contains a detailed discussion of standard methods of carbon–metalloid and carbon–metal bond formation. Houben-Weyl's *Methoden der Organischen Chemie* ⟨74HOU(13/2b)1, 75HOU(13/7)1, 78HOU(13/6)181, 80HOU(13/5)27, 82HOU(13/3a)1⟩ and the appropriate reviews in *Comprehensive Organometallic Chemistry* ⟨82COMC-I(2)1, 82COMC-I(7)265, 82COMC-I(7)515⟩ etc. and *Comprehensive Organic Chemistry* ⟨79COC(3)687⟩ cover parts of the literature up to 1982. The preparation and properties of some polyborylated compounds are also discussed in a monograph by Pelter *et al.* ⟨B-88MI 606-01⟩ and in a book by Pelter ⟨B-87MI 606-02⟩. The chemistry of polylithiated aliphatic hydrocarbons has been reviewed by Maercker and Theis ⟨87TCC(138)1⟩. Discussion of certain aspects of the topic can also be found in *Comprehensive Organic Synthesis* ⟨91COS(1)1, 91COS(1)487⟩ and in a number of other references. ⟨64AOC(2)257, B-67MI 606-01, B-85MI 606-01, B-85MI 606-02, 86MI 606-01, 89MI 606-01, B-89MI 606-02, B-92MI 606-01⟩.

6.06.2 FUNCTIONS CONTAINING AT LEAST ONE METALLOID FUNCTION (AND NO HALOGEN, CHALCOGEN, OR GROUP 15 ELEMENT)

6.06.2.1 Functions Containing Three Metalloids

6.06.2.1.1 Functions bearing three silicons

Trisilylmethyl groups, especially tris(trimethylsilyl)methyl, $(TMS)_3C$, have been widely recognized as versatile building blocks in organic synthesis which have proven their synthetic utility in hundreds of reactions ⟨82JOM(239)93, B-85MI 606-01, 91MI 606-01, 91MI 606-02⟩. These burgeoning applications arise from the unique properties of trisilylmethyl derivatives. Trialkylsilyl groups (R_3Si) can be considered "bulky protons" and have been widely used for the synthesis of sterically protected molecules. Due to the acidity of the Si_3C—H hydrogen (and hence the availability of the corresponding metal derivatives), trisilylmethyl groups can be easily attached to a range of metals and metalloids. Many of the resulting compounds show enhanced stability compared to the corresponding unsubstituted methyl derivatives and thus are easily handled and studied. Lastly, due to the steric as well as to the electronic effects trisilylmethyl groups can stabilize an adjacent carbanionic center or an adjacent carbon–heteroatom double or triple bond.

The main synthetic pathways to 1,1,1-trisilylalkanes are:

(a) "direct synthesis" in the gas phase involving insertion of silicon into a carbon–halogen bond;
(b) reductive silylation of trihalomethanes;
(c) reactions of organometallic carbon nucleophiles with chlorosilanes;

(d) derivatization and rearrangements of the compounds already containing polysilylated C_1 units.

A number of other reactions in which an Si_3C function is created are known, but they are more of academic interest.

(i) Trisilylmethanes and their linear C*-organyl derivatives, $RC(SiX_3)_3$*

Compounds of this type are known with R = H, alkyl, aryl, hetaryl, acyl, and with a variety of substituents on silicon. The parent tris(silyl)methane (**1**) is formed in very low yield via LAH reduction of the corresponding chloro derivative (**2**) ⟨64CB1999, 86ZN(B)1527⟩. This precursor is available, along with other trisilylated methanes such as structures (**3**)–(**6**), in the "direct synthesis" using a copper catalyzed reaction of elemental silicon with chloroform ⟨58CB22⟩ or other halohydrocarbons ⟨85ZC309, 93ZAAC(619)1494⟩.

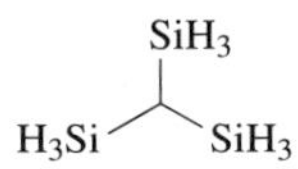
(1)

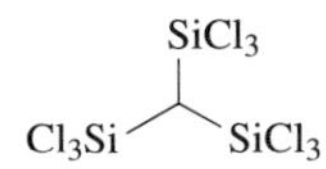
(2)

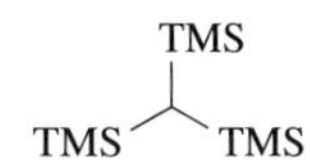
(3)

$HC(SiHCl_2)_n(SiCl_3)_{3-n}$
(**4**)-(**6**); n = 1–3

Attempts were made to obtain tris(silyl)methane by the reaction of silylpotassium, $KSiH_3$, with halomethanes ⟨61JA802, 67IC1751, 75JOM(92)163⟩ and the pyrolysis of methylchlorosilanes ⟨67AG(E)677⟩, but the yields were again in the lower region. More recently, tris(silyl)methane was obtained in high yield by a three-step synthesis presented in Scheme 1 ⟨89CB2115⟩. An *in situ* Grignard reaction of easily accessible chlorophenylsilane with bromoform and magnesium gave tris(phenylsilyl)methane (**7**) in ca. 50% yield. The final conversion of bromo derivative (**8**) into tris(silyl)methane (**1**) has been accomplished through reduction with LAH in a two-phase system employing a phase transfer catalyst. A similar route is suitable for the preparation of the homologous bis(silyl)methane ⟨90CB2087⟩ and tetrakis(silyl)methane ⟨90AG(E)201⟩.

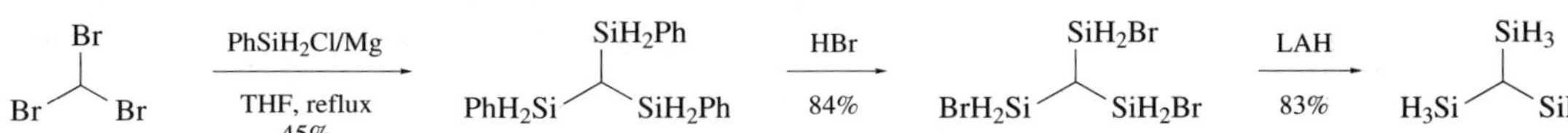

(**7**) (**8**) (**1**)

Scheme 1

Among the methods which can be employed to prepare tris(organosilyl)methanes the most straightforward one involves the Merker–Scott procedure using magnesium or lithium reduction of halohydrocarbons and halosilanes in a polar solvent ⟨63JA2243, 64JOC953, 65JOM(4)98, 70JOM(24)647⟩. The reaction is usually performed by the addition of 1,1,1-trichloro- or tribromoalkane to a haloorganosilane (usually chlorotrimethylsilane) and Mg(Li) in THF. In principle the synthesis provides a general route to symmetrically substituted tris(organosilyl)methanes, but the correct choice of reagents and conditions is essential if good yields of pure compounds are to be obtained. Experimental details for the preparation of tris(TMS)methane (**3**) from chloroform, chlorotrimethylsilane and lithium metal are described in *Inorganic Synthesis* (Equation (1)) ⟨90IS239⟩. A related preparation of the compound $(Me_2PhSi)_3CH$ (70%) has been described starting from the corresponding chlorosilane, bromoform, and *n*-butyllithium ⟨85JCS(P2)729, 93JOM(462)45⟩. The trichloromethyl groups of α,α,α-trichloroethane ⟨63JA2243⟩, 1,1,1-trichlorotoluene ⟨78CB3573⟩, and α,α,α,α′α′α′-hexachloro-*p*-xylene ⟨67AG(E)942⟩ can also be transformed into the tris(TMS)methyl groups.

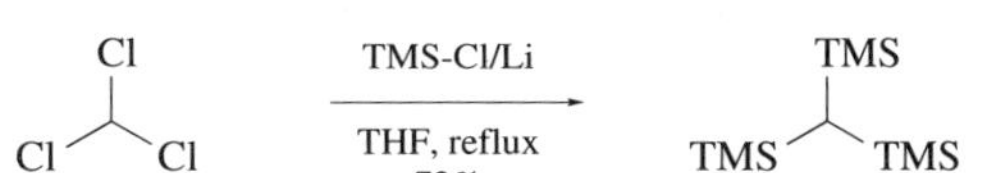

(**3**) (1)

The reductive silylation was shown to tolerate an alkyne function. Thus hexachlorobutyne-2 and chlorotrimethylsilane react with lithium metal to furnish hexa(TMS)butyne-2 (**9**) in 58% yield (Equation (2)) ⟨93JOM(462)31⟩. Remarkably, this compound, owing to the steric shielding of the central π-system, resembles trisilylmethanes in reactivity; its C≡C triple bond cannot be hydrogenated at 200 °C/50 atm in the presence of a palladium–charcoal catalyst.

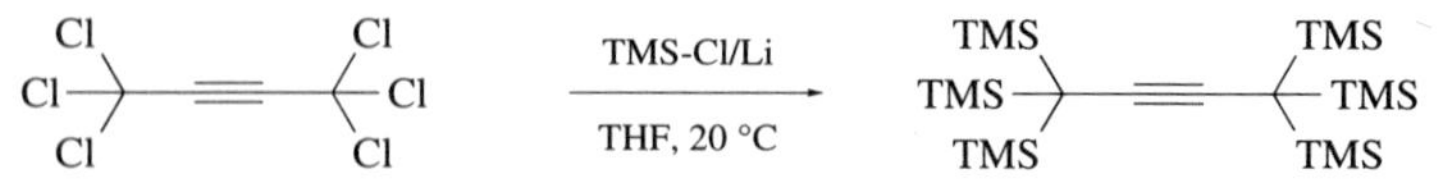

(9) (2)

The partially alkylated halosilanes can also be used to obtain trisilylmethanes. An example is the preparation of tris(dimethylsilyl)methanes (**10**) from dimethylchlorosilane and bromoform or α,α,α-tribromotoluene (Scheme 2). Bromination of silanes (**10**) and the subsequent reactions of the bromosilanes (**11**) with $AgBF_4$, *t*-butylamine, and water illustrate the possible synthetic routes to Si-functionalized trisilylated methanes ⟨83JOM(252)281, 93IC2308⟩.

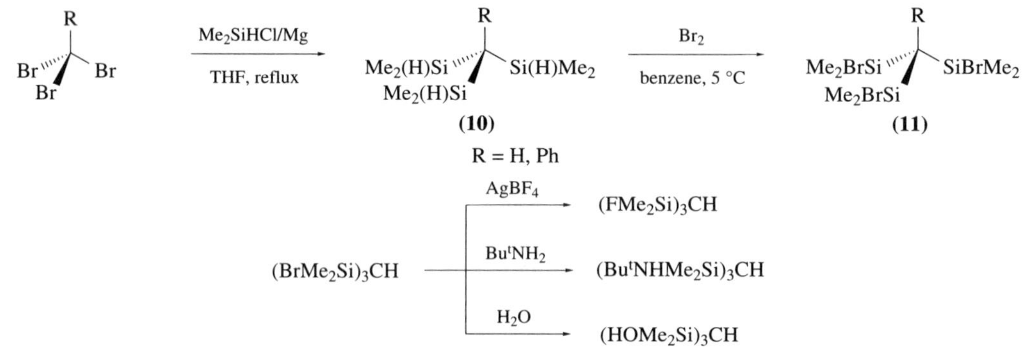

Scheme 2

A potentially very promising method for the build up of trisilylated C-1 units is the direct reductive silylation of carboxylic acid derivatives. However, whereas α-bis(TMS)alkanols can be readily obtained from esters using TMS-Cl/Na in THF ⟨B-81MI 606-01⟩, attempts to realize the direct conversion of esters into trisilylated alkanes has so far had only limited success. For example, hexakis(TMS)-2-butene (**12**) could only be synthesized from dimethyl maleate in 1.5% yield (Equation (3)) ⟨93CB2227⟩.

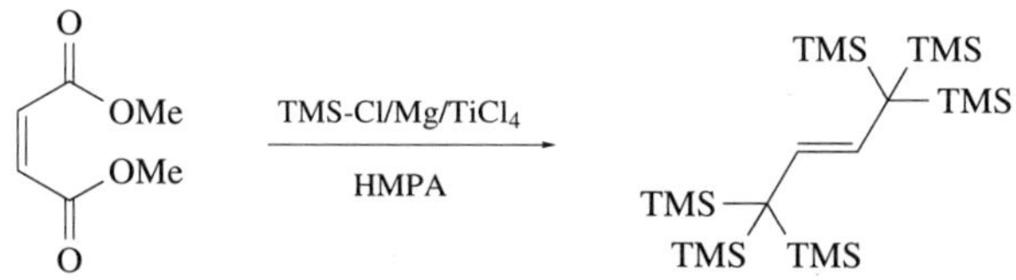

(12) (3)

HMPA = hexamethyl phosphoramide

For the synthesis of trisilylmethanes with different silyl groups attached to the carbon atom the major synthetic pathway is a multistep reaction involving the interaction of a suitable Si-functionalized organometallic carbon nucleophile with a halosilane. Scheme 3 illustrates this methodology ⟨76JCS(D)2268, 82CC1323, 832AAC(497)21, 85S717⟩. Other typical examples are given in Schemes 4–6 ⟨86CB1455, 87AG(E)584, 92ZN(B)805⟩.

Direct incorporation of a trisilylmethyl group into organic molecules may be achieved via organolithium or organomagnesium derivatives. The organolithium reagents, $(R_3Si)_3CLi$, are available by metallation of tris(silyl)methanes with methyllithium in ether/THF ⟨70JOM(24)529, 73JOM(55)209, 83CC827, 84JOM(269)217, 86CC969, 90IS239⟩, but can also be generated at low temperatures from the tris(silyl)methyl halides and lithium metal or via the halogen–metal interconversion with phenyllithium (Table 1) ⟨86CC1043, 92JCS(D)1015, 92OM2938⟩. Tris(TMS)methylmagnesium bromide is formed in 75–90% yields from $(TMS)_3CBr$ (in turn obtained by photobromination of neat $(TMS)_3CH$ at 180–190 °C) and magnesium metal in diethyl ether ⟨92OM2938⟩.

Tris(TMS)methyllithium couples normally with methyl iodide to give the 1,1,1-tris(TMS)ethane. Carbonylation leads to the acid $(TMS)_3CCO_2H$ in 76% yield ⟨70JOM(24)529⟩. The Grignard reagent

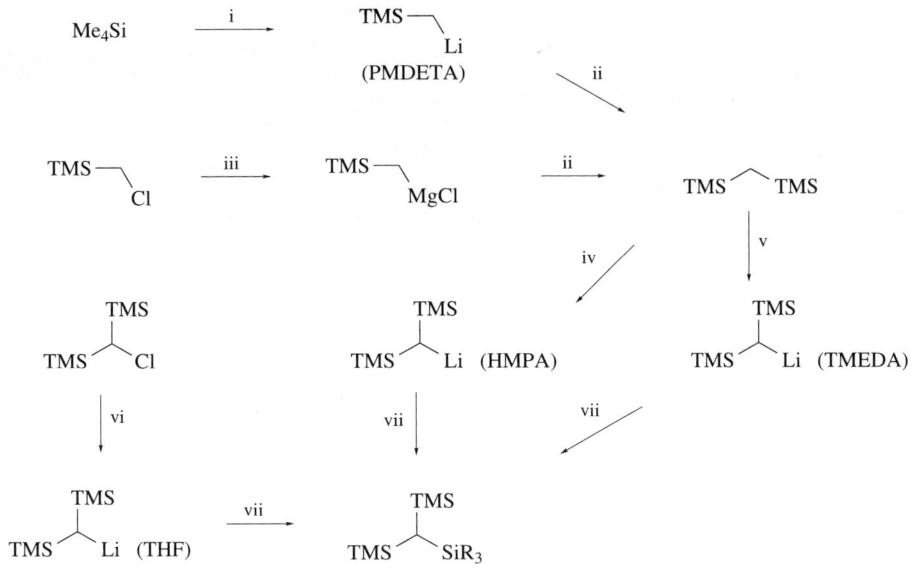

Reagents: i, Bu^nLi/PMDETA, hexane, 20 °C; ii, TMS-Cl, hexane/ether; iii, Mg, ether, 20 °C; iv, Bu^tLi/HMPA, THF, –40 °C, 5 h; v, Bu^nLi/TMEDA, hexane, 25 °C, 12 h; vi, Li, ether, reflux, 24 h; vii, $Me_2PhSiCl$, THF, 20 °C.

(PMDETA = pentamethyldiethylenetriamine)

Scheme 3

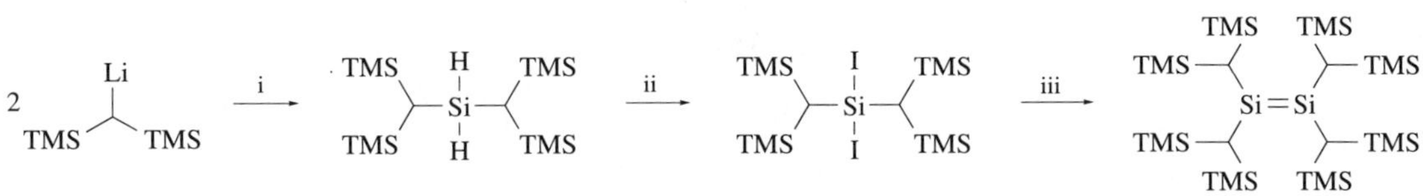

i, SiH_2Cl_2, ether/pentane, –78 °C, 66%; ii, I_2, 100 °C, 24 h, 37%; iii, $C_{10}H_8Li$, THF, –78 °C, 40%

Scheme 4

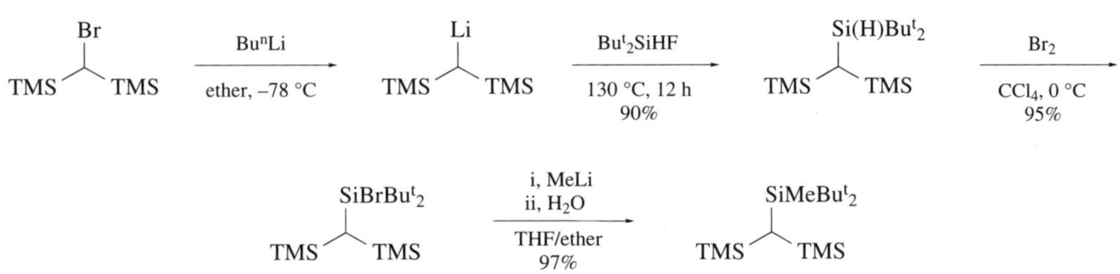

Scheme 5

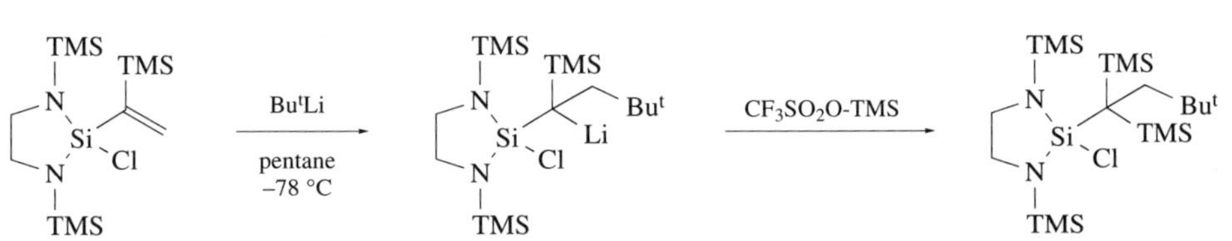

Scheme 6

afforded 1,1,1-tris(TMS)-2-phenylethane (74%) when derivatized with benzyl chloride ⟨92OM2938⟩. Other carbon electrophiles such as nonenolizable aldehydes, ketones, acid chlorides, and epoxides react with $(TMS)_3CLi$ to give the functionalized silanes (Scheme 7) ⟨81JCS(P1)969⟩. However, the method is not completely general; it is limited by the readiness with which $(TMS)_3CLi$ abstracts a proton, if one is available, rather than attacks a carbon ⟨74AG(E)83, 77CB852, 81JCS(P1)969⟩.

Table 1 Preparation of trisilylmethyllithium species.

Substrate	*Reagent*	*Reaction conditions*	*Ref.*
$(TMS)_3CH$	MeLi	Et_2O/THF, reflux, 5 h	90IS239
$(Me_2Ph)_3CH$	MeLi	THF, reflux, 6 h	83CC1390
$(TMS)_3CBr$	Li	ether, 20 °C, 1 h	92OM2938
$(TMS)_3CBr$	PhLi	ether, 0 °C, 3 h	92OM2938
$(MeOMe_2Si)_3CCl$	Bu^nLi	THF, −78 °C	86CC1043, 92JCS(D)1015
$(MeOMe_2Si)_2(TMS)CCl$	Bu^nLi	hexane, −78 °C	92JCS(D)1015
$(MeOMe_2Si)(TMS)_2CCl$	Bu^nLi	THF/ether, −100 °C, 1 h	85JCS(P2)1687, 91JOM(405)149

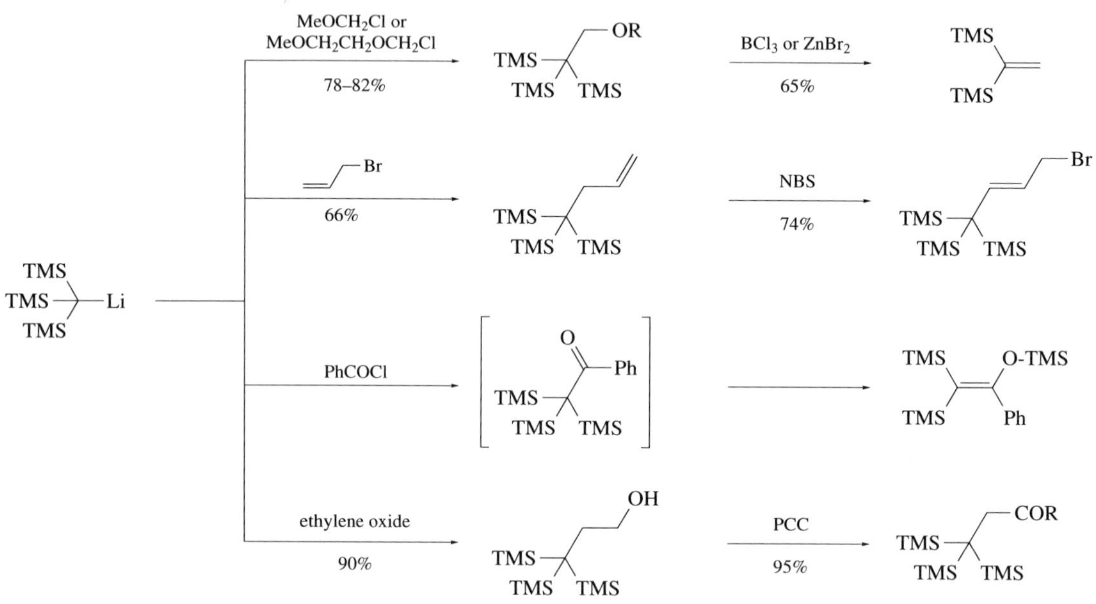

Scheme 7

The reaction of tris(TMS)methyllithium with *O*-ethyl thioformate gave the stable aliphatic thioaldehyde (**13**). This was reduced with sodium borohydride to give the corresponding thiol (**14**) (Scheme 8) ⟨87JA279⟩.

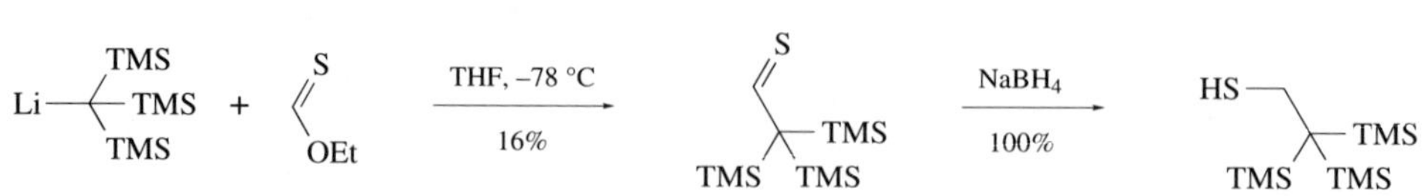

(**13**) (**14**)

Scheme 8

Tris(TMS)methyl-substituted ethene (**15**) is reported to be obtained by the reaction of 1,2-dilithio[tetrakis(TMS)]ethane with paraformaldehyde. The formation of the compound implies 1,2-dianionic rearrangement of the silyl group (Equation (4)) ⟨89JA3748⟩.

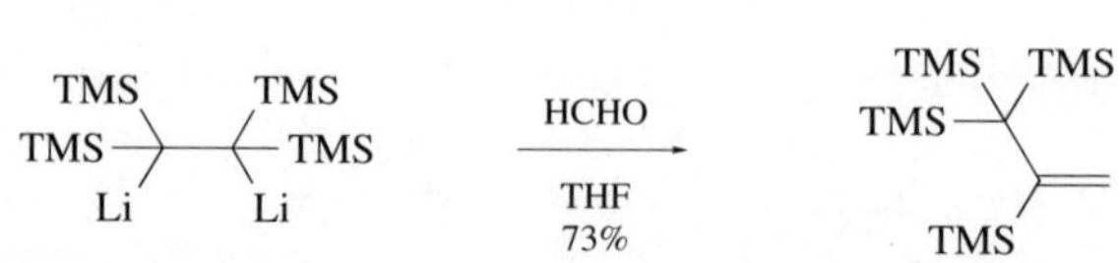

(**15**) (4)

A quite different route to trisilylated methanes and their derivatives involves the use as substrates of low-coordinate silicon derivatives. In principle the insertion of unsaturated silicon compounds of the type $(X_3Si)_2C{=}SiR_2$ into a polar element–hydrogen bond provides a general approach to tris(silyl)methanes. Since the silaethenes are initially obtained from polysilylated methanes, this appears to be a circuitous technique. However, it does lead to the specific polyfunctional compounds which are difficult to prepare by other reactions ⟨84JOM(273)141⟩. Thus alcohols, phenols, thiols, hydrogen halides, and amines add to the silaethenes (**16**) and (**17**) to give the insertion products (Equation (5)). Ionic reagents like PhSLi and $(PhO)_2P(O)OLi$ have been employed in these reactions with equal success ⟨81CB3518, 84JOM(273)141⟩. The 1,1-dimethyl-2,2-bis(TMS)silaethene also inserts into the α-CH bond of pyridine to yield the trisilylmethane (**18**) (Scheme 9) ⟨86JOM(315)9⟩. For all these reactions a nucleophilic attack mechanism can be invoked.

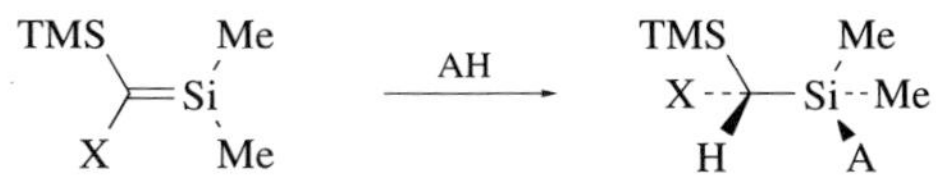

(5)

(**16**) X = TMS

(**17**) X = Bu^t_2MeSi

AH = H_2O, ROH, RSH, R_2NH, HHal

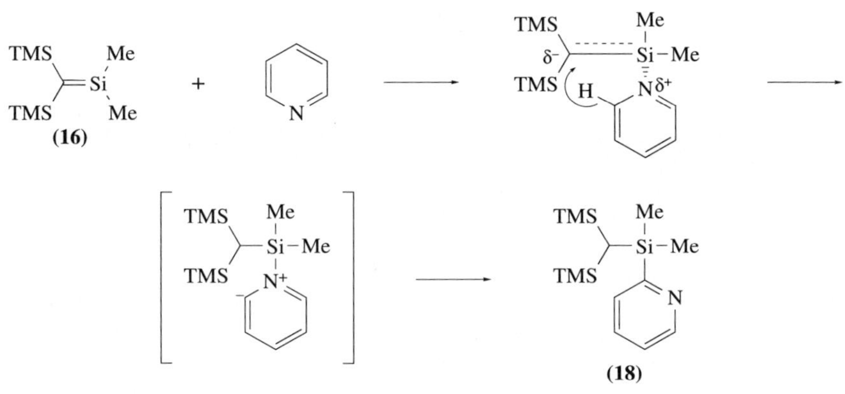

Scheme 9

The formation of trisilylmethane derivatives proceeds equally well when *in situ* generated silaethene is treated with a proton donor reagent ⟨84JOM(273)141, 85CRV419, B-89MI 606-01⟩. For example, trisilylated methanes (**19**) have been isolated in good yield from reacting tris(TMS)methylsilicon halides with methanolic sodium methoxide. The reaction is believed to proceed through an elimination, analogous to E2 eliminations of alkyl halides, involving synchronous attack of MeO^- at a TMS group, liberation of X^-, and formation of the transient silaethene followed by addition of methanol across the C=Si bond (Scheme 10). The compounds $(TMS)_3CSiPhMeF$ and $(TMS)_3CSiPhCl_2$ react analogously to give $(TMS)_2CHSiPhMe(OMe)$ and $(TMS)_2CHSiPhCl(OMe)_2$ ⟨78JOM(157)C50, 80JOM(191)355⟩. It should be noted that in contrast to the silicon species, the germanium compounds $(TMS)_3CGeR_2X$ undergo normal (though very slow) direct nucleophilic substitution at germanium ⟨80JOM(202)157⟩.

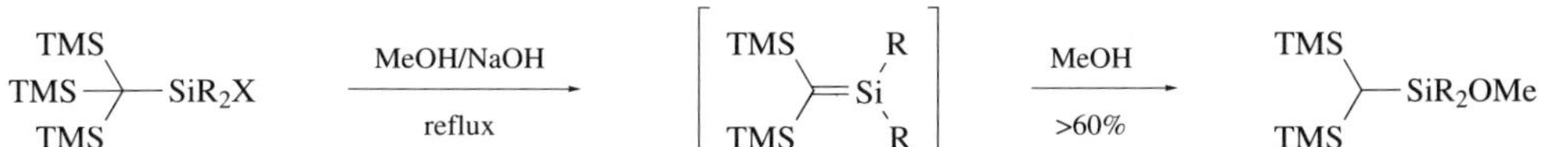

R = Me or Ph, X = Cl, Br or I
R = Ph, X = F

Scheme 10

The ene reactions of silaethenes are also potentially useful for the preparation of functionalized trisilylmethanes. Thus silaethene (**16**) reacts with isobutene at −10 °C to give compound (**20**). A similar reaction occurs between (**16**) and other alkenes ⟨87ZN(B)1062⟩. The rates of these reactions are sensitive to steric hindrance but even the sterically encumbered silaethene (**17**) still adds propene and isobutene ⟨86CB1467⟩. The ene reactions of silaethenes with carbonyl compounds containing a proton in the α-position such as acetone, cyclohexanone, or ethyl acetate are also potentially useful for the preparation of functionalized trisilylmethanes (Scheme 11) ⟨83AG(E)1005, 86CB1467, 86JOM(315)9, 87ZN(B)1062⟩.

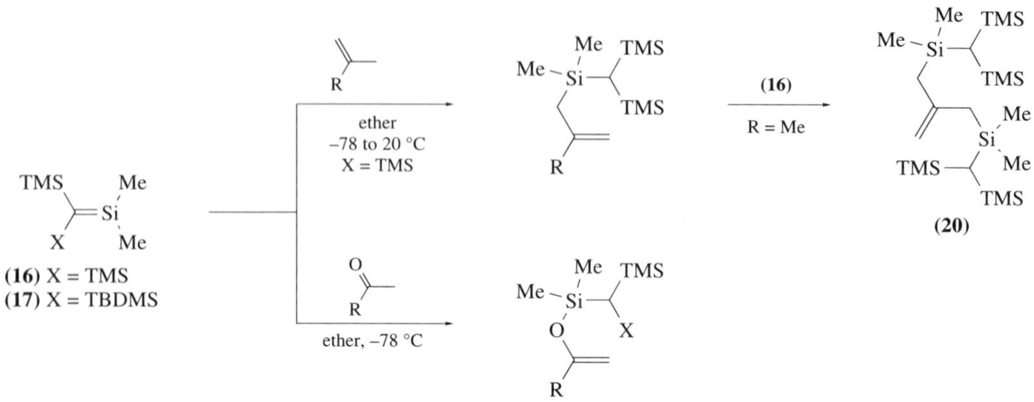

Scheme 11

(ii) Compounds with an Si_3C function as part of one or more ring systems

These can be prepared by a variety of methods of which the main ones are:

(a) reductive silylation of a halohydrocarbon with a suitable halosilane (the Merker–Scott procedure);

(b) metal-induced coupling of chloromethylchlorosilanes;

(c) reaction between a lithiated cyclic carbosilane and a silicon electrophile; and

(d) cycloaddition to multiple carbon–silicon bond bearing silicon substituents at the carbon atom.

Synthesis of the tricyclic carbosilane (**22**) was accomplished by preparing the cyclic carbosilane (**21**) and reacting it with bromoform and lithium metal (Equation (6)) ⟨72ZAAC(390)157⟩. A similar reductive silylation route with silane (**23**) gave the tetrasilabicyclo[2.2.2]octane (**24**) (Scheme 12) ⟨72ZAAC(390)191⟩. 1,3-Disilabicyclo[1.1.0]butane (**26**) has been synthesized by a method involving the low temperature reaction of carbosilane (**25**) with butyllithium in THF (Scheme 13) ⟨81ZAAC(475)87, 84JOM(271)107⟩. From the trisilylmethane (**27**) tris(dimethylchlorosilyl)ethene results in a remarkable yield (>85%) presumably via a silacyclopropane intermediate (Scheme 14) ⟨77ZAAC(430)121⟩.

(21) (22) (6)

Synthesis of cyclocarbosilanes including an Si_3C unit from "carbon anion" precursors appears to be a generally applicable method that is primarily limited by the availability of suitable starting materials ⟨74TCC43⟩. Typically a lithiated cyclocarbosilane is generated in solution and then treated with the appropriate silylating reagent. For example, the conversion of 1,3,5-trisilacyclohexane (**28**) into trisilylated derivative (**29**) has been achieved by lithiation with Bu^nLi/TMEDA complex followed by treatment with a chlorosilane (Scheme 15) ⟨83ZAAC(497)21⟩.

Among the many reactions of unsaturated silicon–carbon compounds which lead to cyclocarbosilanes, several include photochemical or thermal generation of silaethenes and their subsequent conversion into cyclic compounds containing an Si_3C unit. Thus irradiation of a benzene solution of (**30**) with a high-pressure mercury lamp gave 1,2-disilacyclobutane (**31**) in 90% yield (Scheme 16) ⟨79JA1348, 81JA2324⟩.

Both photolysis and thermolysis of bis(TMS)diazomethane produce a carbene which rearranges

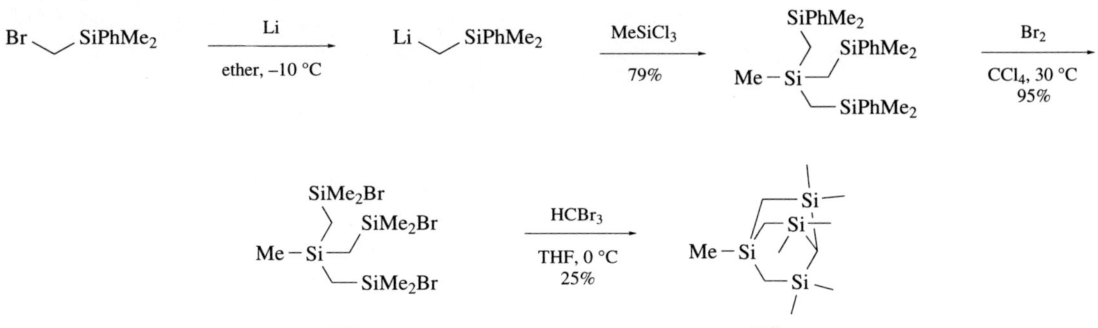

(23) **(24)**

Scheme 12

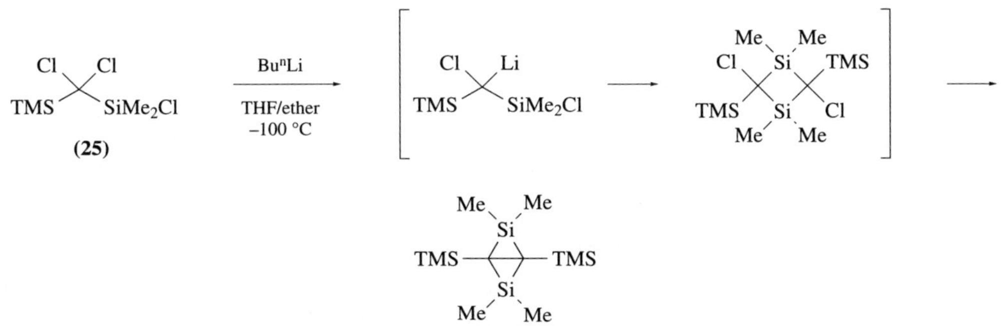

(25) **(26)**

Scheme 13

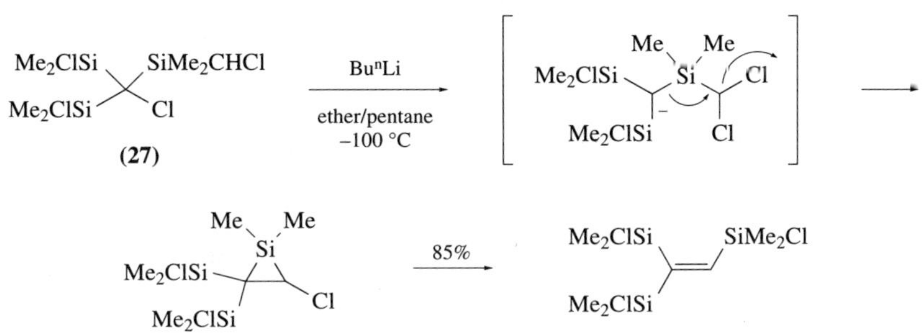

(27)

Scheme 14

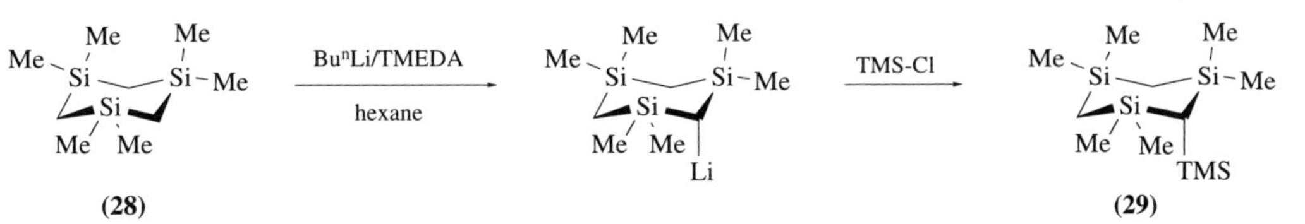

(28) **(29)**

Scheme 15

via methyl migration to silaethene Me(TMS)C═SiMe$_2$. The latter dimerizes to produce as the major products *cis*- and *trans*-1,3-disilacyclobutanes, Me(TMS)C(SiMe$_2$)$_2$C(TMS)Me (46%) ⟨80JA1584⟩.

Decomposition of compound (**32**) in the gas phase at 450°C affords product (**33**) via the corresponding silaethene (Scheme 17) ⟨80JOM(186)309⟩.

Unsaturated systems of the type a═b, a—b═c, and a═b—c═d often combine with the *C*,*C*-disilyl-substituted silaethenes to give [2 + 2]-, [2 + 3]-, and [2 + 4]-cycloadducts containing Si$_3$C units

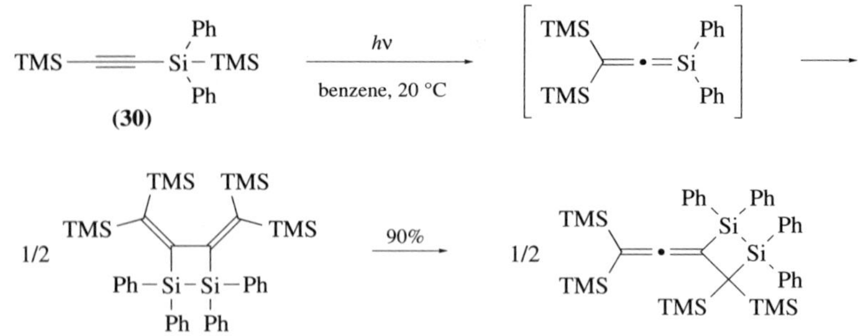

Scheme 16

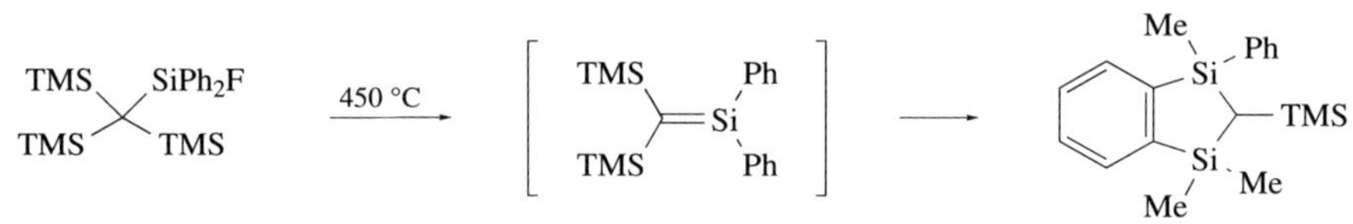

Scheme 17

⟨84JOM(273)141⟩. However, practically all such reactions are only useful for specific derivatives or specific groups of compounds. Characteristic examples are given in Scheme 18. The review by Wiberg contains experimental details for the preparation of compounds **(34)**–**(36)** ⟨84JOM(273)141⟩.

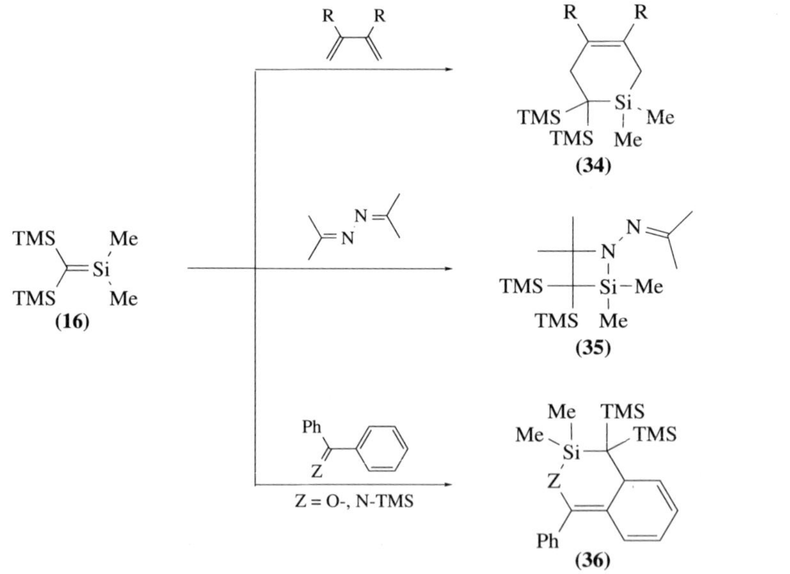

Scheme 18

6.06.2.1.2 Functions bearing three borons

Tri-coordinate boron compounds, BX_3, have a formally vacant orbital in the valence shell and are isoelectronic with the corresponding carbocations, $^{+}CX_3$. For triorganoboranes, BR_3, there is little possibility of populating this orbital by conjugative interaction (apart from hyperconjugation) with the substituents. It is hardly surprising, therefore, that the simple tris(boryl)methanes such as $HC(BH_3)_3$ or $HC(BMe_3)_3$ are highly reactive, unstable compounds. This is also a major reason why compounds of the type $HC(BX_2)_3$ that have actually been isolated, have X groups in which a heteroatom capable of donating π-electron density is bound to boron. The preparation and properties of the species containing a B_3C function vary a great deal with the structural type and character of substituents, and they will therefore be discussed under these headings.

(i) Tris(dihaloboryl)methanes, $RC(BHal_2)_3$

The homologous $C(BHal_2)_nHal_{4-n}$ ($n = 1$–4) were observed in the reaction between various boron halides and carbon vapor generated from a carbon arc. In this manner B_2Cl_4 afforded a mixture of $C(BCl_2)_4$, $ClC(BCl_2)_3$, and $Cl_2C(BCl_2)_2$; and BCl_3 yielded the last two compounds as well as $Cl(Cl_2B)C{=}C(BCl_2)_2$. Compounds $ClC(BCl_2)_3$ and $Cl_2C(BCl_2)_2$ are thermally unstable above $-20\,°C$. From the reaction of B_2F_4 and carbon vapor, $C(BF_2)_4$ was only isolated in a small amount ⟨69JCS(A)1882⟩.

The method for synthesizing tris(dichloroboryl)methane (**37**) is the exchange reaction of tris-(dimethoxyboryl)methane with boron trichloride and a catalytic amount of lithium borohydride (Equation (7)). Reaction conditions were found to be critical; it is important that a large excess of boron trichloride be used and that the crude product be distilled rapidly, since it is particularly unstable to be stored ⟨73IC2472⟩. Reaction of structure (**37**) with boron tribromide failed to lead to pure tris(dibromoboryl)methane, $HC(BBr_2)_3$, but bromine–chlorine exchange did occur and the HCB_3 signal of $HC(BBr_2)_3$ appeared in the NMR spectrum ⟨72MI 606-01, 73IC2472⟩.

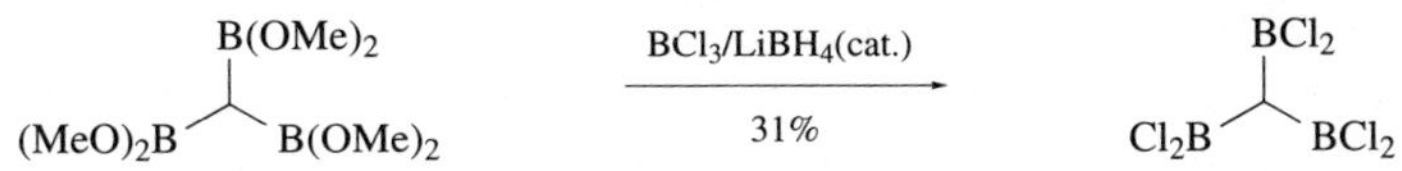

(**37**) (7)

There are several reports in the literature that the addition of boron halides across multiple bonds leads to 1,1-diboryl-substituted alkanes ⟨82HOU(13/3a)1, 91CRV35⟩. Thus, diboron tetrachloride adds twice to alkyne giving $CH(BCl_2)_2CH(BCl_2)_2$, and dehydroboration of alkynes with dichloroborane ethyl etherate ($BHCl_2{\cdot}Et_2O$) in the presence of BCl_3 yields *gem*-diboryl compounds $RCH_2CH(BCl)_2$ ⟨76JA1798⟩. However, despite many unique advantages associated with hydroboration and haloboration, the synthesis of 1,1,1-tris(dihaloboryl)alkanes via alkynes has not yet been realized.

(ii) Tris(dialkoxyboryl)methanes, $RC(B(OAlk)_2)_3$

These compounds have also been called methanetriboronic esters. The compound $HC(B(OMe)_2)_3$ is named in *Chemical Abstracts* as methylidynetrisboronic acid hexamethyl ester.

The preparation and properties of tris(dialkoxyboryl)methanes have been reviewed up to 1976 ⟨75S147, B-77MI 606-01⟩. Tris(dimethoxyboryl)methane (**38**) is readily available on a large scale by direct reaction of chloroform with dimethoxyboron chloride and lithium metal in THF (Equation (8)) ⟨68JA2194, 69JOM(20)19⟩. The procedure has been described in detail; the product is isolated by distillation ⟨75S147⟩. 1,1,1-Trichloroethane and α,α,α-trichlorotoluene react analogously, but attempts to condense polyhalomethanes with bis(dimethylamino)boron chloride, $ClB(NMe_2)_2$, have failed.

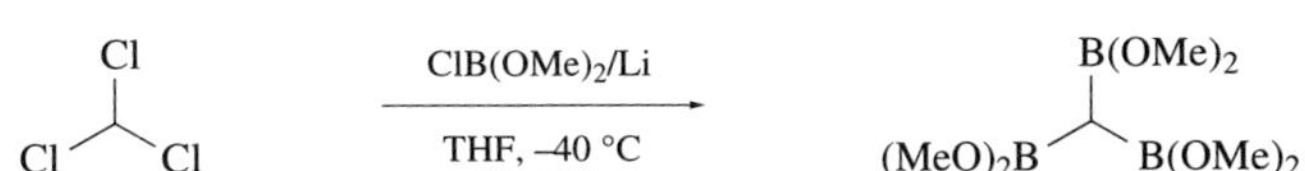

(**38**) (8)

Transesterification of structure (**38**) with ethylene glycol in THF precipitates cyclic boronic ester (**39**) (85%) ⟨75JA5608, 76JOM(110)25⟩. Similarly the cyclic boronic esters (**40**) and (**41**) were formed from the tris(dimethoxyboryl)methane and pinacol or 1,3-propanediol in the presence of a catalytic amount of boron trifluoride etherate ⟨70JOM(24)263, 74JOM(69)45, 75JOM(93)21⟩. These compounds proved to be more stable than their acyclic analogues and better yields of products have been obtained from them in a wide variety of reactions ⟨75S147⟩.

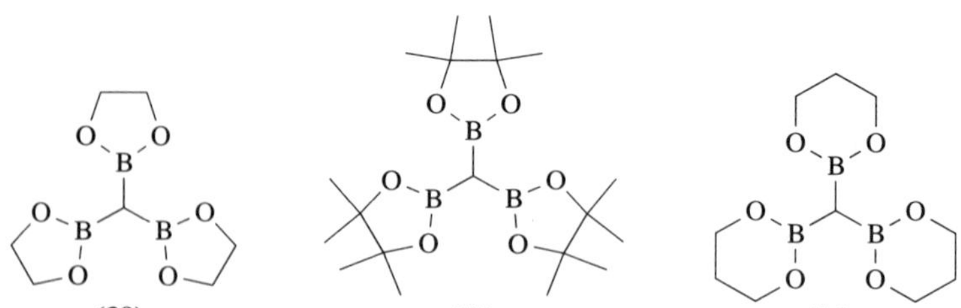

(39) (40) (41)

The transborylation reaction of tris(dimethoxyboryl)methane (**38**) with triethylborane in the presence of "ethylborane" provides a route to the only known 1,1,1-tris(dialkylboryl)methane (**43**). The initially formed mixed triborylmethane (**42**) undergoes transformation to the compound (**43**) on heating at about 100 °C (Scheme 19) ⟨75LA1339⟩. The full scope of this reaction, however, remains to be explored further.

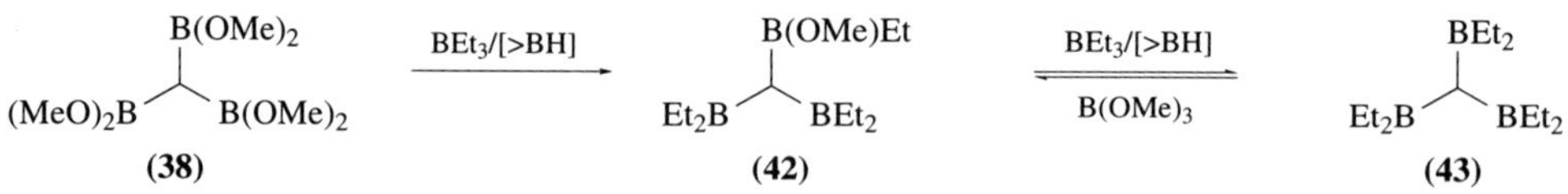

Scheme 19

The alkylation of the anionic species derived from methanetetraboronic esters ⟨73JA5096⟩ by alkyl halides was found to proceed readily, but unfortunately the reaction leads to a mixture of monoalkylated and dialkylated derivatives, as illustrated in Scheme 20 ⟨70JOM(24)263, 75S147⟩. Evidently the initially formed 1,1,1-triborylalkane (**46**) is able to transfer a dimethoxyboryl group to tris(dimethoxyboryl)methide ion (**45**), thus undergoing disproportionation. Condensation of triborylmethide anions with aldehydes and ketones affords alkene-1,1-diboronic esters (**48**), presumably via the unstable cyclic borate anion (**47**) (Scheme 21) ⟨70JOM(21)P6, 74JOM(69)53⟩. The reaction is a general one and tolerates other functional groups, including α-chloro, carbethoxy, or tertiary amino substituents ⟨74JOM(69)63, 75JOM(93)21, 78JOC950, 78JOM(152)1⟩.

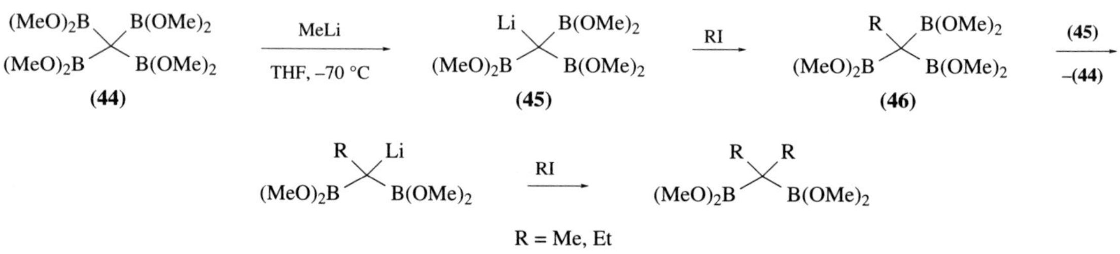

Scheme 20

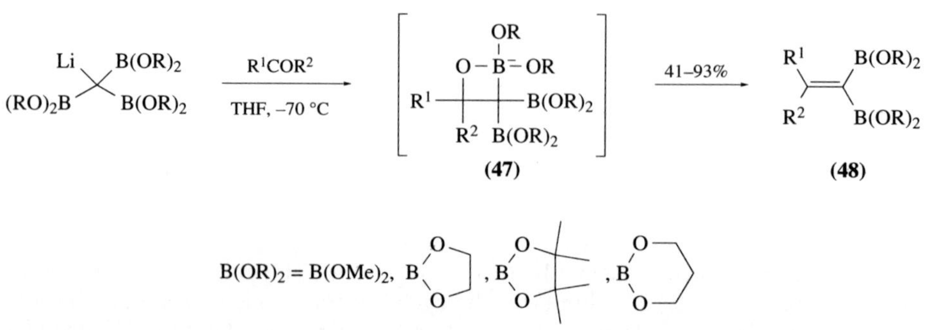

Scheme 21

Finally, it should be noted that metallation of triborylmethide anions with Ph_3ECl, where E = Si, Ge, Sn, or Pb, provides a useful method for the synthesis of compounds $Ph_3EC(B(OR)_2)_3$ ⟨69JA6541, 73JA5096, 73JOM(57)225, 73JOM(57)231, 74JOM(69)63⟩.

(iii) Boracycloalkanes involving a B_3C function

There are several specific methods which are useful for the preparation of boracycloalkanes containing a B_3C unit.

Compounds (**50**) are obtained in up to 50% yield via the isolable 1,1-bis(dialkylboryl)-1-alkenes (**49**) from dialkyl(1-alkynyl)boranes and dialkylboranes (Scheme 22) ⟨64TL1667, 75LA1339, 84HOU(13/3c)162⟩. These have been classified as pentaalkyl-1,5-dicarba-*closo*-pentaboranes(5), but the valency requirements of all atoms can be satisfied without the need to involve electron-deficient bonding ⟨70JA4158, 74JCS(D)665, 91CB2645⟩.

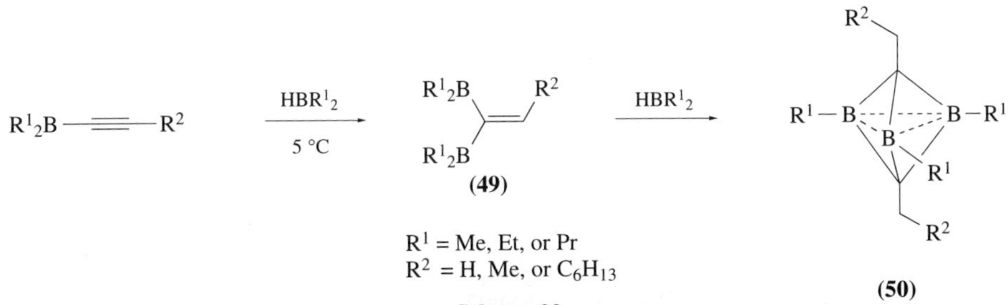

Scheme 22

An efficient synthesis of 2,3,5,6-tetraborabicyclo [2.1.1]hexane (**51**) utilizes the reaction of 1,1,2,2-tetrakis(chloro(diisopropylamino)boryl)ethane with sodium/potassium alloy in benzene (Equation (9)) ⟨90AG(E)292⟩.

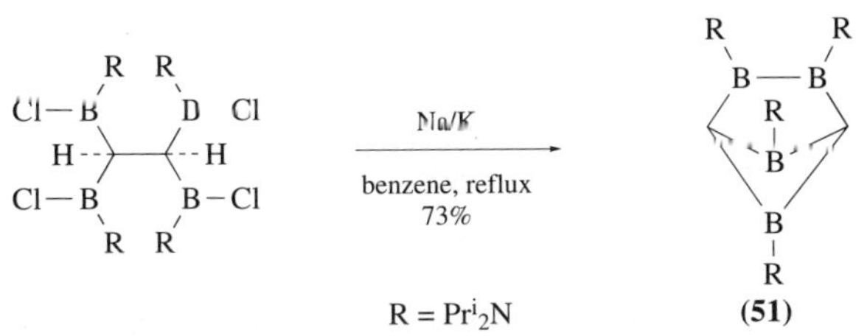

(9)

Sealed-tube pyrolysis of BMe_3 permits isolation of 2,4,6,8,9,10-hexaboraadamantane (**52**) in 25% yield ⟨73CC532, 75JCS(D)148⟩. This boron–carbon cage compound appears to be unique in having a stoichiometry expected of a carborane and yet not possessing the usual carborane type of structure ⟨77JCS(D)136⟩. The ethyl analogue (**53**) is produced by thermal decomposition of the tris-(diethylboryl)methane or polyboryl compounds, themselves obtainable by hydroboration of dialkyl(alkynyl)boranes ⟨75LA1339⟩.

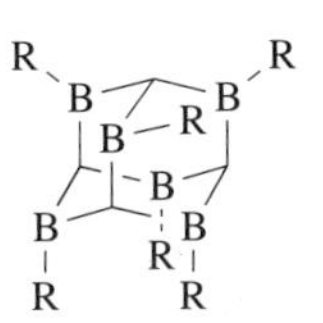

(**52**) R = Me
(**53**) R = Et
(**54**) R = Bu^t
(**55**) R = Cl
(**56**) R =Br
(**57**) R = Me_2N

Pyrolysis of alkyldichloroboranes such as $MeBCl_2$, $Cl_2BCH{=}CHBCl_2$ or $(Cl_2B)_3CH$ at 450°C forms hexachlorohexaboraadamantane (**55**) (5–28%). The synthesis of hexabromohexabora-adamantane (**56**) has been achieved by heating Me_2BBr or $MeBBr_2$ to 520°C. The reactions of structure (**55**) with *t*-butyllithium and diethylamine at −60°C lead to the formation of the corresponding *t*-butyl- (**54**) and dimethylamino- (**57**) derivatives ⟨89JOM(367)19⟩.

Another route to compounds possessing C_4B_6–adamantane structure uses dimerization of the C_2B_3-*closo*-carbaborane (**50**) by treatment with potassium followed by iodine (Equation (10)) ⟨85AG(E)326⟩.

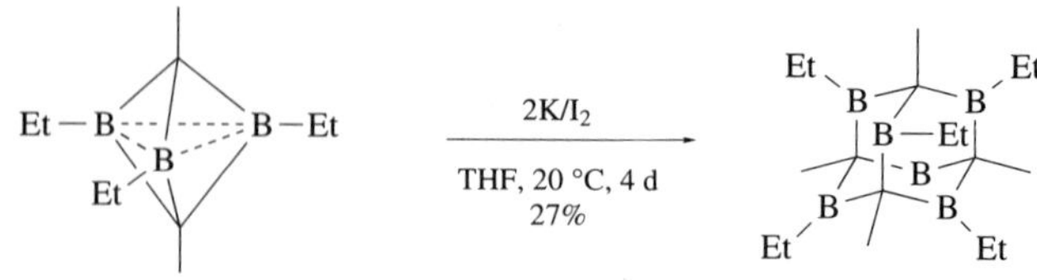

(50) **(58)** (10)

The formation of compounds $C_6H_9{\cdot}x(BNPr^i_2)$ ($x = 1$–6) from the reaction of benzene with dichloro(diisopropylamino)borane and sodium/potassium alloy in 1,2-dimethoxyethane has been confirmed by mass spectroscopic studies. The compound with $x = 3$ was separated from the reaction mixture and identified as 2,8,9-triborabicyclo[3.3.1]nona-3,6-diene-2,8,9-triamine (**59**) (Equation (11)) ⟨88JOM(347)11⟩. Similarly, 1,4,6-triboraspiro[4.4]nona-2.8-diene (**60**) has been isolated from the reaction of benzene with subvalent boron species ⟨90CC741⟩.

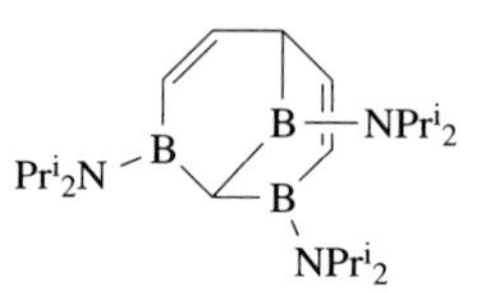

(59)

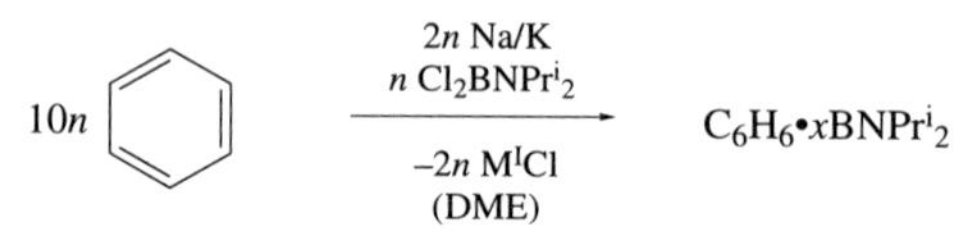

(11)

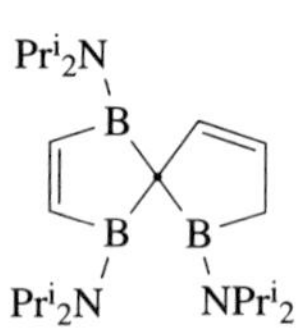

(60)

Addition of a mixture of bis(TMS)butadiyne or 2,5-dimethyl-2,4-hexadiene with $Pr^i_2NBCl_2$ to Na/K alloy in hexane gave the 2,4,5-triborabicyclo[1.1.1]pentane (34%) (Equation (12)) ⟨92CB1807⟩.

(12)

A unique method for the synthesis of compounds containing a B_3C function utilizes low-coordinate boron derivatives. 2-Boranediyl-1,3-diboretanes (**61**) smoothly add ethanol to give compounds of structure (**62**) in quantitative yields (Equation (13)) ⟨89AG(E)781⟩. Despite considerable synthetic potential, this method is clearly limited by the availability of the requisite boron substrates.

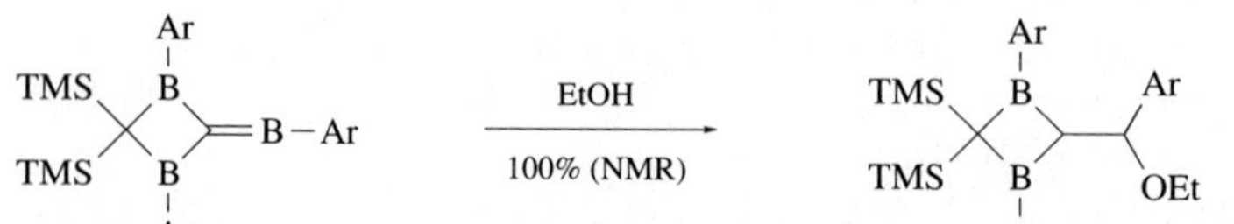

(61) **(62)** (13)

$Ar = 2,4,6\text{-}Me_3C_6H_2,\ 2,3,5,6\text{-}Me_4C_6H$

6.06.2.1.3 Functions bearing three germaniums

In contrast to numerous publications dealing with trisilylmethanes, literature reporting trigermylmethyl derivatives are sparse.

It is curious that straightforward synthetic routes are known for compounds containing Si_3C functions but not for those with functions bearing three germaniums. It appears that many of the methods suitable for the synthesis of mono- and bisgermylalkanes are ineffective or not applicable to trigermylmethanes. For example, although the compound $(Et_3Ge)_2CH_2$ was successfully obtained by the reaction of triethylgermylpotassium with dichloromethane, attempts to prepare $(Et_3Ge)_3CH$ by a similar route have failed, because the metal–halogen exchange reaction, followed by condensation, predominated ⟨67TL1443⟩. The reaction of Ph_3GeNa with chloroform in liquid ammonia was also unsuccessful. Only one of the three chlorine atoms bonded with a carbon is quantitatively replaced by the Ph_3Ge group; a second chlorine is largely substituted while the third is completely replaced by hydrogen ⟨32JA1622, 52JA1418⟩.

The first stable compound with a Ge_3C function, 1,5,5,5-tetrakis(trimethylgermyl)-1,3-pentadiyne (**64**), has been synthesized by condensation of the tetralithium compound from 1,3-pentadiyne with chlorotrimethylgermane (Scheme 23) ⟨73JA3324⟩. Reaction leads to a mixture of two pergermylated isomers, (**63**) (42%) and (**64**) (17%), which were separated by preparative gas chromatography. The method is only of limited value, since polylithiated species are not readily available.

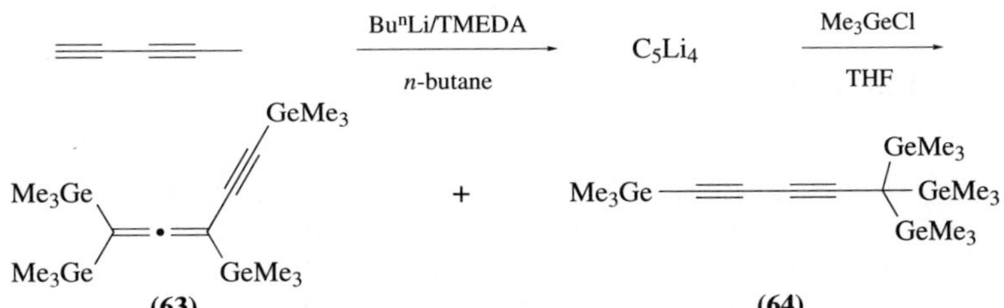

Scheme 23

1,1,1-Tris(triethylgermyl)acetone was reported to form in very low yield from the reaction of triethylgermyldiazoacetone with an equimolar amount of bis(triethylgermyl)mercury in the presence of metallic copper ⟨77JOM(142)155⟩. Although this transformation is of some mechanistic interest, it does not provide a synthetically viable route to trigermylmethyl species.

The more flexible and useful route to 1,1,1-trigermylalkanes involves stepwise base-assisted introduction of an R_3Ge group into compounds containing an activated methyl group. Sato and coworkers have prepared a series of trigermylacetic acid derivatives using this approach ⟨88JOM(354)155, 88OM739, 90OM1325⟩. Some of these reactions are outlined in Schemes 24, 25, and 26. The reaction of the lithium salt of trimethylgermylacetonitrile with bromotrimethylgermane is not straightforward and leads to a mixture of germylated species (Scheme 24). Detailed investigation showed that the trimethylgermylacetonitrile anion readily reacts with Me_3GeCH_2CN via intermolecular anionic rearrangement of the Me_3Ge group to give bis(trimethylgermyl)acetonitrile anion and MeCN. Anion $(Me_3Ge)_2C^-CN$, prepared from structure (**66**) with an equimolar amount of lithium diisopropylamide (LDA), is stable enough at low temperature. However, treatment of (**66**) with a 0.5 molar equivalent of LDA gives a mixture of germylated products ⟨87SC1273⟩.

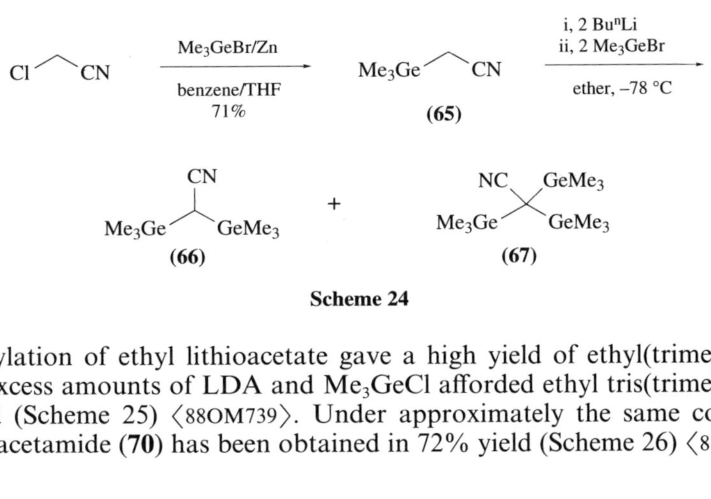

Scheme 24

Trimethylgermylation of ethyl lithioacetate gave a high yield of ethyl(trimethylgermyl)acetate (**68**). The use of excess amounts of LDA and Me_3GeCl afforded ethyl tris(trimethylgermyl)acetate (**69**) in high yield (Scheme 25) ⟨88OM739⟩. Under approximately the same conditions, the tris(trimethylgermyl)acetamide (**70**) has been obtained in 72% yield (Scheme 26) ⟨88JOM(354)155⟩.

(68) **(69)**

Scheme 25

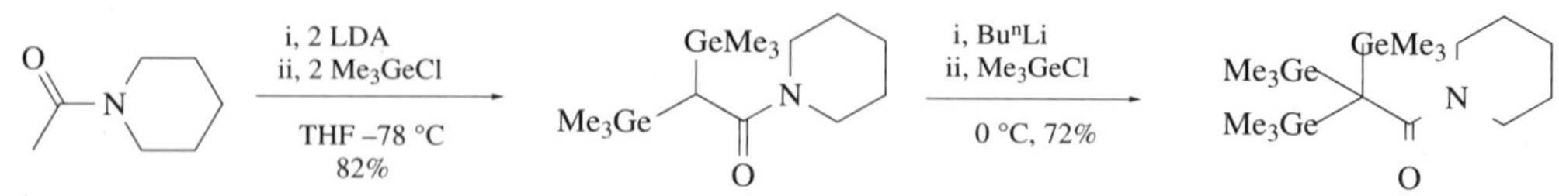

(70)

Scheme 26

The LAH reduction of structure (**69**) gave a 78% yield of 2,2,2-tris(trimethylgermyl)ethanol, and treatment with *n*-butyllithium at room temperature afforded an 86% yield of 1,1,1-tris-(trimethylgermyl)-2-hexanone ⟨88OM739⟩.

6.06.2.1.4 Functions bearing mixed metalloids

Several procedures are available for the synthesis of compounds of the type $RC(E^1X_m)$-$(E^2X_n)(E^3X_{(m)n})$, where E^1, E^2, and E^3 = Si, B, or Ge, but two groups of reactions are most important. The first group includes the reactions of organometallics $RC(M)(E^1X_m)(E^2X_n)$ with inorganic or organoelement halides, $X_{m(n)}E^3Hal$, and the second is the reactions based on unsaturated germanium or boron compounds containing element–carbon double bond, $X_mE^1 = C(E^2X_n)(E^3X_{m(n)})$.

(i) Synthesis based on α-silyl-, boryl-, or germyl-substituted organometallics

Three typical examples illustrating this methodology are shown in Equations (14)–(16) ⟨76JOM(110)25, 78JOM(153)253, 89AG(E)55⟩. The requisite organometallics can be obtained by methods which are discussed in Section 6.06.2.2.

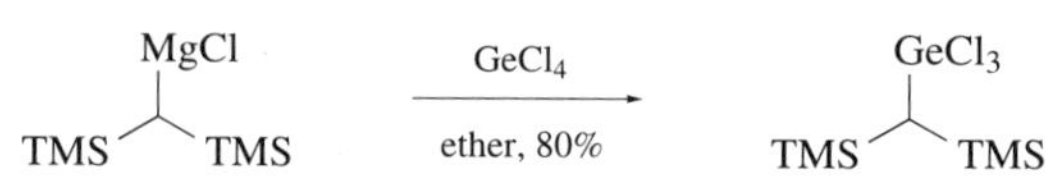

(14)

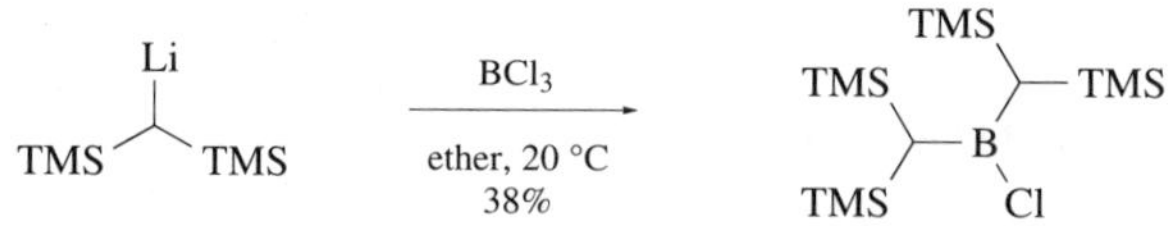

(15)

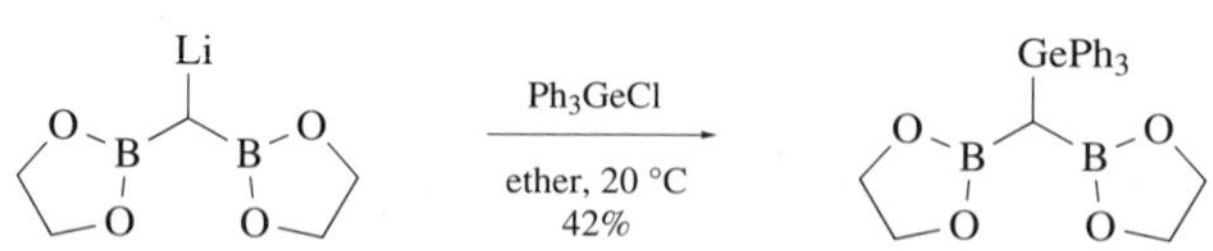

(16)

In principle, the approach provides a general synthetic pathway to compounds containing mixed metalloid functions, but the correct choice of the reactants and reaction conditions is very important. The best results were obtained by the low temperature reaction of a functionalized organolithium or organomagnesium compound with a metalloidal halide in ethereal or THF solution. The group on metalloid which is usually replaced includes Cl, F, RO, or $(TMS)_2N$ ⟨86CB2966, 89CB1057, 91POL1153⟩.

Germanium chlorides react with α-silylated organometallics with retention of the oxidative state of germanium. For example, treatment of $GeCl_2$(diox) or $Ge(N(TMS)_2)_2$ with an equimolar amount of $MgCl(CH(TMS)_2)$ or $Mg(CH(TMS)_2)_2$ in ether leads to the bis(bis(TMS)methy)lgermylene (**71**) in 60–70% yield (Equation (17)) ⟨76CC261, 76JCS(D)2268, 82CC1407, 84CC480, 86JCS(D)1551⟩. The related preparation of bis(TMS)methyl(pentamethylcyclopentadienyl)germylene (**72**) has been described starting from Me_5C_5GeCl and $(TMS)_2CHLi$ (Equation (18)) ⟨86OM1944⟩. It should be noted that bis(TMS)methyl-substituted germylenes such as structures **(71)** and **(72)** and $(TMS)_3CGeCH(TMS)_2$ ⟨91OM1647⟩ represent a unique class of compounds, which can exist as monomers (carbene analogues) or dimers (alkene analogues) depending on the structure and phase state ⟨91CRV311⟩. If an electrophile such as iodine, methyl iodide, acetyl chloride, a diazoalkane, or a diazidosilane is added to the germylene **(72)**, then high yields of the oxidative addition products containing the Ge(IV)$CH(TMS)_2$ unit are obtained ⟨86OM730, 94OM434, 94OM436⟩. A reaction of germylene **(71)** with an alcohol provides the functionalized germane $HGe(CH(TMS)_2)_2OEt$ ⟨81JOM(212)C4⟩.

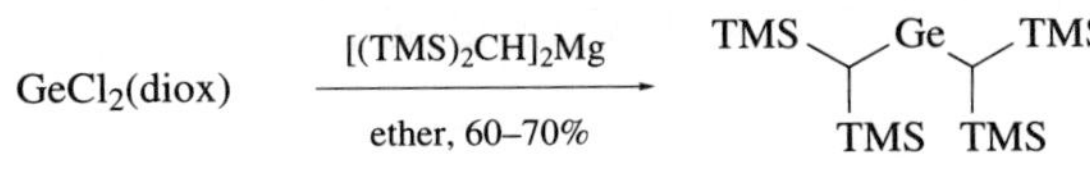

(71) (17)

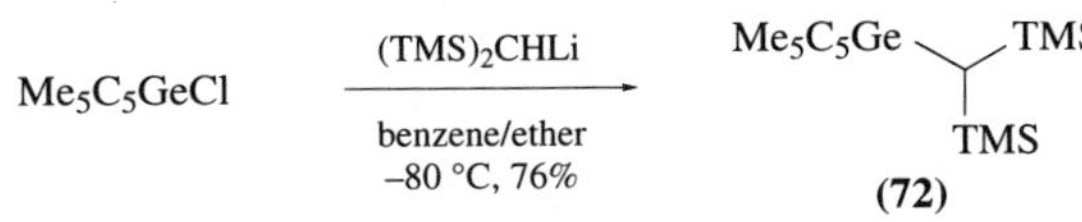

(72) (18)

(ii) Synthesis based on low-coordinate boron and germanium derivatives

C,*C*-Disilyl-substituted methyleneboranes, $XB{=}C(TMS)_2$ ⟨93AG(E)985⟩, and germaethenes, $X_2Ge{=}C(SiR_3)_2$ ⟨84JOM(273)141, 90CRV283⟩, provide access to a variety of compounds containing mixed metalloid functions, but the scope of the reactions is restricted to the preparation of specific polyfunctional compounds.

At a temperature of 500–600 °C methyleneboranes **(73)** and **(74)** undergo rearrangement into cyclic compounds in excellent yield ⟨90CB747⟩. In analogous conditions the methyleneborane **(75)** is transformed to an azasilaboratedine (Scheme 27) ⟨89CB595⟩.

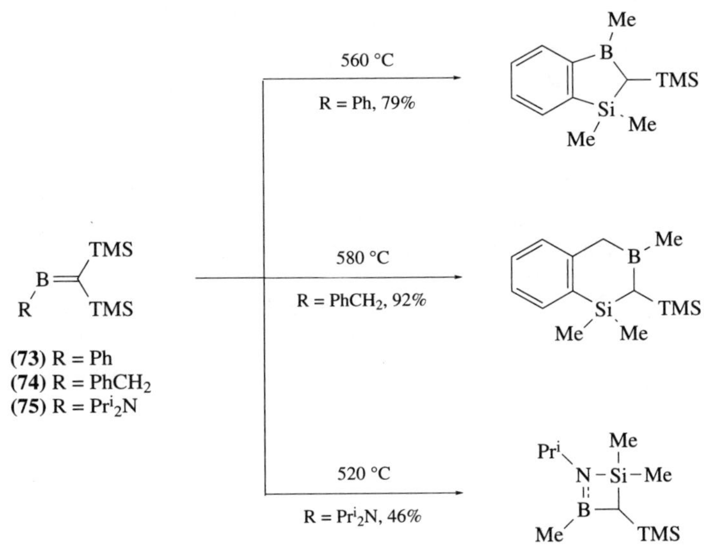

Scheme 27

Treatment of the trisilyl(germyl)methane **(76)** with CsF in DIGLYME leads to a compound of structure **(78)**. The reaction is believed to take place by initial formation of the unstable germaethene **(77)**, which because of steric hindrance, cannot undergo cyclodimerization and looses a TMS group to give a second transient germaethene (Scheme 28) ⟨83IZV959⟩.

(76) **(77)** **(78)**

Scheme 28

A variety of disilyl(boryl)methanes has been prepared by making use of the high reactivity of the B=C double bond towards protic agents and organolithium compounds ⟨93AG(E)985⟩. Some of these reactions are outlined in Scheme 29 ⟨87CB1069, 89CB595⟩. Borinanes (**80**) were obtained from the addition of HCl or $HN(TMS)_2$ to the B=C bonds of boranediylborinanes (**79**) (Equation (19)) ⟨92AG(E)1384⟩. A related preparation of the 1,3-diboretane (**82**) has been described starting from germaethene (**81**) (Equation (20)) ⟨87AG(E)798⟩.

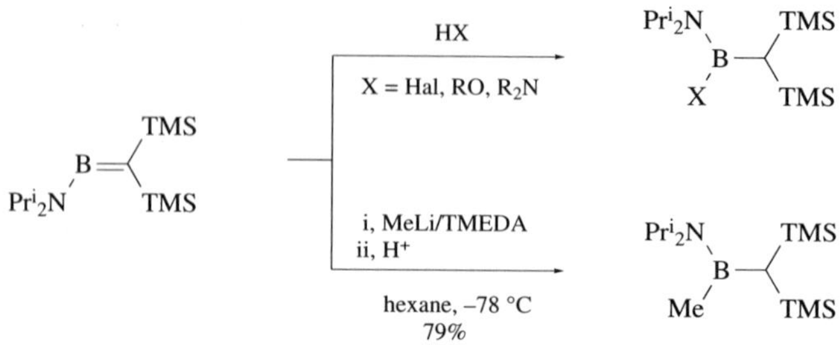

Scheme 29

(**79**) (**80**) (19)

R = Bu^t, 2,3,5,6-tetremethylphenyl
X = Cl, $(TMS)_2N$

(**81**) (**82**) (20)

R = Bu^t, X = $(TMS)_2N$

Germaethene (**83**) generated as a reaction intermediate by thermal elimination of LiCl from Me_2ClGe—C(Li)$(TMS)_2$, can be converted into various compounds containing an Si_2(Ge)C function and in some cases this provides the best route to the new species (Scheme 30) ⟨86CB2966, 86CB2980, 87CB1203⟩. Compound (**85**) has been synthesized by a method involving the low temperature reaction of kinetically stabilized germaethene (**84**) with methanol (Equation (21)) ⟨91POL1153⟩.

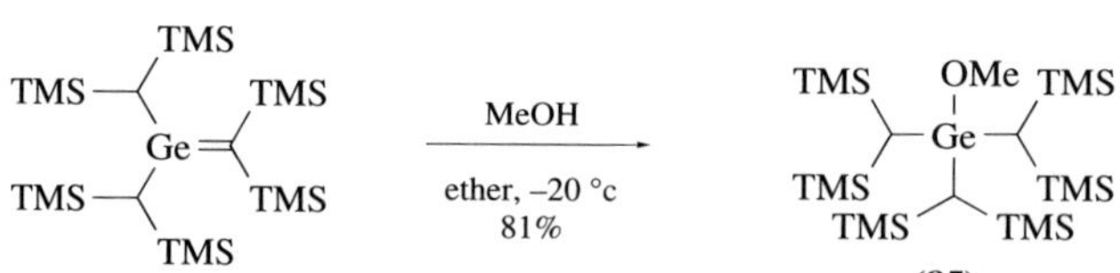

(**84**) (**85**) (21)

(iii) *Miscellaneous*

1,3-Digermacyclobutane (**86**) has been found among the products of "direct synthesis" from TMS(dichloromethyl)silane (Equation (22)) ⟨72ZOB1521⟩. The reaction of 1,3-diborete (**87**) with potassium or lithium in THF leads to compounds $2K^+(\mathbf{87})^{2-}$ or $2Li^+(\mathbf{87})^{2-}$, which are protonated with $(TMS)_2NH$ to yield the 1,3-diboretane (**88**). Reaction of $(\mathbf{87})^{2-}$ with MeI affords the 1,3-diboretane (**89**) (Scheme 31) ⟨89CB1881⟩.

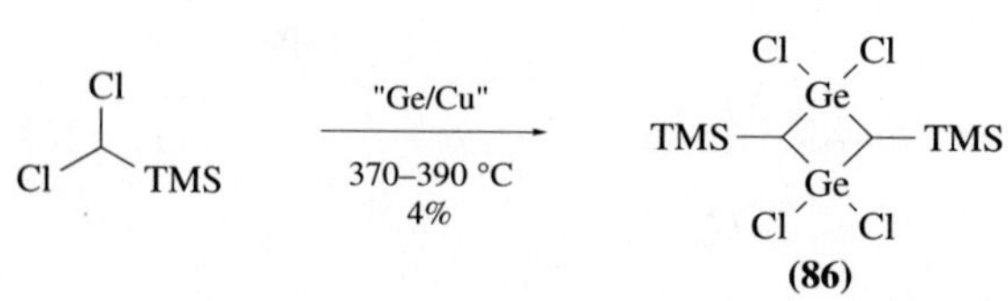

(**86**) (22)

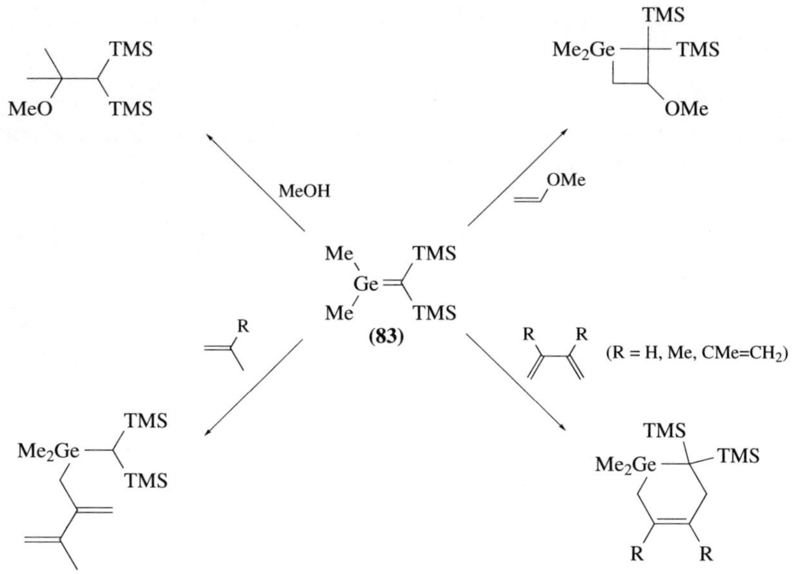

Scheme 30

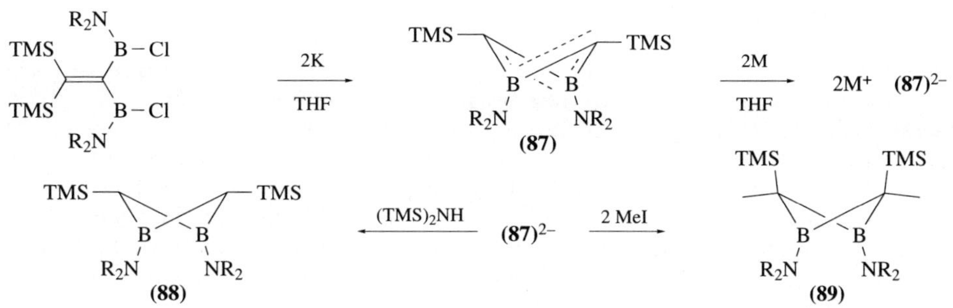

Scheme 31

6.06.2.2 Functions Containing Metalloids and Metals

6.06.2.2.1 Functions bearing silicon(s) and metals(s)

This section is devoted to methods of preparing α-silylated organometallics having the general structure $RC(SiX_3)_n(ML_m)_{3-n}$, where M is a metal, $n = 1$ or 2, and R = H, alkyl, aryl, or hetaryl. These compounds are of special interest due mainly to findings that silylmethyl groups like $(TMS)_2CH$ or $(TMS)_3C$ combine bulkiness with the capability to form strong metal–carbon bonds and that organometallic compounds bearing the polysilylmethyl group often possess unique properties and have unusual structures ⟨B-85MI 606-01, B-91MI 606-01⟩. Although α-metallated organosilane chemistry has been reviewed extensively since the early 1980s ⟨80HOU(13/5)27, B-81MI 606-01, B-83MI 606-01⟩, polysilylated organometallics have received only passing comment within reviews with much wider scope ⟨B-85MI 606-01, 91COS(1)1⟩.

(i) Alkali metal and magnesium derivatives of silylmethanes

In the mid 1990s there are four general methods for the production of compounds of this type. These are:

(a) metal–hydrogen exchange reactions of silylmethanes with basic metal-containing reagents;
(b) oxidative metallation of halo(silyl)methanes;
(c) organometallic addition to vinylsilanes; and
(d) transmetallation of an already α-metallated organosilanes.

Representative examples illustrating the synthesis of alkali metal and magnesium derivatives of silylmethanes which have obvious potential synthetic utility are given in Table 2. A few of the silylmethyl derivatives of the alkali metals display unique aggregate structures and these are discussed by Williard ⟨91COS(1)1⟩.

Table 2 Preparation of trisilylmethyllithium derivatives of the alkali metals and magnesium.

Compound	*Method of preparation*	*Reaction conditions*	*Ref.*
$(TMS)_2CHLi$	$(TMS)_2CH_2 + Bu^tLi$	THF/HMPA (ca. 4:1), $-78\,^\circ C$	77CB852, 80JA1584
$(TMS)_2CHLi$	$(TMS)_2CH_2 + Bu^nLi + TMEDA$	hexane, 12 h, 25°C	82CC1323
$(TMS)_2CHLi$	$(TMS)_2CH_2 + Bu^nLi + PMEDA$	hexane, 0.5 h, 20°C	82CC1323
$(TMS)_2CHLi$	$(TMS)_2CHCl + Li$	ether, reflux	76JCS(D)2268, 80JA1584
$(TMS)_2CHLi$	$(TMS)_3CH + MeOLi$	HMPA, 20°C	73TL4193
$(TMS)(MeOMe_2Si)CHLi$	$(TMS)(MeOMe_2Si)CH_2 + Bu^tLi$	pentane, 20°C	89JOC1784
$(TMS)_2CRLi$[a]	$(TMS)_2CHR + Bu^nLi$	THF/hexane	86CC672
$(TMS)_2CRLi$[a]	$(TMS)_2CHR + Bu^nLi + TMEDA$	hexane	84CC1708
$(TMS)_2CRLi$[a]	$(TMS)_2CHR + Bu^nLi$	hexane/ether, 20°C, 1 h	83CC1419, 84CC1708
$(TMS)_2CRLi$[b]	$(TMS)_2CHR + Bu^nLi + TMEDA$	hexane, 20°C	84JCS(D)1801
$(TMS)_2CHNa$	$(TMS)_3CH + NaOMe$	HMPA, 20°C	73TL4193
$(TMS)_2CHK$	$(TMS)_2CHLi + Bu^tOK$	hexane, 20°C, 16 h	91OM1704
$(TMS)_2CHMgCl$	$(TMS)_2CHCl + Mg$	ether, reflux, 4 h	91JOM(421)175
$[(TMS)_2CR]_2Mg$[a]	$(TMS)_2CHR + MgBu^nBu^s$	heptane	86CC672
$TMS\text{-}CH(MgBr)_2$	$TMSCHBr_2 + Mg(Hg)$	diisopropyl ether	86TL6123

[a] R = 2-pyridyl. [b] R = 4-methylphenyl.

(a) Metal–hydrogen exchange. The methylene group between two silicon atoms shows enhanced reactivity with respect to proton abstraction by strong bases, therefore many bis(silyl)methanes undergo metal–hydrogen exchange on reaction with alkyllithium reagents ⟨77CB852, B-83MI 606-01, 83ZAAC(497)21⟩. In practice, these reactions require strongly coordinating solvents or the use of alkyllithiums in the activating presence of TMEDA or potassium *t*-butoxide. The successful preparation of bis(TMS)methyllithium can be achieved from the bis(TMS)methane using *t*-butyllithium in hexamethylphosphoramide (HMPA) ⟨73TL4193, 74AG(E)83⟩, *t*-butyllithium in THF with HMPA at $-78\,^\circ C$ ⟨77CB852, 80JA1584⟩, Bu^nLi/Bu^tOK reagent in THF solution ⟨91OM551⟩ or the TMEDA complex of *n*-butyllithium in hexane ⟨82CC1323⟩. The last procedure provides one of the simplest routes to $(TMS)_2CHLi$. In addition, a feature of the use of TMEDA is that it enables crystalline monomeric lithium derivatives to be isolated. Silylated product yields for the system $TMS\text{-}CH_2TMS/Bu^nLi/Bu^tOK$ were in the range 78–91% ⟨91OM551⟩. Metallation of the carbosilane $(TMS)_2CHSiMe_2CH_2TMS$ gave, as expected, $(TMS)_2C(Li)SiMe_2CH_2TMS$, since the indicated carbanion is stabilized by three silyl groups. Only monometallation could be achieved for the carbosilane $TMS\text{-}CH_2SiMe_2CH_2SiMe_3$ ⟨91OM551⟩. Poly(dimethylsilene), $(Me_2SiCH_2)_n$, the polymer formally derived from dimethylsilaethene, $Me_2Si{=}CH_2$, has been shown to be metallated by Bu^nLi/Bu^tOK in THF to give a polycarbosilane in which, on average, every fourth CH_2 group in an $SiCH_2Si$ environment is metallated ⟨91OM551⟩.

Methoxydimethylsilyl(TMS)methyl lithium, $HC(Li)(TMS)SiMe_2OMe$, a reagent useful for the direct conversion of aldehydes and ketones to vinylsilanes, is readily formed in hydrocarbon solvent from the appropriate silane and *t*-butyllithium ⟨89JOC1784⟩. Access to the *p*-xylene compound *p*-$MeC_6H_4CH(TMS)_2Li$ is possible by using $Bu^nLi/TMEDA$ in hexane ⟨84JCS(D)1801⟩. However, organolithium tertiary amines reagents failed to yield the dilithium derivative from 1,4-bis(bis-(TMS)methyl)benzene ⟨84JCS(D)1801⟩.

Lithium *t*-butylbis(TMS)acetate, $(TMS)_2C(Li)CO_2Bu^t$, results from the appropriately substituted acetate using LDA in THF at $-78\,^\circ C$ ⟨76TL2737, 77JOC2038⟩. This organolithium compound is a useful synthetic intermediate in the Peterson alkenation reaction ⟨B-83MI 606-01⟩.

Reaction of the pyridine functionalized bis(TMS)methane (**90**) with *n*-butyllithium in hexane yields complex (**91**), whereas metallation of (**90**) with Bu^nLi and TMEDA in hexane affords lithium derivative (**92**). Both compounds contain unprecedented η^3-azaallyl ligand geometries ⟨84CC1708⟩. Treatment of 2-bis(TMS)methylpyridine with Bu^nLi in ether or THF yields binuclear complex (**93**) (Scheme 32) ⟨83CC1419⟩.

A procedure for the preparation of 2-lithio-1,1,3,3,5,5-hexamethyl-1,3,5-trisilacyclohexane by the metallation of *cyclo*-$[Me_2SiCH_2]_3$ with $Bu^nLi/TMEDA$ in hexane has been developed ⟨83ZAAC(497)21, 92OM3464⟩. The yields of monolithium derivative in this synthesis is very good, but further metallation did not occur at 20°C even when excess metallation reagent was used. Similarly 1-TMS-1,3,5-trisilacyclohexane can be selectively metallated at the more sterically available ring proton;

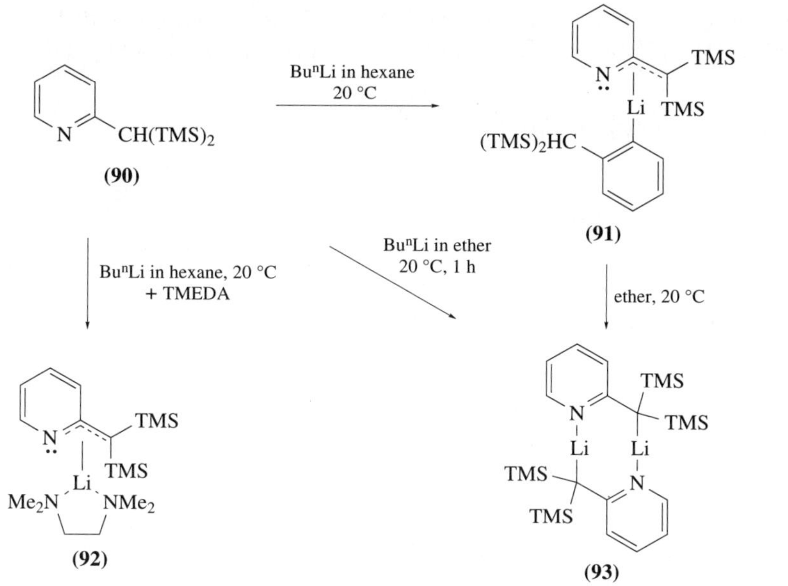

Scheme 32

however, if not immediately quenched the kinetic product (**94**) rearranges to the more stable lithium derivative (**95**) (Scheme 33) ⟨83ZAAC(497)21⟩. In contrast to 1,3,5-trisilacyclohexane, 1,1,3,3-tetramethyl-1,3-disilacyclobutane (**96**) reacts with *n*-butyl, methyl-, and phenyllithium to open the Si_2C_2 ring ⟨75JCS(D)1434, 90OM2677⟩. Even so, the solutions of 2-lithio-1,3-disilacyclobutane (**97**) could be prepared by the action of Bu^tLi/TMEDA on (**96**) in hexane (Scheme 34) ⟨90OM2677⟩. In reactions of (**97**) with TMS-Cl, Me_2HSiCl, Me_3SnCl, *n*-PrI, XCH_2CH_2X (X = Br, I), Me_2S_2, and I_2, the disilacyclobutane ring is retained while the methylene carbon atom is functionalized ⟨90OM2677⟩.

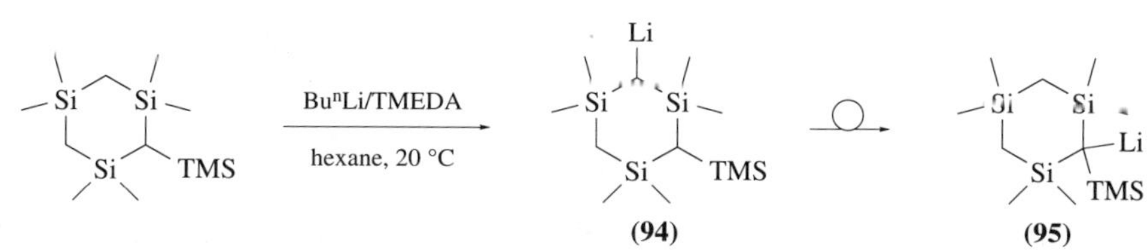

Scheme 33

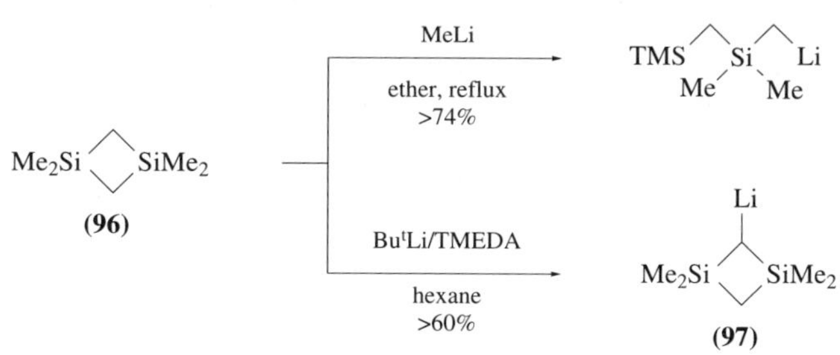

Scheme 34

(b) Oxidative metallation of halo(silyl)methanes. This is a superior method for generating a variety of silyl-substituted organometallics from readily available starting materials. The synthesis is very simple to carry out in the laboratory and yields are normally excellent. For example, chlorobis(TMS)methane reacts with lithium metal in ether to give the corresponding lithium derivative in virtually quantitative yield ⟨76JCS(D)2268⟩. The starting functionalized methane may be readily prepared from CH_2Cl_2, TMS-Cl, and Bu^nLi as shown in Scheme 35 ⟨71JOM(29)389, 81SRI591⟩. In the opinion of the present authors, this route is best for the preparation of $(TMS)_2CHLi$. One limitation of this scheme is that the synthesis of compounds $(TMS)_2C(R)Li$ is difficult or impossible if group

R is methyl or primary alkyl. Thus the reaction of $(TMS)_2C(Me)Cl$ with lithium metal leads to $(TMS)_2C{=}CH_2$ and $(TMS)_2C(H)Me$ instead of the expected lithium species ⟨79ZAAC(448)40⟩.

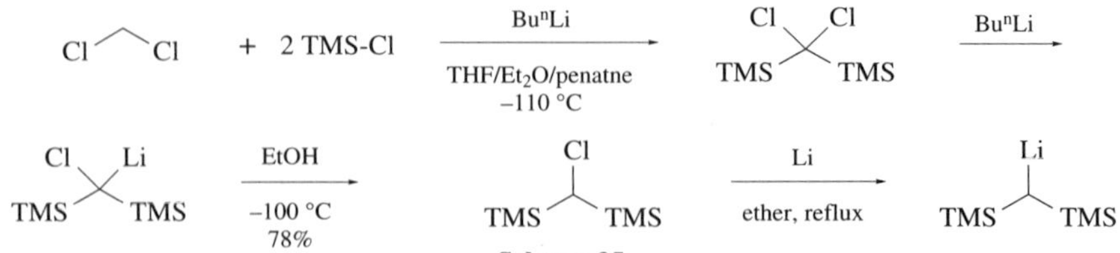

Scheme 35

The preparation of bis(TMS)methylmagnesium halides can be performed in ether or THF using halobis(TMS)methane and magnesium powder ⟨83POL291, 86JCS(D)1551, 91JOM(421)175⟩.

The method can be extended to the preparation of bis(metallated) silylmethanes. Thus, dilithio(TMS)methane (**98**) was obtained by reaction of dichloro(TMS)methane with lithium vapor (Equation (23)) ⟨84CC1664⟩. It should be noted that an alternative approach to this compound is the thermolysis of TMS-CH_2Li ⟨88POL2023⟩. The bulky TMS groups seem to prevent extensive polymerization of the dilithium derivative. In both preparations the only by-products were lithiated methanes and silanes. No higher polymers such as those observed for $(CH_2Li_2)_n$ were seen ⟨84CC1664⟩.

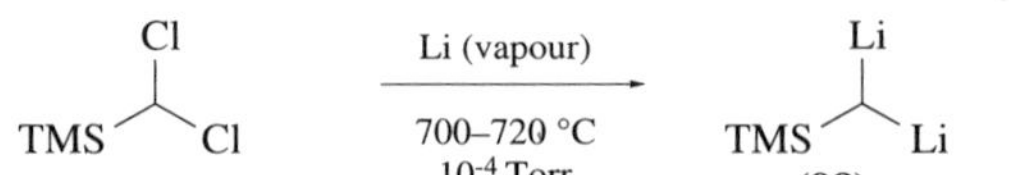

(98) (23)

Bis(bromomagnesio)TMSmethane (**99**) has been prepared directly from the reaction of TMS-$CHBr_2$ and magnesium amalgam in diisopropyl ether (70%) (Equation (24)) ⟨86TL6123⟩. In relation to $CH_2(MgBr)_2$, compound (**99**) is less reactive; this must be due to the anion-stabilizing effect of the TMS group and possibly to steric hindrance.

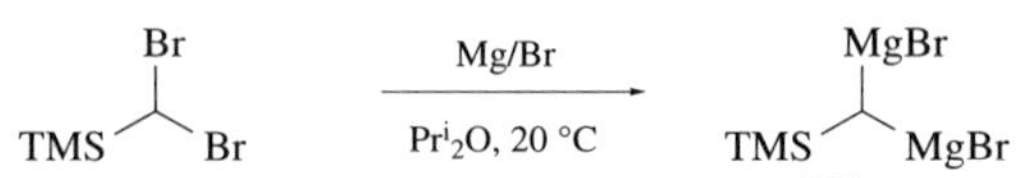

(99) (24)

(c) Organometallic addition to vinylsilanes. While organolithium compounds and Grignard reagents will not, in general, add to isolated C=C double bonds, they will add to the silyl-substituted alkenes to form α-silyllithium or α-silylmagnesium derivatives ⟨52JA4582, 54JOC1278, 70CJC561, 70TL1137⟩. The examples provided in Scheme 36 illustrate the application of the method for the synthesis of lithiated bis(silyl)methanes. Compounds of structure (**100**) are prepared in more than 90% yield ⟨74AG(E)83⟩. The α-lithio derivative (**101**) can be trapped by trimethylsilyl triflate to give compound (**102**), and is a useful silaethene precursor (Scheme 36) ⟨92ZN(B)805⟩.

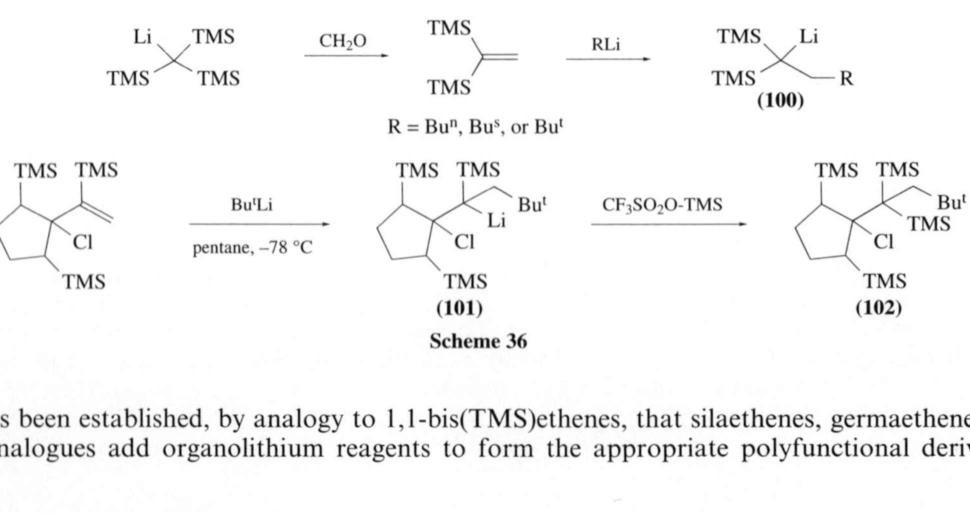

Scheme 36

It has been established, by analogy to 1,1-bis(TMS)ethenes, that silaethenes, germaethenes, and their analogues add organolithium reagents to form the appropriate polyfunctional derivatives

⟨84JOM(273)141, 85OM339, 91AG(E)93⟩. For instance, the lithium species (**104**) was generated as shown in Scheme 37 by addition of phenyllithium to a solution of stannaethene (**103**) ⟨91AG(E)93⟩.

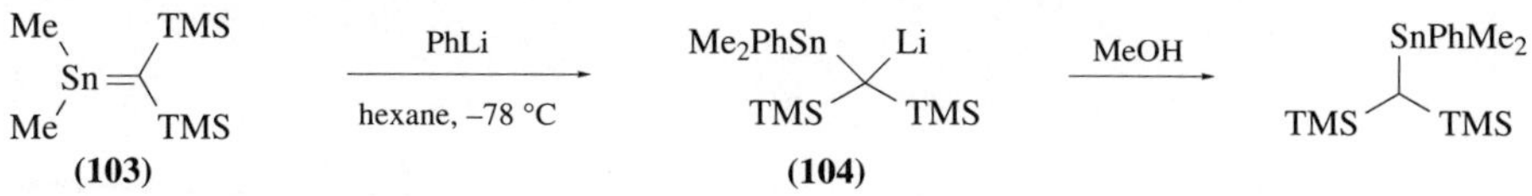

Scheme 37

1,2-Dilithio(tetrakis(TMS))ethane (**105**) results directly from the reaction of tetrakis(TMS) ethylene with excess lithium metal in dry-oxygen-free THF (Equation (25)) ⟨89JA3748⟩. The structure of (**105**) has been unequivocally confirmed by x-ray crystallography and chemical transformations. Treatment of the dilithium compound with H_2O and D_2O cleanly produced tetrakis(TMS)ethane (**106**). The reaction with paraformaldehyde afforded alkene (**107**) (73%) probably via 1,2-anionic rearrangement of the silyl group (Scheme 38). After prolonged standing (9 days), quenching the THF solution of (**105**) with water led to the formation of compounds (**108**) and (**109**) together with small amounts of cyclic compounds (**110**) and (**111**) (Scheme 39) ⟨93CL267⟩. These reactions probably involve intramolecular proton abstraction from a methyl group on silicon followed by subsequent anionic rearrangement.

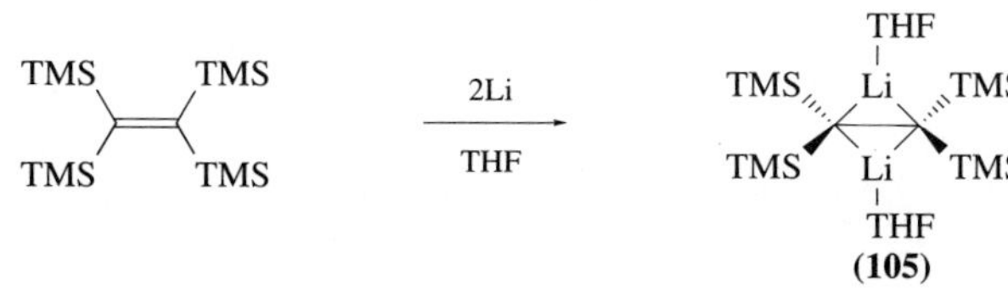

(25)

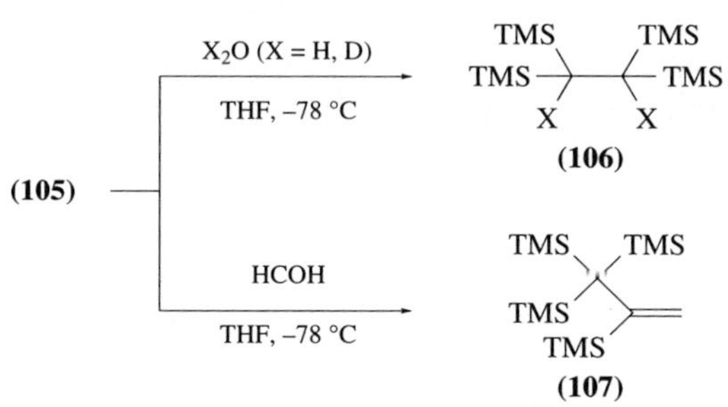

Scheme 38

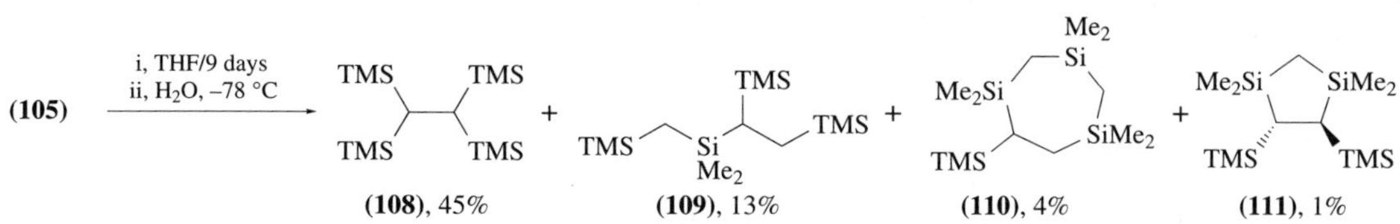

Scheme 39

Similarly, phenyl(tris(TMS)ethylene participates readily in oxidative metallation reaction to give 1,2-dilithiophenyl(tris(TMS))ethane. Hydrolysis of the dilithio derivative leads to the formation of tris(TMS)phenylethane, $Ph(TMS)CHCH(TMS)_2$, in almost quantitative yield ⟨92CL867⟩.

(d) Transmetallation and related reactions. There are several useful syntheses using transmetallation reactions for the preparation of silylmethyl derivatives of alkali and alkaline earth metals.

Bis(bis(TMS)methyl)mercury, which can be prepared from the Grignard reagent and mercury(II) chloride, exchanges mercury with lithium and the heavier alkali metals (Equation (26)) ⟨80POL2023⟩. This procedure has been shown to be a particularly clean method for the generation of α-silyl carbanions. However, the dangers inherent in handling volatile mercury compounds make this route somewhat less than attractive. Bis(TMS)methylpotassium can also be prepared by transmetallation of bis(TMS)methyllithium with potassium *t*-butoxide in hexane (82%) (Equation (27)) ⟨91OM1704⟩.

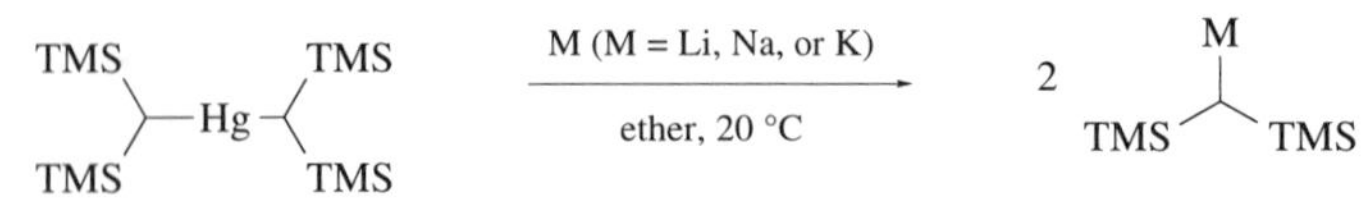

(26)

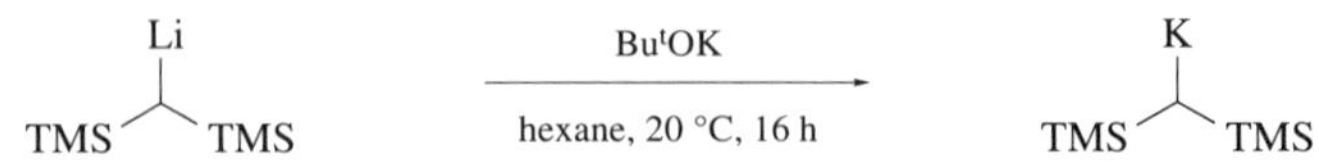

(27)

Also included in this category are desilylation of tris(TMS)methane with sodium methoxide in HMPA (Equation (28)) ⟨73TL4193⟩. The driving force for this reaction may be the formation of a strong Si—O bond of methoxytrimethylsilane. Similar reactions occur with tetrakis(TMS)methane and bis(TMS)methane ⟨73TL4193⟩.

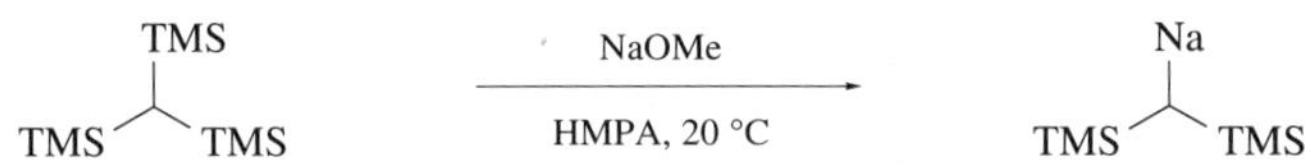

(28)

Dibromo(TMS)methane has been shown to react with a zinc–copper couple in THF to give the dizinc derivative (TMS)CH$(ZnBr)_2$. Transmetallation of the compound with magnesium metal affords the Grignard reagent (TMS)CH$(MgBr)_2$ ⟨72JOM(40)C53⟩.

(ii) Bis(TMS)methyl derivatives of transition metals and group 13 and 14 metals

Practically all compounds of the type $RC(SiX_3)_2(ML_m)$ containing a metal of which the electronegativity is greater than those of lithium and magnesium, have been prepared by transmetallation reactions starting from α-silyl-substituted organolithiums or Grignard reagents and inorganic or organometallic halides. A listing of many of these reactions is given in Table 3. Some specific examples are discussed below.

Lithium bis(TMS)methanide reacts with lanthanum aryloxides (Ln = La or Sm) under ambient conditions to yield $Ln(CH(TMS)_2)_3$—the first structurally characterized neutral homoleptic alkyl of the lanthanide metals ⟨88CC1007⟩.

The conversion of several bis(polymethylcyclopentadienyl)-stabilized lanthanide complexes into bis(TMS)methyl derivatives (**112**) has been achieved by a low-temperature reaction with bis(TMS)methyllithium. Efforts to abstract a proton from the α-carbon atom of the lanthanide complex (**112**; M = Lu) did not yield an alkylidene. Rather, a salt (**113**) was isolated which was assigned a metallacyclobutane structure (Scheme 40) ⟨85JA8103⟩.

The bis(TMS)methylyttrium complex (**114**), which can be prepared from $Cp^*{}_2YCl$ and $(TMS)_2CHLi$, readily undergoes insertion reactions into the σ-yttrium–carbon bond to produce new monomeric complexes, such as (**115**) and (**116**), stabilized by Cp* ligands and $(TMS)_2CH$ group (Scheme 41) ⟨87OM1509⟩.

Compounds (**117**) have been prepared from the appropriate metallocene dichlorides and an equimolar amount of $(TMS)_2CHLi$ in ether. Sodium amalgam in THF under dinitrogen smoothly reduces the bis(TMS)methylzirconium(IV) metallocenes (**117**) to complexes of structure (**118**), containing "side-on" bonded dinitrogen according to Scheme 42 ⟨79JOM(181)25⟩. These results provide a further demonstration that the bulky bis(TMS)methyl ligand may stabilize unusual metal complexes.

The preparation of bis(bis(TMS)methyl)manganese (**119**) was accomplished by treating anhydrous manganese(II) chloride with the Grignard reagent in THF as indicated in Scheme 43. This compound has proved to be a useful agent for transferring the ligand $(TMS)_2CH$ from Mn to another metal site. Thus, the yield of $Pb(CH(TMS)_2)_2$ obtained from structure (**119**) and PbX_2 (X = Cl or $N(TMS)_2$) is certainly superior to that achieved in the direct reaction between $(TMS)_2CHLi$ and PbX_2 ⟨90JOM(394)57⟩.

Mononuclear cuprate (**120**) was prepared as its lithium crown ether salt by addition of the organolithium reagent to CuBr in THF. Its structure consists of a well-separated cation, $(Li(12\text{-crown-}4)_2)^+$, and a cuprate anion (Equation (29)) ⟨85JA4337⟩.

Table 3 Preparation of bis(TMS)methyl derivatives of transition metals and group 13 and 14 metals.

Metal substrate[a]	*Silyl reagent* [$R = (TMS)_2CH$]	*Product*	*Yield* (%)	*Ref.*
$ScCl_3$	RLi	$ScR_3 \cdot 2THF$		74JOM(76)C45
YCl_3	RLi	YR_3		74JOM(76)C45
YCp_2^*Cl	RLi	YCp_2^*R	69	86OM1726
$YCp_2^*(OAr)_2$	RK	$YCp_2^*(OAr)R$	72	94OM69
$La(OAr)_3$	RLi	LaR_3		88CC1007
$[LaCp_2^*Cl_2]^-Li^+$	RLi	$LaCp_2^*R$	60	85JA8091
$CeCp^*(OAr)_2$	RLi	$CeCp^*R_2$		88CC962
$[NdCp_2^*Cl_2]^-Li^+$	RLi	$NdCp_2^*R$	80	85JA8091
$Sm(OAr)_3$	RLi	SmR_3		88CC1007
$[SmCp_2^*Cl_2]^-Li^+$	RLi	$SmCp_2^*R$	45	85JA8091
ECl_3	RLi	$[ER_3Cl][Li(THF)_4]$		78CC140
$YbCl_3$	RLi	$[YbR_3Cl][Li(THF)_4]$		78CC140
$LuCl_3$	RK	$LuR_3(\mu\text{-}Cl)K(OEt)_2$		91OM1704
$[LuCp_2^*Cl_2]^-Li^+$	RLi	$LuCp_2^*R$	64	85JA8091
$TiCl_3(Me_3N)_2$	RLi	TiR_3	6	78JCS(D)734
$TiCp_2Cl_2$	RLi	$TiCp_2R$	63	77JA6645, 78JCS(D)734
$ZrCl_4$	RLi	$ZrClR_3$	45	78JCS(D)734
$ZrCp_2Cl_2$	RLi	$ZrCp_2ClR$	43	77JA6645
$ZrCp_2'Cl_2$	RLi	$ZrCp_2'ClR$	43–79	79JOM(181)25, 81JCS(D)814
$HfCl_4$	RLi	$HfClR_3$	53	78JCS(D)734
$HfCp_2Cl_2$	RLi	$HfCp_2R$	37	77JA6645
$VCl_3(Me_3N)_2$	RLi	VR_3	14	78JCS(D)734
VCp_2Cl_2	RLi	VCp_2R	42	77JA6645
VCp_2Br	RLi	VCp_2R	80	77JA6645
$CrCl_3$	RLi	CrR_3	71	78JCS(D)734
$MnCl_2$	RMgCl	$MnR_2 \cdot THF$	72	90JOM(394)57
CuBr	RLi	[CuBrR]Li		85JA4337
$ZnCl_2$	RLi	ZnR_2	78	91JOM(421)175
$ZnCl_2$	RLi/TMEDA	$[ZnR_3]Li$	73	93ZAAC(619)675
$CdCl_2$	RLi	CdR_2	34	78JOM(153)253
$HgCl_2$	RLi	HgR_2	97	78JOM(153)253
$HgCl_2$	$(TMS)_2CMeLi$	$Hg[C(Me)(TMS)_2]_2$	34	78JOM(153)253
$HgCl_2$	RMgCl	HgR_2	77	66JOM(6)451
$HgCl_2$	R_2Hg	HgClR	72	66JOM(6)451
$HgBr_2$	RLi	HgR_2	16	70JOM(24)647
$HgBr_2$	R_2Hg	HgBrR	36	78JOM(153)253
$AlCl_3$	RLi	$AlClR_2$	47	78JOM(153)253
$AlCl_3$	RLi	AlR_3	69	89ZAAC(579)75
$(AlBrBu^t_2)_2$	RLi	$AlBu^t_2R$	60	92ZAAC(613)67
$(AlBr_2Bu^t)_2$	RLi	$AlBu^tR_2$	78	91ZAAC(595)225
$(AlCl_2)_2CH_2$	RLi	$(AlR_2)_2CH_2$	45	90POL277
$GaCl_3$	RLi	GaR_3	82	80IC3637
Ga_2Br_4	RLi	GaR_3	42	89JOM(364)289
$Ga_2Br_4(diox)_2$	RLi	GaR_2Br	54	92CB1547
$Ga_2Br_4(diox)_2$	RLi	$[GaR_2]_2$	63	89JOM(364)289
$InCl_3$	RLi	InR_3	90	80IC3637
$InPr^iCl_2$	RLi	$InCl(R)Pr^i_2$	61	91ZN(B)1539
$In_2Br_4(TMEDA)_2$	RLi	$[InR_2]_2$	54	89JOM(368)139
$SnCl_2$	RLi	SnR_2	71	86JCS(D)1551, 76JCS(D)2268
$\{SnCl[N(TMS)_2]\}_n$	RLi	SnR_2	71	76JCS(D)2268
$Sn[N(TMS)_2]_2$	RLi	SnR_2	19	76JCS(D)2268
$SnCl_4$	RLi	$SnCl_2R_2$	84	86JCS(D)1551, 82CC1407
$PbCl_2$	RLi	PbR_2	3	76JCS(D)2268

[a]Abbreviations: $Ar = 2,6\text{-}Bu^t_2C_6H_3$, $Cp = \eta\text{-}C_5H_5$, $Cp^* = \eta\text{-}C_5Me_5$, $Cp' = \eta\text{-}C_5H_4X$ (X = H, Me, Et, Pr^i, Bu^t, or TMS).

Studies in the 1980s and early 1990s on the transmetallation reactions of nitrogen-functionalized lithium bis(TMS)methanide (**93**) have yielded a series of molecules with unusual bonding configurations. For example, the reaction of 2-bis(TMS)methylpyridine with butyllithium in THF followed by CuCl, yields a binuclear complex (**121**) in which the metal is not involved in electron-deficient bonding ⟨83CC1419⟩. The silver(I) bis(TMS)methyl derivative (**122**) was prepared by a similar method (Scheme 44) ⟨84CC612⟩. The thermally stable *N*-functionalized bis(TMS)methylcobalt(II) complex has been prepared by reacting the lithium derivative (**93**) with cobalt(II) chloride in ether ⟨93JOM(443)C39⟩. An x-ray structural study has revealed a centrosymmetric molecular skeleton in which a pair of pyridine ligands are *trans*-chelated to the central planar four-

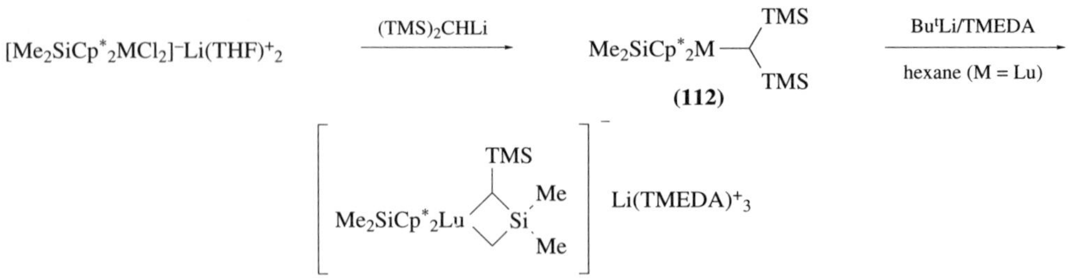

(113)

M = Nd, Sm, or Lu

Scheme 40

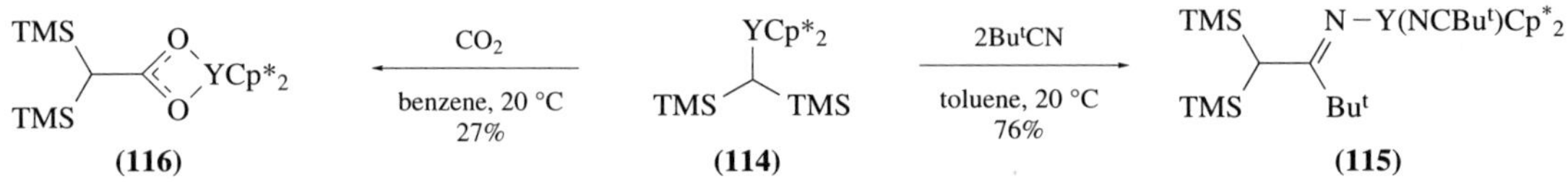

Scheme 41

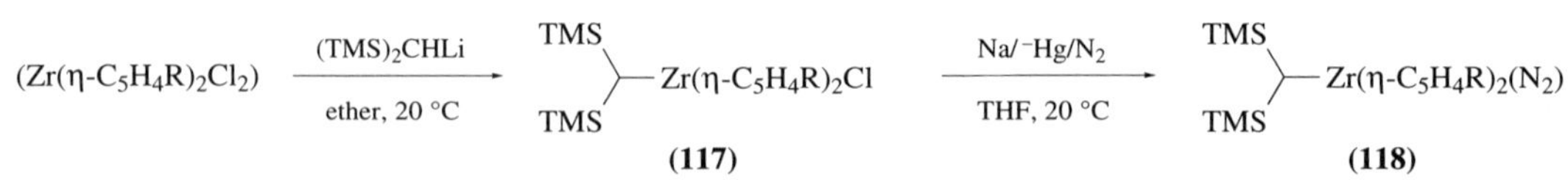

R = H, Me

Scheme 42

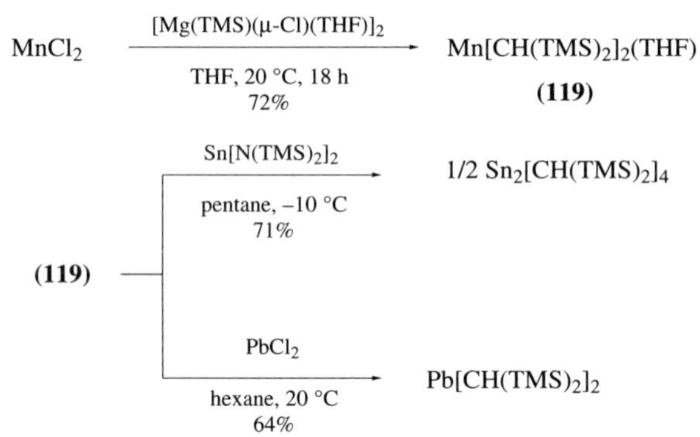

Scheme 43

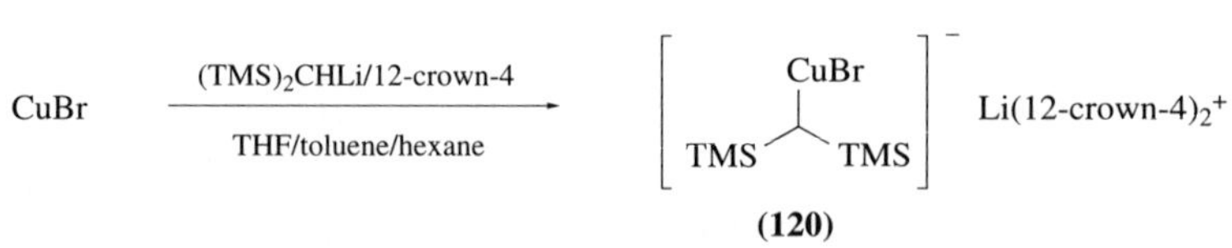

(29)

coordinate cobalt(II) atom. The complexes ($\overline{M\{NC_5H_4C(TMS)_2\text{-}2\}_2}$), where M = Zn, Cd, or Hg, accessible from structure **(93)** and MCl_2, are mononuclear metal(II) alkyls in which the dative metal–nitrogen interactions progressively weaken in the sequence Zn > Cd > Hg ⟨86CC672, 93JCS(D)2653⟩. Monomeric 2-pyridylbis(TMS)methyltin(II) compounds of structures **(123)**–**(125)** were prepared from **(93)** and $SnCl_2$ or $Sn(N(TMS)_2)_2$ (Scheme 45) ⟨88CC336⟩.

Bis(2-pyridylbis(TMS)methyl)stannylene **(124)** upon treatment with aryl azides forms the 1-aza-8-stannabicyclo[3.2.0]octa-2,4,6-triene system **(126)** via stannaimine intermediate (Scheme 46) ⟨93CB2247⟩.

The pentacoordinate aluminum derivative **(127)** is reported to be formed by reaction of lithium bis(TMS)methanide with $AlCl_3$. Compound **(127)** rapidly underwent metal–halogen cleavage with $AlCl_3$ in benzene yielding the salt-like adduct **(128)** (Scheme 47) ⟨87AG(E)681⟩.

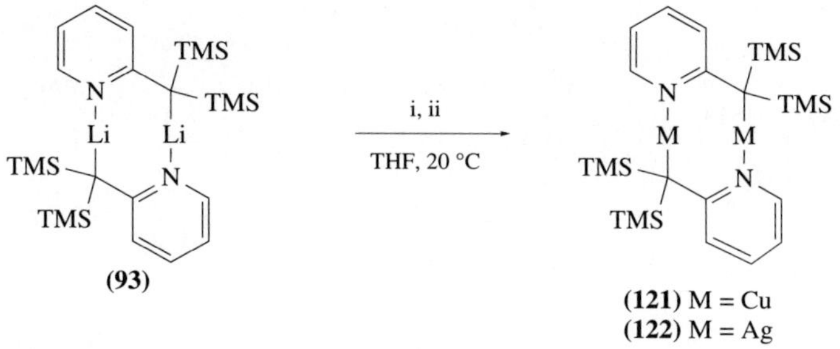

i, BuLi, ether, 1 h at 20 °C; ii, CuCl, THF, 1 h, 20 °C or $AgBF_4$, THF, –78 °C

Scheme 44

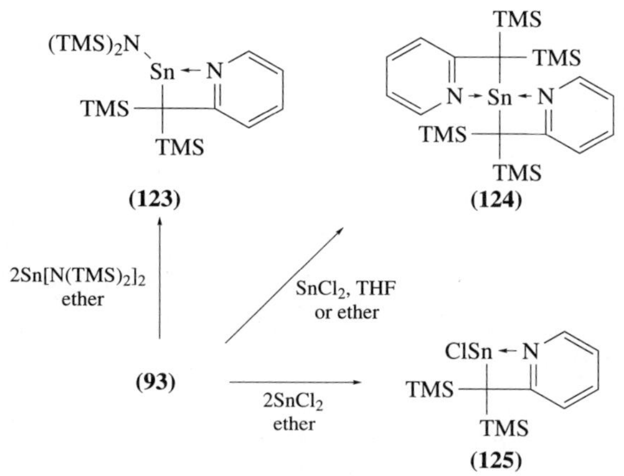

Scheme 45

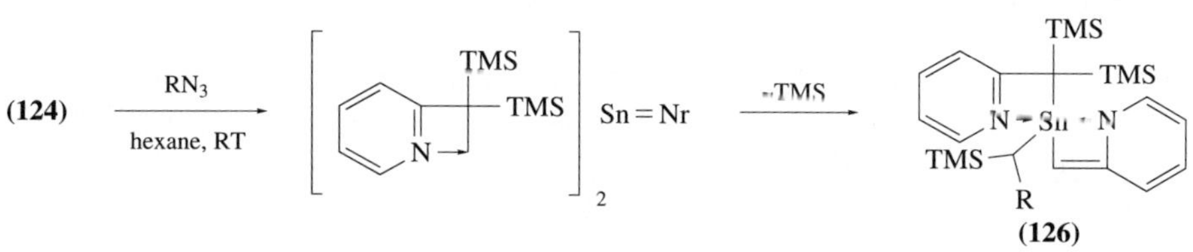

R = 2,6 $Pr^i_2C_6H_3$, 68%, 2,4,6-$Me_3C_6H_2$, 79%

Scheme 46

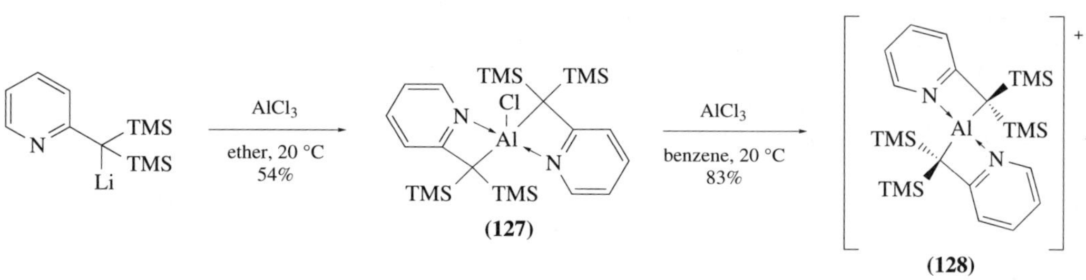

Scheme 47

Efforts to generate a carbanionic species by the reaction of $Al(CH(TMS)_2)_3$ with *t*-butyllithium in the presence of TMEDA have led to 1-sila-3-alanatetane (**129**). The reaction probably involves an attack of the strong base to the bis(TMS)methyl substituent followed by intramolecular addition of a carbanionic center to the coordinatively unsaturated aluminum atom (Equation (30)) ⟨89ZAAC(579)75, 93CB2637⟩. In contrast to $Al(CH(TMS)_2)_3$, a sterically less crowded compound (**130**) reacts with *t*-butyllithium in the presence of *N*,*N*′-dimethylpiperazine (DMP) to give lithium (*μ*-hydrido)trialkylalanate (**131**) (Equation (31)) ⟨91ZAAC(595)225⟩.

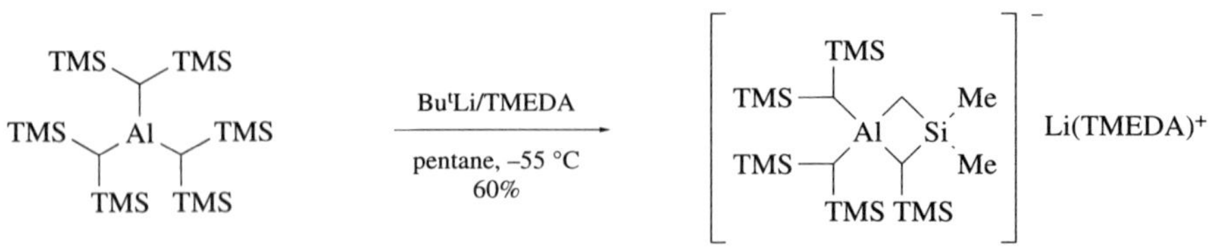

(129)

(30)

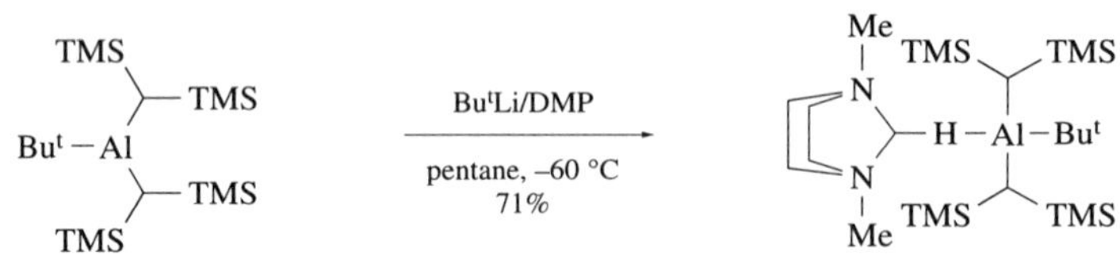

(130) (131)

(31)

A rare example of the double transmetallation reaction is shown in Equation (32). Bis(bromomagnesio)TMS-methane reacts with chlorotrimethylstannane in THF to form bis(trimethylstannyl)TMS-methane (**132**) in almost quantitative yield ⟨86TL6123⟩.

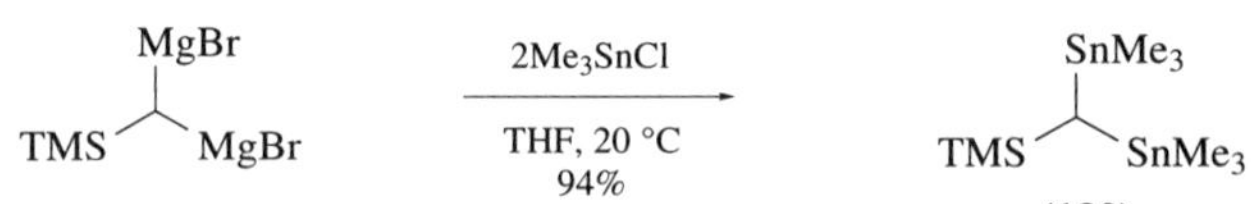

(132)

(32)

6.06.2.2.2 Functions bearing boron(s) and metal(s)

The chemistry and preparation of lithium poly(boryl)methides derived from methanetriboronic esters were reviewed in 1975 ⟨75S147⟩. Since the late 1980s, reviews containing information on the preparation of boron-stabilized carbanions (borylmethide salts) have also appeared ⟨B-87MI 606-02, B-88MI 606-01, 91COS(1)487⟩. The present section considers the preparation of polyheteroatom species of the general formula $RC(BX_2)_n(ML_m)_{3-n}$, where $n = 1$ or 2, M is metal, and R = H, alkyl, or aryl. No methods exist which can be used to prepare all these types of compounds but there are synthetic routes which have some generality.

(i) Synthesis via metal–hydrogen exchange

Alkyllithium reagents and other strong bases can abstract a proton from carbon adjacent to a boryl group generating the metallated borylmethane. In principle, the reaction could provide a general method of synthesis, but in certain cases highly nucleophilic bases react predominantly at the boron atom leading to formation of "ate" complexes (Scheme 48) ⟨65TL3429, 66TL4315, 73S37⟩. Successful metal–hydrogen exchange may be achieved in several ways as follows:

(a) the reagent can be a hindered, nonnucleophilic base;

(b) the groups around boron can be highly branched so that attack on boron is inhibited on steric grounds; or

(c) the electrophilicity of the boron atom may be lowered by the use of heteroatom substituents ⟨91COS(1)487⟩.

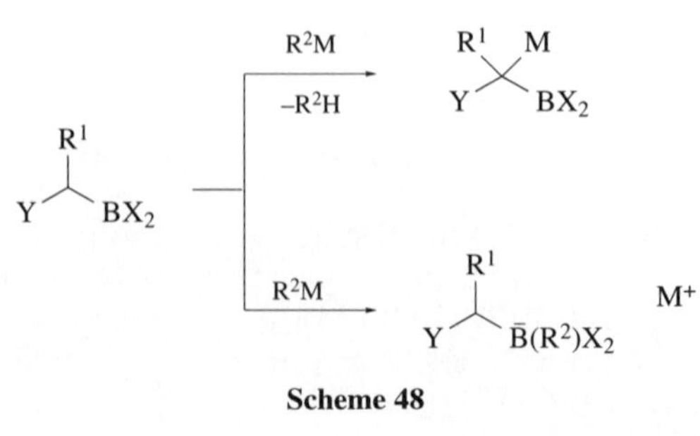

Scheme 48

The example provided in Scheme 49 illustrates the generation of potassium bis(boryl)methanide (**134**) from sterically overcrowded bis(dimesitylboryl)methane ⟨82JCR(S)132⟩. Interestingly, while the reaction of structure (**133**) with lithium dicyclohexylamide gives lithium dimesitylborylmethide, Mes_2BCH_2Li, in preparatively useful yield, *n*-butyllithium attacks at boron to give the borate and *t*-butyllithium gives the hydroborate by *β*-hydrogen transfer ⟨83TL623⟩.

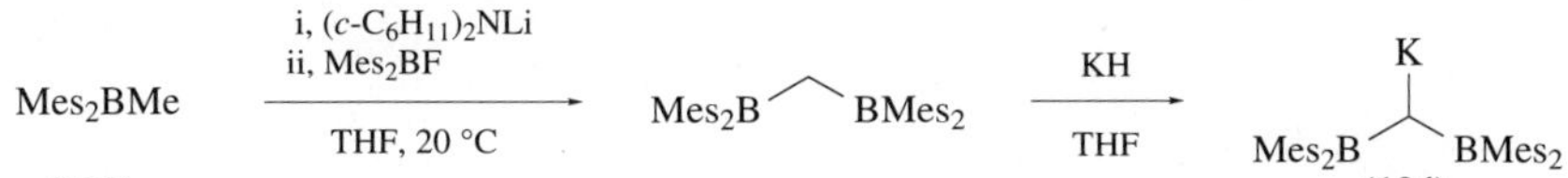

(133) **(134)**

Scheme 49

The deprotonation of bis(dialkoxyboryl)methanes has been accomplished with lithium 2,2,6,6-tetramethylpiperidine (LITMP) in the presence of TMEDA in THF (Equation (33)) ⟨82OM20⟩. However, the reaction is not completely general. Thus, the compound (**136**) undergoes cleavage instead of deprotonation to give lithiated monoborylmethane (**137**) (Equation (34)) ⟨82OM20⟩.

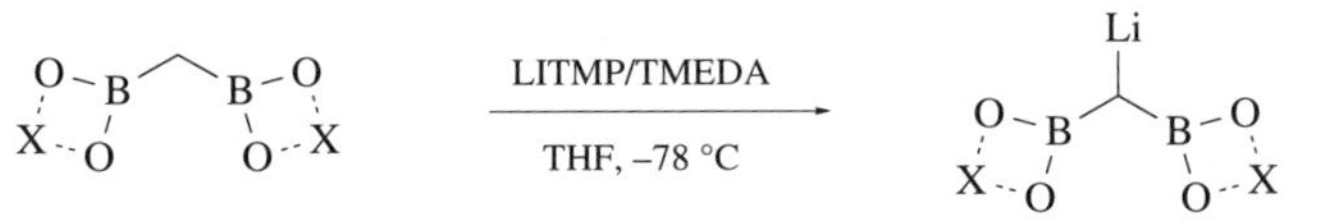

(135) (33)

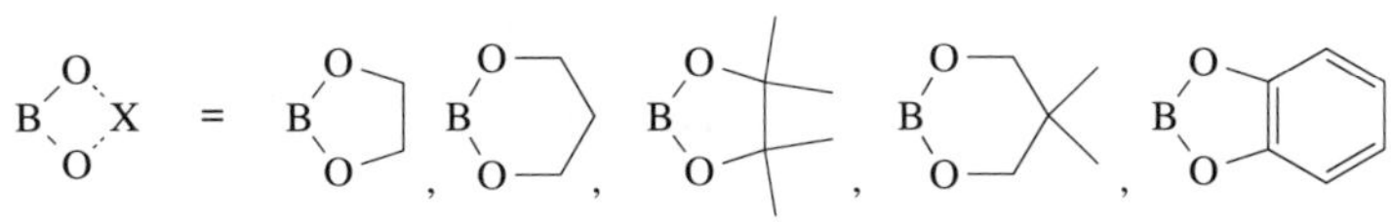

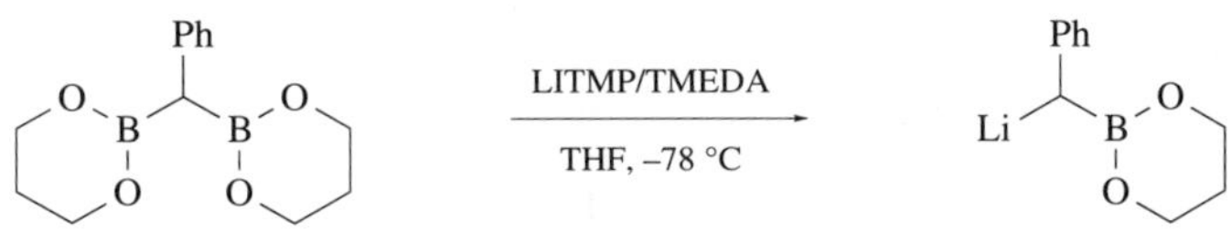

(34)

(136) **(137)**

Diborylmethide salts can be alkylated by alkyl halides to form 1,1-bis(dialkoxyboryl)alkanes, which in turn can be deprotonated and alkylated with a second alkyl halide to form *gem*-diboronic esters, $R^1R^2C(B(OAlk_2))_2$ ⟨82OM20⟩.

(ii) Synthesis via transmetallation reactions

A transmetallation reaction, for the purposes of this section, can be defined as a reaction in which a borylalkyl group is transferred from boron to a metal atom by the reaction of an organometallic compound with a suitable polyborylated methane. The method is widely used for the preparation of lithium tris- and bis(boryl)methides starting from the corresponding polyborylated methanes and organolithium reagents (Equations (35) and (36)) ⟨73JA5096, 74JOM(69)45, 75JA5608, 76JOM(110)25, 76JOM(114)1⟩. In a typical procedure, the tris(dialkoxyboryl)methane and methyllithium are stirred at −78 °C in a polar solvent (normally THF or a mixture of THF and dichloromethane). The reaction time appears to be substrate-dependent, but the reaction is generally complete after about 2.5 h at −78 °C. A borylmethide salt can be isolated or used *in situ* for subsequent derivatization. It should be noted that compounds such as (**135**) and (**138**) are very interesting as reagents for conversion of a carbonyl compound to the homologous aldehyde ⟨B-77MI 606-02, 78JOC950, 80JOC1091⟩.

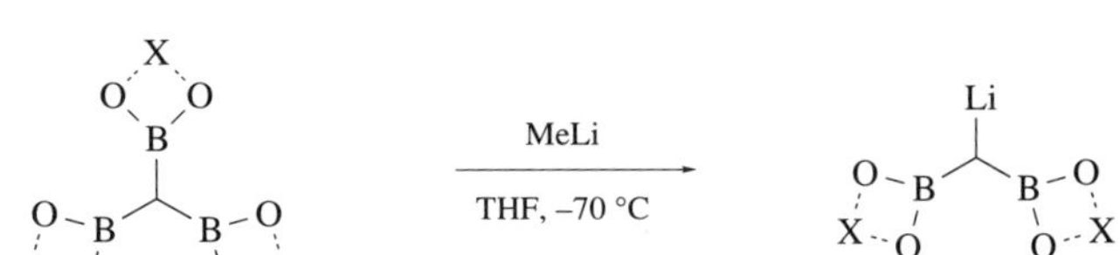

(135) (35)

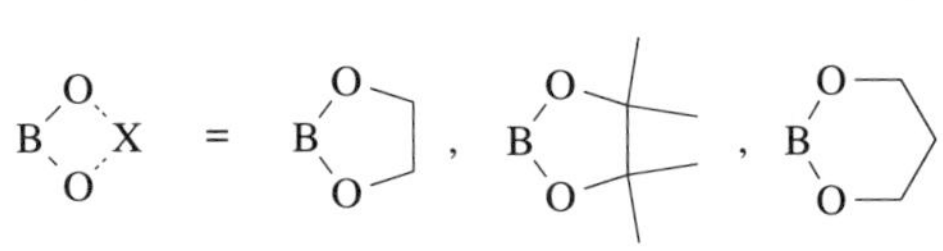

(36)

(138)

The reaction of lithium bis(dialkoxyboryl)methides with organoelement halides, such as Ph_3MCl (M = Ge, Sn, or Pb), provides a general route to bis(boryl)(germyl)methanes and their tin and lead analogues (Equation (37)) ⟨76JOM(110)25⟩. A further extension of this type of chemistry was the successful conversion of the tin compound (**139**) to the methide salt (**140**) stabilized by one boron and one tin atom, and finally to bis(triphenylstannyl)ethylenedioxyborylmethane (**141**) (Scheme 50) ⟨76JOM(110)25⟩. Tris(dialkoxyboryl)methide salts react with Ph_3MCl in a similar fashion to form compound $Ph_3MC(B(OAlk)_2)_3$ ⟨69JA6541, 73JOM(57)225, 73JOM(57)231, 74JOM(69)63⟩.

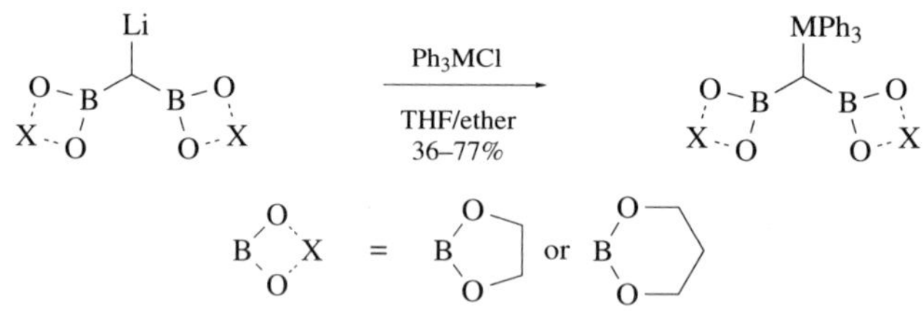

(37)

M = Ge, Sn, or Pb

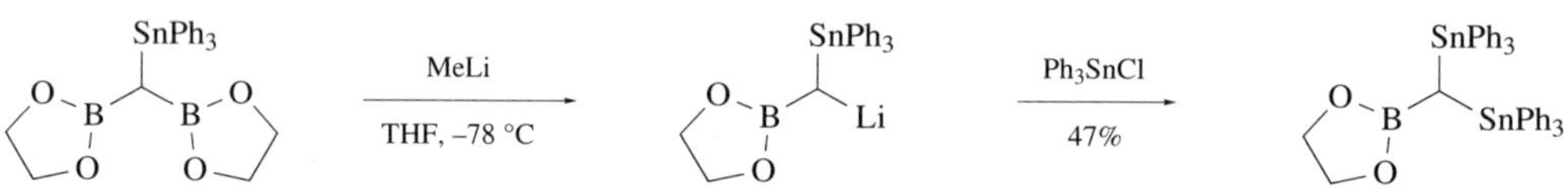

(139) **(140)** **(141)**

Scheme 50

Treatment of tetrakis(dimethoxyboryl)methane with butyllithium or, preferably, lithium methoxide, followed by triphenyltin chloride yields tris(boryl)stannylmethane (**142**), which on further treatment with butyllithium disproportionates to form bis(boryl)bis(stannyl)methane (**143**) (Scheme 51). Reaction of the monotin compound (**142**) with ethylene glycol resulted in the loss of one of the three boron atoms to give (triphenylstannyl)bis(ethylenedioxyboryl)methane (**139**) (Equation (38)). The tin compound (**143**) undergoes similar reaction on treatment with lithium methoxide in methanol, providing an alternative route to (boryl)bis(stannyl)-substituted methanes (Equation (39)) ⟨73JOM(57)225⟩.

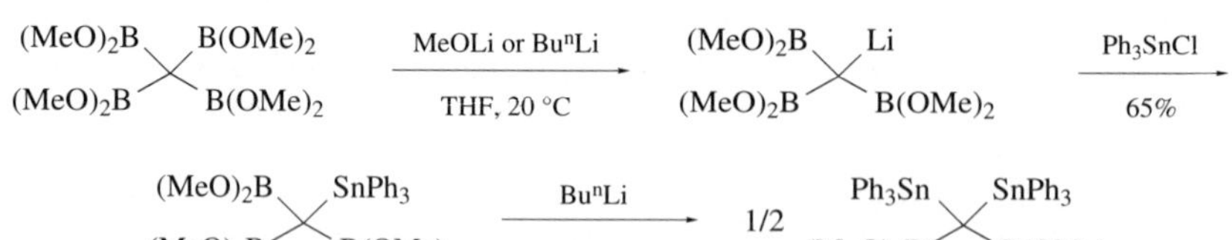

(142) **(143)**

Scheme 51

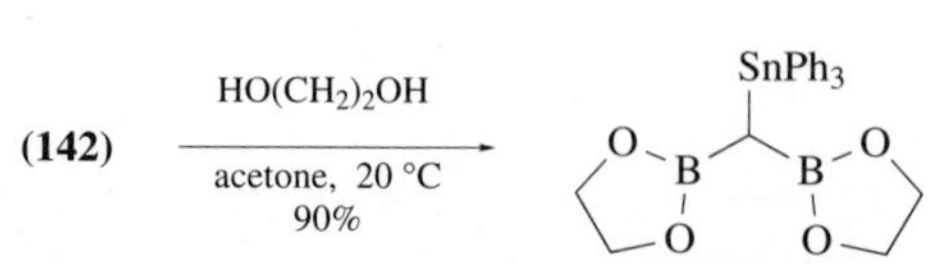

(38)

(139)

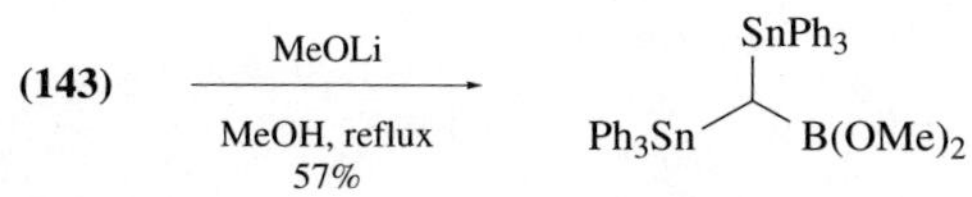

(39)

The reaction of benzylbis(boryl)methane (**144**) with mercuric dichloride and lithium methoxide in anhydrous methanol resulted in the selective replacement of one boron atom by mercury to form the α-metallated borylmethane (**145**) (Equation (40)) ⟨76JOM(114)1⟩. Interaction of boryl(stannyl)-alkanes with alkyllithiums generally leads to cleavage of the tin moiety (Equation (41)) ⟨85OM1690⟩. A useful reaction of dimesitylboryl(TMS)alkanes is that they readily undergo silicon/lithium exchange with LiF to form lithium borylmethides ⟨91COS(1)487⟩.

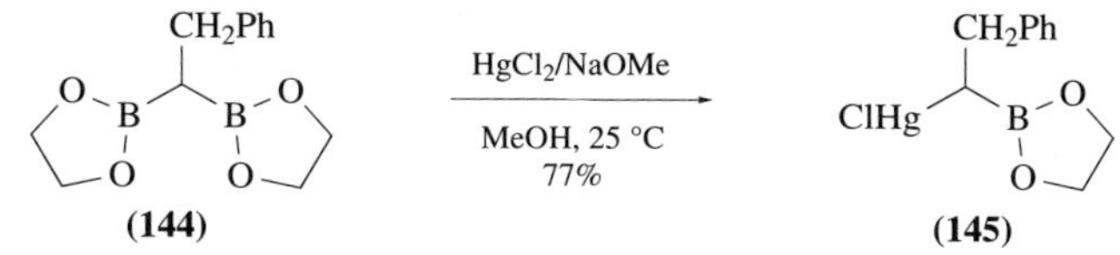

(40)

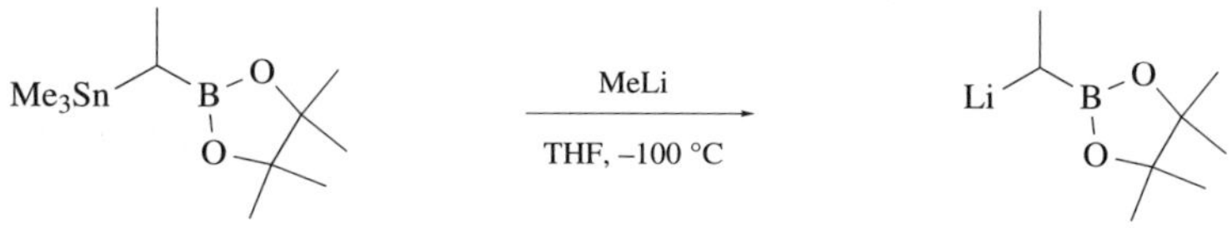

(41)

(iii) Synthesis via addition to vinylboranes and methyleneboranes

The very low yields of lithium borylmethide salts (*ca.* 5%) obtained from the reactions of alkenyldimesitylboranes with organometallic compounds demonstrates that this is not a very useful method for the synthesis of simple α-metallated borylmethanes. However, this approach has proved successful for the preparation of α-metallated silyl(boryl)methanes from 1-TMS-1-dimesitylborylethylenes (see Section 6.06.2.2.4) and several alkali metal polyborylmethides from methyleneboranes ⟨91COS(1)487⟩.

A specific reaction useful for the synthesis of diboriranide (**146**), is shown in Scheme 52. Treatment of (**146**) with the very weak acid *t*-butyl(TMS)amine leads to formation of (**147**) ⟨85AG(E)788⟩.

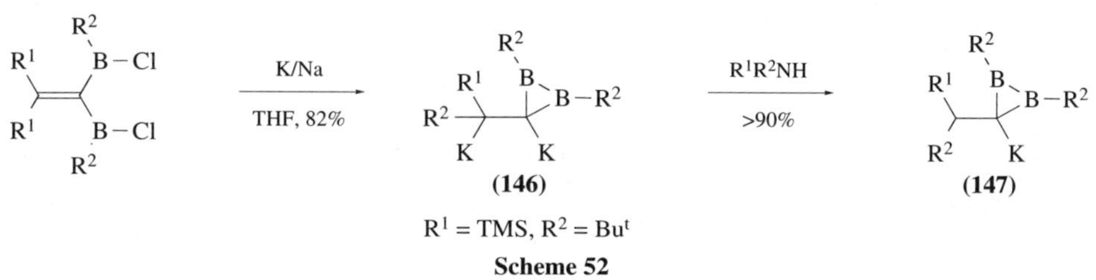

Scheme 52

C-Borylborataalkyne (**148**) with *t*-butyllithium yields 1,3-diborataallene (**149**), which contains a second *t*-butyl group in place of the $(TMS)_2CH$ group. Upon treatment with one equiv. of cyclopentadiene, the 1,3-diborataallene is transformed into the lithium bis(boryl)methide (**150**) (Scheme 53) ⟨88AG(E)1370⟩. Several cyclic compounds, such as (**151**) and (**152**), incorporating a $C(Li)B_2$ unit, have been made by a related 1,2-addition of the C—Li bond to the B=C multiple bond (Equations (42) and (43)) ⟨90AG(E)1030, 92AG(E)1238⟩. Readers are referred to a review by Berndt ⟨93AG(E)985⟩ for more detailed information concerning the synthetic potential of methyleneboranes in carbometallation reactions.

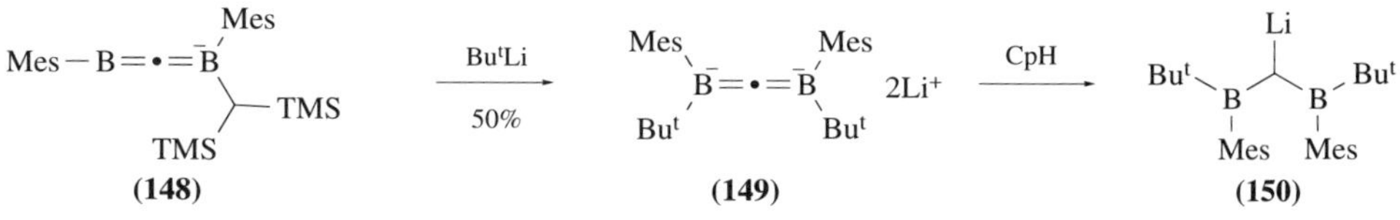

Scheme 53

(151)

Ar = 2,3,5,6-Me_4C_6H (Dur)

(42)

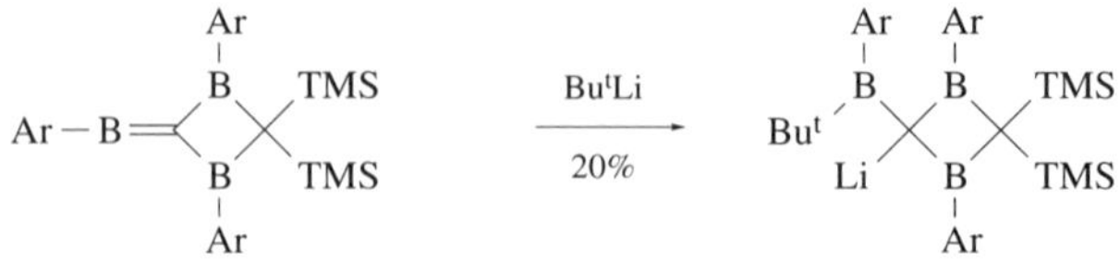

(152)

Ar = Dur, Mes

(43)

A unique reaction leading to the formation of a $C(Sn)B_2$ function has been described by Berndt and co-workers ⟨87AG(E)546⟩. The boranediylborirane (**153**) which, according to calculations, has the nonclassical structure shown in Scheme 54, behaves toward suitable reagents as if it were the carbene. Reaction of (**153**) with a stannylene in pentane led to formation of the stannaethene (**154**) in quantitative yield. Compound (**154**) reacts with HCl to give the 1,3-diboretane (**155**).

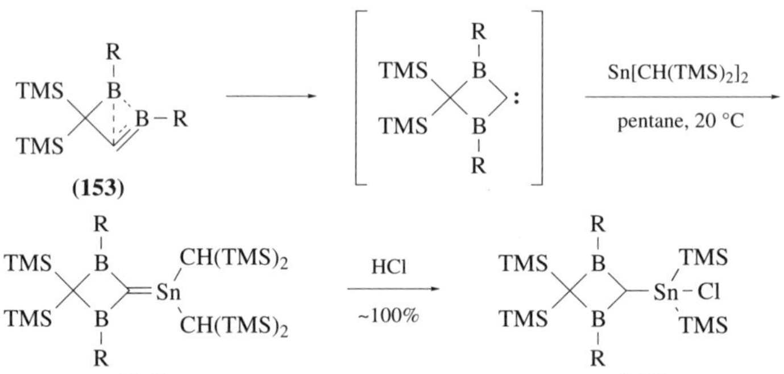

R = Bu^t

Scheme 54

6.06.2.2.3 *Functions bearing germanium(s) and metal(s)*

Isolable examples of compounds of the type $RC(GeX_3)_n(ML_m)_{3-n}$ ($n = 1$ or 2) are virtually unknown. The significant contribution to this class of compounds was made by Barton and Hockman who reported the preparation and synthetic utilization of lithium bis(trimethylgermyl)methide (**156**) ⟨80JA1584⟩. This reagent is formed in good yield by methods similar to those described for the analogous silicon compound (Scheme 55). The synthetic utility of (**156**) has been demonstrated by the reaction with tosyl azide leading to the formation of bis(trimethylgermyl)diazomethane.

The triphenylgermyl group seems to have no acidifying effect; lithiation of $(Ph_3Ge)_2CH_2$ was

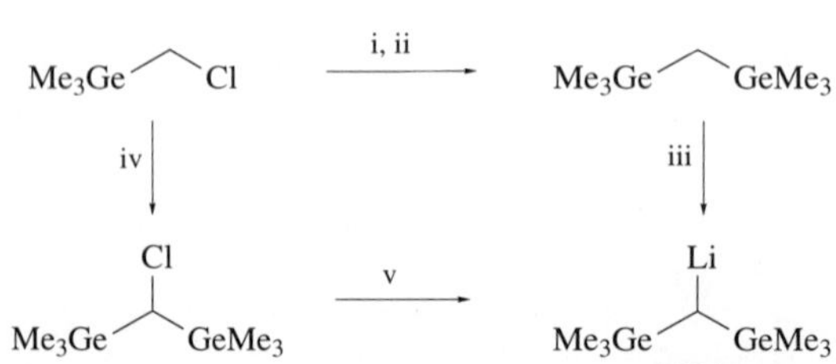

(156)

i, Li, ether, –23 °C; ii, Me_3GeCl; iii, Bu^{t}Li, THF/HMPA, –78 °C; iv, Bu^{s}Li, THF, –78 °C; v, Li, ether

Scheme 55

neither possible with lithium dicyclohexylamide nor with n-butyllithium or t-butyllithium in the presence of HMPA ⟨80TCC(92)109⟩.

Treatment of $(Ph_3Pb)_2CHLi$ with Ph_3GeCl produced the germylbis(stannyl)methane (**157**), which was converted into functionalized lithium methide (**158**) by a transmetallation reaction with phenyllithium (Equations (44) and (45)) ⟨80TL2807⟩.

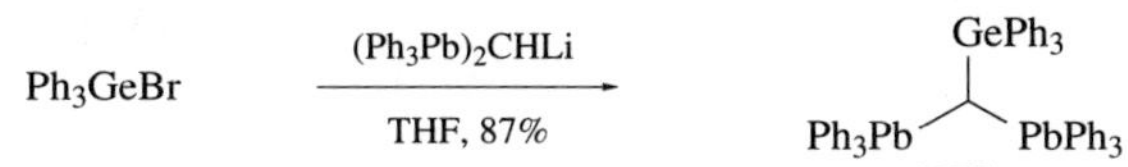

(157) (44)

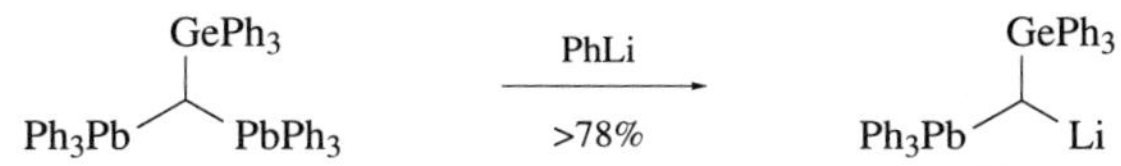

(158) (45)

6.06.2.2.4 *Functions with mixed metalloid(s) and metal(s)*

Only a few totally mixed $RC(E^1X_n)(E^2Y_m)ML_k$ molecules (E^1 and E^2 = Si, Ge, or B; M = metal) have been described. If unsymmetrical compounds such as $X_nE^1CH_2E^2Y_m$ or $X_nE^1CH(Hal)E^2Y_m$ are available, then metal–hydrogen or metal–halogen exchange reactions seem to be the most practical routes to metallated derivatives. Redistribution represents a potential problem and can be avoided by mild reaction conditions and rapid experimental work up procedures.

Several metallating reagents have been examined, but the most satisfactory method for metallation of TMS(ethylenedioxyboryl)methane (**159**) involves the use of LITMP/TMEDA in THF (Scheme 56) ⟨80CC39, 83OM230⟩. In contrast to (**159**), substituted TMS(ethylenedioxyboryl)methanes, $TMSCH(R)BO_2C_2Me_4$ (R = alkyl, aryl) have proved resistant to the usual deprotonation conditions ⟨80JOM(184)C41⟩.

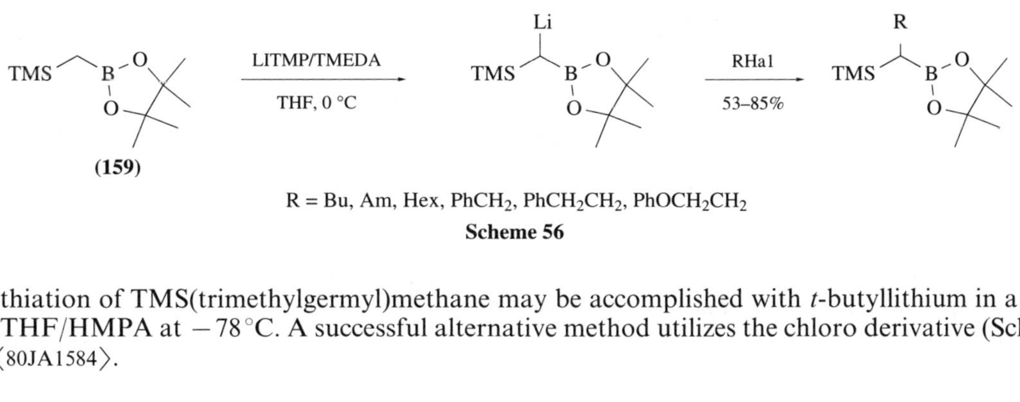

(159)

R = Bu, Am, Hex, $PhCH_2$, $PhCH_2CH_2$, $PhOCH_2CH_2$

Scheme 56

Lithiation of TMS(trimethylgermyl)methane may be accomplished with t-butyllithium in a mixture THF/HMPA at −78 °C. A successful alternative method utilizes the chloro derivative (Scheme 57) ⟨80JA1584⟩.

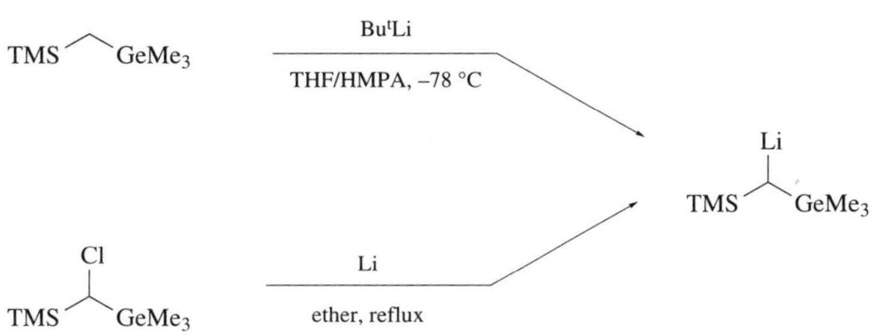

Scheme 57

A versatile approach to metallated polyheteroatom-substituted methanes based on the reactions of tris(triphenylplumbyl)methane has been developed by Kauffmann and co-workers ⟨80TCC109, 80TL2807⟩. Germyl(plumbyl)methide (**158**) undergoes coupling reactions with TMS-Cl to yield the functionalized methane (**160**) (Equation (46)). Transmetallation of silylbis(plumbyl)methanes (**162**) with PhLi gives a new organolithium reagent as shown in Scheme 58 ⟨80TL2807⟩. The required lithium bis(plumbyl)methide (**161**) can be obtained by a transmetallation reaction of $(Ph_3Pb)_3CH$ with PhLi ⟨80TL2807⟩.

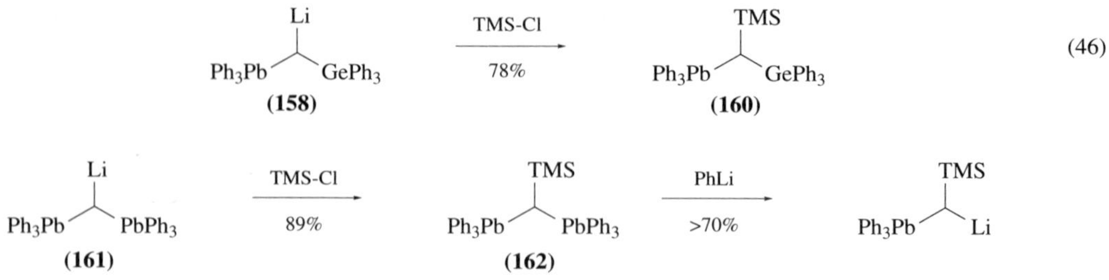

Scheme 58

6.06.3 FUNCTIONS CONTAINING THREE METALS

6.06.3.1 Three Similar Metals

Isolable compounds of the type $RC(ML_m)_3$ (M = metal) are known with M = Li, Hg, Al, Sn, and Pb. Methods of synthesizing polylithiated aliphatic hydrocarbons have been reviewed previously ⟨B-85MI 606-02, 87TCC1⟩. The chemistry and preparation of acyclic compounds with functions containing three heavy main group metals have not been surveyed and only passing comments can be found in appropriate sections of *Methoden der Organischen Chemie* ⟨70HOU(13/4)1, 74HOU(13/2b)1, 78HOU(13/6)181⟩, *Comprehensive Organometallic Chemistry* ⟨82COMC-I(7)265⟩ and in several specialized books ⟨64AOC(2)257, 80TCC(92)109, B-81MI 606-02, B-87MI 606-01⟩.

6.06.3.1.1 Lithium derivatives, $RCLi_3$

Polylithium organic compounds can be prepared by a variety of methods and several have been extensively studied ⟨87TCC(138)1, 92MI 606-02⟩. However, only a few of these methods are readily applicable to the preparation of geminal trilithioalkanes with at least some degree of generality. These are:

(a) reactions of hydrocarbons or halocarbons with lithium vapor;
(b) transmetallation reactions;
(c) metallation of acidic hydrocarbons; and
(d) pyrolysis reactions.

When applicable, transmetallation, especially mercury–lithium exchange, is usually the most convenient and is therefore the method of choice. On the other hand, reactions of hydrocarbon or halocarbon substrates with lithium vapor probably represent the most general approach. Carbon–lithium bonds in polylithiated alkanes are very susceptible to hydrolysis and the handling of any such compound requires the use of vacuum systems and other specialized techniques.

(i) Reactions with lithium vapor

Direct halogen–metal exchange in polyhaloalkanes by treatment with lithium metal, the most popular method for the synthesis of alkyllithium compounds, is only of limited value for the preparation of polylithiated alkanes due to α-elimination of lithium halide after the first step being faster than the halogen–metal exchange ⟨79AG(E)785, 80HCA2046⟩. These difficulties can be overcome by high-temperature reaction of polyhaloalkanes with lithium vapor.

Thus, tetralithiomethane, CLi_4, and hexalithiomethane, C_2Li_6, the first examples of perlithiated alkanes, were prepared by the reaction of lithium vapor with the appropriate perchlorocarbons under vacuum at high temperatures (750–1000 °C) ⟨72CC1078, 83JOM(249)1⟩. The reactions are carried out in a stainless steel reactor consisting of a Knudsen cell containing lithium metal heated in a furnace, an inlet tube, and a liquid nitrogen cold finger. The main advantage of this technique, besides the high reactivity of lithium vapor, is the reduction of rearrangements and secondary reactions from rapid quenching of the products onto the cold finger ⟨87TCC1⟩. In both preparations, depicted in Equations (47) and (48), the main by-products were C_2Li_4 and C_2Li_2. Further investigations showed that the selectivity can be improved by working at 750 °C instead of 850 °C ⟨75CC302, 79JOC2311, 82JA7345⟩. However, the method although optimized, still provides a maximum of 40% CLi_4 together with 58% C_2Li_2 ⟨83JOM(249)1⟩. When chloroform was used as a substrate, trilithiomethane was obtained in 15.5% yield (Equation (49)); the main side products being 39.7% C_2Li_2, 20.1% CLi_4, and 19.1% CH_2Li_2 ⟨82JA7345, 83JOM(249)1⟩. Similarly, the reaction of dichloro(TMS)methane with lithium vapor gave TMS-$CHLi_2$ in only 11% yield (Equation (50)) ⟨84CC1664⟩.

$$CCl_4 + 8Li(g) \xrightarrow{850\,°C} CLi_4 + 4LiCl \qquad (47)$$

$$C_2Cl_6 + 12Li(g) \xrightarrow{850\,°C} C_2Li_6 + 6LiCl \qquad (48)$$

$$CHCl_3 + 6Li(g) \xrightarrow{750\,°C} CHLi_3 + 3LiCl \qquad (49)$$

$$\text{TMS-}CHCl_2 + 4Li(g) \xrightarrow{720\,°C} \text{TMS-}CHLi_2 + 2LiCl \qquad (50)$$

The purest perlithiated ethane (C_2Li_6) was prepared by using diethylmercury instead of hexachloroethane as the starting material ⟨79JA2214⟩. It was also reported that the vapor phase reaction of carbon with atomic lithium yielded perlithiopropyne (C_3Li_4) as the main product ⟨73JA1343⟩.

The major drawback of a technique utilizing reactions between organic substrates and lithium vapor lies in the separation of a mixture of lithium-substituted hydrocarbons. Unfortunately, it has not yet been possible to separate the resulting mixtures, and even separation from the lithium matrix is difficult ⟨83JOM(249)1, 92AG(E)584⟩.

(ii) Transmetallation reactions

In many cases of synthesis of substantial amounts of pure polylithiated alkanes there is no practical alternative to transmetallation reactions ⟨87TCC(138)1⟩. However, since the transmetallation route requires the intermediacy of another organometallic compound, polylithiated alkanes prepared by this method share the same limitations and restrictions that the parent organometallic compound encounters. In practice, mercury–lithium exchange reactions represent one of the most effective routes to the di- and trilithiomethanes because the corresponding organopolymercury compounds are comparatively readily available. Thus, trilithiomethane ($HCLi_3$) can be prepared in good yield by treating tris(chloromercurio)methane with lithium metal in dry THF (Equation (51)) ⟨82MI 606-01⟩. Unfortunately, direct mercury–lithium exchange reaction is not suitable for the synthesis of tetralithiomethane (CLi_4). Treating $C(HgCl)_4$ with lithium metal in diethyl ether resulted in the dimeric products hexalithioethane (C_2Li_6) and tetralithioethylene (C_2Li_4). The reaction is believed to proceed via radical intermediates ⟨87TCC(138)1⟩.

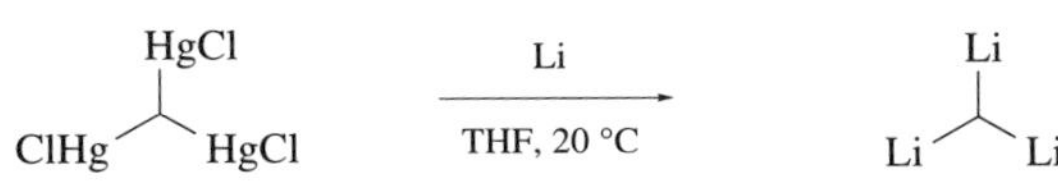

(51)

A variant of the mercury–lithium exchange method is provided by the reaction of organomercury compounds with *n*-butyllithium. Many of the polymercury compounds, including $C(HgCl)_4$ ⟨84AG(E)995, 86OS378⟩ and $CH_2(HgCl)_2$ ⟨83AG(E)733⟩, are useful in this reaction. However, the procedure has not yet been adequately developed for synthesis of trilithiated hydrocarbons.

Transmetallation reactions with several other elements, such as boron and tin, can also be used for the generation of polylithiated organic compounds ⟨80TCC(92)109, 82AG(E)410⟩ but trilithiated hydrocarbons have not yet been prepared this way.

(iii) Metallation of acidic hydrocarbons

Much more easily accessible than the 1,1,1-trilithiated alkanes are certain perlithioalkynes, since alkyllithium reagents can abstract quantitatively not only the hydrogen atoms on *sp*-hybridized carbon atom, but also those of the adjacent methyl group ⟨74MI 606-01, 82AG(E)410, 83T2733⟩. For example, the conversion of propyne into perlithiated derivative can be achieved by reaction with *n*-butyllithium in hexane (Equation (52)) ⟨65JA3788, 69JA6156, 84AG(E)995⟩.

 $\xrightarrow[\text{hexane, 75\%}]{\text{Bu}^n\text{Li}}$ C_3Li_4 (52)

The pattern of reactivity of conjugated diynes closely resembles that of the propyne. Thus, reaction of the 1,3-pentadiyne with excess *n*-butyllithium/TMEDA in hexane gives the perlithiated hydrocarbon C_5Li_4 (Equation (53)) ⟨73JA3324⟩. 2,4-Hexadiyne under the same conditions ⟨76JA8426⟩ or even in the absence of TMEDA ⟨73JCS(P2)599⟩ has been converted into the trilithiated product (Equation (54)). Metallation of 2,4-octadiyne in diethyl ether or THF affords another "lithiocarbon" (Equation (55)) ⟨73JCS(P2)599⟩.

 $\xrightarrow[\text{hexane}]{\text{Bu}^n\text{Li}}$ C_5Li_4 (53)

 $\xrightarrow[\text{hexane}]{\text{Bu}^n\text{Li}}$ MeC_5Li_3 (54)

C_3H_7  $\xrightarrow[\text{ether or THF}]{\text{Bu}^n\text{Li}}$ $C_3H_7C_5Li_3$ (55)

Unlike alkynes, linear and branched alkenes upon treatment with organolithium reagents form only mono- and dilithiated hydrocarbons. For example, propene is dilithiated with *n*-butyllithium/TMEDA in hexane to give the corresponding dianion ⟨75CC877⟩. Isobutylene under these conditions yields the cross-conjugated trimethylenemethane dianion ⟨76T1839⟩. The isomeric 2-butene can be dilithiated to give 1,4-dilithio-2-butene ⟨74JA5640⟩.

(iv) Pyrolysis reactions

The pioneering work in this field is due to Ziegler *et al.* who reported that pyrolysis of halide-free methyllithium resulted in the formation of dilithiomethane in excellent yield (Equation (56)) ⟨55ZAAC(282)345⟩. This procedure is still the method of choice for the preparation of CH_2Li_2, however, it is not the best route to 1,1,1-trilithiated hydrocarbons. Although perlithiopropyne (C_3Li_4) was detected among the products of the thermal decomposition of CH_2Li_2 and CLi_4 (Equations (57) and (58)), it could not be separated from the resulting mixtures ⟨81JA5951, 82JA2637, 85JA5313⟩.

$$\text{MeLi} \xrightarrow{225\,^\circ\text{C}} CH_4 + CH_2Li_2 \qquad (56)$$

$$CH_2Li_2 \xrightarrow[\text{6 min}]{350\,^\circ\text{C}} \underset{20\%}{C_3Li_4} + \underset{80\%}{C_2Li_2} \qquad (57)$$

$$CLi_4 \xrightarrow[\text{8 min}]{225\,^\circ\text{C}} \underset{20\%}{CLi_4} + \underset{40\%}{C_3Li_4} + \underset{30\%}{C_2Li_4} + \underset{10\%}{C_2Li_2} \qquad (58)$$

6.06.3.1.2 *Mercury derivatives, $RC(HgX)_3$*

As in many other cases, transmetallation represents a valuable route to polymercurated methanes. Thus, tris(dimethoxyboryl)methane, readily available via the reaction of chloroform with dimethoxyboron chloride and lithium metal (see Section 6.06.2.1.2), undergoes a rapid reaction with mercuric salts to produce trimercurated methanes (Equation (59)) ⟨80JOM(191)7, 83JOM(243)245⟩.

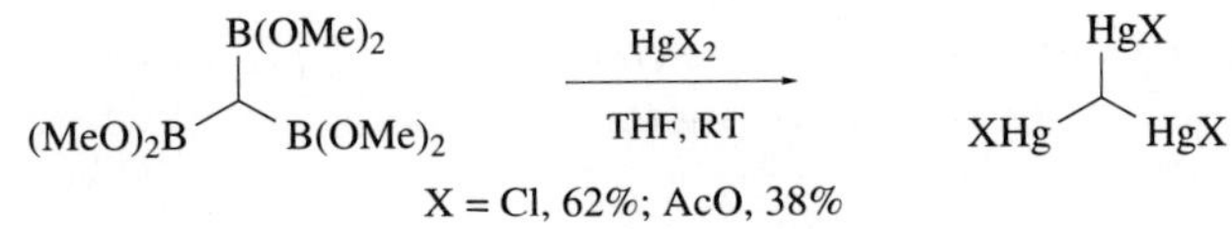

(59)

However, despite the usefulness of transmetallation reactions, this route to trimercurated derivatives has rarely been used, because access is easier via direct mercuration of organic compounds containing an activated methyl group ⟨B-67MI 606-01, 68MI 606-01, 71MI 606-01, 74HOU(13/2b)1, B-81MI 606-02⟩. Treatment of methyl derivatives containing electron-withdrawing groups with a mercuric salt, HgX_2, results in electrophilic mercuration giving a mercury derivative. Use of an excess of HgX_2 may give rise to trimercurated compounds. Typical examples from the literature are given in Table 4. The reactions are normally carried out in aqueous acidic solutions or without solvent at elevated temperatures. For example, acetonitrile reacts with anhydrous mercuric acetate at 150 °C for 20 h in a bomb tube to give the trimercurated product in 93% yield (Table 4, entry 8). Conditions for preparing the mono- and dimercurated derivatives of acetonitrile have also been described ⟨74JPR557⟩.

Table 4 Preparation of 1,1,1-trimercurioalkane derivatives.

Entry	*Compound*	*Method of preparation*	*Yield* (%)	*Ref.*
1	$HC(HgCl)_3$	$HC[B(OMe)_2]_3 + HgCl_2$	62	80JOM(191)7
2	$HC(HgBr)_3$	$HC[B(OMe)_2]_3 + HgBr_2$		83JOM(243)245
3	$HC(HgI)_3$	$HC(HgOAc)_3 + KI$	82	83JOM(243)245
4	$HC(HgOAc)_3$	$HC[B(OMe)_2]_3 + Hg(OAc)_2$	38	83JOM(243)245
5	$HC(HgSCN)_3$	$HC(HgOAc)_3 + KSCN$		83JOM(256)217
6	$HC(HgCN)_3$	$HC(HgOAc)_3 + KCN$	83	83JOM(243)245
7	$HC(HgMe)_3$	$HC[B(OMe)_2]_3 + MeHgOAc$		86JOM(301)1
8	$NCC(HgOAc)_3$	$MeCN + Hg(OAc)_2$	93	74JPR557
9	$NCC(HgMe)_3$	$MeCN + (MeHg)_2O$	95	75ZAAC(415)233
10	$OHCC(HgCl)_3$	$EtOH + HgCl_2/NaOAc$		82JOM(238)327
11	$OHCC(HgBr)_3$	$EtOH + HgBr_2/NaOAc$		82JOM(238)327
12	$OHCC(HgBr)_3$	$AcH + Hg(NO_3)_2$		87JOM(319)1
13[a]	$HO_2CC(HgCl)_3$	'mercuretin' + HCl	95	84JOM(276)1
14[a]	$HO_2CC(HgOAc)_3$	'mercuretin' + AcOH	54	84JOM(276)1
15[a]	$HO_2CC(HgNO_3)_3$	'mercuretin' + HNO_3		86JOM(306)1

[a] "Mercuretin": $AcO[HgC(HgOAc)_2CO_2]_nH$ ($n \sim 0$).

Mercuration of C—H bonds α to carbonyl groups also occurs readily. Thus, reaction of acetic aldehyde with $HgCl_2$ led to the formation of $(ClHg)_3CCHO$ in almost quantitative yield. Mercuration of the aldehyde RCH_2CHO (R = Me, Et) gave $(ClCH_2)_2CRCHO$ (85–88%) ⟨84CCA689⟩. The product obtained by boiling an ethanolic solution of mercuric chloride with sodium acetate ⟨1899CB870⟩ has also been identified as tris(chloromercurio)acetaldehyde (Table 4, entry 10). The bromine analogue was obtained in the same way ⟨82JOM(238)327⟩. The crystal structures of solvates $(ClHg)_3CCHO{\cdot}DMF$ and $(BrHg)_3CCHO{\cdot}DMSO$ have been determined by x-ray diffraction methods ⟨82CSC1571, 82JOM(238)327⟩.

Mercuration of acetaldehyde with mercuric nitrate in aqueous nitric acid always leads to trimercurated acetaldehyde, which separates from the solution in the form of various nitrates whose type and composition depends upon the acid concentration ⟨87JOM(319)1⟩. Addition of an ethanolic solution of acetaldehyde to an aqueous solution of mercuric nitrate gives a hydrated oxonium nitrate of tris(mercurio)acetaldehyde, $(OHg_3CCHO)NO_3{\cdot}H_2O$ ⟨83JOM(253)283⟩.

It has been confirmed that the compounds obtained by melting mercuric acetate, named mercuretin ⟨27JCS2658⟩, is identical with the compound prepared by heating mercuric acetate in acetic anhydride ⟨03LA(329)116⟩. Mercuretin, prepared by both routes, has been identified as the condensation polymer of tris(acetoxymercurio)acetic acid, $AcO(HgC(HgOAc)_2CO_2)_nH$, with n approximately equal to 10. Its hydrolysis in dilute HCl gives tris(chloromercurio)acetic acid as the only Hg-containing product (Table 4, entry 13). Two nitrates of trimercurated acetic acid, obtained from the solution of mercuretin in nitric acid were identified as $[(Hg(H_2OHg)(NO_3Hg)CCO_2)NO_3]$ and $(2(NO_3Hg)_3CCO_2H{\cdot}HNO_3)$ ⟨86JOM(306)1, 91JOM(411)19⟩.

The reaction of 2-ethyl- and 2,4,4,5,5-pentamethyl-1,3-dioxalanium perchlorate with $Hg(OAc)_2$ and $Hg(OCOCF_3)_2$ leads to the formation of mono-, di-, and trimercurated carboxylic acids depending upon the ratio of reagents ⟨81ZOB2241⟩. Mercuration of acetone in acidic aqueous solutions produces at least nine mono-, poly-, and permercurated species, all of which can exist in equilibrium simultaneously ⟨89OM2646⟩.

6.06.3.1.3 *Aluminum derivatives, $RC(AlX_2)_3$*

Geminal trialuminioalkanes are available via bishydroalumination of alkynyldialkylalanes ⟨70HOU(13/4)1⟩. In general, the reaction of terminal alkynes with two molar equiv. of a dialkylaluminum hydride in ether or THF lead to 1,1-dialuminioalkanes (Equation (60)) ⟨60LA(629)222, 66TL6021, 71CC1593⟩. However, in basic solvents such as triethylamine, the reaction of alkynes with one molar equiv. of dialkylaluminum hydride produces metallation rather than hydroalumination ⟨63AG(E)686, 68IZV910⟩. This has been used synthetically to prepare alkynylalanes. Further treatment of the resultant alkynylalanes with excess dialkylaluminum hydride affords trialuminioalkanes (**163**) (Scheme 59) ⟨60LA(629)222, 63BSF1462⟩.

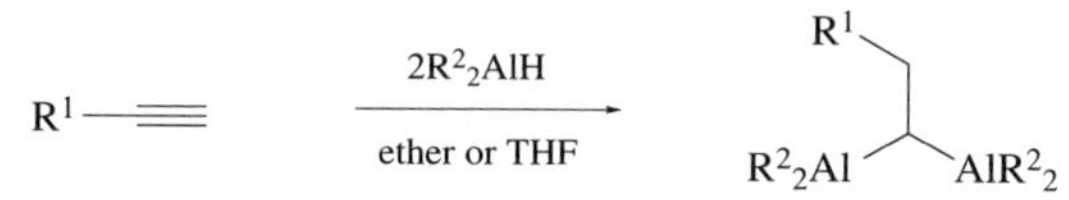

(60)

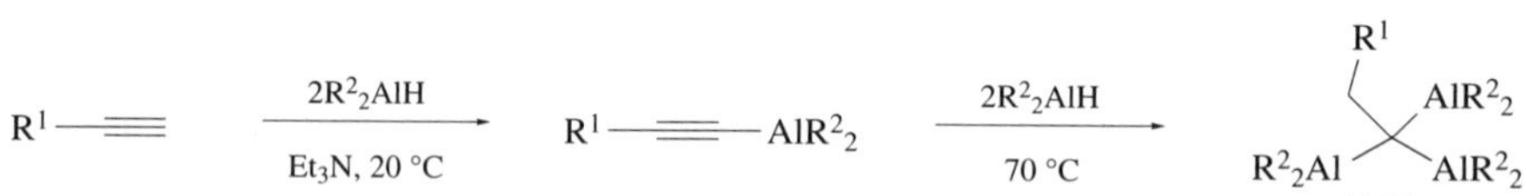

(**163**)

R^1 = Bu^n, R^2 = Et

Scheme 59

Related preparation of 1,1,1-trialuminioalkanes involves bishydroalumination of alkynylalanes derived from the reaction of alkali metal acetylides with dialkylaluminum chlorides (Scheme 60) ⟨60LA(629)222, 68BSF216⟩.

R = Me, Et

Scheme 60

The chemistry of 1,1,1-trialuminioalkanes is not well developed, and only the reactions of compound (**163**) with electron-donor molecules, such as LiH, KF, Et_2O, and pyridine have been reported ⟨83IZV636⟩.

6.06.3.1.4 *Tin and lead derivatives, $RC(MX_3)_3$ (M = Sn or Pb)*

Tris(trimethylstannyl)methane (**164**) can be readily synthesized via interaction of chloroform with trimethylstannyllithium ⟨75OMS(10)18⟩ or via reaction of bromoform with the Grignard reagent obtained from chlorotrimethylstannane and magnesium turning (Scheme 61) ⟨84JOM(266)37⟩. Both procedures give the crude product containing substantial quantities of bis(trimethylstannyl)methane. Distillation and subsequent crystallization from ethanol afford a pure substance in average yield.

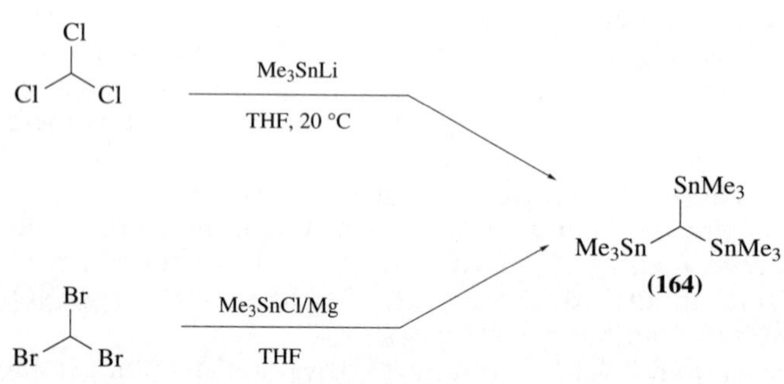

(**164**)

Scheme 61

A more general route to 1,1,1-tristannylalkanes involves the hydrostannation of 1-stannyl-1-alkynes. Trimethyltin hydride has been shown to react with 1-stannyl-1-alkynes in the presence of catalytic amounts of azobisisobutyronitrile (AIBN) to form predominantly 1,1-distannyl-1-alkenes (**165**). The latter undergo a second hydrostannation to give the corresponding 1,1,1-tristannylalkanes (Scheme 62) ⟨83JOM(252)47, 85OM1044⟩.

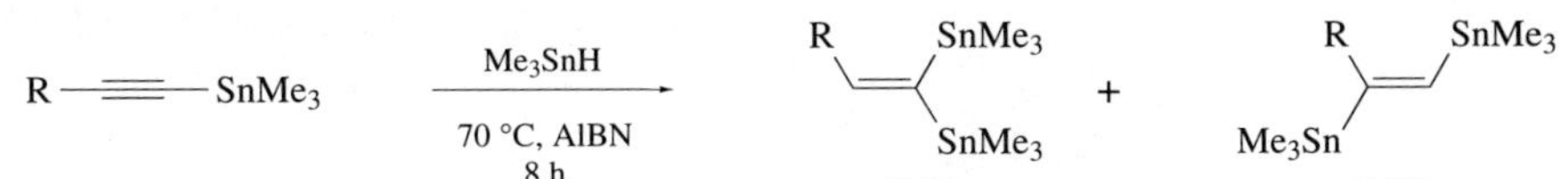

(165) **(166)**

R(ratio **(165)/(166)**, %): Me (97/3), Bu^n (96/4), Bu^t (98/2), Ph (95/5), $MeOCH_2$ (59/41), $PhOCH_2$ (61/39), PhO

(165)
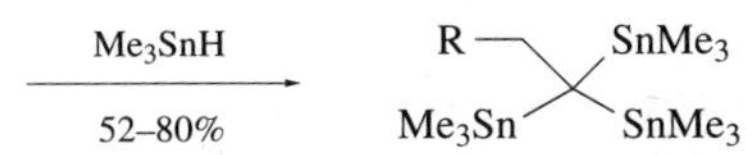

R = Me, Bu^n, Ph, $PhCH_2$, $MeOCH_2$, $PhOCH_2$, PhO

Scheme 62

1,1-Distannyl-2,2-diorganyl-1-alkenes, resulting from the coupling reaction of geminal dibromoalkenes with Me_3SnLi, have also been successfully transformed into the corresponding tristannylalkanes (Scheme 63) ⟨86OM1991⟩.

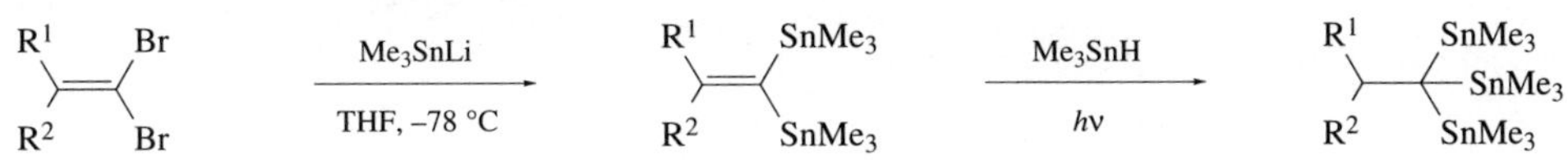

R^1 = Me, CH_2OMe; R^2 = Me, Et, Ph, CH_2CH_2OMe, CH(Me)OMe, CH_2CH_2OEt, CH_2OMe

Scheme 63

Oxymetallation of the alkene derivative (**167**), which is achieved by using Me_3SnOMe in hexane, yields functionalized tristannylalkanes (**168**) (Equation (61)) ⟨91CB503⟩. The reaction is believed to involve the formation of adduct at the boryl group followed by intramolecular rearrangements.

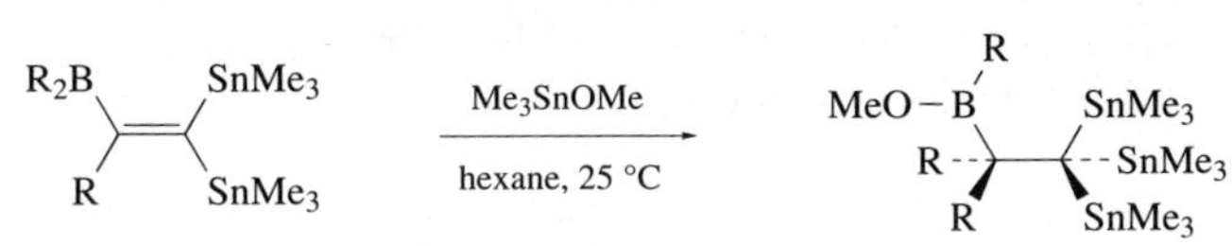 (61)

(167) **(168)**

R = Et

The preparative chemistry of triplumbylmethane derivatives is considerably less rich as compared to the chemistry of tristannylmethanes. The coupling reaction between chloroform and triorganylplumbyllithium is the sole valuable route to tris(triorganylplumbyl)methanes ⟨70JOM(23)471, 85CB380, 91SA(A)849⟩. The method is especially efficient for triphenylplumbyl derivatives since the Ph_3PbLi is readily available from Ph_3PbCl and lithium metal (Equation (62)) ⟨85CB380⟩.

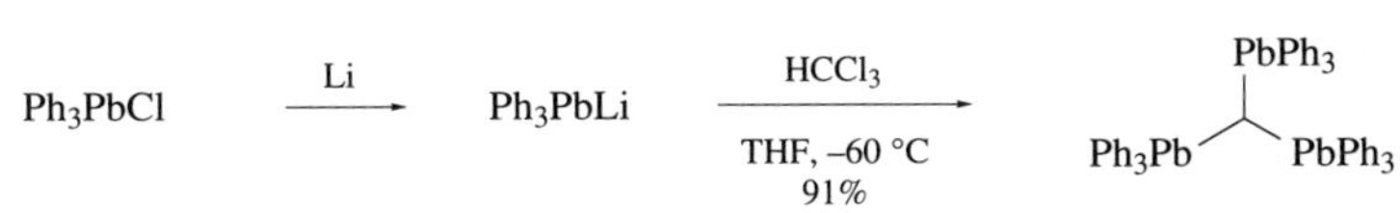 (62)

The following compounds were made analogously; $(Ph_3Pb)_4C$, $(Ph_3Pb)_2CCl_2$, $Ph_3PbCHCl_2$, and $(Ph_3Pb)_2CH_2$. Attempts to prepare $(Ph_3Pb)_3CCl$ and $(Ph_3Pb)_2CHCl$ failed, because of further reaction with triphenylplumbyllithium ⟨70JOM(23)471⟩.

6.06.3.2 Three Dissimilar Metals

Trimetallated alkanes with different metals attached to carbon are rare. Even so, access to these species is of great interest because it is usually difficult to reach such structures by classical routes.

A few of the compounds have been prepared by transmetallation or metallation reactions and examples are described in reviews by Kauffmann ⟨80TCC109, 82AG(E)410⟩.

The conversion of 1,1,1-tris(diethylaluminio)hexane (**163**) into lithiobis(diethylaluminio)hexane has been achieved using reactions with *n*-butyllithium. It was noted that only one diethylaluminio group can be replaced by lithium even when an excess lithium reagent was used (Scheme 63) ⟨83IZV636⟩.

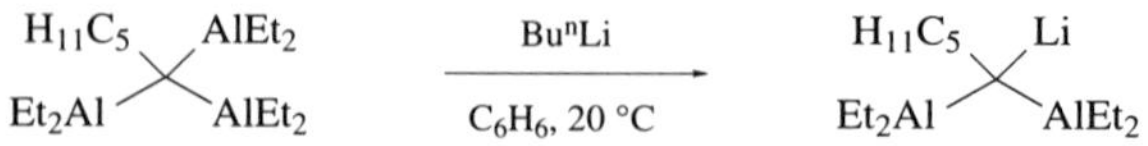

(**163**)

(63)

Compounds with the grouping Sn—C(Li)—Sn are accessible via the reaction of bis(trimethylstannyl)methane with lithium dicyclohexylamide (LDCA) in presence of 1 mol HMPA ⟨79TL501⟩. The nature of the lithium reagent is critical; generally, deprotonation occurs with lithium amides, while transmetallation occurs with alkyllithiums ⟨80TCC109, 90S259⟩. An alternative method for the preparation of lithium bis(stannyl)methide (**169**) consists of treating a tristannylated compound with phenyllithium in ether (Scheme 64). Reaction of lithium derivative (**169**) with Ph_3SnCl gave tris(triphenylstannyl)methane ⟨80TCC(92)109⟩.

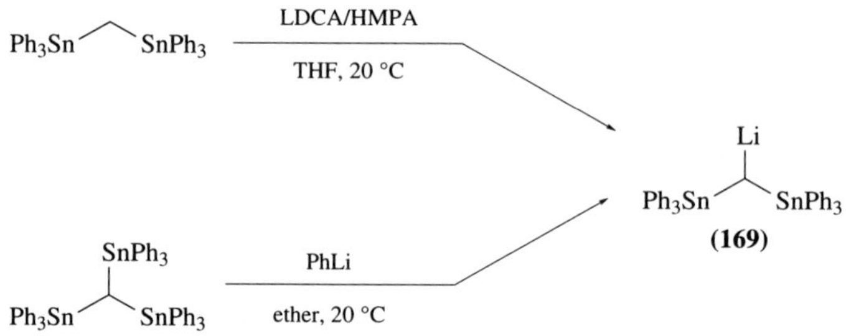

(**169**)

LDCA = lithium dicyclohexylamide

Scheme 64

The lithium bis(triphenylplumbyl)methide (**170**) has been synthesized via several routes; metallation of bis(triphenylstannyl)methane, transmetallation reaction using $(Ph_3Pb)_3CH$ and phenyllithium, or by halogen–metal exchange starting from bromobis(triphenylplumbyl)methane ⟨80TL2807⟩. Lithium derivative (**170**) reacts smoothly with Me_3SnCl to give the expected (stannyl)bis(plumbyl)methane (**171**) (Equation (64)).

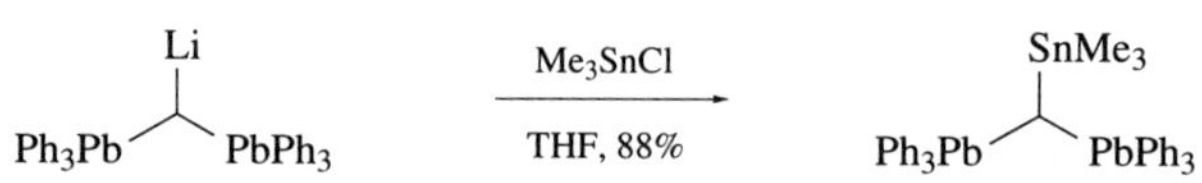

(**170**) (**171**)

(64)

The synthetic potential of compounds of structures (**169**) and (**170**) in reactions with other metal halides remains unexplored.

6.07
Functions Containing Four Halogens or Three Halogens and One Other Heteroatom Substituent

ALEX H. GOULIAEV
Aarhus University, Århus, Denmark

and

ALEXANDER SENNING
Technical University of Denmark, Lyngby, Denmark

6.07.1 TETRAHALOMETHANES—C(Hal)$_4$

The first part of this chapter is confined to a limited number of compounds, they are the 35 possible tetrahalomethanes (Table 1) of which many are commercially available plus a few trifluoromethyl compounds of hypervalent iodine which have been prepared. Most of the tetrahalomethanes have been prepared with the exception of the eight iodine-containing tetrahalomethanes, $CClI_3$, CFI_3, $CBrCl_2I$, CBr_2ClI, CBr_2I_2, $CClFI_2$, $CBrClI_2$ and the chiral CBrClFI. Information on high yielding and selective syntheses of $CBrI_3$, CCl_2I_2 (formed as a by-product) and $CBrFI_2$ (15% as a by-product) is, however, also rather sparse.

Organic perfluoro compounds have many practical applications, such as in refrigerants, as propellants, as fire extinguishers, in the plastics industry as well as in the manufacture of pharmaceuticals, whereas perchloro (apart from CCl_4), perbromo and periodo compounds are more or less only of academic interest.

Methods for introducing fluorine into organic compounds are presented elsewhere in this volume and are not discussed in this chapter. For more information there is a number of excellent books on haloorganic compounds. Organic fluorine compounds have been treated by several authors.

The most recent handbook is Hudlicky's *Chemistry of Organic Fluorine Compounds* ⟨B-92MI 607-01⟩. This book covers all fields of organic fluorine chemistry, including methods for introducing fluorine into organic compounds, reactions, properties, analysis and practical applications of these, as well as synthetic procedures and the synthesis of fluorinating agents. Olah *et al.* ⟨B-92MI 607-02⟩ have published a book covering the synthesis of fluorine compounds, as have Liebman *et al.* ⟨B-88MI 607-01⟩, whereas German and Zemskov mainly cover reagents for the synthesis of fluorine compounds ⟨B-89MI 607-01⟩. Older books include Chambers' *Fluorine in Organic Chemistry* ⟨B-73MI 607-01⟩, Sheppard and Sharts' *Organic Fluorine Chemistry* ⟨B-69MI 607-01⟩, and, finally, Emeléus' *The Chemistry of Fluorine and Its Compounds* ⟨B-69MI 607-02⟩, which covers mainly the inorganic chemistry of fluorides. Organic bromine compounds have been described by Jolles in *Bromine and its Compounds* ⟨B-66MI 607-01⟩.

6.07.1.1 Four Similar Halogens

6.07.1.1.1 Tetrafluoromethane, CF_4

Treatment of metal carbides with fluorine at 20 °C for 90 min or with CoF_3 at 440 °C for 9 h leads to mixtures of various fluorocarbons among which CF_4 predominates. Thus, Cr_2C_3 yields 91% CF_4, as do Al_4C_3 (88%), B_4C (88%), TiC (86%), Fe_3C (86%), CaC_2 (75%) and WC (67%) ⟨59JA806⟩. Fluorination of silicon carbide with F_2 also leads to CF_4 (Equation (1)) ⟨50IS171⟩.

$$M_nC_m + F_2 \xrightarrow{20\,°C,\ 90\ min} CF_4 + C_2F_6 + C_3F_8 + C_4F_{10} \quad (1)$$

relative yields: 68–91% 6–26% 2–14% 0–2%

M, *n*, *m*: see text for specification.

Tetrafluoromethane is furthermore formed, in better than 50% yield, upon fluorination of activated charcoal with F_2 ⟨39JA2962⟩, or upon fluorination of graphite with PF_3, AsF_3, SF_6, CuF_2, FeF_3, VF_5 at 900–1200 °C ⟨55USP2709186, 55USP2709188, 55USP2709190⟩, PF_5 at 2440 °C ⟨58GEP1040011⟩ or HgF_2 at 560–800 °C ⟨55USP2709187⟩.

Fluorination (F_2) of trifluoromethane ⟨37JA1200, 36CB299⟩ or tetrachloromethane at 70 °C in the presence of As gives a relative yield of 74% CF_4 with $CClF_3$ as the main by-product ⟨40JA3477⟩. Treatment of tetrachloromethane ⟨31USP1961622⟩ or tetraiodomethane ⟨48JCS2188⟩, under readily available laboratory conditions, with BrF_3 as fluorinating agent also leads to tetrafluoromethane (Equation (2), Table 2).

Table 1 Bibliographic information on the preparation, melting points and boiling points of the tetrahalomethanes.

Formula	*CA Reg. no.*	*Beilsteins Handbuch der Organischen Chemie*[a]	*Gmelin Handbuch der Anorganischen Chemie*[b]	*M.p.* (°C)	*B.p.* (°C)
CX_4					
CF_4	75-73-0	H 59, I 8, II 11, III 35, IV 26	92/93	−184[c]	−130[c]
CCl_4	56-23-5	H 64, I 12, II 22, III 64, IV 56	139/140	−23[c]	77[c]
CBr_4	558-13-4	H 68, I 17, II 35, III 92, IV 85	254	88–90[c]	190[c]
CI_4	507-25-5	H 74, I 19, II 39, III 104, IV 98	273/283	168[c]	130–140 (1-2 torr)
CF_3X					
$CClF_3$	75-72-9	III 42, IV 34	292/293	−181[c]	−80[c]
$CBrF_3$	75-63-8	III 83, IV 73	306/307	−168[c]	−57 to −58[c]
CF_3I	2314-97-8	III 98, IV 92	314		−22.5[c]
CCl_3X					
CCl_3F	75-69-4	H 64, III 63, IV 54	323	−111 to −110[c]	23.7[c]
$CBrCl_3$	75-62-7	H 67, II 31, III 85, IV 77	339	−6[c]	105[c]
CCl_3I	594-22-9	H 71, III 99, IV 95	283/292, 297/298, 349	−19[a]	142[a]
CBr_3X					
CBr_3F	353-54-8	I 17, III 91, IV 85	284/292, 297, 337/339	−73.6[c]	106–107[c]
CBr_3Cl	594-15-0	H 68, II 35, III 91, IV 85	283/292, 297, 349/351	55[a]	160[a]
CBr_3I	14349-80-5	H 71, IV 96	297/298, 351	35[d](dec.)	
CI_3X					
CFI_3	1495-49-4		297//298		
$CClI_3$	14349-82-7		297/298, 351		
$CBrI_3$	558-16-7	H 74	297/298, 351	113[a]	
CF_2X_2					
$CBrClF_2$	353-59-3	IV 75	352/354, 382/385	−159.5[b,e]	−4[f]
$CClF_2I$	420-49-5	IV 94	383, 385		33[f]
$CBrF_2I$	753-66-2		383, 385		64.5–65.5[g] (746 mmHg)
CCl_2X_2					
CCl_2F_2	75-71-8	H 61, III 48, IV 40	354, 383, 385	−158[c]	−29.8[c]
$CBrCl_2F$	353-58-2	IV 76	354, 383, 385	−106[b]	51–52[f]
CCl_2FI	420-48-4	IV 95	383, 386	−107[b]	90[b]
$CBrCl_2I$	40809-91-4		383, 386		
CBr_2X_2					
CBr_2F_2	75-61-6	I 16, III 87, IV 80	352/354, 375/378	−140[h]	23.8[b,e]
CBr_2Cl_2	594-18-3	H 68, III 88, IV 82	353/354, 379/382	22[a]	135[a]
CBr_2ClF	353-55-9	III 87, IV 82	354, 383, 386		79–80[c]
CBr_2FI	1478-04-2		383, 386		
CBr_2ClI	40809-93-6		383, 386		
CI_2X_2					
CF_2I_2	1184-76-5		354, 378/379		80.5[b,e]
CCl_2I_2	594-23-0		382	85[i]	
CBr_2I_2	14059-90-6		382		
$CClFI_2$	353-49-1		383, 386		
$CBrFI_2$	753-67-3		383, 386		
$CBrClI_2$	40809-94-7		383, 386		
CBrClFI					
CBrClFI (*R*)					
CBrClFI (*S*)					
CBrClFI (±)					
CBrClFI ( )	753-65-1		386		

[a] *Beilsteins Handbuch der Organischen Chemie* ⟨B-18MI 607-01, B-28MI 607-01, B-41MI 607-01, B-59MI 607-01, B-72MI 607-02⟩. [b] *Gmelin Handbuch der Anorganischen Chemie* ⟨B-74MI 607-02⟩. [c] From Aldrich's catalog of chemicals ⟨B-93MI 607-01⟩. [d] From Dehn ⟨09JA1220⟩. [e] Calculated value. [f] From Haszeldine ⟨52JCS4259⟩. [g] From Burton *et al.* ⟨82JFC(20)89⟩. [h] From Miller and Smyth ⟨57JA20⟩. [i] From Höland ⟨1887LA(240)225⟩.

Table 2 Synthesis of fluorotrihalo-, difluorodihalo-, trifluorohalo- and tetrafluoromethanes with BrF_3 and IF_5.

Tetrahalomethane (relative amount 1)	*Relative amount of BrF_3*	*Relative amount of IF_5*	*Temp. (°C)*	*Reaction time*	*Total yield (%)*	*Products[a] (ratio)*
CCl_4	0.33		RT		61	CCl_2F_2 (5) CCl_3F (95)
	0.6		−78 to RT[b]		89	CCl_2F_2 (85) CCl_3F (15)
	0.84–1.06[c]		RT		65–70	CCl_2F_2(15–43) $CClF_3$ (57–85)
CCl_3F	0.8		RT		96	CCl_2F_2
CBr_4	0.33		⩾0		94	CBr_3F (~100) CBr_2F_2 (trace)
	0.66		0		86	CBr_3F (~22) CBr_2F_2 (~88) $CBrF_3$ (trace)
	1[d]		0		94	$CBrF_3$
		0.63	90	3 h	83	CBr_3F (trace) CBr_2F_2 (≈100) $CBrF_3$ (trace) CF_4 (trace)
CI_4[e]	1.92		RT		48	CBr_2F_2 (33) CF_4 (67)
		0.88	RT to 90–100	30 min	65	CIF_3

Source: Banks *et al.* 〈48JCS2188〉. [a] Volatile products were collected in cold traps. [b] Tetrachloromethane was added to solid BrF_3 in carbon dioxide. [c] The reaction was conducted in an autoclave. [d] Bromine trifluoride was added dropwise. [e] Tetraiodomethane was added dropwise over 2 h.

$$CI_4 + BrF_3 \xrightarrow{RT} CBr_2F_2 + CF_4 \qquad (1{:}2) \quad 48\%\ (33{:}67) \tag{2}$$

Tetrafluoromethane can furthermore be made by electrochemical means with HF as the fluorine donor. Thus, electrochemical fluorination of trimethylamine yields mainly tris(trifluoromethyl)-amine, but also some tetrafluoromethane (Equation (3)) 〈52USP2616927〉, whereas electrochemical fluorination of dimethyl sulfide (Equation (4)) yields CF_4 as well as sulfur hexafluoride and trifluoromethylsulfur pentafluoride 〈53JCS2372〉. Electrochemical fluorination of acetic acid, acetyl chloride, acetone, trimethylacetic acid, acetonitrile or methanol in HF with nickel anodes and iron cathodes as well as acetic acid with a KF + 3 HF melt have also been applied in the synthesis of tetrafluoromethane 〈49MI 607-01〉 as has the electrolysis of concentrated trifluoroacetic acid solutions with platinum electrodes (Equation (5)) 〈33BSB102〉.

$$NMe_3 \xrightarrow{\text{HF, electrolysis}} N(CF_3)_3 + CHF_3 + CF_4 + NF_3 \tag{3}$$

$$SMe_2 \xrightarrow[97\%]{\text{HF, electrolysis}} \underset{2\%}{(CF_3)_2SF_4} + \underset{21\%}{CF_3SF_5} + \underset{60\%}{CF_4} + \underset{17\%}{SF_6} \tag{4}$$

$$CF_3CO_2H \xrightarrow{\text{electrolysis}} CF_4 + C_2F_6 + CO_2 + O_2 \tag{5}$$

Other possibilities involve bond breaking in perfluoroalkenes, for example oxidation of tetrafluoroethylene with dioxygen which yields carbon dioxide and tetrafluoromethane (Equation (6)) 〈44USP2351390〉.

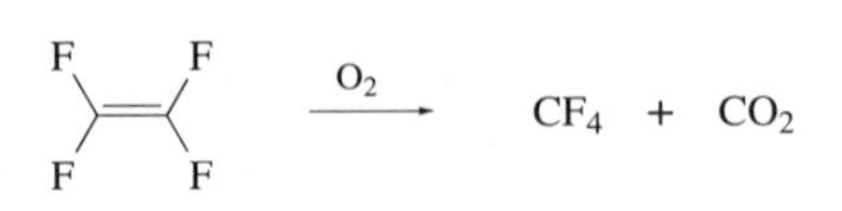

(6)

6.07.1.1.2 *Tetrachloromethane, CCl_4*

An old review on the preparation of tetrachloromethane exists 〈B-48MI 607-01〉.

A major technical preparation uses chlorination of carbon disulfide at 30 °C in the presence of an iron catalyst with formation of tetrachloromethane and disulfur dichloride. The actual chlorinating agent is S_2Cl_2 and the sulfur formed is rechlorinated to S_2Cl_2 (Scheme 1) 〈74MI 607-01, B-93MI 607-03〉.

Tetrachloromethane and disulfur dichloride can be separated by careful distillation in an inert gas stream ⟨75USP3884985⟩.

$$
\begin{aligned}
CS_2 + 3Cl_2 &\xrightarrow[\text{30 °C}]{\text{iron catalyst}} CCl_4 + S_2Cl_2 \\
CS_2 + 2S_2Cl_2 &\longrightarrow CCl_4 + 6S \\
2Cl_2 + 4S &\longrightarrow 2S_2Cl_2 \\
\hline
2CS_2 + 5Cl_2 &\longrightarrow 2CCl_4 + S_2Cl_2 + 2S
\end{aligned}
$$

Scheme 1

Tetrachloromethane is also commercially prepared by chlorination of chloromethane, formed by chlorination of methane or, more commonly, by treatment of methanol with hydrogen chloride ⟨67MI 607-01, 74MI 607-01, 81MI 607-01, B-91MI 607-02, B-93MI 607-02⟩. Another industrial procedure for the manufacture of chloromethanes uses a $CuCl_2/KCl$ melt as catalyst and chlorine source. This melt is regenerated by an oxychlorination procedure which uses the HCl produced in the chlorination reaction. The overall reaction is presented in Equation (7) ⟨B-93MI 607-02⟩.

$$CH_4 + HCl + O_2 \xrightarrow{CuCl_2/KCl \text{ melt}} MeCl + \text{higher chloromethanes} + H_2O \qquad (7)$$

Other processes involve heating of animal charcoal, CO, Cl_2 and PCl_3 at 400 °C at 10 atm in an autoclave with $AlCl_3$ and $FeCl_3$ as catalyst, which gives an 85% conversion to CCl_4 ⟨67MI 607-01⟩.

Pyrolysis of hexachloroethane at 300–420 °C yields a mixture of CCl_4 and CCl_2CCl_2 ⟨50TFS295⟩. Increasing the temperature to 700–800 °C causes further degradation of tetrachloroethylene to tetrachloromethane (Scheme 2) ⟨37GEP68065, 56BRP749408⟩, and subjecting propene to chlorinolysis at 450–550 °C produces a mixture of tetrachloroethylene and CCl_4 the ratio of which ranges from 35 : 65 to 65 : 35 depending on the exact conditions used (Equation (8)) ⟨B-93MI 607-02⟩.

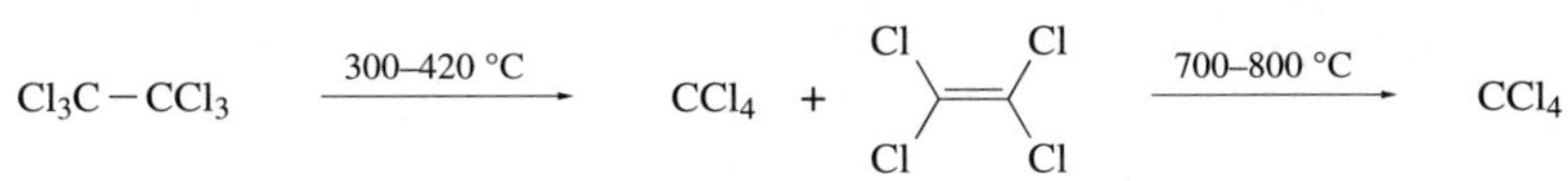

Scheme 2

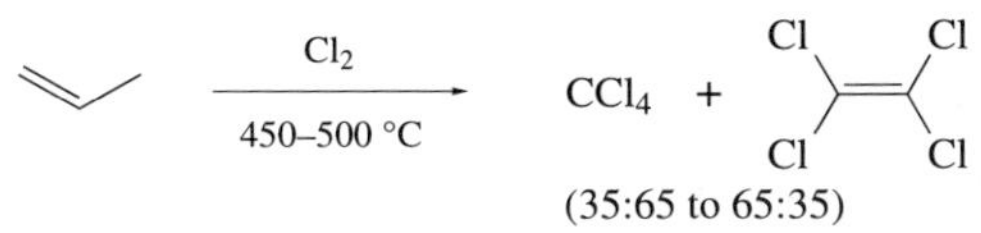

(8)

Thermal decomposition at 600 °C of trichloroacetyl chloride gave a low yield (27%) of tetrachloromethane in admixture with hexachloroethane (Table 3) ⟨39JA435⟩. Under severe conditions (500 °C, in the presence of $FeCl_3$) it is also possible to chlorinate benzene and naphthalene to CCl_4 in high yields (94%) ⟨67MI 607-01⟩. Finally, yields of 90% or more have been achieved by passing thiophosgene with gaseous $MoCl_5$ or WCl_6 at 8 atm over activated charcoal at 430 °C ⟨67MI 607-01⟩.

6.07.1.1.3 Tetrabromomethane, CBr_4

Tetrabromomethane is formed in a relative yield of 94% with an admixture of 6% $CClBr_3$ upon treatment of CCl_4 with $HBr/AlBr_3$; overall yield 87% (Equation (9)) ⟨51USP2553518⟩. It can be made, furthermore, from tribromomethane (or acetone) by oxidation with a sodium hypobromite solution containing excess bromine ⟨32JA2025⟩. This reaction is probably catalyzed by antimony(III) chloride ⟨1870LA(156)60⟩.

$$CCl_4 \xrightarrow{HBr/AlBr_3} \underset{87\%\ (94:6)}{CBr_4 + CClBr_3} \qquad (9)$$

Tetrabromomethane can also be made in a manner similar to the method used in the manufacturing of CCl_4, but with CS_2 and Br_2 ⟨B-93MI 607-02⟩.

Table 3 Synthesis of tetrahalomethanes by thermal decomposition of trihaloacetyl halides or metal haloacetates in the presence of halogen.

Starting material	X_2	*Temp.* (°C)	*Yield* (%)	*Products* (*ratio*)	*Ref.*
CCl_3COCl		600	30	CCl_4(89) CCl_3CCl_3(11)	39JA435
CCl_3COBr		400	15	$CBrCl_3$(67) CCl_3CCl_3(33)	39JA435
CCl_3COI		140	80	$CBrCl_3$(94) CCl_3CCl_3(6)	39JA435
CF_3CO_2Na	I_2[a]	280	61	CF_3I[b]	51JCS584
CF_3CO_2K	I_2[a]	280	55	CF_3I[b]	51JCS584
$(CF_3CO_2)_2Ba$[g]	I_2[a]	280	32	CF_3I[b]	51JCS584
$(CF_3CO_2)_2Hg$[g]	I_2[a]	250	35	CF_3I[b]	51JCS584
$(CF_3CO_2)_2Pb$[g]	I_2[a]	250	26	CF_3I[b]	51JCS584
CF_3CO_2Ag	Cl_2[a]	RT	90	$CClF_3$[b]	51JCS584
	Br_2[a]	50	88	$CBrF_3$[b]	51JCS584
	I_2[a]		89–94	CF_3I[b]	51JCS584
CCl_3CO_2K	Cl_2	110–120	(trace)	CCl_4	1877CB678
	Br_2	110–120	>70	$CBrCl_3$	1877CB678
	I_2	110–120	(trace)	CCl_3I	1877CB678
	ClI	110–120	(trace)	CCl_4	1877CB678
$CClF_2CO_2Ag$[c]	Cl_2[a]	180–260	88	CCl_2F_2[b]	52JCS4259
	Br_2[a]	180–260	91	$CBrClF_2$[b]	52JCS4259
	I_2[a]	180–260	78	$CClF_2I$[b]	52JCS4259
$CBrF_2CO_2Ag$[d]	Br_2[a]	180–260	81	CBr_2F_2[b]	52JCS4259
	I_2[a]	180–260	5	$CBrF_2I$[b]	52JCS4259
CCl_2FCO_2Ag[e]	Cl_2[a]	180–260	63	CCl_3F[b]	52JCS4259
	Br_2[a]	180–260	58	$CBrCl_2F$[b]	52JCS4259
	I_2[a]	180–260	10	CCl_2FI[b]	52JCS4259
$CBrClFCO_2Ag$[f]	Cl_2[a]	180–260	63	$CBrCl_2F$[b]	52JCS4259
	Br_2[a]	180–260	71	CBr_2ClF[b]	52JCS4259

[a] Excess X_2 (1.5 times the amount of metal haloacetate) was used. [b] The products were continously removed by pumping through a cold trap cooled with liquid air. [c] Silver chlorodifluoroacetate was prepared in 65% yield by permanganate oxidation of 1,1,2,3-tetrachloro-3,3-difluoropropene, followed by Ag_2O/P_2O_5 treatment. [d] Bromodifluoroacetic acid was prepared (8% yield) by Swarts' method 03MI 607-01 and converted to the silver salt by Ag_2O/P_2O_5 treatment. [e] Prepared as a mixture of CCl_2FCO_2Ag and $CHClFCO_2Ag$ (33 : 67) by treatment of chlorofluoroacetic acid with 20% excess Cl_2, followed by hydrolysis of the intermediate acid chloride. The silver salts were prepared by treatment of the corresponding acids with Ag_2O/P_2O_5 (57%). [f] Silver bromochlorofluoroacetate was prepared in an overall yield of 23% by treatment of bromofluoroacetic acid with 20% excess Cl_2, followed by hydrolysis of the intermediate acid chloride. The silver salt was prepared by treatment of the corresponding acids with Ag_2O/P_2O_5. [g] Barium trifluoroacetate was prepared from trifluoroacetic acid and $Ba(OH)_2$. Mercury and lead salts were made similarly from trifluoroacetic acid and mercury(II) and lead(II) oxide, respectively.

6.07.1.1.4 Tetraiodomethane, CI_4

Tetraiodomethane may be made in the laboratory by heating CCl_4 with $CaI_2 + 2\,H_2O$ or $LiI + 1.5\,H_2O$ for 5 days at 90–92 °C in an evacuated reaction vessel (yield 50–55%) ⟨15CR(156)652, B-93MI 607-04⟩, by treatment of acetone with KI/NaOCl (sat.) ⟨15CR(156)652⟩ or by treating triiodomethane with potassium hypoiodite under exclusion of light at 80–90 °C ⟨27BSF1251⟩.

The synthesis of tetraiodomethane from acetone or triiodomethane suffers from the fact that only small quantities are preparable and that purification is difficult.

On the other hand, the synthesis from tetrachloromethane and lower alkyl iodides allows the preparation of larger quantities of tetraiodomethane, giving nearly quantitative yields and a high-purity product, for example starting from CCl_4 in the presence of $AlCl_3$ and iodoethane (1 : 0.2 : 4.0) ⟨50IS37⟩ or iodomethane (1 : 0.02 : 4.2) ⟨45JA1642⟩ as the iodine donor. The chloromethane or chloroethane formed during the reaction is distilled off from the reaction vessel (Equations (10) and (11)).

$$\underset{(1:4)}{CCl_4 + MeI} \xrightarrow[40\,°C,\ 2\,h]{AlCl_3\ (cat.)} \underset{92\%\ (purity\ 99\%)}{CI_4 + MeCl} \qquad (10)$$

$$\underset{(1:4)}{CCl_4 + EtI} \xrightarrow[RT,\ 45\,min]{AlCl_3\ (cat.)} \underset{quant.}{CI_4 + EtCl} \qquad (11)$$

Crystalline CI_4 can be prepared by irradiation of a mixture of CF_3I and CO_2 at 30 torr. Carbon dioxide seems, under these conditions, important for the formation of CI_4 ⟨86MI 607-01⟩.

Tetraiodomethane may be purified by recrystallization from chloroform or benzene *in vacuo* ⟨45JA1642, 37JOC76⟩. When stored, tetraiodomethane should, like other iodine containing tetrahalomethanes, be protected from oxygen, water and light ⟨37JOC76⟩.

6.07.1.2 Three Similar and One Different Halogen

6.07.1.2.1 Trifluoromethyl halides

(i) Chlorotrifluoromethane, $CClF_3$

Chlorotrifluoromethane also known as Freon 13, is formed in 95% yield upon treatment of tetrafluoromethane with $SbF_3Cl_2 \cdot 2HF$ at 160°C and 70 atm ⟨47CI(L)427⟩, and furthermore as a by-product (17%) in the preparation of CF_4 from CCl_4 with F_2/As, as mentioned previously ⟨40JA3477⟩.
Chlorotrifluoromethane has, however, been prepared in numerous ways such as by treatment of CCl_4 with fluorine in the presence of iodine or CoF_3 ⟨31ZAAC(201)245⟩, as well as with HF and different oxometal fluoride or metal fluoride catalysts (Equation (12)) ⟨56USP2744147, 56USP2744148, 56USP2745886, 56USP2748177⟩ and with antimony(V) chloride as catalyst at 80–160°C ⟨53USP2658927⟩. Chlorotrifluoromethane is also formed when CCl_2F_2 is kept in contact with oxoaluminum chloride ⟨54USP2694739⟩ or, under more severe conditions, upon treatment of CCl_2F_2 with F_2 in the presence of Hg at 340–370°C ⟨40JA3477⟩.

$$\underset{(1:1.25)}{CCl_4 + HF} \xrightarrow[300\,°C]{AlF_3} \underset{90\%\ (35:60:5)}{CCl_2F_2 + CCl_3F + CClF_3} \qquad (12)$$

Chlorotrifluoromethane also constitutes the main product of the chlorination of graphite in the presence of HF (Equation (13)). This reaction has been optimized and found to be most efficient when conducted at 500°C. The effects of different catalysts have also been investigated ⟨55USP2709184⟩.

$$C \xrightarrow[500\,°C,\ 3\,h]{Cl_2,\ HF} \underset{89\%\ (7:21:53:14:5)}{CCl_3F + CCl_2F_2 + CClF_3 + CF_4 + \text{haloethanes}} \qquad (13)$$

By use of BrF_3 as fluorinating agent chlorotrifluoromethane may be prepared in the laboratory ⟨48JCS2188⟩. Banks *et al.* ⟨48JCS2188⟩ found that slow addition of the fluorinating agent (BrF_3 or IF_5), lowering of the ratio of tetrahalomethane to fluorinating agent as well as low temperature and an efficient condensing system (in order to keep the more volatile intermediates in contact with BrF_3 or IF_5) favors the formation of trifluorohalomethanes and tetrafluoromethane. Bromine trifluoride reacts smoothly, but vigorously with CCl_4 (Equation (14)) and CBr_4 whereas the reaction is *violent* in the case of CI_4. These reactions can be conducted at 0°C. Iodine pentafluoride, on the other hand, is less reactive and the reactions were conducted at 90–100°C (*vide infra*) (Table 2).

$$\underset{(1:0.84\text{–}1)}{CCl_4 + BrF_3} \xrightarrow{RT} \underset{65\text{–}70\%\ (15\text{–}43\text{ to }57\text{–}85)}{CCl_2F_2 + CClF_3} \qquad (14)$$

Other possibilities include fluorination of $CHCl_3$ by the versatile fluorinating agent CF_3OF which yields a mixture of CCl_2F_2 (6%), $CClF_3$ (83%) and CF_4 (11%) ⟨59JA1089⟩. Carbonyl difluoride may by PCl_5 treatment at 265–350°C be converted to a mixture of fluorochloromethanes (Equation (15)) as may phosgene by treatment with $HF/Cl_2/FeCl_3$ at 425°C for 17 h (Equation (16)) ⟨57JA5801⟩. Calcium fluoride/antimony pentafluoride treatment of phosgene at 500°C for 6 h ⟨56USP2757214⟩ leads to a mixture with $CClF_3$ as the main product (Equation (17)).

$$COF_2 + PCl_5 \xrightarrow{\text{autoclave}} CCl_2F_2 + CClF_3 + CCl_4 + COXY + CO_2 \qquad (15)$$

(1:1.25)	265 °C, 16 h	15% (67:33:0:–:–)
(1:2.7)	350 °C, 12 h	35% (14:0:86:–:–)

X, Y = Cl, F

$$COCl_2 + HF + Cl_2 \xrightarrow[\text{425 °C, 17 h, autoclave}]{FeCl_3/C} CCl_2F_2 + CClF_3 + CF_4 \quad (16)$$

(6:6:1) 72% (19:75:6)

$$COCl_2 \xrightarrow[\text{500 °C, 6 h}]{CaF_2/SbCl_5} CClF_3 + CCl_2F_2 + CClF_3 + CF_4 + CCl_4 \quad (17)$$

(65:30:5:trace:trace)

Silver trifluoroacetate when treated with chlorine also forms chlorotrifluoromethane ⟨51JCS584⟩, as does photochlorination of trifluoromethane ⟨37JA1200⟩ and cleavage of hexafluoroethane with Cl_2 at 900 °C ⟨49JA2499⟩.

Isotopically labeled $CClF_2{}^{18}F$ has been prepared by irradiation of a mixture of $CClF_3$ and $^{18}F^{19}F$. $CF_3{}^{18}F$ was the main by-product ⟨79MI 607-01⟩.

(ii) Bromotrifluoromethane, $CBrF_3$

Bromotrifluoromethane may be synthesized in 90% yield under severe conditions (655–680 °C) by treatment of trifluoromethane with bromine ⟨46JA968, 59USP2875254⟩ or in 94% yield under milder conditions by fluorination of tetrabromomethane at 0 °C with BrF_3 being added dropwise (Equation (18), Table 2) ⟨48JCS2188⟩. Furthermore, it can be made from tetrabromomethane with SbF_3/Br_2 at 200 °C ⟨50USP2531372⟩ or in low yield with TiF_4 at 120 °C ⟨58JCS4245⟩. Individual products are separated from the reaction mixture by condensation in cold traps.

$$CBr_4 \xrightarrow[94\%]{BrF_3,\ 0\ °C} CF_3Br \quad (18)$$

Hofmann rearrangement of trifluoroacetamide with sodium hypobromite does not yield trifluoromethylamine, but instead bromotrifluoromethane (Equation (19)) ⟨57JCS30⟩. This reaction is only useful for the generation of $CBrF_3$ and cannot be used for $CClF_3$ and CF_3I ⟨54JA5141⟩. Other degradative processes involve the Hunsdiecker reaction, for example the heating of silver trifluoroacetate and bromine at 65–130 °C with formation of bromotrifluoromethane in 88% yield (Equation (20); Table 3) ⟨51JCS584, 52JA1347⟩, and the heating of trifluoroacetyl bromide to 650 °C causing elimination of carbon monoxide and formation of $CBrF_3$ (Equation (21)) ⟨55USP2704776⟩. Bromination, with elemental bromine, of trifluoroacetic acid at 540 °C (Equation (22)) or trifluoroacetic acid anhydride at 300 °C (Equation (23)) in contact with activated carbon also leads to bromotrifluoromethane ⟨53USP2647933⟩.

$$CF_3CONH_2 \xrightarrow[-NaNCO]{NaOBr} CF_3Br \quad (19)$$

$$CF_3CO_2Ag \xrightarrow{65\text{–}130\ °C} CF_3Br + CO_2 \quad (20)$$

$$CF_3COBr \xrightarrow{650\ °C} CF_3Br + CO \quad (21)$$

$$CF_3CO_2H \xrightarrow[540\ °C]{Br_2,\ \text{activated charcoal}} CF_3Br + CO_2 \quad (22)$$

$$(CF_3CO)_2O \xrightarrow[300\ °C]{Br_2,\ \text{activated charcoal}} CF_3Br + CO_2 \quad (23)$$

Bromotrifluoromethane has also been made by plasma chemical means. The reaction conditions were applied for 4–5 min. Thus, treatment of hexafluoroethane with bromine under the influence of high-frequency electrical discharges (12 MHz, 1 kV) at 4–20 torr gave 73% conversion and a 77% yield of $CBrF_3$. Treatment of CF_3Cl under the same conditions gave only 23% conversion and a yield of 61% ⟨75ZAAC(418)109⟩. As was the case with chlorotrifluoromethane, bromotrifluoromethane is also formed upon bromination of hexafluoroethane at 900 °C ⟨49JA2499⟩.

Bromotrifluoromethane may, furthermore, be prepared by Cl/Br exchange. Thus, treatment of $CClF_3$ with HBr at 400°C in the presence of a $ZnBr_2$/C catalyst leads to $CBrF_3$ in 27% yield ⟨77BEP856233⟩.

(iii) Iodotrifluoromethane, CF_3I

The preparation and reactions of perfluoroiodoalkanes have recently been reviewed by Deev *et al.* ⟨92RCR75⟩.

Synthesis in high yield (91%), of iodotrifluoromethane is possible by the Hunsdiecker reaction, for example from silver trifluoroacetate and iodine, and is used industrially (Equation (24)) ⟨51JA2461, 51JCS584⟩. Sodium, potassium, barium, mercury(II) and lead(II) trifluoroacetate give in general lower yields of CF_3I, but the yield can be improved (80%) when sodium trifluoroacetate is treated with excess iodine in the presence of CuI at 150°C in DMF ⟨67JOC833⟩ or DMSO ⟨89MI 607-02⟩. Other decomposition reactions have also been applied, such as the heating of trifluoroacetyl chloride with KI at 200°C for 6 h (Equation (25)) giving 41% conversion and a yield of 69% iodotrifluoromethane under elimination of CO ⟨58JOC2016⟩, or a procedure where trifluoroacetyl fluoride and lithium iodide at high temperature give CF_3I in 70% yield (Equation (26)) ⟨90CL813⟩.

$$CF_3CO_2Ag \xrightarrow{I_2} \underset{91\%}{CF_3I} + AgI + CO_2 \qquad (24)$$

$$CF_3COCl \xrightarrow[6\ h]{KI,\ 200\ °C} \underset{69\%}{CF_3I} + KCl + CO \qquad (25)$$

$$CF_3COF \xrightarrow{LiI} \underset{70\%}{CF_3I} + LiF + CO \qquad (26)$$

In the laboratory, iodotrifluoromethane may also be conveniently made from the commercially available methyl chlorodifluoroacetate (Equation (27)) which, on the other hand, can be prepared from CF_2CF_2 (Scheme 3). Thus, treatment of $CClF_2CO_2Me$ with equimolar potassium fluoride and iodine at 100–120°C in DMF and in the presence of catalytic amounts of CuI gives a yield of 70–80%. CuI is essential for the high yield which in its absence drops to about 10%. Omission of potassium fluoride results in the formation of chlorodifluoroacetamide in 14% yield as the only identified product. Furthermore, substitution of methyl chlorodifluoroacetate with the bromo analogue gives a yield of 70% upon heating to 80°C in DMF for 5 h ⟨92CC807⟩.

$$ClCF_2CO_2Me + KF + I_2 \xrightarrow[DMF]{CuI\ (cat.),\ 100\text{–}120\ °C,\ 2\text{–}3\ h} \underset{80\text{–}90\%\ (88:12)}{CF_3I + CF_2I_2} + CH_3I + CO_2 + KCl \qquad (27)$$

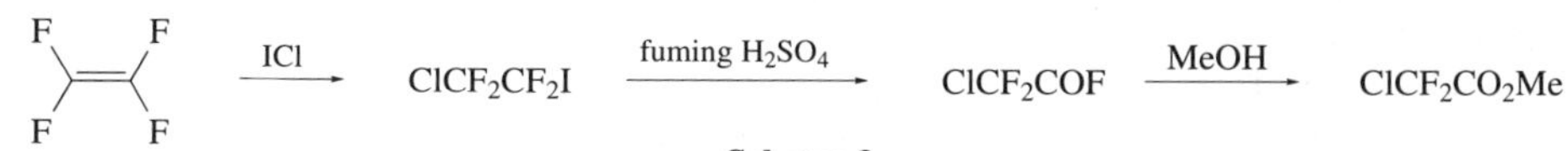

Scheme 3

Other reactions include the reaction of trifluoronitrosomethane with iodine at 50–70°C, with a yield of 80–95% ⟨65IZV1873⟩, and, in low yield, the fluorination of CI_4 with TiF_4 ⟨58JCS4245⟩, or with IF_5 (Table 2) in a yield of 65% ⟨48JCS2188, 49JCS2856⟩. Due to the inferior reactivity of iodine pentafluoride compared to bromine trifluoride, the reaction requires heating at 90–100°C for 30 min.

Furthermore, prolonged UV irradiation of methyl trifluoromethyl sulfide or bis(trifluoromethyl) sulfide in the presence of sources of iodine radicals yields 30–64% iodotrifluoromethane ⟨72JCS(P1)1506, 74JCS(P1)1706⟩.

Iodotrifluoromethane can, like bromotrifluoromethane, be made by plasma chemical means (see the previous section), but only in modest yields. Thus, treatment of hexafluoroethane with iodine under the influence of high frequency electrical discharges (12 MHz, 1 kV) at 4–20 torr gave 30% conversion, but only a yield of 5% CF_3I. Treatment of CF_3Cl under the same conditions gave only 14% conversion but a yield of 28% ⟨75ZAAC(418)109⟩.

(iv) Trifluoromethyliodine(III) compounds, $(CF_3)_nIZ_{3-n}$ (n = 1, 2)

The preparation and isolation of CF_3IF_2 ⟨59AG524⟩, CF_3ICl_2 ⟨89JFC(45)401⟩, $CF_3I(ONO_2)_2$ ⟨74IC2811⟩ and $CF_3I(OCOCF_3)_2$ ⟨76JFC(8)177⟩ have been reported in the literature. Trifluoromethyliodine(III) difluoride, for instance, is prepared by oxidative fluorination (F_2) at −80 °C of trifluoroiodomethane with trichlorofluoromethane as solvent ⟨59AG524⟩.

Bis(trifluoromethyl)iodine derivatives ($n = 2$, Z = ONO_2, $OCOCF_3$, F, Cl) have been prepared by Tyrra and Naumann ⟨91CJC327⟩ by trifluoromethylation of trifluoromethyliodine(III) dichloride, dinitrate, difluoride or bis(trifluoroacetate) with either $Cd(CF_3)_2 \cdot 2$ glyme or $Bi(CF_3)_3$. None of these were, however, isolated, but only detected and investigated by NMR.

6.07.1.2.2 Trichloromethyl halides

(i) Trichlorofluoromethane, CCl_3F

Trichlorofluoromethane has been widely used as refrigerant and propellant and is known as Freon 11. Industrial preparations include Cl/F exchange of HF with CCl_4, in a gas phase reaction at 150 °C with catalysis from aluminum, chromium, nickel or calcium fluoride. The main by-product is Freon 12, CCl_2F_2. Alternatively, the reaction may be conducted in the liquid phase under pressure at 100 °C and with antimony fluoride catalysis ⟨56USP2739989, B-93MI 607-02⟩.

The product of the reaction between tetrachloromethane and HF is highly dependent on the relative amounts of CCl_4 and HF as well as on the temperature. At an HF : CCl_4 ratio of 1 : 9, a temperature of 435 °C and a pressure of 70 atm, the total yield is 95%, consisting of 77% CCl_3F, 18% CCl_4 and 5% CCl_2F_2 (Equation (28)) ⟨47IEC404⟩. In liquid phase reactions antimony chlorides, such as $SbCl_3$, together with a certain amount of chlorine or even better, $SbCl_5$ have been used ⟨47JA947⟩.

$$CCl_4 + HF \xrightarrow[\text{435 °C, 70 atm}]{\text{cat.}} \underset{95\%\ (77:18:5)}{CCl_3F + CCl_4 + CCl_2F_2} \qquad (28)$$

Trichlorofluoromethane is also formed as one of the minor products in the reaction of graphite with Cl_2 and HF at 500 °C as described above ⟨55USP2709184⟩.

Sodium hexafluorosilicate under pressure at 270 °C (Equation (29)) ⟨59AG274⟩, titanium tetrafluoride ⟨58JCS4245⟩, chlorine trifluoride/cobalt trifluoride at 25 °C ⟨53JCS1063⟩, iodine pentafluoride at 30–35 °C ⟨31ZAAC(201)245⟩ and bromine trifluoride (Table 2, Equation (30)) ⟨48JCS2188⟩ are also useful as fluorinating agents for tetrachloromethane.

$$CCl_4 + Na_2SiF_6 \xrightarrow[\text{270 °C, 2 h}]{\text{autoclave}} \underset{\sim 100\%\ (33:66:\text{trace})}{CCl_3F + CCl_2F_2 + CClF_3} + NaCl + SiF_4 \qquad (29)$$

$$\underset{(1:0.33)}{CCl_4 + BrF_3} \xrightarrow{\text{RT}} \underset{61\%\ (5:95)}{CCl_2F_2 + CCl_3F} \qquad (30)$$

In the laboratory the Hunsdiecker reaction may be applied for the synthesis of trichlorofluoromethane (in 63% yield) from silver dichlorofluoroacetate and chlorine (Table 3, Equation (31)) ⟨52JCS4259⟩.

$$\underset{(1:1.5)}{CCl_2FCO_2Ag + Cl_2} \xrightarrow{180\text{–}260\ ^\circ C} \underset{63\%}{CCl_3F} + CO_2 + AgCl \qquad (31)$$

(ii) Bromotrichloromethane, $CBrCl_3$

Treatment of tetrachloromethane with $HBr/AlBr_3$ leads to bromotrichloromethane ⟨34JA1455⟩. Bromotrichloromethane is formed by treatment of trichloromethane with bromine at 225–450 °C ⟨37CR(204)1927, 47JCS674⟩, or in 80–85% yield at 420–450 °C by recycling the material with b.p. below 75 °C ⟨52MI 607-01⟩, and, furthermore, under the influence of light ⟨42JA1342⟩.

Older methods involve heating potassium trichloroacetate with bromine at 120 °C (Equation (32), Table 3) ⟨1877CB678⟩ or heating trichloromethanesulfonyl bromide with ethanol at 100 °C ⟨1869ZC624⟩. Low yields (10%) of bromotrichloromethane have been achieved by heating trichloroacetyl bromide to 400 °C. Most of the trichloroacetyl bromide was recovered (Table 3) ⟨39JA435⟩.

$$Cl_3CCO_2K + Br_2 \xrightarrow[>70\%]{110\text{–}120\ ^\circ C} CBrCl_3 + CO_2 + KBr \qquad (32)$$

The less easily available tetrahalomethanes may, however, also be prepared by base-catalyzed "halogen dance" (Scheme 4) ⟨92MI 607-03, 93BSF599⟩.

Heating chloroform with tetrabromomethane and dibenzoyl peroxide ⟨51USP2553800⟩ or *N*-bromosuccinimide ⟨52MI 607-02⟩ also yields bromotrichloromethane.

$$CBrCl_3 \xrightleftharpoons{\text{base}} CCl_4 + CBr_2Cl_2 + CBr_3Cl + CBr_4$$

$$F_3CCCl_3 + CBr_4 \xrightleftharpoons[\text{RT, 0.5–11 h}]{\text{base/solvent}}$$

$$F_3CCBrCl_2 + F_3CCBr_2Cl + F_3CCBr_3 + CBrCl_3 + CBr_2Cl_2 + CBr_3Cl$$

solvent: DMF, CH_2Cl_2
base: TBAF, NaOPh, NaOH, KOH, NaOEt, $NaNH_2$

Scheme 4

(iii) Trichloroiodomethane, CCl_3I

Good yields (75%) of trichloroiodomethane in admixture with hexachloroethane (16%) are achieved by distillation (140 °C) at atmospheric pressure of trichloroacetyl iodide (Scheme 5, Table 3) ⟨39JA435⟩, prepared from hydrogen iodide and trichloroacetyl chloride. Treatment of bromotrichloromethane with aluminum iodide leads to trichloroiodomethane ⟨53JCS922⟩.

$$Cl_3CCOCl \xrightarrow{HI} Cl_3CCOI \xrightarrow{\substack{\text{distillation}\\ 140\ ^\circ C}} \underset{80\%\ (6:94)}{Cl_3CCCl_3 + CCl_3I}$$

Scheme 5

Trichloromethane treated with iodine and sodium hydroxide at 0 °C for 15 h, gave 10% dichloroiodomethane as well as 1% trichloroiodomethane. The iodinating agent in this medium is sodium hypoiodite generated *in situ*. Use of tetrachloromethane, excess iodine and sodium hydroxide at 0 °C for five days gave trichloroiodomethane in less than 5% yield. An attempt to iodinate tetrachloromethane with NaI in acetone failed ⟨79JCED251⟩.

6.07.1.2.3 *Tribromomethyl halides*

(i) *Tribromofluoromethane, CBr_3F*

Tribromofluoromethane is formed as a coproduct (31%) in the reaction of (dibromofluoromethyl)triphenylphosphonium bromide, formed from triphenylphosphine and $CFBr_3$ in THF, with iodine monobromide in tetraglyme (Scheme 6) ⟨82JFC(20)89⟩, and under controlled reaction conditions, in high yield (94%) upon fluorination of tetrabromomethane with bromine trifluoride (Equation (33), Table 2) ⟨48JCS2188⟩.

$$R_3P + CFXYZ \xrightarrow{\text{solvent}} [R_3PCFXY]^+ Z^-$$

$$[R_3PCFXY]^+ Z^- + X'Y' + KF \xrightarrow{\text{solvent}'} CFXYX'$$

solvent = THF, diethyl ether or triglyme
solvent' = triglyme, tetraglyme or DMF
R = Ph or (NMe_2)
X, Y, Z = F, Cl or Br
X', Y' = Cl, Br, I

Scheme 6

$$\underset{(1:0.33)}{CBr_4 + BrF_3} \xrightarrow{\sim 0\ ^\circ C} \underset{94\%\ (100:\text{trace})}{CBr_3F + CBr_2F_2} \qquad (33)$$

Furthermore, it is available by fluorination of tetrabromomethane with antimony trifluoride (19–92% yield) ⟨18CB669, 57JA5654, 58BSB676, 59JCS13⟩. Birchall and Haszeldine heated tetrabromomethane with antimony trifluoride and bromine (2 : 1 : 0.1) 1 h at 120–130 °C, followed by 30 min at 150–160 °C, with a yield of 65–70% after distillation (Equation (34)) ⟨59JCS13⟩. Silver fluoride has also been applied as fluorinating agent in a distillation apparatus—allowing bromotrifluoromethane to distil out of the reaction mixture ⟨18CB669⟩.

$$\underset{(2:0.1:1)}{CBr_4 + Br_2 + SbF_3} \xrightarrow[\text{ii, 150–160 °C, 0.5 h}]{\text{i, 120–130 °C, 1 h}} \underset{65\text{–}70\%}{CBr_3F} \qquad (34)$$

(ii) *Tribromochloromethane, CBr_3Cl*

Tribromochloromethane is formed as the main product (relative yield 42%) after base-catalyzed "halogen dance" reactions (Scheme 4) ⟨92MI 607-03, 93BSF599⟩. Thus, after treatment of CF_3CCl_3/CBr_4 with KOH in DMF at RT for 4 h, tribromochloromethane is formed with the following by-products: CF_3CCl_2Br (relative yield 32%), CF_3CClBr_2 (21%) and CBr_2Cl_2 (19%) ⟨92MI 607-03⟩.

Dibromochloronitrosomethane may be refluxed with bromine for 3 h giving a 95% yield of tribromochloromethane. Dibromochloronitrosomethane is in turn accessible from $Br_2C{=}NOH$ (Scheme 7) ⟨32CB546⟩.

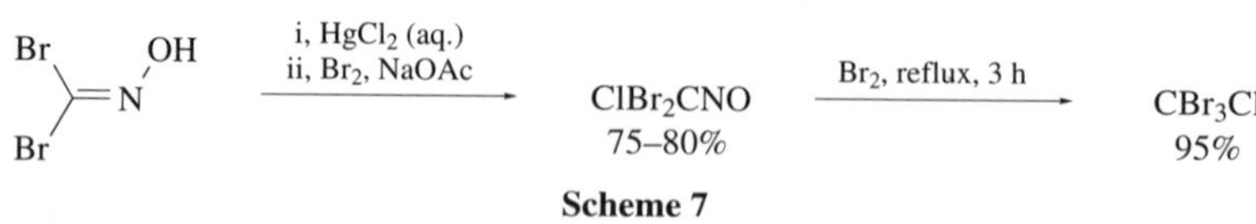

Scheme 7

Treatment of CCl_4 with an equimolar amount of Br_2 for 2 h at 225 °C leads to tribromochloromethane in admixture with bromotrichloromethane and dibromodichloromethane ⟨1892CB188⟩. Trichloromethane heated with bromine for 14 h at 225–275 °C ⟨37CR(204)1927⟩ as well as $HBr/AlCl_3$ treatment of tetrachloromethane gives tribromochloromethane. In the latter reaction, HBr is bubbled into a solution of CCl_4 and $AlCl_3$ and the temperature raised to 90 °C. The product contains 6% tribromochloromethane and 94% tetrabromomethane with an 87% conversion of tetrachloromethane.

(iii) Tribromoiodomethane, CBr_3I

Treatment of $CHBr_3$ with KI, NaOCl and NaOH for 24 h gave mainly $CHBr_2I$ (10%) and only traces (0.1%) of tribromoiodomethane could be detected. The same reaction conditions were also applied to CBr_4 for six days, but again only a very low yield (<1%) of CBr_3I was obtained. On the other hand, when CBr_4 was treated with NaI in acetone for 5 min a yield of more than 30% CBr_3I was observed, although the product was contaminated with $CBrI_3$ and brominated acetone (Equation (35)) ⟨79JCED251⟩. According to an old procedure by Dehn ⟨09JA1220⟩ it should, however, be possible to prepare tribromoiodomethane from tribromomethane and nitrosyl iodide. Dehn reported the product as a solid darkening at 35 °C.

$$CBr_4 \xrightarrow[\text{5 min}]{\text{NaI, acetone}} \underset{\sim 30\%}{CBr_3I} + CBrI_3 + \text{brominated acetone} \quad (35)$$

6.07.1.2.4 *Triiodomethyl halides*

(i) Fluorotriiodomethane, CFI_3 and chlorotriiodomethane, $CClI_3$

Fluorotriiodomethane and chlorotriiodomethane still await preparation, but theoretical discussions of their physical properties are on record ⟨58MI 607-01, 59MI 607-03, 66ZOB1355, 72MI 607-01, 76ZC377, 77MI 607-01, 79JMR559, 83MI 607-01, 91MI 607-01⟩.

(ii) Bromotriiodomethane, $CBrI_3$

Bromotriiodomethane has been prepared by Dehn ⟨09JA1220⟩ by treatment of triiodomethane with NaOBr. It is an unstable solid which liberates iodine upon exposure to light or upon dissolution. Bromotriiodomethane has also been reported as a by-product in the Finkelstein reaction mentioned above, that is the treatment of tetrabromomethane with sodium iodide in acetone (Equation (35)) ⟨79JCED251⟩.

6.07.1.3 Two Similar Halogens

6.07.1.3.1 *Difluoromethylene dihalides*

(i) Bromochlorodifluoromethane, $CBrClF_2$

Bromochlorodifluoromethane may be prepared in 91% yield by the heating of silver chlorodifluoroacetate with bromine (Equation (36), Table 3) ⟨52JCS4259⟩.

$$\underset{(1:1.5)}{CClF_2CO_2Ag + Br_2} \xrightarrow{180\text{–}260\ ^\circ C} \underset{91\%}{CBrClF_2} + CO_2 + AgBr \quad (36)$$

Bromochlorodifluoromethane is formed, but only in very low yield (5%), in the reaction of tris(dimethylamino)(chlorodifluoromethyl)phosphonium chloride, formed from tris(dimethylamino)phosphine and CF_2Cl_2 in triglyme, with bromine in triglyme (Scheme 6) ⟨82JFC(20)89⟩.

Trabalka *et al.* describe in a patent ⟨81MIP68355⟩ the formation of bromochlorodifluoromethane by heating of chlorodifluoromethane with bromine and chlorine at 500–800 °C. Other synthetic methods include the preparation from chlorodifluoromethane, bromine and oxygen with catalysis from chromium trioxide at 500 °C. The yield was 86% and the reaction product consisted of $CBrClF_2$, CBr_2F_2 and $CHClF_2$ (58:7:35; Equation (37)) ⟨59USP2871274⟩. Other sources are bromodifluoromethane by vapor phase chlorination ⟨53USP2639302⟩ or UV irradiation of a mixture of chlorodifluoroiodomethane and bromine ⟨52JCS4259⟩. Finally, treatment of 1,3-dichloro-1,1,3,3-tetrafluoroacetone with bromine at 580–650°C yields bromochlorodifluoromethane ⟨59USP2885450⟩.

$$CHClF_2 + Br_2 + O_2 \xrightarrow[500\ °C]{CrO_3} \underset{86\%\ (58:7:35)}{CBrClF_2 + CBr_2F_2 + CHClF_2} \quad (37)$$

(ii) Chlorodifluoroiodomethane, $CClF_2I$

Chlorodifluoroiodomethane has been prepared by Haszeldine (Equation (38), Table 3) ⟨52JCS4259⟩ in 78% yield by heating silver chlorodifluoroacetate with iodine. This halomethane is, however, unstable and liberates iodine on exposure to O_2 or light ⟨52JCS4259⟩.

$$\underset{(1:1.5)}{CClF_2CO_2Ag + I_2} \xrightarrow{180–260\ °C} \underset{78\%}{CClF_2I} + CO_2 + AgBr \quad (38)$$

Difluorocarbene is generated from methyl chlorodifluoroacetate by treatment with LiCl/HMPA and may, when formed in the presence of I_2 or IBr, be used to prepare chlorodifluoroiodomethane in 15% yield (contaminated with 5% CF_2I_2) and 30% (contaminated with 10% $CBrF_2I$), as shown in Equations (39) and (40), respectively ⟨78JOC2643⟩.

$$CClF_2CO_2Me + LiCl/HMPA + I_2 \xrightarrow{90–95\ °C,\ 48\ h,\ triglyme} \underset{20\%\ (75:25)}{CClF_2I + CF_2I_2} \quad (39)$$

$$CClF_2CO_2Me + LiCl/HMPA + IBr \xrightarrow{90–95\ °C,\ 48\ h,\ triglyme} \underset{40\%\ (75:25)}{CClF_2I + CBrF_2I} \quad (40)$$

Another method is the reaction of $[(Me_2N)_3P(CF_2Cl)]Cl$, formed from tris(dimethylamino)-phosphine and CF_2Cl_2 in triglyme, with dry potassium fluoride and iodine at 0°C for 2 h and then overnight at room temperature. The isolated yield of $CClF_2I$ is 34% (Scheme 6) ⟨82JFC(20)89⟩.

Chlorodifluoroiodomethane can, like bromo- and iodotrifluoromethane, be made by plasma chemical means, but only in modest yields. Thus, treatment of chlorotrifluoromethane with iodine under the influence of high-frequency electrical discharges (12 MHz, 1 kV) at 4–20 torr gave 7% conversion, and a relative yield of 21% ⟨75ZAAC(418)109⟩.

(iii) Bromodifluoroiodomethane, $CBrF_2I$

When difluorocarbene is generated in the presence of IBr, bromodifluoroiodomethane is formed as a by-product (10%) (Equation (40)) ⟨78JOC2643⟩. Another method is the reaction of (bromo-difluoromethyl)triphenylphosphonium bromide, formed from triphenylphosphine and CF_2Br_2 in triglyme, with dry potassium fluoride and iodine in triglyme. The isolated yield of $CBrF_2I$ is 41% (Scheme 6) ⟨83JFC(23)339, 82JFC(20)89⟩.

Bromodifluoroiodomethane is one of the few tetrahalomethanes which could not be prepared except in poor yield (<5%) by the Hunsdiecker method (Table 3).

6.07.1.3.2 Dichloromethylene dihalides

(i) Dichlorodifluoromethane, CCl_2F_2

The industrial preparation of dichlorodifluoromethane, known as Freon 12, involves halogen exchange of CCl_4 with HF, as described above for $CFCl_3$ ⟨B-93MI 607-02⟩.

At a HF : CCl_4 ratio of 4 : 9, a temperature of 490°C and a pressure of 70 atm, the total yield in the reaction between CCl_4 and HF has been reported to be 59%, consisting of 79% CCl_2F_2 and

21% CCl_3F (Equation (41)) ⟨47IEC404⟩, whereas in the presence of $SbCl_3$ and Cl_2 at 110°C and 30 atm, CCl_2F_2 constitutes 90% of the product ⟨B-92MI 607-04⟩. Furthermore, treatment of phosgene with chlorine and anhydrous hydrogen fluoride at 425°C in the presence of $FeCl_3$/C also yields a mixture consisting of CCl_2F_2, $CClF_3$ and CF_4 (Equation (17)) ⟨57JA5801⟩. Trichloromethane does, as mentioned above, also form a CCl_mF_n mixture when treated with CF_3OF ⟨59JA1089⟩, and dichlorodifluoromethane is formed as one of the minor products in the reaction of graphite with Cl_2 and HF at 500°C as described above ⟨55USP2709184⟩.

$$\underset{(9:4)}{CCl_4 + HF} \xrightarrow{490\,°C,\ 70\ atm} \underset{59\%\ (79:21)}{CCl_2F_2 + CCl_3F} \qquad (41)$$

In the laboratory, dichlorodifluoromethane may be prepared in high yield (96%) by fluorination of CCl_3F with BrF_3, or as the main product of the fluorination of CCl_4 with BrF_3 at −78°C→RT in a total yield of 89% (Scheme 8, Table 2). Trichlorofluoromethane is the only by-product ⟨31USP1961622, 48JCS2188⟩.

$$\underset{(1:0.6)}{CCl_4 + BrF_3} \xrightarrow{-78\,°C\ to\ RT} \underset{89\%\ (85:15)}{CCl_2F_2 + CCl_3F}$$

$$\underset{(1:0.8)}{CCl_4 + BrF_3} \xrightarrow{RT} \underset{96\%}{CCl_2F_2}$$

Scheme 8

Treatment of tetrachloromethane with sodium hexafluorosilicate under pressure at 270°C in an autoclave yields a mixture of CCl_nF_{4-n} with dichlorodifluoromethane as the major product (Equation (29)) ⟨59AG274⟩.

Degradative reactions forming CCl_2F_2 include the chlorinolysis at 550°C of 1,1-difluoroethane, formed by treatment of acetylene with HF ⟨B-59MI 607-02⟩ and the heating of silver chlorodifluoroacetate with chlorine (Equation (42), Table 3) ⟨52JCS4259⟩.

$$\underset{(1:1.5)}{CClF_2CO_2Ag + Cl_2} \xrightarrow{180-260\,°C} \underset{88\%}{CCl_2F_2} + CO_2 + AgCl \qquad (42)$$

Dichlorodifluoromethane may finally also be made by electrolysis of chlorodifluoroacetic acid ⟨33BSB102⟩.

(ii) Bromodichlorofluoromethane, $CBrCl_2F$

Bromodichlorofluoromethane has been prepared by Haszeldine ⟨52JCS4259⟩ by use of the Hunsdiecker reaction, in 58% yield by heating silver dichlorofluoroacetate with bromine (Equation (43)) and in 63% yield by heating silver bromochlorofluoroacetate with chlorine (Equation (44), Table 3), and also by bromination of dichlorofluoromethane at 475°C (41% conversion, yield 78%) ⟨56USP2755314, 57JA4159, 59JA2078⟩.

$$\underset{(1:1.5)}{CCl_2FCO_2Ag + Br_2} \xrightarrow{180-260\,°C} \underset{58\%}{CBrCl_2F} + CO_2 + AgBr \qquad (43)$$

$$\underset{(1:1.5)}{CBrClFCO_2Ag + Cl_2} \xrightarrow{180-260\,°C} \underset{63\%}{CBrCl_2F} + CO_2 + AgCl \qquad (44)$$

(iii) Dichlorofluoroiodomethane, CCl_2FI

Dichlorofluoroiodomethane has been prepared in poor yield by heating silver dichlorofluoroacetate with iodine. Dichlorofluoroiodomethane turned out to be unstable to storage at RT. It forms different haloethanes with liberation of iodine ⟨52JCS4259⟩ (Table 3).

It has been prepared in 44% yield (63–73% yield according to Fried and Miller ⟨59JA2078⟩) by a

Finkelstein reaction from bromodichlorofluoromethane/NaI/acetone in a closed vessel for 3 h at 120 °C (Equation (45)) ⟨57JA4159⟩.

$$CBrCl_2F + NaI \xrightarrow[120\ ^\circ C,\ 3\ h]{acetone} \underset{44\text{–}73\%}{CCl_2FI} + NaBr \quad (45)$$

(iv) Bromodichloroiodomethane, $CBrCl_2I$

Bromodichloroiodomethane has not been prepared yet, but theoretical discussions of its physical properties are on record ⟨72MI 607-01, 79JMR559, 91MI 607-01⟩.

6.07.1.3.3 Dibromomethylene dihalides

(i) Dibromodifluoromethane, CBr_2F_2

Dibromodifluoromethane may be prepared in 83% as the only product by fluorination of tetrabromomethane with IF_5 (Equation (46)), as the major product (relative yield 76%) by fluorination with BrF_3, and as a minor product (relative yield 16%) by fluorination of tetraiodomethane with IF_5 (Table 2) ⟨48JCS2188⟩. Fluorination of tetrabromomethane with TiF_4 forms dibromodifluoromethane in low yield as a mixture with bromotrifluoromethane ⟨58JCS4245⟩. Furthermore, CBr_2F_2 is formed in high yield (81%) by the heating of silver bromodifluoroacetate with bromine (Equation (47), Table 3) ⟨52JCS4259⟩.

$$\underset{(1:0.6)}{CBr_4 + IF_5} \xrightarrow{90\ ^\circ C,\ 3\ h} \underset{83\%\ (100:trace:trace:trace)}{CBr_2F_2 + CBr_3F + CBrF_3 + CF_4} \quad (46)$$

$$\underset{(1:1.5)}{CBrF_2CO_2Ag + Br_2} \xrightarrow{180\text{–}260\ ^\circ C} \underset{81\%}{CBr_2F_2} + CO_2 + AgBr \quad (47)$$

Other methods involve vapor phase bromination of difluoromethane at 500 °C (61% conversion; the product is a mixture of $CHBrF_2$ and CBr_2F_2 ⟨53USP2639301⟩) and treatment of difluoromethane with a mixture of bromine and chlorine at 350 °C ⟨53USP2658086⟩. Furthermore, a low yield is obtained from bromochlorodifluoromethane and HBr/Br_2 at 600 °C (66% conversion; the product is a 94 : 6 mixture of $CHBrF_2$ and CBr_2F_2 ⟨56USP2729687⟩), or from tribromofluoromethane, HF and a chromium oxofluoride catalyst at 250 °C ⟨56USP2745886⟩. Reaction between tetrabromomethane, HF and an aluminum oxofluoride catalyst at 230–250 °C, 1,3-dichloro-1,1,3,3-tetrafluoroacetone and bromine at 580–650 °C ⟨59USP2885450⟩ or the heating of tetrabromomethane with AgF (1 : 2), allowing distillation from the reaction mixture to occur ⟨18CB669⟩, all lead to dibromodifluoromethane.

Dibromodifluoromethane is formed, but only in a very low yield (3%), in the reaction of tris(dimethylamino)(chlorodifluoromethyl)phosphonium chloride, formed from tris(dimethylamino)phosphine and CF_2Cl_2 in triglyme, with bromine in triglyme (Scheme 6) ⟨82JFC(20)89⟩.

(ii) Dibromodichloromethane, CBr_2Cl_2

Bromotrichloromethane disproportionates in the presence of tetrabutylammonium fluoride to a mixture of tetrachloromethane, dibromodichloromethane, tribromochloromethane and tetrabromomethane ⟨93BSF599⟩. Other bases and solvent systems have also been tried in such base-catalyzed "halogen dance" reactions. The individual halomethanes were obtained by fractional distillation (Scheme 4) ⟨92MI 607-03⟩.

It is also obtained, in admixture with other halomethanes, from tetrachloromethane treated with bromine trifluoride in the presence of $AlCl_3$ or BBr_3.

Dichloromethane and bromine react under formation of dibromodichloromethane on illumination ⟨42JA1342⟩ or after heating in a sealed vessel at 220–250 °C for more than a week

⟨1887LA(240)192⟩. Furthermore, impure CBr_2Cl_2 is obtained upon treatment of trichloromethane with bromine in a closed reaction vessel at 225–275 °C ⟨37CR(204)1927⟩ or by heating dichlorobromonitrosomethane with bromine (*vide supra*) ⟨32CB546⟩.

(iii) Dibromochlorofluoromethane, CBr_2ClF

Dibromochlorofluoromethane is formed in 71% yield upon heating of silver bromochlorofluoroacetate with bromine (Equation (48), Table 3) ⟨52JCS4259⟩ and by vapor phase bromination of dichlorofluoromethane at 650 °C ⟨56USP2755314⟩.

$$\underset{(1:1.5)}{CBrClFCO_2Ag + Br_2} \xrightarrow{180\text{–}260\ °C} \underset{71\%}{CBr_2ClF} + CO_2 + AgBr \quad (48)$$

(iv) Dibromofluoroiodomethane, CBr_2FI

Dibromofluoroiodomethane is formed as the major product (57%) in the reaction of (dibromofluoromethyl)triphenylphosphonium bromide, formed from triphenylphosphine and CBr_3F in THF, with iodine in tetraglyme. The crude product is, however, contaminated with substantial amounts of other halomethanes (Scheme 6) ⟨82JFC(20)89⟩.

(v) Dibromochloroiodomethane, CBr_2ClI

Dibromochloroiodomethane still awaits preparation, but theoretical discussions of its physical properties are on record ⟨72MI 607-01, 79JMR559, 91MI 607-01⟩.

6.07.1.3.4 Diiodomethylene dihalides

(i) Difluorodiiodomethane, CF_2I_2

Difluorodiiodomethane has been prepared in low yield (27%) by fluorination of CI_4 with HgF_2 in 1,2-dichlorobenzene. Excessive fluorination is minimized by distillation of the CF_2I_2 formed in the reaction ⟨84JOC205, 85DIS(B)840⟩. A better method has, however, been reported by Su *et al.* ⟨92CC807⟩ who discovered a convenient method for obtaining pure CF_2I_2 in fair yield (50–60% isolated yield; 80% by ^{19}F NMR). Treatment of methyl bromodifluoro- or chlorodifluoroacetate with equimolar amounts of KI, I_2 and a catalytic amount of CuI leads to the formation of CF_2I_2 (Equation (27)). This method is, however, somewhat compromized by difficulties in the separation of CH_3I and CF_2I_2. Pure difluorodiiodomethane is obtained from potassium bromodifluoroacetate by heating with equimolar amounts of KI, CuI and I_2 in DMF at 40 °C for 10 h. The use of only a catalytic amount of CuI causes the yield to drop to 10% (Equation (49)) ⟨92CC807⟩.

$$CBrF_2CO_2K + KI + CuI + I_2 \xrightarrow[\ ]{DMF,\ 40\ °C,\ 10\ h} \underset{90\%\ (50\text{–}60\%\ \text{isolated})}{CF_2I_2} + KBr + CO_2 \quad (49)$$

(ii) Dichlorodiiodomethane, CCl_2I_2

Dichlorodiiodomethane has been prepared by Höland in 1887 ⟨1887LA(240)225⟩ by treatment of CH_2Cl_2 with Br_2/I_2 (1 : 4/4) at 100–200 °C for several weeks. Distillation of the crude product gave dichloroiodomethane and dichlorodiiodomethane. Dichlorodiiodomethane melts at 85 °C under liberation of iodine and is probably unstable in solution ⟨1887LA(240)225⟩.

(iii) Bromofluorodiiodomethane, $CBrFI_2$

Bromofluorodiiodomethane is formed as a minor by-product (15%) in the reaction of (dibromofluoromethyl)triphenylphosphonium bromide, formed from triphenylphosphine and CBr_3F in THF, with iodine in DMF (Scheme 26) ⟨82JFC(20)89⟩.

(iv) Dibromodiiodomethane, CBr_2I_2; chlorofluorodiiodomethane, $CClFI_2$; and bromochlorodiiodomethane, $CBrClI_2$

Dibromodiiodomethane, chlorofluorodiiodomethane and bromochlorodiiodomethane still await preparation, but theoretical discussions of their physical properties are on record ⟨58MI 607-01, 59MI 607-03, 66ZOB1355, 72MI 607-01, 76ZC377, 77MI 607-01, 79JMR559, 83MI 607-01, 91MI 607-03⟩.

6.07.1.4 Bromochlorofluoroiodomethane, CBrClFI

Bromochlorofluoroiodomethane, although often mentioned as *the* prototype of a chiral compound, has not been prepared yet. MO calculations indicate that it is the second least stable chiral halomethane judged from the enthalpies of formation ⟨90BCJ1278⟩. The only chiral halomethane which has been prepared is CHBrClF, and it has been shown that this derivative hydrolyzes much faster than other mixed halomethanes ⟨56JA479⟩; this may also be the case with bromochlorofluoroiodomethane.

It would be interesting to try the Hunsdiecker reaction on silver bromochlorofluoroacetate which reacts with both chlorine and bromine as mentioned above. It may, however, turn out to be a very low yielding reaction, as was the case with the treatment of silver bromodifluoroacetate with iodine (Table 3) ⟨52JCS4259⟩.

Theoretical discussions of the physical properties of so far unsynthesized tetrahalomethanes are on record ⟨58MI 607-01, 59MI 607-01, 66ZOB1355, 72MI 607-01, 76ZC377, 77MI 607-01, 79JMR559, 83MI 607-01, 91MI 607-01⟩.

6.07.2 METHANES BEARING THREE HALOGENS

6.07.2.1 Three Halogens and a Chalcogen

Trihalomethoxy as well as trihalomethylthio groups may be regarded as *superhalogens* and behave as such. They are, for example deactivating and *ortho*-, *para*-directing substituents in electrophilic aromatic substitution. As *superhalogens*, they have been used as alternative substituents to circumvent existing patents and to impart increased lipophilicity to compounds relative to their halogen substituted counterparts. For more information see a comparison and discussion on trihalomethyl (and other pseudohalogens) vs. halogen by Haas, who has made a large contribution to this synthetic field ⟨82CZ239, 84MI 607-01⟩.

The subject of trihalomethoxy and trihalomethylthio containing compounds has previously been reviewed by Marhold ⟨83HOU(E4)625, 83HOU(E4)1203⟩, Hocker ⟨83HOU(E4)1368⟩ and McClinton and McClinton ⟨92T6555⟩.

6.07.2.1.1 Three halogens and an oxygen function

(i) Trihalomethanols, $CHal_3OH$

Trifluoromethanol is a rather unstable compound which, above −20 °C, eliminates hydrogen fluoride under formation of carbonyl fluoride. It was therefore not synthesized until 1977, when Seppelt prepared trifluoromethanol by treating trifluoromethyl hypochlorite with HCl at −120 °C ⟨77AG325⟩.

(ii) Trihalomethyl ethers, $CHal_3OR$

Aryl trifluoromethyl ethers may be prepared from the corresponding phenols by treatment with tetrachloromethane and hydrogen fluoride, with the exception of phenols with *ortho*-substituents capable of hydrogen bonding to the hydroxy group. The yields are generally best when electron-withdrawing substituents are present ⟨79JOC2907⟩. Chlorine may be used as such and later removed by hydrogenation ⟨87JA3708⟩. The reaction is catalyzed by antimony trichloride and probably passes through the trichloro derivative ⟨79JOC2907⟩.

Trichloromethyl ethers (aryl and alkyl) can in turn be transformed to trifluoromethyl ethers by treatment with F_2/SnF_2 ⟨91CL1421⟩, HF ⟨86BSF925⟩, $HF/SbCl_5$ ⟨72GEP2129200⟩, SbF_3 ⟨55DOK(105)100⟩ or even better $SbF_3/SbCl_5$ ⟨58JGU2539⟩. Mixed trihalomethyl ethers may be prepared by decreasing the reaction temperature and the amount of HF ($HF/SbCl_3$) used ⟨72GEP2129200, 79JOC2907, 86BSF925⟩.

Rico and Wakselman made mixed trihalomethyl ethers from potassium phenoxides, CBr_2F_2 or $CBrClF_2$ and a catalytic amount of 1-propanethiol kept in DMF at RT for 4 h. Unfortunately, low yields (6–50%) were obtained with aryl difluoromethyl ethers as the major and aryl bromodifluoromethyl ethers as the minor product ⟨81T4209⟩.

Synthesis of aryl trichloromethyl ethers from the corresponding nonhalogenated ether is achieved by photochlorination (70% yield) or by use of Cl_2/PCl_5 (71% yield) ⟨77CI(L)127, 86BSF925⟩.

A carbonyl function can be regarded as a latent CF_2 function. The corresponding transformation is effected by SF_4, sometimes in the presence of HF. Other reagents include the more easily handled alkylsulfur trifluorides, MoF_6/BF_3, COF_2 or the more reactive SeF_4 ⟨B-92MI 607-01⟩. According to Hudlický ⟨B-92MI 607-01⟩ the following order of reactivity is observed with SF_4 as fluorinating agent: alcohols > aldehydes, ketones > acids, amides > esters, acid anhydrides > alkyl halides.

This has been exploited in the synthesis of trifluoromethyl ethers. Thus, treatment of fluoroformic acid esters (FC(=O)OR) with sulfur tetrafluoride leads to trifluoromethyl ethers. This reaction is especially convenient for the synthesis of aryl trifluoromethyl ethers, and only when electron-withdrawing groups are present in the β-position can the aliphatic counterparts be synthesized in satisfactory yields ⟨61JA4860, 64JOC1, 64JOC11⟩. Analogous to this *O*-alkyl chlorothioformic acid esters (ClC(=S)OR) can also be transformed to trifluoromethyl ethers by treatment with MoF_6, in 40–90% yield ⟨73TL2253⟩. Trichloromethyl ethers are prepared similarly by chlorination with Cl_2 of ZC(=S)OR (Z = Cl, SMe) and $[ROC(=S)S]_2$ ⟨55JA1899, 56JA6070, 58JGU2539, 58ZOB2506, 64ZOB1979, 72JOC2651⟩.

A very mild method for the transformation of xanthates or (difluoro(methylthio)methyl) ethers into trifluoromethyl ethers was introduced by Kuroboshi *et al.* They made use of $(HF)_9$/pyridine/1,3-dibromo-5,5-dimethylhydantoin (DBH) as oxidative desulfurization-fluorinating agent and achieved yields of 48–80% ⟨92TL4173⟩.

Electrophilic addition of halogens to alkenes is a standard reaction in organic chemistry, and addition of trihalomethyl hypohalites, which may be considered as a halogen bound to a superhalogen, should therefore be expected to constitute a similar reaction, which is exactly the case. Thus, β-fluoro- or β-chloroalkyl trifluoromethyl ethers may be prepared by addition of trifluoromethyl hypofluorite or hypochlorite to alkenes.

The reaction of trifluoromethyl hypochlorite with unsubstituted alkenes can be controlled at temperatures below $-111\,^{\circ}C$, whereas the reaction with electron-poor alkenes only proceeds at a significantly higher temperature ($\sim 50\,^{\circ}C$). The addition is most probably polar in nature, with Markovnikov regiochemistry, and gives *syn*-addition products. Trifluoromethyl hypofluorite, on the other hand, is more reactive and its reactions have to be conducted at much lower temperature. They are most probably of free-radical nature, that is anti-Markovnikov and lack stereoselectivity. The yields are in most cases quantitative ⟨83JOC242, 86IC376, 91CL1421⟩.

Bis(trifluoromethyl) and bis(trichloromethyl) ether are prepared from dimethyl ether by electrofluorination (HF) ⟨50USP2500388⟩ and photochlorination (Cl_2) ⟨58CT515⟩, respectively.

Trifluoromethoxide is a rather poor nucleophile but, with tris(dimethylamino)sulfonium ($(Me_2N)_3S^+$; TAS^+) as counterion, nucleophilic substitution is possible. Fluorination instead of trifluoromethoxylation may, however, be expected at secondary substitution sites due to the elevated reaction temperature required ⟨85JA4565, 85MI 607-01⟩.

Transformation of an enol group into a difluorohalomethoxy group has been accomplished by reaction with difluorocarbene. Thus, 6-substituted 2-quinoxalones upon treatment with dibromodifluoromethane gave 6-substituted 2-(bromodifluoromethoxy)quinoxalines. Further treatment with poly(hydrogen fluoride)/pyridine yields the corresponding trifluoromethyl ethers (Scheme 9) ⟨92JFC(59)417⟩. A summary of selected methods for the synthesis of trihalomethyl ethers is found in Table 4.

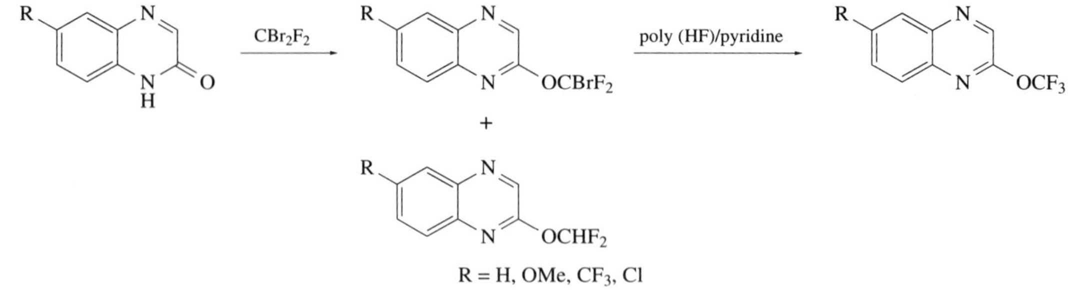

Scheme 9

Table 4 Selected methods for the formation of trihalomethyl ethers.

Starting material	*Reagent(s)*	*Reaction conditions*	*Product(s) (yield) (%)*	*Ref.*
Ar—OH	CCl_4/HF	150 °C, 8 h	Ar—OCF_3 (10–73)	79JOC2907
R—$OCCl_3$ R = aryl or alkyl	HF/$SbCl_5$	140 °C,	R—OCF_3 (75–86)	72GEP2129200
Ar—OMe			Ar—$OCCl_3$	
	$Cl_2/h\nu$	80 °C,	(70)	77CI(L)127
	Cl_2/PCl_5	190–200 °C,	(71)	86BSF925
R—OC(=O)F R = aryl or alkyl	SF_4	100–175 °C, 6 h	R—OCF_3 (18–72)	64JOC1 64JOC11
R—OC(=S)Z Z = Cl, SMe, SSC(=S)OR	Cl_2	0→50 °C[a]	R—$OCCl_3$ (25–90)	(see text)
R = aryl or alkyl	MoF_6	−25–130–190 °C	(40–90)	73TL2253
R—OC(=S)SMe R = aryl or alkyl	Poly-HF/pyridine/DBH[b]	−78–0 °C, 1 h	R—OCF_3 (50–80)	92TL4173
CR_2=CR_2	CF_3OF CF_3OCl	varying	CF_3O—CR_2—CR_2—F CF_3O—CR_2—CR_2—Cl (~quant.)	91CL1421 86IC376 83JOC242
R—OTf R = alkyl	$CF_3O^-TAS^+$[c]		R—OCF_3 : R—F (88, 72 : 28)[d]	85MI 607-01 85JA4565

[a] Depending on the actual compound type. [b] DBH, 1,3-dibromo-5,5-dimethylhydantoin. [c] TAS^+, tris(dimethylamino)sulfonium. [d] This ratio (trifluoromethoxylation vs. fluorination) was achieved at a secondary site. Milder conditions may be used at primary sites which favor trifluoromethoxylation.

(iii) Trihalomethyl esters, $CHal_3OC(=O)R$

Trihalomethyl esters may be considered as carboxylic acid *superhalides* and therefore expected to be very reactive. They may, furthermore, be expected to be thermally unstable and to release carbonyl dihalide with formation of the corresponding carboxylic acid halide, as is the case with trichloromethyl perhalobutanoic acid esters ⟨69CB3127⟩.

Only very few compounds containing a $CHal_3ZC(=Z')R$ (Z, Z′ = O, S, Se, Te) substructure are known. They are in general formed by halogenation of the corresponding methyl esters. Thus, aerosol fluorination of methyl perfluoroadamantaneacetate yields the trifluoromethyl ester plus the acid fluoride in 11% yield ⟨92JOC4749⟩ and low yields were also obtained after electrofluorination of methyl esters ⟨90JFC(50)173⟩, whereas photochlorination of methyl perchloro-3-butenoate at RT gave the corresponding trichloromethyl ester in more than 93% yield ⟨69CB3127⟩, and photochlorination of dimethyl carbonate yields bis(trichloromethyl) carbonate in quantitative yield ⟨83HOU(E4)1368⟩. Trichloromethyl esters are, furthermore, formed by the reaction between an alcohol and diphosgene ($CCl_3C(=O)Cl$) in 62–75% yield ⟨30JPR210, 30JPR81⟩ and obtained in 82–91% yield by photochlorination of methyl chloroformate ⟨80OS(59)195⟩. Bis(trifluoromethyl) oxalate has been obtained in 91% yield by irradiation of a mixture of $(CF_3)_2O_2$ and CO with a low-pressure Hg-lamp ⟨67MI 607-02, 68MI 607-01⟩. Other methods include thermal decomposition of methyl(trifluoromethyl)dioxirane which yields trifluoromethyl acetate (Equation (50)) ⟨91JA7654⟩

and cleavage of amides with CF_3OF at RT which gives *N,N*-dihaloamines and trifluoromethyl esters (Scheme 10) ⟨74JCS(P1)732, 80JFC(15)201⟩.

(50)

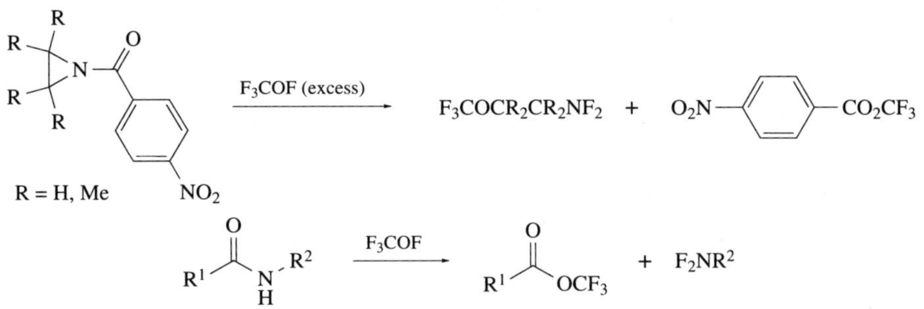

Scheme 10

(iv) *Trihalomethyl hypohalites, $CHal^1{}_3OHal^2$*

The fluorinating agent trifluoromethyl hypofluorite may be prepared in good yields from CO ⟨48JA3986⟩, CO_2 ⟨66IS165⟩ and COF_2 ⟨69JA4432⟩ by treatment with elemental fluorine in the presence of silver(II) fluoride. The hypofluorite is quite thermostable and does not eliminate fluorine unless heated above 275 °C.

Trifluoromethyl hypochlorite is prepared in 99% yield from carbonyl fluoride by treatment with chlorine monofluoride/cesium fluoride ⟨69JA2902⟩.

(v) O-*Trihalomethyl peroxides, $CHal_3OOR$ and trihalomethyl sulfonates, $CHal_3OSO_2R$*

Trifluoromethyl hydroperoxide (CF_3OOH) is prepared in 80% yield by hydrolysis of CF_3OOCOF ⟨71JA3882⟩ or by thermal decomposition of 1-hydroxy-2,2,2-trifluoro-1-(trifluoromethyl)ethyl hydroperoxide (**1**), which is prepared by oxidation of hexafluoroacetone with 90% H_2O_2 ⟨71CC784⟩. Bis(trifluoromethyl) peroxide is prepared in 92% yield from carbonyl fluoride and chlorine trifluoride (5 h at 250 °C) ⟨65USP3202718⟩ or trifluoromethyl hypofluorite ⟨57JA5628⟩. Chloro trifluoromethyl peroxide adds to alkenes (−111 °C→RT, 2→37 h (depending on the alkene type)) in analogy with trifluoromethyl hypochlorite; *β*-chloroalkyl trifluoromethyl peroxides are formed (10–75% yield) (Equation (51)) ⟨75JA13⟩.

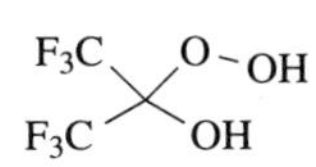

(1)

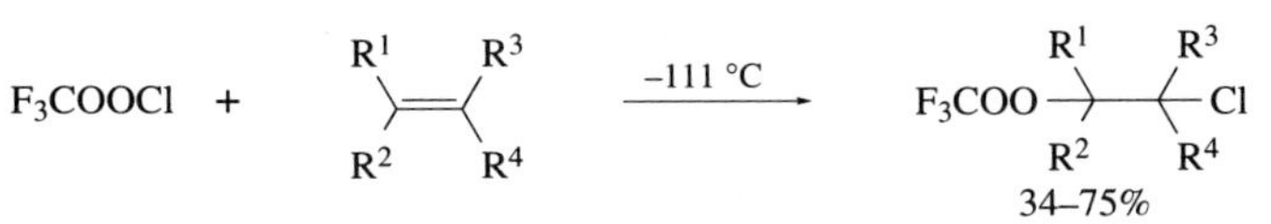

(51)

Trichloromethyl hydroperoxide has also been prepared by photoperoxidation of trichloromethane at RT (Equation (52)) ⟨85AG48⟩.

$$CHCl_3 \xrightarrow[20\ °C]{O_2,\ h\nu} Cl_3COOH \qquad (52)$$

Acylation of trifluoromethyl hydroperoxide is achieved by treatment with an acyl fluoride under dry conditions. Trifluoroacetyl trifluoromethyl peroxide has been prepared in this way (18 h,

RT) in 95% yield ⟨71JA3882⟩. Acetyl trichloromethyl peroxide has been prepared similarly from trichloromethyl hydroperoxide and acetyl chloride ⟨85AG48⟩.

Trifluoromethyl trifluoromethanesulfonate may be prepared from the silver(I) sulfonate and trifluoroiodomethane (200 °C, 24 h) in 86% yield ⟨79TL3865⟩ or from trifluoromethanesulfonyl hypochlorite and trifluorobromomethane (−111 °C, 1 h) in 95% yield ⟨80JA2681⟩. Trichloromethyl trifluoromethanesulfonate has been prepared from mercury(II) trifluoromethanesulfonate and bromotrichloromethane (0 °C, 12 h) in 78% yield ⟨69CB2150⟩.

(vi) N-Trihalomethoxy compounds, $CHal_3ON({=}O)_nR$

A general review of the synthesis of *O*-alkylated hydroxylamines has recently been given by Andree and Kluth ⟨90HOU(E16a)214, 90HOU(E16a)271⟩.

O-(Trifluoromethyl)difluoroformaldehyde oxime has been prepared in excellent yield by dehydrofluorination of CF_3ONHCF_3 with KF. It is thermally stable at RT but dimerizes at RT in the presence of CsF to $CF_3N(OCF_3)CF{=}NOCF_3$ ⟨81JFC(18)441⟩.

Treatment of trichloromethyl hydroperoxide with N_2O_5/$NaHCO_3$ at −25 °C leads to trichloromethyl pernitrate (CCl_3COONO_2) ⟨85AG48⟩.

(vii) Metal trihalomethoxides, $CHal_3OM$

Potassium trifluoromethoxide may be prepared from carbonyl fluoride and potassium fluoride in acetonitrile at 20 °C. This salt is more stable than the corresponding alcohol and can be heated in solution to 80 °C without formation of carbonyl fluoride ⟨65CJC1893⟩.

6.07.2.1.2 Three halogens and a sulfur function

(i) Trihalomethanethiols, $CHal_3SH$

Trifluoromethanethiol is a stable compound whereas the other trihalomethanethiols may be expected to be rather unstable ⟨83HOU(E4)625, 93TL2973⟩. The former has been prepared from mercury(II) trifluoromethanethiolate in 99% yield by treatment with hydrogen chloride ⟨53JCS3219⟩ or in 90% yield by treating (trifluoromethylthio)silane with hydrogen iodide ⟨60JCS3516⟩. The synthesis of trichloromethanethiol has previously been claimed by Connolly and Dyson ⟨34JCS822⟩, but a recent investigation showed this to be in error ⟨93TL2973⟩.

(ii) Trihalomethyl sulfides, $CHal_3SR$

The synthesis of sulfides has previously been reviewed by Gundermann and Hümke ⟨85HOU(E11)158⟩.

Photochlorination of aryl methyl sulfides leads to aryl trichloromethyl sulfides which may be transformed to the corresponding trifluoromethyl analogues by treatment with SbF_3 ⟨37FRP820795, 38BRP479774, 52ZOB2216, 54JGU885, 60JOC60⟩. In the presence of electron-withdrawing groups in the aromatic moiety BF_3 catalysis and higher reaction temperatures are required for the fluorination step ⟨75S721⟩.

Aryl trifluoromethyl sulfides have, furthermore, been prepared from the corresponding aryl halides by direct trifluoromethylthiolation with methyl fluorosulfonyldifluoroacetate (FSO_2CF_2-CO_2Me), CuI and S_8. The choice of solvent and of aryl halide as well as of the metal iodide is of great importance. Yields ranging from 41% to 74% have been reported in HMPA and *N*-methylpyrrolidone whereas no reaction is observed in DMF. Aryl iodides were found to be most reactive and aryl chlorides inactive ⟨93CC918⟩. Replacement of CuI by KI results in the formation of trifluoromethane after work-up ⟨89CC705⟩.

Replacement of aryl bound halogen by a trifluoromethylthio group can also be accomplished with different soft metal trifluoromethanethiolates, CF_3SM, M being Cu(I), Ag(I) or Hg(II).

Thus, CF_3SCu (85–130 °C, 1.5–6 h in DMF, HMPA or *N*-methylpyrrolidone), trifluoromethanethiolates aryl iodides in yields of approximately 75% ⟨85S667⟩. This method is also available for the synthesis of trifluoromethylthio substituted heteroaromatics, and the best yields are achieved with electron-poor systems ⟨75S721⟩. Yields may, furthermore, be improved by use of copper(I) trifluoromethanethiolate on alumina support ⟨90JFC(48)249⟩.

Mercury(II) bis(trifluoromethanethiolate), $(CF_3S)_2Hg$, reacts with alkyl and allyl chlorides with formation of the corresponding trifluoromethyl sulfides ⟨59JA3575, 67JOC2063, 81TL3047⟩.

Silver(I) trifluoromethanethiolate, CF_3SAg, reacts with alkyl halides under formation of alkyl trifluoromethyl sulfides with an insoluble silver halide as a convenient by-product ⟨61JCS2597, 65ZOB1628⟩.

Other sources of trifluoromethanethiolate ions which have been used in nucleophilic aromatic substitution include the *in situ* formation of trifluoromethanethiolate from thiocarbonyl fluoride or bis(trifluoromethyl) trithiocarbonate ($(CF_3S)_2C{=}S$) and calcium, potassium or caesium fluoride ⟨52JCS2198, 85C185, 87JCS(P1)2119, 88JCS(P1)1179⟩.

Electron-rich aromatic systems undergo substitution with trifluoromethanesulfenyl chloride ⟨50JA3529, 53CB557, 64JOC898, 77CB67, 78JFC(11)509⟩ and trichloromethanesulfenyl chloride ⟨63ACS2570⟩, in some cases under free radical conditions, yielding aryl trifluoromethyl sulfides in good yields (generally in excess of 60%) and aryl trichloromethyl sulfides, respectively. Trifluoromethanesulfenyl chloride as well as trifluoromethanesulfenyl fluoride reacts with nucleophiles; with alkenes they form β-halo(trifluoromethylthio)alkanes ⟨62JA3148, 67JOC2063, 87JFC(34)475⟩, and with aryl Grignard reagents, trifluoromethanesulfenyl chloride (at 0°C) forms aryl trifluoromethyl sulfides, which are, however, contaminated with the corresponding aryl chlorides ⟨64JOC895⟩. Trifluoromethanethiol also reacts with alkenes under UV irradiation with formation of the corresponding alkyl trifluoromethyl sulfides ⟨61JA840⟩.

Aromatic potassium thiolates react with bromotrifluoromethane under pressure (2–3 atm, RT, DMF) to give the corresponding trifluoromethyl sulfides (7–75% yield) ⟨84CC793, 85JOC4047⟩. The use of mixed tetrahalomethanes such as dibromodifluoromethane or bromochlorodifluoromethane leads to mixed halomethyl sulfides ⟨81TL1997, 81TL323⟩.

Alkylation of aliphatic and aromatic disulfides may be conducted in DMF/H_2O at RT under radical conditions by treatment of bromotrifluoromethane with sulfur dioxide radical anion (derived from $Na_2S_2O_4/NaO_2SCH_2OH$) and Na_2HPO_4 to neutralize the sulfur dioxide formed during the reaction (31–93% yield) ⟨91CC993⟩. Trifluoroiodomethane reacts similarly with sulfides under UV irradiation, but more slowly, for example, the reaction with dimethyl sulfide (*hv*, 21 days) gives a 92% yield of trifluoromethyl methyl sulfide ⟨72JCS(P1)2180⟩. This reaction has been extended to alkane-, arene- and heteroarenethiols as well ⟨79ZOR396, 79ZOR1245⟩ and may, furthermore, be conducted under phase transfer conditions (52–85% yield) ⟨82JFC(21)365⟩.

N-(Trifluoromethyl)-*N*-nitrosobenzenesulfonamide ($CF_3N(NO)SO_2Ph$, TNS-B), *N*-(trifluoromethyl)-*N*-nitrosotrifluoromethanesulfonamide ($CF_3N(NO)SO_2CF_3$, TNS-Tf), *S*-(trifluoromethyl)-dibenzothiophenium triflate (TBT-Tf) and its seleno analogue (TBSe-Tf) (**2**), and (4-chlorophenyl)-(2,6-dimethylphenyl)trifluoromethylsulfonium hexafluoroantimonate (**3**) have been developed as efficient trifluoromethylating agents and are more easily handled than trifluoroiodomethane gas. Thus, TNS-B reacts with dialkyl sulfides to give the corresponding trifluoromethyl alkyl sulfides in 55–64% yield ⟨82TL3929⟩. TNS-Tf gives generally higher yields of trifluoromethyl sulfides from thiols and dialkyl sulfides after shorter reaction times ⟨86BCJ447⟩. The reaction of sodium alkanethiolates with (**2**) ⟨90TL3579⟩ produces the corresponding trifluoromethyl sulfides in 47–87% yield, as does treatment with (**3**) ⟨84ZOR115⟩.

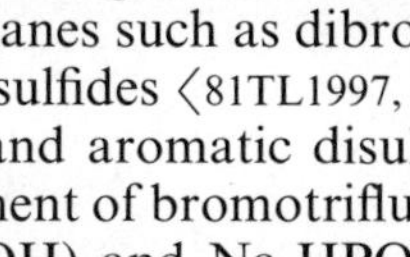

Z = S, Se, Te

(2)

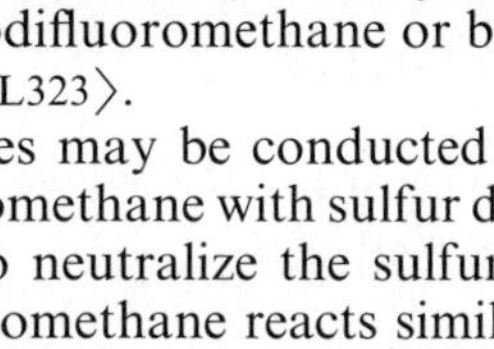

(3)

Direct chlorination of methyl sulfides leads to trichloromethyl sulfides in yields ranging from 56–100%, as does chlorination of alkyl halodithioformates a reaction which also may be used for the formation of dichlorohalomethyl sulfides ⟨39AG457, 59ZOB3786, 60ACS2230, 62ACS117⟩.

Trichloromethyl sulfides have, furthermore, been prepared in 60–76% yield from thiocyanates by treatment with trichloromethyl anions (derived from $CHCl_3/NaOH$) under phase transfer conditions ⟨74S274⟩.

Finally, trihalomethyl sulfides can be interconverted: $CF_3SR \rightarrow CCl_3SR$ (with $AlCl_3$); $CF_3SR \rightarrow$

CBr_3SR (with BBr_3 or $AlBr_3$); $CCl_3SR \rightarrow CF_3SR$ (with KF/18-crown-6, SbF_3 or HF); $CCl_3SR \rightarrow CBr_3SR$ (with HBr). Other fluorinating agents include $AgBF_4$, Hg_2F_2 and $[BnNMe_3]^+F^-$ 〈38USP2108606, 52JA3594, 56CB1160, 67ZOB1770, 71GEP2003143, 77JOC2024, 81TL1997〉.

Kolomeitsev *et al.* have developed a method for the synthesis of trifluoromethyl sulfides from the corresponding alcohols. The alcohol is treated with bis(diethylamino) chloro phosphite which gives the intermediate diethylamino alkyl phosphite ($ROP(NEt_2)_2$) in 80–90% yield. Further treatment with bis(trifluoromethyl) disulfide gives the trifluoromethyl alkyl sulfide in 90–95% yield. The overall yields in this two step procedure are excellent (72–86%) 〈94S145〉. Table 5 presents a summary of selected methods for the formation of trihalomethyl sulfides.

Table 5 Selected methods for the formation of trihalomethyl sulfides.

Starting material	*Reagent(s)*	*Reaction conditions*	*Product (yield) (%)*	*Ref.*
Ar—SMe	Cl_2, *hv*		Ar—$SCCl_3$ (56–100)	(see text)
R—SC(=S)X R = aryl or alkyl X = Cl, F	Cl_2	~40°C,	R—$SCCl_3$ (15–80)	59ZOB3786
Ar—X, X = Br or I Ar = Ph or heteroaromatics	$FSO_2CF_2CO_2Me$, S_8, CuI	100–120°C, 5–8 h	Ar—SCF_3 (41–74)	93CC918, 89CC705
Ar—I, Ar = Ph or pyridine	CF_3YCu Y = S, Se	85–130°C, 5–6 h	Ar—SCF_3 (41–95)	90JFC(48)249, 85S667, 75S721
ArH, ArH = elec. rich aromatics	CF_3SCl, Lewis cat.	60°C, 20 h	Ar—SCF_3 (40–78)	(see text)
$R_2C{=}CR_2$	CF_3SCl, *hv*		CF_3S—CR_2CR_2—Cl (51–83)	67JOC2063, 62JA3148
Ar—MgX	CF_3SCl	0°C	Ar—SCF_3 (~50%)	64JOC895
$R_2C{=}CR_2$	CF_3SH, *hv*		CF_3S—CR_2CR_2—H (56–82)	61JA840
Ar—SK	CF_3Br	2–3 atm, RT, 3 h	Ar—SCF_3 (7–75)	85JOC4047, 84CC793
R—SNa	(**2**) or (**3**)		R—SCF_3 (47–87)	90TL3579, 84ZOR115
R_2S	TNS-B or TNS-Tf[a]		R—SCF_3 (55–64)	86BCJ447, 82TL3929
R_2S, R = aryl or alkyl	CF_3Br, $Na_2S_2O_4$, NaO_2SCH_2OH, Na_2HPO_4	RT	R—SCF_3	91CC993
Ph—$SCCl_3$	SbF_3, (BF_3)		Ph—SCF_3 (70)	(see text)
R—SCX_3, X = F, Cl, Br	(see text)		R—SCY_3 Y = F, Cl, Br	(see text)
R—OH, R = alkyl or benzyl	i, $(Et_2N)_2PCl/NEt_3$; ii, CF_3SSCF_3	i, −40°C, 2 h; ii, −50–20°C, 5 min.	R—SCF_3	94S145

[a] *N*-(Trifluoromethyl)-*N*-nitrosobenzenesulfonamide ($CF_3N(NO)SO_2Ph$, TNS-B), *N*-(trifluoromethyl)-*N*-nitrosotrifluoromethanesulfonamide ($CF_3N(NO)SO_2CF_3$, TNS-Tf).

(iii) Trihalomethyl sulfoxides and trihalomethyl sulfones, $CHal_3S(=O)_nR$

The synthesis of sulfoxides and sulfones has previously been reviewed by Gundermann and Hümke 〈85HOU(E11)665〉 and by Schank 〈85HOU(E11)1129〉.

Partial oxidation of alkyl/aryl trifluoromethyl sulfides with 27% hydrogen peroxide in acetic acid (2 h reflux) yields the corresponding sulfoxides 〈65ZOB1628〉. More drastic oxidation with nitric acid, hydrogen peroxide or chromium trioxide in glacial acetic acid gives the corresponding sulfones 〈52JA3594, 65ZOB1628, 67JOC2063, 87JCS(P1)2119, 93CC918〉.

The transfer of CX_3SO_2 groups is possible by reactions of nucleophiles such as BuLi with electrophilic species like $(CF_3SO_2)_2O$, $(CF_3SO_2)_2NPh$ and CF_3SO_2Cl. In the case of primary or secondary organometallics ditriflation is possible due to the presence of acidic α-protons in the monotriflated products ⟨77JOC3875⟩. Alkynyl trifluoromethyl sulfones have been prepared from sodium alkynides and triflic anhydride in 17–75% yield ⟨92T189⟩.

Monotriflation of aromatic compounds can be carried out as Friedel–Crafts triflation with $(CF_3SO_2)_2O/AlCl_3$, but only aromatic compounds of at least intermediate reactivity such as xylene, toluene, and benzene are triflated ⟨77JOC3875⟩.

Aryl trifluoromethyl sulfones have been prepared in high yields (70–99%) by difluorocarbene insertion into arenesulfonyl fluorides. The reagent mixture used is Me_3SiCF_3 or Me_3SnCF_3 and $(Me_2N)_3S^+Me_3SiF_2^-$(TASF). The yield is unaffected by electronic effects of arene substituents (Equation (53)) ⟨90S1151⟩.

$$ArSO_2F \xrightarrow[\text{THF, 20 °C}]{2Me_3MCF_3,\ \text{TASF}} \underset{70\text{–}99\%}{ArSO_2CF_2F} \qquad (53)$$

$$\text{TASF} = (Me_2N)_3S^+\ Me_3SiF_2^-$$
$$M = Sn, Si$$

Transformation of an arenesulfonamide into an *N*-(trifluoromethyl)azosulfonylarene is effected by treatment of the amide with trifluoronitrosomethane in base (81–92% yield). This intermediate eliminates N_2 on heating to form the corresponding trifluoromethyl phenyl sulfone (~40% yield) (Scheme 11) ⟨82CL1519⟩.

$$F_3CNO + H_2NSO_2Ar \xrightarrow{\text{base}} \underset{81\text{–}92\%}{F_3CN{=}NSO_2Ar} \xrightarrow[-N_2]{\Delta} \underset{\sim 40\%}{F_3CSO_2Ar}$$

Scheme 11

(iv) S-Trihalomethyl thio- and dithioesters, $CHal_3SC({=}Z)R$ ($Z = O$ or S)

Trifluoromethyl thioesters have been prepared by the reaction of mercury(II) bis(trifluoromethanethiolate), $(CF_3S)_2Hg$, with acyl chlorides (40–50 °C, 23–51% yield) ⟨59JA3575⟩. Bis(trifluoromethyl) trithiocarbonate is formed in ~100% yield by CsF treatment at 0 °C of thiocarbonyl difluoride ⟨66AG906, 68CB2609⟩.

Friedel–Crafts (trifluoromethyl)thiothioacylations have been achieved with HF/SbF_5 and $FC({=}S)SCF_3$ ⟨87CB429⟩ with the latter formed from thiocarbonyl fluoride and potassium fluoride ⟨68CB2609⟩ (Scheme 12). The corresponding (trichloromethyl)thiothioacylation (70 °C, 4 h; RT 16 h) has been achieved in 26% yield by treating anthracene with trichloromethyl chlorodithioformate formed in 75% yield from trichloromethanesulfenyl chloride and CS ⟨84JOC3854⟩. Condensation of bis(trifluoromethyl) trithiocarbonate ($(CF_3S)_2C{=}S$) with $(CF_3S)_2C{=}C(SCF_3)_2$ in the presence of a catalytic amount of $HgCl_2$ (3 d at 170 °C) gives 1,1-bis(trifluoromethylthio)-2,2-bis[(trifluoromethylthio)thiocarbonyl]ethene in 10% yield, with bis(trifluoromethyl) disulfide as the main product ⟨77CB916⟩.

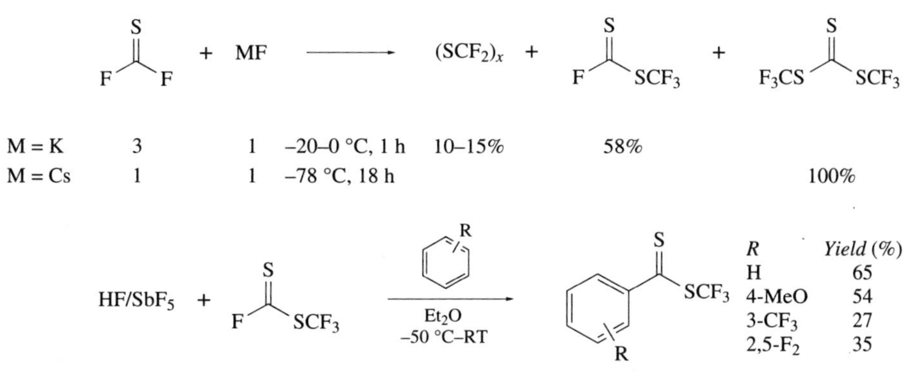

Scheme 12

(v) Trihalomethanesulfenyl, -sulfinyl and -sulfonyl halides; $CHal^1{}_3S(=O)_n(Hal^2)$

The synthesis of sulfenyl, sulfinyl and sulfonyl halides has previously been reviewed by Schubart ⟨85HOU(E11)63⟩, Krauthausen ⟨85HOU(E11)614⟩ and Pawlenko ⟨85HOU(E11)1067⟩. See also the review by Sizov *et al.* on the formation of polyfluoroalkanesulfenyl chlorides ⟨92RCR517⟩. In this section only sulfenyl halides are treated, due to their synthetic importance in the synthesis of trihalomethyl sulfides.

Umpolung of metal trihalomethanethiolates can be achieved by chlorination to the corresponding trihalomethanesulfenyl chlorides ⟨61JCS2597⟩. Bromine and iodine, however, oxidize silver(I) trifluoromethanethiolate to bis(trifluoromethyl) disulfide ⟨61JCS2597⟩.

The industrial preparation of trichloromethanesulfenyl chloride takes place by chlorination of carbon disulfide under conditions which suppress the formation of bis(trichloromethyl) disulfide ⟨71S478, 66FRP1437908⟩.

A general method for the synthesis of trihalomethanesulfenyl halides is to halogenate (Hal_2) thiocarbonyl halides and this method can therefore also be used to prepare mixed trihalomethanesulfenyl halides ⟨59ZOB3792, 62JCS4361, 64AG807, 76CB3432⟩; furthermore, by halogen exchange in trihalomethanesulfenyl halides with $SbF_3/SbCl_5$ (Br→F), HF (Cl→F), NaF (Cl→F), HF/CrOF (Cl→F) or HBr (Cl→Br) ⟨58USP2821554, 59ZOB3401, 59ZOB3784, 60JOC2016, 67GEP1232954⟩.

Still other methods include chlorination of disulfides, for example chlorination of dimethyl disulfide leads to trichloromethanesulfenyl chloride in 80% yield and high purity while photochlorination at RT of bis(trifluoromethyl) disulfide leads to trifluoromethanesulfenyl chloride in 55% yield ⟨53JCS3219, 61USP3014071⟩.

(vi) Trihalomethanesulfenic, -sulfinic and -sulfonic derivatives; $CHal_3S(=O)_nZR$

The trihalomethanesulfenyl, -sulfinyl, and -sulfonyl halides are important starting materials for the synthesis of other trihalomethanesulfenyl, -sulfinyl, and sulfonyl derivatives. For example mercury(II) bis(trifluoromethanethiolate) $[(CF_3S)_2Hg]$, reacts with sulfenyl chlorides with formation of the corresponding trifluoromethyl disulfides ⟨59JA3575⟩. Other methods which allow the formation of mixed trihalomethyl disulfides are the photoreaction between thiocarbonyl halides and trihalomethanesulfenyl halides ⟨68CB2617⟩ and the photoreductive dimerization of trihalomethanesulfenyl chlorides with, for example, CO as reductant ⟨67JINC2819⟩.

Trihalomethanesulfenic acids are, as most other sulfenic acids, unstable. Trichloromethanesulfenic acid for instance eliminates HCl with formation of dichlorosulfine ($Cl_2C{=}S{=}O$) ⟨69CC878⟩, whereas the sulfinic and sulfonic acids are stable compounds.

A preparative method which gives access to halodifluoromethanesulfonic acid derivatives in high yields is the decarbonylative photolysis of $HalC(=O)CF_2SO_2F$ (the reaction is conducted at 80–90°C for Hal = Cl, 50°C for Hal = Br, RT for Hal = I) ⟨79S972⟩.

The synthesis of sulfenic, sulfinic and sulfonic acid derivatives has been reviewed previously by Schubart ⟨85HOU(E11)63⟩, Krauthausen ⟨85HOU614⟩, Pawlenko ⟨85HOU(E11)1055, 85HOU(E11)1084⟩ and Jäger ⟨85HOU(E11)1073⟩ and is not discussed here. See, however, the review by Sizov *et al.* on transformation of polyfluoroalkanesulfenyl chlorides ⟨92RCR517⟩.

The CF_3SO_2 group (in CF_3SO_2Ar) was found by NMR measurements to be even more electron-attracting than the NMe_3 group (in $ArNMe_3{}^+$) ⟨68ZOB2591⟩.

(vii) Metal trihalomethanethiolates; $CHal_3SM$

Metal trihalomethanethiolates are important as precursors of aryl trihalomethyl sulfides (*vide supra*) and their synthesis is therefore mentioned here.

Copper(I) trifluoromethanethiolate, CF_3SCu, is obtained in nearly quantitative yield by heating (50–60°C) CF_3SSCF_3 with copper powder in DMF, HMPA or *N*-methylpyrrolidone for 1–3 h, depending on the activity of the copper powder ⟨85S667⟩. The thiolate can be used *in situ*, but also isolated and stored until needed. The copper(I) thiolate can also be prepared by reaction of mercury(II) bis(trifluoromethanethiolate) with copper powder ⟨59JA3575⟩ or by reaction of $AgSCF_3$ with Cu/CuBr ⟨85S667⟩.

Mercury(II) bis(trifluoromethanethiolate), $(CF_3S)_2Hg$, may be prepared from HgF_2 and carbon disulfide in 72% yield after heating 4 h at 250°C in a steel autoclave ⟨59JA3575⟩ or photochemically in 90% yield from bis(trifluoromethyl) disulfide and metallic mercury ⟨53JCS3219⟩.

Silver(I) trifluoromethanethiolate, CF_3SAg, is formed in 79% yield by the reaction of dry silver(I) fluoride with dry carbon disulfide after heating 12 h at 140°C in a steel autoclave ⟨61JCS2597⟩; it reacts with alkyl halides with formation of alkyl trifluoromethyl sulfides and an insoluble silver halide as a convenient by-product ⟨61JCS2597, 65ZOB1628⟩.

6.07.2.1.3 *Three halogens and a Se or Te function*

(i) Trifluoromethylselenium compounds; CF_3SeR

The chemistry of trifluoromethylselenium compounds has been reviewed and compared with that of the corresponding sulfur compounds ⟨86JFC(32)415⟩.

Aryl trifluoromethyl selenides can be prepared from aryl halides and $(CF_3)SeCu$ in analogy with the synthesis of the corresponding trifluoromethyl sulfides (Equation (54)) ⟨85S667⟩. *Se*-(Trifluoromethyl)dibenzoselenophenium trifluoromethanesulfonate (**2**; Z = Se) can be obtained by fluorination of a mixture of 2-[(trifluoromethyl)seleno]biphenyl with CF_3SO_3H and by treatment of the corresponding sulfoxide with $(CF_3SO_2)_2O$. Compound (**2**) can be used as an electrophilic trifluoromethylation reagent (*vide supra*) ⟨93JA2156⟩. Bis(trifluoromethylseleno)ketene is available by dehydration of bis(trifluoromethylseleno)acetic acid or by dehydrochlorination of the corresponding acid chloride ⟨92CB571⟩.

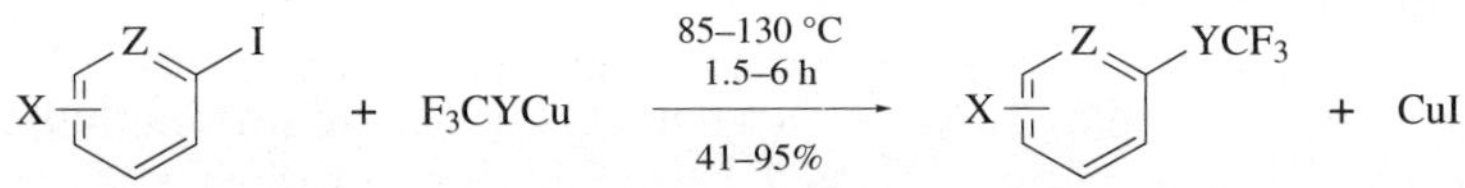

(54)

X = H, Me, NO_2; Y = S, Se; Z = CH, N

Mercury(II) trifluoromethaneselenolate reacts with *S*-trifluoromethyl bromothioformate $[(CF_3)SC({=}O)Br]$ (130°C for 16 h) with formation of $(CF_3S)C({=}O)Se(CF_3)$ ⟨78CB2891⟩. Run overnight at 0°C the same reaction gives a yield of 86% ⟨76CB3432⟩.

Oxidation of an aryl trifluoromethyl selenide with chlorine yields the corresponding aryldichloro(trifluoromethyl)selenane, $ArSeCl_2(CF_3)$ ⟨85S667⟩. Aryl trifluoromethyl selenides can be oxidized with chlorine/water to the corresponding selenoxides and with trifluoroperacetic acid to the corresponding selenones ⟨68ZOB2509⟩. Trifluoromethaneselenonic acid, CF_3SeO_3H was prepared for the first time by Haas and Weiler in 1985 by oxidation of CF_3SeO_2H with $KMnO_4$ under neutral aqueous conditions. Upon attempted purification it decomposes (at concentrations higher than 90%) to CF_4, COF_2, SeO_2 and H_2O ⟨85CB943⟩.

(ii) Trifluoromethyltellurium compounds; CF_3TeR

Mixtures of bis(trifluoromethyl) telluride and dialkyl tellurides are subject to ligand exchange and form alkyl trifluoromethyl tellurides such as $BnTe(CF_3)$ and $Bu^tTe(CF_3)$ ⟨92C78, 92OM2947⟩. Mixtures of $(CF_3)_2Te$ and $(C_6F_5)_2Te$ as well as of $(CF_3)_2Te_2$ and $(C_6F_5)_2Te_2$ undergo scrambling under irradiation to give $(CF_3)Te(C_6F_5)$ and $(CF_3)TeTe(C_6F_5)$, respectively ⟨90JFC(48)207, 93CM1321⟩. An 86% yield of dimethyl(trifluoromethyl)telluronium iodide is obtained in the photoreaction (at −78°C) of Me_2Te and CF_3I ⟨89ZAAC(576)225⟩. *Te*-(Trifluoromethyl)dibenzotellurophenium trifluoromethanesulfonate (**2**; Z = Te) can be obtained by treatment of 2-[(trifluoromethyl) telluro]biphenyl with $(CF_3SO_2)_2O$ and DMSO. Compound (**2**) can be used as an electrophilic trifluoromethylation reagent ⟨93JA2156⟩. Tetrakis(trifluoromethyl)tellurane $(CF_3)_4Te$, which has been obtained from dichlorobis(trifluoromethyl)tellurane and $(CF_3)_2Cd$, reacts with fluoride ions to form $[(CF_3)_4TeF]^-$, with fluoride ion acceptors to form $[(CF_3)_3Te]^+$, and with XeF_2 to form the hypervalent tellurium compound $(CF_3)_4TeF_2$ ⟨92ZAAC(608)69⟩.

6.07.2.2 Three Halogens and a Group 15 Element

6.07.2.2.1 *Three halogens and a nitrogen function*

This subject has previously been reviewed by Marhold ⟨83HOU(E4)625⟩.

(i) *Trihalomethylamines*, $CHal_3NR_2$

The unsubstituted and monosubstituted trihalomethylamines possess more or less salt character and are quite unstable. The phosgeneiminium character makes them prone to hydrolysis (Equation (55)).

(55)

(*a*) *Primary trihalomethylamines*, $CHal_3NH_2$. Trifluoromethylamine has been prepared by treating *N*,*N*-dichloro-*N*-(trifluoromethyl)amine with HCl at −78 °C which gives the hydrochloride. The free base is stable below ~ −30 °C, above which it eliminates HF with formation of difluorophosgeneimine (Scheme 13) ⟨79JA347⟩. Synthesis of trichloromethylamine is achieved by treating ClCN with HCl. The intermediate hydrochloride is unstable, but addition of antimony(V) chloride converts it into the stable hexachloroantimonate salt ⟨64CB1286, 73BSB23⟩.

$$F_3CNCl_2 + 3HCl \xrightarrow{-Cl_2} F_3CNH_3^+\,Cl^- \xrightarrow{\text{amine base}} F_3CNH_2$$

Scheme 13

(*b*) *Secondary trihalomethylamines*, $CHal_3NHR$. Secondary trifluoromethylamines have been prepared from *N*-substituted isocyanide dichlorides ($CCl_2{=}NR$) by reaction with HF. Yields of 69–98% have been achieved with *N*-aryl substituted compounds, whereas the aliphatic counterparts dimerize and/or polymerize (Equation (56)) ⟨63BEP632365⟩. Bis(trifluoromethyl)amine may be synthesized by addition of HF to trifluoromethyl isocyanide difluoride ($CF_3N{=}CF_2$) in 89% yield ⟨59ZOB2159⟩ or by HF treatment of trichloromethyl isocyanide dichloride in 45% yield ⟨55JCS2532, 58JA3604⟩.

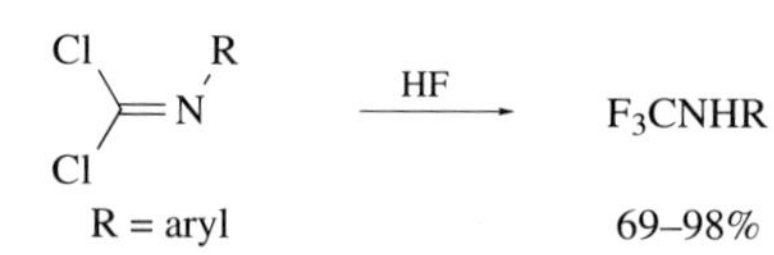

(56)

(*c*) *Tertiary trihalomethylamines*, $CHal_3NR_2$. Tertiary trifluoromethylamines are stable compounds which have been prepared by fluorination of $R_2NC({=}S)SSC({=}S)NR_2$ with either SF_4 or COF_2, or by fluorination of *N*,*N*-disubstituted fluorothioformamides ($FC({=}S)NR_2$) with SF_4. These methods may be applied to both aryl and alkyl trifluoromethylamines (Equation (57)) ⟨61JA3422, 62JA4275, 71ZOR2000⟩.

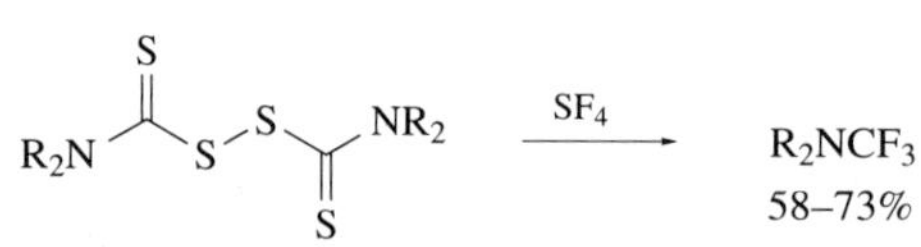

(57)

Tertiary trichloromethylamines have been made similarly by chlorination of *N*,*N*-disubstituted chlorothioformamides with Cl_2 or PCl_5 (~76% yield) ⟨59ZOB3786, 71ZOR2084⟩, by chlorination of $R_2NC({=}S)SC({=}S)NR_2$ with $COCl_2$ (93% for R = Me) ⟨83HOU(E4)625⟩, by chlorination of *N*,*N*-diarylthioformamides ($HC({=}S)NAr_2$) and of alkanesulfonylthioformamides ($RSO_2C({=}S)NR_2$) ⟨82GEP3044216, 72CB2854⟩. *N*-(2,4,6-Trichlorophenyl)-*N*,*N*-bis(trichloromethyl)amine has been prepared by chlorination at 140 °C of the corresponding *N*,*N*-dimethylamine ⟨69LA(730)140⟩. Alkylation of *N*-substituted phosgeneimines with strong alkylating agents such as RO_3SF, followed by Cl^-/FSO_3^--anion exchange, gives the corresponding dichlorophosgeneiminium chlorides ⟨73AG837⟩. Dimethyl(tribromomethyl)amine has been prepared similarly to the trichloromethyl and trifluoromethyl analogues, for example, by bromination with Br_2 of $Me_2NC({=}S)SS(C{=}S)NMe_2$ ⟨72LA(755)145⟩.

Tertiary trifluoromethylamines may alternatively be synthesized from the corresponding trichloromethylamines by treatment with HF or SbF_3 under anhydrous conditions in 58–79% yield ⟨66ZOB1309, 82GEP3044216⟩.

Bis(trifluoromethyl)alkylamines are formed by alkylation of metal (K^+, Cs^+, Hg^{2+}) bis(trifluoromethyl)amides in 44–84% yield ⟨80JCS(P1)2254, 75IZV2279⟩.

N-Bromo- and *N*-iodo-*N*,*N*-bis(trifluoromethyl)amines add to double bonds with formation of β-bromo- and β-iodoalkyl bis(trifluoromethyl)amines, respectively, in 72–98% yield ⟨58JA3604, 65JCS6141, 68JCS(C)796, 80JCS(P1)1544⟩.

Tris(trifluoromethyl)amine has been prepared by treatment of trifluoromethyl isocyanide difluoride with CF_3SF_5 ⟨57JA69⟩ and as a by-product (22% yield) of the reaction between *O,N,N*-tris(trifluoromethyl)hydroxylamine and trifluoromethyl isocyanide difluoride where *N,N,N′,N′*-tetrakis(trifluoromethyl)hydrazine is the main product ⟨67JCS(C)1241⟩.

Mixed *N,N*-dihalo(trihalomethyl)amines (halo equals chlorine or fluorine) have been prepared in 60–75% yield from the corresponding imines by treatment (−20 °C for 3–4 h) with chlorine monofluoride ⟨71IC1635⟩.

(ii) N-*Trihalomethylamides,* $CHal_3NRC({=}O)R$

Isocyanates are used as starting materials for the synthesis of *N*-alkyl-*N*-(trifluoromethyl)-carbamoyl fluorides and of *O*-alkyl-*N*-(trifluoromethyl)carbamates. Thus, trifluoromethyl isocyanate reacts with CsF and subsequently with an alkyl halide, with formation of *N*-alkyl-*N*-(trifluoromethyl)carbamoyl fluorides (Scheme 14) ⟨76IZV209⟩ while the reaction of trifluoromethyl isocyanate with alcohols yields *O*-alkyl-*N*-(trifluoromethyl)carbamates in 66–84% yield (Equation (58)) ⟨59ZOB2157⟩. Aryl isothiocyanates are converted into *N*-aryl-*N*-(trifluoromethyl)carbamoyl fluorides in 43–83% yield by treatment with HgF_2, followed by COF_2/CsF (Equation (59)) ⟨65JA4338⟩.

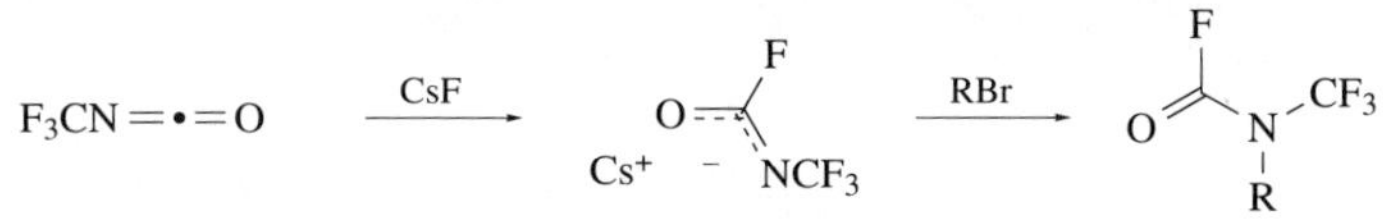

Scheme 14

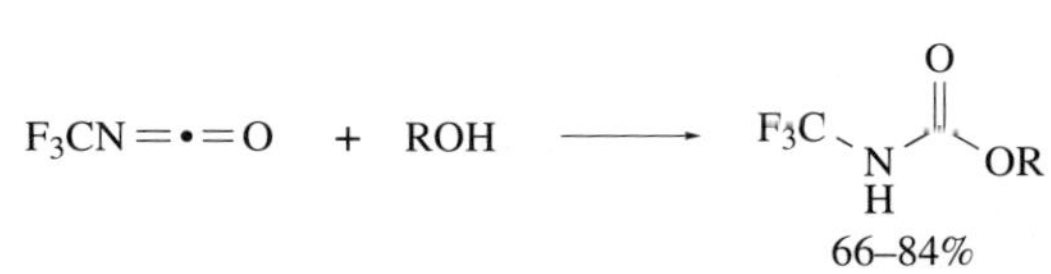

(58)

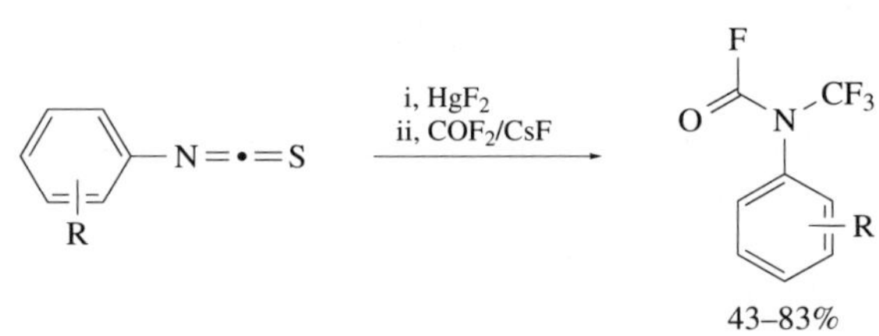

(59)

Hydrolysis of *N,N*-bis(trifluoromethyl)carbodiimide gives *N,N′*-bis(trifluoromethyl)urea ⟨67JCS(C)2302⟩.

(iii) Trihalomethyl isocyanates, $CHal_3N{=}C{=}O$*; trihalomethyl isocyanide dihalides,* $CHal_3N{=}CHal_2$ *and* N-*(trihalomethyl)carbodiimides* $CHal_3N{=}C{=}NR$

Key starting materials in the synthesis of bis(trifluoromethyl)amines are, as mentioned above, trifluoromethyl isocyanide difluoride ($CF_3N{=}CF_2$) and trichloromethyl isocyanide dichloride ($CCl_3N{=}CCl_2$). One high yielding method for the synthesis of these is the chlorination of *N,N*-dimethylcarbamoyl chloride which gives trichloromethyl isocyanide dichloride in 87% yield ⟨63GEP1222917, 70GEP1926659⟩; further treatment with sodium fluoride yields trifluoromethyl isocyanide difluoride which distills off from the reaction mixture (78% yield, Scheme 15) ⟨72CA(77)125952⟩. Other methods of preparation of these compounds are described in Section 6.20.1.

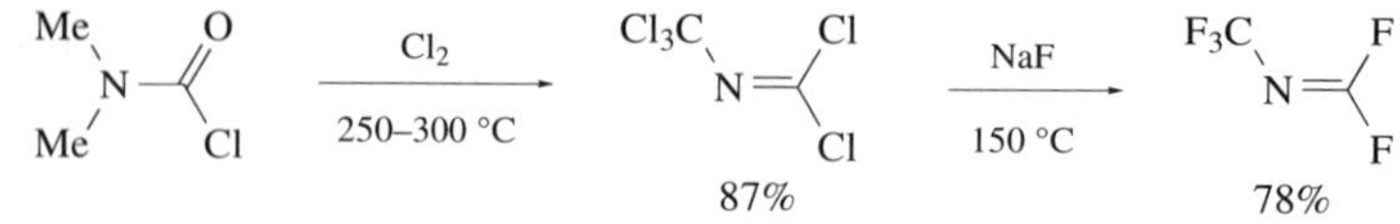

Scheme 15

Other key starting materials in the synthesis of (trihalomethyl)amines are the trihalomethyl isocyanates ($CHal_3N{=}C{=}O$). Trifluoromethyl isocyanate is formed in 75% yield from trifluoroacetyl chloride by treatment with azidotrimethylsilane and thermolysis of the intermediary trifluoroacetyl azide by Curtius amide degradation (Scheme 16) ⟨79CB2158⟩. Trifluoromethyl isothiocyanate is formed in 60% yield upon H_2S treatment of trifluoromethyl isocyanide difluoride, followed by distillation from NaF ⟨69ZOB199⟩. Trichloromethyl isocyanate is unstable and rearranges quantitatively to chloroformyl isocyanide dichloride. The same product is obtained in 89% yield by photochlorination of methyl isocyanate (Scheme 16) ⟨76CA(85)142668⟩. Tribromomethyl isocyanate is formed in 69% yield by NBS bromination of methyl isocyanate (Scheme 16) ⟨82CB860⟩.

N-(Trifluoromethyl)carbodiimides are formed in 46–80% yield by the reaction between trifluoromethyl isocyanide difluoride and an alkyl- or arylamine in the presence of KF/Et_3N ⟨80JFC(15)169⟩.

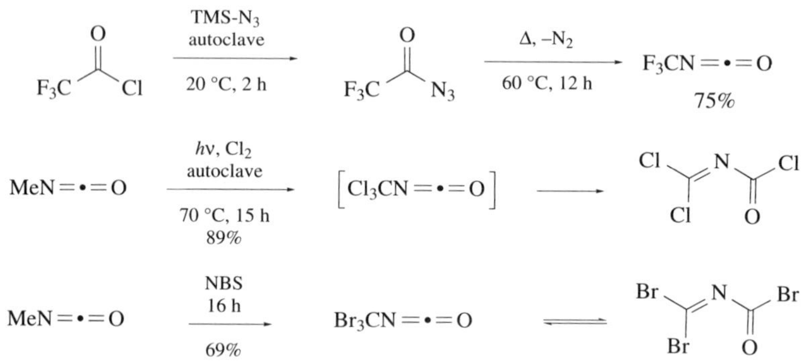

Scheme 16

(iv) N-*Halo(trihalomethyl)amines,* $CHal^1{}_3NRHal^2$

N-Bromo- and *N*-iodo-*N*,*N*-bis(trifluoromethyl)amine are to be considered as *umpoled* amines and thus as key intermediates for the introduction of the bis(trifluoromethyl)amine functionality by reaction with nucleophiles such as alkenes (*vide supra*). They are formed by bromination and iodination of mercury(II) bis(trifluoromethyl)amide in 90% and 67% yield, respectively ⟨58JA3604, 66JCS(A)367⟩. *N*-Chloro-*N*,*N*-bis(trifluoromethyl)amine is similarly formed in 70% yield by chlorination of potassium bis(trifluoromethyl)amide and *N*-fluoro-*N*,*N*-bis(trifluoromethyl)amine is obtained by fluorination of trifluoromethyl isocyanide difluoride with elemental fluorine ⟨81JFC(17)463⟩.

Addition of halogen to *N*-halo isocyanide dihalides ($Hal_2C{=}NHal$) or cyanogen chloride leads to *N*,*N*-dihalo-*N*-(trihalomethyl)amines in 60–97% yield ⟨70CC395, 72LA(755)145, 81JFC(17)463⟩.

(v) N-*(Trihalomethyl)hydroxylamines,* $CHal_3NR^1OR^2$*; trihalonitrosomethanes,* $CHal_3NO$ *and trihalonitromethanes,* $CHal_3NO_2$

Photodimerization of trifluoronitrosomethane, followed by acid hydrolysis and oxidation with silver oxide, lead to bis(trifluoromethyl)aminoxyl radical ⟨57JCS1741⟩ which reacts with alkyl halides with formation of *O*-alkyl-*N*,*N*-bis(trifluoromethyl)hydroxylamines. The *O*-methyl compound was prepared in this way by reaction with iodomethane at RT (Scheme 17) ⟨69JCS(A)431⟩. The bis(trifluoromethyl)aminoxyl radical also reacts with activated positions of alkanes with formation of *N*,*N*-

bis(trifluoromethyl)hydroxylamine and *O*-alkyl-*N*,*N*-bis(trifluoromethyl)hydroxylamines (Scheme 18) ⟨73JCS(P1)1092, 75JCS(D)2225, 75JCS(P1)2033, 80JFC(16)391, 81JCS(P1)455, 81JFC(17)331⟩.

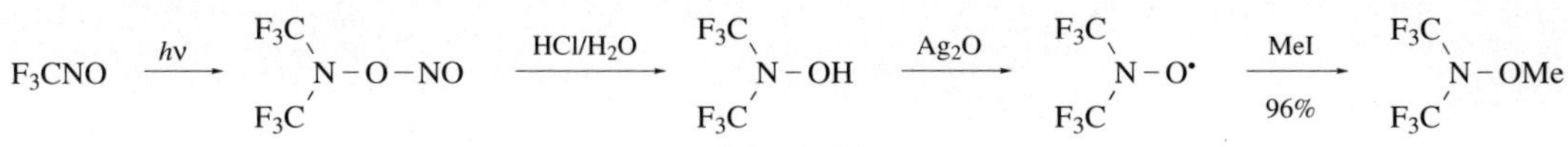

Scheme 17

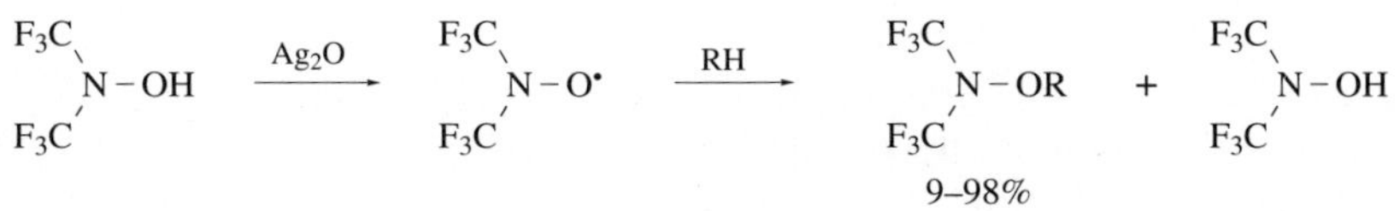

R = CMe_2, CH_2TMS, CH(Ph)CN, CH_2SMe, CH_2Ar

Scheme 18

N-(Trifluoromethyl)hydroxylamine (as its diethyl ether adduct) is formed in 48% yield when trifluoronitrosomethane is treated with potassium hydrogen sulfite in H_2O/Et_2O, followed by hydrolysis (Scheme 19) ⟨82JCS(P1)685⟩. Trifluoronitrosomethane reacts as an ene reaction component, for instance with propene with formation of *N*-(trifluoromethyl)-*N*-allylhydroxylamine in 90% yield (Equation (60)) ⟨80JCS(P1)1960⟩. *O*-Aryl-*N*,*N*-bis(trifluoromethyl)hydroxylamines are prepared in high yield (>80%) from the sodium salt of *N*,*N*-bis(trifluoromethyl)hydroxylamine and an aryl chloride ⟨81JFC(17)85, 81JFC(17)591⟩. *O*-Benzylated derivatives have been prepared in high yield (>80%) from benzyl chlorides and mercury(II) bis[bis(trifluoromethyl)aminoxide] ⟨73JCS(P1)1092⟩.

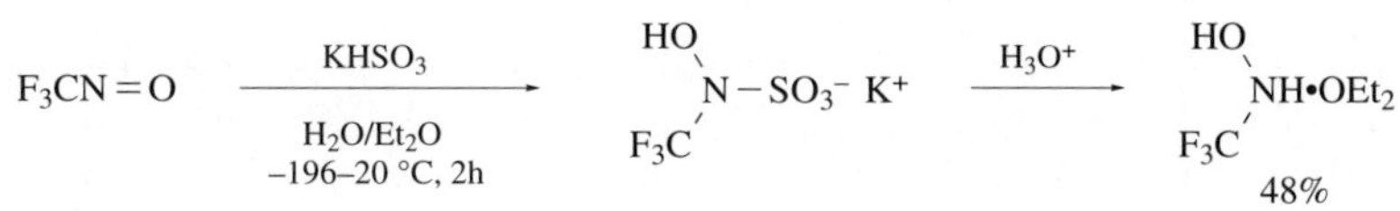

Scheme 19

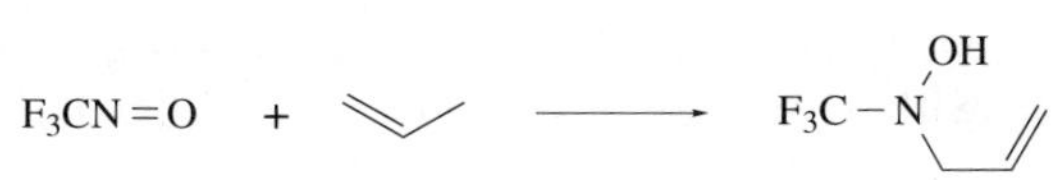 (60)

Trifluoronitrosomethane may be prepared by different methods, for example by free radical substitution of iodotrifluoromethane with NO/Hg/*hv* (89% yield) ⟨54JCS912⟩ or by the heating of $CF_3C(=O)ONO$ (190°C, 56% yield) ⟨62JOC1064⟩ or $CF_3C(=O)NHOH$ (63% yield) ⟨60DOK(132)602⟩. Trichloronitrosomethane has been prepared similarly by the heating of $CCl_3C(=O)NHOH$ (90–95°C, 62% yield) ⟨53JCS2075⟩ and by treatment of CCl_3SO_2Na with HNO_3 (56% yield) ⟨60DOK(132)602⟩.

Nitrotrifluoromethane is available in 49% yield by oxidation of the corresponding nitroso compound with Mn_2O_7 ⟨53JCS2075⟩. Trichloronitromethane (chloropicrin) which is commercially available and useful as an insecticide and as a synthetic building block, can be prepared in a number of ways including the chlorination of nitromethane with strongly alkaline aqueous sodium hypochlorite (90% yield) ⟨44USP2365981⟩.

(vi) N-*(Trihalomethyl)sulfenamides*, $CHal_3NR^1SR^2$

N,*N*-Bis(trifluoromethyl)alkanesulfenamides may be prepared by the reaction of mercury(II) bis(trifluoromethyl)amide with an alkanesulfenyl chloride. *N*,*N*-Bis(trifluoromethyl)methanesulfenamide has been prepared in 96% yield by this method ⟨66JINC1823⟩. Another method is the addition of *N*,*N*-bis(trifluoromethyl)-*S*-(chloro)thiohydroxylamine, formed by the reaction of *N*-chloro-*N*,*N*-bis(trifluoromethyl)amine with elemental sulfur ⟨64USP3121112⟩, to alkenes which leads to *N*,*N*-bis(trifluoromethyl)-2-chloroalkanesulfenamides (Scheme 20) ⟨81JFC(19)91⟩.

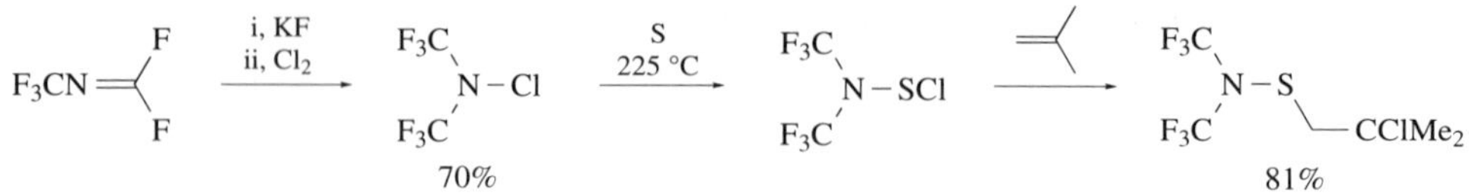

Scheme 20

(vii) Metal (trihalomethyl)amides, $CHal_3NRM$

(Trihalomethyl)amines are poor nucleophiles, but may be converted into stronger nucleophiles by conversion to metal (trihalomethyl)amides. The starting material for metal (trifluoromethyl)-amides is trifluoromethyl isocyanide difluoride ($CF_3N{=}CF_2$) which by reaction with HgF_2 or KF forms the corresponding metal amides ⟨58JA3604, 75IZV2279⟩.

(viii) Miscellaneous compounds

Trifluoronitrosomethane reacts with amines with formation of *N*-trifluoromethyldiazenes ($CF_3N{=}NR$) ⟨61DOK(141)357, 68ZOB709⟩. *N,N*-Bis(dibromomethylene)hydrazine ($Br_2C{=}NN{=}CBr_2$) reacts with silver(II) fluoride with formation of *N,N*-bis(trifluoromethyl)diazene ($CF_3N{=}NCF_3$) in 92% yield, which may be further transformed into the corresponding hydrazine ($CF_3NHNHCF_3$) by treatment with H_2S, HI or PH_3 ⟨62DOK(142)354, 62JA2337, 66JOC3833⟩.

6.07.2.2.2 Three halogens and a phosphorus function

Dialkyl- and diarylphosphines react with tetrachloromethane in the presence of triethylamine to yield the corresponding phosphines $(CCl_3)PR_2$ in 87–95% yield ⟨91PS(55)185⟩ and dichloromethylenephosphines $CCl_2{=}PR$ react with CCl_4 in ether in the presence of $(Et_2N)_3P$ to form bis(trichloromethyl)phosphines $(CCl_3)_2PR$ in 78–81% yield ⟨92ZOB948⟩. Trifluoromethylphosphaalkenes $(CF_3)P{=}CFOR$ have been prepared from the corresponding difluoromethylene compounds ⟨93ZN(B)58⟩. The iminomethylenephosphine $(CF_3)P{=}C{=}NBu^t$ has been prepared from $(CF_3)P{=}CF_2$ and *t*-butylamine and the cyclic diphosphane (**4**) is available from $(CF_3)P{=}CF_2$ and 2,3-dimethyl-2,4-butadiene ⟨91ZN(B)978⟩. Perfluoro-2-phosphapropene upon addition of an alcohol and subsequent treatment with a secondary amine yields the derivative $(CF_3)P{=}C(OR)NR_2$ (R = Me, Et) in 58% yield ⟨91HAC385⟩.

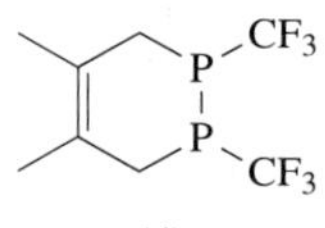

(4)

Trifluoromethanephosphonic acid can be obtained by hydrolysis of diiodo(trifluoromethyl)-phosphine ⟨54JCS3598⟩. With alcohols diiodo(trifluoromethyl)phosphine forms trifluoromethanephosphonic acid monoalkyl esters in 28–59% yield ⟨88ZOB1525⟩. Trimethyl phosphite reacts upon reflux with tetrachloromethane to form dimethyl trichloromethanephosphonate in 91% yield ⟨46ZOB1521⟩. This ester, when refluxed with concentrated hydrochloric acid, is quantitatively converted to trichloromethanephosphonic acid dihydrate ⟨55JA2869⟩.

Azidobis(trifluoromethyl)phosphine reacts with $(CF_3)_2P{-}N{=}PPh_3$ to form $(CF_3)_2P{-}N{=}P{-}(CF_3)_2{-}N{=}PPh_3$ ⟨93JOM(448)219⟩. Aminobis(trichloromethyl)phosphines such as $Me_2NP(CCl_3)_2$ can be oxidized with hydrogen peroxide to the corresponding phosphine oxides and with S_8 to the corresponding phosphine sulfides. With chlorine the corresponding chlorophosphonium chlorides are formed which are dealkylated upon heating to $MeN{=}P(CCl_3)_2$. Aminobis(trichloromethyl)phosphines can be cleaved with HCl to form $ClP(CCl_3)_2$ ⟨93ZOB1240⟩. Dibromo(trifluoromethyl)phosphine $(CF_3)PBr_2$ can be converted to the corresponding (trifluoromethyl)-phosphinylbistriazolide which is useful for the trifluoromethylphosphonation of carbohydrates ⟨93TL149⟩.

6.07.2.2.3 *Three halogens and an As, Sb, or Bi function*

Mixed difluorohalomethylarsines are accessible by treatment of arsenic(III) chloride, bromide or iodide, respectively, with difluorocarbene, generated from $(CF_3)_2Cd \cdot 2MeCN$. Arsenic(III) fluoride reacts with the same cadmium reagent with trifluoromethylation ⟨90ZAAC(588)26⟩. Tris(trifluoromethyl)stilbine ⟨90JFC(46)265⟩ and tris(trifluoromethyl)bismuthine ⟨87JOM(334)323⟩ are accessible from the appropriate element halides and $(CF_3)_2Cd$. Diphenyl(trifluoromethyl)bismuthine and phenylbis(trifluoromethyl)bismuthine have been prepared from the corresponding phenylhalobismuthines and $(CF_3)_2Cd$ ⟨91JOM(417)C47⟩. Bis(trifluoromethyl)chloroarsine yields the corresponding azide when treated with sodium azide ⟨92JST(268)389⟩ and azidobis(trifluoromethyl)arsine reacts with triphenylphosphine to give $(CF_3)_2As—N=PPh_3$ ⟨93JCS(D)663⟩.

6.07.2.3 Three Halogens and a Metalloid

6.07.2.3.1 *Three halogens and a silicon function*

Compounds with an (α- or β-fluoroalkyl)silane structure are prone to rearrangement with elimination of a fluorosilane R_3SiF ⟨59QR233, 64AOC(1)143⟩.

The construction of the CF_3SiR_3 moiety requires several steps such as insertion of difluorosilylene into the C—I bond of trifluoroiodomethane, metathesis of CF_3SiFI with SbF_3 to yield CF_3SiF_2, and subsequent reduction with LAH to trifluoromethylsilane ⟨86JOM(316)41⟩; also bis(trifluoromethyl)silane has been prepared ⟨90JOM(385)207⟩. A more convenient procedure starts from bromotrifluoromethane, tris(diethylamino)phosphine, and an appropriate silicon halide ⟨93OM4930⟩. Tetrakis(trichloromethyl)silane has been prepared in 64% yield by photochlorination of $(CH_2Cl)Me_3Si$. The corresponding chlorination of TMS leads to destruction of the reaction mixture ⟨89ZOB2628⟩.

A reagent prepared from tetrakis(dimethylamino)ethylene and CBr_3F reacts with organochlorosilanes R_nSiCl_{4-n} to give dibromofluoromethylsilanes $(CBr_2F)SiR_nCl_{3-n}$ in 27–54% yield ⟨93OM4930⟩.

6.07.2.3.2 *Three halogens and a boron function*

Many trifluoromethyl boranes are unstable due to the vacant orbital on boron, which causes halide migration from carbon to boron with formation of difluorocarbene. Oxygen or nitrogen(III) ligands as well as tetracoordination of the boron atom increase stability.

Dibutyl(trifluoromethyl)borane, $(CF_3)BBu_2$, is formed in the reaction between potassium dibutylborate(I) and trifluoroiodomethane. The former reacts with boron trifluoride to yield difluoro(trifluoromethyl)borane, CF_3BF_2, ⟨67JA3446⟩. Tris(trifluoromethyl)borane, $(CF_3)_3B$, is an extremely strong Lewis acid and can only be isolated in the shape of adducts with Lewis bases. For example, $(CF_3)_3B \cdot NR^1R^2R^3$ can be obtained from (dialkylamino)dichloroboranes $(R^1R^2N)BCl_2$ and bromotrifluoromethane/tris(diethylamino)phosphine or trifluoroiodomethane/tetrakis(dimethylamino)ethylene and subsequent alkylation of the secondary amine adducts so formed ⟨93JOM(446)25⟩.

In dichloromethane solution aminohaloboranes such as Et_2NBCl_2 and Et_2NBBr_2 react with $CF_3Br/P(NEt_2)_3$ to yield the corresponding $(CF_3)BClNR_2$, $(CF_3)BBrNR_2$, and $(CF_3)_2BNR_2$ ⟨89JOM(378)125⟩.

Dimethylaminobis(trifluoromethyl)borane $(CF_3)_2BNMe_2$ (**5**) inserts into oxiranes to form the corresponding oxaazoniaboratacyclopentanes (**6**) ⟨93ZN(B)935⟩. Compound (**5**) also undergoes ene reactions with alkenes ⟨93JOM(456)19⟩. With diazoalkanes (**5**) forms the corresponding three-membered rings (**7**) ⟨93AG429⟩.

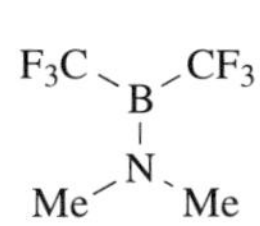

(**5**)

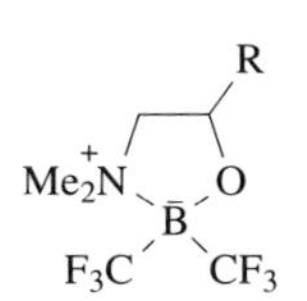

R = Me, CH_2F, CF_3, Et, Bn
(**6**)

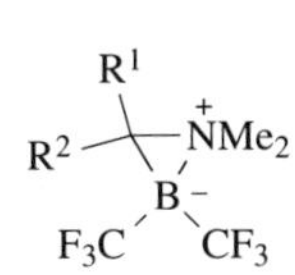

R^1, R^2 = H, Bn, TMS
(**7**)

Hexamethyldistannane reacts metathetically with trifluoroiodomethane to form (trifluoromethyl)trimethylstannane. The latter reacts further with boron trifluoride to generate $[Me_3Sn]^+$ $[(CF_3)BF_3]^-$ which in turn can be converted to the corresponding potassium salt ⟨60JA5298⟩.

6.07.2.3.3 Three halogens and a germanium function

Germanium(II) iodide has been treated with trifluoroiodomethane in an autoclave at 130–135°C and found to give triiodo(trifluoromethyl)germane, $(CF_3)GeI_3$, and diiodobis(trifluoromethyl) germane, $(CF_3)_2GeI_2$. The iodine atoms of the former can be exchanged for chlorine by treatment with silver(I) chloride with formation of $(CF_3)GeCl_3$ ⟨62JA898⟩. The compounds $(CF_3)GeF_3$ and $K_2[(CF_3)GeF_5]$ have been similarly prepared ⟨60JA6228⟩.

Atomic germanium in a solvent slurry reacts with tetrachloromethane to form $(CCl_3)GeCl_3$ and with bromotrichloromethane to form $(CCl_3)GeBr_3$ in low yield ⟨88BCJ3002⟩.

6.07.2.4 Three Halogens and a Metal Function

The chemistry of (trifluoroalkyl)metal compounds has previously been presented in a textbook by Emeléus ⟨B-69MI 607-02⟩ and in reviews by Trichel and Stone ⟨64AOC(1)143⟩. The synthetic methods for the preparation of (perfluoroorgano)metallic compounds have been reviewed by Chambers ⟨B-73MI 607-01⟩, and trifluoromethyl-containing transition metal complexes by Morrison ⟨93AOC(35)211⟩.

Of the different (trifluoromethyl)metal complexes prepared to date, donor [1,2-dimethoxyethanel(glyme) or DMF] stabilized bis(trifluoromethyl)cadmium seems to be the most reliable and efficient trifluoromethylating agent. Other (trifluoromethyl)metal complexes have been prepared, for which some application in the electronic industry may be found, such as the stable gold complexes.

(i) Trihalomethyl alkali and earth alkali metals, $CHal_3M$ ($M = Li, Na, K, Cs, Mg$)

The preparation of α-haloalkyl Grignard reagents has been reviewed by Villiéras ⟨71MI 607-01⟩. The trihalomethyl alkali and earth alkali metals are very difficult to isolate due to the dihalocarbene character of the carbon. They are therefore most often prepared *in situ* and used without isolation. In the case of (trichloromethyl)lithium the position of the equilibrium at −100°C (Equation (61)) lies to the side of (trichloromethyl)lithium ⟨75JA187⟩.

$$Cl_3CLi \rightleftharpoons :CCl_2 + LiCl \qquad (61)$$

Ligand exchange has been attempted in the synthesis of (trifluoromethyl)lithium and -magnesium. When trifluoroiodomethane is treated with methyllithium in diethyl ether, (trifluoromethyl)lithium is formed initially, but this complex decomposes into lithium fluoride and difluorocarbene which dimerizes. This is also the case with magnesium ⟨54JA474, 64AOC(1)143⟩.

According to Köbrich, lithiation of 2,2-diphenyl-1-bromoethene with (dichloromethyl)lithium and subsequent treatment with tetrachloromethane gives (trichloromethyl)lithium and 2,2-diphenyl-1,1-dichloroethene. (Trichloromethyl)lithium has also been prepared *in situ* from butyllithium and tetrachloromethane or trichloromethane and then treated with a number of electrophiles by Hoeg *et al.* ⟨65JA4147⟩. (Trichloromethyl)lithium, -sodium, -potassium and -caesium have, furthermore, been prepared and investigated by IR in an Ar/CCl_4 matrix at 15 K ⟨75JA187⟩.

(Tribromomethyl)lithium should be available by the reaction of butyllithium with tetrabromomethane. No data have, however, been presented ⟨67AG15⟩.

(Trichloromethyl)magnesium chloride has been prepared from isopropylmagnesium chloride and CCl_4 at −115°C in THF or with $CHCl_3$ at −78°C in THF/hexamethylphosphoric triamide (HMPT) (80/20) in 60% and 70% yield, respectively. (Tribromomethyl)magnesium chloride was prepared similarly with CBr_4 at −115°C in THF or with $CHBr_3$ at −95°C in THF/HMPT (80/20) in 50% yield ⟨67BSF1520⟩.

(ii) Trihalomethylaluminum, -gallium, -indium and -thallium, $CHal_3MR_n$, ($M = Al, Ga, In, Tl$) compounds

Tris(trifluoromethyl)gallane has been prepared in 36% yield by Guerra *et al.* by treating gallium tribromide with a slight excess of $Cd(CF_3)_2 \cdot$glyme in CH_2Cl_2. Further treatment with trimethylphosphine or trimethylarsine leads to $[(CF_3)_3GaPMe_3]$ and $[(CF_3)_3GaAsMe_3]$ in quantitative yields ⟨90JOM(390)C73⟩. Naumann *et al.* were able to prepare the following trifluoromethyl group 13 metal complexes $[Ga(CF_3)_2Cl \cdot DMF]$ (in 39% yield), $[Ga(CF_3)_3 \cdot DMF]$ (42%), $[Cd(NCMe)_2][Ga(CF_3)_4]$ (49%), $[In(CF_3)_2Cl \cdot DMF]$ (32%), $[In(CF_3)_3 \cdot 2NCMe]$ (36%) and $[Tl(CF_3)_3 \cdot 2DMF]$ (45%) from $[Cd(CF_3)_2 \cdot D$, D = glyme, diglyme, MeCN] and $GaCl_3$, $InCl_3$ or $TlCl_3$. The reactions were conducted in CH_2Cl_2, MeCN or DMF at −30 to 40°C ⟨91JOM(407)1⟩.

(iii) Trihalomethyltin and -lead, $CHal_3MR_n$ ($M = Sn(II), Sn(IV), Pb(IV)$) compounds

Trifluoromethyl(trimethyl)stannane has been prepared in 81% yield by heating at 80°C for 20 h or irradiation of a mixture of trifluoroiodomethane and hexamethyldistannane ($Me_3SnSnMe_3$). Chlorination of (trifluoromethyl)trimethylstannane leads to (trifluoromethyl)dimethyltin chloride (Scheme 21) ⟨58JOC1565, 60JA1888⟩.

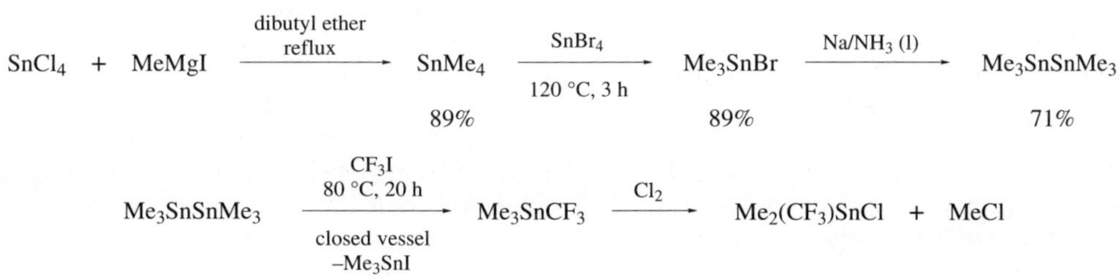

Scheme 21

The (trifluoromethyl)tin halides have been prepared. Thus, when tin tetrabromide is heated for 2 × 15 h at 110°C in an ampoule with $Hg(CF_3)_2$, $(CF_3)_3SnBr$ (8%), $(CF_3)_2SnBr_2$ (16%), and CF_3SnBr_3 (55%) are formed, whereas when $(CF_3)_2Cd \cdot$diglyme is used as trifluoromethylating reagent $(CF_3)_4Sn$ (45%) and CF_3SnBr_3 (10–25%) are collected ⟨92JOM(433)63⟩. Tris(trifluoromethyl)tin iodide may be prepared from tetrakis(trifluoromethyl)stannane; either by treatment with HI or BrI_3 (giving 50–80% yield of $(CF_3)_3SnI$ and 20–50% yield of $(CF_3)_2SnI_2$). (Trifluoromethyl)tin triiodide is formed similarly by HI treatment of trifluoromethyltin tribromide in 94% yield. When these products are treated with AgCl for 3 h at 50°C, the corresponding chlorides are formed in 95% yield. Tris(trifluoromethyl)tin fluoride is formed upon heating (3 h at 100°C) of tetrakis-(trifluoromethyl)stannane ⟨92JOM(433)63⟩.

Mixed (trifluoromethyl)methylstannanes $[(CF_3)_nSnMe_{4-n}]$ have been prepared by ligand exchange between tetramethylstannane and tetrakis(trifluoromethyl)stannane ⟨92JOM(433)63⟩.

Mono-, bis- and tris(trifluoromethyl)stannane may be prepared by reduction of $(CF_3)_nSnX_{4-n}$ with tributylstannane at −50°C for 2 h. The yields were 89%, 92% and 45%, respectively ⟨92JOM(434)159⟩.

A number of (trifluoromethyl)lead complexes have been prepared in a stepwise manner by Eujen and Patorra ⟨92JOM(438)57⟩. Bis(trifluoromethyl)cadmium reacts with trimethyl- and triethyllead bromide in sulfolane at 70°C to give $(CF_3)PbMe_3$ (51% yield) and $(CF_3)PbEt_3$ (56%). Bromination (Br_2) at 0°C or iodination (I_2) at RT gave the corresponding (trifluoromethyl)dialkyllead halides (alkyl = methyl or ethyl) in quantitative yields. The (trifluoromethyl)dialkyllead bromides may be transformed into the bis(trifluoromethyl)dialkyllead complexes ($(CF_3)_2PbMe_2$ (47% yield) and $(CF_3)_2PbEt_2$ (50%)) by treatment with $Cd(CF_3)_2$. Bis(trifluoromethyl)methyllead bromide was then prepared in 75% yield, by bromination at −25°C (72 h) of bis(trifluoromethyl)dimethyllead, which by further reaction with $Cd(CF_3)_2$ gave tris(trifluoromethyl)methyllead in 29% yield. Tetrakis(trifluoromethyl)lead has not been prepared and must be expected to be rather unstable like the analogous tetrahalolead complexes ⟨92JOM(438)57⟩.

(iv) Trihalomethylzinc, -cadmium and -mercury, $CHal_3MR$ ($M = Zn, Cd, Hg(II)$) compounds

Miller *et al.* failed to prepare trihalomethylzinc iodides or bromides from zinc and trihaloiodomethanes or trihalobromomethanes in glyme ⟨57JA4159⟩ whereas Burton and Wiemers did so by treating activated (acid washed) zinc or cadmium with dibromodifluoromethane or bromochlorodifluoromethane in DMF for 2 h at RT. When dichlorodifluoromethane was used, the reaction had to be carried out at 80°C in a sealed ampoule. A yield of 80–95% of (trifluoromethyl)cadmium halide and a yield of 80–85% of (trifluoromethyl)zinc halide was achieved. The (trifluoromethyl)metal halides were not isolated but stored for later use as 1 M solutions in DMF ⟨85JA5014⟩.

Bis(trifluoromethyl)cadmium (11% yield, 99% purity) and bis(trifluoromethyl)zinc (8% yield) were prepared at low temperature (<45°C) by the reaction of the metal with hexafluoroethane in a radio frequency discharge. The "naked", that is nondonor stabilized, compounds $Cd(CF_3)_2$ and $Zn(CF_3)_2$ decompose at 0°C and -40°C, respectively, but may be stabilized with glyme [$Cd(CF_3)_2$·glyme] and pyridine [$Zn(CF_3)_2$·2pyridine] ⟨86JA4103, 91CJC327⟩. The bis(trifluoromethyl)cadmium glyme adduct may also be prepared from dimethylcadmium and bis(trifluoromethyl)mercury in glyme ⟨80CC671, 81JA2995, 86JA832⟩.

Alkali metal tetrakis(trifluoromethyl)cadmates have been prepared and investigated by NMR, but not purified ⟨89JOM(368)131⟩.

The formation of (trifluoromethyl)mercury iodide is accomplished photochemically in 80% yield by treatment of trifluoroiodomethane with mercury at 150°C ⟨48JCS2188⟩. Further treatment of this isolatable intermediate with cadmium amalgam at 120–130°C causes the formation of bis(trifluoromethyl)mercury in 80–90% yield ⟨49JCS2953⟩. Bis(trifluoromethyl)mercury is also formed from mercury(II) oxide and tris(trifluoromethyl)phosphine upon 22 h heating at 100°C in a sealed tube (84% yield) ⟨60JA5759⟩. (Trichloromethyl)mercury(II) chloride and bis(trichloromethyl)mercury have been prepared, by halide substitution, from mercury(II) chloride or mercuric bromide with sodium trichloroacetate. The intermediate mercury(II) trichloroacetate decomposes into the trichloromethyl derivative with elimination of carbon dioxide. The reaction is conducted in diglyme at reflux temperature for 1 h with yields of 69–77%. The actual ratio of sodium trichloroacetate to mercury(II) chloride determines whether the product is bis(trichloromethyl)mercury or the (trichloromethyl) mercury(II) halide ⟨63JOC1129⟩.

(v) Trihalomethylcopper, -silver and -gold, $CHal_3MR_n$ ($M = Cu(I), Ag(I), Au(I), Au(III)$) compounds

The trifluoromethyl complexes of group 11 may be arranged into the following order of stability according to Nair and Morrison: $AuCF_3 > AgCF_3 > CuCF_3$ ⟨89JOM(376)149⟩.

(Trifluoromethyl)copper(I) is probably formed initially in some coupling reactions, for example, when trifluoroiodomethane is heated (130–140°C) with metallic copper in DMF. This intermediate is trapped by aryl iodides with formation of the corresponding trifluoromethylarenes ⟨69TL4095, 70CPB2334⟩. This method has later been improved by Kobayashi *et al.* to include the corresponding reaction with aliphatic halides. They shook trifluoroiodomethane and copper powder in HMPA at 120°C for 2.5 h. After removal of excess copper powder the aliphatic halide was added and the mixture stirred at RT or 70°C under N_2 atmosphere. They were able to trifluoromethylate aliphatic and vinylic halides in 13–81% yield by the (trifluoromethyl)copper(II) iodide formed *in situ* ⟨79TL4071⟩.

The (trifluoromethyl)silver(I) trimethylphosphine complex has been prepared from silver acetate and $Cd(CF_3)_2$·glyme in Et_2O; after 10 min trimethylphosphine was condensed into the reaction vessel. The (trifluoromethyl)silver(I) trimethylphosphine complex, [$(CF_3)AgPMe_3$] was isolated in 36% yield ⟨89JOM(376)149⟩.

(Trifluoromethyl)gold(I) complexes stabilized (air and moisture insensitive) by trimethyl-, triethyl- or triphenylphosphine ($(CF_3)AuL$, $L = PMe_3$, PEt_3 or PPh_3) have been prepared in 65–75% yield from ClAuL ($L = PMe_3$, PEt_3 or PPh_3) and $Cd(CF_3)_2$·glyme in CH_2Cl_2. Bis(trifluoromethyl)mercury was ineffective as a trifluoromethylating agent, even at 150°C ⟨89JOM(376)149, 89OM1498⟩. These gold(I) complexes are oxidized by elemental bromine or iodine with formation of phosphine stabilized (trifluoromethyl)gold(III) dihalides in ~35% yield ($(CF_3)AuX_2L$, $L = PMe_3$, PEt_3 or PPh_3). The analogous addition of trifluoroiodomethane to $(CF_3)AuL$ ($L = PMe_3$ or PEt_3) in CH_2Cl_2 leads to bis(trifluoromethyl)gold(III) iodide in 80% yield (98% *cis*, 2% *trans*) ⟨89JOM(376)149, 89OM1498⟩. The unstable tris(trifluoromethyl)gold(III) complex has been prepared by Guerra *et al.*, from gold vapors and plasma-generated trifluoromethyl radicals at -196°C ⟨86JOM(307)C58⟩, and may also be obtained as the trimethylphosphine stabilized analog [$(CF_3)_3Au(PMe_3)$] by reaction of $(CF_3)_2$-$AuI(PMe_3)$ with $Cd(CF_3)_2$·glyme in the presence of excess trifluoroiodomethane in 80% yield

⟨89OM1498⟩. The organogold(I) complex $[(CF_3)Au(^-C{\equiv}N^+{-}Me)]$ was prepared similarly by Dryden *et al.* in 78% yield ⟨92CM979⟩. Nair and Morrison made $[(CF_3)_3Au(PEt_3)]$ in 20–50% yield from CF_3I and $[(CF_3)Au(PEt_3)]$ ⟨89JOM(376)149⟩.

Bis(trifluoromethyl)gold(III) μ-halide dimers $[Au(CF_3)_2(\mu\text{-}X)]_2$ have also been prepared, by a vapor deposition technique, from gold treated with excess (1 : 100) of either bromotrifluoromethane or trifluoroiodomethane. The products may be recrystallized from pentane and are moderately air and light sensitive. The μ-bromide was isolated in 16% yield wheras the μ-iodide was isolated in 8% yield ⟨90IC3252⟩.

(vi) Miscellaneous trihalomethyl transition metal, $CHal_3MR_n$ (M = transition metal) compounds

A number of platinum complexes containing trifluoromethyl as a ligand have been prepared by ligand exchange from $Pt(CF_3)_nMe_{2-n}NBD$ (NBD, norbornadiene) ⟨93JOM(453)307⟩. This complex is itself made from $PtMe_2NBD$ by trifluoromethyl/methyl exchange with CF_3I ⟨88JOM(342)399, 93JOM(453)299⟩. Trifluoromethyl/methyl exchange has also been used to prepare other platinum complexes containing trifluoromethyl as a ligand ⟨90IC2496⟩.

Trifluoroiodomethane adds to *cis*-$PtMe_2L_2$ ($L = PMe_2Ph$) at 180°C under vacuum to give $PtMe_2L_2(CF_3)I$ with CF_3 and I in axial positions ⟨70IC2556⟩. Hydrido trifluoromethyl complexes of platinum(II) have been prepared by Michelin *et al.* ⟨90OM1449, 89JCS(D)1149⟩.

Other transition metal complexes with trifluoromethyl as ligand have been prepared. These include rhodium ⟨90JOM(388)391, 91CC165⟩, ruthenium ⟨90JOM(395)327, 90JOM(397)209⟩, osmium ⟨90JOM(395)327⟩, iridium ⟨90JOM(394)615⟩, manganese ⟨61JA1598⟩ and cobalt complexes ⟨61JA3593⟩.

6.08
Functions Containing Two Halogens and Two Other Heteroatom Substituents

ANASTASSIOS VARVOGLIS
University of Thessaloniki, Greece

6.08.1 INTRODUCTION

A great variety of functionalities characterize the family of compounds containing functions with two halogens and two other heteroatom substituents; however, the family has few representatives. Heteroatom substituents include O, S, Se, Te, B, N, P, As, Si, Ge, and some metals, and the halogen atoms are F and Cl in most cases. Several members of the family stand on the verge of organic and inorganic chemistry, particularly when relatively small molecules are involved without hydrogen atoms. It is emphasized that the experimental work with elemental fluorine requires special conditions and considerable expertise.

6.08.2 TWO HALOGENS AND TWO CHALCOGEN FUNCTIONS

6.08.2.1 Two Halogens and Two Oxygen Functions

6.08.2.1.1 Difluoro compounds

Addition of fluorine to the carbonyl bond of simple perfluorooxo compounds leads to the formation of either stable adducts (*gem*-fluorohypofluorites) or to products of further transformation. In the case of FC(O)OF (**1**), both types of product may be formed. Thus, the simplest compound of this class, difluorodioxirane (**4**), has been obtained from (**1**) and either fluorine or chlorine fluoride in a flow system using a pelletized caesium fluoride catalyst. After the initial attack of fluoride on carbon, an electron transfer follows between (**2**) and the halogen; the free radical intermediate (**3**) cyclizes to (**4**), or may produce (**5**) when ClF is used (Scheme 1) ⟨93AG(E)905⟩. Bis(fluoroxy)difluoromethane was the unique product obtained from the reaction between FC(O)OF and fluorine, again in the presence of caesium fluoride, in a Monel bomb (Equation (1)) ⟨67JA5161⟩. The same product was also formed from the double addition of fluorine to CO_2, catalyzed by caesium fluoride ⟨67JA1809, 67JA1962⟩. A variation involves the action of fluorine upon sodium trifluoroacetate or sodium oxalate ⟨67JA1811⟩. Another hypofluorite, the peroxide (**7**), was produced by the addition of fluorine to the carbonyl bonds of (**6**) (Equation (2)) ⟨67CC870⟩. In some esters of thionocarbonic acid the C=S group was directly transformed into CF_2 upon treatment with fluorine; in this way, compounds of the type $(RCH_2O)_2CF_2$ were formed (where R is a polynitromethyl group) ⟨89IZV113, 93IZV2513⟩. Similar compounds were obtained by substitution from the corresponding CCl_2 analogues and SbF_5 in SO_2Cl_2.

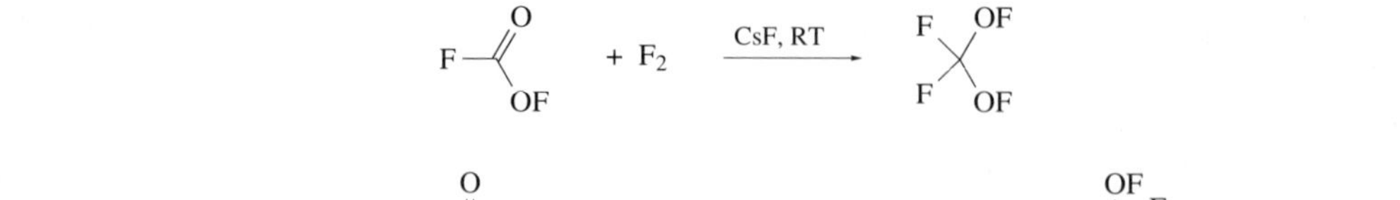

Scheme 1

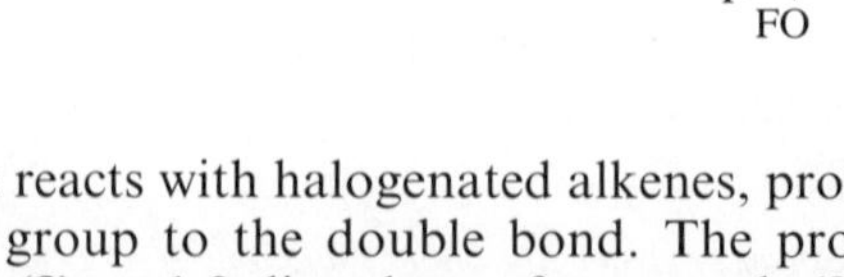

(1)

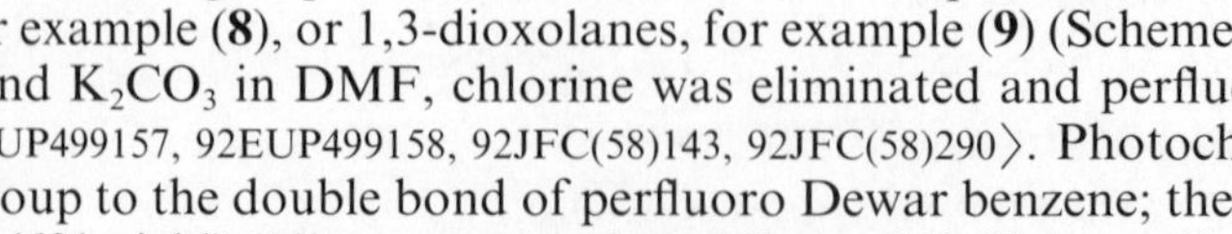

(2)

Bis(fluoroxy)difluoromethane reacts with halogenated alkenes, probably through the free radical (**3**) which transfers its OCF_2O group to the double bond. The products are either halogenated acetals of methanal, for example (**8**), or 1,3-dioxolanes, for example (**9**) (Scheme 2). Upon treatment of the latter with Zn and K_2CO_3 in DMF, chlorine was eliminated and perfluoro-1,3-dioxole was produced ⟨68IC624, 92EUP499157, 92EUP499158, 92JFC(58)143, 92JFC(58)290⟩. Photochemically, $F_2C(OF)_2$ also adds its OCF_2O group to the double bond of perfluoro Dewar benzene; the reaction is complex and the major product (6% yield) (**10**) rearranges thermally to (**11**) (Scheme 3). ⟨79JOC2813⟩. Cyclic thionocarbonates have been converted into 2,2-difluoro-1,3-dioxolanes by fluorinolysis of the C=S function using $Bu_4NH_2F_3$ and *N*-iodosuccinimide ⟨94SL251⟩.

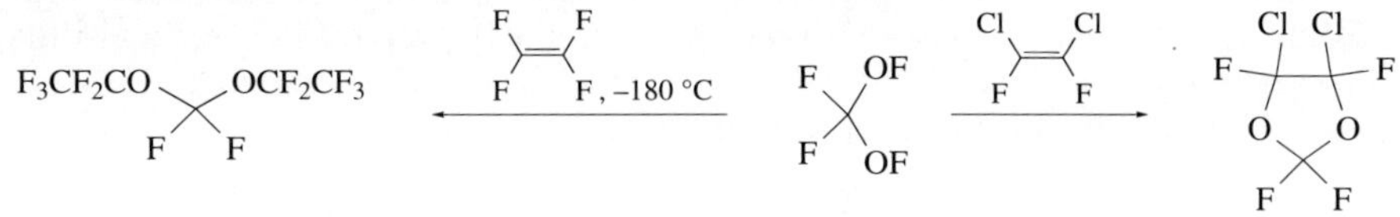

(8) 90% (9)

Scheme 2

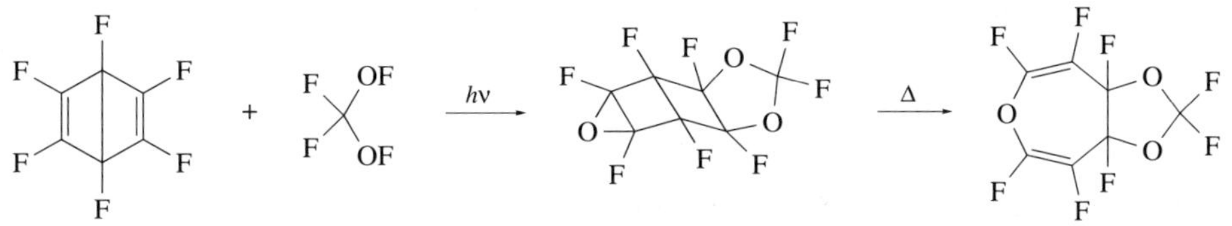

(10) (11)

Scheme 3

Some difluorodialkoxymethanes are produced in a different way, after halogenolysis of the double bond of either $F_3CN{=}C(OR)_2$ or $F_5SN{=}C(OR)_2$ by chlorine fluoride (Equation (3)) ⟨90JFC(48)395⟩. In some cases fluorine can substitute for hydrogen attached to an sp^3 carbon. Perfluorodimethoxymethane and perfluoro-1,3-dioxane—as well as other products—have been reported to result from electrochemical fluorination of methyl 3-methoxypropionate ⟨76ZOR767⟩. The physical properties of such compounds have been described but their preparation has not yet been documented ⟨92MI 608-01, 93MI 608-01⟩. Chemical and electrochemical methods have been used to fluorinate the methanal derivative (**12**) to (**13**) (Equation (4)) ⟨91JFC(55)313⟩.

(3)

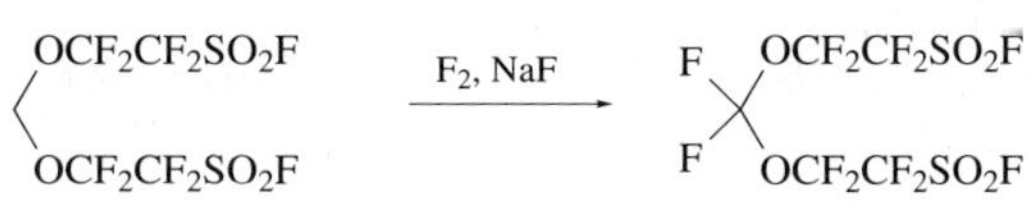

(4)

(12) (13)

The action of fluorine on polymethylene oxide gives mainly a perfluoropolymer, but it also brings about fluorinolysis, with partial fluorination and depolymerization; among the products identified were $F_3COCF_2OCF_2H$ and $HF_2COCF_2OCF_2H$ ⟨81JCS(P1)1321⟩. Chlorine in F_2CCl_2 is substituted by two fluorosulfato groups in its reaction with $ClOSO_2F$; F_2CBr_2 reacts similarly with the peroxy compound FO_2SOOSO_2F. In both cases the product is $F_2C(OSO_2F)_2$ ⟨65IC1828, 74IZV335⟩. Ozonation of tetrafluoroethylene gave rise to an apparently stable moleozonide ⟨68CI(M)197⟩. Singlet oxygen reacted with tetrafluoroethylene at $-50\,^{\circ}C$ to produce peroxidic perfluoroethers, which upon strong and prolonged heating were converted into a mixture containing, *inter alia*, perfluoro-1,3-dioxetane, perfluoro-1,3,5-trioxane, and perfluoro-1,3,5,7-tetraoxocane ⟨72GEP2111696⟩. In another patent, the reaction of the adduct from the reaction between $(F_2N)_2C{=}NF$ and KCN with $F_2C(OF)_2$ is claimed to give a mixture of products, including the peroxide F_3COOCF_2OF ⟨71USP3585218⟩. The unusual compound $F_3COOCF_2OOOCF_3$, bearing both peroxide and trioxide functions, was formed among other products from a complex reaction between $F_2C(OF)_2$ and $CsOCF_3$ ⟨71IC2179⟩.

6.08.2.1.2 Dichloro and dibromo compounds

The number of compounds in this category is few. The compounds can be obtained either by carbonyl (or thiocarbonyl) transformation or by chlorination of a methylene group. Dichlorodiphenoxymethane (**14**) was produced upon heating diphenyl carbonate with PCl_5; similarly, phenylene carbonate gave 2,2-dichloro-benzo-1,3-dioxole (**15**) ⟨61CB544, 82JHC1205⟩. Some diethyl thionocarbonates of the general formula $(RCH_2O)_2C{=}S$ (where R is CF_3, CF_2NO_2 or $CF(NO_2)_2$) were converted into the corresponding CCl_2 compounds using either SO_2Cl_2 or Cl_2 ⟨89IZV113⟩. Photochemical chlorination of 1,3-dioxolane converted it into its perchloro analogue **(16)** ⟨72JOC1458⟩. Some dichlorobis(aryloxy)methanes were similarly prepared via radical-induced chlorination of the formaldehyde diaryl acetals with SO_2Cl_2 or Cl_2 ⟨88S961⟩. Dibromomethylal, $CBr_2(OMe)_2$, was obtained by simply dropping Br_2 into cooled methylal ⟨22AG489⟩.

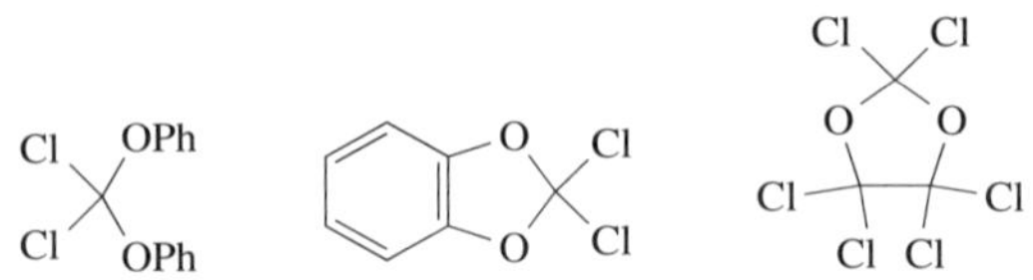

(14) **(15)** **(16)**

6.08.2.2 Two Halogens and Two Sulfur Functions

6.08.2.2.1 Difluoro compounds

The reaction of fluorine with thiocarbonyl compounds usually gives addition products in which sulfur has also been oxidized. Carbon disulfide is thus converted into the isolable compound **(17)** and subsequently into **(18)** (Scheme 4) ⟨77IC2974⟩. The yield of **(18)** increased when caesium fluoride was used as a catalyst; chlorine fluoride reacts in a similar way, leading to the formation of (*trans*-$ClSF_4)_2CF_2$ ⟨83JFC(23)325⟩. The pseudohalogen CF_3SF was also added to carbon disulfide; no oxidation occurred and the double adduct $(CF_3SS)_2CF_2$ was obtained in high yield ⟨92ZN(B)369⟩. The related reaction of $F_2C{=}S$ with CF_3SCl gave the adduct $(F_3CS)CF_2SCl$ ⟨68CB2617⟩. The same product was formed upon addition of ClF to $F_3CSC(S)F$, while a related addition of ClF occurred to the C=S bond of FC(S)SCN ⟨74CZ109⟩. Compound **(17)** reacts with metal fluorides, affording relatively stable anions or cations. For example, with CsF a symmetrically bridged anion was formed, while with AsF_5 in liquid SO_2 the stable sulfonium salt $F_2C(SF_3)SF_2{}^+AsF_6{}^-$ resulted ⟨91CB1353, 92CC1017⟩. When BF_3 in SO_2 was used, **(17)** afforded the solvolysis product **(19)**, as a mixture of two diastereoisomers. With silylated amines, **(19)** was converted into sulfinimides, for example **(20)** or **(21)** (Scheme 5). From the reaction between **(19)** and HCl, the rather unstable $F_2C(S(O)Cl)_2$ was produced, which upon hydrolysis gave the bis(sulfinic acid) $F_2C(SO_2H)_2$. Direct fluorination at an sp^3 carbon has been effected in some cases, for example with **(22)** which was converted into **(23)**; this was hydrolyzed to the sulfonic acid **(25)** through its salt **(24)** (Scheme 6) ⟨90IC4588⟩.

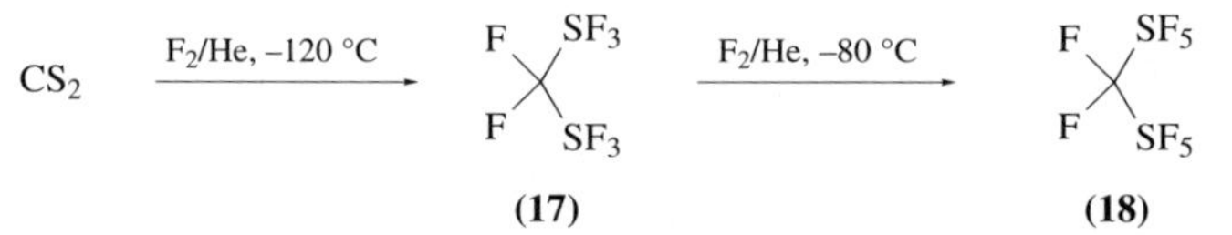

(17) **(18)**

Scheme 4

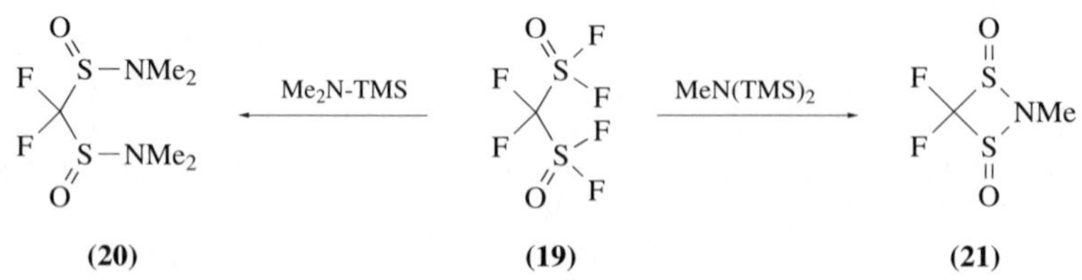

(20) **(19)** **(21)**

Scheme 5

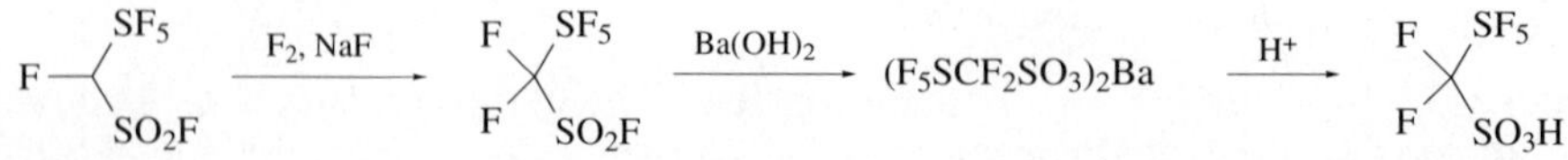

(22) **(23)** **(24)** **(25)**

Scheme 6

Electrochemical methods have been used for the preparation of (**18**) and $F_2C(SO_2F)_2$ ⟨59JA574, 86JFC(32)89⟩. While the latter was formed from $CH_2(SO_2F)_2$ in 80% yield, the former was one of many products into which 1,3,5-trithiane was transformed, for example, $F_3CSF_4CF_2SF_5$, and the trithianic derivative $(CF_2SF_4)_3$. A different approach was used for the preparation of the sulfone, $PhSO_2CF_2SPh$, which was obtained from the reaction of $PhSO_2CHF_2$ and aqueous NaOH in a two-phase system (CH_2Cl_2 with Aliquat 336); the sulfone was oxidized with H_2O_2 to $F_2C(SO_2Ph)_2$ ⟨89JFC(43)53⟩.

A fairly large family of perhalogenated 1,3-dithietanes and their sulfur derivatives is known. Photochemical dimerization of thiophosgene leads in a straightforward way to the formation of (**26**) ⟨33CB567⟩. Upon treatment with SbF_5, this gives rise mainly to compound (**27**), along with some mono- and dichloroperfluoro-1,3-dithietanes (Scheme 7) ⟨65JOC1375⟩. A related method started with compound (**28**), which was converted by strong acids into the isolable salt (**29**); treatment with caesium fluoride afforded the trifluorodithietane (**30**) (Scheme 8) ⟨87CB429⟩. A similar compound to (**29**) is the salt (**31**), which can undergo a variety of transformations at its carbocationic center, as shown in Scheme 9 ⟨90CB1635⟩.

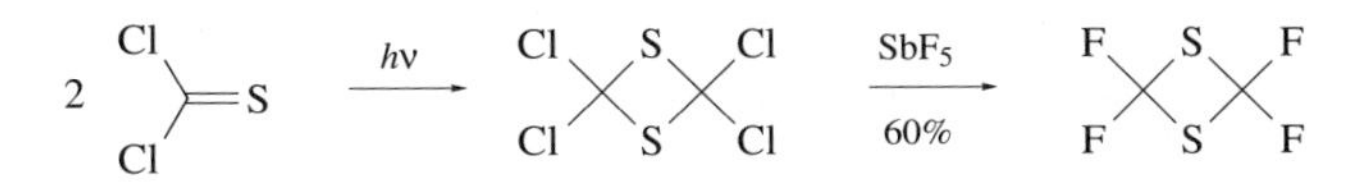

(**26**) (**27**)

Scheme 7

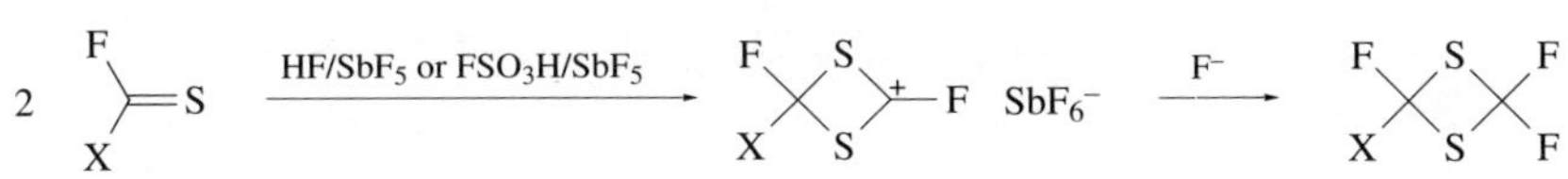

(**28**) X = Cl or SCF3 (**29**) (**30**)

Scheme 8

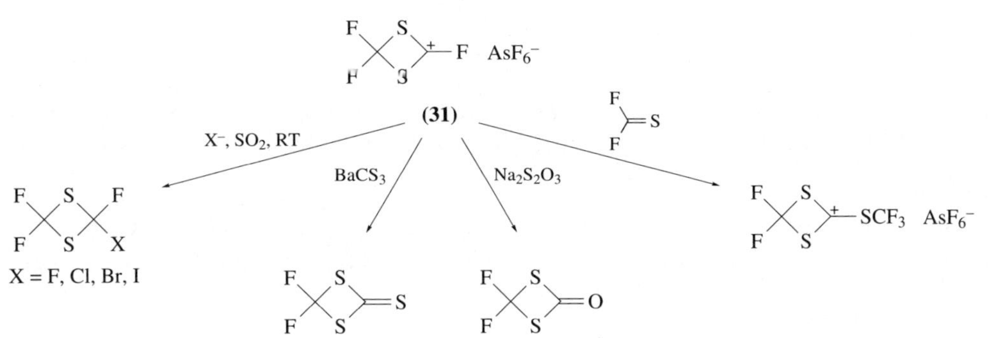

Scheme 9

The majority of the reactions of perfluoro-1,3-dithietane involve oxidation at the sulfur. Depending on the oxidant and the conditions, several products can be obtained using fluorine or xenon difluoride ⟨84JFC(26)359, 89JFC(45)225⟩; chlorine fluoride ⟨73JFC(3)17, 92CB535⟩; chromium trioxide ⟨80AG(E)203⟩; trifluoromethyl hypochlorite or perfluoro-*t*-butyl hypochlorite ⟨77JA4194, 78IC2173, 79JA5949, 81JA406, 86IC275⟩; and trifluoromethaneperoxysulfonic acid ⟨83CB1623⟩. Examples of the oxidation products are the structures (**32**)–(**35**). When (**35**) was heated with chlorine fluoride, it was transformed into the open chain compound $F_3CSO_2CF_2SF_4Cl$ ⟨92CB535⟩.

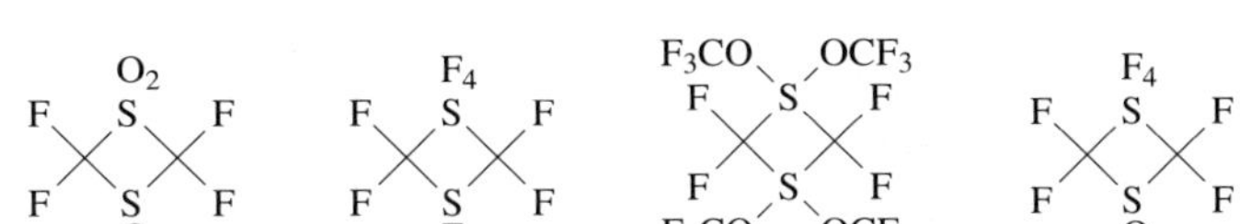

(**32**) (**33**) (**34**) (**35**)

6.08.2.2.2 Dichloro, dibromo, and diiodo compounds

Halogenated 1,3-dithietanes result from the photochemical dimerization of chlorofluorothiophosgene, chlorobromothiophosgene, and dibromothiophosgene ⟨76CB3432, 77CB916, 87CB1499⟩. Tetrabromo-1,3-dithietane was also obtained from its tetrachloro analogue by substitution using BBr_3, in 78% yield ⟨77CB916⟩. These compounds, along with tetrachloro-1,3-dithietane undergo various reactions at either carbon or sulfur, in much the same way as tetrafluoro-1,3-dithietane ⟨33CB567, 77CB916, 83CB1623, 87SUL59, 88JFC(39)329⟩. Some transformations of the parent disulfone 1,1,3,3-tetraoxo-1,3-dithietane (**36**) are of interest: it can be halogenated by aqueous chlorine or bromine, affording tetrachloro or tetrabromo disulfones in high yields (an analogous iodination does not occur). Chlorination was also effected using $C_4F_9SO_2Cl$ and Et_3N. The tetrachlorodisulfone (**37**), upon mild hydrolysis, was converted quantitatively into the sulfonic acid (**38**) (Scheme 10). Trichloro disulfone (**40**) was obtained from the silylated disulfone (**39**) upon treatment with butyllithium and $C_4F_9SO_2Cl/Et_3N$ (Equation (5)) ⟨83CB1285, 85CB2208⟩. Treatment of dibromothiophosgene ($Br_2C{=}S$) with ozone at $-80\,°C$ resulted in its partial trimerization; hexabromo-1,3,5-trithiane was formed in 5% yield ⟨77CB916⟩.

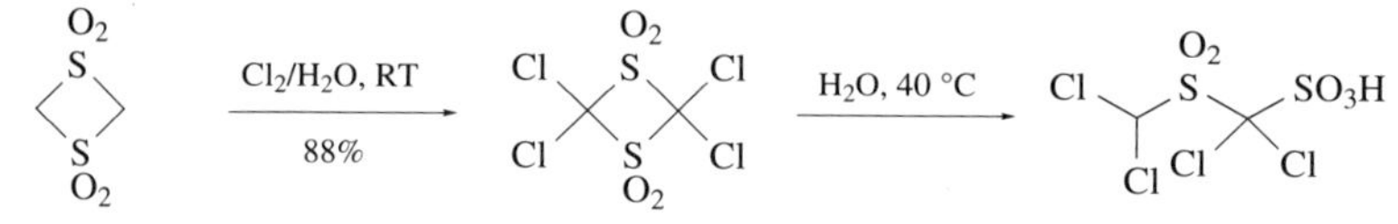

(**36**) (**37**) (**38**)

Scheme 10

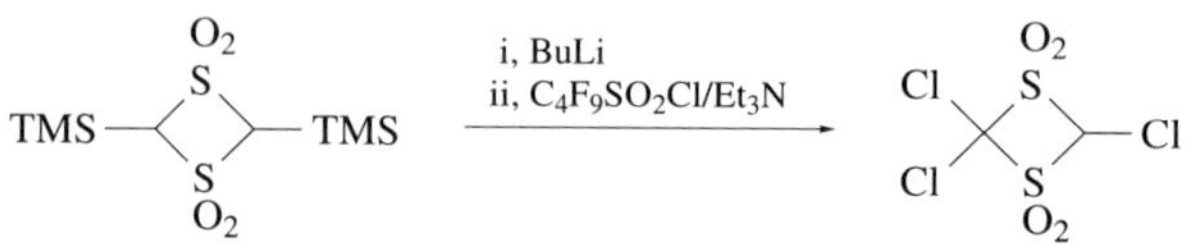

(**39**) (**40**) (5)

Bis(halogenated) open chain *gem*-disulfones exist for all four halogens; some have been known since the end of the nineteenth century, for example $Br_2C(SO_2Et)_2$ ⟨1890CB3226⟩. They may be produced by direct chlorination or bromination of *gem*-disulfones using aqueous halogen; sulfuryl chloride or sulfuryl bromide may also be used ⟨35JA217, 73JOC3358, 75CJC2664, 75JOC1278, 80CJC1996, 88ZOR1327⟩. The trisulfone of 1,3,5-trithiane has been fully chlorinated photochemically, while trithiane itself can be converted directly into the same product, that is, hexachloro-1,1,3,3,5,5-hexaoxo-1,3,5-trithiane (**41**), when heated with aqueous chlorine and MoO_3 ⟨63JPR1, 64ZC457⟩. An interesting transformation of this trisulfone occurred when it was heated with SbF_3 and catalytic amounts of $SbCl_5$ in sulfolane, resulting in the isolation of (**42**). The hexabromo analogue of (**41**) gave the tetrabromo analogue of (**42**) under milder conditions, in dioxane–water. With tertiary amines or phosphines, these trisulfones afforded stable salts such as (**43**) (Scheme 11) ⟨86CB3631⟩. Under certain conditions—when 1,3,5-trithiane is treated with sodium hypochlorite in an alkaline environment—chlorination occurs, but oxidation stops at the disulfone stage and (**44**) is produced ⟨56JCS508⟩. When treated with a tertiary amine, the monosulfone of hexachloro-1,3,5-trithiane was converted to (**45**) ⟨68USP3376314⟩.

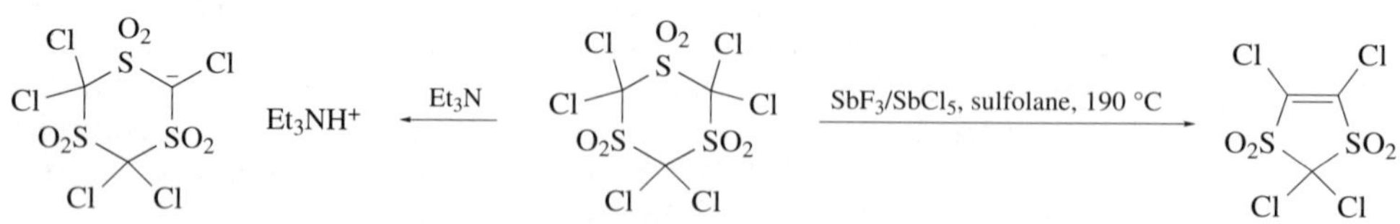

(**43**) (**41**) (**42**) 62%

Scheme 11

The potassium salts of monohalogenated *gem*-disulfones have also been halogenated (Equation (6)) ⟨73JOC3358⟩. Phenyliodonium ylides undergo brominolysis, affording dibromo-*gem*-disulfones ⟨90CC1459, 91CC470⟩. Alternatively, the halosuccinimides NCS, NBS and NIS give the corresponding dihalodisulfones (Equation (7)) ⟨88JCR(S)306⟩. Chlorination of a methylene group in some sulfones bearing another sulfur-containing group has been effected by aqueous chlorine in acetic acid; in

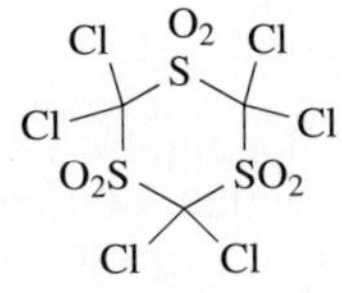

(44)

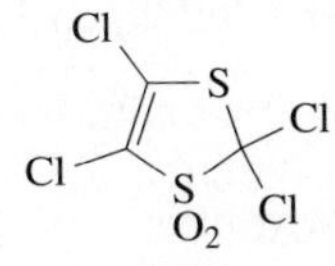

(45)

some cases further transformations led to the isolation of interesting products, as depicted in Scheme 12 ⟨56JCS508, 75CJC2664, 80CJC1996, 83CJC610, 91CJC2127⟩. *N*-Chlorosuccinimide and sulfuryl chloride were used in order to convert $H_2C(SO_2Cl)_2$ into $Cl_2C(SO_2Cl)_2$ ⟨76CZ391⟩.

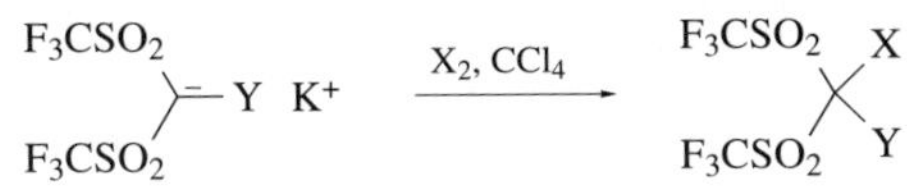

X = Y = Br
X = Y = Cl

(6)

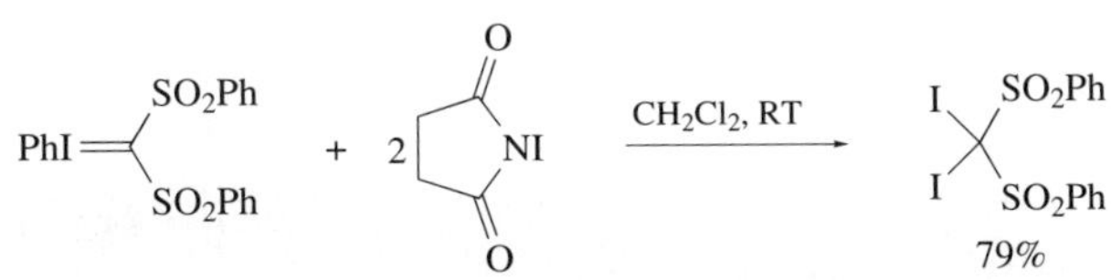

(7)

$MeSO_2CH_2SH$ or $MeSO_2CH_2SO_2OPh$ $\xrightarrow{Cl_2,\ AcOH}$ $MeSO_2CCl_2SO_2OPh$

$MeSO_2CH_2SCH_2Ph$ $\xrightarrow[AcOH]{Cl_2/H_2O}$ $MeSO_2CCl_2SO_2Cl$

4-$ClC_6H_4SCH_2SO_3Na$ $\xrightarrow{NaOCl}$ 4-$ClC_6H_4SO_2CCl_2SO_3Na$

Scheme 12

The transformation of a thiocarbonyl group into CCl_2 can be achieved by chlorination. One example involved photochemical addition of chlorine to carbon disulfide; the dichloromethane bis(sulfuryl chloride), $Cl_2C(SCl)_2$, formed initially was used for the preparation of the heterocycle (**46**) (Scheme 13) ⟨90JCS(P1)509⟩. In another example, the heterocycle (**47**) was heated in a sealed tube with PCl_5 and was converted directly into (**48**) (Equation (8)) ⟨91CB2025⟩.

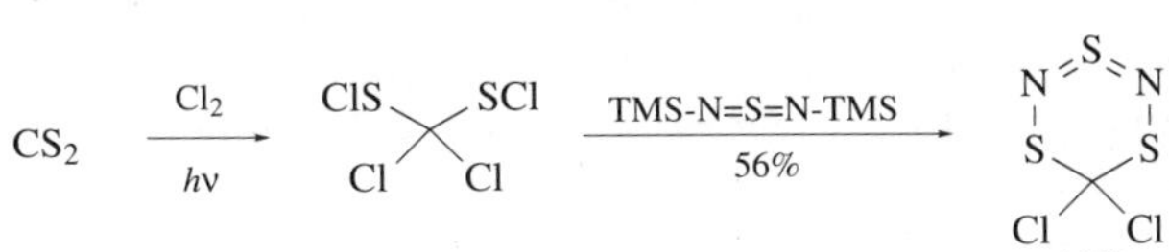

(46)

Scheme 13

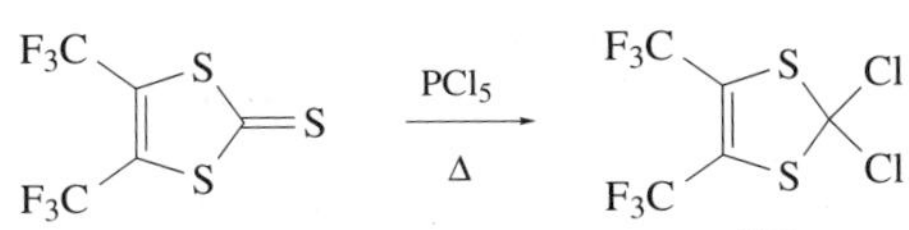

(47) **(48)**

(8)

6.08.2.3 Two Halogens, an Oxygen, and a Sulfur Function

Only one compound of this type is known. It was obtained from the addition of chlorine to either MeOC(S)SMe or MeOC(S)SSC(S)OMe, both of which afforded the same product, $MeOCCl_2SCl$ ⟨53JA4582, 56JA6070⟩.

6.08.2.4 Two Halogens and Other Chalcogen Functions

Tetrafluoro-, tetrachloro-, and tetrabromo-1,3-diselenetanes are known. The fluoro compound **(50)** was obtained by heating polymeric $F_2C{=}Se$, which in turn was formed from the boron compound **(49)** (Scheme 14) ⟨76ZAAC(427)114⟩. Treatment of **(50)** with BCl_3 or BBr_3 gave its tetrachloro or tetrabromo analogues. 1-Trifluoromethyl-1,3,3-trifluoro-1,3-diselenetane was formed from the interaction of $(CF_3CF_2Se)_2Hg$ and $(CF_3Se)_2Hg$, followed by treatment with Et_2AlI ⟨91CB51⟩. Two cyclic ethers—oxetane and tetrahydropyran **(51)**—were converted into **(53)** when treated with the selenoxide **(52)** and acetic anhydride, in a Pummerer-type reaction (Equation (9)) ⟨93TL1311⟩. Tetrafluoro-1,3-ditelluretane was obtained by thermolysis of $Me_3SnTeCF_3$, which initially (at $-196\,°C$) gave the monomer $F_2C{=}Te$ ⟨92C78⟩. The same compound **(54)** was produced from the reaction of $Hg(TeCF_3)_2$ with two equivalents of Et_2AlI; when the monomer $F_2C{=}Te$ was allowed to react with $F_2C{=}Se$, tetrafluoro-1,3-selenatelluretane **(55)** was formed. Compound **(54)** was converted into its tetrachloro analogue with BCl_3 ⟨93JCS(D)2547⟩.

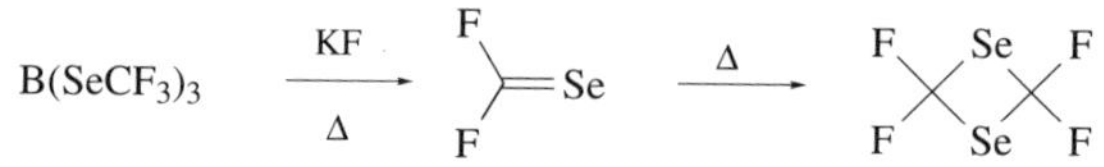

(49) **(50)**

Scheme 14

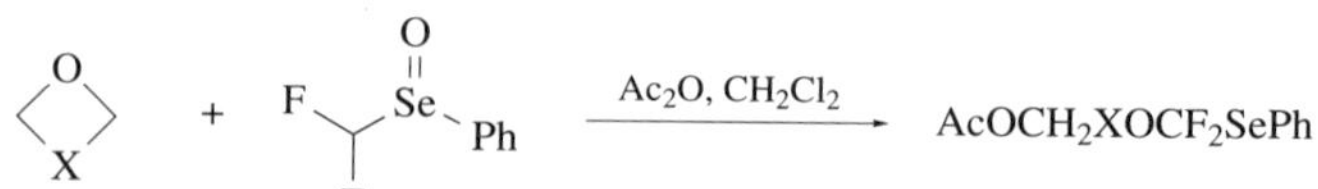

(9)

(51) **(52)** **(53)**

X = CH_2 or $(CH_2)_3$

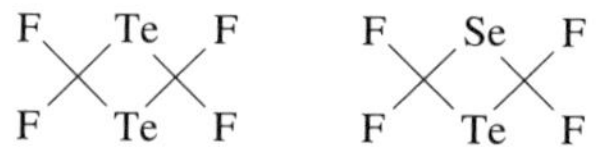

(54) **(55)**

6.08.3 TWO HALOGENS AND ONE CHALCOGEN FUNCTION

6.08.3.1 Two Halogens, a Chalcogen, and a Nitrogen Function

6.08.3.1.1 Oxygen compounds

Perfluoromethyleneimine (*N*-fluorocarbonimidic difluoride) **(56)** is a good precursor for the preparation of several compounds in this category, especially compound **(58)**, through the addition of FSO_2OX **(57)** to its double bond (Equation (10)). The related compound $FSO_2OCF_2N(OSO_2F)F$ was also obtained from **(56)** and the peroxide FO_2SOOSO_2F ⟨83IC805⟩. However, the reaction between $F_2C{=}NCl$ and FSO_2OX led to the formation of a different product, that is, the azo compound $FSO_2OCF_2N{=}NCF_2OSO_2F$ ⟨84JOC3590⟩. Other additions to the double bond of perhalogenated imines include the reactions in Scheme 15 ⟨76JA3529, 83JOC4844, 87AG(E)314, 90JFC(48)395⟩. Some of these compounds were cyclized to form oxaziridines, upon reaction with metal fluorides at a low temperature (Scheme 16) ⟨76JA3529, 79IC919, 80IC1330, 83JOC4844⟩. Direct epoxidation of perfluoroazaalkenes is not normally possible, except with $F_2C{=}NCF_3$ which, with mcpba, gave the corresponding oxaziridine in 50% yield ⟨93JOC4754⟩.

(10)

(56) **(57)** **(58)**

X = F, Cl, Br

Oxaziridine **(59)** underwent cycloaddition reactions with tetrahalogenated ethylenes and also with methyl ketones, affording five-membered heterocycles such as **(60)** and **(61)** (Scheme 17) ⟨82JA4034,

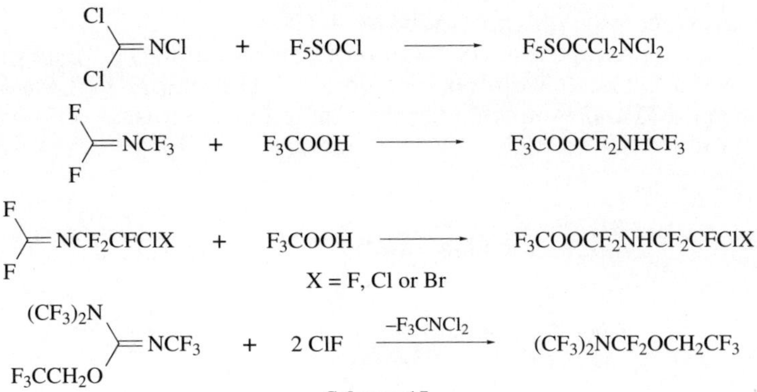

Scheme 15

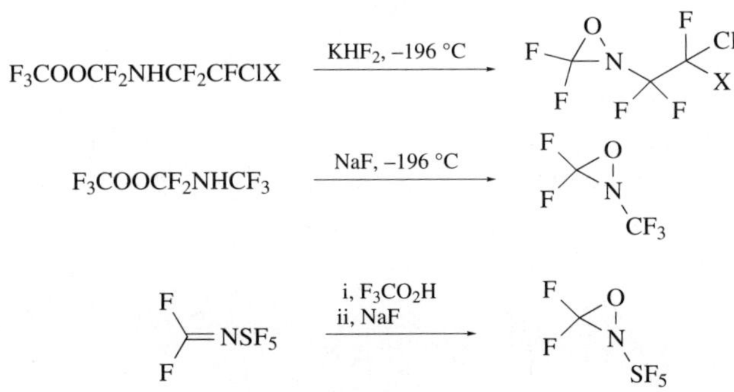

Scheme 16

86JOC4466⟩. The perfluoro oxazolidine **(60)** was also formed in low yield upon pyrolysis of perfluoromorpholine ⟨65JCS6077⟩. The same compound and some related perfluoro-1,3-oxazolidines **(62)**–**(64)** were produced electrochemically from various precursors ⟨56JA5637, 58JA1889, 90JFC(48)257⟩. The potassium and caesium salts of F_2NCF_2OH have been formed upon reaction of difluoroaminocarbonyl fluoride (F_2NCOF) with KF or CsF ⟨67IC1711⟩.

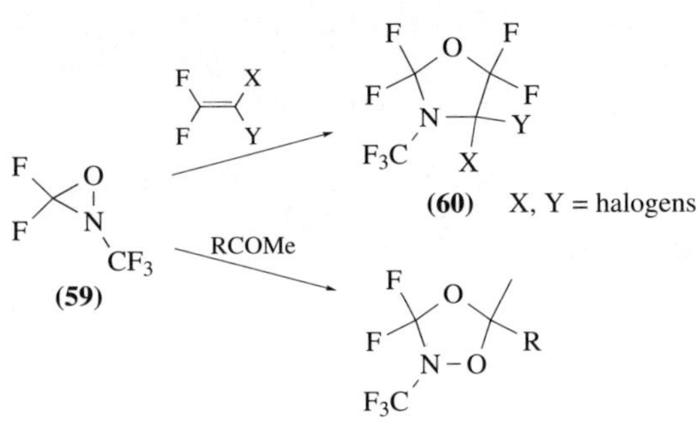

(59) **(60)** X, Y = halogens **(61)**

Scheme 17

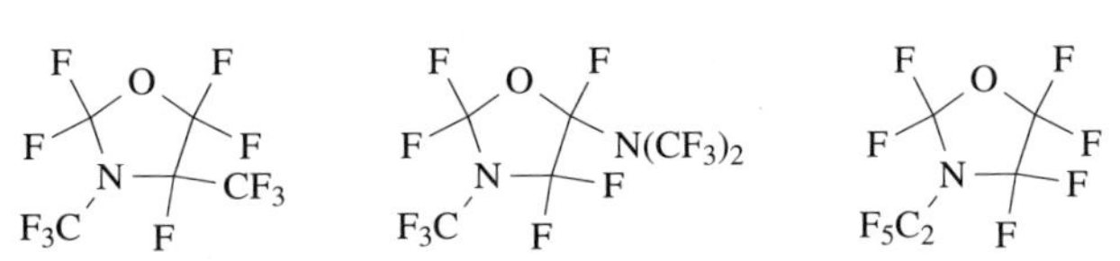

(62) **(63)** **(64)**

6.08.3.1.2 Sulfur compounds

Some unusual reactions have been reported with thiocyanates and fluorine. Methyl thiocyanate was converted into $F_5SCF_2NF_2$ when heated with elemental fluorine ⟨59JA3599⟩. Potassium or silver thiocyanate in the presence of catalytic amounts of CaF_2 gave, among several other products, $FSCF_2NF_2$ ⟨65ZOB1412⟩. An interesting conversion of compound **(65)** occurred upon its treatment with triethylamine, when the zwitterion **(66)** was obtained (Equation (11)) ⟨90AG(E)60⟩. The analogous reaction between **(67)** and quinuclidine **(68)** resulted in the formation of another zwitterion, **(69)** (Equation (12)) ⟨89AG(E)221⟩.

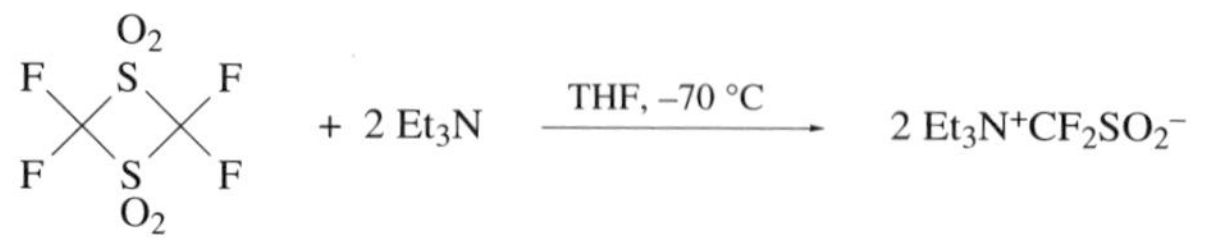

(65) **(66)** (11)

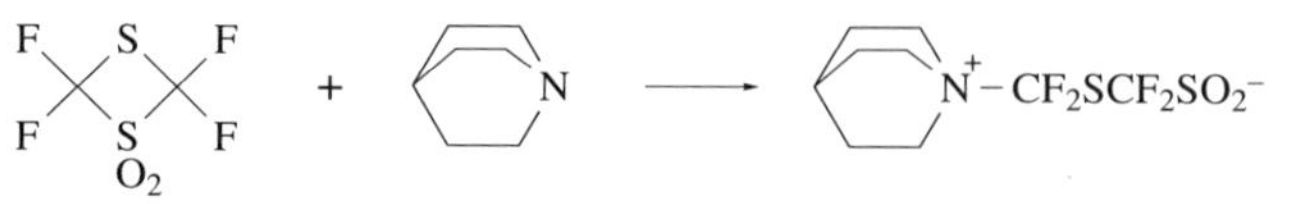

(67) **(68)** **(69)** (12)

Addition of chlorine to **(70)** gave **(71)**. Upon heating this was converted into a mixture of heterocycles **(72)** and **(73)** (Scheme 18) ⟨71CB2732⟩. Compound **(72)** was converted by $Hg(SCF_3)_2$ into F_3CSSCF_2NCS, while **(70)** added photochemically F_3CSCl to afford $F_3CSSCFCl$ (NCS).

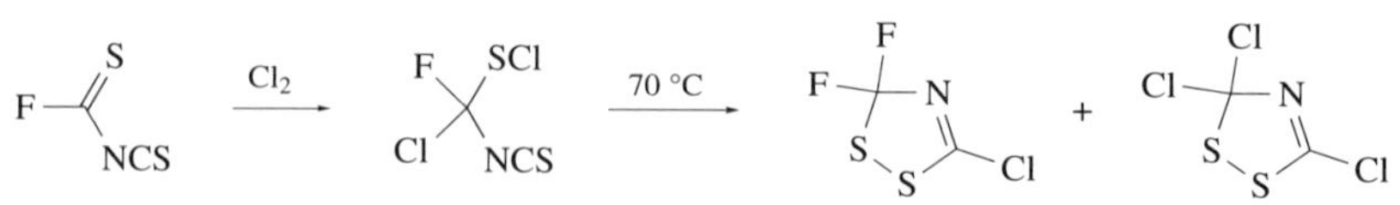

(70) **(71)** **(72)** **(73)**

Scheme 18

6.08.3.2 Two Halogens, a Chalcogen, and Other Functions

Perfluoro-2-phosphapropene, $F_3CP{=}CF_2$, when added to methanol forms F_3CPHCF_2OMe; $C_2F_5P{=}CF_2$ reacts similarly ⟨86ZN(B)149, 90ZN(B)148⟩. The sulfinic salt **(75)** was obtained from **(74)** and sodium sulfite or sodium dithionite; its oxidation with H_2O_2 gave the corresponding sulfonate and, upon acidification, the free acid (Equation (13)) ⟨89JA1773⟩. A similar reaction occurred using $(EtO)_2P(O)CFBr_2$ and $Na_2S_2O_4$ ⟨89CJC1795⟩. The halogenated sulfones **(76)** were converted to **(77)** when their anion attacked PhHgCl (Equation (14)) ⟨74JOM(71)335⟩. Another organometallic mercury compound, $HOHgCI_2SO_3Na$, was obtained from the reaction of HgO and CHI_2SO_3Na ⟨35CB1513⟩.

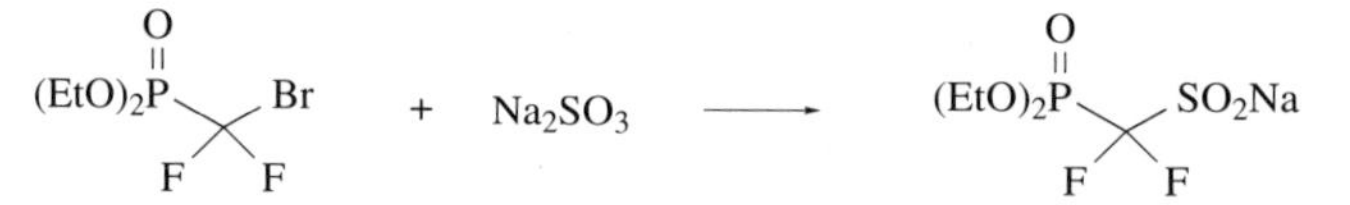

(74) **(75)** (13)

$$PhSO_2CHX_2 + PhHgCl \xrightarrow{Bu^tOK,\ THF,\ -60\ ^\circ C} PhSO_2CX_2HgPh \qquad (14)$$

(76) **(77)** X = Cl, Br

6.08.4 TWO HALOGENS AND TWO GROUP 15 ELEMENT FUNCTIONS

6.08.4.1 Two Halogens and Two Nitrogen Functions

6.08.4.1.1 Diamines and their derivatives

The simplest compound of this class is perfluorodiaminomethane **(79)**, which can be prepared by fluorination either of cyanuric chloride **(78)** or of aminoiminomethane-sulfinic acid **(80)** (Scheme

19) ⟨65ZOB1412, 66JOC4232⟩. The *C*-monochloro analogue of (**79**), that is $FClC(NF_2)_2$, was formed from perfluoroguanidine and $ClNO_2$ ⟨78IZV486⟩. Tetramethylurea was converted into $(Me_2N)_2CF_2$ when treated with COF_2 in the presence of NaF ⟨62JA4275⟩. Fluorine or chlorine fluoride adds to the nitrile function in some perfluorocyanamides, with the formation of perhalogenated methanediamines, as illustrated in Scheme 20 ⟨90JA728, 92IC488⟩. Chlorine fluoride has also been added to the N=C bond of some halogenated imines, in the presence of CsF; for example, (**81**) was converted into (**82**) (Equation (15)) ⟨90JFC(48)395⟩. A related addition of ClF occurred to the C=N bond of (**84**), which was obtained from (**83**) and $LiOR_f$ (where R_f is a perfluoro group) (Equation (16)) ⟨92JFC(57)293⟩. When (**84**) ($n = 1$, $R_f = C(CF_3)_2CH_3$) was treated with $C_6F_5SiMe_3$ and CsF, fluorine at the sp^2 carbon was replaced by the C_6F_5 group ⟨93IC4802⟩.

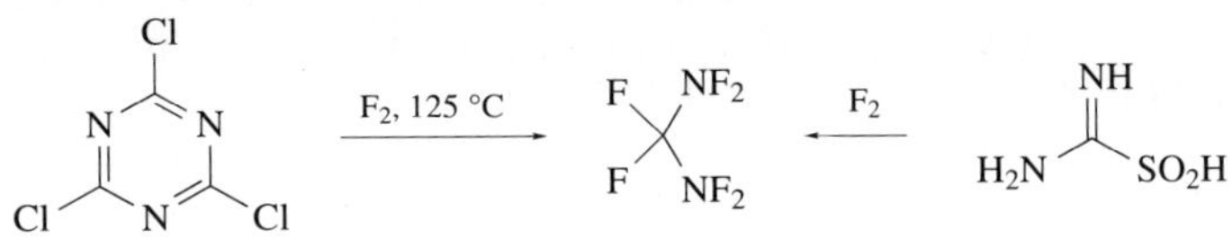

(**78**) (**79**) (**80**)

Scheme 19

$$(CF_3)_2N{-}C{\equiv}N \xrightarrow{F_2,\ CsF} (CF_3)_2NCF_2NF_2$$

$$(CF_3)_2N{-}C{\equiv}N \xrightarrow{ClF} (CF_3)_2NCF_2NCl_2$$

$$F_3CN(F){-}C{\equiv}N \xrightarrow{ClF} F_3CN(F)CF_2NCl_2$$

Scheme 20

$$\underset{(\mathbf{81})}{(CF_3)_2NC(F){=}NCF_3} \xrightarrow{ClF,\ CsF} \underset{(\mathbf{82})}{(CF_3)_2NCF_2NCl(CF_3)} \qquad (15)$$

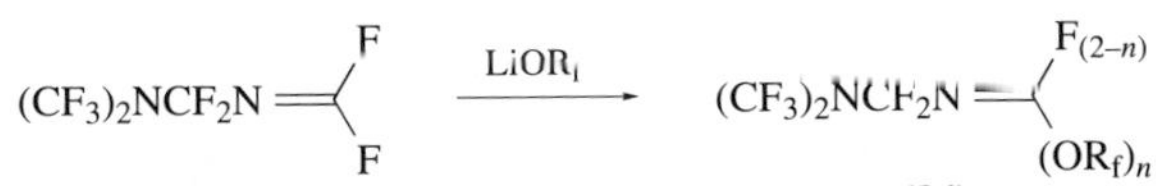

(**83**) (**84**) (16)

Insertion of alkenes and nitriles into the N—Cl bond of (**82**), as well as photolytic reactions of the latter with SF_5Cl and CF_3COCl, occur readily ⟨90JFC(48)395⟩. Several higher halogenated *gem*-diamines were produced in a similar way from (**85**) and halogenated alkenes which inserted into the N—Cl bond thermally; for example (**86**) was obtained using $CF_2{=}CClF$ (Equation (17)) ⟨92IC488⟩. When photolyzed alone (**86**) formed the imine (**83**) in high yield; in the presence of halogenated alkenes, insertion occurred and compounds such as (**87**) were formed (Scheme 21). Photolysis of the simpler *N*-chlorodiamine (**82**) led to the formation of the diazane derivative (**88**) (Equation (18)). Compound (**82**) added to 1,1-difluoroethylene and trifluoroethylene on irradiation forming adducts such as (**89**) (Equation (19)).

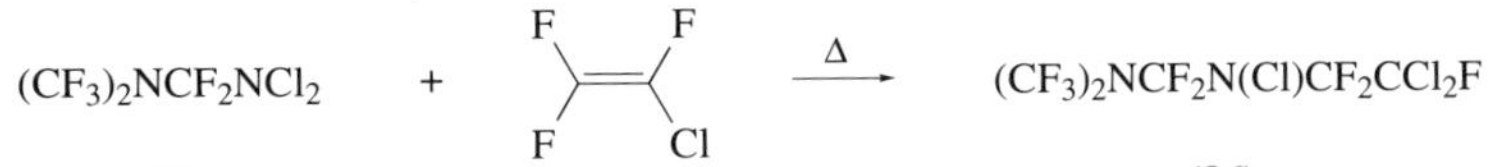

(**85**) (**86**) (17)

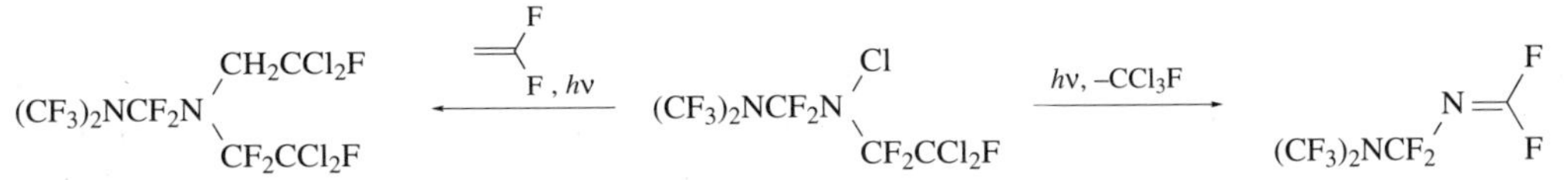

(**87**) (**86**) (**83**)

Scheme 21

(82) (88)

(18)

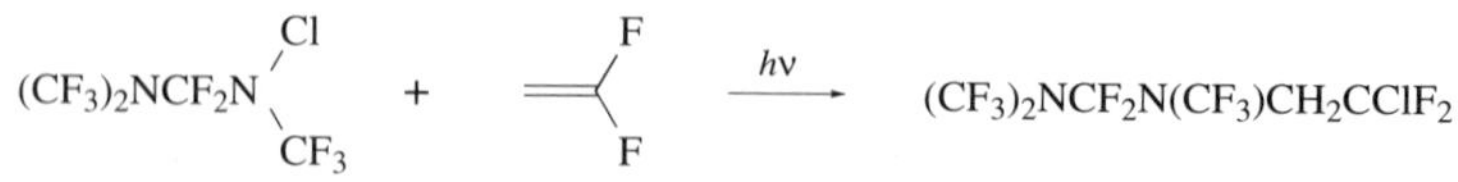

(82) (89)

(19)

Chlorine fluoride has also been added to the C=N bond of a perfluorohydrazone; the adduct, upon irradiation with UV light, was converted into a complex tetrazane having two CF_2 groups flanked by nitrogen functions ⟨89IC3345⟩. A related perfluorohydrazone, **(90)**, added a methyl group and fluorine to its double bond upon reaction with iodomethane and silver fluoride; compound **(91)** was formed (Equation (20)) ⟨92IC4917⟩. Some difluoromethylene *gem*-diamines with alkyl or phenyl groups are also known. A *gem*-diamine with two phenyl groups, $PhNHCF_2NHPh$, was prepared from $PhNHCF_3$ and aniline ⟨59ZOB2169⟩. The bis(azo compound) $O_2NC_6H_4N{=}NCF_2N{=}NC_6H_4NO_2$ was the product from the reaction of difluorodinitromethane with 4-nitroaniline ⟨92IC329⟩. The preparation of the nitramine $F_2NCF_2NO_2$ from $CF_2(OF)_2$ and $F_2NC(F){=}NF$ has been described ⟨68USP3387033⟩. A simple amine derivative, difluoro(bisisocyanato)methane **(93)**, was produced upon thermolysis of **(92)** (Equation (21)) ⟨84JOC4541⟩. Some sulfur-containing amine derivatives have been reported, derived from various simple compounds bearing a cyano group; some examples are illustrated in Scheme 22 ⟨67MI 608-01, 68ZN(B)743, 86CB107⟩.

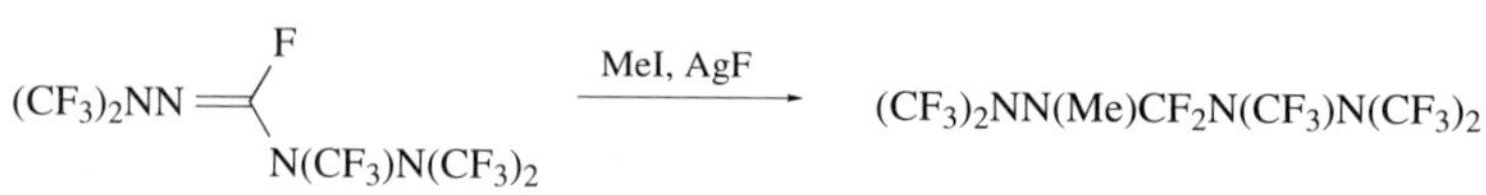

(90) (91)

(20)

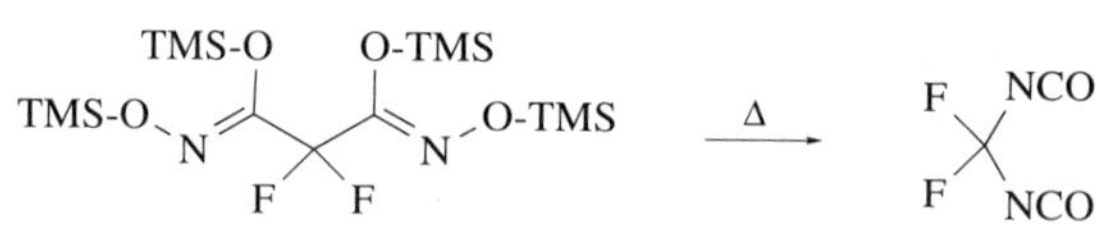

(92) (93)

(21)

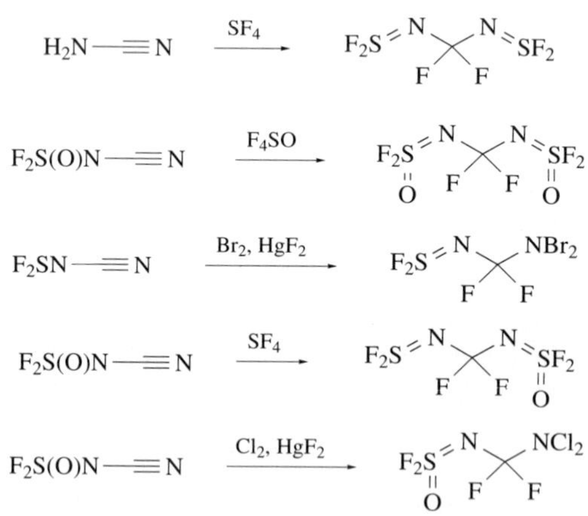

Scheme 22

6.08.4.1.2 *Dinitro compounds*

Di(halo)dinitromethanes of the general formula $X_2C(NO_2)_2$ are known with all four halogens; some mixed halogen compounds have also been reported. Dichloro- and dibromodinitromethanes are formed in several degradative reactions of a variety of precursors (too many to mention in the space available here) by nitric acid. The conversion of 2,4,6-trichloro- and 2,4,6-tribromoaniline into dichloro- and dibromodinitromethane, respectively, is an important preparative method

⟨44JOC419⟩. A more conventional method for the preparation of dibromodinitromethane involved the reaction of sodium dinitromethanide with bromine after treatment with butyllithium, or directly with bromine in THF–HMPA (hexamethylphosphoramide) ⟨89ZOR2490⟩. Phenyliodonium dinitromethylide was converted in high yields into either dichloro- or dibromodinitromethane with the appropriate halogen, NCS or NBS (Equation (22)) ⟨78IZV2348⟩. The very unstable diiododinitromethane was reported to be formed upon acidification of potassium iododinitromethanide ⟨24JCS442⟩. Difluorodinitromethane (**95**) was first prepared from difluorodiazirine (**94**) and N_2O_4 (Equation (23)) ⟨64JHC233⟩. The same substrate (**94**), which on photolysis generates difluorocarbene, gave a 5% yield of (**95**) with NO; the photochemical reaction of difluorodiiodomethane in excess NO afforded a 10% yield of (**95**) ⟨92IC329⟩. The reactions of potassium trinitromethanide with fluorine–sodium fluoride and also with potassium fluoride again led to the formation of (**95**) ⟨68IZV429, 68JOC3073⟩. Difluorodinitromethane was also formed in 38% yield from the reaction of difluoronitroacetic acid with XeF_2; in a similar way, fluorochloronitroacetic acid produced fluorochlorodinitromethane ⟨88IZV466, 88IZV2639⟩. The same compound was obtained when chlorotrinitromethane was treated with CsF in DMF ⟨70IZV2553⟩. Chlorobromodinitromethane was prepared from chlorine and ammonium bromodinitromethanide; the latter was formed from dibromodinitromethane in liquid ammonia (Equation (24)) ⟨44JOC419⟩.

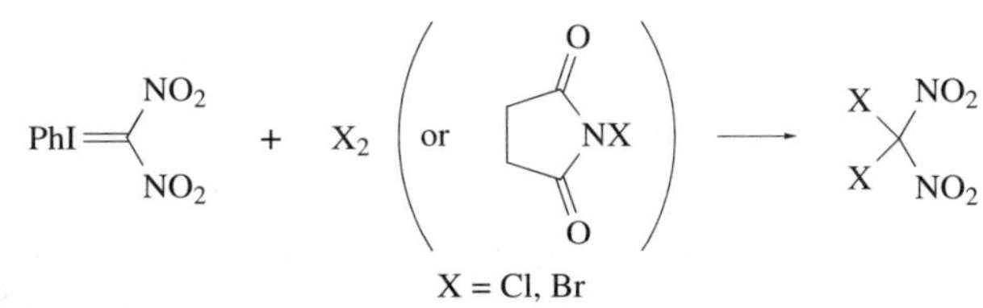

(22)

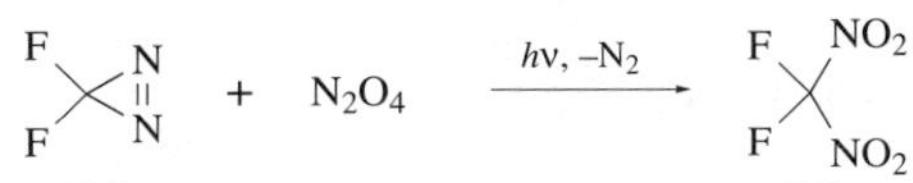

(**94**) (**95**)

(23)

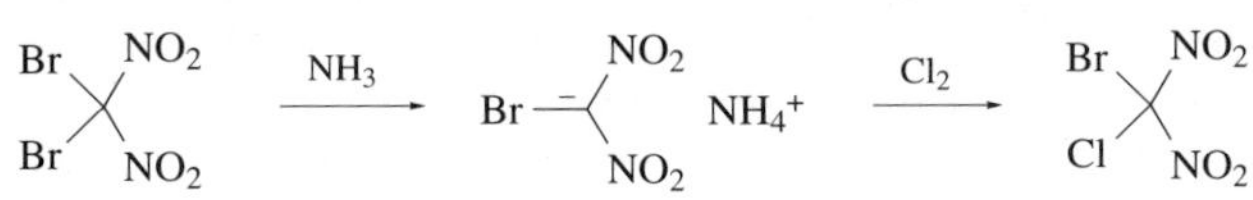

(24)

6.08.4.1.3 Cyclic compounds

Several methods have been developed for the preparation of the important heterocycle difluorodiazirine (**94**). Historically, the first method involved cyclization of $F_2C(NF_2)_2$ by ferrocene ⟨66JHC245⟩. The same substrate was converted more efficiently into (**94**) using tetrabutylammonium iodide (Scheme 23) ⟨67JOC1944, 68JOC1847⟩. Tetrafluoroformamidine (**96**) was also transformed efficiently into (**94**), using either ferrocene or potassium iodide ⟨67JOC4045, 68JOC1847⟩. Tetrafluoroformamidine was isomerized by potassium fluoride at a low temperature to perfluorodiaziridine (**97**), an unstable compound which frequently exploded: with ferrocene this gave (**94**); when heated with caesium fluoride it dimerized to form (**98**) (Scheme 24) ⟨68JOC3489⟩. Curiously, compound (**98**) was claimed in a patent to be a possible heat transfer fluid ⟨92JAP04110383⟩. A unique rearrangement leading to the formation of (**94**) occurred when difluorocyanamide (F_2NCN) was treated with caesium fluoride at room temperature ⟨66IC1455⟩. Chlorofluorodiazirine was obtained from dichlorobis(difluoramino)methane, $Cl_2C(NF_2)_2$, after treatment with boric acid; the precursor was prepared from perfluoroguanidine, HCl and HNO_3 ⟨67JOC1944, 67USP3355492⟩. Some further perhalogenated diaziridines are compounds (**99**)–(**102**). Their preparation involved the use of CsF or KF as catalysts at low temperatures. Compound (**99**) was obtained from $CF_2{=}NF$ or $CF_3N(F)C(F){=}NF$ ⟨83JOC771⟩. Compound (**100**) was formed from $CF_2{=}NCl$; on treatment with mercury in trifluoroacetic acid, chlorine was replaced by hydrogen ⟨84JOC3590⟩. Compound (**101**) was formed from $CF_2{=}NF$ and $CF_2{=}NBr$, which first gave $CF_3N(F)C(F){=}NBr$ ⟨88JOC4443⟩. Compound (**102**) was prepared from $CF_2{=}NF$ and $CF_2{=}NCF_3$ ⟨90JA728⟩.

Fluorine and chlorine—from either ClF or a mixture of chlorine and mercury difluoride—readily add to cyanuric fluoride (**103**) affording compound (**104**) (Equation (25)) ⟨75MI 608-01, 86CB107⟩. Fluorine added to (**103**) in a complex way; in addition to the perfluoro analogue of (**104**), perfluoro-

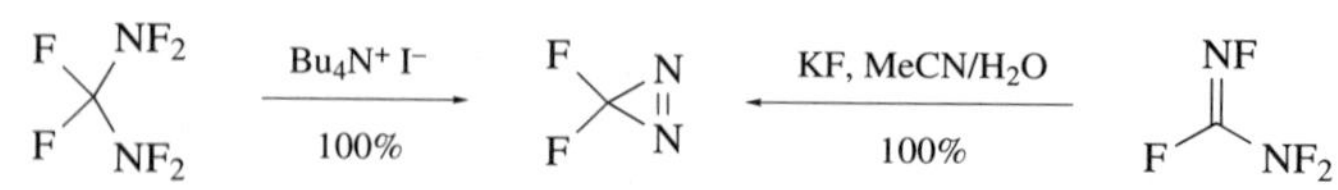

(94) (96)

Scheme 23

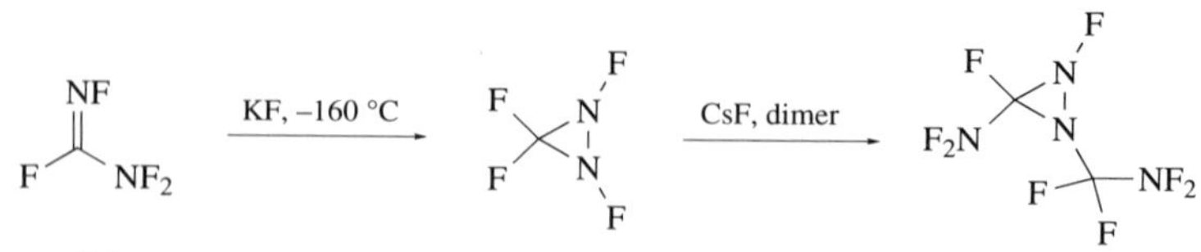

(96) (97) (98)

Scheme 24

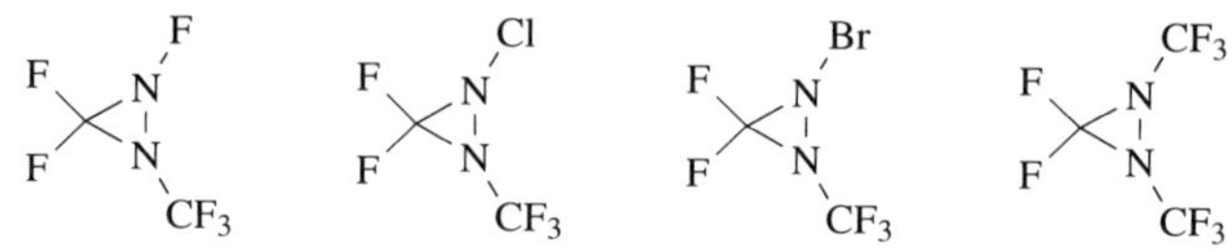

(99) (100) (101) (102)

1,3-diazolidine and noncyclic compounds such as $CF_3N(F)CF_2N(F)CF_3$ were formed ⟨62JA2751⟩. Perfluoro-2,5-diazahexa-2,4-diene (**105**), when heated with CsF, was cyclized to a mixture of isomeric 1,3-diazolines (**106**) and (**107**). In acetonitrile in the presence of CsF, both products gave the anion (**108**), which reacted with alkyl halides affording *N*-alkyl compounds, for example (**109**) with iodomethane. Compounds (**108**), (**106**) and (**107**) also underwent fluoride ion induced reactions with perfluoroalkenes; the main product, when perfluorocyclopentene was used was (**110**) (Scheme 25) ⟨85JCS(P1)1191⟩. The perfluoro analogue of (**109**) was converted to 2,2-dichloro- or 2,2-dibromo-dimethyl-1,3-diazolidine ⟨84JFC(24)457⟩. Another heterocycle (**112**), was formed from (**111**) upon heating with CsF; when irradiated in the presence of CsF, (**111**) was transformed into a complex dimer (Equation (26)) ⟨90HAC167, 92ICA527⟩. Compound (**112**) was also formed upon dimerization of (**106**) and (**107**) ⟨85JCS(P1)1191⟩.

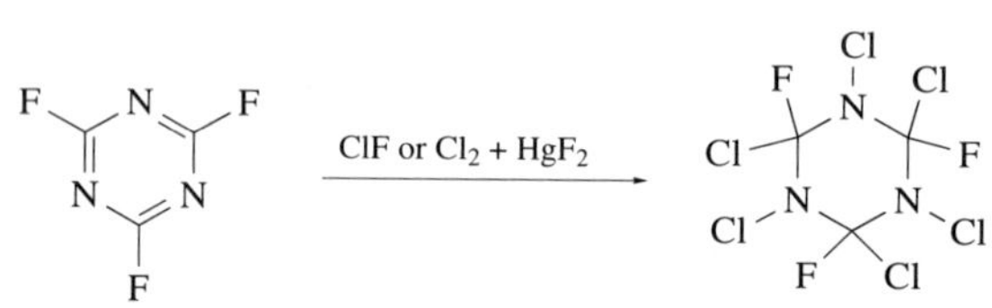

(103) (104) (25)

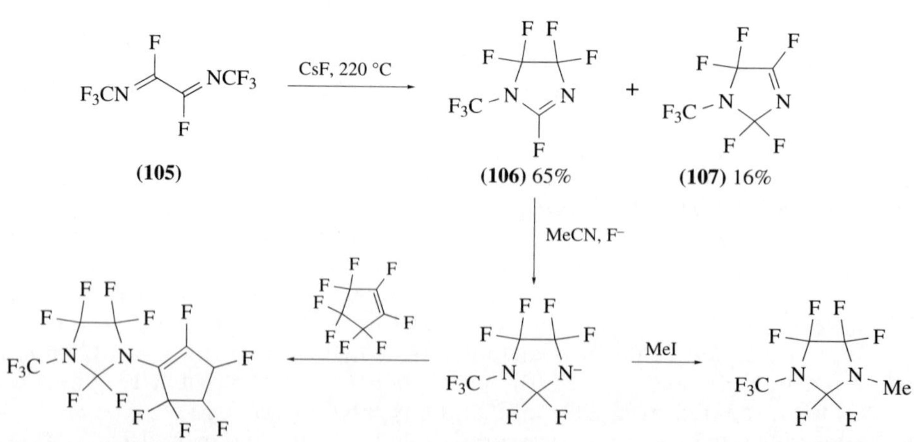

(105) (106) 65% (107) 16%

(110) (108) (109)

Scheme 25

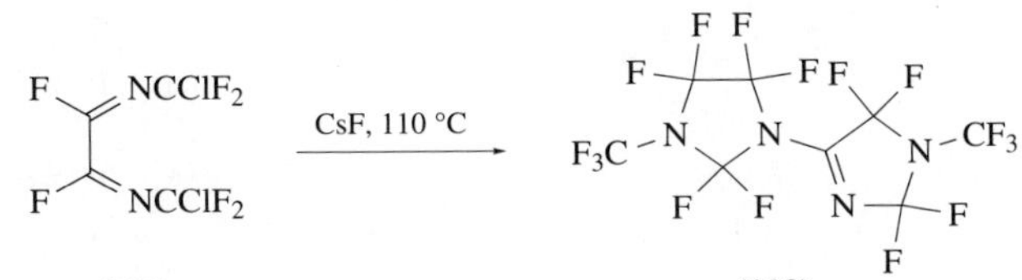

(111) (112)

(26)

6.08.4.2 Two Halogens and Two Phosphorus Functions

6.08.4.2.1 Bis(phosphonates)

The study of these compounds has been intensified since two important discoveries were made: (i) the ability of $Cl_2C(PO_3H_2)_2$ (clodronic acid, (**115**)) to help in the process of bone formation; and (ii) the various physiological properties of nucleotide analogues, especially ATP, in which an oxygen atom in the triphosphate unit is replaced by a difluoromethylene group ⟨78JCE760, 84JCS(P1)1119⟩. Bis(halogenated) methylenebisphosphonic acids are usually prepared through halogenation of the appropriate tetraesters, which are subsequently hydrolyzed. The ester $Cl_2C(PO(OEt)_2)_2$ was obtained in 72% yield by heating $H_2C(PO(OEt)_2)_2$ with PCl_5 at 140 °C ⟨60ZOB1602⟩. However, the cheaper commercially available isopropyl tetraester (**113**) is the preferred starting material. This was converted into its dichloro derivative (**114**) and then hydrolyzed to the acid (**115**) by heating in concentrated hydrochloric acid (Scheme 26) ⟨66BEP672205, 68JOM(13)199, 85JOM(291)145⟩. The dibromo analogues and the relatively unstable diiodo analogues of (**114**) were prepared in a similar way by the hypohalite method ⟨71JOC1835⟩. Several other tetraesters (symmetrical or mixed) of the general formula $X_2C(PO_3R_2)_2$ (where R was an alkyl group and X was chlorine or bromine) were prepared by the hypohalite method; in which careful control of temperature, pH, and reaction time led to high yields ⟨92JCS(P2)835, 92PS(70)183⟩. The ester (**114**) was also converted to the acid (**115**) without hydrolysis by heating in *sym*-tetrachloroethane, with evolution of propene ⟨67JOC4111⟩. Since this acid is highly polar, it was desirable to prepare some partial ester derivatives. This was achieved via two methods: silyl derivatives and selective hydrolysis. The first method was based on the observations that the rate of silylation of methyl-containing mixed tetraesters follows the order methyl > primary alkyl > secondary alkyl, whereas the hydrolysis rate of silyl and alkyl esters follows the order $SiMe_3$ > tertiary alkyl > secondary alkyl > primary alkyl. Taking into account these factors, the preparation of several tri-, di-, and mono-esters was effected. For example, the ester (**116**) was converted first to (**117**) and then to the salt (**118**) in 87% yield (Scheme 27). The second approach was suitable for *n*-alkyl esters, since branched chain alkyl groups in acidic conditions undergo hydrolysis faster than *n*-alkyl groups. Thus, (**119**) was transformed into (**120**) in 44% yield (Scheme 28) ⟨93TL4551⟩.

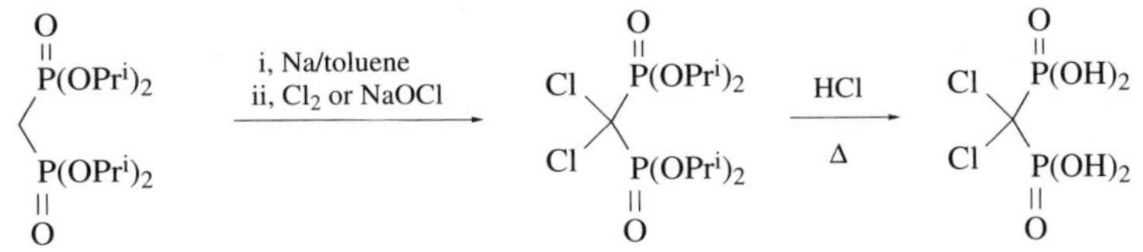

(113) (114) (115)

Scheme 26

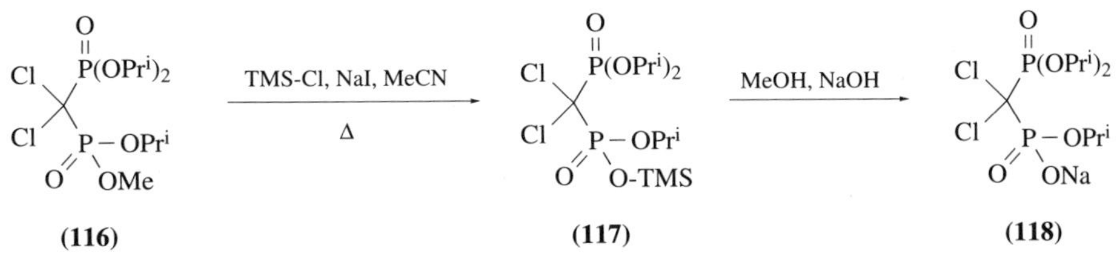

(116) (117) (118)

Scheme 27

Difluoro tetraesters, as well as the free bisphosphonic acid (**122**) which is an isopolar analogue of pyrophosphoric acid, were prepared by several routes. Fluorination of (**121**) with perchloryl fluoride

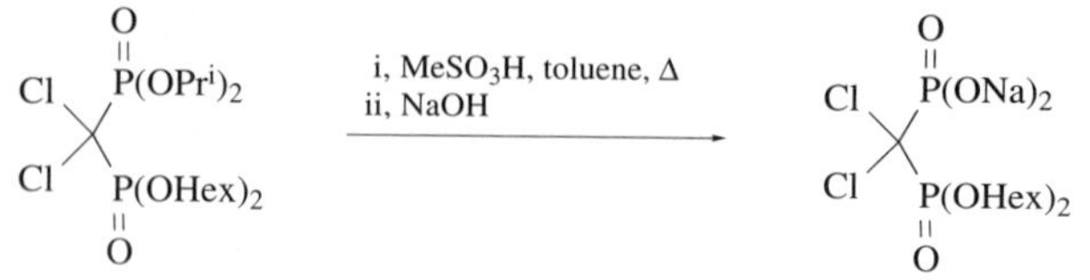

(119) Hex = *n*-C_6H_{13} **(120)**

Scheme 28

in a strongly basic environment gave (**122**) in good yields (R = Et or Pr^i); this was converted with trimethylsilyl bromide into the silyl derivative (**123**), which then was hydrolyzed to the free acid (**124**), isolated as its trisdicyclohexylammonium salt (Scheme 29) ⟨81JOC4573, 82JFC(20)617⟩. Acetyl hypofluorite was also an efficient fluorinating agent ⟨91BMC357⟩. Coupling reactions have also been used; for example, the lithium salt obtained from $HCF_2P(O)(OEt)_2$ and lithium diisopropylamide (LDA) gave, on reaction with $(EtO)_2P(O)Cl$, the tetraester $(EtO)_2P(O)CF_2P(O)(EtO)_2$ in 74% yield ⟨82TL2323⟩. The reaction between $BrF_2CPO(OPr^i)_2$ and $(Pr^iO)_2PONa$ also gave (**122**) (R = Pr^i) ⟨81CC930⟩. The same compound was formed using CBr_2F_2 and $(Pr^iO)_2PONa$; the mixed halogen tetraester $FBrC(PO(OPr^i)_2)_2$ was obtained from $FCH(PO(OPr^i)_2)_2$ and tetraisopropyl pyrophosphate, upon treatment with bromine in aqueous K_2CO_3 ⟨88JOM(340)93⟩. The application of the Michaelis–Becker variation to the preparation of difluoromethylene esters has also been described ⟨80JFC(15)263, 82JFC(20)121⟩, as have other routes to compounds of this type ⟨94JOC2393⟩.

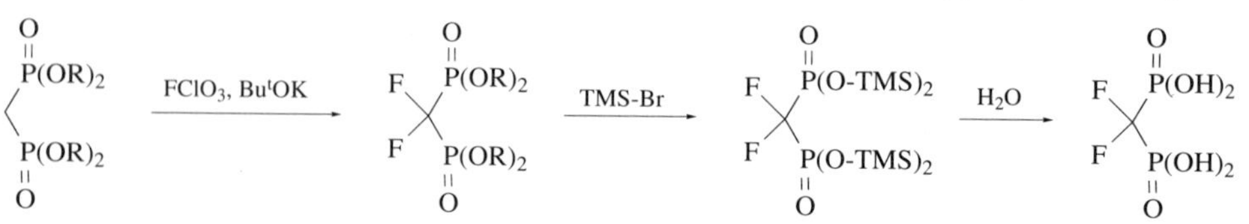

(121) **(122)** **(123)** **(124)**

Scheme 29

Coupling of $X_2C(PO_3H_2)_2$ with 5′-phosphoromorpholidate of adenosine resulted in the formation of (**125**) ⟨84JCS(P1)1119⟩. In the same way, the CF_2 analogue of 3′-azido-3-deoxythymidine triphosphate was obtained ⟨91BMC357⟩. Other nucleotide analogues with a CF_2 group in the place of an oxygen next to the first phosphorus atom, of the general formula (**126**), were prepared from the tris(tetrabutylammonium) salt of $F_2C(PO_3H_2)_2$ and the tosylate of the nucleoside; these initially gave the CF_2 pyrophosphate analogues, which were converted with phosphate into (**126**) ⟨91TL6425⟩. Isopentenyl and geranyl difluoromethylenephosphonates and some nucleotide analogues were also prepared using this methodology ⟨86JOC4768, 87JA5542, 87JOC1794⟩.

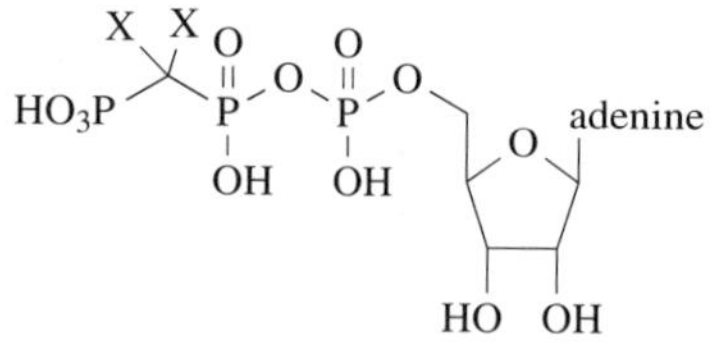

(125)
X = F, Cl

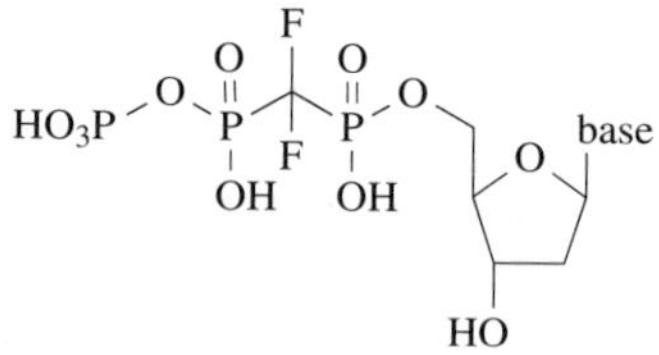

(126)

6.08.4.2.2 *Cyclic compounds*

Reaction of the ylides (**127**) with triphenylphosphine led to the formation of diphosphiranes (**128**) (Equation (27)) ⟨80IJC(B)610⟩. Similar compounds (**130**) were obtained stereospecifically from *trans*-diphosphene (**129**) and dichloro- or dibromocarbenes (Equation (28)) ⟨87T1793⟩. Yields became quantitative when sonication was applied ⟨91TL5965⟩. Bis(trifluoromethyl)phosphine (**131**), upon treatment with $ZnMe_2$ or triethylamine, was converted via $F_3CP{=}CF_2$ to 1,3-diphosphetane (**132**) (Equation (29)) ⟨72CC673, 83IC2573, 85PS(21)349⟩. Thermolysis of $Me_3SnP(CF_3)_2$, however, was more efficient; this initially gave $F_3CP{=}CF_2$, which was found to be fairly stable. Depending on the conditions, the thermolysis products could be either $F_3CP{=}CF_2$ or (**132**) (a mixture of *cis* and *trans* isomers), along with some trimer, that is, 2,4,6-tris(trifluoromethyl)-1,3,5-triphosphorin ⟨84AG(E)710⟩. Similar thermolysis at 0.001 torr (0.13 Pa) of $Me_2NP(SnMe_3)CF_3$ resulted in the formation of the monomer $Me_2NP{=}CF_2$, which underwent polymerization; when the polymer was heated at 500–600 °C, the dimeric compound (**133**) was formed ⟨92ZN(B)321⟩. Treatment of Cl_2PCHCl_2 with triethylamine led to the formation of (**134**) via $ClP{=}CCl_2$ ⟨81ZOB2630⟩. Another heterocycle (**135**) was formed from the reaction of $Cl_2P(S)CF_2P(S)Cl_2$ and butylamine ⟨87CZ176⟩.

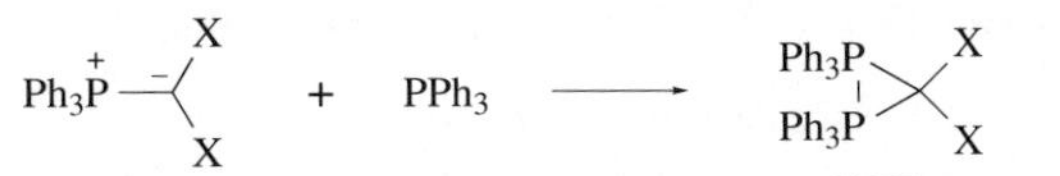

(**127**) X = Cl, Br (**128**) (27)

(**129**) (**130**) (28)

Ar = 2, 4, 6-$Bu^t_3C_6H_2$

X = Cl, Br

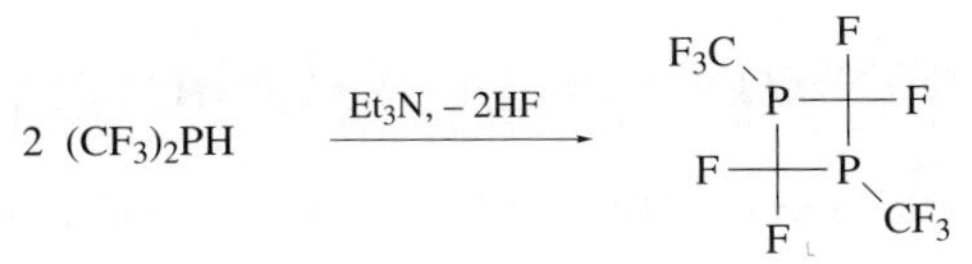

(**131**) (**132**) (29)

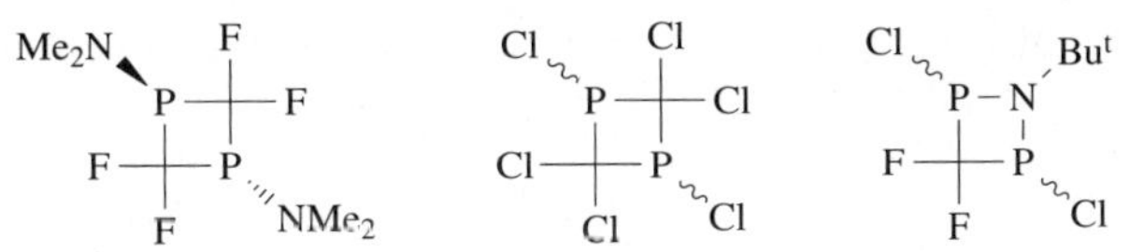

(**133**) (**134**) (**135**)

6.08.4.2.3 *Miscellaneous compounds*

When Me_3GePMe_2 was heated in benzene with $Hg(CF_3)_2$, a mixture of products was formed, including $Me_2PCF_2PMe_2$ ⟨78BSF(2)361⟩. Both isomeric 1,3-diphosphetanes (**132**) reacted with methanol (the *cis* isomer reacted faster), affording $F_3CP(OMe)CF_2P(CF_3)CHF_2$ and other products of further methanolysis ⟨85IC148⟩. The chlorophosphine Cl_2PCHCl_2, when heated with triethylamine and PCl_3, was transformed into $Cl_2PCCl_2PCl_2$ ⟨81ZOB2630⟩. The monomeric $F_2C{=}PCF_3$ reacted with several dialkylphosphines forming (**136**), which upon heating with elemental sulfur afforded (**137**) (Scheme 30) ⟨90ZN(B)299⟩. Compound (**138**) was converted into several derivatives, as illustrated in Scheme 31 ⟨87CZ176⟩. Carbon vapour reacted with PCl_3 forming $Cl_2PCCl_2PCl_2$ ⟨70JCS(A)31⟩. Some chlorinated and fluorinated bis(phosphoranium) salts of the general formula $R_3P^+C^-(X)P^+R_3X^-$ are formed in solution when phosphines are treated with CCl_4, $CFCl_3$, or $CFBr_3$ ⟨77S699, 88JOC366⟩.

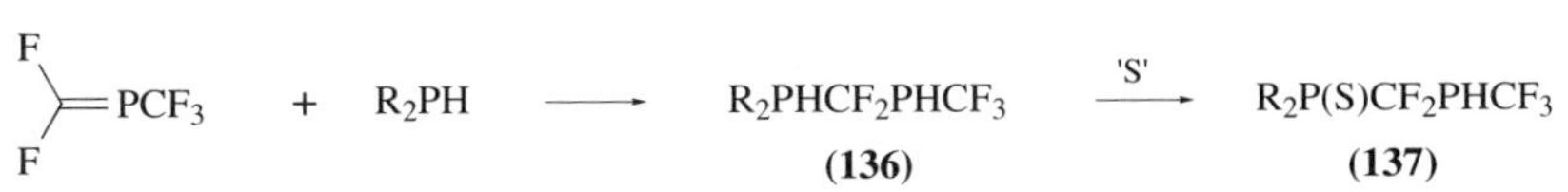

(**136**) (**137**)

Scheme 30

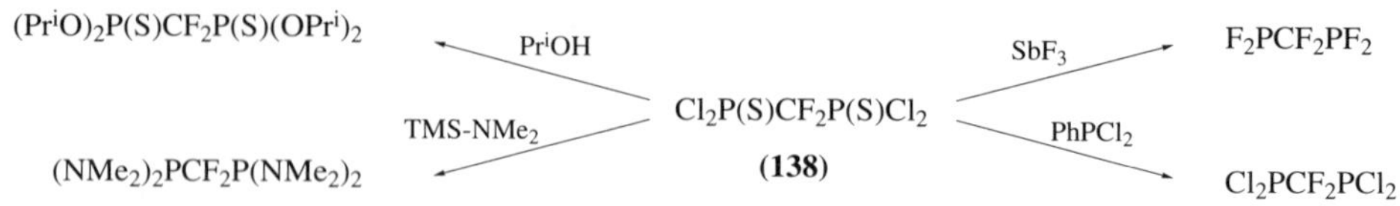

(138)

Scheme 31

6.08.4.3 Two Halogens and Two Other Group 15 Element Functions

The sole compounds in this category are (**140**) and (**141**), which were obtained upon thermolysis of (**139**) (Equation (30)) ⟨84AG(E)710⟩.

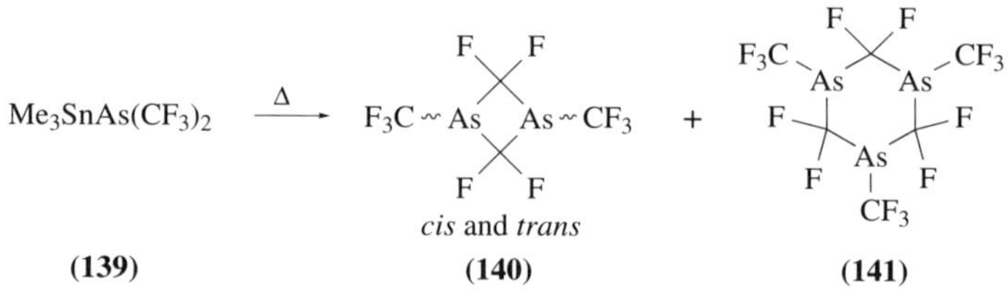

(30)

(139) (140) (141)

6.08.5 TWO HALOGENS AND ONE GROUP 15 ELEMENT FUNCTION

6.08.5.1 Two Halogens, a Phosphorus, and a Metalloid or Metal Function

When reacted with butyllithium in THF solution, diethyl trichloromethylphosphonate (**142**) gives the salt (**143**) which can be converted into (**144**) upon treatment with trimethylsilyl chloride (Scheme 32) ⟨73JOM(59)237⟩. The same salt (**143**) was formed from diethyl chloromethylphosphonate upon treatment with butyllithium and then CCl_4 ⟨75S535⟩. The difluoro analogue of (**143**) was similarly obtained *in situ* from diethyl difluoromethylphosphonate and LDA. With TMS-Cl this afforded TMS-$CF_2P(O)(OEt)_2$ in 87% yield, while with Bu_3SnCl the phosphonate $Bu_3SnCF_2P(O)(OEt)_2$ was obtained in 77% yield ⟨82TL2323⟩. The salt $LiCF_2P(O)Ph_2$ was formed *in situ* from $HCF_2P(O)Ph_2$ and LDA ⟨90TL5571⟩. Several compounds with a P—Ge bond, mainly germylphospholanes, underwent CF_2 insertion into this bond; the mildest way to produce difluorocarbene was the thermal decomposition of $Hg(CF_3)_2$ in benzene ⟨78BSF(2)361⟩. Upon reaction with metallic cadmium the phosphonate (**145**) was converted into the stable salts (**146**) and (**147**) (Equation (31)) ⟨81JFC(18)197⟩. The same phosphonate (**145**) reacted similarly with metallic zinc to afford $BrZnCF_2P(O)(OEt)_2$; after several days in contact with PhHgCl, this gave another stable organometallic compound, $PhHgCF_2P(O)(OEt)_2$ ⟨89JOC613, 90JFC(49)75⟩. The related compound $PhHgCCl_2P(O)(OMe)_2$ was formed from $PhHgCCl_2Br$ and $P(OMe)_3$ ⟨66JOM(5)185⟩. The stable copper compounds BrM-$CuCF_2P(O)(EtO)_2$ (M = Cd, Zn) were produced from the corresponding cadmium or zinc phosphonates and CuBr ⟨92T189⟩. A stable rhodium complex containing the group PCF_2P has been described ⟨90JOM(399)189⟩.

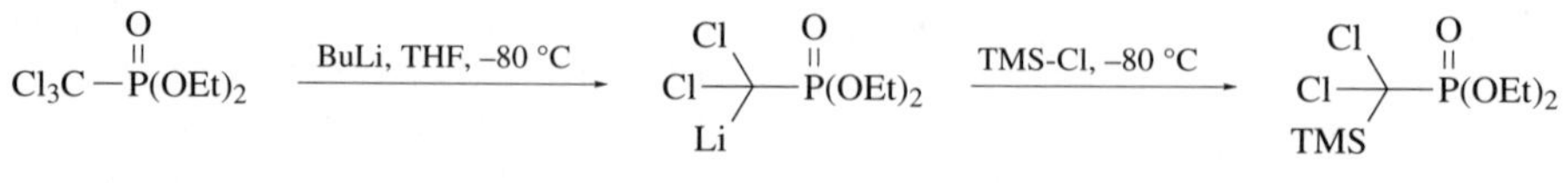

(142) (143) (144)

Scheme 32

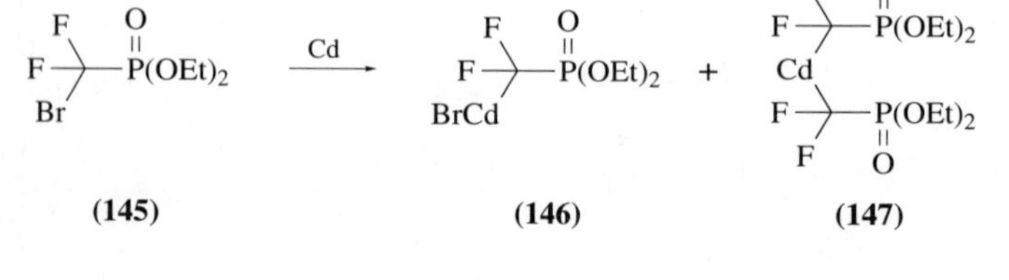

(31)

(145) (146) (147)

6.08.6 TWO HALOGENS AND TWO METALLOID FUNCTIONS

6.08.6.1 Two Halogens and Two Silicon Functions

The chemistry of these compounds is dominated by the work of G. Fritz on carbosilanes; most of his results have been summarized in a book and a review article ⟨B-86MI 608-01, 87AG(E)1111⟩.

6.08.6.1.1 Linear carbosilanes

A simple compound of this class is perchloro-1,3-disilapropane (**148**), which was obtained by photochlorination of the main product resulting from the reaction between chloroform and elemental silicon in the presence of copper ⟨58CB22⟩. It was also produced by photochlorination of $Cl_3SiCH_2SiCl_3$ ⟨57IZV199, 64CB1673⟩. Perchloro-1,3-disilapropane (**148**) readily replaced one or more of its silicon-bound chlorine atoms; for example, upon treatment with SbF_3 or ZnF_2 it was converted to (**149**) ⟨64CB1673, 71AG(E)510, 72ZAAC(391)219⟩. Further transformations were effected with $LiAlH_4$: to (**150**) and to (**151**) (the latter with one equivalent of MeMgCl) (Scheme 33) ⟨71ZAAC(382)9, 71ZN(B)480⟩. Interestingly, MeLi methylated exclusively the carbon atom of compound (**148**). In contrast, the reaction of (**150**) with MeLi was very complex, the major products being MeH_2Si CCl_2SiH_3 and CCl_2-containing 1,3,5-trisilapentanes. The reaction of (**150**) with MeMgCl was similarly complex; however, in the presence of a great excess of MeMgCl, the main product was TMS-CH_2SiHMe_2 ⟨78ZAAC(441)125⟩. Photobromination of $Cl_3SiCH_2SiCl_3$ by BrCl readily afforded $Cl_3SiCBr_2SiCl_3$ ⟨76ZAAC(419)213⟩. A well-studied compound related to (**148**) is (**153**); it has been prepared by several methods. One method was by reaction of (**149**) with MeLi ⟨71ZN(B)480⟩. Another approach involved lithiation of dichloromethane (two equivalents of BuLi) and subsequent reaction with trimethylsilyl chloride ⟨67JCS(C)1470, 72JOC2662⟩. Alternatively, the lithiated (**152**) afforded (**153**), again with TMS-Cl (Scheme 34) ⟨70JOM(23)361⟩. A better yield was achieved upon insertion of dichlorocarbene into the Si—Si bond of (**154**). The dibromo analogue at the central carbon of (**153**) was prepared from TMS-CBr_2Li and TMS-Cl, and also from the complex TMS-$BrP(NEt_2)_3$ and TMS-Br ⟨70JOM(24)647, 70TL4693, 90ZOB709⟩.

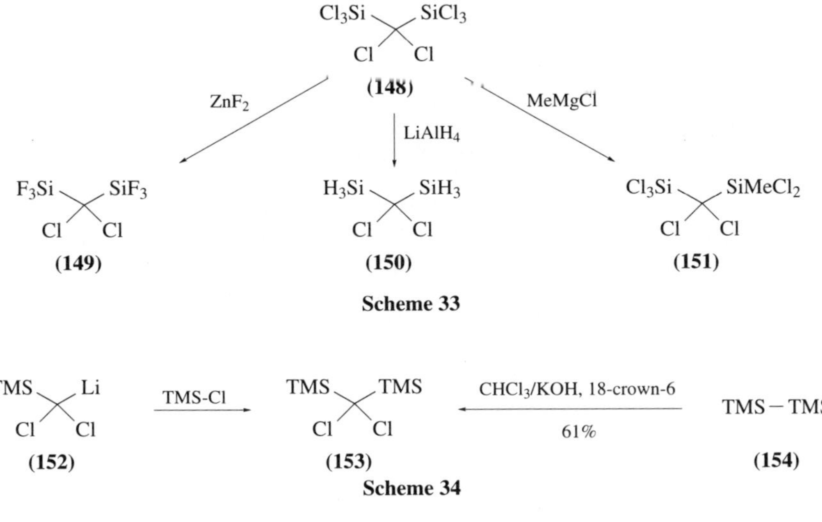

Scheme 33

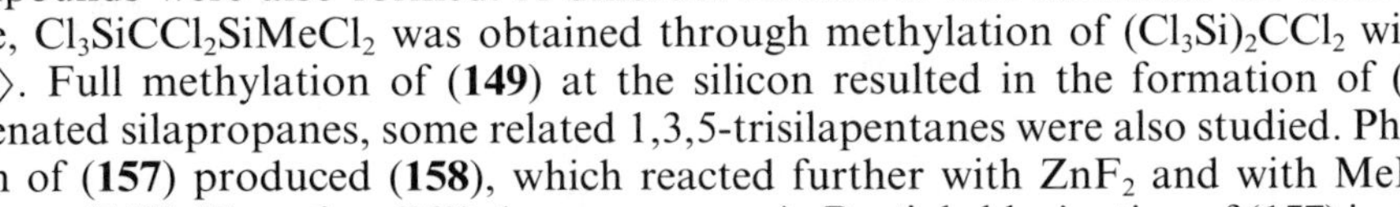

Scheme 34

Partially chlorinated 2,2-dichloro-1,3-disilapropanes such as (**155**) and (**156**) were also obtained from (**152**) and Me_2SiCl_2 or $MeSiCl_3$ (Scheme 35) ⟨79ZAAC(448)55⟩. Starting with $PhMe_2SiCCl_2Li$, similar compounds were also formed. A different method is also accessible for such compounds; for example, $Cl_3SiCCl_2SiMeCl_2$ was obtained through methylation of $(Cl_3Si)_2CCl_2$ with MeMgCl ⟨71ZN(B)480⟩. Full methylation of (**149**) at the silicon resulted in the formation of (**153**). Apart from halogenated silapropanes, some related 1,3,5-trisilapentanes were also studied. Photochemical chlorination of (**157**) produced (**158**), which reacted further with ZnF_2 and with MeMgCl in the same manner as (**148**) (Equation (32)) ⟨78ZAAC(439)161⟩. Partial chlorination of (**157**) is also possible, using sulfuryl chloride and benzoyl peroxide ⟨71ZAAC(382)217⟩. The most interesting reaction of (**158**) was methylation with MeMgCl: when MeMgCl was in large excess (18 : 1), the main products were TMS-C≡C-TMS (47%) and CH_2=C(TMS)CH(TMS)$_2$ (24%) ⟨74ZAAC(404)103⟩. The same reaction was studied with some fluorinated silicon compounds such as $F_3SiCCl_2SiF_2CH_2SiF_3$ and $F_3SiCCl_2SiF_2CHClSiF_3$ ⟨78ZAAC(439)161⟩. An octahedral silicon complex has been described, resulting from the reaction between 2,2′-bipyridyl and $(Cl_3Si)_2CCl_2$ ⟨90CB945⟩.

(155) (152) (156)

Scheme 35

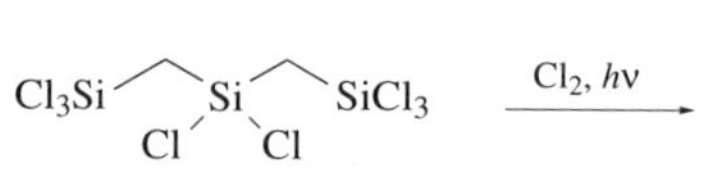

(157) (158) (32)

A fair number of difluorocarbodisilanes are known. Several disilanes of the general formula (**159**) and other related compounds reacted readily with difluorocarbene, produced by thermolysis of Me_3SnCF_3; the products (**160**) resulted from CF_2 insertion into the Si—Si bond (Equation (33)). Some trisilanes also reacted in the same way. For example, $FMe_2Si(SiMe_2)SiMe_2F$, which was converted by single or double CF_2 insertion to $FMe_2SiCF_2SiMe_2SiMe_2F$ or $(FMe_2SiCF_2)_2SiMe_2$. Compound (**160**) (X = F) can react with Grignard and organolithium compounds (MeMgCl, MeLi, PhLi, etc.); one or both silicon-bound fluorine atoms are then replaced by an alkyl or phenyl group, so that with MeLi either TMS-CF_2SiMe_2F or $(TMS)_2CF_2$ is obtained. Other reactions at the Si—F bond in $(FMe_2Si)_2CF_2$ involved $LiAlH_4$, which converted it to $(HMe_2Si)_2CF_2$, and Me_2PLi, which produced $Me_2PMe_2SiCF_2SiMe_2F$ as well as $(Me_2PSiMe_2)_2CF_2$ ⟨83AG(E)730⟩.

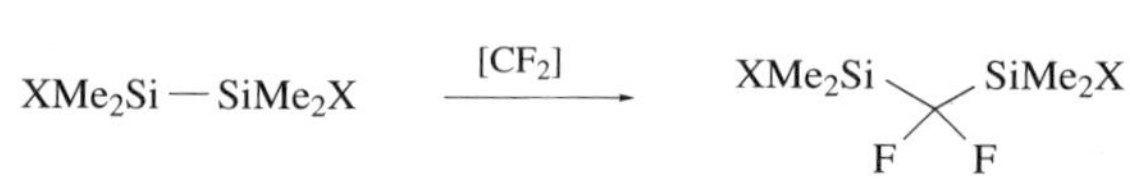

(159) (160) (33)

X = F, Cl, Br, OMe

6.08.6.1.2 *Cyclic carbosilanes*

Reaction of $Cl_3SiCH_2CH_2SiCl_2CH_2Cl$ with magnesium produced 1,3-disilacyclopentane (**161**), which upon photochlorination eventually gave the fully chlorinated compound (**162**) (Scheme 36). An interesting solvent effect was observed in the reaction of (**162**) with an excess of MeMgCl; while in ether several products were formed, the main product (66% yield) being $Cl_2C{=}CClSiMe_2CCl_2$-TMS, in pentane, compound (**163**) was formed exclusively ⟨70ZAAC(372)21, 79ZAAC(458)37⟩. When $MeSiCl_3$ or Me_2SiCl_2 are heated at 700 °C, they are converted to the cyclic carbosilane (**164**), which upon photochlorination affords the perchloro compound (**165**) (Scheme 37) ⟨60ZAAC(303)85⟩. The partially brominated (**166**) was obtained from the photochemical reaction of (**164**) and bromine chloride, whereas the bromo analogue of (**164**) gave the octabromo analogue of (**166**) (Scheme 37) ⟨75ZAAC(419)213⟩. The hexafluoro analogue of (**164**) underwent photochlorination to produce a mixture of 2,2-dichloro-, 2,2,4,4-tetrachloro-, and 2,2,4,4,6,6-hexachloro-hexafluoro trisilacyclohexanes ⟨75ZN(B)965⟩. The reactivity of these compounds is analogous to that already mentioned for their linear analogues, for example (**165**) was converted by $LiAlH_4$ into 2,2,4,4,6,6-hexachlorotrisilacyclohexane ⟨71ZAAC(382)9⟩. Unusual reactivity was, however, noted in the case of MeMgCl; with (**165**), four equivalents of MeMgCl gave the ring-contracted (**167**) in high yield. Further reaction of (**167**) with more MeMgCl afforded several products, the major being (**168**) (Scheme 38) ⟨73ZAAC(395)159, 73ZAAC(399)280⟩. Several partially chlorinated 1,3,5-trisilacyclohexanes are known, some resulting from the interaction of (**165**) with its fully hydrogenated analogue, that is $(H_2SiCH_2)_3$ ⟨71ZAAC(382)217⟩. Their reactivity with MeMgCl is of interest, since different products are obtained in each case depending on the structure of the substrate (Scheme 39). The reactivities of partly brominated trisilacyclohexanes with Grignard and organolithium compounds exhibit some similarity to and a number of differences from their chloro analogues; the most notable difference is the double substitution of carbon-bound bromine by butyl groups when butyllithium is used (Equation (34)). Generally, more results are available in the carbosilane field, with ample discussion and mechanistic details ⟨B-86MI 608-01⟩.

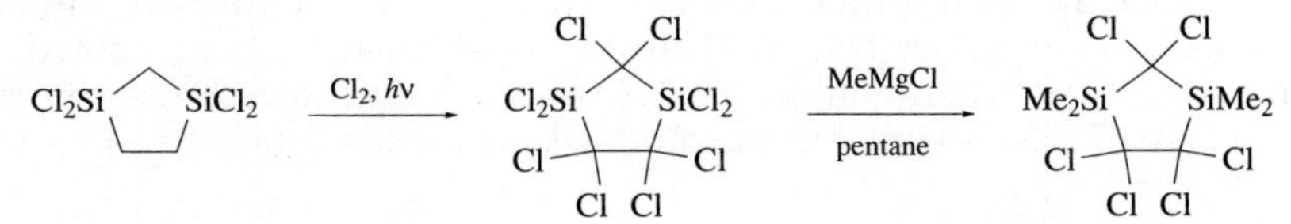

(161) **(162)** **(163)**

Scheme 36

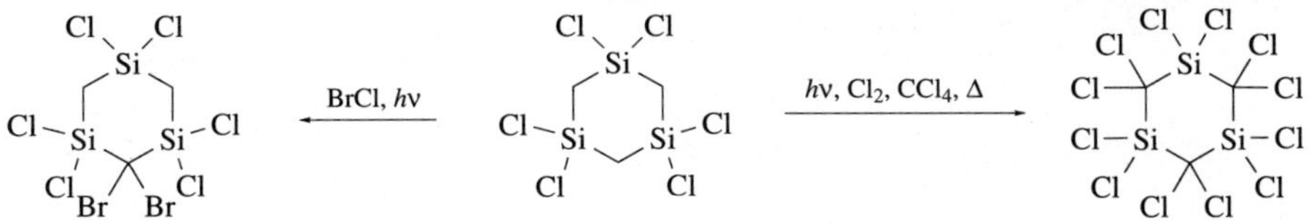

(166) **(164)** **(165)**

Scheme 37

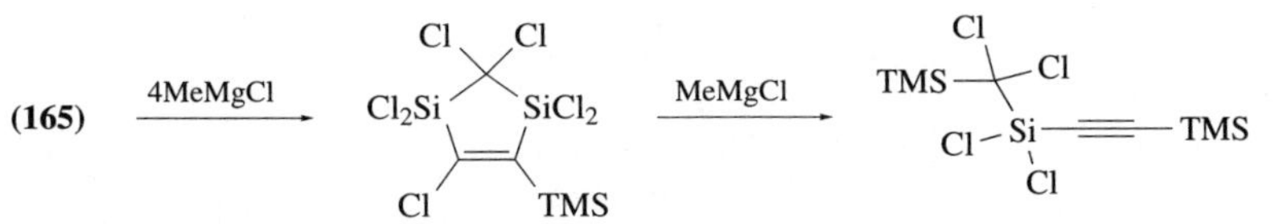

(167) **(168)**

Scheme 38

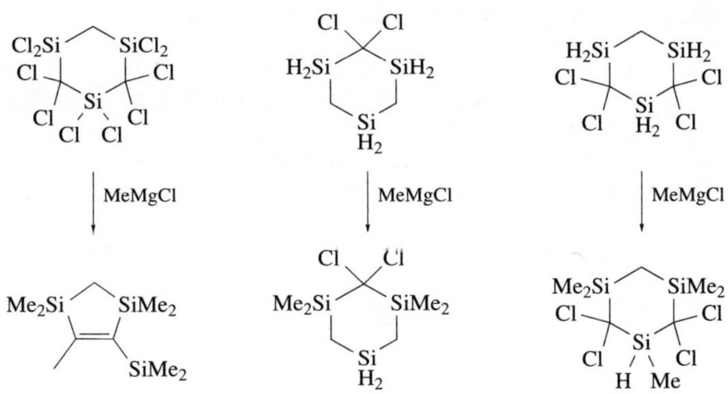

Scheme 39

(34)

6.08.6.2 Two Halogens and Two Boron or Other Metalloid Functions

Two compounds of this type are known. The first is $Cl_2BCCl_2BCl_2$, which was formed when carbon vapour reacted with either BCl_3 or B_2Cl_4 ⟨69JCS(A)1882⟩. The other is $Cl_3GeCCl_2GeCl_3$; this was obtained in a similar way from carbon vapour and $GeCl_4$ ⟨70JCS(A)31⟩.

6.08.7 TWO HALOGENS AND ONE METALLOID FUNCTION

6.08.7.1 Two Halogens, a Silicon, and a Metal Function

Apart from Grignard and organolithium compounds which exist only in solution (Section 6.08.6.1), some stable compounds of this category are known with tin and mercury. They are normally formed from the appropriate organometallic compound coupled with metal or silyl chlorides; in one case, an exchange reaction between organomercurials was noted. Examples are given in Scheme 40 ⟨70JOM(23)361, 70JOM(24)674, 71JOM(27)19⟩. Dichlorocarbene insertion into the Si—Hg and Si—Ge bond also occurs, but the products are further transformed ⟨67JOM(7)P20⟩.

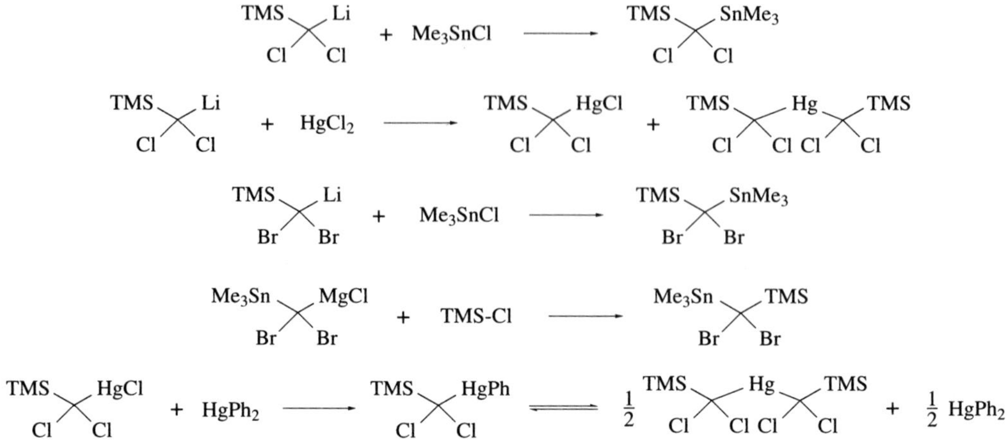

Scheme 40

6.08.8 TWO HALOGENS AND TWO METAL FUNCTIONS

Stable isolable compounds of this category involve tin. Most were prepared by insertion of dihalocarbenes, from organomercurial precursors or chloroform (Equation (35)) ⟨67JA2790, 69JA1954, 91SRI401⟩. Coupling of Me_3SnCBr_2-TMS with Me_3SnCBr_2MgCl also produced $(Me_3Sn)_2CBr_2$ ⟨70JOM(24)647⟩. The theoretically interesting dilithiodifluoromethane was shown by *ab initio* calculations to have a planar lithium bridged structure of CLi_2F^+ F^- ion pair character ⟨85CC5⟩.

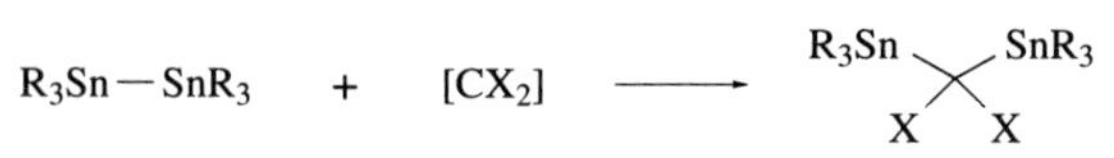

(35)

6.09
Functions Containing One Halogen and Three Other Heteroatom Substituents

ANGELA MARINETTI and PHILIPPE SAVIGNAC
DCPH - Ecole Polytechnique, Palaiseau, France

6.09.1 ONE HALOGEN AND THREE CHALCOGENS

6.09.1.1 One Halogen and Three Oxygen Functions

Acyclic covalent halogenides having the general structure—one halogen and three oxygen functions—are known for chlorine and fluorine. In most cases, the oxygen substituents are polyfluorinated or polynitrated groups.

Poly(nitro)alkyl chloro *ortho*-formates are reactive intermediates in the synthesis of explosive *ortho*-carbonates and are themselves explosive compounds ⟨83USP467715⟩. They have been prepared by chlorination of the corresponding trichloromethyl disulfides (Scheme 1); their reaction with the HF–pyridine complex yields fluoro *ortho*-formates ⟨83JEM95⟩ through a chlorine–fluorine exchange reaction. Highly fluorinated peroxide derivatives—exhibiting a better thermal stability than non-fluorinated analogues—are prepared by CsF-catalyzed fluorination of the corresponding carbonyl compounds (Equation (1)) ⟨68JOC2095, 76JFC(7)501⟩. In both cases, limitation of sample size and suitable safety equipment are required.

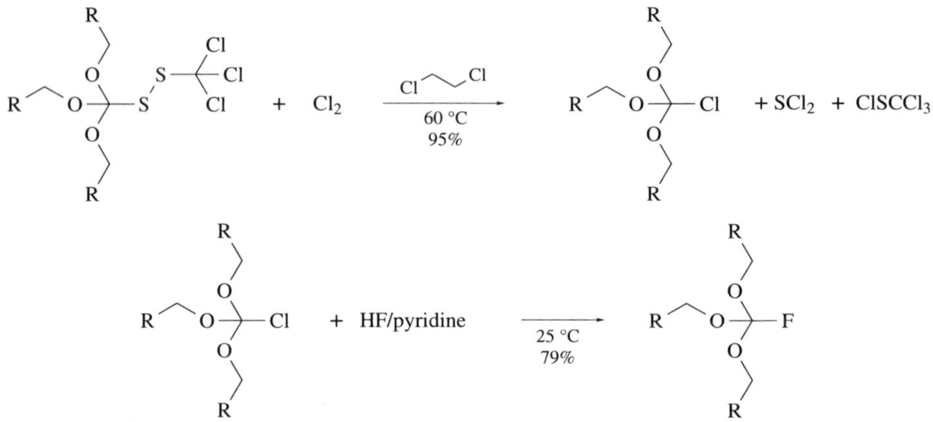

R = $C(NO_2)_2F$

Scheme 1

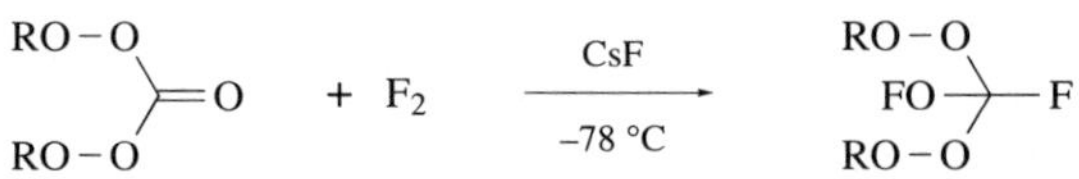

(1)

R = CF_3, 100%
SF_5, 60%

Reaction of carbon tetrachloride with pentafluorophenol in the presence of aluminum chloride, or with bromine fluorosulfate, gives the trisubstituted species in low yields and as a mixture with several by-products (Scheme 2) ⟨68IC434, 88ZOR1513⟩.

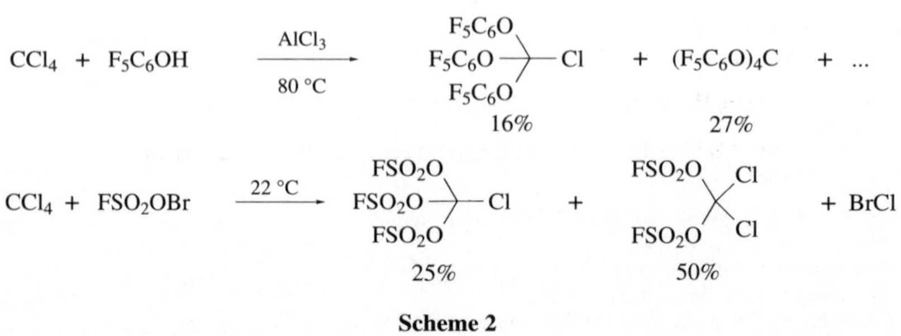

Scheme 2

6.09.1.2 One Halogen, Sulfur, and Oxygen Function

6.09.1.2.1 One halogen and three sulfur functions

(i) Acyclic covalent halogenides

Compounds of this type have been reported with sulfide, oligosulfide, and sulfone substructures. Polyhalogenated derivatives are by far the most numerous. The principal methods of preparation are exemplified below.

(a) From thiocarbonyl compounds and halogens. The oldest method of preparation was described by Hass and Klug ⟨66AG(E)845, 68CB2609⟩. This involves the rapid and quantitative addition of chlorine (or bromine) to bis(trifluoromethyl) trithiocarbonate to give the sulfenyl chloride (Equation (2)). Several examples of halogenations using chlorine and thionyl chloride have been described subsequently (Equation (3) and Table 1). Starting materials are either trithiocarbonates with strong electron withdrawing groups, R^1 and R^2, or sulfone derivatives. Under similar conditions, bromine and iodine failed to react with trithiocarbonate-*S*,*S*-dioxides (entries 2 and 3 in Table 1).

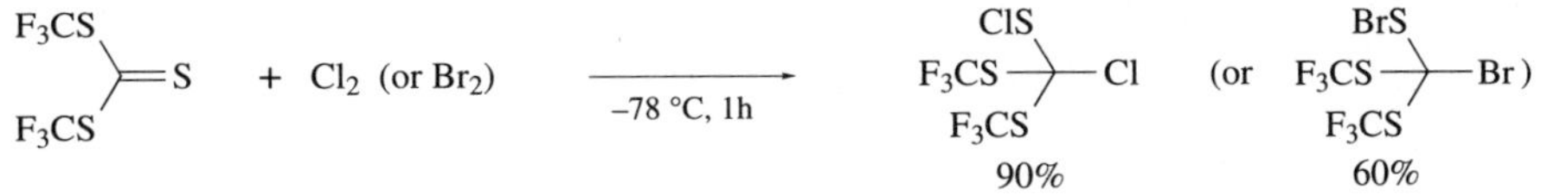

(2)

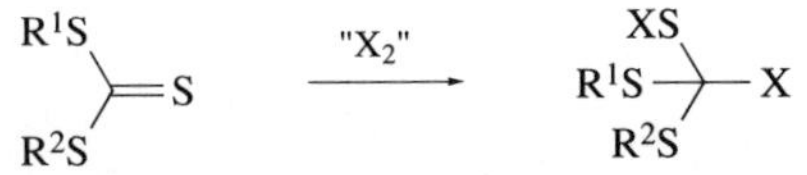

(3)

Table 1 Halogenation of thiocarbonyl compounds.

Entry	R^1	R^2	*Reagents and conditions*	*Yield* (%)	*Ref.*
1	CF_3	CF_3	FCl, −80°C	(mixture)	76CB1976
2	PhS	$PhSO_2$	Cl_2, CCl_4	80	76CB1069
3	PhS	4-Me.$C_6H_4SO_2$	Cl_2, CCl_4	88	76CB1069
4	CN	CN	Cl_2, 10°C, CH_2Cl_2	80	81CB1132
5	Cl_3CS	Cl_3CS	Cl_2, CCl_4	91	84SUL85
6	Cl_3CCCl_2S	Cl_3CCCl_2S	Cl_2, CCl_4	100	85T5145
7	$PhCH_2$	Cl_3CS	SO_2Cl_2, CH_2Cl_2	98	85T5145
8	Me	Me	SO_2Cl_2, −15°C, pentane	53	84SUL241

Chlorination with chlorine or thionyl chloride is a very efficient approach to sulfenyl chlorides: most of the reactions reported in Table 1 are rapid, clean, and quantitative. The sulfenyl chlorides thus obtained are useful starting materials for the synthesis of more elaborate, trisubstituted trithio *ortho*-formates. The most significant examples of nucleophilic substitution reactions of the SCl function by amines, thiols, and amides (Equation (4)) are given in Table 2. The reactivity of the S—Cl bond is generally increased by electron withdrawing substituents R^1 and R^2. Some of the polyfluoro derivatives are claimed to be fungicidal agents, and the synthesis of the phthalimido derivative (entry 2 in Table 2) has been patented ⟨70GEP1908680⟩.

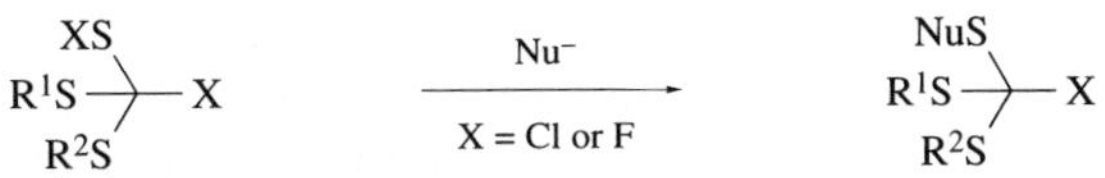

(4)

(b) From thiocarbonyl compounds and sulfenyl halogenides. Thiocarbonates having the general structure $[R(S)_nS]_2C{=}S$ ($n = 1$, 2 and 3) are formed from thallium or barium thiocarbonates by reaction with sulfenyl chlorides (Equation (5) and Table 3). When an excess of sulfenyl chloride is used, the transient thiocarbonate reacts further to give the chlorothio *ortho*-formate in fair to moderate yields (entries 1 and 2 in Table 3). The synthesis of various polyhalogenated *ortho*-formates bearing disulfide or trisulfide units has also been reported ⟨85T5145⟩.

Table 2 Nucleophilic substitution reactions of sulfenyl chlorides.

R[1]	*R*[2]	*Nu*	*Conditions*	*Yield* (%)	*Ref.*
CF_3	CF_3	Et_2NH	petroleum	82	76CB1976
CF_3	CF_3	phthalimide	C_6H_6, Et_3N	54	70GEP1908680, 76CB1976
CF_3	CF_3	K-pyrrole	pentane	66	83CB3325
Cl_3CS	Cl_3CS	Na phthalimide	CH_2Cl_2, H_2O	50	85T5145
CF_3	CF_3	NH_3	−80 °C	67	74CZ109
PhS	4-Me.$C_6H_4SO_2$	morpholine	$CHCl_3$	90	76CB1069
PhS	Ph-SO_2	morpholine	$CHCl_3$	87	76CB1069
PhS	4-Me.$C_6H_4SO_2$	PhSH	$CHCl_3$	75	76CB1069
PhS	4-Me.$C_6H_4SO_2$	Bu^tSH	$CHCl_3$	87	76CB1069
PhS	4-Me.$C_6H_4SO_2$	4-Me.$C_6H_4SO_2Na$	C_6H_6	28	76CB1069
PhS	4-Me.$C_6H_4SO_2$	MeCOSH	CCl_4, 50 °C	95	86T739
CF_3	CF_3	AgCN	20 °C	97	76CB1976
CF_3	CF_3	AgOCN	50 °C	6	76CB1976
CF_3	CF_3	AgSCN	20 °C	66	76CB1976
CF_3	CF_3	AgSeCN	20 °C	79	76CB1976
CN	CN	H_4NSCN	SO_2, −20 °C	92	81CB1132
CF_3	CF_3	$F_2C=S$	hν	30	76CB1976

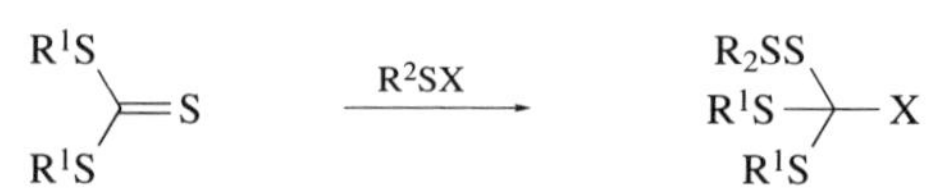

(5)

Table 3 Addition of sulfenyl chlorides to thiocarbonyl compounds.

R[1]	*R*[2]	*Reagents and conditions*	*Yield* (%)	*Ref.*
CF_3S	CF_3	$Tl_2CS_3 + F_3CSCl$	35	76CB3432
Cl_3CS_2	Cl_3CS	$BaCS_3 + Cl_3C\text{-}S_2Cl$, MeCN	10	85T5145
Cl_3CS	Cl_3C	Cl_3CSCl, MeCN, 80 °C		85T5145
Cl_3CCl_2CS	Cl_3CCl_2C	$Cl_3C\text{-}Cl_2CSCl$, MeCN, 80 °C	56	85T5145
FCl_2CCl_2CS	FCl_2CCl_2C	$FCl_2C\text{-}Cl_2C\text{-}SCl$, MeCN, 80 °C		85T5145
Cl_3CS	Cl	SCl_2, 0 °C	100	84SUL85
CF_3	CF_3	F_3CSF, −196 to −60 °C	51	92ZN(B)369

Sulfur dichloride reacts cleanly at low temperature with bis(trichloromethylthio) trithiocarbonate to give the corresponding chlorothio *ortho*-formate in quantitative yield (entry 6 in Table 3). The reaction of bis(trifluoromethyl) thiocarbonate with trifluoromethylsulfenyl fluoride must be performed at very low temperature (−190 °C), because of the much higher reactivity of F_3CSF compared to F_3CSCl. The structures of tris(trichloromethyldithio)- and (trichloromethyltrithio)chloro *ortho*-formate (entries 2 and 3 in Table 3) have been established by x-ray crystallography.

(c) By halogenation of trisulfonylmethanes. Trialkyl-, triaryl-, and trifluoro-sulfonylmethanes are very strong acids; in the presence of bases (Ag_2CO_3 or NaOH) the corresponding silver or sodium salts are formed. They react smoothly with halogens (Cl_2, Br_2, and I_2) (Equation (6) and Table 4). The iodo derivatives are generally less stable than their chloro or bromo analogs. The fluoro derivative could not to be isolated from the reaction of $(F_3CSO_2)_3C^-Ag^+$ with XeF_2 ⟨88IC2135⟩, but $(FSO_2)_3CF$ is obtained from reaction of $(FSO_2)_3CH$ with XeF_2 ⟨80AG(E)942⟩. A few examples of direct halogenation of trisulfonylmethanes in refluxing CCl_4 in the absence of any deprotonating agents have been reported (entries 8–16 in Table 4). Yields are satisfactory.

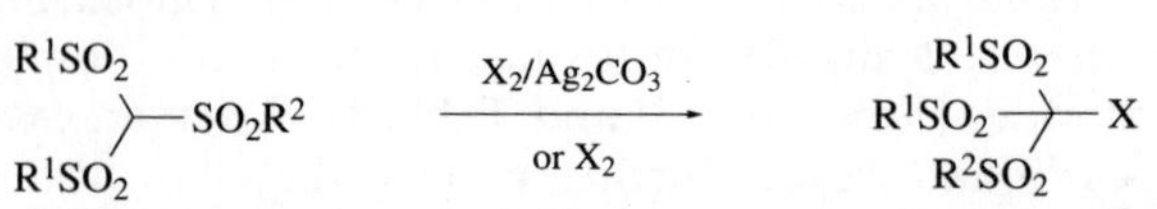

(6)

Only two examples of lithiated chlorobis(sulfonyl)methanes have been reported. They react with

Table 4 Halogenation of trisulfonylmethanes.

R^1	*R^2*	*Reagents and conditions*	*Yield* (%)	*Ref.*
Me	Me	Cl_2, NaOH, H_2O	88	46RTC53
Me	Me	Br_2, NaOH, H_2O	95	46RTC53
F	F	Br_2, Ag_2CO_3	96	80AG(E)942, 83ZOR79
F	F	I_2, Ag_2CO_3	96	80AG(E)942, 83ZOR79
F	F	XeF_2, −10°C, CF_2Cl_2		80AG(E)942
CF_3	CF_3	Cl_2, Ag_2CO_3, 0°C, CH_2Cl_2	90	88IC2135
CF_3	CF_3	Br_2, Ag_2CO_3, 0°C, CH_2Cl_2	94	88IC2135
Et	Me	Cl_2, 70°C, CCl_4–$CHCl_3$	65	76ZOR558
Et	Me	Br_2, 70°C, CCl_4–$CHCl_3$	92	76ZOR558
Ph	Me	Cl_2, 70°C, CCl_4–$CHCl_3$	75	76ZOR558
Ph	Me	Br_2, 70°C, CCl_4–$CHCl_3$	95	76ZOR558
4-Me.C_6H_4	Me	Cl_2, 70°C, CCl_4–$CHCl_3$	70	76ZOR558
4-Me.C_6H_4	Me	Br_2, 70°C, CCl_4–$CHCl_3$	80	76ZOR558
4-Cl.C_6H_4	Me	Cl_2, 70°C, CCl_4–$CHCl_3$	60	76ZOR558
4-Cl.C_6H_4	Me	Br_2, 70°C, CCl_4–$CHCl_3$	90	76ZOR558
4-Me-3-NO_2.C_6H_3	Me	Br_2, 70°C, CCl_4	63	76ZOR2583

phenylsulfenyl chloride at room temperature to afford chloro(phenylthio)bis(sulfonyl)methanes in low to moderate yield (Equation (7)) ⟨72LA(758)132⟩.

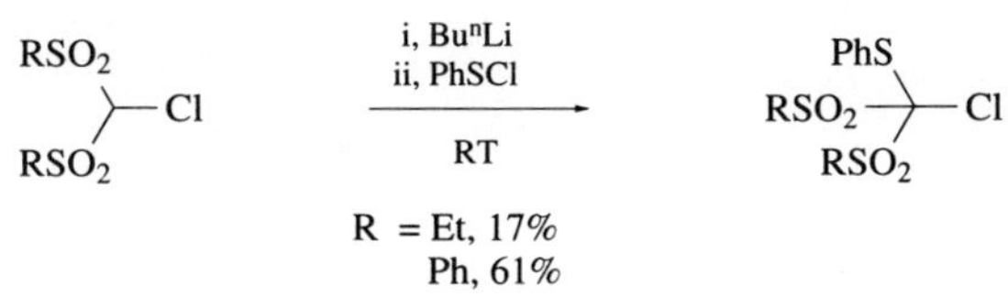

(7)

(ii) Cyclic covalent halogenides

(*a*) *From thiocarbonyl compounds and halogens or sulfenyl halides.* The previously described addition of halogens or sulfenyl halides to trithiocarbonates has been applied efficiently to the synthesis of cyclic derivatives (Scheme 3) ⟨87SUL59, 91CB2025⟩. The four-membered cyclic thiocarbonyl compounds react considerably faster than their acyclic analogues. This reflects the diminished ring strain upon sp^2–sp^3 rehybridization of the C-2 center in the dithietane.

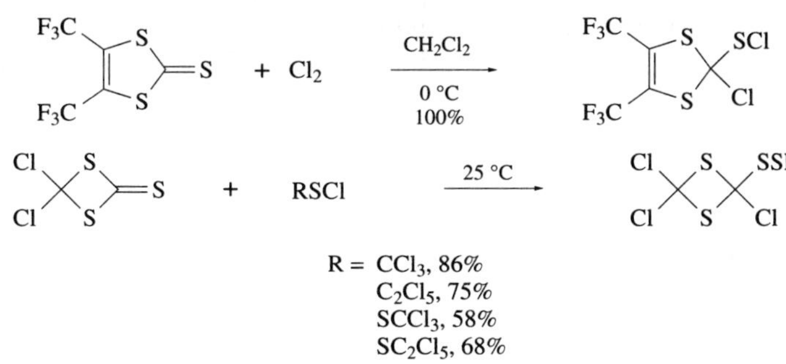

Scheme 3

(*b*) *Fluorination of trithiosubstituted carbocations.* Exchange of the AsF_6- counterion with a fluoride anion in cationic 1,3-dithietane derivatives gives rise to a stable fluorinated compound when R = CF_3, or to an unstable one when R = Me (Equation (8)) ⟨90CB1635⟩. Previously, the same approach had been applied to the synthesis of symmetrical dithietanes from trithiomethyl carbocations (Scheme 4) ⟨87CB429⟩. The first step of the reaction is the dimerization of a fluorodithioformate, induced by the strong acidic medium. Formation of the covalent C—F bond is then promoted by addition of ether, as a solvent, to the reaction mixture.

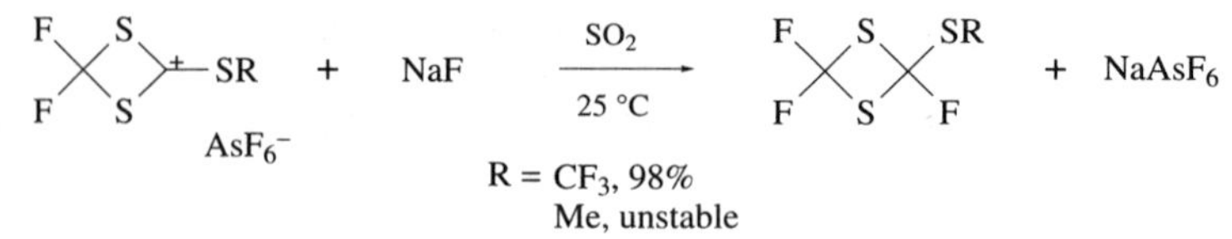

(8)

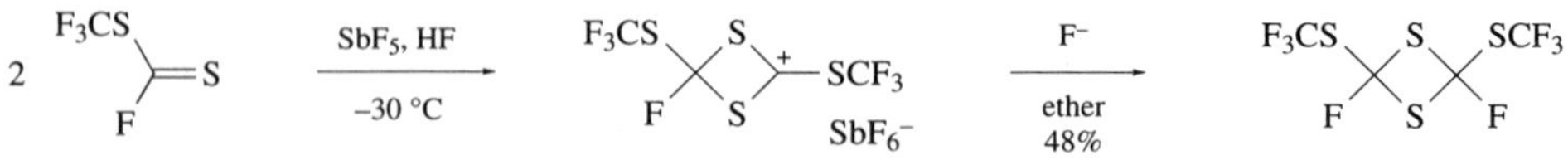

Scheme 4

6.09.1.2.2 One halogen, two sulfur, and one oxygen function

Only two examples of this class of compounds have been isolated and identified: both are derived from carbohydrates and both are symmetrical disulfides resulting from chlorine addition to dithiocarbonates (Equation (9)) ⟨69JOC1642⟩. The final products precipitate from the solution as white solids. They are used as intermediates in the synthesis of sugar chlorothioformates ⟨71CAR(16)145⟩.

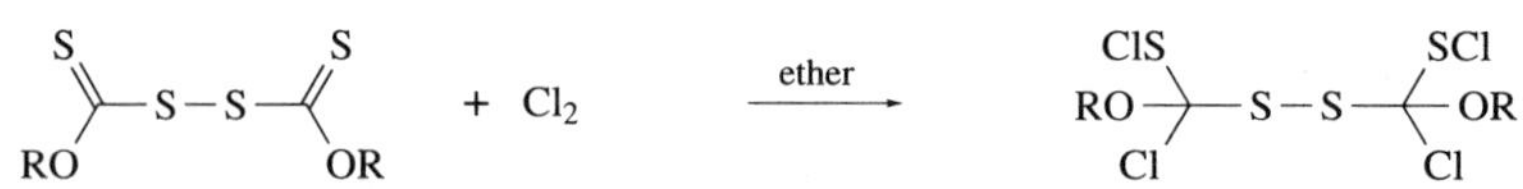

(9)

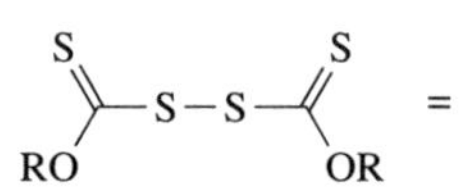

bis (1,2:5,6-di-*O*-isopropylidine-3-*O*-thiocarbonyl-α-D-glucofuranose) disulfide, 100%
bis (methyl 4,6-*O*-benzylidine-2-(and 3-) *O*-thio-carbonyl-α-D-glucopyranoside) disulfide, 35%

Work done in the early 1980s has shown that *O*,*S*-dimethyl dithiocarbonate reacts at 0 °C with sulfuryl chloride to give the expected chlorinated adduct which undergoes a rapid intramolecular rearrangement to afford a methoxydichloro(methyldithio) derivative (Scheme 5) ⟨83TL5683⟩. In the light of this result, a reinterpretation of the earlier literature data on chlorination of thioesters seems to be necessary.

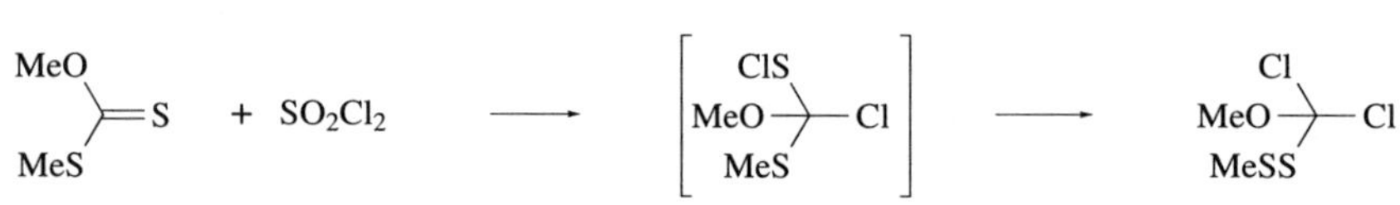

Scheme 5

6.09.1.3 One Halogen, Selenium, and Sulfur Function

The methods available for the synthesis of haloselenothio *ortho*-formates are analogous to those used for the corresponding trithio *ortho*-formates, but with fewer examples. When sulfur and selenium substituents are perhalomethyl, $CF_nCl_{(3-n)}$, chloro *ortho*-formates have been isolated (Equation (10)); the only bromo derivative cited ⟨86ZN(B)413⟩ seems to be unstable. $(CF_3Se)_2C{=}S$ reacts with chlorine at low temperature by addition of halogen to the thiocarbonyl group (Equation (11)) ⟨74CZ511⟩. No experimental details have been given. A report from the mid 1980s concerns the addition of sulfuryl (or selenyl) chlorides to thiocarbonyl derivatives under UV irradiation. Physical and spectroscopical data of the new substances are provided. (Table 5) ⟨86ZN(B)413⟩.

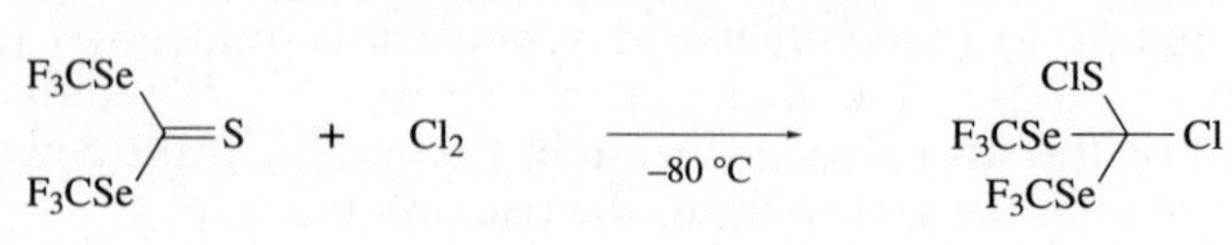

(10)

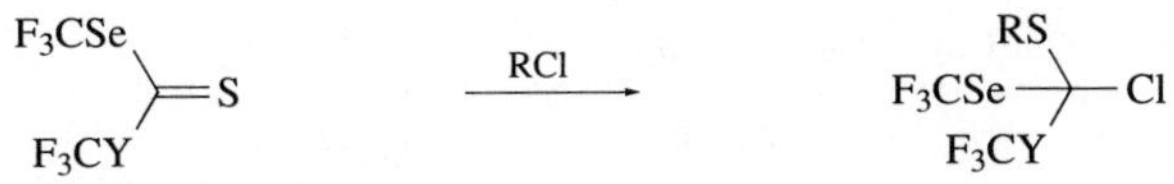
(11)

Table 5 Addition of sulfuryl and selenyl chlorides to thiocarbonyl derivatives.

Entry	*Y*	*Reagents and conditions*	*Yield* (%)
1	Se	F_3CSCl, *hv*	88
2	Se	F_3CSeCl, *hv*	85
3	Se	$F_2ClCSCl$, *hv*	51
4	Se	FCl_2CSCl, *hv*	38
5	S	F_3CSeCl, *hv*	87
6	S	F_3CSCl, *hv*	78

6.09.2 ONE HALOGEN AND TWO CHALCOGENS

6.09.2.1 One Halogen, Two Chalcogens, and a Nitrogen Function

As far as covalent halides are concerned, a few examples of this type have been described, with either oxygen or sulfur as the chalcogens. A unique compound containing two oxygen and a nitrogen functions was obtained by reaction of peroxydisulfuryl difluoride with ClCN under thermal conditions (Equation (12)) ⟨75IC592⟩. The suggested mechanism involves formation of a nitrene and the reaction pathway shown in Scheme 6 is proposed to account for the addition of the $FSO_2O\cdot$ radical to the unsaturated carbon–nitrogen system. This radical addition to ClCN occurs quantitatively and is sufficiently facile to be synthetically useful.

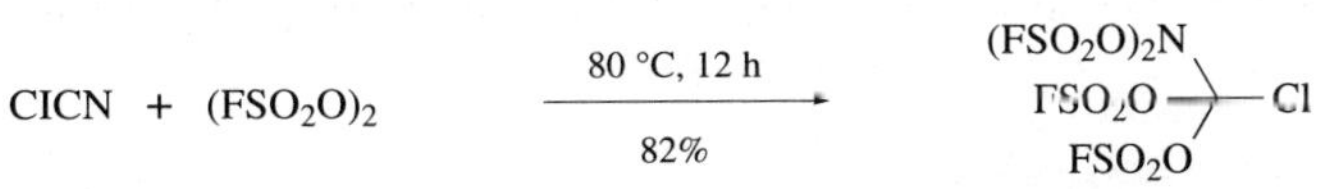
(12)

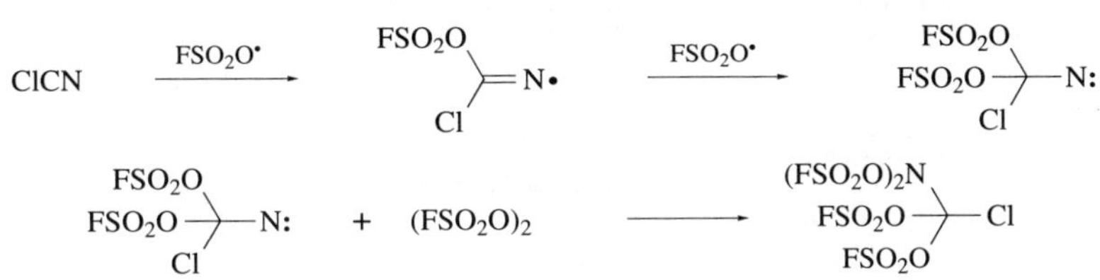

Scheme 6

Bis(3-nitroarylsulfonyl)bromonitromethanes are obtained from bis(arylsulfonyl)(methylsulfonyl) methanes (Scheme 7). After polynitration with HNO_3–H_2SO_4 the methylsulfonyl group is easily removed in basic medium. Two bromination methods, direct bromination with potassium hypobromite and basic cleavage followed by bromine addition, were outlined ⟨76ZOR2583⟩.

Arylazo derivatives of bis(arylsulfonyl)chloromethanes are yellow crystalline compounds which are readily soluble in most organic solvents. They are prepared from bis(arylsulfonyl)chloromethanes in basic medium by reaction with aryldiazonium chlorides (Equation (13)) ⟨68ZOR324⟩.

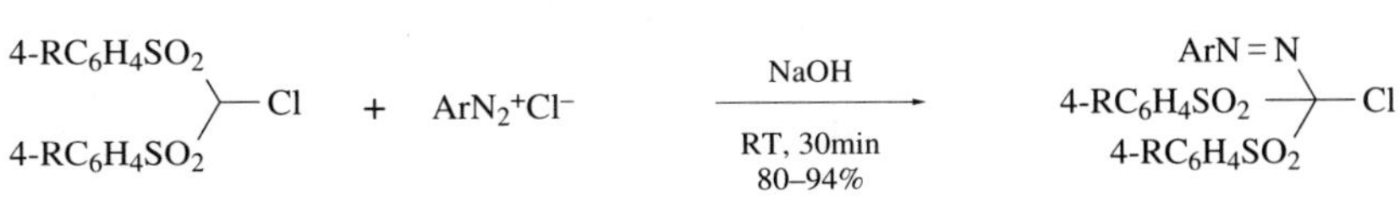
(13)

R = H, Me, Cl, NO_2
Ar = C_6H_5, 4-Me-C_6H_4, 4-Cl-C_6H_4, 4-O_2N-C_6H_4, 4-MeO-C_6H_4, 4-HO_2C-C_6H_4, 2-MeO, 5-Cl-C_6H_3, 2-Me, 5-Cl-C_6H_3

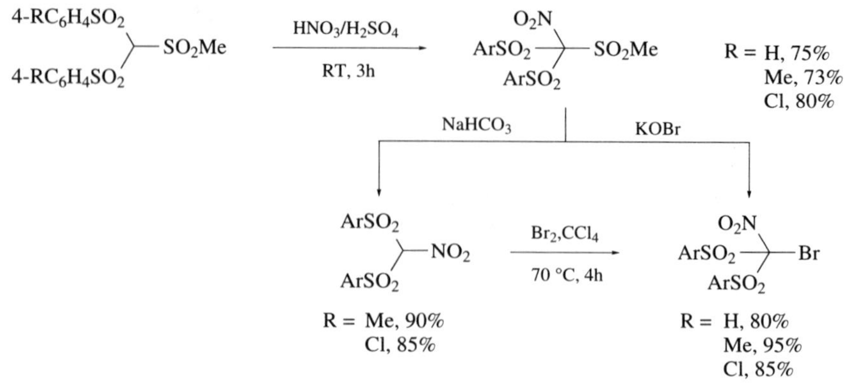

$ArSO_2$ = 4-R, 3-O_2N-C_6H_3

Scheme 7

6.09.2.2 One Halogen, Two Sulfur, and a Phosphorus Function

Triethoxyphosphonium bis(phenylsulfonyl)methylide is one of the few known stable trialkoxyphosphonium ylides, most of which are only reaction intermediates. It reacts as a very weak nucleophile; for example, with bromine it gives the 1-bromomethylphosphonate in very good yield (Scheme 8) ⟨88S913⟩.

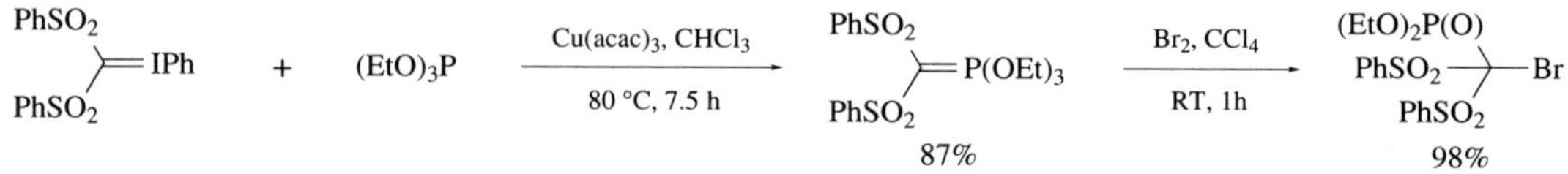

Scheme 8

Mixed sulfonic–phosphonic acids have been synthesized for evaluation as potential electrolytes in fuel cells. The starting material $(EtO)_2P(O)CFBr_2$ was converted into its disodium 1-fluoro-1,1-disulfonylmethylphosphonate salt by repeated reaction with sodium dithionite and hydrogen peroxide (Scheme 9) ⟨89CJC1795⟩. Yields were satisfactory.

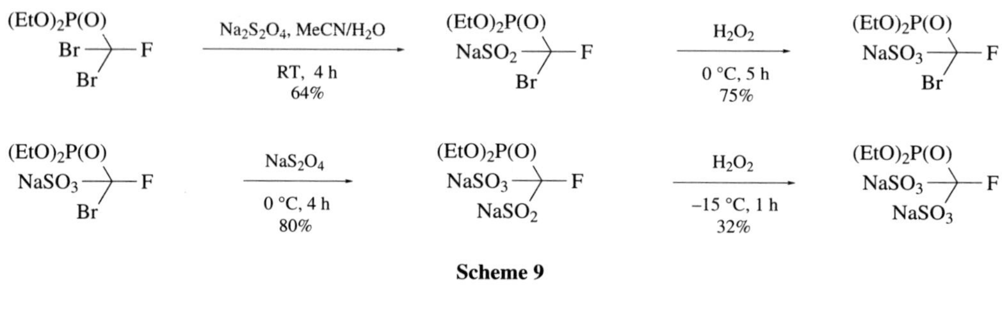

Scheme 9

6.09.2.3 One Halogen, Two Sulfur, and a Metal Function

All the known compounds bearing one halogen, two sulfur, and one metal function on the same carbon atom have the general structure (**1**).

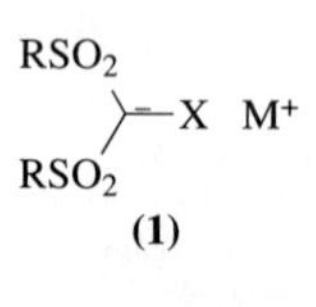

X = Cl,Br
M = Li,Na,K,Ag
R = Ph,CF_3

(**1**)

It has been found that the action of strong bases (for R = Ph) or even weak bases (for R = CF_3) leads to metallation at the C—H bond of bis(sulfonyl)halomethanes (Equation (14)) ⟨72LA(758)132, 73JOC3358, 77ZOR275⟩.

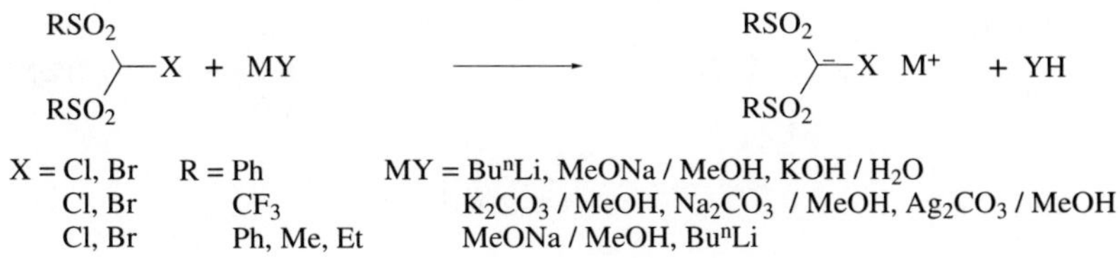

(14)

pK$_a$ values for some monohalodisulfonylmethanes (R = Ph, Me, Et) have been established by potentiometric titration in water–dioxane (1 : 1) ⟨72LA(758)132⟩. The metallohalomethanes are crystalline substances, soluble in water, but insoluble in nonpolar or low polarity organic solvents. They are very stable under ordinary conditions and have been found to be stable up to 250 °C in some cases.

6.09.3 ONE HALOGEN AND ONE CHALCOGEN

6.09.3.1 One Halogen, One Chalcogen, and Two Group 15 Elements

6.09.3.1.1 One halogen, one oxygen, and two nitrogen functions

(i) From perfluoroimines

Addition of alcohols to *N*-fluoroimines gives adducts the stability which depends on the electron withdrawing character of the carbon substituents. If these substituents are F and NF_2 or F and CF_3, the adduct decomposes spontaneously with HF elimination to yield mainly a new fluoroimine (Scheme 10). If both carbon substituents are NF_2 groups, stable adducts are obtained. Interaction of these stable liquid compounds with NOF or NO_2Cl causes replacement of the NHF group by F, through an unstable intermediate bearing an N(F)NO substituent (Equation (15)). The compounds described here are patented as good oxidants ⟨72USP3707555⟩ and are shatteringly explosive under certain conditions. Suitable protective equipment should be used during all phases of work ⟨73JOC1065⟩.

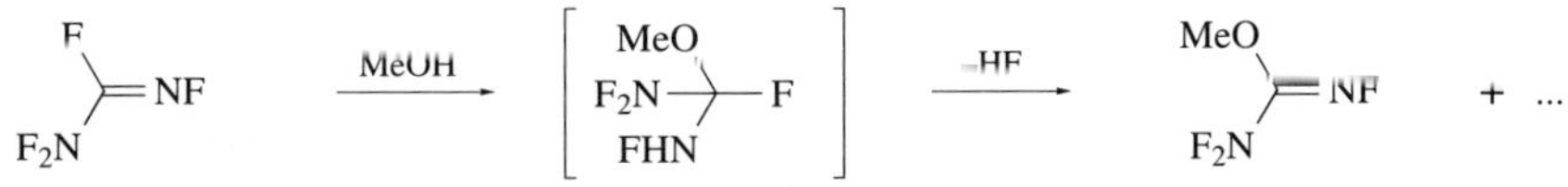

Scheme 10

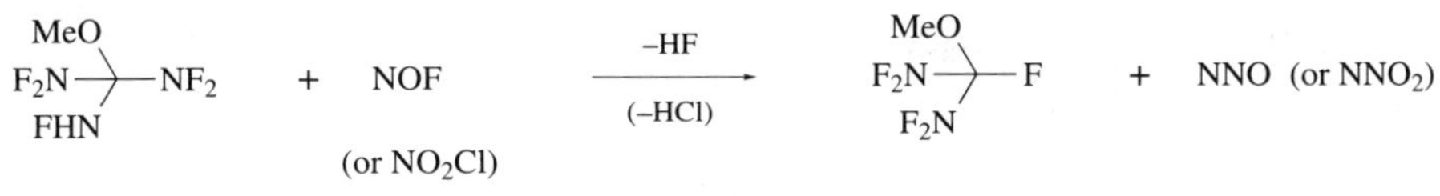

(15)

Oxidation of perfluoroimines by aromatic peroxyacids or H_2O_2 affords oxaziridines (**2**) ⟨89USP4874875, 92EUP496413⟩.

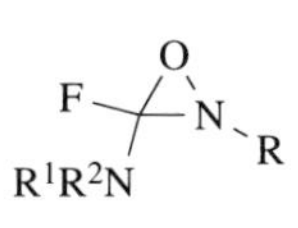

(2)

(ii) From fluorotrinitromethane

In marked contrast to chloro- and bromotrinitromethane, whose reactions with bases and reducing reagents generally lead to nitroform anions, fluorotrinitromethane reacts with certain nucleophiles with formal substitution of a nitro group (Equation (16)) ⟨68JOC3073⟩.

(16)

R = Et, 74%
CF_3CH_2, 55%

(iii) Diazirines from amidinium salts or fluoroimines

In 1965, Graham reported the direct preparation of 3 aryloxy-3-halodiazirines by the action of aqueous NaOCl (or NaOBr) on various isoureas (Equation (17) and Table 6) ⟨65JA4396⟩. Despite the importance of this interconversion, the mechanism remains only partly elucidated. The original suggestions are summarized in Scheme 11. Not all of these intermediates have been isolated and characterized ⟨81JOC5048⟩.

(17)

Table 6 Preparation of 3-halo-, 3-alkoxy-, or aryloxy-diazirines.

R^1	*R^2*	*X*	*Reagents and conditions*	*Yield* (%)	*Ref.*
Me	H	Cl	NaOCl, NaCl, LiCl, 25°C, DMSO	60	65JA4396
Me	H	Cl	NaOCl, NaCl, 25°C, DMSO	64	79JCS(P2)1298
Bu^i	H	Cl	NaOCl, NaCl, 25°C, DMSO	58	79JCS(P2)1298
Ph	$PhSO_3$	Cl	NaOCl, LiCl, −5°C, DMSO–pentane	37	82JOC4177
CD_3	H	Cl	NaOCl, NaCl, LiCl, 25°C, DMSO		86JA99
$Me^{18}O$	H	Cl	NaOCl, NaCl, LiCl, 25°C, DMSO		86JA99
$PhCH_2$	H	Cl	NaOCl, NaCl, LiCl, 25°C, DMSO	57	87TL1969
$PhCH_2$	H	Br	NaOBr, 0°C, DMSO–pentane	35	90JPO694
PhCHD	H	Cl	NaOCl, 15°C, DMSO–pentane		90TL4715
c-$C_3H_5CH_2$	H	Cl	NaOCl, LiCl, 5°C, DMSO–pentane		89TL2473
EtCHMe	H	Cl	NaOCl, LiCl, 10°C, DMSO–pentane		92JA9386

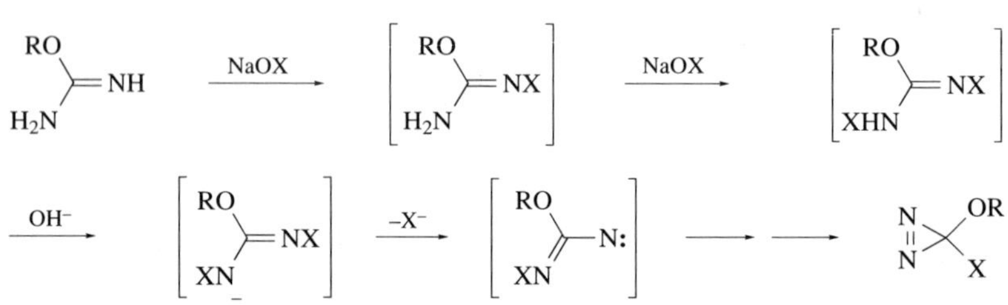

Scheme 11

Diazirines have been widely investigated as sources of numerous halocarbenes. Successive improvement of the synthetic approach, by using *N*-substituted isoureas as starting materials, greatly enlarged the scope and potential of diazirine chemistry. As neat liquids, the diazirines are unpredictably explosive at ambient temperatures, but in solution they can be handled safely. Chloro- or bromoalkoxydiazirines are converted into the corresponding fluoroalkoxydiazirines simply upon stirring with anhydrous tetra-*n*-butylammonium fluoride (tbaf) in acetonitrile (Equation (18)) ⟨83JA6513, 86JOC2168, 86TL419, 90JPO694⟩.

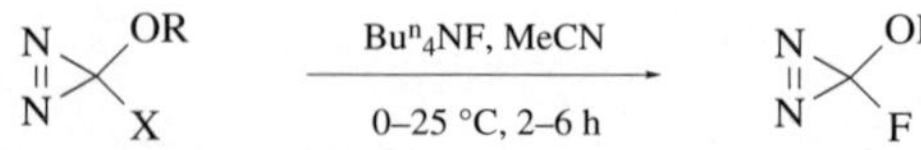

(18)

R = Ph	X = Cl, 55%
Ph	Cl, 39%
Me	Br, -
CH_2Ph	Cl, 40%

The reductive defluorination–cyclization of methoxytrifluoroformamidine in the presence of ferrocene at room temperature represents an alternative approach to 3-fluoromethoxydiazirine (Equation (19)) ⟨73JOC1065⟩.

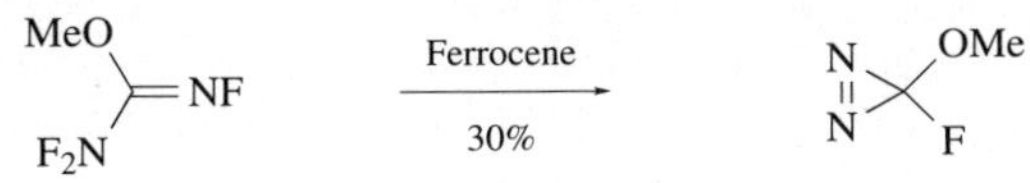

(19)

6.09.3.1.2 *One halogen, one sulfur, and two nitrogen functions*

1-Chloro-1,1-dinitromethyl sulfides or sulfones are the only reported representatives of this class of compounds. They are usually obtained by treatment of the potassium salts of dinitromethyl sulfides or sulfones with chlorine (Equation (20)) ⟨73IZV350, 75IZV2750⟩. The nucleophilicity of the potassium salts is so weak that they do not react even with iodomethane. An alternative route to the analogous sulfide derivatives (Scheme 12) involves the addition of PhSCl to iodonium nitroylides, which proceeds by cleavage of the intermediate iodonium salt ⟨78IZV2348⟩. The iodonium nitroylide is a thermally unstable crystalline compound which decomposes rapidly in solvents like acetone, ethanol, and dichloromethane and explosively after isolation in air or under an inert gas. Substituents such as NO_2 on the benzene ring increase the stability of these dinitroylides.

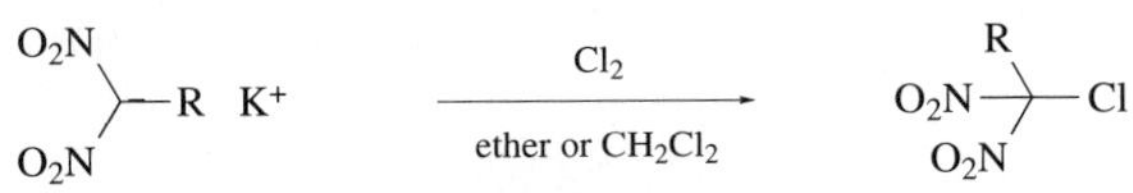

(20)

R = $MeSO_2$, 92%
$2,4\text{-}(O_2N)_2\text{-}C_6H_3SO_2$, 30%
MeS, 90%
$2,4\text{-}(O_2N)_2\text{-}C_6H_3$, 65%

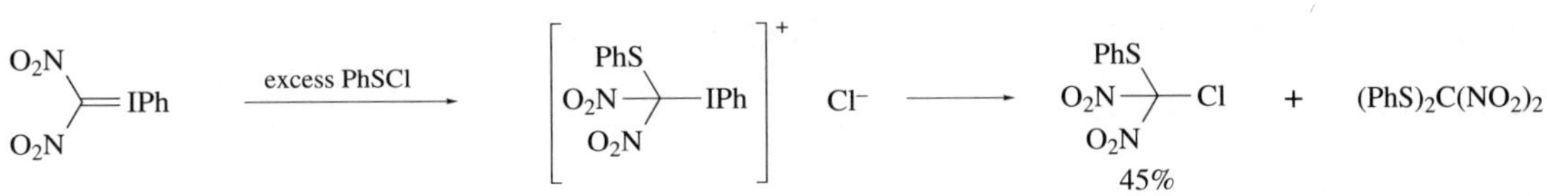

Scheme 12

6.09.3.1.3 *One halogen, one chalcogen, and two phosphorus functions*

The first approach to this kind of compound is a traditional Michaelis–Arbuzov reaction between triethyl phosphite and trichloromethyl phenyl ether (Scheme 13) ⟨70JPR475⟩. Analogous tetraethyl halomethoxymethylene bisphosphonates are obtained by radical halogenation of the corresponding methoxymethylene bisphosphonates (Equation (21)) ⟨84ZOB2504⟩.

Scheme 13

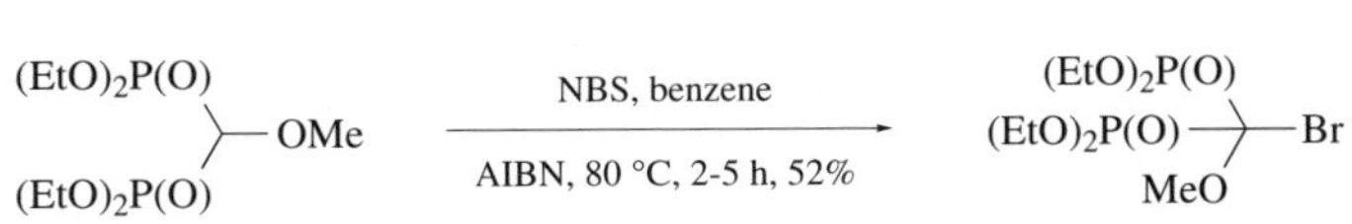

(21)

Unexpectedly, two products were obtained during the reaction of the methylene bisphosphonate salts, $[((RO)_2PO)_2CH]^-M^+$ (M = Li, Na, K), with CF_3SCl. The expected sulfenyl derivative is obtained in a mixture with chloro(trifluoromethylthio)methylene bisphosphonate, which results from a chlorination process (Equation (22)). The use of aluminum trichloride reduces the proportion of chlorinated by-product (16%) with respect to the sulfonylated product (74%) ⟨85JCS(P1)1935⟩. $CClF_2SCl$ and CCl_2FSCl have also been used in analogous reactions.

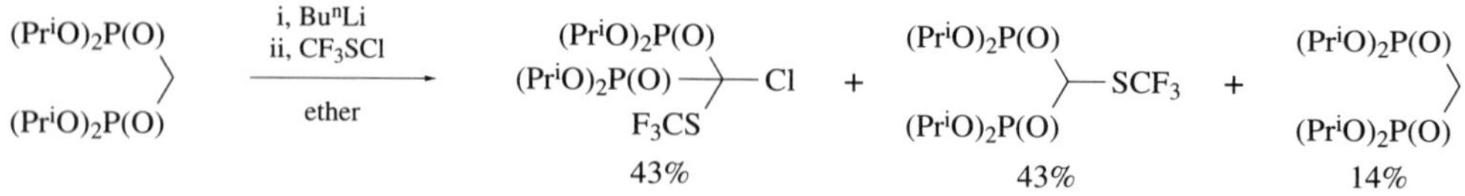

(22)

6.09.3.2 One Halogen, One Sulfur, and Two Silicon Functions

Only one member of this family has been reported in the literature of the late 1980s onwards. It was prepared by NBS bromination of bis(trimethylsilyl)methyl (trimethylsilyl)methyl sulfide (Scheme 14) ⟨87CPB1734⟩. This compound is used as a precursor for the generation of a thiocarbonyl ylide, whose 1,3 cycloaddition to several dipolarophiles leads to tetrahydrothiophene derivatives.

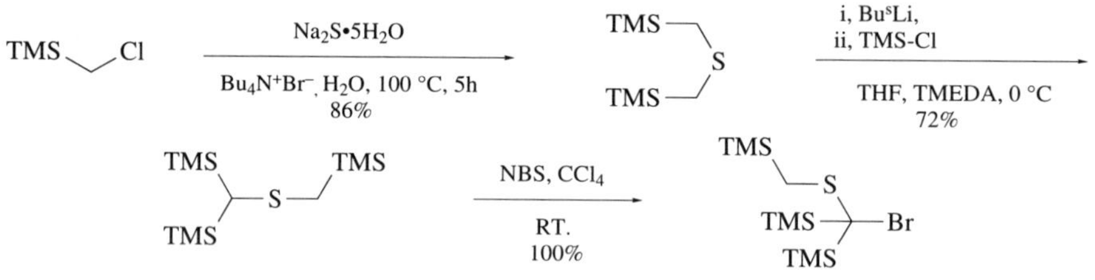

Scheme 14

6.09.4 ONE HALOGEN AND THREE GROUP 15 ELEMENTS

6.09.4.1 One Halogen and Three Nitrogen Functions

6.09.4.1.1 Halotrinitromethanes

Halotrinitromethanes have been extensively studied in view of their use as oxidants in monopropellant fuels and in bipropellant systems with hypergolic fuels ⟨68USP3419625⟩. The best methods for the preparation of halotrinitromethanes are electrophilic halogenations of trinitromethane salts (Equation (23)). A wide range of halogenating agents has been tested in various reaction conditions; the most successful syntheses are collected in Table 7. In entry 1, the reaction conditions have been optimized: aprotic solvents and room temperature (20 °C) are essential factors. Direct fluorination of trinitromethane with fluorine in water (90% yield) or methanol (70% yield) was also reported ⟨68IZV656, 68IZV676⟩ as a satisfactory synthetic method.

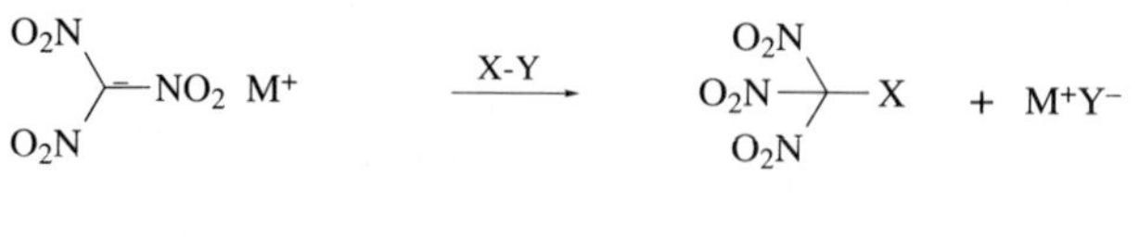

(23)

Table 7 Electrophilic halogenation of nitroform salts.

M	*XY*	*Conditions*	*X*	*Yield* (%)	*Ref.*
K	$FClO_3$	THF, 25 °C, 11 h	F	72	71IZV1487
K	F_2Xe	MeCN, 20 °C	F	78	87ZOR1657
Na	F_2/N_2	H_2O, 0 °C, 1.5 h	F	92	68JOC3080
NH_4	F_2/N_2	H_2O, 0 °C	F	70	68IZV656
Na	NCS	H_2O, 0 °C, 20 min	Cl	99	69IZV2617
K	$ClSO_3F$	Cl_2FC-$CClF_2$, −30–0 °C	Cl	82	76IZV489
Na	NBS	H_2O, 0 °C, 20 min	Br	98	69IZV2617
K	ICl	CCl_4, 10 °C, 30 min	I	71	77IZV2058

When tetranitromethane reacts with anhydrous metal halides in aprotic dipolar solvents one of the nitro groups is replaced by a halogen (Equation (24)) ⟨68USP3419625, 69ZOR1317, 70IZV2553, 71IZV1073, 72IZV1721, 78USP4120904⟩.

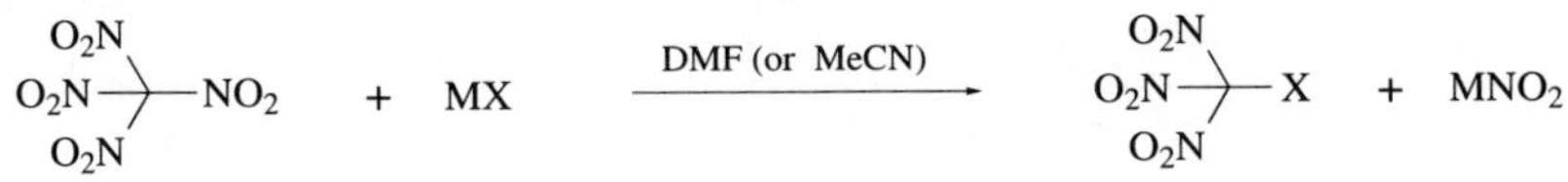

(24)

MX = KF, RbF, 58%
CsF, 49%
$KF/OC(CF_3)_2$, 90%
LiCl, KCl, RbCl, CsCl, 40–60%
LiBr, KBr, 23–30%
$Na(Cl)NSO_2Ph$, 49%

Halotrinitromethanes have also been obtained from arylthio-, arylseleno-, and arylantimono-trinitromethanes by cleavage of the carbon–heteroatom bond, through treatment with halogens, or with sulfuryl chloride under mild conditions (Equation (25)) ⟨74IZV1350, 82IZV161, 85IZV439⟩. Treatment of arylthio- or arylseleno-trinitromethanes with gaseous hydrogen halides (HCl, HF) in the presence of Lewis acids or mineral acids affords the corresponding halotrinitromethanes (Equation (26)). The addition of a Lewis acid promotes the nucleophilic substitution of the chalcogenide group, but yields remain unsatisfactory (1–30%) ⟨89IZV2106⟩.

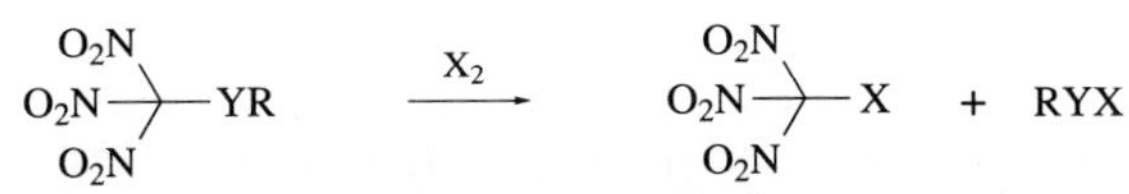

(25)

Y =	X_2 =	R =	Conditions	Yield
S	Cl_2	Ph	DME, 0 °C, 2 h	86%
	SO_2Cl_2	$2,4\text{-}(O_2N)_2\text{-}C_6H_3$	MeCN, 20 °C, 5 min	80%
	Br_2	$2,4\text{-}(O_2N)_2\text{-}C_6H_3$	MeCN, 20 °C, 5 min	78%
Sb	Cl_2	Ph	CH_2Cl_2, 20 °C, 3 h	52%
Se	Cl_2	Ph	CH_2Cl_2, 20 °C, 1h	79%
	Br_2	Ph	CH_2Cl_2, 20 °C, 1h	75%

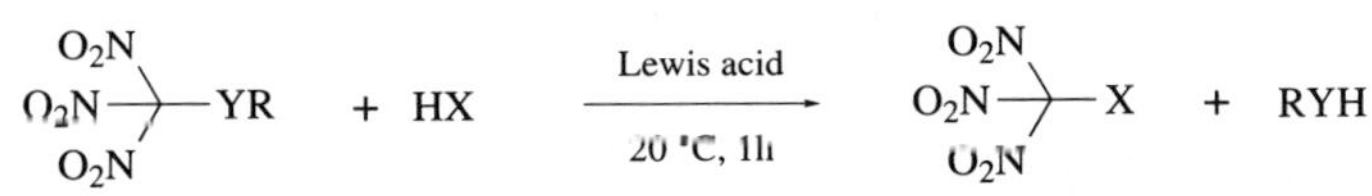

(26)

Y = S, Se X = F, Cl Lewis acid = $AlCl_3$, $SbCl_5$
Solvent = CH_2Cl_2, MeCN, DMSO

Trinitromethyl sulfones were also tested as starting materials for the synthesis of halotrinitromethanes (Equation (27)). The C—SO_2 bond was cleaved efficiently by halogenated electrophiles (X_2, $BrONO_2$, HCl) ⟨77IZV375⟩.

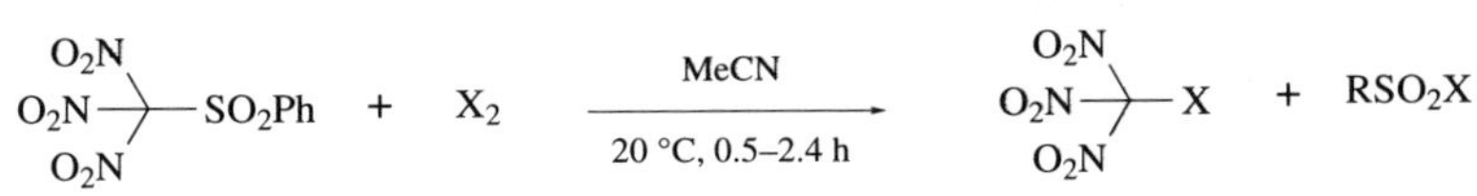

(27)

X_2 = Cl_2, 80%
Br_2, 80%

Electrophiles (X_2 = Cl_2, Br_2, SO_2Cl_2, NCS or NBS) very readily cleave the S—C bond of dimethyl-(trinitromethyl)sulfonium tetrafluoroborate to give the corresponding halotrinitromethanes in fair to good yields (60–80%) (Equation (28)) ⟨77IZV2530⟩. The starting material, obtained by nitration of dimethylsulfonium dinitromethylide with nitronium tetrafluoroborate, is stable for at least 24 h at 20°C. It also reacts under mild conditions with nucleophiles (HCl, HBr) to give the expected halotrinitromethanes. In this case, yields are low (16–33%) because of competing reactions between the product halotrinitromethanes and dimethyl sulfide.

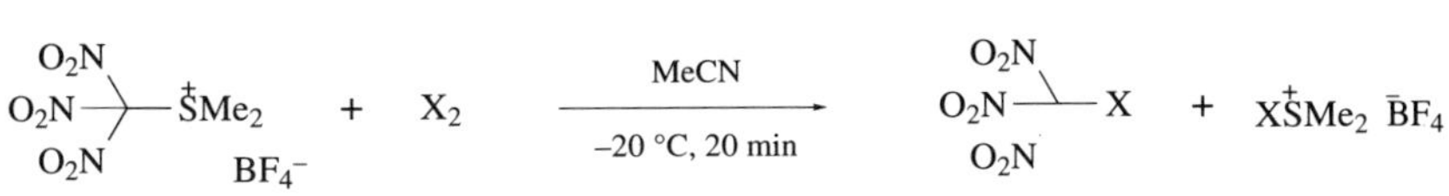

(28)

All the approaches described above employ the halogenation of preformed trinitromethyl moieties. There is only one report of the nitration of a halodinitromethane. Nitryl fluoride was employed as the nitrating agent (Equation (29)). The reaction did not require prior salt formation, but the use of ammonium salts of halodinitromethane slightly increases the yield (to 55%) ⟨74IZV915⟩.

X = F, 47%
Br, 39%

(29)

6.09.4.1.2 *Miscellaneous halonitromethanes*

Bromo- and chlorodinitromethanes react with α,α′-dicarbonylazo compounds to give *N*-(halodinitromethyl)hydrazine derivatives (Equation (30)) ⟨75IZV2838⟩. (Arylazoxy)dinitromethanes, bearing an acidic proton, undergo facile bromination in basic medium to give the (aryloxy)bromodinitromethanes (Equation (31)) ⟨92MC52⟩. Finally, bis(difluoroamino)fluoronitromethane $[(F_2N)_2(NO_2)CF]$ has been identified as a colorless gas ⟨68USP3387033⟩.

(30)

X =	R =
Cl	OMe, 81%
Cl	NHMe, 67%
Br	OEt, 98%
Br	NH_2, 87%

Ar = 2,4-$(O_2N)_2$•C_6H_3, 80%

(31)

6.09.4.1.3 *Halotriaminomethanes*

Several poly(fluoroamino)halomethanes (Scheme 15) have been patented as precursors for bleaching agents, explosives, rocket fuels, and pyrotechnic agents ⟨67JOC3859, 67USP3354217, 68IC1647, 69USP3450762, 72IC418, 72USP3689560, 73JOC1075, 73USP3755404⟩. This compilation is not exhaustive. The various reported syntheses are performed on small scale and generally afford complex mixtures of fluorinated products, including those having the general structure $(R_2N)_3CF$. The reaction between NCl_3 and CCl_4 in the presence of $AlCl_3$ gives a mixture of four compounds, one of them being $(Cl_2C{=}N)_3CCl$ (27%) (Scheme 16) ⟨72RTC331⟩.

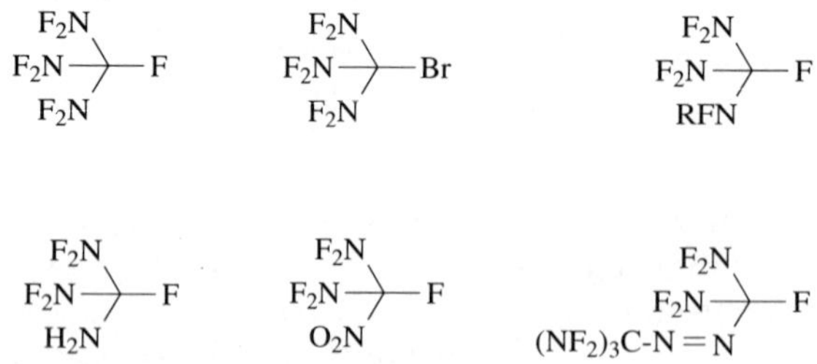

Scheme 15

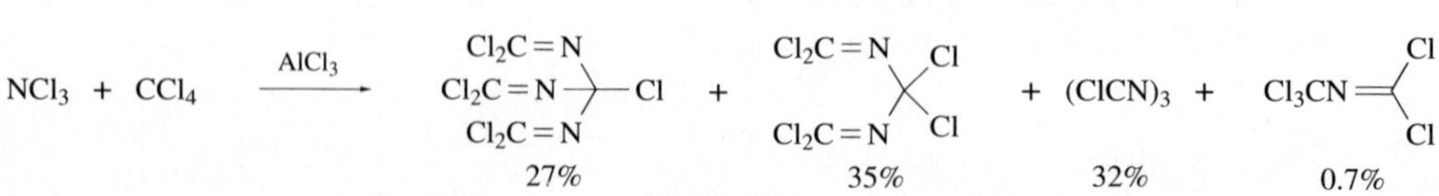

Scheme 16

6.09.4.1.4 *(Difluoroamino)fluorodiazirine*

(Difluoroamino)fluorodiazirine was prepared from tris(difluoroamino)fluoromethane by a reductive defluorination–cyclization reaction with iodide ion in acetonitrile or with ferrocene (17%) (Equation (32)) ⟨67JOC4045, 68JOC1847, 72USP3637663⟩. *Extreme caution should be used when manipulating (difluoroamino)fluorodiazirine because it can explode violently and it is sensitive to phase changes.*

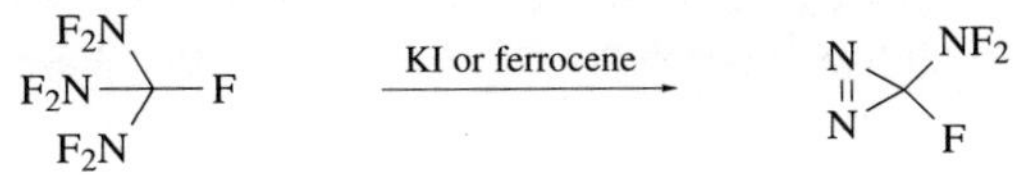

(32)

6.09.4.1.5 *Fluorine-containing diaziridines*

Pentafluoroguanidine undergoes rearrangement to a diaziridine in the presence of alkali metal fluorides (MF) (Equation (33)). CF_4, SiF_4, CO_2, and other impurites were observed in the reaction mixture. Explosions occurred when the volatile diaziridine was condensed or volatilized. The diaziridine is unstable at room temperature and must be stored at $-78\,°C$ ⟨68JOC3489⟩. A similar diaziridine was also obtained by a reductive defluorination–cyclization reaction of bis(difluoramino)[(trifluoromethyl)fluoramino]fluoromethane (Equation (34)) ⟨68JOC1847⟩.

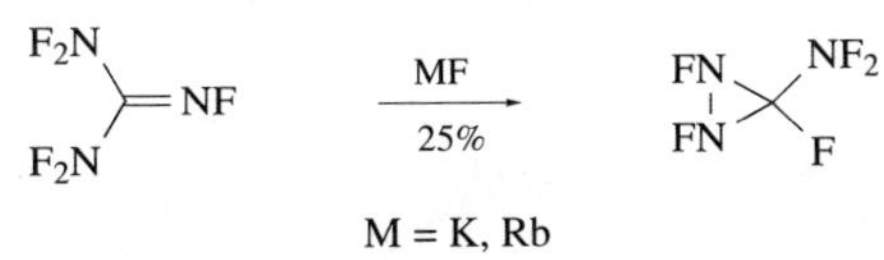

(33)

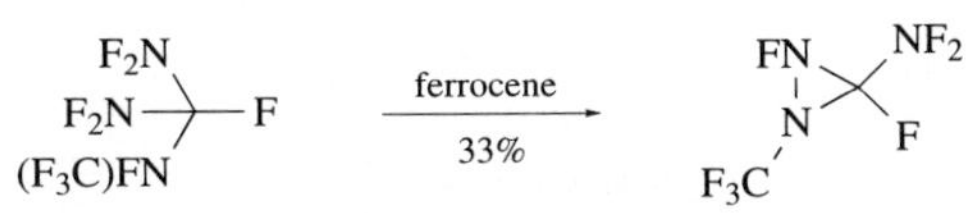

(34)

6.09.4.2 One Halogen, One Nitrogen, and Two Phosphorus Functions

Compounds of this class bear, essentially, two phosphoryl groups and an isocyanate or ammonium function on the carbon atom.

The reaction of trichloromethyl isocyanate with various amounts of triethyl phosphite was studied under several conditions in order to determine the number of chlorine atoms of the trichloromethyl group which are active in the Michaelis–Arbuzov reaction. With optimized stoichiometry of the reagents and reaction conditions, a 33–42% yield of the diphosphonyl substituted derivative is obtained (Equation (35)). In all cases side products are formed ⟨73ZOB544⟩. The analogous reactions with a dialkyl halophosphite and trichloromethyl isocyanate afforded only the corresponding bis(phosphoryl)halomethyl isocyanate in moderate to good yields (Equation (36)). This reaction was extended to 2-chloro-1,3-dioxaphospholanes: as expected, the Michaelis–Arbuzov rearrangement led to ring opening; the yield is very low (6%) ⟨77ZOB2766⟩.

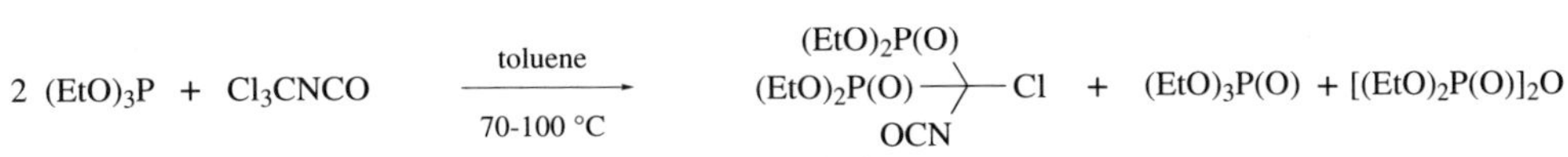

(35)

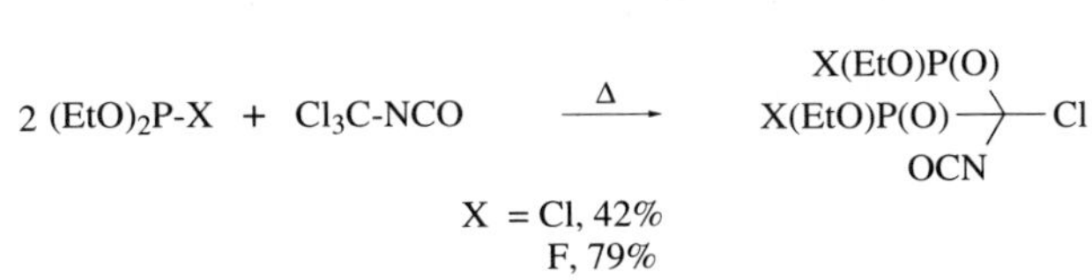

(36)

Dichlorofluoromethyl isocyanate has been used in a reaction with dialkyl chlorophosphites with a reagent ratio of 1 : 2. The main product was the (alkoxychlorophosphonyl)(alkoxyfluorophosphonyl)chloromethyl isocyanate; it resulted from the reaction of the tautomeric form, $Cl_2C{=}NC(O)F$, of the starting isocyanate with the chlorophosphite, followed by a second

Michaelis–Arbuzov reaction (Scheme 17) ⟨88ZOB1516⟩. The same approach was extended to the synthesis of other substituted isocyanates by reaction of monophosphorylated dichloromethyl isocyanates with trialkyl- or dialkyl chlorophosphites (Equation (37)) ⟨75ZOB1965⟩. Tetraethyl (dimethylamino)methylbis(phosphonate) is methylated by dimethyl sulfate; upon addition of gaseous chlorine, the resulting ylide undergoes a chlorination–elimination reaction to give the inner salt of triethyl chloro(trimethylammonium)methylbis(phosphonate) (Scheme 18) ⟨76JPR116⟩. The inner salt was further hydrolyzed by aqueous HCl to the corresponding bisphosphonic acid (63%). Analogous aminophosphonic acids have been patented as herbicides ⟨80JAP8089210⟩.

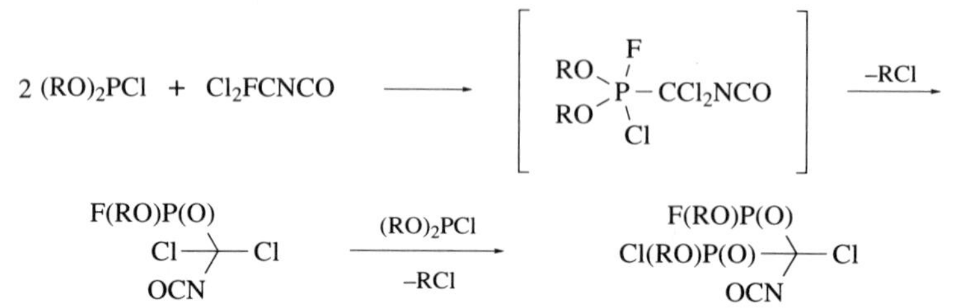

R = Me, 42%
Et, 60%

Scheme 17

(37)

X = EtO, $Z^1 = Z^2 = Cl$, 36%
EtO, $Z^1 = Cl$, $Z^2 = EtO$, 6.5%
Cl, $Z^1 = Z^2 = Cl$, 61%

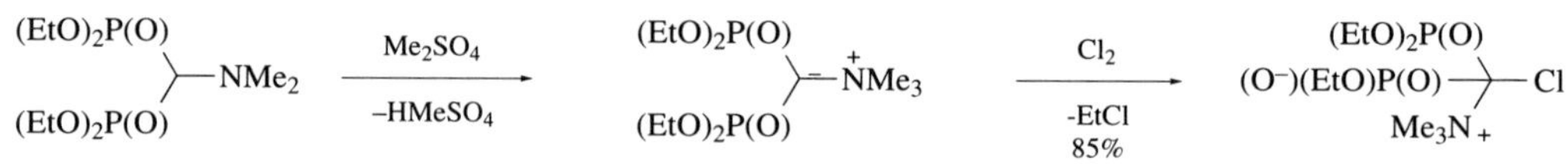

Scheme 18

6.09.4.3 One Halogen and Three Phosphorus Functions

Only one representative of this family is described; it contains one bromine and three diethyl phosphoryl groups (Equation (38)). Carbon tetrabromide reacts with triethyl phosphite under milder conditions than does carbon tetrachloride: in refluxing benzene the reaction is complete after 1 h. The yield of isolated product is 67% ⟨79ZOB1470⟩.

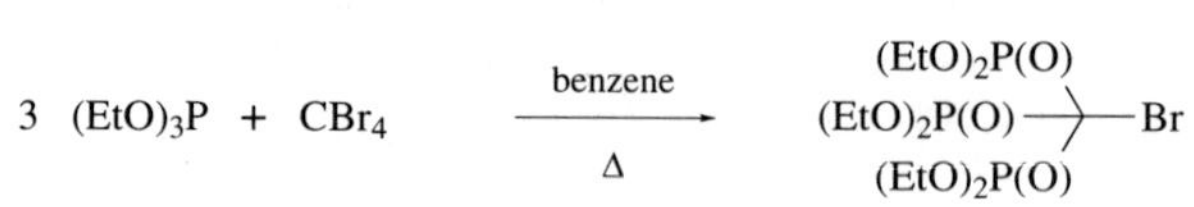

(38)

6.09.5 ONE HALOGEN AND TWO GROUP 15 ELEMENTS

Compounds of this class are essentially carbanions bearing *gem*-dinitro or diphosphorus functions and the halogen substituent; they occur as reaction intermediates.

6.09.5.1 Metal Halodinitromethides

The compounds described here are explosives and may detonate on grinding or impact. Furthermore, fluorodinitro compounds show varying degrees of toxicity and may cause painful burns when brought into contact with the skin. *Consequently, they should be handled with care.*

The potassium bromo- and chlorodinitromethides were prepared from the corresponding 2-halo-2,2-dinitroethanol in basic medium (Scheme 19). Unlike the chloro derivative which is unstable in the free state at room temperature, the bromo derivative can be isolated, recrystallized, and air dried ⟨64JOC3587⟩.

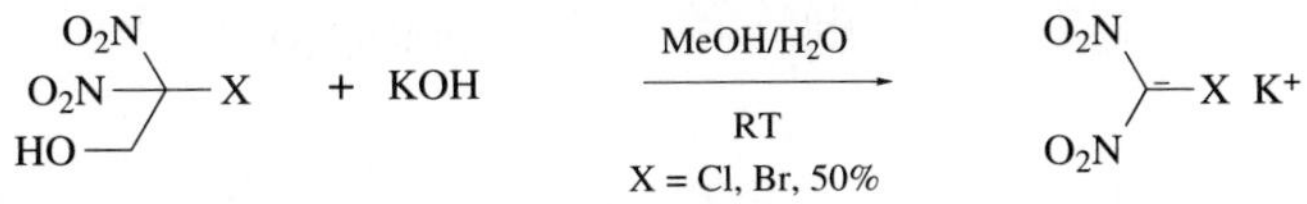

Scheme 19

The fluorodinitromethide was obtained by reaction of fluorotrinitromethane with hydroperoxide ions through displacement of one of the nitro groups (Scheme 20). The fluorinated carbanion is unstable in solution and has been trapped by acid hydrolysis or reacted immediately ⟨68JOC3073⟩. The silver salt of fluorodinitromethane is equally unstable. It is obtained by reaction of silver oxide with fluorodinitromethane and reacts *in situ* with electrophiles to give C—C coupling reactions (Scheme 21) ⟨73S605⟩.

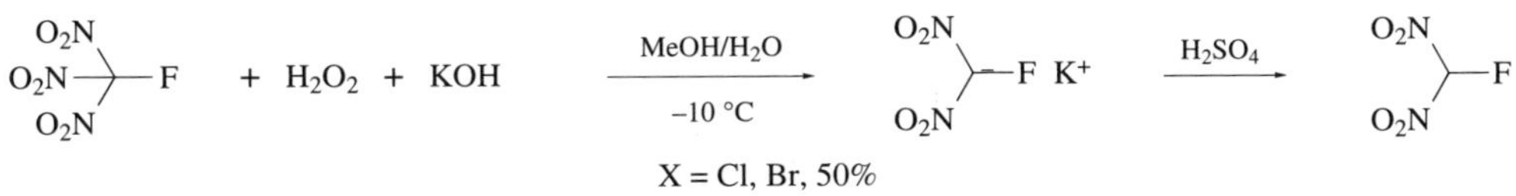

Scheme 20

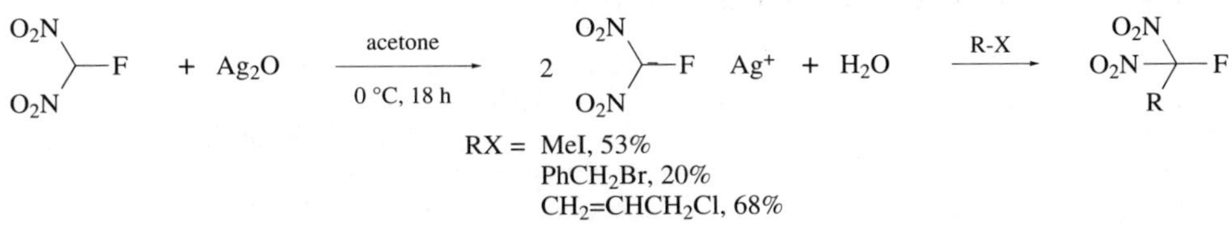

Scheme 21

In contrast to the potassium and silver salts, bis(fluorodinitromethyl)mercury is a white crystalline substance which is stable enough to be stored. It reacts readily at room temperature with KI or KOH through a metal exchange reaction; the potassium salt is formed according to Equation (39). It is slowly hydrolyzed in the presence of moisture. The same mercury derivative is used as a starting material for the synthesis of symmetrical bis(chlorodinitromethyl)mercury by a transmercuration reaction (Equation (40)): the reaction with polynitro compounds bearing a labile hydrogen atom is performed in moist ether at room temperature ⟨67DOK(176)1086⟩.

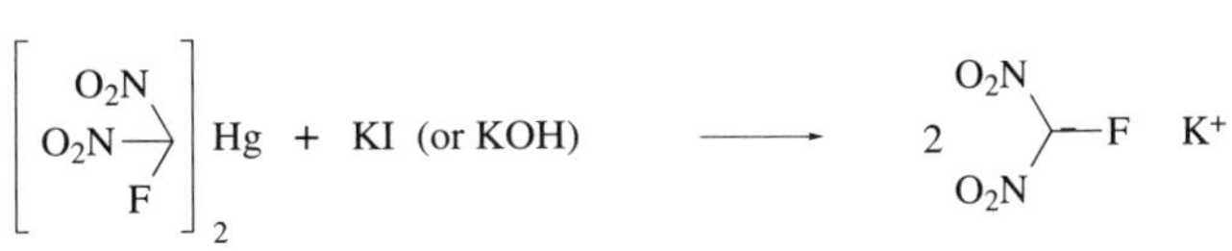 (39)

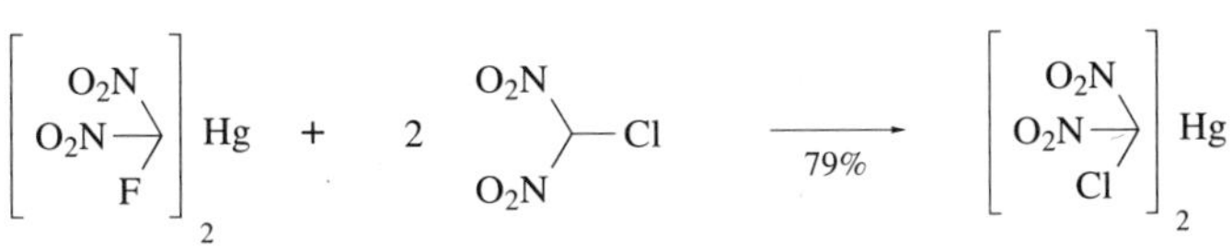 (40)

6.09.5.2 Metal Bis(phosphonyl)halomethides

Halogenated methylenebis(phosphonic acids) have biological applications which are associated with their complexing properties for metal cations (Ca^{2+}, Zn^{2+}, etc.). Alkali metal salts of alkylidenebis(phosphonic acid) are known to improve the cleaning power of soaps and detergents ⟨66BEP672205⟩; 1,1-bis(phosphonyl)-1-halo-1-metal derivatives are useful intermediates for the synthesis of substituted bis(phosphonic acids). The first preparation used a chlorine–lithium exchange between dichloromethylenebis(phosphonate) and butyllithium (Equation (41)) ⟨73JOM(59)237⟩. Other metals (Na, Tl) have been used as counterions ⟨85JOM(291)145, 86JOM(309)C7⟩.

(41)

A second approach uses α-lithio-α-chloromethylphosphonates as starting materials. Condensation with chlorophosphates in the presence of base (lithium diisopropylamide—LDA) affords the target compounds in quantitative yield (Equation (42)) ⟨86JOM(304)283, 87SC1559⟩.

(42)

Only one type of 1-halocarbanion bearing a phosphorus and a nitrogen function is known (Equation (43)). It is used as an alkeneation reagent ⟨88JAP63227595⟩.

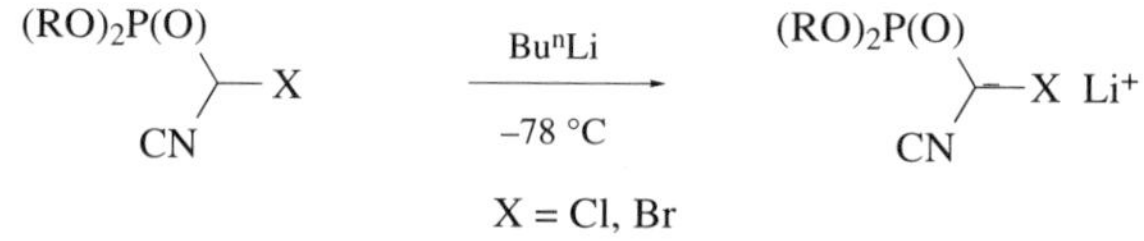

(43)

6.09.6 ONE HALOGEN AND ONE GROUP 15 ELEMENT

Compounds of this class with a phosphorus and two silicon groups or with a phosphorus and one silicon group, together with a lithium substituent, have been described. Most of them were prepared by reaction of dichlorobis(trimethylsilyl)methane with butyllithium and a trivalent chlorophosphane (Scheme 22 and Table 8). The α-chlorophosphanes shown in entries 3, 4, and 5 are stable compounds, while the analogous derivatives described in entries 1, 2, and 6 give spontaneously rearranged products at room temperature. The product shown in entry 7 is stable at room temperature, but it rearranges to a 1-chlorophosphorane at 160°C.

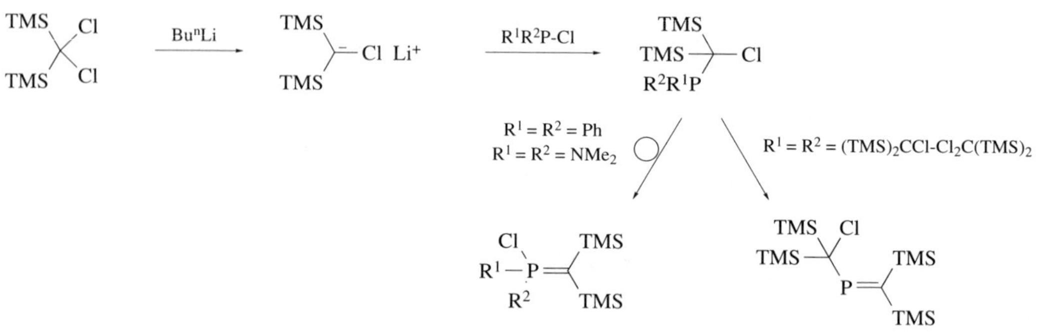

Scheme 22

Table 8 Synthesis of chlorobis(trimethylsilyl)methylphosphanes.

Ent	*R*[1]	*R*[2]	*Chlorophosphane*	*Yield* (%)	*Ref.*
1	Ph	Ph	Ph_2PCl		83CB114
2	NMe_2	NME_2	$(Me_2N)_2PCl$	64[a]	83CB114
3	NMe_2	Cl	Me_2NPCl_2	68	83CB114
4	Ph	Cl	$PhPCl_2$		83CB114
5	Cl	Cl	PCl_3 (1 eq)		83CB114
6	$ClC(TMS)_2$	$ClC(TMS)_2$	PCl_3 (0.33 eq)	54[a]	82TL2017, 86IS117
7	Cl	$(TMS)_2CH$	$(TMS)_2CHPCl_2$		89CB453

Reaction conditions: −110°C, THF/ether/pentane. [a] Yields of final rearranged products.

In a second approach, treatment of an α-lithio-α-fluoromethylphosphonate with bromo- or chlorotrimethylsilane gives a mixture of mono and bis(trimethylsilyl) derivatives because of a facile proton transfer (Equation (44)). These products could not be usefully separated ⟨83CC886, 86JCS(P1)1425⟩.

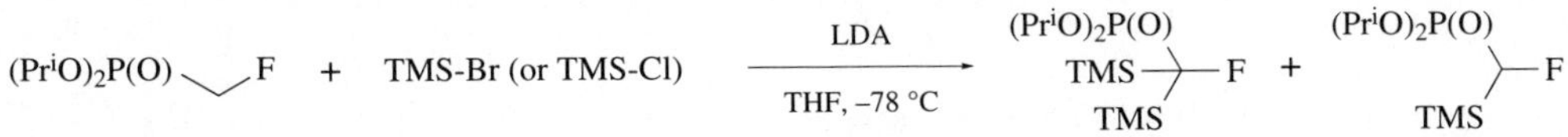

(44)

Only one example of a compound bearing a phosphorus, a silicon, a metal, and a halogen function on the same carbon has been described. It was prepared from diethyl trichloromethylphosphonate by chlorine–lithium exchange in the presence of butyllithium and chlorotrimethylsilane at low temperature (Equation (45)). The compound is obtained in quantitative yield and it can either be protonated by acids or alkylated ⟨88JOM(338)295⟩.

(45)

6.09.7 ONE HALOGEN AND METALLOID FUNCTION (THREE, TWO OR ONE, TOGETHER WITH METALS)

6.09.7.1 One Halogen and Three Metalloid Functions

6.09.7.1.1 Three silicon functions

Organosilanes of this family have been widely investigated and several useful synthetic approaches have been described.

The first one is a carbon–halogen bond forming reaction: tris(trimethylsilyl)methane was halogenated by means of NBS ⟨68CB2815, 71JOM(29)389⟩ or bromine under UV irradiation with or without solvents (Scheme 23) ⟨92OM2938⟩. An alternative route consists of metallation of tris-(trialkylsilyl)methanes followed by halogenation of the resulting anions (Scheme 24) ⟨70JOM(24)529, 86CB1455, 89JOM(366)39⟩. Halogenating agents are tetrabromomethane or dibromodifluoromethane. Crystal structure determinations of the silylated carbanion–THF adducts have been performed ⟨83CC827, 83CC1390⟩.

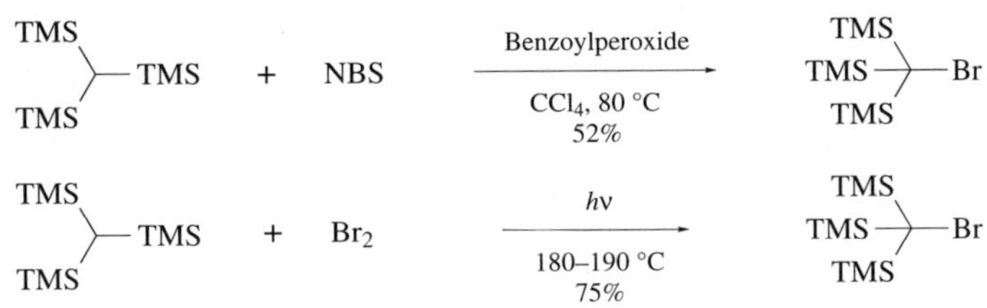

Scheme 23

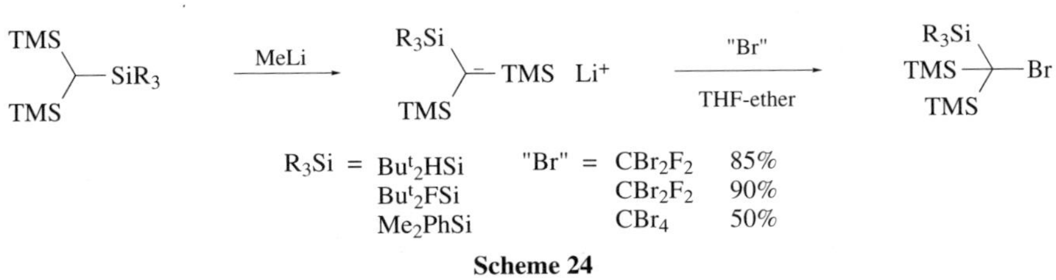

Scheme 24

The most widely and routinely used approach is the silylation of lithiated carbanions derived from *gem*-dihalobis(trialkylsilyl)methanes. While chlorobis(trialkylsilyl)methanes are not suitable reagents for preparation of the corresponding carbanions, dichlorobis- or dibromobis(trialkylsilyl)methanes are readily metallated through halogen–lithium exchange reactions. The organolithium derivatives thus formed react with various chlorosilanes at low temperature to give the halotris(trialkylsilyl)methanes (Scheme 25) ⟨70JOM361, 70JOM529, 70JOM647, 81CB2087, 84ZAAC(510)169, 86JOM(315)C5, 87CB653, 87CC1461, 87JCS(P2)1047, 89CC595⟩. More often, the halotris(trialkylsilyl) methanes were converted directly into new derivatives, either by halogen substitution or by reactions at the silicon substituents.

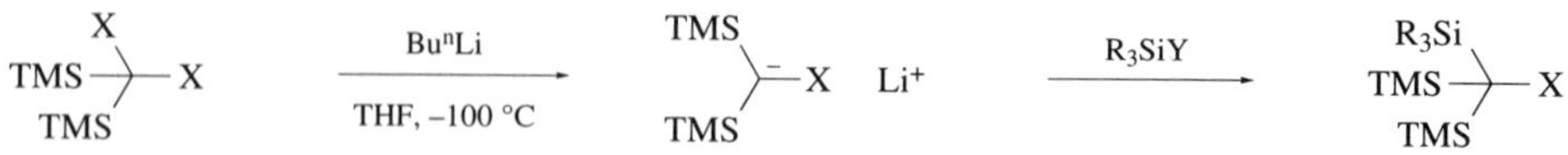

X	R_3SiY	*Yield* (%)
Cl	TMS-Cl	69
		74
Br	TMS-Cl	78
Cl	$Me_2PhSiCl$	-
Br	$Me_2PhSiCl$	95
Br	$(D_3C)_2PhSiCl$	52
Cl	$(CH_2{=}CH)Me_2SiCl$	58
Cl	$Y\text{-}C_6H_4SiMe_2F$ Y = H, 4-MeO, 4-Cl, 3-CF_3	-
Cl	Me_2HSiCl	-
Cl	Et_2HSiCl	-
Cl	Me_2SiF_2	56
Cl	$MePhSiF_2$	44

Scheme 25

Sodium metal may also be used to metallate dibromobis(trimethylsilyl)methane in a coupling reaction with chlorodiphenylsilane. Bromine is then added *in situ* in order to perform bromination at both the carbon and silicon atoms (Scheme 26) ⟨89CB409⟩. Other examples of carbon–silicon bond forming reactions from various starting materials have been described (Equation (46)) ⟨77ZAAC(430)121, 79JOM(178)11⟩. When R^1_3Si is a trichlorosilyl group and the electrophile is iodotrimethylsilane, a significant amount of bis(trichlorosilyl)trimethylsilylmethane (44%) is formed. The same reaction was applied to the silylation of a cyclic six-membered carbosilane (Equation (47)) ⟨79JOM(178)11⟩. The final carbosilane could not be separated from the polymeric side products. The yield was established after reduction of the SiCl and CCl groups with $LiAlH_4$.

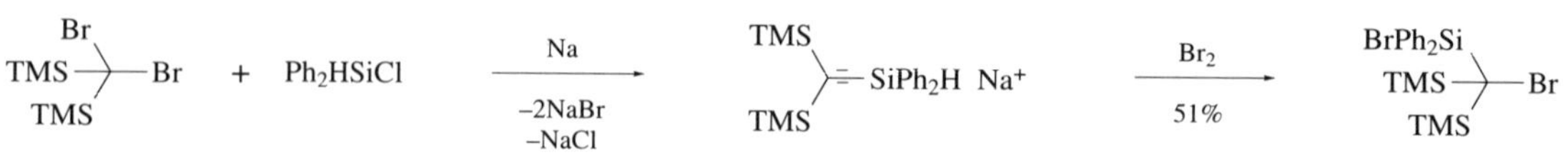

Scheme 26

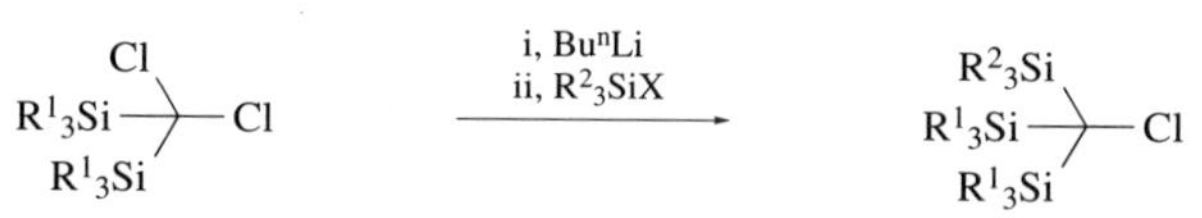

(46)

R^1_3Si	R^2_3SiX	*Yield* (%)
$PhMe_2Si$	$PhMe_2SiCl$	40
Cl_3Si	TMS-I	42
Cl_3Si	TMS-Cl	25

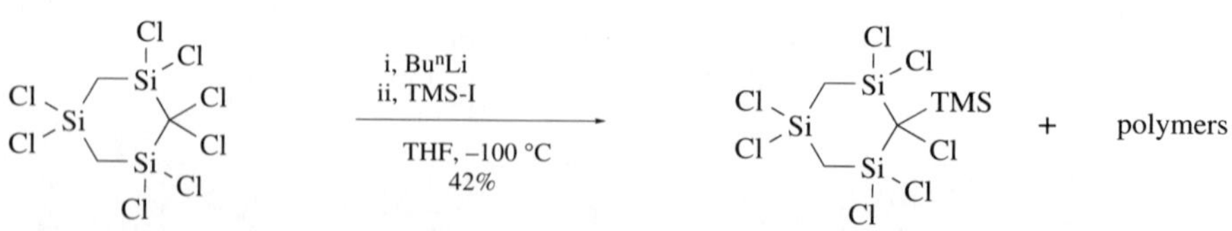

(47)

After reaction between $TMSCCl_2SiMe_2Cl$ and butyllithium at $-100\,°C$ in ether, and subsequent warming to 20 °C, dimeric carbosilanes are formed as the main products (Equation (48)). Purification by HPLC allowed elimination of the by-products and afforded a mixture of the two isomers ⟨84JOM(271)107⟩.

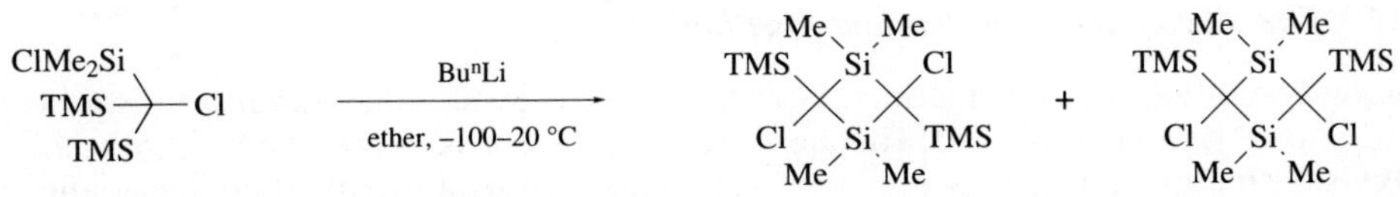

(48)

A great number of bromotris(trialkylsilyl)methane derivatives (Table 9) have been obtained in high yields through further displacement of a phenyl or a hydrogen (Y) by chlorine or bromine in $(YR_2Si)(TMS)_2CBr$, as shown in Equation (49). Bromine substitution reactions at the silicon atom are also effected by various nucleophiles (AgF, MeOH, $AgOP(O)Ph_2$, etc.), as shown in Equation (50) and Table 10.

Table 9 Synthesis of $(XR_2Si)(TMS)_2CBr$ (X = Cl, Br).

R	*Y*	*X*	*Reagents and conditions*	*Yield* (%)	*Ref.*
Me	Ph	Br	Br_2, 60°C, 2 h	96	81CB2087
Me	Ph	Cl	ClI, 70°C, 20 h	74	81CB2087
Me	Ph	I	I_2, 130°C, 8 h	60	81CB2087
CD_3	Ph	Br	Br_2, 60°C	82	87CB653
Ph	H	Br	Br_2	96	89CB409
Ph	H	Cl	Cl_2	88	89CB409
Bu^t	H	Br	Br_2, CCl_4	100	86CB1455

(49)

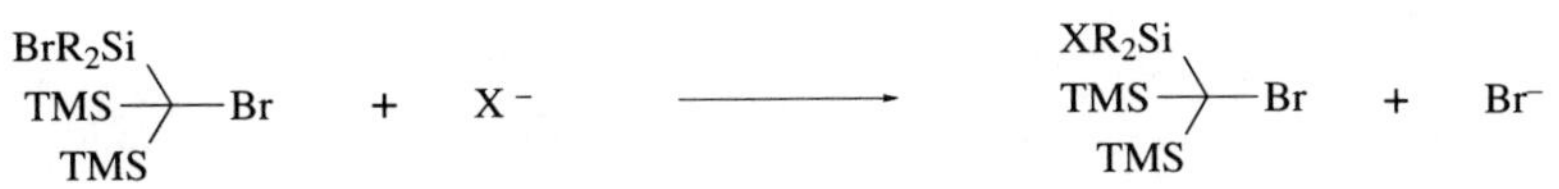

(50)

Table 10 Synthesis of $(XR_2Si)(TMS)_2CBr$.

Ent	*R*	*X*	*Reagents and conditions*	*Yield* (%)	*Ref.*
1	CD_3	F	AgF, 60°C, THF	67	87CB653
2	Me	F	AgF, 66°C, 50 h, THF	48	81CB2087
3	Ph	F	KHF_2, 64°C, 2.5 h, MeOH	89	89CB409
4	Ph	OH	H_2O, 25°C, 72 h, THF	93	89CB409
5	Me	OH	H_2O, 25°C, 1 h, pentane	95	81CB2087
6	Me	MeO	MeOH / Et_3N, 65°C, 10 h	69	81CB2087
7	Ph	MeO	MeOH, 64°C, 6 h	90	89CB409
8	Me	PhO	PhONa, 111°C, 100 h, toluene	87	81CB2087
9	Me	PhS	PhSNa, THF	51	81CB2087
10	Me	4-Me.$C_6H_4SO_2$	AgX, 25°C, 4 h, THF	88	81CB2087
11	Me	4-Me.$C_6H_4SO_2$	AgX, 156°C, 1.5 h,Bu_2O	75	81CB2087
12	Me	$MesSO_3$	AgX, 25°C, 4 h, THF	80	81CB2087
13	Me	$Ph_2P(O)O$	AgX, 25°C, 4 h, THF	85	81CB2087
14	Me	(PhO)PhP(O)O	AgX, 25°C, 4 h, THF	95	81CB2087
15	Me	$(PhO)_2P(O)O$	AgX, 25°C, 4 h, THF	80	81CB2087

Chlorotris(methoxydimethylsilyl)chloromethane can be prepared from chlorotris(trimethylsilyl)methane by addition of a large excess of ICl in tetrachloromethane, followed by a methanol–triethylamine mixture (Scheme 27) ⟨92JCS(D)1015⟩.

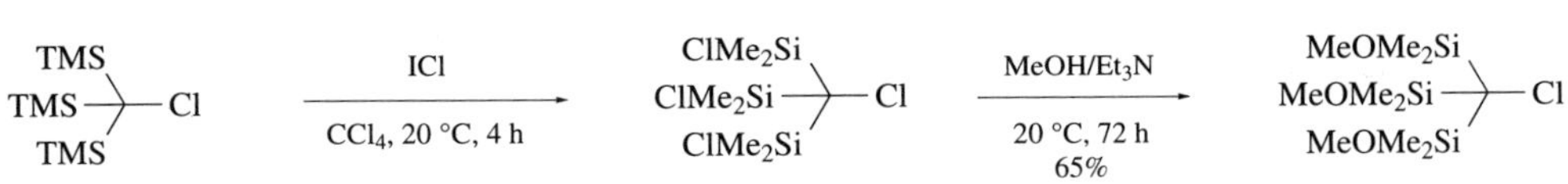

Scheme 27

6.09.7.1.2 *Two silicon and a germanium function*

Compounds containing two silicon functions and one germanium function have been prepared according to the general approaches reported above for the synthesis of halo-tris(trialkylsilyl)methanes (see Section 6.09.7.1.1). Thus, chlorodimethylphenylgermane reacted with bromotris(trimethylsilyl)methyllithium at low temperature to give the corresponding germyldisilyl derivative (Equation (51)) ⟨86CB2966⟩. Cleavage of the phenyl substituent was then performed with bromine (Equation (52)). The colorless, solid product was obtained in good yield. As reported above for silicon (Section 6.09.7.1.1), the bromo function on germanium can participate in further substitution reactions.

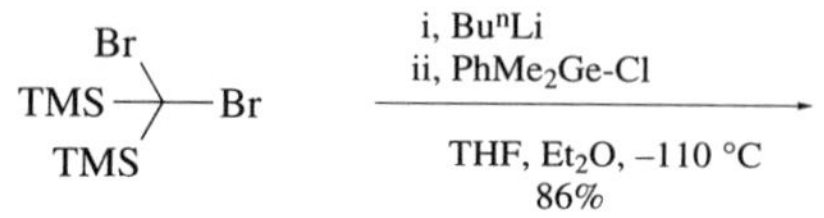

(51)

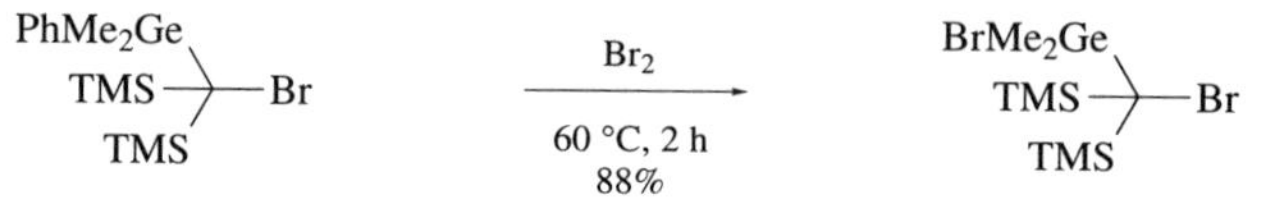

(52)

6.09.7.1.3 *Two silicon and a boron function*

The only example of such compounds is bromo[bromobis(isopropyl)aminoboryl]bis(trimethylsilyl)methane. It is prepared by bromine addition to the corresponding methylidenebis(isopropyl)aminoborane, as shown in Equation (53) ⟨89CB595⟩. The starting material is obtained by the thermolysis of [fluorobis(isopropyl)aminoboryl]tris(trimethylsilyl)methane in the gas phase at 520°C (Equation (54)).

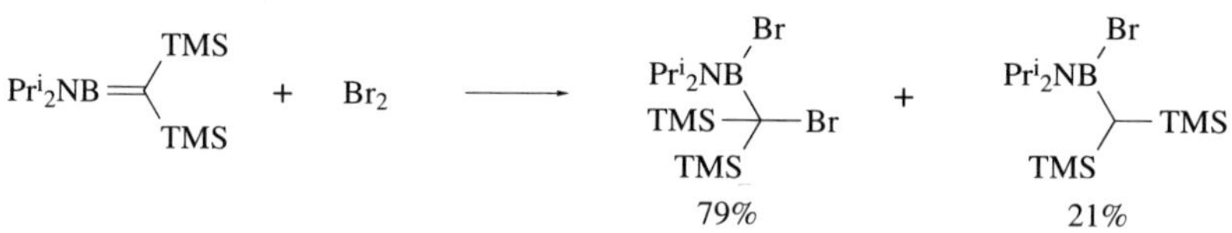

(53)

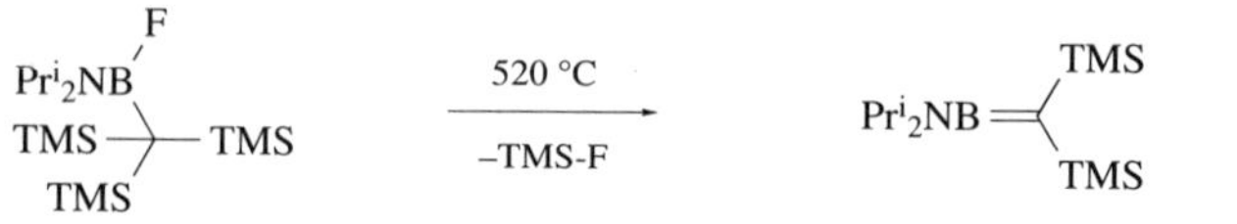

(54)

6.09.7.1.4 *Three boron functions or two boron and a metal function*

Conversion of tetrakis(tetramethylethylenedioxyboryl)methane to the corresponding triborylmethide anion is performed by adding two equivalents of methyllithium at low temperature ⟨74JOM(69)45⟩. The lithiated carbanion is treated *in situ* with bromine in dichloromethane to afford the crystalline bromotriborylmethane (Scheme 28). This last compound reacts with one equivalent of methyllithium to give the bromodiborylmethide anion; however the conversion is incomplete under these conditions. The reaction of the carbanion with chlorotriphenyltin was performed but the final product was not completely characterized (Scheme 29). An analogous iodo derivative $(C_3H_6O_2B)_2(Ph_3Sn)CI$ was more fully characterized ⟨73JOM(57)231⟩. The same approach has been applied to the synthesis of tris(trimethylenedioxyboryl)methane derivatives ⟨73JA5096, 75JOM(93)21⟩. The THF insoluble boryl substituted lithium salt was purified by precipitation, before reaction with bromine (Scheme 30).

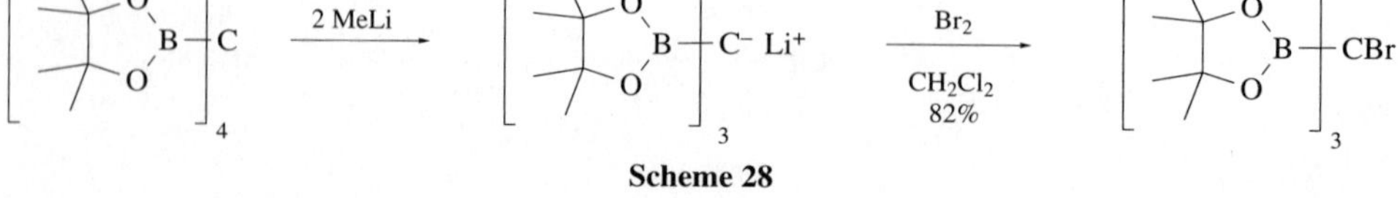

Scheme 28

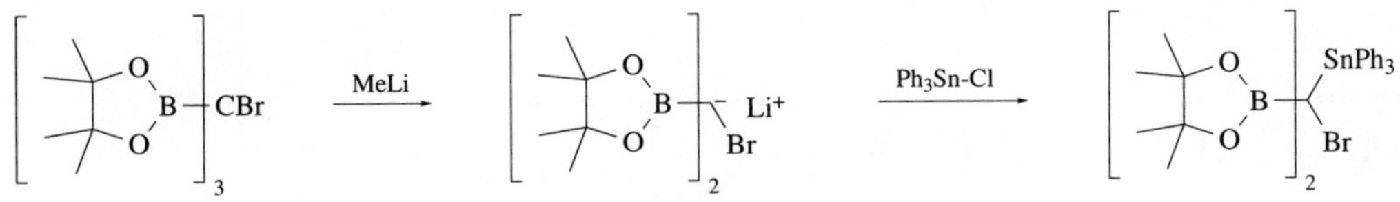

Scheme 29

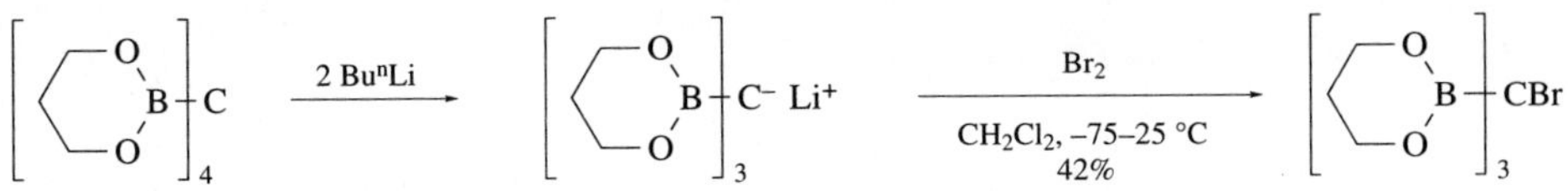

Scheme 30

6.09.7.2 One Halogen, Two Metalloid, and One Metal Function

This class of compounds contains, principally, bis(trimethylsilyl)lithium derivatives which have been studied as potential precursors for metalloid substituted carbenes.

Dihalobis(trimethylsilyl)methanes are readily metallated with butyllithium at low temperature through a halogen–lithium exchange reaction (Equation (55)) ⟨70JOM(23)361, 70JOM(24)647, 70TL4693⟩. The lithium salts are stable at −110 °C. In the absence of a trapping reagent, the salts dimerize to give tetrasubstituted ethylene derivatives ⟨67AG(E)41, 72AG(E)473⟩. They were reacted with silicon, tin, germanium, and phosphorus dihalogenides to afford 1,3-disila-, distanna-, or digerma-cyclobutanes ⟨74JA6237⟩ and bis(methylene)phosphoranes, respectively ⟨82AG(E)80⟩. Preparation of the analogous organomercury reagent was carried out with $(TMS)_2C(Cl)Li$ and mercuric chloride in a 3 : 1 ratio; the yield of $[(TMS)_2ClC]_2Hg$ was 58%. The major by-product has been isolated and characterized as a dimercury compound (Equation (56)) ⟨70JOM(23)361⟩.

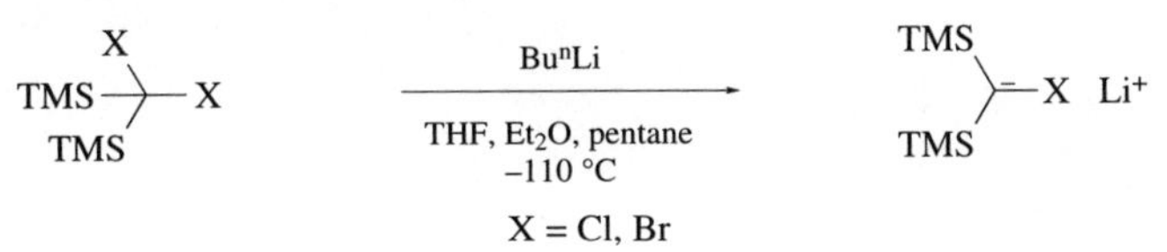

(55)

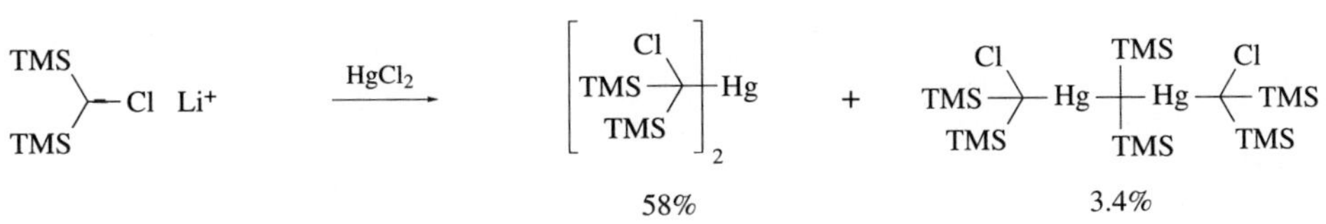

(56)

6.09.8 ACKNOWLEDGEMENTS

In collecting the literature, the authors have benefited greatly from collaboration with Mrs Françoise Girard and Martine Rouyer, who are gratefully acknowledged.

6.10
Functions Containing Four or Three Chalcogens (and No Halogens)

ALEX H. GOULIAEV
Aarhus University, Århus, Denmark

and

ALEXANDER SENNING
Technical University of Denmark, Lyngby, Denmark

6.10.1 TETRACHALCOGENOMETHANES

The compounds discussed here conform to the general formula (**1**), where X is a chalcogen.

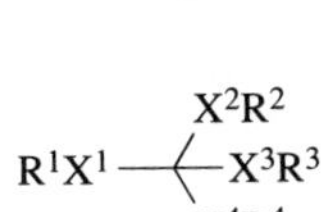

(**1**)

6.10.1.1 Four Similar Chalcogens

Only the fully substituted derivatives of *ortho*-carbonic acid (**2**), tetrathio*ortho*-carbonic acid (**3**), and tetraseleno*ortho*-carbonic acid (**4**) are of preparative significance in the synthesis of tetrachalcogenomethanes with four similar chalcogens, ⟨B-70MI 610-01, B-70MI 610-02⟩. With just one R group equaling hydrogen, the lability of the compounds restricts them to being intermediates, if present at all, in solvolytic reaction sequences, etc. Tetratelluro*ortho*-carbonates (**5**) are unknown. Most known examples of (**1**) are symmetrically substituted; acyclic as well as cyclic and spirocyclic versions of (**1**) have been prepared. Examples of (**1**) where one or several chalcogen atoms (X = S, Se) are in a higher oxidation state than 2 are known, but are rather rare ⟨B-70MI 610-02⟩.

(**2**) (**3**) (**4**) (**5**)

6.10.1.1.1 Four oxygen functions

Derivatives with four oxygen functions are derivatives of *ortho*-carbonic acid, $C(OH)_4$, and have been reviewed extensively ⟨B-70MI 610-01, 83HOU(E4)694⟩. The vast majority are simple *ortho*-esters ((**2**) with R = alkyl or aryl, including cyclic and spirocyclic compounds), but mixed ester *ortho*-esters ((**2**) with R = acyl) have also been described. The available synthetic methods can be classified as involving basic or acidic conditions. In the latter case simple alkanols (which readily form carbenium ions under acidic conditions and are thus subject to C—O bond cleavage) can often not be used as starting materials. An exception are 1-alkanols with strongly electron-withdrawing substituents (such as chlorine, fluorine, or nitro substituents) in the 2-position, where the formation of the corresponding carbenium ions is disfavored. Those synthetic methods which also allow the elaboration of unsymmetrically substituted *ortho*-carbonic acid esters are particularly valuable (Tables 1–5).

The preparation of symmetrical *ortho*-carbonates (**2**) is possible from appropriate C_1 synthons (e.g., tetrachloromethane (**6**), trichloronitromethane (chloropicrin) (**7**), trichloromethanesulfenyl chloride (perchloromethyl thiol) (**8**), trichloroacetonitrile (**9**), trialkoxyacetonitriles (**10**), trichloromethyl isocyanide dichloride (**11**), aryl cyanates (**12**), and trialkoxycarbenium salts (**13**) and alkanols/metal alkoxides or phenols/phenoxides ⟨83HOU(E4)694⟩. While the tetrasubstitution of tetrachloromethane (**6**) generally fails with simple alkoxides, it does proceed satisfactorily with copper(I) phenoxides (Equation (1)) ⟨74TL4409⟩. The catalytic action of iron(III) chloride permits the corresponding synthesis of aliphatic *ortho*-carbonates (**2**) derived from 2-fluoro and/or 2-nitro substituted alkanols under acidic conditions (Equation (2)) ⟨65JOC411⟩.

$$\underset{(\mathbf{6})}{CCl_4} + 4ArOCu \xrightarrow[40\text{–}60\%]{MeCN} \underset{(\mathbf{2})}{C(OAr)_4} \qquad (1)$$

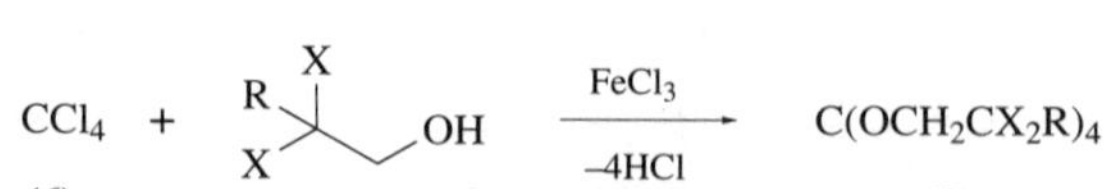

(**6**) (**2**) (2)

Trichloronitromethane (**7**) and trichloromethanesulfenyl chloride (**8**) are more generally applicable as C_1 synthons (Scheme 1) ⟨48JOC265⟩. These reactions fail when applied to branched alkoxides. Trialkoxyacetonitriles (**10**) have been particularly focused upon as precursors of *ortho*-carbonates (**2**) ⟨77S73, 82LA507⟩. Trichloroacetonitrile (**9**) reacts with alkoxides (including branched alkoxides) via the corresponding trialkoxyacetonitriles (**10**) to form *ortho*-carbonates (**2**) (Scheme 2). Both symmetrical ($Alk^1 = Alk^2$) and unsymmetrical ($Alk^1 \neq Alk^2$) tetraalkyl *ortho*-carbonates (**2**) are accessible in this way.

A broadly applicable synthesis of symmetrical tetraaryl *ortho*-carbonates (**2**) (including spirocyclic *ortho*-carbonates derived from dihydroxyarenes) employs trichloromethyl isocyanide dichloride

Table 1 Selected symmetrical acyclic *ortho*-carbonates $C(OR)_4$ (**2**).

R	*Boiling point* (°C/torr)	*Melting point* (°C)	*Ref.*
Me	114	−5.5	77S73
CF_3	20.8		89JOC1990
Et	159		77S73
CCl_3CH_2		131	72S599
$CClF_2CH_2$	80/2.8		65JOC411
$CF(NO_2)_2CH_2$		136	68USP3388147
$C(NO_2)_3CH_2$		163	67USP3306939
CF_3CF_2	80		89JOC1990
Pr	224		48JOC265
Pr^i	70/10		77S73
$CH_2{=}CHCH_2$	100/11		82LA507
$CHF_2CF_2CH_2$	65/0.1		65JOC411
$CF_3CF_2CH_2$	132		89JOC1990
Bu	273		48JOC265
Bu^i	245		37JCS827
$CHF_2(CF_2)_2CH_2$	135/0.05		65JOC411
$Pr^i(CH_2)_2$	81–82/0.0001		77S73
$Me_2CH(CH_2)_2$	168/11		82LA507
Bu^tCH_2		78–79	77S73
$CHF_2(CF_2)_3CH_2$	135/0.05		65JOC411
$Me(CH_2)_5$	184–185/2		49DOK(68)515
c-C_6H_{11}		101–103	77S73
$CHF_2(CF_2)_5CH_2$	170/0.008		65JOC411
$Me(CH_2)_7$	242/2	−32	75FRP2235188
$Me(CH_2)_8$	237–240/1.5		49DOK(68)515
$Me(CH_2)_9$	288–290/2		49DOK(68)515
$Ph(CH_2)_2$	109–110/0.008		B-70MI 610-01
Ph		98	72S599
3-MeC_6H_4		84–85	74TL4409
4-MeC_6H_4		101	74TL4409
2-$Bu^tC_6H_4$		259–260	72S599
2-ClC_6H_4		155	72S599
4-ClC_6H_4		129–130	74TL4409
4-$(O_2N)C_6H_4$		227	72S599
4-$(MeO_2C)C_6H_4$		112	72S599

Table 2 Selected unsymmetrical acyclic *ortho*-carbonates $C(OR^1)_3OR^2$ (**2**).

R^1	R^2	*Boiling point* (°C/torr)	*Ref.*
Me	Bu		82LA507
Me	Bu^s		82LA507
Me	$Et_2N(CH_2)_2$	82–83/15	82LA507
Et	Me	162	77S73
Et	Pr	64–66/10	82LA507
Et	$Et_2N(CH_2)_2$	90–92/10	82LA507
Et	Ph	60–62/0.001	82LA507
Et	4-ClC_6H_4	90–92/0.001	82LA507
Bu	Me		83HOU(E4)694

Table 3 Selected unsymmetrical acyclic *ortho*-carbonates $C(OR^1)_2(OR^2)_2$ (**2**).

R^1	R^2	*Boiling point* (°C/torr)	*Ref.*
Me	Et		82LA507
Me	Bu		83HOU(E4)694
Et	c-C_6H_{11}	93–99/11	82LA507

Table 4 Selected unsymmetrical cyclic *ortho*-carbonates (**2**).

Compound	*Boiling point* (°C/torr)	*Ref.*
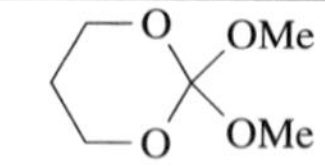	71/15	84S837
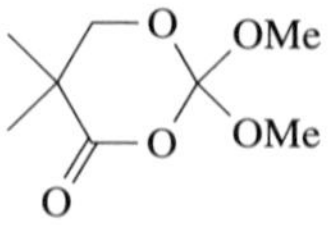	62/0.15	70USP3503993
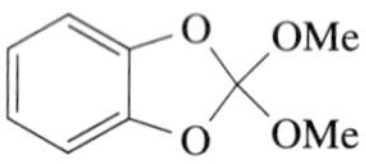	114.5/16	82LA507
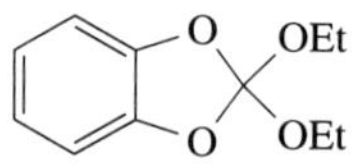	123/15	64LA(675)142
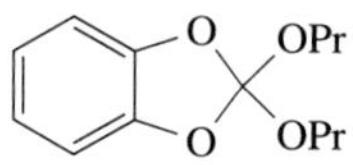	147/12	64LA(675)142

Table 5 Selected spirocyclic *ortho*-carbonates (**2**).

Compound	*Boiling point* (°C/torr)	*Melting point* (°C)	*Ref.*
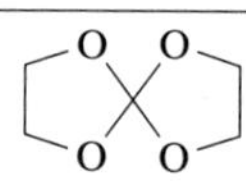		145 121.0–121.5	77S73 92MM3829
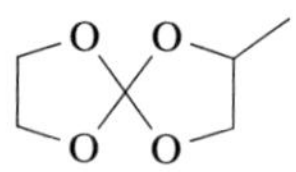			86JAP(K)61289091
	90–94/11		77S73
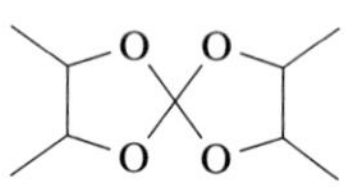		109–110	84S837
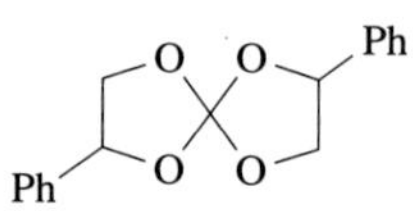		58	92MM3829
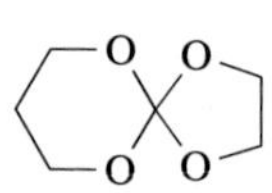	68/1.0		84S837
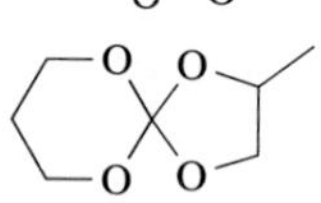	70/5		84S837
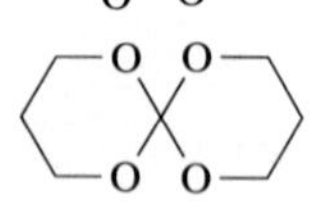		138	82LA507
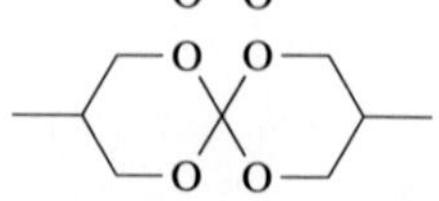		138.5	77S73
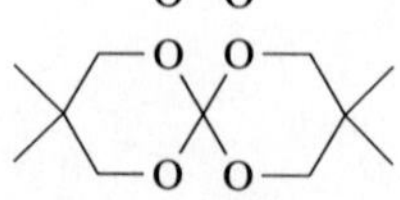		138.5	82LA507

Table 5 (*continued*).

Compound	*Boiling point* (°C/torr)	*Melting point* (°C)	*Ref.*
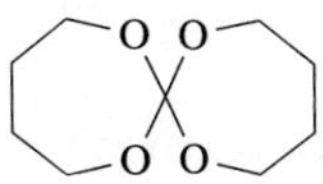		111	84S837
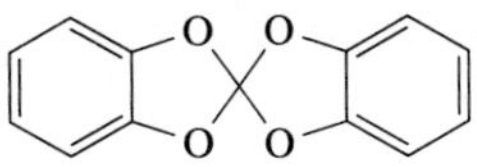		110	67JHC166
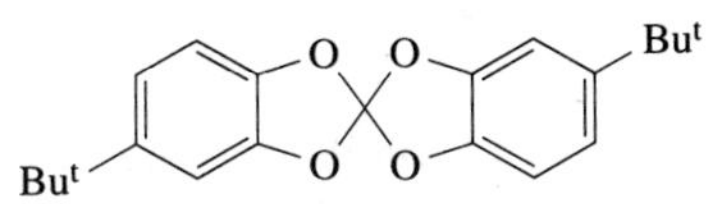		113	72S599
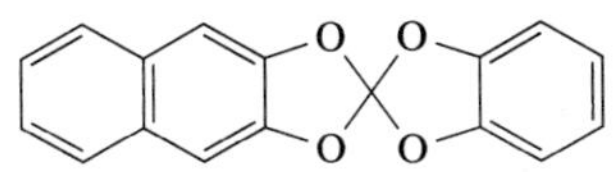		156–157	86JOC370
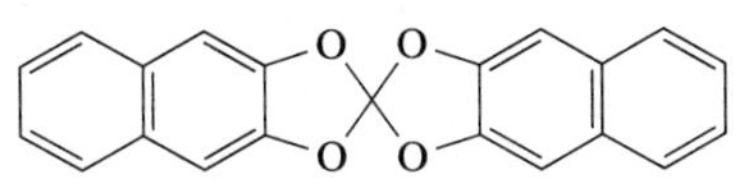		>340	86JOC370
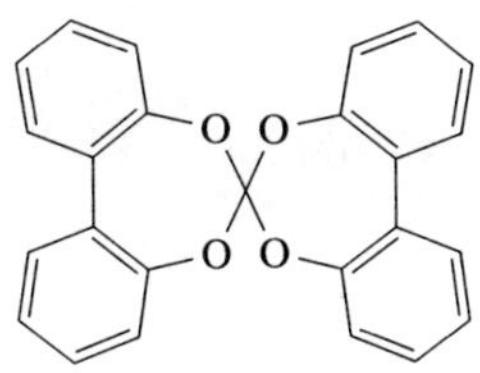		340	72S599
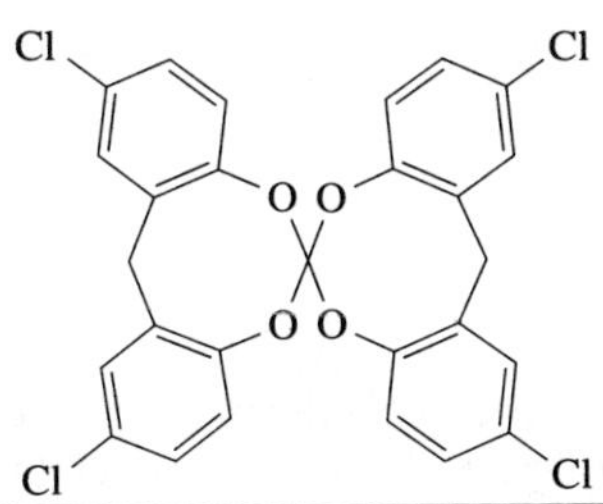		225	72S599

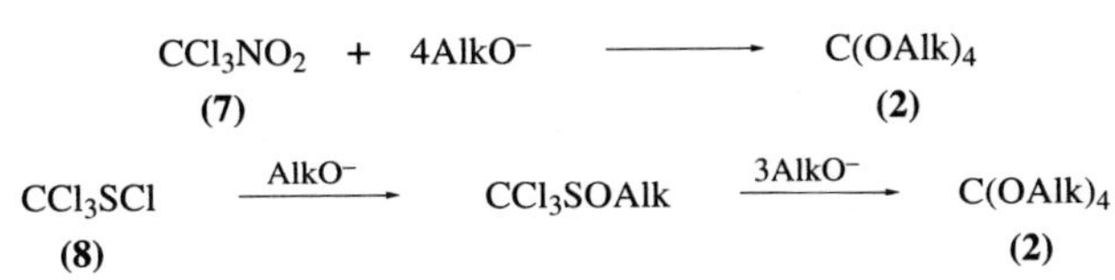

Scheme 1

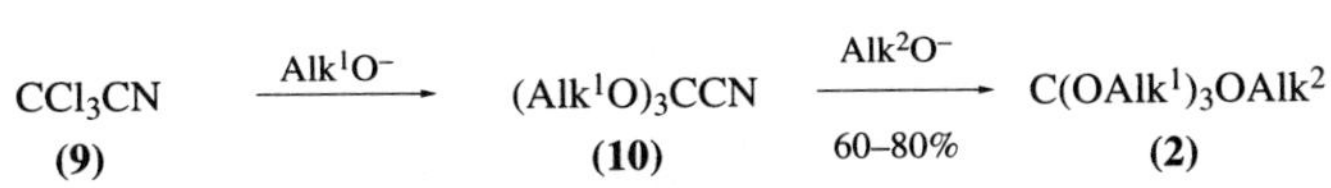

Scheme 2

(N-trichloromethylcarbonimidic dichloride) (**11**) as its C_1 synthon (Equation (3)) ⟨72S599⟩. This method is unsuitable for the preparation of aliphatic *ortho*-carbonates (**2**) (with the exception of tetrakis(2,2,2-trichloroethyl) *ortho*-carbonate), since these are cleaved by the hydrogen chloride evolved.

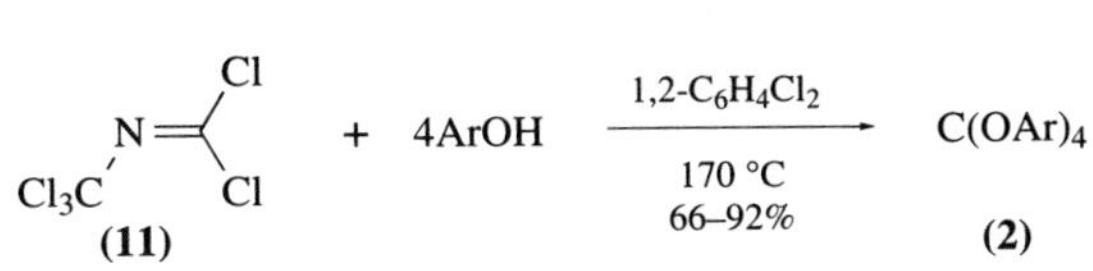

(3)

Aryl cyanates (**12**) react with alkanols under acidic conditions to form tetraalkyl *ortho*-carbonates (**2**) in 20–40% yield (Equation (4)) ⟨65CB3286⟩.

$$\underset{(\mathbf{12})}{ArOCN} + 4AlkOH \xrightarrow[20\text{–}40\%]{H^+} \underset{(\mathbf{2})}{C(OAlk)_4} \qquad (4)$$

Symmetrical tetraalkyl *ortho*-carbonates (**2**) are obtained from trialkoxycarbenium tetrafluoroborates (**13**) and alkoxides (Equation (5)) ⟨56CB2060⟩.

$$\underset{(\mathbf{13})}{(AlkO)_3C^+\ BF_4^-} \xrightarrow{AlkO^-} \underset{(\mathbf{2})}{C(OAlk)_4} \qquad (5)$$

Treatment of thallium(I) alkoxides with carbon disulfide leads to *O*,*O′*-dialkyl thiocarbonates (**14**), which react further with the starting thallium compound to yield tetraalkyl *ortho*-carbonates (**2**), the driving force undoubtedly being the formation of the extremely insoluble thallium(I) sulfide (Scheme 3) ⟨72JOC4198⟩. Dialkoxydibutylstannanes and carbon disulfide react similarly (Scheme 4) ⟨71JOC1176⟩. Thiocarbonates (**14**) can also be similarly desulfurized by treatment with sodium carbonate to yield spirocyclic *ortho*-carbonates (**2**) ⟨67JHC166⟩.

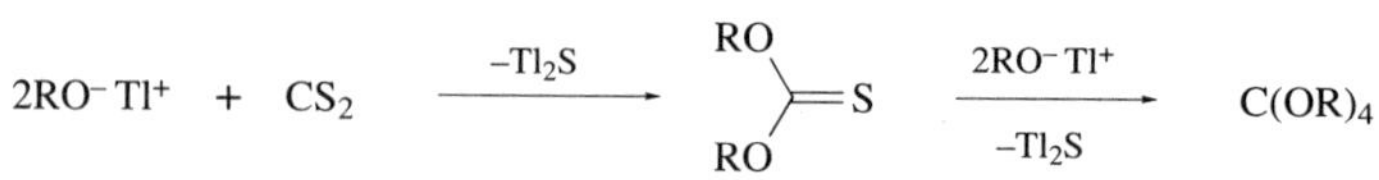

(**14**) (**2**)

Scheme 3

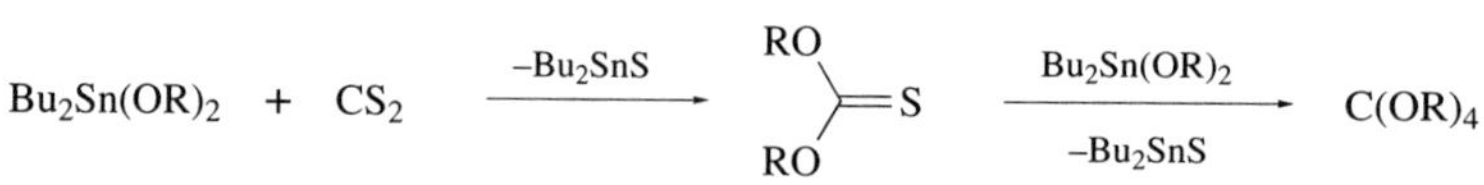

(**14**) (**2**)

Scheme 4

Carbonic acid diesters (**15**) can be converted to the corresponding dichloro derivatives (**16**) with phosphorus pentachloride; these in turn are treated with phenols or phenoxide ions to yield tetra-aryl *ortho*-carbonates (**2**). Unsymmetrical *ortho*-carbonates (**2**) can also be prepared in this way (Scheme 5) ⟨64LA(675)142⟩.

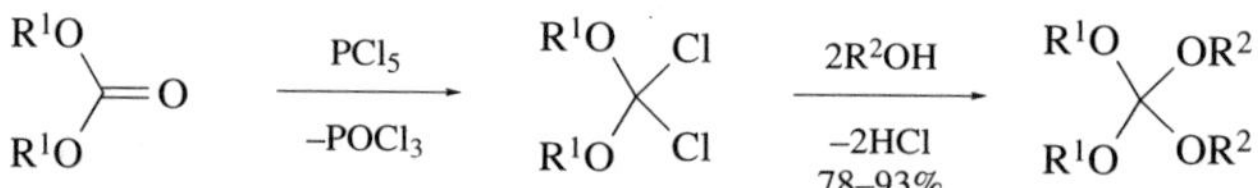

(**15**) (**16**) (**2**)

Scheme 5

Trans-esterification of *ortho*-carbonates (**2**) with alkanols is another method for the preparation of *ortho*-carbonates, even allowing the synthesis of unsymmetrically substituted esters. Spirocyclic *ortho*-carbonates are obtained with α,ω-alkanediols ⟨77S73, 82LA507, 92MM3829⟩. With equimolar amounts of α,ω-diol and tetramethyl *ortho*-carbonate, a half-exchanged monocyclic intermediate can be isolated which, with a second α,ω-diol, can form an unsymmetrical spirocyclic compound (Scheme 6) ⟨84S837⟩. In principle all possible scrambling products of the *trans*-esterification of *ortho*-carbonates $C(OR^1)_4$ with alcohols R^2OH can be obtained (Scheme 7). Spirocyclic and oligospirocyclic *ortho*-carbonates (**2**), including unsymmetrical and substituted versions, have been prepared in considerable variety as monomers for the production of polymers ⟨92JAP(K)04164085⟩. Typical examples are 1,4,6,9-tetraoxaspiro[4.4]nonane, 1,4,6,10-tetraoxaspiro[4.5]decane, 1,5,7,11-tetraoxaspiro[5.5]undecane, 1,5,7,12-tetraoxaspiro[5.6]dodecane, 1,6,8,13-tetraoxaspiro[6.6]tridecane, and 8,10,19,20-tetraoxatrispiro[5.2.2.5.2.2]heneicosane. *Trans*-esterification of spirocyclic *ortho*-carbonates with diols to yield new spirocyclic *ortho*-carbonates (**2**) has also been reported ⟨86JAP(K)61289091⟩.

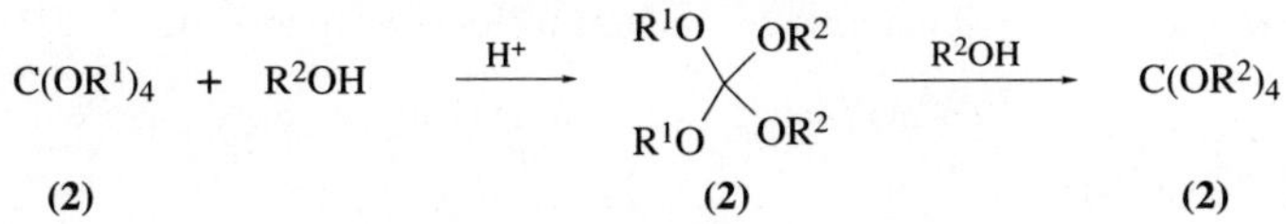

Scheme 6

$$C(OAlk^1)_4 \xrightarrow{Alk^2OH} C(OAlk^1)_3OAlk^2 \xrightarrow{Alk^2OH} C(OAlk^1)_2(OAlk^2)_2 \xrightarrow{Alk^2OH}$$

(2) (2) (2)

$$COAlk^1(OAlk^2)_3 \xrightarrow{Alk^2OH} C(OAlk^2)_4$$

(2) (2)

Scheme 7

Simple *ortho*-carbonates such as tetramethyl, tetraethyl, and tetrapropyl *ortho*-carbonate can be fluorinated with helium-diluted fluorine in a cryogenic reactor to yield the corresponding perfluorinated *ortho*-esters. The hydrogen fluoride formed during the fluorination is scavenged with solid sodium fluoride to prevent cleavage of the carbon–oxygen bonds (Equation (6)) ⟨87JFC(37)327, 89JOC1990⟩.

$$\underset{\substack{\textbf{(2)}\\ n\,=\,1,\,2,\,3}}{C(OC_nH_{2n+1})_4} \xrightarrow[\substack{-120\ ^\circ C\\ 18\text{–}56\%}]{F_2/NaF} \underset{\textbf{(2)}}{C(OC_nF_{2n+1})_4} \qquad (6)$$

6.10.1.1.2 Four sulfur functions

Compounds (**3**), including a considerable number of heterocyclic and spiroheterocyclic tetrathio-*ortho*-carbonic acid esters such as 2,2-dithio substituted 1,3-dithioles and 1,3-dithianes, were briefly reviewed in 1991 ⟨91SUL247⟩. Tetrathio*ortho*-carbonic acid and its tetrasodium salt are known, but they are without preparative importance. The simplest type of organic compounds fitting the above description are the symmetrical tetrathio*ortho*-carbonates $C(SR)_4$, available from the corresponding alkali metal thiolate and a C_1 synthon such as tetrachloromethane (**6**) (Scheme 8) ⟨83HOU(E4)705⟩. With trichloromethanesulfenyl chloride (**8**) the corresponding reaction leads to tetrathiomethane units containing symmetrical disulfides $(RS)_3CSSC(SR)_3$ (**17**) (R = Et, Pr, Bu, Bu^t) in the case of aliphatic thiols (Equation (7)) and trisulfides $(RS)_3CSSSC(SR)_3$ (R = Ph, 2-ClC_6H_4, 4-ClC_6H_4, 3-MeC_6H_4, 4-MeC_6H_4, 2,4,6-$Me_3C_6H_2$, 4-$Bu^tC_6H_4$) in the case of aromatic thiols ⟨52RTC1071⟩. Examples of (**3**) are shown in Tables 6–9.

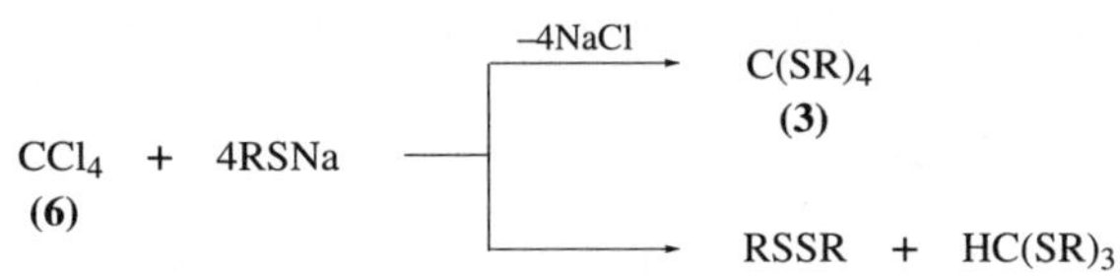

Scheme 8

$$\underset{\textbf{(8)}}{CCl_3SCl} + RSNa \xrightarrow{-RSSR} \underset{\textbf{(17)}}{(RS)_3CSSC(SR)_3} \qquad (7)$$

Tris(methylthio)carbenium cations react with excess thiols, RSH, to give the corresponding tetrathio*ortho*-carbonates $C(SR)_4$ (Equation (8)) ⟨68JOC3333⟩. Unsymmetrical tetrathio*ortho*-carbonates $(R^1S)_2C(SMe)SR^2$ can be obtained from 2-(methylthio)-1,3-dithiolium cations and thiolate anions R^2S^- (Equation (9)) ⟨64ZC384⟩. *N*,*N*′-Dinitroso-*S*-alkyl- or -arylisothioureas (**19**), obtained *in situ* from isothioureas (**18**) and nitrous acid, have been reported to yield unsymmetrical or symmetrical tetrathio*ortho*-carbonates (**3**) with thiols and symmetrical tetrathio*ortho*-carbonates with base (Scheme 9 and Equations (10) and (11)). The impossibility of obtaining tetra-*t*-butyl tetrathio*ortho*-carbonate $C(SBu^t)_4$ in this way has been stressed. This is probably not due to problems with the method, but rather to prohibitive steric congestion ⟨83HOU(E4)705⟩.

Table 6 Selected symmetrical acyclic tetrathio*ortho*-carbonates $C(SR)_4$ (**3**).

R	*Boiling point* (°C/torr)	*Melting point* (°C)	*Ref.*
Me	127/11	66	88S22
CF_3	82/70		67JOC2063
Et		35.5	72RTC1117
Pr	123/0.16		68JOC3333
Pr^i		61.4	72RTC1117
c-C_6H_{11}		169	81JAP(K)56125326
$Me(CH_2)_5CHMe$			86MI 610-01
$Me(CH_2)_{10}$			86MI 610-01
$Me(CH_2)_{11}$			86MI 610-01
$Me(CH_2)_{17}$			73GEP1694210
Ph		159	72CB3280
4-BrC_6H_4		211	67BSF3233
4-ClC_6H_4		212.6–213.6	72CB3280
4-FC_6H_4		169	74CC634
2-MeC_6H_4		126.3–127.8	72CB3905
4-MeC_6H_4		147	74CC634
4-$MeOC_6H_4$		156	74CC634
2-naphthyl		136	68JOC3333

Table 7 Selected unsymmetrical acyclic tetrathio*ortho*-carbonates $C(SR^1)_3SR^2$ (**3**).

R^1	R^2	*Boiling point* (°C/torr)	*Melting point* (°C)	*Ref.*
Pr^i	Me	85–86/0.2		72CB3280
Ph	Me		95.2–95.8	72CB487
Ph	CF_3			69ZOB1755
Ph	4-ClC_6H_4		191	13LA(396)1
2-MeC_6H_4	4-ClC_6H_4		193	13LA(396)1

$$[(MeS)_3C]^+ [BF_4]^- + 4RSH \xrightarrow[-H[BF_4]]{-3MeSH} \underset{\mathbf{(3)}}{C(SR)_4} \qquad (8)$$

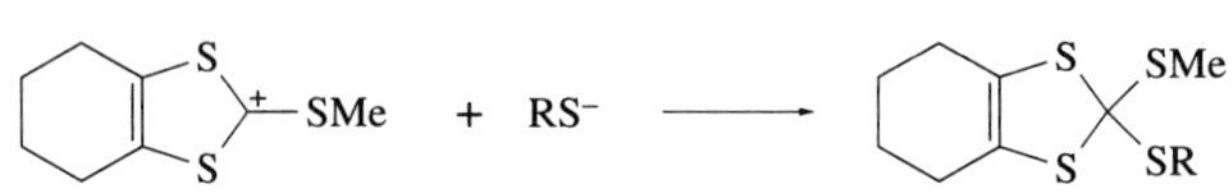

(3)

(9)

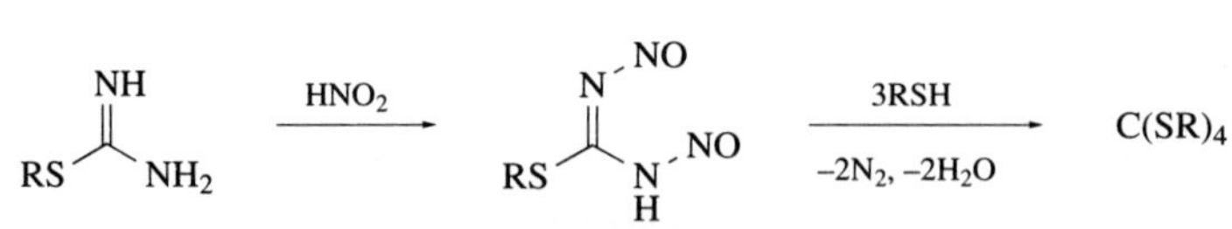

(18) (19) (3)

Scheme 9

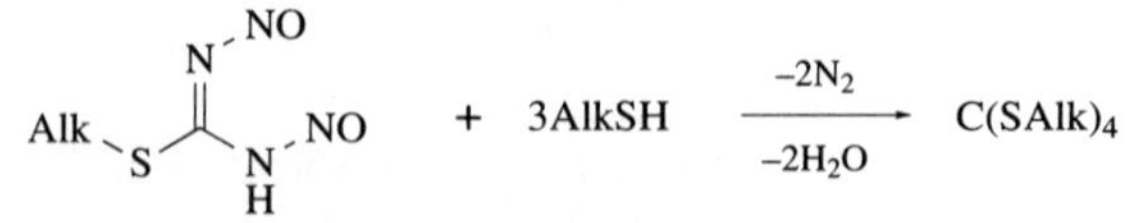

(19) (3)

(10)

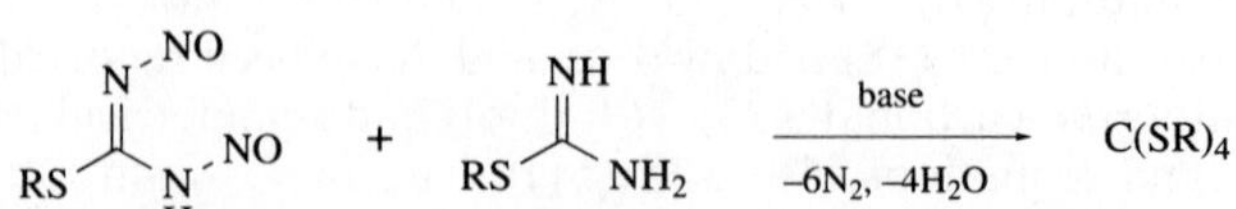

(19) (18) (3)

(11)

Table 8 Selected cyclic tetrathio*ortho*-carbonates (**3**).

Compound	*Melting point* (°C)	*Ref.*
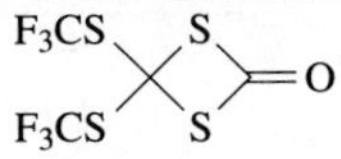	(boiling point 50 °C/10 torr)	77CB916
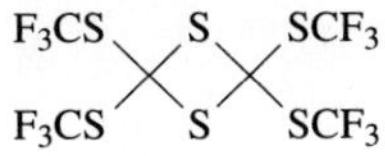	(boiling point 44 °C/5 torr)	77CB916
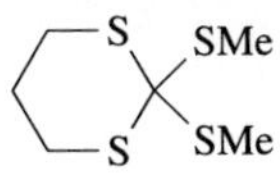	32.7	67AG(E)442
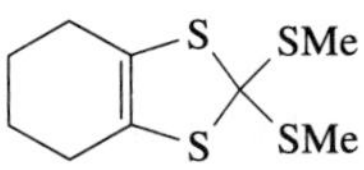	38	64ZC384
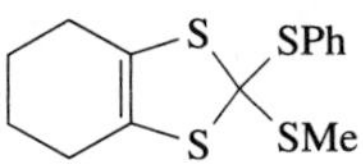	63	64ZC384
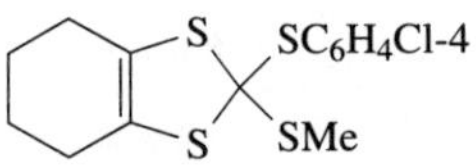	88	64ZC384
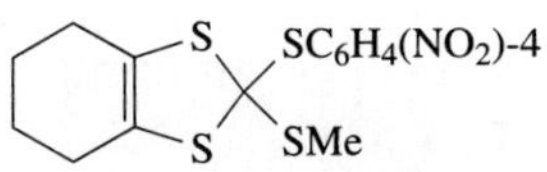	93	64ZC384
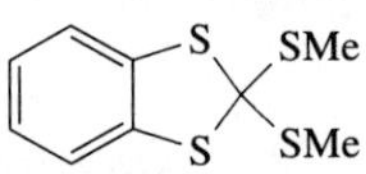	21.5 (boiling point 88 °C/0.6 torr)	72CB3280
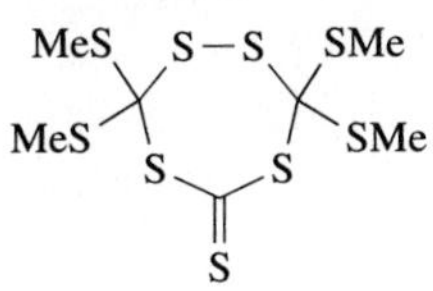 **(30)**		75RTC1
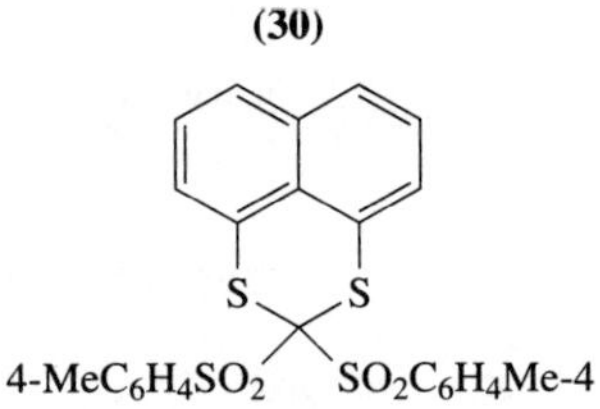 **(35)**	148.0–148.5	72BCJ960

Dithio- ⟨62RTC1009⟩ and trithiomethanide ⟨72CB3280⟩ ions can be sulfenylated with dialkyl or diaryl disulfides to yield the corresponding symmetrical or mixed tetrathio*ortho*-carbonates (**3**) (Equations (12)–(15)).

$$Na^{+}\,{}^{-}[CH(SEt)_2] \; + \; EtSSEt \xrightarrow{NaNH_2/NH_3} \underset{\mathbf{(3)}}{C(SEt)_4} \qquad (12)$$

$$Li^{+}\,{}^{-}[C(SMe)_3] \; + \; MeSSMe \xrightarrow{-MeSLi} \underset{\mathbf{(3)}}{C(SMe)_4} \qquad (13)$$

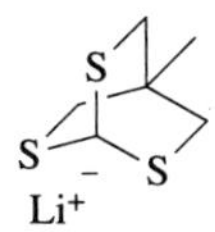 + MeSSMe —(−MeSLi)→ 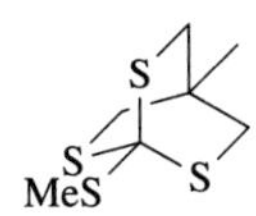 **(3)** (14)

Table 9 Selected spirocyclic tetrathio*ortho*-carbonates (**3**).

Compound	*Melting point* (°C)	*Ref.*
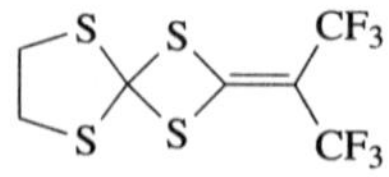	85.5–86.5	70JOC3470
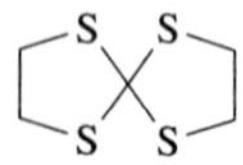	142–143	67JOC2567
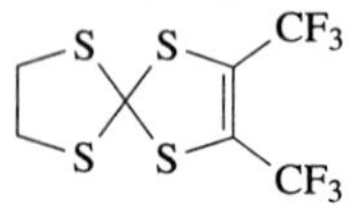 **(32)**	209	91CB2025
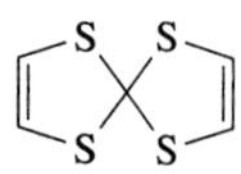		88EUP293749
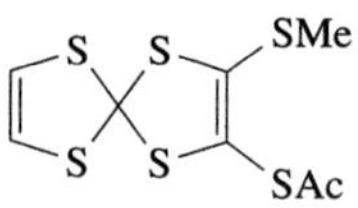		83MI 610-01
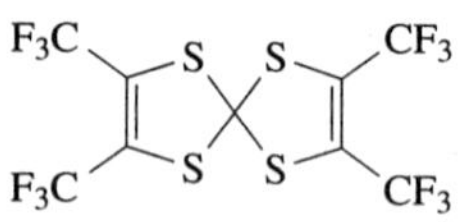 **(33)**	44	91CB2025
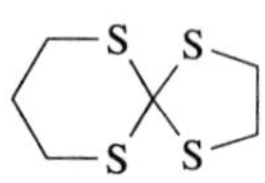	130	67JOC2567
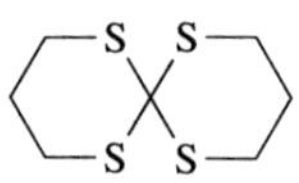	120	67JOC2567
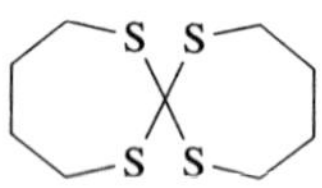	165	70JHC201
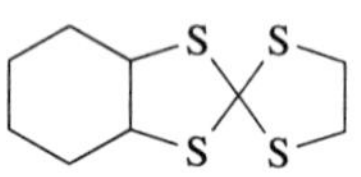	117.5–120.0	75JCS(P1)270
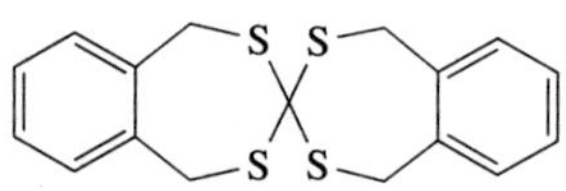		82BCJ1106
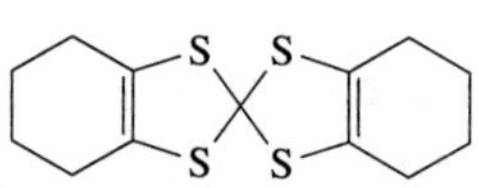 **(27)**	163	76JPR(318)127
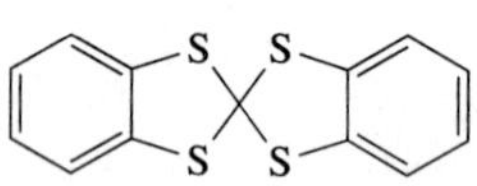	120–121	61E566
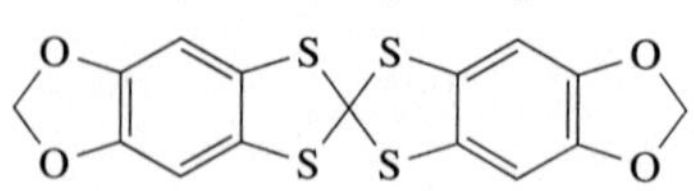	237–238	65LA(689)179
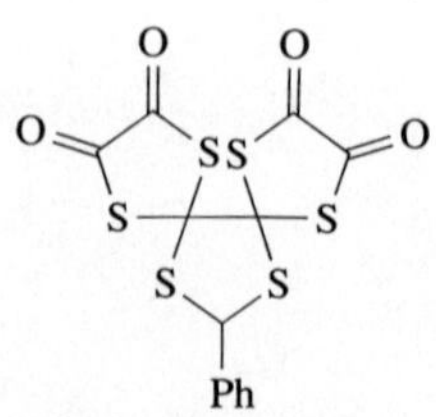	160.5–161.0 (dec.)	67CB767

Table 9 (*continued*).

Compound	*Boiling point* (°C)	*Ref.*
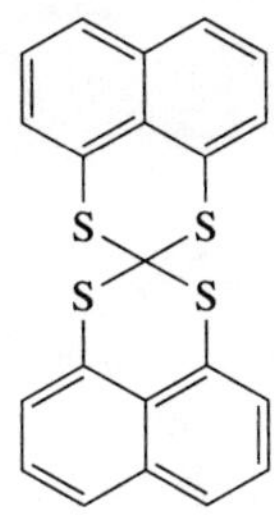	187–188	86JOC370

$$2Li^+ \ ^-[CH(SR^1)_2] + 2R^2SSR^2 \xrightarrow[-2R^2SLi]{-CH_2(SR^1)_2} \underset{(\mathbf{3})}{C(SR^1)_2(SR^2)_2} \qquad (15)$$

The cyclic trithiocarbonate 1,3-dithiane-2-thione (**20**) has been treated with a secondary amine to form the corresponding spirocyclic tetrathio*ortho*-carbonate (**3**) (Equation (16)) ⟨62JOC4068⟩. 1,3-Dithiolane-2-thione (**23**) reacts with 2-phenyloxirane (**55**; R = Ph) to yield the corresponding monosubstituted 1,4,6,9-tetrathiaspiro[4.4]nonane as a by-product (cf. Scheme 19) ⟨91HCA1500⟩.

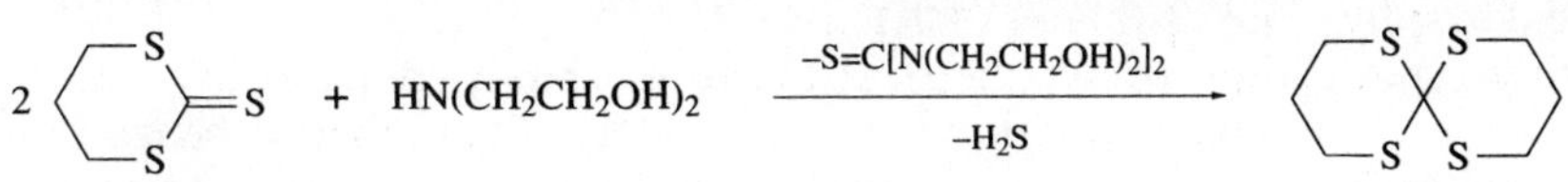
(20) (3) (16)

Similarly, cyclic trithiocarbonates such as (**20**) and cyclic (*N*-cyanoimino)dithiocarbonates (**21**) react with α,ω-alkanedithiols to form the corresponding symmetrical or unsymmetrical spirocyclic compounds (**3**) (Scheme 10 and Equation (17)) ⟨67JOC2567, 72USP3652256⟩.

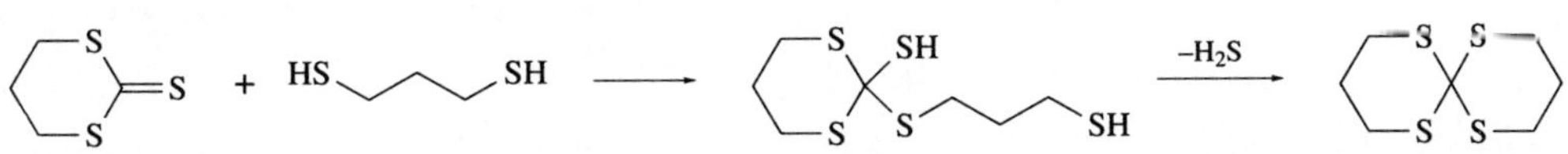
(20) (3) (3)

Scheme 10

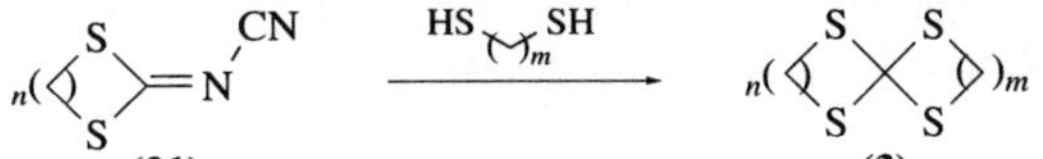
(21) (3) (17)

Carbon disulfide has been subjected to double insertion of an α-thioxo carbene leading to a symmetrical spirocyclic compound (Equation (18)) ⟨61E566⟩.

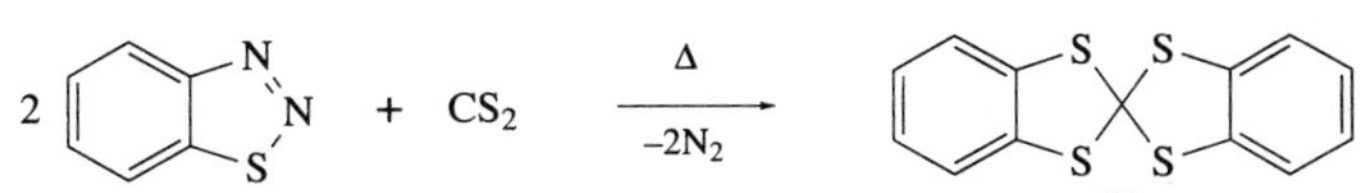
(3) (18)

The [2 + 2] addition of a thioketene such as bis(trifluoromethyl)thioketene (**22**) to a cyclic trithiocarbonate such as 1,3-dithiolane-2-thione (**23**) has been shown to yield the corresponding unsymmetrical spirocyclic compound (Equation (19)) ⟨70JOC3470⟩.

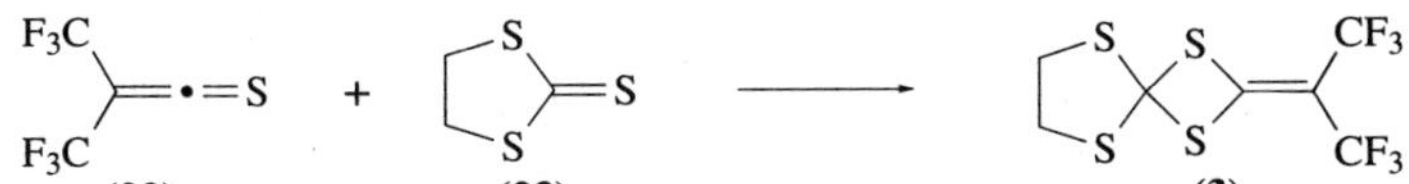
(22) (23) (3) (19)

The 2-azido-1,3-dithiole (**24**) decomposes in boiling dioxan with formation of the corresponding spirocyclic tetrathio*ortho*-carbonate (**27**). The corresponding carbene (**25**) (formed by loss of hydrazoic acid) and 1,2-cyclohexanedithione (**26**) (formed by loss of nitrogen and hydrogen cyanide) have been invoked as putative intermediates (Scheme 11). A more rational synthesis of (**27**) from sodium 1,2-cyclohexenedithiolate (**29**) and a trithiocarbenium salt (**28**) (Equation (20)) has been developed by the same author ⟨76JPR127⟩.

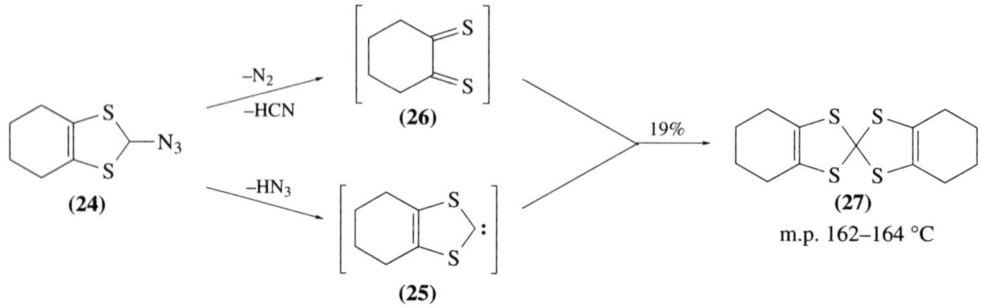

Scheme 11

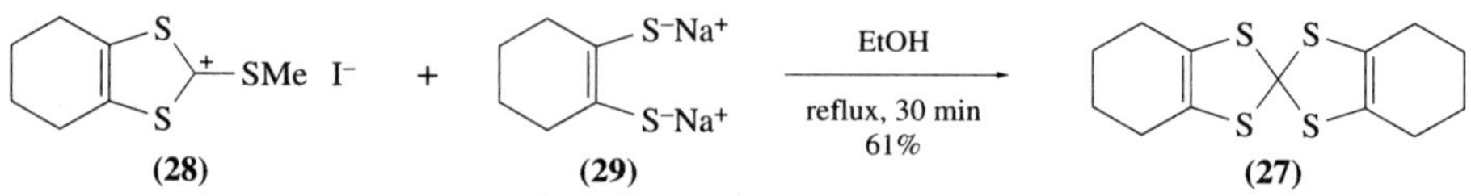

(20)

The obscure reaction of methyl chlorodithioformate with methylmagnesium iodide furnishes the tetrathiepane derivative (**30**) (Table 8) ⟨75RTC1⟩.

Finally, *trans*-esterification of symmetrical tetrathio*ortho*-carbonates (**3**) with thiols, including α,ω-alkanedithiols, is possible (Equation (21)) ⟨70JHC201⟩. A number of benzo annellated 1,4,6,9-tetrathiaspiro[4.4]nonanes with polymerizable alkenyl side chains have been prepared in this way ⟨89JAP(K)01182311⟩. Interestingly, there is no record of the corresponding *trans*-esterification of *ortho*-carbonates (**2**) with thiols.

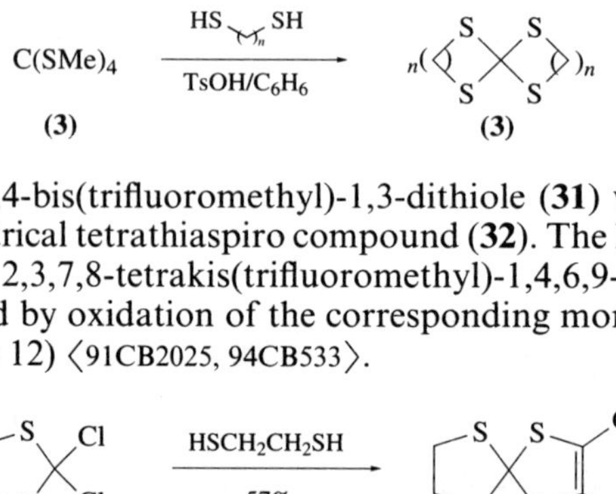

(21)

Treatment of 2,2-dichloro-3,4-bis(trifluoromethyl)-1,3-dithiole (**31**) with 1,2-ethanedithiol leads to the corresponding unsymmetrical tetrathiaspiro compound (**32**). The latter reacts with 1,1,1,3,3,3-hexafluoro-2-butyne to yield 2,3,7,8-tetrakis(trifluoromethyl)-1,4,6,9-tetrathiaspiro[4.4]nona-2,7-diene (**33**), which is also formed by oxidation of the corresponding monocyclic sulfine (**34**) with *m*-chloroperbenzoic acid (Scheme 12) ⟨91CB2025, 94CB533⟩.

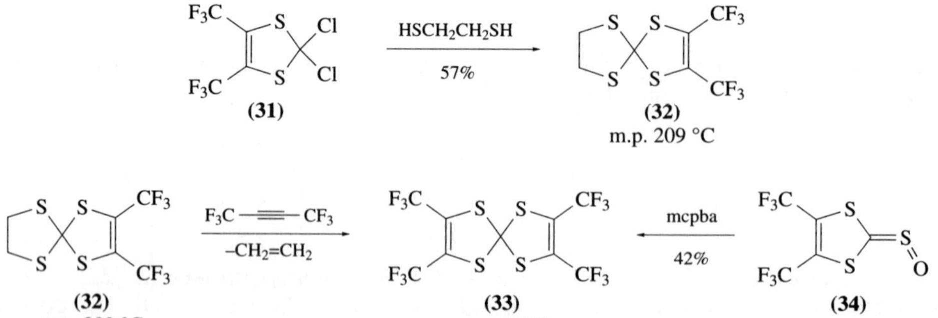

Scheme 12

Of the many theoretically possible partially or fully oxidized derivatives of tetrathio*ortho*-carbonates, only a few have been prepared. Partially ⟨33RTC923, 33RTC1039⟩ and fully ⟨11LA(384)322⟩ *S*-brominated derivatives of symmetrical compounds are known (Scheme 13). The corresponding *S*-iodination of tetramethyl tetrathio*ortho*-carbonate stops after the introduction of three *S*-iodo functions. Additional iodine does not attack the fourth sulfur atom, instead it attacks the iodide ions to form triiodide ions (Scheme 14).

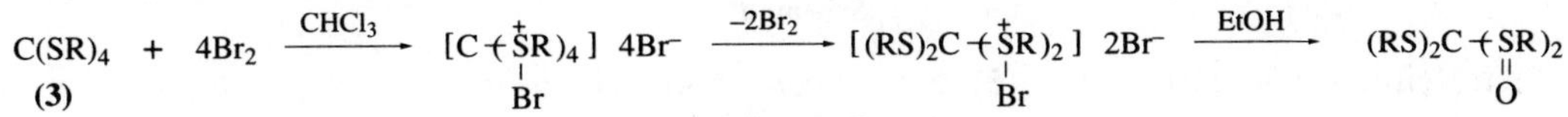

Scheme 13

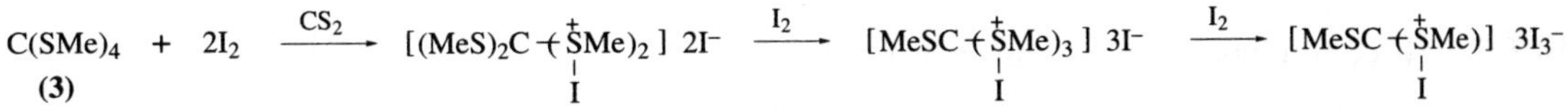

Scheme 14

The disulfoxide tetraphenyl tetrathio*ortho*-carbonate *S,S'*-dioxide has been prepared by solvolysis of the corresponding *S,S,S',S'*-tetrabromo derivative (Scheme 13) ⟨33RTC1039⟩.

Disulfones $(R^1S)_2C(SO_2R^2)_2$ ⟨48RTC894, 52RTC409, 72LA(758)132, 82CC1183⟩ and trisulfones $(R^1S)C(SO_2R^2)_2SO_2R^3$ ⟨B-70MI 610-01⟩ are also accessible (Scheme 15). The disulfone (**35**) (Table 8) has been prepared in 88% yield by copper(II)-catalyzed copyrolysis of the corresponding cyclic disulfide and di(*p*-tolylsulfonyl)diazomethane ⟨72BCJ960⟩.

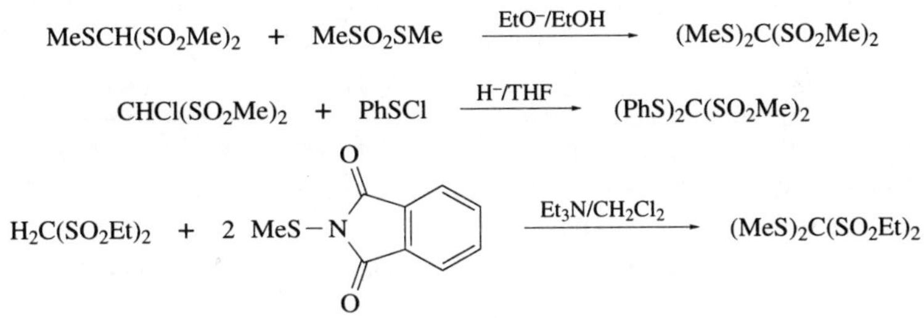

Scheme 15

The failure in a number of attempts to obtain octaoxides $C(SO_2R)_4$ of tetrathio*ortho*-carbonates (**3**) has been attributed to the fact that they must be regarded as mixed anhydrides of two extremely strong acids, the trisulfone $CH(SO_2R)_3$ and the sulfonic acid RSO_3H, and thus subject to facile hydrolysis.

The synthesis of the sulfenic acid derivatives tris[(trifluoromethyl)thio]methanesulfenyl chloride $(CF_3S)_3CSCl$ and *N,N*-diethyl-tris[(trifluoromethyl)thio]methanesulfenamide $(CF_3S)_3CSNEt_2$ is particularly intriguing (Scheme 16) ⟨76CB1976⟩.

$$[(CF_3)S]_2CCl\text{–}S\text{–}NR_2 \; + \; Hg[S(CF_3)]_2 \longrightarrow [(CF_3)S]_3C\text{–}S\text{–}NR_2 \xrightarrow{HCl} [(CF_3)S]_3CSCl$$

Scheme 16

6.10.1.1.3 *Four selenium or tellurium functions*

The chemistry of tetraseleno*ortho*-carbonic esters (**4**) is not well developed; only two compounds, tetrakis(trifluoromethyl) tetraseleno*ortho*-carbonate $C[Se(CF_3)]_4$ and tetraphenyl tetraseleno*ortho*-carbonate $C(SePh)_4$, have been synthesized so far. The former is prepared by reaction of mercury(II) trifluoromethaneselenolate (**36**) with tetrabromomethane; a contamination with 5% unreacted tetrabromomethane could not be removed (Equation (22)) ⟨84T4963⟩. The latter is obtained by selenenylation of tris(phenylseleno)methanide ion with diphenyl diselenide (Scheme 17) ⟨72CB511⟩. While this is the only reported application of this synthetic concept, it should also be useful for the preparation of unsymmetrical tetraseleno*ortho*-carbonates (**4**).

$$\underset{(\mathbf{36})}{[(CF_3)Se]_2Hg} \; + \; CBr_4 \xrightarrow{65\,^{\circ}C} \underset{(\mathbf{4})}{C[Se(CF_3)]_4} \qquad (22)$$

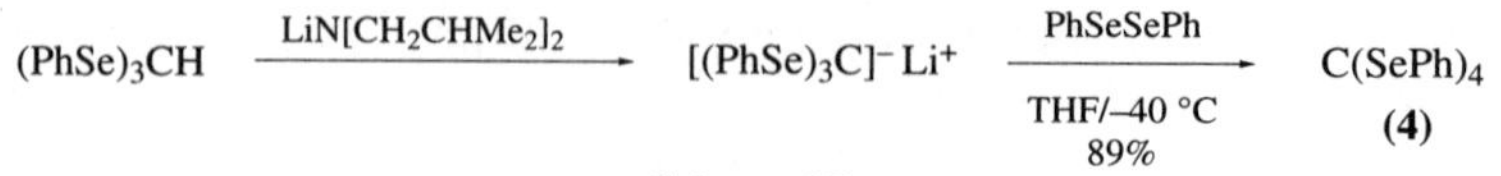

Scheme 17

No tetratelluro*ortho*-carbonates (**5**) have been reported so far ⟨B-86MI 610-02, B-87MI 610-01, 90HOU(E12/b)1⟩. The reason that their synthesis does not even seem to have been attempted must be their expected instability, with the possible exception of perfluoroalkyl tetratelluro*ortho*-carbonates.

6.10.1.2 Three Similar and One Different Chalcogen

This category theoretically consists of thio*ortho*-carbonates (**37**), seleno*ortho*-carbonates (**38**), telluro*ortho*-carbonates (**39**), trithio*ortho*-carbonates (**40**), selenotrithio*ortho*-carbonates (**41**), tellurotrithio*ortho*-carbonates (**42**), triseleno*ortho*-carbonates (**43**), triselenothio*ortho*-carbonates (**44**), triselenotelluro*ortho*-carbonates (**45**), tritelluro*ortho*-carbonates (**46**), tritellurothio*ortho*-carbonates (**47**), and selenotritelluro*ortho*-carbonates (**48**). None of the above selenium- and/or tellurium-containing compounds have yet been reported ⟨B-86MI 610-02, B-86MI 610-03, B-87MI 610-01, 90HOU(E12/b)1⟩.

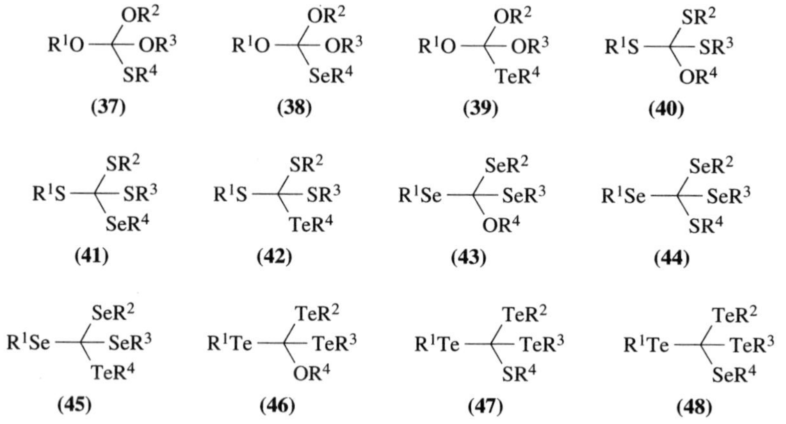

6.10.1.2.1 Trioxygen-substituted methyl chalcogens

The vast majority of known thio*ortho*-carbonates (**37**) are cyclic *ortho*-esters; the simple compound $(MeO)_3CSMe$ has, however, been prepared ⟨83HOU(E4)703, 90MI 610-01⟩. Thio*ortho*-carbonic acid tetraalkyl esters (**37**) have been postulated as intermediates in the preparation of tetraalkyl *ortho*-carbonates (**2**) from trichloromethanesulfenyl chloride (**8**) and sodium alkoxides (cf. Scheme 1) ⟨48JOC265⟩. Disulfides of the type $(RO)_3CSSCCl_3$ (**49**) and $(RO)_3CSSC(OR)_3$ (**50**) are formed concomitantly upon treatment of 2-fluoro- and/or 2-nitroalkanols with trichloromethanesulfenyl chloride (**8**) and base under phase-transfer conditions (Scheme 18 with, for example, $R = (O_2N)_2CFCH_2$) ⟨83JOC3354⟩.

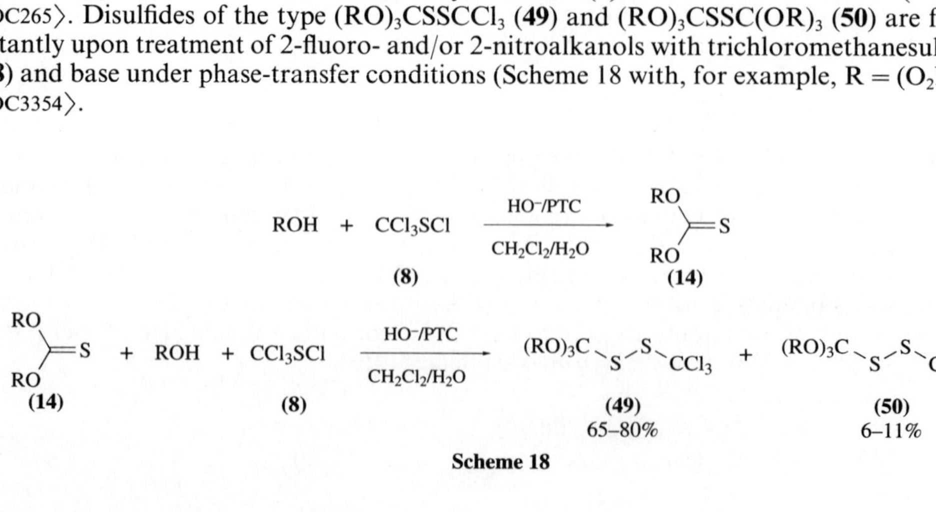

Scheme 18

O,O-Disubstituted thiocarbonates (**14**) react with nitrile oxides such as (**51**) to give the corresponding cyclic *ortho*-ester such as (**52**) (Equation (23)) ⟨61AG656, 72CB2815⟩.

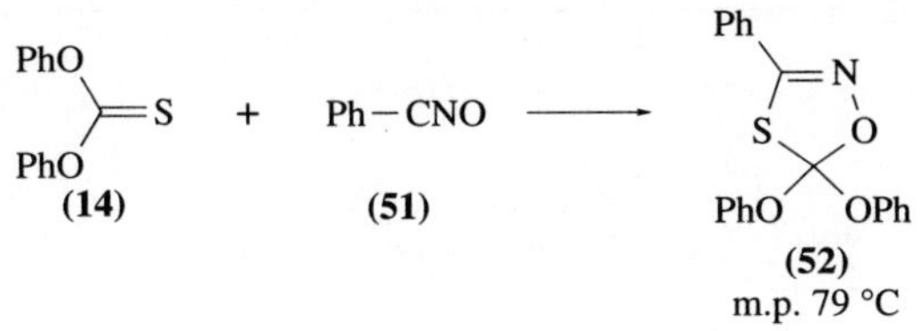

(23)

The oxidation of 1,3-benzodioxole-2-thione (**53**) with lead(IV) acetate to yield the disulfide (**54**) is hardly a general reaction (Equation (24)) ⟨67JCS(C)807⟩.

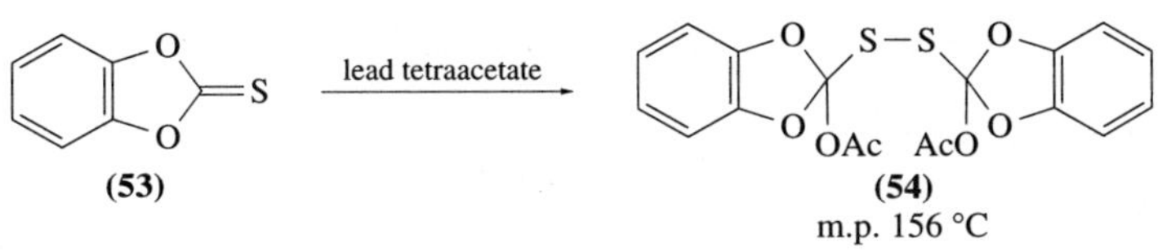

(24)

6.10.1.2.2 Trisulfur-substituted methyl chalcogens

The chemistry of trithio*ortho*-carbonates (**40**) is strongly dominated by cyclic compounds ⟨83HOU(E4)704⟩. See Table 10 for examples.

Spirocyclic trithio*ortho*-carbonates (**40**) are accessible from 1,3-dithiole-2-thione (**56**) and 1,3-dithiolane-2-thione (**23**), and 2-substituted oxiranes (**55**), both possible isomers being formed in total yields of 40–60% (Scheme 19) ⟨91HCA1500⟩.

Acyclic ⟨61AG656, 72CB2815, 91TL4329⟩ as well as cyclic ⟨67BSF2239⟩ trithiocarbonates (**57**) and *C*-sulfonyldithioformates (**60**) add nitrile oxides (**58**) to give the corresponding 1,4,2-oxathiazoles (**59**) and (**61**), respectively (Scheme 20) ⟨76CB1069⟩. Nitrosoalkenes such as 1-nitroso-1-phenylethene (**62**) react with trithiocarbonates such as diphenyl trithiocarbonate to form the corresponding 4*H*-1,5,2-oxathiazines such as (**63**) (Equation (25)) ⟨85TL2131⟩.

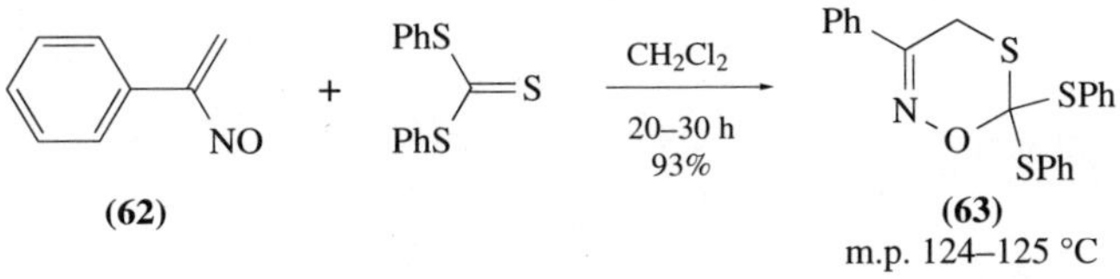

(25)

Pyrolysis of *O,O′*-alkanediyl *S,S′*-dimethyl bis(dithiocarbonates) (**64**) yields the corresponding *O,S*-alkanediyl *S,S*-dimethyl trithio*ortho*-carbonates (**65**) (Scheme 21) ⟨78S286⟩. Carbohydrate derivatives containing cyclic trithio*ortho*-carbonate groups have been prepared in the same way ⟨76S449, 79CAR(74)127⟩; cf. ⟨76JCS(P1)2112⟩.

1-Oxa-4,6,9-trithiaspiro[4.4]non-2-ene derivatives (**67**) are accessible by *S*-alkylation of 1,3-dithiolane-2-thione (**23**) and subsequent deprotonation to the corresponding thiocarbonyl ylide (**66**), which finally rearranges (Scheme 22) ⟨72BCJ1797⟩.

The trithiocarbenium salt (**68**) can be treated with simple primary or secondary alkanols to yield the corresponding trithio*ortho*-carbonates (**69**) (Equation (26)) ⟨64ZC384⟩.

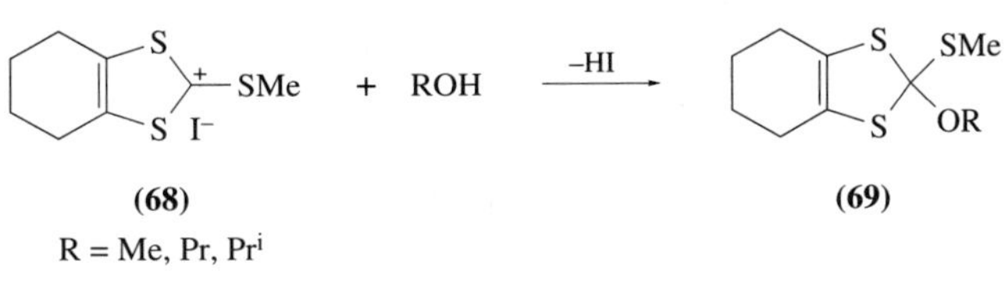

(26)

Table 10 Selected cyclic trithio*ortho*-carbonates (**40**).

Compound	*Boiling point* (°C/torr)	*Ref.*
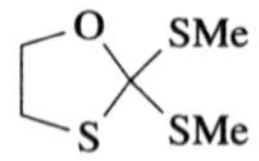	110–112/4	78S286
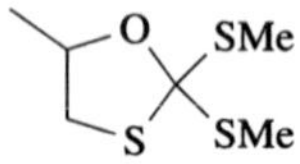	(melting point 29 °C)	78S286
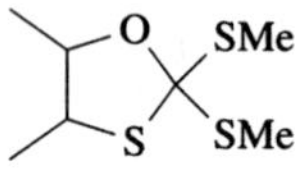	140–142/8	78S286
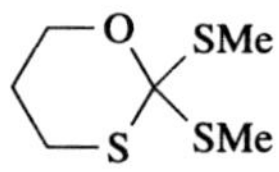		78S286
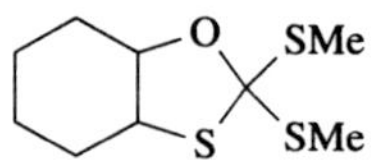	190–193/2	78S286
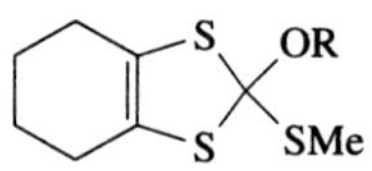 **(69)**		64ZC384
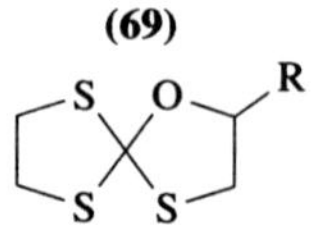		91HCA1500
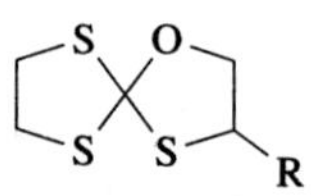		91HCA1500
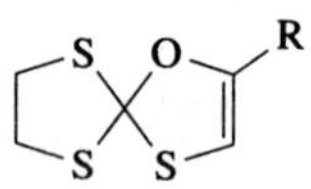		91HCA1500
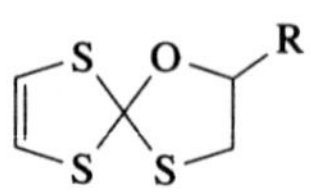		91HCA1500
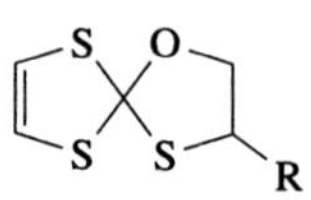		91HCA1500

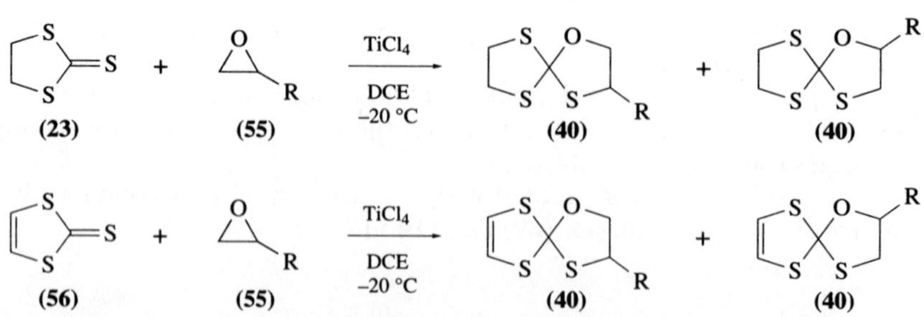

R = Me, CH_2Cl, Et, Bu, CH_2COBu^t, Ph
DCE = $ClCH_2CH_2Cl$

Scheme 19

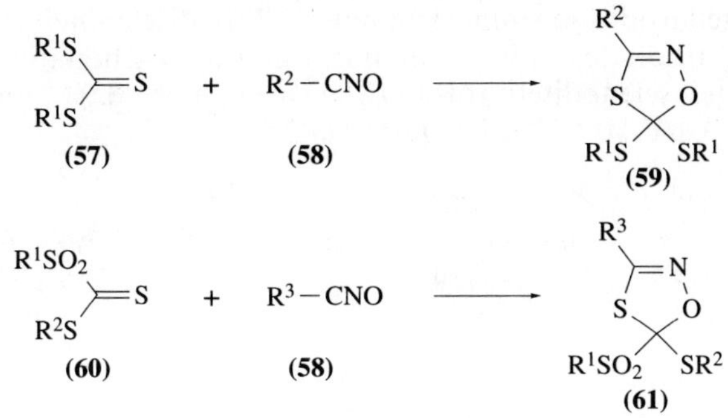

Scheme 20

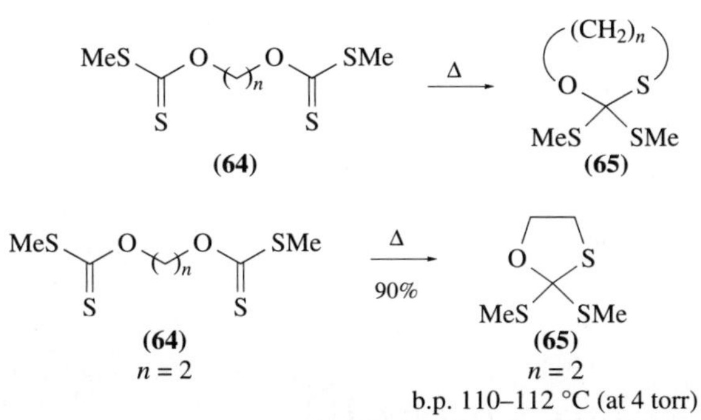

Scheme 21

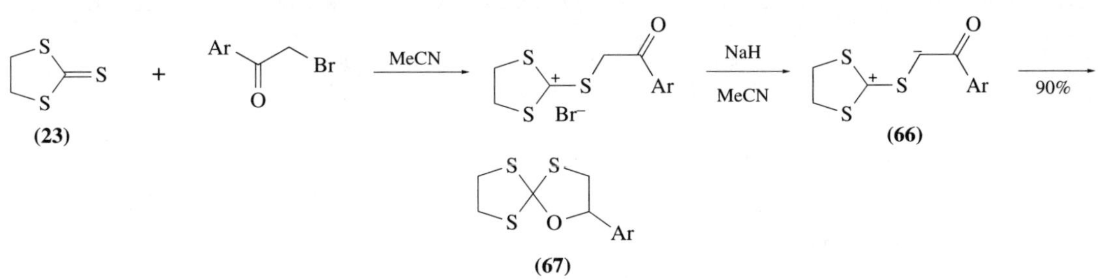

Scheme 22

Bis(trifluoromethyl)thioketene (**22**) adds to *O*,*S*-dialkyl dithiocarbonates (**70**) to form the corresponding 1,3-dithietanes (**71**) in high yields (Equation (27)) ⟨70JOC3470⟩.

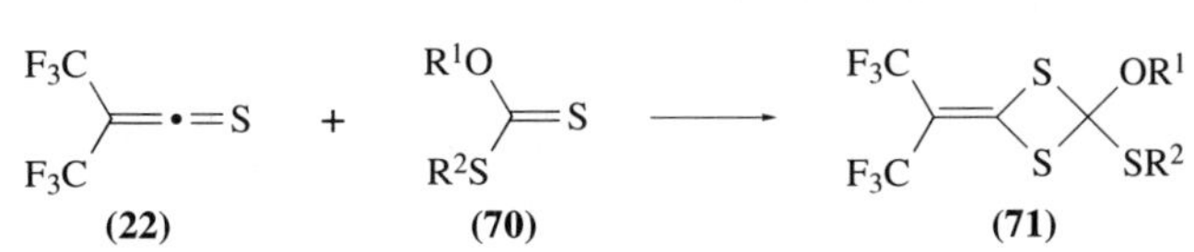

(27)

6.10.1.2.3 *Triselenium- or tritellurium-substituted methyl chalcogens*

No compounds answering this description appear to be on record, cf. ⟨B-86MI 610-02, B-86MI 610-03, B-87MI 610-01, 90HOU(E12/b)1⟩.

6.10.1.3 Two Similar and Two Different Chalcogens

This category theoretically includes selenothio*ortho*-carbonates (**72**), tellurothio*ortho*-carbonates (**73**), selenotelluro*ortho*-carbonates (**74**), selenodithio*ortho*-carbonates (**75**), tellurodithio*ortho*-

carbonates (**76**), selenotellurodithio*ortho*-carbonates (**77**), diselenotelluro*ortho*-carbonates (**78**), diselenotellurothio*ortho*-carbonates (**79**), ditellurothio*ortho*-carbonates (**80**), selenoditelluro-*ortho*-carbonates (**81**), and selenoditellurothio*ortho*-carbonates (**82**), none of which are known ⟨B-86MI 610-02, B-86MI 610-03, B-87MI 610-01, 90HOU(E12/b)1⟩.

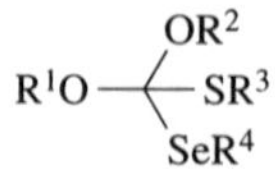

(72)

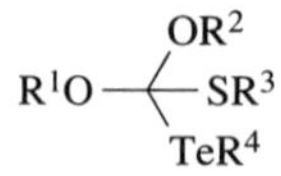

(73)

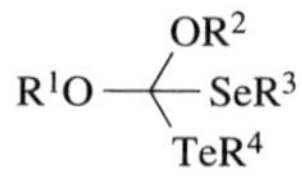

(74)

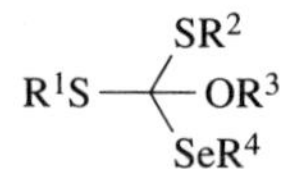

(75)

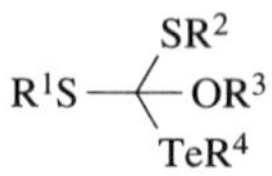

(76)

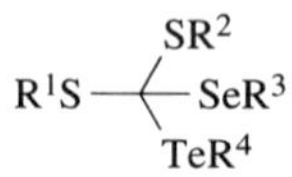

(77)

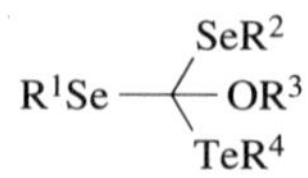

(78)

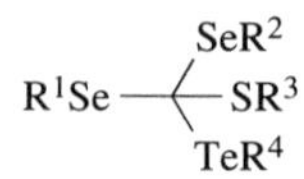

(79)

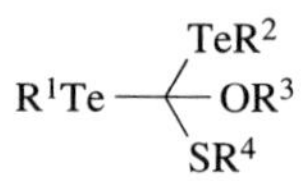

(80)

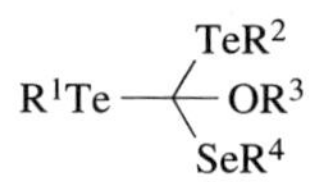

(81)

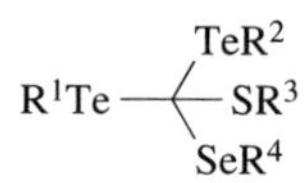

(82)

6.10.1.3.1 Dioxygen-substituted methylene dichalcogens

No examples of selenothio*ortho*-carbonates (**83**), tellurothio*ortho*-carbonates (**84**), or seleno-telluro*ortho*-carbonates (**85**) are known ⟨B-86MI 610-02, B-86MI 610-03, B-87MI 610-01, 90HOU(E12/b)1⟩.

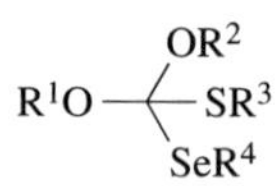

(83)

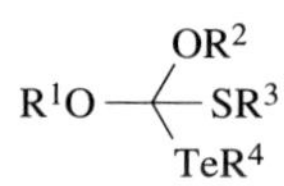

(84)

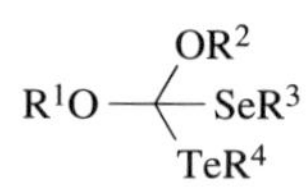

(85)

6.10.1.3.2 Disulfur-substituted methylene dichalcogens

2,2-Dichloro-1,3-benzodioxole (**86**) reacts with benzenethiolate to give the dithio*ortho*-carbonate 2,2-bis(phenylthio)-1,3-benzodioxole (**87**) (Equation (28)) ⟨64LA(675)142⟩. Bis(trifluoromethyl) thioketene (**22**) undergoes cycloaddition with *O*,*O*-disubstituted thiocarbonates (**14**) to yield the corresponding 1,3-dithietanes (**88**) (Equation (29)) ⟨70JOC3470⟩.

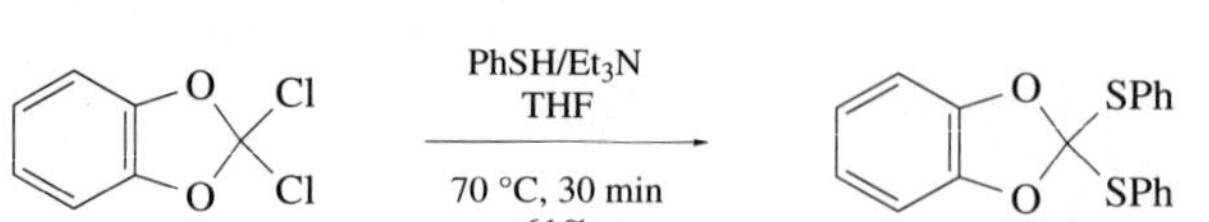

(86) **(87)** m.p. 98.5–99.5 °C (28)

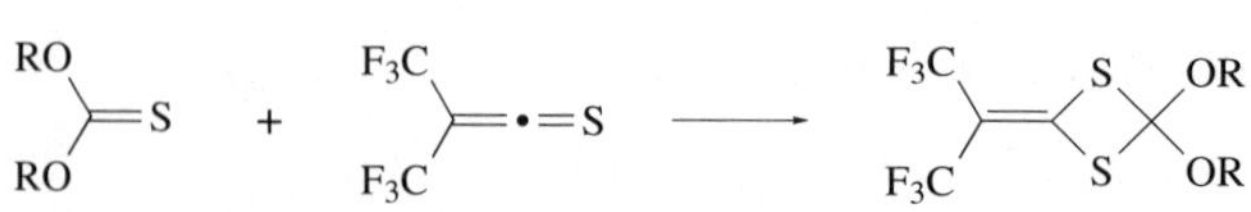

(14) **(22)** **(88)** (29)

The mercury(II) salt of *N*,*N*-bis(trifluoromethyl)hydroxylamine, $[(CF_3)_2NO]_2Hg$, reacts with thiophosgene with the formation of the transient corresponding *O*,*O*-disubstituted thiocarbonate, which in turn dimerizes to 2,2,4,4-tetrakis[bis(trifluoromethyl)aminoxy]-1,3-dithietane ⟨84JFC(24)485⟩.

Cyclic trithiocarbonates (**90**) react with 3,4,5,6-tetrachloro-*o*-benzoquinone (*o*-chloranil) (**89**) to form spirocyclic dithio*ortho*-carbonates (**91**) (Equation (30)) ⟨73ZC465⟩.

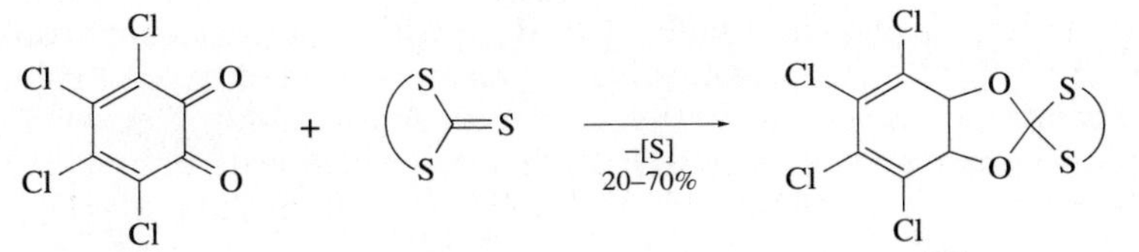

(89) (90) (91) (30)

Aryl *N,N*-dimethyldithiocarbamates with electron-withdrawing substituents in the aryl group such as **(92)** react with methanolic sodium nitrite to form cyclic bis(dithio*ortho*-carbonates) linked via a peroxide group such as **(93)** (Equation (31)) ⟨79JOC267⟩.

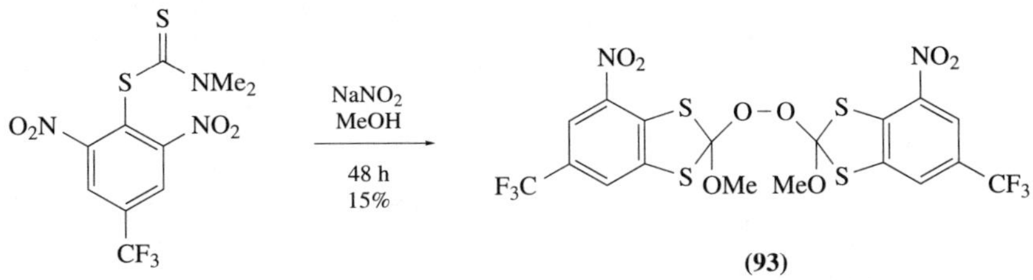

(92) (93) m.p. 177.0–178.5 °C (31)

Tellurodithio*ortho*-carbonates **(94)** and selenotellurodithio*ortho*-carbonates **(95)** are unknown ⟨B-86MI 610-02, B-86MI 610-03, B-87MI 610-01, 90HOU(E12/b)1⟩.

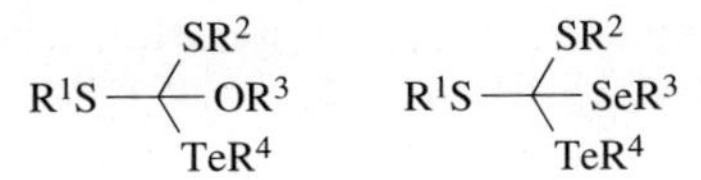

(94) (95)

6.10.1.3.3 *Diselenium- or ditellurium-substituted methylene dichalcogens*

No selenium and/or tellurium compounds fitting this description appear to bc on record ⟨B-86MI 610-02, B-86MI 610-03, B-87MI 610-01, 90HOU(E12/b)1⟩.

6.10.2 TRICHALCOGENOMETHANES

6.10.2.1 Methanes Bearing Three Oxygens and a Group 15 Element, Metalloid or Metal Function

Ethyl *N,N*-dimethylcarbamate upon sequential treatment with triethyloxonium tetrafluoroborate and sodium ethoxide yields 36% of the corresponding acetal $Me_2NC(OEt)_3$ ⟨61LA(641)1⟩. *N,N,N′,N′*-Tetrasubstituted ureas upon treatment with phosgene yield the corresponding formamidinium chlorides which form the corresponding urea acetals with ethoxide. The latter react with ethanol to give $R_2NC(OEt)_3$ ⟨64CB1232⟩. Corresponding methyl compounds $R_2NC(OMe)_3$ have also been prepared. These compounds are remarkably stable to hydrolysis. Transesterification is possible ⟨83HOU(E4)710⟩, for example of $Me_2NC(OMe)_3$ with isopropanol to yield $Me_2NC(OPr^i)_3$ ⟨77S73⟩.

N-Methyl-*N,O*-bis(tributylstannyl)-2-aminoethanol **(96)** reacts with 1,3-dioxolane-2-thione **(97)** to form the corresponding spirocyclic compound **(98)** (Equation (32)) ⟨74JOM(72)103⟩.

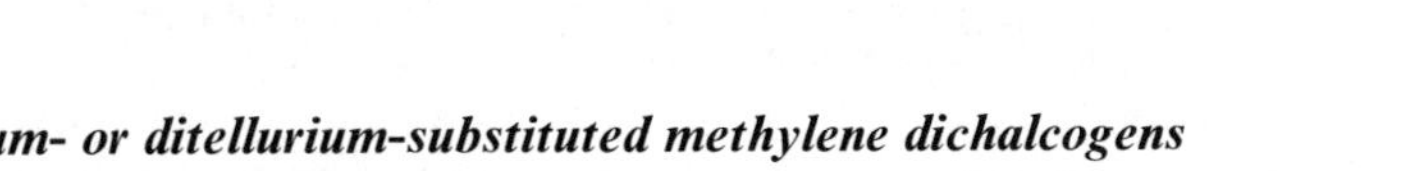

(96) (97) (98) (32)

Trialkoxycarbenium ions (typically obtained by alkylation of carbonates) react with sodium dialkylphosphites to form the corresponding phosphonates in 59–68% yield, for example **(99)** (Equation (33)) ⟨88ZOB2167, 88ZOB2168⟩. The same products are accessible in 62–80% yield by reaction of symmetrical *ortho*-carbonates $C(OR^1)_4$ with dialkylphosphorous acid anhydrides $(R^2O)_2P—O—P(OR^2)_2$ ⟨67ZOB2137, 88ZOB1927⟩. The reaction of trialkoxycarbenium ions with lithium dialkylphosphides gives (trialkoxymethyl)phosphines such as (**100**) (Equation (34)) ⟨88ZOB1447⟩. Dioxymethanephosphonates such as (**101**) upon electrochemical methoxylation yield the corresponding *ortho*-esters such as (**102**) in 38–72% yield (Equation (35)) ⟨87S44⟩.

(33)

(99)

(34)

(100)

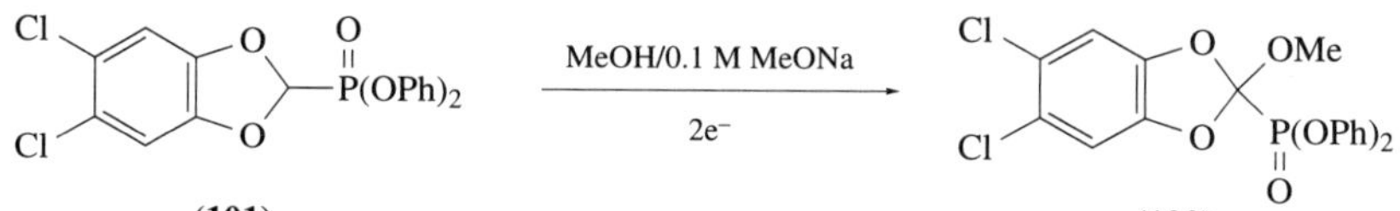

(35)

(101) **(102)**

6.10.2.2 Methanes Bearing Three Sulfurs and a Group 15 Element, Metalloid or Metal Function

Methyl *N*,*N*-dimethyldithiocarbamate after *S*-methylation with dimethyl sulfate and subsequent treatment with sodium ethanethiolate forms $Me_2N—C(SMe)_2SEt$ ⟨70BCJ3528⟩. Aminotrithio-*ortho*-formates such as (**104**; Z = S) have been prepared from trithiocarbenium salts and imidazoles ⟨61LA(641)1⟩. 2-Alkylthio-1,3-dithiolylium cations (e.g., **103**; Z = S) react correspondingly (Equation (36)) ⟨70TL481⟩. Heterocyclic aminotrithio*ortho*-formates such as (**106**) are accessible by double addition of 2-alkylthiiranes (**105**) to methyl isothiocyanate (Equation (37)) ⟨89NKK63⟩. A similar reaction of 2-iodoalkyl isothiocyanates with sulfur nucleophiles has been reported ⟨81JCS(P1)52⟩. Methyl isothiocyanate reacts with undiluted bis(trifluoromethyl)thioketene (**22**) in a 1 : 3 ratio to form the spirocyclic aminotrithio*ortho*-formate (**107**) (Equation (38)) ⟨78JOC2500⟩. In an obscure reaction between 5-chloro-2-nitroaniline and 1,3-thiazolidine-2-thione (**108**), the corresponding bicyclic aminotrithio*ortho*-formate (**109**) is formed (Equation (39)) ⟨77USP4031228⟩. 3,5-Dimethylthiazolo[2,3-*b*]thiazolium bromide (**110**), upon heating with aqueous base, forms the amino-trithio*ortho*-formate (**111**) (Equation (40)) ⟨67JOC2074⟩. 2,3,5,6-Tetrahydrothiazolo[2,3-*b*] thiazolium bromide (**112**) reacts with sodium benzenethiolate in ethanol to form the bicyclic aminotrithio*ortho*-formate (**113**) (Equation (41)); analogous compounds have been prepared in a similar manner ⟨71JCS(C)103⟩. The betaine (**115**) formed by reaction of benzo-1,3-dithiole-2-thione (**114**) and triethyl phosphite has been reported to interact with 1-(1-propenyl)piperidine to form the aminotrithio*ortho*-formate (**116**) (Scheme 23) ⟨74CB3155⟩. The corresponding reaction with piperidine yields the derivative (**117**) (Scheme 24) ⟨74CB3155⟩. Thiobenzoyliminodithiocarbonates such as (**118**) have been shown to react with diphenyldiazomethane with 1,4-cycloaddition to form two stereoisomers of the spiro heterocycle (**119**) (Equation (42)), the reaction probably taking place by way of a thiirane intermediate ⟨78BCJ301⟩.

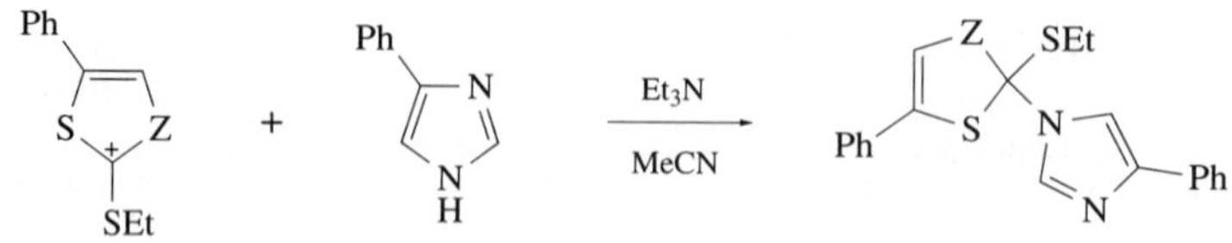

(36)

(103)
Z = S, Se

(104)

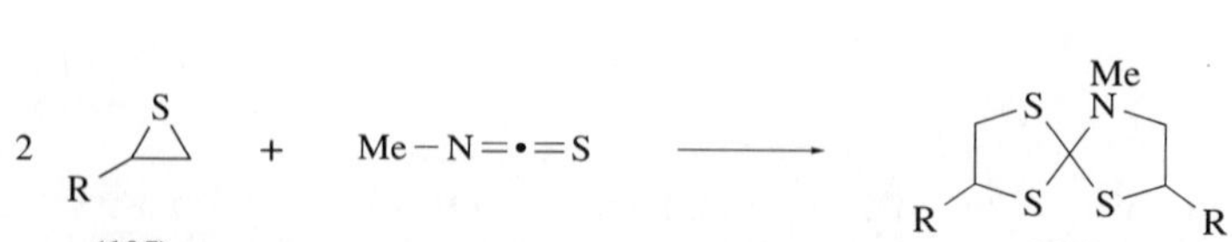

(37)

(105)
R = Me, Et

(106)

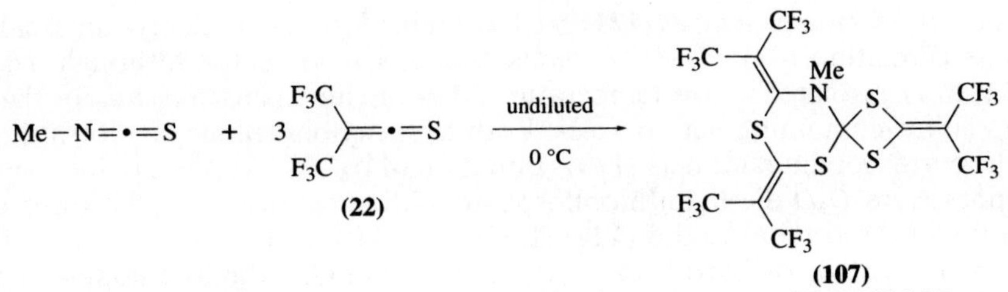

m.p. 121.5–122.5 °C (38)

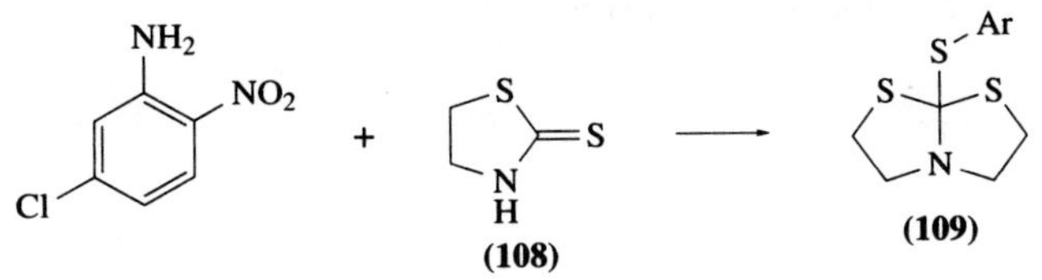

(39)

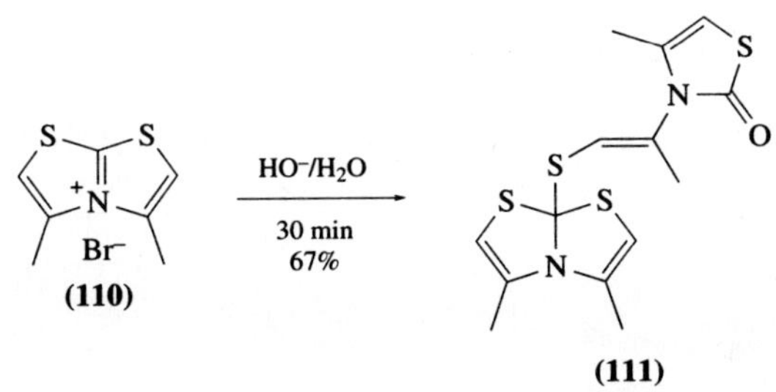

m.p. 194–194.5 °C (dec.) (40)

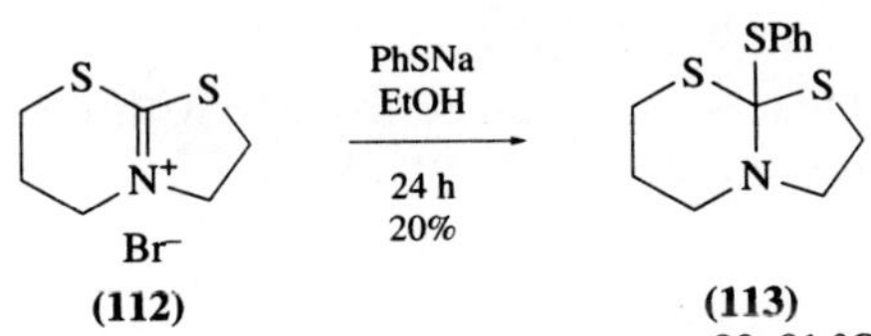

m.p. 80–81 °C (41)

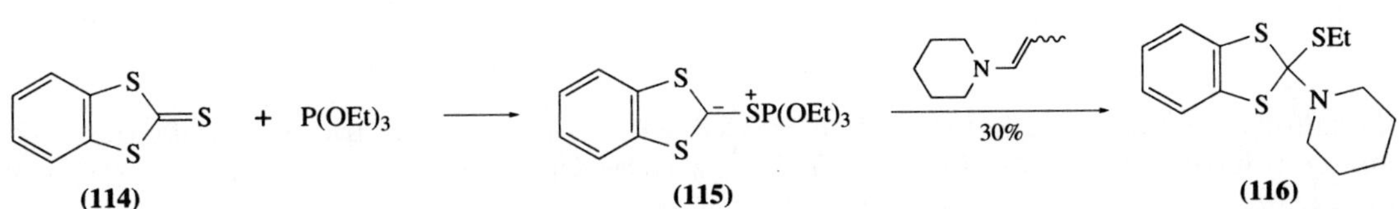

b.p. 190 °C (at 0.6 torr)

Scheme 23

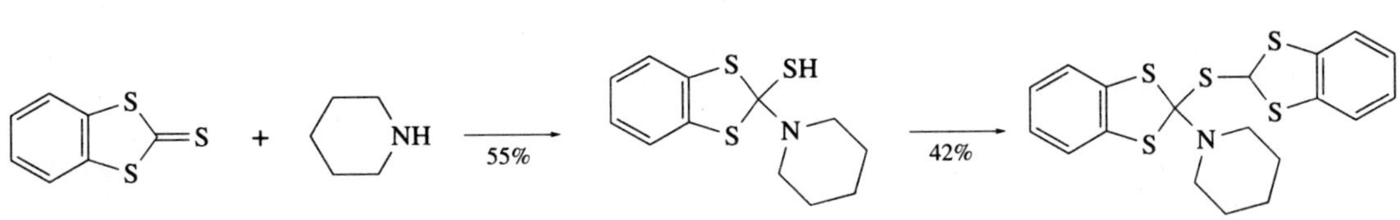

m.p. 158 °C

Scheme 24

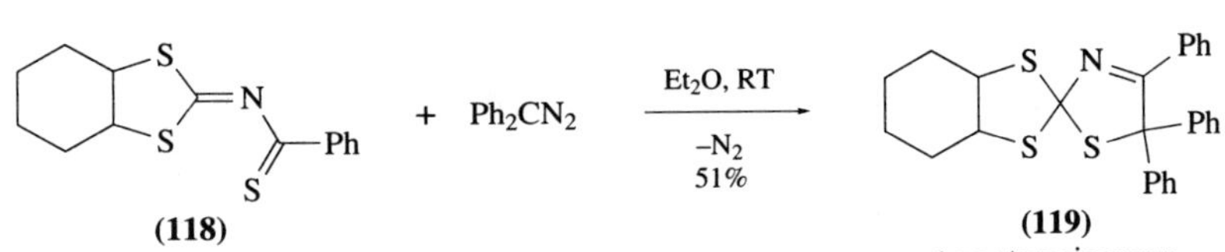

two stereoisomers (42)

Azidotrithio*ortho*-formates such as (**121**) can be obtained from trithiocarbenium salts such as (**120**) and azide ions (Equation (43)) ⟨65ZC386, 76JPR(318)127, 81JCS(P1)618⟩. Although azidotrithio*ortho*-formates (**122**) can be isolated at low temperatures they are mostly used *in situ* for the elaboration of nitrogen- and sulfur-containing heterocycles. With triphenylphosphine, azidotrithio*ortho*-formates (**122**) form the corresponding adducts (**123**) (Equation (44)) ⟨76JPR127⟩. Trichloronitromethane (**7**) reacts with potassium *O*,*O*-diethyldithiophosphate with substitution of the three chlorine atoms to yield the nitrotrithio*ortho*-formate (**124**) (Equation (45)) ⟨60JAP6018199⟩. A similar reaction of trichloromethyl isocyanide dichloride (**11**) with ammonium *O*,*O*-diethylthiophosphate leads to the [(dichloromethylene)amino]trithio*ortho*-formate (**125**) (Equation (46)) ⟨63BEP627486⟩. Analogous reactions of trichloromethyl isocyanide dichloride (**11**) with sodium *N*,*N*-dimethyl- and *N*,*N*-diethyldithiocarbamate have been reported ⟨68BRP1097237⟩.

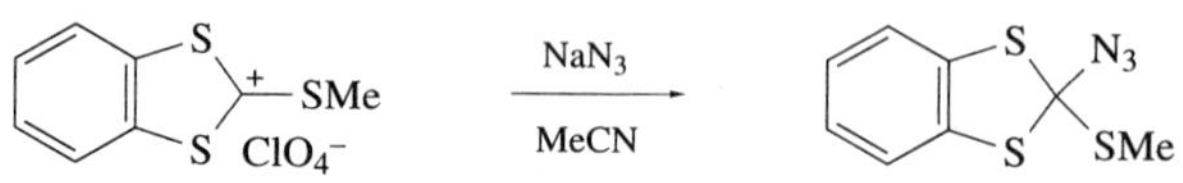

(**120**) (**121**) (43)

(**122**) (**123**) (44)

$$Cl_3CNO_2 + 3(EtO)_2PS_2K \longrightarrow (O_2N)C[SP(S)(OEt)_2]_3 \quad (45)$$

(**7**) (**124**)

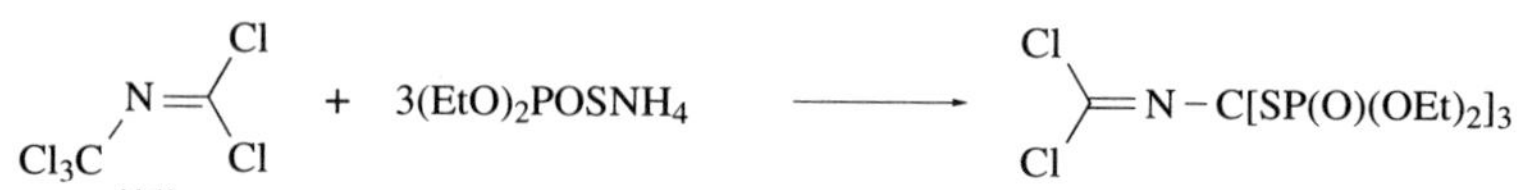
(**11**) (**125**) (46)

Bis(phenylthio)sulfine (**126**) reacts with 2-diazopropane with 1,3-dipolar addition across the C=S bond to form the corresponding thiadiazole derivative (**127**) (Equation (47)) ⟨73TL3589⟩.

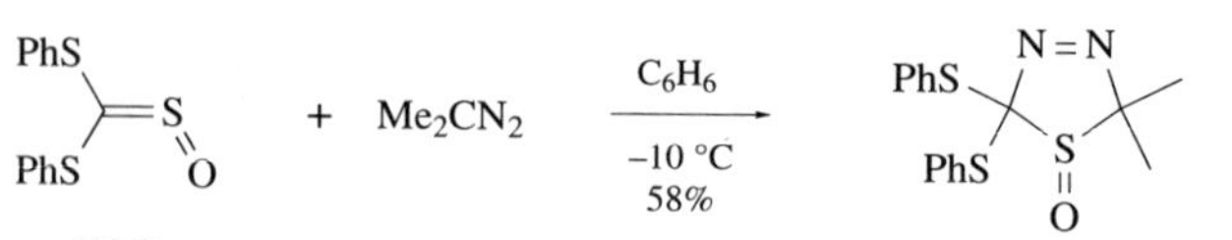

(**126**) (**127**) m.p. 55 °C (dec.) (47)

Trisulfonylmethanes $CH(SO_2R)_3$ when treated with nitric acid/sulfuric acid yield the corresponding *C*-nitro compounds $(O_2N)C(SO_2R)_3$. In cases where R is an electron-rich aryl group, aromatic nitration takes place simultaneously ⟨76ZOR2583⟩.

Trithiosubstituted methanephosphonic acid diesters (**131**) have been prepared in 64–73% yield by addition of alkanethiols to phosphonodithioformates (**128**), followed by subsequent *S*-alkylation and deprotonation of the asymmetrical disulfides (**129**) formed to the corresponding ylides (**130**), which then rearrange to (**131**) (Scheme 25, where LDA is lithium diisopropylamide) ⟨92JOC4507⟩. Methyl diethylphosphonodithioformate (**132**) yields the symmetrical sulfide (**133**) in poor yield when subsequently treated with ethylmagnesium bromide and diethyl disulfide (Scheme 26) ⟨89TL3415⟩. 3,4-Dicyano-1,3-dithiole-2-thione (**134**) reacts with trimethyl phosphite in boiling toluene to give the corresponding 2-methylthio substituted 2-phosphonic acid dimethyl ester (**135**) (Equation (48)) ⟨78JOC595⟩.

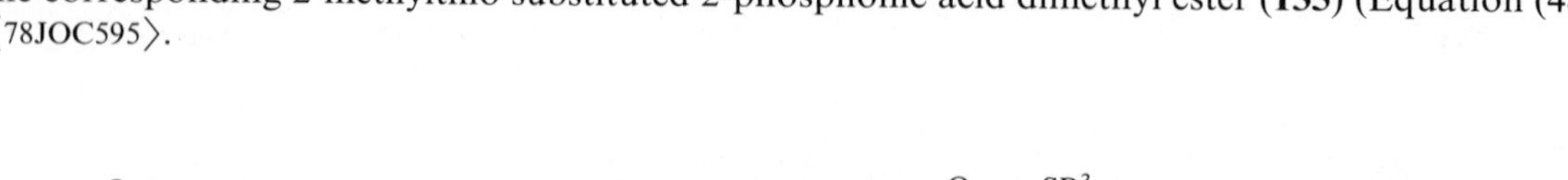

(**128**) (**129**) (**130**) (**131**)

Scheme 25

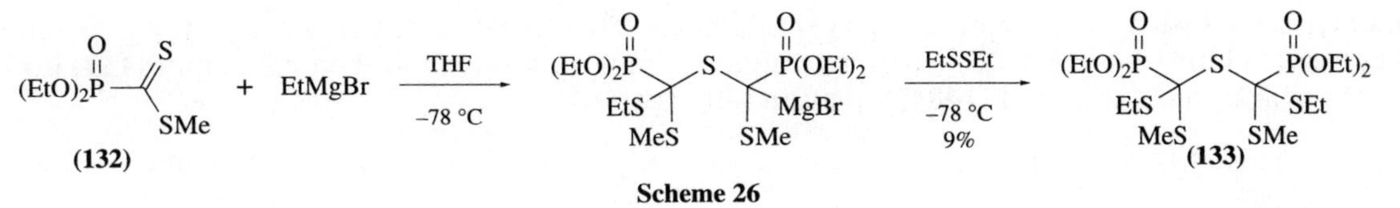

Scheme 26

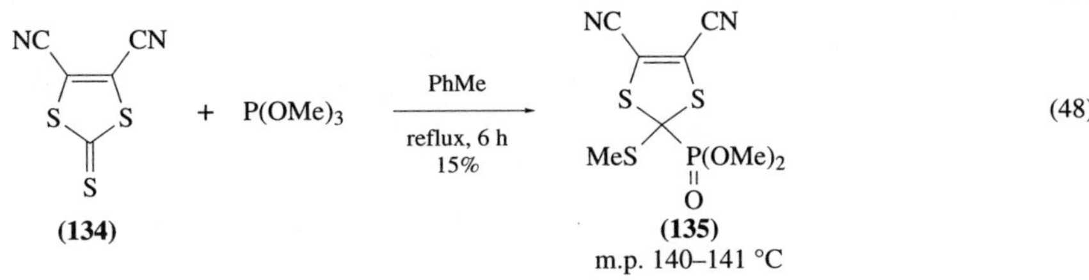

The anions of trithio*ortho*-formates (including oligothiaadamantanes) can be silylated with trimethylchlorosilane to yield (trimethylsilyl)trithio*ortho*-formates $(Me_3Si)C(SR)_3$ ⟨67AG(E)442, 72CB487, 77CB2880⟩. The anion of tri-*t*-butyl trithio*ortho*-formate $CH(Bu^tS)_3$, however, resists trimethylsilylation ⟨77CB2880⟩. Methanetrisulfonyl trifluoride and trimethylchlorosilane form (trimethylsilyl)methanetrisulfonyl trifluoride $(Me_3Si)C(SO_2F)_3$ ⟨83ZOR79⟩.

The lithium salts of trithio*ortho*-formates, $(RS)_3CLi$, are important intermediates in a number of synthetic procedures. Lithiated trithio*ortho*-formates (including mixed compounds)—obtained by metalation of a trithio*ortho*-formate $CH(SR)_3$ ⟨75CC216⟩, by cleavage of a tetrathio*ortho*-carbonate $C(SR)_4$ with butyllithium ⟨72CB487⟩, or by addition of an organolithium compound to a trithiocarbonate ⟨84TL991⟩—play an important role as *in situ* synthetic intermediates, that is as acyl anion equivalents ⟨72CB3280, 72CB3892, 91S347, 91TL6329, 92BMC1607⟩ and in the synthesis of tetrathio*ortho*-carbonates (**3**) (see Section 6.10.1.1.2). In a "one-pot" reaction, trithiocarbenium salts $[C(SR^2)_2SR^1{}_2]X$ can be reduced with sodium borohydride and subsequently metalated with methyllithium to yield mixed lithiotrithio*ortho*-formates $(R^1S)_2(R^2S)CLi$ ⟨93T3035⟩. Isotopically labeled lithiotrithio*ortho*-formates $(RS)_3CLi$ have also been prepared ⟨84HCA1083⟩. In solution, lithiotrithio*ortho*-formates are in equilibrium with the corresponding dithiocarbene and lithium thiolate ⟨67AG(E)443, 72CB487, 72CB3280, 78CB3644⟩. Silylation of lithiotrithio*ortho*-formates leads to the corresponding trialkylsilyl derivatives, for example triphenyl (trimethylsilyl)trithio*ortho*-formate, $(Me_3Si)C(SPh)_3$.

The lithium salt of tris(trifluoromethylsulfonyl)methane, $[(CF_3)SO_2]_3CLi$, is a key electrolyte in nonaqueous batteries ⟨93JAP(K)05062690⟩.

Sodiotrithioformates, $(RS)_3CNa$, are analogously prepared from trithio*ortho*-formates and sodium amide in liquid ammonia ⟨85HOU(E5)3⟩. Like their lithium counterparts they are used *in situ* for synthetic purposes.

6.10.2.3 Methanes Bearing Three Seleniums or Three Telluriums and a Group 15 Element, Metalloid or Metal Function

Triseleno*ortho*-formates such as trimethyl triseleno*ortho*-formate, $CH(SeMe)_3$ ⟨79JOM(177)1, 81TL4009⟩, and triphenyl triseleno*ortho*-formate, $CH(SePh)_3$ ⟨72CB511⟩, have been lithiated with lithium diisopropylamide and other metalation reagents. The lithio derivatives are useful as synthetic acyl anion equivalents.

No tellurium compounds of the above description are on record ⟨B-86MI 610-02, B-87MI 610-01, 90HOU(E12/b)1⟩.

6.10.2.4 Methanes Bearing Three Dissimilar Chalcogens and a Group 15 Element, Metalloid or Metal Function

2-Alkylthio-1,3-thiaselenolylium cations (e.g., (**103**; Z = Se)) probably react with imidazoles to form (**104**; Z = Se) (Equation (36)) ⟨70TL481⟩. Subsequent treatment of 3-methyl-1,3-benzothiazol-2-one (**136**) with triethyloxonium tetrafluoroborate and sodium ethoxide leads to 2,2-diethoxy-3-

methyl-2,3-dihydro-1,3-benzothiazole (**137**) (Scheme 27) ⟨61LA(641)1⟩; the corresponding reaction between 2-chloro-3-methyl-1,3-benzothiazolium tetrafluoroborate (**138**) and resorcinol (**139**) leads to the spirocyclic compound (**140**) (Equation (49)) ⟨68CB1137⟩.

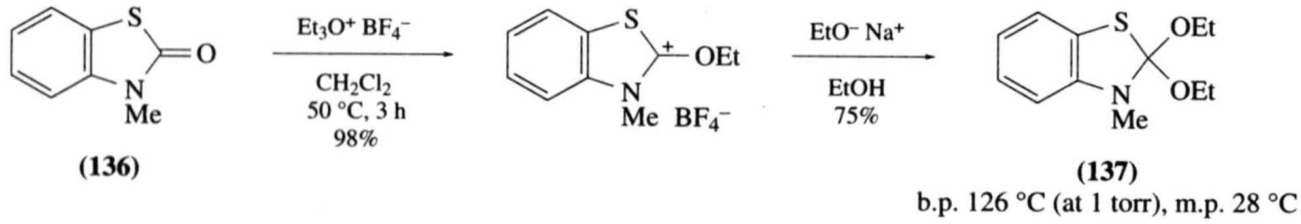

Scheme 27

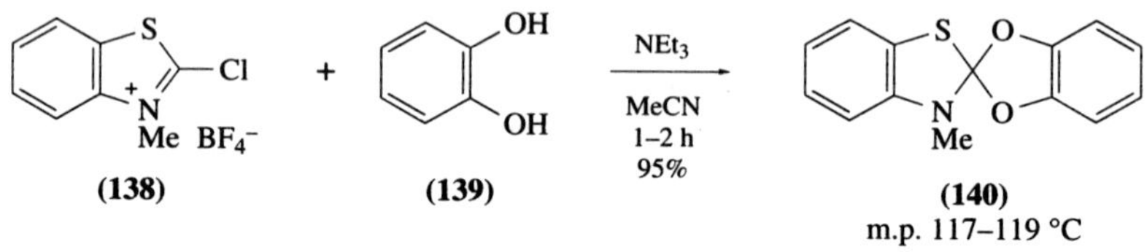

(49)

6.11
Functions Containing Two or One Chalcogens (and No Halogens)

WOLFGANG PETZ
Gmelin-Institut der Max-Planck-Gesellschaft, Frankfurt, Germany

and

FRANK WELLER
Fachbereich Chemie der Universität Marburg, Germany

6.11.1 INTRODUCTION

The compounds in Chapter 6.11 are discussed as follows: first those compounds with a group 15 element at the carbon atom; second those compounds with metalloids (other main group elements); and finally transition metal functions. Compounds with dissimilar atoms at the carbon atom are discussed after compounds with similar atoms.

6.11.2 DICHALCOGENOMETHANES

6.11.2.1 Methanes Bearing Two Similar Chalcogens

Dichalcogenomethanes with similar chalcogens and further atoms other than chalcogens or halogens have only been described with the O_2C, S_2C, and Se_2C units; compounds with the Te_2C unit could not be found.

6.11.2.1.1 Two oxygens and a group 15 element and/or a metalloid and/or a metal function

Up to 1995, compounds with the $(RO)_2C$ unit bonded to group 15 elements other than N have not yet been described. Compounds in which this unit is connected to metalloids or metal functions could also not be found.

(i) Compounds with the O_2CN_2 core

Compounds with the O_2CN_2 core are urea acetals and the chemistry of these compounds and other *O*- and *N*-functional orthocarbonic acid derivatives has been reviewed by Kantlehner *et al.* ⟨77S73⟩.

No common method exists to prepare all these types of compounds, but there are some general procedures. As illustrated in Scheme 1, Meerwein has reported that various dialkylaminoethoxycarbenium tetrafluoroborates can be reversibly converted into the acetals (**1**) by treatment with anhydrous NaOEt in acetonitrile ((**1a**), 56%; (**1c**), 65%; (**1d**), high); with BF_3; reconversion into the water-soluble salts occurs quantitatively. The preparation of the acetals (**1**) cannot be performed in alcohol because in this solvent one NMe_2 group is readily replaced to give $R_2NC(OEt)_3$ compounds ⟨61LA(641)1⟩.

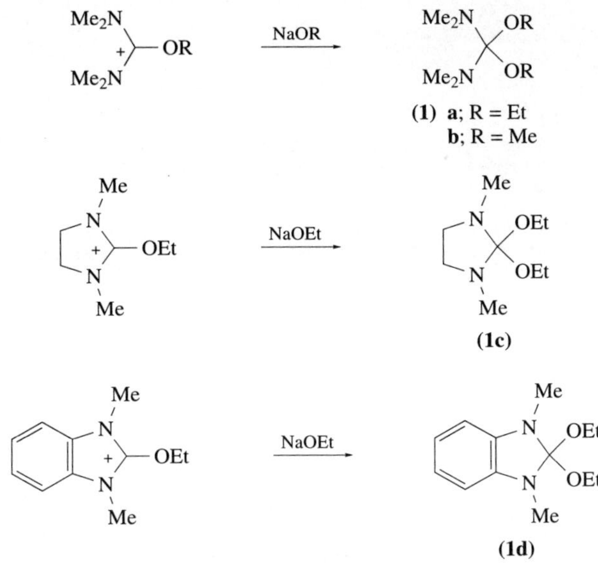

Scheme 1

Compound (**1a**) and the corresponding methoxy derivative $(Me_2N)_2C(OMe)_2$ (**1b**) are obtained by reacting NaOR in a similar procedure with $[(Me_2N)_2CCl]Cl$ ⟨64CB1232⟩, $(Me_2N)_2CF_2$ ⟨77S73⟩, or $(Me_2N)_2C(OR)CN$ ⟨82LA507, 90LA965⟩; the compounds are also formed as a side product during the preparation of the triamine $(Me_2N)_3COR$ in a one-pot reaction from the formamidinium chloride $[(Me_2N)_2CCl]Cl$ and $HNMe_2$ (to give $[C(NMe_2)_3]Cl$ and $[H_2NMe_2]Cl$) followed by treating the mixture with NaOR in THF; the alcohol formed reacts further with the triamine to give $(Me_2N)_2C(OMe)_2$ or $(Me_2N)_2C(OEt)_2$, respectively ⟨79LA2089⟩. The reactions are shown in Scheme 2.

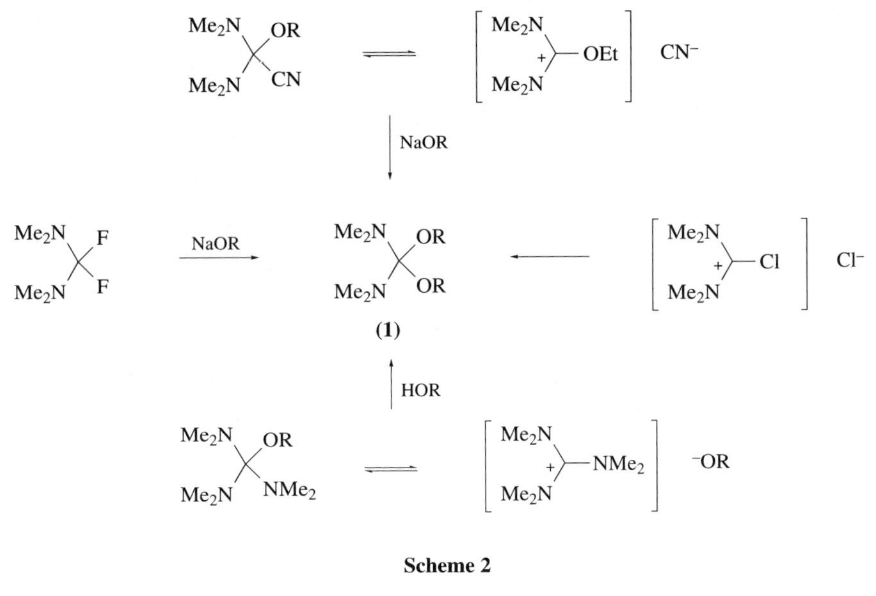

Scheme 2

The first report about acetals with the O_2CN_2 core appeared in 1959 in a short note by Stachel. As depicted in Scheme 3, alcoholysis of the cyclic *N,N′*-diacylurea dichloride derivative, which has been obtained from oxalyl chloride and diisopropylcarbodiimide, produces the corresponding acetals (**1**) ((**1e**), R = Me, m.p. 125°C; (**1f**), R = Et, m.p. 147°C). With ethylene glycol the stable spiroacetal (**1g**) (m.p. 245°C, dec.) is formed ⟨59AG246⟩.

Whereas the methods in Schemes 1 to 3 comprise alcoholysis of diaminodihalogenomethanes, acetals can also be produced by aminolysis of dioxodihalogenomethanes. Thus, 2,2-dichloro-1,3-benzodioxole in ether reacts with piperidine ⟨64LA(675)142⟩ or morpholine ⟨72G558⟩ with cooling to

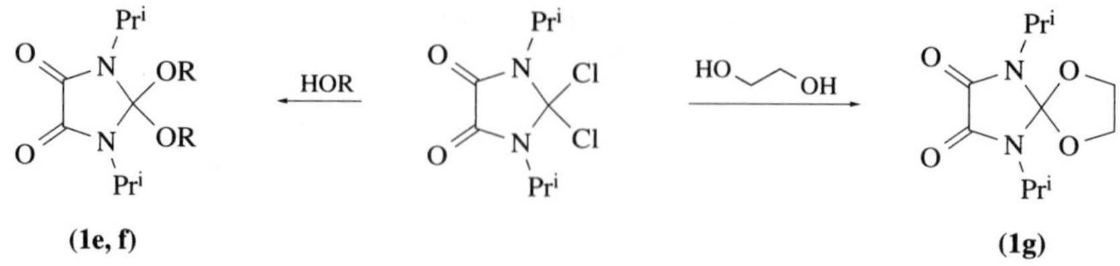

Scheme 3

give the corresponding acetals (**2a**) and (**2b**) in 84% and 72% yield, respectively ((**2a**), m.p. 84–85 °C; (**2b**), m.p. 120–122 °C) (Scheme 4).

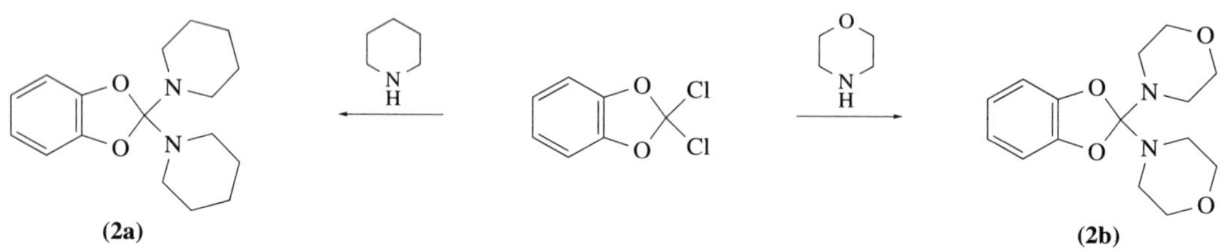

Scheme 4

A more complicated procedure for the preparation of acetals is based on the extrusion of $(Bu_3Sn)_2S$ during the reaction of *N*,*N*′-bis(tributylstannyl)-*N*,*N*′-dimethylethylenediamine with ethylene thionocarbonate in chloroform to form the spiroacetal (**3**) in 87% yield (b.p. 79–89 °C/4 mm Hg). Analogously, the symmetrical acetal (**4**) (b.p. 41 °C/1 mm Hg) was obtained in 90% yield by reacting the bifunctional organotin compounds *O*,*N*-bis(tributylstannyl)-*N*-methylethanolamine with 3-methyl-2-oxazolidinethione. The compounds have been purified by distillation ⟨70MI 611-01⟩ (Scheme 5).

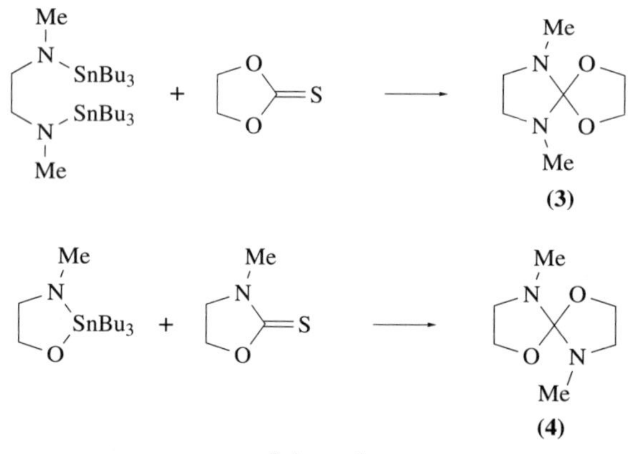

Scheme 5

The structure of the anhydronucleoside (**5**) has been confirmed by x-ray structure analysis. The compound was obtained in 65% yield by treatment of the epoxide (a mixture of isomers) with BF_3 etherate in THF at room temperature. Compound (**6**) was crystallized by slow evaporation of acetone–acetonitrile solution ⟨81CC780⟩.

6.11.2.1.2 Two sulfurs and a group 15 element and/or a metalloid and/or a metal function

Most of the compounds in this section are based on the coordination chemistry of the fragment S_2CPR_3 in which the two sulfur atoms and the carbon atom interact with a mononuclear or dinuclear transition metal fragment in an allyl-like manner. Compounds with the S_2CPMo, S_2CPW, and S_2CPMn core have been described. The coordination mode of the S_2CPR_3 ligand has been studied ⟨93OM4267⟩.

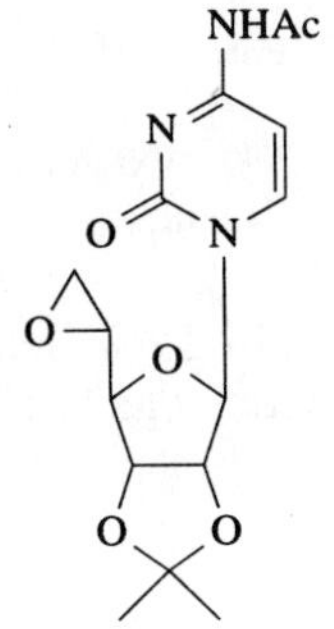

(5)

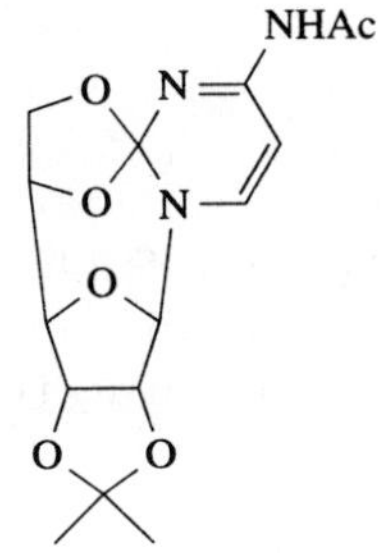

(6)

(i) Compounds with the S_2CN_2 core

There are only few compounds described in the literature with the S_2CN_2 core and because of the different natures of the compounds a common method of preparation for these compounds cannot be formulated.

The starting materials for the preparation of $(PhS)_2C(NO_2)_2$ **(7)** are various highly reactive ylides; these are treated with benzensulfenyl chloride as shown by pathways (a)–(d) in Scheme 6. Thus, the ylide $Me_2S^+—C^-(NO_2)_2$ in acetonitrile or dichloromethane (pathway (a)) with 2 moles of PhSCl at room temperature produces **(7)** (colorless crystals, m.p. 61–62°C) in 60% or 40% yield, respectively. It was assumed that the salt **(8)** is formed in the first step, which is further converted into **(7)** with the liberation of Me_2SCl_2. With the ylide $Ph_2S^+—C^-(NO_2)_2$ under similar conditions (pathway (b)), **(7)** was obtained in 92% yield 〈75IZV2621, 77IZV139〉. Similarly, the selenium dinitro ylide $Ph_2Se^+—C^-(NO_2)_2$ was reacted (pathway (c)) to give **(7)** in 85% yield with removal of the selenium as Ph_2SeCl_2. Purification was achieved by extraction of the dried reaction mixture with hot heptane followed by cooling with dry ice 〈78IZV1091〉. The iodonium ylide $PhI^+—C^-(NO_2)_2$ in ether produces **(7)** in only 20% yield (pathway (d)) along with $PhSC(NO_2)_2Cl$ (45%); the compounds were separated by preparative TLC 〈76IZV2640, 78IZV2348〉.

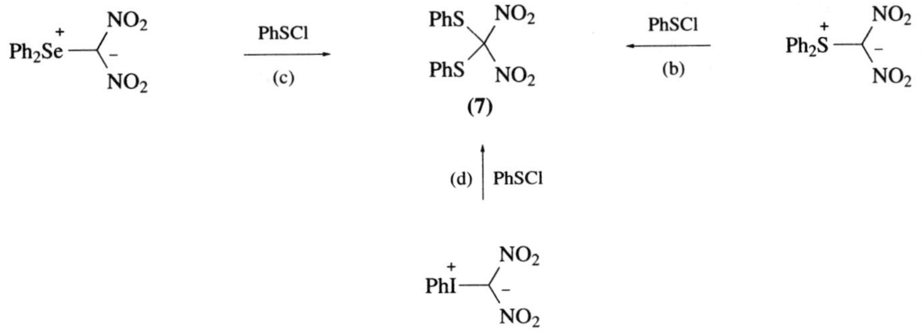

Scheme 6

Thiourea derivatives such as $S{=}C(NHAc)_2$ containing electron-acceptor substituents on nitrogen undergo a radical *S*-arylation by phenyl radicals (generated from *N*-nitrosoacetanilide) to generate the radical $[PhSC(NHAc)_2]^{\bullet}$. Decomposition produces further radicals from which **(9)** (m.p. 100–105°C) could be isolated as a product of recombination in 16% yield (Equation (1)). The reaction was carried out in acetone followed by addition of alcohol 〈82IZV460〉.

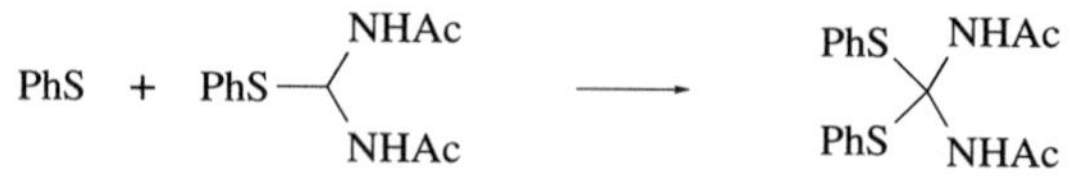

(9)

(1)

Yellow crystals of (**10**) (m.p. 145–146°C) have been obtained in 21% yield by reacting 2-methylcyclopentanone with CS_2 in aqueous ammonia at 0°C. The product was purified by recrystallization from acetic acid–water ⟨73JCS(P1)1009⟩ (Equation (2)).

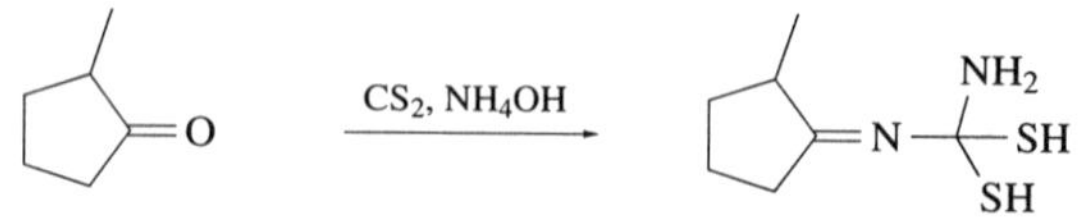

(10)

(2)

Compound (**11**) was shown by NMR spectroscopic studies to be formed when the trimerization product of 2 moles of MeN=C=S and 1 mole of MeN=C=O was allowed to react with ethylene oxide as depicted in Equation (3) ⟨63BAU1384⟩. The preparation of (**12**) has also been described ⟨68MI 611-01⟩.

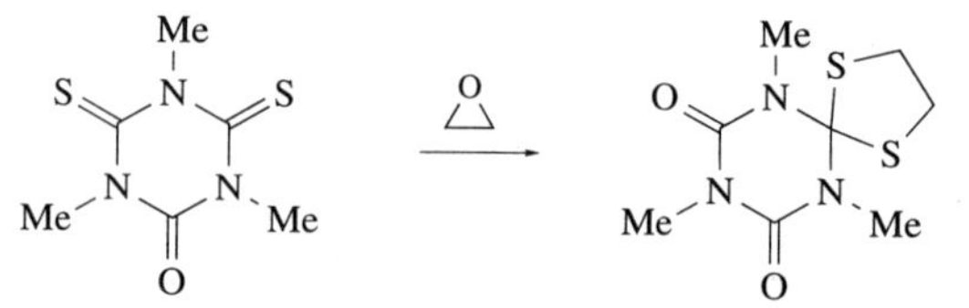

(11)

(3)

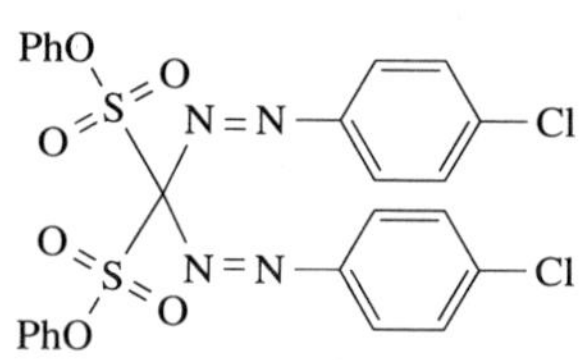

(12)

The cyclic cationic cobalt carbene complex (**13**) reacts with CS_2 in acetone (2 h at 60°C) to the dark brown dimetallaspirocarbene complex (**14**) as depicted in Equation (4); the structure of the compound was confirmed by x-ray analysis. The dication was obtained in 58% yield as the PF_6 salt and crystallizes with 0.5 mole of acetone ⟨85CB3151⟩.

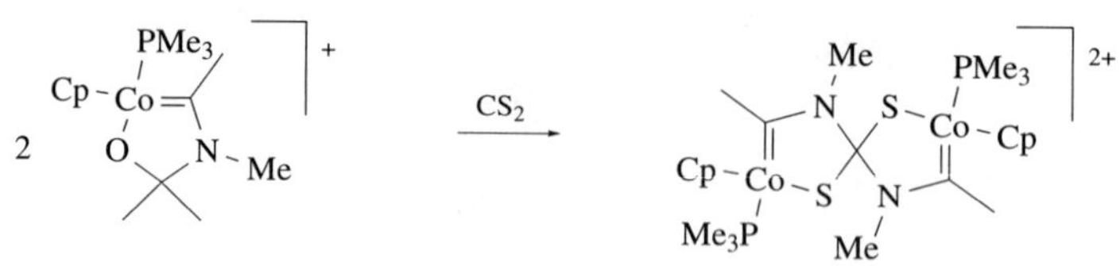

(13) (14)

(4)

(ii) Compounds with the S_2CP_2 core

There are only three quite different compounds known with the S_2CP_2 core. The 1,3-dimethyl-2,2,4,4-tetrakis(trimethylsilylsulfano)-1,3-diphosphetane (**15**) (m.p. 110°C) has been prepared in 79% yield from the reaction of methylbis(trimethylsilyl)phosphine with CS_2 in 1,2-dimethoxyethane at −30°C. In the first step CS_2 inserts into the starting material and the insertion product dimerizes to (**15**). The compound crystallizes from cyclopentane at −78°C as shown in Equation (5) ⟨84ZAAC(517)75⟩.

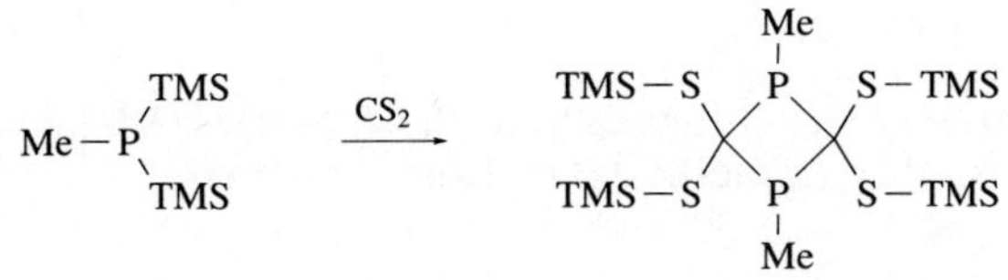

(15) (5)

The dithioacetal **(16)** is formed in 95% yield by sulfenation of the corresponding methylene compound with $MeSSO_2Me$. The starting materials were adsorbed on Al_2O_3–KF and left at room temperature for 1 h according to Equation (6). The compound was extracted with dichloromethane and purified by chromatography ⟨92SC1359⟩.

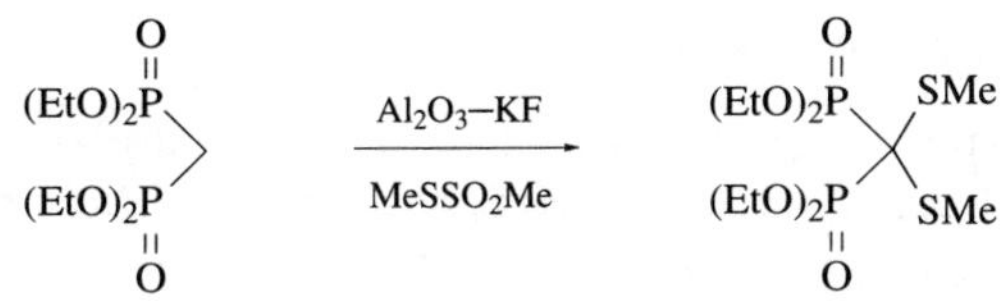

(16) (6)

Green crystals of the cationic complex **(17)** form in 40% yield from a hot mixture of MeC-$(CH_2PEt_2)_3$ (triphos) and $Fe(BF_4)_2 \cdot 6\ H_2O$ in dichloromethane–ethanol when CS_2 is passed through the solution ⟨80AG1055, 81JOM(218)81⟩.

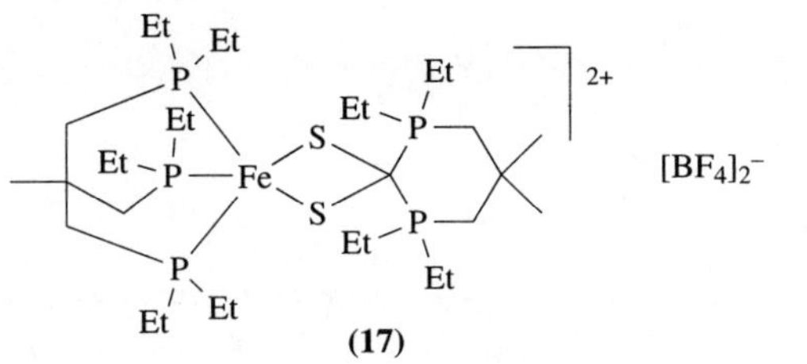

(17)

(iii) Compounds with the S_2CSi_2 core

The compounds of the type $(R^1S)_2C(SiMe_2R^2)_2$ **(18)** have been prepared by three related methods. The methods described in Table 1 start from various lithium salts. Method 1 is the reaction of $(TMS)_2CLi_2$ with 4 moles of $MeSSO_2Me$ in at $-78\,^{\circ}C$ in THF ⟨88TL5237⟩. Method 2 is the reaction of the corresponding $(RS)_2CH_2$ with BuLi followed by addition of RMe_2SiCl to give first $(RS)_2CH$-$(SiMe_2R)$, which is again treated with BuLi/RMe_2SiCl ⟨90CL1411⟩. A similar procedure generates the heterocycles **(18e)** and **(18f)** ⟨67JA431, 67JA434⟩. Method 3 is the reaction of $(TMS)_2C(SPh)Li$ with PhSSPh in THF ⟨84JOC168⟩. The compound **(18a)** forms also as an equimolar mixture with $(TMS)_2C{=}O$ during the reaction of $(TMS)_3CSMe$ with mcpba at $-78\,^{\circ}C$ in dichloromethane ⟨86TL5985⟩.

Table 1 Preparation of $(R^1S)_2C(SiR^2{}_3)_2$, compounds **(18a)**–**(18f)**.

(18)	*R*	*SiR*$^1{}_3$	*M.p.* (°C)	*Method*	*Yield* (%)	*Ref.*
a	Me	TMS		1	85	88TL5237
b	Me	$SiMe_2Ph$		2	60	90CL1411
c	Et	TMS		2		90CL1411
d	Ph	TMS		2, 3	72, 61	84JOC168
e	$—(CH_2)_3—$	TMS	29.6–30	2	96	67JA434
f	$—(CH_2)_3—$	$SiPh_3$	217	2		67JA431

The compound $(TMS)_2C(SH)SOEt$ was proposed to be an intermediate during the thermal decomposition of $(TMS)_2CHSSOEt$ to give finally TMS-C(S)SEt via a sila-Pummerer type rearrangement ⟨85TL2259⟩.

(iv) Compounds with the S_2CGe_2 core

The germanium derivative (**19**), analogous to the heterocycles (**18e**), has been prepared in a similar two-step procedure as described in method 2 in Table 1 ⟨67JA431⟩.

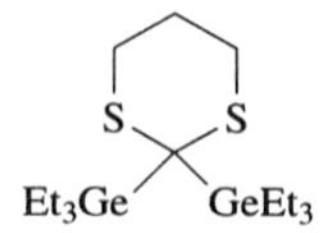

(19)

(v) Compounds with the S_2CSn_2 core

One compound of this type is described in the literature. Colorless crystals of $(MeS)_2C(SnMe_3)_2$ (**20**) (m.p. 91–92°C) have been obtained in 90% yield by addition of $ClSnMe_3$ to a solution of $(MeS)_2(Me_3Sn)CLi$ in THF at −78°C. The compound was purified by vacuum distilation ⟨77CB841⟩.

(vi) Compounds with the S_2CNa_2 core

Only one compound with this core, $(PhOSO_2)_2CNa_2$, has been mentioned in the literature ⟨68MI 611-01⟩; nothing is reported about the nature of the C—Na bond.

(vii) Compounds with the S_2CNSi core

The compound $(MeS)_2C(N{=}CHPh)TMS$ (**21**) has been obtained by the reaction of $PhCH_2N{=}C(SMe)_2$ with lithium diisopropylamide (LDA) in THF at −78°C followed by addition of TMS-Cl together with the silylated product $PhCH_2N{=}C(SMe)(SCH_2TMS)$. The compound ratio varies between 30/70 and 75/25 as monitored by NMR spectroscopy and depends on the time when TMS-Cl was added to the reaction mixture; chromatography over silica gel with petroleum ether/benzene produces (**21**) in 8% yield ⟨75AG449, 80LA1751⟩.

(viii) Compounds with the S_2CPSi core

Mikolajczyk and co-workers describe the formation of $(MeS)_2C(P({=}O)(OEt)_2)(TMS)$ (**22**), when the salt $[(MeS)_2C(P({=}O)(OEt)_2)]Li$ is treated with TMS-Cl in THF at −78°C. The compound was characterized by a ^{31}P NMR signal at $\delta = 37.4$ ppm ⟨89S101⟩.

(ix) Compounds with the S_2CSiGe core

The only compound with this core has been reported by Brook *et al.* In a procedure described for the heterocycles (**18e**), (**18f**), and (**19**), the 2,2-disubstituted 1,3-dithiane (**23**) has been obtained in two steps by lithiation of the 2-germyl-1,3-dithiane followed by addition of TMS-Cl (method 2 for (**18**)) ⟨67JA431⟩.

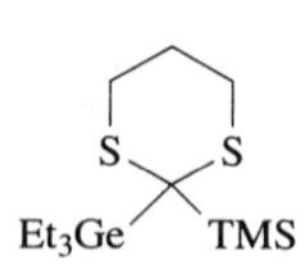

(23)

(x) Compounds with the S_2CSiLi core

The structure of compounds with this core is not reported. Ionic ($[(RS)_2C(SiR_3)]Li$) or covalent formulation is possible. Reich reported a low-temperature multi-NMR study involving

$(PhS)_2C(SiMe_2Ph)Li$ (and similar compounds). In THF–ether the compounds are present as contact ion pairs, whereas in hexamethylphosphoric triamide (HMPTA) weakening of the C—Li bond occurs with formation of separated ion pairs ⟨93AG1489⟩.

The compounds have generally been obtained by treating the corresponding $(RS)_2C(SiR_3)H$ with BuLi or a lithium amide in a polar solvent such as THF or diethyl ether. Thus, a solution of $(MeS)_2C(TMS)Li$ is obtained by addition of $(MeS)_2C(TMS)H$ to a solution of LDA in a mixture of THF, hexane, and HMPTA at −78 °C ⟨77CB841⟩. Similarly, addition of BuLi to a THF solution at −70 to −50 °C of the appropriate $(RS)_2C(SiR_3)H$ species can also be achieved ⟨73CB2277⟩.

In most cases the compounds have been generated *in situ* and used as nucleophiles to produce an element–carbon bond in the reaction with element–halogen compounds. The following species have been used in this manner: $(MeS)_2C(TMS)Li$ ⟨73CB2277, 75AG37, 83JA7719, 85TL3031, 89TL7033, 90CL1411⟩, $(PhS)_2C(TMS)Li$ ⟨84JOC168⟩, $(MeS)_2C(SiMe_2Ph)Li$, $(EtS)_2C(TMS)Li$, $(EtS)_2C(SiMe_2Ph)Li$⟨90CL1411⟩, $(MeS)_2C(SiMe_2Bu^t)Li$ ⟨85TL1903⟩, $(SCH_2CH_2CH_2S)C(TMS)Li$, and $(SCH_2SCH_2S)C(TMS)Li$ ⟨73CB2277⟩.

(xi) Compounds with the S_2CSnLi core

The compounds $(MeS)_2C(SnPh_3)Li$, $(MeS)_2C(SnBu_3)Li$, and $(MeS)_2C(SnMe_3)Li$ have been obtained by reacting $(MeS)_2CHLi$ with the corresponding R_3SnCl compounds followed by treating of the resulting $(MeS)_2CHSnR_3$ with LDA in THF–HMPTA at −78 °C ⟨75AG37⟩. The compound $(MeS)_2C(SnMe_3)Li$ has similarly been prepared in a THF–hexane–HMPTA mixture ⟨77CB841⟩. In all cases the compounds were used in solution for further reactions.

(xii) Compounds with the S_2CPCr core

The compound $(PhS)_2C(PPh_3)Cr(CO)_5$ **(24)** was presented at a symposium without further details of its preparation ⟨75JOM(94)229⟩. The compound can be considered to result from the coordination of the ylide $(PhS)_2$ C=PPh$_3$ to the 16-electron fragment $Cr(CO)_5$ via the carbon atom.

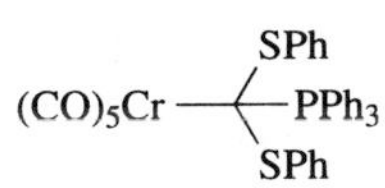

(24)

(xiii) Compounds with the S_2CPMo core

The coordination chemistry of the adduct S_2CPR_3 (R = Me, Et, cyclohexyl (Cy)) towards various molybdenum compounds has been explored by the group of Miguel, Carmona and others. This ligand coordinates in a η^3-pseudoallylic manner to one or two transition metals or to one transition metal and a main group element with formation of a S_2CPMo core.

The first compound in this series, the dimeric species $\{Mo(CO)_2(S_2CPEt_3)(PEt_3)\}_2$ **(25)**, has been detected by Bianchini and prepared either from $Mo(CO)_3(C_7H_8)$ and S_2CPEt_3 or from irradiating the η^1-bonded complex $(CO)_5MoS_2CPEt_3$ in 60% and 40% yield, respectively ⟨82OM778⟩. The addition of $SnBuCl_3$ to **(25)** leads to **(26)** (69% yield) in which the two sulfur atoms form bridges to the tin atom ⟨92AG81⟩; the reactions are summarized in Scheme 7.

Red crystals of **(27)** have been obtained by two routes. Starting from $MoHCl(CO)_2(PMe_3)_3$ the complex is formed in low yield with CS_2 by elimination of HCl. The compound is also formed in an unusual reaction on heating a solution of $Mo(C_2S_4)(C_2H_4)(PMe_3)_3$ at 50 °C under 2 atm CO. Alkylation leads to the cationic species **(28)** as depicted in Scheme 8 ⟨92JCS(D)2307, 93IC5569⟩.

The related mononuclear compound **(29)** is produced in 75% yield by insertion of CS_2 into the Mo—P bond of $MoCl(NO)(PMe_3)_4$. Replacement of Cl and one PMe_3 by the chelating ligand $[S_2CNMe_2]^-$ results in the formation of the yellow complex **(30)** in about 50% yield as shown in Scheme 9 ⟨89IC2120⟩.

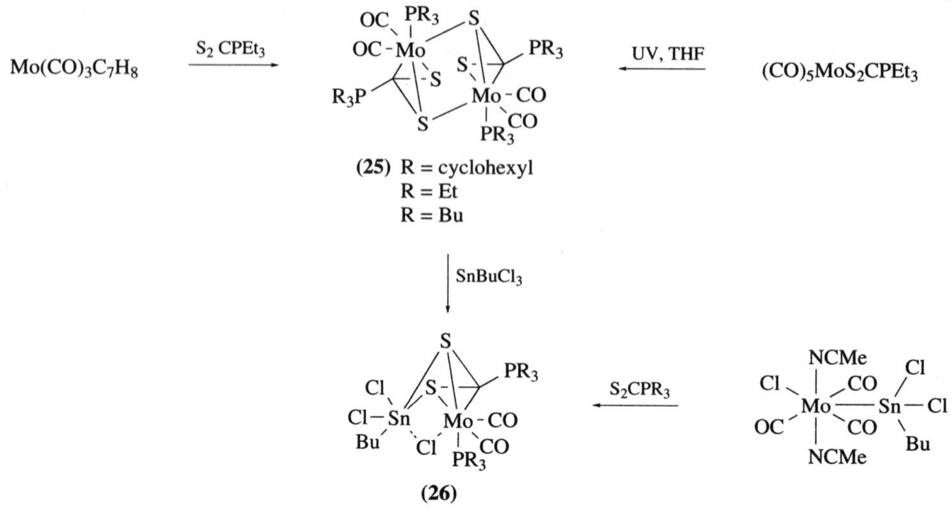

Scheme 7

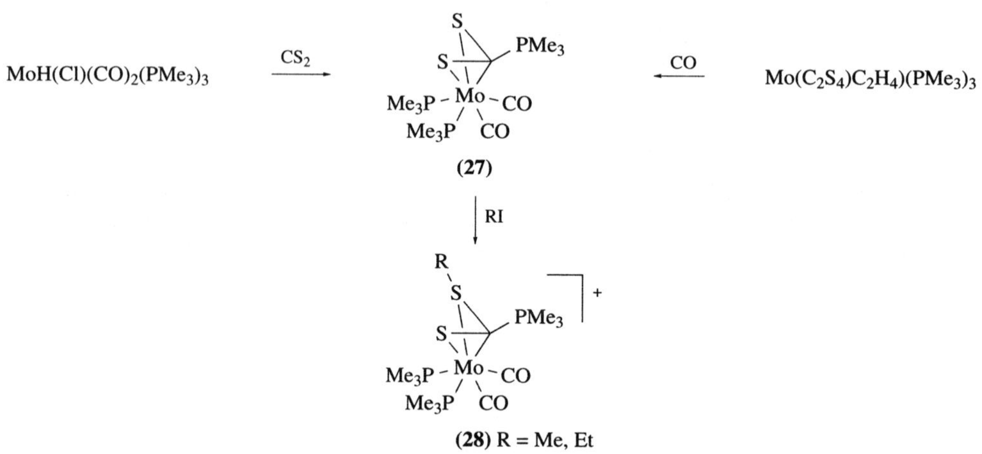

Scheme 8

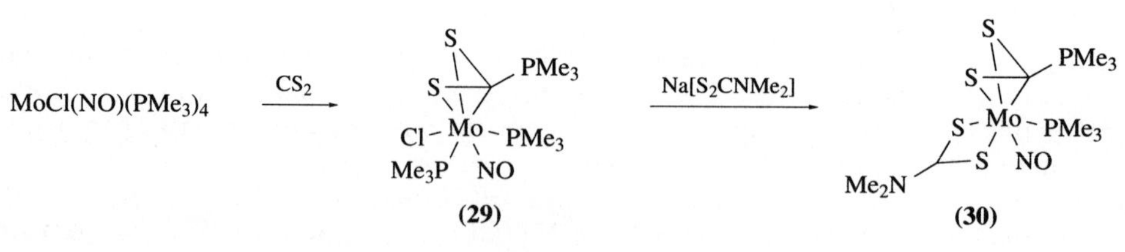

Scheme 9

Dinuclear compounds, in which the S_2CPR_3 group is bridging to another transition metal via the sulfur atoms, have been realized with Mn-, Re-, and Ru-containing fragments. Thus, reaction of $(CO)_3(Br)Mn(S_2CPR_3)$ with $Mo(CO)_3(NCMe)_3$ in THF produces the purple–red complex (**31**) in quantitative yields as shown in Scheme 10 (M = Mn). One CO group at the Mo atom can be replaced by monodentate phosphines to give (**32**) ⟨91JOM(420)C12, 93OM1394⟩; whereas the action of chelating ligands R_2PEPR_2 (E = O, R = Et; E = CH_2, R = Me, Ph) produce red (**33**) in 70–80% yield ⟨93OM1394⟩. Reduction of (**31**) with Na—Hg in THF produces the anion (**34**), probably with formation of a Mo—Mn bond ⟨93OM2888⟩. The analogous Re-containing compounds (**31**–**34**; M = Re) have been obtained ⟨94JOM(467)231⟩.

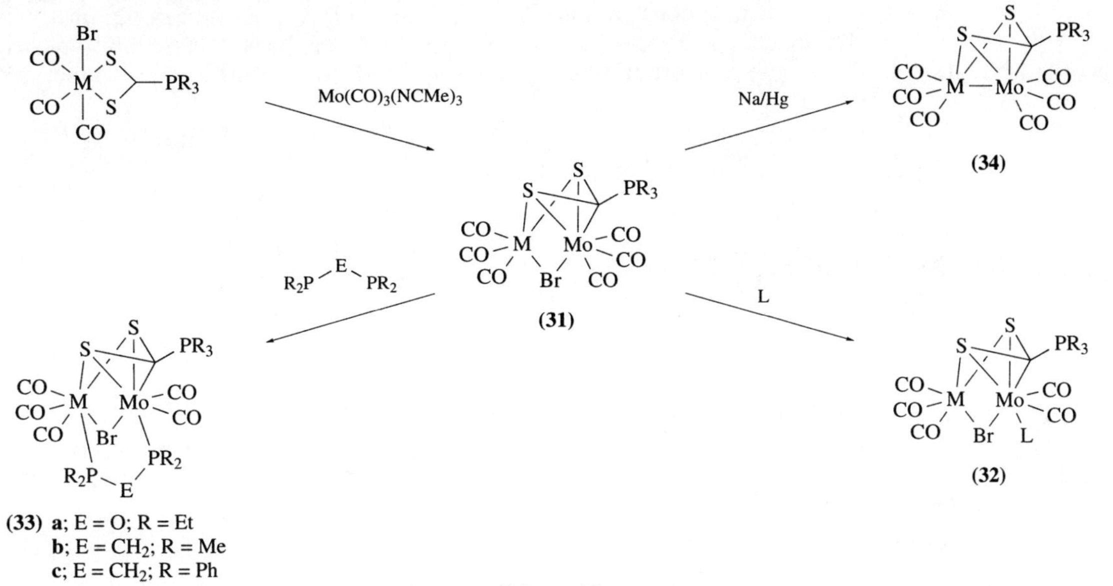

Scheme 10

If the starting complex $Mo(CO)_3(NCMe)_3$ is treated with the cationic species $[(\eta\text{-}Me_6C_6)ClRuC(S_2CPCy_3)]^+$ in THF, (**35**) is formed instantaneously as depicted in Scheme 11. This complex undergoes substitution of one CO group to afford the cationic dicarbonyls (**36**); all compounds have been prepared as the PF_6^- salts ⟨92POL2713⟩.

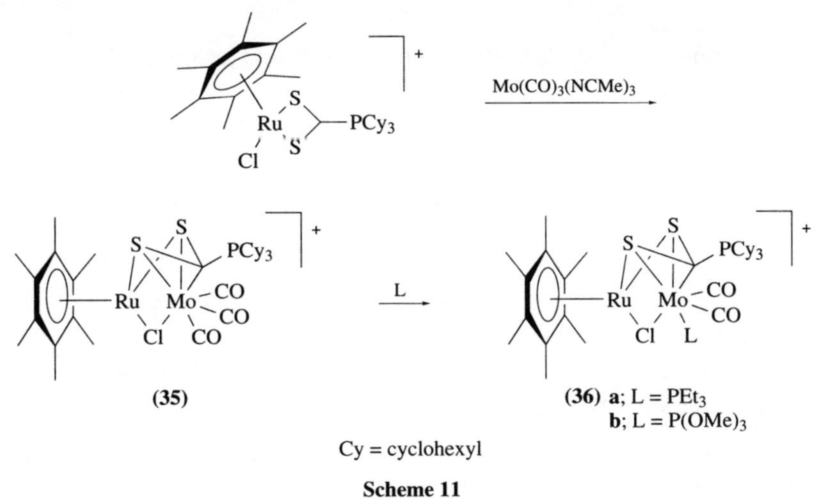

Scheme 11

A further series of compounds with the S_2CPMo core is depicted in Equation (7). The reaction of $(\eta^3\text{-allyl})(CO)_2BrM(\eta^2\text{-}S_2CPR_3)$ (M = Mo, W; R = cyclohexyl, Pr^i) with $Mo(CO)_3(NCMe)_3$ in dichloromethane produces a series of four compounds of the type (**37**) with M′ = Mo, M = Mo or W in yields up to 90% ⟨94OM1336⟩.

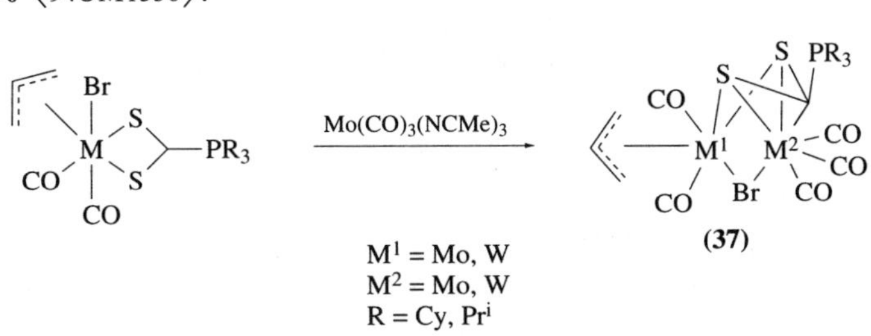

(7)

(xiv) Compounds with the S_2CPW core

There are only few compounds with this core which resemble those of the corresponding Mo compounds. Thus, the series of four compounds (**37**) in Equation (7) have also been obtained with M′ = W ⟨94OM1336⟩. The reaction sequence shown in Scheme 11 under similar conditions with $W(CO)_3(NCMe)_3$ leads to the corresponding cation (**38**) and the substitution product (**39**) ⟨92POL2713⟩.

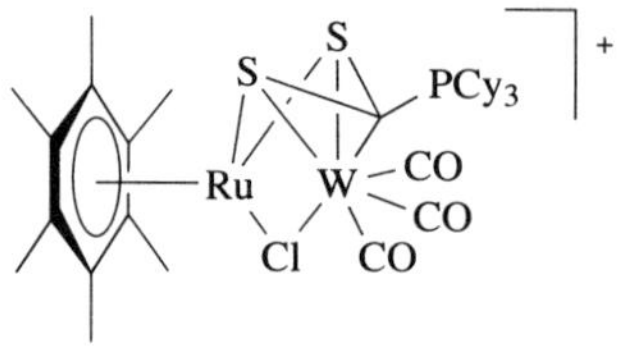

(**38**) Cy = cyclohexyl

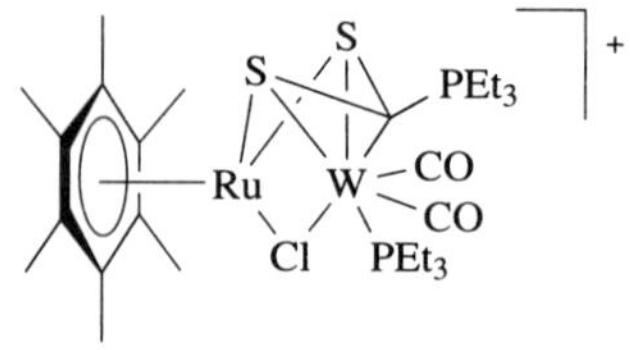

(**39**)

Starting with $WHCl(CO)_2PMe_3$, the red complex (**40**) is formed in 80% yield by reacting with CS_2; HCl is eliminated. This compound can be transferred into the cationic species (**41**) in 50% yield upon alkylation with MeI; the corresponding reaction of the analogous Mo compounds is depicted in Scheme 8 ⟨93IC5569⟩.

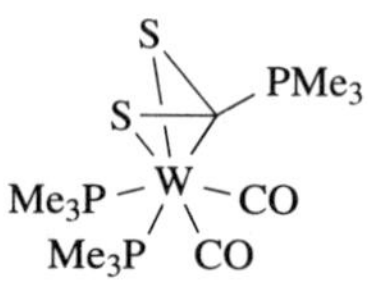

(**40**)

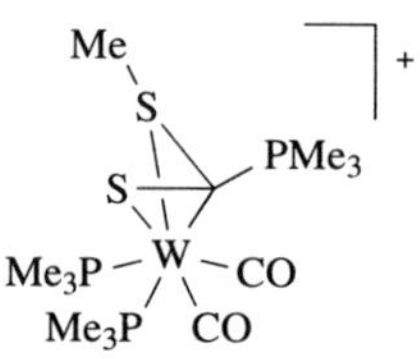

(**41**)

The tungsten ylide compound $(PhS)_2C(PPh_3)W(CO)_5$ (**42**) was presented at a symposium without further details of preparation ⟨75JOM(94)229⟩; it is the tungsten analogue of the chromium complex (**24**).

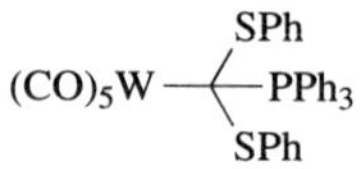

(**42**)

(xv) Compounds with the S_2CPMn core

Compounds with this core are only known with the S_2CPR_3 ligand bonded in a η^3-manner and bridging two transition metals. Starting with the ionic compound (**34**) (M = Mn) in which the ligand is η^3-bonded to the Mo atom, the reaction with $ClSnPh_3$ in THF induces a change in the coordination mode and leads to the red compounds (**43**) (R = cyclohexyl, Pr^i) in 70–80% yield ⟨93OM2888⟩.

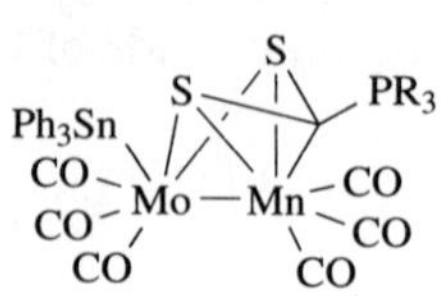

(**43**)

The heterobimetallic compound $ReMn(CO)_6S_2CPCy_3$ (**44**), in which the $Mo(CO)_3SnPh_3$ fragment of (**43**) is replaced by the isoelectronic $Re(CO)_3$ fragment, is obtained in 54% yield either by the reaction of $(CO)_3BrMn(\eta^2\text{-}S_2CPCy_3)$ or the cation $[(CO)_4Mn(\eta^2\text{-}S_2CPCy_3)]^+$ with $Na[Re(CO)_5]$ in

THF or by the reaction of $(CO)_3BrRe(\eta^2\text{-}S_2CPCy_3)$ with $Na[Mn(CO)_5]$ as shown in Scheme 12 ⟨91OM384⟩.

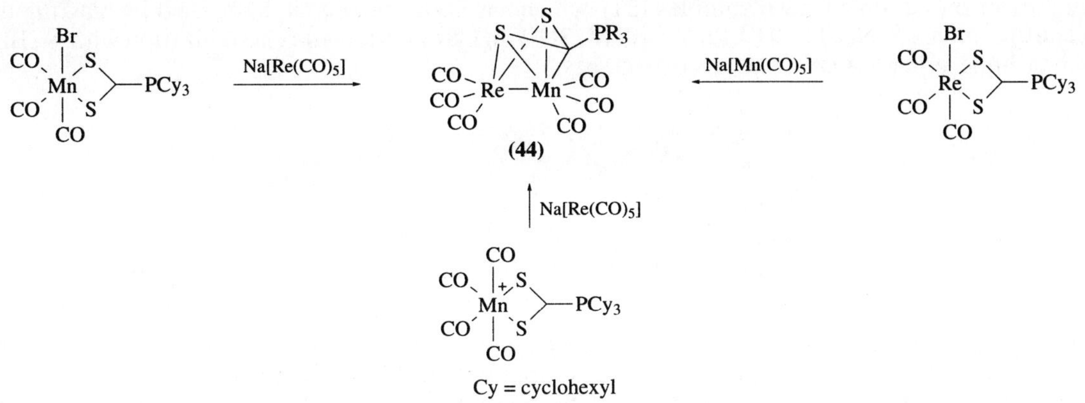

Scheme 12

The preparation of a series of homobinuclear compounds starting from (**45**) is summarized in Scheme 13. The basic compound (**45**) can be prepared by several routes. Thus, similar to its Re analogue (**44**), it forms by the action of $Na[Mn(CO)_5]$ on $(CO)_3BrMn(\eta^2\text{-}S_2CPCy_3)$ or the cationic species $[(CO)_4Mn(\eta^2\text{-}S_2CPCy_3)]^+$ ⟨91OM384⟩. Yields of about 80–90% are obtained when $Mn_2(CO)_{10}$ is refluxed with S_2CPCy_3 or $S_2CP(Pr^i)_3$ in toluene or dichloromethane ⟨87CC472, 91OM1683⟩. A stepwise CO substitution of (**45**) by monodentate ligands or by the chelating ligand $Ph_2PCH_2PPh_2$ generates the derivatives (**46**)–(**48**) ⟨87CC472, 91OM1683⟩. Reduction of (**45**) with $Li[BHEt_3]$ leads to the intermediate lithium salt (**49**) which, on alkylation, can be transformed into the orange hydrido compounds (**50**) ⟨91OM3005⟩.

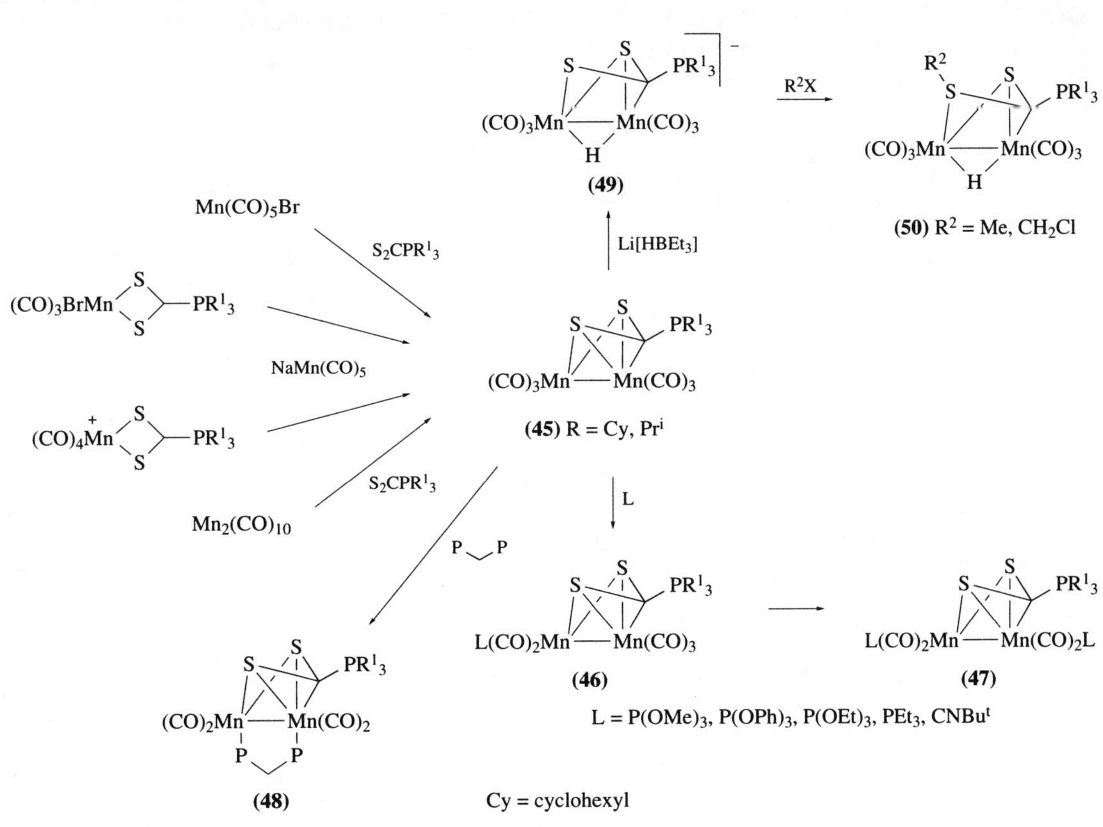

Scheme 13

(xvi) Compounds with the S_2CPNi core

Two types of compounds with the S_2CPNi core have been described. Bianchini reported the structure of the cationic brown complex (**51**), which has been obtained in 86% yield by reacting the nickel(0) complex $NiN(CH_2CH_2PPh_2)_3$ with $MeSO_3F$ in CS_2 solution and crystallization with $NaBH_4$ from acetone–ether ⟨83JOM(246)C13, 84JOM(270)251⟩.

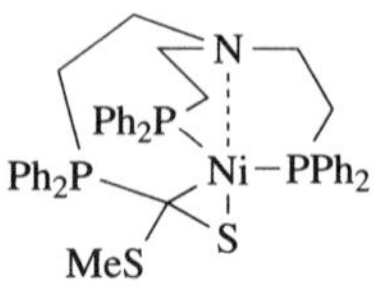

(**51**)

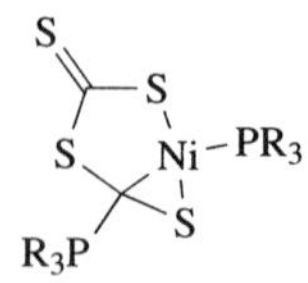

(**52**)

The second type was prepared by Ibers and co-workers. Thus, addition of 2 equivalents of S_2CPR_3 to a solution of $Ni(cod)_2$ in THF produced the compounds (**52**) (R = Me, Et) containing two S_2CPR_3 units under head-to-tail dimerization of two CS_2 molecules and movement of one PR_3 molecule to the nickel atom ⟨83IC411⟩. The compound (**52**) with R = Me has also been obtained according to Equation (8) by reacting the nickel acyclic compound with CS_2 in ether ⟨88OM2577⟩.

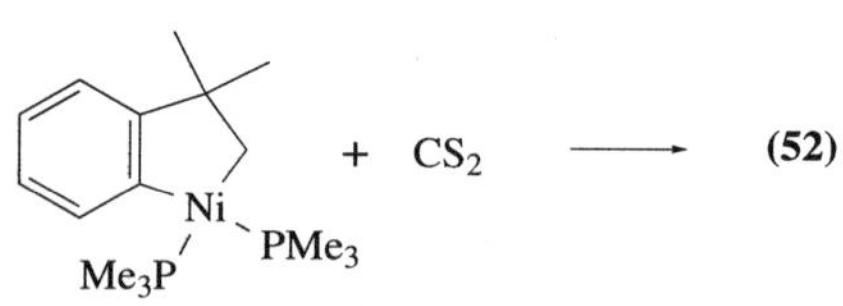 (8)

(xvii) Compounds with the S_2CPPt core

One type of compound with the S_2CPPt core has been reported. Thus, *S*-alkyl (triphenylstannyl)dithioformates $Ph_3SnC(S)SR$ (R = Me, CH_2Ph, allyl) react in benzene solution with one equivalent of the Pt(0) complex $(PPh_3)_2Pt(\eta\text{-}CH_2{=}CH_2)$ to give the corresponding pale yellow (**53**) in quantitative yields as shown in Equation (9). The molecular structure of the methyl derivative is reported ⟨83IC3700⟩.

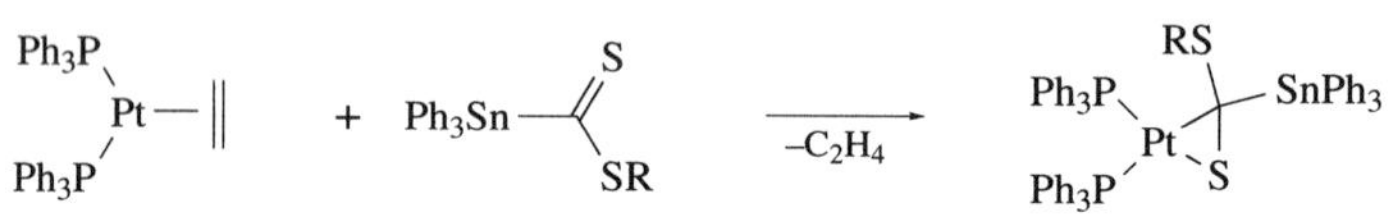 (9)

(xviii) Compounds with the S_2CFe_2 core

Various different types of compounds have been described. Seyferth reported the preparation of (**54**) in a two-step reaction starting from a dithioformate ester–$Fe_2(CO)_6$ complex which contains an acidic hydrogen atom. Deprotonation with LDA followed by treatment with iodomethane produce orange (**54**; $R^1 = R^2 =$ Me) in 77% yield ⟨83OM1696⟩. The starting lithiated species (having a S_2CLiFe core) rearranges to a salt-like product which on alkylation with R′X compounds can be transformed into (**54**), in which different organic groups are located on sulfur as depicted in Scheme 14. The compounds can be viewed as complexes of $Fe_2(CO)_6$ with the six-electron sulfonium ylide donor $R_2S{=}C{=}S$ ⟨87OM283⟩.

A related complex has been prepared by Busetto and co-workers. The complex (**55**) with four iron atoms, in which the CS_2 group functions as an eight-electron donor, was prepared by reacting FpC(S)SFp (Fp = $CpFe(CO)_2$) with excess $Fe_2(CO)_9$ in toluene. Among others, the dark green compound (**55**) could be separated by column chromatography in 9% yield ⟨86G101⟩.

A series of compounds in which a carbene ligand $C(SR)_2$ functions as a four-electron donor forming a S_2CFe_2 core has been described by Angelici and co-workers. The cationic carbene complex $[Cp(CO)(MeCN)Fe{=}C(SR)_2]^+$ reacts in THF with $[(CO)_3FeNO]^-$ to give the violet compounds (**57**) in 40–75% yield. The route outlined in Scheme 15 can also be extended to produce compounds

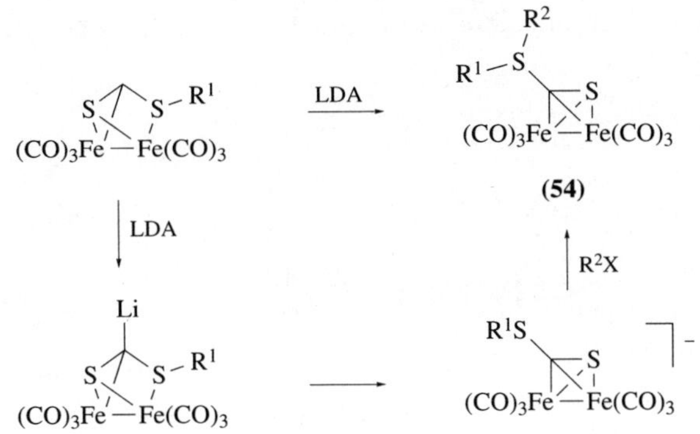

R^1 = Me, Et
R^2 = Me, Et, allyl

Scheme 14

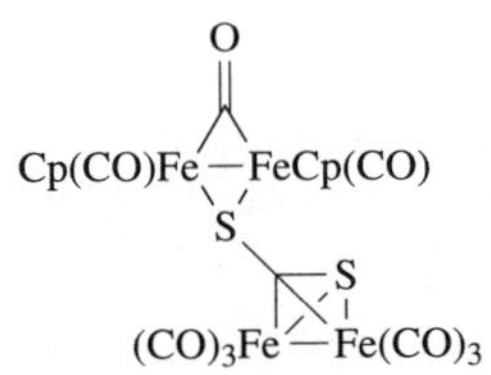

(55)

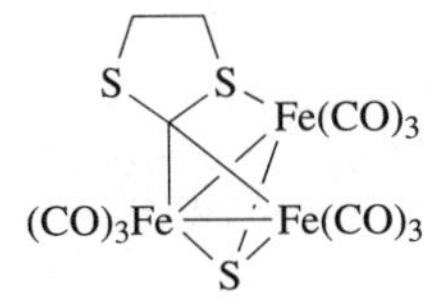

(56)

with the S_2CFeCo and S_2CRuCo core. With PEt_3, the complex (**57**) containing the $C(SMe)_2$ ligand is transformed into the black complex (**58**) (m.p. 102–104°C) in 93% yield ⟨86IC2877⟩. One 50-electron trinuclear complex with the S_2CFe_2 core is also described. As depicted in the structure of (**56**), the carbene ligand of (**57**) can also serve as a μ^3-4-electron donor. Compound (**56**) was one of the compounds obtained from irradiating a mixture of $Fe(CO)_5$ with the trithiocarbonate complex $(CO)_5Cr—S{=}CS(CH_2)_2S$ in THF; column chromatography yields black crystals (m.p. 120°C) in about 14% yield ⟨84JOM(262)69⟩.

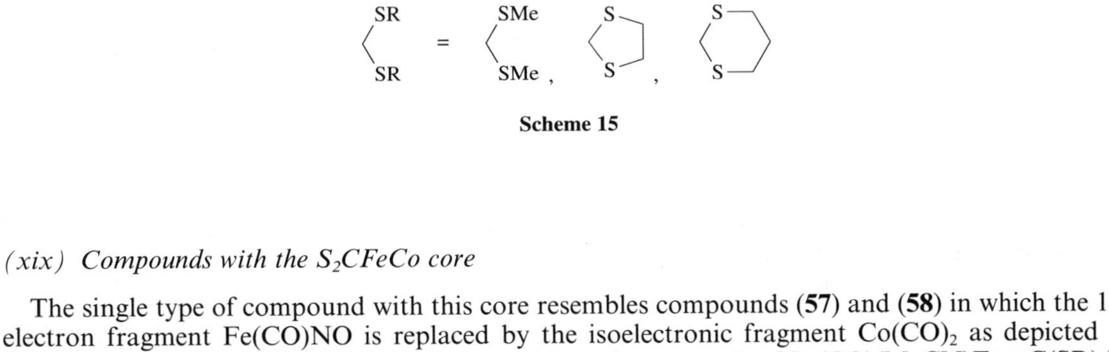

Scheme 15

(*xix*) *Compounds with the* S_2CFeCo *core*

The single type of compound with this core resembles compounds (**57**) and (**58**) in which the 13-electron fragment Fe(CO)NO is replaced by the isoelectronic fragment $Co(CO)_2$ as depicted in Scheme 16. A solution of the appropriate cationic carbene complex $[Cp(CO)(MeCN)Fe{=}C(SR)_2]^+$ in THF was combined with an equimolar solution of $Na[Co(CO)_4]$. Red to black solids (**59**) were obtained in 60–70% yield by column chromatography on silica gel. The compounds (**59**) can be converted into the phosphine-substituted derivatives (**60**) in similar yields as reported for the iron analogues by reacting with PEt_3 in dichloromethane ⟨86IC2877⟩. The chemistry of **(59)** has been described ⟨84OM1038, 85OM1226⟩.

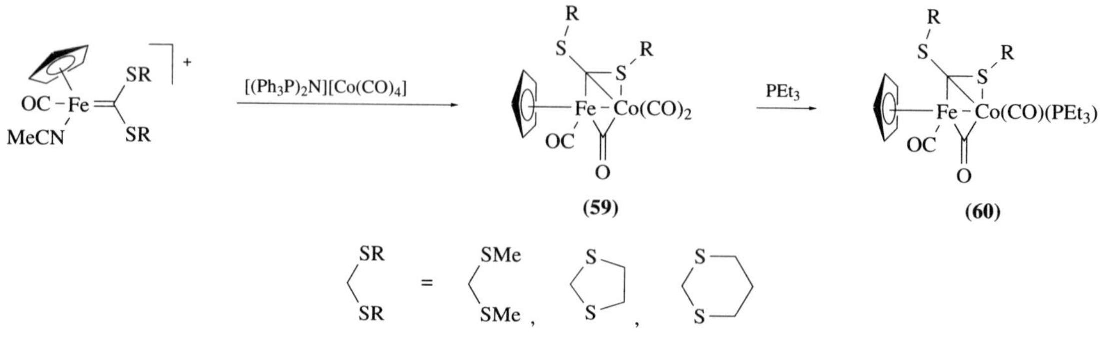

Scheme 16

(xx) Compounds with the S_2CCoRu core

The compound $Cp(CO)_2Ru(\mu\text{-}C(SMe)_2)Co(CO)_2$ **(61)** has been prepared analogously to **(59)** in Scheme 16 but starting with the cationic carbene complex $[Cp(CO)_2(MeCN)Ru{=}C(SMe)_2]^+$; red crystals of **(61)** (m.p. 110–112°C) are produced in 62% yield ⟨86IC2877⟩.

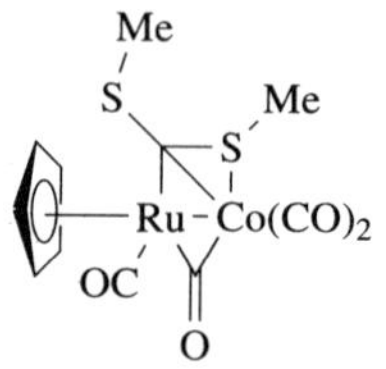

(61)

6.11.2.1.3 Two seleniums or two telluriums and a group 15 element and/or a metalloid and/or a metal function

(i) Compounds with the Se_2CSi_2 core

Only one compound with this core has been described. DuMont and co-workers reported the formation of the yellow triselenacyclopentane derivative **(62)** (m.p. 159°C) in less than 5% yield by reacting a solution of $(TMS)_3CSeSeSeC(TMS)_3$ in dioxane with copper powder for 3 days at 80–90°C; the main product is $(TMS)_3CSeSeC(TMS)_3$ ⟨90CB2325⟩.

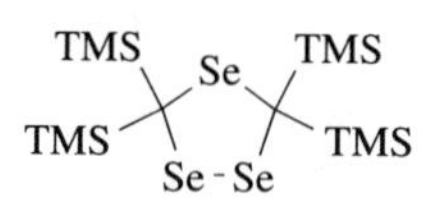

(62)

(ii) Compounds with the Se_2CSiLi core

The compound $(PhSe)_2C(TMS)Li$ has been mentioned as being formed *in situ* by the consecutive reaction of $(PhSe)_2CH_2$ with LDA followed by treatment with TMS-Cl. Further reactions with ketones are described ⟨77CB852⟩.

(iii) Compounds with the Se_2CPPd core

The red–violet to blue–violet compounds **(63a)**–**(63c)**, containing two Se_2CPPd centers, form in about 80% yield upon the reaction of $(Ph_3P)_2Pd(\eta^2\text{-}CSe_2)$ with various methyl substituted phosphine ($PR_3 = PMe_3$, PMe_2Ph, $PMePh_2$) ligands as depicted in Equation (10). The compounds show decomposition points (DTA) of 151, 159, and 116°C, respectively ⟨85ZN(B)1351⟩. The structure of one derivative ($PR_3 = PMe_3$) has been confirmed by an x-ray diffraction study ⟨86JOM(311)63⟩.

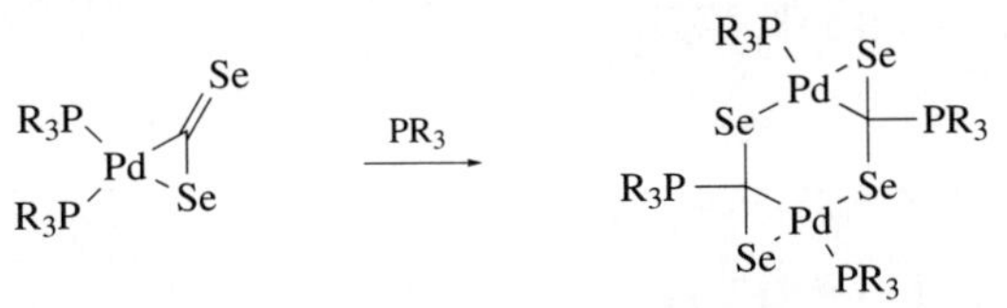

(10)

(63) **a**; $PR_3 = PMe_3$
b; $PR_3 = PMe_2Ph$
c; $PR_3 = PMePh_2$

6.11.2.2 Methanes Bearing Two Dissimilar Chalcogens

6.11.2.2.1 Oxygen, sulfur and two functions derived from a group 15 element, metalloid and/or a metal

(i) Compounds with the $OSCN_2$ core

Compounds with this core are restricted to the species **(64)**–**(66)** and there is no common method of preparation. The spirocyclic compound **(64)** was obtained when the *N*-TMS ester of trithioisocyanuric acid was reacted with ethylene oxide in the presence of tetramethylethylammonium bromide. The compounds have been characterized by ^1H NMR spectroscopy; no further details of preparation are available ⟨63BAU1384⟩. The compound **(65)** has been described ⟨75YGK136⟩. The compound **(66)** (m.p. 166–168°C) has been obtained in 91% yield by reacting 4-(3-methylthioureido)phenyl *N*-methoxycarbonyl-*N*-methylcarbamate with dimethyl sulfate in acetone ⟨75MI 611-01⟩. The radical **(67)** was reported to be formed upon x-ray irradiation of hydrous carnidazole at 77 K; addition of a radiogenic NO_2 radical to the C=S bond was postulated ⟨92C130⟩.

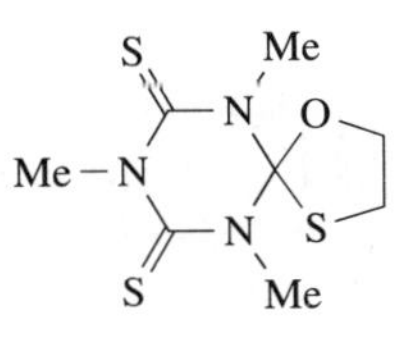

(64)

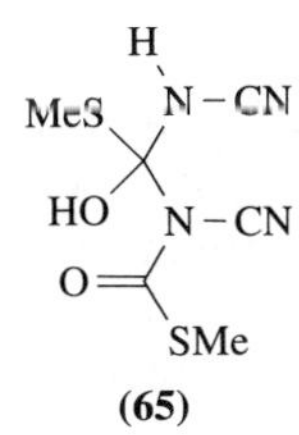

(65)

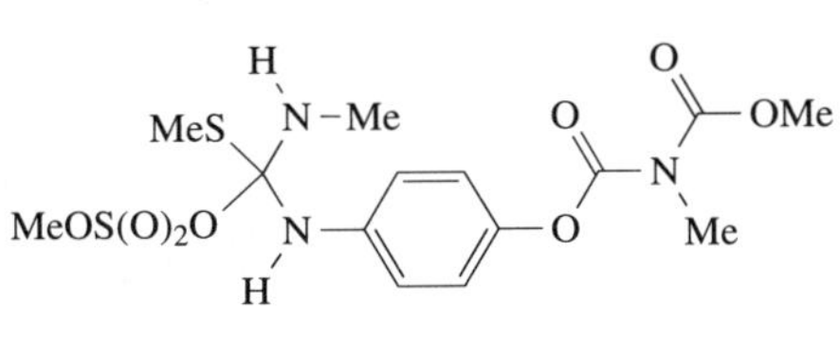

(66)

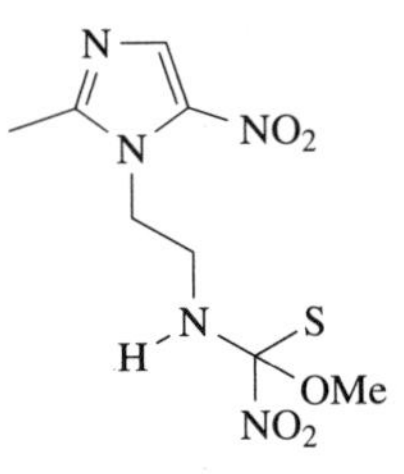

(67)

(ii) Compounds with the $OSCSi_2$ core

The preparation of the *O-Si* hemithioacetals **(68)** and **(69)** have been reported by Ricci *et al.* Thus, treatment of $(TMS)_2CHSMe$ with *n*-buthyllithium in ether at −60°C followed by addition of bis(trimethylsilyl) peroxide (BTMSPO) for oxysilylation produced a 1 : 1 mixture of unreacted starting material and **(68)**. Minor amounts of **(69)** were detected by GC–MS analysis upon reacting $(TMS)_2CHSMe$ with mcpba in dry dichloromethane at −78°C ⟨86TL5985⟩.

(68) R = TMS
(69) R = $OCOC_6H_4Cl$-3

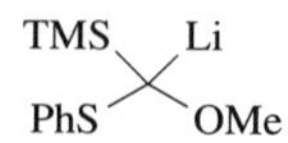

(70)

(iii) Compounds with the OSCSiLi core

Lithiation of the *O,S*-acetal MeO(PhS)CHTMS with *s*-butyllithium in the presence of TMEDA at −78°C generates **(70)** ⟨86JOC879⟩. A similar preparation with *n*-butyllithium in THF was described ⟨83SC985⟩. The lithiated acetal has not been isolated and was used for the conversion of ketones into ketene-*O,S*-acetals.

6.11.2.2.2 *Oxygen and selenium, or oxygen and tellurium and two functions derived from a group 15 element, metalloid and/or a metal*

Compounds with OSeC or OTeC fragments with adjacent group 15 elements, metalloids or transition metals have not yet been described.

6.11.2.2.3 *Sulfur and selenium, or sulfur and tellurium and two functions derived from a group 15 element, metalloid and/or a metal*

(i) Compounds with the SSeCSiLi core

According to low-temperature ^{1}H NMR studies, the compound PhS(PhSe)CLiSiMe_2Ph is present in THF and also in THF–HMPTA solution as a separated ion pair with diastereomeric TMS groups up to −20°C. At about 7°C in THF–ether, the signals coalesce; a rotation barrier around the C—S bond was suggested ⟨93AG1489⟩.

(ii) Compounds with the SSeCPPd core

The only three compounds in this series are reported by Werner *et al.* As with their S_2 analogues and in a manner related to Equation (10), various $Pd(PR_3)_4$ compounds ($PR_3 = PMe_3$, PMe_2Ph, $PMePh_2$) add CSSe in ether to give the dimeric red compounds **(71a)**–**(71c)** in about 85% yield as shown in Equation (11). From spectroscopic data the authors could not decide between the two possible isomers involving a Se or a S bridge of the Pd—C bond ⟨85ZN(B)1351⟩.

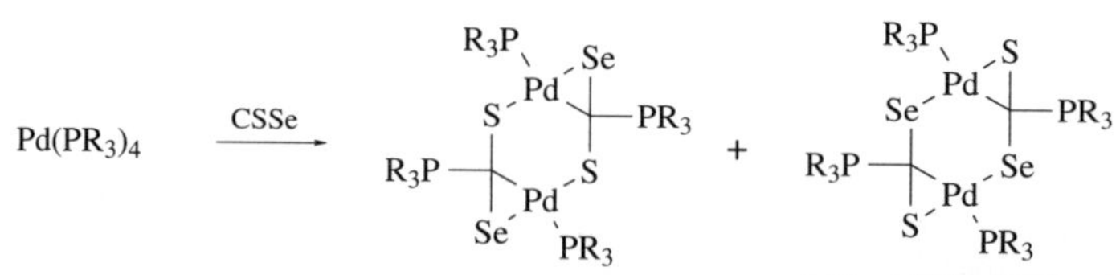

(11)

(71) **a**; $PR_3 = PMe_3$
b; $PR_3 = PMe_2Ph$
c; $PR_3 = PMePh_2$

6.11.3 MONOCHALCOGENOMETHANES

The majority of compounds in this section contain a μ_3-CER carbyne ligand (E = chalcogen) bridging a triangular face of a transition metal cluster compound. R can also be a 16- or 17-electron transition metal fragment, for example $C_5H_5Fe(CO)_2$, $Mn(CO)_5$, etc. In this case the CE unit can be considered as μ_4-bridging. Not included are cluster compounds in which only the 16-electron ligand CE coordinates at an appropriate trinuclear cluster in a μ_3-manner. Many compounds in this series also arise from the coordination of a E═CXY compound at a transition metal in an η^2-manner.

The bonding of a CER fragment to main group elements is restricted to few examples bearing the OCN_3 core. Thus, no compounds with the $ECSi_3$, ECP_3, or related units have been described.

The compounds are arranged in a manner such that the CER fragment (E = chalcogen) is bonded to three identical main group elements, mixed main group elements, mixed main and group transition metal elements, three equal transition metal elements, and different transition metal elements.

6.11.3.1 Methanes Bearing One Oxygen Function and Three Functions Derived From the Group 15 Element, Metalloid and/or a Metal

6.11.3.1.1 Compounds with the OCN_3 core

Two types of compounds with the OCN_3 core have been described. Kantlehner *et al.* reported the preparation of the carbonic orthoamide derivatives **(72a)**–**(72c)** (R = Me, Et, Pr^i) which form upon addition of the corresponding alcohol-free NaOR to $[C(NMe_2)_3]Cl$ in THF solution. Fractional distillation produced **(72a)** (m.p. 40–42 °C), **(72b)** (b.p. 60–75 °C at 10 torr), and **(72c)** (b.p. 39–40 °C at 0.1 torr) in 54%, 39% and 64% yields, respectively ⟨79LA2089⟩. Compounds **(72a)** and **(72b)** have also been obtained by treating $C(NMe_2)_4$ with an equimolar amount of the corresponding HOR as depicted in Scheme 17 ⟨76JOM(105)C19⟩.

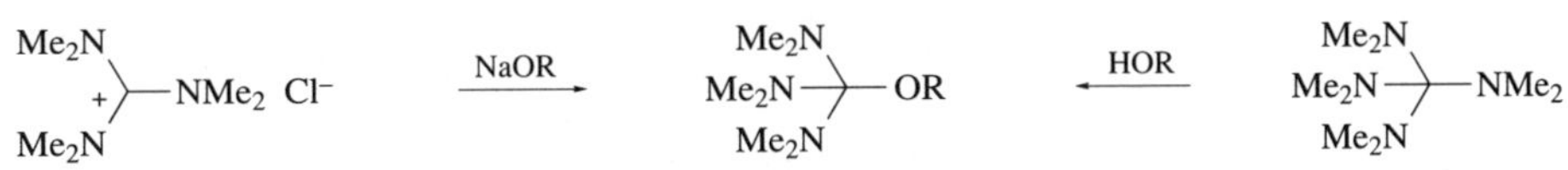

(72) **a**; R = Me
b; R = Et
c; R = Pr^i

Scheme 17

Spirocyclic compounds in this series form as 3 : 1 adducts of various alkyl isocyanides with aminoaziridines in acetonitrile (Scheme 18). The compounds **(73a)** (m.p. 175–182 °C) and **(73b)** (m.p. 161–162 °C) are formed in 93% and 64% yield, respectively. Protonation of **(73a)** with aqueous HCl results in the formation of the cationic compound **(74)** in 43% yield ⟨81LA264⟩.

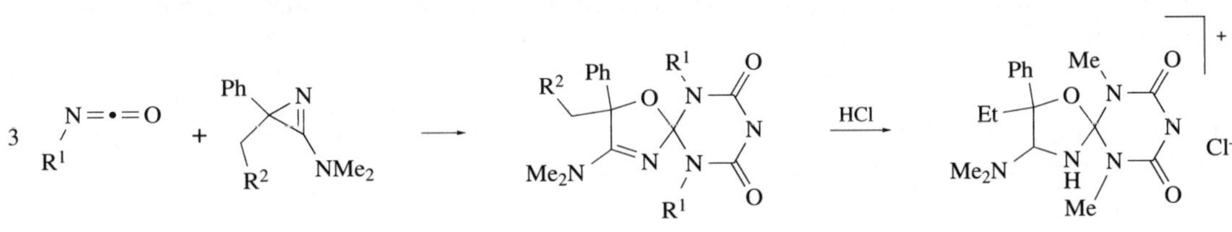

(73) **a**; R^1 = Me; R^2 = Me
b; R^1 = Et; R^2 = H

(74)

Scheme 18

The spirocyclic compound **(75)** is probably formed as an intermediate similar to Scheme 5, when $Me_3SnNMeCH_2CH_2NMeSnMe_3$ is allowed to react with the appropriate ethylene thionocarbonate; the compound could not be isolated ⟨70MI 611-01⟩.

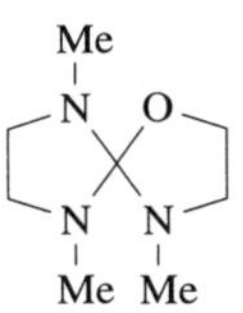

(75)

6.11.3.1.2 Compounds with the OCNSiTa core

Compounds with this core can formally be considered as adducts of pyridine derivatives at η^2-bonded tantalum silaacyl compounds via the acyl carbon atom. The starting unstable silaacyl complex $C_5(Me)_5Cl_3Ta(\eta^2\text{-}C(O)TMS)$ has been obtained by CO insertion into the Ta—Si bond of $C_5(Me)_5Cl_3TaTMS$. The orange compounds (**76**) precipitate from pentane solution in about 90% yield upon addition of the appropriate pyridine derivative NR_2 ((**76a**), NR_2 = pyridine, m.p. 95–98 °C; (**76b**), $NR_2 = NC_5H_4Me\text{-}4$; (**76c**) $NR_2 = NC_5H_3Me_2\text{-}2{,}6$) as follows from Scheme 19. Donor exchange occurs in (**76a**) with the stronger base 4-methylpyridine to give (**76b**) and excess pyridine-d_5 liberates 1 equiv. of pyridine from (**76a**) with incorporation of the deuterated donor ⟨89JA149⟩.

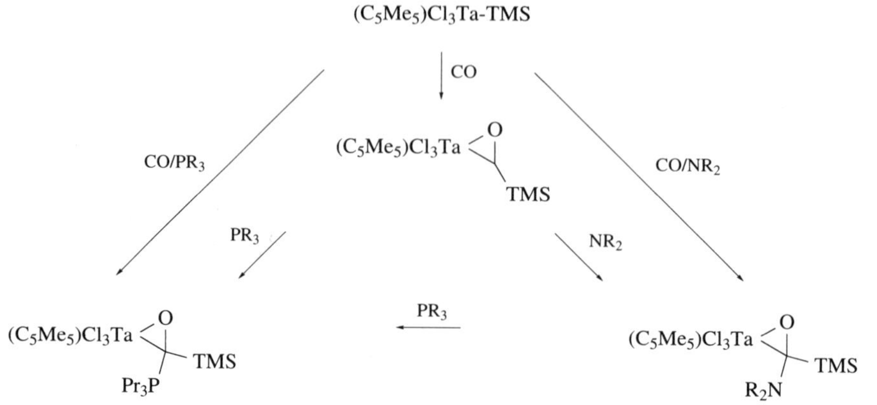

(**77**) **a**; R = Me
b; R = Et
c; R = OMe

(**76**) **a**; $NR_2 = NC_5H_5$
b; $NR_2 = NC_5H_4Me\text{-}p$
c; $NR_2 = NC_5H_3Me_2\text{-}o,o$

Scheme 19

6.11.3.1.3 Compounds with the OCPSiTa core

Closely related to the compounds with the OCNSiTa core are those with the OCPSiTa core which were prepared in a manner analogous to the pyridine derivatives (**76**) as depicted in Scheme 19. Starting again with $C_5(Me)_5Cl_3TaTMS$, addition of PR_3 under CO pressure (0.55 MPa; 80 psi) produced the compounds (**77**) in about 85–90% yield as pale yellow powders. However, solutions of (**77**) in benzene or diethyl ether are intensely colored ((**77a**), blue; (**77b**), deep purple; (**77c**), dark red) and decomposition occurs within a few hours. PMe_3 converts (**76a**) into (**77a**) in 90% yield ⟨89JA149⟩. The structure of (**77b**) was established by an x-ray study ⟨89JA149, 90AX2374⟩.

6.11.3.1.4 Compounds with the OCFe₃ core

This section includes compounds in which the μ_3-bonded COR group bridges the face of a triangular cluster or one face of a tetrahedral cluster, contributing three electrons to the number of 48 or 60 CVE. Not included are butterfly clusters with 62 CVE, in which the carbon atom is bonded to four Fe atoms ⟨85JA8136, 85JA8147⟩.

The neutral orange–red biscarbyne complexes (**78**) (**a**, R = Me) can be prepared either by alkylation of the anion $[Fe_3(CO)_9(\mu_3\text{-}MeCO)]^-$ with excess $MeSO_3F$ in dichloromethane at elevated temperature in about 10% yield ⟨83CC1557⟩ or in not specified yield (**b**, R = Et) by alkylation of the anion $[Fe_3(CO)_9(\mu_3\text{-}CMe)(\mu_3\text{-}CO)]^-$ with $[Et_3O]BF_4$ ⟨85OM1436⟩. Replacement of CO in (**78b**) by PPh_3 leads to (**79**) in about 95% yield. Both compounds can also be generated from the corresponding $\mu_3\text{-}\eta^2$ alkyne complexes as depicted in Scheme 20 ⟨85OM1436⟩.

The dark red ferrocenyl derivative (**80**) forms in 3% yield along with bisalkylidyne complexes (**81a**) and (**81b**), and (**78a**) by reacting ferrocenyllithium with $Fe_2(CO)_9$ in THF at room temperature; the compounds could be separated by column chromatography over silica gel ⟨90ZN(B)447⟩.

The bisalkylidyne complex (**81a**) in which two μ_3-COMe groups are symmetrically bonded to a trinuclear cluster was earlier prepared in about 15% yield by reduction of $Fe_3(CO)_{12}$ with *n*-butyllithium and subsequent treatment of the reaction mixture with $[Me_3O]BF_4$ or in a similar procedure from $HFe_3(\mu_2\text{-}COMe)(CO)_{10}$ in 22% yield using *t*-butyllithium ⟨88OM812⟩.

Anionic species with one μ_3-COR and one μ_3-CO function have also been isolated. One route starts with the trinuclear dianion $[Fe_3(CO)_{11})]^{2-}$, which reacts with the electrophiles $MeSO_3F$, $EtSO_3F$ and MeCOCl to produce the red compounds (**82a**)–(**82c**) (R = Me, Et, MeCO) obtained as the $[(Ph_3P)_2N]$ salts in 73%, 43% and 86% yield, respectively ⟨79IC1236⟩. A similar reaction using

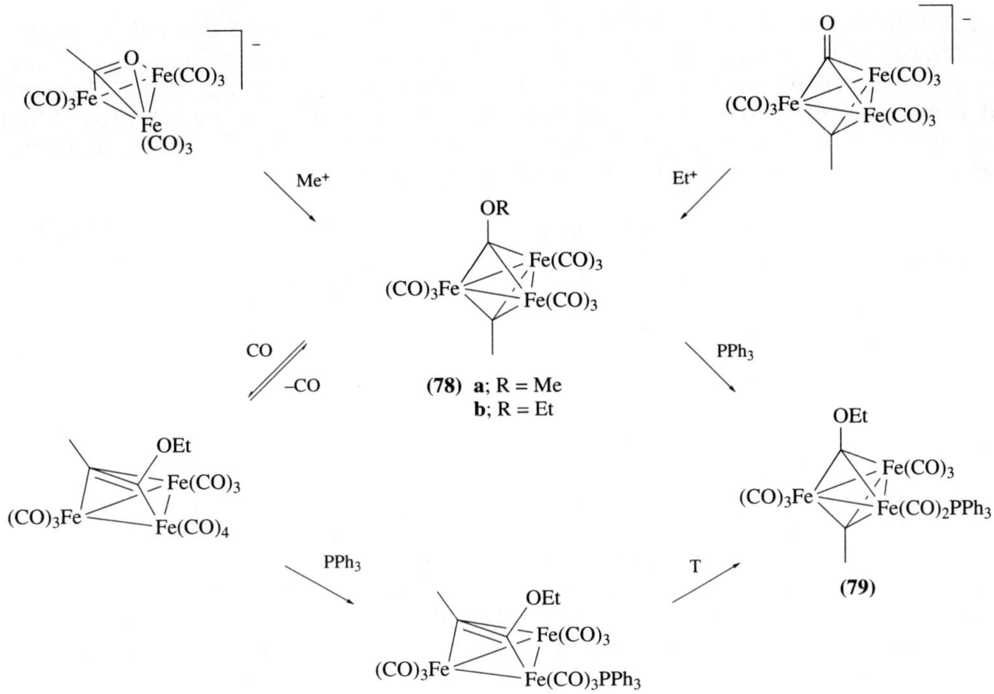

(78) **a**; R = Me
b; R = Et

(79)

Scheme 20

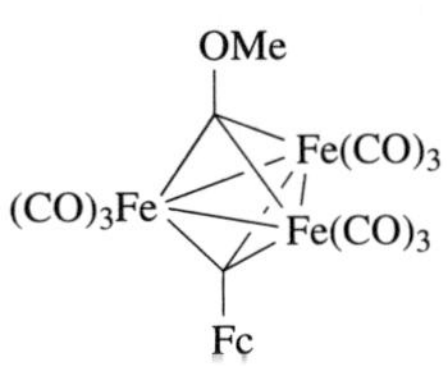

(80)

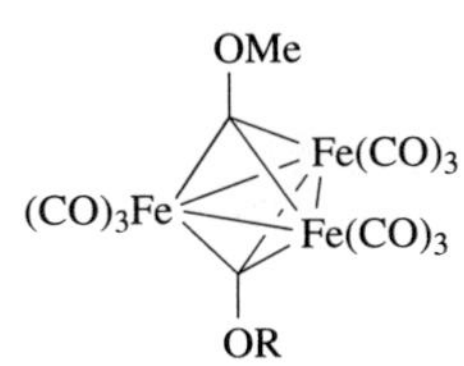

(81) **a**; R = Me
b; R = O(CH$_2$)$_4$OMe

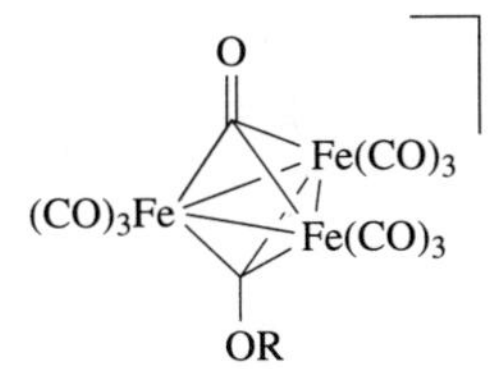

(82) **a**; R = Me
b; R = Et
c; R = COMe
d; R = COCH=CHMe
e; R = COCH=CHPh
f; R = CH$_2$OMe

unsaturated acyl halides as electrophiles gave **(82d)** and **(82e)** in low yields ⟨89JOM(368)199⟩. The other route comprises the reaction of $[Fe_2(CO)_8]^{2-}$ with ClCH$_2$OMe in THF to produce the red salt **(82f)** (R = CH$_2$OMe), which was isolated in 22% yield as the NEt$_4$ salt ⟨81CC189, 81CC2496⟩.

The brown complex **(83)** with one Cp ring coordinated at one Fe atom of the trinuclear core has only been obtained as a by-product during the formation of heterodinuclear and trinuclear compounds. The structure was confirmed by an x-ray determination ⟨87OM819⟩. The complex forms in 3.4% yield upon reacting HFe$_3$(μ-COMe)(CO)$_{10}$ with {Ni(μ-CO)(Cp$_2$}$_2$ ⟨86OM1103⟩ or in about 5% yield from (cyclooctene)$_2$Fe(CO)$_3$ and Cp(CO)Fe(μ-CO)(μ-COMe)Mn(CO)(η^5-CH$_3$C$_5$H$_4$), with the main product being a complex with a OCFe$_2$Mn core ⟨87JA2843⟩.

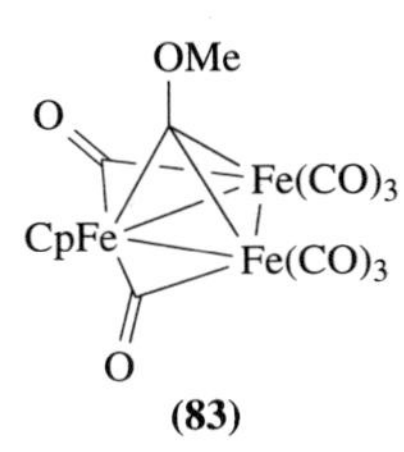

(83)

The unstable hydrido complex (**84**) was mentioned as being formed from hydrogenation of the cluster $HFe_3(\mu\text{-}COMe)(CO)_{10}$; however, its stable dark purple $SbPh_3$ derivative (**85**) can be isolated in 22% yield if the reaction is carried out in the presence of excess of the ligand ⟨83OM219⟩. The related complex (**86**) in which the three hydrogen atoms, serving as one-electron donors, are replaced by the three-electron donor Bi atom is obtained in 65% yield by alkylation of the anion $[BiFe_3(CO)_{10}]^-$ with $MeSO_3CF_3$ in CH_2Cl_2 solution ⟨86IC2472⟩.

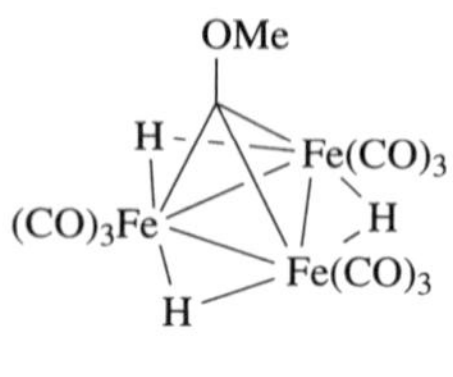

(84)

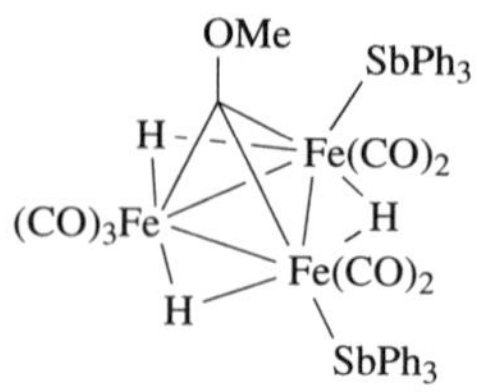

(85)

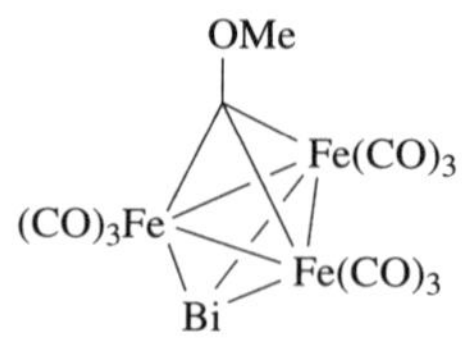

(86)

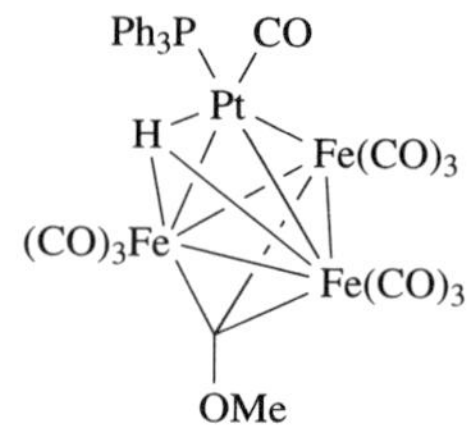

(87)

The trinuclear hydrido cluster $HFe_3(\mu\text{-}COMe)(CO)_{10}$ gave black crystals of the heteronuclear tetrahedral cluster $Fe_3Pt(\mu_3\text{-}H)(\mu_3\text{-}COMe)(CO)_{10}(PPh_3)$ (**87**) in about 50% yield when it was allowed to react with $Pt(C_2H_4)_2PPh_3$ at room temperature in ether ⟨82CC51⟩.

The anionic tetranuclear black purple cluster $[Fe_4(\mu\text{-}CO)(\mu_3\text{-}COMe)(CO)_{11}]^-$ (**88**) was obtained by alkylation of the dianion $[Fe_4(CO)_{13}]^{2-}$ with either $[Me_3O]^+[BF_4]^-$ ⟨80CC780, 80CC781⟩ or $MeSO_3F$ ⟨80CC778, 82JA5621⟩, both in dichloromethane, in about 55% yield; the compound was isolated as the $[(Ph_3P)_2N]$ salt.

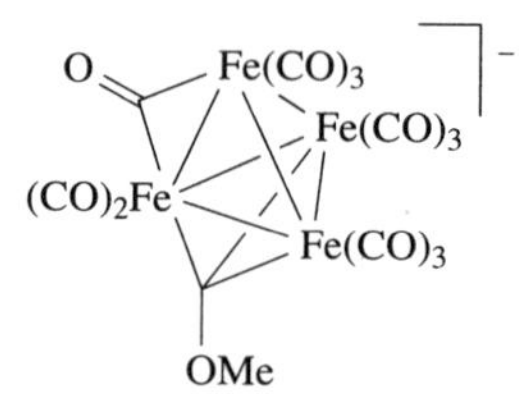

(88)

6.11.3.1.5 *Compounds with the* $OCCo_3$ *core*

This section contains two types of compounds with this core. A few derivatives of the trinuclear cluster $Cp_3Co_3(\mu_3\text{-}COR)_2$ are known, whereas the majority of compounds are based on the trinuclear cluster $Co_3(CO)_9(\mu_3\text{-}COR)$, in which R represents various organic substituents including main group or transition metal elements. Starting material for the second series is mainly $Co_2(CO)_8$ or the covalent trinuclear cluster $Co_3(CO)_9(\mu_3\text{-}COLi)$.

Thus, the reaction of $Co_2(CO)_8$ with the trialkylborane adducts R_3NBH_3 produces the black compounds (**89a**) and (**89b**) in moderate yields ⟨68IC2265⟩. A similar treatment of $Co_2(CO)_8$ in THF with the adducts Et_3NBX_3 (X = Cl, Br) gives (**89c**) and (**89d**) ⟨72JOM(46)149, 75JOM(102)109, 76ZN(B)342⟩. $Co_2(CO)_8$ is also the starting material for the preparation of the red siloxy compound (**90c**), which can be obtained in 54% yield by reacting with 1-hydrosilatrane in THF ⟨93IC5883⟩, whereas red–black (**90a**) (m.p. 108 °C dec.) and (**90b**) (m.p. 91–93 °C) have been obtained by the reaction of $(CO)_9Co_3(\mu_3\text{-}COLi)$ with the corresponding $ClSiR_3$ in diethyl ether in 44% and 20% yield, respectively ⟨70JOM(24)C61⟩.

Various compounds with Ti, Zr or Hf ((**91**)–(**93**)) in the apical position have also been prepared. Thus, one equivalent of $(CO)_9Co_3(\mu_3\text{-}COLi)$ reacts with Cp_2MCl_2 (M = Ti, Zr, Hf) in toluene to produce the deep red to black colored cluster compounds (**91**) (**a**, m.p. 134 °C, 57% yield; **b**, m.p. 138–141 °C, 72% yield; **c**, m.p. 139–142 °C, 88% yield). Two equivalents of $(CO)_9Co_3(\mu_3\text{-}COLi)$ in a similar procedure leads to the disubstituted compounds (**92**) (**a**, m.p. 158–160 °C, 39% yield; **b**, m.p. 145 °C, 35% yield; **c**, m.p. 145 °C, 65% yield). Treatment of (**91**) with solid NaOH in toluene results in the formation of the oxo-bridged derivatives (**93**) (**a**, m.p. 165 °C, 36% yield; **b**, m.p. 160–164 °C, 15% yield; **c**, m.p. 156–160 °C, 19% yield) ⟨78CB1603⟩. Compound (**91a**) can also be obtained from $Co_2(CO)_8$ and Cp_2TiCl_2 ⟨76JOM(113)67⟩. Black crystals of (**94**) are formed in 23% yield from $Na[Co(CO)_4]$ and $CpTiCl_3$ in toluene ⟨78CB1239⟩. The chromium-substituted complex (**95**), which is extremely sensitive to atmospheric oxygen and moisture, forms in about 60% yield by treatment of $Co_2(CO)_8$ with $Cp_2Cr_2(\mu\text{-}OCMe_3)_2$ ⟨90JOM(384)279⟩, while the red complex (**96**) is the product of the reaction of $Co_2(CO)_8$ with one equivalent of pyridine in similar yield ⟨87AG681⟩. All the main group and transition metal-substituted compounds are summarized in Scheme 21.

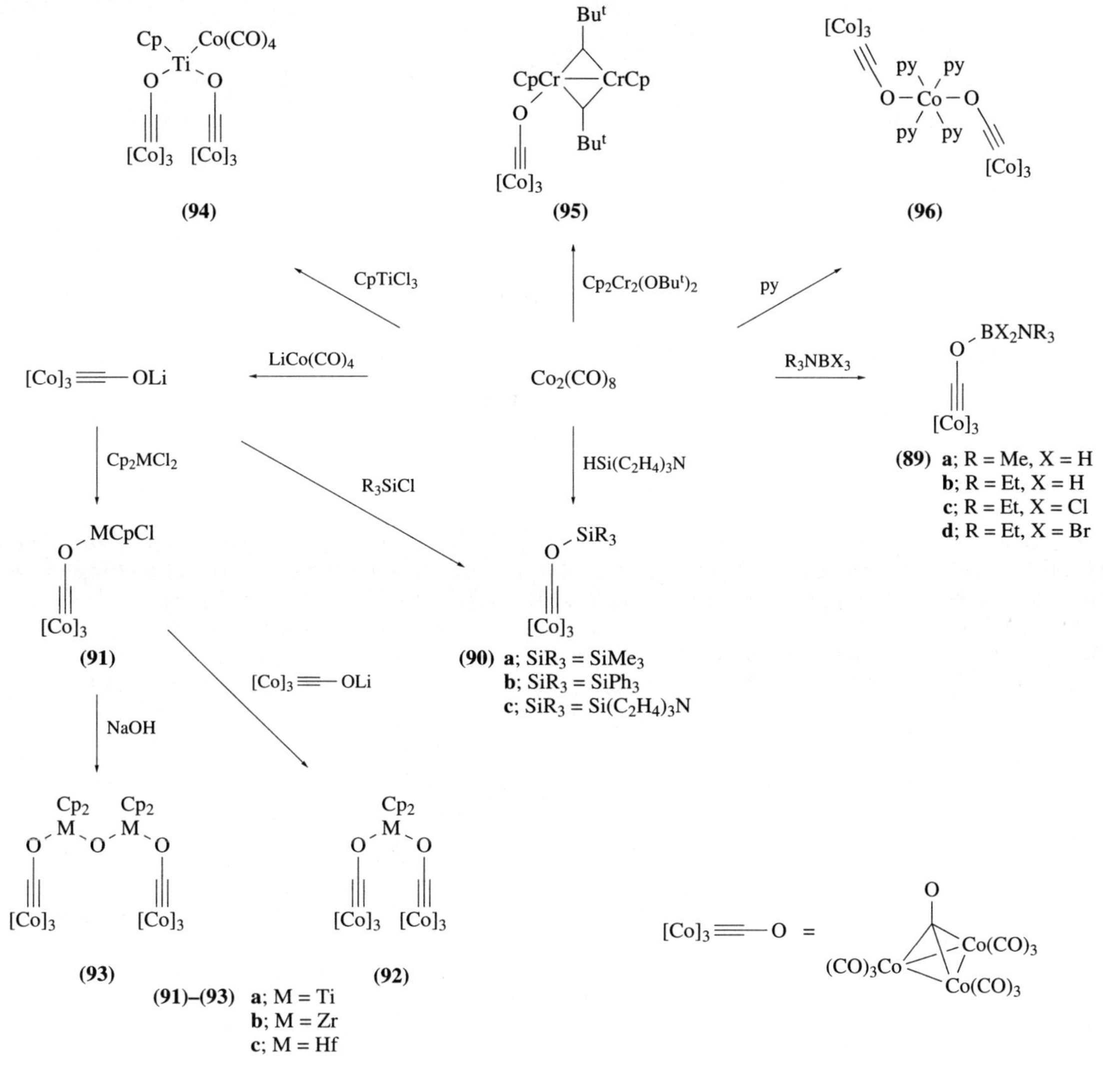

Scheme 21

The starting material for many compounds in Scheme 21, black crystals of $(CO)_9Co_3(\mu_3\text{-}COLi)$, have been obtained from $LiCo(CO)_4$ and $Co_4(CO)_{12}$ or $Co_2(CO)_8$ in ether as Et_2O ⟨79CC396⟩ or $Pr^i{}_2O$ adducts ⟨80AG411⟩. The compound can be freed from ether upon reacting with Ph_2O ⟨81AG94⟩. Addition of dry HCl in hexane gives red crystals of the corresponding acid $(CO)_9Co_3(\mu_3\text{-}COH)$ (**97**), which rapidly decomposes at room temperature ⟨79CC397⟩; the acid can be stored under CO or Ar at liquid-nitrogen temperature ⟨81AG94⟩. With NEt_3 the dark violet adduct $(CO)_9Co_3(\mu_3\text{-}COH \cdot NEt_3)$ is obtained ⟨81AG215⟩. Acetylation of $(CO)_9Co_3(\mu_3\text{-}COLi)$ with MeCOBr in benzene at room temperature gives $(CO)_9Co_3(\mu_3\text{-}COCOMe$ ((**98**), m.p. 123°C) in 67% yield ⟨76CB3339⟩. The alkyl derivatives (**99**) have been obtained by different routes. The red-black methyl derivative (**99a**) can be obtained from $Co_2(CO)_8$ either by reacting with $MeOCHCl_2$ in THF in 27% yield ⟨73JOM(50)265⟩ or by reacting with $Fe_3(H)(\mu\text{-}COMe)(CO)_{10}$ in 48% yield ⟨87OM819⟩. The consecutive addition of LiPh/ether at −60°C and Et_3OBF_4 in CH_2Cl_2 at −10°C to a solution of $Co_2(CO)_8$ does not form a Fischer-type carbene complex but leads to the formation of (**99b**) (the ethyl derivative) in about 4% yield ⟨83M851⟩. The organic species are shown in Scheme 22.

The second type of compound with the Co_3CO core comprises two species and has been reported by the group of Floriani. The compounds originate from trimerization of a CpCo fragment which

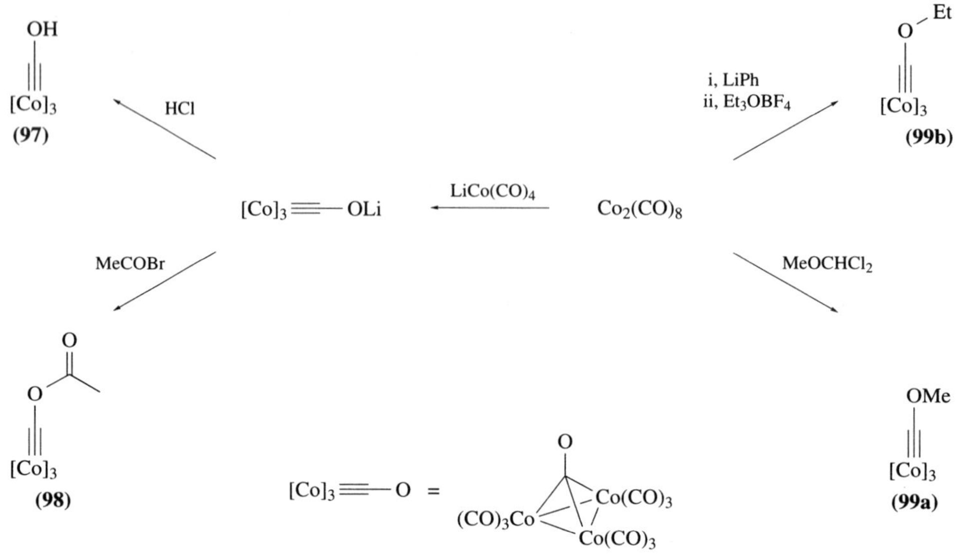

Scheme 22

forms from $CpCo(CH_2{=}CH_2)_2$ upon losing the labile bonded ethylene molecules. Thus, black crystals of (**100**) crystallized in 48% yield from a toluene solution when the ethylene complex was allowed to react with $Cp_2Ti(CO)_2$; in contrast to the other μ_3-COTi species discussed above with Ti(IV), the titanium atom has the oxidation number 3. Reaction of (**100**) with excess TMS-Cl in THF replaces both Cp_2Ti fragments by TMS groups with formation of red-violet (**101**) in about 56% yield ⟨86AG283, 88NJC621⟩. The formation of the compounds is depicted in Scheme 23.

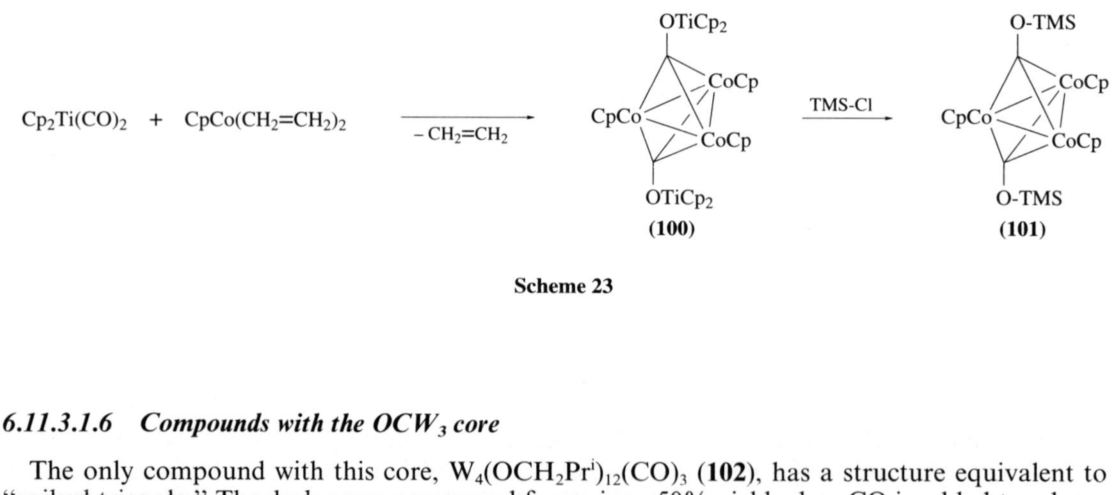

Scheme 23

6.11.3.1.6 *Compounds with the OCW_3 core*

The only compound with this core, $W_4(OCH_2Pr^i)_{12}(CO)_3$ (**102**), has a structure equivalent to a "spiked triangle." The dark green compound forms in ~50% yield when CO is added to a hexane solution of the butterfly cluster $W_4(OCH_2Pr^i)_{12}$ for 24 h; five bridging OR groups ($R = CH_2Pr^i$) stabilize the W—W bonds and the two CO groups are *cis* at one basal W atom ⟨89JA7283, 90JOM(394)265⟩.

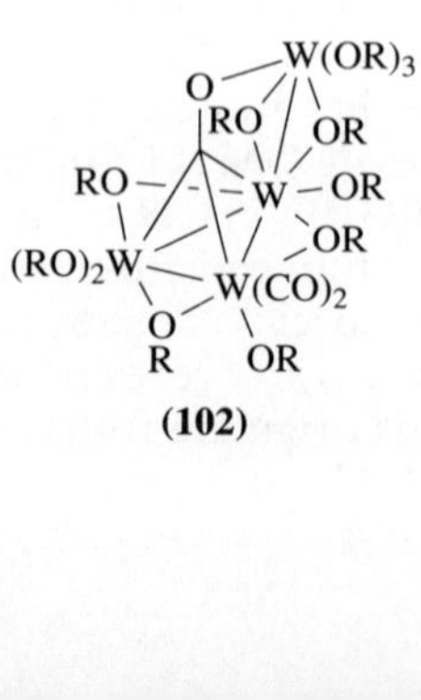

(102)

6.11.3.1.7 Compounds with the OCRu$_3$ core

Compounds with this core comprise homometallic and heterometallic species, both with the same μ_3-COMe ligand. The homometallic compounds are collected in Scheme 24.

The Ru analog $H_3Ru_3(\mu_3\text{-COMe})(CO)_9$ (**103**) of the Fe cluster (**84**) has been obtained as bright orange crystals in 93% yield upon heating the trinuclear precursor $HRu_3(\mu\text{-COMe})(CO)_{10}$ in hexane in the presence of H_2 ⟨85JOM(287)357⟩. Replacement of three CO groups in the basal plane by the tripod ligand $MeC(CH_2PPh_2)_3$ gives the yellow (**104**) in about 50% yield with a face-capped structure. The introduction of the smaller tridentate ligand $HC(PPh_2)_3$ under the same conditions gives orange (**105**), in which the ligand behaves as a bidentate bridge spanning two equatorial sites on two Ru atoms ⟨91OM2384⟩. PPh_3 replaces one or two CO groups to give (**106**) and (**107**) in 30% and 14% yield, respectively ⟨89OM1270⟩. 1,3-Cyclohexadiene behaves as a 4-electron donor to one Ru atom and replaces one CO and H_2 and reacts in the presence of diethyl fumarate to give (**108**) in low yield ⟨83OM1179⟩. Deprotonation of (**103**) with K-Selectride/THF gives the anion (**109**), which can be isolated as the $[(Ph_3P)_2N]$ salt ⟨89OM1270⟩.

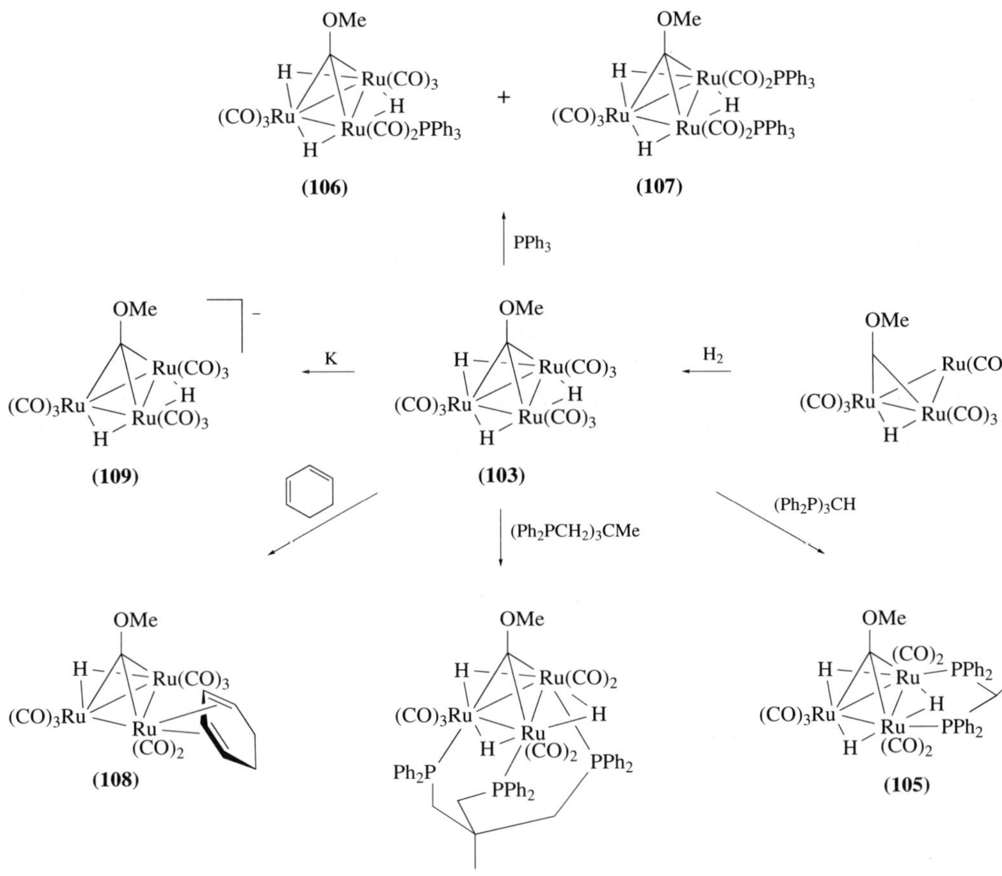

Scheme 24

A series of heterometallic compounds are summarized in Scheme 25. Some compounds are derived from a homometallic trinuclear species by replacement of one or more bridging H atoms by the isolable $AuPR_3$ group or a similar one-electron donating group. Thus, treatment of the basic compound (**103**) with $AuMePPh_3$ under mild conditions produces three orange compounds in which one (**111**), two (**112**), or all (**113**) bridging H atoms of (**103**) are replaced by the $AuPPh_3$ unit ⟨82CC773, 83JCS(D)2599, 83JOM(249)273, 89OM1270⟩. A similar reaction starting from the anion (**109**) and the corresponding $ClMPPh_3$ produces (**111**) (M = Au), (**114**) (M = Ag) and (**115**) (M = Cu) in about 65–75% yield; these compounds can be converted into (**116**)–(**118**) by treating with PPh_3 in refluxing CH_2Cl_2 in yields greater than 90% ⟨89OM1270⟩. The anion (**109**) reacts with $[Rh(CO)_3(PPh_3)_2]^-$ in methanol at $-20\,^{\circ}C$ to form the red air-stable tetranuclear complex (**119**) in 80% yield ⟨89CC1029⟩.

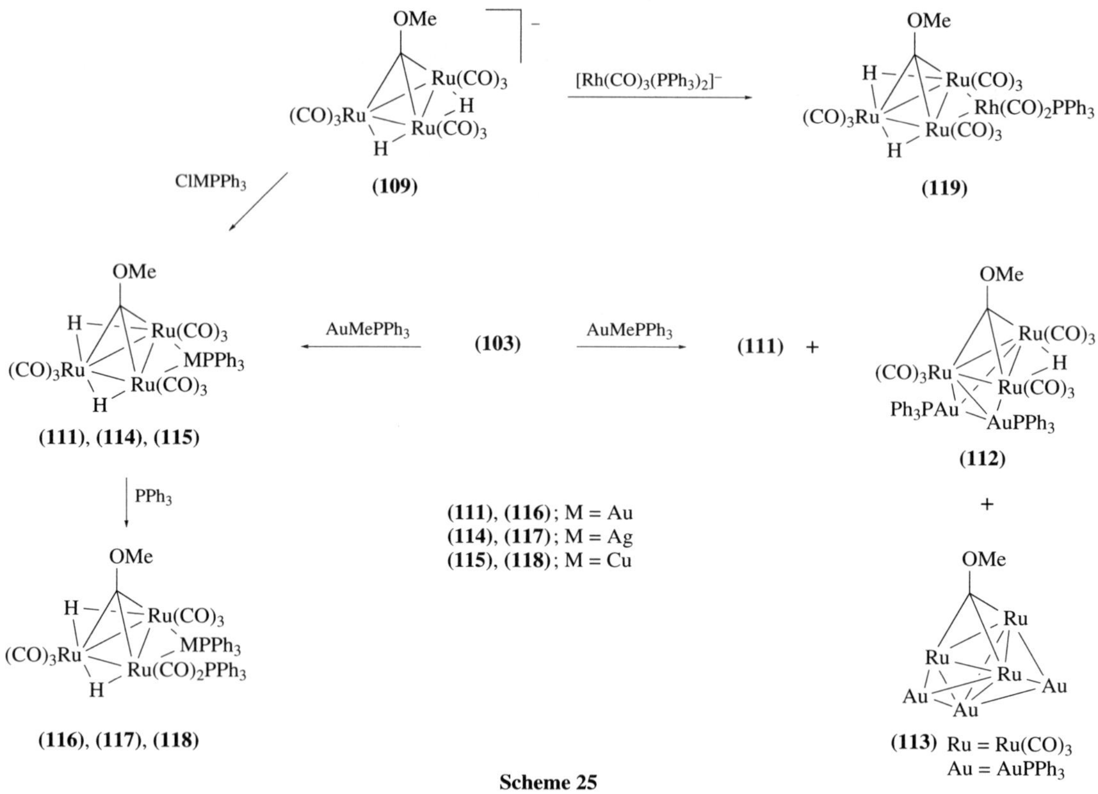

Scheme 25

The red compound (**120**) with three transition metals is obtained in about 5% yield by reducing the tetrahedral cluster $Ru_3RhH_2(CO)_{10}(PPh_3)(\mu\text{-COMe})$ in THF with $K[BHBu_3]$ followed by addition of solid $Au(PPh_3)Cl$ in dichloromethane at $-90\,^{\circ}C$ ⟨91JCS(D)1017⟩; the reaction is depicted in Equation (12).

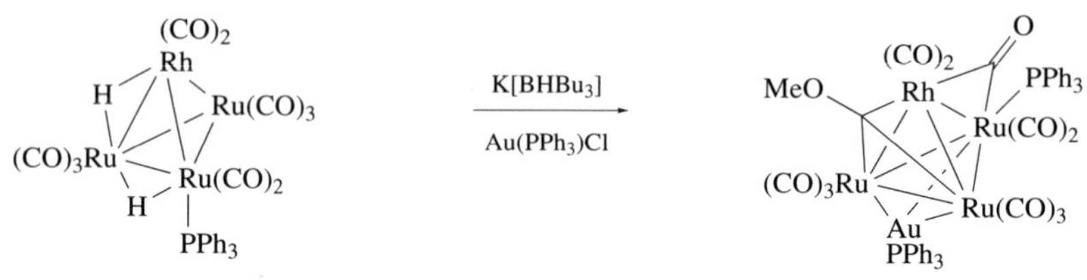

(**120**) (12)

The only compound with a μ_3-CORu ligand spanning a Ru_3 plane is described by Geoffroy and co-workers. The complex (**121**) forms either in about 5% yield by reacting the cluster $Ru_3(CO)_9(\mu_3\text{-CO})(\mu_3\text{-NPh})$ with diphenylacetylene or in 49% by reacting the starting cluster with one of the other reaction products, a metallapyrrolidone complex as depicted in Scheme 26 ⟨86OM2561⟩.

6.11.3.1.8 Compounds with the OCOs$_3$ core

Closely related to the iron cluster (**84**) and its ruthenium derivative (**103**) is the corresponding osmium compound (**122**). It is similarly obtained in about 93% yield by heating $HOs_3(\mu\text{-COMe})(CO)_{10}$ with H_2 in hexane for 2 h ⟨83OM219⟩; as depicted in Equation (13), the reaction is reversible and carbonylation regenerates the starting complex ⟨85JOM(287)357⟩.

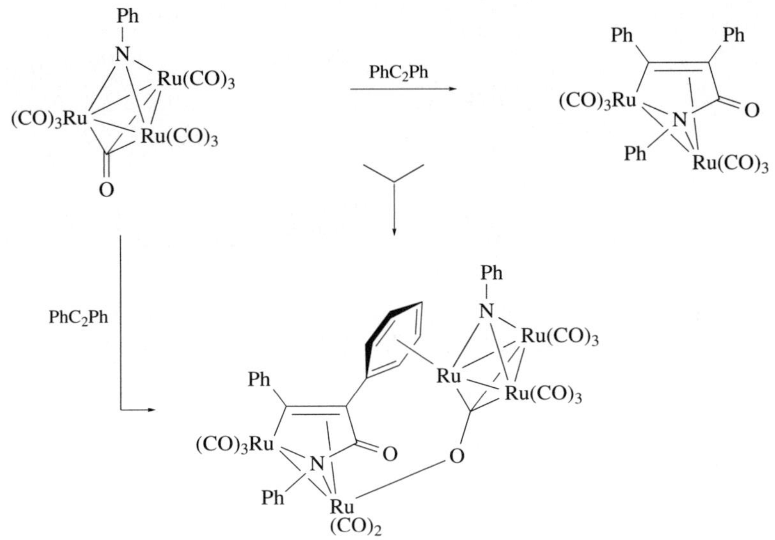

(121)

Scheme 26

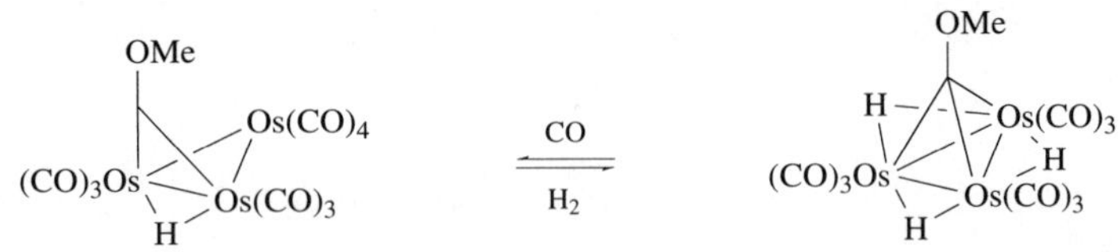

(13)

(122)

The related compound (**123**) with a boroxine center could be isolated as a pale yellow precipitate upon hydroboration of $(\mu\text{-H})_2Os_3(CO)_{10}$ by THF·BH_3 in dichloromethane solution, and the structure was confirmed by an x-ray structure determination ⟨84CC392⟩.

(123)

A series of compounds with two bridging carbyne ligands have been described by Shapley and co-workers as shown in Scheme 27. Thus, sequential addition of phenyllithium and $MeSO_3CF_3$ in a Fischer-type manner to an ethereal solution of $HOs_3(\mu\text{-COMe})(CO)_{10}$ at 0°C produces orange–red air-stable crystals of (**124**) (m.p. 137–139°C) in 41% yield. Protonation with HSO_3CF_3 retains the framework and gives the anionic complex (**125**); the proton bridges one Os—Os bond. CO substitution with excess of PPh_3 gives a mixture of the orange–red compounds (**126**) (20% yield) and (**127**) (54% yield), which could be separated by TLC techniques ⟨86OM1757⟩. Treatment of (**124**) sequentially with phenyllithium and $MeSO_3CF_3$ provided red crystals of the Fischer-type carbene complex (**128**) in 66% yield; the structure was confirmed by an x-ray analysis ⟨89JOM(371)257⟩.

Small amounts (10%) of the cluster (**129**) ($Os_6(CO)_{17}(py)_2$) were formed, when $Os_6(CO)_{18}$ was treated with excess pyridine (py) in dichloromethane; the major product was the dianion $[Os_5(CO)_{15}]^{2-}$; the yield can be enhanced to 20% by addition of Me_3NO to the reaction mixture. An x-ray analysis confirmed the structure with one osmium atom in a "spike" arrangement ⟨84CC1089⟩.

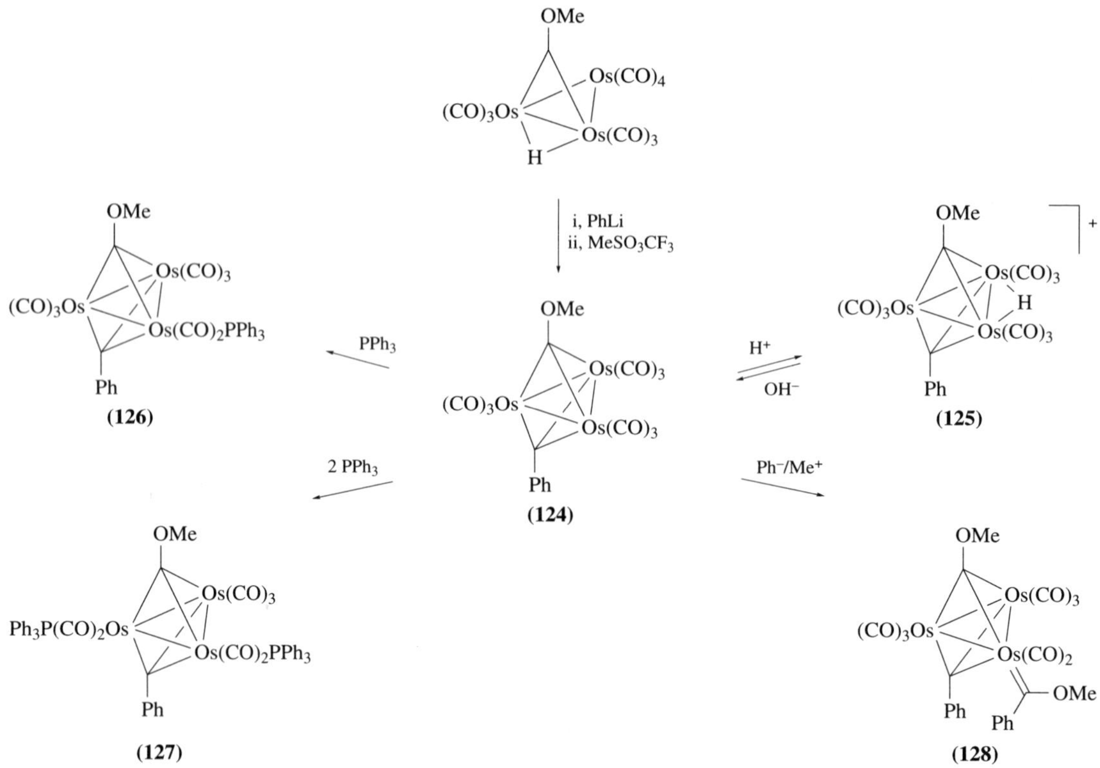

Scheme 27

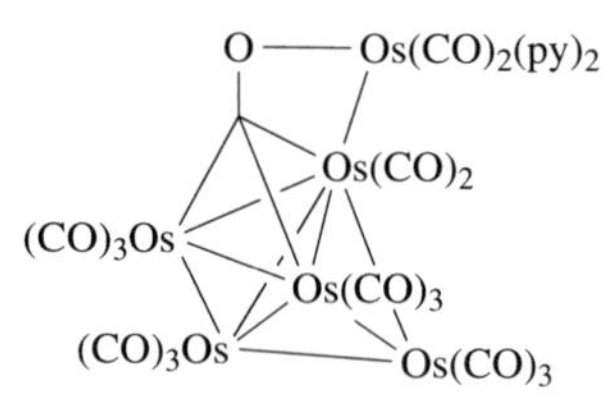

(129)

py = pyridine

6.11.3.1.9 *Compounds with the OCFe$_2$Ni core*

Only one compound with this core has been described. Blue–green crystals of $Fe_2Ni(\mu_3\text{-}COMe)(\mu_3CO)(Cp)$ **(130)** were obtained in 45% yield when $HFe_3(\mu\text{-}COMe)(CO)_{10}$ was allowed to react with $\{Ni(CO)Cp\}_2$ in toluene on heating in a Carius tube. The compound was separated by chromatography and the structure was confirmed by an x-ray analysis ⟨86OM1103⟩.

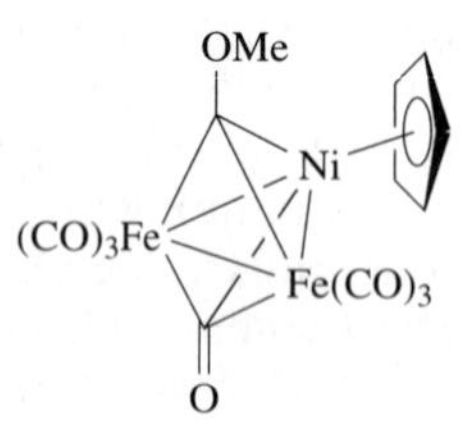

(130)

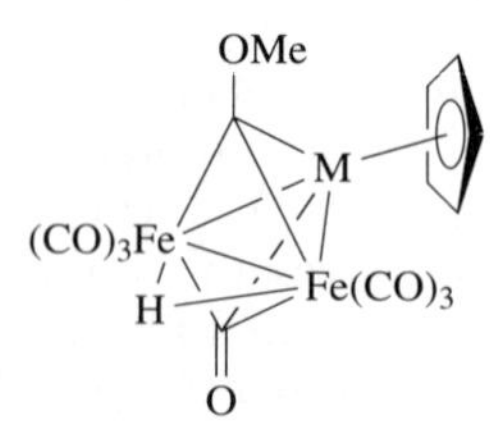

(131) **a**; M = Co
b; M = Rh

6.11.3.1.10 *Compounds with the OCFe$_2$Co core*

The related complex (**131a**; M = Co) is derived from (**130**) by replacement of the Ni atom by the isoelectronic CoH unit. The green–brown complex in which the hydrogen atom bridges the Fe—Fe bond has similarly been obtained in 57% yield by reacting $HFe_3(\mu\text{-}COMe)(CO)_{10}$ with $Co(CO)_2Cp$. Reaction of (**131a**) with $MeAuPPh_3$ leads to the replacement of hydrogen by $AuPPh_3$ with result of black crystals of (**132**) in about 63% yield ⟨86OM1103⟩. As shown by the same group, $HgPh_2$ in toluene at 90 °C similarly produces (**133a**; M = Co) as the sole product ⟨87CC147⟩.

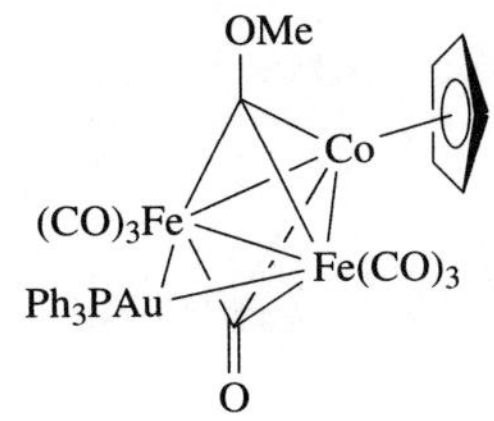

(132)

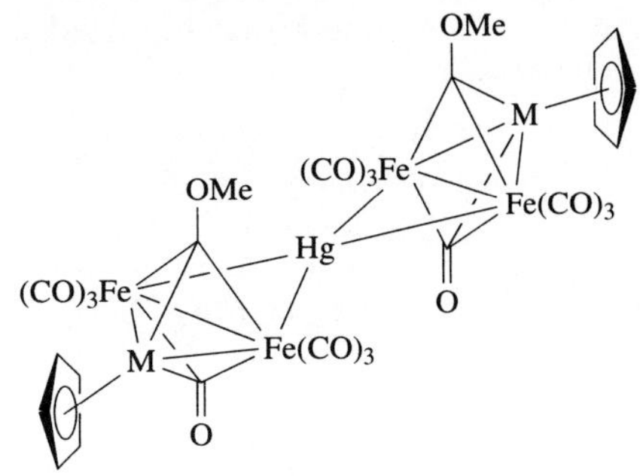

(133) **a**; M = Co
b; M = Rh

6.11.3.1.11 Compounds with the $OCFe_2Rh$ core

The preparation of (**131b**; M = Rh) is carried out in a manner similar to its cobalt derivative by reacting $HFe_3(\mu\text{-}COMe)(CO)_{10}$ with $Rh(CO)_2Cp$; black crystals are obtained in 65% yield ⟨86JOM(310)67⟩. The analogous reaction with $HgPh_2$ leads to (**133b**; M = Rh); however, the solid-state structure shows bridging of the mercury atom the Fe—Rh bonds. In solution a polytopal rearrangement takes place with the mercury atom migrating around the Fe_2Rh triangle ⟨87CC147⟩.

6.11.3.1.12 Compounds with the $OCFe_2Mn$ core

The heterotrinuclear cluster (**134**) has been prepared in 80% yield from $(cyclooctene)_2Fe(CO)_3$ and $Cp(CO)Fe(\mu\text{-}CO)(\mu\text{-}COMe)Mn(CO)(\eta^5\text{-}MeC_5H_4)$; a minor product is a similar complex with a $OCFe_3$ core ⟨87JA2843⟩. As depicted in Scheme 28, addition of excess diazomethane ⟨87JA2843⟩ or diazoethane ⟨88OM794⟩ to (**134**) stereospecifically yields the μ_3-methoxycarbyne, μ_2-alkylidene cluster (**135**) in 15% (**a**, R = H) or 62% (**b**, R = Me) yield, respectively. The ethylidene cluster (**135b**; R = Me) decomposes at room temperature within 8–10 days to give small amounts of the isomer (**135c**) by ethylidene "rotation." Further decomposition products comprise other dinuclear and trinuclear species ⟨88OM794⟩.

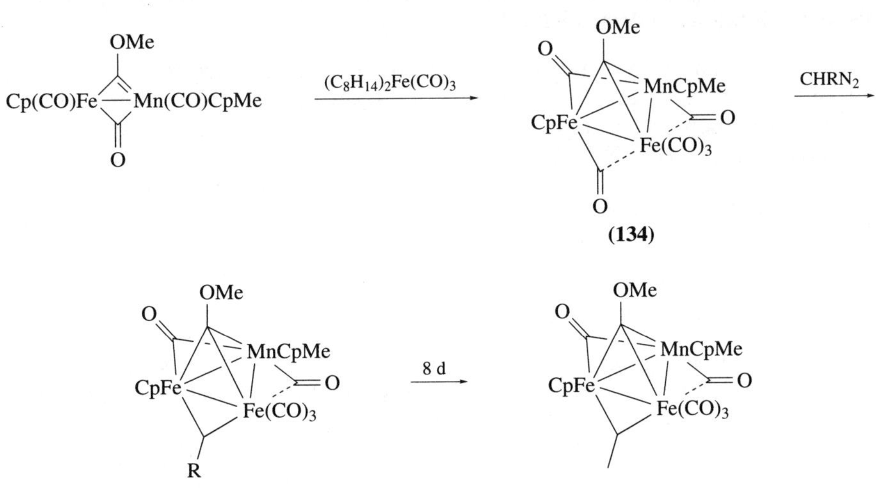

(134)

(135) **a**; R = H
b; R = Me

(135c)

Scheme 28

6.11.3.1.13 Compounds with the $OCRu_2W$ core

The tetranuclear species (**136**) (**a**, R = H; **b**, R = Me), in which the μ-COMe group bridges one heteronuclear face of the tetrahedral cluster have been prepared by heating a mixture of $Ru_3(CO)_{10}(\mu$-COMe)H and the corresponding hydrido complex $Cp'W(CO)_3H$ ($Cp' = C_5H_5$, C_5Me_5) in toluene for 1 h ⟨90JCS(D)3033⟩. The same group reported a similar reaction with $CpW(CO)_3C{\equiv}CPh$, which leads to (**137**) ⟨91OM2485⟩.

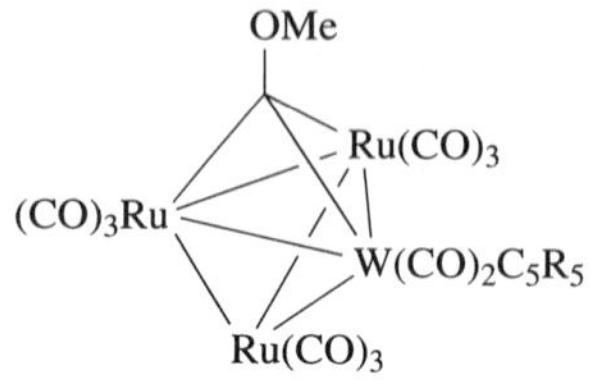

(136)

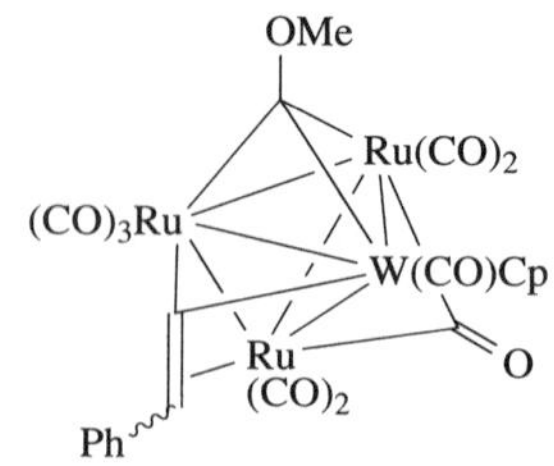

(137)

6.11.3.1.14 Compounds with the OCW_2Ru core

With one equivalent of $CpW(CO)_3H$ the tetranuclear complex (**137**) is converted into the pentanuclear complex (**138**) in which the COMe group bridges a W_2Ru face in a μ_3-manner. The reaction has been performed in refluxing toluene to give red crystals in 24% yield ⟨91OM2485⟩.

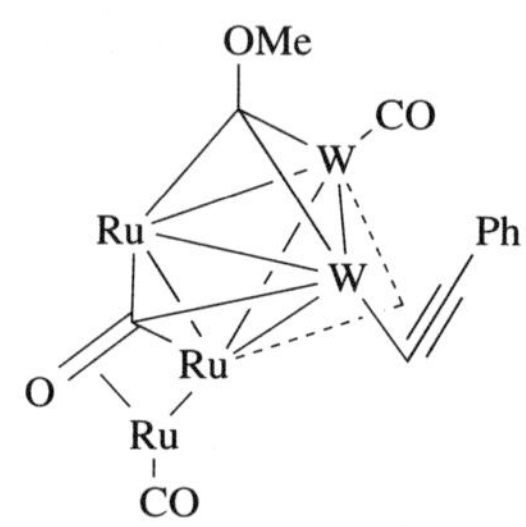

(138) Ru = $Ru(CO)_2$
W = WCp

6.11.3.2 Methanes Bearing One Sulfur Function and Three Functions Derived from the Group 15 Element, Metalloid and/or a Metal

The compounds in which two or three different elements are attached to the SC carbon atom are represented by only a few examples. Main group compounds consist with few exceptions of species with the $SC(TMS)_3$ fragment. The majority of samples in this section, however, are based on Co_3 cluster compounds with a μ_3-C=S ligand or similar clusters in which one or two Co atoms are replaced by other transition metals; addition of an electrophile at the C=S sulfur atom gives a μ_3-methylidine ligand which contributes three electrons to the cluster.

6.11.3.2.1 Compounds with the SCN_3 core

A series of compounds of the general type $RSC(NO_2)_3$, in which the NO_2 groups are bound to the carbon atom via the nitrogen atoms, have been described by a Russian group. The compounds have been obtained by the reaction of $KC(NO_2)_3$ with the appropriate RSCl in dimethoxyethane or ether at about 0 °C. Recrystallization from chloroform–hexane produced 2,4-$(NO_2)_2C_6H_3SC(NO_2)_3$ (m.p. 68–69 °C) in 50% yield and $PhSC(NO)_3$ (m.p. 12–13 °C) in 93% yield, respectively ⟨73IZV350⟩. $MeSC(NO_2)_3$ was similarly obtained in 93% yield but purified by distillation (b.p. 40 °C at 1 mm Hg) ⟨73IZV350, 75IZV1669⟩. The compounds 4-$(NO_2)C_6H_4SC(NO_2)_3$ (m.p. 125–127 °C) and 4-$MeOC_6H_4SC(NO_2)_3$ (m.p. 36–39 °C) have been prepared in the same manner in dichloromethane solution in 87% and 70% yield, respectively ⟨74IZV1350⟩. $PhSC(NO_2)_3$ is also produced in 20% yield by the reaction of $K(C(NO_2)_3$ with $[PhS][BF_4]$ ⟨75IZV1669⟩. Reactions with hydrogen halides have been described ⟨89IZV2106⟩.

6.11.3.2.2 Compounds with the $SCSi_3$ core

The key compound for this series is $HC(TMS)_3$, which can be converted into $HSC(TMS)_3$ (**139**) in 46% yield by treatment with MeLi and then with elemental sulfur after extractive purification with methanol–KOH followed by sublimation ⟨85TL1091, 85TL2259⟩. A similar procedure with $HC(SiMe_2Ph)_3$ generates pale yellow crystals of $HSC(SiMe_2Ph)_3$ (**140**) in 10% yield, melting at 201–203 °C ⟨87IC1488⟩. DuMont and others obtained several sulfur-rich species using $HC(TMS)_3$ or $HSC(TMS)_3$ as starting materials as shown in Scheme 29. Thus, treatment of (**139**) with SCl_2 gives yellow (**141**) in 45% yield. This compound is also produced along with (**142**) by oxidation of (**139**) with air ⟨85TL1091, 93CB1355⟩. Desulfurization of the tetrasulfane (**142**) to (**141**) can be performed with elemental mercury. The corresponding disulfane (**143**) forms as a mixture with the methyl derivative (**144**) ⟨93CB1355⟩. The latter can also be obtained from $(TMS)_3CLi$ and Me_2S ⟨90JOM(398)65⟩. Starting with a lithiated 2-[[(trimethylsilyl)methyl]thio]tetrahydropyran and TMS-Cl the compound (**145**) was prepared in 83% yield ⟨85JA6729⟩ (Equation (14)).

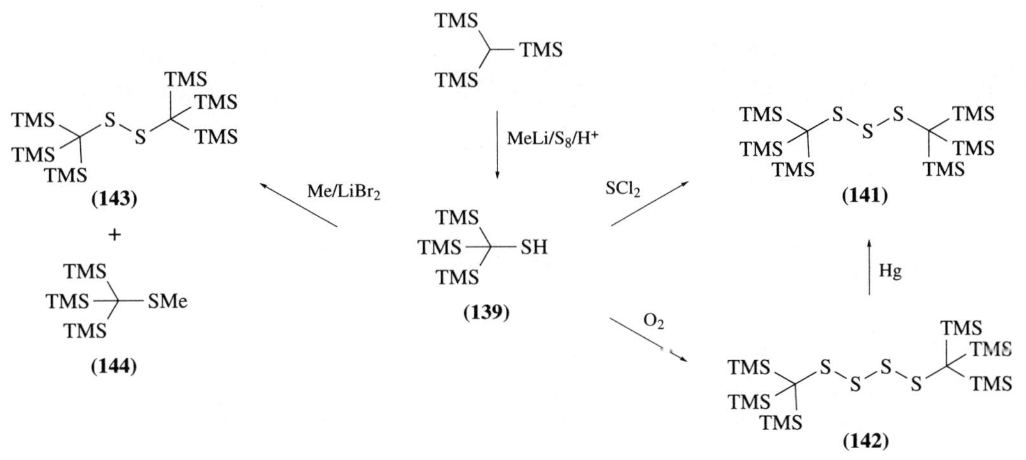

Scheme 29

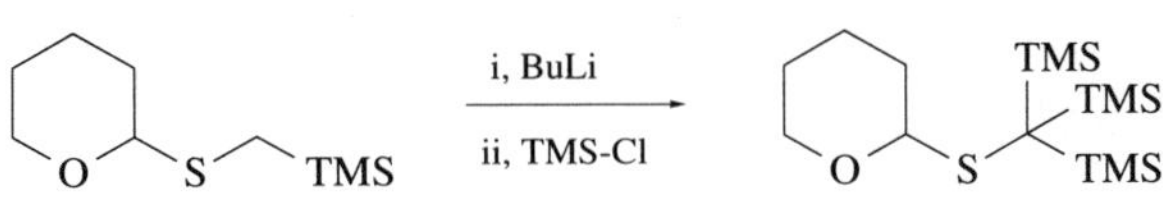

(14)

Several *S*-metalated derivatives have been prepared by the groups of Power and Lappert and the compounds are summarized in Scheme 30; starting materials are the hydrogen compounds (**139**) and (**140**) or the corresponding lithium salts. The pale yellow trimeric and tetrameric silver compounds (**146**) (66% yield) and (**147**) (92% yield, m.p. 235 °C) are the result of the reaction with NEt_3 and $AgNO_3$ in refluxing benzene or acetonitrile, respectively; the compounds contain six- and eight-membered AgS rings ⟨87IC1488⟩. A similar tetrameric gold compound (**148**) has been prepared in 69% yield from the lithium compound and AuCl, whereas Ph_3PAuCl produces the monomeric compound (**149**) (m.p. 204–207 °C) as colorless crystals in 69% yield ⟨93IC5126⟩. The reaction of (**139**) with $M[N(TMS)_2]_2$ compounds (M = Sn, Pb) produces various dimeric or polymeric species (**150**)–(**152**) containing the $SCSi_3$ core; the polymeric lithium salt was obtained with butyllithium in hexane ⟨85CC1776⟩. The lithium salt $LiSC(TMS)_3 \cdot 3.5THF$ was reported to be a dimer with a square Li_2S_2 core and was prepared in the presence of small amounts of THF ⟨85CC1674⟩.

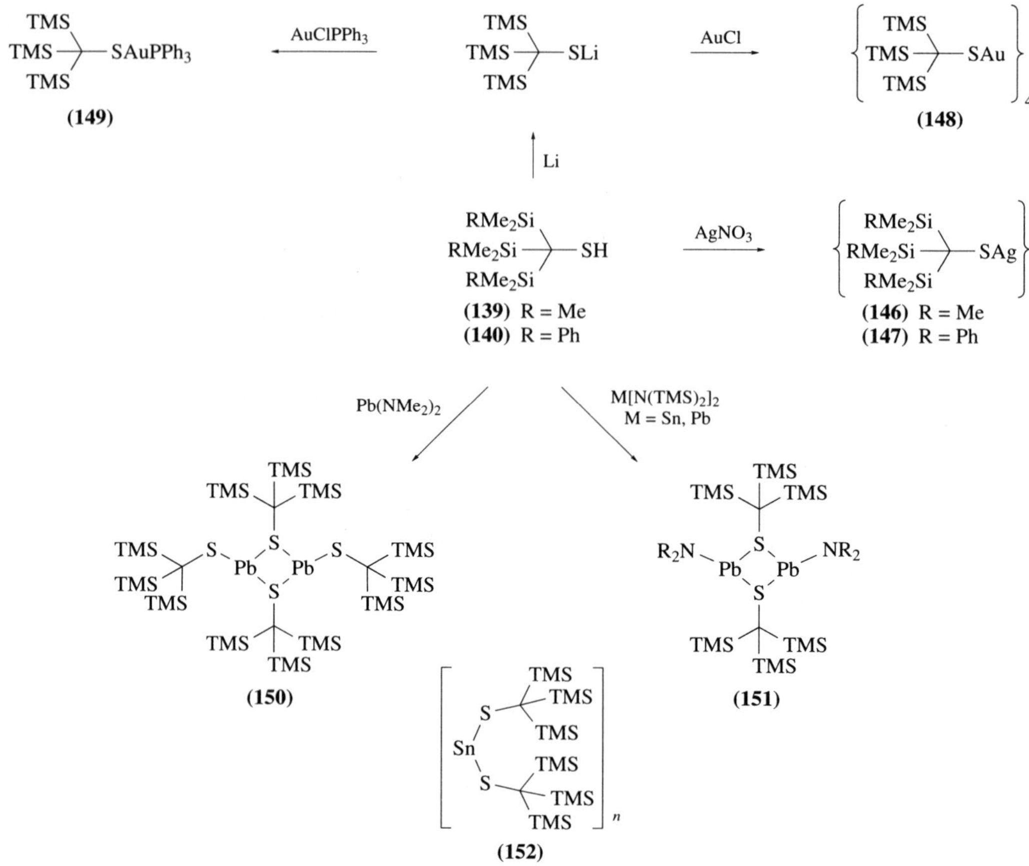

Scheme 30

6.11.3.2.3 Compounds with the $SCSi_2P$ core

Two different compounds with this core have been described. The reaction of MesP=C(TMS)$_2$ (Mes = mesityl) with $\frac{1}{4}$ S_8 leads to an inseparable mixture of MesP(S)=C(TMS)$_2$ and the thiaphosphirane MesP(=S)(μ-S)C(TMS)$_2$ (**153**). The mixture is completely converted into (**153**) by addition of a second equivalent of sulfur (Scheme 31) ⟨84CC698⟩. An additional compound with the $SCSi_2P$ core, (**154**), has been described as very water sensitive and has been prepared by reacting (EtO)$_2$P(O)CSMe(TMS)Li with TMS-Cl in THF at $-78\,°C$ ⟨89S101⟩.

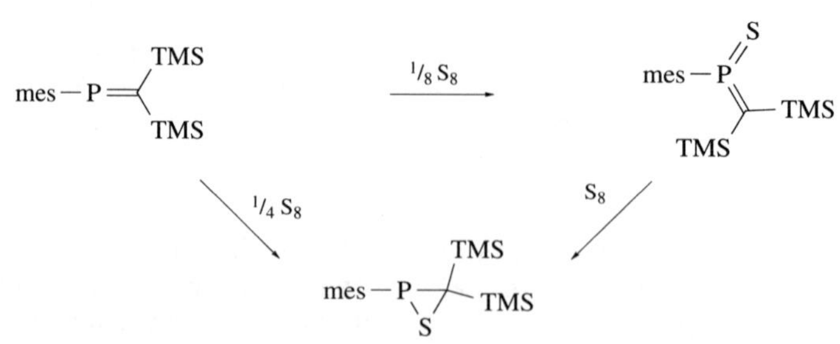

Scheme 31

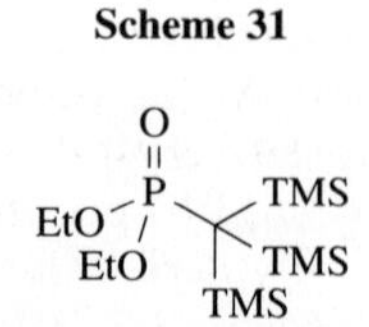

(154)

6.11.3.2.4 Compounds with the SCPSiLi core

$(EtO)_2P(O)CSMe(TMS)Li$ results from lithiation of $(EtO)_2P(O)CHSMe(TMS)$ with Bu^nLi in THF at −78 °C and was not isolated. The species reacts with ketones and aldehydes and can be used to transfer the $(EtO)_2P(O)CSMe(TMS)$ fragment to various substrates ⟨89S101⟩.

6.11.3.2.5 Compounds with the $SCSi_2Li$ core

$PhSCLi(TMS)_2$ (**155**; R = Ph) is obtained *in situ* by dropwise addition of butyllithium in hexane to $PhSCH(TMS)_2$ in THF at −78 °C. This and similar lithium compounds have been formulated as ionic species and have been used to replace the oxygen atom of various epoxides by the carbene PhS(TMS)C to give cyclopropanes ⟨89TL7033⟩. The methyl derivative has been obtained similarly in 100% yield ⟨77CB852⟩.

(155)

6.11.3.2.6 Compounds with the SCN_2Fe core

The reaction of $Fe_2(CO)_9$ with tetramethylthiourea in ether for 24 h followed by chromatography in Fluorosil leads to $Fe_2(CO)_6(SC(NMe_2)_2)$ (**156**) in 2.6% yield along with $Fe_3(CO)_9S_2$ (0.1%) and $(Me_2N)_2C{=}S{-}Fe(CO)_4$ (6.2%) (Equation (15)) ⟨74IC225⟩.

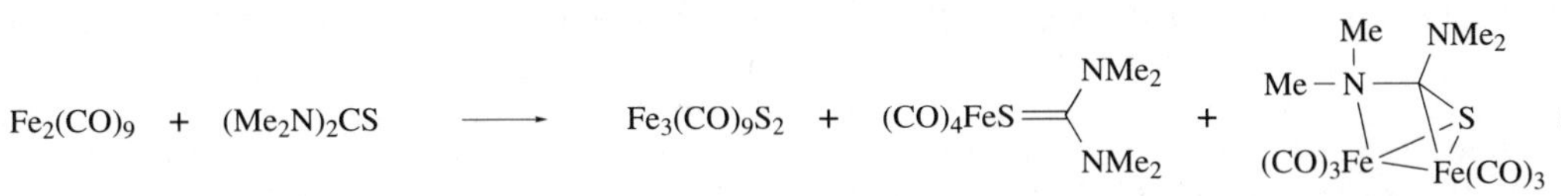

(15)

(156)

6.11.3.2.7 Compounds with the SCNSnPt core

Compounds of the general type $(PPh_3)_2Pt(SCSnPh_3NR_2)$ (**157**; $NR_2 = NMeH$, $N(CH_2)_4$) are prepared by the reaction of the corresponding $Ph_3SnC(S)NR_2$ with $(PPh_3)_2PtC_2H_4$ in benzene for 24 h. The pale yellow products were precipitated with hexane and recrystallized from benzene–hexane. The yields were approximately 50%; by-products are the corresponding Sn-bonded species $(PPh_3)_2PhPtSn(Ph)_2C(S)NR_2$ (Equation (16)) ⟨83IC3700⟩. The compounds can also be considered in terms of an η^2-coordination of $S{=}C(SnPh_3)NR_2$ at the Pt fragment.

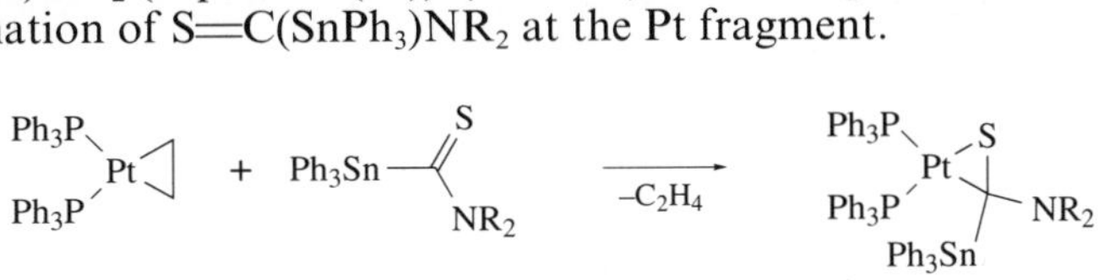

(16)

(157)

NR_2 = NMeH, 

6.11.3.2.8 Compounds with the SCSiSnLi core

The compounds $RSCLi(TMS)(SnMe_3)$ (**158**; R = Me, Ph) have been obtained in 100% yield by lithiation of the corresponding $RSCH(TMS)(SnMe_3)$ with LDA in THF–HMPTA ⟨77CB852⟩.

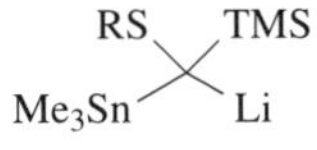

(158)

6.11.3.2.9 Compounds with the SCFe$_3$ core

Only one compound of this type has been described. The tetranuclear compound $Fe_4(CO)_{11}(CS)S$ (**159**) was obtained as one of the products from the reaction of $Fe_3(CO)_{12}$ with CS_2 in hexane solution at 80°C under 10 atm CO–Ar pressure. Both the compounds in Equation (17) have been formed in about 4% yield ⟨80CC812⟩.

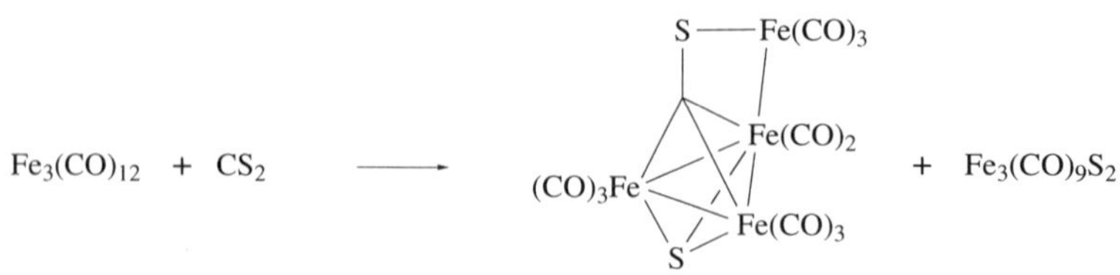

(17)

(159)

6.11.3.2.10 Compounds with the SCCo$_3$ core

The compounds with this core are derived from $Co_3Cp_3(\mu_3\text{-}CS)(\mu_3\text{-}S)$ and the hypothetical cluster anion $[Co_3(CO)_9(\mu_3\text{-}CS)]^-$; both are electron precise cluster compounds. Of the two sulfur atoms of the first species, only that at the μ_3-C=S ligand is nucleophilic enough to react with a variety of electrophilic reagents Y to produce compounds of the general type $Co_3Cp_3(\mu_3\text{-}CSY)(\mu_3\text{-}S)$; the corresponding $Co_3(CO)_9(\mu_3\text{-}CSY)$ compounds are formed in a complex reaction from $Co_2(CO)_8$. The electrophile Y can be a cationic group R^+ or a 16-electron transition metal Lewis acid as shown in Scheme 32.

Starting from $Co_3Cp_3(\mu_3\text{-}CS)(\mu_3\text{-}S)$, the alkylating agents RI (R = Me, Et, Pr^i) produce the black salts $[Co_3Cp_3(\mu_3\text{-}CSR)(\mu_3\text{-}S)]I$ (**160a–c**) from a THF solution in about quantative yields. The reaction with Pr^{i}I has only been conducted in an NMR tube. A decreasing reactivity order MeI > EtI > Pr^{i}I was found, indicated by an increasing reaction time in the order 1 h, 5 h, 5 days, respectively. Black crystals of $Co_3Cp_3(\mu_3\text{-}CSCr(CO)_5)(\mu_3\text{-}S)$ (**161**) have been obtained in 35% yield by treating a freshly prepared THF solution of $(CO)_5CrTHF$ with $Co_3Cp_3(\mu_3\text{-}CS)(\mu_3\text{-}S)$; the complex could be purified by recrystallization from THF–hexane ⟨79AG663, 80CB1654⟩.

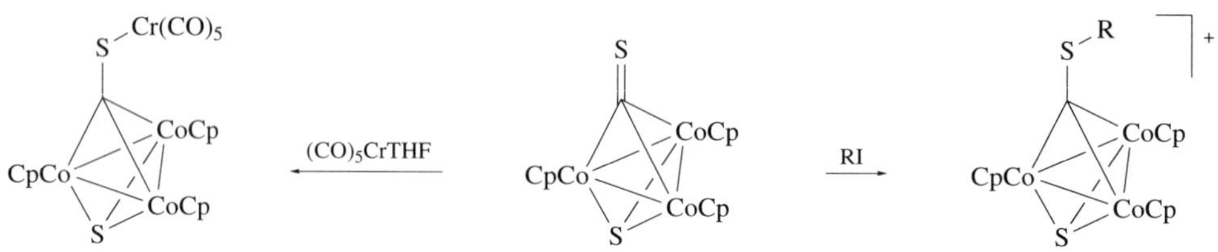

(161)

(160) **a**; R = Me
b; R = Et
c; R = Pr^i

Scheme 32

The reaction of CS_2 with $Co_2(CO)_8$ results in the formation of more than eight different sulfur-containing products, similar to that with $Fe_3(CO)_{12}$. Separation by column chromatography or by thin-layer chromatography gave compounds which could be isolated in low yields and identified by x-ray diffraction analysis. Reaction time, solvents, and the workup procedure all affect the product distribution found. Thus, a 4:1 mixture of the components in light petroleum gave (**162**) after a reaction time of 8 h and purification by thin-layer chromatography ⟨83CC1613⟩; additionally, isomers have also been isolated ⟨82IC3781⟩. Wei described the preparation of (**163**) and unidentified products using the same or a slightly modified procedure ⟨84IC2973⟩; some of the products are summarized in Scheme 33. Earlier results concerning these compounds have been reported from the group of Marko and co-workers ⟨68JOM(11)207⟩.

6.11.3.2.11 Compounds with the SCFe$_2$Co core

The 50-electron cluster compounds $Fe_2CoCp_2(CO)_5CSR$ (**164**) (**a**; R = Me; **b**; R = CH_2Ph) can be prepared by the reaction of the corresponding bridging thiocarbyne complexes $[Cp(CO)Fe(\mu\text{-}CO)$

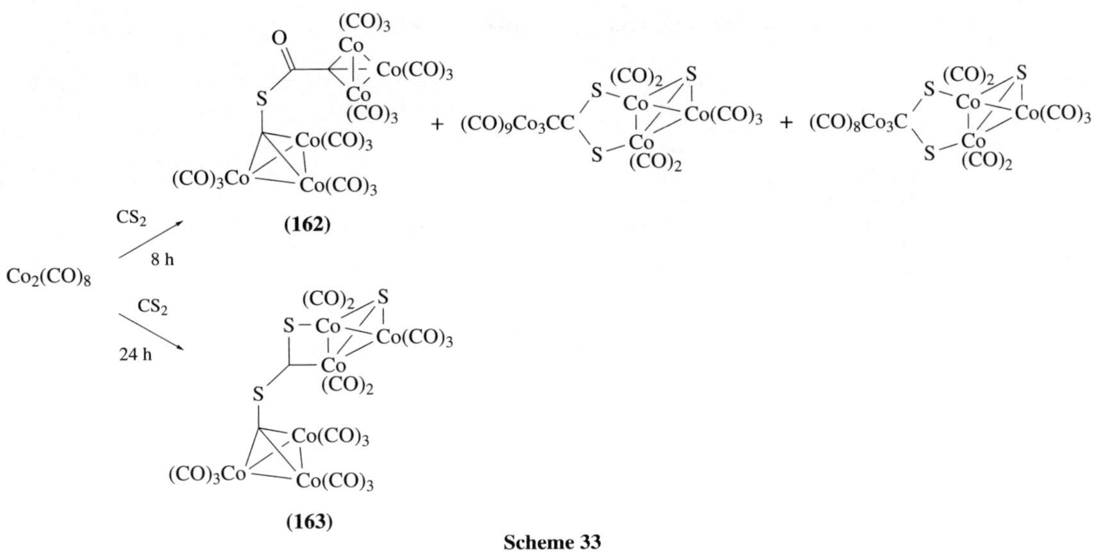

Scheme 33

(μ-SR)]PF_6 with excess Na[$Co(CO)_4$] (prepared *in situ* by combining $Co_2(CO)_8$ with finely ground NaOH) in THF under UV photolysis at room temperature. Extraction of the neutral cluster compounds with benzene and recrystallization from dichloromethane–hexane produces brownish-purple crystals. Whereas (**164a**) is isolated in a yield of about 80%, the similar route to (**164b**) gives the compound in 28% yield but requires photolysis at 12°C. Phosphines replace one CO group at the Co atom when the reaction is carried out in dichloromethane to give (**165**) and (**166**) in good yields (Table 2, entries 1 to 6) as illustrated in Scheme 34 ⟨88JOM(394)183⟩.

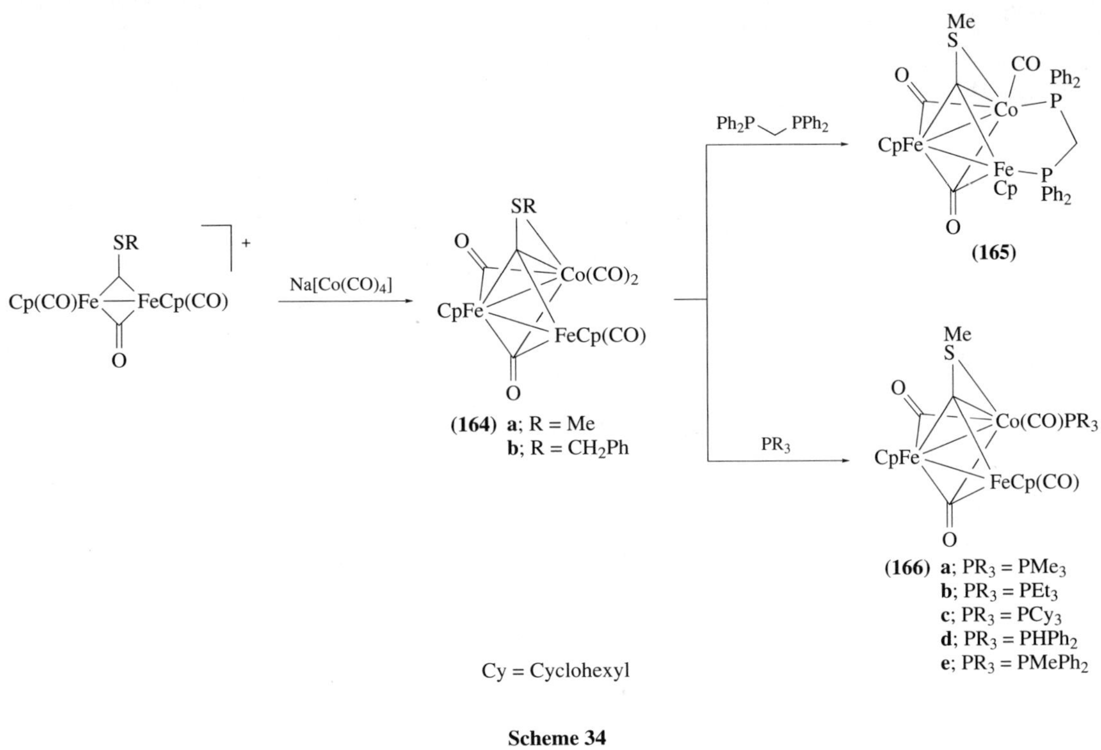

Scheme 34

6.11.3.2.12 Compounds with the $SCCo_2W$ core

Only one set of compounds of this type has been described. The 48-electron cluster compounds (**167a**)–(**167e**) (Table 3) have been obtained by addition of $Co_2(CO)_8$ to an equilibrium mixture of

Table 2 Preparation of **(165)** and **(166)** from (**164**; R = Me) by CO/phosphine exchange.

Entry	*Product*	*Phosphine*	*Decomp. temp.* (°C)	*Time* (h)	*Yield* (%)
1	**(166a)**	PPh_3	108	9	93
2	**(166b)**	PEt_3	95	1.5	68
3	**(166c)**	PCy_3	95	12	54
4	**(166d)**	$PHPh_2$	90	5	63
5	**(166e)**	$PMePh_2$		5	
6	**(165)**	$Ph_2PCH_2PPh_2$	90	45	57

Table 3 Properties of the compounds **(167a)**–**(167e)**.

(167)	R^3	R^1	R^2	*Yield* (%)	*M.p.* (°C)
a	Ph	CO_2Me	CO_2Me	81	58
b	Me	CO_2Me	CO_2Me	66	
c	Ph	CF_3	CF_3	39	
d	Me	CF_3	CF_3	45	
e	Me	CO_2Me	H	80	124

the two isomers of the carbene complex (diphos)$(CO)_3W{=}CS_2C_2R^1R^2$ in dichlormethane solution. After 12 h reaction time the compounds could be precipitated with hexane. From **(167e)** two isomers were formed according to the orientation of the R^1CCR^2 unit. The yields of the compounds are given in Scheme 35 ⟨88JOM(394)183⟩.

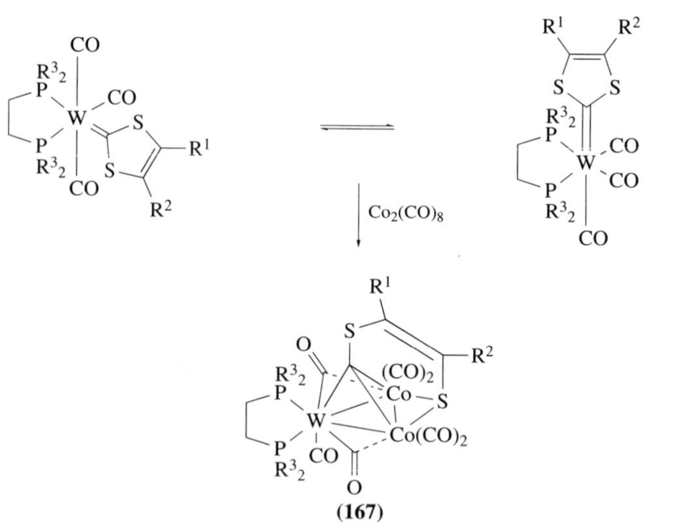

(167)

Scheme 35

6.11.3.3 Methanes Bearing One Selenium or One Tellurium Function and Three Functions Derived from the Group 15 Element, Metalloid and/or a Metal

The compounds bearing the SeC or TeC unit, with few exceptions, are restricted to samples with the TMS group and thus contain the tris(trimethylsilyl)methyl function, $(TMS)_3C$. This section refers mainly to work of the research groups of Sladky, duMont, and Arnold.

6.11.3.3.1 Compounds with the $SeCN_3$ core

Only one compound in this series has been described, by a Russian group in connection with compounds containing the related $SC(NO_2)_3$ fragment. The compound $PhSeC(NO_2)_3$ has been prepared by addition of $K[C(NO_2)_3]$ to a solution of PhSeCl in CH_2Cl_2 solution. KCl was removed and the resulting dried oil was triturated with hexane; the reaction could also be carried out with PhSeBr and the yield of the compound was about 100% in each case ⟨82IZV161⟩. Reactions of the compound with various M—Hal compounds have been described ⟨89IZV2106⟩.

6.11.3.3.2 Compounds with the $SeCSi_3$ core

This section contains compounds of the general type $(TMS)_3CSeX$ in which the selenium atom is further bonded to either hydrogen, alkyl, aryl, halogen, chalcogen, or gold. The $(TMS)_3C$ group is abbreviated below as Tsi. A list of compounds prepared by the groups of Sladky, duMont and Arnold along with conditions and yields are collected in Table 4 and a survey of reactions is shown in Schemes 36–38.

Table 4 Preparation and properties of compounds with the Si_3CSe core.

Entry	*Structure*	*Color* (m.p.; °C)	*Preparation*	*Yield* (%)	*Ref.*
1	**(171)**	Red (204)	TSiLi + Se; H_2O, O_2		85CC1800
2			**(175)** + Se/I_2	95	93ZAAC(619)1693
3	**(175)**	Red (203)	**(171)** + Cu; 90 °C, THF	46	90CB2325
4	**(176)**	Red–brown (192)	**(171)** + Br_2	95–98	93ZAAC(619)1693
5	**(178)**	Yellow–brown (201)	**(171)** + SO_2Cl_2	96	93ZAAC(619)1693
6	**(177)**	Black–violet	**(171)** + I_2; toluene	79	88CB2109
7	**(179)**	Yellow (123)	**(178)** + Me_3SiCN, CH_3CN	89	93ZAAC(619)1693
8	**(180)**	Orange (140)	**(178)** + KSCN in CH_2Cl_2	64	93ZAAC(619)1693
9	**(181)**	Yellow (94)	**(177)** + S; toluene	5	93ZAAC(619)1693
10			**(175)** + S/I_2 in toluene	48	
11	**(182)**		**(177)** + Se; toulene	5	93ZAAC(619)1693
12	**(183)**		**(177)** + S; toluene	10	93ZAAC(619)1693
13	**(184)**	Colorless	**(175)** + MeLi; THF	53	89CB2279
14	**(185)**	Colorless	**(175)** + PhLi	43	89CB2279
15	**(174)**	Colorless (85)	**(175)** + TSiLi/THF	73	89CB2279
16	**(170)**	Colorless (70)	**(175)** + TSiLi/THF	8	89CB2279
17		Pale yellow	**(169)**(THF+)HSO_3CF_3	93	93JA6777
18	**(169)**	Light orange	TSiLi + Se; DMF	73	93JA6777
19	**(172)**	Colorless (dec.)	**(169)** + AuCl; THF	65	93IC5126
20	**(173)**	Colorless (184–189)	**(169)** + Ph_3PAuCl; ether	75	93IC5126
21			**(170)** + $Ph_3PAuN(SiMe_3)_2$	76	

TsiLi has been reported to undergo insertion of two equivalents of selenium to give TsiSeSeLi (**168**) in THF solution ⟨85CC1800⟩ and insertion of one equivalent of selenium to give (**169**) in the presence of dimethoxyethane, which on protonolysis leads to the hydride (**170**) ⟨93JA6777⟩; addition of H_2O and oxidation with O_2 give bis[tris(trimethylsilyl)methyl]triselenide, TsiSeSeSeTsi (**171**) ⟨85CC1800⟩ (Scheme 36).

The middle selenium atom of (**171**) can be removed to give TsiSeSeTsi (**175**), which has been used as starting material for a variety of other TsiSeR compounds ⟨89CB2279, 90CB2325, 93ZAAC1693⟩.

$$\underset{\mathbf{(170)}}{\text{TsiSeH}} \xleftarrow{\text{H}^+\text{/ether}} \underset{\mathbf{(169)}}{\text{TsiSeLi•DME}} \xleftarrow{\text{Se/DME}} \text{TsiLi} \xrightarrow{\text{Se}} \underset{\mathbf{(168)}}{\text{TsiSeSeLi}} \xrightarrow{\text{H}_2\text{O}_2\text{/O}_2} \underset{\mathbf{(171)}}{\text{TsiSeSeSeTsi}} \xrightarrow{\text{Cu}} \underset{\mathbf{(175)}}{\text{TsiSeSeTsi}}$$

Scheme 36

The gold compounds (**172**) and (**173**) have been obtained by treatment of TsiLi · DME with AuCl or $AuClPPh_3$, respectively, as depicted in Scheme 37 ⟨93IC5126⟩.

The production of TsiSeH (**170**) and TsiSeC_4H_7O (**174**) from TsiSeSeTsi (**175**) in THF was reported to be a result of a single electron-transfer reaction (SET) via the TsiSe· radical ⟨89CB2279⟩.

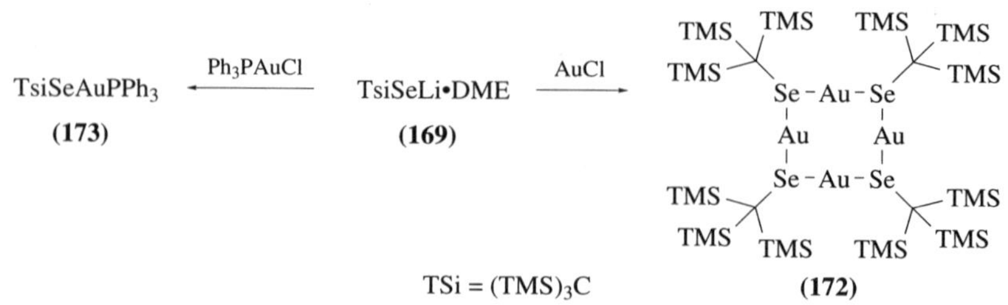

Scheme 37

As shown in Scheme 38 the Se—Se bond in the diselenide (**175**) can be cleaved by various oxidizing agents. While Br_2 ⟨90CB2325⟩ and I_2 ⟨88CB2109⟩ lead to the corresponding halides (**176**) and (**177**), respectively, the chloride (**178**) is obtained by reacting with SO_2Cl_2; the introductions of pseudohalogen CN^- or SCN^- can by achieved by treating the chloride with the appropriate TMS-X ⟨93ZAAC(619)1693⟩. A reversible reaction with elemental sulfur or selenium generates a mixture of further chalcogen-rich derivatives (**181**–**183**) which can also be obtained starting from (**175**) ⟨93ZAAC1693⟩.

The action of LiR (R = Me, Ph) on the diselenide (**175**) generates the species TsiSeMe (**184**) and TsiSePh (**185**), respectively ⟨93ZAAC(619)1693⟩.

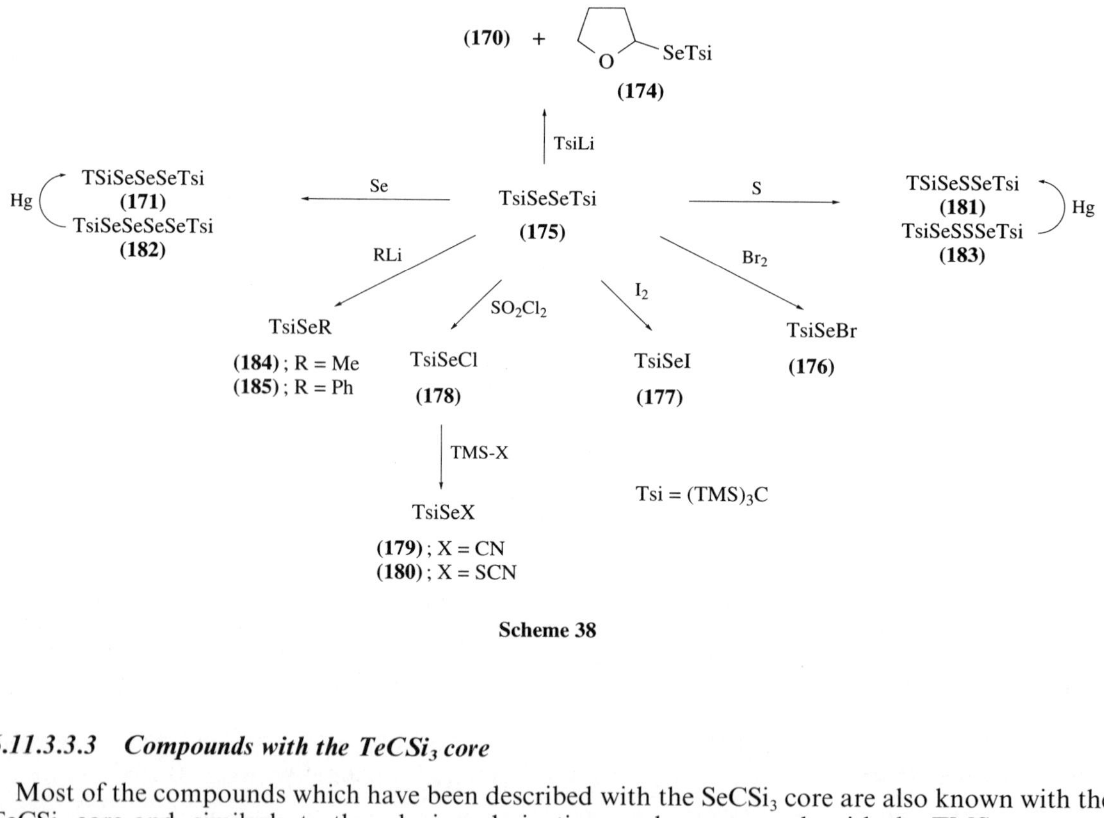

Scheme 38

6.11.3.3.3 Compounds with the TeCSi$_3$ core

Most of the compounds which have been described with the SeCSi_3 core are also known with the TeCSi_3 core and, similarly to the selenium derivatives, only compounds with the TMS group are known. Thus, compounds of the general type $(TMS)_3$CTeX in which the tellurium atom is further bonded to hydrogen, alkyl, aryl, halogen, chalcogen, or some transition metals have been prepared by the same working groups as reported for the corresponding selenium compounds. A list of compounds along with conditions and yields are collected in Table 5 and a survey of reactions is shown in the Schemes 39–40.

As with selenium, insertion of two equivalents of tellurium into the C—Li bond of TsiLi produces TsiTeTeLi (**186**), which can also be obtained by addition of tellurium to the monotelluride (**187**). Oxidation of the lithium salt (**186**) with H_2O/O_2 generates the tritelluride TsiTeTeTeTsi (**188**) from

Table 5 Preparation and properties of compounds with the Si_3CTe core.

Entry	*Structure*	*Color* (m.p., °C)	*Preparation*	*Yield* (%)	*Ref.*
1	**(186)**		**(187)** + Te		85CC1800
2	**(187)**	Orange (141–145)	THF-adduct; TsiLi + Te, THF	56	93JA6777
3			**(189)** + Li		85CC1800
4	**(188)**	Red–black	From decomp. of **(190)**		85JOM(295)C1
5			**(186)** + H_2O/O_2	60	85CC1800
6	**(189)**		**(187)** + Se		85CC1800
7	**(190)**		**(186)** + $ClP(Bu^t)_2$; not isol.		85JOM(295)C1
8	**(191)**		**(187)** + $SeCl_2$		85CC1800
9	**(192)**	Red	**(191)** + Hg in pentane		85CC1800
10			**(189)** + Se, ultrasound 30 h	95	91OM2101
11	**(193)**		**(187)** + $ClP(Bu^t)_2$		85JOM(295)C1
12	**(194)**		From decomp. of **(193)**		85JOM(295)C1
13	**(195)**	Yellow (153–154)	**(187)** + H^+ in ether	81	93JA6777
14	**(196)**	Colorless	**(187)** + AuCl	42	93IC5126
15	**(197)**	Yellow	**(187)** + $ClAuPPh_3$	65	93IC5126
16	**(198)**		**(187)** + Cp_2TiCl_2; unstable		93JA545
17	**(199)**	Red	**(187)** + Cp_2ZrCl_2	57	93JA545
18	**(200)**	Orange	**(189)** + S, ultrasound 30 h	95	91OM2101
19	**(201)**	Blue–black (110)	**(189)** + SO_2Cl_2	95	89CB1255
20	**(202)**	Blue–black (110)	**(189)** + Br_2	95	89CB1255
21	**(203)**	Blue–black (110)	**(189)** + I_2	95	89CB1255
22	**(204)**	Bright yellow (80)	Tsi_2Te_2 + RLi in THF	61	89CB2279
23			**(203)** + MeLi in THF	85	89CB1255
24	**(205)**	Bright yellow (47)	Tsi_2Te_2 + RLi in THF	57	89CB2279
25			**(203)** + PhLi in THF	80	89CB1255
26	**(206)**	Yellow (140)[a]	**(203)** + AgCN[b]	72	91CB1131
27	**(207)**	Dark red (125)[a]	**(203)** + AgSCN[b]	87	91CB1131
28	**(208)**	Red (130)[a]	**(203)** + AgSeCN[b]	82	91CB1131
29	**(209)**	Red (~120)[a]	**(203)** + AgNCO[b]	73	91CB1131
30	**(210)**	Deep red (135)[a]	**(203)** + AgN_3[b]		91CB1131
31			**(203)** + TlOEt/TMS-N_3[b]	84	
32	**(211)**	Dark red (145)[a]	**(203)** + Ag_2NCN[b]	87	91CB1131
33	**(212)**	Orange (140)[a]	**(203)** + K_2NSN[b]	92	91CB1131
34	**(213)**	Colorless (69)	**(189)** + TSiLi in THF	76	89CB2279

[a] With decomposition. [b] In benzene.

which the middle tellurium atom can be removed with mercury to give **(189)**; the latter compound is also formed by oxidation of the lithium salt **(187)** with elemental Se ⟨85CC1800⟩. Reaction of **(186)** with $ClP(Bu^t)_2$ produces the ditellurophosphine **(190)** ⟨85JOM(295)C1⟩. The reactions are summarized in Scheme 39.

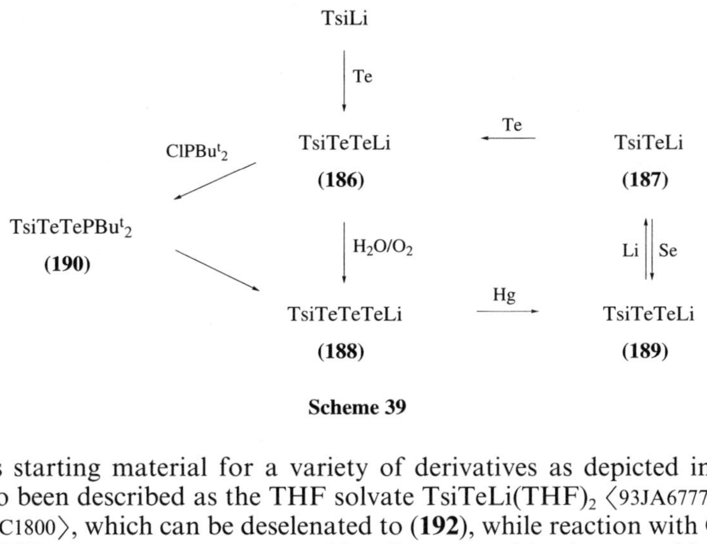

Scheme 39

TsiTeLi **(187)** is starting material for a variety of derivatives as depicted in Scheme 40. This compound has also been described as the THF solvate $TsiTeLi(THF)_2$ ⟨93JA6777⟩. Action of $SeCl_2$ leads to **(191)** ⟨85CC1800⟩, which can be deselenated to **(192)**, while reaction with $ClP(Bu^t)_2$ gives the tellurophosphine **(193)**. At room temperature this tellurophosphine is not stable and converts into

(194) and $Te[P(Bu^t)_2]_2$ ⟨85JOM(295)C1⟩. With triflic acid and the THF solvate **(187)** the corresponding acid **(195)** is obtained ⟨93JA6777⟩.

The THF solvate **(187)** is also the starting material for some transition metal derivatives. With AuCl and Ph_3PAuCl the tetranuclear species **(196)** and the mononuclear compound **(197)**, respectively, have been prepared; the structure of the tetranuclear compound is identical to the structure of the corresponding selenium compound **(172)** ⟨93IC5126⟩. LiCl elimination proceeds also with Cp_2MCl_2 (M = Ti, Zr), to give the $Cp_2M(TeTsi)_2$ compounds **(198)** and **(199)**, respectively ⟨93JA545⟩.

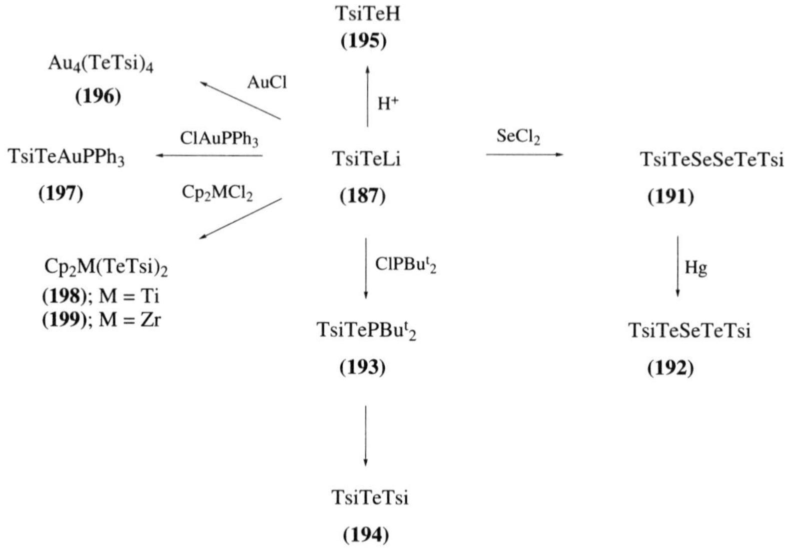

Scheme 40

The chemistry of TsiTeTeTsi **(189)** is summarized in Scheme 41. Ultrasonic activation in the presence of elemental sulfur or selenium leads to the insertion of one chalcogen molecule to give $(TsiTe)_2E$ (**200**; E = S), (**192**; E = Se) ⟨91OM2101⟩.

The tellurohalogenides **(201)**–**(203)** have been prepared by the reaction of **(189)** with SO_2Cl_2, Br_2, or I_2, respectively. The chloride **(201)** can be alkylated with MeLi to give **(204)** or arylated with PhLi or PhMgBr to produce **(205)** ⟨89CB1255⟩; both compounds can also be prepared from Tsi_2Te_2 and the corresponding RLi ⟨89CB2279⟩. Halogen exchange in the iodide **(203)** with various AgX compounds leads to the compounds **(206)**–**(210)**. A similar reaction with two equivalents of the iodide and Ag_2NCN or Ag_2NSN generates the carbodiimide **(211)** and the sulfur diimide **(212)**, respectively. The compound TsiTeOEt was mentioned in a dissertation ⟨91CB1131⟩. Finally, the addition of TsiLi to a solution of **(189)** leads to the incorporation of a THF molecule to produce air-stable **(213)** (m.p. 69 °C) in about 76% yield ⟨89CB2279⟩.

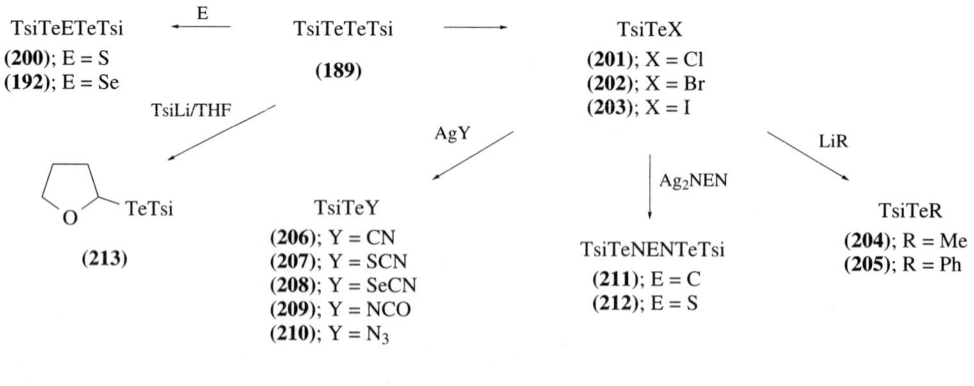

Scheme 41

6.12
Functions Containing at Least One Group 15 Element (and No Halogen or Chalcogen)

DUNCAN CARMICHAEL, ANGELA MARINETTI and PHILIPPE SAVIGNAC
DCPH–Ecole Polytechnique, Palaiseau, France

6.12.1 METHANES BEARING FOUR GROUP 15 ELEMENTS

6.12.1.1 Four Similar Group 15 Element Functions

6.12.1.1.1 Four nitrogen functions

Cyclic, acyclic and spirocyclic compounds bearing four nitrogen functions are known. In all cases, at least two of the nitrogen functions are identical, and totally symmetrical compounds also exist. Examples of this class of compound, where the tetracoordinated carbon is bound to the nitrogen atom of amino, dialkyl or diarylamino, fluoroamino, amido, hydrazino, nitro, and isocyano functions, have been prepared. No single method has been shown to be compatible with every one of these functions, but a number of syntheses are reasonably general. These are detailed below.

(i) By nucleophilic exchange of chlorine atoms

(a) From 2,2-dichloro-4,5-imidazolidinediones and analogues. The 2,2-dichloro-4,5-imidazolidinediones react with primary or secondary amines to give the corresponding tetraaminomethanes, by successive replacement of both chlorine atoms. Dehydrohalogenation may be effected by an excess of the amine itself, or by the addition of bases such as triethylamine or K_2CO_3 (Equation (1) and Table 1).

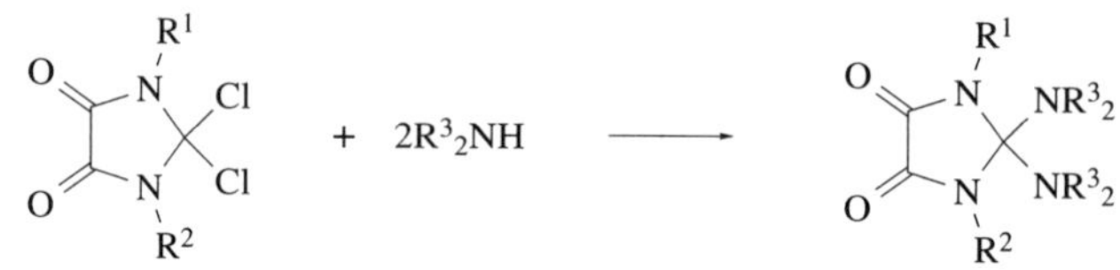

(1)

Table 1 Nucleophilic exchange of chlorine atoms with amines on 2,2-dichloro-4,5-imidazolidinediones.

R^1, R^2	*Reagents*	*Conditions*	*Yield* (%)	*Ref.*
Me_2CH	morpholine	Dioxane, Et_3N, 25°C, 2 h	28	70CB766
Me_2CH	*N*-Me-piperazine	Dioxane, Et_3N, 25°C, 2 h	63	70CB766
Me_2CH	pyrrolidine	Dioxane, 4 equiv. amine, 25°C	46	73CB2315
Me_2CH	piperidine	Dioxane, 4 equiv. amine, 25°C	81	73CB2315
$4\text{-}NO_2C_6H_4$	morpholine	Dioxane, 4 equiv. amine, 25°C	75	73CB2315
$2\text{-}MeC_6H_4$	morpholine	Dioxane, 4 equiv. amine, 25°C	87	73CB2315
C_6H_{11}	morpholine	Dioxane, 4 equiv. amine, 25°C	63	73CB2315

The benzylthiolate leaving group may also be used in place of chloride ⟨78JPR335⟩. Diamines lead to spirocyclic derivatives (Equation (2) and Table 2). 2,2′-Spiroimidazolidine-4′,5′-dione derivatives have been patented as stabilizers for color photographic materials ⟨76GEP2460330⟩.

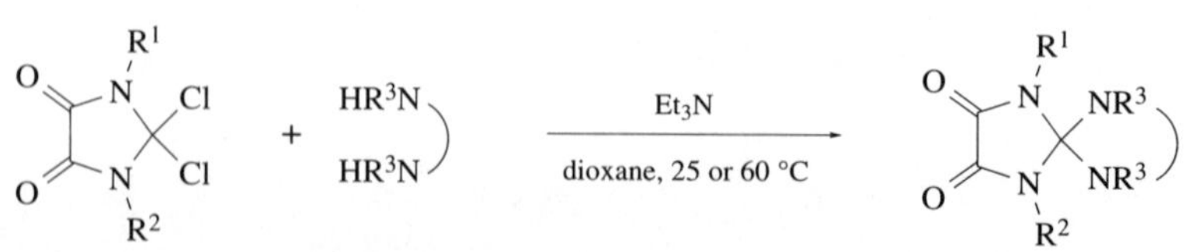

(2)

A related reaction of 2,2-dichloro-2,6-dihydro-4,6-pyrimidinediones with simple or chelating amines proceeds through a substitution of both chlorine atoms to give open or spirocyclic tetraamines, respectively (Equation (3)). Yields are moderate to good (Table 3) ⟨88CZ382⟩.

Table 2 Nucleophilic exchange of chlorine atoms with diamines on 2,2-dichloro-4,5-imidazolidinediones.

R¹, R²	*Reagents*	*Yield (%)*	*Ref.*
Me_2CH	$(MeNHCH_2)_2$	45	70CB766
C_6H_{11}	$(MeNHCH_2)_2$	52	70CB766
Me, C_6H_{11}	$(MeNHCH_2)_2$	11	70CB766
Me_2CH	1,2-$(NH_2)_2C_6H_4$	76	70CB766
c-C_6H_{11}	1,2-$(NH_2)_2C_6H_4$	92	70CB766
4-MeC_6H_4	1,2-$(NH_2)_2C_6H_4$	65	70CB766
H, Ph	1,2-$(NH_2)_2C_6H_4$	67	78JPR335
H, 2-ClC_6H_4	1,2-$(NH_2)_2C_6H_4$	55	78JPR335
H, PhCO	1,2-$(NH_2)_2C_6H_4$	78	78JPR335
H, $PhCH_2$	1,2-$(NH_2)_2C_6H_4$	57	78JPR335
Me_2CH	triphenylguanidine	38	70ZC30
c-C_6H_{11}	triphenylguanidine	15	70ZC30

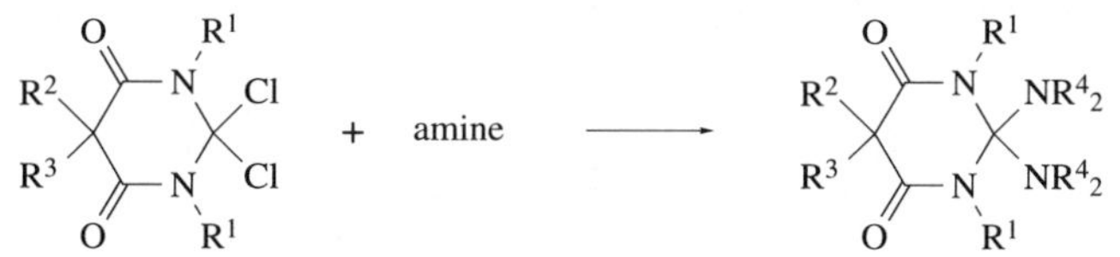

(3)

Table 3 Nucleophilic exchange of chlorine atoms with amines on 2,2-dichloro-2,6-dihydro-4,6-pyrimidinediones.

R¹, R², R³	*Reagents*	*Conditions*	*Yield (%)*
c-C_6H_{11}; Cl; Cl	pyrazole	THF, 25°C	79
c-C_6H_{11}; Cl; Cl	phenoxazine	THF, 25°C	36
c-C_6H_{11}; Cl; Cl	$(PhNHCH_2)_2$	THF, 25°C	44
c-C_6H_{11}; Cl; Cl	4-Ph-1,2-$(NH_2)_2C_6H_3$	THF, 25°C	36
Ph; Ph; Cl	phenoxazine	THF, 25°C	78
Ph; Ph; Cl	1,2-$(NH_2)_2C_6H_4$	THF, 25°C	23
Me_2CH; Cl; Cl	tetramethyl-2,2′-diimidazole	THF, Et_3N, 25°C	34
c-C_6H_{11}, Me, Me	$(PhNHCH_2)_2$	THF, 25°C	44
c-C_6H_{11}, Me, Me	homopiperazine	THF, K_2CO_3, 25°C	18

(b) From 2-chloro,2-aminoiminium salts. Tetrakis(dimethylamino)methane, the first tetraaminomethane described in the literature, was synthesized in 70% yield by allowing tetramethylchloroformamidinium chloride to react with lithium dimethylamide in benzene at room temperature (Equation (4)) ⟨66JA2885⟩. Homologous tetraaminomethanes, prepared analogously, have been patented as aminating agents, herbicides, insecticides, bactericides, and catalysts for urethane polymerization ⟨70USP3551418⟩.

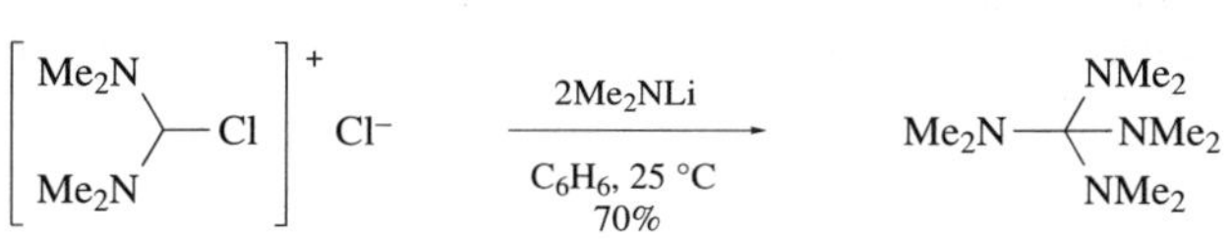

(4)

The preparation of a dibenzoanellated tetraaminomethane from 2-chloro-1,3-dimethyl-benzoimidazolium tetrafluoroborate and *N,N*-dimethyl-*ortho*-phenylenediamine has been reported (Equation (5)) ⟨68CB1137⟩.

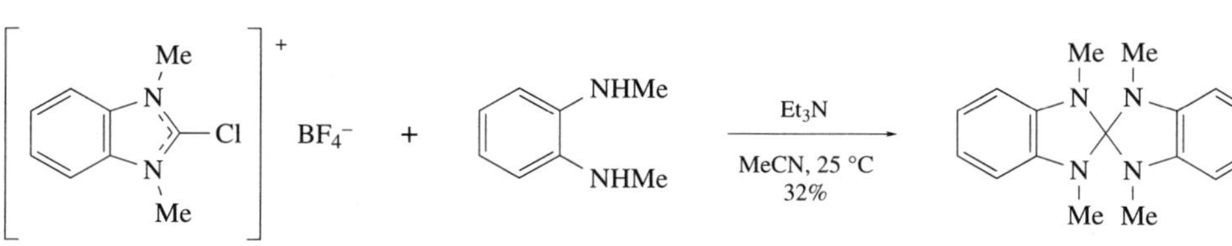

(5)

This synthetic methodology has been extended to the synthesis of the tetraazamethanes indicated in Scheme 1. The properties of these compounds were subsequently investigated by photoelectron spectroscopy ⟨86JOC370⟩.

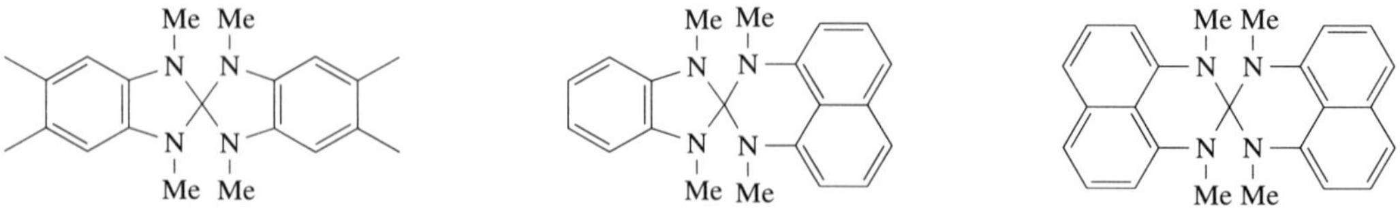

Scheme 1

(c) Tetrakis(1-pyrazolyl)methane from CCl_4. Tetrakis(1-pyrazolyl)methanes have been widely investigated as polydentate ligands in transition metal organometallic complexes. The parent compound was first synthesized in low yield (12%) from carbon tetrachloride and an alkali-metal pyrazolide ⟨70JA5118⟩. The conversion was improved slightly (20%) under solid–liquid phase transfer catalysis conditions (Equation (6)) ⟨84OPP299⟩. Nonetheless, these syntheses remain less efficient than a two-step procedure starting from phosgene and sodium pyrazolide salts (see Section 6.12.1.1.1.(v)).

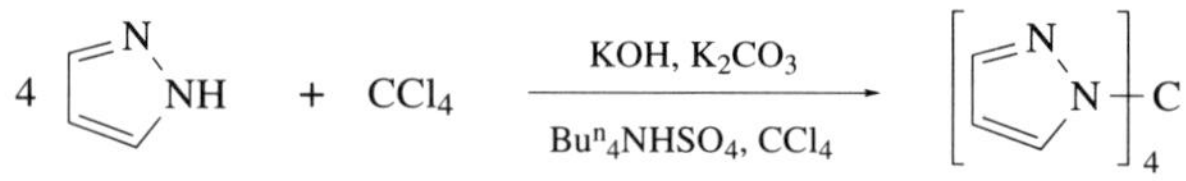

(6)

(d) Tetrakis(difluorosulfoximidoyl)methane from CCl_4. The title compound may be prepared directly from CBr_4 and the corresponding organomercury reagent, but the reaction gives a number of products and yields are poor: the desired compound was only detected by mass spectroscopy. A more practical synthesis involves the interaction of carbon tetrachloride with bis(difluorosulfoximidoyl)mercury and treatment of the resulting intermediate with tris(difluorosulfoximidoyl)borane. Nonetheless, the yield remains low (6%) (Scheme 2) ⟨79CB1189⟩.

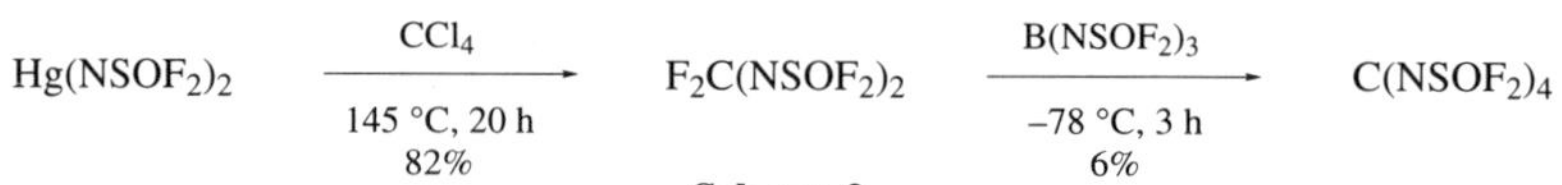

Scheme 2

(ii) By addition to carbon–nitrogen double bonds

A wide variety of NH functionalities add to activated carbon–nitrogen double bonds. Thus, ammonia, 4-cyanoaniline, 4-trifluoromethylaniline, hydrazine, trifluoroacetylhydrazine and succinimide react with pentafluoroguanidine at low temperatures to give the corresponding tetraazamethanes (Scheme 3) ⟨73JOC1075⟩. In the case of relatively basic amines, such as NH_3, Bu^nNH_2 and Me_2NH, spontaneous elimination of HNF_2 occurs below room temperature, to give a polyfluoroguanidine having a different substitution pattern. *These reactions are dangerous, and should be performed in polar solvents in order to reduce the risks of explosion: suitable protective equipment should be used during all phases of work.*

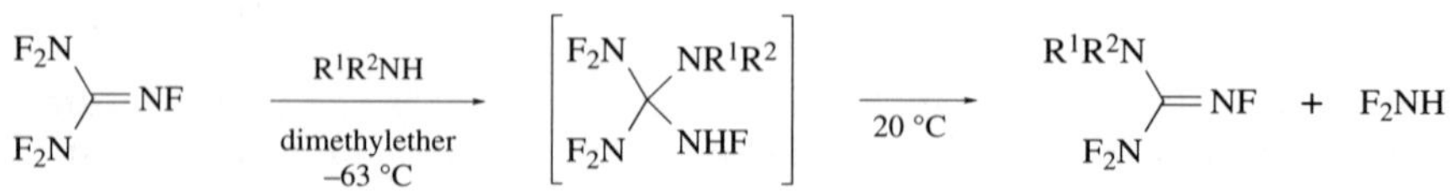

Scheme 3

The first-formed tetraaminomethanes (Scheme 3) are useful synthetic intermediates. Bis(difluoroamino)(fluoroamino)aminomethane may be treated with elemental fluorine at low temperature to give, amongst other products, the highly explosive tetrakis(difluoroamino)methane ⟨72USP3689560, 72USP3663621, 73JOC1075⟩, $C(F_2N)_4$ which has been patented as a rocket-propellant oxidizer (Equation (7)) ⟨72USP3699094, 73USP3755404⟩.

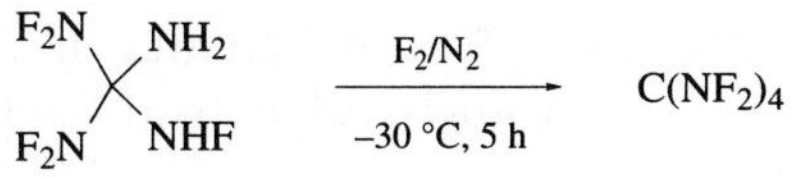
(7)

Similarly, ammonia adds to [bis(difluoroamino)fluoromethylazo]trifluoroformamidine giving an unstable adduct, which undergoes low temperature fluorination. Fluoropentakis(difluoroamino) azomethane is formed in low yield by this method (Scheme 4) ⟨72IC418⟩.

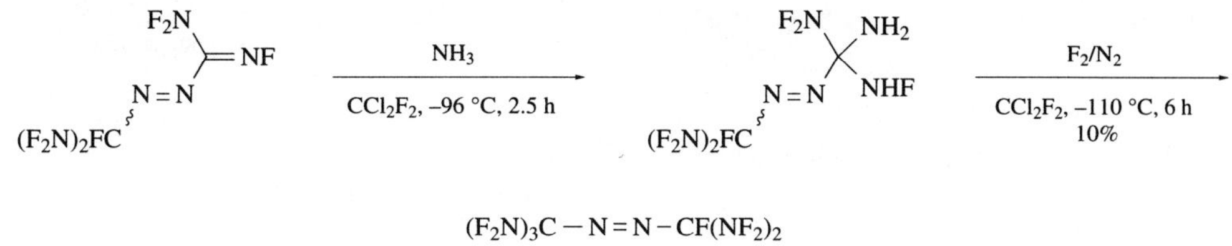

Scheme 4

Pentafluoroguanidine may be subjected to an analogous addition of isocyanic acid, which gives bis(difluoroamino)fluoroaminomethyl isocyanate. *This isocyanate and the other derivatives described in this section are reported to be very powerful explosives which are extremely sensitive to impact, friction, and perhaps temperature changes. Quantities as small as* 100 *mg should be regarded as dangerous.* The first formed isocyanate product may react with a further equivalent of isocyanic acid to give the cyclic derivative shown in Equation (8) ⟨73JOC1080⟩.

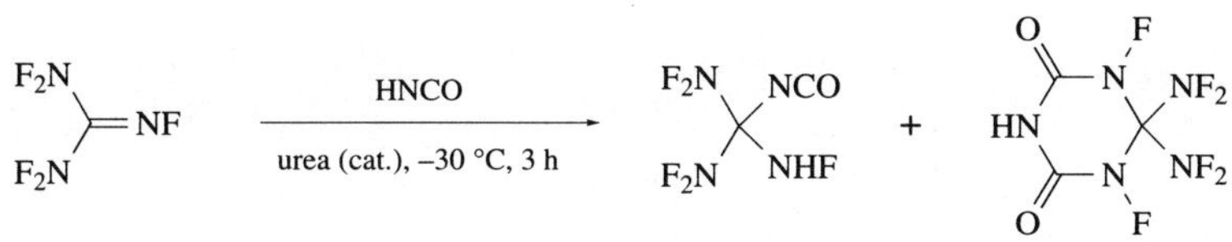
(8)

Other tetraazamethanes may be obtained from bis(difluoroamino)fluoroaminomethyl isocyanate. Treatment with elemental fluorine alone permits substitution of hydrogen by fluorine without a competing reaction at the isocyanate moiety. However, fluorination in the presence of sodium fluoride gives tetrakis(difluoroamino)methane (92%) (Scheme 5) ⟨73JOC1088, 73USP3733360⟩.

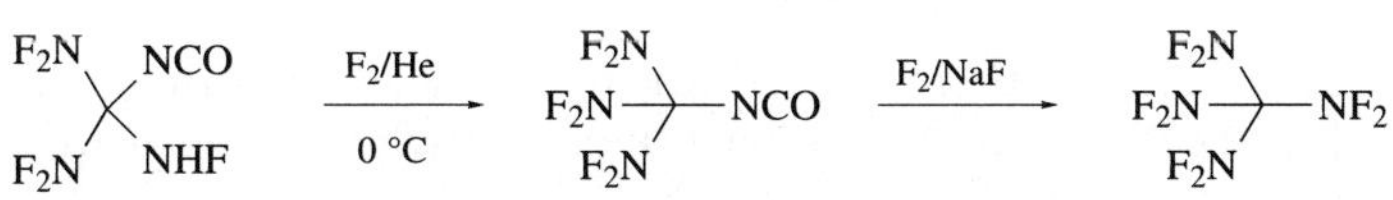

Scheme 5

Addition of ethanol to bis(difluoroamino)fluoroaminomethyl isocyanate results in the transformation of the NCO function into the corresponding ethylcarbamate ⟨73JOC1080⟩. Tris (difluoroamino)methyl isocyanate also undergoes addition of alcohols, water and amines at the isocyanate function (Table 4) ⟨73JOC1083⟩.

Table 4 Addition of water, alcohols and amines at the isocyanate function.

Isocyanate	*Reagents and conditions*	*Products*	*Yield* (%)
$(F_2N)_2(HNF)C(NCO)$	EtOH, $-196/25\,^\circ C$	$(F_2N)_2(NHF)C(NHCO_2Et)$	
$(F_2N)_3C(NCO)$	$HOCH_2CH_2OH$, $25\,^\circ C$	$((F_2N)_3CNHCO_2CH_2)_2$	70
$(F_2N)_3C(NCO)$	$H_2C{=}CHCH_2OH$, $25\,^\circ C$	$(F_2N)_3CNHCO_2CH_2CH{=}CH_2$	91
$(F_2N)_3C(NCO)$	ether, NH_3, $-196/25\,^\circ C$	$(F_2N)_3CNHCONH_2$	73
$(F_2N)_3C(NCO)$	ether, $0.3NH_3$, $-196/25\,^\circ C$	$((F_2N)_3CNHCO)_2NH$	50
$(F_2N)_3C(NCO)$	$(H_2N)_2$, MeCN, $-196/25\,^\circ C$	$((F_2N)_3CNHCONH)_2$	51
$(F_2N)_3C(NCO)$	H_2O	$(F_2N)_3CNH_2$	
$(F_2N)_3C(NCO)$	$(F_2N)_3CNH_2/Ph_3PO$	$((F_2N)_3CNH)_2CO$	

Although most of the compounds described in this section are high explosives, some related methodologies have been applied to more classical synthetic procedures.

Thus, 1,3-bis(methoxycarbonyl)-*S*-methyl isothiourea (BMMTU), a widely used reagent for the preparation of benzimidazole-2-carbamates from *ortho*-phenylenediamines, reacts in slightly acid media with 2-aminobenzamides, to give 2,2-biscarbamates in yields ranging from 56% to 75% (Scheme 6). This reaction permits the synthesis of potential medicinal products, fungicides and antiparasitic agents ⟨90S151⟩.

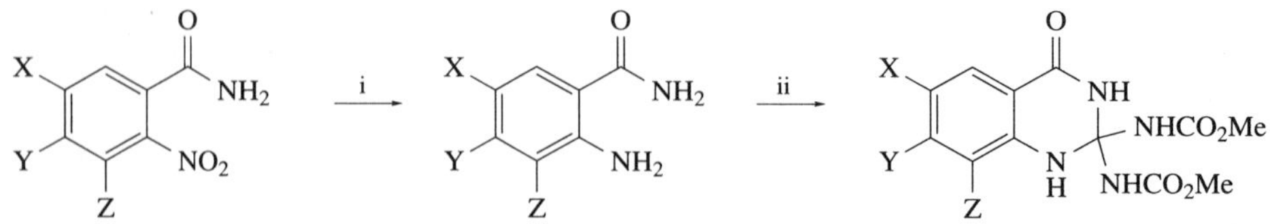

i, Raney - Ni, N_2H_4, MeOH, 25 °C, 3 h

ii, $MeO_2CN{=}C(SMe)NHCO_2Me$, MeOH, $MeCO_2H$, reflux, 2 h

X	*Y*	*Z*	*Yield* (%)
H	H	H	61
OMe	OMe	H	75
OMe	OMe	OMe	56

Scheme 6

The reaction of two equivalents of isocyanate with pentasubstituted guanidines gives a cycloadduct having an *s*-triazine structure. This reaction does not appear to be completely general. Thus, pentamethylguanidine reacts with both phenyl and methyl isocyanate, but *N*-phenyltetramethylguanidine undergoes reaction with methyl isocyanate but not with phenyl isocyanate. With *N*-phenyltetramethylguanidine and an equimolar mixture of the two isocyanates, an unsymmetrically substituted *s*-triazine is produced (Equation (9)) ⟨68TL5037⟩.

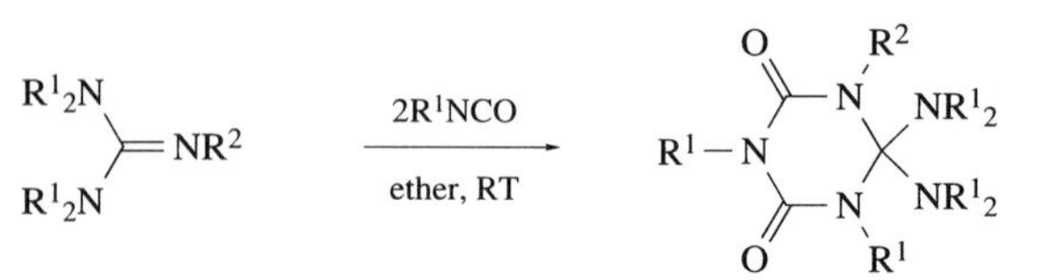

(9)

R^1, R^2 = Me, C_6H_5

(iii) By addition to nitrogen–nitrogen double bonds

Azodicarboxylate esters react either with trinitromethane in dioxane ⟨75IZV2838⟩, or with its mercury salts in water ⟨77IZV1563⟩, to give (trinitromethyl)hydrazine derivatives. Nitroform reacts analogously with azocarboxamides (Scheme 7).

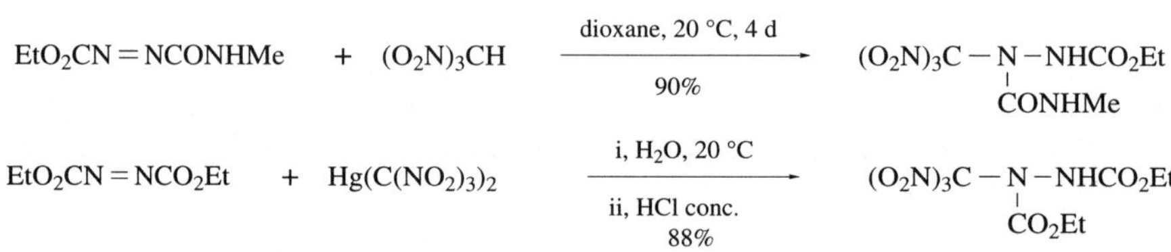

$$EtO_2CN{=}NCONHMe + (O_2N)_3CH \xrightarrow[90\%]{\text{dioxane, 20 °C, 4 d}} (O_2N)_3C{-}N(CONHMe){-}NHCO_2Et$$

$$EtO_2CN{=}NCO_2Et + Hg(C(NO_2)_3)_2 \xrightarrow[\text{ii, HCl conc. } 88\%]{\text{i, } H_2O\text{, 20 °C}} (O_2N)_3C{-}N(CO_2Et){-}NHCO_2Et$$

Scheme 7

(iv) From carbanions bearing two nitrogen substituents

Ammonium trinitromethanide reacts readily with nitryl fluoride to form tetranitromethane in excellent yield. Trinitromethane itself is ionized sufficiently in acetonitrile to interact directly with nitryl fluoride but the yield is slightly poorer in this case (Equation (10)) ⟨71IZV1594, 74IZV915⟩.

$$(O_2N)_3C^- NH_4^+ \xrightarrow[\text{MeCN, } 90\%]{NO_2F} C(NO_2)_4 \qquad (10)$$

Reaction of bis(benzotriazolyl)methane with *n*-butyllithium and tosyl azide gave the *gem*-diazido compound in moderate yield: the expected monoazido compound appears to be more reactive than the starting material. A further Staudinger phosphorylation reaction with triphenylphosphine gave a product, presumed to be the corresponding bis(triphenylphosphoranylidene) compound, which was too unstable to be isolated (Scheme 8) ⟨90JCR(S)330⟩.

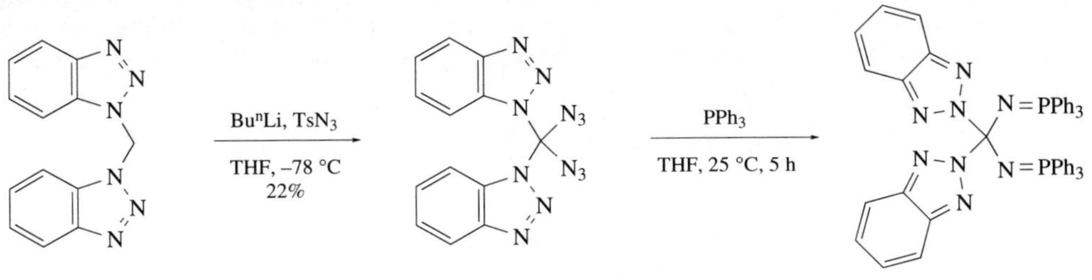

Scheme 8

(v) Miscellaneous syntheses

This section comprises a number of unrelated synthesis of which two are particularly important. The first constitutes the most effective preparation of tetranitromethane, which is obtained in approximately 60% yield by the reaction of anhydrous nitric acid with acetic anhydride below 10 °C (Equation (11)) ⟨55OSC(3)803⟩. The product is a high explosive, and should be purified only by steam distillation. The synthesis has been adapted to the preparation of the labelled compound $C(NO_2{}^{15})_4$ in 40% overall yield from sodium nitrate ⟨90MI 612-01⟩.

$$4(MeCO)_2O + 4HNO_3 \xrightarrow[10\,°C]{} C(NO_2)_4 + 7MeCO_2H + CO_2 \quad (11)$$

The second important synthesis produces tetrakis(1-pyrazolyl)methanes, which have been prepared through a two-step process from the appropriate pyrazolide salt and phosgene. Thermolysis of the intermediate 1,1′carbonyldipyrazole at 190 °C in the presence of cobalt(II) chloride induces an autocondensation reaction which gives the tetrakis(1-pyrazolyl)methanes in a total yield of 50% (Scheme 9) ⟨73CJC2448⟩.

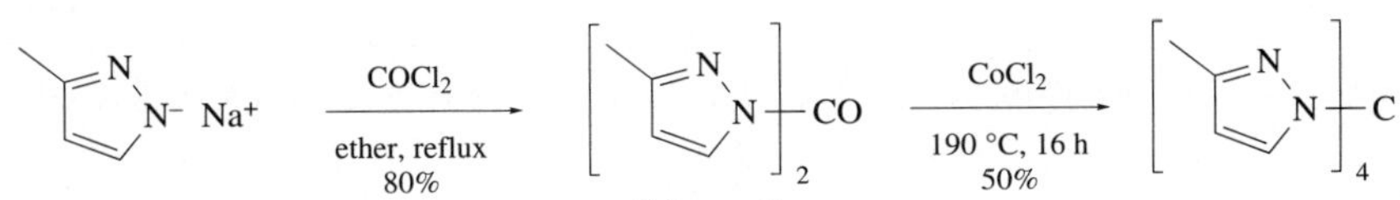

Scheme 9

Interaction of 1,4,7,10-tetraazacyclododecane and ethyl orthocarbonate in acidic ethanol gives solutions which contain a tricyclic guanidinium ion-amino compound: the reaction proceeds in nearly quantitative yield. The corresponding covalent tetracyclic tetraazatridecane may be obtained by neutralization, extraction into benzene, and subsequent drying by azeotropic distillation (Scheme 10). The conformation of the two macrocycles, and the equilibrium which relates them, was studied by NMR spectroscopy ⟨74T1769⟩.

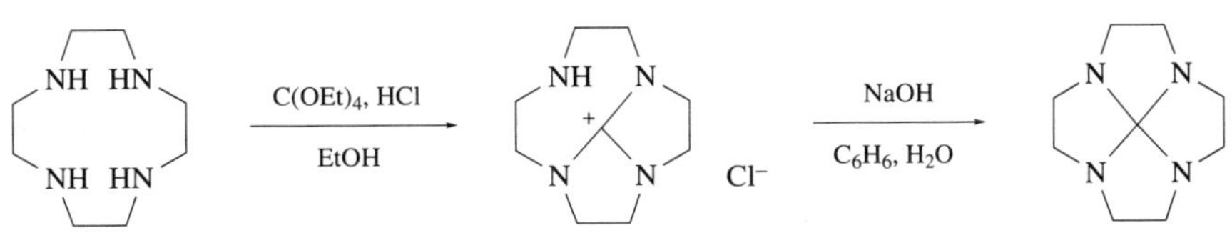

Scheme 10

6.12.1.1.2 *Four phosphorus functions*

Very few papers dealing with the synthesis of this class of compounds have appeared. Nonetheless, a number of patents describing the use of methanetetraphosphonic acid and its derivatives in detergents ⟨75USP3892676⟩ or as chelating agents for heavy metal ions ⟨84USP4440646⟩ have been disclosed.

Chlorodimethylphosphine displaces all four chlorine atoms of carbon tetrachloride, to give tetrakis(dichlorodimethylphosphoranyl)methane in 82% yield. The pentacoordinated phosphorus derivative was converted into the corresponding oxide upon basic hydrolysis (80% yield) (Scheme 11). The reaction is not general and fails for reagents such as PCl_3 and $MePCl_2$, presumably because of their lower electron density at the phosphorus atom ⟨67ZOB2513⟩.

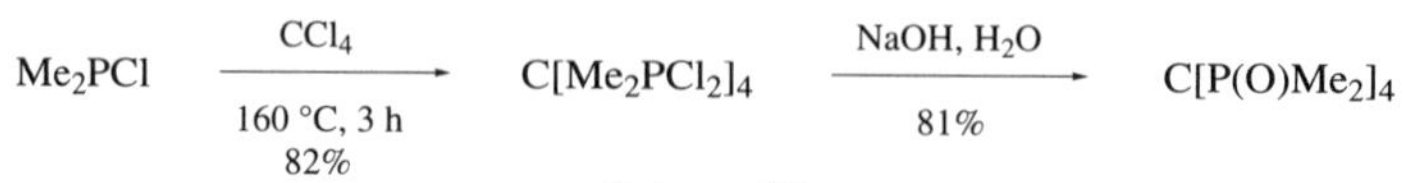

Scheme 11

The only spiro[2.2]tetraphosphapentane which is known to date was prepared by the reaction of carbon tetrachloride with 1,2-dipotassium di-*tert*-butyldiphosphide, in 2% yield after chromatography on alumina. Of the two *anti*-isomers which are observed, the sterically more hindered one converts slowly into the favored isomer at room temperature (Scheme 12) ⟨83AG(E)632⟩.

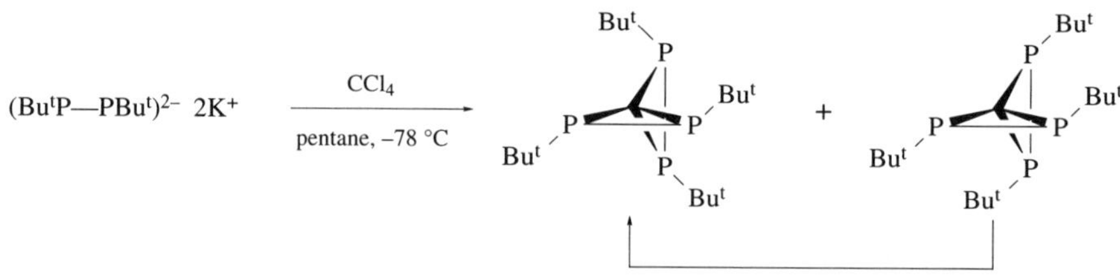

Scheme 12

6.12.1.2 Three Similar and One Different Group 15 Element Functions

6.12.1.2.1 Three nitrogen functions

Only two compounds bearing three nitrogen functions are known: both are cyclic compounds containing three nitrogens and a phosphorus atom. Both syntheses involve the participation of a phosphorus substituted C—N double bond, and in each case two of the nitrogen atoms are included in the ring.

In the first case, the interaction of diethyl phosphite with diphenylcarbodiimide gave an intermediate amidinophosphinic acid which underwent a [2 + 2] cycloaddition reaction with phenyl isocyanate. The final product is obtained in 65% yield (Scheme 13) ⟨81ZOB1035⟩.

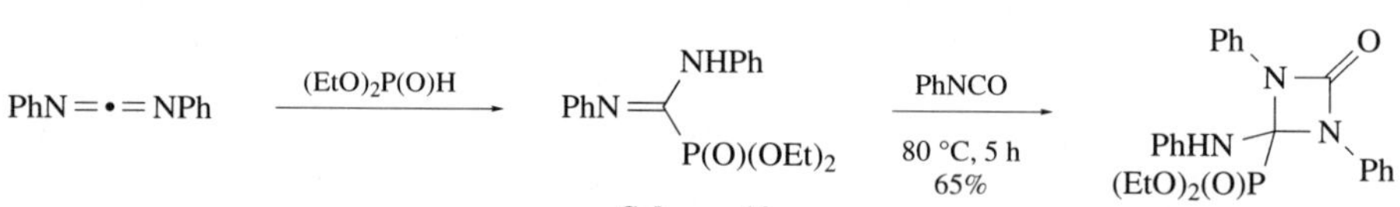

Scheme 13

In the second case, two equivalents of *C*-phosphorylated *C*-chlorohydrazones are reacted with the potassium salt of alanine. Both the amino and the carboxylate groups act as nucleophiles resulting in substitution of the chlorine atom of the phosphorylated substrate. The linear product is converted, through the steric pressure, into the cyclic tautomer by addition of the N—H function to the C=N double bond (Scheme 14) ⟨90ZOB1980⟩.

6.12.1.2.2 Three phosphorus functions

The only representatives of this class of compound bear one nitrogen and three phosphorus functions. All the synthetic methods involve at least one Michaelis–Arbuzov type reaction as an intermediate step.

Reactions between triethyl phosphite and trichloromethylamines or trichloromethyl isocyanate afford tris(diethoxyphosphinyl)methylamines or isocyanates, respectively (Equation (12) and Table 5).

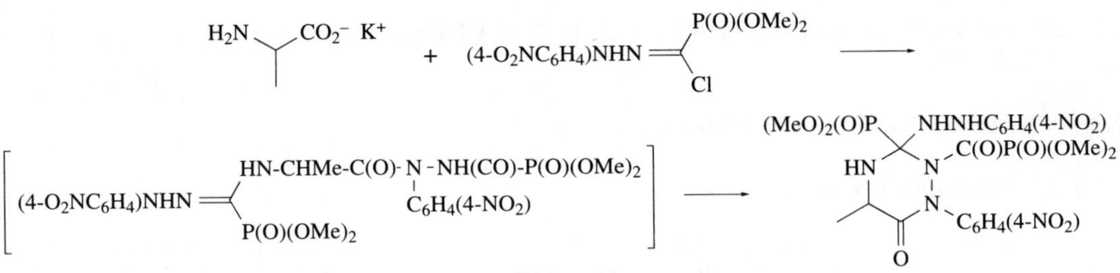

Scheme 14

$$3(EtO)_3P \; + \; ZCX_3 \; \longrightarrow \; ZC[P(O)(OEt)_2]_3 \qquad (12)$$

Table 5 Reactions of triethyl phosphite with trichloromethylamines or isocyanate.

ZCX_3	*Conditions*	*Yield* (%)	*Ref.*
$(Me_2N)CCl_3$	CH_2Cl_2, −40°C, 1 h	71	72ZOB1169
(1-morpholino)CCl_3	CH_2Cl_2, −40°C, 1 h	77	72ZOB1169
$(OCN)CCl_3$	toluene, 90°C, 1.5 h	56	73ZOB544
$(OCN)CBr_3$	toluene, 90°C, 1.5 h	23	89ZOB571
$(Me_2N)CCl_3$	CH_2Cl_2, −80/25°C		74GEP2237879

Unsymmetrically substituted tris(phosphoryl)methylanilines were obtained transiently through a two-step synthesis starting from *N*-phenylcarbonimidic dichloride. An initial Michaelis–Arbuzov reaction between a trialkyl phosphite and the dichloroimine led to the corresponding bis(phosphoryl)imine which was then reacted with various $R_2P(O)H$ compounds: addition of the P—H function to the C═N double bond afforded the tris(phosphoryl)methylanilines. These compounds, which were originally thought to be stable ⟨72JPR87⟩, rearrange spontaneously by migration of one of the phosphoryl groups from the carbon to the nitrogen atom (Scheme 15) ⟨72JPR969, 86JPR231⟩.

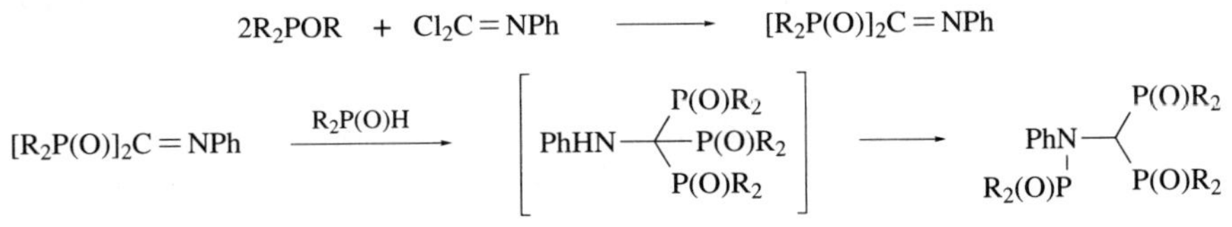

Scheme 15

6.12.1.3 Two Similar and Two Different Group 15 Element Functions

6.12.1.3.1 Two nitrogen functions

Only one report of compounds bearing two nitrogen functions exists.

Treatment of a benzene solution of amidines with chlorobis(diethoxyphosphonyl)methyl isocyanate in the presence of triethylamine readily yields cyclic *s*-triazin-2(1*H*)-ones (Equation (13)) ⟨75ZOB2093⟩.

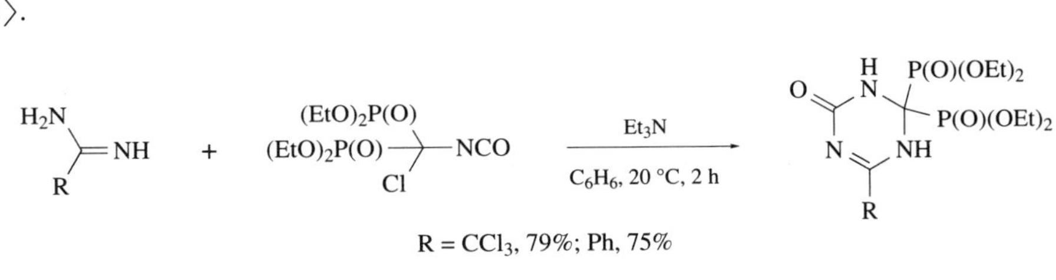

(13)

6.12.1.3.2 Two phosphorus functions

See Section 6.12.1.3.1.

6.12.2 METHANES BEARING THREE GROUP 15 ELEMENTS AND A METALLOID OR A METAL

6.12.2.1 Three Similar Group 15 Elements

6.12.2.1.1 Three nitrogen functions

Compounds bearing three nitrogen functions essentially comprise derivatives of trinitromethane. In cases where they are not explicitly described as stable, they are probably best treated as explosive and should be handled with care.

The most widely studied reagent of this class is bis(trinitromethyl)mercury which may be easily prepared, in 80% yield, by reaction of HgO with trinitromethane in ether at room temperature (Equation (14)) ⟨60IZV505⟩.

$$HC(NO_2)_3 + HgO \xrightarrow[15\ \text{min}]{\text{ether}} Hg[C(NO_2)_3]_2 \quad (14)$$

The interaction of this reagent with a wide range of ethers, aliphatic and aromatic nitro compounds, nitramines and halocarbons has been investigated. The molecular complexes which are formed have been studied by IR and NMR spectroscopy. A number of lead references may be consulted ⟨68IZV2839, 69IZV2304, 72MI 612-01⟩.

Bis(trinitromethyl)mercury undergoes metathesis reactions with R_2Hg species to afford the mixed derivatives $RHgC(NO_2)_3$ (Equation (15)). In the case where R = Bu^n, the direct reaction of trinitromethane with dibutylmercury in ethanol leads to the same product ⟨67JOM(9)5⟩.

$$HgR_2 + Hg[C(NO_2)_3]_2 \xrightarrow[\text{reflux, 6 h}]{\text{EtOH}} RHgC(NO_2)_3 \quad (15)$$

R = Ph, 86%; Bu^n, 50%

The related aryl(trinitromethyl)mercury may also be prepared from diphenylmercury and tetranitromethane in a wide range of solvents such as MeCN, sulfolane and chloroform, probably by a radical mechanism (Equation (16)) ⟨74ZOR1793⟩.

$$HgPh_2 + C(NO_2)_4 \xrightarrow[80\ °C,\ 10\ \text{h};\ 46\%]{\text{sulfolane}} PhHgC(NO_2)_3 \quad (16)$$

Bis(trinitromethyl)mercury reacts with either piperidine or chlorodialkylamines to give (trinitromethyl)mercuric amides in very high yields (Equation (17)) ⟨67IZV2708⟩.

$$Hg[C(NO_2)_3]_2 + R_2NX \xrightarrow[0\ °C,\ 3\ \text{h}]{\text{ether}} R_2NHgC(NO_2)_3 \quad (17)$$

R_2NX = *N*-chloropiperidine, 92%; *N*-chlorodimethylamine, 91%; piperidine (2 equiv.), 30%

The adduct of 2,2′-azobis(isobutyronitrile) with $Hg(C(NO_2)_3)_2$ reacts as a mercurating agent towards 2,4-dioxypentane to afford 3-(2,4-dioxopentyl)trinitromethylmercury (Equation (18)) ⟨77IZV1563⟩.

$$Hg[C(NO_2)_3]_2 + MeCOCH_2COMe \xrightarrow{\text{AIBN}} (MeCO)_2CHHgC(NO_2)_3 \quad (18)$$

Insertion of alkenes into the C—Hg bond of bis(trinitromethyl)mercury has been studied. Cyclohexene undergoes *trans*-addition whilst norbornene gives an *exo-cis* product, as found in the case of more classical mercuration reactions (Equation (19)) ⟨68IZV1638⟩.

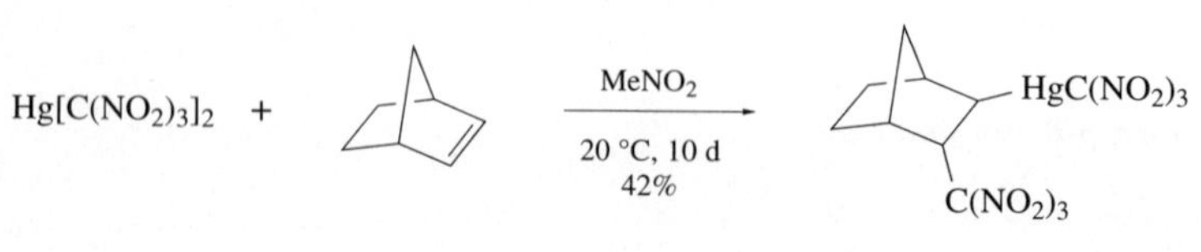

(19)

A similar addition to 3,5-dinitrocyclopentene has been reported ⟨70IZV2108⟩ but the stereochemistry of the addition was not defined.

The reaction with allyltrimethylsilane is relatively complex: after the first mercuration, the elimination of trimethylsilane generates a new site of unsaturation which undergoes a second addition of the organomercury reagent (Equation (20)) ⟨73RZC1243⟩.

(20)

Several metal salts of trinitromethane, $M^{+}\,^{-}C(NO_2)_3$, have been prepared: their stability depends quite strongly upon the counterion. Alkaline metals and ammonium salts are reported to decompose exothermically, without explosion, upon heating ⟨77IZV1965⟩. The specific impulse properties of compounds such as $Me_2AlC(NO_2)_3$ and $LiC(NO_2)_3$ have been measured ⟨71USP3562309⟩, and $Me_3SiC(NO_2)_3$ decomposes violently ⟨93TL1859⟩. The alkali metal salts (K, Rb, Cs) and ammonium compounds of the trinitromethanide anion may be prepared by reaction of trinitromethane with the corresponding hydroxide ⟨77IZV1965⟩, whilst the Al, Mg, B, Be and Li salts were prepared from the metal alkyls and a stoichiometric amount of $HC(NO_2)_3$ or $BrC(NO_2)_3$ in pentane ⟨71USP3562309⟩. The silver salt may be prepared *in situ* from $HC(NO_2)_3$ and silver oxide in MeCN ⟨63T(S)177⟩. It reacts with chlorodiphenylstibine to afford diphenyl(trinitromethyl)stibine (75–80% yield), which is also accessible less efficiently in 30–50% yield from the potassium salt (Equation (21)) ⟨85IZV439⟩.

$$MC(NO_2)_3 + Ph_2SbCl \xrightarrow[20\ ^\circ C,\ 4\ h]{CH_2Cl_2} Ph_2SbC(NO_2)_3 \qquad (21)$$

$$M = K, Ag$$

Finally, two potassium salts of a (dinitromethyl)hydrazine, $K^{+}\,^{-}C(NO_2)_2N(COR)NHCOR$, have been very briefly described, but no synthetic details were given ⟨75IZV2838⟩.

6.12.2.1.2 Three phosphorus functions

A wide variety of methanes having a lithium or sodium atom and three phosphorus substituents have been prepared as anionic tripod ligands for coordination chemistry. Variants having tri- and pentavalent phosphorus atoms are known, and syntheses have been devised which permit the targeting of ligands containing any desired combination of oxygen and sulfur-bound phosphorus atoms. The synthesis of the alkali-metal derivatives has been accomplished by proton abstraction from the parent methane, HtrisXYZ, with various reagents. Selected examples are given in Table 6 (Equation (22)). The lithium salts containing phosphorus(V) atoms are fairly air stable and can be stored for several months without change. An x-ray crystal structure determination of $LiC(PMe_2)_3$ shows that the lithium atom is bound to both of the phosphorus lone pairs and to the carbon atom ⟨84CC569⟩. This implies that the charge is delocalized over the phosphorus and the carbon atoms and indicates that reactivity towards electrophiles is complex.

(22)

Table 6 Metalation reactions of HtrisXYZ.

HTrisXYZ	*Reagents and conditions*	*Yield* (%)	*Ref.*
$HC[P(O)Ph_2]_3$	LiOMe, CH_2Cl_2, 25 °C, 5 h	78	86IC2699
$HC[P(O)Ph_2]_3$	NaH, THF		90JA7984
$HC[P(O)Ph_2]_2[P(S)Ph_2]$	LiOMe, CH_2Cl_2, 25 °C, 5 h	82	86IC2699
$HC[P(O)Ph_2][P(S)Ph_2]_2$	LiOMe, CH_2Cl_2, 25 °C, 5 h	73	86IC2699
$HC[P(S)Ph_2]_3$	LiOMe, CH_2Cl_2, 25 °C, 5 h	72	86IC2699, 82CC930
$HC[P(S)Me_2]_3$	Bu^{t}Li, THF, −20/25 °C, 5 h	76	82CB818
$HC(PMe_2)_3$	Bu^{t}Li, pentane, 25 °C, 10 h	73	79ZN(B)1178
$HC[PPh_2]_3$	LiOMe, MeOH	>66	86IC2699

The analogous Hg "trisS_3" derivatives are prepared directly by reaction of the parent methane "HtrisS_3" with the corresponding metal halides in ethanol solution: the cadmium analogues may be prepared similarly in the presence of a mild base such as triethylamine (Equation (23)) ⟨85IC2889⟩. In the solid state, the complexes are bound through the sulfur atoms, but there is some circumstantial evidence for coordination to carbon in solution.

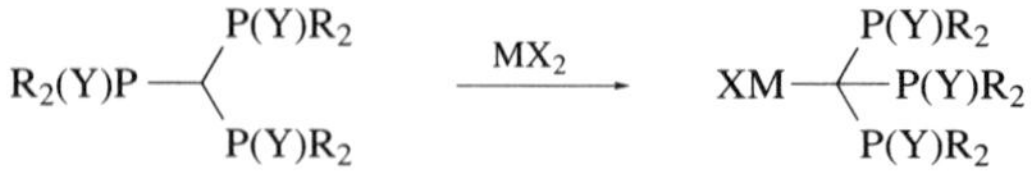

(23)

M	*3P(Y)R$_2$*	*X*	*Yield* (%)
Hg	3P(S)Ph$_2$	Cl	92
Hg	P(S)Ph$_2$[P(S)Me$_2$]$_2$	Cl	85
Cd	3P(S)Ph$_2$	Cl	94
Cd	P(S)Ph$_2$[P(S)Me$_2$]$_2$	Cl	56

6.12.3 METHANES BEARING TWO GROUP 15 ELEMENTS AND METALLOID AND/OR METAL FUNCTIONS

6.12.3.1 Two Similar Group 15 Elements

6.12.3.1.1 Two nitrogen functions

There appears to be only one compound of this general class which has featured in the literature. Lithiation of *N*-(benzotriazol-1-ylmethyl) pyrrole with *n*-butyllithium and subsequent addition of chlorotrimethylsilane gave a mixture of products which included the bis(trimethylsilyl) derivative (Equation (24)) ⟨89JHC829⟩. The reported yield was relatively low (36%), but it appears that the method could be synthetically useful if two equivalents of *n*-butyllithium and chlorotrimethylsilane were used.

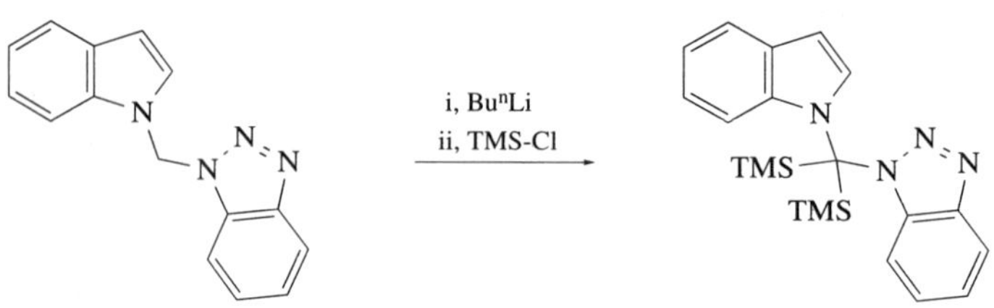

(24)

6.12.3.1.2 Two phosphorus functions

A few compounds in which a carbon atom bears two phosphorus and two silicon functions have been described. All have trimethylsilyl substituents, but the phosphorus moieties may be either phosphines (PR_2), phosphonium salts ($PR_3{}^+$), phosphorus ylides ($=PR_2$) or phosphaalkenes ($R_2C{=}P$).

The earliest report of such a compound concerned a bisphosphonium salt, prepared by addition of two equivalents of chlorotrimethylsilane to a symmetrical dimethyltetraphenylcarbo-diphosphorane (Equation (25)). This phosphonium salt is relatively unstable and decomposes within a few minutes ⟨76ZN(B)721⟩.

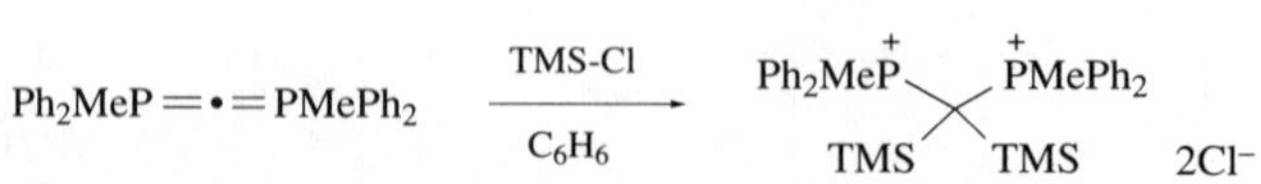

(25)

A second example involves the addition of a chlorophosphaalkene to a phenyl- or alkylnyllithium reagent, which gives a phosphinomethyl substituted phosphaalkene. It is proposed that the reaction proceeds through the intermediacy of a lithium bis(trimethylsilyl) phosphinomethanide which reacts with a second equivalent of the chlorophosphaalkene (Scheme 16) ⟨86CB2609⟩.

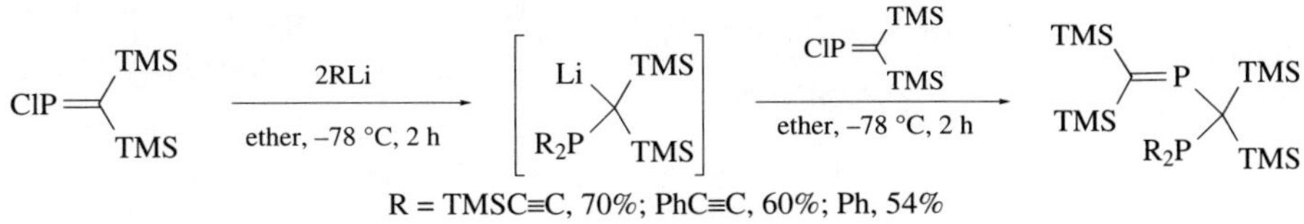

Scheme 16

In the case where R = TMS-C≡C, a subsequent pyrolysis gave a bis(trimethylsilyl)-substituted diphosphirane in 42% yield: this reaction proceeds through an intramolecular [2 + 2] cycloaddition involving the P—C double bond and one of the C—C triple bonds, and a subsequent rearrangement (Equation (26)) ⟨86CB2609⟩.

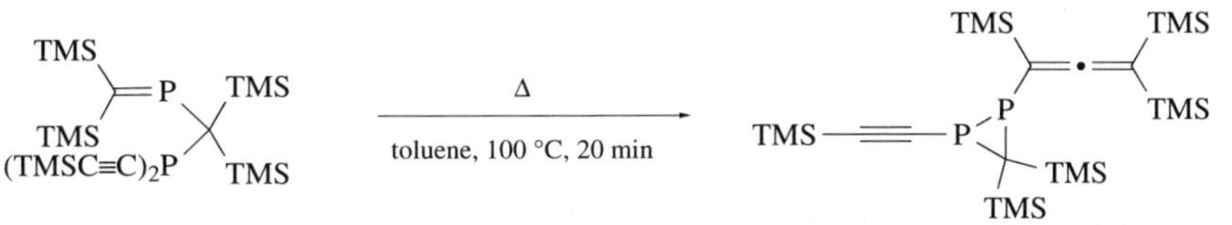 (26)

The last compound is a phosphino-phosphorane derivative, which is prepared in poor yield by means of a zirconium-mediated synthesis (Scheme 17) ⟨93OM2757⟩.

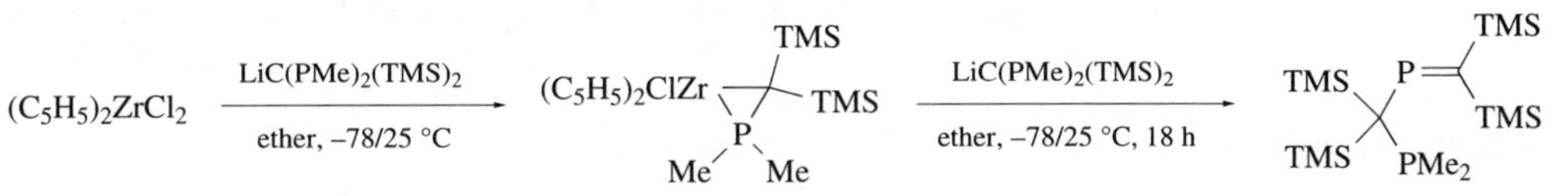

Scheme 17

Compounds containing two phosphorus, a silicon and a lithium function as substituents at the central carbon are also known. Bis(phosphino)methanides and a bis(thiophosphinyl)methanide have been prepared by metalation with *n*-butyllithium of the corresponding trisubstituted methanes (Equation (27)) ⟨88ZN(B)1416, 92CB1333⟩.

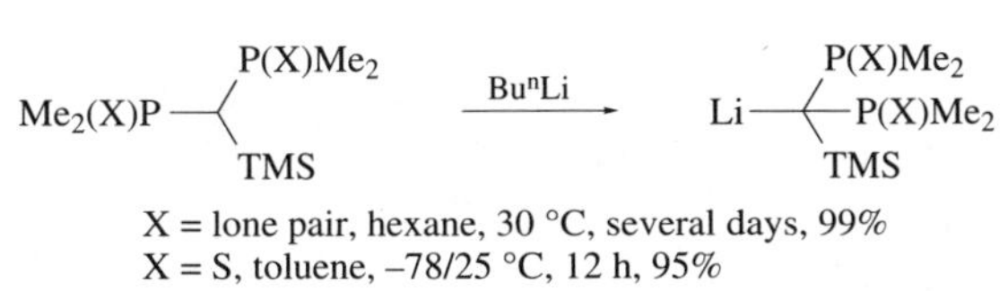 (27)

The mercuration of methylenediphosphonate compounds which contain an acidic hydrogen is known. Reaction of mercuric acetate with a methylenediphosphonate in refluxing tetrahydrofuran provokes bis(mercuration) at the carbon atom with the loss of acetic acid. The product undergoes redistribution reactions with other organomercurials (Scheme 18) ⟨73JOM(59)231⟩

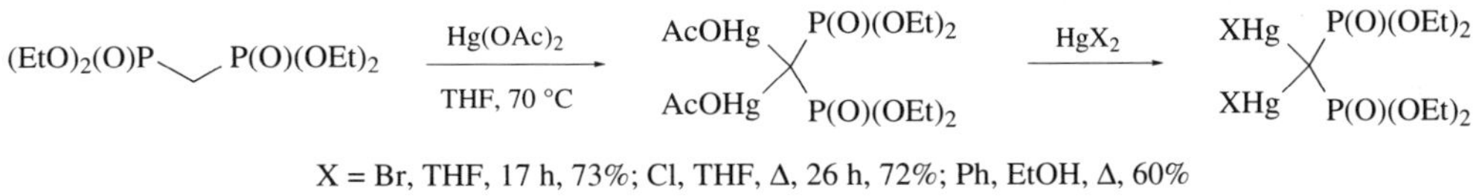

Scheme 18

Methylenediphosphines react in a similar fashion with a mixture of mercuric acetate and the DMSO adduct of mercuric triflate, $Hg(DMSO)_6(OSO_2CF_3)_2$. The first, monomercurated, product reacts further with mercuric acetate to give an unstable dimercurated compound which could not be isolated (Equation (28)) ⟨84OM1916⟩. It should be noted that such a mercuration reaction failed for a related tris(phosphino)methane ⟨86JOM(301)269⟩.

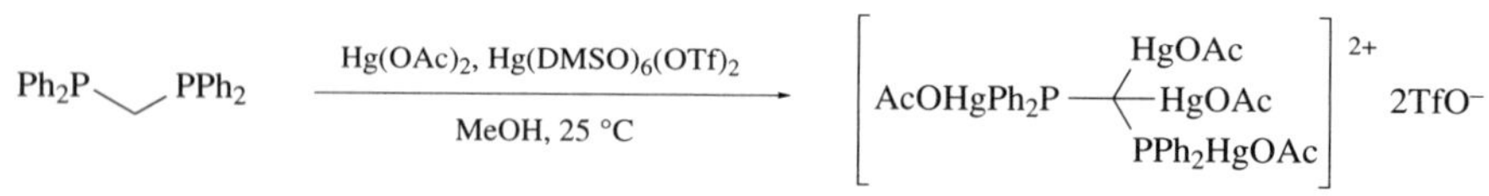

(28)

6.12.4 METHANES BEARING ONE GROUP 15 ELEMENT AND METALLOID AND/OR METAL FUNCTIONS

6.12.4.1 Nitrogen Functions

A computer search of the *Chemical Abstract Data Base* has not revealed any compound belonging to this subgroup.

6.12.4.2 Phosphorus Functions

6.12.4.2.1 One phosphorus and three silicon functions

The most important compounds bearing one phosphorus and three silicon functions are methanes which carry one phosphorus and three trimethylsilyl functions. Many of these compounds may be prepared directly from PCl_3, or the corresponding chlorophosphine, and tris(trimethylsilyl)methyllithium (Equation (29)) according to the conditions outlined in Table 7. The experimental procedure employed for this transformation appear to be quite critical to the success of the reaction. A similar series of reactions which involve the related reagent $(PhMe_2Si)_3CLi$ has been described: the reaction with PCl_3 proceeds in 61% yield ⟨89JOM(366)39⟩.

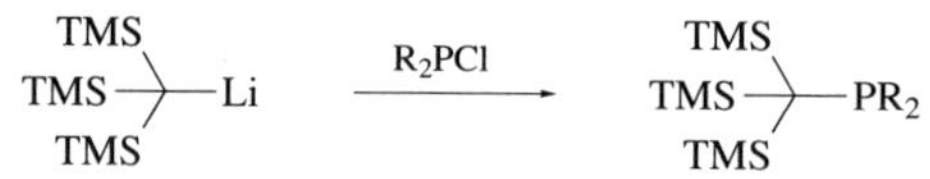

(29)

Table 7 Reactions of tris(trimethylsilyl)methyllithium with chlorophosphanes.

R_2PX	*Conditions*	*Yield (%)*	*Ref.*
PCl_3	THF, 0°C, 2 h	65	80ZC153
	THF-ether, 20°C, 2 h	65	89JOM(366)39
	THF, 0/25°C, 2 h	67	90IS235
$MePCl_2$, Bu^tPCl_2, Et_3CPCl_2, $PhPCl_2$, $TMSCH_2PCl_2$, 2,4,6-Me_3-$C_6H_2PCl_2$	THF, 66°C, 8 h	>70	80ZC419, 81ZAAC(473)85
Ph_2PCl	ether, 40°C, 2 h	48	83JCS(D)905

Treatment of tris(trimethylsilyl)methyllithium with (trimethylsilyl)aminodifluorophosphines causes the replacement of one of the fluorine atoms by tris(trimethylsilyl)methyl group. Small quantities of tris(trimethylsilyl)methyldifluorophosphine are also formed. Upon heating, these compounds are easily transformed in phosphaalkenes through elimination of fluorotrimethylsilane, presumably because of sterical pressure (Scheme 19) ⟨84ZN(B)1500⟩.

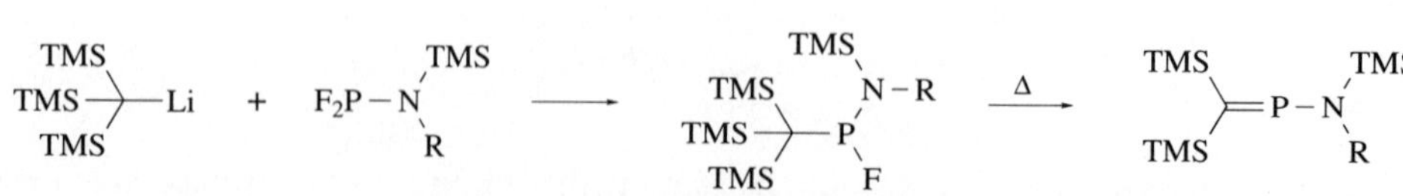

R = adamantyl, 46%; Bu^t, 79%; mesityl, 60%; TMS, 48%

Scheme 19

The reactivity of tris(trimethylsilyl)methyllithium towards electrophiles is complicated by exchange reactions. Thus, this lithium reagent reacts with phosphorus tribromide to give principally bromotris(trimethylsilyl)methane, and only traces of dibromo(tris(trimethylsilyl)methyl)phosphine. If required, dibromo(tris(trimethylsilyl)methyl)phosphine may be prepared efficiently by reaction of BBr_3 with the corresponding dichloro derivative. The compound $(TMS)_3CPCl_2$ has also been used for the synthesis of $(TMS)_3CPH_2$, $(TMS)_3CPH_2 \cdot BH_3$ and $(TMS)_3CP(O)Cl_2$ by reaction with $LiAlH_4$, $NaBH_4$ and $AgNO_3$ respectively ⟨89JOM(366)39⟩.

Reduction of $(TMS)_3CPCl_2$ with Bu^{t}Li, sodium naphthalenide or $(TMS)_3CLi$ gives the bis(tris(trimethylsilyl)methyl)diphosphene (Equation (30)) ⟨82TL4941, 82JA5820⟩.

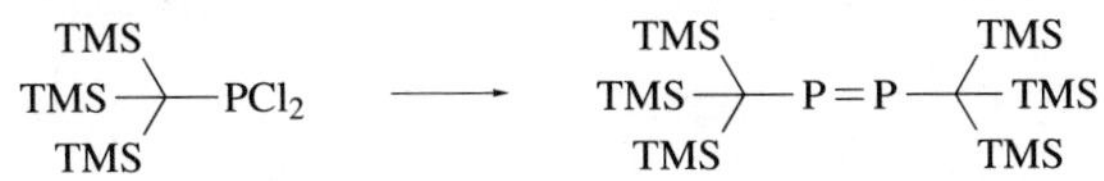

(30)

A similar interaction of $(TMS)_3CPCl_2$ with lithium trimethylsilylphosphides leads to silyl-substituted diphosphines. The triphenylsilyl derivative is stable but its trimethylsilyl analogue evolves to give a compound having a tetraphosphabicyclobutane structure (Scheme 20) ⟨89OM2490⟩.

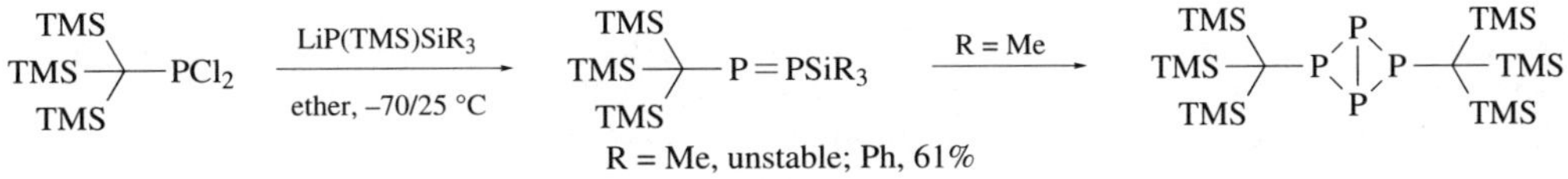

Scheme 20

The diphosphene $(TMS)_3CP{=}PC(TMS)_3$ undergoes further transformations at the phosphorus–phosphorus double bond. These include ozonolysis ⟨84CC1622⟩, [2 + 1] cycloadditions with carbenes ⟨89CC593⟩ or isonitriles ⟨92ZAAC(617)53⟩ and reactions with HCl ⟨86JCS(D)1801⟩: each of these transformations leaves the tetrasubstituted carbon atom unchanged.

A reaction producing a carbon atom bearing one phosphorus and three silicon functions through addition of a phosphorus–oxygen bond to a transient silaethene is known. The addition of $(PhO)_2P(O)OSiMe_3$ to the silicon carbon double bond is favored over a competing insertion of the C=Si function into the O—Si linkage (Scheme 21) ⟨81CB3505⟩.

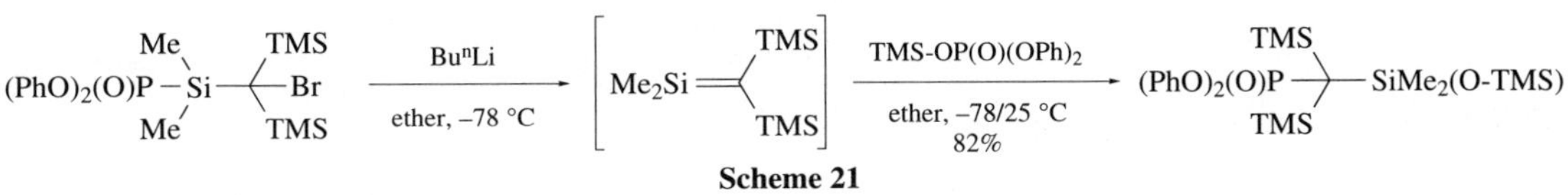

Scheme 21

Finally, the reaction of iodotrimethylsilane with bis(trimethylsilyl)methylene trimethylphosphorane leads to an iodophosphonium compound in 85% yield (Equation (31)) ⟨70CB3448⟩.

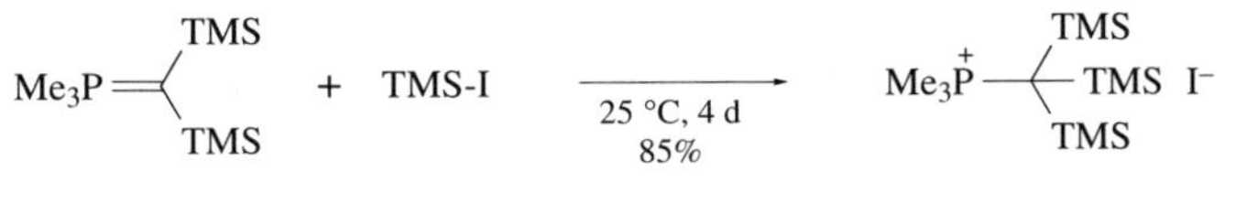

(31)

6.12.4.2.2 *One phosphorus, one metal and two silicon functions*

Most of the compounds which fall within this section are organolithium reagents, but a number of organoaluminum compounds are also known.

Bis(trimethylsilyl)phosphinomethanes, $R_2PCH(TMS)_2$, may be metalated by Bu^{n}Li to give the corresponding organolithium reagents (Equation (32)) (R = Ph ⟨84CB2063⟩; R = Me ⟨88ZN(B)1416⟩). A crystal structure determination of bis(trimethylsilyl)dimethylphosphinylmethyllithium shows a dimeric structure for its TMEDA adduct. In accordance with the ambidentate reactivity which these compounds show towards electrophiles, each lithium atom is bound to carbon and to phosphorus ⟨84CB2063⟩.

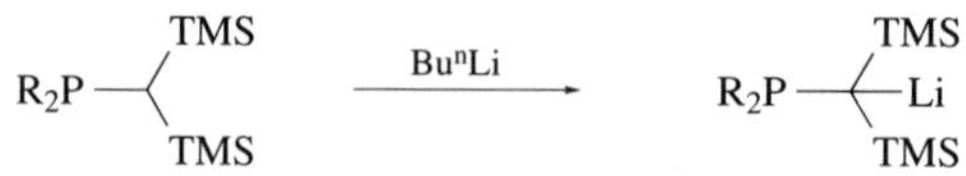

(32)

R = Me, TMEDA, hexane, 25 °C, 7 d, 76%
R = Ph, DME, hexane, 25 °C, 12 h, 95%

Bis(trimethylsilyl)methylphosphonates and -phosphine oxides are also metalated by LDA and methyllithium, respectively. Yields are good to excellent (Equation (33)) ⟨84JOC5087, 87JOM(323)135⟩. The metalation of dialkylmethyl phosphonates by three equivalents of LDA and subsequent treatment with two equivalents of chlorotrimethylsilane also provides a simple and efficient route to the same carbanions (X = MeO, 68%; EtO, 85%; Pr^iO, 25%) ⟨87JOM(323)135⟩.

(33)

X	*RLi*	*Conditions*	*Yield (%)*
EtO	LDA	THF, –50 °C	>91
MeO	LDA	THF, –50 °C	>92
Ph	MeLi	0 °C	

Lithium to aluminum exchange reactions of bis(trimethylsilyl)dimethylphosphinomethyllithium have been reported. The reactions are generally complex and give cyclic structures. A number of examples are given in Scheme 22 ⟨90(CC)1621, 91(CC)466, 91OM2884⟩.

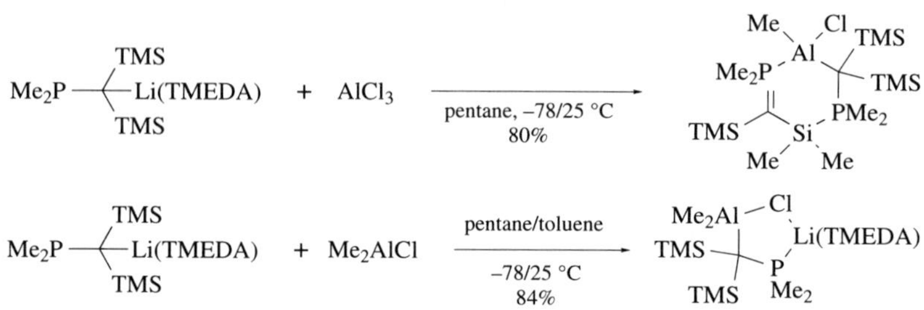

Scheme 22

6.12.4.3 Arsenic or Antimony Functions

Tris(trimethylsilyl)methyldichloroarsane, which forms the starting point for the preparation of each arsenic containing compound of this general type, may be prepared from $AsCl_3$ and $(TMS)_3CLi$ (Equation (34)) ⟨83TL2769⟩. The conditions employed were similar to those used for the preparation of its phosphorus analogue ⟨80ZC153⟩.

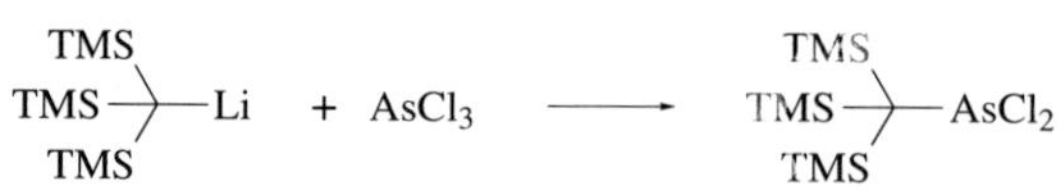

(34)

The chlorine atoms of $(TMS)_3CAsCl_2$ may be substituted either by a methoxyl group or by an allenyl function. Subsequent transformations give a dimer which results from the head-to-tail dimerization of the As—C double bonds of two 1-arsabutatriene units (Scheme 23) ⟨90TL6331⟩.

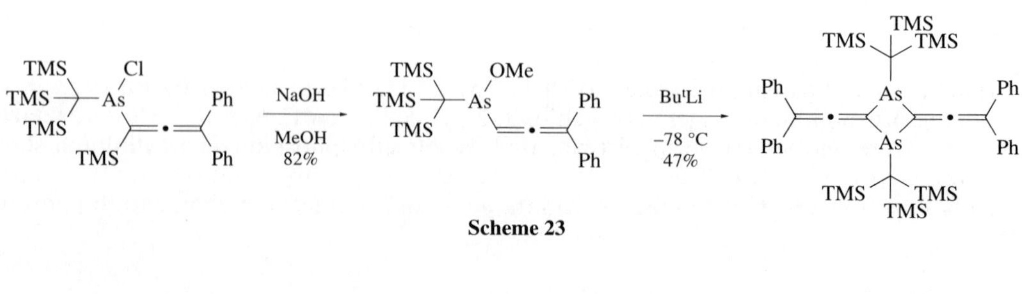

Scheme 23

Reduction of $(TMS)_3CAsCl_2$ by *t*-butyllithium ⟨83TL2769⟩ or organometallic complexes ⟨86ZN(B)191⟩ leads to a tris(trimethylsilyl)methyl-substituted diarsene, whose crystal structure ⟨85JCS(D)383⟩, electrochemical properties ⟨87JCS(D)249⟩, oligomerization reactions ⟨84JCR(S)274⟩, and sulfuration ⟨83TL2769⟩ have been described (Equation (35)).

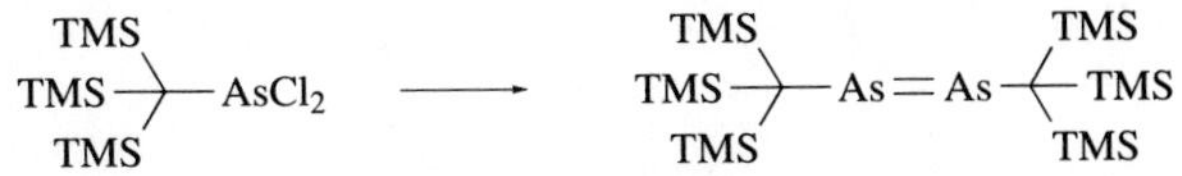

(35)

Reduction of a mixture of $(TMS)_3CAsCl_2$ and $(TMS)_3CPCl_2$ by *t*-butyllithium gave the phosphaarsene $(TMS)_3CAs{=}PC(TMS)_3$, and the symmetrical diarsene and diphosphene in unspecified yield ⟨83TL3625⟩.

Finally, the synthesis of the highly photosensitive $(TMS)_3CSbCl_2$ from $(TMS)_3CLi$ and $SbCl_3$ has been described briefly, but no experimental details were given ⟨85JA8211⟩.

6.12.5 ACKNOWLEDGEMENTS

In collecting the literature, the authors have benefited greatly from collaboration with Ms Françoise Girard, who is gratefully acknowledged.

6.13
Functions Containing at Least One Metalloid (Si, Ge or B) and No Halogen, Chalcogen or Group 15 Element; Also Functions Containing Four Metals

PAUL D. LICKISS
Imperial College of Science, Technology and Medicine, London, UK

6.13.1 METHANES CONTAINING AT LEAST ONE METALLOID (AND NO HALOGEN, CHALCOGEN OR GROUP 15 ELEMENT)

There are a huge number of potential substitution patterns at tetravalent carbon covered by the title of this section. Besides the elements specifically excluded in the title of the section and the lanthanides, the actinides and Fr, Ra and Tc—there are 47 elements which could be considered as substituents to be covered in the section. There are 230 300 possible combinations ($(4+n-1)!/4!(n-1)!$ for $n=47$) of these 47 elements at four coordinate carbons (not including a further 178 365 optical isomers generated by having four different elements as substituents). About 60 i.e., 0.026% are discussed below. Thus it is clear that there is still enormous scope for the synthetic chemist in the preparation of such organometallic compounds. It has been impossible to carry out a comprehensive search of the literature for all the possible functions covered under this heading, but it is hoped that the majority of important functions have been found. In some cases, carbon substituted by four metals may better be regarded as an interstitial carbide and more the province of the inorganic chemist; such species have been omitted.

6.13.1.1 Methanes Bearing Four Metalloid Functions

There are a large number of compounds known containing a carbon substituted by four silicon functions; such compounds may be prepared in a wide variety of ways. In view of the number of tetrasilylmethanes known, surprisingly, there seems to be no analogous germanium substituted species, although this is presumably due to a lack of synthetic effort rather than any inherent instability. A relatively small number of methanes substituted by four boron functions have been prepared, and there is also a range of compounds containing either a variety of metalloid functions or mixed metalloid and metal functions.

6.13.1.1.1 Four similar metalloid functions

(i) Four Si functions

(a) Tetrasilylmethanes from in situ *coupling reactions.* A range of compounds containing the Si_4C function has been prepared by *in situ* coupling reactions in which most, if not all of the four Si—C bonds are formed in the same reaction. Simple compounds of the type $(R^1{}_2R^2Si)_4C$ can be made by coupling reactions using different metals and different halomethanes according to the general reaction (Equation (1)). The range of precursors used for this type of synthesis and the products obtained are given in Table 1.

$$R^1{}_2R^2SiCl + \text{metal} + \text{tetrahalomethane} \longrightarrow (R^1{}_2R^2Si)_4C + \text{metal halide} \quad (1)$$

Table 1 Reagents used in and products formed from *in situ* coupling reactions.

Chlorosilane	*Metal*	*Halomethane*	*Si_4C Product*	*Yield (%)*	*Ref.*
TMSCl	Mg	CBr_4	$(TMS)_4C$	28	64JOC953
TMSCl	Mg	$CBrCl_3$	$(TMS)_4C$	20	64JOC953
TMSCl	Mg	CBr_2Cl_2	$(TMS)_4C$	30	64JOC953
TMSCl	Li	CBr_4	$(TMS)_4C$	28	65JOM(4)98
TMSCl	Li	CCl_4	$(TMS)_4C$	33–66.5	64MI 613-01, 65JOM(4)98
Me_2SiHCl	Mg	CBr_4	$(Me_2HSi)_4C$	55–60	64JOC953, 85JOM(294)305
$PhMe_2SiCl$	Mg	CCl_4	$(PhMe_2Si)_4C$	11	70JOM(24)89
PhH_2SiCl	Mg	CBr_4	$(PhH_2Si)_4C$	*ca.* 45	90AG(E)201
$MeSiCl_3$	Mg	CBr_4	$(MeCl_2Si)_4C$		86ZN(B)1527
$SiCl_4$	Mg	CBr_4	$(Cl_3Si)_4C$		86ZN(B)1527

The *in situ* tetrasilylation reactions may be written as simple sequential metallation–silylation steps, but actually the mechanism is likely to be rather more complicated as rapid ligand exchange between TMS, Br and Li has been shown to occur even at low temperatures in the $TMSCl/CBr_4/Li$ system ⟨70TL4693⟩. Sequential lithiation–silylation reactions can also be carried out using CH_2Cl_2

and Bu^nLi in a 1 : 4.5 ratio, quenching with TMSCl to give $(TMS)_4C$ in 36% yield ⟨72JOC2662⟩. The coupling of CLi_4, either as a solid or in solution, with TMSCl gives a 5% yield of $(TMS)_4C$ ⟨72CC1078, 84AG(E)995⟩. The tetrasilylation of CCl_4 can also be achieved electrochemically, silylation using TMSCl and an aluminium anode giving $(TMS)_4C$ in 60% yield along with di- and trisilylated products ⟨88JOM(358)31⟩. The coupling reaction between (**1**) and CCl_4 in the presence of lithium (Equation (2)) affords the heptasila[4.4.4]propellane (**2**); again all four Si—C bonds are formed in a single reaction. This reaction presumably proceeds in a stepwise manner with sequential formation of chloromethyl lithium reagents which react in turn with the four Si—Br groups in the organosilicon precursor. However, there may be rapid exchanges between Li, Cl, Br and silyl groups also taking place as mentioned above for the $TMSCl/CBr_4/Li$ system. Perhaps not surprisingly for such a complicated reaction, the yield is low (5%) ⟨76ZAAC(419)2⟩.

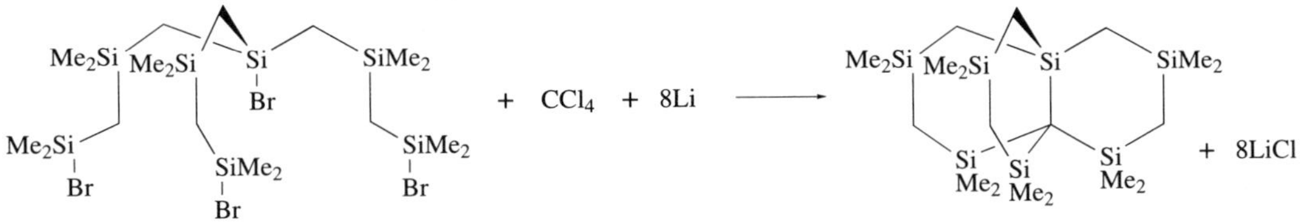

(**1**) (**2**) (2)

The phenylsilylmethane $(PhH_2Si)_4C$ (see Table 1 for its synthesis) is of interest as it is the precursor to $(BrH_2Si)_4C$, via Si—Ph bond cleavage by HBr, which may then be reduced under phase transfer conditions using $LiAlH_4$ to give $(H_3Si)_4C$, the Si/C inverse of the well-known Me_4Si ⟨90AG(E)201⟩. This silane cannot be prepared from $(Cl_3Si)_4C$ as both trisilylmethyl and Cl_3Si^- anions are good leaving groups and attempted reductions lead to Si—C bond cleavage. Interest in $(H_3Si)_4C$ and other H_3Si substituted methanes stems from their potential as chemical vapour deposition precursors to thin films of amorphous hydrogenated silicon carbide ⟨92JOM(429)1⟩.

(b) Formation by reaction between a trisilyllithiomethane and a halo- or pseudohalosilane. The reactions between tris(trimethylsilylmethyl)lithium (for its synthesis see 6.13.1.2.1 (i) (a)) and halo- or pseudohalosilanes have been used to prepare a large number of tetrasilylmethanes via the simple coupling reaction outlined in Equation (3).

$$(TMS)_3CLi + R^1R^2R^3SiX \longrightarrow (TMS)_3CSiR^1R^2R^3 + LiX \quad (3)$$

The range and yields of compounds prepared by the method given in Equation (3) are given in Table 2. The formation of such compounds is greatly affected by steric factors; for example, addition of more than one equiv. of the lithium reagent $(TMS)_3CLi$ ('TsiLi') to a di-, tri- or tetrahalosilane does not lead to the substitution of more than one Tsi group for a halide; i.e., it does not seem to be possible to attach more than one Tsi group to any one Si atom. It should also be noted that, for the synthesis of the most sterically hindered compounds, improved yields are obtained if silyl fluorides are used rather than silyl chlorides (presumably again for steric reasons); for example, TsiLi reacts with Ph_2SiCl_2 and with Ph_2SiF_2 to give $TsiSiPh_2X$ products in 8 and 76% yield, respectively; see Table 2.

A variety of other trisilylmethyllithium reagents that are less symmetrical than TsiLi have also been prepared (see 6.13.1.2.1.(i)(a)), and they react in a similar way with halosilanes to give tetrasilylmethanes (Table 3). The variety of functional groups that may be present in the trisilylmethyllithium reagents is greater than might be expected considering the reactivity of normal alkyllithium reagents with functional organosilanes. This unusual compatibility of a functional group on silicon and an alkyllithium function is presumably due to the fact that the reagents are often prepared at very low temperatures and also because steric constraints are likely to hinder the attack of one bulky molecule on another.

Tetrasilylmethane products may also be formed in the reactions of disilylmethane derivatives with lithium or organolithium reagents. The reaction of $(TMS)_2CCl_2$ with four equivalents of lithium in the presence of Pr^iSiF_3 at room temperature gives $(TMS)_2C(SiPr^iF_2)_2$ in 48% yield ⟨85JOM(291)277⟩. Alternatively, $(TMS)_2CLi_2$ can be prepared (see 6.13.1.3.1) at room temperature and it reacts with TMSCl to give $(TMS)_4C$ in 31% yield ⟨88TL5237⟩. The reactions of $(TMS)_2CClLi$ with halosilanes often gives mixtures containing tetrasilylmethane products as well as the expected trisilylmethanes (Equation (4)) ⟨84ZAAC(510)169⟩.

Table 2 Products obtained from the reaction between $(TMS)_3CLi$ (TsiLi) and halo- or pseudohalosilanes $R^1R^2R^3SiX$.

RR^1R^2SiX	*Product*	*Yield (%)*	*Ref.*
TMSCl	TsiTMS	57–85	70JOM(24)529, 77JOM(142)39, 92OM2938
TMSCN	TsiTMS	78	86JOC3545
$Me_2Si(N_3)_2$	$TsiSiMe_2N_3$		82AG(E)443
SiH_2Cl_2	$TsiSiH_2Cl$		93TH61301
$MeSiHCl_2$	TsiSiMeHCl	19–55	70JOM(24)529, 79JCR(S)12
$MeSiF_3$	$TsiSiMeF_2$	69	85ZN(B)1023
SiF_4	$TsiSiF_3$	68–74	84ZAAC(510)175, 86JOM(309)247
Bu^tSiF_3	$TsiSi(Bu^t)F_2$	52–71	84ZAAC(510)175, 86JOM(309)247
Pr^iSiF_3	$TsiSi(Pr^i)F_2$	67	87JOM(323)1
Me_2SiF_2	$TsiSiMe_2F$	65	87JOM(323)1
Et_2SiF_2	$TsiSiEt_2F$	40	70JOM(24)529
$Pr^i_2SiF_2$	$TsiSi(Pr^i)_2F$	62	87JOM(323)1
$PhSiF_3$	$TsiSiPhF_2$	67	84ZAAC(510)175
Me_2SiHCl	$TsiSiMe_2H$	87	70JOM(24)529
$SiCl_4$	$TsiSiCl_3$	78	70JOM(24)529
$Me_2SiCl(OMe)$	$TsiSiMe_2OMe$	60	70JOM(24)529
$Si(OPr^i)Cl_3$	$TsiSiCl_2(OPr^i)$	40	70JOM(24)529
Me_2SiCl_2	$TsiSiMe_2Cl$	82	70JOM(24)529
Et_2SiCl_2	$TsiSiEt_2Cl$	48	70JOM(24)529
EtMeSiHCl	TsiSiEtMeH	60	70JOM(24)529
$PhSiCl_3$	$TsiSiPhCl_2$	75	70JOM(24)529
$PhSiHCl_2$	TsiSiPhHCl	17	70JOM(24)529
Ph_2SiCl_2	$TsiSiPh_2Cl$	8	70JOM(24)529
Ph_2SiF_2	$TsiSiPh_2F$	76	70JOM(24)529
PhMeSiHCl	TsiSiPhMeH	80	70JOM(24)529
$PhMeSiF_2$	TsiSiPhMeF	74	70JOM(24)529
$Ph_2SiF(OMe)$	$TsiSiPh_2OMe$	9	70JOM(24)529
$PhMe_2SiF$	$TsiSiMe_2Ph$	13	81JOM(221)13
$Me_2(CH_2{=}CH)SiCl$	$TsiSiMe_2(CH{=}CH_2)$	55	85JOM(291)25
$Me_2(CH_2{=}CHCH_2)SiCl$	$TsiSiMe_2(CH_2CH{=}CH_2)$	38	85JOM(291)25
$Me_2(PhC{\equiv}C)SiF$	$TsiSiMe_2(C{\equiv}CPh)$	16	85JOM(291)25
$Me_2(PhCH_2)SiF$	$TsiSiMe_2(CH_2Ph)$	20	85JOM(291)25
$(CH_2{=}CH)_2SiCl_2$	$TsiSi(CH{=}CH_2)_2Cl$	51	86JCS(P2)1289
$(Cl_3Si)_2O$	$(TsiSiCl_2)_2O$	15.5	90JOM(398)59

Tsi = $(Me_3Si)_3C$.

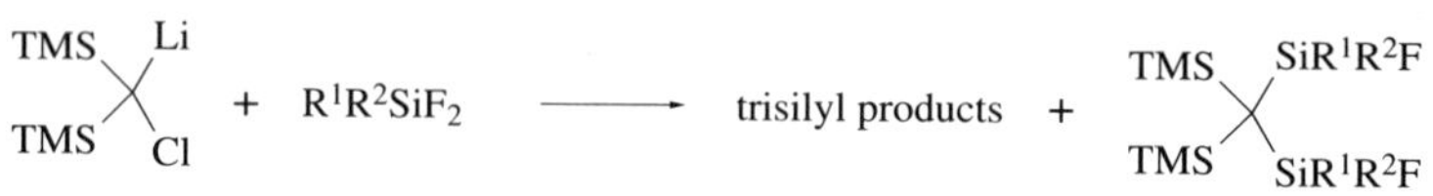

(4)

$R^1 = R^2 = F$, 41%

$R^1 = F$, $R^2 = Me$, 36%

$R^1 = Me$, $R^2 = Ph$, 36%

$R^1 = F$, $R^2 = Ph$, 29%

$R^1 = F$, $R^2 = Bu^t$, 71%

(c) Formation via *addition reactions of silenes.* Cyclic compounds containing an Si_4C function in the ring are formed in similar reactions to those shown in Equation (4); for example, in Scheme 1 the initial metallation gives rise to a lithium reagent that can readily eliminate a salt to give a highly reactive silene (Si═C containing species) which rapidly dimerises in a head-to-tail manner to give **(3)** in 42% yield ⟨84ZAAC(510)169⟩. Similarly, the reaction between $(TMS)_2CBrLi$ and Me_2SiCl_2 gives a 30% yield of **(4)**, presumably via initial formation of $(TMS)_2CBr(SiMe_2Cl)$, which undergoes Li/Br exchange with the original lithium reagent to give $(TMS)_2CLi(SiMe_2Cl)$, which then eliminates LiCl. Finally, the silene $(TMS)_2C{=}SiMe_2$ which is formed, dimerises ⟨76JOM(116)257⟩.

A range of trisilylmethyllithium species $(TMS)_2CLi(SiMe_2X)$ (X = F, Cl, Br, I, PhS, Ts, TsO, MsO, Ph_2PO_2, Ph_2PO_3, Ph_2PO_4) have been prepared (see 6.13.1.2.1.(i)(a) for details) which, in the absence of any trapping reagent, (cf. reactions in Table 3) eliminate LiX at various temperatures ranging from about −100 °C to above room temperature to give the silene $(TMS)_2C{=}SiMe_2$ which rapidly dimerises to give the disilacyclobutane **(4)** ⟨81CB2087, 81CB3505, 81CB3518⟩. (For reactions

Table 3 Tetrasilylmethanes from the reaction of miscellaneous trisilyllithiomethanes with halosilanes.

Lithiomethane	*Halosilane*	*Product*	*Yield* (%)	*Ref.*
$(PhMe_2Si)_3CLi$	$SiCl_4$	$(PhMe_2Si)_3CSiCl_3$		92JOM(427)9
$(PhMe_2Si)_3CLi$	SiH_2Cl_2	$(PhMe_2Si)_3CSiH_2Cl$	59	93TH61301
$(PhMe_2Si)_3CLi$	$MeSiHCl_2$	$(PhMe_2Si)_3CSiMeHCl$	83	85JCS(P2)729
$(PhMe_2Si)_3CLi$	Me_2SiHCl	$(PhMe_2Si)_3CSiMe_2H$	75	85JCS(P2)729
$(p\text{-}MeC_6H_4SiMe_2)_3CLi$	Me_2SiHCl	$(p\text{-}MeC_6H_4SiMe_2)_3CSiMe_2H$		94JOM(466)35
$(MeOMe_2Si)_3CLi$	Me_2SiHCl	$(MeOMe_2Si)_3CSiMe_2H$	89	86CC1043, 92JCS(D)1015
$(MeOMe_2Si)_3CLi$	Me_2SiCl_2	$(MeOMe_2Si)_3CSiMe_2Cl$	44	92JCS(D)1015
$(MeOMe_2Si)_3CLi$	Ph_2SiClH	$(MeOMe_2Si)_3CSiPh_2H$	65	92JCS(D)1015
$R_2C(SiMe_2Ph)Li$[a]	Me_2SiHCl	$R_2C(SiMe_2Ph)(SiMe_2H)$	62	93JCS(P2)59, 94JOM(466)35
$R_2C(SiMe_2Ph)Li$	Et_2SiHCl	$R_2C(SiMe_2Ph)(SiEt_2H)$	54	93JCS(P2)59
$R_2C(SiMe_2Ph)Li$	Et_2SiF_2	$R_2C(SiMe_2Ph)(SiEt_2F)$	30	93JCS(P2)59
$R_2C(SiMe_2Ph)Li$	Et_2SiCl_2	$R_2C(SiMe_2Ph)(SiEt_2Cl)$	9.5	93JCS(P2)59
$R_2C(SiMe_2C_6H_4Y)Li$[b]	Me_2SiHCl	$R_2C(SiMe_2C_6H_4Y)(SiMe_2H)$		94JOM(466)35
$R_2C(SiMe_2OMe)Li$	TMSCl	$R_3C(SiMe_2OMe)$		81CB2087
$R_2C(SiMe_2OMe)Li$	Me_2SiHCl	$R_2C(SiMe_2OMe)(SiMe_2H)$	62	86JCS(P2)1363
$R_2C(SiMe_2OMe)Li$	Me_2SiCl_2	$R_2C(SiMe_2OMe)(SiMe_2Cl)$	45	85JCS(P2)1687
$R_2C(SiMe_2OMe)Li$	Ph_2SiHCl	$R_2C(SiMe_2OMe)(SiPh_2H)$	41	87JCS(P2)891
$RC(SiMe_2OMe)_2Li$	Me_2SiCl_2	$RC(SiMe_2OMe)_2(SiMe_2Cl)$		92JCS(D)1015
$R_2C(SiMe_2H)Li$	Et_2SiCl_2	$R_2C(SiMe_2H)(SiEt_2Cl)$	82	93JCS(P2)391
$R_2C(SiMe_2H)Li$	Me_2SiCl_2	$R_2C(SiMe_2H)(SiMe_2Cl)$	75	93JOM(451)45
$R_2C(SiMe_2Cl)Li$	Me_2SiCl_2	$R_2C(SiMe_2Cl)_2$	31	76JOM(116)257
$R_2C(SiMe_2Bu^n)Li$	TMSCl	$R_3CSiMe_2Bu^n$		81CB2087
$R_2C(SiMe_2N_3)Li$	Me_2SiHCl	$R_2C(SiMe_2N_3)(SiMe_2H)$	40	93JOM(451)45
$R_2C(SiEt_2H)Li$	Me_2SiCl_2	$R_2C(SiEt_2H)(SiMe_2Cl)$	65	93JCS(P2)391
$R_2C(SiMe_2CH{=}CH_2)Li$	$Me_2Si(CH{=}CH_2)Cl$	$R_2C(SiMe_2CH{=}CH_2)_2$	60	87JCS(P2)1047
$R_2C(SiMe_2CH{=}CH_2)Li$	Me_2SiHCl	$R_2C(SiMe_2CH{=}CH_2)(SiMe_2H)$	67	87JCS(P2)1047
$R_2C(SiMe_2CH{=}CH_2)Li$	Me_2SiCl_2	$R_2C(SiMe_2CH{=}CH_2)(SiMe_2Cl)$	66	87JCS(P2)1047
$R_2C(SiMe_2CH{=}CH_2)Li$	TMSCl	$R_3CSiMe_2(CH{=}CH_2)$	88	87JCS(P2)1047
$R_2C(SiMe_2CH{=}CH_2)Li$	Et_2SiCl_2	$R_2C(SiMe_2CH{=}CH_2)(SiEt_2Cl)$	54	87JCS(P2)1047
$R_2C(SiMe_2CH{=}CH_2)Li$	Et_2SiHCl	$R_2C(SiMe_2CH{=}CH_2)(SiEt_2H)$	65	87JCS(P2)1047
$R_2C(SiMe_2CH{=}CH_2)Li$	Ph_2SiHCl	$R_2C(SiMe_2CH{=}CH_2(SiPh_2H)$	58	87JCS(P2)1047
$R_2C(SiMe_2CH{=}CH_2)Li$	PhMeSiHCl	$R_2C(SiMe_2CH{=}CH_2)(SiPhMeH)$	69	87JCS(P2)1047
$R_2C(SiMe_2CH_2SiMe_3)Li$	TMSCl	$R_3CSiMe_2CH_2TMS$	90	91OM551
(**32**)	TMSCl	(**33**)		83ZAAC(497)119
(**35**)	$ClSi(CH_2TMS)_3$	(**36**)		83ZAAC(497)119

[a] R = TMS. [b] Y = *p*-OMe, *p*-Me, *p*-Cl or *m*-CF_3.

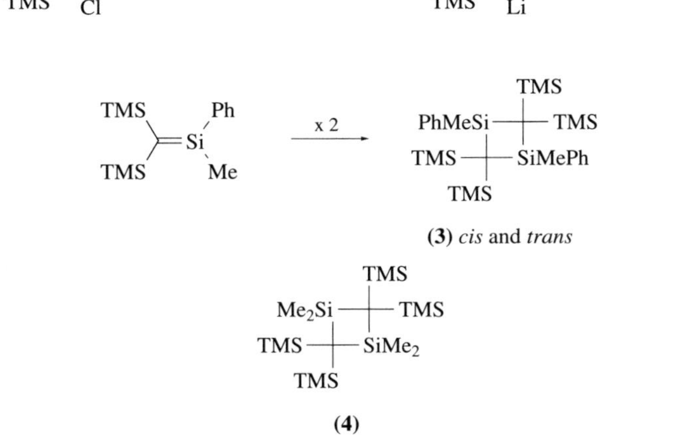

(**4**)

Scheme 1

involving the thermally generated silene $(Cl_3Si)_2C{=}SiCl_2$ see 6.13.1.1.1(i)(d).) Addition of various organosilanes to a reactive Si=C bond can also lead to formation of the Si_4C function. Thus, addition of $(PhO)_2P(O)TMS$ to the silene $(TMS)_2C{=}SiMe_2$ (**5**) gives $(TMS)_3CSiMe_2OP(O)(OPh)_2$ ⟨81CB3505⟩ while addition of TMSX gives $(TMS)_3CSiMe_2X$ (X = Cl, OMe, NMe_2 or OPh) ⟨81CB3505⟩ and addition of $Ph_2C{=}NSiMe_3$ affords (**6**), which when heated at 169 °C rearranges to give $(TMS)_3CSiMe_2N{=}CPh_2$ (Scheme 2) ⟨81CB3518, 87ZN(B)1055, 87ZN(B)1062⟩. The silene (**5**) also reacts with SiF_4 (but not with $SiCl_4$) to give $(TMS)_2C(SiF_3)(SiMe_2F)$ ⟨84JOM(273)141⟩. Addition of PhCN to two equivs. of the silene (**5**) gives (**7**) which rearranges to give (**8**) which reacts with methanol at 160 °C to give (**9**) (Scheme 2) ⟨81CB3518, 87ZN(B)1062, 87ZN(B)1055⟩.

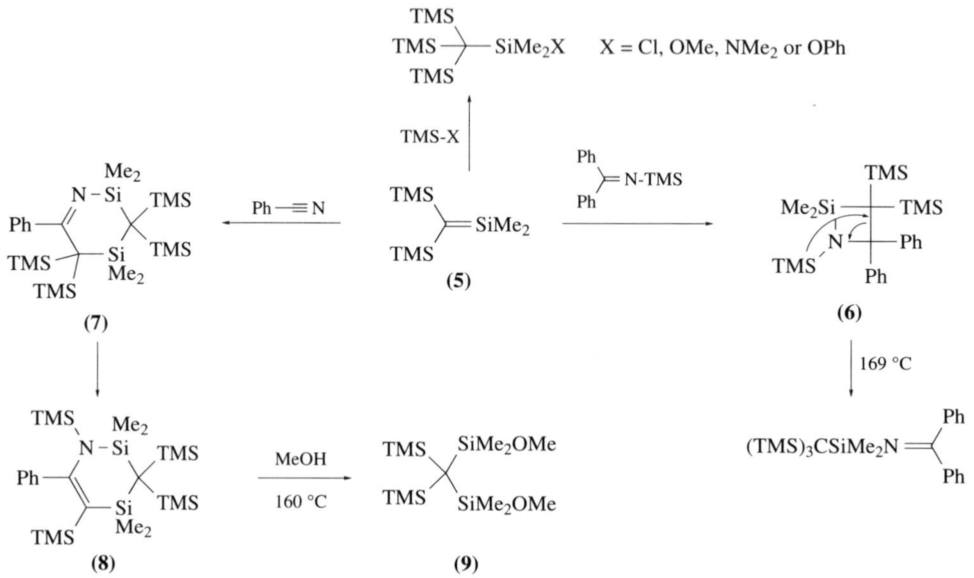

Scheme 2

Intramolecular elimination of LiBr may also lead to cyclic products (**10**) from simple Si—C bond formation (Scheme 3) ⟨81ZAAC(481)60⟩. Alternatively, elimination of LiBr from (**11**; $n = 0$) presumably forms silene (**12**), which dimerises to give (**13**) (Scheme 3) ⟨81ZAAC(481)60⟩. The organometallic syntheses of carbosilanes are reviewed in ⟨74TCC43⟩.

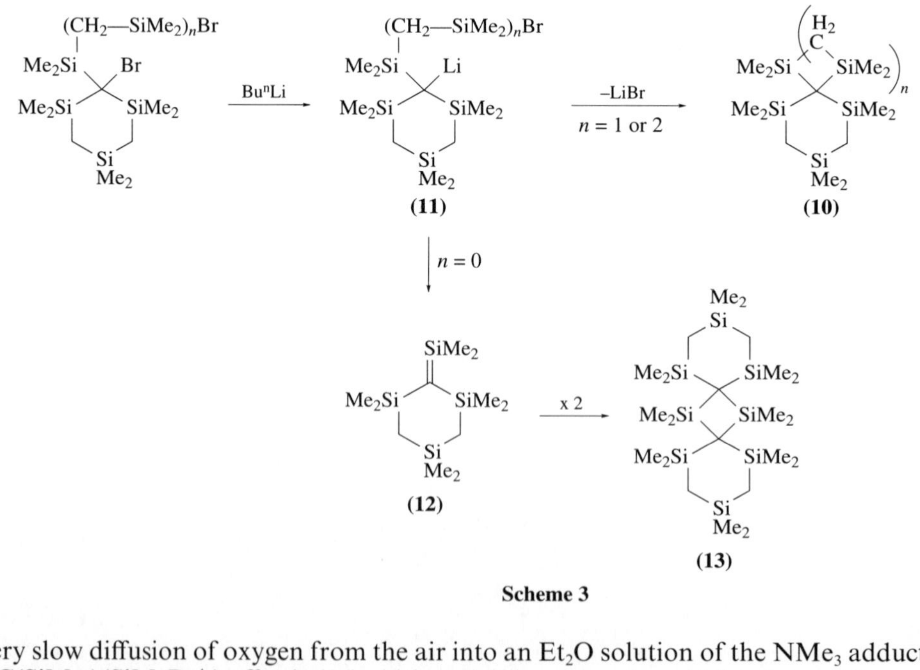

Scheme 3

The very slow diffusion of oxygen from the air into an Et_2O solution of the NMe_3 adduct of silene $Me_2Si{=}C(SiMe_3)(SiMeBu^t_2)$ affords (**14**). Although it is unclear how (**14**) is formed it may again be via an initial head-to-tail dimerisation of the silene and then further reaction with the oxygen ⟨89ZN(B)796⟩.

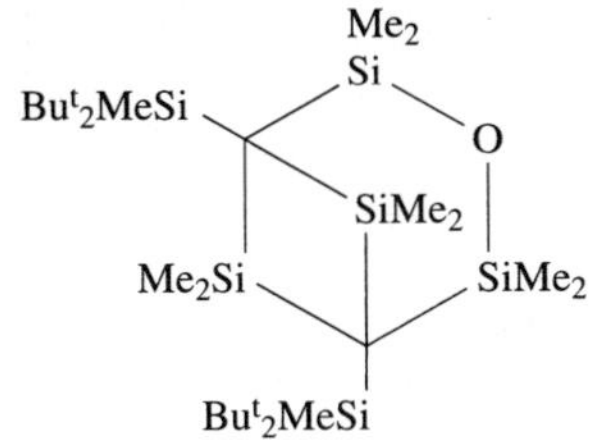

(14)

(d) Thermolytic or photolytic methods. The perchlorotetrasilylmethane $(Cl_3Si)_4C$ is a high melting product from a 'direct synthesis' type of reaction in which CCl_4 is passed over an Si/Cu mixture (in a manner reminiscent of the preparation of methylchlorosilanes from MeCl and Si/Cu) at temperatures above 300 °C. This again is likely to be a stepwise reaction (as in the *in situ* couplings above) in which C—Cl groups are replaced sequentially by $SiCl_3$ groups ⟨59CB1018⟩. Similar reactions involving $(Cl_3Si)_3CCl$ in $SiCl_4$ or Si_2Cl_6 over Si/Cu at high temperatures also yield $(Cl_3Si)_4C$ in up to 44% yield. A surprising coproduct in this type of reaction is **(15)** which could be formed from the dimerisation of the silene $(Cl_3Si)_2C{=}SiCl_2$ ⟨61EGP22169, 63CB2894, 84ZAAC(512)131⟩. Compound **(15)** is also formed in 26% yield when $(Cl_3Si)_3CCl$ is passed over a 10 : 1 Fe/Cu mixture at 350 °C ⟨64CB1111⟩. The disilylmethane $(Cl_3Si)_2CCl_2$ reacts with Si/Cu at 300–350 °C to give $(Cl_3Si)_4C$ **(15)** and the spirocyclic compounds (**16**; $n = 1$ to 3) ⟨87AG(E)1111⟩. Thermolysis of **(17)** similarly affords the spirocyclic compound (**18**) (Equation 5) ⟨87AG(E)1111, 93ZAAC(619)1494⟩.

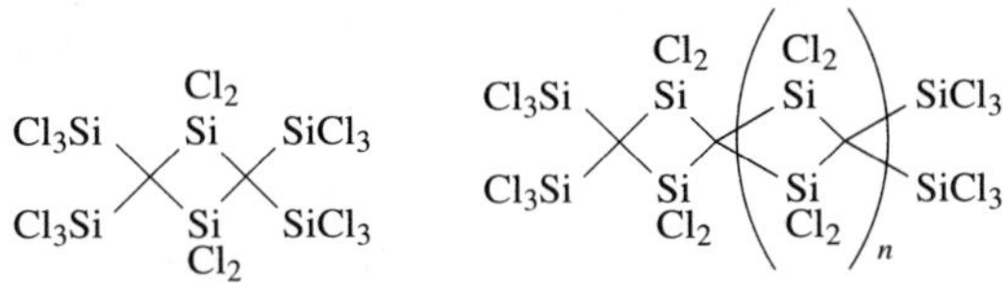

(15) **(16)** n = 1, 2 or 3

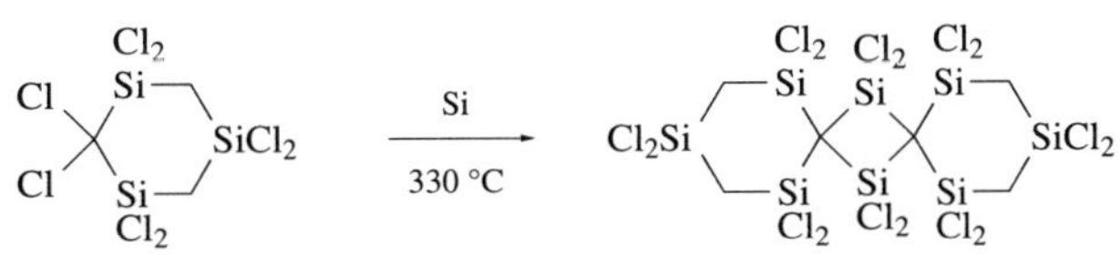

(17) **(18)** (5)

The thermolysis of 1,2-disilacyclobutanes (**19**) with exocyclic double bonds causes rearrangement to a 1,3-disilacyclobutane system (**20**) or (**21**) containing an Si_4 substituted carbon ⟨79JA1348, 82JOM(236)7, 88CL1441⟩. The two types of product (**20**) and (**21**) may arise from different mechanistic pathways; product (**20**) from a concerted 1,2-silyl shift and product (**21**) by breakdown of (**19**) into two silenes, one of which can undergo rapid ligand exchange (which is thought to occur in a number of other related silenes) to give (**22**) which then reacts with the silatriene in a head-to-tail fashion to give (**21**). Alternatively, the two silenes may recombine, without rearrangement, in a head-to-tail fashion (Scheme 4) ⟨79JA1348, 82JOM(236)7, 88CL1441⟩.

The gas phase pyrolysis of Me_4Si at 700 °C affords a large number of compounds, many of which are carbosilanes containing the 1,3,5,7-tetrasilaadamantane framework. However, some that are formed in low yield contain Si_3C_3 rings having a boat conformation such as (**23**) and (**24**) ⟨73AG(E)654, 74ZAAC(404)1⟩. The pyrolytic preparation of the carbosilanes is not very convenient as it leads to the formation of a large number of products which have to be separated by distillation, fractional crystallisation, or high-performance liquid chromatography (HPLC) ⟨84ZAAC(512)93⟩. The synthesis of carbosilanes by pyrolysis or organometallic synthetic routes has been reviewed ⟨87AG(E)1111⟩, and discussed in detail ⟨65MI 613-01⟩.

Photolysis (or thermolysis for R = Me) of the bis(silyldiazomethyl) compounds (**25**), in the absence of trapping agents, gives rise to low yields of (**26**) (a much higher yield is obtained by thermolysis) presumably via dicarbene (**27**) which undergoes the well-known silylcarbene to silene rearrangement to give (**28**) which cyclises in a predominantly head-to-tail fashion to give (**26**) (Scheme 5) ⟨91JA7790⟩.

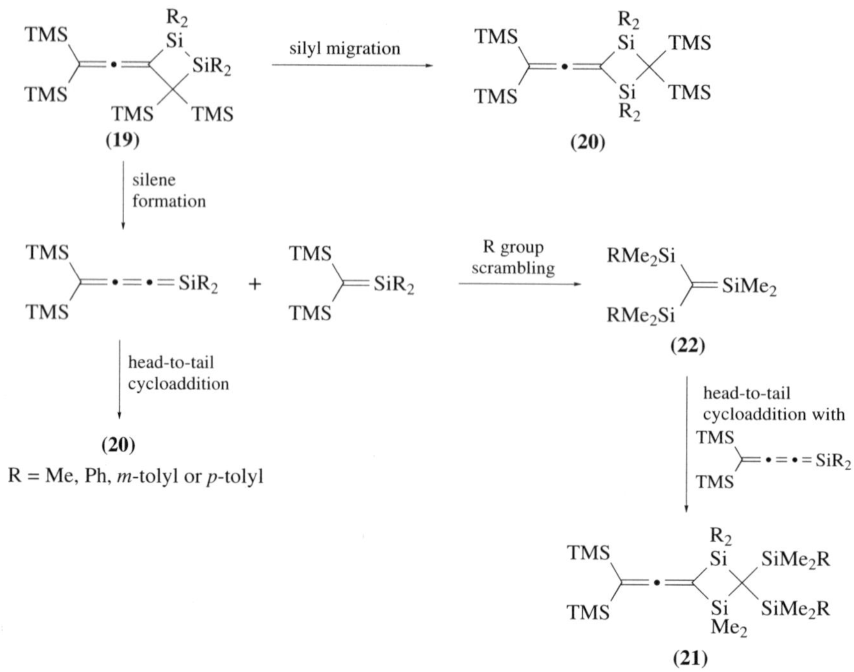

Scheme 4

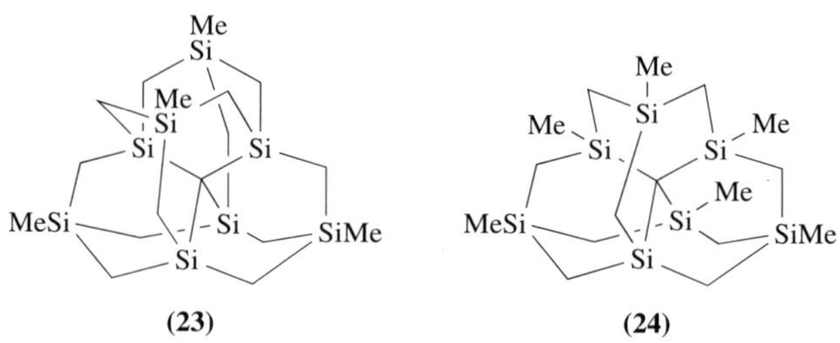

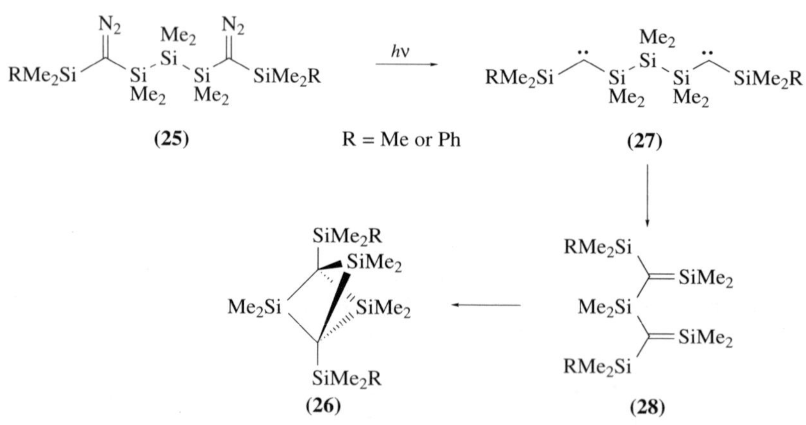

Scheme 5

(e) Transition metal catalysed reactions. The palladium catalysed reactions between isocyanides and oligosilanes in the presence of 1,1,3,3-tetramethylbutyl isocyanide gives products (**29**) from tetrasilanes ((**30**) from the hexasilane TMS$(SiMe_2)_4$TMS) in which both the C to N triple bond and three Si—Si bonds have been broken. The mechanism of this surprising reaction is unclear (the reaction works, but less well, in the absence of the butyl isocyanide) but the first step is thought to involve insertion of the RNC into a terminal Si—Si bond of the oligosilane (Equation (6)) ⟨89(CC)1494, 91JA8899⟩.

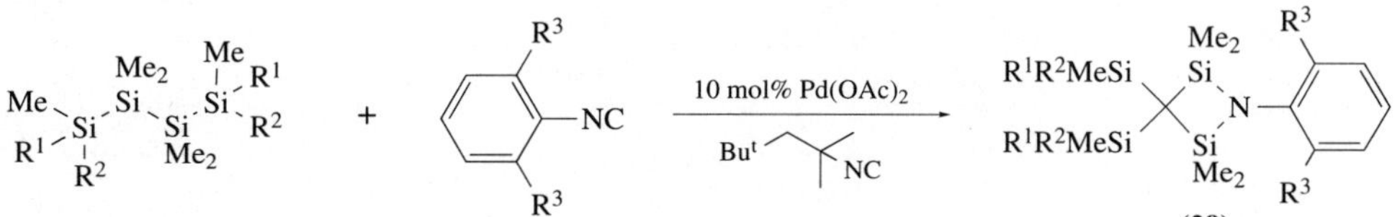

(6)

(29)

$R^1 = R^2 = R^3 = Me$, 40%
$R^1 = R^2 = Me$, $R^3 = Pr^i$, 45%
$R^1 = R^2 = Ph$, $R^3 = Me$, 31%
$R^1 = R^2 = Ph$, $R^3 = Pr^i$, 62%

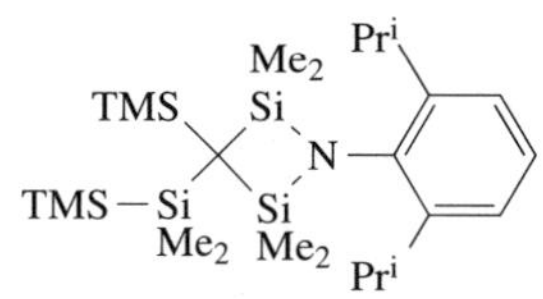

(30)

(f) Rearrangements of carbosilanes caused by $AlBr_3$. A range of cyclic compounds containing Si_4C functions has been prepared by treatment of a carbosilane containing an Si_3C or Si_4C centre, to which is attached a carbosilane chain, with $AlBr_3$. The reactions proceed at between room temperature and 80 °C and usually lead to Si—C bond cleavage and formation of Me_4Si as a by-product as the ring is formed. Thus, $(TMS)_3C(SiMe_2CH_2)_2TMS$ (prepared from $(TMS)_3CSiMe_2Cl$, see Table 2) initially affords cyclic compound (**31**), which reacts further with $AlBr_3$ to give additional Si_4C substituted products (Scheme 6) ⟨83ZAAC(497)21, 83ZAAC(497)119, 83ZAAC(497)134⟩.

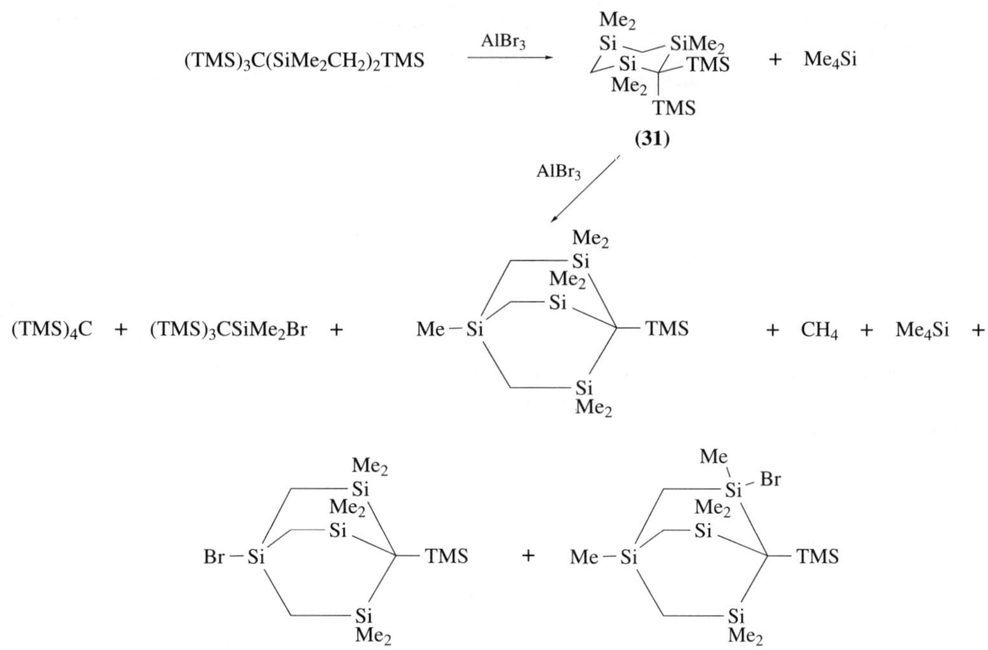

Scheme 6

Cyclic trisilylmethyllithium reagents may also be used as precursors to complicated carbosilanes; thus (**32**) reacts with TMSCl to give (**33**) (see Table 3), which reacts with $AlBr_3$ to give (**34**), and lithium reagent (**35**) can be used to prepare the cyclic compound (**36**) which contains an Si_4C group (see Table 3) (Scheme 7) ⟨83ZAAC(497)119⟩. Lithium compounds (**62**) which are formed by a rearrangement (see Section 6.13.1.2.1(i)(a)) also react with TMSCl to give tetrasilyl substituted products (**37**) ⟨77ZAAC(430)137⟩.

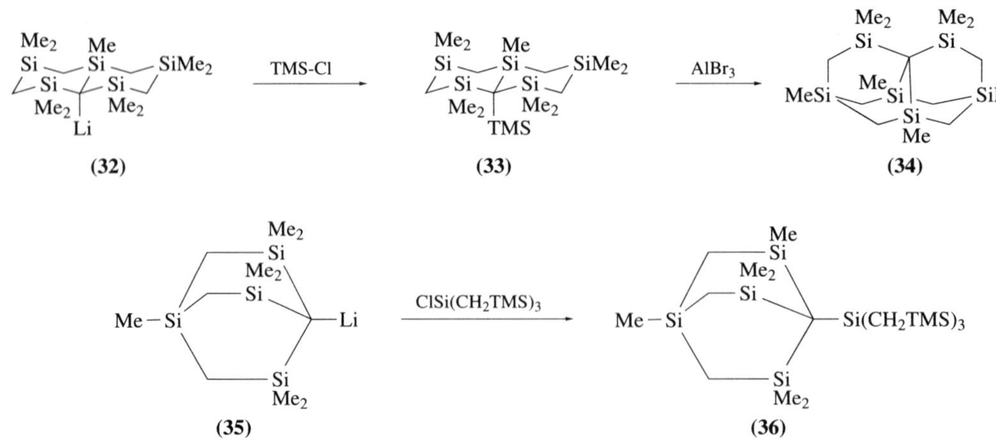

Scheme 7

(ii) Four Ge functions

The abundance of compounds containing the Si_4C group discussed in the previous section might suggest that at least some similar Ge_4C containing compounds should be known. However, there do not seem to have been any compounds containing this grouping prepared. There would not appear to be any inherent reason why such compounds should not be stable; *ab initio* calculations on $(H_3Ge)_4C$ suggest a greater stability than $(H_3Si)_4Si$, a known stable (albeit highly reactive) compound ⟨81TCA285⟩. The reasons for the rarity of tetragermylmethanes are probably the higher costs involved for germanium compounds, the relative lack of interest in organogermanium chemistry generally when compared with organosilicon chemistry, and some of the preparative routes used for the silicon compounds not being readily applicable to corresponding germanium species. It should, however, be possible to prepare compounds containing the Ge_4C group by, for example, a route involving sequential metallations of a halomethane followed by reaction with an organogermanium halide.

(iii) Four B functions

Although the synthesis of such compounds does not seem to have been achieved—calculations on the structures of $C(BH_2)_4$ and the spirocycle (**38**) have been carried out. Both compounds, despite the presence of electropositive substituents and/or small rings, were found to prefer the usual tetrahedral geometry at carbon rather than a planar geometry ⟨76JA5419⟩.

Compounds containing the CB_4 function can be prepared by an *in situ* coupling in a manner similar to that for CSi_4 functions as described above. Thus the reaction (Equation (7)) gives the methanetetraboronic ester in 50–60% yield and can be carried out easily on a scale giving 120–150 g product ⟨68JA2194, 69JOM(20)19⟩. The preparation of $[^{13}C]C(B(OMe)_2)_4$ starting from $[^{13}C]CCl_4$ has been reported to occur in a similar manner ⟨91AG(E)1488⟩. Disproportionation of $HC(B(OMe)_2)_3$ in the presence of Et_3B also occurs to give low yields of $C(B(OMe)_2)_4$ ⟨75LA1339⟩. Reactions of $C(B(OMe)_2)_4$ with the α, ω-alkanediols $HO(CH_2)_nOH$ ($n = 2$ or 3) gives products (**39**) which have been found to be more synthetically useful than the methyl ester ⟨73JOM(57)231⟩. The pinacol ester $C(BO_2C_2Me_4)$ is similarly prepared from the methyl ester and pinacol ⟨74JOM(69)45⟩.

(**37**) R = H or TMS

(**38**)

(**39**)

(**40**) R = Me; X = Cl
R = H; X = F

$$4ClB(OMe)_2 + CCl_4 + 8Li \xrightarrow{THF} C[B(OMe)_2]_4 + 8LiCl \quad (7)$$

Carbon vapour produced from arcing electrodes reacts with B_2Cl_4 to give $C(BCl_2)_4$ as a white crystalline solid, presumably formed by insertion of C atoms into the B—B bond of B_2Cl_4. The reaction of C atoms with B_2F_4 gives low yields of $C(BF_2)_4$ in a similar manner ⟨69JCS(A)1882⟩.

6.13.1.1.2 *Three similar and one different metalloid function*

(i) Three Si functions and one Ge function

The reactions of trisilyllithiomethanes with germanium halides proceed in a manner similar to those described above for silicon halides and the expected products from lithium halide elimination are formed. Thus $(TMS)_2(MeOMe_2Si)CLi$ and Me_3GeBr give $(TMS)_2(MeOMe_2Si)CGeMe_3$ in 89% yield ⟨87JCS(P2)775⟩. The disilylgermyllithiomethane $(TMS)_2(Me_3Ge)CLi$ reacts similarly with Me_2SiHCl to afford $(TMS)_2(Me_3Ge)CSiMe_2H$ in 46% yield ⟨87JCS(P2)779⟩. The reaction of TsiLi with Me_3GeBr, Me_2GeCl_2, Et_2GeBr_2, $GeCl_4$, $EtGeCl_3$, Et_2GeCl_2 or Ph_2GeCl_2 gives $TsiGeMe_3$ (70–80%), $TsiGeMe_2Cl$ (60%), $TsiGeEt_2Br$ (58%), $TsiGeCl_3$, $TsiGeEtCl_2$, $TsiGeEt_2Cl$ and $TsiGePh_2Cl$, respectively ⟨70JOM(24)529, 79JCR(S)12, 79BAU2222, 80JOM(202)157⟩.

In a manner reminiscent of the silicon analogues described above, a germanium halide may also add across the reactive double bond of a silene, i.e., $(TMS)_2C{=}SiMe_2$ reacts with Me_3GeCl to give $(TMS)_2C(GeMe_3)(SiMe_2Cl)$ in 85% yield ⟨87ZN(B)1062⟩.

In an Si_3Ge substituted methane the germanium may also be divalent rather than the more usual tetravalent. For example, the reaction between TsiLi and Me_5C_5GeCl gives $TsiGeMe_5C_5$ in 62% yield ⟨91OM3838⟩ and in a transmetallation reaction TsiLi reacts with the bulky germylene $C_5Me_5GeCH(TMS)_2$ to give $TsiGeCH(TMS)_2$ in 65% yield ⟨91OM1647⟩.

(ii) Three Si functions and one B function

The reaction of TsiLi with $B(OMe)_3$ does not proceed in the simple manner expected, but use of three equivs. of $B(OMe)_3$ affords, after an aqueous workup, a mixture of $TsiB(OMe)_2$, TsiB (OMe)OH and $TsiB(OH)_2$ ⟨86JOM(308)261⟩, while a nonaqueous workup allows $TsiB(OMe)_2$ to be isolated as product in good yield ⟨89JCS(D)447, 89CB1057⟩. Surprisingly, the reaction between TsiLi and BF_3 in Et_2O/THF does not afford $TsiBF_2$ but rather $TsiB(F)O(CH_2)_4Tsi$ in 55% yield in which a ring-opened THF molecule has been incorporated into the product. The mechanism by which this compound is formed is not known in detail; the product is unusual as reactions between TsiLi and other metal or metalloid halides generally give products derived from the simple elimination of a lithium halide ⟨82JOM(235)265⟩, and the reaction between TsiLi and Ph_2BBr does afford $TsiBPh_2$ ⟨84JOM(272)1⟩. In contrast to the TsiLi analogue, the reaction between $(PhMe_2Si)_3CLi$ and $BF_3 \cdot Et_2O$ in THF does give the expected product, $(PhMe_2Si)_3CBF_2$, in 85% yield ⟨89JCS(D)447⟩. The lower reactivity of $(PhMe_2Si)_3CLi$ compared with TsiLi towards the BF_3/THF system and towards other metalloid halides is probably associated with their difference in structure, the compound containing phenyl being a molecular species and TsiLi ionic (see 6.13.1.2.1(i)(a) for further details of these organolithium compounds).

Addition of BF_3 across the reactive double bond of the silene $(MeBu^t_2Si)(TMS)C{=}SiMe_2$ gives $(MeBu^t_2Si)(TMS)C(BF_2)(SiMe_2F)$ in about 70% yield ⟨83AG(E)1005, 86CB1467⟩. Surprisingly, this product is also formed in 65% yield from the reaction between $(TMS)_2CLi(SiBu^t_2F)$ and BF_3, presumably via rearrangement to $(TMS)(FMe_2Si)CLi(SiBu^t_2Me)$ and subsequent formation of the silene to which the BF_3 adds ⟨86CB1467⟩. Although the reaction between TsiLi and BF_3 is complicated, the reaction between TsiLi and $(TMS)_2NBF_2$ affords $TsiB(F)N(TMS)_2$ ⟨85AG(E)324⟩. The reaction between $PhCH_2B(OMe)_2$ and TsiLi in a 1 : 1 ratio gives $TsiB(OMe)CH_2Ph$ in 46% yield ⟨90CB747⟩ and the reaction between TsiLi and Bu^tBF_2 affords $TsiB(F)Bu^t$ in 78% yield ⟨89CB1057⟩. TsiLi also reacts with $(TMS)RNBF_2$ (R = Me, Pr^i, TMS, Bu^t_3) to give TsiB(F)NR(TMS) in yields of 59% (R = Me) and 83% (R = Pr^i) and with dichloro(2,2,6,6-tetramethylpiperidino)borane to give chloro(2,2,6,6-tetramethylpiperidino)-tris(trimethylsilyl)methylborane (**40**; R = Me, X = Cl) in 32% yield ⟨86CB1117⟩. Such compounds readily eliminate TMSF to give TsiB≡NR species (R = Bu^t or TMS). Similarly TsiLi reacts with $Pr^i_2N{=}BF_2$ to give TsiB(F)NPr^i_2 in 78% yield ⟨87CB1069⟩. The reaction between TsiLi and (2,6-dimethylpiperidino)difluoroborane gives the fluoroborane (**40**) (R = H, X = F) as an oil in 71% yield ⟨89CB595⟩.

(iii) Three Ge functions and one Si or B function. Also three B functions and one Si or Ge function

No compounds containing these functions seem to have been prepared. This is not surprising for the case of the CGe_3 compounds when it is taken into account that there are neither CGe_4 functions as described in 6.13.1.1.1.(ii) nor CGe_3M species. Again this lack of compounds is not likely to be due to a problem of inherent instability of such species but rather a lack of interest in their synthesis. The lack of CB_3Si or CB_3Ge functions is more surprising as both the CB_4 function and various CB_3M functions are known (see 6.13.1.2.1(ii) below). The synthesis of these mixed metalloid compounds could probably be easily achieved in a manner similar to those, for example, of CB_3Sn functions; i.e., by reaction between a CB_3Li function and a silicon or germanium halide such as TMSCl or Ph_3GeCl.

6.13.1.1.3 Two similar and two different metalloid functions

(i) Two Si and two Ge functions

When it is considered that there are a large number of compounds containing the CSi_4 function, and no compounds containing the CGe_4 function known, it can be expected that there will only be a handful of compounds known which contain the CSi_2Ge_2 function. Such compounds have been prepared using methods that have been widely used for the synthesis of CSi_4 functions, and many more species could no doubt be prepared in similar ways. The reaction between the highly reactive dilithiomethane $(TMS)_2CLi_2$ and Me_3GeCl gives the expected $(TMS)_2C(GeMe_3)_2$ in 80% yield ⟨88TL5237⟩. This compound is also produced as a by-product (in up to 30% yield) from the low temperature reaction between $(TMS)_2CCl_2$,Bu^{n}Li and Me_3GeBr (($(TMS)_2C(GeMe_3)Cl$ being the desired product) ⟨87JCS(P2)779⟩. The reaction between Me_2GeBr_2 and two equivs. of $(TMS)_2CBrLi$ at low temperature gives compounds (**41**) and (**42**) in 17 and 4.4% yields respectively. These products presumably arise from lithium/halogen exchange after the initial C—Ge bond formation as shown in Scheme 8 ⟨76JOM(116)257⟩. Wiberg has shown independently that a variety of lithium compounds $(TMS)_2C(GeMe_2X)Li$ (X = F, Br, OMe, etc.; see 6.13.1.2.2(i)) eliminate LiX to give germene (**43**) which does then dimerise ⟨86CB2966, 86CB2980⟩.

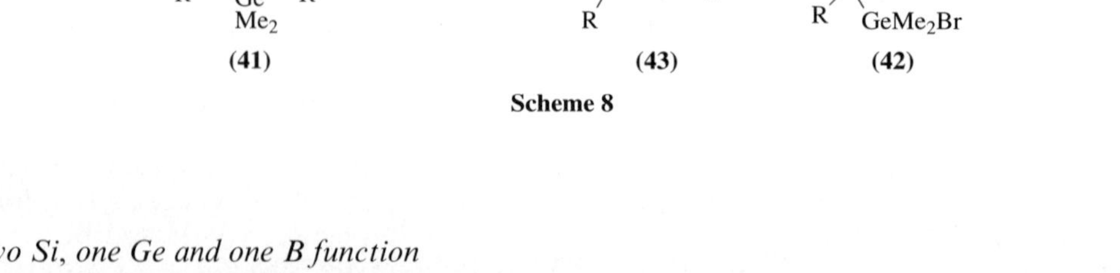

Scheme 8

(ii) Two Si, one Ge and one B function

Compounds containing the CSi_2GeB function do not appear to be known. However, they are likely to be readily available from the reaction of one of the several known CSi_2GeLi containing compounds with a boron halide or $B(OMe_3)$.

(iii) Two Si and two B functions

As for the case of the CSi_2Ge_2 function there are very few compounds known containing the CSi_2B_2 group. The reaction of the unsaturated boron compound (**44**) with $RBBr_2$ species gives products (**45**) (43% yield) and (**46**) (53% yield) that are the result of a formal addition across the C=B bond (Scheme 9) ⟨89CB595⟩.

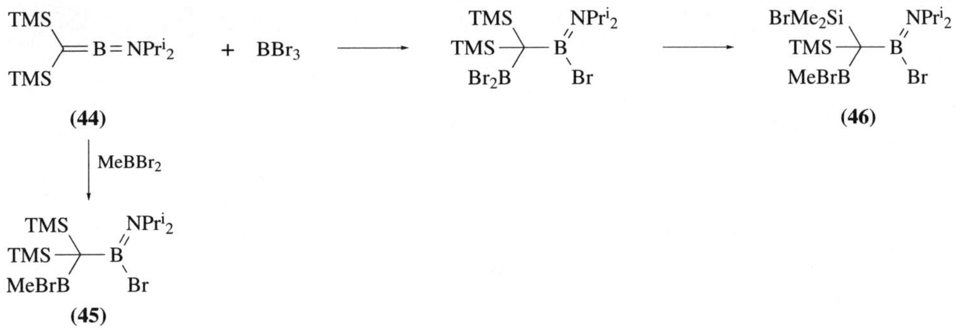

Scheme 9

The reactions of the dilithium reagent (**47**) give a variety of cyclic compounds containing the CSi_2B_2 function (see also Equation (16)) which appear to arise via an intramolecular cyclisation and a double substitution at the triply bonded carbon (Scheme 10) ⟨89AG(E)781, 89AG(E)784⟩.

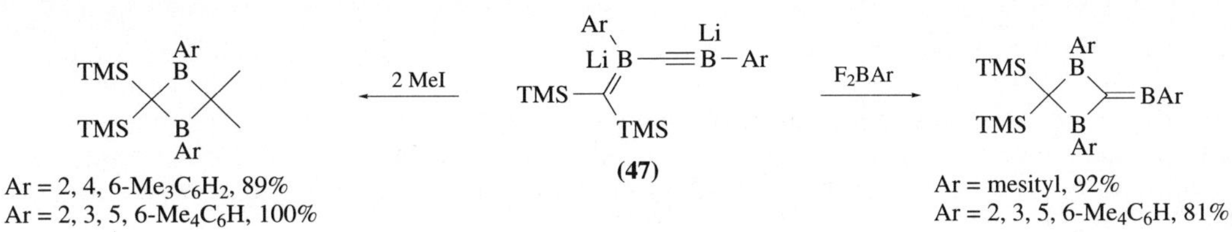

Scheme 10

The unsaturated borane (**83**) (see 6.13.1.2.2(ii)) reacts with Ar_2BBF_2 to give the cyclic species (**48**) which rearranges slowly to give (**49**) (Equation (8)). This has been structurally characterised by x-ray crystallography ⟨91AG(E)594⟩.

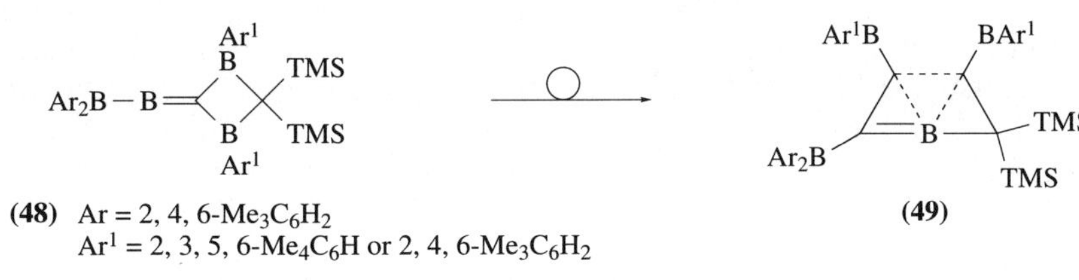

(8)

Treatment of the dianion (**50**) with Ph_3PAuCl or CH_2Br_2 gives compounds (**51**), (**52**) and (**53**) in 22, 30, and 53% yields respectively, all of which seem to have been formed via a silyl migration and a geminal substitution (Scheme 11) ⟨86AG(E)1112, 89CB1881⟩.

(iv) Two Ge and two B functions. Also two Ge, one Si and one B function. Also two B, one Si and one Ge function

As might be expected from the paucity of CSi_2Ge_2 and CSi_2B_2 containing species, there appear to be have been no compounds prepared containing the CGe_2B_2, CGe_2SiB, or CB_2SiGe functions. Again there is not likely to be any good reason why such compounds should not be made. Reactions between appropriately substituted lithiomethanes and halometalloids should readily afford the required functions.

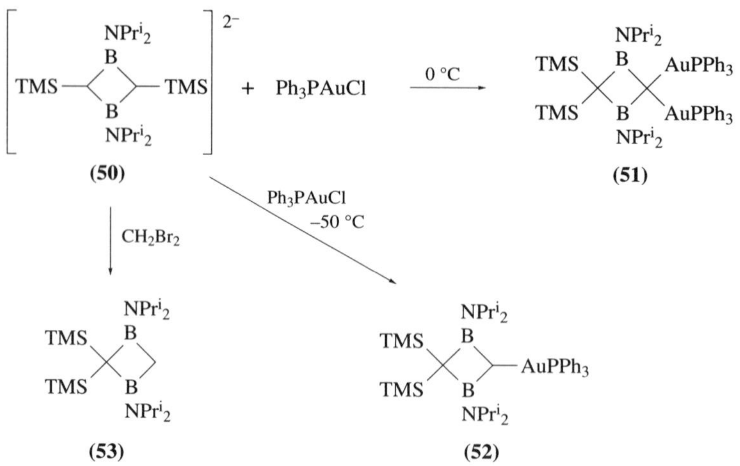

Scheme 11

6.13.1.2 Methanes Bearing Three Metalloid Functions and a Metal Function

6.13.1.2.1 Three similar metalloid functions

(i) Three Si functions

(a) Three Si and one group 1 metal function. The $(TMS)_3C$ (Tsi) group has been attached to a wide variety of metals and metalloids usually using the lithium reagent TsiLi. The bulk of the Tsi group means that molecules containing it often have novel structures or contain elements in unusual oxidation states. The size of the Tsi group has been discussed in terms of a 'cone angle' (a term more commonly used in phosphine chemistry) and when attached to Si, Ge, or Sn it has a cone angle of 208°, 196°, and 190° respectively ⟨91MI 613-01⟩.

The metallation of TsiH by MeLi in THF/Et_2O occurs readily to give a dark, red–brown solution of TsiLi in about 6 h at the reflux temperature or 20 h at room temperature ⟨70JOM(24)529, 77CB852⟩. Improvements to this preparation have been reported. These involve removal of the Et_2O from the reaction solution by distillation, thus allowing a higher reflux temperature and complete metallation in 2 h; removal of any residual MeLi from the reaction mixture by addition of TMSOMe or TMSOEt which reacts with MeLi but not TsiLi; determination of the extent of reaction by treatment of an aliquot of the reaction solution with TMSCl and determining the $(TMS)_4C$–$(TMS)_3CH$ ratio by ^1H NMR spectroscopy ⟨84JOM(269)217⟩. The TsiLi produced in this way is remarkably stable for an alkyllithium reagent, a 0.15 M solution in THF having a half-life in refluxing THF of 70 h ⟨70JOM(24)529⟩. The ease of preparation, relative stability (and hence relative ease of handling) and the steric requirements of the bulky Tsi group have made TsiLi a useful and widely applicable reagent in the preparation of unusual organometallic compounds many of which are described elsewhere in this chapter.

Metallation of TsiH does not occur with $Bu^nLi/Et_2O/THF$, $Bu^nLi/C_5H_{12}/TMEDA$, Bu^tLi/C_5H_{12}, or Bu^tLi/THF. As these are more powerful metallating agents than MeLi, steric effects would seem to play an important role in these reactions ⟨73JOM(55)209⟩. The addition of the base hexamethylphosphoric triamide (HMPT) to reaction solutions does, however, promote metallation and TsiLi is formed in good yield from TsiH using $Bu^tLi/THF/HMPT$ as the metallating agent at $-78\,°C$ ⟨85CB2493⟩. (See below for the reaction between TsiH and $Bu^tLi/C_5H_{12}/TMEDA$ at room temperature.) When prepared in THF from TsiH and MeLi, the TsiLi may be isolated as a crystalline solid in 65% yield. Characterisation by x-ray crystallography shows that, in the solid state, the lithium compound is not simply TsiLi, but that it actually has an unusual ate composition $(Tsi_2Li)(Li(THF)_4)$ ⟨83CC827⟩. In THF or toluene solution the unusual ionic form is also predominant, but at low temperatures and low concentration a second form thought to be $TsiLi(THF)_n$ ($n = 1$ or 2) can be detected by NMR spectroscopy. A different form, thought to be an aggregate of some kind, is also observed in toluene. At higher temperatures the two forms interconvert rapidly on the NMR timescale and the ^1H NMR signal for the TMS groups is commonly seen as a broad singlet at 90 MHz or two singlets at 300 MHz at room temperature ⟨93JCS(D)3259⟩.

Preparation of TsiLi may also be achieved by reaction of TsiBr with Bu^nLi (at $-75\,°C$), PhLi or with lithium metal ⟨77JOM(142)39, 92OM2938⟩. These methods give good yields which may be determined by derivatisation with TMSCl, but the method involving the metallation of TsiH by MeLi is to be preferred in most cases as TsiH is much more readily available than TsiBr. TsiLi may also be prepared by treatment of TsiSPh with Linaphth/THF at $-78\,°C$; however, this is again not a very convenient route as the phenylthio starting material is much more difficult to prepare than TsiH. However, this route may be of use on occasions when a complete absence of halogen in the system is required as the methods involving transmetallation using, for example, MeLi or BuLi usually have some residual halogen present from the lithium reagents preparation ⟨84JOC168⟩.

TsiLi as a TMEDA adduct $(Tsi_2Li)(LiTMEDA_2)$ is formed in 64% yield by metallation of TsiH in hexane in the presence of TMEDA over 80 h at room temperature ⟨87CB1069⟩. The reaction between TsiH and MeLi (evidently containing some LiCl from the MeLi preparation) in the presence of $Me_2N(CH_2)_2NMe(CH_2)_2NMe_2$ (PMDETA) again gives a TsiLi-like species. Again crystallography shows that an ate species is present in the solid state, this time containing a novel, linear, chlorine centred cation as shown in **(54)** ⟨86CC969⟩. The reaction between $TsiHgBu^n$ and Bu^nLi at 65–70 °C in the absence of solvent gives the highly reactive, solvent-free TsiLi which has been characterised by x-ray crystallography and found to be a bridged dimer $(TsiLi)_2$ in the solid state. Although this synthesis of TsiLi requires several more steps than the others described, the absence of solvent, particularly THF—which can cause side reactions when the lithium reagent is used—may mean that this is the method of choice if a coordinated solvent is likely to be troublesome ⟨91AG(E)324⟩.

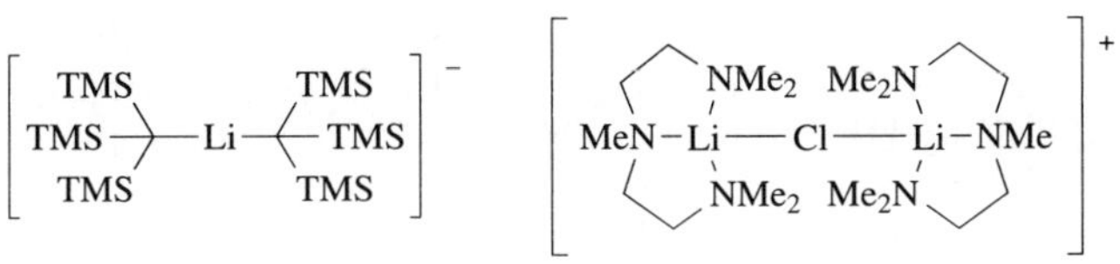

(54)

The metallation of $(PhMe_2Si)_3CH$ with MeLi proceeds in a similar manner to TsiH to give $(PhMe_2Si)_3CLi$ which can be isolated as a solid in 67% yield. This reagent is generally less reactive than TsiLi (it does not react with TMSCl or Me_2SiCl_2) and has been found to have a monomeric structure **(55)** in the solid state ⟨83CC1390, 85JCS(P2)729⟩.

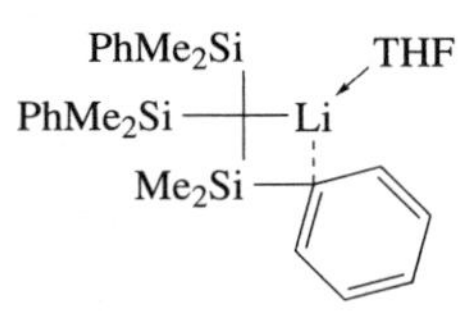

(55)

The metallation of several other Si_3CH functional trisilylmethanes by MeLi or Bu^nLi has also been reported to occur easily and to give preparatively useful reagents. For example, compounds (**32**) and (**35**) are formed from the corresponding C—H compounds by treatment with MeLi/THF and with Bu^nLi/TMEDA respectively ⟨83ZAAC(497)119⟩. The lithium reagent $(TMS)_2(PhMe_2Si)CLi$ is formed from the metallation of $(TMS)_2(PhMe_2Si)CH$ with MeLi in refluxing THF ⟨87CC1461⟩, and trisilylmethane (**56**) is also metallated by MeLi to give (**57**) (Equation (9)) ⟨83ZAAC(497)21⟩. The trisilylmethane derivative (**58**) may also be metallated by MeLi in THF/Et_2O (Equation 10) to give lithium reagent (**59**) but little chemistry has been carried out with the reagent ⟨88JOM(341)109⟩.

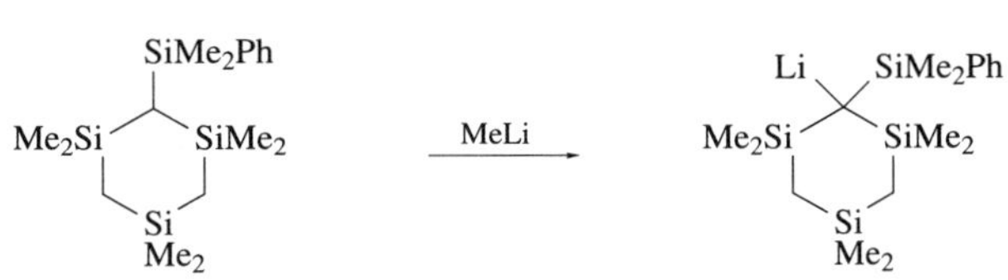

(56) **(57)** (9)

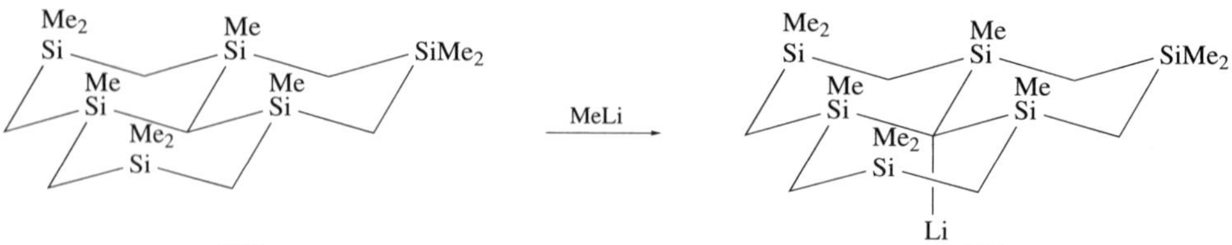

(58) **(59)** (10)

The stabilising effect of three α-silyl groups can be seen in the reactions of species **(60)** with Bu^nLi/TMEDA. The initial reaction products are the HSi_2C carbanions **(61)** but these rearrange to give the more stable Si_3C carbanions **(62)** (Equation (11)) ⟨77ZAAC(430)137⟩.

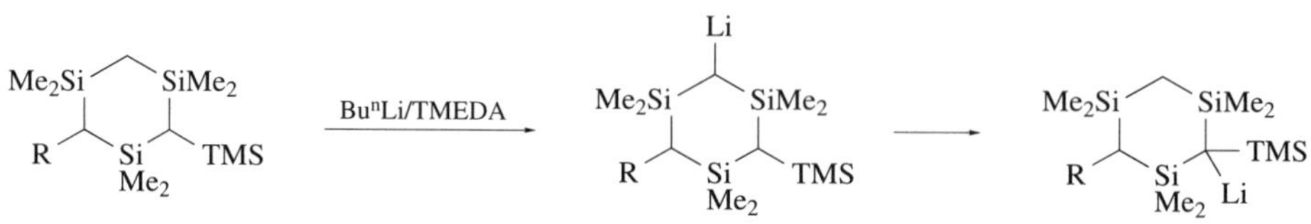

(60) R = H or TMS **(61)** **(62)** (11)

More reactive metallating agents such as Bu^tLi/TMEDA and Bu^nLi/Bu^tOK react readily with $(TMS)_2CHSiMe_2CH_2TMS$ to give $(TMS)_2CLiSiMe_2CH_2TMS$ in >90% yield ⟨91OM551⟩. This is, perhaps, surprising as although the carbanion is stabilised by three α-silyl groups (rather than two if the CH_2 group had been metallated) reaction of the closely related $(TMS)_3CH$ with Bu^tLi/C_5H_{12}/TMEDA gives $(TMS)_2CHSiMe_2CH_2Li$ in which metallation of a methyl group rather than the central methine CH has occured ⟨73JOM(55)209⟩. The highly sterically hindered fluorosilane $Bu^t_2FSiCH(TMS)_2$ reacts with MeLi in THF at room temperature over a period of a week to give the alkyllithium reagent $Bu^t_2FSiCLi(TMS)_2$ which rearranges over a period of weeks to give $Me_2FSiCLi(TMS)(SiBu^t_2Me)$ which has been characterised by x-ray crystallography ⟨87OM35⟩. The metallation of the central carbon by MeLi rather than substitution of F for Me on the silicon is surprising and is presumably due to the severe steric hindrance at the silicon bearing the fluorine together with the fact that the carbanion formed is stabilised by three α-silyl groups ⟨83AG(E)1005, 84JOM(271)381, 86CB1455⟩.

The formation of Si_3LiC functions may also be achieved from Si_3XC substituted methanes (X = halogen) using either other lithium reagents or lithium metal. A range of substituted aryl silanes $(TMS)_2(YC_6H_4Me_2Si)CCl$ (Y = H, *p*-OMe, *p*-Me, *p*-Cl, *m*-CF_3) reacts with Bu^nLi at −110°C to give the corresponding lithium reagents $(TMS)_2(YC_6H_4Me_2Si)CLi$ which react readily with chlorosilanes to give Si_4C containing compounds ⟨94JOM(466)35⟩. A range of $(TMS)_2(XMe_2Si)CLi$ species may be prepared according to the general Equation (12). Such compounds are useful precursors to $(TMS)_2C{=}SiMe_2$ via LiX elimination (which takes place readily except when X = MeO, PhO, Bu^n or Ph which are stable) and the lithium reagents are not, therefore, usually isolated as pure compounds ⟨77AG(E)328, 81CB2087, 81CB3505, 81CB3518, 86JOM(315)9⟩. The reaction between $(TMS)_2(MeOMe_2Si)CCl$ and Bu^nLi in THF/hexane at −80°C gives $(TMS)_2(MeOMe_2Si)CLi$ which reacts with silyl halides to give CSi_4 functions ⟨87JCS(P2)891⟩. The reagent can similarly be prepared at −110°C in THF/Et_2O/pentane ⟨92JOM(437)41⟩. The phenyl containing compound $(TMS)_2(PhMe_2Si)CLi$ can be prepared from the corresponding chloride using Bu^nLi at −110°C ⟨92JOM(437)41⟩. The vinylsilane $(TMS)_2(CH_2{=}CHMe_2Si)CCl$ and Bu^nLi at −100°C in THF/Et_2O/pentane gives the organolithium compound $(TMS)_2(CH_2{=}CHMe_2Si)CLi$ which reacts with various organosilicon halides to give tetrasilylmethanes ⟨87JCS(P2)381, 87JCS(P2)1047⟩. The silyl hydride $(TMS)_2(HMe_2Si)CCl$ reacts with Bu^nLi at −110°C in THF/Et_2O/pentane to give $(TMS)_2(HMe_2Si)CLi$ ⟨92JOM(437)41⟩.

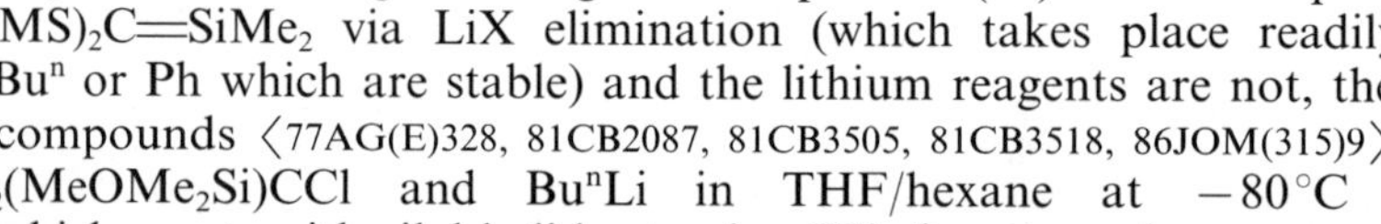

(12)

X = F, Cl, Br, I, OMe, OPh, Ph, Bu^n, Ph_2, PO_2, etc.

The reaction of $(TMS)_2CBr(SiPh_2X)$ (X = F or Br) with PhLi gives the corresponding lithium reagents $(TMS)_2CLi(SiPh_2X)$ (again substitution of Ph for X does not occur). The Si—H compounds

$(TMS)_2CM(SiPh_2H)$ (M = Li or Na) are also formed in good yield from the reaction between $(TMS)_2CBr_2$, Ph_2SiHCl and M in THF at −78 °C. Treatment of $(TMS)_2CBr(SiPh_2Br)$ with two equivs. of RLi or treatment of $(TMS)_2CLi(SiPh_2Br)$ with a second equiv. of RLi affords further trisilyllithium reagents $(TMS)_2CLi(SiPh_2R)$ (R = Me, Bu^n, Ph or OPh) ⟨89CB409⟩. The metallation of a trisilylmethylhalide may also be accomplished by reaction with lithium metal. Thus, $(PhMe_2Si)_3CCl$ reacts with two equivs. of lithium to give $(PhMe_2Si)_3CLi$ and LiCl ⟨77ZAAC(430)121⟩. Lithium reagent $(ClMe_2Si)_2CLi(SiMe_2CCl_2H)$ is probably formed as an intermediate (which rapidly eliminates LiCl) on treatment of $(ClMe_2Si)_2CCl(SiMe_2CCl_2H)$ with Bu^{n}Li ⟨77ZAAC(430)121⟩.

The methoxysilane $(MeOMe_2Si)_3CCl$ reacts with Bu^{n}Li at −78 °C to give the lithium compound $(MeOMe_2Si)_3CLi$ which, in the solid state, is actually a dimer, $[(LiC(SiMe_2OMe)_3)_2]$ having the unusual structure (**63**) where the lithium atoms are solvated intramolecularly by the methoxy groups rather than by solvent molecules ⟨86CC1043⟩.

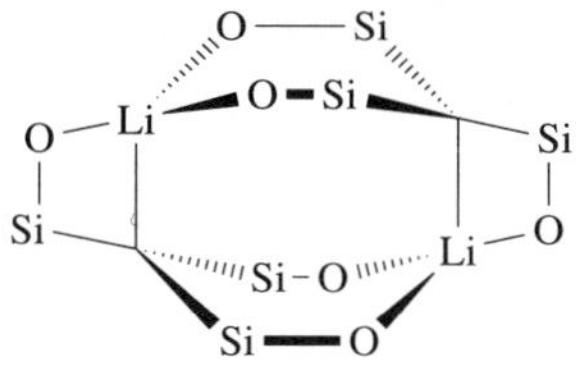

(63)

Coordination around the central carbon in $[\{LiC(SiMe_2OMe)_3\}_2]$. Methyl groups have been omitted for clarity.

The reaction between TsiH and methylsodium in Et_2O in the presence of TMEDA gives a dialkylsodiate $(Na(TMEDA)(Et_2O)(NaTsi_2)$ containing a linear $(Tsi—Na—Tsi)^-$ anion in a manner similar to the metallation of TsiH by MeLi described above ⟨94AG(E)1268⟩. A sodium reagent formulated as TsiNa, which can be derivatised with MeI to give TsiMe, appears to be formed in low yield in the reaction between $(TMS)_4C$ and NaOMe/hexamethylphosphoramide(HMPA) ⟨73TL4193⟩. The alkylpotassium compounds TsiK and $(PhMe_2Si)_3CK$ may both be made in two ways; either by treating the analogous lithium reagents with Bu^{t}OK, or by treating the parent hydrocarbons with MeK, the latter being preferred ⟨94OM753⟩.

(b) Three Si and one group 2 metal function. The highly reactive source of magnesium (Mg(anthracene)(THF)) reacts with TsiCl in THF at 0 °C to give the Grignard reagent TsiMgCl in 90% yield ⟨88JOC3134⟩. The analogous bromide can also be prepared in 75% yield (determined by double titration) from the reaction between TsiBr and Mg in Et_2O ⟨92OM2938⟩. The reaction between magnesium etherate and TsiLi in a 1 : 1 ratio gives complex (**64**) (characterised by x-ray crystallography) ⟨88JCS(D)381⟩. When (**64**) is heated it decomposes to give the lithium-free derivative Tsi_2Mg which x-ray crystallography reveals has the linear structure (**65**) ⟨88JCS(D)381, 89CC273⟩. Although this compound would appear to have potential as a mild alternative to TsiLi for the introduction of the Tsi group into compounds it is rather unreactive. For example, it does not react readily with CO_2, MeI, Me_3COCl or TMSCl. This is presumably because of the steric protection of the C—Mg bonds by the bulky alkyl groups ⟨89CC273, 94JOM(480)199⟩. The reaction between TsiBr and Mg in THF gives complex (**66**). Reactions of this type are complicated by the presence of $MgBr_2$ which arises from the reaction of Mg with $BrCH_2CH_2Br$ used to activate the Mg ⟨94JOM(469)129⟩.

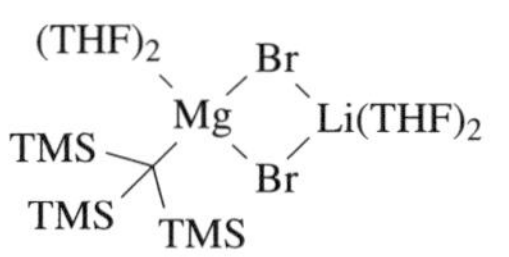

(64)

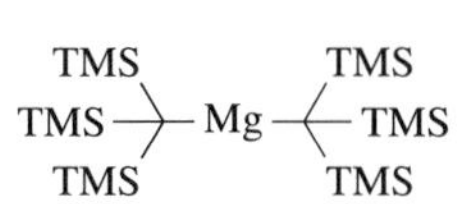

(65)

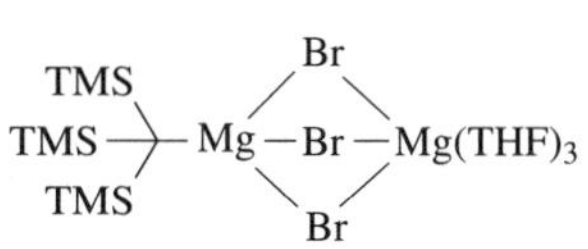

(66)

(c) Three Si and one group 12 metal function. The reaction of about 1.5 equivs. of TsiLi with $ZnCl_2$, $CdCl_2$, or $HgCl_2$ affords the Tsi_2M species in 42, 22 and 35% yields respectively ⟨80JOM(190)101⟩. A better yield is obtained in the reaction between two equivs. of TsiLi (evidently containing some residual MeLi from the metallation of TsiH) and anhydrous $ZnCl_2$ which affords a mixture of Tsi_2Zn (61%) and TsiZnMe (12%) as crystalline solids ⟨91JOM(421)175⟩. These compounds are remarkably thermally and chemically stable because of the high degree of steric hindrance provided by the two bulky Tsi groups to the reactive M—C bonds ⟨80JOM(190)101⟩. The reaction of TsiLi with ZnX_2 in a 1 : 1 stoichiometry affords $[((TMS)_3C)_nX_{n+1}]Li.2THF \cdot Et_2O$ species (X = Cl, $n = 2$; X = Br, n = 1); the chloride reacts with a variety of lithium reagents to give TsiZnR compounds (R = Me, Bu^n, $(TMS)_2CH$, Tsi, $(TMS)_2N$, etc.) ⟨94JOM(469)135⟩.

The reaction between $(PhMe_2Si)_3CLi$ and anhydrous $ZnCl_2$ in THF gives a product formulated as $Li(THF)_2((PhMe_2Si)_3CZnCl_2)$, probably having structure (**67**). When heated, (**67**) affords a sublimate $(PhMe_2Si)_3CZnCl$ which is dimeric in the solid state (as are the analogous Cd and Hg complexes formed in a similar way) and which can be hydrolysed to give the corresponding hydroxide $(PhMe_2Si)_3CZnOH$ ⟨86CC908, 93JOM(462)45⟩. In the solid state this unusual alkylzinc hydroxide has the OH bridged dimeric structure (**68**) ⟨86CC908⟩. Reaction of the lithium reagent $(TMS)_2(HMe_2Si)CLi$ with MBr_2 affords $((TMS)_2(HMe_2Si)C)_2M$ (M = Zn, Cd or Hg) species in 18 and 20% yields for the M = Cd and Hg compounds respectively. The Zn compound reacts with various electrophiles such as Br_2 and CF_3CO_2H to give compounds that are the products from reactions with the Si—H groups rather than the C—Zn bonds ⟨87JOM(320)137, 90CC1471⟩. The reaction of $ZnBr_2$ with two equivs. of the lithium reagents $(TMS)_2(XMe_2Si)CLi$ afford $((TMS)_2(XMe_2Si)C)_2Zn$ in yields of 78 and 50% for X = H and X = OMe respectively ⟨92JOM(437)41⟩. Similarly the reactions of $CdCl_2$ with $(TMS)_2(XMe_2Si)CLi$ reagents afford $((TMS)_2(XMe_2Si)C)_2Cd$ in yields of 40, 66, and 30% respectively for X = H, OMe and Ph ⟨92JOM(437)41⟩. The reaction of two equivs. of $(TMS)_2(MeOMe_2Si)CLi$ with $HgBr_2$ at $-110\,^{\circ}C$ gives $((TMS)_2(MeOMe_2Si)C)_2Hg$ in 30% yield ⟨91JOM(405)149⟩.

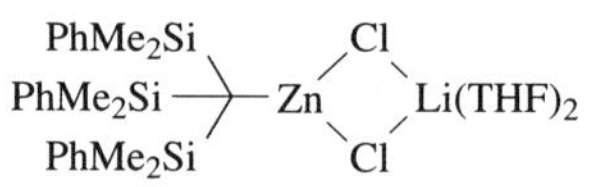

(67)

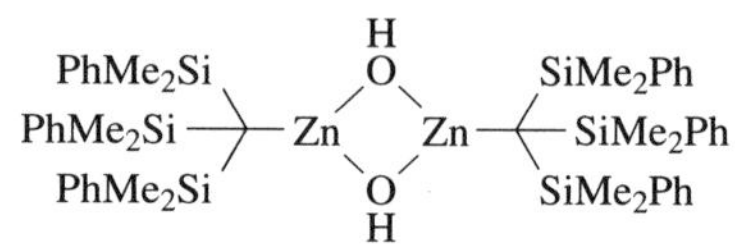

(68)

The reaction of one equiv. of TsiLi and one of $CdBr_2$ in Et_2O/THF affords, after crystallising from cyclohexane, a 21% yield of the trimeric bridged complex $(Li(THF)_4)((CdTsi)_3(\mu\text{-}Br)_3(\mu_3\text{-}Br))$, the anion of which has the structure (**69**), which is based on a cube with one corner missing. A second product (**70**) may be obtained from the reaction in 13% yield by extraction of the reaction residue with heptane. The mechanism by which (**70**) is formed is not known in detail but it is likely to involve some accidental hydrolysis and formation of LiOTMS ⟨88JCS(D)381⟩. The reaction of slightly less than one equiv. of TsiLi and $CdCl_2$ gives some Tsi_2Cd and a 50% yield of $TsiCdCl_2Li(THF)$. On sublimation, the complex breaks down to give TsiCdCl ⟨88JCS(D)381⟩. X-ray crystallography shows that the lithium salt and the alkylcadmium chloride have, in fact, the dimeric and tetrameric structures (**71**) and (**72**) ⟨88CC1389⟩. A similar complex, $Li(THF)_2CdBr_2(C(SiMe_2Ph)_3)$, is formed in the reaction between $(PhMe_2Si)_3CLi$ and $CdBr_2$; this also breaks down on heating to give $(PhMe_2Si)_3CdBr$ ⟨88JCS(D)381⟩. This complex, which readily forms a hydrate when crystallised from wet THF, has also been shown to have a dimeric structure (**73**) ⟨88CC1389⟩.

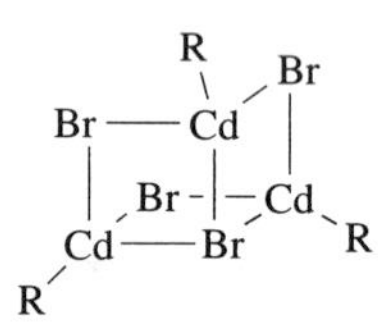

R = $(TMS)_3C$

(69)

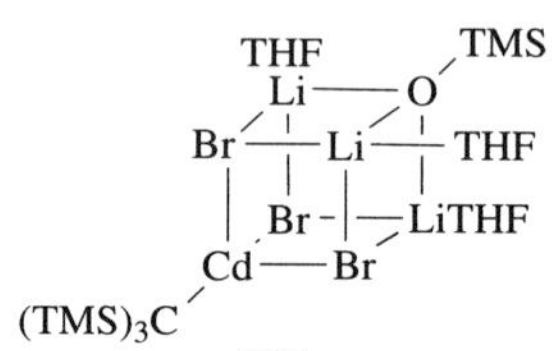

(70)

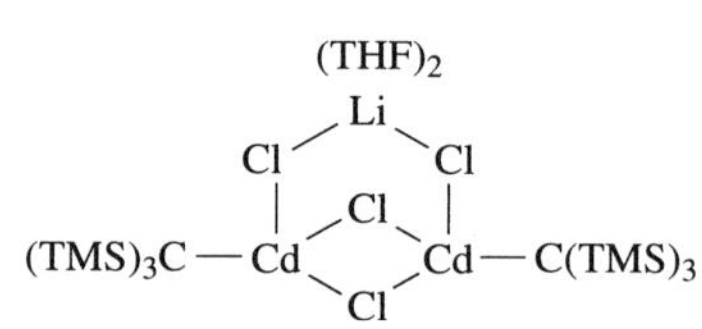

(71)

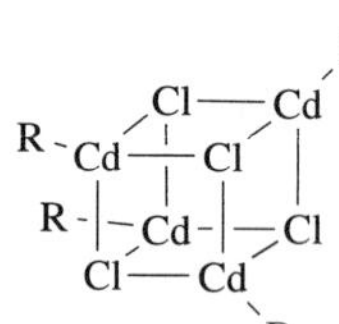

R = $C(TMS)_3$

(72)

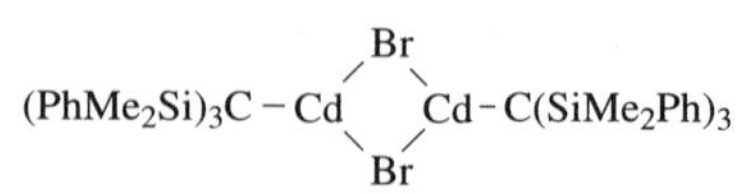

(73)

The reaction between one equiv. of TsiLi and $HgCl_2$ in refluxing THF/Et_2O affords Tsi_2Hg in a 10% yield (20% yield is obtained if $HgBr_2$ is used instead of the chloride) as a colourless solid, while reaction of TsiLi with HgBrMe gives TsiHgMe in 42% yield ⟨77JCR(S)116⟩. The reaction between one equiv. of TsiLi and $HgBr_2$ at 0°C affords TsiHgBr in a yield of 73% ⟨91AG(E)324⟩. This bulky alkylmercury bromide reacts with Grignard reagents to give TsiHgR species (R = Me, Pr^i, Bu^t or Ph) and the chloride $(PhMe_2Si)_3CHgCl$ with lithium reagents or $PhCH_2MgCl$ to give $(PhMe_2Si)_3CHgR$ compounds (R = Me, Bu, Ph, Tsi or $PhCH_2$) ⟨95UP 613-01⟩. The bromide TsiHgBr is also formed in the reaction between $(Fe(CO)_4(HgTsi)_2)$ (obtained from the reaction between TsiHgCl and

($(Fe(CO)_4)^{2-}$) and either $HgBr_2$ or ($Fe(CO)_4(HgBr)_2$) ⟨79JCS(D)767⟩. The reaction of $HgBr_2$ with one equiv. of the lithium reagent $TMS_2(H_2C{=}CHMe_2Si)CLi$ affords $TMS_2(H_2C{=}CHMe_2Si)CHgBr$ in 25% yield ⟨87JOM(320)137⟩. The reaction of $HgCl_2$ and slightly less than two equivs. of TsiLi in refluxing Et_2O has also been reported to give a 70% yield of Tsi_2Hg and between $HgCl_2$ and slightly less than two equivs. of TsiMgBr a 82% yield of the same dialkylmercury species. This is one of the few examples known of the use of a reagent other than TsiLi for the introduction of the Tsi group ⟨92OM2938⟩.

(d) Three Si and one group 13 metal function. A low yield of $TsiAlCl_2$ is obtained from the reaction between TsiLi and $AlCl_3$ in THF/Et_2O. The low yield is probably due to the THF reacting and forming compounds of the type $TsiAl(Cl)O(CH_2)_4Tsi$ as seen in the case of the reaction with BF_3 (See 6.13.1.1.2(ii)) ⟨83JOM(249)23⟩. (See below for the formation of a further TsiAl derivative formed by reduction of a TsiTl species.)

The reaction of $Ga_2Br_4 \cdot 2$ dioxane with TsiLi coordinated to ether solvents (THF or Et_2O) leads only to the isolation of TsiH but if solvent-free TsiLi is used (see 6.13.1.2.1(i)(a)) then the unusual tetragallium tetrahedrane (**74**; M = Ga) is formed as an air-stable solid in 56% yield. The indium analogue of (**74**; M = In) is similarly formed from TsiLi and $In_2Br_4 \cdot 2$ TMEDA ⟨92AG(E)1364⟩. The bulky lithium reagents $(PhMe_2Si)_3CLi$ and TsiLi react with $GaCl_3$ to give products of formula $Li(THF)_2GaCl_3R$ (R = $(PhMe_2Si)_3C$ or Tsi) in 92 and 54% yields respectively. For the case when R = $(PhMe_2Si)_3C$ x-ray crystallography has shown that the structure of the alkylgallium complex has the chlorine bridged structure (**75**) ⟨87JCS(D)747⟩. The reaction between an excess of TsiLi (evidently containing some residual MeLi) and $GaCl_3$ gives a solid thought to be TsiGa(OH)Me which is presumably formed by hydrolysis of TsiGaClMe in the work up ⟨83JOM(249)23⟩.

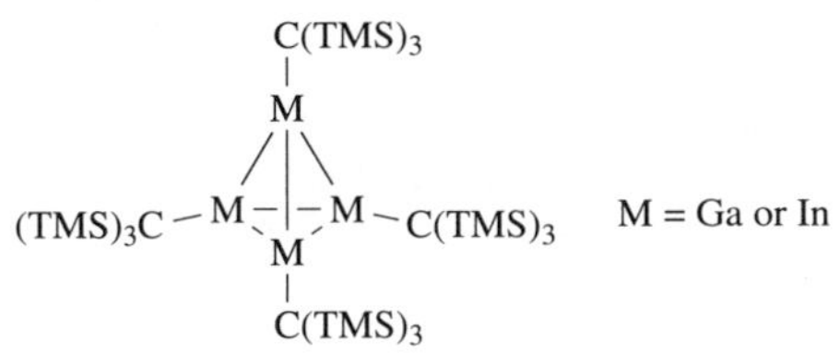

(74)

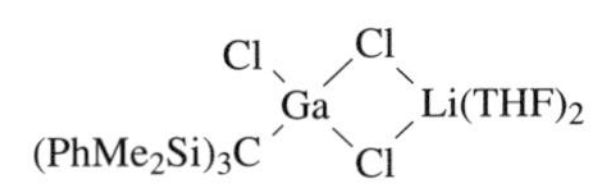

(75)

The reaction between $InCl_3$ and TsiLi gives a complicated alkylindium halide (**76**) containing coordinated LiCl. The indium complex has been reduced using $LiAlH_4$ to give a complex metal hydride (**77**) which when hydrolysed gives a cage compound ($O(TsiIn)_4(OH)_6$) having the structure (**78**) ⟨86CC908, 87JCS(D)747⟩.

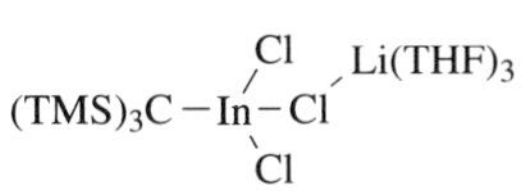

(76)

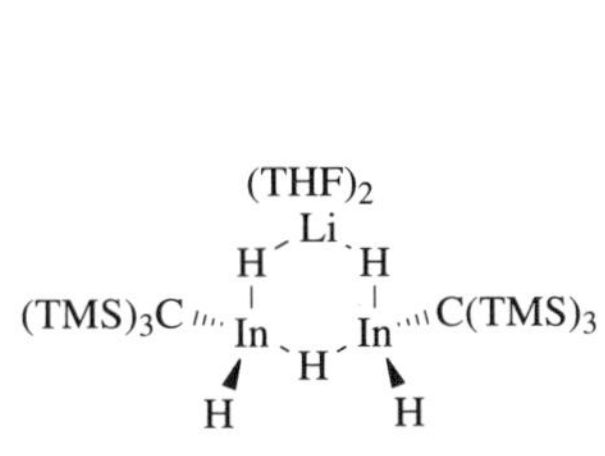

(77)

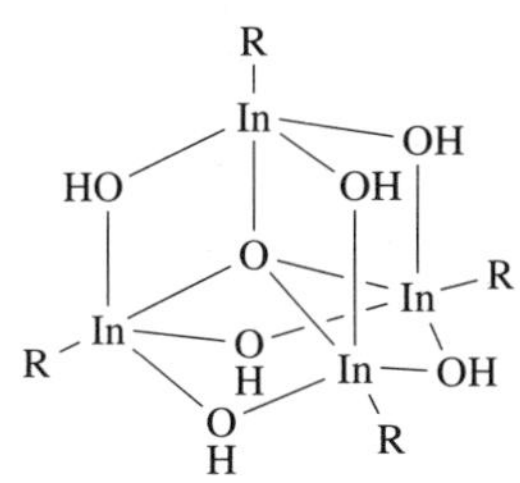

(78)

The reaction between $TlCl_3$ and TsiLi gives a compound formulated as $Li(THF)TlCl_3C(TMS)_3$ in 45% yield but it is not clear whether this compound has a structure containing two bridging Cls as in the gallium complex (**75**) above, only one bridging Cl as in the indium compound (**76**), or some other structure. It is likely, however, that at least two Cls bridge to the Li in order to satisfy the coordination sphere of the Li present. Reduction of the thallium compound with $LiAlH_4$ leads to transfer of the Tsi group and the formation of $Li(THF)_3AlH_3Tsi$ ⟨87JCS(D)747⟩.

(e) Three Si and one group 14 metal function. The reactions between TsiLi and Me_3SnCl, Ph_3SnCl, $(PhCH_2)_2SnCl_2$, $SnCl_4$, $EtSnCl_3$, Me_2SnCl_2, Ph_2SnCl_2, or $SnBr_4$ afford $TsiSnMe_3$ (80%), $TsiSnPh_3$ (34%), $TsiSn(CH_2Ph)_2Cl$ (41%), $TsiSnCl_3$, $TsiSnEtCl_2$, $TsiSnMe_2Cl$, $TsiSnPh_2Cl$ and $TsiSnBr_3$ (containing some impurities) respectively ⟨79BAU2222, 90JCS(D)2643⟩. Treatment of $(Me_3Sn)_4C$ with one equiv. of MeLi followed by one equiv. of TMSCl gives a low yield of $TsiSnMe_3$ together with a variety of other tetrametallomethanes ⟨85JOM(291)179⟩. The reactive silene $(TMS)_2C{=}SiMe_2$ undergoes addition with Me_3SnCl to give $(TMS)_2C(SiMe_2Cl)(SnMe_3)$

⟨87ZN(B)1062⟩. The reaction of $(PhMe_2Si)_3CLi$ with Me_2SnCl_2 in THF/Et_2O gives $(PhMe_2Si)_3CSnMe_2Cl$ in 80% yield ⟨87JOM(325)117, 88JOM(355)33⟩. This is in contrast to the lack of reaction between $(PhMe_2Si)_3CLi$ and Me_2SiCl_2, the longer C—Sn bond length presumably lowering the degree of steric hindrance at the central carbon. The reaction of TsiLi with Me_2SnCl_2 and Ph_2SnCl_2 gives $TsiSnMe_2Cl$ and $TsiSnPh_2Cl$ in 75 and 40% yields respectively ⟨88JOM(355)33⟩. The reaction of the related lithium reagent $(TMS)_2(Me_3Sn)CLi$ with Me_2SiCl_2, TMSCl or Me_2SiHCl gives $(TMS)_2(Me_3Sn)CSiMe_2X$ species (X = Cl, Me, or H) in 75, 84, and 60% yields respectively ⟨88JOM(355)33⟩.

The reactions of TsiLi with lead halides proceed in a manner similar to those for the tin compounds. Thus reactions with Me_3PbCl, Et_3PbCl and Ph_3PbCl afford $TsiPbMe_3$ (79%), $TsiPbEt_3$ (59%) and $TsiPbPh_3$ (50%) respectively ⟨82ICA(58)149⟩. The reaction between two equivs. of TsiLi and Ph_2PbCl_2 does not lead to formation of Tsi_2PbPh_2 (presumably because of the severe steric repulsion between two Tsi groups at a tetrahedral centre) but rather to the unexpected dilead compound $TsiPbPh_2PbPh_3$ in 29% yield ⟨93JCS(D)2891⟩.

(f) Three Si and one transition metal function. The reaction between 1 equiv. of TsiLi and $MnCl_2$ gives a product $(Li(THF)_4)(Tsi_3Mn_3Cl_4THF)$ (**79**) in 60% yield which may be regarded as an 'alkylmanganese chloride'. Reaction of TsiLi with $CoCl_2$ is thought to give a similar cobalt containing species ⟨85CC534, 88JCS(D)381⟩. The monomeric two-coordinate manganese complex Tsi_2Mn is formed in ca. 65% yield as a pale yellow solid from the reaction between 2 equiv. TsiLi and $MnCl_2$ in THF ⟨85CC1380⟩. The preparation of Tsi_2Ni has been reported but few details of the compound are available ⟨71SAP7004922⟩.

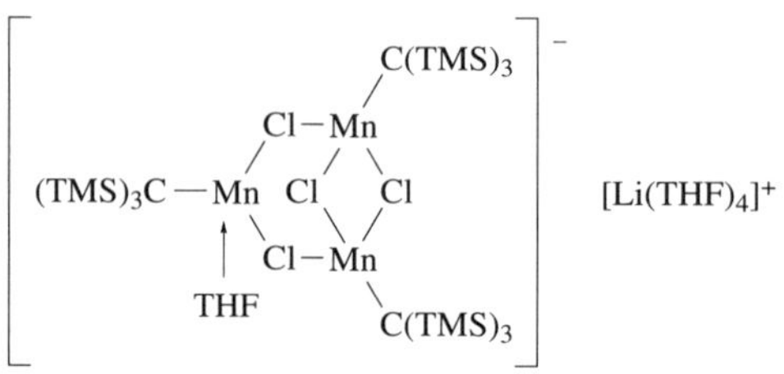

(79)

The Gilman reagent $(Li(THF)_4)(CuTsi_2)$ may be isolated as a solid in 26% yield from the reaction between TsiLi and CuI. As in the case of other Tsi_2M species there is a linear Tsi—M—Tsi arrangement in the solid state ⟨84JOM(263)C23⟩. The reaction between TsiLi and AgI in THF gives the ate complex $(Li(THF)_4)(Tsi_2Ag)$ as a colourless solid in 70% yield which has been structurally characterised ⟨84CC870⟩. The reaction of TsiLi with LAuCl(L = Et_3P, Ph_3P or Ph_3As) gives TsiAuL complexes as stable white crystalline solids ⟨78JCR(S)170⟩.

(ii) Three Ge functions

As has been seen in sections above there is a general lack of compounds containing several germanium atoms attached to the same carbon and there do not appear to be any compounds containing the Ge_3MC grouping known. Such compounds should be readily available via the routes used to prepare the many analogous Si_3MC containing functions as described above.

(iii) Three B functions

The CB_4 substituted compound $C(B(OMe)_2)_4$ reacts with bases MeLi, LiOMe or Bu^nLi to give a compound that may be formulated as $C(B(OMe)_2)_3Li$ which reacts with Ph_3SnCl or Ph_3SnBr to give $C(B(OMe)_2)_3SnPh_3$ in 52 and 65% yield respectively ⟨69JA6541, 73JOM(57)225, 74JOM(69)53⟩. The lithium compound also reacts with Me_3SnCl to give $C(B(OMe)_2)_3SnMe_3$, but rapid disproportionation in the case of similar lead species means that $C(B(OMe)_2)_3PbPh_3$ cannot be isolated from the reaction between the lithium reagent and Ph_3PbCl ⟨69JA6541, 73JOM(57)225⟩. The pinacol ester $C(BO_2C_2Me_4)_4$ also reacts with MeLi to give $C(B_2C_2Me_4)_3Li$ in a similar manner ⟨74JOM(69)53⟩. These lithium reagents would no doubt react with a wide range of other metal and metalloid halides allowing a large number of currently unknown CB_3M functions to be prepared.

6.13.1.2.2 Other mixed metalloid functions

(i) Two Si, one Ge and one metal function

Compounds of the general type $(TMS)_2C(XMe_2Ge)CBr$ (X = F, Br, OMe, OPh, SPh, $(PhO)_2PO_2$, etc.) react with Bu^nLi or PhLi at low temperature (−78 to −120°C) to give lithium compounds $(TMS)_2(XMe_2Ge)CLi$, many of which are thermally unstable, eliminating LiX to give the germene $(TMS)_2C{=}GeMe_2$ in a manner analogous to that of the trisilyl substituted compounds (see 6.13.1.2.1(i)) ⟨86CB2966, 86CB2980⟩. The metallation of $(TMS)_2(Me_3Ge)CCl$ can similarly be achieved by Bu^nLi at −110°C to give $(TMS)_2(Me_3Ge)CLi$ which is relatively thermally stable ⟨87JCS(P2)779⟩. This lithium compound would almost certainly react with many metal and metalloid halides to give a wide range of CSi_2GeM species if required.

The reaction of $(TMS)_2(FMe_2Ge)CBr$ with $Bu^t{}_3SiNa$ gives a quantitative yield of the cyclic species (**41**) (see 6.13.1.1.3(i)). This presumably arises by dimerisation of the germene $(TMS)_2C{=}GeMe_2$ formed by the elimination of NaF from the highly reactive $(TMS)_2(FMe_2Ge)CNa$ intermediate ⟨86CB2966, 86CB2980⟩.

(ii) Two Si, one B and one metal function

The reduction of (**80**) with four equivs. of lithium (Scheme 12) affords the unsaturated boron compound (**81**) which may also be written as its tautomer (**82**). Mono protonation affords a compound analogous to (**81**) containing the $(TMS)_2CH$ group which suggests that one of the lithium atoms was associated with the Si_2B substituted carbon ⟨88AG(E)961⟩. The structure of the 2,3,5,6-Me_4C_6H analogue of (**81**) and (**82**) in the solid state shows that the structure is actually better represented by (**83**) ⟨90AG(E)1032⟩.

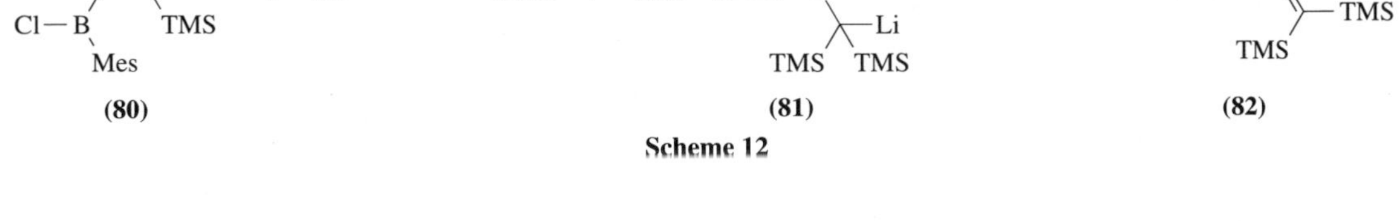

Scheme 12

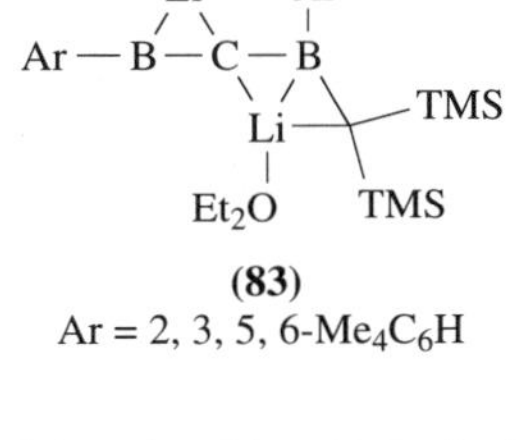

(**83**)
Ar = 2, 3, 5, 6-Me_4C_6H

The addition of lithium to $(TMS)_2C{=}BAr$ (Ar = mesityl or 2,3,5,6-tetramethylphenyl) gives a radical anion which dimerises to give (**84**). X-ray crystallography shows that the compound (**84**) (with Ar = mesityl) has short C—Li and short B—Li distances. The reaction of (**85**; Ar = 2,3,5,6-tetramethylphenyl) (see 6.13.1.1.3(iii)) with Bu^tLi gives (**86**) (Equation (13)) in which there is again an L-aryl interaction ⟨90AG(E)1030⟩.

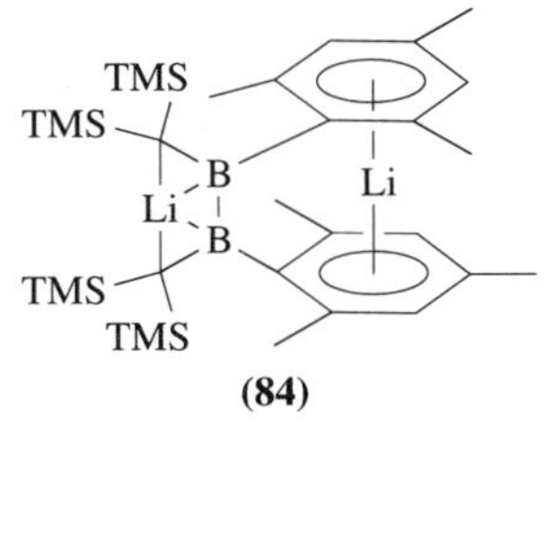

(**84**)

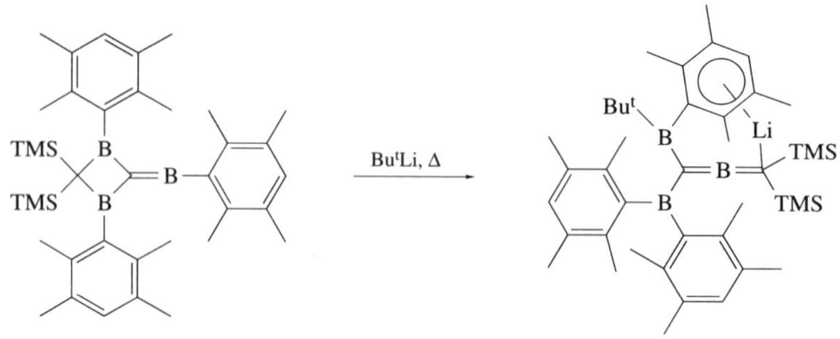

(85) **(86)**

(13)

(iii) One Si, one Ge, one B and one metal function. Also two Ge, one Si and one metal function. Also two Ge, one B and one metal function. Also Two B, one Si and one metal function. Also two B, one Ge and one metal function

Very few examples of compounds containing these functions seem to have been prepared. Compounds containing such functions should, however, be available from synthetic routes described above in this section. In particular, such functions should be available from the reactions between suitably substituted lithium reagents and appropriate metal halides.

The reaction of (**87**; R = Pr^i_2N) with two equivs. of potassium gives (**88**; M = K, R = Pr^i_2N) in which the ring carbons can be considered to have B_2KSiC substitution ⟨86AG(E)1112⟩. The related lithium complex (**88**; M = Li, R = Bu^t) can be prepared in a similar way (Equation (14)), although its actual structure is a 'sandwich' with four lithium atoms between two puckered rings and a complicated coordination pattern between the 4 Li, 2 B and 2 ring carbon atoms ⟨86AG(E)1111⟩.

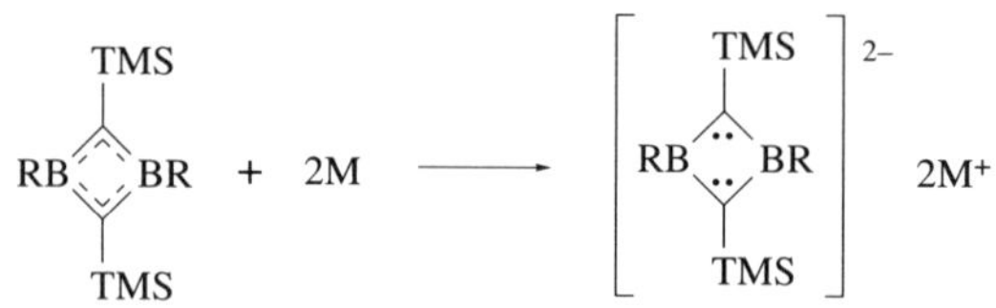

(14)

(87) **(88)**

R = NPr^i_2, M = K
R = Bu^t, M = Li

6.13.1.3 Methanes Bearing Two Metalloid and Two Metal Functions

6.13.1.3.1 Two Si and two metal functions

The reaction between lithium vapour and $(TMS)_2CCl_2$ gives $(TMS)_2CLi_2$ in 26% yield as determined by reaction with D_2O ⟨84CC1664⟩. The dilithium reagent is also produced, together with $(TMS)_2CH_2$, on the thermolysis of (TMS)CHLi at 160 °C ⟨88PO2023⟩. A more practical synthesis of $(TMS)_2CLi_2$ is from the reaction between $(TMS)_2CCl_2$ and four equivs. of LiDBB (DBB = 4,4′-di-*t*-butylbiphenyl). Relatively little synthetic work seems to have been done with this interesting dilithium reagent but it has been shown to react with Me_3SnCl to give $(TMS)_2C(SnMe_3)_2$ ⟨88TL5237⟩ and it is likely that it would react with other metal halides to give further novel Si_2M_2C substituted compounds.

The reaction of (**47**) (Scheme 10) with Me_3SnCl not only gives compounds containing the CSi_2B_2 function but also the CSi_2Sn_2 function (see 6.13.1.1.3(iii)) ⟨89AG(E)781, 89AG(E)784⟩. For a further example of a compound containing the CSi_2Li_2 function, see the CSi_2BLi containing structure (**83**) ⟨90AG(E)1032⟩.

The di-Grignard reagent $(TMS)_2C(MgBr)_2$ may be prepared in several ways (Scheme 13), the best probably being the route using just Mg metal and the halomethane. This route gives a yield of about

60% but the di-Grignard reagent is rather unreactive towards electrophiles and gives relatively low yields of ditin compounds (presumably via halogen exchange) $(TMS)_2C(SnMe_3)(SnMe_2X)$ (X = Cl or Br) when treated with Me_3SnCl ⟨89TL6195⟩. The low reactivity of the di-Grignard reagent is thought to be due to the shielding of the carbanionic centre by the four relatively bulky groups (2 × TMS and 2 × $MgBr(THF)_2$) attached to it when crystallised from THF/hexane ⟨92JA7302⟩.

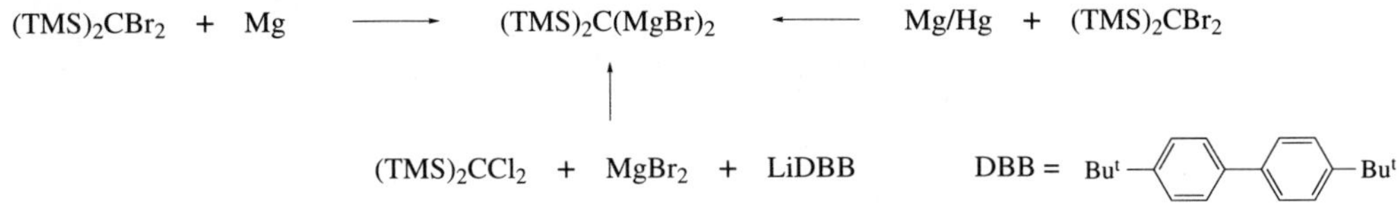

Scheme 13

The Al_2Si_2C species (**89**) and (**90**) are both predicted to have planar geometries at carbon and to be stable in the gas phase or in an inert matrix. It is not clear, however, how such unusual species could be prepared ⟨91CC1536⟩.

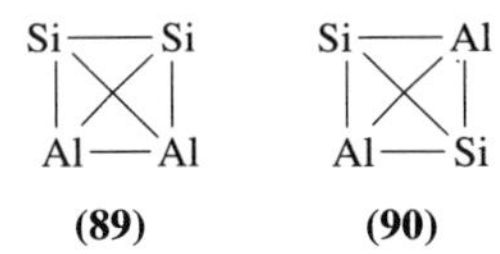

The reaction between $(TMS)_2CBrLi$ and Me_2SnCl_2, Me_2SiCl_2, or Me_2GeBr_2 in a 2:1 ratio gives the cyclobutane species (**91**) in 20, 36 and 17% yields respectively. Although the mechanism of formation of these rings was thought to be via a series of transmetallation steps, more recent work suggests that the doubly bonded species $(TMS)_2C{=}MMe_2$ which dimerise readily in a head-to-tail fashion may well be the source of the rings. Acyclic species (**92**) and (**93**) are formed in the tin case by treatment with Br_2 and MeLi (Scheme 14) ⟨74JA6237, 76JOM(116)257⟩.

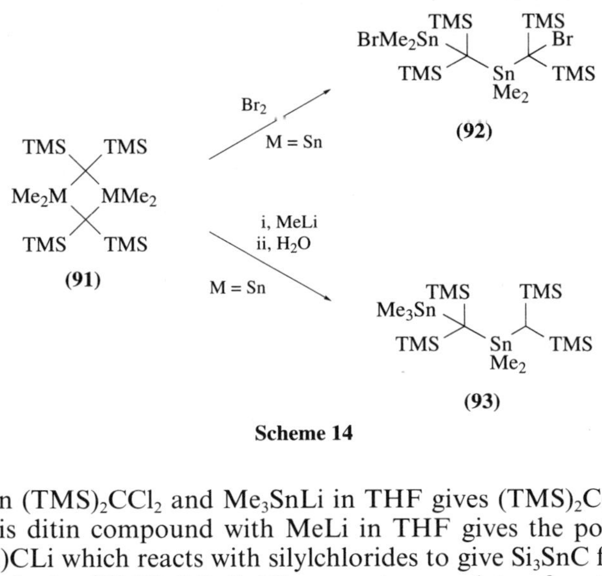

Scheme 14

The reaction between $(TMS)_2CCl_2$ and Me_3SnLi in THF gives $(TMS)_2C(SnMe_3)_2$ in 76% yield. Further reaction of this ditin compound with MeLi in THF gives the potentially useful lithium reagent $(TMS)_2(Me_3Sn)CLi$ which reacts with silylchlorides to give Si_3SnC functions and could, no doubt, be used to attach the $(TMS)_2(Me_3Sn)C$ group to a variety of metals thus increasing the number of Si_2SnMC functions known ⟨87CC1183, 88JOM(355)33⟩. The ditin compound is also formed in 23% yield as one component of a complicated mixture formed from the reaction of $(Me_3Sn)_4C$ with one equiv. of MeLi followed by one equiv. of TMSCl ⟨85JOM(291)179⟩.

Treatment of $(TMS)_2(BrMe_2Sn)CBr$ with PhLi at −110°C in Et_2O affords $(TMS)_2(BrMe_2Sn)CLi$ which readily eliminates LiBr to give the highly reactive unsaturated $(TMS)_2C{=}SnMe_2$; this dimerises to give the distannacyclobutane in good yield ⟨91AG(E)93⟩.

The major product from the reaction between $(TMS)_2CClLi$ and $HgCl_2$ is not the expected $((TMS)_2CCl)_2Hg$, but the Hg_2Si_2C functional species (**94**) (38% yield) which is presumably formed from the desired product via transmetallation reactions ⟨70JOM(23)361⟩.

A transition metal complex (**95**) containing a carbon substituted by two silicon and two tungsten atoms is formed in low yield in the reaction between WCl_4 and Na/Hg in DME followed by treatment with three equivs. of $(TMS)_2NLi$. The mechanism by which (**95**) is formed is unknown ⟨88POL2049⟩.

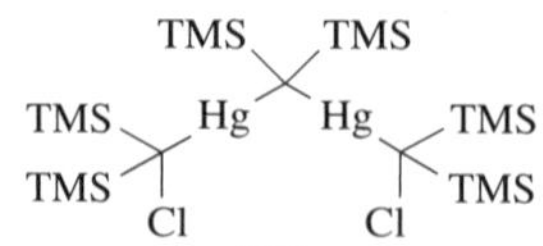

(**94**)

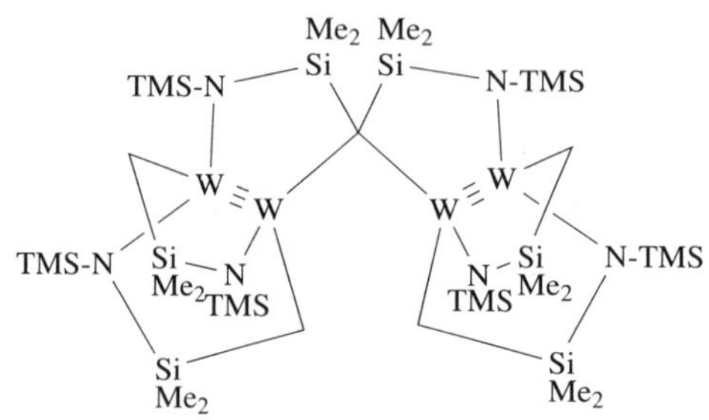

(**95**)

6.13.1.3.2 *Two Ge and two metal functions*

There do not appear to be any compounds containing this function known but they could, no doubt, be prepared in similar ways to the analogous Si_2M_2C containing species described above in 6.13.1.3.1.

6.13.1.3.3 *Two B and two metal functions*

The CB_3Sn substituted compound $C(B(OMe)_2)_3SnPh_3$ rapidly undergoes base-catalysed disproportionation to give a high yield of $C(B(OMe)_2)_2(SnPh_3)_2$ together with $C(B(OMe)_2)_4$. This type of disproportionation occurs even more readily for the lead analogue and only $C(B(OMe)_2)_2(PbPh_3)_2$ is isolated in attempts to prepare $C(B(OMe)_2)_3PbPh_3$ ⟨69JA6541, 73JOM(57)231⟩. The propanediolboronic esters (**96**) react with BuLi to give lithium compounds (**97**) which react with metal halides Ph_3MCl to give compounds (**98**) containing M_2B_2C functions in good yields (Scheme 15). The lithium reagents (**97**) could presumably be used to prepare a wider range of M_2B_2C functions by treatment with various metal halides ⟨73JOM(57)231⟩.

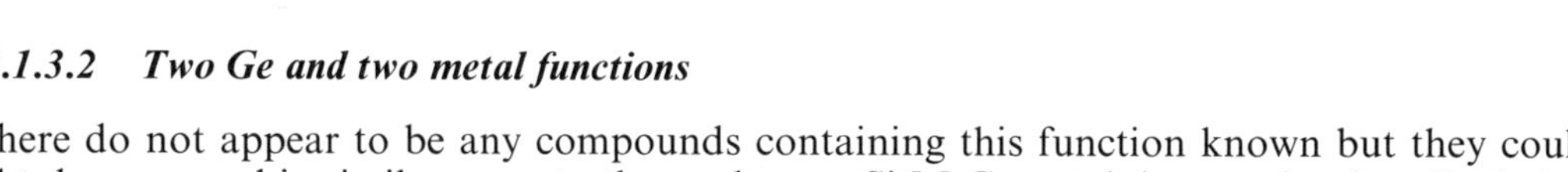

(**96**) (**97**) (**98**)

M = M' = Sn, 84%; M = M' = Pb, 39%; M = Sn, M' = Pb, 79%

Scheme 15

Addition of Bu^tLi to the unsaturated boron species (**99**) leads to formation of a dilithium salt in 50% yield which x-ray crystallography shows to have a diboraallene structure (**100**) in the solid state (Equation (15)) ⟨88AG(E)1370⟩. The dilithium reagent has been used to prepare several other compounds containing multiply bonded boron atoms ⟨90AG(E)399⟩.

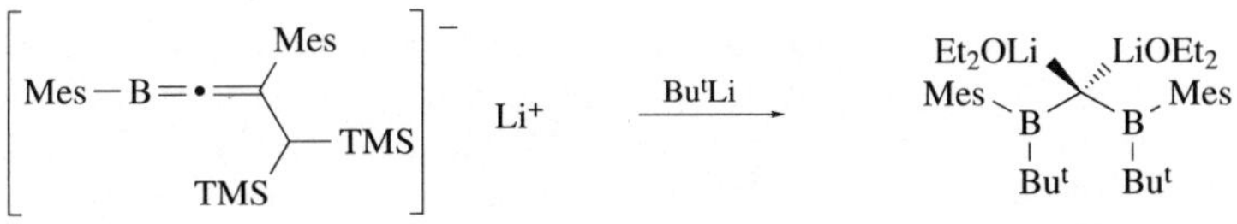

(15)

(99) **(100)**

The reaction between a dilithium reagent and two equivs. of Me_3SnCl (Equation (16)) affords a B_2Sn_2C substituted species (Ar = mesityl, 93% Ar = 2,3,5,6-Me_4C_6H, 98%). Presumably treatment of the dianion with other metal halides would lead to further B_2M_2C containing compounds ⟨89AG(E)784⟩.

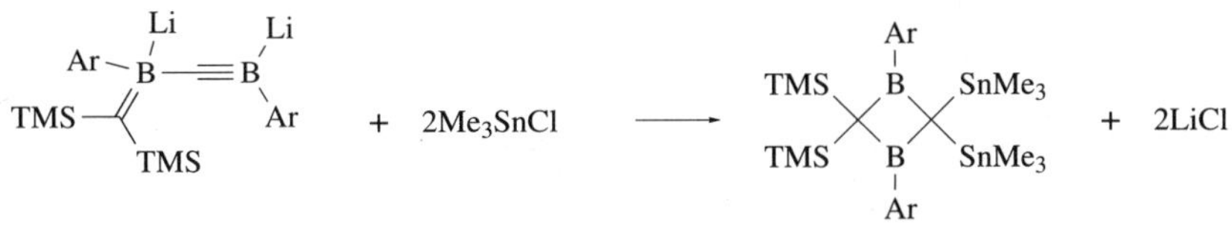

(16)

(47)

Ar = 2, 4, 6-$Me_3C_6H_2$, 93%
Ar = 2, 3, 5, 6-Me_4C_6H, 98%

Treatment of the dianion (**50**) with Ph_3PAuCl gives compounds (**51**) in 22% yield, containing both B_2Si_2C and Au_2B_2C groupings (see 6.13.1.1.3(iv)) which seem to have been formed via a silyl migration and a geminal substitution ⟨86AG(E)1112, 89CB1881⟩.

Calculations on the B_2Li_2C functional compound (**101**) have shown that a planar coordination at C is energetically preferable to a tetrahedral configuration but the compound does not seem to have been made ⟨76JA5419, 87T1019⟩.

(101)

6.13.1.3.4 Other combinations of two metalloids and two metal functions

There are very few, if any, compounds known containing mixed metalloids and two metal functions. However, they should be available by the routes described above for related compounds containing two of the same metalloid.

6.13.1.4 Methanes Bearing One Metalloid and Three Metal Functions

6.13.1.4.1 One Si and three metal functions

The main component of the mixture formed on treating $(Me_3Sn)_4C$ with MeLi followed by one equiv. of TMSCl is $(Me_3Sn)_3CTMS$, but this does not seem to have been isolated as a pure compound ⟨85JOM(291)179⟩. A similar lead compound, $(Me_3Pb)_3CTMS$, is formed in the reaction between $(Me_3Pb)_3CLi$ (see 6.13.2.2) and TMSCl ⟨91SA(A)849⟩. Reactions between the lead substituted lithium reagent and other halosilanes would no doubt allow a range of $(Me_3Pb)_3CSiR_3$ compounds to be prepared.

The reaction between trihalomethanes and $Co_2(CO)_8$ gives trinuclear complexes containing μ_3-carbyne ligands, e.g., (**102**) in Equation (17). This reaction can be extended to include silyl substituted trihalomethanes; thus, $Me_2SiHCCl_3$, $TMSCX_3$ (X = Cl or Br), or $PhMe_2SiCCl_3$ react with $Co_2(CO)_8$ to give complexes (**102**) ($R^1R^2R^3$ = Me_2OH (after some hydrolysis in the workup), Me_3 and $PhMe_2$) ⟨73JOM(50)265, 79JOM(178)227⟩. Unfortunately, silyltrihalomethanes are difficult to make and an alternative route to the carbyne species is to treat methylidyne or bromomethylidyne tricobaltnonacarbonyl complexes with silanes $R^1R^2R^3SiH$ as shown by the general Equation (17) ⟨77JA5209,

79JOM(178)227, 81JOM(221)257⟩. The route using the μ_3-CH species is probably the better one as it does not generate HBr which can cause unwanted side reactions such as Si—Ph bond cleavage at the silicon. Once formed, the complexes (**102**) can undergo a variety of reactions at silicon without disrupting the cluster framework thus enabling a wider range of compounds to be easily prepared ⟨77JA5209⟩.

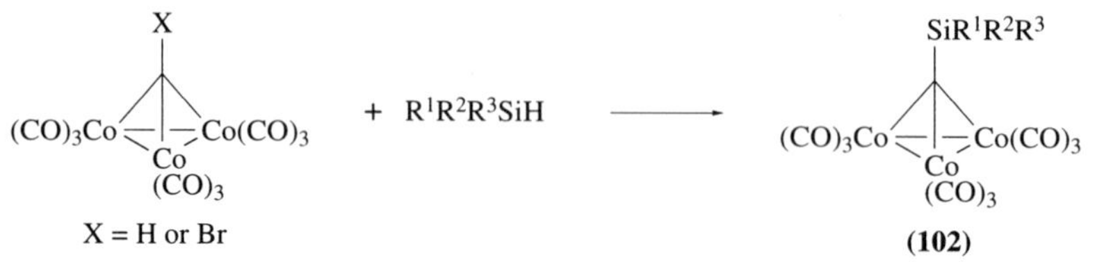

(**102**)

(17)

The reaction of silicon hydrides that are themselves within a transition metal complex with $HCCo_3(CO)_9$ has also given compounds containing an $SiCo_3C$ grouping (see Scheme 16) ⟨92JOM(437)323⟩.

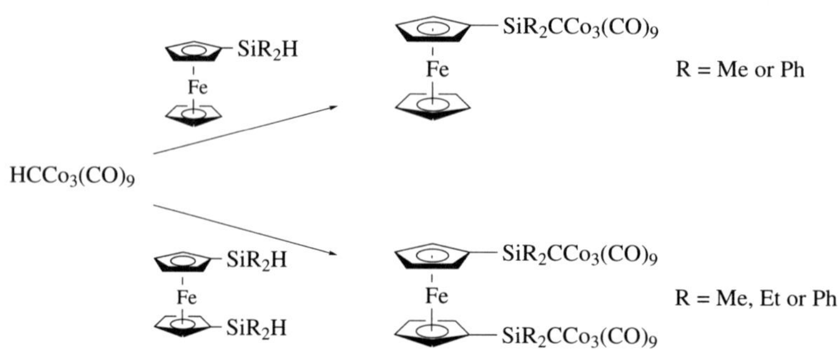

Scheme 16

Cobalt bridged carbyne complexes of the type shown above in Equation (17) undergo metal substitution reactions giving mixed metal clusters as shown in Scheme 17. Alternatively, mixed metal clusters may be prepared from mixed metal carbyne clusters (Scheme 17) ⟨90JOM(381)261⟩.

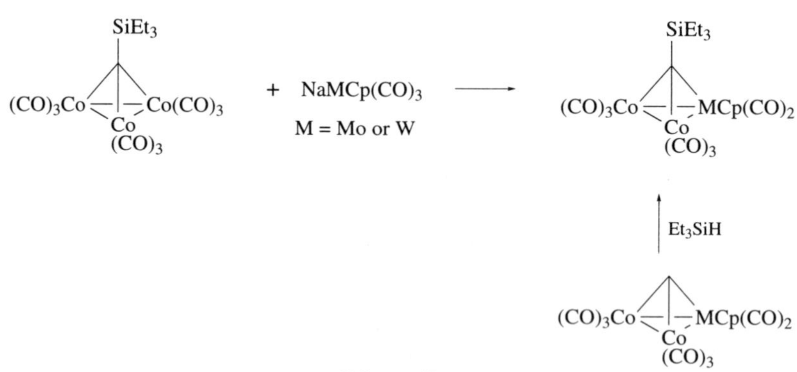

Scheme 17

The cobalt mediated cleavage of alkynes and concomitant build-up of clusters seems to be a general reaction and a wide range of compounds have been prepared using this type of methodology. Thus various alkynes and dialkynes react with either $(\eta^5\text{-}C_5H_5)Co(CO)_2$ or $(\eta^2\text{-}C_2H_4)_2Co(\eta^5\text{-}C_5H_5)$ to give tricobalt cluster species containing μ_3-$CSiMe_3$ ligands as outlined in Equation (18) ⟨81JOM(217)105, 82JOC3192, 86OM394⟩. Such reactions are clearly complicated and often give low yields of the compounds shown in Equation (17) together with numerous other products although, in some cases, the intermediates may be isolated and fully characterised.

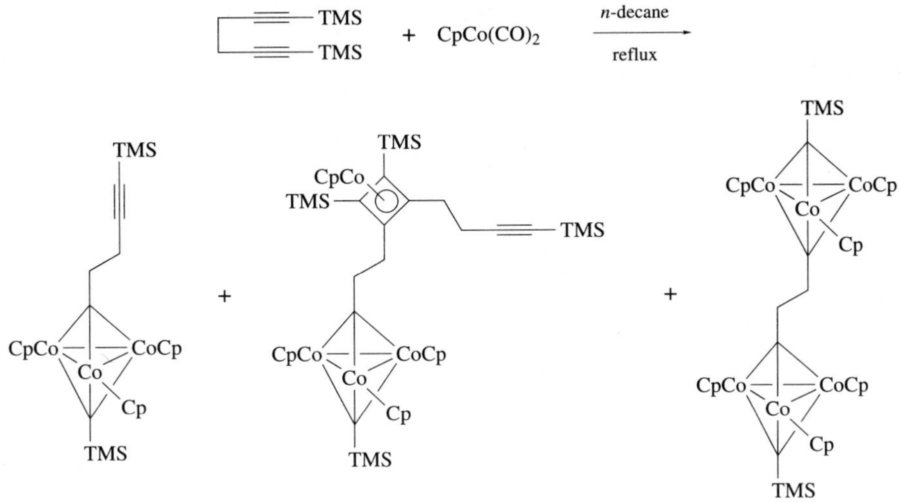

(18)

The reaction between $CpCo(CO)_2$ and TMSC≡CC≡CTMS gives structure (**103**) ⟨79JA2768⟩. The hexametallic complexes (**104**) rearrange on flash vacuum pyrolysis at 550 °C to give the isomers (**105**) containing two μ_3-CTMS ligands (Equation (19)) ⟨83JA1384⟩.

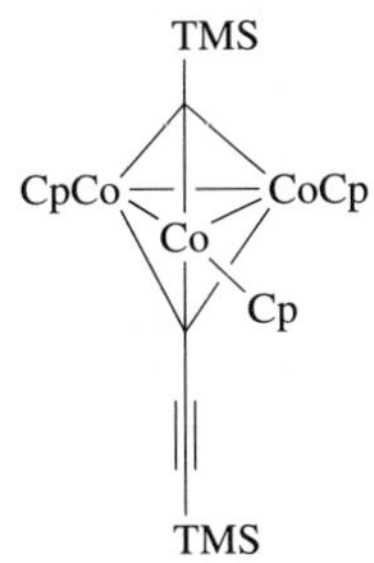

(103)

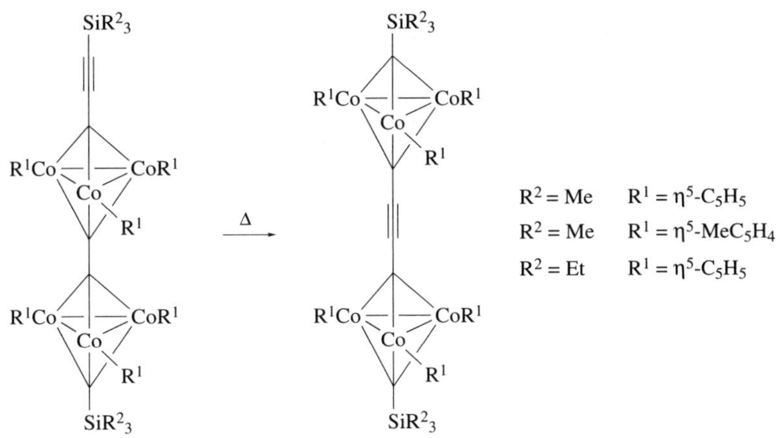

(104) **(105)** (19)

Cluster formation also occurs when nickelocene is treated with alkyllithium reagents. Thus $TMSCH_2Li$ gives the carbyne bridged cluster (**106**) and (**107**) (if excess lithium reagent is used) as a black–violet solid in 15% yield which may also be obtained from the reaction between nickelocene and the nickel alkyl complex (**108**) (Scheme 18) ⟨79JOM(178)371⟩.

The bimetallic asymmetrically bridging alkylidyne complex (**109**) reacts with low valent transition metal fragments to form a range of trimetallic species, (**110**), (**111**) and (**112**) in good yield containing a μ_3-$CSiPh_3$ ligand which bridges three different transition metals (Scheme 19) ⟨88JOM(355)231⟩.

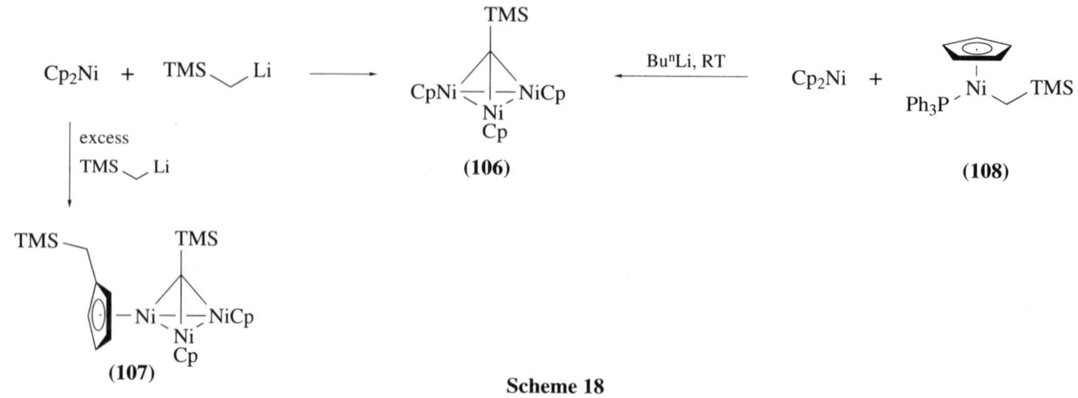

Scheme 18

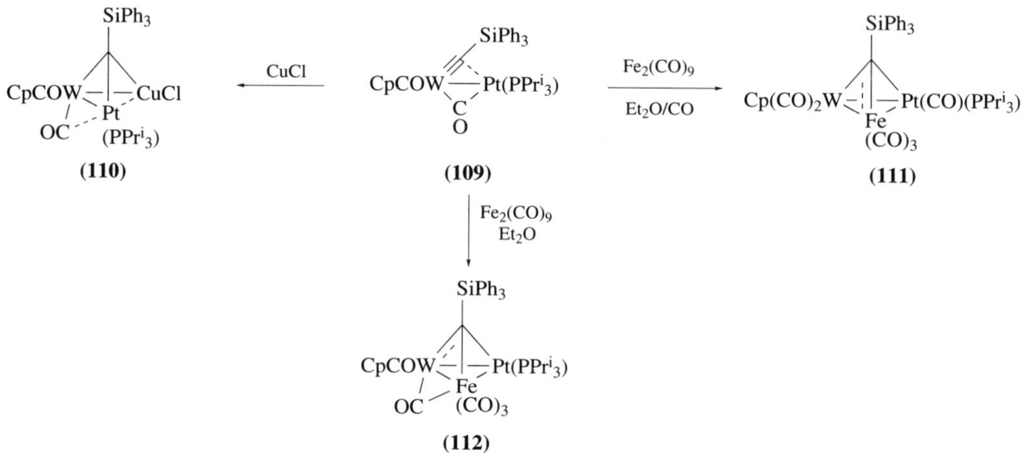

Scheme 19

6.13.1.4.2 One Ge and three metal functions

Germanes, R_3GeH, react in a similar way to the silanes discussed above in 6.13.1.4.1 with the methylidyne cluster $HCCo_3(CO)_9$ to give germyl substituted methylidynes. For example, germanes Ph_3GeH, MePh(1-naphth)GeH, Et_3GeH, $Bu^n{}_3GeH$, Et_2ClGeH and $(PhCH_2)_2ClGeH$ all give the appropriate $R_3GeCCo_3(CO)_9$ species with structure similar to the silicon analogues (**102**) above (Equation (17)) ⟨77JA5209, 79JOM(178)227, 80JOM(188)329, 81JOM(221)257⟩.

6.13.1.4.3 One B and three metal functions

There seem to be few compounds known containing this type of function although they may be available by treatment of a methylidyne cluster with R_2BH species in the same way that the silyl and germyl substituted carbyne complexes described above are made.

6.13.2 METHANES BEARING FOUR METAL FUNCTIONS

6.13.2.1 Methanes Bearing Four Similar Metals

Calculations on the structures of CLi_4 and $C(BeH)_4$ have been carried out in order to ascertain whether the small electropositive substituents might promote planar rather than tetrahedral geometry at carbon. In both cases tetrahedral geometry is still preferred ⟨76JA5419⟩.

The synthesis of CLi_4 may be achieved in several ways. The original synthesis involved the reaction of CCl_4 with lithium vapour at 750 °C giving a 40.5% yield as determined by deuterolysis to give CD_4 ⟨83JOM(249)1⟩. More convenient and apparently higher yielding syntheses involve the reaction between $C(HgEt)_4$ and Bu^tLi or those between $C(HgCl)_4$ and either Bu^tLi or lithium metal ⟨84AG(E)995⟩. Tetralithiomethane reacts in a variety of complicated ways with simple derivatising agents and its synthetic use is therefore probably limited.

The tetrastannylmethane $(Me_3Sn)_4C$ is formed as a readily sublimable solid from the reaction between Me_3SnLi and either $CHCl_3$ or CCl_4 in THF ⟨75OMS(10)18, 84JOM(266)37⟩. The compound has also been prepared enriched in ^{13}C at the central carbon by use of $[^{13}C]CCl_4$ in a similar reaction ⟨79JOM(172)293⟩. The reaction between CCl_4 and four equivs. of Me_3SnNa affords only 18% $(Me_3Sn)_4C$ at room temperature but 41% if carried out at −23 °C ⟨76JOM(122)171⟩. The reaction between $(Me_3Sn)_4C$ and one equiv. of MeLi followed by one equiv. of Me_3PbCl affords a mixture of tetrametallomethanes including a very low yield of $(Me_3Pb)_4C$ ⟨85JOM(291)179⟩. The tetrastannylmethane also reacts with various electrophilic reagents to give other substituted Sn_4C compounds as a result of cleavage of the Sn—Me bonds only ⟨84JOM(266)37⟩. The reaction between four equivs. of Ph_3PbLi and CCl_4 affords the stable crystalline solid $(Ph_3Pb)_4C$. The compound is sparingly soluble in organic solvents and only decomposes at 292–294 °C ⟨65RTC43⟩.

The reaction of ethanol in alkaline solution with mercuric oxide gives a species $C_2Hg_6(OH)_6$ known as 'ethane hexamercarbide'. This material dissolves in aqueous solutions of carboxylic acids to give crystalline compounds $C(HgOCOR)_4$ (R = Me or CF_3) which have been characterised by x-ray crystallography and which react with MeSH in H_2O to give $C(HgSMe)_4$ ⟨74CC646, 77ZN(B)1022⟩. Tetramercuriomethanes may also be prepared from $C(B(OMe)_2)_4$ by treatment with various mercury compounds. Thus, reactions with $Hg(OAc)_2$ and with EtHgOAc afford $C(HgOAc)_4$ and $C(HgEt)_4$ respectively ⟨70JA231, 84AG(E)995⟩. Reactions between the acetate $C(HgOAc)_4$ and HF, halide or cyanide ions gives the compounds $C(HgX)_4$(X = F, Cl, Br, I, or CN) ⟨78JOM(153)1, 79ZN(B)390, 84AG(E)995⟩.

6.13.2.2 Methanes Bearing Three Similar and One Different Metal Function

The reaction between $(Me_3Sn)_4C$ and one equiv. of MeLi followed by one equiv. of Me_3PbCl affords a mixture of tetrametallomethanes including $(Me_3Sn)_3CPbMe_3$ and $(Me_3Pb)_3CSnMe_3$ in yields of 44 and 8% yields respectively. Unfortunately, although the yield of the $PbSn_3C$ compound is reasonable, it is likely to be difficult to separate such a mixture and the reaction is, therefore, of little synthetic use ⟨85JOM(291)179⟩. The lithium reagent $(Me_3Sn)_3CLi$ may be observed by NMR spectroscopy from the reaction between $(Me_3Sn)_4C$ and MeLi in Et_2O ⟨81JMR(44)54⟩. A similar lead-substituted methyllithium $(Me_3Pb)_3CLi$ may be prepared by treating $(Me_3Pb)_3CH$ with MeLi in THF ⟨91SA(A)849⟩ in the same way as $(TMS)_3CLi$ is prepared from $(TMS)_3CH$ (6.13.1.2.1(i)(a)). The lead substituted compound could presumably be used in the same way as the silicon substituted lithium reagent to prepare a wide variety of compounds containing the $(Me_3Pb)_3C$ group by reaction with suitable metal or metalloid halides.

6.13.2.3 Methanes Bearing Two Similar and Two Different Metal Functions

The reaction between $(Me_3Sn)_4C$ and one equiv. of MeLi followed by one equiv. of Me_3PbCl affords a mixture of tetrametallomethanes including a 26% yield of $(Me_3Pb)_2C(SnMe_3)_2$ ⟨85JOM(291)179⟩.

6.13.2.4 Methanes Bearing Four Different Metal Functions

Although there are a potential 135751 different combinations (and their optical isomers) of 43 different metals at a tetrahedral carbon centre very few seem to have been prepared. A comprehensive search of the literature for such a large number of functional groups is clearly very difficult to carry out, and it is quite possible that many such groupings have been missed in compiling this chapter. It is hoped that omission of such groupings will not be serious for the organic chemist.

6.13.3 METHANES BEARING MORE THAN FOUR METALLOID OR METAL FUNCTIONS

The structures of magnesium ⟨91AOC(32)147⟩, lithium ⟨85AOC(24)354⟩ and Na, K, Rb, or Cs ⟨87AOC(27)169⟩ compounds are usually complicated in the solid state often having carbon with a coordination number greater than four. Such species will not be further described here and are fully discussed in the reviews referred to.

There are a variety of carboranes known in which the cage carbons have a metal or metalloid substituent. Such compounds usually contain carbon with a formal coordination number greater than four and are, therefore, strictly outside the coverage of this section, however a few representative compounds are given below. The reaction between $C_2B_{10}H_{12}$ with BuLi affords (**113**) which reacts with various mercury salts to give compounds of type (**114**) (Equation (20)) ⟨67JOM(7)385, 68BAU414, 73JGU848⟩. Such compounds will not be discussed further here but they have been the subject of several reviews (see for example, ⟨90AOC(30)99, 92CRV225, 92CRV251⟩ and references therein).

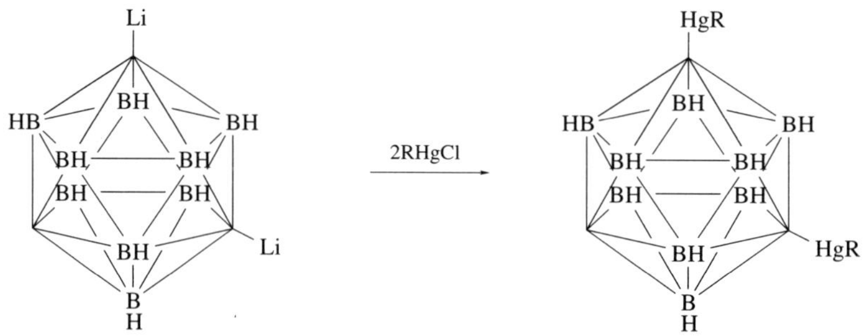

(113) **(114)** (20)

Carbon may have six-coordinate octahedral geometry in complexes containing dicationic ions such as in (**115**) formed according to Equation (21). Such compounds may be characterised by x-ray crystallography and the central carbon observed by [^{13}C]CNMR spectroscopy ⟨88AG(E)1544, 91AG(E)1488⟩.

$$C[B(OMe)_2]_4 + 6LAuCl + 6CsF \longrightarrow 6CsCl + 2B(OMe)_3 +$$

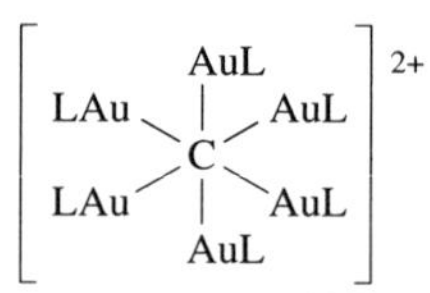

$$[MeOBF_3]^{-}_{2} \quad (21)$$

L = Ph_3 or $Ph_2(p\text{-}Me_2NC_6H_4)P$ **(115)**

6.14
Functions Containing a Carbonyl Group and at Least One Halogen

GEOFFREY E. GYMER and
SUBRAMANIYAN NARAYANASWAMI
Pfizer Central Research, Sandwich, UK

6.14.1 CARBONYL HALIDES WITH TWO SIMILAR HALOGENS

Compounds of formula $X_2C{=}O$, where X = F, Cl, Br or I, are currently named in *Chemical Abstracts* as carbonic dihalides. Other names commonly applied to these compounds are carbonyl dihalides, carbonyl halides and halophosgenes. The latter derives from the historical name phosgene, which is still the term most commonly used to describe carbonic dichloride.

Methods of synthesis have been previously reviewed; the review by Hagemann ⟨83HOU(E4)1⟩ is particularly important because it also covers the remit of Section 6.14.2 and includes some preparative details. This area is covered comprehensively in *Beilstein* up to the end of 1980. By far the most important member of this class is carbonic dichloride, which will henceforth be referred to by its more common name, phosgene. Phosgene is an important industrial chemical used mainly in the manufacture of polyurethanes and polycarbonates. Although the organization of this section retains the logical sequence by commencing with discussion of the synthesis of carbonic difluoride in Section 6.14.1.1, it may be prudent of the reader also to take note of some aspects discussed in Section 6.14.1.2, concerning the preparation and safe handling of phosgene, and possible ways of avoiding its use.

6.14.1.1 Carbonic Difluoride, COF_2

Carbonic difluoride, also called carbonyl difluoride and fluorophosgene, is a stable, colourless gas, b.p. −83 °C. It is commercially available from several suppliers; the industrial preparation probably involves the reaction of carbon monoxide with elemental fluorine. In common with all carbonyl dihalides, it is extremely toxic. In addition, hydrolysis and other nucleophilic attack leads to the formation of hydrogen fluoride, with all its attendant hazards. The suitability of glass apparatus for the preparation or reaction of carbonic difluoride should be carefully considered. If significant quantities of hydrogen fluoride could be generated during the reaction or work-up procedure, then failure of glass apparatus may occur with potentially disastrous results. Gaseous products can be freed of hydrogen fluoride by passage through a copper tube packed with sodium fluoride pellets ⟨60IS155⟩. Glass is inert to pure carbonic difluoride.

Three prinicipal routes to carbonic difluoride have been described, together with several miscellaneous preparations (Scheme 1).

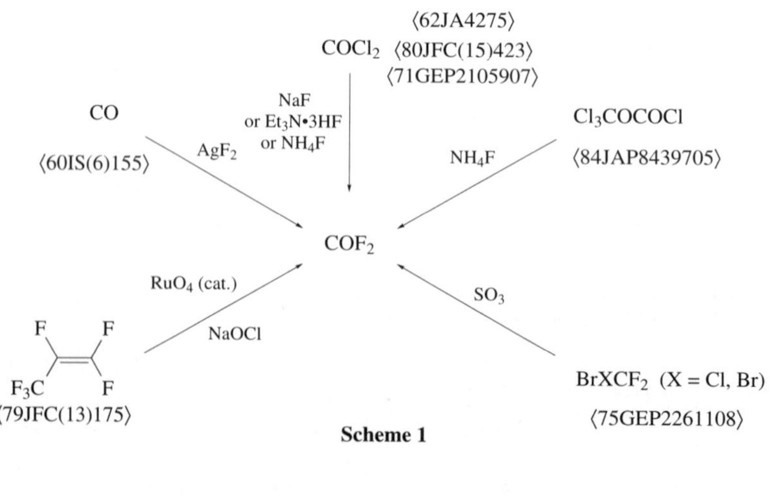

Scheme 1

6.14.1.1.1 *From carbonic dichloride (phosgene)*

Transhalogenation of phosgene with various fluorine donors is a well-described method of preparing carbonic difluoride.

The early work of Steinkopf and Herold led to the preparation of impure carbonic difluoride in low yield. They achieved this by warming a mixture of arsenic trifluoride and phosgene ⟨20JPR79⟩. Emeléus and Wood devised a substantially improved procedure using antimony trifluoride; they obtained carbonic difluoride of 90–95% purity, free from hydrogen chloride, from which it cannot be separated by distillation ⟨48JCS2183⟩. This method was further refined by Haszeldine and Iserson, who added chlorine to the mixture to form pentavalent antimony salts, allowing them to obtain carbonic difluoride of 95% purity in 80–85% yield ⟨57JA5801⟩. These reactions were performed at quite high pressures and temperatures, and these methods were superseded by the work of Tullock and co-workers, who were able to prepare carbonic difluoride of 95% purity in 70–80% yield through the reaction of phosgene and sodium fluoride in acetonitrile at 30–35 °C and atmospheric pressure ⟨62JA4275, B-75MI 614-01⟩. This is an excellent preparative method, only equalled by the alternative of Franz, who has used the triethylamine trishydrogen fluoride complex and phosgene in acetonitrile at room temperature to give carbonic difluoride in quantitative yield, containing carbon dioxide as the only impurity ⟨80JFC(15)423⟩. A similar method employing ammonium hydrogen fluoride as the fluorine source is also reported to give carbonic difluoride in 90% yield, together with 8% carbonic chloride fluoride ⟨71GEP2105907⟩. Another potentially useful route involves substitution of trichloromethyl chloroformate for phosgene using similar reaction conditions, thus avoiding to some extent the dangers associated with the handling of phosgene ⟨84JAP8439705⟩. Because of the extreme toxicity of phosgene, alternatives are described in Section 6.14.1.2.8.

Two gas phase reactions leading to carbonic difluoride are worthy of mention, since they can be run as continuous flow procedures, suitable for industrial production. In the first, thionyl fluoride (from thionyl chloride and sodium fluoride) and phosgene are passed as a 4 : 1 mixture over iron(III) chloride at 400 °C to give carbonic difluoride in 96% yield ⟨80EGP139936⟩. Many other catalysts have been examined, and the SOFCl and $SOCl_2$ formed during the reaction can be recycled by treatment with NaF to regenerate thionyl fluoride. The second method involves the preparation of carbonic chloride fluoride by passing phosgene over calcium fluoride at 200–550 °C, and then passing it over activated carbon, when metathesis to carbonic difluoride and phosgene takes place ⟨88EUP253527⟩. In these last two preparations it should be borne in mind that at the elevated temperatures employed, phosgene is at least partly dissociated to carbon monoxide and chlorine, and thus fluorination of carbon monoxide may be the predominant reaction pathway.

Finally, Glemser and Biermann have prepared carbonic difluoride in quantitative yield by passing a 1 : 1 mixture of phosgene and nitrogen trifluoride through a nickel tube heated to 400 °C ⟨67CB2484⟩. These workers were fully aware that fluorination of carbon monoxide produced by thermal decomposition of phosgene could explain the reaction course observed. The same workers have shown that fluorination of CO with NF_3 does in fact yield COF_2 in nearly quantitative yield ⟨67CB1184⟩.

6.14.1.1.2 *From carbon monoxide*

The fluorination of carbon monoxide with elemental fluorine is probably the preferred method of production of carbonic difluoride on an industrial scale, since it is amenable to continuous flow technology ⟨B-89MI 614-01⟩. As with many reactions employing elemental fluorine, this does not appear to be a straightforward process ⟨34ZAAC(221)154⟩; however, it had already been found to be satisfactory as a preparative procedure by the 1940s ⟨44MI 614-01, B-63MI 614-01⟩. In the latter preparation, carbon monoxide is actually burnt in an atmosphere of electrolytically generated fluorine, and Kwasnik refers to the risk of violent explosion if the procedure is incorrectly conducted. This method has now been improved by introducing carbon monoxide into the liquid HF electrolyte used for the fluorine preparation ⟨70JAP7026611⟩. The purity of the COF_2 produced is apparently improved when potassium fluoride is added to the electrolyte ⟨69USP3461050⟩ when COF_2 is the sole product.

A well-described laboratory procedure by Tullock and co-workers exists for the fluorination of carbon monoxide by a more amenable method. Carbon monoxide and helium are mixed together and passed through a copper tube containing silver(II) fluoride. An exothermic reaction ensues, to give COF_2 contaminated with HF, which is removed by passing the effluent gases through sodium fluoride pellets. Upon condensation, the COF_2 is obtained in 70–85% yield and greater than 99%

purity. This method avoids the low-temperature fractional distillation usually necessary to obtain pure product ⟨60IS155⟩.

Fluorination of carbon monoxide with sulfur hexafluoride under the influence of IR radiation has also been claimed as a simultaneous preparation of carbonic difluoride and sulfur tetrafluoride ⟨81CZP208251⟩. A less attractive but high-yielding method involves passing an equimolar mixture of carbon monoxide and NF_3 through a nickel tube heated to 450–520 °C, to give COF_2 in 90–95% yield based on NF_3 ⟨67CB1184⟩. NF_3 is prepared by electrolysis of molten ammonium fluoride hydrofluoride, so this is not normally an accessible laboratory method ⟨66CB371⟩.

Olah and Kuhn prepared carbonic difluoride as a pure side-product in 61% yield when in pursuit of carbonic bromide fluoride by the exothermic reaction of bromine trifluoride with carbon monoxide ⟨56JOC1319⟩. This method is well described, but is probably unacceptably hazardous ⟨B-63MI 614-01⟩.

6.14.1.1.3 Oxidative methods

Oxidation of terminal 1,1-difluoroalkenes—in practice usually perfluoroalkenes—leads to carbonic difluoride. One method stands out as being particularly suitable for laboratory use, namely catalytic ruthenium tetroxide oxidative cleavage. Commercially available perfluoropropene can be oxidized by a catalytic quantity of ruthenium tetroxide in Freon 113 with gradual addition of peracetic acid, periodic acid or sodium hypochlorite to regenerate the catalyst. The products are TFA and COF_2, both in high yield ⟨79JFC(13)175⟩. It is reasonable to speculate that if tetrafluoroethylene was used in the reaction, the sole product would be COF_2, possibly constituting a very simple and controllable small-scale laboratory preparation. Oxidation of tetrafluoroethylene with oxygen at 200–450 °C has been used as a very high-yielding industrial preparation ⟨65NEP6509518⟩, which can be made more controllable by dilution with Freons ⟨74URP424809⟩. Carbonic difluoride is also produced in quantitative yield from perfluoropropene and oxygen under IR radiation, in the presence of SF_6 as a radiation absorber ⟨90CZP264358⟩.

Terminal 1,1-difluoroalkenes can also be ozonised to give carbonic difluoride with high implied yields; however, this has not yet been developed as a synthetic route to COF_2 ⟨80JA7572, 89BAU1570⟩. Finally, carbonic difluoride can be prepared by oxidation of chloro- or bromodifluoromethane with oxygen at 200–500 °C, a method suitable for an industrial flow process ⟨89EUP310255⟩.

6.14.1.1.4 Other methods

The reaction of fluorohalomethanes with sulfur trioxide provides a versatile route to all carbonic fluoride halides. Carbonic difluoride is thus available by reaction of either dibromodifluoromethane or bromochlorodifluoromethane with SO_3. The presence of catalysts such as sulfuric acid or mercury(I) or mercury(II) salts is beneficial, and is in fact necessary for efficient reaction of the second substrate. In this manner, dropwise addition of dibromodifluoromethane to SO_3 at 35–44 °C gave 63% carbonic difluoride based on starting material consumed ⟨74GEP2261108⟩. The scope of this reaction in fact extends beyond that encompassed above, as will be demonstrated in subsequent sections of this chapter, making it probably the most general laboratory method available.

Fluorination of carbonyl sulfide also provides a route to carbonic difluoride. When carbonyl sulfide is passed through a copper tube packed with cobalt trifluoride maintained at 200 °C, the effluent gases comprise only COF_2 and SF_6, which are readily separated by low-temperature fractionation ⟨52JA5792⟩. In a second example, carbonyl sulfide diluted with helium is introduced into an electrolyte consisting of sodium fluoride in anhydrous hydrogen fluoride. Electrolysis produces carbonic difluoride, 81%, and sulfur hexafluoride, 99% ⟨76JAP7630034⟩.

Lastly, in a patent to the Dow Chemical Company, heating a mixture of calcium fluoride, titanium dioxide and carbon at 2800–3500 °C under argon gives rise to carbonic difluoride as the only volatile product. This industrial process uses very cheap starting materials which may compensate for the large of amount of energy consumed; it can be operated continuously, and various other fluoride sources and oxides are also satisfactory ⟨67USP3322823⟩.

6.14.1.2 Carbonic Dichloride (Phosgene), $COCl_2$

Named in *Chemical Abstracts* as carbonic dichloride, $COCl_2$ is more commonly known as phosgene. Other commonly encountered names are carbonic acid dichloride, carbonyl dichloride and carbonyl chloride. It is an important building block for the chemical industry. It is a colourless gas, b.p. 8 °C, of legendary toxicity ⟨B-93MI 614-01⟩. In 1988 the consumption of phosgene in the USA alone was 1.6×10^9 pounds (7.3×10^8 kg), of which approximately 90% was absorbed by the manufacturers of polyurethane plastics ⟨B-89MI 614-01⟩. It is thus widely available, and is supplied either as a liquified gas in a pressurized cylinder, or in solution, most commonly as a 20% by weight solution in toluene.

A comprehensive review of the methods of preparation of phosgene would not in the authors' opinion be of great value since the vast majority of methods are of no synthetic importance; phosgene is so widely available and so toxic that its preparation in the laboratory should be severely questioned. Exceptions to this are the production of minute quantities of isotopically labelled material, which are covered in detail later in this section. Instead the authors intend to devote part of this section to informing readers on the safe use of phosgene and its replacement by less hazardous alternatives.

Phosgene preparation has been reviewed up to the end of 1925 ⟨20JCS1410, 27CRV109⟩. Subsequent reviews relating to phosgene appear to pay little attention to its synthesis ⟨82KO(17)416, 83HOU(E4)1⟩. This is not surprising since the current method of phosgene manufacture has been known since 1878 ⟨1878G233⟩.

6.14.1.2.1 Phosgene toxicity

In order to handle phosgene in a safe manner, an appreciation of its toxicity is necessary. Phosgene is an insidious poison. The initial effects of irritation to the eyes and mucous membranes often disappear upon removal from exposure; however, within 24 h, pulmonary oedema, due to lung tissue acylation, commonly develops. This is often fatal. At high exposures (>200 ppm), phosgene can pass into the blood, and death from congestive heart failure follows. From animal and, more sadly, human experiments the following generalization emerges: LC_{50} (lethal concentration to 50% of recipients) = 570 ppm min. This implies that 10 ppm exposure for 1 h or 100 ppm for 6 min would exceed the LC_{50}. In the event of phosgene exposure, the casualty should be at once removed from exposure, kept warm and rested. Artificial respiration and external cardiac massage should be employed where necessary. Mouth to mouth resuscitation can be employed without undue risk to the operator. Removal of contaminated clothing and copious irrigation of the skin and eyes with water are also necessary. In all cases the casualty should be removed to hospital and kept under observation. Medical treatment may involve corticosteroids, administration of oxygen and other symptomatic measures ⟨B-93MI 614-02⟩.

6.14.1.2.2 Handling phosgene

Phosgene should always be used in a fume hood with powerful extraction. Two persons should be present during the entire period that exposure could occur. A method of monitoring phosgene concentration is advisable. A simple method is to tape phosgene indicator papers to the fume hood sill. These are prepared by soaking filter papers in a 10% alcoholic solution of equal parts of *p*-dimethylaminobenzaldehyde and diphenylamine, and allowing to dry. These yellow papers turn deep orange in the presence of dangerously high concentrations of phosgene ⟨B-89MI 614-02⟩. A Draeger pump specific for phosgene is a more sophisticated safeguard. Reactions quite often require an excess of phosgene, or fail to proceed, in both cases leaving residual phosgene at the end of the experiment. This must be purged from the reaction mixture by a vigorous stream of nitrogen, the effluent gas being scrubbed with sodium hydroxide solution. Phosgene may still be present after this operation, so reaction solvent subsequently evaporated off should also be freed from phosgene before being discarded. This could be achieved by addition of a reactive amine or alcohol, which would convert any remaining phosgene to a urea or carbonate.

Several excellent alternatives to phosgene have been developed; these will be reviewed in Section 6.14.1.2.8. However, sometimes phosgene provides the simplest, cleanest way of accomplishing a desired reaction.

6.14.1.2.3 *Preparation of phosgene—an overview*

Phosgene was first prepared by the interaction of chlorine and carbon monoxide in the presence of sunlight, by John Davy (the brother of Sir Humphrey Davy) in 1812 ⟨20JCS1410⟩. The name is derived from two Greek words meaning 'light' and 'I produce'. By 1878, Paternò replaced photoinduction by passing the gases over active carbon in the form of animal charcoal ⟨1878G233⟩. This method is still employed today for the manufacture of billions of kilograms of phosgene each year. Currently the process is run at 50°C and 5–10 atm in the presence of activated carbon to produce phosgene at a cost of less than 50 cents per pound, or 25 cents per kilogram ⟨B-89MI 614-01⟩. In 1869 the only other preparation to have withstood the test of time was discovered by Schützenberger ⟨1869BSF198⟩. This involves reaction of carbon tetrachloride with sulfur trioxide. The reaction is conveniently done in oleum, a solution of sulfur trioxide in sulfuric acid ⟨19CR(169)17⟩. Armstrong reported that hexachloroethane also produced phosgene under the same conditions ⟨1870CB730⟩. It was also discovered that the use of sulfuric acid without the addition of sulfur trioxide still led to the production of phosgene ⟨27CRV109⟩. The oleum process was used to produce phosgene in Italy during the First World War, but could not compete with the continuous catalytic process discovered by Paternò. Many other methods have been developed, often alternative ways of combining chlorine and carbon monoxide, with chlorine present as a metallic chloride ⟨28G443, 30CB1221⟩ or nonmetallic chloride such as SCl_2 ⟨68FRP1526958⟩. Many catalysts have also been considered as alternatives to activated carbon in the Paternò process ⟨82MI 614-01⟩. Several methods involving the oxidation of chloroform or carbon tetrachloride were also developed, using oxygen ⟨24MI 614-01⟩ or chromium reagents ⟨1893CB1990⟩. Pyrolysis of various compounds containing chlorine, carbon and oxygen has also often led to the production of phosgene, notably oxalyl chloride, which when heated in a sealed vessel at 340°C for 70 h gave phosgene in quantitative yield together with carbon monoxide ⟨13CB1426⟩. A more recent example involves the decomposition of trichloroacetic acid at 300–400°C ⟨72URP345094⟩. Other substrates were also pyrolysed to give phosgene in this patent.

6.14.1.2.4 *Laboratory methods suitable for preparing phosgene*

Should the laboratory preparation of phosgene be undertaken, great care must be taken that safe handling practices are adopted. It is well to remember that the most hazardous part of any process is often the dismantling of apparatus and disposal of residues. The reader is urged to consult Section 6.14.1.2.2. Some examples are shown in Scheme 2.

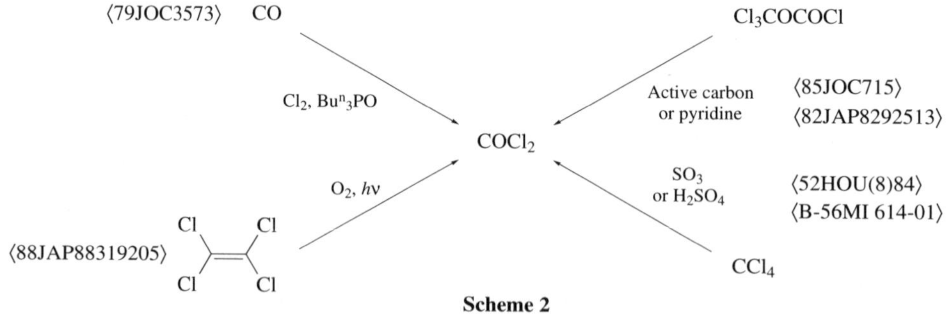

Scheme 2

(i) *Preparation of phosgene from carbon tetrachloride and sulfuric acid*

In this method, carbon tetrachloride is dropped into 100% sulfuric acid containing 2% by weight ignited kieselguhr maintained at 120–130°C. The evolved phosgene is absorbed into toluene, in which it is extremely soluble, whilst the hydrogen chloride produced passes through ⟨B-56MI 614-01⟩.

(ii) From carbon tetrachloride and oleum

This method is essentially that of Grignard and Urbain ⟨19CR(169)17⟩; however, it is more accessible through *Houben-Weyl* ⟨52HOU(8)84⟩. When 45% oleum is heated to 78 °C and treated dropwise with carbon tetrachloride, phosgene is evolved, which is passed through sulfuric acid and collected (92%).

(iii) From trichloromethyl chloroformate, $Cl_3COCOCl$

The use of trichloromethyl chloroformate as a phosgene alternative will be dealt with in detail in Section 6.14.1.2.8. However, dropwise addition of trichloromethyl chloroformate to carbon tetrachloride containing a trace of pyridine gives, after 30 min, 90% conversion to phosgene, with no chloroformate remaining ⟨82JAP8292513⟩. The authors suggest that the less carcinogenic solvent dichloromethane should be substituted for carbon tetrachloride in this preparation.

A solution of trichloromethyl chloroformate in THF is also rapidly decomposed to phosgene by the addition of activated charcoal ⟨85JOC715⟩.

6.14.1.2.5 Other laboratory preparations of phosgene

Four methods follow which are possibly suitable for the small-scale preparation of phosgene. Three of these are less attractive because of special conditions employed (irradiation, pressure) or lack of detail. The fourth is a demonstration experiment designed for educational purposes and may not be of synthetic utility.

(i) Phosphine oxide-catalysed reaction of chlorine and carbon monoxide

Workers at the Ube Chemical Industries Company have published a series of patents in which phosgene is produced in high yield (>90%) by the phosphine oxide- or sulfide-catalysed reaction of chlorine with carbon monoxide under pressure in carbon tetrachloride, tetrachloroethane, chlorobenzene or dichlorobenzene ⟨78GEP2811310, 80JAP80136115, 80JAP80162415⟩. An example of this work has appeared in a more accessible source ⟨79JOC3573⟩. This reaction was carried out under a pressure of 5 kg cm^{-2} of carbon monoxide with tri-*n*-butylphosphine oxide as a catalyst.

(ii) From paraformaldehyde, carbon tetrachloride and aluminum trichloride

Carbon tetrachloride, paraformaldehyde and aluminum chloride in the mole ratio 1 : 0.5 : 0.5 were warmed together at 35–40 °C for 2–2.5 h, then heated under reflux for 0.5–1 h to give phosgene in quantitative yield after careful decomposition with ammonium hydroxide ⟨68ZPK1380⟩. The quantity of ammonium hydroxide used is obviously crucial, since phosgene reacts rapidly with ammonia.

(iii) Photochemical oxidation of tetrachloroethylene

Air or oxygen is bubbled through tetrachloroethylene whilst irradiation is carried out using light with a wavelength of 200–290 nm. This method is claimed to be very controllable and capable of producing phosgene in high yield ⟨88JAP88319205⟩.

(iv) Oxidation of chloroform or carbon tetrachloride using a platinum metal catalyst

In experiments designed to demonstrate the preparation, quantification and reactions of phosgene, chloroform or carbon tetrachloride is oxidised with oxygen in the presence of platinum metal. The resulting phosgene is assayed by reaction with sodium iodide and titration, and converted to crystal violet dye with dimethylaniline and aluminum chloride ⟨78MI 614-01⟩.

6.14.1.2.6 *Purification of phosgene*

Phosgene from cylinders is approximately 99% pure, the remainder comprising of carbon monoxide, carbon dioxide, air, hydrogen chloride and water. Laboratory phosgene prepared by some routes may also contain chlorine. Glemser suggests that evaporation of 20% of the volume of phosgene at room temperature and pressure, and low-temperature vacuum distillation, with fractionation of the remainder leads to a very pure product ⟨B-63MI 614-02⟩. Chlorine in phosgene can be detected by bubbling the gas through mercury. Discolouration indicates the presence of chlorine, which can be removed by passage through two washbottles containing cottonseed oil ⟨63OSC(4)524⟩. A suitable method for removing bromine from carbonic bromide fluoride involves passing the gas through irradiated trichloroethylene ⟨73AG(E)918⟩. This method may also be suitable for removing chlorine from phosgene.

6.14.1.2.7 *Preparation of isotopically labelled phosgene*

Only phosgene labelled at the carbon atom with ^{11}C, ^{13}C or ^{14}C will be covered in this section, although other phosgenes labelled at oxygen and chlorine are known. The authors feel that labelling at carbon is the most relevant, particularly for the preparation of labelled drug molecules, which is the major use of labelled phosgenes.

(i) Carbon-11-labelled phosgene

Two European groups have prepared ^{11}C-labelled phosgene by several routes, resulting in different specific activities. This isotope is a positron emitter with a half-life of only 22 min and is valuable for the study of interactions of drug molecules with receptors by positron emission tomography, and for nuclear medicine imaging. It is obviously necessary to prepare and use the $[^{11}C]COCl_2$ rapidly, and all methods use flow techniques to sequentially execute the various steps involved. Early work employed bombardment of nitrogen by a cyclotron proton beam in the presence of oxygen to produce either $[^{11}C]CO$ ⟨78MI 614-02⟩ or $[^{11}C]CO_2$ ⟨77MI 614-01, 78MI 614-03⟩ by the nuclear reaction $^{14}N(p\alpha)^{11}C$. The latter was then reduced by zinc at 400°C to $[^{11}C]CO$. The $[^{11}C]CO$ was then either photochemically united with chlorine ⟨78MI 614-02, 81MI 614-01⟩, or chlorinated over platininum(IV) chloride ⟨77MI 614-01, 83MI 614-01⟩ at 290–430°C to produce $[^{11}C]COCl_2$. By the photolytic method, the specific activity was below 100 mC μmol^{-1} 20 min after the reaction commenced ⟨78MI 614-02⟩. This was improved to 400–500 mC μmol^{-1} by employing the platininum(IV) chloride route ⟨83MI 614-01⟩. However, more recently a new route involving the production of $[^{11}C]CH_4$ by bombardment of nitrogen in the presence of hydrogen has been developed. The $[^{11}C]CH_4$ is then chlorinated by copper(II) chloride at 380°C, and oxidized to $[^{11}C]COCl_2$ over iron filings ⟨87MI 614-01⟩. The specific activity 30 min after the commencement of the process was 1400–1600 mC μmol^{-1}. Many medically useful molecules have been prepared from $[^{11}C]COCl_2$, including urea, pimozide, diphenylhydantoin, dimethyloxazolidinedione and ketanserin ⟨83MI 614-01⟩.

(ii) Carbon-13-labelled phosgene

Great insight can be obtained into small molecule–enzyme interactions through ^{13}C NMR studies. For this reason, many ^{13}C-labelled drug molecules have been synthesized, and several of these have employed $[^{13}C]COCl_2$ in their preparation. Two methods of preparation have been used, both scaled-down versions of standard phosgene preparations.

(a) From irradiation of ^{13}CO and chlorine. Stoichiometric amounts of chlorine and $[^{13}C]CO$ were condensed into an evacuated flask and externally cooled with liquid nitrogen. The flask was then allowed to warm up and was irradiated for 3 h. A quantitative yield of $[^{13}C]COCl_2$ was claimed; however, subsequent reaction gave phosgene-derived products in only 70% yield ⟨90MI 614-01⟩.

(b) From carbon tetrachloride and sulfuric acid. Oleum was added dropwise to 98% sulfuric acid until a yellow colour persisted, to give 100% sulfuric acid. A small quantity of freshly ignited Celite was added, followed by $[^{13}C]CCl_4$. The mixture was heated to 140°C, and the $[^{13}C]COCl_2$ evolved was collected in dry toluene cooled in an ice–salt bath, whilst the hydrogen chloride produced was

allowed to escape through a calcium chloride drying tube. Although the gain in weight of toluene suggested almost quantitative conversion to [^{13}C]$COCl_2$, subsequent reaction with methanol gave phosgene-derived products in only 60% yield ⟨87MI 614-02⟩.

(iii) Carbon-14-labelled phosgene

In all recent applications [^{14}C]$COCl_2$ has been purchased from one of the established sources. However, the methods used above for the preparation of [^{11}C]$COCl_2$ and [^{13}C]$COCl_2$ should also be suitable for [^{14}C]$COCl_2$, starting from [^{14}C]CO_2, [^{14}C]CO, [^{14}C]CH_4 or [^{14}C]CCl_4. An early preparation of [^{14}C]$COCl_2$ from [^{14}C]CO and chlorine by UV irradiation has been reported ⟨48JA1968⟩.

6.14.1.2.8 Phosgene alternatives

There has been rising pressure to avoid the use of isocyanates, and their precursor phosgene, in recent years. Since over 90% of world consumption of phosgene is in the preparation of isocyanates for polyurethane manufacture, and a further 7% is destined for polycarbonate production, if these plastics could be formed by alternative routes the use of phosgene worldwide would virtually cease. As will be discussed later in this section, the majority of phosgene alternatives suitable for the preparation of fine chemicals are not only expensive, but are generally derived from phosgene. For productions purposes, the alternatives must be cheap and must be prepared in a manner which avoids the use of hazardous reagents. The candidate which fulfils these criteria is dimethyl carbonate, prepared from methanol, oxygen and carbon monoxide in the presence of a copper catalyst at less than \$1 per pound (\$2 per kilogram). A discussion on the replacement of phosgene by dimethyl carbonate, and on other nonphosgene routes to isocyanates, carbonates and urethanes was published in 1989 ⟨B-89MI 614-01⟩. Several laboratory alternatives to phosgene will now be discussed.

(i) Trichloromethyl chloroformate (diphosgene), $Cl_3COCOCl$

All compounds bearing the —$OCCl_3$ leaving group are potential phosgene sources, and this should be borne in mind during their use and subsequent disposal. Under many reaction conditions, trichloromethyl chloroformate reacts with nucleophiles in such a way that the phosgene formed during the first acylation step is destroyed by a second molecule of the nucleophilic species at a more rapid rate than the original attack on trichloromethyl chloroformate; thus, 0.5 mol of the reagent is equivalent to 1.0 mol of phosgene, the latter being present only in minute concentration during the reaction. This leads to its common name, 'diphosgene' (Scheme 3). However, there are several circumstances which lead to rapid and total conversion of trichloromethyl chloroformate into phosgene. These include high temperature ⟨1887JPR99⟩ or contact with iron(III) oxide ⟨19JPC498⟩, activated charcoal ⟨85JOC715⟩ or pyridine ⟨82JAP8292513⟩. It must be assumed that many other commonly encountered substances will also catalyse this transformation. Thus, trichloromethyl chloroformate, b.p. 128 °C, is a stable and easy to handle alternative to phosgene; it is widely and commercially available, or readily prepared by chlorination of methyl chloroformate ⟨79OS(59)195⟩. In many transformations, yields are comparable to those achieved with phosgene; however, in some instances significantly lower yields are reported ⟨80JOC4059⟩. In a paper on the preparation of chiral oxazolidinones, Pridgen *et al.* list many applications for which trichloromethyl chloroformate has proved satisfactory ⟨89JOC3231⟩. The authors suggest that whilst the use of this reagent will normally be much less hazardous than the use of phosgene, the same high standard of safety precautions must be taken since there is a risk that considerable and unexpected concentrations of phosgene may be present during and after reaction.

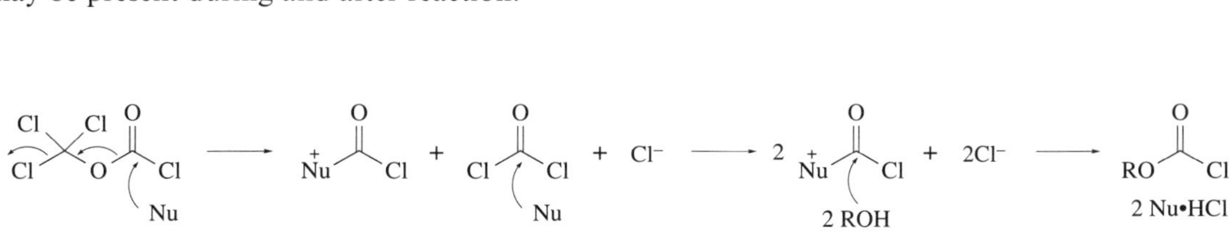

Scheme 3

(ii) Bis(trichloromethyl) carbonate (triphosgene), $Cl_3COCO_2CCl_3$

The use of bis(trichloromethyl) carbonate as a phosgene alternative is a logical extension of the work described above. This compound is a stable crystalline solid, m.p. 81–83 °C, commercially available, or readily prepared by chlorination of dimethyl carbonate ⟨87AG(E)894⟩. It has the advantage as a solid of being easy to weigh, thus simplifying quantitative additions. As might be predicted, 1 mol of reagent is equivalent to 3 mol of phosgene, although in practice this stoichiometry does not always lead to maximum yields (Scheme 4). Whilst no specific references to the extensive decomposition of bis(trichloromethyl) carbonate to phosgene have been discovered, its similarity to trichloromethyl chloroformate strongly suggests that the same safety precautions should be employed.

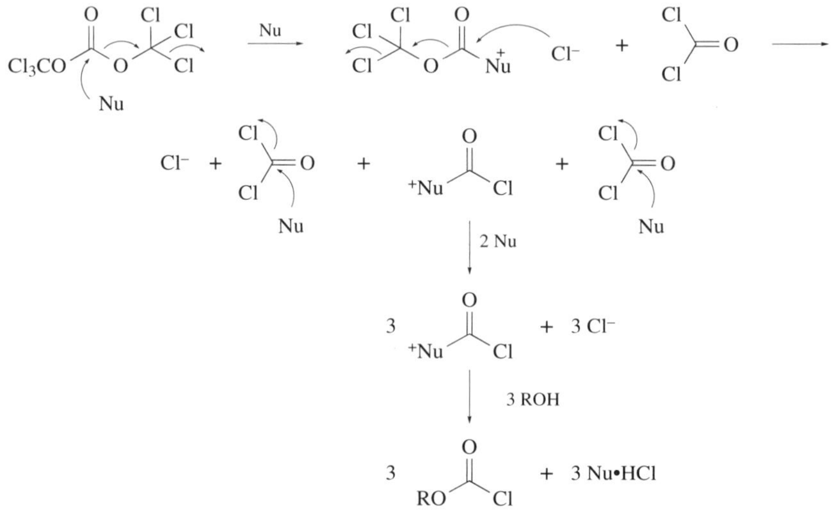

Scheme 4

Eckert and Forster have described a range of applications for bis(trichloromethyl) carbonate, encompassing most of the reactions commonly achieved by phosgene. The yields achieved with strictly stoichiometric quantities of triphosgene range from 68% to 94% ⟨87AG(E)894⟩. Subsequently, its use in the preparation of *N*-carboxyanhydrides of α-amino acids ⟨88TL5859⟩, α-chloro-chloroformates ⟨89TL2033⟩ and quinazolinediones ⟨91SC285⟩, all in good yields, has been reported.

(iii) N,N′-*Carbonyldiimidazole*

N,N′-Carbonyldiimidazole (**1**) appears an excellent substitute for phosgene when its role is to insert a carbonyl function between two nucleophilic groups to complete a five- or six-membered ring. The nucleophilic groups can be amines, enamines, alcohols, enols, thiols or enethiols, and the heterocycles formed may be either aromatic or nonaromatic; the examples are numerous, and the yields normally high. This application is exemplified by formation of the cyclic carbonate from *cis*-cyclohexane-1,2-diol in 94% yield ⟨75SC47⟩, and by benzimidazolones from *o*-phenylenediamines at room temperature ⟨56AG754, 65JHC41⟩ in yields of 80–90%. The use of phosgene to perform this ring closure often requires high temperatures. Examples of most other transformations more commonly employing phosgene are also reported for carbonyldiimidazole; a brief selection follows.

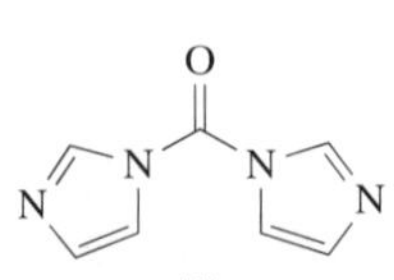

(1)

Symmetrical and unsymmetrical ureas can be prepared by reaction with a single amine ⟨56AG754⟩, or sequentially with two different amines ⟨91JOC891⟩. Carbamates can result from sequential reaction with an alcohol and an amine ⟨91JOC891⟩. Aldoximes are converted into nitriles in quantitative yield ⟨82SC25⟩. Isocyanates may be prepared from amines in good yield ⟨61AG66⟩. By preparing the bismesylate salt of carbonyldiimidazole, formamides can be dehydrated to isonitriles ⟨82JCR(S)79⟩.

(iv) Carbonates as phosgene alternatives

The role of dimethyl carbonate as an industrial substitute for phosgene has already been alluded to. Such relatively unactivated molecules are used in laboratory synthesis, but the reaction conditions are often quite harsh; for example, 3-aminoalkanols and diethyl carbonate give high yields of the cyclic carbamates in the presence of sodium methoxide at 120°C ⟨80LA122⟩. Many more activated carbonates have been developed. Commercially available *N,N′*-disuccinimidyl carbonate (**2**) is a very useful reagent; a recent example of sequential reaction with complex alcohols and a complex amine to give good yields of carbamates by workers from Merck, Sharp & Dohme is worthy of mention ⟨92TL2781⟩. Bis(2-pyridyl) carbonate (**3**) also performs similar reactions, and has been exploited by the same workers to form carbamates. In this case, even tertiary alcohols give excellent yields providing the alkoxide salt is used ⟨91TL4251⟩. Examples of high-yielding ring formation by carbonyl insertion have also been reported for this reagent ⟨86H(24)1625⟩. Ueda *et al.* report that *N,N′*-(carbonyldioxy)bisbenzotriazole (**4**) may have similar utility ⟨83S908⟩.

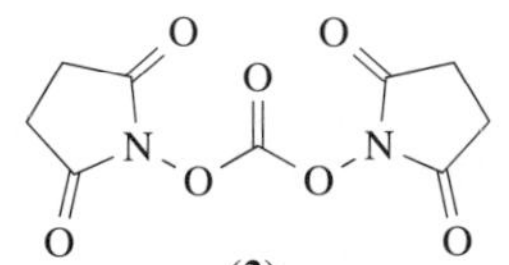
(2)

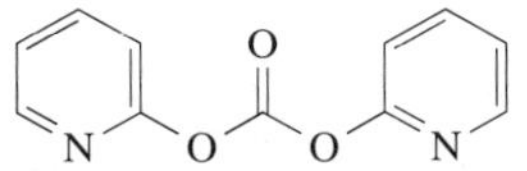
(3)

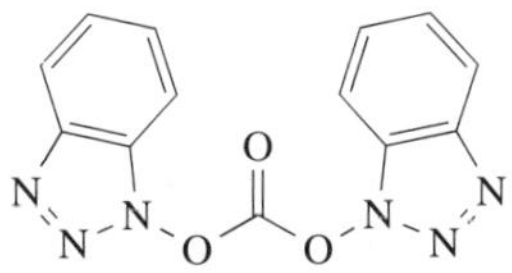
(4)

(v) Chloroformates as phosgene alternatives

The role of tricholoromethyl chloroformate as a phosgene substitute has already been discussed. Several other chloroformates also fulfil this role. Amongst these, commercially available 4-nitrophenyl chloroformate is very useful. The mixed carbonates formed by its reaction with alcohols are often stable to flash chromatography, yet are sufficiently reactive to form carbamates with amines at room temperature ⟨92EUP486948⟩. Further applications of 4-nitrophenyl chloroformate and several other bifunctional chloroformates have been reported ⟨93S103⟩.

6.14.1.3 Carbonic Dibromide, $COBr_2$

Carbonic dibromide, also commonly called bromophosgene, carbonyl bromide and carbonyl dibromide, is a colourless liquid, b.p. 65–67°C ⟨73T3309⟩. It is stable in the absence of light at −10°C; however, at higher temperature or in the presence of light, dissociation into bromine and carbon monoxide takes place. For this reason, relatively little has been published on its preparation and reactions. The first synthesis of pure carbonic dibromide was carried out by von Bartal in 1905 ⟨06LA(345)334⟩. It is to his credit that the yield and purity of $COBr_2$ produced by his method have not been significantly improved. The most accessible description of this preparation is that reported by Fodor and co-workers ⟨73T3309⟩. Carbon tetrabromide was melted and heated to 125°C, and concentrated sulfuric acid was added dropwise, with vigorous stirring. The crude distillate was freed from bromine by the passage of ethylene. Redistillation gave pure carbonic dibromide in 68% yield. Storage of this material, even at low temperature, is not advisable, since any decomposition can lead to a buildup of pressure because of the formation of carbon monoxide. As with all the carbonic halides, the extreme toxicity of carbonic dibromide dictates that all procedures should be carried out in a fume hood with good extraction.

Several other preparations of carbonic dibromide have been attempted. The thermal decomposition of oxalyl bromide at 150–155°C with subsequent cooling of the autoclave to −80°C before opening has been reported to give carbonic dibromide in a distilled yield of 20% ⟨13CB1426⟩. Attempts to prepare carbonic dibromide by passing carbon monoxide over heated bromides of platinum and gold led only to evolution of bromine, presumably by dissociation of carbonic dibromide at the elevated temperatures used ⟨30CB1221⟩. Finally, carbon tetrabromide has been oxidised photochemically to carbonic dibromide by molecular oxygen in the presence of bromine as a sensitizer ⟨37CB1080⟩. This experiment was not designed to be of synthetic utility, but might warrant further investigation.

6.14.1.4 Carbonic Diiodide, COI_2

The preparation of carbonic diiodide has never been successfully demonstrated. Theoretical calculations of its thermodynamic properties ⟨82MI 614-02⟩ and of its electronic structure ⟨90MI 614-02⟩ have been reported. The dissociation temperature of carbonic dibromide is approximately −10 °C ⟨73T3309⟩, and for carbonic fluoride iodide, below −20 °C ⟨B-63MI 614-01⟩; thus, carbonic diiodide would be expected to dissociate at temperatures well below this. Attempted preparation from carbon monoxide and platinum tetraiodide led to liberation of iodine at 135 °C, well below the thermal decomposition temperature of platinum tetraiodide. This probably indicates the formation and instantaneous dissociation of carbonic diiodide ⟨30CB1221⟩. In a less brutal series of experiments, Staudinger and Anthes concluded that the dissociation temperature of carbonic diiodide was below −80 °C ⟨13CB1426⟩. This result was based on the formation of iodine by the reaction of oxalyl chloride and sodium iodide at low temperature.

6.14.2 CARBONYL HALIDES WITH TWO DISSIMILAR HALOGENS

6.14.2.1 One Fluorine and Chlorine, Bromine or Iodine Function

6.14.2.1.1 Carbonic chloride fluoride, COClF

Carbonic chloride fluoride is a stable, colourless gas, b.p. −47 °C. In company with all other carbonic dihalides, it is extremely toxic. The production of hydrogen fluoride on hydrolysis is an added risk. Some suggestions concerning the safe handling of this class of compounds can be found in Section 6.14.1.2.2.

Other names which have been used for this compound are carbonyl chloride fluoride, carbonyl chlorofluoride, chlorocarbonyl fluoride and chlorofluorophosgene. It is not commercially available; however, it is of some importance in the synthesis of fluoroformates, which are a valuable class of intermediates.

Several methods are available for its preparation (Scheme 5).

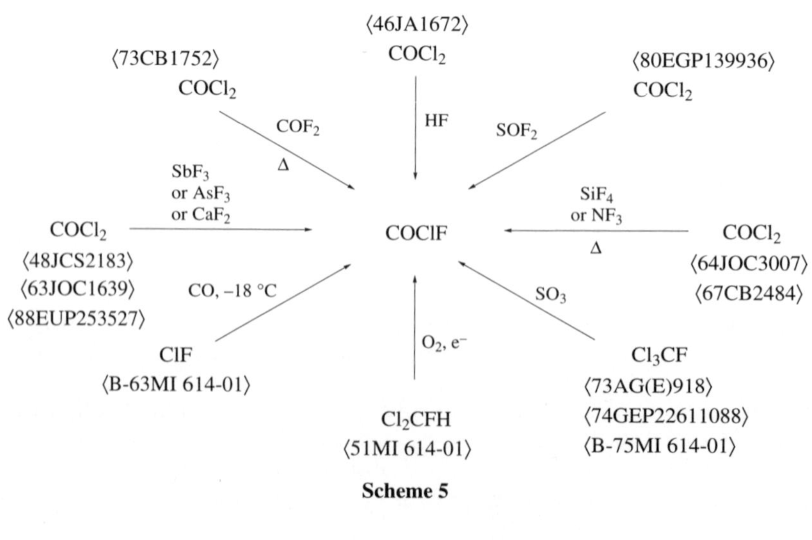

Scheme 5

(i) From carbonic dichloride (phosgene)

The first successful preparation of carbonic chloride fluoride was by Emeléus and Wood, who heated phosgene with antimony trifluoride in a 4 : 1 molar ratio, with the addition of some antimony pentachloride, at 135 °C for 1 h. The condensed product was purified by low-temperature distillation, and was isolated in 70% yield ⟨48JCS2183⟩. Upon repeating this work, Haszeldine and Iserson reported a purity of 95%, with carbon dioxide as the major contaminant ⟨57JA5801⟩. In an extension of this work, Christie and Pavlath substituted arsenic trifluoride for antimony trifluoride, achieving a yield of over 90% ⟨65JOC1639, 83HOU(E4)1⟩. An industrial gas phase process has also been patented by ICI in which phosgene is passed over calcium fluoride at 200–550 °C to give carbonic chloride fluoride as the major product ⟨88EUP253527⟩. Several nonmetallic fluorinating agents have also been successfully employed. Christie and Pavlath passed a 1 : 1 molar ratio of phosgene and silicon tetrafluoride through a quartz tube heated at 420 °C to give carbonic chloride fluoride in quantitative yield, with more than 40% conversion at a single pass ⟨64JOC3007⟩. Glemser and Biermann obtained carbonic chloride fluoride in 65% yield by passing nitrogen trifluoride and phosgene in a 1 : 2 molar ratio through a metal tube heated at 310 °C. The only other product formed was carbonic difluoride, with 82% conversion of phosgene. Higher temperatures led exclusively to carbonic difluoride ⟨67CB2484⟩. Finally, by an industrial gas phase process it is claimed that the reaction of thionyl fluoride with phosgene over a Lewis acid catalyst at 300–400 °C can be made to yield either carbonic difluoride or carbonic chloride fluoride by varying molar ratios and conditions. The thionyl fluoride could be regenerated by reaction of the sulfur-containing by-products with sodium fluoride ⟨80EGP139936⟩. Carbonic chloride fluoride has also been prepared in 50% yield by reaction of phosgene with anhydrous hydrogen fluoride at 80 °C ⟨46JA1672⟩.

(ii) From sulfur trioxide and fluorotrichloromethane

Preparations of carbonic chloride fluoride from phosgene inevitably involve either gas streams, usually at high temperature, or the use of pressure vessels. For this reason, its production from fluorotrichloromethane and oleum or sulfur trioxide is much more suitable for laboratory use. Four minor variations of this method are reported by Siegemund, giving yields from 60% to 83%. The reaction appears quantitative, but total conversion of the perhalogenomethane is not achieved. The highest yield resulted from slow treatment of sulfur trioxide with fluorotrichloromethane ⟨73AG(E)918⟩; however, methods involving oleum may be more simple to carry out ⟨73AG(E)918, 74GEP2261108, B-75MI 614-01⟩.

(iii) From carbon monoxide and chlorine monofluoride

Carbonic chloride fluoride is formed in 85–90% yield by the combination of excess carbon monoxide and chlorine monofluoride at −18 °C ⟨B-63MI 614-01⟩. Small quantities of carbonic difluoride and phosgene are also formed in this reaction, but can readily be removed by low-temperature distillation.

(iv) Metathesis from carbonic difluoride and phosgene

During experiments designed to prepare various perhalomethanes, Haszeldine and Iserson frequently found carbonic chloride fluoride amongst the products isolated from the co-pyrolysis of carbonic difluoride and phosgene over carbon powder at temperatures of 400–450 °C ⟨57JA5801⟩. Subsequently, Jäckh and Sundermeyer have proposed this as a convenient synthetic method, claiming a 30% yield of carbonic chloride fluoride from heating a 1 : 1 molar ratio of phosgene and carbonic difluoride ⟨73CB1752⟩. Unfortunately, no details are given; however, in laboratories used to the safe handling and distillation of gases, this preparation, which uses two commercially available starting materials, may be worth pursuing.

(v) By photochemical oxidation of dichlorofluoromethane

A mechanistic investigation of the chlorine sensitised photochemical oxidation of dichlorofluoromethane by molecular oxygen has been reported by Schumacher ⟨51MI61401⟩. Whilst not suitable as a preparative method in this form, the high yield (95%) could prove attractive to laboratories equipped to carry out photochemistry on a synthetic scale.

6.14.2.1.2 *Carbonic bromide fluoride, COBrF*

Carbonic bromide fluoride is a stable, colourless gas, b.p. −21 °C. It is also commonly called bromofluorophosgene, carbonyl bromofluoride and carbonyl bromide fluoride. It must be assumed to have the high toxicity associated with all phosgene analogues. Carbonic bromide fluoride has been prepared by two methods.

(i) From bromine trifluoride and carbon monoxide

Bromine trifluoride is maintained at a temperature between 8 °C and 30 °C whilst a stream of carbon monoxide is passed through it. This process must be carefully controlled, since bromine trifluoride solidifies at 8 °C, and the exothermic reaction can become explosive above 30 °C. Fractional condensation separates the carbonic bromide fluoride from carbonic difluoride, which is formed in an equimolar amount. Redistillation gives pure carbonic bromide fluoride in 85% yield ⟨56JOC1319, B-75MI 614-01⟩. A minor variation of this preparation is reported to give COBrF in greater than 90% yield ⟨B-63MI 614-01⟩.

(ii) From tribromofluoromethane and sulfur trioxide

Slow addition of tribromofluoromethane to sulfur trioxide at 40 °C yields carbonic bromide fluoride (64%), contaminated with a considerable amount of sulfur dioxide. This method also produces an equimolar amount of bromine, which can be removed by passing the evolved gases through irradiated trichloroethylene before condensation ⟨73AG(E)918, 74GEP2261108⟩.

6.14.2.1.3 *Carbonic fluoride iodide, COFI*

Carbonic fluoride iodide is a colourless gas, b.p. −20.6 °C ⟨B-63MI 614-01⟩ or +23.4 °C ⟨B-75MI 614-01⟩. The great discrepancy between the two reported boiling points may be due to the instability of this rarely reported compound. Above −20 °C, decomposition occurs with the liberation of iodine. The only published synthesis involves the agitation of iodine pentafluoride under 120 atm pressure of carbon monoxide for 7 d. The major products are carbonic difluoride and iodine; however, carbonic fluoride iodide can be isolated in 12% yield by low-temperature distillation at reduced pressure ⟨B-63MI 614-01, B-75MI 614-01⟩.

6.14.2.2 One Chlorine and Bromine or Iodine Function

6.14.2.2.1 *Carbonic bromide chloride, COBrCl*

Carbonic bromide chloride is a colourless gas, b.p. 35–37 °C, first prepared by Besson in 1895 by the reaction of phosgene with boron tribromide at 150 °C ⟨1895BSF444⟩. Von Bartal repeated this work ⟨06LA(345)334⟩, and also reported a preparation from phosgene and aluminum tribromide ⟨07MI 614-01⟩. The only other published account of the synthesis of this compound is by Garino, who claimed that carbonic bromide chloride is formed by aerial oxidation of chlorobromoiodomethane ⟨26G847⟩.

6.14.2.2.2 Carbonic chloride iodide, COClI

Carbonic chloride iodide has never been reported, and, by extrapolation from the stabilities of other carbonic dihalides, its decomposition temperature would be considerably below $-20\,^{\circ}C$.

6.14.2.3 One Bromine and One Iodine Function

6.14.2.3.1 Carbonic bromide iodide, COBrI

Carbonic bromide iodide has also never been reported, and would be expected to be unstable at temperatures considerably below $-20\,^{\circ}C$.

6.14.3 CARBONYL HALIDES WITH ONE HALOGEN AND ONE OTHER HETEROATOM FUNCTION

6.14.3.1 One Halogen and One Oxygen Function

Carbonyl halides with one halogen and one oxygen function are named in *Chemical Abstracts* as esters of the respective carbonohalidic acid; however, they are almost universally referred to as halogenoformates in the chemical literature, and this nomenclature will be adopted in this section. Other names sometimes encountered, and exemplified by ethyl chloroformate, are ethyl carbonochloridate, ethyl chloroformic ester, ethoxycarbonyl chloride and ethyl chlorocarbonate. The free parent halogenoformic acids are unknown, but their esters are generally stable. Chloroformates are important in the preparation of many fine chemicals, particularly carbamates, which are commonly used as amine-protecting groups in peptide chemistry. Fluoroformates are also used to prepare such carbamates, especially when the oxygen substituent is a tertiary alkyl group, where the improved thermal stability relative to the corresponding chloroformate is an advantage. One potentially valuable attribute of fluoroformates is their stability in dipolar aprotic solvents such as DMF and DMSO; under these conditions, chloroformates react exothermically to form complex products. Thus, where insolubility is a problem, alkoxycarbonylation with fluoroformates in a polar medium provides an alternative. This has been demonstrated by the protection of carbohydrates as their carbonate derivatives ⟨90JOC1851⟩. Another important use of fluoroformates is in the conversion of alcohols to their respective fluorides by catalysed or thermally induced loss of carbon dioxide from the alcohol fluoroformate ⟨55JA5033, 65JOC4104⟩. The carbonyl function of fluoroformates can also be fluorinated with sulfur tetrafluoride to give the stable —OCF_3 moiety (Scheme 6) ⟨92GEP4023107⟩.

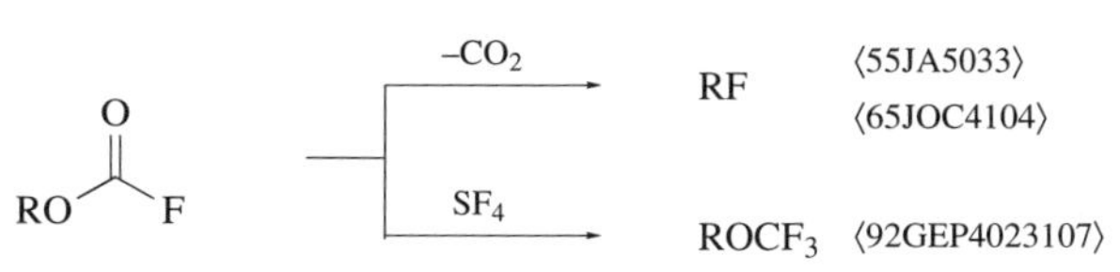

Scheme 6

The preparation of chloroformates has been comprehensively reviewed up to 1982 ⟨64CRV645, 83HOU(E4)9⟩; fluoroformates have been similarly treated ⟨B-72MI 614-01, 83HOU(E4)9⟩.

Bromoformates have been rarely reported, and most aliphatic iodoformates are unstable; some examples of aromatic iodoformates and aliphatic iodoformates with stabilising features will be discussed.

The sulfur analogues of haloformates and halothiolformates are also well known and are covered in Section 6.14.3.2. However, the analogous sellenium and tellurium compounds have not been described.

6.14.3.1.1 Fluoroformate esters, ROCOF

Two general methods are available for the preparation of fluoroformates; preparations of α-fluoroalkyl fluoroformates and α-chloroalkyl fluoroformates are covered separately. The synthesis of several miscellaneous unusual fluoroformates is also included.

(i) Preparation from alcohols or phenols and carbonic halide fluorides

This method is not very attractive as a general laboratory method since the carbonic halide fluorides are highly toxic gases. In addition, only carbonic difluoride is commercially available. Although not usually reported, carbonates would be formed as by-products unless a large excess of the phosgene analogue is used. This would not hinder isolation, however, which is normally carried out by distillation.

Emeléus and Wood originally developed this method, exploring the use of both carbonic difluoride and carbonic chloride fluoride in the presence of pyridine to prepare ethyl fluoroformate and phenyl fluoroformate ⟨48JCS2183⟩. It is notable in this report that whilst ethyl fluoroformate could be prepared by merely bubbling carbonic difluoride through a solution of ethanol and pyridine in toluene, a similar experiment with phenol led to diphenyl carbonate in quantitative yield. However, phenyl fluoroformate was isolated in 50% yield when phenol and excess carbonic difluoride were heated in benzene solution to 100 °C in an autoclave. This suggests that in the presence of pyridine, phenyl fluoroformate reacts much more rapidly with phenol than ethyl fluoroformate reacts with ethanol. In reactions involving carbonic chloride fluoride, only fluoroformates were formed—no chloroformates were detected. Emeléus and Wood also state that the fluoroformates they prepared were more thermally stable and less rapidly hydrolysed than their chloroformate counterparts. A patent from Merck describes the preparation of fluoroformates from primary and secondary alcohols and carbonic difluoride under pressure at 170 °C ⟨92GEP4023107⟩. A well-described method for the preparation of *t*-butyl fluoroformate by Ugi and Wackerle involves the generation of carbonic chloride fluoride from fluorotrichloromethane and oleum, and immediate reaction with *t*-butanol and triethylamine in ether ⟨75S598⟩. The reported yield exceeds 90%; *t*-butyl fluoroformate is sufficiently stable to be stored, in contrast to the corresponding chloroformate, which is stable only at low temperature, rapidly decomposing to *t*-butyl chloride and carbon dioxide. This compound is thus a useful reagent for amine protection. Voelter and co-workers have also exploited the same method for the preparation of fluoroformates for use as amine protecting group precursors. Thus, Adpoc fluoride (**5**) was prepared from 2-(1-adamantyl)propan-2-ol as a crystalline solid in 95% yield ⟨80CC1265⟩, and *t*-Bumeoc fluoride (**6**) as an unstable oil in 71% yield ⟨83LA248⟩. Finally, Oláh and Kuhn have prepared alkyl fluoroformates in good yield from primary and secondary alcohols by reaction with carbonic bromide fluoride ⟨56JOC1319⟩. This method is technically very demanding, and in the authors' view should not be attempted. A compilation of fluoroformates prepared from carbonic dihalides up to 1979 has been published ⟨83HOU(E4)9⟩.

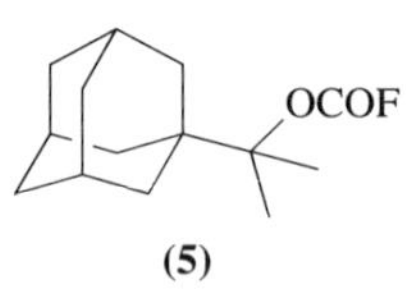

(5)

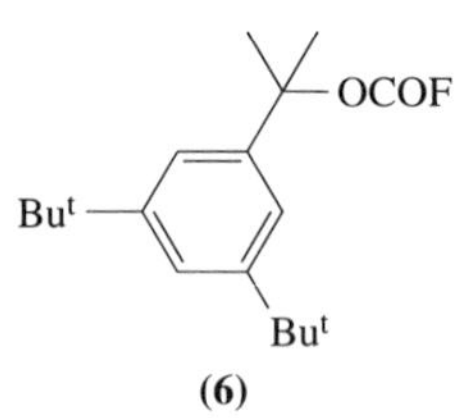

(6)

(ii) Preparation from chloroformates, carbonates or carbamates and a fluoride source

(a) From chloroformates. In cases where the chloroformate is readily available, several excellent methods exist for transhalogenation to the corresponding fluoroformate. The early work of Goswami and Sarker ⟨33JIC537⟩, extended by Nakanishi *et al.* ⟨55JA3099⟩ using thallium(I) fluoride, has been superseded, and it is recommended that this method should not be attempted. By using potassium fluoride and a catalytic 18-crown-6 phase transfer agent, Olofson and Cuomo were able to prepare aryl and primary and secondary alkyl fluoroformates at room temperature in yields of 80–95%. Yields were diminished by conducting the exchange at elevated temperatures, due to

catalysed decomposition of the products to fluorides with the liberation of carbon dioxide ⟨79JOC1016⟩. A similar method has used tetrabutylammonium fluoride or triethylbenzylammonium chloride and potassium fluoride to prepare primary and secondary alkyl fluoroformates in yields ranging from 42% to 92% ⟨82SC513⟩. In 1986, Japanese workers reported that a carefully dried mixture of calcium and potassium fluorides effected 95% conversion of ethyl chloroformate to fluoroformate at room temperature in acetonitrile ⟨86CC793⟩. These methods appear attractive because of their simplicity and high yields. Other methods involving chloroformates and a fluoride source have also been reported ⟨83HOU(E4)9⟩.

(b) From 1-chloroethyl carbonates. Olofson and co-workers have developed a very general and high yielding preparation of fluoroformates by the reaction of potassium fluoride with 1-chloroethyl carbonates of most alcohols or phenols, catalysed by 18-crown-6 ⟨90JOC1847⟩. The reaction proceeds by fluoride attack on the carbonate to give the fluoroformate and the 1-chloroethoxide anion, which eliminates chloride, yielding acetaldehyde (Scheme 7). Removal of this volatile product ensures the reverse reaction does not occur. The use of 1,2,2,2-tetrachloroethyl carbonates in this reaction also leads to high yields of fluoroformates ⟨86EUP176142⟩. The extra labour involved in preparing the 1-chloroalkyl carbonates will in many cases be amply rewarded by the versatility of this reaction; the scope includes carbonates derived from tertiary and benzylic alcohols.

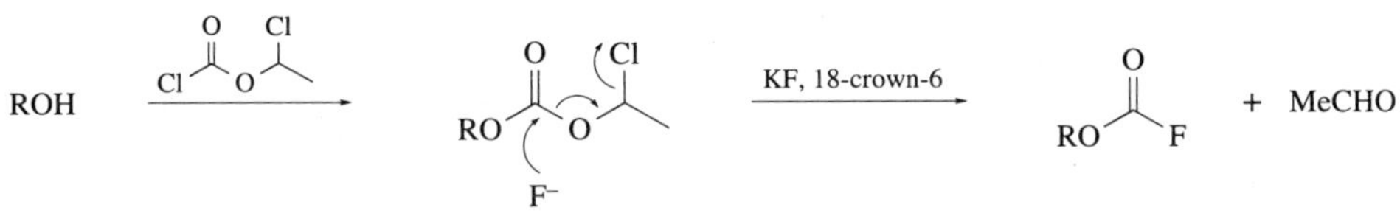

R = *n*-octyl, Bu^t, Ph, $PhCH_2$

Scheme 7

(c) From alkyl carbamates. Olah and Welch have devised a simple method for preparing primary, secondary and tertiary alkyl fluoroformates in yields usually between 50% and 80% by the diazotisation of alkyl carbamates in commercially available pyridinium poly(hydrogen fluoride) (Scheme 8). Isolation is simply achieved by extraction with ether and removal of the solvent, followed by distillation ⟨74S654⟩.

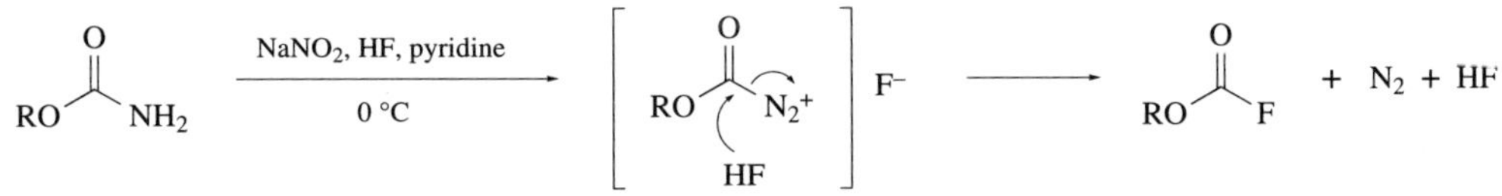

R = Me, Pr^i, Bu^t

Scheme 8

(iii) Preparation of haloalkyl fluoroformates

The reaction of carbonic difluoride with ketones proceeds by addition of its elements across the carbonyl group to yield α-fluoroalkyl fluoroformates. These can be stable compounds; in the case of cyclohexanone the adduct can be isolated in 78% yield as a distillable liquid. Not surprisingly, thermal or boron trifluoride-catalysed decomposition of these adducts yields *geminal* difluoro derivatives and carbon dioxide (Scheme 9) ⟨62JA4275⟩. In a more complex manner, reaction of carbonic difluoride with DMSO yields fluoromethyl fluoroformate in 38% yield. This reaction probably proceeds by an initial Pummerer rearrangement, but the mechanism of the subsequent steps is not clear ⟨85JFC(28)219⟩.

The formation of ω-chloroalkyl fluoroformates by the reaction of carbonic chloride fluoride with

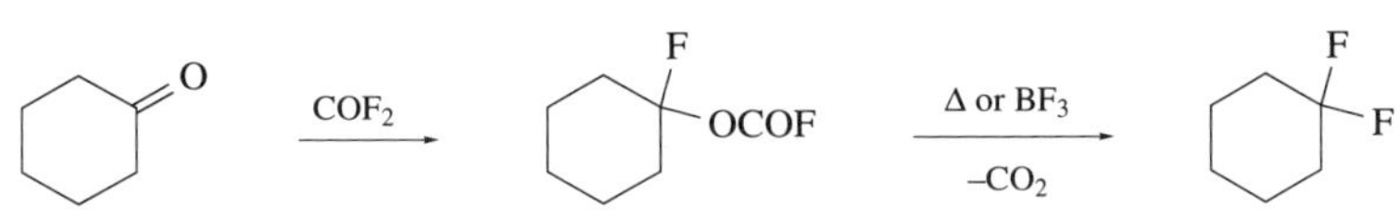

Scheme 9

cyclic ethers has also been reported. Excellent yields are obtained with oxirane, oxetane and tetrahydrofuran in the presence of tri-*n*-butylamine. Tetrahydropyran also gives the corresponding product, but in much lower yield (Equation (1)) ⟨65JOC1639⟩. Carbonic difluoride does not follow the same reaction course with oxirane, instead giving 2-(trifluoromethoxy)ethyl fluoroformate, $CF_3O(CH_2)_2OCOF$. Larger ring sizes do not appear to have been studied ⟨64JOC11⟩.

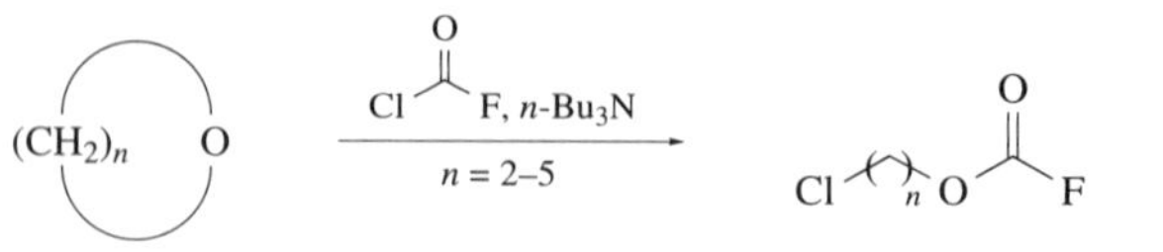

(1)

(iv) Miscellaneous fluoroformates

Trifluoromethyl fluoroformate, CF_3OCOF, was formed in poor yield (15%) from the reaction of carbonic difluoride with cyanogen fluoride at low temperature. A possible mechanism has been proposed ⟨81JFC(18)259⟩. This compound has also been prepared in high yield by the insertion of carbon monoxide into trifluoromethyl hypofluorite under photochemical conditions ⟨65CC241⟩.

The final three preparations in this section are included to exemplify unusual fluoroformates which have been reported. The authors suggest that these preparations are too hazardous to be undertaken except in laboratories where the handling of unpredictably explosive compounds is routine.

Bis(fluoroformyl) peroxide has been studied by several groups as a radical source. It is prepared by mixing streams of oxygen, carbon monoxide and fluorine in precise proportions. Powerful explosions result if the conditions deviate from those reported ⟨72IC2531⟩. Reaction of bis(fluoroformyl) peroxide with a caesium fluoride for a prolonged period at −40 °C gave trifluoromethyl fluoroformyl peroxide in 49% yield, b.p. −16 °C. A mechanism has been suggested ⟨72IC2531⟩. A low yield of the primitive fluoroformate fluorocarbonyl hypofluorite can be prepared from bis(fluoroformyl) peroxide and fluorine under UV irradiation and purified by fractional distillation, m.p. −141 °C, b.p. −55 °C ⟨67JA5161⟩. Reaction of fluorocarbonyl hypofluorite with sulfur tetrafluoride under irradiation leads to a low yield of the unusual pentafluorosulfur fluoroformate ⟨67JA5161⟩. These reactions are summarised in Scheme 10.

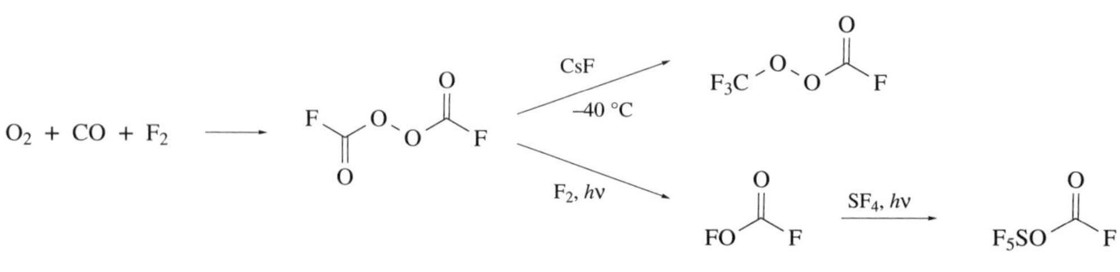

Scheme 10

6.14.3.1.2 Preparation of chloroformate esters, ROCOCl

(i) From alcohols and phosgene

Chloroformates are most commonly prepared from an alcohol or phenol and phosgene, under a variety of conditions. This method of synthesis has been thoroughly reviewed up to 1982 ⟨52HOU(8)101, 64CRV645, 83HOU(E4)9⟩. The authors will, therefore, restrict discussion to some general comments and recent modifications or extensions. The articles cited above contain references to numerous examples of aliphatic ⟨64CRV645⟩ and aromatic ⟨64CRV645, 83HOU(E4)9⟩ chloroformates made by this method. Although the use of phosgene can often be avoided by the use of various mixed carbonates or phosgene equivalents (see Section 6.14.1.2.8), it is sometimes the best or most convenient method for the preparation of chloroformates on a laboratory scale. Strict adherence to safe working procedures must be practised at all times, especially during workup procedures, since an excess of phosgene is often used (see Section 6.14.1.2.2).

Primary and secondary alcohols can react with phosgene to form chloroformates in the absence of a base. An excess of phosgene is often employed, which, together with low temperatures (usually room temperature or below), reduces the formation of carbonates formed by reaction of a further molecule of alcohol with the initially formed chloroformate. On a laboratory scale, commercially available 20% phosgene in toluene is a safe and convenient source of this reagent. Many solvents have been used, the only requirement being that they do not react rapidly with phosgene, or with the chloroformate produced. DMF and DMSO cannot be used for this reason (see Section 6.14.3.1); however, high yields have been recorded in the presence of considerable concentrations of water ⟨74MIP61411⟩. The use of a molar equivalent of a tertiary base, most often triethylamine, pyridine or dimethylaniline, is often advantageous, enabling the quantity of phosgene used to be reduced, and allowing reaction in the presence of acid-sensitive groups which would be otherwise destroyed by the hydrogen chloride produced as formation of the chloroformate proceeds. The addition of a tertiary base, or preformation of the alcoholate anion, is generally essential for the successful preparation of tertiary chloroformates; they are also often thermally unstable, undergoing quite clean decomposition to the tertiary alkyl halide and carbon dioxide at or below room temperature ⟨81JOC1720⟩. This occurs to a greater or lesser degree with retention of configuration in the case of optically active alcohol substrates ⟨82JOC3521, 86JOC1180⟩. This complication can be avoided by substituting the corresponding tertiary fluoroformate, which is much more thermally stable (see Section 6.14.3.1).

Phenols do not generally react with phosgene at room temperature in the absence of an acid acceptor. Good yields can, however, be achieved in chlorobenzene at 126 °C with the addition of a catalytic quantity of stearyltrimethylammonium chloride. The reaction will proceed slowly even at room temperature ⟨65CI(L)791⟩. In the presence of a base such as dimethylaniline, reaction is often rapid and complete at 0 °C ⟨67JOC300⟩. Examples of the use of a two-phase system using concentrated aqueous sodium hydroxide as the base are well described in *Houben-Weyl* ⟨83HOU(E4)9⟩.

A few examples will now be discussed that illustrate a particular point, or which may prove to be useful extensions of the reaction of alcohols with phosgene.

Tertiary chloroformates bearing an electron-withdrawing group on the α carbon atom are thermally stable and apparently more resistant to hydrolysis than their unsubstituted analogues ⟨71JOU1943, 78AG(E)361⟩. Notwithstanding, a satisfactory preparation of *t*-butyl chloroformate has been reported. By working at low temperature it proved possible to prepare *t*-butyl chloroformate from potassium *t*-butoxide and phosgene in a distilled yield of 61%. This material showed no evidence of decomposition after 2 months of storage at −25 °C ⟨81JOC1720⟩. At the other extreme, primary and secondary alkyl chloroformates have been prepared from the corresponding alcohols and phosgene in good yield in the vapour phase under continuous flow conditions, using a reaction time of 3–4 s at 100–250 °C ⟨51JA3796⟩. This method might form the basis of an industrial process. The problem of completely removing the tertiary base and its hydrochloride from chloroformates which cannot be distilled because of scale or instability may have been resolved by Dreiding and Egli, who employed poly(4-vinylpyridine) as an acid acceptor to prepare chloroformate (**7**) in excellent yield (Equation (2)) ⟨86HCA1442⟩. In this case, the use of triethylamine caused partial decomposition of the product. During the preparation of a specialised amine-protecting group in the (a-fluorenyl)methoxycarbonyl (Fmoc) family, Ramage and Raphy found that the modified fluorenylmethyl chloroformate could not be prepared by reaction of phosgene with the hydroxymethylfluorene in the presence of base because of spontaneous fragmentation to the corresponding alkene. This was overcome by converting the alcohol into the *O*-trimethylsilyl ether, which with phosgene gave the chloroformate under neutral conditions (Scheme 11) ⟨92TL385⟩. Finally, chloroformate preparation using phosgene generated *in situ* has been reported. In a method which may be of industrial significance, Japanese workers treated butanol with chlorine and carbon monoxide in the presence of phosphine oxides at room temperature at 60 kg cm^{-2} pressure to give a 93% yield of butyl chloroformate ⟨79JAP7961121⟩. On a laboratory scale, should the preparation of a chloroformate using diphosgene (see Section 6.14.1.2.8) prove unsatisfactory, addition of activated charcoal to the reaction will decompose the diphosgene completely to phosgene in solution. This method has been used to advantage in the preparation of carbamoyl chlorides, and should also be applicable to chloroformates ⟨85JOC715, 90JMC2101⟩.

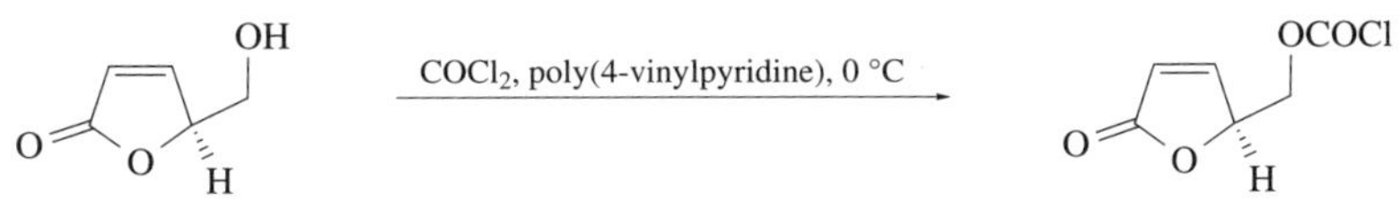

(**7**)

(2)

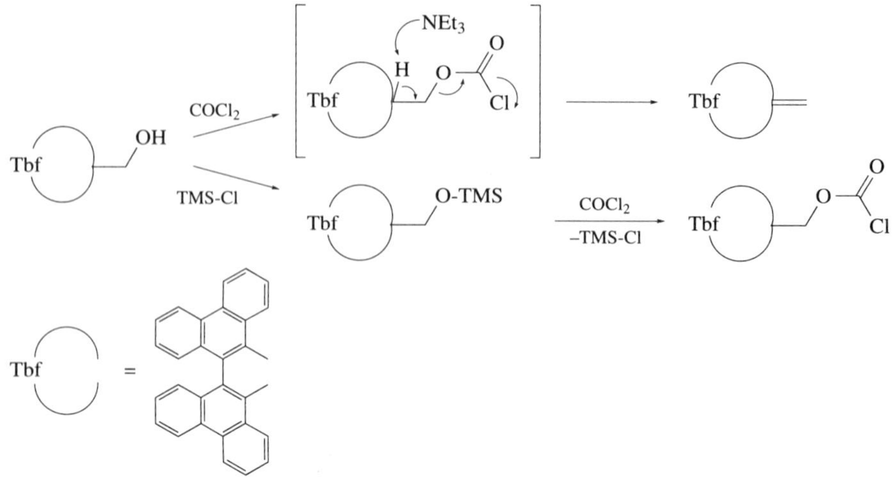

Scheme 11

A series of chiral chloroformates has been prepared for the separation of optically active amines by derivatization and reverse-phase liquid chromatography. Formation from the respective alcohols with phosgene in the presence of triethylamine proceeded in all cases without detectable racemization ⟨90JC593⟩. Some representative structures, **(8)**, **(9)** and **(10)**, are shown. Finally, the chloroformate of cyclic peroxide 3-hydroxymethyldioxetane **(11)** has been prepared with phosgene and pyridine in the standard way. In common with many less exotic chloroformates, it is stable to flash chromatography on silica ⟨86S330⟩.

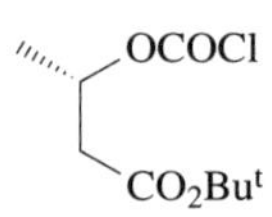

(8)

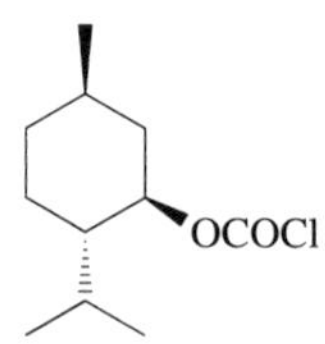

(9)

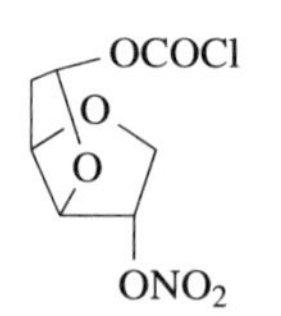

(10)

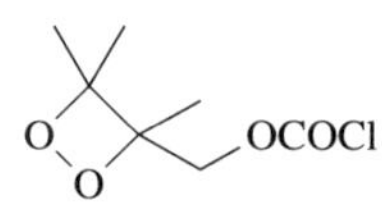

(11)

(ii) From alcohols and diphosgene or triphosgene

The roles of diphosgene and triphosgene as alternatives to phosgene have already been explored (see Section 6.14.1.2.8). Konakahora *et al.* have described the preparation of several sensitive chloroformates using these reagents, and they discuss the effect of tertiary base nucleophilicity on the yield obtained ⟨93S103⟩. A potentially useful modification of this method has been developed by Japanese workers, who were able to prepare the chloroformate **(12)** in high yield using an *in situ* two-step procedure. Initial treatment of the alcohol with diphosgene and a tertiary base gave a large amount of the corresponding carbonate. This was avoided by first treating the alcohol with two equivalents of diphosgene in the absence of base to give a 1 : 1 mixture of chloroformate and carbonate. Addition of a catalytic amount of pyridine to this mixture converted the carbonate to chloroformate, leading to its formation in an overall 78% yield (Scheme 12) ⟨91TL4371⟩. The use of diphosgene and triphosgene in chloroformylation has been reviewed ⟨89MI 614-04, 90MI 614-03⟩.

(iii) By chlorination of ROC(=X)Y compounds (X, Y = S or O)

The chlorination of a variety of carbonates, xanthates and thionoesters leads to chloroformates ⟨64CRV645⟩. These methods are generally not of great synthetic importance because the starting materials are usually less available than the corresponding alcohol, which is more commonly used for chloroformate synthesis. In instances where attempts to form a chloroformate led only to the symmetrical carbonate, treatment of the latter with thionyl chloride or phosphorus pentachloride may offer a useful method of recovering the situation (Equation (3)). Certain alcohols cannot readily

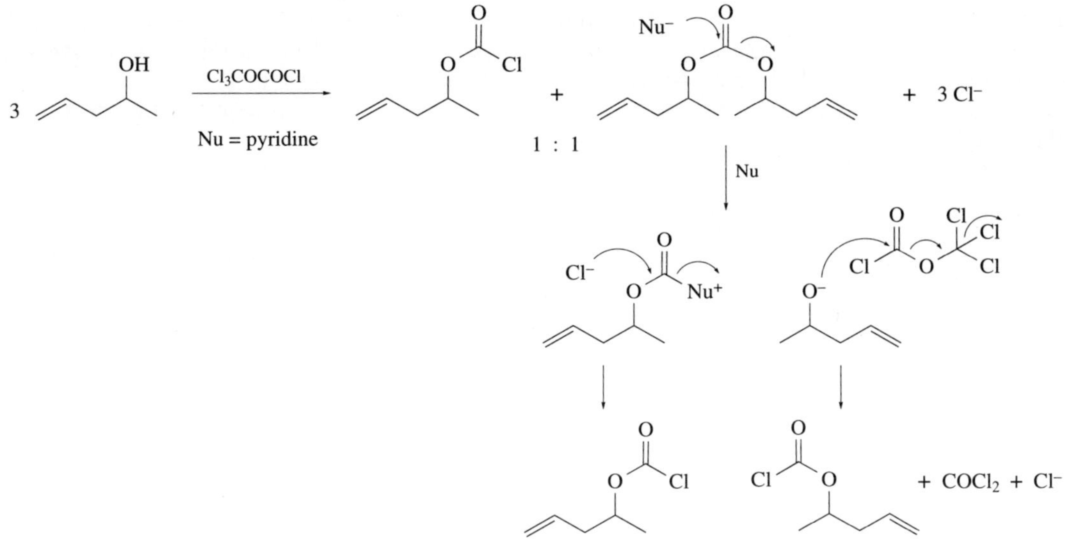

(12)

Scheme 12

be converted to chloroformates with phosgene in the presence of base because of chemical instability. These include alcohols with a β electron-withdrawing group. A high-yielding method of preparing chloroformates from alcohols of this type has been devised by Gilligan and Stafford, who first converted the alcohol to *O*-alkyl *S*-ethyl thiocarbonate with commercially available *S*-ethyl carbonochloridothioate in the presence of iron(III) chloride. This product is then treated with sulfuryl chloride to provide the chloroformate in high yield (Scheme 13) ⟨79S600⟩. Some of the alcohols described in this report contained β-nitro or β,β-dinitro functions; compounds of this type present a serious explosion hazard and should only be handled in laboratories equipped to deal with such eventualities. However, this method is probably quite general; it could also prove valuable in avoiding the symmetrical carbonate by-products which often reduce yields in chloroformate synthesis.

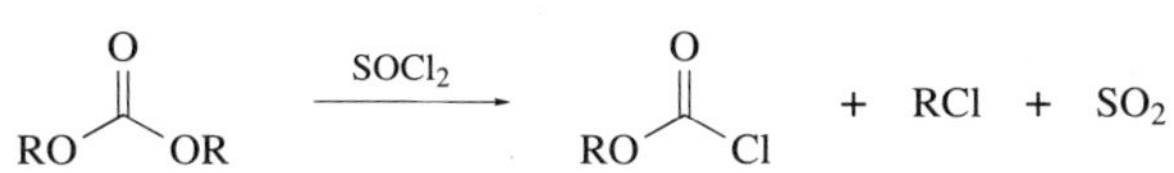

(3)

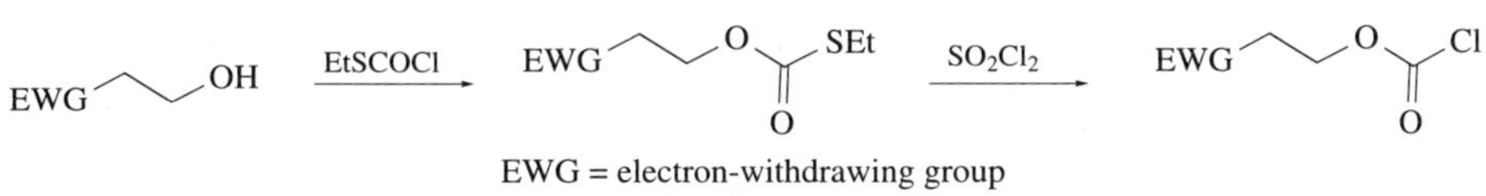

EWG = electron-withdrawing group

Scheme 13

Another use of this method is described by Folkmann and Lund in the preparation of acyloxymethyl chloroformates as precursors of pharmaceutical pro-drugs. In this case, chlorination of the intermediate thiocarbonate (**13**) is best achieved by sulfuryl chloride at room temperature in the presence of $BF_3 \cdot Et_2O$; chlorine will also perform this cleavage (Equation (4)) ⟨90S1159⟩.

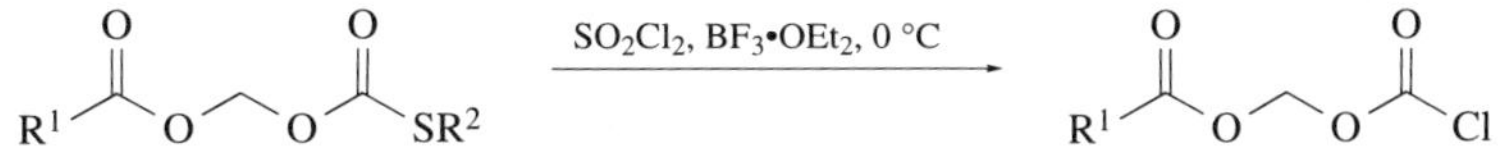

(4)

(13)

(iv) By electrophilic substitution of chloroformates containing aromatic rings

A variety of aromatic electrophilic substitutions have been carried out on compounds in which the chloroformate group is already present. In the search for new amine-protecting groups in the

Fmoc family, Carpino brominated (9-fluorenyl)methyl chloroformate in the presence of iron(III) chloride to give the 2,7-dibromo analogue in 37% yield ⟨80JOC4250⟩. Nitration of 4-methylphenyl chloroformate is reported to give 4-methyl-3-nitrophenyl chloroformate (87%) ⟨80GEP2920386⟩. Phenyl chloroformate has been converted into its 4-sulfonic acid analogue by sulfur trioxide in a similarly high yield ⟨91EUP415473⟩.

(v) Other methods

Hypochlorites bearing electron-withdrawing groups undergo carbon monoxide insertion into the O—Cl bond at low temperatures to give chloroformates in excellent yield. Thus, a series of perfluoroalkyl hypochlorites gave perfluoroalkyl chloroformates in quantitative yield ⟨69TL723⟩. Chlorine trifluoromethanesulfonate (**14**) also undergoes a similar reaction to give the mixed anhydride (**15**) in 82% yield (Equation (5)) ⟨82JFC(19)227⟩.

$$\underset{\textbf{(14)}}{ClOSO_2CF_3} \xrightarrow{CO,\ -110\text{–}0\ ^\circ C} \underset{\textbf{(15)}}{ClCO_2SO_2CF_3} \qquad (5)$$

Electrophilic carbonylation of several substrates has been performed by treating an equimolar mixture of bromine and antimony pentachloride with carbon monoxide in liquid sulfur dioxide to yield the chloro-oxocarbonium ion (**16**), which can be captured by ethanol to give ethyl chloroformate (Scheme 14) ⟨73TL2287⟩.

$$Br_2 + SbCl_5 + CO \xrightarrow{\text{liquid } SO_2} [BrCO]^+ [SbCl_5Br]^- \longrightarrow \underset{\textbf{(16)}}{[ClCO]^+ [SbCl_4Br_2]^-} \xrightarrow{ROH} ROCOCl + H^+ [SbCl_4Br_2]^-$$

Scheme 14

Two methods of preparing methyl chloroformate from palladium bis(methoxycarbonyl) complexes have been reported ⟨93JOM(451)243⟩. Since in these complexes the methoxycarbonyl ligand is obtained by treatment of a simple palladium complex with carbon monoxide in methanol, and the chlorine atom for the decomposition step is supplied by copper(II) chloride or chlorine, it is possible that this could form the basis of a catalytic preparation.

(vi) Preparation of halogenated chloroformates

(a) α-Haloalkyl chloroformates. Two important compounds in this class are commercially available: trichloromethyl chloroformate, which has been already discussed as a phosgene alternative (see Section 6.14.1.2.8), and α-chloroethyl chloroformate, which has found use in the *N*-dealkylation of tertiary amines ⟨84JOC2081⟩. The preparation of mono-α-chloroalkyl chloroformates can be carried out most conveniently by the reaction of phosgene ⟨84JOC2081⟩, or a phosgene equivalent ⟨89TL2033⟩, with an aldehyde at or below room temperature (Scheme 15). Formaldehyde ⟨83GEP3241568⟩ and benzaldehyde ⟨83HOU(E4)9⟩ also undergo this reaction, but it has not been reported for ketones. Alternatively, chloroformates bearing α hydrogen atoms can be readily chlorinated photochemically. Whilst this is an excellent method of preparing trichloromethyl chloroformate ⟨79OS(59)195⟩, and presumably other fully α-chloro-substituted chloroformates, attempts to prepare partially α-chlorinated compounds result in the formation of mixtures. Thus, partial chlorination of methyl chloroformate yields both chloromethyl chloroformate and dichloromethyl chloroformate, which can be isolated in moderate yield by fractional distillation. These compounds can be freed from associated trichloromethyl chloroformate by treatment with charcoal, when only the latter decomposes to phosgene ⟨90S1159⟩. An alternative radical chlorination employing sulfuryl chloride can also be used to prepare α-chloroethyl chloroformate in 51% yield ⟨91JAN1083⟩.

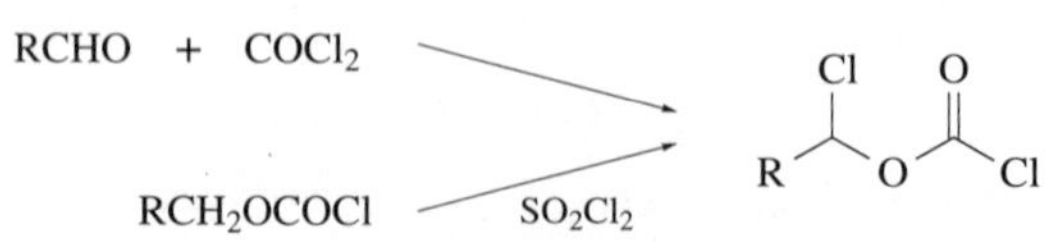

R = H, alkyl, Ph

Scheme 15

(b) β-Chloroalkyl chloroformates. The preparation of β-chloroalkyl chloroformates is well described ⟨83HOU(E4)9⟩. The most general method involves pyridine-catalysed addition of phosgene to an oxirane, first described by Malinovsky and Modyantseva ⟨53JGU229⟩. The reaction usually proceeds regiospecifically with unsymmetrical oxiranes to give the product with the chlorine on the less substituted carbon atom (Equation (6)) ⟨53JGU229, 57JCS2735⟩. Several β-chloroalkyl chloroformates have also been prepared by addition of chlorine to chloroformates bearing allyl or propargyl groups ⟨59JAP594264⟩. Treatment of ethylene carbonate with phosphorus(V) chloride also gives β-chloroethyl chloroformate in high yield ⟨61CB544⟩.

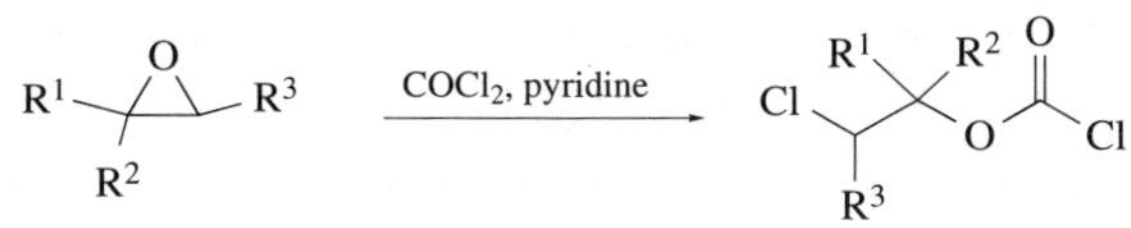

(6)

(vii) Preparation of enol chloroformates, $R^1R^2C{=}CR^3OCOCl$

The simplest member of this class, vinyl chloroformate, was the first to be prepared, by pyrolysis of ethylene glycol bis(chloroformate). This has been well described by Lee, who was able to achieve a moderate yield using a flow process ⟨65JA3943⟩. An earlier report describing 2-propenyl chloroformate ⟨34JA2007⟩ has been reinvestigated by Olofson *et al.*, who could not reproduce the work ⟨78JOC752⟩. This group has subsequently carried out the majority of work in this area. The first general synthesis emerged after failure to trap enolates prepared by many methods with phosgene. The reaction of ethyl 2-propenyl ether with mercury(II) oxide gave di(acctonyl)mercury(II) in quantitative yield. Reaction with phosgene at 0–25 °C led to 2-propenyl chloroformate in 86% distilled yield (Scheme 16) ⟨78JOC752⟩. Another method of some generality has since been developed. Initially, Olofson and co-workers developed a simple and convenient synthesis of 2,2-dichlorovinyl chloroformate by zinc dehalogenation of the adduct between phosgene and trichloroacetaldehyde (Scheme 17) ⟨90JOC2240⟩. Subsequently they have increased the scope of this method to the preparation of other enol chloroformates ⟨90JOC5982⟩.

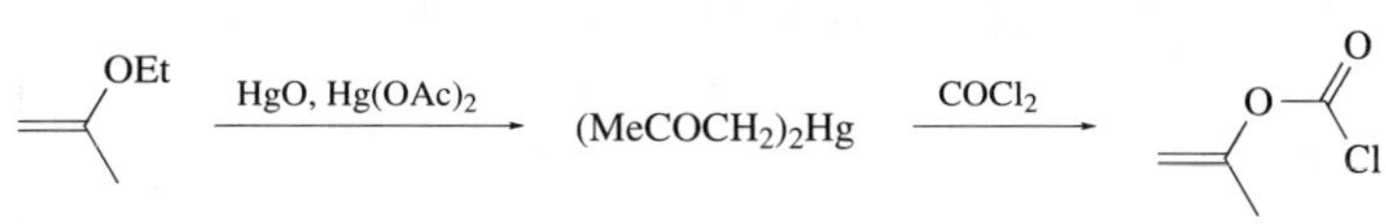

Scheme 16

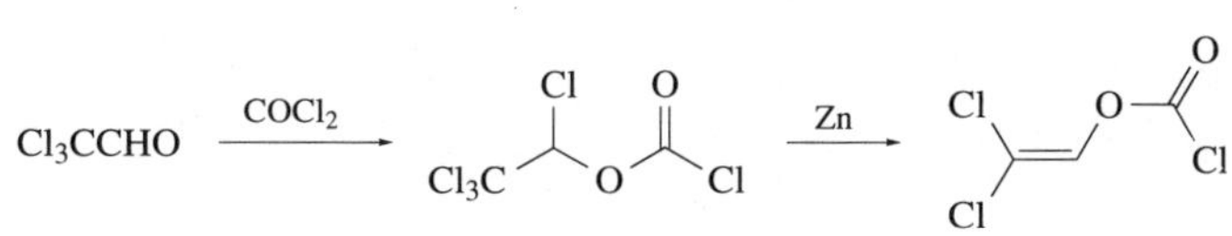

Scheme 17

The corresponding imidate chloroformates (**17**), derived from secondary amides and phosgene, have also been reported (Equation (7)) ⟨83HOU(E4)9⟩.

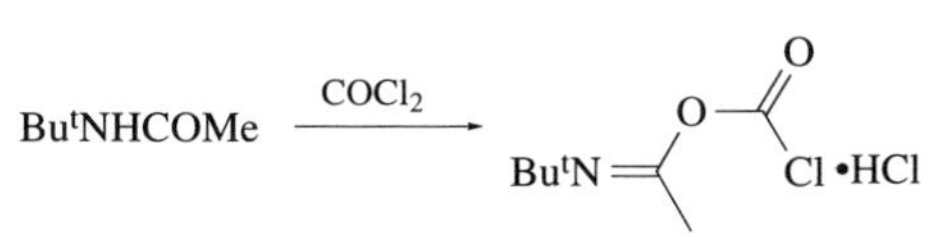

(7)

(17)

(viii) Preparation of oxime chloroformates, $R^1R^2C{=}NOCOCl$

Chloroformates derived from oximes and their tautomeric *C*-nitroso compounds have been used as intermediates in the preparation of oxime carbonates, which have been suggested as *t*-butoxycarbonylating agents ⟨77BCJ718⟩ and as precursors to photolabile amine-protecting groups ⟨89TL1901⟩. The oxime chloroformates were prepared in the standard way from the oxime and phosgene or trichloromethyl chloroformate with pyridine or *N,N*-dimethylaniline, in inert solvents (Equation (8)).

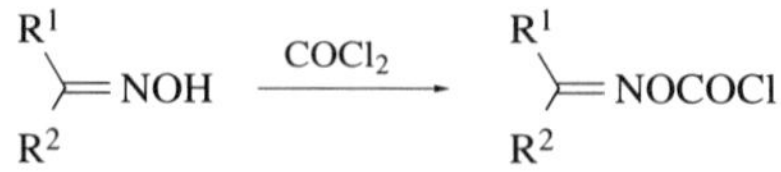

(8)

Conversion of the nitroso compound (**18**) to a mixture of (*E*)- and (*Z*)-oxime chloroformates (**20**) was accomplished by first treating (**18**) with TMS-Cl and zinc to give the silyl protected oxime (**19**), followed by subsequent reaction with phosgene (Scheme 18) ⟨89IZV2849⟩.

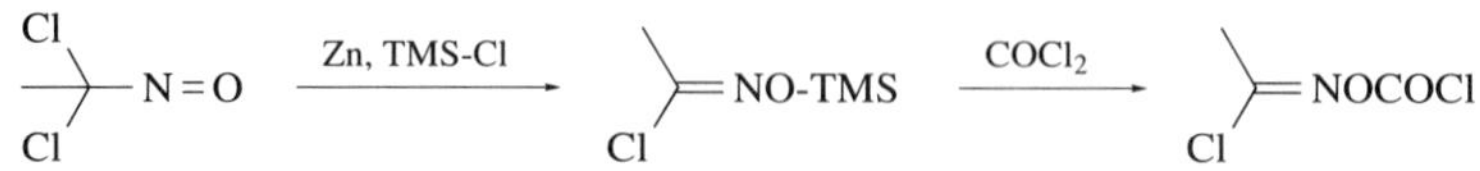

(18) **(19)** **(20)**

Scheme 18

(ix) Formation of chloroformic acid anhydrides

Anhydrides of chloroformic acid with both carboxylic and sulfonic acids have been reported. They are prepared at low temperatures, and are very reactive. Amino acids protected as their *N*-benzyloxycarbonyl derivatives form anhydrides on reaction with phosgene and triethylamine at −70 °C ⟨57HCA604⟩. These anhydrides react with alcohols at −70 °C to form esters. Reaction of unprotected amino acids with phosgene or bis(trichloromethyl) carbonate (triphosgene) without base at room temperature yields the cyclic amino acid carboxyanhydrides. Since the amino group of the amino acid is protonated to a great extent under these conditions, it is possible that a similar anhydride formed by reaction of the carboxylate with phosgene is initially formed, followed by ring closure onto the amine function (Scheme 19) ⟨88TL5859⟩.

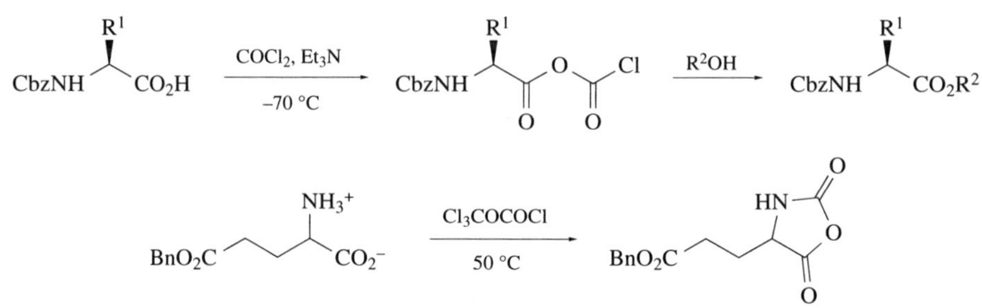

Scheme 19

The anhydride of chloroformic acid and trifluoromethanesulfonic acid is formed by insertion of carbon monoxide into the O—Cl bond of chlorine trifluoromethanesulfonate at low temperature (see Equation (5)) ⟨82JFC(19)227⟩.

(x) Miscellaneous unusual chloroformates

The ^{17}O NMR spectrum of the peroxy chloroformate (**21**) has been reported. This compound was prepared from *t*-butyl hydroperoxide, presumably by reaction with phosgene, and is stable at room temperature ⟨87TL6443⟩.

$Bu^tOOCOCl$

(21)

Also reported is the reaction of phosgene with $Hg(OSeF_5)_2$ to give stable chlorocarbonyl pentafluoroselenate, $ClCOOSeF_5$ ⟨81ZAAC(472)26⟩. The corresponding fluorocarbonyl derivative has been known for some time ⟨73IC769⟩, but was prepared by the reaction of $FOSeF_5$ with carbon monoxide.

6.14.3.1.3 *Preparation of bromoformate esters, ROCOBr*

Bromoformates are poorly represented in the chemical literature. Rosenmund and Döring prepared several alkyl bromoformates and benzyl bromoformate by the reaction of the appropriate alcohols with carbonic dibromide, and reported that they tended to decompose within a few days ⟨28AP(266)277⟩. A much more convenient method involves halogen exchange between the corresponding chloroformate and anhydrous hydrogen bromide in methylene chloride, catalysed by tetrabutylammonium bromide. In this manner, very high distilled yields of phenyl bromoformate and 2,2,2-trichloroethyl bromoformate were achieved ⟨88S407⟩.

6.14.3.1.4 *Preparation of iodoformates, ROCOI*

Stable iodoformates have only been prepared in the last 20 years. In general, they require some steric or electronic feature to render them sufficiently stable to allow their isolation. Thus, attempts to form the iodoformate from cholesteryl chloroformate and sodium iodide led to 3-iodocholesterol in 50% yield, presumably by decomposition of the unstable intermediate iodoformate ⟨67JOC2633⟩. Hoffmann and Iranshahi report that attempts to isolate iodoformates by halogen exchange were unsuccessful with unhindered primary, secondary or benzyl chloroformates ⟨84JOC1174⟩. The first stable iodoformate was prepared by irradiation of 9-triptycyl hydrogen oxalate (**22**) in the presence of iodine and mercury(II) oxide. The 9-triptycyl iodoformate (**23**) proved to be very stable to loss of carbon dioxide, only decomposing to give 9-iodotriptycene (**24**) at 260 °C (Scheme 20). This would be expected for an ionic reaction occurring at the bridgehead carbon ⟨71JCS(C)3766⟩. Subsequently, a range of carefully chosen iodoformates was prepared in high yield from the corresponding chloroformates by halogen exchange with sodium iodide in acetonitrile (Table 1). They could be stored at −20 °C over copper powder. Phenyl iodoformate was sufficiently stable to be distilled at 68 °C and 1 torr, and underwent reaction with copper(I) cyanide to give the corresponding cyanoformate in good yield ⟨84JOC1174⟩.

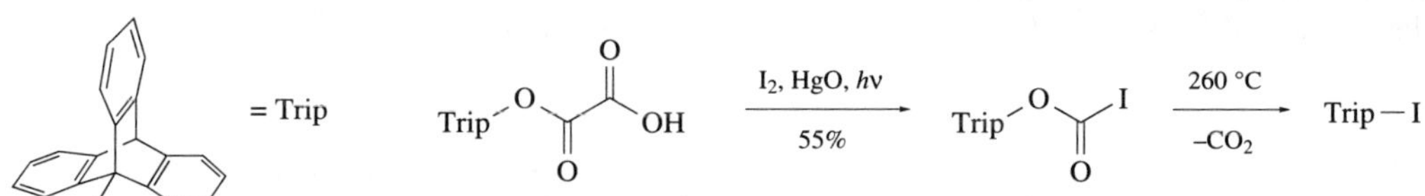

(22) **(23)** **(24)**

Scheme 20

Table 1 Formation of iodoformates from chloroformates.

ROCOCl $\xrightarrow{\text{NaI, MeCN}}$ ROCOI

R	*Temperature* (°C)	*Yield* (%)
CH_2=CH	25	80
Ph	70	85
Bu^tCH_2	70	75
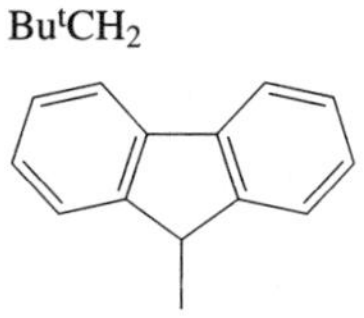	25	64

6.14.3.2 One Halogen and One Sulfur Function

Care should be exercised when searching or naming compounds of this class, RSCOHal, because confusion can sometimes arise over the nature of the connectivity. It is not always clear whether the acid component is connected to the alcohol or thiol by a C—S or C—O bond. This ambiguity is circumvented by the *Chemical Abstracts* nomenclature; the title compounds are named as carbonohalidothioic acid *S*-esters. Alternative names are formic acid halothio-*S*-esters or halothiolformate esters. For convenience, the last name will be used throughout this work. However, many examples also exist where the sulfur atom is connected to atoms other than an alkyl or aryl carbon; these can no longer be regarded as halothiolformates, and are named appropriately as they occur, for example chlorocarbonylsulfenyl chloride, ClCOSCl.

6.14.3.2.1 *Fluorothiolformate esters, RSCOF*

Several general methods of synthesis of fluorothiolformate esters are described by Heywang ⟨83HOU(E4)9⟩; latterly, publication in this area has been dominated by Haas and Della Védova ⟨91ZAAC(600)145⟩, however their work has emphasized inorganic and spectroscopic aspects, so it will not be covered in depth here.

(i) Preparation from carbonic dihalides and thiols or thiophenols

Thiols and thiophenols react with carbonic chloride fluoride and carbonic bromide fluoride in the presence of base to give fluorothiolformate esters in good yields ⟨65JOC1317⟩. It is also reported that trifluoromethyl disulfide, CF_3SSH, reacts with carbonic difluoride in the presence of potassium fluoride at $-25\,^\circ C$ to give CF_3SSCOF in 55% yield ⟨76JA6545⟩. If the scope of this reaction can be extended to include thiols and thiophenols this could be a more convenient route, since carbonic difluoride is commercially available.

(ii) Preparation from chlorothiolformates

The treatment of chlorothiolformate esters with hydrogen fluoride at room temperature gives excellent yields of fluorothiolformate esters by halogen exchange ⟨65JOC1317⟩.

(iii) Preparation from fluorocarbonylsulfenyl chloride and bromide, FCOSHal

Fluorocarbonylsulfenyl halides react with a wide range of nucleophiles by displacement of the halogen from the sulfenyl halide function ⟨67AG(E)705, 69CB2718, 70AG(E)466, 73JFC(3)383, 91SA(A)1619, 91ZAAC(600)145⟩. Some of these reactions are summarized in Scheme 21. Fluorocarbonylsulfenyl chloride is itself prepared by treating chlorocarbonylsulfenyl chloride with antimony trifluoride in sulfolane at 90°C ⟨69CB2718⟩, whilst fluorocarbonylsulfenyl bromide is prepared from fluorocarbonylsulfenyl chloride by reaction with trimethylsilyl bromide ⟨91ZAAC(600)145⟩. A comparison of the reactivities of chlorocarbonylsulfenyl chloride and fluorocarbonylsulfenyl chloride with amines shows that chlorocarbonylsulfenyl chloride initially reacts at the carbonyl group, whilst fluorocarbonylsulfenyl chloride reacts at the sulfur atom ⟨73JFC(3)383⟩.

(iv) Rearrangement of aryl fluorothionoformates

As the key step in the preparation of thiophenols from phenols, aryl fluorothionoformates are thermally rearranged to fluorothiolformates at temperatures of 200–500°C (Equation (9)) ⟨88USP4754072⟩.

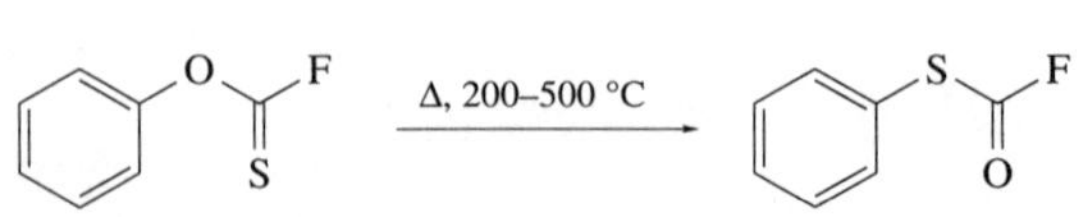

(9)

(v) Reaction of FCOSCl with electron-rich aromatic systems

Fluorocarbonylsulfenyl chloride reacts with pyrrole in the presence of pyridine at $-20\,^\circ C$ to give 2-pyrrolyl fluorothiolformate in 79% yield ⟨77CB67⟩; however, whilst chlorocarbonylsulfenyl

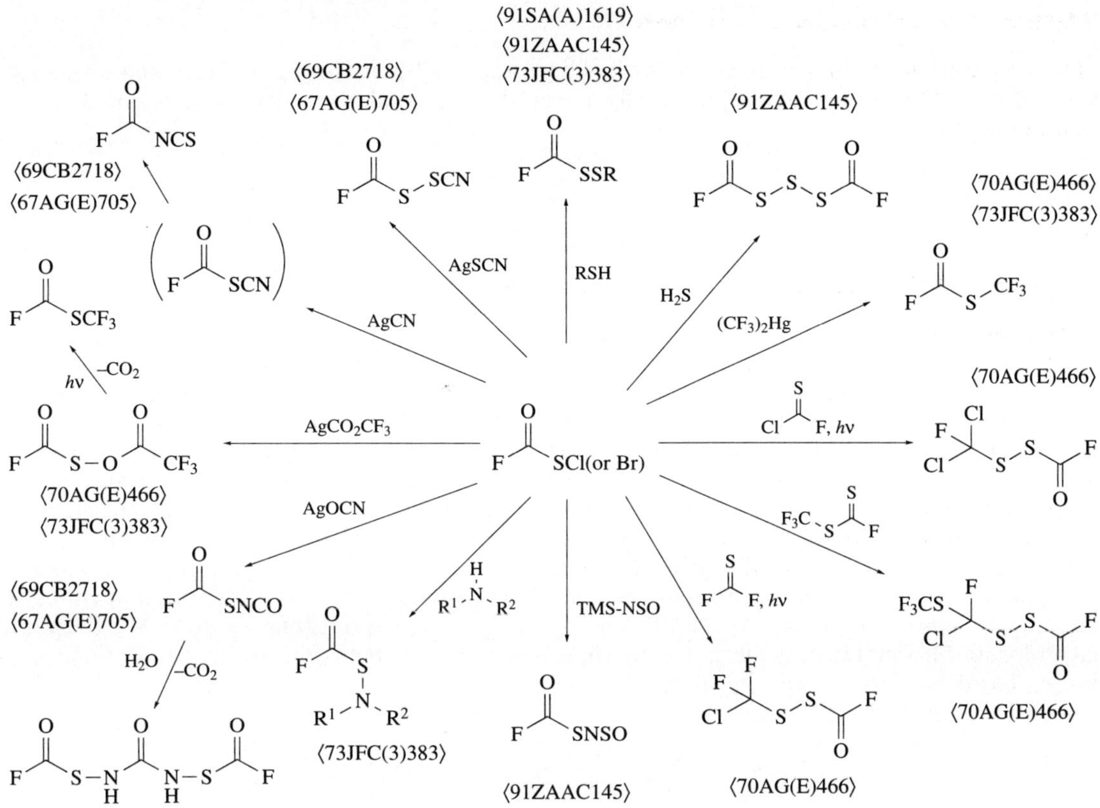

Scheme 21

chloride undergoes the analogous reaction with thiophene in the presence of tin(IV) chloride, the fluorocarbonyl analogue is prepared by reaction of fluorocarbonylsulfenyl chloride with thienylmagnesium bromide only in poor yield (Equation (10)) ⟨76CB2475⟩.

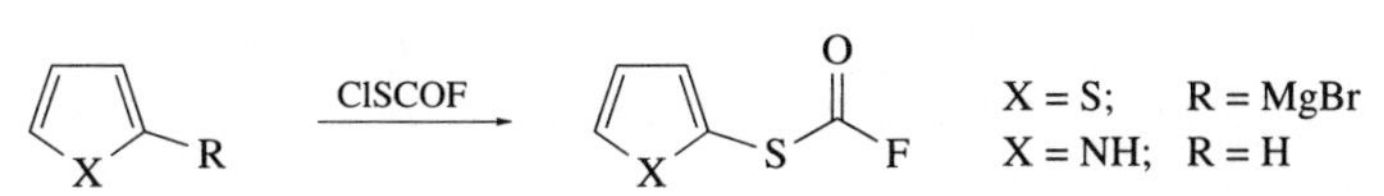

(10)

(vi) Preparation of β-fluoroalkyl fluorothiolformates

Compounds of the structure **(25)** have been reported to undergo chlorine–fluorine exchange with spontaneous rearrangement in the presence of hydrogen fluoride at −10 °C to give β-fluoroalkyl fluorothiolformates (Equation (11)) ⟨88JFC(40)365⟩. The reaction appears quite general, but yields vary in the range 30–60%.

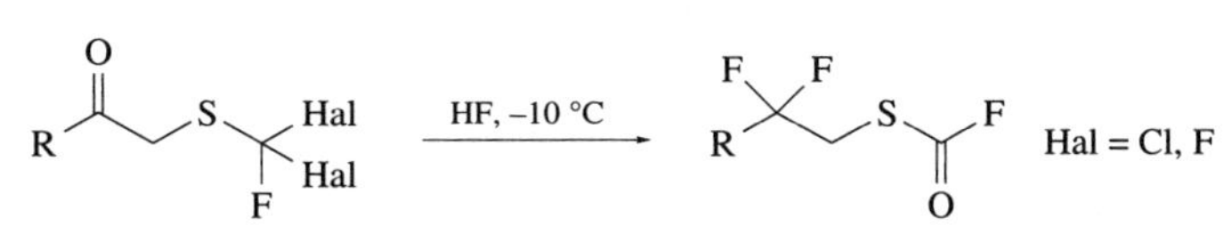

(25) (11)

(vii) Other methods

Pentafluorosulfur carbonyl fluoride F_5SCOF has been prepared in low yield (10–15%), by the irradiation of oxalyl fluoride in the presence of disulfur decafluoride, F_5SSF_5, or pentafluorosulfur hypofluorite F_5SOF. It appears to be a water-sensitive, stable gas (b.p. −10 °C, approximately); the structure is supported by extensive spectroscopic and analytical evidence ⟨68JA3954⟩.

6.14.3.2.2 Chlorothiolformate esters, RSCOCl

The preparation of chlorothiolformates has been thoroughly reviewed by Heywang up to 1982 ⟨83HOU(E4)9⟩. This section will thus briefly cover the principal methods, supplemented by any relevant material.

(i) Preparation from thiols or thiophenols and phosgene

The overwhelming majority of alkyl or aryl chlorothiolformates are prepared from the corresponding thiol or thiophenol or its sodium salt, and phosgene ⟨83HOU(E4)9⟩. If the sodium salt is not used, active carbon, or an organic base such as triethylamine ⟨88EJM(23)561⟩, pyridine ⟨77JOC3686⟩ or dimethylaniline ⟨71BCJ2515⟩ is often added, but the reaction will usually proceed in the absence of base ⟨84ZAAC(508)136⟩. Many literature examples employ a considerable excess of phosgene, often many times the theoretical amount necessary. There is no evidence to suggest that this practice leads to increased yields providing a base is present, and it should be vigorously discouraged on the grounds of safety. The replacement of phosgene by trichloromethyl chloroformate has also been successfully achieved ⟨85JHC1643⟩. In this report, sodium hydride was used to prepare heterocyclic thiol sodium salts *in situ*. This is usually more convenient than forming the salt using aqueous sodium hydroxide followed by precipitation and drying ⟨73BCJ1269⟩. It appears that chlorothiolformates may be more thermally stable than their chloroformate counterparts; *t*-butyl chlorothiolformate can be distilled at 43 °C and stored for several weeks at 4 °C without signs of decomposition ⟨77JOC3686⟩.

(ii) Preparation from sulfenyl chlorides and carbon monoxide

Alkyl and aryl sulfenyl chlorides are converted to chlorothiolformate esters in high yield when treated with carbon monoxide at pressures of 70–400 bars. Electronegatively substituted sulfenyl chlorides such as 2,4-dinitrobenzenesulfenyl chloride and trichloromethanesulfenyl chloride failed to react under these conditions ⟨68CC527⟩.

(iii) Preparation from thiiranes and phosgene

Ring opening of thiiranes with phosgene under pyridine catalysis gives β-chloroalkyl chlorothiolformate esters in a manner analogous to the corresponding reactions with oxiranes (Equation (12)) ⟨83HOU(E4)9⟩; however, the regioselectivity may not be the same.

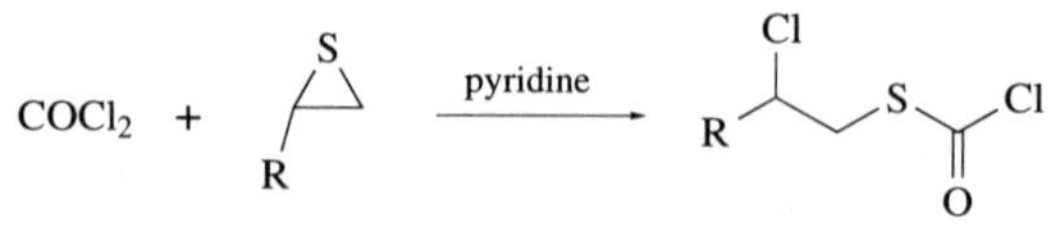

(12)

(iv) Preparation by controlled hydrolysis of trichloromethanethiol derivatives

Trichloromethanesulfenyl chloride can be hydrolysed in good yield to chlorocarbonylsulfenyl chloride, ClCOSCl, by treatment with sulfuric acid containing the calculated quantity of water ⟨64GEP1224720⟩. This method has also been employed in the preparation of bis(chlorocarbonyl) disulfide from bis(trichloromethyl) disulfide ⟨73CL1315⟩; however, other workers have had difficulty reproducing this result ⟨83JOC4750⟩. A milder hydrolysis using dilute hydrochloric acid in refluxing acetone enabled the preparation of (**27**) from (**26**) in 70% yield ⟨86SUL93⟩. Previous attempts to prepare this compound by addition of chlorocarbonylsulfenyl chloride to cyclohexene resulted in the formation of (**29**) through the intermediacy of the thiirane (**28**) (Scheme 22) ⟨70AG(E)54⟩.

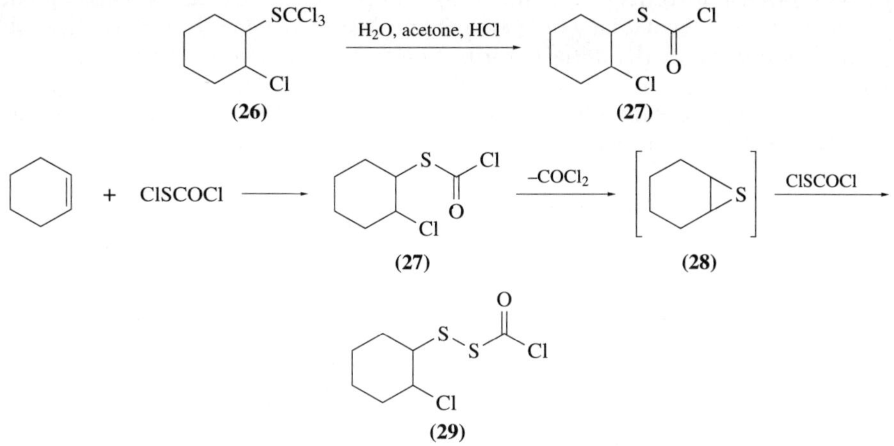

Scheme 22

(v) Preparation from alkoxydichloromethanesulfenyl derivatives

The thermal ⟨81LA1244⟩ or Lewis acid-catalysed ⟨64FRP1372971⟩ elimination of alkyl chlorides from a range of alkoxydichloromethanesulfenyl compounds (**30**) provides a general route to certain classes of chlorocarbonylsulfenyl derivatives. This area has been extensively covered in a series of publications by Barany and co-workers. This work is summarized in Scheme 23 ⟨83JOC4750, 83TL5683, 84JCS(P1)2615, 84JOC1043, 86JOC1866⟩. This reaction has been used for the efficient preparation of [^{18}O]chlorocarbonylsulfenyl chloride ⟨84MI 614-02⟩.

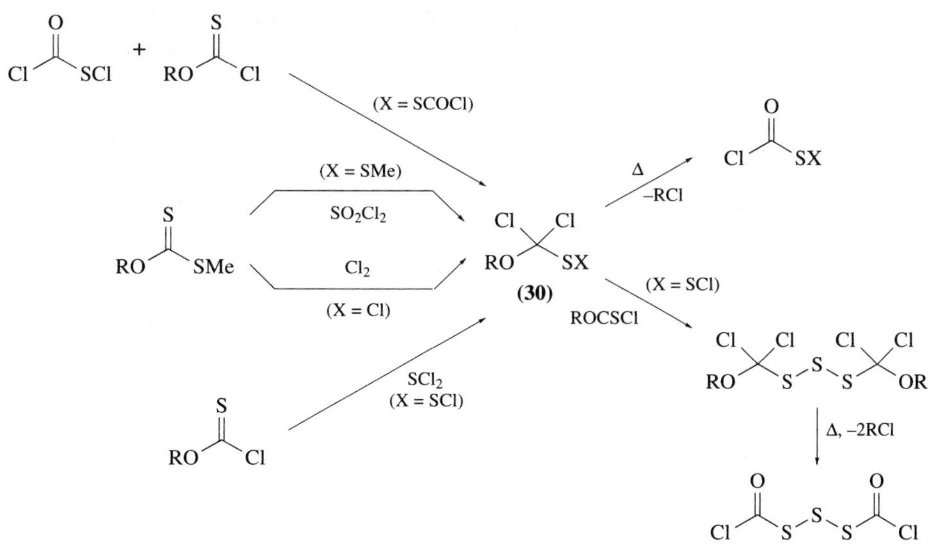

Scheme 23

(vi) Miscellaneous reactions leading to chlorocarbonylsulfenyl derivatives

Chlorocarbonylsulfenyl chloride reacts with silver(I) cyanide to give chlorocarbonyl thiocyanate, in 65% yield ⟨69CB2718⟩; several other nucleophiles also react preferentially at sulfur to give the products shown in Scheme 24 ⟨70AG(E)54⟩. In the case of trifluoroacetamide, further reaction to the

2-oxo-1,3,4-oxathiazole occurs ⟨87JFC(36)429⟩. In the reaction of bis(bistrifluoromethylaminoxy)mercury(II) (**31**) with thiophosgene, migration of the bis(trifluoromethyl)amino groups from oxygen to sulfur occurs, leading to (**32**) in 61% yield (Scheme 25) ⟨84JFC(24)485⟩. Reaction of thiophosgene with bis(trifluoromethyl) sulfine (**33**) also proceeds through a rearrangement to give (**35**), possibly through the intermediacy of (**34**) (Scheme 26) ⟨85CB4553⟩. In the reaction of the allenic alcohol (**36**) with thiophosgene, the initially formed carbonochloridothionic acid *O*-ester undergoes a [3,3] sigmatropic rearrangement to the vinylic chlorothiolformate (**37**) ⟨92AG(E)866⟩. This is an example of the rearrangement of an allylic ester (Scheme 27), and thus may be very general; however, no examples appear in *Houben-Weyl* ⟨83HOU(E4)9⟩.

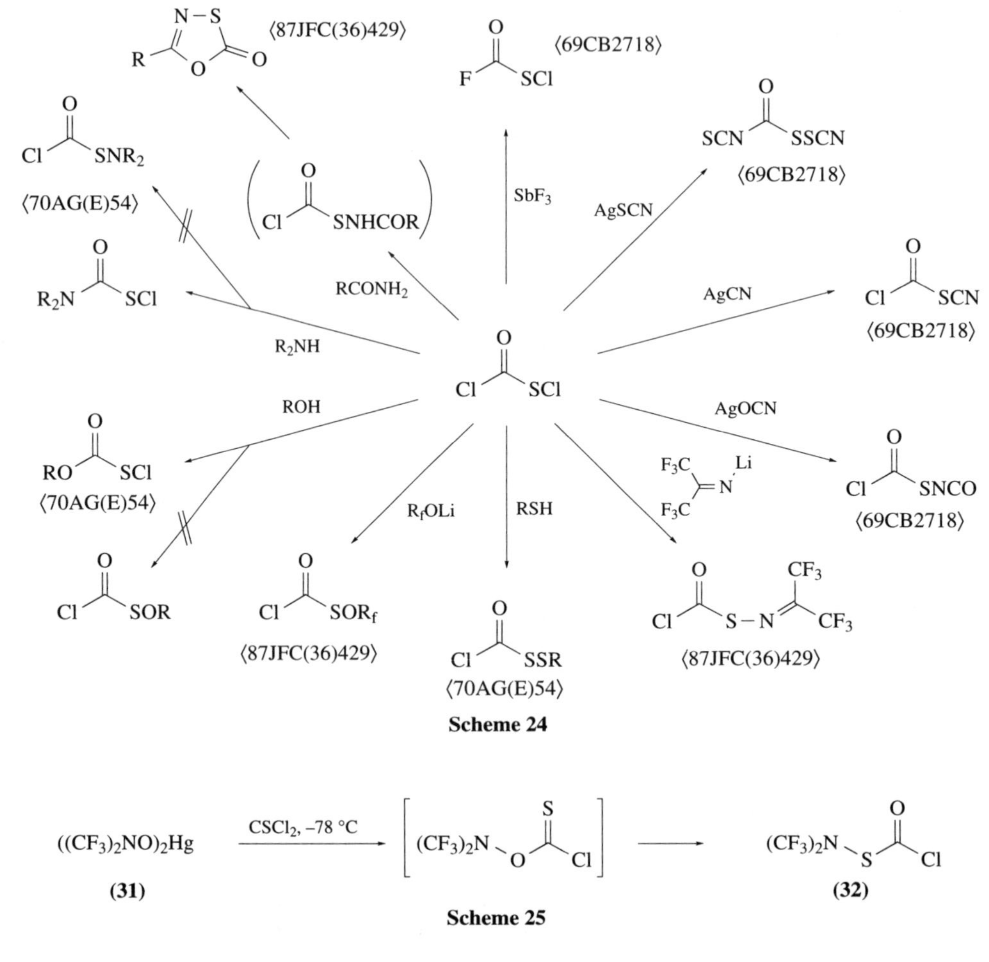

Scheme 24

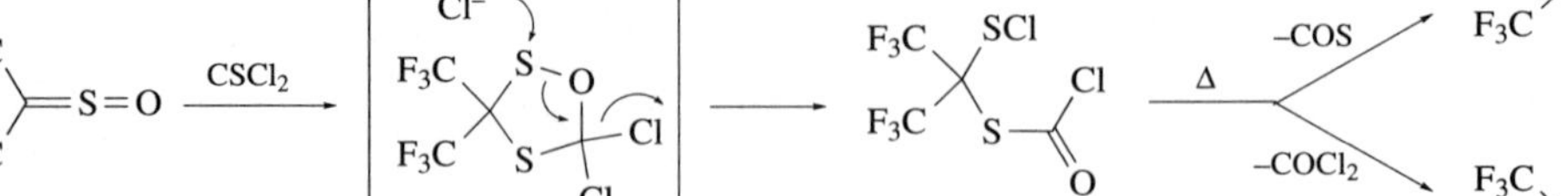

(31) **(32)**

Scheme 25

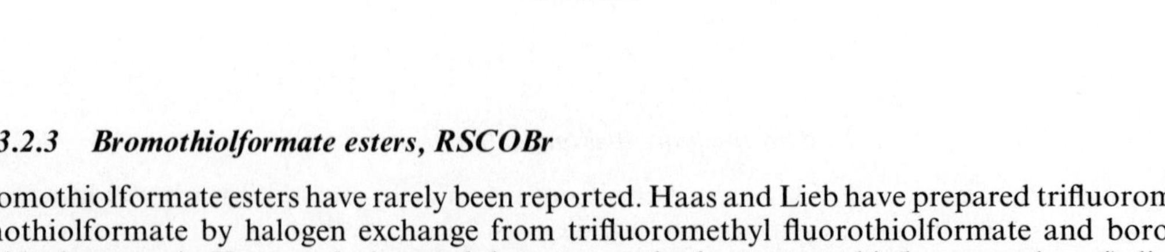

(33) **(34)** **(35)**

Scheme 26

6.14.3.2.3 *Bromothiolformate esters, RSCOBr*

Bromothiolformate esters have rarely been reported. Haas and Lieb have prepared trifluoromethyl bromothiolformate by halogen exchange from trifluoromethyl fluorothiolformate and boron tribromide ⟨78CB2891⟩; Haas *et al.* also used the same method to prepare bis(bromocarbonyl) disulfide

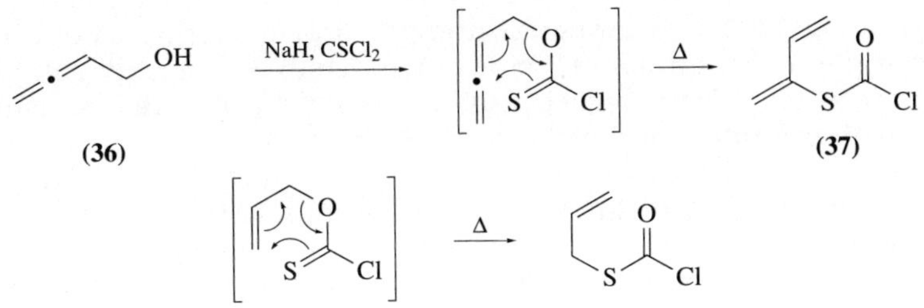

(36) **(37)**

Scheme 27

from bis(fluorocarbonyl) disulfide ⟨73JFC(3)383⟩. A poorly characterized example of carbon monoxide insertion into methanesulfenyl bromide to give methyl bromothiolformate is also reported ⟨83HOU(E4)9⟩. This method is analogous to the reaction of sulfenyl chlorides with carbon monoxide and should proceed under milder conditions in view of the lower S—Br bond strength.

6.14.3.2.4 Iodothiolformate esters, RSCOI

This class of compound does not appear to have been reported.

6.14.3.3 One Halogen and One Nitrogen Function

Compounds of the structure HalCONR^{1}R^2 are referred to in *Chemical Abstracts* as substituted carbamic halides. Alternative names are halogenoformic amides, carbamyl halides and carbamoyl halides. The last name will be adopted in this work, in line with common usage. Other classes of compound incorporating the HalCON functionality also exist, notably those where the nitrogen atom is part of a pseudohalide (e.g., isocyanate) or is part of an imide group. These compounds are variously named, often as halogenocarbonyl derivatives. The entire scope of this section is comprehensively reviewed in *Houben-Weyl* up to 1982: carbamoyl halides by Heywang ⟨83HOU(E4)36⟩; bis(halogenocarbonyl) amines, HalCONRCOHal, by Hagemann ⟨83HOU(E4)1019⟩; halogenocarbonylcarbamic acid esters and amides, HalCONRCOX (X = O, N, S) by Kraatz ⟨83HOU(E4)1023⟩ and Baasner ⟨83HOU(E4)1078⟩; halogenocarbonylisocyanide dihalides and derivatives, HalCON═CHal$_2$, by Hagemann ⟨83HOU(E4)1170⟩; HalCON═CHalX (X = O, N, S) by Salzburg ⟨83HOU(E4)1177⟩; HalCONRCHal$_3$ and derivatives by Marhold ⟨83HOU(E4)1203⟩; halogenocarbonyl isocyanates, HalCONCO, by Hagemann ⟨83HOU(E4)1234⟩; and halogenocarbonyl ureas, HalCONRCONRX (X = O, N, S), by Lieb ⟨83HOU(E4)1275⟩ and Botta ⟨83HOU(E4)1313⟩. The present authors do not hope or wish to incorporate this vast, well-researched body of knowledge into this review. Their aim is to introduce readers to the main synthetic methods employed in this area, incorporating some more recent references, where they exist.

6.14.3.3.1 Carbamoyl fluorides, R^{1}R^{2}NCOF, and other N*-fluorocarbonyl compounds*

The most important members of this class are carbamoyl fluorides bearing one further substituent on the nitrogen atom, RNHCOF. Because of their thermal and chemical stability relative to the corresponding carbamoyl chlorides, they are used as intermediates in the synthesis of important fungicides and insecticides where nitrogen substitution can be carried out, even in the presence of tertiary bases, without elimination to the isocyanate. These *N,N*-disubstituted carbamoyl fluorides can then be further reacted with amines or alcohols to give, respectively, ureas and carbamates.

(i) Preparation from carbonic fluoride halides

Although preparation from carbonic fluoride halides is analogous to the very versatile reaction of phosgene with amines under base-catalysed conditions to form carbamoyl chlorides, it is not a reliable or accessible general method for the preparation of carbamoyl fluorides.

Carbonic chloride fluoride reacts predictably with diphenylamine to produce diphenylcarbamoyl fluoride in good yield ⟨80EUP52842⟩; however, carbonic chloride fluoride is not commercially available, so this method is not very attractive. On the other hand, carbonic difluoride is commercially available; however, its reaction with amines does not usually result in the formation of carbamoyl fluorides; reaction with piperidine or dimethylamine at room temperature leads to oxidative fluorination to form the *N*-fluoro derivative in reasonable yield, with the formation of carbon monoxide ⟨84CC416⟩. The same authors did isolate the carbamoyl fluoride (**39**) in 70% yield from the reaction of (**38**) with carbonic difluoride (Equation (13)). However, some classes of nitrogen compound are fluorocarbonylated by carbonic difluoride. Secondary amides give moderate yields of *N*-fluorocarbonyl derivatives; however, it is difficult to prevent concomitant conversion of the amide carbonyl function to a *gem*-difluoride ⟨62JA4275⟩. The *N*-methoxycarbonyl-*N*-trimethylsilylhydrazine (**40**) is converted to the *N*-fluorocarbonyl derivative (**41**) at −20 °C in quantitative yield (Equation (14)) ⟨82ZN(B)81⟩. This may indicate that compounds with less basic nitrogen atoms are more likely to lead to *N*-fluorocarbonyl products. Benzophenone imine is also reported to react with carbonic difluoride to give *N*-fluorocarbonylation in 40% yield ⟨68ZOR720⟩, whilst at elevated temperatures *N*-aryldifluoroazomethines (**42**) add the elements of carbonic difluoride under fluoride catalysis to give *N*-fluorocarbonyl-*N*-trifluoromethylanilines (**43**) in good yield (Scheme 28) ⟨65JOC4338⟩. In a similar reaction, carbonic difluoride adds to the iminofluoride from HF and HCN to give $CF_2HNHCOF$ in 70% yield ⟨62JA4275⟩. Phenyl isocyanate also adds carbonic difluoride under caesium fluoride catalysis to give *N*,*N*-bis(fluorocarbonyl)aniline in 95% yield ⟨62JA4275⟩. Finally, in a complex reaction, carbonic difluoride reacts with cyanogen fluoride to give bis(trifluoromethyl)carbamoyl chloride, $(CF_3)_2NCOF$, in low yield ⟨81JFC(18)259⟩. This compound can also be prepared in high yield by the fluorination of dimethylcarbamoyl chloride.

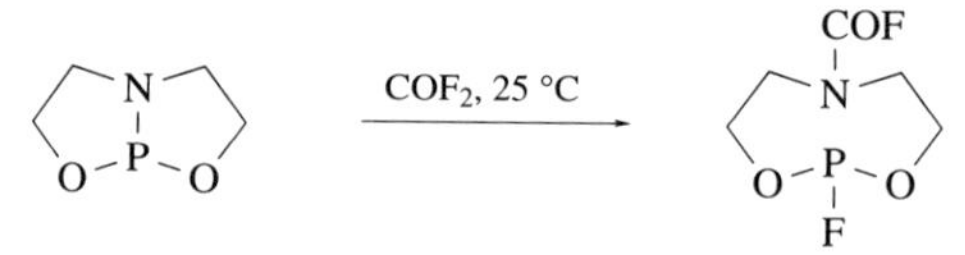

(38) (39) (13)

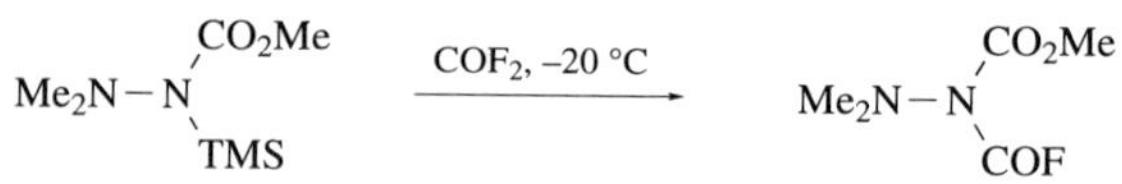

(40) (41) (14)

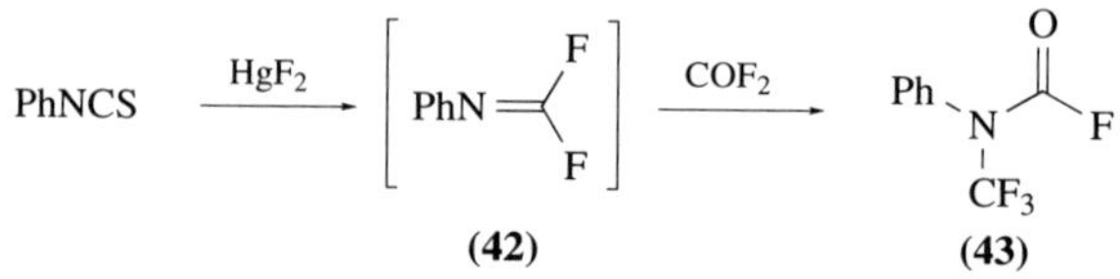

(42) (43)

Scheme 28

(ii) *Preparation from carbamoyl halides by halogen–fluorine exchange*

N-Carbonyl chlorides of many types of are readily prepared; thus, they make ideal starting materials for the preparation of the corresponding *N*-carbonyl fluorides. Several methods have been reported. The use of potassium fluoride in sulfolane requires high temperatures, and is thus limited in its application; any aliphatic chlorine atoms present are also likely to be exchanged ⟨78GEP2706683⟩. The use of the phase transfer catalyst 18-crown-6 enables the exchange to take place in high yield at room temperature, either in the presence or the absence of solvent ⟨79JOC1016⟩. A more recent report from Japanese workers suggests that a carefully dried mixture of potassium and calcium fluorides can effect quantitative exchange at 50 °C in acetonitrile ⟨86CC793⟩. Although reported examples are restricted to the preparation of *N*,*N*-disubstituted carbamoyl fluorides, the last two methods may be applicable to *N*-monosubstituted derivatives, unless the basic nature of potassium fluoride precludes this.

Predictably, carbamoyl bromides are also reported to undergo exchange ⟨78GEP2706683⟩; however, this is of little synthetic use, due to their relative inaccessibility.

(iii) Preparation from alkyl or aryl isocyanates and hydrogen fluoride or other fluoride sources

The method of choice for the preparation of carbamoyl fluorides bearing one further substituent on nitrogen involves addition of hydrogen fluoride to an isocyanate. Alkyl and aryl isocyanates react with hydrogen fluoride in ether at room temperature to give quantitative yields of the corresponding *N*-monosubstituted carbamoyl fluorides ⟨45JCS864⟩. By conducting the reaction at −80 °C, hydrogen fluoride will also add to cyanic acid, HOCN, to give the unsubstituted carbamoyl fluoride, H_2NCOF, which is a stable solid, m.p. 47 °C ⟨40CB177⟩. A more convenient, but lower yielding, method uses the commercially available, stable pyridinium poly(hydrogen fluoride) as both a solvent and hydrogen fluoride source, providing both alkyl and aryl carbamoyl fluorides in 40–60% yield ⟨79JOC3872⟩.

(iv) Preparation from other carbamoyl fluorides

The thermal and base stability of monosubstituted carbamoyl fluorides, RNHCOF, relative to their carbamoyl chloride counterparts renders them excellent intermediates for the preparation of further carbamoyl fluorides by reaction at the nitrogen atom with a range of electrophilic reagents. The reaction with a wide variety of sulfenyl chlorides in the presence of a tertiary base has been thoroughly investigated because the products can be transformed into a range of fungicidal and insecticidal carbamates by reaction with hydroxyl compounds or amines (Scheme 29).

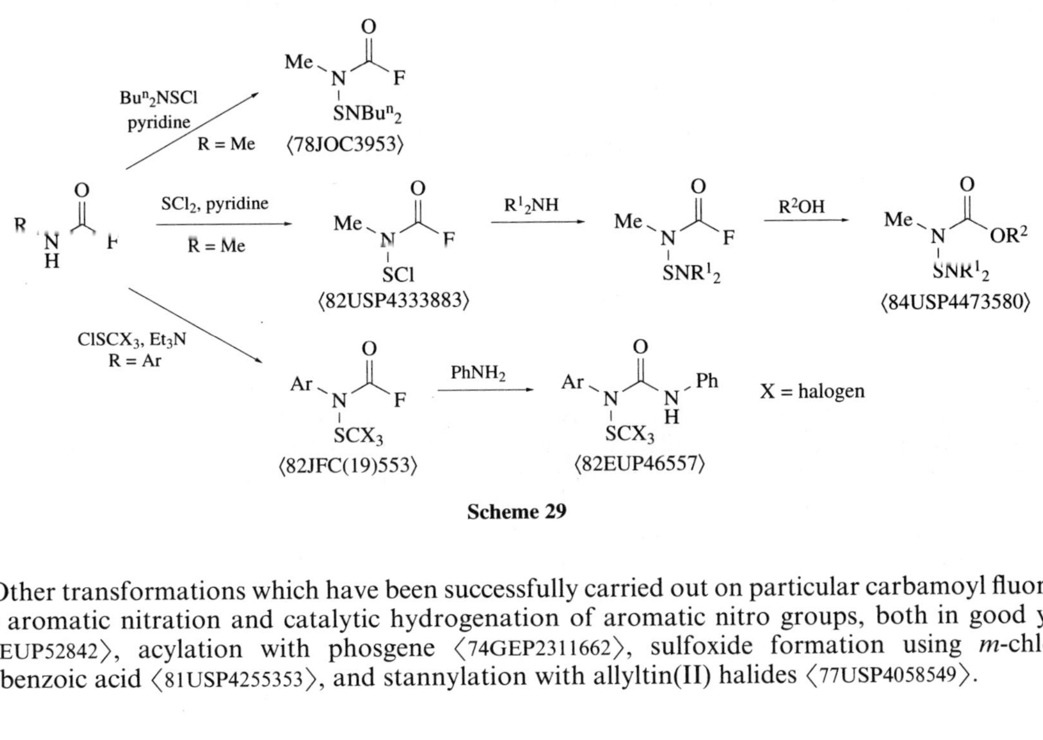

Scheme 29

Other transformations which have been successfully carried out on particular carbamoyl fluorides are aromatic nitration and catalytic hydrogenation of aromatic nitro groups, both in good yield ⟨80EUP52842⟩, acylation with phosgene ⟨74GEP2311662⟩, sulfoxide formation using *m*-chloroperbenzoic acid ⟨81USP4255353⟩, and stannylation with allyltin(II) halides ⟨77USP4058549⟩.

(v) Carbamoyl fluorides arising from fluorination

Carbamoyl fluorides are often formed when appropriate substrates, notably isocyanates, are fluorinated. However, in common with many other fluorinations of organic compounds, the structure of the products can rarely be predicted. Fluorination of *n*-propyl isocyanate with fluorine in Freon 11 can give *n*-propyl-*N*-fluorocarbamoyl fluoride in up to 40% yield. Further studies show that the product does not arise from addition of fluorine to the C═N of the isocyanate (Scheme 30) ⟨67JOC1633⟩. In two related reactions, the fluorination of trifluoromethyl isocyanate ⟨79CB2158⟩ and trifluoroacetyl isocyanate ⟨86JFC(34)251⟩ with xenon difluoride led to good yields of the fluorocarbonylhydrazines (**45**) and (**46**), respectively. Their formation can be explained as arising from elimination of xenon from intermediates of structure (**44**) (Scheme 31). Trifluoromethyl isocyanate reacts with several other fluoride sources to give a range of products (Scheme 32).

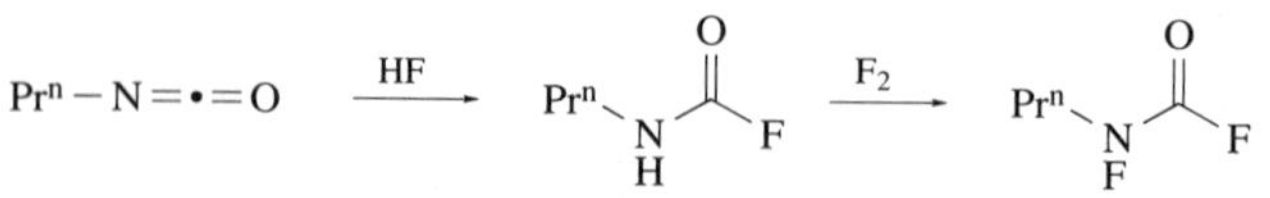

Scheme 30

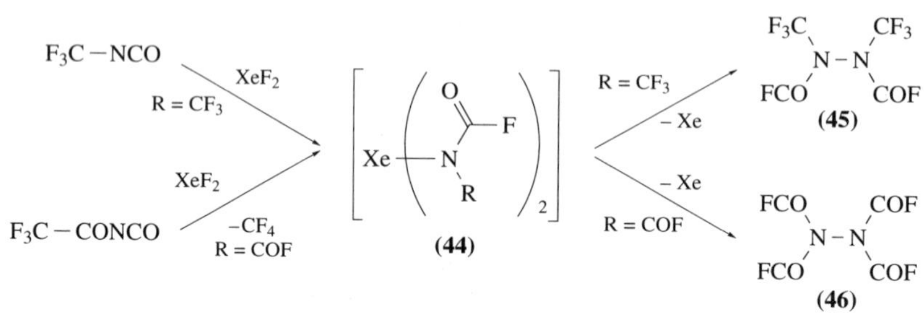

Scheme 31

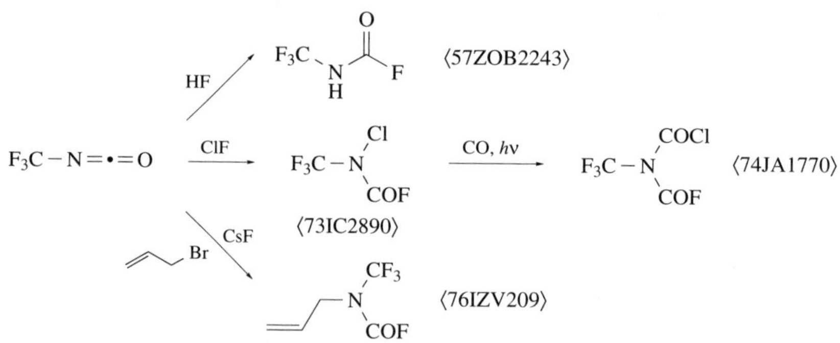

Scheme 32

Electrochemical fluorination of dimethylcarbamoyl chloride, Me_2NCOCl, gives almost exclusively the perfluorinated product, $(CF_3)_2NCOF$ ⟨58JOC1576⟩. When five- or six-membered cyclic ureas (**47**) are fluorinated in water, ω-difluoroaminoalkylcarbamoyl fluorides (**48**) are formed in poor yield (Equation (15)) ⟨70JA2096⟩. Finally, a range of (fluorocarbonyl)imidosulfur compounds $R^1R^2S{=}NCOF$ have been prepared; fluorination of CF_3SNCO at low temperature leads to the formal 1,3-addition of fluorine, to give $CF_3FS{=}NCOF$ in good yield ⟨83CB1257⟩. Compounds of the same class are formed in almost quantitative yield by the reaction of sulfur tetrafluoride ⟨86IS10⟩ or substituted analogues ⟨91CB2411⟩ with silyl isocyanates (Scheme 33).

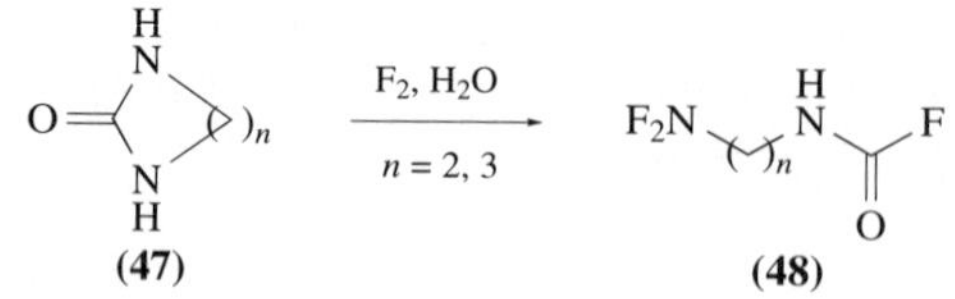

(**47**) (**48**) (15)

(vi) *Carbamoyl fluorides from trifluoromethylamines, $R^1R^2NCF_3$*

N,N-Disubstituted formamides can be fluorinated by silicon tetrafluoride in the presence of potassium fluoride to give trifluoromethylamines, $R^1R^2NCF_3$. Where R^1 and R^2 are alkyl groups, addition of these compounds to crushed ice results in the carbamoyl fluorides R^1R^2NCOF in over 70% yield ⟨83JFC(23)207⟩. Perfluoroalkyl tertiary amines bearing at least one CF_3 group, $R_f^1R_f^2NCF_3$, are converted to the carbamoyl fluorides $R_f^1R_f^2NCOF$ in 38–62% yield by 30% oleum, preferably with catalysis by $MoCl_5$. Where more than one NCF_3 group is present, only one is converted to give the monofluorocarbonyl derivative ⟨89JFC(45)293⟩.

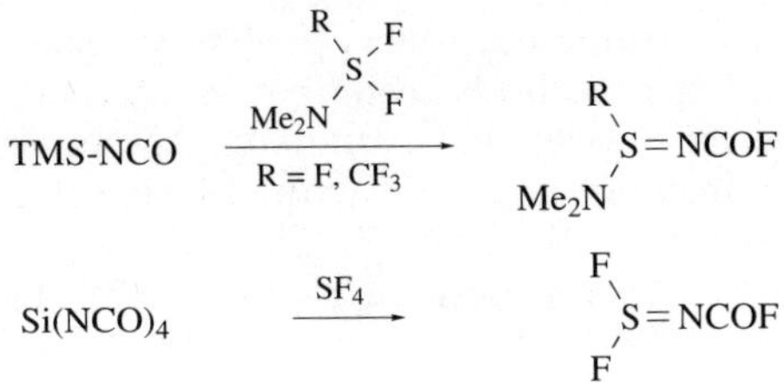

Scheme 33

(vii) Miscellaneous reactions yielding carbamoyl fluorides

A variety of N-unsubstituted aziridines have been shown to react with trifluoromethyl hypofluorite, CF_3OF, at $-40\,^{\circ}C$ to give 50–100% yields of aziridine-N-carbonyl fluorides (**49**). These can rearrange on storing at room temperature to give high yields of β-fluoroalkyl isocyanates (**50**) in high yields (Scheme 34) ⟨80JFC(15)37⟩.

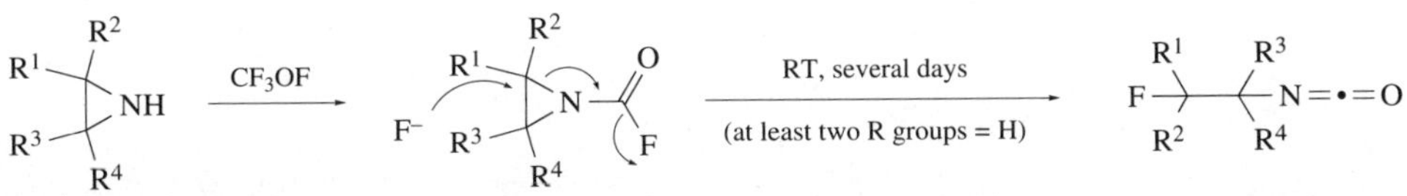

(49) **(50)**

R^1, R^2, R^3, R^4 = H, Me, Ph

Scheme 34

The perfluorinated oxaziridine (**51**) reacts with a wide range of nucleophiles, in the presence of sodium fluoride, generally giving fluorocarbonyl derivatives (**52**) in varying yields by attack at the nitrogen atom. Reactions with diethylamine and potassium thiocyanate did not follow this course, and sodium azide and sodium cyanate failed to react (Scheme 35) ⟨79JFC(14)289⟩. N-Monosubstituted carbamoyl fluorides have been prepared by the reaction of an alcohol or an alkene with cyanogen chloride and HF (Equation (16)) ⟨73GEP2342860⟩. Finally, a report claiming the formation of F_2NCOF in high yield by isomerisation of trifluoronitrosomethane, CF_3NO ⟨36CB684⟩, has been reinvestigated and found to be incorrect; the product appears to be a mixture of trifluoronitromethane and hexafluoroazoxymethane (**53**) ⟨54JCS919⟩.

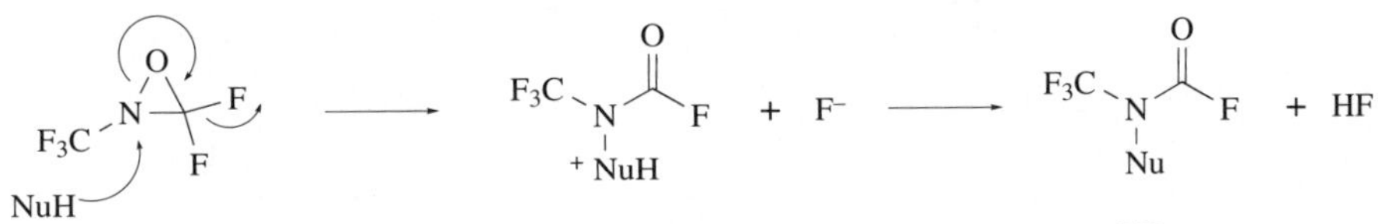

(51) **(52)**

NuH = MeOH (80%), Pr^iOH (88%), Bu^tOH (93%), $MeCO_2H$ (32%), EtSH (30%); and KCN (40%)

Scheme 35

(16)

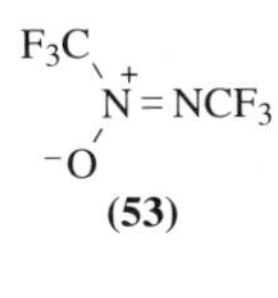

(53)

(viii) Preparation of fluorocarbonyl isocyanate

Whilst the chemistry of chlorocarbonyl isocyanate has been extensively explored, and has been the subject of several reviews (see Section 6.14.3.3.2), the reactions of fluorocarbonyl isocyanate do not appear to have been similarly examined; this is probably because no convenient synthesis yet exists. Until the early 1990s, all the methods available involved the use of COF_2, often at high temperatures. However it has been shown that commercially available chlorocarbonyl isocyanate will undergo halogen exchange in moderate yield with chlorine fluoride or xenon difluoride. The methods available are summarized in Scheme 36. Several preparations of fluorocarbonyl isothiocyanate, FCONCS, have also been described (Scheme 37). These are mainly analogous to those described for fluorocarbonyl isocyanate; formation by halogen exchange has not yet been investigated.

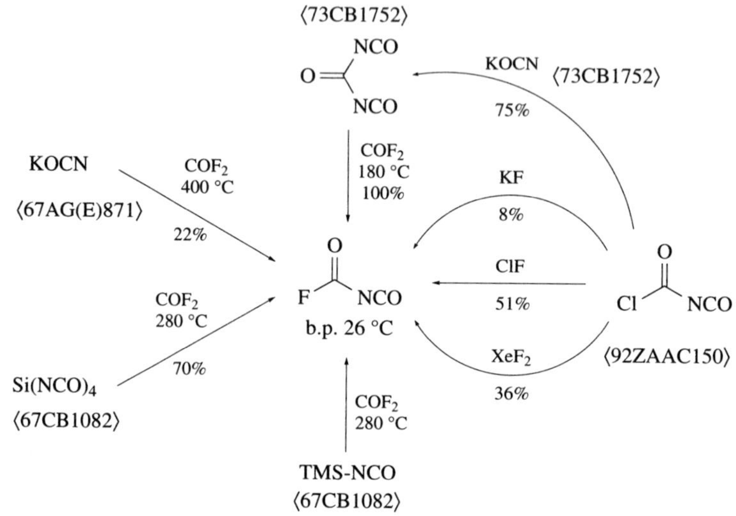

Scheme 36

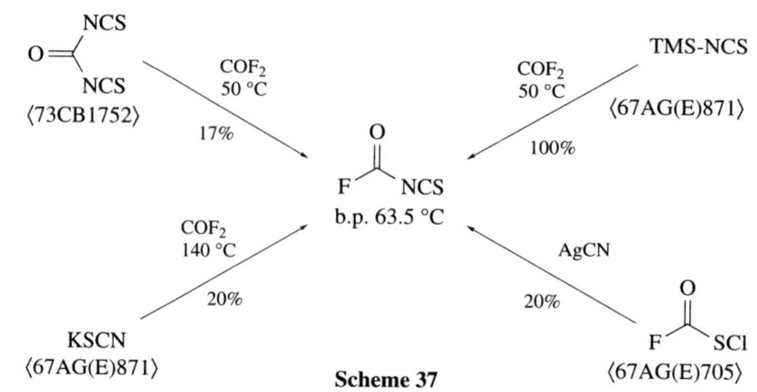

Scheme 37

6.14.3.3.2 *Carbamoyl chlorides, R^1R^2NCOCl, and other* N-*chlorocarbonyl compounds*

Carbamoyl chlorides are quite often rather unreactive towards nucleophiles, and can often be purified by chromatography and recrystallised from alcohols without decomposition. Perhaps because of their relative lack of reactivity, or, in 1993, the discovery of their carcinogenic potential ⟨B-93MI 614-01⟩, their use in synthesis has been quite limited. Carbamoyl chlorides bearing one further substituent on the nitrogen atom are important intermediates in the preparation of isocyanates, but are not usually isolated. Carbamoyl chloride itself, H_2NCOCl, is well described but of limited stability; it has found use in the amidation of aromatic systems under Friedel–Crafts conditions ⟨B-64MI 614-01⟩. Some classes of *N*-chlorocarbonyl compounds are very reactive; when the nitrogen is part of an aromatic ring system, reaction with alcohols is often instantaneous.

This section encompasses a diversity of compound types. The authors have ordered the contents by reagent and/or reaction type, rather than by the structure of the product formed. The exception to this is the final subsection, which is devoted to the synthesis of chlorocarbonyl pseudohalogenides. This area, up to 1982, is covered in great depth in *Houben-Weyl*, volume E4; this work includes many tables of specific compounds and synthetic details.

(i) Methods involving the use of phosgene

Most compounds with a nitrogen atom bearing a hydrogen or TMS function, or which can tautomerize to yield a nitrogen atom bearing a hydrogen, will react with phosgene to give various *N*-chlorocarbonyl compounds. This is by far the most commonly encountered method of preparing such compounds. The use of phosgene can often be avoided by substituting one of the phosgene alternatives, diphosgene or triphosgene; these are treated separately in the next subsection. What follows is an attempt to make some general comments for various classes of nitrogen compounds, which are summarized in Scheme 38–42. Compounds which react abnormally with phosgene are discussed at the end of this subsection.

(a) Reaction of phosgene with secondary amines, R^1R^2NH, to give R^1R^2NCOCl. Since reactions of nucleophiles with phosgene proceed with the elimination of hydrogen chloride, it might be expected that the reaction of phosgene with basic secondary amines should proceed to the stage where 50% of the amine is converted to product, whilst the remainder is converted to the non-nucleophilic amine hydrochloride. This is in fact the case at temperatures near to ambient and below ⟨85CB2294⟩; however, at reflux in toluene with an excess of phosgene, total conversion to the carbamoyl chloride can occur in excellent yield ⟨88MI 614-01⟩. The addition of a tertiary base as an HCl acceptor not only allows full utilisation of the secondary amine, but also enables the quantity of phosgene to be reduced. Stoichiometric amounts of phosgene can sometimes lead to yields approaching 95% ⟨89JCS(P1)1727⟩; however, the use of 1.5–2.0 equivalents is more usual. As the quantity of phosgene is reduced, the possibility of symmetrical urea formation increases. This is rarely specifically mentioned as a problem in the literature; however, isolated yields of carbamoyl chlorides are often only moderate. An internal tertiary basic centre obviates the need for the addition of a further HCl acceptor in some cases ⟨87IJC(B)748⟩; the success of this reaction depends on the tertiary centre hydrochloride of the starting material being to some extent soluble in the reaction medium, so the use of a more polar solvent may be beneficial in this case.

In the reaction of secondary amines with phosgene, anion formation using sodium hydride can also lead to high yields of carbamoyl chlorides ⟨89MI 614-03⟩; this method is potentially useful when separation of the product from the tertiary base hydrochloride could be problematical. Alternatively, the use of an *N*-trimethylsilylamine as the amine component should allow smooth reaction with phosgene, volatile trimethylsilyl chloride being the only by-product ⟨69ZOB2598⟩. *N,N*-Disubstituted carbamoyl chlorides can be very stable; examples exist of reactions in trifluoroacetic acid and water leaving the carbamoyl chloride unchanged ⟨90S1065⟩.

(b) Reaction of phosgene with primary amines, RNH_2, and ammonia, to give RNHCOCl and H_2NCOCl. Phosgene reacts with primary amines to give *N*-monosubstituted carbamoyl chlorides, RNHCOCl. On treatment with tertiary bases, or thermally, under reflux in benzene or toluene, these carbamoyl chlorides eliminate hydrogen chloride to give isocyanates in high yield. Because of the uncertainty involved in isolating the carbamoyl chlorides due to their instability, the preferred substrate for further reactions is usually the isocyanate; the reaction of nucleophiles with isocyanates gives rise to the same product as the corresponding *N*-monosubstituted carbamoyl chloride, and often with the advantage that no acid acceptor is necessary (Scheme 38). Mono- or bisilyl-substituted primary amines react with phosgene at 0°C to give isocyanates directly, with no isolable carbamoyl chloride intermediates ⟨69ZOB2598⟩.

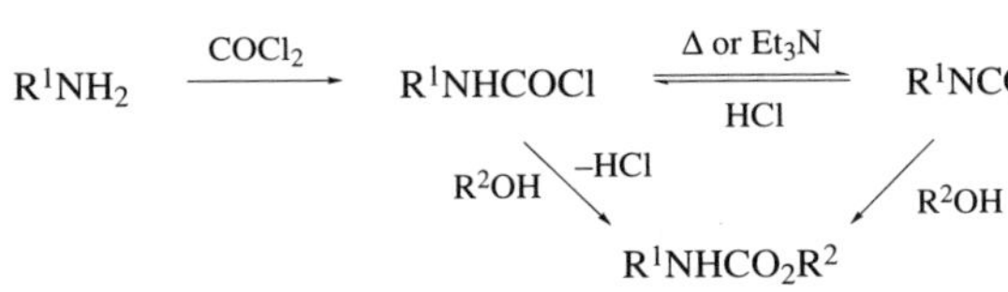

Scheme 38

Primary amines react cleanly with phosgene in the gas phase at 275°C to give 70–90% yields of *N*-monosubstituted carbamoyl chlorides; ammonia undergoes the same reaction at 400–500°C to give carbamoyl chloride ⟨49AG183, 50JA1888⟩. Although secondary amines also undergo this reaction, it is most unlikely to be the best method of preparation of any particular *N,N*-disubstituted carbamoyl chloride.

(c) Reaction of phosgene with tertiary amines. The initial reaction of phosgene with tertiary amines gives rise to acylammonium salts ⟨92JOC5136⟩. These will often eliminate an alkyl chloride on standing for 24–72 h at room temperature to give *N,N*-disubstituted carbamoyl chlorides. On hydrolysis, the unstable *N*-carboxylic acids lose carbon dioxide to give secondary amines (Scheme

39) ⟨75LA2227, 83JHC1477, 85AF217, 89AF539⟩. This dealkylation can proceed in good yield, and is a useful alternative to the Von Braun reaction, which utilises cyanogen bromide to accomplish the same transformation ⟨20CB601⟩. Pyridine was thought to form a stable bispyridinium salt (**54**) with phosgene ⟨56CB2562⟩. Reinvestigation has led to reassignment of the structure as (**55**) ⟨88JOC6145⟩.

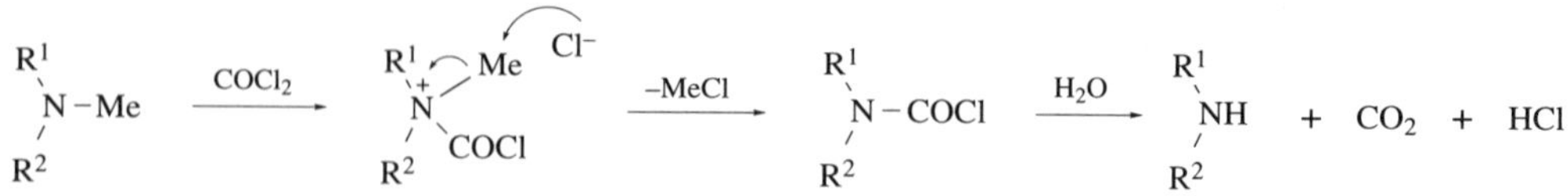

Scheme 39

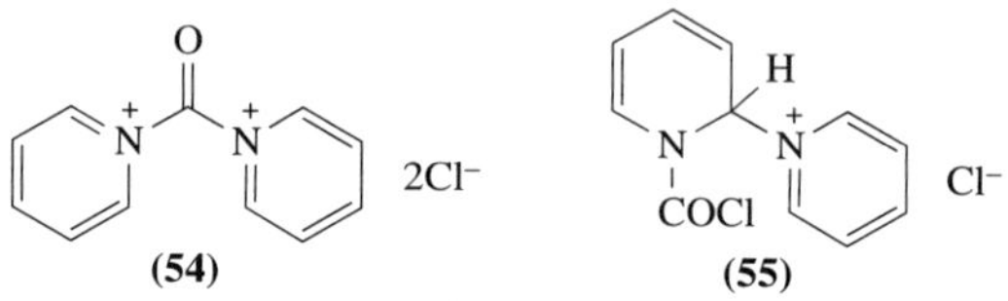

(54) **(55)**

(d) Reaction of phosgene at heteroaromatic nitrogen atoms. Aromatic heterocycles which carry a hydrogen atom on a cyclic nitrogen, or which can tautomerize to do so, react with phosgene to give *N*-chlorocarbonyl derivatives; amongst these are imidazole ⟨79LA1756⟩, benzimidazole ⟨77S704⟩ and 5-aryl-1,3,4-oxadiazol-2(3*H*)-ones ⟨89JHC231⟩. Indole reacts with phosgene to give the unexpected product (**56**) ⟨84IJC(B)986⟩. *N*-Trimethylsilyl heterocycles also react with phosgene to give similar products; examples include imidazoles and pyrazoles ⟨80ZOB875⟩, and 1,2,4-triazolo[4,3-*a*]pyridin-3(2*H*)-one ⟨83BCJ2969⟩. In the latter example, the *N*-trimethylsilyl derivative (**57**) and the parent heterocycle (**58**) react with phosgene to form different *N*-chlorocarbonyl compounds (Scheme 40).

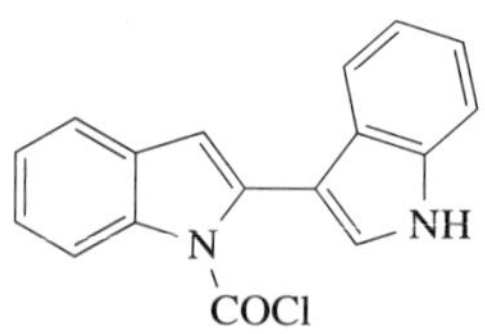

(56)

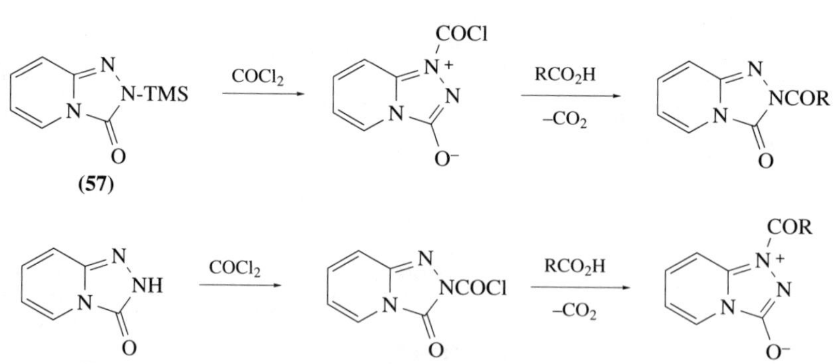

Scheme 40

These heterocyclic *N*-chlorocarbonyl compounds are usually very reactive towards nucleophiles.

(e) Reaction of other N—H compounds with phosgene. Many other classes of N—H-containing compounds also react with phosgene with the elimination of HCl to give *N*-chlorocarbonyl derivatives. These include oximes, hydrazines, ureas, guanidines, carbazates and sulfamoyl chlorides. Imines, amides and *N*-silyl amides also react, but the initial products can sometimes rearrange. Compounds which have tautomeric forms containing N—H bonds can also react to give the *N*-chlorocarbonyl derivatives of that tautomer; *N*-substituted iminoethers (**59**) follow this path,

unless the oxygen substituent is a silyl group (**60**), when they give the product expected from their amide tautomer. These reactions are summarized in Scheme 41.

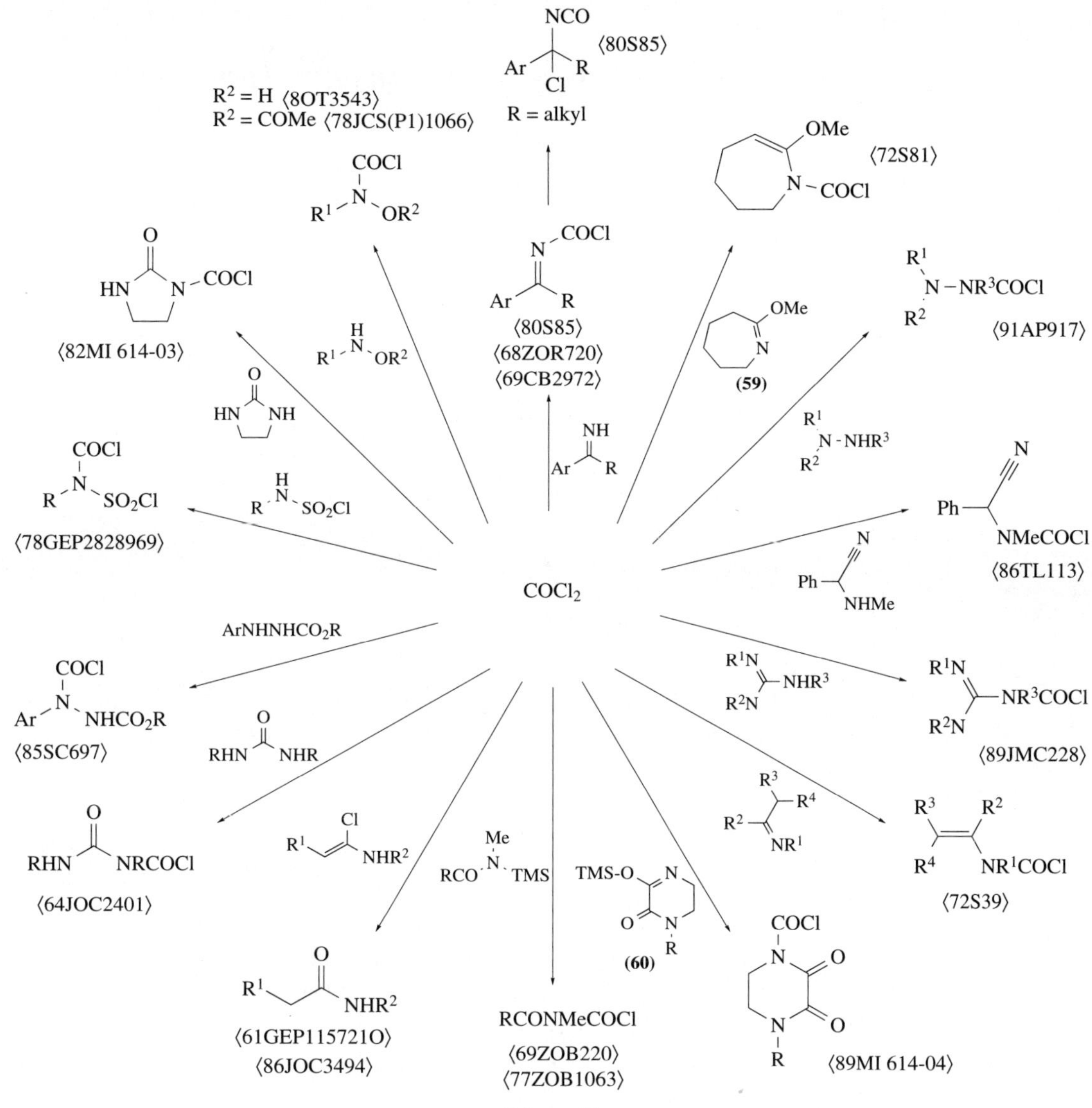

Scheme 41

(f) Reactions of compounds containing C—N multiple bonds with phosgene. Many compounds which contain carbon multiply bonded to nitrogen initially add phosgene across the carbon–nitrogen bond to give α-chloroalkyl *N*-chlorocarbonyl derivatives. If these adducts possess the capacity to eliminate HCl, this is likely to occur. This behaviour is found in suitably substituted imines. Where elimination of HCl cannot occur, the initial adducts are stable, and can be useful bifunctional intermediates in heterocyclic synthesis. Certain imines, carbodiimides, nitriles and cyanamides fall into this category. These reactions are summarized in Scheme 42.

(g) Abnormal reactions of N—H compounds with phosgene. *N*,*N*-Diphenylformamidine (**61**) reacts with two molecules of phosgene, exemplifying at the same time the two modes of reaction discussed in the two previous subsections. Although the initial product (**62**) cannot eliminate HCl readily, it is extremely sensitive to hydrolysis, and it is the initial hydrolysis product (**63**) which eliminates HCl to form *N*-chlorocarbonyl-*N*-phenylformamide (**64**) (Scheme 43) ⟨86JOC4483⟩.

The rearrangement of ω-alkoxyalkyl- and ω-phenoxyalkylcarbamoyl chlorides (**65**) to the ω-chloroalkyl carbamates (**66**) proceeds even at room temperature when $n = 2$ or 3. More vigorous heating leads to the cyclic carbamates (**67**) ⟨75JCS(P1)1836⟩ (Scheme 44).

When 3-hydroxymethylpiperidine reacts with phosgene as its PTSA salt the expected chloroformate (**68**) is formed. If the chloroformate free base is liberated at $-60\,^{\circ}C$ with triethylamine,

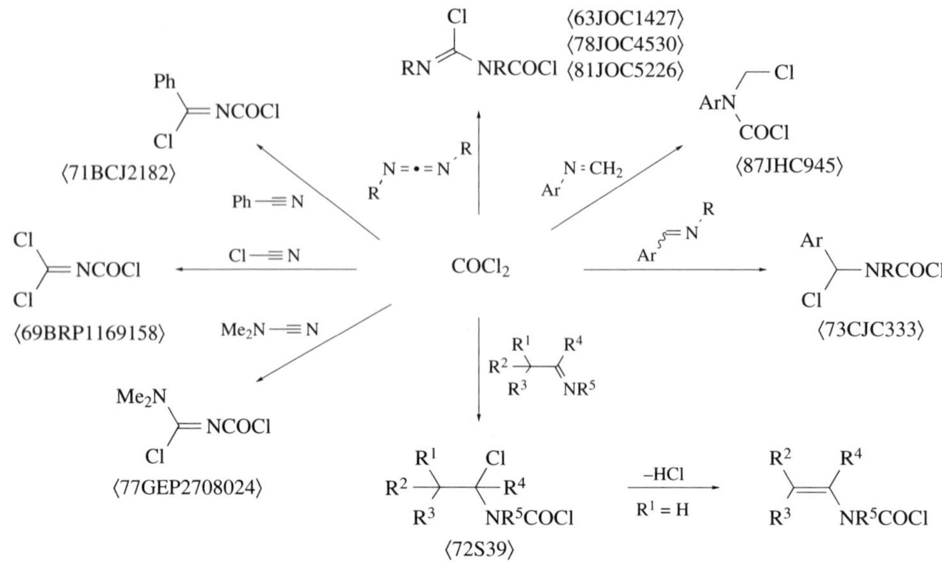

Scheme 42

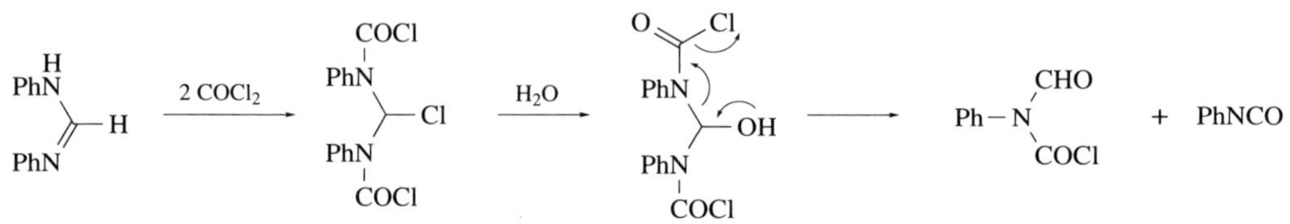

(61) **(62)** **(63)** **(64)**

Scheme 43

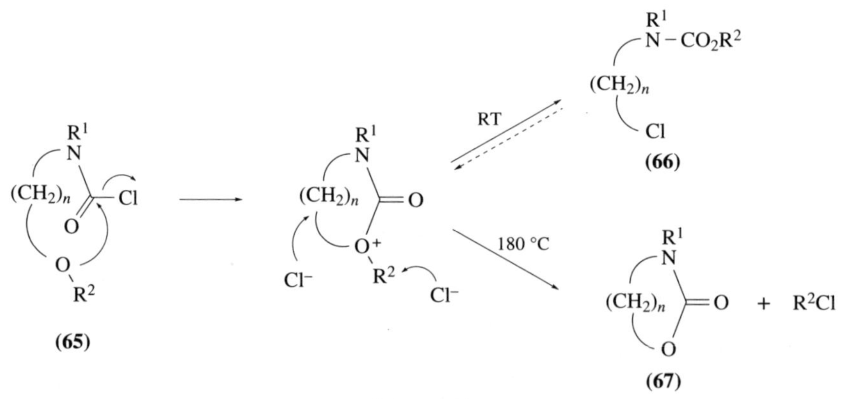

(65) **(66)** **(67)**

Scheme 44

migration of the chlorocarbonyl group from oxygen to nitrogen occurs, to give the stable carbamoyl chloride (**69**) in high yield. The same reaction course was observed for 3-hydroxypiperidine. Compounds (**68**) and (**69**) readily cyclise in the presence of triethylamine to give the bicyclic carbamate (**70**) (Scheme 45) ⟨80JOC5325⟩.

The *N*-chlorocarbonyl derivatives of the *O*-benzylhydroxylamine (**71**) and proline (**73**) cyclise readily to give the corresponding anhydrides (**72**) ⟨92TL2629⟩ and (**74**) ⟨91JOC751⟩ (Scheme 46).

(ii) Methods involving diphosgene or triphosgene

The role of trichloromethyl chloroformate (diphosgene) and bis(trichloromethyl) carbonate (triphosgene) as alternatives to the use of phosgene has already been discussed (see Section 6.14.1.2.8). In the preparation of carbamoyl chlorides both reagents react to give the same product as would have been anticipated if phosgene had been employed to carry out the transformation (Scheme 47).

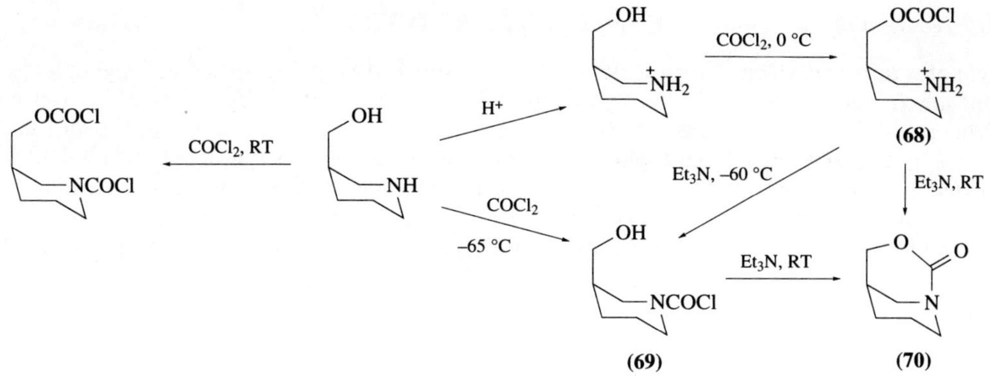

Scheme 45

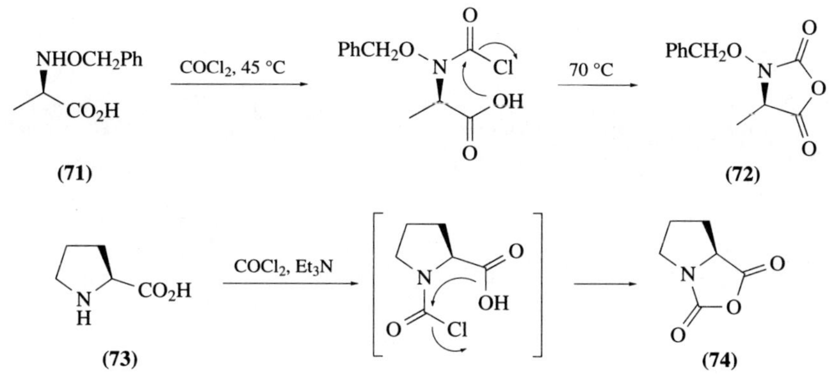

Scheme 46

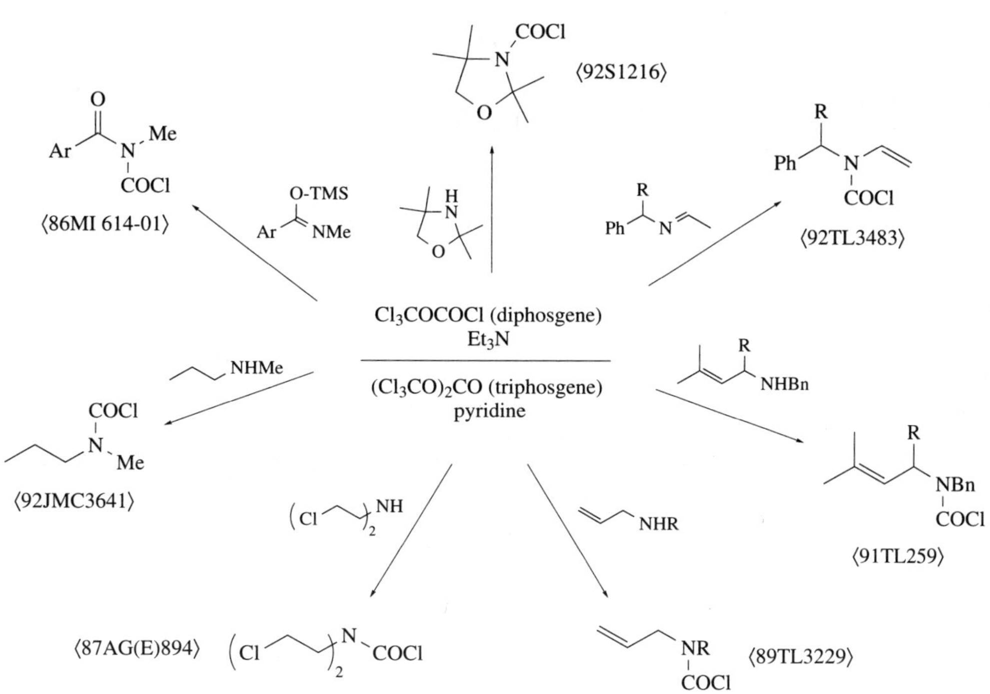

Scheme 47

(iii) The reaction of isocyanates with X—Cl to give RNXCOCl

Isocyanates add hydrogen chloride, acyl chlorides, sulfur dichloride, other sulfenyl chlorides and phosgene across the C=N bond to give *N*-carbonyl chlorides. Vinyl isocyanates add two molecules of hydrogen chloride to give reactive bifunctional electrophiles **(75)** ⟨80AG(E)201⟩, and carbonyl isocyanates R^1CONCO react with phosphorus(V) chloride to give **(76)** ⟨65TL2423, 66CB239⟩, the unsaturated analogues of **(75)**. Reaction with boron trichloride leads to dimerisation and formation of *N*-chlorocarbonyl ureas **(77)** ⟨70AG(E)372⟩. These reactions are summarized in Scheme 48. Many of these products are useful intermediates for the synthesis of heterocycles; several examples are illustrated.

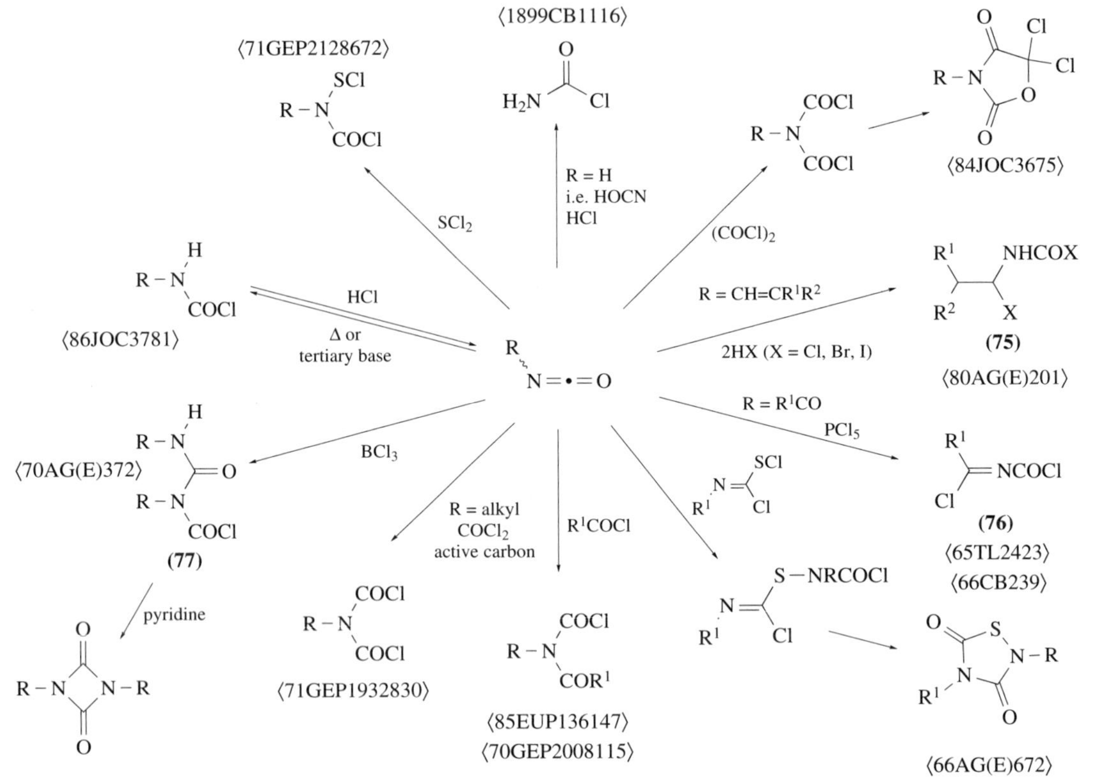

Scheme 48

(iv) Methods involving reaction with carbon monoxide

The *N*-chloro derivatives of primary and secondary alkylamines can be carbonylated by carbon monoxide at 60 atm in the presence of palladium metal or palladium(II) salts. The products are *N*-monosubstituted carbamoyl chlorides (30–40%) and *N*,*N*-disubstituted carbamoyl chlorides (62–99%), respectively (Scheme 49) ⟨71JOC858, 92MI 614-01⟩. The iron tricarbonyl complex **(78)** adds the isobutyronitrile anion. Oxidation of the reaction mixture with copper(II) chloride produced the carbamoyl chloride **(79)** in 72% yield (Equation (17)) ⟨88MI 614-02⟩.

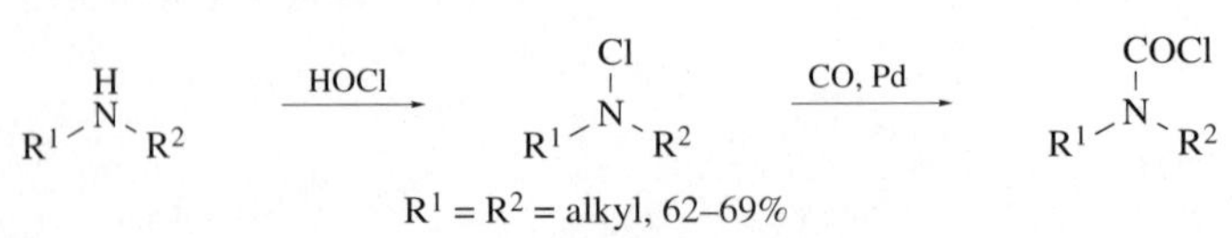

Scheme 49

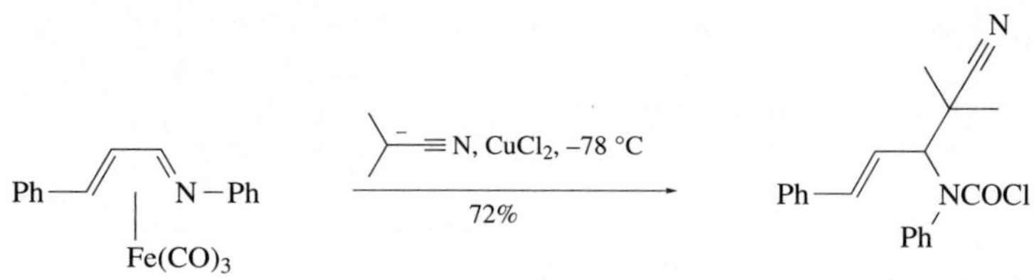

(17)

(78) **(79)**

(v) Preparation from formamides, R^1R^2NCHO

Several methods of converting formamides into carbamoyl chlorides have been reported, employing various chlorinating mixtures, and usually proceeding in 60–70% distilled yields. These methods are summarized in Table 2.

Table 2 Methods for the conversion of formamides into carbamoyl chlorides.

$$R^1R^2NCHO \longrightarrow R^1R^2NCOCl$$

R¹	*R²*	*Conditions*	*Yield (%)*	*Ref.*
Me	Me	SO_2Cl_2, 180 °C		62AG861
Me	Me	SO_2Cl_2, pyridine, CH_2Cl_2, RT	62	68CB113
Me	Me	SCl_2, pyridine, CH_2Cl_2	69	68CB113
$-(CH_2)_4-$		SCl_2, pyridine, CH_2Cl_2	77	68CB113
$-(CH_2)_5-$		SCl_2, pyridine, CH_2Cl_2	59	68CB113
$-(CH_2)_2O(CH_2)_2-$		SCl_2, pyridine, CH_2Cl_2	66	68CB113
Me	n-C_8H_{18}	PCl_3, $SOCl_2$	70	71CB969
Me	Ph	PCl_3, $SOCl_2$	62	71CB969

(vi) Chlorination of carbamic acid esters, thiolesters and related compounds

Carbamic acid esters can be cleaved by various chlorinating agents to yield *N*-chlorocarbonyl compounds. This provides a method of converting primary or secondary amines to carbamoyl chlorides in good yield without the use of phosgene or a phosgene equivalent (Table 3). In a variant of this reaction, chlorine cleaves 3,5-dioxo-1,2,4-dithiazolidines to yield bis(chlorocarbonyl)amines (Equation (18)) ⟨70S542⟩; in a similar reaction, the compound (**80**) produces chlorocarbonyl isocyanide dichloride (**81**) (Equation (19)) ⟨71CB2732⟩.

Table 3 Cleavage of carbamic acid esters.

$$R^1R^2NH \longrightarrow R^1R^2NCOX \xrightarrow{\text{reagent}} R^1R^2NCOCl$$

R¹	*R²*	*X*	*Reagent*	*Yield (%)*	*Ref.*
1° or 2° alkyl	1° or 2° alkyl	OEt	$POCl_3$	42–79	87SC1887
Me	Me	O-TMS	PCl_5	83	81MI 614-02
Ph	Me	O-TMS	PCl_5	80	91OM366
$(NO_2)_2CFCH_2$	$(NO_2)_2CFCH_2$	SEt	SO_2Cl_2	85	82JCED97
$(NO_2)_3CCH_2$	H	SEt	PhSCl (or Cl_2)	84	85JOC5879

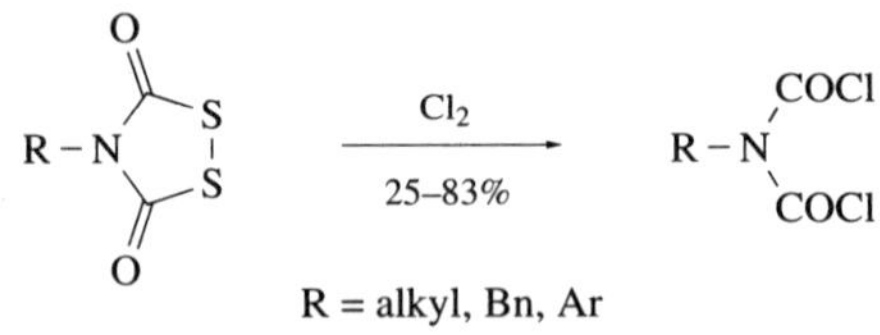

(18)

R = alkyl, Bn, Ar

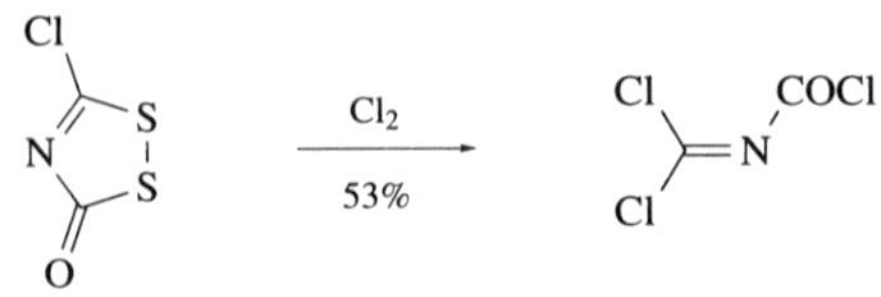

(19)

(80) **(81)**

(vii) Other methods of preparing N-*chlorocarbonyl compounds*

Photochemical chlorination of *N*-methyloctafluoropyrrolidine leads to the trichloromethyl derivative (**82**) in 95% yield, which can be converted to the *N*-chlorocarbonyl compound (**83**) by treatment with oleum at room temperature in 84% yield ⟨83JFC(22)521⟩. *N*-Trichloromethyl-3,6-dihydro-1,2-oxazines (**84**) can also be hydrolysed to the corresponding *N*-chlorocarbonyl compounds (**85**) in good yield; however, only brief treatment with water was necessary to accomplish this transformation ⟨89CJC2153⟩.

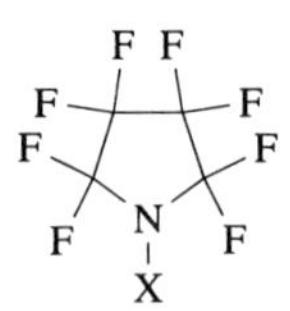

(82) X = CCl_3
(83) X = COCl

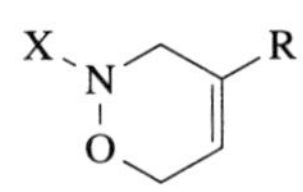

(84) X = CCl_3
(85) X = COCl

Certain strained heterocycles are cleaved by hydrogen chloride to give *N*-chlorocarbonyl products; thus 1,3-diazetidine-2,4-dione (**86**) provides the urea derivative (**87**) in 91% yield (Equation (20)) ⟨66CB3103⟩, whilst diaziridinones (**88**) lead to hydrazine derivatives (**89**) in good yield (Equation (21)) ⟨69JOC2254⟩.

(20)

(86) **(87)**

(21)

(88) **(89)**

Chlorocarbonyl isocyanate can undergo cycloadditions to multiple bonds to produce *N*-chlorocarbonyl derivatives; in this way, the enamine (**90**) gave the *N*-chlorocarbonylazetidine (**91**) in 86% yield (Equation (22)) ⟨77ZOR290⟩, and nitrile oxides (**92**) gave 1,2,4-oxadiazol-5-ones (**93**), also in high yields (Equation (23)) ⟨88S994⟩. Pyrrole forms a stable *N*-carboxylic acid, which reacts normally with oxalyl chloride or the Ghosez reagent to form unstable *N*-chlorocarbonylpyrrole ⟨84CC630, 87JOC2319⟩.

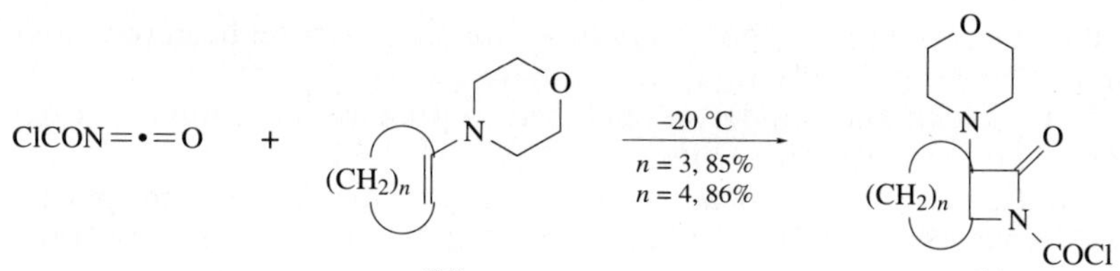

(90) (91) (22)

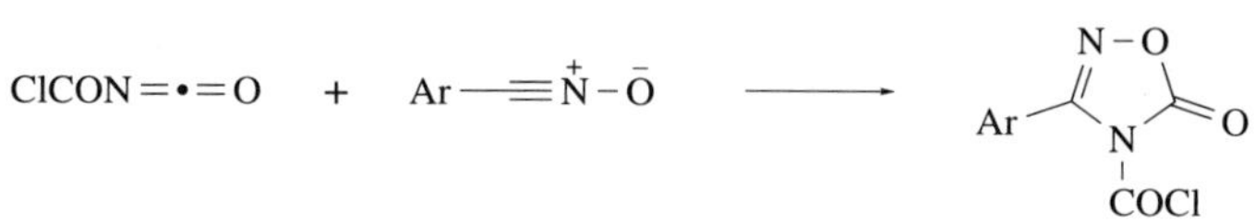

(92) (93) (23)

(viii) Preparation of chlorocarbonyl isocyanate and isothiocyanate

Chlorocarbonyl isocyanate is a thermally stable, very reactive bifunctional molecule, and is of considerable importance, particularly in the preparation of heterocyclic compounds. It is a commercially available colourless liquid, b.p. 64°C, which can be stored indefinitely in the absence of moisture.

It was first prepared in 1969 by Gottardi and Henn in poor yield by the photolysis of chlorine isocyanate ⟨69M1860⟩. Several other methods have been reported, involving high-temperature reactions of phosgene with isocyanates or cyanates ⟨73CB1752⟩; however, only one method, due to Hagemann, has emerged as both high yielding and economical. In this synthesis, inexpensive methyl isocyanate or cyanogen chloride is converted in high yield to *N*-chlorocarbonyl isocyanide dichloride, which is partially hydrolysed with methanesulfonic acid under carefully controlled conditions to give *N*-chlorocarbonyl isocyanate in over 90% yield (Scheme 50) ⟨73AG(E)999⟩. The chemistry of chlorocarbonyl isocyanate has been extensively reviewed ⟨77AG(E)743, 90H(31)1377, 93T3227⟩.

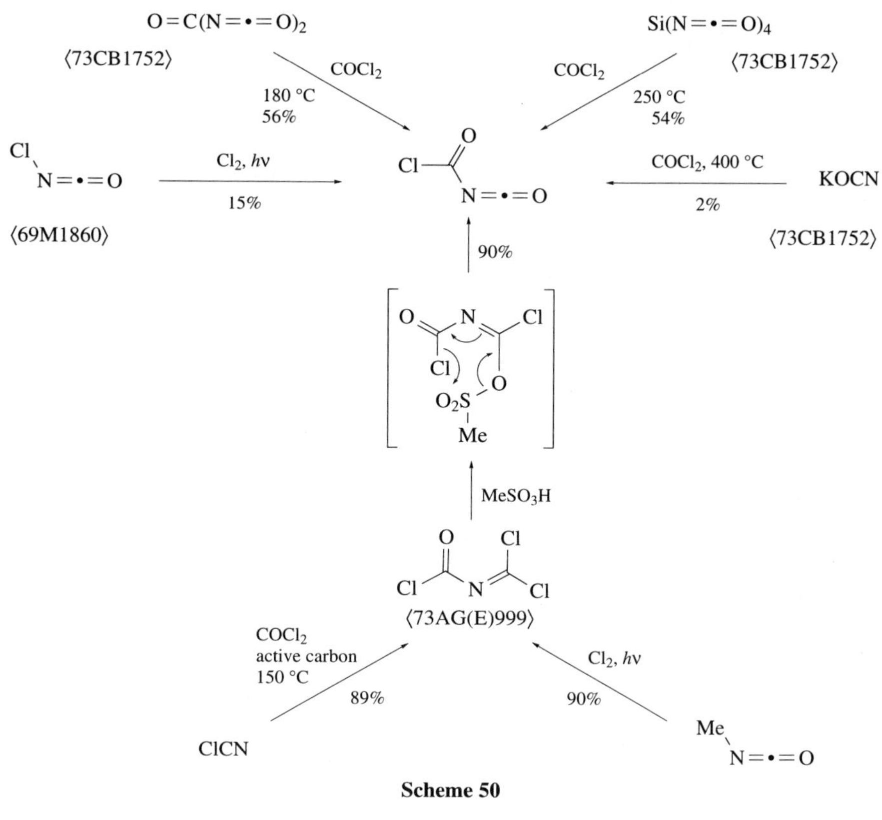

Scheme 50

The sulfur analogue, *N*-chlorocarbonyl isothiocyanate, may have been prepared as early as 1960 by the reaction of phosgene with lead(II) thiocyanate ⟨60AP(293)150⟩; however, full characterisation was not carried out. Jäckh and Sundermeyer failed to prepare this compound from phosgene and potassium thiocyanate, probably due to its instability at the high temperatures employed ⟨73CB1752⟩. A further preparation from chlorocarbonyl isocyanide dichloride and phosphorus(V) sulfide ⟨69AG(E)20⟩ has been disputed by Bunnenberg and Jochims, who failed to repeat this work; however, these workers finally successfully prepared *N*-chlorocarbonyl isothiocyanate in high yield by thermal elimination of carbon monoxide from oxalyl chloride isothiocyanate at 85°C in the presence of active carbon (Scheme 51) ⟨81CB1746⟩.

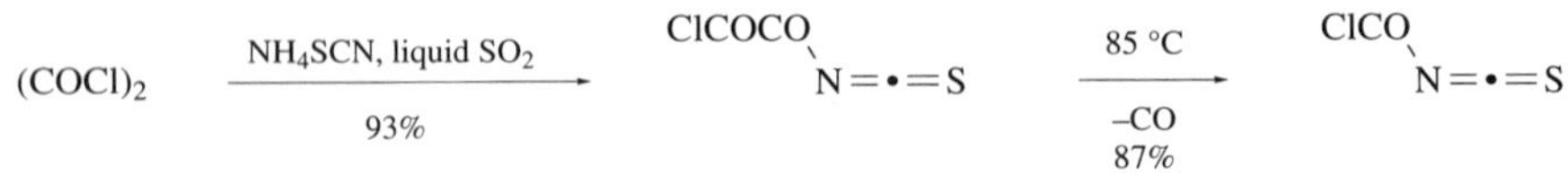

Scheme 51

6.14.3.3.3 *Carbamoyl bromides, R¹R²NCOBr, and other* N-*bromocarbonyl compounds*

Although carbamoyl bromides have been known for well over 100 years, their appearance in the literature is quite rare, and they do not seem to have any uses unique to themselves.

(i) Preparation of N-*monosubstituted carbamoyl bromides from isocyanates and hydrogen bromide*

This is the only important general synthetic method for *N*-monosubstituted carbamoyl bromides, RNHCOBr; it can also be used to prepare the parent unsubstituted carbamoyl bromide, H_2NCOBr, which can be isolated as a pure solid, m.p. 27°C, but is of very limited thermal stability. Examples of compounds prepared by this route are shown in Table 4. Both *N*-aryl- and *N*-alkylcarbamoyl bromides can be prepared by this method; α,β-unsaturated isocyanates add a further molecule of hydrogen bromide to form α-bromoalkylcarbamoyl bromides (Equation (24)) ⟨80AG(E)201⟩.

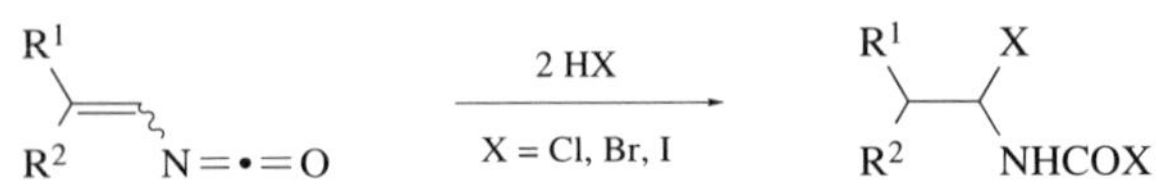

(24)

Table 4 Reaction of isocyanates with hydrogen bromide.

R–N=•=O + HBr ⟶ RNHCOBr

R	*Ref.*
H	40CB177
Me	64CB3162
Et	1866BSF435
Ph	1895MI 614-01

(ii) Preparation of N,N-*disubstituted carbamoyl bromides*

(a) Preparation from carbamoyl chlorides and hydrogen bromide. Dimethylcarbamoyl chloride undergoes halogen exchange with hydrogen bromide at room temperature without solvent to give dimethylcarbamoyl bromide in 90% distilled yield ⟨59CCC760⟩. Presumably this method could be extended to the preparation of a wide range of *N,N*-disubstituted carbamoyl bromides.

(b) Preparation from N,N-*disubstituted formamides.* *N*-Formylpiperidine is converted into piperidine-*N*-carbonyl bromide in 22% yield by the reaction of phosphorus(III) chloride and thionyl bromide ⟨71CB969⟩. In common with the preceding method, this is likely to be of some generality for preparing *N,N*-disubstituted carbamoyl bromides.

(c) Preparation from secondary amines and carbonic dibromide. Although this method has an

obvious analogy with the preparation of *N,N*-disubstituted carbamoyl chlorides from secondary amines and phosgene, it has not been used to prepare the bromine analogues. This is principally because carbonic dibromide is not readily accessible.

(iii) Other methods of preparing N-*bromocarbonyl compounds*

In common with many other isocyanates, fluorocarbonyl isocyanate, FCONCO adds hydrogen bromide to give iminodicarboxylic acid bromide fluoride, FCONHCOBr, as a hygroscopic solid that is unstable in air ⟨73CB1752⟩. Radical bromination of methyl isocyanate initially gives tribromomethyl isocyanate, Br_3CNCO, which rapidly equilibrates with its isomer, bromocarbonyl isocyanide dibromide, $Br_2C{=}N\text{-}COBr$. Radical bromination of neopentyl isocyanate follows the same course, resulting in a tautomeric mixture of Bu^tCBr_2NCO and $Bu^tCBr{=}NCOBr$ ⟨82CB860⟩. A similar tautomerism is observed when benzophenone imine, $Ph_2C{=}NH$ reacts with oxalyl bromide, with loss of carbon monoxide, to give a mixture of $Ph_2C{=}NCOBr$ and $Ph_2BrCNCO$ ⟨71ZOR2229⟩.

Bromine will cleave thiol ester functions in the same manner as observed for chlorine; thus, the cyclic disulfide (**94**) and thiolcarbamate (**95**) both yield bis(bromocarbonyl)methylamine on treatment with bromine (Scheme 52) ⟨68FRP1517379, 75GEP2351556⟩. Formation of *N*-bromocarbonyl compounds from the corresponding *N*-fluorocarbonyl compounds by halogen exchange has also been reported; thus, (**96**) on treatment with aluminum(III) bromide gave the bromocarbonyl derivative (**97**) (Equation (25)) ⟨73GEP2131789⟩. Treatment of 4-bromo-3-arylsydnones bearing an *ortho*-carbonyl function in the aryl group (**98**) with bromine or hydrogen bromide leads to 1-bromocarbonylindazoles (**99**). The reaction tolerates a wide variety of carbonyl functions, and usually proceeds in high yield (Equation (26)) ⟨93TL239⟩.

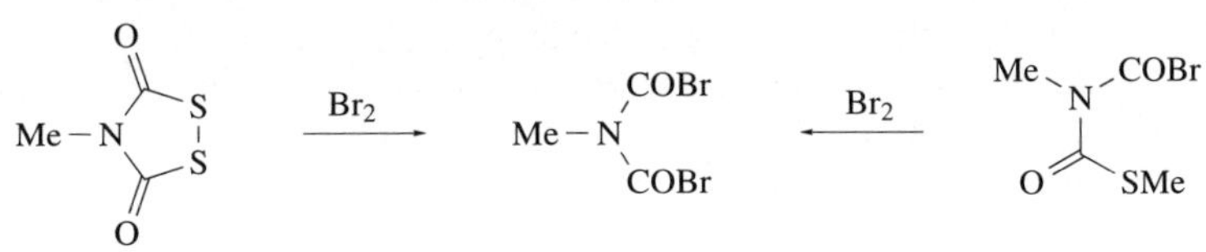

(**94**) (**95**)

Scheme 52

$$\underset{\textbf{(96)}}{Cl_2FCSNRCOF} \xrightarrow{AlBr_3} \underset{\textbf{(97)}}{Cl_2FCSNRCOBr} \qquad (25)$$

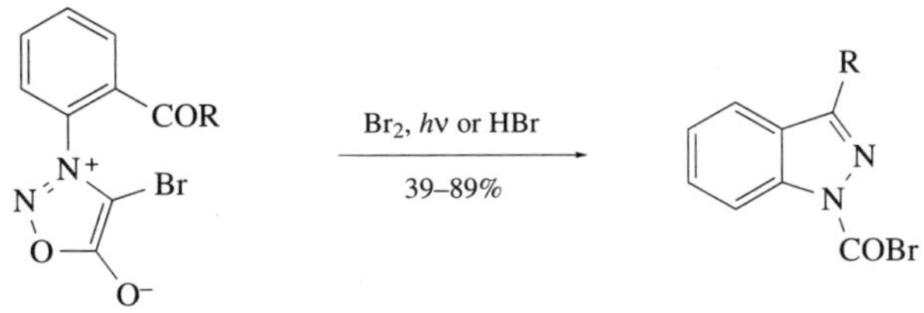

(26)

(**98**) (**99**)

R = H, Me, CH_2Br, CBr_3, Ph, OMe, OEt, NHMe

(iv) N-*Bromocarbonyl isocyanate and isothiocyanate*

N-Bromocarbonyl isocyanate, BrCONCO, has been prepared by the reaction of commercially available chlorocarbonyl isocyanate with boron tribromide ⟨91ZAAC(600)145, 92SA(A)1179⟩. The sulfur analogue, bromocarbonyl isothiocyanate, was prepared 10 years previously by the same route as chlorocarbonyl isothiocyanate, by thermal decarbonylation of oxalyl bromide thiocyanate, BrCO-CONCS, at 85 °C in the presence of active carbon in 46% yield. The compound is a distillable oil, stable at −18 °C for more than 1 week ⟨81CB1746⟩.

6.14.3.3.4 *Carbamoyl iodides, R[1]R[2]NCOI*

Only one instance of relatively stable carbamoyl iodides has been reported. When α,β-unsaturated isocyanates are treated with two equivalents of hydrogen iodide, α-iodoalkylcarbamoyl iodides are formed (Equation (27)) ⟨79EUP7498, 80AG(E)201⟩. Thus, it appears that the stability of carbamoyl iodides is such that more examples could perhaps be prepared. By analogy with iodoformates, ROCOI (see Section 6.14.3.1.4), similar stabilising features could be incorporated to enable isolation.

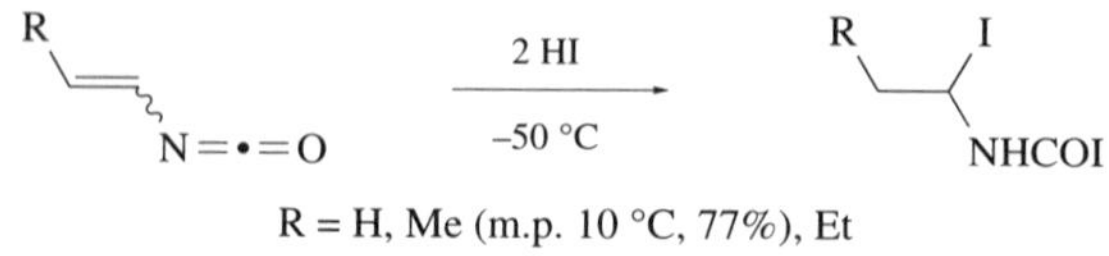

(27)

6.14.3.4 One Halogen and One Phosphorus Function

6.14.3.4.1 *Chlorocarbonyl derivatives of phosphorus(III)*

The reaction of trimethylsilylphosphines with one equivalent of phosgene gives rise to intermediate chlorocarbonyl phosphorus(III) compounds. In the case of *t*-butyltrimethylsilylphosphine, Bu[t]PHTMS, this intermediate, Bu[t]PHCOCl (**100**), is stable at low temperatures, but loses carbon monoxide on warming. Treatment with base leads to the phosphaketene Bu[t]P=C=O, which is stable below $-60\,^{\circ}C$, and adds hydrogen chloride to regenerate the chlorocarbonyl derivative (**100**) ⟨83TL2639⟩. The analogous reaction with *t*-butylbis(trimethylsilyl)phosphine gives the similar intermediate (**101**); however, this loses TMS-Cl at $-90\,^{\circ}C$ to give the same phosphaketene. When Bu[t]P(TMS)$_2$ is treated with two equivalents of phosgene, the reaction follows a different course, through the chlorocarbonyl compound (**102**), which was characterised at $-40\,^{\circ}C$ by ^{31}P NMR spectroscopy, to *t*-butyldichlorophosphine, Bu[t]PCl$_2$ ⟨83CB109⟩. These transformations are summarized in Scheme 53. Mesitylbis(trimethylsilyl)phosphine (**103**) reacts with phosgene via the unisolated chlorocarbonyl intermediate (**104**), which eliminates TMS-Cl to afford the stable phosphaketene (**105**) (Scheme 54) ⟨83AG(E)785⟩.

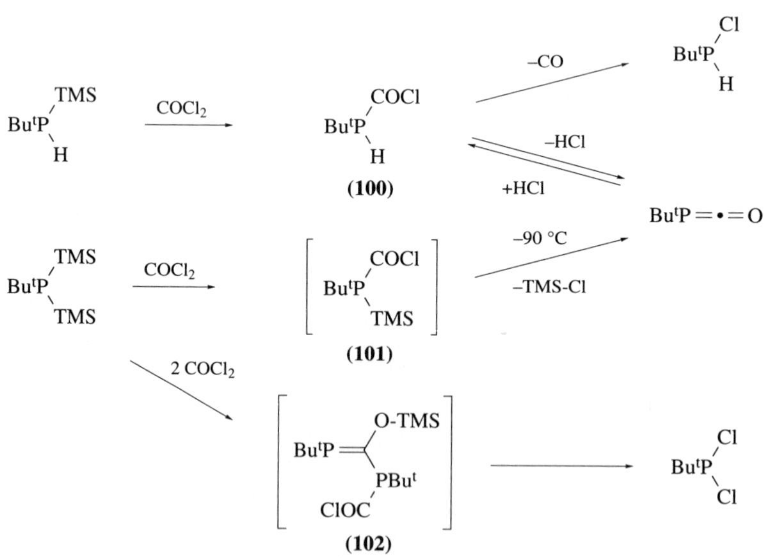

Scheme 53

6.14.3.4.2 *Chlorocarbonyl derivatives of phosphorus(V)*

Only one compound of this type has been reported in the literature; (dimethoxyphosphinyl)formyl chloride (MeO)$_2$P(O)COCl (**106**) was characterized as a stable, distillable liquid, prepared by the

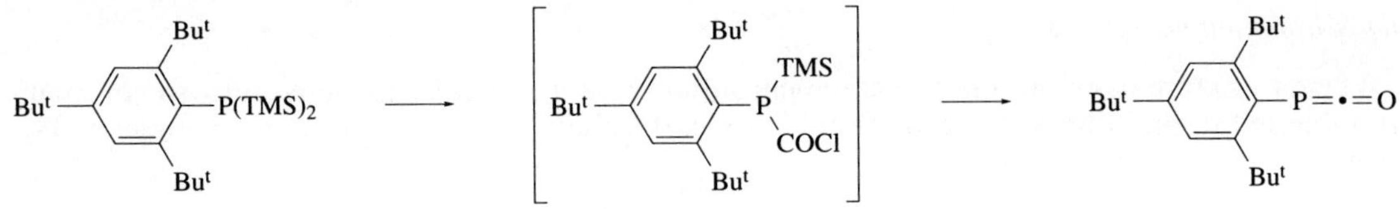

(103) (104) (105)

Scheme 54

Arbuzov reaction of trimethyl phosphite with phosgene (Equation (28), R = Cl) ⟨57IZV48⟩. In view of the fact that a simple, functionalised derivative such as (**106**) would be expected to have been the subject of further investigations, this report must be treated with some caution; however, the Arbuzov reaction of trimethyl phosphite with aliphatic acid chlorides to yield stable α-keto-phosphonic acid dimethyl esters has also been described (Equation (28), R = Me) ⟨45BAU364⟩, so the analogous reaction with phosgene to yield (**106**; R = Cl) would not be surprising.

$$(MeO)_3P + RCOCl \xrightarrow[R = Cl,\ Me]{} (MeO)_2P(=O)COR \quad (28)$$

(**106**)

6.14.3.5 One Halogen and One As, Sb or Bi Function

Preparations of these classes of compound have not been reported.

6.14.3.6 One Halogen and One Metalloid (B, Si or Ge) Function

Preparations of these classes of compound have not been reported.

6.14.3.7 One Halogen and One Metal Function

Metal complexes containing a metal–halocarbonyl bond have been reported in all three transition metal series, but not for the lanthanides or actinides. The stability of these complexes varies considerably, the most stable being isolable solids, unaffected by air; most are intermediates of very limited stability, or unexpected products of little importance.

6.14.3.7.1 Halocarbonyl complexes of first transition series metal (iron and chromium)

(i) Iron complexes

Pentacarbonyl iron has been reported to form an iodocarbonyl iron intermediate on reaction with iodine ⟨68JOM(13)411⟩. Subsequently, tricarbonyl(cyclohexadienyl)iron cation has been shown to give a similar iodocarbonyl complex (**107**) on treatment with tetraethylammonium iodide in acetone or methyl nitrite with the exclusion of light. It can be isolated in admixture with a second product, where the cyclohexadiene ring has been iodinated. It is stable in the absence of light and has an iodocarbonyl C═O stretching frequency in the IR spectrum at 1735 cm^{-1} (Equation (29)) ⟨86JOM(312)C21⟩.

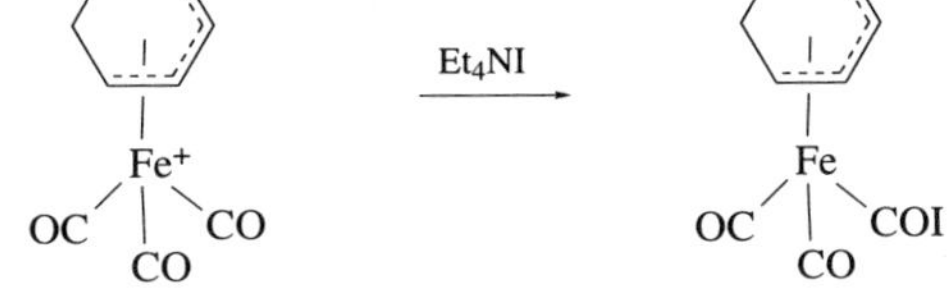

(**107**) (29)

(ii) Chromium complexes

When a mixture of tricarbonyl(benzene)chromium and a sterically hindered nitroso compound are subjected to photolysis in carbon tetrachloride, the chlorocarbonyl chromium complex (**108**) is formed (Equation (30)) ⟨84MI 614-01⟩.

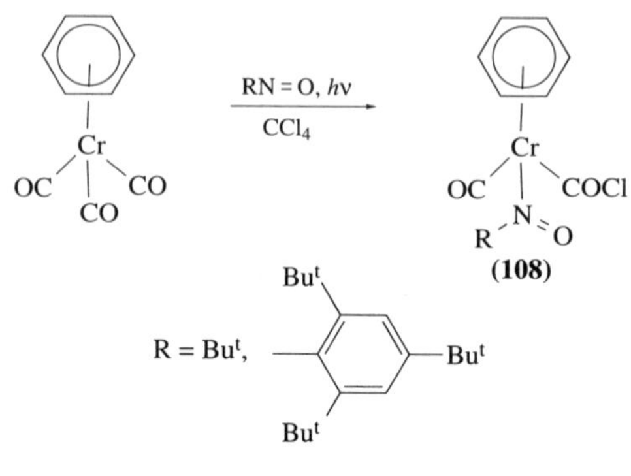

(30)

(**108**)

6.14.3.7.2 *Halocarbonyl complexes of second transition series metals (ruthenium and rhodium)*

(i) Ruthenium complexes

When $Ru_3(CO)_{12}$ and bis(triphenylphosphine)nitrogen chloride are stored in THF for 1 h at room temperature, the dark red–brown solution formed contains the chlorocarbonyl ruthenium complex (**109**), which is not isolated, but can be characterized by IR spectroscopy, the chlorocarbonyl C═O stretching frequency occurring at 1776 cm^{-1} ⟨87JA6015⟩.

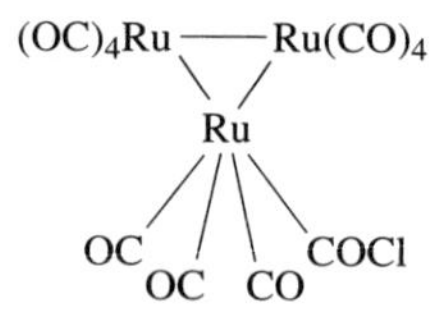

(**109**)

(ii) Rhodium complexes

A stable chlorocarbonyl rhodium complex is formed when (**110**) is treated with oxalyl chloride in toluene at room temperature (Equation (31)). The yellow precipitate of (**111**) is stable in air ⟨69JOM(19)161⟩. The chlorocarbonyl C═O stretching frequency is at 1680 cm^{-1}. A second stable complex (**112**) is formed from the reaction of bis(cyclooctene)chlororhodium with acrolein and 1,2-bis(diphenylphosphino)ethane ⟨74NKK732⟩.

(**110**) (**111**)

(31)

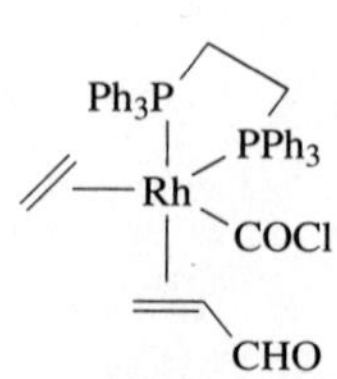

(**112**)

Another chlorocarbonylrhodium complex appears in *Chemical Abstracts*; however, this is a case of ambiguous naming, as the chlorine and carbonyl functions are individually bonded to rhodium ⟨68IS99, 72CL483⟩.

6.14.3.7.3 Halocarbonyl complexes of third transition series metals (rhenium and iridium)

(i) Rhenium complexes

Bromination of tricarbonyl(cyclopentadienyl)rhenium(I) leads to an intermediate bromocarbonyl rhenium complex (**113**) which can decarbonylate to the final product (**114**) (Scheme 55) ⟨87IZV1671⟩.

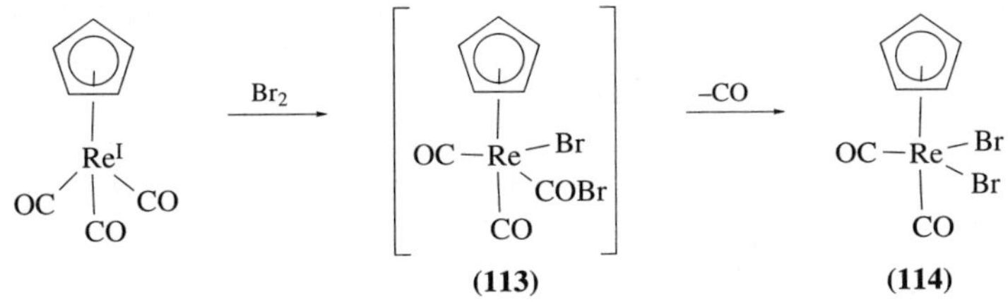

Scheme 55

(ii) Iridium complexes

A series of fluorocarbonyl iridium(III) complexes have been prepared by the reaction of xenon difluoride with a range of pentacoordinate iridium(I) cations. Most of these complexes are stable at room temperature. Extensive structural studies have been carried out, including an x-ray characterisation. The fluorocarbonyl C═O stretching frequency for (**115**) is at 1756 cm^{-1}. This work is summarised in Scheme 56 ⟨88CC529, 93JCS(D)1031⟩.

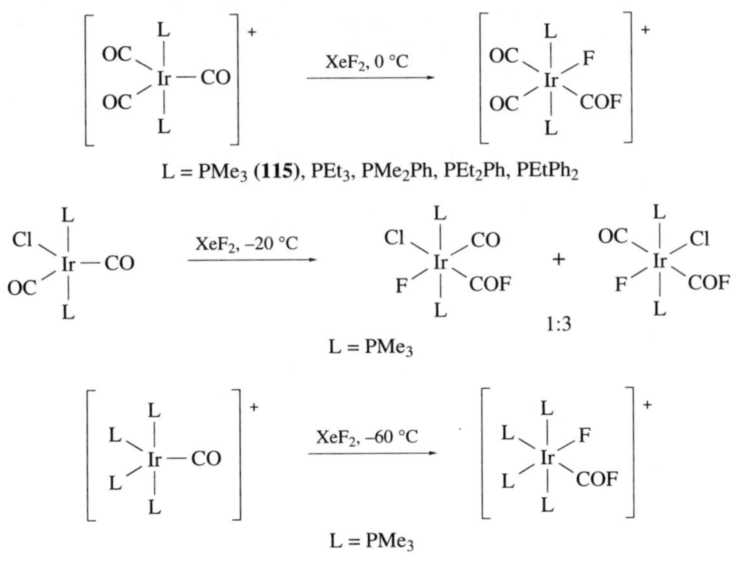

Scheme 56

6.15
Functions Containing a Carbonyl Group and at Least One Chalcogen (but No Halogen)

HEINER ECKERT and ALFONS NESTL
Technischen Universität München, Garching, Germany

6.15.1 CARBONYL CHALCOGENIDES WITH TWO SIMILAR CHALCOGEN FUNCTIONS

6.15.1.1 Two Oxygen Functions

6.15.1.1.1 Dialkyl carbonates from phosgene and substitutes

Carbonic acid is principally a difunctional carboxylic acid, therefore there are many preparative methods which provide access to a great variety of its organic derivatives and which differ widely in rate and selectivity of reactions and reagents.

Phosgene ⟨64CRV645, 73CRV75, B-91MI 615-01⟩, as the formal carboxylic acid dichloride of carbonic acid, is a highly reactive reagent which affords high turnovers and good yields. Thus, both symmetrical and unsymmetrical dicarbonates, the latter via chloroformates, can easily be produced.

Generally phosgene is blown into the reacting alcohol at temperatures of about 50–80 °C.

The reaction runs at lower temperatures in the presence of acid acceptors such as amines or inorganic hydroxides. It can also be accelerated catalytically with *N*,*N*-dialkylamides or quaternary ammonium salts, particularly the adduct of two moles of pyridine and one mole of phosgene, which can also be isolated. Instead of alcohols and phenols, the corresponding alcoholates and phenolates can be reacted with phosgene.

The manufacture of diethyl carbonate from phosgene and ethanol with a purity of 99.9% is described in a patent ⟨86JAP61118349⟩.

The phosgenation of alcohols can be carried out in a one-step as well as in a two-step process (Scheme 1).

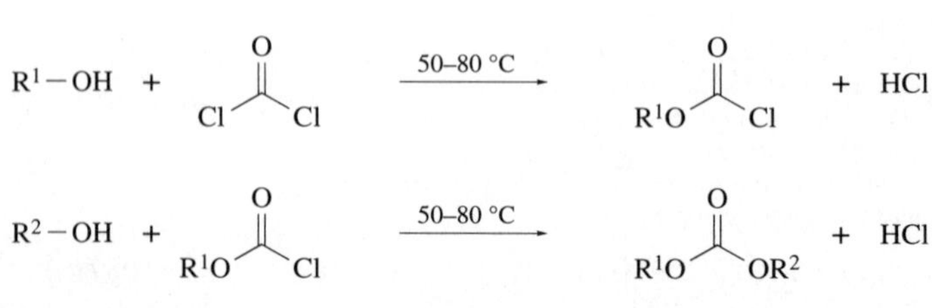

Scheme 1

The advantages of the two-step process for producing diethyl carbonate in technical plants have been described ⟨94JAP06041019⟩. The greatest advantage of this process is formation of the intermediate chloroformate, which allows the preparation of a great number of mixed esters of carbonic acid in a well-defined way.

In a similar manner to that described for reactions of phosgene, chloroformate esters react either with alcohols under reflux or in the presence of amines such as dimethylaniline or pyridine. Here also crystalline adducts of one mole of chloroformate ester and two moles of pyridine can be isolated. Alkali alcoholates also react with success.

A useful building block for palladium-catalyzed carbon–carbon bond formation by conversion of the carbonic acid ester into a carbon acid ester is methyl propargyl carbonate, which can be prepared from methyl chloroformate and propargyl alcohol in high yield (Equation (1)) ⟨93S1109⟩. Even the lithium acetylide derived from this building block is stable under the reaction conditions required for further C—C connections, as shown in a stereoselective synthesis of the side chain of glaucosterol ⟨87T5315⟩.

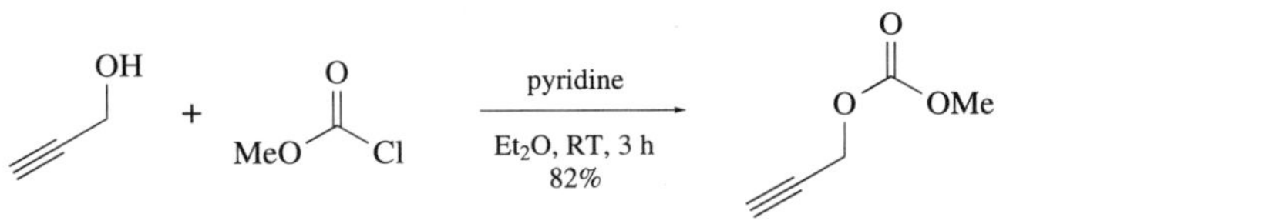

(1)

Similar substituted 2-alkyl carbonates are used in a synthesis of 1,2-dien-4-ynes ⟨91JOM(417)305⟩. A structurally related carbonate with a terminal ethynyl group is prepared according to Equation (2). It is a key intermediate in the preparation of (+/−)-pentalenolactone (*E*)-methyl ester ⟨91HCA465⟩. Reduction of geranylacetone to the corresponding alcohol and conversion into the methyl carbonate affords a juvenile hormone analogue (III) ⟨88ZN(B)1038⟩.

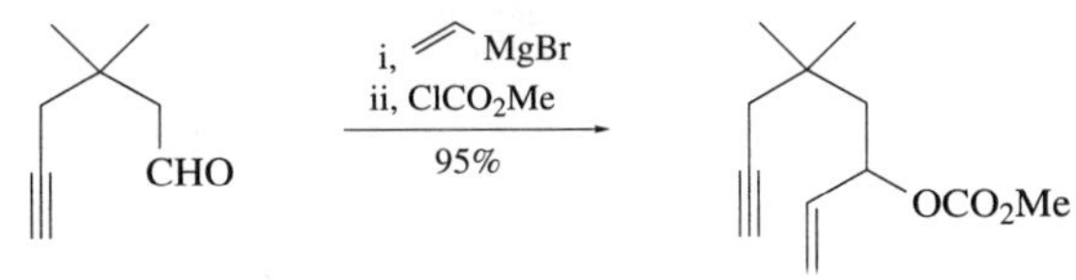

(2)

The conditions required to form the carbonate from an alcohol and methyl chloroformate are so mild that allenes are stable and not subject to rearrangement as shown in Equation (3) ⟨90TL5629⟩. Optically active alcohols react with retention of configuration (Equation (4)) ⟨88JOC4419⟩. Nine different alkyl allyl carbonates were prepared from allyl chloroformate and the corresponding racemic alcohols with final racemic resolution ⟨93T10725⟩.

(3)

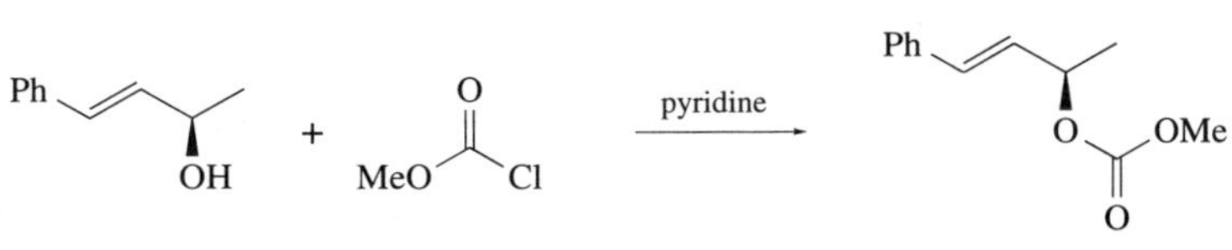

(4)

In 1990 Olofson and co-workers reported the first synthesis of *O*-butadienyl carbonates derived from α,β-unsaturated aldehydes. Treatment of crotonaldehyde with $KOBu^t$ in THF at −78 °C affords the enolate, which is converted to the carbonate by treatment with the chloroformate (Scheme 2). If the aldehyde bears a substituent in the α-position the (*E*)-butadienyl carbonates are formed exclusively ⟨90TL1405⟩.

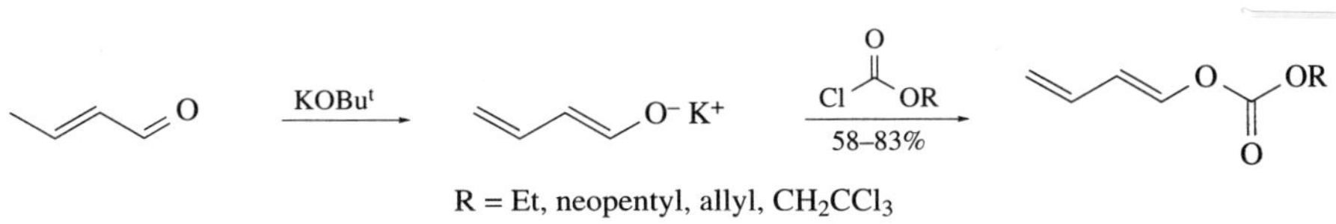

Scheme 2

Trialkylsilyl enol ethers react well with fluoroformates in the presence of a catalytic amount of $PhCH_2NMe_3{}^+F^-$ (btaf). The enolcarbonate from 2-methylcyclohexanone is formed in this way in good yield as illustrated in Equation (5) ⟨80TL819⟩.

(5)

btaf = benzyltrimethylammonium fluoride

This method cannot be used for enolate or silyl enol ethers derived from aldehydes because of aldol or Michael condensation with another molecule of aldehyde. To solve this problem, aldehydes are treated with HF in the presence of an 18-crown-6 catalyst. The enolates are generated and can be trapped efficiently as formed with fluoroformates to give vinylic carbonates. Fluoroformates can be prepared by halide exchange from the chloroformates and two equivalents of KF (Scheme 3) ⟨90JOC1⟩.

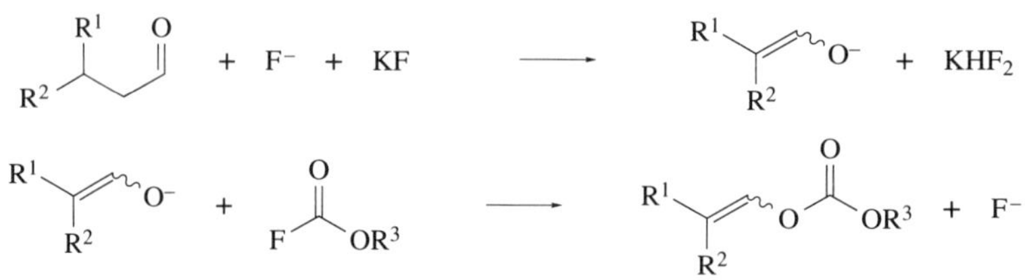

Scheme 3

In all the reactions so far described, phosgene is the basic chemical used for the preparation of carbonates in a direct way as well as being used in the synthesis of the chloroformates. To avoid the difficulties associated with the toxicity of phosgene, substitutes for phosgene have been developed ⟨B-91MI 615-01⟩. The liquid trichloromethyl chloroformate (diphosgene) was introduced by Kurita and co-workers and by Ugi and co-workers ⟨76JOC2070, 77AG(E)259⟩. Much safer in storage, transportation and handling is the crystalline bis(trichloromethyl) carbonate (triphosgene), which was introduced by Eckert in 1987 ⟨87AG(E)894, 90MI 615-02⟩. Since then, triphosgene has become well-established for industrial use. Bis(trichloromethyl) carbonate itself is a dialkyl carbonate and is prepared by radical chlorination of dimethyl carbonate in excellent yield up to 99%.

Carbonate esters ^{14}C-labeled at the carbonyl group can be synthesized by a method utilizing the readily available (^{14}C)phosgene, which is first converted to an isolable alkyl or aryl chloroformate and subsequently reacted with the appropriate alcohol to give the corresponding ester ⟨86MI 615-01⟩.

6.15.1.1.2 Dialkyl carbonates from inorganic carbonates

Carbonic acid esters can be prepared in phase-transfer catalyzed reactions from primary alkyl halides and a mixture of dry potassium hydrogen carbonate and dry potassium carbonate in nonpolar solvents. The conversion is ineffective in the absence of hydrogen carbonate and catalyst. This method uses methyltrioctylammonium chloride as catalyst and toluene or petroleum ether as solvent, forming symmetric dialkyl carbonates in a one-step process with alkyl groups C_6–C_{16} in yields of 67–86%. The unsymmetrical benzyl hexyl carbonate is obtained by reacting dry potassium hydrogen carbonate with benzyl bromide followed by reaction with dry potassium carbonate and hexyl bromide, yielding 34% of product ⟨81CB1210⟩. In a related publication the effect of variation of the phase-transfer catalyst is described ⟨84JOC1122⟩.

Dialkyl carbonates are also easily prepared by the heterogeneous reaction of solid potassium carbonate with alkyl bromides in DMF or DMSO in the presence of organostannyl compounds such as hexabutyldistannoxane or chlorotributylstannane. A mixed catalytic system consisting of a tributylstannyl compound and 18-crown-6 was effective even in less polar solvents.

Well-known phase-transfer catalysts such as 18-crown-6 or benzyltriethylammonium chloride were slightly effective but not sufficiently so for this heterogeneous reaction. It was found by the authors that the reaction was distinctly accelerated by the addition of triorganostannyl compounds, especially tributylstannyl compounds, while the tetrabutyl and dibutyl compounds were not active. DMF and DMSO were good solvents for this reaction. Alkyl bromides were more effective reagents than the corresponding chlorides or iodides (Equation (6)) ⟨81CL749⟩.

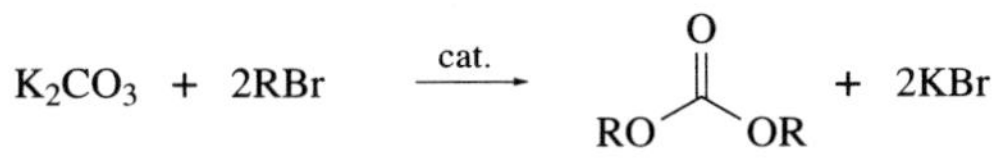

(6)

6.15.1.1.3 Dialkyl carbonates from urea derivatives

A laboratory method for dialkyl carbonates makes use of 1,1′-carbonylbis(4-benzylidene-1,4-dihydropyridine) as a reagent (Equation (7)) ⟨80S485⟩. The required activation energy for this reaction is delivered by the aromatization energy from the 1,4-dihydropyridine system to the 4-substituted pyridine.

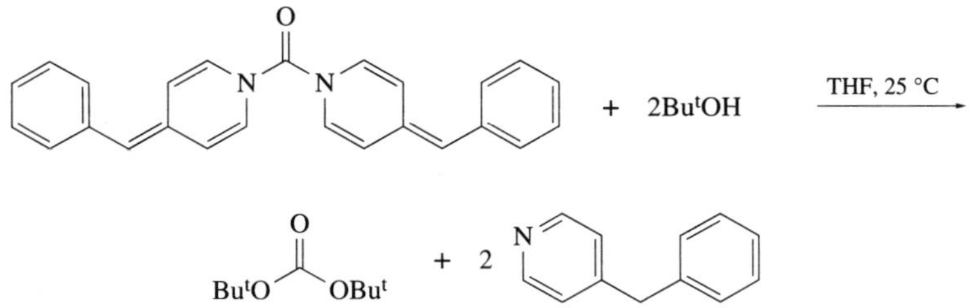

(7)

6.15.1.1.4 Dialkyl carbonates from carbon oxides and alcohols

Strong bases readily absorb carbon dioxide. This fact is exploited in the synthesis of carbonates with at least one tertiary alkyl group. In a two-step process the sodium alcoholate of *t*-butanol reacts with carbon dioxide to form the stable sodium *t*-butyl carbonate which reacts with methyl or ethyl iodide to the corresponding mixed carbonates (Scheme 4) ⟨71ZOR273⟩.

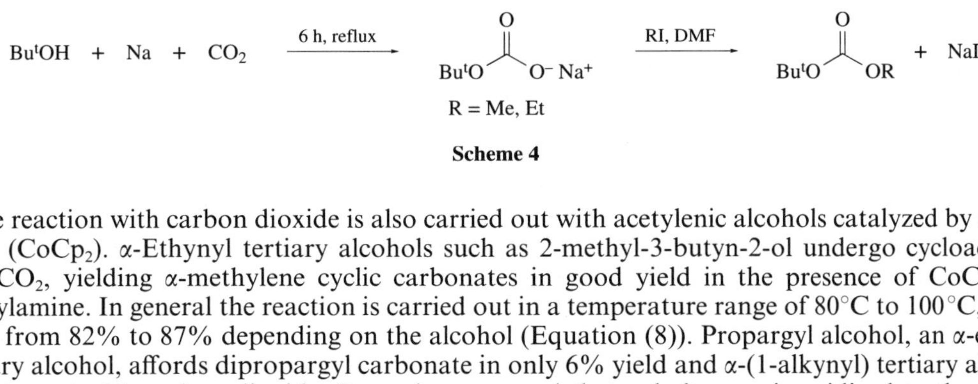

Scheme 4

The reaction with carbon dioxide is also carried out with acetylenic alcohols catalyzed by cobaltocene ($CoCp_2$). α-Ethynyl tertiary alcohols such as 2-methyl-3-butyn-2-ol undergo cycloaddition with CO_2, yielding α-methylene cyclic carbonates in good yield in the presence of $CoCp_2$ and triethylamine. In general the reaction is carried out in a temperature range of 80°C to 100°C, giving yields from 82% to 87% depending on the alcohol (Equation (8)). Propargyl alcohol, an α-ethynyl primary alcohol, affords dipropargyl carbonate in only 6% yield and α-(1-alkynyl) tertiary alcohols do not react with carbon dioxide. It can be presumed that cobaltocene is oxidized to the cobaltocenium ion $[CoCp_2]^+$ in the presence of water or alcohol if air is present.

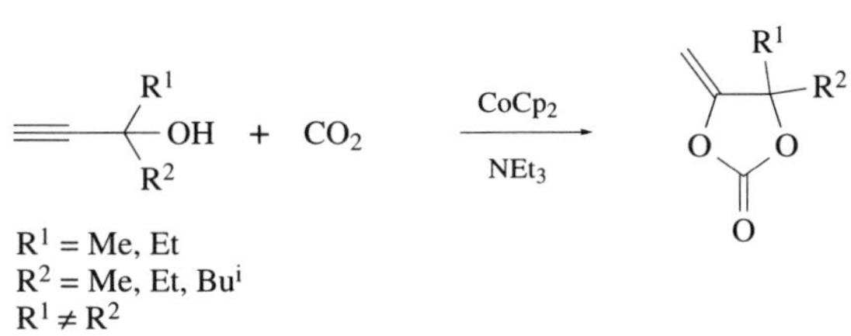

(8)

Another method for obtaining symmetrical carbonates from CO_2 and alcohols uses triphenylphosphane and diethyl azodicarboxylate (dead). The authors obtained dipentyl carbonate in 88%

yield by this route, di-*s*-butyl carbonate in 80% yield, bis(2-ethylhexyl) carbonate in 74% yield, and diallyl carbonate in 75% yield (Equation (9)) ⟨82JOC5209⟩.

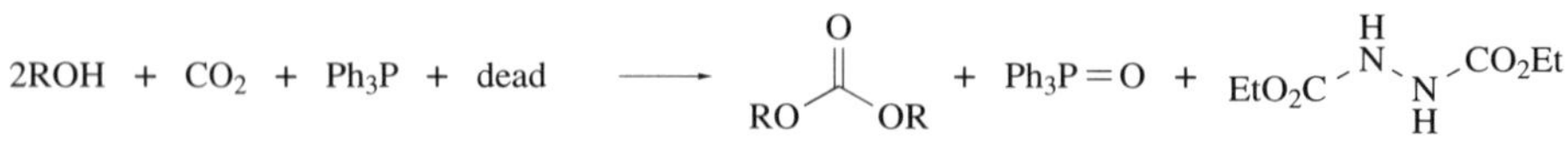

(9)

dead = diethyl azodicarboxylate

The first step of the reaction is the formation of monoalkyl carbonate ion from the alcohol and CO_2. The presence of triethylamine increases the amount of the ion. The monoalkyl carbonate then undergoes cyclization in the presence of cobaltocene as shown in Scheme 5 ⟨87BCJ1204⟩.

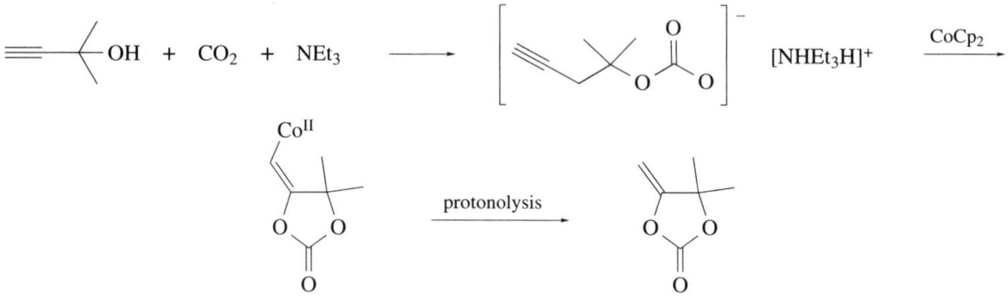

Scheme 5

Carbon monoxide alkoxylation reactions have been described. CO is a low-cost product obtained in many industrial processes and is a good ligand for transition metals; transition metal catalyzed reactions are therefore obviously possible. In the late 1960s, Saegusa and co-workers described a method for producing dimethyl carbonate. This preparation is carried out by reaction between cupric methoxide and carbon monoxide in pyridine as solvent. It may be assumed that MeOCOCuOMe is first formed by the insertion of carbon monoxide into the copper–oxygen linkage; this then decomposes to produce dimethyl carbonate (Scheme 6) ⟨68TL831⟩. This reaction depends strongly on the temperature and reaches its optimum at 70 °C.

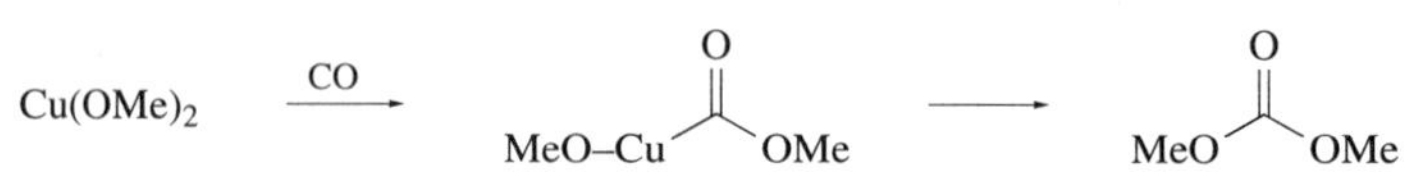

Scheme 6

In a similar process, carbonylation of 1-propanol and 1-butanol in the presence of equivalent amounts of $CuCl_2$ under a pressure of 20–100 atm of CO at 60 °C yielded in the corresponding carbonates dipropyl carbonate (82%) and dibutyl carbonate (64%). At temperatures above 100 °C the proportion of the corresponding alkyl formates increases distinctly ⟨73IZV807⟩. Nefedov and Sergeeva in a series of three publications describe access to symmetrical and unsymmetrical carbonates in one- or two-step processes using mercury(II) acetate. The reactions are carried out in an autoclave with a CO pressure of 100 atm and temperatures of 200–220 °C. The formation of the carbonates is described from the following alcohols, the yield of the carbonate being given in parentheses: methanol (34%), ethanol (37%), 1-propanol (44%), 2-propanol (31%), 1-butanol (51%), 2-methyl-1-propanol (48%), 3-methyl-1-butanol (38%), 1-hexanol (44%), and 1-heptanol (42%) (Scheme 7) ⟨72IZV1635, 72IZV2733, 73IZV562⟩.

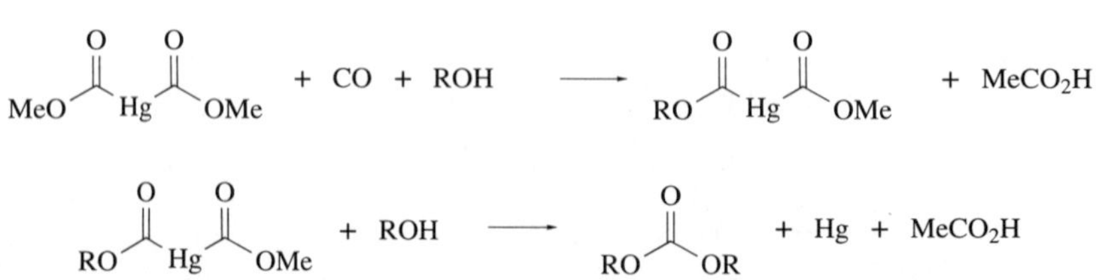

Scheme 7

All the methods described above use transition metal(II) salts as oxidizing agents. Another reagent for the oxidation is selenium. Using a temperature of 20°C and 1 atm of CO, carbonates from methanol, ethanol, 1-propanol, and 1-butanol were prepared in yields ranging from 94% to 99%. Benzyl alcohol affords the corresponding carbonate under these conditions in 76% yield. With secondary alcohols such as 2-propanol and cyclohexanol, and with *t*-butanol, yields are 6% to 16% under these conditions. THF is effectively used as solvent (Scheme 8) ⟨71TL4885⟩. This approach is achieved more conveniently in a catalytic cycle using oxygen as the oxidizing agent. For the preparation of dialkyl carbonates derived from primary alcohols under conditions as described above, reactions according to Scheme 9 are achieved in excellent yields. The intervention of carbonyl selenide as an intermediate in the catalytic cycle offers a reasonable interpretation (path a), although there could be another path, in which the intermediacy of carbonyl selenide is not required (path b) ⟨75BCJ108⟩.

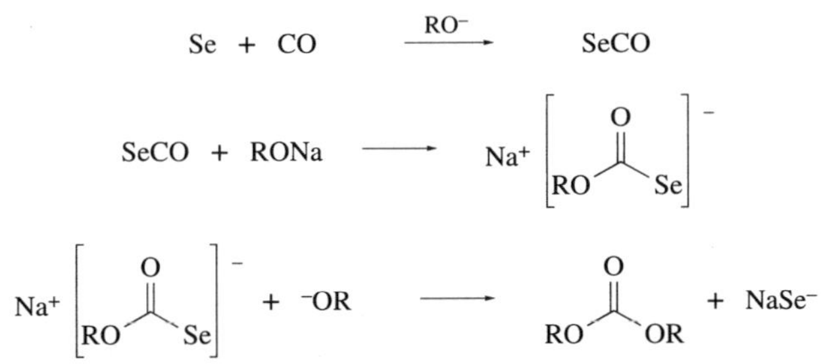

Scheme 8

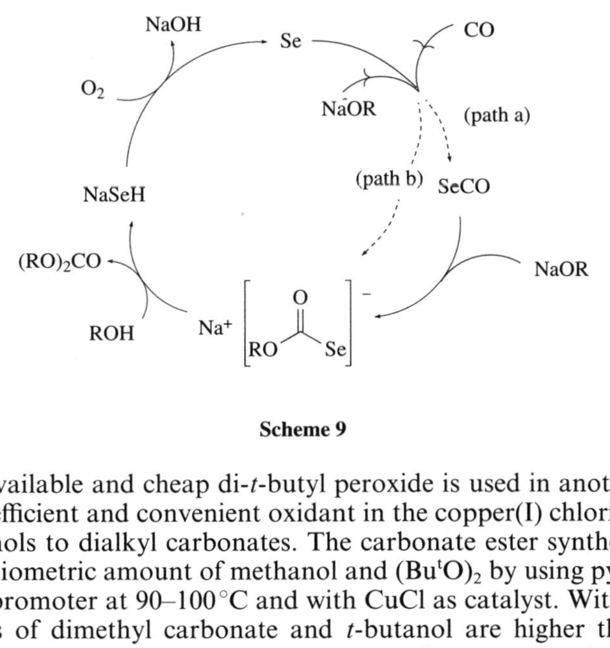

Scheme 9

The commercially available and cheap di-*t*-butyl peroxide is used in another catalytic cycle. Di-*t*-butyl peroxide is an efficient and convenient oxidant in the copper(I) chloride-catalyzed oxidative carbonylation of alcohols to dialkyl carbonates. The carbonate ester synthesis can be carried out efficiently with a stoichiometric amount of methanol and $(Bu^tO)_2$ by using pyridine or a substituted pyridine as a catalyst promoter at 90–100°C and with CuCl as catalyst. With 2,6-dimethylpyridine (92°C, 50 atm), yields of dimethyl carbonate and *t*-butanol are higher than 90% (Scheme 10) ⟨87CC410⟩.

For the preparation of dimethyl carbonate the best combination is carbon monoxide, methanol, and air; a good yield of 89% is obtainable. The catalytic system for this purpose is $MnCl_2$–$CuCl_2$–LiCl (0.5 : 2.5 : 2.5) (Equation (10)) ⟨77IZV363⟩.

Patents for producing dimethyl carbonate, used commercially in the synthesis of triphosgene, and diethyl carbonate appeared during 1992 and 1993. The catalysts are mostly copper(I) salts, sometimes promoted by amines such as trialkylamines, 2-hydroxypyridine, dipyridyl, imidazole and phenanthroline; good results can also be achieved using $Co(acac)_2$ as catalyst. One patent offers the possibility of producing carbonates from C_1 to C_{10} alkanols and from C_3 to C_6 cycloalkanols ⟨92JAP04054156, 92JAP04108765, 92JAP04356446, 93EUP528498, 93EUP534545, 93JAP05255201, 93JAP05310644⟩. Carbonates are also produced as described in a large group of patents using similar metal catalysts to those given above and additionally the nitrous acid ester from the alcohol which is introduced into the carbonate ⟨93EUP558996, 93EUP559001, 93EUP559212, 93GEP4134688, 93JAP05255197, 94EUP581240⟩.

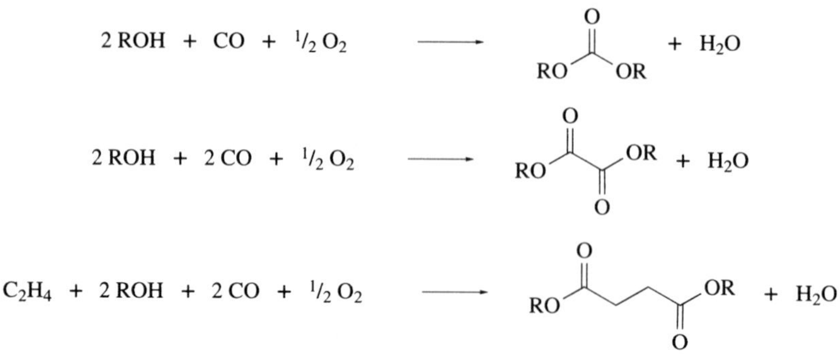

Scheme 10

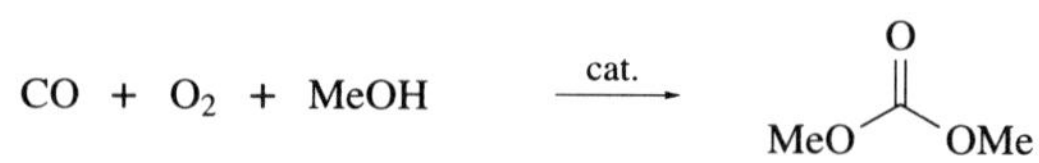

(10)

6.15.1.1.5 *Dialkyl carbonates from formates and ketones*

It is reported by Kondo *et al.* that selenium acts as an unusual oxidizing agent of formates in the presence of alkoxides to afford dialkyl carbonates in excellent yields at room temperature under nitrogen (Scheme 11, reaction 1), and the facile oxidation of the NaSeH thus formed to Se with molecular oxygen (reaction 2) initiates a catalytic reaction (reaction 3). Formation of the dialkyl carbonates is not observed when formates are allowed to react with sodium alkoxide under the same conditions in the absence of Se (Scheme 11) ⟨74TL803⟩.

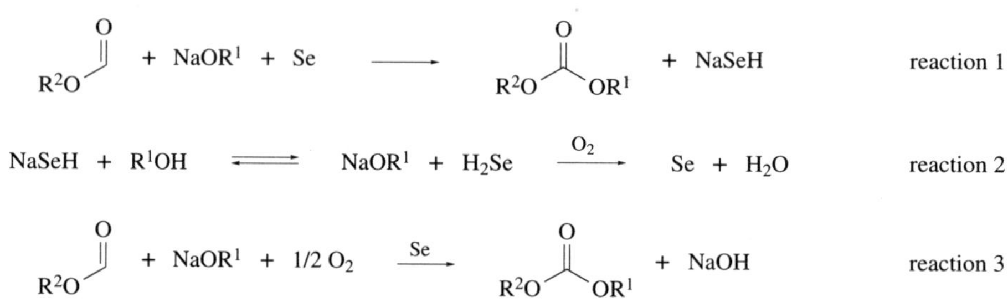

Scheme 11

In contrast to the classical Baeyer–Villiger oxidation of ketones to esters, no simple methodology exists for the double oxidation of ketones to carbonates. It is reported that the formal equivalent of a double Baeyer–Villiger reaction is easily accomplished under mild conditions by oxidation of diethyl ketals with peroxycarboxylic acid. This facile double oxidation of ketals to orthocarbonates provides an efficient method for the removal of a carbonyl function from a ketone. In this reaction the diethyl carbonate is formed in a yield of 50% with 18% of the orthocarbonate (Equations (11) and (12)) ⟨82JA1769⟩.

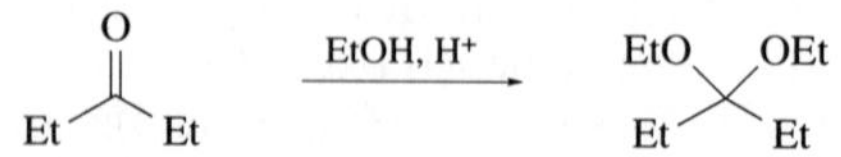

(11)

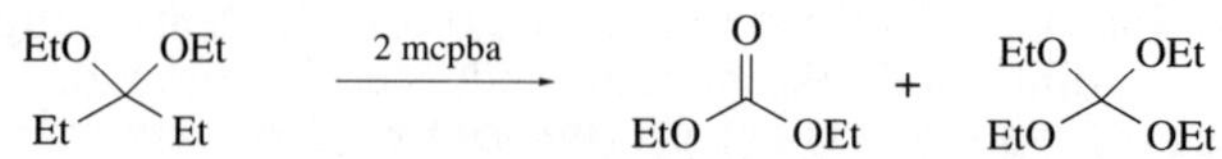

(12)

6.15.1.1.6 Cyclic carbonates and transesterification

Cyclic carbonates are obtained most simply by reacting phosgene with ethane-1,2-diol at room temperature (Equation (13)) ⟨24JCS(125)2259⟩. Another method makes use of sodium hydrogencarbonate as carbonyl source and 2-chloroethanol (Equation (14)) ⟨28GEP516281⟩. A further gas-phase process uses ethylene oxide and carbon dioxide to produce the cyclic carbonate over charcoal at a temperature of 210 °C (Equation (15)) ⟨39GEP740366⟩.

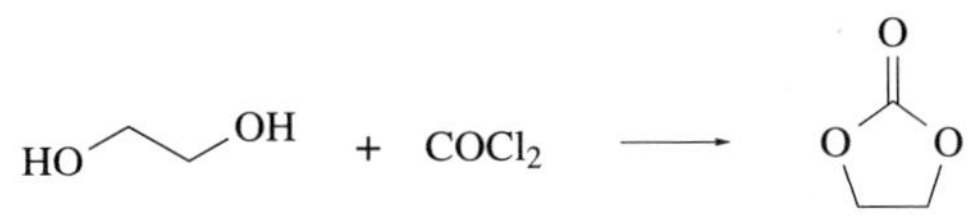

(13)

(14)

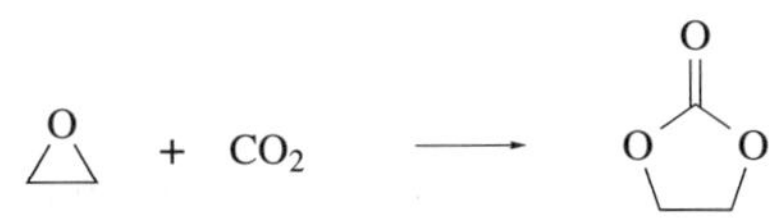

(15)

A viable alternative to the traditional method of producing dimethyl carbonate is its generation through transesterification of ethylene carbonate with methanol (Equation (16)). Dimethyl carbonate is useful as a gasoline octane enhancer, methylating agent and urethane precursor. This reaction is a classical ester exchange, subject to the general rules of acid and base catalysis. Bases are generally more effective, the equilibrium being reached more rapidly with bases which have comparable pK_b values. Effective soluble bases include alkali metal alkoxides, carbonates, bicarbonates and phenoxides; particularly effective are sodium and potassium bicarbonate ⟨91JMOC389⟩. Transesterification is also used industrially to produce dimethyl carbonate and diethyl carbonate as shown in recent patents. Tertiary amines, nitrogen heterocycles such as dbu, and quaternary ammonium groups fixed on basic anion exchangers are used as catalysts ⟨88JAP63238043, 93EUP543234, 94EUP583789⟩.

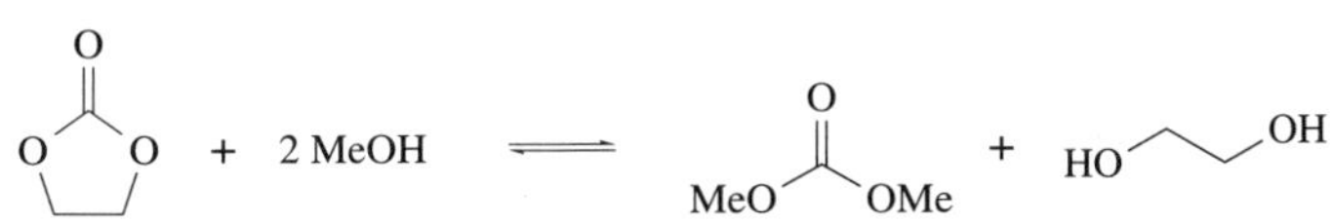

(16)

Laufer and co-workers have investigated the difference between the use of phosgene and triphosgene to generate a cyclic carbonate derived from a catechol derivative. Treatment of equimolar amounts of 4-nitrocatechol and dmap in THF with excess phosgene (20% in toluene) consistently produces *o*-(4-nitrophenylene) carbonate as pale yellow needles in 40–45% yield. Doubling the amount of dmap did not significantly alter the yield of *o*-(4-nitrophenylene) carbonate, but replacing phosgene by 2/3 mole of triphosgene per mole of 4-nitrocatechol and raising the temperature to 50–60 °C improved the yield to 81% (Equation (17)) ⟨89OPP771⟩.

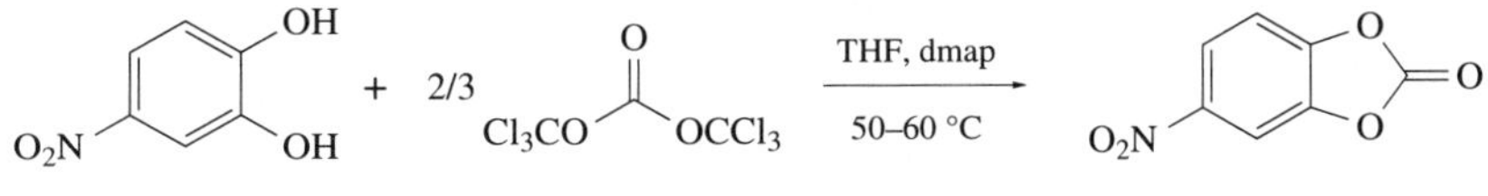

(17)

dmap = dimethylaminopyridine

The proper choice of protecting groups is critical in the synthesis and derivatization of prostaglandins. During the course of the authors' research it was necessary to produce 1,3-hydroxyprostanoid intermediates with a base-labile protecting group. The reaction with triphosgene gave the 1,3-cyclic carbonate of the prostaglandins (**1**) (Equation (18)) ⟨93TL395⟩.

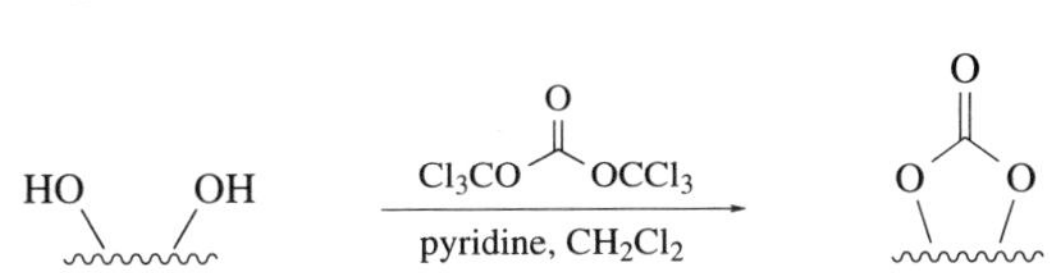

(18)

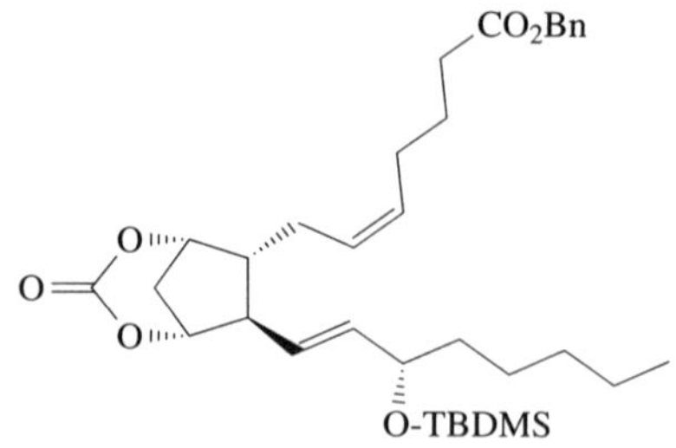

(1)

Kang *et al.* have described the regioselective protection of 1,2,3-, 1,2,4-, and 1,2,5-triols as 5-membered cyclic carbonates with triphosgene. The substituted 1,2,3-triol (2*S*,3*S*)-4-benzyloxy-1,2,3-butanetriol, (*S*)-1,2,4-butanetriol and (2*R*,3*S*)-5-tetradecene-1,2,3-triol all react with triphosgene to afford the 1,2-cyclic carbonates (Equations (19)–(21)). (4*S*,5*S*)-6-Benzyloxy-1,4,5-hexanetriol also reacts in a regioselective manner to form the 4,5-cyclic carbonate (Equation (22)) ⟨94SC305⟩.

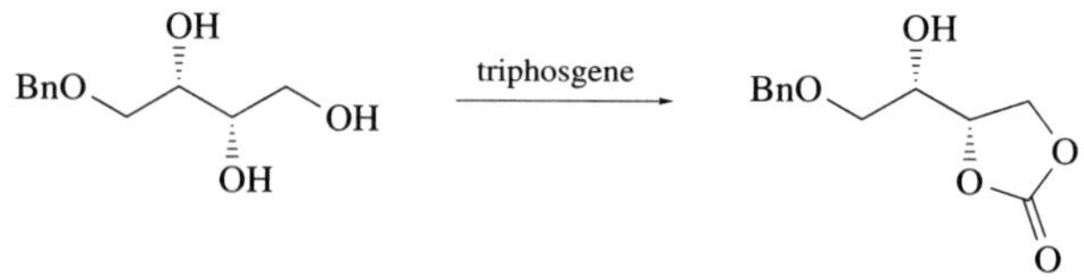

(19)

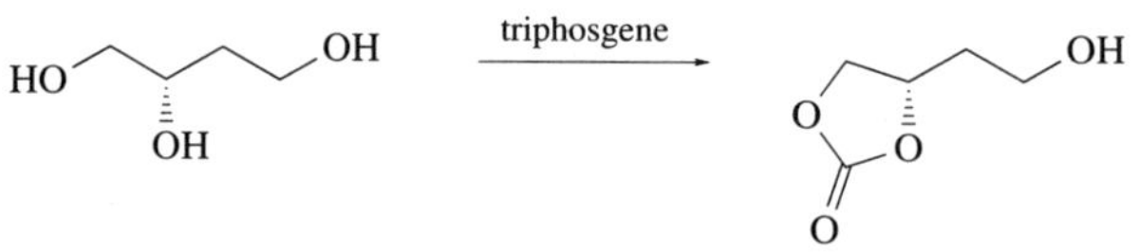

(20)

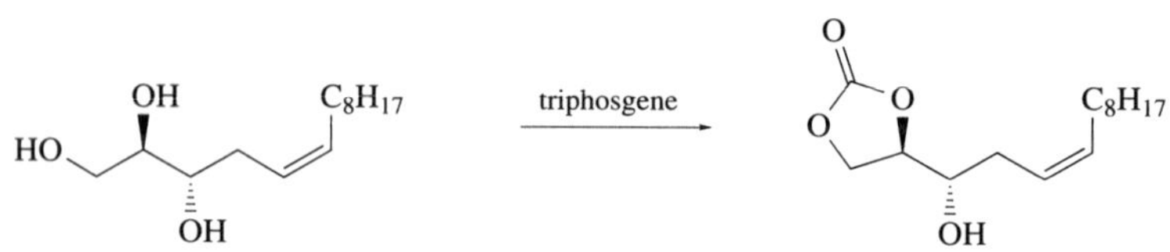

(21)

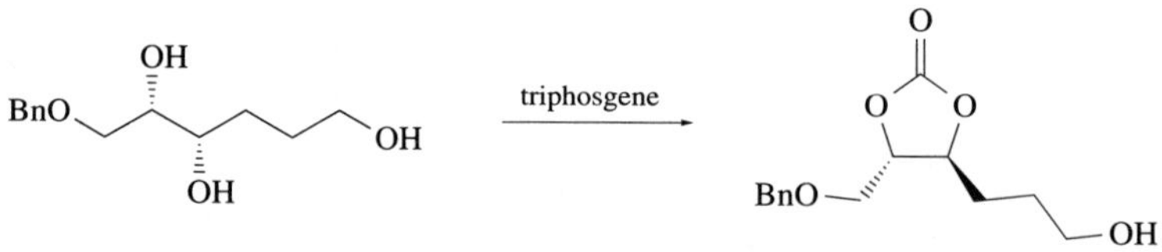

(22)

6.15.1.1.7 *Dialkyl carbonates by iodolactonization*

Iodocyclization of a series of homoallylic *t*-butyl carbonates is an efficient and moderately *erythro* stereoselective method for the functionalization of homoallylic alcohols with 1,3-relative asymmetric induction, as shown in the example in Equation (23) ⟨82JOC4013⟩.

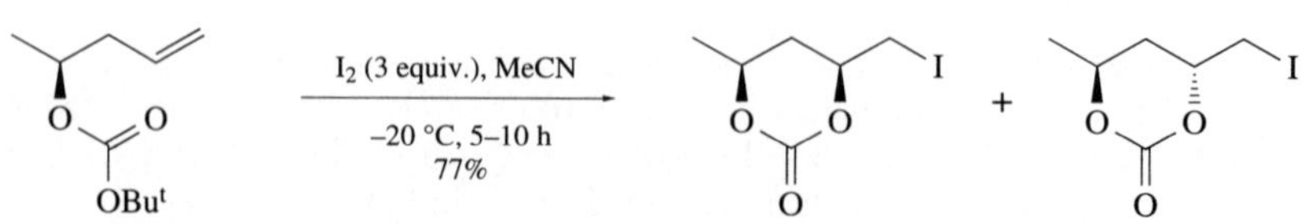

(23)

erythro 10 : *threo* 1

In a variation of the method, the iodolactonization is performed on lithium alkenyl carbonates, prepared in quantitative yield by bubbling CO_2 through a THF solution of lithium alkoxides at room temperature for 1 h. The iodolactonization reaction is carried out in homogenous THF solution at room temperature by adding 2.2 equivalents of I_2 dissolved in THF to the carbonate ⟨82JOC4626⟩.

An important intermediate in the total synthesis of nonactin is obtained by iodolactonization of 1,7-octadien-(*S*)-4-ol-4-*t*-butylcarbonate to yield 1-iodo-7-octene-2,4-diol 2,4-cyclic carbonate in a *cis*/*trans* ratio of 6.5 : 1. Further reduction of the iodo group leads to 7-octene-2,4-diol 2,4-cyclic carbonate in 55% yield ⟨84JA5304⟩. The same reaction has also been used to solve problems in very complex syntheses of natural products ⟨93JOC3703⟩.

6.15.1.1.8 Acyl carbonates

A well-established coupling reagent in peptide chemistry is the tetraethylammonium salt of an acid, usually an *N*-terminal protected amino acid or a peptide, and a chloroformate (Equation (24)). The advantage of this acyl carbonate as a coupling reagent in peptide chemistry is that it can form a peptide bond in high yield and without racemization. The by-products are gaseous carbon dioxide and the corresponding low-boiling alcohol, which can easily be separated from the product. In contrast, the by-products generated in some other methods are often difficult to remove. This coupling technique using acyl carbonates proved its superiority in the synthesis of encephalin analogues ⟨79HCA398⟩.

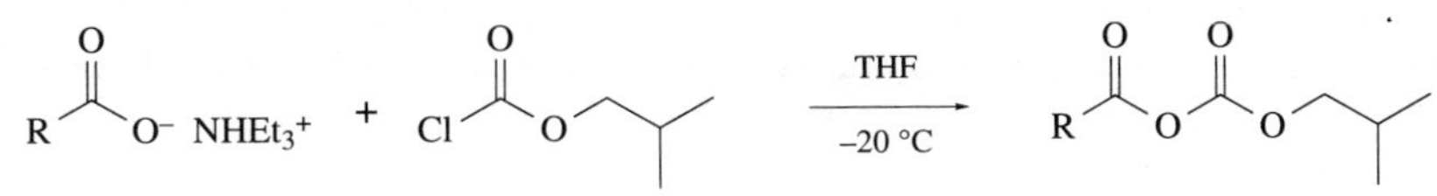

(24)

6.15.1.1.9 Carbonates by electrochemistry

Carbonic acid esters can be produced by electrolysis of carbon monoxide and the corresponding alcohol in the presence of a halide electrolyte. The latter plays a catalytic role in carbonate formation. The efficiency of the process depends, among other variables, on the starting alcohol ⟨78MI 615-01⟩. Electrolytic carbonylation of methanol in the gas phase under atmospheric pressure at 343 K is described by Otsuka *et al.* The anode in this process is doped with $PdCl_2$ and $CuCl_2$. The desired product, dimethyl carbonate, is produced at 0.80 V ⟨94CL495⟩. Several carbonates are also formed in low yield by the anodic oxidation of aqueous potassium butyrate ⟨89JCR(S)15⟩.

6.15.1.1.10 Carbonates via ozonolysis

Ozonolysis of tetramethoxyethylene leads via the ozonide to dimethyl carbonate (20–40%) (Scheme 12). Higher temperatures and a lower initial concentration of tetramethoxyethylene result in increased yields of dimethyl carbonate ⟨88CJC2234⟩. Diethyl carbonate has also been reported as a product of ozonolysis of a ketene diacetal ⟨92CB2493⟩.

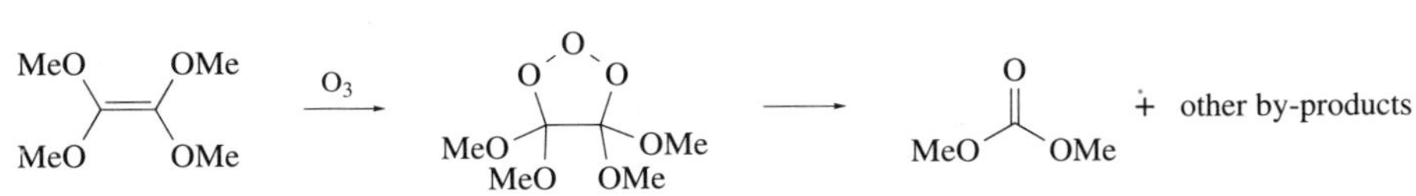

Scheme 12

6.15.1.1.11 Carbonates by miscellaneous methods

Diethyl carbonate is obtained in 12% yield by the reaction of difluorodinitromethane with sodium ethoxide and final hydrolysis with H_2SO_4 (Equation (25)) ⟨82JGU131⟩. Treatment of carbon disulfide with diethoxytin yields diethyl carbonate over several steps (xanthate, orthothiocarbonate) (Scheme 13) ⟨78BCJ3549⟩.

(25)

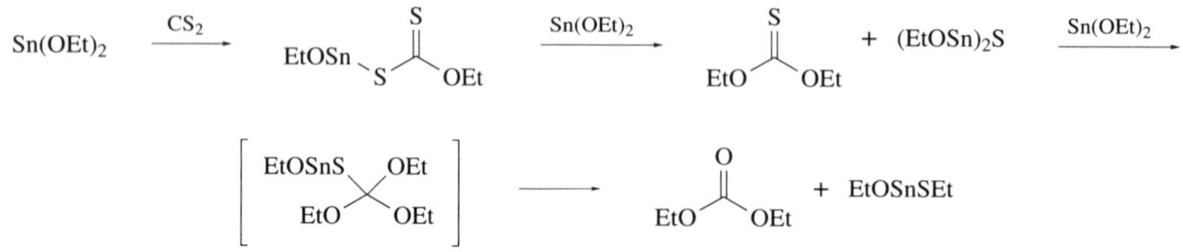

Scheme 13

6.15.1.1.12 *Monoalkyl carbonate complexes*

The tertiary phosphine-coordinated palladium(0) complexes Pd(styrene)L_2 ($L = PMe_3$, PMe_2Ph, $PMePh_2$) react readily with allylic carbonates (methyl 2-methylallyl carbonate and allyl ethyl carbonate) in THF to afford cationic (π-allyl)palladium complexes having an alkyl carbonate anion $[\pi\text{-allylPdL}_2]^+[OCOOR]^-$ (R = Me, Et). These complexes are extremely moisture-sensitive and react readily with water to give the corresponding hydrogencarbonate. Preparation of the corresponding (π-allyl)platinum carbonates leads to analogous products (Scheme 14, Equation (26)) ⟨92OM171⟩.

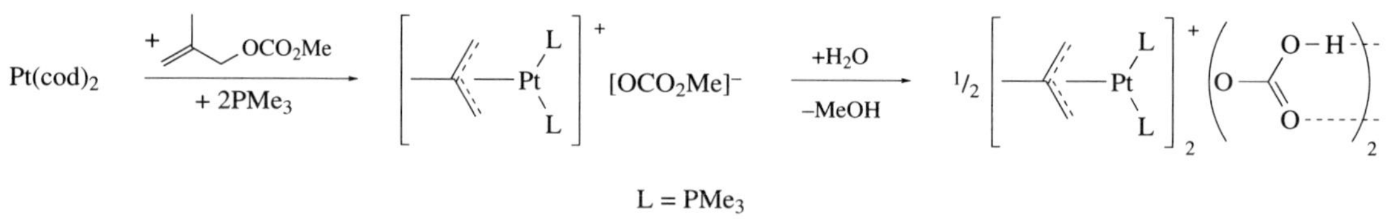

Scheme 14

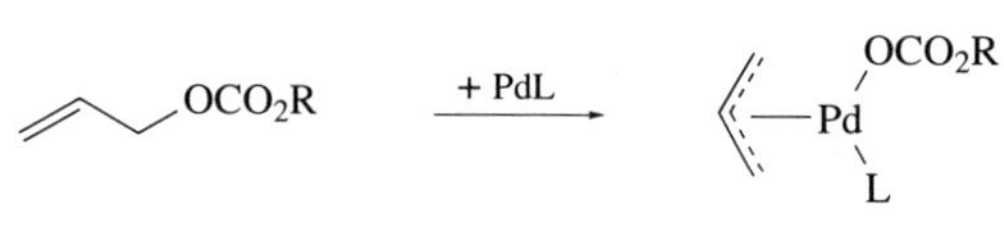

(26)

6.15.1.1.13 *Polycarbonates*

Interfacial polycondensation is currently used for the industrial production of polycarbonates. Bisphenol A is reacted with phosgene at 20–40 °C in a two-phase mixture consisting of an aqueous alkaline phase and an immiscible organic phase. The reaction is described in Equation (27).

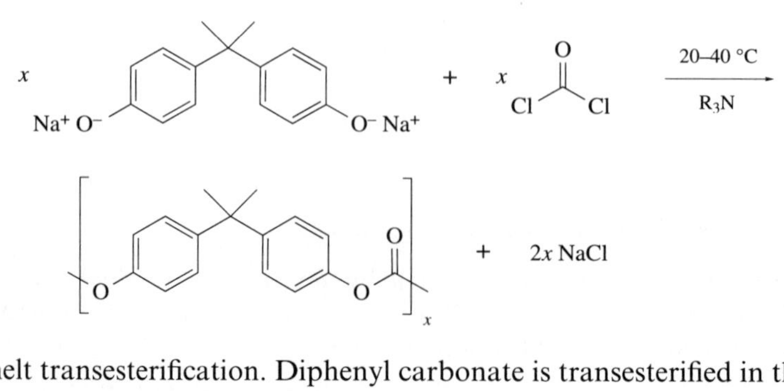

(27)

Another method is melt transesterification. Diphenyl carbonate is transesterified in the melt with bisphenol A to form polycarbonates. During this process phenol is removed by distillation (Equation (28)) ⟨B-92MI 615-01⟩.

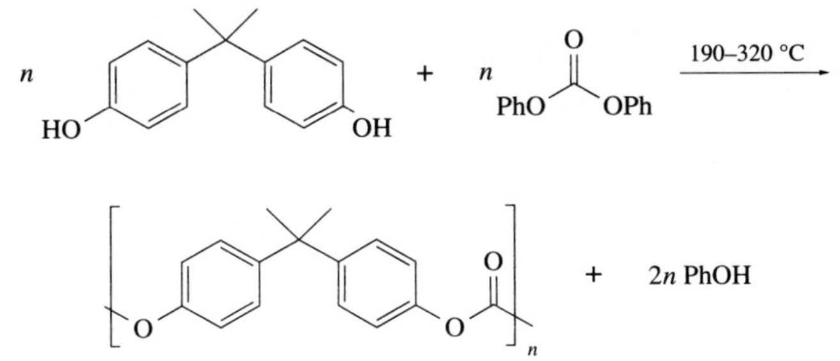

(28)

As previously described diphenyl carbonate can be obtained from dialkyl carbonates. The polycarbonate is prepared using high-purity starting materials (diphenyl carbonate, bisphenol A), improved catalyst systems, high-viscosity reactors, and an alternative process route ⟨91AG(E)1598⟩. Such a process has been described in a patent ⟨93EUP561363⟩.

Triphosgene has recently received attention as a substitute for phosgene owing to the development of ecologically more favorable, phosgene-free production processes. The same polycarbonate can be obtained by reaction of bisphenol A with triphosgene in the presence of triethylamine as described by Eckert ⟨87AG(E)894⟩.

6.15.1.2 Two Sulfur Functions

6.15.1.2.1 Dialkyl dithiocarbonates from phosgene and COS

Symmetrical *S,S′*-dialkyl dithiocarbonates are usually synthesized by reacting phosgene with thiols or their sodium salts (Equation (29)) ⟨1872JPR(2)477, B-62MI 615-01⟩.

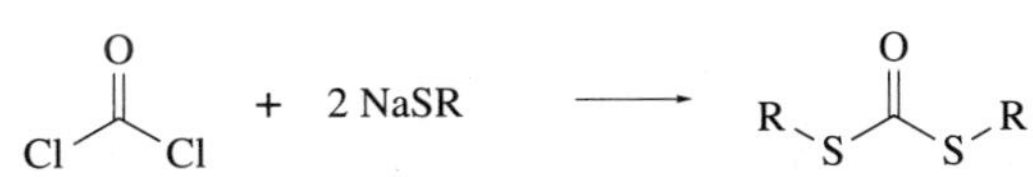

(29)

Carbon oxysulfide is a good alternative to phosgene for making symmetrical *S,S′*-dialkyl dithiocarbonates. Carbon oxysulfide is prepared by triethylamine catalysis from carbon monoxide and elemental sulfur. With triethylamine and water the reactive species triethylammonium dithiocarbonate is formed. The latter is alkylated by alkyl halides in 63–100% yield (Scheme 15) ⟨90CL811⟩. This preparative method is straightforward and convenient; it avoids the use of toxic phosgene and of thiols.

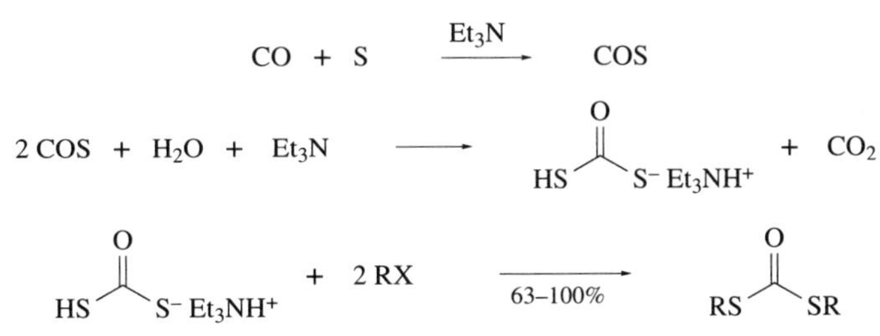

X = Cl, Br, I
R = Bn, Et, Bu^n, Hex^n, Oct^n, $CH_2{=}CHCH_2$, 4-$MeC_6H_4CH_2$, 3,4-$Me_2C_6H_3CH_2$

Scheme 15

6.15.1.2.2 Dialkyl dithiocarbonates by thione–thiol rearrangement

Previously described synthetic routes restrict the variety of possible products to the symmetrical ones. For mixed dialkylated dithiocarbonates R^1SCOSR^2 ($R^1 \neq R^2$), a common method is the thione–thiol rearrangement (Schoenberg rearrangement), which makes use of an *O*,*S*-dialkyl xanthate R^1OCSSR^2 (Equation (30)). The alcohols which are used for this xanthate synthesis vary from simple saturated ⟨81JOC3141⟩ and unsaturated ⟨72CPB2348, 92JOC2523⟩ alkanols up to complex precursors of natural products, for example (+)-milbemycin-β_3 ⟨82JA4708, 85CC1326⟩, angelasidine A ⟨88TL4957⟩, and kaurenoide ⟨81JCS(P1)1293⟩. Xanthates can also be prepared by methylation of potassium *O*-alkyl xanthates by dimethyl sulfate (alkyl = C_1–C_8 functions) in the presence of sodium hydrogencarbonate in water ⟨81S149⟩.

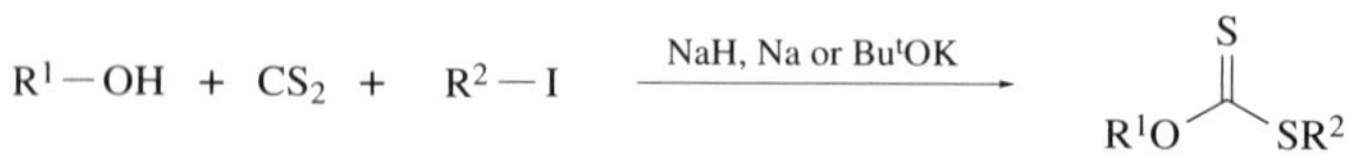

(30)

The thione–thiol rearrangement (Equation (31)) is a formal interchange of the thione sulfur and the oxygen from the ester component and is principally thermally activated. It occurs in conjunction with the thermal elimination and is catalyzed by acids such as TFA ⟨81JOC3141⟩, by amino compounds (for example tricaprylmethylammonium chloride) ⟨81S149⟩, 4-piperidinopyridine ⟨92OPP200⟩, pyridine *N*-oxide ⟨87H(26)2583, 89CPB576⟩, 4-dimethylaminopyridine *N*-oxide ⟨88H(27)2327⟩, and by other species including $BF_3 \cdot Et_2O$ ⟨78CPB3807⟩, $AlCl_3$ ⟨73CPB604⟩, 2,4,6-trinitroalkoxybenzenes ⟨74TL4479⟩, palladium, platinum, rhodium, iridium ⟨86IJ250⟩, and phase-transfer catalysts ⟨80S375⟩. Yields of 58–100%, mostly in the range 70–80%, are obtained.

Two-phase reactions use quarternary ammonium salts as phase-transfer catalysts. However, these conditions proved inadequate in the case in which the *O*-alkyl group was not methyl and when both alkyl groups were methyl, because side reactions (hydrolysis and subsequent reactions in the main) were in competition with the rearrangement to *S*,*S*′-dialkyl dithiocarbonates. In this cases the tricaprylmethylammonium salts are used with advantage ⟨80S375, 81S149⟩.

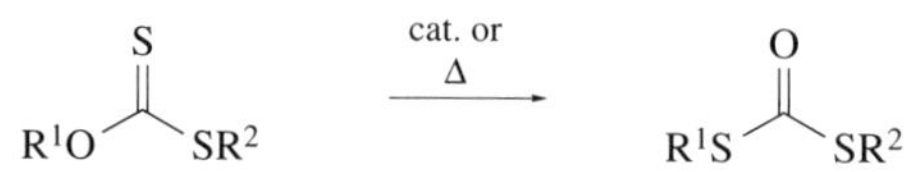

(31)

6.15.1.2.3 *Dialkyl dithiocarbonates by [3,3]-sigmatropic rearrangement*

These rearrangements represent important organic transformations, especially because of the high stereochemical control that accompanies them. They are induced thermally. The following examples of [3,3] sigmatropic rearrangements afford *S*,*S*-dialkyl dithiocarbonates by intramolecular alkyl migration (Equation (32)) ⟨72CPB2348⟩. This is a convenient method for the syntheses of dithiocarbonates (**2**), (**3**) and (**4**) from myrtenol, *trans*-pinocarveol and perillyl alcohol, respectively ⟨85JOC118⟩. Intermediate dithiocarbonates are useful in the preparation of methyl *ent*-17,17,17-trifluorokaur-15-en-19-oate and *ent*-16,16-difluoro-17-norkauran-19-oic acid ⟨81JCS(P1)1293⟩. The first synthesis of angelasidine A was accomplished in eight steps starting from farnesol. The quarternary carbon atom of the agelasidine A was constructed by the successful application of the hetero-Claisen allyl xanthate rearrangement. This methodology provides the basis for a general and efficient route to the agelasidinine skeleton ⟨88TL4957⟩.

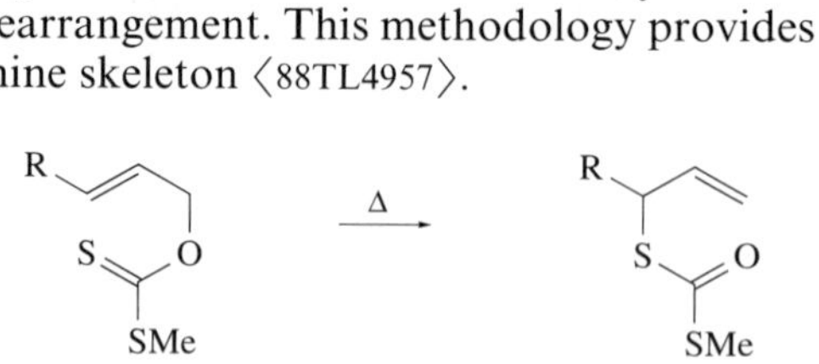

(32)

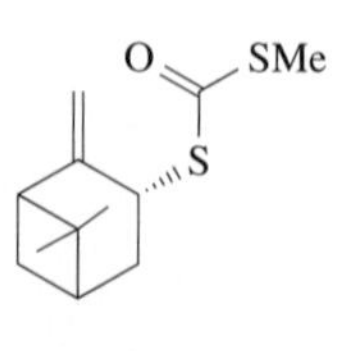

(2)

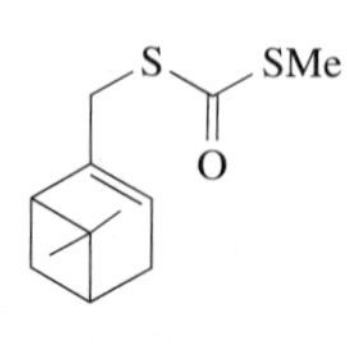

(3)

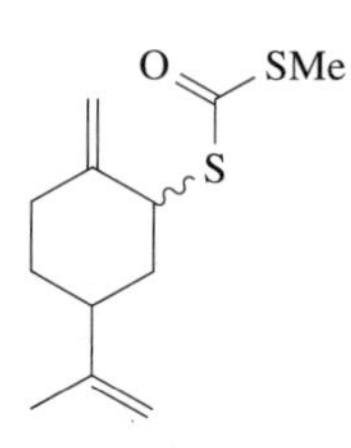

(4)

An important step in the total synthesis of milbemicin-β_3 is also a hetero-Claisen rearrangement (Equation (33)) ⟨82JA4708, 85CC1326⟩. This natural product has enormous potential as a broad-spectrum antiparasitic agent.

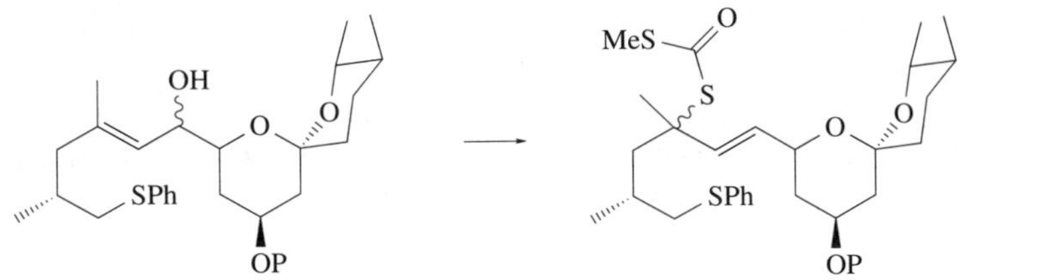

(33)

The next two examples deal with transfer of chirality, since [3,3]-sigmatropic rearrangement is an important tool for regioselective transformation of allylic alcohols to allylically rearranged products. This is demonstrated in the asymmetric synthesis of thiotetronic acids, which is based upon an allyl xanthate to dithiocarbonate rearrangement (Equation (34)) ⟨87CC1228⟩. For this purpose β-cyclodextrin is used as a catalyst to give the product in good yield and enantiomeric excess of 46% (Equation (35)) ⟨91TL7557⟩. β-Cyclodextrin provides a chiral environment of an inclusion complex with the xanthates to induce chirality. The application of [3,3]-sigmatropic rearrangements in the late stages of synthetic sequences has, however, been limited by the generally elevated temperatures that are required to induce reaction. For this and other reasons a number of attempts have been made to find catalysts for these sigmatropic rearrangements. Notable among the attempts is the work of Overman ⟨84AG(E)579⟩, who observed that Hg(II) and Pd(II) complexes caused accelerations of the order of 10^{10} over the thermal uncatalyzed Cope rearrangements. The overall stereochemistry of the reaction was similar to that of the corresponding uncatalyzed rearrangement ⟨82JA7225⟩. An example of this method is shown in Equation (36) ⟨86IJ250⟩.

(34)

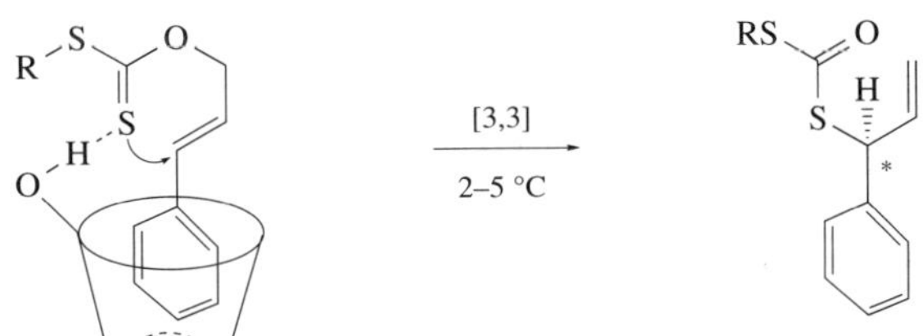

(35)

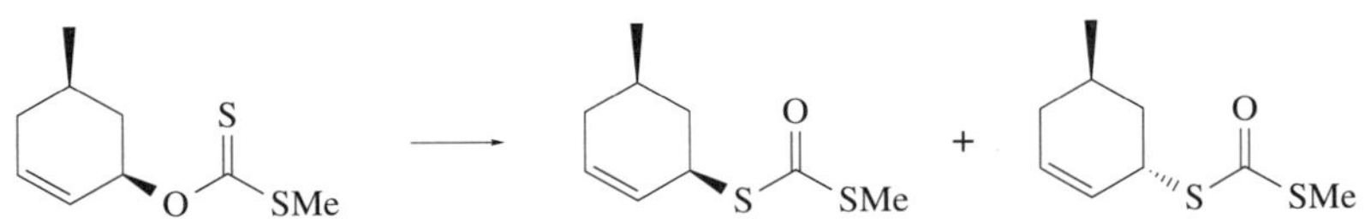

(36)

6.15.1.2.4 *Other methods*

Besides these classical methods, some less common synthetic approaches have been published.

Sigmatropic rearrangements normally introduce the sulfur into position three of an alkene, but there is an example in which sulfur is transferred from position 6 to position 1 of a terminal alkene, accompanied by the formation of a carbon–carbon bond (ring closure). The authors propose a free radical mechanism for this isomerization of an *S*-methyl hex-5-enyl xanthate to an *S*-cyclopentylmethyl *S*′-methyl dithiocarbonate (Equation (37)) ⟨93JOC2894⟩.

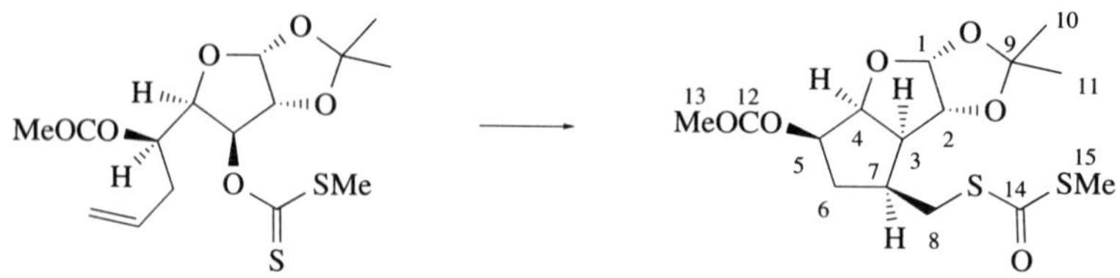

(37)

A variation of thione–thiol rearrangement starting with *O*-alkyl *S*-methyl xanthates affords bis(methylthio carbonyl) poly-sulfanes (**5**) via sulfenyl chloride intermediates ROC(SCl)(Cl)SMe ⟨84JCS(P1)2615, 83JOC4750⟩.

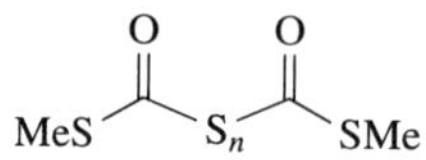

(**5**) $n = 1$ to 6

A method which makes no use of rearrangement and works under very mild thermal conditions is the oxygenation of the lithio derivatives of α-phosphoryl dithiocarbonates (Equation (38)) ⟨84TL2489⟩.

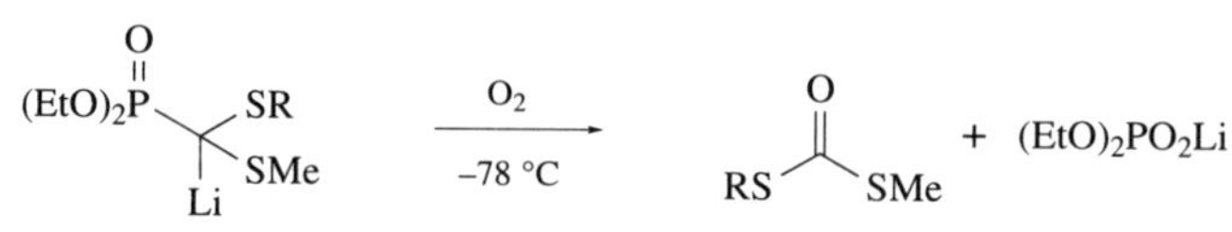

(38)

R = Me, Et

β-Amino-substituted dithiocarbonates are prepared in a simple HBr-catalyzed hydrolysis of substituted thiazolines (Equation (39)) ⟨64JOC2442⟩.

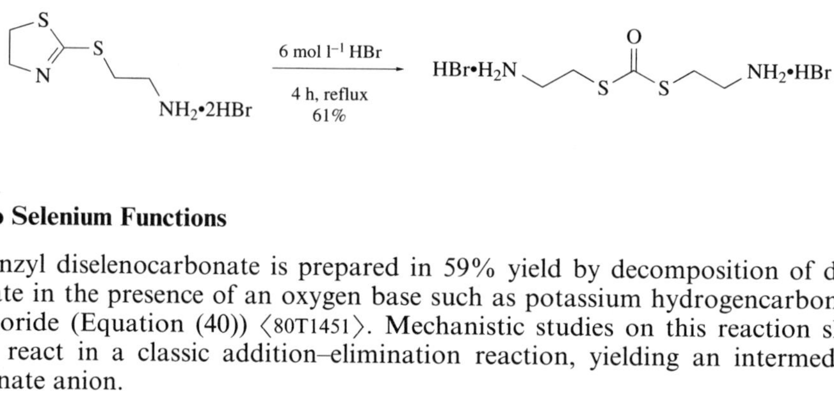

(39)

6.15.1.3 Two Selenium Functions

Se,Se′-Dibenzyl diselenocarbonate is prepared in 59% yield by decomposition of dibenzyl triselenocarbonate in the presence of an oxygen base such as potassium hydrogencarbonate, and of mercury dichloride (Equation (40)) ⟨80T1451⟩. Mechanistic studies on this reaction showed that oxygen bases react in a classic addition–elimination reaction, yielding an intermediate benzyl diselenocarbonate anion.

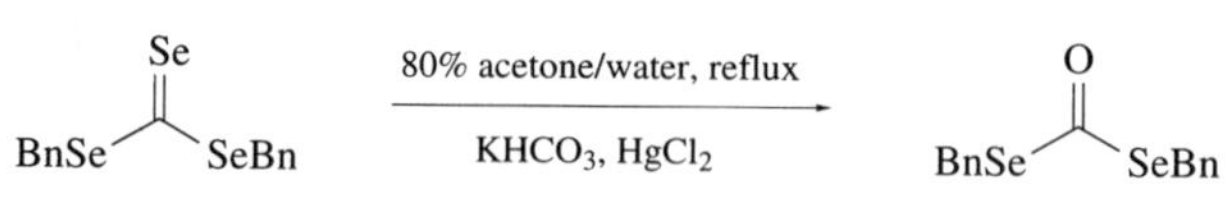

(40)

6.15.2 CARBONYL CHALCOGENIDES WITH TWO DISSIMILAR CHALCOGENIDE ATOM FUNCTIONS

6.15.2.1 Oxygen and Sulfur Functions

6.15.2.1.1 Dialkyl thiocarbonates from activated carbonates

The dominant method for producing dialkyl thiocarbonates and bis(dialkoxycarbonyl) sulfides uses alkyl chloroformates and sulfides.

Bis(methoxycarbonyl) sulfide can be prepared by reacting methyl chloroformate and lithium sulfide in THF to afford the product in 51% yield. Commercial anhydrous Li_2S does not dissolve in THF and has to be prepared freshly (Equation (41)) ⟨79T2329, 78CC838⟩.

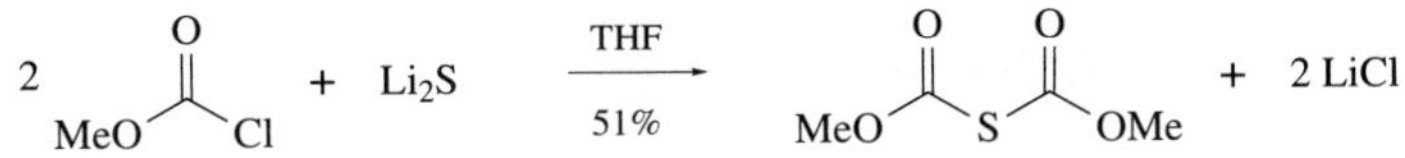

(41)

These compounds may also be obtained in very high yields by the reaction of alkyl chloroformates with sodium sulfide nonahydrate in a solid–liquid system consisting of dichloromethane, the above-mentioned salt, and hexadecyltributylphosphonium bromide as phase-transfer catalyst. A typical procedure consists of the slow addition of 10 mol% excess of powdered sodium sulfide to a well-stirred solution of the alkyl chloroformate and hexadecyltributylphosphonium bromide (5 mol% with respect to the sulfide) in dichloromethane at 0 °C. Bis(alkoxycarbonyl) sulfides are not formed or formed only in low yields without the catalyst. With the catalyst, the high yields obtained under mild conditions, the use of cheap and common laboratory reagents, the absence of side reactions, and the simple workup, make this procedure useful for preparation of bis(alkoxycarbonyl) sulfides (Equation (42)) ⟨82SC897⟩.

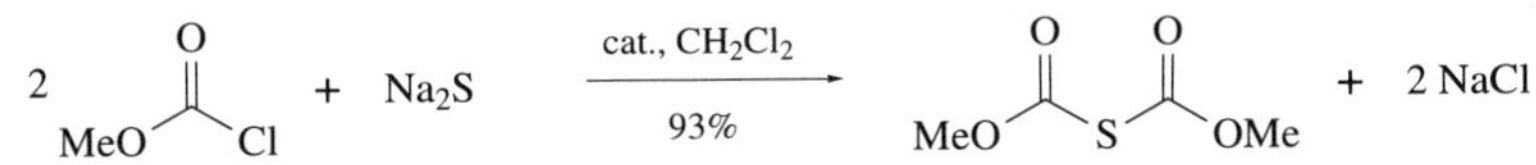

(42)

cat. = $C_{16}H_{33}(C_4H_9)_3P^+ Br^-$

The above methods deal with the reaction of alkyl chloroformates with inorganic sulfides affording *S*-acyl thiocarbonates. By using thiols instead of inorganic sulfides, mixed alkyl thiocarbonates can be formed.

An obvious route to unsymmetric alkyl derivatives of thiocarbonic acid uses alkyl chloroformates, which are obtained from the corresponding alcohol and phosgene. Because of the toxicity of phosgene, chloroformates can also be prepared with advantage using triphosgene.

The previously described alkyl chloroformates react with alkanethiols to give alkyl thiocarbonates. The reaction can be used to obtain branched and unsymmetric dialkyl thiocarbonates (Equation (43)) ⟨76BSF501⟩. Taylor and co-workers have synthesized *S*-ethyl *O*-methyl thiocarbonate, *S*-isopropyl *O*-methyl thiocarbonate, and *S*-butyl *O*-methyl thiocarbonate for use in mechanistic studies of thermal eliminations. They all are prepared by reacting methyl chloroformate and the corresponding thiol in the presence of pyridine to trap the HCl evolved (Equation (44)) ⟨83JCS(P2)291⟩.

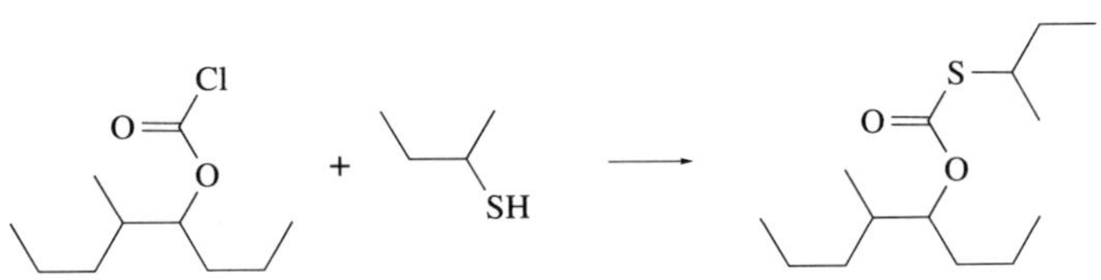

(43)

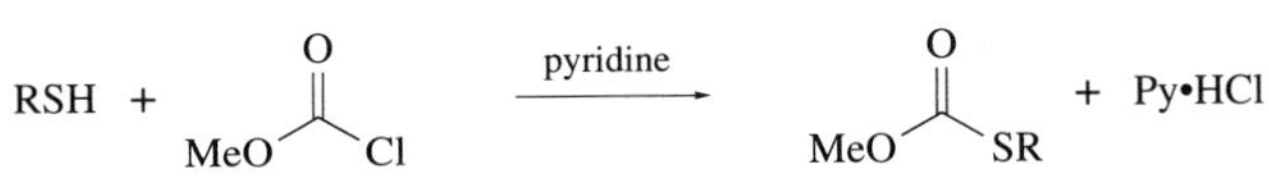

(44)

R = Et, Pr^i, Bu

2-Thiomethyl-2-phenyl-1,3-indanedione is also transformed to the corresponding thiocarbonates as described by Marshalkin *et al.* by using methyl and phenyl chloroformates in the presence of the base triethylamine, as outlined in Equation (45) ⟨85ZOR1670⟩.

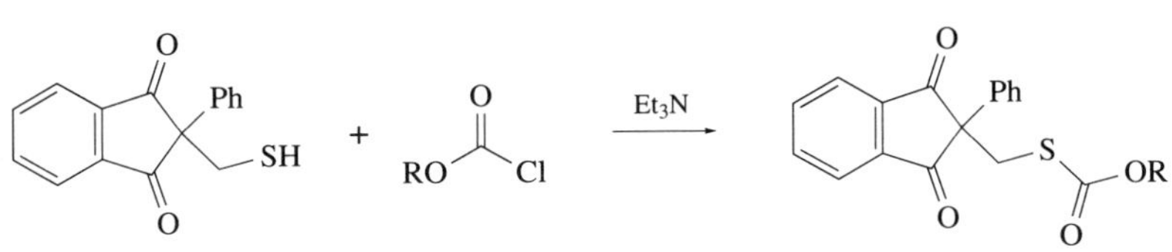

(45)

R = Me, Ph

Polyethyleneoxydiisothiocyanates can be obtained by thermolysis of the thioacyl *O*-ethyl thiocarbonate, which can easily be prepared by reacting ethylene dithiocarbamate, ethyl chloroformate, and triethylamine (Scheme 16) ⟨85ZOB2100⟩.

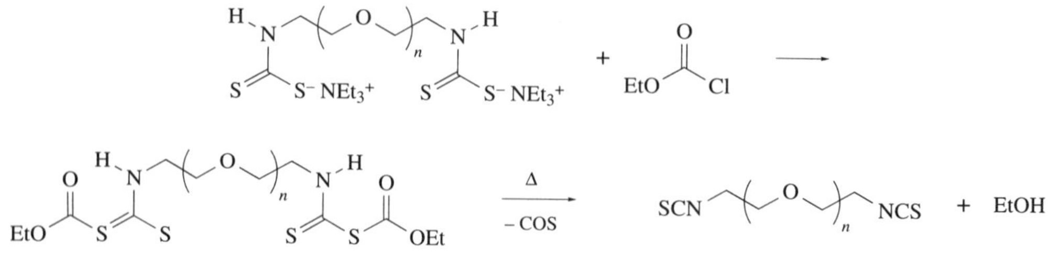

n = 1, 2, 3, 4

Scheme 16

The above examples all describe a simple way to obtain the alkyl thiocarbonates. This class of compounds is also used, for example, in the synthesis of juvenile hormone analogues ⟨88ZN(B)1038⟩, and as intermediates in prodrug synthesis ⟨90S1159⟩. *S*-Ethoxycarbonyl *O*-ethyl dithiocarbonate can be obtained by the Holmberg method, which consists of the reaction of sodium *O*-ethyl dithiocarbonate with ethyl chloroformate using acetone or ethanol as solvent (Equation (46)) ⟨85PS(25)91⟩.

(46)

2-Propoxycarbonyl thiocyanate can be prepared by the reaction of isopropyl chloroformate with a rhodanide anion ⟨88PS(39)257⟩. Ethyl(methoxycarbonylthio)phenylarsine sulfide (**6**) can be prepared as a by-product in a yield of 27% by reacting ethylmethylphenylarsine sulfide with methyl chloroformate in boiling benzene ⟨76JGU2441⟩.

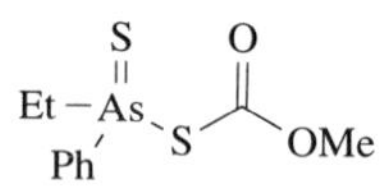

(6)

Esters of *O*,*O*-dialkyl dithiophosphoric acids are of marked interest because of their exceptional insecticidal properties and low mammalian toxicity. These esters can be prepared by reaction of *O*,*O*-diethyl dithiophosphoric acid salts with methyl chloroformate, isopropyl chloroformate, or benzyl chloroformate to afford the corresponding diethyl *S*-alkoxy-carbonyl dithiophosphate in yields from 62% to 80% depending on the reaction conditions. The best results are obtained in benzene at 80 °C, but yields are also acceptable when the reaction is carried out in acetonitrile at 20 °C (Equation (47)) ⟨86CI(L)140⟩.

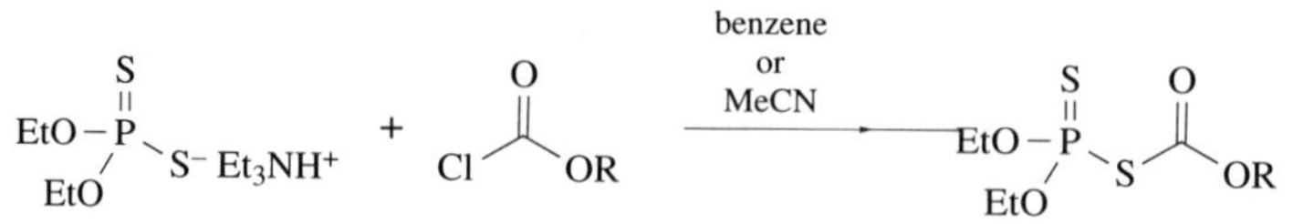

(47)

R = Me, Et, Pr^i, Bn

S-Vinyl *O*-*t*-butyl thiocarbonate is obtained (60%) from *S*-(*β*-chloroethyl) chlorothioformate by reaction with potassium *t*-butoxide at −30 °C. The chloroformate can be prepared by reacting phosgene with thiirane at −10 °C (Scheme 17) ⟨77MI 615-02⟩.

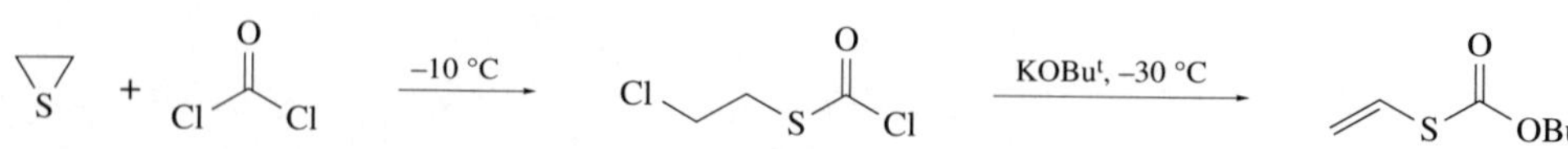

Scheme 17

Thiocarbonates are also important as nonisoprenoid juvenile hormone analogues. For example, 2-(3-phenoxyphenoxy)ethanol is condensed with ethyl chloroformate, giving *S*-ethyl *O*-2-(3-phenoxyphenoxy)ethyl thiocarbonate (Equation (48)) ⟨89ZN(B)983⟩.

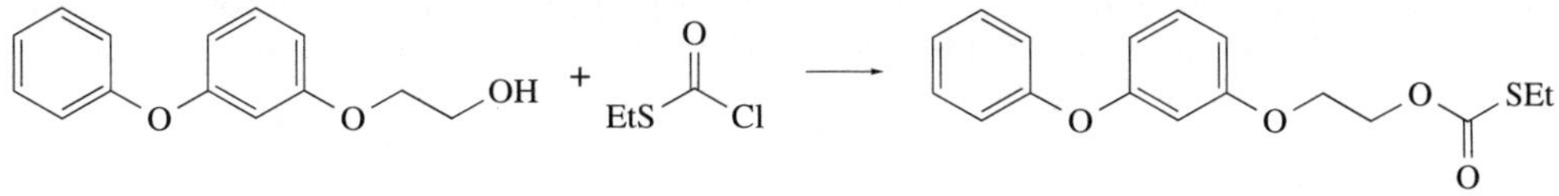

(48)

Ethyl chloroformate also reacts with nitro alcohols in the presence of anhydrous iron(III) chloride to form the respective *O*-alkyl *S*-ethyl thiocarbonates. The reaction proceeds vigorously at ambient temperature, is complete in a few minutes, and is essentially quantitative (Equation (49)) ⟨79S600⟩. It is remarkable that iron(III) chloride exhibits no catalytic effect with other alkoxycarbonyl chlorides.

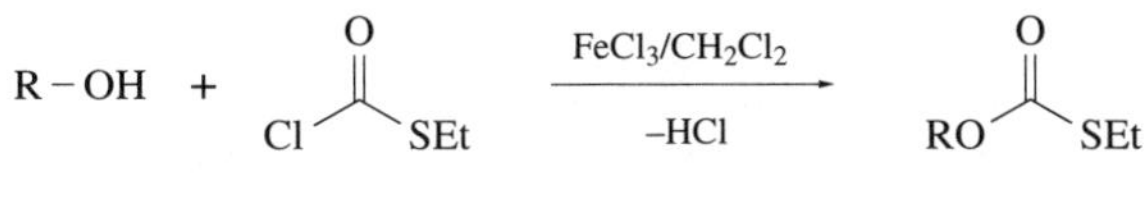

(49)

$R = (O_2N)_3CCH_2$, $FC(NO_2)CH_2$, $MeC(NO_2)_2CH_2$, $O_2NCH_2CH_2$

A substitute for an *S*-alkyl thiochloroformate is an *S*-alkyl thiocarbonylimidazolium chloride. Such nitrogen-containing compounds, for example tertiary amines or imines, are very good acylating co-reagents ⟨78T3105⟩. Another activation of the alkyl thioformate group is provided by the corresponding bromothioformate, an example of which is easily derived from trifluoromethyl thiofluoroformate by halogen interchange. This species reacts well with sodium cyanide to afford the trifluoromethyl thiocyanoformate (Equation (50)) ⟨78CB2891⟩.

(50)

In the case of tertiary alkyl groups the corresponding chloroformate is unstable. In these cases activation is provided by the carbonic acid anhydride. Both thiophenol and butanethiol react rapidly with di-*t*-butyl dicarbonate under phase-transfer conditions using the anhydrous potassium carbonate 18-crown-6 system, and afford good yields of the corresponding *t*-butoxycarbonyl derivatives (Equation (51)). In contrast, the reaction with aqueous sodium hydroxide and tetrabutylammonium hydrogensulfate does not lead to product, because the thiocarbonates are immediately hydrolyzed back to the starting thiols ⟨85CJC153⟩.

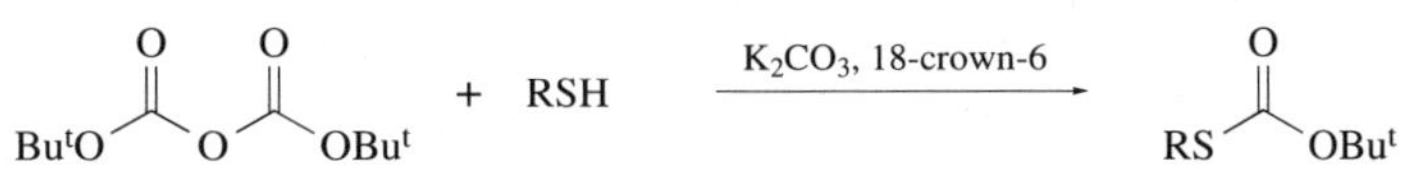

(51)

R = Ph, 96%; Bu^n, 90%

Cyclic thiocarbonates are of especial interest in the chemistry of sugars. Methyl 2,6-di-*O*-methyl-3,4-*O*-thiocarbonyl *β*-D-galactoside is added to a solution of methyl iodide and propylene oxide sealed in a glass ampoule and heated to 80 °C for 14 h to afford the iodothiocarbonates in 99% yield (Equation (52)) ⟨87AJC795⟩. The importance of thiocarbonates formed by reaction with chloroformates is also documented in several patents ⟨81USP224776, 83MIP8303603, 91MIP9110639, 93GEP4140446⟩.

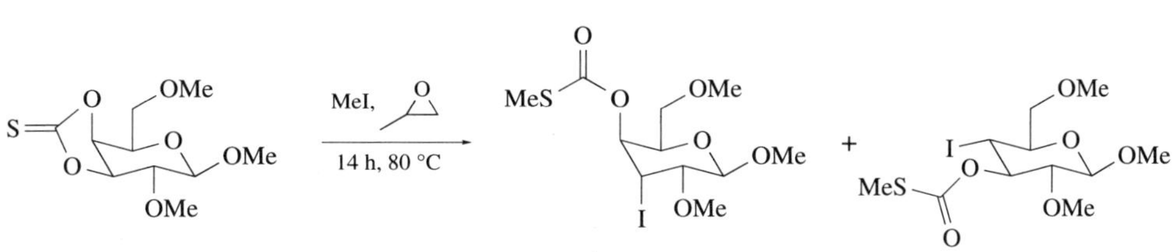

(52)

6.15.2.1.2 *Dialkyl thiocarbonates from alkoxycarbonyl sulfenyl chlorides*

Alkoxycarbonyl sulfenyl chlorides are highly reactive electrophilic thiocarboxylating reagents. Methoxycarbonyl sulfenyl chloride is an established reagent. This reagent can be used to form *S*-aryl thiocarbonates from aromatic compounds in the presence of Lewis acids such as $AlCl_3$ or BF_3 in methylene chloride or carbon disulfide (Equation (53)). These thiocarbonates hydrolyze in methanolic potassium hydroxide to the aryl mercaptans in yields of 90% ⟨77ZC411⟩.

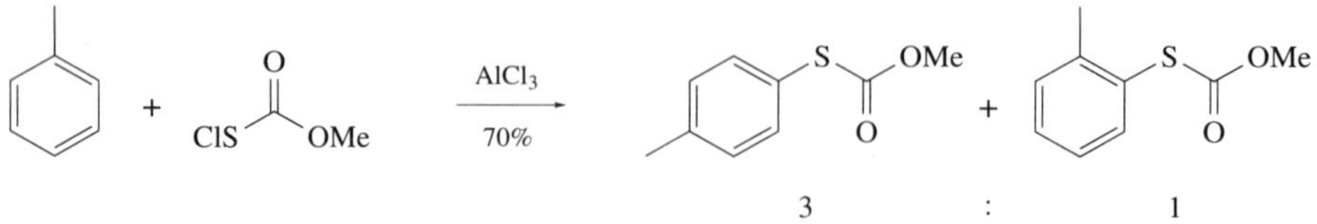

(53)

α-Sulfenylated ketones are also available in a regioselective α-methoxycarbonyl sulfenylation of various ketones or aldehydes employing methoxycarbonyl sulfenyl chloride. Addition of the corresponding ketones or aldehydes to one equivalent of the reagent in chloroform at room temperature for 24 h provides α-methoxycarbonyl sulfenylated carbonyl compounds in good yields. As an example, the reaction of nonan-4-one is shown in Equation (54) ⟨92JOC1053⟩.

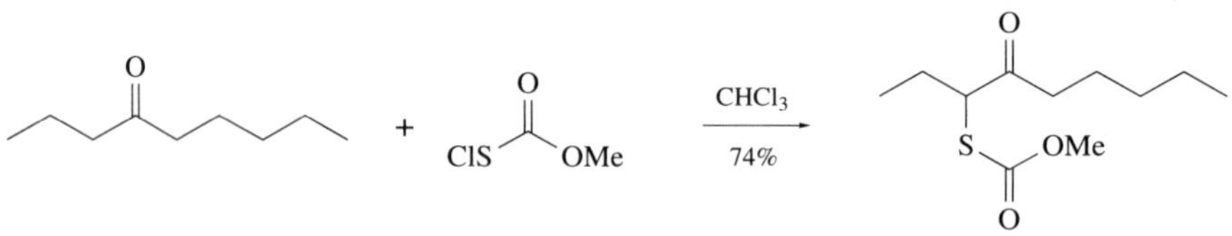

(54)

Alkanesulfenyl methyl thiocarbonates can also be obtained. Reaction of benzyl thiol with methoxycarbonyl sulfenyl chloride in methanol yields the disulfide (Equation (55)) ⟨80JOC271⟩.

(55)

The reagent methoxycarbonyl sulfenyl chloride is easily prepared by reacting methanol with chlorocarbonyl sulfenyl chloride ⟨88JHC193⟩.

6.15.2.1.3 *Dialkyl thiocarbonates from orthoformates*

O,*S*-Diethyl thiocarbonate is prepared by reaction of 2-ethoxy-1,3-oxathiolane with *t*-butyl peroxide as radical initiator (Equation (56)). The key step is hydrogen atom abstraction by the initiator. Under same reaction conditions, butyl diethoxymethyl sulfide is converted into *S*-butyl *O*-ethyl thiocarbonate (Equation (57)) ⟨81ZOR1539, 85ZOR2206, 86JOU270⟩. Lauryl peroxide has also been used as initiator instead of *t*-butyl peroxide ⟨87JGU365⟩.

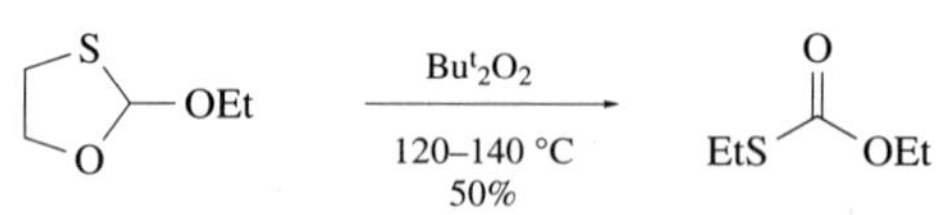

(56)

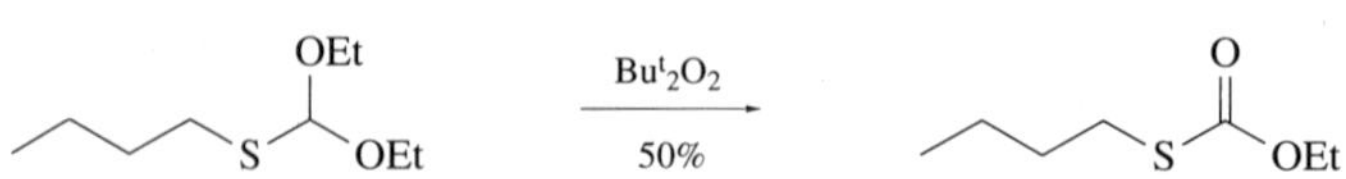

(57)

6.15.2.1.4 *Dialkyl thiocarbonates from ethyl xanthate potassium salt*

In research on model systems for disulfide prodrug formation, dithiocarbonates are often intermediates. They are obtained by the reaction of an alkyl iodide and ethyl xanthate potassium salt. The thiocarbonate from an alkyl isoalloxazine is obtained in good yield (Equation (58)) ⟨91JMC2049⟩.

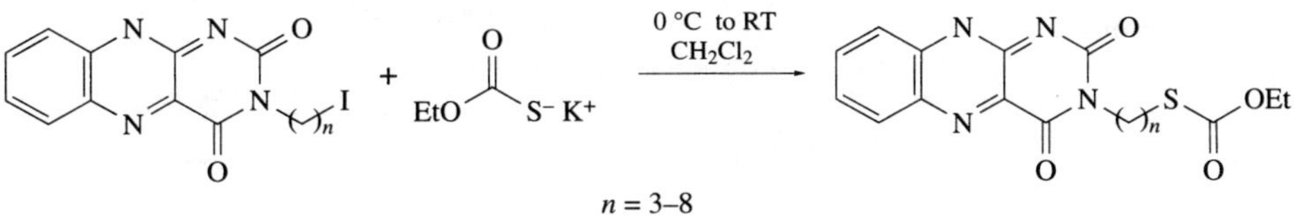

(58)

6.15.2.1.5 *Miscellaneous thiocarbonates from carbonothioic acid salts*

In reactions analogous to the oxidation of xanthates to dixanthogens and the oxidation of dixanthates to tetraxanthogens, monothiocarbonates and bis(monothiocarbonates) can be oxidized to the corresponding disulfides. The oxidation of potassium monothiocarbonates is performed rapidly and quantitatively using iodine (Equation (59)) ⟨73AJC755⟩.

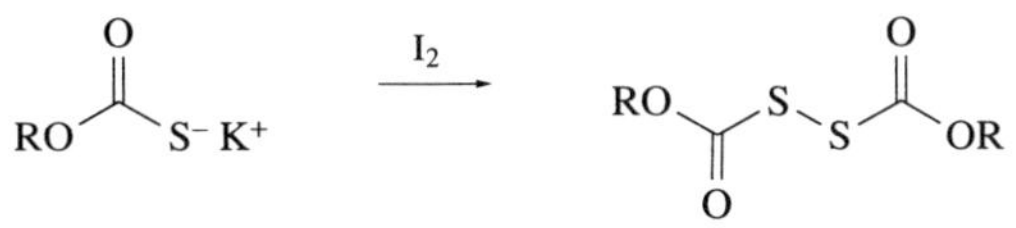

(59)

Potassium monothiocarbonates such as potassium *O*-ethyl thiocarbonate can act as sulfur nucleophiles and undergo addition to electrophiles such as epoxides or alkyl halides (Equations (60) and (61)) ⟨83JOC5033⟩.

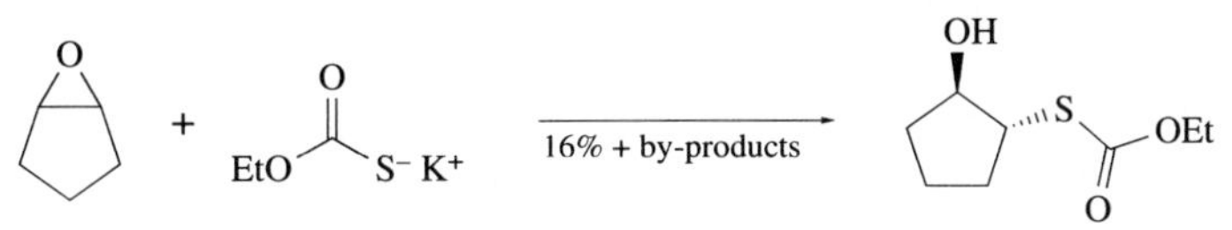

(60)

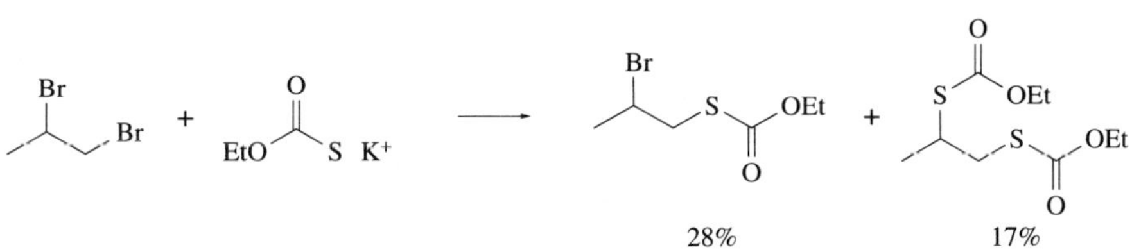

(61)

It is known that the polarizability of chalocogens depends on their position in the periodic table. In the case of charged species the counterion also plays an important role. If the counterion is changed to a tertiary amine, particularly dbu, thiocarbonates are obtained in optimum yields by *S*-alkylation of the ammonium salts (Equations (62) and (63) and Table 1). The reagent itself is easily prepared by reacting carbon monoxide, elemental sulfur, and an alcohol in an autoclave in the presence of dbu (Equation (62)) ⟨88TL4767⟩.

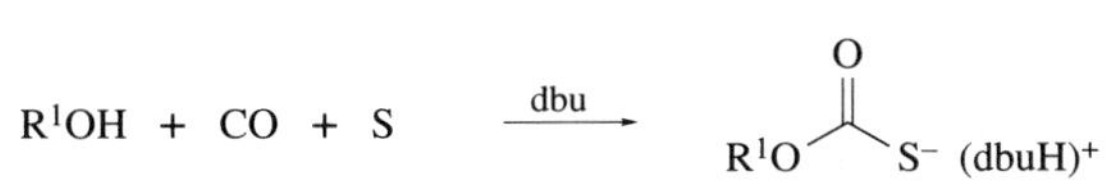

(62)

dbu = 1,5-diazabicyclo[5.4.0]undec-5-ene

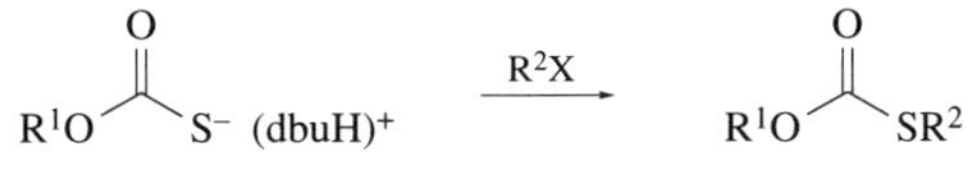

(63)

The nature of the alkylating agent is also important in this type of reaction. If iodomethane is used, excellent yields of products are usually obtained ⟨90ZOB1826⟩. The potassium *O*-alkyl thiocarbonate, used as above, can also be prepared by a selenium-catalyzed reaction of alcohol, carbon monoxide, and sulfur to react in a second step with alkyl halides to afford the desired product ⟨90TL4773⟩.

Using a potassium *O*-alkyl thiocarbonate, unsymmetrical dialkoxycarbonyl sulfides are obtained in a reaction with *O*-alkyl chloroformates (Equation (64)) ⟨90S825⟩.

Table 1 Formation of unsymmetrical thiocarbonates from alcohols and alkyl halides (Equations (62) and (63)).

Alcohol	*Alkyl halide*	*Product*	*Yield*[a] (%)
Bu^nOH	$PhCH_2Br$	$Bu^nOC(O)SCH_2Ph$	86
	$CH_2{=}CHCH_2Br$	$Bu^nOC(O)SCH_2CH{=}CH_2$	100
	Bu^nI	$BuOC(O)SBu^n$	75
MeOH	$PhCH_2Br$	$MeOC(O)SCH_2Ph$	74
EtOH	$PhCH_2Br$	$EtOC(O)SCH_2Ph$	93
Pr^nOH	$PhCH_2Br$	$Pr^nOC(O)SCH_2Ph$	86
Pr^iOH	$PhCH_2Br$	$Pr^iOC(O)SCH_2Ph$	82
n-$C_{10}H_{21}OH$	$PhCH_2Br$	n-$C_{10}H_{21}OC(O)SCH_2Ph$	81
$PhCH_2OH$	$PhCH_2Br$	$PhCH_2OC(O)SCH_2Ph$	83
$CH_2{=}CHCH_2OH$	$PhCH_2Br$	$CH_2{=}CHCH_2OC(O)SCH_2Ph$	71
$CH_3OCH_2CH_2OH$	$PhCH_2Br$	$CH_3OCH_2CH_2OC(O)SCH_2Ph$	87
Bu^tOH	$PhCH_2Br$	$Bu^tOC(O)SCH_2Ph$	52

[a] Isolated yields based on alcohol used.

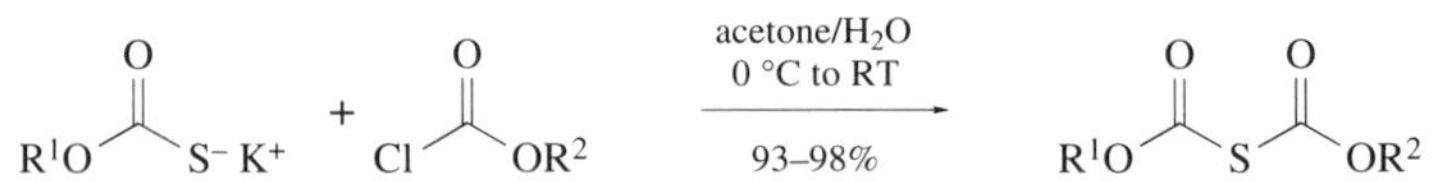

(64)

6.15.2.1.6 *Other methods*

Ozonolysis of 1-methoxyvinyl aryl sulfones leads to *S*-aryl *O*-methyl thiocarbonate *S,S*-dioxides in good yield. The ozonolysis is carried out in ethyl acetate in the presence of tetracyanoethylene (TCNE) (Equation (65)) ⟨76S408⟩.

(65)

Carbamoyldisulfides are obtained by reacting *O*-alkyl chlorothioformates with chlorocarbonyl sulfenyl chloride in the presence of phenylmethylamine followed by hydrolysis (Equation (66)) ⟨86JOC1866⟩.

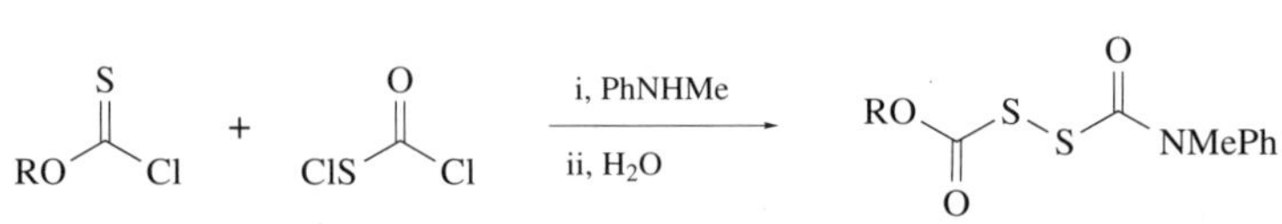

(66)

6.15.2.1.7 *Alkoxycarbonyl protecting groups for sulfur-containing amino acids*

The sulfur terminus of cysteine is protected reversibly by easily removable protecting groups of the thiocarbonate type. Cysteine can also be protected simultaneously at the thiol group and at the amino function (Equation (67)). To prevent oxidation to the disulfide, the sodium salt of the thiol group is dissolved in degassed oxygen-free water ⟨90MI 615-01⟩. The allyloxycarbonyl residue can also be introduced as a side chain protection for the SH group via the chloroformate (Equation (68)) ⟨93MI 615-01⟩. Other protecting groups such as isobutyloxycarbonyl are similarly introduced with success to form the thiocarbonate from sulfur-containing amino acids ⟨92JC241⟩.

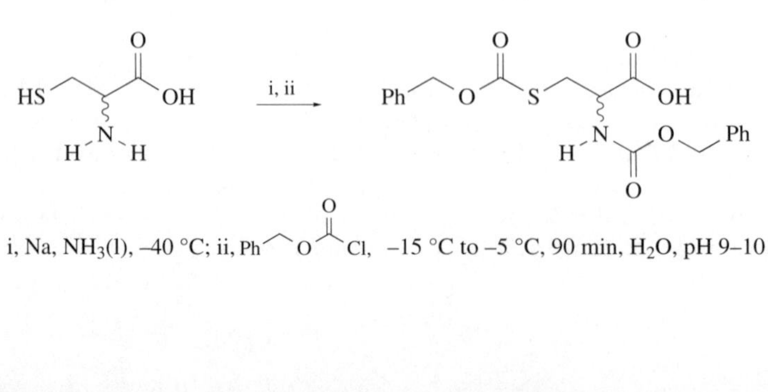

(67)

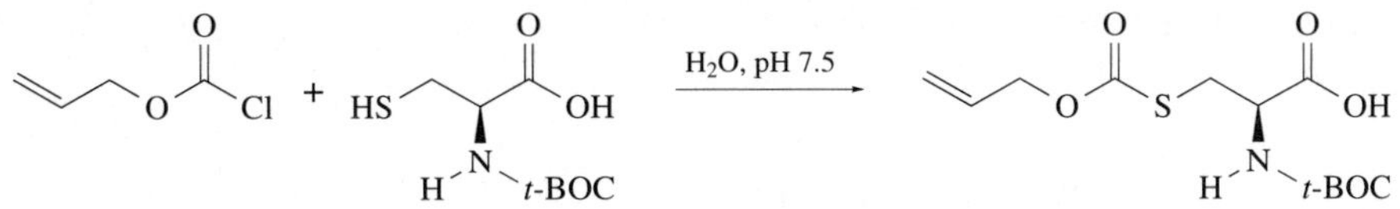

(68)

t-BOC = *t*-butoxycarbonyl

6.15.2.2 Oxygen and Selenium or Tellurium Functions

Selenocarbonates are prepared in a convenient one-pot synthesis as described by Segi *et al.* The reaction of bis(trimethylsilyl) selenide with Bu^nLi followed by alkylation with an alkyl halide gives alkyl silyl selenides. The repetition of a similar one-pot reaction with another mole of Bu^nLi, and the appropriate chloroformate yields unsymmetrical selenocarbonates. By this route *Se*-hexyl *O*-ethyl selenocarbonate, *Se*-benzyl *O*-ethyl selenocarbonate and *Se*-hexyl *O*-phenyl selenocarbonate have been prepared in yields from 33% to 75% (Equations (69)–(72)) ⟨89CL1009⟩.

$$\text{TMS-SeMe}_3 \xrightarrow[0\,^\circ\text{C, 30 min, THF}]{Bu^nLi} \text{LiSe-TMS} \qquad (69)$$

$$\text{LiSe-TMS} + R^1X \xrightarrow{30\,^\circ\text{C, 2 h}} R^1\text{Se-TMS} \qquad (70)$$

$$R^1\text{Se-TMS} \xrightarrow[0\,^\circ\text{C, 30 min}]{Bu^nLi} R^1\text{SeLi} \qquad (71)$$

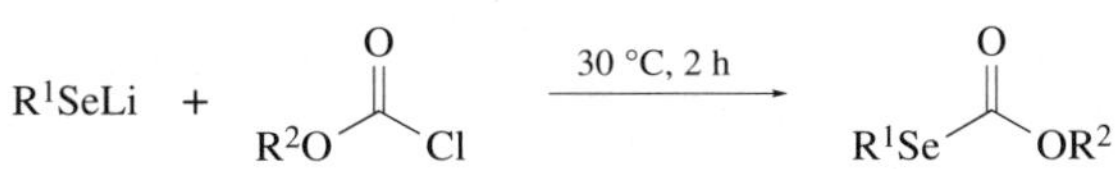

(72)

$R^1 = C_6H_{13}$, $PhCH_2$
R^2 = Et, Ph

Alternatively, the reaction of Bu^nLi and selenium gives lithium *n*-butyl selenide. The latter reacts in a further step with ethyl chloroformate to produce the selenocarbonate in 15% yield (Scheme 18) ⟨75BCJ108⟩. Instead of the lithium salt, a sodium alkyl selenide can be used. It is prepared from an alkyl selenol and elemental sodium. The subsequent step is the reaction with alkyl chloroformates as described above (Scheme 19) ⟨84ZAAC(513)183⟩.

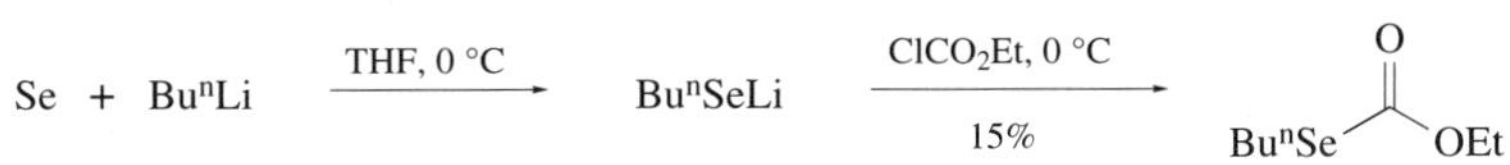

Scheme 18

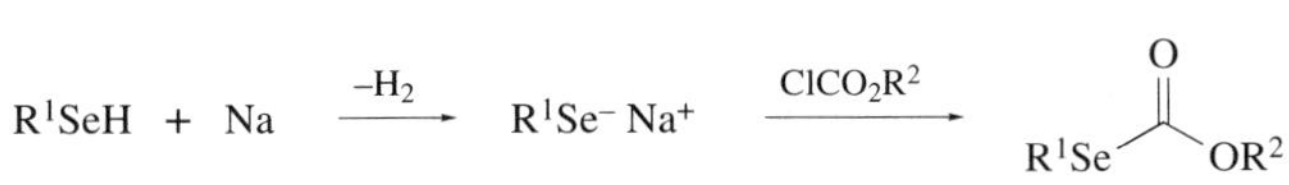

Scheme 19

Selenocarbonates with long-chain alkyl groups are prepared by reacting didecyl diselenide, sodium borohydride, and methyl chloroformate. The tellurocarbonates can be obtained in the same way ⟨91JCS(P2)501⟩.

Tellurodicarbonic acid diesters are obtained as reasonably stable oily compounds by reacting alkyl chloroformates with sodium telluride, prepared *in situ* from sodium hydride and tellurium in dry DMF, under phase-transfer conditions. The yields of diesters range from 39% to 69% (Equation (73)) ⟨89BCJ2117⟩.

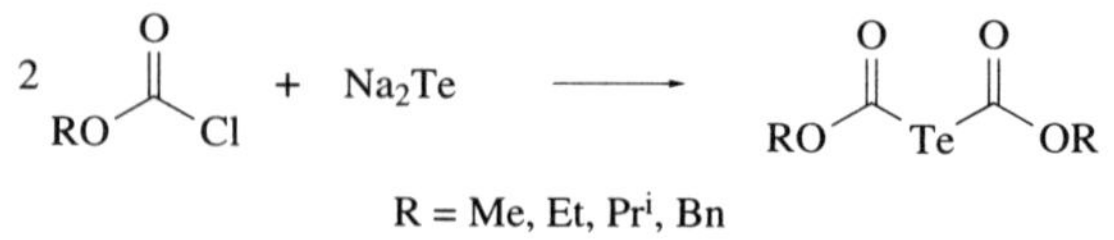

(73)

6.15.2.3 Other Dissimilar Chalcogenide Functions

6.15.2.3.1 Oxygen and selenium functions

Se-Decyl *O*-methyl carbonate is prepared by reacting elemental selenium, hydrazine hydrate, and 1-bromodecane to yield didecyl selenide, which reacts in a further step with methyl chloroformate to afford the product in 69% yield (Scheme 20) ⟨91JCS(P2)501⟩.

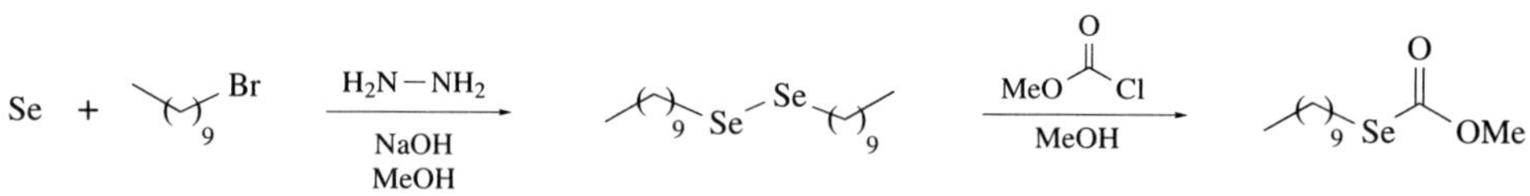

Scheme 20

6.15.2.3.2 Oxygen and tellurium functions

Te-Decyl *O*-methyl carbonate can be prepared (30%) from elemental tellurium by its reduction with sodium borohydride and alkylation with 1-bromodecane, followed by further reaction with methyl chloroformate (Equation (74)) ⟨91JCS(P2)501⟩. Tellurodicarbonic acid diesters can be synthesized by reacting alkyl chloroformates with sodium telluride in the presence of tetrabutylammonium bromide. The products tellurodicarbonic acid dimethyl-, diethyl-, diisopropyl-, and dibenzyl esters are obtained in reasonable yields of 39–69% ⟨89BCJ2117⟩ (Equation (75)).

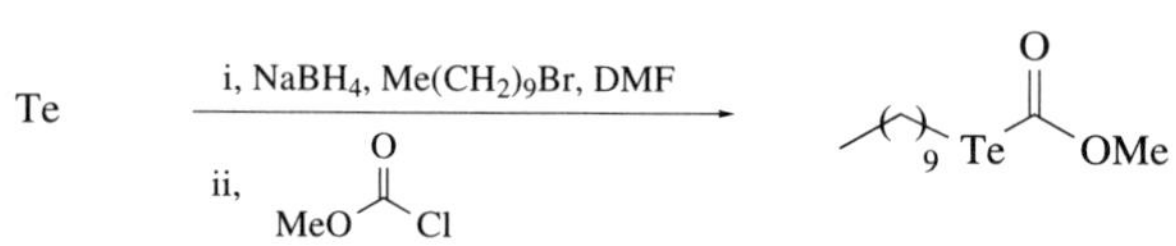

(74)

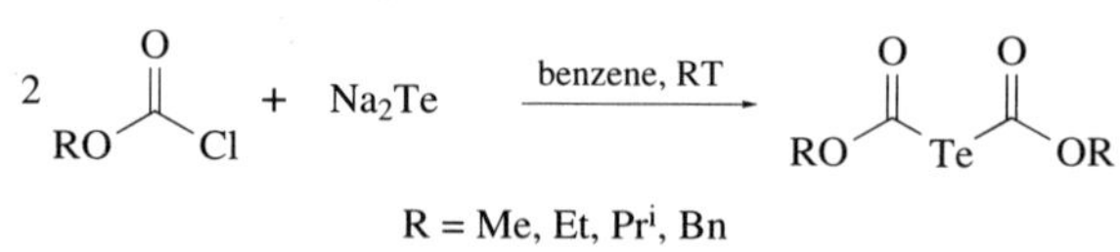

(75)

6.15.2.3.3 Sulfur and selenium functions

S,Se-Dialkyl thioselenocarbonates are prepared in an experimental route starting with the alkylselenol, as described by Sturm and Gattow. Sodium ethyl selenide undergoes reaction with *S*-ethyl thiocarbonate to afford *S,Se*-diethyl thioselenocarbonate in 46% yield (Equation (76)) ⟨84ZAAC(514)120⟩.

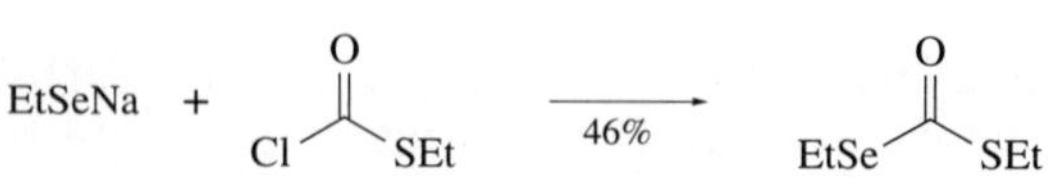

(76)

S,Se-Dialkyl thioselenocarbonates are also prepared using mercury reagents. A typical procedure uses trifluoromethyl bromothioformate and bis(trifluoromethylseleno)mercury. The mixture is allowed to react to 130 °C over 16 h. *S,Se*-Bis(trifluoromethyl) thioselenocarbonate is obtained in 91% yield (Equation (77)) ⟨78CB2891⟩.

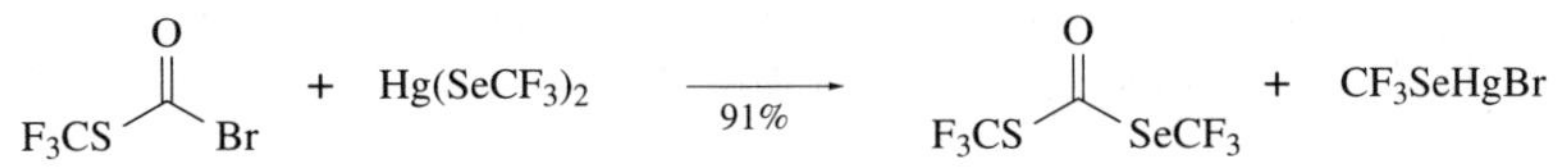

(77)

6.15.3 CARBONYL CHALCOGENIDES WITH A CHALCOGEN FUNCTION AND ONE OTHER HETEROATOM FUNCTION

6.15.3.1 Oxygen and Nitrogen Functions

Urethanes are carbonic acid derivatives containing a carbonyl function directly connected to an alkoxy function and an amino function. They are important for producing pharmaceuticals and polymers, and therefore many preparative methods exist.

The formal constitution allows a great deal of variety in the alkoxy component as well as the basic amino function. In general two synthetic approaches to urethanes can be distinguished; the first method involves a carbonic acid derivative of an alcohol or of a phenol reacting with ammonia or an amine; the second method involves a carbonic acid derivative of ammonia or an amine reacting with an alcohol or a phenol.

6.15.3.1.1 Urethanes from chloroformates

Chloroformates are easily prepared using the readily available triphosgene bis(trichloromethyl)carbonate ⟨87AG(E)894⟩. Chloroformates react easily with ammonia or amines. *n*-Butyl chloroformate reacts in concentrated aqueous ammonia to afford the corresponding urethane, the excess ammonia absorbing the hydrochloric acid (Equation (78)) ⟨20JCS708⟩. This reaction can also be carried out in organic solvents such as ether or toluene. Of course, half of the ammonia, and therefore half of the amine, is lost as its hydrochloride; to prevent this, a tertiary amine, sodium carbonate or sodium hydroxide is used to absorb the HCl. For example, ethyl *N*-aziridinylurethane is prepared by reacting ethyleneimine and ethyl chloroformate in the presence of triethylamine (Equation (79)) ⟨50LA(566)229⟩.

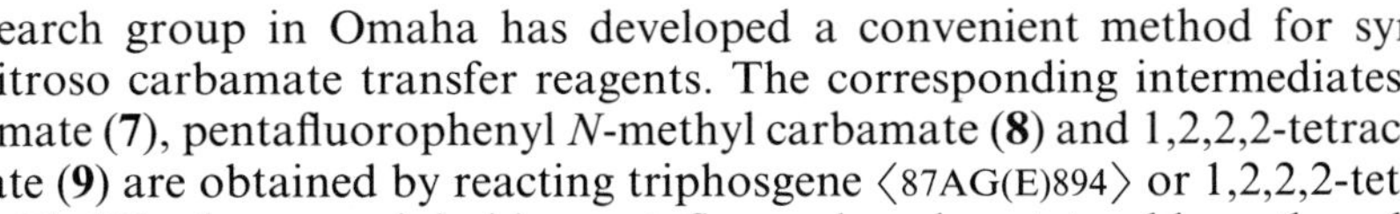

(78)

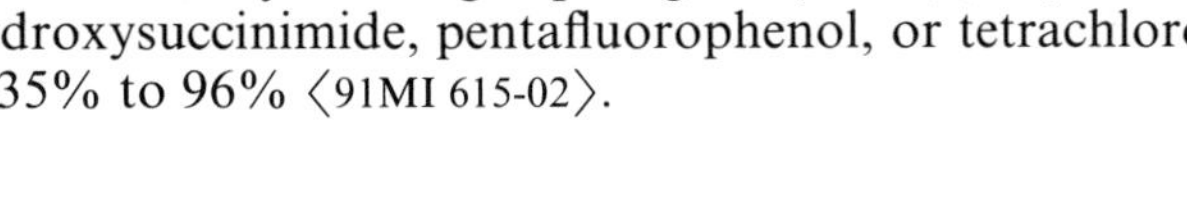

(79)

A cancer research group in Omaha has developed a convenient method for synthesizing *N*-[^{3}H]methyl *N*-nitroso carbamate transfer reagents. The corresponding intermediates succinimidyl *N*-methyl carbamate (**7**), pentafluorophenyl *N*-methyl carbamate (**8**) and 1,2,2,2-tetrachloroethyl *N*-methyl carbamate (**9**) are obtained by reacting triphosgene ⟨87AG(E)894⟩ or 1,2,2,2-tetrachloroethyl chloroformate with *N*-hydroxysuccinimide, pentafluorophenol, or tetrachloroethanol with methylamine. Yields vary from 35% to 96% ⟨91MI 615-02⟩.

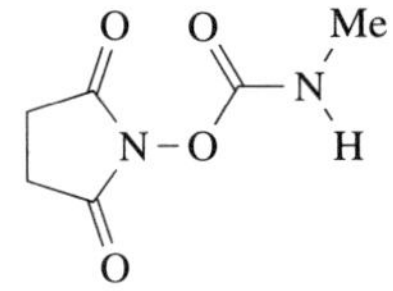

(**7**)

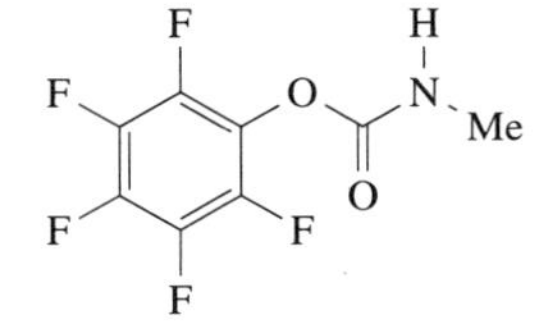

(**8**)

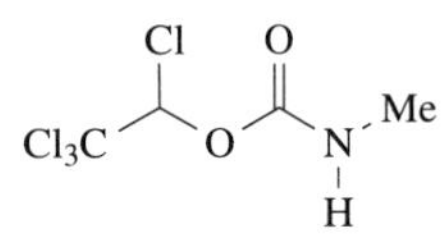

(**9**)

6.15.3.1.2 Urethanes from dialkyl carbonates

Urethanes are also available from dialkyl carbonates by partial aminolysis. This procedure is advantageous if mild reaction conditions are necessary. The components, for example, diethyl carbonate and ethylenediamine, are mixed and heated gently to afford ethyl *N*-2-aminoethylcarbamate (Equation (80)) ⟨34GEP676049⟩.

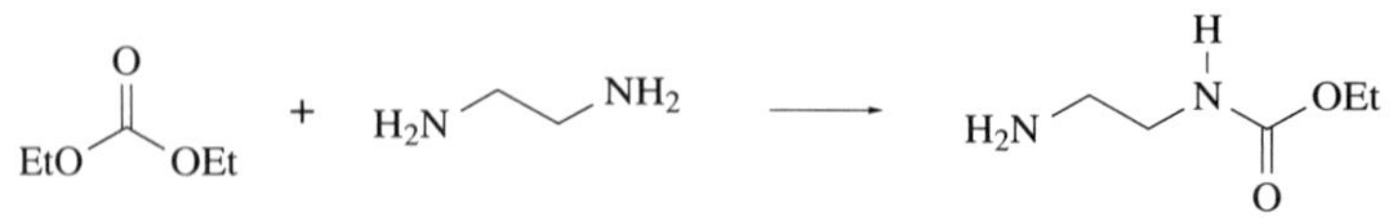
(80)

Urethanes are also available from cyclic carbonates. Einhorn obtained catechol carbonic acid diethylamide from catechol carbonate and diethylamine (Equation (81)) ⟨1898LA(300)145⟩.

(81)

Aminolysis of ethyl isopropenyl carbonate with 2-phenylethylamine affords *O*-ethyl *N*-2-phenylethylurethane in excellent yield. The advantages of this method are the mild reaction conditions, the high yield, the absence of an additional base, and the formation of acetone as the only by-product (Equation (82)) ⟨78TL3737⟩.

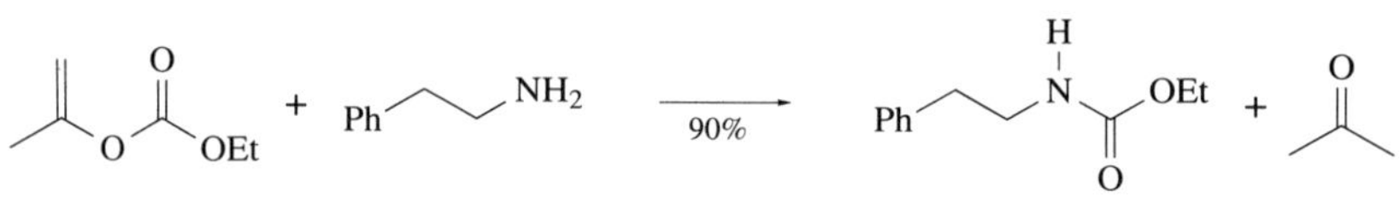
(82)

6.15.3.1.3 Urethanes from urea

Urea is suitable as a cheap starting material for *N*-unsubstituted urethanes. If urea and alcohols are heated at high temperature, the corresponding urethanes are obtained in good yields. In this way carbamic acid benzyl ester can be obtained in yields of 87% by heating benzyl alcohol and urea in the presence of $Zn(OAc)_2$ for 8 h at 150 °C (Equation (83)) ⟨46ZN(B)518⟩. Urea salts, especially urea nitrates, can also be used to obtain the urethanes ⟨1899GEP114396⟩.

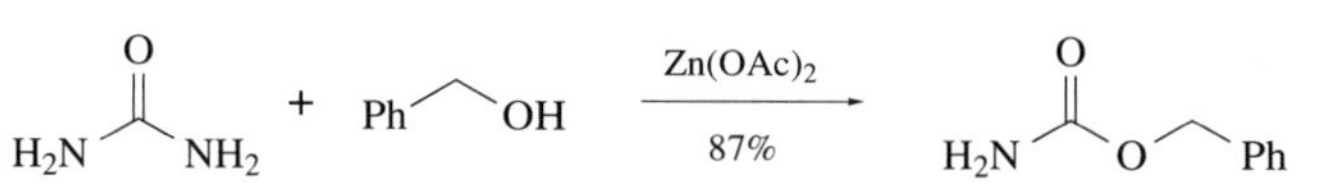
(83)

6.15.3.1.4 Urethanes from isocyanates

Isocyanates react easily and mostly quantitatively with alcohols, particularly primary alcohols, to produce urethanes (Equation (84)). However phenolic hydroxy functions react slowly and only in fair yields. For this reason, urethanes containing phenolic groups are prepared more advantageously by the chloroformate route ⟨B-31MI 615-01⟩.

(84)

Urethanes can also be formed if Hofmann degradation, Curtius degradation, or Lossen rearrangement is performed in the presence of alcohols or phenols.

The cephalosporin derivative (**10**) was prepared in a study of the effect of varying the substituent at C-3. The urethane is prepared by reaction of the 3-desacetyl cephalosporin derivative with benzyl isocyanate (88% yield) ⟨92JMC3731⟩.

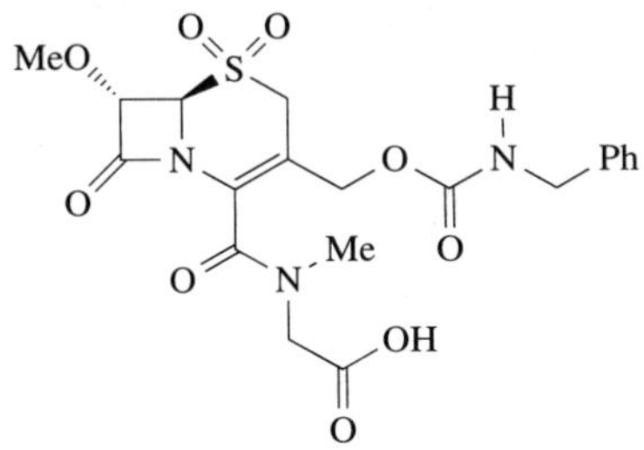

(10)

6.15.3.1.5 *Urethanes from isocyanides*

If isobutyraldehyde and cyclohexyl isocyanide are added to a solution of *n*-butylamine in methanol which is saturated with carbon dioxide, and the solution is then allowed to stand for 20 h at about 20 °C, the urethane can be isolated in almost quantitative yield (Scheme 21) ⟨B-71MI 615-01⟩. An analogous four-component condensation (4CC) is observed with benzylamine, *n*-butyraldehyde, and cyclohexyl isocyanide.

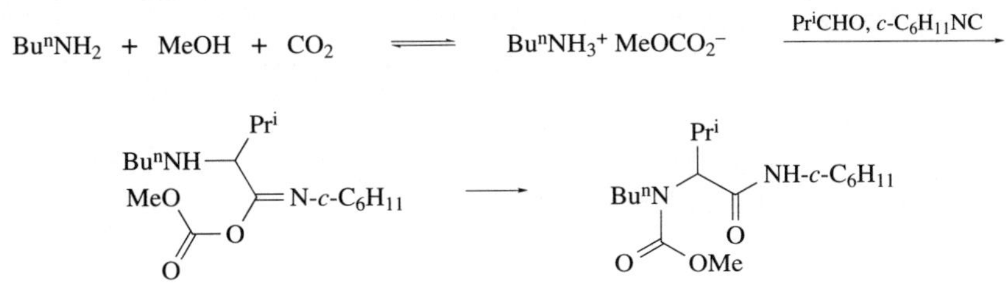

Scheme 21

6.15.3.1.6 *Oxazolidones (cyclic urethanes)*

Easy access to oxazolidones is given by reacting ethanolamine with potassium cyanate in an aqueous solution to afford 2-hydroxyethylurea, which cyclizes on distillation to the desired product (Scheme 22) ⟨03CB1280⟩. The hydroxyethylurea can also be prepared using methods described above.

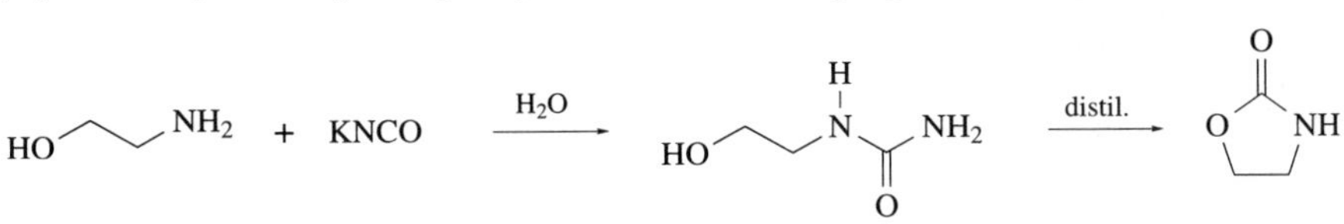

Scheme 22

2-Bromoethoxycarbonyl-protected amino acids, glycine or L-valine, can generate 2-oxazolidone *N*-acetic acid (**11**) or 2-oxazolidone *N*-isovaleric acid (**12**) under very strongly alkaline conditions, like 1 mol l^{-1} sodium hydroxide, as described by Eckert and Ugi ⟨79LA278⟩.

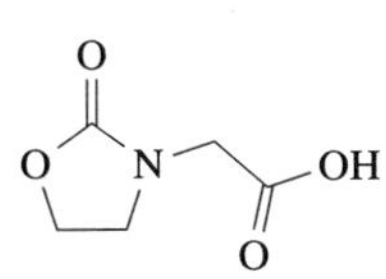

(11)

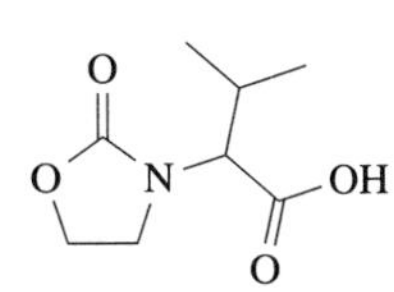

(12)

Oxazolidin-2-ones can be prepared in a simple one-step procedure using triphosgene ⟨87AG(E)894⟩. L-Serine reacts with 1/3 equivalent of triphosgene at room temperature in dioxan and sodium hydroxide to yield 67% of 4-carboxyoxazolidin-2-one (Equation (85)) ⟨93SC2839⟩. 2-Amino-5-methoxyphenol can similarly be converted into 6-methoxybenzoxazolinone (Equation (86)) ⟨89S875⟩.

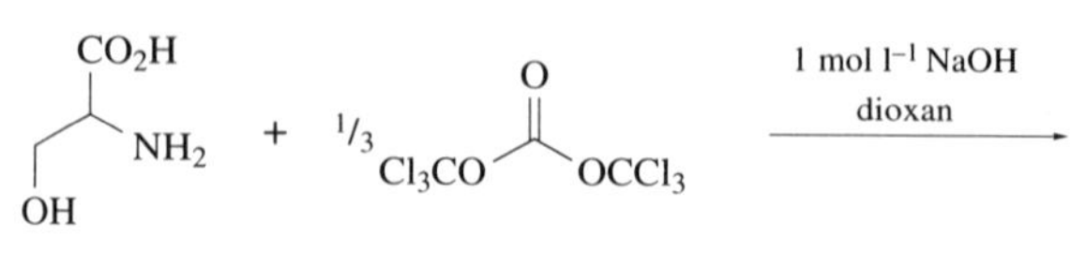

(85)

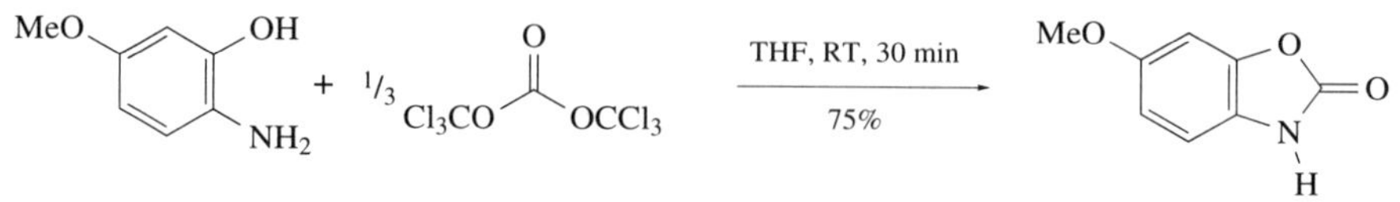

(86)

6.15.3.1.7 *N-Carboxy α-amino acid anhydrides*

N-Carboxy α-amino acid anhydrides (NCAs), or Leuchs anhydrides ⟨06CB857, 74HOU(15/2)187⟩, constitute a special category of mixed anhydrides which achieve both amino group protection and carboxylate activation of α-amino acids simultaneously. The apparent advantage of the concurrent amine protection and carboxylate activation in NCAs is, however, counterbalanced by their high reactivity. These reagents are sensitive to moisture and are prone to polymerization. NCAs can be prepared by a facile one-pot reaction at room temperature. In a typical reaction a *N-t*-BOC-amino acid and triphosgene are stirred in ethyl acetate at room temperature. Triethylamine addition to the solution is accompanied by an instantaneous precipitation of triethylammonium chloride to afford an intermediate, which reacts with the loss of carbon dioxide within 2–20 h depending on the amino acid to give the desired product. By this method the NCA of glycine is obtained in a yield of 83% (Scheme 23) ⟨92JOC2755⟩.

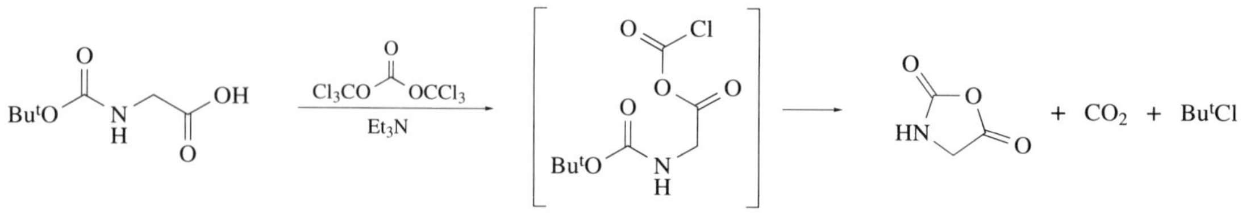

Scheme 23

Daly and co-workers describe a preparative route which also uses triphosgene. Treatment of a suspension of L-phenylalanine in anhydrous tetrahydrofuran with 1/3 equivalent of triphosgene at 40–50 °C leads to a completely homogeneous solution affording L-phenylalanine NCA within 3 h in a yield of 83% (Equation (87)) ⟨88TL5859⟩.

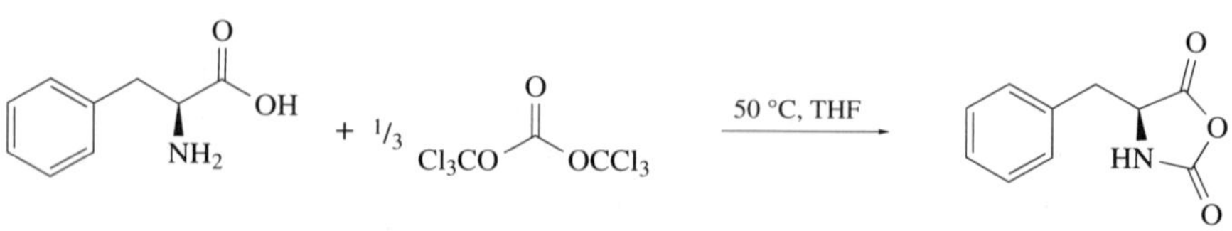

(87)

6.15.3.1.8 *Urethanes as protection for amino functions*

To establish chemoselectivity it is often necessary to block amino functions, particularly in peptide and nucleotide chemistry, by reversibly protecting them, mainly by protective groups of the urethane type. These can easily be introduced as well as cleaved by standard methods ⟨74HOU(15/1)46, B-82MI

615-02⟩. Usually benzyloxycarbonyl- (Z-) ⟨32CB1192⟩, *t*-butoxycarbonyl- (*t*-BOC-) ⟨57JA6180⟩, and fluorenylmethoxycarbonyl- (FMOC-) ⟨70JA5748⟩ residues are applied. In some cases, increased selectivity combined with particularly mild reaction conditions is achieved by special methods. Examples include reagents with β-haloalkyl groups ⟨76AG(E)681, 79LA278⟩ particularly the 2,2,2-trichloro-*t*-butoxycarbonyl- (TCBOC-) residue ⟨78AG(E)361⟩.

A comprehensive survey of carbamate protection of amino functions is available ⟨B-91MI 615-03⟩. The following examples illustrate recent uses of carbamate protection in synthesis.

In a synthetic approach to the alkaloid lycorine the amino function of an intermediate is protected by an ethoxycarbonyl residue, which is introduced by ethyl chloroformate in the presence of triethylamine ⟨90H(30)839⟩.

Water-soluble analogues of the important antitumor alkaloid camptothecin have been developed by SmithKline Beecham Pharmaceuticals. The amino group in a polyfunctional intermediate is protected by the Z-residue, introduced with benzyl chloroformate, as shown in Equation (88) ⟨91JMC98⟩.

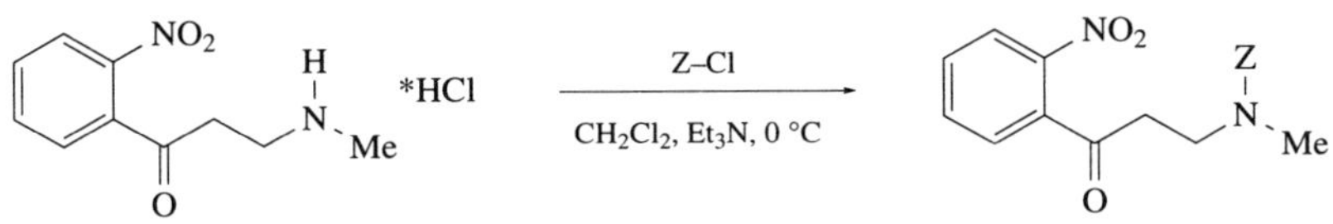

(88)

The problem of guanidino function protection has been taken up by Dodd and Kozikowski in the conversion of alcohols to guanidines. The latter are blocked by Z- or *t*-BOC-groups. Thus *N,N′*-bis(benzyloxycarbonyl)guanidine or *N,N′*-bis(*t*-butoxycarbonyl)guanidine react as nucleophiles in the Mitsunobu protocol with several different alcohols according to Equation (89), affording the corresponding alkylguanidine in excellent yields, in most cases more than 95% ⟨94TL977⟩.

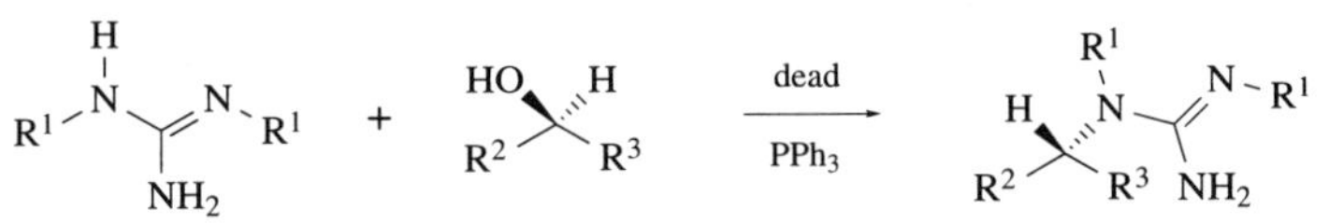

(89)

R^1 = *t*-BOC or Z

As part of a total synthesis of thymosin β_4, an optimized conventional synthesis of the *C*-terminal tridecapeptide fragment (31–43) is described by Link and Voelter ⟨93ZN(B)1000⟩. The *N*-terminal α-amino group is protected by Z- and the ε-amino functions of lysines by *t*-BOC-residues, as shown in the following sequence (31–43) of thyrosin β_4: Z-Lys(t-BOC)-Glu(O-Bu^t)-Thr(Bu^t)-Ile-Glu(O-Bu^t)-Gln-Glu(O-Bu^t)-Lys(t-BOC)-Gln-Ala-Gly-Glu(O-Bu^t)-Ser(Bu^t)-OBu^t.

t-BOC-Group protection is also useful in an efficient procedure for the synthesis of phosphoropeptides by the *t*-BOC mode solid phase method. A phosphothreonine-containing peptide related to the EGF receptor protein was synthesized by this methodology ⟨93CL1401⟩.

For alkylating both ends of tetraamines, reversible masking of all amino groups is necessary and can be performed by reaction with *t*-BOC-O-*t*-BOC ⟨91JMC569⟩.

Previously described protecting groups, Z, BOC, and FMOC residues are also used in the pharmaceutical and other industries as amino function blocking groups, as shown in some patents from the 1990s ⟨93EUP560730, 93EUP547699, 93MIP9305026⟩.

An efficient Leu-enkephalin synthesis using highly lipophilic ferrocenylmethyl (Fem), *t*-BOC, and TCBOC residues has been performed by Eckert *et al.* ⟨91ZN(B)339⟩. Introduction of the TCBOC-residue into the building block H-Fem-Gly-OMe is carried out with 2,2,2-trichloro-*t*-butyl chloroformate, yielding 73% product, as shown in Equation (90). The chloroformate TCBOC-Cl is easily obtained by reaction of 2,2,2-trichloro-*t*-butanol with triphosgene, as described by the same author ⟨87AG(E)894⟩.

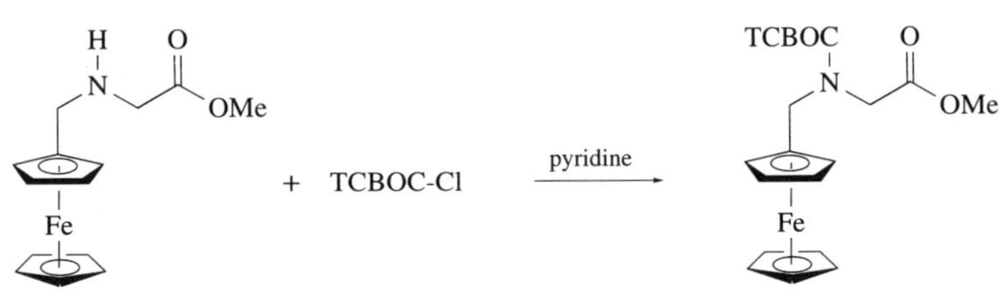

(90)

TCBOC = Cl_3CCMe_2OCO

The extremely selective and mild cleavage of β-haloalkoxycarbonyl groups makes them exceptionally useful in semisyntheses of highly sensitive penicillin and cephalosporin derivatives, as recently demonstrated by Eckert ⟨90ZN(B)1715⟩. Introduction of the 2,2-dibromopropoxycarbonyl group by means of its chloroformate at D-phenylglycine produces 90% of the urethane.

The TCBOC-residue is also suitable for amino function protection in nucleotide chemistry. Ugi and co-workers prepared all possible *N*-protected nucleosides with TCBOC-residue, as exemplified by the nucleic acid bases *N*-TCBOC-adenine **(13)** and *N*-TCBOC-uracil **(14)** ⟨83T2207, 85ACS(B)761⟩. A di-TCBOC-adenine derivative is also described by Ugi in the direct synthesis of cyclic AMP derivatives ⟨93ACS125⟩.

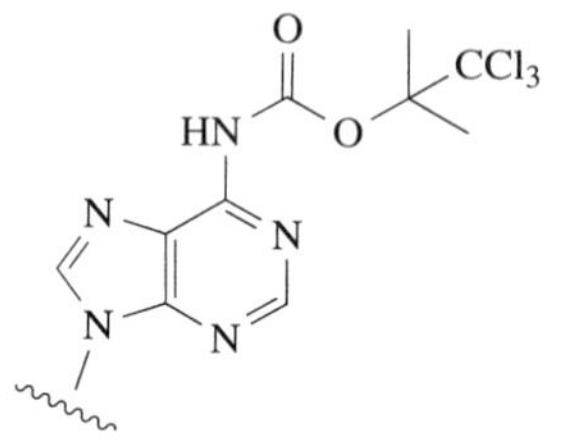

(13)

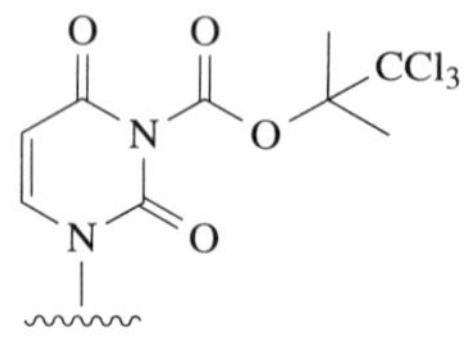

(14)

An example of another special protective group for amino function protection is the 2-trimethylsilylethoxycarbonyl residue TMS-CH_2CH_2OCO (TEOC). It proved useful in the key step of a highly stereoselective three component coupling in the first total synthesis of the antitumor antibiotic (+)-hitachimycin. The urethane intermediate is obtained by reaction with the corresponding chloroformate in excellent yield, according to Equation (91) ⟨92JA8008⟩.

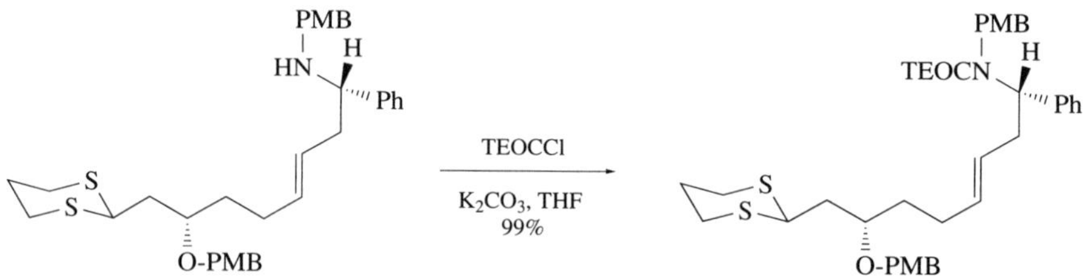

(91)

TEOC = 2-trimethylsilylethoxycarbonyl
PMB = *p*-methoxybenzyl

6.15.3.1.9 *Polyurethanes*

Polyurethanes are an important class of polymers ⟨B-89MI 615-01⟩. They are produced by polyaddition of diisocyanates with diols. Typical representative monomers are hexamethylene-1, 6-diisocyanate, phenylene-1,4-diisocyanate, diphenylmethane-4,4′-diisocyanate, and naphthalene-1,5-diisocyanate as the isocyanate component and 1,4-butanediol, 1,4-bis(3-hydroxypropyl)benzene, 1,4-bis(2-hydroxyethoxy)benzene, 2,2-bis[4-(2-hydroxyethoxy)phenyl]propane as the diol component. Patent literature in this field provides information on advances ⟨92GEP4215648, 92GEP4219418, 92JAP05339226, 92JAP05339542, 92JAP06033008⟩.

6.15.3.1.10 *Carbazates*

Generally carbazates can be divided into two groups: monoacylated and bisacylated derivatives. They are obtained simply by reacting alkyl or aryl chloroformates with hydrazine hydrate, or with alkyl- or arylhydrazine derivatives. Monocarbazates are formed exclusively by carrying out the above reactions in organic solvents such as chloroform or acetone under ice cooling ⟨83HOU(E4)273, 91TL10053, 93JA8898⟩ (Equation (92)).

(92)

Instead of chloroformates, unsymmetrically substituted *O*-alkyl *O'*-phenyl carbonates can be used; the phenoxy group activates the carbonate ester ⟨68HCA622, 84CJC574⟩ Equation (93)).

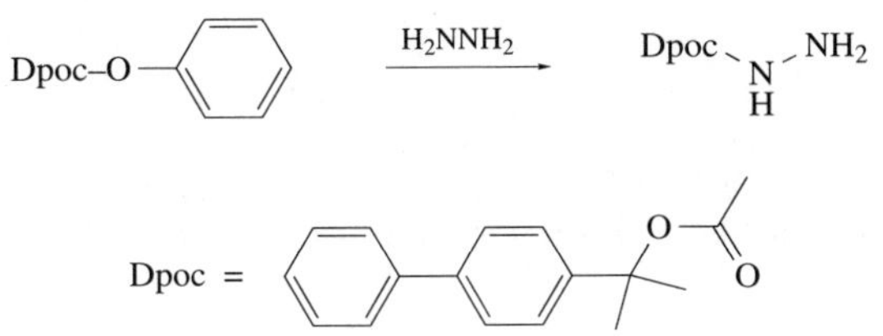

(93)

Another method for preparing monoacylated carbazates makes use of dialkyl dicarbonates; thus, 1-ethoxycarbonyl-1-methylhydrazine is formed by reacting methylhydrazine and diethyl carbonate ⟨83HOU(E4)274⟩ (Equation (94)).

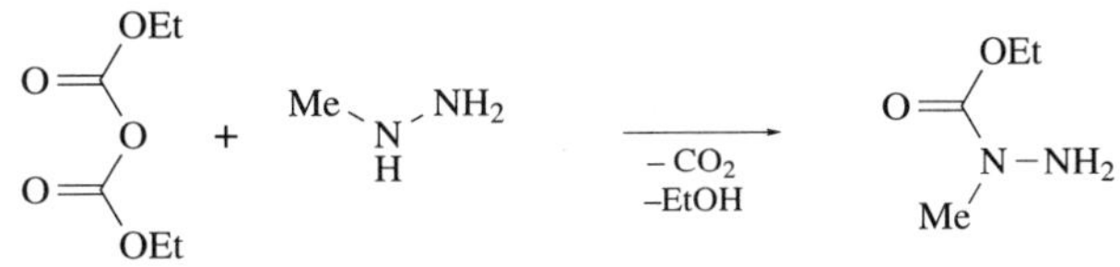

(94)

Bisacylated hydrazines, particularly the symmetrical ones, are obtained very easily by reacting chloroformates and hydrazine hydrate in aqueous solution ⟨66CB2039, 83HOU(E4)279⟩.

In aqueous solution dialkyl dicarbonates also react with hydrazine hydrate to give *N,N'*-bis-(alkoxycarbonyl)hydrazines. Also, mixed anhydrides prepared from potassium alcoholate, carbon dioxide, and dialkyl chlorophosphate react with hydrazine to give bisacylated hydrazines ⟨83HOU(E4)279⟩.

An acylated carbazate derivative is generated in the reaction of a diaziridinone with the sodium salt of malonitrile in addition to a pyrazoline and a tetraazaspirononane derivative ⟨92JOC7359⟩.

6.15.3.1.11 *Azidoformates*

Azidoformate esters, N_3CO_2R, can be prepared by either of the following general routes: (1) the diazotization of the corresponding carbazates NH_2NHCO_2R; (2) the reaction of chloroformates $ClCO_2R$ with sodium azide or a similar reagent ⟨83HOU(E4)284⟩. The chloroformate can be generated *in situ* from an alcohol and phosgene; this is illustrated by the conversion of a 17-hydroxysteroid into the corresponding chloroformate (Equation (95)) ⟨91JCS(P1)37⟩. Alkylcarbonic diethylphosphoric anhydrides, $RCOCO_2P(O)(OEt)_2$, have been used as alternatives to chloroformates; for example, a convenient preparation of *t*-butyl azidoformate, BOC-N_3, is the reaction of *t*-butylcarbonic diethylphosphoric anhydride with potassium azide ⟨88OSC(6)207⟩.

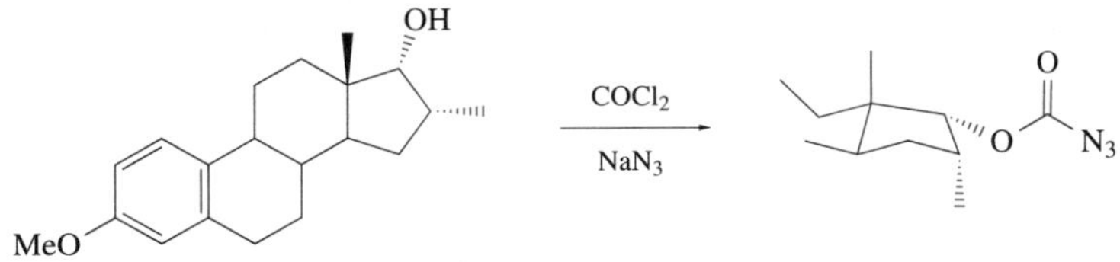

(95)

6.15.3.2 Oxygen and Phosphorus Functions

The main interest in phosphoryl *O*-alkyl carbonates is based on the potential biological activity of these phosphorus compounds ⟨88JMC1831, 90BBR(171)458, 93MI 615-03⟩. In particular, AZT phosphonates have been investigated as possible prodrugs in HIV-research ⟨90BBR(172)288, 91MI 615-02, 93MI 615-02⟩.

6.15.3.2.1 *Phosphinecarboxylates from alkali metal phosphides and carbon dioxide*

Kuchen and co-workers have prepared the lithium and sodium salts of diphenylphosphinecarboxylic acid from the easily available lithium and sodium diphenylphosphides and carbon dioxide ⟨91PS(60)287, 93PS(83)65⟩ (Equation (96)).

(96)

These compounds rapidly decompose in protic media, undergoing decarboxylation and formation of diphenylphosphine. Treatment of the phosphinecarboxylates with TMS-Cl leads to the trimethylsilyl ester, and with dimethyl sulfate to the methyl ester, whereas reaction with alkyl iodides gave only the dialkyldiphenylphosphine with liberation of carbon dioxide (Scheme 24).

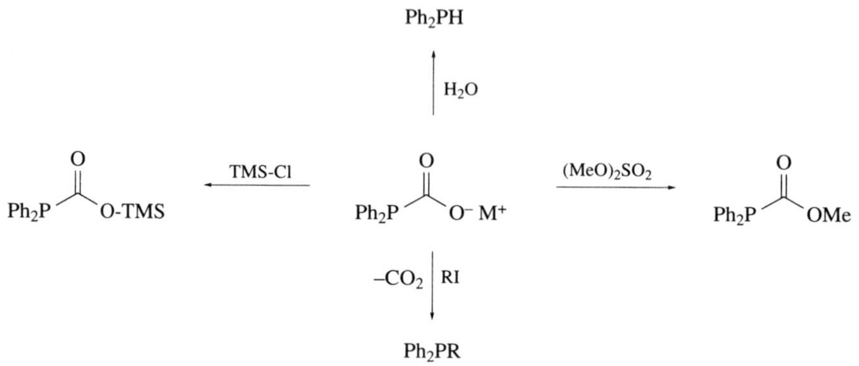

Scheme 24

Bubner and Balszuweit synthesized methyl di-*n*-butoxyphosphinecarboxylate oxide, $(BUO)_2P(O)CO_2Me$, in a similar way starting from sodium di-*n*-butyl phosphite ⟨87ZAAC(546)142⟩.

6.15.3.2.2 *Phosphinecarboxylates by the Arbuzov reaction and related methods*

Reaction of alkyl chloroformates with monoalkylphosphines affords monoalkylphosphinecarboxylates; for example, phenyl phenylphosphinecarboxylate is obtained in 88% yield ⟨81ZN(B)910⟩ (Equation (97)).

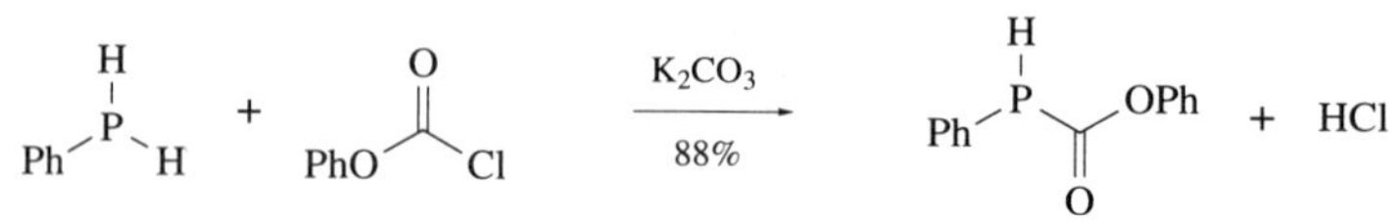

(97)

Alkyl dialkoxyphosphinecarboxylate oxides are prepared by reacting alkyl chloroformates and trialkyl phosphites in a typical Arbuzov reaction ⟨90BBR(171)288⟩ (Equation (98)). If alkoxy(methyl) phosphinecarboxylate oxides are desired, they can easily be prepared from dialkoxyphosphinecarboxylates by addition of methyl iodide ⟨88PS(35)329⟩ (Scheme 25).

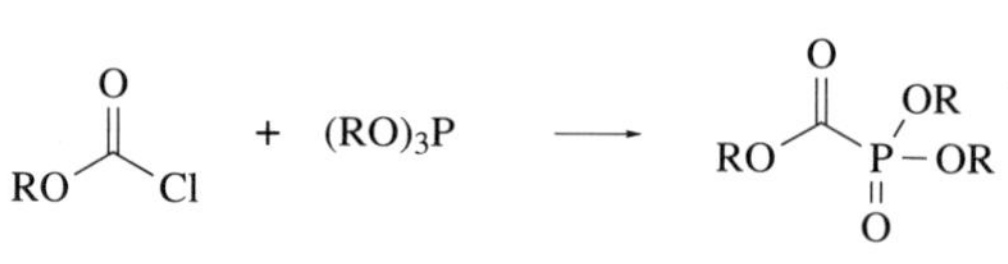

(98)

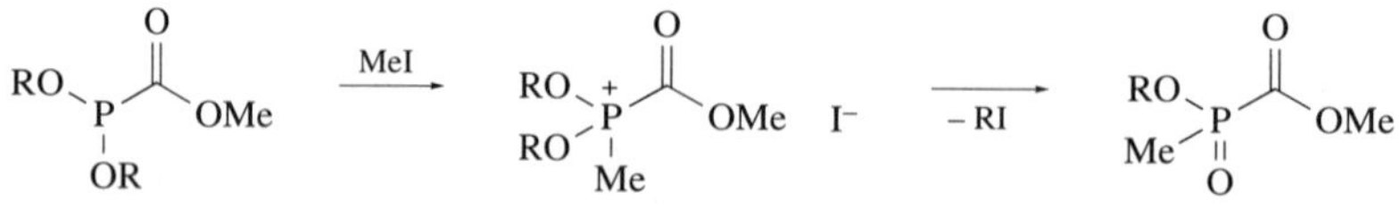

Scheme 25

Similarly, Prishchenko and co-workers have used methyl chloroformate instead of methyl iodide to obtain the phosphinedicarboxylate ⟨87JGU1484, 87JGU1260, 88PS(35)329⟩ (Equation (99)).

(99)

Iyer *et al.* produced bis(trimethylsilyloxy)phosphinecarboxylates by treatment of tris(trimethylsilyl) phosphite with alkyl chloroformates ⟨89TL7141⟩ (Scheme 26). In the same way, alkyl(trimethylsilyl)phosphinecarboxylates are prepared by the use of alkylbis(trimethylsiloxy)phosphines ⟨86ZAAC(542)37⟩. A great variety of silylated phosphinecarboxylates have been prepared by Issleib *et al.* ⟨85ZAAC(530)16⟩. These silylated compounds are suitable for further derivatization.

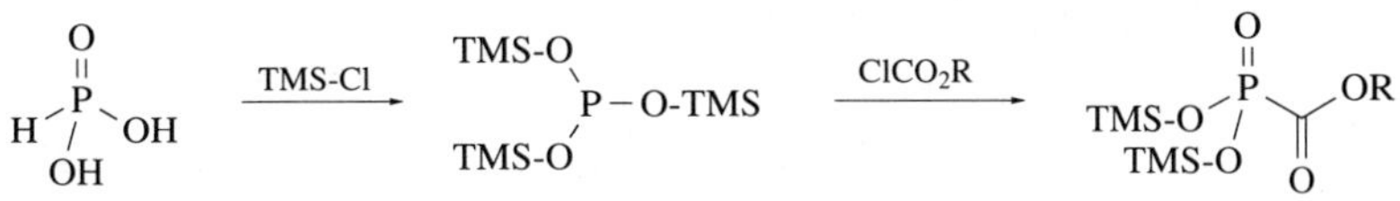

R = Me, Et, Ph, 2,4-$Cl_2C_6H_3$

Scheme 26

Acylation of the silylated phosphinosaccharide shown in Equation (100) by benzyl chloroformate is the best method for introducing the carboxylate function at phosphorus ⟨89CAR(194)209⟩.

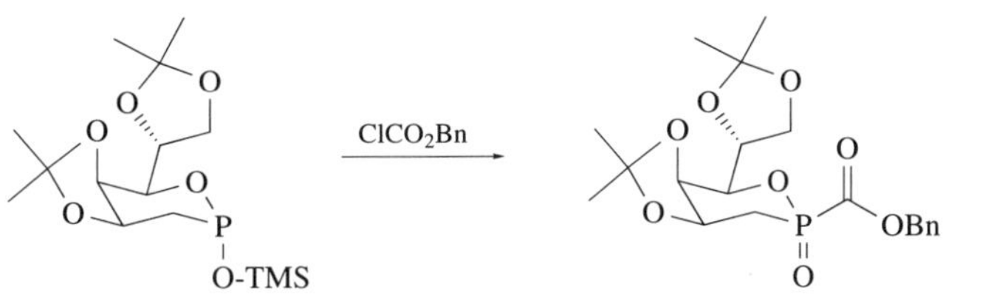

(100)

6.15.3.3 Oxygen and Other Heteroatom Functions

6.15.3.3.1 Oxygen and boron functions

Trimethylamine-carboxyborane is obtained by activating the cyano group in trimethylamine-cyanoborane by ethylation with triethyloxonium tetrafluoroborate, followed by hydrolysis of the resulting nitrilium salt with water (Scheme 27).

$$Me_3NBH_2CN + Et_3O^+ BF_4^- \xrightarrow[\Delta]{CH_2Cl_2} Me_3NBH_2CNEt^+ BF_4^- + Et_2O \xrightarrow{2\ H_2O} Me_3NBH_2CO_2H + EtNH_3^+ BF_4^-$$

Scheme 27

Similar amine-carboxyboranes are obtained by the same route ⟨79JINC1223, 84IC3063, 84IC4322, 88JOM(344)29, 89IS(25)79, 90IC554⟩.

Esterification of the alkylamine-carboxyboranes can be performed in different ways. Reacting trimethylamine-carboxyborane with methanol using dcc as coupling reagent yields 82% of the corresponding methyl ester ⟨86S833, 89IS79⟩ (Equation (101)).

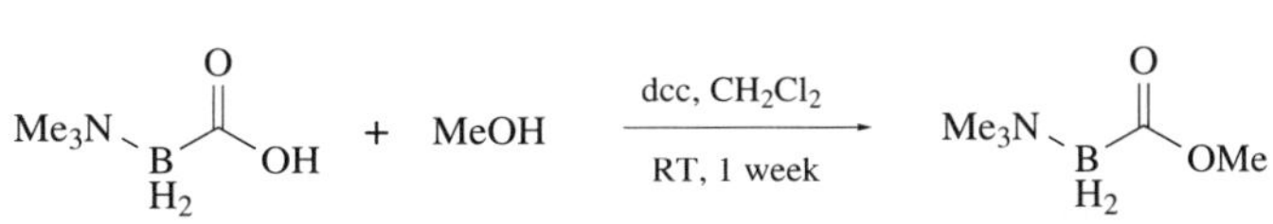

(101)

dcc = dicyclohexylcarbodiimide

To obtain the esters in good yields, alkylamine-carboxyboranes can be treated with the corresponding chloroformate of the alcohol. For example, trimethylamine-carboxy borane methyl ester is obtained (84%) by reacting trimethylamine-carboxyborane and methyl chloroformate with triethylamine ⟨89IS(25)79⟩ (Equation (102)).

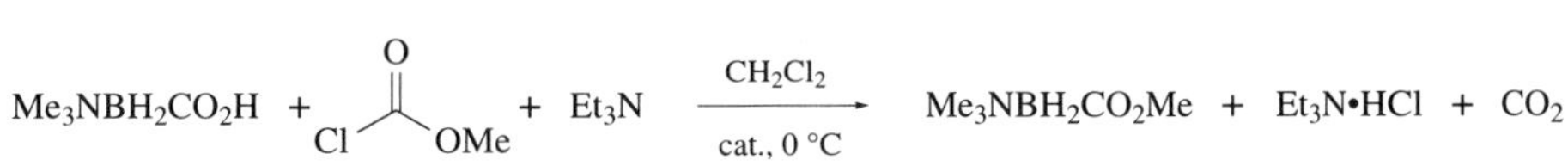

(102)

The ester can also be prepared by reacting trimethylamine-carboxyborane with trimethyl orthoformate in the presence of boron trifluoride etherate ⟨92S380⟩ (Equation (103)). Another route also starting with trimethylamine-carboxyborane uses *N,N'*-carbonyldiimidazole as the activating reagent to obtain trimethylamine-imidazole-carbonylborane, which then reacts with sodium ethoxide to give the corresponding ethyl ester in a yield of 42% ⟨85ZN(C)344⟩.

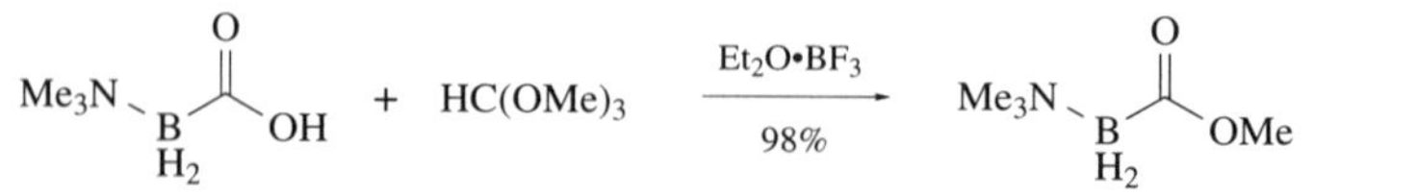

(103)

A method that differs from the previously described routes has been developed by Miller, who has successfully synthesized the 2-carboxylic acid derivative of 1,1,4,4-tetramethyl-1,4-diazonia-2,5-diboratacyclohexane ⟨91IC2228⟩ (Scheme 28).

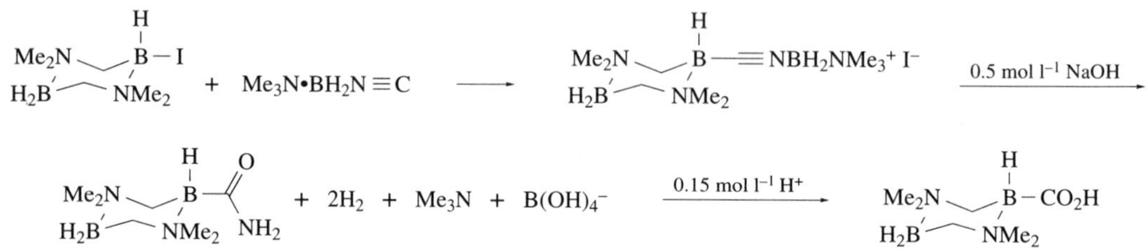

Scheme 28

Alkylphosphine-carboxyboranes can be obtained in a similar manner. Triphenylphosphine-carboxyborane and tricyclohexylphosphine-carboxyborane both are obtained by reacting the corresponding cyanoborate with triethyloxonium tetrafluoroborate followed by hydrolysis with water to yield the products in 30% and 75% yields, respectively ⟨81JINC457, 92AP(325)267⟩ (Equation (104)).

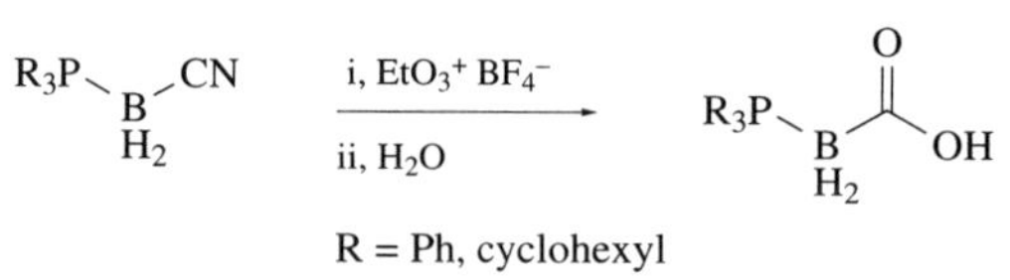

(104)

Triethyl phosphite-carboxyborane is obtained by a different hydrolysis procedure starting from triethyl phosphite-cyanoborane as outlined in Scheme 29 ⟨91T6915⟩.

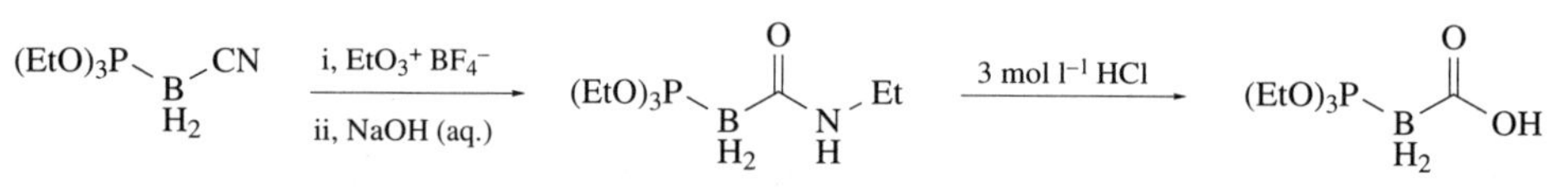

Scheme 29

6.15.3.4 Sulfur and Nitrogen Functions

6.15.3.4.1 Thiocarbamates from chlorothioformates and amines

The synthesis and use of disubstituted thiocarbamates has been well-documented in the literature ⟨B-70MI 615-01, B-77MI 615-01, B-82MI 615-01⟩.

Simple thiocarbamates, such as *N,S*-diethyl thiocarbamate and *S*-ethyl *N*-phenyl thiocarbamate, are readily obtained from the reaction of the appropriate amine with ethyl chlorothioformate (Equation (105)) ⟨85JOC5879⟩.

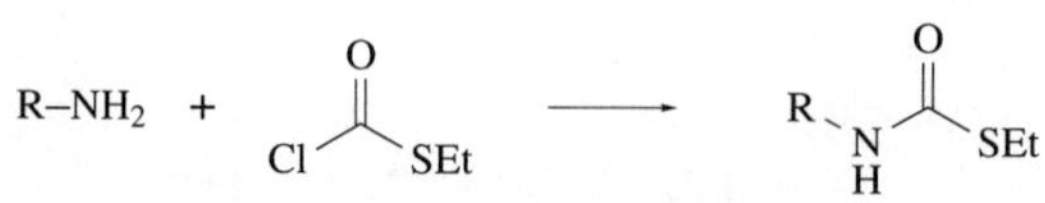

(105)

6.15.3.4.2 Thiocarbamates from carbamoyl chlorides and thiols

Another method is based on the treatment of thiols with carbamoyl chlorides in the presence of bases or alkali metal thiolates. In an example reported in 1989, Threadgill and Gledhill converted *N*-*t*-BOC-cysteine methyl ester into the *S*-(*N*,*N*-dimethylcarbamoyl)cysteine derivative in excellent yield by reaction with dimethylcarbamoyl chloride and pyridine (Equation (106)) ⟨89JOC2940⟩.

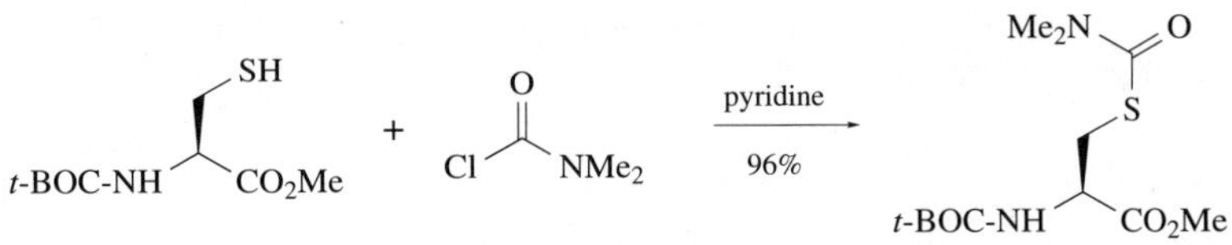

(106)

A special synthesis starting from dialkylcarbamoyl chlorides and episulfides in equimolar amounts at 100–120 °C in the presence of triethylamine or pyridine as catalyst leads to *S*-2-chloroalkyl *N*,*N*-dialkyl thiocarbamates (in analogy to similar epoxide reactions) in yields of 50–80% ⟨68ZOR1661⟩ (Equation (107)).

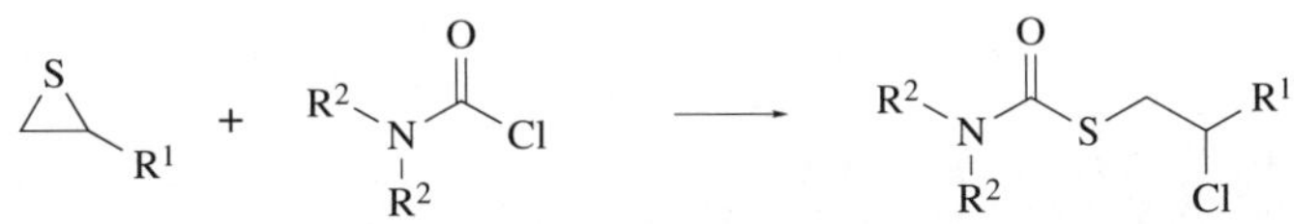

(107)

D'Amico and Schafer reported an alternative synthesis of disubstituted thiocarbamates by using potassium alkyl or benzyl dithiocarbonates ⟨80PS(8)301⟩. Treatment of these dithiocarbonates with disubstituted carbamoyl chlorides affords disubstituted thiocarbamates (Scheme 30).

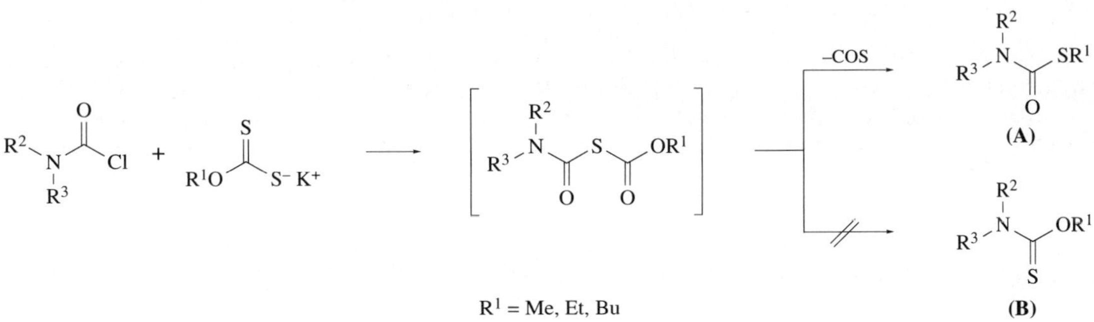

Scheme 30

In all probability a mixed anhydride is formed but immediately decomposes to give carbonyl sulfide and the disubstituted thiocarbamate (**A**). The proposed nucleophilic displacement mechanism is depicted in Scheme 31.

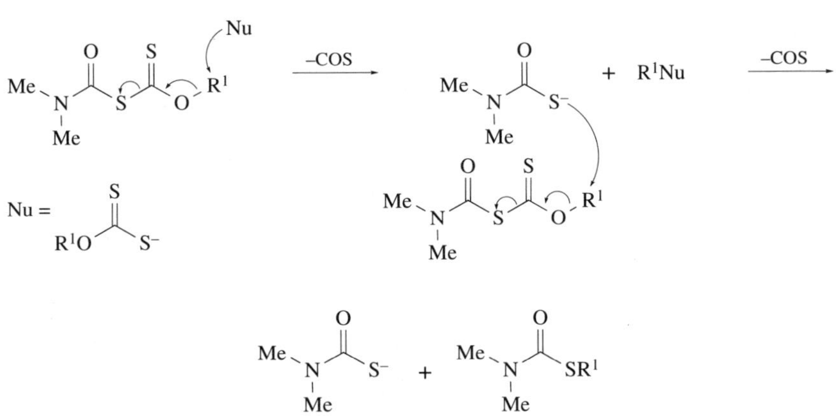

Scheme 31

6.15.3.4.3 *Thiocarbamates from isocyanates and thiols*

A widely used route to thiocarbamates is the reaction of isocyanates with thiols (Equation (108)). In comparison with alcohols, thiols are less reactive. Thus the use of higher temperatures or of catalysts such as triethylamine ⟨57JA366⟩, dabco, diacetoxydibutyl stannate ⟨79GEP2921130⟩, or triton B (benzyltrimethylammonium hydroxide) ⟨80AP(313)995, 81AP(314)315⟩ is necessary.

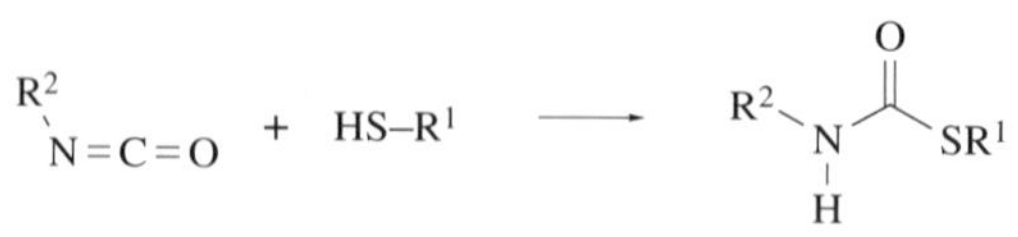

(108)

Threadgill and Gledhill synthesized *S*-(*N*-alkylcarbamoyl) derivatives of cysteine, cysteinylglycine and glutathione by this route ⟨89JOC2940⟩. Thus *N*-*t*-BOC-cysteine methyl ester was carbamoylated smoothly using the appropriate isocyanate in dichloromethane (Equation (109)).

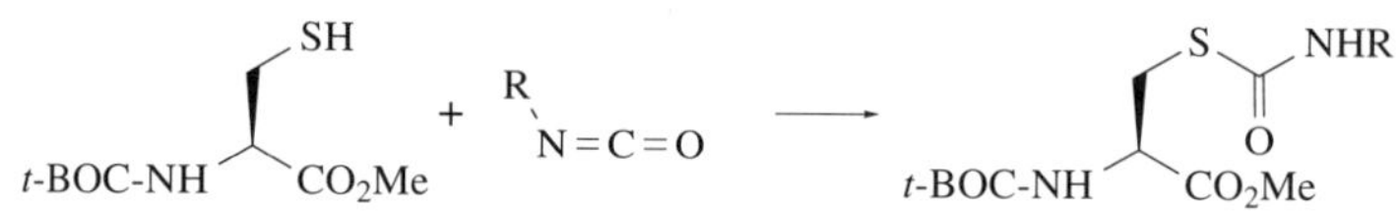

(109)

Cysteinylglycine and glutathione derivatives were prepared in the same way. Many of the deprotected *S*-carbamoylamino acids and peptides are metabolites of the corresponding *N*-alkyl-formamides in rodents and in humans.

6.15.3.4.4 *Thiocarbamates from* 1-*chlorothioformimidates*

Hydrolysis of chlorothioformimidates is another way of preparing thiocarbamates ⟨66AG210⟩ (Equation (110)). The methods of preparation of the precursors are described in Section 6.20.1.3.2.

In an anhydrous medium these alkylimino chloromethyl sulfides give, after addition of carboxylic acid and base, *N*-acylated thiocarbamates (Scheme 32) ⟨82JCS(P1)2813⟩. These are formed by acyl migration from oxygen to nitrogen.

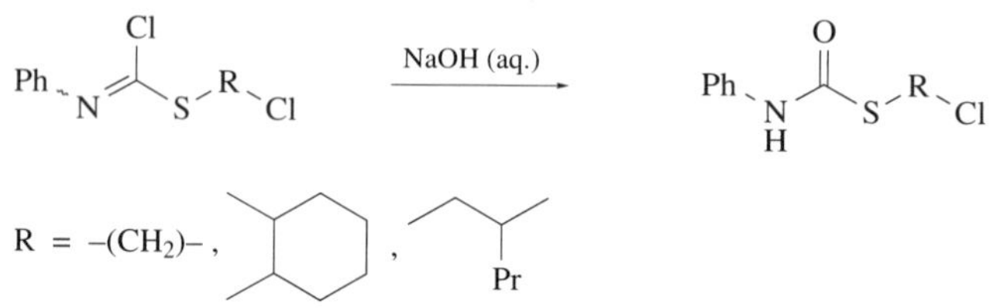

(110)

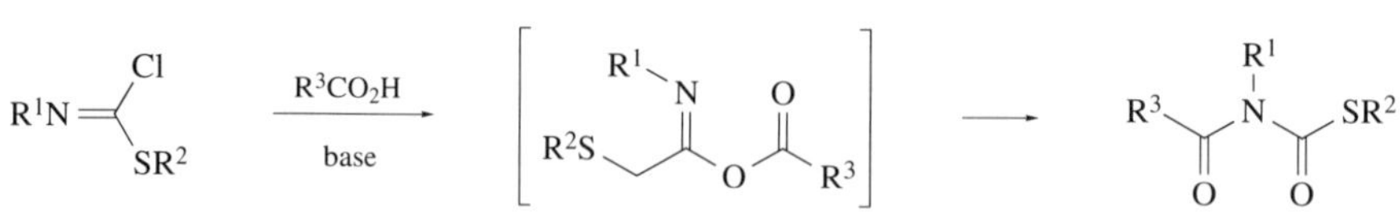

Scheme 32

6.15.3.4.5 *Thiocarbamates from alkylamide salts, carbon monoxide, and sulfur*

An efficient synthesis of *S*-alkyl thiocarbamates was developed by Mizuno *et al.* ⟨91TL6867⟩. The key step of this synthetic method is conversion of a carbamoyllithium species, generated from a lithium dialkylamide and carbon monoxide, into the lithium thiocarbamate by addition of elemental sulfur. In the subsequent step, *S*-alkylation with alkyl halides affords various *S*-alkyl thiocarbamates in yields of 55–99%. The reaction provides a useful method for the synthesis of *S*-alkyl thiocarbamates because of the mild reaction conditions, good yields and easy availability of the reagents (Scheme 33).

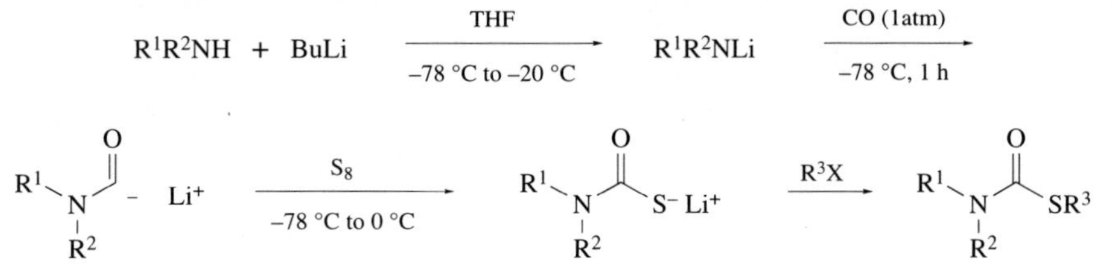

Scheme 33

In an earlier work, Mizuno prepared ammonium thiocarbamates in the presence of selenium as catalyst to obtain *S*-alkyl thio-carbamates ⟨89AG(E)452⟩.

The use of selenium as a catalyst for the synthesis of *S*-alkyl thiocarbamates was described by Koch ⟨75TL2087⟩, who treated primary amines with disulfides and carbon monoxide in the presence of selenium (Equation (111)). The conditions required are mild and aromatic amines can be used provided that triethylamine is added as a cocatalyst. In the catalyzed system a selenocarbamate intermediate may be involved similarly to the method described by Mizuno and co-workers (Equation (112)) ⟨89AG(E)452⟩.

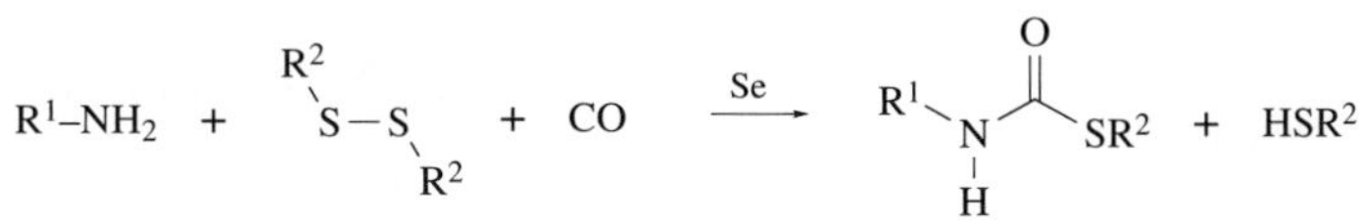

(111)

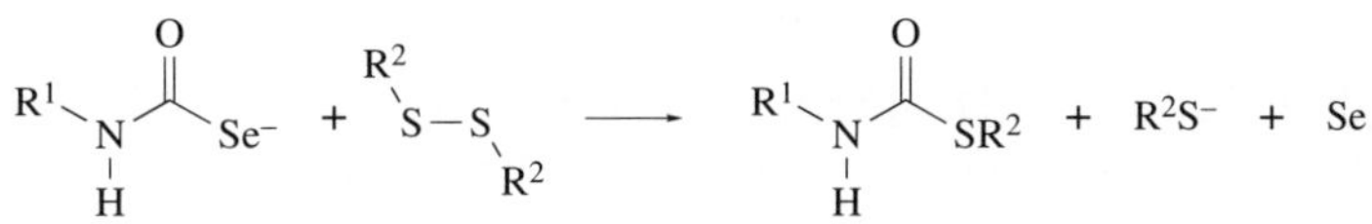

(112)

6.15.3.4.6 *Thiocarbamates by [3,3]-sigmatropic rearrangement*

Various thiocarbamates were prepared by Koch. An elegant path to *S*-aryl thiocarbamates is the thermal rearrangement (Newman–Kwast rearrangement) of *O*-aryl thiocarbamates (Equation (113)).

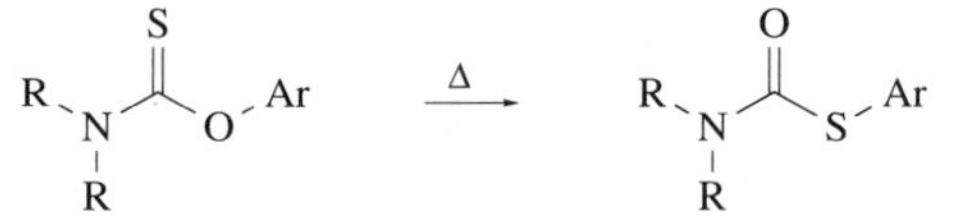

(113)

This reaction is formulated as an intramolecular four-membered ring process ⟨71BCJ1393⟩, whereas thermal treatment of *O*-allyl thiocarbamates to S-allylthiocarbamates is described as a [3,3]-sigmatropic rearrangement (Equation (114)) ⟨79CL1361⟩.

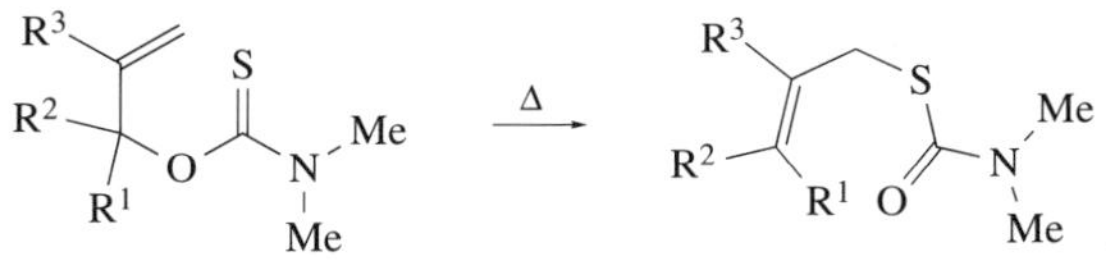

(114)

The thiocarbamates are derived from allylic alcohols after treatment with sodium hydride and *N*,*N*-dimethylthiocarbamoyl chloride. The distillation of this mixture provides the requisite allylic thiocarbamate. The thiocarbamates have been used by Mimura and co-workers as key intermediates for the stereoselective synthesis of trisubstituted alkenes and of various types of α,β-unsaturated carbonyl compounds ⟨79CL1361⟩.

For the synthesis of *N*,*N*-disubstituted thiocarbamates it is also possible to use COS addition to secondary amines and alkyl halides in aqueous sodium hydroxide (Equation (115)) ⟨80PS(8)301⟩. This method requires the reaction to be carried out at low temperatures (0–10 °C). An excess (30–40%) of the secondary amine must be employed in order to minimize the hydrolysis of carbonyl sulfide to carbon dioxide and hydrogen sulfide.

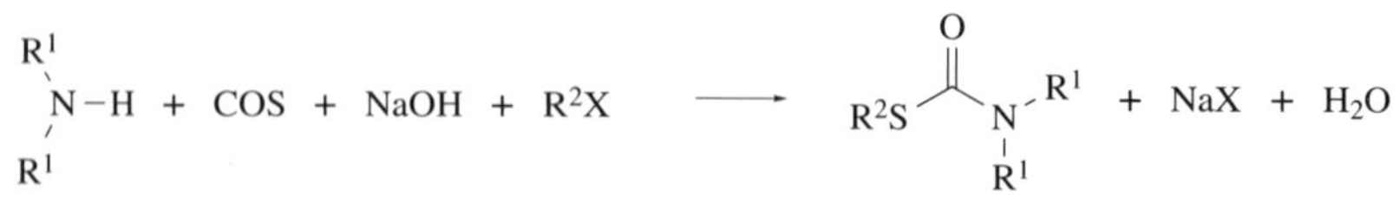

(115)

6.15.3.4.7 *Thiocarbamates from 2-dialkyliminium-1,3-oxothiolane iodides*

When working with 2-alkylimino-1,3-oxathiolanes, Hoppe and Follmann discovered a new method for the preparation of *S*-vinyl thiocarbamates ⟨77AG(E)462⟩. By elimination of HI from 2-dialkyliminium-1,3-oxathiolane iodides, *S*-vinyl thiocarbamates are formed (Scheme 34).

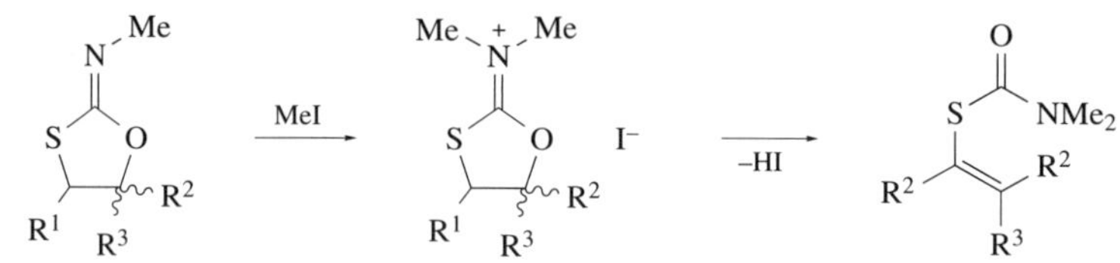

Scheme 34

6.15.3.4.8 *Thiocarbamates as amine protective groups*

The thiol-labile dithiasuccinoyl (Dts) N^{α}-amino protecting group and *N*-thiocarboxy anhydrides of α-amino acids are reported to have certain advantages for peptide synthesis ⟨68JA3254⟩. These amino function protecting groups are available from the appropriate ethoxythiocarbonyl derivatives of amino acids (Scheme 35). A convenient procedure for the synthesis of ethoxythiocarbonyl derivatives was developed by Barany ⟨78JOC2930⟩.

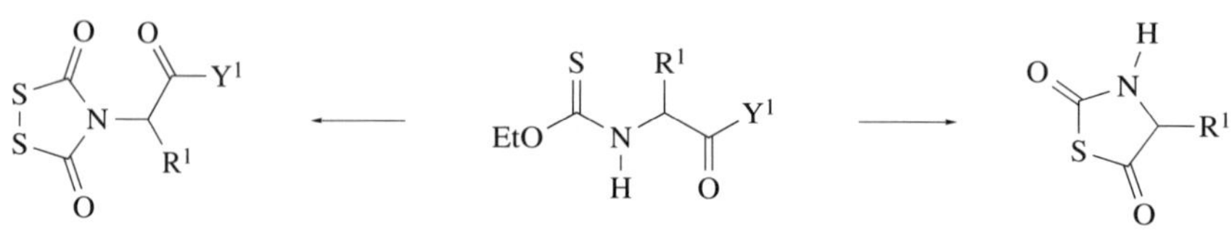

Y^1 = –OH, $–OR^2$, $–NHCH(R^3)COY^2$

Scheme 35

6.15.3.4.9 *Thiocarbamates by other methods*

Akiba and Inamoto developed a synthesis of *S*-alkyl thiocarbamates in which the conversion of an imino group into a carbonyl group is achieved via a nitrosimine ⟨73CC13⟩. A mixture of the *S*-alkylisothiourea or its hydrobromide and isopentyl nitrite in benzene was heated at 50–60 °C to obtain the *S*-alkyl thiocarbamate in good yield (Scheme 36).

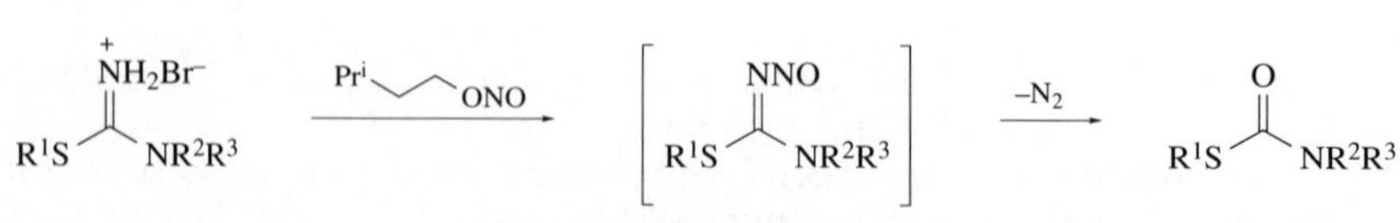

R^1 = Bn, Et; R^2 = Me, Et, Ph; R^3 = Ph

Scheme 36

The synthesis of rather exotic thiocarbamates is described by Schroll and Barany ⟨86JOC1866⟩. The reaction of alkoxydichloromethyl chlorocarbonyl disulfides with an excess of *N*-methylaniline leads to the postulated key intermediate (**A**) in Scheme 37, which offers three pathways to different thiocarbamates.

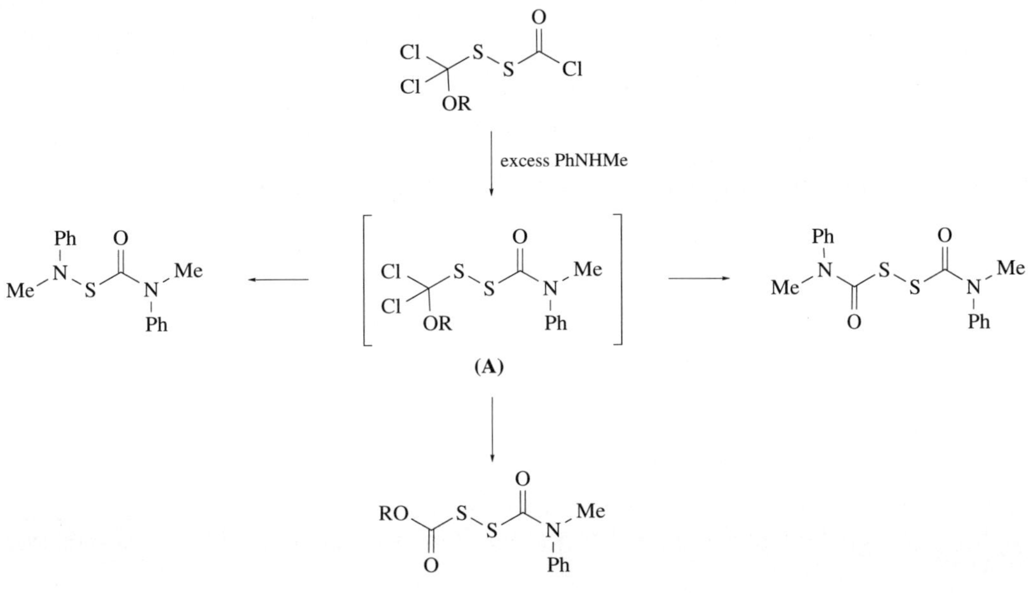

Scheme 37

6.15.3.5 Sulfur and Phosphorus Functions

It is known that *O*-alkylthiocarbonyl compounds undergo alkyl migration from oxygen to sulfur. The ethyl ester of dibutylphosphonous acid reacts with alkyl chlorothioformates to yield phosphinyl alkylthioformates (Equation (116)). The reaction of the components is accompanied by a thione–thiol rearrangement which occurs at 10–35 °C; thus, the expected phosphinyl thionoformates are not obtained. Two mechanisms of the rearrangement are possible at the stage of formation of the intermediate product of the Arbuzov reaction. In the first case, a four-centered cyclic transition state is formed, similar to that in the thione–thiol rearrangement. In a second variant the authors propose that the phosphorus atom becomes pentavalent, forming a P—S bond in a three-membered ring. Stabilization is due to the transfer of the alkyl cation to the sulfur atom ⟨88ZOB26⟩.

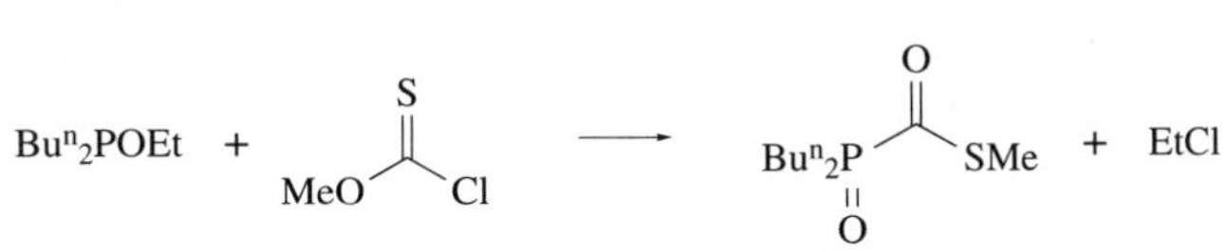 (116)

Analogous compounds from the patent literature are important as plant growth regulators, and are obtained from trialkyl phosphites and chlorothioformates ⟨74USP3849102, 75GEP2435407⟩.

6.15.3.6 Other Mixed Systems

6.15.3.6.1 Selenium and nitrogen functions

Selenocarbamates can be obtained as described by Kondo and co-workers by reaction of an amine, carbon monoxide, and elemental selenium, followed by treatment with an alkyl halide, in good to excellent yields ⟨89AG(E)452⟩. *Se*-*n*-Butyl *N*,*N*-diethyl selenocarbamate is prepared quantitatively by this procedure using *n*-butyl iodide (Equation (117)).

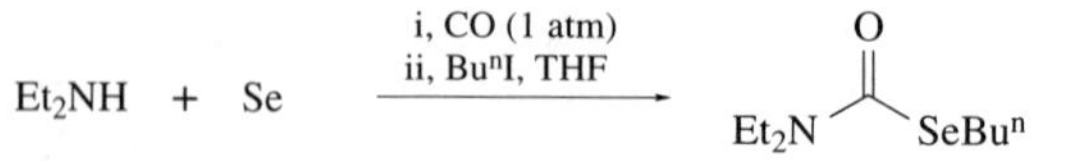

(117)

A different route to the same selenocarbamate uses *N*,*N*-diethylcarbamoyl chloride and lithium *n*-butylselenide ⟨79S597⟩ (Equation (118)).

(118)

4-Methyl-2-phenyl-1,2,4-selenadiazolidine-3,5-dione is obtained by pyrolyzing 2-methyl-4-phenyl allophanoyl selenocyanate with elimination of HCN ⟨82TL2833⟩ (Equation (119)).

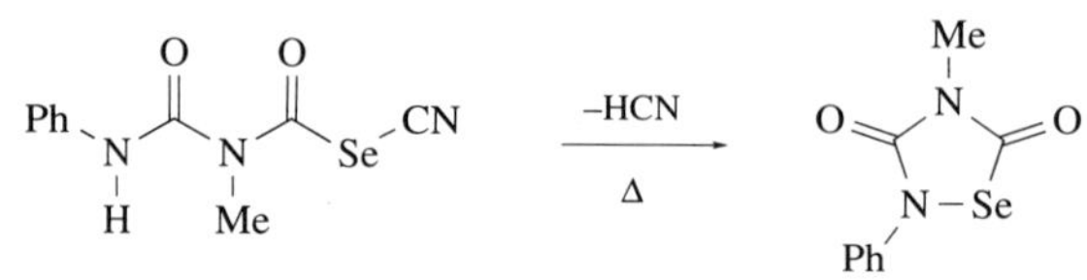

(119)

6.15.3.6.2 *Tellurium and nitrogen functions*

A tellurocarbamate, *Te*-butyl *N*,*N*-diethyl tellurocarbamate is synthesized by reacting metallic tellurium with *n*-butyllithium at 0°C and further reaction with diethylcarbamoylchloride (Scheme 38) ⟨90SC703⟩.

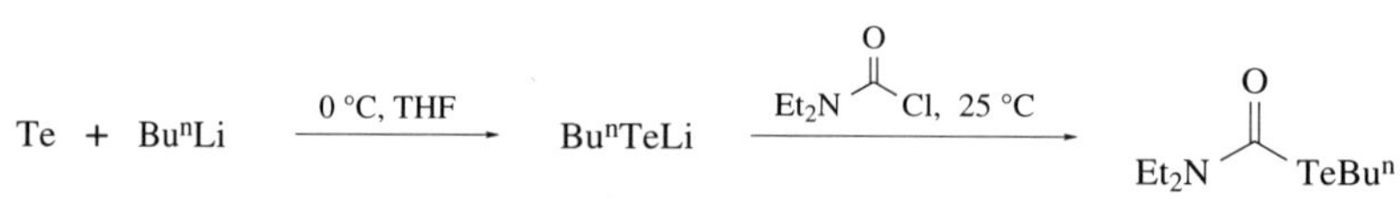

Scheme 38

Another convenient method to obtain unsymmetrical tellurocarbamates uses bis(*N*,*N*-dimethylcarbamoyl) ditelluride, which is prepared in 58% yield by treating DMF with sodium metal and elemental tellurium under argon ⟨85CL1671⟩. Reduction of this reagent by sodium borohydride and further reaction of the resulting monotelluride with *n*-hexyl iodide affords *Te*-*n*-hexyl *N*,*N*-dimethyl tellurocarbamate in an excellent yield of 92% (Equation (120)) ⟨92CL1389⟩.

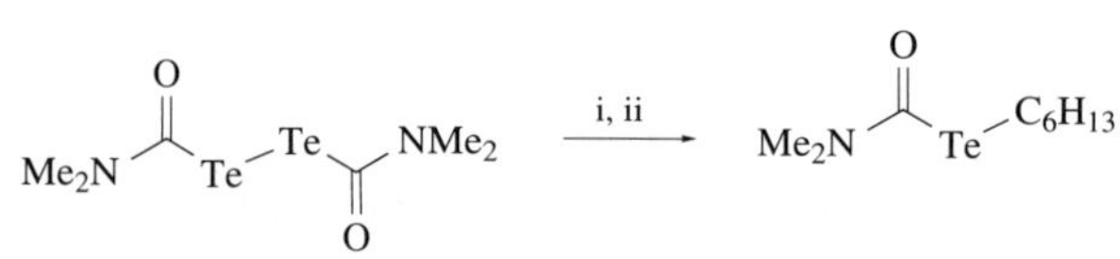

(120)

i, $NaBH_4$ (2.2 equiv.), Et_2O, –50 °C; ii, *n*-$C_6H_{13}I$, EtOH, reflux

6.16
Functions Containing a Carbonyl Group and Two Heteroatoms Other Than a Halogen or Chalcogen

ANTHONY F. HEGARTY and LEO J. DRENNAN
University College Dublin, Republic of Ireland

6.16.1 FUNCTIONS CONTAINING AT LEAST ONE NITROGEN FUNCTION (AND NO HALOGENS OR CHALCOGENS)

6.16.1.1 Carbonyl Derivatives with Two Nitrogen Functions

6.16.1.1.1 Acyclic ureas

(i) Synthesis of urea

Urea (**3**), the parent compound of carbonyl derivatives with two nitrogen functions, may be prepared by the action of ammonia (**1**) on hydrogen isocyanate (**2**) (Equation (1)) or by using ammonia instead of a substituted amine in the reactions outlined in the subsequent sections ⟨83HOU(E4)334⟩. Urea is an important starting material for the synthesis of substituted ureas. As a weak base it can be used as a catalyst in chemical reactions.

(1) (2) (3) (1)

(ii) Reaction of amines/imines with phosgene

Phosgene (**4**) may be attacked by amines or imines to yield *N*-alkylated symmetrical ureas, in all cases. Primary amines require temperatures >100 °C to form substituted ureas successfully; at temperatures <100 °C the isocyanate may be formed. Equation (2) outlines the reaction of a secondary amine (**5**) with phosgene (**4**) to give the *N*,*N*,*N'*,*N'*-tetrasubstituted urea (**6**). The reaction takes place in inert solvents and the generated hydrogen chloride reacts with excess amine. This equation applies equally for primary amines. For R^1 = 2-nitrophenyl, R^2 = H a yield of 81% has been reported ⟨57JGU2590, 57JGU2882, 82JPR227⟩. For tertiary amines (**7**) the reaction involves the loss of an alkyl chloride rather than hydrogen chloride. Benzyl chloride is lost more easily than alkyl chloride but aryl chlorides are more difficult to remove (Equation (3))) ⟨34BSF244, 46MI 616-01, B-71MI 616-01, 73OSC(5)201, 81JOC3011⟩. Imines (**9**) react similarly (Equation (4)), usually involving the loss of hydrogen chloride, which can be reacted with excess imine or an added tertiary amine (e.g., pyridine) ⟨79JPR827⟩.

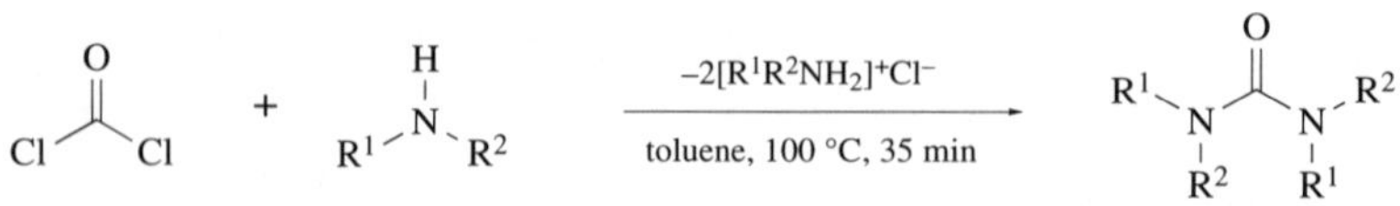

(4) (5) (6) (2)

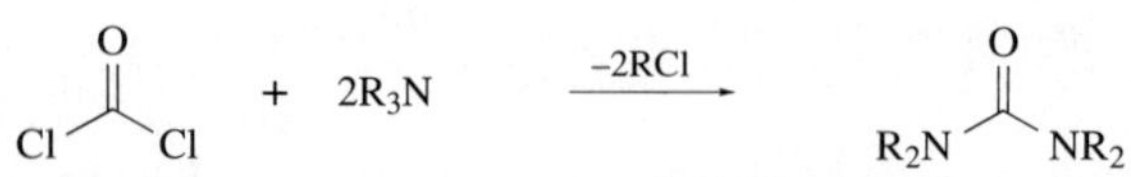

(4) (7) (8) (3)

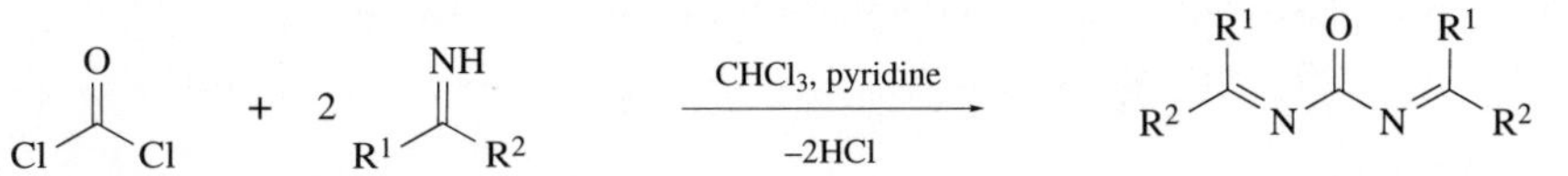

(4) (9) (10)

(4)

(iii) Reaction of amines with S*-chlorothiocarbonyl chloride*

The reaction of 1*H*-benzotriazole (**12**) with *S*-chlorothiocarbonyl chloride (**11**) in dichloromethane at 20 °C for 3 h results in the formation of 1-(1*H*-benzotriazolocarbonylthio)-1*H*-benzotriazole (**13**) which, on heating at 30 °C, yields 1,1′-carbonyl-bisbenzotriazole (**14**) (62%) with loss of sulfur (Scheme 1) ⟨79LA1756⟩.

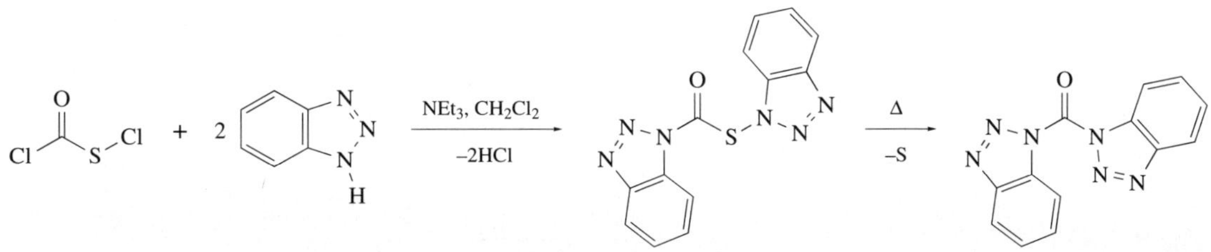

(11) (12) (13) (14)

Scheme 1

(iv) Reaction of amines with carbamic acid chlorides

This synthesis allows the possibility of forming unsymmetrical ureas. Again the reaction involves the formation of hydrogen chloride which may be precipitated as an amine hydrochloride salt either by using an excess of base or by adding a tertiary amine (Equation (5)) ⟨36JCS1273, 67JMC541, 70JOC843, 72GEP2206366, 82JOU1958⟩. For $R^1 = Et$, $R^2 = Ph$, $R^3 = H$, $R^4 = H$ the reaction involves the bubbling of ammonia gas (**16**) ($R^3 = R^4 = H$) through a solution of *N*-ethyl-*N*-phenylcarbamic acid chloride (**15**) and results in the precipitation of ammonium chloride and ~100% yield of the urea (**17**) ⟨36JCS1273⟩. The major drawback of this method is that compound (**15**) is carcinogenic, so other synthetic pathways should be employed if possible.

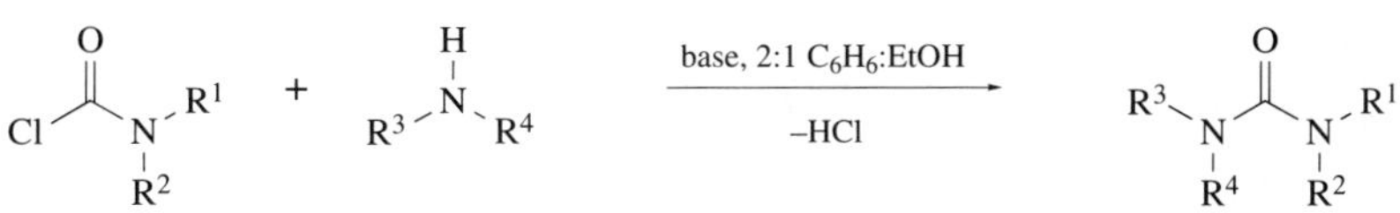

(15) (16) (17)

(5)

(v) Reaction of amines with carbamic acid esters

The sodium methoxide-catalysed reaction of amines (**19**) with carbamic acid esters (**18**) at 165 °C results in the formation of *N*-substituted ureas. For $R^1 = Et$, $R^2 = Ph$, $R^3 = H$, $R^4 = Bu$, $R^5 = Bu$ the product *N*,*N*-dibutyl-*N*′-phenylurea has been isolated in 95% yield (Equation (6)) ⟨68T2367, 72CL971⟩. This method works well if the ester group is a good leaving group such as phenoxy,

4-nitrophenoxy or 2,4,5-trichlorophenoxy. Both symmetrical and unsymmetrical ureas can be synthesized by this method.

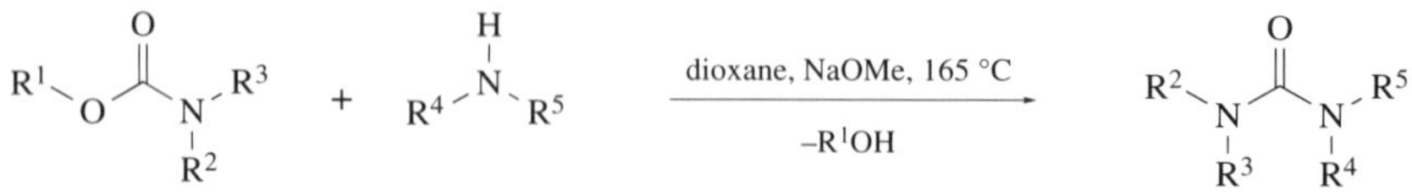

(18) (19) (20) (6)

(vi) Reaction of amines with carbonic acid esters

High temperatures and pressures must be used to effect the reaction of diphenyl carbonate (**21**) with dimethylamine (**22**), but the yield of *N,N,N′,N′*-tetramethylurea (**23**) is 73% (Equation (7)) ⟨63AG1059, 66JMC444, 73GEP2241471, 78JHC1231⟩.

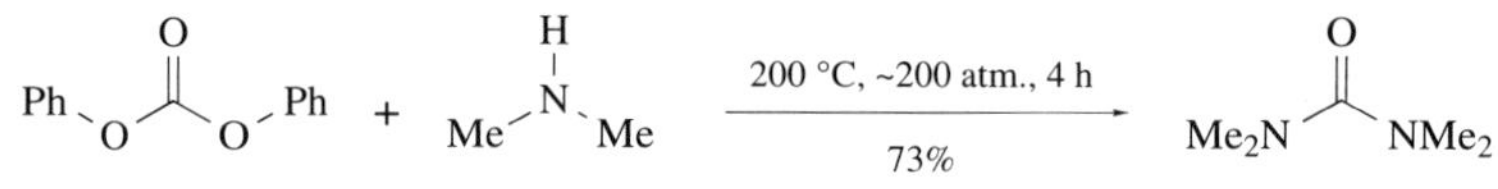

(21) (22) (23) (7)

(vii) Reaction of amines with 4,6-diphenylthieno[3,4-d]-1,3-dioxol-2-one 5,5-dioxide

The following is a general method for the preparation of tetrasubstituted symmetrical or unsymmetrical ureas in very high yields in which the leaving group can be recovered for re-use. The cyclic carbonate (**24**) was reacted in THF with an equimolar quantity of amine (**25**) at 20°C for 30 min. Following addition of an equimolar quantity of the second amine (**26**) and reflux for 20 min, the tetrasubstituted urea (**27**) was isolated in overall 79–98% yield (Scheme 2) ⟨79CB727⟩.

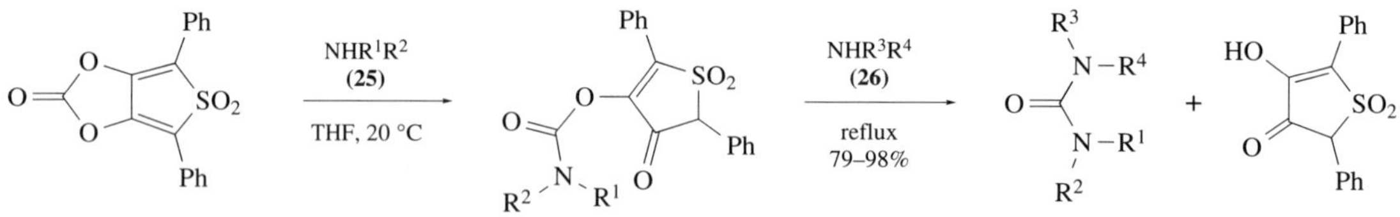

(24) (27)

Scheme 2

(viii) Reaction of amines with urea

This is a general method for the synthesis of *N*-mono- or *N,N*-disubstituted ureas. Alkylamines (**29**) or hydrochloride salts of alkylamines (**30**) react with urea (**28**) in a replacement reaction involving the loss of ammonia or ammonium halide and yielding the required substituted urea (**31**) in high yield (Equation (8)). The *N,N*-disubstituted urea contaminant can be removed by recrystallization. The above reaction using cyclohexylamine yields *N*-cyclohexylurea in 96% yield ⟨83GEP3135111⟩.

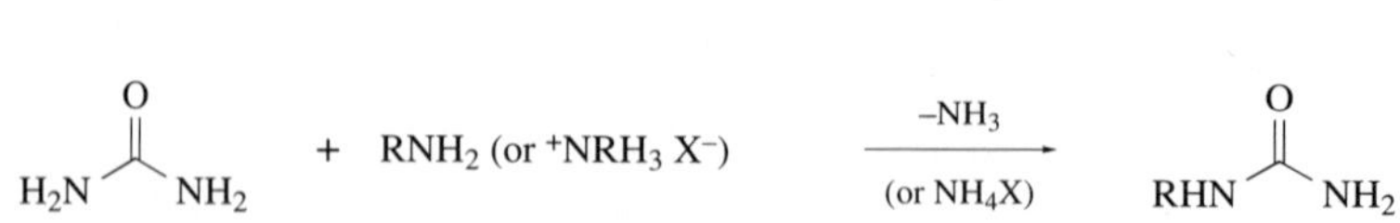

(28) (29) (30) (31) (8)

(ix) Oxidation of thioureas

Thioureas (**32**) may be oxidized to the corresponding ureas (**33**) by a variety of reagents: manganese dioxide in dichloromethane ⟨78BCJ1245⟩, *t*-butyl hypochlorite in carbon tetrachloride ⟨83T1729⟩, sodium nitrite in 4 mol l^{-1} hydrochloric acid ⟨82T1163⟩, or mercuric acetate in dichloromethane (Equation (9)) ⟨90JHC2215, 91TL4791⟩.

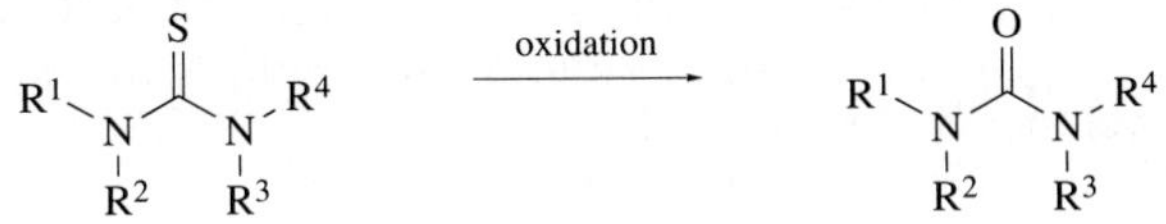

(32) (33) (9)

(x) *Addition to isoureas*

Isoureas (**34**) may be cleaved by the addition of HX (where X = OH, OR, NHR) in a nucleophilic fashion to yield substituted ureas (**35**) as shown (Equation (10)) ⟨74YGK936, 81ACH57⟩.

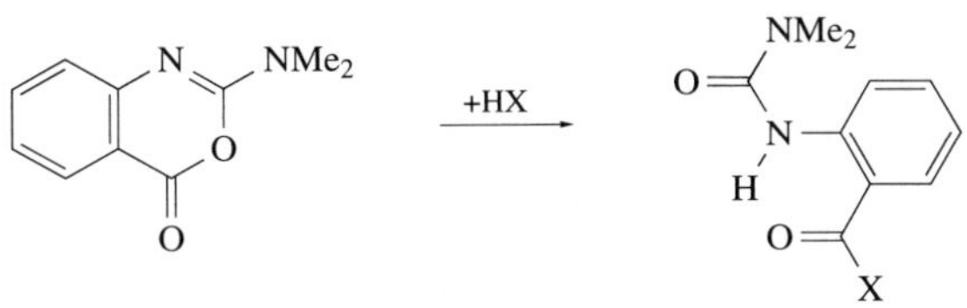

(34) X = OH, OR, NHR (35) (10)

(xi) *Preparation from* ortho-*carbonic acid derivatives*

The hydrolysis of bis(dialkylamino)dichloromethanes (**36**) results in the isolation of ureas (**37**) (Scheme 3) ⟨78IZV220⟩.

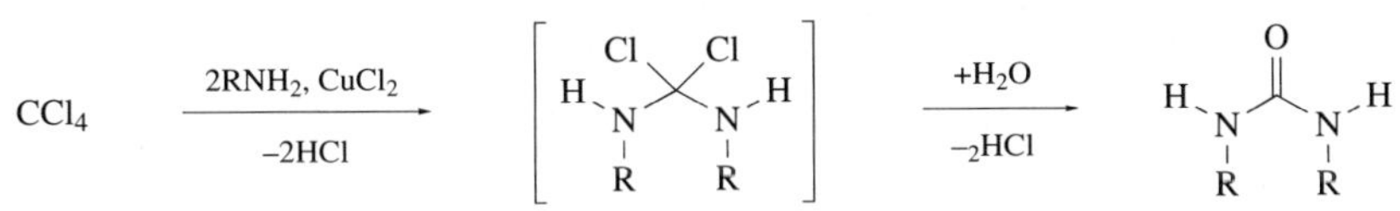

(36) (37)

Scheme 3

(xii) *Addition of amines to carbon oxide sulfide*

This is a two-step reaction which is carried out in inert solvents such as benzene, toluene, methanol, propanol or butanol and involves an intermediate which need not be purified before the subsequent reaction. *N*,*N*-Dimethyl-*N*′-phenylurea (**42**) was prepared by the action of dimethylamine (**39**) on carbon oxide sulfide (**38**) in toluene and the subsequent reaction of aniline (**41**) with the intermediate (**40**) formed. The temperature was maintained at 20 °C for 55 min then brought to 82 °C over 10 min before maintaining it at 82–89 °C for 4 h to effect complete reaction. The isolated yield was 81% (Scheme 4) ⟨79GEP2742158⟩.

(38) (39) (40) (41)

Scheme 4

(xiii) Addition of amines, imines or water to isocyanates or alkali metal cyanates

For this reaction, increasing the electron-withdrawing power on the isocyanate results in an increase in the reactivity towards the nucleophile. *N'*-Cyclohexyl-N-[(2-methoxycarbonyl)phenyl] urea (**45**) can be prepared by the reaction of equimolar amounts of the methyl ester of 2-aminobenzoic acid (**44**) and cyclohexyl isocyanate (**43**) at reflux for 20 min in petroleum ether (b.p. 90–100 °C) using a catalytic amount of triethylamine. The isolated yield was 94% (Equation (11)) ⟨61JOC5238⟩. This reaction type is known for a large variety of amines and isocyanates ⟨26JA1736, 49JA2297, 57JA5717, 57LA(609)83, 62JOC2622, 63JMC669, 79JMC28, 79JOU1077, 80JCED292, 80JOC886, 82CPB534, 83CPB41, 90H(31)1393, 90JOC3699, 91IZV2593, 92S297⟩.

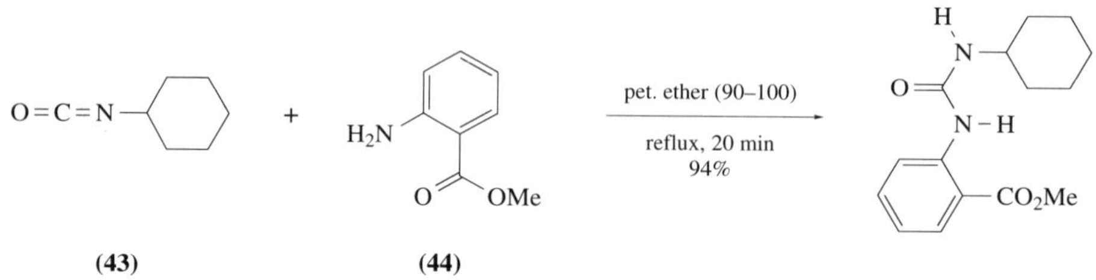

(11)

(**43**) (**44**) (**45**)

N,N'-Diphenyl-*N*-(1-phenyl-1-propenyl)urea (**48**) can be prepared by reacting equimolar amounts of propiophenone phenylimine (**47**) with phenyl isocyanate (**46**) for 2 h at ~130 °C. The reported yield is 77% (Equation (12)) ⟨80CB2499⟩. This reaction is also generally applicable ⟨86CB2553⟩. The reaction of water with alkyl isocyanates (**49**) leads to the production of primary amines (**50**) which then react with excess alkyl isocyanate (**49**) in the usual way to form symmetrical disubstituted ureas (**51**) (Scheme 5). The yield of *N,N'*-didodecylurea made in this manner has been reported as 91% ⟨65MI 616-01⟩.

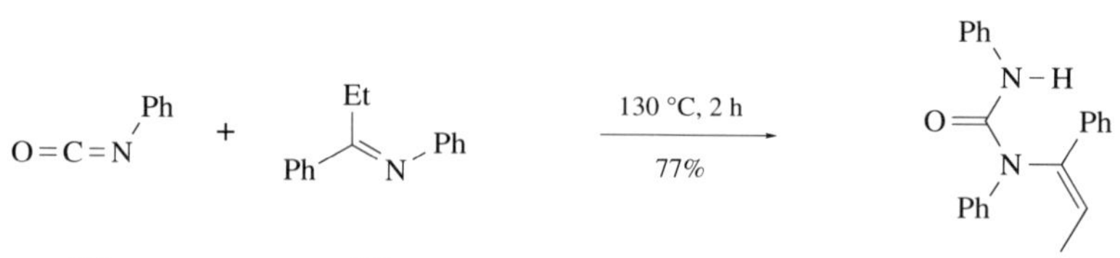

(12)

(**46**) (**47**) (**48**)

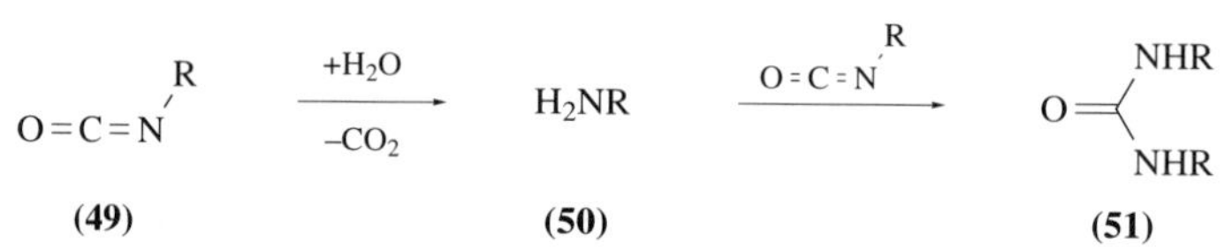

(**49**) (**50**) (**51**)

Scheme 5

Alkali metal cyanates (**52**) react with amines (**53**) to form ureas (**54**) (Equation (13)). In the synthesis of *N*-heptylurea, heptylamine, ice, ice-water and 5 mol l^{-1} hydrochloric acid were heated to 70–80 °C before the sodium cyanate was added portionwise. After 2–4 h two phases separated which, following isolation, yielded 86–88% of the desired monosubstituted urea ⟨63OSC(4)515⟩.

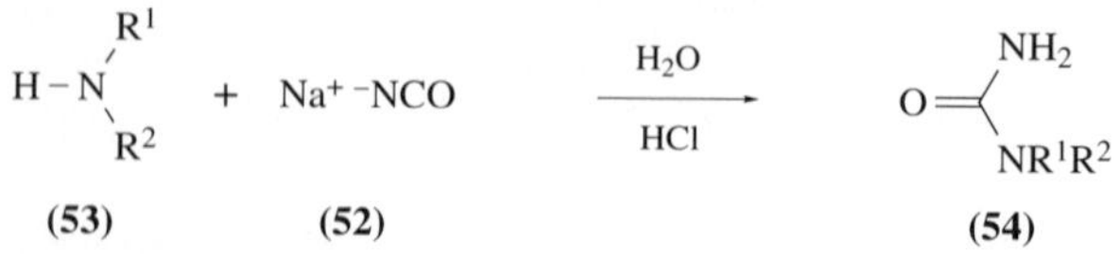

(13)

(**53**) (**52**) (**54**)

(xiv) Hydrolysis of cyanamides

Ureas (**56**) are also formed by the action of water on cyanamides (**55**) in a reaction which can be catalysed by either acids or bases (Equation (14)) ⟨66CB2937, 81GEP2855882⟩. The cyanamides decompose to the corresponding amides on heating to 175 °C with concentrated hydrochloric acid. The best yields are obtained by using basic hydrogen peroxide under mild conditions.

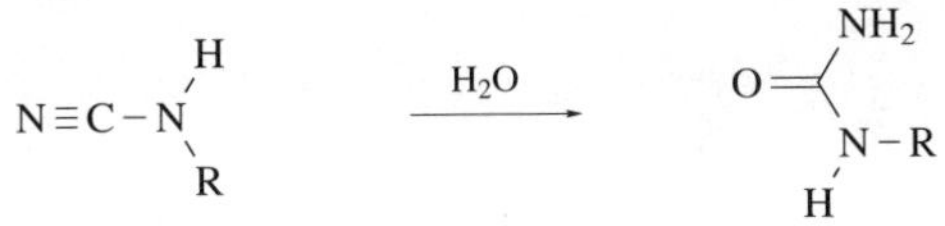

(55) (56) (14)

(xv) *Oxidative amination of carbon monoxide*

Two equivalents of amine may attack carbon monoxide to form ureas (**57**), provided an oxidation process also occurs (Equation (15)). In the synthesis of *N,N′*-bis(ethoxycarbonylmethyl)urea, carbon monoxide is bubbled through a solution of glycine ethyl ester, triethylamine and catalytic selenium in THF at 20 °C for 4 h. The isolated yield was 98% ⟨79S735⟩.

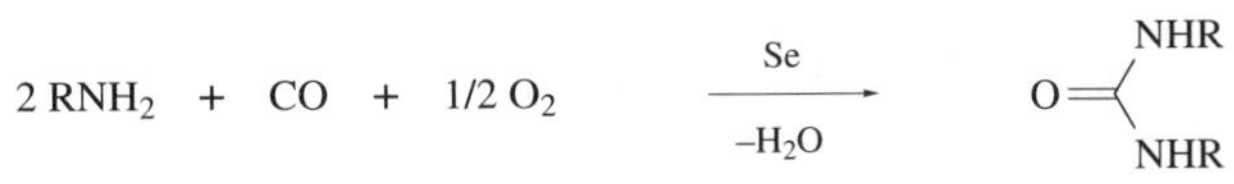

(57) (15)

(xvi) *Reaction of amines with isocyanides*

Ureas may also be synthesized in a redox reaction of mercuric salts (**58**) and amine (**59**) on isocyanide (**60**) (Equation (16)). For example, *N*-butyl-*N′*-(2,6-dimethylphenyl)urea (**61**) is prepared as shown in 88% yield ⟨72TL4263⟩. The order of decreasing reactivity of the mercuric salts is $Hg(OCOMe_3)_2 > Hg(NO_3)_2 > HgCl_2 \gg Hg(CN)_2$.

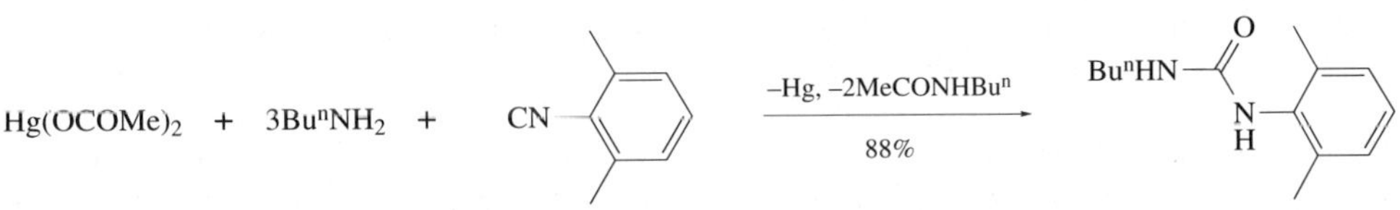

(58) (59) (60) (61) (16)

(xvii) *Hydroxyureas and alkoxyureas*

This class of compounds may be synthesized by the reaction of phosgene with hydroxylamine hydrochloride ⟨66JCS(C)350⟩, with *N*-hydroxylamines ⟨75CZ245⟩ or with *O*-methylhydroxylamine hydrochloride ⟨66JCS(C)350⟩ in dioxane/water in the presence of potassium acetate to yield the symmetrical *N,N′*-dihydroxyurea or *N,N′*-dimethoxyurea.

Carbamic acid chlorides react with hydroxylamine in dioxane, in the presence of triethylamine, to form the corresponding *N,N*-dialkyl-*N′*-hydroxyurea derivatives. The yield of *N,N*-diethyl-*N′*-hydroxyurea prepared by this method is 68%, while the reported yield of the *N,N*-dimethyl derivative is 86% ⟨74AP(307)7⟩.

N-Hydroxyurea has been prepared by the reaction of ethyl carbamate (urethane) with hydroxylamine hydrochloride in alkaline solution at ~20 °C with reported yields of 53–73% (Equation (17)) ⟨73OSC(5)645⟩.

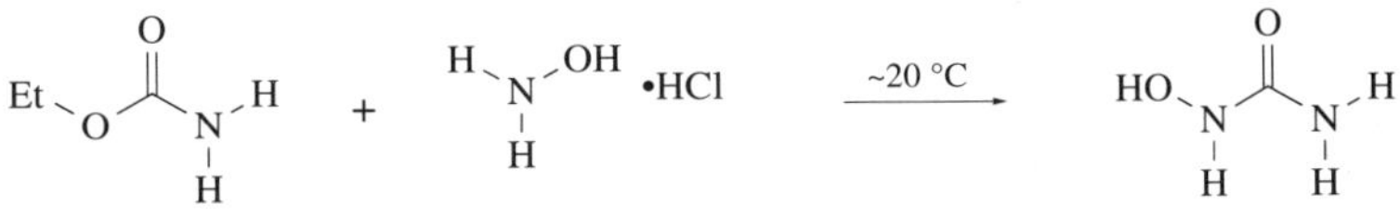

(17)

The addition reaction of isocyanates or alkali metal cyanates with hydroxylamine or hydroxylamine derivatives in benzene, THF, diethyl ether or dichloromethane solution occurs at ~20 °C.

For example, equimolar amounts of *O*-phenylhydroxylamine and methyl isocyanate react at 20 °C over 1 h to yield 92% of the required *N*-methyl-*N*′-phenoxyurea (**62**) (Equation (18)) ⟨83S471⟩.

(18)

(62)

(xviii) Acylureas

The *N*-acyl derivative of urea can be prepared by the synthesis of the *O*-acyl derivative followed by an internal $O \rightarrow N$ rearrangement of the acyl group ⟨79COC(2)1094⟩. For example, the *O*-acyl derivative is formed in the reaction of acryloyl chloride and urea. The $O \rightarrow N$ rearrangement occurs in inert solvents such as dioxane, THF or chloroform at 20–40 °C in the presence of triethylamine or sodium carbonate to give *N*-acryloylurea in 53.5% yield ⟨68CB2419⟩.

The reaction of acetamide with half an equivalent of bromine in base gives an isocyanate intermediate which reacts further with acetamide to form *N*-acetyl-*N*′-methylurea in 80–84% yield ⟨43OSC(2)462⟩.

Other methods which may be employed in the synthesis of acylureas include the reaction of phosgene with acylamines ⟨B-65MI 616-02⟩, the reaction of *N*-acylcarbamidic acid esters with amines ⟨B-65MI 616-02⟩, the hydrolysis of *N*-acylcyanamides ⟨68USP3539569, 68USP3575975⟩, the reaction of isocyanides with *N*-bromosuccinimide or with *N*-bromoacetamide ⟨63BCJ1314⟩, the hydrolysis of *N*-(1-chloro-1-alkenyl)urea ⟨72AG993⟩, the addition of carboxylic acids to carbodiimides ⟨65JOC2849, 66JA1024, 67CRV107⟩ and the reaction of *N*-hydroxyformamides with acyl chlorides ⟨79CZ175⟩.

(xix) Nitrosoureas

The synthesis of nitrosoureas may be carried out, in good yield, by reacting an *N*-nitrosocarbamic acid ester with an amine under mild conditions ⟨82JMC178⟩.

N-Methyl-*N*-nitroso-*N*′-(2-phenylethyl)urea (**63**) has been prepared by the reaction of a solution of *N*-methyl-*N*′-(2-phenylethyl)urea in formic acid and water with an aqueous solution of sodium nitrite at 0–5 °C over 30 min (Equation (19)). The isolated yield of the *N*-nitrosourea derivative is reported as 84% ⟨63JMC669, 66JA3798⟩.

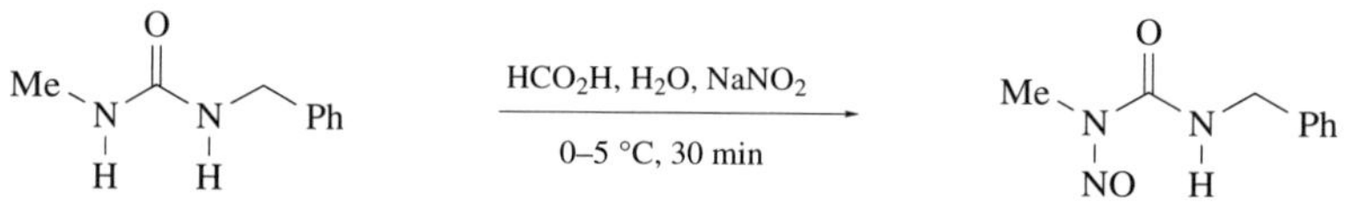

(19)

(63)

(xx) Azodicarbonyl amides (bis(aminocarbonyl)diazines)

Members of this class of compounds have been prepared by the oxidation of the corresponding *N*,*N*′-bis(aminocarbonyl)hydrazines or by the reaction of bis(alkoxycarbonyl)diazines with primary or secondary aliphatic amines ⟨66AG376, 82AHC(30)1⟩. The latter has been used to produce *N*,*N*,*N*′,*N*′-tetramethylazodicarboxamide in 79% yield ⟨73JOC1652⟩.

6.16.1.1.2 Cyclic ureas

(i) Reaction of diamines with phosgene

Cyclic ureas are the products in this and subsequent reactions. Diamines (**64**) do not react with phosgene (**67**) to form the required cyclic urea (**69**) directly but need to be activated. Therefore, the diamine (**64**) is treated with hexamethyldisilazane (HMDS) (**65**) under reflux for 10 h to yield 88% ($n = 4$) of the bis(trimethylsilylamino)alkane (**66**). The reaction of (**66**) with phosgene (**67**) occurs in toluene at 0 °C in 1 h, facilitated by triethylamine. For $n = 3$, the yield is 60%; for $n = 4$, the yield is 75%. The 1,3-bis(trimethylsilyl)-2-oxo product (**68**) is readily hydrolysed in ~100% yield to form the cyclic urea product (**69**) (Scheme 6) ⟨60CB2810, 75JCS(P1)212, 77AJC455⟩. Polymeric ureas may be formed even if the reaction mixture is very dilute.

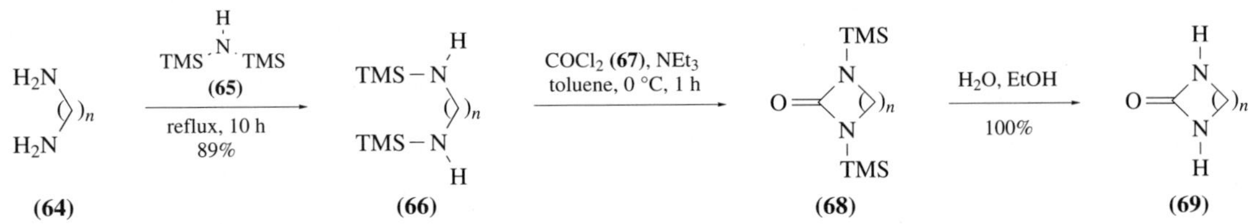

Scheme 6

(ii) Reaction of imines with chloroformate esters

The one-step synthesis of 2-oxo-1,3,4,5-tetraarylimidazolidine (**72**) has been reported via the reaction of two equivalents of arylaldehyde arylimine (**70**) with sodium in ether (3 h reflux) followed by the addition of ethyl chloroformate (**71**) to the filtrate of the filtered initial reaction mixture and 1 h reflux (Equation (20)). The reaction yields 60% of a racemic mixture of *d*- and *l*- products and 28% of the *meso*-compound. The diastereomers can be separated by recrystallization ⟨70JOC2099, 71JOC2516, 80IJC(B)653, 80S1001⟩.

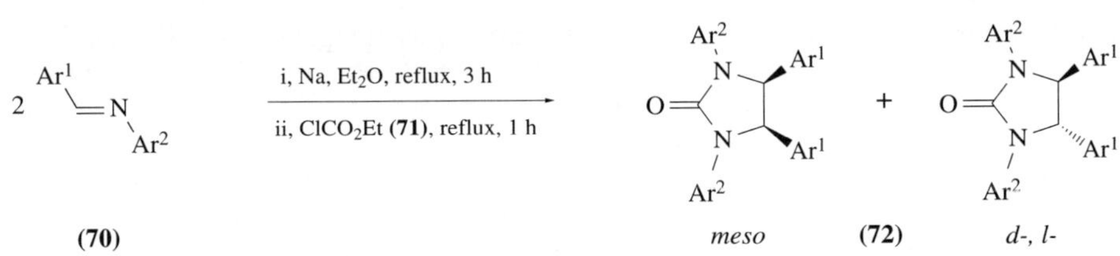

(20)

(iii) Reaction of diamines with dialkyl carbonates

This reaction usually requires vigorous conditions but the reaction of 1,2-diaminoethane (**74**) with bis(succinimido)carbonate (**73**) occurs readily at 20 °C in good yield (90%) to form 2-oxoimidazolidine (**75**) (Equation (21)) ⟨82SC213⟩.

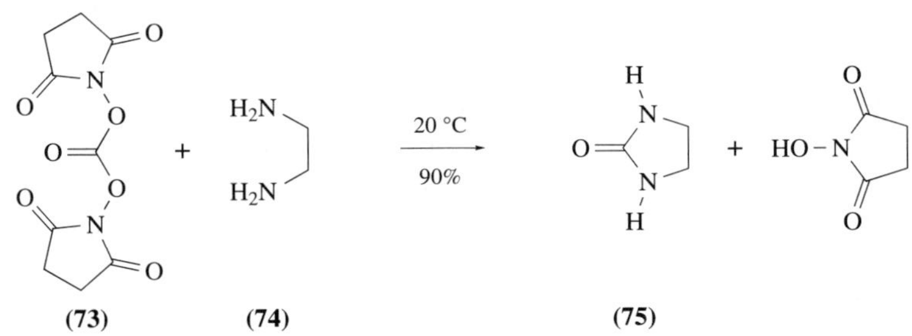

(21)

(iv) Pyrolysis of bis(carboxyalkylamino)alkanes

An excellent yield (~100%) has been reported for the thermal conversion of 1,4-bis(carboxyalkylamino)alkanes (**76**) into 1,3-dimethyl-2-oxoimidazolidine (**77**) (Equation (22)) ⟨79AG(E)503⟩.

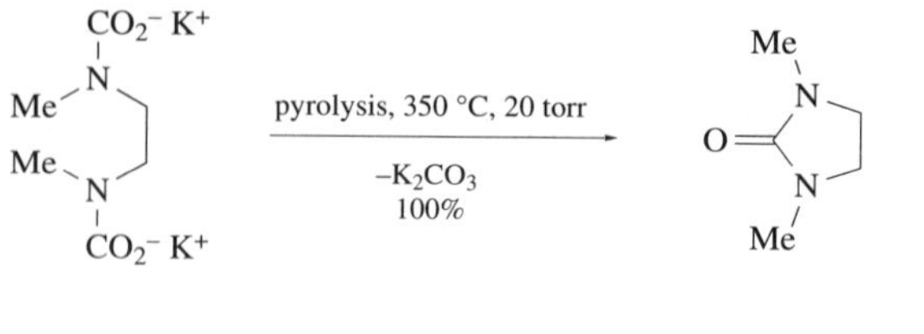

(22)

(76) **(77)**

(v) Reaction of urea with aldehydes and ketones

The condensation of aldehydes and ketones with urea can lead to a large variety of mono- and polycyclic ureas. Cyclic ureas such as 3,7-dioxo-2,4,6,8-tetraazabicyclo[3.3.0]octane (**80**) may be synthesized by the reaction of two equivalents of urea (**78**) with aldehydes such as glyoxal (**79**) (Equation (23)) ⟨1877LA(189)159, 52USP2592565⟩.

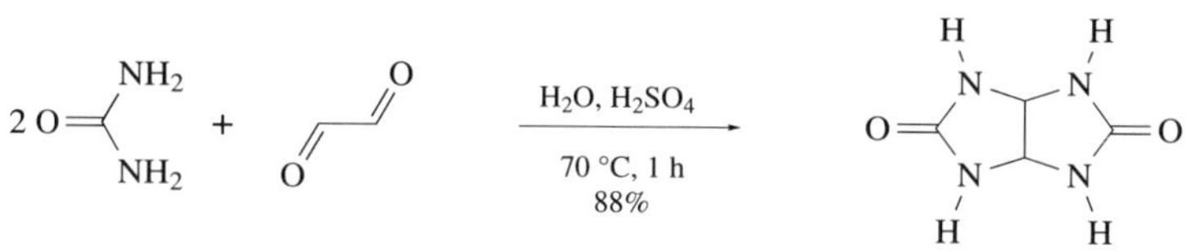

(23)

(78) **(79)** **(80)**

(vi) Cyclization of N-*(ω-haloalkyl)ureas*

Ureas may be cyclized readily if there is a suitable leaving group on the *N*-alkyl chain. The synthesis of 1,3-diphenyl-2-oxoimidazolidine (**82**) ($R^1 = R^2 = Ph$, $R^3 = R^4 = H$) was carried out by reacting 10 mol l^{-1} sodium hydroxide in ethanol for 15 min at reflux with 1-(2-chloroethyl)-1,3-diphenylurea (**81**) ($R^1 = R^2 = Ph$, $R^3 = R^4 = H$). The yield was reported as 46% (Equation (24)) ⟨73JHC639⟩. This procedure is generally applicable for various R groups, leading to a wide variety of cyclic ureas (**82**) ⟨51JOC1829, 53CJC896, 53JA1120, 65CJC1798, 70JOC843, 71JHC557, 79JOU1077⟩.

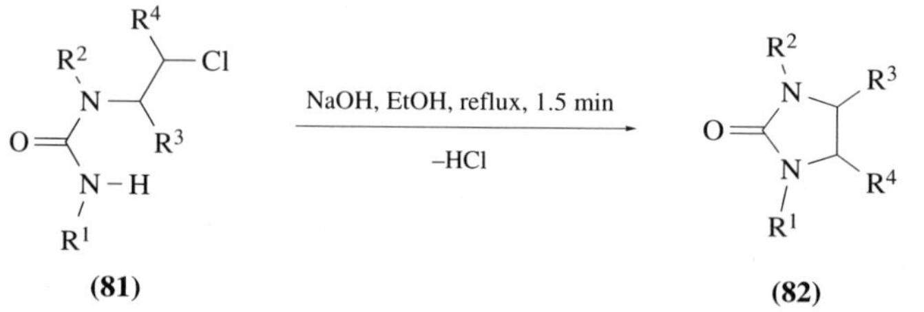

(24)

(81) **(82)**

(vii) Addition of diamines to alkali metal cyanates

Diamines (**83**) add to alkali metal cyanates (**84**) to yield cyclic ureas (**85**) (Equation (25)) ⟨66JMC852⟩. The reaction gives yields of 32–88% for various R groups (R = H, Me, CF_3, SMe, Cl, Br) and is again applicable to other similar systems ⟨45JA2079, 47JA3150, 57JA5710, 73JMC901⟩.

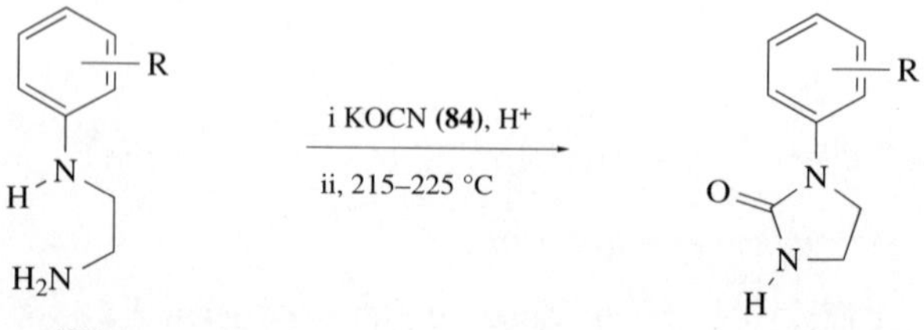

(25)

(83) **(85)**

(viii) Cyclization of N-*(2-propynyl)ureas*

N-(2-Propynyl)ureas (**86**) are cyclized in good yield at mild temperatures by the action of base to form cyclic ureas (**87**) (Equation (26)) ⟨62JOC3079, 63JOC991, 64JOC1851, 78JOC61⟩.

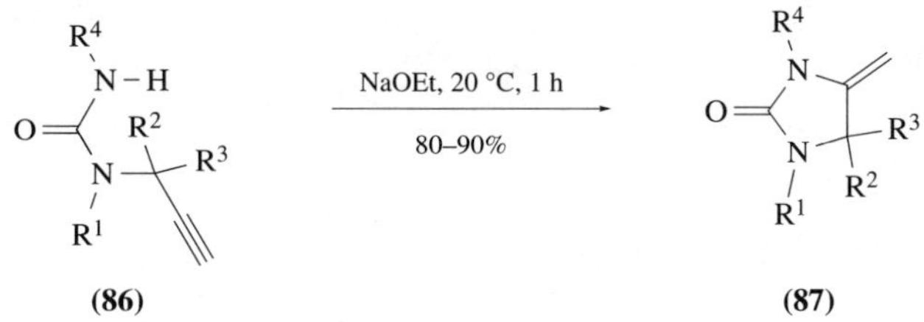

(26)

(86) **(87)**

(ix) Cyclization of N-*haloureas*

3-Membered ring cyclic ureas can be synthesized in good yield from *N*-haloureas. The cyclization is carried out by potassium in pentane or by potassium *t*-butoxide in *t*-butanol. For example, the reaction of bis(*t*-butyl)urea (**88**) in *t*-butanol with *t*-butyl hypochlorite yields the *N*-halourea (**89**), which may then be cyclized by the addition of potassium *t*-butoxide to form the cyclic urea (**90**) in an overall 90% yield (Scheme 7) ⟨69JOC2254⟩. This reaction also works for other *N*-substituted ureas ⟨69JOC2263, 72LA(765)94, 78JOC922⟩.

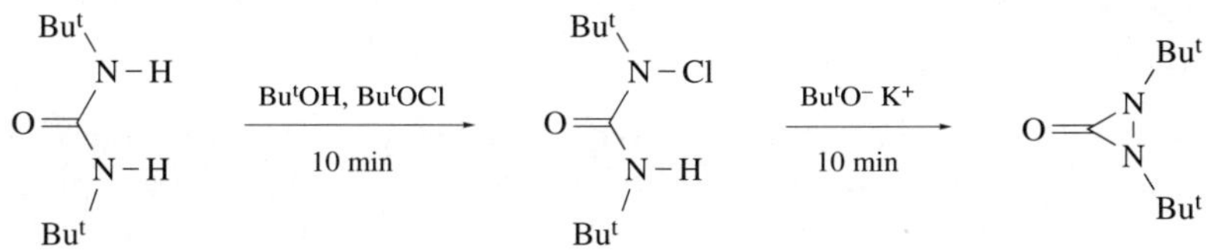

(88) **(89)** **(90)**

Scheme 7

(x) Oxidation of cyclic thioureas

Cyclic thioureas (**91**) can be oxidized to the corresponding cyclic ureas (**92**) by the action of a range of oxidizing agents such as potassium permanganate ⟨77BSB663⟩, hydrogen peroxide ⟨56CB343, 58JA6409, 81JMC1089⟩ or dimethyl selenoxide ⟨78JOC2132⟩ (Equation (27)).

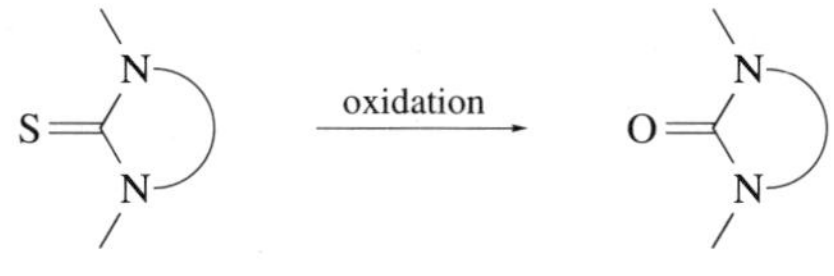

(27)

(91) **(92)**

(xi) Addition of diamines to carbon oxide sulfide

The synthesis of 2-oxo-perhydropyrimidine (**95**) involves the reaction of 1,3-diaminopropane (**93**) and carbon oxide sulfide in 50% ethanol at 67°C for 20 min followed by the addition of HCl and heating at 84°C for 1.5 h. The isolated yield of (**95**) was 81% ⟨79USP4154931⟩. The reaction goes through an intermediate (**94**) (Scheme 8).

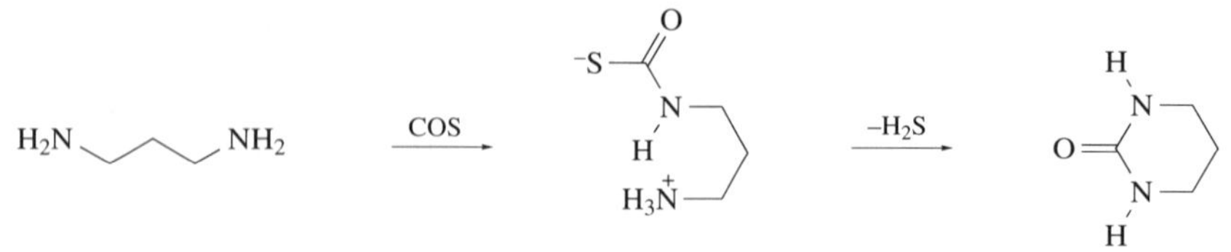

(93) (94) (95)

Scheme 8

(xii) Reaction of isocyanates and aziridines

The reaction of isocyanates (**96**) with aziridines (**97**) proceeds smoothly at 100°C in the presence of lithium halides to yield cyclic ureas (**98**). For R^1 = benzoyl and R^2 = 2-phenylethyl the yield of 1-benzoyl-2-oxo-3-(2-phenylethyl)imidazolidine (**98**) has been reported as 59% (Equation (28)) ⟨66LA(698)180⟩. Alkyl and aryl isocyanates also trimerize in a competing reaction.

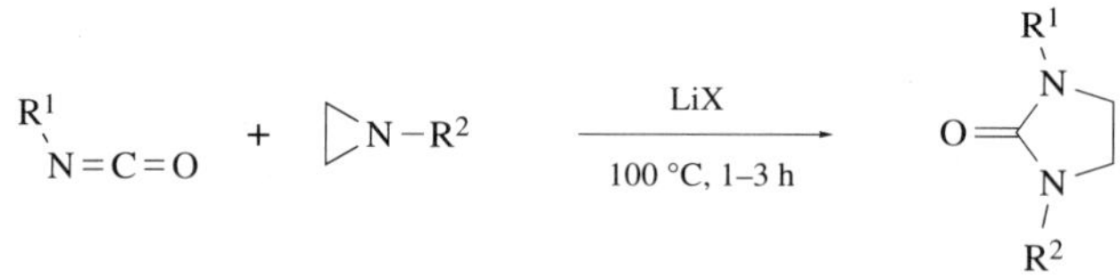

(96) (97) X = Cl, Br (98) (28)

(xiii) Reaction of carbodiimides and oxiranes

This reaction is catalysed by alkali metal hydroxides, tertiary amines or lithium chloride. A cyclic urea (**101**) has been synthesized in the reaction of phenyl oxirane (**100**) with diphenylcarbodiimide (**99**) at 200°C under nitrogen for 7 h and in the presence of lithium chloride. The reported yield was 93% (Equation (29)) ⟨61CB3287⟩.

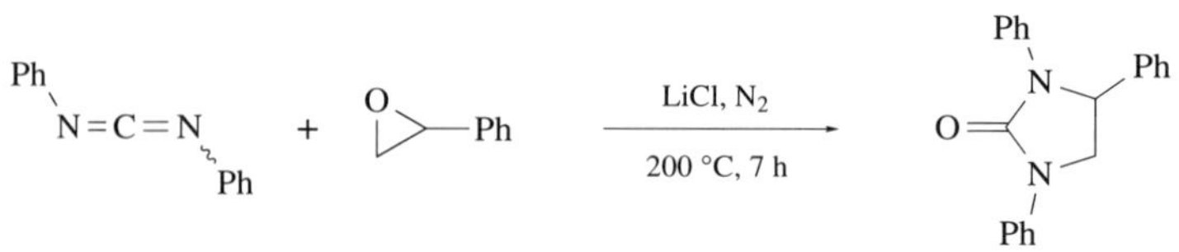

(99) (100) (101) (29)

(xiv) Reaction of diamines with carbon monoxide

The reaction of diamines (**103**) with carbon monoxide (**102**) takes place using various conditions, depending on the chain length, to yield the cyclic ureas (**104**). Either sulfur in methanol or selenium in THF can be used for the reaction and the conditions vary from 40–120°C, and from 4–50 atm pressure to give yields of 58–98% (Equation (30)) ⟨71JA6344⟩.

(102) (103) (104) (30)

(xv) Ring expansion of cyclic amidoximes

The synthesis of 2-oxo-1,3-diazacyclododecane (**106**; $x = 7$) is carried out by Beckmann rearrangement on 2-hydroximinoazacycloundecane (**105**; $x = 7$) in phosphoric acid for 10 min at 120°C, then

50 min at 140–150°C. The reported yield of the cyclic urea (**106**) was 76% (Equation (31)) ⟨57LA(607)67, 78JOC1544⟩. This is a very good method for preparing cyclic ureas containing large rings.

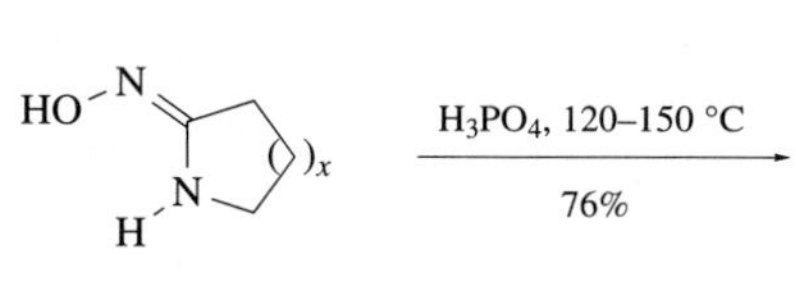

(31)

(**105**) (**106**)

6.16.1.1.3 Carbamoyl azides

Carbamoyl azides, $R^1R^2NCON_3$, are most generally prepared by one of two routes. The first of these is the displacement of chloride from the corresponding carbamoyl chloride, R^1R^2NCOCl, by reaction with sodium azide ⟨70JHC807⟩. The second general route involves the diazotization of the hydrazides $R^1R^2NCONHNH_2$ ⟨86T2677⟩. The second method was used to prepare the highly unstable and explosive carbonic diazide, $CO(N_3)_2$, ⟨1895JPR(2)472⟩ which has subsequently been shown to react with benzene to give *N*-(azidocarbonyl)azepine ⟨66CI(L)1266⟩.

Carbamoyl azides $RNHCON_3$ can also be prepared directly from aldehydes RCHO by oxidation with pyridinium chlorochromate (pcc) in the presence of sodium azide. This method probably involves the intermediacy of an acyl azide which undergoes Curtius rearrangement and further reaction with sodium azide ⟨88SC545⟩.

6.16.1.2 Carbonyl Derivatives with One Nitrogen and One P, As, Sb or Bi Function

6.16.1.2.1 Carbonyl derivatives with one nitrogen and one phosphorus(III) function

The synthesis of carbonyl derivatives with one nitrogen and one phosphorus function, the phosphorus being a secondary phosphine (**109**), is achieved by reacting a phosphine (**107**) with an isocyanate (**108**) (Equation (32)). The reaction has been carried out with $R^1 = R^3 =$ phenyl and $R^2 = (CH_2)_nSH$ ($n = 2$, 3) at room temperature and resulted in an isolated yield of 62–75% ⟨73JPR471⟩. The same reaction with $R^1 =$ Ph or cyclohexyl, $R^2 = CHPhCH_2COR^4$ ($R^4 =$ Me, Ph, Bu^t), $R^3 =$ Ph carried out in ethanol or acetone as solvents gave (**109**) in yields of 56–77% ⟨70JPR366⟩.

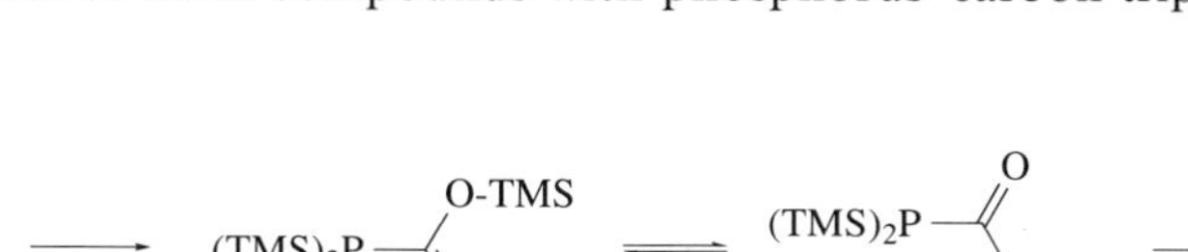

(32)

(**107**) (**108**) (**109**)

Tris(trimethylsilyl)phosphane (**110**) and isopropyl isocyanate (**111**), when stirred in diethylether for 3 days, react to form a compound of the required type (**112b**). However, compound (**112b**) is unstable and reacts further to form compounds with phosphorus–carbon triple bonds (Scheme 9) ⟨89AG(E)53⟩.

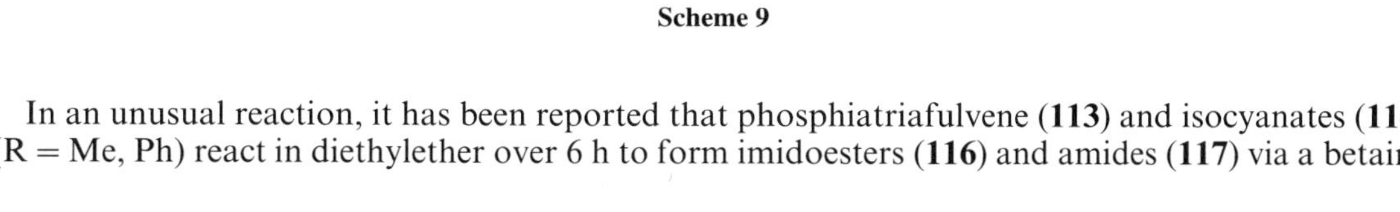

(**110**) (**111**) (**112a**) (**112b**)

Scheme 9

In an unusual reaction, it has been reported that phosphiatriafulvene (**113**) and isocyanates (**114**) (R = Me, Ph) react in diethylether over 6 h to form imidoesters (**116**) and amides (**117**) via a betaine intermediate (**115**) (Scheme 10) ⟨91S1099⟩. If the phosphine does not have a hydrogen substituent, the product of addition to isocyanates may be formed in very good yield and may also be stable ⟨77CB2368⟩.

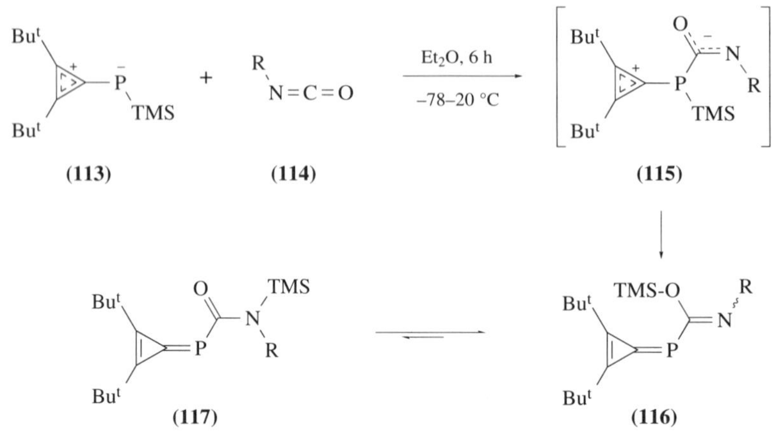

Scheme 10

6.16.1.2.2 *Carbonyl derivatives with one nitrogen and one phosphorus(V) function*

These compounds (**120**) are formed when a phosphine oxide (**118**) reacts with an isocyanate (**119**) (Equation (33)). This has been achieved with $R^1 = R^2 = OMe$ and $R^3 = SO_2NCO$. The reaction in ether at 0°C gave the required product (**120**). However, if the reaction is carried out at 20°C for 1 h a 2:1 product is formed (i.e., $[(MeO)_2P(O)NH]_2SO_2$). This reaction is generally applicable and also works for $R^1 = R^2 =$ Et, Pr, Pr^i, Bu and Bu^i ⟨68USP3401214, 69RZC1443, 73JGU1019⟩. The reaction has also been carried out with $R^1 = R^2 = OMe$ and $R^3 = Ts$ in benzene under reflux for 5 h. The required product (**120**) was isolated. The same authors have synthesized several other compounds (**120**) bearing a variety of arenesulfonyl groups ⟨68USP3413382⟩.

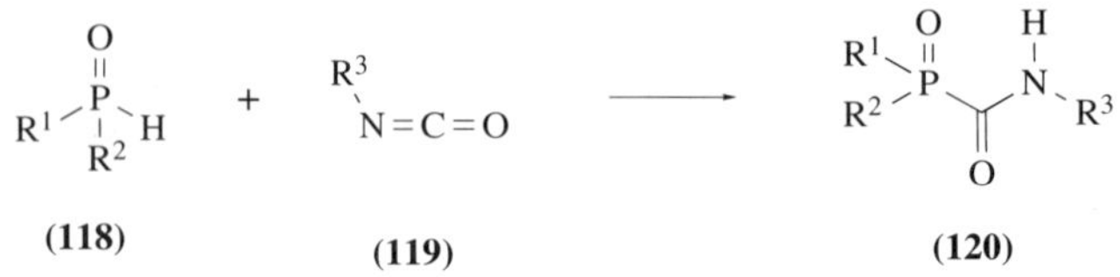

(33)

The synthesis of compounds (**124**) has also been achieved by reacting compounds of type $(TMS\text{-}O)PH(OR^1)$ (**121**) with $OCNCH_2CO_2R^2$ (**122**) in ether at ⩽35°C for 1 h to give (**123**) in quantitative yield; these reacted with aqueous sodium hydroxide at −5°C to give the desired products (**124**) (Scheme 11) ⟨87EGP242812⟩. This method is applicable for $R^1 =$ H, TMS, alkali metal and (substituted) ammonium and $R^2 =$ H, alkyl and an alkali metal.

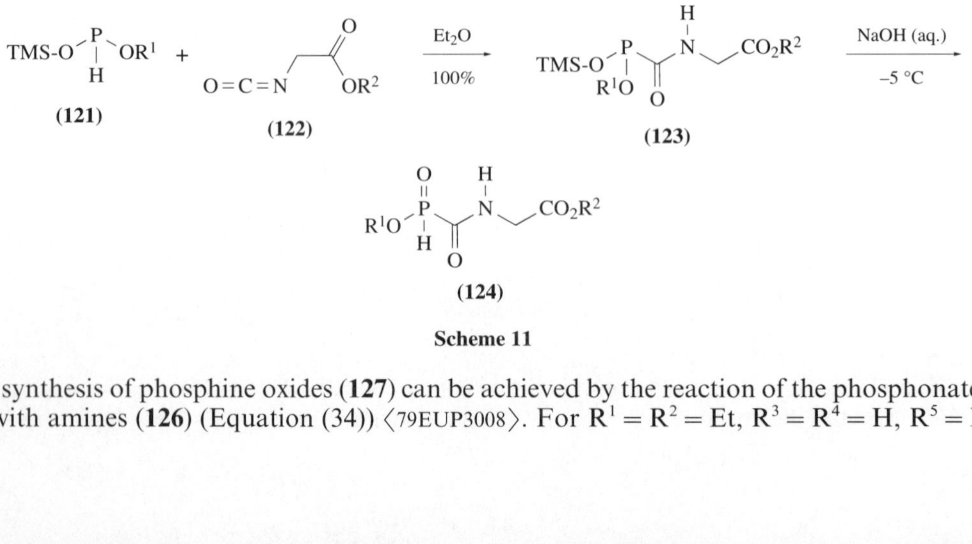

Scheme 11

The synthesis of phosphine oxides (**127**) can be achieved by the reaction of the phosphonate esters (**125**) with amines (**126**) (Equation (34)) ⟨79EUP3008⟩. For $R^1 = R^2 = Et$, $R^3 = R^4 = H$, $R^5 = Me$ the reaction involves the passing of ammonia into the reaction mixture at 40°C for 3 h to yield the required product (**127**). A similar reaction with $R^1 = R^2 = R^5 =$ Me and 1,1,3,3-tetramethylguanidine as the amine (**126**) gave $(MeO)_2P(O)CON{=}C(NMe_2)_2$ as the product ⟨76GEP2553147⟩.

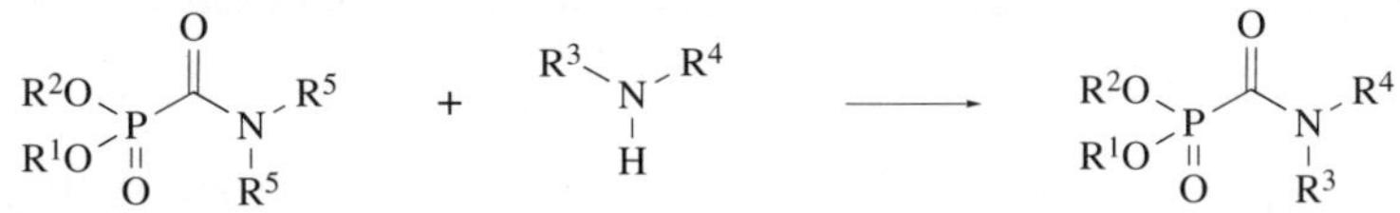

(**125**) (**126**) (**127**) (34)

The reaction of trialkyl phosphite (**128**) and alkyl chloroformate (**129**) at 125°C yields the phosphonate (**130**), which with ammonia, or other amines, gives the required carbamoyl phosphonates (**131**) (Scheme 12). This method works for R^1, R^2 = alkyl, allyl, haloalkyl, alkoxyalkyl and alkenyl; R^3 = Me; R^4 = Me and R^5 = alkyl, alkynyl and hydroxyalkyl ⟨71GEP2040367, 74USP3849102, 77GEP2638754⟩. In an Arbuzov reaction, trimethyl phosphite (**133**) has been incorporated into compound (**132**) to form (**134**) in 98% yield at 20–30°C (Equation (35)) ⟨82EUP43978⟩.

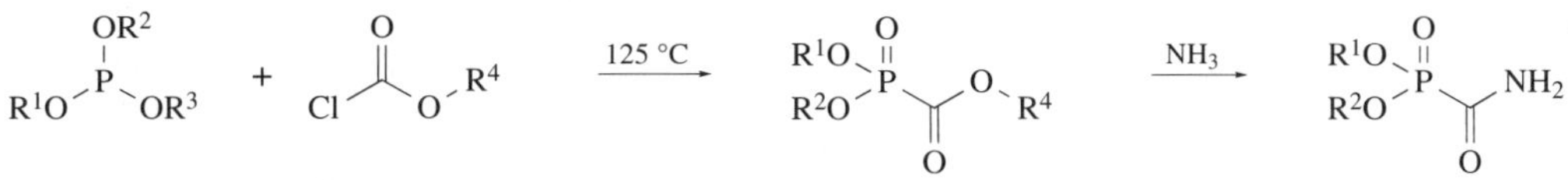

(**128**) (**129**) (**130**) (**131**)

Scheme 12

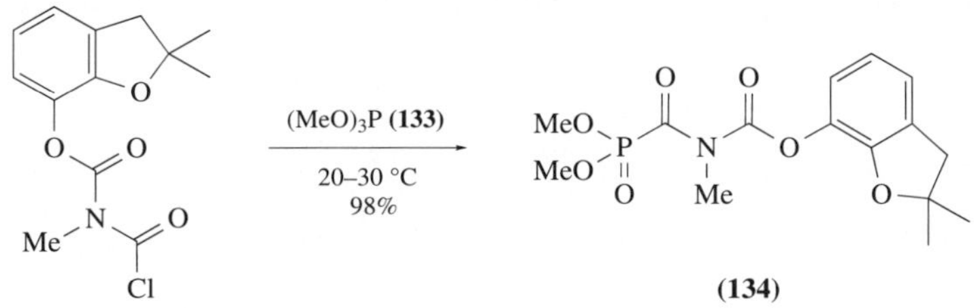

(**132**) (**134**) (35)

6.16.1.2.3 Carbonyl derivatives with one nitrogen and one arsenic function (carbamoylarsines)

The synthesis of carbamoylarsines (**137**) has been carried out by one of two methods; the reaction of alkyl or aryl arsines (**135**) with phenyl isocyanate (**136**) in the presence of dibutyltinacetate to form *N*-phenylcarbamoyl arsines (Equation (36)), or the reaction of lithium arsides (**138**) with cyclohexyl isocyanate (**139**) to form *N*-cyclohexylcarbamoyl arsines (**140**) (Scheme 13). The former reaction does not work for cyclohexyl isocyanate ⟨67LA(709)248⟩.

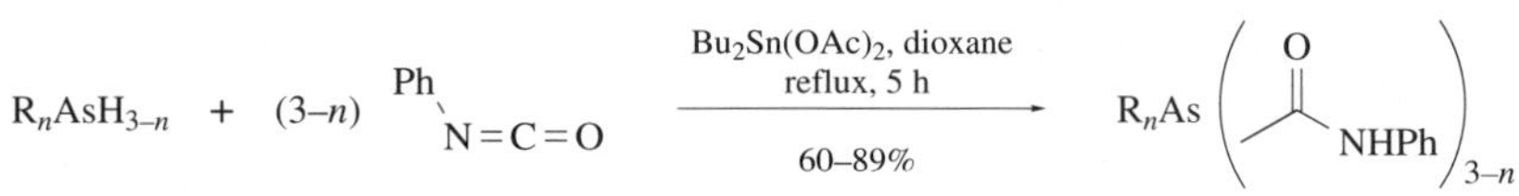

(**135**) (**136**) (**137**) (36)

R = Ph, n = 2; c-C_6H_{11}, n = 2; Ph, c-C_6H_{11}; Ph, n = 1; Bu, n = 1

6.16.1.2.4 Carbonyl derivatives with one nitrogen and one antimony function

No compounds of this type were found in the literature.

6.16.1.2.5 Carbonyl derivatives with one nitrogen and one bismuth function

The synthesis of compounds of this type was achieved by Mishra and Tandon first by forming a 1 : 1 adduct of DMF (**141**) with bismuth trichloride (**142**) and then by reacting this with bases to

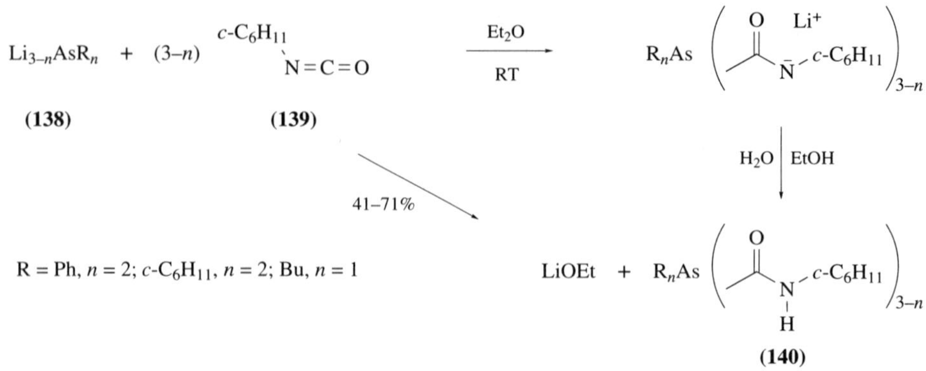

Scheme 13

form the isolated compounds such as (**143**). The bases used were triethylamine, diethylamine, α-picoline and pyridine. Scheme 14 gives an outline of the mechanism when pyridine is used ⟨71IC1896⟩.

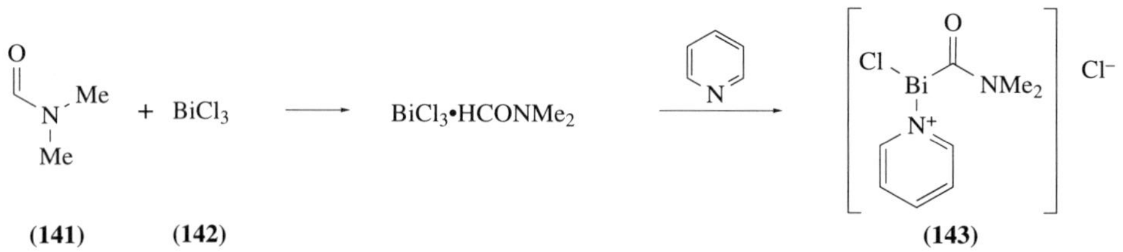

Scheme 14

6.16.1.3 Carbonyl Derivatives with One Nitrogen and One Metalloid (B, Si, Ge) Function

6.16.1.3.1 Carbonyl derivatives with one nitrogen and one boron function

Compounds of this type have been synthesized in many different ways; the reaction of H-Ala-OMe · HCl in dichloromethane with triethylamine, $Me_3N \cdot BH_2CO_2H$ and dicyclohexylcarbodiimide (dcc) gives L-$Me_3N \cdot BH_2CONHCHMeCO_2H$ ⟨89EUP336772⟩. Dipeptide and tripeptide esters can similarly be used ⟨90EJM301⟩. Boron analogues of hydroxamic acids can be made by reacting hydroxylamine hydrochloride with $Me_3N \cdot BH_2CO_2H$ in water to give an 85% yield of $Me_3N \cdot BH_2C(O)NHOH \cdot HCl$ ⟨88IC302⟩; $Me_3N \cdot BH_2CO_2H$ reacts with R^1R^2NH in the presence of dcc in chloroform at room temperature to give $Me_3N \cdot BH_2CONR^1R^2$ ⟨90BCJ3658⟩; cyclic examples may be synthesized (e.g., a carbamoyl analogue (**144**) of the cyanoborane oligomer $(BH_2CN)_x$) by putting $(EtO)_2P(O)CH_2NMe_2 \cdot BH_2CONHEt$ on silica gel ⟨91IC2433⟩. The reaction of phenylaniline methyl ester hydrochloride with trimethylamine-carboxyborane, triphenylphosphine, carbon tetrachloride and triethylamine in acetonitrile for 24 h yields $Me_3N \cdot BH_2CO$-Ph-OMe ⟨92MI 616-01⟩.

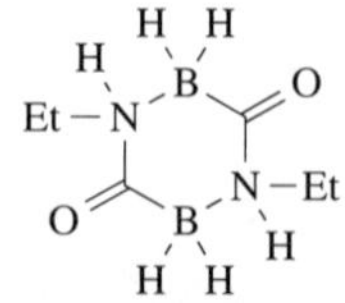

(144)

A common way of making these compounds is exemplified by that reported by Sood *et al.*, which involved the reaction of cyanoborane (**145**) with triethyl phosphite (**146**) to form the triethylphosphite-cyanoborane adduct (**147**) followed by ethylation with triethyloxonium tetrafluoroborate to form the nitrilium ion (**148**). This was readily hydrolysed to triethyl phosphite-*N*-ethylcarbamoylborane (**149**) (Scheme 15) ⟨91T6915⟩. Many other papers make use of the same method on different substrates ⟨76JA5702, 79JINC1223, 81JINC457, 89CC900, 90IC554, 90IC3218, 91IC1046, 92IC4911⟩.

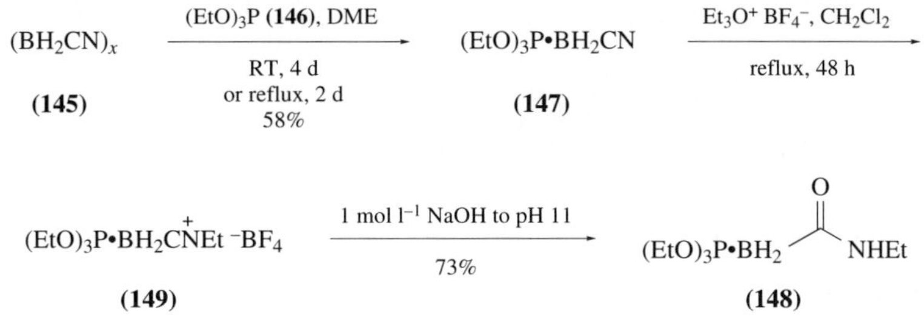

Scheme 15

6.16.1.3.2 *Carbonyl derivatives with one nitrogen and one silicon function (carbamoylsilanes)*

Carbamoylsilanes may be made by the addition of hexamethyldisilathiane (**151**) to *N*,*N*-diethylcarbamoylmercury (**150**) in ether. The product, *N*,*N*-diethylcarbamoylsilane (**152**), however, can decarbonylate (Equation (37)) ⟨69CC462⟩.

$$\underset{(150)}{Hg(CONEt_2)_2} + \underset{(151)}{(TMS)_2S} \xrightarrow{Et_2O} \underset{(152)}{TMS\text{-}CONEt_2} \qquad (37)$$

Hydrosilylation of alkyl isocyanates with triethylsilane in the presence of palladized charcoal or palladium dichloride at 80–130°C for 6 h yields 40–90% of the desired product ($RNHCOSiEt_3$) ⟨77JOM(140)97⟩. Deoxygenation of cyclohexyl isocyanate with $Ph_2(Me_3C)Si^-Li^+$ in an NMR tube at −50°C in THF resulted in the identification and isolation of *N*-cyclohexyl-*t*-butyldiphenylsilylcarboxamide (**153**) ⟨83T2989⟩. A solution of $(TMS)_3SiLi(THF)_3$ in pentane–THF (10:1) added over 2 h to a solution of Me_2NCOCl in dry pentane under dry nitrogen at −30°C then stirred at room temperature yields 75% of $(TMS)_3SiCONMe_2$ ⟨91JOM(403)293⟩.

(153)

6.16.1.3.3 *Carbonyl derivatives with one nitrogen and one germanium function (carbamoylgermanes)*

Carbamoylgermanes may be synthesized by treating lithium *t*-butylamide (**154**) with carbon monoxide (**155**) to form (*t*-butylcarbamoyl)lithium (**156**), which reacts with trimethylgermyl chloride (**157**) to form the required product (**158**) (Scheme 16). Yields are in the region of 55% ⟨71AG(E)339⟩.

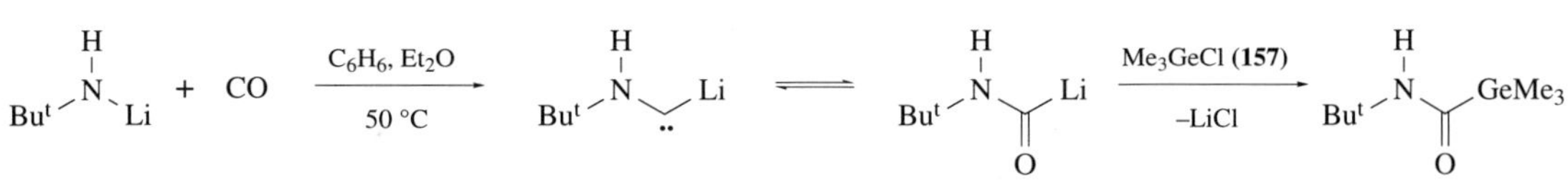

(154) **(155)** **(156)** **(158)**

Scheme 16

Another method for the preparation of carbamoylgermanes is the reaction of triethylgermyllithium (**159**) with amides (**160**) in hexane at −30°C for 5 h followed by hydrolysis. For

R = Me, the yield of *N*,*N*-dimethylamidetriethylgermylcarbonyl acid (**161**) is 45% (Equation (38)) ⟨84IZV1897⟩.

(159) **(160)** R = Me, Et **(161)** (38)

6.16.1.4 Carbonyl Derivatives with One Nitrogen and One Metal Function

6.16.1.4.1 Carbon monoxide insertion reactions

The insertion of carbon monoxide into a previously formed metal–nitrogen bond is the most-studied method of synthesizing this type of compound ⟨87OM372, 88AOC(28)139, 88CRV1059, 89CCR1⟩. The first example of carbon monoxide insertion into a d- or f-element metal to dialkylamide bond giving rise to carbamoyl insertion products was reported by Fagan ⟨81JA2206⟩. The facile insertion of carbon monoxide into chlorobis(pentamethylcyclopentadienyl)uranium and thorium dialkylamido complexes (**162**) led to the formation of the η^2-carbamoyl complexes M[η-$Me_5C_5]_2[\eta^2$-$CONR_2$]Cl (**163**) (Equation (39)). Bis(dialkylamido) compounds (**164**) are more reactive toward carbon monoxide than the chloro analogues (**162**) (Scheme 17). The yield of bis(carbamoyl)M[η-$Me_5C_5]_2[\eta^2$-$CONR_2]_2$ (**166**) is ~30%, but when the solutions are maintained under vacuum at 100 °C, carbon monoxide loss occurs and the bis(carbamoyl) products (**166**) revert to the mono insertion products M[η-$Me_5C_5]_2[\eta^2$-$CONR_2$]NR_2 (**165**).

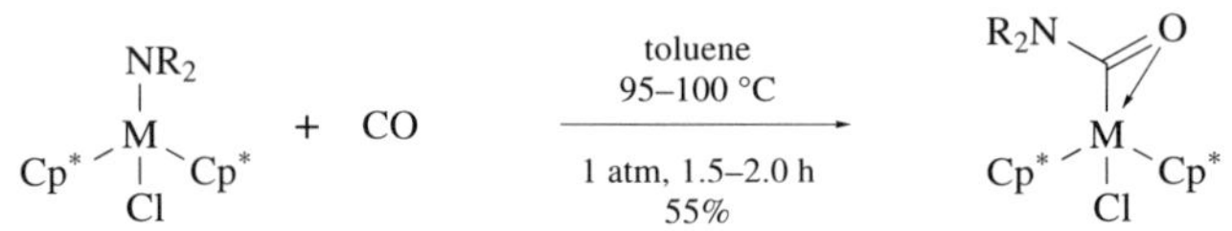

(162) M = Th, U; R = Me, Et **(163)** (39)

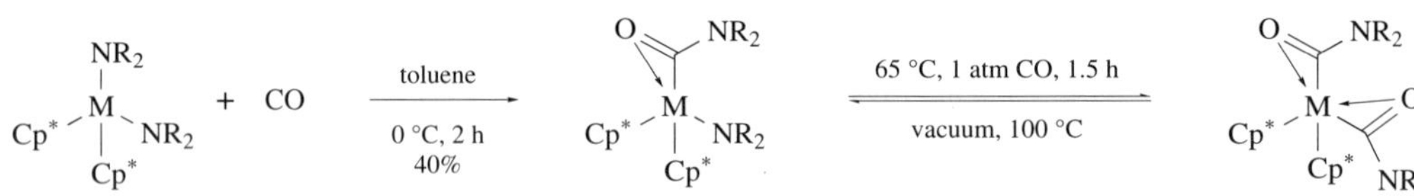

(164) **(165)** **(166)**

M = Th, U; R = Me, Et

Scheme 17

(dppe)PtMe[N(CH_2Ph)(H)] (**167**) reacts with 3 atm of carbon monoxide at 25 °C to give (dppe)PtMe(CONHCH_2Ph) (**168**) in 80% yield (Equation (40)) ⟨85OM939⟩. Similarly, the reaction of (dppe)PtMeNMe_2 (generated *in situ*) with carbon monoxide leads to (dppe)PtMe($CONMe_2$) in 76% isolated yield.

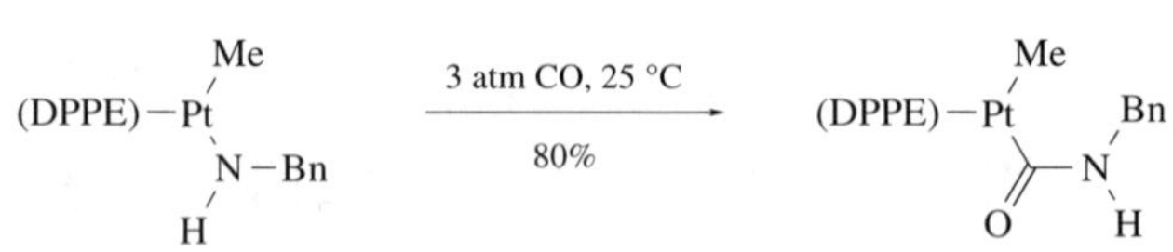

(167) **(168)** (40)

Metals in groups 8, 9 and 10 bearing amide ligands are not well known. Cowan and Trogler have reported the carbon monoxide insertion reaction into the platinum–amide bond of Pt($Ph_2PCH_2CH_2PPh_2$)(Me)[NH(CH_2Ph)] to generate the carbamoyl metal complex Pt($Ph_2PCH_2CH_2PPh_2$)(Me)[CONH(CH_2Ph)] ⟨87OM2451⟩. Carbon monoxide can also be inserted into the ruthenium–amide bond. When carbon monoxide is bubbled through a solution of η^5-CpRu(NH_2)($Cy_2PCH_2CH_2PCy_2$) (Cy = cyclohexyl), the carbamoyl complex η^5-CpRu($CONH_2$)($Cy_2PCH_2CH_2PCy_2$) is isolated ⟨91OM2781⟩. An unusual reaction has been reported involving the insertion of carbon monoxide into the iridium–nitrogen bond of Ir(CO)(NHAr)$(PPh_3)_2$ (**169**) (where Ar = Ph, *p*-C_6H_4Me) at room temperature producing a 90% yield of the carbamoyl–metal complex (**170**) (Equation (41)) ⟨93OM2401⟩.

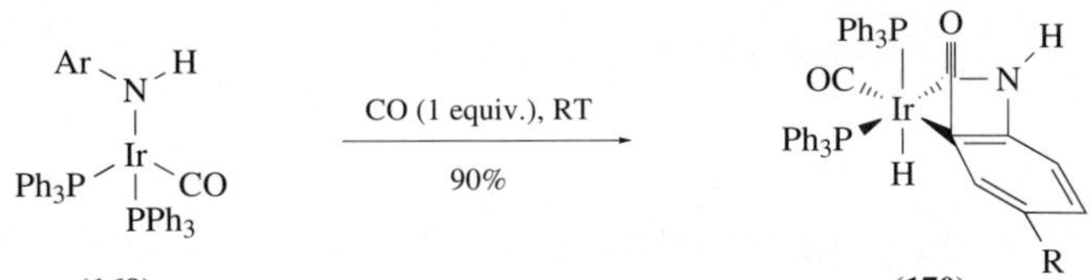

(**169**) (**170**) (41)

6.16.1.4.2 *Nucleophilic attack on metal carbonyl complexes*

Carbamoyl–metal complexes can also be made by nucleophilic attack on metal carbonyls. The direct addition of a nucleophilic nitrogen anion (Nu^-) to a coordinated carbon monoxide ligand (M = Fe, Ru, Os, Co, Re, Mo, W) leads to the required compounds ⟨79JA1627, 82IC1704, 84JA1125, 85JA2355, 85M1103, 85OM478, 85OM1523, 86JOM(312)C21, 87IC526⟩. The attack of the conjugate acid of a nitrogen nucleophile (NuH) on metal carbonyls (M = Pt, Ni, Ir, Fe, Ru, Os, Mn) also results in the preparation of carbamoyl–metal complexes. This method is restricted to strongly activated, usually cationic, metal carbonyl complexes since NuH is always a much weaker nucleophile than Nu^- ⟨75IC2148, 76IC2346, 79IC1165, 80AOC(18)1, 84OM1523, 85OM590⟩. The attack of the nitrogen nucleophile on metal carbonyls (M = Fe, Ni) can be aided by Y groups which make the nitrogen electron rich, thus promoting nucleophilic attack (Y = Li, MgX (X = Hal), HgR, $C(NMe_3)_3$, $HB(OR)_2$, $B(OR)_3$) ⟨83OM1044⟩.

6.16.1.4.3 *Other methods*

Carbamoyl–metal complexes can be synthesized by using an electron-rich transition-metal complex to displace a halide from an organic carbamoyl halide. For example, $FeCp(CO)_2CONR_2$ (R = Me, Et) has been synthesized in this way ⟨63JA1918⟩. $WCp(CO)_2CONHMe$ (**173**) has been made in the reaction of $WCp(CO)_2Me$ (**171**) with hydrogen isocyanate (**172**) (Equation (42)) ⟨72JA3799⟩. *Trans*-$[Pt(PPh_3)_2(CNMe)CONHMe]^+$ (**175**) has been made by the attack of hydroxide ion on *trans*-$[Pt(PPh_3)_2(CNMe)_2]^{2+}$ (**174**) (Equation (43)) ⟨71JA5424⟩.

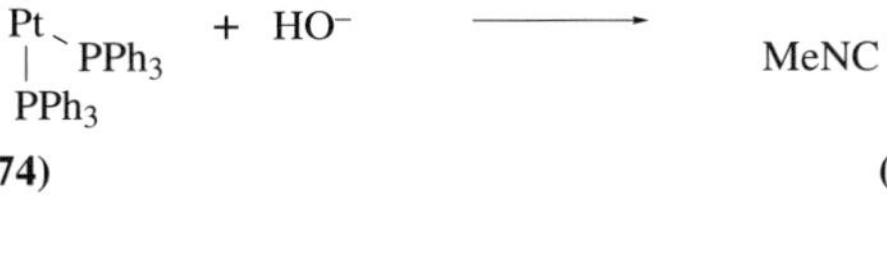

(**171**) (**173**) (42)

(**174**) (**175**) (43)

6.16.2 FUNCTIONS CONTAINING AT LEAST ONE PHOSPHORUS, ARSENIC, ANTIMONY OR BISMUTH FUNCTION (AND NO HALOGEN, CHALCOGEN OR NITROGEN FUNCTIONS)

6.16.2.1 Carbonyl Derivatives with Two P, As, Sb or Bi Functions

Carbonyl bis(diphenylphosphide) (**178**) may be synthesized by adding a solution of diphenyl-(trimethylsilyl)phosgene (**177**) in diethyl ether dropwise to a solution of phosgene (**176**) in diethylether at $-100\,^{\circ}C$ under nitrogen. The reported yield is 40% and involves the production of trimethylsilyl chloride (**179**) (Equation (44)) ⟨73AG(E)842⟩.

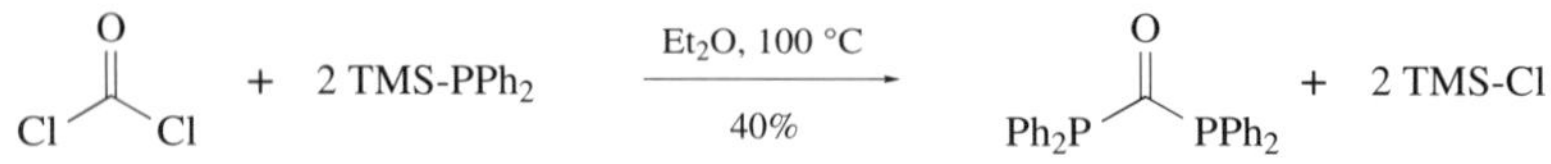

(**176**) (**177**) (**178**) (**179**) (44)

Treatment of $Cl_2P(O)CH{=}CHOEt$ (**180**) with NOCl (**181**) in carbon tetrachloride at $0\,^{\circ}C$ gives a reported 85% yield of $Cl_2P(O)CH(NO)CH(OEt)Cl$ (**182**) which, on ethanolysis, gives a mixture containing Cl_2CHOEt (**183**), $EtO_2CNHCH(OEt)_2$ (**184**), $(EtO)_2P(O)CONHCH(OEt)_2$ (**185**) and $[(EtO)_2P(O)]_2CO$ (**186**) (Scheme 18) ⟨80MI 616-01⟩. Carbonylbis(phosphonates) (**188**) are prepared by alkaline hydrolysis of a salt of a dihalomethylenediphosphonic acid (**187**). Thus, compound (**187**) and sodium hydroxide are refluxed for 1–6 h in water to yield 63% of compound (**188**) (Equation (45)) ⟨67JOC4111, 70USP3497313⟩.

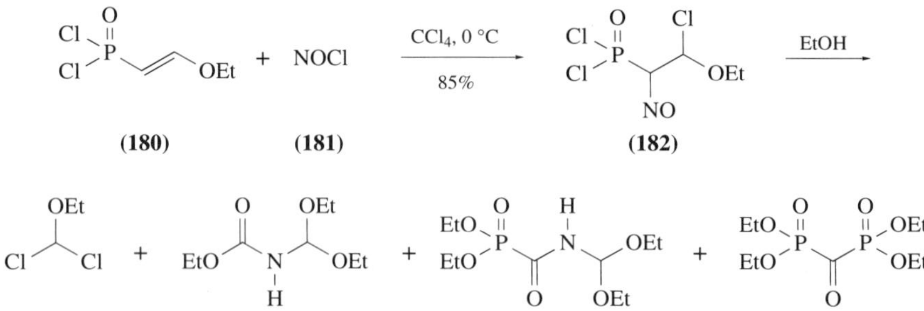

(**180**) (**181**) (**182**)

(**183**) (**184**) (**185**) (**186**)

Scheme 18

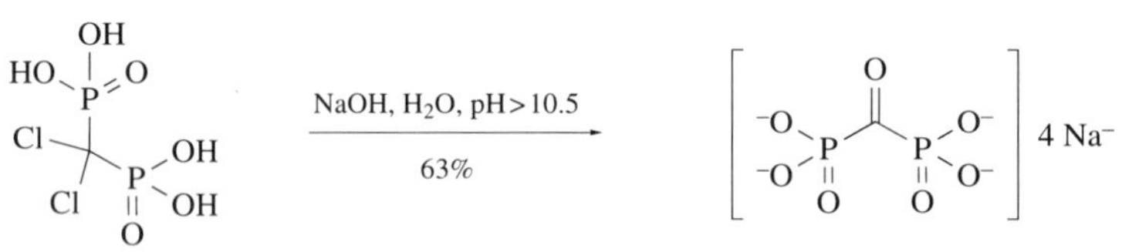

(**187**) (**188**) (45)

Metal carbonyl complexes have also been prepared containing phosphorus ligands bridged by carbonyl groups ⟨90JOM(383)295⟩. The addition of $Pr^i_2NPCl_2$ (**190**) to a solution of $Na_2Fe(CO)_4$ (**189**) in ether at $-78\,^{\circ}C$, followed by 24 h stirring at room temperature, results in the isolation of a 35% yield of $(Pr^i_2NP)_2COFe_2(CO)_6$ (**191**). The amine must be sterically hindered for this reaction to succeed (Equation (46)) ⟨85IC4449, 87JA7764⟩. These compounds may also be prepared by inserting carbon monoxide into a P—P bond. The reaction of the iron complex (**192**) with carbon monoxide at $80\,^{\circ}C$ and 50 atm pressure results in the isolation of the phosphorus-bridging carbonyl derivative (**193**) as the major product (Equation (47)) ⟨86AG(E)755, 86ZN(B)283, 87CB1421⟩.

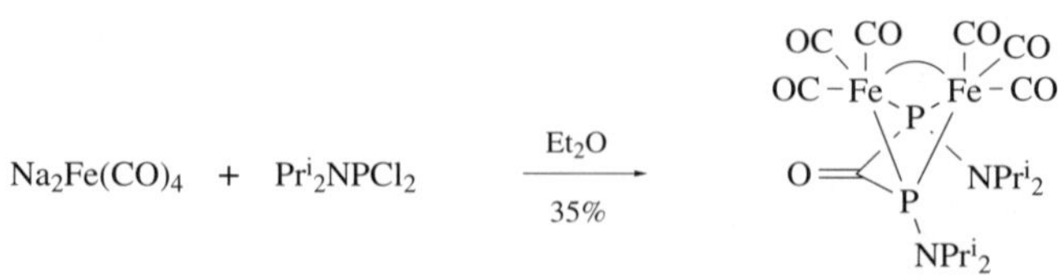

(**189**) (**190**) (**191**) (46)

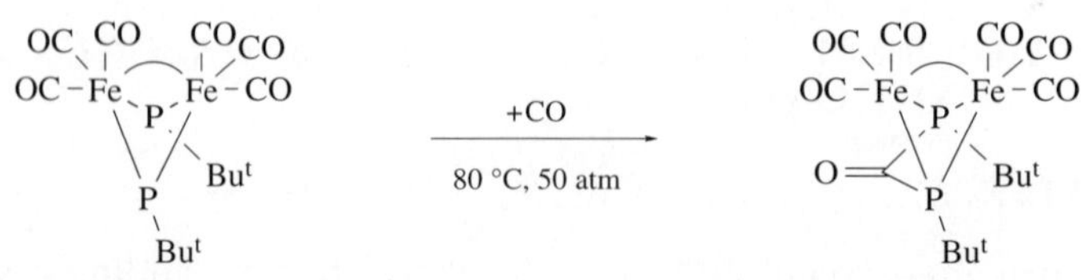

(**192**) (**193**) (47)

Phosphoureas, such as 1,3-bis(trimethylsilyl) substituted diphosphourea derivatives (**194**), may be cyclized to form 1,3-di-*t*-butyl-2-alkyl-1,2,3-triphosphetan-4-ones (**198**) in good yield. The reaction takes place at room temperature in toluene over 12–36 h (Scheme 19) ⟨83CB2371⟩.

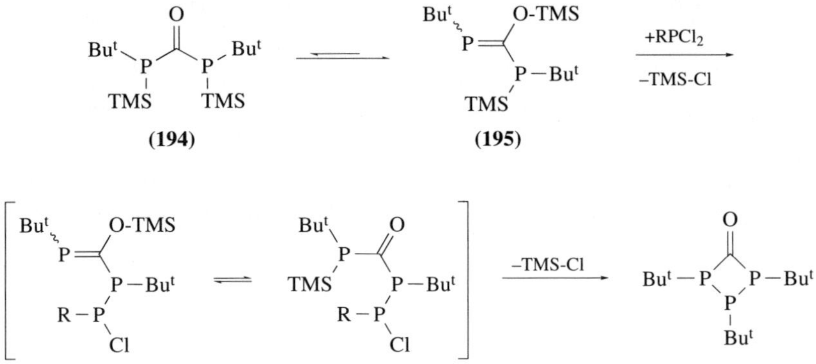

(194) (195) (196) (197) (198)

R = Me, Ph, Bu^t

Scheme 19

No references were found to carbonyl derivatives with two As, Sb or Bi functions.

6.16.2.2 Carbonyl Derivatives with One P, As, Sb or Bi Function Together with a Metalloid or Metal Function

No references were found for carbonyl derivatives either with one P, As, Sb or Bi function together with a metalloid function or with one As, Sb or Bi function together with a metal function. There are, however, reports of carbonyl derivatives with one phosphorus and a metal function. They are made by carbon monoxide insertion into metal–phosphorus bonds. For example, $Cp^*HfCl_2(PBu^t_2)$ (**199**) ($Cp^* = \eta^5$-C_5Me_5) has been shown to react with 1 equivalent of carbon monoxide within minutes at 25 °C to generate the insertion product (**200**) (Equation (48)) ⟨85JA4670, 87JCS(D)2039, 88CRV1059⟩.

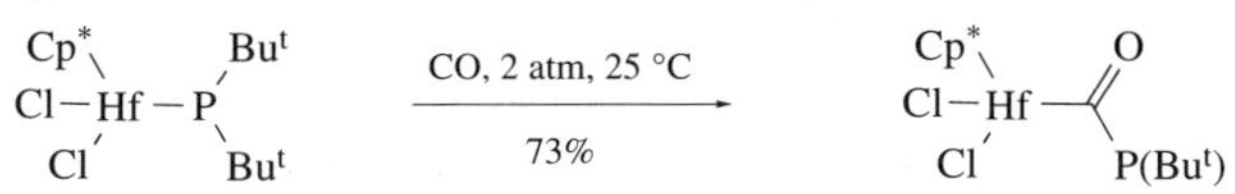

(199) (200) (48)

6.16.3 FUNCTIONS CONTAINING AT LEAST ONE METALLOID FUNCTION (AND NO HALOGEN, CHALCOGEN OR GROUP 5 ELEMENT FUNCTIONS)

6.16.3.1 Carbonyl Derivatives with Two Silicon Functions

Bis(silyl) ketones may be prepared by the oxidation of the corresponding carbinol. For example, bis(triphenylsilyl) ketone (**202**) has been synthesized by oxidizing 1,1-bis(triphenylsilyl)methanol (**201**) with chromium trioxide and sulfuric acid in diethylether at room temperature for 45 min (36% yield); by reacting the methanol derivative (**201**) in ether successively with dcc, pyridinium trifluoroacetate and DMSO and stirring for 12 h at room temperature (26% yield); and by oxidizing the methanol derivative (**201**) in carbon tetrachloride solution with benzoyl peroxide and *N*-bromosuccinimide at 65 °C (6% yield) (Equation (49)) ⟨68CJC2119⟩.

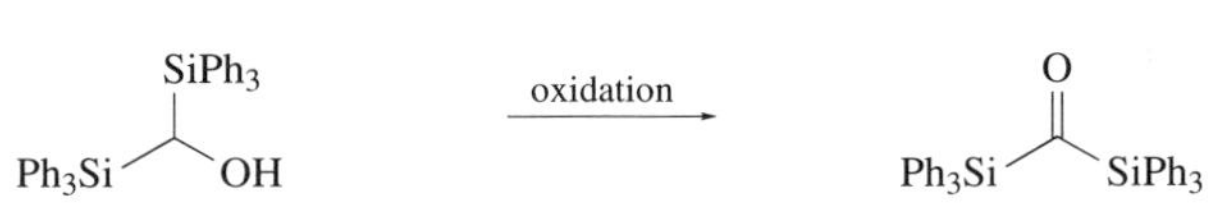

(201) (202) (49)

Bis(trimethylsilyl) ketone (**205**) has been prepared by the reaction of the bis-silylated ylide (**203**) with the freshly prepared adduct formed from ozone and triphenylphosphite (**204**) in toluene at −78 °C for 5 h (Equation (50)) ⟨85AG(E)1068, 92AG(E)1064⟩.

(50)

(203) **(204)** **(205)**

The oxidation of tris(trimethylsilyl)-methylthiomethane (**206**) with *m*-chloroperoxybenzoic acid (mcpba) leads to bis(trimethylsilyl) ketone (**205**) under very mild conditions. A solution of mcpba in dichloromethane is added slowly, under nitrogen, to a solution of the thiomethane derivative (**206**) in dichloromethane at −78 °C. After 10 min stirring and workup, the ketone (**205**) is isolated in 45% yield (Equation (51)) ⟨86TL5985⟩.

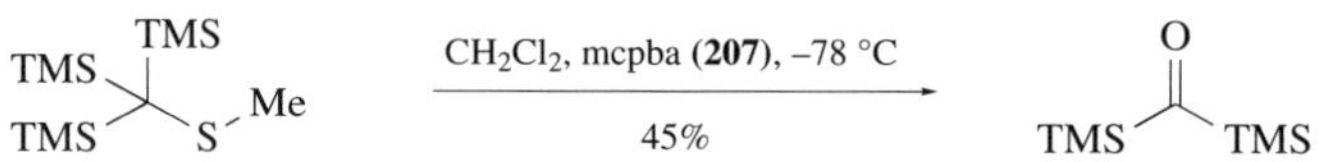

(51)

(206) **(205)**

6.16.3.2 Carbonyl Derivatives with Two Boron Functions

No references were found to carbonyl derivatives with two boron functions.

6.16.3.3 Other Carbonyl Derivatives with Two Dissimilar Metalloid Functions

Germyl silyl ketones have been prepared by hydrolysing the corresponding 2-substituted 1,3-dithianes. For example, triethylgermyl trimethylsilyl ketone (**211**) is synthesized by reacting 2-triethylgermyl-1,3-dithiane (**208**) with *n*-butyllithium in THF over 1–3 h at −20 °C to form the carbanion which readily reacts with trimethylsilyl chloride at 0 °C to form 2-triethylgermyl-2-trimethylsilyl-1,3-dithiane (**209**) in 79% yield. Hydrolysis of the dithiane (**209**) with mercuric chloride (**210**) and cadmium carbonate in 10% water in methanol–THF at 25 °C over 90 min yielded the required ketone (**211**) (Scheme 20) ⟨67JA431⟩.

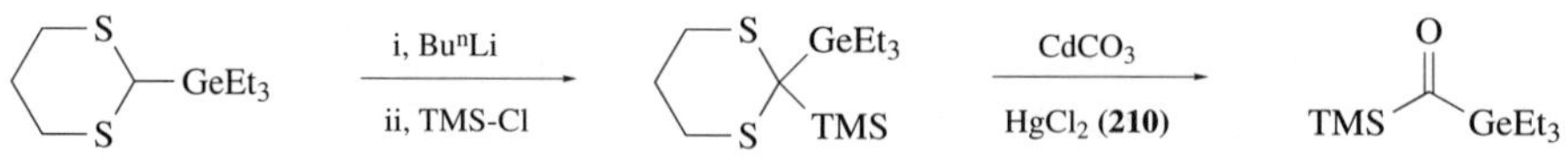

(208) **(209)** **(211)**

Scheme 20

6.16.3.4 Carbonyl Derivatives with One Metalloid and One Metal Function

Transition-metal acylsilanes have been of some interest since the preparation of the first transition-metal-bonded acylsilane, *fac*-Re(CO)$_3$(diphos)[COSiPh$_3$] (**215**). It is prepared by adding a THF solution of Ph$_3$SiLi to a slurry of [Re(CO)$_4$(diphos)][ClO$_4$] (**214**) in THF at 25 °C and reacting for 90 min. The transition-metal-bonded acylsilane (**215**) was isolated as air-stable purple crystals in 25–30% yield (Scheme 21) ⟨76JA4678⟩.

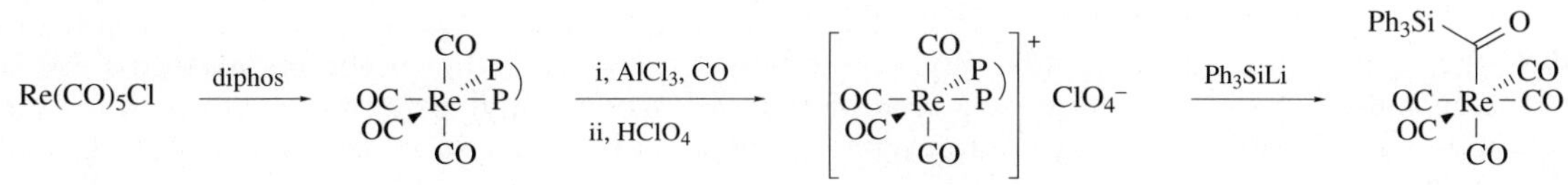

(212) **(213)** **(214)** **(215)**

Scheme 21

Reaction of $Cp_2Zr(TMS)Cl$ (**216**) with carbon monoxide (689 kPa) provides the silaacyl $Cp_2Zr(\eta^2\text{-CO-TMS})Cl$ (**217**) and the first observation of carbon monoxide insertion into a transition metal–silicon bond. The complex (**217**) was isolated in 90% yield. (Equation (52)) ⟨87JA2049⟩.

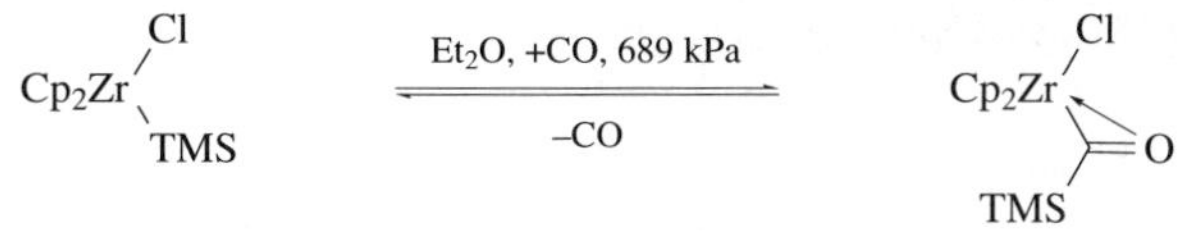

(52)

(216) **(217)**

Pentane solutions of $Cp_2Zr[Si(TMS)_3]TMS$ (**218**) react rapidly with carbon monoxide (689 kPa) to give an orange solution from which the silaacyl $Cp_2Zr(\eta^2\text{-COTMS})[Si(TMS)_3]$ (**219**) can be crystallized in 75% yield (Equation (53)). This reaction is generally applicable and the carbon monoxide insertion reaction of transition-metal silyl compounds has been widely reported ⟨85JA4084, 85JA6409, 86JA5355⟩.

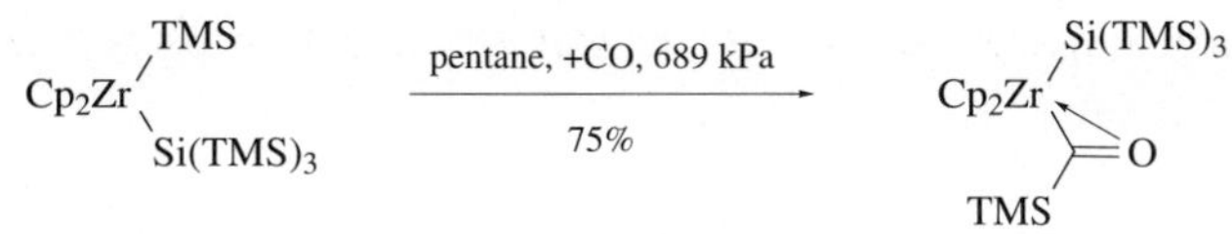

(53)

(218) **(219)**

6.16.4 CARBONYL DERIVATIVES CONTAINING TWO METAL FUNCTIONS

6.16.4.1 Metal Carbonyl Complexes and Fluxionality

Carbon monoxide reacts with transition metals to form transition-metal carbonyl complexes. Where two or more atoms of the transition metal are incorporated into the organometallic complex, the carbonyls adopt either a terminal or a bridged configuration. The energy difference between terminal and bridging carbonyls is usually low, so a fluxionality exists between the configurations. The compound (**220**) contains both terminal and bridging carbonyl ligands ⟨84AOC(23)219⟩. In a simple C-bonded bridge, the carbon monoxide ligand is perpendicular to the metal–metal axis and the carbon is equally bound to both metals, whereas the CO ligand may also be preferentially bound to one or other of the metal atoms in the bridging carbonyl complex. The carbon monoxide ligand is usually bound to two or more metal atoms in one of the ways shown in Figure 1. In (a) the M—C bonds are of equal length; in (b) the M—C bond lengths are significantly different; in (c) and (d) the carbon monoxide ligand is triply bridging, with the obvious differences between the bond lengths.

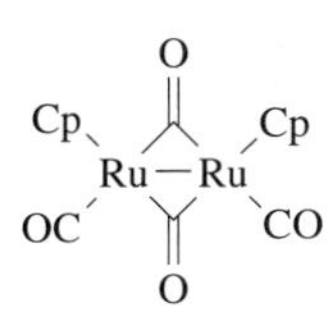

(220)

(a) **(b)** **(c)** **(d)**

Figure 1

If a metal carbonyl complex is synthesized in which there is more than one metal atom, some or all of the carbon monoxide ligands may bridge between two or more of the metal atoms. Some transition metals form only monomer complexes and so carbonyl bridging is impossible. The carbonyl-bridged transition-metal complexes are polymers of metal carbonyls where polymerization is possible.

6.16.4.2 Preparation of Metal Carbonyls

A large number of metal carbonyls is known and the following survey is intended to provide a brief summary of methods of preparation of the more common compounds. For more information, appropriate volumes of *Comprehensive Organometallic Chemistry* and a specialist monograph ⟨B-93MI 616-01⟩ should be consulted.

6.16.4.2.1 Metal carbonyls of titanium, zirconium and hafnium

Titanium hexacarbonyl $[Ti(CO)_6]$ has been produced by condensation of titanium metal vapour with carbon monoxide in a matrix of inert gases at 10–15 K and identified spectroscopically ⟨77IC822⟩. The cyclopentadienyl derivative, $Ti(C_5H_5)_2(CO)_2$, has been prepared by treating $TiCl_2(C_5H_5)_2$ with sodium cyclopentadienide and carbon monoxide under pressure ⟨61JA1287⟩. The first hafnium carbonyl, $(\eta^5\text{-}C_5H_5)_2Hf(CO)_2$, was prepared and studied by Sikora *et al.* ⟨79JA5079⟩. None of the above transition metal carbonyl complexes contains bridging carbon monoxide ligands.

6.16.4.2.2 Metal carbonyls of vanadium, niobium and tantalum

Reduction of vanadium trichloride (VCl_3) by sodium in pyridine under 200 atm of carbon monoxide yields, following workup, the vanadium hexacarbonyl $[V(CO)_6]$ (an odd-electron transition-metal carbonyl complex), which cannot be isolated. The noble gas configuration is attained in the isolable anionic intermediate $[V(CO)_6]^-$ or in the super-reduced 18-electron species $[V(CO)_5]^{3-}$ ⟨81JA6100⟩. A direct synthesis of $V(CO)_6$ and the dimer $V_2(CO)_{12}$ by condensation with carbon monoxide in a matrix of noble gases has been reported ⟨76IC1666⟩. All the transition-metal carbonyl complexes of this group are extremely unstable.

6.16.4.2.3 Metal carbonyls of chromium, molybdenum and tungsten

This is the first group to be discussed in which metal carbonyls have been known for some time and are quite stable. Molybdenum hexacarbonyl, $Mo(CO)_6$, chromium hexacarbonyl, $Cr(CO)_6$, and tungsten hexacarbonyl, $W(CO)_6$, were all prepared for the first time in the early part of the twentieth century ⟨10JCS798, 27BSF1041, 28CR(187)564⟩. The monomers have been made by the reaction of carbon monoxide at 200–250 atm and 200–300 °C with the metals (M = Mo, W) ⟨30GEP531402, 31GEP547025⟩ by reacting anhydrous metal(III) chloride (M = Cr) or metal(V) chloride (M = Mo, W) with phenylmagnesium bromide and carbon monoxide at 4 °C and 1 atm and hydrolysing the reaction products ⟨27BSF1041, 35ZAAC(221)321, 47JA1723, 50IS156⟩ by reacting the metal chlorides with carbon monoxide at 100 atm and 0–10 °C in the presence of zinc or iron powders (M = Mo, W) ⟨40DOK(26)54, 40DOK(26)57⟩.

Chromium hexacarbonyl has also been prepared using a metal–pyridine system as the reducing agent. The reaction of anhydrous pyridine, magnesium powder and small amounts of iodine was used to carbonylate chromium(III) and chromium(II) salts at 130–180 °C and 100–300 atm for 4–12 h ⟨57JA3611, 59G809⟩. Lithium aluminum hydride has also been used as the reducing agent ⟨59MI 616-02⟩.

The most versatile method of preparing the hexacarbonyls of this group is the reductive carbonylation with trialkylaluminum compounds, for which yields of 92%, 76% and 92% have been reported for $Cr(CO)_6$, $Mo(CO)_6$ and $W(CO)_6$, respectively ⟨60JA1325⟩.

Reduction of the hexacarbonyls with a borohydride in liquid ammonia forms dimeric $[M_2(CO)_{10}]^{2-}$. Chromium hexacarbonyl $Cr(CO)_6$ forms chromocene $Cr(\eta^5\text{-}C_5H_5)_2$ on reaction with sodium cyclopentadienide, while under the same conditions the molybdenum and tungsten hexacarbonyls form only $[(\eta^5\text{-}C_5H_5)M(CO)_3]_2$. The chromium analogue of these dimers can also be formed ⟨B-84MI 616-01⟩.

6.16.4.2.4 *Metal carbonyls of manganese, technetium and rhenium*

Decacarbonyldirhenium, $Rh_2(CO)_{10}$ was first prepared in 1941 by Hieber and Fuchs ⟨41ZAAC(248)256⟩. Decacarbonyldimanganese, $Mn_2(CO)_{10}$, was prepared and fully characterized in 1954 by Brimm *et al.* ⟨54JA3831⟩ and the synthesis of decacarbonylditechnetium, $Tc_2(CO)_{10}$, was first described in 1961 ⟨61AG579, 61JA2953, 65ZN(B)1159⟩.

(i) Decacarbonyldimanganese

Decacarbonyldimanganese has been prepared by reacting carbon monoxide at 200 atm at room temperature for 15–17 h with a mixture of magnesium powder, manganese iodide, copper and copper iodide suspended in diethyl ether ⟨54JA3831⟩, by the action of phenylmagnesium bromide or chloride and carbon monoxide at 30 atm on anhydrous $MnCl_2$ in diethyl ether at -20 to 30°C ⟨58USP2822247⟩, and by reducing manganese(II) salts with sodium benzophenone ketyl in THF, carbonylating the resulting mixture with carbon monoxide at 200–700 atm and 65–200°C and hydrolysing and steam distilling the $Mn_2(CO)_{10}$ from the resulting mixture ⟨58JA6167⟩. The reduction of an anhydrous manganese(II) salt with a trialkylaluminum compound dissolved in an ether or benzene in the presence of carbon monoxide under pressure results in a 60% yield of the required compound ⟨58IZV100, 60JA1325, 65IC293⟩.

(ii) Decacarbonylditechnetium

This carbonyl has been prepared by the action of carbon monoxide at 250–350 atm on the heptoxide Tc_2O_7 at 220–275°C for 12–20 h ⟨61AG579, 61JA2953, B-64MI 616-01, 65ZN(B)1159⟩.

(iii) Decacarbonyldirhenium and higher polymers of rhenium hexacarbonyl

Rhenium heptoxide, Re_2O_7, is reduced by carbon monoxide at high temperature and pressure to give $Re_2(CO)_{10}$ (**221**) (Equation (54)) ⟨41ZAAC(248)256⟩. The reaction of $ReCl_5$ or $ReCl_3$ with carbon monoxide at 130°C and 250–280 atm for 8 h using sodium in THF as the reducing agent has yielded 70% of $Re_2(CO)_{10}$ ⟨63JCS1133⟩. The polynuclear tetracarbonylrhenium, $[Re(CO)_4]_n$, has been prepared by reacting Re_2S_7 with carbon monoxide (85 atm) at 200°C in the presence of copper powder.

$$Re_2O_7 + 17\,CO \longrightarrow \underset{(\mathbf{221})}{Re_2(CO)_{10}} + 7\,CO_2 \qquad (54)$$

6.16.4.2.5 *Metal carbonyls of iron, ruthenium and osmium*

(i) Iron carbonyls

The transition-metal carbonyls $Fe_2(CO)_9$ (**222**) and $Fe_3(CO)_{12}$ (**223**) are obtained by the action of sunlight on $Fe(CO)_5$ in glacial acetic acid or in acetic anhydride *in vacuo* ⟨27CB1424, 29MI 616-01⟩. The trimer (**223**) may also be prepared by oxidizing alkaline solutions containing carbonylferrates with manganese dioxide, MnO_2, followed by removal of excess MnO_2, acidification and extraction of the carbonyl with petroleum ether ⟨57ZAAC(289)324⟩. The cyclopentadienyl iron carbonyl dimer (**224**) has been prepared by reacting iron pentacarbonyl and dicyclopentadiene at 135°C in an autoclave. Prolonged reaction of the same reactants produces the tetrameric cluster compound, $[Fe(\eta^5\text{-}C_5H_5)(CO)]_4$, which involves CO groups which are triply bridging ⟨B-78MI 616-01⟩.

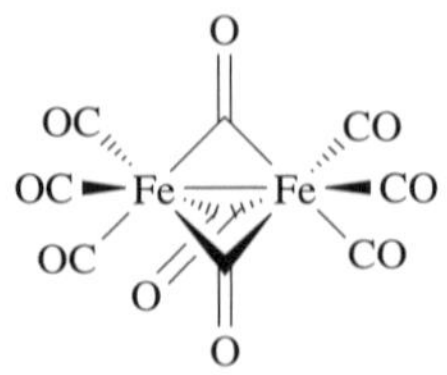

(222)

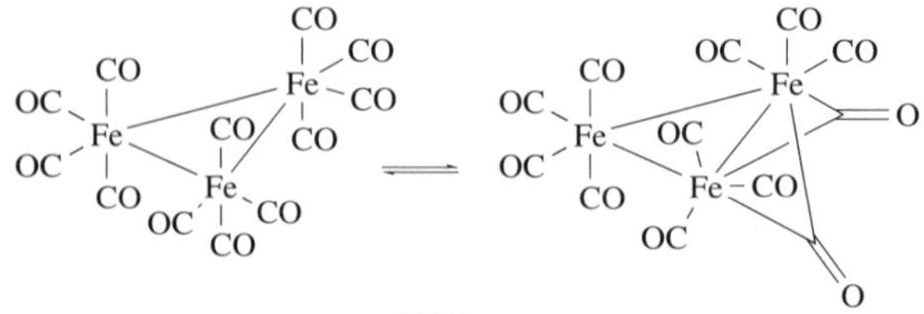

(223)

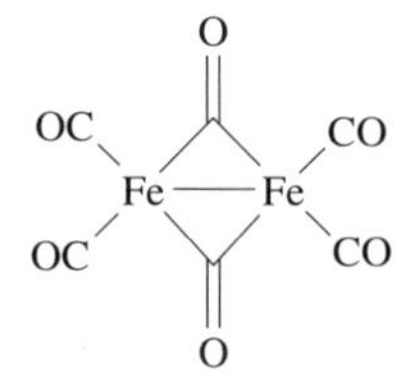

(224)

(ii) Ruthenium carbonyls

Both $Ru(CO)_5$ and $Ru_3(CO)_{12}$ are formed when RuI_3 is mixed with a large excess of silver powder and kept at 170°C for 24 h under carbon monoxide at about 450 atm in a rotating autoclave. The trimer is prepared from the monomer by heating it in hydrocarbons or alcohols for a short time in the presence of light ⟨36ZAAC(226)385⟩.

When ruthenium(III) acetylacetonate reacts with carbon monoxide and hydrogen (3 : 1) under pressure (150–200 atm) at 140–160°C in acetone, benzene or methanol, yields as high as 82% of $Ru_3(CO)_{12}$ are obtained ⟨64CI(M)206⟩.

(iii) Osmium carbonyls

The carbonyls $Os(CO)_5$ and $Os_3(CO)_{12}$ are formed when the halides ($OsCl_3$ and Os_2Br_9) and an oxyiodide are mixed with powdered copper or silver and reacted with carbon monoxide at high pressure (200–300 atm) and temperature (150–300°C) ⟨43MI 616-01⟩.

6.16.4.2.6 Metal carbonyls of cobalt, rhodium and iridium

Mononuclear carbonyls are not formed with this group of transition metals as the elements of this group can only satisfy the 18-electron rule in their carbonyls if metal–metal bonds are present. However, multinuclear carbonyls such as $Co_6(CO)_{16}$, $Rh_6(CO)_{16}$ and $Ir_4(CO)_{12}$ are readily formed ⟨40ZAAC(245)321, 66JA1821, 67CC440⟩.

(i) Cobalt carbonyls

The octacarbonyl $Co_2(CO)_8$ has been prepared by the reaction of carbon monoxide at 30–40 atm and 150°C with finely divided cobalt ⟨10JCS798⟩ or by reacting cobalt halides and sulfides with carbon monoxide in the presence of reducing metals, such as copper ⟨39ZAAC(240)261⟩. By treating $CoSO_4$ or $CoCl_2$ in aqueous ammonia solution with carbon monoxide (95–110 atm) at 120–140°C for 16–18 h a yield of 60% of the required carbonyl may also be obtained ⟨53LA(582)116⟩.

Adkins and Krsek prepared solutions of $Co_2(CO)_8$ by reacting Raney cobalt suspended in diethyl ether with carbon monoxide at about 250 atm and 150°C ⟨48JA383⟩. According to a Japanese patent, Co_2O_3 suspended in benzene is transformed into the carbonyl by reaction with carbon monoxide and hydrogen at 200 atm and 110°C in the presence of some pyridine and preformed $Co_2(CO)_8$ ⟨56JAP10924⟩. Another way to synthesize octacarbonyldicobalt is to use high pressures of carbon monoxide (100–200 atm) at high temperatures (140–300°C) with anhydrous cobalt salts of organic or inorganic acids dissolved or suspended in nonaqueous media, using hydrogen as the reducing agent. The reaction is accelerated by the addition of preformed $Co_2(CO)_8$ ⟨49USP2473993, 49USP2476263, 49USP2477553, 49USP2477554, 60CI(M)133, 60CI(M)137⟩.

The carbonyl $Co_4(CO)_{12}$ (**226**) is prepared by heating $Co_2(CO)_8$ (**225**) to 60°C for 24 h in an inert atmosphere with the evolution of carbon monoxide (Equation (55)) ⟨10JCS798⟩, or by using hydrocarbons as solvents in the reaction ⟨55CI(M)6, 55JA3951⟩. A method has also been reported for the preparation of $Co_4(CO)_{12}$ (**227**) from cobalt(II) ethylhexanoate or cobalt(II) and cobalt(III) acetylacetonates. These are reacted with hydrogen at 30–50 atm and with $Co_2(CO)_8$, giving yields of over 90% (Equation (56)) ⟨59CI(M)132⟩.

$$2\,[Co_2(CO)_8] \xrightarrow{60\,^\circ C} Co_4(CO)_{12} + 4\,CO \quad (55)$$

(**225**) (**226**)

$$2\,(RCO_2)_2Co + 3\,Co_2(CO)_8 + 2\,H_2 \longrightarrow Co_4(CO)_{12} + 4\,RCO_2H \quad (56)$$

(**227**)

(ii) Rhodium carbonyls

When rhodium metal is reacted at 200 °C with carbon monoxide (450 atm) for 15 h, the dimer $Rh_2(CO)_8$ can be isolated. The reaction of anhydrous $RhCl_3$ and carbon monoxide at 200 atm for 15 h in the presence of copper (or silver, cadmium or zinc) results in the isolation of the other polymers. At 50–80 °C, the tetramer $Rh_4(CO)_{12}$ is formed, whereas at 80–230 °C only $Rh_6(CO)_{16}$ is present ⟨43ZAAC(251)96⟩.

(iii) Iridium carbonyls

The reaction of iridium halides, IrX_3, with carbon monoxide at 350 atm for 24–48 h at 100–140 °C in the presence of copper yields a mixture of $Ir_4(CO)_{12}$ and $[Ir(CO)_4]_n$. Separation is based on the solubility of $[Ir(CO)_4]_n$ in ether and carbon tetrachloride relative to that of $Ir_4(CO)_{12}$ ⟨40ZAAC(245)321⟩.

6.16.4.2.7 Metal carbonyls of nickel, palladium and platinum

The first transition-metal carbonyl to be discovered was nickel tetracarbonyl, $Ni(CO)_4$, and it is used to produce metallic nickel. Reduction of $Ni(CO)_4$ leads to a number of polynuclear carbonylate anion clusters such as $[Ni_5(CO)_{12}]^{2-}$ and $[Ni_6(CO)_{12}]^{2-}$ ⟨B-84MI 616-01⟩.

Alkaline reduction of $[PtCl_6]^{2-}$ in an atmosphere of carbon monoxide yields a series of platinum carbonylate anion clusters, $[Pt_3(CO)_6]_n^{\,2-}$. By refluxing salts of the $n = 3$ anion in acetonitrile, the cluster anion $[Pt_{19}(CO)_{22}]^{4-}$ has been obtained ⟨79JA6110⟩.

6.16.4.2.8 Metal carbonyls of copper, silver and gold

Neutral binary carbonyls are not formed by these metals at room temperature, but some have been synthesized by the condensation of copper or silver vapour and carbon monoxide at temperatures of 6–15 K (e.g., $M_2(CO)_6$) ⟨76JA3167⟩.

6.16.4.2.9 Mixed metal carbonyls

These are prepared by reacting a halogenocarbonylmetal with carbonylmetallates of alkali metals (Equations (57), (58), (59), and (60)). The reactions are carried out in an ether, such as THF, and occur very rapidly.

$$NaMn(CO)_5 + ReCl(CO)_5 \longrightarrow NaCl + (CO)_5MnRe(CO)_5 \quad (57)$$

$$NaCo(CO)_4 + MnBr(CO)_5 \longrightarrow NaBr + (CO)_4CoMn(CO)_5 \quad (58)$$

$$NaCo(CO)_4 + Fe(C_5H_5)(CO)_2I \longrightarrow NaI + (CO)_4CoFe(C_5H_5)(CO)_2 \quad (59)$$

$$NaRe(CO)_5 + Fe(C_5H_5)(CO)_2Cl \longrightarrow NaCl + (CO)_5ReFe(C_5H_5)(CO)_2 \quad (60)$$

The anion $[FeCo_3(CO)_{12}]^-$ has been prepared by reacting $Co_2(CO)_8$ with $Fe(CO)_5$ in the presence of acetone ⟨60G1005⟩. Ultraviolet radiation of a hexane solution of $Fe(CO)_5$ and $Mn_2(CO)_{10}$ results in the formation of the mixed metal carbonyl $[(CO)_5Mn]_2Fe(CO)_4$ ⟨65ZN(B)1306⟩. Warming a THF solution of the salt $[Mn(CO)_6][Co(CO)_4]$ to room temperature results in the formation of $(CO)_5MnCo(CO)_4$ ⟨64CB2289⟩.

6.17
Functions Containing a Thiocarbonyl Group and at Least One Halogen; Also at Least One Chalcogen and No Halogen

ERICH KLEINPETER
University of Potsdam, Germany

and

KALEVI PIHLAJA
University of Turku, Finland

6.17.1 FUNCTIONS CONTAINING AT LEAST ONE HALOGEN

Functions containing at least one halogen have been described in a volume of Houben-Weyl ⟨83HOU(E4)407⟩ published in 1983. Thus the literature search for related compounds in Chapter 6.17 is based on *Chemical Abstracts* from 1980 to February 1994. The present text concentrates on literature from the mid 1980s; original references to synthetic methods given in Volume E4 of Houben-Weyl are mentioned only in order to locate them in cases of need. Volume E4 of Houben-Weyl can be strongly recommended because it includes detailed synthetic procedures ⟨83HOU(E4)407, 83HOU(E4)420⟩.

A number of clathrate compounds have proved useful for facilitating the handling of thiophosgene and similar reactive reagents ⟨92TL261⟩.

6.17.1.1 Thiocarbonyl Halides with Two Similar Halogens

6.17.1.1.1 Thiocarbonyl difluoride

(i) From 2,2,4,4-tetrafluoro-1,3-dithietane

Middleton *et al.* ⟨61JA2589, 65JOC1375⟩ prepared thiocarbonyl difluoride $F_2C{=}S$ (a colorless gas boiling at $-54\,^{\circ}C$) ⟨65JOC1375⟩ in high purity and yield (>90%) by a three-step procedure; first thiophosgene was dimerized and the dimer fluorinated with SbF_3. Pyrolysis of the resulting 2,2,4,4-tetrafluoro-1,3-dithietane at 475–500 °C led to thiocarbonyl difluoride in nearly quantitative yield (Scheme 1).

Scheme 1

(ii) Other methods

Thiocarbonyl difluoride was also obtained as a by-product of the reaction of bis(trifluoromethyl) trisulfide with Grignard reagents RMgX (R = Et, Pr^i) ⟨92JFC(59)91, 93JFC(60)85⟩. Middleton *et al.* ⟨65JOC1375⟩ also obtained thiocarbonyl difluoride from tetrafluoroethylene by reaction with sulfur at 450–500 °C. It was obtained in even higher yield (96%) by Marquis ⟨60USP2962529⟩ from chlorodifluoromethane by bubbling it through sulfur at 360 °C. Further $F_2C{=}S$ was obtained from the sulfenyl chloride FCl_2CSCl by reduction with tin and concentrated HCl but in lower yield (47.5%) ⟨59ZOB3792⟩.

6.17.1.1.2 Thiocarbonyl dichloride (thiophosgene)

(i) From trichloromethanesulfenyl chloride and other perchloromethyl derivatives containing sulfur

One of the most useful methods for the synthesis of thiophosgene is the reduction of perchloromethyl sulfenylchloride with H_2S at 110–114°C; this gives thiophosgene in 96% yield ⟨74S26⟩. A number of other reducing agents have been used ⟨83HOU(E4)407⟩, including tin and concentrated HCl (Equation (1)) ⟨89JAP03060418⟩.

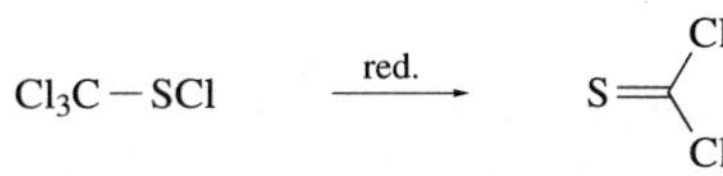
(1)

Alternatively, perchloromethyl derivatives containing sulfur can be thermally decomposed ⟨65GEP1219455⟩. A number of synthetic procedures are published which optimize the thiophosgene yield; in particular, aqueous H_2SO_3 in the presence of catalysts (S_2Cl_2, SCl_2, alkali metal iodide and/or I_2) proved useful ⟨89JAP03037110, 89JAP03060417, 89JAP03060418, 89JAP03197310, 91JAP03060419⟩.

(ii) From trichloromethyl thiol

Thiophosgene, a red liquid boiling at 73.5°C ⟨83HOU(E4)407⟩ was also prepared by reducing trichloromethyl thiol with SO_2 in the presence of KI and S_2Cl_2 ⟨85JAP62113712, 88JAP01257116⟩. When H_2S was used in place of S_2Cl_2, the yields obtained were higher than 97% (Equation (2)) ⟨86JAP62176910⟩.

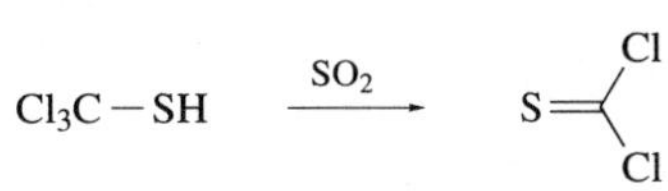
(2)

(iii) From carbon monosulfide

Thiophosgene can be obtained from carbon monosulfide CS (continuously produced by dissociation of CS_2 in a high-frequency discharge at 0.1 torr (mm Hg)) ⟨74IC1778⟩ by reaction with Cl_2 at room temperature ⟨67AG649, 68ZAAC(361)180⟩.

6.17.1.1.3 Thiocarbonyl dibromide

(i) From carbon monosulfide

Carbon monosulfide, continuously produced by dissociation of CS_2 in a high-frequency discharge (0.1 torr (mm Hg)), ⟨74IC1778⟩ reacts at room temperature with Br_2 giving thiocarbonyl dibromide ⟨67AG649, 68ZAAC(361)180⟩. This compound is an orange–red liquid boiling at 142–144°C ⟨65JOC1375⟩.

(ii) Other methods

Thiocarbonyl dibromide was obtained from $F_2C{=}S$ in 97% yield ⟨65JOC1375⟩ by addition of anhydrous hydrogen bromide; thiol intermediates eliminate the hydrogen fluoride (Equation (3)). Thiocarbonyl dibromide could also be obtained from thiophosgene but only in 31% yield ⟨81CB829⟩; the halogens were exchanged in this case with BBr_3 at 60–65°C.

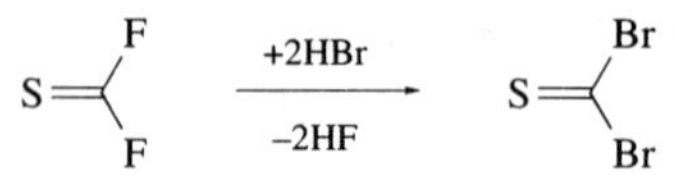

(3)

6.17.1.1.4 Thiocarbonyl diiodide

(i) From carbon monosulfide

Thiocarbonyl diiodide can be synthesized in a similar way to thiophosgene and thiocarbonyl dibromide, from carbon monosulfide by reaction with I_2 ⟨67AG649, 68ZAAC(361)180⟩. Although the diiodide could not be isolated, it was identified by IR spectroscopy ($\nu_{C=S} = 1062$ cm^{-1}; $\nu_{C-I} = 602$ cm^{-1}) ⟨67AG649, 68ZAAC(361)180⟩.

6.17.1.1.5 Sulfoxides of thiocarbonyl halides (sulfines) with two similar halogens

(i) Difluorosulfine, $F_2C{=}SO$

(a) From 1,1,4,4-tetrafluorodithietane. It was much more difficult to obtain this compound than the corresponding dichlorosulfine $Cl_2C{=}SO$ (see below). 2,2,4,4-Tetrafluoro-1,3-dithietane was easily oxidized with trifluorotriacetic acid (in dichloromethane at 0°C) to 2,2,4,4-tetrafluoro-1,3-dithietane-1-oxide; however, to get the corresponding 1,3-dioxide, Henn and Sundermeyer ⟨88JFC(39)329⟩ only succeeded by employing trifluoromethanetrisulfonic acid as a new oxidizing agent (Scheme 2); pyrolysis at 480°C and 0.01 torr (mm Hg) yields $F_2C{=}SO$ quantitatively. Difluorosulfine was detected only by mass spectrometry and decomposes spontaneously at −100°C to $F_2C{=}O$ and sulfur.

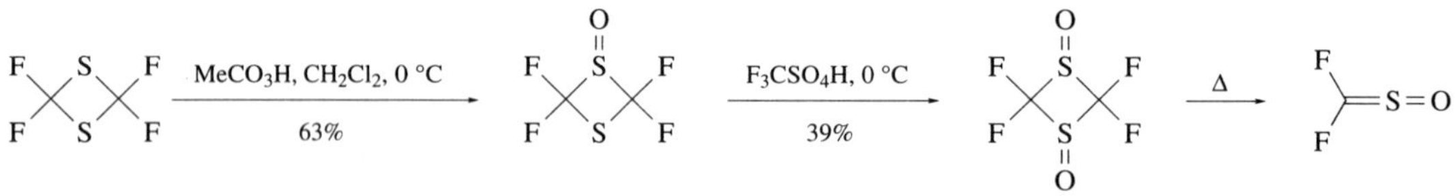

Scheme 2

(b) From allyl difluorochloromethyl sulfoxide. $F_2C{=}SO$ could also be obtained by pyrolysis of $F_2ClCS(O)CH_2CH{=}CH_2$ at 300–400°C ⟨86CB269⟩; again, it was identified by mass spectrometry.

(ii) Dichlorosulfine $Cl_2C{=}SO$

The synthesis, structure, analysis, and chemistry of thiophosgene-*S*-oxide (dichlorosulfine) have been reviewed ⟨92SUL275⟩.

(a) From 2,2,4,4-tetrachloro-1,3-dithietane. 2,2,4,4-Tetrachloro-1,3-dithietane was oxidized with trichlorotriacetic acid (in dichloromethane at 0°C), and the 1-oxide thus obtained further oxidized to the corresponding 1,3-dioxide ⟨83CB1623⟩. The latter compound is cleaved quantitatively to dichlorosulfine by means of vacuum pyrolysis at 480°C and 0.5 torr (mm Hg) (Scheme 3).

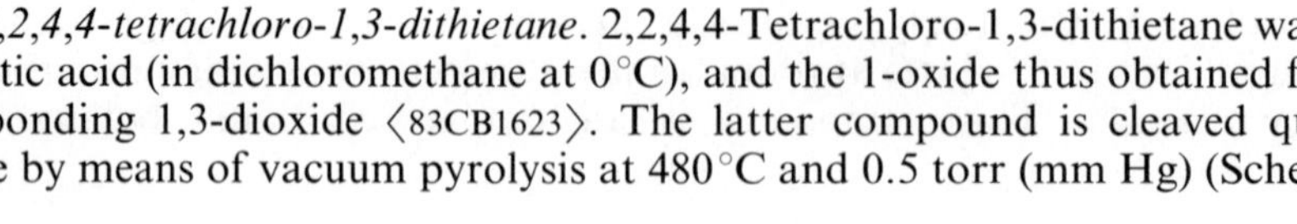

Scheme 3

(b) From thiophosgene. Thiophosgene-*S*-oxide was synthesized from thiophosgene in 32% yield by oxidation with mcpba (Equation (4)) ⟨69TL4461⟩. It is obtained as a yellow liquid (b.p. 34–36 °C at 25 torr (mm Hg)).

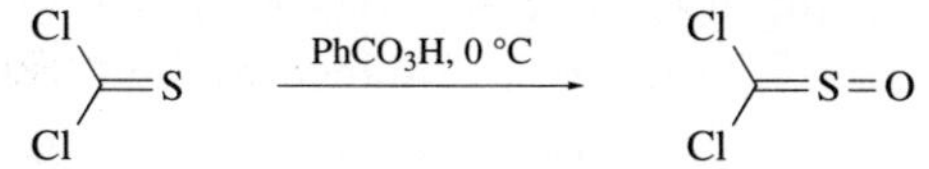

(4)

(c) From trichloromethanesulfenyl chloride. $Cl_2C{=}SO$ was also obtained by hydrolysis of trichloromethanesulfenyl chloride Cl_3CSCl in ca. 30% yield (slightly dependent on the reaction conditions) and was found to be a thermally unstable compound ⟨69CC878⟩.

(d) From allyl trichloromethyl sulfoxide. $Cl_2C{=}SO$ was obtained in 49% yield by pyrolysis of $Cl_3CS(O)CH_2CH{=}CH_2$ at 300–400 °C ⟨86CB269⟩.

(iii) Dibromosulfine $Br_2C{=}SO$

Dibromosulfine (a reddish liquid) was described for the first time by Schork and Sundermeyer ⟨85CB1415⟩. Tetrabromo-1,3-dithietane was oxidized with trifluorotriacetic acid in dichloromethane in two steps to the corresponding 1,3-dioxide; the pyrolysis of this 1,3-dioxide gave dibromosulfine in 70% yield (Scheme 4).

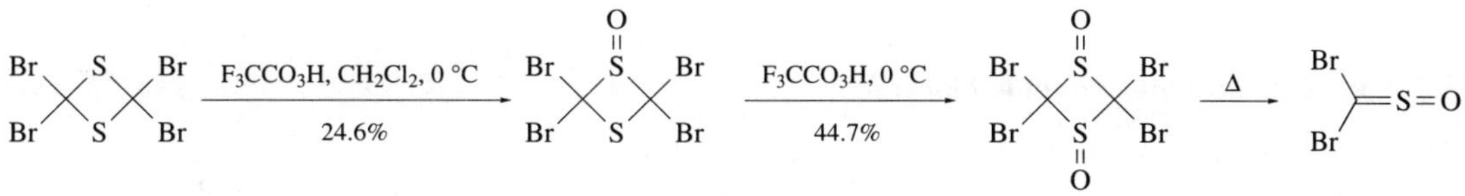

Scheme 4

The IR spectra of $F_2C{=}SO$, $Cl_2C{=}SO$ and $FClC{=}SO$ (Section 6.17.1.2.4) were obtained at 10–25 K in an argon matrix on a CsI window ⟨86SA(A)1281⟩ in order to differentiate the sulfoxides from the corresponding dihalocarbonyls and thiocarbonyl compound. Characteristic vibrations are given in Table 1 ⟨86SA(A)1281⟩.

Table 1 Characteristic frequencies of halocarbonyls, thiocarbonyls, and the corresponding sulfoxides (wavenumbers in cm^{-1}).

	$X^1X^2C{=}S$	$X^1X^2C{=}O$	$X^1X^2C{=}SO$
$X^1 = X^2 = Cl$	1130(νC—S)	1817(νC—O)	1169(νC—S)
	785($\nu_{ass}CCl_2$)	837($\nu_{ass}CCl_2$)	946($\nu_{ass}CCl_2$)
$X^1 = F$, $X^2 = Cl$	1257(νC—S)	1868(νC—O)	1297, 1246(νC—S)
	1014(νC—F)	1095(νC—F)	1131, 1145(νC—S)
	612(νC—Cl)	776(νC—Cl)	640(νC—Cl)?
			1050, 1082(νS—O)
$X^1 = X^2 = F$	1354(νC—S)	1930(νC—O)	1373(νC—S)
	1180($\nu_{ass}CF_2$)	1244($\nu_{ass}CF_2$)	1296($\nu_{ass}CF_2$)
			1118(νS—O)

6.17.1.2 Thiocarbonyl Halides with Two Dissimilar Halogens

6.17.1.2.1 Thiocarbonyl chloride fluoride

(i) From dichlorofluoromethanesulfenyl chloride

Thiocarbonyl chloride fluoride $FClC{=}S$ (a yellow liquid boiling at 7 °C) can be readily synthesized from the sulfenyl chloride FCl_2CSCl in 87% yield by reduction with tin and concentrated HCl ⟨59ZOB3792⟩.

(ii) Other methods

FClC=S was obtained as a by-product of the F_2C=S synthesis via the pyrolysis of 2-chloro-2,4,4-trifluoro-1,3-dithietane and could be readily separated from F_2C=S by distillation ⟨65JOC1375⟩. Also, benzenethiosulfonic acid *S*-dichlorofluoromethyl ester can be thermally decomposed (at 170–250 °C) (Equation (5)); the yields obtained were satisfactory ⟨83HOU(E4)407⟩.

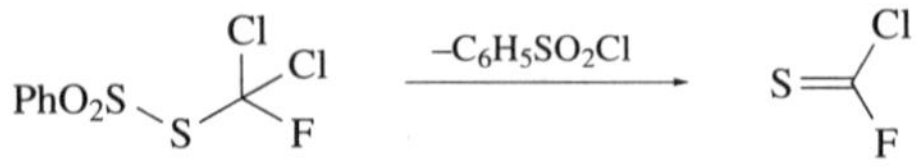

(5)

6.17.1.2.2 *Thiocarbonyl bromide fluoride*

Thiocarbonyl bromide fluoride, FBrC=S, a yellow liquid boiling at 4–8 °C and 100 torr (mm Hg), can be obtained in 34% yield from thiocarbonyl chloride fluoride by halogen exchange with BBr_3 at 60–65 °C ⟨81CB829⟩. FBrC=S spontaneously decomposes at room temperature.

6.17.1.2.3 *Thiocarbonyl bromide chloride*

Thiocarbonyl chloride bromide, ClBrC=S, a red liquid boiling at 47 °C and 80 torr (mm Hg), has been obtained from thiophosgene in 14% yield by halogen exchange with BBr_3 ⟨76CB3432, 81CB829⟩.

6.17.1.2.4 *Sulfoxides of thiocarbonyl halides (sulfines) with two dissimilar halogens*

(i) Chlorofluorosulfine ClFC=SO

Chlorofluorosulfine was synthesized from 2,4-dichloro-2,4-difluoro-1,3-dithietane by oxidation with trifluoroperacetic acid at −10 °C to the 1,3-dioxide; pyrolysis at 450 °C of the latter compound gives ClFC=SO nearly quantitatively. This could not be isolated but was identified by on-line mass spectrometry and photoelectron spectroscopy, respectively (Scheme 5) ⟨87CB1499, 88JFC(39)329⟩. The isomerism of the chlorofluorodithietanes was studied by ^{19}F NMR spectroscopy and x-ray structure analysis ⟨87CB1499⟩. The two isomers of chlorofluorosulfine (Equation (6)) were identified by IR and PE spectroscopy and, synthetically, by cycloaddition to 1,3-cyclopentadiene ⟨87CB1499⟩.

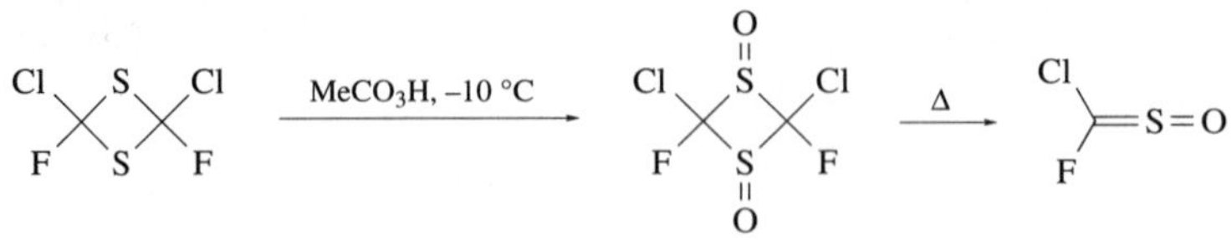

Scheme 5

(6)

ClFC=SO was also obtained by pyrolysis of $FCl_2CS(O)CH_2CH$=CH_2 at 300–400 °C; it was identified by means of mass spectrometry ⟨86CB269⟩.

6.17.1.3 Thiocarbonyl Halides with One Halogen and one Other Heteroatom Function

6.17.1.3.1 Fluorothioformates ROC(F)=S

Thiocarbonyl chloride fluoride reacts with alcohols at low temperature but without any solvent leading selectively to the alkyl fluorothioformates ROC(F)=S (R = Me, Et, Pr^i, Ph) ⟨59ZOB3792⟩.

6.17.1.3.2 Chlorothioformates ROC(Cl)=S

(i) From phenols and naphthols

Phenols and naphthols readily react with thiophosgene in the presence of a base to give aryl chlorothioformates in good yield (Equation (7)) ⟨62JOC4509, 65CB2063, 65LA(681)64, 67AG(E)281, 83HOU(E4)407⟩. Aryl chlorothioformates (in addition to those given in ⟨83HOU(E4)407⟩) which were synthesized in this way are listed in Table 2. Hydrocarbons and chlorohydrocarbons were used as solvents. Phenyl chlorothioformate is a yellow liquid boiling at 91 °C and 10 torr (mm Hg) ⟨59ZOB3792⟩. In a number of cases, the aryl chlorothioformates were not isolated but used for the synthesis of thiocarbonic acid derivatives ⟨81JA932, 82EUP62834, 83JA4059, 87USP4754072⟩. In some cases thiophosgene was produced directly in the reaction mixture (see above) (a) by chlorination of CS_2 ⟨85JAP62120358, 90JAP03220173⟩ from which the corresponding phenyl chlorothioformates were obtained in 93–96% yield; (b) by reduction of CCl_3CSCl ⟨85JAP61229860, 85JAP61229861, 88JAP2009858⟩, the aryl chlorothioformates being obtained in 63% yield and the corresponding naphthyl derivatives in 74% yield; and (c) by reducing CCl_3SH with SO_2 ⟨86GEP3616009⟩.

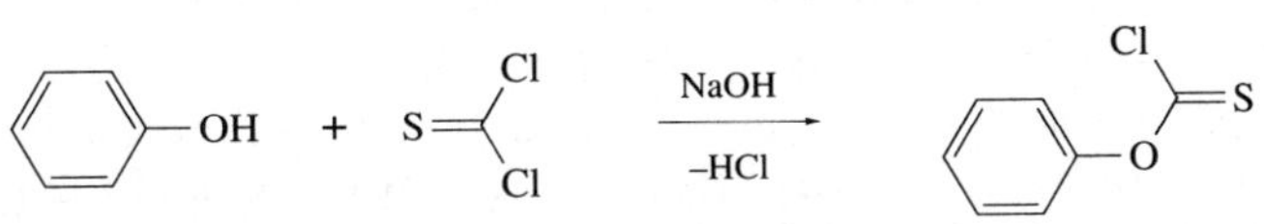

(7)

Table 2 Aryl chlorothioformates synthesized from thiophosgene and phenols or napthols.

Aryl—OH	*Base*	*Yield* (%)	*Ref.*
Phenol	NaOH/H_2O	95.0	91JAP03218336
			91JAP03193759
m-tert-butyl-phenol	NaOH/H_2O		93JAP05058990
m-tert-butyl-phenol	NaOH/H_2O	98.9	88JAP01301654
m-R^1-*p*-R-phenol	NaOH/H_2O		83JAP60155152
R^1 = halogen; R = alkyl $\geqslant$ 2			
5,6,7,8-tetrahydro-naphthol-2	NaOH/H_2O	89.9	90JAP04066566
5,6,7,8-tetrahydro-naphthol-2	NaOH/H_2O	80.9	90JAP04128263
			90JAP05032616
			87JAP01102058
5,6,7,8-tetrahydro-naphthol-2	NaOH/H_2O/$Na_2S_2O_4$	81.0	87JAP01075462
C_6Cl_5—OH	NaOH	99.0[a]	85SUL61
2,4,6-Trimethyl-phenol	NaOH	81.0[b]	85SUL61
2,6-di-OMe—C_6H_3ONa		48[c]	90CL83

[a] Colorless solid; m.p. 108–112 °C. [b] Oily product; m.p. 40–43 °C. [c] In THF at 0 °C.

The esters ROC(Cl)=S (R = succinimido, phthalimido, norborn-5-ene-2,3-dicarboximido) proved to be useful reagents for introducing the thiocarbonyl group into organic compounds ⟨88EGP273832⟩.

(ii) From alcohols, alcoholates, and alkoxytrimethylsilanes

Propargyl chlorothioformate (a light yellow liquid boiling at 41 °C and 8 torr (mm Hg) was obtained (52% yield) as above from propargyl alcohol and thiophosgene with sodium hydride as the base (Equation (8)) ⟨92AG(E)866⟩; the corresponding allene, CH_2=C=CHCH$_2$OC(Cl)=S, could not be isolated owing to spontaneous [3,3]-sigmatropic rearrangement to CH_2=C(CH=CH_2)SC(Cl)=O ⟨92AG(E)866⟩.

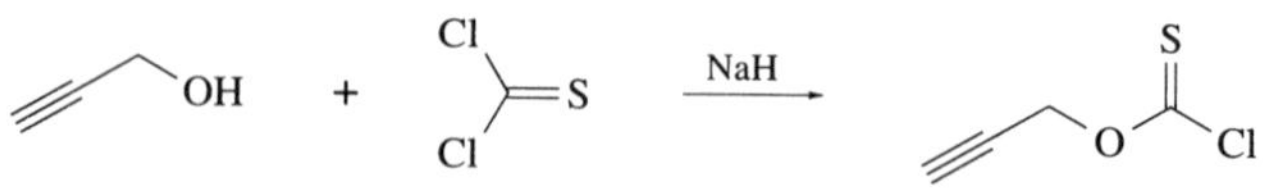

(8)

Alkyl chlorothioformates have been synthesized in two different ways. The first is the reaction of thiophosgene with a potassium alkoxide in the corresponding alcohol ROH (or in tetrahydrofuran) between −65 °C and 15 °C ⟨86S760⟩. EtOC(Cl)=S was prepared from thiophosgene in 93% yield when the reaction mixture was kept for 1 h at −60 °C in ethanol and tetrachloromethane (Equation (9)) and other alkyl chlorothioformates were similarly isolated in high yield. When sodium alkoxides RONa (R = Et, Pr^n) were used, the corresponding alkyl chlorothioformates (R = Et, Pr^n) were obtained in approximately 50% yield ⟨84ZAAC(508)136⟩. As an alternative, the reaction of alkoxytrimethylsilanes with thiophosgene, which does not require basic media, was tried ⟨86S760⟩; the yields, were, however, no higher than 35% (Equation (10)).

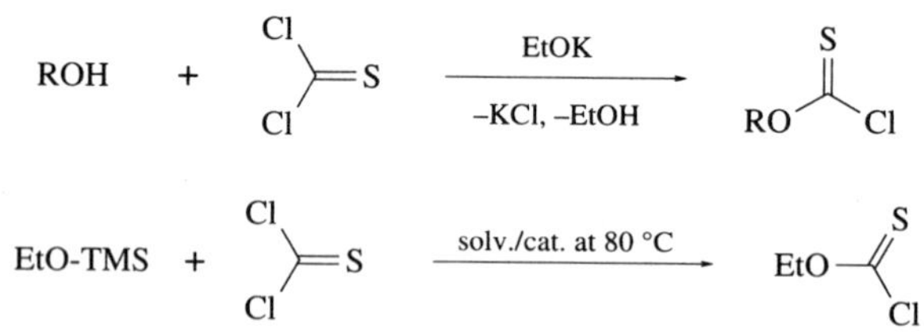

(9)

(10)

The second method is the chlorolysis of bis(alkoxythiocarbonyl) disulfides with Cl_2 or $SOCl_2$ ⟨57GEP1018054, 65CB2059, 83JOC4750⟩. However, the alkyl chlorothioformates are obtained in lower yield than from the first method, the reagents must be very clean, and the products are difficult to purify by distillation (Equation (11)) ⟨86S760⟩.

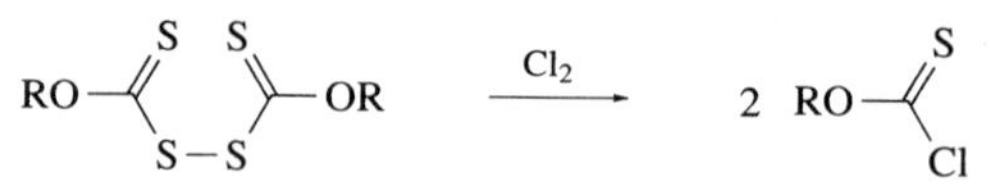

(11)

6.17.1.3.3 *Fluorodithioformates RSC(F)=S*

(i) From chlorodithioformates

The alkyl fluorodithioformates RSC(F)=S (R = Et, Pr^n) have been synthesized in yields of 40% and 60%, respectively, from the corresponding chlorodithioformates with KF under solid–liquid phase-transfer catalysis conditions (18-crown-6 in acetonitrile under argon); spectroscopic data was obtained ⟨84ZAAC(509)67⟩.

(ii) Other methods

The reaction of thiols RSH (R = Me, Et) with FClC=S, but without any solvent, is an alternative for the synthesis of alkyl fluorodithioformates ⟨59ZOB3792⟩. Perfluoromethyl fluorodithioformate CF_3SC(F)=S can be synthesized from trifluoromethyl thiol CF_3SH with anhydrous NH_3 ⟨55JCS3871⟩ (yield 40%) or with KF ⟨68CB2609⟩ (yield 58%) as HF acceptors. The yield increases to ca. 96% if FClC=S and $Hg(SCF_3)_2$ are used at room temperature ⟨72CB820⟩; CF_3SC(F)=S is obtained in this way as a yellow liquid boiling at 43 °C ⟨68CB2609⟩ and it can be readily converted into CF_3SC(Cl)=S (yield 87%) with BCl_3 ⟨76CB1976⟩.

6.17.1.3.4 *Chlorodithioformates RSC(Cl)=S*

(i) From thiols

Alkyl chlorodithioformates RSC(Cl)=S (R = Et, Pr^n, Pr^i, Bu^n) were synthesized in yields of 55% to 70% from the corresponding thiols RSH and thiophosgene in dry CS_2 at room temperature ⟨84ZAAC(508)136⟩. The compounds could be readily stored at −20 °C under argon; spectroscopic data was obtained.

(ii) From carbon monosulfide

The insertion of carbon monosulfide into the SCl bond of sulfenyl chlorides was used for the synthesis of alkyl and aryl chlorodithioformates. However, the production of CS needs special care and equipment ⟨74IC1778⟩. In this way, trichloromethyl chlorodithioformate (an orange oil boiling at 98–100 °C and 14 torr (mm Hg)) was obtained from trichloromethanesulfenyl chloride in 75% yield (Equation (12)) ⟨84JOC3854⟩ and phenyl chlorodithioformate PhSC(Cl)=S from phenyl chlorodithioformate in 73% yield ⟨84JA263⟩. Chlorodithioformates which have been prepared by this method are S=C(Cl)SSC(Cl)=S (50% yield), $Cl_3CSCCl_2SC(Cl)$=S (49%) ⟨86SUL203⟩, and a number of other RSSC(Cl)=S derivatives ⟨86ACS(B)609⟩: R = COMe, an orange oil (50%), R = COCl, a dark red oil boiling at 55–56 °C and 0.04 torr (mm Hg) (53%), R = CCl_3, an orange oil boiling at 78–80 °C and 0.04 torr (mm Hg) (57%), and R = C_2Cl_5, an orange oil boiling at 108–109 °C and 0.03 torr (mm Hg) (19%) ⟨86ACS(B)609⟩.

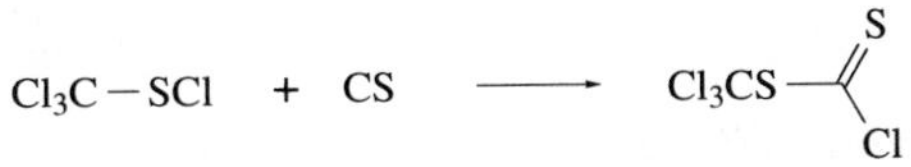

(12)

S=C(Cl)CH_2CH_2SC(Cl)=S (a light yellow oil, 38% yield) and 2,5-bis(chlorothiocarbonylthio)-1,3,4-thiadiazole (yellow needles melting at 69 °C, 100%) were obtained in the same way (Equation (13)) ⟨91SUL143⟩.

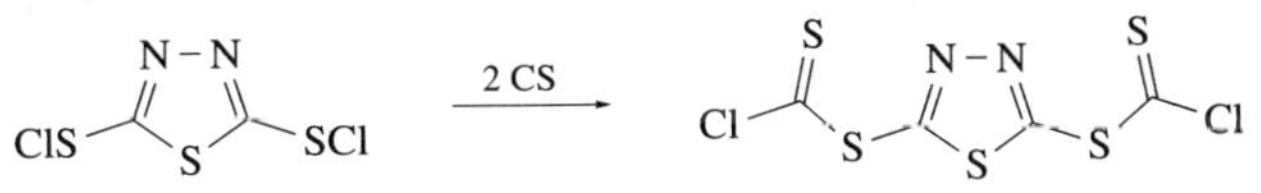

(13)

(iii) Other methods

Methyl chlorodithioformate MeSC(Cl)=S was obtained as the major product from $(MeS)_2C(Cl)SCl$ (with H_2O, 58%; with aq. KI, 55%; with MeSH, 47%), from $MeSCCl_2SSMe$ (with H_2O, 81%; with aq. KI, 72%; with MeSH, 60%) and from $MeSCCl_2SCl$ (with MeSH, 36%), respectively ⟨84SUL241⟩.

Alkali metal chlorodithioformates MSC(Cl)=S were prepared from the alkali metal chlorides (yields: M = Na, 45%, K, 23%, Rb, 35%, and Cs, 31%) and CS_2 in the presence of NaOH pellets ⟨83ZAAC(502)7⟩. Ethyl chlorodithioformate EtSC(Cl)=S was obtained from KSC(Cl)=S and EtI but only in 12% yield ⟨83ZAAC(502)7⟩. Diazocarbonyl compounds can also give chlorodithioformates with thiophosgene (Equation (14)) ⟨64LA(674)124⟩.

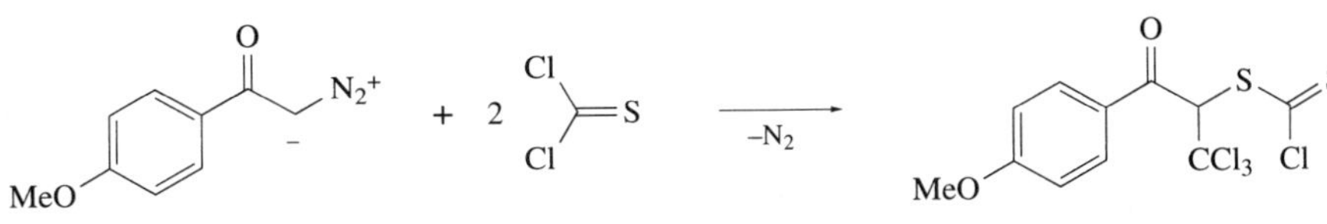

(14)

(iv) Specific methods for aryl chlorodithioformates

The reaction of pentachlorobenzenesulfenyl chloride and sodium trithiocarbonate gave the pentachlorophenyl chlorodithioformate in 46% yield (Equation (15)) ⟨85T5145⟩.

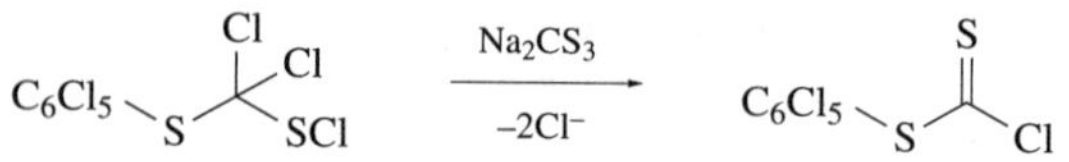

(15)

Aryl chlorodithioformates were also obtained by the reaction of arenediazonium chlorides with CS_2 under Sandmeyer conditions (copper powder or Cu(I)Cl at room temperature) (Equation (16)) ⟨79ZC289⟩ and, in the usual manner, from thiophosgene and arenethiols (Equation (17)) ⟨85SUL61⟩. This last method gave 4-nitrobenzenethio- ⟨73CB1487⟩ and 2-naphthalenethio- derivatives ⟨75JCS(P2)916⟩. Pentafluorobenzenethiocarbonic acid chloride was obtained (60%) by Haas and Kempf ⟨84T4963⟩ from bis(pentafluorobenzenethio)mercury and thiophosgene (Equation (18)).

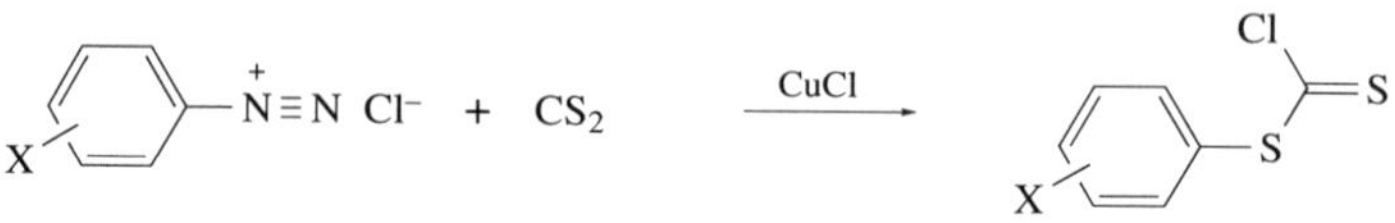

(16)

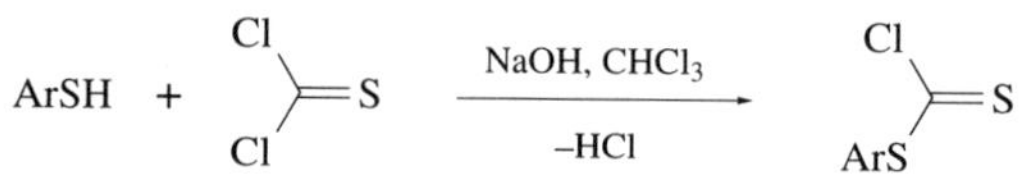

(17)

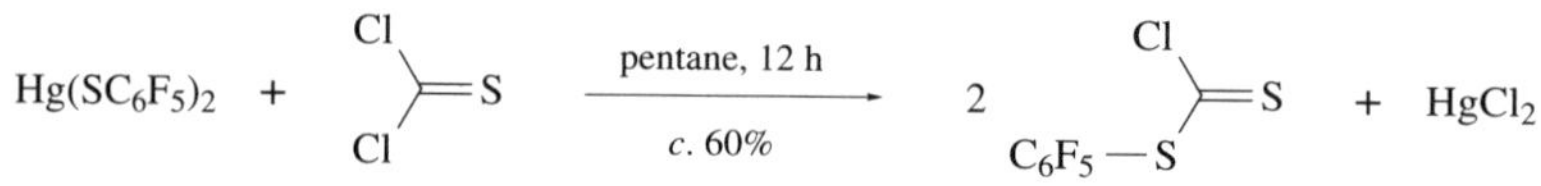

(18)

6.17.1.3.5 *Bromodithioformates RSC(Br)═S*

(i) From thiols

Alkyl bromodithioformates RSC(Br)═S were obtained in 31% yield (R = Et) and 44% yield (R = Pr^n) by treating the corresponding thiol RSH with Br_2C═S in ether at room temperature under argon. The deep red liquids could be stored at −20 °C under argon ⟨84ZAAC(509)61⟩; spectroscopic data was obtained.

(ii) Other methods

Trifluoromethyl bromodithioformate $F_3CSC(Br)$═S, a red liquid boiling at 67 °C and 150 torr (mm Hg), was obtained in 83% yield from $F_3CSC(F)$═S by halogen exchange with BBr_3 at 35–40 °C; IR, ^{19}F NMR, UV, and MS data was obtained ⟨76CB3432⟩.

6.17.1.3.6 *Fluoroselenothioformates RSeC(F)═S*

Trifluoromethyl fluoroselenothioformate $F_3CSeC(F)$═S, a liquid boiling at 57–58 °C, was prepared from $Hg(SeCF_3)_2$ and FClC═S at −78 °C; the ^{19}F chemical shifts (36.3 and −107.6 ppm), the IR and MS data of the compound were measured ⟨76ZAAC(427)114⟩.

6.17.1.3.7 *Chloroselenothioformates RSeC(Cl)═S*

(i) From alkaneselenols

In a similar way to the corresponding alkyl chlorodithioformates, the selenoesters RSeC(Cl)═S (R = Et, Pr^n, yellow viscous oils) were synthesized from the corresponding alkaneselenols and thiophosgene in dry CS_2 as the solvent at room temperature in yields of 63–76% ⟨84ZAAC(509)61⟩.

(ii) Other methods

$F_3CSeC(Cl){=}S$ was synthesized from $F_3CSeC(F){=}S$ in 97% yield by halogen exchange with BCl_3 ⟨76ZAAC(427)114⟩.

6.17.1.3.8 *Bromoselenothioformates RSeC(Br)=S*

$F_3CSeC(Cl){=}S$, a liquid boiling at 54°C and 50 torr (mm Hg), was synthesized photochemically (UV irradiation for 4 h) from F_3CSeBr in 51% yield and by halogen exchange with BBr_3 from $F_3CSeC(F){=}S$ at 40°C in 74% yield ⟨86ZN(B)413⟩.

6.17.1.3.9 *Sulfines derived from $CF_3SeC(Cl){=}S$ and $CF_3SeC(Br){=}S$*

Fockenberg and Haas ⟨86ZN(B)413⟩ obtained the *S*-oxides of $CF_3SeC(Cl){=}S$ and $CF_3SeC(Br){=}S$, respectively, by oxidation with mcpba. $CF_3SeC(Cl){=}SO$, a yellow liquid boiling at 47°C and 10 torr (mm Hg), was obtained in 56% yield and $CF_3SeC(Br){=}SO$, a yellow liquid boiling at 60°C and 10 torr (mm Hg), in 25% yield.

6.17.1.3.10 *Thiocarbamoyl fluorides $R_2NC(F){=}S$*

N,N-Diethylthiocarbamoyl fluoride $Et_2NC(F){=}S$ was obtained from diethylamine and thiocarbonyl chloride fluoride in 45% yield ⟨63CJC1643, 64LA(590)123, 70LA(733)195, 70LA(735)158, 75LA1025⟩. *N,N*-Dimethylthiocarbamoyl fluoride $Me_2NC(F){=}S$ was also prepared by treating $CF_2{=}CFR$ ($R = F$, Cl, CF_3) with tetramethylthiuramide sulfide at 130–135°C ⟨81URP804634⟩. *N*-(Fluorothiocarbonyl)imidosulfuryl difluoride $F_2S{=}NC(F){=}S$ was synthesized by reaction of $Si(NCS)_4$ with SF_4 ⟨77MIP53536⟩.

6.17.1.3.11 *Thiocarbamoyl chlorides $R_2NC(Cl){=}S$*

(i) From secondary amines and related compounds

Thiocarbamoyl chlorides can be synthesized from thiophosgene and secondary amines (Equation (19)) ⟨63CJC1643, 64LA(590)123, 70LA(735)158, 75LA1025⟩. Data published after 1980 is given in Table 3. In two cases ⟨85JAP61233662, 85JAP61233663⟩, thiophosgene, which is added to aromatic amines, was produced directly in the reaction mixture from trichloromethyl thiol and SO_2 ⟨85JAP61233662⟩ or $NaHSO_3$ ⟨85JAP61233663⟩. Higher yields were obtained by replacing secondary amines with trimethylsilylamines (Equation (20)) ⟨80T539⟩ and the chlorothiocarbamates produced in this way proved easier to purify. The TMS derivative $RN(TMS)C(Cl){=}S$ was also obtained ⟨86ZOB1547⟩.

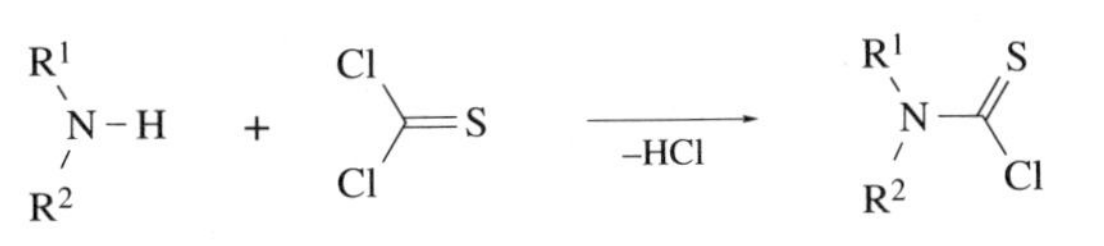

(19)

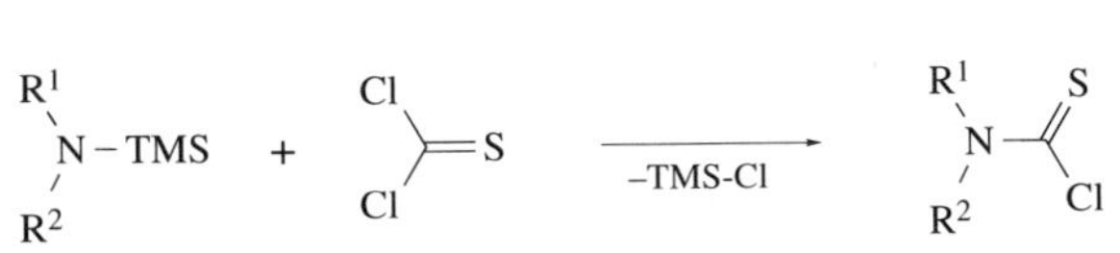

(20)

Triphenylphosphine *N*-trimethylsilylimide and triphenylphosphiniminium chloride were similarly added to thiophosgene (Scheme 6) ⟨87UKZ395⟩.

Table 3 Chlorothiocarbamates as synthesized from secondary amines or amides and thiophosgene.

Chlorothiocarbamate	*Solvent*	*m.p.*	*Yield (%)*	*Ref.*
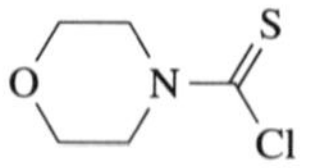 colorless crystals	dry ether	170 °C	62.5	80CB1898
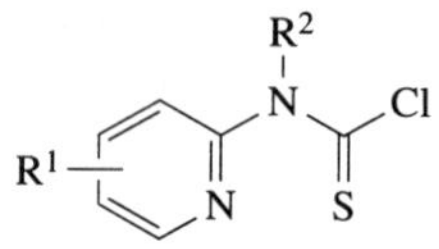 R^1 = lower alkyl R^2 = halogen, lower alkyl, lower alkoxy	pat.			83JAP60056958
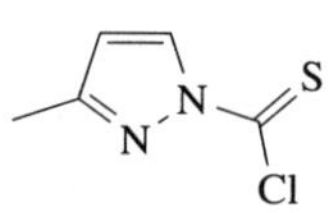	ether			85EGP256705
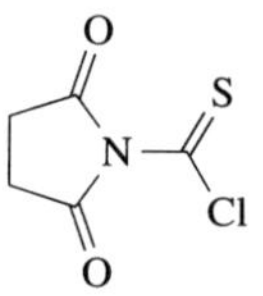 (succinimido, phthalimido and norborn-5-ene-2,3-dicarboximido analogues have also been synthesized)	THF/benzene	62–64 °C		88EGP273832
N,N'-Me,Ph–C(Cl)=S	$SO_2/CCl_3SH/CCl_4/KI$		45.1	85JAP61233662
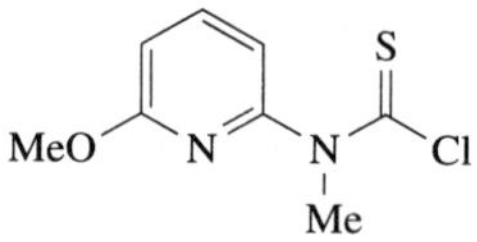	$NaHSO_3/CCl_3SH/CCl_4/KI$		30	85JAP61233663

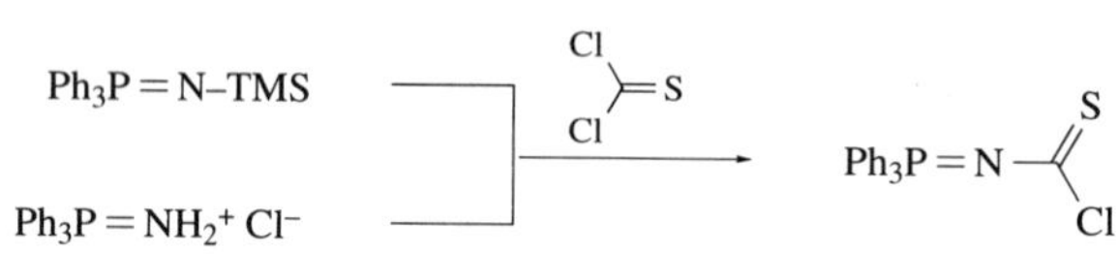

b.p. 182–184 °C

Scheme 6

(ii) From tetraalkylthiuramide disulfides

Another generally used method for the preparation of thiocarbamoyl chlorides is the chlorination of tetraalkylthiuramide disulfides (Table 4 and Equation (21)). Chlorine ⟨89JAP3020256, 92MI 617-01⟩, SO_2Cl_2 ⟨83JOC1449, 88ZOB1489, 89IZV909, 90SC2769⟩ and S_2Cl_2 ⟨89EUP355578⟩ have been used as the chlorinating agent. *N,N*-Dimethylthiocarbamoyl chloride was also synthesized from the monosulfide by reaction with phosgene (Equation (22)) ⟨92MI 617-02⟩.

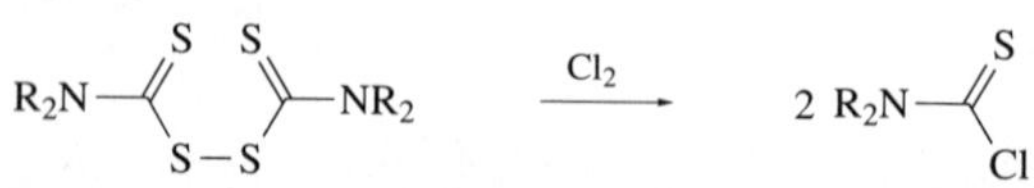 (21)

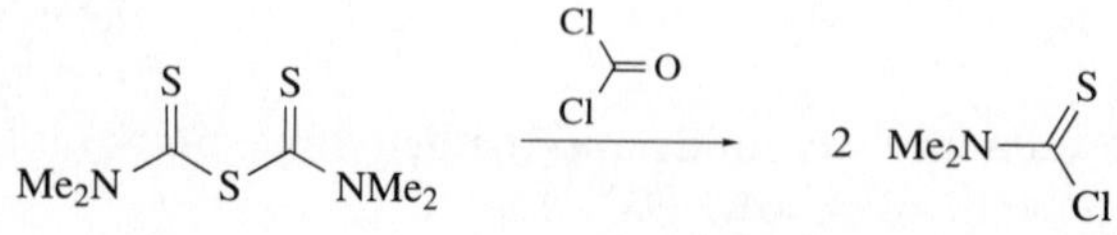 (22)

Table 4 Chlorothiocarbamates synthesized by chlorination of thiuramide disulfides.

Chlorothiocarbamate	*Chlorination*	*b.p.* (°C)	*Yield* (%)	*Ref.*
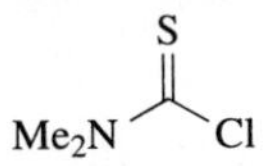	SO_2Cl_2	96/17 torr	83.2	89IZV909 88ZOB1489
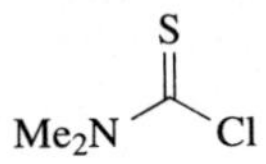	SO_2Cl_2		94	83JOC1449
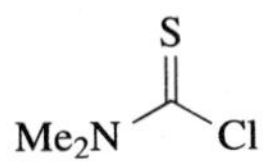	SO_2Cl_2	43 (m.p.)	97	90S2769
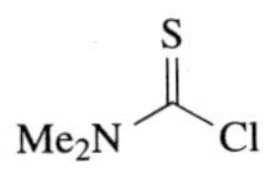	S_2Cl_2		99	89EUP355578
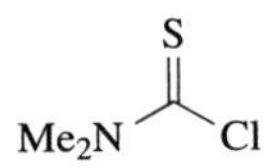	Cl_2		92	92MI-617-01
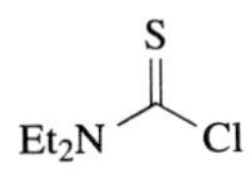	SO_2Cl_2	70/13 torr	87	89IZV909 88ZOB1489
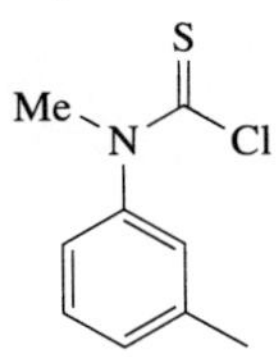	Cl_2	pat.		89JAP03020256

(*iii*) *From thioformamides*

The third synthetic method, which is a general one for thiocarbamoyl chlorides, is the chlorination of thioformamides (Equation (23)) ⟨83HOU(E4)407⟩. SCl_2 ⟨70LA(733)195, 89CB113⟩, Cl_2 ⟨70LA(733)195⟩ and SO_2Cl_2 have been used as the chlorinating agent. The thioformamides can be synthesized from the corresponding formamides with Lawesson's reagent (Scheme 7) ⟨85JAP62114966⟩.

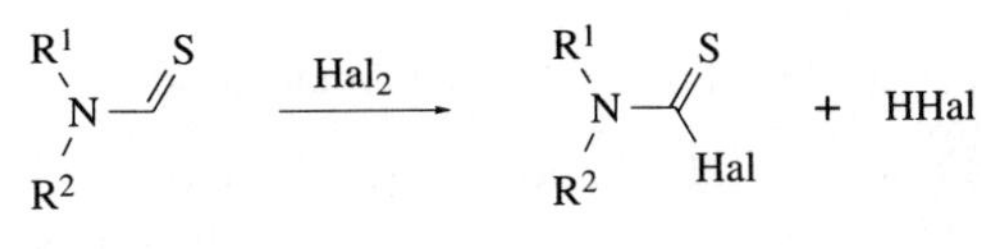

(23)

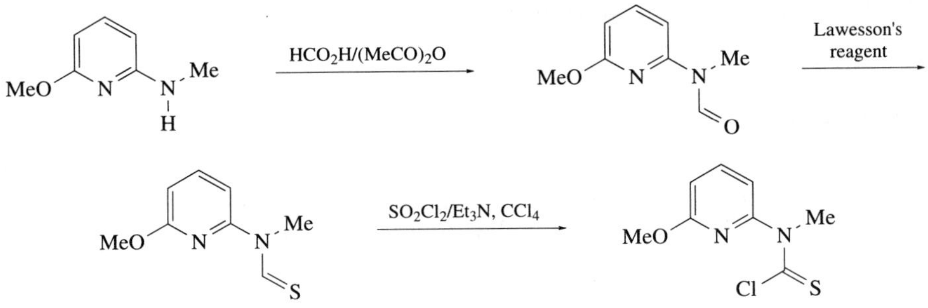

Scheme 7

(*iv*) *Other methods*

N,N-Dimethylthiocarbamoyl chloride was obtained as a by-product in the synthesis of alkyl isocyanates RN=C=O from RNHC(OR′)=S and $Cl_2C{=}N^+Me_2Cl^-$ ⟨92ZOR607⟩. The stable gold chloride complexes $ClAuS{=}C(Cl)NMe_2$ and $Cl_3AuS{=}C(Cl)NMe_2$ have been isolated and characterized ⟨92POL893⟩.

(v) Methods for thiocarbazic acid chlorides ($R^1_2NNR^2C(Cl){=}S$)

Thiocarbazic acid chlorides $R^1R^2NN(R^3)C(Cl){=}S$ were prepared from the corresponding trialkylhydrazines and thiophosgene ⟨72CB2854, 91AP(324)917⟩ (Table 5 and Equation (24)). Thiocarbazic acid chlorides were employed for the thiocarbazoylation of thiazine-2-ones and thiazolidine-2-ones ⟨91LA405⟩.

Table 5 Thiacarbazic acid chlorides, $R^1R^2N{-}N(R^3){-}C(Cl){=}S$, obtained from thiophosgene and the corresponding trialkyl hydrazine.

R[1]	*R*[2]	*R*[3]	*m.p.*	*Yield* (%)	*Ref.*
Me	Me	Me	64–65 °C	69	72CB2854
morpholino		Me	93 °C	54	91AP(324)917
piperidino		Me	41–42 °C	13	91AP(324)917
Me	Me	cyclohexano	30 °C	29	91AP(324)917

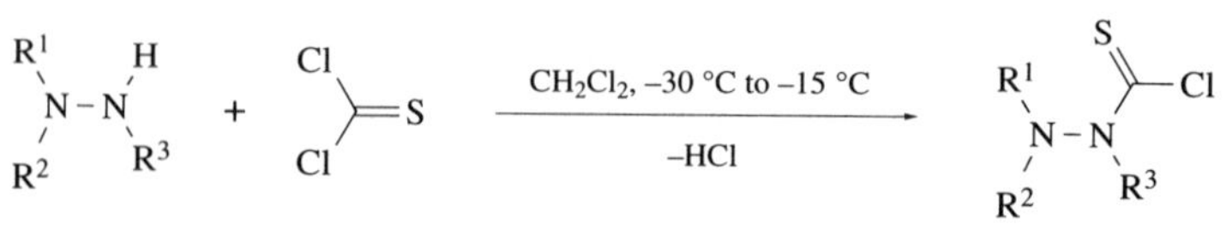

(24)

6.17.1.3.12 *Thiocarbamoyl bromides $R_2NC(Br){=}S$*

Thiocarbamoyl bromides can be synthesized from the corresponding thioformamides by reaction with bromine (Equation (23)) ⟨70LA(733)195, 72LA(755)145⟩. For example, $MePhNC(Br){=}S$ was formed in 78% yield and $EtPhNC(Br){=}S$ in 83% yield.

6.17.2 FUNCTIONS CONTAINING AT LEAST ONE CHALCOGEN FUNCTION (AND NO HALOGEN)

6.17.2.1 Thionocarbonates (*O,O*-Diesters of Thiocarbonic Acid)

The direct reaction of alcohols and phenols with thiocarbonylating agents usually leads to thionocarbonates in good yield ⟨83HOU(E4)420⟩. Thiophosgene and chlorothionoformates are most often used for thiocarbonylation; several other reagents are used in specific cases.

(i) From thiophosgene

In these reactions thiophosgene is allowed to react with a hydroxy compound in the presence of a base (Equation (25)); bases used include potassium carbonate ⟨69JOC3011⟩, pyridine ⟨55JA2479, 62JOC4509, 80CB750⟩ and 4-dimethylaminopyridine (dmap) ⟨82TL1979, 83TL865⟩. 1,2-Diols (Equation (26)) ⟨82TL1979⟩ and enols also react easily ⟨80CB750, 83HOU(E4)420⟩.

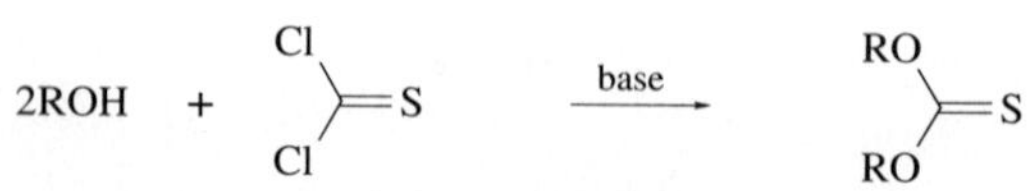

(25)

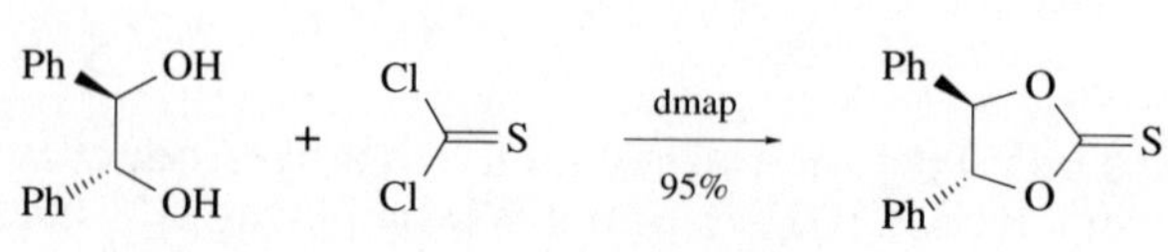

(26)

The above method is successful with 2-fluoroalcohols, an example being the formation of bis(2-fluoro-2,2-dinitroethyl) thionocarbonate (Equation (27)); it is unsuitable for nitroalcohols such as 2,2-dinitropropanol and 2,2,2-trinitroethanol ⟨83HOU(E4)420, 83JCED131⟩.

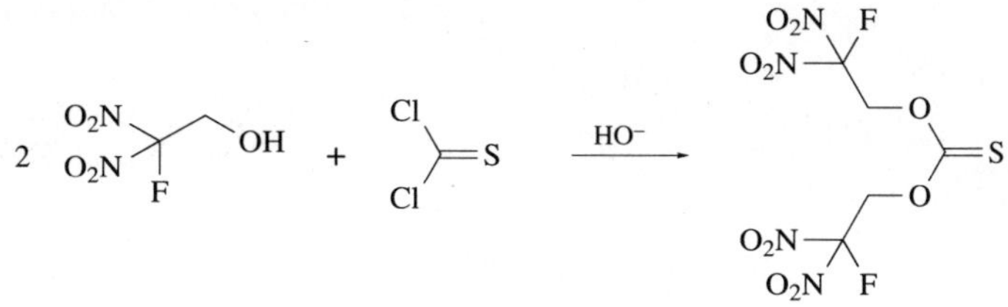

(27)

(ii) From chlorothionoformates

This is the most general method for the preparation of mixed thionocarbonates ⟨83HOU(E4)420⟩. After protection of the 3- and 5-hydroxyl groups of methyl shikimate, the 4-hydroxyl group can be transformed into a thionocarbonate by reaction with phenyl chlorothionoformate in the presence of an equivalent amount of dmap (Equation (28)) ⟨85TL4941⟩. Similarly the 2′-hydroxy group of 1-(3′,5′-*O*-(1,1,3,3-tetraisopropyldisiloxane-1,3-diyl)-*β*-D-ribofuranosyl)-1*H*-benzimidazole was transformed into a thionocarbonate by reaction with dmap and *p*-tolyl chlorothionoformate (Equation (29)) ⟨87HCA138⟩.

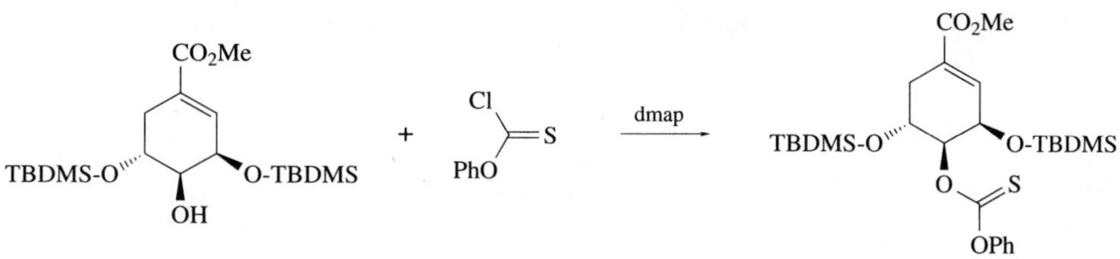

(28)

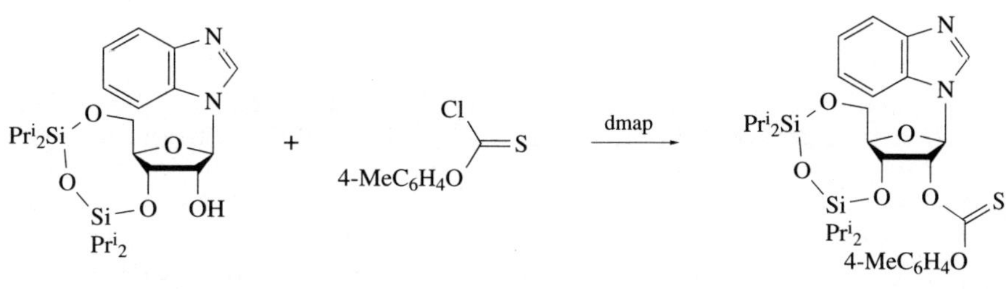

(29)

Sometimes a hydroxy derivative is metallated before treatment with an aryl (or alkyl) chlorothionoformate; for example, by adding methyllithium and then ClC(=S)OPh to a solution of 1-*O*-benzyl-2,3-isopropylidene-L-lyxose in THF at −78 °C, the corresponding thionocarbonate was obtained in 90% yield (Equation (30)) ⟨85T4079⟩.

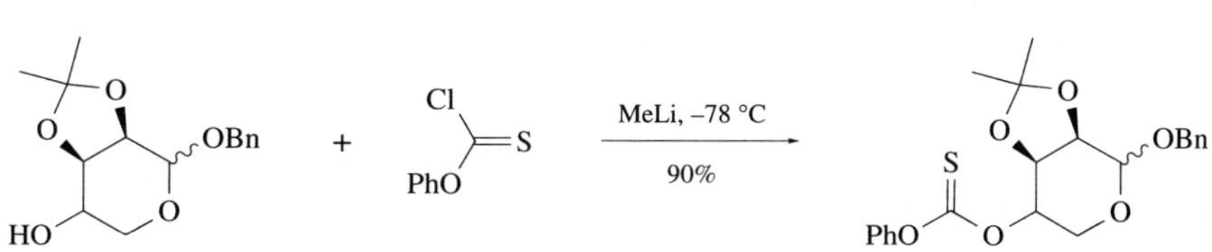

(30)

The reaction of pyranosides which do not possess *cis*-vicinal hydroxyl groups (e.g. Me α-D-glc, Me *β*-D-xyl) with dibutyltin oxide and then with phenyl chlorothionoformate gives monothionocarbonates in good yield (Scheme 8) ⟨86CPB430⟩. They, in turn, can be used for preparing the deoxy compounds. Pyranosides with *cis*-vicinal hydroxyl groups (e.g. Me α-D-gal, Me-*β*-D-ara) give cyclic thionocarbonates instead (Scheme 8).

The reaction of a sulfone with butyllithium and then with an aldehyde at −40 °C gives the lithium salt of a *β*-hydroxysulfone, which can be used *in situ* for the synthesis of the corresponding thionocarbonate (Scheme 9) ⟨91TL2703⟩. When diethyl difluoromethylphosphonate is metallated

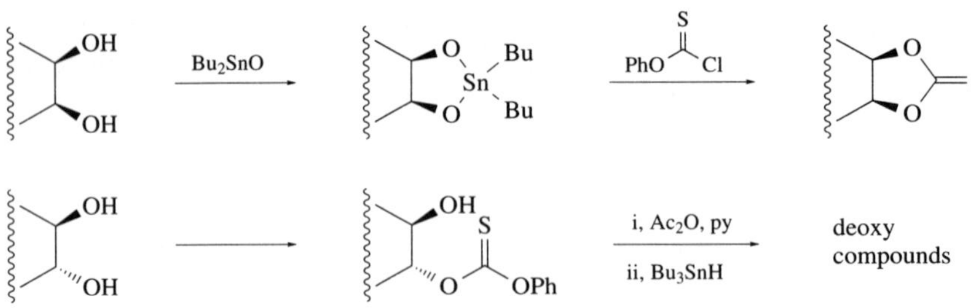

Scheme 8

with lithium diisopropylamide (LDA) in THF at −78 °C and then reacted with a series of aldehydes, intermediate alkoxy adducts are obtained which give thionocarbonates in 71–90% yield with phenyl chlorothionoformate (Equation (31)) ⟨92TL1839⟩. These thionocarbonates can then be used for the synthesis of 1,1-difluoroalkylphosphonates.

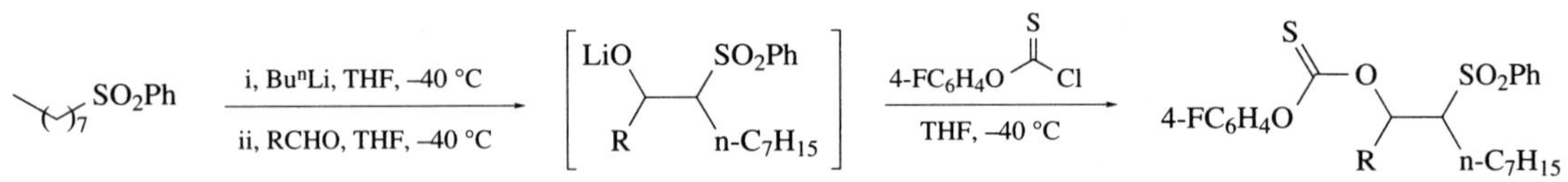

Scheme 9

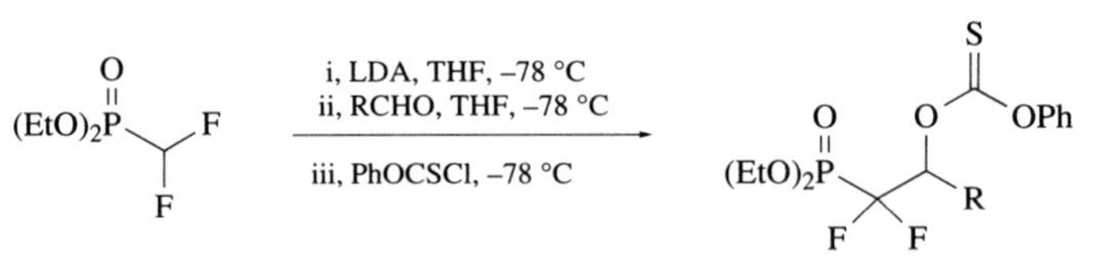

(31)

Diol monothionocarbonates can be prepared as illustrated in Scheme 10. 5-Oxopentyl phenyl thionocarbonate ⟨91CPB1659, 92TL2543⟩ and 5-oxohexyl phenyl thionocarbonate ⟨92T9433⟩ were prepared in the usual manner by treating the respective hydroxy compound with phenyl chlorothionoformate.

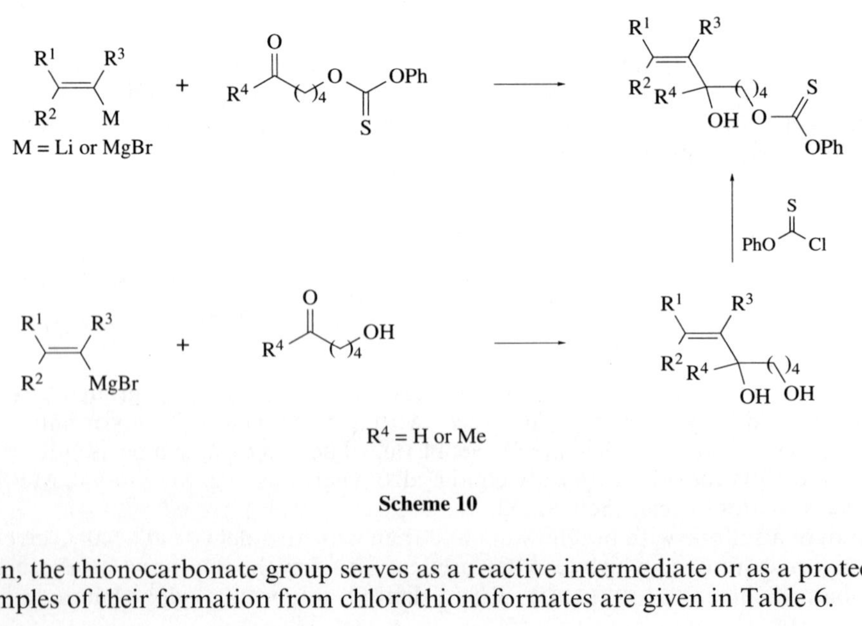

Scheme 10

Very often, the thionocarbonate group serves as a reactive intermediate or as a protective group; further examples of their formation from chlorothionoformates are given in Table 6.

Table 6 Further examples of thionocarbonates prepared by treating a hydroxy compound with a chlorothionoformate.

Thionocarbonate	*Chlorothionoformate*	*Base*	*Yield (%)*	*Ref.*
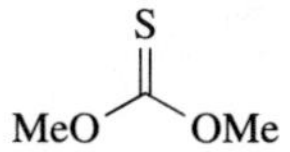	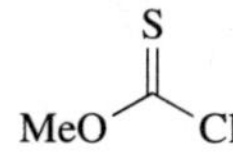	MeONa	72	83JOC4750
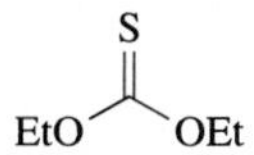	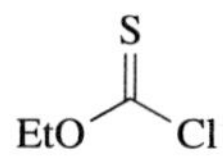	EtONa	40–46	83JOC4750
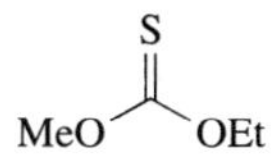	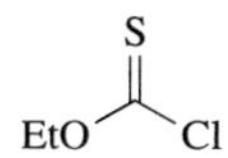	MeONa	28	83JOC4750
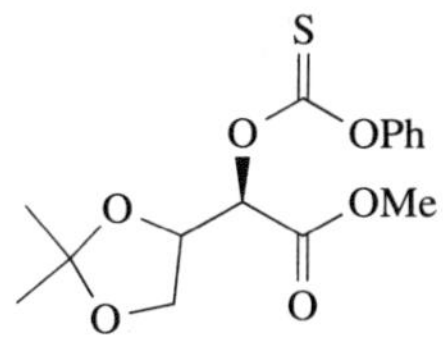	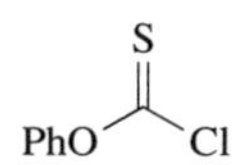	pyridine	84	88S226
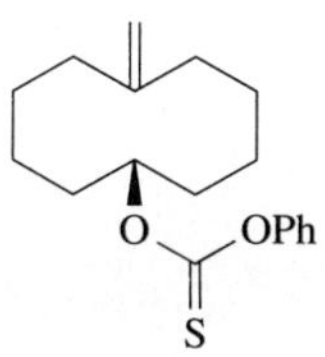	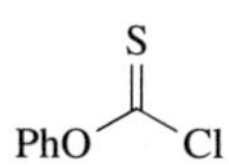	pyridine	62	91T121
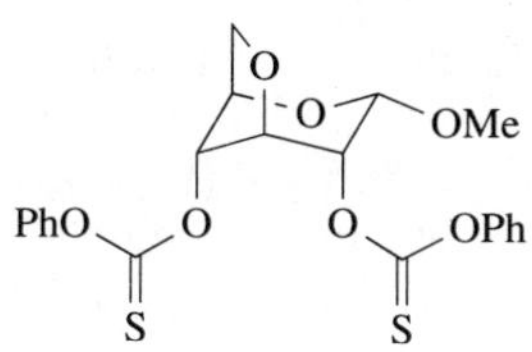	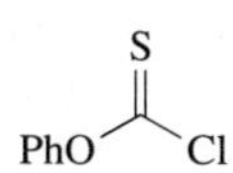	pyridine	43	91T6381
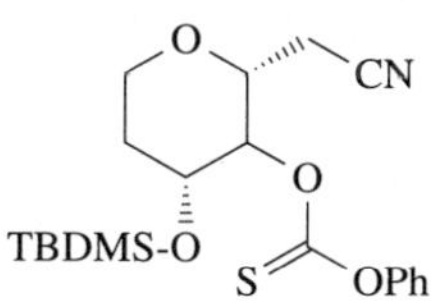	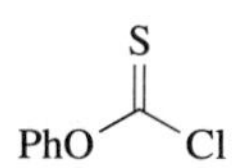	dmap	89	91JCS(P1)787
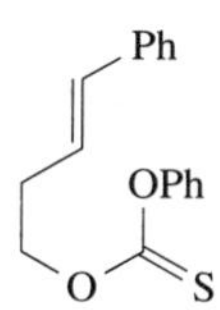	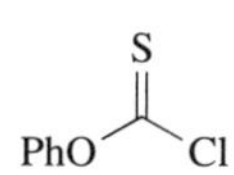	pyridine	92	92JOC6803

(iii) From thiocarbonylbis(azoles)

Thiocarbonylbis(azoles) react with two alcohol molecules or with diols forming acyclic or cyclic thionocarbonates, respectively ⟨83HOU(E4)420⟩. Thiocarbonyldiimidazole is most often used but 1,1′-thiocarbonylbis(1,2,4-triazole) can also serve as the thiocarbonylating reagent ⟨83HOU(E4)420⟩. The latter proved to be very effective in preparing thionocarbonates from 2,2-dinitropropanol and from 2,2-difluoro-2-nitroethanol. The same approach can also be applied to the preparation of unsymmetrical thionocarbonates (Scheme 11) ⟨83JCED131⟩.

An example of the formation of a thionocarbonate from a 1,2-diol and thiocarbonyldiimidazole is shown in Equation (32). Several cyclic thionocarbonates were prepared from various 1,2-diols and thiocarbonyldiimidazole in order to be used as initiators for radical-based cyclization reactions ⟨87TL5973⟩. Catechol has also been used as a diol with thiocarbonyldiimidazole, the thionocarbonate (1,3-benzoxadiole-2-thione) being formed in 82% yield ⟨88JOC2263⟩. Mixed thionocarbonates have been prepared by the successive displacement of imidazole from thiocarbonyldiimidazole by two different alcohols ⟨92JOC6803⟩.

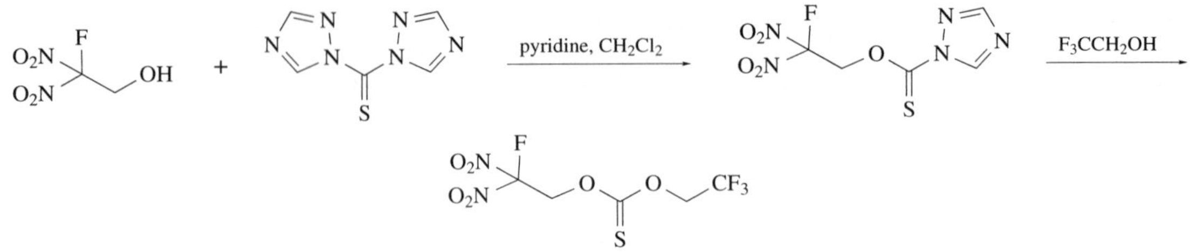

Scheme 11

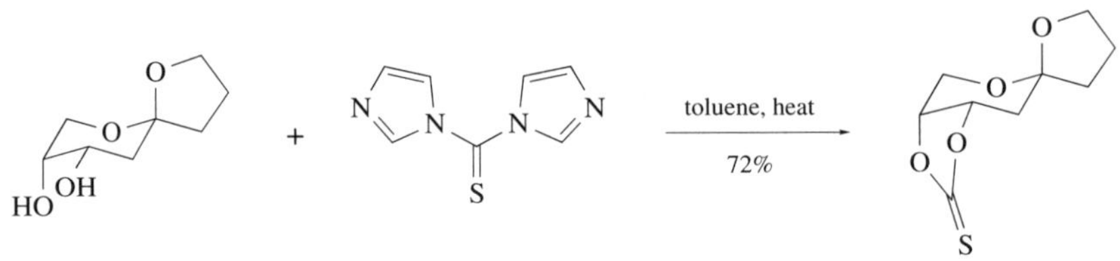

(32)

(iv) Other methods

Several cyclic thionocarbonates have been prepared by cyclization reactions; two examples are shown in Equations (33) and (34) ⟨92CPB2279, 93JOC2894⟩. The reaction shown in Equation (34), and the substitution process shown in Equation (35) ⟨92T6883⟩, involve intermediate thioacyl radicals. By adding NaSH to a stirred mixture of methyl 2-*O*-methyl-*β*-L-arabinopyranoside and Viehe's salt, methyl 2-*O*-methyl-3,4-*O*-thiocarbonyl-*β*-L-arabinopyranoside was isolated in 73% yield (Equation (36)) ⟨82AJC1709⟩; other diols were similarly converted into thionocarbonates. An interesting route to diaryl thionocarbonates proceeds via Cu^I complexes which react with carbon disulfide (Scheme 12) ⟨88TL827⟩.

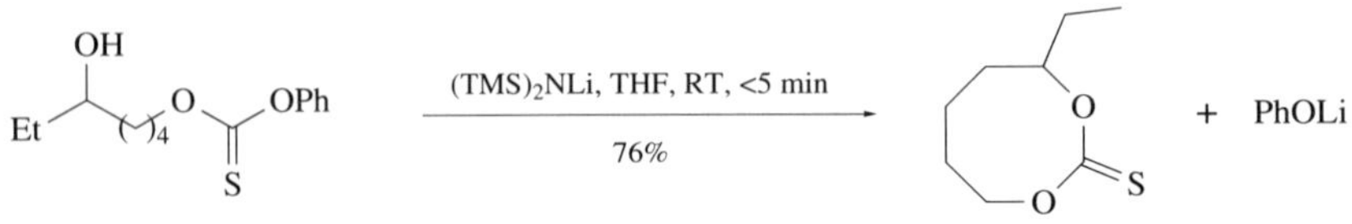

(33)

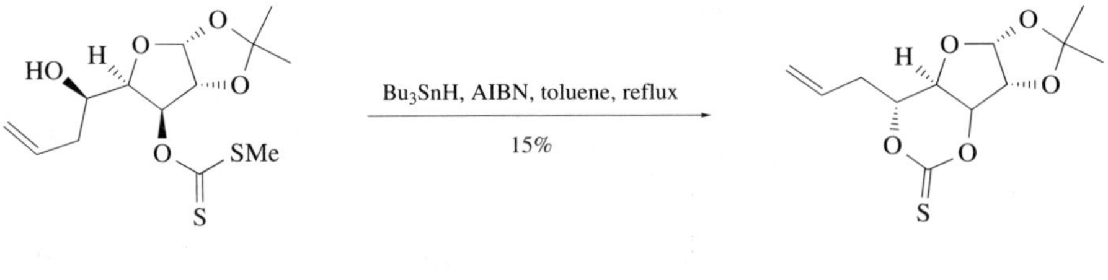

(34)

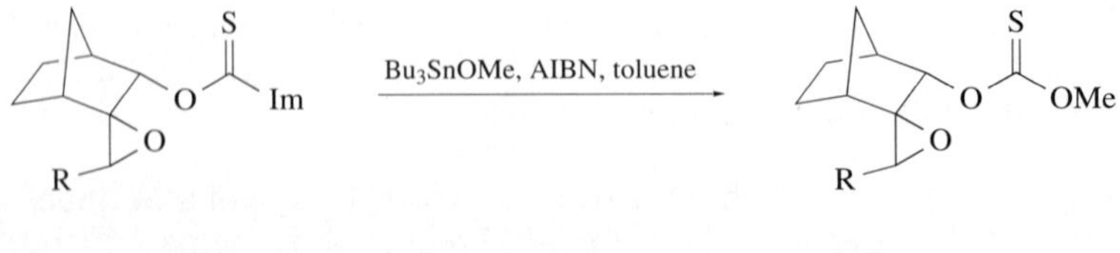

(35)

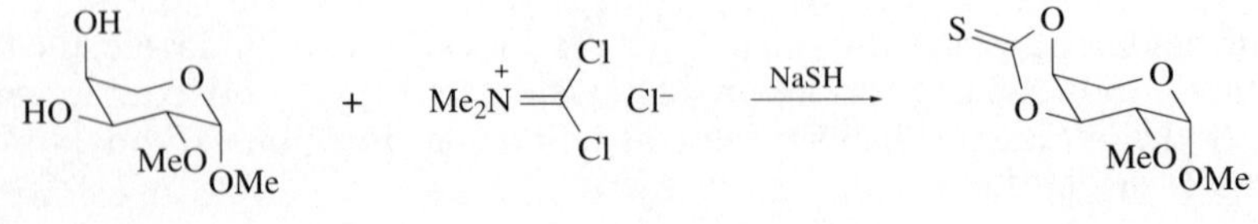

(36)

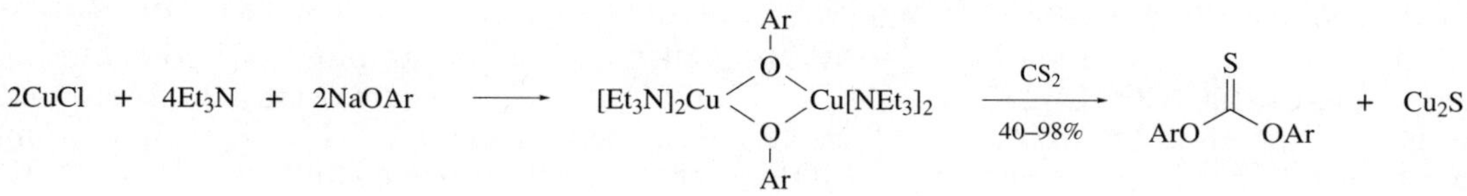

Scheme 12

6.17.2.2 Dithiocarbonates (*O,S*-Diesters of Dithiocarbonic Acid)

Dithiocarbonic acid itself is unstable and decomposes spontaneously to H_2S and COS. However, its *O*-alkyl esters (xanthates) and their salts are relatively stable.

6.17.2.2.1 Salts of O*-alkyl esters of dithiocarbonic acid*

The potassium salts can be easily prepared from alcohols, carbon disulfide and KOH using the appropriate alcohol as the solvent (Equation (37)) ⟨60CB3056, 77S873, 83HOU(E4)420⟩. These salts are very often used to synthesize other derivatives of dithiocarbonic acids. For example, the reaction of 3-(2-hydroxyethyl)- or 3-(3-hydroxypropyl)benzothiazolin-2-ones with potassium hydroxide and excess carbon disulfide afforded the corresponding dithiocarbonates (Scheme 13) ⟨88JHC1601⟩ which were converted by aqueous ammonium persulfate to the corresponding disulfides or by chloroalkanes to the *S*-alkyl derivatives.

 (37)

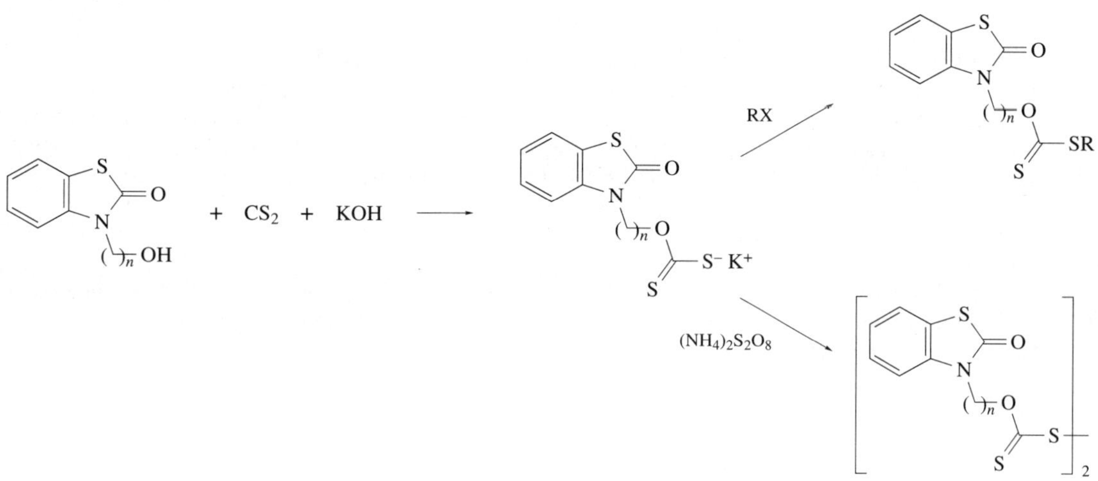

Scheme 13

6.17.2.2.2 O,S*-Diesters of dithiocarbonic acids*

(i) From salts of O*-esters of dithiocarbonic acids*

A salt $M^{+-}SC(=S)OR$ (M = K or Na) can be converted into the corresponding *O,S*-diester by reaction with a chloroalkane or a similar reagent. This is exemplified in Scheme 13; further examples are shown in Equations (38) and (39). By heating equimolar amounts of 2-chloromethylpyridine or 2-chloromethylquinoline with potassium *O*-ethyl dithiocarbonate for 2 h the corresponding dithiocarbonates were obtained (Equation (38)) ⟨89JPR439⟩. Similarly, the reaction of 3-chloromethylbenzothiazolin-2-thione with sodium *O*-alkyl dithiocarbonates (alkyl = Me, Et) gave the corresponding esters (Equation (39)) ⟨89JHC1245, 89JPR439⟩. The preparation of 41 compounds of the type $(EtOC(S)SCH_2C_6H_4)_2Z$ ($Z = (CH_2)_{1-6}$, O, S or SO_2) from potassium *O*-ethyl dithiocarbonate and the corresponding chloro compound has been described ⟨90MI 617-03⟩. The reaction of bromotriphenylmethane with potassium *O*-ethyl dithiocarbonate in various conditions affords *O*-ethyl *S*-triphenylmethyl dithiocarbonate in 79–94% yield ⟨90JCS(P1)2009⟩. *S*-Methyl *O*-trimethylsilyl dithiocarbonate was obtained (38%) from sodium *O*-trimethylsilyl dithiocarbonate and chloromethane ⟨84ZAAC(515)182⟩. The corresponding *S*-ethyl derivative was obtained similarly in 41% yield with iodoethane.

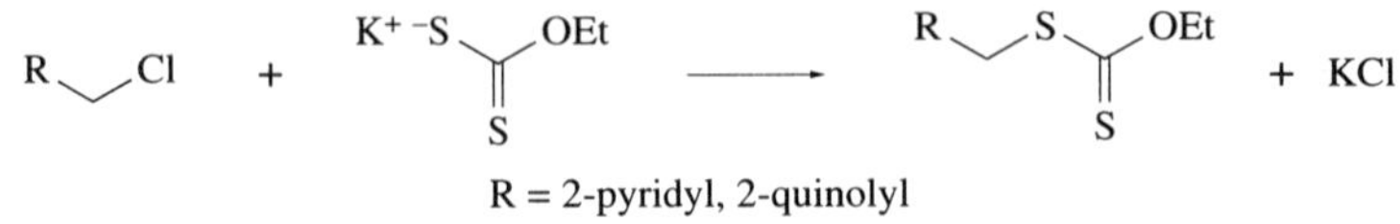

(38)

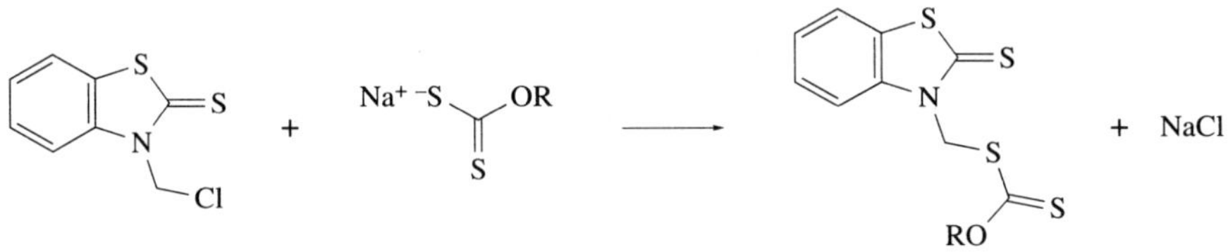

(39)

Protected glycosyl bromides and chlorides can be converted into the corresponding *S*-glycosyl *O*-ethyl dithiocarbonates with complete anomeric inversion in high yield ⟨91TL7453, 92MI 617-04⟩; an example ⟨92MI 617-04⟩ is shown in Equation (40).

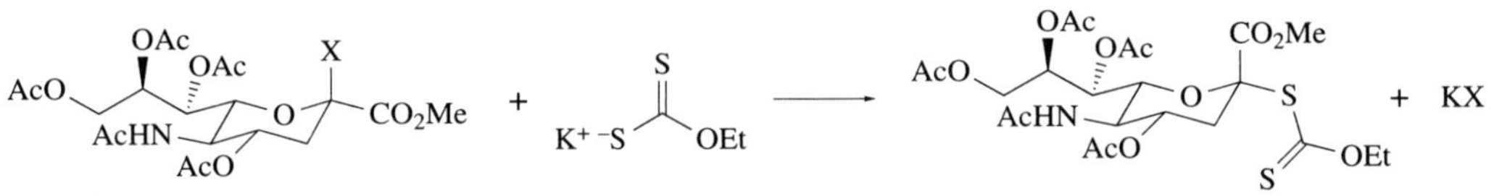

(40)

α-Haloketones and α-haloketoximes undergo displacement easily. For example, the reaction of 2-chlorocyclopentanone with potassium *O*-isopropyl dithiocarbonate in acetone led to the formation of the corresponding dithiocarbonate ⟨93H(35)77⟩, and 3-bromo-5-methylhexan-2-one reacted with potassium *O*-ethyl dithiocarbonate in acetone to give the dithiocarbonate in 86% yield ⟨87JHC581⟩. Chloroacetaldehyde reacts with potassium *O*-ethyldithiocarbonate in an analogous way to give the dithiocarbonate in 65% yield ⟨92HCA907⟩. There are many other examples of displacement of halide from α-halocarbonyl compounds ⟨82JIC882, 84MI 617-01, 85JIC164, 86MI 617-04, 88MI 617-01⟩. Sodium *O*-methyl dithiocarbonate and α-chloroacetophenone oxime reacted to give the displacement product (as a mixture of *syn* and *anti* isomers) in 87% yield ⟨87S349⟩. This was converted into 3-phenyl-4*H*-1,5,2-oxathiazine-6-thione by heating with tosyl isocyanate (Scheme 14). Related reactions of α-haloketoximes have been described ⟨85TL5943⟩.

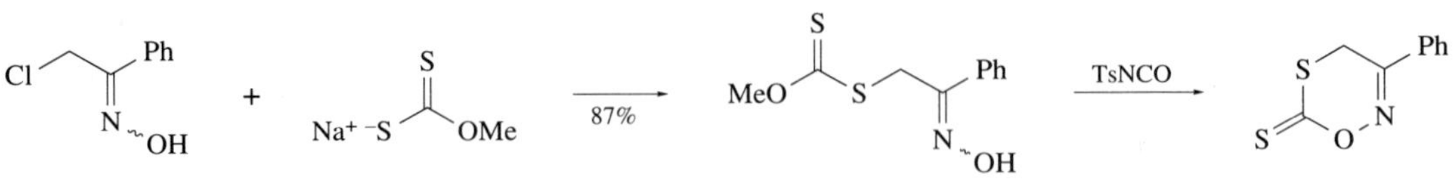

Scheme 14

Carboxylic acid chlorides react readily with potassium *O*-alkyl dithiocarbonates. Acetyl chloride reacts with potassium *O*-ethyl and *O*-methyl dithiocarbonate to give the corresponding anhydrides MeCOSC(=S)OR in high yield ⟨81LA1244⟩. Similar reactions of 2-furoyl chloride ⟨85MI 617-03⟩ and of phthaloyl chloride (Equation (41)) ⟨87JIC194⟩ have been described. Alkyl chloroformates can be used to prepare anhydrides of dithiocarbonic acids ⟨78JOC2930⟩. Thus, *O*-butyl *S*-ethoxycarbonyl dithiocarbonate was prepared by slow addition of potassium *O*-butyl dithiocarbonate to an excess of ethyl chloroformate in acetone at 0°C ⟨83JCS(P1)2641⟩. Phthaloylglycyl chloride reacted with potassium *O*-alkyl dithiocarbonates to give the corresponding *O*-alkyl *S*-phthaloylglycyl dithiocarbonates. Irradiation of the latter in benzene with a mercury lamp gave the corresponding alkyl dithiocarbonates (Scheme 15) ⟨85IJC(B)685⟩. Analogous reactions, involving the extrusion of COS from dithiocarbonic anhydrides, have been reported ⟨89TL4367⟩.

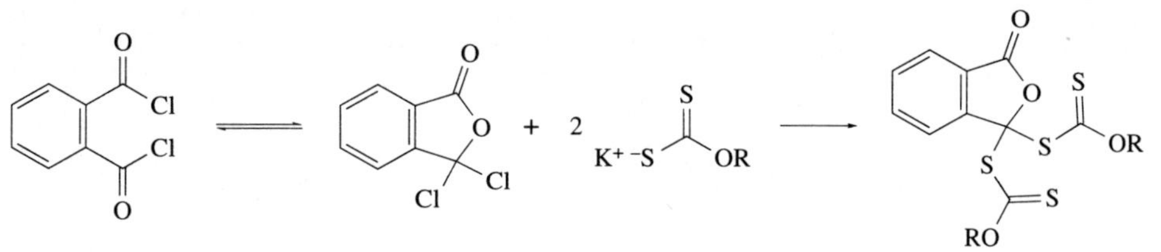

(41)

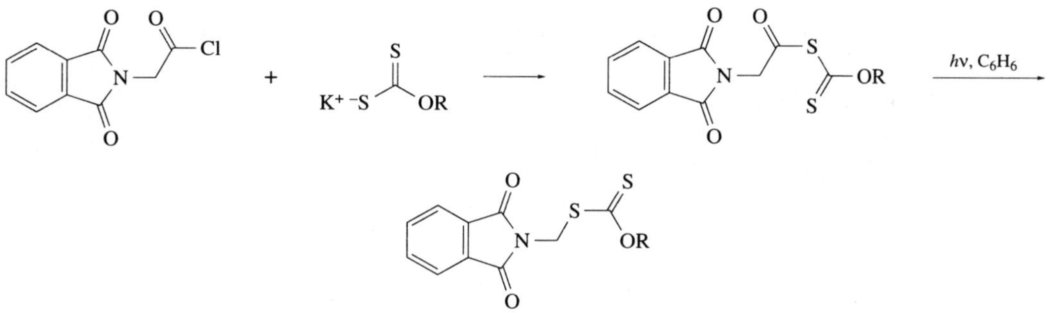

Scheme 15

An alcohol can be converted to a dithiocarbonate by allowing it to react with phosgene and potassium *O*-ethyl dithiocarbonate. The products have been used as condensing agents for the preparation of peptides ⟨85KPS844⟩. When irradiated with visible light, these compounds give the corresponding *S*-alkyl dithiocarbonates (Scheme 16) ⟨90JA2034⟩. The radical intermediates in these reactions can also undergo cyclization, as shown in Scheme 17. Irradiation of *S*-5-hexenyloxy *O*-ethyl dithiocarbonate converted it smoothly in refluxing heptane into *S*-cyclopentylmethyl *O*-ethyl dithiocarbonate whereas *S*-3-butenyloxy *O*-ethyl dithiocarbonate gave a lactone.

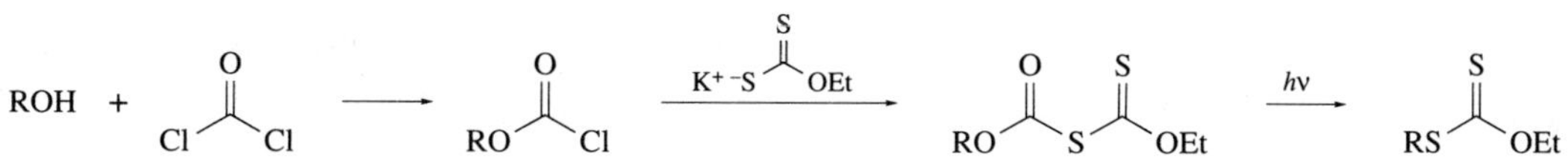

Scheme 16

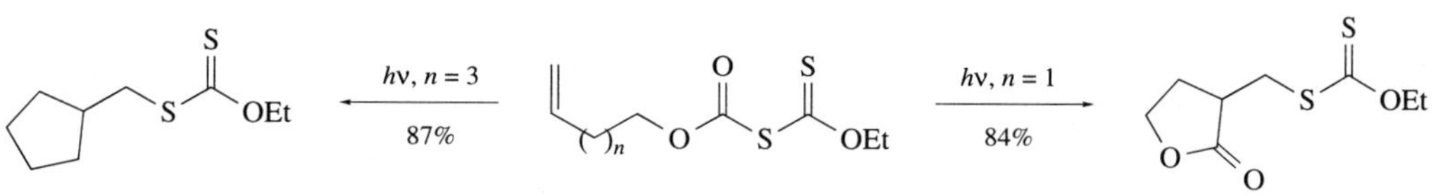

Scheme 17

Several other types of activated halides have been shown to react with *O*-alkyldithiocarbonate salts. 2-Chloro-4,5-dihydroimidazole reacts with two moles of potassium *O*-alkyl dithiocarbonates giving the corresponding diesters $[ROC({=}S)]_2S$ and imidazoline-2-thione ⟨92JCS(P1)47⟩. Displacement of chloride from 1-chloro-1,3,3-triphenylallene by potassium *O*-ethyl dithiocarbonate has also been reported ⟨89BCJ967⟩. The reaction of triphenyltelluronium chloride with the equivalent amount of sodium *O*-alkyl dithiocarbonate gives the corresponding triphenyltelluronium dithiocarbonate (Equation (42)). The same product can be obtained by reacting a triphenyltelluronium alkoxide with carbon disulfide. The reaction of freshly prepared dimethyltellurium diiodide with a sodium *O*-alkyl dithiocarbonate leads to dimethyltellurium bis(*O*-alkyl dithiocarbonates) (Equation (43)), which can also be obtained by inserting carbon disulfide into dimethyltellurium bis(alkoxides) ⟨85ZAAC(525)127⟩. *S*-(Alkoxycarbonyl) *O*-alkyl dithiocarbonates have been prepared by the addition of a solution of a potassium *O*-alkyl dithiocarbonate to that of an *O*-alkyl chlorothiocarbonate (Equation (44)) ⟨80AQ183, 83JCS(P1)2641⟩.

(42)

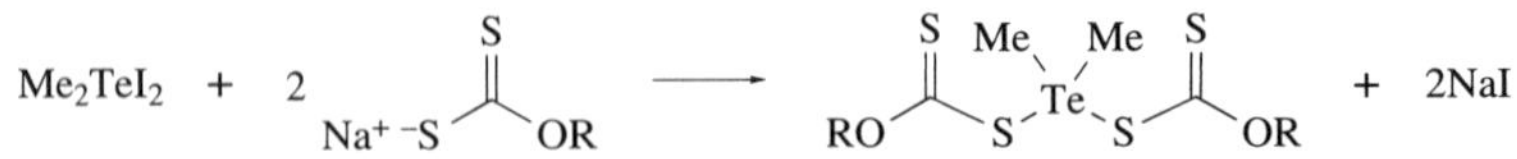
(43)

(44)

Bis(methoxythiocarbonyl)sulfanes have been prepared by a variety of different methods as shown in Scheme 18 ⟨87ZAAC(552)181⟩. Several analogous preparations have been reported ⟨86JOC1866⟩. Methoxydichloromethanesulfenyl chloride, when added to a suspension of potassium *O*-methyl dithiocarbonate in $CDCl_3$ at 5°C, gave the dithiocarbonate which underwent a further displacement reaction with *N*-methylaniline (Scheme 19) ⟨90JOC1475⟩. [Methoxy(thiocarbonyl)](chlorocarbonyl)disulfane, $MeOC(=S)S_2COCl$, is obtained (41%), together with bis[methoxy (thiocarbonyl)]sulfane, when ClSCOCl is added to a suspension of potassium *O*-methyl dithiocarbonate in chloroform ⟨86JOC1866⟩.

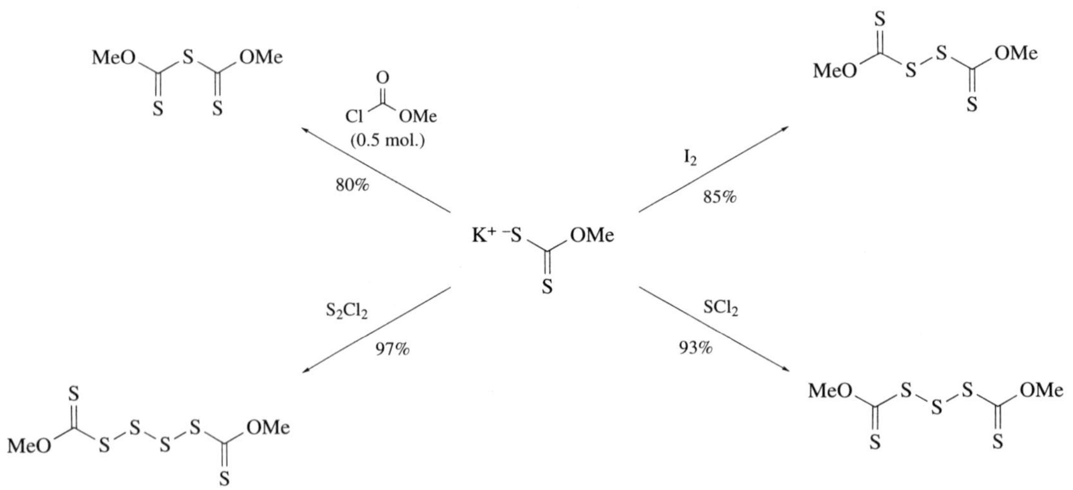

Scheme 18

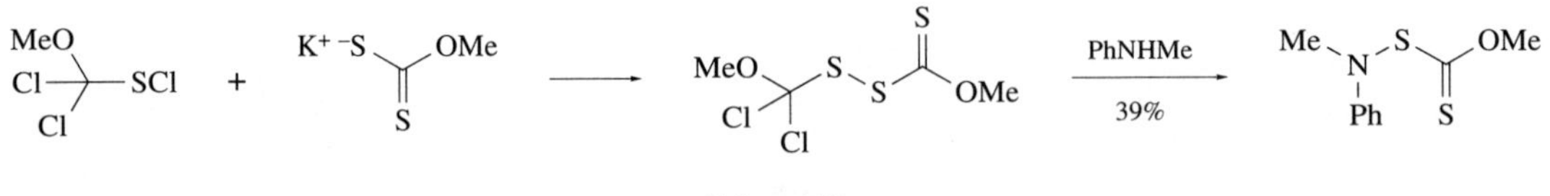

Scheme 19

Salts of *O*-alkyl dithiocarbonates can act as nucleophiles towards a variety of reagents other than halides. Methoxythiocarbonyl methylsulfane can be synthesized by the reaction of potassium *O*-methyl dithiocarbonate with the *S*-methyl ester of methanethiosulfonic acid (Equation (45)) ⟨87ZAAC(550)109⟩. In basic conditions, and in the presence of excess KSC(=S)OMe, a further reaction can occur leading to the formation of a symmetrical disulfane. When potassium *O*-isopropyl dithiocarbonate reacts with (*Z*)-4-phenylbut-3-enyl methanesulfonate in the presence of a phase transfer catalyst *O*-isopropyl (*Z*)-S-4-phenylbut-3-enyl dithiocarbonate is obtained (Equation (46)) ⟨88JCS(P1)1517⟩.

(45)

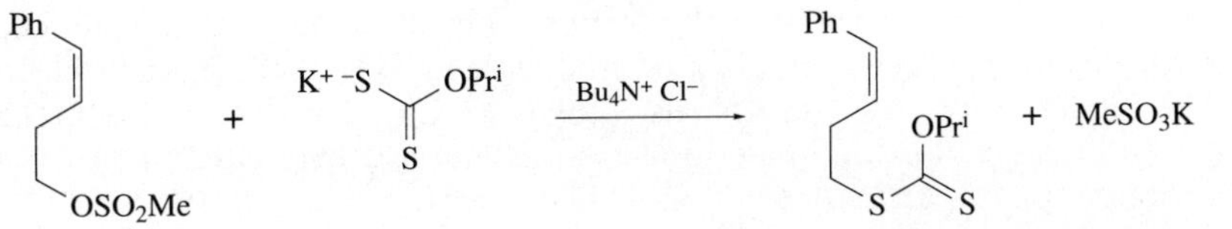

(46)

A mixture of 3-*O*-acetyl-2-azido-4-(benzyloxycarbonyl)amino-2,4,6-trideoxy-D-galactopyranosyl nitrate and the minor talopyranoside was obtained by treating a D-galactal with sodium azide and cerium(IV) nitrate in acetonitrile at −25 °C ⟨92MI 617-05⟩. A reaction of the above mixture with potassium *O*-ethyl dithiocarbonate in acetonitrile in turn gave *O*-ethyl *S*-(3-*O*-acetyl-2-azido-4-(benzyloxycarbonyl)amino-2,4,6-trideoxy-*β*-D-galactopyranosyl) dithiocarbonate (Scheme 20) ⟨92MI 617-05⟩. Several similar reaction sequences have been reported ⟨88LA75, 91T5149, 93HCA995⟩.

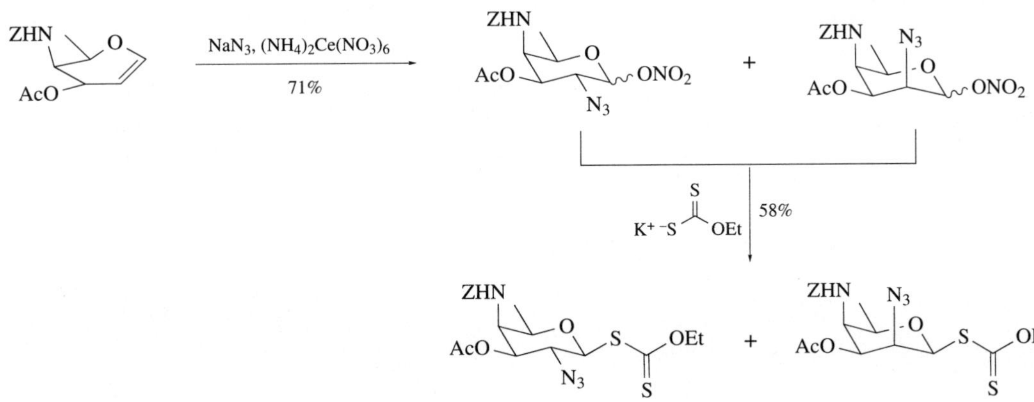

Scheme 20

The addition of *V. harveyi* oxidoreductase (NADH oxidized by riboflavin) to an anaerobic solution of potassium *O*-ethyl dithiocarbonate, 11-deoxydaunomycin, and NADH led to two dithiocarbonates (7*S* and 7*R*) in an 85 : 15 ratio (Equation (47)) ⟨83JA7187⟩. The cyclization shown in Equation (48) is initiated by nucleophilic opening of the three-membered ring by potassium *O*-ethyl dithiocarbonate ⟨88ZOR117⟩.

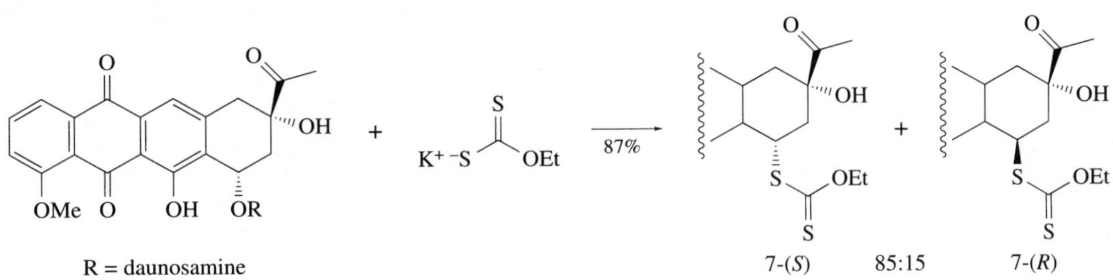

(47)

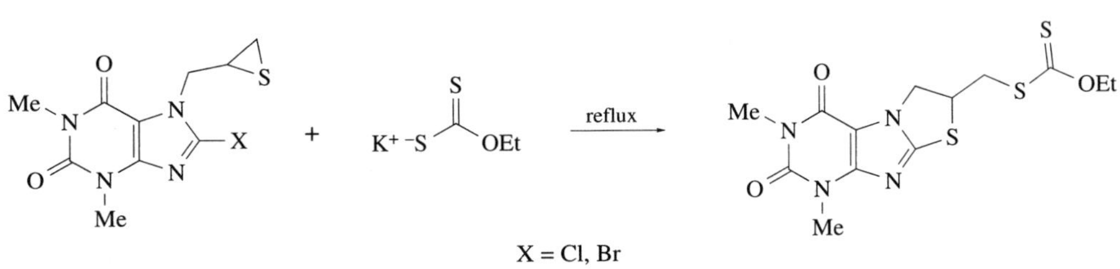

(48)

(ii) From carbon disulfide

It is possible to prepare *O*,*S*-dialkyl dithiocarbonates directly from alcohols by reaction with a base, carbon disulfide and a haloalkane. This is mechanistically equivalent to using salts of *O*-esters of dithiocarbonic acids (6.17.2.2.2(i)), but experimentally more convenient, and the procedure has been widely used. In a typical procedure, the sodium salt of the alcohol is produced by its reaction with sodium hydride and a catalytic amount of imidazole in THF; carbon disulfide and a haloalkane are then added sequentially. An example is shown in Equation (49) ⟨92JA4111⟩; many others, encompassing a wide range of primary alcohols ⟨82JOC132, 89CC1265, 91JOC2849, 92JOC6803⟩, secondary alcohols ⟨79JA6116, 81CC756, 81JCS(P1)2363, 82HCA371, 82YGK352, 83JOC4750, 87CC1228, 88CAR(181)253, 90TL4931, 91T121, 91T6381, 92TL5261, 93CAR(242)281, 93CAR(246)119, 93JOC2894, 93JOC3798⟩, tertiary alcohols ⟨93TL2733⟩ and phenols ⟨86T2329⟩, have been described.

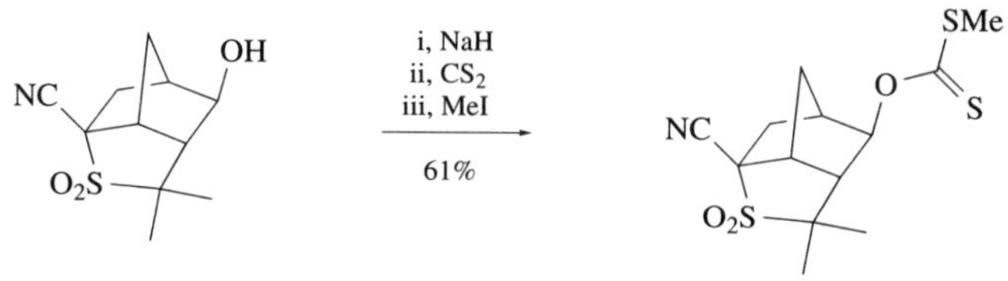

(49)

Sometimes butyllithium is used in place of sodium hydride for the preparation of dithiocarbonates directly from alcohols. An example is shown in Equation (50) ⟨85T4079⟩; other examples have been described ⟨87BCJ1853, 88M127⟩. Phase-transfer catalysts have also been used with aqueous sodium hydroxide as the base ⟨88M127, 89SC547⟩.

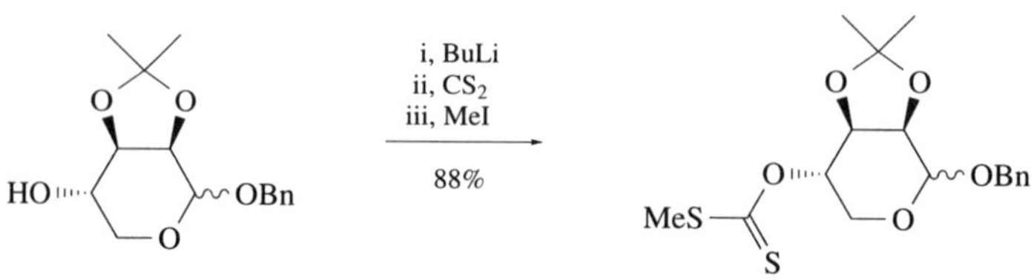

(50)

An extension of the methodology involves the generation of a carbanion using butyllithium and its aldol addition to an aldehyde; the lithium salt of the aldehyde then reacts further with carbon disulfide and iodomethane ⟨88TL6125, 91TL2703⟩. An example of the reaction sequence is shown in Scheme 21 ⟨91TL2703⟩.

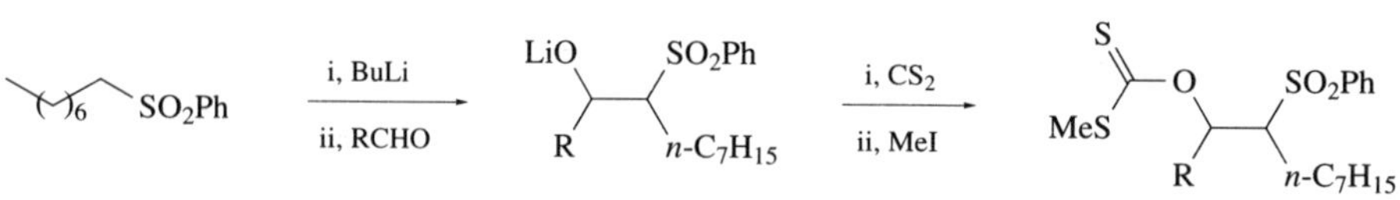

Scheme 21

By treating carboxylic esters with lithium diisopropylamide (LDA) at 0 °C, lithium enolates were generated. The latter were treated with carbon disulfide followed by a haloalkane to produce *O*-(1-alkoxy-2,2-dialkylvinyl) *S*-alkyl dithiocarbonates in very good yield (Scheme 22) ⟨89CC684, 89JOC5603⟩.

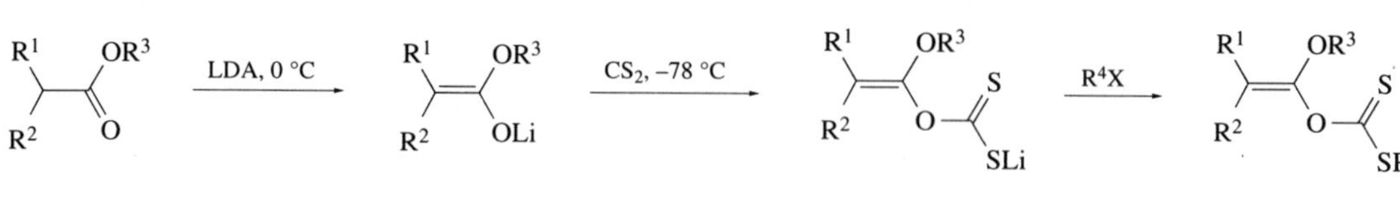

Scheme 22

(iii) From thiocarbonyl chlorides

O,S-Dialkyl dithiocarbonates can be prepared by the displacement of chloride from alkoxythiocarbonyl chlorides by thiolate anions. For example, *S*-ethyl *O*-methyl dithiocarbonate was obtained by adding sodium ethanethiolate to a solution of methoxythiocarbonyl chloride in ether at a rate to maintain a mild reflux (Equation (51)) ⟨83JOC4750⟩. Silver salts have also been used for preparations of this type ⟨86JOC4488⟩, and imidazolides ROC(=S)Im (Im = 1-imidazolyl) have been used in place of alkoxythiocarbonyl chlorides ⟨86T2329⟩.

(51)

S-(Alkoxycarbonyl) *O*-alkyl dithiocarbonates have been prepared in high yield by the addition of a solution of a potassium *O*-alkyl monothiocarbonate to that of an *O*-alkyl chlorothiocarbonate in aqueous acetone (Equation (52)) ⟨85PS(25)91⟩. In the presence of a phase transfer catalyst and aqueous sodium thiosulfate, methoxythiocarbonyl chloride gives bis(methoxythiocarbonyl) sulfide (74%) (Equation (53)) ⟨83JOC4750⟩.

(52)

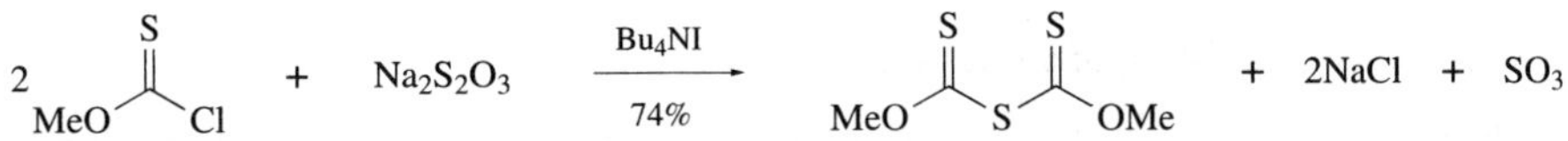

(53)

An alternative procedure is to displace chloride from dithiocarbonyl chlorides such as PhSC(=S)Cl with alkoxide anions ⟨91T121⟩.

(iv) By radical addition to alkenes

Examples of intramolecular addition of radicals ROC(=S)S· to carbon–carbon double bonds have already been illustrated in Scheme 17. Intermolecular radical addition reactions have also been reported ⟨88CC308, 89H(28)171⟩; two examples are shown in Equations (54) and (55).

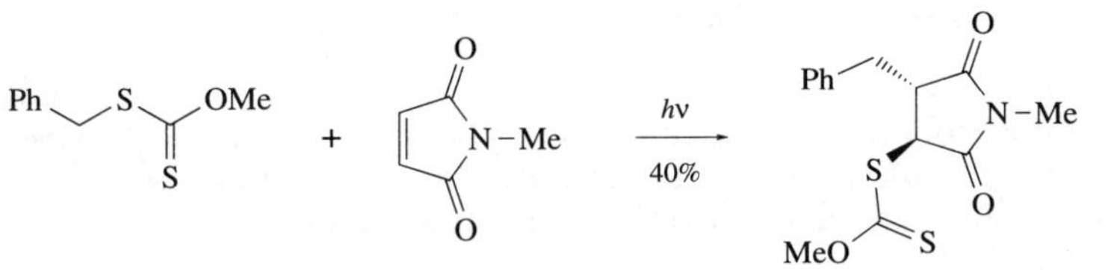

(54)

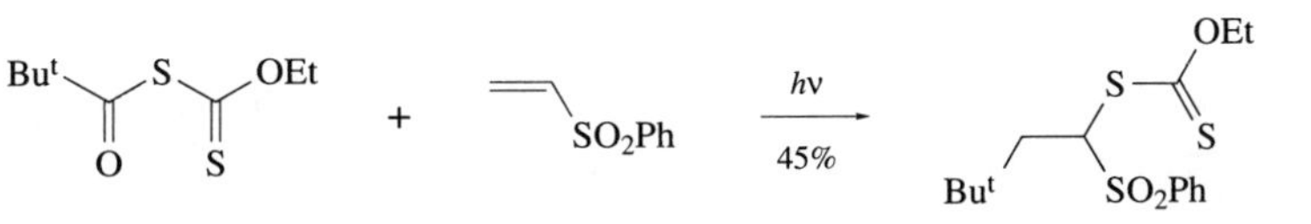

(55)

(v) From other dithiocarbonates

The radical reaction of bromoalkanes with *O*-ethyl triphenyltin dithiocarbonate was initiated by tributyltin hydride and 2,2′-azobisisobutyronitrile (AIBN) or by visible light. Dithiocarbonates were obtained in good yield ⟨92JA7909⟩. The reaction sequence is illustrated in Scheme 23 for a radical which undergoes cyclization before capture; the sequence depends on the ability of triphenyltin radicals to add reversibly to dithiocarbonates.

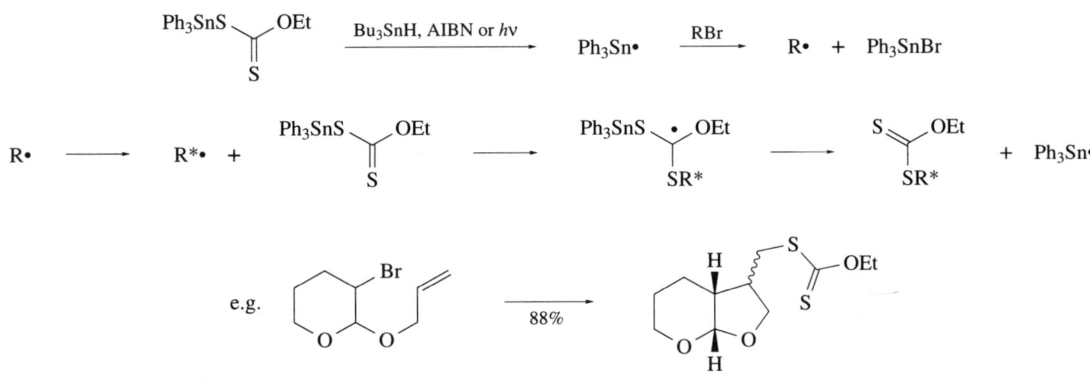

Scheme 23

O-Ethyl *S*-(2-hydroxy-4-oxo-4-phenylbutyl) dithiocarbonate and related esters were prepared by allowing lithium enolates of ketones to react with *O*-ethyl *S*-(2-oxoethyl) dithiocarbonate ⟨92HCA907⟩. An example is shown in Equation (56).

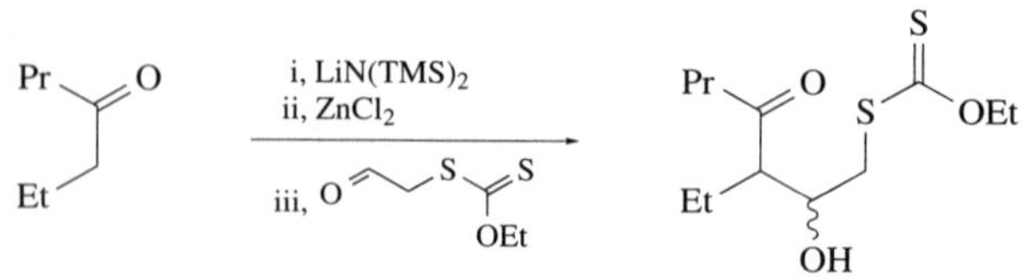

(56)

6.17.2.3 Derivatives of Trithiocarbonic Acid

Trithiocarbonic acid, $(HS)_2C{=}S$, can be generated from barium trithiocarbonate by reaction with dilute HCl at 0 °C ⟨63ZAAC(321)143⟩. It is, however, stable only at −78 °C; otherwise it decomposes to hydrogen sulfide and carbon disulfide. Salts of trithiocarbonic acid are obtained by the reaction of carbon disulfide with metal sulfides or metal hydrogen sulfides ⟨83HOU(E4)420⟩. A 1,5-diazabicyclo[5.4.0]undec-5-ene (dbu) salt was obtained by heating a mixture of water, CS_2, and dbu in molar ratios 2 : 3 : 4 ⟨86CE311, 87NKK1408⟩.

6.17.2.3.1 Salts of monoesters of trithiocarbonic acid

The reaction of aliphatic or aromatic thiolate salts with carbon disulfide gives salts of the monoesters of trithiocarbonic acid (Equation (57)) ⟨70JA3342, 70MI 617-01, 72JA1563, 80JOC175, 93POL13⟩. In a few cases lithium salts have been generated by fragmentation of 1,3-dithiole derivatives ⟨81TL5195, 82BCJ339⟩; an example ⟨81TL5195⟩ is shown in Scheme 24.

(57)

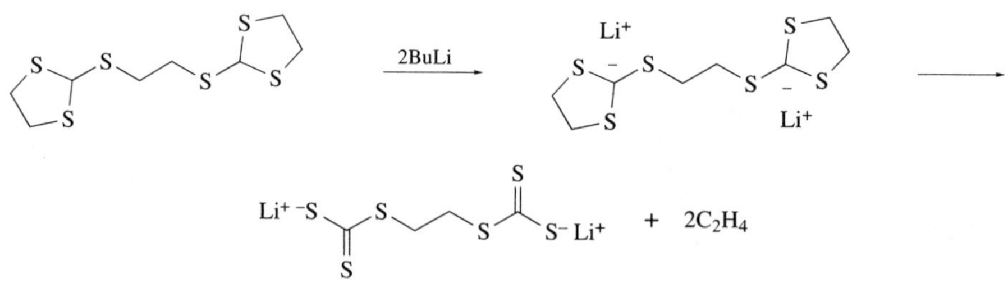

Scheme 24

6.17.2.3.2 Diesters of trithiocarbonic acid

(i) From thiophosgene

Diesters of trithiocarbonic acid can be prepared in very good yield from thiophosgene and salts of thiols and thiophenols (Equation (58)) ⟨61JOC4047, 83HOU(E4)420⟩. The reaction of thiophosgene with dithiolates gives cyclic trithiocarbonic acid esters (Equation (59)) ⟨77CC660, 83HOU(E4)420⟩.

(58)

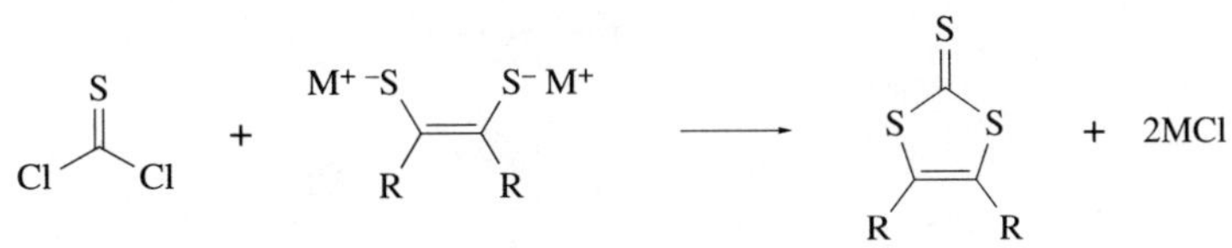
(59)

(ii) From S-*esters of dithiocarbonic acid chloride*

The starting acid chlorides can be isolated as intermediates from the reaction of thiophosgene with thiols or thiophenols ⟨83HOU(E4)420⟩; they form mixed esters when allowed to react further with another alkanethiol or thiophenol (Equation (60)) ⟨61JOC4047, 71CC314⟩.

(60)

(iii) From salts of monoesters of trithiocarbonic acid

The salts are excellent nucleophiles, and they are alkylated by haloalkanes and a range of other alkylating agents ⟨83HOU(E4)420⟩. The reaction can also be conveniently carried out by generating the salt *in situ* from a thiol, carbon disulfide, and a base ⟨86S894⟩. Phase transfer catalysts have been used to achieve one-pot conversions of thiols to dithiocarbonates ⟨87BCJ435⟩.

An example of thc alkylation reaction (which was used as a step in a penem synthesis) is shown in Equation (61) ⟨81TL3485⟩. There are several other similar examples of the preparation of trithiocarbonates derived from azetidinones ⟨81CPB3158, 87H(25)123, 92JOC4352⟩. A reaction of potassium 1,2-ethanebis(trithiocarbonate) with HCl gave 1,2-ethanebis(trithiocarbonic acid), $HSCSSCH_2CH_2SCSSH$. Further reaction with iodoalkanes gave $RSC(=S)SCH_2CH_2SC(=S)SR$ (R = Me, 86%; R = Et, 82%) ⟨85ZAAC(530)101⟩. An example of the use of a leaving group other than a halide is provided by the reaction of potassium *S*-methyl trithiocarbonate with the *S*-methyl ester of methanethiosulfonic acid, $MeSO_2SMe$, which gave methyl methylthiothiocarbonyl disulfane, $MeSC(=S)S_2Me$ ⟨87ZAAC(550)109⟩.

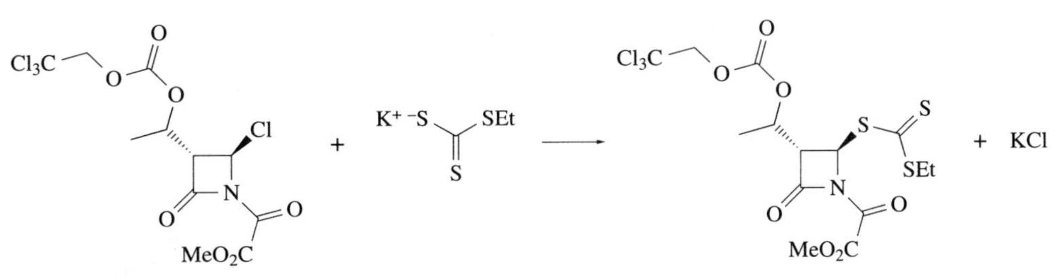
(61)

Many electrophiles other than alkylating agents have also been used, and some of these are illustrated in Scheme 25. Reactions of the salts are shown with 2-chloro-3,5-dinitropyridine ⟨81JHC1581⟩, methyl chlorodithioformate, iodine, sulfur monochloride, and sulfur dichloride ⟨87ZAAC(545)125, 87ZAAC(554)172⟩ and trichloromethanesulfenyl chloride ⟨85T5145⟩. The reaction of $RPCl_2$ or R_2PCl (R = Me, Pr^i, Ph) with $NaSCSSR^1$ (R^1 = Et, Pr^i) gave $RP(SCSSR^1)_2$ and $R_2PSCSSR^1$, respectively ⟨88PS(35)93⟩. Triphenyltellurium chloride reacted with sodium *S*-alkyl trithiocarbonates giving the corresponding *S*-alkyl *S*-triphenyltellurium trithiocarbonates in high yield ⟨88ZAAC(556)189⟩.

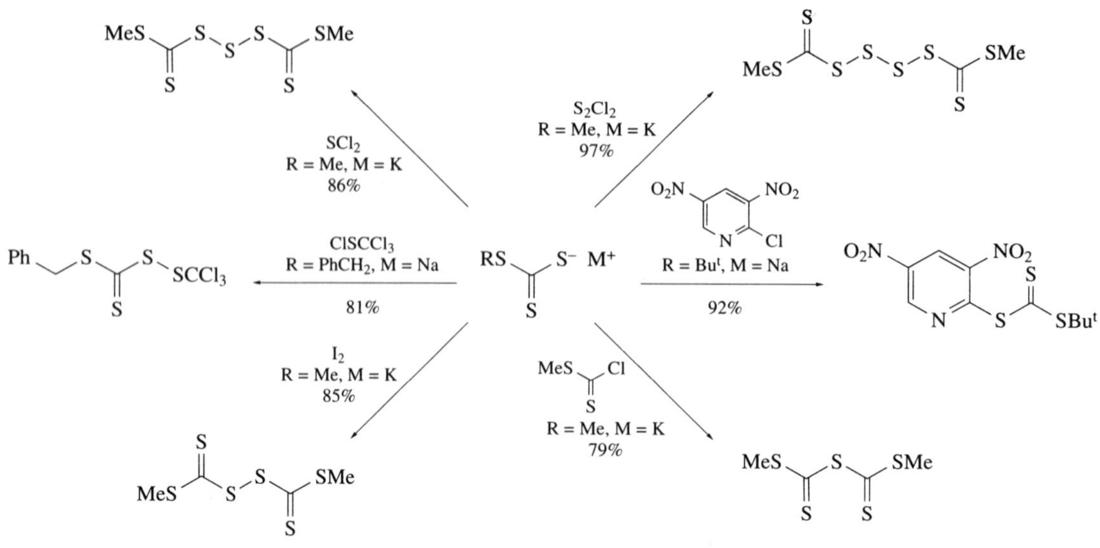

Scheme 25

(iv) From trithiocarbonate salts

Sodium, potassium, and barium salts of trithiocarbonic acid react with alkylating agents giving symmetrical dialkyl trithiocarbonates (Equation (62)) ⟨80CPB110, 82BCJ1174, 83HOU(E4)420⟩. These esters were also synthesized in high yield in the presence of a phase transfer catalyst ⟨86S894, 88S22, 88SC1531⟩. For example, carbon disulfide and 33% aqueous sodium hydroxide were stirred together, and haloalkanes were then added in the presence of tetrabutylammonium hydrogensulfate (Scheme 26) ⟨88SC1531⟩. The dbu trithiocarbonate salt also reacts with alkylating agents to give dialkyl trithiocarbonates in good yields ⟨86CE311, 87NKK1408⟩. The method can be used to prepare cyclic trithiocarbonates; for example, *cis*-3,5-dibromocyclopentene reacted with sodium trithiocarbonate in DMF to give the *cis*-3,4-cyclopenteno-1,3-dithiolane-2-thione (Equation (63)) ⟨93H(35)77⟩.

(62)

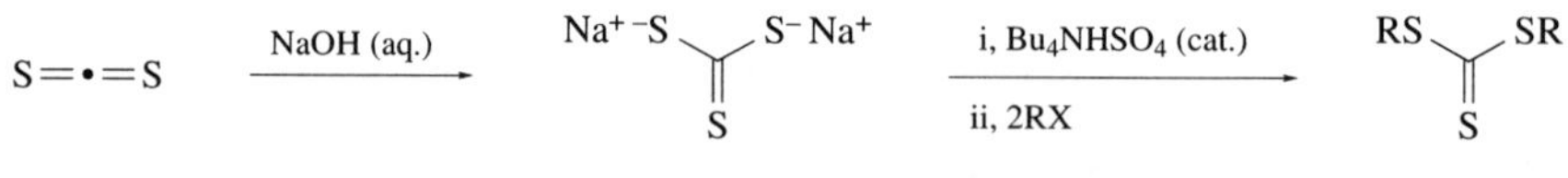

Scheme 26

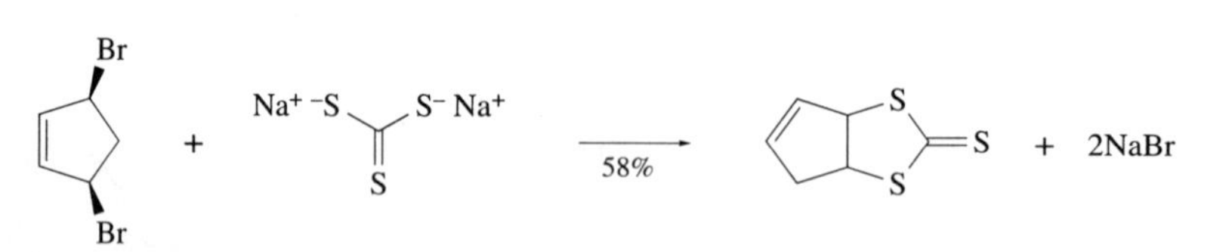

(63)

Other types of electrophile can be used to functionalize the trithiocarbonate salts. For example, the reaction of sodium trithiocarbonate with trichloromethanesulfenyl chloride gave the crystalline $(Cl_3CS_2)_2CS$, which on oxidation gave the sulfine $(Cl_3CS_2)_2CS{=}O$ ⟨84SUL85⟩. The compounds $(RS_2)_2C{=}S$ can be prepared by treating a sulfenyl chloride ($R = CCl_3$, CCl_3CCl_2, CCl_2FCCl_2 or $CClF_2CCl_2$) with an excess of freshly prepared aqueous sodium trithiocarbonate in dichloromethane, or with anhydrous barium trithiocarbonate suspended in acetonitrile ⟨85T5145⟩. Oxidation of these compounds with mcpba gave the sulfines (Scheme 27).

When thioureas are prepared by three-component reactions of amines, carbon tetrachloride, and sodium sulfide, the reactions proceed exothermically in the presence of a phase-transfer catalyst ⟨83LA1839⟩. Thiophosgene is an intermediate in this conversion, and its phase-transfer catalyzed transformation into carbon disulfide and trithiocarbonate can be demonstrated. When the amine

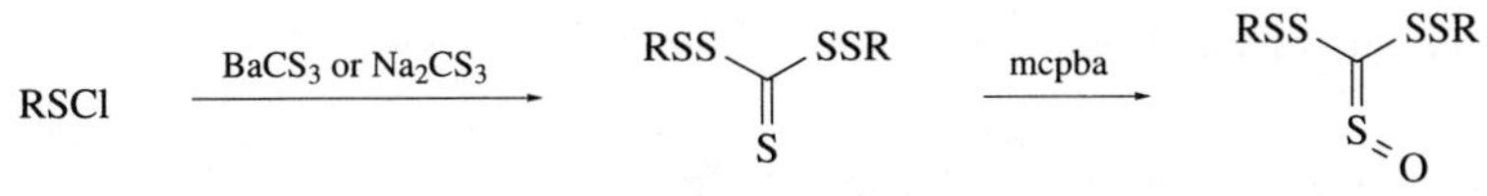

Scheme 27

was replaced by dibromomethane, benzyl chloride or α,β-dibromostyrene, the products included 1,3-dithietane-2-thione (5%), dibenzyl trithiocarbonate (45%), and 4-phenyl-1,3-dithiolane-2-thione (50%), respectively ⟨83LA1839⟩. An analogous reaction, in which 4-chloromethyl-1,3-dioxolane is the electrophile, has been reported ⟨91JPR139⟩.

(v) Methods leading to cyclic trithiocarbonates

Five-membered cyclic trithiocarbonates are accessible by a variety of routes ⟨65JA934, 75JOC3052, 76JOC626, 81TL5195⟩ and several of the general routes described above can be used (e.g. those in Equations (59) and (63)). The following are some specific methods.

Several cyclic trithiocarbonates have been prepared from the corresponding dithio-*N*,*N*-dimethyliminium salts by reaction with hydrogen sulfide or with sodium hydrogensulfide ⟨92CC1410, 92KGS1117, 93H(35)77⟩. An example of the process is shown in Equation (64) ⟨93H(35)77⟩. A different route to cyclic trithiocarbonates is shown in Equation (65). A suspension of tetraethylammonium tetrathiooxalate in acetonitrile was stirred with carbon disulfide at ambient temperature until the former had dissolved. Chloromethane was then bubbled through the solution. The precipitate obtained consisted of 75–90% of 4,5-bis(methylthio)-1,3-dithiole-2-thione and 10–25% of dimethyl trithiocarbonate ⟨86ACS(B)593⟩.

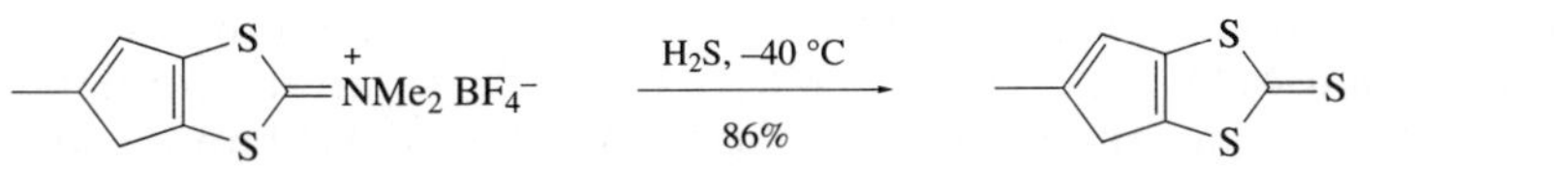

(64)

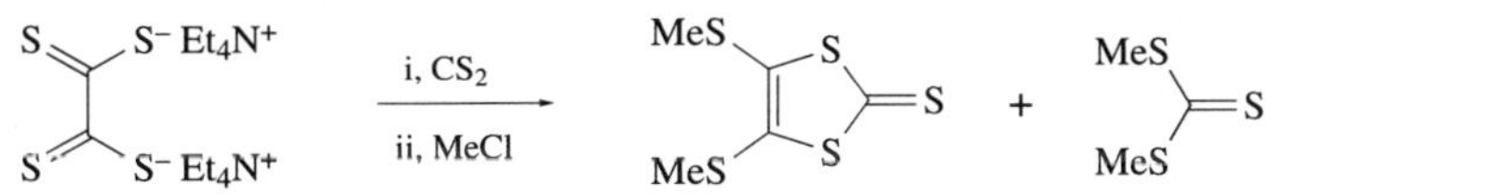

(65)

2-Pyridyl- and 2-quinolyl dithiocarbonates cyclocondense with carbon disulfide to yield 4,5-disubstituted 1,3-dithiole-2-thiones in 15 to 60% yield (Scheme 28) ⟨89JPR(331)439⟩.

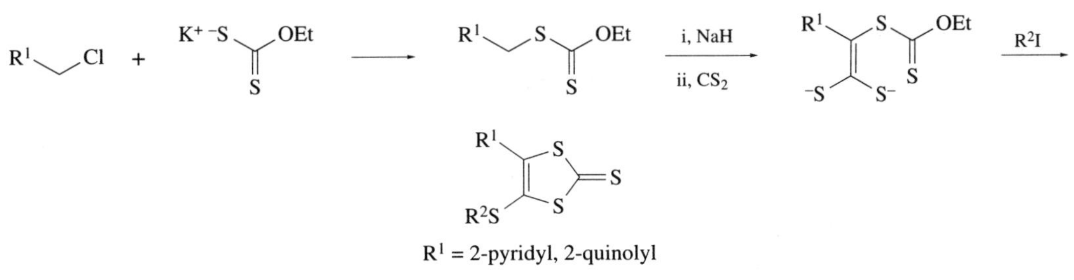

R^1 = 2-pyridyl, 2-quinolyl

Scheme 28

6.17.2.4 Derivatives of Thiocarbamic Acid

Thiocarbamic acid has the structure $H_2NC(=S)OH$. The free acids $R^1R^2NC(=S)OH$ decompose to COS and the corresponding amine when their preparation is attempted, but their salts can be prepared by treating COS with amines in the presence of an alkali metal hydroxide. Alkylation or acylation of these salts usually gives *S*-esters of thiocarbamic acid ⟨81S622, 81ZAAC(481)126⟩. The *O*-esters are isolable but when heated those bearing a dialkylamino group can isomerize to *S*-esters ⟨66JOC3980, 69JOC3604, 73JOC2106⟩; those with a monoalkylamino function can decompose to isothiocyanates and alcohols ⟨83HOU(E4)420⟩.

6.17.2.4.1 *O*-Esters of thiocarbamic acids

(i) From O-*alkyl or* O-*aryl chlorothioformates*

The reaction of *O*-alkyl or *O*-aryl chlorothioformates with amines is a general method for the preparation of *O*-esters of thiocarbamic acids ⟨83HOU(E4)420⟩; the following are examples of the method.

N,N,O-Trimethyl thiocarbamate was prepared from dimethylamine and *O*-methyl chlorothioformate (Equation (66)) ⟨87CB987⟩. 1,2,3,4-Tetrahydro-2-(methoxythiocarbonyl)-*β*-carboline-3-carboxylic acid was obtained by adding an ethereal solution of *O*-methyl chlorothioformate and aqueous sodium hydroxide to an aqueous solution of sodium 1,2,3,4-tetrahydro-*β*-carboline-3-carboxylate (Equation (67)) ⟨87CPB2840⟩. 4-(Nitrophenylamino)aniline reacted with *O*-phenyl chlorothioformate and pyridine in acetone to give the corresponding *O*-phenyl thiocarbamate ⟨88MI 617-02⟩. The amino group of 4-cyano-3-methylthiopyrazol-5-amine was similarly substituted ⟨92PHA251⟩.

(66)

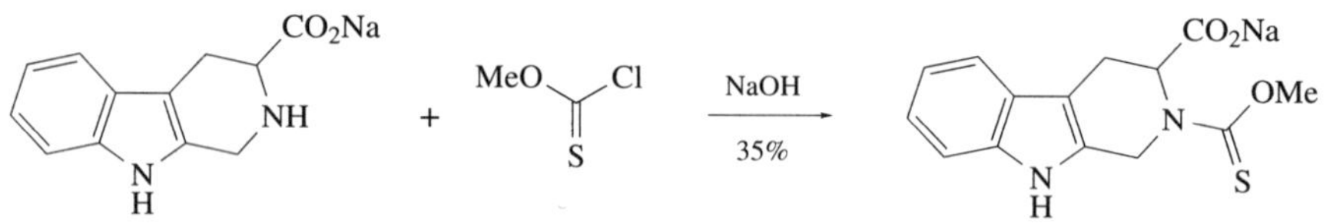

(67)

The reaction of an *O*-aryl chlorothioformate (one equivalent) with an alkyl or aryl bis(trimethylsilylamine) (three equivalents) in ether at room temperature gave the corresponding thiocarbamate in good yield (Scheme 29) ⟨85PS(21)291⟩. If the chlorothioformates were used in excess, the products were the corresponding *N*-alkyl or *N*-aryl bis(*O*-arylthionocarbonates) (Scheme 29).

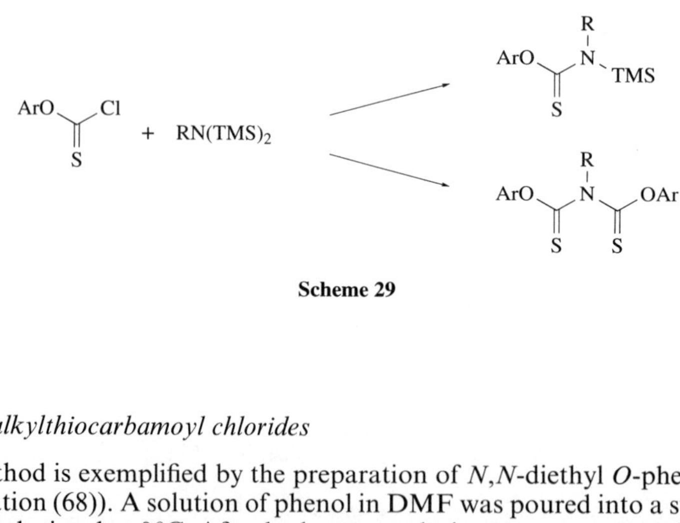

Scheme 29

(ii) From N,N-*dialkylthiocarbamoyl chlorides*

This general method is exemplified by the preparation of *N,N*-diethyl *O*-phenyl dithiocarbamate from phenol (Equation (68)). A solution of phenol in DMF was poured into a suspension of sodium hydride in DMF and stirred at 0°C. After hydrogen evolution was over, *N,N*-diethylthiocarbamoyl chloride was added and the mixture was heated for 1 h at 80°C. After separation, 76% of *N,N*-diethyl *O*-phenyl thiocarbamate was obtained ⟨66JOC3980, 92S112⟩. Similar preparations have been reported from *N,N*-dialkylcarbamoyl chlorides and the following: substituted phenols ⟨59JA714, 91CPB1939⟩, 1-naphthol ⟨92S112⟩, 3-hydroxypyridine ⟨92S112⟩, methyl 3-hydroxythiophene-2-carboxylate ⟨84S172⟩, (*R*)-(+)-1,1′-binaphthol ⟨93JOC1748⟩, other binaphthols ⟨91PS(63)51⟩ and calix-[4]arenes ⟨90CC1432⟩.

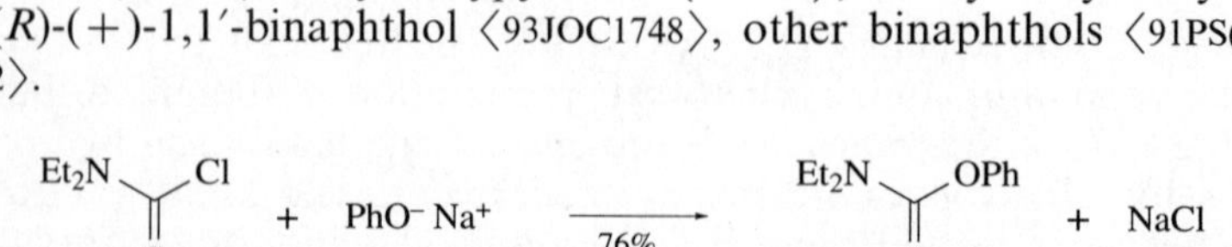

(68)

3-Methyl-2-[(dimethylthiocarbamoyl)oxy]-3-methylcyclopent-2-en-1-one and related compounds

⟨86JOC4741⟩ can be prepared by adding lithium hydroxide solution at room temperature to a stirred solution of the ketone in chloroform, and then a solution of *N*,*N*-dimethylthiocarbamoyl chloride in chloroform (Equation (69)) ⟨87JOC5630⟩. Cyclohexane-1,2-diones have been used as the starting materials for similar preparations ⟨88JOC1110⟩.

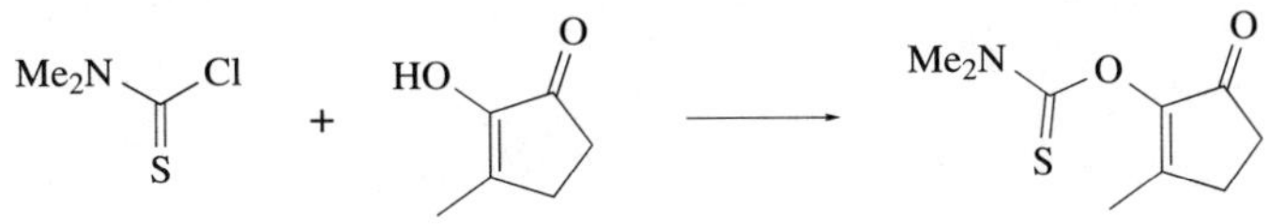

(69)

The thiocarbamoyl chloride can be generated *in situ* and then reacted with the appropriate alcohol or phenol: an example of this approach is shown in Scheme 30 ⟨90OPP128⟩.

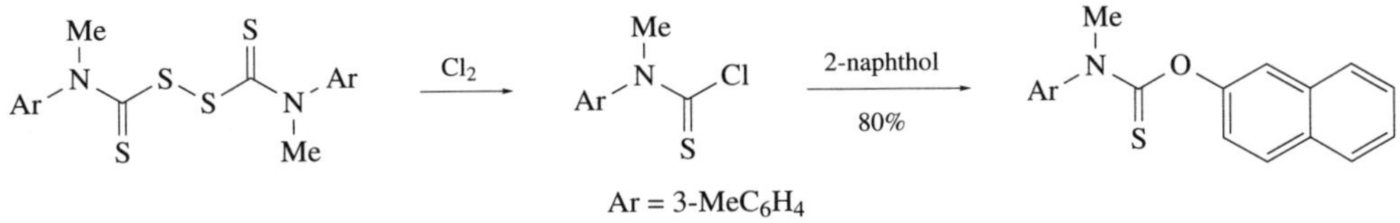

Scheme 30

(iii) From N,N′*-thiocarbonyldiimidazole and related compounds*

A mild procedure for the functionalization of alcohols is to heat the alcohol with *N*,*N*′-thiocarbonyldiimidazole in a solvent such as dichloromethane or THF (Equation (70)). This procedure was used to convert several epoxyalcohols into the thiocarbonylimidazolides in high yield ⟨81JCS(P1)2363⟩, and there are several other examples of similar procedures ⟨75JCS(P1)1574, 85T4079, 87JA609, 92JOC6803, 92T6883, 92T8031, 92TL5261⟩.

(70)

O,*N*-Dibenzyl *N*-methyl thiocarbamate was prepared by adding sodium hydride to a solution of 1-(*N*-benzyl *N*-methyl thiocarbamoyl)imidazole and benzyl alcohol in acetonitrile (Equation (71)) ⟨88JOC2263⟩. 1-[(Benzyloxy)thiocarbonyl]imidazole was obtained from *N*,*N*′-thiocarbonyldiimidazole, benzyl alcohol, and sodium hydride; this compound also gave *O*,*N*-dibenzyl *N*-methyl thiocarbamate when benzylamine was added ⟨88JOC2263⟩.

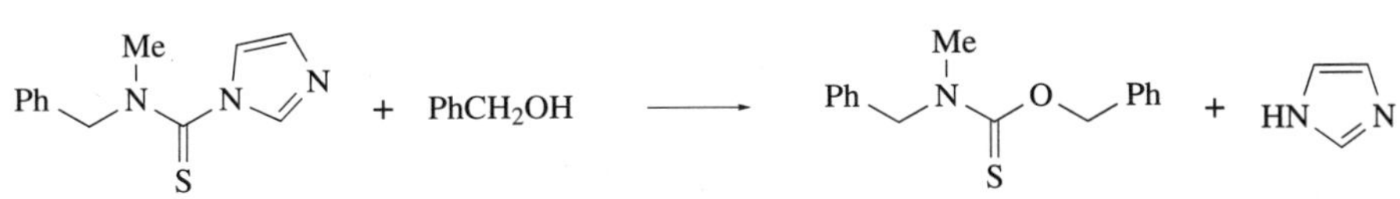

(71)

(iv) From aminoalcohols and carbon disulfide and related methods

Several fused tetrahydro-1,3-oxazine-2-thiones were prepared from aminoalcohols (as in the example shown in Scheme 31) by cooling a solution of the aminoalcohol in aqueous potassium hydroxide to 0 °C, and then mixing the solution with a solution of carbon disulfide in dioxane. Aqueous lead(II) nitrate was then added and lead sulfide precipitated. The oxazine-2-thiones were obtained by extraction with ethanol ⟨83T1829⟩. An alternative procedure is to add thiophosgene to a solution of the aminoalcohol and triethylamine ⟨83T1829⟩. There are other reports of the preparation of a variety of oxazine-2-thiones from aminoalcohols by similar methods ⟨82H(19)1191, 83JHC1181, 84JHC1373, 85S1149, 85T1353, 90MRC1045⟩.

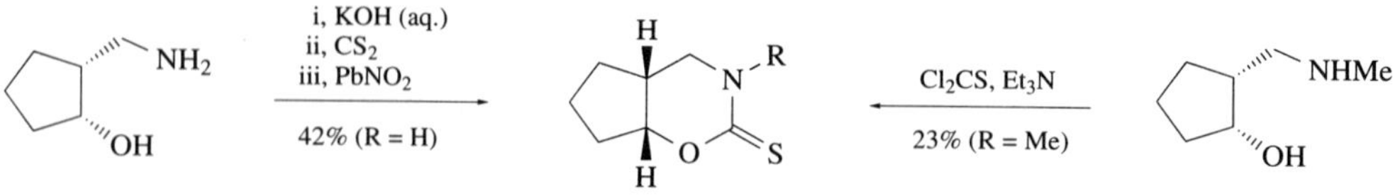

Scheme 31

1,3,4-Oxadiazole-2-thiones were obtained by reacting the respective hydrazides with carbon disulfide in the presence of potassium hydroxide (Equation (72)) ⟨93JMC1090, 93JMC1802⟩.

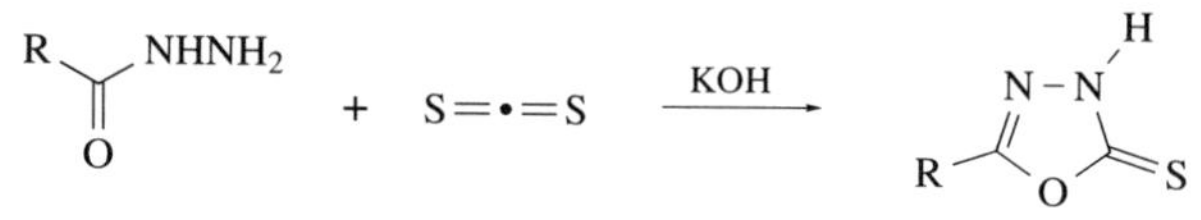

(72)

(v) From isothiocyanates

The addition of alcohols to isothiocyanates (Equation (73)) is a general method for the preparation of O-alkyl thiocarbamates; many examples of the procedure have been reported ⟨83JHC1223, 83ZOR92, 84CCC1577, 84ZOR2158, 85JPR1007, 85KGS339, 85MI 617-01, 85S423, 86MI 617-02, 87CCC1764, 89CPB1249, 89MI 617-01, 90JHC407, 90JHC643, 90JOC5230, 90MI 617-02, 90ZOR1422, 91S265, 93OPP83, 93TL3745⟩. Others include the preparation of the N-pentafluorosulfanyl derivative $SF_5NHC({=}S)OMe$ (87%) ⟨84CB1707⟩ and the reaction of hydroxypyridones with alkyl isothiocyanates ⟨87MI 617-02, 91PJS422⟩. A cyclic trimeric isothiocyanato phosphazene reacts with alcohols (R = Me, Et, Pr, Bu, Pr^i) giving first (at 25 °C) a species in which three of the six thiocyanato groups have reacted, a higher temperature (66 °C) being necessary to bring about reaction of all the isothiocyanato groups (Scheme 32) ⟨91IC1776⟩. Butadien-1-yl thiocyanate undergoes a Diels–Alder reaction with dienophiles, and the cycloadducts were intercepted by reaction with ethanol (Equation (74)) ⟨86HCA1898⟩. The hydroxyl group of fluorenone oxime adds to benzoyl isothiocyanate to give the thiocarbamate ⟨85CC1405⟩. O,O-Dialkyl imidodicarbonothioates $R^1OC({=}S)NHC({=}S)OR^2$ can be obtained by reaction of O-alkyl carbonochlorodithioates $R^1OC({=}S)Cl$ with sodium isothiocyanate and treatment of the resultant alkoxythiocarbonyl isothiocyanates with alcohols R^2OH ⟨85S975⟩.

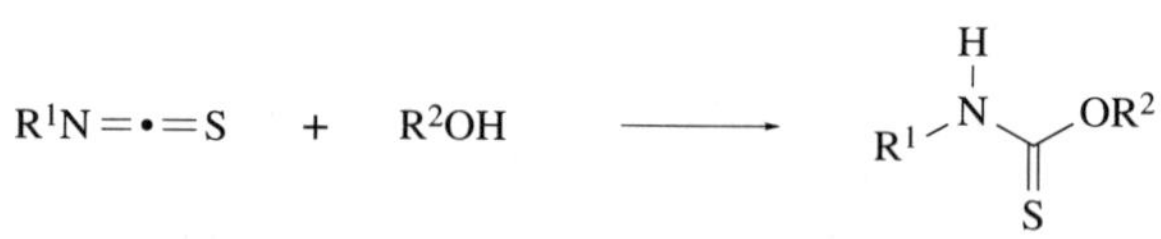

(73)

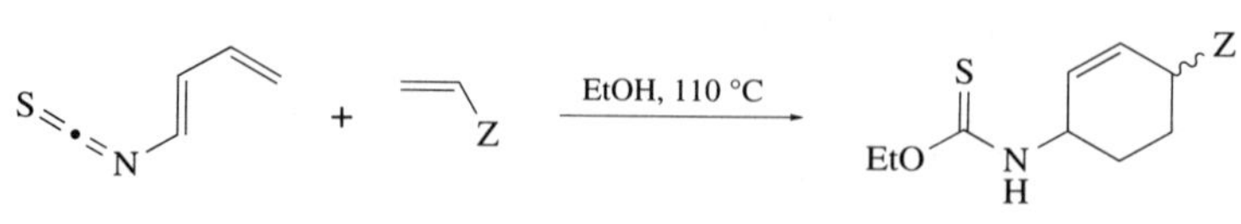

(74)

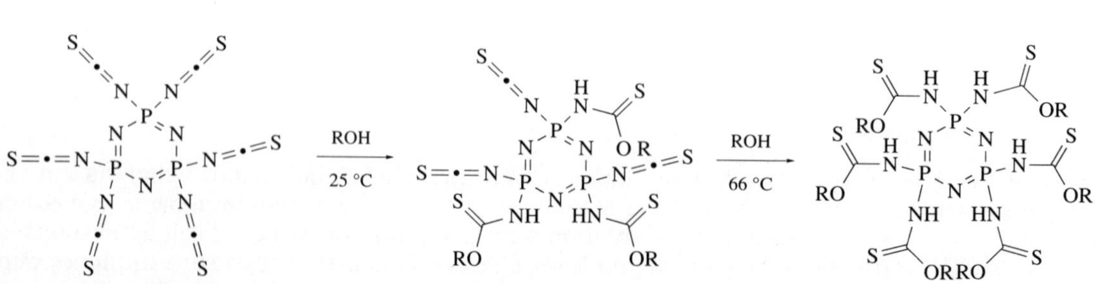

Scheme 32

The reaction of phosgene with one or two equivalents of ammonium thiocyanate gives thiocarbonyl chloride thiocyanate (9%) or thiocarbonyl dithiocyanate (73%), respectively ⟨81CB1132⟩. The latter compound reacts with alcohols to give *O*-alkyl imidodi(thiocarbonic acids), $[ROC(=S)]_2NH$. Carbonyl isocyanate isothiocyanate, SCNC(=O)NCO can be obtained by condensing carbonyl chloride isocyanate with ammonium thiocyanate ⟨81CB2064⟩. This reacts further with an alcohol giving *O*-alkyl (isothiocyanatocarbonyl) carbamates, ROC(=O)NHC(=O)NCS. These can react further with a second equivalent of an alcohol (which can be the same alcohol or a different one) to give the esters ROC(=O)NHC(=O)NHC(=S)OR ⟨81CB2064⟩.

(vi) From O-alkyl thiocarbamic acids and their salts

O-Methyl thiocarbamic acid, $MeOC(=S)NH_2$, can be obtained in 67% yield by treating the corresponding potassium salt in water with chloroacetic acid ⟨81ZAAC(481)126⟩. *O*-Ethyl thiocarbamic acid, $EtOC(=S)NH_2$ was obtained (46–50%) from dry ammonium thiocyanate in ethanol at 12–15°C after adding sulfuric acid ⟨82CB1252⟩. It reacts with oxalyl chloride giving a mixture of ethoxy(thiocarbonyl) isocyanate, EtOC(=S)NCO and its dimer (Scheme 33). Ethoxy(thiocarbonyl) isocyanate reacts further with alcohols and amines as shown in Scheme 33 ⟨82CB1259⟩.

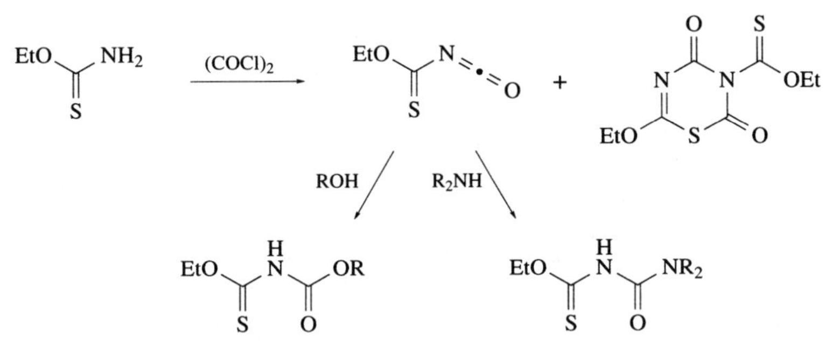

Scheme 33

(vii) Other methods

O-Ethyl thiocarbamates were prepared by the reaction of potassium *O*-ethyl dithiocarbonate with amines ⟨89JOC2978⟩. When *N*-arylbenzimidoyl chlorides were added to various potassium *O*-alkyl dithiocarbonates, *O*-alkyl esters of the corresponding *N*-aryl-*N*-thiobenzoyl thiocarbamic acids were obtained (Equation (75)) ⟨81ZC438⟩. Other imidoyl chlorides reacted similarly ⟨85ZC219⟩. A rearrangement is also involved in the alkylation of some dithiocarbamates (Equation (76)) ⟨88JCS(P1)813⟩. Alkylation of the potassium salts $K^{+}\,{}^{-}SC(=S)O(CH_2)_nOC(=S)S^{-}\,K^{+}$ gave dialkyl bis(dithiocarbonates), which on reaction with amines RNH_2 gave the bis(thiocarbamates) $RNHC(=S)O(CH_2)_nOC(=S)NHR$ ⟨82MI 617-03⟩.

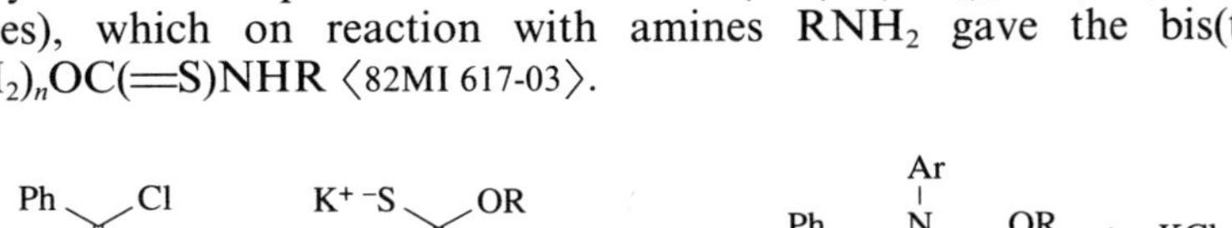

(75)

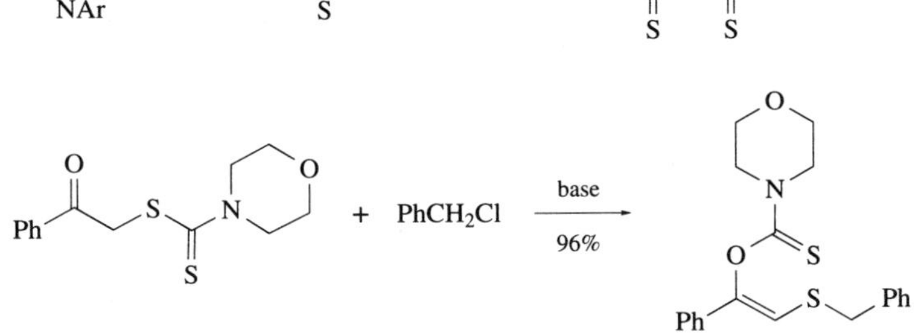

(76)

N,*N*-Dimethyl-*N*′-phenylthiourea, $PhNHC(=S)NMe_2$, reacts with ethyl bromocyanoacetate to give *O*-ethyl *N*-phenylthiocarbamate, EtOC(=S)NHPh ⟨82JCS(P1)653⟩. The reaction of 2,3-dihydro-

6-methoxy-2-thioxo-4*H*-1,3,5-thiadiazine-4-one with dibenzylamine resulted in the formation of a dithiocarbamate by ring cleavage (Equation (77)) ⟨81CB2075⟩.

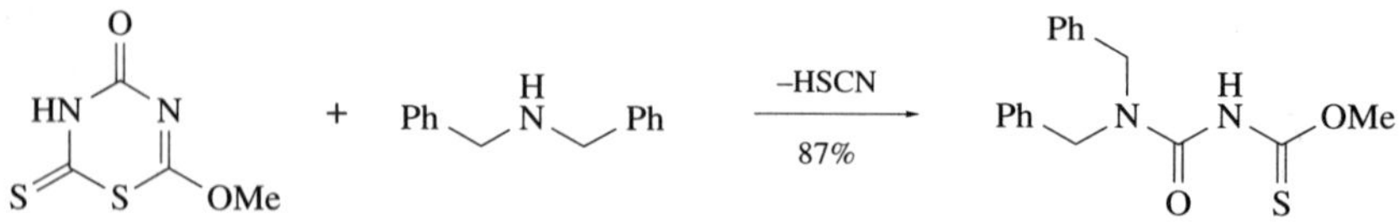

(77)

Several routes to dithiocarbamates involve the displacement of sulfur functions by amino functions. Some of these are illustrated in Equation (78) ⟨90ZC90⟩, Equation (79) ⟨87CPB156⟩, Equation (80) ⟨83JCS(P1)2011⟩, and Equation (81) ⟨86JOC1866⟩. The reaction of dithiocarbonate esters with *N*,*N*-dimethylhydrazine leads to mixtures of thionocarbamates and thionocarbazates whereas alkoxythiocarbonylimidazoles give cleanly the latter ⟨84JCS(P1)1005⟩.

(78)

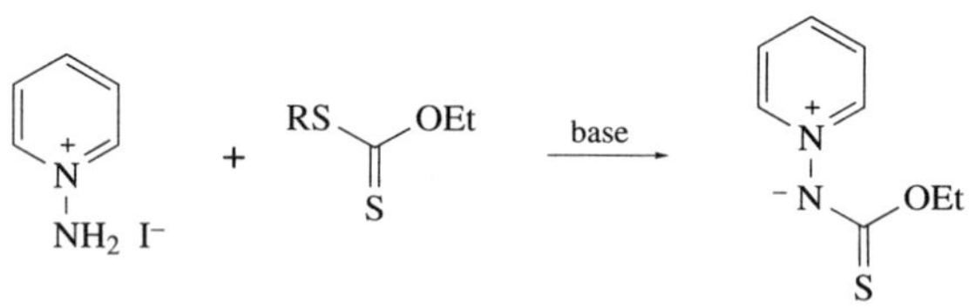

(79)

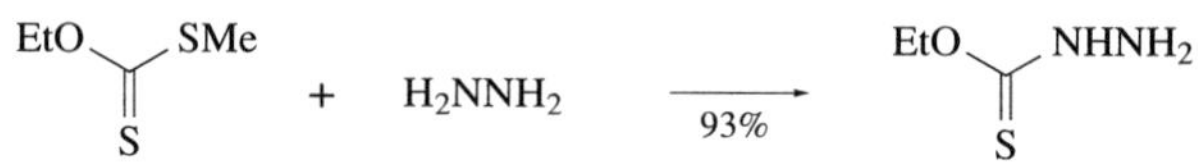

(80)

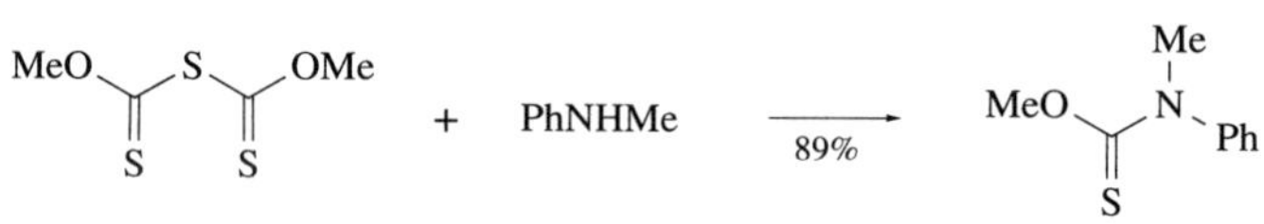

(81)

6.17.2.5 Derivatives of Dithiocarbamic Acid

The parent compound, dithiocarbamidic acid, can be obtained as an unstable crystalline solid from the reaction of ammonium dithiocarbamate and concentrated HCl at 0°C. Other dithiocarbamic acids, $R^1R^2NC(=S)SH$, are also unstable compounds which can be generated from their salts with HCl ⟨B-62MI 617-01, B-77MI 617-01⟩; they decompose to amines R^1R^2NH and carbon disulfide ⟨83HOU(E4)420⟩.

6.17.2.5.1 Salts of dithiocarbamic acids

Alkylammonium dithiocarbamates are prepared from carbon disulfide and two moles of the amines in solvents such as acetone or ethanol ⟨B-62MI 617-02, 79JINC1277, 80ZC101, 83HOU(E4)420⟩. When the reaction is carried out in aqueous sodium hydroxide or potassium hydroxide, the sodium or potassium dithiocarbamates are formed ⟨51RTC917, 51RTC949, 65JMC174⟩. Formamide reacts with carbon disulfide in the presence of sodium hydroxide to give sodium *N*-formyl dithiocarbamate, $Na^{+-}SC(=S)NHCHO$ ⟨85ZAAC(522)145⟩. The reaction of hydrazine hydrate with carbon disulfide and methanolic potassium hydroxide gives potassium dithiocarbazate, $K^{+-}SC(=S)NHNH_2$ ⟨85ZAAC(531)101⟩. With a 2:1 ratio of carbon disulfide to hydrazine, the product is the salt $2K^{+-}SC(=S)NHNHC(=S)S^-$ ⟨85ZAAC(531)82⟩. Carbon disulfide reacts with methylhydrazine at the substituted nitrogen atom ⟨89S923⟩ whereas it reacts with arylhydrazines at the unsubstituted nitrogen atom ⟨92CCC1699⟩. Lithium dithiocarbamates were obtained from secondary amines by deprotonation of the amine using butyllithium in THF at 0°C followed by the addition of carbon disulfide in THF at 0 °C ⟨91S637⟩. This procedure is particularly useful with cyclic secondary amines such as indoline since the salt can be further deprotonated at C-2 by *s*-butyllithium ⟨94S983⟩ and this allows electrophiles to be introduced at C-2.

6.17.2.5.2 Esters of dithiocarbamic acids

(i) From dithiocarbamates

The *S*-alkylation of alkali metal or ammonium dithiocarbamates by haloalkanes and related compounds is the most general method for the preparation of esters of *N*-alkyl or *N*-aryl dithiocarbamates (Equation (82)). The method is fully documented in earlier reviews ⟨B-62MI 617-01, B-62MI 617-02, 83HOU(E4)420⟩, and there are many more recent examples. A wide range of electrophiles has been used; these include haloalkanes, alkyl nitrates, α-haloketones, α-haloesters, activated heteroaryl halides, aromatic diazonium salts, acyl halides, and activated alkenes. A selection of examples from the 1980s is given below.

(82)

The standard method for *S*-methyl esters is the reaction of dithiocarbamate salts with iodomethane ⟨82ZAAC(488)94⟩. *S*-Alkyl esters, $(H_2N)_2C{=}NC({=}S)SR$, of guanidinothioformic acid were prepared from its potassium salt and iodoalkanes (e.g., R = Me, 83%; R = Bn, 43%) ⟨82ZAAC(485)203⟩. *S*-Alkyl-*N*-formyl-*N*-methyl dithiocarbamates, OHCN(Me)C(=S)SR, were obtained from the potassium salt of the corresponding acid in an analogous manner ⟨85ZAAC(527)130⟩. The *S*-alkyl esters, S=CHNHC(=S)SR, of *N*-thioformyl dithiocarbamic acid were prepared by reaction of iodoalkanes with tetra-*n*-butyl *N*-thioformyl dithiocarbamate ⟨85ZAAC(525)112⟩. Several examples of the formation of dithiocarbamates from disaccharides by displacement of bromide have been described ⟨91CAR(216)271⟩; one of these is shown in Equation (83). 2-Chloromethylquinoline and other halomethyl-substituted heterocycles have also been converted into dithiocarbamates by displacement of halide ⟨87BCJ1807, 89JPR(331)439⟩.

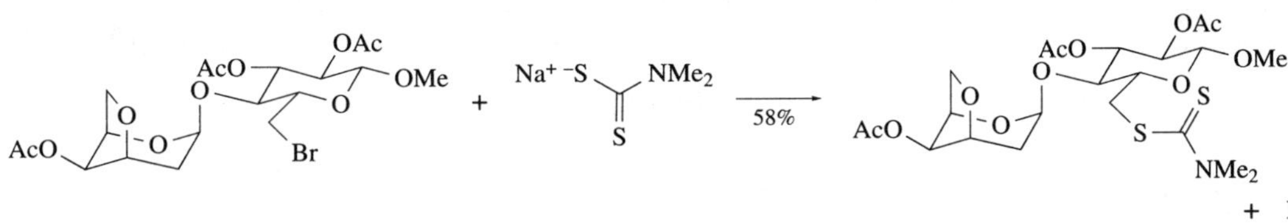

(83)

There are many examples of the displacement of halide from α-halocarbonyl compounds by dithiocarbamate anions. These include reactions of 1-(chloroacetyl)piperidine ⟨81UKZ26⟩, 2-(chloroacetamido)benzothiazole ⟨82FES205⟩, 2-(chloromethyl)benzimidazole ⟨85IJC(B)1298⟩, and of several chloroacetyl substituted heterocycles ⟨85MI 617-04, 85IJC(B)580, 86MI 617-02, 87MI 617-01, 89JIC60, 92AF1453⟩. The reaction of chloroacetonitrile with sodium *N*,*N*-dimethyl dithiocarbamate in the presence of a phase-transfer catalyst, tetrabutylammonium iodide, has been reported ⟨88BCJ2203⟩. The reaction of α-bromocarboxylic acids with sodium or potassium *N*,*N*-dialkyl dithiocarbamate is illustrated in Equation (84) ⟨91JOC5684⟩. The ester (−)-menthyl chloroacetate ⟨85MI 617-01⟩, α-chloroalkylnitrosamines (RN(NO)CH_2Cl) ⟨87LA583⟩, and chloromethyl oximes ⟨87S349⟩ have been used in similar displacement reactions. α-Haloketones undergo displacement of halide easily ⟨85H(23)3099⟩, and some of the dithiocarbamates formed in this type of process have been reduced by baker's yeast to chiral alcohols, as illustrated in Scheme 34 ⟨88BCJ3205⟩.

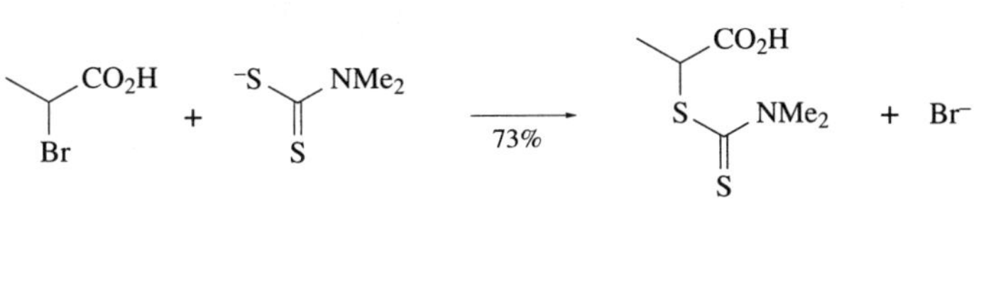

(84)

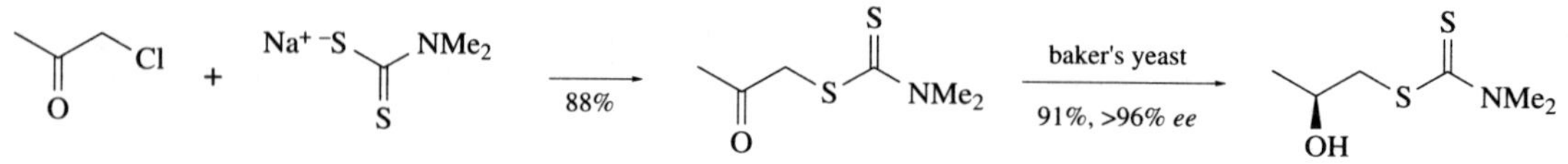

Scheme 34

Carbon electrophiles other than halides can be used: there are examples of the formation of dithiocarbamates from epoxides ⟨84AP(317)1042⟩ and from tosylates ⟨88S224, 92JA10573⟩. Dithiocarbamates can also be formed by conjugate addition reactions or by conjugate addition–elimination reactions with activated alkenes ⟨84MI 617-02⟩. An example of conjugate addition followed by cyclization is illustrated in Scheme 35 ⟨91M1047⟩, and related reactions of 4-benzylideneoxazolones have been described ⟨92S919⟩. 4-Chloro-3-nitrocoumarin reacts with potassium *N*,*N*-dialkyl dithiocarbamates in acetone by displacement of the nitro group, whereas chloride is displaced in the reaction with potassium *N*-phenyl dithiocarbamate (Scheme 36) ⟨92JHC383⟩. Examples of addition–elimination reactions of α,β-unsaturated sulfones ⟨81KGS907, 84ZOR2114⟩ include the preparation of sulfolene derivatives (Equation (85)) ⟨81KGS907⟩.

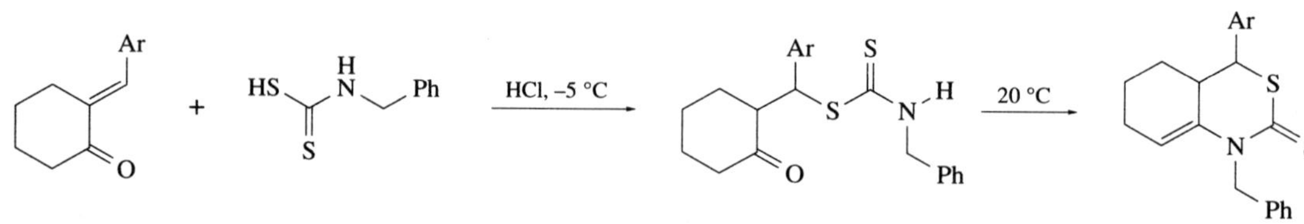

Scheme 35

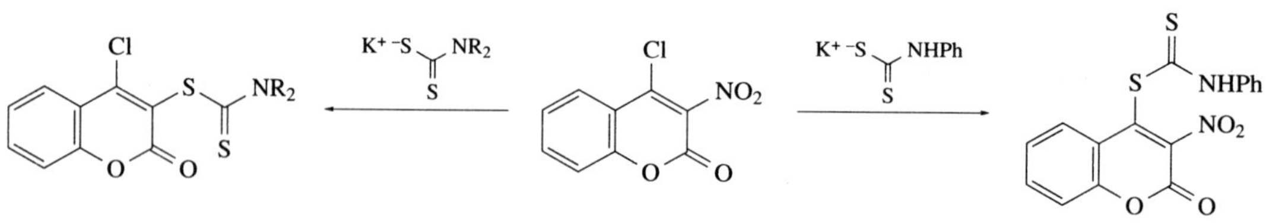

Scheme 36

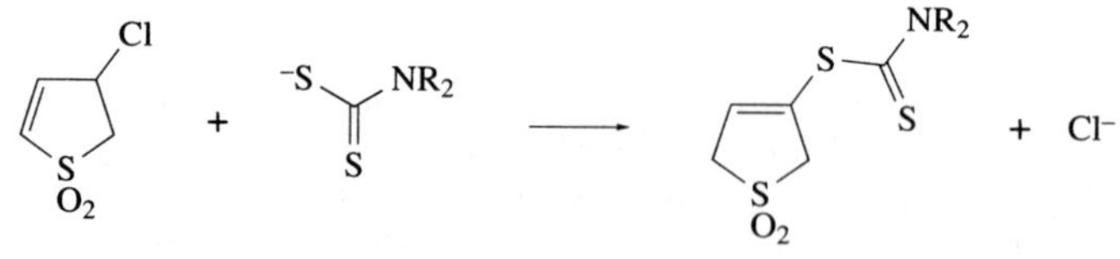

(85)

Acid chlorides react readily with dithiocarbamate salts ⟨81LA1388, 84MI 617-04, 88ZOR2019⟩. Aryl α-chlorooxime *O*-methanesulfonates also react to give isolable but unstable dithiocarbamates which can be cyclized to dithiazolium salts with fluoroboric acid (Scheme 37) ⟨85CC1641⟩. Acetylene reacts with amines and carbon disulfide to give either mono- or bis(dithiocarbamates) (Scheme 38) depending upon the temperature and the reaction conditions ⟨81ZOR2269⟩.

S-Aryl dithiocarbamates have been prepared from sodium dithiocarbamates and diaryliodonium salts (Equation (86)) ⟨87JCS(P1)2759, 87JOC4117⟩.

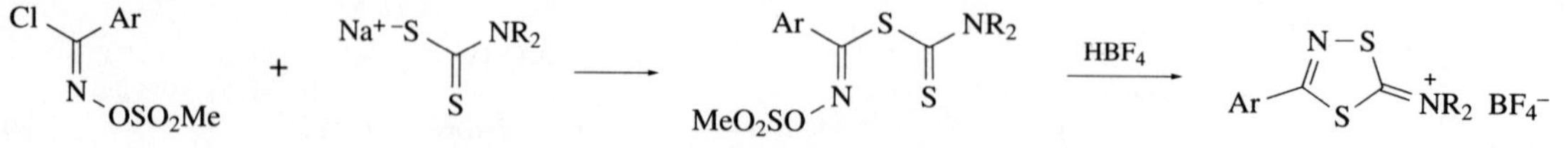

Scheme 37

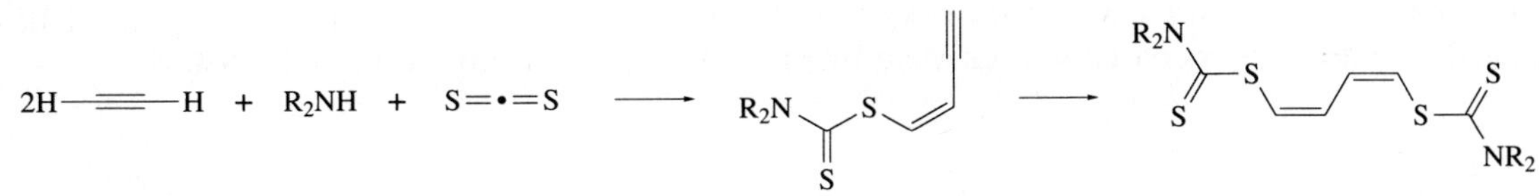

Scheme 38

(86)

There are several examples of the use of heteroatomic electrophiles for the substitution of dithiocarbamate salts. For example, salts $M^{+-}SC(=S)NMe_2$ react with iodine to give $(SC(=S)NMe_2)_2$ and with SCl_2 and S_2Cl_2 to give the corresponding tri- and tetrasulfanes, respectively ⟨88ZAAC(556)141⟩. Other iodine coupling reactions are reported ⟨85ZAAC(524)111, 85ZAAC(527)125⟩. Reaction with $MeSO_2SMe$ gives $MeS_2C(=S)NMe_2$ ⟨87ZAAC(551)191⟩. Substitution reactions using phosphorus halides ⟨81ZOB530, 82MI 617-02, 83JIC806⟩ and tellurium halides ⟨88JCS(D)2363⟩ have also been reported. TMS esters are formed with TMS-Cl ⟨88JOC2263⟩.

(ii) From carbon disulfide

The reaction between amines and carbon disulfide to give dithiocarbamates was described in (i) above. Procedures in which dithiocarbamate *S*-esters are made in one pot from carbon disulfide, an amine, and an electrophile are therefore not essentially different from those already described. However, examples of the procedure are given here since the method is experimentally convenient.

A typical procedure is described for the preparation of *N*-aryl *S*-methyl dithiocarbamates from anilines ⟨81S961⟩. The compounds were obtained by adding aqueous sodium hydroxide and carbon disulfide to a vigorously stirred solution of the aniline derivative in DMSO at room temperature. Iodomethane was then added a with ice cooling. The mixture was poured into water, and the precipitated product was isolated and recrystallized from ethanol. Similar procedures have been described for *N*-aryl *S*-nonyl dithiocarbamates ⟨82JMC557⟩ and for *S*-allyl *N*-aryl dithiocarbamates ⟨91S147⟩. Other procedures have been described for the preparation of *N,N,S*-trialkyl dithiocarbamates ⟨84MI 617-03, 88S775, 91S637⟩ and for the formation of dithiocarbamates from chloroacetic acid and carbon disulfide ⟨82MI 617-01⟩.

The reaction of sodium *N*-formyl dithiocarbamate, $Na^{+-}SC(=S)NHCHO$, obtained from formamide, carbon disulfide and sodium hydride, with iodomethane or iodoethane in ether gives the respective *S*-alkyl *N*-formyl dithiocarbamates ⟨85ZAAC(524)117⟩. Acyl dithiocarbazates RC(=O)NHNHC(=S)SR were prepared from acylhydrazines by reaction with carbon disulfide and aqueous potassium hydroxide, followed by alkylation with an iodoalkane ⟨82JMC557⟩. Methylhydrazine has also often been used in this type of preparation ⟨72ICA11, 78JINC451, 88S690⟩. The reaction of methylhydrazine, carbon disulfide and potassium hydroxide with 1,2-bis(bromomethyl)benzene gave the bis(*N*-methyl dithiocarbazate) (Equation (87)) ⟨91POL823⟩. Dibromomethane has been used as the electrophile in the formation of 1,1-bis(dithiocarbamates) ⟨91ANC1295⟩.

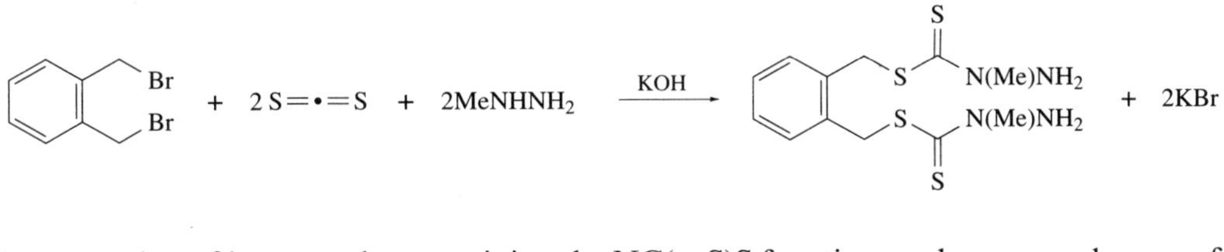

(87)

A large number of heterocycles containing the NC(=S)S function are known, and many of them are prepared using carbon disulfide. A few representative examples are given below.

Reactions of *β*-lactones with dithiocarbamates, generated *in situ* from amines and carbon disulfide in the presence of a base, lead to 1,3-thiazine-2-thiones by way of intermediate dithiocarbamates

(Scheme 39) ⟨81AP(314)587, 82AP(315)103, 84AP(317)297, 85AP(318)848, 86AP(319)1064, 89IJC(B)439⟩. 1,3-Thiazoline-2-thiones were prepared in high yield by the route shown in Equation (88) ⟨89M871⟩.

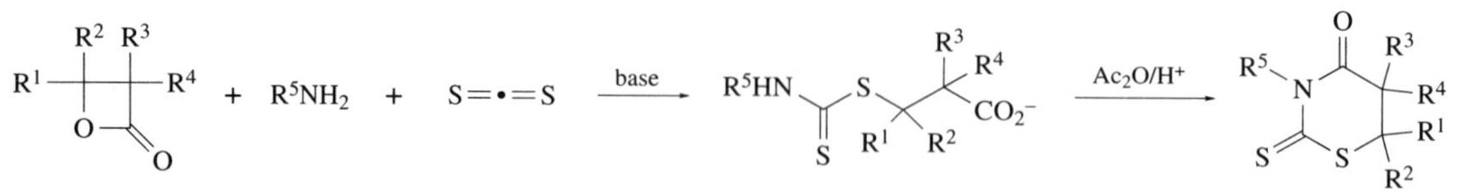

Scheme 39

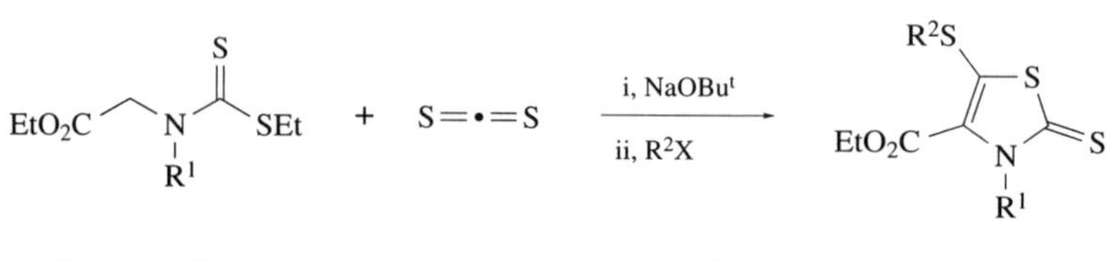

(88)

The oxidation by iodine of potassium *N*-methyl *N*-thioformyl dithiocarbamate, $K^{+-}SC({=}S)NMeC({=}S)H$, obtained in 80% yield from *N*-methylthioformamide, carbon disulfide, and potassium hydroxide at $-15\,^{\circ}C$, gives *N*-methyl-1,2,4-dithiazol-3,5-dithione (Equation (89)) ⟨85ZAAC(528)168⟩. Similarly, 1,3,4-oxadiazol-2-ylmethyl *N*-aryl dithiocarbamates undergo oxidative cyclization with thionyl chloride or iodine to yield 5-aryl-1,3,4-oxadiazolo[3,2-*d*][1,3,4]thiadiazine-6(5*H*)-thiones (Equation (90)) ⟨89IJC(B)439⟩. Various 1,3,4-thiadiazole-2(3*H*)-3-thiones have been prepared from hydrazine derivatives and carbon disulfide ⟨93JMC1090, 93JMC1802⟩. 5-Phenyl-1,2,4-dithiazole-3-thione was obtained from its sodium salt formed from sodium hydride, thiobenzamide, and carbon disulfide in dry pyridine after treating the latter with HCl ⟨85ZAAC(525)112⟩. The reaction of a primary alkylamine and carbon disulfide with 1,4-dibromobutane-2,3-dione also results in the formation of a heterocycle, 4,4′-bis(3-alkyl-4-hydroxy-thiazolidine-2-thione) ⟨89SUL123⟩.

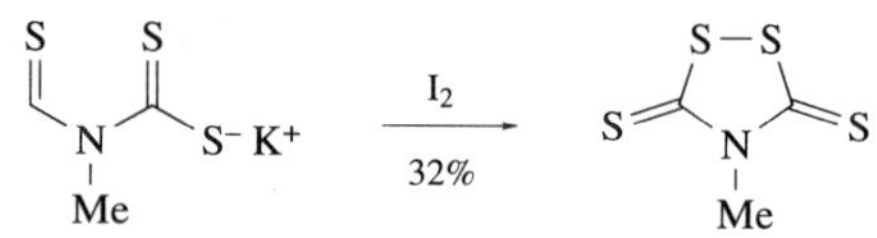

(89)

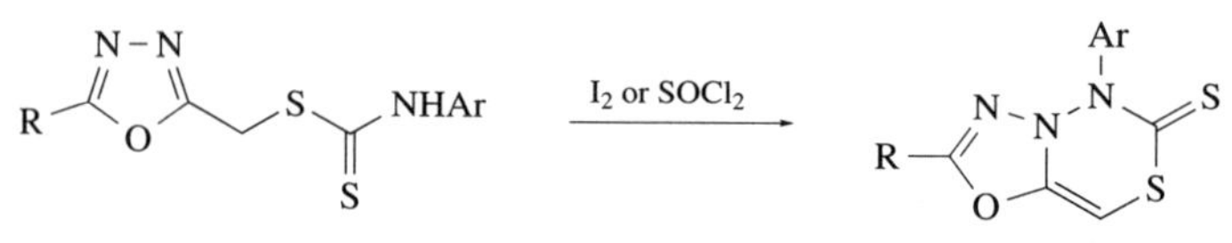

(90)

Many heterocyclic amines have been functionalized on nitrogen by reaction in basic conditions with carbon disulfide and an alkylating agent. For example, carbon disulfide was reacted with the lithium salt of imidazoline 2-thione in THF, and iodomethane was then added to give *S*-methyl 2-methylthio-4,5-dihydroimidazole-1-carbodithioate (Equation (91)) ⟨81JCS(P1)2499, 81S908⟩. The reaction of 5-amino-1,2,4-triazoles with carbon disulfide (0.1 mole), potassium hydroxide (0.1 mole), and alkylating agents results in substitution of either the exocyclic amino group or a ring nitrogen atom, depending on the amount of base used ⟨90JHC1249⟩. 2-Aminothiazole has similarly been converted into a dithiocarbamate ⟨81ZOR191⟩. The reactions of pipecolic acid, 1,2,3,4-tetrahydroisoquinoline-3-carboxylic acid, and 1,2,3,4-tetrahydro-*β*-carboline-3-carboxylic acid with carbon disulfide in the presence of sodium hydroxide followed by alkylation with methyl iodide readily gave the corresponding dithiocarbamates ⟨87CPB2840⟩.

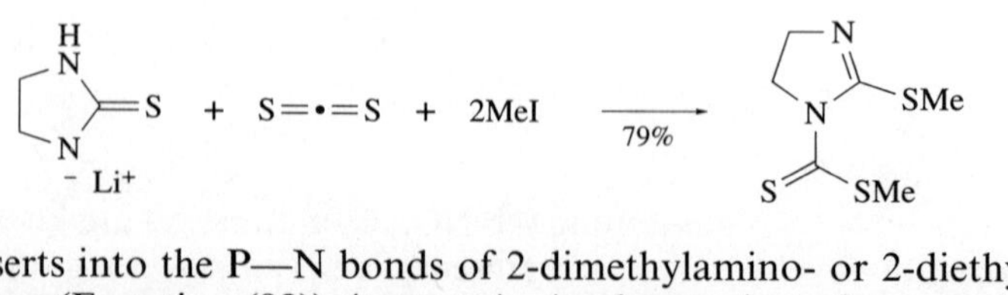

(91)

Carbon disulfide inserts into the P—N bonds of 2-dimethylamino- or 2-diethylamino-3-phenyl-1,3,2-oxazaphospholanes (Equation (92)) ⟨81ZOB28⟩. Analogous insertion reactions of carbon disulfide into P—N bonds ⟨83ZOB785⟩ and into As—N bonds ⟨89ZOB2520⟩ have been reported.

Formamide reacts with carbon disulfide in the presence of sodium hydroxide to yield crude sodium *N*-formyl dithiocarbamate, NaSC(=S)NHCHO ⟨85ZAAC(22)145⟩.

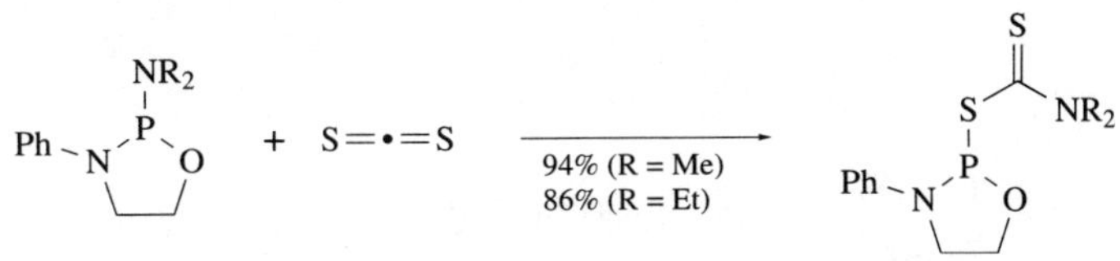

(92)

Dithiocarbamic acids derived from amines and carbon disulfide add to vinyl ethers and thioethers (Equation (93)) ⟨86ZOR2145, 87ZC24⟩. Additions to alkoxyallenes have also been reported ⟨90ZOR1364⟩. Hydroxybenzyl dialkyldithiocarbamates were prepared in 45–89% yields from the corresponding phenols by aminomethylation with formaldehyde and secondary amines followed by thiocarbamoylation with carbon disulfide ⟨86MI 617-03⟩. Bis(dithiocarbamates) have been prepared by treating diamines with CS_2. Thus the reaction of *m*-phenylenediamine with carbon disulfide and ammonium hydroxide gave the bis(dithiocarbamate) in 70% yield ⟨86MI 617-05⟩. Dialkylamines react similarly ⟨87ZPK1829⟩. The reaction of ethynyl ketones, ArCOC≡CTMS, with amines and carbon disulfide in MeCN gave the adducts ArCOCH=C(TMS)SC(=S)NR_2 ⟨90ZOB602⟩.

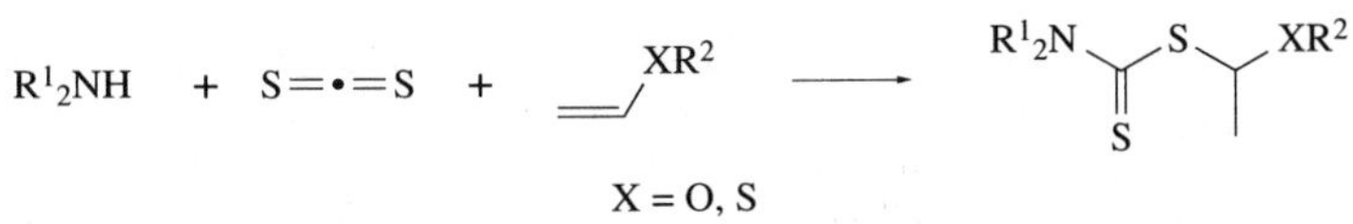

(93)

(iii) From isothiocyanates

Two general procedures for obtaining dithiocarbamates from isothiocyanates are outlined in Equation (94) and in Scheme 40. The addition of thiols to isothiocyanates is represented by the reaction of benzenethiol and phenyl isothiocyanate to give 73% of *S*-benzyl *N*-phenyl dithiocarbamate ⟨81CCC1970, 84CCC1577⟩ and by the addition of methanethiol to pentafluorosulfanyl isothiocyanate ⟨84CB1707⟩. The general procedure of Scheme 40 is illustrated by the reaction of benzoyl chloride, potassium isothiocyanate, and alkanethiols ⟨81ZAAC(476)7, 82JMC557⟩. The addition of potassium isothiocyanate and benzenethiol or methanethiol to a solution of 3-chlorobenzo[*b*]thiophene-2-carbonyl chloride similarly gives the dithiocarbamates ⟨84JPR633⟩. In an analogous reaction, *S*-ethyl and *S*-propyl isothiocyanodithioformic acids are obtained from RSC(=S)Cl (R = Et, Pr) and potassium isothiocyanate in the presence of phase-transfer catalyst, 18-crown-6 ⟨84ZAAC(512)231⟩.

(94)

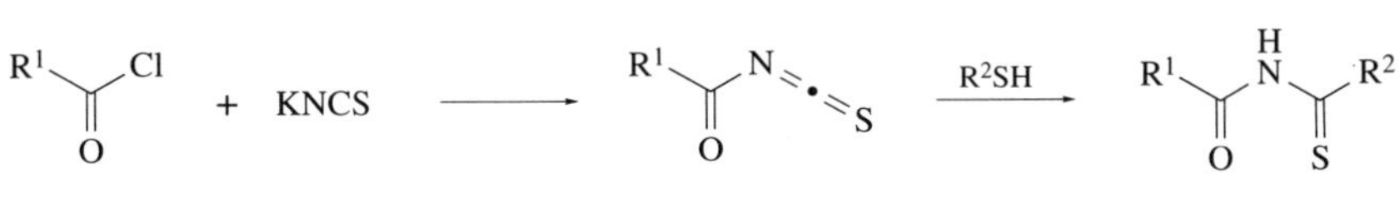

Scheme 40

Carbonyl diisothiocyanate reacts with a range of nucleophiles to give cyclic dithiocarbamates (Scheme 41) ⟨81CB1132, 81CB2075⟩. Analogous reactions of carbonyl isocyanate isothiocyanate have been described ⟨81CB2064⟩. *N*-(2,6-Dimethylphenyl)-*N*-isothiocyanatomethyl chloroacetamide reacts with aqueous sodium hydrogen sulfide giving, after stirring for 1 h, 2-(2,6-dimethylphenyl)-6-oxo-2-thioxo-hexahydro-1,3,5-thiadiazepine in 86% yield ⟨86S817⟩.

Several 1,3-thiazolidine-2-thiones were prepared from 1-iodo-2-isothiocyanatocycloalkane and sodium sulfides; an example is shown in Equation (95) ⟨81JCS(P1)52⟩. 1,3-Thiazine-2-thiones are formed by the cyclization of appropriate dithiocarbamates. For instance, treatment of PhCH=CRC(=O)NCS (R = H, Me, Ph) with NaSH in methanol gave dithiocarbamates H_2NC(=S)SCHPhCHRCO_2Me which then cyclized to 2-thioxo-1,3-thiazin-4-ones ⟨90MI 617-01⟩.

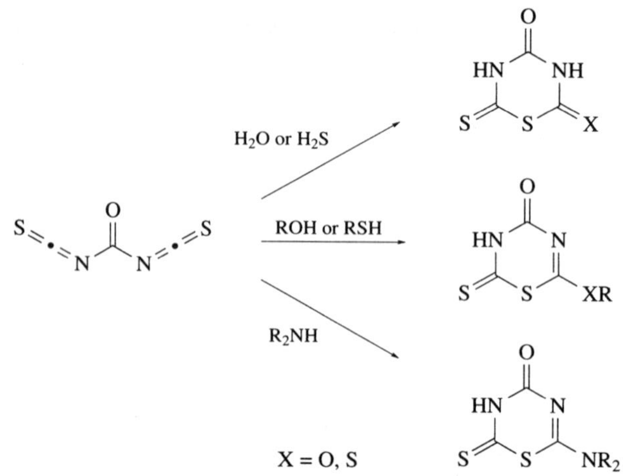

Scheme 41

Similarly, the condensation of $RCOCHR^1CR^2R^3NCS$ with NaHS gave $RCOCHR^1CR^2R^3NHC(=S)S^-Na^+$ which reacted with iodomethane to afford the corresponding *S*-methyl dithiocarbamates. With mineral acids the latter give the corresponding dithiocarbamic acids and their ring tautomers, thiazinethiones ⟨91KGS416⟩. 2-Thioxo-6-nitro-2,3-dihydro-4*H*-1,3-benzothiazin-4-one can be obtained by the reaction of 2-chloro-5-nitrobenzoyl isothiocyanate with sodium hydrogen sulfide ⟨92MI 617-03⟩.

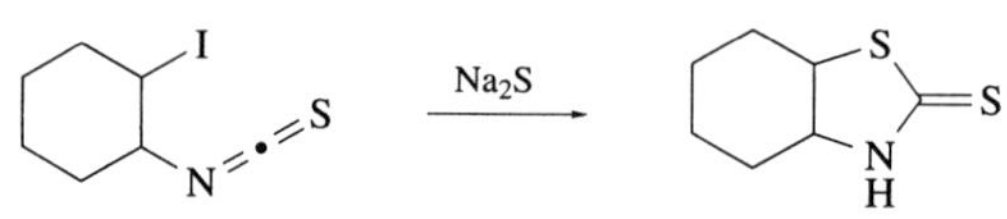

(95)

(iv) From thiocarbamoyl chlorides

Several dithiocarbamates were prepared by stirring a mixture of a thiophenol, dimethylthiocarbamoyl chloride, potassium *t*-butoxide, and DMF at room temperature, as illustrated in Equation (96) ⟨91CPB1939⟩. The reaction of benzimidazole-2-thiol with *N*,*N*-diethylthiocarbamoyl chloride in the presence of anhydrous potassium carbonate in refluxing THF afforded the corresponding dithiocarbamates ⟨85JHC1269⟩. Reactions of thiazolidine-2-thione with thiocarbamoyl or thiocarbazic acid chlorides led to *S*-substitution products (Scheme 42) ⟨92AP(325)173⟩.

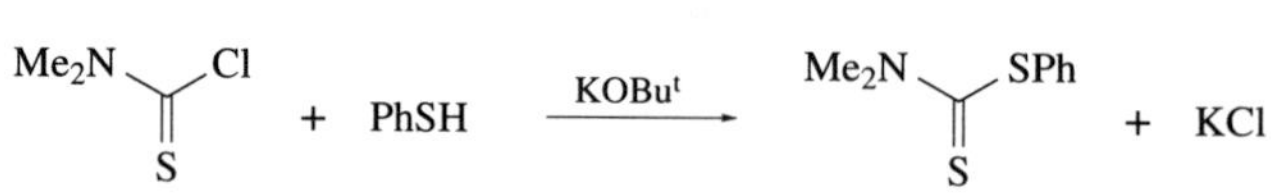

(96)

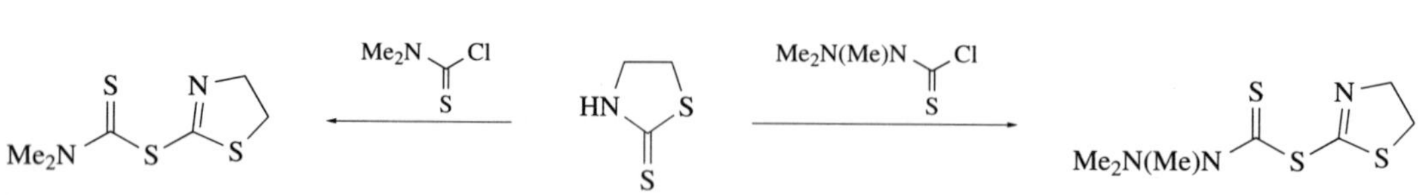

Scheme 42

(v) From thiuram disulfides

N,*N*-Dimethyl dithiocarbamates can be prepared from thiuram disulfides as shown in Equation (97) ⟨93S483⟩. This approach has been used with other aryllithium derivatives ⟨87JHC749⟩. Compounds of the type $R^1_2Te[SC(=S)NR^2_2]_2$ were obtained by reaction of dimethyltellurium with thiuram disulfides ⟨88ZAAC(556)179⟩. Sodium cyclopentadienide cleaves thiuram disulfides: after acidification, a mixture of tautomeric *S*-cyclopentadienyl dithiocarbamates was isolated ⟨93H(35)77⟩.

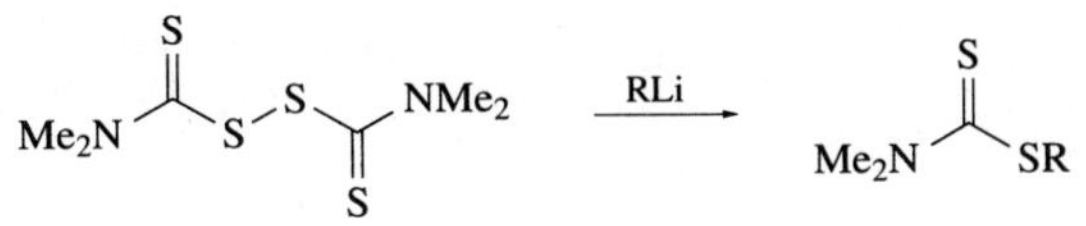

(97)

(vi) Other methods

S-Methyl *N,N*-dimethyl dithiocarbamate was prepared from dimethylamine and MeSC(=S)Cl ⟨87CB987⟩. Similarly, condensation of the latter with methylphenylamine gave the *N*-methyl-*N*-phenyldithiocarbamate ⟨83JOC4750⟩. A few thiazolidine-2-thiones were prepared from *vic*-iodoalkanecarbamates, potassium *S*-ethyl dithiocarbonate, and aqueous sodium hydroxide ⟨85H(23)1181⟩. A method of preparing *N*-hydroxydithiocarbamates is illustrated in Equation (98) ⟨85TL5943⟩. 8-Hydrazino-5-chloro-3,4-dihydro-2,6-dimethylbenzopyran reacts readily with allyl chlorodithioformates in the presence of diisopropylethylamine (Hünig's base) to give the corresponding allyl hydrazinecarbodithioates ⟨88JOC38⟩.

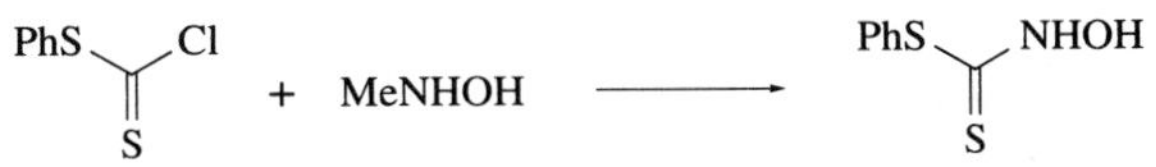

(98)

6.18 Functions Containing a Thiocarbonyl Group Bearing Two Heteroatoms Other Than a Halogen or Chalcogen

JOSÉ BARLUENGA, EDUARDO RUBIO and MIGUEL TOMÁS
Universidad de Oviedo, Spain

6.18.1 THIOCARBONYL DERIVATIVES CONTAINING AT LEAST ONE NITROGEN FUNCTION (AND NO HALOGEN OR CHALCOGEN FUNCTIONS)

6.18.1.1 Thiocarbonyl Derivatives with Two Nitrogen Functions

The synthesis of thioureas and their derivatives have been reviewed previously ⟨79COC(3)373, 83HOU(E4)407⟩. This chapter has been organized according to the principal reagents used as primary sources for the synthesis of these compounds.

6.18.1.1.1 From isothiocyanates

Isothiocyanates are, probably, the most powerful starting materials for preparing thioureas and derivatives. Thus, in a general manner, alkyl or aryl isothiocyanates react with ammonia, primary amines, and secondary amines to give 1-substituted, 1,3-disubstituted and trisubstituted thioureas, respectively ⟨55CRV181⟩. This reaction generally takes place with good yields and works better in polar solvents such as diethyl ether, ethanol, water, or acetone (Equation (1)). One of the main advantages of the method is its versatility, allowing the use of reagents with a wide range of substitutents, both in the amine and in the isothiocyanate components.

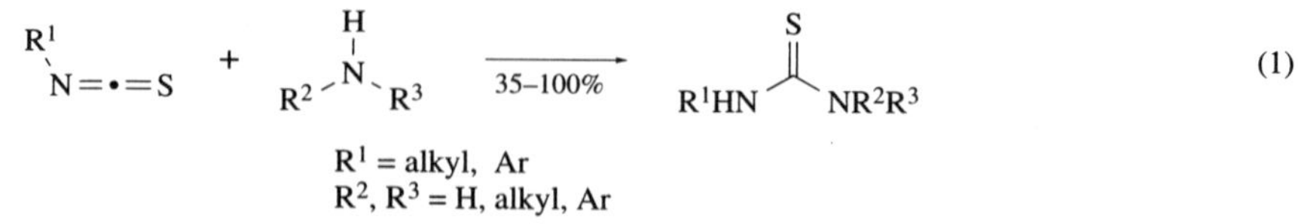

The amine component can be extensively functionalized. Thus, ω-aminoalcohols and aminophenols react chemoselectively in boiling chloroform with isothiocyanates through the amino group exclusively (Equation (2)) ⟨75JMC90⟩. Aminothiols also react with phenyl isothiocyanate through the amino group, while alkyl isothiocyanates produce a 1 : 1 mixture of isomers (**1**) and (**2**) resulting from competitive addition of the amino and the thiol group. If the hydrochloride derivative is used instead of the free amine, the reaction leads to the exclusive formation of the dithiocarbamate derivative (**2**) (Scheme 1) ⟨63JOC3140⟩.

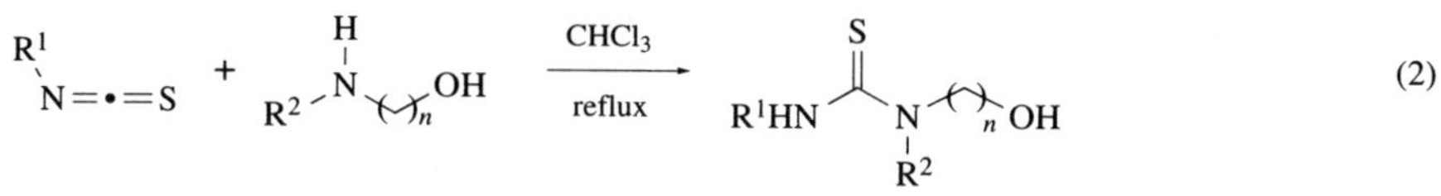

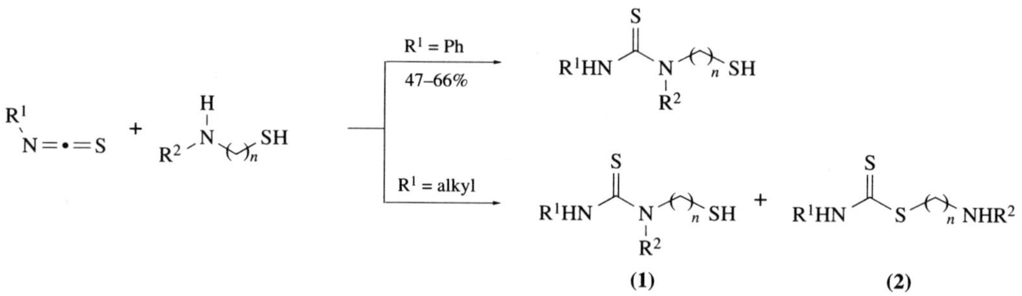

Scheme 1

Hydroxylamine and substituted hydroxylamines react with alkyl and aryl isothiocyanates to give the corresponding hydroxythiourea derivatives (**3**) in variable yields; the best results are obtained with aromatic isothiocyanates and *N*- or *O*-alkylhydroxylamines, while isothiocyanates having bulky substituents, such as *t*-butyl, give rise to lower yields (Equation (3), Table 1) ⟨1897JPR71, 66PHA23, 69ACS324⟩.

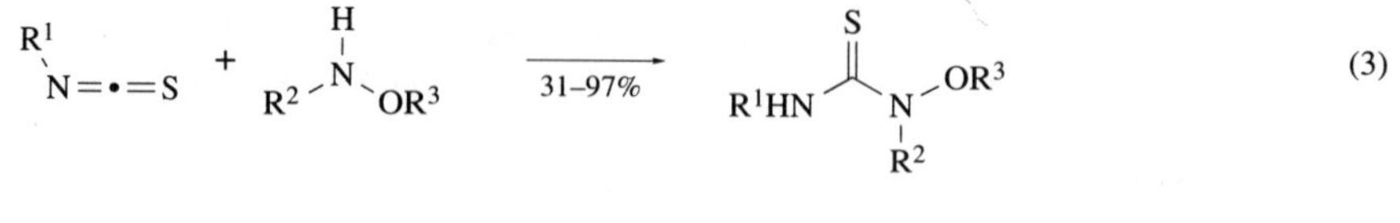

Table 1 Thioureas (**3**) from hydroxylamines and isothiocyanates.

R^1	R^2	R^3	*Yield* (%)	*Ref.*
Me	H	Me	52	66PHA23
Bu^n	H	H	48	66PHA23
Bu^t	H	H	31	70JMC377
Ph	Me	H	97	61AP765
Ph	H	Et	60	69ACS324

Steroids containing fused heterocycles, such as the thiazole ring, have been the subject of attention due to the influence exerted by the heterocyclic component on their pharmacological behavior. This type of compound has been prepared stereospecifically by employing a general method of ring expansion of *N*-thioacylaziridines. Thus, steroidal 2,3-aziridines react with phenyl isothiocyanate to give unstable thiourea intermediates which, with sodium iodide, undergo ring expansion to give the final derivatives (**4**) and (**5**) (Scheme 2) ⟨80JCS(P1)766⟩.

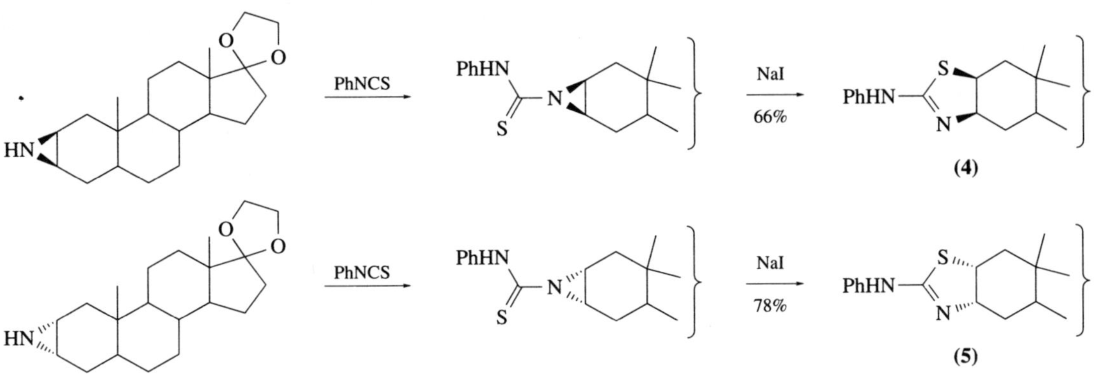

Scheme 2

The N—Si bond of trimethylsilylamines has also been found to insert into the C═N bond of isothiocyanates. The reaction takes place in ether at room temperature to give *N*-silylated thioureas, which can in turn be easily hydrolyzed to the corresponding derivatives (Scheme 3) ⟨79LA263⟩.

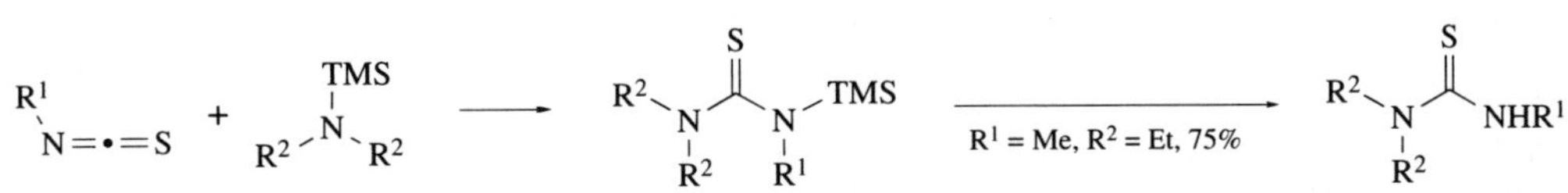

Scheme 3

In a similar way, trimethylstannylthioureas have been prepared in essentially quantitative yield from phenyl isothiocyanate and stannylamines ⟨65JCS2157⟩. This aminostannylation reaction is effected under mild conditions and appears to involve a cyclic transition state as suggested by the low activation parameters found (Scheme 4).

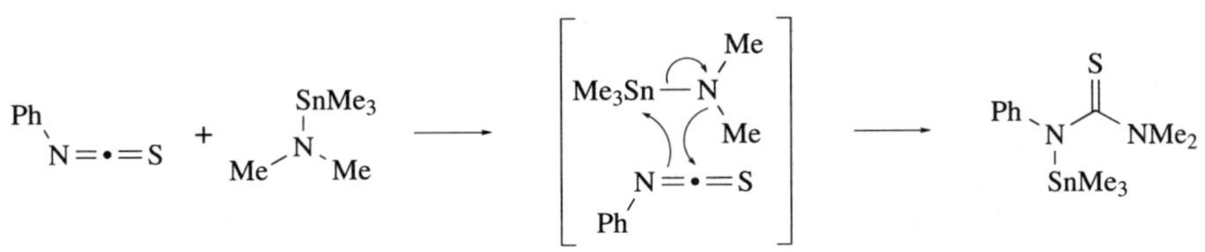

Scheme 4

The N—H bond of imines also reacts with isothiocyanates. For instance, diphenylmethyleneamine reacts with phenyl isothiocyanate to give methylenethioureas (**6**), which have the structure of 2-amino-3-aza-1-thiabutadienes ⟨90JOC1721⟩. The same compounds are obtained by reacting methyl and phenyl isothiocyanate with *N*-tributylstannyldiphenylmethyleneamine ⟨72JCS(P1)130⟩. The *N*-phenyl derivative is stable and may also be prepared from bis(tributyltin) oxide and *N*-phenyl-*N'*-diphenylmethyleneaminothiourea in boiling benzene. Methyl isothiocyanate, however, undergoes addition at the carbon–sulfur double bond and requires hydrolysis to yield the thiourea derivative. The related reaction of *N*-silylated methyleneamines is also known and represents a more general entry into this type of compound ⟨72JCS(P1)1678, 89S228⟩. Another approach to these compounds involves the reaction of secondary amines with 1-bromoalkyl isothiocyanates; the reaction takes place in ether at low temperature, and the brominated intermediate is transformed with loss of hydrogen bromide into the azathiabutadiene (Scheme 5) ⟨79CB1956⟩.

An interesting example of the insertion of a nitrogen–metalloid bond into isothiocyanates is the

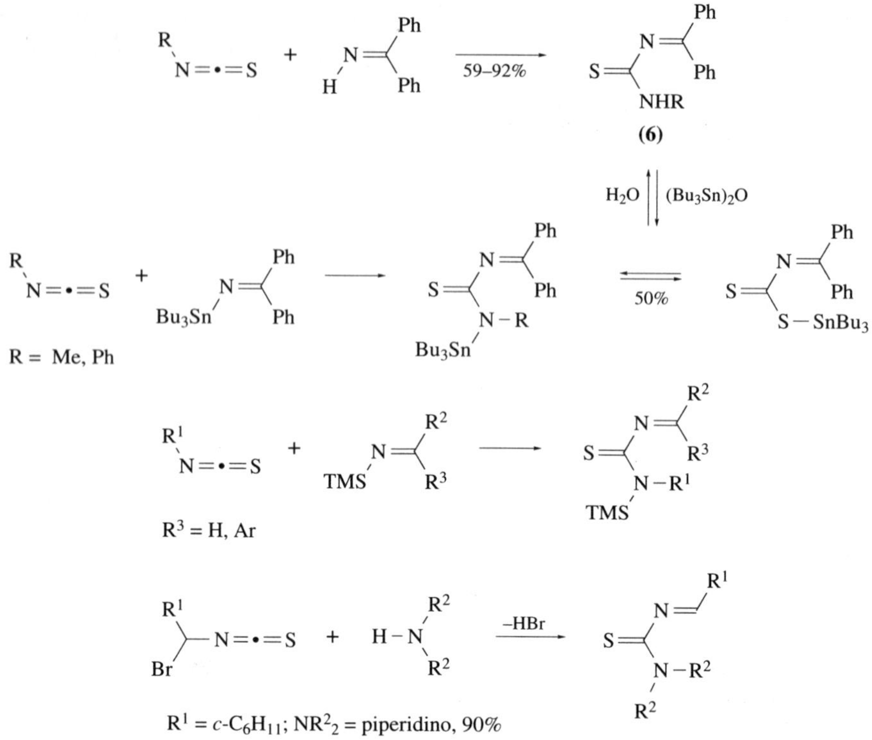

Scheme 5

reaction of diazaboracyclohexane with phenyl isothiocyanate which results in the formation of the rare eight-membered boratriaza heterocycle (**7**) (Equation (4)) ⟨76JOM(110)15⟩.

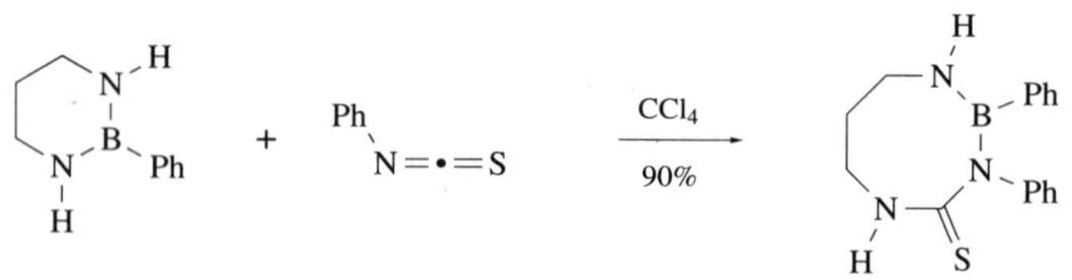

(**7**) (4)

Functionalized isothiocyanates appear to be adequate starting materials for the synthesis of *N*-unsubstituted thioureas for which the preparation seems to be otherwise difficult or requires cumbersome procedures. For instance, alkanoyl isothiocyanates, easily available from acid chlorides and lead(II) thiocyanate, react with secondary amines in acetone to yield *N*-alkanoylthioureas which are in turn deprotected with aqueous base ⟨55OSC(3)773, 87JCS(P1)1153⟩. Thioureas of this sort are also available by reaction of alkoxycarbonyl isothiocyanates with amines followed by acid hydrolysis of the carbamate moiety (Scheme 6) ⟨73JPR144⟩.

Silicon(IV) isothiocyanate reacts in benzene with secondary amines to give, after solvolysis with a mixture of water and isopropanol, 1,1-disubstituted thioureas ⟨73OSC(4)801⟩. Phosphorus derivatives, such as 2-isothiocyanatobenzo-1,3,2-dioxophosphole, react with secondary amines to furnish the corresponding addition compounds (**8**); further treatment with aqueous sodium hydroxide allows the removal of the phosphorus group yielding 1,1-disubstituted thioureas (Scheme 7) ⟨71LA(743)167⟩.

In what constitutes a special reaction, aliphatic and aromatic isothiocyanates dimerize on heating in DMSO to afford *N*,*N*′-symmetrically substituted thioureas (Equation (5)) ⟨74S289⟩.

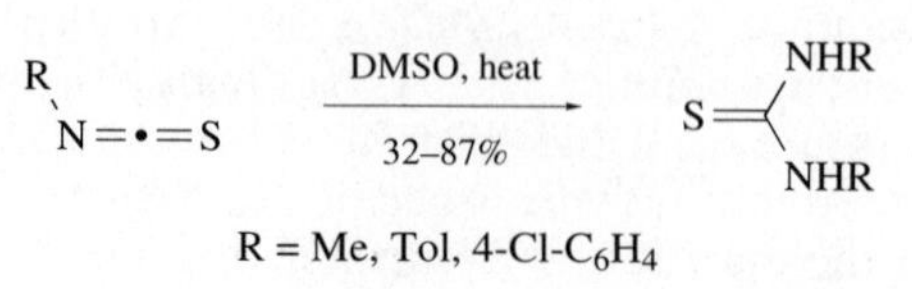

(5)

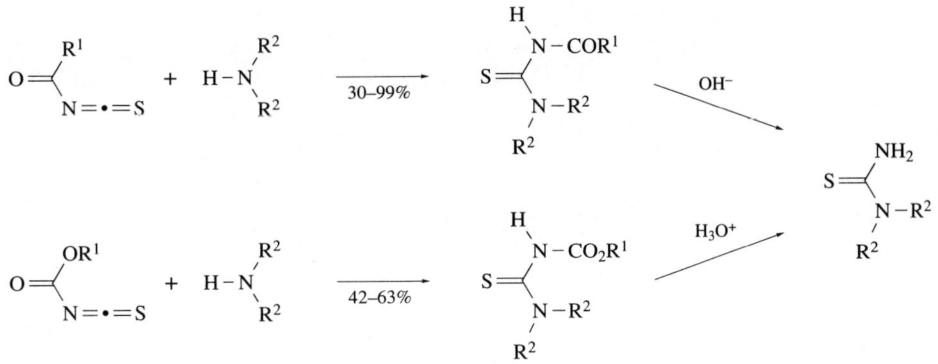

Scheme 6

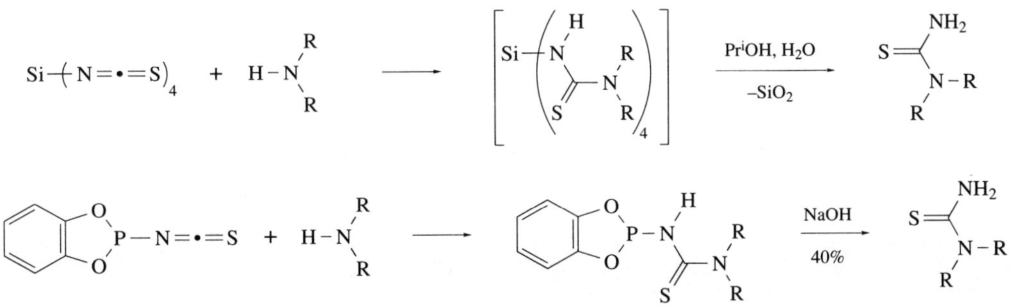

(8)

Scheme 7

The synthesis of heterocyclic derivatives of thioureas has been reviewed by Mukerjee and Ashare ⟨91CRV1⟩. The strategies developed are based on the idea of the counter-attack reagent, since the thiourea adduct resulting from the addition of one equivalent of amine to an isothiocyanate might still undergo a subsequent intramolecular cyclization if adequate functionalization is present. Amino acids and their derivatives, amino aldehydes, amino ketones, amino oximes, hydrazines, and hydrazides are among the many reagents employed for this purpose. Thus, a wide variety of five- and six-membered heterocycles has been easily prepared (Equation (6)).

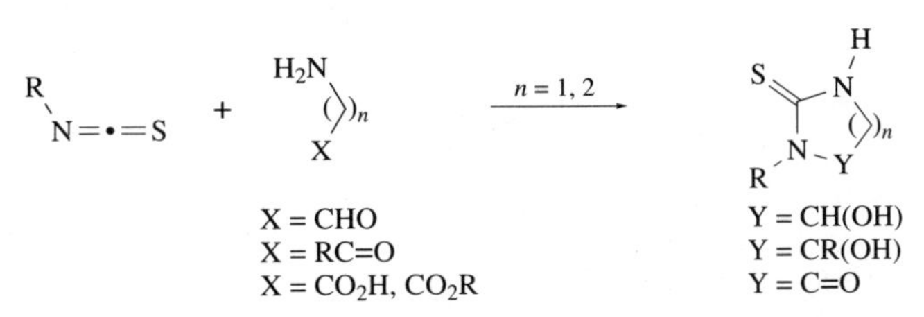

(6)

Thiosemicarbazides are also prepared in a general manner from isothiocyanates. Thus, alkyl and aryl isothiocyanates react exothermically with mono-, di-, or trisubstituted hydrazines, as well as with hydrazine itself, to give thiosemicarbazides in good yield. When unsymmetrical mono- and disubstituted hydrazines are employed both of the two possible regioisomers (**9**) and (**10**) are actually obtained, although the thiosemicarbazides resulting from the attack of the less-substituted nitrogen predominate in the case of monosubstituted hydrazines (Scheme 8) ⟨61ZOB3726, 70MI 618-01⟩.

Another entry into thiosemicarbazides is provided by the reaction of isothiocyanates, prepared from azides and carbon disulfide, with triphenylphosphine and hydrazine. However, the yields obtained through this method are only moderate (Scheme 9) ⟨92T7505⟩.

Functionalized hydrazines have also been used as suitable precursors of thiourea derivatives. For instance, acylhydrazines ⟨68JOC851, 72TL2939, 80JHC1369⟩ and sulfonylhydrazines ⟨70JMC334⟩ react regioselectively through the free amino group with isothiocyanates in ethanol to give thiosemicarbazides in good yields (Equation (7)).

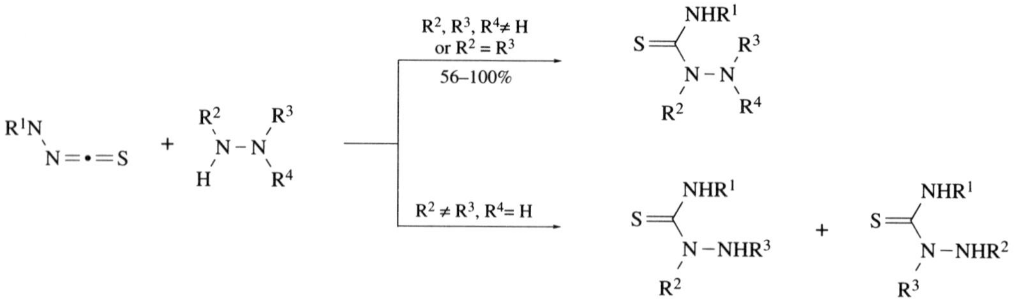

(9) (10)

Scheme 8

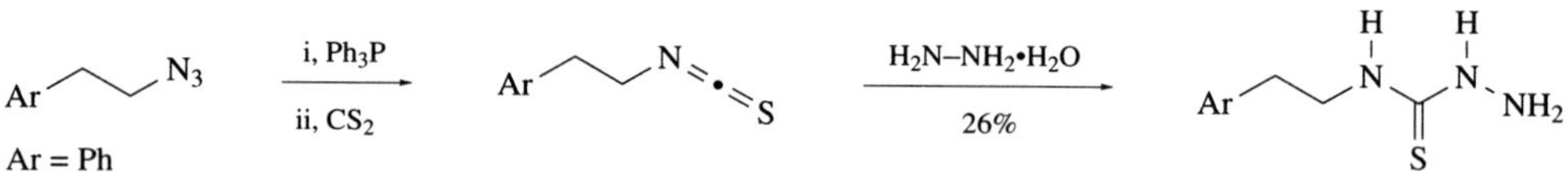

Scheme 9

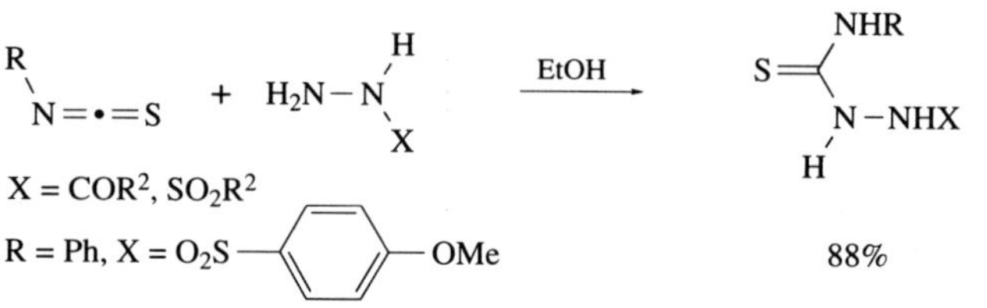

(7)

Thiosemicarbazides are also available from amino isothiocyanates. In this case, reaction of 3-aminopyridine with substituted amino isothiocyanates represents another route to 1,1,4-trisubstituted thiosemicarbazides (Equation (8)) ⟨68ACS1898⟩.

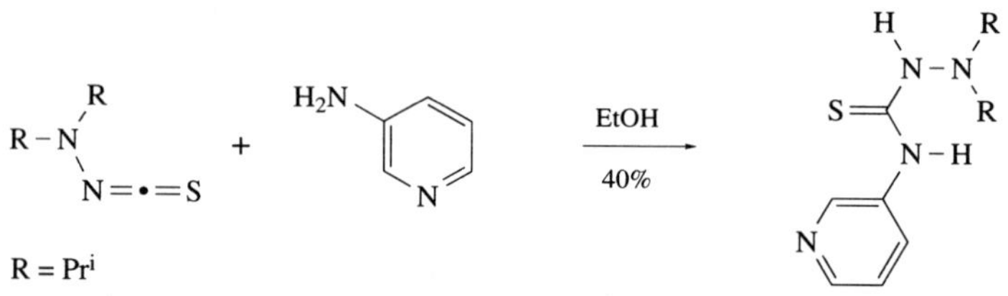

(8)

It is also possible to obtain cyclic derivatives of thiosemicarbazides from properly functionalized isothiocyanates following the same pattern described above for thioureas. Thus, hydrazine reacts with 2-acetoxyvinyl isothiocyanate in refluxing ether to give 4,5-dihydro-1,2,4-triazine-3(2*H*)-thione (**11**) (Equation (9)) ⟨79JCR(S)240⟩.

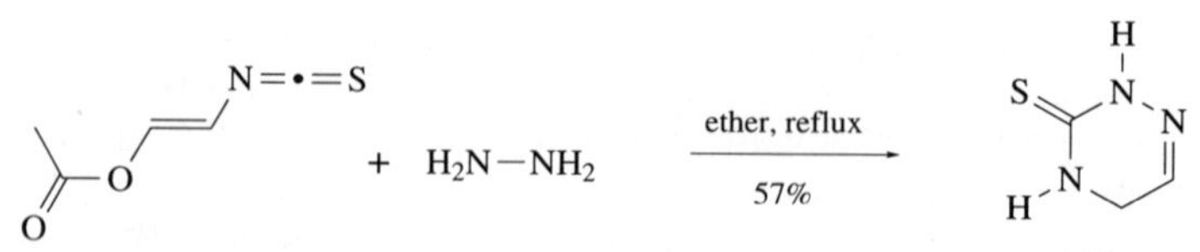

(9)

(11)

6.18.1.1.2 *From carbon disulfide*

Symmetrical thioureas are synthesized inexpensively by reacting primary amines with carbon disulfide in the presence of a catalyst, such as sulfur, hydrogen peroxide, hydroxide ion, iodine, or pyridine. The reaction appears to involve an ammonium dithiocarbamate intermediate which upon loss of hydrogen sulfide and addition of a second equivalent of amine yields 1,3-disubstituted thioureas. This reaction pathway is supported by the formation of unsymmetrical thioureas when the reaction is carried out in the presence of ammonia or other amine (Scheme 10) ⟨55CRV181⟩.

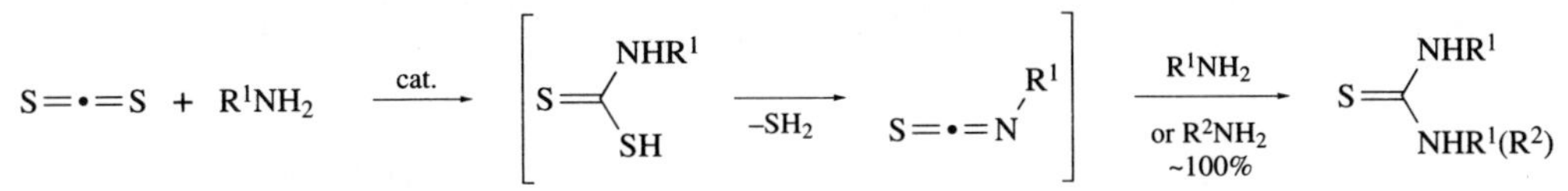

Scheme 10

Trisubstituted thioureas are synthesized from carbon disulfide, aliphatic or aromatic primary amines and hexaalkylphosphorus triamides, which transfer one of their dialkylamino groups. The reaction takes place in pyridine giving the thioureas in almost quantitative yield (Equation (10)) ⟨75S384⟩.

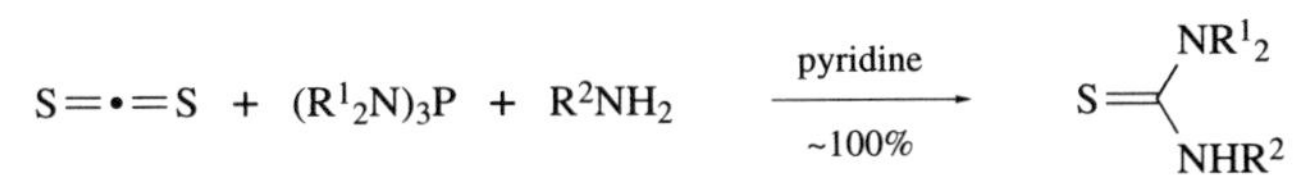 (10)

R^2 = Ph, Bn, *c*-C_6H_{11}

Carbon disulfide condenses with three molar equivalents of dialkylcyanamides at 100 °C and high pressure to give substituted thiocarbamoylimino-1,3,5-thiadiazines (**12**) in good yields. The reaction is thought to proceed through a repeated [2 + 2] cycloaddition–retroaddition cycle (Scheme 11) ⟨90JCS(P1)1218⟩.

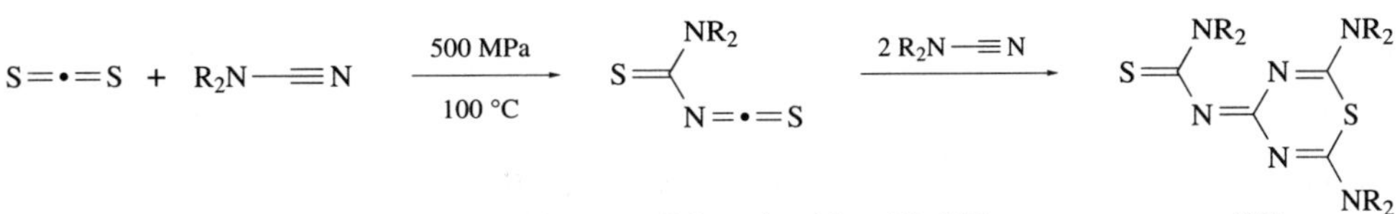

NR_2 = NMe_2, pyrrolidine, piperidine, 55–84% **(12)**

Scheme 11

Hydroxylamines show anomalous behavior towards carbon disulfide in DMF since the formation of monohydroxythioureas (**13**) instead of the expected *N*,*N*′-dihydroxy derivatives has been reported to occur (Equation (11)) ⟨70LA(731)171⟩.

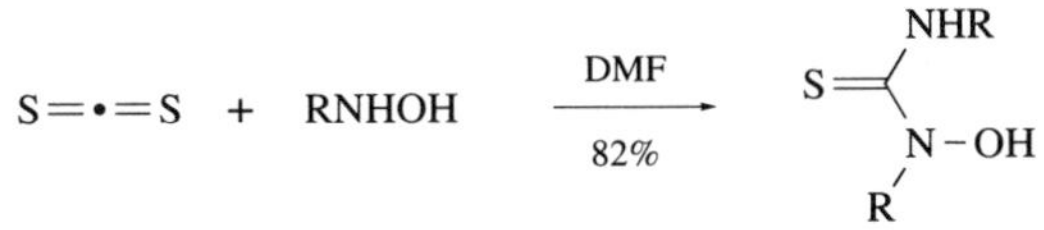 (11)

R = *c*-C_6H_{11} **(13)**

As expected, α,ω-diamines react with carbon disulfide in ethanol to give cyclic thioureas ⟨63FCF554⟩. The method allows the preparation of heterocycles of various ring sizes, including medium ring, macrocyclic, and benzocondensed products as is exemplified in Scheme 12 by the reaction of 1,8-diaminooctane ⟨80TL313⟩ and 2-aminobenzylamine ⟨69JOC3604, 71ZC12⟩ with carbon disulfide. Substituted pyrimidine-2(1*H*)-thiones (**14**) are available in a regioselective fashion by refluxing 4-amino-1-azabutadienes in carbon disulfide (Scheme 12) ⟨79CC675, 82JCS(P1)2149⟩.

2-Thiohydantoins are immediate precursors of imidazothiazoles, potential central nervous system antitumor agents. These thiohydantoins are generally prepared in good yields from α-amino acids or α-amino esters and isothiocyanates. Alternatively, they can be obtained in one pot by reaction of dithiocarbamic esters, generated *in situ* from a primary amine, carbon disulfide and methyl iodide, and α-amino acids (Scheme 13) ⟨83S391⟩.

The Schiff bases derived from aromatic aldehydes and arylamines are precursors of imidazole-2-thione in a reaction sequence that comprises three steps; thus, the sodium cyanide promoted dimerization of the imines in DMF giving the benzil dianil is followed by reduction with sodium in ether and cyclization with carbon disulfide. This method is restricted to all-aryl substituted com-

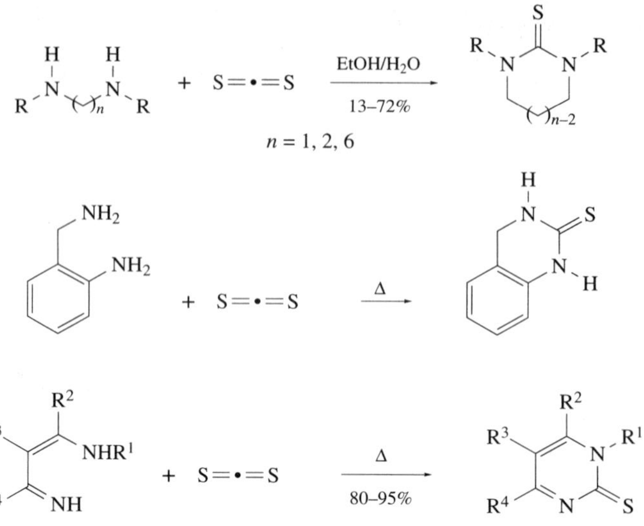

(14)

Scheme 12

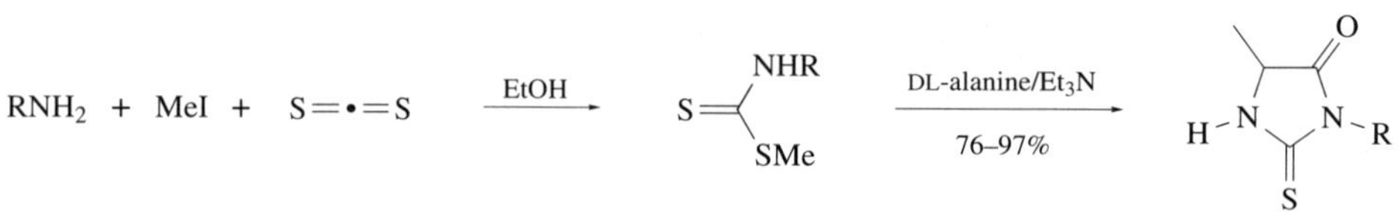

Scheme 13

pounds since only aromatic aldimines undergo cyanide-catalyzed dimerisation to form dianils (Scheme 14) ⟨80S1001⟩.

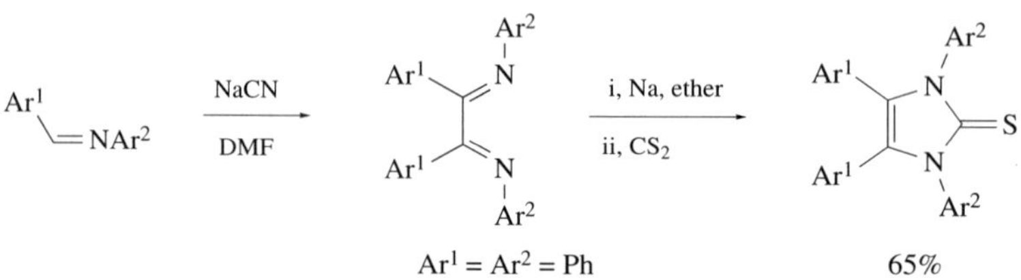

Scheme 14

Carbon disulfide also inserts into the N—Si bond of siladiazoles to give an unstable intermediate (**15**) which readily decomposes to the corresponding thiourea. The reaction takes place at or below room temperature and sometimes requires the use of a high excess of carbon disulfide ⟨91JOM(419)9⟩. Other reactive amines such as bis(tributyltin)ethylene diamine have also been reported to afford cyclic thioureas (Scheme 15) ⟨70MI 618-02⟩.

6.18.1.1.3 From thiophosgene

Substituted thioureas can also be prepared by reaction of thiophosgene with primary and secondary amines. This method is closely related to the one described in Section 6.18.1.1.1, and involves the formation of isothiocyanates and thiocarbamoyl chlorides as the intermediates in the reaction with primary or secondary amines, respectively ⟨55CRV181, 73RCR587⟩. Accordingly, when the reaction is carried out using equimolecular amounts of thiophosgene and secondary amine, the corresponding thiocarbamoyl chloride is isolated; this intermediate can in turn react with another equivalent of a different amine, thus allowing the synthesis of unsymmetrically substituted thioureas (Scheme 16).

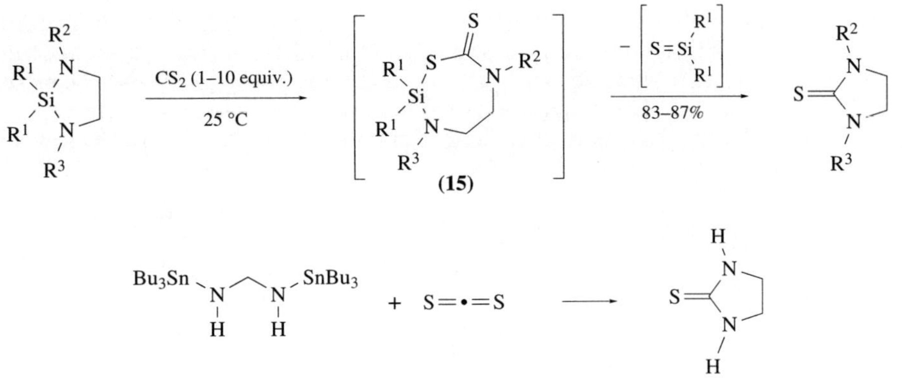

Scheme 15

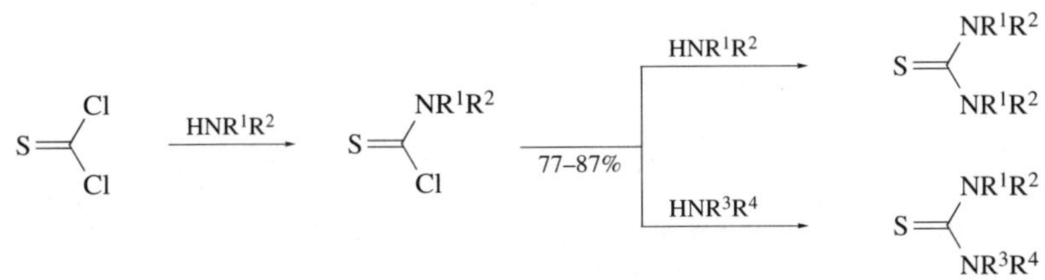

Scheme 16

Thiophosgene also reacts with nitrogen heterocycles to give bis(heterocyclic) thiocarbonyl compounds which have found utility as thiocarbonyl transfer reagents in other syntheses of thiourea derivatives (see Section 6.18.1.1.4). The synthesis of these compounds can be achieved either from N—H containing heterocycles, mostly azoles, or activated derivatives thereof. For instance, *N*-trimethylsilylimidazole reacts with thiophosgene in carbon tetrachloride at room temperature to furnish *N*,*N*′-thiocarbonylbis(imidazole) (Equation (12)) ⟨78JOC337, 79LA1756⟩.

(12)

R = H, TMS

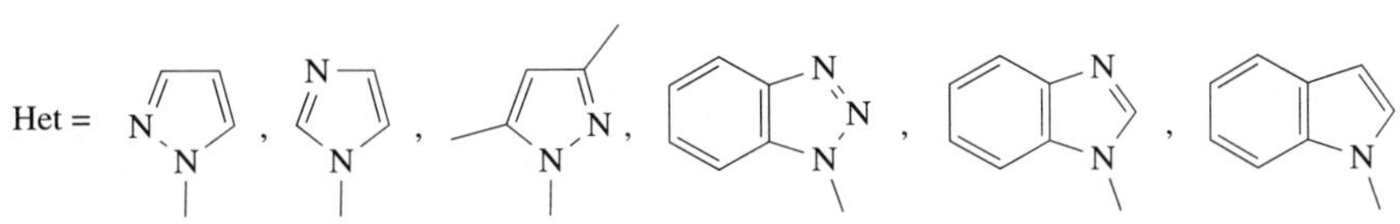

As in the case of carbon disulfide, aliphatic diamines produce cyclic thioureas ⟨75JCS(P1)212, 75JMC913, 80JMC1261⟩. 1,3-Dimethyl-2,2-diethyl-1,3-diaza-2-germacyclopentane reacts with thiophosgene at room temperature to give 1,3-dimethylimidazolidine-2-thione (**16**) in 78% yield (Equation (13)) ⟨75JOM(88)C35, 91JOM(419)9⟩.

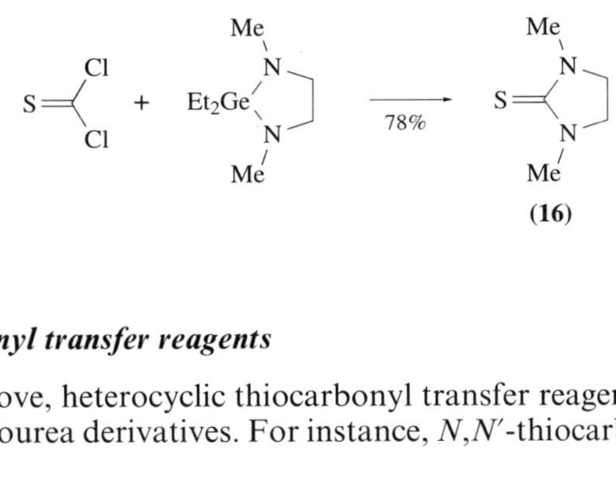

(13)

(16)

6.18.1.1.4 From thiocarbonyl transfer reagents

As has been indicated above, heterocyclic thiocarbonyl transfer reagents have been used increasingly for the synthesis of thiourea derivatives. For instance, *N*,*N*′-thiocarbonyl-bis(imidazole) reacts with two equivalents of aliphatic or aromatic primary amines in CH_2Cl_2 or chloroform at room temperature to give 1,3-disubstituted thioureas in nearly quantitative yield ⟨78JOC337, 80JCS(P1)766⟩. Secondary amines such as diethylamine expel just one imidazole group to furnish the mixed thiourea (**17**) (Scheme 17) ⟨62LA(657)98⟩. The other azoles studied, while usually less reactive, may in certain situations offer some advantages due to their higher stability to moisture and the low solubility of the free heterocycle recovered (Scheme 17) ⟨78JOC337⟩.

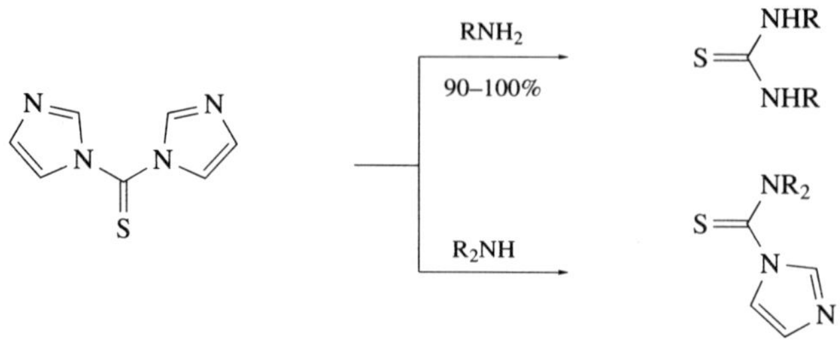

Scheme 17

N,N′-Thiocarbonylbis(3,5-dimethyl)pyrazole has been reported to give *N,N′*-diphenylthiourea and benzimidazoline-2-thione on reaction with aniline and 1,2-diaminobenzene, respectively. When 1,4-diaminobenzene is used the reaction results in the formation of a thiourea derived polymer. Aliphatic primary diamines show differing behavior depending on the length of the methylene chain. Thus, while 1,3-diaminopropane and 1,4-diaminobutane form the bis(3,5-dimethyl-1-pyrazolylthioformyl)diamines (DMP = 3,5-dimethylpyrazolyl), 1,5-diaminopentane produces a thermoplastic polymer (Scheme 18) ⟨61LA(646)96⟩.

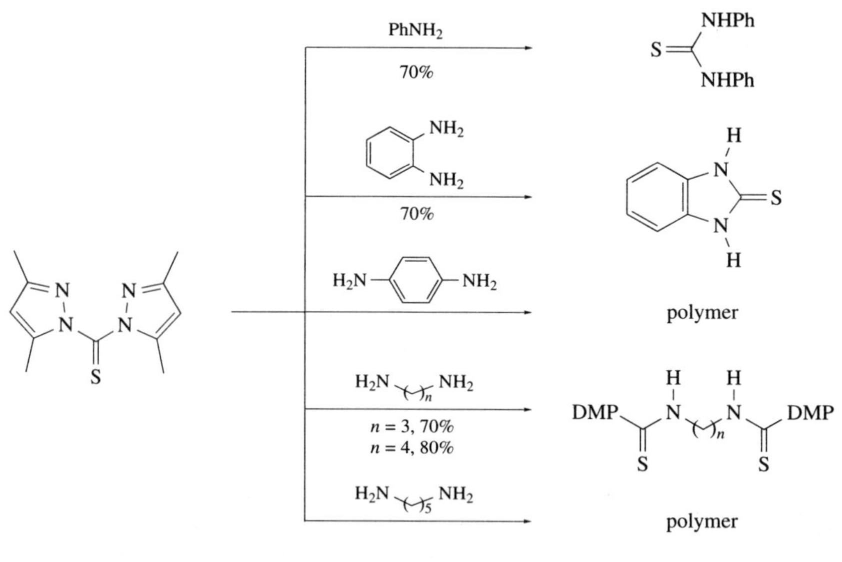

Scheme 18

A particularly useful preparation of thiourea derivatives takes advantage of the *in situ* formation of thiocarbonylbis(imidazole) from imidazole, carbon disulfide and a 2-halothiazolium salt (molar ratio 2 : 2 : 1), thus avoiding the use of thiophosgene. The utility of this *in situ* generated thiocarbonylbis(imidazole) has been demonstrated in the preparation of unsymmetrical thioureas (**18**) and thiosemicarbazides (**19**) by stepwise addition of two different amines or an amine and hydrazine hydrate, respectively (Table 2, Scheme 19) ⟨88JOC2263⟩.

6.18.1.1.5 From ureas

The direct conversion of the carbonyl moiety into a thiocarbonyl function is possible in general with some phosphorus derivatives, such as P_4S_{10} and the popular Lawesson's reagent. Some examples

Table 2 Thioureas (**18**) and thiosemicarbazides (**19**) from thiocarbonylbis(imidazole).

R¹R²NH	*R³R⁴NH*	*Yield (%)*
$PhCH_2MeNH$	NH_3	58
PhMeNH	NH_3	75
Morpholine	NH_3	48
$PhCH_2MeNH$	$NH_2NH_2 \cdot xH_2O$	41
PhMeNH	$NH_2NH_2 \cdot xH_2O$	74
Morpholine	$NH_2NH_2 \cdot xH_2O$	52
PhMeNH	Piperidine	88

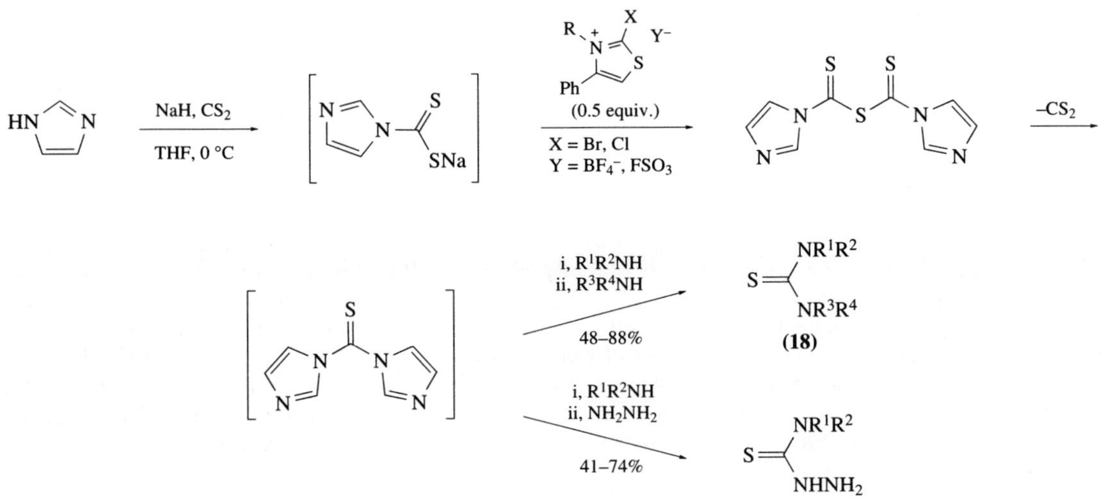

(**18**)

(**19**)

Scheme 19

involving P_4S_{10} and tetrasubstituted thioureas have been described. The yields vary from moderate to poor ⟨71LA(746)92⟩, a notable exception being the transformation of imidazolidin-2-one into the thione (**20**) which has been achieved in nearly quantitative yield (Equation (14)).

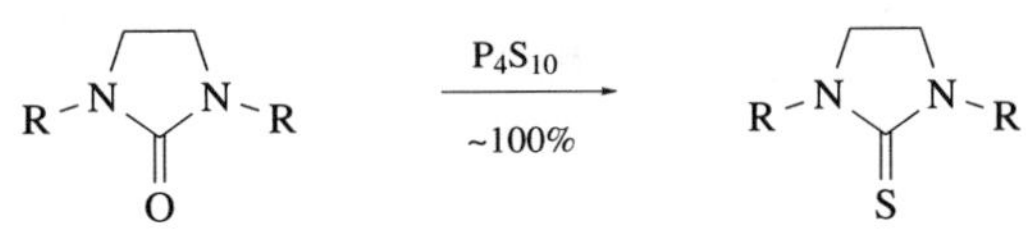

(**20**) (14)

High yields are obtained with some heterocyclic derivatives. Thus, 1,4,6-trisubstituted 2(1*H*)-pyrimidones are converted into the corresponding pyrimidine-2-thiones (**21**) by refluxing in benzene with Lawesson's reagent. In this case the yields vary from poor to quantitative (Equation (15)) ⟨82H(19)2283⟩.

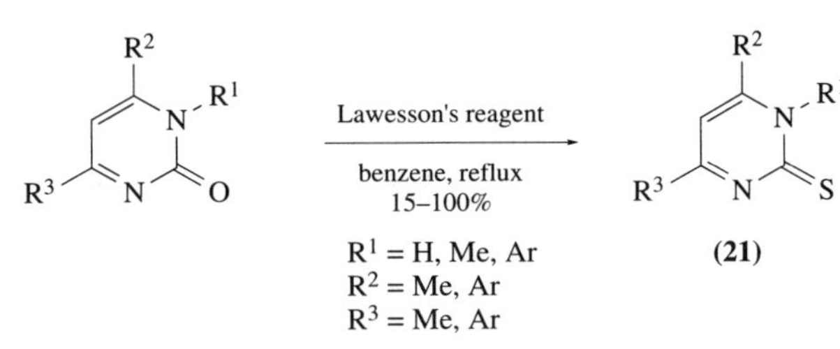

(**21**) (15)

6.18.1.1.6 *Miscellaneous methods*

A nearly quantitative conversion of symmetrical thioureas into unsymmetrical thioureas, by exchange of one amino group, has been described ⟨93TL6447⟩. The process is accomplished by refluxing 1,3-diphenylthiourea with different types of amines and triethylamine in acetonitrile giving high yields of 1-phenylthiourea derivatives (**22**); the reaction also works, though in lower yields, in other polar solvents such as methanol, ethanol, THF, or DMF. The procedure has also been shown to take place under phase transfer catalysis, thus allowing aqueous solutions of low boiling point amines to be used (Equation (16)). Sodium cyanothioformate reacts with primary diamines in refluxing butanol to afford heterocyclic derivatives (**23**) in good yields (Equation (17)) ⟨55MI 618-01, 81JPR41⟩.

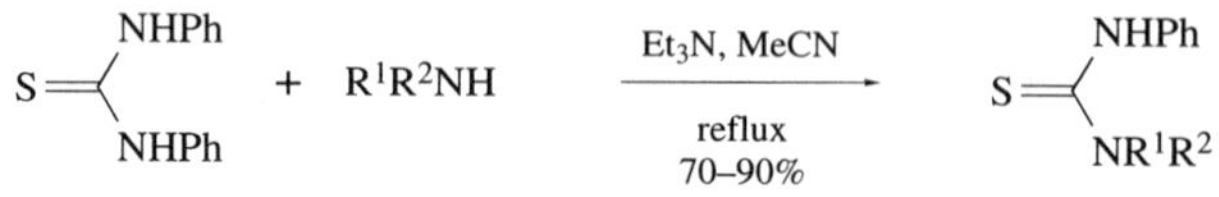

(16)

(**22**)

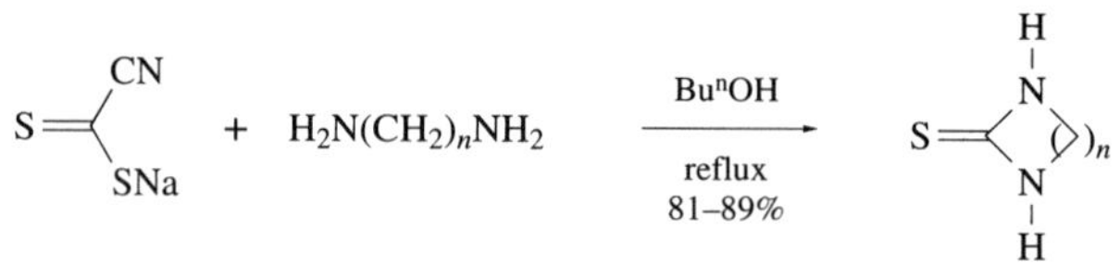

(17)

$n = 1, 2$ (**23**)

Thiourea *S,S*-dioxide derivatives have found application in the reduction of different compounds, such as dinitrodiaryl derivatives ⟨88AJC995⟩, organoselenium and tellurium halides, and oxides ⟨88SC301⟩ as well as azobenzenes ⟨88MI 618-01⟩. Thiourea itself can be quantitatively converted into its dioxide by treatment with H_2O_2 and sodium acetate in water at 5°C ⟨90MI 618-01⟩. The *S,S*-dioxides derived from monosubstituted thioureas are stable and can be obtained in a similar way by treatment with two equivalents of hydrogen peroxide in ethanol at low temperature ⟨69LA(722)80⟩. When the thiourea bears bulky substituents at both nitrogen atoms, the oxidation process stops at the *S*-oxide stage ⟨69LA(722)52⟩. The oxidation of the thioxo function of 1-aryl thiosemicarbazides with hydrogen peroxide leads to the *S*-oxide derivative but it is also accompanied with the additional oxidation of the hydrazo group to the azo function (Scheme 20) ⟨66TL3695⟩.

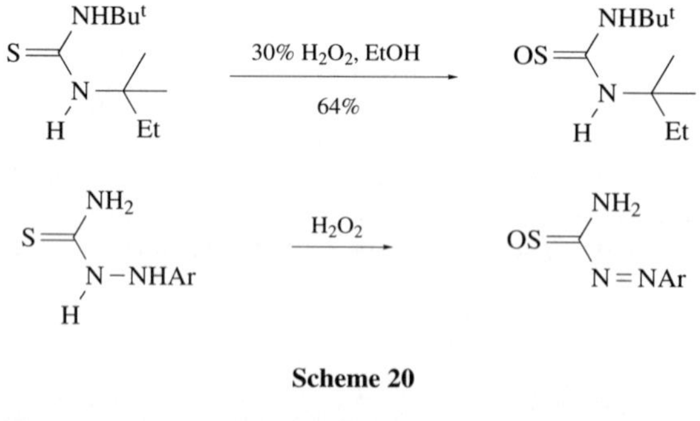

Scheme 20

6.18.1.2 Thiocarbonyl Derivatives with One Nitrogen and One Phosphorus Function

Phosphinothioformamides, also named thiocarbamoylphosphines, are generally used as ligands in organometallic chemistry. The related compounds containing nitrogen and one atom of group 15 other than phosphorus (As, Sb, Bi) are not known.

6.18.1.2.1 *From isothiocyanates*

Isothiocyanates are the most valuable source for preparing phosphinothioformamides. Typically, the P—H bond of phosphines adds to isothiocyanates to give the title compounds (Equation (18)) ⟨87ZN(B)77⟩. Thioalkyl phosphines, prepared from sodium phosphines and chloroalkyl thiols, react with PhNCS to give the thioalkylphosphinothioformamide derivatives which can undergo a subsequent intramolecular cyclization to the thiocarbonylphosphazine (**24**) (Scheme 21) ⟨73JPR471⟩. Bis(ethylthio)alkylphosphines undergo reduction of the phosphorus–sulfur bond with dibutyltin dihydride giving rise to the corresponding species with a P—H bond, which yields alkylthiophosphinothioformamides upon reaction with phenyl isothiocyanate (Scheme 22) ⟨88ZOB1468⟩. The reaction of phosphonate and thiophosphonate isothiocyanates with secondary phosphines results in the formation of the corresponding phosphonates and thiophosphonates (Equation (19)) ⟨92PS(73)81⟩.

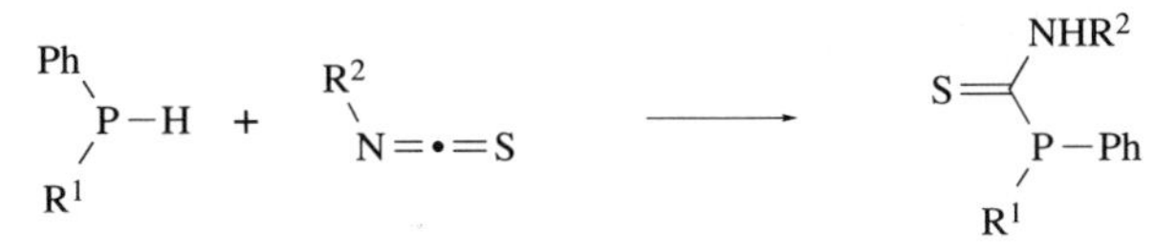

(18)

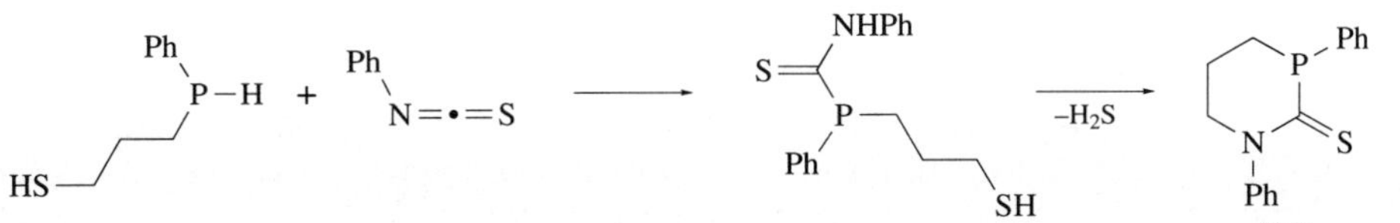

Scheme 21 **(24)**

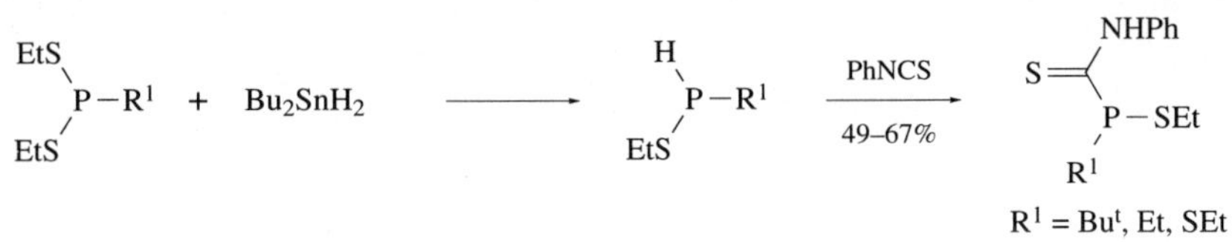

Scheme 22

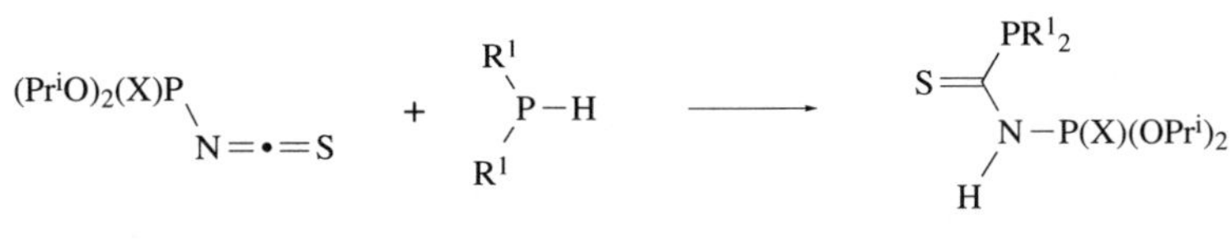

(19)

The P—H bond of phosphorus(V) derivatives also adds to isothiocyanates to yield phosphinothioformamides (Equation (20)) ⟨69BCJ2975, 87ZN(B)860⟩. Chiral phosphinothioformamides (**25**) are obtained when using isothiocyanates prepared from optically active amines, such as (*S*)- and (*R*)-phenylethylamine (Equation (21)) ⟨86ZN(B)1142⟩. (*N*-Alkylthiocarbamoyl)phosphonic acid esters (**26**) have been prepared and shown to be versatile precursors of α-substituted methyl phosphonate derivatives which display interesting biological activity. Their preparation cannot be achieved by the Arbuzov-type reaction of trialkyl phosphites and isothiocyanates. However, when dialkyl phosphates are used instead, the reaction works well and addition of the P—H bond to the C=N double bond of the isothiocyanate takes place. The process is base catalyzed and the reaction is usually run in the presence of alkoxide ion, except for compounds sensitive to these reagents such as phosphonates; in these cases, triethylamine is a good alternative (Equation (22)) ⟨75ZOB1484, 82JOC3012⟩. The related (*N*-alkylthiocarbamoyl)thiophosphonic acid esters have been prepared in a similar way by starting from the corresponding thiophosphonic esters (Equation (23)) ⟨83JOC3966⟩.

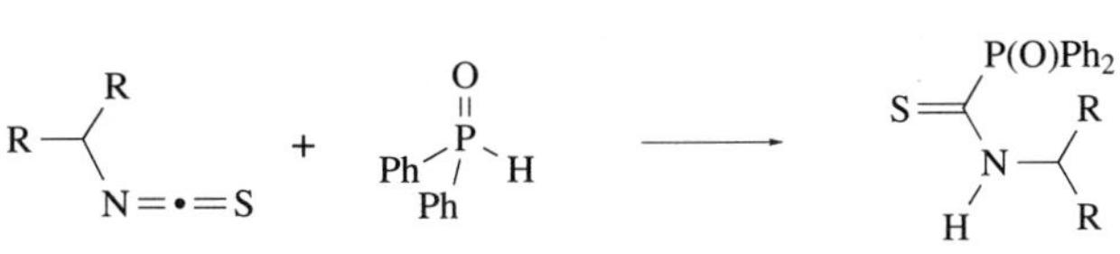

(20)

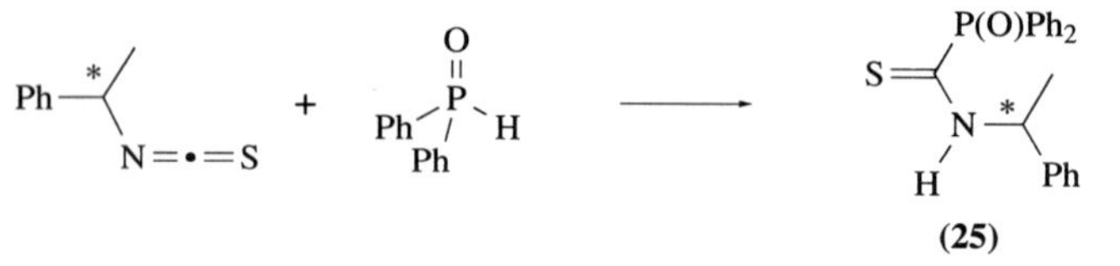

(21)

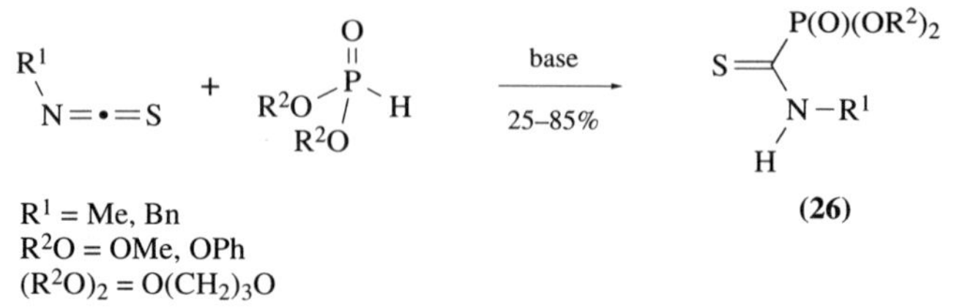

(22)

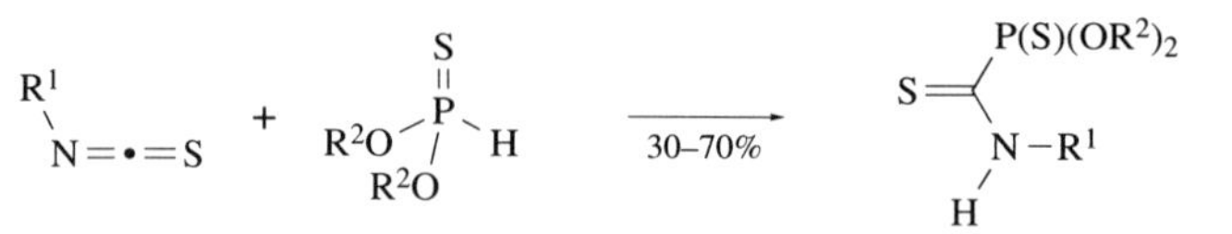

(23)

Trimethylsilyl tetraethylphosphorodiamidite reacts with phenyl isothiocyanate yielding, after hydrolysis of the N—Si bond, the thiocarbamoylphosphonate derivative (**27**) (Scheme 23) ⟨76IZV455⟩. Thiocarbamoylphosphonic acids have also been prepared by reaction of tris(trimethylsilyl) phosphite with alkyl and aryl isothiocyanates. The exothermic reaction takes place very rapidly at room temperature to give an adduct which upon treatment with aniline in methanol furnishes the monoanilinium salt of thiocarbamoylphosphonic acid (**28**) in almost quantitative yield (Scheme 24) ⟨79TL3013⟩.

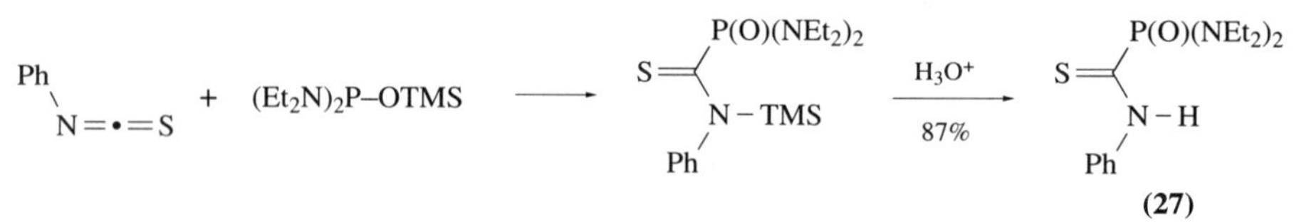

Scheme 23

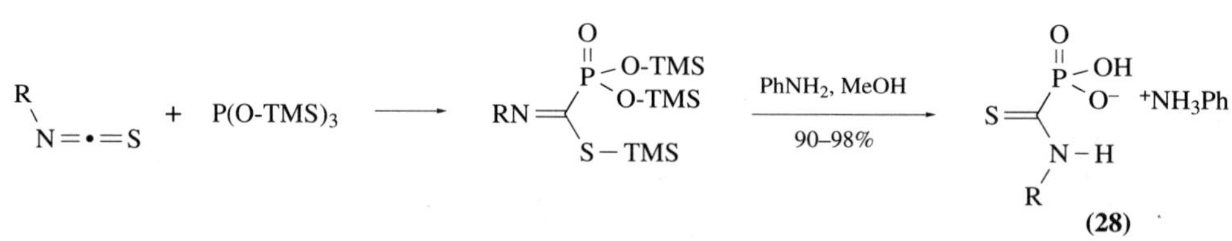

Scheme 24

Alkyl and aryl isothiocyanates react with trimethylsilyldiphenylphosphine at 80°C to give the expected *N*-silylated derivatives arising from addition of the P—Si bond, which undergo easy methanolysis producing diphenylphosphinothioformamides (**29**) (Scheme 25) ⟨68JCS(A)1105⟩. The insertion of isothiocyanates into the P—Si bond has also been described in several other phosphorus derivatives; thus, bis(trimethylsilyl)phosphines give similar insertion products with phenyl isothiocyanate (Equation (24)) ⟨85ZAAC(520)120, 85ZAAC(520)139⟩. The P—Si bond of trimethylsilyl phosphatriafulvenes also inserts into the C=N bond of isothiocyanates yielding compounds (**30**) through the betaine intermediate shown (Scheme 26) ⟨91S1099⟩. Silylphosphines and germylphosphines react at room temperature, in almost quantitative yield, with equimolar amounts of phenyl isothiocyanate to give the corresponding phosphinothioformamide derivatives (**31**). The reaction proceeds with retention of the configuration at the silicon or germanium centers suggesting a four-center mechanism (Equation (25)) ⟨79TL3507⟩.

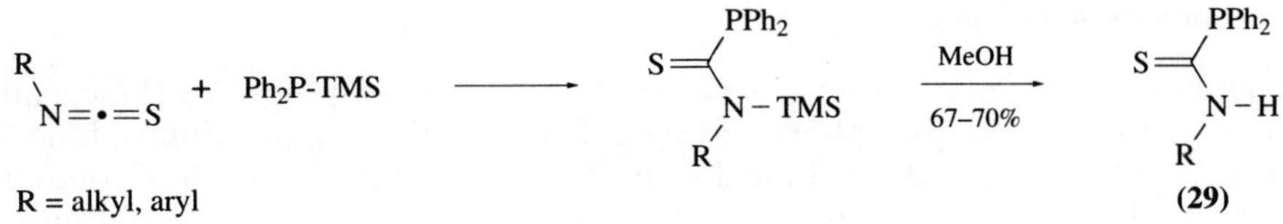

(29)

Scheme 25

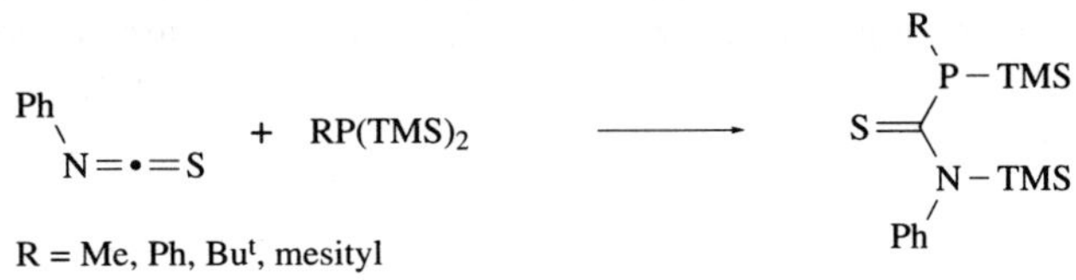

(24)

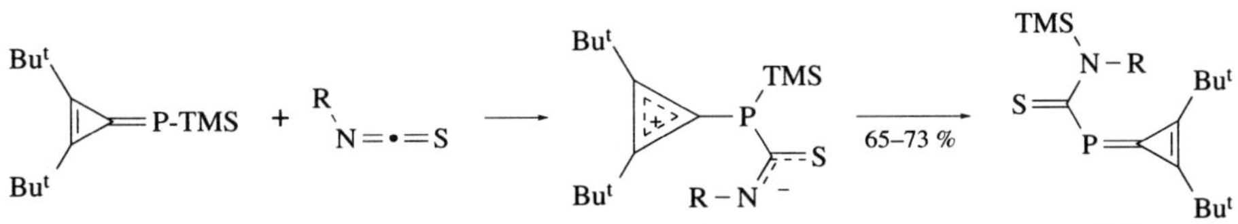

(30)

Scheme 26

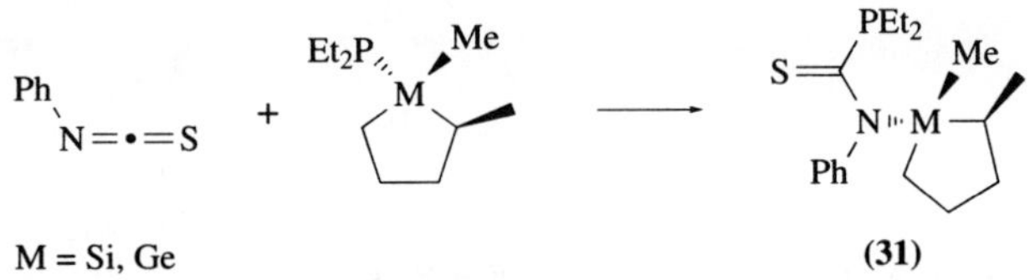

(31)

(25)

Phosphorus(III) acid amides react rapidly with isothiocyanates to form an addition intermediate which, upon hydrolysis, furnishes phosphinothioformamides (**32**) (Equation (26)) ⟨77IZV1177⟩. Also, phosphorus(V) acid *t*-butyl esters and amides react with phenyl isothiocyanate to give phosphinothioformamides (Equation (27)) ⟨90ZOB563⟩.

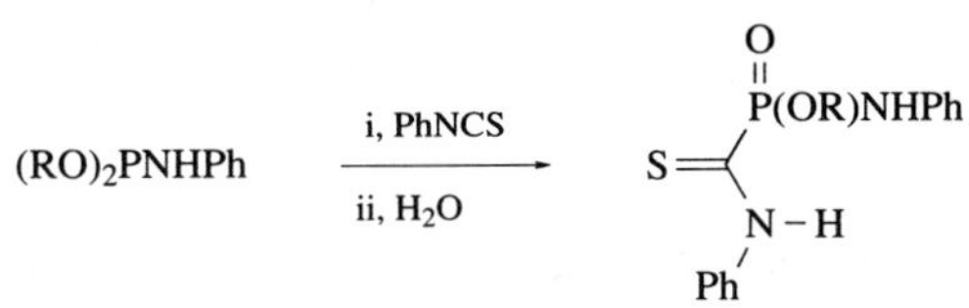

(32)

(26)

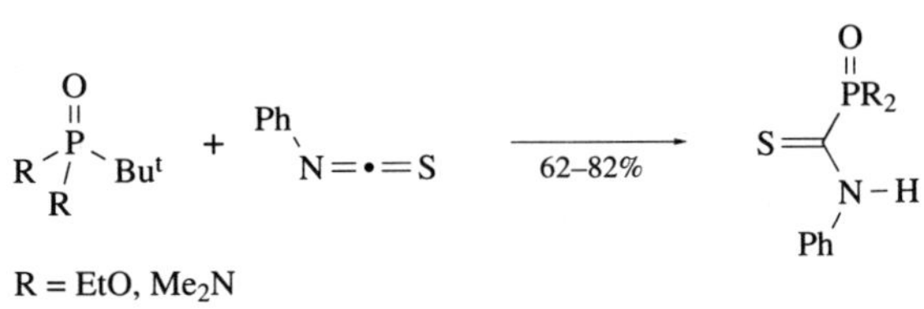

(27)

N-Thiophosphonate phosphinothioformamides (**33**) can be prepared by reaction of dialkoxythiophosphoryl isothiocyanates with sodium dialkylphosphites in ethanol. However, the yield in this reaction is very low, and heating in benzene does not improve the synthesis of (**33**), but results in phosphonate–thiophosphate rearrangement (Equation (28)) ⟨93ZOB633⟩.

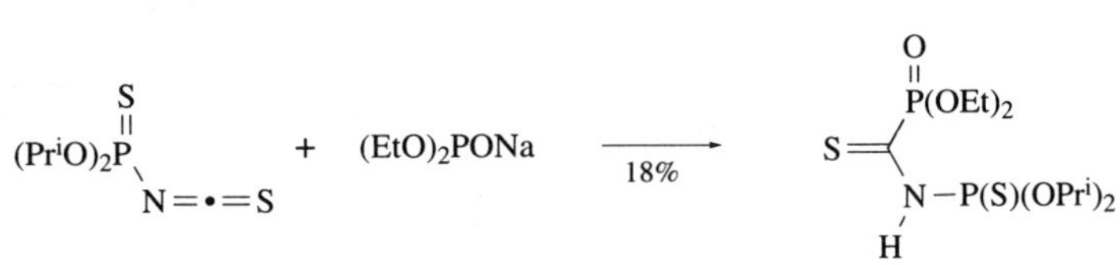

(33)

(28)

6.18.1.2.2 *From halothioamides*

Chlorothioamides are the main alternative to isothiocyanates for preparing thiocarbonyl derivatives containing nitrogen and phosphorus atoms. Thus, lithium diphenylphosphide reacts with *N*-alkyl- and *N*-arylthioformamidoyl chlorides to give diphenylphosphinothioformamides (**34**) in 54–73% yield ⟨85CB227⟩. Good yields are also obtained in the reaction of lithium trimethylsilylphosphides with thioformamidoyl chlorides to give the trimethylsilylphosphinothioformamide derivatives (**35**) ⟨84ZAAC(518)21⟩. Phosphorus(III) esters react with dialkylamides of chlorothioformic acid to give phosphinothioformamides (**36**) in moderate to good yields (Scheme 27) ⟨88ZOB1489⟩.

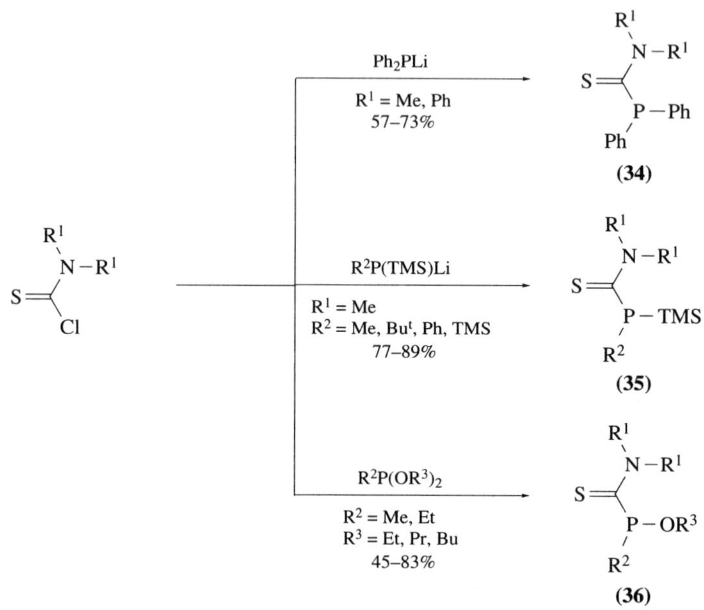

Scheme 27

6.18.1.2.3 *From thiophosphinoyldithioformates*

Substituted thiophosphinoylthioformamides (**37**) are obtained by aminolysis reaction of thiophosphinoyldithioformates. The reaction works well for ammonia and primary aliphatic amines with moderate steric hindrance (Equation (29)) ⟨70ACS1094⟩.

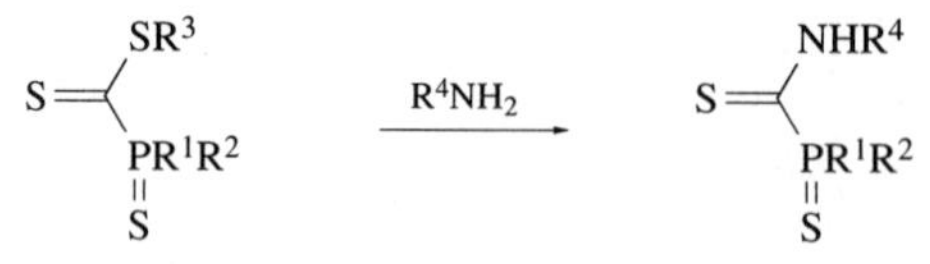

(29)

6.18.2 FUNCTIONS CONTAINING AT LEAST ONE METALLOID FUNCTION

Thiocarbonyl compounds bearing metalloid functions are almost unknown. Two different approaches for the synthesis of bis(trimethylsilyl)thioketone have been described up to now, and some controversy has been generated concerning the stability of this compound. Block *et al.* proposed in 1985 the intermediacy of bis(trimethylsilyl) thioketone (**38**) in the serendipitous synthesis of alkyltrimethylsilyldithioformates from alkanesulfenic acids. This bis(trimethylsilyl) thioketone was prepared by pyrolysis of the corresponding thiosulfinate and trapped *in situ* with 2,3-dimethylbutadiene to give the corresponding Diels–Alder cycloadduct. In the absence of the trapping agent, the major product is the trimethylsilyldithioformate (**39**) (Scheme 28) ⟨85TL2259, 88T281⟩.

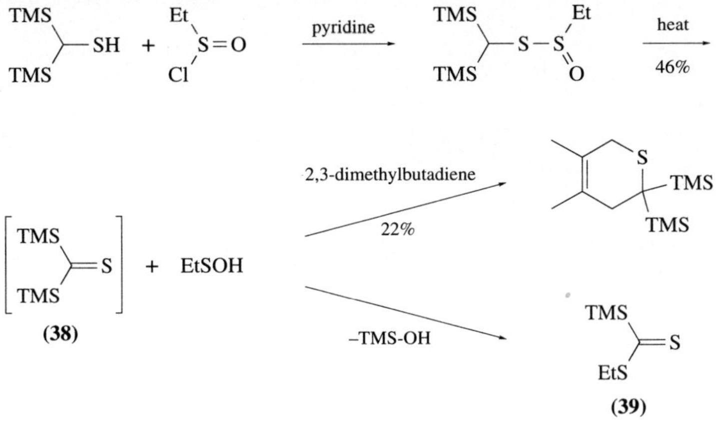

Scheme 28

Ricci *et al.* almost simultaneously reported the first synthesis of bis(trimethylsilyl) thioketone (**38**) starting from tris(trimethylsilyl)methane. Thus, the treatment of the latter with one equivalent of MeLi followed by addition of elemental sulfur and acidic hydrolysis leads to tris(trimethylsilyl)methanethiol which is transformed into tris(trimethylsilyl)methanesulfenyl bromide by reaction with MeLi and bromine. Finally, heating of the sulfenyl bromide results in loss of bromotrimethylsilane and formation of the target compound (**38**) (Scheme 29) ⟨85TL1091⟩.

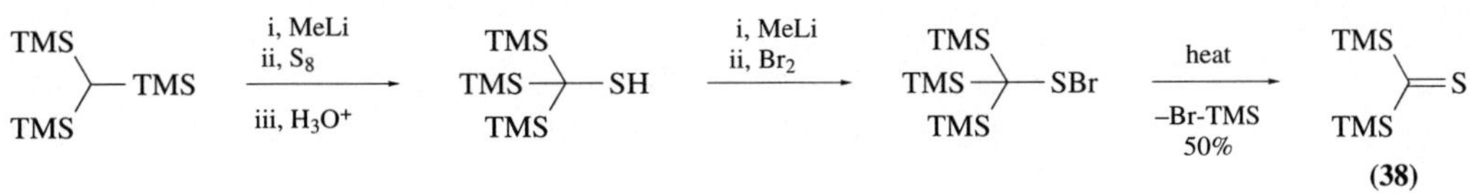

Scheme 29

6.19
Functions Containing a Selenocarbonyl or Tellurocarbonyl Group—SeC(X)X′ and TeC(X)X′

FRANK S. GUZIEC, Jr. and LYNN JAMES GUZIEC
New Mexico State University, Las Cruces, NM, USA

6.19.1 OVERVIEW

Initially it should be noted that examples of selenocarbonyl analogues of well-known carbonyl-based functional groups are much less common than the corresponding oxygen and sulfur compounds. The related tellurocarbonyl compounds are even more rare. This scarcity is probably due to a number of factors. Selenocarbonyl and tellurocarbonyl compounds are much less stable than their oxygen and sulfur analogues, presumably owing to poor overlap in the π bonds of the selenium and tellurium compounds. These compounds are also prone to lose volatile selenium and tellurium species and are therefore often unpleasant to work with and potentially toxic. The early literature associated with selenocarbonyl or tellurocarbonyl compounds is fraught with errors. Even the nomenclature describing these analogues is often inconsistent, leading to difficulties in searching the literature (e.g., the term 'selenone' is often used incorrectly to designate a carbon–selenium double bond instead of the accepted term 'selone'). A number of reports dealing with these problems have been published ⟨B-73MI 619-01, B-73MI 619-02, B-86MI 619-01, B-87MI 619-01, B-87MI 619-02⟩. On a positive note it should also be stated that advances in the availability of novel reagents for the introduction of selenium and tellurium into organic molecules and progress in spectroscopic characterization may make further examples of these compounds more common in the near future.

6.19.2 SELENO- AND TELLUROCARBONYL FUNCTIONS CONTAINING AT LEAST ONE ATTACHED HALOGEN

Compounds containing a halogen directly attached to a selenocarbonyl or tellurocarbonyl moiety are quite rare. In addition, very little has been reported on the reactions of these compounds. Selenocarbonyl difluoride (**1**) is quite unstable to dimerization in solution and to polymerization in the presence of Lewis acids. It can be obtained by pyrolysis of the dimer at 360°C (Equation (1)), or on a preparative scale by treatment of aluminum triiodide with $Hg(SeCF_3)_2$ (Equation (2)) ⟨80ZN(B)526⟩. Its tellurium analogue (**2**) can be prepared similarly using diethylaluminum iodide as a Lewis acid (Equation (3)) ⟨91CC1378⟩. This compound is also highly reactive and prone to dimerization.

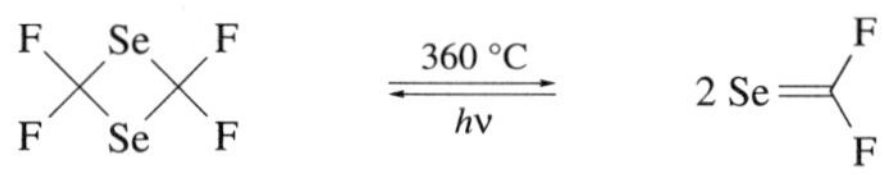

(1)

(1)

(2)

(3)

(2)

Selenophosgene (**3**) can be prepared by vacuum pyrolysis of 2,2,4,4-tetrachloro-1,3-diseletane (Equation (4)), but is stable only below −130°C ⟨81ZN(B)1261⟩.

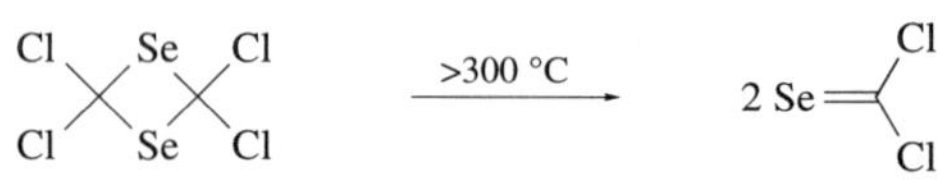

(4)

(3)

A single example of a monosubstituted selenoacyl halide has been reported. It is prepared starting from selenocarbonyl difluoride (Equation (5)) ⟨80ZN(B)526⟩.

(5)

6.19.3 SELENOCARBONYL FUNCTIONS CONTAINING AT LEAST ONE ATTACHED CHALCOGEN (AND NO HALOGENS)

Note that no tellurocarbonyl compounds of this class have been reported to date.

6.19.3.1 Dialkoxy-substituted Selenocarbonates, $(RO)_2C{=}Se$

Oxygen-substituted selenocarbonates are not particularly well represented in the literature, and generally appear only as minor by-products in the preparations of selenium-containing compounds

⟨88AJC549⟩. Treatment of sugar-derived *cis*-vicinal diols with phosgeneiminium chloride (**4**), followed by sodium hydrogen selenide treatment, afforded the corresponding cyclic selenocarbonates (**5**) in good yield (Equation (6)) ⟨88AJC549⟩. This reaction is reported to fail in the case of catechol, affording instead the open-chain selenourethane (**6**) (Equation (7)) ⟨88AJC549⟩.

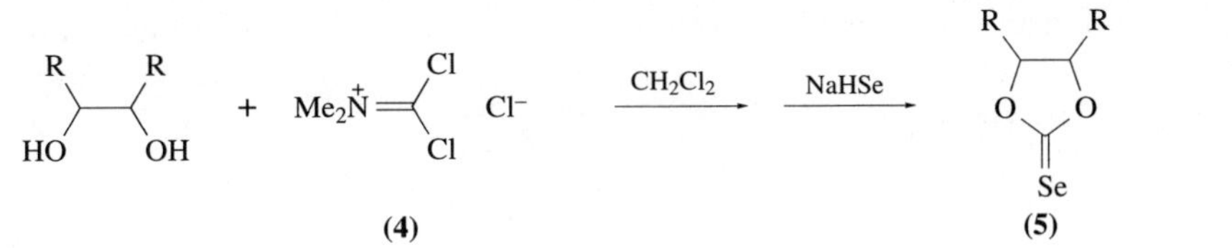

(6)

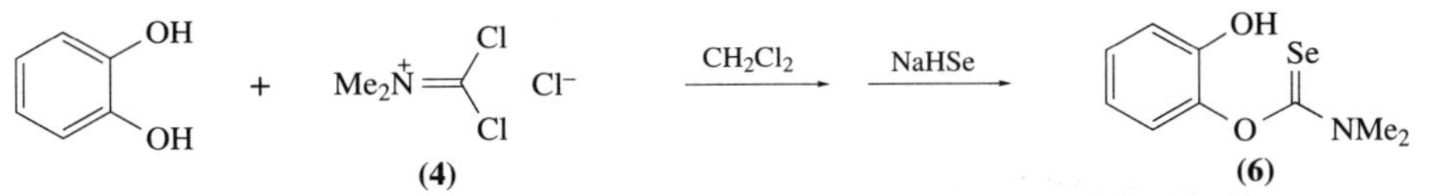

(7)

6.19.3.2 Dithio-substituted Selenocarbonates $(RS)_2C{=}Se$

Cyclic dithio-substituted selenocarbonates are important intermediates in the preparation of tetrathiafulvalenes. A general route to these compounds involves alkylation of dithiocarbamate anions (**7**), readily obtained by reaction of secondary amines with carbon disulfide in the presence of base, followed by cyclization in strong acid to the corresponding iminium salts (**8**). Treatment of this isolable salt with NaHSe or H_2Se affords the desired cyclic selenocarbonate (**9**) (Scheme 1) ⟨89JCS(P1)1068⟩.

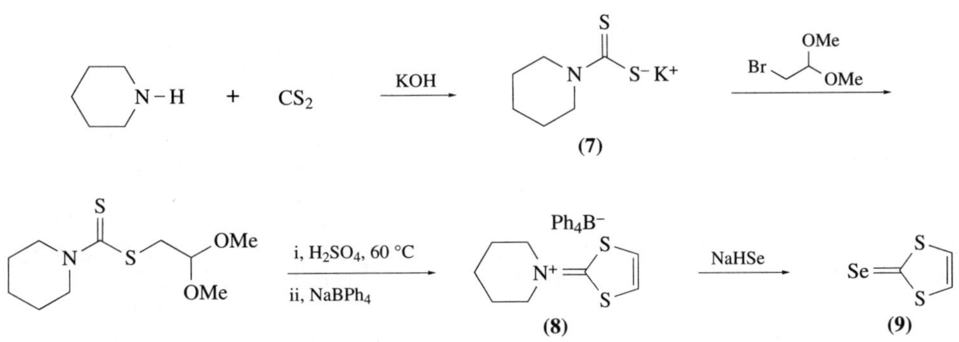

Scheme 1

The previously mentioned phosgeneiminium chloride route to dialkoxy-substituted selenocarbonates can also be used in the preparation of dithio-substituted derivatives. Treatment of butanethiol with Viehe's salt (**4**), followed by addition of sodium hydrogen selenide, afforded the dithioselenocarbonate (**10**) in good yield (Scheme 2). Cyclic dithioselenocarbonate derivatives (**11**) and (**12**) have also been prepared using this method ⟨88AJC549⟩.

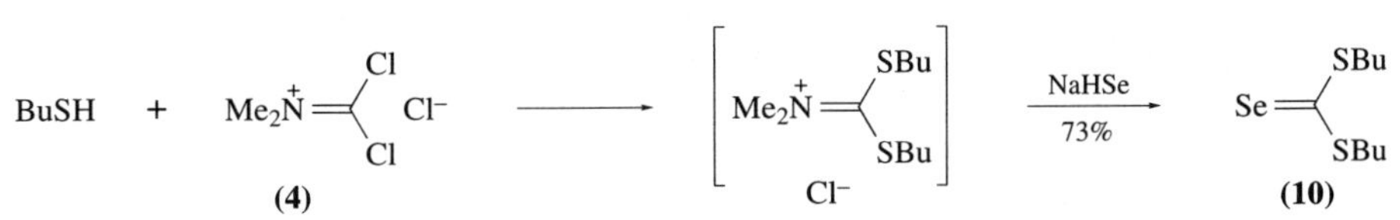

Scheme 2

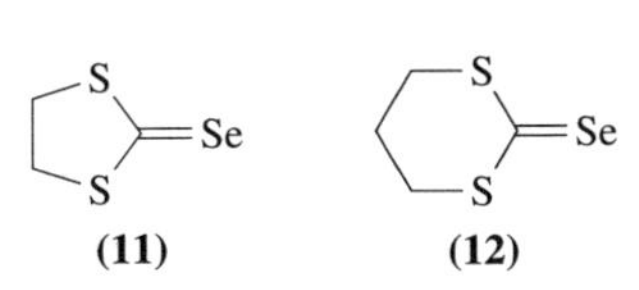

Complex selenocarbonates can be prepared by direct metallation of 4,5-unsubstituted selenocarbonates. For example, consecutive treatment of the cyclic unsaturated dithioselenocarbonate (**9**) with lithium diisopropylamide (LDA), selenium, zinc chloride, and tetrabutylammonium bromide affords a diselenolate salt that can be acylated to the complex selenocarbonate (**13**) (Scheme 3) ⟨92JOM(427)213⟩.

Similar metallation reactions have been reported for other selenocarbonate derivatives.

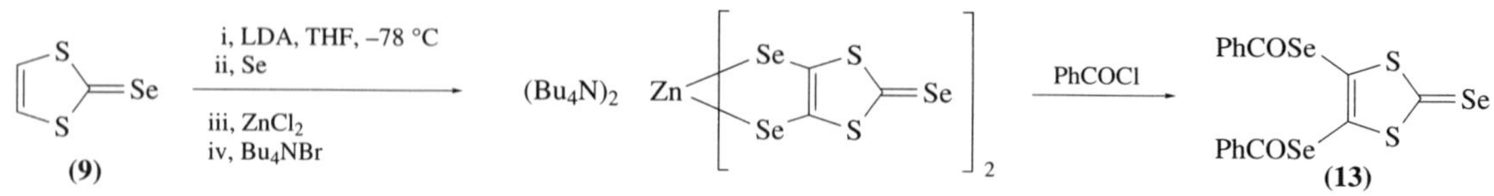

Scheme 3

6.19.3.3 Selenocarbonyl Functions Flanked by Two Selenium Atoms, $(RSe)_2C{=}Se$

Compounds of this type are generally referred to as triselenocarbonates in the literature. A variety of preparations of this class of compounds, particularly the cyclic triselenocarbonates, have been reported owing to their importance as intermediates in the preparation of tetraselenafulvalenes. The parent acid, triselenocarbonic acid (**14**), can be prepared by acidification of barium triselenocarbonate (Equation (8)). It is unstable even at low temperatures, liberating hydrogen selenide and polymeric selenium-containing materials. The chemistry of this and related selenium-substituted carbonic acids has been reviewed ⟨68AG(E)868, B-73MI 619-03⟩.

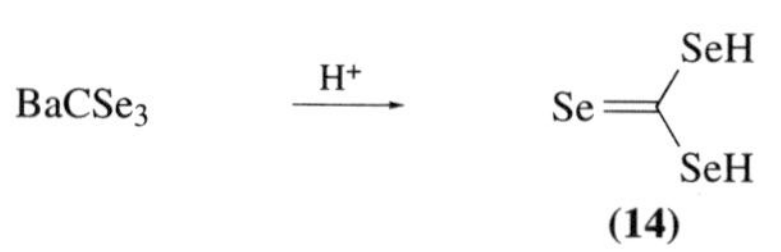

(8)

A number of triselenocarbonate preparations begin with carbon diselenide as a starting material. (It should be noted that carbon diselenide is a remarkably unpleasant reagent and should be used only with great care ⟨B-73MI 619-04⟩.) Reaction of sodium acetylide with carbon diselenide in the presence of selenium, followed by acidification, affords the cyclic unsaturated triselenocarbonate (**15**) (Scheme 4) ⟨77JA5909⟩. Electrochemical reduction of carbon diselenide, followed by alkylation of the intermediate dianion, can be used to prepare complex selenium-substituted triselenocarbonates such as (**16**) (Scheme 5) ⟨76CC148, 83CC235⟩. Heating 1,2,3-benzoselenadiazole in refluxing xylene in the presence of a two-fold excess of carbon diselenide affords benzo-1,3-diselena-2-selone (**17**) in 69% yield (Equation (9)) ⟨83CC295⟩.

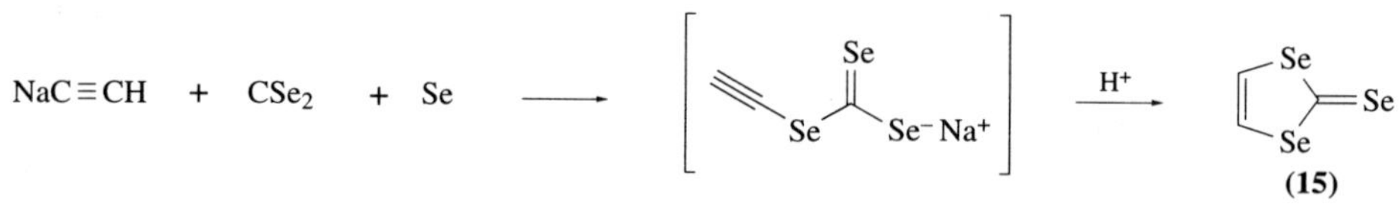

Scheme 4

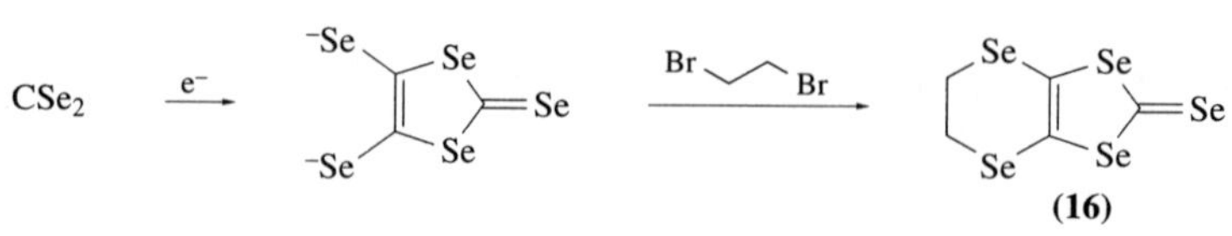

Scheme 5

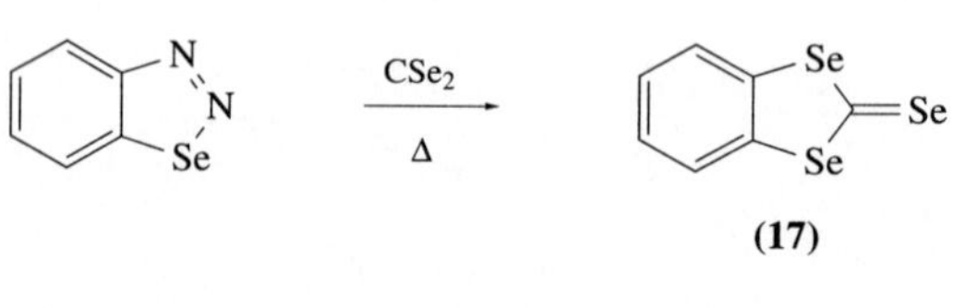

(9)

Treatment of alkyl halides with carbon diselenide and potassium hydroxide in dimethyl sulfoxide affords a variety of both cyclic and acyclic triselenocarbonates (**18**) in moderate yields (Equation (10)) ⟨67ACS1981⟩. It should be noted that the expected direct preparation of acyclic triselenocarbonates via reaction of a selenolate with carbon diselenide, followed by alkylation, fails to afford the desired product (Equation (11)) ⟨B-73MI 619-05⟩.

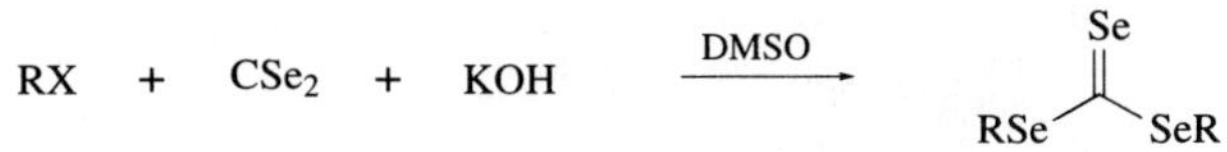

(10)

(18)

$R = -(CH_2)_2-, -(CH_2)_3-, -CH_2Ph, -CH_2CO_2H, -Me$

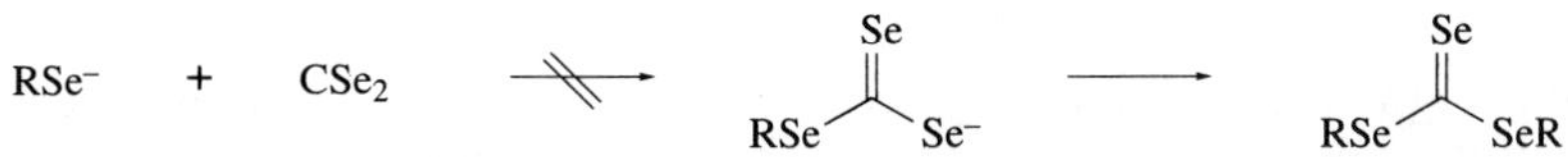

(11)

Reaction of cyclic iminium intermediates such as (**19**) with hydrogen selenide provides a convenient route to substituted unsaturated triselenocarbonates (Scheme 6 ⟨75JOC746⟩, Scheme 7 ⟨84CC89⟩, Scheme 8 ⟨77CC505⟩ and Scheme 9 ⟨80CC866A, 80CC866B⟩). The required iminium salts can be obtained by cyclization of diselenocarbamates. The latter two approaches have the advantage that they avoid the use of carbon diselenide as a source of selenium.

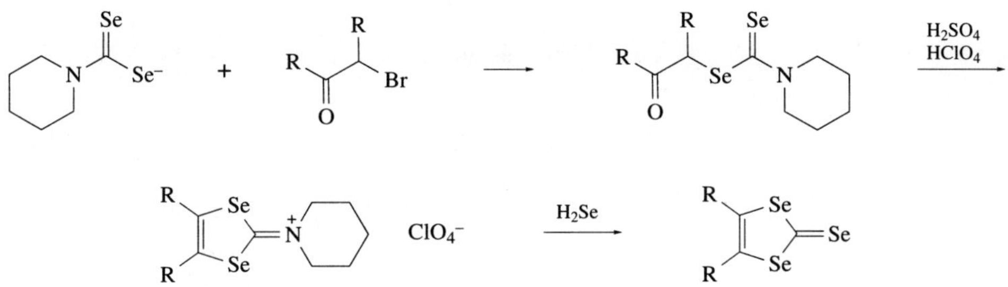

(19)

Scheme 6

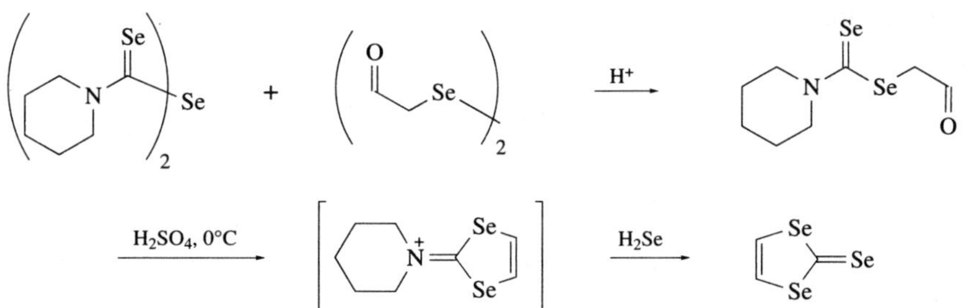

(15)

Scheme 7

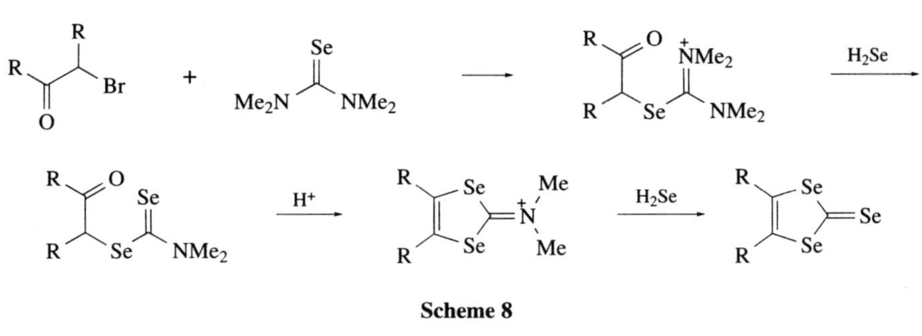

Scheme 8

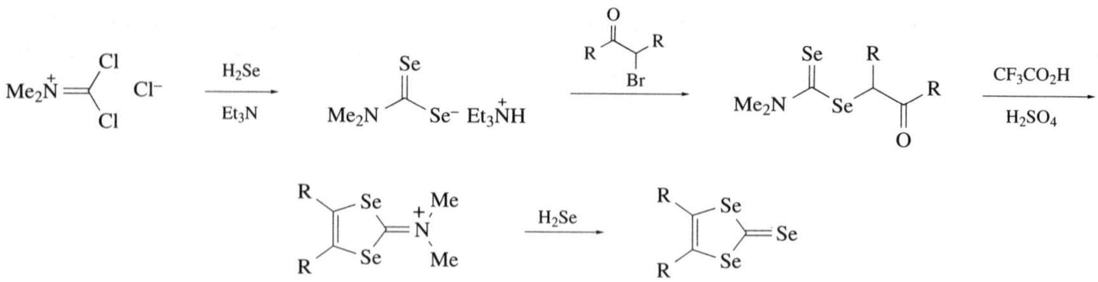

Scheme 9

Thiocarbonyl compounds can similarly be converted to the corresponding selenocarbonyl derivatives by alkylation followed by hydrogen selenide treatment (Scheme 10) ⟨83CC294⟩. The triselenocarbonate moiety has been incorporated into more complex structures by an extrusion–cycloaddition sequence (Equation (12)) ⟨89CC1520⟩. The previously described metallation route to other selenocarbonates (Scheme 3) has also been used for functionalization of cyclic 4,5-unsaturated triselenocarbonates ⟨92JOM(427)213⟩.

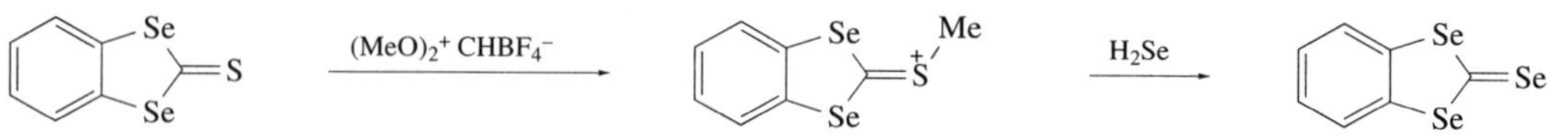

Scheme 10

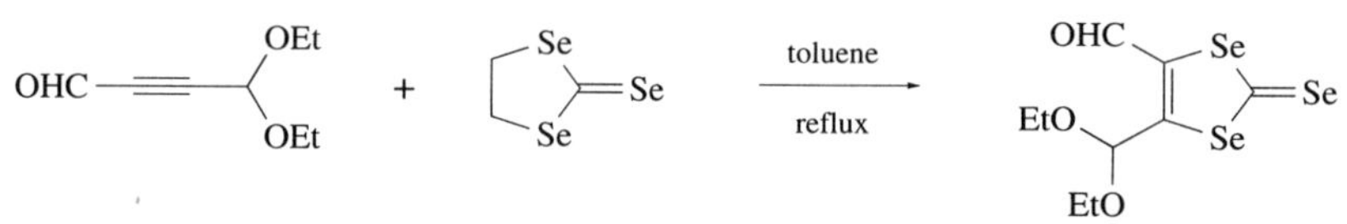

(12)

Bis(trifluoromethyl) triselenocarbonate (**20**) has been prepared from 2,2,4,4-tetrafluoro-1,2-selenetane and $B(SeCF_3)_3$ (Equation (13)) ⟨80ZN(B)526⟩. Another less general triselenocarbonate preparation involves the alkylation of barium triselenocarbonate (Equation (14)) ⟨71CB1429⟩; however, whether the structure of the product obtained is in fact a triselenocarbonate appears to be questionable ⟨B-73MI 619-03⟩.

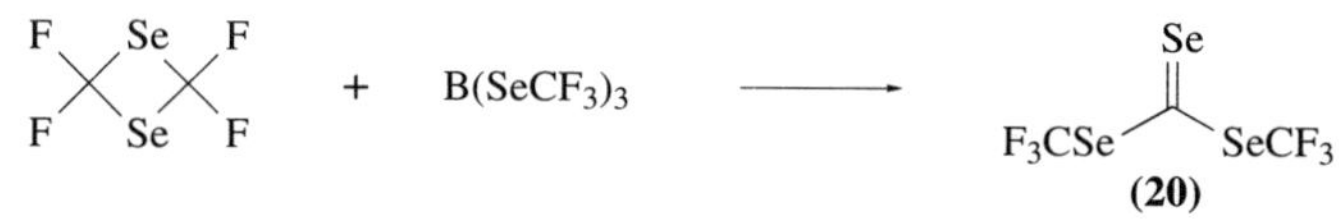

(13)

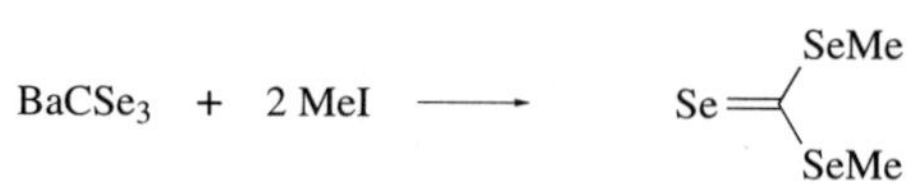

(14)

6.19.3.4 Selenocarbonyl Functions Flanked by Two Different Chalcogen Atoms, RX(C=Se)YR

Compounds of this type are best prepared from intermediate diselenoxanthate salts. For example, treatment of carbon diselenide with alkoxides and alkylation affords *O*-alkyl diselenocarbonates (**21**) (Scheme 11) ⟨70ACS2061⟩. *S*-Alkyl derivatives (**22**) can be similarly prepared using thiolates (Scheme 12) ⟨70ACS2055⟩.

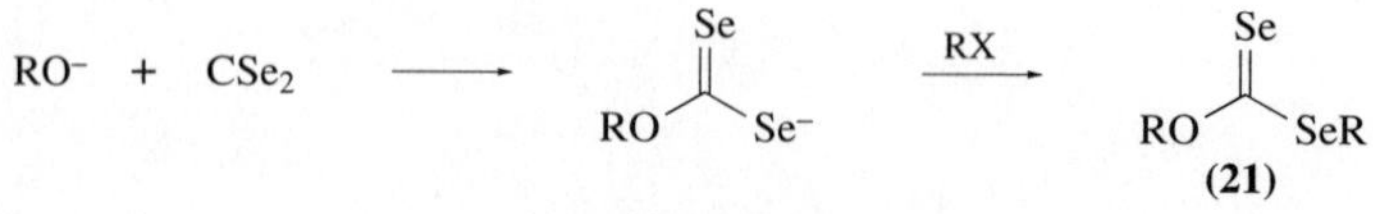

Scheme 11

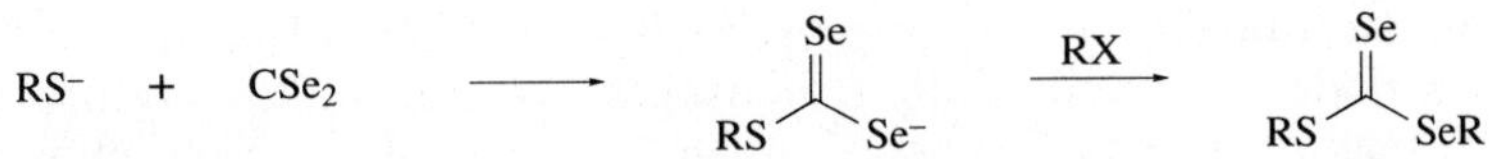

(22)

Scheme 12

The cyclic selenocarbonate (**23**) can be prepared by treatment of 2-hydroxythiophenol with Viehe's salt, followed by addition of sodium hydrogen selenide (Equation (15)). The acyclic derivative (**24**) could be similarly prepared by consecutive treatment of Viehe's salt with phenol and butanethiol, followed by sodium hydrogen selenide treatment (Equation (16)) ⟨88AJC549⟩.

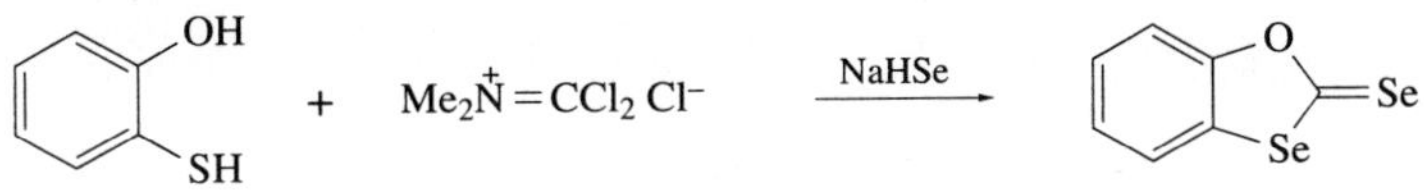

(23) (15)

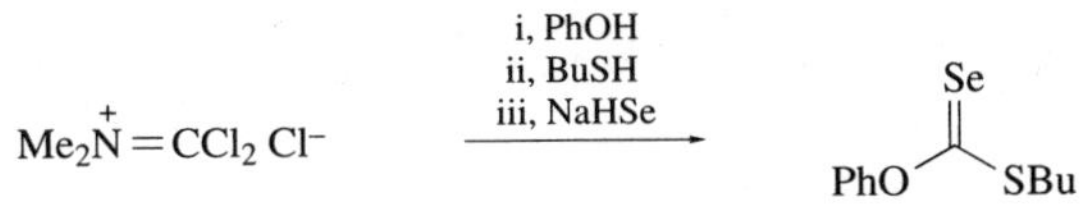

(24) (16)

Heterocyclic organometallic selenocarbonate derivatives such as (**25**) have been prepared by addition of osmium–formaldehyde complexes to carbon diselenide (Equation (17)) ⟨83JOM(244)C53⟩. Mixed selenocarbonates can also be metallated to more complex derivatives as previously described (Scheme 3) ⟨92JOM(427)213⟩.

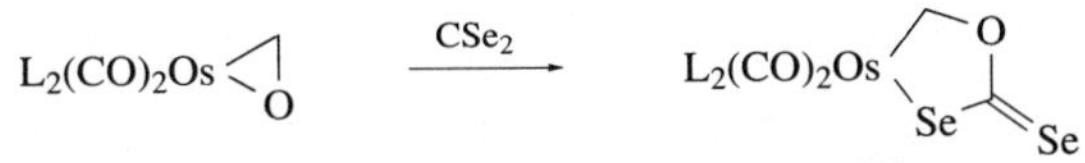

(25) (17)

6.19.3.5 Selenocarbamates RO(C═Se)NHR, RS(C═Se)NHR, and RSe(C═Se)NHR

Monoselenocarbamates can be prepared by a variety of methods. Treatment of an alcohol with an isoselenocyanate affords the corresponding selenourethane (**26**) in good yield (Equation (18)) ⟨72BCJ2937⟩. Addition of hydrogen selenide to an alkyl cyanate affords primary monoselenourethanes (**27**) (Equation (19)) ⟨66ACS2091⟩. These compounds can also be prepared by ammonolysis or hydrazinolysis of diselenocarbonates (Equations (20), (21)) ⟨64ACS2417, 70ACS2061, 72ANY(192)101⟩.

(26) (18)

(27) (19)

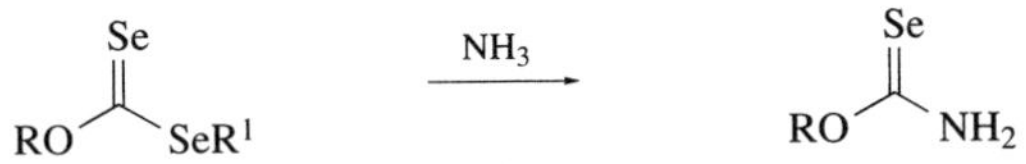

(20)

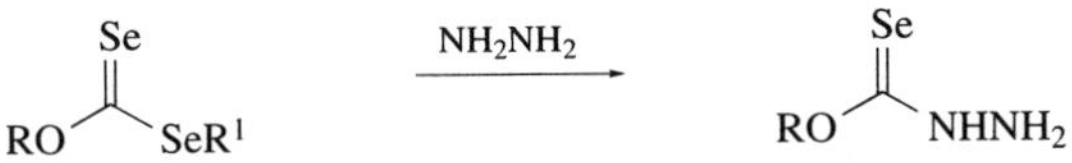

(21)

An aryl selenocarbamate (**6**) was the unexpected major product in the previously described attempt to prepare a cyclic dioxygen-substituted selenocarbonate (Equation (7)).

Selenothiocarbamates (**28**) can readily be prepared by alkylation of thioureas, followed by treatment of the intermediate salts with sodium hydrogen selenide under slightly acidic conditions (Scheme 13) ⟨69JOC3549⟩. Under basic conditions the corresponding selenoureas are obtained (see below, Equation (28)).

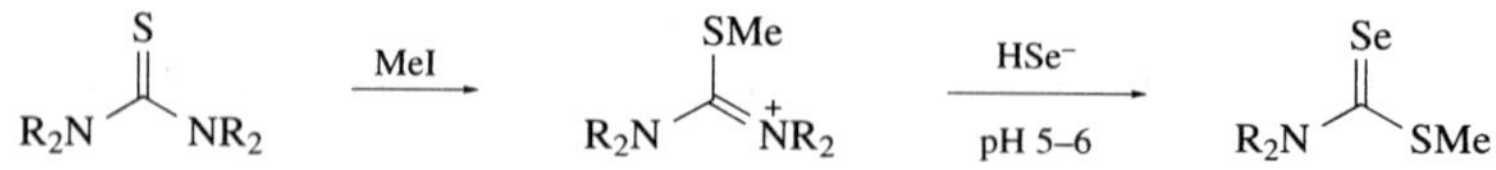

Scheme 13

Reactions of carbon diselenide with secondary amines afford salts of diselenocarbamic acids (**29**) (Equation (22)) ⟨67ACS2904, 68ZC230, 70ACS351, 72IJS(A)133⟩. Related salts can be prepared by the reaction of phosgeneiminium chloride with hydrogen selenide in the presence of triethylamine (see above, Scheme 9). This route avoids the use of carbon diselenide ⟨80CC866A, 80CC866B, 86BCJ1741⟩. Such salts can readily be alkylated to the corresponding diselenocarbamates (**30**) (Equation (23)) ⟨61JCS2922, 86BCJ1741⟩.

$$2\,R_2NH + CSe_2 \longrightarrow R_2NCSe_2^- \; R_2NH_2^+ \quad \textbf{(29)} \qquad (22)$$

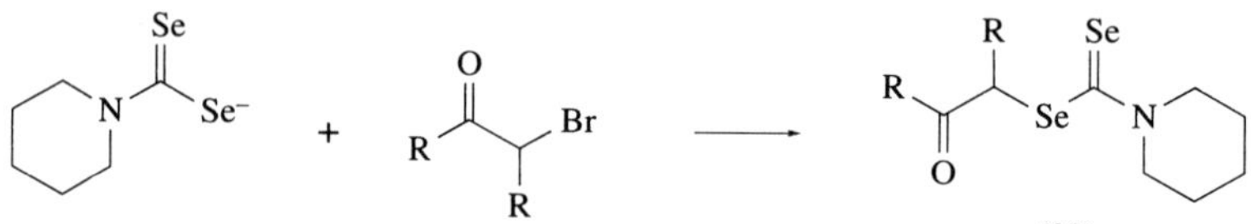

(23)

Salts of dialkyl selenocarbamates under mild oxidative conditions afford mixtures of bis(selenocarbamoyl) selenides (**32**) and the corresponding triselenides (**33**) (Scheme 14) ⟨61JCS2922, 70ACS959, 82S771⟩. This reaction proceeds through an intermediate unstable diselenide (**31**) that is in equilibrium with these species. The monoselenide (**32**) can be converted to the triselenide (**33**) upon treatment with elemental selenium. Conversely, treatment of the triselenide with triethyl phosphite affords the monoselenide (Equation (24)) ⟨82S771⟩.

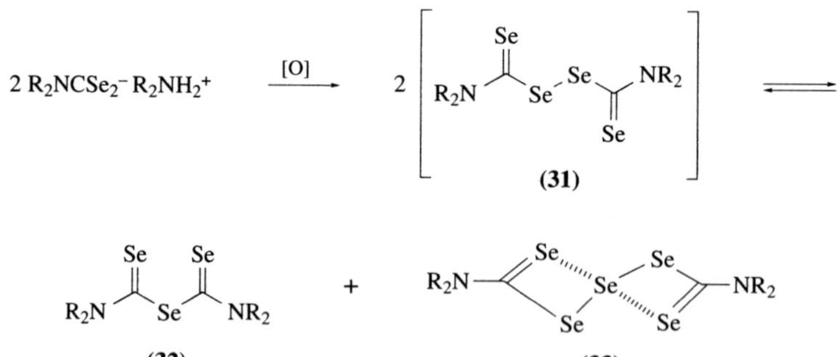

Scheme 14

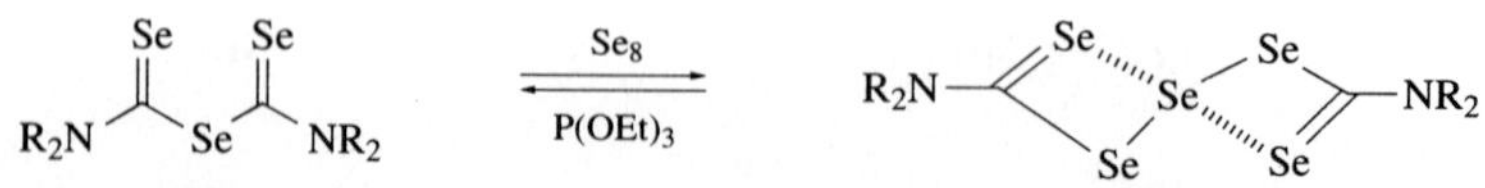

(24)

Hydrazines react with carbon diselenide to afford diselenocarbazate salts (**34**) (Equation (25)) ⟨66ACS2742⟩. Nickel complexes of these compounds can be prepared but appear to have eneselenolate structures such as (**35**) ⟨66ACS2742⟩. The free diselenocarbazic acids (**36**), which can be prepared by acidification of the simple salts (Scheme 15), are reported in some cases to be more stable than the corresponding dithiocarbazic acids ⟨70ACS959⟩.

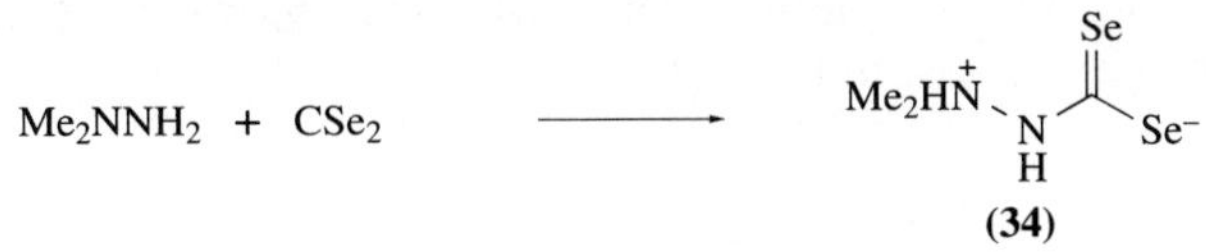

(25)

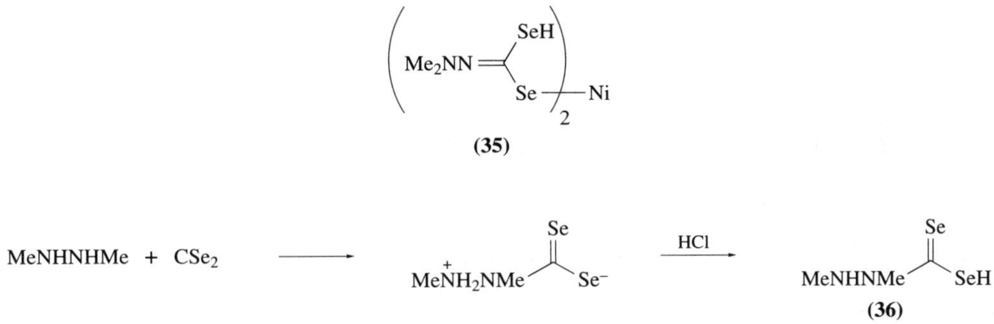

Scheme 15

Heterocyclic selenocarbamate derivatives—oxazolidine-2-selones (**37**)—have been prepared by a variety of routes. Mercuric chloride-promoted reaction of 1,2-amino alcohols with carbon diselenide readily affords this class of compounds in moderate yields (Equation (26)) ⟨80JHC571, 89H(28)269⟩. Another route to these compounds that avoids the problems associated with the use of carbon diselenide is outlined in Scheme 16 ⟨91JCS(P1)2495⟩. This route allows a large-scale preparation of chiral oxazolidineselones such as (**38**), useful as NMR shift reagents. A more direct route to these compounds involves metallation and direct selenation of the corresponding readily available oxazolines (**39**) (Equation (27)) ⟨91JOC6733⟩.

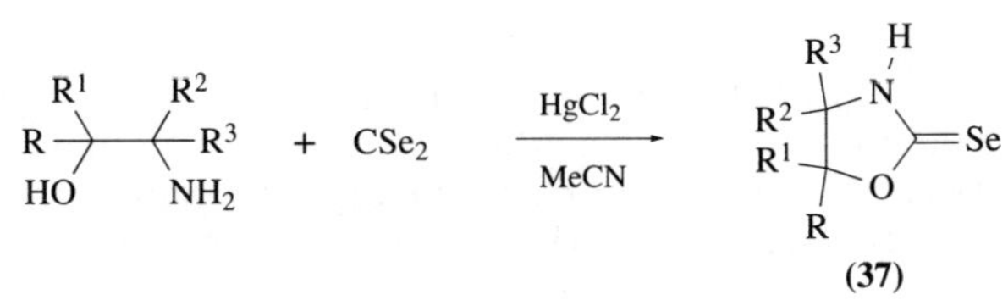

(26)

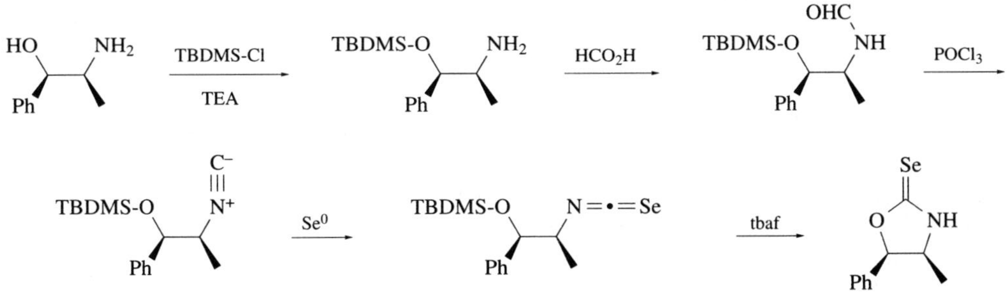

TBDMS = *t*-butyldimethylsilyl

Scheme 16

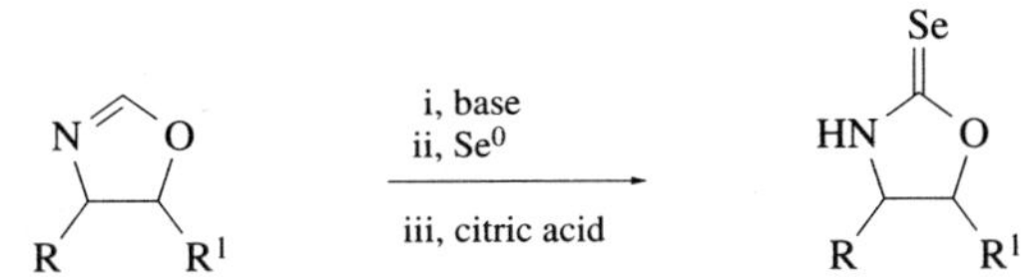

(27)

6.19.3.6 Phosphorus-substituted Selenocarbonyl Derivatives

Reaction of trialkylphosphines with carbon diselenide affords moderately stable zwitterionic selenocarbonyl compounds (**40**) (Equation (28)) ⟨63ACS549, 70QRS45⟩. Complex reaction mixtures were obtained using triphenylphosphine.

R=Me, Et, Pr (**40**) (28)

6.19.4 FUNCTIONS CONTAINING AT LEAST ONE NITROGEN FUNCTION (AND NO HALOGEN OR CHALCOGEN FUNCTIONS)

6.19.4.1 Selenoureas $(R_2N)_2C{=}Se$

Selenoureas are probably the most widely studied selenocarbonyl derivatives. The early chemistry associated with the preparation and reactions of this class of compounds has been reviewed ⟨B-73MI 619-06⟩.

Selenoureas can be prepared using a variety of general methods. Thioureas can be converted to selenoureas (**41**) via an alkylation–displacement sequence, provided the pH is kept in the range 8–9 (Scheme 17) ⟨63BSB149, 69JOC3549⟩. Reaction at a more acidic pH affords the selenocarbamate instead (see Scheme 13). This route provides a convenient method for the preparation of both tri- and tetrasubstituted selenoureas.

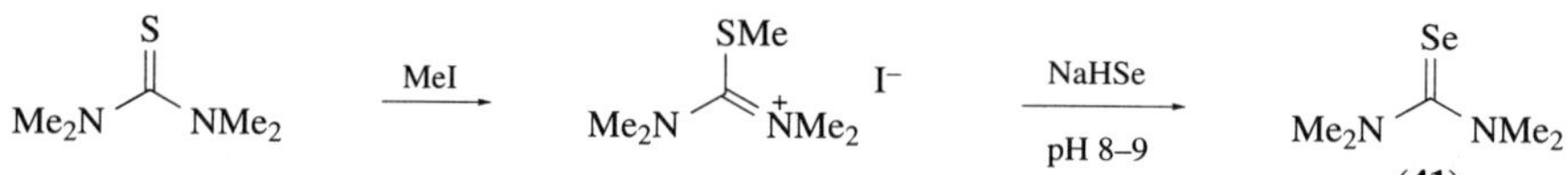

(**41**)

Scheme 17

Reaction of isoselenocyanates with ammonia, primary amines, or secondary amines provides another general route to mono-, di-, or trisubstituted selenoureas (Equation (29)) ⟨59M41, 63JOC1642, 63ZC348, 67CB1373, 67CB1459, 80S60⟩. Reaction of a primary ammonium salt with selenocyanate ion also affords the monosubstituted selenourea (Equation (30)) ⟨59G693⟩.

(29)

(30)

Carbodiimides react with hydrogen selenide to afford 1,3-disubstituted selenoureas (Equation (31)) ⟨53JOC292⟩. Cyanamides also react with H_2Se to afford unsubstituted, mono- or 1,1-disubstituted selenoureas (Equation (32)) ⟨47JA1833, 51JA1864, 53JOC292, 63OSC(4)359⟩.

(31)

(32)

Cyanamide and cyanamide salts are reported to react with phosphorus pentaselenide in aqueous solution, presumably via formation of H_2Se, to afford selenourea (Equation (33)) ⟨91EGP291752⟩. Phosphorus pentaselenide is reported to convert urea into selenourea in only poor yield (Equation

(34)) ⟨66ACS281⟩. *N*,*N*-Dimethylcyanamide is also reported to react with another selenating reagent, bis(trimethylsilyl) selenide, to afford 1,1-dimethylselenourea in low yield (Equation (35)) ⟨90CL1403⟩.

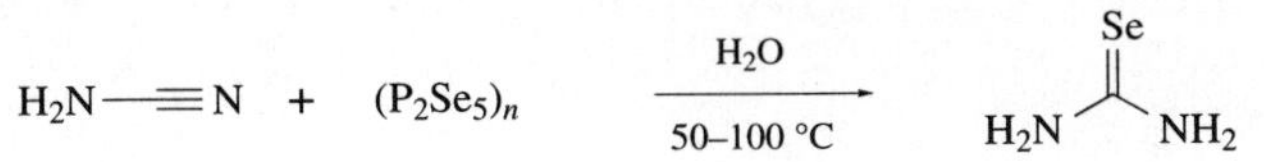
(33)

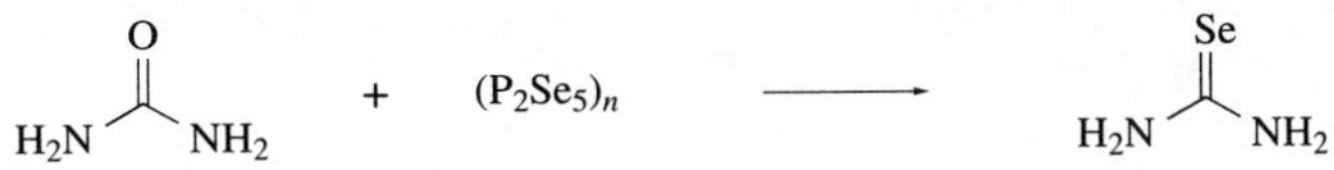
(34)

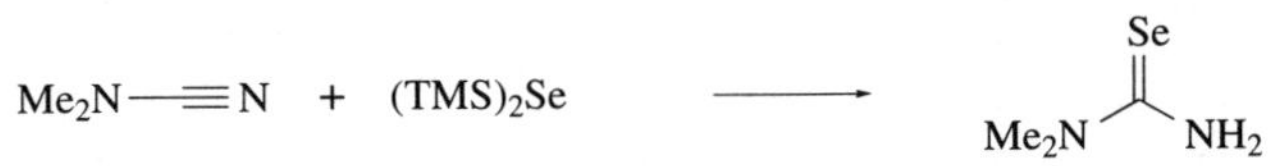
(35)

Treatment of dibenzyl triselenocarbonate (**42**) with excess primary amine affords symmetrical 1,3-disubstituted selenoureas, presumably via an isoselenocyanate intermediate (**43**) (Scheme 18) ⟨72ANY(192)101⟩. Treatment of carbon diselenide with excess primary amine also affords a 1,3-disubstituted selenourea ⟨36CB1356⟩. This reaction also probably involves an isoselenocyanate intermediate (Scheme 19).

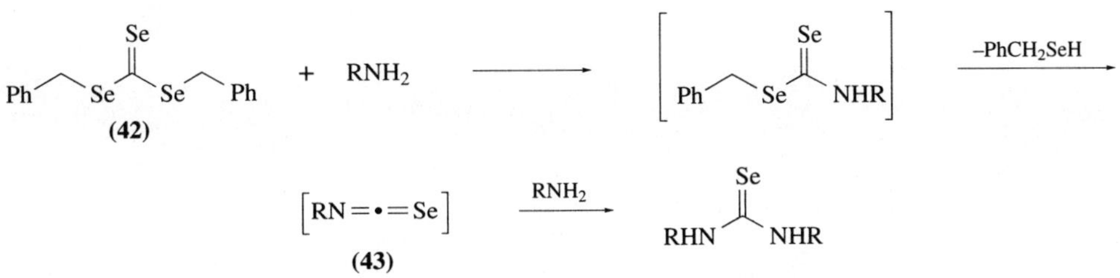

Scheme 18

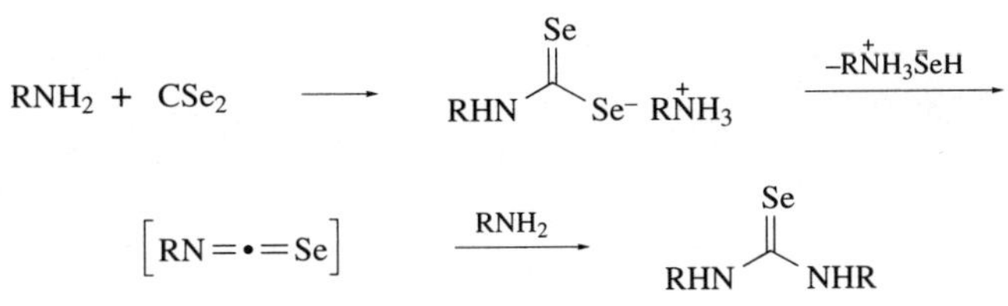

Scheme 19

Tetrasubstituted selenoureas can be prepared from selenocarbamoyl selenides or triselenides (**44**) (see Scheme 14) by treatment with secondary amines at elevated temperature in the presence of oxygen (Equation (36)) ⟨82S771⟩.

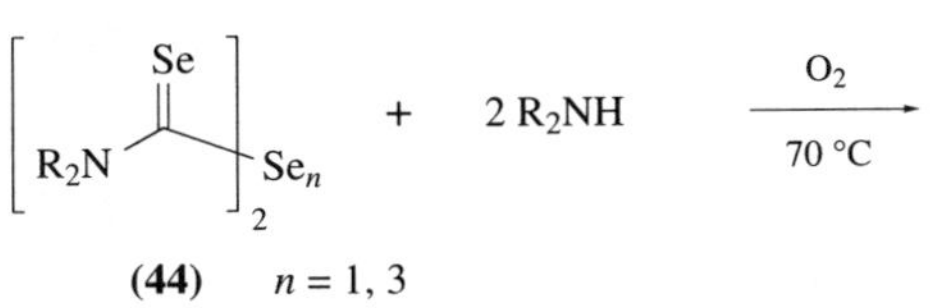
(36)

(**44**) $n = 1, 3$

Selenobiurets (**45**) can be prepared from the corresponding thiocarbonyl analogs by an alkylation–selenation procedure similar to that used for the preparation of simple selenoureas (Equation (37)).

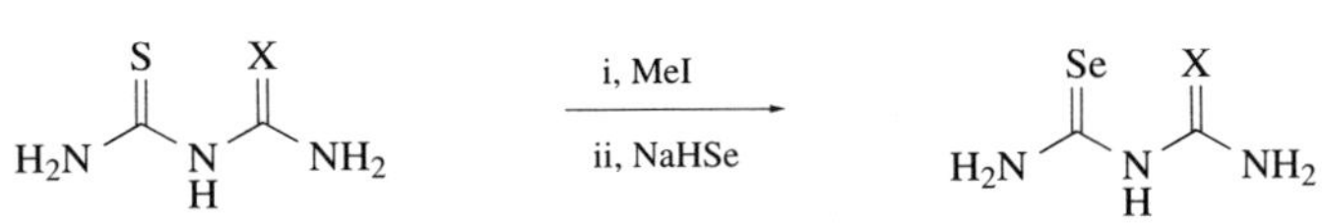
(37)

(**45**) X = O, S, Se

Heterocyclic selenoureas such as selenopyrimidines (**46**) and selenobarbiturates (**47**) have been prepared from selenourea and β-dicarbonyl compounds (Equations (38) and (39)) ⟨56JA5292, 59JA6270⟩.

(46) (38)

(47) (39)

Metallation of *N*-methylimidazole at low temperature followed by selenation with elemental selenium affords the selenium analogue (**48**) of the antithyroid drug MMI (Equation (40)) ⟨94JOC4691⟩.

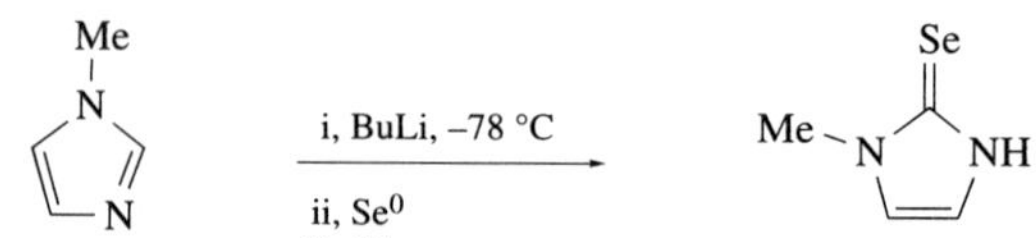

(48) (40)

Selenosemicarbazides (**49**) can be prepared by reaction of an isoselenocyanate with a hydrazine (Equation (41)) ⟨62BSB541, 66ACS278, 67CB1373⟩. These compounds can be converted to the corresponding selenosemicarbazones ⟨56JA97, 62BSB541, 66ACS278, 67CB1373⟩.

(49) (41)

Other biologically interesting complex selenocarbonyl derivatives have also been prepared via displacements using selenourea. These include the selenium analogue of the benzodiazepin drug "TIBO" (**50**) (Equation (42)) ⟨91JMC3187⟩ and selenoguanosine derivatives (**51**) and (**52**) (Scheme 20) ⟨91TL4823⟩. The latter compound is formed via a novel polyhetero-Claisen rearrangement. Finally, tris(methylphenylamino)methane (**53**) reacts with elemental selenium to afford the corresponding selenourea in good yield (Equation (43)) ⟨87CB1581⟩.

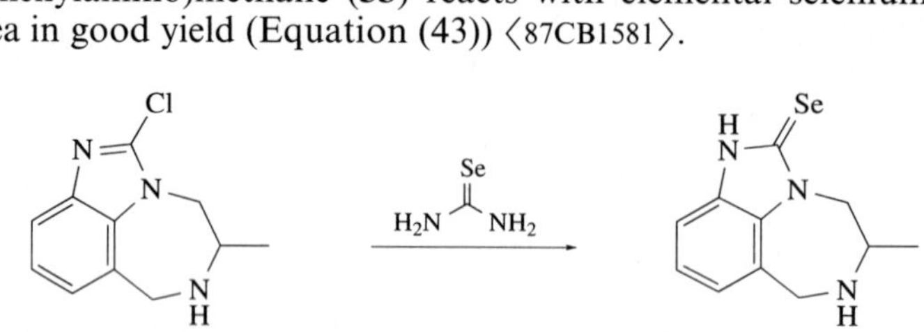

(50) (42)

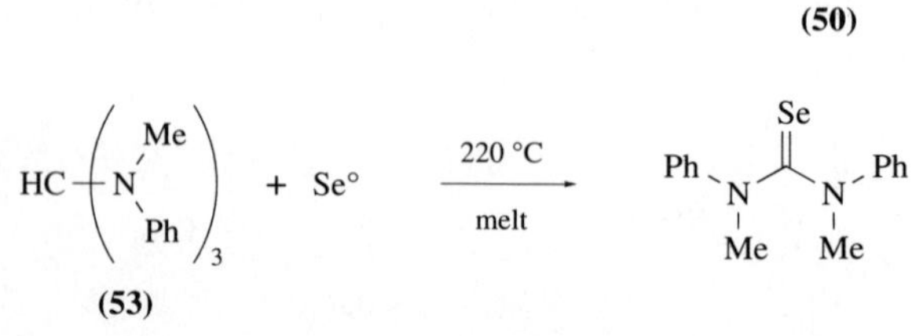

(53) (43)

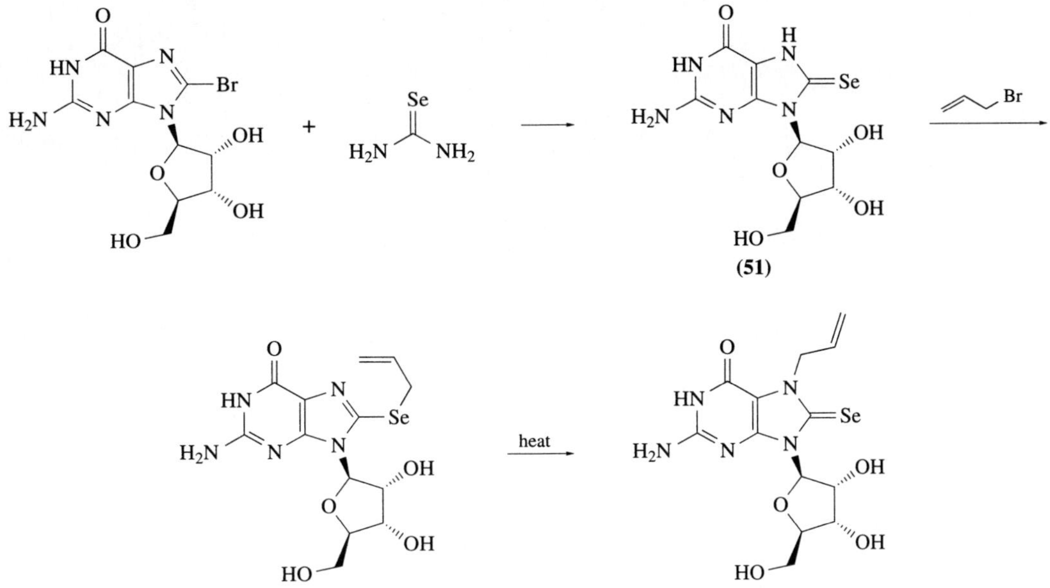

(51) (52)

Scheme 20

6.19.4.2 Telluroureas $(R_2N)_2C{=}Te$

Cyclic telluroureas (**55**) have been prepared by tellurium-promoted cleavage of the tetra-aminoethylene derivatives such as (**54**) (Equation (44)) ⟨80CC635⟩. A variety of metal penta-carbonyl complexes of these compounds (**57**) have also been prepared by treatment of the telluroureas with the metal isonitrile pentacarbonyl complex (**56**) (Equation (45)) ⟨80CC635⟩.

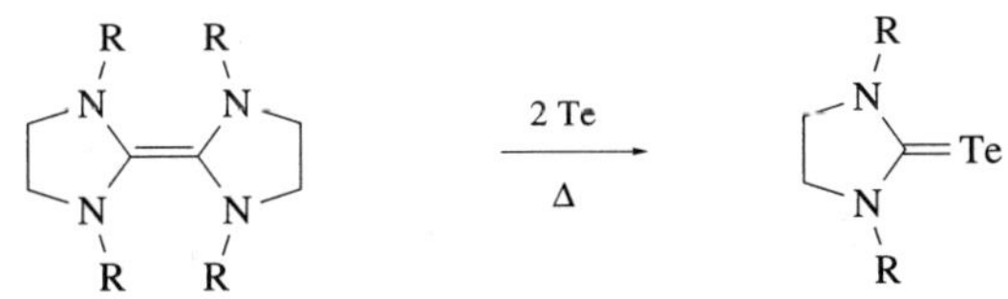

(54) R = Me, Et (55) (44)

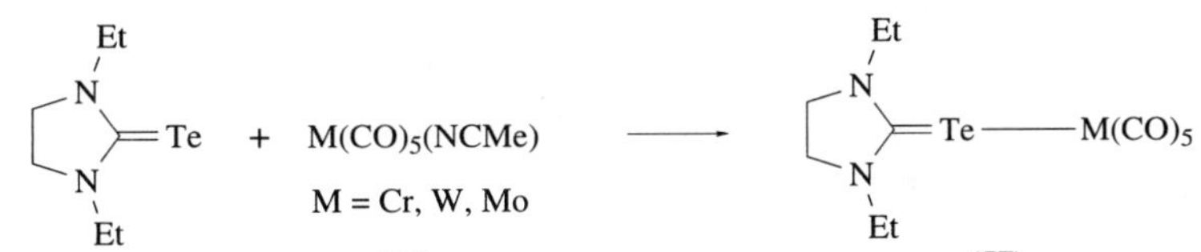

(56) (57) (45)

Tellurium is also reported to react with 1,3-dimethylbenzimidazoline at elevated temperature to afford the corresponding cyclic tellurourea (**58**) (Equation (46)) ⟨69ZOB941⟩.

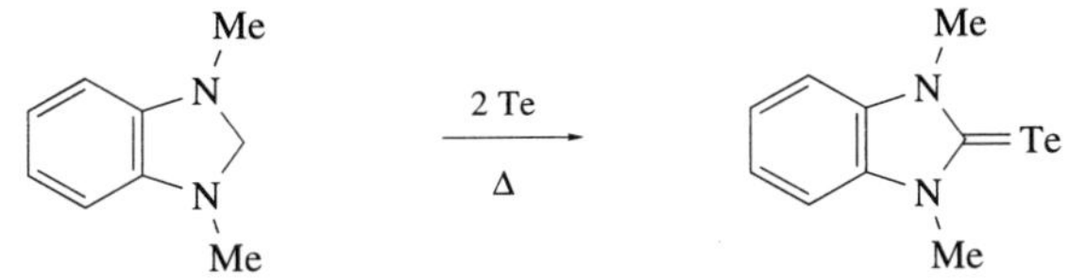

(58) (46)

6.20
Functions Containing an Iminocarbonyl Group and at Least One Halogen; Also One Chalcogen and No Halogen

THOMAS L. GILCHRIST

University of Liverpool, UK

6.20.1 FUNCTIONS CONTAINING AT LEAST ONE HALOGEN

6.20.1.1 Iminocarbonyl Halides with Two Similar Halogen Functions

In the mid-1990s these iminocarbonyl halide compounds XYC=NR (X, Y = halogen) are named in *Chemical Abstracts* as carbonimidic dihalides; they are also commonly called isocyanide dihalides. Methods of synthesis of compounds of this type have been reviewed previously. The most important of these reviews are by Ulrich ⟨B-68MI 620-01⟩, by Ulrich and Richter ⟨B-77MI 620-01⟩, by Kühle *et al.* ⟨67AG(E)649, B-71MI 620-01⟩ and by Kühle ⟨83HOU(E4)522⟩. The reviews by Ulrich and those by Kühle and colleagues all contain experimental details for the preparation of some compounds of this class. Several specific compounds containing fluorine are covered in appropriate sections of the *Gmelin Handbooks* on perfluorohaloorganic compounds ⟨B-79MI 620-01, B-80MI 620-01, B-81MI 620-01⟩.

6.20.1.1.1 *Carbonimidic difluorides, $F_2C{=}NR$*

Carbonimidic difluoride compounds have been reported with R = H, D, alkyl, perhaloalkyl, aryl, hetaryl, acyl- and with a variety of heteroatomic substituents on nitrogen. No methods exist which can be used to prepare all these types of compound but there are methods which have some generality, these are listed below. Methods which are specific to derivatives of one type are described for the compounds concerned.

(i) From trifluoromethylamines, F_3CNHR

Trifluoromethylamines serve as precursors to carbonimidic difluorides of various types, including the parent compound (R = H). Conditions and reagents used to bring about the dehydrofluorination vary depending upon the substituent, but low temperatures are desirable because several types of carbonimidic difluoride dimerise easily at ambient temperatures. Triethylamine and potassium fluoride have been used most often as co-reagents.

Examples of carbonimidic difluorides which have been prepared by this method are listed in Table 1.

Table 1 Carbonimidic difluorides formed by dehydrofluorination of trifluoramines, F_3CNHR.

R	*Reagents and conditions*	*Yield (%)*	*Ref.*
H	NEt_3, 195–263 K		88CC105
CF_3	NaF, 140°C	70	59JGU2662
Ph	KF, 140°C		59JGU2662 B-71MI 620-01
C_6H_4Cl-4	NEt_3, MeCOCl, 40°C	38	B-71MI 620-01
OMe, OBu^t, OCF_3	KF, −196–24°C		81JFC(18)441
OCF_2CF_3	KF, −196–24°C	92	90EUP353743

(ii) From carbonimidic dichlorides

Some carbonimidic dichlorides have been converted into the corresponding difluorides. Trifluoroamines can be intermediates, so that in such examples the method is an extension of method (i). This is illustrated by the reaction of *N*-arylcarbonimidic dichlorides with hydrogen fluoride to give *N*-aryltrifluoroamines which are then dehydrofluorinated by heating with potassium fluoride (Scheme 1) ⟨66AG(E)848, B-71MI 620-01⟩. Aromatic carbonimidic dichlorides can also be converted into the difluorides by heating with an alkali metal fluoride in a high boiling inert solvent; for example, *N*-pentachlorophenylcarbonimidic difluoride is obtained from the dichloride and potassium fluoride in 68% yield ⟨B-71MI 620-01⟩. Related preparations of compounds $F_2C{=}NXF_5$ (X = S or Te) have been described starting from the corresponding carbonimidic dichlorides (Scheme

2) (X = S: ⟨76IC14, 88IC570⟩, X = Te: ⟨85IC4171⟩) and the azine $F_2C{=}NN{=}CF_2$ has been prepared by heating the corresponding tetrabromide with silver fluoride ⟨62JOC2589⟩. $F_2C{=}NCF_3$ is prepared in good yield by heating $Cl_2C{=}NCCl_3$ with an excess of sodium fluoride in sulfolane ⟨72GEP2101107⟩.

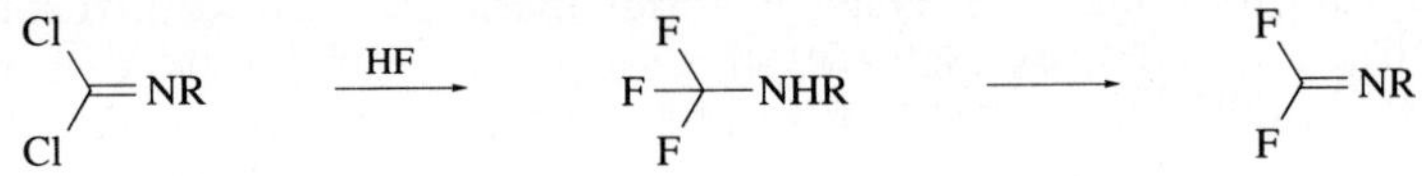

Scheme 1

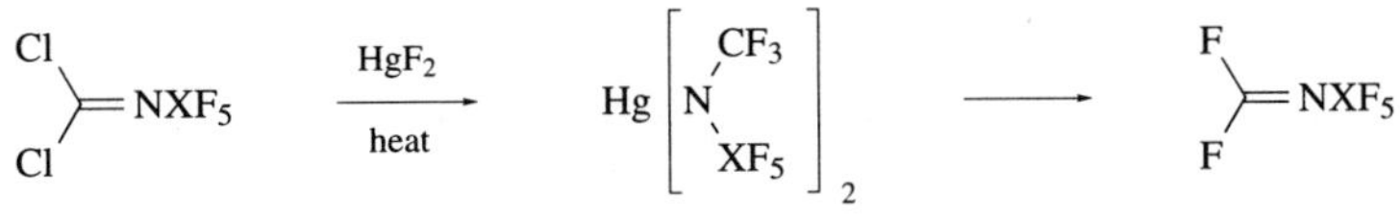

Scheme 2

(iii) From isothiocyanates

The conversion of several aryl isothiocyanates into carbonimidic difluorides has been achieved by reaction with mercury(II) fluoride and other metal fluorides ⟨65JA4338⟩. *N*-Ethylcarbonimidic difluoride was also prepared (37%) by this method.

(iv) From isocyanides

The reaction of isocyanides RNC (R = Bu^t, Ph, 2-ClC_6H_4, and 2-FC_6H_4) with fluorine at −78 °C leads to the formation of the corresponding carbonimidic difluorides ⟨80TL4893⟩.

(v) Dehalogenation methods

The best methods for the preparation of the *N*-halocarbonimidic difluorides start from nitriles. An efficient synthesis of the *N*-chloro compound is shown in Scheme 3 ⟨83JOC4844⟩ and the *N*-fluoro analogue can be prepared in a related way by dechlorination of $ClCF_2NClF$ with mercury in trifluoroacetic anhydride ⟨81JOC1277, 84IC2188⟩. Several *N*-perfluoroalkyl compounds, including $F_2C{=}NCF_3$ and $F_2C{=}NCF_2CF_2N{=}CF_2$, have been made by a related photochemical dechlorination method ⟨90IC571⟩. *N*-Bromocarbonimidic difluoride is prepared from cyanogen fluoride, bromine and potassium fluoride ⟨84JA4266, 88JOC4443⟩. The reaction of $(CF_3)_2NX$ (X = Br or Cl) with iron pentacarbonyl leads to the formation of $F_2C{=}NCF_3$ in high yield ⟨65JCS5774⟩.

A related reductive method has provided a route to carbonimidic difluorides (**1**) (Equation (1)). The reaction of chlorodifluoronitrosomethane with diethyl phosphite gave (**1**) (42%) ⟨69JGU1694⟩.

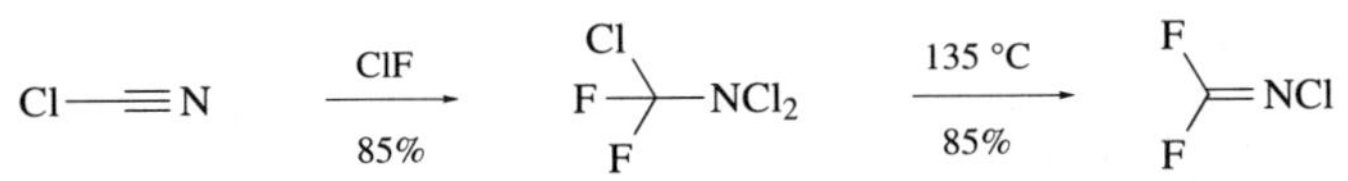

Scheme 3

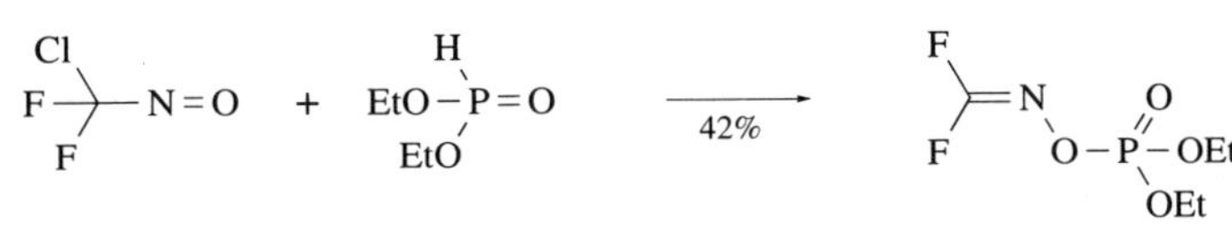 (1)

(**1**)

(vi) Specific methods

Some methods which are useful for specific derivatives or for specific groups of compounds are discussed under this heading.

The hydrolysis of trifluoromethyl isocyanate in triethylamine has been used to prepare carbonimidic difluoride; $F_2C{=}ND$ has also been made in this way ⟨92ZN(B)351⟩. Among the many methods which can be used to prepare *N*-trifluoromethylcarbonimidic difluoride ⟨B-79MI 620-01⟩, several involve the vapour phase pyrolysis of copolymers of trifluoronitrosomethane and perhaloalkenes ⟨55JCS1881, 64JCS1351, 65JCS6209, 91MI 620-01⟩. The compound can also be formed by decomposition of $(CF_3)_2NOSN$ ⟨86JFC(34)46⟩. The vapour phase pyrolysis of oxazetidines (**2**); $R = CF_3$, C_2F_5 and C_3F_7) also provides a route to the corresponding perfluoroalkylcarbonimidic difluorides (Equation (2)) ⟨56JCS3416, 64JCS1351, 64JCS4066⟩. The imine $(CF_3)_2C{=}NCF_3$ is isomerised to $F_2C{=}NCF(CF_3)_2$ by heating with caesium fluoride as a catalyst ⟨91JFC(53)181⟩.

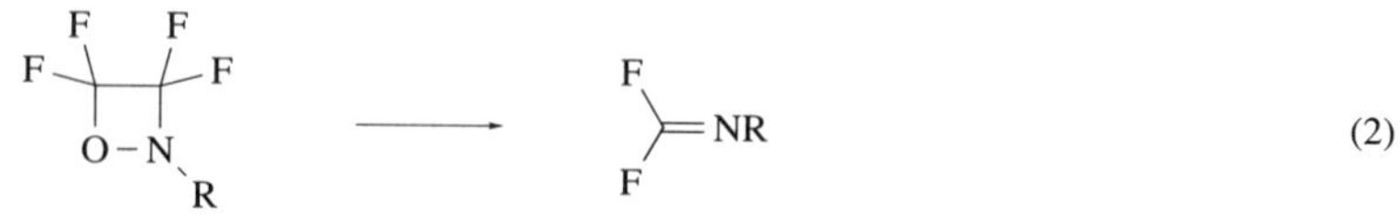

(2) (2)

The pyrolysis of difluorodiazirine provides a route to $F_2C{=}NN{=}CF_2$ ⟨68JHC41, 84USP4446068⟩. Perhaloalkenes can be inserted into the N—N bond of this azine by irradiation of the mixture to give new carbonimidic difluorides (for example, the compound $F_2C{=}NCF_2CF_2N{=}CF_2$ from tetrafluoroethylene) ⟨67JA3868⟩. The hydrazone (**3**) has been isolated from the photolysis of the azoalkane (**4**); the reaction sequence shown in Scheme 4 can account for its formation ⟨90JOC1503⟩.

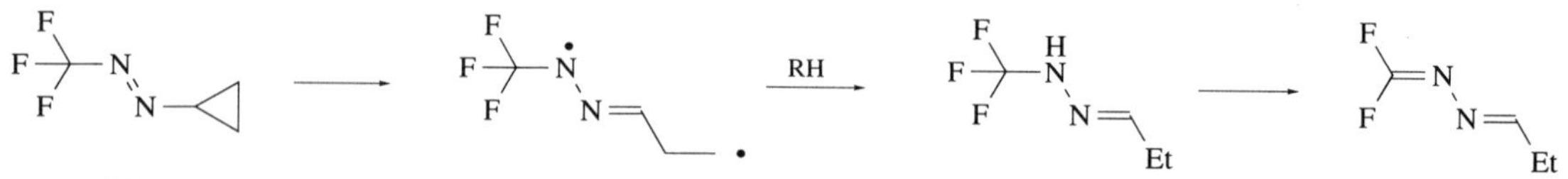

(4) **(3)**

Scheme 4

6.20.1.1.2 *Carbonimidic dichlorides, $Cl_2C{=}NR$*

The dichlorides are by far the most numerous of the carbonimidic dihalides; a few compounds of this class have been identified as natural products ⟨77JA7367, 78TL1391, 78TL1395⟩ and there are several well-established methods of synthesis of dichlorides bearing a wide variety of functional groups on nitrogen. The review by Kühle ⟨83HOU(E4)522⟩ provides a valuable survey of the literature up to 1982; this and the earlier review by Kühle *et al.* ⟨B-71MI 620-01⟩ contain experimental details of preparations and tables of examples which are not readily accessible from other sources. The major methods of preparation are detailed below and, where appropriate, examples of their use are quoted, particularly when the original paper gives experimental details.

(i) From isothiocyanates

The oldest method of preparation is from isothiocyanates: it is known for alkyl-, aryl-, acyl-, phosphoryl- and sulfonyl-substituted carbonimidic dichlorides. The chlorination of alkyl and acyl isothiocyanates is usually carried out with chlorine in an inert solvent and the products can sometimes be isolated in good yield, for example R = Me, 85% ⟨59JGU2131⟩; R = Ph, 96% ⟨B-71MI 620-01⟩; R = *β*-tetraacetylglucosyl, 92% ⟨87AG(E)359⟩. The method is not completely general: the products can be unstable (as with 2-nitrophenyl isothiocyanate) or they can undergo further reaction with chlorine (e.g., reaction of phenyl isothiocyanate with an excess of chlorine in chloroform results in further chlorination at the 4-position of the ring ⟨64JOC1613⟩). The course of the reaction of chlorine with isothiocyanates is shown in Scheme 5, the intermediate sulfenyl chlorides (**5**) having been isolated in a few cases.

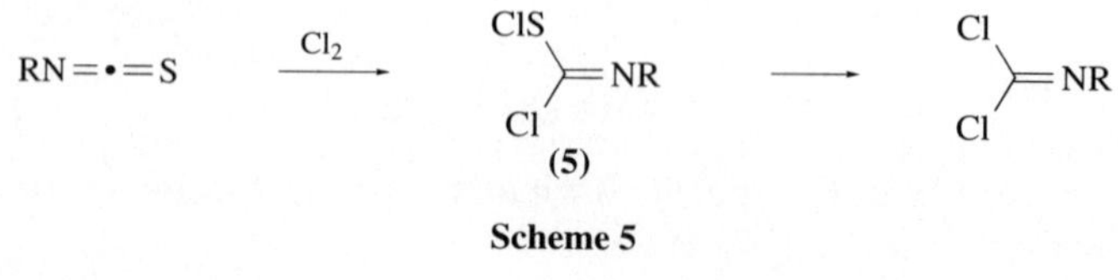

(5)

Scheme 5

Acyl isothiocyanates can be chlorinated in much the same way, with chlorine in an inert solvent ⟨B-71MI 620-01, 75JCS(P2)1046⟩. A Lewis acid catalyst facilitates the reaction and both aluminium(III) chloride and titanium(IV) chloride have been used. An example of the effect of a catalyst is shown in Scheme 6. Chlorination of carbonyl diisothiocyanate (**6**) in the absence of a catalyst gave the carbonimidic dichloride (**7**) but in the presence of iodine both functional groups were chlorinated to give (**8**) ⟨82CB3587⟩. The thiocyanato group of carbonyl isocyanate isothiothiocyanate (**9**) has also been selectively chlorinated to give compound (**10**) in good yield; the isocyanate function can then be selectively hydrolysed ⟨81CB2064⟩.

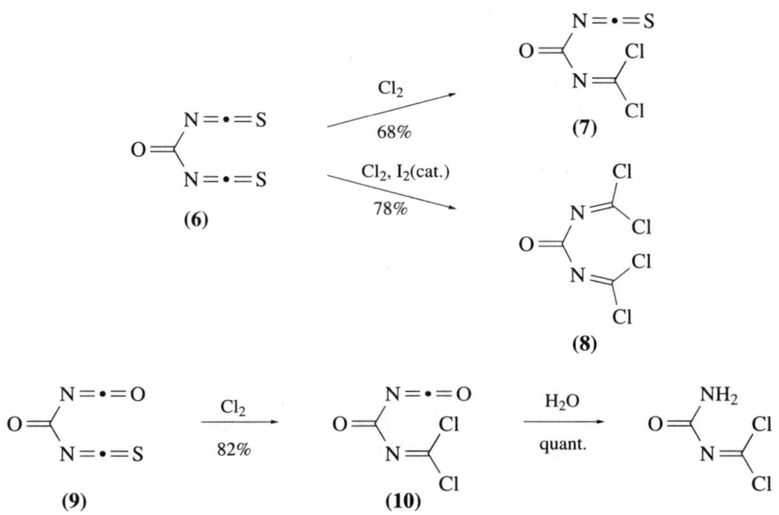

Scheme 6

Phosphoryl isothiocyanates are similarly readily chlorinated but the chlorination of sulfonyl isothiocyanates is more sluggish. An alternative is to use related reagents such as the dithiocarbamate salts (**11**) ⟨B-71MI 620-01⟩ or the corresponding bismethyl compounds (**12**) ⟨83AQ(C)115⟩, both readily available from the corresponding sulfonamide. Both are more readily chlorinated than the isothiocyanate. The benzothiazolyl-substituted carbonimidic dichloride (**13**) has been obtained (65%) in an analogous manner by chlorination of the corresponding bis(methylthio) compound ⟨84S520⟩. The final product of chlorination of trimethylsilyl isothiocyanate is the sulfenyl chloride $Cl_2C{=}NSCl$ ⟨B-71MI 620-01⟩.

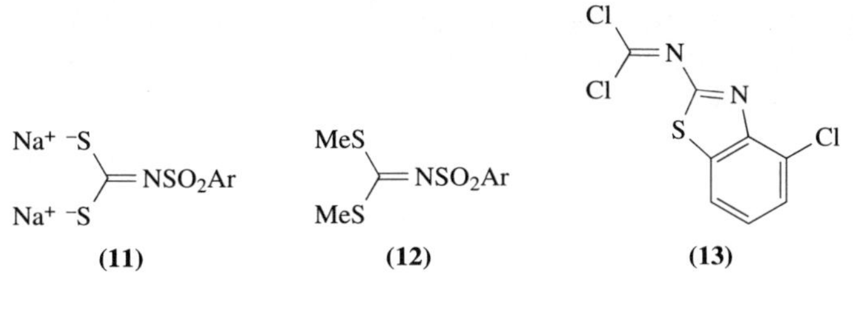

(**11**) (**12**) (**13**)

(ii) From isocyanides

The addition of chlorine to isocyanides in the cold is also a long-established method for the preparation of carbonimidic dichlorides. It has been used mainly for alkyl derivatives, for example $R = Bu^t$, 52% ⟨B-71MI 620-01⟩, $R = TsCH_2$, 67% ⟨81JHC1127⟩ and $R = PhCH_2$, 68% ⟨86T2677⟩. Aryl derivatives can also be prepared by this method and because of the mild reaction conditions it can be useful for the preparation of compounds which are sensitive to further chlorination ⟨87CB421⟩. A specific preparation based on an isocyanide is the formation of chromium carbene complexes by reaction of coordinated trichloromethyl isocyanide with secondary amines (Equation (3)) ⟨88AG(E)1344, 89CB1907⟩.

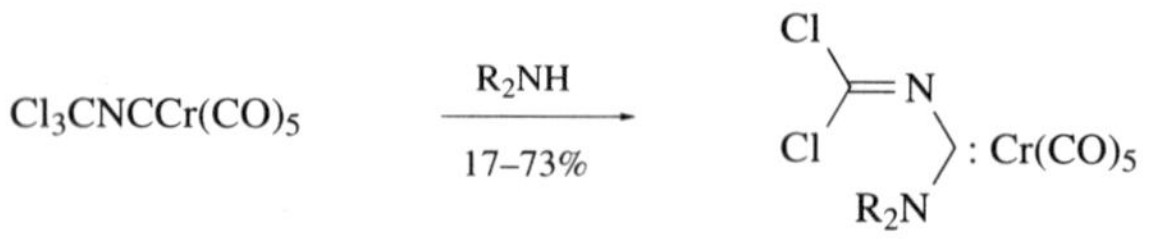
(3)

(iii) From formanilides

Many *N*-arylcarbonimidic dichlorides have been prepared in good yield from the corresponding formanilides by chlorination in thionyl chloride as a solvent (Equation (4)) ⟨62AG(E)647, B-71MI 620-01⟩. Sulfuryl chloride is most often used as the chlorinating agent. The method is most efficient for aryl derivatives with deactivating groups on the ring which inhibit ring chlorination. Examples from more recent literature include preparation of the 3-trifluoromethylphenyl derivative ⟨87JCS(P1)1069⟩, the 4-cyanophenyl derivative ⟨92SC1191⟩ and the 2,6-dichloro-4-nitrophenyl derivative ⟨87JMC1241⟩.

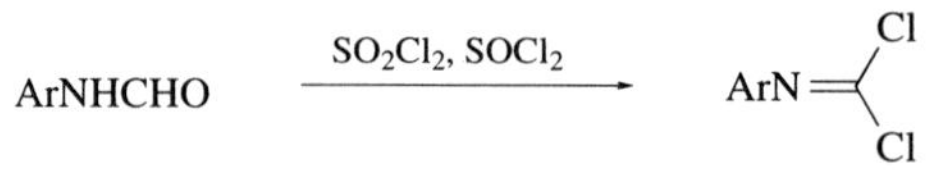
(4)

An example of the use of an analogous procedure for the preparation of an aliphatic carbonimidic dichloride is the preparation of $Cl_2C{=}NCH_2CH_2CN$ by chlorination of the corresponding formamide ⟨79GEP2753204, 79GEP2754604⟩.

(iv) From isocyanates

The reaction of several aliphatic isocyanates with chlorinating agents such as phosphorus pentachloride is reported to give *N*-alkylcarbonimidic dichlorides. For example, methyl isocyanate ⟨B-68MI 620-01⟩ and butyl isocyanate ⟨B-71MI 620-01⟩ have been converted into the corresponding dichlorides (in 83% and 50% yields) by reaction with a mixture of phosphorus pentachloride and phosphorus oxychloride. The method is not a useful one for aryl derivatives because at the temperatures required for reaction phosgene is eliminated. However, there are examples of successful chlorination of phosphoryl and sulfonyl isocyanates ⟨B-68MI 620-01⟩.

Some α-chloroalkyl isocyanates exist in tautomeric equilibrium with carbonimidic dichlorides. For example, trichloromethyl isocyanate is in equilibrium with the chloroformyl derivative **(14)** (Scheme 7) and reaction of the mixture with antimony pentafluoride gave the fluoride **(15)** as the major product ⟨88ZOB1516⟩. Compound **(14)** is reported as the product of radical chlorination of methyl isocyanate ⟨85GEP3337939⟩; it can be prepared in the laboratory by Curtius rearrangement of trichloroacetyl azide ⟨93T3227⟩.

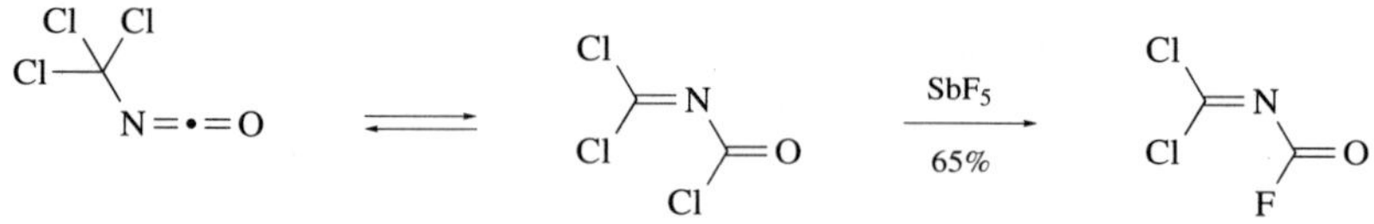
(14) **(15)**

Scheme 7

(v) From cyanogen chloride

Cyanogen chloride is a very useful starting material for the preparation of a variety of carbonimidic dichlorides. Examples of several types which can be prepared from cyanogen chloride are given in Table 2. Reaction of cyanogen chloride with chlorine in the presence of alkenes leads to the formation of *N*-(β-chloroalkyl)carbonimidic dichlorides (Equation (5)) ⟨69JPR15, 70S20⟩. These 'three component' reactions may go by way of $Cl_2C{=}NCl$ since this compound has been shown independently to add to the double bond of alkenes ⟨69AG(E)606⟩. A tetramer of cyanogen chloride, prepared in the presence of hydrogen chloride and DMF, has also been shown to be a carbonimidic dichloride (**16**) ⟨B-71MI 620-01⟩.

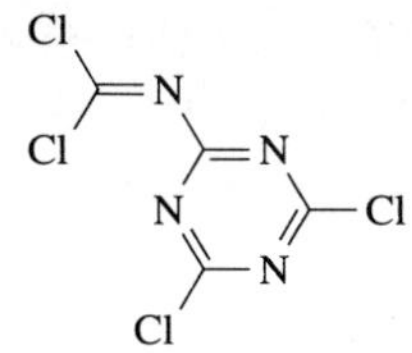

(16)

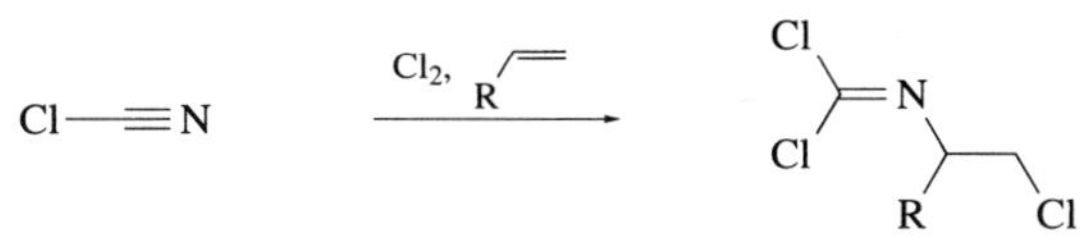

(5)

Table 2 Carbonimidic dichlorides prepared from cyanogen chloride, ClCN.

R	*Reagents and conditions*	*Yield* (%)	*Ref.*
CH_2OMe	$MeOCH_2Cl$, 50 °C, 5 atm	10	66AG(E)845
CH_2CH_2Cl	Cl_2, $H_2C{=}CH_2$, 0 °C	19	69JPR15
$CH(Cl)CH_2Cl$	Cl_2, $H_2C{=}CHCl$, $FeCl_3$	90	70S20
Bu^t	Cl_2, Bu^tCl, $FeCl_3$	91	73T297
$MeC{=}CHCl$	Cl_2, $H_2C{=}C(Cl)Me$, $FeCl_3$	60	70S20
$COCH_2Cl$	$ClCH_2COCl$, 50 °C, 5 atm	64	66AG(E)845
COPh	PhCOCl, 50 °C, 5 atm	20	66AG(E)845
Cl	Cl_2, 60 °C, carbon	73	69AG(E)606, 87JFC(37)29
SCl	SCl_2, 0 °C	27	58JCS764
$S(CF_3)SF_4$	CF_3SF_4Cl, *hv*	40	76IC14
TeF_5	TeF_5Cl, *hv*	32	85IC4171
$N(CF_3)_2$	$(CF_3)_2NCl$, *hv*	100	66JCS(A)933, 92IC488
$N(CF_3)C_2F_5$	$CF_3(C_2F_5)NCl$, *hv*	60	89IC3345

(vi) Other chlorination methods

There are several useful examples of the formation of carbonimidic dichlorides by chlorination of amines and amides, but none of the reactions is general. Reaction of *N*,*N*-dimethylaniline with chlorine under carefully controlled conditions leads to the formation of the carbonimidic dichloride (**17**) in high yield ⟨62AG(E)632⟩. Other examples of chlorination routes are shown in Scheme 8 ⟨69LA(730)133, 78CB1619, 83MI 620-01⟩.

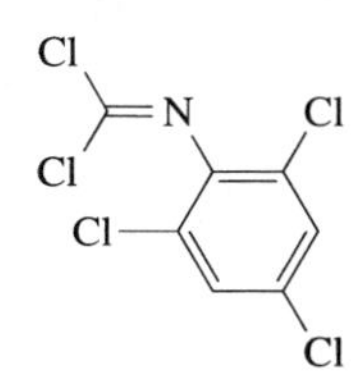

(17)

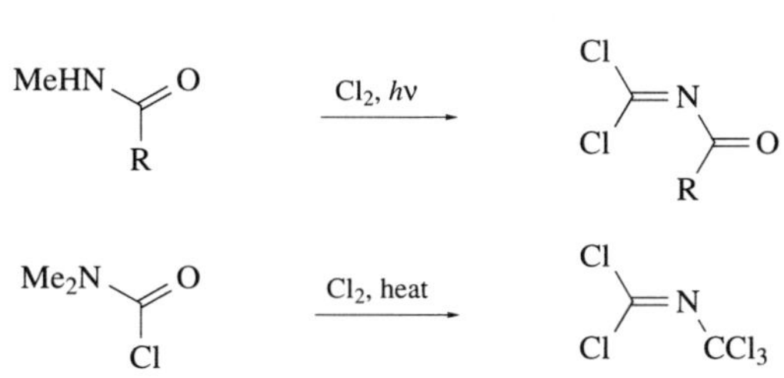

Scheme 8

An important chlorination reaction is the reaction of glyoxylic acid aldoxime to give *N*-hydroxy-carbonimidic dichloride (Scheme 9). The chlorination has been carried out with chlorine ⟨88SC1171⟩ and also with *N*-chlorosuccinimide or *t*-butyl hypochlorite ⟨89LA985⟩. (This oxime can also be prepared by reductive methods ⟨B-71MI 620-01⟩; e.g., it has been prepared by the reduction of trichloronitromethane with tin and HCl ⟨85USP4558160⟩.) The chlorination of glyoxylic acid aldoxime has an analogy in the formation of the hydrazone (**18**), a possible mechanism for which is shown ⟨72JOC2005⟩.

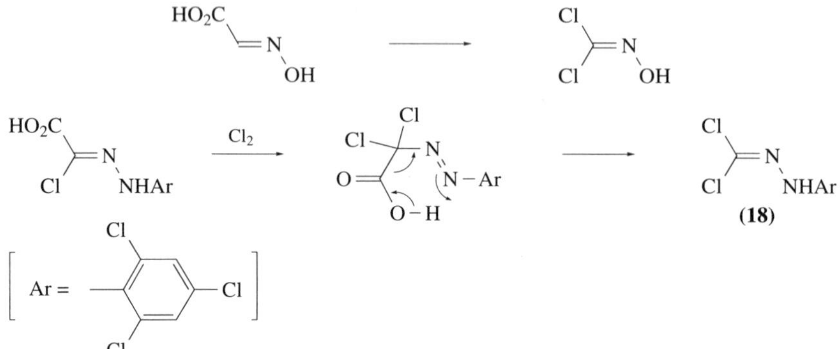

Scheme 9

(vii) Methods involving introduction of the CCl_2 fragment

Dichlorocarbene can be transferred to a variety of nitrogen compounds to produce carbonimidic dichlorides ⟨72TL3827, 73ZOR1759, B-77MI 620-01⟩. Some of these reactions are outlined in Scheme 10 ⟨68CC968, 71JOC1786, 73JA2982, 83JPR787, 92TL2339⟩. The reaction of azidooctane with dichlorocarbene appears to be capable of extension to the preparation of other carbonimidic dihalides.

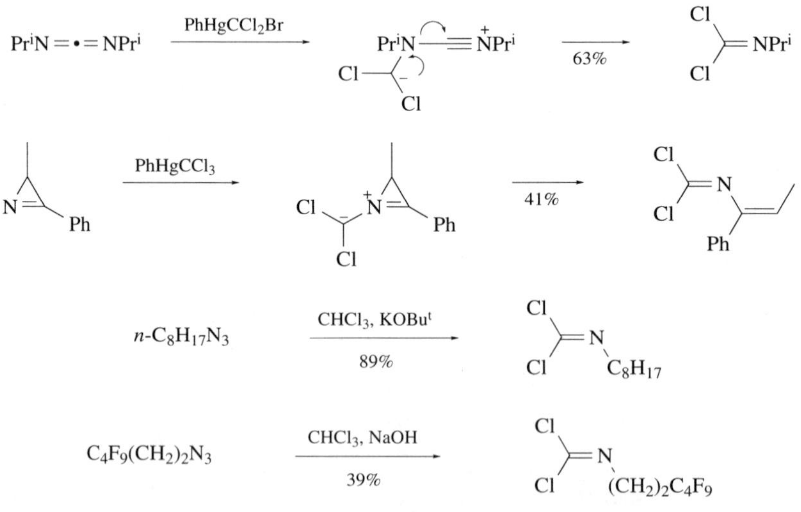

Scheme 10

There are also some useful specific syntheses based on the formal transfer of CCl_2 from tetrachloromethane. The hydrazone (**19**) has been synthesised by methods involving the low temperature reaction of tetrachloromethane with the azosilane (**20**) and with salts of silylated hydrazines ⟨71CB3989, 76ZN(B)1317, 86JOM(301)15, 91ZAAC(605)131⟩. These reactions probably involve dichlorocarbene as an intermediate since compound (**19**) is also formed from (**20**) and a different source of dichlorocarbene (Equation (6)).

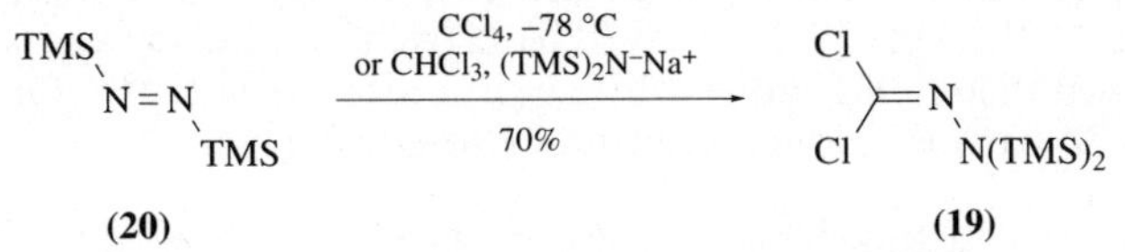

(6)

Some perfluoroaryl- and perchloroaryl-carbonimidic dichlorides can be synthesised by the reaction of the perhaloamine with tetrachloromethane and aluminium(III) chloride ⟨83JFC(22)439, 83KGS798⟩ and the same procedure has been used to convert pentafluorophenylhydrazine into the hydrazone (**21**) (Equation (7)) ⟨86ZOR1297⟩. *N*-Phenylcarbonimidic dichloride is one of the products of photolysis of mixtures of aniline and tetrachloromethane or chloroform ⟨89ZN(B)1589⟩ and pentafluorophenylcarbonimidic dichloride is produced by the vapour phase pyrolysis of chloroform with pentafluoroaniline ⟨77JFC(9)505⟩.

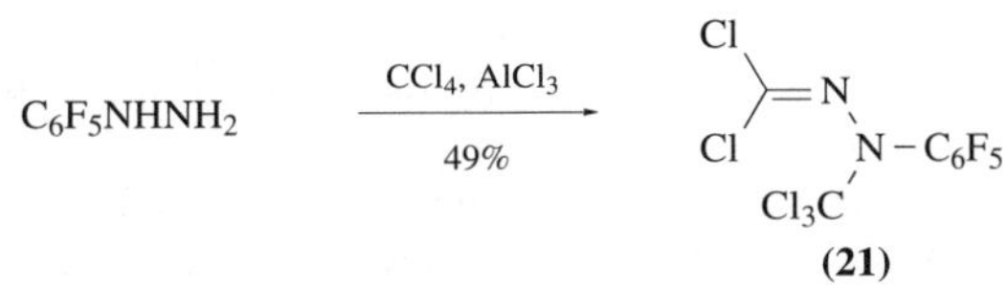

(7)

(viii) From other carbonimidic dichlorides

Carbonimidic dichlorides with reactive functional groups attached to nitrogen can be converted into other carbonimidic dichlorides and in some cases this provides the best route to the new species. The addition of *N*-chlorocarbonimidic dichloride to carbon–carbon double bonds has been referred to earlier; this compound also reacts with sulfur, in a process catalysed by chloride ions ⟨74TL1687⟩ and with dimethyl or diphenyl disulfide (Scheme 11) ⟨74TL1691⟩. The sulfenyl chloride $Cl_2C{=}NSCl$, like other sulfenyl halides, adds readily to alkenes: the adduct with cyclohexene has been isolated in 80% yield ⟨58JCS764⟩. New acylcarbonimidic dichlorides can also be formed from $Cl_2C{=}NCOCl$, an example being the generation of the phosphorus-containing species (**22**) (Equation (8)) ⟨78JGU1512⟩.

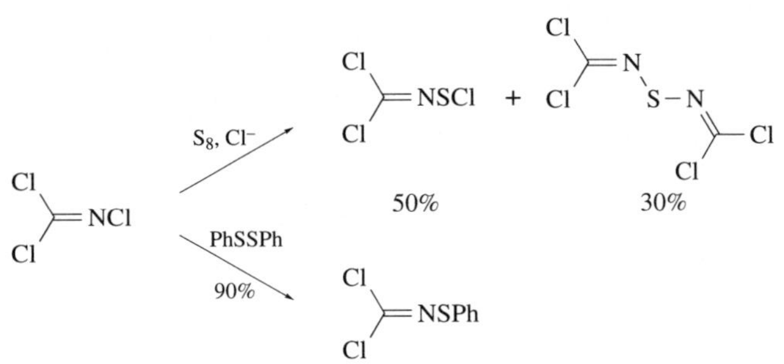

Scheme 11

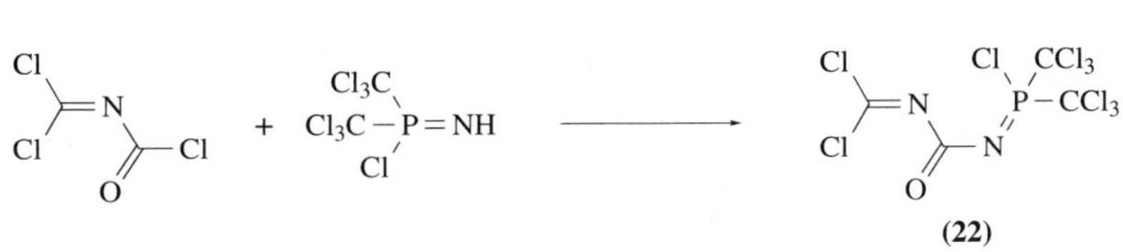

(8)

(ix) Methods for dichloroiminium cations

Salts of the type (**23**; R^1, R^2 = alkyl) are very useful synthetic intermediates. The methods of preparation and their chemistry have been reviewed by Janousek and Viehe ⟨B-76MI 620-01⟩ and by Marhold ⟨83HOU(E4)655⟩. No new general preparative methods have been reported since the publication of these reviews. The most important methods are all of a similar type, involving the chlorination of compounds of general structure (**24**; X = H, Cl, SR or SO_2R) (Equation (9)) ⟨71ZOR2084, 73ZOR39, B-76MI 620-01⟩. Salts derived from cyclic secondary amines have been synthesised by reaction with carbon disulfide followed by chlorination ⟨89SC2825⟩. On heating to 140 °C they undergo ring opening to carbonimidic dichlorides (Equation (10)).

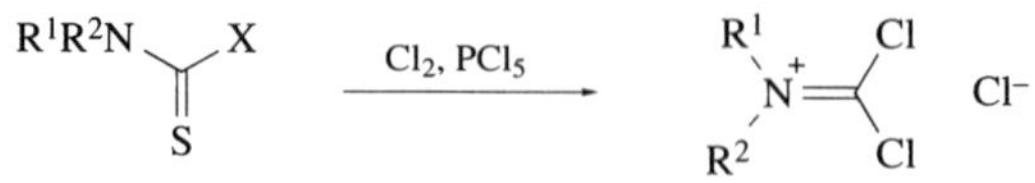

(9)

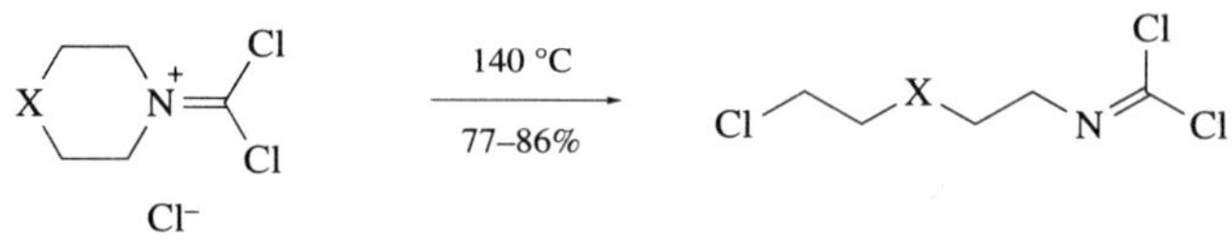

(10)

There are a few examples of the preparation of dichloroiminium salts (**23**) by *N*-alkylation of carbonimidic dichlorides ⟨B-76MI 620-01, 92ZAAC(617)136⟩ but this is not a general method since the carbonimidic dichlorides are poor nucleophiles. The compound $Cl_2C{=}NCl$ can be protonated in a superacid medium ⟨92ZAAC(617)136⟩. The parent dichloroiminium cation $Cl_2C{=}NH_2^+$ has been generated from cyanogen chloride and hydrogen chloride and both tetrachloroferrate and hexachloroantimonate salts have been isolated ⟨73T297⟩. A review of conjugated iminium salts (**25**), which are prepared from a variety of chloroiminium salts and nitriles R^1CN, includes examples of dichloroiminium salts (**25**; R^1 = Cl) ⟨88S655⟩.

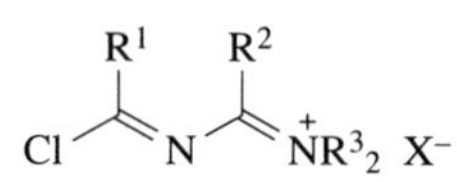

(25)

A compound which can formally be regarded as a coordinated dichloroiminium cation is the dichlorodiazomethane complex (**26**), which has been generated from a structurally related diimide complex by CCl_2 transfer ⟨80CC879, 81JCR(S)276⟩. The salt $[(CO)_5Cr{=}C{=}N^+{=}CCl_2]\ AlCl_4^-$ has been isolated from the reaction of the isocyanide complex $(CO)_5CrCNCCl_3$ with aluminium trichloride ⟨86OM2187⟩.

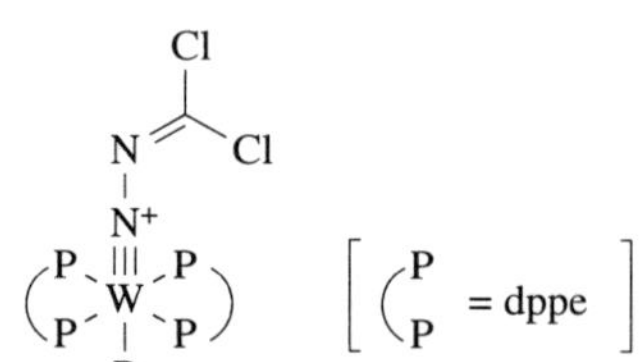

(26)

6.20.1.1.3 *Carbonimidic dibromides, $Br_2C{=}NR$*

The methods available for the synthesis of carbonimidic dibromides are nearly all analogous to methods used for the corresponding dichlorides, but with fewer examples.

The most straightforward method of preparation is the addition of bromine in the cold to isocyanides, a method which was first described in 1878 ⟨1878BSF(2)185⟩. Experimental details have been given for the preparation (76%) of *N*-phenylcarbonimidic dibromide by this method ⟨B-71MI 620-01⟩ and the thermally unstable *N*-cyclohexyl analogue has been prepared in the same way. The low stability of simple *N*-alkyl derivatives can preclude their full characterisation. Similarly, $Br_2C{=}NCN$, prepared by the addition of bromine to isocyanogen, is unstable and polymerises easily ⟨93JOC6930⟩.

N-4-Bromophenylcarbonimidic dibromide is reported to be formed by bromination of phenylthioformamide (Scheme 12) ⟨61MI 620-01⟩. Other methods for the formation of *N*-aryl derivatives which are analogous to those used for carbonimidic dichlorides are also illustrated in Scheme 12 ⟨79BSF(2)520, 88CB1445⟩.

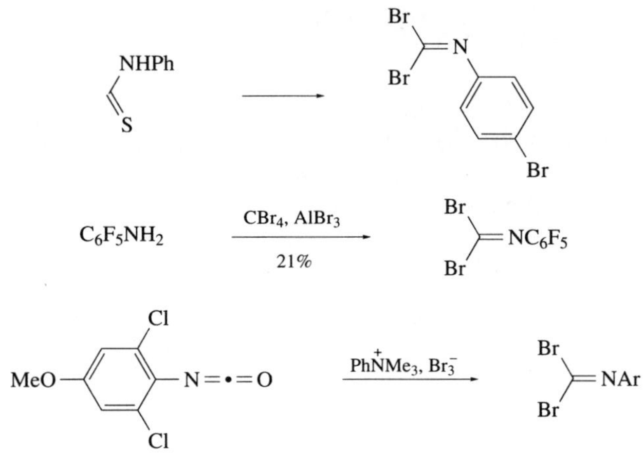

Scheme 12

The bromination of methyl isocyanate with *N*-bromosuccinimide leads to the formation of tribromomethyl isocyanate (**27a**) (60%) ⟨82CB860⟩. Like the corresponding trichloromethyl isocyanate this compound exists in equilibrium with a carbonimidic dihalide isomer (**27b**) (Equation (11)). *N*-Trifluoromethylcarbonimidic dibromide has been prepared by the reaction of the amine $(CF_3)_2NH$ with boron tribromide ⟨84JFC(24)523⟩.

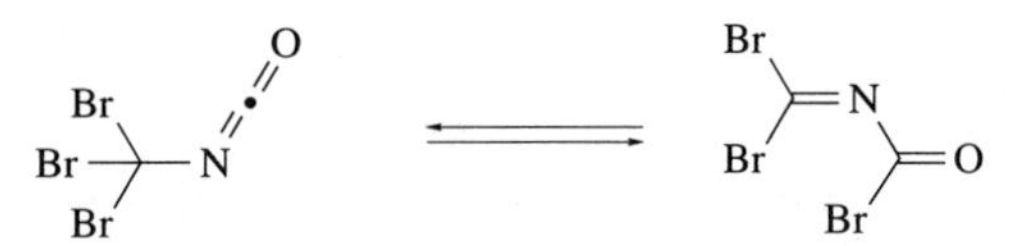

(27a) **(27b)** (11)

The best route to the oxime (**28**), a useful precursor to bromonitrile oxide, is the bromination of glyoxylic acid aldoxime (Equation (12)). This reaction has been carried out using *N*-bromosuccinimide ⟨89LA985⟩ and with bromine in water ⟨90JOC3045, 92TL3113⟩. The preparation has been performed on a 500 g scale ⟨84TL487⟩. The oxime can be acylated on oxygen to give other carbonimidic dibromides ⟨87EUP232012⟩. There are also related bromination routes to the arylhydrazones (**29**) ⟨72JOC2005, 76CI(L)164⟩ and to the *N*,*N*-dimethylhydrazone ⟨91ZOR533⟩.

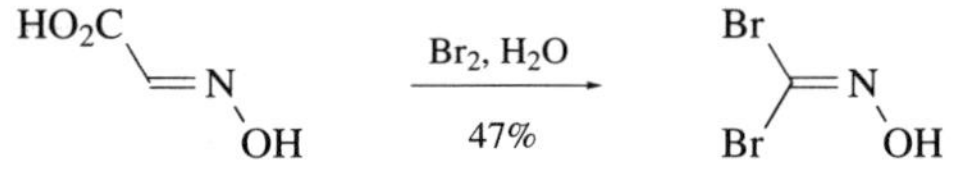

(28) (12)

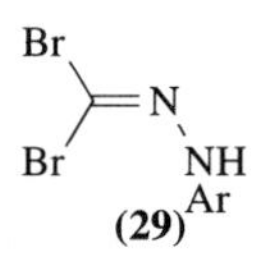

(29)

The bromination of tetrazolylhydrazones (**30**) has provided a route to several azines (**31**) which have been isolated in moderate to good yield (Equation (13)) ⟨64CI(L)1757, 79JCS(P1)724⟩. The tetrabromoazine (**32**) has been prepared by an analogous route ⟨B-71MI 620-01⟩.

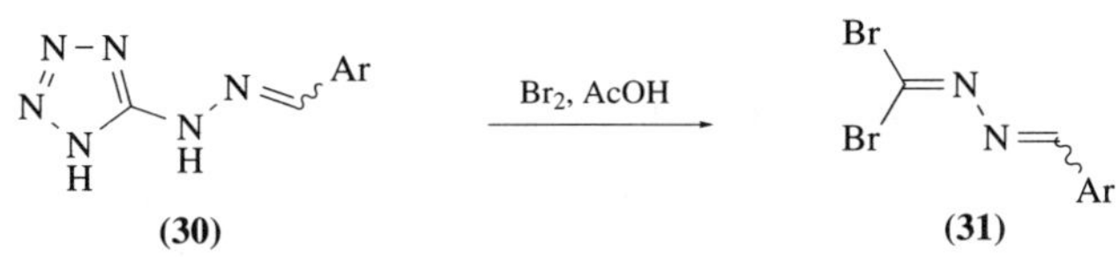

(30) **(31)** (13)

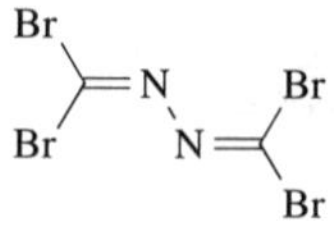

(32)

Dibromoiminium salts $Br_2C{=}NR_2{}^+Br^-$ (R = Me or Et) have been prepared by routes analogous to those used for the dichloroiminium salts ⟨74ZOR449, 88AJC563⟩. Thus, the dimethyliminium salt was formed by bromination of the disulfide (**33**) (Equation (14)).

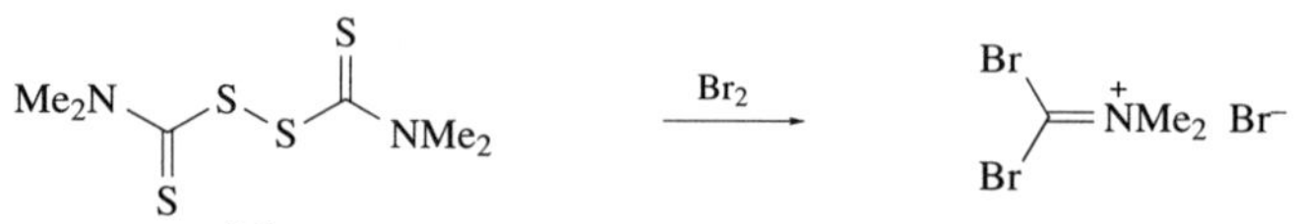

(14)

6.20.1.1.4 *Carbonimidic diiodides, $I_2C{=}NR$*

Isolable examples of this class of compounds are virtually unknown. The likely reason is their tendency to dissociate to iodine and the isocyanide. For example, *N*-alkylcarbonimidic dichlorides react with potassium iodide with the liberation of iodine ⟨59JGU2131⟩; the probable explanation is that carbonimidic diiodides are formed as intermediates but then dissociate (Scheme 13).

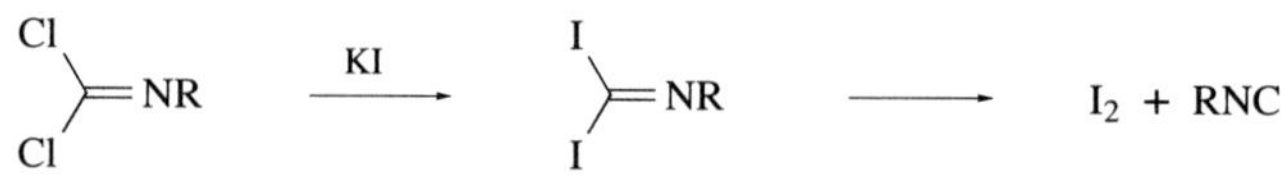

Scheme 13

Diiodoformaldoxime (**34**) has been isolated in good yield from the reaction of mercury or sodium fulminate with iodine ⟨31LA(489)7, 32CB65⟩. The iminium salt $I_2C{=}NMe_2{}^+I^-$ has been prepared (80%) by the reaction of the corresponding dichloroiminium chloride with iodomethane ⟨79ZOR215, 81ZOR180⟩.

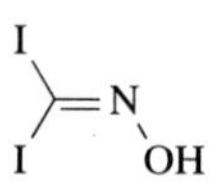

(34)

6.20.1.2 Iminocarbonyl Halides with Two Dissimilar Halogen Functions

There are relatively few examples of carbonimidic dihalides with different halogens attached to carbon. Most contain fluorine, with chlorine present as the second halogen.

6.20.1.2.1 *Carbonimidic chloride fluorides, ClFC=NR*

ClFC=NH has been suggested as an intermediate in the reaction of HF with cyanogen chloride, but it has not been detected directly ⟨69HCA812⟩. Several methods of preparation of ClFC=NF have been described and the (*E*)- and (*Z*)-isomers have been identified ⟨70JA3665⟩. The methods of preparation are summarised in Scheme 14 ⟨67JGU1343, 69JGU1301, 70JA3665, 73JOC1075⟩.

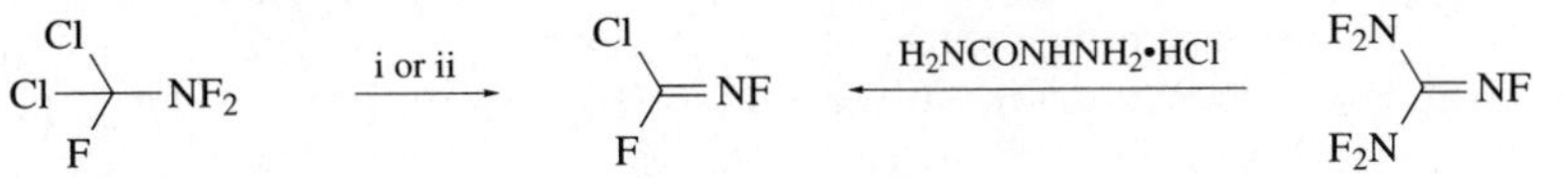

i, Hg, 100 °C, 52%(*E*), 39%(*Z*); ii, $Fe(CO)_5$, 10 °C, 43%

Scheme 14

$ClFC{=}NCF_3$ has been produced by the high temperature pyrolysis of 2,3-dichloroperfluoro-*N*-methylazetine and of copolymers of trifluoronitrosomethane with chlorofluoroalkenes such as trifluorochloroethylene ⟨65JCS6209, 72GEP2101107⟩. This compound is also one of the products obtained from the reaction of trifluoromethyl isocyanide with chlorine fluoride ⟨85IC4665⟩. Reductive routes to the azine (**35**) ⟨71ZOR2267⟩ and to the oxime (**36**) ⟨60JGU4029⟩ have been described and these are outlined in Scheme 15.

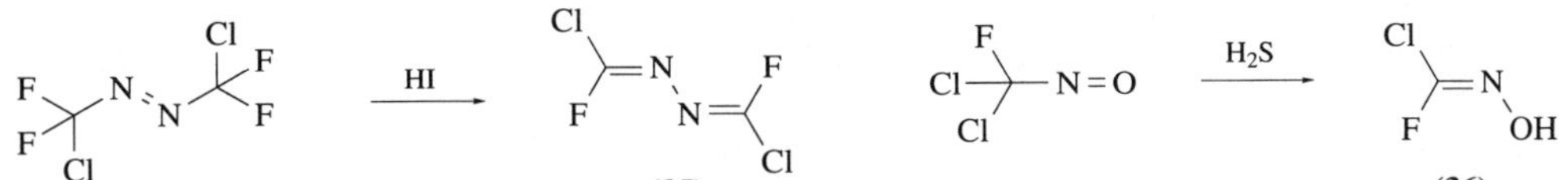

(**35**) (**36**)

Scheme 15

Chlorofluoroformaldoxime (**36**) is not isolable; it is converted into a crystalline polymer when the solution is concentrated. However, several series of phosphorus-substituted oximes have been isolated and characterised from the reaction of dichlorofluoronitrosomethane with phosphites and other phosphorus(III) reagents. Two examples, one resulting from oxidative addition to a cyclic phosphorus(III) reagent without ring opening ⟨72JGU293⟩ and the other resulting from ring opening ⟨69JGU1235⟩, are shown in Scheme 16; many related reactions have been described ⟨69JGU1692, 70JGU540, 72JGU799, 72JGU803, 88KFZ143, 89ZOB1455⟩.

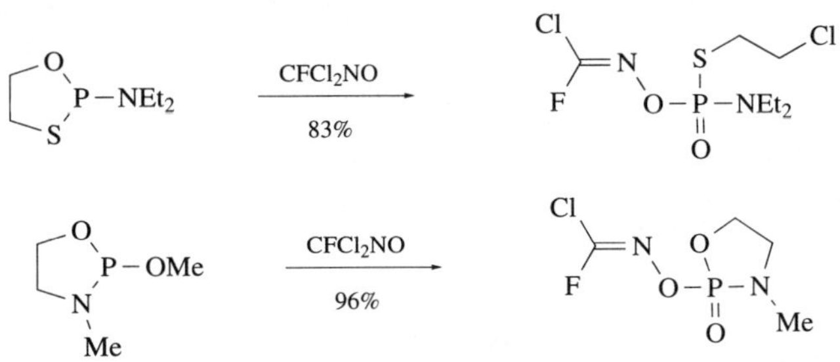

Scheme 16

6.20.1.2.2 *Carbonimidic bromide fluorides, BrFC═NR*

A few of these compounds have been described. Photolysis of a mixture of N_2F_4 and fluorotribromomethane gave $BrFC{=}NF$ as a mixture of (*E*)- and (*Z*)-isomers ⟨66IC1795⟩. Reaction of tetrabromoformaldazine (**32**) with silver fluoride gave a mixture containing $BrFC{=}NN{=}CF_2$ (the major component), $BrFC{=}NN{=}CBrF$ and $BrFC{=}NN{=}CBr_2$ ⟨66JOC3833⟩.

6.20.1.2.3 *Carbonimidic bromide chlorides and carbonimidic chloride iodides, ClXC═NR (X = Br or I)*

There are a few salts which contain these functional groups. The salts $ClXC{=}NH_2{}^+SbCl_6{}^-$ (X = Br and I) have been isolated from the reaction of cyanogen bromide and cyanogen iodide with antimony(V) chloride and HCl ⟨64CB1286⟩ and the crystal structure of the salt formed from cyanogen bromide has been determined ⟨93ZN(B)19⟩. There are also examples of salts of the type (**25**; R^1 = Br) ⟨88S655⟩.

6.20.1.3 Iminocarbonyl Halides with One Halogen and One Other Heteroatom Function

Compounds having this general structure are known with fluorine, chlorine, bromine and iodine as the heteroatom. The second heteroatom is most commonly oxygen, sulfur or nitrogen. The methods of preparation of the oxygen, sulfur and nitrogen compounds have been reviewed by Ulrich ⟨B-68MI 620-01⟩ and by Kühle ⟨69AG(E)20, 83HOU(E4)543⟩. Similar displacement reactions of dihaloiminium ions are reviewed by Janousek and Viehe ⟨B-76MI 620-01⟩ and by Marhold ⟨83HOU(E4)655⟩.

6.20.1.3.1 Iminocarbonyl fluorides with one other heteroatom function

(i) Iminocarbonyl fluorides with an oxygen function, $F(R^1O)C{=}NR^2$

The parent compound of this type ($R^1 = R^2 = H$) is named in *Chemical Abstracts* as carbonofluoridoimidic acid. A few of these compounds have been prepared by the selective displacement of fluoride from carbonimidic difluorides by oxygen nucleophiles. The reaction can be carried out on *N*-perfluoroalkylcarbonimidic difluorides with alkoxides (Scheme 17) ⟨76IZV2381, 92JFC(57)293⟩. Nitric acid has also been used as the nucleophile; dropwise addition of concentrated nitric acid at −78 °C gave the nitrate (**37**) ⟨68JGU37⟩. An example of a related method in which the oxygen nucleophile is generated *in situ* is shown in Scheme 18 ⟨82POL129⟩. Two specific methods based on different approaches are also shown. Thermolysis of the diazirine (**38**) at 80 °C gave the carbene dimer as the major product, but the azine (**39**) as a minor product (13%) (Equation (15)) ⟨86TL419⟩. Compounds of this class have also been formed by thermal rearrangement of perfluorinated oxaziridines; a *C*-perfluoroalkyl substituent of the oxaziridine migrates to oxygen in the rearrangement ⟨93JCS(P1)505⟩. A reductive dechlorination method provided a route to compound (**40**) (as a mixture of the two isomers) in 79% yield (Equation (16)) ⟨87AG(E)134⟩.

(**37**)

Scheme 17

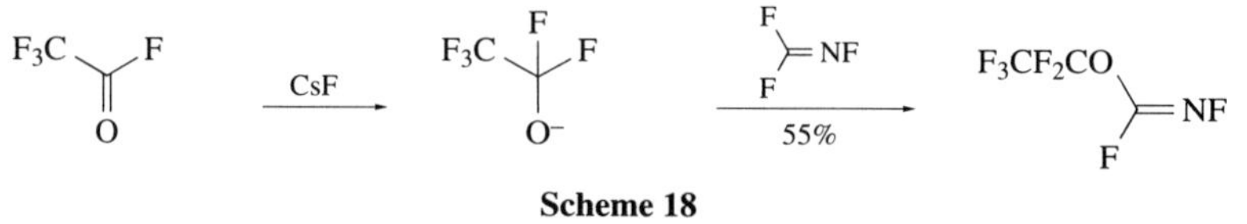

Scheme 18

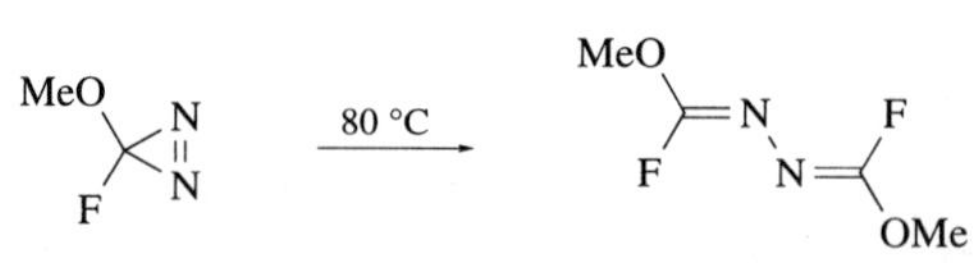

(**38**) (**39**) (15)

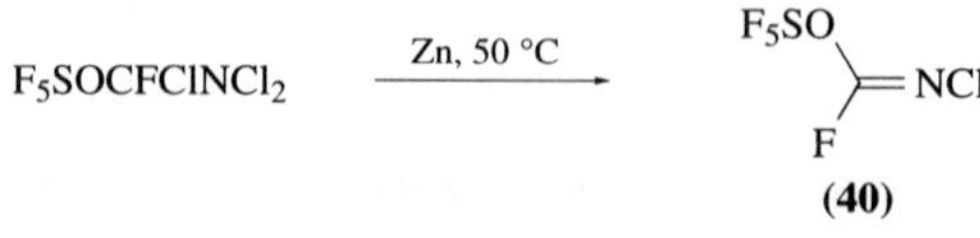

(**40**) (16)

(ii) Iminocarbonyl fluorides with a sulfur function, $F(R^1S)C{=}NR^2$

Such compounds can be prepared in a manner analogous to those with an oxygen function, by displacement of one fluoride of carbonimidic difluorides using a thiol as the nucleophile ⟨69JGU183⟩.

(iii) Iminocarbonyl fluorides with a nitrogen function, $F(R^1R^2N)C{=}NR^3$

These compounds are described in *Chemical Abstracts* as carbamimidic fluorides. The addition of ammonia ⟨83IZV694⟩ or primary alkylamines ⟨80JFC(15)169⟩ to $F_2C{=}NCF_3$ leads to compounds of this type having the general structure (**41**) in good yield. With an excess of the carbonimidic difluoride 2 : 1 adducts (**42**) can be isolated ⟨83MI 620-02⟩. Trimethylsilyl azide reacts with $F_2C{=}NCF_3$ to give the α-azidoimide $F(N_3)C{=}NCF_3$ which shows no tendency to cyclise to the corresponding tetrazole ⟨85IZV700⟩. The carbonimidic difluoride $F_2C{=}NCF_3$ also reacts with antimony(V) fluoride, either to give a cyclic trimer ⟨83JFC(22)175⟩ or, in liquid sulfur dioxide, to give the salt $(CF_3)_2NC(F){=}N^+{=}CFN(CF_3)_2SbF_6^-$ ⟨84JFC(26)321⟩.

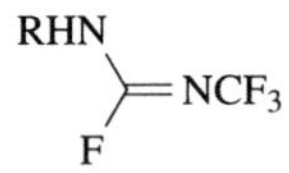

(**41**)

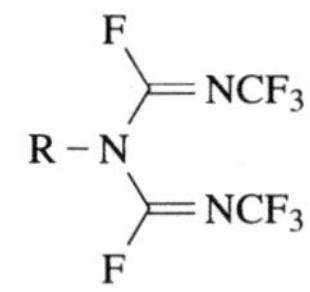

(**42**)

The most numerous examples of compounds of this class are formal dimers (**43**) of carbonimidic difluorides. In the presence of fluoride ions the carbonimidic difluorides can dimerise by the route shown in Scheme 19. This is in effect a special case of nucleophilic displacement of fluoride. Some examples of dimers formed in this manner are listed in Table 3. Related reactions are the combination of $F_2C{=}NF$ with $F_2C{=}NBr$ to form a 'mixed dimer' (**44**) ⟨88JOC4443⟩ and with $F_4S{=}NF$ to give

Table 3 Dimers (**43**) of carbonimidic difluorides $F_2C{=}NR$.

R	*Ref.*
F	82POL129, 83JOC771, 90JA728
Bu^t	80TL4893
CF_3	87JFC(37)259, 76IZV212
Ph	66AG(E)848, 80TL4893
OCF_3	81JFC(18)441
TeF_5	85IC4171
$N(CF_3)C_2F_5$	89IC3345

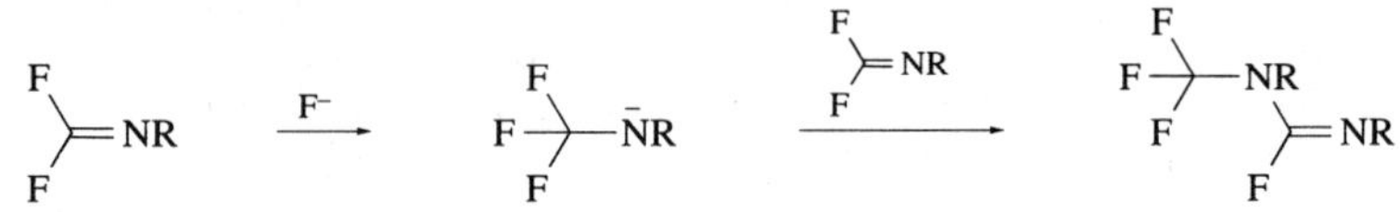

(**43**)

Scheme 19

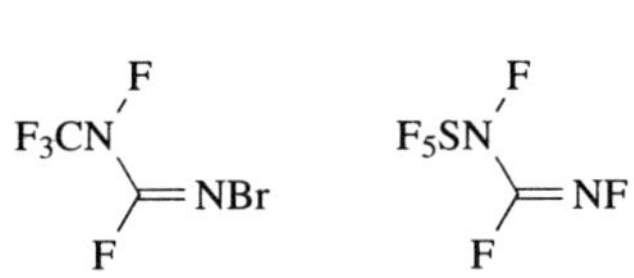

(**44**) (**45**)

the 1 : 1 adduct (**45**) ⟨84IC2188⟩.

Some useful specific preparations of compounds of this class have been described. The *N*-bromoimide (**46**) has been prepared in 90% yield from $(CF_3)_2NCN$, bromine and caesium fluoride ⟨90JA728⟩ and compound (**47**) in 84% yield from fluorocyclohexane, cyanogen chloride and hydrogen fluoride ⟨69HCA812⟩. The reaction of TeF_5NHCF_3 with potassium fluoride gives the imide (**48**) (49%) ⟨85IC4171⟩. The azoimide (**49**; $R = CF_2NF_2$) has been obtained in low yield by thermolysis of perfluoroaminodiazirine ⟨67JHC389⟩ and the related compound (**49**; R = C(F)=NF) by reaction of the azo compound (**50**) with carbon monoxide at 180 °C ⟨72IC418⟩. The perfluoro compound $F_2NCF{=}NF$ is produced in low yield by fluorination of guanylurea ⟨67JOC3859⟩. A potentially more general method of synthesis of this class of compounds which has not been exploited is the exchange of the chlorine of the corresponding imidoyl chlorides for fluorine by the action of hydrofluoric acid ⟨69AG(E)20⟩.

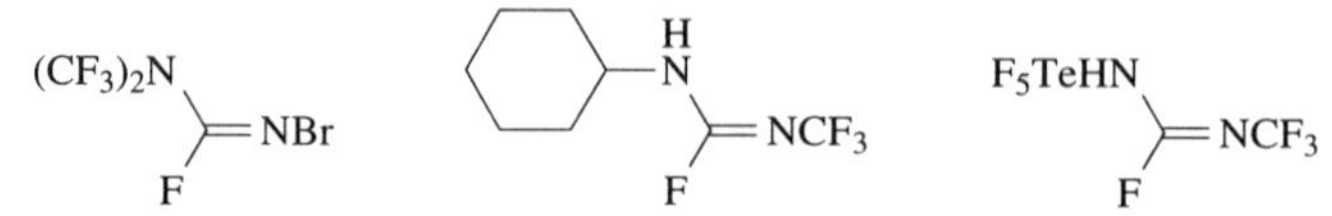

(**46**) (**47**) (**48**)

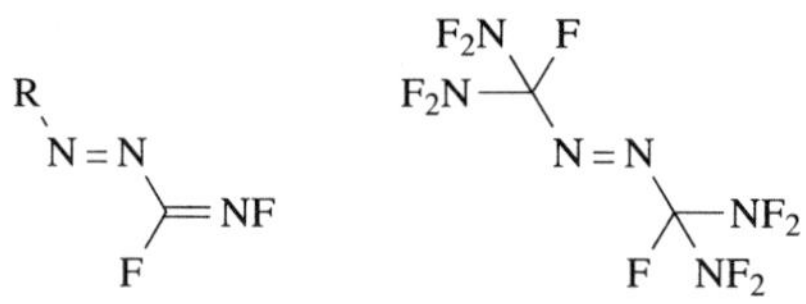

(**49**) (**50**)

(iii) Iminocarbonyl fluorides with a phosphorus function

The reaction of triethyl phosphite or trimethyl phosphite with $F_2C{=}NCF_3$ leads to the formation of the adducts $(RO)_3P(F)C(F){=}NCF_3$ in good yield ⟨77IZV2379, 81IZV2632⟩.

6.20.1.3.2 *Iminocarbonyl chlorides with one other heteroatom function*

(i) Iminocarbonyl chlorides with an oxygen function, $Cl(R^1O)C{=}NR^2$

There are two principal methods for the preparation of compounds of this type (named in *Chemical Abstracts* as derivatives of carbonochloridoimidic acid). The first, and older, method is the selective displacement of one chloride from a carbonimidic dichloride by an oxygen nucleophile, usually an alkoxide or a phenoxide ion. Several examples of reactions of this type are described by Kühle ⟨69AG(E)20, 83HOU(E4)543⟩. The method recommended for reaction of alkoxides with *N*-arylcarbonimidic dichlorides is to use a two-phase system of chlorobenzene and aqueous alkali; this prevents further displacement of chloride. Examples of this method of preparation include those for R^2 = alkyl ⟨82TL3539⟩, aryl ⟨75CB2290, 82TL3539, 84ZOR1197⟩ and SF_5 ⟨82JFC(19)411⟩.

Reaction with phenoxides has been performed with equimolar quantities of reagents in a polar solvent such as dioxane ⟨69AG(E)20⟩. An example is the selective displacement of chloride from the dichloride (**51**) with phenol in the presence of triethylamine (Equation (17)) ⟨75JCS(P2)1046, 77ZC172⟩.

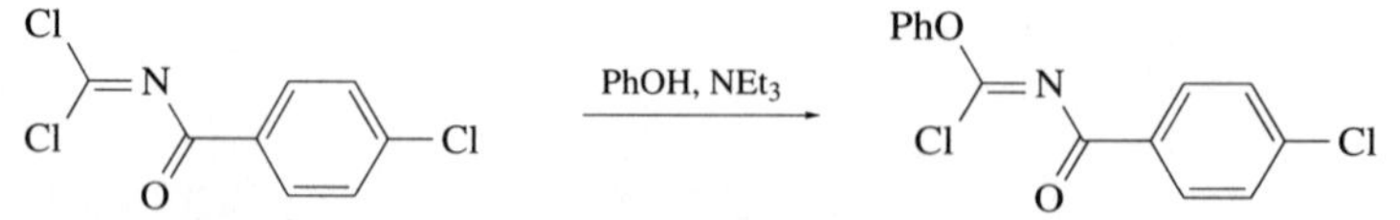

(17)

(**51**)

The second general method of preparation of these compounds is the reaction of aryl cyanates with acyl chlorides (Equation (18)) and with other active chlorine compounds such as sulfenyl chlorides. Several examples of this reaction are described in a review by Grigat ⟨72AG(E)949⟩. The range of examples has subsequently been extended. With phosphorus pentachloride at −50 °C the salts (**52**) were formed in high yield ⟨72JGU97⟩. The pyrone derivatives (**53**) have been isolated in low yield from the reactions of aryl cyanates with malonyl chloride (which behaves as the pyrone acid chloride) ⟨85CB1371⟩ and the 'mixed' imidoyl chlorides (**54**) were isolated in moderate yield from the reaction of $Cl_2C{=}NCOCl$ with aryl cyanates ⟨81ZOR398⟩. Cyanates also react with chloroiminium salts to give salts of the type (**25**) ⟨88S655⟩.

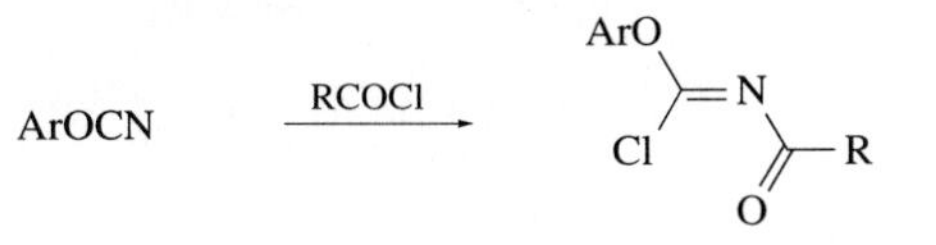

(18)

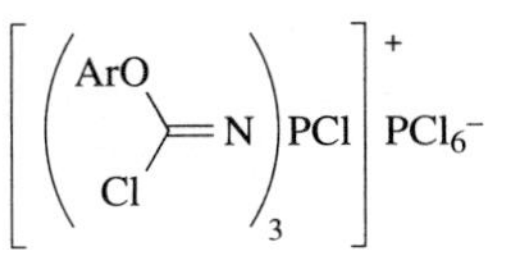

(**52**)

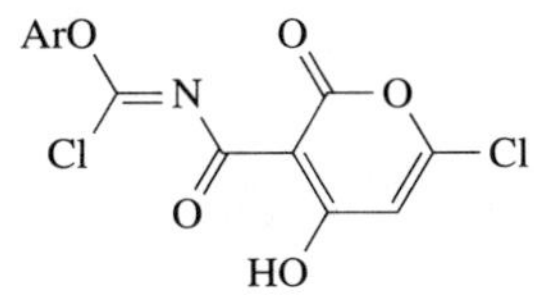

(**53**)

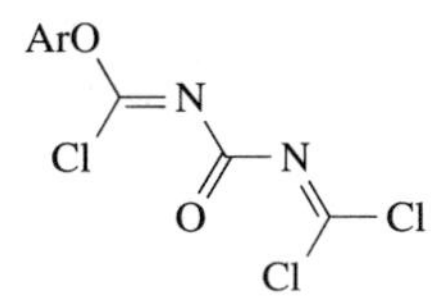

(**54**)

In principle the reaction of carbamate esters $R^1NHCO_2R^2$ with phosphorus pentachloride could provide a third general method of synthesis, but such reactions normally go further and yield isocyanates. In the reaction of F_5SNHCO_2Ph with phosphorus pentachloride some of the imidoyl halide $F_5SN{=}C(Cl)OPh$ was isolated as well as the isocyanate ⟨83JFC(23)593⟩.

(ii) Iminocarbonyl chlorides with a sulfur function, $Cl(R^1S)C{=}NR^2$

These compounds have also been called 1-halothioformimidates. The parent compound ($R^1 = R^2 = H$) is named as carbonochloridimidothioic acid in *Chemical Abstracts*. The four principal methods of preparation are exemplified below.

(a) From carbonimidic dichlorides. The displacement of chloride from carbonimidic dichlorides by sulfur nucleophiles has been used as a method of preparation for a few types of compounds of this type, including *N*-aryl, aroyl and phosphoryl derivatives ⟨69AG(E)20, 75CB2290, 75JCS(P2)1046, 84JOC528⟩. For example, the imidoyl chloride (**55**) was isolated (60%) from the reaction of the corresponding carbonimidic dichloride with phenylmethanethiol and triethylamine ⟨84JOC528⟩. A related reaction is the preparation of the heterocycle (**56**) from the imidoyl chloride (**57**) and sulfur (Equation (19)) ⟨76GEP2451635⟩.

(**55**)

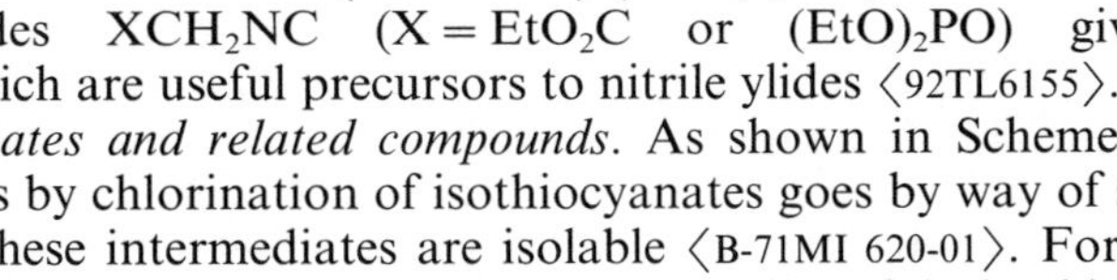

(19)

(**57**) (**56**)

(b) From isocyanides. The addition of sulfenyl chlorides to isocyanides provides a general route to compounds of this class ⟨62AG(E)647, 84T1075, 85JOC771⟩. With bis(trimethylsilyl)amino isocyanide a range of sulfenyl chlorides has been used (Scheme 20) ⟨77ZN(B)1003⟩. The addition of methanesulfenyl chloride to isocyanides XCH_2NC ($X = EtO_2C$ or $(EtO)_2PO$) gives the compounds $MeSC(Cl){=}NCH_2X$ which are useful precursors to nitrile ylides ⟨92TL6155⟩.

(c) From isothiocyanates and related compounds. As shown in Scheme 5 the preparation of carbonimidic dichlorides by chlorination of isothiocyanates goes by way of sulfenyl chloride intermediates and some of these intermediates are isolable ⟨B-71MI 620-01⟩. For example, the sulfenyl chlorides (**58**) have been isolated in good yield from the reaction of the isothiocyanates with chlorine ⟨78CB698⟩. Chlorination of aryl isothiocyanates in the presence of thiols RSH leads to the formation of the compounds (**59**) ⟨90JHC1191⟩. A related high yielding preparative method is the reaction of compounds (**60**; $X = COAr$ or SO_2Ar) with sulfuryl chloride to give the chlorides (**61**) ⟨75AP(308)379, 84CZ404, 91JPR187⟩. Thio- and dithiocarbamic esters PhNHCXSR ($X = O$ or S) have also been

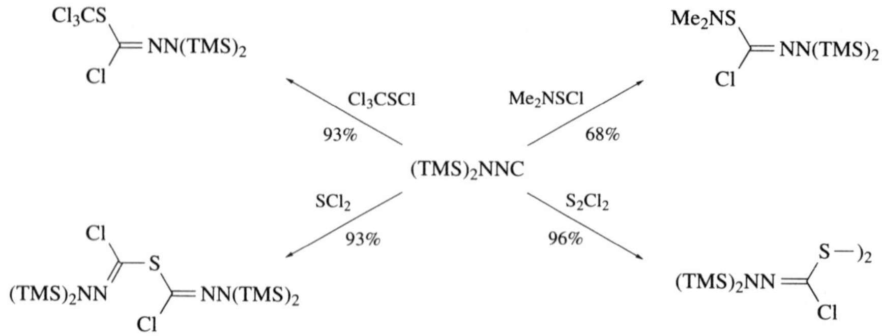

Scheme 20

chlorinated in high yield by means of the Appel reagent (triphenylphosphine and tetrachloromethane) to give compounds of this type ⟨76CB810⟩.

(58) **(59)** **(60)** **(61)**

(d) From thiocyanates. The reaction of cyanates with acyl chlorides described by Grigat ⟨72AG(E)949⟩ has an analogy in the reaction of methyl or ethyl thiocyanate with oxalyl chloride ⟨78LA1704⟩. With 2 mol of methyl thiocyanate and in the presence of boron trifluoride the imidoyl chloride (**62**) was formed (66%). The *N*-chloro compounds (**63**) have been isolated from the reaction of thiocyanates with iodine monochloride ⟨82MI 620-02⟩. By analogy with the reaction shown in Equation (5) the three-component reaction of alkyl thiocyanates with chlorine and alkenes gave the adducts (**64**) in moderate yield ⟨69JPR15⟩. Ethyl thiocyanate is also reported to react with hydrogen chloride to give the 2:2 adduct (**65**) (40%) ⟨73BCJ303⟩. Salts of the type (**25**) are formed from thiocyanates and imidoyl chlorides ⟨88S655, 92T8271⟩. The methods of preparation of salts $R^1SC(Cl){=}NR^2{}_2{}^+X^-$ have been reviewed by Kantlehner ⟨79MI 620-02⟩.

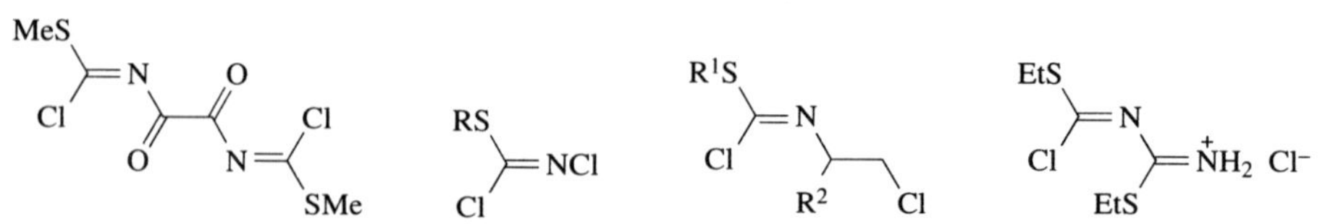

(62) **(63)** **(64)** **(65)**

(iii) Iminocarbonyl chlorides with a selenium function, $Cl(R^1Se)C{=}NR^2$

An example of a compound with this functional group is the selenazinone (**67**), which was formed by cyclisation of the selenocyanate (**66**) with HCl ⟨68AG(E)464⟩.

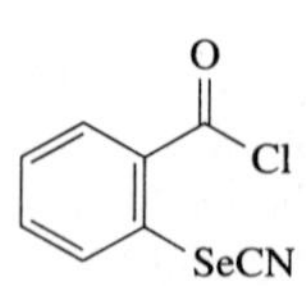

(66)

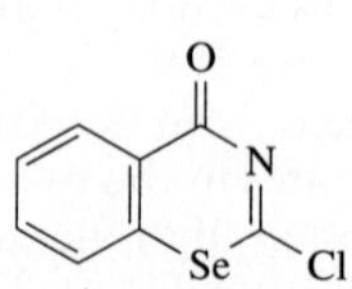

(67)

(iv) Iminocarbonyl chlorides with a nitrogen function, $Cl(R^1R^2N)C{=}NR^3$

The current *Chemical Abstracts* name for the parent compound ($R^1 = R^2 = R^3 = H$) is carbamimidic chloride. This compound is still unknown but its salts and many substituted carbamimidic chlorides have been described. There are five principal methods of preparation.

(a) From carbonimidic dichlorides and dichloroiminium salts. In principle the reaction of carbonimidic dichlorides with amines provides a general route to carbamimidic chlorides, but the reaction must be carried out under carefully controlled conditions in order to be a useful synthetic method. For example, displacement of chloride from *N*-aroylcarbonimidic dichlorides takes place in high yield with dialkyltrimethylsilylamines at low temperature (Equation (20)) ⟨89LA931⟩. There are several other useful examples of displacement of chloride from carbonimidic dichlorides by secondary amines ⟨80CC879, 83KGS798, 88AG(E)1344, 88JFC(40)217, 89CB1907⟩ one of which was shown to lead specifically to the (*Z*)-isomer ⟨93JPO319⟩. Primary amines tend to react further to give guanidines. *N*-Phenylcarbonimidic dichloride reacts with cyclic tertiary amines to give quaternary ammonium salts (Scheme 21) which either demethylate ($n = 5$ and 6) or undergo ring opening ($n = 2$–4) above room temperature ⟨74TL3765⟩. Triazinones (**68**) have been prepared by the reaction of the carbonimidic dichloride (**8**) with arylamines ⟨82CB3587⟩.

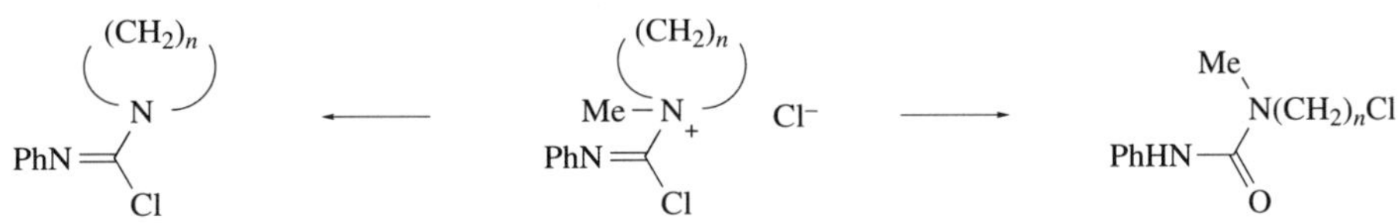

Scheme 21

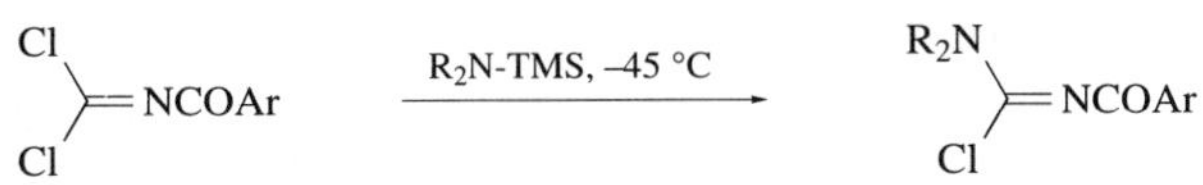

(20)

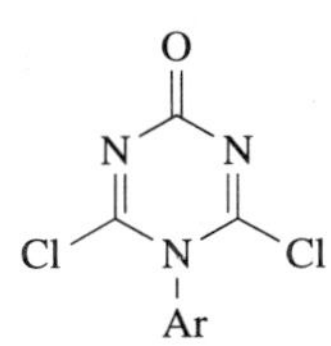

(68)

The major route to *N*,*N*-dimethylcarbamimidic chlorides (**69**) is provided by the displacement of chloride from the iminium salt (**70**) by amines, dichloroamines and other nitrogen nucleophiles (Scheme 22) ⟨B-76MI 620-01⟩. This reaction provides a wide range of compounds having the general structure (**69**). Many of these compounds are described in the review by Janousek and Viehe ⟨B-76MI 620-01⟩; representative examples of reactions of the salt (**70**) with reagents of the general type RNH_2 from this review and from later references are given in Table 4.

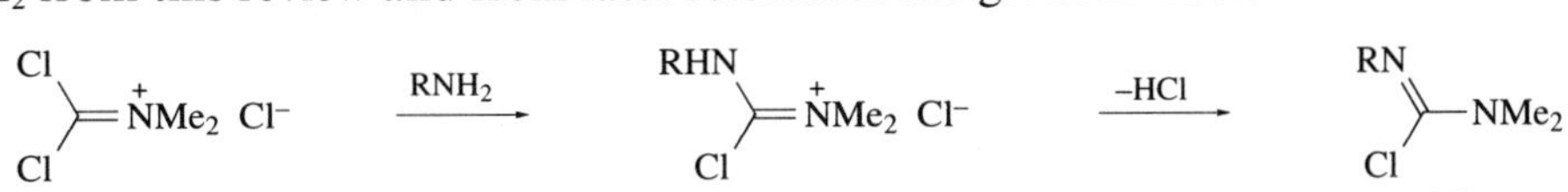

Scheme 22

N,*N*-Dichlorourethanes ⟨75ZOR71⟩ and *N*,*N*-dichlorosulfonamides ⟨73ZOR43, 78ZOR1841⟩ also both react with the iminium salt (**70**) to give compounds of general structure (**69**), with the elimination of chlorine. A variety of other nitrogen compounds react with the salt (**70**): some examples are shown in Scheme 23 ⟨74ZOR36, 75C209, 78BSB391, 79AG(E)333, 84ZOB1110, 86T6645⟩.

(b) From thioureas. Several types of thiourea can be converted in good yield into carbamimidic chlorides by reaction with an equimolar amount of phosgene ⟨B-68MI 620-01⟩. An example is the preparation of the *N*-*p*-toluenesulfonylimide (**71**) (88%) from the thiourea. A variant of this method is provided by the reaction of *N*-acetylthiourea with phosphorus pentachloride, which gave the imide (**72**) (71%) ⟨90ZOB706⟩. *N*-Sulfonylcarbamimidic chlorides have also been obtained by chlorination of *S*-methylisothioureas ⟨75ZOR2243, 80CP1076132⟩.

Table 4 Carbamimidic dichlorides (**69**) from the iminium salt (**70**) and RNH_2.

R	*Yield* (%)	*Ref.*
$C_6H_4NO_2$-4	80	71AG(E)573
C_6H_4CN-2	quant.	84TL1557
CO_2Et	72	B-76MI 620-01
$CH{=}C(CN)_2$		82UKZ395
OCH_2Ph	91	B-76MI 620-01
SO_2Ph	68	82LA2105
$N(Me)C_6H_3(NO_2)_2$-2,4	88	80JCS(P2)867

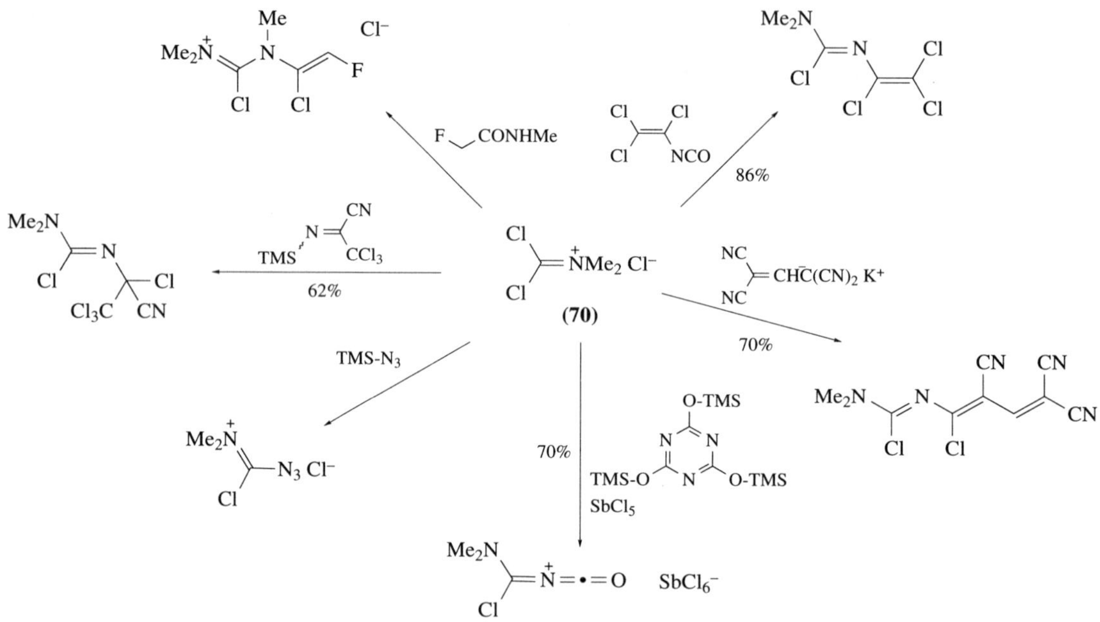

Scheme 23

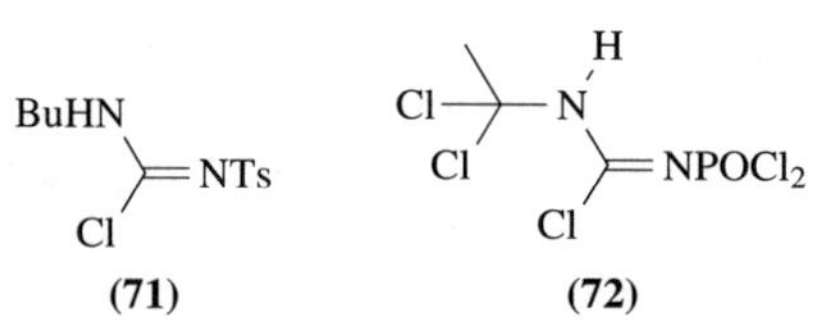

(*c*) *From ureas.* Ureas can also be chlorinated with phosgene, but attack on oxygen, rather than on nitrogen, which is the favoured reaction only with ureas having secondary or tertiary alkyl substituents on nitrogen ⟨B-68MI 620-01⟩. An example of the reaction is shown in Equation (21) ⟨86JOC1719⟩. There are also several examples of chlorinations of this type which have been carried out with the triphenylphosphine–tetrachloromethane reagent ⟨73CB2093, 74CB698, 76CB1643, 76CB2921, 79CB640, 80CB1095⟩ and with phosphorus(V) chloride ⟨91ZOB643, 92ZOB306, 92ZOB324⟩.

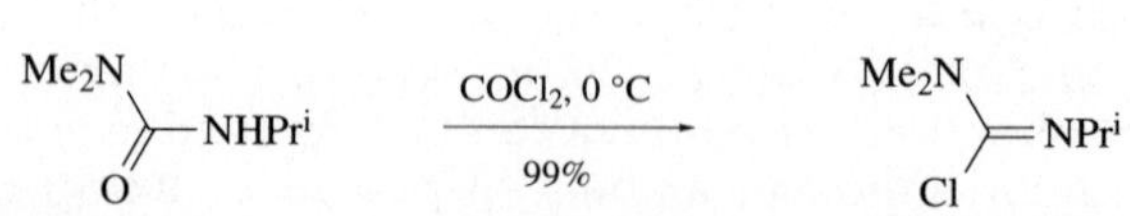

(21)

(*d*) *From cyanamides.* Dialkylcyanamides add to electron deficient aroyl chlorides to give carbamimidic chlorides, generally in good yield. The reaction was first described by Bredereck and Richter ⟨66CB2454⟩ but was later greatly expanded, especially by Ried *et al.* ⟨80S619⟩. The method is exemplified by the preparation of the imidoyl chloride (**73**) (96%) from 4-nitrobenzoyl chloride and *N*-cyanopyrrolidine (Equation (22)). The reaction has been performed with a variety of aroyl chlorides ⟨80CB2583, 80S619⟩ and other activated acyl chlorides ⟨80S619, 81MI 620-02, 86LA1997⟩. Other activated halogen compounds also react with dialkylcyanamides to give carbamimidic chlorides: examples of reaction with sulfenyl chlorides ⟨80S619, 87CZ339⟩, sulfinyl chlorides ⟨80S619⟩, diarylimidoyl chlorides ⟨87CC99⟩, chloroiminium salts ⟨84CC1105, 88S655, 89S918, 92T8271, 92JHC1551, 93H(36)1521⟩, thionyl chloride ⟨74ZOR1000⟩, sulfur dichloride ⟨85IC2453⟩, sulfuryl chloride ⟨73ZOR633⟩ and phosphorus trichloride ⟨78JGU1078⟩ are known. Some of these are illustrated in Scheme 24 ⟨89S918⟩.

(**73**)

(22)

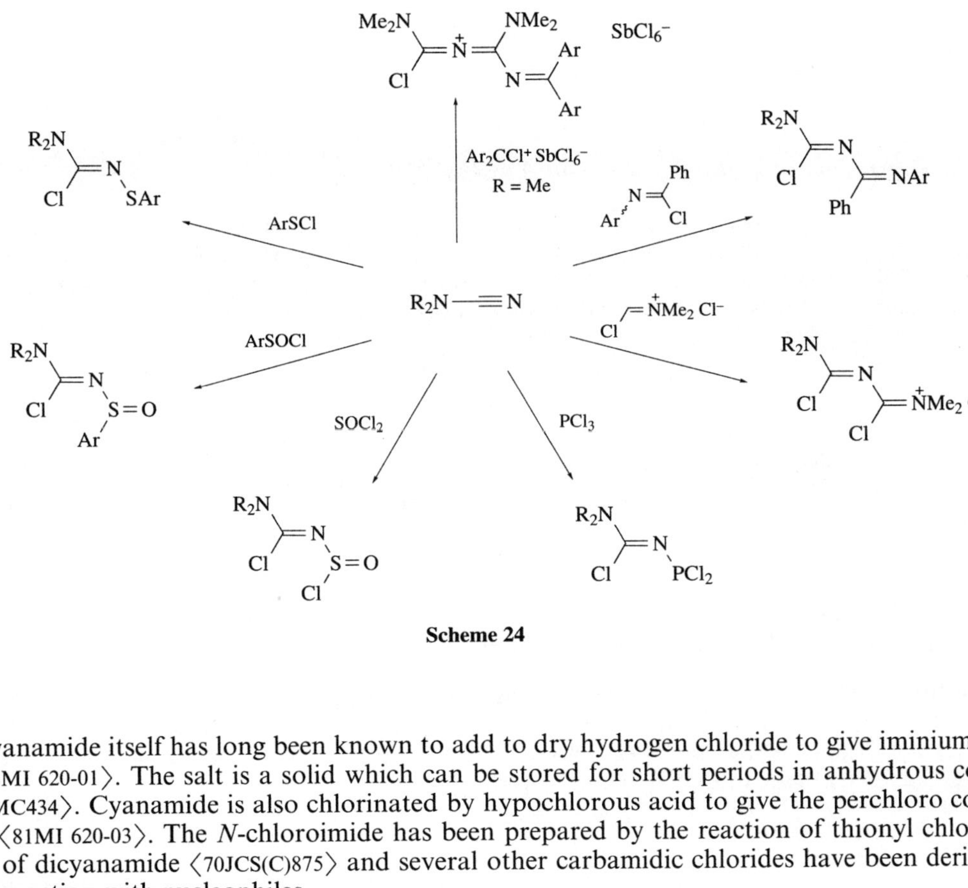

Scheme 24

Cyanamide itself has long been known to add to dry hydrogen chloride to give iminium salt (**74**) ⟨B-68MI 620-01⟩. The salt is a solid which can be stored for short periods in anhydrous conditions ⟨90JMC434⟩. Cyanamide is also chlorinated by hypochlorous acid to give the perchloro compound (**75**) ⟨81MI 620-03⟩. The *N*-chloroimide has been prepared by the reaction of thionyl chloride with salts of dicyanamide ⟨70JCS(C)875⟩ and several other carbamidic chlorides have been derived from it by reaction with nucleophiles.

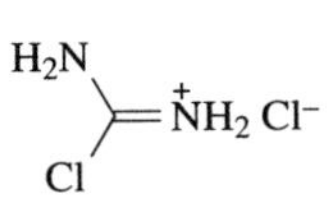

(**74**)

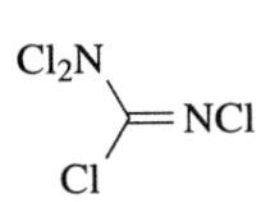

(**75**)

(e) From carbodiimides. Diisopropylcarbodiimide and other carbodiimides with secondary alkyl substituents have been shown to react with activated halogen compounds to give carbamimidic

chlorides. The reaction has been described with phosgene and thiophosgene by Ulrich ⟨B-68MI 620-01⟩. The reaction with trichlorotriazine is shown in Equation (23) ⟨82KGS121⟩; the same type of reaction takes place with benzoyl chloride ⟨85AP(318)1057⟩, aliphatic acid chlorides including chloroacetyl chloride ⟨66CB3155, 77JOC3220⟩, squaric acid dichloride ⟨85AP(318)992⟩ and hydrogen chloride ⟨89TL2379⟩. The reaction with phosphorus pentachloride leads to the formation of six-coordinate phosphorus species (**76**) with partial CN double bonds and with two equivalent phosphorus–nitrogen bonds ⟨68AG(E)299, 69AG(E)455, 90IC5081⟩.

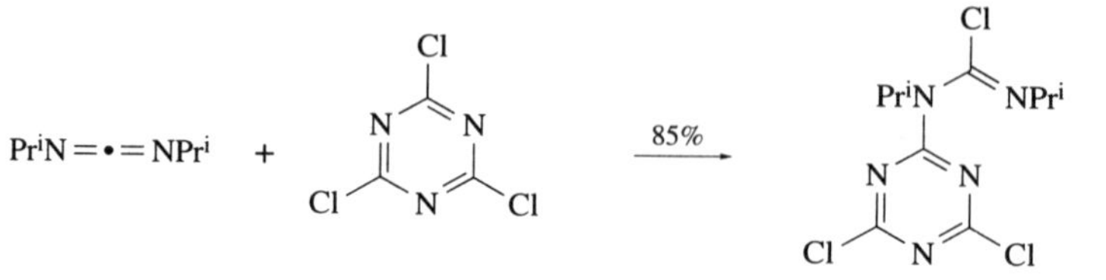

(23)

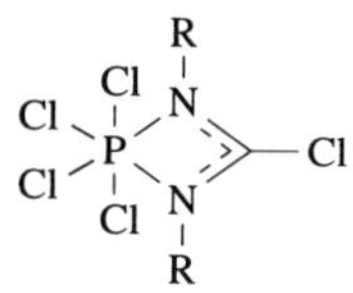

(76)

(v) Iminocarbonyl chlorides with a phosphorus function

There are a few compounds containing this functional group. The phosphonates (**77**) are formed in good yield by the reaction of the isocyanate $Cl_2P(O)CCl_2NCO$ with alcohols ⟨71ZOB2155⟩. The chlorooxime (**78**) is the major product of the nitrosation of compound (**79**) ⟨90ZOB223⟩; it is a precursor to the phosphorus substituted nitrile oxide (**80**) ⟨93ZOB637⟩.

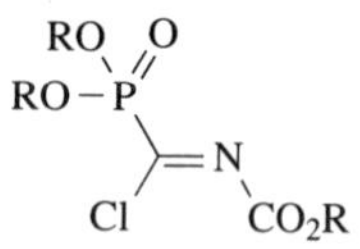

(77)

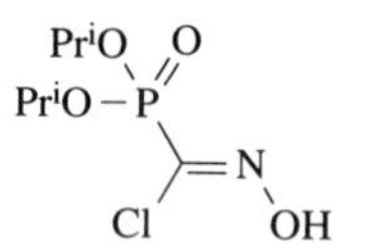

(78)

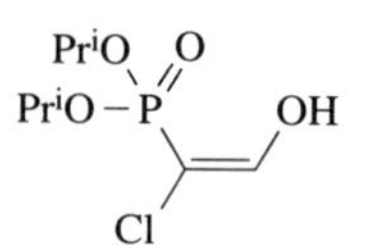

(79)

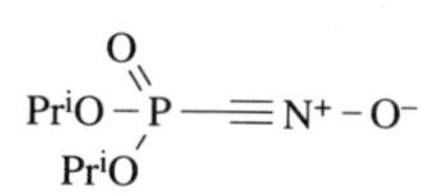

(80)

(vi) Iminocarbonyl chlorides with a metalloid or a metal function

The insertion of alkyl isocyanides into boron–halogen bonds of BX_3 or into metal–chlorine bonds has provided a route to compounds of this type. The reactions are summarised in Scheme 25 ⟨69M1823, 75JOM(101)C1, 77JCS(D)2015, 79JOM(181)69⟩. The unstable adducts formed with titanium(IV) chloride are useful synthetic intermediates ⟨88CB507⟩.

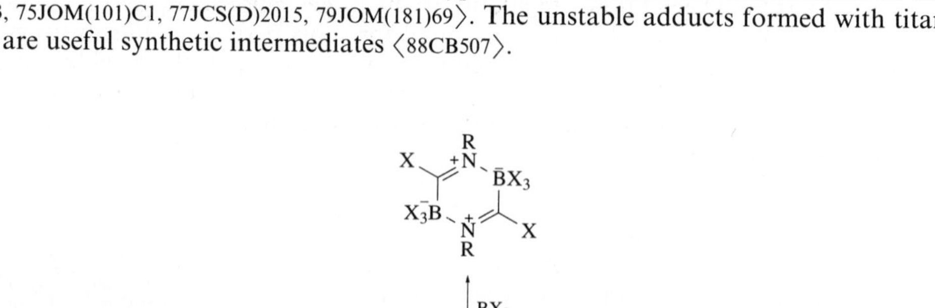

Scheme 25

6.20.1.3.3 *Iminocarbonyl bromides and iodides with one other heteroatom function*

Several examples of compounds of this type have been described with either sulfur or nitrogen as the second heteroatom attached to carbon. The synthetic methods are analogous to some of those used for the corresponding iminocarbonyl chlorides.

Alkyl thiocyanates RSCN have been found to react with bromine and with iodine to give 1 : 1 adducts having the structure (**81**; X = Br or I) ⟨82MI 620-01, 82MI 620-02⟩. The addition of SF_5Br to trifluoromethyl isocyanide gave the imidoyl bromide (**82**) (62%) ⟨85IC4665⟩. Compounds of this type which have been prepared by nucleophilic displacement are the hydrazones (**83**), which are derived from the corresponding dibromo compounds by reaction with thiophenolate anions ⟨72JCS(P2)1050⟩ and the iminium salt (**84**), which is formed from the chloroiminium salt by reaction with iodomethane ⟨81ZOR180⟩.

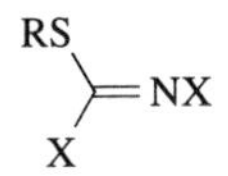
(**81**)

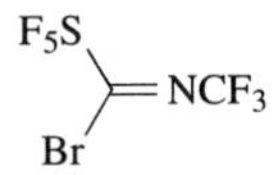
(**82**)

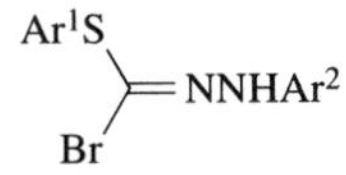
(**83**)

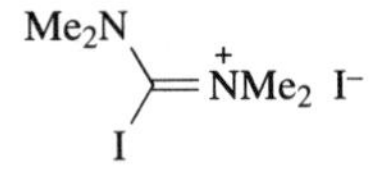
(**84**)

Addition reactions of cyanamides have also been used to prepare carbamimidic bromides and iodides. Cyanamide reacted with hydrogen bromide and hydrogen iodide to give the salts (**85**) in good yield ⟨70GEP1915668⟩ and the addition of hydrogen bromide to diethylcyanamide or to *N*-cyanopyrrolidine also gave the corresponding bromoiminium salts ⟨88JCS(P1)899⟩. Similarly the iminium salts (**86**; X = Br and I) were formed from the sodium salt of the corresponding cyanamide by addition of hydrogen bromide or hydrogen iodide ⟨76CPB26⟩. The squaric acid derivatives (**87**; X = Br and I) have been prepared by addition of the corresponding halides to *N*-cyanopyrrolidine ⟨80S619⟩.

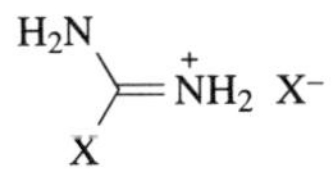
(**85**) X = Br or I

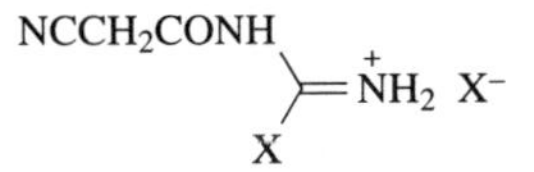
(**86**)

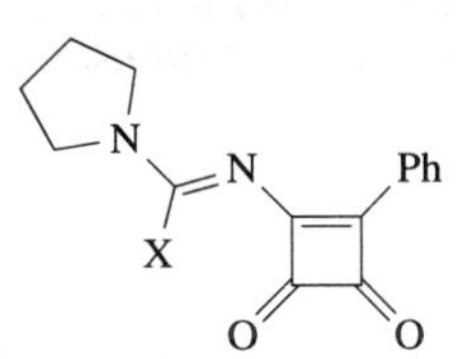
(**87**)

Examples of imidoyl halides with a phosphorus function are provided by the hydrazone (**88**) (which is prepared by a method analogous to that used for the oxime (**78**)) ⟨89ZOB714⟩ and by the diazo compounds (**89**; X = Br and I, R = Ph and OMe) which are prepared by halogenation of the corresponding (**89**; X = H) ⟨79LA1002⟩.

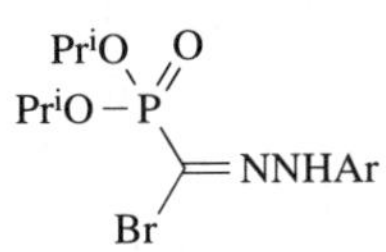
(**88**)

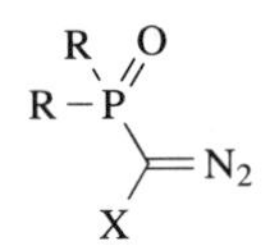
(**89**)

6.20.2 FUNCTIONS CONTAINING AT LEAST ONE CHALCOGEN (AND NO HALOGENS)

6.20.2.1 Iminocarbonyl Compounds with Two Similar Chalcogen Functions

6.20.2.1.1 *Iminocarbonyl compounds with two oxygen functions*

The parent compound of this class, $(HO)_2C{=}NH$, is named in *Chemical Abstracts* as carbonimidic acid. Derivatives are also called iminocarbonates. The most extensive review of the methods of preparation of these compounds is by Kühle ⟨83HOU(E4)561⟩.

(i) From carbonimidic halides and related compounds

The reaction of carbonimidic dichlorides with sodium alcoholates or phenolates in a 1 : 2 molar ratio, and sometimes with alcohols in excess, provides a general route to carbonimidic diesters **(90)** (Equation (24)) ⟨69AG(E)20⟩. Examples include the formation of the diethoxyiminium salt **(91)** from the corresponding dichloroiminium salt and ethanol ⟨66ZAAC(344)113⟩, the diester **(92)** from the carbonimidic dichloride and sodium methoxide ⟨81JHC1127⟩, dimethyl *N*-cyclohexyliminocarbonate from the carbonimidic dichloride and sodium methoxide ⟨87CB339⟩ and the polymer **(93)** from *N*-phenylcarbonimidic dichloride and bisphenol A ⟨91MM2302⟩. *N*-Arylcarbonimidic dichlorides have also been used to trap the intermediates formed in the electrochemical reduction of diaryl-1,2-diketones, giving the cyclic iminocarbonates **(94)** ⟨94TL2365⟩.

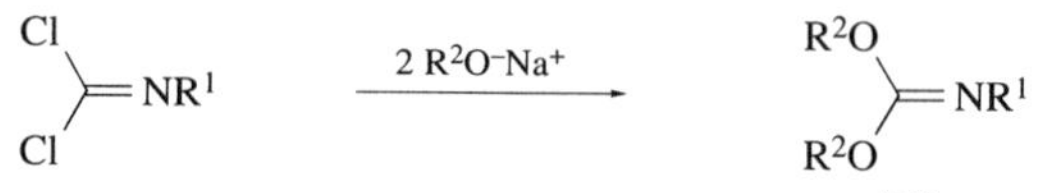

(90) (24)

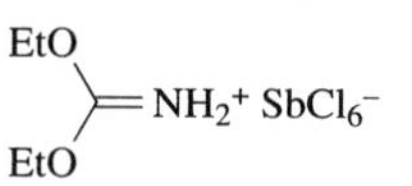

(91)

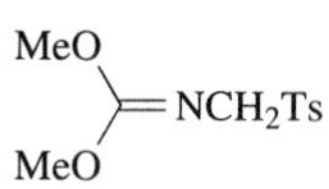

(92)

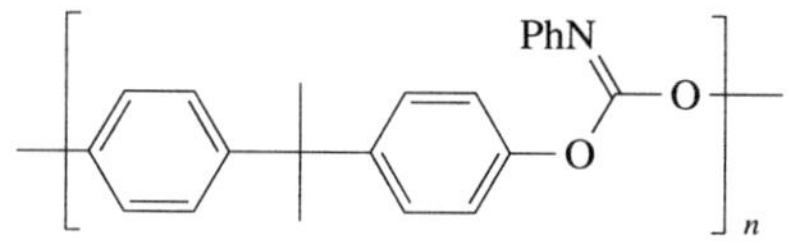

(93)

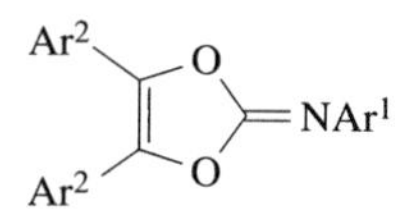

(94)

Carbonimidic diesters have also been prepared by displacement of fluoride from the corresponding carbonimidic difluorides ⟨90JFC(48)395, 92JFC(57)293⟩. For example, compound **(95)** was isolated (70%) from the reaction of the corresponding difluoride with lithium 2,2,2-trifluoroethoxide ⟨90JFC(48)395⟩. Instead of halide, 1,2,4-triazolide can also act as a leaving group for the preparation of carbonimidic diesters ⟨73JPR640⟩.

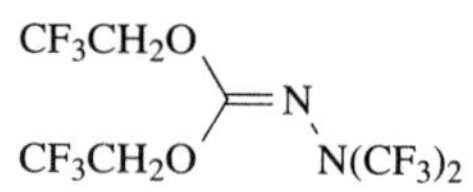

(95)

(ii) From cyanogen halides, cyanates and isothiocyanates

N-Unsubstituted carbonimidic esters have been prepared by nucleophilic addition of alcohols to cyanogen chloride, to cyanogen bromide or to cyanate esters ⟨83HOU(E4)561⟩. The addition to cyanate esters allows esters to be prepared which have two different oxygen substituents (Equation (25)). An example is the preparation of the methyl 4-tolyl ester ($R^1 = Me$, $R^2 = 4\text{-}MeC_6H_4$) (95%) from 4-tolyl cyanate and methanol ⟨73JPR289⟩. A useful method for the preparation of $(PhO)_2C{=}NH$ is the reaction of potassium cyanide with two equivs. of phenyl cyanate: cyanide displaces phenoxide from one mole of phenyl cyanate and this then adds to the second mole to give the iminocarbonate in high yield ⟨65CB3662⟩. Iminocarbonates can also be obtained from isothiocyanates, an example being the preparation of 2-(*N*-phenylimino)-1,3-dioxolane (93%) from phenyl isothiocyanate and $Bu_2Sn(OCH_2CH_2O)$ ⟨77BCJ3271⟩.

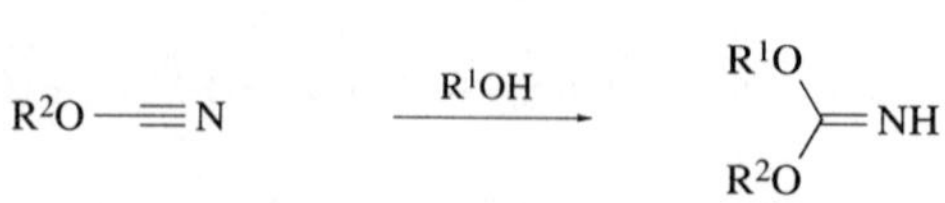

(25)

(iii) From dichlorobis(aryloxy)methanes and orthocarbonates

Dichlorodiphenoxymethane is a useful intermediate for the preparation of carbonimidic acid diphenyl esters (Equation (26)). It has been shown to react with cyanamide to give the *N*-cyano compound (R = CN) (90%) ⟨82JHC1205⟩. An analogous route has been used to prepare the aminosulfonyl compound (R = SO_2NH_2) (67%) ⟨91S753⟩.

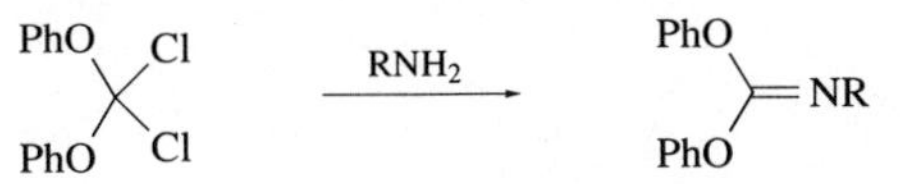
(26)

Reactions of primary amines with tetraethyl orthocarbonate and with trialkyl orthoformates have also been used to prepare iminocarbonates ⟨70BCJ187, 77MI 620-02⟩. Triethyl orthocarbonate also reacts with arenesulfonamides to give the *N*-arenesulfonyl iminocarbonates in good yield ⟨63JOC2902⟩.

(iv) By functional group interconversion on nitrogen

For several types of *N*-substituted carbonimidic diesters the most convenient method of preparation is from another carbonimidic diester (Equation (27)). Examples of these reactions are summarised in Table 5. The simplest type is from *N*-unsubstituted compounds which are halogenated, alkylated or acylated, but in other cases the imidoyl nitrogen of the product is derived from the reagent. Thus, the conversion of dimethyl *N*-phenyl carbonimidate into the *N*-tosyl compound takes place by a [2 + 2] cycloaddition–cycloreversion process ⟨87CB339⟩.

(27)

Table 5 Carbonimidic diesters prepared by exchange of *N*-substituents (Equation (27)).

R[1]	*R*[2]	*R*[3]	*Reagent*	*Yield* (%)	*Ref.*
Et	H	Cl	Cl_2	58	36CB2358
Et	H	$CHPh_2$	$Ph_2CHNH_3{}^+Cl^-$	33	87CB1271
Ph	H	CH_2OMe	$ClCH_2OMe$	50	87CB1271
Me	H	$(CF_2Cl)_2COH$	$(CF_2Cl)_2CO$	83	62USP3115513
Et	H	COMe	MeCOCl	87	77ACH77
Et	H	-COCO-	$(COCl)_2$	88	77ACH77
Et	H	CO_2Et	$ClCO_2Et$	68	86CB3236
Et	H	CSNHPh	PhNCS	22	75ZOR2223
Et	H	PPh_2	Ph_2PCl	> 80	71CB1199
2-$NO_2C_6H_4$	H	SO_2NHMe	$ClSO_2NHMe$	82	85TL4149
2-$NO_2C_6H_4$	H	-SO-	$SOCl_2$	77	87S170
Et	H	NHTs	$TsNHNH_2$	54	63PCS370
Me	H	CN	H_2NCN	73	80EUP14064
Et	Cl	OH	NH_2OH	80	13CB2447
Et	Cl	4-$NO_2C_6H_4SCl_2$	4-$NO_2C_6H_4SCl$	83	85ZOR1281
Me	Ph	Ts	TsNCO	91	87CB339

(v) By O*-alkylation of alkoxycarbonyl functions*

Carbamate esters can be alkylated on oxygen with trialkyloxonium tetrafluoroborates, and this reaction provides a route to some carbonimidic diesters ⟨87CB339⟩. Thus, ethyl *N*-methylcarbamate, EtO_2CNHMe, gave diethyl *N*-methyliminocarbonate (53%) when alkylated with triethyloxonium tetrafluoroborate. Cationic species can also be produced by this method: some examples are shown in Scheme 26 ⟨87CB1271, 92CB2487⟩. There are also examples of *N*-(trimethylsilyl)carbamates which exist in equilibrium with *O*-trimethylsilyl isomers ⟨76IZV2547⟩.

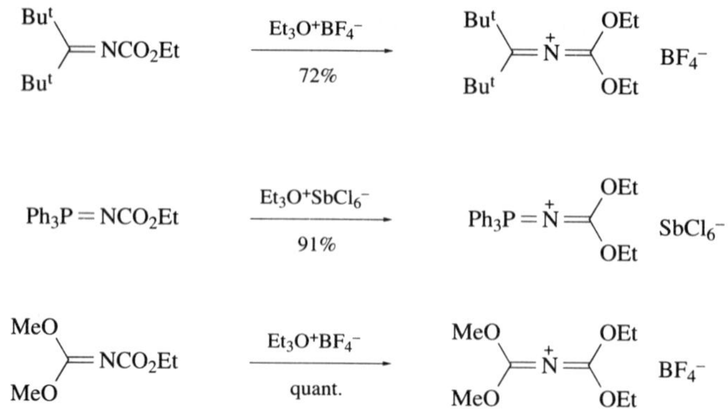

Scheme 26

(vi) Other methods

Azodicarboxylic esters are useful partners in Diels–Alder reactions. They normally act as dienophiles but they can sometimes act as heterodienes ⟨83CJC1213, 89JA2995⟩; for example, 3,4-dihydro-2*H*-pyran and dibenzyl azodicarboxylate reacted to give the cycloadduct (**96**) (70%) ⟨89JA2995⟩.

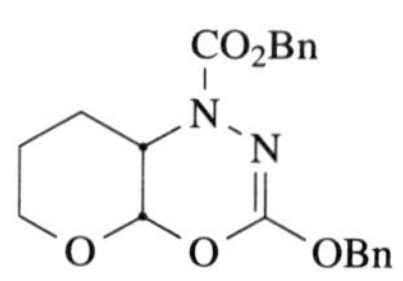

(96)

Hydrazine sulfate is reported to react with bis(trimethylsilyl)amine and carbon dioxide to give the compound (**97**) as a separable component of a mixture with *N*-trimethylsilyl isomers ⟨88ZOB393, 91ZOB1202⟩.

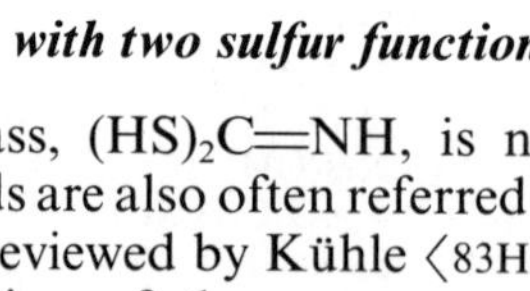

(97)

6.20.2.1.2 *Iminocarbonyl compounds with two sulfur functions*

The parent compound of this class, $(HS)_2C{=}NH$, is named in *Chemical Abstracts* as carbonimidodithioic acid. The compounds are also often referred to as iminodithiocarbonates. Methods of preparation have previously been reviewed by Kühle ⟨83HOU(E4)579⟩.

Most of the methods of preparation of these compounds are analogous to those used for iminocarbonates; however, because of the relative ease with which sulfur can be alkylated, alkylation methods are more important than for the oxygen compounds.

(i) From carbonimidic halides

This method of preparation has not been used extensively. Carbonimidic dichlorides react with thiols or thiolate anions with displacement of one (Section 6.20.1.3.2) or both chloride ions (Scheme 27). The reaction is successful with carbonimidic dichlorides bearing a range of substituents on nitrogen ⟨83HOU(E4)579⟩. Different sulfur substituents can be introduced by successive displacement; for example, compound (**98**; $R^3 = Ph$) was prepared (68%) from the intermediate (**99**; $R^1 = Ts$, $R^2 = Me$) and sodium thiophenolate ⟨86ZC204⟩. Successive displacement of bromide by different sulfur nucleophiles has also been carried out with a carbonimidic dibromide $Br_2C{=}NN{=}CHAr$

⟨72JCS(P2)1050⟩. The carbonimidic difluoride $F_2C{=}NCF_3$ reacts in an analogous way with thiols ⟨69JGU183⟩.

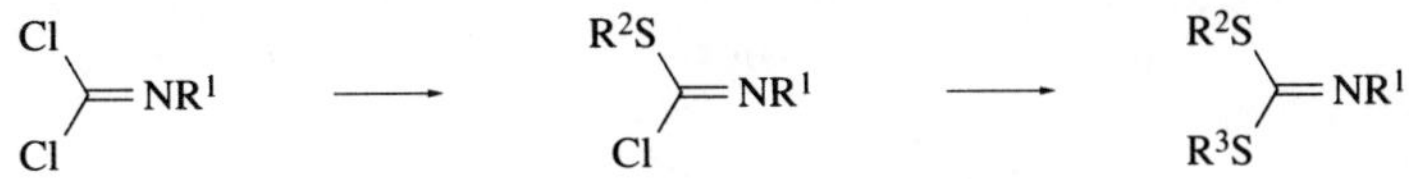

Scheme 27

(ii) From cyanogen chloride and thiocyanates

Cyanogen chloride reacts with dithiols such as ethane-1,2-dithiol and benzene-1,2-dithiol in the presence of HCl to give hydrochloride salts of *N*-unsubstituted iminodithiocarbonates; an example is the salt (**100**) derived (64%) from propane-1,3-dithiol ⟨83HOU(E4)579⟩. Alkyl thiocyanates also react with thiols to give compounds of this type, including those in which the sulfur substituents are different ⟨83HOU(E4)579⟩.

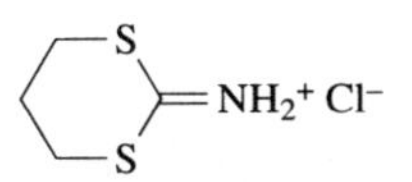

(100)

(iii) Methods involving S*-alkylation*

A variety of activated amino compounds can be converted into *S*,*S*-dialkyl iminodithiocarbonates by reaction with a base, carbon disulfide and a haloalkane (usually iodomethane) (Scheme 28) ⟨83HOU(E4)579⟩. The method was used initially to convert sulfonamides into the dithioesters $RSO_2N{=}C(SMe)_2$ ⟨66CB2885, 91S220⟩ but it also works well with cyanamides ⟨77ZAAC(434)115, 84CCC2285⟩, with aromatic primary carboxamides ⟨79S554⟩ and with heteroaromatic amines including 2-aminothiazoles ⟨91JCR(S)160⟩ and 2-aminobenzothiazoles ⟨82S590⟩. *S*-Alkylation and *S*-acylation of the intermediates shown in Scheme 28 can also be used as a method of preparation. Dithiocarbamates $R^1NHCS_2R^2$ can be alkylated in high yield to give iminodithiocarbonates bearing either identical or different substituents on sulfur ⟨82LA231, 83HOU(E4)579, 83S375, 85S891⟩. Dithiocarbamates $R^1_2NCS_2R^2$ can also be alkylated on sulfur to give salts. The salt $(Me_2S)_2C{=}NMe_2{}^+I^-$ has been prepared (80%) in this way; it can also be made by *N*-alkylation of $(Me_2S)_2C{=}NMe$ ⟨66CB3268⟩. The tosylhydrazones (**101**), which are useful as precursors to bis(alkylthio)carbenes, can similarly be made by alkylation of $TsNHNHCS_2Me$ ⟨72CC354, 87S267⟩ or of the potassium salt $TsNHNHCS_2{}^-K^+$ ⟨66LA(694)44⟩. Reaction of this salt with α-chloroketones results in the formation of the cyclic iminodithiocarbonates (**102**) in good yield ⟨89S132⟩. An intramolecular *S*-alkylation occurs when the dithiocarbamate (**103**) is treated with isocyanates; the thiadiazoles (**104**) are formed in high yield ⟨84JOC1703⟩. Similarly, attempts to prepare compound (**106**) by *S*-methylation of the salt (**105**; $X = NHCS_2Me$) resulted in its formation in poor yield, the major product being 2,5-bis(methylthio)-1,3,4-thiadiazole resulting from intramolecular *S*-acylation ⟨86ZAAC(533)99⟩.

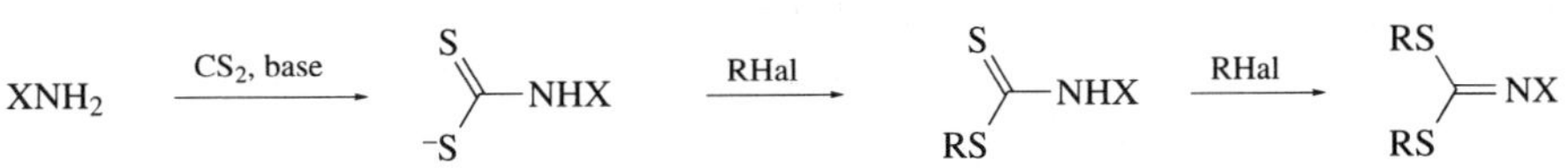

Scheme 28

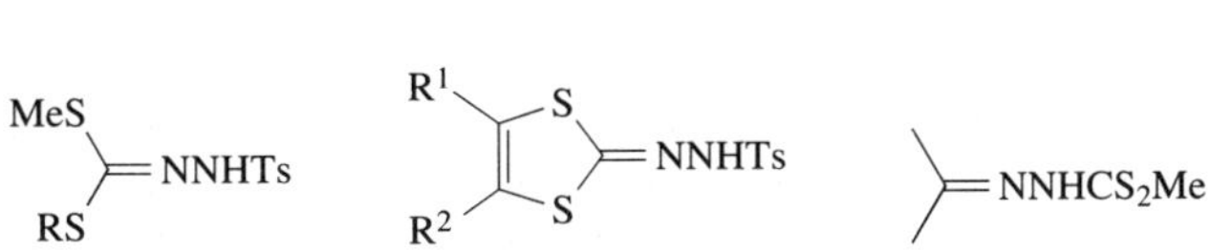

(101) **(102)** **(103)**

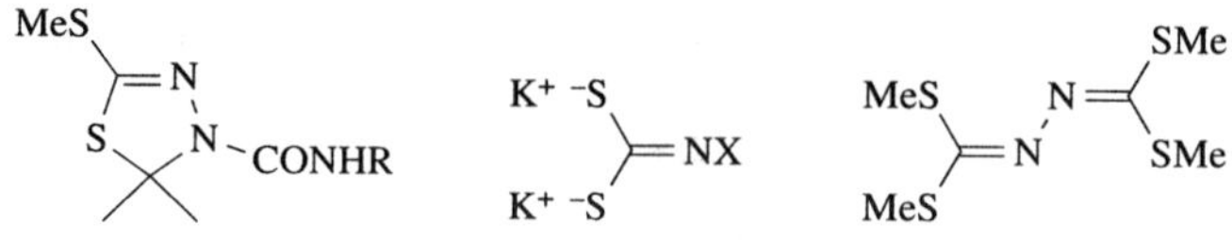

(104) (105) (106)

The salt (**105**; X = CN) can also be protonated by treatment with HCl in ether at −20 °C; however, protonation occurs predominantly on nitrogen to give a zwitterionic product ⟨77ZAAC(434)110⟩.

(iv) By functional group interconversion on nitrogen

The NH group of iminodithiocarbonates can be substituted by a variety of electrophiles ⟨83HOU(E4)579⟩, including isocyanates, acyl chlorides, sulfamoyl chlorides ⟨85TL1105⟩ and aromatic sulfenyl chlorides. *N*-Hydroxyiminodithiocarbonates can be made using hydroxylamine. The reactions are analogous to those of iminocarbonates shown in Equation (27) and Table 5.

(v) Other methods

Diazosulfones (R = Ph, Ar or Et) have been prepared by diazo transfer (Equation (28)) using tosyl azide and a base ⟨64CB735, 94T3195⟩.

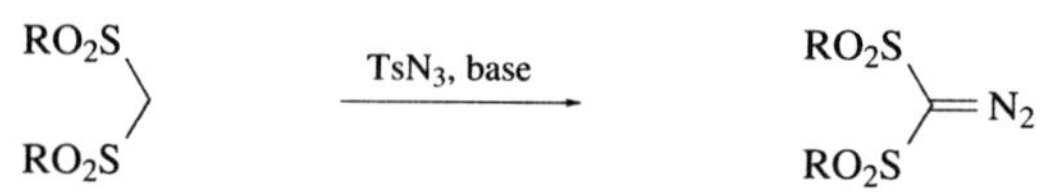

(28)

6.20.2.1.3 *Iminocarbonyl compounds with two selenium functions*

A few compounds with this functional group are known ⟨71JPR804, 72BCJ489, 80CC866⟩. One example is the salt (**107**), which was prepared from $Cl_2C{=}NMe_2{}^+Cl^-$ by successive reaction with hydrogen selenide and triethylamine, then 3-bromobutan-2-one and trifluoroacetic acid ⟨80CC866⟩.

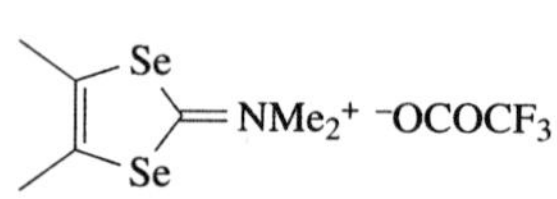

(**107**)

6.20.2.2 Iminocarbonyl Compounds with Two Dissimilar Chalcogen Functions

6.20.2.2.1 *Iminocarbonyl compounds with one oxygen and one sulfur function*

The parent compound of this class, HO(HS)C=NH, is named as carbonimidothioic acid in *Chemical Abstracts*. Most of the methods for the preparation of derivatives are analogous to those used for compounds with two oxygen or two sulfur functions.

(i) From carbonimidic halides

Carbonimidothioic esters (**108**) bearing a variety of substituents have been prepared by displacement of chloride from carbonimidic chlorides and appropriate oxygen or sulfur nucleophiles (Scheme 29) ⟨64JGU4149, 83HOU(E4)561⟩. Both chloride functions of *N*-benzenesulfonylcarbonimidic dichloride can be displaced by 2-thiolethanol and similar bidentate nucleophiles; thus, the oxathiolane (**109**) is prepared in good yield ⟨67AP(300)553⟩. The reaction is probably a general one: it has also been reported for the *N*-phenyl- ⟨64JGU4149⟩ and the *N*-benzoyl- ⟨66CB1912⟩ carbonimidic dichlorides.

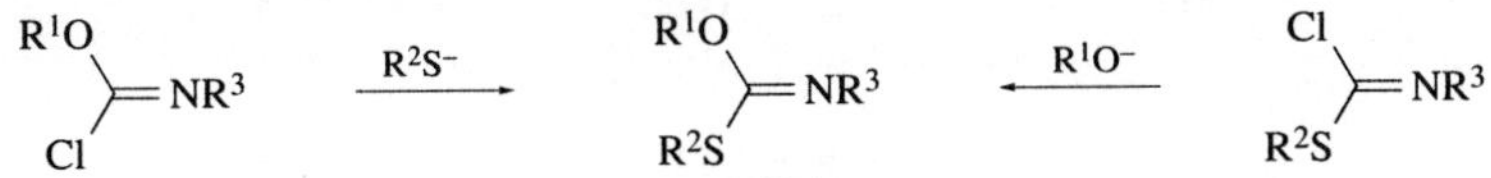

(**108**)

Scheme 29

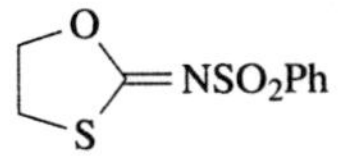

(**109**)

(ii) *From cyanates and thiocyanates*

N-Unsubstituted compounds of this class are most easily prepared either by the addition of thiols to cyanates or from alcohols and thiocyanates (Scheme 30) ⟨64CB3022, 83HOU(E4)561⟩. Cyanogen chloride can also be used as a starting material, with 2-thiolethanol and related compounds, for the preparation of cyclic derivatives such as (**110**). An alternative method of preparation of the salt (**110**) is the reaction of HSCN with ethylene oxide and HCl ⟨58HCA377⟩.

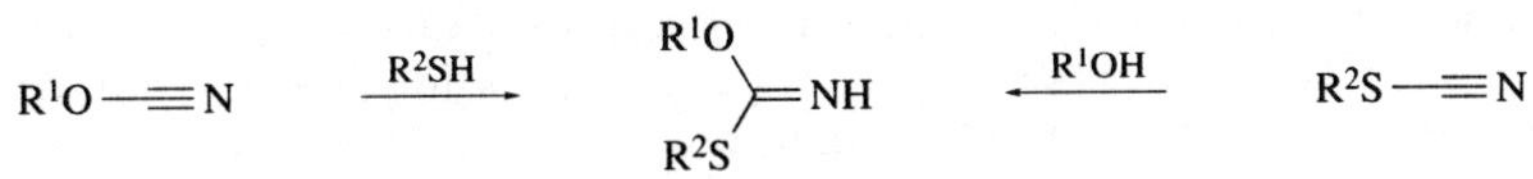

Scheme 30

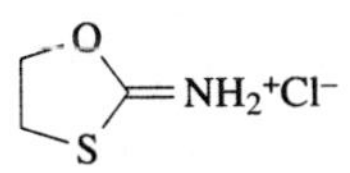

(**110**)

(iii) S-*Alkylation methods*

A simple method for the preparation of compounds of this class is, in principle, the *S*-alkylation of anions (**111**) derived from *O*-alkyl thiocarbamates ⟨73JCS(P1)2644⟩. In practice the most efficient way of bringing about the reaction is often to generate the anions *in situ* from isothiocyanates and alkoxide ions (Scheme 31) ⟨75ZOR1628, 92TL1025⟩. A variant of this reaction is the reaction of acyl isothiocyanates with propargyl alcohol: the intermediates (**112**) undergo spontaneous cyclisation to give the cyclic iminothiocarbonates (**113**) ⟨73CPB62⟩. An intramolecular alkylation also accounts for the formation of the dihydrothiazole (**114**) (72%) from $Br(CH_2)_2NCS$ and phenoxide anions ⟨84CCC295⟩. Oxiranes ⟨78JOC3732⟩ and nitrile oxides ⟨65T1537⟩ can also add across the C═S bond of isothiocyanates.

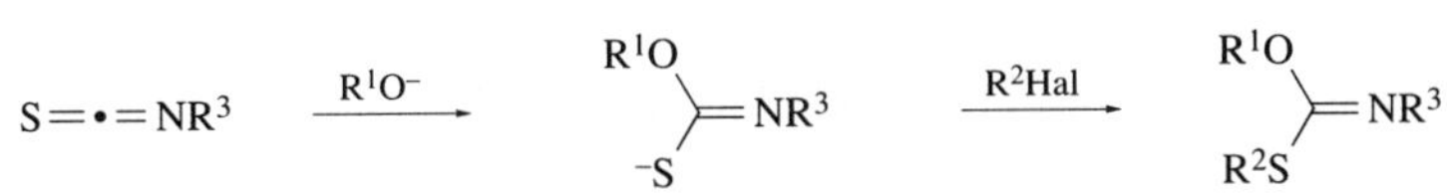

(**111**)

Scheme 31

The procedure illustrated in Scheme 28 for the preparation of iminodithiocarbonates has been applied to the preparation of compounds such as (**115**) by using the sodium salt of cyanamide and

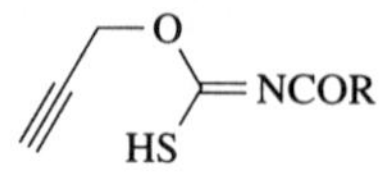
(112)

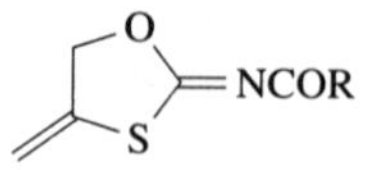
(113)

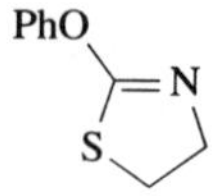
(114)

COS in place of CS_2 ⟨83ZAAC(501)157, 83ZAAC(501)176⟩. *O*-Acyl compounds can also be prepared by successive *S*-alkylation and *O*-acylation of the intermediate formed from sodium cyanamide and COS ⟨83ZAAC(501)169⟩.

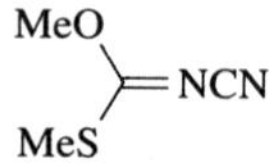
(115)

(iv) By functional group interconversion on nitrogen

Reactions of the type shown in Equation (27) and Table 5 are also possible with this class of compounds, although fewer examples have been reported ⟨83HOU(E4)561⟩. *N*-Acylation can be carried out using acyl chlorides and sulfonation by sulfonyl chlorides ⟨90GEP3841185⟩. Hydroxylamine reacts with salts such as (**110**) to give the corresponding hydroxyimino derivatives, and potassium cyanate introduces the $CONH_2$ group on to nitrogen ⟨58HCA377⟩.

(v) Other methods

The diester (**116**) has been prepared (70%) by heating phenyl isothiocyanate with dimethyl formamide diethyl acetal, $Me_2NCH(OEt)_2$ ⟨77CB37⟩. A good route to compounds (**108**; $R^3 = CN$) is the reaction of the dithioesters (**117**) with cyanamide and a base, followed by *S*-alkylation ⟨70AP(303)625, 84JHC61⟩. The salts (**118**) have been obtained from the iminodithiocarbonates (**119**) and potassium hydroxide ⟨87PS(29)1⟩.

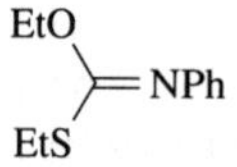
(116)

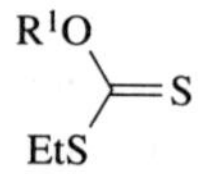
(117)

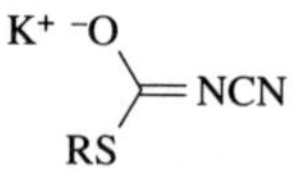
(118)

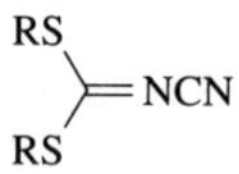
(119)

6.20.2.2.2 *Iminocarbonyl compounds with one oxygen or sulfur and one selenium or tellurium function*

Several cyclic structures incorporate functional groups of these types. The salts (**120**) have been prepared by acid catalysed cyclisation of α-selenocyanoketones, $R^1COCH_2R^2SeCN$ ⟨89EGP270530⟩. Analogous thiaselenoliminium salts, such as (**121**), are formed when the thiaselenazines (**122**) are treated with HI ⟨84S667⟩. The imines (**123**; X = NPh or NCO_2Et) are conveniently prepared from the corresponding thiones (X = S) by reaction with azidobenzene or with ethyl azidoformate ⟨80JHC117⟩.

Compound (**123**; X = NPh) has also been prepared from phenyl isothiocyanate and the anion PhC≡CSe⁻ ⟨71JPR804⟩. Reaction of the tellurium compound (**124**) with bromine gave a solid product which was formulated as (**125**) but it was too unstable to characterise fully ⟨81JOM(217)329⟩.

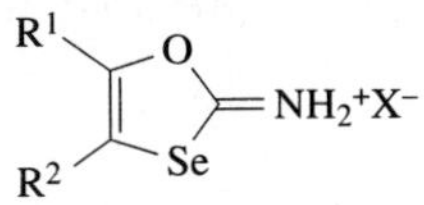

(120)

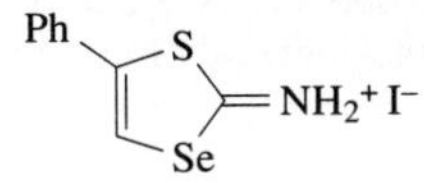

(121)

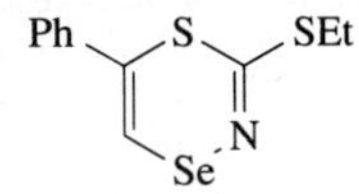

(122)

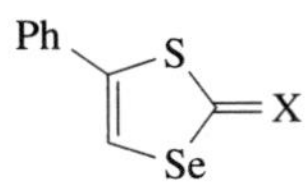

(123)

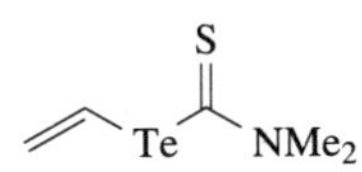

(124)

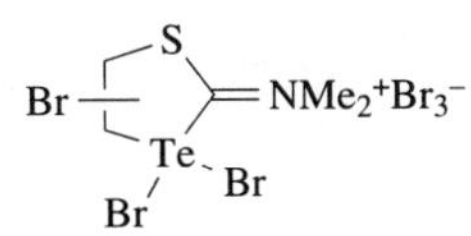

(125)

6.20.2.3 Iminocarbonyl Compounds with One Chalcogen and One Other Heteroatom Function

6.20.2.3.1 Iminocarbonyl compounds with one oxygen and one nitrogen function

These compounds, commonly called isoureas and named in *Chemical Abstracts* as derivatives of carbamimidic acid, $H_2N(HO)C{=}NH$, are widely represented in the literature. The most comprehensive survey of methods of preparation is by Kühle ⟨83HOU(E4)587⟩. A review by Mathias ⟨79S561⟩ on isoureas as alkylating agents also contains a survey of methods for their preparation.

(i) From carbonimidic halides, esters and thioesters

One of the most general methods of preparation of isoureas (**126**) is the displacement of a leaving group on the imidoyl carbon of carbonimidic halides, esters and thioesters by a nucleophile. The types of reaction which have been used are outlined in Scheme 32.

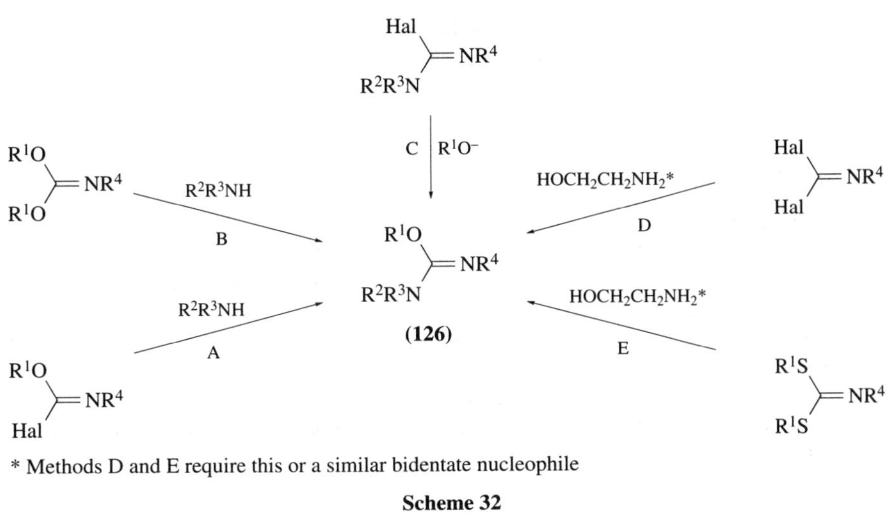

* Methods D and E require this or a similar bidentate nucleophile

Scheme 32

Method A, the displacement of halide from haloformamidic esters, is a useful one when appropriately substituted starting materials are available. An example is the formation of (**126**; $R^1 = R^2 = R^3 = Me$, $R^4 = Ph$) in high yield from methyl *N*-phenylchloroformamidate and dimethylamine ⟨83HOU(E4)587⟩. The zwitterionic triazolones (**127**) are also formed by a reaction of this type from $PhO(Cl)C{=}NCOCl$ and 1,1,disubstituted hydrazines ⟨84JOC2404⟩. The displacement of an oxygen function by an amine (method B) has been used more widely. Several iminocarbonates, especially those with an electron-withdrawing substituent R^4 on nitrogen, participate in this reaction readily. Some compounds of this type which react with primary and secondary amines are listed in Table 6.

Table 6 Carbonimidic diesters $(R^1O)_2C{=}NR^4$ useful for preparation of isoureas (method B in Scheme 32).

R[1]	*R*[4]	*Ref.*
Et	Ts	63JOC2902
Et	CN	67CB2604
Me	CN	85OPP256, 91OPP721
Ph	CN	87AF1008, 87JHC275, 88AF7, 89JOC1062
Ph	COPh	87AF1003
Ph	SO_2NH_2	91S753
Ph	$SO_2C_6H_4Cl$-2	88GEP(O)3634929

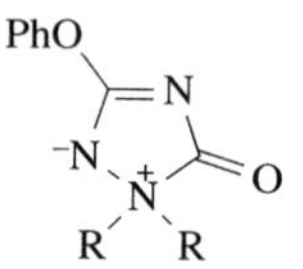

(127)

Formamidinium chlorides bearing a variety of substituents on nitrogen react with alcoholates or phenolates to give isoureas (method C) ⟨83HOU(E4)587, 86GEP(O)3504453⟩. Fluoride can also be displaced, as in the formation of the isourea (**128**) from the corresponding formamidinium fluoride ⟨84JFC(26)321⟩. Methods D and E require the introduction of both nitrogen and oxygen functions; both are therefore essentially restricted to the formation of cyclic isoureas from 2-aminoethanol and related nucleophiles ⟨83HOU(E4)587, 85BCJ3379⟩.

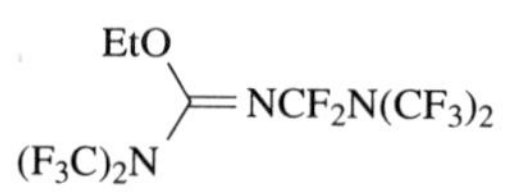

(128)

(ii) From cyanogen halides, cyanates and cyanamides

The reaction of cyanate esters with amines leads to isoureas, as shown in Scheme 33. Strongly basic aliphatic primary and secondary amines give 1 : 1 adducts ⟨83HOU(E4)587⟩ although in the presence of an excess of the cyanate, primary amines can give 2 : 1 adducts ⟨64CB3027⟩. Hydroxylamine adds to phenyl cyanate to give a 1 : 1 adduct ⟨82JOC4177⟩. Cyanogen chloride and cyanogen bromide can give isoureas with suitable bidentate nucleophiles such as 2-aminophenols. A more widely used method is the addition of alcohols to cyanamides. This approach is illustrated by the formation of the isourea (**129**) (68%) from dimethylcyanamide and but-3-en-2-ol ⟨91S86⟩ and of (**130**) (81%) from benzyl alcohol, cyanamide and *p*-toluenesulfonic acid ⟨87TL1969, 90TL4715⟩. A related reaction is the conversion of the nitrilium salt (**131**) (Ad = 1-adamantyl) into the isouronium salt (**132**) by reaction with methanol ⟨84CB502⟩.

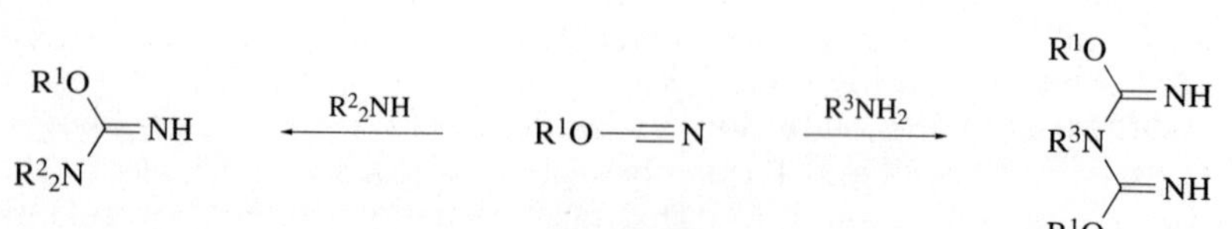

Scheme 33

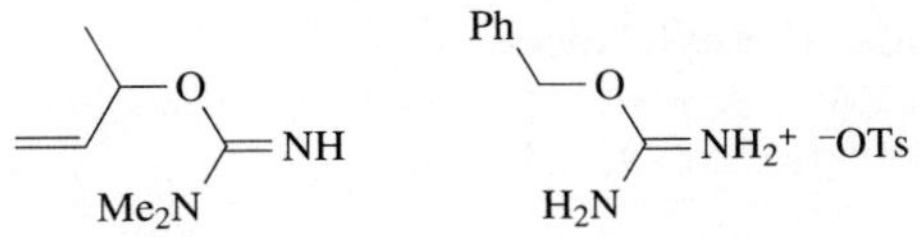

(129) (130)

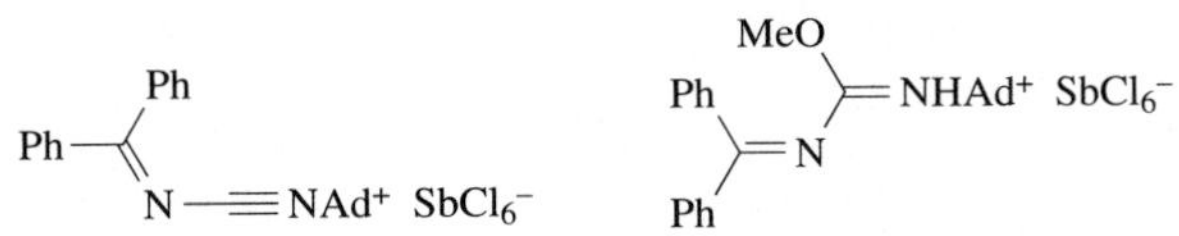

(131) (132)

(iii) From carbodiimides

N,N′-Disubstituted isoureas can be formed by the addition of alcohols or phenols to carbodiimides. The reaction can be catalysed by bases but also goes well in the presence of copper(I) chloride, which allows even hindered alcohols to react. Thus, a range of alcohols can be added to diisopropylcarbodiimide in the presence of CuCl to give the isoureas ROC(=NPr^i)NHPr^i in high yield ⟨87TL4445⟩. The reaction is useful because it converts the oxygen function into a good leaving group. This application is illustrated by the reaction of (*S*)-octan-2-ol and dcc, which gave the isourea with retention of configuration (Scheme 34); the leaving group can then be displaced with inversion of configuration ⟨91S465⟩.

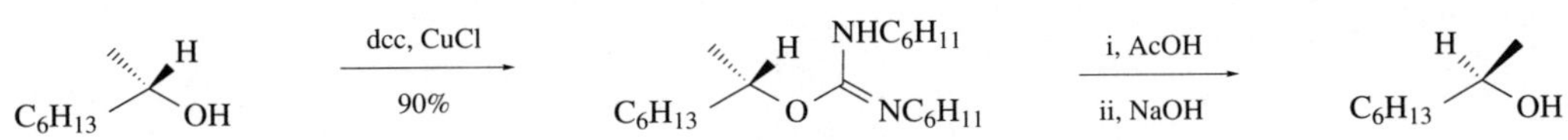

Scheme 34

(iv) By functional group interconversion on nitrogen

Isoureas with one or more hydrogen atoms attached to nitrogen can be functionalised using acyl chlorides and alkanesulfonyl chlorides ⟨83HOU(E4)587⟩. Other types of electrophilic substitution have been reported; thus, MeOC(=NH)NH_2 can be selectively mono- and dichlorinated by reaction with sodium hypochlorite ⟨81JOC5048⟩ and the isourea (**133**) reacted with thionyl chloride to give thiatriazine *S*-oxide (**134**) ⟨85TL1105⟩. There are also several examples of intramolecular *N*-alkylation of isoureas ⟨88CC1175, 89CC452⟩.

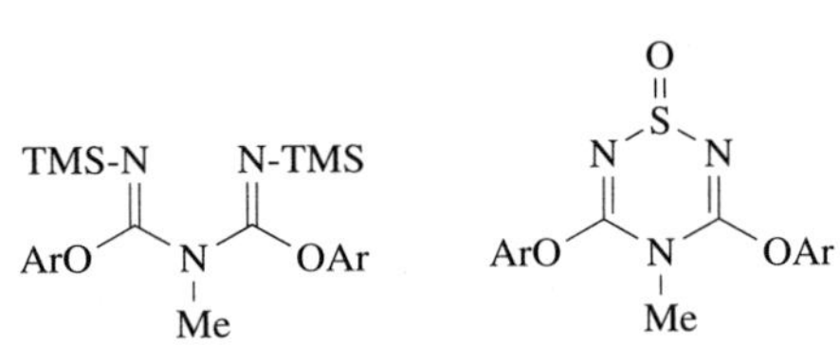

(133) Ar = 4-MeC$_6$H$_4$ **(134)**

(v) Other methods

Isoureas can be prepared by *O*-alkylation of ureas using trialkyloxonium tetrafluoroborates ⟨87CB339⟩. Some examples of intramolecular *O*-alkylation have also been reported ⟨83HOU(E4)587,

89CC452⟩. Dimethyl formamide diethyl acetal reacts with *N*-chloroamides to give isoureas RCON=C(OMe)NMe_2 ⟨70CB256⟩. A series of cyclic isoureas has been prepared using a 1,3-dipolar cycloaddition route (Scheme 35) ⟨91TL5987⟩.

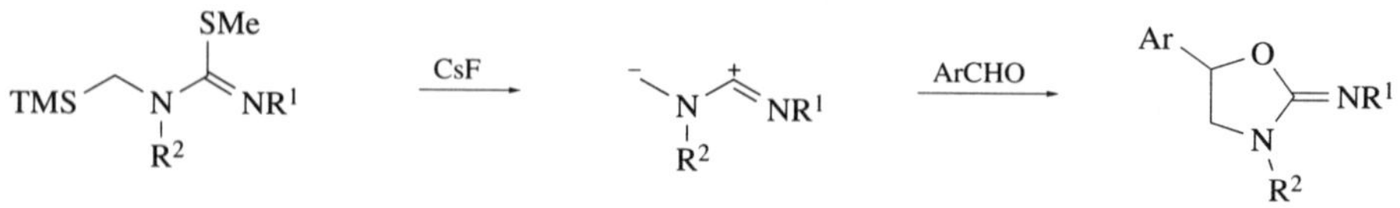

Scheme 35

6.20.2.3.2 *Iminocarbonyl compounds with one oxygen and one phosphorus function*

Compounds of this class have been prepared by nucleophilic displacement of chloride from carbonimidic chlorides with a phosphorus function ⟨71ZOB2155, 90ZOB222⟩. An example is the arylhydrazone (**135**), which was prepared from the corresponding chloro compound, sodium sulfite and HCl ⟨90ZOB222⟩.

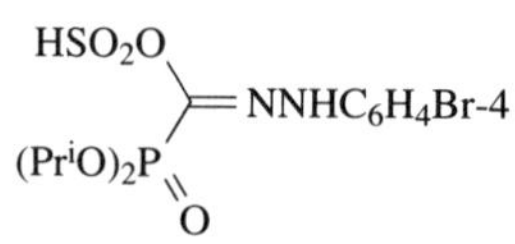

(135)

6.20.2.3.3 *Iminocarbonyl compounds with one oxygen and one metalloid or metal function*

Oxazoles bearing a trimethylsilyl or a trimethylstannyl function at the 2-position are representatives of this type ⟨87JOC3413⟩. The dihydrooxazole (**136**) was prepared (70%) from the oxazolinyllithium intermediate (**137**) and trimethyltin chloride, but attempts to form the corresponding trimethylsilyl derivative led only to opening of the ring ⟨87S693⟩. The corresponding boron substituted species (**138**) were generated as intermediates when (**137**) was treated with trialkylboranes ⟨83SC367⟩.

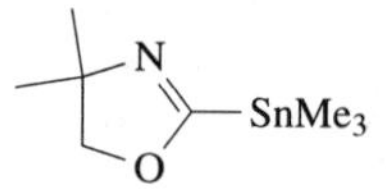

(136)

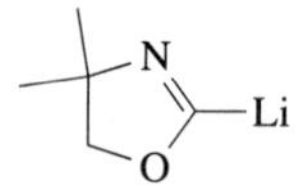

(137)

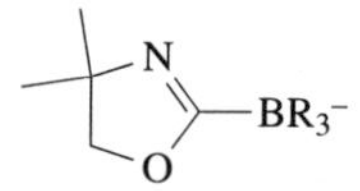

(138)

6.20.2.4 Iminocarbonyl Compounds with One Sulfur and One Other Heteroatom Function

6.20.2.4.1 *Iminocarbonyl compounds with one sulfur and one nitrogen function*

Compounds of this class are usually referred to as isothioureas; the parent compound (HS)(H_2N)C=NH is named as carbamidothioic acid in *Chemical Abstracts*. There are many compounds of this type in the literature and by far the most common method of preparation is from thioureas, by alkylation or other electrophilic substitution on sulfur. More highly functionalised compounds can be obtained from thioureas by successive substitution on sulfur and on nitrogen. A review by Kühle ⟨83HOU(E4)597⟩ provides a comprehensive survey of methods of preparation. The major methods are summarised below.

(i) From carbonimidic halides

Carbonimidic dichlorides have been used only infrequently as starting materials for isothioureas, because of the problems associated with introducing two nucleophiles successively. Compounds of the type RS(PhNH)C=NSO_2Ph have been prepared in moderate yield by successive displacement of chloride by thiols and by aniline ⟨85M651⟩. When both nucleophiles are in the same molecule, as with 2-aminothiophenol, the method works well. If the precursor imidoyl chloride already contains the sulfur function, replacement of chloride by a nitrogen nucleophile occurs readily ⟨91JPR187⟩. The nucleophile is normally a primary or secondary amine but sodium thiocyanate has been used to prepare compound (**139**) from the corresponding chloride ⟨86JOC4043⟩; benzophenone imine, Ph_2C=NH, has also been used in this type of displacement ⟨89JOC1185⟩. The chloride of chloroformamides is similarly readily replaced by sulfur nucleophiles to give isothioureas.

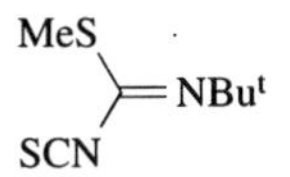

(139)

(ii) From cyanamides, carbodiimides or thiocyanates

The addition of sulfur nucleophiles, usually thiols, to cyanamides or to carbodiimides produces isoureas (Scheme 36). The sulfur function of thiophosphoric diesters $(RO)_2P(=S)OH$ also adds to carbodiimides but the isothioureas so formed are unstable and readily rearrange to thioureas ⟨77JOC3629⟩. Isothioureas unsubstituted on nitrogen can also be prepared by the addition of amines to thiocyanates (Scheme 36).

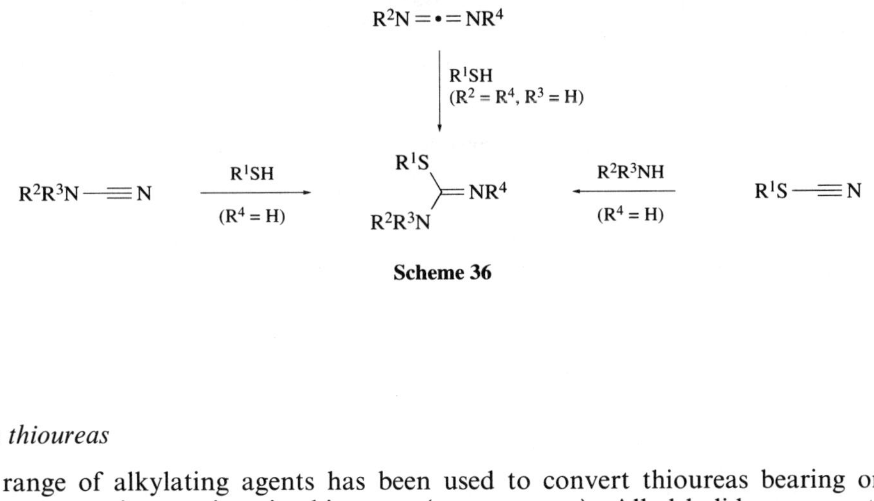

Scheme 36

(iii) From thioureas

A wide range of alkylating agents has been used to convert thioureas bearing one or more hydrogen atom on nitrogen into isothioureas ⟨83HOU(E4)597⟩. Alkyl halides are most commonly used ⟨88S460⟩ but there are many examples of the use of other types of alkylating agent. Even the tertiary bromide 1-bromoadamantane reacted with thiourea to give the isothiourea (**140**; Ad = 1-adamantyl) in 21% yield ⟨92CCC1947⟩. It is also possible to introduce aryl substituents on sulfur by using activated aryl halides or arenediazonium salts as electrophiles. Thioureas can be phosphorylated on sulfur by reaction with a variety of phosphorus electrophiles ⟨90ZOB798, 92IZV425⟩.

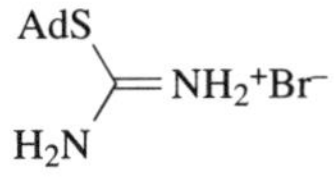

(140)

(iv) From iminodithiocarbonates

Compounds of the type $(MeS)_2C{=}NX$ (X = CN, COR, SO_2R, etc.) react readily with nitrogen nucleophiles (including amines and sulfonamides). Usually one of the sulfur functions can be displaced selectively to give an isothiourea, although the use of nucleophilic amines in excess can lead to the formation of guanidines. Diamines can give either guanidines or bis(isothioureas) (**141**), depending on the dilution ⟨93BCJ148⟩. The reaction has proved to be widely used for the preparation of *N*-cyano substituted thioureas ⟨84CPB4893, 88JOC3120⟩.

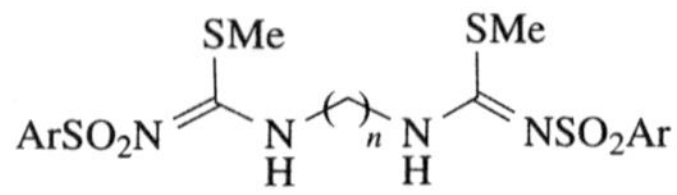

(141)

(v) By substitution on nitrogen

Isothioureas with a free NH group can be acylated by a wide range of reagents including acyl chlorides, isocyanates and isothiocyanates ⟨83HOU(E4)597⟩.

(vi) Cycloaddition methods

A [2 + 2] cycloaddition–cycloreversion method of preparing isothioureas from arenesulfonyl isocyanates and dithiocarbamates is illustrated in Scheme 37. Conjugated isothioureas (**142**) ⟨88CPB4755⟩ act as dienes and undergo [4 + 2] cycloaddition to sulfene ($H_2C{=}SO_2$] to give the thiadiazine dioxides (**143**) ⟨87TL2641⟩.

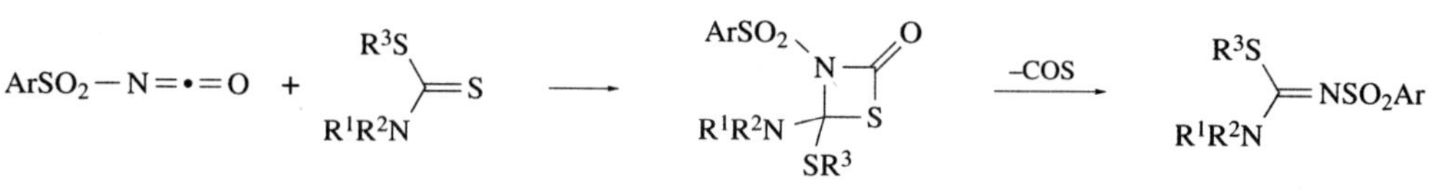

Scheme 37

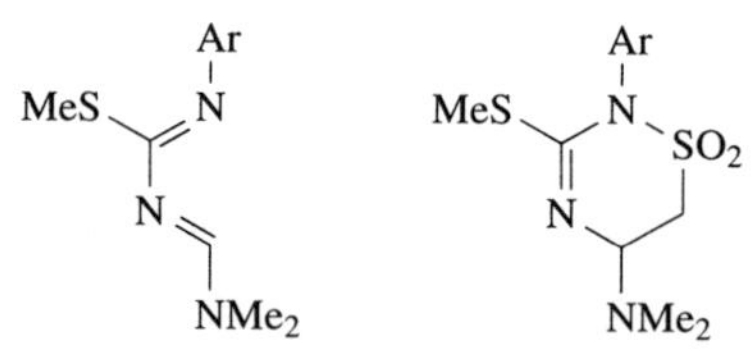

(142) **(143)**

6.20.2.4.2 *Iminocarbonyl compounds with one sulfur and one phosphorus function*

Diazo compounds (**144**) ⟨82JOC1284⟩ and (**145**) ⟨94T3195⟩ bearing a sulfur and a phosphorus substituent have been prepared in good yield from the corresponding methylene compounds by diazo transfer using tosyl azide and a base.

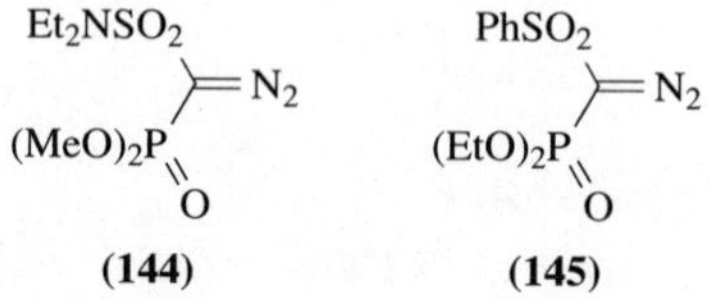

(144) **(145)**

6.20.2.4.3 Iminocarbonyl compounds with one sulfur and one metalloid or metal function

2-Trimethylsilylthiazole (**146**) provides an example of this class of compounds ⟨90PAC643⟩. Dihydrothiazoles bearing both trimethylsilyl and trimethylstannyl substituents are also known ⟨90H(31)1213, 93H(36)473⟩. The route to these compounds is illustrated in Scheme 38. The boron containing species (**147**) has been generated from methyl isocyanide ⟨72LA(755)67⟩ but it is unstable and reacts further to give a zwitterionic 2 : 1 adduct (Scheme 39).

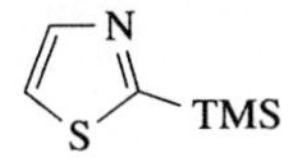

(146)

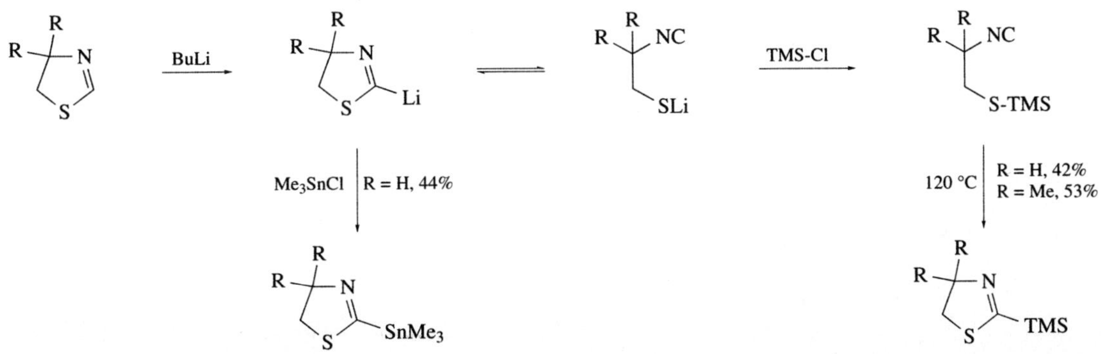

Scheme 38

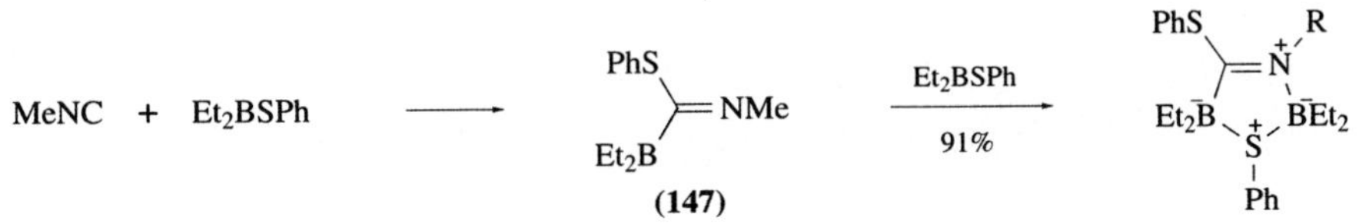

Scheme 39

6.20.2.5 Iminocarbonyl Compounds with One Selenium and One Other Heteroatom Function

Compounds of the type $R^1Se(R^2R^3N)C{=}NR^4$ are isoselenoureas. Although not as numerous as isothioureas, they can be prepared by analogous methods. The most common method of preparation is the alkylation of selenoureas ⟨62JOC2899, 78JHC473, 86NJC51, 89MI 620-01⟩.

An intramolecular addition of an amine to a selenocyanide leads to the formation of a cyclic selenourea (Scheme 40). The selenocyanide was generated *in situ* from 2-aminoethylselenosulfuric acid and cyanide ⟨65JOC2454⟩. An intermediate isoselenocyanate is formed by the ring opening of the azirine (**148**) by carbon diselenide (Scheme 41) and its reaction with dimethylamine then iodomethane gives the isoselenourea (**149**) ⟨82CB2516⟩.

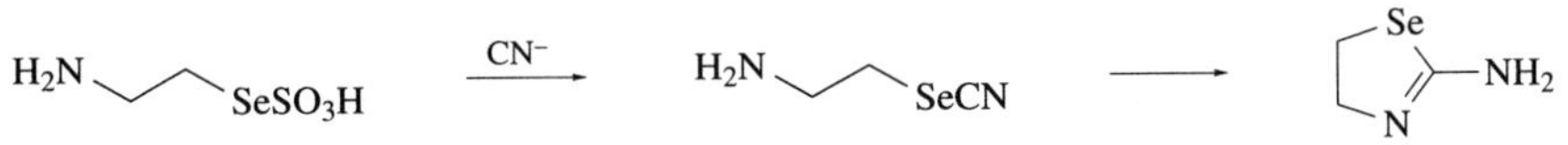

Scheme 40

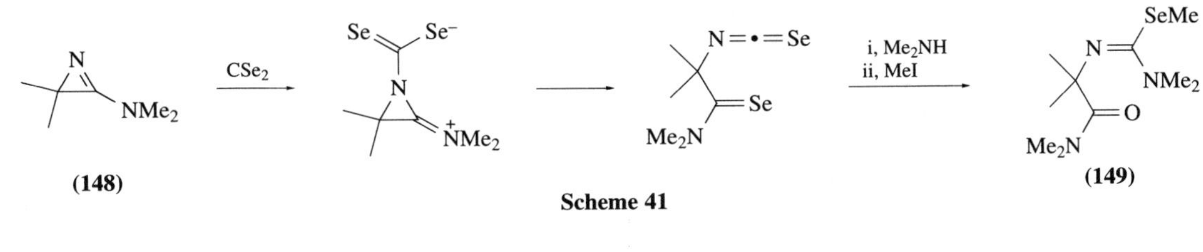

Scheme 41

6.21
Functions Containing an Iminocarbonyl Group and Any Elements Other Than a Halogen or Chalcogen

IAN A. CLIFFE
Wyeth Research (UK), Taplow, UK

6.21.1 IMINOCARBONYL DERIVATIVES CONTAINING AT LEAST ONE NITROGEN FUNCTION (AND NO HALOGEN OR CHALCOGEN FUNCTIONS)

6.21.1.1 Iminocarbonyl Derivatives with Two Nitrogen Functions

Iminocarbonyl derivatives with two nitrogen functions are called guanidines. Common methods of preparation of this class of compounds are shown in outline in Scheme 1. The following sections are ordered by type of substituent on the imino N^2-atom.

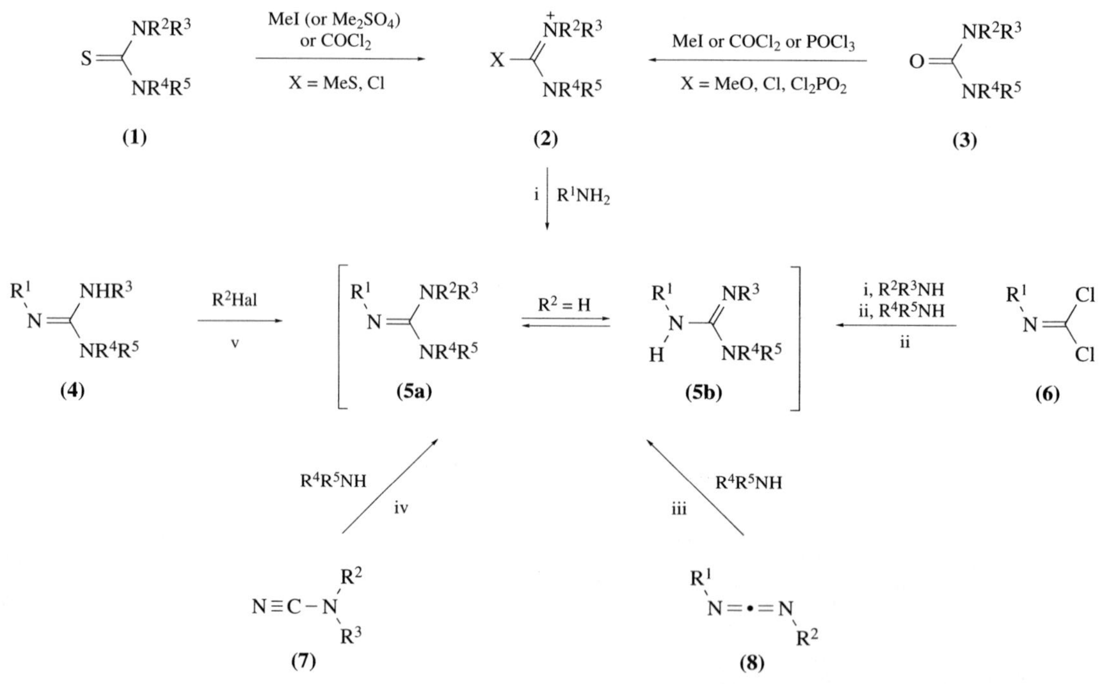

i, *S*-methylisothiouronium, *O*-methylisouronium, *O*-phosphorylisouronium or chloroformamidinium salts; ii, carbonimidic dichlorides; iii, carbodiimides; iv, cyanamides; v, guanidines

Scheme 1

6.21.1.1.1 N-*Unsubstituted iminocarbonyl derivatives*

The preparation of guanidine (**5a**; R^1–R^5 = H) ⟨83HOU(E4)608, 91MI 621-01⟩ and biguanide (**5a**; R^1–R^4 = H, R^5 = C(=NH)NH_2) ⟨59JA3728, 61CRV313⟩ will not be discussed here.

(i) N-*Unsubstituted iminocarbonyl derivatives from amidinium salts and urea derivatives*

The reaction of amines with *S*-methylisothiouronium salts (**2**; X = MeS) produces guanidinium salts in high yields (Scheme 1, *i*). Common variants of this classic reaction, known as the Rathke guanidine synthesis ⟨1884CB297⟩, involve the use of *O*-methylisouronium, *O*-phosphorylisouronium and chloroformamidinium (Vilsmeier) salts in place of *S*-methylisothiouronium salt. The method produces a wide range of substituted guanidines and is particularly suitable when ammonia and primary alkylamines are used as nucleophiles ⟨B-93MI 621-01⟩. The reaction with secondary amines is generally less facile than with primary ⟨63JMC275⟩, whilst hindered amines such as *t*-butylamine fail to react ⟨64CPB946⟩. 1-*t*-Butylguanidine (**5b**; $R^1 = Bu^t$, R^3–R^5 = H) may, however, be prepared in 50% yield by using the more reactive zwitterionic aminoiminomethane sulfonic acid (**2**; X = SO_3^-, R^2–R^5 = H) as the electrophile. This reagent converts primary amines to guanidine derivatives in high yields under mild conditions at room temperature ⟨88TL3183, 92JCS(P1)3179⟩ and has the advantage over the Rathke procedure of not producing unpleasant methanethiol as a by-product. Thermally stable *N*-substituted aminoiminomethanesulfonic acids (**2**; X = SO_3^-, R^2 = alkyl, R^3–R^4 = H, alkyl) are prepared by direct oxidation of thioureas (**1**; R^2 = alkyl, R^3–R^4 = H, alkyl) with peracids in the absence ⟨86S777⟩ or presence ⟨86JOC1882⟩ of sodium molybdate as catalyst. Vilsmeier salts are also more reactive than isothiouronium salts and the highly hindered 1,1,3,3-tetra-isopropylguanidine (**5a**; R^1 = H, R^2–R^5 = Pr^i) is obtained in 81% yield by treating the Vilsmeier salt (**2**; X = Cl, R^2–R^5 = Pr^i) with ammonia ⟨82JCS(P1)2085⟩.

Other variants of the Rathke synthesis utilise heteroaromatic species as leaving groups. 3,5-Dimethyl-1*H*-pyrazole-1-carboxamidine nitrate (**2**; X = 3,5-Me_2-pyrazol-1-yl, R^2–R^5 = H) ⟨53JA4053, 58CJC1541⟩ and 1*H*-pyrazole-1-carboxamidine hydrochloride (**2**; X = pyrazol-1-yl, R^2–R^5 = H) ⟨92JOC2497⟩ both appear to be superior to *S*-methylisothiouronium sulfate as guanylating agents for the preparation of monoalkylguanidines (**5b**; R^1 = alkyl, R^3–R^5 = H). The second reagent is the more reactive of the two, producing products which can be purified by simple crystallisation, but both reagents are poor for preparing guanidine derivatives of sterically hindered, more basic secondary amines. However, the use of heteroaromatic leaving groups has been extended by the finding that *N,N′*-bisurethane-protected carboxamidines are a very reactive species (Scheme 2; R^1 = *t*-butoxycarbonyl (*t*-BOC) $PhCH_2OCO$): in reactions with weak nucleophiles such as 2,2,2-trifluoroethylamine or aniline they give high yields of guanidines after deprotection (using TFA for R^1 = *t*-BOC and catalytic hydrogenation for R^1 = $PhCH_2OCO$). This observation indicates that the electrophilicity of neutral diacylated derivatives is greater than that of protonated unacylated analogues ⟨93TL3389⟩. Pyrazole-containing guanylating agents show utility as reagents for the conversion of ornithine to arginine in solid-phase peptide synthesis ⟨93SC657⟩.

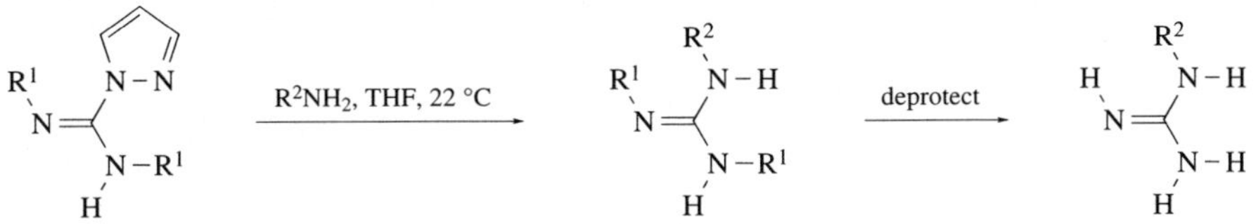

Scheme 2

N-Unsubstituted iminocarbonyl derivatives may be prepared directly and in good yields by the treatment of thiourea derivatives (**1**) with ammonia in the presence of zinc(II) or preferably lead(II) salts (e.g., PbO) with concomitant formation of metal(II) sulfides ⟨68JMC1268⟩.

1,3-Dialkylguanidines are prepared from 1-alkylthiourea derivatives by formation of a 1-alkyl-3-*t*-BOC-thiourea, reaction with an alkylamine in the presence of a water-soluble carbodiimide (WSC), and deprotection (Scheme 3) ⟨92TL5933⟩. 1-Alkylguanidines are obtained in an analogous fashion from 1,3-di-*t*-BOC-thiourea.

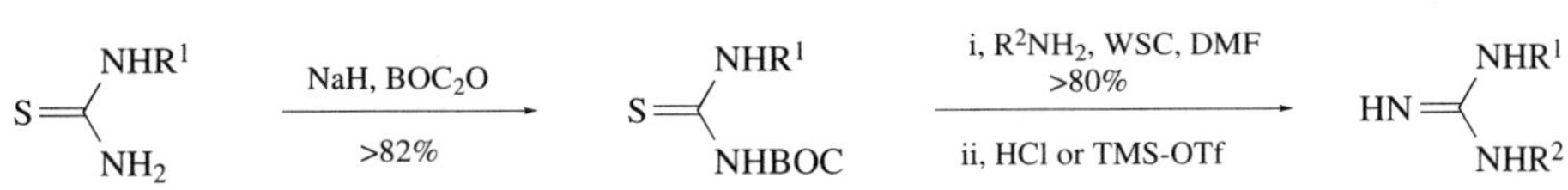

Tf = trifluoromethanesulfonyl

Scheme 3

(ii) N-*Unsubstituted iminocarbonyl derivatives from cyanamides*

The preparation of 1-alkyl- and 1,1-dialkylguanidines (**5a**; R^1–R^3 = H, R^4, R^5 = H, alkyl) by the reaction of amine salts with cyanamide (**7**; $R^2 = R^3$ = H) in aqueous or alcoholic solution gives low yields (0–20%) of products (Scheme 1, *iv*). The reaction only works well when metal (e.g., calcium) cyanamides are fused with alkylammonium salts ⟨58CJC1541⟩. A far more important and widely used preparative reaction is that of a mono- or dialkylamine base or salt with a monoalkylcyanamide (**7**; R^2 = H, R^3 = alkyl), which gives 1,1-di and 1,1,2-trialkylguanidines (**5a**; $R^1 = R^2$ = H, $R^3 = R^4$ = alkyl, R^5 = H, alkyl) in high yields over a wide temperature range (0–180 °C) ⟨68JMC1129, 83HOU(E4)608⟩.

1-Cyano-3*H*-guanidines such as 1-cyano-1,3,3-triethylguanidine are obtained when *N*-alkyldicyanamides are treated with dialkylamines at room temperature (Scheme 4; $R^1 = R^2$ = Et) but, on standing, they decompose into a mixture of mono- and dialkylcyanamides ⟨73TL3653⟩.

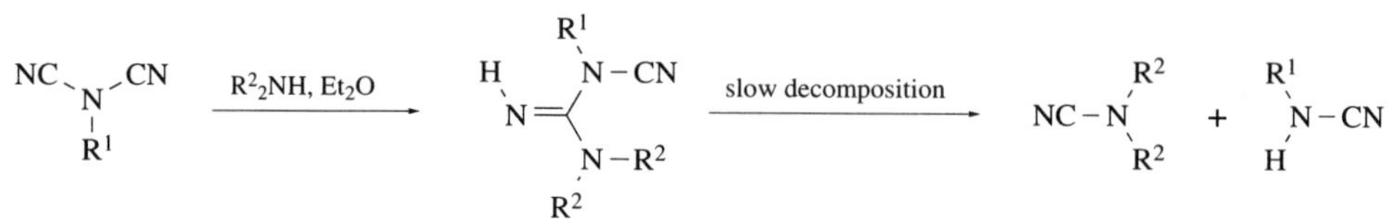

Scheme 4

(iii) N-*Unsubstituted iminocarbonyl derivatives from guanidines*

The alkylation of unsubstituted guanidine with simple alkyl halides usually gives multicomponent mixtures (Scheme 1, *v*). The few reports of good yields of monoalkylated products include the reaction of guanidine with tosylates in the presence of sodium hydride (Equation (1)) ⟨65JMC446⟩ and the reaction of 2-(arylamino)imidazolines with alkyl halides (Equation (2)), which gives exocyclic *N*-alkylation with Na_2CO_3 or NEt_3 as base, and endocyclic *N*-alkylation with NaH ⟨80JMC1217⟩.

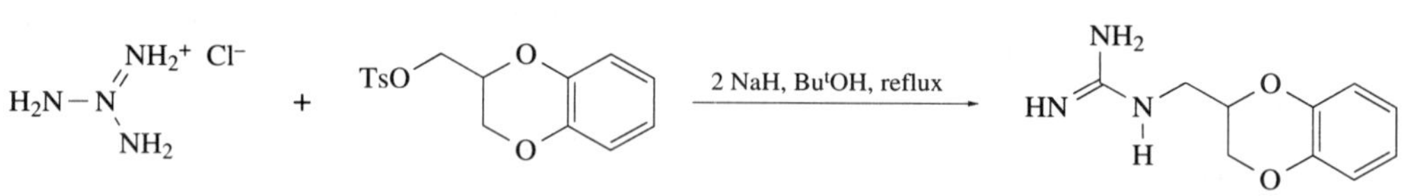

(1)

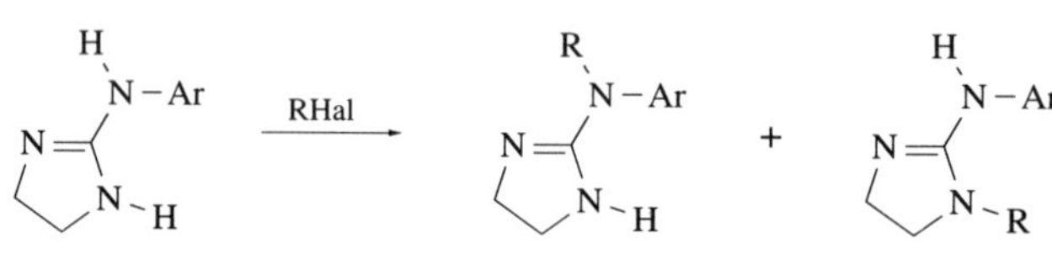

(2)

The reaction of guanidines with aryl halides occurs only when the aromatic group is either activated by electron-withdrawing substituents ⟨64RTC1305⟩ or is itself an electron-deficient heterocycle ⟨62JOC2504⟩ (Scheme 1, *v*; R^1, R^3–R^5 = H, R^2 = 2,4-$(NO_2)_2C_6H_3$ or 3-NO_2-pyridin-2-yl, respectively). An alternative method involving the electrophilic substitution of benzene with hydroxyguanidine-*O*-sulfonate requires Lewis acid catalyst activation (Equation (3)) ⟨67ZN(B)820⟩.

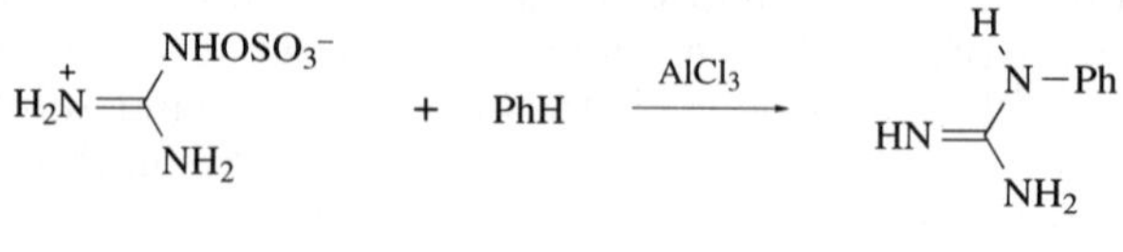

(3)

6.21.1.1.2 N-*Alkyliminocarbonyl derivatives*

(i) N-*Alkyliminocarbonyl derivatives from amidinium salts and urea derivatives*

There are many examples of the Rathke synthesis of *N*-alkylguanidines from *S*-alkylisothiouronium salts (**2**; X = SMe, SEt) and product yields are usually excellent (Scheme 1, *i*) ⟨83HOU(E4)608, 93MI 621-01⟩. Preparation from ureas may also be carried out conveniently via

O-alkylisouronium (**2**; X = OMe) ⟨58CJC1541, 69JMC712⟩ or *O*-phosphorylisouronium salts (**2**; X = Cl_2PO_2) ⟨87CJC626⟩, but the preferred method, via the reactive chloroformamidinium (Vilsmeier) salts (**2**; X = Cl), permits the preparation of sterically hindered pentaalkylguanidines (**5a**; R^1–R^5 = alkyl) ⟨84LA108, 85LA2178, 90T1839⟩ and highly hindered analogues (e.g., **5a**; R^1 = Bu^t, R^2–R^5 = Pr^i) ⟨82JCS(P1)2085⟩. The direct reaction of thioureas (**1**) with alkylamines in hot ethanol produces good yields of guanidines (**5a**; e.g., R^1 = methyl, R^2–R^5 = H) in the presence of lead oxide ⟨69JMC558⟩.

(ii) N-*Alkyliminocarbonyl derivatives from carbonimidic dihalides*

Tri-, tetra- and pentaalkylguanidines are obtained from the reaction of carbonimidic dichloride (**6**) with two molar equivs of the same amine or one each of two different amines ⟨69AG(E)20⟩ (Scheme 1, *ii*). An example of the reaction of a carbonimidic difluoride is provided by perfluoro-2-azapropene ($CF_3N{=}CF_2$), which reacts with various dialkylamines such as dimethylamine ⟨78MI 612-01⟩ and TMS diethylamide ⟨92IC488⟩ to give 2-trifluoromethylguanidines (**5a**; R^1 = CF_3, R^2–R^5 = Me, Et).

(iii) N-*Alkyliminocarbonyl derivatives from carbodiimides*

The reactions of *N,N′*-dialkylcarbodiimides (**8**; R^1, R^2 = alkyl) with ammonium or mono- and dialkylammonium salts at room temperature give high yields of di-, tri- or tetraalkylguanidines (**5a**; R^1 = R^2 = alkyl, R^3 = H, R^4, R^5 = H, alkyl) (Scheme 1, *iii*) ⟨62JA3673⟩.

Carbodiimides of type $Bu^tN{=}C{=}NCHR^1_2$ (**9**; R^1 = Bu^t, NMe_2) react with NBS to afford unstable alkylidenecyanamidinium bromides (**10**; R^1 = Bu^t, NMe_2, X = Br) which undergo von Braun elimination of Bu^tBr to give alkylidenecyanamides (**12**; R^1 = Bu^t, NMe_2) (Scheme 5). Treating the alkylidenecyanamide (**12**; R^1 = Bu^t, NMe_2) with *t*-butyl chloride and antimony pentachloride produces the stable hexachloroantimonate (**10**; R^1 = Bu^t, NMe_2, X = $SbCl_6$) in quantitative yield and this reacts with amines to give alkylideneuronium salts (**11**; R^1 = Bu^t, NMe_2) ⟨84CB502⟩.

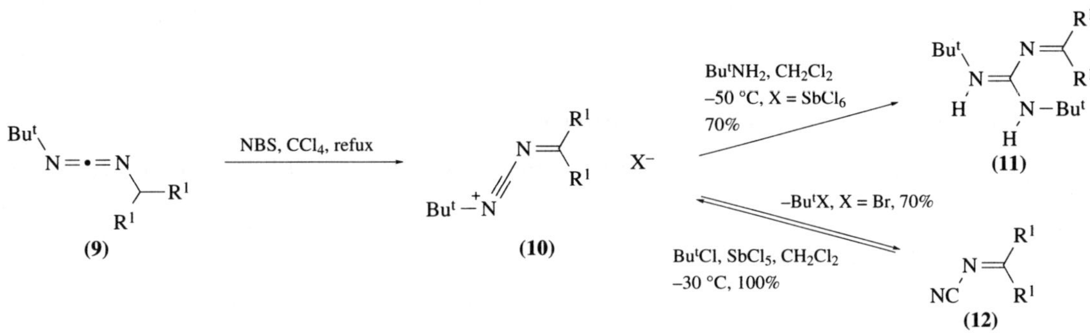

Scheme 5

(iv) N-*Alkyliminocarbonyl derivatives from cyanamides*

Reactions of amines or amine salts with alkylcyanamides at room temperature give high yields of guanidines (Scheme 1, *iv*) ⟨83HOU(E4)608⟩. A particularly mild method performed at very low temperatures involves the reaction of cyanamidinium hexachloroantimonates with ammonia or mono- or dialkylamines (Scheme 6; R^2 = R^3 = alkyl, R^4, R^5 = H, alkyl) ⟨84CB1161⟩.

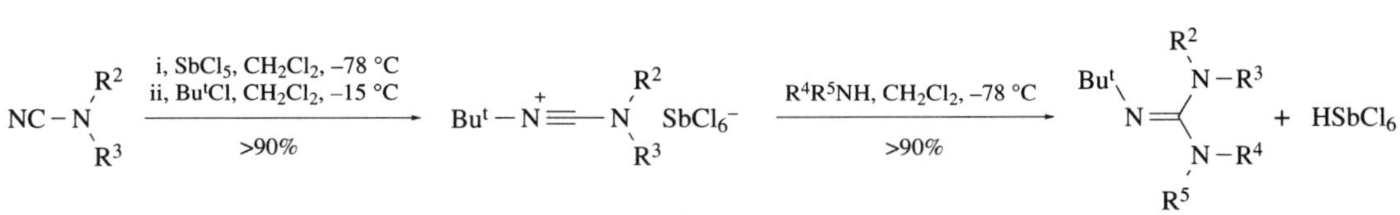

Scheme 6

(v) N-*Alkyliminocarbonyl derivatives by miscellaneous methods*

Complexes of mercury(II) chloride and *t*-butyl isocyanide react with an excess of mono- and dialkylamines to give guanidine derivatives and mercury through a redox decomposition reaction (Equation (4); $R^2 = Bu$, $R^3 = H$ or $R^2 = R^3 = Et$). The reaction with weak nucleophiles (e.g., aniline) is enhanced by the addition of strong bases (e.g., triethylamine) and a mechanism for the reaction has been proposed ⟨75JOM(94)333⟩.

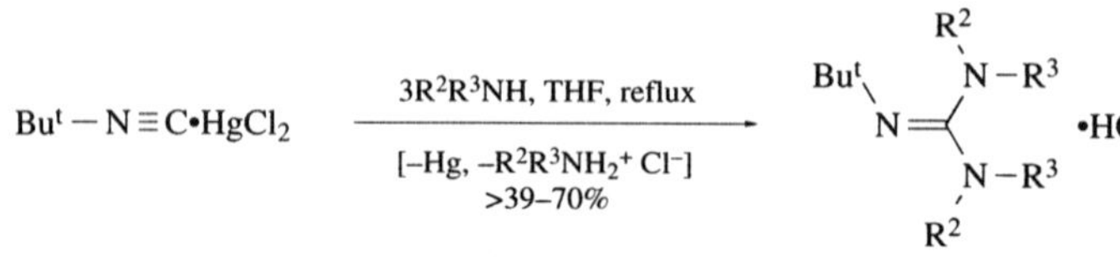

(4)

Lithium aluminum hydride reduction of various alkyl-substituted acylguanidines (**5a**; R^1 = acyl, R^2–R^5 = H, alkyl) gives alkylguanidines in yields ranging from 51% to 62%. The rate of the reaction depends on the number of NH atoms, and yields are improved by using excess reducing agent in THF at room temperature ⟨77JOC3608⟩.

6.21.1.1.3 N-*Alkenyliminocarbonyl derivatives*

(i) N-*Alkenyliminocarbonyl derivatives from amidinium salts*

Ketone imines react with Vilsmeier salts to give amino-substituted enimines (Equation (5); R = H, Me) ⟨75C514⟩.

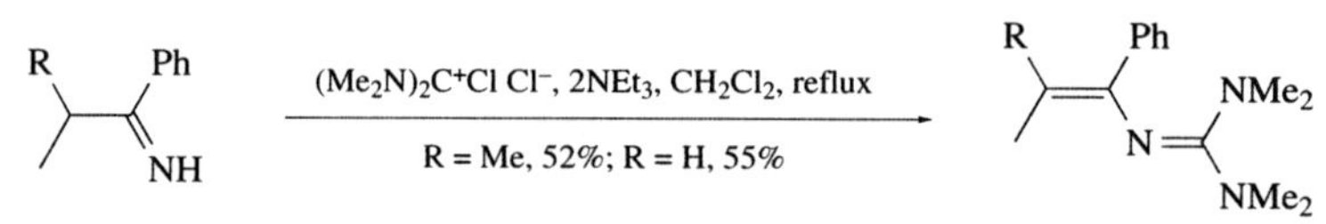

(5)

(ii) N-*Alkenyliminocarbonyl derivatives from guanidines*

1,1,3,3-Tetramethylguanidine reacts with isobutyraldehyde or with dimethyl acetylenedicarboxylate to give amino-substituted enimines (Scheme 7). The former reaction fails with aldehydes bearing one or no α-substituent, and yields in the latter reaction are often poor owing to the base-sensitive nature of the acetylene ⟨75C514⟩.

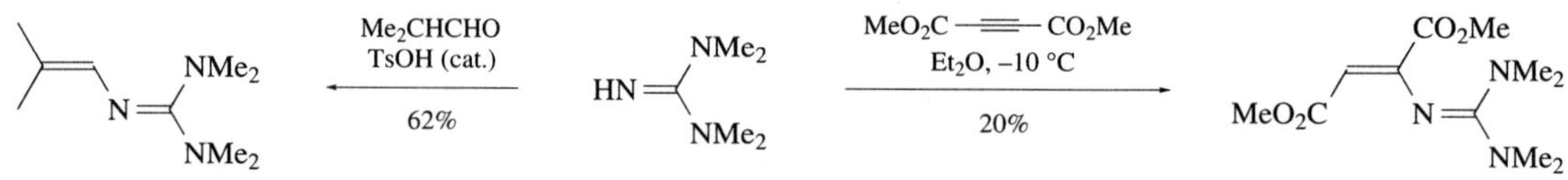

Scheme 7

6.21.1.1.4 N-*Aryliminocarbonyl derivatives*

The methods of synthesis of *N*-aryliminocarbonyl derivatives (i.e., 2-arylguanidines) are essentially the same as those used for 2-alkylguanidines.

(i) N-*Aryliminocarbonyl derivatives from amidinium salts and urea derivatives*

Although anilines are less nucleophilic than alkylamines they will react with *S*-alkylisothiouronium salts (Scheme 1, *i*) to give 2-arylguanidines (**5a**; R^1 = aryl) ⟨66BSF73⟩, and high yields of 2-phenyl-1,1,3,3-tetrabutylguanidine (**5a**; R^1 = Ph, R^2–R^5 = Me) have been obtained from the reaction of aniline with chloroformamidinium (**2**; X = Cl, R^2–R^5 = Me, yield = 84%) ⟨90T1839⟩ and *O*-phosphorylisouronium salts (**2**; X = Cl_2PO_2, R^2–R^5 = Me, yield = 46%) ⟨61CB2278⟩. 2-(Arylimino)tetrahydroimidazoles such as clonidine (**5a**; R^1 = 2,6-$Cl_2C_6H_3$, R^2 = R^4 = H, R^3R^5 = $(CH_2)_2$) are prepared by the reaction of (a) *S*-methylisothiouronium salts (**2**; X = SMe, R^2 = aryl, R^3–R^5 = H) with ethylenediamine ⟨78RTC51⟩; (b) sulfonic acid derivatives (**2**; X = SO_3^-, R^2 = R^4 = H, R^3R^5 = $(CH_2)_2$) with anilines ⟨86JOC1882⟩; and (c) the complex formed from 1-acetylimidazolidinone and phosphorus oxychloride (**2**; X = Cl_2PO_2, R^2 = H, R^4 = MeCO, R^3R^5 = $(CH_2)_2$) with anilines followed by deacylation ⟨80AF1733⟩.

The transimination reaction of *N*-arylphosphimides with thioureas gives 2-arylguanidines in high yield (Equation (6)) ⟨91MIP54636⟩.

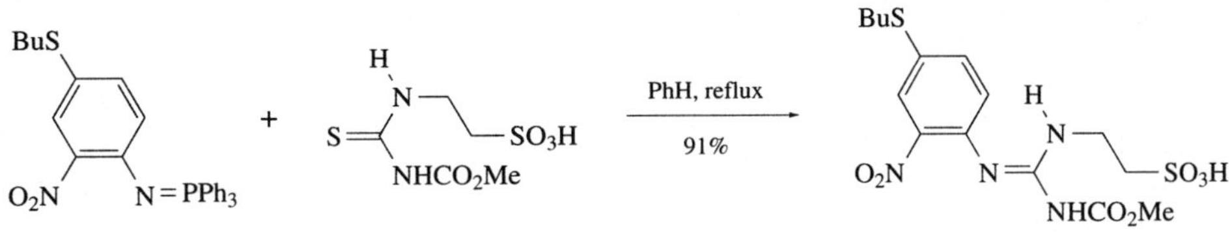

(6)

(*ii*) N-*Aryliminocarbonyl derivatives from carbonimidic dichlorides*

The stepwise replacement of the two chlorine atoms of the *N*-arylcarbonimidic dichloride (**6**; R^1 = 4-NCC_6H_4) by two different monoalkylamines gives 1,3-dialkyl-2-arylguanidines (**5a**; R^1 = 4-NCC_6H_4, R^2–R^4 = alkyl, R^3 = R^5 = H) in high yields (Scheme 1, *ii*) ⟨92SC1191⟩. The reaction of *N*-phenylcarbonimidic dichloride with TMS-dialkylamides gives high yields of intermediate chloroformamidines which react further, only in the case of sterically unhindered chloroformamidines (Scheme 8; R^1 = Me) and TMS-dimethylamide, to give 1,1,3,3-tetramethyl-2-phenylguanidine (R^1 = R^2 = Me) ⟨69TL1421⟩. Organothallium amides (Me_2TlNMe_2) undergo related reactions ⟨73ZC192⟩. The reaction of *N*-(2,4-dichlorophenyl)carbonimidic dichloride (**6**; R^1 = 2,4-$Cl_2C_6H_3$) with ethylenediamine gives clonidine (**5a**; R^1 = 2,6-$Cl_2C_6H_3$, R^2 = R^4 = H, R^3R^5 = $(CH_2)_2$) ⟨78RTC51⟩.

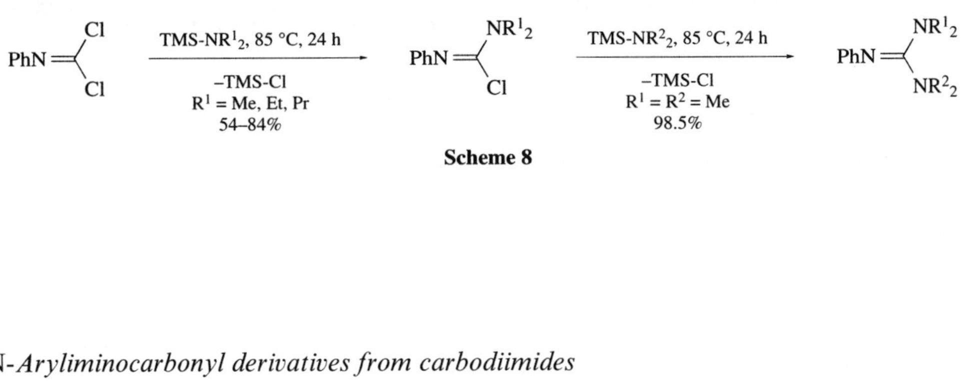

Scheme 8

(*iii*) N-*Aryliminocarbonyl derivatives from carbodiimides*

2-Aryl-1,3-disubstituted guanidines are prepared in good yields from *N,N′*-disubstituted carbodiimides and anilines in the absence of solvent with or without heating; in refluxing toluene; or in DMF at room temperature in the presence of sodium hydride (Scheme 1, *iii*) ⟨80JMC13⟩.

(*iv*) N-*Aryliminocarbonyl derivatives from cyanamides*

Arylcyanamides (**7**; R^2 = H, R^3 = aryl) react with ammonia and primary amines (Scheme 1, *iv*) in methanol at 100°C in a steel bomb to give 2-arylguanidines (**5b**; R^1 = R^4 = H, R^3 = aryl, R^5 = H, alkyl) in reasonable yields (42–85%) ⟨75JMC90⟩.

(v) N-*Aryliminocarbonyl derivatives by miscellaneous methods*

Triarylguanidines are obtained in good yield by the reaction of anilines with 2,2-dichloro-1,3-benzodioxole (Equation (7)) ⟨64LA(675)142⟩.

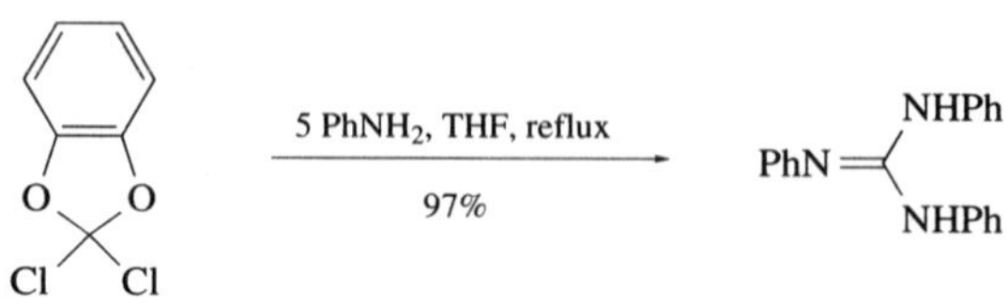

(7)

6.21.1.1.5 N-*Alkynyliminocarbonyl derivatives*

2-(Phenylethynyl)-1,1,3,3-tetramethylguanidine is formed as a transient intermediate in the flash photolysis of phenylguanidinocyclopropenones (Scheme 9). An acylguanidine is the final product in aqueous solution ⟨91AG(E)1356⟩.

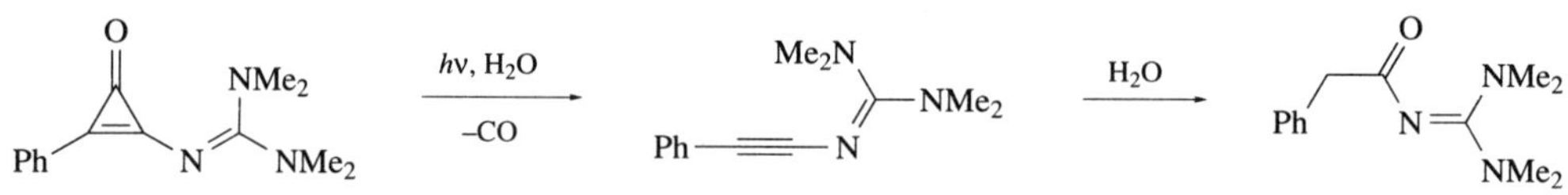

Scheme 9

6.21.1.1.6 N-*Acyliminocarbonyl derivatives*

The tautomerism of monoacylguanidines has been studied ⟨61CJC1017⟩.

(i) N-*Acyliminocarbonyl derivatives from* S-*methylisothiourea*

Acylation of the free base of *S*-methylisothiouronium salts (**2**; X = SMe, R^2 = alkyl, R^3–R^5 = H) with acid chlorides followed by treatment of the resulting *S*-methyl-*N*-acylthioureas with appropriate monoalkylamines gives 1-acyl-2,3-dialkylguanidines (**5a**; $R^1 = R^2$ = alkyl, $R^3 = R^4$ = H, R^5 = acyl) ⟨75AF1477⟩.

(ii) N-*Acyliminocarbonyl derivatives from cyanamides*

The reaction of *N*,*N*-dialkylcyanamides with acid chlorides gives acylated chloroformamidinium salts which on reaction with secondary amines produce acylated guanidines (Scheme 10; R = Me, Et, allyl) ⟨81MI 621-01⟩.

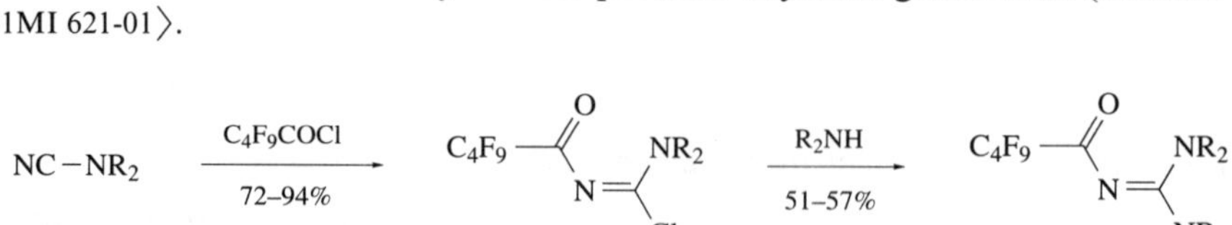

Scheme 10

(iii) N-*Acyliminocarbonyl derivatives from guanidines*

Unsubstituted and monosubstituted guanidines are acylated by condensation of the requisite guanidine free base with the appropriate acid ester ⟨61CJC1017, 75AF1477⟩. The preparation of di- and trisubstituted guanidines takes longer and in these cases acylation of guanidines with acid chlorides or anhydrides is the normal practice ⟨68JOC552⟩, although mixtures of products invariably result ⟨75AF1477⟩. Di- and triacylated guanidines have been deacylated selectively to give monoacylguanidines with either ethanol or a quaternary ammonium hydroxide ion-exchange resin ⟨77JOC3608⟩.

2-Phenyl-1,1,3,3-tetramethylguanidine undergoes exchange reactions with benzoyl and thiobenzoyl isocyanates to give high yields of acylated or thioacylated guanidines (Equation (8)) ⟨78LA1543⟩.

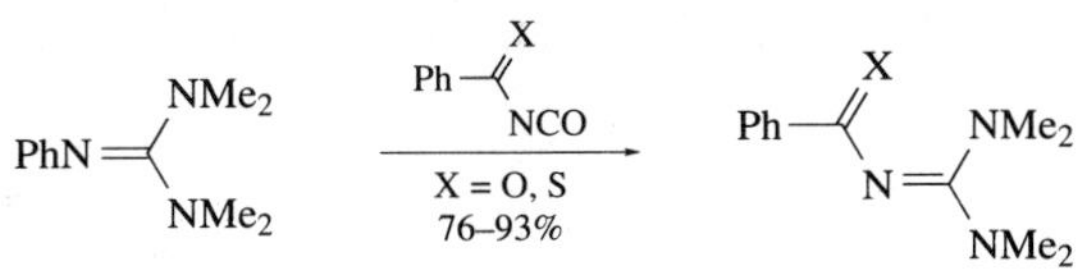

(8)

2-(Thiocarbamoyl)guanidines are produced by the reaction of 1,1,3,3-tetramethylguanidine with *N*,*N*-dimethylthiocarbamoyl chloride ⟨72IJS(A)105⟩ or isothiocyanates (Equation (9)) ⟨82PIA441⟩.

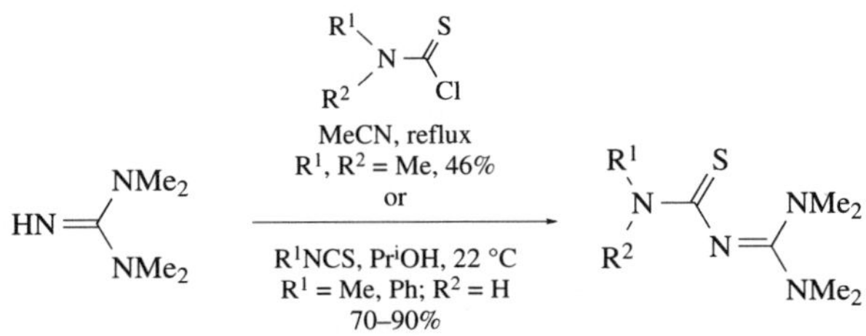

(9)

2-Amidinoureas (**5a**; $R^1 = ArNHCO$) are obtained from the reaction of guanidines with aryl isocyanates (ArNCO) and/or carbamoyl chlorides (ArNRCOCl) ⟨78AF1435⟩.

(iv) N-*Acyliminocarbonyl derivatives by miscellaneous methods*

Carbonic acid *ortho*-amides react with acid amides to give acylguanidines (Equation (10); R = H, Et, Ph) ⟨79LA2096⟩.

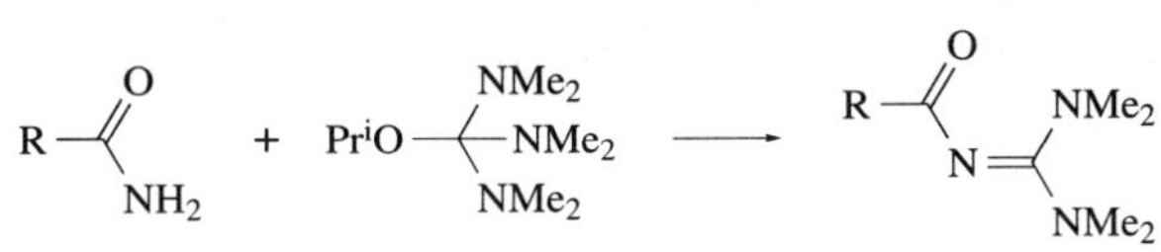

(10)

6.21.1.1.7 N-*Cyanoiminocarbonyl derivatives*

(i) N-*Cyanoiminocarbonyl derivatives from thioureas and their derivatives*

Cimetidine (**16**; R = 2-((5-methylimidazol-4-yl)methylthio)ethyl) (Scheme 11) is an example of an *N*-cyanoiminocarbonyl derivative which has been prepared by several methods. These methods include the reaction of amine (**13**) with methyl isothiocyanate to give the thiourea (**14**), which is treated with lead cyanamide (*i*); the reaction of amine (**13**) with *N*-cyano-*N′*,*S*-dimethylisothiourea (*ii*); and the reaction of amine (**13**) with dimethylcyanodithioimidocarbonate followed by treatment of the intermediate *S*-methylisothiourea (**15**) with methylamine (*iii*) ⟨77JMC901⟩. The Rathke reaction is acid catalysed and so the use of the weak base *N*-cyano-*N′*,*S*-dimethylisothiourea in (*ii*) leads to a low yield of product. This limitation has been circumvented by the room temperature reaction of *N*-aryl- or *N*-alkyl-*N′*-cyanothioureas (**1**; $R^2 = R^4 = H$, $R^3 = CN$, R^5 = aryl, alkyl) with monoalkylamines in DMF in the presence of a water-soluble carbodiimide, which gives 2-cyanoguanidines (**5b**; R^1 = alkyl, $R^3 = CN$, $R^4 = H$, R^5 = aryl, alkyl) in a single step in high yields (>65%) ⟨89TL7313⟩. The reaction works well with sterically hindered *t*-alkylamines and *N*,*N*-dialkylamines and is an improvement over the earlier procedure using dicyclohexylcarbodiimide(dcc) ⟨84SC1275⟩. The room temperature reaction of cyanamide (R^1NH_2, $R^1 = CN$) with a range of *N*-(4-pyridyl)thioureas (**1**;

$R^2 = R^5 = H$, $R^3 = $ 4-pyridyl, $R^4 = $ Ph, neopentyl, etc.) in the presence of dcc also gives 2-cyano-1-(4-pyridyl)guanidines (**5b**; $R^1 = CN$, $R^3 = $ 4-pyridyl, $R^4 = $ Ph, neopentyl, etc., $R^5 = H$) in good yields (>40%) ⟨78JMC773⟩.

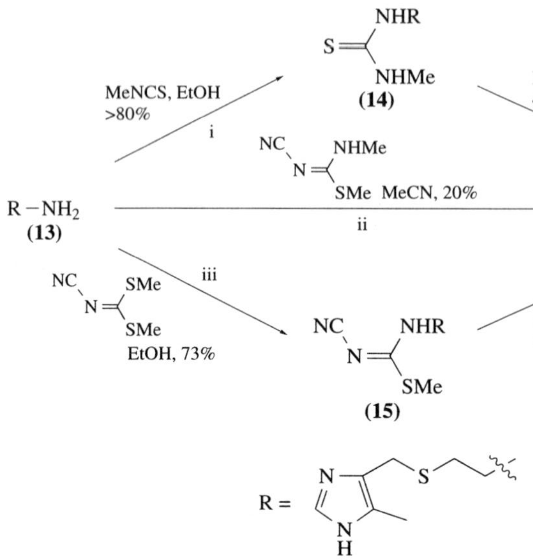

Scheme 11

(ii) N-*Cyanoiminocarbonyl derivatives from carbodiimides*

The reaction of cyanamide (H_2NCN) with carbodiimides gives 2-cyanoguanidines in high yields (Scheme 1, *iii*). The reaction with hindered carbodiimides (**8**; $R^1 = $ aryl, $R^2 = t$-alkyl) in the absence of solvent occurs smoothly in the presence of catalytic diisopropylethylamine, whilst reaction with less-hindered carbodiimides is exothermic ⟨78JMC773⟩.

(iii) N-*Cyanoiminocarbonyl derivatives from cyanamides*

The synthesis of monosubstituted 2-cyanoguanidines (**5b**; $R^1 = R^5 = H$, $R^3 = CN$, $R^4 = $ alkyl) is accomplished in reasonable yields by the condensation of sodium (or lithium) dicyanamide (**7**; $R^2 = Na$, $R^3 = CN$) with monoalkylammonium salts in either refluxing *t*-butanol or water (Scheme 1, *iv*) ⟨68JMC811⟩.

A cyanoguanidine is formed from the reaction of *N*-methyl-*N*-phenylcyanamide with sodium (Equation (11)). The reaction is highly dependent on the nature of the nitrogen substituents and proceeds via an electron transfer mechanism involving the formation of a catalytic quantity of $PhMeN^-$ ⟨91CJC861⟩. The same product is formed in a multistep sequence from the reaction of *N*-methyl-*N*-phenylcyanamide with phenyllithium in 58% yield ⟨71CJC2315⟩.

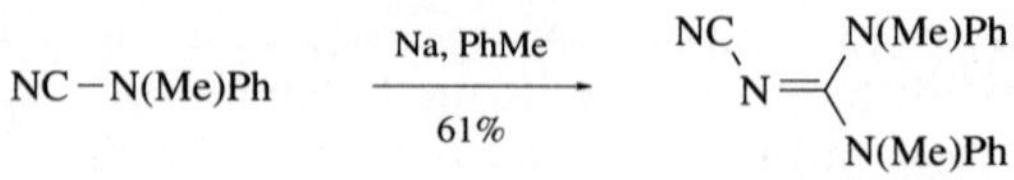

(11)

6.21.1.1.8 N-*Haloiminocarbonyl derivatives*

The most common method of preparation of *N*-haloiminocarbonyl derivatives (*N*-haloguanidines) is the elemental halogenation of appropriately substituted guanidines. It should be noted that many *N*-haloguanidines are unstable and/or explosive and consequently most halogenations in the laboratory are carried out on a small scale and with appropriate shielding and protective equipment. The issue of tautomerism and double bond isomerism in unsymmetrical compounds has been studied ⟨67TL3851⟩.

Perfluoroguanidine is produced by the reaction of guanidine salts with elemental fluorine either in aqueous solution ⟨75USP3968158⟩ or in the absence of solvent but in the presence of alkali metal fluorides as diluents ⟨67JOC1662, 67JOC3859⟩ (Equation (12)). The latter technique has been extended to the direct fluorination of cyanoguanidine in the presence of a large amount of sodium fluoride (Equation (13)) ⟨68JOC2522⟩. In a similar way, guanidine nitrate has been chlorinated to give a 37% yield of perchloroguanidine using a two-phase system consisting of water and chloroform in the presence of finely divided marble and NaCl at $-10\,^{\circ}C$ (Equation (14)) ⟨81MI 621-02⟩. Fluorination of monoalkoxycarbonylguanidines gives tetrafluoro derivatives (Equation (15)), whilst fluorination of 1,3-bis(alkoxycarbonyl)guanidines gives not only the expected fluorinated product, but also produces some fluorinolysis of the alkoxycarbonyl groups so that bis(alkoxycarbonyl)fluoroguanidines and alkoxycarbonyltetrafluoroguanidines are both formed (Equation (16)) ⟨69JOC2840⟩. 2-Chloro- and 2-bromotetramethylguanidines are prepared by the direct halogenation of 1,1,3,3-tetramethylguanidine (Equation (17); Hal = Cl, Br) ⟨66JOC1426⟩.

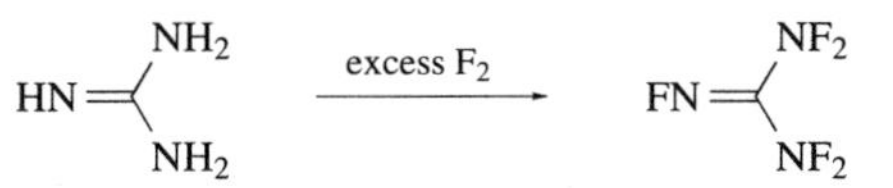

(12)

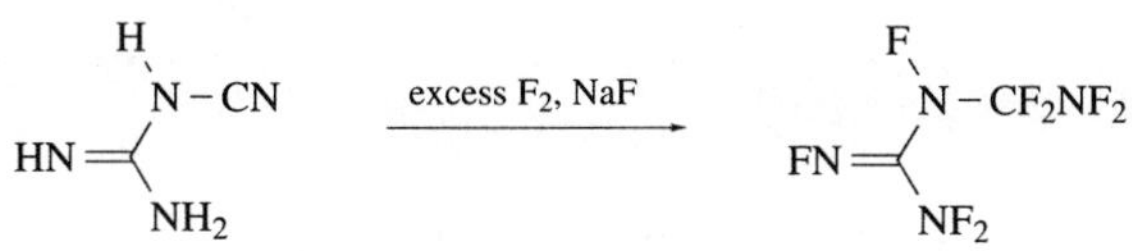

(13)

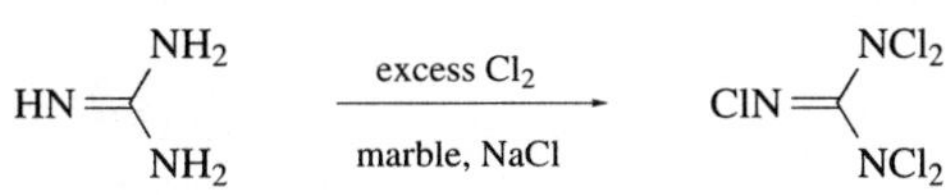

(14)

(15)

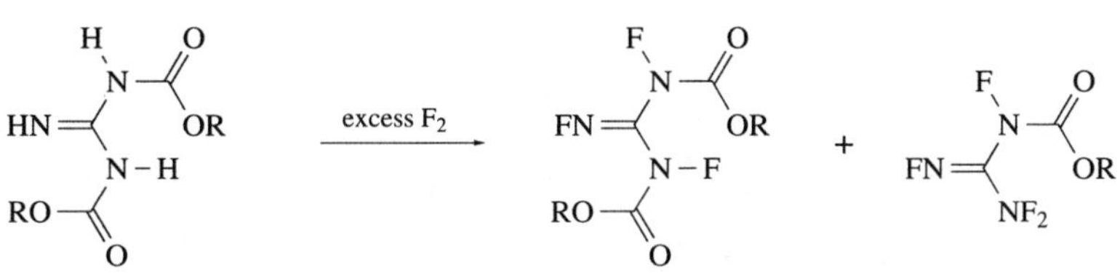

(16)

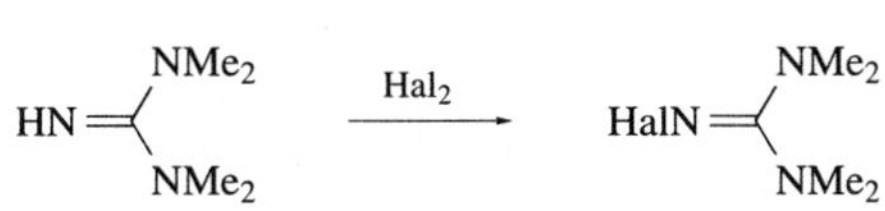

(17)

2-Chloro-1,1-dialkylguanidines may be prepared by an oxidation route in which the appropriate unchlorinated guanidine salt is reacted with sodium hypochlorite in a two-phase solvent system at low temperature (Equation (18); X = Cl, NO_3, SO_4). 1-Aryl-2-chloroguanidines (Equation (18); R^1 = Ph, R^2 = H) are obtained in low yields (< 10%) by this route owing to oxidative side reactions; however, if the possibility of the interfering aromatic quinone-type intermediates is eliminated (because, e.g., R^1 = Ph, R^2 = Me), then 2-chloro derivatives are obtained in high yields ⟨65ZN(B)1165⟩.

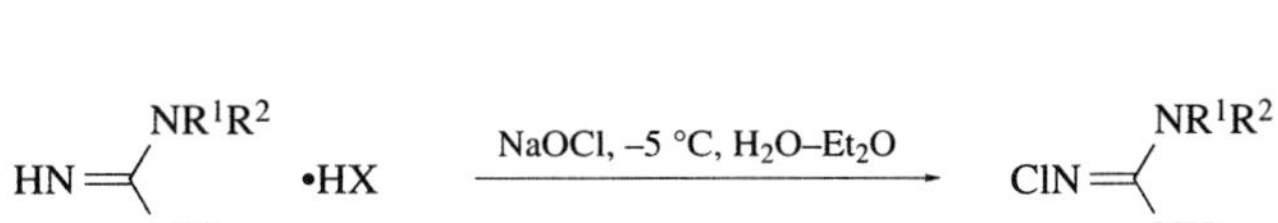

(18)

An example where a 2-haloguanidine is not prepared by the formation of a nitrogen—halogen bond is the reaction of an amine with an *N*-fluoroimine: the reaction of electron-deficient pentafluoroguanidine with alkyl- or arylamines at low temperatures gives an adduct which on warming loses difluoroamine to yield a trifluorinated product (Scheme 12). The rate of reaction and the type of product formed depends upon the nucleophilicity of the amino component ⟨73JOC1075⟩.

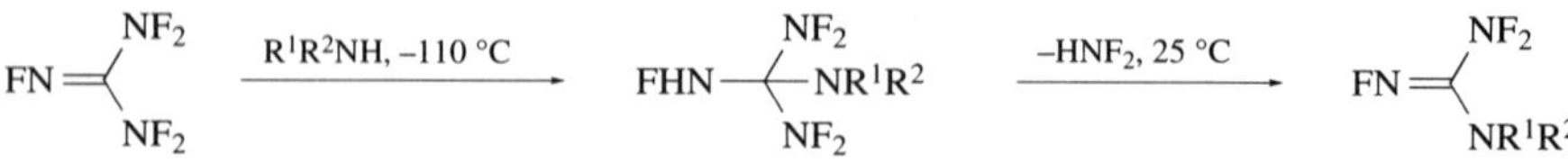

Scheme 12

6.21.1.1.9 N-*Chalcogenoiminocarbonyl derivatives*

(i) Oxygen derivatives

The most common methods of synthesis of hydroxyguanidines involve the reactions of *N*-substituted or unsubstituted hydroxylamines with *S*-alkylisothioureas ⟨67ZN(B)820, 74JOC1166⟩ or chloroformamidines ⟨68JA6846⟩, carbodiimides ⟨72CB1709, 73ZC58⟩, and cyanamides (Scheme 13, (*i*)–(*iii*), respectively) ⟨66JCS(C)2031, 70CC806⟩.

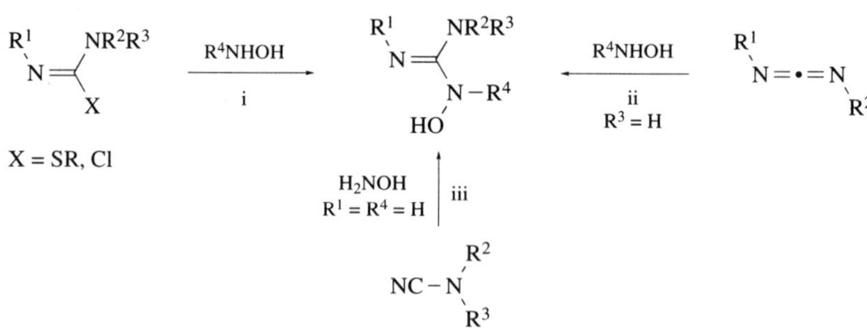

i, *S*-alkylisothioureas or chloroformamidines; ii, carbodiimides; iii, cyanamides

Scheme 13

Tetrasubstituted hydroxyguanidines are best prepared from chloroformamidinium chlorides via reaction with *O*-(THP)hydroxylamine followed by removal of the protecting group with acid (Equation (19)). Cyclic tri- and tetrasubstituted hydroxyguanidines are prepared by the reaction of phosgene-*O*-(THP)oxime with a diamine, followed by removal of the protecting group (Equation (20)) ⟨76JOC3253⟩. Other methods for the formation of hydroxyguanidine-like compounds employ the reduction of nitro precursors. Dinitroformaldehyde oxime is formed by the reduction of trinitromethane in an acid medium on a mercury cathode (Equation (21)) ⟨75MI 621-01, 76MI 621-01⟩, and 1,1′-(hydroxyimidoyl)bis(2-aryl)diazenes are produced when 3-nitroformazans are reduced by H_2S in aqueous alcoholic ammonia solution at 0°C (Equation (22)) ⟨80MI 621-01⟩.

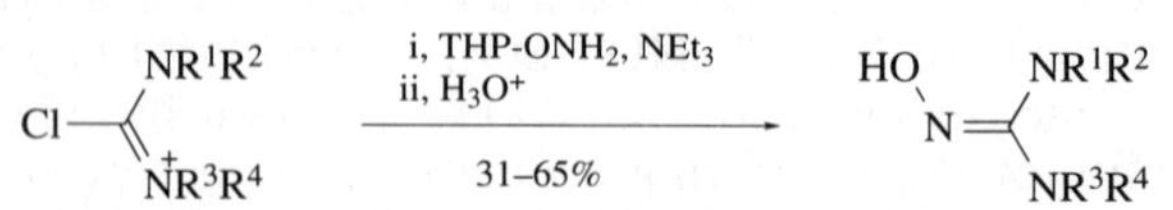

(19)

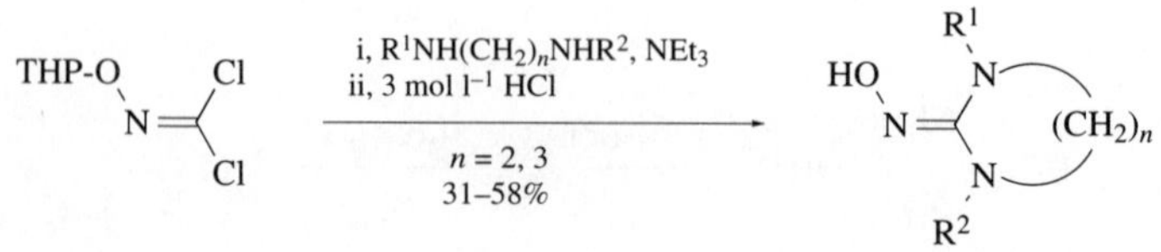

(20)

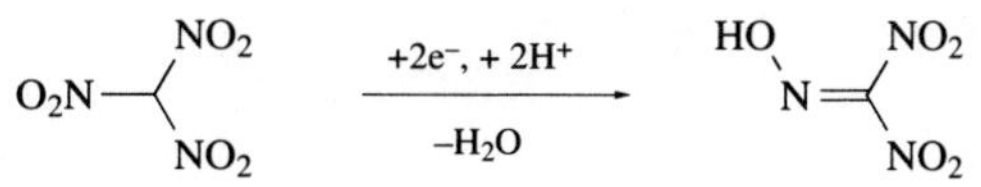

(21)

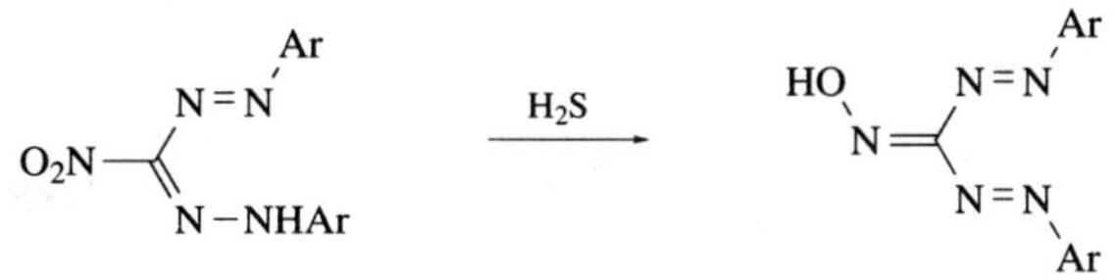

(22)

Alkoxyguanidines are prepared by methods analogous to those used for the hydroxy derivatives above but with *O*-alkylhydroxylamines in place of hydroxylamines. They are also obtained by alkylation of hydroxyguanidines with alkyl bromides or iodides (Scheme 14) ⟨74JPR434⟩. Acetoxy- and benzoyloxyguanidines are prepared by acylation of hydroxyguanidines (Scheme 14) ⟨70BAP569, 74JPR434⟩.

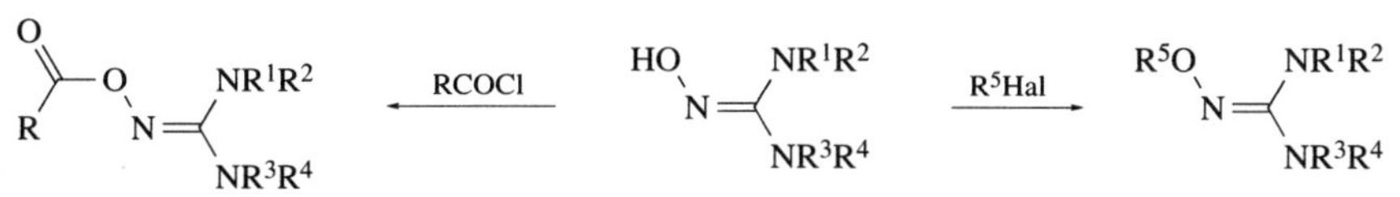

Scheme 14

(ii) Sulfur derivatives

Pentafluorosulfenylguanidines are obtained from the reaction of $SF_5N{=}CCl_2$ either with dialkylamines ⟨82JFC(19)411⟩ or with more nucleophilic dialkylaminosilanes (Equation (23); R = alkyl; X = H, TMS) ⟨92IC488⟩. They are also formed by the reaction of guanidine N—H bonds with sulfenyl chlorides ⟨59CB2563⟩.

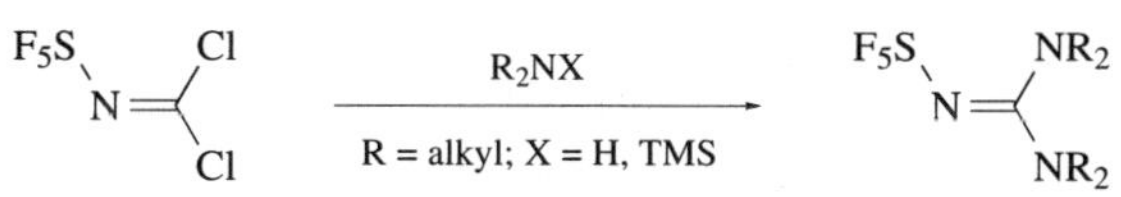

(23)

Several methods for the synthesis of sulfonylguanidines are known. The simplest method employs the condensation of an arylsulfonyl chloride with guanidine (Equation (24)) ⟨40USP2218490⟩.

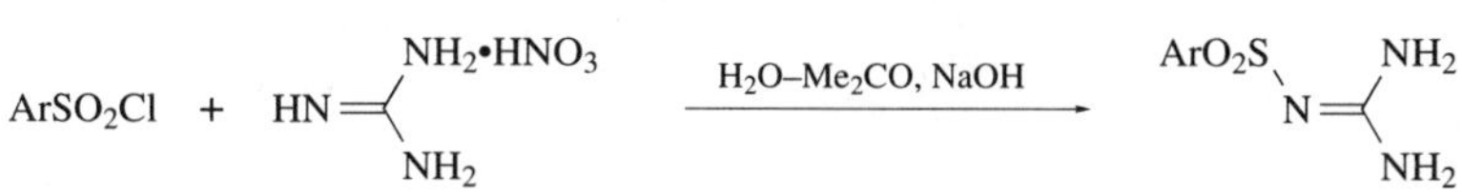

(24)

The Rathke guanidine synthesis using S-methylisothiouronium salts tends to give low yields of sulfonylguanidines (Equation (25); X = SMe) ⟨73JOC1591⟩, and so the more reactive amidinium chlorides are usually employed (Equation (25); X = Cl, R^1 = Ar, R^2–R^5 = alkyl ⟨73JOC1591, 79ZC250⟩; X = Cl, R^1 = Me, R^2 = Bu^i, R^3 = H, R^4 = R^5 = alkyl ⟨67AP(300)567⟩; and X = Cl R^1 = R^2 = Ph, R^3–R^5 = H ⟨74ZOR42⟩). Amidinium chlorides will react with a range of nitrogen nucleophiles other than amines, such as sodium azide and hydrazine, to give sulfonylcarbamimidic azides and *N*-((dialkylamino)hydrazinomethylene)sulfonamides (Scheme 15; R^2, R^3 = Me, Ph). Sulfonylcarbamimidic azides react with triphenylphosphane to give iminophosphoranes ⟨82LA2105⟩. The leaving-group ability of cyanide has been exploited in a sequence involving the room temperature reaction of 4-chloro-5-Ts-1,2,3-dithiazole with secondary amines to give *N*,*N*-dialkylcyanoformamidines which on heating with excess secondary amine give 2-Ts-tetraalkylguanidines (Scheme 16) ⟨92TL4963⟩. The aryl- or alkylsulfonyl group can be attached to the nucleophile instead of the electrophile as in the reaction of arylsulfonamides with *S*-alkylisothioureas (Equation (26)) ⟨67AG(E)862⟩. (Alkylaminosulfonyl)guanidines have been prepared by the reaction of *N*,*N*-dialkyl-*N'*-chlorosulfonylchloroformamidines with primary or secondary amines (Equation (27)) ⟨88JPR900⟩. *N*-Sulfonylguanidines are also prepared by the reaction of amines with either *S*,*S*-dimethyl-*N*-arylsulfonyliminodithiocarbonimidate ⟨66CB2885⟩ or *N*-Ts-carbonimidic dichloride (Equation (28); X = MeS, Cl) ⟨73JOC1591, 77AP820, 79JOC4536⟩.

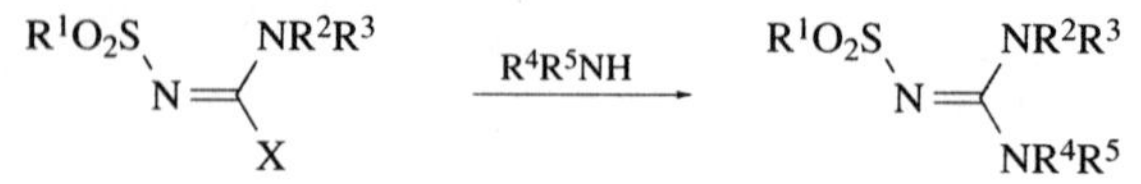

(25)

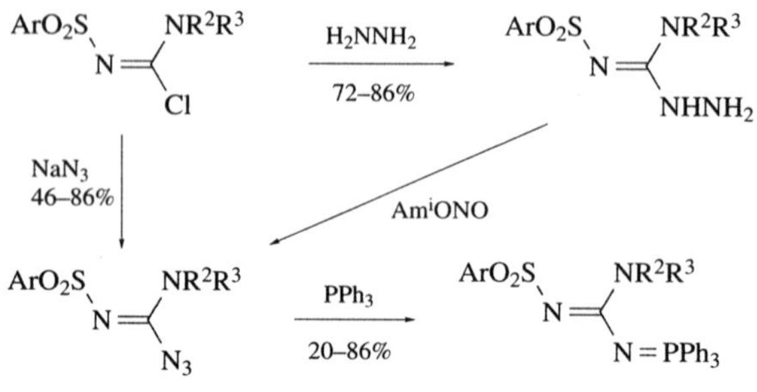

Scheme 15

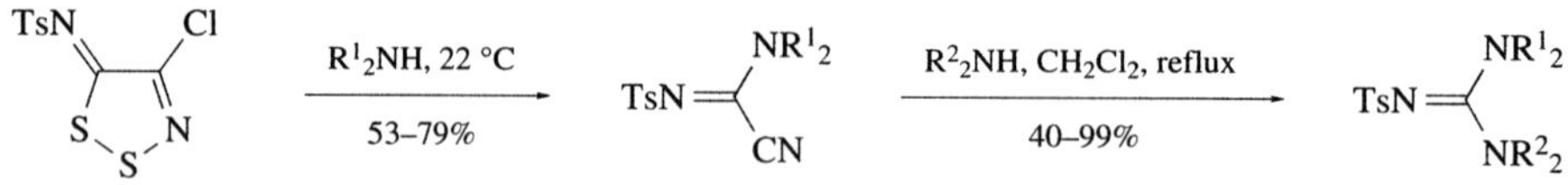

Scheme 16

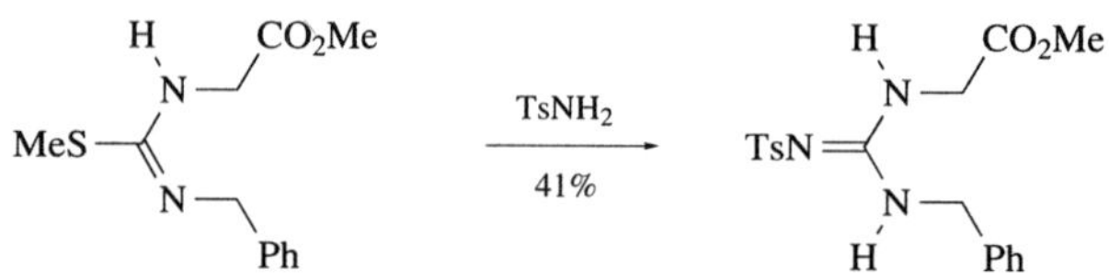

(26)

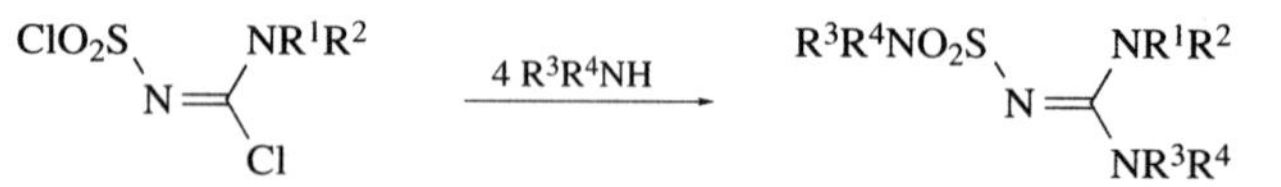

(27)

(28)

A further method for the preparation of sulfonylguanidines involves the reaction of either *N*-sulfonyl-*N'*-alkylcarbodiimides with alkylamines (Scheme 17, *path* (*a*); $R^1 = Ph$, $R^2 = Bu^t$, $R^3R^4 = (CH_2)_4$) ⟨67AP(300)567⟩ or *N*,*N'*-dialkylcarbodiimides with sulfonamides (Scheme 17, *path* (*b*)) ⟨70JCS(C)1429⟩.

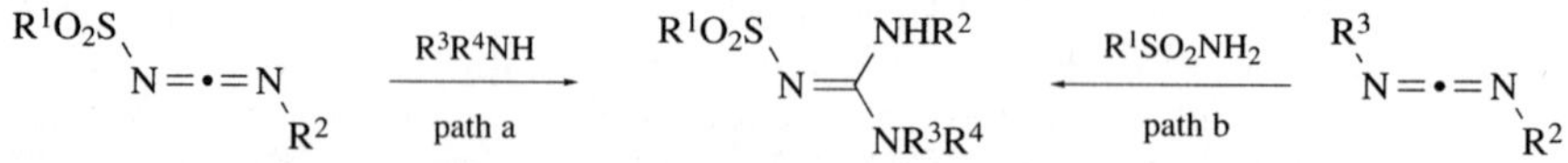

Scheme 17

Finally, sulfonylguanidines may by prepared by [2 + 2] cycloaddition reactions (Scheme 18) involving sulfonyl isocyanates and guanidines ($Ar = 4\text{-}MeC_6H_4$, X = C, Y = O, Z = NPh) ⟨68AG(E)291⟩; sulfonyl isocyanates and thioureas ($Ar = 4\text{-}MeC_6H_4$, X = C, Y = O, Z = S) ⟨91CPB1939, 92MI 621-02⟩; and *N*-sulfinylsulfonamides and thioureas (Ar = Me, X = S, Y = O, Z = S) ⟨67ACS1293⟩. In a related reaction, the crystalline bis(imino)thiazetidine, formed from the reaction of methyl-*t*-butylcarbodiimide with tosyl isothiocyanate, undergoes a reaction with excess dimethylamine at 0 °C to give a tosylguanidine (Scheme 19) ⟨81BSB63⟩.

The anion of trinitromethane reacts with soft electrophiles like phenylsulfenyl chloride at the carbon atom to give *C*-substituted trinitromethane derivatives (Scheme 20; M = K, Ag, piperi-

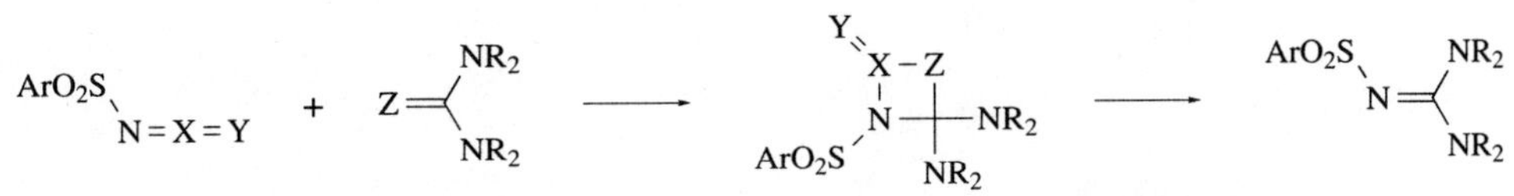

Scheme 18

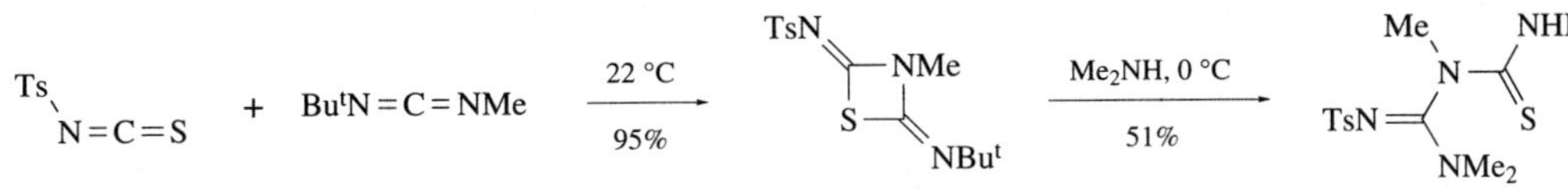

Scheme 19

dinium) ⟨74MI 621-02⟩ and with hard electrophiles like phenylsulfenyl tetrafluoroborate ⟨75IZV1669⟩, acetyl chloride ⟨73IZV819⟩, or TMS-Cl ⟨73ZOR896⟩ at the oxygen atom to give *aci*-nitro derivatives (R = PhS, Ac, TMS, respectively).

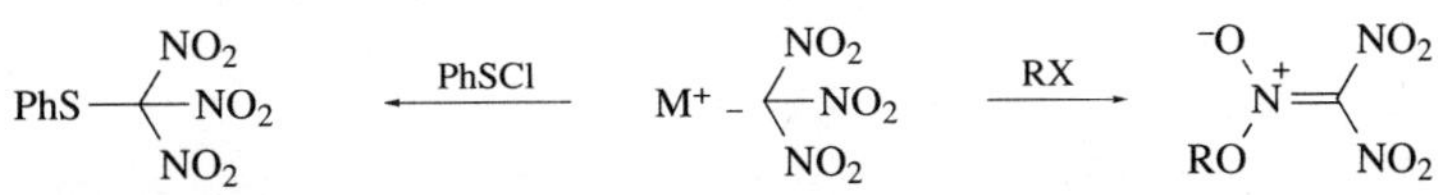

Scheme 20

6.21.1.1.10 N-*Aminoiminocarbonyl derivatives*

(i) Alkyl and aryl derivatives

The methods of synthesis of *N*-aminoiminocarbonyl derivatives (*N*-aminoguanidines) are analogous to those used for guanidines and will not be discussed in detail. Scheme 21 shows the general methods of preparation from amidinium salts and hydrazines (*i*) ⟨90T3897⟩; from thioureas and hydrazines (*ii*) ⟨00CB1058⟩; from *S*-alkylisothiosemicarbazide (salts) and amines (*iii*) ⟨70JHC689⟩; from cyanamides and hydrazines (*iv*) ⟨70CC806⟩; from carbodiimides and hydrazines (*v*) ⟨66AP709⟩; from *S*-alkylthiourea (salts) and hydrazines (*vi*) ⟨62LA(651)89⟩; and from *N*-aminocarbonimidic dichlorides and amines ⟨72JOC2005⟩ or the more reactive silylated amines, e.g., TMS-NMe_2 ⟨92IC488⟩ (*vii*). These methods will not be discussed further. All the theoretically possible methylated aminoguanidines ($H_2NNRC({=}NR_2)NR_2$, where R = Me or H) have been prepared ⟨63JMC283⟩.

(ii) Imino, nitro, nitroso and azido derivatives

A wide range of imino and nitro derivatives of guanidines have been prepared by the reaction of aryl diazonium salts with nitroalkyl derivatives (the Bamberger reaction, ⟨26LA(446)260⟩. 1,5-Diaryl-3-nitroformazans (**18**) are readily obtained by this procedure (Scheme 22), and any tendency to tar formation is minimised by keeping the temperature below 0 °C, waiting long enough for the nitrogen oxides to escape after diazotization, and employing pure nitromethane and freshly distilled arylamines ⟨36JCS1693, 43JA2390, 57ZOB2134, 87POL13⟩. The reaction of nitromalonyl aldehyde anion (**19**) with aryl diazonium salts gives analogous products (Scheme 22; X = Cl, BF_4) ⟨65ZOR735, 72ZOR332⟩. Dinitroformaldehyde hydrazones have also been prepared from nitroalkyl derivatives by the condensation of hydrazines with tetranitromethane at low temperatures (Scheme 23; $R^1 = R^2$ = Me, Et, Ph) ⟨79IZV813, 91CL569⟩ and by the diazo coupling reaction of arylhydrazines with dinitromethane (Scheme 23; R^1 = H, R^2 = 2-4-dinitrophenyl (dnp) ⟨77ACS(B)589⟩. The reaction of aromatic diazo compounds with malonic acid instead of a nitroalkyl derivative (Equation (29)) gives 1,5-diaryl-3-arylazoformazans ⟨74MI 621-01⟩ via a 1,5-diarylformazan intermediate ⟨62MI 621-01, 64ZOB2855⟩.

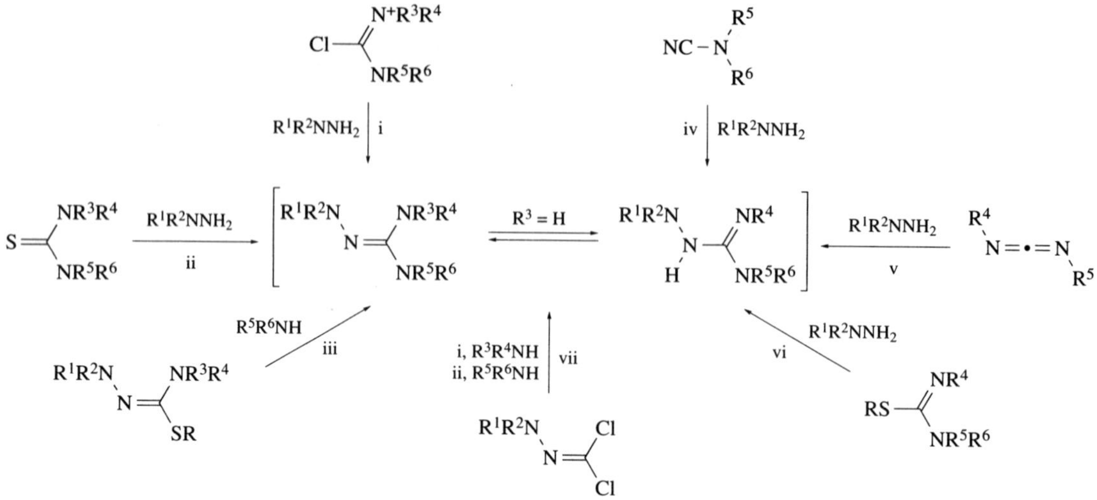

i, amidinium salts; ii, thioureas; iii, *S*-alkylisothiosemicarbazides; iv, cyanamides and hydrazines; v, carbodiimides; vi, *S*-alkylthioureas; vii, dichlorocarbiimides

Scheme 21

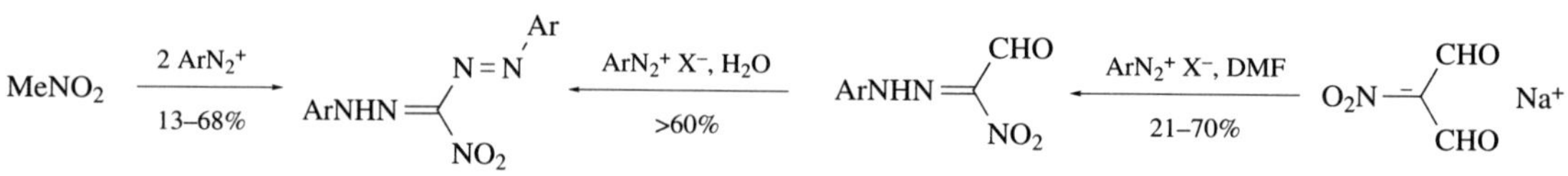

(17) **(18)** **(20)** **(19)**

Scheme 22

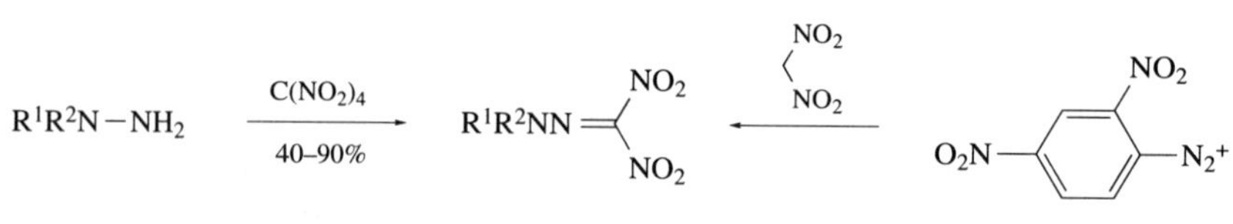

Scheme 23

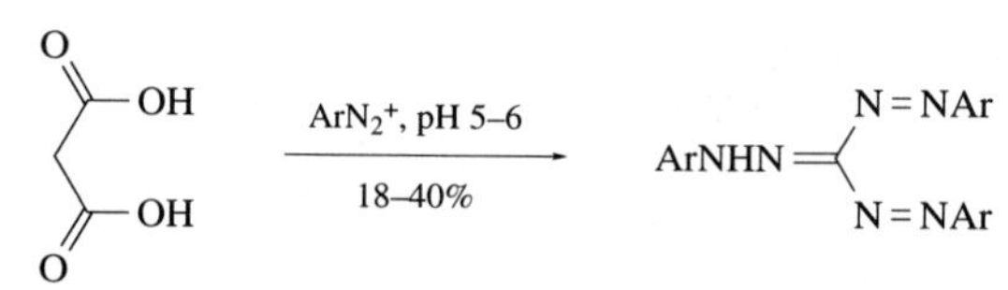

(29)

1,5-Diaryl-3-aminoformazans are prepared by nucleophilic displacement reactions of amines with 3-chloro- ⟨63ZOB113⟩ or 3-nitro-1,5-diarylformazans (Equation (30); $X = Cl$, NO_2, R^1, $R^2 = H$, alkyl) ⟨85JHC813⟩. In a similar fashion, *N*-aryl-1-nitromethanehydrazonyl azides are prepared by the reaction of sodium azide with bromo derivatives in aqueous alcoholic media (Equation (31)) ⟨72ZOR39⟩, and a wide range of (heteroarylmethyl)nitroguanidines have been prepared in the patent literature using the reaction of *S*-methyl-*N*-nitroisothioureas with amines (e.g., Equation (32)) ⟨91EUP425978⟩.

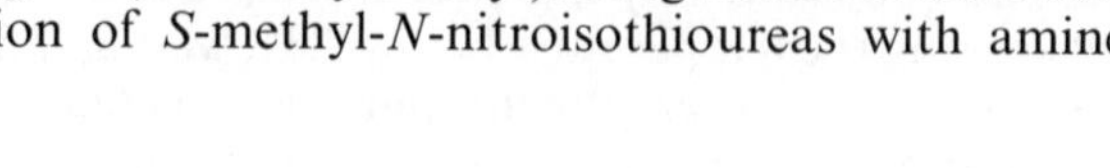

(30)

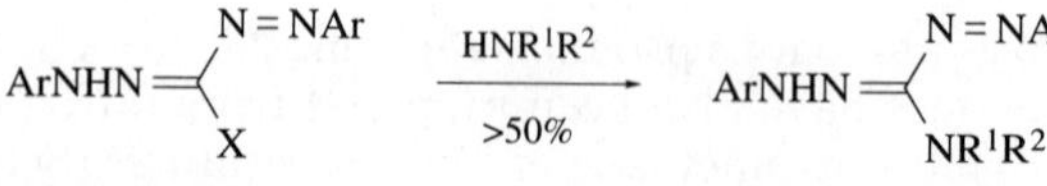

(31)

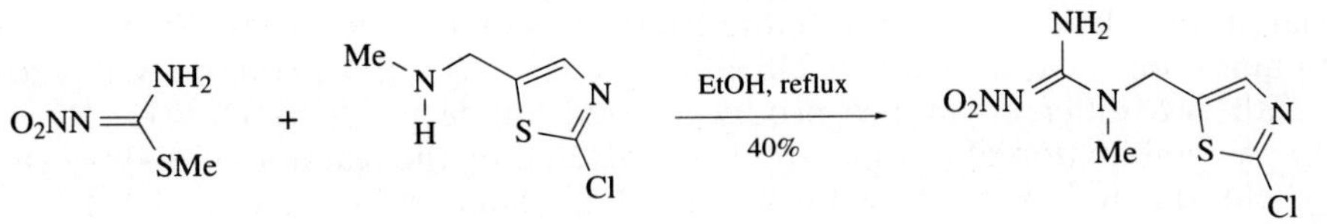
(32)

A kinetic study indicates that the nitration of *syn*-tetraethylguanidine with nitric acid is 700 times slower than the nitration of guanidine and occurs in sulfuric acid but not in acetic acid (Equation (33)) ⟨57CJC527⟩.

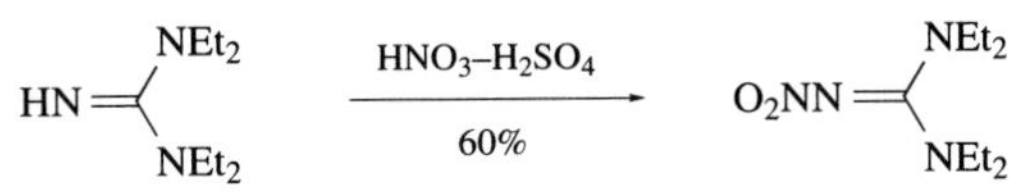
(33)

Most 2-nitrosoguanidines are unstable and decompose to give the corresponding ureas and nitrogen; however, the 1,1,3,3-tetraphenyl derivative obtained by nitrosylation of 1,1,3,3-tetraphenylguanidine is a stable solid at 20 °C (Scheme 24) ⟨74BCJ935⟩. 3-Nitrosoformazans are obtained by the reduction of 1,5-diaryl-3-nitroformazans with H_2S (Equation (34)) ⟨80MI 621-01⟩.

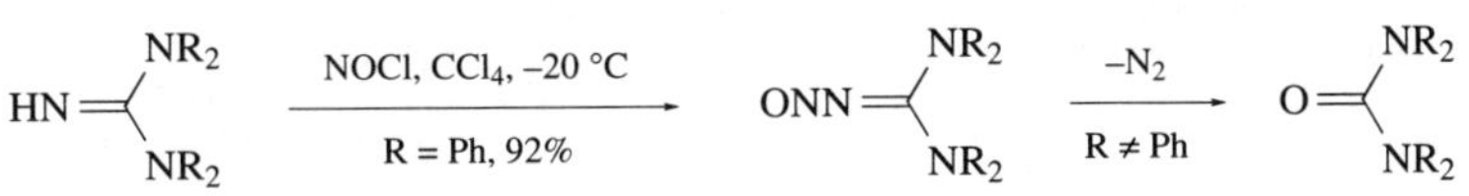

Scheme 24

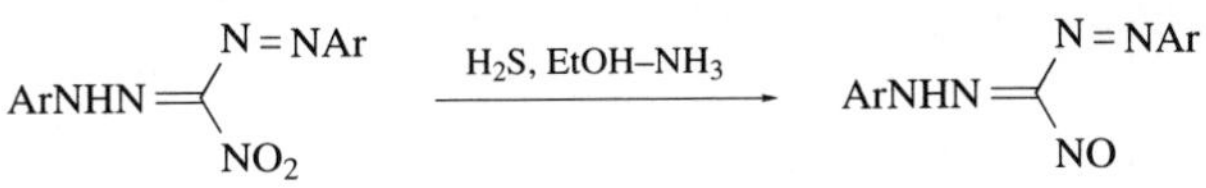
(34)

6.21.1.1.11 N-*P*, N-*As*, N-*Sb and* N-*Bi iminocarbonyl derivatives*

(i) Phosphorus(III) derivatives

The few examples of this class of compounds are iminophosphinoguanidines which have been made by reaction of iminochlorophosphines with silylated guanidines (Equation (35)) ⟨91ZOB401⟩. Tris(trichlorophosphoranediylamino)carbenium salts are prepared by the photolysis of tri-azidocarbenium hexachloroantimonate in PCl_3 (Equation (36)) ⟨79AG(E)693⟩ via a mechanism which is analogous to the Staudinger reaction of organic azides with phosphanes.

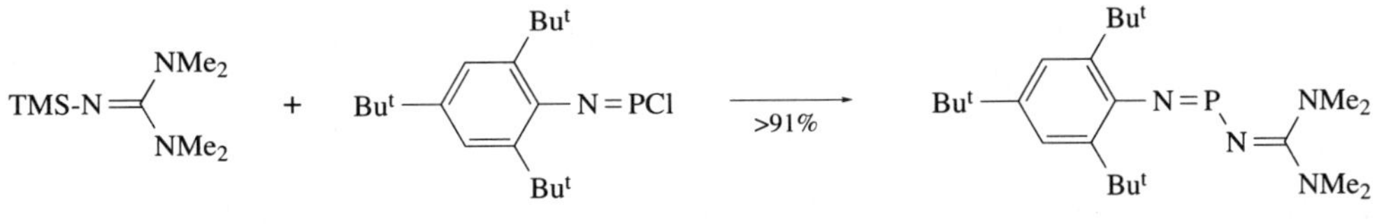
(35)

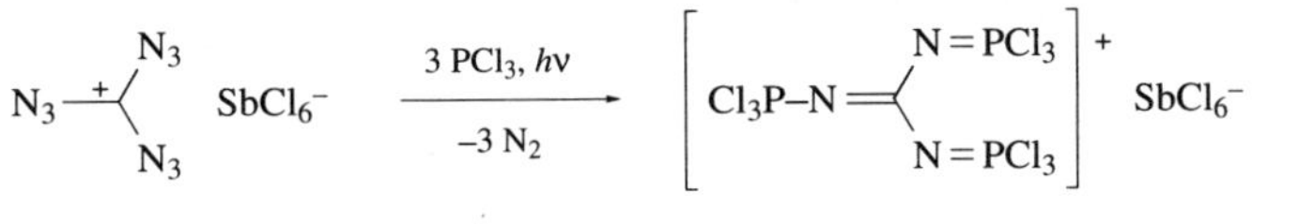
(36)

(ii) Phosphorus(V) derivatives

Two major synthetic methods exist for this class of compound. The first route entails formation of the amide bond by the reaction of phosphoryl chlorides with guanidines to give, e.g., guanidinylphosphoric diamides (Equation (37)) ⟨67JMC118, 67JMC273⟩. The second route involves the formation of two guanidinyl C—N bonds by the reaction of dichloromethylenephosphoramidic derivatives with dialkylamines (Equation (38)): thus, reactions with excess dialkylamines produce bis(dialkylamino)methylenephosphoramidic esters (X = O, $R^1 = R^2$ = Me, Et, allyl) ⟨69IZV119⟩ and phosphoramidothioic acid derivatives (X = S, $R^1 = R^2$ = Me, Et) ⟨74OMR(6)494, 74ZOB538⟩, whilst reactions with two different amines produce mixed amide products (X = O, R^1 = Ph, R^2 = Et) ⟨66ZOB461⟩. A combination of both methods is provided by the reaction of dialkylcyanamides with phosphoryl chloride followed by treatment with dialkylamines (Scheme 25; R^1, R^2 = various alkyl groups) ⟨74ZOB1264⟩.

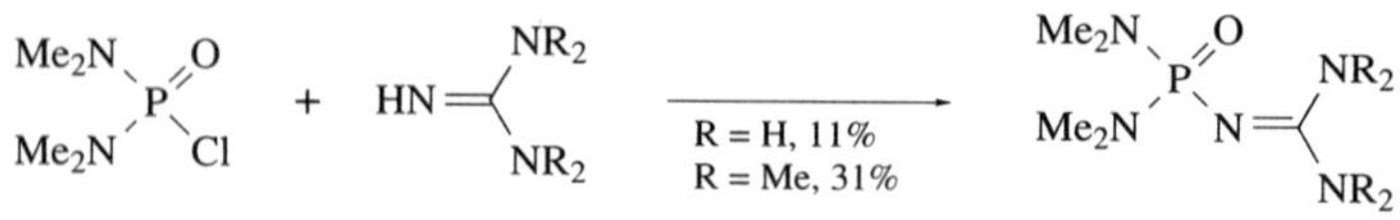

(37)

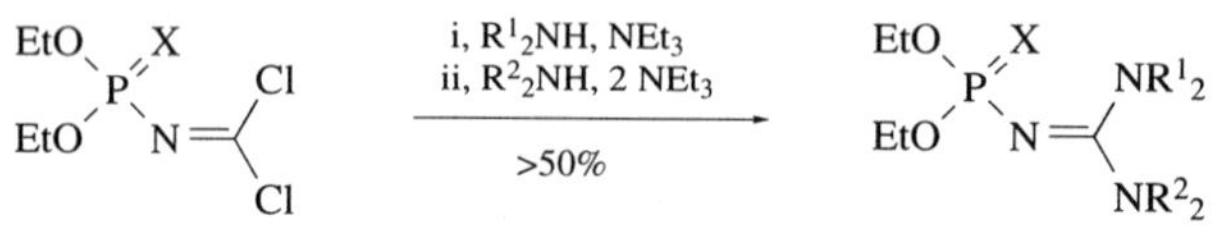

(38)

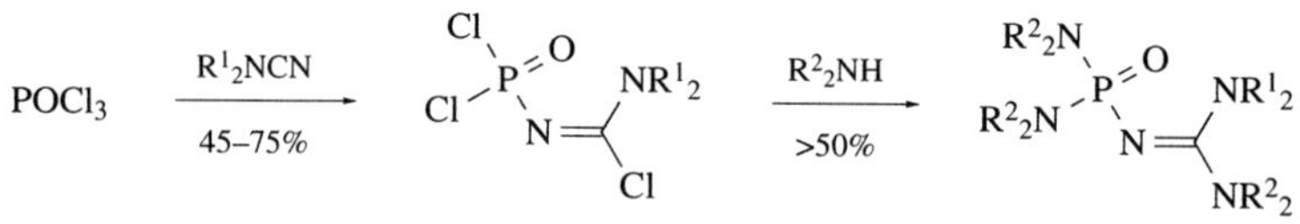

Scheme 25

(iii) As, Sb and Bi derivatives

The single report of guanidines attached to group 15 elements other than N or P refers to dative-bonded complexes of guanidines and bismuth or antimony trihalides ⟨88EUP273458⟩.

6.21.1.1.12 N-*Si*, N-*Ge and* N-*B iminocarbonyl derivatives*

N-Silylated guanidines are prepared by treating the corresponding guanidine with a chlorosilane in the presence of a base (Equation (39)) ⟨75ZOR762⟩. In the absence of added base, a stable guanidinium salt $((Me_2N)_2CNHTMS)^+$ Hal^- is formed ⟨88JOM(339)241⟩.

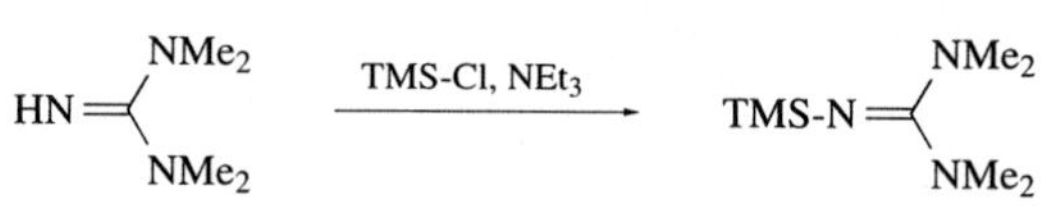

(39)

The only example of *N*-germylated or *N*-borated guanidine derivatives are triarylboron complexes (Equation (40); R = H, Ph) ⟨78ZOB811, 81ZOB880⟩.

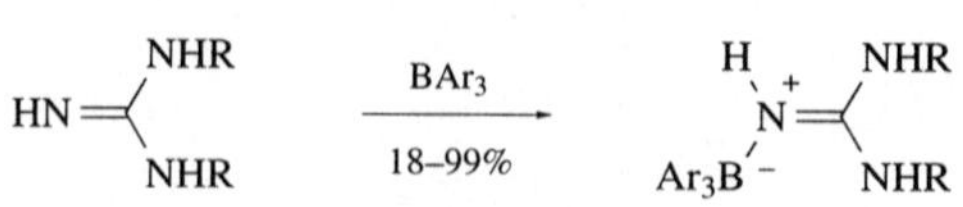

(40)

6.21.1.2 Iminocarbonyl Derivatives with One Nitrogen and One P, As, Sb or Bi Function

The majority of compounds in this class contain a phosphorus function. Very few compounds contain arsenic functions, and none contain antimony or bismuth functions. The compounds will be discussed in the order of the substituent on the imino nitrogen atom.

6.21.1.2.1 N-*Alkylimino derivatives with one P or As function*

The starting materials for *N*-alkyl- and *N*-arylimino derivatives are either chloroimine analogues or carbodiimides. The few examples of *N*-alkylimino derivatives include compound (**22**; R = Bu), prepared by a Michaelis–Arbuzov reaction of the carbamimidic chloride (**21**) with trimethyl phosphite (Equation (41)) ⟨62BEP612225⟩, and the amidinoarsine (**24**; $R^1 = C_6H_{11}$, Ph, $R^2 = C_6H_{11}$, R^3 = H, Et), formed from the attack of alkali metal organoarsides on *N,N'*-dialkylcarbodiimides (**23**; $R^1 = C_6H_{11}$, Ph) (Equation (42)) followed by hydrolysis or alkylation of the intermediate (lithioamidino)arsines ⟨68JOM(13)363⟩.

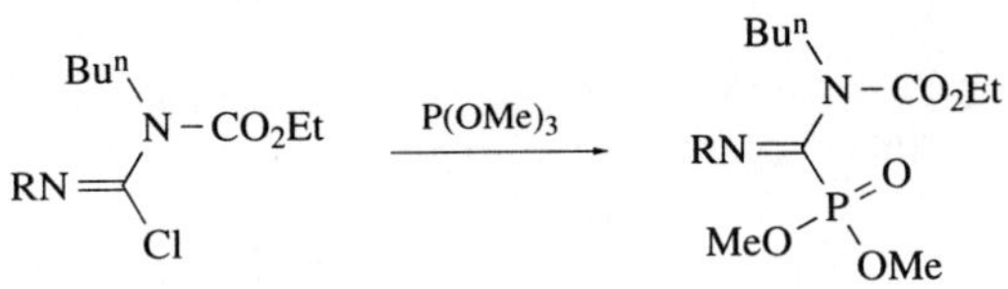

(**21**) (**22**) (41)

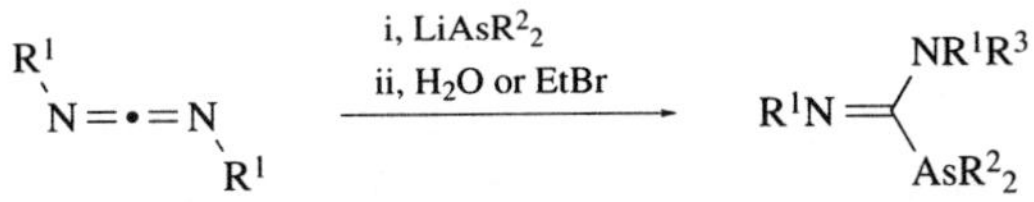

(**23**) (**24**) (42)

6.21.1.2.2 N-*Arylimino derivatives with one P function*

(i) *Amidinophosphonates*

Amidinophosphonates (**27**) have been obtained by three different methods (Scheme 26). The first method involves the reversible reaction of *N,N'*-diphenylcarbodiimide (**25**) with the phosphite triester (**26**) to give the amidinophosphonates (**27**; Ar = Ph, R^1 = Me, Et, Bu, Pr^i, R^2 = Ph, R^3 = TMS) ⟨90ZOB778⟩. The second method is the direct aminolysis of imino chlorides (**28**), which has been used to give a wide range of amidinophosphonates (**27**; Ar = Ph, 4-ClC_6H_4, 3-MeC_6H_4, 4,2-Cl(Me)C_6H_3, R^1 = Me, Et, Pr^i, R^2 = H, R^3 = Me, Et, Pr, Pr^i, Bu, Bu^t, C_6H_{13}, C_6H_{11} or $R^2 = R^3$ = Me, Et, Pr, Pr^i) ⟨77GEP2547512⟩. The third method yields the amidinophosphonates (**27**; Ar = Ph, R^1 = Me, R^2 = Ph, R^3 = COP(O)(OMe)$_2$) via an exothermic Michaelis–Arbuzov reaction of trialkyl phosphites and the amidino chlorides (**29**) ⟨62BEP612225⟩.

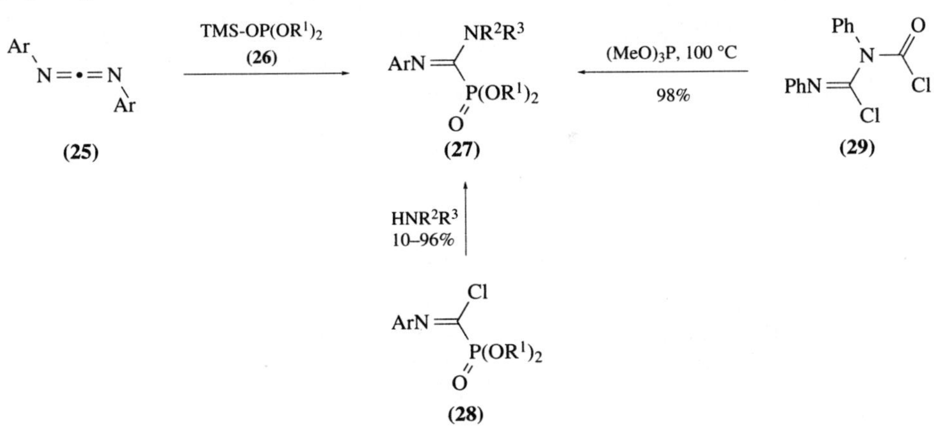

(**25**) (**27**) (**29**) (**28**)

Scheme 26

(ii) *Alkylideneamidophosphonates*

In a variant of the trialkyl phosphite and carbodiimide reaction, treatment of triethyl phosphite with the chloroalkylcarbodiimide (**30**) gives the alkylideneamidophosphonate (**31**) (Equation (43)) ⟨78ZOB1425⟩.

(43)

(iii) Amidinophosphines

Phosphines which contain P—H or P—Si bonds will add across the C═N bond of carbodiimides with transfer of hydrogen or silicon from phosphorus to nitrogen (Scheme 27); thus, the reaction of *N,N'*-diphenylcarbodiimide (**32**) with monophenylphosphine gives the bis(amidino)phosphine (**33**); with diphenylphosphine (**34**; $R^1 = R^2 = Ph$, $R^3 = H$) the product is the mono(amidino)phosphine (**35**; $R^1 = R^2 = Ph$, $R^3 = H$); and with tris(TMS)phosphine (**34**; $R^1 = R^2 = R^3 = TMS$) it is the trisilylated amidinophosphine (**35**; $R^1 = R^2 = R^3 = TMS$) ⟨82IZV1426⟩. Treatment of compound (**35**; $R^1 = H$, TMS, $R^2 = R^3 = TMS$) with methanol gives the desilylated product (**37**). Reaction of compound (**35**; $R^1 = R^2 = R^3 = TMS$) with *N,N'*-diphenylcarbodiimide or reaction of excess compound (**32**) with tris(TMS)phosphine gives the product (**36**) ⟨81SRI279⟩. Silicon appears to be transferred in preference to hydrogen because the reaction of bis(TMS)phosphine (**34**; $R^1 = H$, $R^2 = R^3 = TMS$) gives the product (**35**; $R^1 = H$, $R^2 = R^3 = TMS$).

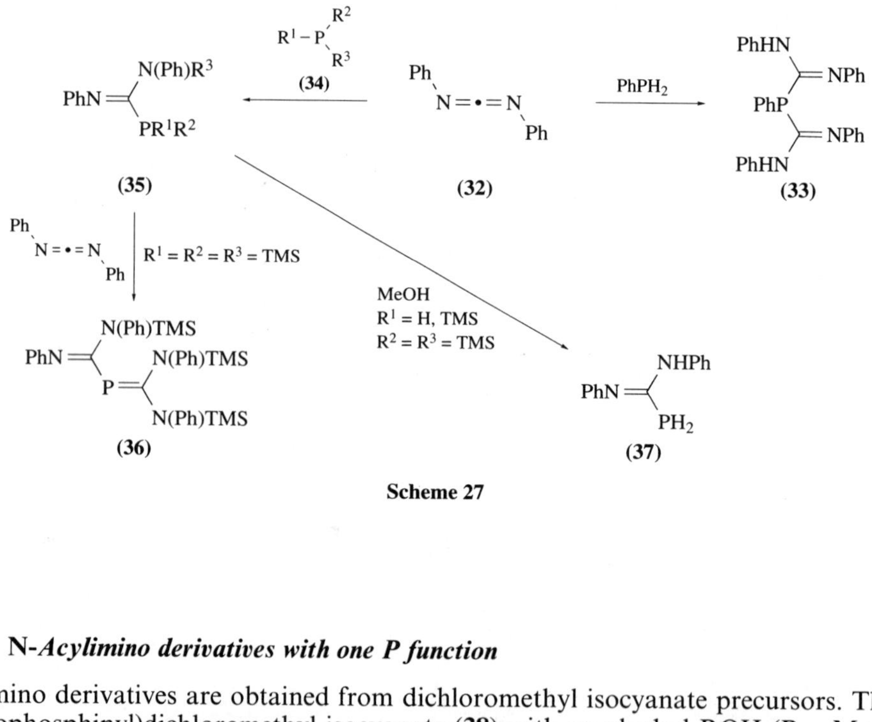

Scheme 27

6.21.1.2.3 N-*Acylimino derivatives with one P function*

N-Acylimino derivatives are obtained from dichloromethyl isocyanate precursors. The reaction of (dichlorophosphinyl)dichloromethyl isocyanate (**38**) with an alcohol ROH (R = Me, Et) in the cold gives the chloroimine (**39**; R = Me, Et) which on treatment with diethylamine gives the *N*-acylimino compound (**40**; R = Me, Et) (Scheme 28) ⟨71ZOB2155⟩. Similarly, the *N*-phosphinylcarbonylimino derivative (**42**, R = Me, Et) is obtained by treating the isocyanate (**41**; R = Me, Et) with triethylphosphite (Equation (44)) ⟨73ZOR1815⟩.

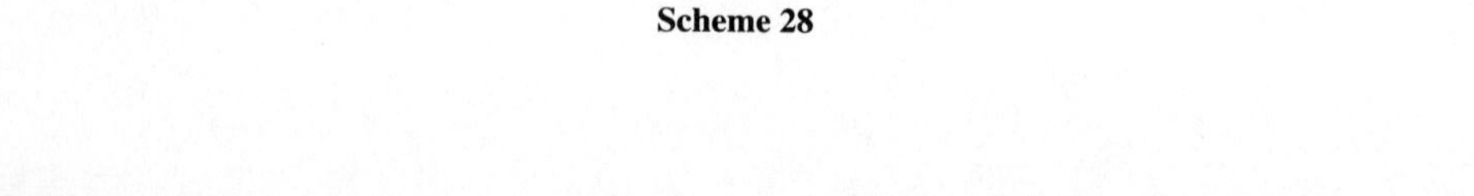

Scheme 28

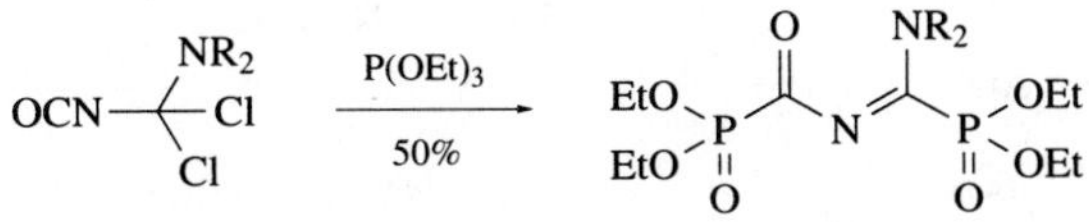

(41) (42) (44)

6.21.1.2.4 Hydrazono derivatives with one P function

Hydrazono derivatives are obtained by several different reactions: the reaction of chlorohydrazones with nitrogen nucleophiles; the 1,3-addition reaction of amines and nitrile amines; the treatment of diazonium salts with α-phosphineacetyl derivatives; alkylation reactions; and oxidation reactions.

N'-Aryl-*C*-(dialkoxyphosphoryl)formamidrazones (**44**; $X = NR^2_2$, Ar = 4-$NO_2C_6H_4$, 4-HalC_6H_4, 4-MeC_6H_4, Hal = F, Cl, Br, I, R^1 = OMe, OEt, OPr^i, R^2 = Me, Ph) are prepared in a yield of 29–99% by the treatment of chlorohydrazono derivatives (**43**) with aqueous solutions of amines (Equation (45)) ⟨90ZOB1293, 90ZOB2694⟩. Phosphonic acid diamides (**44**; X = NMe_2, Ar = 4-$NO_2C_6H_4$, R^1 = NEt_2) are prepared in a similar fashion ⟨90ZOB1986⟩, whilst treating compound (**43**; Ar = 4-$NO_2C_6H_4$, R^1 = OPr^i) with sodium azide or sodium nitrite gives the azido- and nitrohydrazones (**44**; X = N_3, NO_2, Ar = 4-$NO_2C_6H_4$, R^1 = OPr^i) ⟨90ZOB1980⟩.

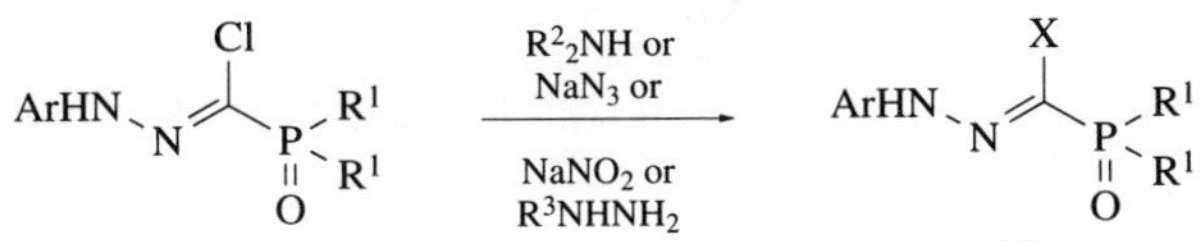

(43) (44) (45)

N'-Phosphineformamidrazone (**46**) is obtained by the 1,3-addition of diisopropylamine to the nitrile imine (**45**) (Equation (46)) ⟨91CB1739⟩. Related 1,3-additions of monoalkylamines to nitrile imines have been reported ⟨92CC1274⟩.

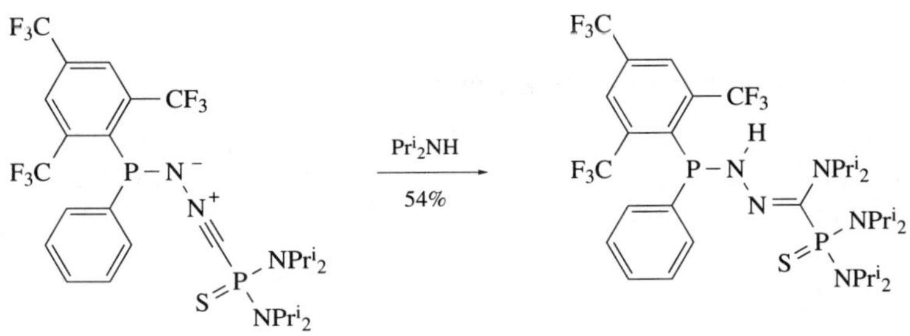

(45) (46) (46)

An interesting solvent effect is observed in the preparation of (arylhydrazono)(arylazomethyl) phosphonates (**50**; Ar = Ph, 4-MeOC_6H_4, 4-$NO_2C_6H_4$, 4-ClC_6H_4, 4-BrC_6H_4, 4-IC_6H_4, R = Me, Et, Pr^i). The reaction of the phosphinylaldehyde (**47**) with diazonium salts (**48**; X = Cl, BF_4) gives the phosphonate (**50**) as the major product in pyridine, and compound (**49**) as the major product in water (Equation (47)) ⟨86ZOB1427, 87ZOB2467⟩. Phosphonate (**50**; Ar = 4-$NO_2C_6H_4$, R = Pr^i) has also been prepared by the reaction of compound (**43**; Ar = 4-$NO_2C_6H_4$, R^1 = Pr^i) with 4-nitrophenylhydrazine (Equation (45); R^3 = 4-$NO_2C_6H_4$) ⟨90ZOB1288⟩. The coupling reaction of the phosphineacetic acid salt (**51**) with 2 molar equivalents of diazonium salt (**52**) gives the bis(arylazo)methylenephosphine (**53**), which is isolated as the tetrafluoroborate salt (**54**; Ar = Ph, 4-$NO_2C_6H_4$, 4-$Me_2NC_6H_4$, 2-MeOC_6H_4) (Scheme 29) ⟨62ZN(B)782⟩.

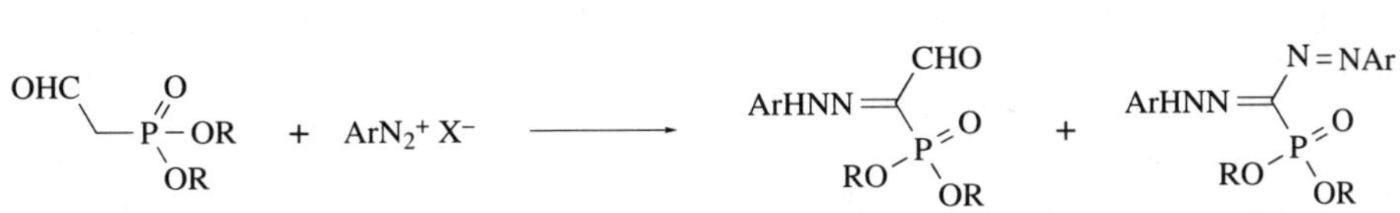

(47) (48) (49) (50) (47)

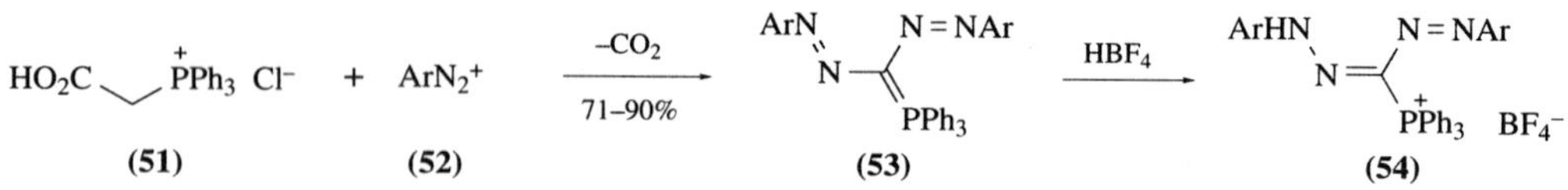

Scheme 29

Other preparations of compounds in this class include the isolation of the inner salt (**56**) from the alkylation of the formamidrazone (**55**) with methyl iodide (Equation (48)) ⟨91ZOB776⟩, and oxidation of the amidrazone (**57**; R = Bu^t) with silver(I) oxide to give the *N*-(arylazo)(dialkoxyphosphoryl)methylene-*t*-butylamine (**58**; R = Bu^t) (Equation (49)) ⟨93S59⟩. The (arylazo)imines (**58**) cannot be made with an N^3-methylene group (e.g., R = Me, $PhCH_2$) because the products of such reactions tautomerise to give *N*-alkylideneamide arylhydrazones which then cyclise to form dihydro-1,2,4-triazoles.

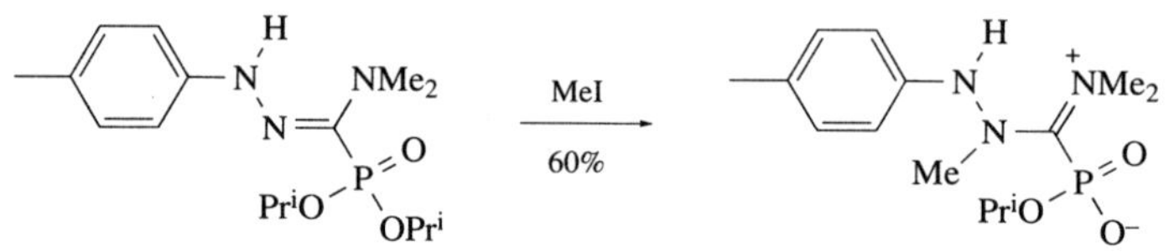

(48)

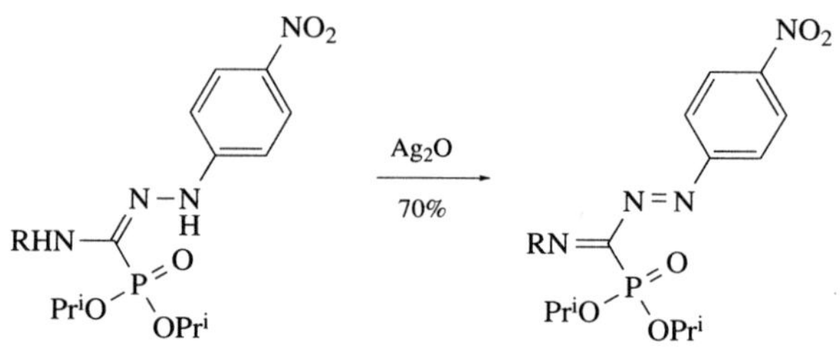

(49)

6.21.1.2.5 *Diazonium derivatives with one P function*

The diazonitromethyl derivatives (**60**; R^1 = Ph, R^2 = OMe; R^1 = R^2 = Ph, OMe, OEt) and the nitrate side products (**61**) are obtained by treating the diazomethylphosphonates (**59**) with dinitrogen pentoxide (Equation (50)) ⟨79LA1002⟩.

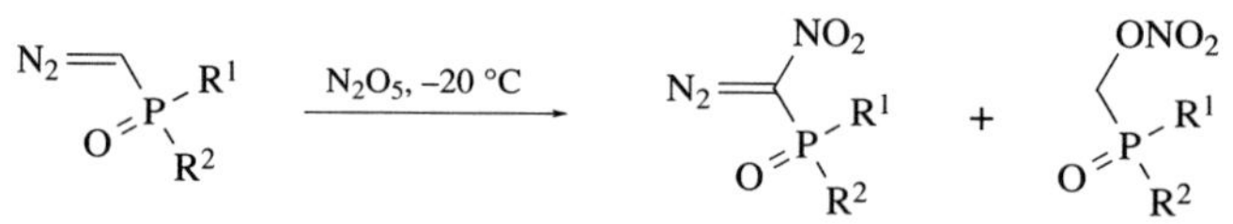

(50)

6.21.1.2.6 N,N-*Dialkyliminium derivatives with one P function*

2-Phosphaallylic salts and phosphorylated amidinium salts may be prepared from chloroamidinium salts or phosphaalkenes.

2-Phosphaallylic salt (**68**; R^2 = Me, X = Cl) is formed in high yield by the reaction of tris(TMS)phosphane with two molar equivalents of the imidoyl chloride (**66**; R^2 = Me) (Scheme 30) ⟨83AG(E)545, 84AG(E)903⟩. The analogous compound (**68**; R^2 = Et, X = CF_3SO_3), along with other unidentified products, is formed by rearrangement of the 1,3-tetraalkylformamidinium-substituted cyclotetraphosphane (**67**) produced by cyclisation of the cationic diphosphene (**64**; R^1_2N = Pr^i_2N, 2,2,6,6-tetramethylpiperidinyl; R^2 = Et; X = CF_3SO_3). Diphosphene cation (**64**) is obtained by the reaction of the chlorophosphine (**63**) with TMS trifylate or by the condensation of phosphaalkene (**62**) with chlorophosphenium cation, followed by valence isomerism of the first-formed phosphaalkene–phosphenium cation (**65**) ⟨91TL2775⟩. The 2-phosphaallylic salt (**68**; R^2 = Me, Et, X = Cl) is also prepared by leaving benzene solutions of the phosphaalkene (**63**) standing at room temperature for several hours.

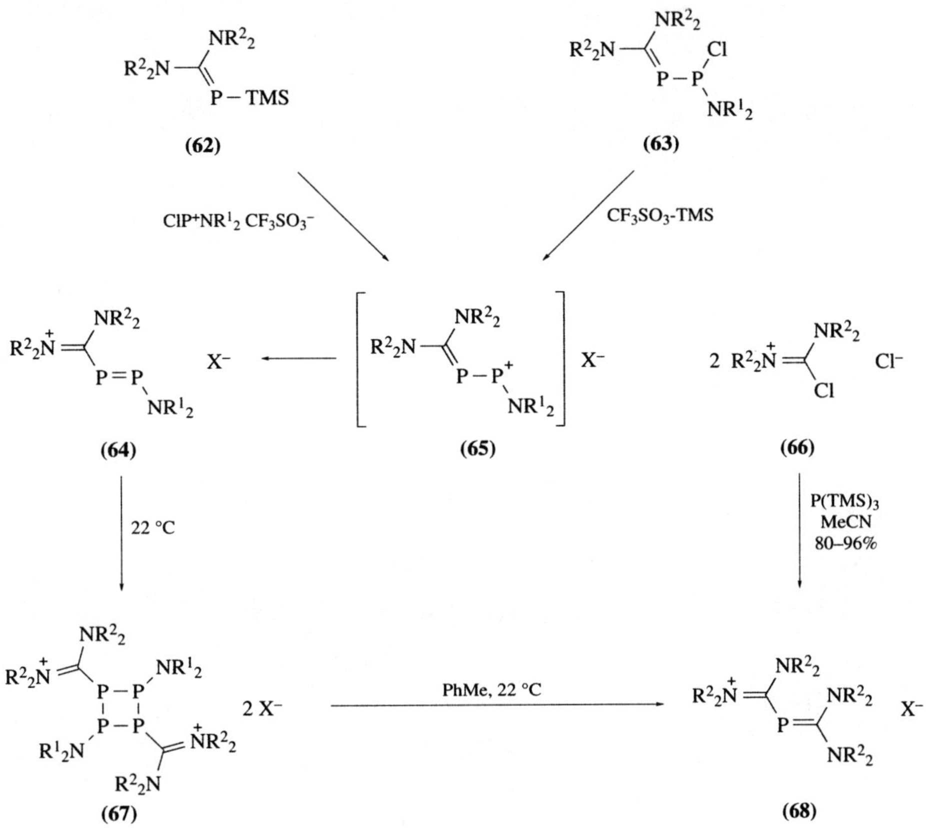

Scheme 30

The synthesis of phosphorylated amidinium salts is shown in Scheme 31. Reaction of the cation (**71**; $R^2 = Me$) with triethyl phosphite gives the phosphonic anhydride inner salt (**73**; $R^2 = Me$) ⟨72JOC2730⟩ via the monophosphorylated amidinium salt intermediate (**69**; $R^2 = Me$) which can also, depending upon reaction conditions, give the betaine (**70**; $R^2 = Me$) as the major product ⟨72JCS(D)2267, 78JPR389⟩. The reaction of the cation (**71**; $R^2 = Me$) with an excess of diethyl phenylphosphonite gives the salt (**76**; $R^2 = Me$), which is converted to the stable inner salt (**72**; $R^2 = Me$) by gentle warming ⟨72JOC2730⟩. In contrast, ethyl and isopropyl diphenylphosphinite (**74**; $R^3 = Et$, Pr^i) have only one displaceable alkyl group and undergo a standard Michaelis–Arbuzov reaction with cation (**71**) to give the salt (**75**; $R^2 = Me$) ⟨72JOC2730, 78JPR389⟩. The inner salt (**72**; $R^2 = Me$) is also prepared by oxidation of the electron-rich phosphaalkene (**77**; $R^2 = Me$) with ozone ⟨86AG(E)457⟩.

6.21.1.3 Iminocarbonyl Derivatives with One Nitrogen and One Metalloid Function

This class of compounds consists largely of boron-containing compounds. There appear to be no germanium-containing compounds and only one example of a silanecarboximidamide.

6.21.1.3.1 Silicon derivatives

Silanecarboximidamide (**79**) is prepared from the reaction of *N*,*N*′-diphenylcarbodiimide (**32**) and bis(TMS)mercury (**78**) (Equation (51)) ⟨72JOM(42)293⟩.

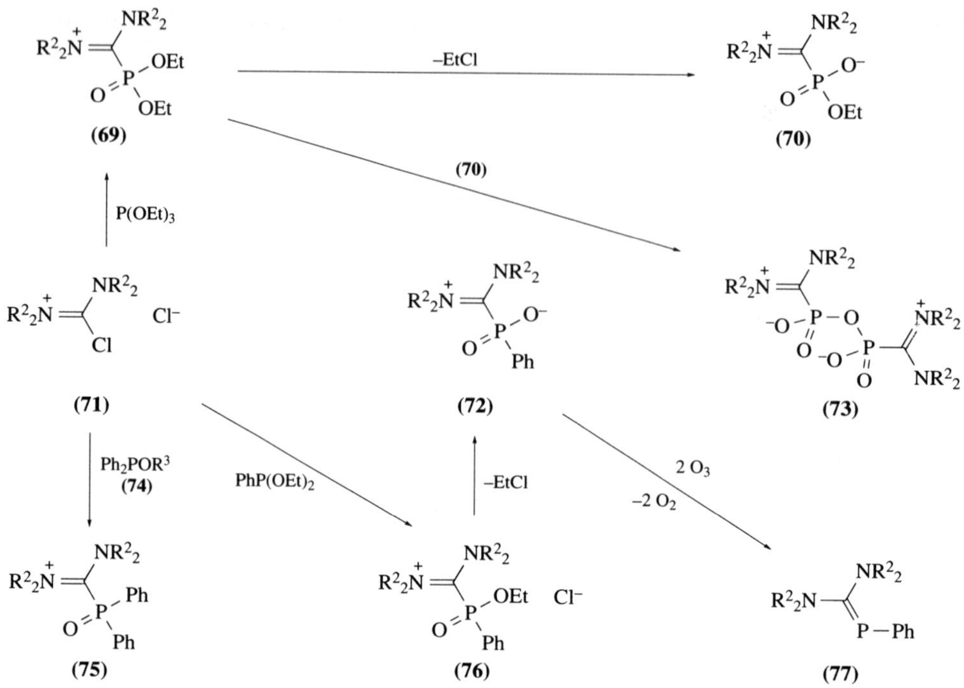

Scheme 31

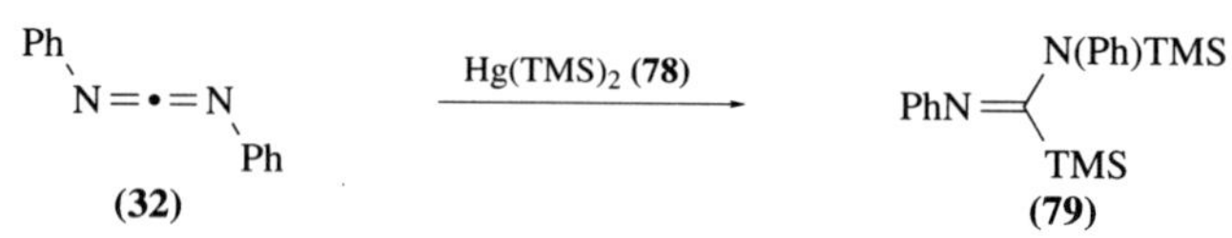

(51)

6.21.1.3.2 *Boron derivatives*

Borane carboximidamides (**82**) are crystalline compounds which are stabilised by internal donor–acceptor bonding ⟨82IZV2370⟩. Compounds (**82**; $R^1 = Ph$, $R^2 = R^3 = Me, Pr^i$, $R^4 = Bu, Pr^i$; $R^1 = Me$, $R^2 = R^3 = Ph$, $R^4 = Bu$) ⟨79DOK(245)121⟩ and (**82**; $R^1 = Ph$, $R^2 = R^4 = Pr^i$, $R^3 = H$) ⟨82IZV2370⟩ are prepared by the exothermic reaction of phenyl isocyanide with either a dialkylboryl derivative (**81**) or a boraneamidine (**83**; $R^2 = Pr^i$), respectively (Scheme 32). The mechanism of the latter reaction may involve the intermediate (**84**), which undergoes an electrocyclic reaction. The sterically hindered boraneamidine (**83**; $R^2 = Bu^t$) gives the related product (**82**; $R^1 = Ph$, $R^2 = H$, $R^3 = Bu^t$, $R^4 = Pr^i$) via the insertion of phenylisonitrile into the nitrogen–boron bond of the compound (**83**) instead of an electrocyclic reaction ⟨82IZV2370⟩. The reaction of the 2-(dialkylboryl)aminopyridine (**85**) with phenyl isocyanide gives the product (**86**) by the electrocyclic and not the insertion mechanism (Equation (52)) ⟨87IZV1139⟩.

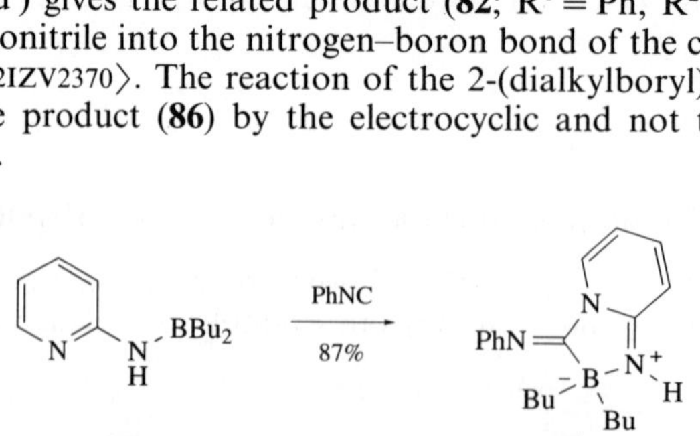

(52)

The only examples of *N'*-alkylboranecarboximidamides reported in the literature are carboximidamides (**87**; $R^1 = Bu^t$, $R^2 = NMe_2$; $R^1 = Pr^i$, $R^2 = Me$) formed in very low yields as side products of other reactions ⟨80ZAAC(486)99, 82JOM(231)191⟩.

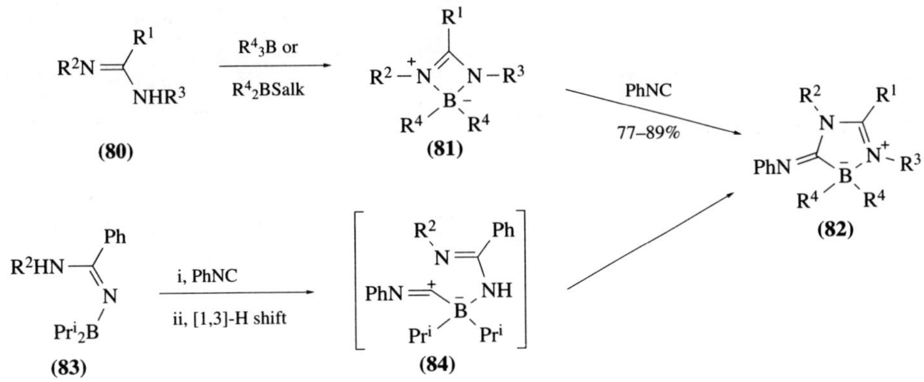

Scheme 32

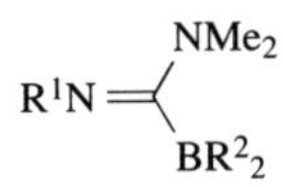

(87)

6.21.1.4 Iminocarbonyl Derivatives with One Nitrogen and One Metal Function

Most of the compounds in this section are carbene complexes of Group 18 transition metals that are formed by nucleophilic reactions of amines with isocyanide ligands of transition-metal complexes.

6.21.1.4.1 Group 18 transition-metal derivatives

Azetidine reacts with the coordinated isocyanide ligand of the neutral complex (*cis*-**88**; M = Pd, Pt, R = Bu^t, 4-$MeOC_6H_4$) and the cationic complex (*trans*-**90**; M = Pd, Pt) in THF to form the corresponding diaminocarbene derivatives (**89** and **91**) (Equations (53) and (54)) ⟨90JCS(D)1197⟩. Azetidine also reacts with the bis(isocyanide) complexes (*cis*-**92**; R = Bu^t, 4-$MeOC_6H_4$), converting one or two ligands into acyclic carbenes (**93**; R = Bu^t and **94**; R = 4-$MeOC_6H_4$) (Equation (55)). The isocyanide complexes (*cis*-**95**; M = Pd, Pt) react in THF with two equivs of azetidine to afford the cationic acyclic diaminocarbene complexes (**96**) containing a metal-coordinated azetidine ligand (Equation (56)) ⟨90JCS(D)1197⟩. Hydrido alkyl carbene complexes (**98**; R = CH_2CN, CH_2CF_3) are prepared by reaction of azetidine with hydrido alkyl isocyanide complexes (**97**) (Scheme 33). These latter complexes react with triphenylphosphine with elimination of HR to give the complex (*trans*-**99**) by a nonreductive elimination pathway ⟨90OM1449⟩.

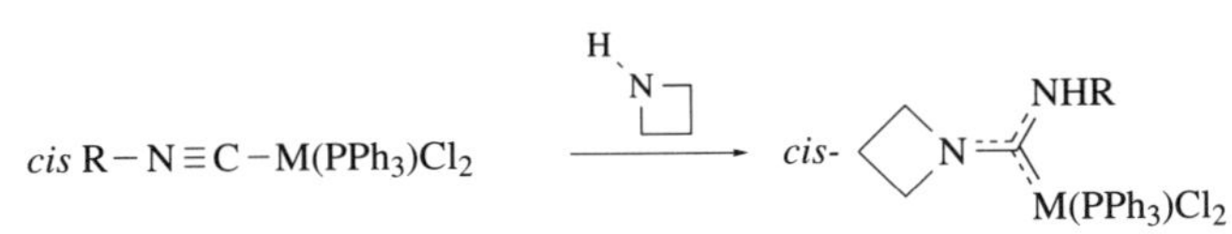

(88) **(89)** (53)

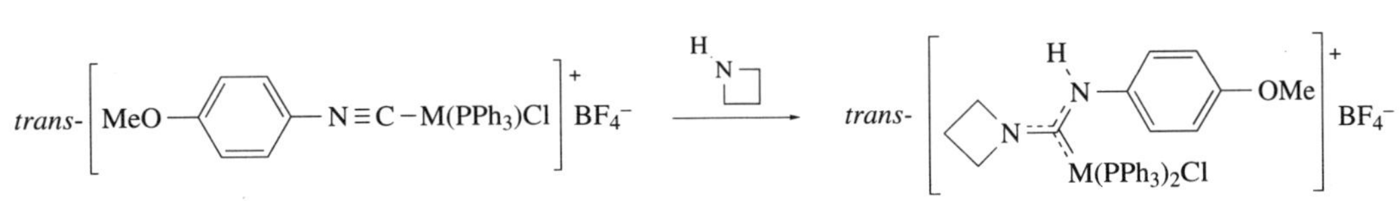

(90) **(91)** (54)

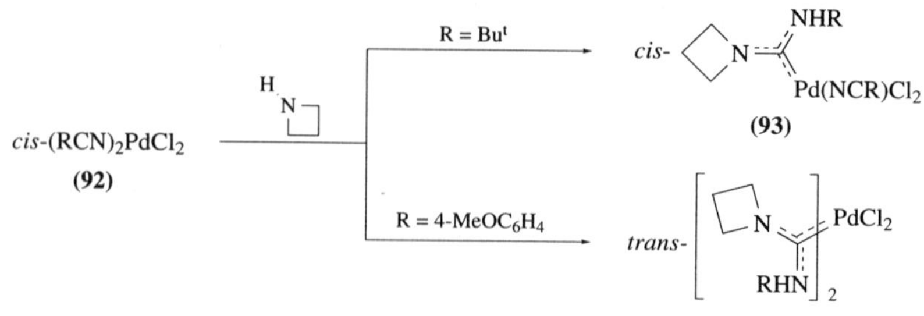

(92) (93) (94) (55)

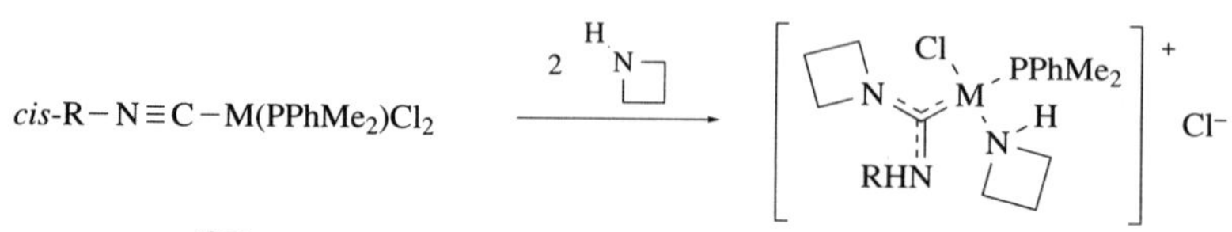

(95) (96) (56)

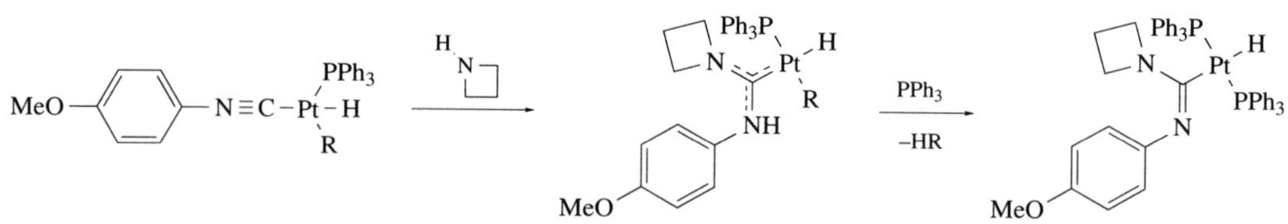

(97) (98) (99)

Scheme 33

Dimethylamine and diethylamine have been used in place of azetidine in reactions with the tetrakis(t-butyl isocyanide)rhodium(I) salt (**100**) to give crystalline 1 : 1 adducts (**101**; R = Me, Et) containing a σ-bonded amidinium cation (Equation (57)) ⟨72JCS(D)1303⟩.

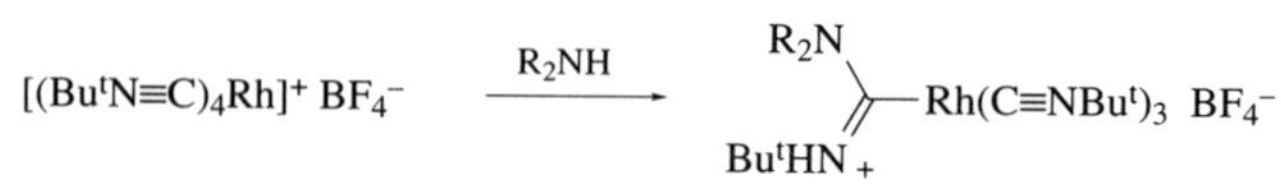

(100) (101) (57)

6.21.1.4.2 *Other metal derivatives*

The few examples of non group 18 transition metal derivatives include the trialkylstannyl- and trialkylplumbylformamidines (**104**; M = Sn, Ar = 4-MeC_6H_4, $R^1 = R^2$ = Me; M = Pb, Ar = Ph, R^1 = Bu, R^2 = Et) obtained by the 1,1-addition of a metal amide (**102**) to an aryl isocyanide (**103**) (Equation (58)) ⟨66TL3423, 68JOM(14)327⟩ and the bis(diazonitromethyl)mercury (**106**) prepared by the reaction of nitrodiazomethane (**105**) with mercury(II) oxide (Equation (59)) ⟨71LA(753)143⟩.

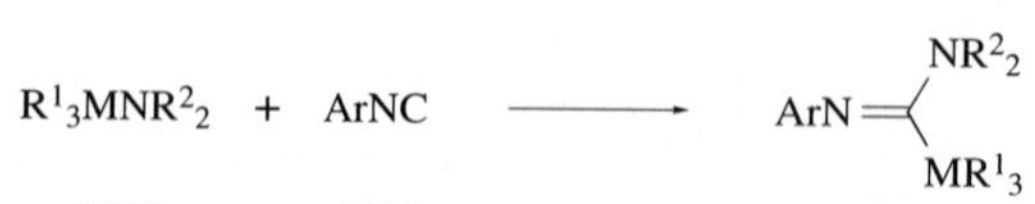

(102) (103) (104) (58)

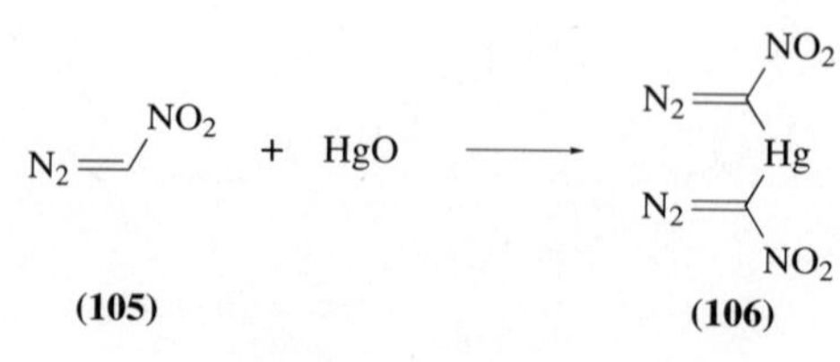

(105) (106) (59)

6.21.2 IMINOCARBONYL DERIVATIVES CONTAINING AT LEAST ONE P, As, Sb OR Bi FUNCTION (AND NO HALOGEN, CHALCOGEN OR NITROGEN FUNCTIONS)

6.21.2.1 Iminocarbonyl Derivatives with One P, As, Sb or Bi Function and One P, As, Sb or Bi Function

Although a number of iminocarbonyl derivatives containing two phosphorus functions exist, there are no examples where the first function is phosphorus and the second function is arsenic, antimony or bismuth. Iminocarbonyl derivatives in which the first function is arsenic, antimony or bismuth are known only for organometallic diazomethanes.

6.21.2.1.1 Iminocarbonyl derivatives with one P function and one P, As, Sb or Bi function

(i) Bis(phosphino)iminocarbonyl derivatives

Low-coordination phosphorus compounds such as the methylenephosphines (*R**,*S**-**107**) and (*R**,*R**-**108**) in which the carbon atom of the iminocarbonyl group is connected to two 3-valent phosphorus atoms, are relatively unstable and have been studied very little ⟨82AG(E)448⟩. However, (diazomethylene)bis(phosphonous diamides) (**111**; R = Pr^i) are more stable and are formed in good yield by the addition of the chlorophosphane (**110**; R = Pr^i) to the lithium salt of the (bisphosphanyl)diazomethane (**109**; R = Pr^i) (Equation (60)) ⟨86JA7868, 88JA2663, 89JOM(372)201, 90JA6277⟩. Minor modification to one of the reagents dramatically alters the course of the reaction, and treatment of compound (**109**; R = C_6H_{11}) with compound (**110**; R = C_6H_{11}) gives a mixture of the diazo derivative (**111**; R = C_6H_{11}) and the nitrile imine (**112**; R = C_6H_{11}) in a 4:21 ratio ⟨92JA6059⟩.

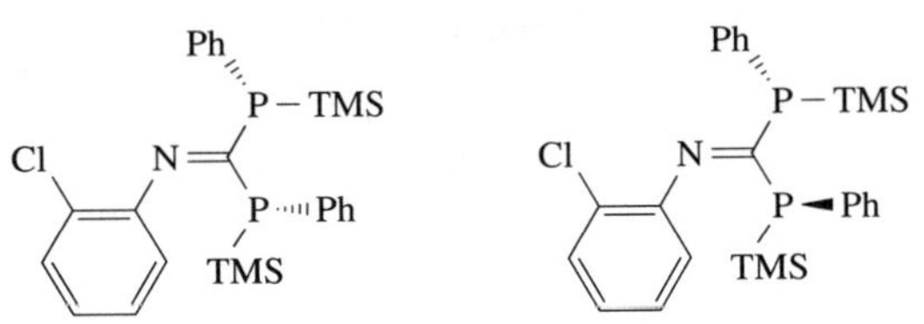

(107) **(108)**

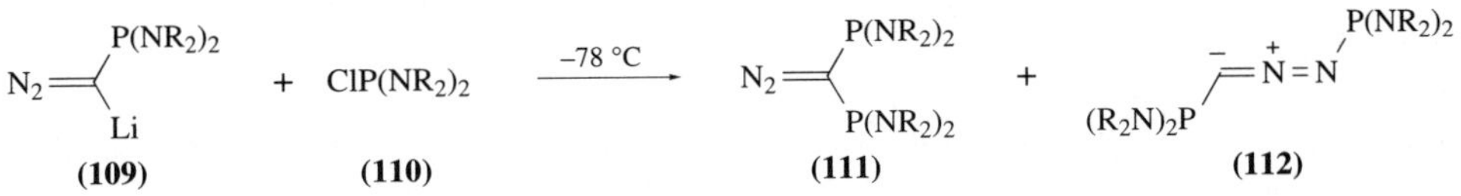

(109) **(110)** **(111)** **(112)** (60)

(ii) Bis(phosphinyl)iminocarbonyl derivatives

The long list of compounds containing 5-valent phosphorus atoms includes diphenyl-, dialkoxy-, diamino-, alkoxyamino- and alkoxyfluorophosphinyl derivatives, and phosphinothioyl derivatives. The methods of synthesis of all these compounds are essentially the same.

(a) Bis(phosphinyl)iminocarbonyl derivatives from carbonimidic dichlorides and organophosphorus reagents. The reaction of carbonimidic dichlorides and organophosphorus compounds is the most popular method of synthesis of bis(phosphinyl)iminocarbonyl derivatives. Gross *et al.* have prepared a large number of (arylcarbonimidoyl)bisphosphonic acid esters by the reaction of a carbonimidic dichloride with $(EtO)_2P(O)R$ (Equation (61); R = OEt ⟨72EGP93766⟩; R = Me or Et ⟨72JPR87⟩. *N*-(Bis(diphenylphosphinyl)methylene)arylamines are produced in a similar fashion with $Ph_2P(O)(OR)$ (Equation (62); R = Me, Et) ⟨72EGP93766, 72JPR87⟩.

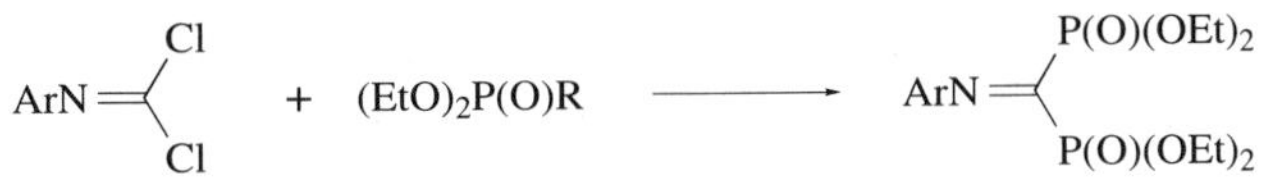

(61)

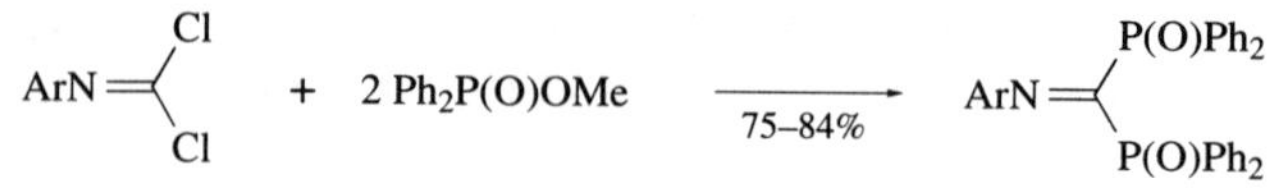

(62)

The Michaelis–Arbuzov reaction of trichloromethyl isocyanate (**113**) with alkyldihalo-, dialkylhalo- and trialkyl phosphites has been studied extensively by Shokol *et al.* (Scheme 34). Treatment of trichloromethyl isocyanate with two molar equivs of $ClP(OEt)_2$ in the presence of $FeCl_3$ as catalyst gives an intermediate (**114**) which can be reacted with a variety of nucleophiles; for example, ethanol gives the carbamate (**116**; R = OEt) ⟨75ZOB1965⟩. In a similar fashion, reaction of trichloromethyl isocyanate with two molar equivs of $(EtO)_3P$ gives a related intermediate (**115**) which can be converted to either the ester (**116**; R = EtO) or urea (**116**; R = PhNH) by reaction with ethanol or aniline, respectively ⟨73ZOB544⟩. Phenylsulfonylimino derivatives (**119**; R = Me, Et) are also prepared by the Michaelis–Arbuzov reaction (Equation (63)) ⟨66CB1252⟩.

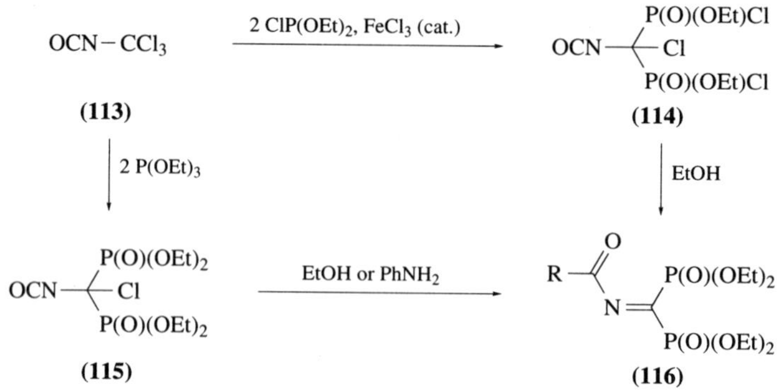

Scheme 34

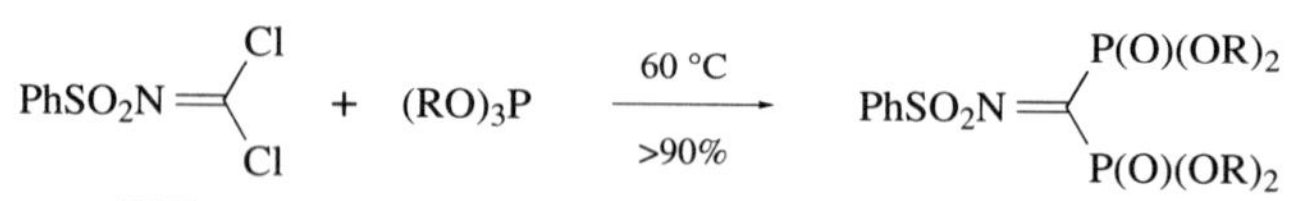

(63)

(117) **(118)** **(119)**

(b) Bis(phosphinyl)iminocarbonyl derivatives from carbonimidic dichlorides by metal–halogen exchange. *N*-(Bis(diphenylphosphinyl)methylene)arylamines are produced from carbonimidic dichlorides by metal–halogen exchange with $Me_2TlP(O)Ph_2$ (Equation (64)) ⟨73ZC192⟩.

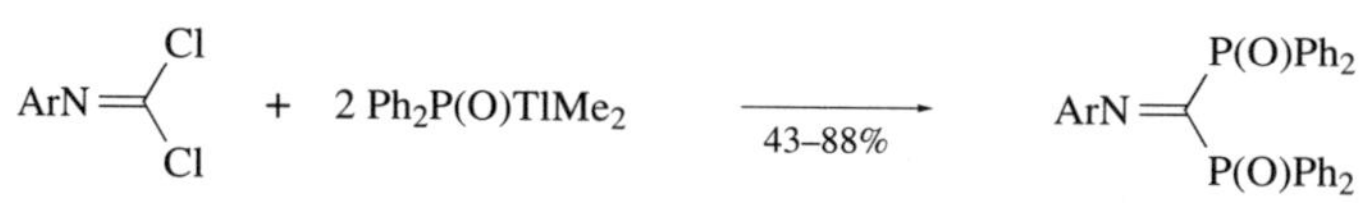

(64)

(c) Bis(phosphinyl)iminocarbonyl derivatives from tosyl azides. (Diazomethylene)bisphosphonates are synthesised by reaction of tosyl azide with the corresponding C—H active methylene precursor in the presence of potassium *t*-butoxide (Equation (65); X = O, R = OMe, OEt) ⟨68CB3734, 68TL3171, 92JOC178⟩. The yield in the reaction is improved by using 2-naphthalenesulfonyl azide in place of tosyl azide and by separating the silica-sensitive product from the insoluble sulfonamide by-product by filtration rather than chromatography ⟨91S405⟩. Bis(diphenylphosphinothioyl)imino derivatives are made by an analogous reaction (Equation (65); X = S, R = Ph) ⟨75MI 621-02, 75OMS(10)1021, 79T181⟩. Mixed diesters of (diazomethylene)bisphosphonates (**121**) are obtained in excellent yield by treating the potassium salt of the methylenebisphosphonate (**120**) with tosyl azide in THF (Scheme 35) ⟨72S351, 77JA1267, 82JOC1284⟩. The methyl ester group of (**121**) is hydrolysed with careful temperature control to the phosphonate salt (**122**) via a *trans*-esterification reaction with TMS-Br ⟨82JOC1284⟩. The α-diazophosphonate intermediate (**123**) may be isolated by extraction from CCl_4 solution into aqueous base or buffer followed by evaporation. A study of the stability and photochemical behaviour of α-diazophosphonates has been carried out by Bartlett *et al.* ⟨82CC536, 82JOC1284⟩.

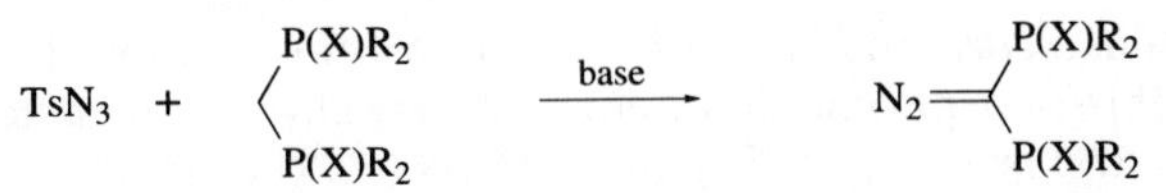

(65)

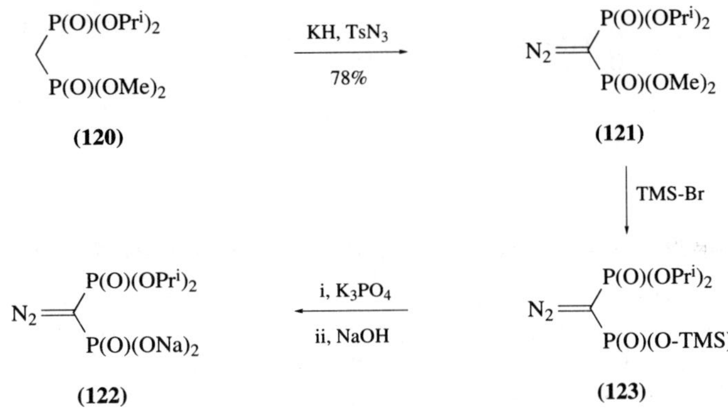

Scheme 35

(d) Bis(phosphinyl)iminocarbonyl derivatives from diazometallic salts. α-Diazophosphonothioic diamides are made from diazolithium salts (Scheme 36). Addition of chlorophosphane derivatives (**125**; $R^1 = Ph, NMe_2$) to the lithium salt of the thioxophosphoranyldiazomethane (**124**; $R^2 = NPr^i{}_2$) leads to the diazo compounds (**126**; $R^1 = Ph, NMe_2, R^2 = NPr^i{}_2$). In contrast, the chlorophosphanes (**125**; $R^1 = Bu^t, NPr^i{}_2$) react with compound (**124**; $R^2 = NPr^i{}_2$) to give the stable nitrile imines (**127**; $R^1 = Bu^t, NPr^i{}_2, R^2 = NPr^i{}_2$). In the same way, chlorophosphane (**125**; $R^1 = NPr^i{}_2$) also reacts with compound (**124**; $R^2 = Bu^t$) to give the nitrile imine (**127**; $R^1 = NPr^i{}_2, R^2 = Bu^t$). Nitrile imine (**127**; $R^1 = R^2 = NPr^i{}_2$) rearranges on heating into the isomeric diazo derivatives (**126**; $R^1 = R^2 = NPr^i{}_2$). Reaction of the diazo compounds (**126**; $R^1 = Ph, NMe_2, R^2 = NPr^i{}_2$) with elemental sulfur gives the bis(thioxophosphoranyl)diazo derivatives (**128**; $R^1 = Ph, NMe_2, R^2 = NPr^i{}_2$) ⟨90JA6277⟩. The chemistry of phosphinothioyl derivatives (**126**; $R^1 = R^2 = NPr^i{}_2$; $R^1 = NPr^i{}_2, R^2 = Bu^t$; and $R^1 = Ph, NMe_2, R^2 = NPr^i{}_2$) has been described by Bertrand and co-workers ⟨88AG(E)1350, 88JA2663, 90JA6277, 91CB1739⟩.

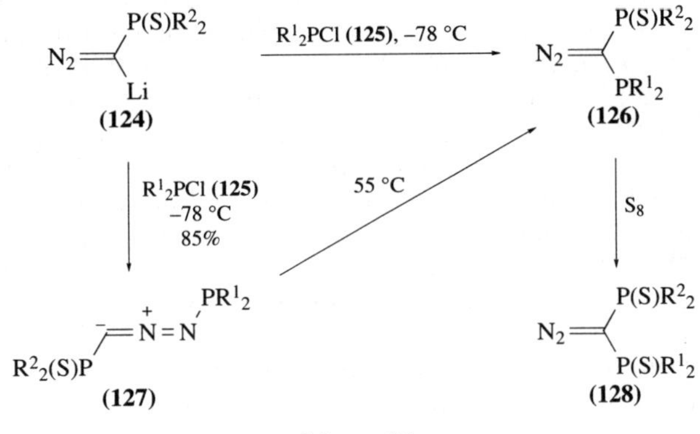

Scheme 36

The diazosilver derivative (**129**) reacts with the chlorophosphine (**130**) to give the unstable (bis(TMS)methylene)phosphino)diazomethylphosphine oxide (**131**), which undergoes an immediate 1,5-cyclisation and a TMS shift at low temperature to give the phosphadiazole (**132**) (Scheme 37) ⟨89S511⟩.

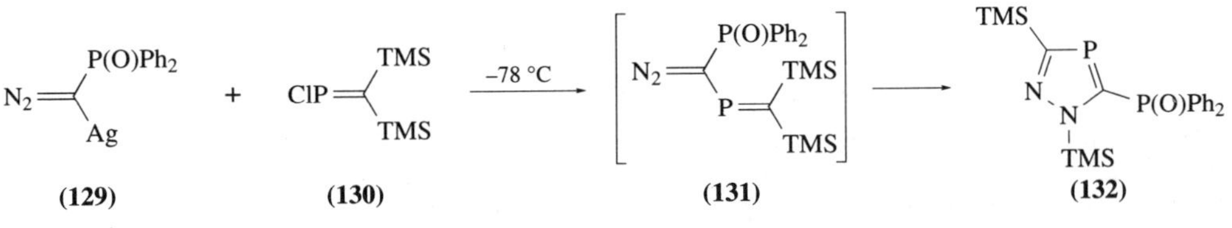

Scheme 37

(e) Bis(phosphinyl)iminocarbonyl derivatives from arylisocyanates. The only known example of a *P,P′*-(carbonimidoyl)bis(phosphonic diamide) is prepared from the reaction of a phosphorus(III) acid anhydride with an aryl isocyanate (Equation (66)) ⟨82ZOB2187⟩.

(66)

6.21.2.1.2 *Iminocarbonyl derivatives with one As, Sb or Bi function and another As, Sb or Bi function*

The only representatives of this class of compound are the bis(3-valent) organometallic diazomethanes (**135**; M = As, Sb, Bi) prepared by treating arsino-, stibino- and bismuthino dimethylamides (**134**; M = As, Sb, Bi) with diazomethane (**133**) (Equation (67)) ⟨75JOM(93)339⟩. Mixed organometallic derivatives arise from two step reactions ⟨77JOM(132)359⟩.

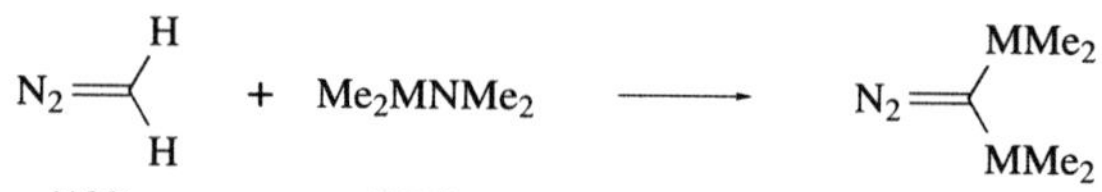

(67)

(133) **(134)** **(135)**

6.21.2.2 Iminocarbonyl Derivatives with One P, As, Sb or Bi Function and One Si, Ge, or B Function

6.21.2.2.1 *Iminocarbonyl derivatives with one P function and one Si, Ge or B function*

(i) Silicon derivatives

All of the compounds in this class are phosphorus-containing silyldiazomethane derivatives. They are often prepared by the reaction of the lithium salt of a diazomethane derivative with a stoichiometric amount of the requisite chlorophosphane or chlorosilane (Scheme 38). Thus, dialkylphosphanylsilyldiazomethanes (**137**; X = lone pair, $R^1R^2R^3Si$ = TMS, $R^4 = R^5 = Bu^t$) ⟨90JA6277⟩ and dialkylaminophosphanylsilyldiazomethanes (**137**; X = lone pair, $R^1R^2R^3Si$ = TMS, $R^4 = R^5$ = dialkylamino) ⟨92BSF367⟩ are prepared readily from chlorophosphanes (**136**) at low temperature; however, the phosphavinyldiazoalkanes (**137**; X = lone pair, $R^1R^2R^3Si$ = TMS, R^4R^5 = :C(TMS)$_2$, :C(TMS)Ph) are only stable at −78 °C and on warming undergo rearrangement reactions to give phosphadiazoles ⟨89S511⟩. Silylated α-diazo phosphonates (**137**; X = O, $R^1R^2R^3Si$ = TMS, TBDMS, $SiPr^i_3$, $R^4 = R^5$ = OMe, OEt) ⟨79LA1002, 85JOM(290)33⟩ and phosphonothioic diamides (**137**; X = S, $R^1R^2R^3Si$ = TMS, $SiPh_3$, $R^4 = R^5 = NPr^i_2$) ⟨88JA2663, 91JOC1801⟩ are prepared from lithiated α-diazo phosphonates (**138**) by reaction with silyl electrophiles (Z = Cl, CF_3SO_3). The phosphonothioic diamides are formed via nitrile imine intermediates ($R^1R^2R^3SiN{=}N^+{=}C^-P(S)NPr^i_2$), which are the only products isolated in reactions with bulky electrophiles, (e.g., Pr^i_3SiCl).

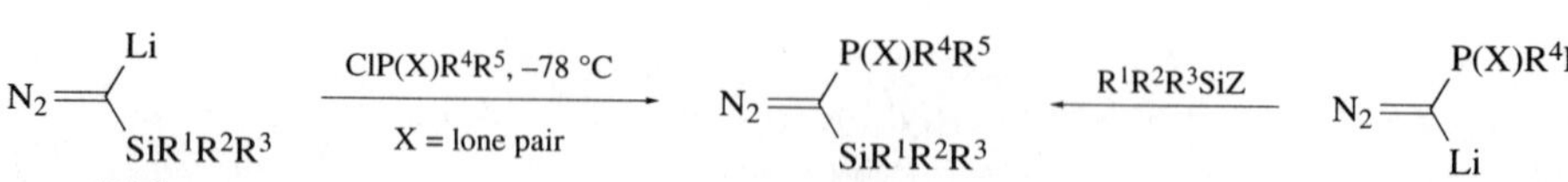

(136) **(137)** **(138)**

Scheme 38

Dialkylaminophosphanylsilyldiazomethanes (**137**; X = lone pair, $R^1R^2R^3Si$ = TMS, $R^4 = R^5$ = dialkylamino) are precursors of donor-substituted 'stabilised' carbenes ⟨91AG(E)674, 91NJC393, 92BSF367⟩. The diisopropylamino analogue (**139**) on flash thermolysis gives the stable phosphanylcarbene (**140**) (Scheme 39), which undergoes a [2 + 3] cycloaddition followed by a ring-opening reaction with either N_2O to give the diazo phosphonic diamide (**142**; X = O) in a yield of

60% ⟨91NJC393⟩ or TMS-N_3 at $-78\,°C$ to 4°C to form the diazo compound (**142**; X = TMS-N) ⟨91AG(E)674⟩. The reaction with N_2O may be considered a formal oxidation of compound (**139**).

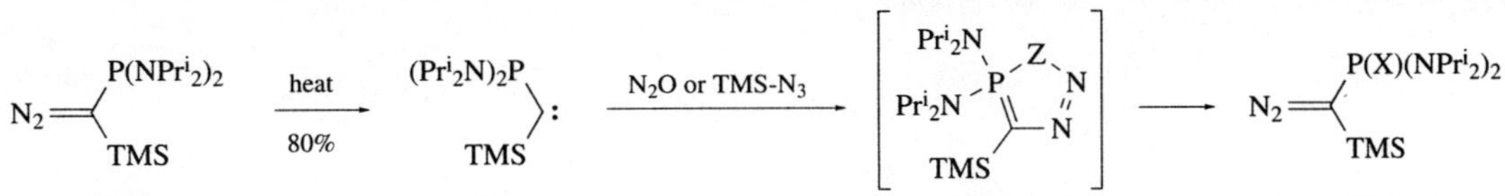

(139) **(140)** **(141)** **(142)**

Scheme 39

α-Diazo phosphine sulfides (e.g., **144**; R = Bu^t) ⟨90JA6277⟩ and α-diazo phosphonothioic diamides (e.g., **144**; R = NPr^i_2) ⟨89JOC4426⟩ may be prepared by direct sulfuration of the phosphanyl precursors (**143**) (Equation (68)).

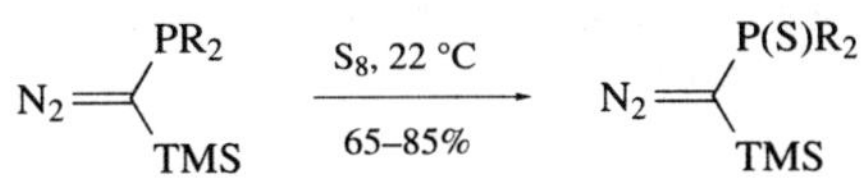 (68)

(143) **(144)**

(ii) Germanium derivatives

There are few reports of germanium derivatives, but preparations involving the use of trialkylgermanium chloride are known. Thus, the α-diazo phosphonothioic diamide triethylgermanium derivative (**145**) is prepared by the same method as the silicon analogue (**142**) ⟨88JA2663⟩. A reaction which exploits the weakness of the Ge—P bond of germylphosphines involves an insertion reaction of phenyl isocyanide (Equation (69)) ⟨72JOM(34)83⟩.

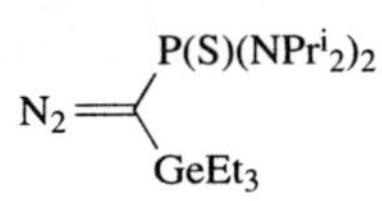

(145)

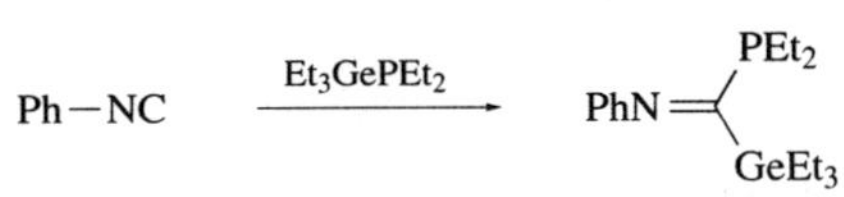 (69)

(iii) Boron derivatives

The only examples of this class are the internal salts (**148**; R = H, F). These are synthesised by oxidative ylidation of the α-diazophosphane (**146**) with CCl_4 followed by reaction with boron-containing Lewis acids (Scheme 40) ⟨91AG(E)1154, 92BSF367⟩.

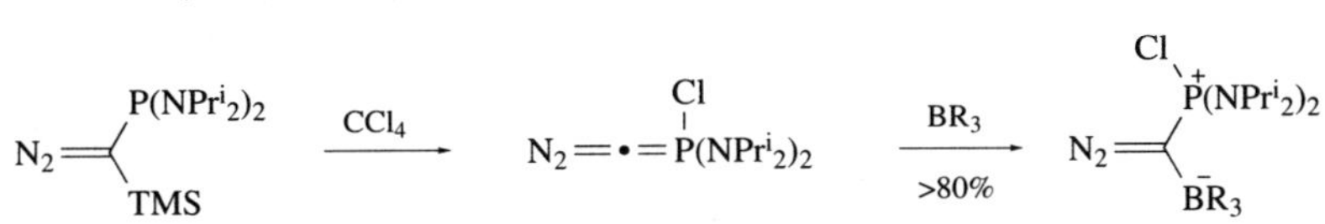

(146) **(147)** **(148)**

Scheme 40

6.21.2.2.2 *Iminocarbonyl derivatives with one As, Sb or Bi function and one Si, Ge or B function*

(α-Diazo(TMS)methyl)dimethyl arsines, stibines and bismuthines (**150**; M^1 = Si, M^2 = As, Sb, Bi, R = Me, n = 3) are prepared by the reaction of (TMS)diazomethanes (**149**) with metal amides

(Equation (70)). Trimethyltin chloride is used as a catalyst for the arsine derivative ⟨80JOM(191)371⟩. The (diazotrimethylgermylmethyl)dimethylarsine (**150**; $M^1 = Ge$, $M^2 = As$, R = Me, n = 3) is prepared in a similar fashion ⟨77JOM(132)359⟩.

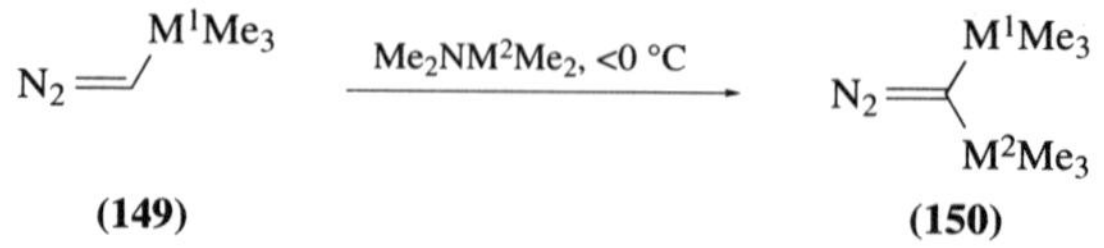

(70)

6.21.2.3 Iminocarbonyl Derivatives with One P, As, Sb and Bi Function and One Metal Function

The only examples of this class are diazomethane derivatives prepared as intermediates for the study of carbenes. Most of the compounds are prepared by metallation or transmetallation of diazomethane precursors.

6.21.2.3.1 Iminocarbonyl derivatives with one P function and one metal function

(i) Group 1 metals

Diazolithium salts (**152**; X = lone pair, $R^1 = R^2 = NPr^i_2$; X = S, $R^1 = R^2 = NPr^i_2$, Bu^t) are prepared at low temperature by treating the corresponding diazomethane precursors (**153**) with BuLi (Scheme 41) ⟨90JA6277, 91OM3205⟩.

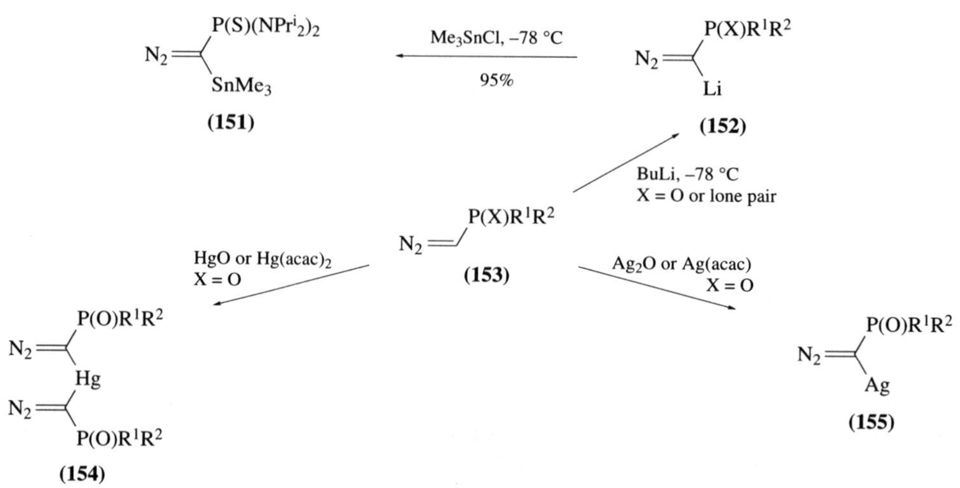

Scheme 41

(ii) Transition metals

The phosphorus-containing diazomethylsilver (**155**; $R^1 = R^2 = OMe$, OEt; $R^1 = OMe$, $R^2 = Ph$) ⟨71LA(748)207, 89MI 621-01⟩ and the bis(diazomethyl)mercury (**154**; $R^1 = R^2 = OMe$, OEt, Ph) ⟨71JOC1379, 71LA(748)207⟩ are prepared by metallation reactions with metal oxides or metal acetylacetonates (Scheme 41). The stannyldiazomethane (**151**) is prepared by transmetallation of compound (**152**; X = S, $R^1 = R^2 = NPr^i_2$) with trimethylstannyl chloride ⟨92JA6059⟩. The crystalline bis(diazomethylene)mercury(II) salt (**157**) is produced by the catalytic decomposition of the diazomethylenephosphorane (**156**) with mercury(II) chloride followed by anion exchange with potassium hexafluorophosphate (Equation (71)) ⟨92BSF367⟩.

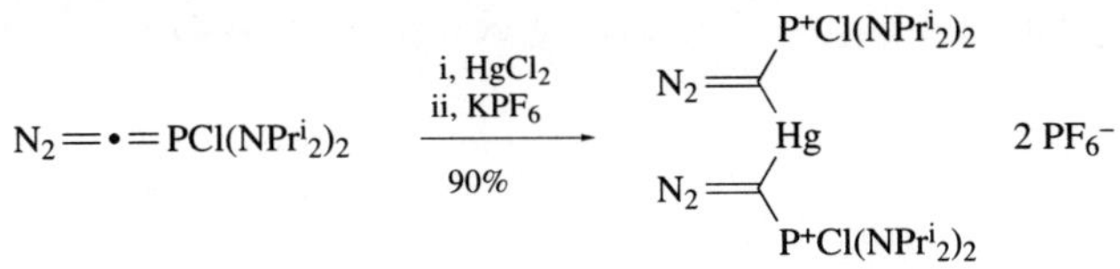

(71)

(156) **(157)**

6.21.2.3.2 *Iminocarbonyl derivatives with one As, Sb and Bi function and one metal function*

Diazomethylarsine derivatives with trimethylstannyl and trimethylplumbyl substituents (**158**; M = Sn, Pb) are prepared by metallation of diazomethylarsines (**159**) with metal amides (Scheme 42). Explosive bis(diazo(dimethylarsino)methyl)mercury (**160**) is produced using $Hg(N(TMS)_2)_2$ as metallating agent ⟨77JOM(132)359⟩.

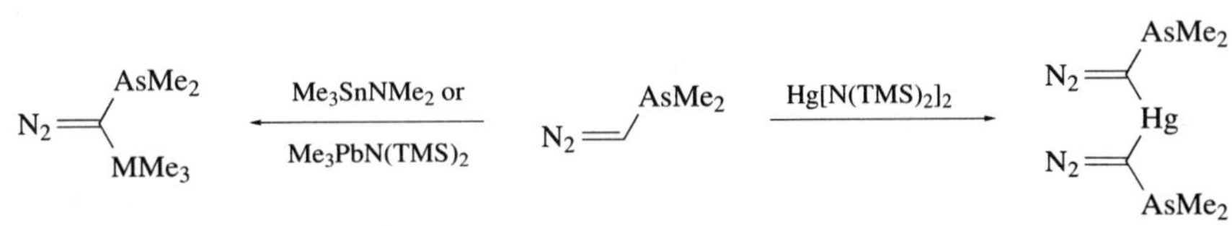

(158) **(159)** **(160)**

Scheme 42

6.21.3 IMINOCARBONYL DERIVATIVES CONTAINING AT LEAST ONE METALLOID FUNCTION (AND NO HALOGEN, CHALCOGEN, OR GROUP 5 ELEMENT FUNCTIONS)

6.21.3.1 Iminocarbonyl Derivatives with Two Metalloid Functions

6.21.3.1.1 N-*Unsubstituted iminocarbonyl derivatives*

The electronic structure of HN=C(TMS)$_2$ has been calculated but its synthesis is not reported ⟨92MI 621-01⟩.

6.21.3.1.2 N-*Alkyl- and* N-*aryliminocarbonyl derivatives*

These compounds are prepared by the insertion of alkyl- or aryl isocyanides into metal–metal bonds or by transmetallation reactions.

N-Alkylbis(silyl)imines are less stable than their *N*-aryl analogues and only one *N*-alkyl-substituted compound has been reported. The *N*-cyclohexyl derivative (**161**; R = C_6H_{11}) is prepared by the insertion reaction of *N*-cyclohexyl isocyanide into the Si—Si bond of a disilane with catalysis by Pd(0) ⟨87TL1293⟩ or Pt(0) ⟨90JAP2209879⟩ (Equation (72), M = Pd, Pt). Sterically-hindered *t*-alkyl isocyanides fail to undergo the reaction. A wide range of *N*-aryl analogues has been prepared by the Pd(0)-catalysed method ⟨87TL1293⟩. The reaction is exothermic with disilanes containing electron-withdrawing substituents. A study of the insertion of aryl isocyanides into the Si—Si linkage of polysilanes indicates that palladium(II) acetate is a better catalyst for the preparation of poly(sila(*N*-substituted)imines) ⟨88JA3692, 91JA8899⟩.

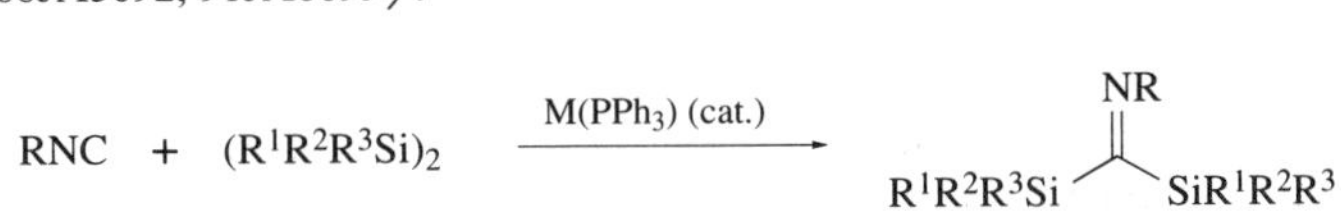

(72)

(161)

(2,6-Xylylimino)bissilanes (**165**; $SiR^1R^2R^3$ = TMS, *t*-butyldimethylsilyl(TBDMS)) are made from (2,6-xylylimino)(TMS)methyllithium (**163**) by treatment with the requisite chlorosilane (Scheme 43) ⟨87JA7888⟩. ((2,6-Xylylimino)trialkylsilyl)metal derivatives are of interest as acyl anion synthetic equivalents; the 2,6-xylylimino copper-containing compound (**164**) is more stable than its 2-tolylimino counterpart.

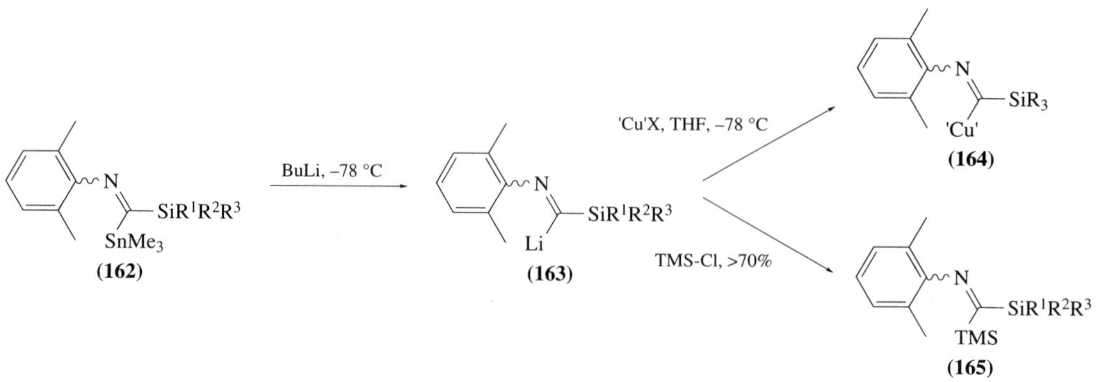

Scheme 43

6.21.3.1.3 N-*Haloiminocarbonyl derivatives*

The electronic structures of the imines HalN=C(TMS)$_2$ have been calculated but their syntheses are not reported ⟨92MI 621-01⟩.

6.21.3.1.4 N-*Aminoiminocarbonyl (diazomethane) derivatives*

The major methods of preparation of this class of compounds are transmetallation, diazo group transfer and cycloaddition reactions.

Examples of metallation reactions include the treatment of silylated ⟨87OM1857⟩ and germylated ⟨70JCS(A)2954⟩ lithiodiazomethanes (**166**; R^1 = TMS, $GeMe_3$) with chlorosilanes and chlorogermanes to give bis(silyl)diazomethanes (**167**; R^1 = TMS, R^2 = TMS, (TMS)$SiMe_2$, (TMS)$_3$Si) and bis(germyl)diazomethanes (**167**; $R^1 = R^2$ = $GeMe_3$), respectively (Scheme 44). Bis(germyl)diazomethane (**167**; $R^1 = R^2$ = $GeMe_3$) is much less stable than its silyl counterpart and has also been prepared by a dimethylamine elimination reaction with diazomethane (**168**) ⟨70JCS(A)2954⟩. This method fails for the less reactive amides of silicon (i.e., $TMSNMe_2$) and boron (i.e., Ph_2BNMe_2) owing to their less ionic character. A further method of preparation of this class of compound involves the transfer of the diazo group from tosyl azide to the carbanion (**169**; $R^1 = R^2$ = TMS, $GeMe_3$) derived from bis(TMS)- or bis(germyl)methanes ⟨80JA1584⟩.

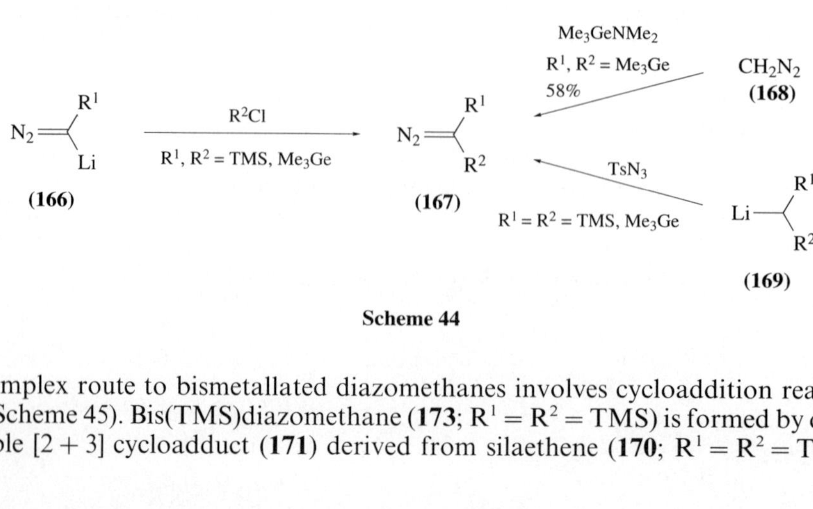

Scheme 44

A more complex route to bismetallated diazomethanes involves cycloaddition reactions of metalloethenes (Scheme 45). Bis(TMS)diazomethane (**173**; $R^1 = R^2$ = TMS) is formed by decomposition of the unstable [2 + 3] cycloadduct (**171**) derived from silaethene (**170**; $R^1 = R^2$ = TMS) and N_2O

⟨88CB1407⟩. A similar reaction involving the [2 + 3] cycloaddition of silaethene (**170**; $R^1 = TMS$, $R^2 = TMS$, TBDMS) with a silyl azide R^3N_3 (R^3 = silyl group) forms siladihydrotriazole (**174**), which undergoes a [2 + 3] cycloreversion to give the product (**173**; $R^1 = TMS$, $R^2 = TMS$, TBDMS) ⟨86CB1467, 86CC591, 87CB1203⟩. Siladihydrotriazoles also rearrange by an alternative mechanism to give bis(silyl-substituted)diazomethanes and this reaction has been used to prepare a wide range of silyl compounds (**175**; R^1–R^3 = silyl group) ⟨86CB1467, 87CB1203, 87CB1213⟩. Similar reactions are observed with the silaoxadiazole (**171**; $R^1 = R^2 = TMS$), which rearranges to give the bis(silyl-substituted)diazomethane (**172**; $R^1 = R^2 = TMS$) ⟨88CB1407⟩.

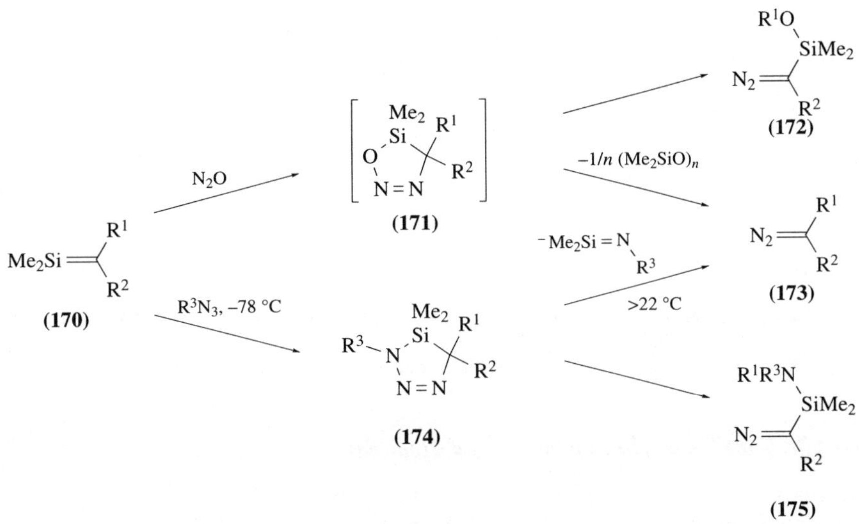

Scheme 45

The use of the germaethene (**177**) and the alkylidenaminoborane (**179**) in reactions analogous to those shown for silaethene in Scheme 45 produces the respective germylsilyldiazomethanes (**176** and **178**) (Scheme 46) ⟨86CB2980, 87CB1203⟩ and the boranylsilyldiazomethane (**180**) (Equation (73)) ⟨89CB595⟩.

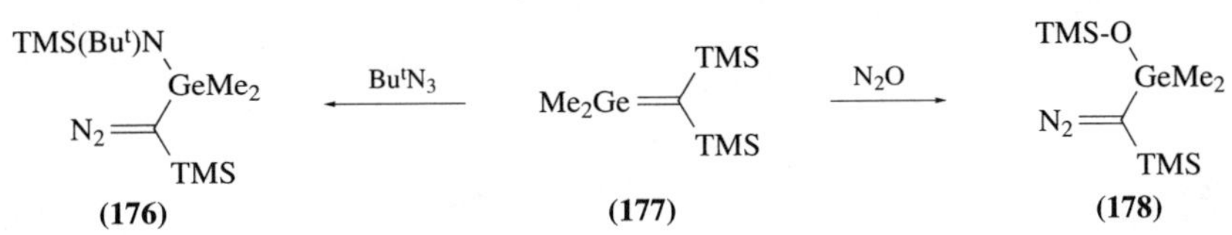

Scheme 46

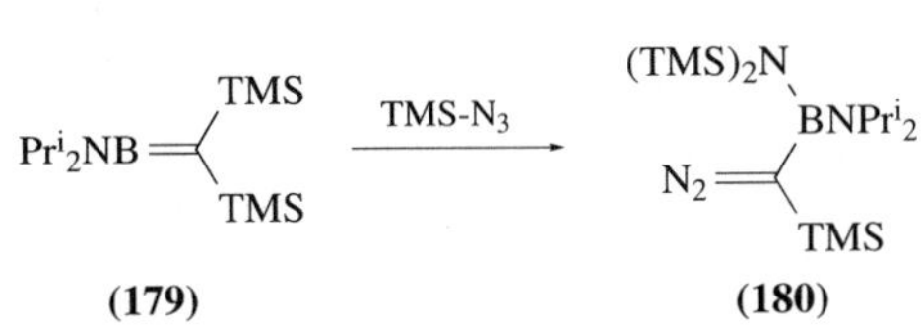

(73)

Finally, diazoalkanes are useful precursors to carbenoid species. The transformation is catalysed by transition metals, and an investigation of the reaction of the anion of (TMS)diazomethane with $M(CO)_5PPh_3$ (M = Cr, W) (Scheme 47) resulted in the isolation of two stable crystalline *N*-bonded diazoalkane–metal complex intermediates (**181**; M = Cr, W) ⟨88CC1598⟩.

6.21.3.1.5 N-*Silyliminocarbonyl derivatives*

N-Silylated bis(TMS)imines (**183**; R = TMS, $(TMS)_2NSiMe_2$) are formed as side products in the reaction of silaethene (**182**) with silyl azides (RN_3) (Scheme 48) ⟨81CB3518, 87CB1203⟩.

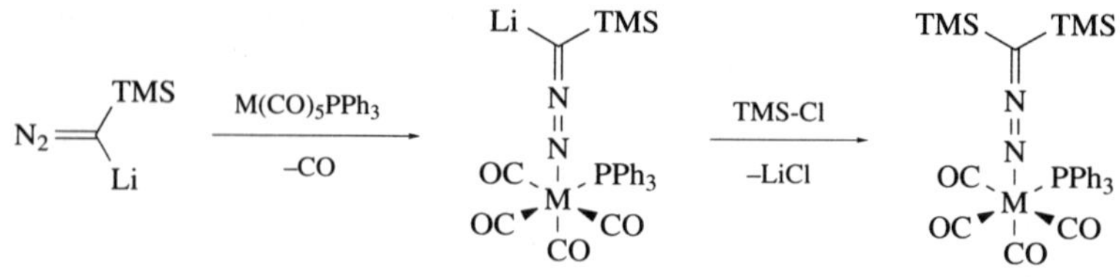

(181)

Scheme 47

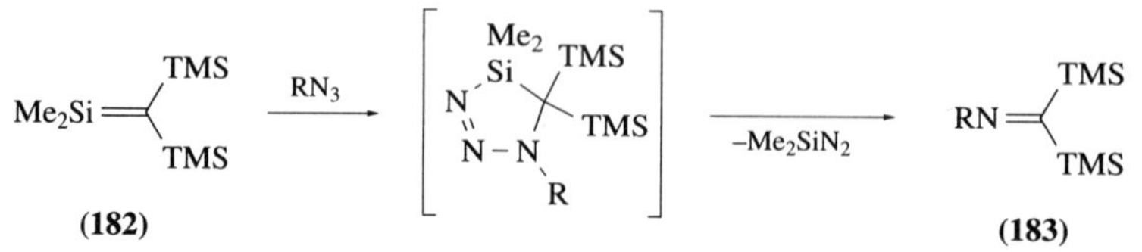

(182) (183)

Scheme 48

6.21.3.2 Iminocarbonyl Derivatives with One Metalloid Function and One Metal Function

The only examples of this class of compound have silicon as the metalloid.

6.21.3.2.1 N-*Alkyl- and* N-*aryliminocarbonyl derivatives*

These compounds are similar to iminocarbonyl derivatives with two metalloid functions (Section 6.21.3.1.2) and are prepared in a similar fashion by the insertion of alkyl- or aryl isocyanides into metal–metal bonds and by transmetallation reactions.

Organosilyl(*N*-alkylimino)stannanes (**186**; $R^1 = R^2 = Me$, $R^3 = Pr^i$, C_6H_{11}, C_6H_{13}, 2-MeC_6H_4; $R^1 = Bu$, $R^2 = Me$, $R^3 = Pr^i$; $R^1 = Me$, $R^2 = Bu^t$, $R^3 = Pr^i$) are produced by the Pd(0)-catalysed insertion of isonitriles into the Si—Sn bond of organosilylstannanes (Equation (74)) ⟨86CC980⟩. The reaction proceeds in competition with the disproportionation of compound (**184**) to give the distannane (**187**) and disilane (**188**). The Pd(0)-catalysed reaction of (TMS)dimethylstannane with the sterically bulky Bu^tNC gives only disproportionation products.

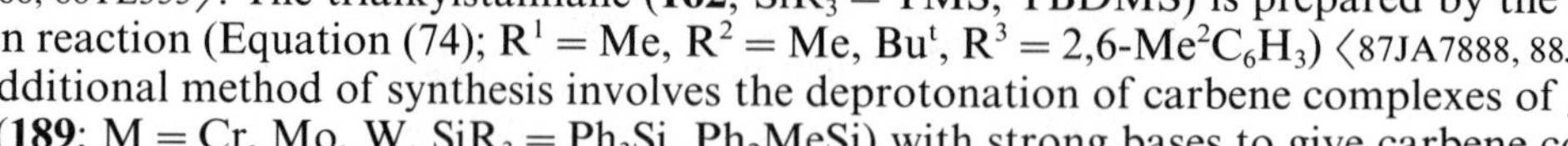

(184) (185) (186) (187) (188) (74)

The lithium derivative (**163**; $SiR_3 = TMS$, TBDMS), generated *in situ* at $-78\,°C$ by transmetallation of the trialkylstannane (**162**; $SiR_3 = TMS$, TBDMS) with BuLi, has been converted into various copper reagents (**165**; $SiR_3 = TMS$) by reaction with $CuBr \cdot SMe_2$ or Cu acetylide (Scheme 43) ⟨87JA7888, 88TL355⟩. The trialkylstannane (**162**; $SiR_3 = TMS$, TBDMS) is prepared by the isonitrile insertion reaction (Equation (74); $R^1 = Me$, $R^2 = Me$, Bu^t, $R^3 = 2,6\text{-}Me^2C_6H_3$) ⟨87JA7888, 88JA3692⟩.

An additional method of synthesis involves the deprotonation of carbene complexes of group 16 metals (**189**; M = Cr, Mo, W, $SiR_3 = Ph_3Si$, Ph_2MeSi) with strong bases to give carbene complexes (**190**) containing formal imino bonds (Equation (75)) ⟨90JOM(385)221⟩.

(189) (190) (75)

6.21.3.2.2 N-*Aminoiminocarbonyl (diazomethane) derivatives*

These compounds are obtained by metallation reactions. Lithium silyldiazomethides (**191**; $SiR_3 = TMS$, tips, TMS-$SiMe_2$) are obtained by treatment of silyldiazomethanes (**192**) with BuLi or

LDA (Scheme 49). The reaction is usually carried out at low temperature (−78 °C) in THF as solvent ⟨85H(23)2371, 91OM3205⟩. The plumbyl- and stannyl(TMS)diazomethanes (**193**; SiR_3 = TMS, M = Pb, Sn) are obtained by reaction of (TMS)diazomethane (**192**) with metal amides ⟨80JOM(191)371⟩. The stannyl(triisopropylsilyl)diazomethane (**196**) is obtained from the reaction of bis(trimethylstannyl)diazomethane (**195**) with Pr^i_3SiCl in a variety of solvents (THF, CH_2Cl_2 or toluene), but the same reaction performed in the absence of solvent gives the bis(triisopropylsilyl)nitrile imine (**194**) (Scheme (50)). A discussion of the scope and mechanism of these reactions has been provided by Bertrand and co-workers ⟨92JA6059⟩.

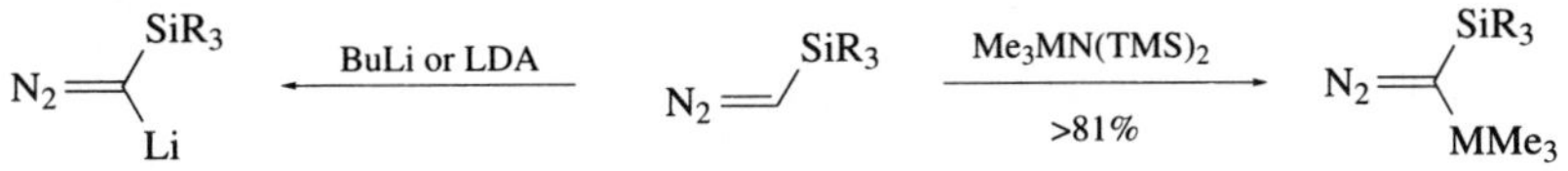

(191) **(192)** **(193)**

Scheme 49

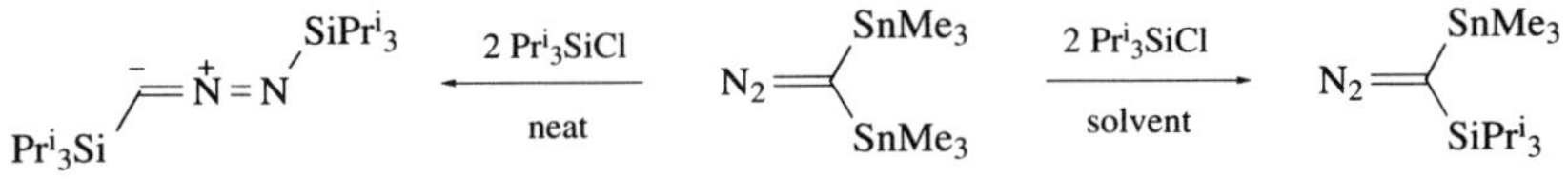

(194) **(195)** **(196)**

Scheme 50

Organic diazoalkanes are useful as precursors to carbenoid species or as 1,3-dipolar cycloaddition reagents in heterocyclic synthesis, and consequently a number of *N*-coordinated diazoalkane transition-metal complexes are known (see, e.g., Scheme 47). *C*-Bonded metal diazoalkane complexes are fewer in number but examples include the crystalline (diazomethyl) trimethylsilanenickel(II) complex (**199**) which decomposes above −25 °C. It is formed by treatment of lithium (TMS)diazomethide (**197**) with the nickel reagent (**198**) in a high-yielding reaction (Scheme 51) ⟨90JA5351⟩. The unstable (diazomethyl)trimethylsilanerhodium(I) complex (**201**) is prepared from the reagent (**200**), also in high yield, and on reaction with iodomethane produces the more stable rhodium(III) diazo complex (**202**) ⟨87OM1822⟩.

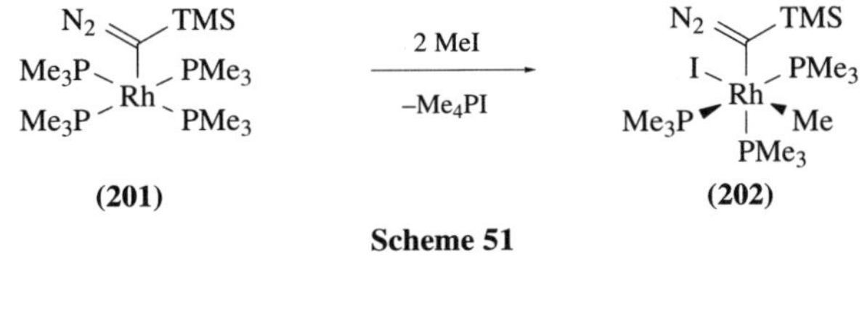

(197) **(199)** **(201)** **(202)**

Scheme 51

6.21.4 IMINOCARBONYL DERIVATIVES CONTAINING TWO METAL FUNCTIONS

There are no examples of this class of compound, if organometallic complexes with bridging isocyanide ligands ⟨82COMC-I(4)523⟩ are excluded.

6.22
Functions Containing Doubly Bonded P, As, Sb, Bi, Si, Ge, B or a Metal

VADIM D. ROMANENKO
Institute of Organic Chemistry, Kiev, Ukraine

and

MICHEL SANCHEZ and LYDIA LAMANDÉ
Université Paul Sabatier, Toulouse, France

6.22.1 FUNCTIONS CONTAINING DOUBLY BONDED P, As, Sb or Bi

6.22.1.1 General Remarks

The chemistry of functions $X^1X^2C{=}E$ (X^1, X^2 = heteroatom substituents, E = P, As, Sb, or Bi) featuring a double bond between carbon and the heavier Group 15 elements is centered largely around phosphorus. The latter forms four types of isolable compounds (**1**–**4**) containing a formal C=P bond.

(1) **(2)** **(3)** **(4)**

In contrast to the phosphorus species the chemistry of their arsenic analogues is in its infancy. Although several structural types of the coordinatively unsaturated arsenic derivatives are known, information about *C*,*C*-diheterosubstituted species $X^1X^2C{=}As(Z)_n(Y)_m$ is very sparse. There is no experimental evidence for double bonding between a X^1X^2C unit and trivalent or tetravalent antimony or bismuth.

6.22.1.2 Dicoordinate Phosphorus and Arsenic Derivatives

The compounds $X^1X^2C{=}EY$ (E = P, As) are currently named in *Chemical Abstracts* as methylenephosphines or methylenearsines. They possess a genuine $(p\text{–}p)\pi$ double bond and can be regarded as the heavier congeners of alkenes. The chemistry of phosphaalkenes is documented by two monographs ⟨B-88MI 622-01, B-90MI 622-01⟩ and by several comprehensive reviews ⟨81AG(E)731, 86ZOB253, 89T6019⟩ that appeared in the last decade. Individual aspects of phosphaalkene chemistry have been covered in numerous review articles. The most important of these are by Becker, Becker and Mundt ⟨83PS(14)267⟩, Kroto ⟨82CSR435⟩, Lutsenko ⟨84ZC345⟩, and Grobe and co-workers ⟨86PS(28)239⟩. The methods of preparation of the low coordinate arsenic compounds have been reviewed by Dianova and Zabotina ⟨91RCR162⟩. This section concerns functions with two heteroatom–carbon bonds, whatever the nature of substituent on the phosphorus or arsenic. Functions of the type $R(X)C{=}EY$ (E = P, As; R = Alk, Ar) are discussed in Chapter 5.23.

Simple heterosubstituted phosphaalkenes and arsaalkenes like $F_2C{=}PH$, $(MeO)_2C{=}PPh$ or $Cl_2C{=}AsPh$ are very unstable. They polymerize rapidly, forming, dependent on the conditions, dimeric or oligomeric compounds $(X^1X^2CEY)_n$ with C—E σ-bonds. Special kinetic or thermodynamic factors are required to increase the stability of $X^1X^2C{=}EY$ species ⟨81AG(E)731, 90MI 622-01⟩. Three general synthetic methods are applied to prepare the $X^1X^2C{=}PY$ functions: (i) the 1,2-elimination reactions of suitable organophosphines, (ii) condensation reactions, and (iii) 1,3-trimethylsilyl migration (Scheme 1). All other methods are of relatively minor importance and are used only in special cases. Table 1 surveys the diversity of the known *C*,*C*-diheterosubstituted phosphaalkenes reported to date with an example from each major class. The list indicates the section in which specific phosphaalkene is covered and also the routes used for its synthesis.

6.22.1.2.1 Dihalomethylenephosphines, $Hal_2C{=}PY$

The synthesis of compounds having this structure has been the subject of extensive investigations, largely because of the ease of making stable molecules of this kind and because of the interesting

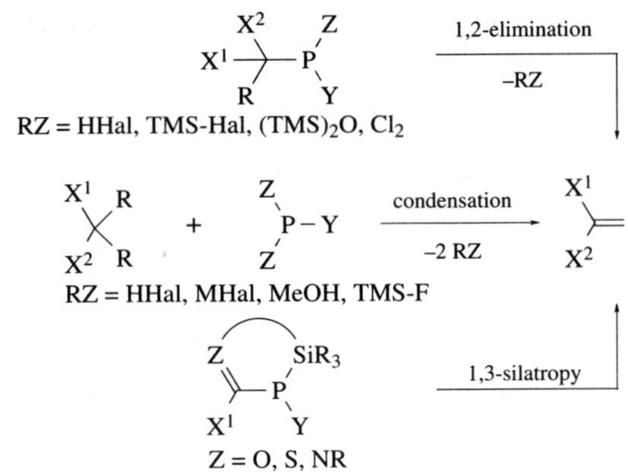

Scheme 1

properties which they possess. These phosphaalkenes allow derivatization at the sp^2-hybridized carbon and they are therefore valuable synthons for the preparation of a large variety of novel functions, such as Hal(X)C=PY, X^1X^2C=PY and XC≡P. Practically all synthetic routes to the dihalomethylenephosphines are based on 1,2-elimination reactions.

As already noted, F_2C=PH is extremely unstable. It has been identified by microwave spectroscopy as a product in the pyrolysis of CF_3PH_2 ⟨76CC513⟩ or in the room temperature reactions between CF_3PH_2 and solid KOH or NaOEt ⟨78JA446, 79CC653⟩. Unlike F_2C=PH, perfluoro-2-phosphapropene (**5**) containing a bulky CF_3 group on phosphorus is stable enough to be isolated in a pure state. This compound was generated for the first time from the reaction of CF_3PH_2 with NaOMe ⟨65JCS6875⟩ or $ZnMe_2$ ⟨81IC3734, 83IC2573⟩. The product obtained by these methods is usually contaminated with other volatile substances and the starting phosphine. In 1984 Grobe and Le Van developed a simple thermolysis procedure according to Equation (1) to produce (**5**) in almost quantitative yield ⟨84AG(E)710, 87PS(30)401⟩. The phosphaalkene formed at 10^{-3} torr is quenched from the gas phase at -196°C. Unreacted starting phosphine is repeatedly fed through the hot zone until the conversion into (**5**) is complete. The kinetic stability observed for the phosphaalkene (**5**) is rather high; thus, in ca. 10% toluene or pentane solutions the dimer is first detectable after about 10 h at 25°C. Therefore, handling in vacuum lines and reactivity studies in organic solvents are possible under ordinary conditions.

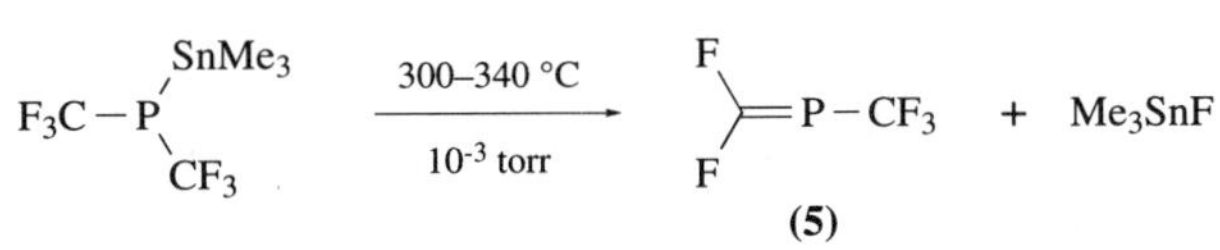

(**5**) (1)

In analogy to (**5**), 2-phospha-1-perfluorobutene (**6**) is obtained from stannylphosphine. The reaction leads to the isomeric phosphaalkenes (**6**) and (**7**) in a molar ratio of 3:1 respectively (Equation (2)) ⟨90ZN(B)148⟩.

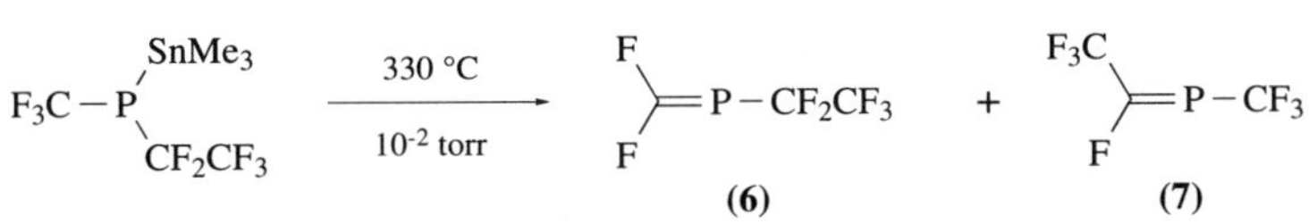

(**6**) (**7**) (2)

Attempts to synthesize perchlorophosphaalkene (**8**) by means of dehydrohalogenation of dichloromethyldichlorophosphine with triethylamine yielded four-membered 1,3-diphosphetane (**9**) ⟨81ZOB2630⟩. Evidence for the transient formation of phosphaalkene (**8**) comes from its trapping with the conjugated dienes to give 2-chlorophosphinines (Scheme 2) ⟨89TL817⟩.

Dichloromethylenephosphines having bulky substituents on phosphorus such as bis(trimethylsilyl)amino-, *N*-trimethylsilyl-*t*-butylamino- or 2,2,6,6-tetramethylpiperidino groups, enjoy high thermal stability. These may be prepared by the reaction of Cl_2CHPCl_2 with the corresponding lithium or sodium amides (Scheme 3). For example, sodium bis(trimethylsilyl)amide reacts with Cl_2CHPCl_2 as a nucleophile and as a dehydrohalogenating agent to give the phosphaalkene (**10**) in good yield ⟨81DOK(256)1401, 81ZOB2630⟩. The reaction of Bu^t(TMS)NLi with Cl_2CHPCl_2 gives rise

Table 1 Known types of *C,C*-diheterosubstituted methylenephosphines.

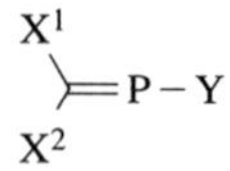

X¹	*X²*	*Method*	*Section number*	*Example*	*Ref.*
F	F	a	6.22.1.2.1	$F_2C{=}PCF_3$	84AG(E)710
Cl	Cl	a	6.22.1.2.1	$Cl_2C{=}PN(TMS)_2$	84ZOB1520
Br	Cl	a	6.22.1.2.1	Br(Cl)C=PAr*	87ZAAC(553)7
Br	Br	a	6.22.1.2.1	$Br_2C{=}PAr^*$	89ZOB1902
I	I	a	6.22.1.2.1	$I_2C{=}PAr^*$	89ZOB1902
RO	F	a, b	6.22.1.2.2	$MeO(F)C{=}PCF_3$	86ZN(B)149
RO	RO	a, b	6.22.1.2.2	$(TMS\text{-}O)_2C{=}PPh$	84TL4447
RS	RS	a, b	6.22.1.2.2	$(Bu^iS)_2C{=}PBu^t$	85ZOB264
R_2N	F	a	6.22.1.2.3	$R_2N(F)C{=}PCF_3$	86ZN(B)149
R_2N	RO	a	6.22.1.2.3	$Me_2N(MeO)C{=}PCF_3$	91HC385
R_2N	RS	b	6.22.1.2.3	$Me_2N(TMS\text{-}S)C{=}PBu^t$	84ZAAC(518)21
R_2N	R_2N	d	6.22.1.2.3	$(Me_2N)_2C{=}PH$	85ZOB1188
R_2P	F	a	6.22.1.2.3	$Me_2P(F)C{=}PCF_3$	86ZN(B)149
R_2P	Cl	e	6.22.1.2.3	[Cl(Ar*)P]ClC=PAr*	89TL177
R_2P	Br	e	6.22.1.2.3	[Br(Ar*)P]BrC=PAr*	89TL177
R_2P	RO	b	6.22.1.2.3	Ar*PH(TMS-O)C=PAr*	84AG(E)619
R_2P	RS	e	6.22.1.2.3	Ar*PH(HS)C=PAr*	84AG(E)970
R_2P	R_2N	e	6.22.1.2.3	$(R_2N)_2P(R_2N)C{=}PNR_2$ (R = Pr^i)	86JA7868
R—P—P—R		d	6.22.1.2.3	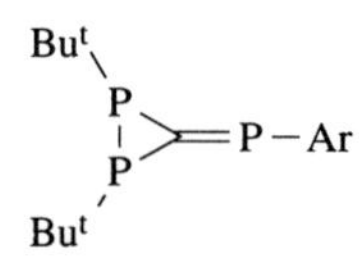 Ar = 2,4-Bu^t_2-4-MeC_6H_2	88CB281
R_3Si	Br	d	6.22.1.2.3	TMS(Br)C=PAr*	84TL4109
R_3Si	R_3Si	a	6.22.1.2.4	$(TMS)_2C{=}PCl$	81ZC357
R_3Ge	R_3Si	e	6.22.1.2.4	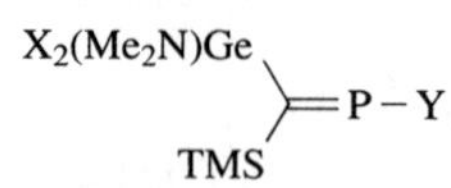 X = $(TMS)_2N$ Y = 2,2,6,6-tetramethyl-piperidino	92IC3493
Li	Cl	d	6.22.1.2.5	Li(Cl)C=PAr*	88CL1733
L_nRe	RO		6.22.1.2.5	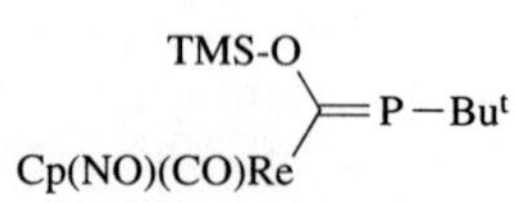	85AG(E)53
L_nPt	Cl	d	6.22.1.2.5	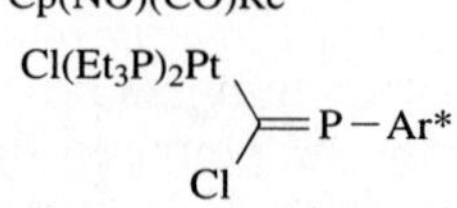	91JA9379

Table 1 (continued)

X¹	*X²*	*Method*	*Section number*	*Example*	*Ref.*
L_nPt	L_nPt	d	6.22.1.2.5	$Cl(Et_3P)_2Pt$ =P−Ar* $Cl(Et_3P)Pt$	91JA9379
R_3Sn	Cl	d	6.22.1.2.5	$Me_3Sn(Cl)C$=PAr*	85TL3551
R_3Sn	Br	d	6.22.1.2.5	$Me_3Sn(Br)C$=PAr*	85TL3551
R_3Sn	R_3Si	d	6.22.1.2.5	$X_2(Me_2N)Sn$ =P−Y TMS X = $(TMS)_2N$ Y = 2,2,6,6-tetramethyl-piperidino	92IC3493

Ar* = 2,4,6-tri-*t*-butylphenyl. [a] 1,2-Elimination. [b] 1,3-Trimethylsilyl migration. [c] Condensation reactions. [d] Reactions at peripheral substituents. [e] Miscellaneous.

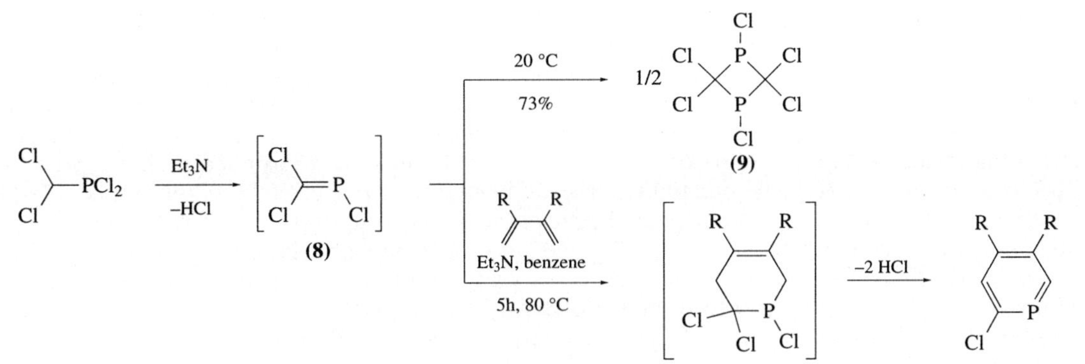

R = H 33%
R = Me 35%

Scheme 2

to the phosphaalkene (**12**) ⟨84ZOB1520⟩. The dibromo-substituted methylenephosphines are also available by this method (**11**), however, the difficulties of synthesizing starting dibromophosphine Br_2CHPBr_2 essentially limit this synthetic methodology.

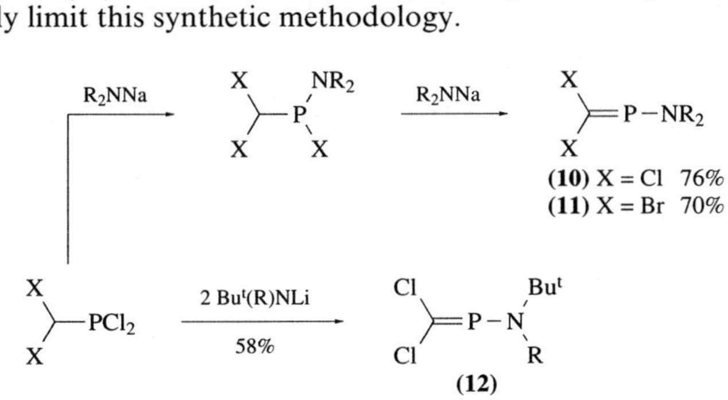

R = TMS; X = Cl, Br

Scheme 3

A simpler synthesis of dichloromethylenephosphines involves dechlorination of the compounds (**13**) with tris(diethylamido)phosphite as shown in Scheme 4 ⟨88ZOB1923⟩. Trichloromethylphosphines (**13**) are accessible through a one-pot synthesis starting from R_2NPCl_2 and the two-component system $CCl_4/(Et_2N)_3P$ ⟨88ZOB482⟩. Treatment of the bulky phosphites R_2NPCl_2 with CCl_4 and Bu^t_3P in a molar ratio of 1 : 1 : 2 allows the preparation of the phosphaalkenes without isolation of the intermediate trichloromethylphosphines ⟨88ZOB1923⟩.

Phosphaalkenes $Hal_2C{=}PPh$ are not stable enough to permit isolation. Methyl, isopropyl or *t*-butyl substituents in the 2-, 4-, and 6-positions of the phenyl ring greatly increase the kinetic stability. Thus, the compound (**14**) is air- and moisture-sensitive but very thermally stable (until

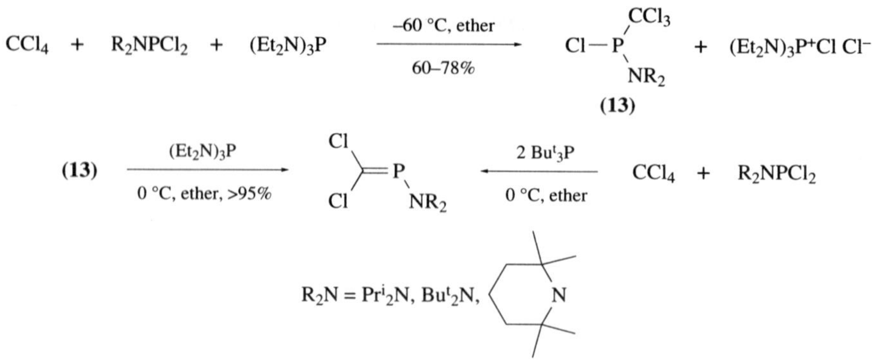

Scheme 4

150 °C); it can be stored without change for months in an inert atmosphere. The best route to (**14**) is the reaction of dichloromesitylphosphine with the system $CCl_4/(Et_2N)_3P$ ⟨92ZOB948⟩. Dihalomethylenephosphines (**15**) ⟨88CB281⟩ and (**16**)–(**19**) which are sterically protected by the 2,4-di-*t*-butyl-4-methylphenyl or 2,4,6-tri-*t*-butylphenyl groups are not sensitive towards air. Difluoro- and dichloromethylenephosphines (**16**) ⟨87ZAAC(553)7⟩ and (**17**) ⟨91CB2677⟩ have been isolated in good yield (64% and 73%) from the reaction of chloro(trihalomethyl)phosphines with *n*-butyllithium (Scheme 5) and dichloromethylenephosphines (**15**) and (**17**) ⟨85TL3551⟩ were isolated in moderate yields from the reactions of trichloromethylphosphines with 1,5-diazabicyclo[5.4.0]undec-5-ene (dbu) (Schemes 6 and 7). Although the latter methods are also applied to the synthesis of dibromo- and diiodomethylenephosphines, the more practical route to the compounds (**17**)–(**19**) starts from dichloro(2,4,6-tri-*t*-butylphenyl)phosphine. This compound reacts with the two-component system $CHal_4/(Et_2N)_3P$ (Hal = Cl, Br) in ether to give phosphaalkenes (**17**) and (**18**) (Scheme 8) ⟨89ZOB1902⟩. Extension of this procedure to the preparation of diiodomethylenephosphines (**19**) was unsuccessful. However, in the presence of tris(*t*-butyl)phosphine, Ar^*PCl_2 (Ar^* = 2,4,6-tri-*t*-butylphenyl) reacted smoothly with CHI_3 to give the compound (**19**) in good yield (Scheme 9).

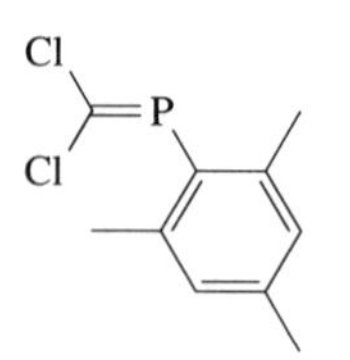

(**14**) 70%

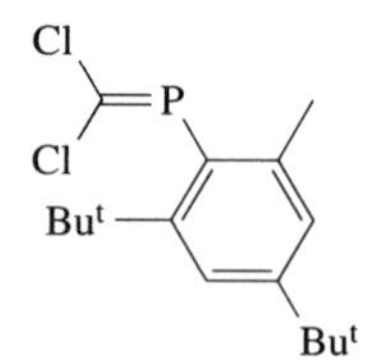

(**15**) 41%

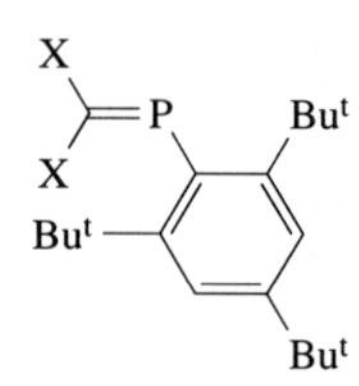

(**16**) X = F 64% ⟨87ZAAC(553)7⟩
(**17**) X = Cl 62% ⟨89ZOB1902⟩; ⟨85TL3551⟩; 73% ⟨91CB2677⟩
(**18**) X = Br 63% ⟨89ZOB1902⟩; ⟨85TL3551⟩; 55% ⟨91CB2677⟩
(**19**) X = I 34% ⟨89ZOB1902⟩; 93% ⟨91CB2677⟩

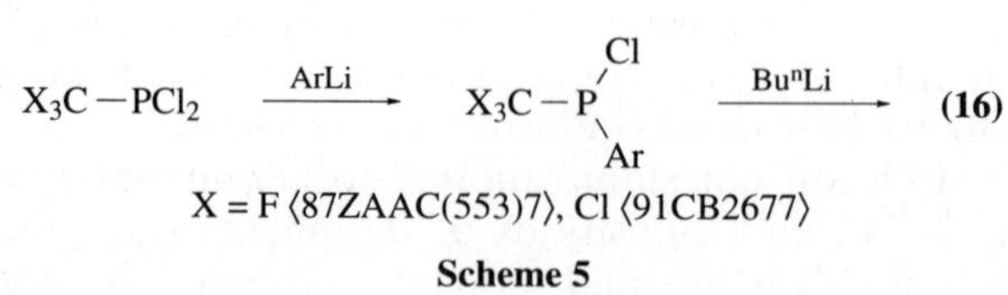

X = F ⟨87ZAAC(553)7⟩, Cl ⟨91CB2677⟩

Scheme 5

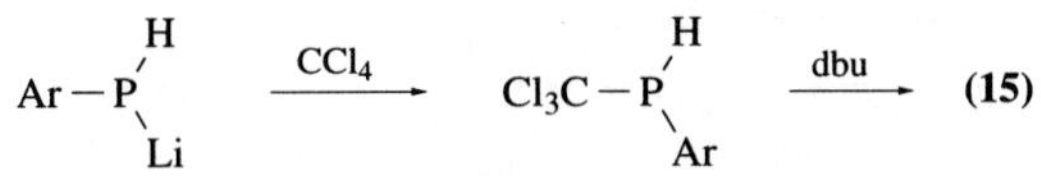

Scheme 6

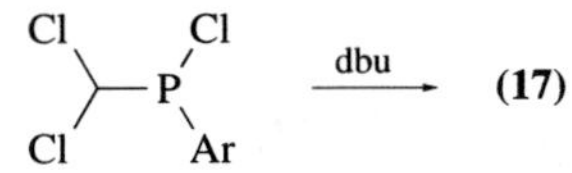

Scheme 7

$ArPCl_2$ + $CHal_4$ + 2 Et_3P ⟶ (**17**) or (**18**) + 2 $Et_3PHalCl$

Scheme 8

$ArPCl_2$ + $2CHI_3$ + 2 Bu^t_3P ⟶ (**19**) + 2 $Bu^t_3P(Cl)I$

Scheme 9

The only known mixed chlorobromomethylenephosphine Cl(Br)C═PAr* has been obtained by addition of bromine to the phosphaalkene H(Cl)C═PAr* (in turn obtained by base-catalyzed condensation of Ar^*PH_2 with trichloromethane ⟨84AG(E)895⟩ followed by dehydrohalogenation of the intermediate phosphine $ClBr_2CHP(Ar^*)Br$ ⟨87ZAAC(553)7⟩).

6.22.1.2.2 Oxygen- and sulfur-substituted methylenephosphines, RO(X)C═PY and RS(X)C═PY

(i) Synthesis by dehydrochlorination reactions

Perfluoro-2-phosphapropene (**5**) reacts with alcohols to yield phosphines (**20**) which can be readily dehydrohalogenated to fluoro(alkoxy)methylenephosphines (**21**). Sulfur-substituted methylenephosphines could not be obtained by such a reaction because addition of thiols to (**5**) leads to compounds having the structure (**22**) (Scheme 10) ⟨86ZN(B)149⟩. In the presence of Pr^i_2NH the phosphines (**20**) react with AlkOH to give the phosphaalkenes $AlkO(RO)C{=}PCF_3$. With Et_2NH, the compound $F_3CP(H)CO_2Me$ undergoes an addition–elimination process yielding the push–pull system $Et_2N(F)C{=}PCO_2Me$. Phosphaalkenes (**21**) react with primary amines $AlkNH_2$ (Alk = Bu^t, Me) with stereoselective formation of the fairly labile phosphaalkenes $AlkNH(RO)C{=}PCF_3$ ⟨93ZN(B)58⟩.

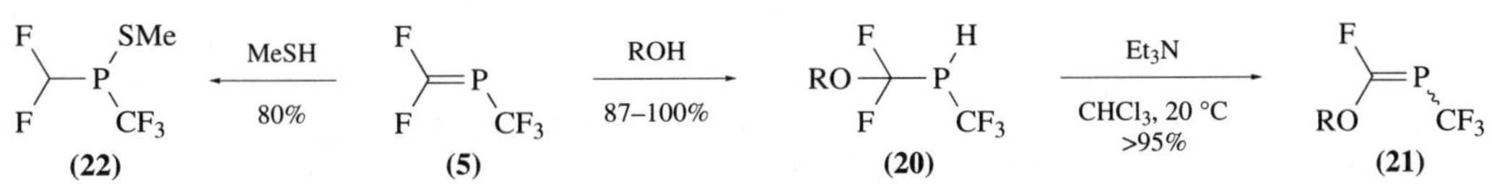

R = Me, Et, Pr^i, $PhCH_2$

Scheme 10

Base-induced dehydrohalogenation is the best method for producing bis(alkylthio)methylenephosphines from readily available chlorophosphines (**23**) or (**25**) ⟨87PS(30)433⟩. For instance, the compound (**23**) underwent dehydrochlorination when treated with an equimolar amount of triethylamine (Equation (3)) ⟨85ZOB264, 85ZOB2797⟩. Similarly, compounds (**26**) may be prepared by treatment of the chlorophosphines (**25**) with $(TMS)_2NLi$ (Equation (4)) ⟨83ZOB473, 85ZOB264⟩. Phosphaalkenes (**24**) are at the borderline of kinetic stability. They are stable in solution up to 0 °C but dimerize to 1,3-diphosphetanes within a few hours at ambient temperatures. The displacement of a chloride from (**24**) by various kinds of nucleophiles such as primary or secondary amines and phosphines has been used for the preparation of novel *P*-functionalized phosphaalkenes ⟨85ZOB264⟩.

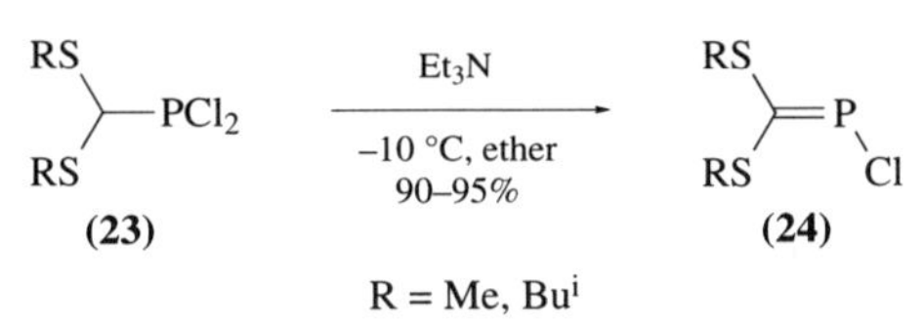 (3)

R = Me, Bu^i

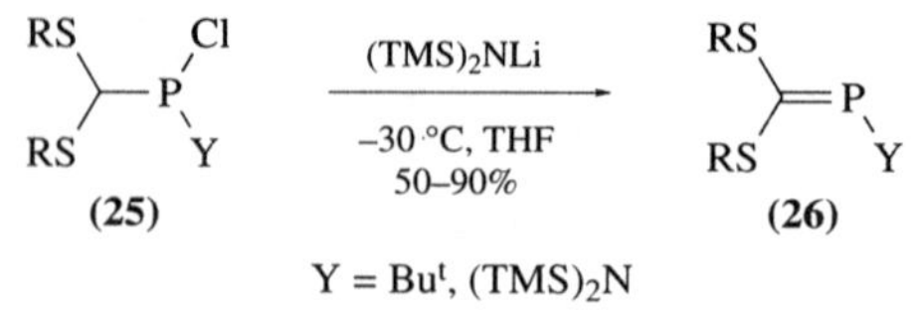
(4)

(ii) Synthesis by reactions including [1,3]-silyl migration

[1,3]-Silyl migration is one of the most useful routes to numerous functions of the type TMSO(X)C=PY or TMSS(X)C=PY. The method is based on the [1,3] migration of a *P*-silyl group to doubly bonded oxygen or sulfur and is usually used in combination with the preceding addition or condensation reactions.

Scheme 11 illustrates the formation of oxygen- and sulfur-substituted functions $X_2C{=}PY$ via insertion of carbon dioxide ⟨84TL4447⟩ or carbon disulfide ⟨80ZAAC(463)144, 84ZAAC(517)75, 84ZAAC(517)89⟩ between the P—Si bond followed by silyl migration. Another possibility to generate thiol-substituted phosphaalkenes starts from lithium silylphosphide Ar*P(Li)TMS (Ar* = 2,4,6-tri-*t*-butylphenyl) which readily reacts with CS_2 forming the lithium sulfidomethylene-phosphine TMS(LiS)C=PAr* ⟨88MI 622-02⟩.

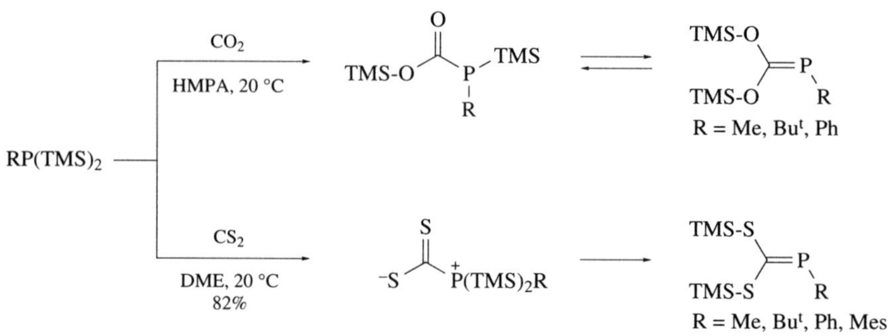
HMPA = hexamethylphosphoramide

Scheme 11

Although the method is successful for some other types of heterocumulenes (*vide infra*), it is not general. Depending upon the nature of the heterocumulene, interaction of the reagents may lead to the phosphaalkene or the reaction may be terminated by the formation of addition product. Aryl isocyanates ⟨79ZAAC(459)87, 69JCS(C)2002⟩, aryl isothiocyanates ⟨85ZAAC(520)120, 85ZAAC(520)139⟩, and diphenylketene ⟨81ZOB2189⟩ react with silylphosphines to give only addition products.

As it will be seen below some trimethylsilyloxy-substituted phosphaalkenes may be obtained from silylphosphines $(TMS)_2PR$ or $(TMS)_3P$ and carbonic acid halides. However, this method does not work in the case of chlorides of the type ROC(O)Cl ⟨80ZOB233⟩ or $R_2NC(S)Cl$ ⟨84ZAAC(518)21⟩.

A variant of the [1,3] silatropic methodology is provided by the reaction of phosphenimidous amides (**27**) with 2-lithium-2-trimethylsilyl-1,3-dithiane. The reaction proceeds via nucleophilic displacement at the dicoordinated phosphorus atom with subsequent silyl migration from carbon to nitrogen (Scheme 12) ⟨84ZOB965⟩.

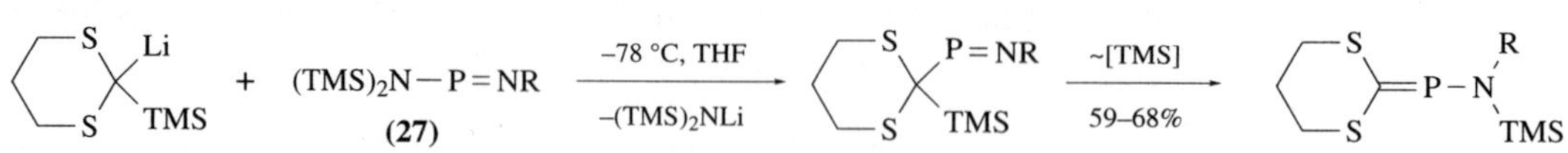
R = Bu^t, TMS

Scheme 12

6.22.1.2.3 *Nitrogen- and phosphorus-substituted methylenephosphines, $R_2N(X)C{=}PY$ and $R_2P(X)C{=}PY$*

(i) From C,C*-dihalomethylenephosphines*

There are several useful examples of the formation of amino-substituted phosphaalkenes starting from trifluoromethylphosphines ⟨93PS(76)265⟩. Thus fluoro(amino)methylenephosphines (**28**) and

(**29**) have been prepared in high yields by reacting CF_3PH_2 or $(CF_3)_2PH$ with Me_2NH or Et_2NH (Equations (5) and (6)) ⟨88CB655, 90CB2317⟩. The reaction of difluoromethylenephosphine $F_2C{=}PCF_3$ (**5**) with two equivalents of R_2NH also gives phosphaalkenes (**29**) via a series of HF elimination and R_2NH addition steps ⟨88CB655⟩. A similar process has been employed for the synthesis of the compound (**30**) (Scheme 13). The transformation can be carried out as a one-pot reaction without isolating the known phosphaalkene (**21**) ⟨91HAC385⟩. An alternative route to compounds of type (**30**) could be the base-catalysed addition of alcohols to (**29**) followed by HF elimination. In practice this route has failed because the aminomethylenephosphines do not react with alcohols.

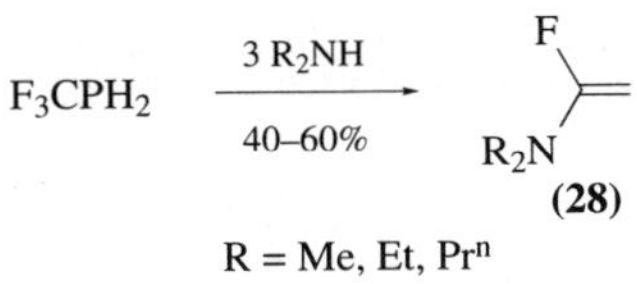

(5)

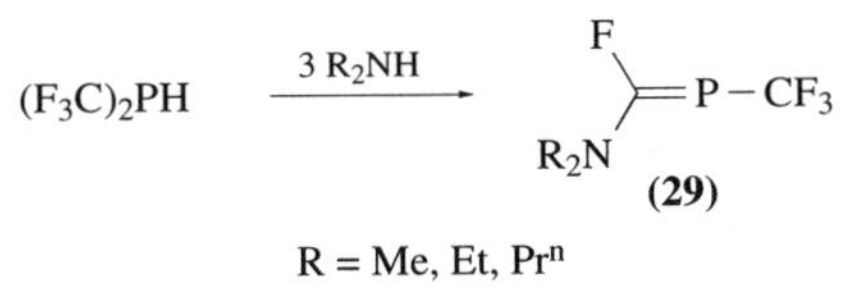

(6)

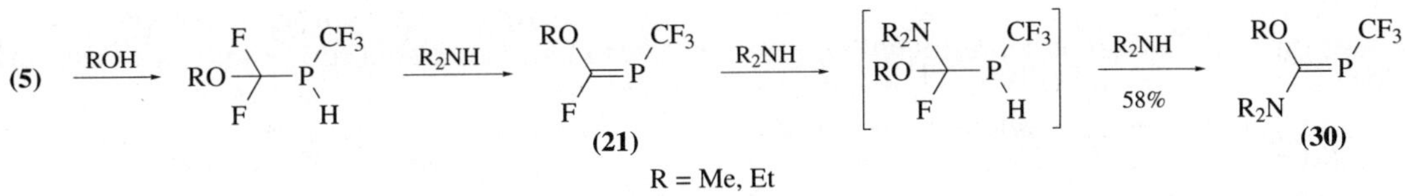

Scheme 13

Bis(amino)methylenephosphines have been suggested as intermediates in the reactions of $Cl_2C{=}PN(TMS)_2$ with $(TMS)_2NLi$ and $Bu^t(TMS)NLi$, but they have not been detected directly ⟨89TL4813⟩. However, phosphino-substituted phosphaalkene (**31**) has been isolated and characterized from the reaction of dichloromethylenephosphine (**15**) with 1,2-dipotassium diorganylphosphides (Equation (7)) ⟨88CB281⟩.

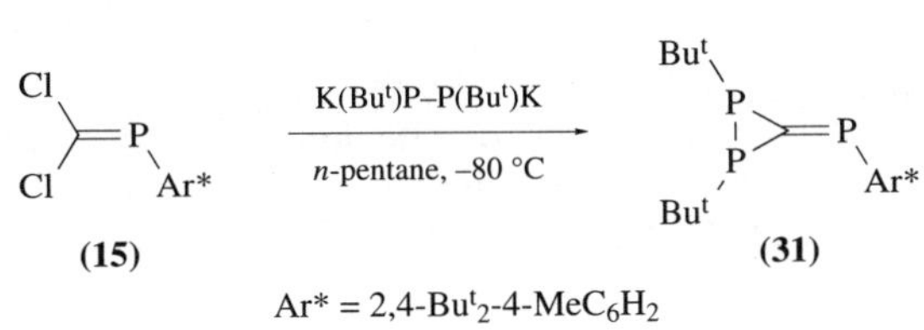

(7)

(ii) Synthesis by condensation reactions

The simplest route for the synthesis of bis(amino)methylenephosphines is the condensation of silylphosphines with functionally substituted dihaloalkanes containing mobile halogen atoms. The formation of phosphaalkene (**32**) from bis(dialkylamino)difluoromethanes and silylphosphines is illustrated in Equation (8). Due to the high affinity of silicon for fluorine, the reaction proceeds under mild conditions. Yields are good to excellent in most examples ⟨82ZOB1925, 90ZOB2238⟩. The use of $P(TMS)_3$ in the reaction makes it possible to obtain *P*-trimethylsilyl-substituted phosphaalkenes (**33**) which are key compounds for the preparation of a broad spectrum *P*-functionalized bis(amino)methylenephosphines (Equation (9)) ⟨83ZOB1672, 85ZOB221, 85ZOB1437, 87ZOB901, 90ZOB2238, 91ZOB401⟩.

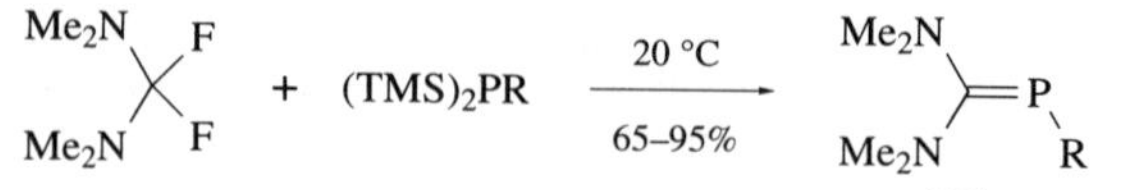

(8)

(32)

R = Bu^t, Ph, 2, 4, 6-$Bu^t_3C_6H_2$

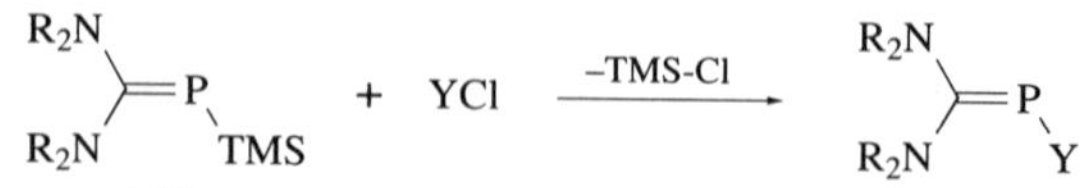

(9)

(33)

Y = Ph_3Ge, 20 °C, benzene, 75% ⟨85ZOB221⟩; Ph_3Sn, 20 °C, benzene, 95%, ⟨85ZOB221⟩; Bu^t_2P, 0 °C, ether, 80% ⟨90ZOB2238⟩; $(TMS)_2C$=P, 0 °C, ether, >90% ⟨87ZOB901⟩; Ar*N=P, 0 °C, ether, 91% ⟨91ZOB401⟩

Unlike the covalent fluorinated analogues, the ionic imidoyl chlorides (**34**) react with tris(trimethylsilyl)phosphine in a 2 : 1 ratio, forming the mesomerically stabilized phosphaalkene (**35**) (Equation (10)) ⟨83AG(E)545, 84AG(E)903, 87PS(30)495⟩.

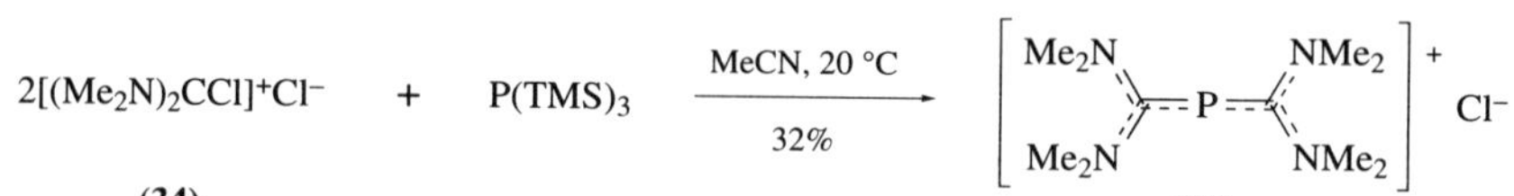

(10)

(34) **(35)**

The successful alternative method for bis(amino)methylenephosphine synthesis involves reactions of phosphines and metal phosphides with highly reactive masked carbonyl compounds. For example, amide acetals condense with arylphosphines forming dialkylamino-substituted phosphaalkenes ⟨80TL1141, 83TL5885⟩. Alkylphosphines are inert with respect to amide acetals. However, sodium phosphides readily react with carbenium tetrafluoroborates giving phosphaalkenes (**32**) (Equation (11)) ⟨81ZC407, 81TL4475⟩. The method is suitable for obtaining the *P*-hydrogen substituted derivative (**32**; R = H) ⟨83ZC99⟩.

(11)

(32)

(iii) Reactions involving trimethylsilyl migrations

There are some useful specific syntheses of the amino- and phosphino-substituted methylenephosphines based on the reactions of silylphosphines such as $RP(TMS)_2$ and $P(TMS)_3$ with carbonic acid halides. Phosgene ⟨79AG(E)469, 83CB109⟩ and isocyanide dichlorides ⟨79AG(E)873, 82AG(E)448, 82CB1617, 83CB1873⟩ were shown to undergo double substitution giving the phosphaalkenes (**36**) and (**37**) (Scheme 14).

(37) **(36)**

Scheme 14

In the reaction of $COCl_2$ with $Bu^tP(TMS)_2$, *t*-butylphosphaketene, $Bu^tP{=}C{=}O$, has been detected as an intermediate by ^{31}P-NMR. When the temperature exceeds −60 °C it dimerizes yielding the corresponding 1,3-diphosphetane. The presence of silylphosphine in excess leads to the phosphaalkene (**36**) ⟨83TL2639⟩. Interaction of benzoylisocyanide dichloride with $PhP(TMS)_2$ yields by halosilane condensation the phosphaalkene (**38**), which then undergoes cyclization and phosphorus to oxygen migration of the silyl group yielding the isomeric 1,3-azaphosphetidine (**39**) with an exocyclic P=C bond (Scheme 15) ⟨82CB2371⟩.

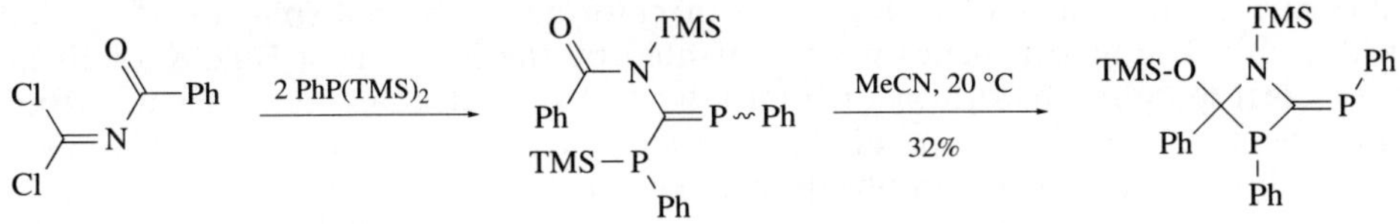

(38) **(39)**

Scheme 15

A number of amino- and phosphino-substituted methylenephosphines have been prepared by addition of silylphosphines to heterocumulenes. Examples of the successful application of this method are shown in Equations (12) ⟨80JOM(192)33, 81SRI279⟩ and (13) ⟨86TL1661, 88ZAAC(556)7⟩. The synthesis of (**40**) was achieved from 1,2-bis[bis(trimethylsilyl)phosphino]benzene and diphenylcarbodiimide ⟨85ZAAC(529)216⟩. The treatment of silyldiphosphines TMS(R)PP(R)TMS by two moles of 2,4,6-tri-*t*-butylphenylphosphaketene, Ar*P=C=O, leads to compounds (**41**) ⟨86CB2748⟩. The addition of Ar*P(H)TMS ⟨84AG(E)619⟩ and Bu^t(RO)C=PR (R = TMS) ⟨86TL1661⟩ to the phosphaketene Ar*P=C=O with the formation of the corresponding phosphaalkenes has been reported.

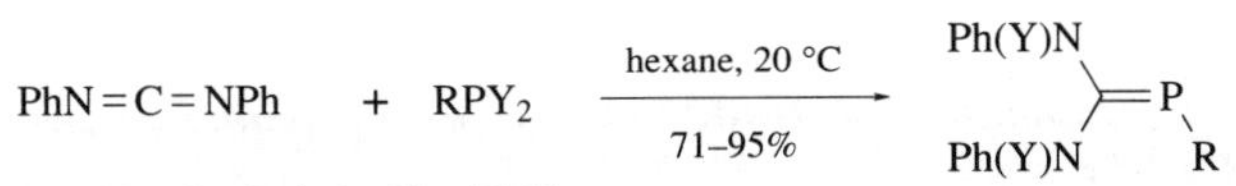 (12)

R = Me, Cy, Bu^t, Ar; Y = TMS

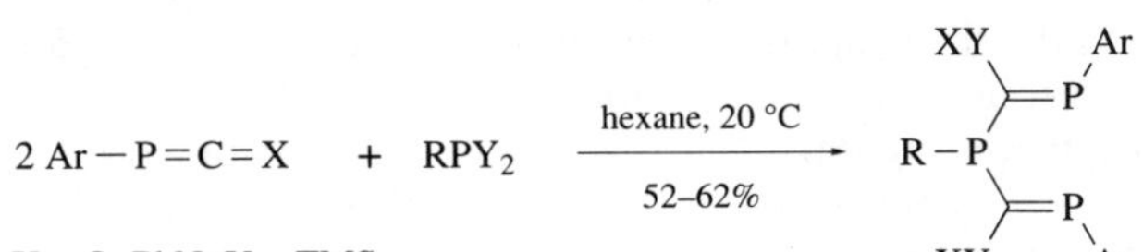 (13)

X = O, PhN; Y = TMS

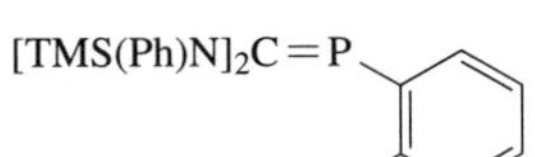

(40)

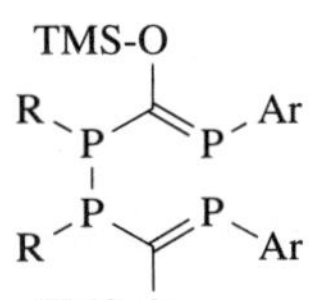

(41)

Silylphosphine (**42**) containing the very bulky 2,4,6-tri-*t*-butylphenyl group reacts with thiophosgene at −78 °C to yield phosphaalkene (**43**) instead of the expected phosphathioketene. The latter can be regenerated from (**43**) by photolysis. Its reaction with 2,4,6-tri-*t*-butylphenylphosphine provides a convenient route to functionalized phosphaalkene (**44**) (Scheme 16) ⟨84AG(E)970⟩.

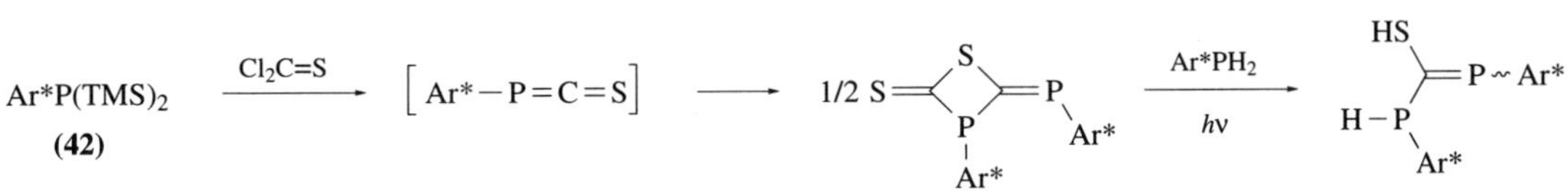

(43) **(44)**

Ar* = 2, 4, 6-$Bu^t_3C_6H_2$

Scheme 16

As noted previously, alkyl- and aryl isocyanates react with silylphosphines to give the addition products which exist in the form of acylphosphides ⟨79ZAAC(459)87⟩. In contrast to simple isocyanates, a reaction between α-methoxybenzyl isocyanate and Bu^tP(TMS)$_2$ led to the corresponding acylphosphide which underwent 1,2-elimination of TMS-OMe and [1,3] trimethylsilyl shift from phosphorus to oxygen, to afford the functionalized phosphaalkene ⟨84ZOB715⟩.

(iv) Intramolecular rearrangements

Diphosphiranes (**45**) formed in the reaction of diphosphenes with halocarbenes readily undergo a photoinduced rearrangement with migration of chlorine to the phosphorus atom ⟨87PS(30)495, 89CC593, 90JOC5750⟩. The halocarbenes were generated by the reaction of Bu^{t}OK or BuLi with an excess of the corresponding haloform or alternatively from tetrahalomethanes and BuLi at low temperatures (Scheme 17). The scope of this method is restricted to 1,3-diphosphapropenes (**46**) containing very bulky substituents on phosphorus.

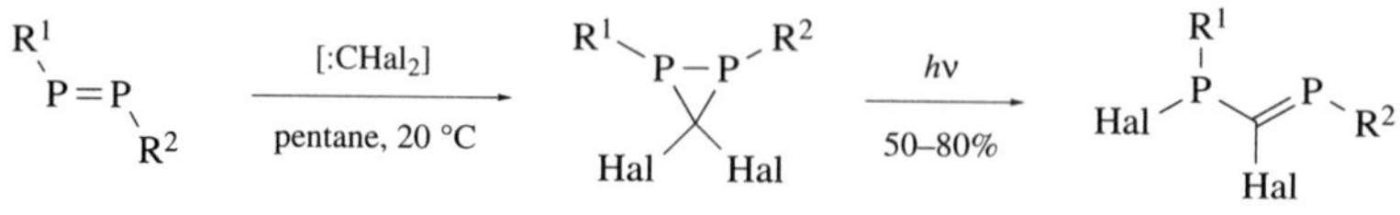

(**45**) (**46**)

R^1, R^2 = 2, 4, 6-Bu$^t_3C_6H_2$

Scheme 17

6.22.1.2.4 *Silicon- and germanium-substituted methylenephosphines, $R_3Si(X)C{=}PY$ and $R_3Ge(X)C{=}PY$*

Starting in 1981 (with the synthesis of the first *C*-silylated phosphaalkene ⟨81ZAAC(473)85⟩, approximately forty stable silicon- and germanium-substituted methylenephosphines have been synthesized so far, the stability of which is due primarily to the presence of bulky silyl or germyl substituents on the carbon atom ⟨88MI 622-01, 90MI 622-01⟩. Two principal methods for the preparation of the compounds include 1,2-elimination reactions and derivatization of the other types of low-coordinated phosphorus derivatives.

(i) Synthesis by elimination reactions

Several *P*-substituted bis(trimethylsilyl)methylenephosphines were obtained by dehydrochlorination of the corresponding chlorophosphines (Scheme 18) ⟨81TL2159, 81TL4957, 85OM339⟩. Base-induced hydrogen halide elimination requires both strong and bulky bases to prevent nucleophilic substitution at the phosphorus atom or addition of the base to the P═C bond. The best reagents meeting these conditions are dbu and 1,4-diazabicyclo[2.2.2]octane (dabco); triethylamine in most cases was not effective.

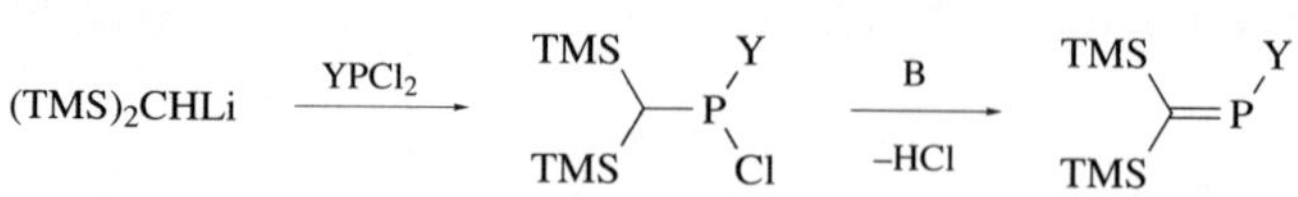

Y = Cl, Br ⟨81TL2159⟩, 2, 4, 6-$Me_3C_6H_2$ ⟨85OM339⟩; B = dabco, dbu

Scheme 18

The synthesis of bis(trimethylsilyl)methylenephosphines is sometimes more simple where the key stage is the thermal elimination of chlorotrimethylsilane. The reaction is promoted by the energy gained from Cl—Si bond formation and the reduction of steric hindrance at the phosphorus atom by cleavage of the bulky silyl group. The latter circumstance is essential because sterically unhindered halophosphines, containing the Hal—P—C—TMS skeleton, are thermally quite stable. For instance, among the compounds $TMSCH_2PCl_2$, $(TMS)_2CHPCl_2$ and $(TMS)_3CPCl_2$ only the last one splits off chlorosilane under sufficiently gentle conditions (150 °C, 0.01 torr), giving phosphaalkene (**47**) ⟨81ZC357⟩. The syntheses of compounds (**48**) and (**49**) are examples of the use of chlorosilane elimination methodology to obtain *P*-alkyl- and *P*-aryl-substituted phosphaalkenes (Scheme 19) ⟨81ZAAC(473)85⟩.

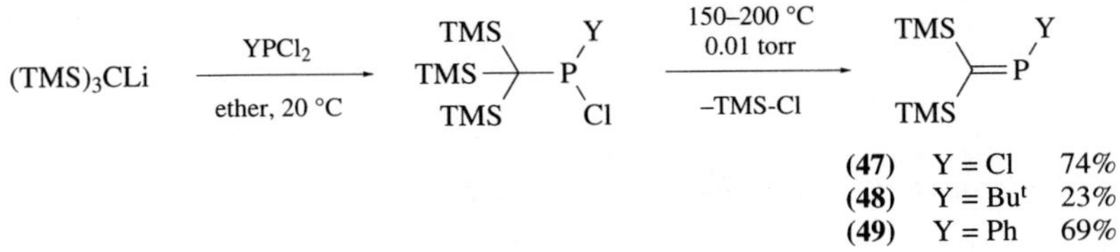

Scheme 19

By analogy with the 1,2-dihalogen elimination of vicinal dihaloalkanes with electropositive metals, the dechlorination of *P*-chloro-α-chloroalkylphosphines with lithium provides an additional tool for phosphaalkene synthesis ⟨81TL4957⟩. The generality of this approach is restricted, however, by difficulties in the preparation of the starting material.

(ii) Derivatization and rearrangements

Treatment of the highly hindered phosphaalkene (**50**) with bromine leads to the compound (**51**) (Equation (14)) ⟨84TL4109⟩. When *N*-bromosuccinimide was utilized as a halogenating reagent, dibromomethylenephosphine was obtained ⟨85TL3551⟩.

(**50**) (**51**) (14)

$Ar^* = 2,4,6\text{-}Bu^t_3C_6H_2$

P-Chloro-bis(trimethylsilyl)methylenephosphine (**47**) has a reactivity which is comparable to that of the secondary chlorophosphines (R_2PCl). It is therefore a key compound for the preparation of *P*-functionalized derivatives such as alkoxy-, alkylthio-, amino- and phosphino-substituted derivatives ⟨81AG(E)731, 82AG(E)219, 88IC784, 89S511⟩. *P*-Fluoro-*C*,*C*-bis(trimethylsilyl)methylenephosphine was obtained by exchange reactions with AgF ⟨84ZOB2800⟩, and $AgBF_4$ ⟨87ZAAC(545)7⟩. Chlorine substitution at the two-coordinate phosphorus atom in (**47**) by bromine and iodine with TMS-Br and TMS-I proceeds as readily as with chlorophosphines ⟨84ZOB2800⟩. The treatment of phosphaalkene (**47**) with Bu^t_2AsLi gives a *P*-arsino-substituted phosphaalkene ⟨85ZOB1862⟩. The coupling reactions of (**47**) with $(R_2N)_2C{=}PTMS$ ⟨87ZOB901⟩ and $Ph_3P{=}CH_2$ ⟨89TL6869⟩ lead to the corresponding diphosphabutadienes. Even the selective replacement of the halogen by alkyl and aryl groups with organolithium and Grignard reagents can be accomplished under mild conditions ⟨84TL4109, 84JA7015, 85ZOB2214⟩. The reaction of (**47**) with pentamethylcyclopentadienyllithium (Cp*Li) in hexane at room temperature affords moderate yields of the phosphaalkene $(TMS)_2C{=}PCp^*$ ⟨85C277⟩. *P*-(Ethynyl)phosphaalkenes were obtained in good yield by reacting (**47**) with alkynyl Grignard reagents ⟨84CB2693⟩. Treatment of (**47**) with alkali metallates of molybdenum and tungsten yields the bis(trimethylsilyl)methylenephosphines in which metal–ligand fragments are bonded to the phosphorus ⟨85CC1687, 86OM593⟩. Another synthetic approach to the compounds $(TMS)_2C{=}P{-}ML_n$ was developed by Niecke and co-workers ⟨85C277, 87CC10⟩. It is based on cleavage of the P—C bond of the phosphaalkene $(TMS)_2C{=}P{-}Cp^*$ with complexes of the type $[M(CO)_3(MeCN)_3]$ (M = Mo, W).

There are several useful methods for the synthesis of silyl- and germyl-substituted phosphaalkenes based on the reactions of other types of low-coordinate phosphorus compounds, but none of the reactions is general ⟨87ZOB1433⟩. For example, the phosphaalkenes (**53**) have been synthesized from the aminoiminophosphines (**52**) by a reaction involving the nucleophilic displacement at the two-coordinate phosphorus atom with subsequent [1,3]-silyl rearrangement (Scheme 20) ⟨84PS(19)189⟩. The mixed *C*-germyl-*C*-silyl-substituted phosphaalkenes, $X^1X^2_2Ge(TMS)C{=}PY$, have been recently obtained by reacting a stable phosphinocarbene $[TMS(X^1YP)C:]$ ($X^1 = Me_2N$, Y = 2,2,6,6-tetramethylpiperidino) and germanediyls $[GeX^2_2]$ ($X^2 = (TMS)_2N$ or Ar*NH) ⟨92IC3493⟩.

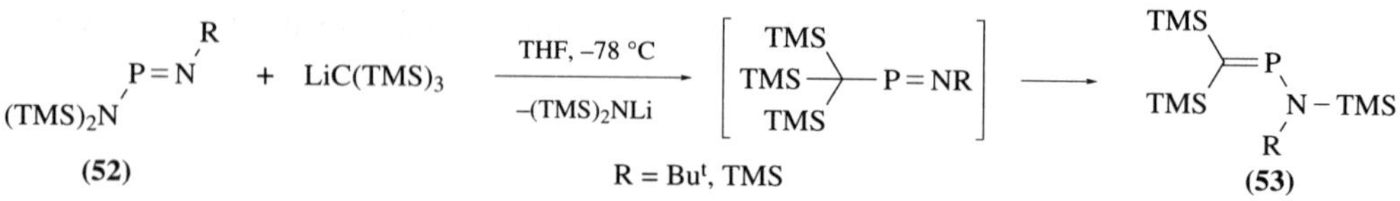

Scheme 20

6.22.1.2.5 *Metallated methylenephosphines, $L_nM(X)C{=}PY$*

Halogen–lithium exchange reactions provide the most general method for the preparation of metallated methylenephosphines. Thus, the C—Cl bond in dichloromethylenephosphine (**17**) is selectively cleaved by *n*-butyllithium at $-80\,^{\circ}C$ in a mixture of ether and toluene ⟨85TL3551⟩. The same lithio(chloro)methylenephosphine can be produced via the reaction of Br(Cl)C=PAr* with $(TMS)_2PLi$ ⟨87ZAAC(545)7⟩. Alternatively, the lithiated phosphaalkene (**54**) has been generated in near quantitative yield from phosphaalkene H(Cl)C=PAr* and *t*-butyllithium at $-78\,^{\circ}C$ in THF ⟨88CL1733⟩. The successive treatment of (**17**) with butyllithium followed by an electrophile such as TMS-Cl, Me_3GeCl or Me_3SnCl gives trimethylsilyl-, trimethylgermyl- or trimethylstannyl-substituted methylenephosphines (**55**) (Scheme 21) ⟨85TL3551, 91CB2677⟩. At room temperature lithiomethylenephosphine (**54**) is unstable and easily splits off lithium chloride to yield the phosphaalkyne Ar*CP ⟨88CL1733⟩. A similar transformation has been described for *C,C*-dichloromethylene-*P*-(2,2,6,6-tetramethylpiperidino)phosphine ⟨89ZOB2133⟩.

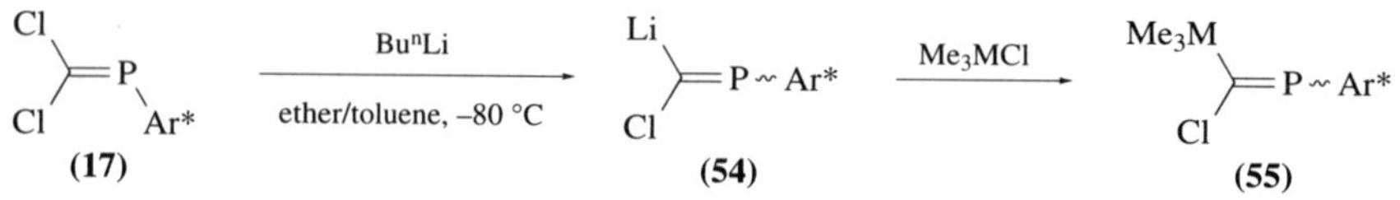

M = Si, Ge, Sn

Scheme 21

The silatropic route to the P=C bond formation represents an alternative strategy for the synthesis of *C*-metallated methylenephosphines ⟨88CRV1327⟩. For example, treatment of the complex $[Re(CO)_2(NO)(\eta^5\text{-}C_5Me_5)][BF_4]$ with LiP(R)TMS yields the phosphaalkenyl complexes (**56**). The reaction involves formation of phosphino-carbonyl complexes via nucleophilic addition of phosphide to a CO ligand prior to the [1,3]-silyl migration from phosphorus to oxygen (Scheme 22) ⟨85AG(E)53⟩. Cationic Fe and Ru complexes of the type $[M(\eta^5\text{-}C_5Me_5)(CO)_3]^+$ react with lithium phosphide LiP(Ar*)TMS analogously ⟨86CB1857⟩.

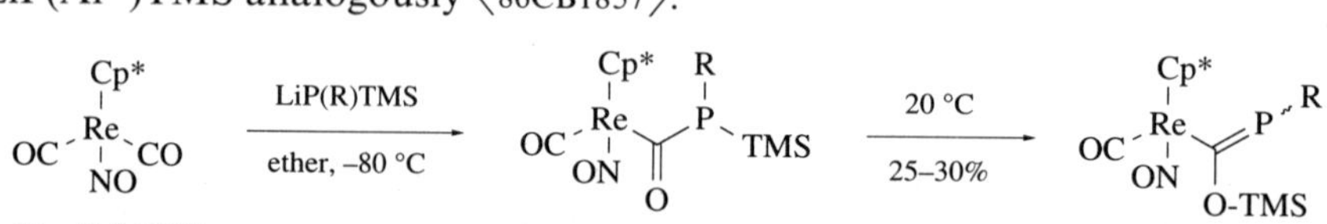

R = Bu^t, TMS

(55)

Scheme 22

Another approach to metallated methylenephosphines is based on the oxidative insertion of transition-metal complexes into the C—Hal bond of the *C*-halomethylenephosphine. Angelici and co-workers used this method for preparing phosphaalkenyl–platinum complexes (**57**) and (**58**) containing Pt—C σ-bond (Scheme 23) ⟨91JA9379⟩. Under similar conditions, reaction of the phosphaalkene (**17**) with $Pd(PPh_3)_4$ affords the phosphaalkyne, Ar*CP. The transformation was interpreted as a multi-step process including the generation of the unstable palladium-substituted methylenephosphine, $Cl(L_3Pd)C{=}PAr^*$ ($L = PPh_3$), and its subsequent conversion into phosphaisocyanide which isomerizes to phosphaalkyne ⟨92TL2981, 93OM4265⟩.

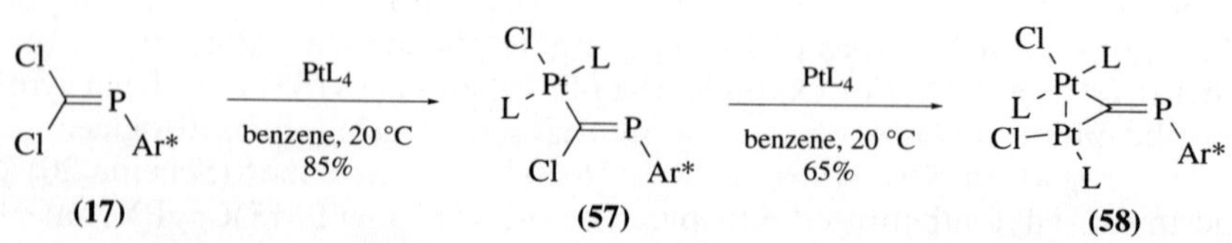

Ar* = 2,4,6-$Bu^t_3C_6H_2$; L = PEt_3

Scheme 23

6.22.1.2.6 C,C-diheterosubstituted methylenearsines, $X_2C{=}AsY$

Only sparse information is available concerning heterosubstituted arsaalkenes ⟨91RCR162⟩. This neglect stems largely from the ease of their oligomerization and polymerization which not only makes them difficult to isolate, but hitherto has precluded their use in preparative chemistry.

Difluoromethylenearsine, $F_2C{=}AsCF_3$, obtained by pyrolysis of a stannylarsine, $Me_3SnAs(CF_3)_2$, has been described by Grobe and Duc Le Van ⟨84AG(E)710⟩. Despite its high reactivity, the arsaalkene could be unequivocally characterized by its ^{19}F NMR spectrum at −110°C and finally identified by its dimerization products.

A dialkylamino group on sp^2-hybridized carbon was found to stabilize methylenearsines $R_2N(R)C{=}AsAr$ ⟨92PS(66)257⟩. However, no attempts have so far been made to obtain *C,C*-bis(amino)-substituted arsaalkenes. An unexpected reaction leading to the formation of the stable tricyclic macroheterocycle (**60**) containing Si(N)C=As fragment was observed when silaarsene (**59**) was treated with 1,6-diisocyanohexane (Equation (15)) ⟨93CC1585⟩.

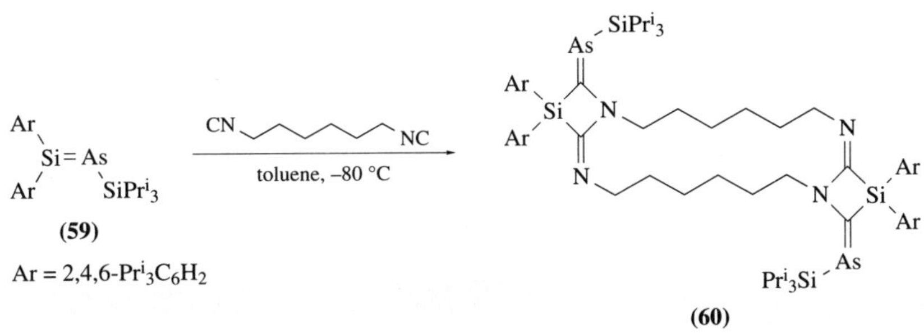

(15)

6.22.1.3 Tricoordinate Phosphorus Derivatives

6.22.1.3.1 Stabilized $[X_2C{=}PY_2]^+$ species

Stabilized $[X_2C{=}PY_2]^+$ species are the heavier congeners of iminium ions $[X_2C{=}NR_2]^+$, and are named as methylenephosphonium ions. They have been investigated only briefly, due primarily to their low stability. After several years of speculation about these species ⟨84JPC1981, 86JA5395, 87CC1399⟩ Bertrand and co-workers were able to prepare, isolate and characterize the first stable methylenephosphonium salt (**62**) starting from the electron rich diazo compound (**61**) according to Scheme 24 ⟨89JA6853⟩. These same workers discovered that the diazo compound (**61**) reacts with triethylboron in toluene at −80°C to give the borane-carbene adduct which can be regarded as a methylenephosphonium type compound ⟨93CC1354⟩.

Grutzmacher *et al.* suggested the more general route to the compounds $[X_2C{=}PY_2]^+A^-$ based on chloride ion abstraction from *P*-chlorinated ylides as illustrated in Scheme 25 ⟨91AG(E)709, 93CC673⟩. The main side process is oxidation of the substrates by the system Al_2Cl_6–CH_2Cl_2 to radical cations $[(TMS)_2CP(Cl)R_2]^{\cdot +}$ and their subsequent transformation into chlorophosphonium salts $[(TMS)_2CHP(Cl)R_2]^+AlCl_4^-$ ⟨93PS(76)21⟩. When *P*-chloro-*P,P*-bis(dialkylamino) substituted ylides react with Al_2Cl_6 in dichloromethane solution, the side reaction now predominates and the only reaction product is the phosphonium salt ⟨93PS(76)21⟩.

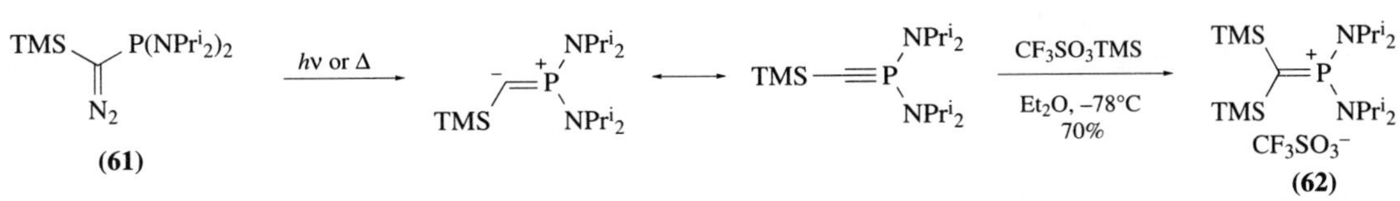

Scheme 24

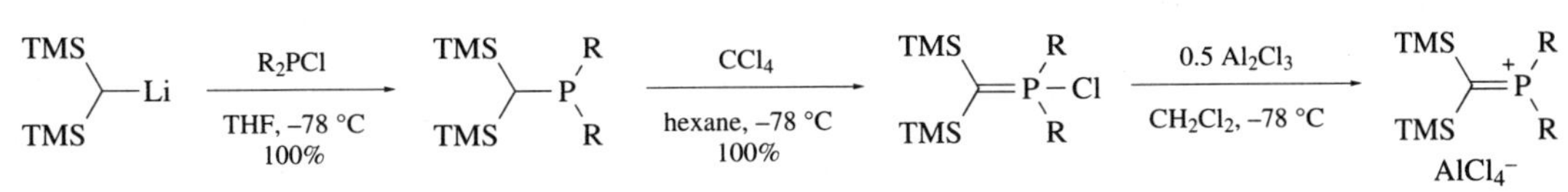

Scheme 25

6.22.1.3.2 Functions with a phosphorus-metal σ-donor bond, $X_2C{=}P(ML_n)Y$

Phosphaalkene transition-metal complexes have been reviewed in 1985 by Scherer ⟨85AG(E)924⟩, in 1981 and 1989 by Appel and Knoll ⟨81AG(E)731, B-89MI 622-01⟩, and in 1988 by Nixon ⟨88CRV1327⟩.

Generally, three synthetic methods can be applied to prepare η^1(P)-coordinated methylenephosphines: (i) complexation of a phosphaalkene which already possesses the P=C double bond, (ii) the so-called 'phospha-Wittig' synthesis ⟨90OM793, 92ACR90⟩, and (iii) an approach based on the reactions of terminal phosphinidene complexes ⟨87AG(E)275⟩ (Scheme 26). However, practically all known *C,C*-diheterosubstituted phosphaalkene complexes $X_2C{=}P(ML_n)Y$ were obtained by starting with compounds in which a phosphorus-carbon double bond already exists. Table 2 summarizes some typical examples. The formulation of the complexes was based on mass spectral and NMR data and in many cases was confirmed by single-crystal x-ray crystallographic studies.

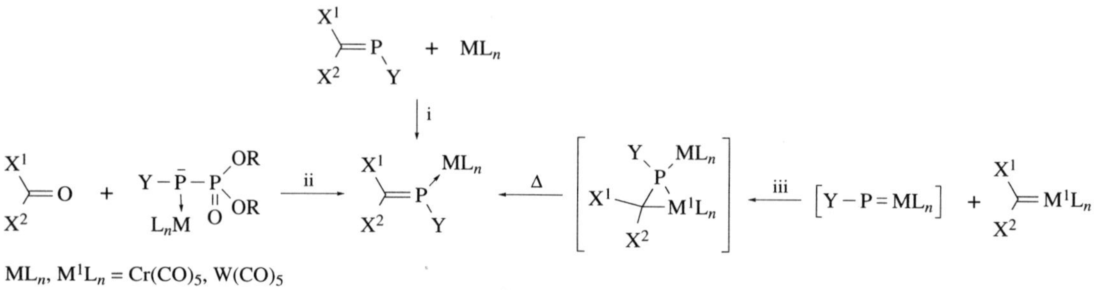

Scheme 26

Table 2 η^1-Metal complexes derived from *C,C*-diheterosubstituted phosphaalkenes.

Phosphaalkene, *X¹*	$X^1X^2C{=}PY$ *X²*	*Y*	*Reagent*	*Complex*	*Ref.*
F	F	F_3C	$Cr(CO)_5(CH_2Cl_2)$	$F_2C{=}P[Cr(CO)_5]CF_3$	88CB655
F	Me_2N	F_3C	$Cr(CO)_5(THF)$	$F(Me_2N)C{=}P[Cr(CO)_5]CF_3$	88CB655
EtO	Me_2N	F_3C	$Cr(CO)_5(THF)$	$EtO(Me_2N)C{=}P[Cr(CO)_5]CF_3$	91HC385
TMS	TMS	$(TMS)_2N$	$Fe(CO)_5$	$(TMS)_2C{=}P[Fe(CO)_4]N(TMS)_2$	84OM1132
TMS	TMS	Ar*	$Fe(CO)_5$	$(TMS)_2C{=}P[Fe(CO)_4]Ar^*$	84TL4109
TMS	TMS	Cl	$[PtCl_2(PEt_3)]_2$	$(TMS)_2C{=}P[(PEt_3)Cl_2]Cl$	92ZOB474
TMS	TMS	Cl	$RhCl(PPh_3)_2$	$(TMS)_2C{=}P[RhCl(PPh_3)_2]Cl$	89JOM(368)C29
TMS	TMS	Me_5C_5	$AgSO_3CF_3$	$[(TMS)_2C{=}P(Ag)C_5Me_5]^+CF_3SO_3^-$	85C277

Ar* = 2,4,6-tri-*t*-butylphenyl

6.22.1.3.3 $\sigma^3\lambda^5$-Methylenephosphoranes, $X_2C{=}P({=}Z)Y$

Stable $\sigma^3\lambda^5$-methylenephosphoranes have been reported with Z = R_2C, O, S, Se and RN. The hetero substituents on the carbon atom (X) are almost completely restricted in the known examples to silyl groups.

General aspects of the chemistry of $\sigma^3\lambda^5$-phosphoranes were covered in a recent monography ⟨92MI 622-01⟩ and several reviews ⟨81TCC71, 82HOU(E1)583, 86PS(26)327, B-90MI 622-01⟩. Tetrakis(trimethylsilyl)-substituted bis(methylene)phosphoranes, $[(TMS)_2C{=}]_2PY$, can be obtained directly from organodichlorophosphines and lithium bis(trimethylsilyl)chloromethanide (*vide infra*). The synthesis of other types of $\sigma^3\lambda^5$-methylenephosphoranes is based on 1,1-oxidative addition reactions to dicoordinated phosphorus derivatives. Both methylenephosphines and iminophosphines can be employed as substrates for the synthesis of methylene(imino)phosphoranes. Since the stable oxo-, thioxo-, and selenoxo-phosphines [YP=O(S,Se)] are unknown, the methods for preparing corresponding phosphoranes consist of the reactions of methylenephosphines with elemental oxygen, sulfur or selenium.

(i) From organodichlorophosphines

The bis(methylene)phosphoranes of type $[TMS)_2C{=}]_2PY$ can be obtained by reaction of the lithiated chlorobis(trimethylsilyl)methane with dichlorophosphines ⟨82AG(E)80, 86CB535, 88PS(36)147⟩. The synthesis is a three-step process as shown in Scheme 27. All *P*-substituted bis(methylene)phosphoranes bearing four silyl groups are thermally stable, soluble in nonpolar solvents, and somewhat volatile. Attempts to apply this approach to the synthesis of bis(methylene)-arsoranes were unsuccessful. Unlike the analogous phosphoranes, bis(methylene)-arsoranes are unstable and isomerize within a short period of time into the stable arsiranes ⟨85AG(E)419⟩.

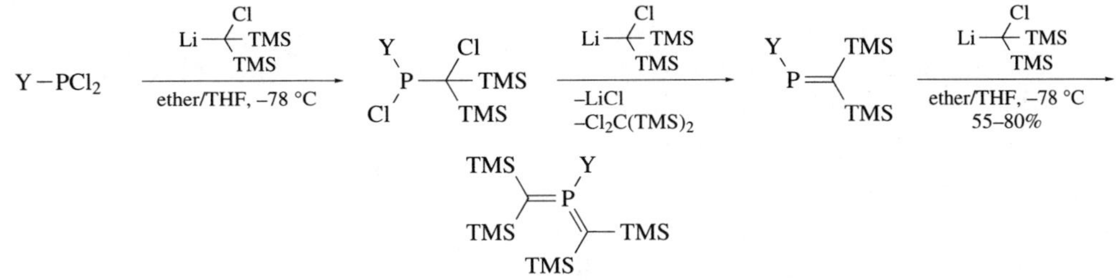

(63)

Y = Me_2N, PhS, Ph_2CH, $C_2H_5CH(Me)$

Scheme 27

The reaction of PCl_3 with three equivalents of $(TMS)_2CClLi$ leads to the stable bis(methylene)chlorophosphorane (**63**; Y = Cl) which is a key substance for the synthesis of other *P*-functionalized derivatives ⟨82TL2017⟩. Another procedure to obtain this compound involves the reaction of bis(methylene)methoxyphosphorane $[(TMS)_2C{=}]_2POMe$ with BCl_3. The bromo- and iodo-substituted compounds are also available by this method ⟨84ZC384⟩.

(ii) From methylenephosphines and iminophosphines

As shown in Scheme 28 the action of some carbenoids on phosphaalkenes affords monomeric bis(methylene)phosphoranes of the type (**64**) ⟨86CB1977⟩. The same approach has been employed for the synthesis of some methylene(imino)phosphoranes. For example, when aminoiminophosphines, $R_2N{-}P{=}NR$, ⟨83ZOB693⟩ or *P*-organoiminophosphines, RP=NR, ⟨91ZOB79⟩ reacted with $(TMS)_2CClLi$ in THF at $-100\,°C$, the corresponding methylene(imino)phosphoranes were isolated in yields of 55–80%. The mechanism postulated for the reactions includes addition of the organolithium derivative to the P=N bond followed by 1,2-elimination of lithium chloride.

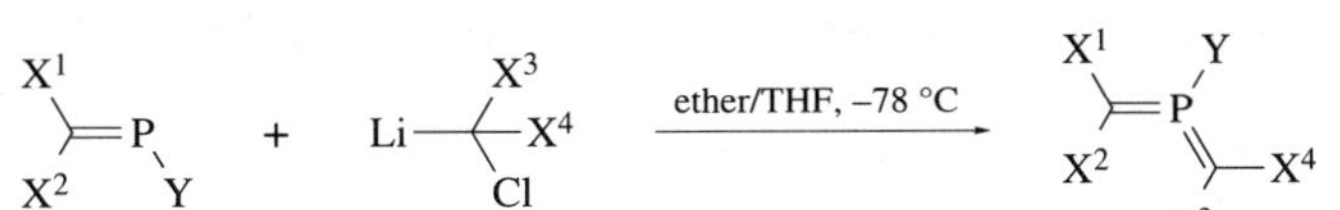

(64)

Entry	*Y*	*X*[1]	*X*[2]	*X*[3]	*X*[4]
1	$(TMS)_2N$	Cl	Cl	TMS	TMS
2	Et_2N	Cl	Cl	TMS	TMS
3	$Pr^i{}_2N$	TMS	TMS	H	H
4	Bu^t	H	Bu^t	TMS	TMS
5	2, 4, 6-$Me_3C_6H_2$	Ph	Ph	TMS	TMS
6	2, 4, 6-$Me_3C_6H_2$	TMS	TMS	Ph	Ph

Scheme 28

As an alternative to the synthesis of methylene(imino)phosphoranes from iminophosphines the reverse process, namely the transfer of an imino group to methylenephosphine, may be employed for the preparation of these compounds. Besides organic azides, the synthetic application of which in low coordinated phosphorus chemistry is widely discussed in references ⟨B-90MI 622-01, 92MI 622-01⟩, *N*-silylated chloroamines have been used as oxidants for methylenephosphines (Scheme 29). The reaction was found to proceed via oxidative addition of the chloroamine to give a *P*-halogenated ylide, which is further converted into the methylene(imino)phosphorane by elimination of chlorotrimethylsilane ⟨89ZOB2131⟩.

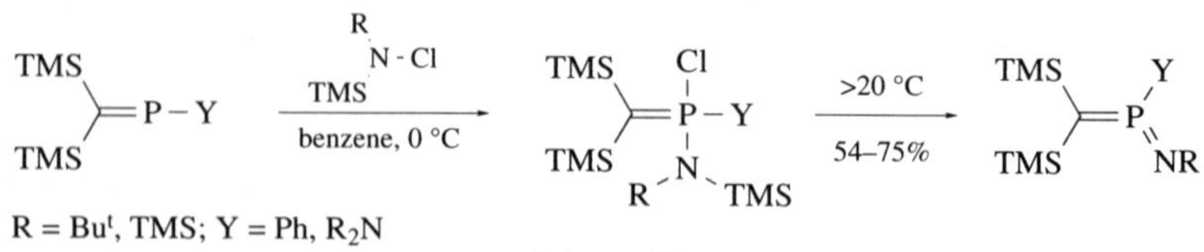

Scheme 29

The methylene(oxo-, thioxo- or selenoxo)phosphoranes, $X_2C{=}P({=}Z)R$ (Z = O, S or Se), have been obtained by oxidation of the phosphaalkenes with ozone, sulfur or selenium. These tricoordinate pentavalent phosphorus species are stable only in the presence of sterically demanding groups at the P=C bond. Thus, among the compounds $X^1X^2C{=}P({=}O)Y$ (X^1,X^2 = heteroatom substituents) the methylene(oxo)-phosphorane $(TMS)_2C{=}P({=}O)Ar^*$ (Ar* = 2,4,6-tri-*t*-butylphenyl) is the only known stable compound ⟨84TL4109⟩. The range of phosphoranes $X^1X^2C{=}P({=}S;Se)Y$ is somewhat larger than that of methylene(oxo)phosphoranes; however, the possibilities for the synthesis of stable examples of these compounds are relatively limited ⟨81CC72, 84CC698, 84TL4109⟩.

6.22.1.4 Tetracoordinate Phosphorus Derivatives

Tetracoordinate phosphorus derivatives of the type $X^1X^2CPY_3$ (methylenephosphoranes or phosphonium ylides) have received considerable attention. Their structure and chemical properties have been reviewed by several authors as part of a wider discussion of phosphonium ylide chemistry ⟨82RCR1, 82HOU(E1)616, 83RCR1096, 91RCR391, 91COS(6)171⟩. This survey covers only the major developments in the synthesis of *C,C*-diheterosubstituted ylide functions; readers are referred to publications cited in the text for a more thorough treatment of other aspects of methylenephosphorane chemistry.

Structurally $\sigma^4\lambda^5$-methylenephosphoranes may be represented as hybrids of covalent, $X^1X^2C{=}PY_3$ (ylene), and dipolar, $X^1X^2C^-{-}P^+Y_3$ (ylide), structures. Hetero substituents at the ylide carbon atom influence the stability of the compounds markedly since σ-donor and π-acceptor groups (e.g., electropositive elements) stabilize ylides, whereas π-donors and σ-acceptors (e.g., electronegative elements) destabilize them ⟨79JA7169⟩. Lifetimes of nonstabilized ylides vary from several hours at ambient temperature to short periods at low temperatures.

Synthetic routes to the *C,C*-heterosubstituted phosphonium ylides are based on the following procedures: (i) the direct introduction of a methylene group (X^1X^2C) into tervalent phosphorus compounds (PY_3); (ii) the synthesis from phosphonium salts; (iii) the oxidative ylidation of tertiary phosphines containing a mobile hydrogen atom at the α-carbon atom and behaving as CH acids; (iv) the rearrangement of α-haloalkylphosphines into *P*-halogenated ylides. The other methods of synthesis are used only in individual instances, but they sometimes have an advantage over standard methods.

6.22.1.4.1 Dihalosubstituted ylides, $Hal_2C{=}PY_3$

C,C-Dihalomethylenephosphoranes are valuable synthons for the Wittig alkenation reactions ⟨62JA1745, 89CRV863⟩. Nevertheless, the number of such compounds which have been isolated and characterized is rather limited. Owing to their low stability and high reactivity dihalomethylene ylides are usually generated *in situ* and quenched with a carbonyl compound.

Barton and co-workers utilized such an *in situ* generation–capture methodology for the reactions of tertiary phosphines (especially triphenylphosphine) with sodium chlorodifluoroacetate and carbonyl compounds (Scheme 30) ⟨67JOC1311, 68JOC1854, 71JFC(1)347, 72JA820⟩. Difluoromethylenetributylphosphorane, $F_2C{=}PBu^n_3$, has also been prepared *in situ* via the reaction of tributylphosphine with sodium chlorodifluoroacetate in *N*-methylpyrrolidone ⟨65JOC1027⟩.

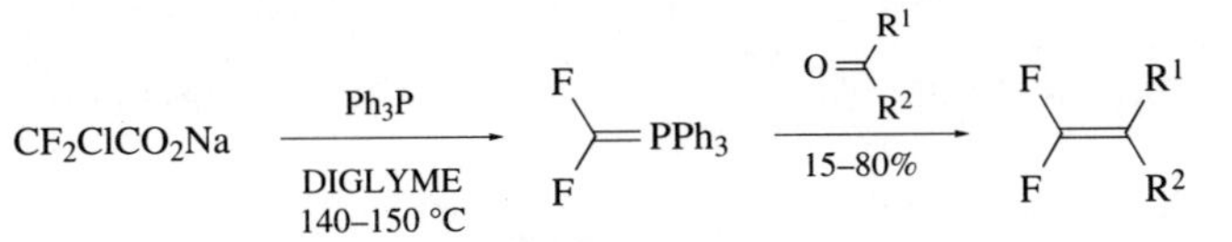

Scheme 30

Triphenylphosphine assisted decomposition of mercurials, $PhHgCFCl_2$ ⟨76CAR(46)9⟩ and $PhHgCX_2Br$ ($X = Cl$, Br) ⟨65JOM(3)337⟩, provided a route to a series of methylenephosphoranes (**66**). Mechanistic studies suggested the intermediacy of an ylide–mercuric halide complex (**65**) which dissociates to afford the reactive methylenephosphorane species (Scheme 31).

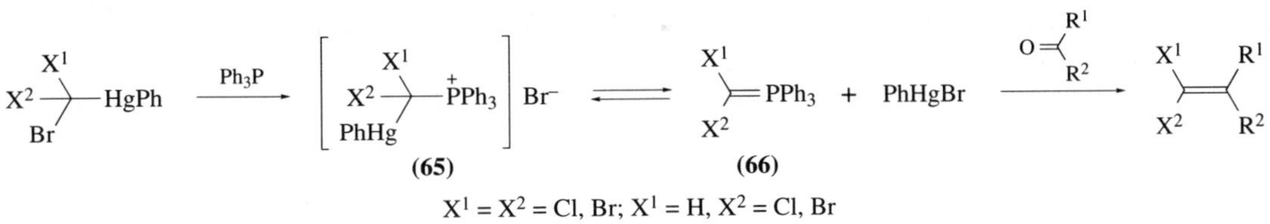

Scheme 31

The well known *in situ* synthesis of dihalomethylenephosphoranes involves the reaction of haloform with potassium *t*-butoxide in the presence of tertiary phosphines. This method has proved especially successful for the formation of dihalomethylenetriphenylphosphoranes (Scheme 32) ⟨62JA854⟩. Dichlorocarbene can also be generated from chloroform and 50% aq. sodium hydroxide solution under phase transfer conditions ⟨84T1523⟩. Another efficient route to dihalomethylene ylides is based on the dehalogenation of phosphonium salts with Group 2B metals ⟨79JOM(169)123, 77JFC(10)131, 75JOC2796⟩ or suitable tervalent phosphorus derivatives ⟨82HOU(E1)616⟩. The most direct procedure consists of interaction of phosphines with tetrahalomethanes (Scheme 33). Evidence has been provided that the initial step of the reaction sequence is the formation of an ion pair formed via attack of phosphorus on a halogen. Subsequent recombination of the ion pair (in the absence of a trapping agent) and dehalogenation of the resultant phosphonium salt results in the formation of the ylide and the dihalophosphorane. The reaction is very simple to carry out in the laboratory and is a good general preparative method. In particular, using this approach Appel and Veltman were able for the first time to isolate dichloromethylene-triphenylphosphorane in a pure state ⟨77TL399⟩. From tertiary phosphines and mixed tetrahalomethanes difluoro- and dichloromethylenephosphoranes are formed preferentially ⟨75TL3789, 76JA556⟩. The addition of zinc dust proved to be favorable in the synthesis of the difluoro ⟨79CL983, 81TL1421⟩, dibromo ⟨72TL3769, 83TL3387⟩, and diiodomethylenephosphoranes ⟨85CC296⟩. Subsequent improvements include the use of activated magnesium instead of zinc ⟨92TL683⟩. The method was also successful for a generation *in situ* of highly reactive tris(dialkylamido)phosphonium ylides containing a dihalomethylene group ⟨77TL1239, 80S554, 84JOC706⟩.

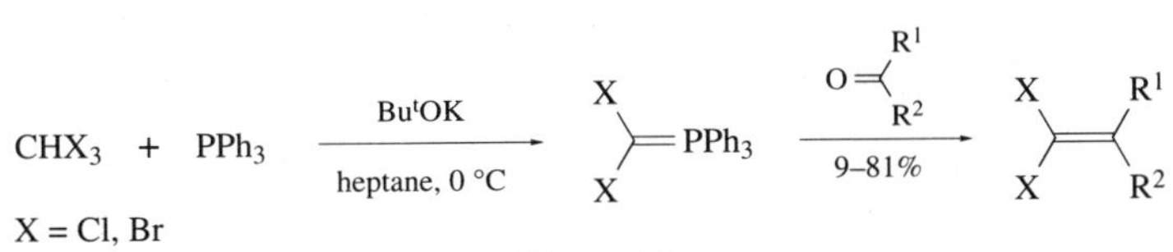

Scheme 32

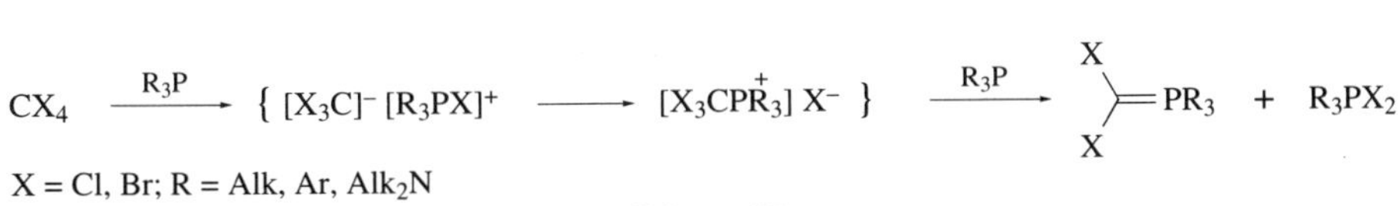

Scheme 33

In contrast to tertiary phosphines, dialkyl- and diarylchlorophosphines do not react with tetrahalomethanes due to the low nucleophilicity of the phosphorus atom. However, di-*t*-butylfluorophosphine can be converted into the ylides (**67**) by the reaction with tetrachloro- and tetrabromomethane ⟨88ZOB2164⟩. The compounds (**67**) are remarkably stable and can be recrystallized from pentane (Equation (16)).

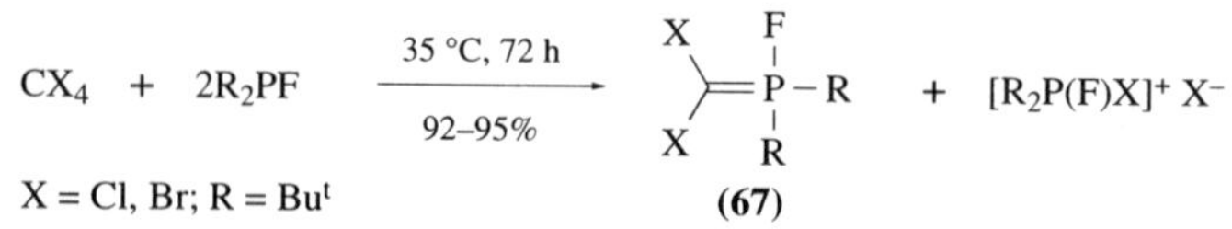

(16)

Dichloromethylene(amino)difluorophosphoranes (**69**) have been produced in good yields from the phosphorane (**68**) by the reaction with amines followed by dechlorination with hexaethyltriamidophosphite (Scheme 34) ⟨90ZOB2814⟩. The best method for preparing dichloromethylenebis(amino)fluorophosphoranes (**70**) is based on the reaction of tetraalkyldiamidofluorophosphites with carbon tetrachloride. These ylides may be distilled and kept in an inert atmosphere without significant decomposition ⟨87ZOB2794⟩. *P*-Chlorinated analogues of the ylide (**70**) can not be obtained by this route since they readily add tetraalkyldiamidochlorophosphites to give the phosphonium salts (**71**) ⟨87ZOB2141⟩. However, the dechlorination of phosphonium salt (**72**) with hexamethyltriamidophosphite proceeded quite smoothly and enabled the ylide (**73**) to be prepared in high yield (Scheme 35) ⟨89ZOB959⟩.

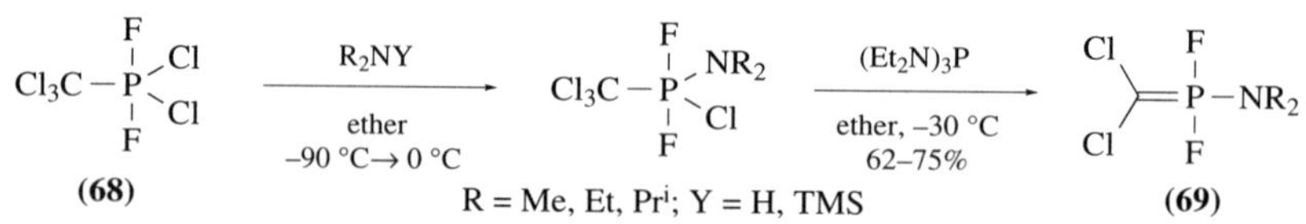

Scheme 34

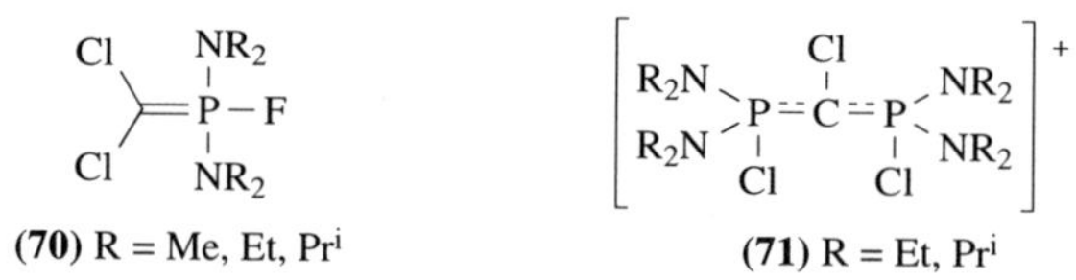

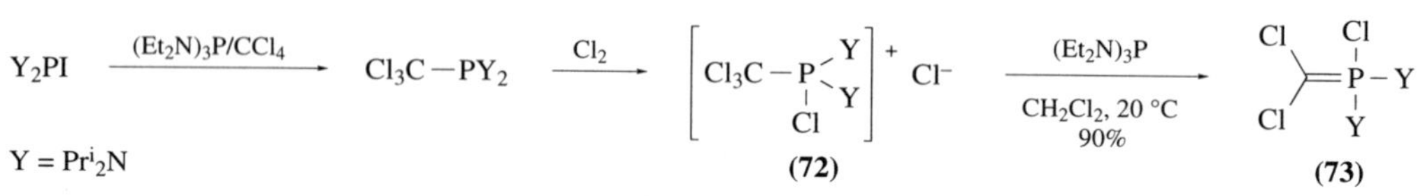

Scheme 35

A convenient method for the synthesis of P—Cl ylides containing a dihalomethylene group involves 1,2(C to P)-halotropic rearrangement. Lutsenko and co-workers obtained bis(dimethylamino)trichloromethylphosphine (**74**), which is stable at room temperature, by an exchange reaction of dichloro(trichloromethyl)phosphine with dimethylamine. The product was purified by distillation in vacuum and was then converted into the *P*-chloro ylide (**75**) by boiling in dichloromethane (Scheme 36) ⟨85ZOB1194⟩. It was shown that the rearrangement is strongly catalyzed by the phosphonium salt $[(Me_2N)_2P(Cl)CCl_3]^+Cl^-$ ⟨88ZOB1461⟩. Similar chlorotropic rearrangement was observed for trichloromethylphosphine (**76**). The latter, in the absence of solvent, in 2 h at room temperature is converted in 70% yield into ylide (**77**) (Scheme 37) ⟨89ZOB1904⟩.

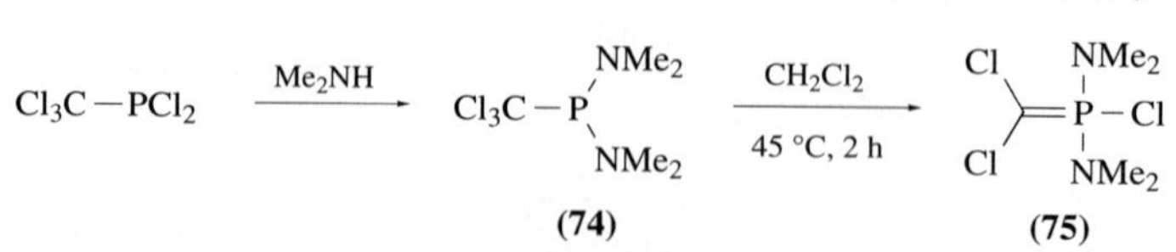

Scheme 36

Other methods for the synthesis of dihalo ylides are restricted to examples including substitution of a proton at the ylide carbon atom. For example, exchange of hydrogen atoms at the ylide carbon atom in the compounds $R(H)C{=}P(F)(NEt_2)_2$ (R = H, Alk) by chlorine occurs on reaction with carbon tetrachloride in ether at -10°C in yields of 70–85% ⟨89ZOB330⟩. *P*-Halo ylides formed by

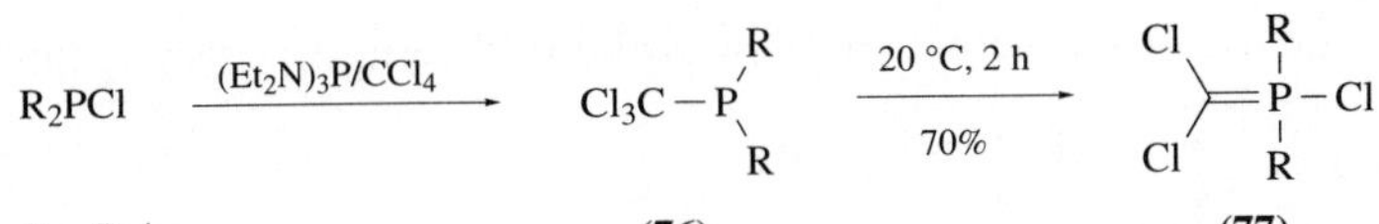

Scheme 37

reaction of tervalent phosphorus compounds with carbon tetrahalides may also react with a second molecule of CCl_4 or CBr_4 exchanging a hydrogen atom at the α-carbon atom by an atom of halogen ⟨88ZOB491⟩. The preparative applicability of this approach, however, seems to be limited.

6.22.1.4.2 Oxygen-, sulfur-, and selenium-substituted ylides, RE(X)C═PY₃ (E = O, S, or Se)

Phosphorus ylides carrying at least one oxygen at the α-carbon atom are very unstable at room temperature and can only be generated and detected *in situ* ⟨61CB1373, 62CB2514, 64TL3323⟩. Ylides of the type $(RO)_2C{=}PY_3$ have not yet been characterized. *Ab initio* calculations show that substituents like OH stabilize singlet carbenes but not the phosphonium ylides. Therefore C—O substituted methylenephosphoranes have an enhanced tendency to dissociate, in agreement with experimental observations ⟨86CB1331⟩.

The synthetic chemistry of C—S substituted phosphorus ylides is considerably richer than for their oxygen analogues. Direct incorporation of $(RS)_2C$ units into tervalent phosphorus compounds may be achieved by the carbenoid method. One of the earliest *in situ* syntheses involved the thermal decomposition of *p*-toluenesulfonylhydrazone salts (**78**) in THF containing a severalfold excess of triphenylphosphine (Scheme 38) ⟨64TL245⟩. The reaction is not applicable to arylthiomethylenephosphoranes. Subsequently, Seebach and co-workers suggested an excellent one-pot synthesis of bis(phenylthio)methylene ylide (**80**) based on the reaction of tris(phenylthio)methyllithium (**79**) with phosphines (Scheme 39) ⟨72CB487⟩. The reaction pathway was rationalized by assuming the existence of an equilibrium between a carbenoid and a carbene. Tris(phenylseleno)methyllithium reacts with triphenylphosphine similarly to give the bis(phenylseleno)methylenephosphorane in 65% yield ⟨72CB511⟩. Comparison with the analogous sulfur ylide shows that PhSe groups stabilize adjacent carbanionic centers nearly as well as PhS groups do.

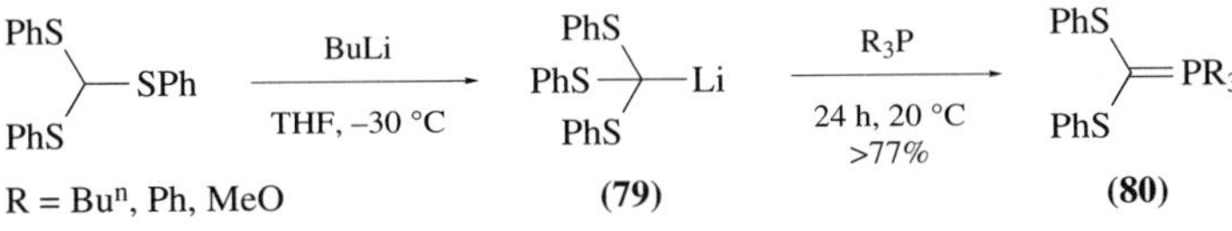

Scheme 38

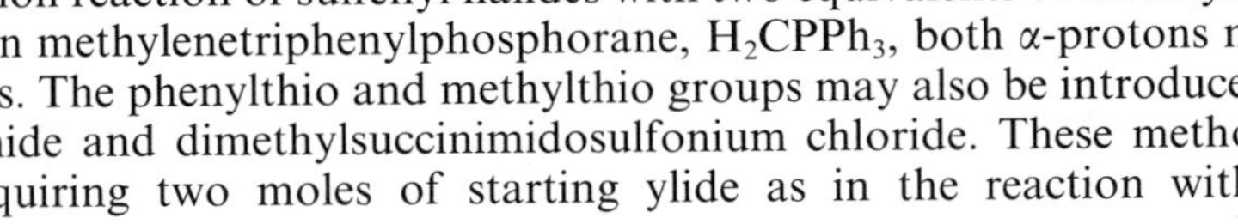

Scheme 39

Methylenephosphoranes containing alkylthio or arylthio functionality may be readily synthesized by the transylidation reaction of sulfenyl halides with two equivalents of an alkylidenephosphorane ⟨82HOU(E1)616⟩. In methylenetriphenylphosphorane, H_2CPPh_3, both α-protons may be substituted by sulfenyl groups. The phenylthio and methylthio groups may also be introduced by *N*-methyl-*N*-phenylthioacetamide and dimethylsuccinimidosulfonium chloride. These methods avoid the disadvantage of requiring two moles of starting ylide as in the reaction with sulfenyl halides ⟨91COS(6)171⟩.

In a similar fashion, seleno-substituted ylides can be synthesized using areneselenyl chlorides or bromides ⟨76JOM(114)281, 79CB355⟩. Mixed phenylseleno(phenylthio)methylenephosphorane (**82**) has been prepared from chloromethyl phenyl sulfide via phenylthiomethylenephosphorane (**81**) (Scheme 40) ⟨83CB1955⟩.

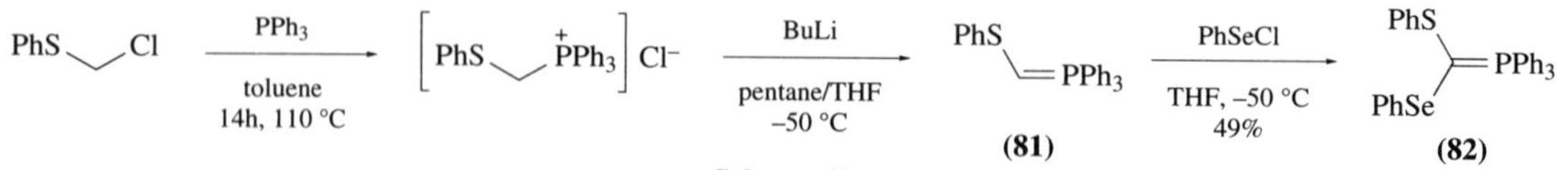

Scheme 40

Unlike bis(phenylthio)methylenetributylphosphorane (**80**; R = Bu^n) which is stable enough to be isolated in pure state, ylides of the type (**84**) can only be generated *in situ*. These species have received increased attention since they were shown to present great synthetic potential in the synthesis of derivatives and analogues of bis(ethylenedithio)tetrathiafulvalene ⟨86T1209, 93PS(74)279⟩. The most simple synthetic route to ylides (**84**) is shown in Scheme 41. Deprotonation of phosphonium salts (**83**) yields an ylide which can be trapped in good yield with a carbonyl compound to afford a dithiafulvalene ⟨76TL3695, 78JOC369, 83TL3469, 91S26⟩. Taking into account the availability of starting materials and the usually high yields in all steps, this is the method of choice for the synthesis of ylides (**84**) in cases where R^1 and R^2 equal H, alkyl, or a condensed π-donor ring. However, this sequence of reactions cannot be employed for the preparation of ylides in which R^1 and R^2 are electron-withdrawing groups. The latter are available from the reaction of carbon disulfide-tri-*n*-butylphosphine adduct with activated carbon–carbon multiple bonds ⟨71JA4961, 75CC960, 79JOC930⟩. For example, when the CS_2–Bu_3P adduct is treated with a mixture of dimethyl alkynedicarboxylate and fluoroboric acid etherate at −65°C, the initially produced phosphorane (**85**) is trapped by protonation, and the resulting phosphonium salt (**86**) can be isolated in yields up to 72%. Under aprotic conditions, salt (**86**) can be used for the *in situ* generation of the unstable ylide (**85**) (Scheme 42) ⟨79JOC930⟩. The same approach has proved successful for the preparation of selenium analogues of the ylide (**85**) ⟨84TL4227⟩.

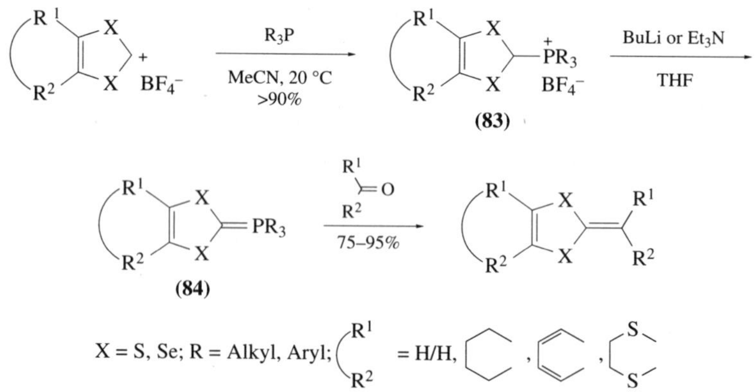

Scheme 41

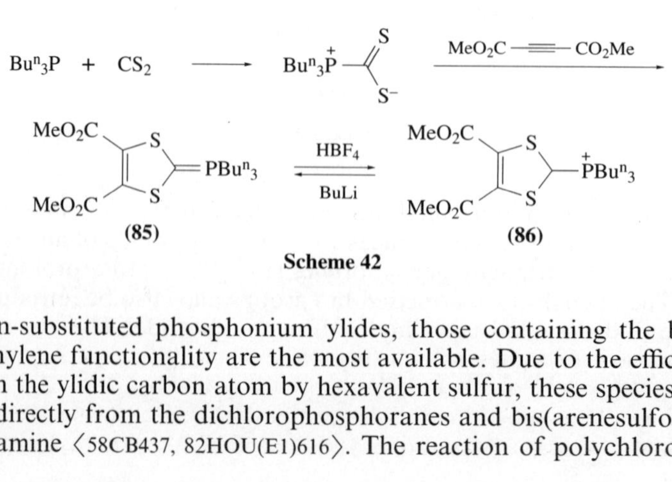

Scheme 42

Among chalcogen-substituted phosphonium ylides, those containing the bis(arenesulfonyl or alkanesulfonyl)methylene functionality are the most available. Due to the efficient stabilization of a negative charge on the ylidic carbon atom by hexavalent sulfur, these species are very stable and can be synthesized directly from the dichlorophosphoranes and bis(arenesulfonyl)methanes in the presence of triethylamine ⟨58CB437, 82HOU(E1)616⟩. The reaction of polychlorophosphoranes with activated methylene compounds affords bis(arenesulfonyl)methylene ylides (**87**) with both one and two chlorine atoms at the phosphorus atom (Scheme 43) ⟨77ZOB2390, 82ZOB1538⟩.

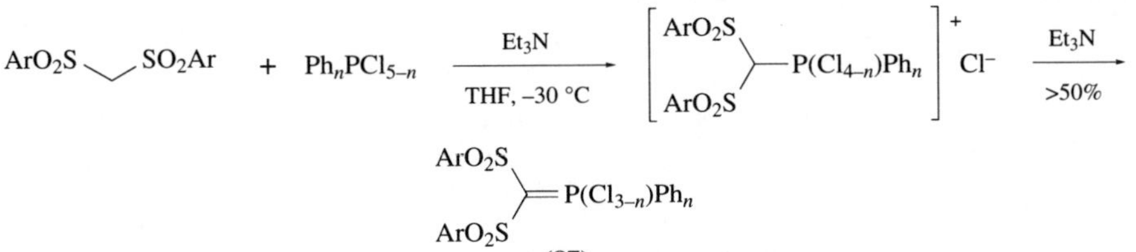

(**87**)

n = 1, 2; Ar = 4-XC_6H_4 (X = H, Cl, Me, MeO)

Scheme 43

Bis(benzenesulfonyl)halomethanes, being comparatively strong CH-acids, react with chlorodiphenylphosphine in the presence of triethylamine with the formation of tertiary α-haloalkylphosphines (**88**). The latter rearrange smoothly into *P*-halogenated ylides (**89**), which are isolated in good yield as stable crystalline compounds and are used as reagents in various chemical conversions (Scheme 44) ⟨77ZOR275⟩. A successful alternative approach to the ylide species such as (**89**) utilizes reaction of bis(arenesulfonyl)methylenephosphines, $(ArSO_2)_2CHPY_2$, with carbon tetrahalides ⟨79ZOB104⟩. Electronegative arenesulfonyl groups at the α-carbon atom reduce the reactivity of the tervalent phosphorus atom towards the polyhaloalkanes. However the reaction rate may in this case be increased by the addition to the reaction medium of tertiary amines.

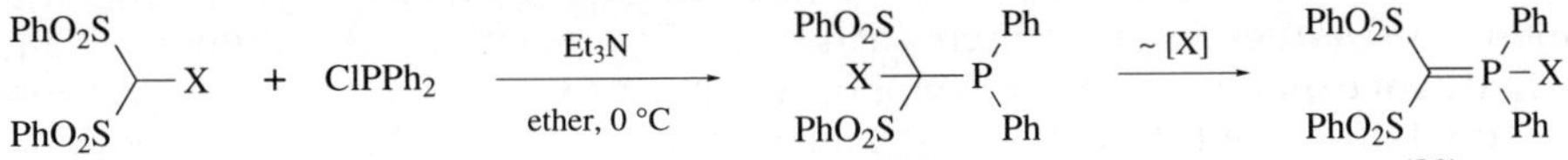

X = Cl, Br (**88**) (**89**)

Scheme 44

Sulfonyl stabilized methylenephosphoranes are also available from simple ylides, $R(H)CPPh_3$ (R = H, alkyl, aryl) and sulfonic acid halides in a transylidation reaction. However, this method is far from straightforward. Aromatic and aliphatic sulfonyl fluorides are most suitable for the introduction of a sulfonyl group into the α-position of methylenephosphoranes. Aliphatic sulfonyl chlorides give only poor yields of alkanesulfonyl substituted ylides. With aromatic sulfonyl chlorides halogenation and sulfenation may occur instead of sulfonation ⟨72RTC37, 74JOC2728⟩.

Hadjiarapoglou and Varvoglis have reported a more direct and general approach to bis(arenesulfonyl or alkanesulfonyl)methylenephosphoranes based on transylidation with phenyliodonium ylides (Equation (17)) ⟨88S913⟩. The reaction is catalyzed by cupric acetonylacetonate and most likely involves the initial complexing of the reactants with the Cu atom, followed by transylidation. Among ylides obtained by this method tris(ethoxy)phosphonium ylide (**90**, $X^1 = X^2 = PhSO_2$, Y = EtO) is of special interest because it is one of the few known stable trialkoxyphosphonium ylides ⟨88S913⟩. As a further development of this work the synthesis of ylides containing bis(perfluoroalkanesulfonyl)methylene functionality has been achieved by reaction of phenyliodonium bis(perfluoroalkanesulfonyl)methide with phosphines ⟨93JFC(60)175⟩.

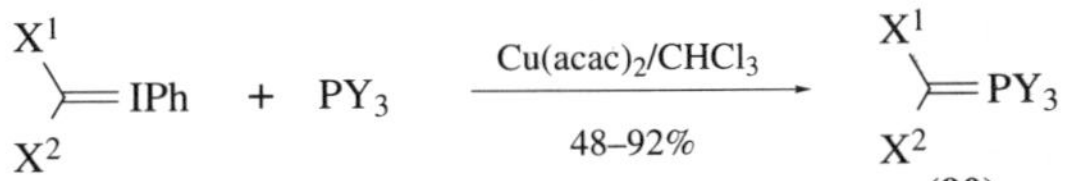

(**90**) (17)

X^1, X^2 = $PhSO_2$, 4-$MeC_6H_4SO_2$, $MeSO_2$, $-SO_2(CH_2)_3SO_2-$; Y = Ph, OEt

Certain *P*-functionalized phosphonium ylides may be synthesized starting from bis(arenesulfonyl)methylphosphine oxides, $(ArSO_2)_2CHP(O)R_2$, which exist in solution in equilibrium with the corresponding methylenephosphoranes ⟨83RCR1096⟩. For example, the reactions of the title phosphine oxides with phosphorus pentachloride and diazomethane lead to the formation of the corresponding *P*-chloro and *P*-methoxy phosphonium ylides ⟨82ZOB1538⟩. It should be noted that *P*-alkoxy substituted phosphonium ylides readily rearrange into phosphonates with migration of the alkyl group to the ylide carbon atom (the ylide version of the Pishchimuka reaction ⟨75CB2465⟩). However, since the electron-withdrawing arenesulfonyl substituents reduce the nucleophilicity of the ylide carbon, the compounds $(ArSO_2)_2C{=}P(OR)R_2$ are considerably more stable than their *C*-alkylated or arylated analogues. The ylide $(PhSO_2)_2C{=}P(OMe)Ph_2$ was reported to be stable on heating to 200 °C ⟨79ZOB104⟩.

6.22.1.4.3 *Nitrogen-, phosphorus-, arsenic- and antimony-substituted ylides, $R_2E(X)C{=}PY_3$ (E = N, P, As or Sb)*

Ab initio calculations predict thermal instability for the aminomethylenephosphorane $H_2N(H)C{=}PH_3$ with respect to dissociation to phosphine PH_3 and singlet carbene $[H_2N(H)C:]$ ⟨86CB1331⟩. In accordance with the theoretical data simple *C*-amino-substituted phosphonium ylides are extremely labile compounds. Few examples of this class have been isolated and fully characterized.

The first stable methylenephosphoranes with a dialkylamino group at the ylide carbon atom were prepared by Grobe and co-workers in 1989 ⟨89NJC363⟩. It was reported that fluorophosphaalkenes (**29**) reacted with tertiary phosphines according to Equation (18) to give the phosphorus ylide (**91**). The rate of reaction under discussion strongly depends on both substituents R^1 and R^2 and is mainly influenced by steric effects. This is demonstrated by the fact that phosphaalkenes (**29**) do not react with tri-*t*-butylphosphine and triphenylphosphine. The ylides (**91**) show a surprising thermal stability: slow decomposition occurs only at temperatures above 50 °C. A 30% solution of the compound (**91**; $R^1 = Me$, $X = Me_2N$) in toluene does not change even on heating to 80 °C for several hours. The unusual stability of the ylides (**91**) has been explained by electron delocalization in the planar $P^{III}CNP^{V}$ skeleton aided by the electron withdrawing substituents CF_3 and F on the tervalent phosphorus atom. As a further development of this work the synthesis of ylides (**92**) and (**93**) with other π-donor substituents has been achieved by reaction of difluoro- and fluoro-(alkoxy)methylenephosphines with trialkylphosphines ⟨92CB567⟩. Compounds (**92**) and (**93**) are stable up to about 10 °C, but decompose at higher temperatures yielding in the case of the C—F substituted ylide difluorotrimethylphosphorane as the main product. Similar to the ylide (**91**) the fluorine and oxygen substituted ylides owe their existence to the electron withdrawing effect of the $F_3C(F)P$ unit which overrides the destabilizing influence of the fluorine or alkoxy substituents on the ylide carbon atom.

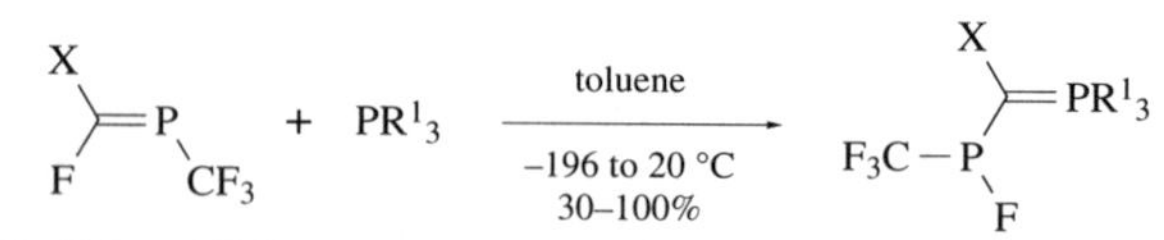

(18)

(**1**); (**21**) and (**29**) (**91**)–(**93**)

$X = R^2N$ (**29, 91**), F (**5, 92**), R^2O (**21, 93**); R^1, R^2 = Me, Et

At least in a very formal sense, one can consider the compound (**95**) produced by reaction of trimethylphosphine with cationic carbyne complex (**94**) as a *C*-dialkylamino-substituted phosphonium salt, although it is obvious that the real bonding pattern of this cationic half ylide is much closer to the mesomerically stabilized zwitterionic structure (Scheme 45) ⟨78CB2451⟩.

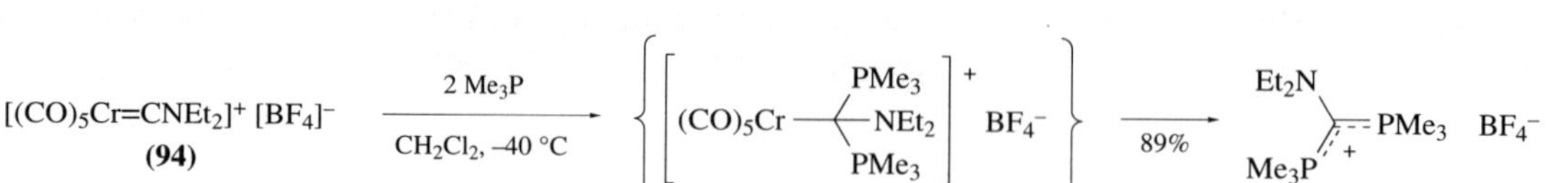

(**94**) (**95**)

Scheme 45

Unlike nitrogen, the heavier Group 15 elements, especially phosphorus, exert a stabilizing effect on the ylide functions and consequently numerous phosphino substituted phosphonium ylides have been synthesized and thoroughly investigated. These will, however, despite their abundance, be treated briefly here because of the availability of a series of review articles that provide quick access to the original papers published up to 1989 ⟨83RCR1096, 91COS(6)171, 91RCR391⟩.

The most general synthesis of phosphino substituted phosphonium ylides is based on transylidation methodology, investigated by Schmidbaur and co-workers in the early 1970s ⟨70AG(E)77, 71CB150⟩. This route in generalized form is outlined in Scheme 46. Ylides carrying at least one hydrogen atom at the α-carbon atom react with chlorodialkyl- or chlorodiarylphosphines with the formation of α-substituted phosphonium salts, from which phosphino-substituted ylides are generated by deprotonation. If the reaction provides a salt whose acidity is greater than that of the starting ylide precursor, a second mole of the starting ylide reacts with the zwitterionic intermediate in a transylidation reaction ⟨91COS(6)171⟩. A large variety of phosphino substituted ylides have been made by this method ⟨B-72MI 622-01, 91RCR391⟩. The synthesis of arsenic- and antimony-substituted ylides may be carried out analogously. From methylenetriphenylphosphorane and phosphorus, arsenic, and antimony halogen compounds the corresponding disubstituted ylides such as **(96)**–**(98)** are available via double transylidation ⟨66LA(699)40, 70JPR456, 84OM38, 84ZN(B)1456⟩.

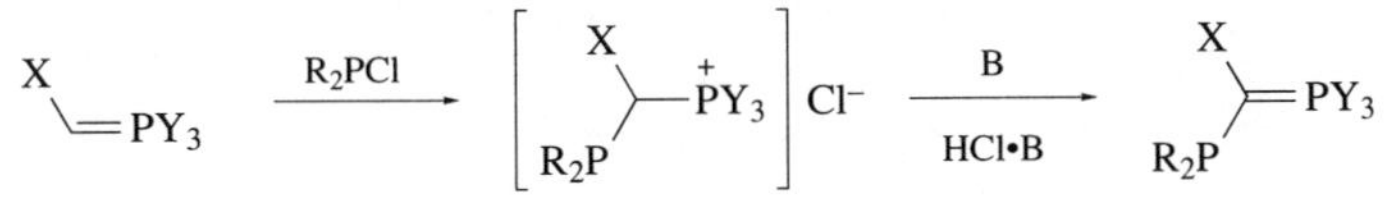

Scheme 46

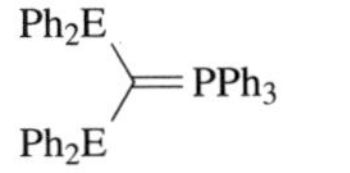

(96) E = P ⟨66LA(699)40, 70JPR(312)456⟩
(97) E = As ⟨84ZN(B)1456⟩
(98) E = Sb ⟨84OM38⟩

Reactions of ylide anions with electrophiles lead directly to functionalized ylides ⟨87AG(E)79, 87TL2111⟩. For instance, the ylide **(99)** can be metallated at the methyl group by LiMe, $NaNH_2$ or KH to give partially solvated products **(100)**. The reaction of the latter with chlorodiphenylphosphine leads to the new ylides **(101)** and **(102)** (Scheme 47) ⟨83CB1386⟩. Pure **(101)** is obtained from (**100**; M = Li) and chlorodiphenylphosphine in the presence of TMEDA.

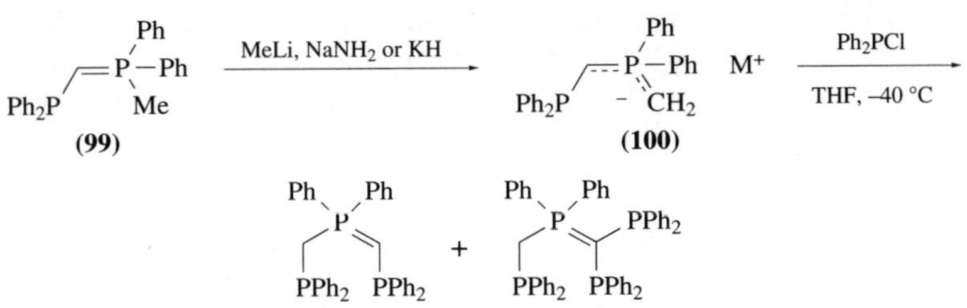

Scheme 47

The convenient procedure for the synthesis of methylenephosphoranes bearing phosphorus, arsenic or antimony at the ylidic carbon atom is condensation of trimethylsilyl-substituted ylides with chlorophosphines, chloroarsines and chlorostibines. The reactions proceed in a 1 : 1 ratio of starting reagents and are followed by elimination of chlorotrimethylsilane. This method permits access to compounds which are often difficult to obtain by other routes and produces salt-free functionalized ylides in good yield. The silylated precursors, representing the starting materials for the syntheses are readily available with a large variety of substituents (cf. Section 6.22.1.4.4). The reactions in both Equation (19) and Scheme 48 constitute examples from the literature which illustrate the synthetic potential of the method.

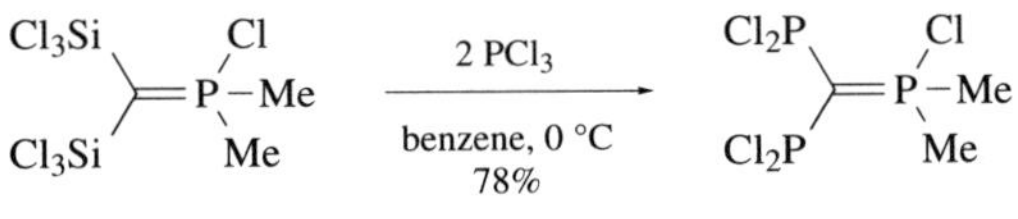

(19)

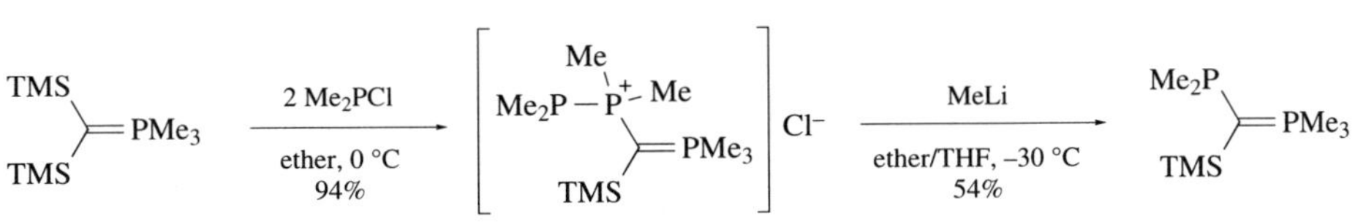

Scheme 48

Ylides bearing $\sigma^3\lambda^5$ and $\sigma^2\lambda^3$ phosphorus at the anionic centre are stable enough to be isolated. A methylenephosphorane bearing two-coordinate phosphorus at the ylidic carbon atom was obtained by reaction of phosphonium salt (**103**) with three equivalents of sodium bis(trimethylsilyl)amide in THF at −78°C. Compound (**104**) is probably formed via [1,3]-silyl migration. According to an x-ray analysis, this ylide can be described as a hetero allyl anion with a three-center-four-electron bond C1—P1—N1. Oxidative addition of sulfur or selenium results in the formation of the imino(thioxo)phosphorane (**105**) or imino(selenoxo)phosphorane (**106**) (Scheme 49) ⟨91CB329⟩.

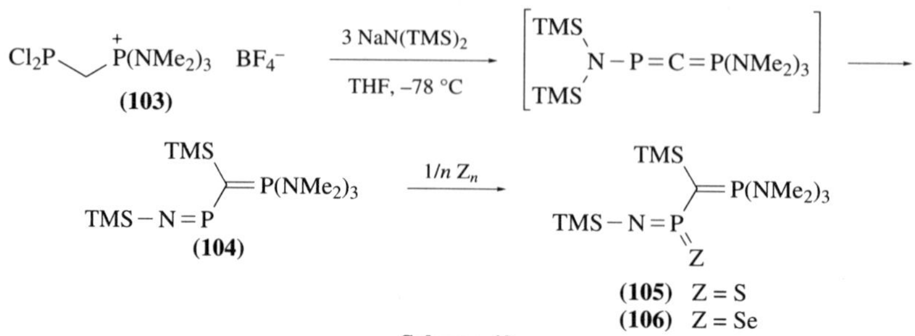

Scheme 49

6.22.1.4.4 *Silicon-, germanium- and boron-substituted ylides, $R_3E(X)C{=}PY_3$ (E = Si or Ge) and $R_2B(X)C{=}PY_3$*

Trialkylsilyl groups exhibit a pronounced stabilizing effect on ylidic carbanions; therefore, numerous silicon substituted phosphonium ylides are available and widely used in preparative chemistry. A few review articles serve very well as introductions to this field ⟨75ACR62, 91RCR391, 91COS(6)171⟩. Germanium- and boron-substituted phosphonium ylides are considerably less studied.

A wide variety of α-silylated methylenephosphoranes have been synthesized from simple methylenephosphoranes and chloroalkyl(aryl)silanes (especially TMS-Cl) by transylidation ⟨82AG(E)545, 85AG(E)1068, 92S787⟩. From methylenetrialkyl(aryl)phosphoranes both mono- or bis-silylated products may result (Scheme 50) ⟨67CB1032, 92CB1053⟩. Apart from the above mentioned silicon halogen compounds, siletanes, 1,3-disiletanes and the highly strained hexamethylsiliranes have been reacted with methylenephosphoranes to give the corresponding silicon substituted ylides ⟨91COS(6)171⟩. The introduction of germanium substituents may be carried out analogously or starting from ylide anions (Scheme 51) ⟨67CB1032, 77CB677⟩. Mixed silicon/germanium or silicon/tin substituted ylides were synthesized by reacting the monosilylmethylenephosphorane (**107**) with germyl or stannyl chlorides (Equation (20)) ⟨67CB1032⟩.

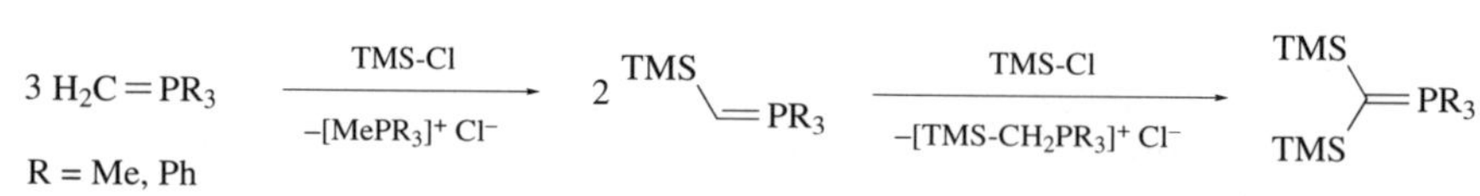

Scheme 50

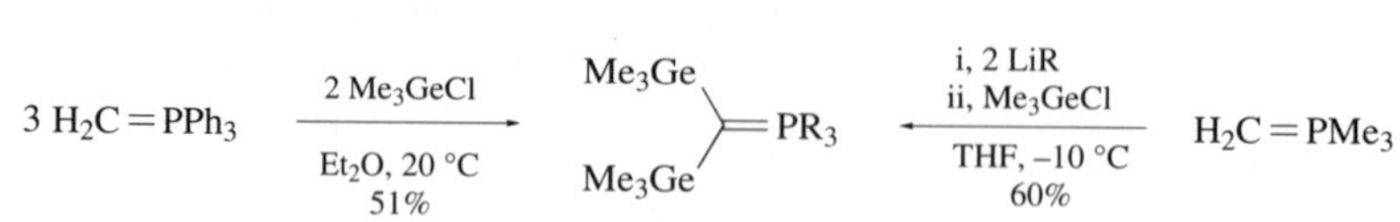

Scheme 51

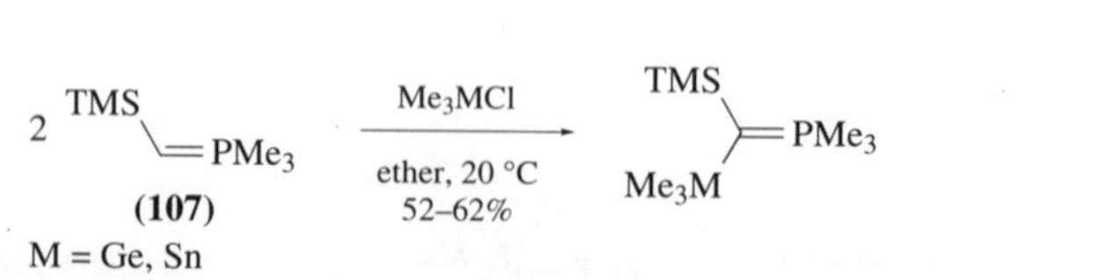

(20)

The reaction of methylenetrimethylphosphorane with dichlorodimethylsilane leads via a twofold transylidation to the formation of Si-bridged bisphosphorane (**108**) (Scheme 52) ⟨70CB97⟩. At elevated temperatures and in the presence of excess ylide as a catalyst, the four-membered ring undergoes an isomerization through ring expansion (**108**)–(**109**) ⟨74AG(E)540⟩. Methylenetriphenylphosphorane reacts with 1,2-dichlorotetramethyldisilane to give a methylenephosphorane (**110**) with two exocyclic ylide functions (Equation (21)) ⟨70AG(E)737⟩.

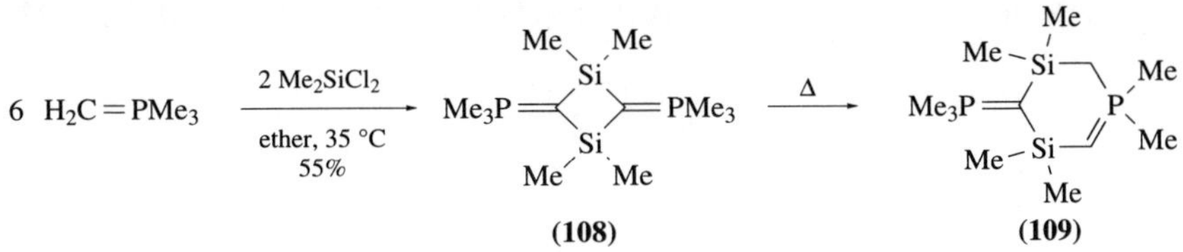

Scheme 52

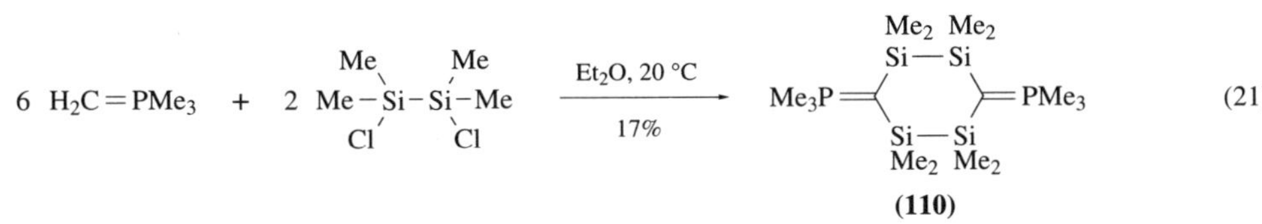

A moderately important process for the preparation of silyl ylides is transsilylation. This method provides a means of introducing more complicated substituents, starting from simple trimethylsilyl ylides. Equation (22) gives some selected examples. As a rule these reactions proceed sufficiently rapidly even at room temperature to give pure products in high yields; the only by-product formed in the reactions is the readily separable chlorotrialkylsilane ⟨71CB150, 70AG(E)77⟩.

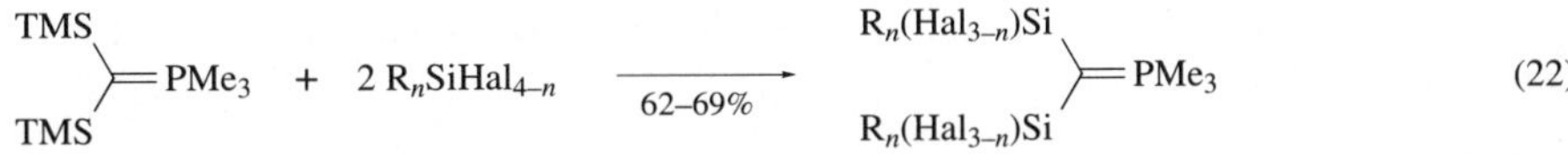

R_nSiHal_{4-n} = Me_2SiCl_2, $MeSiCl_3$, $SiCl_4$, Me_2SiBr_2, $SiBr_4$, $Me(ClCH_2)SiCl_2$, $ClCH_2SiCl_3$, $MeSiHCl_2$, $HSiCl_3$, Me_2SiFCl

The silicon substituted *P*-chloro ylides are available by 1,2(C to P) halotropic rearrangements of trimethylsilyl substituted α-haloalkylphosphines. The latter may in turn be prepared by reacting silylalkylphosphines with tetrahalomethanes or condensation of perchlorinated carbosilanes with lithio- and silylphosphines. The interaction of compounds of the type (**111**) with carbon tetrahalides (CCl_4, $CBrCl_3$, CBr_4) is usually carried out in dichloromethane at room temperature or below 0 °C. Yields of ylides obtained in this way are usually very high (Equation (23)) ⟨79CB1068⟩. However, in certain cases tervalent phosphorus compounds containing trimethylsilyl groups on the α-carbon atom react with carbon tetrahalides with elimination of trihalotrimethylsilylmethane instead of haloform. It should be noted that ylide (**112**) containing a trimethylsilyl group at the sp^2-hybridized carbon atom is readily desilylated by excess of carbon tetrachloride to give carbodiphosphorane (**113**) (Scheme 53) ⟨79CB648⟩.

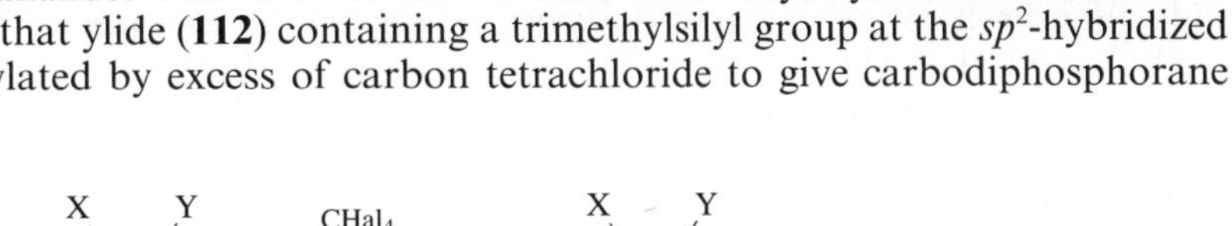

Hal = Cl, Br; X = TMS, Ph, H; Y = Ph, R_2N, RO

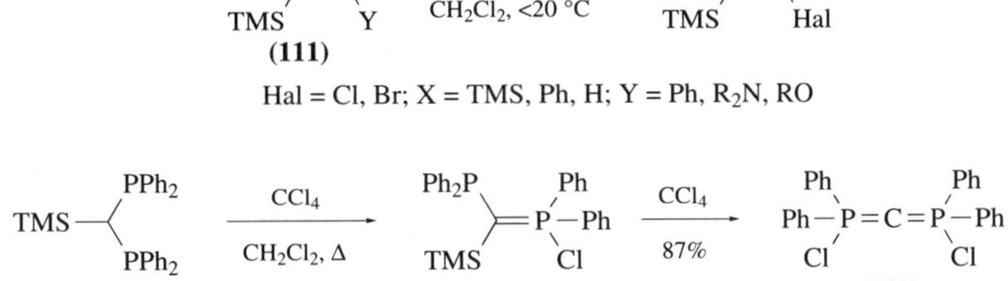

Scheme 53

The α,α-disilyl substituted *P*-chloro ylides have been prepared in good yields via the reaction of chlorocarbosilanes with lithio- or silylphosphines (Schemes 54 and 55). If *C*,*C*-dihalo-1,3,5-trisilacyclohexanes are used the major products are methylenephosphoranes (**114**) with exocyclic ylide functions (Equation (24)). These are stable crystalline substances or liquids distillable in vacuum ⟨81ZAAC(472)45, 84ZAAC(511)95⟩.

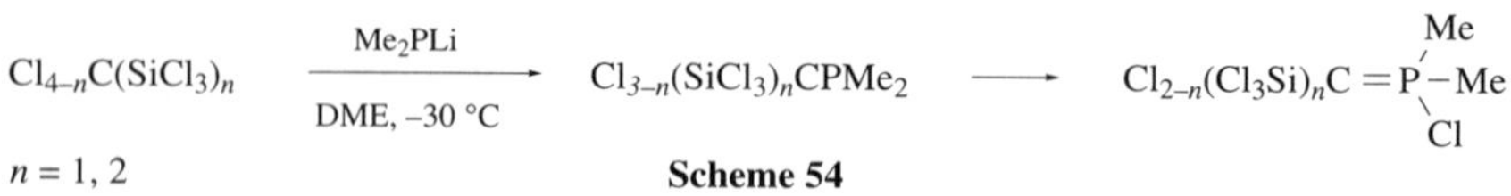

Scheme 54

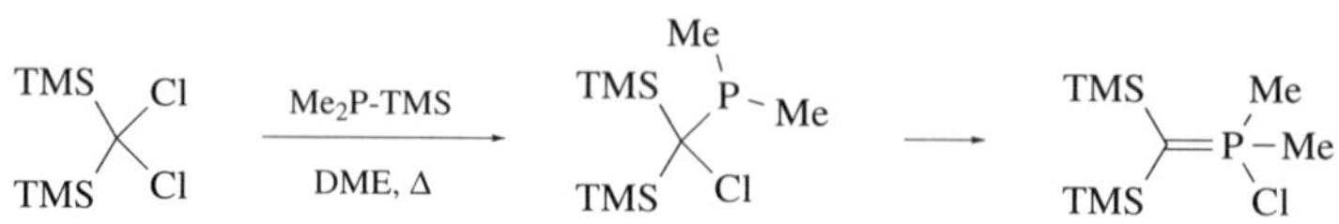

Scheme 55

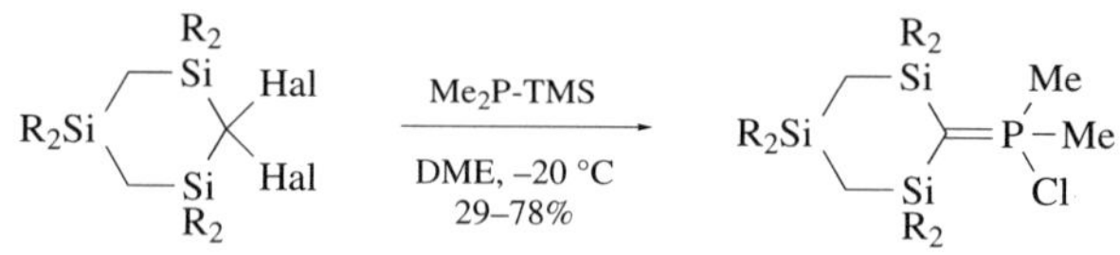

Hal = Cl, Br; R = F, Cl, Me **(114)** (24)

P-Chloro ylides (**115**) containing trichlorosilyl groups on the α-carbon selectively exchange the chlorine atom on phosphorus for a dimethylphosphine group on interaction with lithium dimethylphosphide; the trichlorosilyl groups are unaffected by this. In addition, during the reaction of ylide (**115**) with Me(TMS)PLi, the P—P ylide (**116**) is formed which reacts with a second molecule of the P—Cl ylide and is converted into a double ylide with a P—P bond (**117**) (Scheme 56) ⟨84ZAAC(518)14⟩.

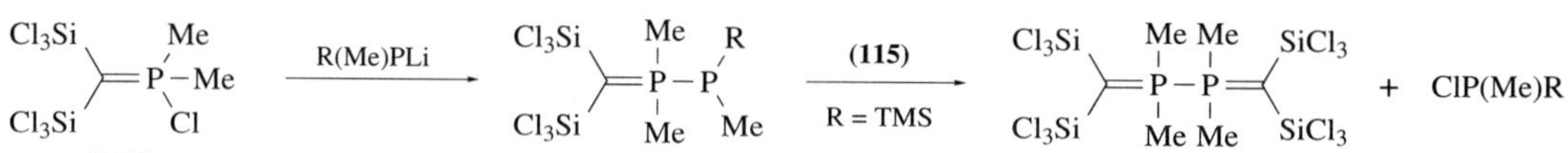

(115) **(116)** **(117)**

R = Me, TMS

Scheme 56

It has been reported that photolysis of phosphinodiazomethane (**118**) in benzene at room temperature in the presence of an excess of chlorotrimethylsilane gives the ylide (**119**) in almost quantitative yield ⟨88JA6463⟩ (Scheme 57). The generality of this approach is restricted by possible difficulties in the preparation of the requisite phosphinocarbenes.

Dialkylborylalkylidenetriphenylphosphoranes have been studied rarely. They are available via transylidation starting from simple alkylidenephosphoranes and chlorodialkyl- or dichloroalkylboranes ⟨84AG(E)381, 86AG(E)559⟩. The synthesis of the mixed silicon–boron substituted ylide (**120**) follows the reaction pathway analogous to the approach to disilylated methylenephosphorane (**119**) (Scheme 58) ⟨93CC1354⟩.

The first *C*-gallyl-substituted phosphorus ylide, $(R_2N)_2P(Me){=}C(TMS)GaMe_2$, was recently prepared by the reaction of a stable phosphinocarbene with $GaMe_3$ ⟨94AG(E)578⟩.

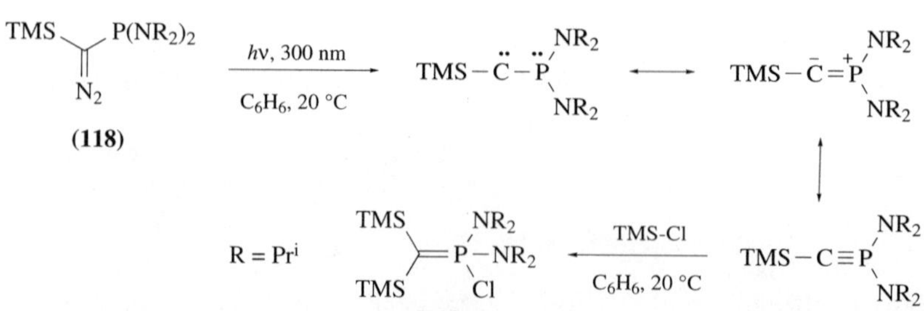

(119)

Scheme 57

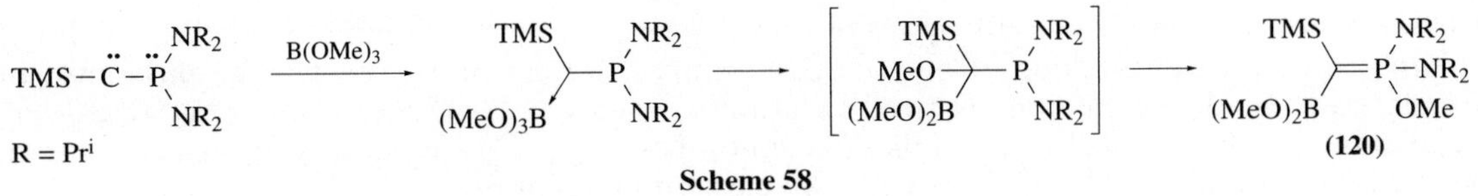

Scheme 58

6.22.1.4.5 *Metal substituted ylides, $L_nM(X)C{=}PY_3$*

Whereas the chemistry of the metal complexes of phosphorus ylides (**121**) has been the subject of intensive investigation and has developed into a rapidly growing field of research with important applications ⟨B-82MI 622-01, 83AG(E)907, 83CCR1⟩, the chemistry of metal substituted ylides (**122**) has been investigated relatively little. Moreover, most of the examples that have been studied in detail are derived from the simple methylenephosphoranes, H_2CPY_3.

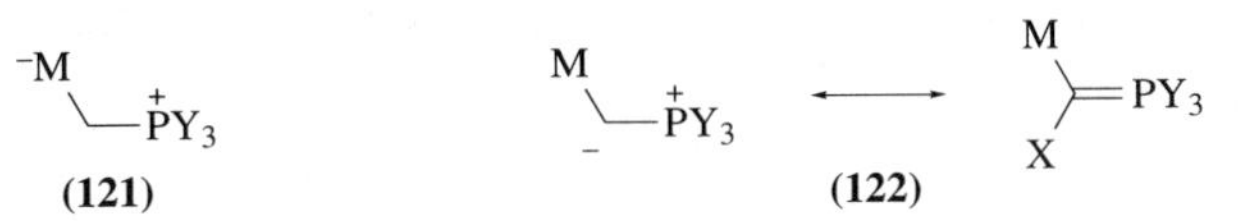

Under this heading, only the synthesis of functions of the type (**122**) which formally arise from substitution of a hydrogen atom at the ylidic carbon by a metal atom will be considered. Metal complexes of phosphonium ylides will not be discussed.

(i) Ylides with a Li—C bond

The synthesis of lithiated ylides by direct substitution of a proton at the α-carbon atom seems to be restricted to alkylidenephosphoranes.

In 1982 Corey and co-workers reported that α-lithiomethylenetriphenylphosphorane (**126**) is available by reaction of the corresponding ylide with *t*-butyllithium in THF solution at low temperature ⟨82JA4724⟩. The same lithium substituted ylide can be generated by reaction of methyltriphenylphosphonium bromide with 2 equivalents of *s*-butyllithium in ether. However, examination of the reactions by means of low temperature NMR spectroscopy showed that the interaction of the methylenetriphenylphosphorane with lithium alkyls includes replacement of a phenyl group at the phosphorus atom by an alkyl group with formation of the ylide (**123**) (route *a*) and metallation of a benzene ring with formation of the ylide (**124**) (route *b*). Spectral characteristics of the lithiated ylide obtained by the exchange reaction of bromomethylenephosphorane (**125**) with *t*-butyllithium in hexane at −75 °C differed from those for the lithiated ylide prepared by direct metallation ⟨83C10, 84TL4097, 85TL1623⟩. To explain these results it was proposed that lithium methylide (**126**) exists in a metallotropic tautomeric equilibrium with the ylide (**124**) lithiated in the benzene ring (Scheme 59) ⟨85TL555⟩.

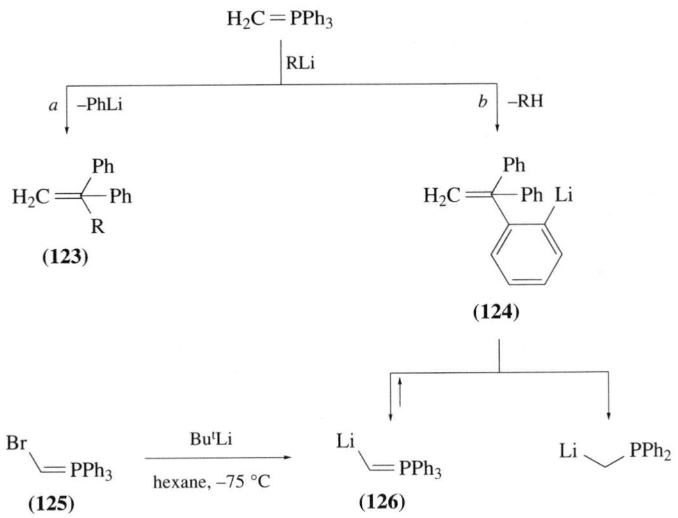

Scheme 59

The reaction of methylenetrialkylphosphoranes with equimolar amounts of methyl- or butyllithium proceeds with lithiation of one of alkyl groups. The use of excess of organolithium compounds leads to the formation of twofold lithiated ylides (**127**) (Scheme 60) ⟨81JOM(209)C10⟩. Metallation of silylated methylenetrialkylphosphoranes (e.g., TMS(H)C=PMe_3) again occurs in the alkyl groups and affords products with two carbanionic centers.

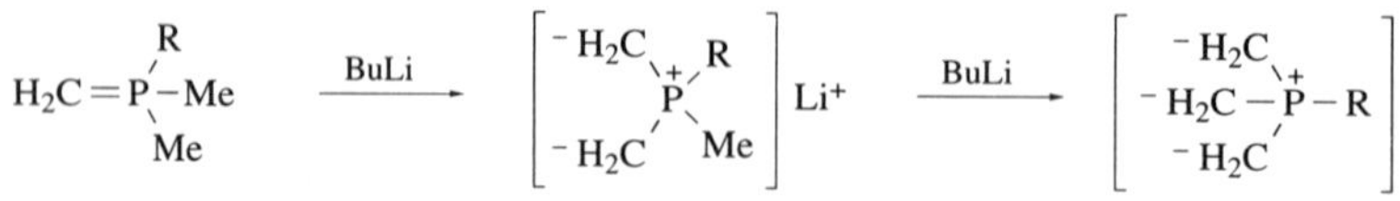

(**127**)

Scheme 60

The monomeric phosphonium ylides α-substituted by Group I metals other than lithium or by alkaline earth metals have not been described.

(ii) Ylides with a transition metal–carbon bond

The first attempts to obtain phosphonium ylides with a transition metal atom in place of the ylide hydrogen atom met with little success since the simplest reactions (e.g., of metal halides with H_2CPR_3) yield only insoluble products which are difficult to characterize ⟨77ZN(B)858⟩. Kaska and co-workers have successfully exploited the stabilizing influence of the η^5-cyclopentadienyl group to effect substitution of a proton at the ylide carbon atom by a $(\eta^5\text{-}C_5H_5)_2ZrCl$ group. Bis(η^5-cyclopentadienyl)chlorozirconylmethylenetriphenylphosphorane (**128**) was isolated as yellow-orange needles in 75% yield. In a similar fashion $(\eta^5\text{-}C_5H_5)_2HfCl_2$ reacts with two equivalents of methylenetriphenylphosphorane in THF for 14 days at room temperature to give hafnium-substituted ylide (**129**) (Equation (25)) ⟨80ZN(B)1289⟩.

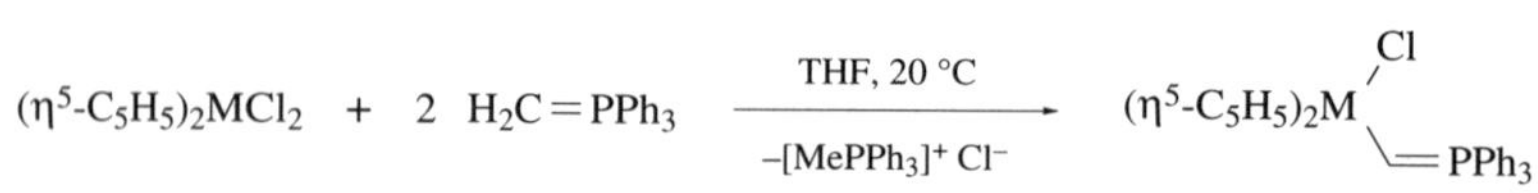

(25)

(**128**) M = Zr
(**129**) M = Hf

Subsequently the series of complexes of Group 4 metals with bridging ylide ligands has been described. Thus, when the methylenephosphorane (**130**) containing bulky diethylamino groups on the phosphorus atom was allowed to react with $TiCl_4$, the bis(phosphoranylidene) dititanacyclobutane (**131**) has been isolated in over 70% yield (Equation (26)) ⟨86AG(E)574⟩. Analogously, reaction of the ylide H_2CPMe_3 with $TiCl_2(NMe_2)_2$ affords the 1,3-bis(dimethylamino)dititanacyclobutane ⟨77ZN(B)858⟩. In both reactions the ylides are deprotonating in the manner of a transylidation, giving rise to the formation of a phosphonium salt as a by-product. When an excess of methylenetris(dimethylamino)phosphorane was treated with $TiCl_2(NMe_2)_2$ according to the same procedure (by employing the phosphorus methylide as base as well as nucleophile), compound (**132**) having different terminal ligands on titanium was obtained in 3% yield. The addition of 2,4,6-trimethylpyridine to an equimolar mixture of ylide and $TiCl_2(NMe_2)_2$ provided the metallacycle (**133**) in 17% yield (Scheme 61) ⟨93AG(E)554⟩. Use of $TiCl_3(NMe_2)$ in place of $TiCl_2(NMe_2)_2$ afforded (**133**) in the same isolated yield. Compound (**133**) underwent ligand exchange reactions at titanium without disruption of the metallacycle core. Thus, the addition of 2.5 equivalents of $Ti(NMe_2)_4$ to (**133**) at ambient temperature afforded (**132**) in quantitative yield. It should be noted that because of their two oxophilic centers, the metal-substituted ylides of type (**133**) may be regarded as a synthetic equivalent for a carbon atom for the selective production of allenes from unhindered aromatic aldehydes. For example, metallacycle (**133**) reacted immediately with an excess of 4-methylbenzaldehyde or 4-methoxybenzaldehyde in THF solution to give the corresponding 1,3-diarylallenes in 43% and 40% isolated yield, respectively, based on an expected two equivalents of allene per metallacycle (Equation (27)). The compound (**133**) underwent the same reaction with 4-methylbenzaldehyde in 20% yield ⟨93AG(E)554⟩.

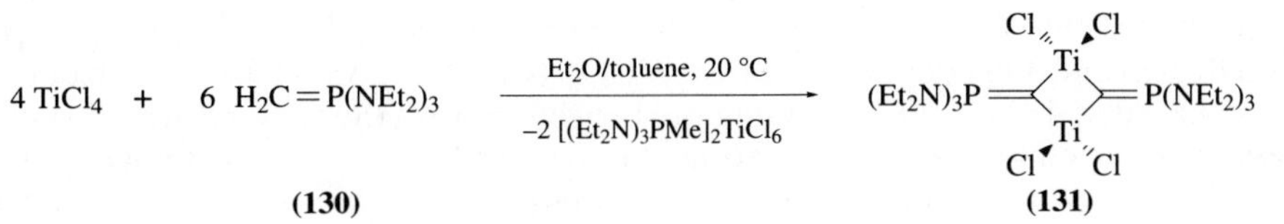

(26)

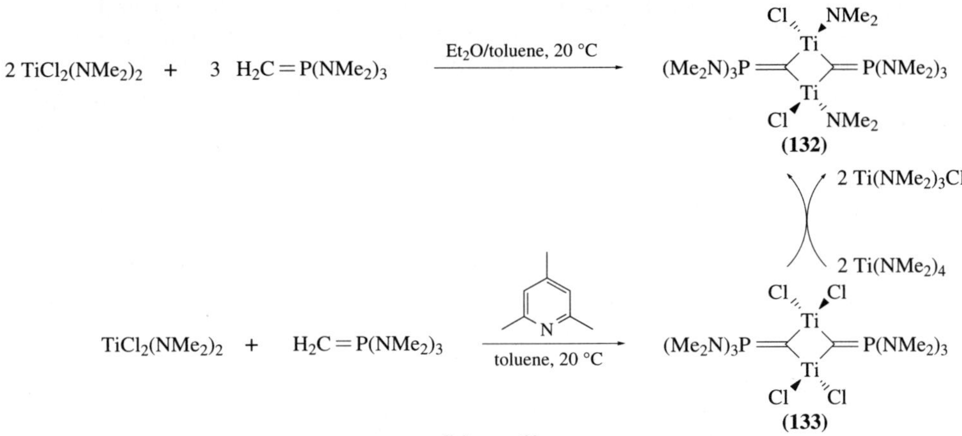

Scheme 61

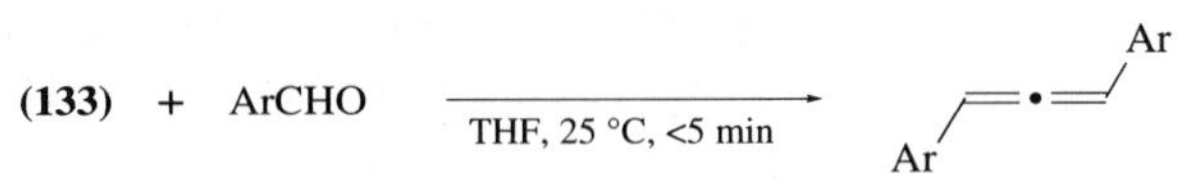

(27)

Ar = *p*-MeC_6H_4, *p*-$MeOC_6H_4$

1,3-Dizirconiacyclobutane containing exocyclic ylide functions has been synthesized from $[Me_3CCH_2]_4Zr$ and excess H_2CPMe_3 ⟨83OM154⟩.

Tin- and lead-substituted ylides are available from methylenephosphoranes and stannyl or plumbyl chlorides via transylidation reactions (cf. Section 6.22.1.4.4) ⟨77CB677⟩.

6.22.1.5 Tetracoordinate Arsenic, Antimony and Bismuth Derivatives

In comparison with the phosphonium ylides ($X^1X^2CPY_3$) relatively little information is available regarding the synthesis of *C,C*-diheterosubstituted arsonium, stibonium and bismuthonium ylides ($X^1X^2CEY_3$) ⟨B-70MI 622-01, 82AOC(20)115, 82COMC-I(2)681, 87CSR45⟩. The size difference between carbon and arsenic, antimony and bismuth atoms leads to relatively inefficient $p\pi$–$d\pi$ overlap between the C-sp^2 orbitals and the large and more diffuse *d*-orbitals of the heteroatom (E), and to decreased electrostatic interaction across the ylide P^+—E^- bond. Consequently, arsonium, and especially stibonium and bismuthonium ylides, are in general more dipolar than their phosphorus analogs and have a tendency to decompose both in solution and in the solid state unless there is electronic stabilization. A significant increase in stability of the ylides $X^1X^2CEY_3$ is observed, however, if electron withdrawing groups (X^1, X^2) are conjugated with the ylidic carbon atom. For example, bis(arenesulfonyl)-substituted arsonium ylides are thermodynamically stable at ambient temperature and may be kept in air without significant decomposition. Stibonium ylides need anhydrous conditions and are frequently not isolated but used *in situ*. Bismuthonium ylides are the most unstable of all Group 5B ylides.

The methods available for the synthesis of arsonium, stibonium and bismuthonium ylides are nearly all analogous to methods used for the corresponding phosphonium ylides.

6.22.1.5.1 C,C-Diheterosubstituted arsonium ylides, $X_2C{=}AsY_3$

Among the above ylides the only well characterized and studied compounds are bis(arenesulfonyl)methylenetriarylarsoranes. The most simple synthesis of these compounds is based on the reaction of dichloroarsoranes with compounds having acidic methylene groups, first investigated by Horner and Oediger in the late 1950s ⟨58CB437, 59LA(627)142⟩. For example, $(PhSO_2)_2CAsPh_3$ is prepared in 49% yield by heating bis(benzenesulfonyl)methane with Ph_3AsCl_2 in benzene in the presence of triethylamine. The strong stabilizing influence of the $PhSO_2$ groups upon the C=As bond is clearly indicated by the remarkable stability of the triphenylarsonium bis(benzenesulfonyl)methylide towards oxygen and water over a period of days. The structure of this ylide was confirmed by x-ray crystallography ⟨88JCS(P2)1829⟩.

Variants of the above method include reaction of $(ArSO_2)_2CH_2$ with triphenylarsine oxide in refluxing acetic anhydride ⟨73T1697⟩. The reaction presumably proceeds with initial formation of an acetoxyarsonium cation $[Ph_3AsOAc]^+$. This then reacts with a carbanion $[(ArSO_2)_2CH]^-$ to form an arsonium salt which is readily converted into an ylide by loss of a proton. The method is limited to the preparation of highly stabilized arsonium ylides since it is necessary that the methylene reactant should be a fairly strong carbon acid. Almost all examples of this approach have utilized triphenyl substituted arsenic derivatives.

Silyl substituted arsonium ylides can be prepared by the salt method based on the reaction of an arsonium salt, usually obtained from a functionalized alkyl halide and an arsine, with a suitable base. Thus triphenylarsonium trimethylsilylmethylide, $TMSCHAsPh_3$, has been prepared in good yield by treatment of the corresponding α-trimethylsilyl substituted onium salt with butyllithium in ethyl ether ⟨65IC1458⟩. The compound is also available from the reaction of the simple ylide $H_2CAsPPh_3$ with chlorotrimethylsilane followed by dehydrochlorination of the so formed arsonium salt with phenyllithium ⟨84TL4425⟩. The ylide $TMSCHAsPh_3$ readily undergoes desilylation by trimethylsilanol ⟨68IC168⟩. Similar reaction of the ylide with $TMSOGeMe_3$ leads to a monogermyl substituted arsonium ylide which disproportionates in solution to give $(Me_3Ge)_2CAsPh_3$ (72%) ⟨77CB677⟩.

The interaction of quaternary arsonium salt (**134**) with $NaNH_2$ in liquid ammonia occurs as outlined in Scheme 62 to give ylide (**135**). Treatment of the latter with Ph_2AsCl affords diarsenylmethylenearsorane (**136**) as the product of a transylidation reaction ⟨85OM344, 87CB1281⟩.

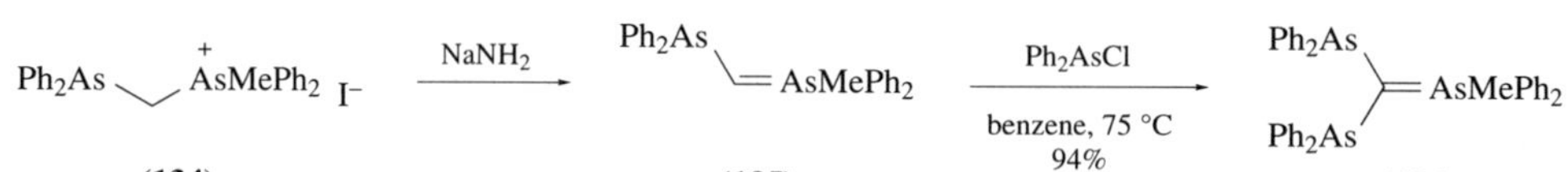

(**134**) (**135**) (**136**)

Scheme 62

Another method for the synthesis of bis(arenesulfonyl)methylenearsoranes consists of thermal decomposition of diazo compounds in the presence of an arsine (Equation (28)). The initial methodology consisted of heating a mixture of the diazo compound with an excess of the trapping agent, without solvent, above the decomposition point of the diazo compound. Subsequent improvement includes the use of bis(hexafluoroacetylacetonato)copper, $Cu(hfacac)_2$, as a homogeneous catalyst ⟨82T3355, 88S319⟩.

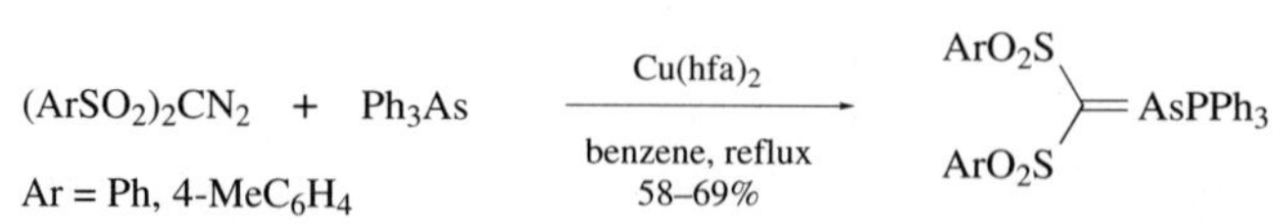

(28)

Related to the conversion of diazo compounds into arsonium ylides is the formation of arsonium ylides by thermal or photolytical decomposition of iodonium ylides. For example, by analogy with the synthesis of phosphonium ylides shown in Equation (17), arsonium ylides can be obtained from phenyliodonium bis(arenesulfonyl)methylides and triphenylarsine in yields of 59–72% ⟨88S913⟩.

6.22.1.5.2 *Stibonium and bismuthonium ylides bearing heterosubstituents, $X_2C{=}EY_3$ (E = Sb or Bi)*

Isolable examples of the above compounds, with the exception of bis(arenesulfonyl)-substituted ylides, are virtually unknown. The likely reason is their low stability. Use of copper hexafluoroacetylacetonate as a catalyst has enabled the ylides $(ArSO_2)_2CSbPh_3$, Ar = Ph (78%) and $4\text{-}MeC_6H_4$ (77%) to be obtained from triphenyl-antimony and the diazo compounds $(ArSO_2)_2CN_2$ in boiling benzene ⟨86T3887⟩. These stibonium ylides are stable for long periods if kept in a dry atmosphere but in solution they are readily hydrolysed if any moisture is present. Being dipolar molecules the ylides are insoluble in alkanes and ether. Attempts to prepare stibonium ylides from dichlorotriphenylantimony always led to ylides contaminated with $(Ph_3SbCl)_2O$ ⟨88JCS(P2)1829⟩.

The bismuthonium ylide $(PhSO_2)_2CBiPh_3$ has been made by the same method ⟨88S319⟩. Although reasonably stable as a solid it decomposed rapidly in solution and hence could not be purified ⟨88S319⟩. It was also reported that this ylide can be generated by interaction of $(PhSO_2)_2CHNa$ with Ph_3BiCl_2 or Ph_3BiO (90BCJ950). It should be noted here that stibonium and bismuthonium ylides have specific chemical properties which little resemble those of phosphonium ylides. Thus they do not take part in Wittig reactions, even with very reactive aldehydes ⟨86T3887, 88JCS(P2)1829⟩.

6.22.2 FUNCTIONS CONTAINING A DOUBLY BONDED METALLOID

6.22.2.1 Tricoordinate Silicon and Germanium Derivatives

Despite the intense activity on multiply bonded silicon and germanium species ⟨85CRV419, B-89MI 622-02, 90CRV283, 90JOM(400)121, 94CCR427⟩ there is little information on *C,C*-diheterosubstituted functions $X^1X^2C{=}EY_2$ (E = Si or Ge). Among the latter only disilicon-substituted (X^1, $X^2 = R_3Si$) silaethenes have been studied in considerable detail by Wiberg and co-workers ⟨84JOM(273)141⟩ over the past two decades. The most successful method for the generation of these compounds involves 1,2-elimination from α-lithiated silanes carrying a good leaving group on the silicon atom.

6.22.2.1.1 Disilylsubstituted silaethenes, $(R_3Si)_2C{=}SiY_2$

A synthetic approach to disilylsubstituted silaethenes is shown in Scheme 63. Generally, the 1,2-elimination of LiY from the silanes (**137**) is reversible. With poor leaving groups, such as Y = MeO or PhO, the equilibrium disfavors the silaethene. In the case of good leaving groups, such as Hal, $ArSO_2$ or $(PhO)_2PO_2$, the equilibrium lies to the right ⟨84JOM(273)141, 89CB409⟩. The rate of elimination of LiY from (**137**) in ether obeys first-order kinetics and increases for substituent Y in the sequence $Ph_2PO_2 < F < SPh < Ph_2PO_4 < I < Br < Cl < p\text{-}MeC_6H_4SO_3$ ⟨81CB3505⟩. The solvent also has an important influence on the results of these reactions. In general, the silaethenes tend to be formed as intermediates in ether solutions, while in tetrahydrofuran they are not. In a less basic solvent such as pentane, silaethene formation would be favored even more, but the lithiation step then proceeds too slowly. Thus, a solvent of intermediate basicity like diethyl ether is the best compromise medium for the formation of silaethenes.

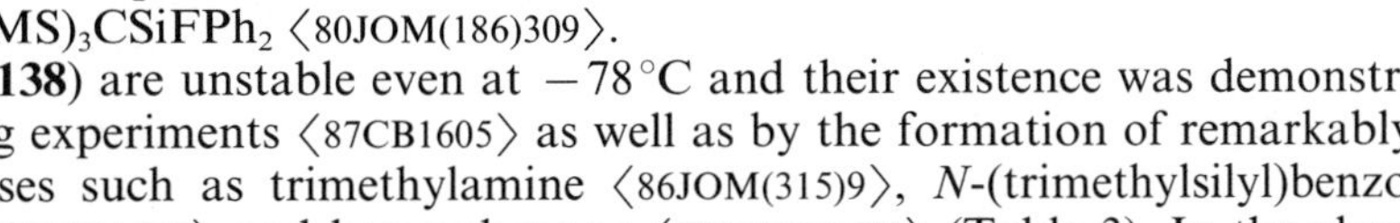

Scheme 63

Thermal elimination of fluorotrimethylsilane represents an alternative to the elimination reactions in solution. For example, the silaethene $(TMS)_2C{=}SiPh_2$ can also be generated by the vapor phase pyrolysis of $(TMS)_3CSiFPh_2$ ⟨80JOM(186)309⟩.

Silaethenes (**138**) are unstable even at −78 °C and their existence was demonstrated by characteristic trapping experiments ⟨87CB1605⟩ as well as by the formation of remarkably stable adducts with Lewis bases such as trimethylamine ⟨86JOM(315)9⟩, *N*-(trimethylsilyl)benzophenone imine ⟨87ZN(B)1055, 87ZN(B)1062⟩ and benzophenone ⟨88ZN(B)1468⟩ (Table 3). In the absence of trapping reagents, the formation of dimers or rearrangement products was observed ⟨81CB2087, 81CB3505, 81CB3518⟩. Note that some of the derivatives of (**138**) obtained by cycloaddition of unsaturated compounds, for example, *N*-(trimethylsilyl)benzophenone imine, can be used as a 'store' for silaethenes (**138**) since they decompose easily to the starting materials by thermal cycloreversion ⟨91CB1981⟩.

An increase in the bulkiness of one of the substituents through the use of a TBDMS group led to an isolable silaethene (**140**) ⟨83AG(E)1005⟩. Transformation of $[(TMS)_2CLi{-}SiFBu^t_2 \cdot 4THF]$ (**139**) into (**140**) proceeds in diethyl ether in the presence of chlorotrimethylsilane at room temperature and leads to a monotetrahydrofuran adduct of (**140**) (45%) ⟨83AG(E)1005⟩. Pure crystalline silaethene is obtained from the adduct by removing THF by azeotropic distillation with benzene. It is kinetically

Table 3 Generation and trapping of transient silaethenes $R^1(TMS)C{=}SiR^2_2$ (**138**).

Silaethene *R*[1]	*R*[2]	*Precursor*	*Trapping reagent or detected product*	*Ref.*
TMS	Me	$(TMS)_2(Me_2YSi)CLi$ Y = Ph_2PO_2, Ph_2PO_3, Ph_2PO_4, *p*-$MeC_6H_4SO_3$	2,3-dimethylbutadiene, TMSN=NTMS and $TMSN_3$; dimer	77AG(E)328
TMS	Me	$(TMS)_3CSiMe_2Y$ Y = Cl, Br, I	MeOH/MeONa	78JOM(157)C50, 80JOM(191)355
TMS	Ph	$(TMS)_3CSiPh_2Y$ Y = Cl, Br, I	MeOH/MeONa	80JOM(191)355
TMS	Me	$(TMS)_2(Me_2YSi)CLi$ Y = F, Cl, Br, I, PhS, *p*-$MeC_6H_4SO_3$, $MeSO_3$, Ph_2PO_2, Ph_2PO_3, Ph_2PO_4	2,3-dimethylbutadiene; dimer	81CB2087, 81CB3505
TMS	Me	$(TMS)_2(Me_2YSi)CLi$ Y = $(PhO)_2PO_2$	TMS-OMe, TMS-Cl, TMSN=NTMS, RN_3, N_2O, Ph_2CO, PhCN, Ph_2C = NTMS, butadiene, isoprene, 2,3-dimethyl-butadiene and isobutene	81CB3518
TMS	Ph	$(TMS)_3CSiPh_2F$	rearrangement products	80JOM(186)309
$MeBu^t_2Si$	Me	$(MeBu^t_2Si)(TMS)(Me_2FSi)CLi$	1,3-butadiene	86CB1455
TMS	Me	Adduct of silaethene with Me_3N	Bu^tN_3, Ph_2CNTMS, acetone, isobutene	86JOM(315)9
TMS	Me	$(TMS)_2(Me_2YSi)CLi$ Y = F, Br, Ph_2PO_4	benzophenone	88ZN(B)1468
TMS	Me	$(TMS)_2(Me_2YSi)CLi$ Y = F, Br	Ph_2C=NTMS	87ZN(B)1055, 87ZN(B)1062
TMS	Me	adduct of silaethene with Ph_2CNTMS	organic dienes and trans-piperylene	86CB3498, 86CB1605, 91CB1981
TMS	Ph	$(TMS)_2(Ph_2YSi)CLi$ Y = H, Hal, OMe, OPh, Bu, Ph	RLi, Ph_2CNTMS, 2,3-dimethylbutadiene	89CB409

stable at ambient temperature and decomposes slowly at 60 °C. The ^1H and ^{13}C NMR spectra of the silaethene (**140**) indicate a rapid intramolecular methyl exchange (Scheme 64) ⟨85AG(E)229, 86CB1467⟩. X-ray-structure determinations are now available for free silaethene (**140**) ⟨85AG(E)229, 87OM32⟩, its donor adducts with THF ⟨84JOM(271)381⟩, and fluoride anion (as its $Li(12\text{-crown-}4)_2]^+$ salt) ⟨87OM35⟩.

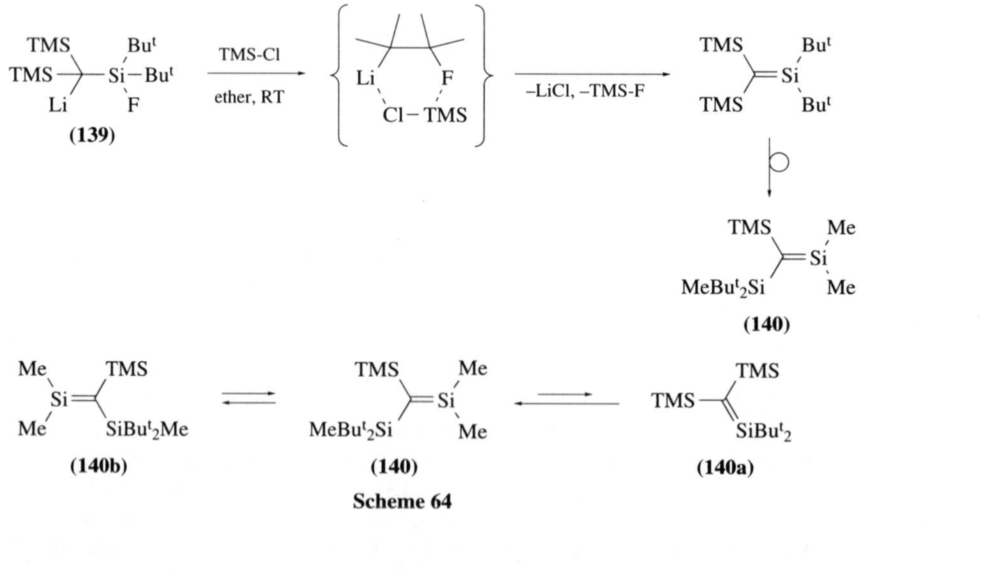

Scheme 64

6.22.2.1.2 C,C-*Diheterosubstituted germaethenes*, $X_2C{=}GeY_2$

The formation of transient germaethene (**141**) was postulated in the reaction of chlorodiphenyl[tris(trimethylsilyl)methyl]germane with cesium fluoride ⟨83IZV959⟩. It can be speculated that the precursor of (**142**) is the germaethene (**141**), which loses a TMS group to give a second unstable germaethene followed by cyclodimerization to give (**142**) (Scheme 65).

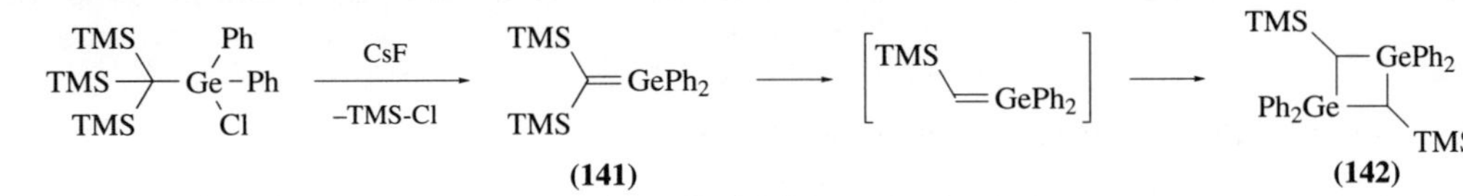

Scheme 65

A more established process for the generation of germaethenes is based on the formation of the C═Ge double bond by the salt elimination method (Scheme 66) ⟨86CB2966⟩. For example, thermal decomposition of germyldisilylmethanes (**143**) in diethyl ether at −110 °C to 100 °C (depending on X) leads to transient germaethene (**144**). The latter has been identified by its chemical reactivity: both by insertion into the O—H bond of alcohols ⟨86CB2980⟩ and by ene reactions ⟨86CB2966, 87CB1203⟩. Various cycloaddition reactions have also been observed: [2 + 2] cycloadditions are obtained with the C═C bond of CH_2═CHOMe and the C═O bond of ketones ⟨86CB2980⟩, [2 + 3] cycloadducts are obtained with azides and N_2O_5 ⟨87CB1203⟩, and [2 + 4] cycloadducts with dienes ⟨86CB2980, 87CB1203⟩. The unwanted side reactions can be reduced by adding a sufficient excess of trapping agent, by the slow generation of the germaethene, and by increasing the reaction temperature. If no trapping agents are added during the elimination of LiX from (**143**) the formation of 'head to tail dimer' is observed. Under special conditions (high temperatures or reactive substituents) germaethenes also stabilize themselves by rearrangement ⟨84JOM(273)141⟩. By comparison with the silaethene $(TMS)_2C$═$SiMe_2$ the Lewis acidity of germaethene (**144**) is weaker and its double bond is less polar ⟨86CB2980⟩.

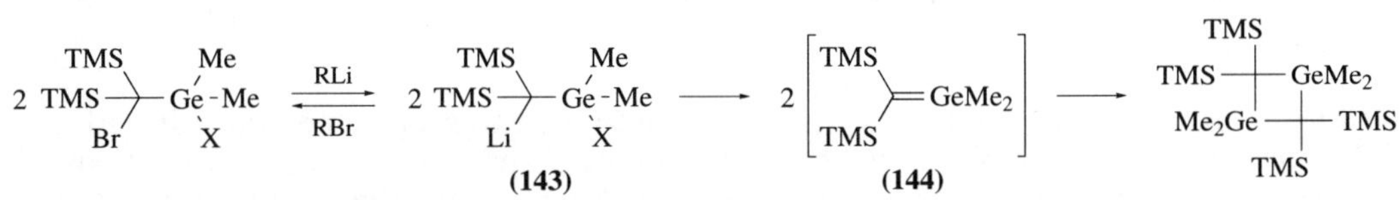

X = F, Br, OMe, OPh, OC_6F_5, SPh, Ph_2PO_4, Ph_2PO_2

Scheme 66

So far, evidence for a stable *C*,*C*-diheterosubstituted germaethene is scarce. The only isolated compounds with three heteroatoms attached to sp^2-hybridized carbon are germaethenes (**148**) and (**149**) obtained by the reaction between the electrophilic cryptocarbene (**145**) and the stable germylenes (**146**) and (**147**) (Scheme 67) ⟨87AG(E)798, 87PAC1011⟩.

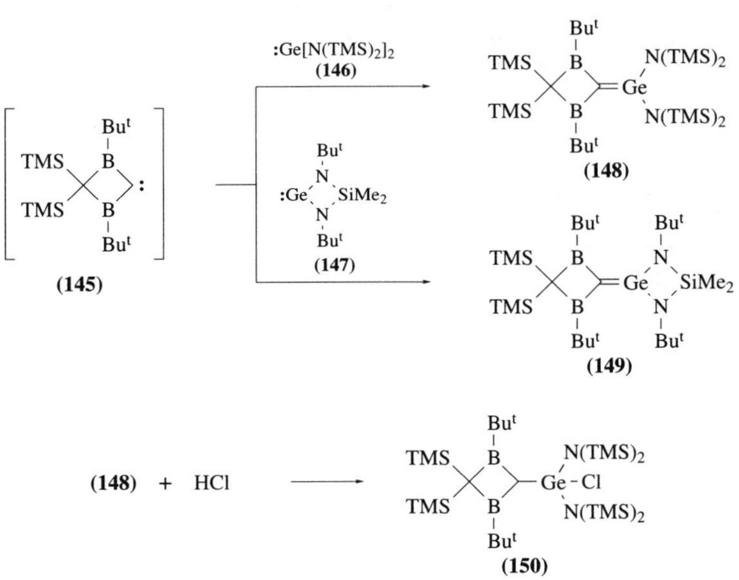

Scheme 67

The yellow crystals (**148**) and (**149**) very slowly decolorize on exposure to air and melt without decomposition. The presence of the C=Ge double bond in (**148**) has been confirmed by x-ray diffraction. Germaethene (**148**) adds HCl with quantitative formation of the 1,3-diboretane (**150**).

6.22.2.2 Functions Incorporating a Doubly Bonded Boron

Like the silaethenes and the germaethenes, compounds with a C=B double bond are stable only when they are sterically shielded by large substituents. In addition, electronic stabilization of the electron-deficient boron center is necessary either through π–π delocalization (classical methyleneboranes (**151**), or borataallenes (**152**) or through σ–π interaction of the neighboring C—X bonds with the dicoordinate boron atom (nonclassical methyleneboranes with the three-center, two-electron 3c-2e bonds, (**153**)) (Scheme 68) ⟨85MI 622-01, 87MI 622-01, 93AG(E)985⟩.

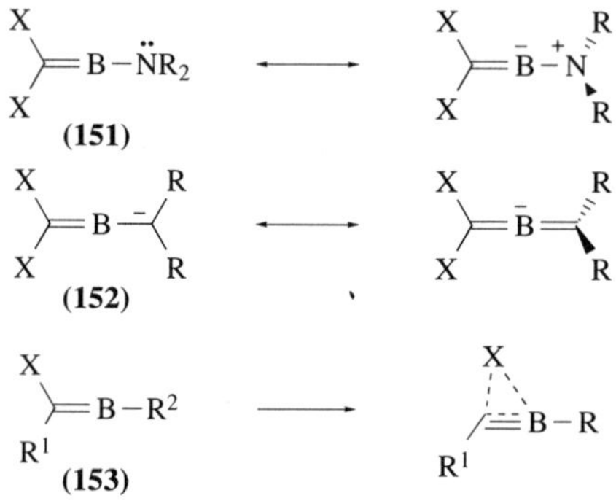

Scheme 68

A few standard and many unique reactions have been used to prepare compounds with a C=B bond. For reason of space, emphasis will be placed on the preparatively important methods for the synthesis of *C,C*-diheterosubstituted methyleneboranes. Some *C*-monoheterosubstituted derivatives are described because their reactions are useful for the preparation of functionalized methyleneboranes. There are also a number of specialized methods for the formation of a C=B bond which, however, do not seem to have general applicability. For further details a comprehensive review on this subject should be consulted ⟨93AG(E)985⟩. It should also be noted that the reactivity of methyleneboranes resembles that of vinyl cations and also that of silaethenes. The tricoordinate silicon atom is related to the dicoordinate boron atom through the diagonal relationship in the periodic system of elements ⟨90AG(E)401, 90CB747⟩.

6.22.2.2.1 *Methyleneboranes, $X_2C{=}B{-}Y$*

(i) 1,2-Elimination reactions

C,C-Disilyl-substituted methyleneboranes can be synthesized by the elimination of fluorotrimethylsilane or methoxytrimethylsilane from tris(trimethylsilyl)methylboranes at temperature between 470 °C and 560 °C (Equation (29)). This method has proved successful for the preparation not only of the amino(methylene)boranes (**156**) ⟨87CB1069⟩ and (**157**) ⟨89CB595⟩ but also the methyleneboranes (**154**) and (**155**) with methyl and *t*-butyl groups on the boron atoms ⟨89CB1057⟩. Compounds (**155**)–(**157**) are remarkably stable, whereas methyleneborane (**154**) rapidly dimerizes to the corresponding 1,3-diboretane. X-ray structural data is available for the compounds (**155**) ⟨89CB1057⟩ and (**156**) ⟨87CB1069⟩.

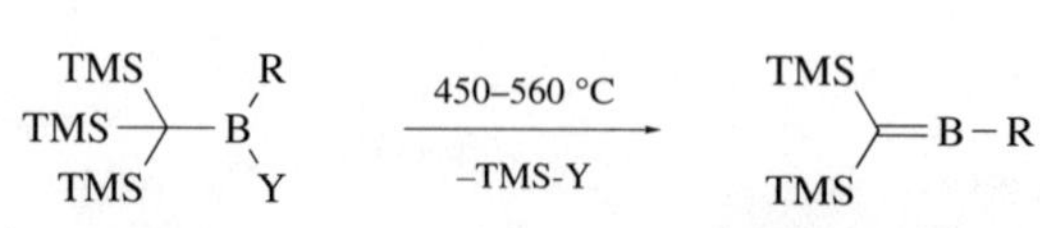

(29)

Y = F, MeO

(**154**) R = Me, 73%
(**155**) R = Bu^t, 54%
(**156**) R = Pr$^i_2$N, 67%
(**157**) R = 2,2,6,6-tetramethylpiperidino

Attempts to prepare the halo(methylene)boranes (**158**) via elimination of TMSOMe from $(TMS)_3CB(OMe)Hal$ resulted in the rearrangement products (**160**) (Scheme 69) ⟨90CB747⟩. The authors supposed that the anticipated compounds (**158**) underwent a double [1,3] shift of the Me and Hal groups, via the intermediates (**159**). The boranes (**160a,b**) cyclodimerize at the B=C bond at 25°C and 69°C, respectively. Benzophenone reacts easily with (**160**) to produce the cyclic 1,2-oxaboretanes ⟨90CB747⟩.

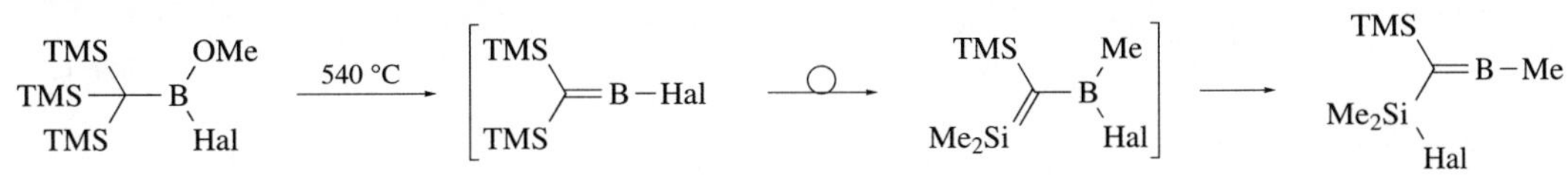

(158) **(159)** **(160a)** Hal = Cl **(160b)** Hal = Br

Scheme 69

(ii) Cycloreversion reactions

The thermal cycloreversion of 1-bora-2,4-cyclohexadienes ⟨85AG(E)1065⟩, 1,3-diboretanes ⟨89AG(E)784⟩ and 1,2-dihydroboretes ⟨90AG(E)401⟩ is of considerable potential for the generation of methyleneboranes. For example, upon melting (148°C and 177°C, respectively) 1,3-diboretanes (**161a,b**) furnish the readily volatile methyleneboranes (**162a,b**), which can be separated by condensation (yield ca. 35%) from mixtures of predominantly higher boiling products (Equation (30)) ⟨89AG(E)784⟩. The bis(stannyl)methyleneborane (**163b**) has been detected by multinuclear NMR spectroscopy ⟨90AG(E)401⟩.

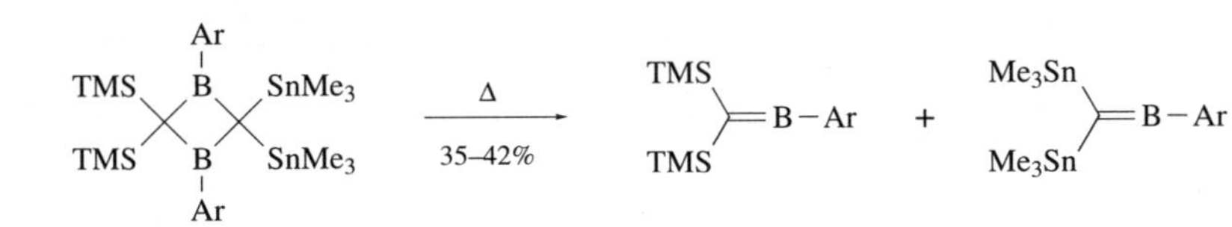

(30)

(161a) Ar = 2, 4, 6-$Me_3C_6H_2$
(161b) Ar = 2, 3, 5, 6-Me_4C_6H **(162a, b)** **(163a, b)**

The thermal cycloreversion of 1,2-dihydroborete (**164**) takes place at 120°C (reverse reaction to the formation of (**164**)). Distillation into CD_2Cl_2 cooled to −80°C afforded a solution of (**165**), which was investigated spectroscopically and trapped in a subsequent reaction step by 4-*t*-butylpyridine (Scheme 70) ⟨90AG(E)401⟩.

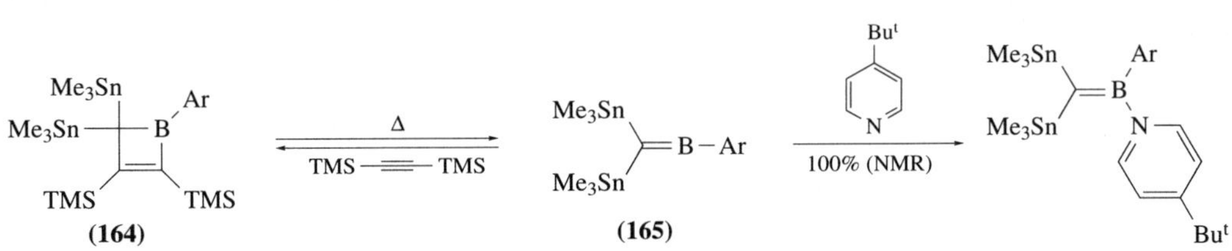

(164) **(165)**

Ar = 2,3,5,6-Me_4C_6H

Scheme 70

(iii) Miscellaneous reactions

Berndt and co-workers obtained the methyleneboranes (**168**) from an attempt to trap an intermediate in the reductive dimerization of diboranes (**166**) by addition of bis(trimethylsilyl)acetylene ⟨82JOM(234)C17, 84ZN(B)1042, 90AG(E)398⟩. They found that the diboranes (**166**) themselves reacted with the alkyne, and 1,1-diborylalkenes (**167**) could be reduced by a sodium/potassium alloy

⟨88AG(E)961⟩, magnesium under ultrasonication ⟨88ZN(B)801⟩ or Bogdanovic'-magnesium in diethyl ether ⟨90AG(E)398⟩ to yield boriranylideneboranes (**168**) (Scheme 71).

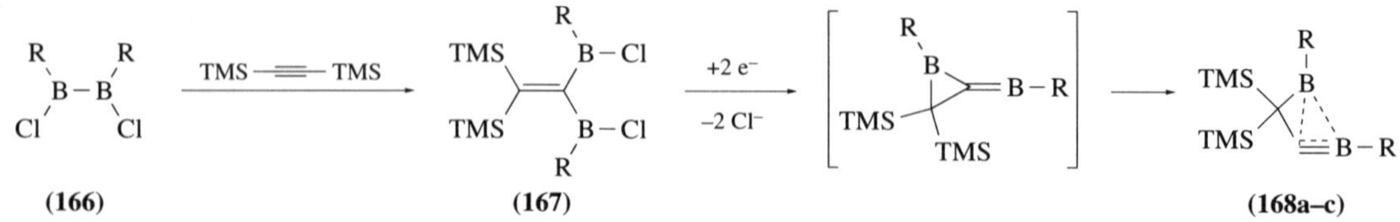

R = 2,3,5,6-Me_4C_6H (**a**), 2,4,6-$Me_3C_6H_2$ (**b**), Bu^t (**c**)

Scheme 71

Reaction of (**168a**) with mesityl(trimethylsilyl)acetylene results in the formation of [boryl (silyl)methylene]borane (**169**). Similarly, treatment of (**168a**) with bis(trimethylstannyl)acetylene affords [boryl(stannyl)methylene]borane (**170**) (Scheme 72) ⟨93AG(E)985⟩.

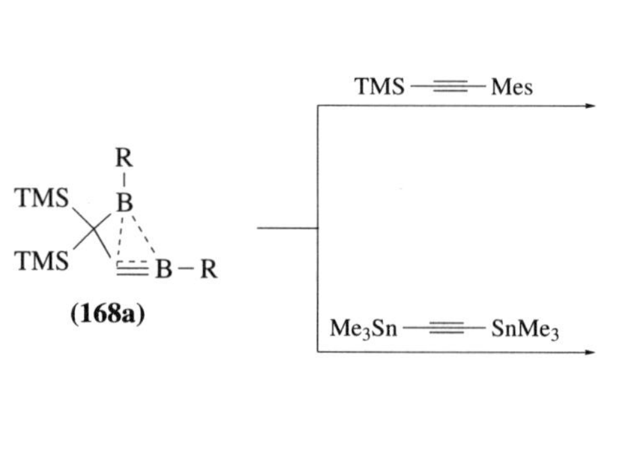

Scheme 72

The C—C bond of the three-membered ring in the compounds (**168**) is reductively cleaved by lithium in ether to give the borataalkynes (**171**) ⟨88AG(E)961⟩. These compounds can also be obtained by reaction of the 1,1-diborylalkenes (**167**) with an excess of lithium in diethyl ether (Scheme 73) ⟨89AG(E)781⟩. The interaction of the borylborataalkynes with electrophiles has provided a route to several new methyleneboranes. Thus, the reaction of (**171**) with Me_3SnCl gave the [boryl(stannyl)methylene]borane (**172**) ⟨90ZN(B)290⟩. The action of aryldifluoroboranes, RBF_2, on (**171**) resulted in formation of the stable (diborylmethylene)boranes (**173**) ⟨89AG(E)781⟩. In the same fashion the borataalkynes (**171**) react with 1,1-diaryl-2,2-difluorodiborane, Dur_2BBF_2, to form the orange-red *B*-boryl(1,3-diboretane-2-ylidene)boranes (**174**) ⟨91AG(E)594⟩.

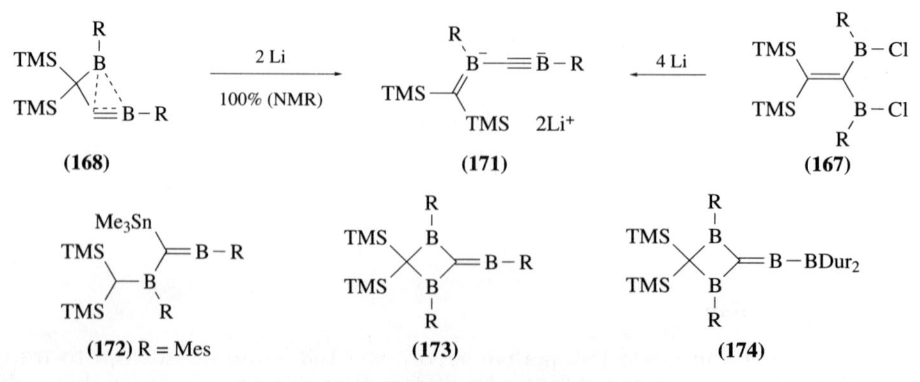

R = 2,4,6-$Me_3C_6H_2$ (Mes), 2,3,5,6-Me_4C_6H (Dur)

Scheme 73

6.22.2.2.2 2-Borataallenes, [X$_2$C=B=CY$_2$]$^-$

Among the methods which can be used to prepare 2-borataallenes the most important one involves the thermal isomerization of *C*-borylboriranides ⟨92AG(E)1238⟩. 2-Borataallene (**177**) is formed together with the 1,3-diboretanide (**176**) by heating (110 °C, 3 h) a toluene solution of the boriranide (**175**). In this reaction, the C—C bond of the three-membered ring is cleaved, and for each product a different duryl group migrates from a boron atom to the carbon atom located between the two boron atoms (Scheme 74) ⟨93AG(E)985⟩.

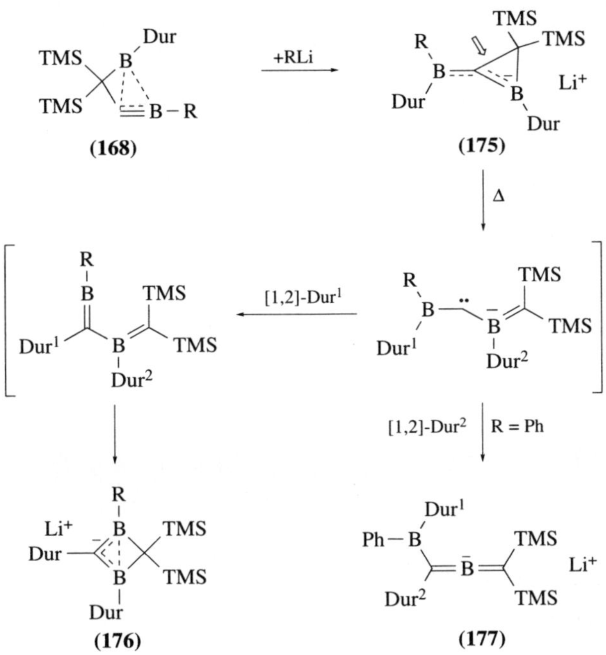

Dur = 2,3,5,6-Me$_4$C$_6$H

Scheme 74

The 2-borataallenes (**180**) are accessible through thermal isomerization of 2-boryl-1,3-diboretanides (**178**) (Scheme 75). Presumably, this reaction initially involves a ring-opening to give (**179**), a process typical of 1,3-diboretanes. Then, one aryl group undergoes a [1,3] migration from a tri- to the dicoordinate boron atom. Compound (**178**) is accessible from the methyleneborane (**173**) and *t*-butyllithium (20%) ⟨90AG(E)1030⟩.

The high reactivity of the boron–carbon double bond in the methyleneboranes (**162**) has also

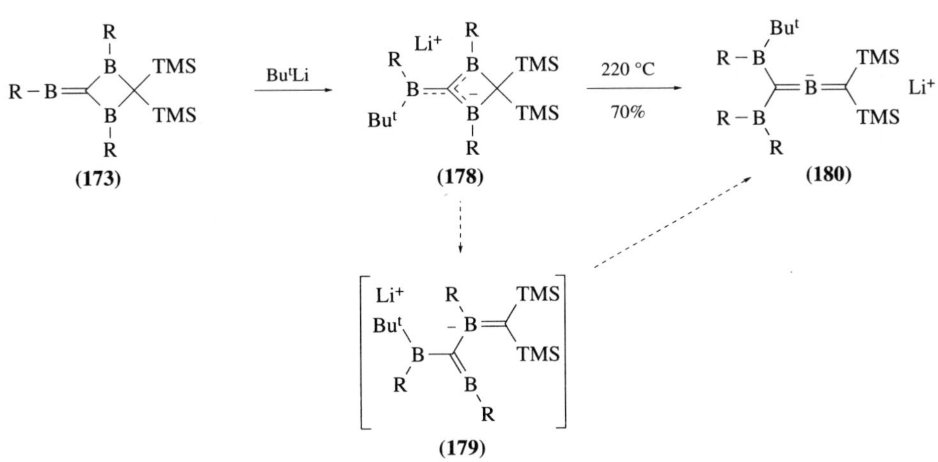

R = 2,3,5,6-tetramethylphenyl

Scheme 75

been exploited in the synthesis of 2-borataallenes and 2,3-diboratabutadienes (Scheme 76). The methyleneborane (**162a**) reacts with lithium in toluene at 25°C to give the 2,3-diboratabutadiene (**182a**), the symmetrical dimer of its radical anion (**181a**). Under the same conditions, (**162b**) yields the 2-borataallene (**183b**) (the unsymmetrical dimer of the radical anion (**181b**)) as major product (56%) together with (**182b**) as minor product (16%) ⟨90AG(E)1030⟩.

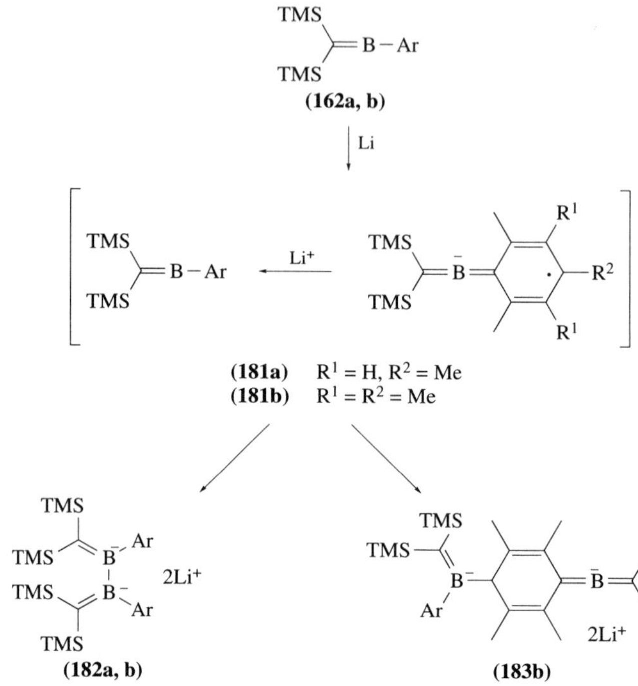

Ar = 2, 4, 6-$Me_3C_6H_2$ (**a**); 2, 3, 5, 6-Me_4C_6H (**b**)

Scheme 76

The reactions of 2-borataallenes with electrophiles have been used as a method for preparation of a few types of heterosubstituted methyleneboranes, including *C,C*-diboryl derivatives. For example, 1,1-diboryl-2-borataallenes (**180**) can be protonated with cyclopentadiene to give the (diborylmethylene)boranes. In all cases the electrophiles attack the carbon atom bonded to the two trimethylsilyl groups regioselectively, with the result that only (diborylmethylene)boranes and not the isomeric (disilylmethylene)boranes result ⟨93AG(E)985⟩.

6.22.3 FUNCTIONS INCORPORATING A DOUBLY BONDED METAL

6.22.3.1 Transition Metal Carbene Complexes

For the purpose of this survey, transition metal *C,C*-diheterosubstituted carbene complexes will be defined as the species of the formula $X^1X^2C{=}ML_n$ which formally contain a double bond between carbon and the transition metal. Systematic (IUPAC) nomenclature for these compounds uses the suffix 'ylidene', the ligand being regarded as neutral with respect to the metal oxidation state; thus the complex $[(OC)_5Cr{=}C(OMe)Cl]$ is called pentacarbonyl[chloro(methoxy)methylidene] chromium(0), but trivially is chloro(methoxy)carbenepentacarbonylchromium(0).

Complexes containing heterosubstituted carbene ligand were first prepared by Fischer in the 1960s ⟨64AG(E)580, 76AOC(14)1⟩. Since then this field has experienced explosive growth and continues to expand at a rapid rate ⟨72CRV545, 73CSR99, 80MI 622-01, 89CRV1703⟩. A large number of methods have been developed, providing access to a great variety of carbene complexes involving nearly all the transition metals. Various aspects of this chemistry have been reviewed extensively in recent years ⟨86AOC(25)121, 88CRV1293, 91SL381, 91CSR503⟩. Therefore, only most important synthetic examples will be selected here from the abundance of material available ⟨B-84MI 622-01, 89MI 622-03⟩.

6.22.3.1.1 *Dihalocarbene complexes, $Hal_2C{=}ML_n$*

The first transition metal dihalocarbene complex of the type $Hal_2C{=}ML_n$ was prepared in 1977 by reaction of tetraphenylporphyrinoiron(II), [(TPP)Fe^{II}], with CCl_4 in the presence of excess iron powder ⟨77CC648, 88CSR1121⟩. In 1978 the difluorocarbene complex of molybdenum, $[(\eta^5\text{-}C_5H_5)(CO)_3Mo{=}CF_2]^+$, was detected in solution by ^{19}F and ^{13}C NMR spectroscopy ⟨78JOM(153)67⟩. In 1980 a stable dichlorocarbene complex of osmium(I) was described ⟨80JA1206⟩ and since then a large number of dihalocarbene complexes has been prepared and thoroughly characterized ⟨83MI 622-01⟩. The review by Brothers and Roper ⟨88CRV1293⟩ provides a valuable survey of the literature up to 1987.

Several synthetic routes to $Hal_2C{=}ML_n$ species have been developed, but each one is appropriate for only a limited number of transition metal substrates. The most general method for the preparation of the dihalocarbene complexes starts from metal trihalomethyl derivatives, $Hal_3C{-}ML_n$.

(i) Synthesis from trihalomethyl complexes

The chemistry of trifluoromethyl derivatives of the transition metals has been the focus of much attention ⟨93AOC(35)211⟩. As a result, many transition metal trifluoromethyl species are known, and these offer the possibility of modification of the CF_3 ligand to form $Hal_2C{=}ML_n$ carbene complexes. In principle an electrophilic attack on coordinated CF_3 ligand using H^+, TMS^+, BF_3 or SbF_5 as fluoride anion abstracting agents lead to difluorocarbene complexes (Scheme 77). Thus, the ruthenium(II) complex (**184**), when treated with anhydrous HCl or TMS-Cl as fluoride abstracting reagents forms the difluorocarbene product (**185**) (Scheme 78) ⟨82JOM(234)C9⟩. The attack of SbF_5 on the CF_3 group in $CpMo(CF_3)(CO)_3$ generates the cationic difluorocarbene complex $[F_2C{=}Mo(Cp)(CO)_3]SbF_6$ ⟨78JOM(153)67⟩. Likewise, $[F_2C{=}Mn(CO)_5]BF_4$ is obtained from $Mn(CO)_5(CF_3)$ and BF_3 ⟨85OM1830⟩.

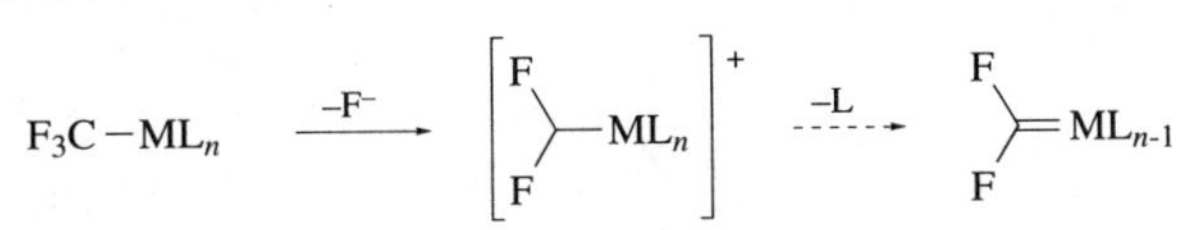

Scheme 77

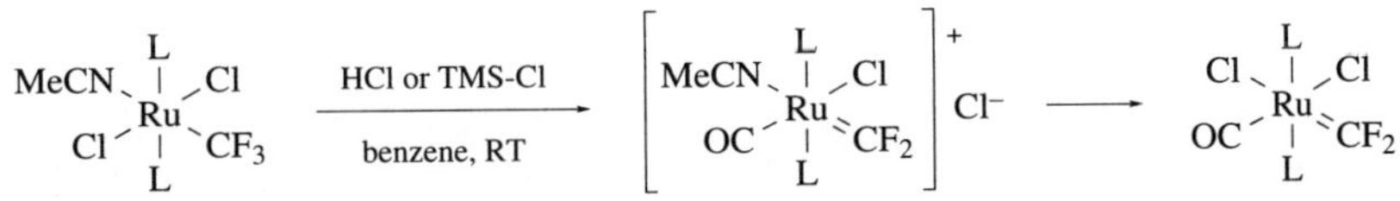

(**184**)
L = PPh3

(**185**)

Scheme 78

Reaction of trifluoromethyl complexes with $BHal_3$ (Hal = Cl, Br, I) converts F_3CML_n to Hal_3CML_n. The use of excess $BHal_3$ resulted in both halide exchange and stabilization of the cationic dihalocarbene complexes. For example, treatment of (**186**) with excess BCl_3 produced the stable dichlorocarbene complex (**187**) which was fully characterized, including an x-ray crystal structure study. An analogous reaction was successfully used in the preparation of dibromocarbene complex (**188**) (Equation (31)) ⟨84OM314, 85OM1830, 88CRV1293⟩. Trifluoromethyl species $[F_3C(Cp)Fe(CO)(PPh_3)]$ ⟨85OM1830⟩ and $[F_3C(Cl)Ru(CO)_2(PPh_3)_2]$ ⟨86AOC(25)121⟩ also react with BCl_3 to give dichlorocarbene complexes.

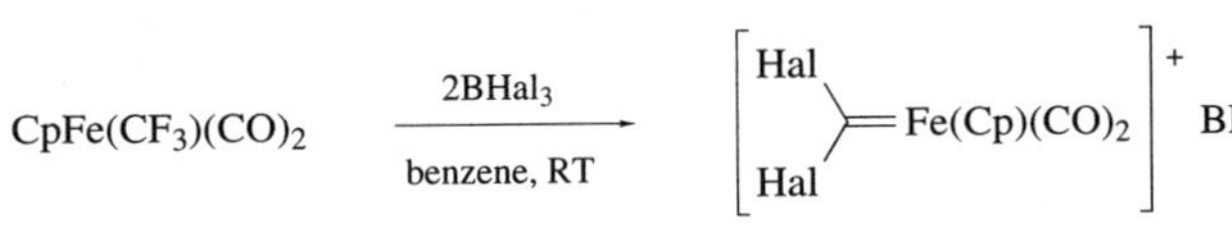

(31)

(**186**)

(**187**) Hal = Cl
(**188**) Hal = Br

(ii) From reactions with $Cd(CF_3)_2$ and $Hg(CCl_3)_2$

Roper and co-workers found that the highly reactive complex $[Cd(CF_3)_2{\cdot}GLYME]$ ⟨86IS52⟩ can be used for CF_2 transfer to a suitable metal substrate ⟨86AOC(25)121⟩. A good example of this method is the synthesis of difluorocarbene ruthenium and osmium complexes (**189**) (Equation (32)) ⟨86JOM(300)167, 83CC719⟩. The oxidative addition product $F_3CRu(CdCF_3)(CO)_2(PPh_3)_2$ is a likely intermediate in this reaction. Other zerovalent ruthenium and osmium complexes, $RuCl(NO)(PPh_3)_2$, $OsCl(NO)(PPh_3)_3$ and $Os(CO)(CS)(PPh_3)_3$, react similarly, giving $F_2C{=}MCl(NO)(PPh_3)_2$ (M = Ru, Os) and $F_2C{=}Os(CO)(CS)(PPh_3)_2$, respectively ⟨88CRV1293⟩.

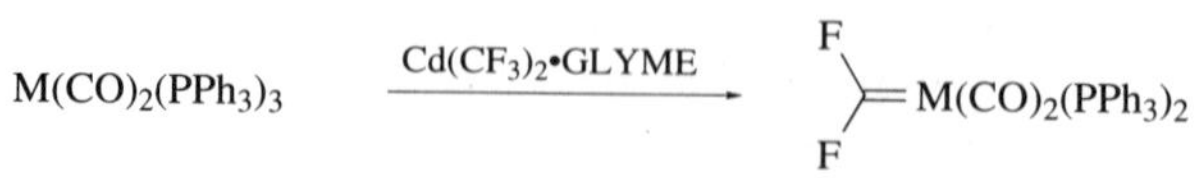

(32)

(**189**) M = Ru, Os

The cadmium reagent was also successfully used for CF_2 transfer to a d^8 iridium center. An example is the preparation of the difluorocarbene complex $F_2C{=}Ir(I)(CO)(PPh_3)_2$ from $IrI(CO)(PPh_3)_2$. At higher temperatures, reaction of $[Cd(CF_3)_2 \cdot glyme]$ with Vaska's complex or with $F_3CIr(CO)(PPh_3)_2$ results in transfer of a CF_2 group and also, for the former substrate, replacement of chloride with CF_3 (Equation (33)) ⟨88CRV1293⟩.

(33)

The mercury compound $Hg(CF_3)_2$ oxidatively adds across one Hg—C bond to $Ru(CO)_2(PPh_3)_2$ producing the stable complex $F_3CRu(HgCF_3)(CO)_2(PPh_3)_2$, containing both F_3C and F_3CHg ligands ⟨82JOM(234)C9⟩. However, ruthenium ⟨82JOM(233)C59⟩, osmium ⟨80JA1206⟩ and iridium ⟨82JOM(236)C7⟩ d^6 complexes react with $Hg(CCl_3)_2$ to give the dichlorocarbene species (Equations (34) and (35)); the probable explanation is that Cl_3C—M derivatives are formed as intermediates but then lose a phosphine ligand, $HCCl_3$ and mercury. The osmium and iridium complexes have been structurally characterized ⟨82JOM(236)C7⟩.

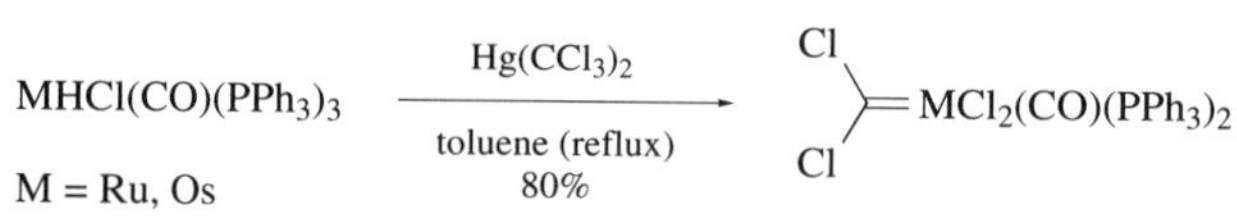

(34)

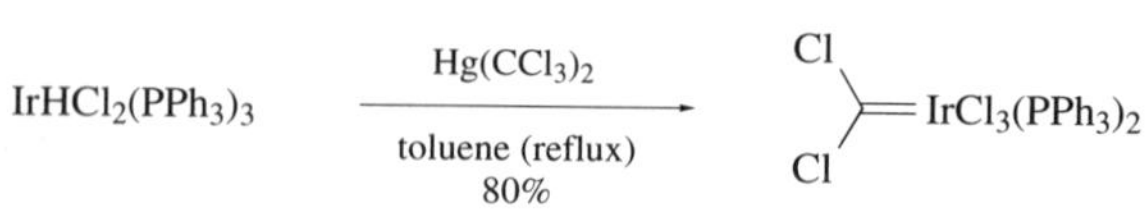

(35)

(iii) Modification of dichlorocarbene complexes

The displacement of chloride from a Cl_2C ligand by other nucleophiles can, in some cases, be used for the preparation of new dihalocarbene complexes. Thus, by analogy with the conversion of RCOBr to RCOF upon reaction with $[Cd(CF_3)_2 \cdot glyme]$ the reaction of the osmium complex (**190**) with the cadmium reagent includes exchange of a chloro for a fluoro substituent in a metal coordination sphere (Equation (36)) ⟨84JOM(269)C55⟩. The use of $AgClO_4$–MeCN to abstract a labile chloride from a carbene complex, with concomitant coordination of acetonitrile or CO, is a useful route to several ruthenium and osmium cationic complexes (Structures (**191**)–(**193**); Equation (37)) ⟨84JOM(269)C55, 88CRV1293⟩.

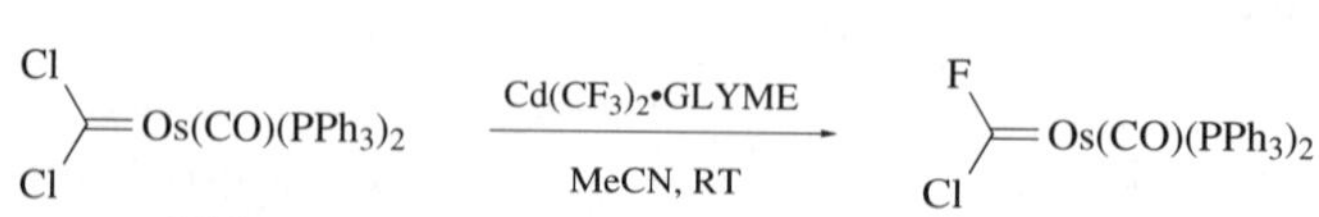

(36)

(**190**)

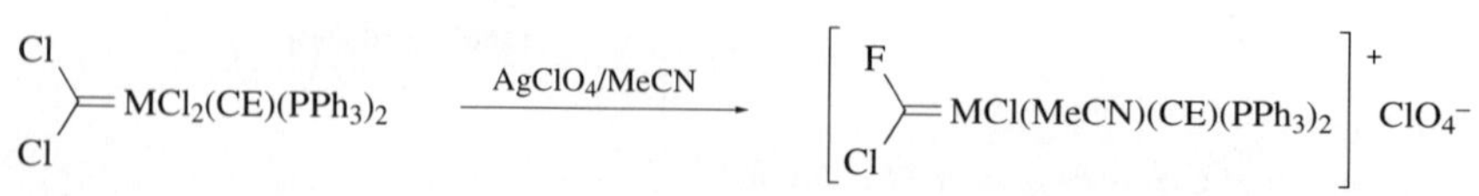

(37)

(**191**) M = Ru, E = O
(**192**) M = Os, E = O
(**193**) M = Os, E = S

6.22.3.1.2 *Oxygen-, sulfur- and nitrogen-substituted carbene complexes, RE(X)C═ML$_n$ (E = O or S) and R$_2$N(X)CML$_n$*

Since very authoritative reviews of the *O*-, *S*- and *N*-substituted transition metal carbene complexes exist ⟨80MI 622-01, B-84MI 622-01, 91SL381⟩ only a brief survey of some of the more general synthetic approaches will be provided here. Basically, three strategies are used for the preparation of the title compounds: (i) modification of a coordinated (noncarbene) carbon ligand, (ii) modification of a carbene or carbyne ligand coordinated to a metal, (iii) reactions of transition metal complexes with organic carbene precursors. The various types of synthesis have differed widely and it will be convenient to deal with each separately.

(i) From lithium acyl metallates

The route from lithium acyl metallates is the most useful and general approach to the preparation of oxygen-substituted carbene complexes from noncarbene precursors. In this method a carbonyl ligand is converted into an alkoxy- or aryloxycarbene ligand by successive addition of a nucleophile (**194**) and an electrophile (Scheme 79) ⟨76AOC(14)1, B-84MI 622-01⟩. Various metal carbonyls have been used as precursors to carbene complexes, including $W(CO)_6$, $Cr(CO)_6$, $Mo(CO)_6$, $Mn_2(CO)_{10}$, $Tc_2(CO)_{10}$, $Re_2(CO)_{10}$, $Fe(CO)_5$ and $Ni(CO)_4$. These compounds are arranged in order of decreasing stability of the carbene product ⟨80MI 622-01⟩.

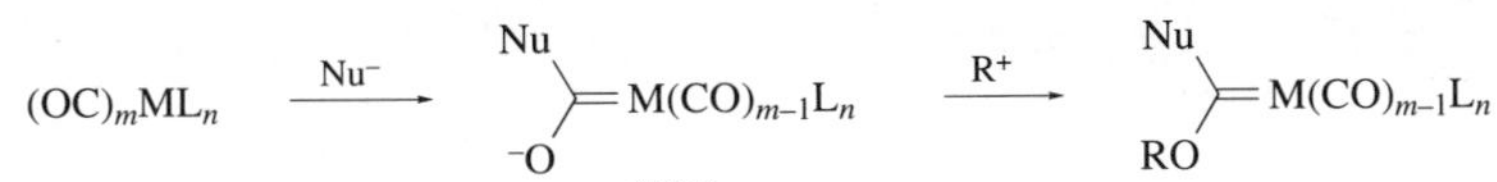

(194)

Scheme 79

Alkoxycarbene complexes are normally synthesized by the original Fischer procedure, involving the reaction of an organolithium reagent with a metal carbonyl, followed by alkylation of the resulting 'ate' complex. Alkylation can be performed with oxonium salts ⟨69CB1495⟩, methyl fluorosulfonate ⟨73SRI249⟩ or methyl triflate ⟨90TL2529⟩. More recently it was shown that the use of phase transfer conditions sufficiently activate the acyl metallate to allow alkylation by readily accessible alkyl halides. The procedure is preparatively useful for accessing chromium containing carbenes, but yields of the analogous molybdenum and tungsten containing species are only $<10\%$ ⟨93OM2806⟩. Chlorosilanes have been used to generate silyloxycarbene complexes ⟨77CB2574, 68JOM(12)P1⟩. A wide range of other heterosubstituted carbene complexes has been synthesized via the metal carbonyl route ⟨B-84MI 622-01, 89MI 622-03⟩. Selected examples are presented in Table 4.

(ii) Electrophilic addition to coordinated acyl, thioacyl and imidoyl ligands

Alkylation of anionic acyl metallate is the second step in the classic Fischer synthesis of carbene complexes. A related synthetic technique involves electrophilic addition to certain neutral acyl complexes ⟨74JOM(80)C35, 74JOM(81)C7, 75BCJ3691, 76JOM(112)227, 78JOM(159)73⟩. Selected reactions are presented here to illustrate the scope of this type of transformation.

The adducts of alkoxy(dialkylamino)carbenes with Hg^{2+} are synthesized by methylation of bis(carbamoyl)mercury compounds with trimethyl- or triethyloxonium fluoroborate (Equation (38)) ⟨67AG(E)560⟩. Thioacyl platinum complex reacts in a similar fashion (Equation (39)) ⟨75IC1513⟩. The conversion of several acyl complexes into *C*,*C*-dioxygen-substituted cyclic carbene ligands has been achieved by intramolecular alkylation reactions (Equations (40) and (41)) ⟨74JCS(D)351, 74JCS(D)1189, 76JOM(113)C45⟩.

$$\mathrm{Hg(CONR_2)_2} \xrightarrow[\mathrm{CH_2Cl_2,\ RT}]{\mathrm{Me_3O^+\,BF_4^-}} \{[\mathrm{MeO(R_2N)C]_2Hg}\}^{2+}\ \ 2\mathrm{BF_4^-} \qquad (38)$$

Table 4 Selected examples illustrating the synthesis of diheterosubstituted metal carbene complexes via acylmetallates.

Reaction	*Yield* (%)	*Ref.*
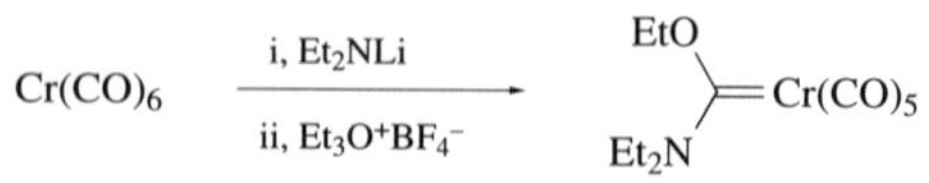	20	76JOM(118)C33 70AG(E)309
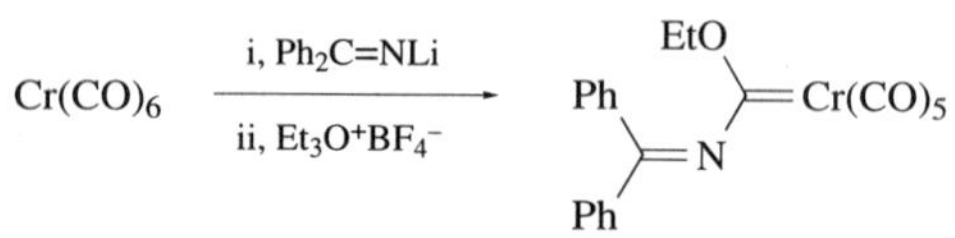	40	74JCS(D)1494
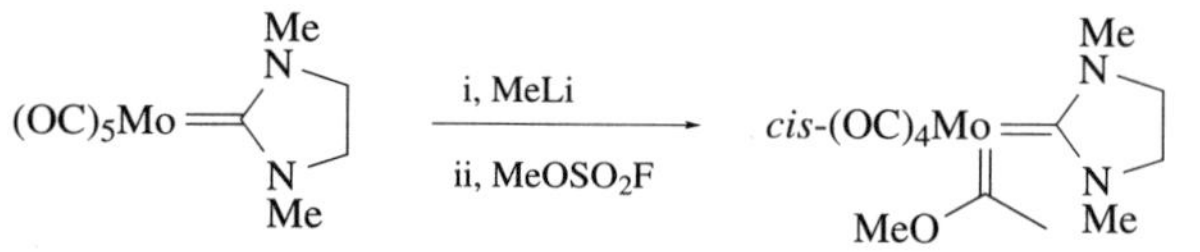	50	77JCS(D)1272
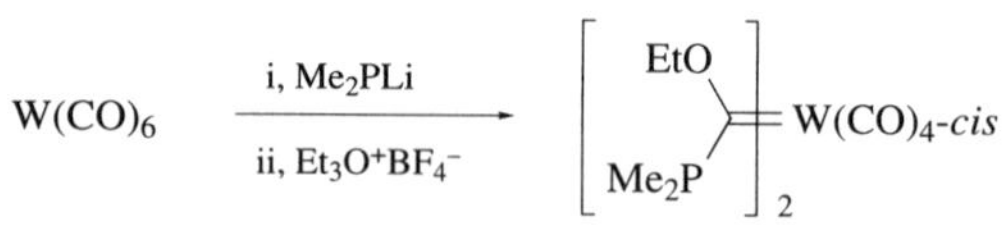	1.5	72CB2558
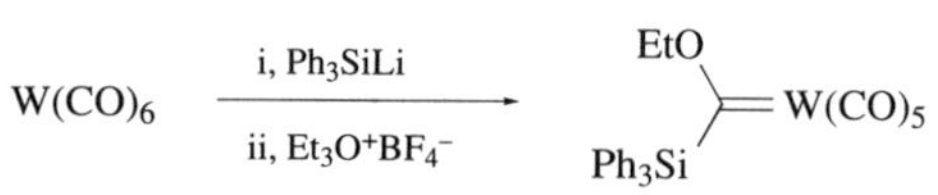	15.4	77CB3467 76JOM(113)C31
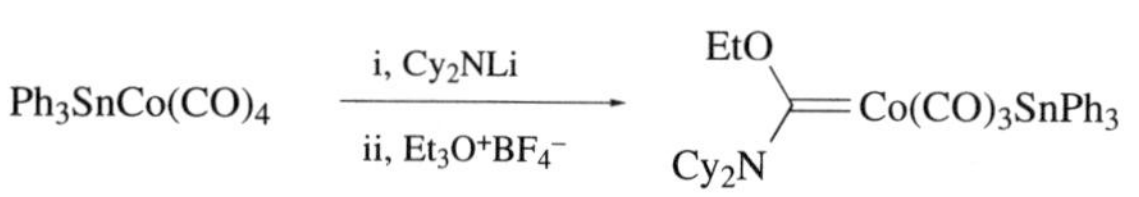	75	88JOM(355)437
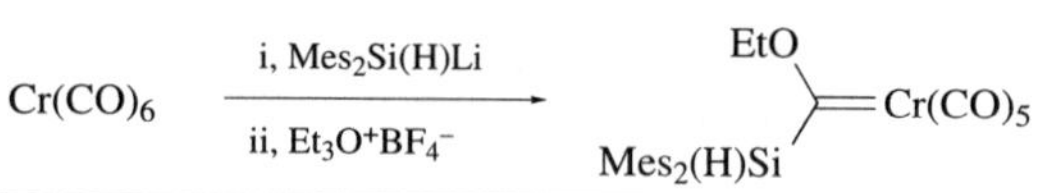	70	93JOM(459)55

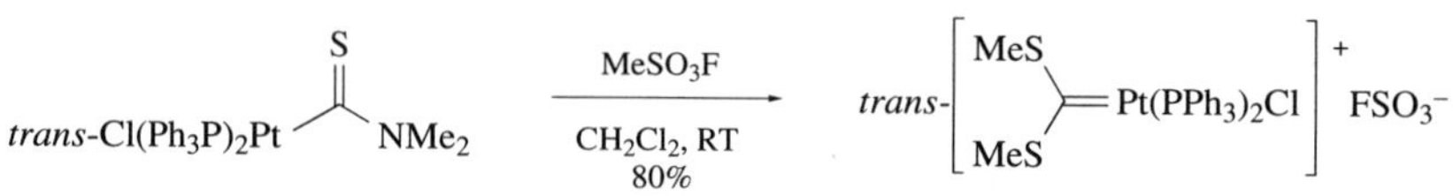

(39)

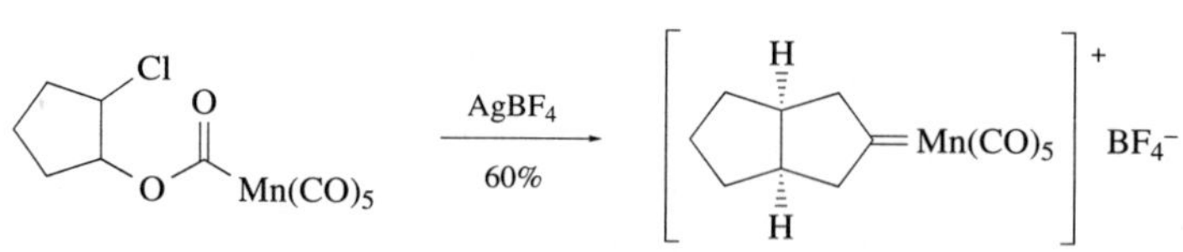

(40)

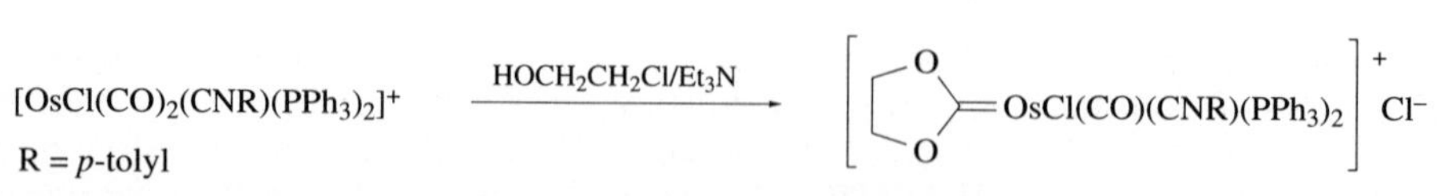

(41)

Some dithiocarbene complexes can be obtained via η^2-CS_2 adducts of transition metals. An example is alkylation of the osmium complex (**195**) with MeI (Equation (42)) ⟨75JOM(90)C34⟩. Cyclic dithio- and disilenocarbene ligands are formed when dihaloalkanes are used as the electrophilic reagents in these reactions (Equation (43)) ⟨76JOM(107)C37⟩.

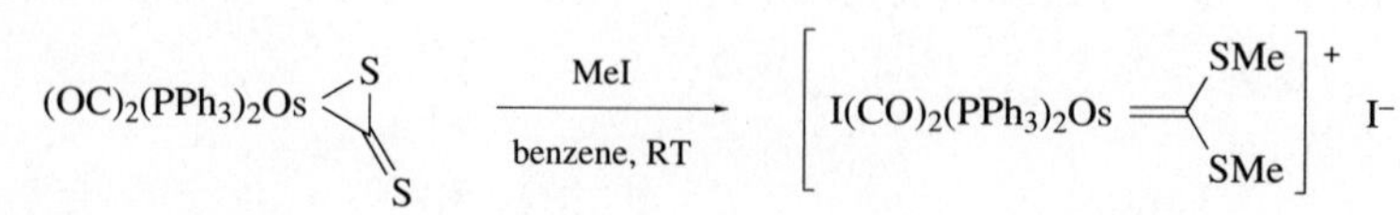

(42)

(**195**)

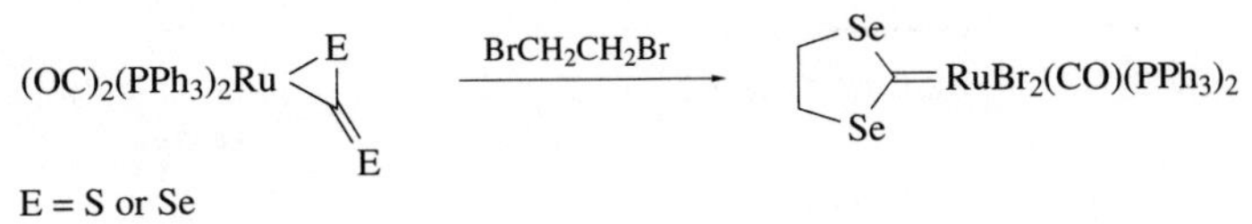

(43)

(iii) Nucleophilic addition to coordinated isocyanides

Coordinated isocyanides react with nucleophiles to yield carbene complexes. Platinum(II) and palladium(II) species have been most extensively investigated, and the range of nucleophilic reagents employed in these reactions has included alcohols, amines and thiols. The typical examples are shown in Equations (44) and (45) ⟨69CC1322, 71JCS(A)21, 74JOM(81)C7⟩. The method has been extended to gold(I) ⟨73G373⟩, mercury(II) ⟨69JOM(16)275, 70JOM(25)255⟩, iron(II) ⟨69CC423, 70JOM(24)205, 72JA417⟩, nickel(II) ⟨74JCS(D)357⟩, ruthenium(II) ⟨73JA4769, 74IC921⟩ and osmium(II) ⟨75JOM(91)C61⟩ compounds. Examples from more recent literature include preparation of chromium carbene complexes by reaction of coordinated trichloromethylisocyanide ⟨88AG(E)1344, 89CB1907⟩ and trifluoromethylisocyanide ⟨90CB751⟩ with secondary amines and synthesis of neutral five-membered cyclic diamino-, aminothio-, and aminooxycarbene derivatives of palladium(II) and platinum(II) starting from the isocyanide complexes *cis*-$Cl_2(PPh_3)M(CNR)$ ⟨88IC2809⟩.

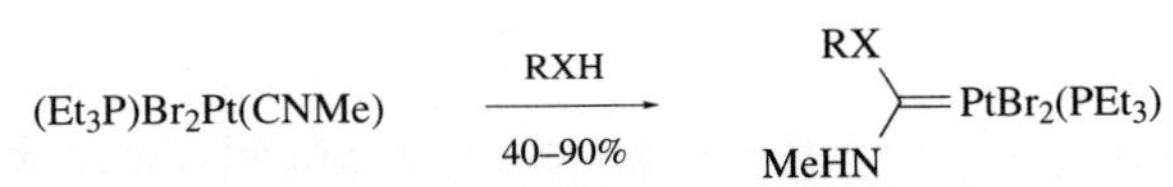

(44)

RXH = MeOH, Pr^{n}OH, $PhNH_2$, Bu^sNH_2

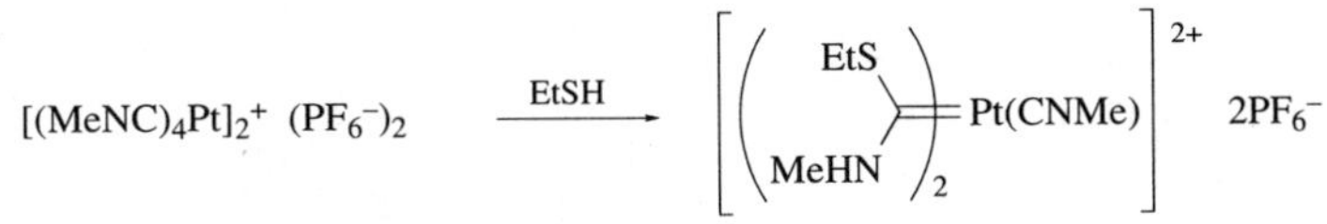

(45)

(iv) From organic carbene precursors

A wide range of *C,C*-dinitrogen substituted carbene complexes has been synthesized via scission of electron-rich alkenes ⟨80MI 622-01⟩. The early works dealt with platinum complexes where the carbene could be introduced via halogen-bridge cleavage or ligand displacement ⟨72CRV545, 75JOM(100)139⟩. The first carbene complex **(196)** prepared in this way is shown in Equation (46) ⟨71CC400⟩. This reaction, which is normally carried out in refluxing xylene, was generalized. Other carbene complexes have been obtained by this procedure including the complexes of chromium(0) ⟨77JCS(D)2160⟩, molybdenum(0) ⟨77JCS(D)1272⟩, tungsten(0) ⟨77JCS(D)1283⟩, ruthenium(I) ⟨72CC927, 77MI 622-01⟩, iridium(I) ⟨74JCS(D)1827⟩, osmium(II) ⟨78JCS(D)826, 78JCS(D)837⟩, and ruthenium(II) ⟨76CC644, 77JCS(D)2172, 78JCS(D)837, 79JCS(D)1929⟩.

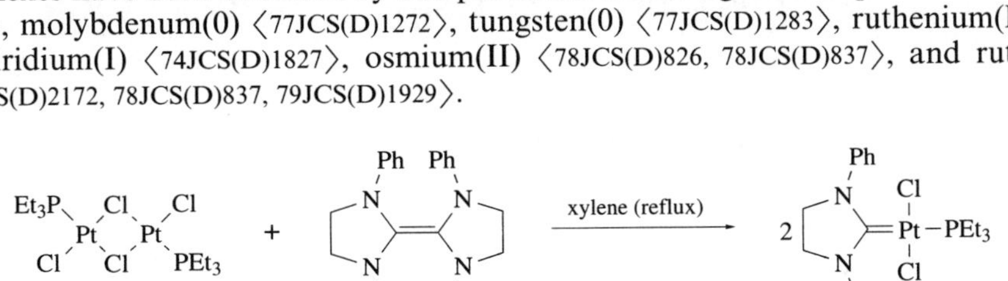

(196)

(46)

A related synthetic technique involves reactions of transition metal complexes with electron rich *gem*-dichlorides in which the C—Cl bonds have appreciable ionic character ⟨80MI 622-01⟩. Two examples are shown in Equations (47) ⟨72CC851⟩ and (48) ⟨75CC929⟩. The procedure is significant in that it allows the preparation of carbene complexes in unusually high oxidation states ⟨74JCS(D)1591, 78JCS(D)348⟩. Imidazolium, oxazolium and thiazolium salts have been used to obtain certain carbene complexes ⟨75JCS(D)939, 86AOC(25)121⟩. Although the products of these reactions are often difficult to prepare by other methods, this technique is of limited applicability.

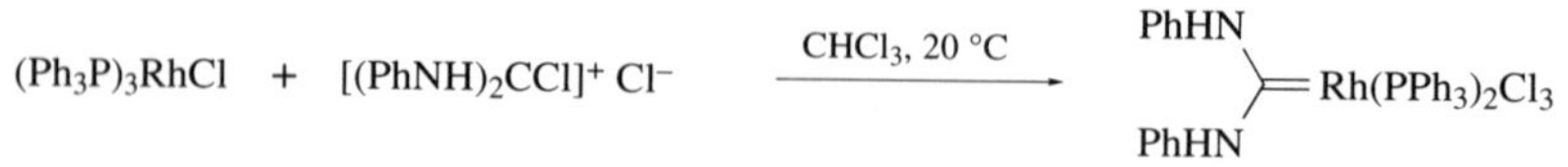
(47)

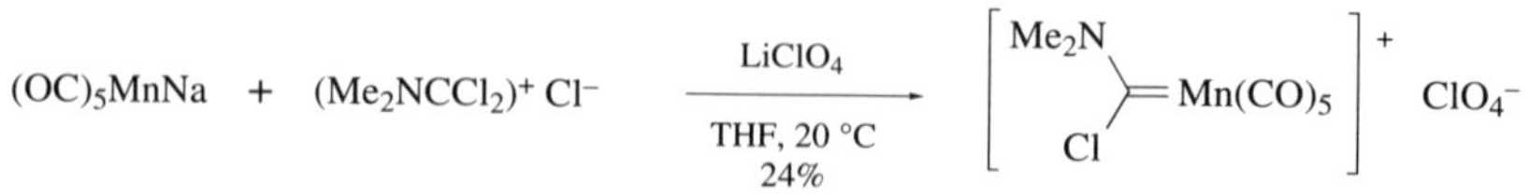
(48)

The homoleptic bis(carbene) adducts of silver(I) and copper(I) are available directly from the reaction of the stable nucleophilic carbene 1,3-dimesitylimidazol-2-ylidene and the corresponding metal triflate ⟨93OM3405⟩.

(v) Syntheses involving functionalization of the carbene ligand

The nucleophilic substitution reactions at the carbene carbon have found wide application in the modification of carbene ligands ⟨80MI 622-01, B-84MI 622-01⟩. The alkoxycarbene complexes, which are obtainable directly from the carbonylmetal compounds, can be used as organometallic ester equivalents ⟨91SL381⟩. Thiols, primary and secondary amines, and organolithium compounds react by substituting for the alkoxy group, leading to heteroatom-stabilized carbene complexes. Reactions normally afford excellent yields, and products are easily separated. An example is the preparation of the carbene complex **(197)** by aminolysis of an alkoxycarbene substrate (Equation (49)) ⟨76JOM(113)C31, 77CB3467⟩.

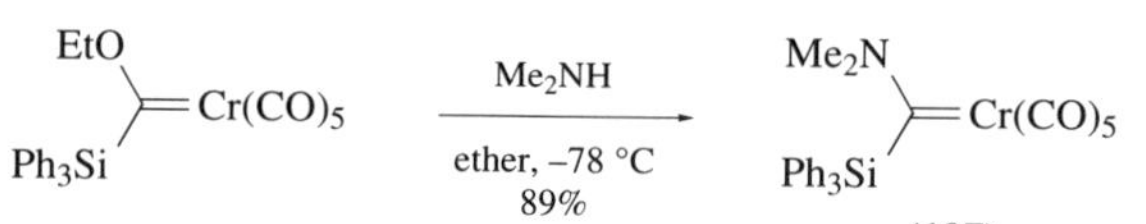
(197)
(49)

Halide displacement from the dihalocarbene ligands by nitrogen, oxygen and sulfur based nucleophiles frequently leads to the formation of new heteroatom-substituted carbene complexes ⟨80MI 622-01, 88CRV1293⟩. Equations (50)–(52) illustrate the scope of this method. Primary alkyl- and arylamines react with dihalocarbene complexes to give products containing isocyanide ligands ⟨82JOM(233)C59, 82JOM(234)C9, 84OM314, 88JOM(338)393⟩. The mechanism of isocyanide formation may involve successive HCl loss from a Cl(RNH)C=ML_n intermediate. The reaction of the ruthenium complex [$Cl_2C{=}RuCl_2(CO)(PPh_3)_2$] with ethanolamine includes the formation of the isocyanide complex and its subsequent cyclization into an aminoalkoxycarbene complex ⟨82JOM(233)C59⟩.

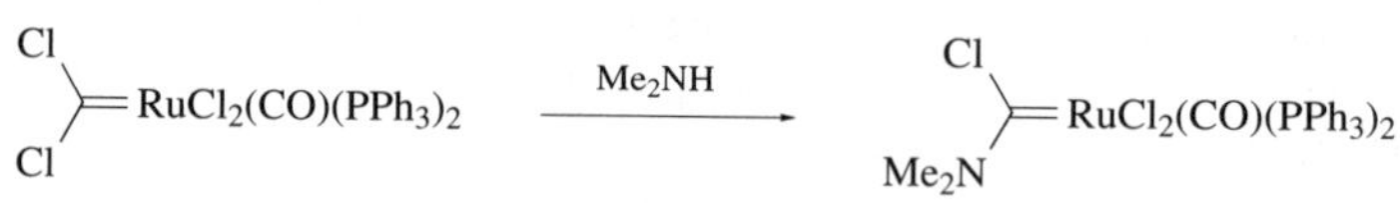
(50)

(51)

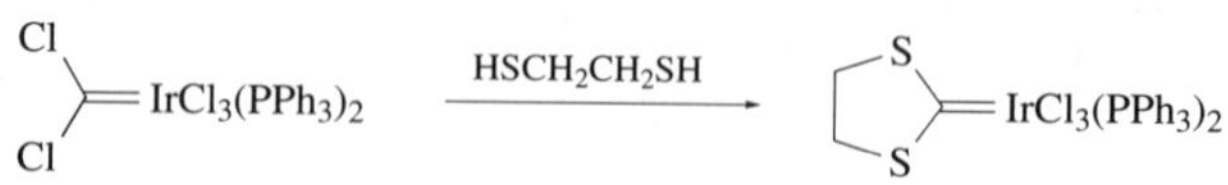
(52)

(vi) Nucleophilic addition to carbyne complexes

Carbyne complexes can be converted into carbene complexes by the nucleophilic addition reactions ⟨77JOM(129)197⟩. In some cases this route leads to transition metal carbenes that are inaccessible by any of the other synthetic methods. Two examples, one resulting from nucleophilic addition to the carbyne complexes (**198**) fluoride anion ⟨76AG(E)616⟩ and the other resulting from addition of lithium thiocyanate ⟨77JOM(128)C49, 78CB3542⟩, are shown in Scheme 80. Many related reactions have been described ⟨91AOC(32)227⟩.

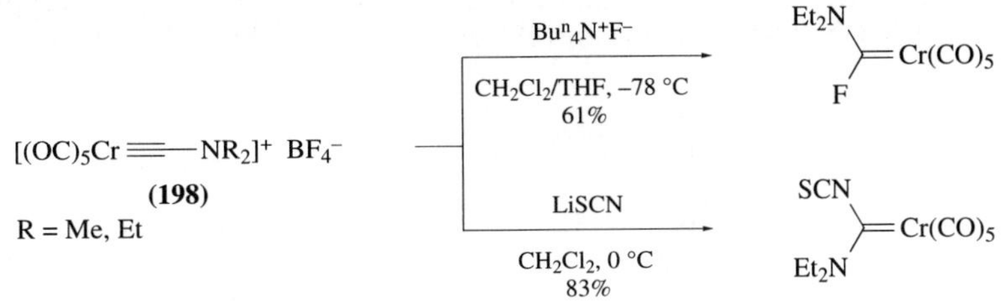

Scheme 80

(vii) Miscellaneous

Open chain Fischer type carbene complexes $(CO)_5Cr{=}C(OMe)CH_2R^1$ react easily with *N,N*-dimethylformamide–dialkyl acetal (dmf–daa). When $R^1 = H$, the carbene complex $(CO)_5Cr{=}C(OMe)CH{=}CHNMe_2$ is formed in high yield; however, when R^1 = alkyl, the reaction products are dialkoxy complexes $(CO)_5Cr{=}C(OMe)OR^2$, in which the OR^2 group comes from the dmf–daa ⟨93OM2994⟩. The latter reaction represents a new and general method for the synthesis of chromium dialkoxycarbene complexes.

Various neutral and cationic gold(I) amino(thio)carbene complexes were obtained in satisfactory yields on addition of lithiated 4-methylthiazole to gold(I) chloride compounds and subsequent protonation or alkylation of the products formed ⟨90CC1722⟩. Similarly, the first monocarbene complex of copper(I) has recently been synthesized by alkylation of a thiazolyl cuprate(I), $[Li_n CuCl(dmt)_m]$ (dmt = 4,5-dimethylthiazolyl), with CF_3SO_3Me ⟨94AG(E)672⟩.

6.22.3.1.3 *Silicon-substituted carbene complexes, $R_3Si(X)C{=}ML_n$*

The chemistry of silyl-substituted Fischer-type carbene complexes was reviewed by Schubert ⟨88JOM(358)215⟩. Compared with the rich chemistry of transition metal carbene complexes having electron-donating substituents at the carbene carbon center, compounds $R_3Si(X)C{=}ML_n$ have been investigated only briefly. The alkoxy(silyl)carbene complexes $R_3Si(EtO)C{=}ML_n$ (M = Cr, Mo or W) can be prepared by standard methods from metal carbonyls and alkali-metal silyls, followed by alkylation with $EtO_3{}^+BF_4{}^-$ ⟨77CB3467, 93JOM(459)55⟩. Several alkylthio(silyl)carbene complexes were obtained via the nucleophilic displacement of an alkoxy group from a complexed silyl-substituted carbene ligand ⟨86CB2900⟩. For example, $Ph_3Si(EtO)CW(CO)_5$ reacts with EtSH in the presence of disulfide (Et_2S_2) to give $Ph_3Si(EtS)CW(CO)_5$ in almost quantitative yield.

Reaction of alkoxy(silyl)carbene complexes with ammonia, primary and secondary amines yields amino(silyl)carbene complexes if the amine is not too bulky ⟨86OM173, 93JOM(459)55⟩. With bulky amines (e.g., $HNPr^i_2$, $HNCy_2$) no aminolysis takes place. When the sterically demanding amines $HNEt_2$, $HNBu^nMe$ or $HN(CH_2Ph)Me$ are used, monoalkylamino-substituted carbene complexes $R_3Si(AlkNH)CM(CO)_5$ are formed instead, owing to cleavage of one of the *N*-alkyl substituents ⟨90JOM(385)221⟩.

Depending on the group X thermolysis of silyl-substituted carbene complexes $R_3Si(X)CM(CO)_5$ results in fragmentation of the carbene ligand (X = OAlk or NHAlk), formation of stable 16-electron carbene complexes $R_3Si(Alk_2N)CML_n$, or formation of ketenes $Ph_3Si(X)C{=}C{=}O$ (X = RO, RS) ⟨88JOM(355)243, 89JOM(373)203, 90JOM(385)221⟩. The stabilities of the carbene complexes $R_3Si(Alk_2N)CM(CO)_4$ decrease in the order W > Mo ~ Cr for a given R_3Si group, and in order $Ph_3Si > MePh_2Si > Me_2PhSi$ for a given metal ⟨88JOM(358)215⟩.

6.22.3.2 Functions With a Formal Tin–Carbon Double Bond

Isolable compounds of doubly bonded tin are very rare ⟨90CRV283, 92OM2748⟩. Up to 1995 only two examples of the stable heterosubstituted stannaalkenes have been described. The Berndt compounds (**199**) were synthesized by the reaction between stannylenes and the cryptocarbene (**145**) (Equation (53)) ⟨87AG(E)546⟩. The structure of (**199a**) has been confirmed by x-ray crystallography.

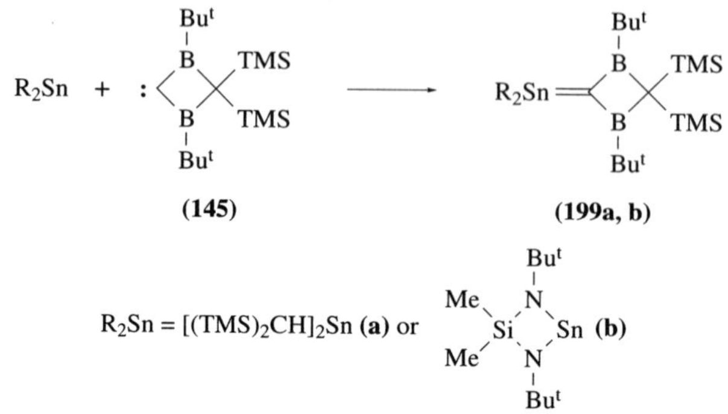

(53)

The chemistry of stannaalkenes is not well developed, and only the addition of HCl across the tin–carbon double bond has been performed.

6.23
Tricoordinated Stabilized Cations and Radicals, ^+CXYZ and $^{\bullet}CXYZ$

THOMAS L. GILCHRIST
University of Liverpool, UK

6.23.1 CARBOCATIONS BEARING THREE HETEROATOM FUNCTIONS

The aim of this chapter is to draw together the literature on the methods of preparation of salts in which the positively charged component is, at least formally, a carbocation with three attached heteroatom functions. Many salts of this type are isolable and several methods of preparation have been discussed in earlier chapters of this volume. The carbocations have been classified according to the types of heteroatom attached to carbon. This review also contains a brief discussion of carbon-centered radicals with three heteroatom functions. None of these species is isolable but a few such radicals can be classified as 'persistent' ⟨76ACR13⟩.

6.23.1.1 Introduction

Tricoordinate carbocations are well established as reaction intermediates but special stabilizing factors are necessary to permit salts containing such cations to be isolated. Heteroatom substituents can provide the necessary stabilization by electron donation from a lone pair into the vacant *p*-orbital on carbon. Nitrogen is the most effective of the common heteroatoms at this type of electron pair donation, to the extent that the presence of just one nitrogen substituent can enable salts of the formal carbocation to be isolated. The structure of such a cation can be represented as a resonance hybrid of carbenium ion (**1a**) and iminium ion (**1b**) forms. On the basis of x-ray crystal structure data and spectroscopic evidence it is clear that the structures are much more accurately represented as the iminium ions (**1b**) containing a formal double bond, and with the positive charge largely centered on nitrogen. When the formal carbocation bears more than one heteroatom substituent there are further possibilities for charge delocalisation.

(**1a**) carbenium ion (**1b**) iminium ion

This chapter includes examples of methods of preparation of salts in which the cations contain a tricoordinate carbon atom with three attached heteroatoms. The positive charge is delocalised, to a greater or lesser extent, over the heteroatom substituents and the bonds to the central carbon have considerable π-bonding character. The types of structure discussed include some species which are not normally regarded as stabilized carbocations, such as dichloroiminium cations and cationic metal carbene complexes; examples of these types are included to illustrate the range of possible substituents.

6.23.1.2 Cations Bearing Three Halogens

The salts $^{+}CX_3SbF_5X^{-}$ ($X = Cl$, Br and I) have been generated at $-78\,^{\circ}C$ by reaction of the appropriate tetrahalides CX_4 with antimony pentafluoride in SO_2ClF ⟨89JA8020⟩. The salts are stable in solution below $-50\,^{\circ}C$ and spectroscopic data were obtained. The corresponding trifluoromethyl cation could not be generated in this way. It has been suggested that the π-electron donor ability of fluorine is more than offset by its electron-withdrawing inductive effect: calculations predict that the trifluoromethyl cation should be less stable than the other trihalomethyl cations ⟨91CC975⟩.

6.23.1.3 Cations Bearing Halogen and Chalcogen Functions

There are few examples of isolable salts of this type. The salt (**2**) bearing one fluorine and two sulfur functions, and analogues with one chlorine or one bromine substituent, have been prepared from tetrafluoro-1,3-dithietane as shown in Scheme 1 ⟨85CB4997, 90CB1635⟩. X-ray data has been obtained for (**2**) ⟨85CB5007⟩ and a series of related salts has been studied spectroscopically ⟨87CB429⟩.

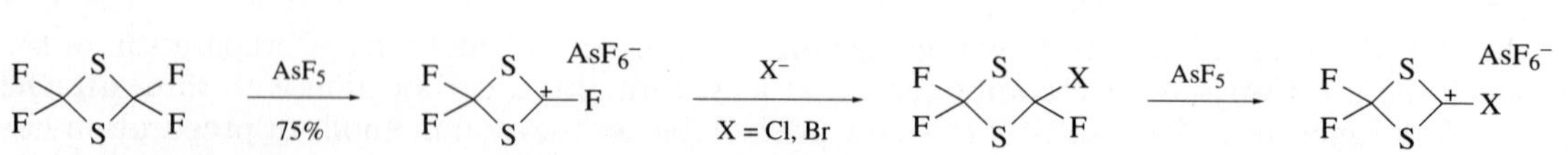

Scheme 1

6.23.1.4 Cations Bearing Halogen, Other Elements and (Possibly) Chalcogen Functions

6.23.1.4.1 Cations bearing two halogens and one nitrogen function

Dichloroiminium salts, and in particular *N,N*-dimethyldichloroiminium chloride (**3**), are important reagents with a wide range of synthetic applications. Dihaloiminium salts with bromine or iodine substituents are known but have been little studied. Stable difluoroiminium salts have not been described. The chemistry of dihaloiminium salts has been reviewed by Janousek and Viehe ⟨B-76MI 623-01⟩ and by Marhold ⟨83HOU(E4)655⟩.

The salt $Cl_2C{=}NH_2^+SbCl_6^-$ has been prepared from cyanogen chloride, HCl and antimony pentachloride; compounds $BrClC{=}NH_2^+SbCl_6^-$ and $ClIC{=}NH_2^+SbCl_6^-$ are prepared analogously from cyanogen bromide and cyanogen iodide ⟨64CB1286⟩. Salts of this type can also be prepared by protonation or alkylation of carbonimidic dichlorides (isocyanide dichlorides) ⟨72CB3050, 83HOU(E4)655, 92ZAAC(617)136⟩; an example is the protonation of $Cl_2C{=}NCl$ in a superacid medium (HF and SbF_6) ⟨92ZAAC(617)136⟩. The method of choice for preparing *N,N*-dialkyldichloroiminium salts is, however, usually based on the chlorination of thiocarbonyl chlorides or thiocarbamates ⟨73ZOR39, 83HOU(E4)655⟩. The preparation can be carried out in one pot starting from a secondary amine, carbon disulfide and a chlorinating agent such as chlorine or phosphorus pentachloride; thus, the salt (**4**) was prepared (91%) from pyrrolidine ⟨89SC2825⟩. The dibromo compound (**5**) was obtained in a similar way, by reaction of the disulfide (**6**) with bromine (Equation (1)) ⟨74ZOR449⟩. The corresponding diiodo compound (**7**) was prepared from the salt (**3**) in good yield by heating with three equivalents of iodomethane ⟨79ZOR215⟩.

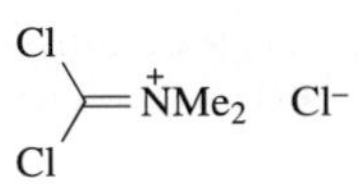

(3)

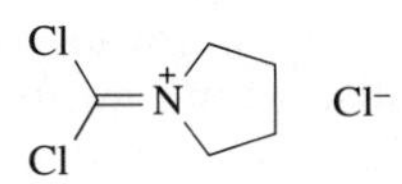

(4)

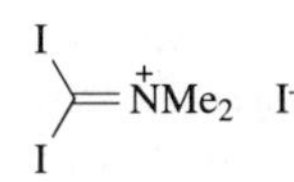

(7)

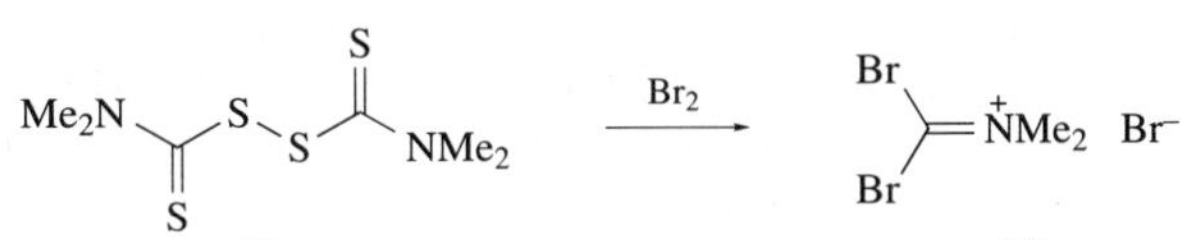

(6) **(5)** (1)

6.23.1.4.2 Cations bearing two halogens and one other heteroatom function

Several transition metal complexes exist in which a dihalocarbene acts as a ligand ⟨88CRV1293⟩. In some of these the metal complex bears an overall positive charge and the species can formally be regarded as a stabilized carbocation bearing three heteroatoms. Structural studies have shown that there is considerable back donation of electrons from the metal to the dihalocarbene ligand in these complexes, so the metal to carbon bond is best regarded as a double bond.

Cationic complexes containing the CF_2 ligand can be formed from trifluoromethyl complexes by abstraction of fluoride using antimony pentafluoride or boron trifluoride. The complex $[CpMo(CF_2)(CO)_3]^+SbF_6^-$ was detected in solution at low temperature from the reaction of antimony pentafluoride with $CpMoCF_3(CO)_3$ ⟨78JOM(153)67⟩. The iron complex $[CpFe(CF_2)(CO)PPh_3]^+BF_4^-$, prepared in a similar manner, was fully characterised ⟨85OM1830⟩. The corresponding dichlorocarbene complex $[CpFe(CCl_2)(CO)PPh_3]^+BF_4^-$ was prepared by halide exchange, with BCl_3 as the Lewis acid ⟨85OM1830⟩. Ruthenium(II) and osmium(II) complexes of the type $[M(CClX)(MeCN)(CO)(PPh_3)_2]^+Y^-$ (X = Cl or F) are formed by abstraction of a chloride ligand from neutral carbene complexes by silver(I) salts in acetonitrile ⟨84JOM(269)C55, 88CRV1293⟩.

6.23.1.4.3 Cations bearing one halogen, one chalcogen and one nitrogen function

The reaction of *N,N*-dimethyldichloroiminium chloride (**3**) with phenol or with thiophenol results in the formation of cations of this type (Scheme 2) ⟨B-76MI 623-01⟩. Cations (**8**) with sulfur substituents can also be prepared by chlorination of dithiocarbamate esters ⟨B-79MI 623-02⟩. Thiocyanates RSCN (R = Me, Ph) react with HBr to form isolable salts $[RSC(Br)NH_2]^+Br^-$ ⟨64CB3162⟩ and the reaction of aryl cyanates with phosphorus pentachloride leads to the formation of the salts (**9**) in high yield ⟨72JGU97⟩.

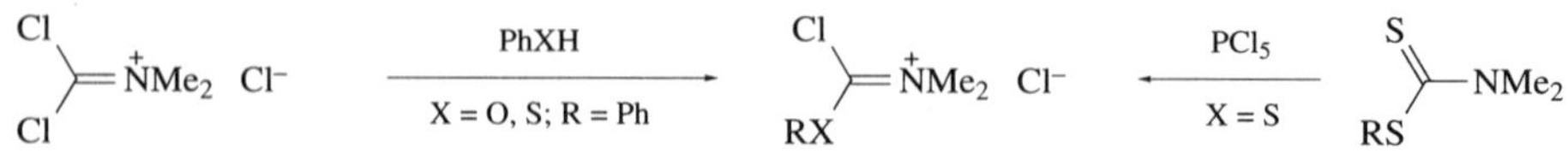

(**3**) (**8**)

Scheme 2

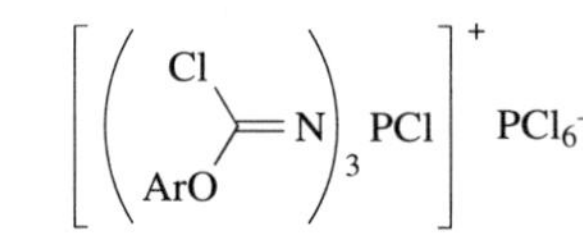

(**9**)

6.23.1.4.4 Cations bearing one halogen and two nitrogen functions

Salts of this type are well represented in the literature; many are stable, isolable solids. Those in which the two nitrogen functions are dialkylamino groups are the best known and the methods of preparation of chloroformamidinium salts has been reviewed ⟨B-79MI 623-01⟩. The simplest salts of this class are derived from cyanamides by reaction with HCl or HBr. Cyanamide reacts with HCl to give the salt $H_2NC(Cl){=}NH_2{}^+Cl^-$ as a solid which is stable for a month at room temperature in anhydrous conditions ⟨90JMC434⟩. Dialkylcyanamides react with HBr in a similar way ⟨88JCS(P1)899⟩. Carbodiimides react with 2 mol HCl to give salts $[(RNH)_2CCl]^+Cl^-$ ⟨B-79MI 623-01⟩ and it is possible to form a crystalline fluorine substituted salt (**10**) by reaction of $(CF_3)_2NC(F){=}NCF_3$ with HF/AsF_5 below $-30\,^\circ C$ ⟨87JFC(37)259⟩. Dimethylcyanamide has also been used as the starting material for more complex salts; examples include the sulfur-containing species (**11**) formed by reaction with sulfur dichloride ⟨85IC2453⟩ and the conjugated iminium salts (**12**) which were produced from *N*-arylbenzamides, phosphorus oxychloride and perchloric acid ⟨87CC99⟩.

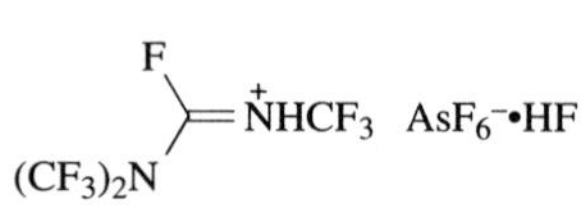

(**10**)

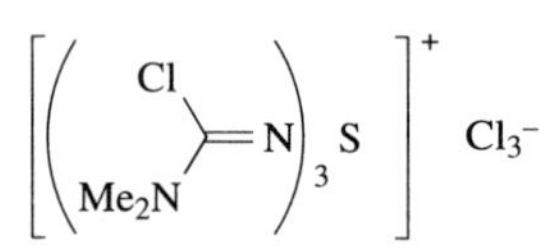

(**11**)

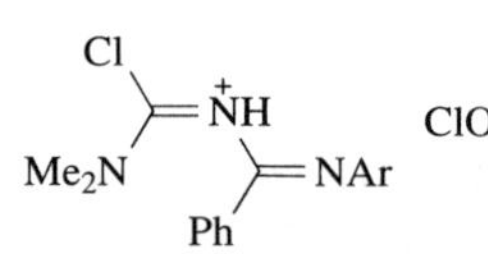

(**12**)

The most general method for the preparation of chloroformamidinium salts is the chlorination of ureas or thioureas using phosgene or similar reagents ⟨B-79MI 623-01⟩. An example of the reaction, the chlorination of tetramethylurea, is shown in Equation (2) ⟨79S339⟩. Another potentially general method is the reaction of dichloroiminium salts such as (**3**) with nitrogen nucleophiles but this is not always successful since the reactions often proceed further, particularly when basic amines are used as nucleophiles ⟨B-76MI 623-01⟩. An example of a salt which has been prepared by this method is (**13**), which was prepared from (**3**) and azidotrimethylsilane ⟨75C209⟩.

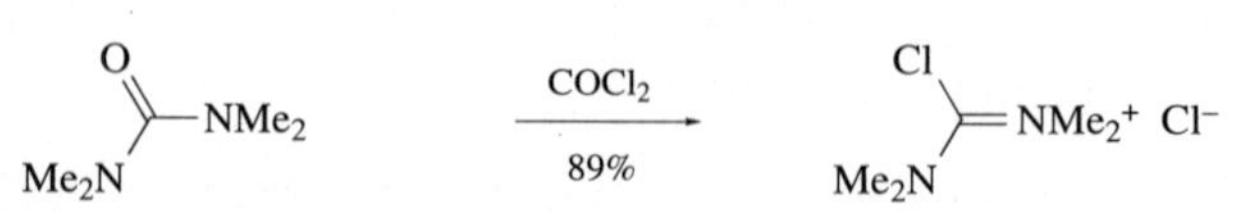

(2)

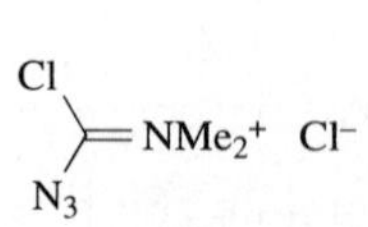

(**13**)

6.23.1.5 Cations Bearing Three Chalcogen Functions

6.23.1.5.1 *Three oxygen functions*

Tris(alkoxy)methylium salts, $(RO)_3C^+BF_4^-$, are isolable species. A review of the early work on their preparation and properties is available ⟨B-71MI 623-01⟩ and they are referred to in more recent reviews on carbenium ions ⟨87CSR75, 93AG(E)767⟩. The best method for their preparation is the reaction of the *ortho*-carbonate $C(OR)_4$ with boron trifluoride etherate and HBF_4; papers which provide good experimental details are available for the preparation of the triethoxy- ⟨87JOM(328)249⟩ and the trimethoxy- ⟨89SC2307⟩ substituted carbenium tetrafluoroborates. Hexachloroantimonate salts can be made in an analogous manner. The salt $(EtO)_3C^+BF_4^-$ has also been prepared by *O*-alkylation of diethyl carbonate ⟨81LA70⟩. The parent cation $^+C(OH)_3$ has been generated in a superacid medium from di-*t*-butyl carbonate ⟨93AG(E)767⟩.

6.23.1.5.2 *Three sulfur functions*

There are several examples of the preparation of stable salts of carbenium ions bearing three sulfur functions. The methods for their preparation are analogous to those used for the tris(alkoxy)carbenium ions but, since it is much easier to alkylate sulfur than oxygen, a wider range of alkylating agents can be used.

The most convenient method of preparation is usually *S*-alkylation of trithiocarbonate diesters (Equation (3)) ⟨67TL2747, 88S22⟩. This method has been reviewed briefly ⟨66AHC(7)39⟩ and there are many more recent examples, including the preparation of some species in which the substituents provide little stabilization to the cation. Some examples (salts (**14**) ⟨78JOC678⟩, (**15**) ⟨92JOC1696⟩ and (**16**) ⟨90CB1635⟩) are shown. Less commonly, tetrathio-*ortho*-carbonates are used as starting materials ⟨67TL2747, 94CB597⟩; the stable salt (**17**) was prepared by this method ⟨94CB597⟩. Trithiocarbonates have also been *S*-arylated using arenediazonium tetrafluoroborates ⟨82JPR669⟩.

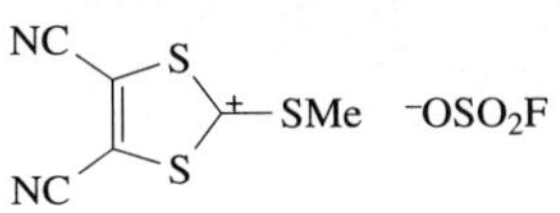

(14)

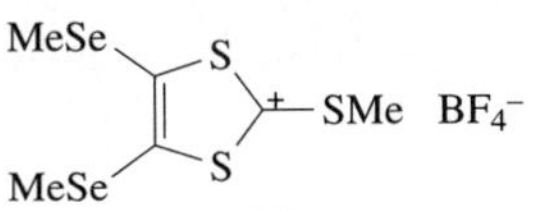

(15)

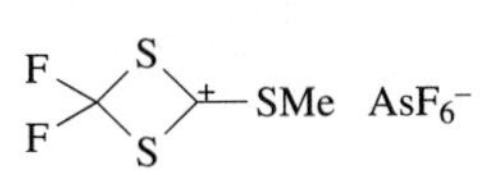

(16)

$(F_3CS)_3C^+$ AsF_6^-

(17)

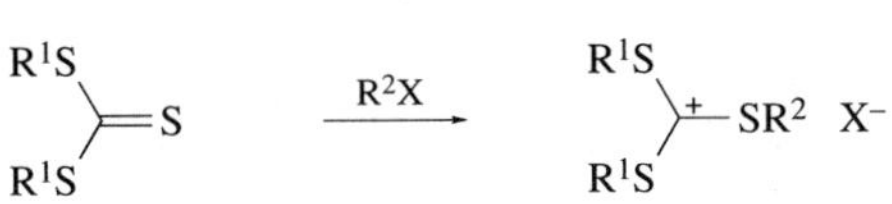

(3)

6.23.1.5.3 *Three selenium functions*

The salt (**18**) was generated by *Se*-alkylation of the corresponding triselenocarbonate ⟨75TL1259⟩. Salts containing mixed sulfur and selenium substituents were also generated.

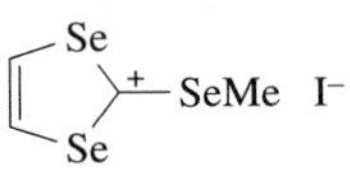

(18)

6.23.1.6 Cations Bearing Chalcogen and Nitrogen Functions

6.23.1.6.1 *Two oxygen and one nitrogen function*

In principle, salts of this type are readily available from dichloroiminium salts and alcohols. This reaction is successful with the salt (**3**) and phenol, catechol and pinacol ⟨B-76MI 623-01⟩; thus, (**19**) was prepared from (**3**) and pinacol. With simple alcohols the chloride salts are unstable, chloride acting as a nucleophile to displace one of the oxygen functions. The problem is overcome with nonnucleophilic counterions; for example, the salt $(EtO)_2C{=}NH_2{}^+SbCl_6{}^-$ can be isolated ⟨66ZAAC(344)113⟩.

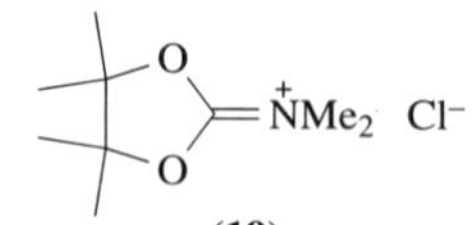

(19)

Another method of preparation is *O*-alkylation of urethanes. *N,N*-Dimethylurethane was ethylated with triethyloxonium tetrafluoroborate to give (**20**) ⟨61LA(641)1⟩ and the same method was used to prepare the salt (**21**) from $(MeO)_2C{=}NCO_2Et$ ⟨92CB2487⟩.

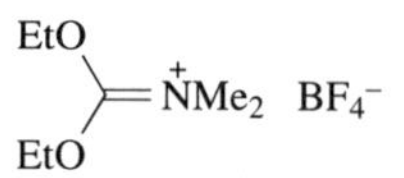

(20)

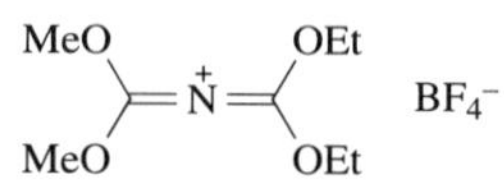

(21)

6.23.1.6.2 *Two sulfur and one nitrogen function*

These salts are most often prepared by *S*-alkylation of dithiocarbamates ⟨74BCJ398, 85S891⟩; for example, the salt (**22**) was prepared (85%) by methylation of the cyclic dithiocarbamate with dimethyl sulfate and subsequent reaction with sodium perchlorate (Equation (4)) ⟨74BCJ398⟩. Double alkylation of dithiocarbamate salts is also possible: thus, the salt (**23**) was derived from $Me_2NCS_2{}^-Na^+$ and 1,2-bis(bromomethyl)benzene ⟨74BCJ398⟩. An alternative method of preparation is *N*-alkylation of iminodithiocarbonates $(R^1S)_2C{=}NR^2$ ⟨68T6485⟩.

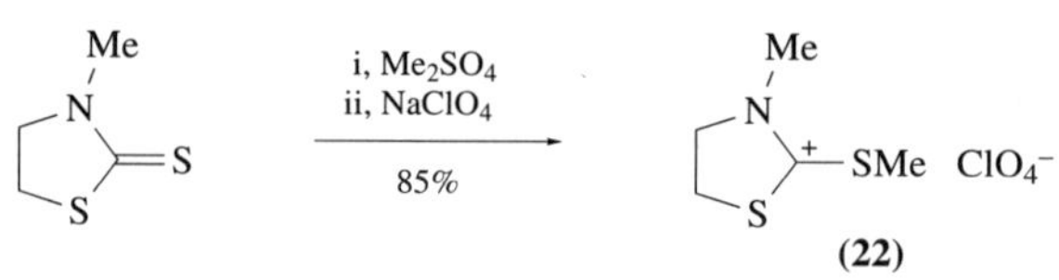

(22) (4)

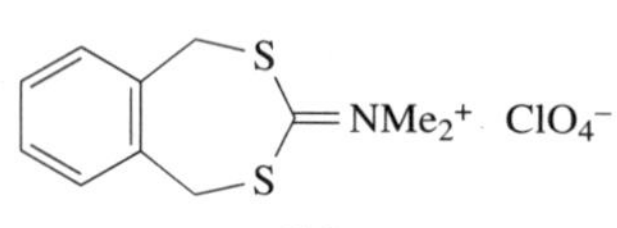

(23)

6.23.1.6.3 *Two selenium and one nitrogen function*

Two examples of salts in which the cations bear these functional groups are (**24**) and (**25**). Both were prepared from salts containing the anion $Me_2NCSe_2{}^-$, by alkylation with 1,3-dibromopropane and 3-bromobutan-2-one, respectively ⟨72BCJ489, 80CC866⟩.

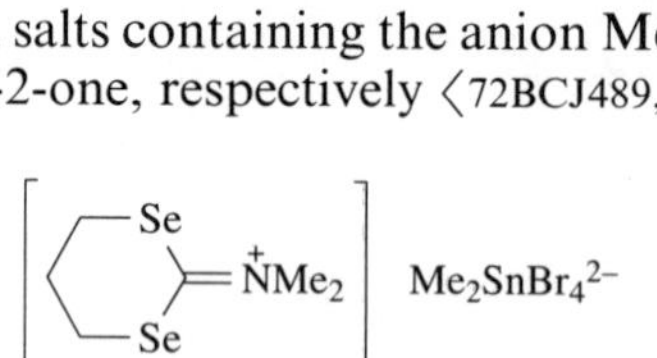

(24)

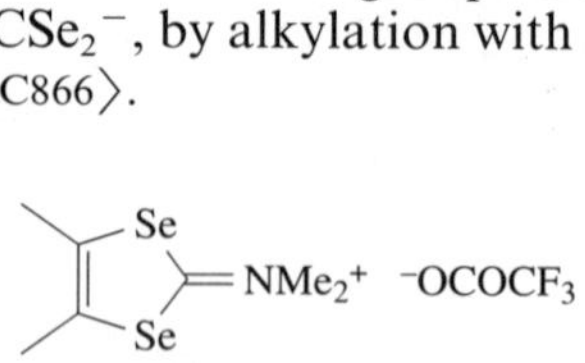

(25)

6.23.1.6.4 *Two different chalcogen and one nitrogen function*

Iminothiocarbonate salts unsubstituted on nitrogen can be prepared either by *S*-alkylation of thiourethanes or by the addition of alcohols and HCl to thiocyanates (Scheme 3) ⟨83HOU(E4)677⟩.

The dithiocarbamidic ester (**26**) can be cyclised by *S*-methylation and the salt (**27**) can be isolated in good yield after the addition of sodium perchlorate ⟨79S182⟩.

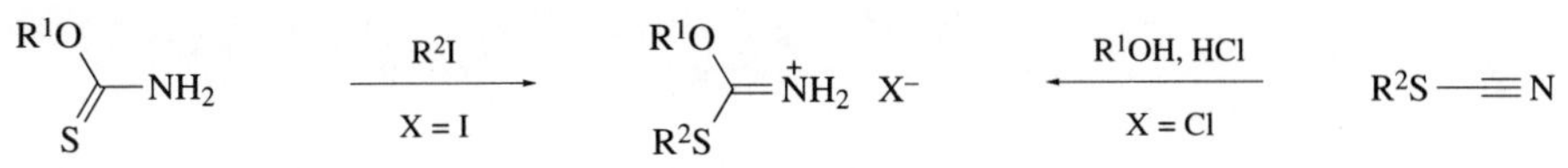

Scheme 3

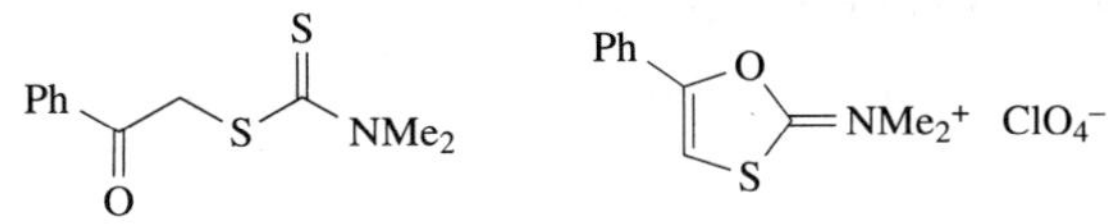

(**26**) (**27**)

Several salts of cations bearing sulfur, selenium and nitrogen functions have been isolated. An example is the salt (**28**); this was obtained (in a 53% yield) by ring contraction of the six-membered heterocycle (**29**) which was brought about by reaction with HI ⟨84S667⟩.

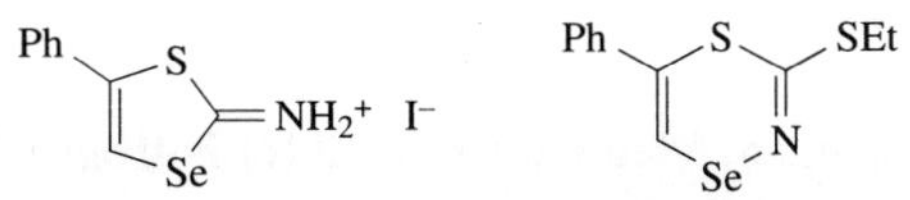

(**28**) (**29**)

6.23.1.6.5 *One oxygen and two nitrogen functions*

Tetraalkylureas are alkylated on oxygen by trialkyloxonium tetrafluoroborates; for example, the salt $[(Me_2N)_2COEt]^+BF_4^-$ was prepared in high yield from tetramethylurea and triethyloxonium tetrafluoroborate ⟨61LA(641)1⟩; trimethyloxonium tetrafluoroborate has also been used ⟨71LA(743)1⟩. With some tetraalkyureas, *N*-alkylation can compete. An alternative procedure for the preparation of salts is the displacement of chloride from chloroformamidinium cations by alcohols or alkoxides. The salt (**30**) was prepared in this way (in a 57% yield) from $[(Me_2N)_2CCl]^+Cl^-$ ⟨71LA(743)1⟩. An interesting example of this reaction is the preparation of the salt (**31**) (Equation (5)) ⟨67CB3319⟩. Salts unsubstituted on nitrogen are obtained by the addition of alcohols to cyanamide; thus, $[BnOC(NH_2)_2]^+$ ^-OTs was prepared (in an 81% yield) from benzyl alcohol, cyanamide and *p*-toluenesulfonic acid in dry chloroform ⟨87TL1969⟩.

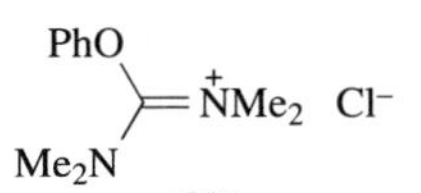

(**30**)

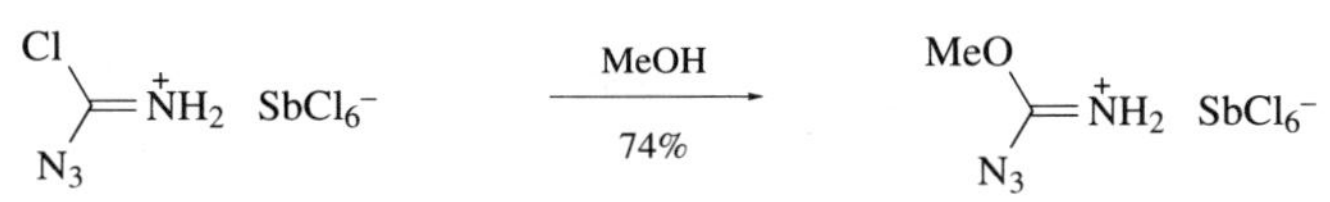

(**31**) (5)

6.23.1.6.6 *One sulfur and two nitrogen functions*

Thioureas are readily alkylated on sulfur by a wide range of electrophiles to give stable isothiouronium salts (**32**) (Equation (6)) ⟨83HOU(E4)690⟩. This is by far the most important method of preparation of these salts. Thioureas can also be *S*-arylated by means of arenediazonium salts. By analogy with the preparation of the salt (**30**), *S*-arylisothiouronium salts can also be prepared by the reaction of chloroformamidinium salts with thiophenols ⟨69LA(727)228⟩.

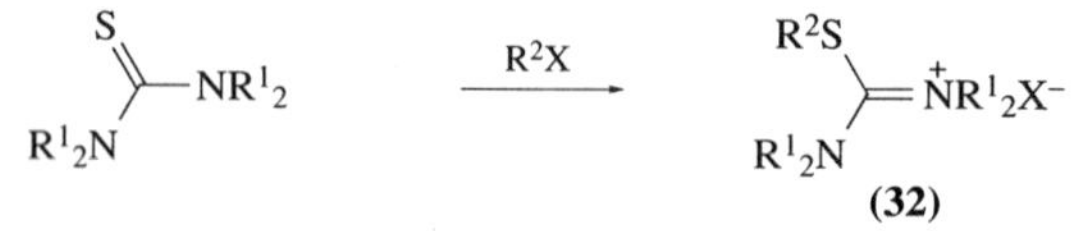

(6)

6.23.1.6.7 *One selenium and two nitrogen functions*

The reaction of selenoureas with alkylating agents is analogous to that of thioureas and stable salts (**33**) can be isolated ⟨62JOC2899, 86NJC51⟩.

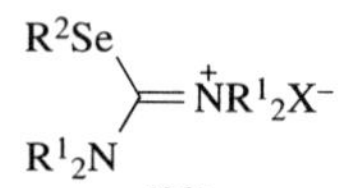

(33)

6.23.1.7 Cations Bearing Chalcogen, Metal and (Possibly) Nitrogen Functions

There are two general methods for the preparation of cationic carbene complexes of this class: (i) alkylation of a ligand on sulfur or oxygen and (ii) nucleophilic addition to an isocyanide ligand. An example of a preparation of the first type is the formation of the mercury complex shown in Equation (7) ⟨67AG(E)560⟩. The platinum complex (**34**) was similarly prepared by *S*-methylation of a thioamide function ⟨75IC1513⟩. The complex (**35**) was formed by alkylation of both sulfur atoms of a CS_2 ligand by 1,2-dibromoethane ⟨76JOM(107)C37⟩. Several related preparations have been reported ⟨74JCS(D)351, 75JCS(D)939, 75JOM(90)C34⟩. The second approach is illustrated by the preparation of the platinum complex (**36**) by addition of ethanethiol to isocyanide ligands of (**37**) ⟨72IC2069⟩. Attempts to add alcohols to coordinated isocyanides are generally unsuccessful, although an intramolecular addition has been reported ⟨74AG(E)599⟩.

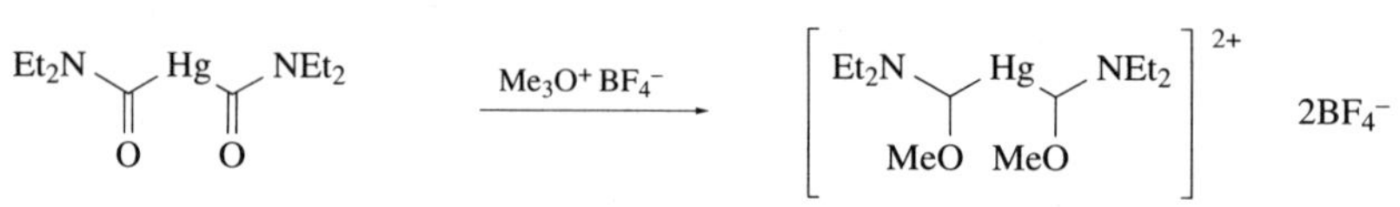

(7)

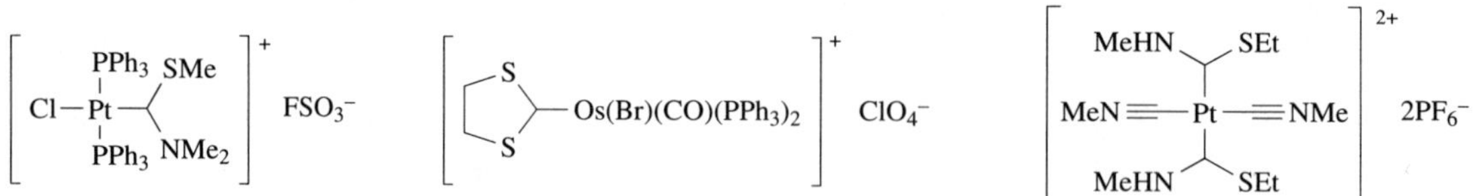

(34) **(35)** **(36)**

$[Pt(CNMe)_4]^{2+}\ 2PF_6^-$

(37)

6.23.1.8 Cations Bearing Three Nitrogen Functions

Methods for the preparation of guanidines have been described in Chapter 6.21. Guanidines are strong bases and excellent nucleophiles, so that a great many guanidinium salts have been prepared by protonation or alkylation of guanidines. Even guanidines bearing electron-withdrawing substituents can be alkylated: an example is the preparation of the guanidinium salt (**38**) (in a 68%

yield) by alkylation of *N*-cyano-*N'*,*N'*,*N'*,*N'*-tetramethylguanidine with triethyloxonium tetrafluoroborate ⟨84CB502⟩.

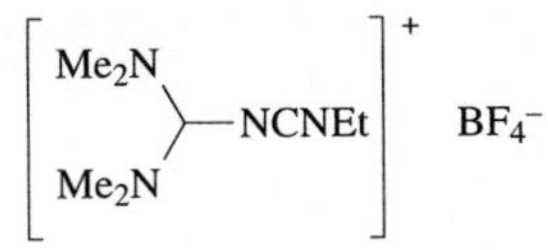

(38)

Salts (**39**) and (**40**) with cations bearing one and two azido functions have been prepared by displacement of chloride from the corresponding chloro-substituted cations ⟨67CB3275⟩. Tris (azido)carbenium hexachloroantimonate (**41**) is isolated in high yield from the reaction of tetrachloroantimony(V) azide, $(SbCl_4N_3)_2$, with carbon tetrachloride ⟨66AG(E)841, 67CB3725⟩. This salt reacts with PCl_3 ⟨79AG(E)693⟩ and with PBr_3 ⟨80ZAAC(468)165⟩ to give the new salts (**42**; X = Cl or Br).

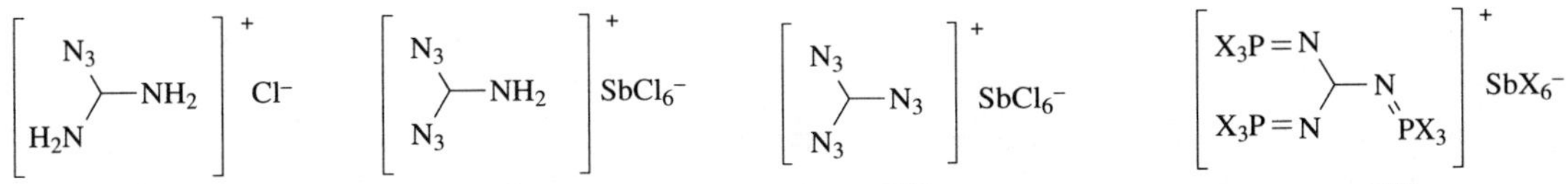

(39) **(40)** **(41)** **(42)**

6.23.1.9 Cations Bearing Nitrogen and Other Element Functions

A few salts have been described in which the cation bears two nitrogen and one phosphorus function ⟨72JOC2730, 78JPR389⟩. The most stable of these is the crystalline salt (**43**) which was prepared as shown in Equation (8) ⟨78JPR389⟩. The salt (**44**), described by the authors as a 'cationic half ylide', has also been prepared ⟨78CB2451⟩. There are a number of cationic carbene complexes of this class known, of which (**45**) is an example ⟨93OM3405⟩; this was prepared from the stable carbene and silver triflate.

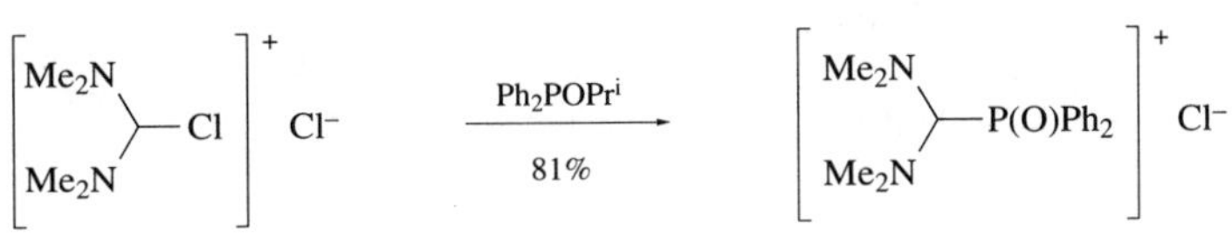

(8)

(43)

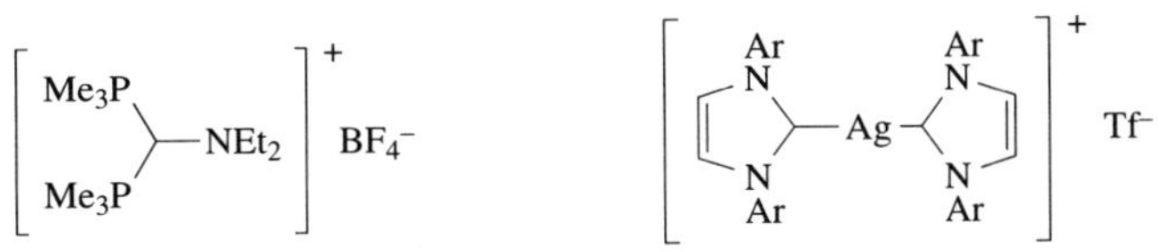

(44) **(45)**

6.23.1.10 Cations Bearing Phosphorus and Silicon Functions

The salt (**46**) has been prepared as shown in Equation (9); it is a colourless solid which is unstable above −30 °C ⟨91AG(E)709, 93PS(76)21⟩. Crystal structure determinations carried out on closely related salts show that both phosphorus and carbon are trigonal planar: the salt can be regarded as the phosphorus analogue of an iminium salt, with stabilization of the carbenium ion by the lone pair on phosphorus. The salt (**47**) has also been isolated but in this case the nitrogen substituents on phosphorus bear much of the positive charge ⟨89JA6853⟩.

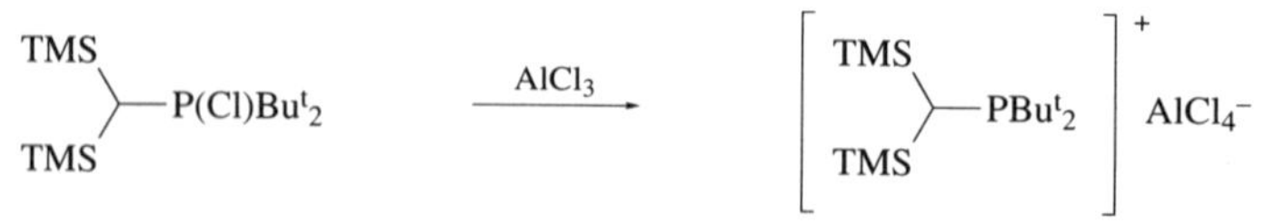

(9)

(**46**)

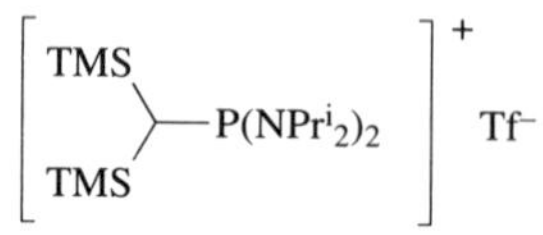

(**47**)

6.23.2 CARBON-CENTERED RADICALS BEARING THREE HETEROATOM FUNCTIONS

No carbon-centered radicals bearing three heteroatom functions can be regarded as 'stable' in that they cannot be isolated and handled, but a few such radicals have significant lifetimes (ranging from a few seconds to a few hours) in solution. Such radicals have been described as 'persistent' ⟨76ACR13⟩. These radicals bear silicon, phosphorus or sulfur substituents and their increased lifetime can be ascribed to steric factors rather than to electron delocalization.

The tris(trimethylsilyl)methyl radical $^{\bullet}C(TMS)_3$ can be generated by decomposition of $[(TMS)_3C]_2Hg$ ⟨70CC559⟩ or by reduction of $(TMS)_3CI$ ⟨91CC1608⟩. This radical is long-lived in solution. A number of carbon radicals bearing three sulfur functions can also be described as persistent, among them $^{\bullet}C(SCF_3)_3$, which is generated reversibly from its dimer at room temperature ⟨79JA6282⟩, and $^{\bullet}C(SCF_3)_2SC_6F_5$, which is produced from its dimer at 140–190 °C ⟨84T4963⟩. A number of other radicals bearing three sulfur functions or two sulfur and one silicon function can be produced by thermal dissociation of their dimers ⟨77CB2880⟩. A series of persistent radicals (**48**) has been produced by the reaction of the dithioesters (**49**; X = O or S) with a wide range of radicals $R^{\bullet}$, including $MeS^{\bullet}$, $Me^{\bullet}$ and $Ph_3Pb^{\bullet}$ ⟨93JA8444⟩.

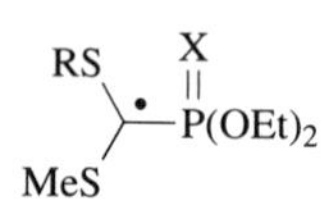

(**48**)

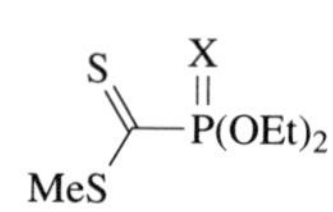

(**49**)

References

EXPLANATION OF THE REFERENCE SYSTEM

Throughout this work, references are designated by a number–lettering coding of which the first two numbers denote tens and units of the year of publication, the next one to three letters denote the journal, and the final numbers denote the page. This code appears in the text each time a reference is quoted; the advantages of this system are outlined in the Introduction. The system has been used previously in "Comprehensive Heterocyclic Chemistry," eds A. R. Katritzky and C. W. Rees, Pergamon, Oxford, 1984 and is based on that used in the following two monographs: (a) A. R. Katritzky and J. M. Lagowski, "Chemistry of the Heterocyclic N-Oxides," Academic Press, New York, 1971; (b) J. Elguero, C. Marzin, A. R. Katritzky and P. Linda, "The Tautomerism of Heterocycles," in "Advances in Heterocyclic Chemistry," Supplement 1, Academic Press, New York, 1976.

The following additional notes apply:

1. A list of journal codes in alphabetical order, together with the journals to which they refer, is given immediately following these notes. Journal names are abbreviated throughout using the CASSI (Chemical Abstracts Service Source Index) system.
2. Each volume contains all the references cited *in that volume*; no separate lists are given for individual chapters.
3. The list of references is arranged in order of (a) year, (b) journal in alphabetical order of journal code, (c) part letter or number if relevant, (d) volume number if relevant, (e) page number.
4. In the reference list the code is followed by (a) the complete literature citation in the conventional manner and (b) the number(s) of the page(s) on which the reference appears, whether in the text or in tables, schemes, etc.
5. For nontwentieth-century references the year is given in full in the code.
6. For journals which are published in separate parts, the part letter or number is given (when necessary) in parentheses immediately after the journal code letters.
7. Journal volume numbers are *not* included in the code numbers unless more than one volume was published in the year in question, in which case the volume number is included in parentheses immediately after the journal code letters.
8. Patents are assigned appropriate three-letter codes.
9. Frequently cited books are assigned codes.
10. Less common journals and books are given the code "MI" for miscellaneous with the whole code for books prefixed by the letter "B-".
11. Where journals have changed names, the same code is used throughout, e.g. CB refers to both *Chem. Ber.* and to *Ber. Dtsch. Chem. Ges.*

Journal Codes

AAC	Antimicrob. Agents Chemother.
ABC	Agric. Biol. Chem.
AC	Appl. Catal.
AC(P)	Ann. Chim. (Paris)
AC(R)	Ann. Chim. (Rome)
ACH	Acta Chim. Acad. Sci. Hung.

ACR	Acc. Chem. Res.
ACS	Acta Chem. Scand.
ACS(A)	Acta Chem. Scand., Ser. A
ACS(B)	Acta Chem. Scand., Ser. B
AF	Arzneim.-Forsch.
AFC	Adv. Fluorine Chem.
AG	Angew. Chem.
AG(E)	Angew. Chem., Int. Ed. Engl.
AHC	Adv. Heterocycl. Chem.
AHCS	Adv. Heterocycl. Chem. Supplement
AI	Anal. Instrum.
AJC	Aust. J. Chem.
AK	Ark. Kemi
AKZ	Arm. Khim. Zh.
AM	Adv. Mater. (Weinheim, Ger.)
AMLS	Adv. Mol. Spectrosc.
AMS	Adv. Mass. Spectrom.
ANC	Anal. Chem.
ANL	Acad. Naz. Lncei
ANY	Ann. N. Y. Acad. Sci.
AOC	Adv. Organomet. Chem.
AP	Arch. Pharm. (Weinheim, Ger.)
APO	Adv. Phys. Org. Chem.
AQ	An. Quim.
AR	Annu. Rep. Prog. Chem.
AR(A)	Annu. Rep. Prog. Chem., Sect. A
AR(B)	Annu. Rep. Prog. Chem., Sect. B
ARP	Annu. Rev. Phys. Chem.
ASI	Acta Chim. Sin. Engl. Ed.
ASIN	Acta Chim. Sin.
AX	Acta Crystallogr.
AX(A)	Acta Crystallogr., Part A
AX(B)	Acta Crystallogr., Part B
B	Biochemistry
BAP	Bull. Acad. Pol. Sci., Ser. Sci. Chim.
BAU	Bull. Acad. Sci. USSR, Div. Chim. Sci.
BBA	Biochim. Biophys. Acta
BBR	Biochim. Biophys. Res. Commun.
BCJ	Bull. Chem. Soc. Jpn.
BEP	Belg. Pat.
BJ	Biochem. J.
BJP	Br. J. Pharmacol.
BMC	Bioorg. Med. Chem. Lett.
BP	Biochem. Biopharmacol.
BPJ	Br. Polym. J.
BRP	Br. Pat.
BSB	Bull. Soc. Chim. Belg.
BSF	Bull. Soc. Chim. Fr.
BSF(2)	Bull. Soc. Chim. Fr., Part 2
C	Chimia
CA	Chem. Abstr.
CAN	Cancer
CAR	Carbohydr. Res.
CAT	Chim. Acta Turc.
CB	Chem. Ber.

CBR	Chem. Br.
CC	J. Chem. Soc., Chem. Commun.
CCA	Croat. Chem. Acta
CCC	Collect. Czech. Chem. Commun.
CCR	Coord. Chem. Rev.
CE	Chem. Express
CEN	Chem. Eng. News
CHE	Chem. Heterocycl. Compd. (Engl. Transl.)
CHEC	Comp. Heterocycl. Chem.
CI(L)	Chem. Ind. (London)
CI(M)	Chem. Ind. (Milan)
CJC	Can. J. Chem.
CJS	Can. J. Spectrosc.
CL	Chem. Lett.
CLY	Chem. Listy
CM	Chem. Mater.
CMC	Comp. Med. Chem.
COC	Comp. Org. Chem.
COMC-I	Comp. Organomet. Chem., 1st edn.
COS	Comp. Org. Synth.
CP	Can. Pat.
CPB	Chem. Pharm. Bull.
CPH	Chem. Phys.
CPL	Chem. Phys. Lett.
CR	C. R. Hebd. Seances Acad. Sci.
CR(A)	C. R. Hebd. Seances Acad. Sci., Ser. A
CR(B)	C. R. Hebd. Seances Acad. Sci., Ser. B
CR(C)	C. R. Hebd. Seances Acad. Sci., Ser. C
CRAC	Crit. Rev. Anal. Chem.
CRV	Chem. Rev.
CS	Chem. Scr.
CSC	Cryst. Struct. Commun.
CSR	Chem. Soc. Rev.
CT	Chem. Tech.
CZ	Chem.-Ztg.
CZP	Czech. Pat.
DIS	Diss. Abstr.
DIS(B)	Diss. Abstr. Int. B.
DOK	Dokl. Akad. Nauk SSSR
DP	Dyes Pigm.
E	Experientia
EC	Educ. Chem.
EF	Energy Fuels
EGP	Ger. (East) Pat.
EJM	Eur. J. Med. Chem.
EUP	Eur. Pat.
FCF	Forschr. Chem. Forsch.
FCR	Fluorine Chem. Rev.
FES	Farmaco Ed. Sci.
FOR	Forschr. Chem. Org. Naturst.
FRP	Fr. Pat.
G	Gazz. Chim. Ital.
GAK	Gummi Asbest Kunstst.
GEP	Ger. Pat.
GEP(O)	Ger. Pat. Offen.

GSM	Gen. Synth. Methods
H	Heterocycles
HAC	Heteroatom Chem.
HC	Chem. Heterocycl. Compd.
HCA	Helv. Chim. Acta
HOU	Methoden Org. Chem. (Houben-Weyl)
HP	Hydrocarbon Process
IC	Inorg. Chem.
ICA	Inorg. Chim. Acta
IEC	Ind. Eng. Chem. Res.
IJ	Isr. J. Chem.
IJC	Indian J. Chem.
IJC(A)	Indian J. Chem., Sect. A
IJC(B)	Indian J. Chem., Sect. B
IJM	Int. J. Mass Spectrom. Ion Phys.
IJQ	Int. J. Quantum Chem.
IJS	Int. J. Sulfur Chem.
IJS(A)	Int. J. Sulfur Chem., Part A
IJS(B)	Int. J. Sulfur Chem., Part B
IS	Inorg. Synth
IZV	Izv. Akad. Nauk SSSR Ser. Khim.
JA	J. Am. Chem. Soc.
JAN	J. Antibiot.
JAP	Jpn. Pat.
JAP(K)	Jpn. Kokai
JBC	J. Biol. Chem.
JC	J. Chromatogr.
JCC	J. Coord. Chem.
JCE	J. Chem. Ed.
JCED	J. Chem. Eng. Data
JCI	J. Chem. Inf. Comput. Sci.
JCP	J. Chem. Phys.
JCPB	J. Chim. Phys. Physico-Chim. Biol.
JCR(M)	J. Chem. Res. (M)
JCR(S)	J. Chem. Res. (S)
JCS	J. Chem. Soc.
JCS(A)	J. Chem. Soc. (A)
JCS(B)	J. Chem. Soc. (B)
JCS(C)	J. Chem. Soc. (C)
JCS(D)	J. Chem. Soc., Dalton Trans.
JCS(F1)	J. Chem. Soc., Faraday Trans. 1
JCS(F2)	J. Chem. Soc., Faraday Trans. 2
JCS(P1)	J. Chem. Soc., Perkin Trans. 1
JCS(P2)	J. Chem. Soc., Perkin Trans. 2
JCS(S2)	J. Chem. Soc. (Suppl. 2)
JEC	J. Electroanal. Chem. Interfacial Electrochem.
JEM	J. Energy Mater.
JES	J. Electron. Spectrosc.
JFA	J. Sci. Food. Agri.
JFC	J. Fluorine Chem.
JGU	J. Gen. Chem. USSR (Engl. Transl.)
JHC	J. Heterocycl. Chem.
JIC	J. Indian Chem. Soc.
JINC	J. Inorg. Nucl. Chem.
JLC	J. Liq. Chromatogr.

JMAS	J. Mat. Sci.
JMC	J. Med. Chem.
JMOC	J. Mol. Catal.
JMR	J. Magn. Reson.
JMS	J. Mol. Sci.
JOC	J. Org. Chem.
JOM	J. Organomet. Chem.
JOU	J. Org. Chem. USSR (Engl. Transl.)
JPC	J. Phys. Chem.
JPJ	J. Pharm. Soc. Jpn.
JPO	J. Phys. Org. Chem.
JPP	J. Pharm. Pharmacol.
JPR	J. Prakt. Chem.
JPS	J. Pharm. Sci.
JPS(A)	J. Polym. Sci., Polym. Chem., Part A
JPU	J. Phys. Chem. USSR (Engl. Transl.)
JSC	J. Serbochem. Soc.
JSP	J. Mol. Spectrosc.
JST	J. Mol. Struct.
K	Kristallografiya
KFZ	Khim. Farm. Zh.
KGS	Khim. Geterotsikl. Soedin.
KO	Kirk-Othmer Encyc.
KPS	Khim. Prir. Soedin.
L	Langmuir
LA	Liebigs Ann. Chem.
LC	Liq. Cryst.
LS	Life Sci.
M	Monatsh. Chem.
MAC	Macromol. Chem.
MC	Mendeleev Chem. J. (Engl. Transl.)
MCLC	Mol. Cryst. Liq. Cryst.
MI	Miscellaneous [book/journal]
MIP	Miscellaneous Pat.
MM	Macromolecules
MP	Mol. Phys.
MRC	Magn. Reson. Chem.
N	Naturwissenschaften
NAT	Nat.
NEP	Neth. Pat.
NJC	Nouv. J. Chim.
NKK	Nippon Kagaku Kaishi (J. Chem. Soc. Jpn.)
NKZ	Nippon Kagaku Zasshi
NZJ	N. Z. J. Sci. Technol.
OCS	Organomet. Synth.
OM	Organometallics
OMR	Org. Magn. Reson.
OMS	Org. Mass Spectrom.
OPP	Org. Prep. Proced. Int.
OR	Org. React.
OS	Org. Synth.
OSC	Org. Synth., Coll. Vol.
P	Phytochemistry
PA	Polym. Age
PAC	Pure Appl. Chem.

PAS	Pol. Acad. Sci.
PB	Polym. Bull.
PC	Personal Communication
PCS	Proc. Chem. Soc.
PHA	Pharmazi
PHC	Prog. Heterocycl. Chem.
PIA	Proc. Indian Acad. Sci.
PIA(A)	Proc. Indian Acad. Sci., Sect. A
PJC	Pol. J. Chem.
PJS	Pak. J. Sci. Ind. Res.
PMH	Phys. Methods Heterocycl. Chem.
PNA	Proc. Natl. Acad. Sci. USA
POL	Polyhedron
PP	Polym. Prepr.
PRS	Proceed. Roy. Soc.
PS	Phosphorus Sulfur
QR	Q. Rev., Chem. Soc.
QRS	Quart. Rep. Sulfur. Chem.
QSAR	Quant. Struct. Act. Relat. Pharmacol. Chem. Biol.
RC	Rubber Chem. Technol.
RCM	Rapid Commun. Mass Spectrom.
RCP	Rec. Chem. Prog.
RCR	Russ. Chem. Rev. (Engl. Transl.)
RHA	Rev. Heteroatom Chem.
RJ	Rubber J.
RP	Rev. Polarogr.
RRC	Rev. Roum. Chim.
RS	Ric. Sci.
RTC	Recl. Trav. Chim. Pays-Bas
RZC	Rocz. Chem.
S	Synthesis
SA	Spectrochim. Acta
SA(A)	Spectrochim. Acta, Part A
SAP	S. Afr. Pat.
SC	Synth. Commun.
SCI	Science
SL	Synlett
SM	Synth. Met.
SR	Sulfur Reports
SRI	Synth. React. Inorg. Metal-Org. Chem.
SS	Sch. Sci. Rev.
SST	Org. Compd. Sulphur, Selenium, Tellurium [R. Soc. Chem. series]
SUL	Sulfur Letters
SZP	Swiss Pat.
T	Tetrahedron
T(S)	Tetrahedron, Suppl.
TA	Tetrahedron Asymmetry
TAL	Talanta
TCA	Theor. Chim. Acta
TCC	Top. Curr. Chem.
TCM	Tetrahedron, Comp. Method
TFS	Trans. Faraday Soc.
TH	Thesis
TL	Tetrahedron Lett.
TS	Top. Stereochem.

UK	Usp. Khim.
UKZ	Ukr. Khim. Zh. (Russ. Ed.)
UP	Unpublished Results
URP	USSR Pat.
USP	US Pat.
WCH	Wiadom. Chem.
YGK	Yuki Gosei Kagaku Kyokaishi
YZ	Yakugaku Zasshi (J. Pharm. Soc. Jpn.)
ZAAC	Z. Anorg. Allg. Chem.
ZAK	Zh. Anal. Khim.
ZC	Z. Chem.
ZN	Z. Naturforsch.
ZN(A)	Z. Naturforsch., Teil A
ZN(B)	Z. Naturforsch., Teil B
ZOB	Zh. Obshch. Khim.
ZOR	Zh. Org. Khim.
ZPC	Hoppe–Seyler's Z. Physiol. Chem.
ZPK	Zh. Prikl. Khim.

VOLUME 6 REFERENCES

1854LA(92)346 A. W. Williamson and G. Kay; *Justus Liebigs Ann. Chem.*, 1854, **92**, 346. 68
1854PRS135 A. W. Williamson and G. Kay; *Proceed. Roy. Soc.*, 1854, **7**, 135. 68

1866BSF435 M. H. Gal; *Bull. Soc. Chim. Fr.*, 1866, **6**, 435. 452

1869BSF198 M. D. Schützenberger; *Bull. Soc. Chim. Fr.*, 1869, **12**, 198. 412
1869ZC624 O. Loew; *Z. Chem.*, 1869, 624. 221

1870CB730 R. Gerstl; *Ber. Dtsch. Chem. Ges.*, 1870, **3**, 730. 412
1870LA(156)60 T. Bolas and C. E. Groves; *Justus Liebigs Ann. Chem.*, 1870, **156**, 60. 215

1872JPR(2)477 F. Salomon; *J. Prakt. Chem. (2)*, 1872, **5**, 477. 471

1877CB185 S. Gabriel; *Ber. Dtsch. Chem. Ges.*, 1877, **10**, 185. 81
1877CB678 J. H. van't Hoff; *Ber. Dtsch. Chem. Ges.*, 1877, **10**, 678. 216, 221
1877LA(189)159 H. Schiff; *Justus Liebigs Ann. Chem.*, 1877, **189**, 159. 508

1878BSF(2)185 M. Tscherniak; *Bull. Soc. Chim. Fr., Part 2*, 1878, **30**, 185. 610
1878CB2265 M. Dennstedt; *Ber. Dtsch. Chem. Ges.*, 1878, **11**, 2265. 81
1878G233 E. Paternò; *Gazz. Chim. Ital.*, 1878, **8**, 233. 411, 412

1879CB115 A. Deutsch; *Ber. Dtsch. Chem. Ges.*, 1879, **12**, 115. 68

1882CB2685 F. Tiemann; *Ber. Dtsch. Chem. Ges.*, 1882, **13**, 2685. 69

1883CB352 A. Pinner; *Ber. Dtsch. Chem. Ges.*, 1883, **16**, 352. 70
1883CB1643 A. Pinner; *Ber. Dtsch. Chem. Ges.*, 1883, **16**, 1643. 70

1884CB297 B. Rathke; *Ber. Dtsch. Chem. Ges.*, 1884, **17**, 297. 641

1887JPR99 W. Hentschel; *J. Prakt. Chem.*, 1887, **36**, 99. 415
1887LA(240)192 M. Arnhold; *Justus Liebigs Ann. Chem.*, 1887, **240**, 192. 227
1887LA(240)225 R. Höland; *Justus Liebigs Ann. Chem.*, 1887, **240**, 225. 213, 227

1890CB1414 E. Laves; *Ber. Dtsch. Chem. Ges.*, 1890, **23**, 1414. 91
1890CB3226 E. Stauffer; *Ber. Dtsch. Chem. Ges.*, 1890, **23**, 3226. 254

1892CB188 A. Besson; *Ber. Dtsch. Chem. Ges.*, 1892, **25**, 188. 222
1892CB347 E. Laves; *Ber. Dtsch. Chem. Ges.*, 1892, **25**, 347. 90, 91
1892CB361 E. Laves; *Ber. Dtsch. Chem. Ges.*, 1892, **25**, 361. 90

1893CB1990 H. Erdmann; *Ber. Dtsch. Chem. Ges.*, 1893, **26**, 1990. 412

1895BSF444 A. Besson; *Bull. Soc. Chim. Fr.*, 1895, **13**, 444. 420
1895JPR(2)472 T. Curtius and K. Heidenreich; *J. Prakt. Chem. (2)*, 1895, **52**, 472. 511
1895MI 614-01 F. Lengfeld and J. Stieglitz; *Am. Chem. J.*, 1895, **17**, 98. 452

1897JPR71 E. Beckmann; *J. Prakt. Chem.*, 1897, **56**, 71. 570, 571

1898LA(300)145 A. Einhorn; *Justus Liebigs Ann. Chem.*, 1898, **300**, 145. 484

1899CB1116 L. Gattermann; *Ber. Dtsch. Chem. Ges.*, 1899, **32**, 1116. 448
1899GEP114396 Farbenfabrik Bayer, *Ger. Pat.* 114 396 (1899). 484

00CB1058 M. Busch and P. Bauer; *Ber. Dtsch. Chem. Ges.*, 1900, **33**, 1058. 653

03CB1280 L. Knorr and P. Rössler; *Ber. Dtsch. Chem. Ges.*, 1903, **36**, 1280. 485
03LA(329)116 J. Sand and F. Singer; *Justus Liebigs Ann. Chem.*, 1903, **329**, 116. 207
03MI 607-01 F. Swarts; *Chem. Zentr.*, 1903, **II**, 709. 216

05CB561 A. E. Tschitschibabin; *Ber. Dtsch. Chem. Ges.*, 1905, **38**, 561. 77

06CB857 H. Leuchs; *Ber. Dtsch. Chem. Ges.*, 1906, **39**, 857. 486
06LA(345)334 A. von Bartal; *Justus Liebigs Ann. Chem.*, 1906, **345**, 334. 417, 420

07CB1740 B. Holmberg; *Ber. Dtsch. Chem. Ges.*, 1907, **40**, 1740. 83, 84, 91
07LA(353)131 B. Holmberg; *Justus Liebigs Ann. Chem.*, 1907, **353**, 131. 83, 84
07MI 614-01 A. von Bartal; *Z. Anorg. Chem.*, 1907, **56**, 49. 420

09JA1220 W. M. Dehn; *J. Am. Chem. Soc.*, 1909, **31**, 1220. 213, 223

10JCS798 L. Mond, H. Hirtz and M. D. Cowap; *J. Chem. Soc.*, 1910, 798. 522, 524

11BSF1308 J. I. Iotsitch and F. Kochlelov; *Bull. Soc. Chim. Fr.*, 1911, **10**, 1308. 77
11CB3235 J. Houben and K. M. L. Schultze; *Ber. Dtsch. Chem. Ges.*, 1911, **44**, 3235. 84
11LA(384)322 F. Arndt; *Justus Liebigs Ann. Chem.*, 1911, **384**, 322. 306

12CB2942 J. Houben; *Ber. Dtsch. Chem. Ges.*, 1912, **45**, 2942. 84

13CB1426 H. Staudinger and E. Anthes; *Ber. Dtsch. Chem. Ges.*, 1913, **46**, 1426. 412, 417, 418
13CB2447 J. Houben and E. Schmidt; *Ber. Dtsch. Chem. Ges.*, 1913, **46**, 2447. 625
13LA(396)1 F. Arndt; *Justus Liebigs Ann. Chem.*, 1913, **396**, 1. 302

14JPR415 H. J. Prins; *J. Prakt. Chem.*, 1914, **89**, 415. 4

15CR(156)652 M. Lantenois; *C. R. Hebd. Séances Acad. Sci.*, 1915, **156**, 652. 216

18CB669 H. Rathsburg; *Ber. Dtsch. Chem. Ges.*, 1918, **51**, 669. 222, 226
B-18MI 607-01 B. Prager, P. Jacobsen, P. Schmidt and D. Stern; in "Beilsteins Handbuch der Organischen Chemie," 4th edn., Hauptwerk, Deutsche Chemische Gesellschaft, Springer-Verlag, Berlin, 1918. 118

19CR(169)17 V. Grignard and E. Urbain; *C. R. Hebd. Seances Acad. Sci.*, 1919, **169**, 17. 412, 413
19JPC498 H. P. Hood and H. R. Murdock; *J. Phys. Chem.*, 1919, **23**, 498. 415

20CB601 J. von Braun and K. Räth; *Ber. Dtsch. Chem. Ges.*, 1920, **53**, 601. 444
20JCS708 F. D. Chattaway and E. Saerens; *J. Chem. Soc.*, 1920, **117**, 708. 483
20JCS1410 R. H. Atkinson, C. T. Heycock and W. J. Pope; *J. Chem. Soc.*, 1920, **117**, 1410. 411, 412
20JPR79 W. Steinkopf and J. Herold; *J. Prakt. Chem.*, 1920, **101**, 79. 409

21JCS1222 C. K. Ingold and W. J. Powell; *J. Chem. Soc.*, 1921, **119**, 1222. 69

22AG489 F. Feist; *Angew. Chem.*, 1922, **35**, 489. 252

23CB979 H. Pauly and H. Schanz; *Ber. Dtsch. Chem. Ges.*, 1923, **56**, 979. 27
23JCS2882 D. L. Hammick; *J. Chem. Soc.*, 1923, 2882. 28

24JA2090 J. E. Driver; *J. Am. Chem. Soc.*, 1924, **46**, 2090. 69
24JCS442 R. A. Gotts and L. Hunter; *J. Chem. Soc.*, 1924, **125**, 442. 261
24JCS(125)2259 C. F. Allpress and W. Maw; *J. Chem. Soc.*, 1924, **125**, 2259. 467
24MI 614-01 E. Biesalski; *Z. Angew. Chem.*, 1924, **37**, 314 (*Chem. Abstr.*, 1924, **18**, 2480). 412

25JPR125 F. Mauthner; *J. Prakt. Chem.*, 1925, **110**, 125. 27

26BSB412 F. Swarts; *Bull. Soc. Chim. Belg.*, 1926, **35**, 412. 68
26G847 M. Garino and E. Teofili; *Gazz. Chim. Ital.*, 1926, **56**, 847. 420
26JA1736 H. E. French and A. F. Wirtel; *J. Am. Chem. Soc.*, 1926, **48**, 1736. 504

26JCS2263	W. R. H. Hurtley and S. Smiles; *J. Chem. Soc.*, 1926, 2263.	83
26LA(446)260	E. Bamberger, R. Padova and E. Ormerod; *Justus Liebigs Ann. Chem.*, 1926, **446**, 260.	653
27BSF1041	A. Job and A. Cassal; *Bull. Soc. Chim. Fr.*, 1927, **41**, 1041.	522
27BSF1251	J. F. Durand; *Bull. Soc. Chim. Fr.*, 1927, **41**, 1251.	216
27CB1424	E. Speyer and H. Wolf; *Ber. Dtsch. Chem. Ges.*, 1927, **60**, 1424.	523
27CRV109	G. M. Dyson; *Chem. Rev.*, 1927, **4**, 109.	411, 412
27JCS2658	J. E. Marsh and R. J. F. Struthers; *J. Chem. Soc.*, 1927, 2658.	207
28AP(266)277	K. W. Rosenmund and H. Döring; *Arch. Pharm.* (*Weinheim, Ger.*), 1928, **266**, 277.	431
28CR(187)564	A. Job and J. Rouvillois; *C. R. Hebd. Seances Acad. Sci.*, 1928, **187**, 564.	522
28GEP516281	G. Steimmig and M. Wittwer, *Ger. Pat.* 516 281 (1928) (*Chem. Abstr.*, 1931, **25**, 1840).	467
28G443	L. Belladen, M. Noli and A. Sommariva; *Gazz. Chim. Ital.*, 1928, **58**, 443.	412
28JA516	P. P. T. Sah; *J. Am. Chem. Soc.*, 1928, **50**, 516.	70
B-28MI 607-01	F. Richter; in "Beilsteins Handbuch der Organischen Chemie," 4th edn., Erstes Ergänzungswerk, Deutsche Chemische Gesellschaft, Springer-Verlag, Berlin, 1928.	213
29MI 616-01	G. Eyber; *Z. Phys. Chem.*, 1929, **144A**, 1.	523
30CB1221	W. Manchot and G. Lehmann; *Ber. Dtsch. Chem. Ges.*, 1930, **63**, 1221.	412, 418
30GEP531402	M. Naumann; *Ger. Pat.* 531 402 (1930) (*Chem. Abstr.*, 1931, **25**, 5523).	522
30JPR81	V. V. Nekrasov and N. N. Mel'nikov; *J. Prakt. Chem.*, 1930, **126**, 81.	230
30JPR210	V. V. Nekrasov and N. N. Mel'nikov; *J. Prakt. Chem.*, 1930, **126**, 210.	230
30RTC1107	H. J. Backer; *Recl. Trav. Chim. Pays-Bas*, 1930, **49**, 1107.	91
31GEP547025	M. Naumann; *Ger. Pat.* 547 025 (1931) (*Chem. Abstr.*, 1932, **26**, 3632).	522
31JCS2637	D. T. Gibson; *J. Chem. Soc.*, 1931, 2637.	91
31LA(489)7	L. Birckenbach and K. Sennewald; *Justus Liebigs Ann. Chem.*, 1931, **489**, 7.	612
B-31MI 615-01	H. Meyer; "Analyse und Konstitutionsermittlung organischer Verbindungen," Springer, Berlin, 1931, p. 367.	484
31USP1961622	H. S. Nutting and P. S. Petrie (Dow Chem. Co.); *US Pat.*, 1961622 (1931) (*Chem. Abstr.*, 1934, **28**, 4746).	212, 225
31ZAAC(201)245	O. Ruff and R. Keim; *Z. Anorg. Allg. Chem.*, 1931, **201**, 245.	217, 220
32CB65	G. Endres; *Ber. Dtsch. Chem. Ges.*, 1932, **65**, 65.	612
32CB546	L. Birchenbach and K. Sennewald; *Ber. Dtsch. Chem. Ges.*, 1932, **65**, 546.	222, 227
32CB1192	M. Bergmann and L. Zervas; *Ber. Dtsch. Chem. Ges.*, 1932, **65**, 1192.	487
32JA1622	C. A. Kraus and H. S. Nutting; *J. Am. Chem. Soc.*, 1932, **54**, 1622.	185
32JA2025	W. H. Hunter and D. E. Edgar; *J. Am. Chem. Soc.*, 1932, **54**, 2025.	215
32JA2964	P. P. T. Sah and T. S. Ma; *J. Am. Chem. Soc.*, 1932, **54**, 2964.	68
33BSB102	F. Swarts; *Bull. Soc. Chim. Belg.*, 1933, **42**, 102.	214, 225
33CB567	A. Schönberg and A. Stephenson; *Ber. Dtsch. Chem. Ges.*, 1933, **66**, 567.	253, 254
33JA3851	H. W. Post and E. R. Erickson; *J. Am. Chem. Soc.*, 1933, **55**, 3851.	68, 79
33JCS306	D. W. Cowie and D. T. Gibson; *J. Chem. Soc.*, 1933, 306.	90
33JIC537	H. C. Goswami and P. B. Sarker; *J. Indian Chem. Soc.*, 1933, **10**, 537.	422
33MI 603-01	P. P. T. Sah; *J. Chinese Chem. Soc.*, 1933, **1**, 100 (*Chem. Abstr.*, 1933, **27**, 5729).	70
33RTC437	H. J. Backer and P. L. Stedehouder; *Recl. Trav. Chim. Pays Bas*, 1933, **52**, 437.	81, 91
33RTC923	H. J. Backer and P. L. Stedehouder; *Recl. Trav. Chim. Pays-Bas*, 1933, **52**, 923.	306
33RTC1039	H. J. Backer and P. L. Stedehouder; *Recl. Trav. Chim. Pays-Bas*, 1933, **52**, 1039.	306, 307
34BSF244	A. Wahl; *Bull Soc. Chim. Fr.*, 1934, **1**, 244.	500
34GEP676049	K. W. Rosenmund, *Ger. Pat.* 676 049 (1934) (*Chem. Abstr.*, 1939, **33**, 6529).	484
34JA1455	H. G. Vesper and G. K. Rollefson; *J. Am. Chem. Soc.*, 1934, **56**, 1455.	221
34JA2007	M. P. Matuszak; *J. Am. Chem. Soc.*, 1934, **56**, 2007.	429
34JCS822	J. M. Connolly and G. M. Dyson; *J. Chem. Soc.*, 1934, 822.	232
34ZAAC(221)154	O. Ruff and G. Miltschitzky; *Z. Anorg. Allgem. Chem.*, 1934, **221**, 154.	409
35CB1513	A. Binz and B. Hughes; *Ber. Dtsch. Chem. Ges.*, 1935, **68**, 1513.	258
35CB2151	H. Scheibler and M. Depner; *Ber. Dtsch. Chem. Ges.*, 1935, **68B**, 2151.	69
35JA217	E. P. Kohler and M. Tishler; *J. Am. Chem. Soc.*, 1935, **57**, 217.	254
35JA2480	L. G. S. Brooker and F. L. White; *J. Am. Chem. Soc.*, 1935, **57**, 2480.	70
35RTC249	H. J. Prins; *Recl. Trav. Chim. Pays-Bas*, 1935, **54**, 249.	4
35RTC307	H. J. Prins and F. J. W. Engelhard; *Recl. Trav. Chim. Pays-Bas*, 1935, **54**, 307.	4
35ZAAC(221)321	W. Hieber and E. Romberg; *Z. Anorg. Allg. Chem.*, 1935, **221**, 321.	522
36BCJ157	H. Suginome and S. Umezawa; *Bull. Chem. Soc. Jpn.*, 1936, **11**, 157.	49
36CB299	O. Ruff; *Ber. Dtsch. Chem. Ges.*, 1936, **69**, 299.	212
36CB684	O. Ruff and M. Giese; *Ber. Dtsch. Chem. Ges.*, 1936, **69**, 684.	441
36CB1356	H. G. Grimm and H. Metzger; *Ber. Dtsch. Chem. Ges.*, 1936, **69**, 1356.	597
36CB2358	J. Houben and R. Zivadinovitsch; *Ber. Dtsch. Chem. Ges.*, 1936, **69**, 2358.	625
36JA529	F. Beyerstedt and S. M. McElvain; *J. Am. Chem. Soc.*, 1936, **58**, 529.	74
36JCS1273	E. N. Abrahart; *J. Chem. Soc.*, 1936, 1273.	501

36JCS1693 F. D. Chattaway, J. G. N. Drewitt and G. D. Parkes; *J. Chem. Soc.*, 1936, 1693. 653
36ZAAC(226)385 W. Manchot and W. J. Manchot; *Z. Anorg. Allg. Chem.*, 1936, **226**, 385. 524

37CB1080 W. Koblitz, H. Meissner and H. J. Schumacher; *Ber. Dtsch. Chem. Ges.*, 1937, **70**, 1080. 418
37CR(204)1927 J. Lecomte, H. Volringer and A. Tchakirian; *C. R. Hebd. Séances Acad. Sci.*, 1937, **204**, 1927. 221, 222, 227
37FRP820795 (I. G. Farbenindustrie AG); *Fr. Pat.* 820795 (1937) (*Chem. Abstr.*, 1938, **32**, 3422). 232
37GEP68065 M. Mugdan (Consortium für Elektrochemische Industrie); *Ger. Pat.* 680659 (1937). 215
37JA1200 A. L. Henne; *J. Am. Chem. Soc.*, 1937, **59**, 1200. 212, 218
37JA1273 F. Beyerstedt and S. M. McElvain; *J. Am. Chem. Soc.*, 1937, **59**, 1273. 80
37JA2266 F. Beyerstedt and S. M. McElvain; *J. Am. Chem. Soc.*, 1937, **59**, 2266. 74
37JCS827 J. M. Connolly and G. M. Dyson; *J. Chem. Soc.*, 1937, 827. 297
37JOC76 M. S. Kharasch, W. G. Alsop and F. R. Mayo; *J. Org. Chem.*, 1937, **2**, 76. 217

38BRP479774 (I. G. Farbenindustrie AG); *Br. Pat.* 479774 (1938) (*Chem. Abstr.*, 1938, **32**, 5226). 232
38USP2108606 F. Müller, O. Scherer and W. Schumacher (IG-Farben); *US Pat.* 2108606 (1938) (*Chem. Abstr.*, 1938, **32**, 2958). 234

39AG457 I. G. Scherer; *Angew. Chem.*, 1939, **52**, 457. 233
39GEP740366 K. Vierling, *Ger. Pat.* 740 366 (1939) (*Chem. Abstr.*, 1945, **39**, 2294). 467
39JA435 J. H. Simons, T. K. Sloat and A. C. Meunier; *J. Am. Chem. Soc.*, 1939, **61**, 435. 215, 216, 221
39JA2962 J. H. Simons and L. P. Bloch; *J. Am. Chem. Soc.*, 1939, **61**, 2962. 212
39JCS781 P. Dyson and D. L. Hammick; *J. Chem. Soc.*, 1939, 781. 28
39ZAAC(240)261 W. Hieber, H. Schulten and R. Marin; *Z. Anorg. Allg. Chem.*, 1939, **240**, 261. 524

40CB177 M. Linhard and K. Betz; *Ber. Dtsch. Chem. Ges.*, 1940, **73**, 177. 439, 452
40DOK(26)54 K. A. Kocheshkov, A. N. Nesmeyanov, M. M. Nadj, I. M. Rossinskaya and L. M. Borissova; *Dokl. Akad. Nauk SSSR*, 1940, **26**, 54 (*Chem. Abstr.*, 1940, **34**, 6183). 522
40DOK(26)57 K. N. Anisimov and A. N. Nesmeyanov; *Dokl. Akad. Nauk SSSR*, 1940, **26**, 57 (*Chem. Abstr.*, 1940, **34**, 6183). 522
40JA3477 J. H. Simons, R. L. Bond and R. E. McArthur; *J. Am. Chem. Soc.*, 1940, **62**, 3477. 212, 217
40USP2218490 American Cyanamid Co.; *US Pat.* 2 218 490 (1940) (*Chem. Abstr.*, 1941, **35**, 1185). 651
40ZAAC(245)321 W. Hieber and H. Lagally; *Z. Anorg. Allg. Chem.*, 1940, **245**, 321. 524, 525

41CB1667 H. Böhme and R. Marx; *Ber. Dtsch. Chem. Ges.*, 1941, **74**, 1667. 90
41JA3476 A. L. Henne and F. W. Haeckl; *J. Am. Chem. Soc.*, 1941, **63**, 3476. 2
41JA3478 A. L. Henne, A. M. Whaley and J. K. Stevenson; *J. Am. Chem. Soc.*, 1941, **63**, 3478. 16
B-41MI 607-01 F. Richter; "Beilsteins Handbuch der Organischen Chemie," 4th edn., Zweites Ergänzungswerk, Deutsche Chemische Gesellschaft, Springer-Verlag, Berlin, 1941. 213
41ZAAC(248)256 W. Hieber and H. Fuchs; *Z. Anorg. Allg. Chem.*, 1941, **248**, 256. 523

42JA1342 E. G. Bohlman and J. E. Willard; *J. Am. Chem. Soc.*, 1942, **64**, 1342. 221, 226
42JA1825 S. M. McElvain and J. W. Nelson; *J. Am. Chem. Soc.*, 1942, **64**, 1825. 70, 77
42JA1963 S. M. McElvain and P. M. Walters; *J. Am. Chem. Soc.*, 1942, **64**, 1963. 74, 80
42JA2515 J. Malkeil and J. P. Mason; *J. Am. Chem. Soc.*, 1942, **64**, 2515. 27
42JA2525 S. M. McElvain, H. I. Anthes and S. H. Shapiro; *J. Am. Chem. Soc.*, 1942, **64**, 2525. 69, 74, 80

43JA613 S. Winstein and R. E. Buckles; *J. Am. Chem. Soc.*, 1943, **65**, 613. 71
43JA2390 D. M. Hubbard and E. W. Scott; *J. Am. Chem. Soc.*, 1943, **65**, 2390. 653
B-43MI 603-01 H. W. Post; "The Chemistry of the Aliphatic Ortho Esters," Reinhold, New York, 1943. 68, 77, 81
43MI 616-01 W. Hieber and H. Stallman; *Z. Electrochem.*, 1943, **49**, 288. 524
43OSC(2)462 E. D. Amstutz and R. R. Myers; *Org. Synth., Coll. Vol.*, 1943, **2**, 462. 506
43ZAAC(251)96 H. Lagally; *Z. Anorg. Allg. Chem.*, 1943, **251**, 96. 525

44JCS74 W. S. Rapson, D. H. Saunder and E. T. Stewart; *J. Chem. Soc.*, 1944, 74. 25
44JOC419 F. C. Schmidt, C. E. Sunderling and Q. P. Cole; *J. Org. Chem.*, 1944, **9**, 419. 261
44MI 614-01 S.-C. Li; *J. Chin. Chem. Soc.*, 1944, **11**, 14 (*Chem. Abstr.*, 1945, **39**, 1099). 409
44USP2351390 A. F. Benning and J. D. Park; *US Pat.* 2351 390 (1944) (*Chem. Abstr.*, 1944, **38**, 5228). 214
44USP2365981 J. B. Tindall (Commercial Solvents Corp.); *US Pat.* 2365981, 1944 (*Chem. Abstr.*, 1946, **40**, 3126). 241

45BAU364 M. I. Kabachnik and P. A. Rossiiskaya; *Bull. Acad. Sci. USSR, Div. Chim. Sci.*, 1945, 364 (*Chem. Abstr.*, 1946, **40**, 4688). 455
45JA650 S. M. McElvain and B. Fajardo-Pinzon; *J. Am. Chem. Soc.*, 1945, **67**, 650. 74
45JA1642 H. Soroos and J. B. Hinkamp; *J. Am. Chem. Soc.*, 1945, **67**, 1642. 216, 217
45JA2079 R. Duschinsky and L. A. Dolan; *J. Am. Chem. Soc.*, 1945, **67**, 2079. 508
45JCS864 G. D. Buckley, H. A. Piggott and A. J. E. Welch; *J. Chem. Soc.*, 1945, 864. 439
45USP2389153 J. D. Kendall; *US Pat.* 2 389 153 (1945) (*Chem. Abstr.*, 1946, **40**, 1540). 84

46JA968 T. J. Brice, W. H. Pearlson and J. H. Simons; *J. Am. Chem. Soc.*, 1946, **68**, 968. 218
46JA1672 J. H. Simons, D. F. Herman and W. H. Pearlson; *J. Am. Chem. Soc.*, 1946, **68**, 1672. 419
46JA1922 S. M. McElvain, R. E. Kent and C. L. Stevens; *J. Am. Chem. Soc.*, 1946, **68**, 1922. 70, 80
46MI 616-01 R. P. Lastovskii; *J. Appl. Chem. (USSR)*, 1946, **19**, 440. 500

46RTC53 H. J. Backer; *Recl. Trav. Chim. Pays-Bas*, 1946, **65**, 53. 91, 275
46ZN(B)518 A. M. Paquin; *Z. Naturforsch., Teil B*, 1946, **1**, 518. 484
46ZOB1521 G. Kamai and L. P. Egorova; *Zh. Obshch. Khim.*, 1946, **16**, 1521 (*Chem. Abstr.*, 1947, **41**, 5439). 242

47CI(L)427 W. B. Whalley; *Chem. Ind. (London)*, 1947, **66**, 427. 217
47IEC404 E. T. McBee, H. B. Hass, L. W. Frost and Z. D. Welch; *Ind. Eng. Chem. Res.*, 1947, **39**, 404. 220, 225
47JA947 E. T. McBee, R. O. Bolt, P. J. Graham and R. F. Tebbe; *J. Am. Chem. Soc.*, 1947, **69**, 947. 220
47JA1100 M. S. Kharasch, E. V. Jensen and W. H. Urry; *J. Am. Chem. Soc.*, 1947, **69**, 1100. 5
47JA1723 B. B. Owen, J. English Jr., H. G. Cassidy and C. V. Dundon; *J. Am. Chem. Soc.*, 1947, **69**, 1723. 522
47JA1820 A. L. Henne and P. Trott; *J. Am. Chem. Soc.*, 1947, **69**, 1820. 16
47JA1833 R. E. Dunbar and E. P. Painter; *J. Am. Chem. Soc.*, 1947, **69**, 1833. 596
47JA3150 R. Duschinsky, L. A. Dolan, L. O. Randall and G. Lehmann; *J. Am. Chem. Soc.*, 1947, **69**, 3150. 508
47JCS674 F. R. Atherton and A. R. Todd; *J. Chem. Soc.*, 1947, 674. 221

48JA383 H. Adkins and G. Krsek; *J. Am. Chem. Soc.*, 1948, **70**, 383. 524
48JA1968 J. L. Hutson and T. H. Norris; *J. Am. Chem. Soc.*, 1948, **70**, 1968. 415
48JA2268 W. E. Mochel, C. L. Agre and W. E. Hanford; *J. Am. Chem. Soc.*, 1948, **70**, 2268. 80
48JA3986 K. B. Kellogg and G. H. Cady; *J. Am. Chem. Soc.*, 1948, **70**, 3986. 231
48JCS687 J. D. Kendall and J. R. Majer; *J. Chem. Soc.*, 1948, 687. 84
48JCS2183 H. J. Emeléus and J. F. Wood; *J. Chem. Soc.*, 1948, 2183. 409, 419, 422
48JCS2188 A. A. Banks, H. J. Emeléus, R. N. Haszeldine and V. Kerrigan; *J. Chem. Soc.*, 1948, 2188. 212, 214, 217, 218, 219, 220, 222, 225, 226, 246
48JOC265 H. Tieckelmann and H. W. Post; *J. Org. Chem.*, 1948, **13**, 265. 296, 297, 308
B-48MI 607-01 E. H. Huntress; "Organic Chlorine Compounds," John Wiley & Sons, New York, 1948, p. 576. 214
48RTC884 H. J. Backer and G. J. de Jong; *Recl. Trav. Chim. Pays-Bas*, 1048, **67**, 884. 90
48RTC894 H. J. Backer; *Recl. Trav. Chim. Pays-Bas*, 1948, **67**, 894. 307

49AG183 H. Hopff and H. Ohlinger; *Angew. Chem.*, 1949, **61**, 183. 443
49AK(1)231 E. Samén; *Ark. Kemi*, 1949, **1**, 231. 91
49DOK(68)515 B. A. Arbuzov and T. G. Shavsha; *Dokl. Akad. Nauk SSSR*, 1949, **68**, 515 (*Chem. Abstr.*, 1950, **44**, 392). 297
49JA298 A. L. Henne and W. G. Finnegan; *J. Am. Chem. Soc.*, 1949, **71**, 298. 16
49JA2297 R. A. Henry and W. M. Dehn; *J. Am. Chem. Soc.*, 1949, **71**, 2297. 504
49JA2499 T. J. Brice, W. H. Pearlson and J. H. Simons; *J. Am. Chem. Soc.*, 1949, **71**, 2499. 218
49JCS2856 R. N. Haszeldine; *J. Chem. Soc.*, 1949, 2856. 219
49JCS2948 H. J. Emeléus and R. N. Haszeldine; *J. Chem. Soc.*, 1949, 2948. 2
49JCS2953 H. J. Emeléus and R. N. Haszeldine; *J. Chem. Soc.*, 1949, 2953. 246
49JOC747 D. D. Coffman, M. S. Raasch, G. W. Rigby, P. L. Barrick and W. E. Hanford; *J. Org. Chem.*, 1949, **14**, 747. 2
49LA(562)229 A. Bertho; *Justus Liebigs Ann. Chem.*, 1949, **562**, 229. 105
49MI 603-01 R. Barre and B. Ladouceur; *Can. J. Res.*, 1949, **B27**, 61 (*Chem. Abstr.*, 1949, **43**, 5365). 77
49MI 607-01 J. H. Simons; *J. Electrochem. Soc.*, 1949, **95**, 47 (*Chem. Abstr.*, 1949, **43**, 2876). 214
49USP2473993 W. F. Gresham and J. V. E. Hardy; *US Pat.* 2 473 993 (1949) (*Chem. Abstr.*, 1949, **43**, 9398). 524
49USP2476263 C. H. McKeever; *US Pat.* 2 476 263 (1949) (*Chem. Abstr.*, 1949, **43**, 9397). 524
49USP2477553 C. H. McKeever; *US Pat.* 2 477 553 (1949) (*Chem. Abstr.*, 1949, **43**, 8107). 524
49USP2477554 C. H. McKeever; *US Pat.* 2 477 554 (1949) (*Chem. Abstr.*, 1949, **43**, 8108). 524

50ACS397 E. Samén; *Acta Chem. Scand.*, 1950, **4**, 397. 91
50DOK(70)231 B. A. Arbusov and E. K. Yuldasheva; *Dokl. Akad. Nauk SSSR*, 1950, **70**, 231 (*Chem. Abstr.*, 1950, **44**, 3 759). 78
50IS37 R. E. McArthur and J. H. Simons; *Inorg. Synth.*, 1950, **3**, 37. 216
50IS156 B. B. Owen, J. English, Jr., H. G. Cassidy and C. V. Dundon; *Inorg. Synth.*, 1950, **3**, 156. 522
50IS171 H. F. Priest; *Inorg. Synth.*, 1950, **3**, 171. 212
50JA1661 S. M. McElvain and J. T. Venerable; *J. Am. Chem. Soc.*, 1950, **72**, 1661. 69
50JA1888 R. J. Slocombe, E. E. Hardy, J. H. Saunders and R. L. Jenkins; *J. Am. Chem. Soc.*, 1950, **72**, 1888. 443
50JA2213 J. Harmon, T. A. Ford, W. E. Hanford and R. M. Joyce; *J. Am. Chem. Soc.*, 1950, **72**, 2213. 5
50JA3529 C. M. Buess and N. Kharasch; *J. Am. Chem. Soc.*, 1950, **72**, 3529. 233
50JA3642 K. E. Rapp, R. L. Pruett, J. T. Barr, C. T. Bahner, J. D. Gibson and R. H. Lafferty, Jr.; *J. Am. Chem. Soc.*, 1950, **72**, 3642. 46
50JA5012 E. D. Bergmann, D. Ginsberg and D. Lavie; *J. Am. Chem. Soc.*, 1950, **72**, 5012. 25
50JA5329 N. S. Marans and R. P. Zelinski; *J. Am. Chem. Soc.*, 1950, **72**, 5329. 145
50LA(566)229 H. Bestian; *Justus Liebigs Ann. Chem.*, 1950, **566**, 229. 483
50RTC1223 H. J. Backer and H. Bos; *Recl. Trav. Chim. Pays-Bas*, 1950, **69**, 1223. 114
50TFS295 F. S. Dainton and K. J. Irvin; *Trans. Faraday Soc.*, 1950, **46**, 295. 215
50USP2500388 J. H. Simons (Minnesota Mining & Mfg. Co.); *US Pat.* 2500388 (1950) (*Chem. Abstr.*, 1950, **44**, 5236). 229
50USP2531372 H. Waterman (Minnesota Mining & Mfg. Co.); *US Pat.* 2531372 (1950) (*Chem. Abstr.*, 1951, **45**, 1705). 218

51JA1864 L. C. King and R. J. Hlavacek; *J. Am. Chem. Soc.*, 1951, **73**, 1864. 596
51JA2233 S. M. McElvain and B. E. Tate; *J. Am. Chem. Soc.*, 1951, **73**, 2233. 70
51JA2461 M. Hauptschein and A. V. Grosse; *J. Am. Chem. Soc.*, 1951, **73**, 2461. 219
51JA3796 J. H. Saunders, R. J. Slocombe and E. E. Hardy; *J. Am. Chem. Soc.*, 1951, **73**, 3796. 425
51JA3807 S. M. McElvain, S. B. Mirviss and C. L. Stevens; *J. Am. Chem. Soc.*, 1951, **73**, 3807. 74
51JA5168 R. G. Jones; *J. Am. Chem. Soc.*, 1951, **73**, 5168. 39
51JCS584 R. N. Haszeldine; *J. Chem. Soc.*, 1951, 584. 32, 216, 218, 219
51JCS1145 B. R. Brown, D. L. Hammick and B. H. Thewlis; *J. Chem. Soc.*, 1951, 1145. 28
51JOC1829 A. F. McKay and R. O. Braun; *J. Org. Chem.*, 1951, **16**, 1829. 508
51MI 614-01 H. J. Schumacher; *Anales de la Asociución Quimica Argentina*, 1951, **39**, 159 (*Chem. Abstr.*, 1953, **46**, 5439). 409
51M1008 R. Riemschneider; *Monatsh. Chem.*, 1951, **82**, 1008. 27
51OSC(1)258 W. E. Kaufmann and E. E. Dreger; *Org. Synth.*, *Coll. Vol.* 1, 1951, 258. 68
51RTC733 H. J. Backer; *Recl. Trav. Chim. Pays-Bas*, 1951, **70**, 733. 114
51RTC892 H. J. Backer; *Recl. Trav. Chim. Pays-Bas*, 1951, **70**, 892. 114
51RTC917 H. L. Klöpping and G. J. M. van der Kerk; *Recl. Trav. Chim. Pays-Bas*, 1951, **70**, 917. 560
51RTC949 H. L. Klöpping and G. J. M. van der Kerk; *Recl. Trav. Chim. Pays-Bas*, 1951, **70**, 949. 560
51USP2553518 D. E. Lake and A. A. Asadorian (The Dow Chemical Company); *US Pat.* 2553518 (1951) (*Chem. Abstr.*, 1952, **46**, 2561). 215
51USP2553800 J. P. West and L. Schmerling (Universal Oil Products Co.); *US Pat.* 2553800 (1951) (*Chem. Abstr.*, 1952, **46**, 2561). 221

52CB901 H. Schlenk; *Chem. Ber.*, 1952, **85**, 901. 27
52CB949 B. Witkop, J. B. Patrick and H. M. Kissman; *Chem. Ber.*, 1952, **85**, 949. 108
52HOU(8)84 S. Petersen; *Methoden Org. Chem.* (*Houben-Weyl*), 1952, **8**, 84. 413
52HOU(8)101 S. Petersen; *Methoden Org. Chem.* (*Houben-Weyl*), 1952, **8**, 101. 424
52JA554 E. R. Alexander and H. M. Busch; *J. Am. Chem. Soc.*, 1952, **74**, 554. 78
52JA848 M. Hauptschein, C. S. Stokes and A. V. Grosse; *J. Am. Chem. Soc.* 1952, **74**, 848. 32
52JA849 M. Hauptschein, R. L. Kinsman and A. V. Grosse; *J. Am. Chem. Soc.* 1952, **74**, 849. 32
52JA1347 M. Hauptschein, E. A. Nodiff and A. V. Grosse; *J. Am. Chem. Soc.*, 1952, **74**, 1347. 218
52JA1418 F. B. Smith and C. A. Kraus; *J. Am. Chem. Soc.*, 1952, **74**, 1418. 185
52JA1974 M. Hauptschein, C. S. Stokes and A. V. Grosse; *J. Am. Chem. Soc.* 1952, **74**, 1974. 32
52JA2292 J. D. Park, D. M. Griffin and J. R. Lacher; *J. Am. Chem. Soc.*, 1952, **74**, 2292. 36
52JA3594 W. E. Truce, G. H. Birum and E. T. McBee; *J. Am. Chem. Soc.*, 1952, **74**, 3594. 16, 44, 45, 234
52JA3855 B. Witkop and J. B. Patrick; *J. Am. Chem. Soc.*, 1952, **74**, 3855. 108
52JA4582 L. F. Cason and H. G. Brooks; *J. Am. Chem. Soc.*, 1952, **74**, 4582. 192
52JA5792 G. A. Silvey and G. H. Cady; *J. Am. Chem. Soc.*, 1952, **74**, 5792. 410
52JCS2198 G. A. R. Brandt, H. J. Emeléus and R. N. Haszeldine; *J. Chem. Soc.*, 1952, 2198. 233
52JCS3490 R. N. Haszeldine; *J. Chem. Soc.*, 1952, 3490. 5
52JCS4259 R. N. Haszeldine; *J. Chem. Soc.*, 1952, 4259. 213, 216, 220, 223, 224, 225, 226, 227, 228
52JCS4423 R. N. Haszeldine; *J. Chem. Soc.*, 1952, 4423. 2
52MI 601-01 F. Smith, M. Stacey, J. C. Tatlow, J. K. Dawson and B. R. J. Thomas; *J. Appl. Chem.*, 1952, **2**, 97. 9
52MI 607-01 L. I. Zakharkin; *Sint. Org. Soedin. Sbornik*, 1952, **2**, 18 (*Chem. Abstr.*, 1954, **48**, 547). 221
52MI 607-02 O. O. Orazi, N. M. I. Mercere, R. A. Corral and J. Meseri; *An. Asoc. Quim. Arg.*, 1952, **40**, 91 (*Chem. Abstr.*, 1953, **47**, 8667). 221
52RTC409 H. J. Backer; *Recl. Trav. Chim. Pays-Bas*, 1952, **71**, 409. 307
52RTC1071 H. J. Backer and E. Westerhuis; *Recl. Trav. Chim. Pays-Bas*, 1952, **71**, 1071. 301
52USP2592565 M. T. Harvey; *US Pat.* 2 592 565 (1952) (*Chem. Abstr.*, 1953, **47**, 601). 508
52USP2616927 E. A. Kauck and J. H. Simons (Minnesota Mining & Mfg. Co.); *US Pat.* 2616927 (1952) (*Chem. Abstr.*, 1953, **47**, 8771). 214
52ZOB2216 L. B. Yagupol'skii and A. I. Kiprianov; *Zhur. Obshch. Khim.*, 1952, **22**, 2216 (*Chem. Abstr.*, 1953, **47**, 4771). 232

53CB557 H. Brintzinger and S. M. Langheck; *Chem. Ber.*, 1953, **86**, 557. 233
53CB790 H. Stetter and K. H. Steinacker; *Chem. Ber.*, 1953, **86**, 790. 80
53CJC896 J. P. Picard and A. F. McKay; *Can. J. Chem.*, 1953, **31**, 896. 508
53JA671 C. W. Whitehead; *J. Am. Chem. Soc.*, 1953, **75**, 671. 104
53JA1120 A. F. McKay, W. G. Hatton and G. W. Taylor; *J. Am. Chem. Soc.*, 1953, **75**, 1120. 508
53JA1668 D. S. Tarbell and A. H. Herz; *J. Am. Chem. Soc.*, 1953, **75**, 1668. 83, 84
53JA3987 S. M. McElvain and C. L. Aldridge; *J. Am. Chem. Soc.*, 1953, **75**, 3987. 70
53JA4053 F. L. Scott, D. G. O'Donovan and J. Reilly; *J. Am. Chem. Soc.*, 1953, **75**, 4053. 641
53JA4582 I. B. Douglass and C. E. Osborne; *J. Am. Chem. Soc.*, 1953, **75**, 4582. 255
53JCS922 R. N. Haszeldine; *J. Chem. Soc.*, 1953, 922. 5, 221
53JCS1063 J. F. Ellis and W. K. R. Musgrave; *J. Chem. Soc.*, 1953, 1063. 220
53JCS1548 R. N. Haszeldine and K. Leedham; *J. Chem. Soc.*, 1953, 1548. 2
53JCS2075 R. N. Haszeldine; *J. Chem. Soc.*, 1953, 2075. 241
53JCS2372 A. F. Clifford, H. K. El-Shamy, H. J. Emeléus and R. N. Haszeldine; *J. Chem. Soc.*, 1953, 2372. 214
53JCS3219 R. N. Haszeldine and J. M. Kidd; *J. Chem. Soc.*, 1953, 3219. 232, 236
53JCS3607 R. N. Haszeldine and E. G. Walaschewski; *J. Chem. Soc.*, 1953, 3607. 64
53JCS3761 R. N. Haszeldine; *J. Chem. Soc.*, 1953, 3761. 5

53JGU229 M. S. Malinovsky and N. M. Medyansteva; *J. Gen. Chem. USSR* (*Engl. Transl.*), 1953, **23**, 229. 429
53JOC292 R. A. Zingaro, F. C. Bennett, Jr. and G. W. Hammar; *J. Org. Chem.*, 1953, **18**, 292. 596
53JOC328 M. S. Kharasch, E. Simon and W. Nudenberg; *J. Org. Chem.*, 1953, **18**, 328. 5
53LA(581)133 H. Böhme and H. J. Gran; *Justus Liebigs Ann. Chem.*, 1953, **581**, 133. 44, 91
53LA(582)116 W. Reppe; *Justus Liebigs Ann. Chem.*, 1953, **582**, 116. 524
53MI 603-01 T. R. Rix and J. F. Arens; *Koninkl. Ned. Akad. Wetenschap., Proc.*, 1953, **B56**, 364 (*Chem. Abstr.*, 1955, **49**, 2299). 74
53USP2639301 R. P. Ruh and R. A. Davis (Dow Chem. Co.); *US Pat.* 2639301 (1953) (*Chem. Abstr.*, 1954, **48**, 3382). 226
53USP2639302 R. P. Ruh and R. A. Davis (Dow Chem. Co.); *US Pat.* 2639302 (1953) (*Chem. Abstr.*, 1954, **48**, 3382). 223
53USP2647933 J. D. La Zerte, W. H. Pearlson and E. A. Kauck (Minnesota Mining & Mfg. Co.); *US Pat.* 2647933 (1953) (*Chem. Abstr.*, 1954, **48**, 7622). 32, 218
53USP2658086 R. P. Ruh and R. A. Davis (Dow Chem. Co.); *US Pat.* 2658086 (1953) (*Chem. Abstr.*, 1954, **48**, 12787). 226
53USP2658927 W. Kwasnik (Farbenfabr. Bayer AG); *US Pat.* 2658927 (1953) (*Chem. Abstr.*, 1954, **48**, 12787). 217

54CB205 H. Stetter and K. H. Steinacker; *Chem. Ber.*, 1954, **87**, 205. 80
54JA474 O. R. Pierce, E. T. McBee and G. F. Judd; *J. Am. Chem. Soc.*, 1954, **76**, 474. 244
54JA3831 E. O. Brimm, M. A. Lynch Jr. and W. J. Sesny; *J. Am. Chem. Soc.*, 1954, **76**, 3831. 523
54JA5141 D. R. Husted and W. L. Kohlhase; *J. Am. Chem. Soc.*, 1954, **76**, 5141. 218
54JA5736 S. M. McElvain, E. R. Degginger and J. D. Behun; *J. Am. Chem. Soc.*, 1954, **76**, 5736. 75
54JCS912 J. Jander and R. N. Haszeldine; *J. Chem. Soc.*, 1954, 912. 241
54JCS919 J. Jander and R. N. Haszeldine; *J. Chem. Soc.*, 1954, 919. 441
54JCS923 R. N. Haszeldine and B. R. Steele; *J. Chem. Soc.*, 1954, 923. 2
54JCS3598 F. W. Bennett, H. J. Emeléus and R. N. Haszeldine; *J. Chem. Soc.*, 1954, 3598. 242
54JGU885 L. M. Yagupol'skii and M. S. Marenets; *J. Gen. Chem. USSR* (*Engl. Transl.*), 1954, **24**, 885. 232
54JOC1278 L. F. Cason and H. G. Brooks; *J. Org. Chem.*, 1954, **19**, 1278. 192
54LA(587)226 H. P. Kaufmann, A. Seher and P. Hagedorn; *Justus Liebigs Ann. Chem.*, 1954, **587**, 226. 69
54USP2694739 J. R. Pailthorp (du Pont de Nemours & Co); *US Pat.* 2694739 (1954) (*Chem. Abstr.*, 1955, **49**, 12526). 217

55AG374 H. Meerwein; *Angew. Chem.*, 1955, **67**, 374. 71
55CI(M)6 G. Natta, R. Ercoli and S. Castellano; *Chim. Ind.* (*Milan*), 1955, **37**, 6. 524
55CRV181 D. C. Schroeder; *Chem. Rev.*, 1955, **55**, 181. 569, 575
55DOK(105)100 L. M. Yagupol'skii, *Dokl. Akad. Nauk SSSR*, 1955, **105**, 100 (*Chem. Abstr.*, 1956, **50**, 11270). 229
55JA509 W. von E. Doering and L. K. Levy; *J. Am. Chem. Soc.*, 1955, **77**, 509. 82, 83, 91
55JA768 P. Tarrant and A. M. Lovelace; *J. Am. Chem. Soc.*, 1955, **77**, 768. 5
55JA1899 I. B. Douglass and F. J. Marascia; *J. Am. Chem. Soc.*, 1955, **77**, 1899. 229
55JA2479 H. R. Al-Kazimi, D. S. Tarbell and D. Plant; *J. Am. Chem. Soc.*, 1954, **77**, 2479. 540
55JA2869 I. S. Bengelsdorf and L. B. Barron; *J. Am. Chem. Soc.*, 1955, **77**, 2869. 242
55JA3099 S. Nakanishi, T. C. Meyers and E. V. Jensen; *J. Am. Chem. Soc.*, 1955, **77**, 3099. 422
55JA3801 R. M. Roberts, T. D. Higgins, Jr. and P. R. Noyes; *J. Am. Chem. Soc.*, 1955, **77**, 3801. 78
55JA3951 R. A. Friedel, I. Wender, S. L. Shufler and H. W. Sternberg; *J. Am. Chem. Soc.*, 1955, **77**, 3951. 524
55JA5033 S. Nakanishi, T. C. Meyers and E. V. Jensen; *J. Am. Chem. Soc.*, 1955, **77**, 5033. 421
55JA5601 S. M. McElvain and G. R. McKay; *J. Am. Chem. Soc.*, 1955, **77**, 5601. 75
55JA6644 H. Gilman and R. G. Wilder; *J. Am. Chem. Soc.*, 1955, **77**, 6644. 69
55JCS1881 D. A. Barr and R. N. Haszeldine; *J. Chem. Soc.*, 1955, 1881. 604
55JCS2532 D. A. Barr and R. N. Haszeldine; *J. Am. Chem. Soc.*, 1955, 2532. 238
55JCS3871 R. N. Haszeldine and J. M. Kidd; *J. Chem. Soc.*, 1955, 3871. 534
55JOC1573 J. G. Erickson; *J. Org. Chem.*, 1955, **20**, 1573. 70
55MI 618-01 S. Kodama, S. Fukushima and S. Nose; *Bull. Chem. Soc. Jpn.*, 1955, **58**, 211 (*Chem. Abstr.*, 1955, **49**, 13685). 580
55OSC(3)773 R. L. Frank and P. V. Smith; *Org. Synth., Coll. Vol.*, 1955, **3**, 773. 572
55OSC(3)803 P. Liang; *Org. Synth. Coll. Vol.*, 1955, **3**, 803. 365
55USP2704776 J. D. LaZerte, W. H. Pearlson and E. A. Kauck; *US Pat.* 2704776 (1955) (*Chem. Abstr.*, 1956, **50**, 2650). 218
55USP2709184 E. L. Muetterties (du Pont de Nemours & Co.); *US Pat.* 2709184 (1955) (*Chem. Abstr.*, 1956, **50**, 6498). 217, 220, 225
55USP2709186 M. W. Farlow and E. L. Muetterties (du Pont de Nemours & Co.); *US Pat.* 2709186 (1955) (*Chem. Abstr.*, 1956, **50**, 6499). 212
55USP2709187 M. W. Farlow and E. L. Muetterties (du Pont de Nemours & Co.); *US Pat.* 2709187 (1955) (*Chem. Abstr.*, 1956, **50**, 6499). 212
55USP2709188 M. W. Farlow and E. L. Muetterties (du Pont de Nemours & Co.); *US Pat.* 2709188 (1955) (*Chem. Abstr.*, 1956, **50**, 6499). 212
55USP2709190 M. W. Farlow and E. L. Muetterties (du Pont de Nemours & Co.); *US Pat.* 2709190 (1955) (*Chem. Abstr.*, 1956, **50**, 6499). 212
55ZAAC(282)345 K. Ziegler, K. Nagel and M. Patheiger; *Z. Anorg. Allg. Chem.*, 1955, **282**, 345. 206

56ACS1006 B. Smith; *Acta Chem. Scand.*, 1956, **10**, 1006. 78, 79

56AG754 H. A. Staab; *Angew. Chem.*, 1956, **68**, 754. 416
56BRP749408 Société d'Electrochimie, d'Electro-Metallurgie et des Aciéries Electriques d'Ugine; *Br. Pat.* 749408 (1956) (*Chem. Abstr.*, 1956, **50**, 16058). 215
56BSF1441 H. Normant and J. Ficini; *Bull. Soc. Chim. Fr.*, 1956, 1441. 27
56CB314 B. Helfrerich and K. Weis; *Chem. Ber.*, 1956, **89**, 314. 71
56CB343 R. Mecke, Jr. and R. Mecke; *Chem. Ber.*, 1956, **89**, 343. 509
56CB1160 F. Boberg, G. Winter and G. R. Schultze; *Chem. Ber.*, 1956, **89**, 1160. 234
56CB2060 H. Meerwein, P. Borner, O. Fuchs, H. J. Sasse, H. Schrodt and J. Spille; *Chem. Ber.*, 1956, **89**, 2060. 73, 97, 300
56CB2562 C. Scholtissek; *Chem. Ber.*, 1956, **89**, 2562. 444
56CB2578 W. Ried and H. Keller; *Chem. Ber.*, 1956, **89**, 2578. 27
56CI(M)371 M. DeMalde, F. Minisci, U. Pallini, E. Volterra and A. Quilico; *Chem. Ind. (Milan)*, 1956, **38**, 371. 6
56JA97 H. G. Mautner and W. D. Kumler; *J. Am. Chem. Soc.*, 1956, **78**, 97. 598
56JA479 J. Hine, A. M. Dowell, Jr. and J. E. Singley, Jr.; *J. Am. Chem. Soc.*, 1956, **78**, 479. 33, 81, 228
56JA4450 R. L. McConnell and H. W. Coover, Jr.; *J. Am. Chem. Soc.*, 1956, **78**, 4450. 118
56JA4458 D. Elmore, J. Ogle, P. Toseland and W. Flechtner; *J. Am. Chem. Soc.*, 1956, **78**, 4458.
56JA5292 H. G. Mautner; *J. Am. Chem. Soc.*, 1956, **78**, 5292. 598
56JA5637 J. A. Young, T. C. Simmons and F. W. Hoffmann; *J. Am. Chem. Soc.*, 1956, **78**, 5637. 257
56JA6070 I. B. Douglass and G. H. Warner; *J. Am. Chem. Soc.*, 1956, **78**, 6070. 229, 255
56JAP10924 Y. Iwanaga, T. Mori and T. Yoshida; *Jpn. Pat.* 10924 (1956) (*Chem. Abstr.*, 1958, **52**, 17641). 524
56JCS508 W. V. Farrar; *J. Chem. Soc.*, 1956, 508. 254, 255
56JCS3416 D. A. Barr and R. N. Haszeldine; *J. Chem. Soc.*, 1956, 3416. 604
56JOC1319 G. A. Oláh and S. J. Kuhn; *J. Org. Chem.*, 1956, **21**, 1319. 410, 420, 422
56M323 W. Klementschitz and K. Gitschthaler; *Monatsh. Chem.*, 1956, **87**, 323. 39
B-56MI 614-01 A. I. Vogel; "Practical Organic Chemistry," 3rd edn., Longman, London, 1956, p. 185. 412
56USP2729687 J. D. Sterling (du Pont de Nemours & Co.); *US Pat.* 2729687 (1956) (*Chem. Abstr.*, 1956, **50**, 11360). 226
56USP2739989 C. M. Barringer, B. F. Harvey, J. M. Quattlebaum and J. D. Sterling, Jr. (du Pont de Nemours & Co.); *US Pat.* 2739989 (1956) (*Chem. Abstr*, 1956, **50**, 10757). 220
56USP2744147 W. N. Milks (Dow Chem. Co.); *US Pat.* 2744147 (1956) (*Chem. Abstr.*, 1957, **51**, 1240). 217
56USP2744148 R. P. Ruh and R. A. Davis (Dow Chem. Co.); *US Pat.* 2744148 (1956) (*Chem. Abstr.*, 1957, **51**, 1240). 217
56USP2745886 R. P. Ruh and R. A. Davis (Dow Chem. Co.); *US Pat.* 2745886 (1956) (*Chem. Abstr.*, 1957, **51**, 1245). 217, 226
56USP2748177 C. B. Miller and J. D. Calfee (Allied Chem. & Dye Corp.); *US Pat.* 2748177 (1956) (*Chem. Abstr.*, 1957, **51**, 455). 217
56USP2755314 R. J. Reid, E. S. Hanson and R. J. Pikna (Firestone Tire & Rubber Corp.); *US Pat.* 2755314 (1956) (*Chem. Abstr.*, 1957, **51**, 2016). 225, 227
56USP2757214 E. L. Muetterties (du Pont de Nemours & Co); *US Pat.* 2757214 (1956) (*Chem. Abstr.*, 1957, **51**, 2016). 217

57BRP757893 R. N. Haszeldine; *Br. Pat.* 757893 (1957) (*Chem. Abstr.* 1957, **51**, 12126). 32
57BRP767779 C. W. Suckling and J. Raventos; *Br. Pat.* 767779 (1957) (*Chem. Abstr.*, 1957, **51**, 15546). 16
57CB2460 R. Pfleger and A. Jager; *Chem. Ber.*, 1957, **90**, 2460. 113
57CJC527 M. W. Kirkwood and G. F. Wright; *Can. J. Chem.*, 1957, **35**, 527. 655
57GEP1002355 I. Lerch and A. Kottler; *Ger. Pat.* 1002355 (1957) (*Chem. Abstr.*, 1959, **53**, 21814).
57GEP1018054 K. Sasse; *Ger. Pat.*, 1018054 (1957) (*Chem. Abstr.*, 1960, **54**, 5480). 534
57HCA604 M. Brenner, J. P. Zimmermann, P. Quitt, W. Schneider and A. Hartmann; *Helv. Chim. Acta*, 1957, **15**, 604. 430
57IZV48 M. I. Kabachnik and P. A. Rossiiskaya; *Izv. Akad. Nauk SSSR Ser. Khim.*, 1957, 48 (*Chem. Abstr.*, 1957, **51**, 10366). 455
57IZV199 V. F. Mironov and V. A. Ponomarenko; *Izv. Akad. Nauk SSSR Ser. Khim.*, 1957, 199 (*Chem. Abstr.*, 1957, **51**, 11272). 267
57JA20 R. C. Miller and C. P. Smyth; *J. Am. Chem. Soc.*, 1957, **79**, 20. 213
57JA69 R. Dresdner; *J. Am. Chem. Soc.*, 1957, **79**, 69. 239
57JA366 E. Dyer and J. F. Glenn; *J. Am. Chem. Soc.*, 1957, **79**, 366. 494
57JA3351 R. E. Bell and R. E. Herfert; *J. Am. Chem. Soc.*, 1957, **79**, 3351. 63
57JA3611 G. Natta, R. Ercoli, F. Calderazzo and A. Rabizzoni; *J. Am. Chem. Soc.*, 1957, **79**, 3611. 522
57JA4159 W. T. Miller, Jr., E. Bergman and A. H. Fainberg; *J. Am. Chem. Soc.*, 1957, **79**, 4159. 64, 225, 226, 245
57JA4708 L. Zeldin and H. Shechter; *J. Am. Chem. Soc.*, 1957, **79**, 4708. 170
57JA5493 J. Hine and J. J. Porter; *J. Am. Chem. Soc.*, 1957, **79**, 5493. 45, 69
57JA5628 R. S. Porter and G. H. Cady; *J. Am. Chem. Soc.*, 1957, **79**, 5628. 231
57JA5654 M. Hellmann, E. Peters, W. J. Pummer and L. A. Wall; *J. Am. Chem. Soc.*, 1957, **79**, 5654. 222
57JA5710 G. DeStevens and A. Halamandaris; *J. Am. Chem. Soc.*, 1957, **79**, 5710. 508
57JA5717 T. M. Laakso and D. D. Reynolds; *J. Am. Chem. Soc.*, 1957, **79**, 5717. 504
57JA5801 R. N. Haszeldine and H. Iserson; *J. Am. Chem. Soc.*, 1957, **79**, 5801. 217, 225, 409, 419
57JA6180 G. W. Anderson and A. C. McGregor; *J. Am. Chem. Soc.*, 1957, **79**, 6180. 487
57JCS30 D. A. Barr and R. N. Haszeldine; *J. Chem. Soc.*, 1957, 30. 218
57JCS1741 R. N. Haszeldine and B. J. H. Mattinson; *J. Chem. Soc.*, 1957, 1741. 240
57JCS2735 J. I. Jones; *J. Chem. Soc.*, 1957, 2735. 429

57JGU2590 D. F. Kutepov and N. S. Rozanova; *J. Gen. Chem. USSR* (*Engl. Transl.*), 1957, **27**, 2590. 500

57JGU2882 D. F. Kutepov and N. S. Rozanova; *J. Gen. Chem. USSR* (*Engl. Transl.*), 1957, **27**, 2882. 500

57LA(607)67 H. Behringer and H. Meier; *Justus Liebigs Ann. Chem.*, 1957, **607**, 67. 511

57LA(609)83 H. A. Staab; *Justus Liebigs Ann. Chem.*, 1957, **609**, 83. 504

57MI 603-01 N. Barbulescu and W. Adlef; *Rev. Chim.* (*Bucharest*), 1957, **9**, 605 (*Chem. Abstr.*, 1958, **52**, 8948). 71

57RTC949 Y. A. Sinnema and J. F. Arens; *Recl. Trav. Chim. Pays-Bas*, 1957, **76**, 949. 110

57RTC969 L. Heslinga, G. J. Katerberg and J. F. Arens; *Recl. Trav. Chim. Pays-Bas*, 1957, **76**, 969. 41

57ZAAC(289)324 W. Hieber and G. Brendel; *Z. Anorg. Allg. Chem.*, 1957, **289**, 324. 523

57ZOB2134 P. S. Pel'kis, R. G. Dubenko and L. S. Pupko; *Zh. Obshch. Khim.*, 1957, **27**, 2134 (*Chem. Abstr.*, 1958, **52**, 6230). 653

57ZOB2243 N. N. Yarovenko, S. P. Motornyĭ, L. I. Kirensakya and A. S. Vasil'eva; *Zh. Obshch. Khim.*, 1957, **27**, 2243 (*Chem. Abstr.*, 1958, **52**, 8052). 440

58BSB676 Y. Desirant; *Bull. Soc. Chim. Belg.*, 1958, **67**, 676. 222

58CB22 R. Müller and G. Seitz; *Chem. Ber.*, 1958, **91**, 22. 173, 267

58CB437 L. Horner and H. Oediger; *Chem. Ber.*, 1958, **91**, 437. 698, 708

58CB650 H. Baganz and L. Domascke; *Chem. Ber.*, 1958, **91**, 650. 79

58CB806 H. Baganz and K. E. Krüger; *Chem. Ber.*, 1958, **91**, 806. 41, 46

58CB1942 A. Rieche, E. Schmitz and E. Beyer; *Chem. Ber.*, 1958, **91**, 1942. 79

58CB2664 E. Kaspar and R. Wiechert; *Chem. Ber.*, 1958, **91**, 2664. 25

58CCC452 Z. Arnold and F. Sorm; *Collect. Czech. Chem. Commun.*, 1958, **23**, 452. 57

58CJC1541 R. A. B. Bannard, A. A. Casselman, W. F. Cockburn and G. M. Brown; *Can. J. Chem.*, 1958, **36**, 1541. 641, 642, 643

58CT515 A. Rieche and H. Gross; *Chem. Tech.*, 1958, **10**, 515 (*Chem. Abstr.*, 1961, **55**, 4350). 229

58GEP1040011 J. T. Denison, M. W. Farlow, E. Muetterties and G. H. Whipple (du Pont de Nemours & Co.); *Ger. Pat.* 1040011 (1958) (*Chem. Abstr.*, 1960, **54**, 17264). 212

58HCA377 T. Wagner-Jauregg and M. Häring; *Helv. Chim. Acta*, 1958, **41**, 377. 629, 630

58IZV100 L. I. Zakharkin, V. V. Gavrilenko and O. Y. Okhlobystin; *Izv. Akad. Nauk SSSR, Ser. Khim.*, 1958, 100 (*Chem. Abstr.*, 1958, **52**, 11 736). 523

58JA851 M. Hauptschein, M. Braid and A. H. Fainberg; *J. Am. Chem. Soc.*, 1958, **80**, 851. 5

58JA1247 R. M. Roberts, J. Corse, R. Boschan, D. Seymour and S. Winstein; *J. Am. Chem. Soc*, 1958, **80**, 1247. 74

58JA1889 J. A. Young and R. D. Dresdner; *J. Am. Chem. Soc.*, 1958, **80**, 1889. 257

58JA3002 J. Hine and K. Tanabe; *J. Am. Chem. Soc.*, 1958, **80**, 3002. 38

58JA3604 J. A. Young, S. N. Tsoukalas and R. D. Dresdener; *J. Am. Chem. Soc.*, 1958, **80**, 3604. 238, 239, 240, 242

58JA3915 S. M. McElvain and D. H. Clemens; *J. Am. Chem. Soc.*, 1958, **80**, 3915. 70

58JA4282 J. Hine and F. P. Prosser; *J. Am. Chem. Soc.*, 1958, **80**, 4282. 33

58JA6167 R. D. Closson, L. R. Buzbee and G. G. Ecke; *J. Am. Chem. Soc.*, 1958, **80**, 6167. 523

58JA6409 H. K. Hall, Jr. and A. K. Schneider; *J. Am. Chem. Soc.*, 1958, **80**, 6409. 509

58JCS764 R. G. R. Bacon, R. S. Irvin, J. M. Pollack and A. D. E. Pullin; *J. Chem. Soc.*, 1958, 764. 607, 609

58JCS2939 J. W. Dale, H. J. Emeléus and R. N. Haszeldine; *J. Chem. Soc.*, 1958, 2939. 49

58JCS4245 H. J. Emeléus and G. S. Rao; *J. Chem. Soc.*, 1958, 4245. 218, 219, 220, 226

58JGU2539 N. N. Iarovenko and A. S. Vasil'eva; *J. Gen. Chem. USSR* (*Engl. Transl.*), 1958, **28**, 2539. 229

58JOC322 M. Hauptschein, A. H. Fainberg and M. Braid; *J. Org. Chem.*, 1958, **23**, 322. 2

58JOC1565 C. G. Krespan and V. A. Engelhardt; *J. Org. Chem.*, 1958, **23**, 1565. 245

58JOC1576 J. A. Young and R. D. Dresdner; *J. Org. Chem.*, 1958, **23**, 1576. 440

58JOC2016 C. G. Krespan; *J. Org. Chem.*, 1958, **23**, 2016. 32, 219

58JPR60 H. Scheibler and M. Depner; *J. Prakt. Chem.*, 1958, **7**, 60. 69

58JPR70 H. Scheibler, U. Faas and B. Hadji-Walassis; *J. Prakt. Chem.*, 1958, **7**, 70. 105

58LA(616)1 F. Boberg, G. Winter and J. Moos; *Justus Liebigs Ann. Chem.*, 1958, **616**, 1. 44

58MI 607-01 P. G. Maslov and Yu. P. Maslov; *Khim. i Tekhnol. Topliv i Masel*, 1958, **3**, 50 (*Chem. Abstr.*, 1959, **53**, 1910). 223, 228

58USP2821554 J. N. Ospenson (California Spray-Chemical Co.); *US Pat.* 2821554 (1958) (*Chem. Abstr.*, 1958, **52**, 10145). 236

58USP2822247 V. Hnizda; *US Pat.* 2 822 247 (1958) (*Chem. Abstr.*, 1958, **52**, 10 520). 523

58USP2853531 C. S. Cleaver; *US Pat.* 2 853 531 (1957) (*Chem. Abstr.*, 1959, **53**, 5135). 69

58ZOB2506 N. N. Yarovenko; *Zh. Obshch. Khim.*, 1958, **28**, 2506. 229

59AG246 H.-D. Stachel; *Angew. Chem.*, 1959, **71**, 246. 321

59AG274 J. Dahmlos; *Angew. Chem.*, 1959, **71**, 274. 220, 225

59AG524 M. Schmeisser and E. Scharf; *Angew. Chem.*, 1959, **71**, 524. 220

59CB329 H. Bredereck, R. Gompper, H. Rempfer, K. Klemm and H. Keck; *Chem. Ber.*, 1959, **2**, 329. 139, 140

59CB1018 R. Müller and H. Beyer; *Chem. Ber.*, 1959, **92**, 1018. 383

59CB2563 J. Goerdeler and J. Vitt; *Chem. Ber.*, 1959, **92**, 2563. 651

59CB3170 H. Baganz and L. Domaschke; *Chem. Ber.*, 1959, **92**, 3170. 39

59CCC760 Z. Arnold; *Collect. Czech. Chem. Commun.*, 1959, **24**, 760. 452

59CCC4048 Z. Arnold; *Collect. Czech. Chem. Commun.*, 1959, **24**, 4048. 51

59CI(L)1216 E. K. Fields and J. M. Sandri; *Chem. Ind.* (*London*), 1959, 1216. 51

59CI(M)132 R. Ercoli, P. Chini and M. Massi-Mauri; *Chim. Ind.* (*Milan*), 1959, **41**, 132. 525

59G693 M. Giua and R. Bianco; *Gazz. Chim. Ital.*, 1959, **89**, 693. 596

59G809 R. Ercoli, F. Calderazzo and G. Bernardi; *Gazz. Chim. Ital.*, 1959, **89**, 809. 522

59G1511 R. Scarpati and G. Speroni; *Gazz. Chim. Ital.*, 1959, **89**, 1511. 76
59HCA1653 H. H. Bosshard, R. Mory, M. Schmid and H. Zollinger; *Helv. Chim. Acta*, 1959, **42**, 1653. 52
59HCA1659 H. H. Bosshard and H. Zollinger; *Helv. Chim. Acta*, 1959, **42**, 1659. 57
59JA574 R. D. Dresdner and J. A. Young; *J. Am. Chem. Soc.*, 1959, **81**, 574. 253
59JA806 W. C. Schumb and J. R. Aronson; *J. Am. Chem. Soc.*, 1959, **81**, 806. 212
59JA1089 J. A. C. Allison and G. H. Cady; *J. Am. Chem. Soc.*, 1959, **81**, 1089. 217, 225
59JA2078 J. H. Fried and W. T. Miller, Jr., *J. Am. Chem. Soc.*, 1959, **81**, 2078. 18, 225
59JA2579 S. M. McElvain and P. L. Weyna; *J. Am. Chem. Soc.*, 1959, **81**, 2579. 76
59JA3575 E. H. Man, D. D. Coffman and E. L. Muetterties; *J. Am. Chem. Soc.*, 1959, **81**, 3575. 233, 235, 236
59JA3599 J. A. Attaway, R. H. Groth and L. A. Bigelow; *J. Am. Chem. Soc.*, 1959, **81**, 3599. 257
59JA3728 S. L. Shapiro, V. A. Parrino and L. Freedman; *J. Am. Chem. Soc.*, 1959, **81**, 3728. 640
59JA4801 G. W. Parshall, D. C. England and R. V. Lindsey, Jr., *J. Am. Chem. Soc.*, 1959, **81**, 4801. 58
59JA6270 H. G. Mautner and E. M. Clayton; *J. Am. Chem. Soc.*, 1959, **81**, 6270. 598
59JAP594264 T. Hoga; *Jpn. Pat.* 594264 (1959) (*Chem. Abstr.*, 1960, **54**, 299). 429
59JCS13 J. M. Birchall and R. N. Haszeldine; *J. Chem. Soc.*, 1959, 13. 222
59JGU1115 L. Z. Soborovskii and N. F. Baina; *J. Gen. Chem. USSR* (*Engl. Transl.*), 1959, **29**, 1115. 58
59JGU2125 N. N. Yarovenko and M. A. Raksha; *J. Gen. Chem. USSR* (*Engl. Transl.*), 1959, **29**, 2125. 54
59JGU2131 K. A. Petrov and A. A. Neimysheva; *J. Gen. Chem. USSR* (*Engl. Transl.*), 1959, **29**, 2131. 604, 612
59JGU2662 K. A. Petrov and A. A. Neimysheva; *J. Gen. Chem. USSR* (*Engl. Transl.*), 1959, **29**, 2662. 602
59LA(621)8 F. Boberg, G. Winter and G. R. Schultze; *Justus Liebigs Ann. Chem.*, 1959, **621**, 8. 44, 47
59LA(623)103 N. Kreutzkamp and G. Cordes; *Justus Liebigs Ann. Chem.*, 1959, **623**, 103. 159
59LA(623)183 K. Bodebenner; *Justus Liebigs Ann. Chem.*, 1959, **623**, 183. 73
59LA(627)142 L. Horner and H. Oediger; *Justus Liebigs Ann. Chem.*, 1959, **627**, 142. 708
59M41 M. Lipp, F. Dallacker and I. Meier zu Köcker; *Monatsh. Chem.*, 1959, **90**, 41. 596
59MI 604-01 R. Scarpati, G. Del Re and T. Maone; *Rend. Acad. Sci. Fis. Mat.* (*Soc. Nazl. Sci., Napoli*), 1959, **26**, 20 (*Chem. Abstr.*, 1961, **55**, 11423). 106
B-59MI 607-01 F. Richter, R. Ostertag, G. Ammerlahn, M. Baumann, E. Behrle and M. Kobel; in "Beilsteins Handbuch der Organischen Chemie," 4th edn., Drittes Ergänzungswerk, Deutsche Chemische Gesellschaft, Springer-Verlag, Berlin, 1959. 213
B-59MI 607-02 A. K. Barbour; *Int. Symp. Fluorine Chem.*, Birmingham, 1959. From M. Hudlický; "Chemistry of Organic Fluorine Compounds," Pergamon, Oxford, 1961, ref. 35. 225
59MI 607-03 B. F. Yudin and G. A. Khachkuruzov; *Trudy Gosudarst. Inst. Priklad. Khim.*, 1959, **42**, 132 (*Chem. Abstr.*, 1961, **55**, 19448). 223
59MI 616-02 A. N. Nesmeyanov, E. P. Mikheev, K. N. Anisimov and N. P. Filimonova; *Zh. Neorg. Khim.*, 1959, **4**, 1958 (*Chem. Abstr.*, 1959, **54**, 11 795). 522
59PCS229 W. M. Wagner; *Proc. Chem. Soc.*, 1959, 229. 25
59QR233 J. J. Lagowski; *Q. Rev., Chem. Soc.*, 1959, **8**, 233. 243
59RTC354 L. C. Rinzema, J. Stoffelsma and J. F. Arens; *Recl. Trav. Chim. Pays-Bas*, 1959, **78**, 354. 83, 84
59UK484 G. A. Shvekhgeimer, N. F. Pyatakov and S. S. Novikov; *Usp. Khim.*, 1959, **28**, 484 (*Chem. Abstr.*, 1959, **53**, 16951). 144
59USP2871274 R. P. Ruh and R. A. Davis (Dow Chem. Co.); *US Pat.* 2871274 (1959) (*Chem. Abstr.*, 1959, **53**, 12173). 223
59USP2875254 F. J. Gradishar; *US Pat.* 2875254 (1959) (*Chem. Abstr.*, 1959, **53**, 14000). 218
59USP2885450 C. Woolf (Allied Chem. Corp.); *US Pat.* 2885450 (1959) (*Chem. Abstr.*, 1959, **53**, 16961). 224, 226
59ZOB942 N. N. Yarovenko, V. N. Shemanina and G. B. Gazieva; *Zh. Obshch. Khim.*, 1959, **29**, 942 (*Chem. Abstr.*, 1960, **54**, 2 158). 49
59ZOB2157 S. P. Motornyi, L. I. Kirenskaya and N. N. Yarovenko; *Zh. Obshch. Khim.*, 1959, **29**, 2157. 239
59ZOB2159 K. A. Petrov and A. A. Neimysheva; *Zh. Obshch. Khim.*, 1959, **29**, 2159. 238
59ZOB2169 K. A. Petrov and A. A. Neimysheva; *Zh. Obshch. Khim.*, 1959, **29**, 2169 (*Chem. Abstr.*, 1960, **54**, 10 912). 260
59ZOB3401 K. A. Petrov and A. A. Neimysheva; *Zh. Obshch. Khim.*, 1959, **29**, 3401. 236
59ZOB3784 N. N. Yarovenko and A. S. Vasil'eva; *Zh. Obshch. Khim.*, 1959, **29**, 3784. 236
59ZOB3786 N. N. Yarovenko and A. S. Vasil'eva; *Zh. Obshch. Khim.*, 1959, **29**, 3786. 233, 234, 238
59ZOB3792 N. N. Yarovenko and A. S. Vasil'eva; *Zh. Obshch. Khim.*, 1959, **29**, 3792 (*Chem. Abstr.*, 1960, **54**, 19 479). 236, 528, 533, 534

60ACS2230 A. Senning and S.-O. Lawesson; *Acta Chem. Scand.*, 1960, **14**, 2230. 233
60AFC(1)166 M. Stacey and J. C. Tatlow; *Adv. Fluorine Chem.*, 1960, **1**, 166. 20
60AG836 H. Eilingsfeld, M. Seefelder and H. Weidinger; *Angew. Chem.*, 1960, **72**, 836. 51
60AG956 H. Gold; *Angew. Chem.*, 1960, **72**, 956. 105
60AP(293)150 H. P. Kaufmann; *Arch. Pharm.* (*Weinheim, Ger.*), 1960, **293**, 150. 452
60BRP823519 Dow Corning Corp.; *Br. Pat.* 823519 (1959) (*Chem. Abstr.*, 1960, **54**, 7558). 18
60CB1398 H. Bredereck, R. Gompper, F. Effenberger, H. Keck and H. Heise; *Chem. Ber.*, 1960, **93**, 1398. 140
60CB2810 L. Birkofer, H. P. Kühltau and A Ritter; *Chem. Ber.*, 1960, **93**, 2810. 507
60CB3056 F. Drawert, K. H. Reuther and F. Born; *Chem. Ber.*, 1960, **93**, 3056. 545
60CI(M)133 P. Chini; *Chim. Ind.* (*Milan*), 1960, **42**, 133. 524
60CI(M)137 P. Chini; *Chim. Ind.* (*Milan*), 1960, **42**, 137. 524
60CRV54 M. L. Bender; *Chem. Rev.*, 1960, **60**, 54. 104
60DOK(131)98 V. F. Mironov, N. G. Dzhurinskaya and A. D. Petrov; *Dokl. Akad. Nauk SSSR*, 1960, **131**, 98 (*Chem. Abstr.*, 1960, **54**, 11 977). 62
60DOK(132)602 I. L. Knunyants and G. A. Sokol'skii; *Dokl. Akad. Nauk SSSR*, 1960, **132**, 602 (*Chem. Abstr.*, 1960, **54**, 24366). 241

60G1005 P. Chini, L. Colli and M. Peraldo; *Gazz. Chim. Ital.*, 1960, **90**, 1005. 526

60GEP1084733 K. Bodenbenner and O. Bayer; *Ger. Pat.* 1 084 733 (1960) (*Chem. Abstr.*, 1961, **55**, 20 963). 73

60IS155 M. W. Farlow, E. H. Man and C. W. Tullock; *Inorg. Syn.*, 1960, **6**, 155. 408, 410

60IZV231 I. L. Knunyants, B. L. Dyatkin, L. S. German and E. P. Mochalina; *Izv. Akad. Nauk SSSR, Ser. Khim.*, 1960, 231 (*Chem. Abstr.*, 1960, **54**, 20 871). 69

60IZV505 S. S. Novikov, T. I. Godovikova and V. A. Tartakovskii; *Izv. Akad. Nauk. SSSR, Ser. Khim.*, 1960, 505 (*Chem. Abstr.*, 1960, **54**, 22 559). 368

60IZV1783 S. S. Novikov, L. I. Khmelnitskii and O. V. Lebedev; *Izv. Akad. Nauk SSSR Ser. Khim.*, 1960, 1783 (*Chem. Abstr.*, 1961, **55**, 19832). 150

60IZV1828 B. P. Federov and F. M. Stoyanovich; *Izv. Akad. Nauk SSSR, Ser. Khim.*, 1960, 1828 (*Chem. Abstr.*, 1969, **55**, 14 298). 83

60IZV1901 M. F. Shostakovskii, A. V. Bogdanova, G. I. Plotnikova and A. N. Dolgikh; *Izv. Akad. Nauk SSSR, Ser. Khim.*, 1960, 1901 (*Chem. Abstr.*, 1961, **55**, 16 407). 83

60JA543 W. R. Hasek, W. C. Smith and V. A. Engelhardt; *J. Am. Chem. Soc.*, 1960, **82**, 543. 19, 39, 52

60JA1325 H. E. Podall, J. H. Dunn and H. Shapiro; *J. Am. Chem. Soc.*, 1960, **82**, 1325. 522, 523

60JA1398 J. Hine, A. D. Ketley and K. Tanabe; *J. Am. Chem. Soc.*, 1960, **82**, 1398. 69

60JA1888 H. C. Clark and C. J. Willis; *J. Am. Chem. Soc.*, 1960, **82**, 1888. 245

60JA3091 W. T. Miller, Jr., J. H. Fried and H. Goldwhite; *J. Am. Chem. Soc.*, 1960, **82**, 3091. 18

60JA5116 D. C. England, L. R. Melby, M. A. Dietrich and R. V. Lindsey, Jr.; *J. Am. Chem. Soc.*, 1960, **82**, 5116. 46

60JA5298 R. D. Chambers, H. C. Clark and C. J. Willis; *J. Am. Chem. Soc.*, 1960, **82**, 5298. 244

60JA5759 J. E. Griffiths and A. B. Burg; *J. Am. Chem. Soc.*, 1960, **82**, 5759. 246

60JA6118 J. Hine and J. J. Porter; *J. Am. Chem. Soc.*, 1960, **82**, 6118. 81

60JA6181 D. C. England, M. A. Dietrich and R. V. Lindsey, Jr.; *J. Am. Chem. Soc.*, 1960, **82**, 6181. 15

60JA6228 H. D. Kaesz, J. R. Phillips and F. G. A. Stone; *J. Am. Chem. Soc.*, 1960, **82**, 6228. 65, 244

60JAP6018199 T. Yamaguchi, T. Yamamoto and K. Sonobe (Nippon Chemical Industrial Co. Ltd); *Jpn. Pat.* 6018199 (1960) (*Chem. Abstr.*, 1961, **55**, 21468). 316

60JCS3516 A. J. Downs and E. A. V. Ebsworth; *J. Chem. Soc.*, 1960, 3516. 232

60JGU4029 N. N. Yarovenko and S. P. Motornyi; *J. Gen. Chem. USSR* (*Engl. Transl.*), 1960, **30**, 4029. 613

60JOC60 E. A. Nodiff, S. Lipschutz, P. N. Craig and M. Gordon; *J. Org. Chem.*, 1960, **25**, 60. 232

60JOC105 C. G. Krespan; *J. Org. Chem.*, 1960, **25**, 105. 64

60JOC993 A. Demiel; *J. Org. Chem.*, 1960, **25**, 993. 41

60JOC2009 T. G. Miller and J. W. Thanassi; *J. Org. Chem.*, 1960, **25**, 2009. 38

60JOC2016 C. W. Tullock and D. D. Coffman; *J. Org. Chem.*, 1960, **25**, 2016. 236

60JOC2069 H. Feuer and T. J. Kucera; *J. Org. Chem.*, 1960, **25**, 2069. 145, 147

60LA(629)222 G. Wilke and H. Müller; *Justus Liebigs Ann. Chem.*, 1960, **629**, 222. 208

60LA(632)38 H. Meerwein, K. Bodenbenner, P. Borner, F. Kunert and K. Wunderlich; *Justus Liebigs Ann. Chem.*, 1960, **632**, 38. 73

60MI 604-01 B. P. Fedorov and F. M. Stoyanovich; *Izv Akad. Nauk SSSR, Otdel. Khim. Nauk.*, 1960, 1828 (*Chem. Abstr.*, 1961, **55**, 14 298). 111

60RTC888 H. D. A. Tigchelaar-Lutjeboer, H. Bootsma and J. F. Arens; *Recl. Trav. Chim. Pays-Bas*, 1960, **79**, 888. 74

60USP2962529 D. M. Marquis; *US Pat.* 2 962 529 (1960) (*Chem. Abstr.*, 1961, **55**, 7285). 528

60ZAAC(303)85 G. Fritz, D. Habel and G. Teichmann; *Z. Anorg. Allg. Chem.*, 1960, **303**, 85. 268

60ZOB1602 K. A. Petrov, F. L. Maklyaev and N. K. Bliznyuk; *Zh. Obshch. Khim.*, 1960, **30**, 1602 (*Chem. Abstr.*, 1961, **55**, 1414). 263

61AG66 H. A. Staab and W. Benz; *Angew. Chem.*, 1961, **73**, 66. 417

61AG265 R. Huisgen, W. Mack and E. Anneser; *Angew. Chem.*, 1961, **73**, 265.

61AG579 W. Hieber and C. Herget; *Angew. Chem.*, 1961, **73**, 579. 523

61AG656 R. Huisgen, W. Mack and E. Anneser; *Angew. Chem.*, 1961, **73**, 656. 309

61AP765 G. Zinner; *Arch. Pharm.* (*Weinheim, Ger.*), 1961, **294**, 765. 570, 571

61CB538 H. Gross and A. Rieche; *Chem. Ber.*, 1961, **94**, 538. 69, 82

61CB544 H. Gross, A. Rieche and E. Höft; *Chem. Ber.*, 1961, **94**, 544. 43, 251, 429

61CB1373 G. Wittig and M. Schlosser; *Chem. Ber.*, 1961, **94**, 1373. 697

61CB2278 H. Bredereck and K. Bredereck; *Chem. Ber.*, 1961, **94**, 2278. 645

61CB2594 H. Gold and O. Bayer; *Chem. Ber.*, 1961, **94**, 2594. 55

61CB3109 H. Bohme and F. Soldan; *Chem. Ber.*, 1961, **94**, 3109. 138

61CB3287 K. Gulbins and K. Hamann, *Chem. Ber.*, 1961, **94**, 3287. 510

61CCC3059 Z. Arnold and A. Holy; *Collect. Czech. Chem. Commun.*, 1961, **26**, 3059. 54

61CJC1017 R. Greenhalgh and R. A. B. Bannard; *Can. J. Chem.*, 1961, **39**, 1017. 646

61CRV313 P. Rây; *Chem. Rev.*, 1961, **61**, 313. 640

61DOK(141)357 S. P. Makarov, A. Y. Yakubovich, V. A. Ginsburg, A. S. Filatov, M. A. Englin, N. F. Privezentseva and T. Y. Nikoforova; *Dokl. Akad. Nauk. USSR*, 1961, **141**, 357 (*Chem. Abstr.*, 1962, **56**, 11425). 242

61E566 R. Huisgen and V. Weberndörfer; *Experientia*, 1961, **17**, 566. 304, 305

61EGP22169 R. Müller and W. Müller; *Ger.* (*East*) *Pat.* 22169, (1961) (*Chem. Abstr.*, 1963, **58**, 2471). 383

61GEP1157210 J. H. Ottenheym and J. W. Garritsen; *Ger. Pat.* 1157 210 (1961) (*Chem. Abstr.*, 1964, **60**, 6756).

61JA802 M. A. Ring and D. M. Ritter; *J. Am. Chem. Soc.*, 1961, **83**, 802. 173

61JA840 J. F. Harris, Jr. and F. W. Stacey; *J. Am. Chem. Soc.*, 1961, **83**, 840. 46, 233, 234

61JA1287 J. G. Murray; *J. Am. Chem. Soc.*, 1961, **83**, 1287. 522

61JA1598 W. R. McClennen; *J. Am. Chem. Soc.*, 1961, **83**, 1598. 247

61JA1767	W. T. Miller, Jr., W. Frass and P. R. Resnick; *J. Am. Chem. Soc.*, 1961, **83**, 1767	18
61JA2383	M. Hauptschein and M. Braid; *J. Am. Chem. Soc.*, 1961, **83**, 2383.	3
61JA2495	M. Hauptschein, M. Braid and A. H. Fainberg; *J. Am. Chem. Soc.*, 1961, **83**, 2495.	2
61JA2589	W. J. Middleton, E. G. Howard and W. H. Sharkey; *J. Am. Chem. Soc.*, 1961, **83**, 2589.	528
61JA2953	J. C. Hileman, D. K. Huggins and H. D. Kaesz; *J. Am. Chem. Soc.*, 1961, **83**, 2953.	523
61JA3422	R. J. Harder and W. C. Smith; *J. Am. Chem. Soc.*, 1961, **83**, 3422.	238
61JA3535	R. B. Kaplan and H. Shechter; *J. Am. Chem. Soc.*, 1961, **83**, 3535.	150
61JA3593	R. B. King, P. M. Treichel and F. G. A. Stone; *J. Am. Chem. Soc.*, 1961, **83**, 3593.	247
61JA4460	H. Hart and R. W. Fish; *J. Am. Chem. Soc.*, 1961, **83**, 4460.	29
61JA4860	W. A. Sheppard; *J. Am. Chem. Soc.*, 1961, **83**, 4860.	229
61JCS2597	H. J. Emeléus and D. E. MacDuffie; *J. Chem. Soc.*, 1961, 2597.	233, 236, 237
61JCS2922	D. Barnard and D. T. Woodbridge; *J. Chem. Soc.*, 1961, 2922.	594
61JCS3779	R. D. Chambers, W. K. R. Musgrave and J. Savory; *J. Chem. Soc.*, 1961, 3779.	2, 3
61JCS3825	H. Goldwhite, R. N. Haszeldine and R. N. Mukherjee; *J. Chem. Soc.*, 1961, 3825.	64
61JOC51	H. Shechter and H. L. Cates, Jr.; *J. Org. Chem.*, 1961, **26**, 51.	145
61JOC391	H. Feuer and U. E. Lynch-Hart; *J. Org. Chem.*, 1961, **26**, 391.	147
61JOC2202	T. Mukaiyama, K. Tonooka and K. Inoue; *J. Org. Chem.*, 1961, **26**, 2202.	74
61JOC4047	H. C. Godt, Jr. and R. E. Wann; *J. Org. Chem.*, 1961, **26**, 4047.	552, 553
61JOC4315	H. L. Yale and J. T. Sheehan; *J. Org. Chem.*, 1961, **26**, 4315.	104
61JOC4734	D. J. Glover and M. J. Kamlet; *J. Org. Chem.*, 1961, **26**, 4734.	170
61JOC5238	B. Taub and J. B. Hino; *J. Org. Chem.*, 1961, **26**, 5238.	504
61LA(641)1	H. Meerwein, W. Florian, N. Schon and G. Stopp; *Justus Liebigs Ann. Chem.*, 1961, **641**, 1.	104, 105, 313, 314, 318, 321, 730, 731
61LA(641)21	H. Bohme and J. Roehr; *Justus Liebigs Ann. Chem.*, 1961, **641**, 21.	111
61LA(646)96	W. Ried and B. M. Beck; *Justus Liebigs Ann. Chem.*, 1961, **646**, 96.	578
61LA(648)21	H. Böhme and J. Roehr; *Justus Liebigs Ann. Chem.*, 1961, **648**, 21.	48, 82, 89
61MI 620-01	M. B. Antia and M. I. Pandit; *Vikram. J. Vikram. Univ.*, 1961, **5**, 106 (*Chem. Abstr.*, 1963, **59**, 3797).	611
61USP2991315	C. W. Plummer; *US Pat.* 2991315 (1956) (*Chem. Abstr.*, 1962, **56**, 2330e).	150
61USP3014071	C. H. Hoyt and J. Kamlet (Crown Zellerbach Co.); *US Pat.* 3014071 (1961) (*Chem. Abstr.*, 1962, **56**, 8566).	236
61ZOB3726	P. S. Pel'kis and M. Z. Peretyazhko; *Zh. Obshch. Khim.*, 1961, **31**, 3726.	573
62ACS117	A. Senning and S.-O. Lawesson; *Acta Chem. Scand.*, 1962, **16**, 117.	233
62AG861	E. Kühle; *Angew. Chem.*, 1962, **74**, 861.	449
62AG(E)331	H. Bredereck, F. Effenberger and G. Simchen; *Angew. Chem., Int. Ed. Engl.*, 1962, **1**, 331.	106
62AG(E)592	H. von Brachel and R. Merten; *Angew. Chem., Int. Ed. Engl.*, 1962, **1**, 592.	105
62AG(E)632	H. Holtschmidt; *Angew. Chem., Int. Ed. Engl.*, 1962, **1**, 632.	607
62AG(E)647	E. Kühle; *Angew. Chem., Int. Ed. Engl.*, 1962, **1**, 647.	606, 617
62BEP612225	Farbenfabr. Bayer A.-G.; *Belg. Pat.* 612225 (1962) (*Chem. Abstr.* 1962, **57**, 15154).	657
62BSB541	C. Collard-Charon, R. Huls and M. Renson; *Bull. Soc. Chim. Belg.*, 1962, **71**, 541.	598
62CB1859	H. Böhme and R. Neidlein; *Chem. Ber.*, 1962, **95**, 1859.	44, 95
62CB2514	G. Wittig, W. Böll and K.-H. Krück; *Chem. Ber.*, 1962, **95**, 2514.	697
62CCC2886	Z. Arnold and A. Holy; *Collect. Czech. Chem. Commun.*, 1962, **27**, 2886.	54
62DOK(142)354	V. A. Ginsburg, A. Y. Yakubovich, A. S. Filatov, G. E. Zelenin, S. P. Makarov, V. A. Shpanskii, G. P. Kotel'nikova, L. F. Sergienko and L. L. Martynova; *Dokl. Akad. Nauk. USSR*, 1962, **142**, 354 (*Chem Abstr.*, 1962, **57**, 4518).	242
62HOU(5/3)735	R. Stroh; *Methoden Org. Chem.*, (*Houben-Weyl*), 1962, **5/3**, 735.	28
62IZV272	S. S. Novikov, V. A. Tartakovskii, T. I. Godovikova and B. G. Gribov; *Izv. Akad. Nauk SSSR Ser. Khim.*, 1962, 272 (*Chem. Abstr.*, 1962, **57**, 12522).	149
62JA854	A. J. Speziale and K. W. Ratts; *J. Am. Chem. Soc.*, 1962, **84**, 854.	695
62JA898	H. C. Clark and C. J. Willis; *J. Am. Chem. Soc.*, 1962, **84**, 898.	244
62JA1312	R. Rabinowitz and R. Marxus; *J. Am. Chem. Soc.*, 1962, **84**, 1312.	
62JA1745	F. Ramirez, N. B. Desai and N. Mc Kelvie; *J. Am. Chem. Soc.*, 1962, **84**, 1745.	694
62JA1751	J. Hine, R. P. Bayer and G. G. Hammer; *J. Am. Chem. Soc.*, 1962, **84**, 1751.	88
62JA2337	W. J. Chambers, C. W. Tullock and D. D. Coffman; *J. Am. Chem. Soc.*, 1962, **84**, 2337.	242
62JA2751	J. B. Hynes and L. A. Bigelow; *J. Am. Chem. Soc* 1962, **84**, 2751.	262
62JA3148	J. F. Harris, Jr.; *J. Am. Chem. Soc.*, 1962, **84**, 3148.	46, 47, 233, 234
62JA3673	T. W. Campbell, J. J. Monagle and V. S. Foldi; *J. Am. Chem. Soc.*, 1962, **84**, 3673.	643
62JA4275	F. S. Fawcett, C. W. Tullock and D. D. Coffman; *J. Am. Chem. Soc.*, 1962, **84**, 4275.	52, 53, 238, 259, 409, 423, 438
62JCS2290	E. A. V. Ebsworth, H. J. Emeléus and N. Welcman; *J. Chem. Soc.*, 1962, 2290.	49
62JCS4361	A. J. Downs; *J. Chem. Soc.*, 1962, 4361.	236
62JOC780	L. A. Kaplan and M. J. Kamlet; *J. Org. Chem.*, 1962, **27**, 780.	144
62JOC1064	C. W. Taylor, T. J. Brice and R. L. Wear; *J. Org. Chem.*, 1962, **27**, 1064.	241
62JOC1455	H. Feuer and W. A. Swarts; *J. Org. Chem.*, 1962, **27**, 1455.	147
62JOC1482	E. R. Bissel and G. C. Shaw; *J. Org. Chem.*, 1962, **27**, 1482.	2
62JOC2504	J. A. Carbon and S. H. Tabata; *J. Org. Chem.*, 1962, **27**, 2504.	642
62JOC2589	H. Schroeder, R. Rätz, W. Schnabel, H. Ulrich, E. Kober and C. Grundmann; *J. Org. Chem.*, 1962, **27**, 2589.	603
62JOC2622	E. C. Taylor and R. V. Ravindranathan; *J. Org. Chem.*, 1962, **27**, 2622.	504
62JOC2858	H. C. Brown and R. Pater; *J. Org. Chem.*, 1962, **27**, 2858.	83
62JOC2899	S.-H. Chu and H. G. Mautner; *J. Org. Chem.*, 1962, **27**, 2899.	637, 732

62JOC3079	P. J. Stoffel and A. J. Speziale; *J. Org. Chem.*, 1962, **27**, 3079.	509
62JOC3584	C. G. Krespan and C. M. Langkammerer; *J. Org. Chem.*, 1962, **27**, 3584.	49, 58
62JOC3664	D. H. Clemens, E. Y. Shropshire and W. D. Emmons; *J. Org. Chem.*, 1962, **27**, 3664.	105, 138, 140
62JOC3686	A. G. Cook and E. K. Fields; *J. Org. Chem.*, 1962, **27**, 3686.	51
62JOC3995	C. G. Krespan and W. R. Brasen; *J. Org. Chem.*, 1962, **27**, 3995.	46
62JOC4068	T. P. Johnston, C. R. Stringfellow, Jr. and A. Gallagher; *J. Org. Chem.*, 1962, **27**, 4068.	305
62JOC4509	D. L. Garmaise, A. Uchiyama and A. F. McKay; *J. Org. Chem.*, 1962, **27**, 4509.	533, 540
62LA(651)89	S. Hünig and F. Müller *Justus Liebigs Ann. Chem.*, 1962, **651**, 89.	653
62LA(657)98	H. A. Staab and G. Walther; *Justus Liebigs Ann. Chem.*, 1962, **657**, 98.	578
B-62MI 615-01	E. E. Reid; "Organic Chemistry of Bivalent Sulfur," Chemical Publishing Co., New York, 1962, vol. 4, p. 170; and references therein.	471
B-62MI 617-01	E. E. Reid; "Organic Chemistry of Bivalent Sulfur," Chemical Publishing Co., New York, 1962, Vol. 4.	560, 561
B-62MI 617-02	G. D. Thorn and R. A. Ludwig; "The Dithiocarbamates and Related Compounds," Elsevier, Amsterdam, 1962.	560, 561
62MI 621-01	H. Wahl and M.-T. Le Bris; *Recent Progr. Chem. Nat. Synth. Coloring Matters Related Fields*, 1962, 507 (*Chem. Abstr.*, 1963, **59**, 7401).	653
62NKZ936	Y. Urata; *Nippon Kagaku Zasshi*, 1962, **83**, 936 (*Chem. Abstr.*, 1963, **59**, 5045).	32
62RTC1009	A. Fröhling and J. F. Arens; *Recl. Trav. Chim. Pays-Bas*, 1962, **81**, 1009.	84, 86, 303
62T79	C. O. Parker, W. D. Emmons, H. A. Rolewicz and K. S. McCallum; *Tetrahedron*, 1962, **17**, 79.	150
62TL701	M. M. Shemyakin, V. K. Antonov, A. M. Shkrob, Y. N. Sheinker and L. B. Senyavina; *Tetrahedron Lett.*, 1962, 701.	104
62USP3115513	F. W. Stacey; *US Pat.* 3115513 (1962) (*Chem. Abstr.*, 1964, **60**, 6754).	625
62ZN(B)782	G. Märkl; *Z. Naturforsch., Teil B*, 1962, **17**, 782.	659
62ZOB745	I. I. Lapkin and N. I. Panova; *Zh. Obshch. Khim.*, 1962, **32**, 745 (*Chem. Abstr.*, 1963, **58**, 5 554).	84
63ACS549	K. A. Jensen and P. H. Nielsen; *Acta Chem. Scand.*, 1963, **17**, 549.	596
63ACS2570	A. Senning; *Acta Chem. Scand.*, 1963, **17**, 2570.	233
63AFC(3)181	A. K. Barbour, L. J. Belf and M. W. Buxton; *Adv. Fluorine Chem.*, 1963, **3**, 181.	16, 38
63AG296	H. Brechbühler, H. Buchi, E. Hatz, J. Schreiber and A. Eschenmoser; *Angew. Chem.*, 1963, **75**, 296.	104
63AG1059	A. Lüttringhaus and H.-W. Dirksen; *Angew. Chem.*, 1963, **75**, 1059.	502
63AG(E)686	P. Binger; *Angew. Chem., Int. Ed. Engl.*, 1963, **2**, 686.	208
63AJC722	T. N. Bell, B. J. Pullman and B. O. West; *Aust. J. Chem.*, 1963, **16**, 722.	49
63BAU1384	G. A. Razuvaev, A. N. Egorochkin, V. S. Etlis and A. P. Sineokov; *Bull. Acad. Sci. USSR, Div. Chim. Sci.*, 1963, 1384.	324, 335
63BCJ1314	M. Okano, Y. Ito, T. Shono and R. Oda; *Bull. Chem. Soc. Jpn.*, 1963, **36**, 1314.	506
63BEP627486	H. Malz, H. Holtschmidt, E. Kühle and O. Bayer (Farbenfabriken Bayer AG); *Belg. Pat.* 627486 (1963) (*Chem. Abstr.*, 1964, **60**, 10557).	316
63BEP632635	E. Klauke and E. Kühle (Bayer AG); *Belg. Pat.* 632365 (1963) (*Chem. Abstr.*, 1964, **61**, 6952).	
63BSB149	C. Collard-Charon and M. Renson; *Bull. Soc. Chim. Belg.*, 1963, **72**, 149.	596
63BSF1462	G. Wilke and W. Schneider; *Bull. Soc. Chim. Fr.*, 1963, 1462.	208
63CB1387	H. Gross and J. Gloede; *Chem. Ber.*, 1963, **96**, 1387.	39
63CB1505	H. Bredereck, F. Effenberger and H. J. Treiber; *Chem. Ber.*, 1963, **96**, 1505.	140
63CB2671	H. Eilingsfeld, M. Seefelder and H. Weidinger; *Chem. Ber.*, 1963, **96**, 2671.	70
63CB2894	R. Müller and W. Müller; *Chem. Ber.*, 1963, **96**, 2894.	383
63CCC2040	Z. Arnold and A. Holy; *Collect. Czech. Chem. Commun.*, 1963, **28**, 2040.	104
63CCC2047	Z. Arnold; *Collect. Czech. Chem. Commun.*, 1963, **28**, 2047.	54
63CJC1643	E. Lieber, C. N. R. Rao, C. B. Laroyer and J. P. Trivedi; *Can. J. Chem.*, 1963, **41**, 1643.	537
63FCF554	J. F. Willems; *Fortschr. Chem. Forsch.*, 1963, **4**, 554.	575
63G942	R. Scarpati and D. Sica; *Gazz. Chim. Ital.*, 1963, 942 (*Chem. Abstr.*, 1964, **60**, 12005).	107
63GEP1146892	H. Gold; *Ger. Pat.* 1146892 (1963) (*Chem. Abstr.*, 1963, **59**, 10070).	105
63GEP1222917	W. Zecher and H. Holtschmidt (Bayer AG); *Ger. Pat.* 1222917 (1963) (*Chem. Abstr.*, 1965, **63**, 8327).	239
63HCA2306	A. Hofmann, H. Ott, R. Griot, P. A. Stadler and A. J. Frey; *Helv. Chim. Acta*, 1963, **46**, 2306.	104
63IZV1946	I. L. Knunyants, L. S. German and I. N. Rozhkov; *Izv. Akad. Nauk SSSR Ser. Khim.*, 1963, 1946 (*Chem. Abstr.*, 1964, **60**, 5325).	54
63JA1918	R. B. King; *J. Am. Chem. Soc.*, 1963, **85**, 1918.	517
63JA2243	R. L. Merker and M. J. Scott; *J. Am. Chem. Soc.*, 1963, **85**, 2243.	173
63JCS1083	G. M. Burch, H. Goldwhite and R. N. Haszeldine; *J. Chem. Soc.*, 1963, 1083.	58
63JCS1133	A. Davison, J. A. McCleverty and G. Wilkinson; *J. Chem. Soc.*, 1963, 1133.	523
63JCS1268	H. J. Emeléus and N. Welcman; *J. Chem. Soc.*, 1963, 1268.	49
63JCS1527	A. J. Burn, J. I. G. Cadogan and P. J. Bunyan; *J. Chem. Soc.*, 1963, 1527.	154
63JCS1887	M. Asscher and D. Vofsi; *J. Chem. Soc.*, 1963, 1887.	7
63JMC275	J. H. Short, U. Biermacher, D. A. Dunnigan and T. D. Leth; *J. Med. Chem.*, 1963, **6**, 275.	641
63JMC283	E. G. Podrebarac, W. H. Nyberg, A. F. French and C. C. Cheng; *J. Med. Chem.*, 1963, **6**, 283.	653
63JMC669	T. P. Johnston, G. S. McCaleb and J. A. Montgomery; *J. Med. Chem.*, 1963, **6**, 669.	504, 506
63JOC112	J. T. Maynard; *J. Org. Chem.*, 1963, **28**, 112.	18

63JOC991 N. Shachat and J. J. Bagnell, Jr.; *J. Org. Chem.*, 1963, **28**, 991. 509
63JOC1129 T. J. Logan; *J. Org. Chem.*, 1963, **28**, 1129. 246
63JOC1427 H. Ulrich and A. A. R. Sayigh; *J. Org. Chem.*, 1963, **28**, 1427. 445
63JOC1642 J. S. Warner; *J. Org. Chem.*, 1963, **28**, 1642. 596
63JOC2902 R. F. Meyer; *J. Org. Chem.*, 1963, **28**, 2902. 632
63JOC3140 A. F. Ferris and B. A. Schutz; *J. Org. Chem.*, 1963, **28**, 3140. 570, 571
63JPR1 Z. El-Hewehi and D. Hempel; *J. Prakt. Chem.*, 1963, **22**, 1. 254
B-63MI 614-01 W. Kwasnik; in "Handbook of Preparative Inorganic Chemistry," ed. G. Brauer, Academic Press, New York, 1963, p. 206. 409, 410, 418, 419, 420
B-63MI 614-02 O. Glemser; in "Handbook of Preparative Inorganic Chemistry," ed. G. Brauer, Academic Press, New York, 1963, p. 650. 414
63OR(13)91 C. Walling and E. S. Huyser; *Org. React.*, 1963, **13**, 91. 5
63OSC(4)359 F. C. Bennett, Jr. and R. A. Zingaro; *Org. Synth. Coll. Vol.*, 1963, **4**, 359. 596
63OSC(4)515 B. B. Kehm and C. W. Whitehead; *Org. Synth., Coll. Vol.*, 1963, **4**, 515. 504
63OSC(4)524 M. W. Farlow; *Org. Syn., Coll. Vol.*, 1963, **4**, 524. 414
63PCS370 R. J. Crawford and R. Raap; *Proc. Chem. Soc.*, 1963, 370. 625
63PCS374 O. J. Kovács, G. Schneider and L. K. Láng; *Proc. Chem. Soc.*, 1963, 374. 71
63PIA(A)45 R. P. Narain and R. C. Mehrotra; *Proc. Indian Acad. Sci., Sect. A*, 1963, **33**, 45 (*Chem. Abstr.*, 1963, **59**, 5018). 78
63T1661 R. G. Griot and A. J. Frey; *Tetrahedron*, 1963, **19**, 1661. 104
63T1675 H. Ott, A. J. Frey and A. Hofmann; *Tetrahedron*, 1963, **19**, 1675. 104
63TL439 V. K. Antonov, A. M. Shkrob and M. M. Shemyakin; *Tetrahedron Lett.*, 1963, 439. 104
63T(S)177 G. S. Hammond, W. D. Emmons, C. O. Parker, B. M. Graybill, J. H. Waters and M. F. Hawthorne; *Tetrahedron, Suppl. 1*, 1963, **19**, 177. 144, 145, 369
63T(S)155 A. Wetterholm; *Tetrahedron, Suppl. 1*, 1963, **19**, 155. 144, 150
63USP3076041 R. R. Twelves and V. Weinmayr; *US Pat.* 3076041 (1963) (*Chem. Abstr.*, 1963, **58**, 13792). 2
63USP3092637 M. Brown; *US Pat.* 3 092 637 (1963) (*Chem. Abstr.*, 1963, **59**, 12 764). 105
63ZAAC(321)143 G. Gattow and B. Krebs; *Z. Anorg. Allg. Chem.*, 1963, **321**, 143. 552
63ZC348 E. Bulka and K.-D. Ahlers; *Z. Chem.*, 1963, **3**, 348. 596
63ZOB113 M. O. Lozinskii and P. S. Pel'kis; *Zh. Obshch. Khim.*, 1963, **33**, 113 (*Chem. Abstr.*, 1963, **59**, 612). 654

64ACS2417 K. A. Jensen and A. Holm; *Acta Chem. Scand.*, 1964, **18**, 2417. 593
64AG807 E. Kühle, E. Klauke and F. Grewe; *Angew. Chem.*, 1964, **76**, 807. 236
64AG(E)380 R. W. Hoffman and H. Hauser; *Angew. Chem., Int. Ed. Engl.*, 1964, **3**, 380. 106
64AG(E)513 G. Köbrich, K. Flory and W. Drischel; *Angew. Chem., Int. Ed. Engl.*, 1964, **3**, 513. 25
64AG(E)580 E. O. Fischer and A. Maasböl; *Angew. Chem., Int. Ed. Engl.*, 1964, **3**, 580. 716
64AOC(1)143 P. M. Treichel and F. G. A. Stone; *Adv. Organomet. Chem.*, 1964, **1**, 143. 243, 244
64AOC(2)257 R. Köster; *Adv. Organomet. Chem.*, 1964, **2**, 257. 172, 204
64CB735 F. Klages and K. Bott; *Chem. Ber.*, 1964, **97**, 735. 628
64CB1111 R. Müller and W. Müller; *Chem. Ber.*, 1964, **97**, 1111. 383
64CB1115 R. Müller and W. Müller; *Chem. Ber.*, 1964, **97**, 1115. 61
64CB1232 H. Eilingsfeld, G. Neubauer, M. Seefelder and H. Weidinger; *Chem. Ber.*, 1964, **97**, 1232. 313, 321
64CB1286 E. Allenstein and A. Schmidt; *Chem. Ber.*, 1964, **97**, 1286. 238, 613, 727
64CB1361 W. Jentzsch; *Chem. Ber.*, 1964, **97**, 1361. 52
64CB1673 R. Müller, S. Reichel and C. Dathe; *Chem. Ber.*, 1964, **97**, 1673. 60, 267
64CB1999 E. Amberger and H. D. Boeters; *Chem. Ber.*, 1964, **97**, 1999. 173
64CB2289 T. Kruck and M. Höfler; *Chem. Ber.*, 1964, **97**, 2289. 526
64CB3022 E. Grigat and R. Pütter; *Chem. Ber.*, 1964, **97**, 3022. 629
64CB3027 E. Grigat and R. Pütter; *Chem. Ber.*, 1964, **97**, 3027. 632
64CB3162 E. Allenstein and P. Quis; *Chem. Ber.*, 1964, **97**, 3162. 452, 728
64CCC645 Z. Arnold and M. Kornilov; *Collect. Czech. Chem. Commun.*, 1964, **29**, 645. 104, 105
64CI(L)581 J. Žemlička; *Chem. Ind.* (*London*), 1964, 581. 80
64CI(L)1757 F. L. Scott and D. A. Cronin; *Chem. Ind.* (*London*), 1964, 1757. 611
64CI(M)206 G. Braca, G. Sbrana and P. Pino; *Chim. Ind.* (*Milan*), 1964, **46**, 206 (*Chem. Abstr.*, 1966, **65**, P8409). 524
64CJC1123 W. R. Cullen and N. K. Hota; *Can. J. Chem.*, 1964, **42**, 1123. 60
64CPB946 T. Tsuji and T. Ueda; *Chem. Pharm. Bull.*, 1964, **12**, 946. 641
64CRV19 P. Noble, Jr., F. G. Borgardt and W. L. Reed; *Chem. Rev.*, 1964, **64**, 19. 144, 150
64CRV645 M. Matzner, R. P. Kurkjy and R. J. Cotter; *Chem. Rev.*, 1964, **64**, 645. 421, 426, 460
64CR(259)4051 G. Jeminet and A. Kergomard; *C. R. Hebd. Seances Acad. Sci.*, 1964, **259**, 4051. 81, 90, 91
64FRP1369784 L. G. Anello, H. R. Nychka and C. Woolf; *Fr. Pat.* 1 369 784 (1964) (*Chem. Abstr.*, 1965, **62**, 1570). 16
64FRP1372971 B. Freedman (California Research Corp.); *Fr. Pat.* 1 372 971 (1964) (*Chem. Abstr.*, 1965, **62**, 1363). 435
64GEP1161285 W. Stilze; *Ger. Pat.* 1 161 285 (1964) (*Chem. Abstr.*, 1964, **60**, 9156). 107
64GEP1224720 W. Weiss (Farbenfabriken Bayer AG.); *Ger. Pat.* 1224720 (1964) (*Chem. Abstr.* 1966, **65**, 12112). 434
64JA2961 D. Seyferth, J. Y.-P. Mui and L. J. Todd; *J. Am. Chem. Soc.*, 1964, **86**, 2961. 38
64JCS1351 D. A. Barr, R. N. Haszeldine and V. Matthews; *J. Chem. Soc.*, 1964, 1351. 604
64JCS4066 R. E. Banks, R. N. Haszeldine and H. Sutcliffe; *J. Chem. Soc.*, 1964, 4066. 604
64JGU4149 V. S. Etlis, A. P. Sineokov and G. A. Razuvaev; *J. Gen. Chem. USSR* (*Engl. Transl.*), 1964, **34**, 4149. 628, 629

64JHC233	R. A. Mitsch; *J. Heterocycl. Chem.*, 1964, **1**, 233.	261
64JOC1	W. A. Sheppard; *J. Org. Chem.*, 1964, **29**, 1.	39, 229, 230
64JOC11	P. E. Aldrich and W. A. Sheppard; *J. Org. Chem.*, 1964, **29**, 11.	229, 230, 424
64JOC252	E. R. Bissel; *J. Org. Chem.*, 1964, **29**, 252.	2
64JOC895	W. A. Sheppard; *J. Org. Chem.*, 1964, **29**, 895.	233, 234
64JOC898	S. Andreades, J. F. Harris, Jr., and W. A. Sheppard; *J. Org. Chem.*, 1964, **29**, 898.	233
64JOC953	R. L. Merker and M. J. Scott; *J. Org. Chem.*, 1964, **29**, 953.	173, 378
64JOC1148	W. Reeve and L. W. Fine; *J. Org. Chem.*, 1964, **29**, 1148.	25
64JOC1198	P. Tarrant and E. C. Stump, Jr.; *J. Org. Chem.*, 1964, **29**, 1198.	5
64JOC1613	D. B. Murphy; *J. Org. Chem.*, 1964, **29**, 1613.	604
64JOC1851	N. R. Easton, D. R. Cassady and R. D. Dillard; *J. Org. Chem.*, 1964, **29**, 1851.	509
64JOC2401	H. Ulrich, J. N. Tilley and A. A. R. Sayigh; *J. Org. Chem.*, 1964, **29**, 2401.	445
64JOC2442	T. P. Johnston and A. Gallagher; *J. Org. Chem.*, 1964, **29**, 2442.	474
64JOC2769	H. E. Zaugg and R. W. DeNet; *J. Org. Chem.*, 1964, **29**, 2769.	104
64JOC2773	W. C. Kuryla and D. G. Leis; *J. Org. Chem.*, 1964, **29**, 2773.	74, 75
64JOC3007	K. O. Christie and A. E. Pavlath; *J. Org. Chem.*, 1964, **29**, 3007.	419
64JOC3587	T. N. Hall; *J. Org. Chem.*, 1964, **29**, 3587.	287
64LA(590)123	W. Ried and O. G. Hillenbrand; *Justus Liebigs Ann. Chem.*, 1964, **590**, 123.	537
64LA(674)124	W. Ried and B. M. Beck; *Justus Liebigs Ann. Chem.*, 1964, **674**, 124.	535
64LA(675)142	H. Gross, J. Rusche and H. Bornowski; *Justus Liebigs Ann. Chem.*, 1964, **675**, 142.	298, 300, 312, 321, 646
64MI 613-01	S.-L. Liu; *J. Chinese Chem. Soc., Series II*, 1964, **11**, 163, (*Chem. Abstr.*, **62**, 9166).	378
B-64MI 614-01	G. A. Olah and J. A. Olah; in "Friedel–Crafts and Related Reactions," ed. G. A. Olah, Interscience, New York, 1964, vol. III, pp. 1262–1267.	442
B-64MI 616-01	J. C. Hileman; "Preparative Inorganic Reactions," Interscience, New York, 1964, vol. 1, p. 102.	523
64RTC119	R. M. Pike and A. M. Dewidar; *Recl. Trav. Chim. Pays-Bas*, 1964, **83**, 119.	68
64RTC1305	H. Dolman, H. A. Peperkamp and H. D. Moed; *Recl. Trav. Chim. Pays-Bas*, 1964, **83**, 1305.	642
64T25	A. Mustafa, W. Asker, A. F. A. M. Shalaby, A. H. Harhash and R. Daguer; *Tetrahedron*, 1964, **20**, 25.	114
64T417	S. Oae, W. Tagaki and A. Ohno; *Tetrahedron*, 1964, **20**, 417.	82
64TL47	M. M. Shemyakin, Y. A. Ovchinnikov, V. K. Antonov, A. A. Kiryushkin, V. T. Ivanov, V. I. Shchelokov and A. M. Shkrob; *Tetrahedron Lett.*, 1964, 47.	104
64TL245	D. M. Lemal and E. H. Banitt; *Tetrahedron Lett.*, 1964, 245.	697
64TL1667	R. Köster and G. W. Rotermund; *Tetrahedron Lett.*, 1964, 1667.	183
64TL3323	D. R. Coulsen; *Tetrahedron Lett.*, 1964, 3323.	697
64USP3121084	H. E. Winberg; *US Pat.* 3 121 084 (1964) (*Chem. Abstr.*, 1964, **60**, 13 197).	105
64USP3121112	C. W. Tullock (DuPont Co.); *US Pat.* 3121112 (1964) (*Chem. Abstr.*, 1964, **60**, 13143).	241
64ZC384	E. Fanghänel and R. Mayer; *Z. Chem.*, 1964, **4**, 384.	301, 303, 309, 310
64ZC457	D. Hempel and F. Wolf; *Z. Chem.*, 1964, **4**, 457.	254
64ZOB1979	L. M. Yagupol'skii and V. V. Orda; *Zh. Obshch. Khim.*, 1964, **34**, 1979.	229
64ZOB2855	M. I. Ermakova and I. Y. Postovskii; *Zh. Obshch. Khim.*, 1964, **34**, 2855 (*Chem. Abstr.*, 1964, **61**, 15 993).	653
65AFC(4)50	R. D. Chambers and R. H. Mobbs; *Adv. Fluorine Chem.*, 1965, **4**, 50.	41, 58, 60, 65
65AG427	E. Steininger; *Angew. Chem.*, 1965, **77**, 427.	109
65AG(E)873	K. Dimroth, P. Heinrich and K. Schromm; *Angew, Chem., Int. Ed. Engl.*, 1965, **4**, 873.	73
65AJC1967	G. Crank and F. W. Eastwood; *Aust. J. Chem.*, 1965, **18**, 1967.	104
65BCJ2107	T. Mukaiyama, T. Yamaguchi and H. Nohira; *Bull. Chem. Soc. Jpn.*, 1965, **38**, 2107.	113
65BRP1014252	R. N. Haszeldine and J. M. Birchall; *Br. Pat.* 1014252 (1965) (*Chem. Abstr.*, 1966, **64**, 9633).	32
65CB1078	H. Bredereck, F. Effenberger and G. Simchen; *Chem. Ber.*, 1965, **98**, 1078.	55
65CB2059	D. Martin and W. Mucke; *Chem. Ber.*, 1965, **98**, 2059.	534
65CB2063	P. Reich and D. Martin; *Chem. Ber.*, 1965, **98**, 2063.	533
65CB3286	D. Martin; *Chem. Ber.*, 1965, **98**, 3286.	300
65CB3303	A. Schönberg, B. König and E. Frese; *Chem. Ber.*, 1965, **98**, 3303.	87
65CB3662	D. Martin and S. Rackow; *Chem. Ber.*, 1965, **98**, 3662.	624
65CC241	P. J. Aymonino; *J. Chem. Soc., Chem. Commun.*, 1965, 241.	424
65CI(L)791	R. J. Cotter, M. Matzner and R. P. Kurkjy; *Chem. Ind.* (*London*), 1965, 791.	425
65CJC1798	G. R. Pettit, D. S. Blonda and R. A. Upham; *Can. J. Chem.*, 1965, **43**, 1798.	508
65CJC1893	M. E. Redwood and C. J. Willis; *Can. J. Chem.*, 1965, **43**, 1893.	232
65CJC1918	M. Mazurek and A. S. Perlin; *Can. J. Chem.*, 1965, **43**, 1918.	71
65GEP1219455	E. Klauke, E. Kühle and W. Weiss; *Ger. Pat.*, 1 219 455 (1965) (*Chem. Abstr.*, 1966, **65**, 13 553).	529
65HCA1746	H. Brechbühler, H. Büchi, E. Hatz, J. Schreiber and A. Eshenmoser; *Helv. Chim. Acta*, 1965, **48**, 1746.	104
65HOU(6)361	H. Meerwein; *Methoden Org. Chem.* (*Houben-Weyl*), 1965, **6**, 361.	73
65IC293	F. Calderazzo; *Inorg. Chem.*, 1965, **4**, 293.	523
65IC1458	N. E. Miller; *Inorg. Chem.*, 1965, **4**, 1458.	708
65IC1828	M. Lustig; *Inorg. Chem.*, 1965, **4**, 1828.	251
65IZV1873	B. L. Dyatkin, A. A. Gevorkyan and I. L. Knunyants; *Izv. Akad. Nauk SSSR Ser. Khim.*; 1965, 1873 (*Chem. Abstr.*, 1966, **64**, 1944).	219
65IZV2179	I. A. Ivanova, B. P. Fedorov and F. M. Stoyanovich; *Izv. Akad. Nauk SSSR, Ser. Khim.*, 1965, 2179 (*Chem. Abstr.*, 1966, **64**, 12 538).	111, 140

65JA934 E. J. Corey, F. A. Carey and R. A. E. Winter; *J. Am. Chem. Soc.*, 1965, **87**, 934. 555
65JA3788 R. West, P. A. Carney and I. C. Mineo; *J. Am. Chem. Soc.*, 1965, **87**, 3788. 206
65JA3943 L.-H. Lee; *J. Am. Chem. Soc.*, 1965, **30**, 3943. 429
65JA4147 D. F. Hoeg, D. I. Lusk and A. L. Crumbliss; *J. Am. Chem. Soc.*, 1965, **87**, 4147. 62, 244
65JA4338 W. A. Sheppard; *J. Am. Chem. Soc.*, 1965, **87**, 4338. 239, 603
65JA4396 W. H. Graham, *J. Am. Chem. Soc.*, 1965, **87**, 4396. 55, 280
65JCS32 D. Leaver, D. M. McKinnon and W. A. H. Robertson; *J. Chem. Soc.*, 1965, 32. 113
65JCS2101 R. N. Haszeldine, M. J. Newlands and J. B. Plumb; *J. Chem. Soc.*, 1965, 2101. 61
65JCS2157 T. A. George, K. Jones and M. F. Lappert; *J. Chem. Soc.*, 1965, 2157.
65JCS5774 M. Green and A. E. Tipping; *J. Chem. Soc.*, 1965, 5774. 603
65JCS6077 R. E. Banks and E. D. Burling; *J. Chem. Soc.*, 1965, 6077. 257
65JCS6141 R. N. Haszeldine and A. E. Tipping; *J. Chem. Soc.*, 1965, 6141. 239
65JCS6149 R. E. Banks, M. G. Barlow and R. N. Haszeldine; *J. Chem. Soc.*, 1965, 6149. 54
65JCS6875 H. Goldwhite, R. N. Haszeldine and D. G. Roswell; *J. Chem. Soc.*, 1965, 6875. 679
65JCS7511 N. Welcmān and H. Regev; *J. Chem. Soc.*, 1965, 7511. 49
65JCS7515 N. Welcman and I. Rot; *J. Chem. Soc.*, 1965, 7515. 49
65JHC41 W. B. Wright, Jr., *J. Heterocycl. Chem.*, 1965, **2**, 41. 416
65JMC174 E. van Heyningen and C. N. Brown; *J. Med. Chem.*, 1965, **8**, 174. 560
65JMC446 J. Augstein, S. M. Green, A. M. Monro, G. W. H. Potter, C. R. Worthing and T. I. Wrigley; *J. Med. Chem.*, 1965, **8**, 446. 642
65JOC411 M. E. Hill, D. T. Carty, D. Tegg, J. C. Butler and A. F. Stang; *J. Org. Chem.*, 1965, **30**, 411. 69, 296, 297
65JOC829 C. Temple, Jr., R. L. McKee and J. A. Montgomery; *J. Org. Chem.*, 1965, **30**, 829. 109
65JOC1027 S. A. Fuqua, W. G. Duncan and R. M. Silverstein; *J. Org. Chem.*, 1965, **30**, 1027. 695
65JOC1317 G. A. Olah, S. J. Kuhn and R. E. A. Dear; *J. Org. Chem.*, 1965, **30**, 1317. 432, 432
65JOC1375 W. J. Middleton, E. G. Howard and W. H. Sharkey; *J. Org. Chem.*, 1965, **30**, 1375. 49, 253, 528, 529, 532
65JOC1639 K. O. Christie and A. E. Pavlath; *J. Org. Chem.*, 1965, **30**, 1639. 419, 424
65JOC2454 D. L. Klayman; *J. Org. Chem.*, 1965, **30**, 2454. 637
65JOC2849 Z. G. Hajos, D. R. Parrish and M. W. Goldberg; *J. Org. Chem.*, 1965, **30**, 2849. 506
65JOC4011 R. C. Terrell, T. Ucciardi and J. F. Vitcha; *J. Org. Chem.*, 1965, **30**, 4011. 44, 45, 46
65JOC4104 K. O. Christie and A. E. Pavlath; *J. Org. Chem.*, 1965, **30**, 4104. 421
65JOC4303 A. J. Speziale, L. R. Smith amd J. E. Fedder; *J. Org. Chem.*, 1965, **30**, 4303. 57
65JOM(3)337 D. Seyferth, H. D. Simmons, Jr. and G. Singh; *J. Organomet. Chem.*, 1965, **3**, 337. 695
65JOM(4)98 R. L. Merker and M. J. Scott; *J. Organomet. Chem.*, 1965, **4**, 98. 173, 378
65LA(681)64 W. Walter and K. D. Bode; *Justus Liebigs Ann. Chem.*, 1965, **681**, 64. 533
65LA(689)179 F. Dallacker, E. Kaiser and P. Uddrich; *Justus Liebigs Ann. Chem.*, 1965, **689**, 179. 304
65MI 613-01 G. Fritz, J. Grobe and D. Kummer; *Adv. Inorg. Radiochem. Chem.*, 1965, **7**, 349. 383
65MI 616-01 W. Gerhardt; *Tenside*, 1965, **2**, 101 (*Chem. Abstr.*, 1966, **64**, 12 558); (see W. Gerhardt, *J. Prakt. Chem.*, 1968, **38**, 77). 504
B-65MI 616-02 R. Howe; "Rodd's Chemistry of Carbon Compounds," Elsevier, Amsterdam, 1965, vol. Ic, p. 316. 506
65NEP6412636 Celanese Corp. of America; *Neth. Pat.* 6 412 636 (1965) (*Chem. Abstr.*, 1965, **63**, 16 370). 71
65NEP6509518 E. I. du Pont de Nemours and Co.; *Neth. Pat.* 6 509 518 (1965) (*Chem. Abstr.*, 1966, **64**, 9292). 410
65RTC43 L. C. Willemsens and G. J. M. van der Kerk; *Recl. Trav. Chim. Pays-Bas*, 1965, **84**, 43. 405
65T1537 J. M. Borsus, G. L'Abbe and G. Smets; *Tetrahedron*, 1965, **21**, 1537. 629
65T2059 T. Goto, Y. Kishi, S. Takahashi and Y. Hirata; *Tetrahedron*, 1965, **21**, 2059. 68
65T3537 M. M. Shemyakin, V. K. Antonov, A. M. Shkrob, V. I. Shchelokov and Z. E. Agadzhanyan; *Tetrahedron*, 1965, **21**, 3537. 104
65TL969 G. Köbrich, H. R. Merkle and H. Trapp; *Tetrahedron Lett.*, 1965, 969. 62
65TL2423 R. Neidlein and W. Haussmann; *Tetrahedron Lett.*, 1965, 2423. 448
65TL3429 G. Cainelli, G. Dal Bello and G. Zubiani; *Tetrahedron Lett.*, 1965, 3429. 198
65USP3202718 E. K. Ellingboe and A. L. McClelland; *US Pat.* 3 202 718 (1965) (*Chem. Abstr.*, 1965, **63**, 14706). 231
65USP3214412 M. Brown; *US Pat.* 3 214 412 (1965) (*Chem. Abstr.*, 1966, **64**, 3 542). 105
65ZC386 E. Fanghänel; *Z. Chem.*, 1965, **5**, 386. 316
65ZN(B)1159 W. Hieber, F. Lux and C. Herget; *Z. Naturforsch., Teil B*, 1965, **20**, 1159. 523
65ZN(B)1165 A. Heesing and U. Wernicke; *Z. Naturforsch., Teil B*, 1965, **20**, 1165. 649
65ZOB1412 M. A. Englin, S. P. Makarov, S. S. Dubov and A. Ya. Yakubovich; *Zh. Obshch. Khim.*, 1965, **35**, 1412 (*Chem. Abstr.*, 1965, **63**, 14344). 258, 259
65ZOB1628 V. V. Orda, L. M. Yagupol'skii, V. F. Bystrov and A. U. Stepanyants; *Zh. Obshch. Khim.*, 1965, **35**, 1628 (*Chem. Abstr.*, 1965, **63**, 17861). 233, 234, 237
65ZOR735 L. S. Pupko and P. S. Pel'kis; *Zh. Org. Khim.*, 1965, **1**, 735 (*Chem. Abstr.*, 1965, **63**, 5555). 653

66ACS278 K. A. Jensen, G. Felbert, C. T. Pedersen and U. Svanholm; *Acta Chem. Scand.*, 1966, **20**, 278. 598
66ACS281 K. A. Jensen, G. Felbert and B. Kägi; *Acta Chem. Scand.*, 1966, **20**, 281. 597
66ACS2091 K. A. Jensen, M. Due, A. Holm and C. Wentrup; *Acta Chem. Scand.*, 1966, **20**, 2091. 593
66ACS2742 U. Anthoni; *Acta Chem. Scand.*, 1966, **20**, 2742. 594
66AG210 G. Ottman and H. Hookes, Jr.; *Angew. Chem.*, 1966, **78**, 210. 494
66AG376 E. Fahr and H. Lind; *Angew. Chem.*, 1966, **78**, 376. 506
66AG906 A. Haas and W. Klug; *Angew. Chem.*, 1966, **78**, 906. 235
66AG(E)132 H. Bredereck, F. Effenberger and Th. Brendle; *Angew. Chem., Int. Ed. Engl.*, 1966, **5**, 132. 138

66AG(E)672 G. Ottomann and H. Hooks, Jr.; *Angew Chem., Int. Ed. Engl.*, 1966, **5**, 672. 448
66AG(E)841 U. Müller and K. Dehnicke; *Angew. Chem., Int. Ed. Engl.*, 1966, **5**, 841. 733
66AG(E)845 A. Haas and W. Klug; *Angew. Chem., Int. Ed. Engl.*, 1966, **5**, 845. 273, 607
66AG(E)848 E. Klauke; *Angew. Chem., Int. Ed. Engl.*, 1966, **5**, 848. 602, 615
66AG(E)875 W. Seelinger, E. Aufderhaar, W. Diepers, R. Feinauer, R. Nehring, W. Thier and H. Hellman; *Angew. Chem., Int. Ed. Engl.*, 1966, **5**, 875. 107
66AG(E)894 R. Feinauer; *Angew. Chem., Int. Ed. Engl.*, 1966, **5**, 894. 107
66AHC(7)39 H. Prinzbach and E. Futterer; *Adv. Heterocycl. Chem.*, 1966, **7**, 39. 48, 729
66AP709 R. Neidlein and E. Heukelbach; *Arch. Pharm. (Weinheim, Ger.)*, 1966, **299**, 709. 653
66BCJ2005 T. Mukaiyama and T. Yamaguchi; *Bull. Chem. Soc. Jpn.*, 1966, **39**, 2005. 113
66BCJ2191 K. Inukai, T. Ueda, and H. Muramatsu; *Bull. Chem. Soc. Jpn.*, 1966, **39**, 2191. 46, 47
66BEP672205 Procter & Gamble Co.; *Belgian Pat.* 672205 (1966) (*Chem. Abstr.*, 1967, **67**, 4040). 263, 287
66BSF73 V. I. Cohen, P. Reynaud and R. C. Moreau; *Bull. Soc. Chim. Fr.*, 1966, 73. 644
66CB239 R. Neidlein and W. Haussmann; *Chem. Ber.*, 1966, **99**, 239. 448
66CB371 O. Glemser, J. Schröder and J. Knaak; *Chem. Ber.*, 1966, **99**, 371. 410
66CB793 R. Müller and S. Reichel; *Chem. Ber.*, 1966, **99**, 793. 61
66CB1252 R. Neidlein, W. Haussmann and E. Heukelbach; *Chem. Ber.*, 1966, **99**, 1252. 666
66CB1892 R. W. Hoffmann, J. Schneider and H. Häuser; *Chem. Ber.*, 1966, **99**, 1892. 74
66CB1912 J. Burkhardt, R. Feinauer, E. Gulbins and K. Hamann; *Chem. Ber.*, 1966, **99**, 1912. 629
66CB2039 H. Bock and I. Kroner; *Chem. Ber.*, 1966, **99**, 2039. 489
66CB2454 K. Bredereck and R. Richter; *Chem. Ber.*, 1966, **99**, 2454. 620
66CB2625 H. Gross and J. Rusche; *Chem. Ber.*, 1966, **99**, 2625. 43
66CB2885 R. Gompper and W. Hägele; *Chem. Ber.*, 1966, **99**, 2885. 627, 651
66CB2937 H. Beyer and K. Pommerening; *Chem. Ber.*, 1966, **99**, 2937. 504
66CB3103 A. Haas and P. Schott; *Chem. Ber.*, 1966, **99**, 3103. 450
66CB3155 K. Hartke and E. Palou; *Chem. Ber.*, 1966, **99**, 3155. 622
66CB3268 K. Hartke, E. Schmidt, M. Castillo and J. Bartulin; *Chem. Ber.*, 1966, **99**, 3268. 627
66CB3892 F. Effenberger, R. Gleiter and G. Kiefer; *Chem. Ber.*, 1966, **99**, 3892. 106
66CCC3584 O. Paleta and A. Posta; *Collect. Czech. Chem. Commun.*, 1966, **31**, 3584. 5
66CI(L)1266 L. E. Chapman and R. F. Robbins; *Chem. Ind. (London)*, 1966, 1266. 511
66CJC575 W. Reeve, J. P. Mutchler and C. L. Liotta; 1966, *Can. J. Chem.*, 1966, **44**, 575. 27
66FRP1437908 M. Zbirovsky and J. Masát; *Fr. Pat.* 1437908 (1966) (*Chem. Abstr.*, 1967, **66**, 37431). 236
66G375 R. Scarpati, D. Sica and C. Santacroce; *Gazz. Chim. Ital.*, 1966, **96**, 375. 76
66GEP1156780 H. von Brachel; *Ger. Pat.* 1156780 (1966) (*Chem. Abstr.*, 1966, **60**, 5344). 105
66IC488 J. B. Hynes and T. E. Austin; *Inorg. Chem.*, 1966, **5**, 488. 53
66IC1455 M. D. Meyers and S. Frank; *Inorg. Chem.*, 1966, **5**, 1455. 261
66IC1795 D. H. Dybvig; *Inorg. Chem.*, 1966, **5**, 1795. 613
66IS165 G. H. Cady; *Inorg. Synth.*, 1966, **8**, 165. 231
66IZV98 V. F. Sedova, L. D. Dikanskaya and V. P. Mamaev; *Izv. Akad. Nauk SSSR, Ser. Khim.*, 1966, 98 (*Chem. Abstr.*, 1967, **65**, 13701). 110
66IZV364 V. A. Dorokhov and B. M. Mikhailov; *Izv. Akad. Nauk SSSR, Ser. Khim.*, 1966, 364 (*Chem. Abstr.*, 1966, **64**, 17624). 113
66JA850 H. Weingarten and W. A. White; *J. Am. Chem. Soc.*, 1966, **88**, 850. 141
66JA1024 D. F. DeTar, R. Silverstein and F. F. Rogers, Jr.; *J. Am. Chem. Soc.*, 1966, **88**, 1024. 506
66JA1821 C. H. Wei and L. F. Dahl; *J. Am. Chem. Soc.*, 1966, **88**, 1821. 524
66JA1834 D. Seyferth and B. Prokai; *J. Am. Chem. Soc.*, 1966, **88**, 1834. 61
66JA2885 H. Weingarten and W. A. White; *J. Am. Chem. Soc.*, 1966, **88**, 2885. 361
66JA3798 W. M. Jones and D. L. Muck; *J. Am. Chem. Soc.*, 1966, **88**, 3798. 506
66JA3825 D. M. White and J. Sonnenberg; *J. Am. Chem. Soc.*, 1966, **88**, 3825. 143
66JCS(A)367 R. C. Dobbie and H. J. Emeléus; *J. Chem. Soc. (A)*, 1966, 367. 240
66JCS(A)933 R. C. Dobbie and H. J. Emeléus; *J. Chem. Soc. (A)*, 1966, 933. 607
66JCS(C)350 E. Boyland and R. Nery; *J. Chem. Soc. (C)*, 1966, 350. 505
66JCS(C)2031 J. A. Bee and F. L. Rose; *J. Chem. Soc. (C)*, 1966, 2031. 650
66JHC245 R. A. Mitsch; *J. Heterocycl. Chem.*, 1966, **3**, 245. 261
66JINC1823 H. J. Emeléus and B. W. Tattershall; *J. Inorg. Nucl. Chem.*, 1966, **28**, 1823. 241
66JMC444 C. W. Mosher, E. M. Acton and L. Goodman; *J. Med. Chem.*, 1966, **9**, 444. 502
66JMC852 W. B. Wright, Jr., H. J. Brabander, R. A. Hardy, Jr. and A. C. Osterberg; *J. Med. Chem.*, 1966, **9**, 852. 508
66JOC789 W. Carpenter, A. Haymaker and D. W. Moore; *J. Org. Chem.*, 1966, **31**, 789. 54
66JOC1426 A. J. Papa; *J. Org. Chem.*, 1966, **31**, 1426. 649
66JOC2312 D. Sianesi, A. Pasetti and F. Tarli; *J. Org. Chem.*, 1966, **31**, 2312. 42
66JOC2874 H. Weingarten and W. A. White; *J. Org. Chem.*, 1966, **31**, 2874. 141
66JOC3833 R. A. Mitsch and P. H. Ogden; *J. Org. Chem.*, 1966, **31**, 3833. 242, 613
66JOC3980 M. S. Newman and H. A. Karnes; *J. Org. Chem.*, 1966, **31**, 3980. 555, 556
66JOC4054 N. L. Weinberg and E. A. Brown; *J. Org. Chem.*, 1966, **31**, 4054. 107
66JOC4232 R. J. Koshar, D. R. Husted and R. A. Meiklejohn; *J. Org. Chem.*, 1966, **31**, 4232. 259
66JOM(5)185 D. Seyferth, J. Y.-P. Mui and G. Singh; *J. Organomet. Chem.*, 1966, **5**, 185. 266
66JOM(6)451 M. Kumada and M. Ishikawa; *J. Organomet. Chem.*, 1966, **6**, 451. 195
66LA(694)44 U. Schöllkopf and E. Wiskott; *Justus Liebigs Ann. Chem.*, 1966, **694**, 44. 627
66LA(698)174 R. Feinauer and W. Seeliger; *Justus Liebigs Ann. Chem.*, 1966, **698**, 174. 107
66LA(698)180 E. Gulbins, R. Morlock and K. Hamann; *Justus Liebigs Ann. Chem.*, 1966, **698**, 180. 510
66LA(699)40 K. Issleib and R. Lindner; *Justus Liebigs Ann. Chem.*, 1966, **699**, 40. 701
B-66MI 607-01 Z. E. Jolles (ed.); "Bromine and its Compounds," Ernest Benn Ltd., London, 1966. 212

66PHA23	G. Zinner and R. O. Weber; *Pharmazi*, 1966, **21**, 23.	570, 571
66RTC793	J. W. Scheeren and W. Stevens; *Recl. Trav. Chim. Pays Bas*, 1966, **85**, 793.	79
66TL3423	W. P. Neumann and K. Kühlein; *Tetrahedron Lett.*, 1966, 3423.	664
66TL3695	C. W. Pluijgers and J. Berg; *Tetrahedron Lett.*, 1966, 3695.	580
66TL4315	G. Cainelli, G. Dal Bello and G. Zubiani; *Tetrahedron Lett.*, 1966, 4315.	198
66TL6021	G. Zweifel and R. B. Steele; *Tetrahedron Lett.*, 1966, 6021.	208
66UK1740	V. I. Erashko, S. A. Shevelev and A. A. Fainzil'berg; *Usp. Khim.*, 1966, **35**, 1740 (*Chem. Abstr.*, 1967, **66**, 54660).	144
66UKZ204	F. S. Babichev, F. A. Mikhailenko, V. K. Kibirev and V. A. Bogolyubskii; *Ukr. Khim. Zh. (Russ. Ed.)*, 1966, **32**, 204 (*Chem. Abstr.*, 1966, **64**, 17 749).	104
66USP3239519	H. E. Winberg; *US Pat.* 3 239 519 (1966) (*Chem. Abstr.*, 1966, **64**, 15 854).	104
66ZAAC(344)113	E. Allenstein and A. Schmidt; *Z. Anorg. Allg. Chem.*, 1966, **344**, 113.	624, 730
66ZOB461	G. I. Derkach and N. I. Liptuga; *Zh. Obshch. Khim.*, 1966, **36**, 461 (*Chem. Abstr.*, 1966, **65**, 634).	656
66ZOB728	A. Y. Yakubovich, S. M. Rozenshtein, S. E. Vasyukov, B. I. Tetel'baum and V. I. Yakutin; *Zh. Obshch. Khim.*, 1966, **36**, 728 (*Chem. Abstr.*, 1966, **65**, 8 901).	54
66ZOB1309	L. M. Yagupol'skii and M. I. Dronkina; *Zh. Obshch. Khim.*, 1966, **36**, 1309.	238
66ZOB1355	Yu. P. Maslov, A. S. Stepanov and P. G. Maslov; *Zh. Obshch. Khim.*, 1966, **36**, 1355 (*Chem. Abstr.*, 1967, **66**, 10467).	223, 228
66ZOR1902	K. V. Altukhov and V. V. Perekalin; *Zh. Org. Khim.*, 1966, **2**, 1902 (*Chem. Abstr.*, 1967, **66**, 46354).	149

67ACS1293	A. Senning; *Acta Chem. Scand.*, 1967, **21**, 1293.	652
67ACS1981	L. Henriksen; *Acta Chem. Scand.*, 1967, **21**, 1981.	591
67ACS2904	K. A. Jensen and V. Krishnan; *Acta Chem. Scand.*, 1967, **21**, 2904.	594
67ACS(C)1392	K. Torssell; *Acta Chem. Scand., Ser. C*, 1967, **21**, 1392 (*Chem. Abstr.*, 1968, **68**, 78178).	149
67AG15	G. Köbrich; *Angew. Chem.*, 1967, **79**, 15.	244
67AG649	R. Steudel; *Angew. Chem.*, 1967, **79**, 649.	529, 530
67AG(E)41	G. Köbrich; *Angew. Chem., Int. Ed. Engl.*, 1967, **6**, 41.	293
67AG(E)281	W. Walter and K. D. Bode; *Angew. Chem., Int. Ed. Engl.*, 1967, **6**, 281.	533
67AG(E)442	D. Seebach; *Angew. Chem., Int. Ed. Engl.*, 1967, **6**, 442.	82, 87, 88, 303, 317
67AG(E)443	D. Seebach; *Angew. Chem., Int. Ed. Engl.*, 1967, **6**, 443.	317
67AG(E)560	U. Schöllkopf and F. Gerhart; *Angew. Chem., Int. Ed. Engl.*, 1967, **6**, 560.	719, 732
67AG(E)649	E. Kühle, B. Anders and G. Zumach; *Angew. Chem., Int. Ed. Engl.*, 1967, **6**, 649.	602
67AG(E)677	G. Fritz; *Angew. Chem., Int. Ed. Engl.*, 1967, **6**, 677.	173
67AG(E)705	A. Haas and H. Reinke; *Angew. Chem., Int. Ed. Engl.*, 1967, **6**, 705.	432
67AG(E)862	J. Gante; *Angew. Chem. Int. Ed. Engl.*, 1967, **6**, 862.	651
67AG(E)871	W. Verbeek and W. Sundermeyer; *Angew. Chem., Int. Ed. Engl.*, 1967, **6**, 871.	442
67AG(E)942	H. Bock and H. Alt; *Angew. Chem., Int. Ed. Engl.*, 1967, **6**, 942.	173
67AG(E)949	C. Rüchardt and G. Hamprecht; *Angew. Chem., Int. Ed. Engl.*, 1967, **6**, 949.	71
67AP(300)553	R. Neidlein and W. Haussmann; *Arch. Pharm. (Weinheim, Ger.)*, 1967, **300**, 553.	629
67AP(300)567	R. Neidlein and E. Heukelbach; *Arch. Pharm. (Weinheim, Ger.)*, 1967, **300**, 567.	651, 652
67BCJ2641	T. Mukaiyama, S. Aizawa and T. Yamaguchi; *Bull. Chem. Soc. Jpn.*, 1967, **40**, 2641.	105, 106
67BSF1520	J. Villiéras; *Bull. Chem. Soc. Fr.*, 1967, 1520.	244
67BSF2239	D. Noël and J. Vialle; *Bull. Soc. Chim. Fr.*, 1967, 2239.	309
67BSF3233	G. Jeminet and A. Kergomard; *Bull. Soc. Chim. Fr.*, 1967, 3233.	302
67CB767	A. Schönberg, B. König and E. Singer; *Chem. Ber.*, 1967, **100**, 767.	304
67CB1032	H. Schmidbaur and W. Tronich; *Chem. Ber.* 1967, **100**, 1032.	702
67CB1082	O. Glemser, U. Biermann and M. Fild; *Chem. Ber.*, 1967, **100**, 1082.	442
67CB1184	O. Glemser and U. Biermann; *Chem. Ber.*, 1967, **100**, 1184.	409, 410
67CB1373	E. Bulka, K.-D. Ahlers and E. Tuček; *Chem. Ber.*, 1967, **100**, 1373.	596, 598
67CB1459	E. Bulka, K.-D. Ahlers and E. Tuček; *Chem. Ber.*, 1967, **100**, 1459.	596
67CB2484	O. Glemser and U. Biermann; *Chem. Ber.*, 1967, **100**, 2484.	409, 419
67CB2604	E. Allenstein and R. Fuchs; *Chem. Ber.*, 1967, **100**, 2604.	632
67CB2946	A. Roedig, K. Grohe and W. Mayer; *Chem. Ber.*, 1967, **100**, 2946.	69
67CB3319	A. Schmidt; *Chem. Ber.*, 1967, **100**, 3319.	731
67CB3725	A. Schmidt; *Chem. Ber.*, 1967, **100**, 3725.	733
67CC13	G. Schneider and L. K. Láng; *J. Chem. Soc., Chem. Commun.*, 1967, 13.	73
67CC440	P. Chini; *J. Chem. Soc., Chem. Commun.*, 1967, 440.	524
67CC870	M. Lustig and J. K. Ruff; *J. Chem. Soc., Chem. Commun.*, 1967, 870.	250
67CC1089	D. L. Coffen; *J. Chem. Soc., Chem. Commun.*, 1967, 1089.	85
67CCC3064	A. Holý; *Collect. Czech. Chem. Commun.*, 1967, **32**, 3064.	80
67CCC3888	O. Paleta, A. Posta, V. Bartl and A. Balhar; *Collect. Czech. Chem. Commun.*, 1967, **32**, 3888.	5
67CJC2659	C. E. Holloway and M. H. Gitliz; *Can. J. Chem.*, 1967, **45**, 2659.	
67CRV107	F. Kurzer and K. Douraghi-Zadeh; *Chem. Rev.*, 1967, **67**, 107.	506
67DOK(176)1086	L. V. Okhlobystina, G. Ya. Legin and A. A. Fainzil'berg; *Dokl. Akad. Nauk SSSR*, 1967, **176**, 1086 (*Chem. Abstr.*, 1968, **68**, 78397).	287
67FCR77	S. Nagase; *Fluorine Chem. Rev.*, 1967, **1**, 77.	20
67GEP1232954	O. Scherer, J. Korinth and D. Starck (Hoechst AG); *Ger. Pat.* 1232954 (1967) (*Chem. Abstr.*, 1967, **66**, 65093).	236
67IC1711	G. W. Fraser and J. M. Shreeve; *Inorg. Chem.*, 1967, **6**, 1711.	257
67IC1751	S. Cradock, G. A. Gibbon and C. H. Van Dyke; *Inorg. Chem.*, 1967, **6**, 1751.	173

67IZV693 M. F. Shostakovskii, V. N. Komarov and O. G. Yarosh; *Izv. Akad. Nauk SSSR Ser. Khim.*, 1967, 693 (*Chem. Abstr.*, 1968, **68**, 21998). 60
67IZV2708 V. I. Erashko, S. A. Shevelev and A. A. Fainzil'berg; *Izv. Akad. Nauk. SSSR, Ser. Khim.*, 1967, 2708 (*Chem. Abstr.*, 1968, **69**, 67193). 368
67JA182 T. E. Stevens and W. H. Graham; *J. Am. Chem. Soc.* 1967, **89**, 182. 55
67JA431 A. G. Brook, J. M. Duff, P. F. Jones and N. R. Davis; *J. Am. Chem. Soc.*, 1967, **89**, 431. 127, 132, 325, 326, 520
67JA434 E. J. Corey, D. Seebach and R. Freedman; *J. Am. Chem. Soc.*, 1967, **89**, 434. 127, 325
67JA716 W. H. Graham and J. P. Freeman; *J. Am. Chem. Soc.*, 1967, **89**, 716. 104
67JA1538 D. Seyferth, R. Damrauer and S. S. Washburne; *J. Am. Chem. Soc.*, 1967, **89**, 1538. 60
67JA1809 F. A. Hohorst and J. M. Shreeve; *J. Am. Chem. Soc.*, 1967, **89**, 1809. 250
67JA1811 P. G. Thompson; *J. Am. Chem. Soc.*, 1967, **89**, 1811. 250
67JA1962 R. L. Cauble and G. H. Cady; *J. Am. Chem. Soc.*, 1967, **89**, 1962. 250
67JA2502 G. M. Atkins, Jr. and E. M. Burgess; *J. Am. Chem. Soc.*, 1967, **89**, 2502. 106
67JA2790 D. Seyferth and F. M. Armbrecht, Jr.; *J. Am. Chem. Soc.*, 1967, **89**, 2790. 270
67JA3446 T. D. Parsons, J. M. Self and L. H. Schaad; *J. Am. Chem. Soc.*, 1967, **89**, 3446. 243
67JA3868 P. H. Ogden and R. A. Mitsch; *J. Am. Chem. Soc.*, 1967, **89**, 3868. 604
67JA5161 R. L. Cauble and G. H. Cady; *J. Am. Chem. Soc.*, 1967, **89**, 5161. 250, 424
67JA5724 K. R. Henery-Logan and T. L. Fridinger; *J. Am. Chem. Soc.*, 1967, **89**, 5724. 77
67JCS(C)30 B. C. Ennis, G. Holan and E. L. Samuel; *J. Chem. Soc.* (*C*), 1967, 30. 81
67JCS(C)807 T. J. Adley, A. K. M. Anisuzzaman and L. N. Owen; *J. Chem. Soc.* (*C*), 1967, 807. 309
67JCS(C)1150 A. E. Platt and B. Tittle; *J. Chem. Soc.* (*C*), 1967, 1150. 7
67JCS(C)1241 R. N. Haszeldine and A. E. Tipping; *J. Chem. Soc.* (*C*), 1967, 1241. 239
67JCS(C)1470 W. R. Bamford and B. C. Pant; *J. Chem. Soc.* (*C*), 1967, 1470. 267
67JCS(C)1547 R. S. Davidson, R. A. Sheldon and S. Trippett; *J. Chem. Soc.*, *Perkin Trans. 1*, 1967, 1547. 118
67JCS(C)2263 R. E. Banks, R. N. Haszeldine and V. Matthews; *J. Chem. Soc.* (*C*), 1967, 2263. 54
67JCS(C)2302 P. H. Ogden; *J. Chem. Soc.* (*C*), 1967, 2302. 239
67JGU1343 V. A. Ginsburg and K. N. Smirnov; *J. Gen. Chem. USSR* (*Engl. Transl.*), 1967, **37**, 1343. 613
67JHC166 C. M. S. Yoder and J. J. Zuckerman; *J. Heterocycl. Chem.*, 1967, **4**, 166. 299, 300
67JHC389 R. A. Mitsch, E. W. Neuvar and P. H. Ogden; *J. Heterocycl. Chem.*, 1967, **4**, 389. 616
67JINC2819 B. W. Tattershall and G. H. Cady; *J. Inorg. Nucl. Chem.*, 1967, **29**, 2819. 236
67JMC118 P. H. Terry and A. B. Borkovec; *J. Med. Chem.*, 1967, **10**, 118. 655
67JMC273 W. J. Fanshawe and V. J. Bauer; *J. Med. Chem.*, 1967, **10**, 273. 655
67JMC541 W. E. Coyne and J. W. Cusic; *J. Med Chem.*, 1967, **10**, 541. 501
67JOC300 M. J. Zabik and R. D. Schuetz; *J. Org. Chem.*, 1967, **32**, 300. 425
67JOC833 D. Paskovich, P. Gaspar and G. S. Hammond; *J. Org. Chem.*, 1967, **32**, 833. 32, 219
67JOC1217 R. Filler and R. M. Schure; *J. Org. Chem.* 1967, **32**, 1217. 25
67JOC1311 F. E. Herkes and D. J. Burton; *J. Org. Chem.*, 1967, **32**, 1311. 694
67JOC1633 R. F. Merritt; *J. Org. Chem.*, 1967, **32**, 1633. 439
67JOC1662 R. A. Davis, J. L. Kroon and D. A. Rausch; *J. Org. Chem.*, 1967, **32**, 1662. 649
67JOC1944 R. L. Rebertus, J. J. McBrady and J. G. Gagnon; *J. Org. Chem.*, 1967, **32**, 1944. 261
67JOC2063 J. F. Harris, Jr.; *J. Org. Chem.*, 1967, **32**, 2063. 81, 233, 234, 302
67JOC2074 C. K. Bradsher and W. J. Jones, Jr.; *J. Org. Chem.*, 1967, **32**, 2074. 314
67JOC2166 A. Winston, J. C. Sharp, K. E. Atkins and D. E. Battin; *J. Org. Chem.*, 1967, **32**, 2166. 25, 31
67JOC2567 J. J. D'Amico and R. H. Campbell; *J. Org. Chem.*, 1967, **32**, 2567. 304, 305
67JOC2630 F. M. Beringer and S. A. Galton; *J. Org. Chem.*, 1967, **32**, 2630. 73
67JOC2633 P. N. Kevill and F. L. Weitl; *J. Org. Chem.*, 1967, **32**, 2633. 431
67JOC3859 R. J. Koshar, D. R. Husted and C. D. Wright; *J. Org. Chem.*, 1967, **32**, 3859. 284, 616, 649
67JOC4045 R. L. Rebertus and P. E. Toren; *J. Org. Chem.*, 1967, **32**, 4045. 261, 285
67JOC4111 O. T. Quimby, J. B. Prentice and D. A. Nicholson; *J. Org. Chem.*, 1967, **32**, 4111. 263, 518
67JOM(7)P20 D. Seyferth, R. J. Cross and B. Prokai; *J. Organomet. Chem.*, 1967, **7**, P20. 270
67JOM(7)385 L. I. Zakharkin and L. S. Podvisozkaya; *J. Organomet. Chem.*, 1967, **7**, 385. 406
67JOM(9)5 M. V. Kashutina and O. Yu. Okhlobystin; *J. Organomet. Chem.*, 1967, **9**, 5. 368
67KGS1096 N. N. Vereshchagina, I. Y. Postovskii and S. L. Mertsalov; *Khim. Geterotsikl. Soedin.*, 1967, 1096 (*Chem. Abstr.*, 1968, **69**, 59188). 109
67LA(707)35 H. Gross, G. Engelhardt, J. Freiberg, W. Bürger and B. Costisella; *Justus Liebigs Ann. Chem.*, 1967, **707**, 35. 43
67LA(709)248 A. Tzschach and R. Schwarzer; *Justus Liebigs Ann. Chem.*, 1967, **709**, 248. 513
67M1043 G. A. Wildschut, H. J. T. Bos, L. Brandsma and J. F. Arens; *Monatsh. Chem.*, 1967, **98**, 1043. 88
B-67MI 606-01 L. G. Makarova and A. N. Nesmeyanov; "The Organic Compounds of Mercury", North-Holland, Amsterdam, 1967. 172, 207
67MI 607-01 S. A. Miller; *Chem. Process Eng.*, 1967, **48**, 79. 215
67MI 607-02 E. L. Varetti and P. J. Aymonino; *An. Asoc. Quim. Argent.*, 1967, **55**, 153 (*Chem. Abstr.*, 1968, **69**, 76555). 230
67MI 608-01 O. Glemser and U. Biermann; *Inorg. Nucl. Chem. Lett.*, 1967, **3**, 223 (*Chem. Abstr.*, 1967, **67**, 99612). 260
67OS(47)47 H. Gross, A. Rieche, E. Höft and E. Beyer; *Org. Synth.*, 1967, **47**, 47. 39
67T45 H. Nozaki, S. Fujita, H. Tayaka and R. Noyori; *Tetrahedron*, 1967, **23**, 45. 107
67T4181 Ö. K. J. Kovács, G. Schneider, L. K. Láng and A. Apjok; *Tetrahedron*, 1967, **23**, 4181. 73
67TL1443 E. J. Bulten and J. G. Noltes; *Tetrahedron Lett.*, 1967, 1443. 185
67TL2747 W. P. Tucker and G. L. Roof; *Tetrahedron Lett.*, 1967, 2747. 729
67TL3501 R. W. Hoffmann and H. J. Luthardt; *Tetrahedron Lett.*, 1967, 3501. 107

67TL3851 A. Heesing and G. Maleck; *Tetrahedron Lett*, 1967, 3851. 649

67USP3306939 M. E. Hill (US Dept. of the Navy), *US Pat.* 3306939 (1967) (*Chem. Abstr.*, 1967, **66**, 104676). 297

67USP3316311 C. W. Plummer; *US Pat.* 3316311 (1967) (*Chem. Abstr.*, 1967, **67**, 53650). 145, 150

67USP3322823 H. G. Langer (Dow Chemical Co.); *US Pat.* 3322823 (1967) (*Chem. Abstr.*, 1967, **67**, 55770). 410

67USP3333007 C. S. Scanley; *US Pat.* 3333007 (1967) (*Chem. Abstr.*, 1967, **67**, 90393). 91

67USP3354217 C. D. Burton (Dow Chemical Co.); *US Pat.* 3354217 (1967) (*Chem. Abstr.*, 1968, **68**, 39083). 284

67USP3355492 R. L. Rebertus and J. G. Gagnon (Minnesota Mining and Manufacturing Co.), US Pat. 3355492 (1967) (*Chem. Abstr.*, 1968, **68**, 49031). 261

67ZN(B)820 A. Heesing and R. Peppmöller; *Z. Naturforsch., Teil B*, 1967, **22**, 820. 642, 650

67ZOB152 A. V. Fokin, V. A. Komarov, K. V. Frosina and K. A. Abdulzanieva; *Zh. Obshch. Khim.*, 1967, **37**, 152 (*Chem. Abstr.*, 1967, **66**, 85399). 54

67ZOB1040 M. A. Kadina, V. A. Ponomarenko, A. V. Kessenikh and E. P. Prokof'ev; *Zh. Obshch. Khim.*, 1967, **37**, 1040 (*Chem. Abstr.*, 1968, **68**, 22019). 62

67ZOB1163 V. A. Tartakovskii, G. A. Shvekhgeimer, N. I. Sobtsova and S. S. Novikov; *Zh. Obshch. Khim.*, 1967, **37**, 1163 (*Chem. Abstr.*, 1968, **68**, 22000). 149

67ZOB1770 L. M. Yagupol'skii and N. V. Kondratenko; *Zh. Obshch. Khim.*, 1967, **37**, 1770. 234

67ZOB2137 V. V. Moskva, A. I. Maikova and A. I. Razumov; *Zh. Obshch. Khim.*, 1967, **37**, 2137 (*Chem. Abstr.*, 1968, **68**, 29802). 314

67ZOB2513 B. M. Gladshtein, V. G. Noskov and L. Z. Soborovskii; *Zh. Obshch. Khim.*, 1967, **37**, 2513 (*Chem. Abstr.*, 1968, **69**, 27493). 366

67ZOR1749 A. V. Stavroskaya, T. V. Protopopova and A. P. Skoldinov; *Zh. Org. Khim.*, 1967, **3**, 1749 (*Chem. Abstr.*, 1968, **68**, 12448). 140

68ACS1898 U. Anthoni, C. Larsen and P. H. Nielsen; *Acta Chem. Scand.*, 1968, **22**, 1898. 574

68AG(E)291 H. Ulrich, B. Tucker and A. A. R. Sayigh; *Angew. Chem., Int. Ed. Engl.*, 1968, **7**, 291. 652

68AG(E)299 H. P. Latscha and P. B. Hormuth; *Angew. Chem., Int. Ed. Engl.*, 1968, **7**, 299. 622

68AG(E)391 H. Gross and B. Costisella; *Angew. Chem., Int. Ed. Engl.*, 1968, **7**, 391. 157

68AG(E)463 H. Gross and B. Costisella; *Angew. Chem., Int. Ed. Engl.*, 1968, **7**, 463. 157

68AG(E)464 G. Simchen; *Angew. Chem., Int. Ed. Engl.*, 1968, **7**, 464. 618

68AG(E)868 M. Dräger and G. Gattow; *Angew. Chem., Int. Ed. Engl.*, 1968, **7**, 868. 590

68BAU414 V. I. Stanko, V. I. Bregadze, A. I. Klimova, O. Yu. Okhlobystin, A. N. Kashin, K. P. Butin and I. P. Beletskaya; *Bull. Acad. Sci. USSR. Div. Chim. Sci.*, 1968, 414. 406

68BRP1097237 E. Kühle, H. Holtschmidt, F. Grewe and H. Kaspers (Farbenfabriken Bayer AG); *Br. Pat.* 1097237 (1968) (*Chem. Abstr.*, 1968, **69**, 10111). 316

68BSF216 H. Demarne and P. Cadiot; *Bull. Soc. Chim. Fr.*, 1968, 216. 208

68CB41 H. Bredereck, G. Simchen, S. Rebsdat, W. Kantlehner, P. Horn, R. Wahl, H. Hoffmann and P. Grieshaber; *Chem. Ber.*, 1968, **101**, 41. 104

68CB51 G. Simchen, H. Hoffmann and H. Bredereck; *Chem. Ber.*, 1968, **101**, 51. 138, 140

68CB113 U. Hasserodt; *Chem. Ber.*, 1968, **101**, 113. 449

68CB1137 H. Quast and E. Schmidt; *Chem. Ber.*, 1968, **101**, 1137. 318, 361

68CB1885 H. Bredereck, F. Effenberger, Th. Brendle and H. Muffler; *Chem. Ber.*, 1968, **101**, 1885. 138

68CB2419 F. Merger; *Chem. Ber.*, 1968, **101**, 2419. 506

68CB2609 A. Haas and W. Klug; *Chem. Ber.*, 1968, **101**, 2609. 235, 273, 534

68CB2617 A. Haas and W. Klug; *Chem. Ber.*, 1968, **101**, 2617. 236, 252

68CB2815 H. Bock, H. Seidl and M. Fochler; *Chem. Ber.*, 1968, **101**, 2815. 289

68CB3002 R. Richter; *Chem. Ber.*, 1968, **101**, 3002. 143

68CB3058 H. Bredereck, G. Simchen and H. U. Schenck; *Chem. Ber.*, 1968, **101**, 3058. 138

68CB3734 M. Regitz, W. Anschütz and A. Liedhegener; *Chem. Ber.*, 1968, **101**, 3734. 666

68CB3851 R. W. Hoffmann and H. J. Luthardt; *Chem. Ber.*, 1968, **101**, 3851. 107

68CC527 W. A. Thaler; *J. Chem. Soc., Chem. Commun.*, 1968, 527. 434

68CC968 J. E. Baldwin and J. E. Patrick; *Chem. Commun.*, 1968, 968. 608

68CI(M)197 F. Gozzo and G. Camaggi; *Chem. Ind. (Milan)*, 1968, **50**, 197. 251

68CJC2119 A. G. Brook, P. F. Jones and G. J. D. Peddle; *Can. J. Chem.*, 1968, **46**, 2119. 125, 131, 132, 519

68CJC2233 W. Reeve, J. C. Hoffsommer and P. F. Aluotto; *Can. J. Chem.*, 1968, **46**, 2233. 27

68CR(C)1506 C. Feugeas and D. Olschwang; *C. R. Hebd. Seances Acad. Sci., Ser. C.*, 1968, **266**, 1506. 95

68FRP1517379 G. Zumach and E. Kühle (Bayer AG); *Fr. Pat.* 1517379 (1968) (*Chem. Abstr.*, 1969, **71**, 12848). 453

68FRP1526958 W. V. Bauer (Lummus Co.); *Fr. Pat.* 1526958 (1968) (*Chem. Abstr.*, 1969, **71**, 51789). 412

68G681 R. Scarpati, M. L. Graziano and R. A. Nicolaus; *Gazz. Chim. Ital.*, 1968, 681 (*Chem. Abstr.*, 1968, **69**, 96531). 107

68HCA622 P. Sieber and B. Iselin; *Helv. Chim. Acta*, 1968, **51**, 622. 489

68IC168 H. Schmidbaur and W. Tronich; *Inorg. Chem.*, 1968, **7**, 168. 708

68IC434 D. D. Desmarteau; *Inorg. Chem.*, 1968, **7**, 434. 272

68IC624 F. A. Hohorst and J. M. Shreeve; *Inorg. Chem.* 1968, **7**, 624. 250

68IC1647 J. B. Hynes, T. E. Austin and L. A. Bigelow; *Inorg. Chem.*, 1968, **7**, 1647. 284

68IC2265 F. Klanberg, W. B. Askew and L. J. Guggenberger; *Inorg. Chem.*, 1968, **7**, 2265. 340

68IS99 D. Evans, J. A. Osborn and G. Wilkinson; *Inorg. Synth.*, 1968, **11**, 99. 457

68IZV429 L. T. Eremenko, F. Ya. Natsibullin and I. P. Borovinskaya; *Izv. Akad. Nauk SSSR Ser. Khim.*, 1968, 429 (*Chem. Abstr.*, 1968, **69**, 18508). 261

68IZV431 L. T. Eremenko, F. Y. Natsibullin, I. P. Borovinskaya and N. D. Karpova; *Izv. Akad. Nauk SSSR Ser. Khim.*, 1968, 431 (*Chem. Abstr.*, 1968, **69**, 43324). 55

68IZV447 S. A. Shevelev, V. I. Erashko and A. A. Fainzil'berg; *Izv. Akad. Nauk SSSR Ser. Khim.*, 1968, 447 (*Chem. Abstr.*, 1968, **69**, 59140). 145

68IZV621 V. A. Tartakovskii, A. A. Fainzil'berg, V. I. Gulevskaya and S. S. Novikov; *Izv. Akad. Nauk SSSR Ser. Khim.*, 1968, 621 (*Chem. Abstr.*, 1968, **69**, 106598). 149
68IZV656 A. V. Fokin, L. T. Eremenko, V. S. Galakhov, A. V. Davydov and F. Ya. Natsibullin; *Izv. Akad. Nauk SSSR Ser. Khim.*, 1968, 656. 282
68IZV676 L. T. Eremenko and F. Ya. Natsibullin; *Izv. Akad. Nauk SSSR Ser. Khim.*, 1968, 676. 282
68IZV910 V. V. Gavrilenko, B. A. Palei and L. I. Zakharkin; *Izv. Akad. Nauk. SSSR Ser. Khim.*, 1968, 910 (*Chem. Abstr.*, 1968, **69**, 77310). 208
68IZV912 L. T. Eremenko and F. Y. Natsibullin; *Izv. Akad. Nauk SSSR Ser. Khim.*, 1968, 912 (*Chem. Abstr.*, 1968, **69**, 35315). 56
68IZV1638 V. A. Tartakovskii; N. S. Zefirov and V. N. Chekulaeva; *Izv. Akad. Nauk. SSSR, Ser. Khim.*, 1968, 1638 (*Chem. Abstr.*, 1968, **69**, 87144). 368
68IZV2307 L. T. Eremenko, G. V. Oreshko and L. I. Berezina; *Izv. Akad. Nauk SSSR Ser. Khim.*, 1968, 2307 (*Chem. Abstr.*, 1969, **70**, 28325. 55
68IZV2379 I. S. Ivanova and T. A. Klimova; *Izv. Akad. Nauk SSSR Ser. Khim.*, 1968, 2379 (*Chem. Abstr.*, 1969, **70**, 19685). 147
68IZV2839 A. L. Fridman, T. N. Ivshina, V. A. Tartakovskii and S. S. Novikov; *Izv. Akad. Nauk. SSSR, Ser. Khim.*, 1968, 2839 (*Chem. Abstr.*, 1969, **70**, 78115). 368
68JA2194 R. B. Castle and D. S. Matteson; *J. Am. Chem. Soc.*, 1968, **90**, 2194. 181, 386
68JA2944 D. Seyferth, R. Damrauer, J. Y. P. Mui and T. F. Jula; *J. Am. Chem. Soc.*, 1968, **90**, 2944. 60
68JA3254 R. S. Dewey, E. F. Schoenewaldt, H. Joshua, W. J. Paleveda, Jr., H. Schwam, H. Barkemeyer, B. H. Arison, D. F. Veber, R. G. Denkewalter and R. Hirschmann; *J. Am. Chem. Soc.*, 1968, **90**, 3254. 496
68JA3954 R. Czerepinski and G. H. Cady; *J. Am. Chem. Soc.*, 1968, **90**, 3954. 433
68JA6846 V. J. Bauer, W. Fulmor, G. O. Morton and S. R. Safir; *J. Am. Chem. Soc.*, 1968, **90**, 6846. 650
68JCED437 D. L. Ross, C. L. Coon, M. E. Hill and R. L. Simon; *J. Chem. Eng. Data*, 1968, **13**, 437. 144
68JCS(A)1105 E. W. Abel and I. H. Sabherwal; *J. Chem. Soc. (A)*, 1968, 1105. 582
68JCS(C)796 E. S. Alexander, R. N. Haszeldine, M. J. Newlands and A. E. Tipping; *J. Chem. Soc. (C)*, 1968, 796. 239
68JGU37 S. P. Makarov, M. A. Englin, V. A. Shpanskii and I. V. Ermakova; *J. Gen. Chem. USSR (Engl. Transl.)*, 1968, **38**, 37. 614
68JHC41 P. H. Ogden and R. A. Mitsch; *J. Heterocycl. Chem.*, 1968, **5**, 41. 604
68JMC811 S. M. Gadekar, S. Nibi and E. Cohen; *J. Med. Chem.*, 1968, **11**, 811. 648
68JMC1129 J. H. Short, C. W. Ours and W. J. Ranus; *J. Med. Chem.*, 1968, **11**, 1129. 642
68JMC1268 W. U. Malik, P. K. Srivastava and S. C. Mehra; *J. Med. Chem.*, 1968, **11**, 1268. 641
68JOC552 K. Matsumoto and H. Rapoport; *J. Org. Chem.*, 1968, **33**, 552. 646
68JOC851 R. F. Smith, A. C. Bates, A. J. Battisti, P. G. Byrnes, C. T. Mroz, T. J. Smearing and F. X. Albrecht; *J. Org. Chem.*, 1968, **33**, 851. 573
68JOC1016 B. C. Anderson; *J. Org. Chem.*, 1968, **33**, 1016. 2
68JOC1084 V. A. Pattison, J. G. Colson and R. L. K. Carr; *J. Org. Chem.*, 1968, **33**, 1084. 57
68JOC1246 J. D. Wilson and H. Weingarten; *J. Org. Chem.*, 1968, **33**, 1246. 141
68JOC1247 M. Graff and W. H. Gilligan; *J. Org. Chem.*, 1968, **33**, 1247. 144
68JOC1257 R. A. Wiesboeck and J. K. Ruff; *J. Org. Chem.*, 1968, **33**, 1257. 57
68JOC1847 R. A. Mitsch, *J. Org. Chem.*, 1968, **33**, 1847. 55, 261, 285
68JOC1854 D. J. Burton and F. E. Herkes; *J. Org. Chem.*, 1968, **33**, 1854. 694
68JOC2095 R. L. Talbott; *J. Org. Chem.*, 1968, **33**, 2095. 272
68JOC2504 D. R. Bender and D. L. Coffen; *J. Org. Chem.*, 1968, **33**, 2504. 85
68JOC2522 D. A. Rausch and J. J. Hoekstra; *J. Org. Chem.*, 1968, **33**, 2522. 649
68JOC3073 M. J. Kamlet and H. G. Adolph; *J. Org. Chem.*, 1968, **33**, 3073. 56, 261, 279, 287
68JOC3080 V. Grakauskas and K. Baum; *J. Org. Chem.*, 1968, **33**, 3080. 56, 282
68JOC3333 G. L. Roof and W. P. Tucker; *J. Org. Chem.*, 1968, **33**, 3333. 301, 302
68JOC3489 W. C. Firth, Jr.; *J. Org. Chem.*, 1968, **33**, 3489. 261, 285
68JOC3675 R. A. Mitsch and E. W. Neuvar; *J. Org. Chem.*, 1968, **33**, 3675. 52
68JOC3928 H. Ulrich, B. Tucker, F. A. Stuber and A. A. R. Sayigh; *J. Org. Chem.*, 1968, **33**, 3928. 143
68JOC3931 E. Dyer, T. E. Majewski and J. D. Travis; *J. Org. Chem.*, 1968, **33**, 3931. 143
68JOM(11)207 E. Klumpp, G. Bor and L. Marko; *J. Organomet. Chem.*, 1968, **11**, 207. 352
68JOM(12)P1 E. Moser and E. O. Fischer; *J. Organomet. Chem.*, 1968, **12**, P1. 719
68JOM(13)199 O. T. Quimby, J. D. Curry, D. A. Nicholson, J. B. Prentice and C. H. Roy; *J. Organomet. Chem.*, 1968, **13**, 199. 263
68JOM(13)363 A. Tzschach and R. Schwarzer; *J. Organomet. Chem.*, 1968, **13**, 363. 657
68JOM(13)411 K. Noack; *J. Organomet. Chem.*, 1968, **13**, 411. 455
68JOM(14)327 T. A. George and M. F. Lappert; *J. Organomet. Chem.*, 1968, **14**, 327. 664
68LA(716)207 R. Ohme and E. Schmitz; *Justus Liebigs Ann. Chem.*, 1968, **716**, 207. 70
68LA(720)146 H. Gross and B. Costisella; *Justus Liebigs Ann. Chem.*, 1968, **720**, 146. 43
68MI 606-01 W. Kitching; *Organomet. Chem. Rev.*, 1968, **3**, 35 (*Chem. Abstr.*, 1968, **69**, 18334). 207
68MI 607-01 P. J. Aymonino; *Photochem. Photobiol.*, 1968, **7**, 761 (*Chem. Abstr.*, 1969, **70**, 10834). 230
68MI 611-01 B. G. Boldyrev; *Biol. Aktiv. Soedin.*, 1968, 64 (*Chem. Abstr.*, 1969, **71**, 80/845h). 324, 326
B-68MI 620-01 H. Ulrich; "The Chemistry of Imidoyl Halides," Plenum Press, New York, 1968. 602, 606, 614, 619, 620, 621, 622
68PAC519 R. B. Woodward; *Pure Appl. Chem.*, 1968, **17**, 519. 105
68T2367 Y. Furuya, S. Goto, K. Itoho, I. Urasaki and A. Morita; *Tetrahedron*, 1968, **24**, 2367. 501
68T2767 H. Weingarten; *Tetrahedron*, 1968, **24**, 2767. 139
68T4217 R. Buyle and H. G. Viehe; *Tetrahedron*, 1968, **24**, 4217. 57
68T6485 J. L. Richards, D. S. Tarbell and E. H. Hoffmeister; *Tetrahedron*, 1968, **24**, 6485. 730

68TL497 M. Oishi, M. Ochiai, M. Nagai and Y. Ban; *Tetrahedron Lett.*, 1968, 497. 106
68TL659 C. A. Fyfe; *Tetrahedron Lett.*, 1968, 659. 109
68TL831 T. Saegusa, T. Tsuda, K. Isayama and K. Nishijima; *Tetrahedron Lett.*, 1968, 831. 464
68TL2713 G. Barnáth, K. Kovács and K. L. Láng; *Tetrahedron Lett.*, 1968, 2713. 104
68TL3171 M. Regitz, W. Anschütz, W. Bartz and A. Liedhegener; *Tetrahedron Lett.*, 1968, 3171. 666
68TL5037 R. Richter; *Tetrahedron Lett.*, 1968, 5037. 364
68USP3376314 D. R. Baker, G. E. Lukes and M. B. McClellan, (Stauffer Chemical Co.); US Pat. 3376314 (1968) (*Chem. Abstr.*, 1968, **69**, 59217). 254
68USP3387033 R. L. Talbott and R. J. Koshar (Minnesota Mining and Manufacturing Co.), US Pat. 3387033 (1968) (*Chem. Abstr.*, 1968, **69**, 43417). 260, 284
68USP3388147 M. J. Kamlet, K. G. Shipp and M. E. Hill (US Dept. of the Navy); *US Pat.* 3388147 (1968) (*Chem. Abstr.*, 1968, **69**, 86358). 297
68USP3401214 J. J. Kohler; *US Pat.* 3 401 214 (1968) (*Chem. Abstr.*, 1968, **69**, 106 869). 512
68USP3413382 H. Ulrich; *US. Pat.* 3 413 382 (1968) (*Chem. Abstr.*, 1969, **70**, 58 015). 512
68USP3419625 R. C. Doss (Phillips Petroleum Co.); *US Pat.* 3419625 (1968) (*Chem. Abstr.*, 1969, **70**, 77293). 282, 283
68USP3539569 R. J. Tull and P. I. Pollack; *US Pat.* 3 539 569 (1968) (*Chem. Abstr.*, 1971, **74**, 42 387). 506
68USP3575975 E. J. Cragoe, Jr. and K. L. Shepard; *US Pat.* 3 575 975 (1968) (*Chem. Abstr.*, 1971, **75**, 36 125). 506
68ZAAC(361)180 R. Steudel; *Z. Anorg. Allg. Chem.*, 1968, **361**, 180. 529, 530
68ZC230 B. Lorenz and E. Hoyer; *Z. Chem.*, 1968, **8**, 230. 594
68ZN(B)743 O. Glemser and S. P. von Halasz; *Z. Naturforsch., Teil B*, 1968, **23**, 743. 260
68ZOB709 S. P. Makarov, A. Y. Yakubovich, A. S. Filatov, M. A. Englin and T. Y. Nikiforova; *Zh. Obshch. Khim.*, 1968, **38**, 709. 242
68ZOB2509 L. M. Yagupol'skii and V. G. Voloshchuk; *Zh. Obshch. Khim.*, 1968, **38**, 2509. 237
68ZOB2591 L. M. Yagupol'skii, G. P. Syrova, V. G. Voloshchuk and V. F. Bystrov; *Zh. Obshch. Khim.*, 1968, **38**, 2591 (*Chem. Abstr.*, 1969, **70**, 46677). 236
68ZOB2813 L. M. Yagupol'skii, V. V. Lyalin, V. V. Orda and L. A. Alekseeva; *Zh. Obshch. Khim.*, 1968, **38**, 2813 (*Chem. Abstr.*, 1969, **70**, 77 464). 61
68ZOR231 V. A. Tartakovskii, O. A. Luk'yanov, N. I. Shlykova and S. S. Novikov; *Zh. Org. Khim.*, 1968, **4**, 231 (*Chem. Abstr.*, 1968, **68**, 95740). 149
68ZOR324 R. G. Dubenko, V. M. Neplyuev and P. S. Pel'kis; *Zh. Org. Khim.*, 1968, **4**, 324. 277
68ZOR720 L. I. Samarai, V. P. Belaya, O. V. Vishnevskii and G. I. Derkach; *Zh. Org. Khim.*, 1968, **4**, 720 (*Chem. Abstr.*, 1968, **69**, 2681). 438
68ZOR1661 A. P. Sineokov, G. É. Kholodenko and V. S. Étlis; *Zh. Org. Khim.*, 1968, **4**, 1661 (*Chem. Abstr.*, 1968, **69**, 106 801). 493
68ZPK1380 N. V. Kheifets, V. A. Lopyrev and I. N. Aizenshtadt; *Zh. Prikl. Khim.*, 1968, **41**, 1380 (*Chem. Abstr.*, 1968, **69**, 64234). 413

69ACS324 C. Larsen, U. Anthoni, C. Christophersen and P. H. Nielsen; *Acta Chem. Scand.*, 1969, **23**, 324. 570, 571
69AG(E)20 E. Kühle; *Angew. Chem., Int. Ed. Engl.*, 1969, **8**, 20. 452, 614, 616, 617, 643
69AG(E)450 D. Seebach and N. Peleties; *Angew. Chem., Int. Ed. Engl.*, 1969, **8**, 450. 93, 94
69AG(E)455 M. L. Ziegler and J. Weiss; *Angew. Chem., Int. Ed. Engl.*, 1969, **8**, 455. 622
69AG(E)606 H. Hagemann, D. Arlt and I. Ugi; *Angew. Chem., Int. Ed. Engl.*, 1969, **8**, 606. 607
69BCJ2323 T. Yonezawa, M. Matsumoto, Y. Matsumura and H. Kato; *Bull. Chem. Soc. Jpn.*, 1969, **42**, 2323. 115
69BCJ2975 I. Ojima, K. Akiba and N. Inamoto; *Bull. Chem. Soc. Jpn.*, 1969, **42**, 2975. 581
69BRP1169158 H. Hagemann; *Br. Pat.* 1169158 (1969) (*Chem. Abstr.*, 1970, **72**, 42858). 445
69BSF325 C. Feugeas and D. Olschwang; *Bull. Soc. Chim. Fr.*, 1969, 325. 95, 98
69BSF332 C. Feugeas and D. Olschwang; *Bull. Soc. Chim. Fr.*, 1969, 332. 111
69CB931 R. Richter and W.-P. Trautwein; *Chem. Ber.*, 1969, **102**, 931. 143
69CB1495 E. O. Fischer and R. Aumann; *Chem. Ber.*, 1969, **102**, 1495. 719
69CB2150 M. Schmeisser, P. Sartori and B. Lippsmeier; *Chem. Ber.*, 1969, **102**, 2150. 232
69CB2718 A. Haas and H. Reinke; *Chem. Ber.*, 1969, **102**, 2718. 432, 435
69CB2972 L. I. Samarai, O. W. Wischnewskij and G. I. Derkatsh; *Chem. Ber.*, 1969, **102**, 2972. 445
69CB3127 A. Roedig and W. Wenzel; *Chem. Ber.*, 1969, **102**, 3127. 230
69CC423 P. M. Treichel and J. P. Stenson; *J. Chem. Soc., Chem. Commun.*, 1969, 423. 721
69CC462 G. J. D. Peddle and R. W. Walsingham; *J. Chem. Soc., Chem. Commun.*, 1969, 462. 515
69CC878 J. Silhanek and M. Zbirovsky; *J. Chem. Soc., Chem. Commun.*, 1969, 878. 236, 531
69CC1322 E. M. Badley, J. Chatt, R. L. Richards and G. A. Sim; *J. Chem. Soc., Chem. Commun.*, 1969, 1322. 721
69CJC4347 A. G. Brook and P. J. Dillon; *Can. J. Chem.*, 1969, **47**, 4347. 124
69CPB1924 A. Takamizawa and K. Hirai; *Chem. Pharm. Bull.*, 1969, **17**, 1924. 113
69CPB1931 A. Takamizawa and K. Hirai; *Chem. Pharm. Bull.*, 1969, **17**, 1931. 85
69HCA812 K. O. Alt and C. D. Weis; *Helv. Chim. Acta*, 1969, **52**, 812. 612, 616
69IZV119 M. P. Alimov and P. I. Alimov; *Izv. Akad. Nauk SSSR Ser. Khim.*, 1969, 119 (*Chem. Abstr.*, 1969, **70**, 114 523). 656
69IZV196 N. N. Vorozhtsov, Jr., N. V. Ermolenko, S. A. Mazalov, O. M. Osina, V. E. Platanov, V. S. Tyurin and G. G. Yakobson; *Izv. Akad. Nauk SSSR Ser. Khim.*, 1969, 196. 12
69IZV1188 L. V. Okhlobystina and V. M. Khutoretskii; *Izv. Akad. Nauk SSSR Ser. Khim.*, 1969, 1188 (*Chem. Abstr.*, 1969, **71**, 38 237). 56

69IZV1331 L. T. Eremenko and F. Y. Natsibullin; *Izv. Akad. Nauk SSSR Ser. Khim.*, 1969, 1331 (*Chem. Abstr.*, 1969, **71**, 90743). 55

69IZV1845 L. T. Eremenko and G. N. Nesterenko; *Izv. Akad. Nauk SSSR Ser. Khim.*, 1969, 1845 (*Chem. Abstr.*, 1970, **72**, 2976). 149

69IZV2304 A. L. Fridman, T. N. Ivshina, V. A. Tartakovskii and S. S. Novikov; *Izv. Akad. Nauk. SSSR, Ser. Khim.*, 1969, 2304 (*Chem. Abstr.*, 1970, **72**, 43824). 368

69IZV2309 L. S. Simonenko, I. S. Korsakova and S. S. Novikov; *Izv. Akad. Nauk SSSR Ser. Khim.*, 1969, 2309 (*Chem. Abstr.*, 1970, **72**, 43044). 147

69IZV2566 V. I. Grigos and L. T. Eremenko; *Izv. Akad. Nauk SSSR Ser. Khim.*, 1969, 2566 (*Chem. Abstr.*, 1970, **72**, 66293). 145

69IZV2617 A. L. Fridman, V. P. Ivshin, T. N. Ivshina and S. S. Novikov; *Izv. Akad. Nauk SSSR Ser. Khim.*, 1969, 2617 (*Chem. Abstr.*, 1970, **72**, 54637). 56, 282

69JA1954 D. Seyferth, F. M. Armbrecht, Jr. and B. Schneider; *J. Am. Chem. Soc.*, 1969, **91**, 1954. 270

69JA2902 C. J. Schack and W. Maya; *J. Am. Chem. Soc.*, 1969, **91**, 2902. 231

69JA4432 M. Wechsberg and G. H. Cady; *J. Am. Chem. Soc.*, 1969, **91**, 4432. 231

69JA6156 R. West and P. C. Jones; *J. Am. Chem. Soc.*, 1969, **91**, 6156. 206

69JA6541 D. S. Matteson and G. L. Larson; *J. Am. Chem. Soc.*, 1969, **91**, 6541. 183, 200, 396, 400

69JCS(A)431 H. J. Emeléus, J. M. Shreeve and P. M. Spaziante; *J. Chem. Soc. (A)*, 1969, 431. 240

69JCS(A)1882 J. E. Dobson, P. M. Tucker, F. G. A. Stone and R. Schaeffer; *J. Chem. Soc. (A)*, 1969, 1882. 181, 269

69JCS(C)2002 K. Itoh, M. Fukui and Y. Ishii; *J. Chem. Soc. (C)*, 1969, 2002. 684

69JCS(C)2251 R. L. Jones and C. W. Rees; *J. Chem. Soc. (C)*, 1969, 2251. 141

69JGU183 S. P. Makarov, M. A. Englin, V. A. Shpanskii and I. V. Ermakova; *J. Gen. Chem. USSR (Engl. Transl.)*, 1969, **39**, 183. 614, 627

69JGU1235 Y. L. Kruglyak, S. I. Malekin, I. V. Martynov and O. G. Strukov; *J. Gen. Chem. USSR (Engl. Transl.)*, 1969, **39**, 1235. 613

69JGU1301 V. A. Ginsburg and K. N. Smirnov; *J. Gen. Chem. USSR (Engl. Transl.)*, 1969, **39**, 1301. 613

69JGU1692 Y. L. Kruglyak, S. I. Malekin, Z. I. Khromova and I. V. Martynov; *J. Gen. Chem. USSR (Engl. Transl.)*, 1969, **39**, 1692. 613

69JGU1694 I. V. Martynov, N. F. Privezentseva and Y. L. Kruglyak; *J. Gen. Chem. USSR (Engl. Transl.)*, 1969, **39**, 1694. 603

69JHC729 A. de Souza Gomes and M. M. Joullié; *J. Heterocycl. Chem.*, 1969, **6**, 729. 111, 114

69JMC558 P. N. Bhargava and H. Singh; *J. Med. Chem.*, 1969, **12**, 558. 643

69JMC712 H. W. Geluk, J. Schut and J. L. M. A. Schlatmann; *J. Med. Chem.*, 1969, **12**, 712. 643

69JOC45 H. G. Adolph and M. J. Kamlet; *J. Org. Chem.*, 1969, **34**, 45. 57

69JOC1642 B. S. Shasha, W. M. Doane, C. R. Russell and C. E. Rist; *J. Org. Chem.*, 1969, **34**, 1642. 276

69JOC2254 F. D. Greene, J. C. Stowell and W. R. Bergmark; *J. Org. Chem.*, 1969, **34**, 2254. 450, 509

69JOC2263 F. D. Greene, W. R. Bergmark and J. G. Pacifici; *J. Org. Chem.*, 1969, **34**, 2263. 509

69JOC2840 V. Grakauskas and K. Baum; *J. Org. Chem.*, 1969, **34**, 2840. 649

69JOC3011 F. N. Jones and S. Andreades; *J. Org. Chem.*, 1969, **34**, 3011. 540

69JOC3549 D. L. Klayman and R. J. Shine; *J. Org. Chem.*, 1969, **34**, 3549. 594, 596

69JOC3604 M. S. Newman and F. W. Hetzel; *J. Org. Chem.*, 1969, **34**, 3604. 555, 575

69JOM(16)275 E. Moser and E. O. Fischer; *J. Organomet. Chem.*, 1969, **16**, 275. 721

69JOM(16)33 D. S. Matteson, R. A. Bowie and G. Srivastava; *J. Organomet. Chem.*, 1969, **16**, 33. 61

69JOM(19)161 W. Keim; *J. Organomet. Chem.*, 1969, **19**, 161. 456

69JOM(20)19 R. B. Castle and D. S. Matteson; *J. Organomet. Chem.*, 1969, **20**, 19. 181, 386

69JPR15 J. Beger, K. Günther and J. Vogel; *J. Prakt. Chem.*, 1969, **311**, 15. 606, 607, 618

69JPR577 H. Gross, B. Costisella and L. Haase; *J. Prakt. Chem.*, 1969, **311**, 577. 157

69JPR925 H. Gross and B. Costisella; *J. Prakt. Chem.*, 1969, **311**, 925. 156

69LA(722)52 W. Walter and G. Randau; *Justus Liebigs Ann. Chem.*, 1969, **722**, 52. 580

69LA(722)80 W. Walter and G. Randau; *Justus Liebigs Ann. Chem.*, 1969, **722**, 80. 580

69LA(725)15 W. Funke; *Justus Liebigs Ann. Chem.*, 1969, **725**, 15. 107, 109

69LA(727)228 H. Kessler, H.-O. Kalinowski and C. v. Chamier; *Justus Liebigs Ann. Chem.*, 1969, **727**, 228. 732

69LA(730)133 K. Grohe, E. Degener, H. Holtschmidt and H. Heitzer; *Justus Liebigs Ann. Chem.*, 1969, **730**, 133. 607

69LA(730)140 K. Grohe, E. Klauke, H. Holtschmidt, U. Mitarbeit and H. Heitzer; *Justus Liebigs Ann. Chem.*, 1969, **730**, 140. 51, 238

69M1823 A. Meller and H. Batka; *Monatsh. Chem.*, 1969, **100**, 1823. 622

69M1860 W. Gottardi and D. Henn; *Monatsh. Chem.*, 1969, **100**, 1860. 451

B-69MI 603-01 E. H. Cordes; in "The Chemistry of Carboxylic Acids and Esters," S. Patai (ed.), John Wiley, London, 1969, p. 623. 68

69MI 605-01 H. Bredereck; Dissertation, University of Stuttgart, 1969. 138

B-69MI 607-01 W. A. Sheppard and C. M. Sharts (eds.); "Organic Fluorine Chemistry," W. A. Benjamin Inc., New York, 1969. 212

B-69MI 607-02 H. J. Emeléus; "The Chemistry of Fluorine and Its Compounds," Academic Press, New York, 1969, pp. 45–71. 212, 244

69OS(49)86 U. Schöllkopf, J. Paust and M. R. Patsch; *Org. Synth.*, 1969, **49**, 86. 63

69RTC289 J. W. Scheeren and R. J. F. Nivard; *Recl. Trav. Chim. Pays-Bas*, 1969, **88**, 289. 55, 107

69RTC897 J. W. Scheeren and W. Stevens; *Recl. Trav. Chim. Pays-Bas*, 1969, **85**, 897. 79, 95

69RZC1443 Z. Arnold and B. Fiszer; *Rocz. Chem.*, 1969, **43**, 1443 (*Chem. Abstr.*, 1969, **71**, 124578). 512

69S17 D. Seebach; *Synthesis*, 1969, 17. 98

69T5995 G. Arnold and G. Paal; *Tetrahedron*, 1969, **25**, 5995. 111

69TL173 R. M. Carlson and P. M. Helquist; *Tetrahedron Lett.*, 1969, 173. 134

69TL723 D. E. Young, L. R. Anderson, D. E. Gould and W. B. Fox; *Tetrahedron Lett.*, 1969, 723. 428

69TL1047 R. C. Blume; *Tetrahedron Lett.*, 1969, 1047. 79

69TL1421 K. Itoh, A. Nozawa and Y. Ishii; *Tetrahedron Lett.*, 1969, 1421. 645

69TL2161 G. Ferré and A.-L. Palomo; *Tetrahedron Lett.*, 1969, **26**, 2161. 57

69TL2223 W. H. Graham; *Tetrahedron Lett.*, 1969, 2223. 77

69TL4095 Y. Kobayashi and I. Kumadaki; *Tetrahedron Lett.*, 1969, 4095. 246

69TL4461 B. Zwangenburg, L. Thijs and J. Strating; *Tetrahedron Lett.*, 1969, 4461. 531

69TL5003 W. de Priester and A. P. M. van der Veek; *Tetrahedron Lett.*, 1969, 5003. 43

69USP3450762 D. L. Esmay and W. G. Kottong (Minnesota Mining & Mfg. Co.); *US Pat.* 3450762 (1969) (*Chem. Abstr.*, 1969, **71**, 49355). 284

69USP3461050 W. V. Childs (Phillips Petroleum Co.); *US Pat.* 3461050 (1969) (*Chem. Abstr.*, 1969, **71**, 87225). 409

69USP3471406 E. G. Budnick; *US Pat.* 3471406 (1969) (*Chem. Abstr.*, 1969, **72**, 102091). 155

69ZC201 J. Gloede, L. Hasse and H. Gros; *Z. Chem.*, 1969, **9**, 201. 104

69ZOB199 S. P. Makarov, M. A. Englin, V. A. Shpanskii and I. V. Ermakova; *Zh. Obshch. Khim.*, 1969, **39**, 199. 240

69ZOB220 V. F. Mironov, V. D. Sheludyakov and V. P. Kosyukov; *Zh. Obshch. Khim.*, 1969, **39**, 220. 445

69ZOB941 M. Z. Girshovich and A. V. El'tsov; *Zh. Obshch. Khim.*, 1969, **39**, 941. 599

69ZOB1755 L. M. Yagupol'skii and N. V. Kondratenko; *Zh. Obshch. Khim.*, 1969, **39**, 1755 (*Chem. Abstr.*, 1970, **72**, 3185). 82, 302

69ZOB2598 V. F. Mironov, V. D. Sheludyakov and V. P. Kozyukov; *Zh. Obshch. Khim.*, 1969, **39**, 2598 (*Chem. Abstr.*, 1970, **72**, 66300). 443

69ZOR220 L. M. Andreeva, K. V. Altukhov and V. V. Perekalin; *Zh. Org. Khim.*, 1969, **5**, 220 (*Chem. Abstr.*, 1969, **70**, 106419). 149

69ZOR1313 L. M. Andreeva, K. V. Altukhov and V. V. Perekalin; *Zh. Org. Khim.*, 1969, **5**, 1313 (*Chem. Abstr.*, 1969, **71**, 101213). 149

69ZOR1317 A. L. Fridman, V. P. Ivshin and S. S. Novikov; *Zh. Org. Khim.*, 1969, **5**, 1317. 283

69ZOR2246 K. V. Altukhov, E. V. Ratsino and V. V. Perekalin; *Zh. Org. Khim.*, 1969, **5**, 2246 (*Chem. Abstr.*, 1970, **72**, 66521). 147

70ACS351 M. L. Shankaranarayana; *Acta Chem. Scand.*, 1970, **24**, 351. 594

70ACS959 U. Anthoni, B. M. Dahl, C. Larsen and P. H. Nielsen; *Acta Chem. Scand.*, 1970, **24**, 959. 594, 595

70ACS1094 O. Dahl and O. Larsen; *Acta Chem. Scand.*, 1970, **24**, 1094. 584

70ACS2055 K. A. Jensen and U. Anthoni; *Acta Chem. Scand.*, 1970, **24**, 2055. 592

70ACS2061 K. A. Jensen, P. A. A. Frederiksen and L. Henriksen; *Acta Chem. Scand.*, 1970, **24**, 2061. 592, 593

70AG(E)54 G. Zumach and E. Kühle; *Angew. Chem., Int. Ed. Engl.*, 1970, **9**, 54. 435

70AG(E)77 H. Schmidbaur and W. Malisch; *Angew. Chem., Int. Ed. Engl.* 1970, **9**, 77. 700, 703

70AG(E)309 E. O. Fischer and M. J. Kollmeier; *Angew. Chem., Int. Ed. Engl.*, 1970, **9**, 309. 720

70AG(E)372 H. Helfert and E. Fahr; *Angew. Chem., Int. Ed. Engl.*, 1970, **9**, 372. 448

70AG(E)456 R. Finding and U. Schmidt; *Angew. Chem., Int. Ed. Engl.*, 1970, **9**, 456. 77

70AG(E)466 A. Hass, H. Reinke and J. Sommerhoff; *Angew. Chem., Int. Ed. Engl.*, 1970, **9**, 466. 432

70AP(303)625 K. Hartke and B. Seib; *Arch Pharm.* (*Weinheim, Ger.*), 1970, **303**, 625. 630

70BAP569 C. Belzecki, B. Hintze and S. Kwiatkowska; *Bull. Acad. Pol. Sci., Ser. Sci. Chim.*, 1970, **18**, 569 (*Chem. Abstr.*, 1971, **74**, 99596). 651

70BCJ187 Y. Ohtsuka; *Bull. Chem. Soc. Jpn.*, 1970, **43**, 187. 625

70BCJ3528 T. Nakai and M. Okawara; *Bull. Chem. Soc. Jpn.*, 1970, **43**, 3528. 314

70BSF2013 F. M. Stoyanovich, I. A. Ivanova and B. P. Federov; *Bull. Soc. Chim. Fr.*, 1970, 2013. 111

70CB256 H. Bredereck, G. Simchen and H. Porkert; *Chem. Ber.*, 1970, **103**, 256. 634

70CB639 H. Stetter and E. Reske; *Chem. Ber.*, 1970, **103**, 639. 78, 81

70CB643 H. Stetter and E. Reske; *Chem. Ber.*, 1970, **103**, 643. 79

70CB766 G. Zinner and R. Vollrath; *Chem. Ber.*, 1970, **103**, 766. 360, 361

70CB3448 H. Schmidbaur and W. Malisch; *Chem. Ber.*, 1970, **103**, 3448. 373

70CC395 D. E. Yound, L. R. Anderson and W. B. Fox; *J. Chem. Soc., Chem. Commun.*, 1970, 395. 240

70CC469 G. M. J. Slusarczuk and M. M. Joullié; *J. Chem. Soc., Chem. Commun.*, 1970, 469. 142

70CC559 A. R. Bassindale, A. J. Bowles, M. A. Cook, C. Eaborn, A. Hudson, R. A. Jackson and A. E. Jukes; *J. Chem. Soc., Chem. Commun.*, 1970, 559. 734

70CC806 C. Belzecki, B. Hintze and S. Kwiatkowska; *J. Chem. Soc., Chem. Commun.*, 1970, 806. 650, 653

70CJC561 A. G. Brook, J. M. Duff and D. G. Anderson; *Can. J. Chem.*, 1970, **48**, 561. 192

70CPB784 M. Itoh; *Chem. Pharm. Bull.*, 1970, **18**, 784. 57

70CPB2334 Y. Kobayashi, I. Kumadaki, S. Sato, N. Hara and E. Chikami; *Chem. Pharm. Bull.*, 1970, **18**, 2334. 246

70CR(C)468 J. Ficini, J. Besseyre, J. D'Angelo and C. Barabara; *C. R. Hebd. Seances Acad. Sci., Ser. C*, 1970, **271**, 468. 106

70GEP1908680 E. Kuehle, A. Haas, W. Klug and F. Grewe (Farbenfabriken Bayer AG); *Ger. Pat.* 1908680 (1970) (*Chem. Abstr.*, 1970, **73**, 109522). 273, 274

70GEP1915668 H. Michaud, H. Prietzel and C. Lienhard (Sueddeutsche Kalkstickstoff-Werke AG); *Ger. Pat.* 1915668 (1970) (*Chem. Abstr.*, 1970, **73**, 120129). 623

70GEP1926659 K. Findeisen, H. Holtschmidt, R. Braden and H. Schwarz (Bayer AG); *Ger. Pat.* 1926659 (1970) (*Chem. Abstr.*, 1971, **74**, 41887). 239

70GEP2008115 H. Disselnkötter (Farbenfabriken Bayer AG); *Ger. Pat.* 2008115 (1970) (*Chem. Abstr.*, 1971, **75**, 140530). 448

70HOU(13/4)1 H. Lehmkuhl and K. Ziegler; *Methoden Org. Chem.* (*Houben-Weyl*), 1970, **13/4**, 1. 204, 208

70IC1682 W. R. Cullen, J. R. Sams and M. C. Waldman; *Inorg. Chem.*, 1970, **9**, 1682. 65

70IC2556 H. C. Clark and J. D. Ruddick; *Inorg. Chem.*, 1970, **9**, 2556. 247

70IZV387 V. M. Khutoretskii, L. V. Okhlobystina and A. A. Fainzil'berg; *Izv. Akad. Nauk SSSR Ser. Khim.*, 1970, 387 (*Chem. Abstr.*, 1970, **73**, 3407). 56

70IZV948 A. A. Onischenko and V. A. Tartakovskii; *Izv. Akad. Nauk SSSR Ser. Khim.*, 1970, 948 (*Chem. Abstr.*, 1970, **73**, 34718). 145

70IZV2108 L. I. Khmel'nitskii, T. I. Godovikova, A. V. Igritskaya and S. S. Novikov; *Izv. Akad. Nauk. SSSR, Ser. Khim.*, 1970, 2108 (*Chem. Abstr.*, 1971, **74**, 53956). 369

70IZV2553 G. K. Khisamutdinov, V. I. Slovetskii, M. S. L'vova, O. G. Usyshkin, M. A. Vesprozvannyi and A. A. Fainzilberg; *Izv. Akad. Nauk SSSR Ser. Khim.*, 1970, 2553 (*Chem. Abstr.*, 1971, **74**, 87266). 261, 283

70JA231 D. S. Matteson, R. B. Castle and G. L. Larson; *J. Am. Chem. Soc.*, 1970, **92**, 231. 405

70JA584 E. L. Eliel and F. W. Nader; *J. Am. Chem. Soc.*, 1970, **92**, 584. 79

70JA2096 V. Grakauskas and K. Baum; *J. Am. Chem. Soc.*, 1970, **92**, 2096. 440

70JA3342 D. Coucouvanis, S. J. Lippard and J. A. Zubieta; *J. Am. Chem. Soc.*, 1970, **92**, 3342. 552

70JA3665 L. M. Zaborowski and J. M. Shreeve; *J. Am. Chem. Soc.*, 1970, **92**, 3665. 613

70JA4158 C.-C. S. Cheung, R. A. Beaudet and G. A. Segal; *J. Am. Chem. Soc.*, 1970, **92**, 4158. 183

70JA5118 S. Trofimenko; *J. Am. Chem. Soc.*, 1970, **92**, 5118. 362

70JA5748 L. A. Carpino and G. Y. Han; *J. Am. Chem. Soc.*, 1970, **92**, 5748. 487

70JA7212 C. F. Lane and H. C. Brown; *J. Am. Chem. Soc.*, 1970, **92**, 7212. 61

70JAP7026611 T. Nagase, H. Baba and T. Abe (Japan Bureau of Industrial Technology); *Jpn. Pat.* 7026611 (1970) (*Chem. Abstr.*, 1970, **73**, 120118). 409

70JCS(A)31 M. J. McGlinchey, J. D. Odom, T. Reynoldson and F. G. A. Stone; *J. Chem. Soc.* (*A*), 1970, 31. 265, 269

70JCS(A)2954 M. F. Lappert, J. Lorberth and J. S. Poland; *J. Chem. Soc.* (*A*), 1970, 2954. 672

70JCS(C)875 J. Geevers, J. Th. Hackmann and W. P. Trompen; *J. Chem. Soc.* (*C*), 1970, 875. 621

70JCS(C)1429 A. Lawson and R. B. Tinkler; *J. Chem. Soc.* (*C*), 1970, 1429. 652

70JGU540 I. V. Martynov, L. N. Shitov and E. A. Mordvintseva; *J. Gen. Chem. USSR* (*Engl. Transl.*), 1970, **40**, 540. 613

70JHC201 D. L. Coffen; *J. Heterocycl. Chem.*, 1970, **7**, 201. 304, 306

70JHC689 I. Lalezari and H. Golgolab; *J. Heterocycl. Chem.*, 1970, **7**, 689. 653

70JHC807 T. Kametani, K. Sota and M. Shio; *J. Heterocycl. Chem.*, 1970, **7**, 807. 511

70JMC334 C. Sunkel and H. Gómez; *J. Med. Chem.*, 1970, **13**, 334. 573

70JMC377 G. Clifton, S. R. Bryant and C. G. Skinner; *J. Med. Chem.*, 1970, **13**, 377. 570, 571

70JOC843 J. A. Settepani and G. R. Pettit; *J. Org. Chem.*, 1970, **35**, 843. 501, 508

70JOC846 K. Baum; *J. Org. Chem.*, 1970, **35**, 846. 55

70JOC2054 F. J. Pavlik and P. E. Toren; *J. Org. Chem.*, 1970, **35**, 2054. 42

70JOC2099 H. D. Becker; *J. Org. Chem.*, 1970, **35**, 2099. 507

70JOC2982 T. Susuki and J. Tsuji; *J. Org. Chem.*, 1970, **35**, 2982. 7

70JOC3188 H. G. Adolph; *J. Org. Chem.*, 1970, **35**, 3188. 57

70JOC3470 M. S. Raasch; *J. Org. Chem.*, 1970, **35**, 3470. 304, 305, 311, 312

70JOC3637 H. Hart and J. F. Janssen; *J. Org. Chem.*, 1970, **35**, 3637. 29

70JOC3730 L. R. Anderson, D. E. Young, D. E. Gould, R. Juurik-Hogan, D. Nuechterlein and W. B. Fox; *J. Org. Chem.*, 1970, **35**, 3730. 42, 43

70JOM(21)P6 D. S. Matteson and P. B. Tripathy; *J. Organomet. Chem.*, 1970, **21**, P6. 182

70JOM(23)361 D. Seyferth, E. M. Hanson and F. M. Armbrecht, Jr.; *J. Organomet. Chem.*, 1970, **23**, 361. 267, 270, 293, 399

70JOM(23)471 L. C. Willemsens and G. J. M. Van der Kerk; *J. Organomet. Chem.*, 1970, **23**, 471. 209

70JOM(24)C61 C. D. M. Mann, A. J. Cleland, S. A. Fieldhouse, B. H. Freeland and R. J. O'Brien; *J. Organomet. Chem.*, 1970, **24**, C61. 340

70JOM(24)205 R. J. Angelici and L. M. Charley; *J. Organomet. Chem.*, 1970, **24**, 205. 721

70JOM(24)263 D. S. Matteson and J. R. Thomas; *J. Organomet. Chem.*, 1970, **24**, 263. 181, 182

70JOM(24)529 M. A. Cook, C. Eaborn, A. E. Jukes and D. R. M. Walton; *J. Organomet. Chem.*, 1970, **24**, 529. 174, 289, 380, 387, 390

70JOM(24)647 D. Seyferth, R. L. Lambert, Jr. and E. M. Hanson; *J. Organomet. Chem.*, 1970, **24**, 647. 173, 195, 267, 270, 293

70JOM(24)89 S.-L. Liu and M.-M. Ma; *J. Organomet. Chem.*, 1970, **24**, 89. 378

70JOM(25)255 F. Bonati, G. Minghetti, T. Boschi and B. Crociani; *J. Organomet. Chem.*, 1970, **25**, 255. 721

70JOM(25)385 R. West and G. A. Gornowicz; *J. Organomet. Chem.*, 1970, **25**, 385. 162

70JPR366 K. Issleib and P. V. Malotki; *J. Prakt. Chem.*, 1970, **312**, 366. 511

70JPR456 K. Issleib and H. P. Abicht; *J. Prakt. Chem.*, 1970, **312**, 456. 151, 701

70JPR475 H. Gross and H. Seibt; *J. Prakt. Chem.*, 1970, **312**, 475. 281

70KGS590 S. S. Novikov, L. I. Khmel'nitskii, T. S. Novikova and O. V. Lebedev; *Khim. Geterotsikl. Soedin.*, 1970, 590 (*Chem. Abstr.*, 1970, **73**, 77006). 150

70LA(731)171 R. D. Haugwitz; *Justus Liebigs Ann. Chem.*, 1970, **731**, 171. 575

70LA(733)195 W. Walter and R. F. Becker; *Justus Liebigs Ann. Chem.*, 1970, **733**, 195. 537, 539, 540

70LA(735)158 K. Sasse; *Justus Liebigs Ann. Chem.*, 1970, **735**, 158. 537

70LA(742)128 K-H. Magosch and R. Feinauer; *Justus Liebigs Ann. Chem.*, 1970, **742**, 128. 110

B-70MI 603-01 R. H. DeWolfe; "Carboxylic Ortho Acid Derivatives," Academic Press, New York, 1970. 68, 69, 80, 81, 91

70MI 603-02 C. Feugeas and D. Olschwang; *C.R. Congr. Natl. Soc. Savantes, Sect. Sci.*, 1970, **95**, 129. 95, 98

B-70MI 604-01 R. H. DeWolfe; "Carboxylic Ortho Acid Derivatives," Academic Press, New York, 1970, p. 420. 104, 105

B-70MI 605-01 R. N. Grimes; "Carboranes," Academic Press, New York, 1970, p. 107. 169

B-70MI 605-02	L. A. Kaplan; in "The Chemistry of Nitro and Nitroso Groups, Part 2," ed. H. Feuer, in "The Chemistry of Functional Groups," series ed. S. Patai, Wiley, New York, 1970, p. 289.	144, 145
B-70MI 605-03	R. H. DeWolfe; in "Organic Chemistry," Academic Press, New York, 1970, p. 420.	138
B-70MI 610-01	R. H. De Wolfe; "Carboxylic Ortho Acid Derivatives," Academic Press, New York, 1970, p. 12.	296, 297, 307
B-70MI 610-02	R. H. DeWolfe; "Carboxylic Ortho Acid Derivatives," Academic Press, New York, 1970, p. 348.	296
70MI 611-01	S. Sakai, Y. Asai, Y. Kiyohara, K. Itoh and Y. Ishii; *Organometal. Chem. Syn.*, 1970/1971, **1**, 45 (*Chem. Abstr.*, 1971, **74**, 76483).	322, 337
B-70MI 615-01	R. Wegler; "Chemie der Pflanzenschutz- und Schädlingsbekämpfungsmittel," Springer, Berlin, Heidelberg, New York, 1970, vol. 2, p. 221.	492
70MI 617-01	D. Coucouvanis; *Prog. Inorg. Chem.*, 1970, **11**, 233.	552
70MI 618-01	V. J. Ram; *Indian J. Appl. Chem.*, 1970, **33**, 394 (*Chem. Abstr.*, 1973, **79**, 126038).	573
70MI 618-02	S. Sakai, Y. Asai, Y. Kiyohara, K. Itoh and Y. Ishii; *Organomet. Chem. Synth.*, 1970, **1**, 45 (*Chem. Abstr.*, 1971, **74**, 76483).	576
B-70MI 622-01	G. O. Doak and L. D. Freedman; "Organometallic Compounds of Arsenic, Antimony and Bismuth," Wiley-Interscience, New York, 1970.	707
70QRS45	K. A. Jensen; *Quart. Rep. Sulfur Chem.*, 1970, **5**, 45.	596
70S20	D. Arlt; *Synthesis*, 1970, 20.	606, 607
70S542	G. Zumach and E. Kühle; *Synthesis*, 1970, 542.	449
70TL481	R. Weiss and R. Gompper; *Tetrahedron Lett.*, 1970, 481.	314, 317
70TL1137	T. H. Chan, E. Chang and E. Vinokur; *Tetrahedron Lett.*, 1970, 1137.	192
70TL4693	G. Köbrich and R. von Nagel; *Tetrahedron Lett.*, 1970, 4693.	267, 293, 378
70TL5111	R. A. LeMahieu and R. W. Kierstead; *Tetrahedron Lett.*, 1970, 5111.	72
70USP3497313	O. T. Quimby; *US Pat.*, 3497313 (1970) (*Chem. Abstr.*, 1970, **72**, 132960).	518
70USP3503993	R. C. Blume (E. I. du Pont de Nemours and Co.); *US Pat.* 3503993 (1970) (*Chem. Abstr.*, 1970, **73**, 14860).	298
70USP3551418	H. I. Weingarten and W. A. White (Monsanto Co.); *US Pat.* 3551418 (1970) (*Chem. Abstr.*, 1971, **74**, 54407).	361
70ZAAC(372)21	G. Fritz and P. Schober; *Z. Anorg. Allg. Chem.*, 1970, **372**, 21.	268
70ZC30	G. Zinner and R. Vollrath; *Z. Chem.*, 1970, **10**, 30.	361
70ZOR144	A. I. Burmakov, L. A. Alekseeva and L. M. Yagupol'skii; *Zh. Org. Khim.*, 1970, **6**, 144 (*Chem. Abstr.*, 1970, **72**, 89975).	38, 39
70ZOR189	F. M. Mukhametshin and A. L. Fridman; *Zh. Org. Khim.*, 1970, **6**, 189 (*Chem. Abstr.*, 1970, **72**, 89679).	145
70ZOR2498	A. I. Burmakov, L. A. Alekseeva and L. M. Yagupol'skii; *Zh. Org. Khim.*, 1970, **6**, 2498 (*Chem. Abstr.*, 1971, **74**, 64133).	39
71AG(E)339	P. Jutzi and F.-W. Schröder; *Angew. Chem., Int. Ed. Engl.*, 1971, **10**, 339.	515
71AG(E)510	G. Fritz and M. Berndt; *Angew. Chem., Int. Ed. Engl.*, 1971, **10**, 510.	267
71AG(E)573	H. G. Viehe and Z. Janousek; *Angew. Chem., Int. Ed. Engl.*, 1971, **10**, 573.	620
71AG(E)810	K. H. Magosch and R. Feinauer; *Angew. Chem., Int. Ed. Engl.*, 1971, **10**, 810.	107
71BCJ1393	A. Kaji, Y. Araki and K. Miyazaki; *Bull. Chem. Soc. Jpn.*, 1971, **44**, 1393.	495
71BCJ2182	S. Yanagida, M. Yokoe, M. Ohoka and S. Komori; *Bull. Chem. Soc. Jpn.*, 1971, **44**, 2182.	445
71BCJ2515	T. Inui; *Bull. Chem. Soc. Jpn.*, 1971, **44**, 2515.	434
71BSF556	R. Guglielmetti, E. Davin-Pretelli and J. Metzger; *Bull. Soc. Chim. Fr.*, 1971, 556.	115
71BSF3354	D. Olschwang; *Bull. Soc. Chim. Fr.*, 1971, 3354.	111
71CAR(16)145	B. S. Shasha and W. M. Doane; *Carbohydr. Res.*, 1971, **16**, 145.	276
71CAR(19)139	G. Wulff and W. Krueger; *Carbohydr. Res.*, 1971, **19**, 139.	71
71CB150	H. Schmidbaur and W. Malisch; *Chem. Ber.*, 1971, **104**, 150.	700, 703
71CB873	R. W. Hoffman, U. Bressel, J. Gehlhaus and H. HaÅser; *Chem. Ber.*, 1971, **104**, 873.	106
71CB969	N. Schindler and W. Plöger; *Chem. Ber.*, 1971, **104**, 969.	449, 452
71CB1199	A. Schmidpeter and W. Zeiss; *Chem. Ber.*, 1971, **104**, 1199.	625
71CB1429	M. Dräger and G. Gattow; *Chem. Ber.*, 1971, **104**, 1429.	592
71CB2732	G. Dahms, A. Haas and W. Klug; *Chem. Ber.*, 1971, **104**, 2732.	258, 449
71CB3989	N. Wiberg and W. Uhlenbrock; *Chem. Ber.*, 1971, **104**, 3989.	608
71CC314	N. H. Nilsson, C. Jacobsen and A. Senning; *J. Chem. Soc., Chem. Commun.*, 1971, 314.	553
71CC383	O. L. Chapman and C. L. McIntosh; *J. Chem. Soc., Chem. Commun.*, 1971, 383.	75
71CC400	D. J. Cardin, B. Cetinkaya, M. F. Lappert, Lj. Manojlovic-Muir and K. W. Muir; *J. Chem. Soc., Chem. Commun.*, 1971, 400.	721
71CC784	C. T. Ratcliffe, C. V. Hardin, L. R. Anderson and W. B. Fox; *J. Chem. Soc., Chem. Commun.*, 1971, 784.	231
71CC1593	G. Zweifel, G. M. Clark and R. Lynd; *J. Chem. Soc., Chem. Commun.*, 1971, 1593.	208
71CCC1867	O. Paleta, A. Postra and K. Tesarik; *Collect. Czech. Chem. Commun.*, 1971, **36**, 1867.	5
71CJC2315	H. J. Anderson, N. C. Wang and E. T. P. Jwili; *Can. J. Chem.*, 1971, **49**, 2315.	648
71CPB2222	A. Takamizawa, K. Hirai and T. Ishiba; *Chem. Pharm. Bull.*, 1971, **19**, 2222.	114
71CPB2629	K. Kikugawa and T. Kawashima; *Chem. Pharm. Bull.*, 1971, **19**, 2629.	57
71GEP1932830	H. Hagemann and K. Ley (Farbf. Bayer AG); *Ger. Pat.* 1932830 (1971) (*Chem. Abstr.*, 1971, **74**, 63942).	448
71GEP2003143	E. Klauke, E. Kühle and L. Eue (Bayer AG); *Ger. Pat.* 2003143 (1971) (*Chem. Abstr.*, 1971, **75**, 98343).	234

71GEP2040367 H. C. Bucha, P. Langsdorf, Jr. and A. G. Jelinek; *Ger. Pat.*, 2 040 367 (1971) (*Chem. Abstr.*, 1971, **74**, 112 211). 513

71GEP2105907 B. M. Lichstein and L. Woolf (Allied Chemical Corp.); *Ger. Pat.* 2105907 (1971) (*Chem. Abstr.*, 1971, **75**, 117983). 409

71GEP2128672 G. K. Kohn (Chevron Research Co.); *Ger. Pat.* 2128672 (1971) (*Chem. Abstr.*, 1972, **76**, 45777). 448

71IC1635 R. F. Swindell, L. M. Zaborowski and J. M. Shreeve; *Inorg. Chem.*, 1971, **10**, 1635. 239

71IC1896 A. K. Mishra and K. N. Tandon; *Inorg. Chem.*, 1971, **10**, 1896. 514

71IC2179 D. D. DesMarteau; *Inorg. Chem.*, 1971, **9**, 2179. 251

71IZV1073 G. Kh. Khisamutdinov, V. I. Slovetskii, A. A. Fainzil'berg and M. Sh. L'vova; *Izv. Akad. Nauk SSSR Ser. Khim.*, 1971, 1073. 283

71IZV1487 L. V. Okhlobystina, V. M. Khutoretskii and A. A. Fainzil'berg; *Izv. Akad. Nauk SSSR Ser. Khim.*, 1971, 1487 (*Chem. Abstr.*, 1971, **75**, 109 761). 56, 282

71IZV1505 V. I. Bregadze, T. I. Ivanova, L. A. Leites, L. V. Okhlobystina and O. Yu. Okhlobystin; *Izv. Akad. Nauk SSSR Ser. Khim.*, 1971, 1505 (*Chem. Abstr.*, 1971, **75**, 98610). 170

71IZV1594 R. G. Gafurov, B. S. Fedorov and L. T. Eremenko; *Izv. Akad. Nauk. SSSR, Ser. Khim.*, 1971, 1594 (*Chem. Abstr.*, 1971, **75**, 98 107). 150, 364

71JA2796 H. C. Brown and Y. Yamamoto; *J. Am. Chem. Soc.*, 1971, **93**, 2796. 61

71JA3882 P. A. Bernstein, F. A. Hohorst and D. D. DesMarteau; *J. Am. Chem. Soc.*, 1971, **93**, 3882. 231, 232

71JA4075 J. S. Splitter, T.-M. Su, H. Ono and M. Calvin; *J. Am. Chem. Soc.*, 1971, **93**, 4075. 110

71JA4961 H. D. Hartzler; *J. Am. Chem. Soc.*, 1971, **93**, 4961. 698

71JA5424 P. M. Treichel, W. J. Knebel and R. W. Hess; *J. Am. Chem. Soc.*, 1971, **93**, 5424. 517

71JA6344 N. Sonoda, T. Yasuhara, K. Kondo, T. Ikeda and S. Tsutsumi; *J. Am. Chem. Soc.*, 1971, **93**, 6344. 510

71JCS(A)21 E. M. Badley, J. Chatt and R. L. Richards; *J. Chem. Soc.* (*A*), 1971, 21. 721

71JCS(C)103 A. D. Clark and P. Sykes; *J. Chem. Soc.* (*C*), 1971, 103. 314

71JCS(C)409 G. V. Boyd and A. J. H. Summers; *J. Chem. Soc.* (*C*), 1971, 409. 110

71JCS(C)3766 K. Bartel, A. Goosen and A. Scheffer; *J. Chem. Soc.* (*C*), 1971, 3766. 431

71JCS(C)3833 G. L. Fleming, R. N. Haszeldine and A. E. Tipping; *J. Chem. Soc.* (*C*), 1971, 3833. 54

71JFC(1)347 D. J. Burton and E. A. Zawistowski; *J. Fluorine Chem.*, 1971, **1**, 347. 694

71JHC557 J. P. Chupp; *J. Heterocycl. Chem.*, 1971, **8**, 557. 508

71JMC517 R. C. Terrell, L. Speers, A. J. Szur, J. Treadwell and T. R. Ucciardi; *J. Med. Chem.*, 1971, **14**, 517. 36

71JMC593 L. Speers, A. J. Szur, R. C. Terrell, J. Treadwell and T. U. Ucciardi; *J. Med. Chem.*, 1971, **14**, 593. 36, 43

71JOC858 T. Saegusa, T. Tsuda and Y. Isegawa; *J. Org. Chem.*, 1971, **36**, 858. 448

71JOC1155 R. F. Smith, D. S. Johnson, C. L. Hyde, T. C. Rosenthal and A. C. Bates; *J. Org. Chem.*, 1971, **36**, 1155. 55

71JOC1176 S. Sakai, Y. Kobayashi and Y. Ishii; *J. Org. Chem.*, 1971, **36**, 1176. 300

71JOC1379 D. Seyferth, R. S. Marmor and P. Hilbert; *J. Org. Chem.*, 1971, **36**, 1379. 670

71JOC1786 D. Seyferth, R. Damrauer, H. Shih, W. Tronich, W. E. Smith and J. Y.-P. Mui; *J. Org. Chem.*, 1971, **36**, 1786. 608

71JOC1835 D. A. Nicholson and H. Vaughn; *J. Org. Chem.*, 1971, **36**, 1835. 263

71JOC2262 L. A. Cescon, G. R. Coraor, R. Dessauer, E. F. Silversmith and E. J. Urban; *J. Org. Chem.*, 1971, **36**, 2262. 143

71JOC2516 E. J. MacPherson and J. G. Smith; *J. Org. Chem.*, 1971, **36**, 2516. 507

71JOC2885 C. F. Hobbs and H. Weingarten; *J. Org. Chem.*, 1971, **36**, 2885. 140

71JOC3627 M. K. Meilahn, L. L. Augenstein and J. L. McManaman; *J. Org. Chem.*, 1971, **36**, 3627. 51

71JOC3843 D. A. Nicholson and H. Vaughn; *J. Org. Chem.*, 1971, **36**, 3843. 118

71JOM(26)183 J.-P. Picard, R. Calas, J. Dunoguès and N. Duffaut; *J. Organomet. Chem.*, 1971, **26**, 183. 124

71JOM(27)19 D. Seyferth and E. M. Hanson; *J. Organomet. Chem.*, 1971, **27**, 19. 270

71JOM(29)389 M. A. Cook, C. Eaborn and D. R. M. Walton; *J. Organomet. Chem.*, 1971, **29**, 389. 60, 191, 289

71JOM(32)79 P. Bourgeois, R. Calas, N. Duffaut and J. Dunoguès; *J. Organomet. Chem.*, 1971, **32**, 79. 162

71JOU1943 A. I. Kirillov, L. F. Ivanova and L. A. Tsareva; *J. Org. Chem. USSR* (*Engl. Transl.*), 1971, 1943. 425

71JPR804 H. Spies, K. Gewald and R. Mayer; *J. Prakt. Chem.*, 1971, **313**, 804. 628, 631

71LA(743)1 H. Kessler and H.-O. Kalinowski; *Justus Liebigs Ann. Chem.*, 1971, **743**, 1. 731

71LA(743)167 W. Walter and K. P. Ruess; *Justus Liebigs Ann. Chem.*, 1971, **743**, 167. 572

71LA(746)92 J. Voss; *Justus Liebigs Ann. Chem.*, 1971, **746**, 92. 579

71LA(748)1 M. Regitz, J. Hocker, W. Schössler, B. Weber and A. Liedhegener; *Justus Liebigs Ann. Chem.*, 1971, **748**, 1. 143

71LA(748)207 M. Regitz, A. Liedhegener, U. Eckstein, M. Martin and W. Anschütz; *Justus Liebigs Ann. Chem.*, 1971, **748**, 207. 670

71LA(750)44 H. Gross and B. Costisella; *Justus Liebigs Ann. Chem.*, 1971, **750**, 44. 156, 159

71LA(751)145 J. Hocker and R. Merten; *Justus Liebigs Ann. Chem.*, 1971, **751**, 145. 143

71LA(753)143 U. Schöllkopf and P. Markusch; *Justus Liebigs Ann. Chem.*, 1971, **753**, 143. 664

71MI 606-01 E. Sebe, S. Kitamura, K. Hayakawa and K. Sumino; *J. Chin. Chem. Soc.* (*Taipei*), 1971, **18**, 65 (*Chem. Abstr.*, 1971, **75**, 140949). 207

71MI 607-01 J. Villiéras; *Organometal. Chem. Rev. A*; 1971, **7**, 91 (*Chem. Abstr.*, 1972, **76**, 99733). 244

B-71MI 615-01 I. Ugi; "Isonitrile Chemistry," Academic Press, New York, London, 1971, p. 185. 485

B-71MI 616-01 S. R. Sandler and W. Karo; "Organic Functional Group Preparations," Academic Press, New York, 1971, vol. 2, p. 135. 500

B-71MI 620-01 E. Kühle, B. Anders and G. Zumach; in "Newer Methods of Preparative Organic Chemistry," ed. W. Foerst, Verlag Chemie, Weinheim, 1971, vol. vi, p. 127. 602, 604, 605, 606

B-71MI 623-01 H. Perst; "Oxonium Ions in Organic Chemistry," Verlag Chemie, Weinheim, 1971. 729

71RTC556 D. H. Holsboer, J. W. Scheeren and A. P. M. Van Der Veek; *Recl. Trav. Chim. Pays-Bas*, 1971, **90**, 556. 39, 40

71RTC1123 J. W. Scheeren, J. E. W. Van Melick and R. J. F. Nivard; *Recl. Trav. Chim. Pays-Bas*, 1971, **90**, 1123. 44

71S16 R. Feinauer; *Synthesis*, 1971, 16. 104

71S131 W. Reeve; 1971, *Synthesis*, 1971, 131. 25

71S312 S. Kabusz and W. Tritschler; *Synthesis*, 1971, 312. 105

71S478 H. Mägerlein, G. Meyer and H. D. Rupp; *Synthesis*, 1971, 478. 236

71SAP7004922 M. R. Collier, B. M. Kingston, M. F. Lappert and M. M. Truelock; *S. Afr. Pat.*, ZA 7004922 (1971), (*Chem. Abstr.* 1971, **75**, 129943). 396

71TL4519 D. St. C. Black, R. C. F. Brown and A. M. Wade; *Tetrahedron Lett.*, 1971, 4519. 107

71TL4885 K. Kondo, N. Sonoda and S. Tsutsumi; *Tetrahedron Lett.*, 1971, 4885. 465

71T4533 J. Burdon and I. W. Parsons; *Tetrahedron*, 1971, **27**, 4533. 37

71USP3562309 J. R. Lovett (Esso Research and Engineering Co.); *US Pat.* 3562309 (1971) (*Chem. Abstr.*, 1971, **74**, 142042). 369

71USP3585218 R. L. Talbott (Minnesota Mining and Manufacturing Co.); *US Pat.* 3585218 (1971) (*Chem. Abstr.*, 1971, **75**, 76161). 251

71ZAAC(382)217 G. Fritz and H. Fröhlich; *Z. Anorg. Allg. Chem.*, 1971, **382**, 217. 267, 268

71ZAAC(382)9 G. Fritz and H. Fröhlich; *Z. Anorg. Allg. Chem.*, 1971, **382**, 9. 267, 268

71ZC12 W. Hampel and M. Kapp; *Z. Chem.*, 1971, **11**, 12. 575

71ZN(B)480 G. Fritz, R. Riekens, T. Günther and M. Berndt; *Z. Naturforsch., Teil B*, 1971, **26**, 480. 267

71ZOB2155 V. V. Doroshenko, V. A. Stukalo, V. A. Shokol and B. N. Kozhushko; *Zh. Obshch. Khim.*, 1971, **41**, 2155 (*Chem. Abstr.*, 1972, **76**, 72605). 622, 634, 658

71ZOR30 G. A. Shvekhgeimer and G. A. Mikheichev; *Zh. Org. Khim.*, 1971, **7**, 30 (*Chem. Abstr.*, 1971, **74**, 111526). 145

71ZOR273 V. F. Pozdnev and E. S. Chaman; *Zh. Org. Khim.*, 1971, **7**, 273 (*Chem. Abstr.*, 1971, **75**, 48379). 463

71ZOR473 I. I. Lapkin, N. N. Pavlova and V. I. Proshutinskii; *Zh. Org. Khim*, 1971, **7**, 473. (*Chem. Abstr.* 1971, **75**, 19659). 92

71ZOR1126 A. L. Fridman, F. A. Gabitov and A. D. Nikolaeva; *Zh. Org. Khim.*, 1971, **7**, 1126 (*Chem. Abstr.*, 1971, **75**, 140168). 145

71ZOR1304 I. N. Aizenshtadt and L. I. Bagal; *Zh. Org. Khim.*, 1971, **7**, 1304 (*Chem. Abstr.*, 1971, **75**, 98650). 145

71ZOR1887 I. I. Lapkin, N. S. Zelenina and V. I. Proshutinskii; *Zh. Org. Khim.*, 1971, **7**, 1887 (*Chem. Abstr.*, 1972, **76**, 3487). 82

71ZOR2000 M. I. Dronkina and L. M. Yagupol'skii; *Zh. Org. Khim.*, 1971, **7**, 2000 (*Chem. Abstr.*, 1972, **76**, 13968). 238

71ZOR2084 V. P. Kukhar, V. I. Pasternak and A. V. Kirsanov; *Zh. Org. Khim.*, 1971, **7**, 2084 (*Chem. Abstr.*, 1972, **76**, 13713). 238, 610

71ZOR2161 L. M. Yagupol'skii and E. A. Chaika; *Zh. Org. Khim*, 1971, **7**, 2161 (*Chem. Abstr.*, 1972, **76**, 13985). 69

71ZOR2229 L. I. Samarai, V. I. Gorbatenko, O. V. Vishnevskii and N. F. Radchenko; *Zh. Org. Khim.*, 1971, **7**, 2229 (*Chem. Abstr.*, 1972, **76**, 45885). 453

71ZOR2267 V. A. Ginsburg, M. N. Vasil'eva and L. L. Martynova; *Zh. Org. Khim.*, 1971, **7**, 2267 (*Chem. Abstr.*, 1972, **76**, 45849). 613

72ACS1874 H. J. Bjørnstad and B. Klewe; *Acta Chem. Scand.*, 1972, **26**, 1874. 170

72AG993 Z. Janousek, J. Collard and H. G. Viehe; *Angew. Chem.*, 1972, **84**, 993. 506

72AG(E)473 G. Köbrich; *Angew. Chem., Int. Ed. Engl.*, 1972, **11**, 473. 293

72AG(E)949 E. Grigat; *Angew. Chem., Int. Ed. Engl.*, 1972, **11**, 949. 616, 618

72AG(E)964 J. Hocker and R. Merten; *Angew. Chem., Int. Ed. Engl.*, 1972, **11**, 964. 141

72ANY(192)101 L. Henriksen and E. S. S. Kristiansen; *Ann. N. Y. Acad. Sci.*, 1972, **192**, 101. 593, 597

72BCJ489 K. Tanaka and T. Tanaka; *Bull. Chem. Soc. Jpn.*, 1972, **45**, 489. 628, 730

72BCJ960 S. Tamagaki and S. Oae; *Bull. Chem. Soc. Jpn.*, 1972, **45**, 960. 303, 307

72BCJ1797 Y. Ueno and M. Okawara; *Bull. Chem. Soc. Jpn.*, 1972, **45**, 1797. 309

72BCJ2937 N. Sonoda, G. Yamamoto and S. Tsutsumi; *Bull. Soc. Chem. Jpn.*, 1972, **45**, 2937. 593

72CA(77)125952 E. Klauke, H. Holtschmidt and K. Findeisen (Bayer AG); *Chem. Abstr.*, 1972, **77**, 125952 (DOS 2/101/107 (1971)). 240

72CB487 D. Seebach; *Chem. Ber.*, 1972, **105**, 487. 87, 88, 302, 317, 697

72CB511 D. Seebach and N. Peleties; *Chem. Ber.*, 1972, **105**, 511. 92, 307, 317, 697

72CB820 A. Haas, W. Klug and H. Marsmann; *Chem. Ber.*, 1972, **105**, 820. 534

72CB1340 W. Kantlehner and P. Speh; *Chem. Ber.*, 1972, **105**, 1340. 105, 109

72CB1709 G. Zinner and H. Gross; *Chem. Ber.*, 1972, **105**, 1709. 650

72CB2558 E. O. Fischer, F. R. Kreissl, C. G. Kreiter and E. W. Meineke; *Chem. Ber.*, 1972, **105**, 2558. 720

72CB2815 R. Huisgen and W. Mack; *Chem. Ber.*, 1972, **105**, 2815. 309

72CB2854 N. H. Nilsson, C. Jacobsen, O. N. Sørensen, N. K. Haunsøe and A. Senning; *Chem. Ber.*, 1972, **105**, 2854. 238, 540

72CB3050 A. Schmidt; *Chem. Ber.*, 1972, **105**, 3050. 727

72CB3280 D. Seebach, K.-H. Geiss, A. K. Beck, B. Graf and H. Daum; *Chem. Ber.*, 1972, **105**, 3280. 82, 86, 88, 302, 303, 317

72CB3892 D. Seebach and A. K. Beck; *Chem. Ber.*, 1972, **105**, 3892. 88, 317

72CB3905 D. Seebach, H. B. Stegmann, K. Scheffler, A. K. Beck and K. H. Geiß; *Chem. Ber.*, 1972, **105**, 3905. 302

72CC354 J. E. Baldwin and J. A. Walker; *J. Chem. Soc., Chem. Commun.*, 1972, 354. 627

72CC673 D. K. Kang and A. B. Burg; *J. Chem. Soc., Chem. Commun.*, 1972, 673. 264

72CC851 B. Cetinkaya, M. F. Lappert and K. Turner; *J. Chem. Soc., Chem. Commun.*, 1972, 851. 721

72CC863 A. Bélanger and P. Brassard; *J. Chem. Soc., Chem. Commun.*, 1972, 863. 75

72CC927 D. J. Cardin, M. J. Doyler and M. F. Lappert; *J. Chem. Soc., Chem. Commun.*, 1972, 927. 721

72CC1078 C. Chung and R. J. Ladow; *J. Chem. Soc., Chem. Commun.*, 1972, 1078. 204, 379

72CLY937 O. Paleta and A. Posta; *Chem. Listy*, 1972, **66(9)**, 937 (*Chem. Abstr.*, 1972, **77**, 125780). 4

72CL483 M. Tanaka, Y. Watanabe, T. Mitsudo, K. Yamamoto and Y. Takegami; *Chem. Lett.*, 1972, 483. 457

72CL971 Y. Furuya, K. Itoho, O. Shibata and K. Ohkubo; *Chem. Lett.*, 1972, 971. 501

72CPB2348 K. Harano and T. Taguchi; *Chem. Pharm. Bull.*, 1972, **20**, 2348. 472

72CRV357 C. U. Pittman, Jr., S. P. McManus and J. W. Larsen; *Chem. Rev.*, 1972, **72**, 357. 43

72CRV545 D. J. Cardin, B. Cetinkaya and M. F. Lappert; *Chem. Rev.*, 1972, **72**, 545. 716, 721

72EGP93766 H. Gross and B. Costisella; *Ger. (East) Pat.* 93 766 (1972) (*Chem. Abstr.*, 1973, **78**, 159 850). 665

72G558 S. Cabiddu, E. Marongiu and F. Sotgiu; *Gazz. Chim. Ital.*, 1972, **102**, 558. 321

72GEP2101107 E. Klauke, H. Holtschmidt and K. Findeisen (Bayer AG); *Ger. Pat.* 2101107 (1972) (*Chem. Abstr.*, 1972, **77**, 125952). 603, 613

72GEP2111696 D. Sianesi, R. Fontanelli and A. Grazioli (Montecatini Edison SpA); *Ger. Pat.* 2111696 (1972) (*Chem. Abstr.*, 1972, **76**, 113263). 251

72GEP2129200 J. Hempel and E. Klauke (Bayer AG); *Ger. Pat.* 2129200 (1972) (*Chem. Abstr.*, 1973, **78**, 71714). 229, 230

72GEP2206366 T. Somlo; *Ger. Pat.*, 2 206 366 (1972) (*Chem. Abstr.*, 1972, **77**, 151 532). 501

72HCA75 P. Stütz and A. Stadler; *Helv. Chim. Acta*, 1972, **55**, 75. 97

72HCA1782 A. Samat, R. Guglielmetti and J. Metzger; *Helv. Chim. Acta*, 1972, **55**, 1782. 111

72ICA11 M. A. Ali, S. E. Livingstone and D. J. Phillips; *Inorg. Chim. Acta*, 1972, **6**, 11. 563

72IC418 J. B. Hynes, T. E. Austin and L. A. Bigelow; *Inorg. Chem.*, 1972, **11**, 418. 284, 363, 616

72IC2069 J. S. Miller and A. L. Balch; *Inorg. Chem.*, 1972, **11**, 2069. 732

72IC2531 D. Pilipovich, C. J. Schack and R. D. Wilson; *Inorg. Chem.*, 1972, **11(10)**, 2531. 424

72IJS(A)105 J. E. Oliver and J. B. Stokes; *Int. J. Sulfur Chem., Part A*, 1972, **2**, 105. 647

72IJS(A)133 L. Hendriksen and E. S. S. Kristiansen; *Int. J. Sulfur Chem., Part A*, 1972, **2**, 133. 594

72IZV85 N. S. Vyazankin, O. A. Kruglya, B. I. Petrov, L. S. German, V. R. Polischuk, B. L. Dyatkin, S. R. Sterlin and B. I. Martynov; *Izv. Akad. Nauk SSSR Ser. Khim.*, 1972, 85 (*Chem. Abstr.*, 1972, **77**, 19 766). 62

72IZV490 O. P. Shitov, S. L. Ioffe, V. A. Tartakovskii and S. S. Novikov; *Izv. Akad. Nauk SSSR Ser. Khim.*, 1972, 490 (*Chem. Abstr.*, 1972, **77**, 34217). 148

72IZV1635 B. K. Nefedov, N. S. Sergeeva and Y. T. Eidus; *Izv. Akad. Nauk. SSSR Ser. Khim.*, 1972, 1635 (*Chem. Abstr.*, 1972, **77**, 151 395). 464

72IZV1717 V. A. Kokorekina, S. A. Shevelev, L. G. Feoktistov and A. A. Fainzil'berg; *Izv. Akad. Nauk SSSR Ser. Khim.*, 1972, 1717 (*Chem. Abstr.*, 1972, **77**, 159470). 149

72IZV1721 V. I. Slovetskii, G. Kh. Khisamutdinov and A. A. Fainzil'berg; *Izv. Akad. Nauk SSSR Ser. Khim.*, 1972, 1721. 283

72IZV2603 V. A. Kokorekina, O. P. Shitov, L. G. Feoktistov, S. L. Ioffe, V. A. Tartakovskii and S. S. Novikov; *Izv. Akad. Nauk SSSR Ser. Khim.*, 1972, 2603 (*Chem. Abstr.*, 1973, **78**, 71791). 149

72IZV2733 B. K. Nefedov, N. S. Sergeeva and Y. T. Eidus; *Izv. Akad. Nauk. SSSR Ser. Khim.*, 1972, 2733 (*Chem. Abstr.*, 1973, **78**, 97 074). 464

72JA125 P. von R. Schleyer, W. F. Sliwinski, G. W. Van Dine, U. Schöllkopf, J. Paust and K. Fellenberger; *J. Am. Chem. Soc.*, 1972, **94**, 125. 63

72JA417 A. L. Balch and J. Miller; *J. Am. Chem. Soc.*, 1972, **94**, 417. 721

72JA820 K. J. Klabunde and D. J. Burton; *J. Am. Chem. Soc.*, 1972, **94**, 820. 694

72JA1563 D. F. Lewis, S. J. Lippard and J. A. Zubieta; *J. Am. Chem. Soc.*, 1972, **94**, 1563. 552

72JA3201 H. M. R. Hoffmann, K. E. Clemens, E. A. Schmidt and R. Smithers; *J. Am. Chem. Soc.*, 1972, **94**, 3201. 107

72JA3799 W. Jetz and R. J. Angelici; *J. Am. Chem. Soc.*, 1972, **94**, 3799. 517

72JA6135 G. M. Atkins, Jr. and E. M. Burgess; *J. Am. Chem. Soc.*, 1972, **94**, 6135. 106

72JCS(D)1303 P. R. Branson and M. Green; *J. Chem. Soc., Dalton Trans.*, 1972, 1303. 664

72JCS(D)2267 F. Sanz and J. J. Daly; *J. Chem. Soc., Dalton Trans.*, 1972, 2267. 661

72JCS(P1)34 G. Haran and D. W. A. Sharp; *J. Chem. Soc., Perkin Trans. 1*, 1972, 34. 46

72JCS(P1)130 P. G. Harrison; *J. Chem. Soc., Perkin Trans. 1*, 1972, 130. 571

72JCS(P1)155 R. N. Haszeldine, B. Higginbottom, R. B. Rigby and A. E. Tipping; *J. Chem. Soc., Perkin Trans. 1*, 1972, 155. 45

72JCS(P1)159 R. N. Haszeldine, R. B. Rigby and A. E. Tipping; *J. Chem. Soc., Perkin Trans. 1*, 1972, 159. 45

72JCS(P1)1506 R. N. Haszeldine, R. B. Rigby and A. E. Tipping; *J. Chem. Soc., Perkin Trans. 1*, 1972, 1506. 45, 219

72JCS(P1)1678 I. Matsuda, K. Itoh and Y. Ishii; *J. Chem. Soc., Perkin Trans. 1*, 1972, 1678. 571

72JCS(P1)2180 R. N. Haszeldine, R. B. Rigby and A. E. Tipping; *J. Chem. Soc., Perkin Trans. 1*, 1972, 2180. 233

72JCS(P1)2438 R. N. Haszeldine, R. B. Rigby and A. E. Tipping; *J. Chem. Soc., Perkin Trans. 1*, 1972, 2438. 45

72JCS(P2)1050 J. Donovan, J. Cronin, F. L. Scott and A. F. Hegarty; *J. Chem. Soc., Perkin Trans. 2*, 1972, 1050. 623, 627

72JGU97 V. I. Shevchenko, N. K. Kulibaba and A. V. Kirsanov; *J. Gen. Chem. USSR* (*Engl. Transl.*), 1972, **42**, 97. 616, 728

72JGU293 S. I. Malekin, M. A. Sokalskii, Y. L. Kruglyak and I. V. Martynov; *J. Gen. Chem. USSR* (*Engl. Transl.*), 1972, **42**, 293. 613

72JGU799 S. I. Malekin, V. I. Yakutin, M. A. Sokalskii, Y. L. Kruglyak and I. V. Martynov; *J. Gen. Chem. USSR* (*Engl. Transl.*), 1972, **42**, 799. 613

72JGU803 Y. L. Kruglyak, S. I. Malekin and I. V. Martynov; *J. Gen. Chem. USSR* (*Engl. Transl.*), 1972, **42**, 803. 613

72JHC1087 R. Scarpati and M. L. Graziano; *J. Heterocycl. Chem.*, 1972, **9**, 1087. 107

72JMC604 R. C. Terrell, L. Speers, A. J. Szur, T. Ucciardi and J. F. Vitcha; *J. Med. Chem.*, 1972, **15**, 604. 38, 43

72JMC606 L. Speers, A. J. Szur and R. C. Terrell; *J. Med. Chem.*, 1972, **15**, 606. 38, 43

72JOC152 E. F. Witucki, G. L. Rowley, M. Warner and M. B. Frankel; *J. Org. Chem.*, 1972, **37**, 152. 56

72JOC1458 D. R. Maulding and B. G. Roberts; *J. Org. Chem.* 1972, **37**, 1458. 252

72JOC2005 H. W. Moon; *J. Org. Chem.*, 1972, **37**, 2005. 608, 611, 653

72JOC2651 I. W. Serfaty, T. Hodgins and E. T. McBee; *J. Org. Chem.*, 1972, **37**, 2651. 229

72JOC2662 D. R. Dimmel, C. A. Wilkie and F. Ramon; *J. Org. Chem.*, 1972, **37**, 2662. 267, 379

72JOC2730 G. H. Birum and J. D. Wilson; *J. Org. Chem.*, 1972, **37**, 2730. 661, 733

72JOC2757 R. A. Ellison, W. D. Woessner and C. C. Williams; *J. Org. Chem.*, 1972, **37**, 2757. 86

72JOC4198 S. Sakai, Y. Kuroda and Y. Ishii; *J. Org. Chem.*, 1972, **37**, 4198. 300

72JOM(34)299 B. I. Petrov, O. A. Kruglaya, N. S. Vyazankin, B. I. Martynov, S. R. Sterlin and B. L. Dyatkin; *J. Organomet. Chem.*, 1972, **34**, 299. 62

72JOM(34)83 J. Satgé, C. Couret and J. Escudié; *J. Organomet. Chem.*, 1972, **34**, 83. 669

72JOM(39)C49 J.-P. Picard, P. Bourgeois, J. Dunoguès and R. Calas; *J. Organomet. Chem.*, 1972, **39**, C49. 125

72JOM(40)C53 B. Martel and M. Varache; *J. Organomet. Chem.*, 1972, **40**, C53. 194

72JOM(42)293 G. Neumann and W. P. Neumann; *J. Organomet. Chem.*, 1972, **42**, 293. 661

72JOM(46)149 G. Schmid and V. Bätzel; *J. Organomet. Chem.*, 1972, **46**, 149. 340

72JPR87 H. Gross and B. Costisella; *J. Prakt. Chem.*, 1972, **314**, 87. 367, 665

72JPR969 H. Gross, B. Costisella, L. Brennecke and G. Engelhardt; *J. Prakt. Chem.* 1972, **314**, 969. 367

72JPR145 E. Höft and S. Ganschow; *J. Prakt. Chem.*, 1972, **314**, 145. 109

72JPR969 H. Gross, B. Costisella, L. Brennecke and G. Engelhardt; *J. Prakt. Chem.*, 1972, **314**, 969. 159

72KGS713 A. K. Pan'kov, M. S. Pevzner and L. I. Bagal; *Khim. Geterotsikl. Soedin.*, 1972, 713 (*Chem. Abstr.*, 1972, **77**, 126523). 145

72LA(755)145 W. Walter and R. F. Becker; *Justus Liebigs Ann. Chem.*, 1972, **755**, 145. 238, 240, 540

72LA(755)67 A. Grote, A. Haag and G. Hesse; *Justus Liebigs Ann. Chem.*, 1972, **755**, 67. 637

72LA(758)132 H. Böhme and M. Junga; *Justus Liebigs Ann. Chem.*, 1972, **758**, 132. 275, 278, 279, 307

72LA(765)94 J. Hinz and C. Rüchardt; *Justus Liebigs Ann. Chem.*, 1972, **765**, 94. 509

72MI 602-01 H. S. Eleuterio; *J. Macromol. Sci., Chem.*, 1972, **6**, 1027. 42

72MI 604-01 I. D. Kolpakova, M. I. Kabachnik, T. Y. Medved, R. P. Lastovskii, L. V. Krinitskaya, E. M. Urinovich and V. A. Smirnova; *Khim. Prom.*, 1972, **48**, 576 (*Chem. Abstr.*, 1972, **77**, 152282). 118

72MI 606-01 P. K. Mattschei; Dissertation, Washington State University, 1972 (*Chem. Abstr.*, 1973, **78**, 111399). 181

72MI 607-01 Yu. G. Papulov, L. V. Chulkova, V. P. Levin and A. E. Stepan'yan; *Zh. Strukt. Khim.*, 1972, **13**, 956 (*Chem. Abstr.*, 1973, **78**, 29045). 223, 226, 227, 228

B-72MI 607-02 H. G. Boit, O. Weissbach, M. E. Fernholz, I. Hagel, R. Meister, A. Reichard, E. Schieber, E. Bayer, F. H. Flock, F. Giese and L. Mähler; in "Beilsteins Handbuch der Organischen Chemie," 4th edn., Viertes Ergänzungswerk, Deutsche Chemische Gesellschaft, Springer-Verlag, Berlin, 1972. 213

72MI 610-01 D. G. H. Ballard, J. Myatt and J. F. P. Richter; *J. Appl. Polym. Sci.*, 1972, **16**, 2647 (*Chem. Abstr.*, 1973, **78**, 17287). 297

72MI 612-01 T. N. Ivshina, V. M. Shitkin, S. L. Ioffe, E. T. Lippmaa and M. Y. Magi; *Zh. Strukt. Khim.*, 1972, **13**, 933 (*Chem. Abstr.*, 1973, **78**, 34556). 368

B-72MI 614-01 D. N. Kevill; in "The Chemistry of Acyl Halides," ed. S. Patai, Wiley, London, 1972, p. 433. 421

B-72MI 622-01 H. J. Bestmann and R. Zimmermann; in "Organic Phosphorus Compounds," eds. G. Kosolapoff and L. Maier, Wiley-Interscience, New York, 1972, vol. 3, p.1. 701

72RTC37 A. M. van Leusen, B. A. Reith, A. J. W. Iedema and J. Strating; *Recl. Trav. Chim. Pays-Bas*, 1972, **91**, 37. 699

72RTC331 J. Geevers, W. P. Trompen and J. Th. Hackmann; *Recl. Trav. Chim. Pays-Bas*, 1972, **91**, 331. 284

72RTC349 D. H. Holsboer and A. P. M. Van Der Veek; *Recl. Trav. Chim. Pays-Bas*, 1972, **91**, 349. 46, 48

72RTC1117 J. W. Ypenburg and H. Gerding; *Recl. Trav. Chim. Pays-Bas*, 1972, **91**, 1117. 302

72S32 W. Tritschler and S. Kabusz; *Synthesis*, 1972, 32. 105

72S39 H. Kiefer; *Synthesis*, 1972, 39. 445

72S81 H. Alper and L. Dinkes; *Synthesis*, 1972, 81. 445

72S351 M. Regitz; *Synthesis*, 1972, 351. 666

72S418 S. Kabusz and W. Tritschler; *Synthesis*, 1972, 418. 105

72S599 K. Findeisen, K. Wagner and H. Holtschmidt; *Synthesis*, 1972, 599. 297, 299

72T1965 W. P. Norris; *Tetrahedron*, 1972, **28**, 1965. 105

72TL2939 R. Grashey, M. Baumann and R. Hamprecht; *Tetrahedron Lett.*, 1972, 2939. 573

72TL3769 E. J. Corey and P. L. Fuchs; *Tetrahedron Lett.*, 1972, 3769. 695

72TL3827 Y. Hata and M. Watanabe; *Tetrahedron Lett.*, 1972, 3827. 608

72TL3957 T. Martini; *Tetrahedron Lett.*, 1972, 3957. 106

72TL4217 G. Ege and H. O. Frey; *Tetrahedron Lett.*, 1972, 4217. 57, 105
72TL4263 H. Sawai and T. Takizawa; *Tetrahedron Lett.*, 1972, 4263. 505
72URP345094 M. S. Salakhov, M. M. Guseinov and I. R. Akhverdiev; *USSR Pat.* 345094 (1972) (*Chem. Abstr.*, 1973, **78**, 83842). 412
72USP3637663 R. A. Mitsch (Minnesota Mining & Mfg. Co.); *US Pat.* 3637663 (1972) (*Chem. Abstr.* 1972, **76**, 113195). 285
72USP3652256 J. J. D'Amico (Monsanto Co.); *US Pat.* 3652256 (1972) (*Chem. Abstr.*, 1972, **77**, 62008). 305
72USP3663621 D. L. Esmay (Minnesota Mining & Mfg. Co.); *US Pat.* 3663621 (1972) (*Chem. Abstr.*, 1972, **77**, 74818). 362
72USP3689560 C. D. Wright and J. L. M. Zollinger (Minnesota Mining & Mfg. Co.); *US Pat.* 3689560 (1972) (*Chem. Abstr.*, 1973, **78**, 29216). 284, 362
72USP3699094 W. C. Firth, Jr. and S. Frank (American Cyanamid Co.); *US Pat.* 3699094 (1972) (*Chem. Abstr.* 1973, **78**, 32239). 362
72USP3707555 C. D. Wright and J. L. M. Zollinger (Minnesota Mining & Mfg. Co.); *US Pat.* 3707555 (1972) (*Chem. Abstr.*, 1973, **79**, 78082). 279
72ZAAC(389)119 W. Plöger, N. Schindler, K. Wollmann and K.-H. Worms; *Z. Anorg. Allg. Chem.*, 1972, **389**, 119. 158, 159
72ZAAC(390)157 G. Fritz and M. Hähnke; *Z. Anorg. Allg. Chem.*, 1972, **390**, 157. 178
72ZAAC(390)191 G. Fritz and M. Hähnke; *Z. Anorg. Allg. Chem.*, 1972, **390**, 191. 178
72ZAAC(391)219 G. Fritz, M. Berndt and R. Huber; *Z. Anorg. Allg. Chem.*, 1972, **391**, 219. 267
72ZOB1169 V. P. Kukhar, V. I. Pasternak and A. V. Kirsanov; *Zh. Obshch. Khim.*, 1972, **42**, 1169 (*Chem. Abstr.*, 1972, **77**, 101790). 367
72ZOB1521 T. K. Gar, A. A. Buyakov and V. F. Mironov; *Zh. Obshch. Khim.*, 1972, **42**, 1521 (*Chem. Abstr.*, 1972, **77**, 140216). 188
72ZOR39 L. S. Pupko, A. I. Dychenko and P. S. Pel'kis; *Zh. Org. Khim.*, 1972, **8**, 39 (*Chem. Abstr.*, 1972, **76**, 112852). 654
72ZOR332 Y. V. Maksimov, S. M. Kvitko and V. V. Perekalin; *Zh. Org. Khim.*, 1972, **8**, 332 (*Chem. Abstr.*, 1972, **76**, 126544). 653

73AG837 H. G. Viehe and Z. Janousek; *Angew. Chem.*, 1973, **85**, 837. 238
73AG(E)654 G. Fritz, G. Marquardt and H. Scheer; *Angew. Chem., Int. Ed. Engl.*, 1973, **12**, 654. 383
73AG(E)842 H. J. Becher and E. Langer; *Angew. Chem., Int. Ed. Engl.*, 1973, **12**, 842. 518
73AG(E)918 G. Siegemund; *Angew. Chem., Int. Ed. Engl.*, 1973, **12**, 918. 414, 419, 420
73AG(E)996 K. Burzin and R. Feinauer; *Angew. Chem., Int. Ed. Engl.*, 1973, **12**, 996. 106
73AG(E)999 H. Hagemann; *Angew. Chem., Int. Ed. Engl.*, 1973, **12**, 999. 451
73AJC755 C. N. Murphy and G. Winter; *Aust. J. Chem.*, 1973, **26**, 755. 479
73BCJ303 S. Yanagida, T. Fujita, M. Ohoka, I. Katagiri, M. Miyake and S. Komori; *Bull. Chem. Soc. Jpn.*, 1973, **46**, 303. 618
73BCJ1269 T. Nagasawa, K. Kuroiwa, K. Narita and Y. Isowa; *Bull. Chem. Soc. Jpn.*, 1973, **46**, 1269. 434
73BSB23 R. Fuks and M. A. Hartemink; *Bull. Soc. Chim. Belg.*, 1973, **82**, 23. 238
73CB1487 J. Goerdeler and H. Hohage; *Chem. Ber.*, 1973, **106**, 1487. 536
73CB1752 C. Jäckh and W. Sundermeyer; *Chem. Ber.*, 1973, **106**, 1752. 419, 442, 451, 452, 453
73CB2093 R. Appel, K.-D. Ziehn and K. Warning; *Chem. Ber.*, 1973, **106**, 2093. 620
73CB2277 D. Seebach, M. Kolb and B.-T. Gröbel; *Chem. Ber.*, 1973, **106**, 2277. 327
73CB2315 H. Gross and G. Zinner; *Chem. Ber.*, 1973, **106**, 2315. 360
73CB2523 H. Stetter and J. Bremen; *Chem. Ber.*, 1973, **106**, 2523. 140
73CB3725 H. Bredereck, G. Simchen and H. Hoffmann; *Chem. Ber.*, 1973, **106**, 3725. 111, 115
73CCC66 O. Paleta and J. Konarek; *Collect. Czech. Chem. Commun.*, 1973, **38**, 66. 42
73CC13 K.-Y. Akiba and N. Inamoto; *J. Chem. Soc., Chem. Commun.*, 1973, 13. 496
73CC532 M. P. Brown, A. K. Holliday and G. M. Way; *J. Chem. Soc., Chem. Commun.*, 1973, 532. 183
73CJC333 K. Koyano and C. R. McArthur; *Can. J. Chem.*, 1973, **51**, 333. 445
73CJC2448 K. I. The, L. K. Peterson and E. Kiehlmann; *Can. J. Chem.*, 1973, **51**, 2448. 365
73CL1315 N. Kobayashi, A. Osawa and T. Fujisawa; *Chem. Lett.*, 1973, 1315. 434
73CPB62 M. Nagamo, T. Matsui, J. Tobitsuka and K. Oyamada; *Chem. Pharm. Bull.*, 1973, **21**, 62. 629
73CPB604 T. Kawata, K. Harano and T. Taguchi; *Chem. Pharm. Bull.*, 1973, **21**, 604. 472
73CRV75 H. Babad and A. G. Zeiler; *Chem. Rev.*, 1973, **73**, 75. 460
73CSR99 D. J. Cardin, B. Cetinkaya, M. J. Doyle and M. F. Lappert; *Chem. Soc. Rev.*, 1973, **2**, 99. 716
73G373 F. Bonati and G. Minghetti; *Gazz. Chim. Ital.*, 1973, **103**, 373. 721
73GEP1694210 C. Heuck, O. Mauz, J. Winter and F. Schülde (Farbwerke Hoechst AG); *Ger. Pat.* 1694210 (1973) (*Chem. Abstr.*, 1973, **79**, 54333). 302
73GEP2131789 E. Kühle and E. Klauke (Bayer AG); *Ger. Pat.* 2131789 (1973) (*Chem. Abstr.*, 1973, **78**, 84075). 453
73GEP2241471 L. M. Shindarov, A. S. Galabov, D. S. Antonov, N. A. Neikova, K. A. Davidkov, V. Vasileva, V. B. Kalcheva and D. S. Stoicheva; *Ger. Pat.*, 2241471 (1973) (*Chem. Abstr.*, 1973, **78**, 159294). 502
73GEP2342860 M. Ribi (Ciba-Geigy AG); *Ger. Pat.* 2342860 (1973) (*Chem. Abstr.*, 1974, **81**, 25199). 441
73IC769 W. L. Reichert and G. H. Cady; *Inorg. Chem.*, 1973, **12**, 769. 430
73IC2472 D. S. Matteson and P. K. Mattschei; *Inorg. Chem.*, 1973, **12**, 2472. 181
73IC2890 G. H. Sprenger, K. J. Wright and J. M. Shreeve; *Inorg. Chem.*, 1973, **12**, 2890. 440
73IZV122 O. P. Shitov, S. L. Ioffe, V. A. Tartakovskii, S. S. Novikov, N. N. Rozhdestvenskaya and G. V. Isagulyants; *Izv. Akad. Nauk SSSR Ser. Khim.*, 1973, 122 (*Chem. Abstr.*, 1973, **78**, 147699). 145

73IZV350 V. I. Erashko, A. V. Sultanov, S. A. Shevelev and A. A. Fainzil'berg; *Izv. Akad. Nauk SSSR, Ser. Khim.* 1973, **350** (*Chem. Abstr.*, 1973, **78**, 147 742). 281, 348

73IZV562 B. K. Nefedov, N. S. Sergeeva and Y. T. Eidus; *Izv. Akad. Nauk. SSSR Ser. Khim.*, 1973, 562 (*Chem. Abstr.*, 1973, **79**, 4963). 464

73IZV807 B. K. Nefedov, N. S. Sergeeva and Y. T. Eidus; *Izv. Akad. Nauk. SSSR Ser. Khim.*, 1973, 807 (*Chem. Abstr.*, 1973, **79**, 31 813). 464

73IZV819 B. G. Sankov, V. I. Erashko and S. A. Shevelev; *Izv. Akad. Nauk SSSR Ser. Khim.*, 1973, 819 (*Chem. Abstr.*, 1973, **79**, 41 855). 653

73IZV1149 I. A. Leenson, G. B. Sergeev, O. P. Shitov, S. L. Ioffe and V. A. Tartakovskii; *Izv. Akad. Nauk SSSR Ser. Khim.*, 1973, 1149 (*Chem. Abstr.*, 1973, **79**, 65447). 148

73IZV1424 L. T. Eremenko, I. V. Tselinskii, F. Y. Natsibullin and G. V. Oreshko; *Izv. Akad. Nauk SSSR Ser. Khim.*, 1973, 1424 (*Chem. Abstr.*, 1973, **79**, 91 707). 55

73JA278 M. S. Newman and C. H. Chen; *J. Am. Chem. Soc.*, 1973, **95**, 278. 79

73JA1343 L. A. Shimp and R. J. Lagow; *J. Am. Chem. Soc.*, 1973, **95**, 1343. 205

73JA2982 A. Hassner, J. O. Currie, Jr., A. S. Steinfeld and R. F. Atkinson; *J. Am. Chem. Soc.*, 1973, **95**, 2982. 608

73JA3324 T. Ling Chwang and R. West; *J. Am. Chem. Soc.*, 1973, **95**, 3324. 185, 206

73JA4016 S. Greenberg and J. G. Moffatt; *J. Am. Chem. Soc.*, 1973, **95**, 4016. 71

73JA4769 D. J. Doonan and A. L. Balch; *J. Am. Chem. Soc.*, 1973, **95**, 4769. 721

73JA5096 D. S. Matteson, L. A. Hagelee and R. J. Wilcsek; *J. Am. Chem. Soc.*, 1973, **95**, 5096. 182, 183, 199, 292

73JCS(D)2314 C. D. Desjardins and J. Passmore; *J. Chem. Soc., Dalton Trans.*, 1973, 2314. 49

73JCS(P1)1009 T. Takeshima, T. Miyauchi, N. Fukada, S. Koshizawa and M. Muraoka; *J. Chem. Soc., Perkin Trans. 1*, 1973, 1009. 324

73JCS(P1)1092 R. E. Banks, D. R. Choudhury and R. N. Haszeldine; *J. Chem. Soc., Perkin Trans. 1*, 1973, 1092. 241

73JCS(P1)2272 P. F. Jones, M. F. Lappert and A. C. Szary; *J. Chem. Soc., Perkin Trans. 1*, 1973, 2272. 128

73JCS(P1)2644 P. R. Atkins, S. E. J. Glue and I. T. Kay; *J. Chem. Soc., Perkin Trans. 1*, 1973, 2644. 629

73JCS(P2)599 J. Klein and J. Y. Becker; *J. Chem. Soc., Perkin Trans. 2*, 1973, 599. 206

73JFC(3)17 T. Abe and J. M. Shreeve; *J. Fluorine Chem.*, 1973, **3**, 17. 253

73JFC(3)383 A. Haas, J. Helmbrecht, W. Klug, B. Koch, H. Reinke and J. Sommerhoff; *J. Fluorine Chem.*, 1973, **3**, 383. 432

73JGU848 L. I. Zakharkin, V. N. Kalinin and E. G. Rys; *J. Gen. Chem. USSR* (*Engl. Transl.*), 1973, **43**, 848. 406

73JGU1019 S. S. Gerasimova, M. I. Bakhitov and E. V. Kuznetsov; *J. Gen. Chem. USSR* (*Engl. Transl.*), 1973, **43**, 1019. 512

73JHC639 J. A. Settepani, R. T. Brown and A. B. Borkovec; *J. Heterocycl. Chem.*, 1973, **10**, 639. 508

73JHC791 A. K. Bose, J. L. Fahey and M. S. Manhas; *J. Heterocycl. Chem.*, 1973, **10**, 791. 113

73JMC901 W. L. Matier, D. A. Owens, W. T. Comer, D. Deitchman, H. C. Ferguson, R. J. Seidehamel and J. R. Young; *J. Med. Chem.*, 1973, **16**, 901. 508

73JMC978 M. Hatanaka and T. Ishimaru; *J. Med. Chem.*, 1973, **16**, 978. 77

73JOC1065 J. L. Zollinger, C. D. Wright, J. J. McBrady, D. H. Dybvig, F. A. Fleming, G. A. Kurhajec, R. A. Mitsch and E. W. Nuevar; *J. Org. Chem.*, 1973, **38**, 1065. 279, 281

73JOC1075 C. D. Wright and J. L. Zollinger; *J. Org. Chem.*, 1973, **38**, 1075. 284, 362, 613, 650

73JOC1080 W. C. Firth, Jr., S. Frank and E. J. Schriffert; *J. Org. Chem.*, 1973, **38**, 1080. 363

73JOC1083 W. C. Firth, Jr. and S. Frank; *J. Org. Chem.*, 1973, **38**, 1083. 363

73JOC1088 W. C. Firth, Jr., S. Frank and M. D. Meyers; *J. Org. Chem.*, 1973, **38**, 1088. 363

73JOC1361 M. E. Parham, W. D. McKown, V. Nelson, S. Kajigaeshi and N. Ishikawa; *J. Org. Chem.*, 1973, **38**, 1361. 76, 77

73JOC1591 T. R. Bosin, R. N. Hanson, J. V. Rodricks, R. A. Simpson and H. Rapoport; *J. Org. Chem.*, 1973, **38**, 1591. 651

73JOC1652 E. E. Smissman and A. Makriyannis; *J. Org. Chem.*, 1973, **38**, 1652. 506

73JOC2106 R. E. Hackler and T. W. Balko; *J. Org. Chem.*, 1973, **38**, 2106. 555

73JOC2614 R. Richter and H. Ulrich; *J. Org. Chem.*, 1973, **38**, 2614. 143

73JOC3358 R. J. Koshar and R. A. Mitsch; *J. Org. Chem.*, 1973, **38**, 3358. 254, 278

73JOC3617 J. L. Adcock and R. J. Lagow; *J. Org. Chem.*, 1973, **38**, 3617. 36

73JOM(50)265 D. Seyferth, J. E. Hallgren and P. L. K. Hung; *J. Organomet. Chem.*, 1973, **50**, 265. 341, 401

73JOM(55)209 O. W. Steward and J. S. Johnson; *J. Organomet. Chem.*, 1973, **55**, 209. 174, 390, 392

73JOM(57)225 D. S. Matteson and G. L. Larson; *J. Organomet. Chem.*, 1973, **57**, 225. 183, 200, 396

73JOM(57)231 D. S. Matteson and R. J. Wilcsek; *J. Organomet. Chem.*, 1973, **57**, 231. 183, 200, 292, 386, 400

73JOM(57)423 B. L. Dyatkin, B. I. Martynov, L. G. Martynova, N. G. Kizim, S. R. Sterlin, Z. A. Stumbrevichute and L. A. Fedorov; *J. Organomet. Chem.*, 1973, **57**, 423. 64

73JOM(59)231 D. Seyferth and R. S. Marmor; *J. Organomet. Chem.*, 1973, **59**, 231. 371

73JOM(59)237 D. Seyferth and R. S. Marmor; *J. Organomet. Chem.*, 1973, **59**, 237. 266, 288

73JPR144 H. Hartmann and I. Reuther; *J. Prakt. Chem.*, 1973, **315**, 144. 572

73JPR471 K. Issleib and K.-D. Franze; *J. Prakt. Chem.*, 1973, **315**, 471. 511, 581

73JPR289 D. Martin, A. Berger and H. J. Niclas; *J. Prakt. Chem.*, 1973, **315**, 289. 624

73JPR640 H. G. O. Becker and V. Eisenschmidt; *J. Prakt. Chem.*, 1973, **315**, 640. 624

73LA40 E. Rabe and H.-W. Wanzlick; *Justus Liebigs Ann. Chem.*, 1973, 40. 55

B-73MI 601-01 R. D. Chambers; "Fluorine in Organic Chemistry," Wiley-Interscience, New York, 1973. 16

B-73MI 601-02 R. D. Chambers; "Fluorine in Organic Chemistry," Wiley-Interscience, New York, 1973, p. 163. 18

B-73MI 601-03 R. D. Chambers; "Fluorine in Organic Chemistry," Wiley-Interscience, New York, 1973, p. 30. 20

B-73MI 607-01 R. D. Chambers; "Fluorine in Organic Chemistry," ed. G. A. Olah, Wiley, New York, 1973. 212, 244

B-73MI 619-01 R. J. Shine; in "Organic Selenium Compounds: Their Chemistry and Biology," ed. D. L. Klayman and W. H. H. Günther, Wiley-Interscience, New York, 1973, p. 273. 587

B-73MI 619-02 K. A. Jensen; in "Organic Selenium Compounds: Their Chemistry and Biology," ed. D. L. Klayman and W. H. H. Günther, Wiley-Interscience, New York, 1973, p. 263. 587

B-73MI 619-03 K. A. Jensen; in "Organic Selenium Compounds: Their Chemistry and Biology," ed. D. L. Klayman and W. H. H. Günther, Wiley-Interscience, New York, 1973, p. 269. 590, 592

B-73MI 619-04 W. H. H. Günther; in "Organic Selenium Compounds: Their Chemistry and Biology," ed. D. L. Klayman and W. H. H. Günther, Wiley-Interscience, New York, 1973, p. 59. 590

B-73MI 619-05 K. A. Jensen; in "Organic Selenium Compounds: Their Chemistry and Biology," ed. D. L. Klayman and W. H. H. Günther, Wiley-Interscience, New York, 1973, p. 270. 591

B-73MI 619-06 R. J. Shine; in "Organic Selenium Compounds: Their Chemistry and Biology," ed. D. L. Klayman and W. H. H. Günther, Wiley-Interscience, New York, 1973, p. 281. 596

73OSC(4)801 R. G. Neville and J. J. McGee; *Org. Synth., Coll. Vol.*, 1973, **4**, 801. 572

73OSC(5)201 H. A. Staab and K. Wendel; *Org. Synth., Coll. Vol.*, 1973, **5**, 201. 500

73OSC(5)645 R. Degheneghi; *Org. Synth., Coll. Vol.*, 1973, **5**, 645. 505

73RCR587 V. V. Mozolis and S. P. Iokubaitite; *Russ. Chem. Rev. (Engl. Transl.)*, 1973, **42**, 587. 576

73RZC1243 G. A. Shvekhgeimer, N. I. Sobtsova and A. Baranski; *Rocz. Chem.*, 1973, **47**, 1243 (*Chem. Abstr.*, 1974, **80**, 27321). 149, 369

73S37 G. Zweifel, R. P. Fisher and A. Horng; *Synthesis*, 1973, 37. 198

73S207 H. Stetter and W. Uerdingen; *Synthesis*, 1973, 207. 80

73S423 W. Tritschler and S. Kabusz; *Synthesis*, 1973, 423. 105

73S547 P. Golborn; *Synthesis*, 1973, 547. 116

73S605 C. L. Coon and R. L. Simon; *Synthesis*, 1973, 605. 287

73SRI249 C. P. Casey, C. R. Cyr and R. A. Boggs; *Synth. React. Inorg. Metal.-Org. Chem.*, 1973, **3**, 249. 719

73T297 R. Fuks and W. A. Harteminck; *Tetrahedron*, 1973, **29**, 297. 607, 610

73T1697 I. Gosney and D. Lloyd; *Tetrahedron*, 1973, **29**, 1697. 708

73T3309 B. A. Phillips, G. Fodor, J. Gal, F. Letourneau and J. J. Ryan; *Tetrahedron*, 1973, **29**, 3309. 417, 418

73TL1219 U. Kraatz; *Tetrahedron Lett.*, 1973, 1219. 110

73TL1807 D. Thomas and D. H. Aue; *Tetrahedron Lett.*, 1973, 1807. 107

73TL2253 F. Mathey and J. Bensoam; *Tetrahedron Lett.*, 1973, 2253. 229, 230

73TL2287 M. Yoshimura, T. Namba and N. Tokura; *Tetrahedron Lett.*, 1973, 2287. 428

73TL3589 L. Thijs, A. Wagenaar, E. M. M. van Rens and B. Zwanenburg; *Tetrahedron Lett.*, 1973, 3589. 316

73TL3653 P. H. Benders; *Tetrahedron Lett.*, 1973, 3653. 642

73TL4193 H. Sakurai, Ken-ichiro Nishiwaki and M. Kira; *Tetrahedron Lett.*, 1973, 4193. 190, 194, 393

73TL4511 R. R. Koganty, M. B. Shambhu and G. A. Digenis; *Tetrahedron Lett.*, 1973, 4511. 57

73TL5147 H. Matsumoto, T. Nakano and Y. Nagai; *Tetrahedron Lett.*, 1973, 5147. 7

73USP3733360 W. C. Firth, Jr.; *US Pat.* 3733360 (1973) (*Chem. Abstr.*, 1973, **79**, 41892). 363

73USP3745220 R. C. Terrell and G. L. Moore; *US Pat.* 3 745 220 (1973) (*Chem. Abstr.*, 1973, **79**, 91 596). 74

73USP3755404 W. C. Firth, Jr. and S. Frank (American Cyanamid Co.); *US Pat.* 3755404 (1973) (*Chem. Abstr.*, 1974, **80**, 5389). 284, 362

73ZAAC(395)159 G. Fritz and P. Bottinger; *Z. Anorg. Allg. Chem.*, 1973, **395**, 159. 268

73ZAAC(399)1 K.-H. Worms and H. Kranz; *Z. Anorg. Allg. Chem.*, 1973, **399**, 1. 118

73ZAAC(399)280 G. Fritz and N. Braunagel; *Z. Anorg. Allg. Chem.*, 1973, **399**, 280. 268

73ZC58 G. Voss, E. Fischer and H. Werchan; *Z. Chem.*, 1973, **13**, 58. 650

73ZC192 B. Walther and W. Kolbe; *Z. Chem.*, 1973, **13**, 192. 645, 666

73ZC465 W. Schroth, H. Bahn and G. Huck; *Z. Chem.*, 1973, **13**, 465. 312

73ZOB544 V. A. Shokol, B. N. Kozhushko and A. V. Kirsanov; *Zh. Obshch. Khim.*, 1973, **43**, 544 (*Chem. Abstr.*, 1973, **79**, 42 617). 285, 367, 666

73ZOR39 V. P. Kukhar, V. I. Pasternak and G. V. Psotskaya; *Zh. Org. Khim.*, 1973, **9**, 39 (*Chem. Abstr.*, 1973, **78**, 84304). 610, 727

73ZOR43 V. P. Kukhar, A. M. Pinchuk and M. V. Shevchenko; *Zh. Org. Khim.*, 1973, **9**, 43 (*Chem. Abstr.*, 1973, **78**, 83996). 619

73ZOR269 K. V. Altukhov, E. V. Ratsino and V. V. Perekalin; *Zh. Org. Khim.*, 1973, **9**, 269 (*Chem. Abstr.*, 1973, **78**, 111180). 149

73ZOR633 L. N. Markovskii, Y. G. Shermolovich and V. I. Shevchenko; *Zh. Org. Khim.*, 1973, **9**, 633 (*Chem. Abstr.*, 1975, **82**, 111734). 621

73ZOR896 S. L. Ioffe, M. V. Kashuting, V. M. Shitkin, A. A. Levin and V. A. Tartakovskii; *Zh. Org. Khim.*, 1973, **9**, 896 (*Chem. Abstr.*, 1973, **79**, 53425). 149, 653

73ZOR1759 V. I. Markov and A. E. Polyakov; *Zh. Org. Khim.*, 1973, **9**, 1759 (*Chem. Abstr.*, 1973, **79**, 126175). 608

73ZOR1815 V. P. Kukhar, M. V. Shevchenko and N. A. Kirsanova; *Zh. Org. Khim.*, 1973, **9**, 1815 (*Chem. Abstr.*, 1973, **79**, 137 100). 658

73ZOR2013 A. Y. Zapevalov, I. P. Kolenko and V. S. Plashkin; *Zh. Org. Khim.*, 1973, **9**, 2013 (*Chem. Abstr.*, 1974, **80**, 47 722). 42

73ZOR2346 R. R. Kostikov, A. F. Khlebnikov and K. A. Ogloblin; *Zh. Org. Khim.*, 1973, **9**, 2346 (*Chem. Abstr.*, 1974, **80**, 36 488). 51

74AG(E)83 B. T. Gröbel and D. Seebach; *Angew. Chem., Int. Ed. Engl.*, 1974, **13**, 83. 175, 190, 192

74AG(E)540 W. Malisch and H. Schmidbaur; *Angew. Chem., Int. Ed. Engl.*, 1974, **13**, 540. 703

74AG(E)599 K. Bartel and W. Fehlhammer; *Angew. Chem., Int. Ed. Engl.*, 1974, **13**, 599. 732

74AP(307)7 G. Zinner and G. Isensee; *Arch. Pharm. (Weinheim, Ger.)*, 1974, **307**, 7 (*Chem. Abstr.*, 1974, **80**, 108 148). 505

74BCJ398 T. Nakai, K. Hiratani and M. Okawara; *Bull. Chem. Soc. Jpn.*, 1974, **47**, 398. 730

74BCJ935 K. Akiba, S. Matsunami, C. Eguchi and N. Inamoto; *Bull. Chem. Soc. Jpn.*, 1974, **47**, 935. 655

74BCJ1496 T. Sugawara, H. Iwamura and M. Oki; *Bull. Chem. Soc. Jpn.*, 1974, **47**, 1496. 99

74BCJ2457 M. Oki, T. Sugawara and H. Iwamura; *Bull. Chem. Soc. Jpn.*, 1974, **47**, 2457. 82

74BSF2072 J. F. Normant, O. Reboul, R. Sauvetre, H. Deshayes, D. Masure and J. Villieras; *Bull. Soc. Chim. Fr.*, 1974, 2072. 41

74CB698 R. Appel, K. Warning and K. D. Ziehn; *Chem. Ber.*, 1974, **107**, 698. 620

74CB1931 W. Schössler and M. Regitz; *Chem. Ber.*, 1974, **107**, 1931. 143

74CB3155 G. Scherowsky and J. Weiland; *Chem. Ber.*, 1974, **107**, 3155. 314

74CC166 J. Nakayama; *J. Chem. Soc., Chem. Commun.*, 1974, 166. 99

74CC323 C. J. Brookes, P. L. Coe, D. M. Owen, A. E. Pedler and J. C. Tatlow; *J. Chem. Soc., Chem. Commun.*, 1974, 323. 20

74CC634 J. G. Gourcy, G. Jeminet and J. Simonet; *J. Chem. Soc., Chem. Commun.*, 1974, 634. 302

74CC646 D. Grdenić, B. Kamenar, B. Korpar-Čolig, M. Sikirica and G. Jovanovski; *J. Chem. Soc., Chem. Commun.*, 1974, 646. 405

74CCC1330 A. Posta, O. Paleta, J. Volves and P. Trska; *Collect. Czech. Chem. Commun.*, 1974, **39**, 1330. 5

74CZ109 G. Dahms, G. Diderrich, A. Haas and M. Yazdanbakhsch; *Chem.-Ztg.*, 1974, **98**, 109. 252, 274

74CZ511 A. Haas, B. Koch and N. Welcman; *Chem.-Ztg.*, 1974, **98**, 511. 276

74GEP2237879 N. Schindler, W. Ploeger and G. Haeusler (Henkel und Cie. GmbH); *Ger. Pat.* 2237879 (1974) (*Chem. Abstr.*, 1974, **80**, 121 096). 367

74GEP2261108 G. Siegemund (Farbwerke Hoechst AG); *Ger. Pat.* 2261108 (1974) (*Chem. Abstr.*, 1974, **81**, 120008). 410, 419, 420

74GEP2311662 G. Büttner, E. Klauke and K. Sasse (Bayer AG); *Ger. Pat.* 2311662 (1974) (*Chem. Abstr.*, 1975, **82**, 4013). 439

74HOU(13/2b)1 H. Straub, K. P. Zeller and H. Leditschke; *Methoden Org. Chem. (Houben-Weyl)*, 1974, **13/2b**, 1. 172, 204, 207

74HOU(15/1)46 E. Wünsch; *Methoden Org. Chem. (Houben-Weyl)*, 1974, **15/1**, 46. 486

74HOU(15/2)187 P. Stelzel; *Methoden Org. Chem. (Houben-Weyl)*, 1974, **15/2**, 187. 486

74IC225 H. Alper and A. S. K. Chan; *Inorg. Chem.*, 1974, **13**, 225. 351

74IC921 D. J. Doonan and A. L. Balch; *Inorg. Chem.*, 1974, **13**, 921. 721

74IC1778 K. J. Klabunde, C. M. White and H. F. Efner; *Inorg. Chem.*, 1974, **13**, 1778. 529, 535

74IC2811 K. O. Christe, C. J. Schack and R. D. Wilson; *Inorg. Chem.*, 1974, **13**, 2811. 220

74IZV335 A. V. Fokin, Yu. N. Studnev, A. I. Rapkin and L. D. Kuznetsova; *Izv. Akad. Nauk SSSR Ser. Khim.*, 1974, **71**, 335 (*Chem. Abstr.*, 1974, **81**, 135378). 251

74IZV915 B. S. Fedorov, R. G. Gafurov and L. T. Eremenko; *Izv. Akad. Nauk. SSSR Ser. Khim.*, 1974, 915 (*Chem. Abstr.*, 1974, **81**, 24 945). 284, 364

74IZV1350 V. I. Erashko, A. V. Sultanov, S. A. Shevelev and A. A. Fainzil'berg; *Izv. Akad. Nauk SSSR Ser. Khim.* 1974, 1350 (*Chem. Abstr.*, 1974, **81**, 90 679). 283, 349

74IZV2144 A. M. Krzhizhevskii, Y. A. Cheburkov and I. L. Knunyants; *Izv. Akad. Nauk SSSR Ser. Khim.*, 1974, 2144 (*Chem. Abstr.*, 1975, **82**, 86 040). 54

74JA925 G. A. Olah, M. Nojima and I. Kerekes; *J. Am. Chem. Soc.*, 1974, **96**, 925. 52

74JA1770 G. H. Sprenger and J. M. Shreeve; *J. Am. Chem. Soc.*, 1974, **96**, 1770. 440

74JA3010 H. Taguchi, H. Yamamoto and H. Nozaki; *J. Am. Chem. Soc.*, 1974, **96**, 3010. 25, 31

74JA5640 R. B. Bates, W. A. Beavers, M. G. Greene and J. H. Klein; *J. Am. Chem. Soc.*, 1974, **96**, 5640. 206

74JA6237 D. Seyferth and J. L. Lefferts; *J. Am. Chem. Soc.*, 1974, **96**, 6237. 293, 399

74JCS(D)351 C. H. Game, M. Green, J. R. Moss and F. G. A. Stone; *J. Chem. Soc., Dalton Trans.*, 1974, 351. 719, 732

74JCS(D)357 C. H. Davies, C. H. Game, M. Green and F. G. A. Stone; *J. Chem. Soc., Dalton Trans.*, 1974, 357. 721

74JCS(D)665 T. Onak and E. Wan; *J. Chem. Soc., Dalton Trans.*, 1974, 665. 183

74JCS(D)1189 D. H. Bowen, M. Green, D. M. Grove, J. R. Moss and F. G. A. Stone; *J. Chem. Soc., Dalton Trans.*, 1974, 1189. 719

74JCS(D)1494 M. J. Doyle, M. F. Lappert, G. M. McLaughlin and J. McMeeking; *J. Chem. Soc., Dalton Trans.*, 1974, 1494. 720

74JCS(D)1591 B. Cetinkaya, M. F. Lappert, G. M. McLaughlin and K. Turner; *J. Chem. Soc., Dalton Trans.*, 1974, 1591. 721

74JCS(D)1827 B. Cetinkaya, P. Dixneuf and M. F. Lappert; *J. Chem. Soc., Dalton Trans.*, 1974, 1827. 721

74JCS(P1)22 J. P. Clayton, J. H. C. Nayler, M. J. Pearson and R. Southgate; *J. Chem. Soc., Perkin Trans. 1*, 1974, 22. 111

74JCS(P1)108 R. D. Chambers, R. P. Corbally, T. F. Holmes and W. K. R. Musgrave; *J. Chem. Soc., Perkin Trans. 1*, 1974, 108. 12

74JCS(P1)540 S. I. Mathew and F. Stansfield; *J. Chem. Soc., Perkin Trans. 1*, 1974, 540. 114

74JCS(P1)732 D. H. R. Barton, R. H. Hesse, M. M. Pechet and H. T. Toh; *J. Chem. Soc., Perkin Trans. 1*, 1974, 732. 231

74JCS(P1)1706 R. N. Haszeldine, B. Hewitson and A. E. Tipping; *J. Chem. Soc., Perkin Trans. 1*, 1974, 1706. 219

74JFC(4)107 R. E. A. Dear and E. E. Gilbert; *J. Fluorine Chem.*, 1974, **4**, 107. 46, 47

74JHC529 M. L. Graziano, R. Scarpati and E. Fattorusso; *J. Heterocycl. Chem.*, 1974, **11**, 529. 107

74JHC943 L. Costa, I. Degani, R. Fochi and P. Tundo; *J. Heterocycl. Chem.*, 1974, **11**, 943. 50
74JOC158 D. Seyferth, W. Tronich and H.-M. Shih; *J. Org. Chem.*, 1974, **39**, 158. 51
74JOC1166 H. Maehr and M. Leach; *J. Org. Chem.*, 1974, **39**, 1166. 650
74JOC1298 F. A. Hohorst, J. V. Paukstelis and D. D. DesMarteau; *J. Org. Chem.*, 1974, **39**, 1298. 42
74JOC2728 B. A. Reith, J. Strating and A. M. van Leusen; *J. Org. Chem.*, 1974, **39**, 2728. 699
74JOM(67)341 D. Seyferth, W. Tronich, W. E. Smith and S. P. Hopper; *J. Organomet. Chem.*, 1974, **67**, 341. 38
74JOM(69)45 D. S. Matteson, R. A. Davies and L. A. Hagelee; *J. Organomet. Chem.*, 1974, **69**, 45. 181, 199, 292, 386
74JOM(69)53 D. S. Matteson and P. B. Tripathy; *J. Organomet. Chem.*, 1974, **69**, 53. 182, 396
74JOM(69)63 D. S. Matteson and M. Furue; *J. Organomet. Chem.*, 1974, **69**, 63. 182, 183, 200
74JOM(71)335 D. Seyferth and R. A. Woodruff; *J. Organomet. Chem.*, 1974, **71**, 335. 258
74JOM(72)103 S. Sakai, H. Niimi and Y. Ishii; *J. Organomet. Chem.*, 1974, **72**, 103. 313
74JOM(76)C45 G. K. Barker and M. F. Lappert; *J. Organomet. Chem.*, 1974, **76**, C45. 195
74JOM(80)C35 D. F. Christian and W. R. Roper; *J. Organomet. Chem.*, 1974, **80**, C35. 719
74JOM(81)C7 D. F. Christian, G. R. Clark and W. R. Roper; *J. Organomet. Chem.*, 1974, **81**, C7. 719, 721
74JPR434 M. Gross, P. Held and H. Schubert; *J. Prakt. Chem.*, 1974, **316**, 434. 651
74JPR557 J. W. Grimm, K. C. Röber, G. Oehme, J. Alm, H. Mennenga and H. Pracejus; *J. Prakt. Chem.*, 1974, **316**, 557. 207
74JPR913 B. Mlotkowska, B. Costisella and H. Gross; *J. Prakt. Chem.*, 1974, **316**, 913. 116
74MIP 614-01 I. Szedlak; *Hung. Pat.* 7577 (1974) (*Chem. Abstr.*, 1974, **81**, 3389). 425
B-74MI 601-01 C. M. Starks; "Free Radical Telomerisation," Academic Press, New York, 1974, chaps 5 and 6. 5, 7
74MI 605-01 S. L. Ioffe, L. M. Leont'eva, O. P. Shitov, B. N. Khasapov and V. A. Tartakovskii; *Tezisy Vses. Soveshch. Khim. Nitrosoedinenii*, 1974, **5**, 49 (*Chem. Abstr.*, 1974, **87**, 23357). 169
B-74MI 605-02 G. Simchen; in "Methodicum Chemicum; Vol. 6 C-N Verbindungen," ed. F. Zymalkowski, Georg Thieme, Stuttgart, 1974, p. 765. 137
74MI 606-01 R. West; *Adv. Chem. Ser.*, 1974, **130**, 211 (*Chem. Abstr.*, 1974, **80**, 133505). 206
74MI 607-01 G. T. Austin; *Chem. Eng.*, 1974, **81**, 125 (*Chem. Abstr.*, 1974, **81**, 119292). 214, 215
B-74MI 607-02 G. Kirschstein, D. Koschel and S. Ruprecht; "Gmelin Handbook of Inorganic Chemistry," Kohlenstoff, Teil D2, ed. D. Koschel, Springer-Verlag, Berlin, 1974. 213
74MI 621-01 S. E. Suvorova, Y. M. Kulikov and V. V. Kozlov; *Tezisy Vses. Simp. Org. Sint.: Benzoidnye Aromat. Soedin., 1st*, 1974, 61 (*Chem. Abstr.*, 1977, **87**, 5539). 653
74MI 621-02 A. V. Sultanov, V. I. Erashko, S. A. Shevelev and A. A. Fainzil'berg; *Tezisy Vses. Soveshch. Khim. Nitrosoedinenii, 5th*, 1974, 4 (*Chem. Abstr.*, 1977, **87**, 22647). 653
74NKK732 M. Hidai, K. Ishimi, M. Iwase and Y. Uchida; *Nippon Kagaku Kaishi* (*J. Chem. Soc. Jpn.*) 1974, 732 (*Chem. Abstr.*, 1974, **81**, 37635). 456
74OMR(6)494 C. K. Tseng and A. Mihailovski; *Org. Magn. Reson.*, 1974, **6**, 494. 656
74PS(4)241 H. Gross, B. Costisella and L. Brennecke; *Phosphorus Sulfur*, 1974, **4**, 241. 159
74S26 H. Mägerlein, G. Meyer and H. D. Rupp; *Synthesis*, 1974, 26. 529
74S153 R. H. DeWolfe; *Synthesis*, 1974, 153. 68, 69, 70, 78
74S274 M. Makosza and F. Fedorynski; *Synthesis*, 1974, 274. 233
74S289 J. B. Chattopadhaya and A. V. R. Rav; *Synthesis*, 1974, 289. 572
74S654 G. A. Olah and J. Welch; *Synthesis*, 1974, 654. 423
74T1769 J. E. Richman and H. E. Simmons; *Tetrahedron*, 1974, **30**, 1769. 365
74TCC43 G. Fritz; *Top. Curr. Chem.*, 1974, **50**, 43. 178, 382
74TL803 K. Kondo, N. Sonoda and H. Sakurai; *Tetrahedron Lett.*, 1974, 803. 466
74TL1687 J. Geevers and W. P. Trompen; *Tetrahedron Lett.*, 1974, 1687. 609
74TL1691 J. Geevers and W. P. Trompen; *Tetrahedron Lett.*, 1974, 1691. 609
74TL3765 G. Leclerc, B. Rouot and C. G. Wermuth; *Tetrahedron Lett.*, 1974, 3765. 619
74TL4409 T. H. Chan, J. F. Harrod and P. van Gheluwe; *Tetrahedron Lett.*, 1974, 4409. 296, 297
74TL4479 K. Harano and T. Taguchi; *Tetrahedron Lett.*, 1974, 4479. 472
74URP424809 E. A. Manuilova, L. Sokolov and S. V. Sokolov; *USSR Pat.* 424809 (1974) (*Chem. Abstr.*, 1974, **81**, 123727). 410
74USP3849102 H. C. Bucha and W. P. Langsdorf, Jr.; *US. Pat.* 3 849 102 (1974), (*Chem. Abstr.*, 1974, **82**, 107 540). 497, 513
74YGK936 E. Ichikawa, T. Takizawa and K. Odo; *Yuki Gosei Kagaku Kyokaishi*, 1974, **32**, 936 (*Chem. Abstr.*, 1975, **82**, 112043). 503
74ZAAC(404)1 G. Fritz and G. Marquardt; *Z. Anorg. Allg. Chem.*, 1974, **404**, 1. 383
74ZAAC(404)103 G. Fritz and E. Bosch; *Z. Anorg. Allg. Chem.* 1974, **404**, 103. 267
74ZAAC(404)204 G. Schott and K. Golz; *Z. Anorg. Allg. Chem.*, 1974, **404**, 204. 164
74ZOB538 Z. M. Ivanova, N. I. Gusar and Y. G. Gololobov; *Zh. Obshch. Khim.*, 1974, **44**, 538 (*Chem. Abstr.*, 1974, **80**, 133 551). 656
74ZOB1264 I. M. Kosinskaya, N. P. Pisanenko and V. I. Shevchenko; *Zh. Obshch. Khim.*, 1974, **44**, 1264 (*Chem. Abstr.*, 1974, **81**, 49 759). 656
74ZOR36 V. P. Kukhar and N. G. Pavlenko; *Zh. Org. Khim.*, 1974, **10**, 36 (*Chem. Abstr.*, 1974, **80**, 120187). 619
74ZOR42 E. A. Abrazhanova, L. F. Pronskii and V. N. Sevast'yanov; *Zh. Org. Khim.*, 1974, **10**, 42 (*Chem. Abstr.*, 1974, **80**, 108 144). 651
74ZOR449 V. P. Kukhar, V. I. Pasternak, M. I. Povolotskii and N. G. Pavlenko; *Zh. Org. Khim.*, 1974, **10**, 449 (*Chem. Abstr.*, 1974, **80**, 145300). 612, 727
74ZOR1000 L. N. Markovskii, Y. G. Shermolovich, Y. A. Nuzhdina and V. I. Shevchenko; *Zh. Org. Khim.*, 1974, **10**, 1000 (*Chem. Abstr.*, 1974, **81**, 49238). 621

74ZOR1793 S. A. Shevelev, R. V. Kolesnikov and I. P. Beletskaya; *Zh. Org. Khim.*, 1974, **10**, 1793 (*Chem. Abstr.*, 1975, **82**, 4386). 368

74ZOR2003 A. L. Fridman, V. D. Surkov, F. A. Gabitov, S. I. Gushchin, S. A. Petukhov and F. M. Mukhametshin; *Zh. Org. Khim.*, 1974, **10**, 2003 (*Chem. Abstr.*, 1974, **81**, 135355). 145

74ZOR2031 V. V. Berenblit, Y. P. Dolnakov, V. P. Sass, L. N. Senyushov and S. V. Sokolov; *Zh. Org. Khim.*, 1974, **10**, 2031 (*Chem. Abstr.*, 1975, **82**, 43287). 36

75ACR62 H. Schmidbaur; *Acc. Chem. Res.*, 1975, **8**, 62. 702

75ACR165 F. Minisci; *Acc. Chem. Res.*, 1975, **8**, 165. 7

75AF1477 J. B. Bream, H. Lauener, C. W. Picard, G. Scholtysik and T. G. White; *Arzneim-Forsch.*, 1975, **25**, 1477. 646, 647

75AG37 D. Seebach and R. Buerstinghaus; *Angew. Chem.*, 1975, **87**, 37. 135, 327

75AG449 D. Hoppe; *Angew. Chem.*, 1975, **87**, 449. 326

75AP(308)379 R. Neidlein and R. Bottler; *Arch. Pharm.* (*Weinheim, Ger.*), 1975, **308**, 379 (*Chem. Abstr.*, 1975, **83**, 58372). 617

75BCJ108 K. Kondo, N. Sonoda and H. Sakurai; *Bull. Chem. Soc. Jpn.*, 1975, **48**, 108. 465, 481

75BCJ1030 M. Fukuda, Y. Okamoto and H. Sakurai; *Bull. Chem. Soc. Jpn.*, 1975, **48**, 1030. 158

75BCJ2496 M. Oki, T. Endo and T. Sugawara; *Bull. Chem. Soc. Jpn.*, 1975, **48**, 2496. 82

75BCJ3691 Y. Yamamoto and H. Yamazaki; *Bull. Chem. Soc. Jpn.*, 1975, **48**, 3691. 719

75C209 P. George and H. G. Viehe; *Chimia*, 1975, **29**, 209. 619, 728

75C514 J. Baum, D. Scholz, F. Tataruch and H. G. Viehe; *Chimia*, 1975, **29**, 514. 644

75CB215 J. Hocker and R. Merten; *Chem. Ber.*, 1975, **108**, 215. 141

75CB1142 G. Schwenker and R. Kolb; *Chem. Ber.*, 1975, **108**, 1142. 143

75CB2290 W. Reid, W. Merkel and R. Schweitzer; *Chem. Ber.*, 1975, **108**, 2290. 616, 617

75CC216 A.-R. B. Manas and R. A. J. Smith; *J. Chem. Soc., Chem. Commun.*, 1975, 216. 88, 317

75CC302 L. G. Sneddon and R. J. Lagow; *J. Chem. Soc., Chem. Commun.*, 1975, 302. 205

75CC877 J. Klein and A. Medlik-Balan; *J. Chem. Soc., Chem. Commun.*, 1975, 877. 206

75CC929 A. J. Hartshorn, M. F. Lappert and K. Turner; *J. Chem. Soc., Chem. Commun.*, 1975, 929. 721

75CC960 C. U. Pittman, Jr. and M. Narita; *J. Chem. Soc., Chem. Commun.*, 1975, 960. 121, 698

75CJC529 R. N. Renaud and P. J. Champagne; *Can. J. Chem.*, 1975, **53**, 529. 20

75CJC2664 W. R. Hardstaff, R. F. Langler, J. Leahy and M. J. Newman; *Can. J. Chem.*, 1975, **53**, 2664. 254, 255

75CR(C)563 A. Etienne, G. Lonchambon and J. Roques; *C. R. Hebd. Seances Acad. Sci., Ser. C.*, 1975, **281**, 563. 142

75CZ245 G. Zinner and J. Schmidt; *Chem.-Ztg.*, 1975, **99**, 245. 505

75DOK(221)1331 R. A. Bekker, G. G. Melikyan, B. L. Dyatkin, and I. L. Knunyants; *Dokl. Akad. Nauk SSSR*, 1975, **221**, 1331 (*Chem. Abstr.*, 1975, **83**, 114585). 65

75FRP2235188 S. Lecolier, D. Boutte and J. C. Mondet (Soc. Nat. des Poudres et Explosifs); *Fr. Pat.* 2235188 (1975) (*Chem. Abstr.*, 1975, **83**, 208310). 297

75GEP2351556 H. G. Schlee, K. Sasse and E. Kühle (Bayer AG); *Ger. Pat.* 2351556 (1975) (*Chem. Abstr.*, 1975, **83**, 58466). 453

75GEP2435407 C. L. McIntosh; *Ger. Pat.* 2435407 (1975) (*Chem. Abstr.*, 1975, **83**, 73487). 497

75HOU(13/7)1 G. Bähr and E. Langer; *Methoden Org. Chem.* (*Houben-Weyl*), 1975, **13/7**, 1. 172

75IC592 R. L. Kirchmeier, U. I. Lasouris and J. M. Shreeve; *Inorg. Chem.*, 1975, **14**, 592. 277

75IC1223 K. E. Peterman and J. M. Shreeve; *Inorg. Chem.*, 1975, **14**, 1223. 51, 52

75IC1513 E. D. Dobrzynski and R. A. Angelici; *Inorg. Chem.*, 1975, **14**, 1513. 719, 732

75IC2148 H. Kurosawa; *Inorg. Chem.*, 1975, **14**, 2148. 517

75IZV1669 V. I. Erashko, A. V. Sultanov and S. A. Shevelev; *Izv. Akad. Nauk SSSR Ser. Khim.*, 1975, 1669 (*Chem. Abstr.*, 1975, **83**, 147250). 348, 349, 653

75IZV2279 A. F. Gontar, E. G. Bykhovskaya and I. L. Knunyants; *Izv. Akad. Nauk SSSR Ser. Khim.*, 1975, 2279. 238, 242

75IZV2621 S. A. Shevelev, V. V. Semenov and A. A. Fainzil'berg; *Izv. Akad. Nauk SSSR Ser. Khim.*, 1975, 2621 (*Chem. Abstr.*, 1976, **84**, 58540). 323

75IZV2750 A. V. Sultanov, V. I. Erashko, S. A. Shevelev and A. A. Fainzil'berg; *Izv. Akad. Nauk SSSR Ser. Khim.*, 1975, 2750. 281

75IZV2838 V. I. Markovskii, G. A. Marchenko, S. S. Novikov and V. I. Slovetskii; *Izv. Akad. Nauk. SSSR Ser. Khim.*, 1975, 2838 (*Chem. Abstr.*, 1976, **84**, 73609). 284, 364, 369

75JA13 N. Walker and D. D. DesMarteau; *J. Am. Chem. Soc.*, 1975, **97**, 13. 42, 231

75JA187 D. A. Hatzenbühler, L. Andrews and F. A. Carey; *J. Am. Chem. Soc.*, 1975, **97**, 187. 244

75JA5608 D. S. Matteson, R. J. Moody and P. K. Jesthi; *J. Am. Chem. Soc.*, 1975, **97**, 5608. 181, 199

75JCS(D)148 M. P. Brown, A. K. Holliday and G. M. Way; *J. Chem. Soc., Dalton Trans.*, 1975, 148. 183

75JCS(D)488 C. D. Desjardins, H. L. Paige, J. Passmore and P. Taylor; *J. Chem. Soc., Dalton Trans.*, 1975, 488. 49, 50

75JCS(D)939 M. Green, F. G. A. Stone and M. Underhill; *J. Chem. Soc., Dalton Trans.*, 1975, 939. 721, 732

75JCS(D)1434 A. M. Devine, P. A. Griffin, R. N. Haszeldine, M. J. Newlands and A. E. Tipping; *J. Chem. Soc., Dalton Trans.*, 1975, 1434. 191

75JCS(D)2225 R. N. Haszeldine, D. J. Rogers and A. E. Tipping; *J. Chem. Soc., Dalton Trans.*, 1975, 2225. 241

75JCS(P1)212 R. J. Hayward and O. Meth-Cohn; *J. Chem. Soc., Perkin Trans. 1*, 1975, 212. 507, 577

75JCS(P1)270 R. Okazaki, F. Ishii, K. Okawa, K. Ozawa and N. Inamoto; *J. Chem. Soc., Perkin Trans. 1*, 1975, 270. 304

75JCS(P1)1302 B. T. Golding and D. R. Hall; *J. Chem. Soc., Perkin Trans. 1*, 1975, 1302. 105

75JCS(P1)1574 D. H. R. Barton and S. W. McCombie; *J. Chem. Soc., Perkin Trans. 1*, 1975, 1574. 557

75JCS(P1)1836 R. Banks, R. F. Brookes and D. H. Godson; *J. Chem. Soc., Perkin Trans. 1*, 1975, 1836. 445

75JCS(P1)2033 R. E. Banks, J. M. Birchall, A. K. Brown, R. N. Haszeldine and F. Moss; *J. Chem. Soc., Perkin Trans 1*, 1975, 2033. 241

75JCS(P2)916 A. Tangerman and B. Zwanenburg; *J. Chem. Soc., Perkin Trans. 2*, 1975, 916. 536

75JCS(P2)1046 A. F. Hegarty and K. J. Dignam; *J. Chem. Soc., Perkin Trans. 2*, 1975, 1046. 605, 616, 617

75JFC(5)25 M. S. Toy and R. S. Stringham; *J. Fluorine Chem.*, 1975, **5**, 25. 42

75JFC(5)475 D. D. Denson, G. J. Moore and C. Tamborski; *J. Fluorine Chem.*, 1975, **5**, 475. 63

75JFC(6)37 M. Brandwood, P. L. Coe and J. C. Tatlow; *J. Fluorine Chem.*, 1975, **6**, 37. 43

75JMC90 T. Jen, H. Van Hoeven, W. Groves, R. A. McLean and B. Loev; *J. Med. Chem.*, 1975, **18**, 90. 570, 645

75JMC913 S. Sharma, R. Bindra, R. N. Iyer and N. Anand; *J. Med. Chem.*, 1975, **18**, 913. 577

75JOC746 K. Bechgaard, D. O. Cowan, A. N. Bloch and L. Henriksen; *J. Org. Chem.*, 1975, **40**, 746. 591

75JOC963 M. Hojo and R. Masuda; *J. Org. Chem.*, 1975, **40**, 963. 84, 89

75JOC1278 B. B. Jarvis and H. E. Fried; *J. Org. Chem.*, 1975, **40**, 1278. 254

75JOC2570 H. J. Reich; *J. Org. Chem.*, 1975, **40**, 2570. 136

75JOC2577 M. G. Miles, J. S. Wager, J. D. Wilson and A. R. Siedle; *J. Org. Chem.*, 1975, **40**, 2577. 120

75JOC2796 D. J. Burton and P. E. Greenlimb; *J. Org. Chem.*, 1975, **40**, 2796. 695

75JOC2911 F. Terrier, A.-P. Chatrousse, C. Paulmier and R. Schaal; *J. Org. Chem.*, 1975, **40**, 2911. 97, 100

75JOC3052 P. Beak, J. Yamamoto and C. J. Upton; *J. Org. Chem.*, 1975, **40**, 3052. 555

75JOC3271 J. L. Adcock, R. A. Beh and R. J. Lagow; *J. Org. Chem.*, 1975, **40**, 3271. 37

75JOM(88)C35 G. Dousse, H. Lavayssiere and J. Satgé; *J. Organomet. Chem.*, 1975, **88**, C35. 577

75JOM(90)C34 K. R. Grundy, R. O. Harris and W. R. Roper; *J. Organomet. Chem.*, 1975, **90**, C34. 720, 732

75JOM(91)C61 K. R. Grundy and W. R. Roper; *J. Organomet. Chem.*, 1975, **91**, C61. 721

75JOM(92)163 J. A. Morrison and J. M. Bellama; *J. Organomet. Chem.*, 1975, **92**, 163. 173

75JOM(93)21 D. S. Matteson and L. A. Hagelee; *J. Organomet. Chem.*, 1975, **93**, 21. 181, 182, 292

75JOM(93)339 P. Krommes and J. Lorberth; *J. Organomet. Chem.*, 1975, **93**, 339. 668

75JOM(93)43 G. Deleris, J. Dunogués and R. Calas; *J. Organomet. Chem.*, 1975, **93**, 43. 27

75JOM(94)229 E. Lindner; *J. Organomet. Chem.*, 1975, **94**, 229. 327, 330

75JOM(94)333 H. Sawai and T. Takizawa; *J. Organomet. Chem.*, 1975, **94**, 333. 644

75JOM(100)139 M. F. Lappert; *J. Organomet. Chem.*, 1975, **100**, 139. 721

75JOM(101)C1 B. Crociani, M. Nicolini and R. L. Richards; *J. Organomet. Chem.*, 1975, **101**, C1. 622

75JOM(102)109 V. Bätzel, U. Müller and R. Allman; *J. Organomet. Chem.*, 1975, **102**, 109. 340

75JOM(102)423 C. Furet, C. Servens and M. Pereye; *J. Organomet. Chem.*, 1975, **102**, 423. 26, 31

75LA195 E. Rabe and H.-W. Wanzlick; *Justus Liebigs Ann. Chem.*, 1975, 195. 55

75LA958 W. Reid and L. Kaiser; *Justus Liebigs Ann. Chem.*, 1975, 958. 111

75LA1025 D. Herrmann and D. Kleune; *Justus Liebigs Ann. Chem.*, 1975, 1025. 537

75LA1339 R. Köster, H.-J. Horstschäfer, P. Binger and P. K. Mattschei; *Justus Liebigs Ann. Chem.*, 1975, 1339. 182, 183, 386

75LA2227 R. Banholzer, A. Heusner and W. Schulz; *Justus Liebigs Ann. Chem.*, 1975, 2227. 444

75MI 608-01 R. L. Kirchmeier, G. H. Sprenger and J. M. Shreeve; *Inorg. Nucl. Chem. Lett.*, 1975, **11**, 699 (*Chem. Abstr.*, 1975, **83**, 178188). 261

75MI 611-01 V. Konecny; *Chem. Zvesti*, 1975, **29**, 811 (*Chem. Abstr.*, 1976, **85**, 155046). 335

B-75MI 614-01 W. Kwasnik; in "Handbuch der Präparativen Anorganischen Chemie," ed. G. Brauer, Enke, Stuttgart, 1975. 409, 419, 420

75MI 621-01 V. V. Paramanov, V. P. Petrosyan and V. I. Slovetskii; *Nov. Polyarogr., Tezisy Dokl. Vses. Soveshch. Polyarogr., 6th, 25. Ed. J. Stradins. 'Zinatne': Riga*, USSR., 1975, 72 (*Chem. Abstr.*, 1977, **86**, 23495). 650

75MI 621-02 G. Cauquis, B. Divisia and J. Ulrich; *Colloq. Spectrosc. Int.*, [*Proc.*], *18th, Volume 3, 1000-3. G.A.M.S.: Paris, Fr.*, 1975, 22 (*Chem. Abstr.*, 1977, **87**, 200326). 666

75OMS(10)18 D. R. Dimmel, C. A. Wilke and P. J. Lamothe; *Org. Mass Spectrom.*, 1975, **10**, 18. 208, 405

75OMS(10)1021 G. Cauquis, B. Divisia and J. Ulrich; *Org. Mass Spectrom.* 1975, **10**, 1021. 666

75RTC1 F. C. V. Larsson and S.-O. Lawesson; *Recl. Trav. Chim. Pays-Bas*, 1975, **94**, 1. 303, 306

75RTC209 R. J. Broekema; *Recl. Trav. Chim. Pays-Bas*, 1975, **94**, 209. 74

75S38 J. Nakayama; *Synthesis*, 1975, 38. 99

75S147 D. S. Matteson; *Synthesis*, 1975, 147. 181, 182, 198

75S272 S. Kabusz, W. Tritschler and A. Lienemann; *Synthesis*, 1975, 272. 105

75S384 N. Yamazaki, T. Tomioka and F. Higashi; *Synthesis*, 1975, 384. 575

75S436 J. Nakayama; *Synthesis*, 1975, 436. 82

75S527 J. W. Scheeren and P. M. op den Brouw; *Synthesis*, 1975, 527. 39

75S535 P. Savignac, J. Petrova, M. Dreux and P. Coutrot; *Synthesis*, 1975, 535. 266

75S598 L. Wackerle and I. Ugi; *Synthesis*, 1975, 598. 422

75S660 S. Cabiddu, S. Melis, L. Bonsignore and M. T. Cocco; *Synthesis*, 1975, 660. 95

75S721 L. M. Yagupol'skii, N. V. Kondratenko and V. P. Sambur; *Synthesis*, 1975, 721. 232, 233, 234

75S785 W. J. Stec, K. Lesiak and M. Sudol; *Synthesis*, 1975, 785. 122

75SC47 J. P. Kutney and A. H. Ratcliffe; *Synth. Commun.*, 1975, 47. 416

75TL899 H. Matsumoto, T. Nikaido and Y. Nagai; *Tetrahedron Lett.*, 1975, 899. 7

75TL1259 E. M. Engler and V. V. Patel; *Tetrahedron Lett.*, 1975, 1259. 93, 101, 729

75TL2087 P. Koch; *Tetrahedron Lett.*, 1975, 2087. 495

75TL3789 D. G. Naae, H. S. Kesling and D. J. Burton; *Tetrahedron Lett.*, 1975, 3789. 695

75USP3883377 C. M. Wright; *US Pat.* 3883377 (1975) (*Chem. Abstr.*, 1975, **83**, 149779). 151

75USP3884985 K. P. Hoffman (Stauffer Chemical Co.); *US Pat.* 3884985 (1975) (*Chem. Abstr.*, 1975, **83**, 82222). 215

75USP3892676 E. G. Budnick (Plains Chemical Development Co.); *US Pat.* 3892676 (1975) (*Chem. Abstr.*, 1975, **83**, 117635r). 155, 366

75USP3914265 W. J. Middleton; *US Pat.* 3914265 (1975) (*Chem. Abstr.*, 1976, **84**, 42635). 10

75USP3968158 Rohm and Haas Co.; *US Pat.* 3 968 158 (1976) (*Chem. Abstr.*, 1976, **85**, 159 473). 649

75YGK136 T. Suyama and K. Odo; *Yuki Gosei Kagaku Kyokai*, 1975, **33**, 136 (*Chem. Abstr.*, 1975, **83**, 78 529). 335

75ZAAC(415)233 F. Weller; *Z. Anorg. Allg. Chem.*, 1975, **415**, 233. 207

75ZAAC(418)109 M. Schmeisser, W. Heuser and E. Danzmann; *Z. Anorg. Allg. Chem.*, 1975, **418**, 109. 218, 220, 224

75ZAAC(418)208 M. Schmidt and E. Weissflog; *Z. Anorg. Allg. Chem.*, 1975, **418**, 208. 127, 132, 134

75ZN(B)965 G. Fritz and M. Portner; *Z. Naturforsch., Teil B*, 1975, **30**, 965. 268

75ZOB1450 A. N. Pudovik, V. I. Nikitina, M. G. Zimin and N. L. Vostretsova; *Zh. Obshch. Khim.*, 1975, **45**, 1450 (*Chem. Abstr.*, 1975, **83**, 179217). 160

75ZOB1484 V. V. Alekseev and M. S. Malinovskii; *Zh. Obshch. Khim.*, 1975, **45**, 1484 (*Chem. Abstr.*, 1975, **83**, 179 221). 581

75ZOB1965 V. A. Shokol, B. N. Kozhushko and A. V. Gumenyuk; *Zh. Obshch. Khim.*, 1975, **45**, 1965 (*Chem. Abstr.*, 1976, **84**, 17 498). 286, 666

75ZOB2093 V. A. Shokol, B. N. Kozhushko and A. M. Kolesnikov; *Zh. Obshch. Khim.*, 1975, **45**, 2093 (*Chem. Abstr.*, 1976, **84**, 17 499). 367

75ZOR71 V. P. Kukhar and M. V. Shevchenko; *Zh. Org. Khim.*, 1975, **11**, 71 (*Chem. Abstr.*, 1975, **83**, 9 148). 619

75ZOR762 E. M. Dorokhova, E. S. Levchenko and N. P. Pel'kis; *Zh. Org. Khim.*, 1975, **11**, 762 (*Chem. Abstr.*, 1975, **83**, 42 988). 656

75ZOR1628 V. L. Dubina, V. P. Pedan, N. I. Levchenko, E. A. Klimkovich, M. I. Shashalova and L. I. Pribor; *Zh. Org. Khim.*, 1975, **11**, 1628 (*Chem. Abstr.*, 1975, **83**, 192765). 629

75ZOR1672 V. M. Belous, L. A. Alekseeva and L. M. Yagupol'skii; *Zh. Org. Khim.*, 1975, **11**, 1672 (*Chem. Abstr.*, 1975, **83**, 178 464). 39

75ZOR2223 V. P. Kukhar and V. I. Pasternak; *Zh. Org. Khim.*, 1975, **11**, 2223 (*Chem. Abstr.*, 1976, **84**, 30308). 625

75ZOR2243 E. A. Abrazhanova, V. N. Sevast'yanov and L. F. Pronskii; *Zh. Org. Khim.*, 1975, **11**, 2243 (*Chem. Abstr.*, 1976, **84**, 58866). 619

76ACR13 D. Griller and K. U. Ingold; *Acc. Chem. Res.*, 1976, **9**, 13. 726, 734

76ACS(B)837 P. Wolkoff and S. Hammerum; *Acta Chem. Scand., Ser. B*, 1976, **30**, 837. 111, 114, 115

76AG(E)616 E. O. Fischer, W. Kleine and F. R. Kreissl; *Angew. Chem., Int. Ed. Engl.*, 1976, **15**, 616. 723

76AG(E)681 H. Eckert and I. Ugi; *Angew. Chem., Int. Ed. Engl.*, 1976, **15**, 681. 487

76AOC(14)1 E. O. Fischer; *Adv. Organomet. Chem.*, 1976, **14**, 1. 716, 719

76BCJ553 K. Arai and M. Oki; *Bull. Chem. Soc. Jpn.*, 1976, **49**, 553. 48, 82

76BCJ1888 T. Abe, S. Nagase and H. Baba; *Bull. Chem. Soc. Jpn.*, 1976, **49**, 1888. 36

76BCJ1996 C. U. Pittman, Jr. and M. Narita; *Bull. Chem. Soc. Jpn.*, 1976, **49**, 1996. 121

76BCJ3567 J. Nakayama, K. Fujiwara and M. Hoshino; *Bull. Chem. Soc. Jpn.*, 1976, **49**, 3567. 113

76BSF501 C. P. Pinazzi, J.-C. Rabadeux and A. Pleurdeau; *Bull. Soc. Chim. Fr.*, 1976, 501. 475

76CA(85)142668 H. Hagemann, H. Schwarz and F. Doering (Bayer AG); *DOS* 2503168 (1976) (*Chem. Abstr.*, 1976, **85**, 142668). 240

76CAR(46)9 J. M. J. Tronchet, D. Schwarzenbach and F. Barbalat-Ray; *Carbohydr. Res.* 1976, **46**, 9. 695

76CAR(51)C13 T. Ogawa and M. Matsui; *Carbohydr. Res.*, 1976, **51**, C13. 71

76CB810 R. Appel and K. Giesen; *Chem. Ber.*, 1976, **109**, 810. 618

76CB1069 S. Holm, J. A. Boerma, N. H. Nilsson and A. Senning; *Chem. Ber.*, 1976, **109**, 1069. 274, 275, 309

76CB1239 M. Schmidt and E. Weissflog; *Chem. Ber.*, 1976, **109**, 1239. 132

76CB1643 W. Ried and R. Schweitzer; *Chem. Ber.*, 1976, **109**, 1643. 620

76CB1976 A. Haas and M. Yazdanbakhsch; *Chem. Ber.*, 1976, **109**, 1976. 274, 275, 307, 534

76CB2351 M. Kamel, W. Kimpenhaus and J. Buddrus; *Chem. Ber.*, 1976, **109**, 2351. 38

76CB2475 A. Haas and V. Hellwig; *Chem. Ber.*, 1976, **109**, 2475. 433

76CB2921 W. Ried, N. Kothe, R. Schweitzer and A. Höhle; *Chem. Ber.*, 1976, **109**, 2921. 620

76CB3339 V. Bätzel and G. Schmid; *Chem. Ber.*, 1976, **109**, 3339. 341

76CB3432 G. Diderrich and A. Haas; *Chem. Ber.*, 1976, **109**, 3432. 236, 237, 254, 274, 532, 536

76CC148 E. M. Engler, D. C. Green and J. Q. Chambers; *J. Chem. Soc., Chem. Commun.*, 1976, 148. 590

76CC261 D. E. Goldberg, D. H. Harris, M. F. Lappert and K. M. Thomas; *J. Chem. Soc., Chem. Commun.*, 1976, 261. 186

76CC513 M. J. Hopkinson, H. W. Kroto, J. F. Nixon and N. P. C. Simmons; *J. Chem. Soc., Chem. Commun.*, 1976, 513. 679

76CC644 P. B. Hitchcock, M. F. Lappert and P. L. Pye; *J. Chem. Soc., Chem. Commun.*, 1976, 644. 721

76CI(L)164 M. M. El-Abadelah; *Chem. Ind. (London)*, 1976, 164. 611

76CL43 W. D. Woessner; *Chem. Lett.*, 1976, 43. 88

76CL891 S. Sakai, T. Fujinami, K. Kosugi and K. Matsnaga; *Chem. Lett.*, 1976, 891. 73

76CPB26 T. Hirayama, M. Kamada, H. Tsurumi and M. Mimura; *Chem. Pharm. Bull.*, 1976, **24**, 26. 623

76CS122 K. Olsson, L. Adolfsson and R. Andersson; *Chem. Scr.*, 1976, **10**, 122. 85, 89

76CZ391 M. Fild and H. P. Rieck; *Chem.-Ztg.*, 1976, **100**, 391. 255

76GEP2451635 G. Beck, H. Heitzer and H. Holtschmidt (Bayer AG); *Ger. Pat.* 2451635 (1976); (*Chem. Abstr.*, 1976, **85**, 46648). 617

76GEP2460330 K. Lohmer, A. Von Koenig, S. Schuetz and J. Stoltefuss (Agfa-Gevaert AG); *Ger. Pat.* 2460330 (1976) (*Chem. Abstr.*, 1977, **86**, 148 775). 360

76GEP2553147 C. L. McIntosh; *Ger. Pat.*, 2 553 147 (1976) (*Chem. Abstr.*, 1976, **85**, 192 885). 513

76IC14 S.-L. Yu and J. M. Shreeve; *Inorg. Chem.*, 1976, **15**, 14. 603, 607

76IC1666 T. A. Ford, M. Huber, W. K. Klotzbücher, M. Moskovits and G. A. Ozin; *Inorg. Chem.*, 1976, **15**, 1666. 522

76IC2346 M. Wada and K. Oguro; *Inorg. Chem.*, 1976, **15**, 2346. 517

76IZV209 A. F. Gontar, E. G. Bykhovskaya and I. L. Knunyants; *Izv. Akad. Nauk SSSR Ser. Khim.*, 1976, 209 (*Chem. Abstr.*, 1976, **84**, 150134). 239, 440

76IZV212 A. F. Gontar, E. G. Bykhovskaya and I. L. Knunyants; *Izv. Akad. Nauk SSSR Ser. Khim.*, 1976, 212 (*Chem. Abstr.*, 1976, **84**, 150135). 615

76IZV455 E. S. Batyeva, Y. N. Girfanova, G. U. Zamaletdinova and A. N. Pudovik; *Izv. Akad. Nauk SSSR Ser. Khim.*, 1976, 455 (*Chem. Abstr.*, 1976, **84**, 180344). 582

76IZV489 A. V. Fokin, Y. N. Studnev, A. I. Rapkin, L. D. Kuznetsova and V. A. Komarov; *Izv. Akad. Nauk SSSR Ser. Khim.*, 1976, 489 (*Chem. Abstr.*, 1976, **85**, 20539). 56, 282

76IZV858 D. P. Deltosova and N. P. Gambaryan; *Izv. Akad. Nauk SSSR Ser. Khim.*, 1976, 858 (*Chem. Abstr.*, 1976, **85**, 123732). 106

76IZV2381 A. F. Gontar, E. G. Bykhovskaya, Y. A. Sizov and I. L. Knunyants; *Izv. Akad. Nauk SSSR Ser. Khim.*, 1976, 2381 (*Chem. Abstr.*, 1977, **86**, 54921). 614

76IZV2547 S. L. Ioffe, A. S. Shashkov, A. L. Blyumenfel'd, L. M. Leont'eva, L. M. Makarenkova, O. B. Belkina and V. A. Tartakovskii; *Izv. Akad. Nauk SSSR Ser. Khim.*, 1976, 2547 (*Chem. Abstr.*, 1977, **86**, 121426). 65

76IZV2640 V. V. Semenov, S. A. Shevelev and A. A. Fainzil'berg; *Izv. Akad. Nauk SSSR Ser. Khim.*, 1976, 2640 (*Chem. Abstr.*, 1977, **86**, 139517). 323

76JA1798 H. C. Brown and N. Ravindran; *J. Am. Chem. Soc.*, 1976, **98**, 1798. 181

76JA3167 D. McIntosh and G. A. Ozin; *J. Am. Chem. Soc.*, 1976, **98**, 3167. 525

76JA3529 E. R. Falardeau and D. D. DesMarteau; *J. Am. Chem. Soc.*, 1976, **98**, 3529. 256

76JA4678 J. R. Anglin and W. A. G. Graham; *J. Am. Chem. Soc.*, 1976, **98**, 4678. 520

76JA5419 J. B. Collins, J. D. Dill, E. D. Jemmis, Y. Apeloig, P. von R. Schleyer, R. Seeger and J. A. Pople; *J. Am. Chem. Soc.*, 1976, **98**, 5419. 386, 401, 404

76JA5702 B. F. Spielvogel, L. Wojnowich, M. K. Das, A. T. McPhail and K. D. Hargrave; *J. Am. Chem. Soc.*, 1976, **98**, 5702. 515

76JA6545 C. A. Burton and J. M. Shreeve; *J. Am. Chem. Soc.*, 1976, **98**, 6545. 432

76JA8426 W. Priester and R. West; *J. Am. Chem. Soc.*, 1976, **98**, 8426. 206

76JAP7630034 S. Nagase, H. Baba, K. Kodaira and T. Abe (Agency of Industrial Sciences & Technology); *Jpn. Pat.* 7630034 (1976) (*Chem. Abstr.*, 1977, **86**, 48582). 410

76JAP(K)76108012 Y. Omura, S. Aihara, F. Mori, Y. Tamai, T. Hosogai, Y. Fujita, F. Wada, T. Nishida and K. Itoi; *Jpn. Kokai*, 76108012 (1976) (*Chem. Abstr.*, 1977, **86**, 120800). 70

76JCS(D)2268 P. J. Davidson, D. H. Harris and M. F. Lappert; *J. Chem. Soc., Dalton Trans.*, 1976, 2268. 174, 186, 190, 191, 195

76JCS(P1)1048 P. H. J. Ooms, J. W. Scheeren and R. J. F. Nivard; *J. Chem. Soc., Perkin Trans. 1*, 1976, 1048. 75

76JCS(P1)1178 R. N. Haszeldine, B. Hewitson and A. E. Tipping; *J. Chem. Soc., Perkin Trans. 1*, 1976, 1178. 46, 47

76JCS(P1)2112 D. H. R. Barton, R. V. Stick and R. Subramanian; *J. Chem. Soc., Perkin. Trans. 1*, 1976, 2112. 309

76JFC(7)261 C. Lau and J. Passmore; *J. Fluorine Chem.*, 1976, **7**, 261. 49

76JFC(7)501 M. J. Hopkinson and D. D. Desmarteau; *J. Fluorine Chem.*, 1976, **7**, 501. 272

76JFC(8)177 D. Naumann and J. Baumanns; *J. Fluorine Chem.*, 1976, **8**, 177. 220

76JGU2441 L. B. Ionov, A. P. Korovyakov and S. S. Molodtsov; *J. Gen. Chem. USSR* (*Engl. Transl.*), 1976, **46**, 2441. 476

76JHC205 M. L. Graziano and R. Scarpati; *J. Heterocycl. Chem.*, 1976, **13**, 205. 107

76JOC626 F. J. Goodman and J. O. Chambers; *J. Org. Chem.*, 1976, **41**, 626. 555

76JOC813 K. T. Potts, J. Baum, S. K. Datta and E. Houghton; *J. Org. Chem.*, 1976, **41**, 813. 115

76JOC1112 D. F. Sullivan, D. I. C. Scopes, A. F. Kluge and J. A. Edwards; *J. Org. Chem.*, 1976, **41**, 1112. 113

76JOC1285 V. L. Heasley, D. R. Titterington and T. L. Rold; *J. Org. Chem.*, 1976, **41**, 1285. 54

76JOC2070 K. Kurita, T. Matsumura and Y. Iwakura; *J. Org. Chem.*, 1976, **41**, 2070. 462

76JOC3253 S. D. Ziman; *J. Org. Chem.*, 1976, **41**, 3253. 650

76JOC3794 H. Yamanaka, J. Kikui and K. Termura; *J. Org. Chem.*, 1976, **41**, 3794. 51

76JOC3975 F. A. Carey, O. D. Dailey, Jr., O. Hernandez and J. R. Tucker; *J. Org. Chem.*, 1976, **41**, 3975. 128

76JOM(105)C19 W. Petz; *J. Organomet. Chem.*, 1976, **105**, C19. 337

76JOM(107)C37 T. J. Collins, K. R. Grundy, W. R. Roper and S. F. Wong; *J. Organomet. Chem.*, 1976, **107**, C37. 720, 732

76JOM(110)15 K.-D. Müller and U. W. Gerwarth; *J. Organomet. Chem.*, 1976, **110**, 15. 572

76JOM(110)25 D. S. Matteson and P. K. Jesthi; *J. Organomet. Chem.*, 1976, **110**, 25. 181, 186, 199, 200

76JOM(112)227 D. F. Christian, H. C. Clark and R. F. Stepaniak; *J. Organomet. Chem.*, 1976, **112**, 227. 719

76JOM(113)C31 E. O. Fischer, H. Hollfelder, F. R. Kreissl and W. Uedelhoven; *J. Organomet. Chem.*, 1976, **113**, C31. 720, 722

76JOM(113)C45 K. R. Grundy and W. R. Roper; *J. Organomet. Chem.*, 1976, **113**, C45. 719

76JOM(113)67 G. Schmid, V. Bätzel and B. Stutte; *J. Organomet. Chem.*, 1976, **113**, 67. 340

76JOM(114)1 D. S. Matteson and P. K. Jesthi; *J. Organomet. Chem.*, 1976, **114**, 1. 199, 201

76JOM(114)281 N. Petragnani, R. Rodrigues and J. V. Comasseto; *J. Organomet. Chem.*, 1976, **114**, 281. 697

76JOM(116)257 D. Seyferth and J. L. Lefferts; *J. Organomet. Chem.*, 1976, **116**, 257. 380, 381, 388, 399

76JOM(118)C33 E. O. Fischer, K. Scherzer and F. R. Kreissl; *J. Organomet. Chem.*, 1976, **118**, C33. 720

76JOM(122)171 H. G. Kuivila and F. V. DiStefano; *J. Organomet. Chem.*, 1976, **122**, 171. 405

76JPR116 H. Gross, B. Costisella, Th. Gnauk and L. Brennecke; *J. Prakt. Chem.*, 1976, **318**, 116. 286

76JPR127 E. Fanghänel; *J. Prakt. Chem.*, 1976, **318**, 127. 113, 304, 306, 316

76JPR272 H. Gross, L. Brennecke and B. Costisella; *J. Prakt. Chem.*, 1976, **318**, 272. 159

76KFZ35 A. L. Fridman, V. S. Zalesov, L. O. Kon'shina, N. A. Kobolov, K. V. Dolbilkin, I. K. Moiseev, T. A. Mratkhuzina and P. G. Belyaev; *Khim. Farm. Zh.*, 1976, **10**, 35 (*Chem. Abstr.*, 1976, **85**, 46054). 145

B-76MI 601-01 M. Hudlicky; "Chemistry of Organic Fluorine Compounds," Ellis Horwood, Chichester, 1976, chap. 5. 5

B-76MI 601-02 M. Hudlicky; "Chemistry of Organic Fluorine Compounds," Ellis Horwood, Chichester, 1976, p. 685. 9

B-76MI 601-03 M. Hudlicky; "Chemistry of Organic Fluorine Compounds," Ellis Horwood, Chichester, 1976. 16

B-76MI 602-01 C. Jutz; in "Advances in Organic Chemistry," eds. H. Böhme and H. G. Viehe, Wiley, New York, 1976, vol 9, part 1, p. 225. 51

B-76MI 620-01 Z. Janousek and H. G. Viehe; in "Iminium Salts in Organic Chemistry," ed. H. Böhme and H. G. Viehe, Wiley, New York, 1976, part 1, p. 343. 609, 610, 614, 619, 620

76MI 621-01 M. Hartmann and G. Geschwend; *Faserforsch. Textiltech.*, 1976, **27**, 475 (*Chem. Abstr.*, 1976, **85**, 193 217). 650

B-76MI 623-01 Z. Janousek and H. G. Viehe; in "Iminium Salts in Organic Chemistry" Part 1, ed. H. Böhme and H. G. Viehe, Wiley, New York, 1976, p. 343. 727, 728, 729

76S408 G. Ferdinand and K. Schank; *Synthesis*, 1976, 408. 480

76S449 G. Descotes and A. Faure; *Synthesis*, 1976, 449. 309

76S551 H. Yoshida, M. Yoshikane, T. Ogata and S. Inokawa; *Synthesis*, 1976, 551. 87, 99

76S798 F. De Angelis, A. Gambacorta and R. Nicoletti; *Synthesis*, 1976, 798. 141

76T1839 J. Klein, A. Medlik-Balan, A. Y. Meyer and M. Chorev; *Tetrahedron*, 1976, **32**, 1839. 206

76T2295 M. Corallo and Y, Pietrasanta; *Tetrahedron*, 1976, **32**, 2295. 7

76T2625 D. W. Nagel, K. G. R. Pachler, P. S. Steyn, R. Vleggaar and P. L. Wessels; *Tetrahedron*, 1976, **32**, 2625. 138

76TL477 M. Mikolajczyk, B. Costisella, S. Grzejszczak and A. Zatorski; *Tetrahedron Lett.*, 1976, 477. 119

76TL1591 I. Kuwajima, T. Sato, N. Minami and T. Abe; *Tetrahedron Lett.*, 1976, 1591. 125

76TL2737 S. L. Hartzell and M. W. Rathke; *Tetrahedron Lett.*, 1976, 2737. 190

76TL3563 H. Gotthardt and S. Nieberl; *Tetrahedron Lett.*, 1976, 3563. 114

76TL3695 K. Ishikawa, K. Akiba and N. Inamoto; *Tetrahedron Lett.*, 1976, 3695. 120, 698

76USP397691 W. J. Middleton; *US Pat.* 3976691 (1976) (*Chem. Abstr.*, 1977, **86**, 29054) 10

76ZAAC(419)2 G. Fritz and H. J. Vogt; *Z. Anorg. Allg. Chem.*, 1976, **419**, 2. 379

76ZAAC(419)213 G. Fritz and I. Arnason; *Z. Anorg. Allg. Chem.*, 1976, **419**, 213. 267

76ZAAC(427)114 A. Haas, B. Koch and N. Welcman; *Z. Anorg. Allg. Chem.*, 1976, **427**, 114. 256, 536, 537

76ZC377 G. Kauschka and L. Kolditz; *Z. Chem.*, 1976, **16**, 377. 223, 228

76ZN(B)342 V. Bätzel; *Z. Naturforsch., Teil B*, 1976, **31**, 342. 340

76ZN(B)721 M. S. Hussain and H. Schmidbaur; *Z. Naturforsch., Teil B*, 1976, **31**, 721. 370

76ZN(B)1317 N. Wiberg and G. Hübler; *Z. Naturforsch., Teil. B*, 1976, **31**, 1317. 608

76ZOR558 V. N. Neplyuev, T. A. Sinenko and P. S. Pel'kis; *Zh. Org. Khim.*, 1976, **12**, 558. 275

76ZOR767 V. V. Berenblit, V. P. Sass, L. N. Senyushov and Yu. K. Starobin; *Zh. Org. Khim.*, 1976, **12**, 767 (*Chem. Abstr.*, 1976, **85**, 5135). 251

76ZOR1287 L. M. Yagupol'skii, V. M. Belous and L. A. Alekseeva; *Zh. Org. Khim.*, 1976, **12**, 1287 (*Chem. Abstr.*, 1976, **85**, 77 797). 39, 43

76ZOR1377 R. A. Bekker, G. G. Melikyan, B. L. Dyatkin and I. L. Knunyants; *Zh. Org. Khim.*, 1976, **12**, 1377 (*Chem. Abstr.*, 1976, **85**, 159 779). 65

76ZOR1798 V. M. Belous, K. D. Litvinova, L. A. Alekseeva and L. M. Yagupol'skii; *Zh. Org. Khim.*, 1976, **12**, 1798 (*Chem. Abstr.*, 1976, **85**, 159 593). 39

76ZOR2213 Y. L. Yagupol'skii, B. K. Kerzhner and L. M. Yagupol'skii; *Zh. Org. Khim.*, 1976, **12**, 2213 (*Chem. Abstr.*, 1977, **86**, 71 870). 52

76ZOR2583 V. M. Neplyuev and T. A. Sinenko; *Zh. Org. Khim.*, 1976, **12**, 2583 (*Chem. Abstr.*, 1977, **86**, 120923). 275, 277, 316

77ACH77 Z. Bende, I. Bitter and Z. Csürös; *Acta Chim. Acad. Sci. Hung.*, 1977, **93**, 77. 625

77ACS(B)589 J. Sire; *Acta Chem. Scand., Ser. B*, 1977, **31**, 589. 653

77AG325 K. Seppelt; *Angew. Chem.*, 1977, **89**, 325. 228

77AG(E)259 G. Skorna and I. Ugi; *Angew. Chem., Int. Ed. Engl.*, 1977, **16**, 259. 462

77AG(E)328 N. Wiberg and G. Preiner; *Angew. Chem., Int. Ed. Engl.*, 1977, **16**, 328. 392, 709, 710

77AG(E)462 D. Hoppe and R. Follmann; *Angew. Chem., Int. Ed. Engl.*, 1977, **16**, 462. 496

77AG(E)743 H. Hagemann; *Angew. Chem., Int. Ed. Engl.*, 1977, **16**, 743. 451

77AG(E)854 P. Gölitz and A. de Meijere; *Angew. Chem., Int. Ed. Engl.*, 1977, **16**, 854. 20

77AJC455 J. A. MacLaren; *Aust. J. Chem.*, 1977, **30**, 455. 507

77AP820 R. Neidlein and U. Askani; *Arch. Pharm.* (*Weinheim, Ger.*), 1977, **310**, 820. 651

77BCJ718 M. Itoh, D. Hagiwana and T. Kamiya; *Bull. Chem. Soc. Jpn.*, 1977, **50**, 718. 429

77BCJ3271 S. Sakai, H. Niimi, Y. Kobayashi and Y. Ishii; *Bull. Chem. Soc. Jpn.*, 1977, **50**, 3271. 624

77BEP856233 H. Boehm and K. H. Hellberg; *Belg. Pat.* 856233 (1977) (*Chem. Abstr.*, 1978, **88**, 190061). 219

77BSB663 N. De Kimpe, R. Verhe, L. De Buyck and N. Schamp; *Bull. Soc. Chim. Belg.*, 1977, **86**, 663. 509

77CB37 M. Reiffen and R. W. Hoffmann; *Chem. Ber.*, 1977, **110**, 37. 630

77CB67 A. Haas and U. Niemann; *Chem. Ber.*, 1977, **110**, 67. 233, 432

77CB677 H. Schmidbaur, J. Eberlein and W. Richter; *Chem. Ber.*, 1977, **110**, 677. 702, 707, 708

77CB841 R. Buerstinghaus and D. Seebach; *Chem. Ber.*, 1977, **110**, 841. 128, 135, 326, 327

77CB852 B.-T. Gröbel and D. Seebach; *Chem. Ber.*, 1977, **110**, 852. 175, 190, 334, 351, 390

77CB916 G. Diderrich, A. Haas and M. Yazdanbakhsch; *Chem. Ber.*, 1977, **110**, 916. 235, 254, 303

77CB2368 R. Appel and M. Montenarh; *Chem. Ber.*, 1977, **110**, 2368. 512

77CB2574 E. O. Fischer, T. Selmayr, F. R. Kreissl and U. Schubert; *Chem. Ber.*, 1977, **110**, 2574. 719

77CB2880 R. Schlecker, U. Henkel and D. Seebach; *Chem. Ber.*, 1977, **110**, 2880. 317, 734

77CB3467 E. O. Fischer, H. Hollfelder, P. Friedrich, F. R. Kreissl and G. Huttner; *Chem. Ber.*, 1977, **110**, 3467. 720, 722, 723

77CC259 G. E. Gerhardt and R. J. Lagow; *J. Chem. Soc., Chem. Commun.*, 1977, 259. 36

77CC505 P. Shu, A. N. Block, T. F. Carruthers and D. O. Cowan; *J. Chem. Soc., Chem. Commun.*, 1977, 505. 591

77CC648 D. Mansuy, M. Lange, J. C. Chottard, P. Guerin, P. Morliere, D. Brault and M. Rougee; *J. Chem. Soc., Chem. Commun.*, 1977, 648. 717

77CC660 W. P. Krug, A. N. Bloch and D. O. Cowan; *J. Chem. Soc., Chem. Commun.*, 1977, 660. 552

77CI(L)127 R. Louw and P. W. Franken; *Chem. Ind.* (*London*), 1977, 127. 229, 230

77CJC3136 C. D. Desjardins and J. Passmore; *Can. J. Chem.*, 1977, **55**, 3136. 49

77CR(C)321 A. Etienne, G. Lonchambon and P. Giraudeau; *C. R. Hebd. Seances Acad. Sci., Ser. C.*, 1977, **285**, 321. 143

77FCR39 O. Paleta; *Fluorine Chem. Rev.*, 1977, **8**, 39. 4

77GEP2547512 Bayer AG; *Ger. Pat.* 2 547 512 (1977) (*Chem. Abstr.* 1977, **87**, 135 913). 657

77GEP2638754 T. J. Hernandez; *Ger. Pat.*, 2 638 754 (1977) (*Chem. Abstr.*, 1977, **86**, 166 392). 513

77GEP2708024 S. Tocker (Du Pont De Nemours, E. I., and Co.); *Ger. Pat.* 2708024 (1977) (*Chem. Abstr.*, 1977, **87**, 201 591).

77IC822 R. Busby, W. Klotzbücher and G. A. Ozin; *Inorg. Chem.*, 1977, **16**, 822. 522

77IC2974 L. A. Shimp and R. J. Lagow; *Inorg. Chem.*, 1977, **16**, 2974. 46, 252

77IZV136 V. I. Erashko, S. A. Shevelev and A. A. Fainzil'berg; *Izv. Akad. Nauk SSSR Ser. Khim.*, 1977, 136 (*Chem. Abstr.*, 1977, **86**, 170770). 145

77IZV139 S. A. Shevelev, V. V. Semenov and A. A. Fainzil'berg; *Izv. Akad. Nauk SSSR Ser. Khim.*, 1977, 139 (*Chem. Abstr.*, 1977, **86**, 170 789). 56, 323

77IZV363 Y.-P. Yang and A. L. Lapidus; *Izv. Akad. Nauk. SSSR Ser. Khim.*, 1977, 363 (*Chem. Abstr.*, 1977, **86**, 171 868). 465

77IZV375 V. I. Erashko, A. V. Sultanov, S. A. Shevelev and A. A. Fainzil'berg; *Izv. Akad. Nauk SSSR Ser. Khim.*, 1977, 375. 283

77IZV1177 A. N. Pudovik, E. S. Batyeva, E. N. Ofitserov, O. G. Sinyashin and N. V. Ivasyuk; *Izv. Akad. Nauk SSSR Ser. Khim.*, 1977, 1177 (*Chem. Abstr.*, 1977, **87**, 102 413). 583

77IZV1563 G. A. Marchenko, V. I. Markovskii, V. I. Slovetskii and S. S. Novikov; *Izv. Akad. Nauk. SSSR Ser. Khim.*, 1977, 1563 (*Chem. Abstr.*, 1977, **87**, 201 669). 364, 368

77IZV1965 K. V. Titova, E. I. Kolmakova and V. Ya. Rosolovskii; *Izv. Akad. Nauk. SSSR Ser. Khim.*, 1977, 1965 (*Chem. Abstr.*, 1977, **87**, 193 036). 369

77IZV2058 A. L. Fridman and S. S. Novikov; *Izv. Akad. Nauk SSSR Ser. Khim.*, 1977, 2058 (*Chem. Abstr.*, 1978, **88**, 22 043). 56, 282

77IZV2379 A. F. Gontar, N. A. Til'kunova, E. G. Bykhovskaya and I. L. Knunyants; *Izv. Akad. Nauk SSSR, Ser. Khim.*, 1977, 2379 (*Chem. Abstr.*, 1978, **88**, 23064). 616

77IZV2530 S. A. Shevelev, V. V. Semenov and A. A. Fainzil'berg; *Izv. Akad. Nauk SSSR Ser. Khim.*, 1977, 2530. 283

77JA1267 P. A. Bartlett and K. P. Long; *J. Am. Chem. Soc.*, 1977, **99**, 1267. 666

77JA4194 T. Kitazume and J. M. Shreeve; *J. Am. Chem. Soc.*, 1977, **99**, 4194. 253

77JA5209 D. Seyferth and C. L. Nivert; *J. Am. Chem. Soc.*, 1977, **99**, 5209. 401, 402, 404

77JA5909 E. M. Engler, B. A. Scott, S. Etemad, T. Penney and V. V. Patel; *J. Am. Chem. Soc.*, 1977, **99**, 5909. 590

77JA6645 J. L. Atwood, G. K. Barker, J. Holton, W. E. Hunter, M. F. Lappert and R. Pearce; *J. Am. Chem. Soc.*, 1977, **99**, 6645. 195

77JA7367 S. Wratten and D. J. Faulkner; *J. Am. Chem. Soc.*, 1977, **99**, 7367. 604

77JCR(S)116 F. Glockling, N. S. Hosmane, V. B. Mahale, J. J. Swindall, L. Magos and T. J. King; *J. Chem. Res.* (*S*), 1977, 116. 394

77JCS(D)136 I. Rayment and H. M. M. Shearer; *J. Chem. Soc., Dalton Trans.*, 1977, 136. 183

77JCS(D)1272 M. F. Lappert, P. L. Pye and G. M. McLaughlin; *J. Chem. Soc., Dalton Trans.*, 1977, 1272. 720, 721

77JCS(D)1283 M. F. Lappert and P. L. Pye; *J. Chem. Soc., Dalton Trans.*, 1977, 1283. 721

77JCS(D)2015 M. Behnam-Dahkordy, B. Crociani and R. L. Richards; *J. Chem. Soc., Dalton Trans.*, 1977, 2015. 622

77JCS(D)2160 P. B. Hitchcock, M. F. Lappert and P. L. Pye; *J. Chem. Soc., Dalton Trans.*, 1977, 2160. 721

77JCS(D)2172 M. F. Lappert and P. L. Pye; *J. Chem. Soc., Dalton Trans.*, 1977, 2172. 721

77JCS(P1)1162 A. Banerji, J. C. Cass and A. R. Katritzky; *J. Chem. Soc., Perkin Trans. 1*, 1977, 1162. 143

77JCS(P1)1898 J. A. Miller and D. Stewart; *J. Chem. Soc., Perkin Trans. 1*, 1977, 1898. 118

77JCS(P1)2263 J. I. Grayson and S. Warren; *J. Chem. Soc., Perkin Trans. 1*, 1977, 2263. 121

77JFC(9)279 K. Omori, S. Nagase, H. Baba, K. Kodaira and T. Abe; *J. Fluorine Chem.*, 1977, **9**, 279. 51

77JFC(9)505 T. I. Savchenko, T. D. Petrova, V. E. Platonov and G. G. Yakobson; *J. Fluorine Chem.*, 1977, **9**, 505. 609

77JFC(10)131 M. H. J. Van Hamme and D. J. Burton; *J. Fluorine Chem.*, 1977, **10**, 131. 695

77JHC881 K. Takahashi and K. Mitsuhashi; *J. Heterocycl. Chem.*, 1977, **14**, 881. 28

77JMC901 G. J. Durant, J. C. Emmett, C. R. Ganellin, P. D. Miles, M. E. Parsons, H. D. Prain and G. R. White; *J. Med. Chem.*, 1977, **20**, 901. 647

77JOC565 C. Wakselman and T. Nguyen; *J. Org. Chem.*, 1977, **42**, 565. 63

77JOC679 G. B. Butler, K. H. Herring, P. L. Lewis, V. V. Sharpe, III and R. L. Veazey; *J. Org. Chem.*, 1977, **42**, 679. 76

77JOC1543 K. Hirai, H. Sugimoto and T. Ishiba; *J. Org. Chem.*, 1977, **42**, 1543. 85

77JOC1667 P. J. Stang and D. P. Fox; *J. Org. Chem.*, 1977, **42**, 1667. 127

77JOC1773 H. J. Reich and S. K. Shah; *J. Org. Chem.*, 1977, **42**, 1773. 130

77JOC2024 R. M. DeMarinis and W. M. Bryan; *J. Org. Chem.*, 1977, **42**, 2024. 234

77JOC2038 D. F. Sullivan, R. P. Woodbury and M. W. Rathke; *J. Org. Chem.*, 1977, **42**, 2038. 190

77JOC3128 H. W. Scheeren, R. W. M. Aben, P. H. J. Ooms and R. J. F. Nivard; *J. Org. Chem.*, 1977, **42**, 3128. 75

77JOC3220 W. T. Brady and R. A. Owens; *J. Org. Chem.*, 1977, **42**, 3220. 622

77JOC3608 J. F. Stearns and H. Rapoport; *J. Org. Chem.*, 1977, **42**, 3608. 644, 647

77JOC3629 M. Mikolajczyk, P. Kielbasiński and Z. Goszczyńska; *J. Org. Chem.*, 1977, **42**, 3629. 635

77JOC3686 R. L. Stanley and D. S. Tarbell; *J. Org. Chem.*, 1977, **42**, 3686. 434

77JOC3875 J. B. Hendrickson and K. W. Bair; *J. Org. Chem.*, 1977, **42**, 3875. 235

77JOM(128)C49 E. O. Fischer, W. Kleine, F. R. Kreissl, H. Fischer, P. Friedrich and G. Hüttner; *J. Organomet. Chem.*, 1977, **128**, C49. 723

77JOM(129)197 E. O. Fischer, P. Stückler and F. R. Kreissl; *J. Organomet. Chem.*, 1977, **129**, 197. 722

77JOM(132)359 E. Glozbach and J. Lorberth; *J. Organomet. Chem.*, 1977, **132**, 359. 668, 670, 671

77JOM(140)97 I. Ojima and S.-I. Inaba; *J. Organomet. Chem.*, 1977, **140**, 97. 515

77JOM(141)35 C. Couret, J. Satge and J.-P. Picard; *J. Organomet. Chem.*, 1977, **141**, 35. 131

77JOM(142)155 O. A. Kruglaya, I. B. Fedot'eva, B. V. Fedot'ev, I. D. Kalikhman, E. I. Brodskaya and N. S. Vyazankin; *J. Organomet. Chem.*, 1977, **142**, 155. 185

77JOM(142)39 D. Seyferth, J. L. Lefferts and R. L. Lambert, Jr.; *J. Organomet. Chem.*, 1977, **142**, 39. 380, 391

77JPR17 B. Mlotkowska, H. Gross, B. Costisella, M. Mkolajczyk, S. Grzejszczak and A. Zatorski; *J. Prakt. Chem.*, 1977, **319**, 17. 48

77MIP53536 Yeda Research and Development Co. Ltd.; *Israeli Pat.* 53 536 (1977) (*Chem. Abstr.*, 1981, **97**, 127 026). 537

77MI 603-01 S. Tanimoto, T. Miyake and M. Okano; *Bull. Inst. Chem. Res. Kyoto Univ.*, 1977, **55**, 276 (*Chem. Abstr.*, 1978, **88**, 62 315). 95

B-77MI 606-01 D. S. Matteson; in "Gmelin Handbuch der Anorganischen Chemie," eds K. Niedenzu and K.-C. Buschbeck, Springer, Berlin, 1977, vol. 48, part 16, p. 37. 181

B-77MI 606-02 M. Fieser and L. F. Fieser; "Reagents for Organic Synthesis," Wiley, New York, 1977, vol. 6, p. 329. 199

77MI 607-01 N. I. Nikolaev; *Zh. Fiz. Khim.*, 1977, **51**, 2136 (*Chem. Abstr.*, 1977, **87**, 183809). 223

77MI 614-01 D. Roeda and B. Van Zanten; *J. Labelled Compd. Radiopharm.*, 1977, **13**, 284 (*Chem. Abstr.*, 1977, **87**, 61929). 414

B-77MI 615-01 R. Wegler; "Chemie der Pflanzenschutz- und Schädlingsbekämpfungsmittel," Springer, Berlin, Heidelberg, New York, 1977, vol. 5, pp. 21, 411. 492

77MI 615-02 W. H. Daly, C. G. Overberger, C.-D. S. Lee and T. J. Pacansky; *Macromol. Synth.*, 1977, **6**, 23. 476

B-77MI 617-01 G. Gattow and W. Behrendt; "Carbon Sulfides and Their Inorganic and Complex Chemistry," Georg Thième Verlag, Stuttgart, 1977. 560

B-77MI 620-01 H. Ulrich and R. Richter; in "The Chemistry of Cyanates and Their Thio Derivatives," ed. S. Patai, Wiley, Chichester, 1977, p. 969. 602, 608

77MI 620-02 A. V. Eremeev, V. G. Andrianov and I. P. Piskunova; *Latv. PSR Zinat. Akad. Vestis, Kim. Ser.*, 1977, 371 (*Chem. Abstr.*, 1977, **87**, 133848). 625

77MI 622-01 M. F. Lappert and P. L. Pye; *J. Less-Common Met.*, 1977, **54**, 191 (*Chem. Abstr.*, 1977, **87**, 201682u). 721

77OS(56)8 P. Stütz and P. A. Stadler; *Org. Synth.*, 1977, **56**, 8. 86

77S73 W. Kantlehner, B. Funke, E. Haug, P. Speh, L. Kienitz and T. Maier; *Synthesis*, 1977, 73. 296, 297, 298, 300, 313, 320, 321

77S181 S. E. Dinizo and D. S. Watt; *Synthesis*, 1977, 181. 119

77S699 R. Appel and W. Morbach; *Synthesis*, 1977, 699. 265

77S704 I. Butula and L. Curkovic; *Synthesis*, 1977, 704. 444

77S873 I. Degani, R. Fochi and M. Santi; *Synthesis*, 1977, 873. 545

77TL399 R. Appel and V. Veltmann; *Tetrahedron Lett.*, 1977, 399. 695

77TL885 A. van der Gen, C. G. Kruse, N. L. J. M. Broekhof and A. Wijsman; *Tetrahedron Lett.*, 1977, 885. 48, 120

77TL1065 J. A. Miller and D. Stewart; *Tetrahedron Lett.*, 1977, 1065. 118

77TL1239 W. G. Salmond; *Tetrahedron Lett.*, 1977, 1239. 695

77TL2987 M. J. Gallagher and H. Honegger; *Tetrahedron Lett.*, 1977, 2987. 116

77TL4315 B. Boutevin and E. B. Dongala; *Tetrahedron Lett.*, 1977, **49**, 4315. 7

77UK685 V. G. Granik, A. M. Zhidkova and R. G. Glushkov; *Usp. Khim*, 1977, **46**, 685 (*Chem. Abstr.*, 1977, **87**, 21 650). 104

77USP4020091 E. G. Budnick; *US Pat.* 4020091 (1977) (*Chem. Abstr.*, 1977, **87**, 135912). 151, 155

77USP4021459 M. J. Green; *US Pat.* 4 021 459 (1977) (*Chem. Abstr.*, 1977, **87**, 102 537). 80

77USP4031228 C. C. Beard (Syntex (USA) Inc.); *US Pat.* 4031228 (1977) (*Chem. Abstr.*, 1977, **87**, 102333). 314

77USP4058549 T. D. J. D'Silva (Union Carbide Corp.); *US Pat.* 4058549 (1977) (*Chem. Abstr.*, 1978, **88**, 50329). 439

77ZAAC(430)121 G. Fritz and U. Finke; *Z. Anorg. Allg. Chem.*, 1977, **430**, 121. 178, 290, 393

77ZAAC(430)137 G. Fritz and U. Finke; *Z. Anorg. Allg. Chem.*, 1977, **430**, 137. 385, 392

77ZAAC(434)110 G. Gattow and K. Klaeser; *Z. Anorg. Allg. Chem.*, 1977, **434**, 110. 628

77ZAAC(434)115 G. Gattow and K. Klaeser; *Z. Anorg. Allg. Chem.*, 1977, **434**, 115. 627

77ZC172 W. Schroth, R. Spitzner and J. Herrmann; *Z. Chem.*, 1977, **17**, 172 (*Chem. Abstr.*, 1977, **87**, 102276). 616

77ZC411 W. Schroth, M. Hassfeld, W. Schiedewitz and C. Pfotenhauer; *Z. Chem.*, 1977, **17**, 411. 478

77ZN(B)858 H. Schmidbaur, W. Scharf and H.-J. Füller; *Z. Naturforsch., Teil B*, 1977, **32**, 858. 706

77ZN(B)1003 N. Wiberg and G. Hübler; *Z. Naturforsch., Teil. B*, 1977, **32**, 1003. 617

77ZN(B)1022 D. Breitinger and W. Morell; *Z. Naturforsch., Teil B*, 1977, **32**, 1022. 405

77ZOB321 V. A. Shokol and B. N. Kozhushko; *Zh. Obshch. Khim.*, 1977, **47**, 321 (*Chem. Abstr.*, 1977, **86**, 190108). 160

77ZOB1063 V. D. Sheludyakov, A. S. Tkachev, S. V. Sheludyakova and V. F. Mironov; *Zh. Obshch. Khim.*, 1977, **47**, 1063. 445

77ZOB2390 O. I. Kolodyazhnyi; *Zh. Obshch. Khim.*, 1977, **47**, 2390 (*Chem. Abstr.*, 1978, **88**, 23068). 699

77ZOB2766 B. N. Kozhushko, A. V. Gumenyuk and V. A. Shokol; *Zh. Obshch. Khim.*, 1977, **47**, 2766. 285

77ZOR275 O. I. Kolodyazhnyi and V. P. Kukhar; *Zh. Org. Khim.*, 1977, **13**, 275 (*Chem. Abstr.*, 1977, **87**, 6100). 278, 699

77ZOR290 L. A. Lazukina, V. I. Gorbatenko, L. F. Lur'e and V. P. Kukhar; *Zh. Org Khim.*, 1977, **13**, 290 (*Chem. Abstr.*, 1977, **87**, 39405). 450

77ZOR943 K. A. Petrov, N. A. Tikhonova and N. A. Shchekotikhina; *Zh. Org. Khim.*, 1977, **13**, 943 (*Chem. Abstr.*, 1977, **87**, 52696). 80

77ZOR1559 F. A. Gabitov, A. L. Fridman and V. A. Zakharov; *Zh. Org. Khim.*, 1977, **13**, 1559 (*Chem. Abstr.*, 1977, **87**, 133837). 145

77ZOR1857 R. R. Kostikov, A. F. Khlebnikov and K. A. Ogloblin; *Zh. Org. Khim.*, 1977, **13**, 1857 (*Chem. Abstr.*, 1978, **88**, 6061). 51

77ZOR2008 V. G. Voloshchuk, V. N. Boiko and L. M. Yagupol'skii; *Zh. Org. Khim.*, 1977, **13**, 2008 (*Chem. Abstr.*, 1978, **88**, 6483). 49

78AF1435 G. H. Douglas, J. Diamond, W. L. Studt, G. N. Mir, R. L. Alioto, K. Auyang, B. J. Burns, J. Cias and P. R. Darkes; *Arzneim.-Forsch.*, 1978, **28**, 1435. 647

78AG(E)361 H. Eckert, M. Listl and I. Ugi; *Angew. Chem., Int. Ed. Engl.*, 1978, **17**, 361. 425, 487

78BCJ301 M. Ooka, A. Kitamura, R. Okazaki and N. Inamoto; *Bull. Chem. Soc. Jpn.*, 1978, **51**, 301. 314

78BCJ1245 T. C. Sharma, N. S. Sahni and A. Lal; *Bull. Chem. Soc. Jpn.*, 1978, **51**, 1245. 502

78BCJ2391 I. Kawajima, N. Minami, T. Abe and T. Sato; *Bull. Chem. Soc. Jpn.*, 1978, **51**, 2391. 125

78BCJ3549 I. Wakeshima, Y. Saitoh and I. Kijima; *Bull. Chem. Soc. Jpn.*, 1978, **51**, 3549. 469

78BSB391 J. Gorissen and H. G. Viehe; *Bull. Soc. Chim. Belg.*, 1978, **87**, 391. 619

78BSB693 E. Francotte, R. Verbruggen, H. G. Viehe, M. van Meerssche, G. Germain and J. P. Declerq; *Bull. Soc. Chim. Belg.*, 1978, **87**, 693. 74

78BSF(2)361 J. Escudié, C. Couret and J. Satgé; *Bull. Soc. Chim. Fr., Part 2*, 1978, 361. 265, 266

78CB698 R. Neidlein, P. Leinberger, A. Gieren and B. Dederer; *Chem. Ber.*, 1978, **111**, 698. 617

78CB921 E. Allenstein and F. Sille; *Chem. Ber.*, 1978, **111**, 921. 52

78CB1239 G. Schmid, B. Stutte and R. Boese; *Chem. Ber.*, 1978, **111**, 1239. 340

78CB1603 B. Stutte, V. Bätzel, R. Boese and G. Schmid; *Chem. Ber.*, 1978, **111**, 1603. 340

78CB1619 H. Reimlinger, F. Billiau and R. Merényi; *Chem. Ber.*, 1978, **111**, 1619. 607

78CB2451 F. R. Kreissl, K. Eberl and W. Kleine; *Chem. Ber.*, 1978, **111**, 2451. 700, 733

78CB2891 A. Haas and M. Lieb; *Chem. Ber.*, 1978, **111**, 2891. 237, 436, 477, 483

78CB3542 E. O. Fischer, W. Kleine, G. Kreis and F. R. Kreissl; *Chem. Ber.*, 1978, **111**, 3542. 723

78CB3573 H. Bock, W. Kaim and H. E. Rohwer; *Chem. Ber.*, 1978, **111**, 3573. 173

78CB3644 M. Nitsche, D. Seebach and A. K. Beck; *Chem. Ber.*, 1978, **111**, 3644. 317

78CC140 J. L. Atwood, W. E. Hunter, R. D. Rogers, J. Holton, J. McMeeking and R. Pearce; *J. Chem. Soc., Chem. Commun.*, 1978, 140. 195

78CC285 M. Sekine and T. Hata; *J. Chem. Soc., Chem. Commun.*, 1978, 285. 118

78CC838 J. A. Gladysz, V. K. Wong, and B. S. Jick; *J. Chem. Soc., Chem. Commun.*, 1978, 838. 475

78CPB3807 K. Komaki, T. Kawata, K. Harano and T. Taguchi; *Chem. Pharm. Bull.*, 1978, **26**, 3807. 472

78GEP2706683 S. Thym, K. H. Koenig and G. Hamprecht (BASF AG); *Ger. Pat.* 2706683 (1978) (*Chem. Abstr.*, 1979, **90**, 22357). 438

78GEP2811310 M. Masaki, S. Fuzimura (Ube Industries Ltd.); *Ger. Pat.* 2811310 (1978) (*Chem. Abstr.*, 1979, **90**, 6537). 413

78GEP2828969 G. Hamprecht, A. Parg and K. H. König (BASF AG); *Ger. Pat.* 2828969 (1978) (*Chem. Abstr.*, 1980, **92**, 214885). 445

78HOU(13/6)181 G. Bähr and S. Pawlenko; *Methoden Org. Chem.* (*Houben-Weyl*), 1978, **13/6**, 181. 172, 204

78IC2173 T. Kitazume and J. M. Shreeve; *Inorg. Chem.*, 1978, **17**, 2173. 253

78IJ129 J. W. Thompson, J. L. Howell and A. F. Clifford; *Israel J. Chem.*, 1978, **17**, 129. 10

78IZV220 I. Weisz, F. Havelka and B. K. Nefedov; *Izv. Akad. Nauk SSSR Ser. Khim.*, 1978, 220 (*Chem. Abstr.*, 1978, **88**, 136272). 503

78IZV468 Y. A. Davidovich, L. A. Pavlova and S. V. Rogozhin; *Izv. Akad. Nauk SSSR Ser. Khim.*, 1978, 468 (*Chem. Abstr.* 1978, **88**, 190647). 95

78IZV486 A. V. Fokin, A. M. Gukov, Yu. N. Studnev, V. A. Komarov and A. T. Uzun; *Izv. Akad. Nauk SSSR Ser. Khim.*, 1978, 486 (*Chem. Abstr.*, 1978, **88**, 190051). 259

78IZV1091 S. A. Shevelev, V. V. Semenov and A. A. Fainzil'berg; *Izv. Akad. Nauk SSSR Ser. Khim.*, 1978, 1091 (*Chem. Abstr.*, 1978, **89**, 146358). 323

78IZV2348 V. V. Semenov, S. A. Shevelev and A. A. Fainzil'berg; *Izv. Akad. Nauk SSSR Ser. Khim.*, 1978, 2348 (*Chem. Abstr.*, 1979, **90**, 54589). 261, 281, 323

78JA446 H. W. Kroto, J. F. Nixon, N. P. C. Simmons and N. P. C. Westwood; *J. Am. Chem. Soc.*, 1978, **100**, 446. 679

78JA1325 D. S. Matteson and K. Arne; *J. Am. Chem. Soc.*, 1978, **100**, 1325. 124, 135

78JA1722 R. J. Lagow, R. Eujen, L. L. Gerchman and J. A. Morrison; *J. Am. Chem. Soc.*, 1978, **100**, 1722. 62

78JCE760 M. D. Francis and R. L. Centner; *J. Chem. Ed.*, 1978, **55**, 760. 263

78JCR(S)41 A. Kotynski and W. J. Stec; *J. Chem. Res.* (*S*), 1978, 41. 122

78JCR(S)170 F. Glockling and V. B. Mahale; *J. Chem. Res.* (*S*), 1978, 170. 396

78JCS(D)348 A. J. Hartshorn, M. F. Lappert and K. Turner; *J. Chem. Soc., Dalton Trans.*, 1978, 348. 721

78JCS(D)734 G. K. Barker, M. F. Lappert and J. A. K. Howard; *J. Chem. Soc., Dalton Trans.*, 1978, 734. 195

78JCS(D)826 P. B. Hitchcock, M. F. Lappert and P. L. Pye; *J. Chem. Soc., Dalton Trans.*, 1978, 826. 721

78JCS(D)837 M. F. Lappert and P. L. Pye; *J. Chem. Soc., Dalton Trans.*, 1978, 837. 721

78JCS(P1)202 C. J. Brookes, P. L. Coe, A. E. Pedler and J. C. Tatlow; *J. Chem. Soc., Perkin Trans. 1*, 1978, 202. 20

78JCS(P1)1066 P. E. Alewood, S. A. Hussain, T. C. Jenkins, M. J. Perkins, A. H. Sharma, N. P. Y. Siew and P. Ward; *J. Chem. Soc., Perkin Trans. 1*, 1978, 1066. 445

78JFC(11)509 A. Haas and U. Niemann; *J. Flourine Chem.*, 1978, **11**, 509. 233

78JFC(12)1 T. Abe, K. Kodaira, H. Baba and S. Nagase; *J. Fluorine Chem.*, 1978, **12**, 1. 39

78JFC(12)359 T. Abe and S. Nagase; *J. Fluorine Chem.*, 1978, **12**, 359. 39

78JGU1078 V. I. Shevchenko, N. P. Pisanenko and I. M. Kosinskaya; *J. Gen. Chem. USSR* (*Engl. Transl.*), 1978, **48**, 1078. 621

78JGU1512 E. S. Kozlov, L. G. Dubenko and V. I. Gorbatenko; *J. Gen. Chem. USSR* (*Engl. Transl.*), 1978, **48**, 1512. 609

78JHC473 A. Shafiee; *J. Heterocycl. Chem.*, 1978, **15**, 473. 637

78JHC893 K. Takahashi, K. Takeda and K. Mitsuhashi; *J. Heterocycl. Chem.*, 1978, **15**, 893. 28

78JHC1231 K. Iwata and S. Hara; *J. Heterocycl. Chem.*, 1978, **15**, 1231. 502

78JINC451 M. A. Ali and S. C. Teoh; *J. Inorg. Nucl. Chem.*, 1978, **40**, 451. 563

78JMC773 H. J. Petersen, C. K. Nielsen and E. Arrigoni-Martelli; *J. Med. Chem.*, 1978, **21**, 773. 648

78JOC61 S. K. Chiu, M. Dube, L. Keifer, S. Szilagyi and J. W. Timberlake; *J. Org. Chem.*, 1978, **43**, 61. 509

78JOC337 C. Larsen, K. Steliou and D. N. Harpp; *J. Org. Chem.*, 1978, **43**, 337. 577, 578

78JOC369 N. C. Gonella and M. P. Cava; *J. Org. Chem.*, 1978, **43**, 369. 698

78JOC595 S. Yoneda, T. Kawase, M. Inaba and Z. Yoshida; *J. Org. Chem.*, 1978, **43**, 595. 316

78JOC678 N. F. Haley; *J. Org. Chem.*, 1978, **43**, 678. 729

78JOC752 R. A. Olofson, B. A. Bauman and D. J. Wancowicz; *J. Org. Chem.*, 1978, **43**, 752. 429

78JOC922 P. E. McGann, J. T. Groves, F. D. Greene, G. M. Stack, R. J. Majeste and L. M. Trefonas; *J. Org. Chem.*, 1978, **43**, 922. 509

78JOC950 D. S. Matteson, M. S. Biernbaum, R. A. Bechtold, J. D. Campbell and R. J. Wilcsek; *J. Org. Chem.*, 1978, **43**, 950. 182, 199

78JOC1346 M. K. Meilahn, D. K. Olsen, W. J. Brittain and R. T. Anders; *J. Org. Chem.*, 1978, **43**, 1346. 51

78JOC1459 K. G. Taylor and J. B. Simons; *J. Org. Chem.*, 1978, **43**, 1459. 107

78JOC1544 H. Ulrich, B. Tucker and B. Richter; *J. Org. Chem.*, 1978, **43**, 1544. 511

78JOC1727 R. J. De Pasquale; *J. Org. Chem.*, 1978, **43**, 1727. 52

78JOC1734 H. Matsumoto, T. Nakano, K. Takasu and Y. Nagai; *J. Org. Chem.*, 1978, **43**, 1734. 7

78JOC2132 M. Mikolajczyk and J. Luczak; *J. Org. Chem.*, 1978, **43**, 2132. 509

78JOC2500 M. S. Raasch; *J. Org. Chem.*, 1978, **43**, 2500. 314

78JOC2518 M. Mikolajczyk, A. Zatorski, S. Grzejszczak, B. Costisella and W. Midura, *J. Org. Chem.*, 1978, **43**, 2518. 119

78JOC2643 G. A. Wheaton and D. J. Burton; *J. Org. Chem.*, 1978, **43**, 2643. 224

78JOC2930 G. Barany, B. W. Fulpius and T. P. King; *J. Org. Chem.*, 1978, **43**, 2930. 496, 546

78JOC3485 V. Grakauskas and A. M. Guest; *J. Org. Chem.*, 1978, **43**, 3485. 55

78JOC3732 M. Baudy, A. Robert and A. Foucaud; *J. Org. Chem.*, 1978, **43**, 3732. 629

78JOC3953 C. E. Hatch, III; *J. Org. Chem.*, 1978, **43**, 3953. 439

78JOC4235 M. J. Taschner and G. A. Kraus; *J. Org. Chem.*, 1978, **43**, 4235. 134

78JOC4367 M. S. Newman and P. K. Sujeeth; *J. Org. Chem.*, 1978, **43**, 4367. 28

78JOC4530 D. K. White and F. D. Greene; *J. Org. Chem.*, 1978, **43**, 4530. 445

78JOM(149)C10 D. Van Ende, W. Dumont and A. Krief; *J. Organomet. Chem.*, 1978, **149**, C10. 136

78JOM(149)167 K. Kellner, B. Seidel and A. Tzschach; *J. Organomet. Chem.*, 1978, **149**, 167. 161

78JOM(152)1 A. Mendoza and D. S. Matteson; *J. Organomet. Chem.*, 1978, **152**, 1. 182

78JOM(153)1 D. Grdenić, M. Sikirica and B. Korpar-Čolig; *J. Organomet. Chem.*, 1978, **153**, 1. 405

78JOM(153)253 S. Al-Hashimi and J. D. Smith; *J. Organomet. Chem.*, 1978, **153**, 253. 186, 195

78JOM(153)67 D. L. Reger and M. D. Dukes; *J. Organomet. Chem.*, 1978, **153**, 67. 717, 727

78JOM(157)C50 C. Eaborn, D. A. R. Happer, K. D. Safa and D. R. M. Walton; *J. Organomet. Chem.*, 1978, **157**, C50. 177, 709, 710

78JOM(159)73 T. J. Collins and W. R. Roper; *J. Organomet. Chem.*, 1978, **159**, 73. 719

78JPR255 H. Gross, I. Keitel and B. Costisella; *J. Prakt. Chem.*, 1978, **320**, 255. 120

78JPR335 H. J. Block, E. Fischer and G. Rembarz; *J. Prakt. Chem.*, 1978, **320**, 335. 360, 361

78JPR389 L. Riesel, U. Franke, B. Costisella and H. Gross; *J. Prakt. Chem.*, 1978, **320**, 389. 661, 733

78LA16 J. Hocker and R. Merten; *Justus Liebigs Ann. Chem.*, 1978, 16. 156

78LA1543 E. Schaumann, E. Kausch and E. Rossmanith; *Justus Liebigs Ann. Chem.*, 1978, 1543. 647

78LA1704 W. Böhmer and D. Herrmann; *Justus Liebigs Ann. Chem.*, 1978, 1704. 618

78MI 604-01 H. J. Reich, P. M. Gold and F. Chow; *Symp. Pap.–11th IUPAC Int. Symp. Chem. Nat. Prod.*, 1978, **3**, 133 (*Chem. Abstr.*, 1980, **92**, 111 172). 134

78MI 605-01 K. V. Altukhov and I. E. Titova; *Metody Sint., Str. Khim. Prevrashch. Nitrosoedin, Gertsenovskii Chteniya, 31st*, 1978, 30 (*Chem. Abstr.*, 1979, **91**, 20624). 144

78MI 612-01 N. A. Til'kunova, A. F. Gontar, E. G. Bykhovskaya, V. A. Petrov and I. L. Knunyants; *Zh. Vses. Khim. O va*, 1978, **23**, 712 (*Chem. Abstr.*, 1979, **90**, 120 972). 643

78MI 614-01 H. Kern; *Prax. Naturwiss., Chem.*, 1978, **27**(2), 29 (*Chem. Abstr.*, 1978, **88**, 151494). 413

78MI 614-02 G. A. Brinkman, I. Hass-Lisewska, J. T. Veenboer and L. Lindner; *Int. J. Appl. Radiat. Isot.*, 1978, **29**, 701 (*Chem. Abstr.*, 1979, **90**, 68538). 414

78MI 614-03 D. Roeda, C. Crouzel and B. Van Zanten; *Radiochem. Radioanal. Lett.*, 1978, **33**, 175 (*Chem. Abstr.*, 1978, **89**, 42380). 414

78MI 615-01 D. Cipris and I. L. Mador; *J. Electrochem. Soc.*, 1978, **125**, 1954 (*Chem. Abstr.*, 1979, **90**, 71 695). 469

B-78MI 616-01 E. A. K. Von Gustorf, F.-W. Grevels and I. Fischler; "The Organic Chemistry of Iron," Academic Press, New York, 1978, vol. 1, p. 673; 1981, vol. 2, p. 340. 524

78MIP87868 K. Rudnicki, Z. Kubica, E. Grzywa, J. Wojciechowski, E. Chromniak and K. Ertel; *Polish Pat.* 87868 (1978) (*Chem. Abstr.*, 1978, **88**, 191052). 156

78PJC71 W. Dmowski and A. Kolinski, *Pol. J. Chem.*, 1978, **52**, 71. 19

78RTC51 P. B. M. W. M. Timmermans, P. A. Van Zwieten and W. N. Speckamp; *Recl. Trav. Chim. Pays-Bas*, 1978, **97**, 51. 645

78S283 J. W. Scheeren, H. J. M. Goossens and A. W. H. Top; *Synthesis*, 1978, 283. 77, 78

78S286 A. Faure and G. Descotes; *Synthesis*, 1978, 286. 309, 310

78S309 D. L. Comins, A. F. Jacobine, J. L. Marshall and M. M. Turnbull; *Synthesis*, 1978, 309. 120

78T197 R. W. Rendell and B. Wright; *Tetrahedron*, 1978, **34**, 197. 37

78T3105 G. Le Corre, E. Guibé-Jampel and M. Wakselman; *Tetrahedron*, 1978, **34**, 3105. 477

78TL363 T. Hata, A. Hashizume, M. Nakajima and M. Sekine; *Tetrahedron Lett.*, 1978, 363. 119

78TL1391 S. J. Wratten, D. J. Faulkner, D. Van Engen and J. Clardy; *Tetrahedron Lett.*, 1978, 1391. 604

78TL1395 S. J. Wratten and D. J. Faulkner; *Tetrahedron Lett.*, 1978, 1395. 604

78TL3737 Y. Tamura, J. Haruta, S. Okuyama and Y. Kita; *Tetrahedron Lett.*, 1978, **39**, 3737. 484

78TL3971 S. Halazy, J. Lucchetti and A. Krief; *Tetrahedron Lett.*, 1978, 3971. 94

78USP4098831 F. D. Marsh; *US Pat.* 4098831 (1978) (*Chem. Abstr.* 1979, **90**, 22546). 28

78USP4120904 D. Pilipovich, L. R. Grant, Jr. and R. D. Wilson (US Air Force); *US Pat.* 4120904 (1978) (*Chem. Abstr.*, 1979, **90**, 54461). 283

78ZAAC(439)161 G. Fritz and W. H. Dresel; *Z. Anorg. Allg. Chem.*, 1978, **439**, 161. 267, 268

78ZAAC(441)125 G. Fritz and K-H. Schmid; *Z. Anorg. Allg. Chem.*, 1978, **441**, 125. 267

78ZC452 J. Heinicke and A. Tzschach; *Z. Chem.*, 1978, **18**, 452. 161

78ZN849 E. Lindner and H. Lesiecki; *Z. Naturforsch.*, 1978, **33**, 849.

78ZOB811 G. A. Yuzhakova, R. P. Drovneva, M. I. Vakhrin and I. I. Lapkin; *Zh. Obshch. Khim.*, 1978, **48**, 811 (*Chem. Abstr.*, 1978, **89**, 43 548). 656

78ZOB1425 V. I. Gorbatenko, N. V. Mel'nichenko and L. I. Samarai; *Zh. Obshch. Khim.*, 1978, **48**, 1425 (*Chem. Abstr.*, 1978, **89**, 109 784). 657

78ZOR1841 V. P. Kukhar, V. I. Pasternak and M. V. Shevchenko; *Zh. Org. Khim.*, 1978, **14**, 1841 (*Chem. Abstr.*, 1979, **90**, 22234). 619

78ZOR2229 V. A. Buevich and N. Z. Nakova; *Zh. Org. Khim.*, 1978, **14**, 2229 (*Chem. Abstr.*, 1979, **90**, 71 719). 74

79AG663 H. Werner and K. Leonhard; *Angew. Chem.*, 1979, **91**, 663. 352

79AG(E)333 D. Van Broeck, Z. Janousek, R. Merényi and H. G. Viehe; *Angew. Chem., Int. Ed. Engl.*, 1979, **18**, 333. 619

79AG(E)469 R. Appel and V. Barth; *Angew. Chem., Int. Ed. Engl.*, 1979, **18**, 469. 686

79AG(E)484 H. H. Karsch, B. Zimmer-Gasser, D. Neugebauer and U. Schubert; *Angew. Chem., Int. Ed. Engl.*, 1979, **18**, 484. 151

79AG(E)503 B. J. Barker, J. Rosenfarb and J. A. Caruso; *Angew. Chem., Int. Ed. Engl.*, 1979, **18**, 503. 507

79AG(E)615 F. Huys, R. Merényi, Z. Janousek, L. Stella and H. G. Viehe; *Angew. Chem., Int. Ed. Engl.*, 1979, **18**, 615. 52

79AG(E)693 U. Müller, I. Lorenz and F. Schmock; *Angew. Chem., Int. Ed. Engl.*, 1979, **18**, 693. 655, 733

79AG(E)785 H. Siegel, K. Hiltbrunner and D. Seebach; *Angew. Chem., Int. Ed. Engl.*, 1979, **18**, 785. 204

79AG(E)873 R. Appel, V. Barth, F. Knoll and I. Ruppert; *Angew. Chem., Int. Ed. Engl.*, 1979, **18**, 873. 686

79BAU2222 I. V. Borisova, N. N. Zemlyanski, N. D. Kolosova, Yu. A. Ustynyuk and I. P. Beletskaya; *Bull. Acad. Sci. USSR. Div. Chim. Sci.*, 1979, 2222. 387, 395

79BSF(2)520 B. Rouot and G. Leclerc; *Bull. Soc. Chim. Fr., Part 2*, 1979, 520. 611

79CAR(74)127 A. Faure, B. Kryczka and G. Descotes; *Carbohydr. Res.*, 1979, **74**, 127 (*Chem. Abstr.*, 1980, **92**, 6873). 309

79CB355 G. Saleh, T. Minami, Y. Ohshiro and T. Agawa; *Chem. Ber.*, 1979, **112**, 355. 123, 697

79CB640 W. Ried and H.-E. Erle; *Chem. Ber.*, 1979, **112**, 640. 620

79CB648 R. Appel, K. Geisler and H.-F. Schöler; *Chem. Ber.*, 1979, **112**, 648. 164, 703

79CB727 H. Schmidt, O. Hollitzer, A. Seewald and W. Steglich; *Chem. Ber.*, 1979, **112**, 727. 502

79CB1068 R. Appel and H.-F. Schöler; *Chem. Ber.*, 1979, **112**, 1068. 703

79CB1189 K. Horn, H. Schachner and W. Sundermeyer; *Chem. Ber.*, 1979, **112**, 1189. 362

79CB1873 K. Friedrich and M. Zamkanei; *Chem. Ber.*, 1979, **112**, 1873. 111

79CB1956 A. Q. Hussein and J. C. Jochims; *Chem. Ber.*, 1979, **112**, 1956. 571

79CB2158 W. Lutz and W. Sundermeyer; *Chem. Ber.*, 1979, **112**, 2158. 240, 439

79CC100 C. A. Brown, and A. Yamaichi; *J. Chem. Soc., Chem. Commun.*, 1979, 100. 128

79CC128 R. A. Jacob, L. L. Gerchman, T. J. Juhlke and R. J. Lagow; *J. Chem. Soc., Chem. Commun.*, 1979, 128. 21

79CC396 G. Fachinetti; *J. Chem. Soc., Chem. Commun.*, 1979, 396. 341

79CC397 G. Fachinetti; *J. Chem. Soc., Chem. Commun.*, 1979, 397. 341

79CC485 H. Matsumura, T. Yano, M. Ueyama, K. Tori and W. Nagata; *J. Chem. Soc., Chem. Commun.*, 1979, 485. 111

79CC653 H. Eshtiagh-Hosseini, H. W. Kroto and J. F. Nixon; *J. Chem. Soc., Chem. Comm.*, 1979, 653. 679

79CC675 J. Barluenga, M. Tomás, V. Rubio and V. Gotor; *J. Chem. Soc., Chem. Commun.*, 1979, 675. 575

79CC964 R. D. Chambers, C. G. P. Jones, G. Taylor and R. L. Powell; *J. Chem. Soc., Chem. Commun.*, 1979, 964. 18

79CJC2617 R. N. Renaud and P. J. Champagne, M. Savard; *Can. J. Chem.*, 1979, **57**, 2617. 20

79CL983 S. Hayashi, T. Nakai, N. Ishikawa, D. J. Burton, D. J. Naae and H. S. Kesling; *Chem. Lett.*, 1979, 983. 695

79CL1361 T. Mimura, Y. Kimura and T. Nakai; *Chem. Lett.*, 1979, 1361. 495

79COC(2)1094 A. F. Hegarty; *Comp. Org. Chem.*, 1979, **2**, 1094. 506

79COC(3)373 F. Duus; *Comp. Org. Chem.*, 1979, **3**, 373. 569

79COC(3)687 A. Pelter and K. Smith; *Comp. Org. Chem.*, 1979, **3**, 687. 172

79CZ175 G. Zinner, H.-G. Schecker and W. Heuer; *Chem.-Ztg.*, 1979, **103**, 175. 506

79DOK(245)121 V. A. Dorokhov, L. I. Lavrinovich and B. M. Mikhailov; *Dokl. Akad. Nauk SSSR*, 1979, **245**, 121 (*Chem. Abstr.*, 1979, **91**, 57 081). 662

79EUP3008 A. J. E. Helgstrand, K. N. G. Johansson, A. Misiorny, J. O. Noren and G. B. Stening; *Eur. Pat.* 3008 (1979) (*Chem. Abstr.*, 1979, **92**, 76 662). 512

79EUP7498 K.-H. König, H. G. Oeser and K.-H. Feuerherd (BASF AG); *Eur. Pat.* 7498 (1979) (*Chem. Abstr.* 1980, **93**, 25928). 454

79GEP2742158 H. Hintermaier, H. Michaud and M. Obinger; *Ger. Pat.* 2 742 158 (1979) (*Chem. Abstr.*, 1979, **91**, 20 147). 503

79GEP2753204 G. Beck, G. Dankert and F. Doering (Bayer AG); *Ger. Pat.* 2753204 (1979) (*Chem. Abstr.*, 1979, **91**, 91 664). 606

79GEP2754604 G. Beck and H. Heitzer (Bayer AG); *Ger. Pat.* 2754604 (1979) (*Chem. Abstr.*, 1979, **91**, 666). 606

79GEP2824028 H. Hagen and W. Kantlehner; *Ger. Pat.* 2824028 (1979) (*Chem. Abstr.*, 1980, **93**, 71 541). 140

79GEP2921130 U. Schirmer, K. H. König, B. Wuerzer and G. Retzlaff; *Ger. Pat.* 2 921 130 (1979) (*Chem. Abstr.*, 1981, **95**, 6870). 494

79H(12)637 M. A. Abou-Gharbia and P. H. Doukas; *Heterocycles*, 1979, **12**, 637. 51

79HCA398 J. Pless, W. Bauer, F. Cardinaux, A. Closse, D. Hauser, R. Huguenin, D. Roemer, H.-H. Buescher and R. C. Hill; *Helv. Chim. Acta*, 1979, **62**, 398. 469

79IC919 A. Sekiya and D. D. DesMarteau; *Inorg. Chem.*, 1979, **18**, 919. 256

79IC1165 D. B. Dell'Amico, F. Calderazzo and G. Pelizzi; *Inorg. Chem.*, 1979, **18**, 1165. 517

79IC1236 H. A. Hodali and D. F. Shriver; *Inorg. Chem.*, 1979, **18**, 1236. 338

79IZV813 I. S. Shvarts, M. M. Krayushkin, V. V. Sevost'yanova and V. N. Yarovenko; *Izv. Akad. Nauk SSSR Ser. Khim*, 1979, 813 (*Chem. Abstr.*, 1979, **91**, 19 810). 653

79IZV1911 A. V. Fokin and A. N. Voronkov; *Izv. Akad. Nauk SSSR Ser. Khim.*, 1979, 1911 (*Chem. Abstr.*, 1979, **91**, 192 744). 54

79JA347 G. Klöter and K. J. Seppelt; *J. Am. Chem. Soc.*, 1979, **101**, 347. 238

79JA1348 M. Ishikawa, T. Fuchikami, M. Kumada, T. Higuchi and S. Miyamoto; *J. Am. Chem. Soc.*, 1979, **101**, 1348. 178, 383

79JA1627 N. Grice, S. C. Kao and R. Pettit; *J. Am. Chem. Soc.*, 1979, **101**, 1627. 517

79JA2214 L. A. Shimp and R. J. Lagow; *J. Am. Chem. Soc.*, 1979, **101**, 2214. 205

79JA2768 J. R. Fritch, K. P. C. Vollhardt, M. R. Thompson and V. W. Day; *J. Am. Chem. Soc.*, 1979, **101**, 2768. 403

79JA4732 A. I. Meyers, D. M. Roland, D. L. Comins, R. Henning, M. P. Fleming and K. Shimizu; *J. Am. Chem. Soc.*, 1979, **101**, 4732. 134

79JA5079 D. J. Sikora, M. D. Ruasch, R. D. Rogers and J. L. Atwood; *J. Am. Chem. Soc.*, 1979, **101**, 5079. 522

79JA5949 Q.-C. Mir, K. A. Lawrence, R. W. Shreeve, D. P. Babb and J. M. Shreeve; *J. Am. Chem. Soc.*, 1979, **101**, 5949. 253

79JA6110 D. M. Washecheck, E. J. Wucherer, L. F. Dahl, A. Ceriotti, G. Longoni, M. Manassero, M. Sansoni and P. Chini; *J. Am. Chem. Soc.*, 1979, **101**, 6110. 525

79JA6116 K. Tatsuta, K. Akimoto and M. Kinoshita; *J. Am. Chem. Soc.*, 1979, **101**, 6116. 550

79JA6282 A. Haas, K. Schlosser and S. Steenken; *J. Am. Chem. Soc.*, 1979, **101**, 6282. 734

79JA6638 H. J. Reich, F. Chow and S. K. Shah; *J. Am. Chem. Soc.*, 1979, **101**, 6638. 100

79JA7169 G. Trinquier and J.-P. Malrieu; *J. Am. Chem. Soc.*, 1979, **101**, 7169. 694

79JA7640 A. Sekiya and D. D. DesMarteau; *J. Am. Chem. Soc.*, 1979, **101**, 7640. 53

79JAP79283 Y. Maki; *Jpn. Pat.* 79283 (1979) (*Chem. Abstr.*, 1980, **92**, 6416). 12

79JAP7961121 S. Nakatomi and Y. Shintoku (Ube Industries Ltd.); *Jpn. Pat.* 7961 121 (*Chem. Abstr.*, 1979, **91**, 174838). 425

79JCED251 J. R. Kennedy, T. M. Vickrey, G. R. Somayajulu and B. J. Zwolinski; *J. Chem. Eng. Data*, 1979, **24**, 251 (*Chem. Abstr.*, 1979, **91**, 56256). 221, 223

79JCR(S)12 F. Glockling, P. Harriott and W.-K. Ng; *J. Chem. Res. (S)*, 1979, 12. 380, 387

79JCR(S)240 A. W. Faull and R. Hull; *J. Chem. Res. (S)*, 1979, 240. 574

79JCS(D)767 F. Glockling, V. B. Mahale and J. J. Sweeney; *J. Chem. Soc., Dalton Trans.*, 1979, 767. 395

79JCS(D)1929 P. B. Hitchcock, M. F. Lappert, P. L. Pye and S. Thomas; *J. Chem. Soc., Dalton Trans.*, 1979, 1929. 721

79JCS(P1)724 E. Cawkill, W. D. Ollis, C. A. Ramsden and G. P. Rowson; *J. Chem. Soc., Perkin Trans. 1*, 1979, 724. 611

79JCS(P2)1298 N. P. Smith and I. D. R. Stevens; *J. Chem. Soc., Perkin Trans. 2*, 1979, 1298. 280

79JFC(13)175 C. G. Guizard and H. Cheradame; *J. Fluorine Chem.*, 1979, **13**, 175. 410

79JFC(13)519 T. Abe and S. Nagase; *J. Fluorine Chem.*, 1979, **13**, 519. 37

79JFC(14)289 A. Sekiya and D. D. Desmarteau; *J. Fluorine Chem.*, 1979, **14**, 289. 441

79JFC(15)353 T. Abe, E. Hayashi, H. Baba, K. Kodaira and S. Nagase; *J. Fluorine Chem.*, 1980, **15**, 353. 37

79JINC1223 B. F. Spielvogel, F. Harchelroad Jr. and P. Wisian-Neilson; *J. Inorg. Nucl. Chem.*, 1979, **41**, 1223. 491, 515

79JINC1277 K. J. Cavell, J. O. Hill and R. J. Magee; *J. Inorg. Nucl. Chem.*, 1979, **41**, 1277. 560

79JMC28 R. K. Y. Zee-Cheng and C. C. Cheng; *J. Med. Chem.*, 1979, **22**, 28. 504

79JMR559 G. R. Somayajulu, J. R. Kennedy, T. M. Vickrey and B. J. Zwolinski; *J. Magn. Reson.*, 1979, **33**, 559. 223, 226, 227, 228

79JOC267 K. Rasheed and J. D. Warkentin; *J. Org. Chem.*, 1979, **44**, 267. 313

79JOC420 J.-P. Picard, R. Calas, J. Dunoguès, N. Duffaut, J. Gerval and P. Lapouyade; *J. Org. Chem.*, 1979, **44**, 420. 124

79JOC930 M. Sato, N. C. Gonnella and M. P. Cava; *J. Org. Chem.*, 1979, **44**, 930. 121, 698

79JOC1016 J. Cuomo and R. A. Olofson; *J. Org. Chem.*, 1979, **44**, 1016. 423, 438

79JOC1708 G. G. I. Moore; *J. Org. Chem.*, 1979, **44**, 1708. 45

79JOC1847 C. G. Kruse, A. Wijsman, and A. van der Gen; *J. Org. Chem.*, 1979, **44**, 1847. 48

79JOC1883 D. L. J. Clive and S. M. Menchen; *J. Org. Chem.*, 1979, **44**, 1883. 92

79JOC2311 L. A. Shimp and R. J. Lagow; *J. Org. Chem.*, 1979, **44**, 2311. 205

79JOC2813 M. S. Toy and R. S. Stringham; *J. Org. Chem.*, 1979, **44**, 2813. 250

79JOC2907 A. E. Feiring; *J. Org. Chem.*, 1979, **44**, 2907. 229, 230

79JOC2967 M. Mikolajczyk, S. Grzejszczak, A. Chefczynska and A. Zatorski; *J. Org. Chem.*, 1979, **44**, 2967. 119

79JOC3573 M. Masaki, N. Kakeya and S. Fujimura; *J. Org. Chem.*, 1979, **44**, 3573. 413

79JOC3872 G. A. Olah, J. T. Welch, Y. D. Vankar, M. Nojima, I. Kerekes and J. A. Olah; *J. Org. Chem.*, 1979, **44**, 3872. 439

79JOC4279 D. L. J. Clive and S. M. Menchen; *J. Org. Chem.*, 1979, **44**, 4279. 92

79JOC4536 R. A. Houghten, R. A. Simpson, R. N. Hanson and H. Rapoport; *J. Org. Chem.*, 1979, **44**, 4536. 651

79JOM(169)123 M. J. Van Hamme and D. J. Burton; *J. Organomet. Chem.*, 1979, **169**, 123. 695

79JOM(172)293 T. N. Mitchell and M. El-Behairy; *J. Organomet. Chem.*, 1979, **172**, 293. 405

79JOM(177)1 D. Van Ende, A. Cravador and A. Krief; *J. Organomet. Chem.*, 1979, **177**, 1. 94

79JOM(177)137 A. Ekouya, R. Calas, J. Dunoguès, C. Biran and N. Duffaut; *J. Organomet. Chem.*, 1979, **177**, 137. 125, 162

79JOM(178)11 G. Fritz and S. Wartanessian; *J. Organomet. Chem.*, 1979, **178**, 11. 290

79JOM(178)227 D. Seyferth, C. N. Rudie and M. O. Nestle; *J. Organomet. Chem.*, 1979, **178**, 227. 401, 402, 404

79JOM(178)371 B. L. Booth and G. C. Casey; *J. Organomet. Chem.*, 1979, **178**, 371. 403

79JOM(181)25 J. Jeffery, M. F. Lappert and P. I. Riley; *J. Organomet. Chem.*, 1979, **181**, 25. 194

79JOM(181)69 M. Behnam-Dehkordy, B. Crociani, M. Nicolini and R. L. Richards; *J. Organomet. Chem.*, 1979, **181**, 69. 622

79JOU1077 Y. G. Bal'on and R. N. Moskaleva; *J. Org. Chem. USSR* (*Engl. Transl.*), 1979, **15**, 1077. 504, 508

79JPR827 E. Fanghänel and A. M. Richter; *J. Prakt. Chem.*, 1979, **321**, 827. 500

79LA263 W. Walter, H. Kubel and H.-W. Lüke; *Justus Liebigs Ann. Chem.*, 1979, 263. 571

79LA278 H. Eckert and I. Ugi; *Justus Liebigs Ann. Chem.*, 1979, 278. 485, 487

79LA1002 M. Regitz, B. Weber and U. Eckstein; *Justus Liebigs Ann. Chem.*, 1979, 1002. 623, 660, 668

79LA1456 E. V. Dehmlow and K. Franke; *Justus Liebigs Ann. Chem.*, 1979, 1456. 141

79LA1756 W. Walter and M. Radke; *Justus Liebigs Ann. Chem.*, 1979, 1756. 444, 501, 577

79LA2089 W. Kantlehner, L. Kienitz, H. Jaus and H. Bredereck; *Justus Liebigs Ann. Chem.*, 1979, 2089. 321, 337

79LA2096 W. Kantlehner, H. Jaus, L. Kienitz and H. Bredereck; *Justus Liebigs Ann. Chem*, 1979, **12**, 2096. 647

B-79MI 601-01 H. C. Fielding; in "Organofluorine Chemicals and their Industrial Applications," ed. R. E. Banks, Ellis Horwood, Chichester, 1979, p. 214. 3

B-79MI 602-01 W. Kantlehner; in "Advances in Organic Chemistry," eds. H. Böhme and H. G. Viehe, Wiley, New York, 1979, vol. 9, part 2, p. 5. 51

B-79MI 602-02 W. Kantlehner; in "Advances in Organic Chemistry," eds. H. Böhme and H. G. Viehe, Wiley, New York, 1979, vol. 9, part 2, p. 65. 51

B-79MI 602-03 W. Kantlehner; in "Advances in Organic Chemistry," eds. H. Böhme and H. G. Viehe, Wiley, New York, 1979, vol. 9, part 2, p. 181. 51, 57

B-79MI 602-04 W. Kantlehner; in "Advances in Organic Chemistry," eds. H. Böhme and H. G. Viehe, Wiley, New York, 1979, vol. 9, part 2, p. 279. 51

B-79MI 602-05 W. Kantlehner; in "Advances in Organic Chemistry," eds. H. Böhme and H. G. Viehe, Wiley, New York, 1979, vol. 9, part 2, p. 321. 51, 57

B-79MI 604-01 S. Patai (ed.); "The Chemistry of Acid Derivatives," Supplement B, Wiley, Chichester, 1979. 104, 105

B-79MI 604-02 G. Simchen; in "Advances in Organic Chemistry," eds. H. Böhme and H. G. Viehe, Wiley, New York, 1979, vol. 9, part 2, p. 393. 115

B-79MI 605-01 W. Kantlehner; in "The Chemistry of Acid Derivatives," ed. S. Patai, Supplement B, part 1, Wiley, Chichester, 1979, p. 533. 138

79MI 607-01 G. A. Brinkman and J. Visser; *Int. J. Appl. Radiat. Isot.*, 1979, **30**, 517 (*Chem. Abstr.*, 1979, **91**, 164730) 218

B-79MI 620-01 A. Haas; "Gmelin Handbook of Inorganic Chemistry" 8th edn, "Perfluorohalogenoorganic Compounds of the Main Group Elements," Part 7, ed. D. Koschel, Springer, Berlin, 1979. 602, 604

B-79MI 623-01 W. Kantlehner; in "Iminium Salts in Organic Chemistry" Part 2, ed. H. Böhme and H. G. Viehe, Wiley, New York, 1979, p. 143. 728

B-79MI 623-02 W. Kantlehner; in "Iminium Salts in Organic Chemistry" Part 2, ed. H. Böhme and H. G. Viehe, Wiley, New York, 1979, p. 173. 728

79OS(59)195 K. Kurita and Y. Iwakura; *Org. Synth.*, 1979, **59**, 195. 415, 428
79PS(7)203 P. Jutzi and O. Lorey; *Phosphorus Sulfur*, 1979, **7**, 203. 134
79S81 A. N. Pudovik and I. V. Konovalova; *Synthesis*, 1979, 81. 118
79S182 Y. Ueno and M. Okawara; *Synthesis*, 1979, 182. 731
79S223 G. Aimo, I. Degani and R. Fochi; *Synthesis*, 1979, 223. 97
79S241 J. Liebscher and H. Hartmann; *Synthesis*, 1979, 241. 51
79S311 K. Baum and A. M. Guest; *Synthesis*, 1979, 311. 57
79S339 W. Kantlehner and U. Greiner; *Synthesis*, 1979, 339. 728
79S342 W. Kantlehner, T. Maier and P. Speh; *Synthesis*, 1979, 342. 138
79S554 M. Sato, N. Fukada, M. Kurauchi and T. Takeshima; *Synthesis*, 1979, 554. 627
79S561 L. J. Mathias; *Synthesis*, 1979, 561. 631
79S597 K. Kondo, M. Takarada, S. Murai, N. Sonoda; *Synthesis*, 1979, 597. 498
79S600 W. H. Gilligan and S. L. Stafford; *Synthesis*, 1979, 600. 427, 477
79S735 K. Kondo, K. Murata, N. Miyoshi, S. Murai and N. Sonoda; *Synthesis*, 1979, 735. 505
79S972 N. D. Volkov, V. P. Nazaretian and L. M. Yagupol'skii; *Synthesis*, 1979, 972. 236
79T181 B. Divisia; *Tetrahedron*, 1979, **35**, 181. 666
79T1675 R. F. Abdulla and R. S. Brinkmeyer; *Tetrahedron*, 1979, **35**, 1675. 104
79T2329 J. A. Gladysz, V. K. Wong and B. S. Jick; *Tetrahedron*, 1979, **35**, 2329. 475
79TL501 T. Kauffmann, B. Altepeter, K.-J. Echsler, J. Ennen, A. Hamsen and R. Joussen; *Tetrahedron Lett.*, 1979, 501. 170, 210
79TL513 W. Hartmann and H.-G. Heine; *Tetrahedron Lett.*, 1979, 513. 108
79TL1473 J. T. Martz, G. W. Gokel and R. A. Olofson; *Tetrahedron Lett.*, 1979, **17**, 1473. 25
79TL1827 W. S. Murphy and S. Wattanasin; *Tetrahedron Lett.*, 1979, 1827. 127
79TL2925 J. W. Scheeren, F. J. M. Dahmen and C. G. Bakker; *Tetrahedron Lett.*, 1979, 2925. 75
79TL3013 M. Sekine, H. Yamagata and T. Hata; *Tetrahedron Lett.*, 1979, 3013. 582
79TL3375 G. W. Gokel, H. M. Gerdes, D. E. Miles, J. M. Hufnal and G. A. Zerby; *Tetrahedron Lett.*, 1979, 3375. 126, 135
79TL3507 J. Escudié, C. Couret, J. Dubac, J. Cavezzan, J. Satgé and P. Mazerolles; *Tetrahedron Lett.*, 1979, 3507. 582
79TL3865 Y. Kobayashi, T. Yoshida and I. Kumadaki; *Tetrahedron Lett.*, 1979, 3865. 232
79TL4071 Y. Kobayashi, K. Yamamoto and I. Kumadaki; *Tetrahedron Lett.*, 1979, 4071. 22, 246
79TL4475 M. Sekine, M. Nakajima, A. Kume and T. Hata; *Tetrahedron Lett.*, 1979, 4475. 125
79USP4154931 R. H. Richter, B. W. Tucker and H. Ulrich; *US Pat.* 4 154 931 (1979) (*Chem. Abstr.*, 1979, **91**, 57 068). 509
79ZAAC(448)40 G. Fritz and W. Himmel; *Z. Anorg. Allg. Chem.*, 1979, **448**, 40. 192
79ZAAC(448)55 G. Fritz and W. Himmel; *Z. Anorg. Allg. Chem.*, 1979, **448**, 55. 267
79ZAAC(457)209 K.-H. Worms and H. Blum; *Z. Anorg. Allg. Chem.*, 1979, **457**, 209. 159
79ZAAC(457)214 K.-H. Worms, H. Blum and H.-U. Hempel; *Z. Anorg. Allg. Chem.*, 1979, **457**, 214. 158
79ZAAC(458)37 G. Fritz and J. Mittag; *Z. Anorg. Allg. Chem.*, 1979, **458**, 37. 268
79ZAAC(459)87 G. Becker and O. Mundt; *Z. Anorg. Allg. Chem.*, 1979, **459**, 87. 684, 687
79ZC250 R. Moll; *Z. Chem.*, 1979, **19**, 250. 651
79ZC289 H. Viola and R. Meyer; *Z. Chem.*, 1979, **19**, 289. 536
79ZN(B)390 D. Breitinger, W. Morell and K. Grabetz; *Z. Naturforsch., Teil B*, 1979, **34**, 390. 405
79ZN(B)999 H. A. Klein; *Z. Naturforsch., Teil B*, 1979, **34**, 999. 113
79ZN(B)1171 H. H. Karsch; *Z. Naturforsch., Teil B*, 1979, **34**, 1171. 151
79ZN(B)1178 H. H. Karsch; *Z. Naturforsch., Teil B*, 1979, **34**, 1178. 369
79ZOB104 O. I. Kolodyazhnyi; *Zh. Obshch. Khim.*, 1979, **49**, 104 (*Chem. Abstr.*, 1979, **90**, 204 186). 699, 700
79ZOB1025 V. P. Kukhar and E. I. Sagina; *Zh. Obshch. Khim.*, 1979, **49**, 1025 (*Chem. Abstr.*, 1979, **91**, 74 676). 58
79ZOB1470 V. P. Kukhar, and E. I. Sagina; *Zh. Obshch. Khim.*, 1979, **49**, 1470. 286
79ZOR215 V. P. Kukhar, V. I. Pasternak and M. V. Shevchenko; *Zh. Org. Khim.*, 1979, **15**, 215 (*Chem. Abstr.*, 1979, **90**, 167966). 612, 727
79ZOR396 V. N. Boiko, T. A. Dashevskaya, G. M. Schupak and L. M. Yagupol'skii; *Zhur. Org. Khim.*, 1979, **15**, 396 (*Chem. Abstr.*, 1979, **91**, 20439). 233
79ZOR1245 V. N. Boiko, T. A. Dashevskaya, N. V. Ignart'ev and L. M. Yagupol'skii; *Zhur. Org. Khim.*, 1979, **15**, 1245 (*Chem. Abstr.*, 1979, **91**, 157378). 233
79ZOR1561 N. V. Kondratenko, V. I. Popov, A. A. Kolomeistov, I. D. Sadekov, V. I. Minkin and L. M. Yagupol'skii; *Zh. Org. Khim.*, 1979, **15**, 1561 (*Chem. Abstr.*, 1979, **91**, 174 948). 50

80AF1733 H.-J. May; *Arzneim.-Forsch.*, 1980, **30**, 1733. 645
80AG411 H.-N. Adams, G. Fachinetti and J. Strähle; *Angew. Chem.*, 1980, **92**, 411. 341
80AG1055 C. Bianchini, A. Meli, A. Orlandini and L. Sacconi; *Angew. Chem.*, 1980, **92**, 1055. 325
80AG(E)201 K.-H. König, C. Reitel and K.-H. Feuerherd; *Angew. Chem., Int. Ed. Engl.*, 1980, **19**, 201. 448, 452, 454
80AG(E)203 R. Seelinger and W. Sundermeyer; *Angew. Chem., Int. Ed. Engl.*, 1980, **19**, 203. 253
80AG(E)942 G. Klöter, H. Pritzkow and K. Seppelt; *Angew. Chem., Int. Ed. Engl.*, 1980, **19**, 942. 275
80AHC(27)151 N. Lozac'h and M. Stavaux, *Adv. Heterocycl. Chem.*, 1980, **27**, 151. 48
80AJC287 M. J. Gallagher and H. Honegger; *Aust. J. Chem.*, 1980, **33**, 287. 116
80AOC(18)1 H. Behrens; *Adv. Organomet. Chem.*, 1980, **18**, 1. 517
80AP(313)995 W. Hanefeld; *Arch. Pharm. (Weinheim, Ger.)*, 1980, **313**, 995. 494
80AQ183 M. Herrera, V. M. Ruiz, R. Tapia, J. A. Valderrama and J. C. Vega; *An. Quim., Ser. C*, 1980, **76**, 183 (*Chem. Abstr.*, 1981, **94**, 139 181). 547
80BCJ3687 M. Tamaru, S. Horito and J. Yoshimura; *Bull. Chem. Soc. Jpn.*, 1980, **53**, 3687. 72

80CB750	W. Ried, O. Bellinger and G. Oremek; *Chem. Ber.*, 1980, **113**, 750.	540
80CB1095	J. Goerdeler and S. Raddatz; *Chem. Ber.*, 1980, **113**, 1095.	620
80CB1654	H. Werner, K. Leonhard, O. Kolb, E. Röttinger and H. Vahrenkamp; *Chem. Ber.*, 1980, **113**, 1654.	352
80CB1898	K. Hartke, T. Kissel, J. Quante and R. Matusch; *Chem. Ber.*, 1980, **113**, 1898.	538
80CB2499	J. Goerdeler and C. Lindner; *Chem. Ber.*, 1980, **113**, 2499.	504
80CB2583	W. Ried, G. Oremek and R. Guryn; *Chem. Ber.*, 1980, **113**, 2583.	620, 621
80CC39	D. S. Matteson and D. Majumdar; *J. Chem. Soc., Chem. Commun.*, 1980, 39.	203
80CC207	M. P. Atkins, B. T. Golding, D. A. Howes and P. J. Sellars; *J. Chem. Soc., Chem. Commun.*, 1980, 207.	79
80CC635	M. F. Lappert, T. R. Martin and G. M. McLaughlin; *J. Chem. Soc., Chem. Commun.*, 1980, 635.	599
80CC671	L. J. Krause and J. A. Morrison; *J. Chem. Soc., Chem. Commun.*, 1980, 671.	246
80CC778	E. M. Holt, K. Whitmire and D. F. Shriver; *J. Chem. Soc., Chem. Commun.*, 1980, 778.	340
80CC780	K. Whitmire, D. F. Shriver and E. M. Holt; *J. Chem. Soc., Chem. Commun.*, 1980, 780.	340
80CC781	P. A. Dawson, B. F. G. Johnson, J. Lewis and P. R. Raithby; *J. Chem. Soc., Chem. Commun.*, 1980, 781.	340
80CC812	B. F. G. Johnson, J. Lewis and P. R. Raithby; *J. Chem. Soc., Chem. Commun.*, 1980, 812.	352
80CC866	F. Wudl and D. Nalewajek; *J. Chem. Soc., Chem. Commun.*, 1980, 866.	628, 730
80CC879	H. M. Colquhoun and T. J. King; *J. Chem. Soc., Chem. Commun.*, 1980, 879.	610, 619
80CC1265	H. Kalbacher and W. Voelter; *J. Chem. Soc., Chem. Commun.*, 1980, 1265.	422
80CC866A	F. Wudl and D. Nalewajek; *J. Chem. Soc., Chem. Commun.*, 1980, 866.	591, 594
80CC866B	L.-Y. Chiang, T. O. Poehler, A. N. Bloch and D. O. Cowan; *J. Chem. Soc., Chem. Commun.*, 1980, 866.	591, 594
80CCC3488	M. Hajek, P. Silhavy and J. Malek; *Collect. Czech. Chem. Commun.*, 1980, **45**, 3488.	7
80CCC3502	M. Hajek, P. Silhavy and J. Malek; *Collect. Czech. Chem. Commun.*, 1980, **45**, 3502.	7
80CJC485	W. Reeve, J. R. McKee, R. Brown, S. Lakshmanan and G. A. McKee; *Can. J. Chem.*, 1980, **58**, 485.	27
80CJC1996	T. P. Ahern, R. F. Langler and R. L. McNeil; *Can. J. Chem.*, 1980, **58**, 1996.	254, 255
80CPB110	T. Yamaguchi, M. Watanabe, T. Taguchi and M. Kojima; *Chem. Pharm. Bull.*, 1980, **28**, 110.	554
80CPB2987	Y. Konda, M. Onda, A. Hirano and S. Omura; *Chem. Pharm. Bull.*, 1980, **28**, 2987.	138
80CP1076132	V. Dittrich, W. Toepfl and O. Kristiansen (Ciba-Geigy A.-G.); *Can. Pat.* 1076132 (1980) (*Chem. Abstr.*, 1980, **93**, 239056).	619
80EGP139936	G. Furcht and R. Scholz (Veb Chemiewerk Nuenchritz); *Ger. (East) Pat.* 139936 (1980) (*Chem. Abstr.*, 1980, **93**, 167666).	409, 419
80EUP14064	G. E. Robinson; *Eur. Pat.* 14064 (1980) (*Chem. Abstr.*, 1981, **94**, 83632).	625
80EUP52842	B. Baasner, H. Hagemann and E. Klauke (Bayer AG); *Eur. Pat.* 52842 (1980) (*Chem. Abstr.*, 1982, **97**, 162620).	438, 439
80GEP2920386	H. Scheuermann and D. Schneider (BASF AG); *Ger. Pat.* 2920386 (1980) (*Chem. Abstr.*, 1981, **94**, 191912).	428
80GEP3008081	R. Nishiyama, K. Fujikawa, I. Yokomichi, Y. Tsujii and S. Nishimura; *Ger. Pat.* 3008081(1980) (*Chem. Abstr.*, 1981, **94**, 30577).	9
80H(14)271	M. V. Lakshmikantham and M. P. Cava; *Heterocycles*, 1980, **14**, 271.	101
80HCA1190	U. Burger and F. Dreier; *Helv. Chim. Acta*, 1980, **63**, 1190.	170
80HCA1958	K. Seckinger; *Helv. Chim. Acta*, 1980, **63**, 1958.	143
80HCA2046	D. Seebach, H. Siegel, J. Gabriel and R. Hässig; *Helv. Chim. Acta*, 1980, **63**, 2046.	204
80HOU(13/5)27	S. Pawlenko; *Methoden Org. Chem. (Houben-Weyl)*, 1980, **13/5**, 27.	172, 189
80IC1330	A. Sekiya and D. D. Desmarteau; *Inorg. Chem.*, 1980, **19**, 1330.	256
80IC3637	A. J. Carty, M. J. S. Gynane, M. F. Lappert, S. J. Miles, A. Singh and N. J. Taylor; *Inorg. Chem.*, 1980, **19**, 3637.	195
80IJC(B)610	N. M. Kansal, S. Verma, R. S. Mishra and M. M. Bokadia; *Indian J. Chem., Sect. B*, 1980, **19**, 610 (*Chem. Abstr.*, 1981, **94**, 15817).	264
80IJC(B)653	G. Prasad, G. Singh and K. N. Mehrotra; *Indian J. Chem., Sect. B*, 1980, **19**, 653.	507
80IZV2356	Y. S. Shvetsov, V. D. Cherepinskii-Malov, A. N. Shirshov, V. S. Reznik and V. G. Andrianov; *Izv. Akad. Nauk SSSR Ser. Khim.*, 1980, 2356 (*Chem. Abstr.*, 1981, **94**, 121447).	57
80JA1206	G. R. Clark, K. Marsden, W. R. Roper and L. J. Wright; *J. Am. Chem. Soc.*, 1980, **102**, 1206.	717, 718
80JA1584	T. J. Barton and S. K. Hockman; *J. Am. Chem. Soc.*, 1980, **102**, 1584.	179, 190, 202, 203, 672
80JA2681	Y. Katsuhara and D. D. Des Marteau; *J. Am. Chem. Soc.*, 1980, **102**, 2681.	232
80JA6161	T. N. Salzmann, R. W. Ratcliffe, B. G. Christensen and F. A. Bouffard; *J. Am. Chem. Soc.*, 1980, **102**, 6161.	128
80JA7572	J. W. Agopovich and C. W. Gillies; *J. Am. Chem. Soc.*, 1980, **102**, 7572.	410
80JA7579	M. K. Kaloustian and F. F. Khouri; *J. Am. Chem. Soc.*, 1980, **102**, 7579.	96
80JAP80136115	Ube Chemical Industries Co. Ltd.; *Jpn. Pat.* 80136615 (1980) (*Chem. Abstr.*, 1981, **95**, 6475).	413
80JAP80162415	Ube Industries Ltd.; *Jpn. Pat.* 80162415 (1980) (*Chem. Abstr.*, 1981, **95**, 42330).	413
80JAP8089210	Nissan Chemical Industries; *Jpn. Pat.* 8089210 (1980) (*Chem. Abstr.*, 1981, **94**, 11638).	286
80JAP(K)8083727	Nippon Synthetic Chemical Industry Co., Ltd.; *Jpn. Kokai* 8083727 (1980) (*Chem. Abstr.*, 1981, **94**, 120888).	70
80JCED292	R. S. Bezwada and S. S. Stivala; *J. Chem. Eng. Data*, 1980, **25**, 292.	504
80JCS(D)2428	M. J. S. Gynane, A. Hudson, M. F. Lappert and P. P. Power; *J. Chem. Soc., Dalton Trans.*, 1980, 2428.	165, 167, 168
80JCS(P1)756	R. P. Houghton and A. D. Morgan; *J. Chem. Soc., Perkin Trans. 1*, 1980, 756.	79

80JCS(P1)766 M. M. Campbell and R. C. Craig; *J. Chem. Soc., Perkin Trans. 1*, 1980, 766. 578

80JCS(P1)1544 T. W. Hart, R. N. Haszeldine and A. E. Tipping; *J. Chem. Soc., Perkin Trans. 1*, 1980, 1544. 239

80JCS(P1)1960 M. G. Barlow, R. N. Haszeldine and K. W. Murray; *J. Chem. Soc., Perkin Trans. 1*, 1980, 1960. 241

80JCS(P1)2254 M. G. Barlow, R. N. Haszeldine and C. Simon; *J. Chem. Soc., Perkin Trans. 1*, 1980, 2254. 238

80JCS(P1)2678 W. S. Murphy and S. Wattanasin; *J. Chem. Soc., Perkin Trans. 1*, 1980, 2678. 127

80JCS(P1)2755 Y. Kobayashi, K. Yamamoto, T. Asai, M. Nakano and I. Kumadaki; *J. Chem. Soc., Perkin Trans. 1.*, 1980, 2755. 14

80JCS(P2)867 A. F. Hegarty, M. T. McCormack, K. Brady, G. Ferguson and P. J. Roberts; *J. Chem. Soc., Perkin Trans. 2*, 1980, 867. 620

80JFC(15)37 M. Seguin, J. C. Adenis, C. Michaud and J. J. Basselier; *J. Fluorine Chem.*, 1980, **15**, 37. 441

80JFC(15)169 I. L. Knunyants, A. F. Gontar, N. A. Tilkunova, A. S. Vinigradov and E. G. Bykhovskaya; *J. Fluorine Chem.*, 1980, **15**, 169. 240, 615

80JFC(15)201 M. Seguin, J. C. Adenis, C. Michaud and J. J. Basselier; *J. Fluorine Chem.*, 1980, **15**, 201. 231

80JFC(15)263 D. J. Burton and R. M. Flynn; *J. Fluorine Chem.* 1980, **15**, 263. 264

80JFC(15)423 R. Franz; *J. Fluorine Chem.*, 1980, **15**, 423. 409

80JFC(16)391 R. E. Banks, J. M. Birchall, R. N. Haszeldine and S. N. Nona; *J. Fluorine Chem.*, 1980, **16**, 391. 241

80JFC(17)65 H. Burger, R. Eujen, H. Niepel and G. Pawelke; *J. Fluorine Chem.*, 1980, **17**, 65. 51

80JHC117 A. Shafiee, M. Vosooghi and R. Asgharian; *J. Heterocycl. Chem.*, 1980, **17**, 117. 630

80JHC571 F. A. Devillanova and G. Verani; *J. Heterocycl. Chem.*, 1980, **17**, 571. 595

80JHC1369 P. J. Kothari, M. A. Mehlhoff, S. P. Singh, S. S. Parmar and V. I. Stenberg; *J. Heterocycl. Chem.*, 1980, **17**, 1369. 573

80JMC13 S. Rachlin, E. Bramm, I. Ahnfelt-Rønne and E. Arrigoni-Martelli; *J. Med. Chem.*, 1980, **23**, 13. 645

80JMC1217 H. Stahle, H. Daniel, W. Kobinger, C. Lillie and L. Pichler; *J. Med. Chem.*, 1980, **23**, 1217. 642

80JMC1261 J. A. Grosso, D. E. Nichols, M. B. Nichols and G. K. W. Yim; *J. Med. Chem.*, 1980, **23**, 1261. 577

80JOC175 N. F. Haley and M. W. Fichtner; *J. Org. Chem.*, 1980, **45**, 175. 552

80JOC271 D. N. Harpp and A. Granata; *J. Org. Chem.*, 1980, **45**, 271. 478

80JOC740 R. Breslow and P. S. Pandey; *J. Org. Chem.*, 1980, **45**, 740. 80, 84

80JOC886 J. K. Snyder and L. M. Stock; *J. Org. Chem.*, 1980, **45**, 886. 504

80JOC1091 D. S. Matteson and R. J. Moody; *J. Org. Chem.*, 1980, **45**, 1091. 199

80JOC2024 J. Nakayama, K. Fujiwara and M. Hoshino; *J. Org. Chem.*, 1980, **45**, 2024. 113

80JOC2096 H. Hart, S.-M. Chen and S. Lee; *J. Org. Chem.*, 1980, **45**, 2096. 68

80JOC2236 A. P. Kozikowski and Y.-Y. Chen; *J. Org. Chem.*, 1980, **45**, 2236. 98

80JOC3713 C. Larsen and D. N. Harpp; *J. Org. Chem.*, 1980, **45**, 3713. 114

80JOC3957 J. Elzinga and H. Hogeveen; *J. Org. Chem.*, 1980, **45**, 3957. 7

80JOC4038 N. J. Curtis and R. S. Brown; *J. Org. Chem.*, 1980, **45**, 4038. 104

80JOC4059 A. Efraty, I. Feinstein, L. Wackerle and A. Goldman; *J. Org. Chem.*, 1980, **45**, 4059. 415

80JOC4250 L. A. Carpino; *J. Org. Chem.*, 1980, **45**, 4250. 428

80JOC5214 W. Reeve and R. Tsuk; *J. Org. Chem.*, 1980, **45**, 5214. 25

80JOC5325 H. K. Hall, Jr. and A. El-Shekeil; *J. Org. Chem.*, 1980, **45**, 5325. 446

80JOM(184)C41 D. S. Matteson and D. Majumdar; *J. Organomet. Chem.*, 1980, **184**, C41. 203

80JOM(186)309 C. Eaborn, D. A. R. Happer, P. B. Hitchcock, S. P. Hopper, K. D. Safa, S. S. Washburne and D. R. M. Walton; *J. Organomet. Chem.*, 1980, **186**, 309. 179, 709, 710

80JOM(188)329 D. Seyferth and H. P. Withers Jr.; *J. Organomet. Chem.*, 1980, **188**, 329. 404

80JOM(190)101 C. Eaborn, N. Retta and J. D. Smith; *J. Organomet. Chem.*, 1980, **190**, 101. 393

80JOM(191)355 C. Eaborn, D. A. R. Happer and K. D. Safa; *J. Organomet. Chem.*, 1980, **191**, 355. 177, 709, 710

80JOM(191)371 E. Glozbach and J. Lorberth; *J. Organomet. Chem.*, 1980, **191**, 371. 670, 675

80JOM(191)7 D. K. Breitinger, K. Geibel, W. Kress and R. Sendelbeck; *J. Organomet. Chem.*, 1980, **191**, 7. 206, 207

80JOM(192)33 K. Issleib, H. Schmidt and H. Meyer, *J. Organomet. Chem.*, 1980, **192**, 33. 687

80JOM(202)157 C. Eaborn and N. P. Y. Siew; *J. Organomet. Chem.*, 1980, **202**, 157. 387

80LA122 V. Jäger, V. Buss and W. Schwab; *Liebigs Ann. Chem.*, 1980, 122. 417

80LA873 H. Gotthardt, S. Nuberl and J. Donecke; *Justus Liebigs Ann. Chem.*, 1980, 873. 114

80LA1216 H. Hagen, R.-D. Kohler and H. Fleig; *Liebigs Ann. Chem.*, 1980, 1216. 69

80LA1677 W. Kantlehner and H.-D. Gutbrod; *Liebigs Ann. Chem.*, 1980, 1677. 70

80LA1751 D. Hoppe and L. Beckman; *Liebigs Ann. Chem.*, 1980, 1751. 326

80MI 603-01 S. Nannessian and J. Banoub; *Methods Carbohydr. Chem.*, 1980, **8**, 237. 71

80MI 605-01 E. M. Maksyutov, I. A. Leenson, G. B. Sergeev and V. I. Slovetskii; *Vestn. Mosk. Univ., Ser. 2: Khim.*, 1980, **21**, 262 (*Chem. Abstr.*, 1980, **93**, 123589). 147

80MI 605-02 A. D. Nikolaeva, I. L. Tsentsiper and V. S. Perekhod'ko; *Izv. Vyshh. Uchebn. Zaved., Khim. Tekhnol.*, 1980, **23**, 152 (*Chem. Abstr.*, 1980, **93**, 45896). 145

B-80MI 606-01 E. Negishi; "Organometallics in Organic Synthesis," Wiley, New York, 1980. 172

80MI 616-01 V. M. Ismailov, Kh. Ya. Al-Khaidar and A. N. Guliev; *Azerb. Khim. Zh.*, 1980, 58 (*Chem. Abstr.*, 1981, **95**, 97911). 518

B-80MI 620-01 A. Haas; "Gmelin Handbook of Inorganic Chemistry" 8th edn, "Perfluorohalogenoorganic Compounds of the Main Group Elements," Part 8, ed. D. Koschel, Springer, Berlin, 1980. 602

80MI 621-01 K. C. Kalia, A. Kumar and M. Singla; *Indian Chem. Manuf.*, 1980, **18**, 11 (*Chem. Abstr.*, 1981, **94**, 174500). 650, 655

80MI 622-01 F. J. Brown; *Progr. Inorg. Chem.*, 1980, **27**, 1. 716, 719, 721, 722

80MM252 Y. Yokoyama, A. B. Padias, F. DeBlauwe and H. K. Hall, Jr.; *Macromolecules*, 1980, **13**, 252. 79
80OS(59)195 K. Kurita and Y. Iwakura; *Org. Synth.*, 1980, **59**, 195. 230
80PJC145 D. Buza and W. Gradowska; *Pol. J. Chem.*, 1980, **54**, 145. 111
80PS(8)301 J. J. D'Amico and T. Schafer; *Phosphorus Sulfur*, 1980, **8**, 301. 493, 495
80PS(9)123 S. O. Grim and E. D. Walton; *Phosphorus Sulfur*, 1980, **9**, 123. 151
80S60 V. I. Cohen; *Synthesis*, 1980, 60. 596
80S85 V. I. Gorbatenko and L. I. Samarai; *Synthesis*, 1980, 85. 445
80S127 M. Mikolajczyk, P. Balczewski and S. Grzejszczak; *Synthesis*, 1980, 127. 120, 121, 123
80S375 I. Degani, R. Fochi and V. Regondi; *Synthesis*, 1980, 375. 472
80S485 E. Anders and W. Will; *Synthesis*, 1980, 485. 463
80S554 W. G. Taylor; *Synthesis*, 1980, 554. 695
80S619 W. Ried, H. Dietschmann and H.-E. Erle; *Synthesis*, 1980, 619. 620, 621, 623
80S1001 K. N. Mehrotra and G. Singh; *Synthesis*, 1980, 1001. 507, 576
80T539 S. J. Scorini and A. Senning; *Tetrahedron*, 1980, **36**, 539. 537
80T1451 F. A. Devillanova and G. Verani; *Tetrahedron*, 1980, **36**, 1451. 474
80T3543 H. G. Aurich and H. Czepluch; *Tetrahedron*, 1980, **36**, 3543. 445
80TCC109 T. Kauffmann; *Top. Curr. Chem.*, 1980, **92**, 109. 203, 204, 205, 210
80TL313 A. V. Bogatsky, N. G. Lukyanenko and T. I. Kirichenko; *Tetrahedron Lett.*, 1980, **21**, 313. 575
80TL819 R. A. Olofson and J. Cuomo; *Tetrahedron Lett.*, 1980, 819. 462
80TL1141 H. Oehme, E. Leissring and H. Meyer; *Tetrahedron Lett.*, 1980, **21**, 1141. 686
80TL1357 T. Tsunoda, M. Suzuki and Y. Noyori; *Tetrahedron Lett.*, 1980, 1357. 124
80TL2807 A. Rensing, K.-J. Echsler and T. Kauffmann; *Tetrahedron Lett.*, 1980, **21**, 2807. 203, 204, 210
80TL4893 I. Ruppert; *Tetrahedron Lett.*, 1980, **21**, 4893. 603, 615
80ZAAC(462)130 G. Becker and O. Mundt; *Z. Anorg. Allg. Chem.*, 1980, **462**, 130. 160
80ZAAC(463)144 G. Becker, G. Gresser and W. Uhl; *Z. Anorg. Allg. Chem.*, 1980, **463**, 144. 684
80ZAAC(468)165 U. Müller and F. Schmock; *Z. Anorg. Allg. Chem.*, 1980, **468**, 165. 733
80ZAAC(468)99 W. Maringgele; *Z. Anorg. Allg. Chem.*, 1980, **468**, 99. 662
80ZC101 M. Mortag and K. Möckel; *Z. Chem.*, 1980, **20**, 101. 560
80ZC153 K. Issleib, H. Schmidt and C. Wirkner; *Z. Chem.*, 1980, **20**, 153. 372, 374
80ZC419 K. Issleib, H. Schmidt and C. Wirkner; *Z. Chem.*, 1980, **20**, 419. 372
80ZN(B)526 A. Darmadi, A. Haas and B. Koch; *Z. Naturforsch., Teil B*, 1980, **35**, 526. 588, 592
80ZN(B)1289 J. C. Baldwin, N. L. Keder, Ch. E. Strouse and W. C. Kaska; *Z. Naturforsch., Teil B*, 1980, **35**, 1289. 706
80ZOB233 O. I. Kolodyazhnyi and V. P. Kukhar; *Zh. Obshch. Khim.*, 1980, **50**, 233, (*Chem. Abstr.*, 1980, **92**, 181291). 684
80ZOB875 V. D. Sheludyakov, S. V. Sheludyakova, M. G. Kutnetsova, N. N. Silkina and V. F. Mironov; *Zh. Obshch. Khim.*, 1980, **50**, 875 (*Chem. Abstr.*, 1980, **93**, 114404). 444
80ZOB1897 A. P. Marchenko, V. V. Miroshnichenko, G. N. Kodian and A. M. Pinchuk; *Zh. Obsch. Khim.*, 1980, **50**, 1897 (*Chem. Abstr.*, 1980, **93**, 239512). 14

81ACH57 I. Bitter, L. Szöcs and L. Töke; *Acta Chim. Acad. Sci. Hung.*, 1981, **107**, 57. 503
81AG94 H.-N. Adams, G. Fachinetti and J. Strähle; *Angew. Chem.*, 1981, **93**, 94. 341
81AG215 G. Fachinetti, L. Balocchi, F. Secco and M. Venturini; *Angew. Chem.*, 1981, **93**, 215. 341
81AG(E)131 E. Niecke, W. W. Schoeller and D.-A. Wildbredt; *Angew. Chem., Int. Ed. Engl.*, 1981, **20**, 131. 168
81AG(E)475 S. Murai, R. Sugise and N. Sonoda; *Angew. Chem., Int. Ed. Engl.*, 1981, **20**, 475. 7
81AG(E)581 L. Roesch, G. Altnau and W. H. Otto; *Angew. Chem., Int. Ed. Engl.*, 1981, **20**, 581. 125
81AG(E)585 J.-C. Pommelet, C. Nyns, F. Lahousse, R. Merényi and H. G. Viehe; *Angew. Chem., Int. Ed. Engl.*, 1981, **20**, 585. 44
81AG(E)647 M. R. C. Gerstenberger and A. Haas; *Angew. Chem., Int. Ed. Engl.*, 1981, **20**, 647. 51, 54
81AG(E)731 R. Appel, F. Knoll and I. Ruppert; *Angew. Chem., Int. Ed. Engl.*, 1981, **20**, 731. 678, 689, 692
81AHC(28)1 R. D. Chambers and C. R. Sargent; *Adv. Heterocycl. Chem.*, 1981, **28**, 1. 9, 10
81AP(314)315 W. Hanefeld; *Arch. Pharm.* (*Weinheim, Ger.*), 1981, **314**, 315. 494
81AP(314)587 W. Hanefeld, G. Glaeske and P. Schulze-Weisschu; *Arch. Pharm.* (*Weinheim, Ger.*), 1981, **314**, 587. 564
81BSB63 G. L'abbe, D. Sorgeloos, S. Toppet, G. S. D. King and V. M. Luc; *Bull. Soc. Chim. Belg.*, 1981, **90**, 63. 652
81CB829 A. Haas and J. Mikolajczak; *Chem. Ber.*, 1981, **114**, 829. 529, 532
81CB1132 R. Bunnenberg and J. C. Jochims; *Chem. Ber.*, 1981, **114**, 1132. 274, 275, 559, 565
81CB1210 M. Lissel and E. V. Dehmlow; *Chem. Ber.*, 1981, **114**, 1210. 462
81CB1746 R. Bunnenberg and J. C. Jochims; *Chem. Ber.*, 1981, **114**, 1746. 452, 453
81CB2064 R. Bunnenberg and J. C. Jochims; *Chem. Ber.*, 1981, **114**, 2064. 559, 565, 605
81CB2075 R. Bunnenberg and J. C. Jochims; *Chem. Ber.*, 1981, **114**, 2075. 560, 565
81CB2087 N. Wiberg, G. Preiner and O. Schieda; *Chem. Ber.*, 1981, **114**, 2087. 290, 291, 380, 381, 392, 709, 710
81CB3505 N. Wiberg, G. Preiner, O. Schieda and G. Fischer; *Chem Ber.*, 1981, **114**, 3505. 373, 380, 382, 392, 709, 710
81CB3518 N. Wiberg, G. Preiner and O. Schieda; *Chem. Ber.*, 1981, **114**, 3518. 177, 380, 382, 392, 673, 709, 710
81CC72 E. Niecke and D.-A. Wildbredt; *J. Chem. Soc., Chem. Commun.*, 1981, 72. 694
81CC189 W.-K. Wong, G. Wilkinson, A. M. R. Galas, M. B. Hursthouse and M. Thornton-Pett; *J. Chem. Soc., Chem. Commun.*, 1981, 189. 339
81CC756 G. Mehta and A. V. Reddy; *J. Chem. Soc., Chem. Commun.*, 1981, 756. 550

81CC780 S. David, G. de Sennyey, C. Pascard and J. Guilhem; *J. Chem. Soc., Chem. Commun.*, 1981, 780. 322

81CC822 B. F. Bonini, G. Mazzanti, S. Sarti, P. Zanirato and G. Maccagnani; *J. Chem. Soc., Chem. Commun.*, 1981, 822. 127

81CC930 G. M. Blackburn, D. A. England and F. Kolkmann; *J. Chem. Soc., Chem. Commun.*, 1981, 930. 264

81CC969 M. Sekine, A. Kume and T. Hata; *J. Chem. Soc., Chem. Commun.*, 1981, 969. 125

81CC2496 W.-K. Wong, G. Wilkinson, A. M. Galas, M. B. Hursthouse and M. Thornton-Pett; *J. Chem. Soc., Chem. Commun.*, 1981, 2496. 339

81CCC1970 J. Mindl, J. Sulzer and M. Vecera; *Collect. Czech. Chem. Commun.*, 1981, **46**, 1970. 565

81CL375 J. Yoshimura, S. Horito and H. Hashimoto; *Chem. Lett.*, 1981, 375. 72

81CL749 T. Fujinami, S. Sato and S. Sakai; *Chem. Lett.*, 1981, 749. 463

81CL1169 J. Tsuji, K. Sato and H. Nagashima; *Chem. Lett.*, 1981, 1169. 7

81CL1337 T. Kitazume and N. Ishikawa; *Chem. Lett.*, 1981, 1337. 65

81CL1719 K. Matsui, E. Tobita, M. Ando and K. Kondo; *Chem. Lett.*, 1981, 1719. 14

81CPB3158 T. Hayashi, A. Yoshida, N. Takeda, S. Oida, S. Sugawara and E. Ohki; *Chem. Pharm. Bull.*, 1981, **29**, 3158. 553

81CZP208251 J. Pola and J. Vitek; *Czech Pat.* 208251 (1981) (*Chem. Abstr.*, 1982, **96**, 183681). 410

81DOK(256)1401 I. F. Lutsenko, A. A. Prischchenko, A. A. Borisenko and Z. S. Novikova; *Dokl. Akad. Nauk SSSR*, 1981, **256**, 1401 (*Chem. Abstr.*, 1981, **95**, 81109). 679

81GEP2855882 T. Papenfuhs and K. Gengnagel; *Ger. Pat.*, 2855882 (1981) (*Chem. Abstr.*, 1981, **94**, 15414). 504

81HCA357 B. Weidmann, L. Widler, A. G. Olivero, C. D. Maycock and D. Seebach; *Helv. Chim. Acta*, 1981, **64**, 357. 134

81IC1 A. Sekiya and D. D. DesMarteau; *Inorg. Chem.*, 1981, **20**, 1. 53

81IC3734 A. B. Burg; *Inorg Chem.*, 1981, **20**, 3734. 679

81IZV161 V. I. Erashko, O. M. Sazonova, A. A. Tishaninova and A. A. Fainzil'berg; *Izv. Akad. Nauk SSSR Ser. Khim.*, 1981, 161.

81IZV2146 O. A. Luk'yanov, T. G. Mel'nikova, E. G. Kashirskaya and V. I. Tartakovskii; *Izv. Akad. Nauk SSSR Ser. Khim.*, 1981, 2146 (*Chem. Abstr.*, 1982, **96**, 34511). 145

81IZV2632 A. F. Gontar, E. G. Bykhovskaya, A. S. Vinogradov and I. L. Knunyants; *Izv. Akad. Nauk SSSR Ser. Khim.*, 1981, 2632 (*Chem. Abstr.*, 1982, **96**, 51430). 616

81JA406 D. Schomburg, Q. C. Mir and J. M. Shreeve; *J. Am. Chem. Soc.*, 1981, **103**, 406. 253

81JA932 M. J. Robins and J. S. Wilson; *J. Am. Chem. Soc.*, 1981, **103**, 932. 533

81JA2206 P. J. Fagan, J. M. Manriquez, S. H. Vollmer, C. S. Day, V. W. Day and T. J. Marks; *J. Am. Chem. Soc.*, 1981, **103**, 2206. 516

81JA2324 M. Ishikawa, D. Kovar, T. Fuchikami, K. Nishimura, M. Kumada, T. Higuchi and S. Miyamoto; *J. Am. Chem. Soc.*, 1981, **103**, 2324. 178

81JA2995 L. J. Krause and J. A. Morrison; *J. Am. Chem. Soc.*, 1981, **103**, 2995. 246

81JA5951 L. A. Shimp, J. A. Morrison, J. A. Gurak, J. W. Chinn, Jr. and R. J. Lagow; *J. Am. Chem. Soc.*, 1981, **103**, 5951. 206

81JA6100 J. E. Ellis, K. L. Fjare and T. G. Hayes; *J. Am. Chem. Soc.*, 1981, **103**, 6100. 522

81JA6164 R. A. Moss, W. Guo, D. Z. Denney, K. N. Houk and N. G. Rondan, *J. Am. Chem. Soc.*, 1981, **103**, 6164. 55

81JAP(K)56125326 Z. Yoshida (Asahi Kagaku Kogyo K. K.); *Jpn. Kokai* 56125326 (1981) (*Chem. Abstr.*, 1982, **96**, 34840). 302

81JCR(S)276 H. M. Colquhoun; *J. Chem. Res. (S)*, 1981, 276 (*J. Chem. Res. (M)*, 3416). 610

81JCS(D)814 M. F. Lappert, P. I. Riley, P. I. W. Yarrow, J. L. Atwood, W. E. Hunter and M. J. Zaworotko; *J. Chem. Soc, Dalton Trans.*, 1981, 814. 195

81JCS(P1)52 R. C. Cambie, G. D. Mayer, P. S. Rutledge and P. D. Woodgate; *J. Chem. Soc., Perkin Trans. 1*, 1981, 52. 314, 565

81JCS(P1)455 R. E. Banks, J. M. Birchall, R. N. Haszeldine, R. A. Hughes, S. N. Nona and C. W. Stephens; *J. Chem. Soc., Perkin Trans. 1*, 1981, 455. 241

81JCS(P1)618 J. Nakayama, M. Ochiai, K. Kawada and M. Hoshino; *J. Chem. Soc., Perkin Trans. 1*, 1981, 618. 113, 316

81JCS(P1)969 I. Fleming and C. D. Floyd; *J. Chem. Soc., Perkin Trans. 1*, 1981, 969. 175

81JCS(P1)1293 B. E. Cross, A. Erasmuson and P. Filippone; *J. Chem. Soc., Perkin Trans. 1*, 1981, 1293. 472

81JCS(P1)1321 G. E. Gerhardt and R. J. Lagow; *J. Chem. Soc., Perkin Trans. 1*, 1981, 1321. 36, 251

81JCS(P1)2363 D. H. R. Barton, R. S. H. Motherwell and W. B. Motherwell; *J. Chem. Soc., Perkin Trans. 1*, 1981, 2363. 550, 557

81JCS(P1)2499 M. Yokoyama, K. Motozawa, E.-i. Kawamura and T. Imamoto; *J. Chem. Soc., Perkin Trans. 1*, 1981, 2499. 564

81JFC(17)85 R. E. Banks, A. K. Brown, R. N. Haszeldine, A. Kenny and A. E. Tipping; *J. Fluorine Chem.*, 1981, **17**, 85. 241

81JFC(17)331 R. E. Banks, J. A. Bernardin, R. N. Haszeldine, B. Justin and A. Vavayannis; *J. Fluorine Chem.*, 1981, **17**, 331. 241

81JFC(17)345 A. B. Cowell and C. Tamborski; *J. Fluorine Chem.*, 1981, **17**, 345. 11

81JFC(17)463 A. Sekiya and D. D. DesMarteau; *J. Fluorine Chem.*, 1981, **17**, 463. 240

81JFC(17)591 C. M. Irvin and A. E. Tipping; *J. Flourine Chem.*, 1981, **17**, 591. 241

81JFC(18)93 A. I. Ayi, M. Remli and R. Guedj; *J. Fluorine Chem.*, 1981, **18**, 93. 52, 54

81JFC(18)197 D. J. Burton, R. Takei and S. Shinya; *J. Fluorine Chem.*, 1981, **18**, 197. 58, 266

81JFC(18)259 H. Schachner and W. Sundermeyer; *J. Fluorine Chem.*, 1981, **18**, 259. 424, 438

81JFC(18)281 A. Marhold and E. Klauke; *J. Fluorine Chem.*, 1981, **18**, 281. 9

81JFC(18)441 W. Y. Lam and D. D. DesMarteau; *J. Fluorine Chem.*, 1981, **18**, 441. 232, 602, 615

81JFC(19)91 C. F. Service and A. E. Tipping; *J. Fluorine Chem.*, 1981, **19**, 91. 241

81JHC1127 H. A. Houwing and A. M. van Leusen; *J. Heterocycl. Chem.*, 1981, **18**, 1127. 605, 624

81JHC1477 V. Grakauskas and A. H. Albert; *J. Heterocycl. Chem.*, 1981, **18**, 1477. 149

81JHC1581 K. Rasheed and J. D. Warkentin; *J. Heterocycl. Chem.*, 1981, **18**, 1581. 553

81JINC457 P. Wisian-Neilson, M. A. Wilkins, F. C. Weigel, C. J. Foret and D. R. Martin; *J. Inorg. Nucl. Chem.*, 1981, **43**, 457. 492, 515

81JMC1089 Chau-DerLi, S. L. Mella and A. C. Sartorelli; *J. Med. Chem.*, 1981, **24**, 1089. 509

81JMR(44)54 W. Biffar, T. Gasparis-Ebeling, H. Nöth, W. Storch and B. Wrackmeyer; *J. Magn. Reson.*, 1981, **44**, 54. 405

81JOC886 R. A. McClelland, S. Gedge and J. Bohonek; *J. Org. Chem.*, 1981, **46**, 886. 80

81JOC1277 A. Sekiya and D. D. DesMarteau; *J. Org. Chem.*, 1981, **46**, 1277. 603

81JOC1512 W. S. Johnson, B. Frei and A. S. Gopalan; *J. Org. Chem.*, 1981, **46**, 1512. 134

81JOC1571 K. Ozaki, Y. Yamada and T. Oine; *J. Org. Chem.*, 1981, **46**, 1571. 143

81JOC1720 J. C. Howard; *J. Org. Chem.*, 1981, **46**, 1720. 425

81JOC3011 R. Richter and H. Ulrich; *J. Org. Chem.*, 1981, **46**, 3011. 500

81JOC3141 M. W. Fichtner and N. F. Haley; *J. Org. Chem.*, 1981, **46**, 3141 472

81JOC4573 C. E. McKenna and P. D. Shen; *J. Org. Chem.*, 1981, **46**, 4573. 264

81JOC5048 R. A. Moss, J. Włostowska, W. Guo, M. Fedorynski, J. P. Springer and J. M. Hirshfield; *J. Org. Chem.*, 1981, **46**, 5048. 55, 280, 633

81JOC5226 R. Richter, B. Tucker and H. Ulrich; *J. Org. Chem.*, 1981, **46**, 5226. 445

81JOM(209)C10 H. Schumann and F. W. Reier; *J. Organomet. Chem.*, 1981, **209**, C10. 706

81JOM(212)C4 M. F. Lappert, S. J. Miles, J. L. Atwood, M. J. Zaworotko and A. J. Carty; *J. Organomet. Chem.*, 1981, **212**, C4. 187

81JOM(217)105 E. R. F. Gesing and K. P. C. Vollhardt; *J. Organomet. Chem.*, 1981, **217**, 105. 402

81JOM(217)329 F. Wudl and D. Nalewajek; *J. Organomet. Chem.*, 1981, **217**, 329. 631

81JOM(218)81 C. Bianchini, A. Meli, A. Orlandini and L. Sacconi; *J. Organomet. Chem.*, 1981, **218**, 81. 325

81JOM(221)13 C. Eaborn, P. B. Hitchcock and P. D. Lickiss; *J. Organomet. Chem.*, 1981, **221**, 13. 380

81JOM(221)257 C. E. J. Combes, R. J. P. Corriu and B. J. L. Henner; *J. Organomet. Chem.*, 1981, **221**, 257. 402, 404

81JPR41 H. U. Kibbel, M. Kücken, E. Peters and H. Weber; *J. Prakt. Chem.*, 1981, **323**, 41. 580

81KGS907 T. E. Bezmenova, G. I. Khaskin, V. I. Slutskii, P. G. Dul'nev, L. N. Zakharov, V. I. Kulishov and Yu. T. Struchkov; *Khim. Geterotsikl. Soedin.*, 1981, 907 (*Chem. Abstr.*, 1981, **95**, 187 136). 562

81LA70 W. Kantlehner, T. Maier and J. J. Kapassakalidis; *Liebigs Ann. Chem.*, 1981, 70. 729

81LA264 E. Schaumann, S. Grabley and G. Adiwidjaja; *Liebigs Ann. Chem.*, 1981, 264. 337

81LA1244 H. Böhme, M. Brinkmann and H.-P. Steudel; *Liebigs Ann. Chem.*, 1981, 1244. 435, 546

81LA1388 W. Hanefeld and G. Glaeske; *Liebigs Ann. Chem.*, 1981, 1388. 562

81MIP68355 C. Trabalka, A. Kecskes and V. Lencu; *Rom. Pat.* 68 355 (1981) (*Chem. Abstr.*, 1982, **96**, 34 521). 223

81MI 603-01 O. B. Chalova, E. A. Kantor, T. K. Kiladze, R. A. Karakhanov and D. L. Rakhmankulov; *Izv. Akad. Nauk Gruz. SSR, Ser. Khim.*, 1981, **7**, 86 (*Chem. Abstr.*, 1981, **95**, 115 410). 70

B-81MI 606-01 E. W. Colvin; "Silicon in Organic Synthesis," Butterworth, London, 1981. 172, 174, 189

B-81MI 606-02 R. N. Butler; in "Synthetic Reagents," ed. J. S. Pizey; Ellis Horwood, Chichester, 1981, vol. 4, p. 92. 204, 207

81MI 607-01 S. Akiyama, T. Hisamoto and S. Mochizuki; *Hydrocarbon Process.*, 1981, **3**, 76 (*Chem. Abstr.*, 1981, **95**, 61 321). 215

81MI 614-01 D. Roeda and G. Westera; *Int. J. Appl. Radiat. Isot.*, 1981, **32**, 931 (*Chem. Abstr.*, 1982, **96**, 51 779). 414

81MI 614-02 J. Volford, D. Knausz, A. Meszticzky, L. Horváth and B. Csákvári; *J. Labelled Compd. Radiopharm.*, 1981, **18**, 555. 449

B-81MI 620-01 A. Haas; "Gmelin Handbook of Inorganic Chemistry" 8th edn, "Perfluorohalogenoorganic Compounds of the Main Group Elements," ed. D. Koschel, Springer, Berlin, 1981, part 9. 602

81MI 620-02 V. N. Lebedev, V. V. Yurechko and T. P. Vishnyakova; *Zh. Vses. Khim. O-va.*, 1981, **26**, 465 (*Chem. Abstr.*, 1981, **95**, 168 502). 621

81MI 620-03 A. V. Davydov and A. E. Kretov; *Zh. Vses. Khim. O-va.*, 1981, **26**, 278 (*Chem. Abstr.*, 1981, **95**, 168 503). 621

81MI 621-01 V. N. Lebedev, V. V. Yurechko and T. P. Vishnyakova; *Zh. Vses. Khim. O va*, 1981, **26**, 465 (*Chem. Abstr.*, 1981, **95**, 168 502). 646

81MI 621-02 A. V. Davydov and A. E. Kretov; *Zh. Vses. Khim. O-va*, 1981, **26**, 478 (*Chem. Abstr.*, 1981, **95**, 168 503). 649

81PS(11)311 L. Maier; *Phosphorus Sulfur*, 1981, **11**, 311. 157, 159

81RTC13 C. G. Bakker, J. W. Scheeren and R. J. F. Nivard; *Recl. Trav. Chim. Pays-Bas*, 1981, **100**, 13. 75

81S53 S. Tanimoto, S. Jo and T. Sugimoto; *Synthesis*, 1981, 53. 120

81S149 I. Degani, R. Fochi and V. Regondi; *Synthesis*, 1981, 149 472

81S380 W. Kantlehner, T. Maier and J. J. Kapassakalidis; *Synthesis*, 1981, 380. 78

81S622 W. Chin-Hsien; *Synthesis*, 1981, 622. 555

81S908 M. Yokoyama, K. Hosi and T. Imamoto; *Synthesis*, 1981, 908. 564

81S961 J. Garin, V. Martinez, J. Mayoral, E. Melendez and F. Merchan; *Synthesis*, 1981, 961. 563

81SRI279 K. Issleib, H. Schmidt and C. Wirkner; *Synth. React. Inorg. Metal-Org. Chem.*, 1981, **11**, 279 (*Chem. Abstr.*, 1981, **95**, 62 317). 658, 687

81SRI591 A. H. Cowley and R. A. Kemp; *Synth. React. Inorg. Metal-Org. Chem.*, 1981, **11**, 591. 191

81T4209 I. Rico and C. Wakselman; *Tetrahedron*, 1981, **37**, 4209. 229

81TCA285 J. J. Toman, A. A. Frost, S. Topiol, S. Jacobson and M. A. Ratner; *Theor. Chim. Acta*, 1981, **58**, 285. 386

81TCC71 M. Regitz and G. Maas; *Top. Curr. Chem.*, 1981, **97**, 71. 692

81TL323 I. Rico and C. Wakselman; *Tetrahedron Lett.*, 1981, **22**, 323. 233

81TL871 T. Shono, H. Ohmizu, S. Kawakami, S. Nakano and N. Kise; *Tetrahedron Lett.*, 1981, **22**, 871. 26

81TL1421 M. Suda; *Tetrahedron Lett.*, 1981, **22**, 1421. 695

81TL1997 M. Suda and C. Hino; *Tetrahedron Lett.*, 1981, **22**, 1997. 233, 234

81TL2005 K. Fuji, M. Ueda, K. Sumi and E. Fujita; *Tetrahedron Lett.*, 1981, 2005. 98, 101, 132, 134

81TL2159 R. Appel and A. Westerhaus; *Tetrahedron Lett.*, 1981, **22**, 2159. 688

81TL3047 M. Hanack and A. Kühnle; *Tetrahedron Lett.*, 1981, **22**, 3047. 233

81TL3097 M. Mikolajczyk, S. Grzejszczak and K. Korbacz; *Tetrahedron Lett.*, 1981, **22**, 3097. 123

81TL3485 V. M. Girijavallabhan, A. K. Ganguly, S. W. McCombie, P. Pinto and R. Rizvi; *Tetrahedron Lett.*, 1981, **22**, 3485. 553

81TL4009 J.-N. Denis, S. Desauvage, L. Hevesi and A. Krief; *Tetrahedron Lett.*, 1981, **22**, 4009. 88, 94, 317

81TL4287 M. Isobe, Y. Ichikawa and T. Goto; *Tetrahedron Lett.*, 1981, **22**, 4287. 130

81TL4365 G. R. Weisman, V. B. Johnson and M. B. Coolidge; *Tetrahedron Lett.*, 1981, **22**, 4365. 140

81TL4475 K. Issleib, E. Leissring and H. Meyer; *Tetrahedron Lett.*, 1981, **22**, 4475. 686

81TL4957 R. Appel, J. Peters and A. Westerhaus; *Tetrahedron Lett.*, 1981, **22**, 4957. 688, 689

81TL5195 K. Hatanaka, S. Tanimoto, T. Oida and M. Okano; *Tetrahedron Lett.*, 1981, **22**, 5195. 552, 555

81UKZ26 N. K. Maksudov and M. A. Safaev; *Ukr. Khim. Zh* (*Russ. Ed.*), 1981, 26 (*Chem. Abstr.*, 1982, **96**, 85 383). 561

81URP804634 I. L. Knunyants, L. S. German, S. R. Sterlin and B. L. Tumanski; *USSR Pat.* 804 634 (1981) (*Chem. Abstr.*, 1981, **95**, 24 314). 537

81USP224776 W. H. Gilligan and S. L. Stafford; *US Pat.* 224 776 (1981) (*Chem. Abstr.* 1981, **96**, 34 590). 477

81USP4255353 A. D. Wolf (Du Pont De Nemours, E. I., and Co.); *US Pat.* 4255353 (1981) (*Chem. Abstr.*, 1981, **94**, 191 733). 439

81ZAAC(472)26 P. Huppmann, D. Lentz and K. Seppelt; *Z. Anorg. Allg. Chem.*, 1981, **472**, 26. 430

81ZAAC(473)85 K. Issleib, H. Schmidt and C. Wirkner; *Z. Anorg. Allg. Chem.*, 1981, **473**, 85. 372, 688, 689

81ZAAC(474)7 E. Allenstein and G. Schrempf, *Z. Anorg. Allg. Chem.*, 1981, **474**, 7. 51

81ZAAC(475)18 R. Appel, J. Peters and R. Schmitz; *Z. Anorg. Allg. Chem.*, 1981, **475**, 18. 164

81ZAAC(475)87 G. Fritz, S. Wartanessian, E. Matern, W. Hönle and H. G. von Schnering; *Z. Anorg. Allg. Chem.*, 1981, **475**, 87. 178

81ZAAC(476)7 L. Beyer, R. Kirmse, J. Stach, R. Szargan and E. Hoyer; *Z. Anorg. Allg. Chem.*, 1981, **476**, 7. 565

81ZAAC(481)60 G. Fritz and W. Speck; *Z. Anorg. Allg. Chem.*, 1981, **481**, 60. 382

81ZAAC(481)126 K. Klaeser and G. Gattow; *Z. Anorg. Allg. Chem.*, 1981, **481**, 126. 555, 559

81ZC357 K. Issleib, H. Schmidt and C. Wirkner; *Z. Chem.*, 1981, **21**, 357. 680, 688

81ZC407 H. Oehme, E. Leissring and H. Meyer; *Z. Chem.*, 1981, **21**, 407. 686

81ZC438 G. Barnikow and A. A. Martin; *Z. Chem.*, 1981, **21**, 438. 559

81ZN(B)910 R. Thamm and E. Fluck; *Z. Naturforsch., Teil B*, 1981, **36**, 910. 490

81ZN(B)1261 A. Darmadi, A. Haas, H. Willner and H. Schnöckel; *Z. Naturforsch., Teil. B*, 1981, **36**, 1261. 588

81ZOB28 M. A. Pudovik, L. K. Kibardina, I. A. Aleksandrova, V. K. Khairullin and A. N. Pudovik; *Zh. Obshch. Khim.*, 1981, **51**, 28 (*Chem. Abstr.*, 1981, **95**, 150 542). 564

81ZOB530 M. A. Pudovik, L. K. Kibardina, I. A. Aleksandrova, V. K. Khairullin and A. N. Pudovik; *Zh. Obshch. Khim.*, 1981, **51**, 530 (*Chem. Abstr.*, 1981, **95**, 17 354). 563

81ZOB880 G. A. Yuzhakova, I. I. Lapkin, R. P. Drovneva and M. V. Mazilkina; *Zh. Obshch. Khim.*, 1981, **51**, 880 (*Chem. Abstr.*, 1981, **95**, 79 780). 656

81ZOB1035 M. I. Bakhitov, V. V. Zharkov, E. V. Kuznetsov and F. L. Kligman; *Zh. Obshch. Khim.*, 1981, **51**, 1035 (*Chem. Abstr.*, 1981, **95**, 96503). 366

81ZOB1842 O. P. Shvaika and V. F. Lipnitskii; *Zh. Obshch. Khim.*, 1981, **51**, 1842 (*Chem. Abstr.*, 1982, **96**, 20035). 115

81ZOB2189 O. I. Kolodyazhnyi and V. P. Kukhar; *Zh. Obshch. Khim.*, 1981, **51**, 2189, (*Chem. Abstr.*, 1982, **96**, 20167). 684

81ZOB2241 V. I. Boev and A. V. Dombrovskii; *Zh. Obshch. Khim.*, 1981, **51**, 2241 (*Chem. Abstr.*, 1982, **96**, 122929). 208

81ZOB2630 A. A. Prishchenko and I. F. Lutsenko; *Zh. Obshch. Khim.*, 1981, **51**, 2630, (*Chem. Abstr.*, 1982, **96**, 104395). 265, 679

81ZOR180 V. P. Kukhar, V. I. Pasternak, I. V. Shevchenko, M. B. Shevchenko, A. P. Marchenko and Yu. P. Makovetskii; *Zh. Org. Khim.*, 1981, **17**, 180 (*Chem. Abstr.*, 1981, **95**, 6397). 612, 623

81ZOR191 I. M. Bazazova, R. G. Dubenko and P. S. Pel'kis; *Zh. Org. Khim.*, 1981, **17**, 191 (*Chem. Abstr.*, 1981, **95**, 7129). 564

81ZOR398 V. I. Gorbatenko and L. F. Lur'e; *Zh. Org. Khim.*, 1981, **17**, 398 (*Chem. Abstr.*, 1981, **95**, 97313). 617

81ZOR1539 N. A. Batyrbaev, V. V. Zorin, S. S. Zlotskii and D. L. Rakhmankulov; *Zh. Org. Khim.*, 1981, **17**, 1539 (*Chem. Abstr.*, 1982, **96**, 6626). 478

81ZOR2269 L. V. Sokolyanskaya, A. N. Vavilova, A. N. Volkov and B. A. Trofimov; *Zh. Org. Khim.*, 1981, **17**, 2269 (*Chem. Abstr.*, 1982, **96**, 103 605). 562

82AG(E)80 R. Appel, J. Peters and A. Westerhaus; *Angew. Chem., Int. Ed. Engl.*, 1982, **21**, 80. 293, 693

82AG(E)219 R. Appel and U. Kündgen; *Angew. Chem., Int. Ed. Engl.*, 1982, **21**, 219. 689

82AG(E)410 T. Kauffmann; *Angew. Chem., Int. Ed. Engl.*, 1982, **21**, 410. 170, 205, 206, 210

82AG(E)443 U. Hartkopf and A. de Meijere; *Angew. Chem., Int. Ed. Engl.*, 1982, **21**, 443. 380

82AG(E)448 R. Appel, H. Förster, B. Laubach, F. Knoll and I. Ruppert; *Angew. Chem., Int. Ed. Engl.*, 1982, **21**, 448. 665, 686
82AG(E)545 H. J. Bestmann and A. Bomhard; *Angew. Chem., Int. Ed. Engl.*, 1982, **21**, 545. 702
82AHC(30)1 C. J. Moody; *Adv. Heterocycl. Chem.*, 1982, **30**, 1. 506
82AJC1709 C. Copeland and R. V. Stick; *Aust. J. Chem.*, 1982, **35**, 1709. 544
82AOC(20)115 H. Yaozeng and S. Yanchang; *Adv. Organomet. Chem.*, 1982, **20**, 115. 707
82AP(315)103 W. Hanefeld, G. Glaeske and H.-J. Staude; *Arch. Pharm.* (*Weinheim, Ger.*) 1982, **315**, 103. 564
82BCJ224 M. Sekine, M. Nakajima, A. Kume, A. Hashizume and T. Hata; *Bull. Chem. Soc. Jpn.*, 1982, **55**, 224. 125
82BCJ339 S. Tanimoto, T. Oida, T. Kokubo and M. Okano; *Bull. Chem. Soc. Jpn.*, 1982, **55**, 339. 552
82BCJ1106 S. Smolinski, M. Balazy, H. Iwamura, T. Sugawara, Y. Kawada and M. Iwamura; *Bull. Chem. Soc. Jpn.*, 1982, **55**, 1106. 304
82BCJ1163 T. Hirao, A. Yamada, K. Hayashi, Y. Ohshiro and T. Agawa; *Bull. Chem. Soc. Jpn.*, 1982, **55**, 1163. 162
82BCJ1174 T. Fujinami, S. Sato, N. Uchida and S. Sakai; *Bull. Chem. Soc. Jpn.*, 1982, **55**, 1174. 554
82CB818 H. H. Karsch; *Chem. Ber.*, 1982, **115**, 818. 151, 369
82CB860 R. Reck and J. C. Jochims; *Chem. Ber.*, 1982, **115**, 860. 240, 453, 611
82CB1252 J. Goerdeler and A. Schulze; *Chem. Ber.*, 1982, **115**, 1252. 559
82CB1259 J. Goerdeler and A. Schulze; *Chem. Ber.*, 1982, **115**, 1259. 559
82CB1617 R. Appel, V. Barth and M. Halstenberg; *Chem. Ber.*, 1982, **115**, 1617. 686
82CB1721 W. Kantlehner, E. Haug, H. Isak, W. Schultz, S. Hippich, R. Baur and H. Hagen; *Chem. Ber.*, 1982, **115**, 1721. 143
82CB2371 R. Appel, M. Halstenberg, F. Knoch and H. Kunze; *Chem. Ber.*, 1982, **115**, 2371. 687
82CB2516 E. Schaumann, H. Nimmesgern, G. Adiwidjaja and L. Carlsen; *Chem. Ber.*, 1982, **115**, 2516. 637
82CB3587 R. Bunnenberg, J. C. Jochims and H. Härle; *Chem. Ber.*, 1982, **115**, 3587. 605, 619
82CB3894 E. V. Dehmlow and W. Broda; *Chem. Ber.*, 1982, **115**, 3894. 32
82CC51 M. Green, K. A. Mead, R. M. Mills, I. D. Salter, F. G. A. Stone and P. Woodward; *J. Chem. Soc., Chem. Commun.*, 1982, 51. 340
82CC286 S. Al-Jibori and B. L. Shaw; *J. Chem. Soc., Chem. Commun.*, 1982, 286. 151
82CC536 P. A. Bartlett and N. I. Carruthers; *J. Chem. Soc., Chem. Commun*, 1982, 536. 666
82CC773 L. W. Bateman, M. Green, J. A. K. Howard, K. A. Mead, R. M. Mills, I. D. Salter, F. G. A. Stone and P. Woodward; *J. Chem. Soc., Chem. Commun.*, 1982, 773. 343
82CC930 S. O. Grim, S. A. Sangokoya, I. J. Colquhoun and W. McFarlane; *J. Chem. Soc., Chem. Commun.*, 1982, 930. 369
82CC1183 J. S. Grossert and P. K. Dubey; *J. Chem. Soc., Chem. Commun.*, 1982, 1183. 90, 91, 307
82CC1323 M. F. Lappert, L. M. Engelhardt, C. L. Raston and A. H. White; *J. Chem. Soc., Chem. Commun.*, 1982, 1323. 174, 190
82CC1407 T. Fjeldberg, A. Haaland, M. F. Lappert, B. E. R. Schilling, R. Seip and A. J. Thorne; *J. Chem. Soc., Chem. Commun.*, 1982, 1407. 186, 195
82CL1453 T. Kitazume and N. Ishikawa; *Chem. Lett.*, 1982, 1453. 22
82CL1519 A. Sekiya and T. Umemoto; *Chem. Lett.*, 1982, 1519. 235
82COMC-I(1)411 T. Onak; *Comp. Organomet. Chem., 1st edn.*, 1982, **1**, 411. 169
82COMC-I(1)543 L. J. Todd; *Comp. Organomet. Chem., 1st edn.*, 1982, **1**, 543. 169
82COMC-I(2)1 D. A. Armitage; *Comp. Organomet. Chem., 1st edn.*, 1982, **2**, 1. 172
82COMC-I(2)681 J. L. Wardell; *Comp. Organomet. Chem., 1st edn.*, 1982, **2**, 681. 707
82COMC-I(4)523 W. P. Fehlhammer and H. Stolzenberg; *Comp. Organometal. Chem., 1st edn*, 1982, **4**, 523. 675
82COMC-I(7)265 E. Negishi; *Comp. Organomet. Chem., 1st edn.*, 1982, **7**, 265. 172, 204
82COMC-I(7)515 P. D. Magnus, T. Sarkar and S. Djuric; *Comp. Organomet. Chem., 1st edn.*, 1982, **7**, 515. 172
82CPB534 T. Morikawa, M. Ozeki, N. Umino, M. Kawamori, Y. Arai and K. Tsujihara; *Chem. Pharm. Bull.*, 1982, **30**, 534. 504
82CSC1571 M. Sikirica and D. Grdenić; *Cryst. Struct. Commun.*, 1982, **11A**, 1571 (*Chem. Abstr.*, 1983, **98**, 207910). 207
82CSR435 H. W. Kroto; *Chem. Soc. Rev.*, 1982, **11**, 435. 678
82CZ239 A. Haas; *Chem.-Ztg.*, 1982, **106**, 239. 228
82EUP43978 G. Heywang, E. Kuehle, I. Hammann and B. Homeyer; *Eur. Pat.* 43 978 (1982) (*Chem. Abstr.*, 1982, **96**, 199 882). 513
82EUP46557 E. Kuehle, W. Brandes and P. E. Frohberger (Bayer AG); *Eur. Pat.* 46 557 (1982) (*Chem. Abstr.*, 1982, **96**, 199346). 439
82EUP62834 E. Winkelmann, W. Dittmar and W. Raether; *Eur. Pat.* 62 834 (1982) (*Chem. Abstr.*, 1982, **98**, 125 682). 533
82EUP65358 G. Whittaker (ICI); *Eur. Pat. Appl.* 65358 (1982) (*Chem. Abstr.* 1983, **98**, 143282). 28
82FES205 M. R. Chaurasia, S. K. Sharma and B. Arora; *Farmaco Ed. Sci.*, 1982, **37**, 205 (*Chem. Abstr.*, 1982, **96**, 162 275). 561
82GEP3044216 B. Baasner, G. M. Petruk, H. Hagemann and E. Klauke (Bayer AG); *Ger. Pat.* 3044216 (1982) (*Chem. Abstr.*, 1982, **97**, 181932). 238
82H(18)201 T. Aoki, K. Minami, T. Kubota, Y. Hamashima and W. Nagata; *Heterocycles*, 1982, **18**, 201. 107
82H(19)1191 G. Stájer, A. E. Szabó, F. Fülöp and G. Bernáth; *Heterocycles*, 1982, **19**, 1191. 558
82H(19)2283 A. Katoh, C. Kashima and Y. Omote; *Heterocycles*, 1982, **19**, 2283. 579
82HCA371 D. Bravetti, R. M. Bettolo and A. Lupi; *Helv. Chim. Acta*, 1982, **65**, 371. 550
82HOU(E1)583 M. Regitz; *Methoden Org. Chem.* (*Houben-Weyl*), 1982, **E1**, 583. 692
82HOU(E1)616 H.-J. Bestmann and R. Zimmermann; *Methoden Org. Chem.* (*Houben-Weyl*), 1982, **E1**, 616. 694, 695, 697, 698

82HOU(13/3a)1	R. Köster; *Methoden Org. Chem.* (*Houben-Weyl*), 1982, **13/3a**, 1.	172, 181
82ICA(58)149	F. Glockling and N. M. N. Gowda; *Inorg. Chim. Acta*, 1982, **58**, 149.	396
82IC1704	A. Mayr, Y. C. Lin, N. M. Baog and H. D. Kaesz; *Inorg. Chem.*, 1982, **21**, 1704.	517
82IC3781	G. Gervasio, R. Rossetti, P. L. Stanghellini and G. Bor; *Inorg. Chem.*, 1982, **21**, 3781.	352
82IZV161	V. I. Erashko, O. M. Sazonova, A. A. Tishaninova and A. A. Fainzil'berg; *Izv. Akad. Nauk SSSR Ser. Khim.*, 1982, 161 (*Chem. Abstr.*, 1982, **96**, 142 370).	116, 283, 355
82IZV460	L. V. Yashkina, B. V. Kopylova, R. G. Gasanov and R. Kh. Freidlina; *Izv. Akad. Nauk SSSR Ser. Khim.*, 1982, 460 (*Chem. Abstr.*, 1982, **96**, 217 413).	323
82IZV1426	A. N. Pudovik, G. V. Romanov and T. Y. Stepanova; *Izv. Akad. Nauk SSSR Ser. Khim*, 1982, 1416 (*Chem. Abstr.*, 1982, **97**, 163 106).	658
82IZV2370	V. A. Dorokhov, L. G. Vorontsova, M. G. Kurella, O. S. Chizhov and B. M. Mikhailov; *Izv. Akad. Nauk SSSR Ser. Khim*, 1982, 2370 (*Chem. Abstr.*, 1983, **98**, 126 181).	662
82IZV2637	M. E. Niyazymbetov and V. A. Petrosyan; *Izv. Akad. Nauk SSSR Ser. Khim.*, 1982, 2637 (*Chem. Abstr.*, 1983, **98**, 125 061).	149
82JA1769	W. F. Bailey and M.-J. Shih; *J. Am. Chem. Soc.*, 1982, **104**, 1769.	466
82JA2637	J. A. Gurak, J. W. Chinn, Jr. and R. J. Lagow; *J. Am. Chem. Soc.*, 1982, **104**, 2637.	206
82JA4034	W. Y. Lam and D. D. DesMarteau; *J. Am. Chem. Soc.*, 1982, **104**, 4034.	256
82JA4708	D. R. Williams, B. A. Barner, K. Nishitani and J. G. Phillips; *J. Am. Chem. Soc.*, 1982, **104**, 4708.	472, 473
82JA4724	E. J. Corey and J. Kang; *J. Am. Chem. Soc.*, 1982, **104**, 4724.	705
82JA5621	E. M. Holt, K. H. Whitmire and D. F. Shriver; *J. Am. Chem. Soc.*, 1982, **104**, 5621.	340
82JA5820	A. H. Cowley, J. E. Kilduff, T. H. Newman and M. Pakulski; *J. Am. Chem. Soc.*, 1982, **104**, 5820.	373
82JA7225	L. E. Overman and E. J. Jacobsen; *J. Am. Chem. Soc.*, 1982, **104**, 7225.	473
82JA7345	F. J. Landro, J. A. Gurak, J. W. Chinn, Jr., R. M. Newman and R. J. Lagow; *J. Am. Chem. Soc.*, 1982, **104**, 7345.	205
82JAP8292513	Hodogaya Chemical Co. Ltd.; *Jpn. Pat.* 8292513 (1982) (*Chem. Abstr.*, 1982, **97**, 181747).	413, 415
82JCED97	W. H. Gilligan and M. E. Sitzmann; *J. Chem. Eng. Data*, 1982, **27**, 97.	449
82JCR(S)79	G. Giesemann, G. Von Hinrichs and I. Ugi; *J. Chem. Res. (S)*, 1982, 79.	417
82JCR(S)132	M. V. Garad and J. W. Wilson; *J. Chem. Res. (S)*, 1982, 132.	199
82JCS(D)2281	R. Davis and I. F. Groves; *J. Chem. Soc., Dalton Trans.*, 1982, 2281.	7
82JCS(P1)653	H. Singh, A. S. Ahuja and N. Malhotra; *J. Chem. Soc., Perkin Trans. 1*, 1982, 653.	559
82JCS(P1)685	R. E. Banks and N. Dickinson; *J. Chem. Soc., Perkin Trans. 1*, 1982, 685.	241
82JCS(P1)2085	D. H. R. Barton, J. D. Elliott and S. D. Géro; *J. Chem. Soc., Perkin Trans. 1*, 1982, **1**, 2085.	641, 643
82JCS(P1)2149	T. Nishio and Y. Omote; *J. Chem. Soc., Perkin Trans. 1*, 1982, 2149.	575
82JCS(P1)2813	R. Galli, O. Palla and F. Gozzo; *J. Chem. Soc. Perkin Trans. 1*, 1982, 2813.	494
82JCS(P1)2871	H. C. Ooi and H. Suschitzky; *J. Chem. Soc., Perkin Trans. 1*, 1982, 2871.	104
82JFC(19)227	K. K. Johri, Y. Katsuhara and D. D. Desmarteau; *J. Fluorine Chem.*, 1982, **19**, 227.	428, 430
82JFC(19)411	J. S. Thrasher and A. F. Clifford; *J. Fluorine Chem.*, 1982, **19**, 411.	616, 651
82JFC(19)553	B. Baasner and E. Klauke; *J. Fluorine Chem.*, 1982, **19**, 553.	439
82JFC(20)89	D. J. Burton, S. Shin Ya and H. S. Kesling; *J. Fluorine Chem.*, 1982, **20**, 89.	213, 222, 223, 224, 226, 227, 228
82JFC(20)121	D. J. Burton, T. Ishihara and R. M. Flynn; *J. Fluorine Chem.*, 1982, **20**, 121.	264
82JFC(20)617	D. J. Burton, D. J. Pietrzyk, T. Ishihara, T. Fonong and R. M. Flynn; *J. Fluorine Chem.*, 1982, **20**, 617.	264
82JFC(21)365	V. I. Popov, V. N. Boiko and L. M. Yagupol'skii; *J. Fluorine Chem.*, 1982, **21**, 365.	233
82JGU131	L. V. Okhlobystina, T. I. Cherkasova and V. A. Tyurikov; *J. Gen. Chem. USSR* (*Engl. Transl.*) 1982, **52**, 131.	469
82JHC1205	R. L. Webb and C. S. Labaw; *J. Heterocycl. Chem.*, 1982, **19**, 1205.	251, 625
82JIC882	Y. D. Kulkarni and B. Kumar; *J. Indian Chem. Soc.*, 1983, **60**, 882 (*Chem. Abstr.*, 1984, **101**, 23 411).	546
82JMC178	J. Martinez, J. Oiry, J. L. Imbach and F. Winternitz; *J. Med. Chem.*, 1982, **25**, 178.	506
82JMC557	M. Uda, K. Toyooka, K. Hori, M. Shibuya, S. Kubota and H. Terada; *J. Med. Chem.*, 1982, **25**, 557.	563, 565
82JOC132	S. J. Cristol and D. G. Seapy; *J. Org. Chem.*, 1982, **47**, 132.	550
82JOC1145	N. H. Andersen, A. D. Denniston and D. A. McCrae; *J. Org. Chem.*, 1982, **47**, 1145.	134
82JOC1284	P. A. Bartlett, N. I. Carruthers, B. M. Winter and K. P. Long; *J. Org. Chem.*, 1982, **47**, 1284.	118, 119, 636, 666
82JOC3012	Z. Tashma; *J. Org. Chem.*, 1982, **47**, 3012.	581
82JOC3192	E. R. F. Gesing; *J. Org. Chem.*, 1982, **47**, 3192.	402
82JOC3521	D. A. Bright, D. E. Mathisen and H. E. Zieger; *J. Org. Chem.*, 1982, **47**, 3521.	425
82JOC4013	P. A. Bartlett, J. D. Meadows, E. G. Brown, A. Morimoto and K. K. Jernstedt; *J. Org. Chem.*, 1982, **47**, 4013.	468
82JOC4177	R. A. Moss, L. A. Perez, J. Włostowska, W. Guo and K. Krogh-Jespersen; *J. Org. Chem.*, 1982, **47**, 4177.	280, 632
82JOC4626	A. Bongini, G. Cardillo, M. Orena, G. Porzi and S. Sandri; *J. Org. Chem.*, 1982, **47**, 4626.	469
82JOC5209	W. A. Hoffmann, III; *J. Org. Chem.*, 1982, **47**, 5209.	464
82JOM(231)191	U. Sicker, A. Meller and W. Maringgele; *J. Organomet. Chem.*, 1982, **231**, 191.	662
82JOM(233)C59	W. R. Roper and A. H. Wright; *J. Organomet. Chem.*, 1982, **233**, C59.	718, 722
82JOM(234)C9	G. R. Clark, S. V. Hoskins and W. R. Roper; *J. Organomet. Chem.*, 1982, **234**, C9.	717, 718, 722
82JOM(234)C17	H. Klusik and A. Berndt; *J. Organomet. Chem.*, 182, **234**, C17.	713
82JOM(235)265	C. Eaborn, N. Retta, J. D. Smith and P. B. Hitchcock; *J. Organomet. Chem.*, 1982, **235**, 265.	387
82JOM(236)C7	G. R. Clark, W. R. Roper and A. H. Wright; *J. Organomet. Chem.*, 1982, **236**, C7.	718

82JOM(236)7	M. Ishikawa, K. Nishimura, H. Ochiai and M. Kumada; *J. Organomet. Chem.*, 1982, **236**, 7.	383
82JOM(238)327	D. Grdenić, B. Korpar-Čolig, M. Sikirica and M. Bruvo; *J. Organomet. Chem.*, 1982, **238**, 327.	207
82JOM(239)93	C. Eaborn; *J. Organomet. Chem.*, 1982, **239**, 93.	172
82JOU1958	L. M. Yagupol'skii, M. I. Drokina and K. N. Bil'dinov; *J. Org. Chem. USSR* (*Engl. Transl.*), 1982, **18**, 1958.	501
82JPR227	R. Beckert and R. Mayer; *J. Prakt. Chem.*, 1982, **324**, 227.	500
82JPR669	E. Fanghänel, K.-H. Kühnemund and A. M. Richter; *J. Prakt. Chem.*, 1982, **324**, 669.	729
82KGS121	E. Fischer; *Khim. Geterotsikl. Soedin.*, 1982, 121 (*Chem. Abstr.*, 1982, **96**, 162659).	622
82KO(17)416	E. E. Hardy; *Kirk-Othmer Encyc.*, 1982, **17**, 416.	411
82LA231	W. Walter and W. Ruback; *Liebigs Ann. Chem.*, 1982, 231.	627
82LA507	W. Kantlehner, T. Maier, W. Löffler and J. J. Kapassakalidis; *Liebigs Ann. Chem.*, 1982, 507.	296, 297, 298, 300, 321
82LA2105	G. Himbert and W. Schwickerath; *Liebigs Ann. Chem.*, 1982, **12**, 2105.	620, 651
B-82MI 601-01	R. Filler and Y. Kobayashi (eds.); "Biomedical Aspects of Fluorine Chemistry," Elsevier Biomedical, New York, 1982.	9
B-82MI 601-02	D. J. Sam, F. D. Marsh, W. B. Farnham and B. E. Smart; in "Current Trends in Organic Synthesis," ed. H. Nozaki, Pergamon, New York, 1982, p. 413.	28
82MI 603-01	B. Trathnigg, G. Hippmann and H. Junek; *Angew. Makromol. Chem.*, 1982, **105**, 1 (*Chem. Abstr.*, 1982, **97**, 92 177).	73
82MI 604-01	E. Magnien, M. Shaikh, J. T. Suh and H. Jones; *J. Labelled Compd. Radiopharm.*, 1982, **19**, 1145.	118
82MI 606-01	V. S. Bleshinskii and S. V. Bleshinskii; *Izv. Akad. Nauk Kirg. SSSR*, 1982, 47 (*Chem. Abstr.*, 1982, **97**, 216 264).	205
82MI 614-01	D. B. Dell'amico and F. Calderazzo; *Congr. Naz. Chim. Inorg.* [*Att.*], 15th (1982), 173 (*Chem. Abstr.*, 1984, **100**, 109 741).	412
82MI 614-02	G. Dittmer and U. Niemann; *Philips J. Res.*, 1982, **37(1–2)**, 1 (*Chem. Abstr.*, 1982, **97**, 116 243).	418
82MI 614-03	H.-B. König, K. G. Metzer, H. A. Offe and W. Schröck; *Eur. J. Med. Chem. – Chim. Ther.*, 1982, **17**, 59.	445
B-82MI 615-01	R. Wegler, "Chemie der Pflanzenschutz- und Schädlingsbekämpfungsmittel," Springer, Berlin, Heidelberg, New York, 1982, vol. 8, p. 158.	492
B-82MI 615-02	H. D. Jakubke and H. Jeschkeit; "Aminosäuren, Peptide, Proteine," Verlag Chemie, Weinheim, 1982.	486
82MI 617-01	V. K. Pandey and H. C. Lohani; *Acta Cienc. Indica Chem.*, 1982, **8**, 77 (*Chem. Abstr.*, 1983, **98**, 125 981).	563
82MI 617-02	F. A. Bakhtiyarova, M. M. Makhamatkhanov and N. Kh. Maksudov; *Uzb. Khim. Zh.*, 1982, 59 (*Chem. Abstr.*, 1983, **98**, 72 256).	563
82MI 617-03	Y. S. Shim, O. Y. Chung, W. J. Kim and M. H. Lee; *Yakhak Hoe Chi*, 1982, **26**, 91 (*Chem. Abstr.*, 1982, **97**, 162 320).	559
82MI 620-01	C. Raby, J. Claude and J. Buxeraud; *Bull. Soc. Pharm. Bordeaux*, 1982, **121**, 26 (*Chem. Abstr.*, 1983, **98**, 53 126).	623
82MI 620-02	C. Raby and J. Buxeraud; *Bull. Soc. Pharm. Bordeaux*, 1982, **121**, 21 (*Chem. Abstr.*, 1983, **98**, 53 125).	618, 623
B-82MI 622-01	L. Weber; in "The Chemistry of the Metal-Carbon Bond," ed. F. R. Hartley and S. Patai, Wiley, Chichester, 1982, p. 91.	705
82OM20	D. S. Matteson and R. J. Moody; *Organometallics*, 1982, **1**, 20.	199
82OM778	C. Bianchini, C. A. Ghilardi, A. Meli, S. Midollini and A. Orlandini; *Organometallics*, 1982, **1**, 778.	327
82OM1114	A. A. Bahsoun, J. A. Osborn, C. Voelker, J. J. Bonnet and G. Lavigne; *Organometallics*, 1982, **1**, 1114.	155
82OM1720	A. H. Cowley, E. A. V. Ebsworth, R. A. Kemp, D. W. H. Rankin and C. A. Stewart; *Organometallics*, 1982, **1**, 1720.	165, 167
82PIA441	S. Rajappa, V. Sudarsanam, V. G. Yadav and B. V. Gaikwad; *Proc. Indian Acad. Sci.*, 1982, **91**, 441 (*Chem. Abstr.*, 1983, **98**, 89 235).	647
82PJC1369	W. Dmowski and M. Kaminski; *Pol. J. Chem.*, 1982, **56**, 1369.	52
82POL129	S.-C. Chang and D. D. DesMarteau; *Polyhedron*, 1982, **1**, 129.	614, 615
82RCR1	V. V. Tyuleneva, E. M. Rokhlin and I. L. Knunyants; *Russ. Chem. Rev.* (*Engl. Transl.*), 1982, **51**, 1.	694
82S1	H. Emde, D. Domsch, H. Feger, U. Frick, A. Götz, H. H. Hergott, K. Hoffman, W. Kober, K. Krägeloh, T. Oesterle, W. Steppan, W. West and G. Simchen; *Synthesis*, 1982, 1.	43
82S590	F. Merchán, J. Garín and F. Meléndez; *Synthesis*, 1982, 590.	627
82S771	L. Henriksen; *Synthesis*, 1982, 771.	594, 597
82S951	A. Marhold and E. Klauke; *Synthesis*, 1982, 951.	28
82SC25	C. Botthegi, G. Chelucci and M. Marchetti; *Synth. Commun.*, 1982, **12**, 25.	417
82SC213	K. Takeda and H. Ogura; *Synth. Commun.*, 1982, **12**, 213.	507
82SC415	C. R. Degenhardt; *Synth. Commun.*, 1982, **12**, 415.	159
82SC513	M. Tordeux and C. Wakselman; *Synth. Commun.*, 1982, **12**, 513.	423
82SC897	S. Juliá, G. Tagle and J. C. Vega; *Synth. Commun.*, 1982, **12**, 897.	475
82T1163	K. A. Jørgensen, A.-B. A. G. Ghattas and S.-O. Lawesson; *Tetrahedron*, 1982, **38**, 1163.	503
82T3355	J. N. C. Hood, D. Lloyd, W. A. MacDonald and T. M. Shepherd; *Tetrahedron*, 1982, **38**, 3355.	708

82TL309 P. Pappalardo, E. Ehlinger and P. Magnus; *Tetrahedron Lett.*, 1982, **23**, 309. 124

82TL913 G. H. Hakimelahi and A. Ugolini; *Tetrahedron Lett.*, 1982, **23**, 913. 122

82TL1557 C. Paulmier and P. Lerouge; *Tetrahedron Lett.*, 1982, 1557. 93

82TL1609 T. Shono, N. Kise, A. Yamazaki and H. Ohmizu; *Tetrahedron Lett.*, 1982, **23**, 1609. 25, 26

82TL1979 E. J. Corey and P. B. Hopkins; *Tetrahedron Lett.*, 1982, 1979. 540

82TL2017 R. Appel and A. Westerhaus; *Tetrahedron Lett.*, 1982, **23**, 2017. 288, 693

82TL2323 M. Obayashi, E. Ito, K. Matsui and K. Kondo; *Tetrahedron Lett.*, 1982, **23**, 2323. 58, 264, 266

82TL2833 H. G. Werchan and G. Dittrich; *Tetrahedron Lett.*, 1982, **23**, 2833. 498

82TL3407 J. N. Denis and A. Krief; *Tetrahedron Lett.*, 1982, 3407. 88, 94

82TL3539 S. Knapp and D. V. Patel; *Tetrahedron Lett.*, 1982, **23**, 3539. 616

82TL3929 T. Umemoto and O. Miyano; *Tetrahedron Lett.*, 1982, **23**, 3929. 233, 234

82TL4801 T. Shono, H. Ohmizu and N. Kise; *Tetrahedron Lett.*, 1982, **23**, 4801. 26

82TL4941 C. Couret, J. Escudié and J. Satgé; *Tetrahedron Lett.*, 1982, **23**, 4941. 373

82UKZ395 N. G. Pavlenko and V. P. Kukhar; *Ukr. Khim. Zh.* (*Russ. Ed.*), 1982, **48**, 395 (*Chem. Abstr.*, 1982, **97**, 23735). 620

82USP4333883 S. J. Nelson (Upjohn Co.); *US Pat.* 4333883 (1982) (*Chem. Abstr.*, 1982, **97**, 109574). 439

82YGK352 K. Fujii, H. Tamura and Y. Kinoshita; *Yuki Gosei Kagaku Kyokaishi*, 1982, **40**, 352 (*Chem. Abstr.*, 1982, **87**, 38536). 550

82ZAAC(485)203 G. Gattow and W. Eul; *Z. Anorg. Allg. Chem.*, 1982, **485**, 203. 561

82ZAAC(485)23 G. Becker, W. Massa, O. Mundt and R. Schmidt; *Z. Anorg. Allg. Chem.*, 1982, **485**, 23. 160

82ZAAC(488)94 G. Gattow and R. Gerner; *Z. Anorg. Allg. Chem.*, 1982, **488**, 94. 561

82ZN(B)81 V. Zürn, W. Schwarz, W. Rozdzinski and A. Schmidt; *Z. Naturforsch., Teil B*, 1982, **37**, 81. 438

82ZOB1538 O. I. Kolodyazhnyi; *Zh. Obshch. Khim.*, 1982, **52**, 1538 (*Chem. Abstr.*, 1982, **97**, 163120). 699

82ZOB1925 L. N. Markovskii, V. D. Romanenko and T. I. Pidvarko; *Zh. Obshch. Khim.*, 1982, **52**, 1925 (*Chem. Abstr.*, 1982, **97**, 216330). 685

82ZOB2187 V. L. Foss, N. V. Lukashev and I. F. Lutsenko; *Zh. Obshch. Khim.*, 1982, **52**, 2187 (*Chem. Abstr.*, 1983, **98**, 126227). 668

82ZOR228 A. E. Sorochinskii, A. M. Aleksandrov and V. P. Kukhar; *Zh. Org. Khim.*, 1982, **18**, 228 (*Chem. Abstr.*, 1982, **96**, 142320). 19

83AG(E)545 A. Schmidpeter, S. Lochschmidt and A. Willhalm; *Angew. Chem., Int. Ed. Engl.*, 1983, **22**, 545. 660, 686

83AG(E)632 M. Baudler and W. Leonhardt; *Angew. Chem., Int. Ed. Engl.*, 1983, **22**, 632. 366

83AG(E)730 G. Fritz and H. Bauer; *Angew. Chem., Int. Ed. Engl.*, 1983, **22**, 730. 268

83AG(E)733 A. Maercker, M. Theis, A. J. Kos and P. von R. Schleyer; *Angew. Chem., Int. Ed. Engl.*, 1983, **22**, 733. 205

83AG(E)785 R. Appel and W. Paulen; *Angew. Chem., Int. Ed. Engl.*, 1983, **22**, 785. 454

83AG(E)907 H. Schmidbaur; *Angew. Chem., Int. Ed. Engl.*, 1983, **22**, 907. 705

83AG(E)1005 N. Wiberg and G. Wagner; *Angew. Chem., Int. Ed. Engl.*, 1983, **22**, 1005. 178, 387, 392, 709

83AQ(C)115 F. L. Merchan Alvarez; *An. Quim., Ser. C*, 1983, **79**, 115 (*Chem. Abstr.*, 1984, **101**, 230445). 605

83BCJ2969 A. Saito and B. Shimizu; *Bull. Chem. Soc. Jpn.*, 1983, **56**, 2969. 444

83CB109 R. Appel and W. Paulen, *Chem. Ber.*, 1983, **116**, 109. 454, 686

83CB114 R. Appel, M. Huppertz and A. Westerhaus; *Chem. Ber.*, 1983, **116**, 114. 288

83CB549 J. Högel, A. Schmidpeter and W. S. Sheldrick; *Chem. Ber.*, 1983, **116**, 549. 156

83CB1257 D. Bielefeldt and A. Haas; *Chem. Ber.*, 1983, **116**, 1257. 440

83CB1285 U. Rheude and W. Sundermeyer; *Chem. Ber.*, 1983, **116**, 1285. 254

83CB1386 H. Schmidbaur and U. Deschler; *Chem. Ber.*, 1983, **116**, 1386. 701

83CB1463 R. W. Saalfrank, P. Schierling and P. Schätzlein; *Chem. Ber.*, 1983, **116**, 1463. 75

83CB1623 M. Eschwey, W. Sundermeyer and D. S. Stephenson; *Chem. Ber.*, 1983, **116**, 1623. 253, 254, 530

83CB1873 R. Appel, F. Knoch, B. Laubach and R. Sievers, *Chem. Ber.*, 1983, **116**, 1873. 686

83CB1955 H. Schmidbaur, C. Zybill, C. Krueger and H. J. Kraus; *Chem. Ber.*, 1983, **116**, 1955. 123, 698

83CB2068 A. Schönberg, E. Singer and W. Stephan; *Chem. Ber.*, 1983, **116**, 2068. 143

83CB2371 R. Appel and W. Paulen; *Chem. Ber.*, 1983, **116**, 2371. 519

83CB3325 D. M. Ceacareanu, M. R. C. Gerstenberger and A. Haas; *Chem. Ber.*, 1983, **116**, 3325. 274

83CB3482 R. W. Saalfrank, P. Schierling and W. Hafner; *Chem. Ber.*, 1983, **116**, 3482. 75

83CCR1 W. C. Kaska; *Coord. Chem. Rev.*, 1983, **48**, 1. 705

83CC140 P. J. Atkins and V. Gold; *J. Chem. Soc., Chem. Commun.*, 1983, 140. 30

83CC235 V. Y. Lee, E. M. Engler, R. R. Schumaker and S. S. P. Parkin; *J. Chem. Soc., Chem. Commun.*, 1983, 235. 590

83CC239 K. J. H. Kruithof, R. F. Schmitz and G. W. Klumpp; *J. Chem. Soc., Chem. Commun.*, 1983, 239. 130, 133

83CC283 P. J. Atkins, V. Gold and W. N. Wassef; *J. Chem. Soc., Chem. Commun.*, 1983, 283. 25, 31

83CC294 K. Lerstrup, M. Lee, F. M. Wiygyl, T. J. Kistenmacher and D. O. Cowan; *J. Chem. Soc., Chem. Commun.*, 1983, 294. 592

83CC295 I. Johannsen, K. Bechgaard, K. Mortensen and C. Jacobsen; *J. Chem. Soc., Chem. Commun.*, 1983, 295. 590

83CC528 A. H. Cowley, J. E. Kilduff, S. K. Mehrotra, N. C. Norman and M. Pakulski; *J. Chem. Soc., Chem. Commun.*, 1983, 528. 167

83CC719 G. R. Clark, S. V. Hoskins, T. C. Jones and W. R. Roper; *J. Chem. Soc., Chem. Commun.*, 1983, 719. 718

83CC827 C. Eaborn, P. B. Hitchcock, J. D. Smith and A. C. Sullivan; *J. Chem. Soc., Chem. Commun.*, 1983, 827. 174, 289, 390

83CC881 A. H. Cowley, J. G. Lasch, N. C. Norman, M. Pakulski and B. R. Whittlesley; *J. Chem. Soc., Chem. Commun.*, 1983, 881. 169
83CC886 G. M. Blackburn and M. J. Parratt; *J. Chem. Soc., Chem. Commun.*, 1983, 886. 289
83CC1171 E. Niecke, M. Leuer, D.-A. Wildbredt and W. W. Schoeller; *J. Chem. Soc., Chem. Commun.*, 1983, 1171. 168
83CC1390 C. Eaborn, P. B. Hitchcock, J. D. Smith and A. C. Sullivan; *J. Chem. Soc., Chem. Commun.*, 1983, 1390. 176, 289, 391
83CC1419 R. I. Papasergio, C. L. Raston and A. H. White; *J. Chem. Soc., Chem. Commun.*, 1983, 1419. 190, 195
83CC1557 W.-K. Wong, K. W. Chiu, G. Wilkinson, A. M. R. Galas, M. Thornton-Pett and M. B. Hursthouse; *J. Chem. Soc., Chem. Commun.*, 1983, 1557. 338
83CC1613 G. Gervasio, R. Rossetti, P. L. Stanghellini and G. Bor; *J. Chem. Soc., Chem. Commun.*, 1983, 1613. 352
83CI(L)205 S. A. Okecha; *Chem. Ind.* (*London*), 1983, 205. 113
83CJC610 T. P. Ahern, M. F. Haley, R. F. Langler and J. E. Trenholm; *Can. J. Chem.*, 1983, **62**, 610. 255
83CJC1213 D. Mackay, D. D. McIntyre, N. J. Taylor and L. L. Wong; *Can. J. Chem.*, 1983, **61**, 1213. 626
83CPB41 M. Miyahara, S. Kamiya and M. Nakadate; *Chem. Pharm. Bull.*, 1983, **31**, 41. 504
83C10 M. Schlosser, Huynh Ba Tuong, J. Respondek and B. Schaub; *Chimia*, 1983, **37**, 10. 705
83GEP3135111 R. Fauss, K. Findeisen, U. J. Döbereiner and J. Uwe; *Ger. Pat.* 3 135 111 (1983) (*Chem. Abstr.*, 1983, **99**, 5259). 502
83GEP3241568 R. A. Olofson and J. T. Martz; *Ger. Pat.* 3241568 (1983) (*Chem. Abstr.*, 1983, **99**, 53164). 428
83HOU(E4)1 H. Hagemann; *Methoden Org. Chem.* (*Houben-Weyl*), 1983, **E4**, 1. 408, 411, 419
83HOU(E4)9 G. Heywang; *Methoden Org. Chem.* (*Houben-Weyl*), 1983, **E4**, 9. 421, 422, 423, 424, 425, 428, 429, 432, 434, 436, 437
83HOU(E4)36 G. Heywang; *Methoden Org. Chem.* (*Houben-Weyl*), 1983, **E4**, 36. 437
83HOU(E4)273 U. Petersen; *Methoden Org. Chem.* (*Houben-Weyl*), 1983, **E4**, 273. 488
83HOU(E4)274 U. Petersen; *Methoden Org. Chem.* (*Houben-Weyl*), 1983, **E4**, 274. 489
83HOU(E4)279 U. Petersen; *Methoden Org. Chem.* (*Houben-Weyl*), 1983, **E4**, 279. 489
83HOU(E4)284 U. Petersen; *Methoden Org. Chem.* (*Houben-Weyl*), 1983, **E4**, 284. 489
83HOU(E4)334 U. Petersen; *Methoden Org. Chem.* (*Houben-Weyl*), 1983, **E4**, 334. 500
83HOU(E4)407 U. Kraatz; *Methoden Org. Chem.* (*Houben-Weyl*), 1983, **E4**, 407. 528, 529, 532, 533
83HOU(E4)420 U. Kraatz; *Methoden Org. Chem.* (*Houben-Weyl*), 1983, **E4**, 420. 528, 540, 541, 543, 545, 552
83HOU(E4)522 E. Kühle; *Methoden Org. Chem.* (*Houben-Weyl*), 1983, **E4**, 522. 602, 604
83HOU(E4)543 E. Kühle; *Methoden Org. Chem.* (*Houben-Weyl*), 1983, **E4**, 543. 614, 616
83HOU(E4)561 E. Kühle; *Methoden Org. Chem.* (*Houben-Weyl*), 1983, **E4**, 561. 623, 624, 628, 629, 630
83HOU(E4)579 E. Kühle; *Methoden Org. Chem.* (*Houben-Weyl*), 1983, **E4**, 579. 626, 627, 628
83HOU(E4)587 E. Kühle; *Methoden Org. Chem.* (*Houben-Weyl*), 1983, **E4**, 587. 631, 632, 633
83HOU(E4)597 E. Kühle; *Methoden Org. Chem.* (*Houben-Weyl*), 1983, **E4**, 597. 634, 635, 636
83HOU(E4)608 E. Kühle; *Methoden Org. Chem.* (*Houben-Weyl*), 1983, **E4**, 608. 640, 642, 643
83HOU(E4)625 A. Marhold; *Methoden Org. Chem.* (*Houben-Weyl*), 1983, **E4**, 625. 228, 232, 237, 238
83HOU(E4)655 A. Marhold; *Methoden Org. Chem.* (*Houben-Weyl*), 1983, **E4**, 655. 610, 614, 727
83HOU(E4)677 J. Hocker and F. Jonas; *Methoden Org. Chem.* (*Houben-Weyl*), 1983, **E4**, 677. 730
83HOU(E4)690 J. Hocker and F. Jonas; *Methoden Org. Chem.* (*Houben-Weyl*), 1983, **E4**, 690. 731
83HOU(E4)694 J. Hocker; *Methoden Org. Chem.* (*Houben-Weyl*), 1983, **E4**, 694. 296, 297
83HOU(E4)703 J. Hocker; *Methoden Org. Chem.* (*Houben-Weyl*), 1983, **E4**, 703. 308
83HOU(E4)704 J. Hocker; *Methoden Org. Chem.* (*Houben-Weyl*), 1983, **E4**, 704. 309
83HOU(E4)705 J. Hocker; *Methoden Org. Chem.* (*Houben-Weyl*), 1983, **E4**, 705. 301
83HOU(E4)710 J. Hocker; *Methoden Org. Chem.* (*Houben-Weyl*), 1983, **E4**, 710. 313
83HOU(E4)1019 H. Hagemann; *Methoden Org. Chem.* (*Houben-Weyl*), 1983, **E4**, 1019. 437
83HOU(E4)1023 U. Kraatz; *Methoden Org. Chem.* (*Houben-Weyl*), 1983, **E4**, 1023. 437
83HOU(E4)1078 B. Baasner; *Methoden Org. Chem.* (*Houben-Weyl*), 1983, **E4**, 1078. 437
83HOU(E4)1170 H. Hagemann; *Methoden Org. Chem.* (*Houben-Weyl*), 1983, **E4**, 1170. 437
83HOU(E4)1177 H. Salzburg; *Methoden Org. Chem.* (*Houben-Weyl*), 1983, **E4**, 1177. 437
83HOU(E4)1203 A. Marhold; *Methoden Org. Chem.* (*Houben-Weyl*), 1983, **E4**, 1203. 228, 437
83HOU(E4)1234 H. Hagemann; *Methoden Org. Chem.* (*Houben-Weyl*), 1983, **E4**, 1234. 437
83HOU(E4)1275 F. Lieb; *Methoden Org. Chem.* (*Houben-Weyl*), 1983, **E4**, 1275. 437
83HOU(E4)1313 A. Botta; *Methoden. Org. Chem.* (*Houben-Weyl*), 1983, **E4**, 1313. 437
83HOU(E4)1368 J. Hocker; *Methoden Org. Chem.* (*Houben-Weyl*), 1983, **E4**, 1368. 228, 230
83IC411 M. G. Mason, P. N. Swepston and J. A. Ibers; *Inorg. Chem.*, 1983, **22**, 411. 332
83IC805 S.-C. Chang and D. D. DesMarteau; *Inorg. Chem.*, 1983, **22**, 805. 256
83IC2573 A. B. Burg; *Inorg. Chem.*, 1983, **22**, 2573. 264, 679
83IC3421 B. Murray, J. Hvoslef, H. Hope and P. P. Power; *Inorg. Chem.*, 1983, **22**, 3421. 168
83IC3700 S. W. Carr, R. Colton, D. Dakternieks, B. F. Hoskins and R. J. Steen; *Inorg. Chem.*, 1983, **22**, 3700. 332, 351
83IZV636 V. V. Gavrilenko, L. A. Chekulaeva, V. A. Antonovich and L. I. Zakharkin; *Izv. Akad. Nauk SSSR Ser. Khim.*, 1983, 636 (*Chem. Abstr.*, 1983, **99**, 5668). 208, 210
83IZV694 I. L. Knunyants, A. F. Gontar and A. S. Vinogradov; *Izv. Akad. Nauk SSSR Ser. Khim.*, 1983, 694 (*Chem. Abstr.*, 1983, **98**, 198169). 615
83IZV959 N. N. Zemlyanskii, I. I. Borisova, V. K. Bel'skii, N. D. Kolosova and I. P. Beletskaya; *Izv. Akad. Nauk SSSR Ser. Khim.*, 1983, 959 (*Chem. Abstr.*, 1983, **99**, 38572). 187, 710
83IZV2655 A. A. Gakh and V. M. Khutoretskii; *Izv. Akad. Nauk SSSR Ser. Khim.*, 1983, 2655 (*Chem. Abstr.*, 1984, **100**, 85558). 145

83JA1384 N. T. Allison, J. R. Fritch, K. P. C. Vollhardt and E. C. Walborsky; *J. Am. Chem. Soc.*, 1983, **105**, 1384. 403
83JA4059 M. J. Robins, J. S. Wilson and F. Hansske; *J. Am. Chem. Soc.*, 1983, **105**, 4059. 535
83JA5506 A. H. Cowley, J. G. Lasch, N. C. Norman and M. Pakulski; *J. Am. Chem. Soc.*, 1983, **105**, 5506. 168
83JA6513 D. P. Cox, R. A. Moss and J. Terpinski; *J. Am. Chem. Soc.*, 1983, **105**, 6513. 280
83JA7187 K. Ramakrishnan and J. Fisher; *J. Am. Chem. Soc.*, 1983, **105**, 7187. 549
83JA7719 R. V. Stevens, J. H. Chang, R. Lapalme, S. Schow, M. G. Schlageter, R. Shapiro and H. N. Weller; *J. Am. Chem. Soc.*, 1983, **105**, 7719. 327
83JAP60056958 Tokyo Soda Mfg. Co. Ltd.; *Jpn. Pat.*, 60 056 958 (1983) (*Chem. Abstr.*, 1985, **103**, 104 859). 538
83JAP60155152 H. Morinaka, A. Nakanishi and Y. Nokaka; *Jpn. Pat.* 60 155 152 (1983) (*Chem. Abstr.*, 1985, **104**, 50 680). 533
83JCED131 W. H. Gilligan; *J. Chem. Eng. Data*, 1983, **28**, 131. 541, 543
83JCS(D)905 C. Eaborn, N. Retta and J. D. Smith; *J. Chem. Soc., Dalton Trans.*, 1983, 905. 165, 372
83JCS(D)2599 L. W. Bateman, M. Green, K. A. Mead, R. M. Mills, I. D. Salter, F. G. A. Stone and P. Woodward; *J. Chem. Soc., Dalton Trans.*, 1983, 2599. 343
83JCS(P1)2011 M. Lempert-Sréter, K. Lempert and J. Möller; *J. Chem. Soc., Perkin Trans. 1*, 1983, 2011. 560
83JCS(P1)2641 M. A. Palominos, J. G. Santos, J. A. Valderrama and J. C. Vega; *J. Chem. Soc., Perkin Trans. 1*, 1983, 2641. 546, 547
83JCS(P2)291 R. Taylor; *J. Chem. Soc., Perkin Trans. 2*, 1983, 291. 475
83JEM95 W. H. Gilligan and M. E. Sitzmann; *J. Energy Mater.*, 1983, **1**, 95, (*Chem. Abstr.*, 1986, **104**, 108991). 272
83JFC(22)105 B. V. Kunshenko, N. N. Muratov, A. I. Burmakov, L. A. Alekseeva and L. M. Yagupol'skii; *J. Fluorine Chem.*, 1983, **22**, 105. 37
83JFC(22)175 H. Bürger, R. Köplin, G. Pawelke and C. Krüger; *J. Fluorine Chem.*, 1983, **22**, 175. 615
83JFC(22)207 W. Dmowski and M. Kaminski; *J. Fluorine Chem.*, 1983, **22**, 207. 39
83JFC(22)439 T. I. Savchenko, I. V. Kolesnikova, T. D. Petrova and V. E. Platonov; *J. Fluorine Chem.*, 1983, **22**, 439. 609
83JFC(22)521 P. L. Coe, A. G. Holton, J. H. Sleigh, P. Smith and J. C. Tatlow; *J. Fluorine Chem.*, 1983, **22**, 521. 450
83JFC(23)123 T. Abe, H. Baba, E. Hayashi and S. Nagase; *J. Fluorine Chem.*, 1983, **23**, 123. 39
83JFC(23)207 W. Dmowski and M. Kaminski; *J. Fluorine Chem.*, 1983, **23**, 207. 440
83JFC(23)325 A. Waterfeld and R. Mews; *J. Fluorine Chem.*, 1983, **23**, 325. 252
83JFC(23)339 D. J. Burton; *J. Fluorine Chem.*, 1983, **23**, 339. 224
83JFC(23)593 J. S. Thrasher and A. F. Clifford; *J. Fluorine Chem.*, 1983, **23**, 593. 617
83JHC1181 G. Stájer, E. A. Szabó, F. Fülöp, G. Bernáth and P. Sohár; *J. Heterocycl. Chem.*, 1983, **20**, 1181. 558
83JHC1223 L. N. Pridgen and G. Miller; *J. Heterocycl. Chem.*, 1983, **20**, 1223. 558
83JHC1477 F. Sauter, P. Stanetty, E. Hetzl and F. Fuhrmann; *J. Heterocycl. Chem.*, 1983, **20**, 1477. 444
83JIC806 G. S. Sodhi and N. K. Kaushik; *J. Indian Chem. Soc.*, 1983, **60**, 806 (*Chem. Abstr.*, 1984, **100**, 174 914). 563
83JOC242 K. K. Johri and D. D. DesMarteau; *J. Org. Chem.*, 1983, **48**, 242. 229, 230
83JOC771 S-C. Chang and D. D. DesMarteau; *J. Org. Chem.*, 1983, **48**, 771. 261, 615
83JOC1449 K. Y. Yen and M. P. Cava; *J. Org. Chem.*, 1983, **48**, 1449. 538, 539
83JOC1776 R. W. Binkley and M. G. Ambrose; *J. Org. Chem.*, 1983, **48**, 1776. 22
83JOC3189 A. Padwa and J. G. MacDonald; *J. Org. Chem.*, 1983, **48**, 3189. 126
83JOC3354 M. E. Sitzmann and W. H. Gilligan; *J. Org. Chem.*, 1983, **48**, 3354. 308
83JOC3966 Z. Tashma; *J. Org. Chem.*, 1983, **48**, 3966. 581
83JOC4750 G. Barany, A. L. Schroll, A. W. Mott and D. A. Halsrud; *J. Org. Chem.*, 1983, **48**, 4750. 434, 435, 474, 534, 550, 551, 567
83JOC4844 Y. Y. Zheng and D. D. DesMarteau; *J. Org. Chem.*, 1983, **48**, 4844. 256, 603
83JOC5033 M. Bean and H. Kohn; *J. Org. Chem.*, 1983, **48**, 5033. 479
83JOM(243)245 D. K. Breitinger, W. Kress, R. Sendelbeck and K. Ishiwada; *J. Organomet. Chem.*, 1983, **243**, 245. 206, 207
83JOM(244)C53 C. E. L. Headford and W. R. Roper; *J. Organomet. Chem.*, 1983, **244**, C53. 593
83JOM(246)C13 C. Bianchini, C. A. Ghilardi, A. Meli and A. Orlandini; *J. Organomet. Chem.*, 1983, **246**, C13. 332
83JOM(248)51 D. A. Bravo-Zhivotovskii, S. D. Pigarev, I. D. Kalikhman, O. A. Vyazankina and N. S. Vyazankina; *J. Organomet. Chem.*, 1983, **248**, 51. 131
83JOM(249)1 F. J. Landro, J. A. Gurak, J. W. Chinn, Jr. and R. J. Lagow; *J. Organomet. Chem.*, 1983, **249**, 1. 204, 205, 404
83JOM(249)23 C. Eaborn, M. N. El-Kheli, N. Retta and J. D. Smith; *J. Organomet. Chem.*, 1983, **249**, 23. 395
83JOM(249)273 L. J. Farrugia, M. J. Freeman, M. Green, A. G. Orpen, F. G. A. Stone and I. D. Salter; *J. Organomet. Chem.*, 1983, **249**, 273. 343
83JOM(252)281 C. Eaborn, P. B. Hitchcock and P. D. Lickiss; *J. Organomet. Chem.*, 1983, **252**, 281. 174
83JOM(252)47 T. N. Mitchell and A. Amamria; *J. Organomet. Chem.*, 1983, **252**, 47. 209
83JOM(253)283 D. Grdenić, M. Sikirica, D. Matkovic-Čalogović and A. Nagl; *J. Organomet. Chem.*, 1983, **253**, 283. 207
83JOM(256)217 D. K. Breitinger and W. Kress; *J. Organomet. Chem.*, 1983, **256**, 217. 207
83JPR787 W. Schroth, H. Kluge, R. Frach, W. Hodek and H. D. Schädler; *J. Prakt. Chem.*, 1983, **325**, 787. 608
83KGS798 L. S. Sologub, A. A. Kisilenko, V. P. Kukhar and S. I. Vdovenko; *Khim. Geterotsikl. Soedin.*, 1983, 798 (*Chem. Abstr.*, 1983, **99**, 122246). 609, 619

83LA248 W. Voelter and J. Müller; *Liebigs Ann. Chem.*, 1983, 248. 422

83LA1839 W. Broda and E. V. Dehmlow; *Liebigs Ann. Chem.*, 1983, 1839. 554, 555

83M813 M. Saljoughian, A. Raisi, E. Alipou and S. Afshar, 1983, *Monatsh. Chem.*, **114**, 813. 25

83M851 E. O. Fischer, J. Schneider and D. Neugebauer; *Monatsh. Chem.*, 1983, **114**, 851. 341

B-83MI 606-01 W. P. Weber; "Silicon Reagents for Organic Synthesis," Springer, Berlin, 1983, p. 59. 172, 189, 190

83MI 607-01 I. B. Sladkov; *Zh. Fiz. Khim.*, 1983, **57**, 2619 (*Chem. Abstr.*, 1984, **100**, 12917). 223, 228

83MI 610-01 S. Wawzonek, H. F. Chang, W. Everett and M. Ryan; *J. Electrochem. Soc.*, 1983, **130**, 803 (*Chem. Abstr.*, 1983, **98**, 224022). 304

83MI 614-01 C. Crouzel, D. Roeda, M. Berridge, R. Knipper and D. Comar; *Int. J. Appl. Radiat. Isot.*, 1983, **34**, 1558 (*Chem. Abstr.*, 1984, **100**, 120478). 414

83MI 620-01 L. R. Nassimbeni, M. L. Niven, D. Sohn and W. Sundermeyer; *S. Afr. J. Chem.*, 1983, **36**, 7 (*Chem. Abstr.*, 1983, **98**, 178395). 607

83MI 620-02 A. S. Vinogradov, A. F. Gontar, E. G. Bykhovskaya and I. L. Knunyants; *Zh. Vses. Khim. O-va.*, 1983, **28**, 352 (*Chem. Abstr.*, 1983, **99**, 175175). 615

83MI 622-01 J. A. Morrison; *Adv. Inorg. Chem. Radiochem.*, 1983, **27**, 293. 717

83MIP8303603 M. Bendly, G. Boda, M. Dienes, J. Csutak, G. Szilagyi and C. Tar; *W. Pat.* 8 303 603 (1983) (*Chem. Abstr.* 1984, **101**, 23 003). 477

83OM154 G. W. Rice, G. B. Ansell, M. A. Modrick and S. Zentz; *Organometallics*, 1983, **2**, 154. 707

83OM219 J. B. Keister, M. W. Payne and M. J. Muscatella; *Organometallics*, 1983, **2**, 219. 340, 344

83OM230 D. S. Matteson and D. Majumdar; *Organometallics*, 1983, **2**, 230. 203

83OM1044 W. Petz; *Organometallics*, 1983, **2**, 1044. 517

83OM1179 M. R. Churchill, L. R. Beanan, H. J. Wasserman, C. Bueno, Z. A. Rahman and J. B. Keister; *Organometallics*, 1983, **2**, 1179. 343

83OM1696 D. Seyferth, G. W. Womack, M. Cowle and B. W. Hames; *Organometallics*, 1983, **2**, 1696. 332

83POL291 H. J. Breunig, W. Kanig and A. Soltani-Neshan; *Polyhedron*, 1983, **2**, 291. 168, 192

83PS(14)267 G. Becker, W. Becker and O. Mundt; *Phosphorus Sulfur*, 1983, **14**, 267. 678

83PS(18)27 J. Navech, J. P. Majoral, A. Meriem and R. Kraemer; *Phosphorus Sulfur*, 1983, **18**, 27. 160

83PS(18)43 R. H. Neilson; *Phosphorus Sulfur*, 1983, **18**, 43. 164

83RCR1096 O. I. Kolodyazhni and V. P. Kukhar; *Russ. Chem. Rev.* (*Engl. Transl.*), 1983, **52**, 1096. 694, 699, 700

83SC367 T. Baba and A. Suzuki; *Synth. Commun.*, 1983, **13**, 367. 634

83S375 J. Garín, E. Meléndez, F. L. Merchán, C. Tejel and T. Tejero; *Synthesis*, 1983, 375. 627

83S391 G. Blotny; *Synthesis*, 1983, 391. 575

83S471 Y. Endo, K. Shudo and T. Okamoto; *Synthesis*, 1983, 471. 506

83S905 W. Kantlehner, P. Speh and H. J. Bräuner; *Synthesis*, 1983, 905. 140

83S908 M. Ueda, H. Oikawa and T. Teshirogi; *Synthesis*, 1983, 908. 417

83SC985 A. E. de Groot and B. J. M. Jansen; *Synth. Commun.*, 1983, **13**, 985. 336

83SUL131 N. Kamigata, H. Sawada, J.-I. Ozaki and M. Kobayashi; *Sulfur Letters*, 1983, **1**, 131. 24

83T1729 M. T. M. El-Wassimy, K. A. Jørgensen and S.-O. Lawesson; *Tetrahedron*, 1983, **39**, 1729. 502

83T1829 G. Stájer, E. A. Szabó, F. Fülöp, G. Bernáth, A. Kálmán, G. Argay and P. Sohár; *Tetrahedron*, 1983, **39**, 1829. 557

83T2207 R. G. K. Schneiderwind and I. Ugi; *Tetrahedron*, 1983, **39**, 2207. 488

83T2733 J. Klein; *Tetrahedron*, 1983, **39**, 2733. 206

83T2989 J. E. Baldwin, A. E. Derome and P. D. Riordan; *Tetrahedron*, 1983, **39**, 2989. 515

83T3073 K. J. H. Kruithof, R. F. Schmitz and G. W. Klumpp; *Tetrahedron*, 1983, **39**, 3073. 130, 133

83TL623 A. Pelter, B. Singaram, L. Williams and J. W. Wilson; *Tetrahedron Lett.*, 1983, **24**, 623. 199

83TL865 K. K. Ogilvie, G. H. Hakimelahi, Z. A. Proba and N. Usman; *Tetrahedron Lett.*, 1983, **24**, 865. 540

83TL1035 B. Le Clef, L. Vermylen, H. G. Viehe, M. Van Meerssche, G. Germain and J. P. Declercq; *Tetrahedron Lett.*, 1983, **24**, 1035. 105

83TL1303 T. A. M. Van Schaik, A. V. Henzen and A. van der Gen; *Tetrahedron Lett.*, 1983, **24**, 1303. 116

83TL2639 R. Appel and W. Paulen; *Tetrahedron Lett.*, 1983, **24**, 2639. 454, 686

83TL2769 C. Couret, J. Escudié, Y. Madaule, H. Ranaivonjatovo and J.-G. Wolf; *Tetrahedron Lett.*, 1983, **24**, 2769. 374, 375

83TL3387 F. Camps, J. Coll, G. Fabriás, A. Guerrero and M. Riba; *Tetrahedron Lett.*, 1983, **24**, 3387. 695

83TL3469 Z.-I. Yoshida, T. Kawase, H. Awaji, I. Sugimoto, T. Sugimoto and Sh. Yoneda; *Tetrahedron Lett.*, 1983, **24**, 3469. 698

83TL3625 J. Escudié, C. Couret, H. Ranaivonjatovo and J.-G. Wolf; *Tetrahedron Lett.*, 1983, **24**, 3625. 375

83TL5481 J. Heinicke and A. Tzschach; *Tetrahedron Lett.*, 1983, **24**, 5481. 119

83TL5571 E. J. Corey and N. Raju; *Tetrahedron Lett.*, 1983, 5571. 71

83TL5683 G. Barany; *Tetrahedron Lett.*, 1983, **24**, 5683. 276, 435

83TL5885 J. Navech, J. P. Majoral and R. Kraemer; *Tetrahedron Lett.*, 1983, **24**, 5885. 686

83USP4419514 L. H. McHendry, M. J. Ricks and R. B. Rogers; *US Pat.* 4419514 (1983) (*Chem. Abstr.* 1984, **100**, 85412). 28

83USP467715 M. E. Sitzmann and W. H. Gilligan (US Navy); *US Pat.* 467715 (1983) (*Chem. Abstr.*, 1984, **100**, 24077). 272

83ZAAC(497)119 G. Fritz, H. Volk, K. Peters, E. M. Peters and H. G. von Schnering; *Z. Anorg. Allg. Chem.*, 1983, **497**, 119. 381, 385, 391

83ZAAC(497)134 G. Fritz and G. Brauch; *Z. Anorg. Allg. Chem.*, 1983, **497**, 134. 385

83ZAAC(497)21 G. Fritz, J. Neutzner and H. Volk; *Z. Anorg. Allg. Chem.*, 1983, **497**, 21. 178, 190, 191, 385, 391

83ZAAC(501)157 H. Hlawatschek and G. Gattow; *Z. Anorg. Allg. Chem.*, 1983, **501**, 157. 630

83ZAAC(501)169 H. Hlawatschek and G. Gattow; *Z. Anorg. Allg. Chem.*, 1983, **501**, 169. 630

83ZAAC(501)176 H. Hlawatschek and G. Gattow; *Z. Anorg. Allg. Chem.*, 1983, **501**, 176. 630

83ZAAC(502)7 B. Sturm and G. Gattow; *Z. Anorg. Allg. Chem.*, 1983, **502**, 7. 535

83ZC99 K. Issleib, E. Leissring, M. Riemer and H. Oehme; *Z. Chem.*, 1983, **23**, 99. 686

83ZOB473 O. I. Kolodiazhnyi, I. V. Shevchenko and V. P. Kukhar; *Zh. Obshch. Khim.*, 1983, **53**, 473 (*Chem. Abstr.*, 1983, **98**, 198353). 683

83ZOB693 L. N. Markovskii, V. D. Romanenko, A. V. Ruban and L. K. Polyachenko; *Zh. Obshch. Khim.*, 1983, **53**, 693 (*Chem. Abstr.*, 1983, **99**, 70840). 693

83ZOB785 S. Kh. Nurtdinov, S. V. Khairullina, V. K. Khairullina, R. R. Shagidullin and A. N. Chernov; *Zh. Obshch. Khim.*, 1983, **53**, 785 (*Chem. Abstr.*, 1983, **99**, 105347). 564

83ZOB1672 L. N. Markovskii, V. D. Romanenko and T. V. Pidvarko; *Zh. Obshch. Khim.*, 1983, **53**, 1672 (*Chem. Abstr.*, 1983, **99**, 195100). 685

83ZOR79 Y. L. Yagupol'skii and T. I. Savina; *Zh. Org. Khim.*, 1983, **19**, 79. 275, 317

83ZOR92 V. S. Gasanov and R. K. Alekperov; *Zh. Org. Khim.*, 1983, **19**, 92 (*Chem. Abstr.*, 1983, **98**, 215269d). 558

84AG(E)381 H.-J. Bestmann and T. Arenz; *Angew. Chem., Int. Ed. Engl.*, 1984, **23**, 381. 704

84AG(E)579 L. E. Overman; *Angew. Chem., Int. Ed. Engl.*, 1984, **23**, 579. 473

84AG(E)619 R. Appel; P. Fölling, B. Josten, M. Siray, V. Winkhaus and F. Knoch; *Angew. Chem., Int. Ed. Engl.*, 1984, **23**, 619. 680, 687

84AG(E)710 J. Grobe and D. Le Van; *Angew. Chem., Int. Ed. Engl.*, 1984, **23**, 710. 265, 266, 679, 680, 691

84AG(E)895 R. Appel, C. Casser, M. Immenkepel and F. Knoch; *Angew. Chem., Int. Ed. Engl.*, 1984, **23**, 895. 683

84AG(E)903 A. Schmidpeter and A. Willhalm; *Angew. Chem., Int. Ed. Engl.*, 1984, **23**, 903. 660, 686

84AG(E)970 R. Appel, P. Fölling, L. Krieger, M. Siray and F. Knoch; *Angew. Chem., Int. Ed. Engl.*, 1984, **23**, 970. 680, 687

84AG(E)995 A. Maercker and M. Theis; *Angew. Chem., Int. Ed. Engl.*, 1984, **23**, 995. 405

84AOC(23)219 C. P. Horwitz and D. F. Shriver; *Adv. Organomet. Chem.*, 1984, **23**, 219. 521

84AP(317)297 W. Hanefeld; *Arch. Pharm. (Weinheim, Ger.)*, 1984, **317**, 297. 564

84AP(317)1042 U. Stanior and M. Wiessler; *Arch. Pharm. (Weinheim, Ger.)*, 1984, **317**, 1042. 562

84BRP2123829 M. B. Frankel and E. R. Wilson; *Br Pat.* 2123829 (1984) (*Chem. Abstr.*, 1984, **100**, 194573). 151

84CB502 J. C. Jochims and M. A. Rahman; *Chem. Ber.*, 1984, **117**, 502. 643, 733

84CB1161 J. C. Jochims, R. Abu-El-Halawa, L. Zsolnai and G. Huttner; *Chem. Ber.*, 1984, **117**, 1161. 643

84CB1707 J. S. Thrasher, J. L. Howell and A. F. Clifford; *Chem. Ber.*, 1984, **117**, 1707. 558, 565

84CB2063 R. Appel, G. Haubrich and F. Knoch; *Chem. Ber.*, 1984, **117**, 2063. 164, 373

84CB2693 R. Appel, C. Casser and F. Knoch; *Chem. Ber.*, 1984, **117**, 2693. 166, 689

84CC89 I. Johannsen, K. Lerstrup, L. Henriksen and K. Bechgaard; *J. Chem. Soc., Chem. Commun.*, 1984, 89. 591

84CC392 S. G. Shore, D.-Y. Jan, W.-L. Hsu, L.-Y. Hsu, S. Kennedy, J. C. Huffman, T.-C. L. Wang and A. G. Marshall; *J. Chem. Soc., Chem. Commun.*, 1984, 392. 345

84CC416 O. D. Gupta and J. M. Shreeve; *J. Chem. Soc., Chem. Commun.*, 1984, 416. 438

84CC480 P. B. Hitchcock, M. F. Lappert, S. J. Miles and A. J. Thorne; *J. Chem. Soc., Chem. Commun.*, 1984, 480. 186

84CC569 H. H. Karsch and G. Müller; *J. Chem. Soc., Chem. Commun.*, 1984, 569. 369

84CC612 R. I. Papasergio, C. L. Raston and A. H. White; *J. Chem. Soc., Chem. Commun.*, 1984, 612. 195

84CC630 M. E. Jung and J. C. Rohloff; *J. Chem. Soc., Chem. Commun.*, 1984, 630. 450

84CC698 M. Caira, R. H. Neilson, W. H. Watson, P. Wisian-Neilson and Z.-M. Xie; *J. Chem. Soc., Chem. Commun.*, 1984, 698. 350, 694

84CC793 C. Wakselman and M. Tordeux; *J. Chem. Soc., Chem. Commun.*, 1984, 793. 233, 234

84CC870 C. Eaborn, P. B. Hitchcock, J. D. Smith and A. C. Sullivan; *J. Chem. Soc., Chem. Commun.*, 1984, 870. 396

84CC1089 B. F. G. Johnson, J. Lewis, M. McPartlin, M.-A. Pearsall and A. Sironi; *J. Chem. Soc., Chem. Commun.*, 1984, 1089. 345

84CC1105 G. V. Boyd, P. F. Lindley and G. A. Nicolaou; *J. Chem. Soc., Chem. Commun.*, 1984, 1105. 621

84CC1622 A. M. Caminade, C. Couret, J. Escudié and M. Koenig; *J. Chem. Soc., Chem. Commun.*, 1984, 1622. 373

84CC1634 Y. Y. C. Yeung Lam Ko, R. Carrié, A. Muench and G. Becker; *J. Chem. Soc., Chem. Commun.*, 1984, 1634. 117

84CC1664 H. Kawa, J. W. Chinn, Jr. and R. J. Lagow; *J. Chem. Soc., Chem. Commun.*, 1984, 1664. 192, 205, 398

84CC1669 P. B. Hitchcock, M. F. Lappert, P. P. Power and S. J. Smith; *J. Chem. Soc., Chem. Commun.*, 1984, 1669. 167

84CC1708 D. Colgan, R. I. Papasergio, C. L. Raston and A. H. White; *J. Chem. Soc., Chem. Commun.*, 1984, 1708. 190

84CCA689 B. Korpar-Čolig, Z. Popović and M. Sikirica; *Croat Chem. Acta*, 1984, **57**, 689 (*Chem. Abstr.*, 1985, **102**, 220968). 207

84CCC295 P. Kristian and J. Gonda; *Collect. Czech. Chem. Commun.*, 1984, **49**, 295. 629

84CCC1577 J. Mindl, V. Sterba, V. Kaderábek, Jr. and J. Klicnar; *Collect. Czech. Chem. Commun.*, 1984, **49**, 1577. 558, 565

84CCC2285 G. D. Krapivin, A. Jurášek, J. Kováč and V. G. Kul'nevich; *Collect. Czech. Chem. Commun.*, 1984, **49**, 2285. 627

84CHEC(3)91 H. Neunhoeffer; *Comp. Heterocycl. Chem.*, 1984, **3**, 91. 109

84CJC574 D. A. Holden; *Can. J. Chem.*, 1984, **62**, 574. 489

84CPB4893 K. Shimada, H. Fujisaki, K. Oketani, M. Murakami, T. Shoji, T. Wakabayashi, K. Ueda, K. Ema, K. Hashimoto and S. Tanaka; *Chem. Pharm. Bull.*, 1984, **32**, 4893. 636

84CZ404 R. Neidlein and U. J. Klotz; *Chem.-Ztg.*, 1984, **108**, 404. 617

84H(22)13 Y. Maki, M. Tanabe, Y. Kojima, M. Sako and K. Hirota; *Heterocycles*, 1984, **22**, 13. 111

84H(22)281 P. A. Jacobi, K. T. Weiss and M. Egbertson; *Heterocycles*, 1984, **22**, 281. 111

84HCA1070 J. Gabriel and D. Seebach; *Helv. Chim. Acta*, 1984, **67**, 1070. 86, 92

84HCA1083 D. Seebach, J. Gabriel and R. Hässig; *Helv. Chim. Acta*, 1984, **67**, 1083. 317

84HCA1734 M. E. Scheller and B. Frei; *Helv. Chim. Acta*, 1984, **67**, 1734. 128

84HOU(13/3c)162 M. Grassberger and R. Köster; *Methoden Org. Chem.* (*Houben-Weyl*), 1984, **13/3c**, 162. 183

84IC2188 B. A. O'Brien and D. D. DesMarteau; *Inorg. Chem.*, 1984, **23**, 2188. 603, 615

84IC2582 A. H. Cowley, J. E. Kilduff, J. G. Lasch, S. K. Mehrotra, N. C. Norman, M. Pakulski, B. R. Whittlesey, J. L. Atwood and W. E. Hunter; *Inorg. Chem.*, 1984, **23**, 2582. 168

84IC2973 C. H. Wei; *Inorg. Chem.*, 1984, **23**, 2973. 352

84IC3063 B. Kemp, S. Kalbag and R. A. Geanangel; *Inorg. Chem.*, 1984, **23**, 3063. 491

84IC4322 B. F. Spielvogel, F. U. Ahmed, G. L. Silvey, P. Wisian-Neilson and A. T. McPhail; *Inorg. Chem.*, 1984, **23**, 4322. 491

84IJC(B)986 M. Husain, M. Husain and N. H. Khan; *Indian J. Chem., Sect. B*, 1984, **23**, 986. 444

84IZV1897 D. A. Bravo-Zhivotovskii, S. D. Pigarev, O. A. Vyazankina, A. S. Medvedeva, L. P. Safronova and N. S. Vyazankina; *Izv. Akad. Nauk SSSR Ser. Khim.*, 1984, 1897 (*Chem. Abstr.*, 1985, **102**, 6710). 516

84JA263 K. J. Klabunde, M. P. Kramer, A. Senning and E. K. Moltzen; *J. Am. Chem. Soc.*, 1984, **106**, 263. 535

84JA1125 D. H. Gibson, K. Owens and T.-S. Ong; *J. Am. Chem. Soc.*, 1984, **106**, 1125. 517

84JA3784 L. Hevesi, S. Desauvage, B. Georges, G. Evrard, P. Blanpain, A. Michel, S. Harkema and G. J. van Hummel; *J. Am. Chem. Soc.*, 1984, **106**, 3784. 92

84JA4266 B. A. O'Brien, J. S. Thrasher, C. W. Bauknight, Jr., M. L. Robin and D. D. DesMarteau; *J. Am. Chem. Soc.*, 1984, **106**, 4266. 603

84JA5304 P. A. Bartlett, J. D. Meadows and E. Ottow; *J. Am. Chem. Soc.*, 1984, **106**, 5304. 469

84JA7015 A. H. Cowley, R. A. Jones, J. G. Lasch, N. C. Norman, C. A. Stewart, A. L. Stuart, J. L. Atwood, W. E. Hunter and H.-M. Zhang; *J. Am. Chem. Soc.*, 1984, **106**, 7015. 689

84JAP59193897 Asai Germanium Research Institute; *Jpn. Pat.* 59 193 897 (1984) (*Chem. Abstr.*, 1985, **103**, 22 788). 132

84JAP5951225 Daikin; *Jpn. Pat.* 5951225 (1984) (*Chem. Abstr.*, 1984, **101**, 54544). 3

84JAP8439705 Hodogaya Chemical Co. Ltd.; *Jpn. Pat.* 8439705 (1984) (*Chem Abstr.*, 1984, **100**, 212478). 409

84JCR(S)274 P. Tisnes, Y. Madaule, J. G. Wolf, C. Couret and J. Escudié; *J. Chem. Res.* (*S*), 1984, 274. 375

84JCS(D)1801 W.-P. Leung, C. L. Raston, B. W. Skelton and A. H. White; *J. Chem. Soc., Dalton Trans.*, 1984, 1801. 190

84JCS(P1)455 G. S. Phull, R. G. Plevey and J. C. Tatlow; *J. Chem. Soc., Perkin Trans. 1*, 1984, 455. 53

84JCS(P1)1005 D. H. R. Barton, J. Boivin, S. Legreneur and W. B. Motherwell; *J. Chem. Soc., Perkin Trans. 1*, 1984, 1005. 560

84JCS(P1)1119 G. M. Blackburn, D. E. Kent and F. Kolkmann; *J. Chem. Soc., Perkin Trans. 1*, 1984, 1119. 263

84JCS(P1)2615 A. W. Mott and G. Barany; *J. Chem. Soc., Perkin Trans. 1*, 1984, 2615. 435

84JCS(P2)1239 P. J. Atkins, V. Gold and R. Marsh; *J. Chem. Soc., Perkin Trans. 2*, 1984, 1239. 30

84JCS(P2)1247 P. J. Atkins, V. Gold and W. N. Wassef; *J. Chem. Soc., Perkin Trans. 2*, 1984, 1247. 25, 31

84JFC(24)457 H. Buerger, R. Koeplin and G. Pawelke; *J. Fluorine Chem.*, 1984, **24**, 457. 262

84JFC(24)485 B. L. Booth, R. F. Browne, R. N. Haszeldine and J. S. Varley; *J. Fluorine Chem.*, 1984, **24**, 485. 312, 436

84JFC(24)523 D. Lentz; *J. Fluorine Chem.*, 1984, **24**, 523. 611

84JFC(25)69 P. Tarrant; *J. Fluorine Chem.*, 1984, **25**, 69. 32

84JFC(26)321 G. Pawelke and H. Bürger; *J. Fluorine Chem.*, 1984, **26**, 321. 615, 632

84JFC(26)359 W. Sundermeyer and M. Witz; *J. Fluorine Chem.*, 1984, **26**, 359. 253

84JFC(26)435 H. Lange and D. J. Naumann; *J. Fluorine Chem.*, 1984, **26**, 435. 64

84JHC61 T. P. Selby and G. E. Lepone; *J. Heterocycl. Chem.*, 1984, **21**, 61. 630

84JHC1373 G. Stájer, A. E. Szabó, F. Fülöp and G. Bernáth; *J. Heterocycl. Chem.*, 1984, **21**, 1373. 558

84JOC168 D. J. Ager; *J. Org. Chem.*, 1984, **49**, 168. 325, 327, 391

84JOC205 S. Elsheimer, W. R. Dolbier, Jr. and M. Murla; *J. Org. Chem.*, 1984, **49**, 205. 227

84JOC528 E. M. Acton and K. J. Ryan; *J. Org. Chem.*, 1984, **49**, 528. 617

84JOC605 T. Cohen and L.-C. Yu; *J. Org. Chem.*, 1984, **49**, 605. 134

84JOC706 T. N. Wheeler; *J. Org. Chem.*, 1984, **49**, 706. 695

84JOC1043 G. Barany and A. W. Mott; *J. Org. Chem.*, 1984, **49**, 1043. 435

84JOC1122 J. A. Cella and S. W. Bacon; *J. Org. Chem.*, 1984, **49**, 1122. 462

84JOC1174 H. M. R. Hoffmann and L. Iranshahi; *J. Org. Chem.*, 1984, **49**, 1174. 431

84JOC1703 Y. Nakayama and Y. Sanemitsu; *J. Org. Chem.*, 1984, **49**, 1703. 627

84JOC2081 R. A. Olofson, J. T. Martz, J.-P. Senet, M. Piteau and T. Malfroot; *J. Org. Chem.*, 1984, **49**, 2081. 428

84JOC2404 K. T. Potts, W. R. Kuehnling and P. Murphy; *J. Org. Chem.*, 1984, **49**, 2404. 631

84JOC3590 Y. Y. Zheng, C. W. Bauknight, Jr. and D. D. DesMarteau; *J. Org. Chem.*, 1984, **49**, 3590. 256, 261

84JOC3675 R. Richter, F. A. Stuber and B. Tucker; *J. Org. Chem.*, 1984, **49**, 3675. 448

84JOC3854 E. K. Moltzen, A. Senning, M. P. Kramer and K. J. Klabunde; *J. Org. Chem.*, 1984, **49**, 3854. 235, 535

84JOC4541 W. J. Middleton; *J. Org. Chem.*, 1984, **49**, 4541. 260

84JOC5087 H. H. Schubert and P. J. Stang; *J. Org. Chem.*, 1984, **49**, 5087. 129, 167, 374

84JOM(260)C75 W. J. Bland, R. Davis and J. L. A. Durrant; *J. Organomet. Chem.*, 1984, **260**, C75. 7

84JOM(262)69 G. J. Kruger, L. Linford, H. G. Raubenheimer and A. A. Chalmers; *J. Organomet. Chem.*, 1984, **262**, 69. 333

84JOM(263)C23 C. Eaborn, P. B. Hitchcock, J. D. Smith and A. C. Sullivan; *J. Organomet. Chem.*, 1984, **263**, C23. 396

84JOM(265)237 E. Lukevics, V. N. Gevorgyan, Y. S. Goldberg, A. P. Gaukhman, M. P. Gavars, J. J. Popelis and M. V. Shymanska; *J. Organomet. Chem.*, 1984, **265**, 237. 124

84JOM(266)37 D. W. Hawker and P. R. Wells; *J. Organomet. Chem.*, 1984, **266**, 37. 208, 405

84JOM(269)C55 S. V. Hoskins, R. A. Pauptit, W. R. Roper and J. M. Waters; *J. Organomet. Chem.*, 1984, **269**, C55. 718, 727

84JOM(269)217 Z. H. Aiube and C. Eaborn; *J. Organomet. Chem.*, 1984, **269**, 217. 174, 390

84JOM(270)251 C. Bianchini, C. A. Ghilardi, A. Meli and A. Orlandini; *J. Organomet. Chem.*, 1984, **270**, 251. 332

84JOM(271)107 G. Fritz and J. Thomas; *J. Organomet. Chem.*, 1984, **271**, 107. 178, 290

84JOM(271)381 N. Wiberg, G. Wagner, G. Müller and J. Riede; *J. Organomet. Chem.*, 1984, **271**, 381. 392, 710

84JOM(272)1 C. Eaborn, M. N. A. El-Kheli, P. B. Hitchcock and J. D. Smith; *J. Organomet. Chem.*, 1984, **272**, 1. 387

84JOM(273)141 N. Wiberg; *J. Organomet. Chem.*, 1984, **273**, 141. 177, 180, 187, 193, 382, 709, 711

84JOM(276)1 D. Grdenić, B. Korpar-Čolig and M. Sikirica; *J. Organomet. Chem.*, 1984, **276**, 1. 207

84JPC1981 L. L. Lohr, H. B. Schlegel and K. Morokuma; *J. Phys. Chem.*, 1984, **88**, 1981. 691

84JPR633 M. Richter, M. Augustin and K. Strauss; *J. Prakt. Chem.*, 1984, **326**, 633. 565

84LA108 W. Kantlehner, E. Haug, W. W. Mergen, P. Speh, T. Maier, J. J. Kapassakalidis, H.-J. Bräuner and H. Hagen; *Liebigs Ann. Chem.*, 1984, **1**, 108. 643

84MI 601-01 B. Huang and W. Huang; *Huaxue Xnebao*, 1984, **42**, 1106 (*Chem. Abstr.*, 1985, **102**, 78312). 32

84MI 603-01 A. Krief and L. Hevesi; *Janssen Chim. Acta*, 1984, **2**, 3 (*Chem. Abstr.*, 1984, **103**, 122504). 91

84MI 607-01 A. Haas; *Adv. Inorg. Chem. Radiochem.*, 1984, **28**, 167 (*Chem. Abstr.*, 1985, **102**, 177998). 228

84MI 614-01 R. G. Gasanov, E. A. Domogatskaya, V. N. Setkina, N. K. Baranetskaya, V. N. Trembovla and B. M. Yavorskii; *Khim. Fiz.*, 1984, **3**, 700 (*Chem. Abstr.*, 1984, **101**, 211355). 456

84MI 614-02 A. W. Mott, S. J. Eastep, U. Slomczynska and G. Barany; *J. Labelled Comp. Radiopharm.*, 1984, **21**, 329. 435

B-84MI 616-01 N. N. Greenwood and A. Earnshaw; "Chemistry of the Elements," Pergamon Press, Oxford, 1984. 523, 525

84MI 617-01 D. Skwarski, H. Sobolewski and A. Sobolewski; *Poznan Rocz. Med.*, 1984, **8**, 109 (*Chem. Abstr.*, 1987, **107**, 197706). 546

84MI 617-02 V. Konecny; *Chem. Zvesti*, 1984, **38**, 523 (*Chem. Abstr.*, 1985, **102**, 45727). 562

84MI 617-03 M. Strube and H. Kunze; *Wiss. Beitr.-Martin-Luther-Univ. Halle-Wittenberg*, 1984, 191 (*Chem. Abstr.*, 1984, **101**, 22966). 563

84MI 617-04 V. Konecny; *Chem. Zvesti*, 1984, **38**, 839 (*Chem. Abstr.*, 1985, **102**, 148710). 562

B-84MI 622-01 K. H. Dotz, H. Fischer, P. Hofmann, F. R. Kreissl, U. Schubert and K. Weiss; "Transition Metal Carbene Complexes," Verlag Chemie, Deerfield Beach, FL, 1984. 716, 719, 722

84MIP521372 M. J. Moreno Lopez and A. Llamas Marcos; *Span. ES* 521372 (1984) (*Chem. Abstr.*, 1987, **107**, 217096). 70

84OM38 H. Schmidbaur, B. Milewski-Mahrla, G. Müller and C. Krüger; *Organometallics*, 1984, **3**, 38. 701

84OM314 T. G. Richmond, A. M. Crespi and D. F. Shriver; *Organometallics*, 1984, **3**, 314. 717, 722

84OM1038 J. R. Matachek, R. J. Angelici, K. A. Schugart, K. J. Haller and R. F. Fenske; *Organometallics*, 1984, **3**, 1038. 334

84OM1132 R. H. Neilson, R. J. Thoma, I. Vickovic and W. H. Watson; *Organometallics*, 1984, **3**, 1132. 692

84OM1523 A. J. Lindsay, S. Kim, R. A. Jacobson and R. J. Angelici; *Organometallics*, 1984, **3**, 1523. 517

84OM1916 M. Lusser and P. Peringer; *Organometallics*, 1984, **3**, 1916. 372

84OPP299 S. Julia, J. M. del Mazo, L. Avila and J. Elguero; *Org. Prep. Proced. Int.*, 1984, **16**, 299. 362

84POL389 A. H. Cowley; *Polyhedron*, 1984, **3**, 389. 167

84PS(19)189 V. D. Romanenko, L. K. Polyachenko and L. N. Markovskii; *Phosphorus Sulfur*, 1984, **19**, 189. 689

84S172 S. Corral, J. Lissavetzky and A. M. Valdeolmillos; *Synthesis*, 1984, 172. 556

84S439 L. Syper and J. Mlochowski; *Synthesis*, 1984, 439. 92

84S520 J. Garín, E. Meléndez, F. L. Merchán and T. Tejero; *Synthesis*, 1984, 520. 605

84S667 H. Poleschner, J. Böttger and E. Fanghänel; *Synthesis*, 1984, 667. 630, 731

84S837 T. Endo and M. Okawara; *Synthesis*, 1984, 837. 298, 299, 300

84SC1275 E. T. Hansen and H. J. Petersen; *Synth. Commun*, 1984, **14**, 1275. 647

84SUL85 H. C. Hansen and A. Senning; *Sulfur Letters*, 1984, **2**, 85. 274, 275, 554

84SUL241 A. W. Mott and G. Barany; *Sulfur Letters*, 1984, **2**, 241. 275, 535

84T1075 G. Morel, E. Marchand, K. H. Nguyen Thi and A. Foucaud; *Tetrahedron*, 1984, **40**, 1075. 617

84T1523 R. Galli, L. Scaglioni, O. Palla and F. Gozzo; *Tetrahedron*, 1984, **40**, 1523. 695

84T4963 A. Haas and K. W. Kempf; *Tetrahedron*, 1984, **40**, 4963. 94, 307, 536, 734

84TL81 T. Umemoto; *Tetrahedron Lett.*, 1984, **25**, 81. 12

84TL303 T. Fuchikami and I. Ojima; *Tetrahedron Lett.*, 1984, **25**, 303. 21

84TL487 D. M. Vyas, Y. Chiang and T. W. Doyle; *Tetrahedron Lett.*, 1984, **25**, 487. 611

84TL991 C. A. Brown, R. D. Miller, C. M. Lindsay and K. Smith; *Tetrahedron Lett.*, 1984, **25**, 991. 317

84TL1557 B. Kokel, G. Menichi and M. Hubert-Habart; *Tetrahedron Lett.*, 1984, **25**, 1557. 620

84TL1849 S. D. Sharma and U. Mehra; *Tetrahedron Lett.*, 1984, **25**, 1849. 115

84TL2021 M. Isobe, Y. Funabashi, Y. Ichikawa, S. Mio and T. Goto; *Tetrahedron Lett.*, 1984, **25**, 2021. 130

84TL2195 I. Ruppert, K. Schlich and W. Volbach; *Tetrahedron Lett.*, 1984, **25**, 2195. 14

84TL2489 M. Mikolajczyk, W. Midura and S. Grzejszczak; *Tetrahedron Lett.*, 1984, 2489. 474

84TL3519 A. H. Cowley, N. C. Norman, C. A. Stewart and B. R. Whittlesey; *Tetrahedron Lett.*, 1984, **25**, 3519. 167
84TL4097 B. Schaub, T. Jenny and M. Schlosser; *Tetrahedron Lett.*, 1984, **25**, 4097. 705
84TL4109 R. Appel and C. Casser; *Tetrahedron Lett.*, 1984, **25**, 4109. 680, 689, 692, 694
84TL4227 Z.-i. Yoshida, H. Awaji and T. Sugimoto; *Tetrahedron Lett.*, 1984, **25**, 4227. 124, 698
84TL4425 Sh. Yanchang, G. Zhengxiang, D. Weiyu and H. Yaozeng; *Tetrahedron Lett.*, 1984, **25**, 4425. 708
84TL4447 R. Appel, B. Laubach and M. Siray; *Tetrahedron Lett.*, 1984, **25**, 4447. 680, 684
84USP4440646 E. G. Budnick (Plains Chemical Development Co.); *US Pat.* 4440646 (1984) (*Chem. Abstr.*, 1984, **101**, 74787). 155, 366
84USP4446068 R. A. Mitsch (US Dept. Army); *US Pat.* 4446068 (1984) (*Chem. Abstr.*, 1984, **101**, 90750). 604
84USP4473580 F. E. Dutton and S. J. Nelson (Upjohn Co.); *US Pat.* 4473580 (1984) (*Chem. Abstr.*, 1985, **102**, 24467). 439
84ZAAC(508)136 B. Sturm and G. Gattow; *Z. Anorg. Allg. Chem.*, 1984, **508**, 136. 434, 534, 535
84ZAAC(509)61 B. Sturm and G. Gattow; *Z. Anorg. Allg. Chem.*, 1984, **509**, 61. 536
84ZAAC(509)67 B. Sturm and G. Gattow; *Z. Anorg. Allg. Chem.*, 1984, **509**, 67. 534
84ZAAC(510)169 S. Pohlmann and U. Klingebiel; *Z. Anorg. Allg. Chem.*, 1984, **510**, 169. 290, 379, 380
84ZAAC(510)175 U. Klingebiel and L. Skoda; *Z. Anorg. Allg. Chem.*, 1984, **510**, 175. 380
84ZAAC(511)95 G. Fritz, W. Schick, W. Hönle and H. G. von Schnering; *Z. Anorg. Allg. Chem.*, 1984, **511**, 95. 704
84ZAAC(511)108 G. Fritz and W. Schick; *Z. Anorg. Allg. Chem.*, 1984, **511**, 108. 165
84ZAAC(512)93 G. Fritz, H. Reuss, K.-P. Wörns and A. Wörsching; *Z. Anorg. Allg. Chem.*, 1984, **512**, 93. 383
84ZAAC(512)131 G. Fritz and A. Wörsching; *Z. Anorg. Allg. Chem.*, 1984, **512**, 131. 383
84ZAAC(512)231 B. Sturm and G. Gattow; *Z. Anorg. Allg. Chem.*, 1984, **512**, 231. 565
84ZAAC(513)183 B. Sturm and G. Gattow; *Z. Anorg. Allg. Chem.*, 1984, **513**, 183. 481
84ZAAC(514)120 B. Sturm and G. Gattow; *Z. Anorg. Allg. Chem.*, 1984, **514**, 120. 482
84ZAAC(515)182 G. Gattow and G. Schubert; *Z. Anorg. Allg. Chem.*, 1984, **515**, 182. 546
84ZAAC(517)75 G. Becker, W. Massa, R. E. Schmidt and G. Uhl; *Z. Anorg. Allg. Chem.*, 1984, **517**, 75. 324, 684
84ZAAC(517)89 G. Becker, O. Mundt and G. Uhl; *Z. Anorg Allg. Chem.*, 1984, **517**, 89. 684
84ZAAC(518)14 G. Fritz and W. Schick; *Z. Anorg. Allg. Chem.*, 1984, **518**, 14. 704
84ZAAC(518)21 G. Becker, W. Becker and G. Uhl; *Z. Anorg. Allg. Chem.*, 1984, **518**, 21. 584, 680, 684
84ZC345 I. F. Lutsenko; *Z. Chem.*, 1984, **24**, 345. 678
84ZC384 R. Appel, K. H. Dunker, E. Gaitzsch and T. Gaitzsch, *Z. Chem.*, 1984, **24**, 384. 693
84ZN(B)1042 H. Klusik, C. Pues and A. Berndt; *Z. Naturforsch. Teil B*, 1984, **39**, 1042. 713
84ZN(B)1456 H. Schmidbaur, P. Nusstein and G. Müller; *Z. Naturforsch.*, *Teil B*, 1984, **39**, 1456. 701
84ZN(B)1500 M. Haase, U. Klingebiel and L. Skoda; *Z. Naturforsch.*, *Teil B*, 1984, **39**, 1500. 372
84ZN(B)1518 H. H. Karsch, L. Weber, D. Wewers, R. Boese and G. Müller; *Z. Naturforsch.*, *Teil B*, 1984, **39**, 1518. 152
84ZOB715 A. D. Sinitsa, N. A. Parchomenko, M. I. Povolotskii and L. N. Markovskii; *Zh. Obshch. Khim.*, 1984, **54**, 515, (*Chem. Abstr.*, 1984, **101**, 111029). 687
84ZOB965 V. D. Romanenko, L. K. Polyachenko and L. N. Markovskii; *Zh. Obshch. Khim.*, 1984, **54**, 965 (*Chem. Abstr.*, 1984, **101**, 130765). 684
84ZOB1110 V. I. Pasternak, M. I. Dronkina, L. A. Lazukina and V. P. Kukhar; *Zh. Obshch. Khim.*, 1984, **54**, 1110 (*Chem. Abstr.*, 1985, **102**, 78347). 619
84ZOB1520 A. A. Prishchenko, A. V. Gromov, Yu. N. Luzikov, A. A. Borisenko, E. I. Lazhko, K. Klaus and I. F. Lutsenko; *Zh. Obshch. Khim.*, 1984, **54**, 1520 (*Chem. Abstr.*, 1985, **102**, 6660). 680, 681
84ZOB1842 A. G. Shipov and Y. I. Baukov; *Zh. Obshch. Khim.*, 1984, **54**, 1842 (*Chem. Abstr.*, 1984, **101**, 230689). 132
84ZOB2504 M. V. Livantsov, M. V. Proskurnina, A. A. Prishchenko and I. F. Lutsenko; *Zh. Obshch. Khim.*, 1984, **54**, 2504. 281
84ZOB2800 L. N. Markovskii, V. D. Romanenko, L. S. Kachkovskaya and M. I. Povolotskii; *Zh. Obshch. Khim.*, 1984, **54**, 2800 (*Chem. Abstr.*, 1985, **103**, 6424). 689
84ZOR115 L. M. Yagupol'skii, N. V. Kondratenko and G. N. Timofeeva; *Zh. Org. Khim.*, 1984, **20**, 115 (*Chem. Abstr.*, 1984, **100**, 191494). 233, 234
84ZOR1197 T. D. Petrova, I. V. Kolesnikova, T. I. Savchenko and V. E. Platonov; *Zh. Org. Khim.*, 1984, **20**, 1197 (*Chem. Abstr.*, 1985, **102**, 24240). 616
84ZOR2114 T. I. Bychkova, M. A. Vasil'eva, L. B. Krivdin and A. V. Kalabina; *Zh. Org. Khim.*, 1984, **20**, 2114 (*Chem. Abstr.*, 1985, **102**, 112545). 562
84ZOR2158 V. S. Gasanov, R. K. Alekperov and Sh. Sh. Ragimova; *Zh. Org. Khim.*, 1984, **20**, 2158 (*Chem. Abstr.*, 1985, **102**, 131650). 558

85ACS(B)761 X.-X. Zhou, I. Ugi and J. Chattopadhaya; *Acta Chem. Scand. Ser. B*, 1985, **39**, 761. 488
85AF217 R. Von Banholzer and K.-H. Pook; *Arzneim.-Forsch.*, 1985, **35**, 217. 444
85AG48 S. Gäb and W. V. Turner; *Angew. Chem.*, 1985, **97**, 48. 231, 232
85AG(E)53 L. Weber and K. Reizig; *Angew. Chem.*, *Int. Ed. Engl.*, 1985, **24**, 53. 680, 690
85AG(E)229 N. Wiberg, G. Wagner and G. Müller, *Angew. Chem.*, *Int. Ed. Engl.*, 1985, **24**, 229. 710
85AG(E)324 M. Haase and U. Klingebiel; *Angew. Chem.*, *Int. Ed. Engl.*, 1985, **24**, 324. 387
85AG(E)326 R. Köster, G. Seidel and B. Wrackmeyer; *Angew. Chem.*, *Int. Ed. Engl.*, 1985, **24**, 326. 183
85AG(E)419 R. Appel, T. Gaitzsch and F. Knoch; *Angew. Chem.*, *Int. Ed. Engl.*, 1985, **24**, 419. 693
85AG(E)788 R. Wehrmann, H. Meyer and A. Berndt; *Angew. Chem.*, *Int. Ed. Engl.*, 1985, **24**, 788. 201
85AG(E)924 O. J. Scherer; *Angew. Chem.*, *Int. Ed. Engl.*, 1985, **24**, 924. 692
85AG(E)1065 G. Maier, J. Henkelmann and H. P. Reisenauer; *Angew. Chem.*, *Int. Ed. Engl.*, 1985, **24**, 1065. 713

85AG(E)1068	A. Ricci, M. Fiorenza, A. Degl'Innocenti, G. Seconi, P. Dembech, K. Witzgall and H. J. Bestmann; *Angew. Chem., Int. Ed. Engl.*, 1985, **24**, 1068.	520, 702
85AOC(24)354	W. N. Setzer and P. von R. Schleyer; *Adv. Organomet. Chem.*, 1985, **24**, 354.	406
85AP(318)848	W. Hanefeld and E. Bercin; *Arch. Pharm.* (*Weinheim, Ger.*), 1985, **318**, 848.	564
85AP(318)992	G. Zinner and J. Grünefeld; *Arch. Pharm.* (*Weinheim, Ger.*), 1985, **318**, 992 (*Chem. Abstr.*, 1986, **104**, 109063).	622
85AP(318)1057	J. Grünefeld and G. Zinner; *Arch. Pharm.* (*Weinheim, Ger.*), 1985, **318**, 1057 (*Chem. Abstr.*, 1986, **104**, 129580).	622
85BCJ3379	N. Fukada, M. Hayashi and Y. Suzuki; *Bull. Chem. Soc. Jpn.*, 1985, **58**, 3379.	632
85BSF1219	P. Lerouge and C. Paulmier; *Bull. Soc. Chim. Fr.*, 1985, 1219.	93
85C277	D. Gudat, E. Niecke, B. Krebs and M. Dartmann; *Chimia*, 1985, **39**, 277.	689, 692
85CB227	U. Kunze, A. Bruns, W. Hiller and J. Mohyla; *Chem. Ber.*, 1985, **118**, 227.	584
85CB380	T. Kauffmann and A. Rensing; *Chem. Ber.*, 1985, **118**, 380.	209
85CB814	R. Appel, F. Knoch and R. Zimmerman; *Chem. Ber.*, 1985, **118**, 814.	166
85CB943	A. Haas and H.-U. Weiler; *Chem. Ber.*, 1985, **118**, 943.	237
85CB1371	W. Ried, J. Nenninger and J. W. Bats; *Chem. Ber.*, 1985, **118**, 1371.	616
85CB1415	R. Schork and W. Sundermeyer; *Chem. Ber.*, 1985, **118**, 1415.	531
85CB2208	U. Rheude and W. Sundermeyer; *Chem. Ber.*, 1985, **118**, 2208.	254
85CB2294	J. Goerdeler and H.-J. Bartsch; *Chem. Ber.*, 1985, **118**, 2294.	443
85CB2493	R. W. Hoffmann, A. Riemann and B. Mayer; *Chem. Ber.*, 1985, **118**, 2493.	390
85CB3151	U. Schubert, B. Heiser, L. Hee and H. Werner; *Chem. Ber.*, 1985, **118**, 3151.	324
85CB3959	G. Herrmann and G. Süss-Fink; *Chem. Ber.*, 1985, **118**, 3959.	143
85CB4553	A. Elsässer and W. Sundermeyer; *Chem. Ber.*, 1985, **118**, 4553.	436
85CB4997	A. Waterfeld and R. Mews; *Chem. Ber.*, 1985, **118**, 4997.	726
85CB5007	J. Antel, K. Harms, P. G. Jones, R. Mews, G. M. Sheldrick and A. Waterfeld; *Chem. Ber.*, 1985, **118**, 5007.	726
85CC5	J. D. Watts and J. G. Stamper; *J. Chem. Soc., Chem. Commun.*, 1985, 5.	270
85CC296	F. Gavina, S. V. Luis, P. Ferrer, A. M. Costero and J. A. Marco; *J. Chem. Soc., Chem. Commun.*, 1985, 296.	695
85CC534	C. Eaborn, P. B. Hitchcock, J. D. Smith and A. C. Sullivan; *J. Chem. Soc., Chem. Commun.*, 1985, 534.	396
85CC1326	R. Baker, M. J. O'Mahony and C. J. Swain; *J. Chem. Soc., Chem. Commun.*, 1985, 1326.	472, 473
85CC1380	N. H. Buttrus, C. Eaborn, P. B. Hitchcock, J. D. Smith and A. C. Sullivan; *J. Chem. Soc., Chem. Commun.*, 1985, 1380.	396
85CC1405	D. P. Hansell and R. F. Hudson; *J. Chem. Soc., Chem. Commun.*, 1985, 1405.	558
85CC1641	F. S. Y. Chan and M. P. Sammes; *J. Chem. Soc., Chem. Commun.*, 1985, 1641.	562
85CC1674	M. Aslam, R. A. Bartlett, E. Block, M. M. Olmstead, P. P. Power and G. E. Sigel; *J. Chem. Soc., Chem. Commun.*, 1985, 1674.	349
85CC1687	D. Gudat, E. Niecke, W. Malisch, U. Hofmockel, S. Quashie, A. H. Cowley, A. M. Arif, B. Krebs and M. Dartmann; *J. Chem. Soc., Chem. Commun.*, 1985, 1687.	689
85CC1776	P. B. Hitchcock, H. A. Jasim, R. E. Kelly and M. L. Lappert; *J. Chem. Soc., Chem. Commun.*, 1985, 1776.	349
85CC1800	F. Sladky, B. Bildstein, C. Rieker, A. Gieren, H. Betz and T. Hübner; *J. Chem. Soc., Chem. Commun.*, 1985, 1800.	355, 357
85CJC153	F. Houlihan, F. Bouchard and J. M. J. Frechet; *Can. J. Chem.*, 1985, **63**, 153.	477
85CL1099	T. Minami, S. Tokumasu, R. Mimasu and I. Hirao; *Chem. Lett.*, 1985, 1099.	122
85CL1671	H. Suzuki, H. Manabe and M. Inouye; *Chem. Lett.*, 1985, 1671.	498
85CRV419	G. Raabe and J. Michl; *Chem. Rev.*, 1985, **85**, 419.	177, 709
85DIS(B)840	S. Elsheimer; *Diss. Abstr. Int. B*, 1985, **46**, 840 (*Chem. Abstr.*, 1986, **104**, 129263).	227
85EGP256705	D. Klemm, M. Hartmann, P. Schumann, G. Geschwend, K. Wermann, I. Seewald, G. Kaestner, G. Schwarz and H.-J. Michel; *Ger.* (*East*) *Pat.*, 256 705 (1985) (*Chem. Abstr.*, 1988, **110**, 116953).	538
85EUP136147	C. K. Rao, S. K. Arora, R. Grover, J. A. Durden and T. D. J. D'Silva; *Eur. Pat.* 136147 (1985) (*Chem. Abstr.*, 1985, **103**, 87518).	448
85GEP3337939	H. Hagemann, M. Lenthe, F. Doering and K. H. Mohrmann (Bayer AG); *Ger. Pat.* 3337939 (1985) (*Chem. Abstr.*, 1985, **103**, 195836).	606
85H(23)1181	E. Brunet, M. C. Carreño and J. L. G. Ruano; *Heterocycles*, 1985, **23**, 1181.	567
85H(23)2371	T. Aoyama, M. Kabeya and T. Shioiri; *Heterocycles*, 1985, **23**, 2371.	675
85H(23)3099	A. R. Katritzky and S. Bayyuk; *Heterocycles*, 1985, **23**, 3099.	561
85HOU(E5)	J. Falbe (ed.); *Methoden Org. Chem.* (*Houben-Weyl*), 1985, **E5**.	36, 43, 51, 57
85HOU(E5)3	G. Simchen; *Methoden Org. Chem.* (*Houben-Weyl*), 1985, **E5**, 3.	68, 81, 95, 104, 138, 317
85HOU(E11)63	R. Schubart; *Methoden Org. Chem.* (*Houben-Weyl*), 1985, **E11**, 63.	236
85HOU(E11)158	K.-D. Gundermann and K. Hümke; *Methoden Org. Chem.* (*Houben-Weyl*), 1985, **E11**, 158.	232
85HOU(E11)614	E. Krauthausen; *Methoden Org. Chem.* (*Houben-Weyl*), 1985, **E11**, 614.	236
85HOU(E11)665	K.-D. Gundermann and K. Hümke; *Methoden Org. Chem.* (*Houben-Weyl*), 1985, **E11**, 665.	234
85HOU(E11)1055	S. Pawlenko; *Methoden Org. Chem.* (*Houben-Weyl*), 1985, **E11**, 1055.	236
85HOU(E11)1067	S. Pawlenko; *Methoden Org. Chem.* (*Houben-Weyl*), 1985, **E11**, 1067.	236
85HOU(E11)1073	H. Jäger; *Methoden Org. Chem.* (*Houben-Weyl*), 1985, **E11**, 1073.	236
85HOU(E11)1084	S. Pawlenko; *Methoden Org. Chem.* (*Houben-Weyl*), 1985, **E11**, 1084.	236
85HOU(E11)1129	K. Schank; *Methoden Org. Chem.* (*Houben-Weyl*), 1985, **E11**, 1129.	234
85IC148	A. B. Burg; *Inorg. Chem.*, 1985, **24**, 148.	265
85IC2453	T. Chivers, J. F. Richardson and N. R. M. Smith; *Inorg. Chem.*, 1985, **24**, 2453.	621, 728

85IC2889 S. O. Grim, P. H. Smith, S. Nittolo, H. L. Ammon, L. C. Satek, S. A. Sangokoya, R. K. Khanna, I. J. Colquhoun, W. McFarlane and J. R. Holden; *Inorg. Chem.*, 1985, **24**, 2889. 370

85IC4171 J. S. Thrasher and K. Seppelt; *Inorg. Chem.*, 1985, **24**, 4171. 603, 607, 615, 616

85IC4449 R. B. King, F.-J. Wu, N. D. Sadanani and E. M. Holt; *Inorg. Chem.*, 1985, **24**, 4449. 518

85IC4665 D. Lentz and H. Oberhammer; *Inorg. Chem.*, 1985, **24**, 4665. 613, 623

85IJC(B)580 C. R. Shankar, A. D. Rao, V. M. Reddy and P. B. Sattur; *Indian J. Chem., Sect. B*, 1985, **24**, 580 (*Chem. Abstr.*, 1986, **104**, 168 431). 561

85IJC(B)685 R. R. Daraji and A. Shah; *Indian J. Chem., Sect. B*, 1985, **24**, 685 (*Chem. Abstr.*, 1986, **104**, 186 267). 547

85IJC(B)1298 B. V. Kumar and V. M. Reddy; *Indian J. Chem., Sect. B*, 1985, **24**, 1298 (*Chem. Abstr.*, 1987, **106**, 119 766). 561

85IZV439 V. I. Erashko and S. A. Shevelev; *Izv. Akad. Nauk. SSSR Ser. Khim.*, 1985, 439 (*Chem. Abstr.*, 1985, **103**, 71423). 161, 283, 369

85IZV659 A. V. Fokin, Y. N. Studnev, A. I. Rapkin, V. G. Chilikin and O. V. Verenikin; *Izv. Akad. Nauk SSSR Ser. Khim.*, 1985, 659 (*Chem. Abstr.*, 1985, **103**, 87 468). 42

85IZV700 A. F. Gontar, E. N. Glotov, A. S. Vinogradov, E. G. Bykhovskaya and I. L. Knunyants; *Izv. Akad. Nauk SSSR Ser. Khim.*, 1985, 700 (*Chem. Abstr.*, 1985, **103**, 87459). 615

85IZV2635 A. V. Kalinin, E. T. Apasov, S. L. Ioffe, V. P. Kozyukov and Vik. P. Kozyukov; *Izv. Akad. Nauk SSSR Ser. Khim.*, 1985, 2635 (*Chem. Abstr.*, 1986, **105**, 191 186). 147

85JA2355 R. J. Trautman, D. C. Gross and P. C. Ford; *J. Am. Chem. Soc.*, 1985, **107**, 2355. 517

85JA4084 T. D. Tilley; *J. Am. Chem. Soc.*, 1985, **107**, 4084. 521

85JA4085 M. Fujita and T. Hiyama; *J. Am. Chem. Soc.*, 1985, **107**, 4085. 26

85JA4337 H. Hope, M. M. Olmstead, P. P. Power, J. Sandell and X. Xu; *J. Am. Chem. Soc.*, 1985, **107**, 4337. 194, 195

85JA4565 W. B. Farnham, B. E. Smart, W. J. Middleton, J. C. Calabrese and D. A. Dixon; *J. Am. Chem. Soc.*, 1985, **107**, 4565. 229

85JA4670 D. M. Roddick, B. D. Santarsiero and J. E. Bercaw; *J. Am. Chem. Soc.*, 1985, **107**, 4670. 519

85JA5014 D. J. Burton and D. M. Wiemers; *J. Am. Chem. Soc.*, 1985, **107**, 5014. 15, 246

85JA5313 H. Kawa, B. C. Manley and R. Lagow; *J. Am. Chem. Soc.*, 1985, **107**, 5313. 206

85JA6409 J. Arnold and T. D. Tilley; *J. Am. Chem. Soc.*, 1985, **107**, 6409. 521

85JA6729 E. Block and M. Aslam; *J. Am. Chem. Soc.*, 1985, **107**, 6729. 349

85JA6993 J. Canceill, L. Lancombe and A. Collet; *J. Am. Chem. Soc.*, 1985, **107**, 6993. 33

85JA8091 G. Jeske, H. Lauke, H. Mauermann, P. N. Swepston, H. Schumann and T. J. Marks; *J. Am. Chem. Soc.*, 1985, **107**, 8091. 195

85JA8103 G. Jeske, L. E. Schock, P. N. Swepston, H. Schumann and T. J. Marks; *J. Am. Chem. Soc.*, 1985, **107**, 8103. 194

85JA8136 C. P. Horwitz, E. M. Holt, C. P. Brock and D. F. Shriver; *J. Am. Chem. Soc.*, 1985, **107**, 8136. 338

85JA8147 C. P. Horwitz and D. F. Shriver; *J. Am. Chem. Soc.*, 1985, **107**, 8147. 338

85JA8211 A. H. Cowley, N. C. Norman, M. Pakulski, D. L. Bricker and D. H. Russell; *J. Am. Chem. Soc.*, 1985, **107**, 8211. 375

85JAP61229860 K. Tsuzuki and T. Uotani; *Jpn. Pat.* 61 229 860 (1985) (*Chem. Abstr.*, 1986, **106**, 101 892). 533

85JAP61229861 K. Tsuzuki and T. Uotani; *Jpn. Pat.* 61 229 861 (1985) (*Chem. Abstr.*, 1986, **106**, 101 891). 533

85JAP61233662 K. Tsuzuki and T. Uotani; *Jpn. Pat.* 61 233 662 (1985) (*Chem. Abstr.*, 1986, **106**, 101 900). 537, 538

85JAP61233663 K. Tsuzuki and T. Uotani; *Jpn. Pat.* 61 233 663 (1985) (*Chem. Abstr.*, 1986, **106**, 101 901). 537, 538

85JAP62113712 K. Tsuzuki, T. Uotani, C. Higuchi and H. Tenma; *Jpn. Pat.* 62 113 712 (1985) (*Chem. Abstr.*, 1987, **107**, 201 478). 529

85JAP62114966 K. Sakai and M. Yamashita; *Jpn. Pat.* 62 114 966 (1985) (*Chem. Abstr.*, 1987, **108**, 94 389). 539

85JAP62120358 K. Tsuzuki, T. Uotani, C. Higuchi and H. Tenma; *Jpn. Pat.* 62 120 358 (1985) (*Chem. Abstr.*, 1987, **107**, 236 245). 533

85JAP(K)60208983 T. Endo, H. Nishida, T. Nakahara and K. Kusumoto; *Jpn. Kokai* 60 208 983 (1985) (*Chem. Abstr.*, 1986, **104**, 109 658). 80

85JCS(D)383 A. H. Cowley, N. C. Norman and M. Pakulski; *J. Chem. Soc., Dalton Trans.*, 1985, 383. 375

85JCS(P1)1191 R. N. Barnes, R. D. Chambers, C. D. Hewitt, M. J. Silvester and E. Klauke; *J. Chem. Soc., Perkin Trans. 1*, 1985, **53**, 1191. 262

85JCS(P1)1935 G. M. Blackburn and T. W. Maciej; *J. Chem. Soc., Perkin Trans. 1*, 1985, 1935. 282

85JCS(P1)2585 C. A. Cornish and S. Warren; *J. Chem. Soc., Perkin Trans. 1*, 1985, 2585. 121

85JCS(P2)729 C. Eaborn and A. I. Mansour; *J. Chem. Soc., Perkin Trans. 2*, 1985, 729. 173, 381, 391

85JCS(P2)1687 C. Eaborn and D. E. Reed; *J. Chem. Soc., Perkin Trans. 2*, 1985, 1687. 176, 381

85JFC(27)309 H. Lange and D. Naumann; *J. Fluorine Chem.*, 1985, **27**, 309. 65, 66

85JFC(28)219 H. G. Lange and J. M. Shreeve; *J. Fluorine Chem.*, 1985, **28**, 219. 423

85JHC813 A. O. Abdelhamid, I. M. Abbas, M. A. Abdallah, A. A. Fahmi and A. S. Shawali; *J. Heterocycl. Chem.*, 1985, **22**, 813. 654

85JHC1269 S. Ram, D. S. Wise and L. B. Townsend; *J. Heterocycl. Chem.*, 1985, **22**, 1269. 566

85JHC1643 A. Ohta, Y. Inagawa and C. Mitsugi; *J. Heterocycl. Chem.*, 1985, **22**, 1643. 434

85JIC164 P. C. Joshi, Jr., K. B. Karnatak, M. M. Sah and A. K. Pant; *J. Indian Chem. Soc.*, 1985, **62**, 164 (*Chem. Abstr.*, 1986, **105**, 6281). 546

85JOC118 F. Chatzopoulos-Ouar and G. Descotes; *J. Org. Chem.*, 1985, **50**, 118. 472

85JOC657 K. Fuji, M. Ueda, K. Sumi, K. Kajiwara, E. Fujita, T. Iwashita and I. Miura; *J. Org. Chem.*, 1985, **50**, 657. 101

85JOC662 K. Fuji, M. Ueda, K. Sumi and E. Fujita; *J. Org. Chem.*, 1985, **50**, 662. 119, 126, 132

85JOC715 R. Katakai and Y. Iizuka; *J. Org. Chem.*, 1985, **50**, 715. 413, 415, 425

85JOC771	G. Morel, E. Marchand and A. Foucaud; *J. Org. Chem.*, 1985, **50**, 771.	617
85JOC3266	T. Cohen and L.-C. Yu; *J. Org. Chem.*, 1985, **50**, 3266.	134
85JOC4047	C. Wakselman and M. Tordeux; *J. Org. Chem.*, 1985, **50**, 4047.	233, 234
85JOC5245	J. M. Masnovi and J. K. Kochi; *J. Org. Chem.*, 1985, **50**, 5245.	147
85JOC5879	M. E. Sitzmann and W. Gilligau; *J. Org. Chem.*, 1985, **50**, 5879.	449, 492
85JOM(280)397	W. J. Bland, R. Davis and J. L. A. Durrant; *J. Organomet. Chem.*, 1985, **280**, 397.	7
85JOM(287)357	L. M. Bavaro and J. B. Keister; *J. Organomet. Chem.*, 1985, **287**, 357.	343, 344
85JOM(290)33	T. Allspach, H. Gümbel and M. Regitz; *J. Organomet. Chem.*, 1985, **290**, 33.	668
85JOM(291)145	D. W. Hutchinson and G. Semple; *J. Organomet. Chem.*, 1985, **291**, 145.	263, 288
85JOM(291)179	T. N. Mitchell and R. Wickenkamp; *J. Organomet. Chem.*, 1985, **291**, 179.	395, 399, 401, 405
85JOM(291)25	R. I. Damja, C. Eaborn and W.-C. Sham; *J. Organomet. Chem.*, 1985, **291**, 25.	380
85JOM(291)277	U. Klingebiel, S. Pohlmann and L. Skoda; *J. Organomet. Chem.*, 1985, **291**, 277.	379
85JOM(294)305	C. Eaborn and P. D. Lickiss; *J. Organomet. Chem.*, 1985, **294**, 305.	378
85JOM(295)C1	F. Sladky, B. Bildstein and D. Obendorf; *J. Organomet. Chem.* 1985, **295**, C1.	357, 358
85JPR1007	M. Vlassa, M. Kezdi and M. Bogdan; *J. Prakt. Chem.*, 1985, **327**, 1007.	558
85KGS339	L. Petrulenis, G. Veinberg, L. I. Kononov and E. Lukevics; *Khim. Geterotsikl. Soedin.*, 1985, 339 (*Chem. Abstr.*, 1985, **103**, 123 222).	558
85KPS844	S. K. Khalikov, S. V. Alieva and M. M. Sobirov; *Khim. Prir. Soedin.*, 1985, 844 (*Chem. Abstr.*, 1986, **105**, 79 339).	547
85LA555	U. Schöllkopf, I. Hoppe and A. Thiele; *Liebigs Ann. Chem.*, 1985, 555.	158
85LA2178	G. Wieland and G. Simchen; *Liebigs Ann. Chem.*, 1985, **11**, 2178.	643
85M651	R. Neidlein and U. J. Klotz; *Monatsh. Chem.*, 1985, **116**, 651.	635
85M1103	M. Tasi, A. Sisak, F. Ungváry and G. Pályi; *Monatsh. Chem.*, 1985, **116**, 1103.	517
85MI 603-01	A. B. Padias, R. Szymanski and H. K. Hall, Jr.; *ACS Symp. Ser.*, 1985, **286**, 313.	68
B-85MI 606-01	C. Eaborn; in "Organosilicon and Bioorganosilicon Chemistry," ed. H. Sakurai, Ellis Horwood, Chichester, 1985, p. 123.	172, 189
B-85MI 606-02	R. O. Bach (ed.); "Lithium. Current Applications in Science, Medicine and Technology," Wiley, New York, 1985, p. 291.	172, 204
85MI 607-01	G. L. Trainor; *J. Carbohydrate Chem.*, 1985, **4**, 545 (*Chem. Abstr.*, 1986, **105**, 115325).	229, 230
85MI 617-01	R. K. Alekperov, V. S. Gasanov and E. S. Guseinova; *Azerb. Khim. Zh.*, 1985, 47 (*Chem. Abstr.*, 1987, **106**, 49 734).	558
85MI 617-02	A. J. Baxi, V. H. Shah and A. R. Parikh; *J. Inst. Chem.* (*India*) 1985, **57**, 165 (*Chem. Abstr.*, 1986, **104**, 102 007).	561
85MI 617-03	J. Tulecki, J. Dudzinska-Usarewicz and R. Przemyslaw; *Pol. J. Pharmacol. Pharm.*, 1985, **37**, 709 (*Chem. Abstr.*, 1987, **106**, 18 279).	546
85MI 617-04	A. D. Rao, C. R. Shankar and V. M. Reddy; *Curr. Sci.*, 1985, **54**, 720 (*Chem. Abstr.*, 1986, **105**, 133 835).	561
85MI 622-01	H. Nöth; *Nova Acta Leopold.*, 1985, **59**, 277.	712
85OM339	Z.-M. Xie, P. Wisian-Neilson and R. H. Neilson; *Organometallics*, 1985, **4**, 339.	165, 167, 193, 688
85OM344	H. Schmidbaur and P. Nusstein; *Organometallics*, 1985, **4**, 344.	708
85OM478	C. F. Barrientos-Penna, A. B. Gilchrist, A. H. Klahn-Oliva, A. J. L. Hanlan and D. Sutton; *Organometallics*, 1985, **4**, 478.	517
85OM590	M. A. Lilga and J. A. Ibers; *Organometallics*, 1985, **4**, 590.	517
85OM939	H. E. Bryndza, W. C. Fultz and W. Tam; *Organometallics*, 1985, **4**, 939.	516
85OM1044	T. N. Mitchell, W. Reimann and C. Nettelbeck; *Organometallics*, 1985, **4**, 1044.	209
85OM1226	N. C. Schroeder, J. W. Richardson, Jr., S.-L. Wang, R. A. Jacobson and R. J. Angelici; *Organometallics*, 1985, **4**, 1226.	334
85OM1436	D. Nuel, F. Dahan and R. Mathieu; *Organometallics*, 1985, **4**, 1436.	338
85OM1523	M. Tasi and G. Pályi; *Organometallics*, 1985, **4**, 1523.	517
85OM1690	D. S. Matteson and J. W. Wilson; *Organometallics*, 1985, **4**, 1690.	201
85OM1830	A. M. Crespi and D. F. Shriver; *Organometallics*, 1985, **4**, 1830.	717, 727
85OPP256	J. Zmitek, B. Jenco, A. Kosak and D. Milivojevic; *Org. Prep. Proced. Int.*, 1985, **17**, 256.	632
85OR(34)319	C.-L. J. Wang; *Org. React.*, 1985, **34**, 319.	10
85PAC1827	D. Bellus; *Pure Appl. Chem.*, 1985, **57**, 1827.	7
85PS(21)291	W. Walter and M. Akram; *Phosphorus Sulfur*, 1985, **21**, 291.	556
85PS(21)349	M. Binnewies, J. Grobe and D. Le Van; *Phosphorus Sulfur*, 1985, **21**, 349.	264
85PS(25)91	M. A. Palominos, J. G. Santos, J. C. Vega and M. A. Martinez; *Phosphorus Sulfur*, 1985, **25**, 91.	476, 551
85RTC177	A. M. Van Leusen, J. Wildeman, J. Moskal and A. W. Van Hemert; *Recl. Trav. Chim. Pays-Bas*, 1985, **104**, 177.	122
85S31	H. G. Thomas and A. Schmitz; *Synthesis*, 1985, 31.	77, 78
85S408	A. Kume, H. Tanimura, S. Nishiyama, M. Sekine and T. Hata; *Synthesis*, 1985, 408.	80
85S423	L. L. Whitfield, Jr. and E. P. Papadopoulos; *Synthesis*, 1985, 423.	558
85S497	H. Suzuki, M. Yoshinaga, K. Takaoka and Y. Hiroi; *Synthesis*, 1985, 497.	45, 49
85S667	N. V. Kondratenko, A. A. Kolomeytsev, V. I. Popov and L. M. Yagupolskii; *Synthesis*, 1985, 667.	232, 234, 236, 237
85S698	Y. Ito, E. Nakajo, K. Sho and T. Saegusa; *Synthesis*, 1985, 698.	127
85S717	R. Anderson; *Synthesis*, 1985, 717.	174
85S762	M. L. Graziano and M. R. Iesce; *Synthesis*, 1985, 762.	76, 77
85S891	M. Yokoyama and H. Hatanaka; *Synthesis*, 1985, 891.	627, 730
85S975	M. A. Martinez and J. C. Vega; *Synthesis*, 1985, 975.	558
85S1149	F. Fülöp, G. Csirinyi, G. Bernáth and J. A. Szabó; *Synthesis*, 1985, 1149.	558
85SC697	K. H. Pilgram; *Synth. Commun.*, 1985, **15**, 697.	445

85SUL61 H. C. Hansen, B. Jensen and A. Senning; *Sulfur Letters*, 1985, **3**, 61. 533, 536

85T393 J. Tsuji, K. Sato and H. Nagashima; *Tetrahedron*, 1985, **41**, 393. 7

85T1353 G. Bernáth, G. Stájer, A. E. Szabó, F. Fülöp and P. Sohár; *Tetrahedron*, 1985, **41**, 1353. 558

85T4079 G. E. Keck, E. J. Enholm, J. B. Yates and M. R. Wiley; *Tetrahedron*, 1985, **41**, 4079. 541, 550, 557

85T4503 J. Y. Nedelec, D. Blanchet, D. Lefort and J. F. Beillmann; *Tetrahedron*, 1985, **41**, 4503. 5

85T5145 H. C. Hansen, A. Senning and R. G. Hazell; *Tetrahedron*, 1985, **41**, 5145. 273, 274, 275, 535, 553, 554

85TL555 E. J. Corey, J. Kang and K. Kyler; *Tetrahedron Lett.*, 1985, **26**, 555. 705

85TL1091 A. Ricci, A. Degl'Innocenti, M. Fiorenza, P. Dembech, N. Ramadan, G. Seconi and D. R. M. Walton; *Tetrahedron Lett.*, 1985, **26**, 1091. 349, 585

85TL1105 S. J. Cousins, B. C. Ross, G. N. Maw and J. D. Michael; *Tetrahedron Lett.*, 1985, **26**, 1105. 628, 633

85TL1175 J. M. Renga and P.-C. Wang; *Tetrahedron Lett.*, 1985, **26**, 1175. 26

85TL1623 B. Schaub and M. Schlosser; *Tetrahedron Lett.*, 1985, **26**, 1623. 705

85TL2131 B. F. Bonini, E. Foresti, G. Maccagnani, G. Mazzanti, P. Sabatino and P. Zani; *Tetrahedron Lett.*, 1985, **26**, 2131. 127, 309

85TL2259 E. Block and M. Aslam; *Tetrahedron Lett.*, 1985, **26**, 2259. 349, 585

85TL2675 T. Mandai, M. Yamaguchi, Y. Nakayama, J. Otera and M. Kawada; *Tetrahedron Lett.*, 1985, **26**, 2675. 126

85TL3031 H. Kawasaki, K. Tomioka and K. Koga; *Tetrahedron Lett.*, 1985, **26**, 3031. 128, 327

85TL3479 T. H. Kim and D. Y. Oh; *Tetrahedron Lett.*, 1985, **26**, 3479. 119

85TL3551 R. Appel, C. Casser and M. Immenkeppel; *Tetrahedron Lett.*, 1985, **26**, 3551. 681, 682, 689, 690

85TL4149 J. D. Michael, B. C. Ross and P. M. Rees; *Tetrahedron Lett.*, 1985, **26**, 4149. 625

85TL4941 D. Desmaele and S. Tanier; *Tetrahedron Lett.*, 1985, **26**, 4941. 541

85TL5001 J. D. Buynak, M. N. Rao, M. Y. Chandrasekaran, E. Haley, P. De Meester and S. C. Chu; *Tetrahedron Lett.*, 1985, **26**, 5001. 128

85TL5243 P. G. Gassman and N. J. O'Reilly; *Tetrahedron Lett.*, 1985, 5243. 63

85TL5943 D. H. R. Barton, D. Crich and P. Potier; *Tetrahedron Lett.*, 1985, **26**, 5943. 546, 567

85USP4558160 W. R. Hydro; *US Pat.* 4558160 (1985) (*Chem. Abstr.*, 1986, **104**, 185991). 608

85ZAAC(520)120 G. Becker, J. Härer, G. Uhl and H.-J. Wessely; *Z. Anorg Allg. Chem.*, 1985, **520**, 120. 582, 684

85ZAAC(520)139 G. Becker, W. Massa, R. E. Schmidt and G. Uhl; *Z. Anorg Allg. Chem.*, 1985, **520**, 139. 582, 684

85ZAAC(522)145 R. Gerner and G. Gattow; *Z. Anorg. Allg. Chem.*, 1985, **522**, 145. 560

85ZAAC(524)111 R. Gerner and G. Gattow; *Z. Anorg. Allg. Chem.*, 1985, **524**, 111. 563

85ZAAC(524)117 R. Gerner and G. Gattow; *Z. Anorg. Allg. Chem.*, 1985, **524**, 117. 563

85ZAAC(525)112 R. Gerner and G. Gattow; *Z. Anorg. Allg. Chem.*, 1985, **525**, 112. 561, 564

85ZAAC(525)127 M. Wieber, E. Schmidt and C. Burschka; *Z. Anorg. Allg. Chem.*, 1985, **525**, 127. 547

85ZAAC(527)125 R. Gerner and G. Gattow; *Z. Anorg. Allg. Chem.*, 1985, **527**, 125. 563

85ZAAC(527)130 R. Gerner and G. Gattow; *Z. Anorg. Allg. Chem.*, 1985, **527**, 130. 561

85ZAAC(528)168 R. Gerner and G. Gattow; *Z. Anorg. Allg. Chem.*, 1985, **528**, 168. 564

85ZAAC(529)216 K. Issleib, H. Schmidt and P. Bergmann; *Z. Anorg Allg. Chem.*, 1985, **529**, 216. 687

85ZAAC(530)16 K. Issleib, W. Mögelin and A. Balszuweit; *Z. Anorg. Allg. Chem.*, 1985, **530**, 16. 491

85ZAAC(530)101 G. Gattow and U. Schubert; *Z. Anorg. Allg. Chem.*, 1985, **530**, 101. 553

85ZAAC(531)82 G. Gattow and S. Lotz; *Z. Anorg. Allg. Chem.*, 1985, **531**, 82. 560

85ZAAC(531)101 R. Gerner and G. Gattow; *Z. Anorg. Allg. Chem.*, 1985, **531**, 101. 560

85ZC219 G. Barnikow, J. Lehner and A. A. Martin; *Z. Chem.*, 1985, **25**, 219. 559

85ZC309 R. Müller; *Z. Chem.*, 1985, **25**, 309. 173

85ZN(B)1023 U. Klingebiel, S. Pohlmann, L. Skoda, C. Lensch and G. M. Sheldrick; *Z. Naturforsch., Teil B*, 1985, **40**, 1023. 380

85ZN(B)1351 H. Werner, M. Ebner and W. Bertleff; *Z. Naturforsch., Teil B*, 1985, **40**, 1351. 335, 336

85ZN(C)344 F. Dallacker, T. Böhmel and W. Müllners; *Z. Naturforsch., Teil C*, 1985, **40**, 344. 492

85ZOB221 L. N. Markovskii, V. D. Romanenko, T. V. Sarina-Pidvarko and M. I. Povolotskii; *Zh. Obshch. Khim.*, 1985, **55**, 221 (*Chem. Abstr.*, 1985, **103**, 6462). 685, 686

85ZOB264 V. P. Kukhar, I. V. Schevchenko and O. I. Kolodyazhnyi; *Zh. Obshch. Khim.*, 1985, **55**, 264 (*Chem. Abstr.*, 1985, **103**, 6443). 680, 683

85ZOB1188 V. D. Romanenko, T. V. Sarina, N. V. Kolotilo and L. N. Markovskii; *Zh. Obshch. Khim.*, 1985, **55**, 1188 (*Chem. Abstr.*, 1986, **104**, 5933). 680

85ZOB1194 A. A. Prishchenko, A. V. Gromov, Y. N. Luzikov, E. I. Lazhko and I. F. Lutsenko; *Zh. Obshch. Khim.*, 1985, **55**, 1194, (*Chem. Abstr.*, 1986, **104**, 109778). 696

85ZOB1437 V. D. Romanenko, T. V. Sarina, M. I. Povolotskii and L. N. Markovskii; *Zh. Obshch. Khim.*, 1985, **55**, 1437 (*Chem. Abstr.*, 1986, **104**, 149000). 685

85ZOB1862 O. I. Kolodyazhnyi, I. V. Shevchenko and V. P. Kukhar; *Zh. Obshch. Khim.*, 1985, **55**, 1862, (*Chem. Abstr.*, 1986, **105**, 24365). 689

85ZOB2100 N. G. Luk'yanenko, V. V. Limich, S. V. Shcherbakov and T. I. Kirichenko; *Zh. Obshch. Khim.*, 1985, **55**, 2100 (*Chem. Abstr.*, 1986, **108**, 42276) 476

85ZOB2214 O. I. Kolodyazhnyi, I. V. Shevchenko and V. P. Kukhar; *Zh. Obshch. Khim.*, 1985, **55**, 2214 (*Chem. Abstr.*, 1987, **106**, 102387). 689

85ZOB2797 O. I. Kolodyazhnyi, I. V. Shevchenko, M. I. Povolotskii, I. E. Boldeskul and V. P. Kukhar; *Zh. Obshch. Khim.*, 1985, **55**, 2797 (*Chem. Abstr.*, 1986, **105**, 226765). 683

85ZOR1281 E. S. Levchenko, T. N. Dubinina, L. V. Budnik and E. A. Romanenko; *Zh. Org. Khim.*, 1985, **21**, 1281 (*Chem. Abstr.*, 1986, **104**, 148441). 625

85ZOR1670 V. N. Marshalkin, T. V. Smirnova and T. I. Marshalkina; *Zh. Org. Khim.*, 1985, **21**, 1670 (*Chem. Abstr.*, 1986, **105**, 97115). 475

85ZOR2206 A. Taganliev, L. Z. Rol'nik, E. V. Pastushenko, S. S. Zlotskii and D. L. Rakhmankulov; *Zh. Org. Khim.*, 1985, **21**, (*Chem. Abstr.*, 1986, **104**, 148038). 478

86ACS(B)593 E. K. Moltzen, A. Senning, R. Grønbæk Hazell and H. Lund; *Acta Chem. Scand., Ser. B.*, 1986, **40**, 593. 555
86ACS(B)609 E. K. Moltzen, B. Jensen and A. Senning; *Acta Chem. Scand., Ser. B.*, 1986, **40**, 609. 535
86AG283 S. Gambarotta, S. Stella, C. Floriani, A. Chiesi-Villa and C. Guastini; *Angew. Chem.*, 1986, **98**, 283. 342
86AG(E)457 A. Schmidpeter, K.-H. Zirzow, A. Willhalm, J. M. Holmes, R. O. Day and R. R. Holmes; *Angew. Chem., Int. Ed. Engl.*, 1986, **25**, 457. 661
86AG(E)559 H. J. Bestmann and T. Arenz; *Angew. Chem., Int. Ed. Engl.*, 1986, **25**, 559. 704
86AG(E)574 H. Schmidbaur, R. Pichl and G. Müller; *Angew. Chem., Int. Ed. Engl.*, 1986, **25**, 574. 706
86AG(E)755 L. Weber, K. Reizig and R. Boese; *Angew. Chem., Int. Ed. Engl.*, 1986, **25**, 755. 518
86AG(E)919 P. Jutzi, U. Meyer, B. Krebs and M. Dartmann; *Angew. Chem., Int. Ed. Engl.*, 1986, **25**, 919. 166
86AG(E)1111 G. Schmidt, G. Baum, W. Massa and A. Berndt; *Angew. Chem., Int. Ed. Engl.*, 1986, **25**, 1111. 398
86AG(E)1112 P. Hornbach, M. Hildenbrand, H. Pritzkow and W. Siebert; *Angew. Chem., Int. Ed. Engl.*, 1986, **25**, 1112. 389, 398, 401
86AOC(25)121 M. A. Gallop and W. R. Roper; *Adv. Organomet. Chem.*, 1986, **25**, 121. 716, 717, 721
86AP(319)1064 W. Hanefeld and Z. E. Günes; *Arch. Pharm.* (*Weinheim, Ger.*), 1986, **319**, 1064. 564
86BCJ447 T. Umemoto and A. Ando; *Bull. Chem. Soc. Jpn.*, 1986, **59**, 447. 233, 234
86BCJ1741 K. Kobayashi, H. Tukada, K. Kikuchi and I. Ikemoto; *Bull. Chem. Soc. Jpn.*, 1986, **59**, 1741. 594
86BCJ3293 T. Oshikawa and M. Yamashita; *Bull. Chem. Soc. Jpn.*, 1986, **59**, 3293. 119
86BSF925 B. Langlois and G. Soula; *Bull. Chim. Soc. Fr.*, 1986, 925. 229, 230
86CB107 M. Geisel and R. Mews; *Chem. Ber.*, 1986, **119**, 107. 260, 261
86CB269 J. Holoch and W. Sundermeyer; *Chem. Ber.*, 1986, **119**, 269. 530, 531
86CB535 R. Appel, E. Gaitzsch, K.-H. Dunker and F. Knoch; *Chem. Ber.*, 1986, **119**, 535. 693
86CB1117 M. Haase, U. Klingebiel, R. Boese and M. Polk; *Chem. Ber.*, 1986, **119**, 1117. 388
86CB1331 H. J. Bestmann, A. J. Kos, K. Witzgall and P. von R. Schleyer; *Chem. Ber.*, 1986, **119**, 1331. 697, 700
86CB1455 N. Wiberg and G. Wagner; *Chem. Ber.*, 1986, **119**, 1455. 174, 289, 291, 392, 709, 710
86CB1467 N. Wiberg and G. Wagner; *Chem. Ber.*, 1986, **119**, 1467. 178, 387, 673, 710
86CB1857 L. Weber, K. Reizig and M. Frebel; *Chem. Ber.*, 1986, **119**, 1857. 690
86CB1977 R. Appel, T. Gaitzsch, F. Knoch and G. Lenz; *Chem. Ber.*, 1986, **119**, 1977. 166, 693
86CB2553 W. Lösel and K.-H. Pook; *Chem. Ber.*, 1986, **119**, 2553. 504
86CB2609 R. Appel, C. Casser and F. Knoch; *Chem. Ber.*, 1986, **119**, 2609. 371
86CB2748 R. Appel, Ch. Porz and F. Knoch; *Chem. Ber.*, 1986, **119**, 2748. 687
86CB2900 J. Kron, H. Hörnig and U. Schubert; *Chem. Ber.*, 1986, **119**, 2900. 723
86CB2966 N. Wiberg and C.-K. Kim; *Chem. Ber.*, 1986, **119**, 2966. 186, 188, 292, 388, 397, 711
86CB2980 N. Wiberg and C.-K. Kim; *Chem. Ber.*, 1986, **119**, 2980. 188, 388, 397, 673, 711
86CB3236 R. Kupfer, E. U. Würthwein, M. Krestel and R. Allmann; *Chem. Ber.*, 1986, **119**, 3236. 625
86CB3631 G. Arens, W. Sundermeyer and H. Pritzkow; *Chem. Ber.*, 1986, **119**, 3631. 254
86CC591 N. Wiberg, K. Schurz, G. Reber and G. Müller; *J. Chem. Soc., Chem. Commun.*, 1986, 591. 673
86CC672 M. J. Henderson, R. I. Papasergio, C. L. Raston, A. H. White and M. F. Lappert; *J. Chem. Soc., Chem. Commun.*, 1986, 672. 190, 196
86CC793 J. Ichihara, T. Matsuo, T. Hanafusa and T. Ando; *J. Chem. Soc., Chem. Commun.*, 1986, 793. 423, 438
86CC908 S. S. Al-Juaid, N. H. Buttrus, C. Eaborn, P. B. Hitchcock, A. T. L. Roberts, J. D. Smith and A. C. Sullivan; *J. Chem. Soc., Chem. Commun.*, 1986, 908. 394, 395
86CC969 N. H. Buttrus, C. Eaborn, P. B. Hitchcock, J. D. Smith, J. G. Stamper and A. C. Sullivan; *J. Chem. Soc., Chem. Commun.*, 1986, 969. 174, 391
86CC980 Y. Ito, T. Bando, T. Matsuura and M. Ishikawa; *J. Chem. Soc., Chem. Commun.*, 1986, 980. 674
86CC1043 N. H. Buttrus, C. Eaborn, S. H. Gupta, P. B. Hitchcock, J. D. Smith and A. C. Sullivan; *J. Chem. Soc., Chem. Commun.*, 1986, 1043. 174, 176, 381, 393
86CE311 Y. Hori, Y. Nagano, S. Miyake, S. Teramoto and H. Taniguchi; *Chem. Express*, 1986, **1**, 311 (*Chem. Abstr.*, 1987, **106**, 66484). 552, 554
86CI(L)140 R. Shabana, M. T. El Wassimy, A. El Badih and A. G. Ghattas; *Chem. Ind.* (*London*), 1986, 140. 476
86CJC732 B. M. Pinto, J. Sandoval-Ramirez, R. D. Sharma, A. C. Willis and F. W. B. Einstein; *Can. J. Chem.*, 1986, **64**, 732. 102
86CPB430 M. E. Hague, T. Kikuchi, K. Kanemitsu and Y. Tsuda; *Chem. Pharm. Bull.*, 1986, **34**, 430. 541
86EUP176412 M. Piteau, J.-P. Senet, P. Wolf, Vu Anh Dang and R. A. Olofson; *Eur. Pat.*, 176412 (1986) (*Chem. Abstr.*, 1986, **105**, 171854).
86G101 L. Busetto, V. Zanotti, V. G. Albano, D. Braga and M. Monari; *Gazz. Chim. Ital.*, 1986, **116**, 101. 332
86GEP3616009 K. Tsuzuki and T. Uotani; *Ger. Pat.* 3616009 (1986) (*Chem. Abstr.*, 1986, **106**, 138090). 533
86GEP(O)3504453 K. Kiehs, B. Würzer and N. Meyer (BASF AG); *Ger. Pat. Offen.* 3504453 (1986) (*Chem. Abstr.*, 1986, **105**, 190692). 632
86H(24)1571 Y. Terao, M. Aono and K. Achiwa; *Heterocycles*, 1986, **24**, 1571. 129
86H(24)1625 S. Kim and Y. K. Ko; *Heterocycles*, 1986, **24**, 1625. 417
86HCA44 M. E. Scheller and B. Frei; *Helv. Chim. Acta*, 1986, **69**, 44. 127
86HCA1378 M. E. Scheller, G. Iwasaki and B. Frei; *Helv. Chim. Acta*, 1986, **69**, 1378. 127

86HCA1442	M. Egli and A. S. Dreiding; *Helv. Chim. Acta.*, 1986, **69**, 1442.	425
86HCA1898	S. Huber, P. Stamouli, T. Jenny and R. Neier; *Helv. Chim. Acta*, 1986, **69**, 1898.	558
86IC275	K. D. Gupta, R. Mews, A. Waterfeld, J. M. Shreeve and H. Oberhammer; *Inorg. Chem.*, 1986, **25**, 275.	253
86IC376	W. A. Kamil, F. Haspel-Hentrich and J. M. Shreeve; *Inorg. Chem.*, 1986, **25**, 376.	229, 230
86IC2472	K. H. Whitmire, C. B. Lagrone and A. L. Rheingold; *Inorg. Chem.*, 1986, **25**, 2472.	340
86IC2699	S. O. Grim, S. A. Sangokoya, I. J. Colquhoun, W. McFarlane and R. K. Khanna; *Inorg. Chem.*, 1986, **25**, 2699.	151, 369
86IC2877	J. R. Matachek and R. J. Angelici; *Inorg. Chem.*, 1986, **25**, 2877.	333, 334
86IJ250	P. R. Auburn, J. Whelan and B. Bosnich; *Isr. J. Chem.*, 1986, **27**, 250.	472, 473
86IS117	R. Appel; *Inorg. Synth.*, 1986, **24**, 117.	288
86IS10	J. S. Thrasher; *Inorg. Synth.*, 1986, **24**, 10.	440
86IS52	R. Eujen; *Inorg. Synth.*, 1986, **24**, 52.	717
86JA99	M. A. Kesselmayer and R. S. Sheridan; *J. Am. Chem. Soc.*, 1986, **108**, 99.	280
86JA832	D. M. Wiemers and D. J. Burton; *J. Am. Chem. Soc.*, 1986, **108**, 832.	15, 246
86JA4103	M. A. Guerra, T. R. Bierschenk and R. J. Lagow; *J. Am. Chem. Soc.*, 1986, **108**, 4103.	246
86JA5355	J. Arnold, T. D. Tilley and A. L. Rheingold; *J. Am. Chem. Soc.*, 1986, **108**, 5355.	521
86JA5395	F. Bernardi, A. Bottoni and A. Venturini; *J. Am. Chem. Soc.*, 1986, **108**, 5395.	691
86JA6683	F. F. Khouri and M. K. Kaloustian; *J. Am. Chem. Soc.*, 1986, **108**, 6683.	96
86JA7868	A. Baceiredo, A. Igau, G. Bertrand, M. J. Menu, Y. Dartiguenave and J. J. Bonnet; *J. Am. Chem. Soc.*, 1986, **108**, 7868.	665, 680
86JAN230	N. Ohi, B. Aoki, T. Shinozaki, K. Moro, T. Noto, T. Nehashi, H. Okazaki and I. Matsunaga; *J. Antibiot.*, 1986, **39**, 230.	447
86JAP61118349	H. Wada and K. Suzuki; *Jpn. Pat.* 61 118 349 (1986) (*Chem. Abstr.*, 1986, **105**, 190 499).	460
86JAP61158986	E. Fujita, Y. Nagao, T. Kumagai and T. Abe; *Jpn. Pat.*, 61 158 986 (1986) (*Chem. Abstr.*, 1987, **106**, 49 872).	116
86JAP62176910	K. Tsuzuki, K. Uotani, C. Higuchi and H. Tenma; *Jpn. Pat.*, 62 176 910 (1986) (*Chem. Abstr.*, 1987, **107**, 157 697).	529
86JAP(K)61289091	H. Taguchi and H. Kato (Toa Gosei Chemical Industry Co., Ltd); *Jpn. Kokai* 61 289 091 (86 289 091) (1986) (*Chem. Abstr.*, 1987, **107**, 97 268).	298, 300
86JCS(D)1551	T. Fjeldberg, A. Haaland, B. E. R. Schilling, M. F. Lappert and A. J. Thorne; *J. Chem. Soc., Dalton Trans.*, 1986, 1551.	186, 192, 195
86JCS(D)1801	A. H. Cowley, J. E. Kilduff, N. C. Norman and M. Pakulski; *J. Chem. Soc., Dalton Trans.*, 1986, 1801.	373
86JCS(P1)195	D. J. Ager; *J. Chem. Soc., Perkin Trans. 1*, 1986, 195.	127, 135
86JCS(P1)1425	G. M. Blackburn and M. J. Parratt; *J. Chem. Soc., Perkin Trans. 1*, 1986, 1425.	289
86JCS(P2)471	V. Gold, G. J. Johnston and W. N. Wassef; *J. Chem. Soc., Perkin Trans. 2*, 1986, 471.	31
86JCS(P2)1289	G. A. Ayoko and C. Eaborn; *J. Chem. Soc., Perkin Trans. 2*, 1986, 1289.	380
86JCS(P2)1363	A. I. Al-Wassil, C. Eaborn and M. N. Romanelli; *J. Chem. Soc., Perkin Trans. 2*, 1986, 1363.	381
86JFC(31)37	T. Umemoto, Y. Kuriu, H. Shuyama, O. Miyano and S. Nakayama; *J. Fluorine Chem.*, 1986, **31**, 37.	12
86JFC(32)89	F. W. Klink, D. J. Wasser and C. C. Liu; *J. Fluorine Chem.*, 1986, **32**, 89.	253
86JFC(32)415	A. Haas; *J. Fluorine Chem.*, 1986, **32**, 415.	237
86JFC(34)46	G. Hartmann, R. Mews, G. M. Sheldrick, R. Anderskewitz, M. Niemeyer, H. J. Emeleus and H. Oberhammer; *J. Fluorine Chem.*, 1986, **34**, 46.	604
86JFC(34)251	W. Sundermeyer and M. Witz; *J. Fluorine Chem.*, 1986, **34**, 251.	439
86JHC81	M. E. Sitzmann and W. H. Gilligan; *J. Heterocycl. Chem.*, 1986, **23**, 81.	145
86JHC553	M. L. Graziano, M. R. Iesce and R. Scarpati; *J. Heterocycl. Chem.*, 1986, **23**, 553.	
86JOC370	R. Gleiter and J. Uschmann; *J. Org. Chem.*, 1986, **51**, 370.	299, 305, 362
86JOC879	S. Hackett and T. Livinghouse; *J. Org. Chem.*, 1986, **51**, 879.	336
86JOC955	Y. Takeuchi, M. Asahina, A. Murayama, K. Hori and T. Koizumi; *J. Org. Chem.*, 1986, **51**, 955.	43, 47, 54, 57
86JOC1180	H. E. Zieger, D. A. Bright and H. Haubenstock; *J. Org. Chem.*, 1986, **51**, 1180.	425
86JOC1719	W. Lwowski, S. Kanemasa, R. A. Murray, V. T. Ramakrishnan, T. K. Thiruvengadam, K. Yoshida and A. Subbaraj; *J. Org. Chem.*, 1986, **51**, 1719.	620
86JOC1866	A. L. Schroll and G. Barany; *J. Org. Chem.*, 1986, **51**, 1866.	435, 480, 497, 548, 560
86JOC1882	C. A. Maryanoff, R. C. Stanzione, J. N. Plampin and J. E. Mills; *J. Org. Chem.*, 1986, **51**, 1882.	641, 645
86JOC2168	R. A. Moss, G. Kmiecik-Lawrynowicz and K. Krogh-Jespersen; *J. Org. Chem.*, 1986, **51**, 2168.	280
86JOC3494	K. Drauz, A. Kleemann, J. Martens and P. Scherberich; *J. Org. Chem.*, 1986, **51**, 3494.	445
86JOC3545	K. Mai and G. Patil; *J. Org. Chem.*, 1986, **51**, 3545.	380
86JOC3781	T. D. J. D'Silva, A. Lopes, R. L. Jones, S. Singhawangcha and J. K. Chan; *J. Org. Chem.*, 1986, **51**, 3781.	448
86JOC3983	D. J. Ager and M. B. East; *J. Org. Chem.*, 1986, **51**, 3983.	135
86JOC4043	G. Morel, E. Marchand, C. Haquin and A. Foucaud; *J. Org. Chem.*, 1986, **51**, 4043.	635
86JOC4466	B. A. O'Brien, W. Y. Lam and D. D. DesMarteau; *J. Org. Chem.*, 1986, **51**, 4466.	257
86JOC4483	E. A. Barsa and R. Richter; *J. Org. Chem.*, 1986, **51**, 4483.	445
86JOC4488	D. Gala, M. Steinman and R. S. Jaret; *J. Org. Chem.*, 1986, **51**, 4488.	550
86JOC4737	Y. Nagao, T. Kumagai, S. Takao, T. Abe, M. Ochiai, Y. Inoue, T. Taga and E. Fujita; *J. Org. Chem.*, 1986, **51**, 4737.	116
86JOC4741	A. A. Ponaras and Ö. Zaim; *J. Org. Chem.*, 1986, **51**, 4741.	557

86JOC4768 V. J. Davisson, A. B. Woodside, T. R. Neal, K. E. Stremler, M. Muehlbacher and C. D. Poulter; *J. Org. Chem.*, 1986, **51**, 4768. 264

86JOM(300)167 W. R. Roper; *J. Organomet. Chem.*, 1986, **300**, 167. 718

86JOM(301)1 J. Mink, D. K. Breitinger, K. Dietrich and W. Kress; *J. Organomet. Chem.*, 1986, **301**, 1. 207

86JOM(301)15 S. K. Vasisht, M. Sood, N. Sood and G. Singh; *J. Organomet. Chem.*, 1986, **301**, 15. 608

86JOM(301)269 M. Lusser and P. Peringer; *J. Organomet. Chem.*, 1986, **301**, 269. 372

86JOM(303)189 J. Klaveness, F. Rise and K. Undheim; *J. Organomet. Chem.*, 1986, **303**, 189. 134

86JOM(304)283 M. P. Teulade, P. Savignac, E. E. Aboujaoude, S. Lietge and N. Collignon; *J. Organomet. Chem.*, 1986, **304**, 283. 288

86JOM(306)1 D. Grdenić, M. Sikirica and D. Matkovic-Čalogović; *J. Organomet. Chem.*, 1986, **306**, 1. 207

86JOM(307)C58 M. A. Guerra, T. R. Bierschenk and R. J. Lagow; *J. Organomet. Chem.*, 1986, **307**, C58. 246

86JOM(308)261 P. D. Lickiss; *J. Organomet. Chem.*, 1986, **308**, 261. 387

86JOM(309)C7 D. W. Hutchinson and G. Semple; *J. Organomet. Chem.*, 1986, **309**, C7. 288

86JOM(309)247 Y. Nakadaira, K. Oharu and H. Sakurai; *J. Organomet. Chem.*, 1986, **309**, 247. 380

86JOM(310)67 L. J. Farrugia; *J. Organomet. Chem.*, 1986, **310**, 67. 347

86JOM(311)63 H. Otto, M. Ebner and H. Werner; *J. Organomet. Chem.*, 1986, **311**, 63. 335

86JOM(312)C1 H. H. Karsch, A. Appelt and G. Hanika; *J. Organomet. Chem.*, 1986, **312**, C1. 169

86JOM(312)C21 G. R. John, L. A. P. Kane-Maguire and R. Kanitz; *J. Organomet. Chem.*, 1986, **312**, C21. 455, 517

86JOM(315)C5 C. Eaborn, P. D. Lickiss, S. T. Najim and M. N. Romanelli; *J. Organomet. Chem.*, 1986, **315**, C5. 290

86JOM(315)9 N. Wiberg and H. Köpf; *J. Organomet. Chem.*, 1986, **315**, 9. 177, 178, 392, 709, 710

86JOM(316)41 H. Beckers, H. Bürger, P. Bursch and I. Ruppert; *J. Organomet. Chem.*, 1986, **316**, 41. 243

86JOU270 A. Taganlyev, L. Z. Rol'nik, E. V. Pastushenko, S. S. Zlotskii and D. L. Rakhmankulov; *J. Org. Chem. USSR* (*Engl. Transl.*), 1986, **22**, 270. 478

86JPR231 H. Gross and B. Costisella; *J. Prakt. Chem.*, 1986, **328**, 231. 367

86LA1997 W. Ried and K. Schöpke; *Liebigs Ann. Chem.*, 1986, 1997. 621

B-86MI 603-01 S. R. Sandler and W. Karo; "Organic Functional Group Preparations," 2nd edn., Academic Press, New York, 1986, vol. 2, p. 48. 68

86MI 604-01 M. Halmann, A. N. Nkansah, M. Peretz, A. K. Singh, D. Vofsi and S. Yanai; *Plant Growth Regul.*, 1986, **4**, 247 (*Chem. Abstr.*, 1986, **105**, 129327). 116

86MI 604-02 D. Y. Kim and D. Y. Oh; *Bull. Korean Chem. Soc.*, 1986, **7**, 486 (*Chem. Abstr.*, 1987, **107**, 198490). 120

86MI 606-01 A. H. Cowley and N. C. Norman; *Prog. Inorg. Chem.*, 1986, **34**, 1. 172

86MI 607-01 Z. Wang, L. Yang, Z. Liu, C. Wu, and L. Ge; *Kexue Tongbao*, 1986, **31**, 1504 (*Chem. Abstr.*, 1987, **106**, 93468). 216

B-86MI 608-01 G. Fritz and E. Matern; *"Carbosilanes,"* Springer-Verlag, Berlin, 1986. 267, 268

86MI 610-01 O. I. Korchev, Yu. I. Puzin and G. V. Leplyanin; *Vysokomol. Soedin., Ser. A*, 1986, **28**, 468 (*Chem. Abstr.*, 1986, **105**, 60971). 302

B-86MI 610-02 S. Patai and Z. Rappoport (eds.); "The Chemistry of Functional Groups—The Chemistry of Organic Selenium and Tellurium Compounds," Wiley, Chichester, 1986, vol. 1. 308, 311, 312, 313, 317

B-86MI 610-03 C. Paulmier; "Selenium Reagents and Intermediates in Organic Synthesis," Pergamon, Oxford, 1986. 308, 311, 312, 313

86MI 615-01 M. N. Nassar, B. J. Agha and G. Digenis; *J. Labelled Comp. Radiopharm.*, 1986, **23**, 667. 462

86MI 617-01 V. H. Shah, A. J. Baxi and A. R. Parikh; *J. Inst. Chem.* (*India*), 1986, **58**, 165 (*Chem. Abstr.*, 1987, **107**, 58 953g. 561

86MI 617-02 A. B. H. Amor and B. Baccar; *J. Soc. Chim. Tunis*, 1986, **2**, 9 (*Chem. Abstr.*, 1987, **107**, 39 709). 558

86MI 617-03 N. S. Pereslegina, G. N. Kuz'mina, E. I. Markova and P. I. Sanin; *Neftekhimiya*, 1986, **26**, 563 (*Chem. Abstr.*, 1987, **107**, 175 623). 565

86MI 617-04 D. Skwarski, H. Sobolewski and Z. Szczepanska; *Poznan Rozs. Med.*, 1986, **10**, 97 (*Chem. Abstr.*, 1991, **114**, 142 807). 546

86MI 617-05 Z. Hu, D. Zheng, J. Zheng and G. Liu; *Yingyong Huaxue*, 1986, **3**, 75 (*Chem. Abstr.*, 1987, **106**, 156 011). 565

B-86MI 619-01 C. Paulmier; "Selenium Reagents and Intermediates in Organic Synthesis," Pergamon, Oxford, 1986, p. 58. 587

86NJC51 M. Brutus, Y. Mollier and M. Stavaux; *Nouv. J. Chem.*, 1986, **10**, 51. 637, 732

86OM173 U. Schubert, W. Hepp and J. Müller; *Organometallics*, 1986, **5**, 173. 723

86OM394 B. Eaton, J. M. O'Connor and K. P. C. Vollhardt; *Organometallics*, 1986, **5**, 394. 402

86OM593 D. Gudat, E. Niecke, A. M. Arif, A. H. Cowley and S. Quashie; *Organometallics*, 1986, **5**, 593. 689

86OM730 P. Jutzi and B. Hampel; *Organometallics*, 1986, **5**, 730. 187

86OM1103 A. A. Aitchison and L. J. Farrugia; *Organometallics*, 1986, **5**, 1103. 339, 346, 347

86OM1726 K. H. den Haan, J. L. de Boer, J. H. Teuben, A. L. Spek, B. Kojić-Prodić, G. R. Hays and R. Huis; *Organometallics*, 1986, **5**, 1726. 195

86OM1757 W.-Y. Yeh, J. R. Shapley, J. W. Ziller and M. R. Churchill; *Organometallics*, 1986, **5**, 1757. 345

86OM1944 P. Jutzi, B. Hampel, M. B. Hursthouse and A. J. Howes; *Organometallics*, 1986, **5**, 1944. 187

86OM1991 T. N. Mitchell and W. Reimann; *Organometallics*, 1986, **5**, 1991. 209

86OM2187 F. Seitz, H. Fischer, J. Riede and J. Vogel; *Organometallics*, 1986, **5**, 2187. 610

86OM2561 S.-H. Han, G. L. Geoffroy and A. L. Rheingold; *Organometallics*, 1986, **5**, 2561. 344

86OS378 A. Maercker and M. Theis; *Organomet. Synth.*, 1986, **3**, 378. 205

86PS(26)327 H. Germa and J. Navech; *Phosphorus Sulfur*, 1986, **26**, 327. 692

86PS(28)239 J. Grobe, D. Le Van, J. Schulze and J. Szameitat; *Phosphorus Sulfur*, 1986, **28**, 239. 678

86RTC278 S. Sankararaman and J. K. Kochi; *Recl. Trav. Chim. Pays-Bas*, 1986, **105**, 278. 147

86RTC286 J. M. Masnovi and J. K. Kochi; *Recl. Trav. Chim. Pays-Bas*, 1986, **105**, 286. 147

86S224 J. P. Chupp, R. C. Grabiak, K. L. Leschinsky and T. L. Neumann; *Synthesis*, 1986, 224. 28

86S330 W. Adam, V. Bhushan, T. Dirnberger and R. Fuchs; *Synthesis*, 1986, 330. 426

86S397 H. J. Bestmann and K. Roth; *Synthesis*, 1986, 397. 75

86S633 M. Taddei and A. Ricci; *Synthesis*, 1986, 633. 134

86S760 M. A. Martinez and J. C. Vega; *Synthesis*, 1986, 760. 534

86S777 A. E. Miller and J. J. Bischoff; *Synthesis*, 1986, 777. 641

86S817 A. Vass and G. Szalontai; *Synthesis*, 1986, 817. 565

86S833 B. F. Spielvogel, F. U. Ahmed and A. T. McPhail; *Synthesis*, 1986, 833. 491

86S894 I. Degani, R. Fochi, A. Gatti and V. Regondi; *Synthesis*, 1986, 894. 553, 554

86SA(A)1281 W. Sander, R. Henn and W. Sundermeyer; *Spectrochim. Acta, Part A*, 1986, **42**, 1281. 531

86SUL93 E. K. Moltzen and A. Senning; *Sulfur Letters*, 1986, **4**, 93. 435

86SUL203 E. K. Moltzen, B. Jensen and A. Senning; *Sulfur Letters*, 1986, **4**, 203. 535

86T553 K-H. G. Brinkhaus, E. Steckhan and D. Degner; *Tetrahedron*, 1986, **42**, 553. 77, 78

86T739 A. Senning, H. C. Hansen, M. F. Abdel-Megeed, W. Mazurkiewicz and B. Jensen; *Tetrahedron*, 1986, **42**, 739. 274

86T1209 A. Krief; *Tetrahedron*, 1986, **42**, 1209. 698

86T1963 E. Juaristi, B. Gordillo and L. Valle; *Tetrahedron*, 1986, **42**, 1963. 120

86T2329 D. H. R. Barton, D. Crich, A. Löbberding and S. Z. Zard; *Tetrahedron*, 1986, **42**, 2329. 550

86T2677 M. Klich and G. Teutsch; *Tetrahedron*, 1986, **42**, 2677. 511, 605

86T3887 C. Glidewell, D. Lloyd and S. Metcalfe; *Tetrahedron*, 1986, **42**, 3887. 708, 709

86T6645 A. Hamed, E. Müller and J. C. Jochims; *Tetrahedron*, 1986, **42**, 6645. 619

86TL113 I. T. Kay and S. E. J. Glue; *Tetrahedron Lett.*, 1986, **27**, 113. 445

86TL419 R. A. Moss, M. Fedorynski, J. Terpinski and D. Z. Denney; *Tetrahedron Lett.*, 1986, **27**, 419. 280, 614

86TL1335 M. Sakamoto, H. Aoyama and Y. Omote; *Tetrahedron Lett.*, 1986, **27**, 1335. 114

86TL1661 R. Appel, P. Fölling, W. Schuhn and F. Knoch; *Tetrahedron Lett.*, 1986, **27**, 1661. 687

86TL2957 G. Märkl and H. Seitz; *Tetrahedron Lett.*, 1986, **27**, 2957. 161

86TL3845 L. M. Weinstock, J. M. Stevenson, S. A. Tomelini, S. H. Pan, T. Utne, R. B. Jobson and D. F. Reinhold; *Tetrahedron Lett.*, 1986, **27**, 3845. 25

86TL4861 D. P. Mathews, J. P. Whitten and J. R. McCarthy; *Tetrahedron Lett.*, 1986, **27**, 4861. 10, 84

86TL5143 M. S. Baird and H. H. Hussain; *Tetrahedron Lett.*, 1986, **27**, 5143. 124

86TL5985 A. Ricci, A. Degl'Innocenti, M. Ancillotti, G. Seconi and P. Dembech; *Tetrahedron Lett.*, 1986, **27**, 5985. 325, 335, 520

86TL6123 B. J. J. van de Heisteeg, G. Schat, M. A. G. M. Tinga, O. S. Akkerman and F. Bickelhaupt; *Tetrahedron Lett.*, 1986, **27**, 6123. 190, 192, 198

86UK1803 L. A. Pavlova, Y. A. Davidovich and S. V. Rogozhin; *Usp. Khim.*, 1986, **55**, 1803 (*Chem. Abstr.*, 1987, **106**, 101 424). 68

86USP4575565 W. F. Goure (Monsanto); *US Pat.* 4 575 565 (1986) (*Chem. Abstr.* 1986, **105**, 6305). 28

86ZAAC(533)99 G. Gattow and S. Lotz; *Z. Anorg. Allg. Chem.*, 1986, **533**, 99. 627

86ZAAC(542)37 K. Issleib and B. Stiebitz; *Z. Anorg. Allg. Chem.*, 1986, **542**, 37. 491

86ZC138 H. Poleschner; *Z. Chem.*, 1986, **26**, 138. 93

86ZC204 M. Hans, H. Dehne and R. Hartwig; *Z. Chem.*, 1986, **26**, 204. 626

86ZN(B)149 J. Grobe, D. Le Van and J. Nientiedt; *Z. Naturforsch., Teil B*, 1986, **41**, 149. 258, 680, 683

86ZN(B)191 H. Lang, G. Huttner, B. Sigwarth, U. Weber, L. Zsolnai, I. Jibril and O. Orama; *Z. Naturforsch, Teil B*, 1986, **41**, 191. 375

86ZN(B)283 R. L. De, D. Wolters and H. Vahrenkamp; *Z. Naturforsch., Teil B*, 1986, **41**, 283. 518

86ZN(B)413 F. Fockenberg and A. Haas; *Z. Naturforsch., Teil B*, 1986, **41**, 413. 276, 537

86ZN(B)1142 U. Kunze, H. Jawad and R. Burghardt; *Z. Naturforsch., Teil B*, 1986, **41**, 1142. 581

86ZN(B)1527 H. Schmidbaur and J. Ebenhöch; *Z. Naturforsch., Teil B*, 1986, **41**, 1527. 173, 378

86ZOB253 L. N. Markovskii and V. D. Romanenko; *Zh. Obshch. Khim.*, 1986, **56**, 253 (*Chem. Abstr.*, 1986, **105**, 172 535). 678

86ZOB1427 B. I. Buzykin, M. P. Sokolov, V. N. Ivanova, B. G. Liorber and V. A. Pavlov; *Zh. Obshch. Khim.*, 1986, **56**, 1427 (*Chem. Abstr.*, 1986, **105**, 226 812). 659

86ZOB1547 M. D. Mizhiritskii and V. O. Reikhsfeld; *Zh. Obshch. Khim.*, 1986, **56**, 1547 (*Chem. Abstr.*, 1987, **106**, 137 656). 537

86ZOB2427 M. V. Livantsov, A. A. Prishchenko and I. F. Lutsenko; *Zh. Obshch. Khim.*, 1986, **56**, 2427 (*Chem. Abstr.*, 1987, **107**, 176 123). 117

86ZOR107 I. I. Lapkin, A. N. Nedugov, P. O. Bulatov, N. N. Pavlova, G. A. Gartman and A. S. Novichkova; *Zh. Org. Khim.*, 1986, **22**, 107 (*Chem. Abstr.*, 1986, **105**, 208 555). 100

86ZOR1297 T. D. Petrova, A. G. Ryabichev, T. I. Savchenko and V. E. Platonov; *Zh. Org. Khim.*, 1986, **22**, 1297 (*Chem. Abstr.*, 1988, **108**, 37 338). 609

86ZOR1393 V. P. Tverdokhlebov, N. Yu. Polyakova and I. V. Tselinskii, *Zh. Org. Khim.*, 1986, **22**, 1393 (*Chem. Abstr.*, 1987, **106**, 214 099). 145

86ZOR1834 A. F. Benda, A. F. Eleev and G. A. Sokol'skii; *Zh. Org. Khim.*, 1986, **22**, 1834 (*Chem. Abstr.*, 1987, **107**, 22 938). 42, 46

86ZOR1842 A. F. Benda, A. F. Eleev, A. F. Ermolov and G. A. Sokol'skii; *Zh. Org. Khim.*, 1986, **22**, 1842 (*Chem. Abstr.*, 1987, **107**, 22 939). 42

86ZOR2145 B. A. Trofimov, S. V. Amosova, N. I. Ivanova, O. A. Tarasova, M. L. Al'pert, G. V. Dmitrieva and M. V. Sigalov; *Zh. Org. Khim.*, 1986, **22**, 2145 (*Chem. Abstr.*, 1988, **108**, 37 214). 565

86ZOR2618 T. D. Ladyzhnikova, K. V. Altukhov and N. A. Solov'ev; *Zh. Org. Khim.*, 1986, **22**, 2618 (*Chem. Abstr.*, 1987, **107**, 175950). 149

87AF1003 A. Buschauer; *Arzneim.-Forsch.*, 1987, **37**, 1003. 632
87AF1008 A. Buschauer; *Arzneim.-Forsch.*, 1987, **37**, 1008. 632
87AG681 G. Fachinetti, G. Fochi, T. Funaioli and P. F. Zanazzi; *Angew. Chem.*, 1987, **99**, 681. 341
87AG(E)79 H. J. Bestmann and M. Schmidt; *Angew. Chem., Int. Ed. Engl.*, 1987, **26**, 79. 701
87AG(E)134 A. Schmuck and K. Seppelt; *Angew. Chem., Int. Ed. Engl.*, 1987, **26**, 134. 614
87AG(E)275 F. Mathey; *Angew. Chem., Int. Ed. Engl.*, 1987, **26**, 275. 692
87AG(E)314 A. Schmuck and K. Seppelt; *Angew. Chem., Int. Ed. Engl.*, 1987, **26**, 314. 256
87AG(E)359 T. Hassel and H. P. Müller; *Angew. Chem., Int. Ed. Engl.*, 1987, **26**, 359. 604
87AG(E)546 H. Meyer, G. Baum, W. Massa, S. Berger and A. Berndt; *Angew. Chem., Int. Ed. Engl.*, 1987, **26**, 546. 202, 724
87AG(E)584 S. Masamune, Y. Eriyama and T. Kawase; *Angew. Chem., Int. Ed. Engl.*, 1987, **26**, 584. 174
87AG(E)681 L. M. Engelhardt, U. Kynast, C. L. Raston and A. H. White; *Angew. Chem., Int. Ed. Engl.*, 1987, **26**, 681. 196
87AG(E)798 H. Meyer, G. Baum, W. Massa and A. Berndt; *Angew. Chem., Int. Ed. Engl.*, 1987, **26**, 798. 188, 711
87AG(E)894 H. Eckert and B. Forster; *Angew. Chem., Int. Ed. Engl.*, 1987, **26**, 894. 416, 447, 462, 471, 483, 486, 487
87AG(E)908 R. Milczarek, W. Rüsseler, P. Binger, K. Jonas, K. Angermund, C. Krüger and M. Regitz; *Angew. Chem., Int. Ed. Engl.*, 1987, **26**, 908. 154
87AG(E)1111 G. Fritz; *Angew. Chem., Int. Ed. Engl.*, 1987, **26**, 1111. 267, 383
87AJC795 J. J. Patroni and R. V. Stick; *Aust. J. Chem.*, 1987, **40**, 795. 477
87AOC(27)169 C. Schade and P. von R. Schleyer; *Adv. Organomet. Chem.*, 1987, **27**, 169. 406
87BCJ435 A. Sugawara, K. Hasegawa, K.-i. Suzuki, Y. Takahashi and R. Sato; *Bull. Chem. Soc. Jpn.*, 1987, **60**, 435. 553
87BCJ1204 Y. Inoue, J. Ishikawa, M. Taniguchi and H. Hashimoto; *Bull. Chem. Soc. Jpn.*, 1987, **60**, 1204. 464
87BCJ1807 S. Tsuboi, S. Mimura, S.-I. Ono, K. Watanabe and A. Takeda; *Bull. Chem. Soc. Jpn.*, 1987, **60**, 1807. 561
87BCJ1853 S. Matsui; *Bull. Chem. Soc. Jpn.*, 1987, **60**, 1853. 550
87BCJ3687 M. Kameyama and N. Kamigata; *Bull. Chem. Soc. Jpn.*, 1987, **60**, 3687. 7, 24
87CB339 E. Schaumann, J. Dietz, E. Kausch and G. C. Schmerse; *Chem. Ber.*, 1987, **120**, 339. 624, 625, 633
87CB421 M. Baudler and J. Simon; *Chem. Ber.*, 1987, **120**, 421. 605
87CB429 A. Haas and W. Wanzke; *Chem. Ber.*, 1987, **120**, 429. 235, 253, 275, 726
87CB653 N. Wiberg and H. Köpf; *Chem. Ber.*, 1987, **120**, 653. 290, 291
87CB987 L. Carlsen and H. Egsgaard; *Chem. Ber.*, 1987, **120**, 987. 556, 567
87CB1069 R. Boese, P. Paetzold and A. Tapper; *Chem. Ber.*, 1987, **120**, 1069. 188, 388, 391, 712
87CB1203 N. Wiberg, P. Karampatses and C.-K. Kim; *Chem. Ber.*, 1987, **120**, 1203. 188, 673, 711
87CB1213 N. Wiberg, P. Karampatses and C.-K. Kim; *Chem. Ber.*, 1987, **120**, 1213. 673
87CB1271 M. Krestel, R. Kupfer, R. Allmann and E.-U. Würthwein; *Chem. Ber.*, 1987, **120**, 1271. 625
87CB1281 H. Schmidbaur and P. Nussstein; *Chem. Ber.*, 1987, **120**, 1281. 708
87CB1357 N. Wiberg, G. Preiner, P. Karampatses, C.-K. Kim and K. Schurz; *Chem. Ber.*, 1987, **120**, 1357. 163
87CB1421 L. Weber, K. Reizig, D. Bungardt and R. Boese; *Chem. Ber.*, 1987, **120**, 1421. 518
87CB1499 R. Henn, W. Sundermeyer and H. Pritzkow; *Chem. Ber.*, 1987, **120**, 1499. 254, 532
87CB1581 A. Schönberg, E. Singer and W. Stephan; *Chem. Ber.*, 1987, **120**, 1581. 598
87CB1605 N. Wiberg, G. Fischer and K. Schurz; *Chem. Ber.*, 1987, **120**, 1605. 709
87CB1707 K. P. Langhans and O. Stelzer; *Chem. Ber.*, 1987, **120**, 1707. 165
87CC10 D. Gudat and E. Niecke; *J. Chem. Soc., Chem. Commun.*, 1987, 10. 689
87CC99 G. V. Boyd, J. Cobb, P. F. Lindley, J. C. Mitchell and G. A. Nicolaou; *J. Chem. Soc., Chem. Commun.*, 1987, 99. 621, 728
87CC147 L. J. Farrugia; *J. Chem. Soc., Chem. Commun.*, 1987, 147. 347
87CC410 G. E. Morris, D. Oakley, D. A. Pippard and D. J. H. Smith; *J. Chem. Soc., Chem. Commun.*, 1987, 410. 465
87CC472 D. Miguel, V. Riera, J. A. Miguel, X. Solans and M. Font-Altaba; *J. Chem. Soc., Chem. Commun.*, 1987, 472. 331
87CC1092 A. R. Barron and A. H. Cowley; *J. Chem. Soc., Chem. Commun.*, 1987, 1092. 166
87CC1183 S. M. Dhaher, C. Eaborn and J. D. Smith; *J. Chem. Soc., Chem. Commun.*, 1987, 1183. 399
87CC1228 M. S. Chambers, E. J. Thomas and D. J. Williams; *J. Chem. Soc., Chem. Commun.*, 1987, 1228. 473, 550
87CC1399 C. Schade, P. von R. Schleyer; *J. Chem. Soc., Chem. Commun.*, 1987, 1399. 691
87CC1461 C. Eaborn, P. D. Lickiss, S. T. Najim and W. A. Stańczyk; *J. Chem. Soc., Chem. Commun.*, 1987, 1461. 290, 391
87CC1701 C. Wakselman and M. Tordeux; *J. Chem. Soc., Chem. Commun.*, 1987, 1701. 11
87CCC1764 P. Kutschy, M. Dzurilla, D. Koscík and P. Kristian; *Collect. Czech. Chem. Commun.*, 1987, **52**, 1764. 558
87CJC626 P. Pruszynski; *Can. J. Chem.*, 1987, **65**, 626. 643
87CL2177 N. Tokitoh, N. Choi and W. Ando; *Chem. Lett.*, 1987, 2177. 129
87CPB156 A. Kakehi, S. Ito, K. Nagata, N. Kinoshita and N. Kakinuma; *Chem. Pharm. Bull.*, 1987, **35**, 156. 560
87CPB1734 Y. Terao, M. Aono, N. Imai and K. Achiwa; *Chem. Pharm. Bull.*, 1987, **35**, 1734. 129, 282

87CPB2840 Y. Saiga, I. Iijima, A. Ishida, T. Miyagishima, T. Oh-ishi, M. Matsumoto and Y. Matsuoka; *Chem. Pharm. Bull.*, 1987, **35**, 2840. 556, 564

87CSR45 D. Lloyd, I. Gosney and R. A. Ormiston; *Chem. Soc. Rev.*, 1987, **16**, 45. 707

87CSR75 U. Pindur, J. Müller, C. Flo and H. Witzel; *Chem. Soc. Rev.*, 1987, **16**, 75. 729

87CZ176 M. Fild and K. H. Reichert; *Chem.-Ztg.*, 1987, **111**, 176. 265

87CZ339 W. Ried and J. Eliadis; *Chem.-Ztg.*, 1987, **111**, 339. 621

87EGP242812 K. Issleib, A. Balszuweit, B. Steibitz, W. Moegelin and J. Reefschlaeger; *Ger. (East) Pat.* 242 812 (1987) (*Chem. Abstr.*, 1987, **107**, 97 121). 512

87EUP232012 A. C. T. Hsu (Rohm and Haas Co.); *Eur. Pat.* 232012 (1987) (*Chem. Abstr.*, 1988, **108**, 5580). 611

87H(25)123 I. Kawamoto, R. Endo, K. Suzuki and T. Hata; *Heterocycles*, 1987, **25**, 123. 553

87H(26)2583 K. Harano, I. Shinohara, M. Murase and T. Hisano; *Heterocycles*, 1987, **26**, 2583. 472

87HCA138 C. Papageorgiou and C. Tamm; *Helv. Chim. Acta*, 1987, **70**, 138. 541

87HCA1320 A. Griesbeck and D. Seebach; *Helv. Chim. Acta*, 1987, **70**, 1320. 79

87IC526 D. J. Taube, A. Rokicki, M. Anstock and P. C. Ford; *Inorg. Chem.*, 1987, **26**, 526. 517

87IC1488 K. Tag, M. Aslam, E. Block, T. Nicholson and J. Zubieta; *Inorg. Chem.*, 1987, **26**, 1488. 349

87IJC(B)748 S. Sharma, V. K. Agarwal, S. K. Dubey, R. N. Iyer, N. Anand, R. K. Chatterjee, S. Chandra and A. B. Sen; *Indian J. Chem., Sect. B*, 1987, **26**, 748. 443

87IZV306 V. A. Petrosyan, M. E. Niyazymbetov and B. V. Lyalin; *Izv. Akad. Nauk SSSR Ser. Khim.*, 1987, 306 (*Chem. Abstr.*, 1987, **106**, 203 993). 56

87IZV808 T. T. Vasil'eva, V. A. Kochetkova, B. V. Nelyubin, V. I. Dostovalova and R. Kh. Freidlina; *Izv. Akad. Nauk. SSSR Ser. Khim.*, 1987, 808 (*Chem. Abstr.*, 1988, **108**, 111 753). 5

87IZV1139 V. A. Dorokhov and E. A. Shagova; *Izv. Akad. Nauk SSSR Ser. Khim*, 1987, 1139 (*Chem. Abstr.*, 1988, **108**, 204 663). 662

87IZV1671 N. E. Kolobova, Z. P. Valueva and E. I. Kazimirchuk; *Izv. Akad. Nauk SSSR Ser. Khim.*, 1987, 1671 (*Chem. Abstr.*, 1988, **109**, 37924). 457

87IZV2872 D. A. Bravo-Zhivotovskii, I. S. Biltueva, T. I. Vakul'skaya, N. S. Vyazankin and M. G. Voronkov; *Izv. Akad. Nauk SSSR Ser. Khim.*, 1987, 2872 (*Chem. Abstr.*, 1988, **109**, 149655). 131

87JA279 R. Okazaki, A. Ishii and N. Inamoto; *J. Am. Chem. Soc.*, 1987, **109**, 279. 176

87JA609 T. V. RajanBabu; *J. Am. Chem. Soc.*, 1987, **109**, 609. 557

87JA2049 B. K. Campion, J. Falk and T. D. Tilley; *J. Am. Chem. Soc.*, 1987, **109**, 2049. 521

87JA2504 K. C. Nicolaou, D. G. McGarry, P. K. Somers, C. A. Veale and G. T. Furst; *J. Am. Chem. Soc.*, 1987, **109**, 2504. 134

87JA2843 R. H. Fong and W. H. Hersh; *J. Am. Chem. Soc.*, 1987, **109**, 2843. 339, 347

87JA3708 G. A. Olah, T. Yamato, T. Hashimoto, J. G. Shih, N. Trivedi, B. P. Singh, M. Piteau and J. A. Olah; *J. Am. Chem. Soc.*, 1987, **109**, 3708. 229

87JA5542 K. E. Stremler and C. D. Poulter; *J. Am. Chem. Soc.*, 1987, **109**, 5542. 264

87JA6015 M. J. Sailor, C. P. Brock and D. F. Shriver; *J. Am. Chem. Soc.*, 1987, **109**, 6015. 456

87JA7764 R. B. King, F.-J. Wu and E. M. Holt; *J. Am. Chem. Soc.*, 1987, **107**, 7764. 518

87JA7888 Y. Ito, T. Matsuura and M. Murakami; *J. Am. Chem. Soc.*, 1987, **109**, 7888. 672, 674

87JAP01075462 Y. Awano and K. Tsuzuki; *Jpn. Pat.* 01 075 462 (1987) (*Chem. Abstr.*, 1989, **111**, 133 808). 533

87JAP01102058 Y. Awano, K. Tsuzuki and K. Katsura; *Jpn. Pat.* 01 102 058 (1987) (*Chem. Abstr.*, 1989, **111**, 153 382). 533

87JAP6261987 J. Odera, M. Kawada and T. Bandai; *Jpn. Pat.* 6 261 987 (1987) (*Chem. Abstr.*, 1987, **107**, 78 069). 126

87JCR(S)362 M. L. Graziano and M. R. Iesce; *J. Chem. Res. (S)*, 1987, 362. 107

87JCS(D)249 A. J. Bard, A. H. Cowley, J. E. Kilduff, J. K. Leland, N. C. Norman, M. Pakulski and G. A. Heath; *J. Chem. Soc., Dalton Trans.*, 1987, 249. 375

87JCS(D)747 J. L. Atwood, S. G. Bott, P. B. Hitchcock, C. Eaborn, R. S. Shariffudin, J. D. Smith and A. C. Sullivan; *J. Chem. Soc., Dalton Trans.*, 1987, 747. 395

87JCS(D)2039 P. Zanella, N. Brianese, U. Castellato, F. Ossola, M. Porchia, G. Rossetto and R. Graziani; *J. Chem. Soc., Dalton Trans.*, 1987, 2039. 519

87JCS(P1)181 G. M. Blackburn, D. Brown, S. J. Martin and M. J. Parratt; *J. Chem. Soc., Perkin Trans. 1*, 1987, 181. 121

87JCS(P1)1069 R. A. W. Johnstone and P. J. Price; *J. Chem. Soc., Perkin Trans. 1*, 1987, 1069. 606

87JCS(P1)1153 J. C. Brindley, J. M. Caldwell, G. D. Meakins, S. J. Plackett and S. J. Price; *J. Chem. Soc., Perkin Trans. 1*, 1987, 1153. 572

87JCS(P1)1515 R. Grigg, J. Devlin, A. Ramasubbu, R. M. Scott and P. Stevenson; *J. Chem. Soc., Perkin Trans. 1*, 1987, 1515. 7

87JCS(P1)2119 W. Dmowski and A. Haas; *J. Chem. Soc., Perkin Trans. 1*, 1987, 2119. 233, 234

87JCS(P1)2647 P. Carisi, G. Mazzanti, P. Zani, G. Barbaro, A. Battaglia and P. Giorgianni; *J. Chem. Soc., Perkin Trans. 1*, 1987, 2647. 127

87JCS(P1)2759 E. Kotali and A. Varvoglis; *J. Chem. Soc., Perkin Trans. 1*, 1987, 2759. 562

87JCS(P2)381 G. A. Ayoko and C. Eaborn; *J. Chem. Soc., Perkin Trans. 2*, 1987, 381. 392

87JCS(P2)775 C. Eaborn and A. K. Saxena; *J. Chem. Soc., Perkin Trans. 2*, 1987, 775. 387

87JCS(P2)779 C. Eaborn and A. K. Saxena; *J. Chem. Soc., Perkin Trans. 2*, 1987, 779. 387, 388, 397

87JCS(P2)891 N. H. Buttrus, C. Eaborn, P. B. Hitchcock, P. D. Lickiss and S. T. Najim; *J. Chem. Soc., Perkin Trans. 2*, 1987, 891. 381, 392

87JCS(P2)1047 G. A. Ayoko and C. Eaborn; *J. Chem. Soc., Perkin Trans. 2*, 1987, 1047. 290, 381, 392

87JFC(34)475 W. Gombler and G. Bollmann; *J. Fluorine Chem.*, 1987, **34**, 475. 233

87JFC(36)429 E. O. John and J. M. Shreeve; *J. Fluorine Chem.*, 1987, **36**, 429. 436

87JFC(37)29 C. M. De Vohringer, E. R. De Staricco and E. H. Staricco; *J. Fluorine Chem.*, 1987, **37**, 29. 607

87JFC(37)171 Q. Chen, Z. Yang and Y. He; *J. Fluorine Chem.*, 1987, **37**, 171. 63

87JFC(37)223 A. Probst, K. Raab, K. Ulmand and K. von Werner; *J. Fluorine Chem.*, 1987, **37**, 223. 3

87JFC(37)259 R. Minkwitz, R. Kerbachi, R. Nass, D. Bernstein and H. Preut; *J. Fluorine Chem.*, 1987, **37**, 259. 615, 728

87JFC(37)327 J. L. Adcock, M. L. Robin and S. Zuberi; *J. Fluorine Chem.*, 1987, **37**, 327. 301

87JGU365 É. A. Berdyeva, D. Kurbanov, Y. K. Khekimov, L. Z. Rol'nik, S. S. Zlotskii and D. L. Rakhmankulov; *J. Gen. Chem. USSR* (*Engl. Transl.*), 1987, **57**, 365. 478

87JGU1260 A. A. Prishchenko, M. V. Livantsov and I. F. Lutsenko; *J. Gen. Chem. USSR* (*Engl. Transl.*), 1987, **57**, 1260. 490

87JGU1484 A. A. Prishchenko, M. V. Livantsov, S. A. Moshnikov and I. F. Lutsenko; *J. Gen. Chem. USSR* (*Engl. Transl.*), 1987, **57**, 1484. 490

87JHC275 R. L. Webb, D. S. Eggleston, C. S. Labaw, J. J. Lewis and K. Wert; *J. Heterocycl. Chem.*, 1987, **24**, 275. 632

87JHC581 G. D. Madding and M. D. Thompson; *J. Heterocycl. Chem.*, 1987, **24**, 581. 546

87JHC749 L. V. Dunkerton, B. C. Barot and A. Nigam; *J. Heterocycl. Chem.*, 1987, **24**, 749. 566

87JHC945 J. J. D'Amico, F. G. Bollinger, C. C. Tung and W. E. Dahl; *J. Heterocycl. Chem.*, 1987, **24**, 945. 445

87JIC194 R. R. Darji and A. Shah; *J. Indian Chem. Soc.*, 1987, **64**, 195 (*Chem. Abstr.*, 1988, **108**, 111 793). 546

87JMC1241 M. Van Dort, R. Neubig and R. E. Counsell; *J. Med. Chem.*, 1987, **30**, 1241. 606

87JOC855 J. M. Fang, B. C. Hong and L.-F. Liao; *J. Org. Chem.*, 1987, **52**, 855. 127

87JOC944 J. M. Wyvratt, G. G. Hazen and L. M. Weinstock; *J. Org. Chem.*, 1987, **52**, 944. 25

87JOC1794 V. J. Davisson, D. R. Davis, V. M. Dixit and C. D. Poulter; *J. Org. Chem.*, 1987, **52**, 1794. 264

87JOC2015 G. Hajos, A. Messmer and T. Koritsanszky; *J. Org. Chem.*, 1987, **52**, 2015. 114

87JOC2319 D. L. Boger and M. Patel; *J. Org. Chem.*, 1987, **52**, 2319. 450

87JOC2657 T. Okuyama, M. Kitano and T. Fueno; *J. Org. Chem.*, 1987, **52**, 2657. 96, 98

87JOC3413 A. Dondoni, G. Fantin, M. Fogagnolo, A. Medici and P. Pedrini; *J. Org. Chem.*, 1987, **52**, 3413. 634

87JOC4117 Z.-C. Chen, Y.-Y. Jin and P. J. Stang; *J. Org. Chem.*, 1987, **52**, 4117. 562

87JOC5630 A. A. Ponaras and Ö. Zaim; *J. Org. Chem.*, 1987, **52**, 5630. 557

87JOM(319)1 D. Grdenić, D. Matkovic-Čalogović and M. Sikirica; *J. Organomet. Chem.*, 1987, **319**, 1. 207

87JOM(320)C27 P. B. Hitchcock, M. F. Lappert and S. J. Smith; *J. Organomet. Chem.*, 1987, **320**, C27. 169

87JOM(320)137 G. A. Ayoko, N. H. Buttrus, C. Eaborn, P. B. Hitchcock and J. D. Smith; *J. Organomet. Chem.*, 1987, **320**, 137. 394, 395

87JOM(323)1 S. Pillmann and U. Klingebiel; *J. Organomet. Chem.*, 1987, **323**, 1. 380

87JOM(323)135 P. Savignac, M.-P. Teulade and N. Collignon; *J. Organomet. Chem.*, 1987, **323**, 135. 164, 374

87JOM(325)117 S. S. Al-Juaid, S. M. Dhaher, C. Eaborn, P. B. Hitchcock and J. D. Smith; *J. Organomet. Chem.*, 1987, **325**, 117. 396

87JOM(325)13 G. E. Carr, R. D. Chambers, T. F. Holmes and D. G. Parker; *J. Organomet. Chem.*, 1987, **325**, 13. 10

87JOM(328)249 G. Picotin and P. Miginiac; *J. Organomet. Chem.*, 1987, **328**, 249. 729

87JOM(334)323 D. Naumann and W. Tyrra; *J. Organomet. Chem.*, 1987, **334**, 323. 243

87JPR1131 D. Raabe and H. H. Hoerhold; *J Prakt. Chem.*, 1987, **329**, 1131 (*Chem. Abstr.*, 1989, **110**, 153796). 29

87JPS(A)3025 B. Boutevin, Y. Pietrasanta and B. Youssef; *J. Polym. Sci., Polym. Chem., Part A*, 1987, **25**, 3025. 7

87LA583 B. Rugewitz-Blackholm and M. Wiessler; *Liebigs Ann. Chem.*, 1987, 583. 561

B-87MI 606-01 M. Pereyre, J.-P. Quintard and A. Rahm; "Tin in Organic Synthesis," Butterworth, London, 1987, p. 149. 172, 204

B-87MI 606-02 A. Pelter; in "Boron Chemistry," ed. S. Hermanek, World Scientific Publishers, Singapore, 1987, p. 416. 172, 198

B-87MI 610-01 S. Patai and Z. Rappoport (eds.); "The Chemistry of Functional Groups—The Chemistry of Organic Selenium and Tellurium Compounds," Wiley, Chichester, 1987, vol. 2. 308, 311, 312, 313, 317

87MI 614-01 P. Landais and C. Crouzel; *Appl. Radiat. Isot.*, 1987, **38**, 297 (*Chem. Abstr.*, 1987, **107**, 58474). 414

87MI 614-02 H. T. A. Cheung, Chau Diem Dieu and E. Lacey; *J. Labelled Compd. Radiopharm.*, 1987, **24**, 879 (*Chem. Abstr.*, 1988, **108**, 186645). 415

87MI 617-01 V. H. Shah, H. Parekh and A. R. Parikh; *J. Inst. Chem.* (*India*), 1987, **59**, 100 (*Chem. Abstr.*, 1988, **108**, 167 391). 561

87MI 617-02 M. A. Butt, R. Perveen, R. Kamal and A. Salam; *Pak. J. Sci. Ind. Res.*, 1987, **30**, 485 (*Chem. Abstr.*, 1988, **108**, 150 261). 558

B-87MI 619-01 F. S. Guziec, Jr.; in "Organoselenium Chemistry," ed. D. Liotta, Wiley-Interscience, New York, 1987, p. 277. 587

B-87MI 619-02 F. S. Guziec, Jr.; in "The Chemistry of Organic Selenium and Tellurium Compounds," ed. S. Patai, Wiley-Interscience, Chichester, 1987, vol. 2, p. 215. 587

87MI 622-01 H. Nöth; in "Proc. 6th Int. Meet. Boron Chem. (IMEBORON)," ed. S. Hermanek, World Scientific, Singapore, NJ, 1987, p. 438. 712

87NKK1408 Y. Hori, Y. Nagano, T. Fukuhara, S. Teramoto and H. Taniguchi; *Nippon Kagaku Kaishi*, (*J. Chem. Soc. Jpn.*) 1987, 1408 (*Chem. Abstr.*, 1988, **109**, 54 270). 552, 554

87NKK1447 Y. Nagao, T. Kumagai, S. Takao, T. Abe, M. Ochiai, T. Taga, Y. Inoue and E. Fujita; *Nippon Kagaku Kaishi* (*J. Chem. Soc. Jpn.*), 1987, 1447. 116

87OM32 N. Wiberg, G. Wagner, J. Riede and G. Müller; *Organometallics*, 1987, **6**, 32. 710

87OM35 N. Wiberg, G. Wagner, G. Reber, J. Riede and G. Müller; *Organometallics*, 1987, **6**, 35. 392, 710
87OM283 D. Seyferth, G. B. Wormack, M. K. Gallagher, M. Cowie, J. P. Fackler, Jr. and A. M. Mazany; *Organometallics*, 1987, **6**, 283. 332
87OM372 P. A. Kongshaug and R. G. Miller; *Organometallics*, 1987, **6**, 372. 516
87OM819 A. A. Aitchison and L. J. Farrugia; *Organometallics*, 1987, **6**, 819. 339, 341
87OM1509 K. H. den Haan, G. A. Luinstra, A. Meetsma and J. H. Teuben; *Organometallics*, 1987, **6**, 1509. 194
87OM1822 M. J. Menu, P. Desrosiers, M. Dartiguenave, Y. Dartiguenave and G. Bertrand; *Organometallics*, 1987, **6**, 1822. 675
87OM1857 A. Sekiguchi and W. Ando; *Organometallics*, 1987, **6**, 1857. 672
87OM2451 R. L. Cowan and W. C. Trogler; *Organometallics*, 1987, **6**, 2451. 516
87PAC1011 A. Berndt, H. Meyer, G. Baum, W. Massa and S. Berger; *Pure Appl. Chem.*, 1987, **59**, 1011. 711
87POL13 A. T. Hutton; *Polyhedron*, 1987, **6**, 13. 653
87PS(29)1 T. Schafer, L. A. Suba, P. G. Ruminski and J. J. d'Amico; *Phosphorus Sulfur*, 1987, **29**, 1. 630
87PS(30)401 J. Grobe, D. Le Van, J. Nientiedt and J. Szameitat; *Phosphorus Sulfur*; 1987, **30**, 401. 679
87PS(30)433 O. I. Kolodyazhnyi, I. V. Schevchenko and V. P. Kukhar; *Phosphorus Sulfur*, 1987, **30**, 433. 683
87PS(30)495 A. Schmidpeter, A. Willhalm, J. Kroner, R. O. Day, J. M. Holmes and R. R. Holmes; *Phosphorus Sulfur*, 1987, **30**, 495. 686, 688
87PS(31)81 H. Ranaivonjatovo, J. Escudié, C. Couret and J. Satgé; *Phosphorus Sulfur*, 1987, **31**, 81. 167
87PS(38)79 S. O. Grim, S. A. Sangokoya, E. Delaubenfels, I. J. Colquhoun and W. McFarlane; *Phosphorus Sulfur*, 1987, **38**, 79. 151
87S44 B. Costisella and I. Keitel; *Synthesis*, 1987, 44. 119, 314
87S169 P. Coutrot, C. Grison and M. Youssefi-Tabrizi; *Synthesis*, 1987, 169. 136
87S170 J. D. Michael; *Synthesis*, 1987, 170. 625
87S267 P. Gosselin; *Synthesis*, 1987, 267. 627
87S349 M. Ishida, K. Ichikawa, M. Asano, H. Nakanishi and S. Kato; *Synthesis*, 1987, 349. 546, 561
87S693 A. Dondoni, G. Fantin, M. Fogagnolo, A. Medici and P. Pedrini; *Synthesis*, 1987, 693. 634
87S990 S. D. Sharma, U. Mehra, J. P. S. Khurana and S. B. Pandhi; *Synthesis*, 1987, 990. 113
87SC1047 M. A. de Jesus, J. A. Prieto L. del Valle and G. L. Larson; *Synth. Commun.*, 1987, **17**, 1047. 26
87SC1273 N. Kanemoto, S. Inoue and Y. Sato; *Synth. Commun.*, 1987, **17**, 1273. 185
87SC1559 P. Savignac, M. P. Teulade, E. E. Aboujaoude and N. Collignon; *Synth. Commun.*, 1987, **17**, 1559. 288
87SC1887 O. Hoshino, K. Saito, M. Ishizaki and B. Umezawa; *Synth. Commun.*, 1987, **17**, 1887. 449
87SUL59 H. C. Hansen, B. Jensen and A. Senning; *Sulfur Letters*, 1987, **6**, 59. 254, 275
87T1019 J. D. Watts and J. G. Stamper; *Tetrahedron*, 1987, 1019. 401
87T1793 G. Etemad-Moghadam, J. Bellan, C. Tachon and M. Koenig; *Tetrahedron*, 1987, **43**, 1793. 264
87T5315 J. P. Genet and J. M. Gandin; *Tetrahedron*, 1987, **43**, 5315. 461
87TCC(138)1 A. Maercker and M. Theis; *Top. Curr. Chem.*, 1987, **138**, 1. 172, 204, 205
87TL1293 Y. Ito, S. Nishimura and M. Ishikawa; *Tetrahedron Lett.*, 1987, **28**, 1293. 671
87TL1969 R. A. Moss, B. K. Wilk and L. M. Hadel; *Tetrahedron Lett.*, 1987, **28**, 1969. 280, 632, 731
87TL2111 H. J. Bestmann and M. Schmidt; *Tetrahedron Lett.*, 1987, **28**, 2111. 701
87TL2147 J. Otera, Y. Niibo and H. Aikawa; *Tetrahedron Lett.*, 1987, **28**, 2147. 126
87TL2641 S. N. Mazumdar, M. Sharma and M. Pal Mahajan; *Tetrahedron Lett.*, 1987, **28**, 2641. 636
87TL2769 D. Sulzle; *Tetrahedron Lett.*, 1987, **28**, 2769. 32
87TL4153 A. M. Kini, S. F. Tytko, J. E. Hunt and J. M. Williams; *Tetrahedron Lett.*, 1987, **28**, 4153. 87
87TL4445 S. P. Collingwood, A. P. Davies and B. T. Golding; *Tetrahedron Lett.*, 1987, **28**, 4445. 633
87TL5121 P. T. Meinke and G. A. Krafft; *Tetrahedron Lett.*, 1987, **28**, 5121. 130
87TL5801 W. P. Dailey; *Tetrahedron Lett.*, 1987, **28**, 5801. 55
87TL5973 F. E. Ziegler and Z.-I. Zheng; *Tetrahedron Lett.*, 1987, **28**, 5973. 543
87TL6121 B. A. Boyd, R. J. Thoma and R. H. Neilson; *Tetrahedron Lett.*, 1987, **28**, 6121. 129
87TL6443 J. J. Barieux and J. P. Schirmann; *Tetrahedron Lett.*, 1987, **28**, 6443. 430
87UKZ395 A. V. Kirsanov, V. A. Zasorina, A. S. Shtepanek and A. M. Pinchuk; *Ukr. Khim. Zh.* (*Russ. Ed.*), 1987, **53**, 395 (*Chem. Abstr.*, 1987, **109**, 110 502). 537
87USP4754072 E. Bay; *US Pat.* 4 754 072 (1987) (*Chem. Abstr.*, 1988, **109**, 230 528). 533
87ZAAC(545)125 W. Eul and G. Gattow; *Z. Anorg. Allg. Chem.*, 1987, **545**, 125. 553
87ZAAC(545)7 D. Gudat, E. Niecke, W. Sachs and P. Rademacher, *Z. Anorg. Allg. Chem.*, 1987, **545**, 7. 689, 690
87ZAAC(546)142 K. Issleib, M. Bubner and A. Balszuweit; *Z. Anorg. Allg. Chem.*, 1987, **546**, 142. 490
87ZAAC(550)109 W. Knoth and G. Gattow; *Z. Anorg. Allg. Chem.*, 1987, **550**, 109. 548, 553
87ZAAC(551)191 W. Knoth and G. Gattow; *Z. Anorg. Allg. Chem.*, 1987, **551**, 191. 563
87ZAAC(552)181 W. Knoth and G. Gattow; *Z. Anorg. Allg. Chem.*, 1987, **552**, 181. 548
87ZAAC(553)7 R. Appel and M. Immenkeppel; *Z. Anorg. Allg. Chem.*, 1987, **553**, 7. 680, 682, 683
87ZAAC(554)172 W. Knoth and G. Gattow; *Z. Anorg. Allg. Chem.*, 1987, **554**, 172. 553
87ZC24 B. A. Trofimov, N. A. Nedolya and V. I. Komel'kova; *Z. Chem.*, 1987, **27**, 24. 565
87ZN(B)77 U. Kunze and R. Tittmann; *Z. Naturforsch.*, *Teil B*, 1987, **42**, 77. 581
87ZN(B)860 U. Kunze and R. Burghardt; *Z. Naturforsch.*, *Teil B*, 1987, **42**, 860. 581
87ZN(B)1055 N. Wiberg, G. Preiner, G. Wagner, H. Koepf and G. Fischer; *Z. Naturforsch. Teil B*, 1987, **42**, 1055. 382, 709, 710
87ZN(B)1062 N. Wiberg, G. Preiner, G. Wagner and H. Koepf; *Z. Naturforsch.*, *Teil B*, 1987, **42**, 1062. 177, 178, 382, 387, 396, 709, 710
87ZOB901 L. N. Markovskii, V. D. Romanenko, L. S. Kachkovskaya, M. I Povolotskii, I. I. Patsanovskii, Yu. Z. Stepanova and E. A. Ishmaeva; *Zh. Obshch. Khim.*, 1987, **57**, 901 (*Chem. Abstr.*, 1988, **108**, 204691). 685, 686, 689

87ZOB1433 L. N. Markovskii, V. D. Romanenko and A. V. Ruban; *Zh. Obshch. Khim.*, 1987, **57**, 1433 (*Chem. Abstr.*, 1988, **108**, 204651). 689

87ZOB2141 A. P. Marchenko, G. N. Koidan, V. A. Oleinik, A. A. Kudryavtsev and A. M. Pinchuk; *Zh. Obshch. Khim.*, 1987, **57**, 2141 (*Chem. Abstr.*, 1988, **109**, 129147). 696

87ZOB2467 V. N. Ivanova, M. P. Sokolov, B. I. Buzykin, V. A. Pavlov, B. G. Liorber, T. V. Zykova, F. S. Shagvaleev, M. V. Alparova and L. F. Chertanova; *Zh. Obshch. Khim.*, 1987, **57**, 2467 (*Chem. Abstr.*, 1989, **110**, 135334). 659

87ZOB2794 G. N. Koidan, V. A. Oleinik, A. P. Marchenko and A. M. Pinchuk; *Zh. Obshch. Khim.*, 1987, **57**, 2794 (*Chem. Abstr.*, 1989, **110**, 114934). 696

87ZOR1657 I. V. Tselinskii, A. A. Mel'nikov and A. E. Trubitsin; *Zh. Org. Khim.*, 1987, **23**, 1657 (*Chem. Abstr.*, 1988, **109**, 6053). 55, 282

87ZOR2624 T. D. Ladyzhnikova, A. A. Mel'nikov, N. A. Solov'ev, I. V. Tselinskii and K. V. Altukhov; *Zh. Org. Khim.*, 1987, **23**, 2624 (*Chem. Abstr.*, 1988, **109**, 110340). 149

87ZPK1829 R. N. Zagidullin; *Zh. Prikl. Khim.*, 1987, **60**, 1829 (*Chem. Abstr.*, 1988, **109**, 92265). 565

88ACS(B)666 L. Eberson, M. Lepisto, M. Finkelstein, S. A. Hart, W. M. Moore and S. D. Ross; *Acta Chem. Scand., Ser. B*, 1988, **42**, 666. 143

88AF7 S. Elz and W. Schunach; *Arzneim.-Forsch.*, 1988, **38**, 7. 632

88AG(E)389 G. Märkl, S. Dietl, M. L. Ziegler and B. Nuber; *Angew. Chem., Int. Ed. Engl.*, 1988, **27**, 389. 154

88AG(E)709 G. Märkl, S. Dietl, M. L. Ziegler and B. Nuber; *Angew. Chem., Int. Ed. Engl.*, 1988, **27**, 709. 154

88AG(E)961 R. Hunold, J. Allwohn, G. Baum, W. Massa and A. Berndt; *Angew. Chem., Int. Ed. Engl.*, 1988, **27**, 961. 397, 714

88AG(E)1344 G. Beck and W. P. Fehlhammer; *Angew. Chem., Int. Ed. Engl.*, 1988, **27**, 1344. 606, 619, 721

88AG(E)1350 M. Granier, A. Baceiredo and G. Bertrand; *Angew. Chem., Int. Ed. Engl.*, 1988, **27**, 1350. 667

88AG(E)1370 M. Pilz, J. Allwohn, R. Hunold, W. Massa and A. Berndt; *Angew. Chem., Int. Ed. Engl.*, 1988, **27**, 1370. 201, 400

88AG(E)1484 M. Regitz and P. Binger; *Angew. Chem., Int. Ed. Engl.*, 1988, **27**, 1484. 152

88AG(E)1544 F. Scherbaum, A. Grohmann, B. Huber, C. Krüger, and H. Schmidbaur; *Angew. Chem., Int. Ed. Engl.*, 1988, **27**, 1544. 406

88AJC549 C. M. Copeland, J. Ghosh, D. P. McAdam, B. W. Skelton, R. V. Stick and A. H. White; *Aust. J. Chem.*, 1988, **41**, 549. 589, 593

88AJC563 D. P. McAdam and R. V. Stick; *Aust. J. Chem.*, 1988, **41**, 563. 612

88AJC995 J. F. K. Wilshire; *Aust. J. Chem.*, 1988, **41**, 995. 580

88AOC(28)139 P. C. Ford and A. Rokicki; *Adv. Organomet. Chem.*, 1988, **28**, 139. 516

88BCJ171 J. D. Meinhart and R. H. Grubbs; *Bull. Chem. Soc. Jpn.*, 1988, **61**, 171. 140

88BCJ2147 M. Kinokshita, M. Taniguchi, M. Morioka, H. Takami and Y. Mizusawa; *Bull. Chem. Soc. Jpn.*, 1988, **61**, 2147. 134

88BCJ2203 M. Yanagawa, O. Moriya, Y. Watanabe, Y. Ueno and T. Endo; *Bull. Chem. Soc. Jpn.*, 1988, **61**, 2203. 561

88BCJ2927 H. Suzuki, H. Takaoka, H. Yamamoto and T. Ogawa; *Bull. Chem. Soc. Jpn.*, 1988, **61**, 2927. 150

88BCJ3002 K. Mochida, Y. Nakai and S. Shiota; *Bull. Chem. Soc. Jpn.*, 1988, **61**, 3002. 244

88BCJ3205 S. Tsuboi, N. Kohara, K. Doi, M. Utaka and A. Takeda; *Bull. Chem. Soc. Jpn.*, 1988, **61**, 3205. 561

88BCJ3321 M. Fujita and T. Hiyama; *Bull. Chem. Soc. Jpn.*, 1988, **61**, 3321. 63

88CAR(181)253 J.-C. Florent, J. Ugghetto-Monfrin and C. Monneret; *Carbohydr. Res.*, 1988, **181**, 253. 550

88CB281 M. Baudler and J. Simon; *Chem. Ber.*, 1988, **121**, 281. 680, 682, 685

88CB507 D. Seebach, G. Adam, T. Gees, M. Schiess and W. Weigand; *Chem. Ber.*, 1988, **121**, 507. 622

88CB655 J. Grobe, Duc Le Van, J. Nientiedt, B. Krebs, M. Dartmann; *Chem. Ber.*, 1988, **121**, 655. 685, 692

88CB1407 N. Wiberg, G. Preiner and K. Schurz; *Chem. Ber.*, 1988, **121**, 1407. 673

88CB1445 D. Lentz, K. Graske and D. Preugschat; *Chem. Ber.*, 1988, **121**, 1445. 611

88CB2109 W. W. Du Mont and I. Wagner; *Chem. Ber.*, 1988, **121**, 2109. 355, 356

88CC105 H. Bürger and G. Pawelke; *J. Chem. Soc., Chem. Commun.*, 1988, 105. 602

88CC308 P. Delduc, C. Tailhan and S. Z. Zard; *J. Chem. Soc., Chem. Commun.*, 1988, 308. 551

88CC336 L. M. Engelhardt, B. S. Jolly, M. F. Lappert, C. L. Raston and A. H. White; *J. Chem. Soc., Chem. Commun.*, 1988, 336. 196

88CC529 A. J. Blake, R. W. Cockman, E. A. V. Ebsworth and J. H. Holloway; *J. Chem. Soc., Chem. Commun.*, 1988, 529. 457

88CC638 J. H. Clark, M. A. McClinton and R. J. Blade; *J. Chem. Soc., Chem. Commun.*, 1988, 638. 15

88CC645 M. Ashwell and R. F. W. Jackson; *J. Chem. Soc., Chem. Commun.*, 1988, 645. 126

88CC962 H. J. Heeres, A. Meetsma and J. H. Teuben; *J. Chem. Soc., Chem. Commun.*, 1988, 962. 195

88CC1007 P. B. Hitchcock, M. F. Lappert, R. G. Smith, R. A. Bartlett and P. P. Power; *J. Chem. Soc., Chem. Commun.*, 1988, 1007. 194, 195

88CC1175 W. Clegg, S. P. Collingwood, B. T. Golding and S. M. Hodgson; *J. Chem. Soc., Chem. Commun.*, 1988, 1175. 633

88CC1389 S. S. Al-Juaid, N. H. Buttrus, C. Eaborn, P. B. Hitchcock, J. D. Smith and K. Tavakkoli; *J. Chem. Soc., Chem. Commun.*, 1988, 1389. 394

88CC1598 M. J. Menu, G. Crocco, M. Dartiguenave, Y. Dartiguenave and G. Bertrand; *J. Chem. Soc., Chem. Commun*, 1988, 1598. 673

88CJC2234 K. R. Kopecky, J. Molina and R. Rico; *Can. J. Chem.*, 1988, **66**, 2234. 469

88CL1441 H. Sakurai, M. Kudo, K. Sakamoto, Y. Nakadaira, M. Kira and A. Sekiguchi; *Chem. Lett.*, 1988, 1441. 383

88CL1733 M. Yoshifuji, T. Niitsu and N. Inamoto; *Chem. Lett.*, 1988, 1733. 680, 690

88CPB4755 T. Iwakawa, H. Tamura, T. Sato and Y. Hayase; *Chem. Pharm. Bull.*, 1988, **36**, 4755. 636

88CRV1059 L. D. Durfee and I. P. Rothwell; *Chem. Rev.*, 1988, **88**, 1059. 516, 519

88CRV1121 R. Guilard and K. M. Kadish; *Chem. Rev.*, 1988, **88**, 1121.

88CRV1293 P. J. Brothers and W. R. Roper; *Chem. Rev.*, 1988, **88**, 1293. 716, 717, 718, 722, 727

88CRV1327 J. F. Nixon; *Chem. Rev.*, 1988, **88**, 1327. 690, 692

88CZ382 W. Ried and H. Nenninger; *Chem. Ztg.* 1988, **112**, 382. 360

88DIS(B)2973 J. C. Easdon; *Diss. Abstr. Int. B*, 1988, **48**, 2973 (PhD Thesis, University of Iowa, 1987) (*Chem. Abstr.*, 1988, **109**, 149 667). 63

88EGP273832 P. Henklein, H. U. Heyne and P. Franke; *Ger.* (*East*) *Pat.* 273 832 (1988) (*Chem. Abstr.*, 1989, **113**, 23 689). 533, 538

88EJM561 J. Marchand-Brynaert, R. Laub, F. De Meester and J.-M. Frêre; *Eur. J. Med. Chem.*, 1988, **23**, 561. 434

88EUP253527 D. P. Ashton and T. A. Ryan (ICI plc); *Eur. Pat.* 253 527 (1988) (*Chem. Abstr.*, 1988, **108**, 115 169). 409, 419

88EUP260979 P. J. Simpson; *Eur. Pat.* 260 979 (1988) (*Chem. Abstr.*, 1988, **109**, 110 844). 80

88EUP273458 Himont, Inc.; *Eur. Pat.* 273 458 (1988) (*Chem. Abstr.* 1989, **110**, 39 875). 656

88EUP293749 A. Aumüller, S. Hünig, P. Erk, H. Meixner and T. Metzenthin (BASF AG); *Eur. Pat.* 293 749 (1988) (*Chem. Abstr.*, 1990, **112**, 7148). 304

88GEP3611522 H. Blum and S. Hemmann; *Ger. Pat.* 3 611 522 (1988) (*Chem. Abstr.*, 1988, **108**, 132 051). 159

88GEP(O)3634929 C. Fest, H. J. Riebel, H. J. Santel, R. R. Schmidt and H. Strang (Bayer AG); *Ger. Pat. Offen.* 3 634 929 (1988) (*Chem. Abstr.*, 1988, **109**, 73 481). 632

88H(27)2327 K. Harano, H. Kiyonaga, S. I. Sugimoto, T. Matsuoka and T. Hisano; *Heterocycles*, 1988, **27**, 2327. 472

88HCA1242 R. Neidlein and F. Lucchesini; *Helv. Chim. Acta*, 1988, **71**, 1242. 100, 124

88IC302 V. M. Norwood and K. W. Morse; *Inorg. Chem.*, 1988, **27**, 302. 514

88IC570 J. S. Thrasher, J. B. Nielsen, S. G. Bott, D. J. McClure, S. A. Morris and J. L. Atwood; *Inorg. Chem.*, 1988, **27**, 570. 603

88IC784 R. J. Thoma, C. A. Prieto and R. H. Neilson; *Inorg. Chem.*, 1988, **27**, 784. 689

88IC2135 L. Turowsky and K. Seppelt; *Inorg. Chem.*, 1988, **27**, 2135. 275

88IC2809 R. Bertani, M. Mozzon and R. A. Michelin; *Inorg. Chem.*, 1988, **27**, 2809. 721

88IZV466 I. V. Martynov, V. K. Brel, V. I. Uvarov, I. A. Pomytkin, N. N. Aleinikov and S. A. Kashtanov; *Izv. Akad. Nauk SSSR Ser. Khim.*, 1988, 466 (*Chem. Abstr.*, 1988, **109**, 109 842). 261

88IZV2639 I. V. Martynov, V. K. Brel, V. I. Uvarov, N. N. Aleinikov and S. A. Kashtanov; *Izv. Akad. Nauk SSSR Ser. Khim.*, 1988, 2639 (*Chem. Abstr.*, 1989, **110**, 231 066). 261

88JA2663 G. Sicard, A. Baceiredo and G. Bertrand; *J. Am. Chem. Soc.*, 1988, **110**, 2663. 665, 667, 668, 669

88JA3692 Y. Ito, T. Matsuura and M. Murakami; *J. Am. Chem. Soc.*, 1988, **110**, 3692. 163, 671, 674

88JA4144 A. E. Wroblewski, J. Applequist, A. Takaya, R. Honzatko, S.-S. Kim, R. A. Jacobson, B. H. Reitsma, E. S. Yeung and J. G. Verkade; *J. Am. Chem. Soc.*, 1988, **110**, 4144. 79

88JA6463 A. Igau, H. Grützmacher, A. Baceiredo and G. Bertrand; *J. Am. Chem. Soc.*, 1988, **110**, 6463. 704

88JA8671 P. T. Meinke and G. A. Krafft; *J. Am. Chem. Soc.*, 1988, **110**, 8671. 130

88JAP01257116 T. Kagawa, T. Uotani and K. Tsuzuki; *Jpn. Pat.* 01 257 116 (1988) (*Chem. Abstr.*, 1989, **112**, 201 687j). 529

88JAP01301654 T. Kagawa, T. Uotani and K. Tsuzuki; *Jpn. Pat.* 01 301 654 (1988) (*Chem. Abstr.*, 1989, **112**, 216 462). 533

88JAP02009858 T. Kagawa, T. Uotani and K. Tsuzuki; *Jpn. Pat.* 02 009 858 (1988) (*Chem. Abstr.*, 1990, **113**, 5923). 533

88JAP63227595 T. Takeuchi, K. Tatsuta, K. Itsushiki, Y. Takahashi and T. Sawa (Microbiochemical Research Foundation); *Jpn. Pat.* 63 227 595 (1988) (*Chem. Abstr.*, 1989, **110**, 135 503). 288

88JAP63238043 M. Tojo and S. Fukuoka, *Jpn. Pat.* 63 238 043 (1988) (*Chem. Abstr.*, 1989, **110**, 74 823). 467

88JAP88319205 M. Takahashi (Asahi Chemical Industry Co. Ltd.); *Jpn. Pat.* 88 319 205 (1988) (*Chem. Abstr.*, 1989, **110**, 138 162). 413

88JCR(S)306 L. Hadjiarapoglou and A. Varvoglis; *J. Chem. Res.* (*S*), 1988, 306. 254

88JCS(D)381 N. H. Buttrus, C. Eaborn, M. N. A. El-Kheli, P. B. Hitchcock, J. D. Smith, A. C. Sullivan and K. Tavakkoli; *J. Chem. Soc., Dalton Trans.*, 1988, 381. 393, 394, 396

88JCS(D)2363 N. Al-Salim, A. A. West and W. R. McWhinnie; *J. Chem. Soc., Dalton Trans.*, 1988, 2363. 563

88JCS(P1)813 M. D. Brown, C. J. Meunier and G. H. Whitham; *J. Chem. Soc., Perkin Trans. 1*, 1988, 813. 559

88JCS(P1)899 F. S. Y. Chan, M. P. Sammes and R. L. Harlow; *J. Chem. Soc., Perkin Trans. 1*, 1988, 899. 623, 728

88JCS(P1)921 G. E. Carr, R. D. Chambers, T. F. Holmes and D. G. Parker; *J. Chem. Soc., Perkin Trans. 1*, 1988, 921. 14

88JCS(P1)1179 W. Dmowski and A. Haas; *J. Chem. Soc., Perkin Trans. 1*, 1988, 1179. 233

88JCS(P1)1517 M. D. Bachi and E. Bosch; *J. Chem. Soc., Perkin Trans. 1*, 1988, 1517. 548

88JCS(P1)3367 E. Dziadulewicz, D. Hodgson and T. Gallagher; *J. Chem. Soc., Perkin Trans. 1*, 1988, 3367. 127

88JCS(P2)1829 G. Ferguson, C. Glidewell, I. Gosney, D. Lloyd, S. Metcalfe and H. Lumbroso; *J. Chem. Soc., Perkin Trans. 2*, 1988, 1829. 708, 709

88JFC(39)329 R. Henn and W. Sundermeyer; *J. Fluorine Chem.*, 1988, **39**, 329. 254, 530, 532

88JFC(39)425 G. A. Hartgraves and D. J. Burton; *J. Fluorine Chem.*, 1988, **39**, 425. 64

88JFC(40)217 I. V. Kolesnikova, T. D. Petrova, V. E. Platonov, V. A. Mikhailov, A. A. Popov and V. A. Savelova; *J. Fluorine Chem.*, 1988, **40**, 217. 619

88JFC(40)365 D. Bielefeldt and E. Klauke; *J. Fluorine Chem.*, 1988, **40**, 365. 433

88JFC(41)185 H. Lange and D. Naumann; *J. Fluorine Chem.*, 1988, **41**, 185. 64

88JFC(41)247 B. R. Langlois; *J. Fluorine Chem.*, 1988, **41**, 247. 38, 45

88JHC193 K. H. Pilgram and L. H. Gale; *J. Heterocycl. Chem.*, 1988, **25**, 193. 478
88JHC907 H. Witzel and U. Pindur; *J. Heterocycl. Chem.*, 1988, **25**, 907. 104
88JHC1601 J. J. D'Amico and F. G. Bollinger; *J. Heterocycl. Chem.*, 1988, **25**, 1601. 545
88JMC1831 H. Griengl, W. Hayden, G. Penn, E. De Clerq and B. Rosenwirth; *J. Med. Chem.*, 1988, **31**, 1831. 489
88JOC38 K. H. Pilgram, R. D. Skiles and D. A. Kleier; *J. Org. Chem.*, 1988, **53**, 38. 567
88JOC366 D. G. Cox and D. J. Burton; *J. Org. Chem.*, 1988, **53**, 366. 265
88JOC1110 A. A. Ponaras, Ö. Zaim, Y. Pazo and L. Ohannesian; *J. Org. Chem.*, 1988, **53**, 1110. 557
88JOC2263 H. Sugimoto, I. Makino and K. Hirai; *J. Org. Chem.*, 1988, **53**, 2263. 543, 557, 563, 578
88JOC2350 T. Kitazume and T. Ikeya; *J. Org. Chem.*, 1988, **53**, 2350. 21
88JOC3120 W. A. Kleschick, J. E. Dunbar, S. W. Snider and A. P. Vinogradoff; *J. Org. Chem.*, 1988, **53**, 3120. 636
88JOC3134 S. Harvey, P. C. Junk, C. L. Raston and G. Salem; *J. Org. Chem.*, 1988, **53**, 3134. 393
88JOC3609 M. Mikolajczyk, P. Graczyk, M. W. Wieczorek, G. Bujacz, Y. T. Struchkov and M. Y. Antipin; *J. Org. Chem.*, 1988, **53**, 3609. 116
88JOC3637 A. Ando, T. Miki and I. Kumadaki; *J. Org. Chem.*, 1988, **53**, 3637. 9
88JOC3766 B. M. Pinto, B. D. Johnston, J. Sandoval-Ramirez and R. D. Sharma; *J. Org. Chem.*, 1988, **53**, 3766. 101, 102
88JOC4419 B. Zhou and Y. Xu; *J. Org. Chem.*, 1988, **53**, 4419. 461
88JOC4443 C. W. Bauknight, Jr. and D. D. DesMarteau; *J. Org. Chem.*, 1988, **53**, 4443. 261, 603, 615
88JOC5668 B. M. Pinto, B. D. Johnston and R. Nagelkerke; *J. Org. Chem.*, 1988, **53**, 5668. 101
88JOC6145 J. A. King, Jr., P. E. Donahue and J. E. Smith; *J. Org. Chem.*, 1988, **53**, 6145. 444
88JOM(338)295 M. P. Teulade and P. Savignac; *J. Organomet. Chem.*, 1988, **338**, 295. 289
88JOM(338)393 G. R. Clark, K. Marsden, C. E. F. Rickard, W. R. Roper and L. J. Wright; *J. Organomet. Chem.*, 1988, **338**, 393. 722
88JOM(339)17 D. Bellus, B. Klingert, R. W. Lang and G. Rihs; *J. Organomet. Chem.*, 1988, **339**, 17. 64
88JOM(339)241 S. C. Chaudhry and D. Kummer; *J. Organomet. Chem.*, 1988, **339**, 241. 656
88JOM(340)93 D. W. Hutchinson and D. M. Thornton; *J. Organomet. Chem.*, 1988, **340**, 93. 264
88JOM(341)109 G. Fritz, M. Volk, M. Straub, H. G. von Schnering, K. Peters and E.-M. Peters; *J. Organomet. Chem.*, 1988, **341**, 109. 391
88JOM(341)293 J. J. Eisch and J. E. Galle; *J. Organomet. Chem.*, 1988, **341**, 293. 131, 133
88JOM(342)399 T. G. Appleton, R. D. Berry, J. R. Hall and D. W. Neale; *J. Organomet. Chem.*, 1988, **342**, 399. 247
88JOM(344)29 B. Györi, I. Lázár and J. Emri; *J. Organomet. Chem.*, 1988, **344**, 29. 491
88JOM(347)11 A. Meller, D. Bromm, W. Maringgele, D. Böhler and G. Elter; *J. Organomet. Chem.*, 1988, **347**, 11. 184
88JOM(348)317 R. C. Cambie, A. D. Erson, A. C. Gourdie, P. S. Rutledge and P. D. Woodgate; *J. Organomet. Chem.*, 1988, **348**, 317. 131
88JOM(354)155 S. Urayama, S. Inoue and Y. Sato; *J. Organomet. Chem.*, 1988, **354**, 155. 185
88JOM(355)33 S. M. Dhaher, C. Eaborn and J. D. Smith; *J. Organomet. Chem.*, 1988, **355**, 33. 396, 399
88JOM(355)231 J. C. Jeffery, M. A. Ruiz and F. G. A. Stone; *J. Organomet. Chem.*, 1988, **355**, 231. 403
88JOM(355)243 U. Schubert, J. Kron and H. Hörnig; *J. Organomet. Chem.*, 1988, **355**, 243. 723
88JOM(355)437 A. C. Filippou, E. Herdtweck and H. G. Alt; *J. Organomet. Chem.*, 1988, **355**, 437. 720
88JOM(358)31 P. Pons, C. Biran, M. Bordeau and J. Dunoguès; *J. Organomet. Chem.*, 1988, **358**, 31. 379
88JOM(358)215 U. Schubert; *J. Organomet. Chem.*, 1988, **358**, 215. 723
88JOM(394)183 W. A. Schenk, D. Kuemmerle and C. Burschka; *J. Organomet. Chem.*, 1988, **394**, 183. 353, 354
88JPR900 H. Schroeder, E. Fischer and M. Michalik; *J. Prakt. Chem.*, 1988, **330**, 900. 651
88KFZ143 L. A. Chepakova, V. K. Brel, G. F. Makheva, V. L. Yankovskaya, B. K. Beznosko and V. Malygin; *Khim. Farm. Zh.*, 1988, **22**, 143 (*Chem. Abstr.*, 1988, **109**, 2876). 613
88LA75 H. Paulsen, W. Rauwald and U. Weichert; *Liebigs Ann. Chem.*, 1988, 75. 549
88LA599 W. Ried and A. von der Eltz; *Liebigs Ann. Chem.*, 1988, 599. 111
88LA605 E. Haug, W. Kantlehner, H. Hagen, P. Speh and H.-J. Bräuner; *Liebigs Ann. Chem.*, 1988, 605. 115
88M127 E. Zbiral, H. H. Brandstetter and E. P. Schreiner; *Monatsh. Chem.*, 1988, **119**, 127. 550
88M1427 F. Sauter, P. Stanetty, W. Sittenthaler and R. Waditschatka; *Monatsh. Chem.*, 1988, **119**, 1427. 25, 31
88MI 601-01 K. Ulm; *Spec. Chem.*, 1988, 418. 32
88MI 603-01 Y. K. Khekimov, D. Kurbanov, A. Taganlyev, T. K. Khodzhalyev, S. Tulegenov and I. Kurbanov; *Izv. Akad. Nau. Turkm. SSR, Ser. Fiz.-Tekh., Khim. Geol. Nauk*, 1988, 107 (*Chem. Abstr.*, 1989, **111**, 57 029). 88
88MI 604-01 I. S. Biltueva, D. A. Bravo-Zhivotovskii, T. I. Vakul'skaya, N. S. Vuazankin and M. G. Voronkov; *Metalloorg. Khim.*, 1988, **1**, 789 (*Chem. Abstr.*, 1989, **111**, 194 918). 131
B-88MI 606-01 A. Pelter, K. Smith and H. C. Brown; "Borane Reagents," Academic Press, New York, 1988. 172, 198
B-88MI 607-01 J. F. Liebman, A. Greenberg and W. R. Dolbier, Jr.; "Fluorine-Containing Molecules, Structure, Reactivity, Synthesis and Applications," VCH Publishers Inc., New York, 1988. 212
88MI 614-01 T. Ohsumi, N. Mito, H. Oshio and T. Itaya; *J. Pesticide Sci.*, 1988, **13**, 71. 443
88MI 614-02 C.-H. Cheng; *J. Chin. Chem. Soc.*, 1988, **35**, 261. 448
88MI 617-01 D. Skwarski, H. Sobolewski and A. Mackowiak; *Acta Pol. Pharm.*, 1988, **45**, 491 (*Chem. Abstr.*, 1990, **112**, 157 811). 546
88MI 617-02 B. Chen, Z. Qiu, Y. Lu and X. Yan; *Yiyao Gongye*, 1988, **19**, 299 (*Chem. Abstr.*, 1989, **110**, 38 705). 556

88MI 618-01 Y. V. Polenov and V. V. Budanov; *Izv. Vyssh. Uchebn. Zaved. Khim. Khim. Tekhnol.* 1988, **31**, 66 (*Chem. Abstr.*, 1989, **110**, 211 874). 580

B-88MI 622-01 L. N. Markovskii, V. D. Romanenko and A. V. Ruban; "The Chemistry of Acyclic Compounds of Dicoordinated Phosphorus," Naukova Dumka, Kiev, 1988, p. 296. 678

88MI 622-02 L. Krieger; Dissertation, University of Bonn, 1988. 684

88NJC621 S. Stella, C. Floriani, A. Chiesi-Villa and C. Guastini; *Nuov. J. Chim.*, 1988, **12**, 621. 342

88OM739 S. Inoue, Y. Sato and T. Suzuki; *Organometallics*, 1988, **7**, 739. 185, 186

88OM794 R. H. Fong and W. H. Hersh; *Organometallics*, 1988, **7**, 794. 347

88OM812 A. A. Aradi, F.-W. Grevels, C. Krüger and E. Raabe; *Organometallics*, 1988, **7**, 812. 338

88OM2577 J. Campora, E. Carmona, E. Gutierrez-Puebla, M. L. Poveda and C. Ruiz; *Organometallics*, 1988, **7**, 2577. 332

88OR(35)513 M. Hudlický; *Org. React.*, 1988, **35**, 513. 10

88OSC(6)207 M. J. Insalaco and D. S. Tarbell, *Org. Synth.*, *Coll. Vol.*, 1988, **6**, 207. 489

88POL2023 H. Kawa, B. C. Manley and R. J. Lagow; *Polyhedron*, 1988, **7**, 2023. 192, 398

88POL2049 D. C. Bradley, H. M. Dawes, M. B. Hursthouse and H. R. Powell; *Polyhedron*, 1988, **7**, 2049. 400

88PS(35)329 I. F. Lutsenko, A. A. Prishchenko and M. V. Livantsov; *Phosphorus Sulfur*, 1988, **35**, 329. 117, 490

88PS(35)93 M. Wieber and B. Bauer; *Phosphorus Sulfur*, 1988, **35**, 93. 553

88PS(36)147 R. Appel, P. Schulte and F. Knoch; *Phosphorus Sulfur*, 1988, **36**, 147. 693

88PS(39)257 V. V. Dragalov and V. E. Lavrov; *Phosphorus Sulfur*, 1988, **39**, 257. 476

88PS(40)1 S. Masson, A. Sene, D. W. Hutchinson and D. M. Thornton; *Phosphorus Sulfur*, 1988, **40**, 1. 123

88RTC641 A. R. Katritzky, R. J. Offerman, P. Cabildo and M. Soleiman; *Recl. Trav. Chim. Pays-Bas*, 1988, **107**, 641. 162

88S22 M. Barbero, S. Cadamuro, I. Degani, R. Fochi, A. Gatti and V. Regondi; *Synthesis*, 1988, 22. 302, 554, 729

88S224 W. Szeja and J. Bogusiak; *Synthesis*, 1988, 224. 562

88S226 K.-C. Luk and C.-C. Wei; *Synthesis*, 1988, 226. 543

88S319 C. Glidewell, D. Lloyd and S. Metcalfe; *Synthesis*, 1988, 319. 708, 709

88S407 J.-P. Senet, G. Sennyey and G. P. Wooden; *Synthesis*, 1988, 407. 431

88S460 C. R. Rasmussen, F. J. Villani, Jr., B. E. Reynolds, J. N. Plampin, A. R. Hood, L. R. Hecker, S. O. Nortey, A. Hanskin, M. J. Costanzo, R. M. Howse, Jr. and A. Molinari; *Synthesis*, 1988, 460. 635

88S562 J. Barluenga, F. López and F. Palacios; *Synthesis*, 1988, 562. 121

88S655 J. Liebscher; *Synthesis*, 1988, 655. 610, 613, 617, 618, 621

88S690 P. Molina, A. Tárraga and A. Espinosa; *Synthesis*, 1988, 690. 563

88S775 H. Ahlbrect and D. Kornetzky; *Synthesis*, 1988, 775. 563

88S913 L. Hadjiarapoglou, A. Varvoglis; *Synthesis*, 1988, 913. 278, 699, 708

88S961 A. Cambanis, E. Bauml and H. Mayr; *Synthesis*, 1988, 961. 252

88S994 K. R. Rao, Y. V. D. Nageswar, A. Gangadhar and P. B. Sattur; *Synthesis*, 1988, 994. 450

88SC301 E. S. Lang and J. V. Comasseto; *Synth. Commun.*, 1988, **18**, 301. 580

88SC545 P. S. Reddy, P. Yadagiri, S. Lumin, D.-S. Shin and J. R. Falck; *Synth. Commun.*, 1988, **18**, 545. 511

88SC1171 D. Chiarino, M. Napoletano and A. Sala; *Synth. Commun.*, 1988, **18**, 1171. 608

88SC1531 A. W. M. Lee, W. H. Chan and H. C. Wong; *Synth. Commun.*, 1988, **18**, 1531. 554

88SC1611 T. H. Kim and D. Y. Oh; *Synth. Commun.*, 1988, **18**, 1611. 121

88T281 E. Block and M. Aslam; *Tetrahedron*, 1988, **44**, 281. 585

88T2063 P. Herczegh, M. Zsély and L. Szilágyi; *Tetrahedron*, 1988, **44**, 2063. 88

88T4135 M. Fujita, M. Obayashi and T. Hiyama; *Tetrahedron*, 1988, **44**, 4135. 26

88T5095 R. W. Saalfrank, W. Hafner, J. Markmann and H.-J. Bestmann; *Tetrahedron*, 1988, **44**, 5095. 75

88TL355 M. Murakami, T. Matsuura and Y. Ito; *Tetrahedron Lett.*, 1988, **29**, 355. 674

88TL827 N. Narasimhamurthy and A. G. Samuelson; *Tetrahedron Lett.*, 1988, **29**, 827. 544

88TL1699 J. Y. Nédélec, H. Ait-Haddou-Mouloud, J. C. Folest and J. Périchon; *Tetrahedron Lett.*, 1988, **29**, 1699. 24

88TL2023 F. Perron, T. C. Gahman and K. F. Albizati; *Tetrahedron Lett.*, 1988, **29**, 2023. 79

88TL2327 H. Fukuda and T. Endo; *Tetrahedron Lett.*, 1988, **29**, 2327. 106

88TL3183 K. Kim, Y.-T. Lin and H. M. Mosher; *Tetrahedron Lett.*, 1988, **29**, 3183. 641

88TL4767 T. Mizuno, I. Nishiguchi, T. Hirashima, A. Ogawa, N. Kambe and N. Sonoda; *Tetrahedron Lett.*, 1988, 4767. 479

88TL4957 Y. Ichikawa; *Tetrahedron Lett.*, 1988, **29**, 4957. 472

88TL5237 N. J. R. Van Eikema Hommes, F. Bickelhaupt and G. W. Klumpp; *Tetrahedron Lett.*, 1988, **29**, 5237. 325, 379, 388, 398

88TL5859 W. H. Daly and D. Poché; *Tetrahedron Lett.*, 1988, **29**, 5859. 416, 430, 486

88TL6125 K. Nozaki, K. Oshima and K. Utimoto; *Tetrahedron Lett.*, 1988, **29**, 6125. 550

88URP1421744 N. M. Golub, B. A. Kirilyuk, I. A. Mel'nitskii, J. K. Kiladze, E. A. Kantor and D. L. Rakhmankulov; *USSR Pat.* 1 421 744 (1988) (*Chem. Abstr.*, 1989, **110**, 192 833). 89

88USP4754072 E. Bay (Stauffer Chemical Co.); *US Pat.* 4754072 (1988) (*Chem. Abstr.*, 1988, **109**, 230 528). 432

88ZAAC(556)7 R. Appel, P. Fölling, B. Josten, W. Schunn, H. W. Wenzel and F. Knoch; *Z. Anorg. Allg. Chem.*, 1988, **556**, 7. 687

88ZAAC(556)141 W. Knoth and G. Gattow; *Z. Anorg. Allg. Chem.*, 1988, **556**, 141. 563

88ZAAC(556)179 M. Wieber and E. Schmidt; *Z. Anorg. Allg. Chem.*, 1988, **556**, 179. 567

88ZAAC(556)189 M. Wieber and E. Schmidt; *Z. Anorg. Allg. Chem.*, 1988, **556**, 189. 553

88ZN(B)75 H. Blum; *Z. Naturforsch., Teil B*, 1988, **43**, 75. 159

88ZN(B)801 H. Meyer, G. Schmidt-Lukasch, G. Baum, W. Massa and A. Berndt; *Z. Naturforsch., Teil B*, 1988, **43**, 801. 714

88ZN(B)1038 J. B. Rodriguez, E. G. Gros and A. M. Stoka; *Z. Naturforsch., Teil B*, 1988, **43**, 1038. 461, 476

88ZN(B)1416 H. H. Karsch, A. Appelt, B. Deubelly, K. Zellner, J. Riede and G. Müller; *Z. Naturforsch., Teil B*, 1988, **43**, 1416. 164, 371, 373

88ZN(B)1468 N. Wiberg, G. Preiner, K. Schurz and G. Fischer; *Z. Naturforsch., Teil B*, 1988, **43**, 1468. 709, 710

88ZOB26 N. I. Buvashkina, L. V. Kovalenko, L. I. Virin and I. Y. Popova; *Zh. Obshch. Khim.*, 1988, **58**, 26. 497

88ZOB393 V. D. Sheludyakov, A. B. Lebeveda, A. V. Kisin, I. S. Nikishina, A. V. Lebedev and A. D. Kirilin; *Zh. Obshch. Khim.*, 1988, **58**, 393 (*Chem. Abstr.*, 1989, **110**, 8273). 626

88ZOB482 V. A. Oleinik, G. N. Koidan, A. P. Marchenko and A. M. Pinchuk; *Zh. Obshch. Khim.*, 1988, **58**, 482 (*Chem. Abstr.*, 1989, **110**, 8296). 681

88ZOB491 O. I. Kolodyazhnyi and D. B. Golokhov; *Zh. Obshch. Khim.*, 1988, **58**, 491 (*Chem. Abstr.*, 1989, **110**, 135337). 697

88ZOB1066 S. V. Nikolaeva, V. V. Zorin, V. I. Larionov, R. S. Musavirov, D. M. Kukovitskii and D. L. Rakhmankulov; *Zh. Obshch. Khim.*, 1988, **58**, 1066 (*Chem. Abstr.*, 1989, **111**, 7468). 127

88ZOB1447 A. A. Prishchenko, M. V. Livantsov, N. V. Boganova and I. F. Lutsenko; *Zh. Obshch. Khim.*, 1988, **58**, 1447 (*Chem. Abstr.*, 1989, **110**, 192946). 314

88ZOB1461 G. N. Koidan, A. P. Marchenko, V. A. Oleinik and A. M. Pinchuk; *Zh. Obshch. Khim.*, 1988, **58**, 1461 (*Chem. Abstr.*, 1989, **110**, 173340). 696

88ZOB1468 O. G. Sinyashin, I. Y. Gorshunov, E. S. Batyeva and A. N. Pudovik; *Zh. Obshch. Khim.*, 1988, **58**, 1468 (*Chem. Abstr.*, 1989, **110**, 212935). 581

88ZOB1489 A. N. Podovik, V. K. Khairullin and M. A. Vasyanina; *Zh. Obsch. Khim.*, 1988, **58**, 1489 (*Chem. Abstr.*, 1988, **110**, 192948). 538, 539, 584

88ZOB1516 B. N. Kozhushko and V. A. Shokol; *Zh. Obshch. Khim.*, 1988, **58**, 1516 (*Chem. Abstr.*, 1989, **110**, 154412). 286, 606

88ZOB1525 A. V. Golovanov, I. G. Maslennikov and A. N. Lavrent'ev; *Zh. Obshch. Khim.*, 1988, **58**, 1525 (*Chem. Abstr.*, 1989, **110**, 192949) 242

88ZOB1923 A. P. Marchenko, G. N. Koidan, V. A. Oleinik and A. M. Pinchuk; *Zh. Obshch. Khim.*, 1988, **58**, 1923 (*Chem. Abstr.*, 1989, **111**, 78139). 681

88ZOB1927 A. A. Prishchenko, M. V. Livantsov, N. V. Boganova and I. F. Lutsenko; *Zh. Obshch. Khim.*, 1988, **58**, 1927 (*Chem. Abstr.*, 1989, **111**, 78141). 314

88ZOB2164 G. N. Koidan, V. A. Oleinik, A. P. Marchenko and A. M. Pinchuk; *Zh. Obshch. Khim.*, 1988, **58**, 2164 (*Chem. Abstr.*, 1989, **111**, 78147). 696

88ZOB2167 V. Yu. Mavrin and V. V. Moskva; *Zh. Obshch. Khim.*, 1988, **58**, 2167 (*Chem. Abstr.*, 1989, **110**, 212942). 156, 313

88ZOB2168 A. A. Prishchenko, M. V. Livantsov, N. V. Boganova and I. F. Lutsenko; *Zh. Obshch. Khim.*, 1988, **58**, 2168 (*Chem. Abstr.*, 1989, **110**, 212943). 313

88ZOR117 Yu. V. Strokin and F. A. Khaliullin; *Zh. Org. Khim.*, 1988, **24**, 117 (*Chem. Abstr.*, 1989, **111**, 134028j). 549

88ZOR644 T. D. Ladyzhnikova, N. A. Solov'ev, K. V. Altukhov, V. V. Perekalin and G. A. Berkova; *Zh. Org. Khim.*, 1988, **24**, 644 (*Chem. Abstr.*, 1988, **109**, 110341). 149

88ZOR1327 V. M. Neplyuev, I. M. Bazavova and M. O. Lozinskii; *Zh. Org. Khim.*, 1988, **24**, 1327 (*Chem. Abstr.*, 1989, **110**, 212716). 254

88ZOR1513 T. D. Petrova, A. G. Ryabichev, T. I. Savchenko, I. V. Kolesnikova and V. E. Platonov; *Zh. Org. Khim.*, 1988, **24**, 1513. 272

88ZOR2019 K. Krus, A. Masias and I. P. Beletskaya; *Zh. Org. Khim.*, 1988, **24**, 2019 (*Chem. Abstr.*, 1989, **111**, 23458). 562

89ACR15 R. A. Moss; *Acc. Chem. Res.*, 1989, **22**, 15. 55

89AF539 M. Wolf-Pflugmann, G. Lambrecht, J. Wess and E. Mutschler; *Arzneim.-Forsch.*, 1989, **39**, 539. 444

89AG617 A. Igau, A. Baceiredo, G. Trinquier and G. Bertrand; *Angew. Chem.*, 1989, **101**, 617. 125

89AG929 D. Hu, H. Schäufele, H. Pritzkow and U. Zenneck; *Angew. Chem.*, 1989, **101**, 929. 154

89AG(E)53 R. Appel and M. Poppe; *Angew. Chem., Int. Ed. Engl.*, 1989, **28**, 53. 511

89AG(E)55 A. Sekiguchi, C. Kabuto and H. Sakurai; *Angew. Chem., Int. Ed. Engl.*, 1989, **28**, 55. 186

89AG(E)221 U. Hartwig, H. Pritzkow, K. Rall and W. Sundermeyer; *Angew. Chem., Int. Ed. Engl.*, 1989, **28**, 221. 258

89AG(E)452 N. Sonoda, T. Mizuno, S. Murakami, K. Kondo, A. Ogawa, J. Ryu and N. Kambe; *Angew. Chem., Int. Ed. Engl.*, 1989, **28**, 452. 495, 497

89AG(E)781 R. Hunold, M. Pilz, J. Allwohn, M. Stadler, W. Massa, P. von R. Schleyer and A. Berndt; *Angew. Chem., Int. Ed. Engl.*, 1989, **28**, 781. 184, 389, 398, 714

89AG(E)784 M. Pilz, M. Stadler, R. Hunold, J. Allwohn, W. Massa and A. Berndt; *Angew. Chem., Int. Ed. Engl.*, 1989, **28**, 784. 389, 398, 401, 713

89AG(E)1013 T. Wettling, J. Schneider, O. Wagner, C. G. Kreiter and M. Regitz; *Angew. Chem., Int. Ed. Engl.*, 1989, **28**, 1013. 153

89AJC1235 D. J. Collins, L. M. Downes, A. G. Jhingran, S. B. Rutschmann and G. J. Sharp; *Aust. J. Chem.*, 1989, **42**, 1235. 72

89BAU1570 N. I. Moiseeva, A. E. Gekhman, E. S. Rumyantsev, V. I. Klimanov and I. I. Moiseev; *Bull. Acad. Sci. USSR. Div. Chim. Sci.*, 1989, **38**, 1570. 410

89BCJ967 T. Kitamura, S. Miyake, S. Kobayashi and H. Taniuchi; *Bull. Chem. Soc. Jpn.*, 1989, **62**, 967. 547

89BCJ2117 H. Suzuki and Y. Nishioka; *Bull. Chem. Soc. Jpn.*, 1989, **62**, 2117. 482

89CAR(194)209 H. Molin, J.-O. Nore'n and A. Claesson; *Carbohydr. Res.*, 1989, **194**, 209. 491

89CB113 U. Hasserodt; *Chem. Ber.*, 1968, **101**, 113. 539

89CB409 N. Wiberg, M. Link and G. Fischer; *Chem. Ber.*, 1989, **122**, 409. 290, 291, 393, 709, 710

89CB453 E. Niecke, M. Leuer and M. Nieger; *Chem. Ber.*, 1989, **122**, 453. 288

89CB595 A. Tapper, T. Schmitz and P. Paetzold; *Chem. Ber.*, 1989, **122**, 595. 187, 188, 292, 388, 389, 673, 712

89CB1057 R. Boese, P. Paetzold, A. Tapper and R. Ziembinski; *Chem. Ber.*, 1989, **122**, 1057. 186, 387, 712

89CB1255 K. Giselbrecht, B. Bildstein and F. Sladky; *Chem. Ber.*, 1989, **122**, 1255. 357, 358

89CB1881 G. Karger, P. Hornbach, A. Krämer, H. Pritzkow and W. Siebert; *Chem. Ber.*, 1989, **122**, 1881. 188, 389, 401

89CB1907 W. P. Fehlhammer and G. Beck; *Chem. Ber.*, 1989, **122**, 1907. 606, 619, 721

89CB2115 R. Hager, O. Steigelmann, G. Müller and H. Schmidbaur; *Chem. Ber.*, 1989, **122**, 2115. 173

89CB2279 B. Bildstein, K. Giselbrecht and F. Sladky; *Chem. Ber.*, 1989, **122**, 2279. 355, 356, 357, 358

89CC246 C. E. McKenna and J. N. Levy; *J. Chem. Soc., Chem. Commun.*, 1989, 246. 117

89CC273 S. S. Al-Juaid, C. Eaborn, P. B. Hitchcock, C. A. McGeary and J. D. Smith; *J. Chem. Soc., Chem. Commun.*, 1989, 273. 393

89CC452 R. Freire, E. I. León, J. A. Salazar and E. Suárez; *J. Chem. Soc., Chem. Commun.*, 1989, 452. 633, 634

89CC593 M. Gouygou, J. Bellan, J. Escudié, C. Couret, A. Dubourg, J.-P. Declercq and M. Koenig; *J. Chem. Soc., Chem. Commun.*, 1989, 593. 373, 688

89CC595 C. Eaborn, K. L. Jones and P. D. Lickiss; *J. Chem. Soc., Chem. Commun.*, 1989, 595. 290

89CC684 S. M. Ali and S. Tanimoto; *J. Chem. Soc., Chem. Commun.*, 1989, 684. 550

89CC705 Q.-Y. Chen and S.-W. Wu; *J. Chem. Soc., Chem. Commun.*, 1989, 705. 232, 234

89CC900 W. J. Mills, L. J. Todd and J. C. Huffman; *J. Chem. Soc., Chem. Commun.*, 1989, 900. 515

89CC1029 J. Evans, P. M. Stroud and M. Webster; *J. Chem. Soc., Chem. Commun.*, 1989, 1029. 343

89CC1046 R. Bartsch, P. B. Hitchcock and J. F. Nixon; *J. Chem. Soc., Chem. Commun.*, 1989, 1046. 154

89CC1265 M. Yamamoto, T. Uruma, S. Iwasa, S. Kohmoto and K. Yamada; *J. Chem. Soc., Chem. Commun.*, 1989, 1265. 550

89CC1494 Y. Ito, M. Suginome, M. Murakami and M. Shiro; *J. Chem. Soc., Chem. Commun.*, 1989, 1494.

89CC1520 M. Sallé, A. Gorgues, J.-M. Fabre, K. Bechgaard, M. Jubault and F. Texier; *J. Chem. Soc., Chem. Commun.*, 1989, 1520. 592

89CC1559 N. Kamigata, T. Fukushima and M. Yoshida; *J. Chem. Soc., Chem. Commun.*, 1989, 1559. 21

89CCR1 M. D. Fryzuk and C. D. Montgomery; *Coord. Chem. Rev.*, 1989, **95**, 1. 516

89CJC1795 D. Su, W. Cen, R. L. Kirchmeier and J. M. Shreeve; *Can. J. Chem.*, 1989, **67**, 1795. 258, 278

89CJC2153 C.-T. Lin and W.-J. Hsu; *Can. J. Chem.*, 1989, **67**, 2153. 450

89CL1009 M. Segi, M. Kato, T. Nakajima, S. Suga and N. Sonoda; *Chem. Lett.*, 1989, 1009. 481

89CPB576 K. Harano, I. Shinohara, S.-I. Sugimoto, T. Matsuoka and T. Hisano; *Chem. Pharm. Bull.*, 1989, **37**, 576. 472

89CPB1249 K. Ukawa, E. Imamiya, H. Yamamoto, K. Mizuno, A. Tasaka, Z.-i. Terashita, T. Okutani, H. Nomura, T. Kasukabe, M. Hozumi, I. Kudo and K. Inoue; *Chem. Pharm. Bull.*, 1989, **37**, 1249. 558

89CRV863 B. E. Maryanoff and A. B. Reitz; *Chem. Rev.*, 1989, **89**, 863. 694

89CRV1703 R. D. Adams; *Chem. Rev.*, 1989, **89**, 1703. 716

89CS81 L. I. Belen'kii, D. B. Brokhovetskii and M. M. Krayushkin; *Chem. Scr.*, 1989, **29**, 81 (*Chem. Abstr.*, 1990, **112**, 55100). 29

89EGP270530 H. Hartmann, M. Weber, T. Sleurich, U. Leuski, H. I. Ayman and U. Simm; *Ger. (East) Pat.* 270530 (1989) (*Chem. Abstr.*, 1990, **112**, 158258). 630

89EUP298553 F. H. Ebetino; *Eur. Pat.* 298553 (1989) (*Chem. Abstr.*, 1989, **110**, 213077). 157

89EUP310255 E. Bay and M. Coates; *Eur. Pat.* 310255 (1989) (*Chem. Abstr.*, 1989, **111**, 9765). 410

89EUP336772 A. T. McPhall, I. H. Hall and B. T. Spielvogel; *Eur. Pat.* 336772 (1989) (*Chem. Abstr.*, 1989, **112**, 158969). 514

89EUP355578 J. Becherer; *Eur. Pat.* 355578 (1989) (*Chem. Abstr.*, 1990, **113**, 114667). 538, 539

89H(28)171 F. Mestre, C. Tailhan and S. Z. Zard; *Heterocycles*, 1989, **28**, 171. 551

89H(28)269 A. Kjaer and I. Skrydstrup; *Heterocycles*, 1989, **28**, 269. 595

89H(28)389 B. M. Pinto, B. Johnston and R. Nagelkerke; *Heterocycles*, 1989, **28**, 389. 100, 102

89IC2120 E. Carmona, E. Guitiérrez-Puebla, A. Monge, P. J. Pérez and L. J. Sánchez; *Inorg. Chem.*, 1989, **28**, 2120. 327

89IC3345 G. Sarwar, R. L. Kirchmeier and J. M. Shreeve; *Inorg. Chem.*, 1989, **28**, 3345. 260, 615

89IJC(B)439 H. Singh, R. Dwivedi and L. D. S. Yadav; *Indian J. Chem., Sect B*, 1989, **28**, 439 (*Chem. Abstr.*, 1990, **112**, 77133). 564

89IS79 B. F. Spielvogel, F. U. Ahmed and A. T. McPhail; *Inorg. Synth.*, 1989, **25**, 79. 491

89IZV113 L. T. Eremenko, G. V. Oreshko and M. A. Fadeev; *Izv. Akad. Nauk SSSR Ser. Khim.*, 1989, 113 (*Chem. Abstr.*, 1989, **111**, 38840). 250, 252

89IZV909 A. N. Pudovik, V. K. Khairullin, M. A. Vasyanina, R. M. Kamalov, O. N. Kataeva, I. A. Litvinov and V. A. Naumov; *Izv. Akad. Nauk SSSR Ser. Khim.*, 1989, 909 (*Chem. Abstr.*, 1989, **111**, 194728). 538, 539

89IZV2106 V. I. Erashko, A. T. Baryshnikov, N. I. Zubanova and A. A. Tishaninova; *Izv. Akad. Nauk SSSR Ser. Khim.*, 1989, 2106 (*Chem. Abstr.*, 1990, **112**, 216337). 283, 349, 355

89IZV2849 V. B. Sokolov, A. N. Chekhlov, T. A. Epishina and I. V. Martynov; *Izv. Akad. Nauk SSSR Ser. Khim.*, 1989, 2849 (*Chem. Abstr.*, 1990, **113**, 5694). 430

89JA149 J. Arnold, T. D. Tilley, A. L. Rheingold, S. J. Geib and A. M. Arif; *J. Am. Chem. Soc.*, 1989, **111**, 149. 338

89JA1773 D. J. Burton, A. S. Modak, R. D. Guneratne, D. Su, W. Cen, R. L. Kirchmeier and J. M. Shreeve; *J. Am. Chem. Soc.*, 1989, **111**, 1773. 258

89JA2995 Y. Leblanc, B. J. Fitzsimmons, J. P. Springer and J. Rokach; *J. Am. Chem. Soc.*, 1989, **111**, 2995. 626

89JA3748 A. Sekiguchi, T. Nakanishi, C. Kabuto and H. Sakurai; *J. Am. Chem. Soc.*, 1989, **111**, 3748. 176, 193

89JA6853 A. Igau, A. Baceiredo, H. Grützmacher, H. Pritzkow and G. Bertrand; *J. Am. Chem. Soc.*, 1989, **111**, 6853. 691, 734

89JA7283 M. H. Chisholm, K. Folting, M. J. Hampden-Smith and C. E. Hammond; *J. Am. Chem. Soc.*, 1989, **111**, 7283. 342

89JA8020 G. A. Olah, L. Heiliger and G. K. Prakash; *J. Am. Chem. Soc.*, 1989, **111**, 8020. 726

89JA8502 V. Nair and G. S. Buenger; *J. Am. Chem. Soc.*, 1989, **111**, 8502. 15

89JA8510 T. R. Doyle and O. Vogl; *J. Am. Chem. Soc.*, 1989, **111**, 8510. 33

89JAN367 D. T. Davies, F. P. Harrington, S. J. Knott and R. Southgate; *J. Antibiot.*, 1989, **42**, 367. 445

89JAP03020256 H. Omura, S. Yoshino, H. Hayashi, T. Kayano and Y. Tanabe; *Jpn. Pat.* 03 020 256 (1989) (*Chem. Abstr.*, 1991, **114**, 250 233). 539

89JAP03037110 T. Kagawa, T. Uotani and K. Tsuzuki; *Jpn. Pat.* 03 037 110 (1989) (*Chem. Abstr.*, 1991, **114**, 250 233). 529

89JAP03060417 T. Kagawa, T. Uotani and K. Tsuzuki; *Jpn. Pat.* 03 060 417 (1989) (*Chem. Abstr.*, 1991, **115**, 52 901). 529

89JAP03060418 T. Kagawa, T. Uotani and K. Tsuzuki; *Jpn. Pat.* 03 060 418 (1989) (*Chem. Abstr.*, 1991, **115**, 52 902). 529

89JAP03197310 T. Kagawa, T. Uotani and K. Tsuzuki; *Jpn. Pat.* 03 197 310 (1989) (*Chem. Abstr.*, 1991, **115**, 235 761). 529

89JAP(K)1182311 M. Ishidoya and T. Nakamichi (Nippon Oils and Fats Co. Ltd.); *Jpn. Kokai* 01 182 311 (1989) (*Chem. Abstr.*, 1990, **112**, 57459). 306

89JCR(S)15 R. Brettle and M. A. Khan; *J. Chem. Res. (S)*, 1989, 15. 469

89JCS(D)447 S. S. Al-Juaid, C. Eaborn, M. N. A. El-Kheli, P. B. Hitchcock, P. D. Lickiss, E. Molla, J. D. Smith and J. A. Zora; *J. Chem. Soc., Dalton Trans.*, 1989, 447. 387

89JCS(D)1149 R. A. Michelin and R. Ros; *J. Chem. Soc., Dalton Trans.*, 1989, 1149. 247

89JCS(P1)283 D. S. Dainter, T. Jackson, A. H. H. Omar, H. Suschitzky, B. J. Wakefield, N. Hughes, A. J. Nelson and G. Varvounis; *J. Chem. Soc., Perkin Trans. 1*, 1989, 283. 28

89JCS(P1)1068 F. S. Guziec, Jr., J. M. Russo, F. F. Torres, G. C. Long and M. R. Tellez; *J. Chem. Soc., Perkin Trans. 1*, 1989, 1068. 589

89JCS(P1)1319 P. P. McCleery and B. Tuck; *J. Chem. Soc., Perkin Trans. 1*, 1989, 1319. 117

89JCS(P1)1727 H. Marley, S. H. B. Wright and P. N. Preston; *J. Chem. Soc. Perkin Trans. 1*, 1989, 1727. 443

89JCS(P1)2385 Q.-Y. Chen and S.-W. Wu; *J. Chem. Soc., Perkin Trans. 1*, 1989, 2385. 15

89JCS(P2)251 W. N. Wassef and N. M. Ghobrial; *J. Chem. Soc., Perkin Trans. 2*, 1989, 251. 25

89JFC(42)429 G. Pawelke; *J. Fluorine Chem.*, 1989, **42**, 429. 14

89JFC(43)53 G. P. Stahly; *J. Fluorine Chem.*, 1989, **43**, 53. 253

89JFC(44)433 Q. Chen and S. Wu; *J. Fluorine Chem.*, 1989, **44**, 433. 38

89JFC(45)86 A. Heaton and R. L. Powell; *J. Fluorine Chem.*, 1989, **45**, 86. 14

89JFC(45)215 A. Battais, B. Boutevin, Y. Pietrasanta, R. Bertocchio and A. Lantz; *J. Fluorine Chem.*, 1989, **45**, 215. 7

89JFC(45)225 R. Henn, W. Sundermeyer, M. Witz and H. Pritzkow; *J. Fluorine Chem.*, 1989, **45**, 225. 253

89JFC(45)293 T. Abe and E. Hayashi; *J. Fluorine Chem.*, 1989, **45**, 293. 440

89JFC(45)401 W. Tyrra and D. Naumann; *J. Fluorine Chem.*, 1989, **45**, 401. 220

89JHC231 R. Milcent, G. Barbier, T. Tzirenstchikow and L. Lebreton; *J. Heterocycl. Chem.*, 1989, **26**, 231. 444

89JHC829 A. R. Katritzky, M. Drewniak-Deyrup, X. Lan and F. Brunner; *J. Heterocycl. Chem.* 1989, **26**, 829. 370

89JHC1245 J. J. D'Amico and F. G. Bollinger; *J. Heterocycl. Chem.*, 1989, **26**, 1245. 546

89JIC60 M. K. Jani, N. K. Undavia and B. Trivedi; *J. Indian Chem. Soc.*, 1989, **66**, 60 (*Chem. Abstr.*, 1990, **112**, 77 052). 561

89JMC228 E. W. Thomas, E. E. Nishizawa, D. C. Zimmermann and D. J. Williams; *J. Med. Chem.*, 1989, **32**, 228. 445

89JOC444 H. Tanaka, S. Yamashita, M. Yamanoue and S. Torii; *J. Org. Chem.*, 1989, **54**, 444. 26, 31

89JOC613 D. J. Burton and L. G. Sprague; *J. Org. Chem.*, 1989, **54**, 613. 58, 266

89JOC1062 P. J. Garratt, C. J. Hobbs and R. Wrigglesworth; *J. Org. Chem.*, 1989, **54**, 1062. 632

89JOC1185 G. Morel, E. Marchand, A. Foucaud and L. Topet; *J. Org. Chem.*, 1989, **54**, 1185. 635

89JOC1290 M. R. Myers and T. Cohen; *J. Org. Chem.*, 1989, **54**, 1290. 128

89JOC1782 J. P. Marino, M.-W. Kim and R. Lawrence; *J. Org. Chem.*, 1989, **54**, 1782. 111

89JOC1784 T. F. Bates and R. D. Thomas; *J. Org. Chem.*, 1989, **54**, 1784. 190

89JOC1990 W.-H. Lin, W. D. Clark and R. J. Lagow; *J. Org. Chem.*, 1989, **54**, 1990. 297, 301

89JOC2940 M. D. Threadgill and A. P. Gledhill; *J. Org. Chem.*, 1989, **54**, 2940. 493, 494

89JOC2978 F. E. Scully, Jr. and T. Ortega; *J. Org. Chem.*, 1989, **54**, 2978. 559

89JOC3231 L. N. Pridgen, J. Prol, Jr., B. Alexander and L. Gillyard; *J. Org. Chem.*, 1989, **54**, 3231. 415

89JOC4042 G. A. Molander and K. Mautner; *J. Org. Chem.*, 1989, **54**, 4042. 133

89JOC4272 R. C. Gatrone; *J. Org. Chem.*, 1989, **54**, 4272. 118

89JOC4426 G. Sicard, H. Gruetzmacher, A. Baceiredo, J. Fischer and G. Bertrand; *J. Org. Chem.*, 1989, **54**, 4426. 669

89JOC5003 J. Otera, Y. Niibo and H. Nozaki; *J. Org. Chem.*, 1989, **54**, 5003. 126

89JOC5603 S. M. Ali and S. Tanimoto; *J. Org. Chem.*, 1989, **54**, 5603. 550

89JOC5613 P. F. Hudrlik, E. L. O. Agwaramgbo and A. M. Hudlik; *J. Org. Chem.*, 1989, **54**, 5613. 125

89JOM(364)289 W. Uhl, M. Layh and T. Hildenbrand; *J. Organomet. Chem.*, 1989, **364**, 289. 195

89JOM(366)39 S. S. Al-Juaid, S. M. Dhaher, C. Eaborn, P. B. Hitchcock, C. A. McGeary and J. D. Smith; *J. Organomet. Chem.*, 1989, **366**, 39. 165, 289, 372, 373

89JOM(367)19 W. Haubold, W. Keller and G. Sawitzki; *J. Organomet. Chem.*, 1989, **367**, 19. 183

89JOM(368)C29 P. B. Hitchcock, M. F. Meidine, J. F. Nixon, H. Wang, D. Gudat, and E. Niecke, *J. Organomet. Chem.*, 1989, **368**, C29. 692

89JOM(368)131 D. Naumann and W. Tyrra; *J. Organomet. Chem.*, 1989, **368**, 131. 246

89JOM(368)139 W. Uhl, M. Layh and W. Hiller; *J. Organomet. Chem.*, 1989, **368**, 139. 195

89JOM(368)199 A. Ishihara, T.-A. Mitsudo and Y. Watanabe; *J. Organomet. Chem.*, 1989, **368**, 199. 339

89JOM(371)257 W.-Y. Yeh, S. R. Wilson and J. R. Shapley; *J. Organomet. Chem.*, 1989, **371**, 257. 345

89JOM(372)201 M. J. Menu, M. Dartiguenave, Y. Dartiguenave, J. J. Bonnet, G. Bertrand and A. Baceiredo; *J. Organomet. Chem.*, 1989, **372**, 201. 665

89JOM(373)203 J. Kron and U. Schubert; *J. Organomet. Chem.*, 1989, **373**, 203. 723

89JOM(376)149 H. K. Nair and J. A. Morrison; *J. Organomet. Chem.*, 1989, **376**, 149. 246, 247

89JOM(378)125 D. J. Brauer, H. Bürger, F. Dörrenbach, G. Pawelke and W. Weuter; *J. Organomet. Chem.*, 1989, **378**, 125. 243

89JPO349 E. Juaristi, A. Flores-Vela, V. Labastida and M. Ordonez; *J. Phys. Org. Chem.*, 1989, **2**, 349. 119

89JPR439 W. Dölling and A. Hildebrandt; *J. Prakt. Chem.*, 1989, **331**, 439. 545, 546

89LA187 E. V. Dehmlow and J. Stutten; *Liebigs Ann. Chem.*, 1989, 187. 32

89LA931 R. Spitzner, J. Andersch and W. Schroth; *Liebigs Ann. Chem.*, 1989, 931. 619

89LA975 M. Müller, E. Meyle, E. F. Paulus, H. Plagge and H. H. Otto; *Liebigs Ann. Chem.*, 1989, 975. 129

89LA985 K. Halling, I. Thomsen and T. B. G. Torssell; *Liebigs Ann. Chem.*, 1989, 985. 608, 611

89M871 P. Kindt, W. Dölling and M. Augustin; *Monatsh. Chem.*, 1989, **120**, 871. 564

B-89MI 601-01 L. German and S. Zemskov (eds.); "New Fluorinating Agents in Organic Chemistry," Springer-Verlag, Berlin, 1989. 16

B-89MI 601-02 A. I. Burmakov, B. V. Kuushenko, L. A. Alekseeva and L. M. Yagupolskii; in "New Florinating Agents in Organic Synthesis," ed. L. German and S. Zemskov, Springer-Verlag, 1989, p. 197. 19

89MI 605-01 M. E. Sitzmann and H. G. Adolph; *Statutory Invent. Regist. US* 644 (*Chem. Abstr.*, 1989, **111**, 214120). 145

B-89MI 606-01 G. Raabe and J. Michl; in "The Chemistry of Organic Silicon Compounds," eds S. Patai and Z. Rappoport, Wiley, Chichester, 1989, vol. 2, p. 1015. 177

B-89MI 606-02 P. Mazerolles; in "Rings, Clusters and Polymers of Main Groups and Transition Elements," ed. H. W. Roesky, Elsevier, Amsterdam, 1989, p. 139. 172

B-89MI 607-01 L. German and S. Zemskov (eds.); "New Fluorinating Agents in Organic Synthesis," Springer-Verlag, Heidelberg, 1989. 212

89MI 607-02 H.-T. Xu and H.-X. Zhai; *Huaxue Shiji*, 1989, **11**, 123 (*Chem. Abstr.*, 1989, **111**, 194056). 219

B-89MI 614-01 H. H. Szmant; "Organic Building Blocks of the Chemical Industry," Wiley, New York, 1989, pp. 97, 125, 138, 146, 162. 409, 411, 412, 415

B-89MI 614-02 "The Merck Index," 11th edn., Merck, Rahway, NJ, USA, 1989, p. 1165. 411

89MI 614-03 B. Ilien, A. Mejean and C. Hirth; *Biochem. Pharm.*, 1989, **38**, 2879. 443

89MI 614-04 T. Kawaguchi; *J. Synth. Org. Chem.*, 1989, **47**, 384. 426

B-89MI 615-01 K. C. Frisch and D. Klempner; in "Comprehensive Polymer Science," ed. G. C. Eastmond, A. Ledwith, S. Russo and P. Sigwalt, Pergamon Press, Oxford, 1989, vol. 5, p. 413. 488

89MI 617-01 B. Khadzhieva, V. Kalcheva, G. Vasilev, B. Gulubov and Z. Dimchev; *Dokl. Bolg. Akad. Nauk*, 1989, **42**, 91 (*Chem. Abstr.*, 1991, **114**, 61973). 558

89MI 620-01 S. Bilinski, L. Bielak, J. Chmielewski, B. Marcewicz-Rojewska and I. Musik; *Acta Pol. Pharm.*, 1989, **46**, 343 (*Chem. Abstr.*, 1991, **114**, 6379). 637

89MI 621-01 H. Tomioka; *Kenkyu Hokoku-Asahi Garasu Kogyo Gijutsu Shoreikai*, 1989, **54**, 63 (*Chem. Abstr.*, 1991, **114**, 163264). 670

B-89MI 622-01 R. Appel and F. Knoll; *Adv. Inorg. Chem.*, 1989, **33**, 259. 692

B-89MI 622-02 G. Raabe and J. Michl; in "The Chemistry of Organic Silicon Compounds," eds. S. Patai and Z. Rappoport, Wiley, Chichester, 1989, vol. 2, p. 1015. 709

89MI 622-03 U. Schubert; "Advances in Metal Carbene Chemistry," Kluwer Academic, Dordrecht, 1989. 716, 719

89NJC363 J. Grobe, D. Le Van and J. Nientiedt; *Nouv. J. Chim.*, 1989, **13**, 363. 700

89NJC891 M. Rahmoune, Y. Y. C. Y. L. Ko, R. Carrié and F. Tonnard; *Nouv. J. Chem.*, 1989, **13**, 891. 156

89NKK63 Y. Taguchi, I. Shibuya and Y. Suhara; *Nippon Kagaku Kaishi*, 1989, 63 (*Chem. Abstr.*, 1989, **110**, 212660). 314

89OM1270 J. Evans, P. M. Stroud and M. Webster; *Organometallics*, 1989, **8**, 1270. 343

89OM1498 R. D. Sanner, J. H. Satcher, Jr. and M. W. Droege; *Organometallics*, 1989, **8**, 1498. 246, 247

89OM2490 A. H. Cowley, P. C. Knüppel and C. M. Nunn; *Organometallics*, 1989, **8**, 2490. 373

89OM2646 F. A. Johnson and W. D. Perry; *Organometallics*, 1989, **8**, 2646. 208

89OPP771 D. A. Laufer, K. Doyle and X. D. Zhang; *Org. Prep. Proced. Int.*, 1989, **21**, 771. 467

89S101 M. Mikolajczyk and P. Balczewski; *Synthesis*, 1989, 101. 129, 326, 350, 351

89S132 S. Bittner, A. Moradpour and P. Krief; *Synthesis*, 1989, 132. 627

89S228 J. Barluenga, M. Tomás, A. Ballesteros and L. A. López; *Synthesis*, 1989, 228. 571

89S463 G. A. Olah, T. Weber, D. R. Bellew and O. Farooq; *Synthesis*, 1989, 463. 45

89S466 C. Gallina and C. Giordano; *Synthesis*, 1989, 466. 25

89S511 W. Schnurr and M. Regitz; *Synthesis*, 1989, 511. 667, 668, 689

89S684 J. Voss and B. Wollny; *Synthesis*, 1989, 684. 96

89S875 D. Sicker; *Synthesis*, 1989, 875. 486

89S918 J. C. Jochims, A. Hamed, T. Huu-Phuoc, J. Hofmann and H. Fischer; *Synthesis*, 1989, 918. 621
89S923 P. Molina, A. Tárraga and A. Espinosa; *Synthesis*, 1989, 923. 560
89SC31 B. Labiad and D. Villemin; *Synth. Commun.*, 1989, **19**, 31. 82
89SC547 A. W. M. Lee, W. H. Chan, H. C. Wong and M. S. Wong; *Synth. Commun.*, 1989, **15**, 547. 550
89SC2307 U. Pindur and C. Flo; *Synth. Commun.*, 1989, **19**, 2307. 729
89SC2825 N. Guillot, B. Marin, Z. Janousek and H. G. Viehe; *Synth. Commun.*, 1989, **19**, 2825. 610, 727
89SUL123 E. Vajda and A. Simon; *Sulfur Letters*, 1989, **10**, 123. 564
89T637 S. Shimizu and M. Ogata; *Tetrahedron*, 1989, **45**, 637. 61
89T1691 K.-D. Ginzel, P. Brungs and E. Steckhan; *Tetrahedron*, 1989, **45**, 1691. 117
89T3487 S. A. Fedoreyev, S. G. Llyin, N. K. Utkina, O. B. Maximov, M. V. Reshetnyak, M. Yu. Antipin and Yu. T. Struchkov; *Tetrahedron*, 1989, **45**, 3487. 138
89T3787 J. G. Dingwall, J. Ehrenfreund and R. G. Hall; *Tetrahedron*, 1989, **45**, 3787. 116
89T6019 L. N. Markovskii and V. D. Romanenko; *Tetrahedron*, 1989, **45**, 6019. 678
89TL15 B. H. Lipshutz, R. Moretti and R. Crow; *Tetrahedron Lett.*, 1989, **30**, 15. 134
89TL109 K. Uneyama, O. Morimoto and H. Nanbu; *Tetrahedron Lett.*, 1989, **30**, 109. 20
89TL177 M. Gouygou, C. Tachon, R. El Ouatib, O. Ramarijaona, G. Etemad-Moghadam and M. Koenig; *Tetrahedron Lett.*, 1989, **30**, 177. 680
89TL219 J.-i. Yoshida, S.-i. Matsunaga and S. Isoe; *Tetrahedron Lett.*, 1989, **30**, 219. 133
89TL817 P. Le Floch and F. Mathey; *Tetrahedron Lett.*, 1989, **30**, 817. 679
89TL1901 G. Church, J. M. Ferland and J. Gauthier; *Tetrahedron Lett.*, 1989, **30**, 1901. 429
89TL2033 M. J. Coghlan and B. A. Caley; *Tetrahedron Lett.*, 1989, **30**, 2033. 416, 428
89TL2379 F. Jung, A. Olivier, D. Boucherot and F. Loftus; *Tetrahedron Lett.*, 1989, **30**, 2379. 622
89TL2473 R. A. Moss, G. J. Ho and B. K. Wilk; *Tetrahedron Lett.*, 1989, **30**, 2473. 280
89TL2873 M. J. Szymonifka and J. V. Heck; *Tetrahedron Lett.*, 1989, **30**, 2873. 129
89TL3159 Y. Takeyama, Y. Ichinose, K. Oshima and K. Utimoto; *Tetrahedron Lett.*, 1989, **30**, 3159. 21
89TL3229 G. B. Gill, G. Pattenden and S. J. Reynolds; *Tetrahedron Lett.*, 1989, **30**, 3229. 447
89TL3415 A. Bulpin, S. Masson and A. Sene; *Tetrahedron Lett.*, 1989, **30**, 3415. 121, 316
89TL4367 J. E. Forbes and S. Z. Zard; *Tetrahedron Lett.*, 1989, **30**, 4367. 547
89TL4501 G. Märkl and W. Hölzl; *Tetrahedron Lett.*, 1989, **30**, 4501. 129
89TL4813 N. Dufour, A.-M. Caminade and J.-P. Majoral; *Tetrahedron. Lett.*, 1989, **30**, 4813. 685
89TL5481 J. F. G. A. Jansen and B. L. Feringa; *Tetrahedron Lett.*, 1989, **30**, 5481. 88
89TL5841 J.-C. Cuevas, P. Patil and V. Snieckus; *Tetrahedron Lett.*, 1989, **43**, 5841. 162
89TL6195 M. Hogenbirk, N. J. R. van Eikema Hommes, G. Schat, O. S. Akkerman, F. Bickelhaupt and G. W. Klumpp; *Tetrahedron Lett.*, 1989, **30**, 6195. 399
89TL6869 A.-M. Caminade, Ch. Roques, N. Dufour, D. Colombo, F. Gonce and J.-P. Majoral; *Tetrahedron Lett.*, 1989, **30**, 6869. 689
89TL7013 C. F. Bigge, J. T. Drummond and G. Johnson; *Tetrahedron Lett.*, 1989, **30**, 7013. 58
89TL7033 E. Schaumann and C. Friese; *Tetrahedron Lett.*, 1989, **30**, 7033. 327, 351
89TL7141 R. P. Iyer, L. R. Phillips, J. A. Biddle, D. R. Thakker, W. Egan, S. Aoki and H. Mitsuya; *Tetrahedron Lett.*, 1989, 7141. 491
89TL7313 K. S. Atwal, S. Z. Ahmed and B. C. O'Reilly; *Tetrahedron Lett.*, 1989, **30**, 7313. 647
89USP4833250 ICI Americas; *US Pat.* 4833250 (1989) (*Chem. Abstr.*, 1991, **114**, 5484). 28
89USP4874875 W. Navarrini and D. D. DesMarteau (Ausimont Srl); *US Pat.* 4874875 (1989) (*Chem. Abstr.*, 1990, **112**, 159140). 279
89ZAAC(576)225 D. Naumann, J. Fischer and B. Wilkes; *Z. Anorg. Allg. Chem.*, 1989, **576**, 225. 237
89ZAAC(579)75 W. Uhl; *Z. Anorg. Allg. Chem.*, 1989, **579**, 75. 195, 197
89ZN(B)796 G. Reber, J. Riede, N. Wiberg, G. Wagner and G. Müller; *Z. Naturforsch., Teil B*, 1989, **44**, 796. 383
89ZN(B)983 J. B. Rodriguez, E. G. Gros and A. M. Stoka; *Z. Naturforsch., Teil B*, 1989, **44**, 983. 477
89ZN(B)1589 W. Boszczyk and T. Latowski; *Z. Naturforsch., Teil B*, 1989, **44**, 1589. 609
89ZOB330 O. I. Kolodyazhnyi; *Zh. Obshch. Khim.*, 1989, **59**, 330, (*Chem. Abstr.*, 1989, **111**, 134333). 696
89ZOB571 E. B. Silina, B. N. Kozhushko and v. A. Shokol; *Zh. Obshch. Khim.*, 1989, **59**, 571 (*Chem. Abstr.*, 1989, **111**, 194895). 367
89ZOB714 B. I. Buzykin, M. P. Sokolov and V. N. Ivanova; *Zh. Obshch. Khim.*, 1989, **59**, 714 (*Chem. Abstr.*, 1989, **111**, 194898). 623
89ZOB959 A. P. Marchenko, V. A. Oleinik, G. N. Koidan and A. M. Pinchuk; *Zh. Obshch. Khim.*, 1989, **59**, 959, (*Chem. Abstr.*, 1989, **111**, 233003). 696
89ZOB1455 L. A. Chepakova, V. K. Brel, I. V. Martynov and I. G. Maslennikov; *Zh. Obshch. Khim.*, 1989, **59**, 1455 (*Chem. Abstr.*, 1990, **112**, 179148). 613
89ZOB1902 G. N. Koidan, A. V. Oleinik, A. P. Marchenko and A. M. Pinchuk; *Zh. Obshch. Khim.*, 1989, **59**, 1902, (*Chem. Abstr.*, 1990, **112**, 98663). 680, 682
89ZOB1904 G. N. Koidan, V. A. Oleinik, A. P. Marchenko and A. M. Pinchuk; *Zh. Obshch. Khim.*, 1989, **59**, 1904, (*Chem. Abstr.*, 1990, **112**, 98664). 696
89ZOB2131 A. V. Ruban, L. S. Kachkovskaya, M. I. Povolovtskii, V. D. Romanenko and L. N. Markovskii; *Zh. Obshch. Khim.*, 1989, **59**, 2131, (*Chem. Abstr.*, 1990, **112**, 98671). 694
89ZOB2133 L. N. Markovskii, G. N. Koidan, A. P. Marchenko, V. D. Romanenko, M. I. Povolotskii and A. M. Pinchuk; *Zh. Obshch. Khim.*, 1989, **59**, 2133, (*Chem. Abstr.*, 1990, **112**, 98672). 690
89ZOB2520 K. I. Kokorev, Sh. Kh. Badrutdinov, L. A. Al'metkina and F. D. Yambushev; *Zh. Obshch. Khim.*, 1989, **59**, 2520 (*Chem. Abstr.*, 1990, **112**, 217107). 564
89ZOB2628 G. V. Motsarev, V. T. Inshakova and A. D. Raskina; *Zh. Obshch. Khim.*, 1989, **59**, 2628 (*Chem. Abstr.*, 1990, **112**, 158329). 243
89ZOR2490 G. V. Nekrasova, E. E. Boldysh, E. S. Lipina and V. V. Perekalin; *Zh. Org. Khim.*, 1987, **25**, 2490 (*Chem. Abstr.*, 1988, **109**, 6053). 54, 261

90AG(E)60 H. Pritzkow, K. Rall, S. Reimann-Andersen and W. Sundermeyer; *Angew. Chem., Int. Ed. Engl.*, 1990, **29**, 60. 258

90AG(E)201 R. Hager, O. Steigelmann, G. Müller, H. Schmidbaur, H. E. Robertson and D. W. H. Rankin; *Angew. Chem., Int. Ed. Engl.*, 1990, **29**, 201. 173, 378, 379

90AG(E)292 A. Krämer, H. Pritzkow and W. Siebert; *Angew. Chem., Int. Ed. Engl.*, 1990, **29**, 292. 183

90AG(E)398 Ch. Wieczorek, J. Allwohn, G. Schmidt-Lukasch, R. Hunold, W. Massa and A. Berndt; *Angew. Chem., Int. Ed. Engl.*, 1990, **29**, 398. 713, 714

90AG(E)399 M. Pilz, J. Allwohn, W. Massa and A. Berndt; *Angew. Chem., Int. Ed. Engl.*, 1990, **29**, 399. 400

90AG(E)401 M. Pilz, H. Michel and A. Berndt; *Angew. Chem., Int. Ed. Engl.*, 1990, **29**, 401. 712, 713

90AG(E)1030 M. Pilz, J. Allwohn, P. Willershausen, W. Massa and A. Berndt; *Angew. Chem., Int. Ed. Engl.*, 1990, **29**, 1030. 201, 397, 715, 716

90AG(E)1032 J. Allwohn, M. Pilz, R. Hunold, W. Massa and A. Berndt; *Angew. Chem., Int. Ed. Engl.*, 1990, **29**, 1032. 397, 398

90AOC(30)99 N. S. Hosmane and J. A. Maguire; *Adv. Organomet. Chem.*, 1990, **30**, 99. 406

90AX2374 A. L. Rheingold; *Acta Crystallogr.*, 1990, **C46**, 2374. 338

90BBR(171)458 C. C. Neto, J. M. Steim, P. S. Sarin, D. K. Sun, N. N. Bhongle, R. K. Piratla and J. G. Turcotte; *Biochem. Biophys. Res. Commun.*, 1990, **171**, 458. 489

90BBR(172)288 A. Rosowsky, J. Saha, F. Fazely, J. Koch and R. M. Ruprecht; *Biochem Biophys. Res. Commun.*, 1990, **172**, 288. 489, 490

90BCJ950 H. Suzuki and T. Murafuji; *Bull. Chem. Soc. Jpn.*, 1990, **63**, 950. 709

90BCJ1278 I. Novak; *Bull. Chem. Soc. Jpn.*, 1990, **63**, 1278. 228

90BCJ3658 M. K. Das, P. Mukherjee and S. Roy; *Bull. Chem. Soc. Jpn.*, 1990, **63**, 3658. 514

90CB747 P. Paetzold, T. Schmitz, A. Tapper and R. Ziembinski; *Chem. Ber.*, 1990, **123**, 747. 187, 387, 712, 713

90CB751 D. Lentz and R. Marschall; *Chem. Ber.*, 1990, **123**, 751. 721

90CB945 D. Kummer, S. C. Chaudhry, T. Debaerdemaeker and U. Thewalt; *Chem. Ber.*, 1990, **123**, 945. 268

90CB1245 R. Streubel and E. Niecke; *Chem. Ber.*, 1990, **123**, 1245. 166

90CB1635 A. Waterfeld; *Chem. Ber.*, 1990, **123**, 1635. 253, 275, 726, 729

90CB2087 J. Zech and H. Schmidbaur; *Chem. Ber.*, 1990, **123**, 2087. 173

90CB2317 J. Grobe, D. Le Van, B. Lüth and M. Hegemann; *Chem. Ber.*, 1990, **123**, 2317. 685

90CB2325 I. Wagner, W.-W. duMont, S. Pohl and W. Saak; *Chem. Ber.*, 1990, **123**, 2325. 334, 355, 356

90CC470 M. R. Bryce, A. J. Moore D. Lorcy, A. S. Dhindsa and A. Robert; *J. Chem. Soc., Chem. Commun.*, 1990, 470. 120, 123

90CC741 A. Meller, D. Bromm, W. Maringgele, A. Heine, D. Stalke and G. M. Sheldrick; *J. Chem. Soc., Chem. Commun.*, 1990, 741. 184

90CC1133 M. B. Gazizov, R. A. Khairullin, R. F. Kadirova, E. S. Lewis and A. M. Kook; *J. Chem. Soc., Chem. Commun.*, 1990, 1133. 116

90CC1432 Y. Ting, W. Verboom, L. C. Groenen, J.-D. van Loon and D. N. Reinhoudt; *J. Chem. Soc., Chem. Commun.*, 1990, 1432. 556

90CC1459 S. Z. Zhu and O. X. Chen; *J. Chem. Soc., Chem Commun.*, 1990, 1459. 254

90CC1471 F. Aigbirhio, C. Eaborn, A. Habtemariam and J. D. Smith; *J. Chem. Soc., Chem. Commun.*, 1990, 1471. 394

90CC1621 H. H. Karsch, K. Zellner and G. Müller; *J. Chem. Soc., Chem. Commun.*, 1990, 1621.

90CC1722 H. G. Raubenheimer, F. Scott, M. Roos and R. Otte; *J. Chem. Soc., Chem. Commun.*, 1990, 1722. 723

90CL83 M. Koreeda and I. A. George; *Chem. Lett.*, 1990, 83. 533

90CL811 T. Mizuno, T Yamaguchi, J. Nishiguchi, T. Okushi and T. Hirashima; *Chem. Lett.*, 1990, 811 471

90CL813 H. Fukaya, T. Abe and E. Hayashi; *Chem. Lett.*, 1990, 813. 219

90CL1403 K. Shimada, S. Hikage, Y. Takeishi and Y. Takikawa; *Chem. Lett.*, 1990, 1403. 597

90CL1411 K. Narasaka, N. Saito, Y. Hayashi and H. Ichida; *Chem. Lett.*, 1990, 1411. 127, 325, 327

90CRV191 M. Regitz; *Chem. Rev.*, 1990, **90**, 191. 153

90CRV283 J. Barrau, J. Escudie and J. Satge; *Chem Rev.*, 1990, **90**, 283. 187, 709, 724

90CZP264358 J. Pola, J. Vitek, Z. Chvatal and V. Dedek; *Czech. Pat.* 264358 (1990) (*Chem. Abstr.*, 1991, **114**, 163550). 410

90CZ176 W. Kantlehner, P. Speh, H. Lehmann, E. Haug and W. W. Mergen; *Chem.-Ztg.*, 1990, **114**, 176. 139

90EJM301 A. Sood, C. K. Sood, B. F. Spielvogel and I. H. Hall; *Eur. J. Med. Chem.*, 1990, **25**, 301. 514

90EUP353743 D. D. DesMarteau and S. P. Kotun (Ausimont Srl); *Eur. Pat.* 353743 (1990) (*Chem. Abstr.*, 1990, **113**, 39916). 602, 602

90GEP3841185 C. Fest (Bayer AG); *Ger. Pat.* 3841185 (1990) (*Chem Abstr.*, 1990, **113**, 190956). 630

90H(30)839 M. E. Jung and S. J. Miller; *Heterocycles*, 1990, **30**, 839. 487

90H(30)855 F. H. Ebetino, C. R. Degenhardt, L. A. Jamieson and D. C. Burdsall; *Heterocycles*, 1990, **30**, 855. 123

90H(31)47 B. F. Bonini, G. Maccagnani, G. Mazzanti, G. Abdel-Lateef, A. Atwa, P. Zani and G. Barbaro; *Heterocycles*, 1990, **31**, 47. 127, 128

90H(31)1213 O. Bortolini, G. Fantin, M. Fogagnolo, A. Medici and P. Pedrini; *Heterocycles*, 1990, **31**, 1213. 637

90H(31)1377 A. Kamal; *Heterocycles*, 1990, **31**, 1377. 451

90H(31)1393 J. H. Musser, S. C. Lewis and R. H. W. Bender; *Heterocycles*, 1990, **31**, 1393. 504

90HAC167 G. Sarwar, R. L. Kirchmeier and J. M. Shreeve; *Heteroatom Chem.*, 1990, **1**, 167. 262

90HOU(E12b)1 K. J. Irgolic; *Methoden Org. Chem. (Houben-Weyl)*, 1990, **E12b**, 1. 308

90HOU(E16a)214 R. Andree and J. F. Kluth; *Methoden Org. Chem. (Houben-Weyl)*, 1990, **E16a**, 214. 232

90HOU(E16a)271 R. Andree and J. F. Kluth; *Methoden Org. Chem.* (*Houben-Weyl*), 1990, **E16a**, 271. 232

90IC554 M. Mittakanti and K. W. Morse; *Inorg. Chem.*, 1990, **29**, 554. 491, 515

90IC571 G. Sarwar, R. L. Kirchmeier and J. M. Shreeve; *Inorg. Chem.*, 1990, **29**, 571. 603

90IC2496 D.-S. Yang, G. M. Bancroft, R. J. Puddephatt and J. S. Tse; *Inorg. Chem.*, 1990, **29**, 2496. 247

90IC3218 M. Mittakanti, M. R. M. D. Charandabi and K. W. Morse; *Inorg. Chem.*, 1990, **29**, 3218. 515

90IC3252 J. L. Margrave, K. H. Whitmire, R. H. Hauge and N. T. Norem; *Inorg. Chem.*, 1990, **29**, 3252. 247

90IC4588 G. L. Gard, A. Waterfeld, R. Mews, J. Mohtasham and R. Winter; *Inorg. Chem.*, 1990, **29**, 4588. 252

90IC5081 D. K. Kennepohl, B. D. Santarsiero and R. G. Cavell; *Inorg. Chem.*, 1990, **29**, 5081. 622

90IJC(B)180 H. Singh, D. Singh and S. Kumar; *Indian J. Chem., Sect. B*, 1990, **29**, 180. 111

90IS235 A. H. Cowley, N. C. Norman and M. Pakulski; *Inorg. Synth.*, 1990, **27**, 235. 372

90IS239 A. H. Cowley, N. C. Norman and M. Pakulski; *Inorg. Synth.*, 1990, **27**, 239. 173, 174, 176

90IZV491 L. S. German, E. I. Mysov and E. G. Ter-Gabrielyan; *Izv. Akad. Nauk SSSR Ser. Khim.*, 1990, 491 (*Chem. Abstr.*, 1990, **113**, 114615). 22

90IZV1620 O. A. Rakitin, V. A. Ogurtsov, T. I. Godovikova and L. I. Khmel'nitskii; *Izv. Akad. Nauk SSSR Ser. Khim.*, 1990, 1620 (*Chem. Abstr.*, 1990, **113**, 190863). 150

90IZV1623 O. A. Rakitin, V. A. Ogurtsov, T. I. Godovikova and L. I. Khmel'nitskii; *Izv. Akad. Nauk SSSR Ser. Khim.*, 1990, 1623 (*Chem. Abstr.*, 1990, **113**, 211464). 150

90IZV1816 S. A. Shevelev, I. L. Dalinger, V. M. Vinogradov and A. A. Fainzil'berg; *Izv. Akad. Nauk SSSR Ser. Khim.*, 1990, 1816 (*Chem. Abstr.*, 1991, **114**, 5765). 145

90JA728 C. W. Bauknight, Jr. and D. D. DesMarteau; *J. Am. Chem. Soc.*, 1990, **112**, 728. 259, 261, 615

90JA1962 J.-i. Yoshida, T.-i. Maekawa, T. Murata, S.-i. Matsunaga and S. Isoe; *J. Am. Chem. Soc.*, 1990, **112**, 1962. 124

90JA2034 J. E. Forbes and S. Z. Zard; *J. Am. Chem. Soc.*, 1990, **112**, 2034. 547

90JA3696 K. C. Nicolaou, D. G. McGarry and P. K. Sommers; *J. Am. Chem. Soc.*, 1990, **112**, 3696. 134

90JA4552 D. R. Williams, P. A. Jass, H. L. A. Tse and R. D. Gaston; *J. Am. Chem. Soc.*, 1990, **112**, 4552. 134

90JA5351 H. König, M. J. Menu, M. Dartiguenave, Y. Dartiguenave and H. F. Klein; *J. Am. Chem. Soc.*, 1990, **112**, 5351. 675

90JA5609 H. J. Reich, R. C. Holtan and C. Bolm; *J. Am. Chem. Soc.*, 1990, **112**, 5609. 135

90JA6277 M. Granier, A. Baceiredo, Y. Dartiguenave, M. Dartiguenave, M. J. Menu and G. Bertrand; *J. Am. Chem. Soc.*, 1990, **112**, 6277. 665, 667, 668, 669, 670

90JA7984 R. S. Tanke and R. H. Crabtree; *J. Am. Chem. Soc.*, 1990, **112**, 7984. 369

90JAP03220173 T. Kagawa, T. Uotani, S. Kuranashi, K. Hirokane and K. Tsuzuki; *Jpn. Pat.* 03 220 173 (1990) (*Chem. Abstr.*, 1991, **116**, 128 384). 533

90JAP04066566 H. Tenma, Y. Awano and K. Tsuzuki; *Jpn. Pat.* 04 066 566 (1990) (*Chem. Abstr.*, 1992, **117**, 48 127). 533

90JAP04128263 H. Tenma, Y. Awano and K. Tsuzuki; *Jpn. Pat.* 04 128 263 (1990) (*Chem. Abstr.*, 1992, **117**, 150 740). 533

90JAP05032616 H. Tenma, Y. Awano and K. Tsuzuki; *Jpn. Pat.* 05 032 616 (1990) (*Chem. Abstr.*, 1993, **119**, 95 121). 533

90JAP2209879 Agency of Industrial Sciences and Technology; *Jpn. Pat.* 2 209 879 (1990) (*Chem. Abstr.* 1991, **114**, 102 384). 671

90JC593 M. Ahnoff, S. Chen, A. Green and I. Grundevik; *J. Chromatogr.*, 1990, **506**, 593. 426

90JCR(S)330 A. R. Katritzky, L. Wrobel and G. P. Savage; *J. Chem. Res.* (*S*), 1990, 330. 365

90JCS(D)1197 R. Bertani, M. Mozzon, F. Benetollo, G. Bombieri and R. A. Michelin; *J. Chem. Soc., Dalton Trans*, 1990, **4**, 1197. 663

90JCS(D)2643 P. Brown, M. F. Mahon and K. C. Molloy; *J. Chem. Soc., Dalton Trans.*, 1990, 2643. 395

90JCS(D)3033 Y. Chi, S.-H. Chu, B.-F. Chen, S.-M. Peng and G.-H. Lee; *J. Chem. Soc., Dalton Trans.*, 1990, 3033. 348

90JCS(P1)19 G. Baccolini and E. Mezzina; *J. Chem. Soc., Perkin Trans. 1*, 1990, 19. 121

90JCS(P1)375 R. C. F. Jones and J. Schofield; *J. Chem. Soc., Perkin Trans. 1*, 1990, 375. 71

90JCS(P1)509 R. M. Bannister and C. W. Rees; *J. Chem. Soc., Perkin Trans. 1*, 1990, 509. 255

90JCS(P1)1218 T. Tsuchiya, M. Yasumoto, I. Shibuya and M. Goto; *J. Chem. Soc., Perkin Trans. 1.*, 1990, 1218. 575

90JCS(P1)2009 P. Huszthy, G. Izsó, K. Lempert, M. Györ and A. Rockenbauer; *J. Chem. Soc., Perkin Trans. 1*, 1990, 2009. 546

90JCS(P1)2031 S. Ma and X. Lu; *J. Chem. Soc., Perkin Trans.* 1, 1990, 2031. 7

90JFC(46)265 D. Naumann and J. Kischkewitz; *J. Fluorine Chem.*, 1990, **46**, 265. 243

90JFC(46)423 H. Sawada, M. Nakayama, M. Yoshida, T. Yoshida and N. Kamigata; *J. Fluorine Chem.*, 1990, **46**, 423. 12

90JFC(47)95 B. Boutevin, Y. Furet, L. Lemanach and F. Vial-Reveillon; *J. Fluorine Chem.*, 1990, **47**, 95. 7

90JFC(47)261 L. S. Chen; *J. Fluorine Chem.*, 1990, **47**, 261. 5

90JFC(48)207 R. Kasemann and D. Naumann; *J. Fluorine Chem.*, 1990, **48**, 207. 237

90JFC(48)249 J. H. Clark, C. W. Jones, A. P. Kybett, M. A. McClinton, J. M. Miller, D. Bishop and R. J. Blade; *J. Fluorine Chem.*, 1990, **48**, 249. 233, 234

90JFC(48)257 T. Abe, E. Hayashi, H. Baba and H. Fukaya; *J. Fluorine Chem.*, 1990, **48**, 257. 257

90JFC(48)395 N. R. Patel, R. L. Kirchmeier and J. M. Shreeve; *J. Fluorine Chem.*, 1990, **48**, 395. 251, 256, 259, 624

90JFC(49)75 L. G. Sprague, D. J. Burton, R. D. Guneratne and W. E. Bennett; *J. Fluorine Chem.*, 1990, **49**, 75. 266

90JFC(50)173 T. Abe, E. Hayashi, H. Fukaya and H. Baba; *J. Fluorine Chem.*, 1990, **50**, 173. 230

90JHC407 U. Urleb, B. Stanovnik and M. Tisler; *J. Heterocycl. Chem.*, 1990, **27**, 407. 558

90JHC643 U. Urleb, B. Stanovnik and M. Tisler; *J. Heterocycl. Chem.*, 1990, **27**, 643. 558

90JHC1191 D. T. W. Chu and A. K. Claiborne; *J. Heterocycl. Chem.*, 1990, **27**, 1191. 617

90JHC1249 L. Pongó, J. Reiter, J. Barkóczy and P. Sohár; *J. Heterocycl. Chem.*, 1990, **27**, 1249. 564

90JHC1953 J. Reisch, C. O. Usifoh and J. O. Oluwadiya; *J. Heterocycl. Chem.*, 1990, **27**, 1953. 109

90JHC2215 H. Inoue; *J. Heterocycl. Chem.*, 1990, **27**, 2215. 503

90JMC434 N. V. Harris, C. Smith and K. Bowden; *J. Med. Chem.*, 1990, **33**, 434. 621, 728

90JMC2101 M. Turconi, M. Nicola, M. G. Quintero, L. Maiocchi, R. Micheletti, E. Giraldo and A. Donetti; *J. Med. Chem.*, 1990, **33**, 2101. 425

90JOC1 R. A. Olofson, V. A. Dang, D. S. Morrison and P. F. DeCusati; *J. Org. Chem.*, 1990, **55**, 1. 462

90JOC1281 G. M. Lee and S. M. Weinreb; *J. Org. Chem.*, 1990, **55**, 1281. 23, 25

90JOC1475 A. L. Schroll, S. J. Eastep and G. Barany; *J. Org. Chem.*, 1990, **55**, 1475. 548

90JOC1503 P. S. Engel and W.-X. Wu; *J. Org. Chem.*, 1990, **55**, 1503. 604

90JOC1721 G. Morel, E. Marchand and A. Foucaud; *J. Org. Chem.*, 1990, **55**, 1721. 571

90JOC1847 Vu Anh Dang, R. A. Olofson, P. R. Wolf, M. D. Piteau and J.-P. G. Senet; *J. Org. Chem.*, 1990, **55**, 1847. 423

90JOC1851 Vu Anh Dang and R. A. Olofson; *J. Org. Chem.*, 1990, **55**, 1851. 421

90JOC2240 M. P. Bowman, R. A. Olofson, J.-P. Senet and T. Malfroot; *J. Org. Chem.*, 1990, **55**, 2240. 429

90JOC3045 P. A. Wade, J. F. Bereznak, B. A. Palfey, P. J. Carroll, W. P. Dailey and S. Sivasubramanian; *J. Org. Chem.*, 1990, **55**, 3045. 611

90JOC3699 T. H. Kim and H. Rapoport; *J. Org. Chem.*, 1990, **55**, 3699. 504

90JOC4757 S. F. Wnuk and M. J. Robins; *J. Org. Chem.*, 1990, **55**, 4757. 46

90JOC5230 M. E. Lewellyn, S. S. Wang and P. J. Strydom; *J. Org. Chem.*, 1990, **55**, 5230. 558

90JOC5750 M. Gouygou, Ch. Tachon, M. Koenig, A. Dubourg, J.-P. Declercq, J. Jaud and G. Etemad-Moghadam; *J. Org. Chem.*, 1990, **55**, 5750. 688

90JOC5982 M. P. Bowman, J.-P. G. Senet, T. Malfroot and R. A. Olofson; *J. Org. Chem.*, 1990, **55**, 5982. 429

90JOM(381)261 H. T. Schacht and H. Vahrenkamp; *J. Organomet. Chem.*, 1990, **381**, 261. 402

90JOM(383)295 R. B. King, F.-J. Wu and E. M. Holt; *J. Organomet. Chem.*, 1990, **383**, 295. 518

90JOM(384)279 S. E. Nefedov, A. A. Pasynskii, I. L. Eremenko, B. Orazsakhatov, V. M. Novotortsev, O. G. Ellert, A. F. Shestakov, A. I. Yanovsky and Yu. T. Struchkov; *J. Organomet. Chem.*, 1990, **384**, 279. 341

90JOM(385)207 H. Beckers and H. Bürger; *J. Organomet. Chem.*, 1990, **385**, 207. 243

90JOM(385)221 W. Hepp and U. Schubert; *J. Organomet. Chem.*, 1990, **385**, 221. 674, 723

90JOM(386)229 R. Davis, J. L. A. Durrant and N. M. S. Khazal; *J. Organomet. Chem.*, 1990, **386**, 229. 7

90JOM(386)321 J. Grobe, D. Le Van and J. Welzel; *J. Organomet. Chem.*, 1990, **386**, 321. 49

90JOM(388)47 R. Tacke, K. Fritsche, A. Tafel and F. Wuttke; *J. Organomet. Chem.*, 1990, **388**, 47. 127

90JOM(388)391 A. K. Burrell, G. R. Clark, J. G. Jeffrey, C. E. F. Rickard and W. R. Roper; *J. Organomet. Chem.*, 1990, **388**, 391. 247

90JOM(390)C73 M. A. Guerra, S. K. Mehrotra, D. W. Dyer and R. J. Lagow; *J. Organomet. Chem.*, 1990, **390**, C73. 245

90JOM(394)57 P. B. Hitchcock, M. F. Lappert, W.-P. Leung and N. H. Buttrus; *J. Organomet. Chem.*, 1990, **394**, 57. 194, 195

90JOM(394)265 M. H. Chisholm, K. Folting, V. J. Johnston and C. E. Hammond; *J. Organomet. Chem.*, 1990, **394**, 265. 342

90JOM(394)615 P. J. Brothers, A. K. Burrell, G. R. Clark, C. E. F. Rickard and W. R. Roper; *J. Organomet. Chem.*, 1990, **394**, 615. 247

90JOM(395)327 M. Herberhold and A. F. Hill; *J. Organomet. Chem.*, 1990, **395**, 327. 247

90JOM(397)51 R. Davis, N. M. S. Khazal and T. E. Bitterwolf; *J. Organomet. Chem.*, 1990, **397**, 51. 7

90JOM(397)209 L. M. Boyd, G. R. Clark and W. R. Roper; *J. Organomet. Chem.*, 1990, **397**, 209. 247

90JOM(398)59 C. Eaborn, P. B. Hitchcock, P. D. Lickiss and A. D. Taylor; *J. Organomet. Chem.*, 1990, **398**, 59. 380

90JOM(398)65 H. Bock, J. Meuret and U. Stein; *J. Organomet. Chem.*, 1990, **398**, 65. 349

90JOM(399)189 J. Grobe, D. Le Van, B. Krebs, M. Dartmann, F. Stone, A. Gordon and J. Szameitat; *J. Organomet. Chem.*, 1990, **399**, 189. 266

90JOM(400)121 J. Satgé; *J. Organomet. Chem.*, 1990, **400**, 121. 709

90JPO694 R. A. Moss and T. Zdrojewski; *J. Phys. Org. Chem.*, 1990, **3**, 694. 280

90LA499 R. Bognár and I. Farkas; *Liebigs Ann. Chem.*, 1990, 499. 71

90LA965 W. Kantlehner and U. Greiner; *Liebigs Ann. Chem.*, 1990, 965. 321

90MI 602-01 T. G. Archibald, N. Nguyen, J. S. Khosrowshahi and K. Baum; *Gov. Rep. Announce. Index (US)* 1990, **90**, Abstr. 035963 (*Chem. Abstr.*, 1991, **115**, 52917). 55

90MI 604-01 S. S. Pochapsky, H. F. VanBrocklin, M. J. Welch and J. A. Katzenellenbogen; *Bioconjugate Chem.*, 1990, **1**, 231 (*Chem. Abstr.*, 1990, **113**, 94071). 134

90MI 605-01 J. F. Weber and M. B. Frankel; *Propellants, Explos., Pyrotech.*, 1990, **15**, 26 (*Chem. Abstr.*, 1990, **112**, 237777). 151

90MI 610-01 J. G. Bowen, M. H. Hockley and J. R. Housley (Boots Co. plc), *Chin. Pat.* 1053921 (1990) (*Chem. Abstr.*, 1992, **116**, 194304). 308

90MI 612-01 W. Skawinski, J. Flisak, A. C. Chung, F. Jordan and R. Mendelsohn; *J. Labelled Comp. Radiopharm.*, 1990, **28**, 1179 (*Chem. Abstr.*, 1991, **114**, 61496). 365

90MI 614-01 W. A. Etzel and S. Berger; *J. Labelled Compd. Radiopharm.*, 1990, **28**, 977 (*Chem. Abstr.*, 1991, **114**, 23643). 414

90MI 614-02 E. A. Sneddon (Daresbury Lab., SERC); *Report* DL/SCI/P-659E (1990) (*Chem. Abstr.*, 1991, **114**, 163262). 418

90MI 614-03 T. Imagawa; *J. Synth. Org. Chem.*, 1990, **48**, 1058. 426

90MI 615-01 P. Parent, F. Leborgne, J. P. Lellouche, J. P. Beaucourt and A. Vanhove; *J. Labelled Compd. Radiopharm.*, 1990, **28**, 633. 480

90MI 615-02 T. Imagawa; *Yuki Gosei Kagaku Ryokai*, 1990, **48**, 1058 (*Chem. Abstr.*, 1991, **114**, 120 928). 462

90MI 617-01 M. Dzurilla, P. Kutschy, D. Koscik, V. Ficeri and R. Kraus; *Chem. Pap.* 1990, **44**, 45 (*Chem. Abstr.*, 1990, **113**, 115 217). 565

90MI 617-02 Sh. A. Musytafaev, V. S. Gasanov, I. A. Dzhafarov and R. K. Alekperov; *Izv. Vyssh. Uchebn. Zaved., Khim. Khim. Tekhnol.*, 1990, **33**, 100 (*Chem. Abstr.*, 1991, **114**, 23 374). 558

90MI 617-03 W. Podkoscielny, B. Tarasiuk, M. Dethloff, W. Kowalewska, H. Maziarczyk, B. Zieba-Zoltowska and J. Gorska-Poczopko, *Przem. Chem.*, 1990, **69**, 495 (*Chem. Abstr.*, 1991, **114**, 184 935). 546

90MI 618-01 T. Zhang; *Huaxue Shiji*, 1990, **12**, 126 78 (*Chem. Abstr.*, 1990, **113**, 230 774). 580

B-90MI 622-01 M. Regitz and O. J. Scherer; "Multiple Bonds and Low Coordination in Phosphorus Chemistry," Thieme Verlag, Stuttgart, 1990. 678, 688, 692, 693

90MRC1045 P. Sohár, G. Bernáth, S. Frimpong-Manso, A. E. Szabó and G. Stájer; *Magn. Reson. Chem.*, 1990, **28**, 1045. 558

90OM793 A. Marinetti, S. Bauer, L. Ricard and F. Mathey; *Organometallics*, 1990, **9**, 793. 692

90OM1325 S. Inoue and Y. Sato; *Organometallics*, 1990, **9**, 1325. 185

90OM1449 R. A. Michelin, R. Bertani, M. Mozzon, L. Zanotto, F. Benetollo and G. Bombieri; *Organometallics*, 1990, **9**, 1449. 247

90OM2677 D. Seyferth, J. L. Robison and J. Mercer; *Organometallics*, 1990, **9**, 2677. 191

90OPP128 N. R. Ayyangar, R. D. Wakharkar, V. H. Deshpande and B. Rai; *Org. Prep. Proced. Int.*, 1990, **22**, 128. 557

90PAC643 A. Dondoni; *Pure Appl. Chem.*, 1990, **62**, 643. 637

90POL277 M. Layh and W. Uhl; *Polyhedron*, 1990, **9**, 277. 195

90PS(47)7 H. Gross, S. Ozegowski and B. Costisella; *Phosphorus Sulfur*, 1990, **47**, 7. 155

90PS(47)261 K. Kosta and A. Kotyński; *Phosphorus Sulfur*, 1990, **47**, 261. 156, 157

90PS(51/52)23 F. H. Ebetino and L. A. Jamieson; *Phosphorus Sulfur*, 1990, **51/52**, 23. 158

90PS(51/52)153 U. Schülke; *Phosphorus Sulfur*, 1990, **51/52**, 153. 158

90PS(54)197 K. A. Jaeggi and T. Winkler; *Phosphorus Sulfur*, 1990, **54**, 197. 160

90S151 J. White and D. C. Baker; *Synthesis*, 1990, 151. 364

90S259 T. Sato; *Synthesis*, 1990, 259. 210

90S327 R. Ferraccioli, C. Gallina and C. Giordano; *Synthesis*, 1990, 327. 25

90S666 A. R. Katritzky, Z. Yang and J. N. Lam; *Synthesis*, 1990, 666. 143

90S825 M. A. Palominos and J. C. Vega; *Synthesis*, 1990, 825. 479

90S1065 P. Kessler, B. Chatrenet, M. Goeldner and C. Hirth; *Synthesis*, 1990, 1065. 443

90S1151 A. A. Kolomeitsev, V. N. Movchun, N. V. Kondratenko and Y. L. Yagupol'skii; *Synthesis*, 1990, 1151. 235

90S1159 M. Folkmann and F. J. Lund; *Synthesis*, 1990, 1159. 427, 428, 476

90SC703 T. Hiiro, T. Mogami, N. Kambe, S.-I. Fujiwara and N. Sonoda; *Synth. Commun.*, 1990, **20**, 703. 498

90SC2769 M. Vilkas and D. Quasmi; *Synth. Commun.*, 1990, **20**, 2769. 538

90SC3303 M. E. Sitzmann and H. G. Adolph; *Synth. Commun.*, 1990, **20**, 3303. 145

90SL217 T. Hirao and Y. Ohshiro; *Synlett.*, 1990, 217. 7

90T1839 G. Barcelo, D. Grenouillat, J.-P. Senet and G. Sennyey; *Tetrahedron*, 1990, **46**, 1839. 643, 645

90T3859 J. Y. Thoraval, W. Nagai, Y. Y. C. Yeung Lam Ko and R. Carrié; *Tetrahedron*, 1990, **46**, 3859. 60

90T3897 N. Guillot, H. G. Viehe, B. Tinant and J. P. Declercq; *Tetrahedron*, 1990, **46**, 3897. 653

90TL123 A.-B. N. Luheshi, R. K. Smalley, P. D. Kennewell and R. Westwood; *Tetrahedron Lett.*, 1990, **31**, 123. 111

90TL1151 A. Bulpin, S. Masson and A. Sene; *Tetrahedron Lett.*, 1990, **31**, 1151. 123

90TL1405 P. F. De Cusati and R. A. Olfson; *Tetrahedron Lett.*, 1990, 1405. 461

90TL1877 C. T. Hewkin and R. F. W. Jackson; *Tetrahedron Lett.*, 1990, **31**, 1877. 126

90TL2529 D. F. Harvey and M. F. Brown; *Tetrahedron Lett.*, 1990, **31**, 2529. 719

90TL3357 M. Zupan and Z. Bregar; *Tetrahedron Lett.*, 1990, **31**, 3357. 10

90TL3579 T. Umemomto and S. Ishihara; *Tetrahedron Lett.*, 1990, **31**, 3579. 22, 233, 234

90TL3859 J. Y. Thoraval, W. Nagai, Y. Y. C. Y. L. Ko and R. Carrié; *Tetrahedron Lett.*, 1990, **46**, 3859. 165

90TL4359 A. J. Barker, M. M. Campbell and M. J. Jenkins; *Tetrahedron Lett.*, 1990, **31**, 4359. 121

90TL4715 R. A. Moss and H. R. Kim; *Tetrahedron Lett.*, 1990, **31**, 4715. 280, 632

90TL4773 T. Mizuno, I. Nishiguchi, T. Hirashima, A. Ogawa, N. Kambe and N. Sonoda; *Tetrahedron Lett.*, 1990, 4773. 479

90TL4931 G. V. M. Sharma and S. R. Vepachedu; *Tetrahedron Lett.*, 1990, **31**, 4931. 550

90TL5571 M. L. Edwards, D. M. Stemerick, E. T. Jarvi, D. P. Matthews and J. R. McCarthy; *Tetrahedron Lett.*, 1990, **31**, 5571. 266

90TL5629 J. Nokami, A. Maihara and J. Tsuji; *Tetrahedron Lett.*, 1990, 5629. 461

90TL6331 G. Märkl and S. Reithinger; *Tetrahedron Lett.*, 1990, **31**, 6331. 374

90TL6831 R. G. Sutherland, C. Zhang and A. Piorko; *Tetrahedron Lett.*, 1990, **31**, 6831. 30

90TL7181 T. Shono, M. Ishifune, O. Ishige, H. Uyama and S. Kashimura; *Tetrahedron Lett.*, 1990, **31**, 7181. 24

90TL7201 B. L. Feringa, O. J. Gelling and L. Meesters; *Tetrahedron Lett.*, 1990, **31**, 7201. 124

90TL7379 A. A. Gakh, A. S. Kiselyov and V. V. Semenov; *Tetrahedron Lett.*, 1990, **31**, 7379. 145

90ZAAC(588)26 L. Riesel, H. Vogt and A. Brückner; *Z. Anorg. Allg. Chem.*, 1990, **588**, 26. 243

90ZC90 A. A. Martin, F. Zenner and G. Barnikow; *Z. Chem.*, 1990, **30**, 90. 560

90ZN(B)148 J. Grobe, M. Hagemann and D. Le Van; *Z. Naturforsch., Teil B*, 1990, **45**, 148. 258, 679

90ZN(B)290 J. Allwohn, R. Hunold, M. Pilz, R.-G. Müller, W. Massa and A. Berndt; *Z. Naturforsch., Teil B*, 1990, **45**, 290. 714

90ZN(B)299 J. Grobe, G. Lange and D. Le Van; *Z. Naturforsch., Teil B*, 1990, **45**, 299. 265

90ZN(B)447 M. Herberhold, W. Feger and U. Thewalt; *Z. Naturforsch., Teil B*, 1990, **45**, 447. 338

90ZN(B)1715 H. Eckert; *Z. Naturforsch., Teil B*, 1990, **45**, 1715. 488

90ZOB222 M. P. Sokolov, B. I. Buzykin and Z. V. Molodykh; *Zh. Obshch. Khim.*, 1990, **60**, 222 (*Chem. Abstr.*, 1990, **113**, 40849). 634

90ZOB223 M. P. Sokolov, B. I. Buzykin and V. A. Pavlov; *Zh. Obshch. Khim.*, 1990, **60**, 223 (*Chem. Abstr.*, 1990, **113**, 40850). 622

90ZOB563 T. K. Gazizov, Y. V. Chugunov and L. K. Sal'keeva; *Zh. Obshch. Khim.*, 1990, **60**, 563 (*Chem. Abstr.*, 1990, **113**, 115416). 583

90ZOB602 A. S. Nakhmanovich, R. V. Karnaukhova, O. G. Yarosh, V. N. Elokhina, I. D. Kalikhman and R. A. Gromkova; *Zh. Obshch. Khim.*, 1990, **60**, 602 (*Chem. Abstr.*, 1990, **113**, 24010). 565

90ZOB706 V. G. Rozinov, M. Y. Dmitrichenko, G. V. Dolgushin and V. I. Donskikh; *Zh. Obshch. Khim.*, 1990, **60**, 706 (*Chem. Abstr.*, 1990, **113**, 78534). 619

90ZOB709 G. N. Koidan, A. P. Marchenko and A. M. Pinchuk; *Zh. Obshch. Khim.*, 1990, **60**, 709 (*Chem. Abstr.*, 1990, **113**, 78535). 267

90ZOB778 R. M. Kamalov, G. M. Makarov, R. G. Islamov, M. A. Pudovik, R. A. Cherkasov and A. N. Pudovik; *Zh. Obshch. Khim.*, 1990, **60**, 778 (*Chem. Abstr.*, 1990, **113**, 132334). 657

90ZOB798 V. N. Zontova, T. I. Koroleva, N. N. Mel'nikov and A. F. Grapov; *Zh. Obshch. Khim.*, 1990, **60**, 798 (*Chem. Abstr.*, 1990, **113**, 172180). 635

90ZOB1288 B. I. Buzykin and M. P. Sokolov; *Zh. Obshch. Khim.*, 1990, **60**, 1288 (*Chem. Abstr.*, 1991, **114**, 24023). 659

90ZOB1293 M. P. Sokolov, B. I. Buzykin and T. A. Zyablikova; *Zh. Obshch. Khim.*, 1990, **60**, 1293 (*Chem. Abstr.*, 1991, **114**, 24024). 659

90ZOB1826 R. F. Mavlyutov, S. M. Kalashnikov, É. M. Kuramshin, A. I. Naimushin and U. B. Imashev; *Zh. Obshch. Khim.*, 1990, **60**, 1826 (*Chem. Abstr.*, 1991, **114**, 61324). 479

90ZOB1980 B. I. Buzykin and M. P. Sokolov; *Zh. Obshch. Khim.*, 1990, **60**, 1980 (*Chem. Abstr.*, 1991, **114**, 102200). 366, 659

90ZOB1986 M. P. Sokolov, B. I. Buzykin and G. V. Mavrin; *Zh. Obshch. Khim.*, 1990, **60**, 1986 (*Chem. Abstr.*, 1991, **114**, 42963). 659

90ZOB2238 M. I. Povolotskii, V. V. Negrebetskii, V. D. Romanenko, V. I. Ivanchenko, T. V. Sarina and L. N. Markovskii; *Zh. Obshch. Khim.*, 1990, **60**, 2238, (*Chem. Abstr.*, 1991, **115**, 8934). 685,686

90ZOB2694 M. P. Sokolov, G. V. Mavrin, B. I. Buzykin, M. N. Miftakhov, R. M. Gainutdinov and R. G. Timergaliev; *Zh. Obshch. Khim.*, 1990, **60**, 2694 (*Chem. Abstr.*, 1991, **115**, 29480). 659

90ZOB2814 A. P. Marchenko, G. N. Koidan, V. A. Schevchenko, M. I. Povolotskii and A. M. Pinchuk; *Zh. Obshch. Khim.*, 1990, **60**, 2814, (*Chem. Abstr.*, 1991, **115**, 29483). 696

90ZOR1129 M. L. Petrov, M. A. Abramov, K. A. Potekhin, Y. T. Struchkov, V. A. Galishev and A. A. Petrov; *Zh. Org. Khim.*, 1990, **26**, 1129 (*Chem. Abstr.*, 1990, **113**, 171955). 116

90ZOR1364 O. A. Tarasova, B. A. Tromifov, N. I. Ivanova, L. M. Sinegovskaya and S. V. Amosova; *Zh. Org. Khim.*, 1990, **26**, 1364 (*Chem. Abstr.*, 1991, **114**, 23374). 565

90ZOR1422 N. A. Nedolya, V. V. Gerasimova and G. I. Sarapulova; *Zh. Org. Khim.*, 1990, **26**, 1422 (*Chem. Abstr.*, 1991, **114**, 42058). 558

91AG(E)93 N. Wiberg and S.-K. Vasisht; *Angew. Chem., Int. Ed. Engl.*, 1991, **30**, 93. 193, 399

91AG(E)207 P. Binger, T. Wettling, R. Schneider, F. Zurmühlen, U. Bergsträsser, J. Hoffmann, G. Maas and M. Regitz; *Angew. Chem., Int. Ed. Engl.*, 1991, **30**, 207. 153

91AG(E)324 W. Hiller, M. Layh and W. Uhl; *Angew. Chem., Int. Ed. Engl.*, 1991, **30**, 324. 391, 394

91AG(E)594 A. Höfner, B. Ziegler, R. Hunold, P. Willershausen, W. Massa and A. Berndt; *Angew. Chem., Int. Ed. Engl.*, 1991, **30**, 594. 389, 714

91AG(E)674 M. Regitz; *Angew. Chem., Int. Ed. Engl.*, 1991, **30**, 674. 668, 669

91AG(E)709 H. Grützmacher and H. Pritzkow; *Angew. Chem., Int. Ed. Engl.*, 1991, **30**, 709. 165, 691, 733

91AG(E)1154 J.-M. Sotiropoulos, A. Baceiredo, K. Horchler v. Locquenghien, F. Dahan and G. Bertrand; *Angew. Chem., Int. Ed. Engl.*, 1991, **30**, 1154. 669

91AG(E)1356 Y. Chiang, A. S. Grant, A. J. Kresge, P. Pruszynski, N. P. Schepp and J. Wirz; *Angew. Chem., Int. Ed. Engl.*, 1991, **30**, 1356. 646

91AG(E)1488 H. Schmidbaur, B. Brachthäuser and O. Steigelmann; *Angew. Chem., Int. Ed. Engl.*, 1991, **30**, 1488. 386, 406

91AG(E)1598 D. Freitag, G. Fengler and L. Morbitzer; *Angew. Chem., Int. Ed. Engl.*, 1991, **30**, 1598. 471

91ANC1295 S. Kamata and K. Onoyama; *Anal. Chem.*, 1991, **63**, 1295. 563

91AOC(32)147 P. R. Markies, O. S. Akkerman, F. Bickelhaupt, W. J. J. Smeets and A. L. Spek; *Adv. Organomet. Chem.*, 1991, **32**, 147. 406

91AOC(32)227 A. Mayr and H. Hoffmeister; *Adv. Organomet. Chem.*, 1991, **32**, 227. 723

91AP(324)917 W. Hanefeld and H. J. von Gösseln; *Arch. Pharm. (Weinheim, Ger.)*, 1991, **324**, 917. 445, 540

91BMC357 D. Hebel, K. L. Kirk, J. Kinjo, T. Kovacs, K. Lesiak, J. Balzarini, E. DeClercq and P. F. Torrence; *Bioorg. Med. Chem. Lett.*, 1991, **1**, 357. 264

91CAR(216)271 M. A. Ali, L. Hough and A. C. Richardson; *Carbohydr. Res.*, 1991, **216**, 271. 561

91CB51 R. Boese, A. Haas and M. Spehr; *Chem. Ber.*, 1991, **124**, 51. 256

91CB329 U. Krüger, H. Pritzkow and H. Grützmacher; *Chem. Ber.*, 1991, **124**, 329. 702

91CB503 B. Wrackmeyer and K. Wagner; *Chem. Ber.*, 1991, **124**, 503. 209

91CB1131 W. Fimml and F. Sladky; *Chem. Ber.*, 1991, **124**, 1131. 357, 358

91CB1353 D. Viets, R. Mews, M. Noltemeyer, H. G. Schmidt and A. Waterfeld; *Chem. Ber.*, 1991, **124**, 1353. 252

91CB1425 M. Conrads and J. Mattay; *Chem. Ber.*, 1991, **124**, 1425. 125

91CB1739 F. Castan, M. Granier, T. A. Straw, A. Baceiredo, K. B. Dillon and G. Bertrand; *Chem. Ber.*, 1991, **124**, 1739. 659, 667

91CB1805 M. Frasch, W. Sundermeyer and J. Waldi; *Chem. Ber.*, 1991, **124**, 1805. 128

91CB1981 N. Wiberg, S. Wagner and G. Fischer; *Chem. Ber.*, 1991, **124**, 1981. 709, 710

91CB2025 C. Maletzko, W. Sundermeyer, H. Pritzkow, H. Irngartinger and U. Huber-Patz; *Chem. Ber.*, 1991, **124**, 2025. 255, 275, 304, 306

91CB2411 M. Erhart, R. Mews, F. Pauer, D. Stalke and G. Sheldrick; *Chem. Ber.*, 1991, **124**, 2411. 440

91CB2645 D. Bromm, U. Seebold, M. Noltemeyer and A. Meller; *Chem. Ber.*, 1991, **124**, 2645. 183

91CB2677 S. J. Goede and F. Bickelhaupt; *Chem. Ber.*, 1991, **124**, 2677. 682, 690

91CC165 C. J. Cable, H. Adams, N. A. Bailey, J. Crosby and C. White; *J. Chem. Soc., Chem. Commun.*, 1991, 165. 247

91CC466 H. H. Karsch, K. Zellner and G. Müller; *J. Chem. Soc., Chem. Commun.*, 1991, 466.

91CC470 J. Hackenberg and M. Hanack; *J. Chem. Soc., Chem. Commun.*, 1991, 470. 254

91CC975 C. H. Reynolds; *J. Chem. Soc., Chem. Commun.*, 1991, 975. 726

91CC979 S. O. Grim and P. B. Kettler; *J. Chem. Soc., Chem. Commun.*, 1991, 979. 151

91CC993 C. Wakselman, M. Tordeux, J. L. Clavel and B. Langlois; *J. Chem. Soc. Chem. Commun.*, 1991, 993. 233, 234

91CC1305 F. Mercier, L. Ricard, F. Mathey and M. Regitz; *J. Chem. Soc., Chem. Commun.*, 1991, 1305. 154

91CC1378 R. Boese, A. Haas and C. Limberg; *J. Chem. Soc., Chem. Commun.*, 1991, 1378. 588

91CC1536 P. von R. Schleyer and A. I. Boldyrev; *J. Chem. Soc., Chem. Commun.*, 1991, 1536. 399

91CC1560 C. Jones, L. M. Engelhardt, P. C. Junk, D. S. Hutchings, W. C. Patalinghug, C. L. Raston and A. H. White; *J. Chem. Soc., Chem. Commun.*, 1991, 1560. 168

91CC1608 C. Eaborn and D. A. R. Happer; *J. Chem. Soc., Chem. Commun.*, 1991, 1608. 734

91CJC327 W. Tyrra and D. Naumann; *Can. J. Chem.*, 1991, **69**, 327. 220, 246

91CJC415 P. Yates, A. Seif-El-Nasr, J. Stanton and J. J. Krepinsky; *Can. J. Chem.*, 1991, **69**, 415. 88

91CJC861 M. Cariou and J. Simonet; *Can. J. Chem.*, 1991, **69**, 861. 648

91CJC2127 J. C. Baum, K. A. Durkin, L. Precedo, S. B. O'Blenes, J. E. Goehl, R. F. Langler, G. K. MacCormack and L. L. Smith; *Can. J. Chem.*, 1991, **69**, 2127. 255

91CL569 A. M. Krishnan, L. T. Wolford and J. H. Boyer; *Chem. Lett*, 1991, 569. 653

91CL1421 A. Sekiya and K. Ueda; *Chem. Lett.*, 1991, 1421. 229, 230

91CL1873 M. Hirano, A. Miyashita, H. Shitara and H. Nohira *Chem. Lett.*, 1991, 1873. 115

91COS(1)1 P. G. Williard; *Comp. Org. Synth.*, 1991, **1**, 1. 172, 189, 190

91COS(1)487 A. Pelter and K. Smith; *Comp. Org. Synth.*, 1991, **1**, 487. 172, 198, 201

91COS(6)171 H. J. Bestmann and R. Zimmermann; *Comp. Org. Synth.*, 1991, **6**, 171. 694, 697, 700, 701, 702

91CPB1659 S. Harusawa, H. Osaki, T. Kurokawa, H. Fujii, R. Yoneda and T. Kurihara; *Chem. Pharm. Bull.*, 1991, **39**, 2543. 542

91CPB1939 T. Iwakawa, H. Tamura, A. Murabayashi and Y. Hayase; *Chem. Pharm. Bull.*, 1991, **39**, 1939. 556, 566, 652

91CRV1 A. K. Mukerjee and R. Ashare; *Chem. Rev.*, 1991, **91**, 1. 573

91CRV35 J. A. Morrison; *Chem. Rev.*, 1991, **91**, 35. 181

91CRV311 W. P. Neumann; *Chem. Rev.*, 1991, **91**, 311. 187

91CSR503 H. Rudler, M. Audouin, E. Chelain, B. Denise, R. Goumont, A. Massoud, A. Parlier, A. Pacreau, M. Rudler, R. Yefsah, C. Alvarez and F. Delgado-Reyes; *Chem. Soc. Rev.*, 1991, **20**, 503. 716

91EGP291752 K. Geisler, A. Kuenzler and E. Bulka; *Ger. (East) Pat.* DD291 752 (1991) (*Chem. Abstr.* 1991, **115**, 207 516). 596

91EUP415473 S. A. Madison, L. M. Ilardi and H. Cerfontain (Unilever NV; Unilever plc); *Eur. Pat.* 415473 (1991) (*Chem. Abstr.*, 1991, **114**, 228528). 428

91EUP421322 H. Aoyama, T. Yasuhara, S. Kono and S. Koyama (Daikin Ind.); *Eur. Pat. Appl.* 421322 (1991) (*Chem. Abstr.* 1991, **115**, 70901). 4

91EUP425798 Agro-Kanesho Co. Ltd., Jpn.; *Eur. Pat. Appl.* 425978 (1991) (*Chem. Abstr.*, 1991, **115**, 280003).

91EUP450584 S. Kumai and A. Wada (Asahi Glass); *Eur. Pat. Appl.* 450584 (1991) (*Chem. Abstr.*, 1992, **116**, 83205). 32

91EUP461563 K. Raab (Hoechst); *Eur. Pat.* 461563 (1991) (*Chem. Abstr.*, 1992, **116**, 128158). 32

91GEP3937567 K. Raab (Hoechst); *Ger. Pat.* 3937567 (1991) (*Chem. Abstr.* 1991, **115**, 28684). 32

91HAC385 J. Grobe, Duc Le Van, U. Althoff, B. Krebs and M. Dartmann and R. Gleiter; *Heteroatom Chem.*, 1991, **2**, 385. 242, 685

91HCA465 W. Oppolzer, J.-Z. Xu and C. Stone; *Helv. Chim. Acta*, 1991, **74**, 465. 461

91HCA1500 V. Oremus, A. Linden and H. Heimgartner; *Helv. Chim. Acta*, 1991, **74**, 1500. 305, 309, 310

91IC1046 W. J. Mills, C. H. Sutton, M. W. Baize and L. J. Todd; *Inorg. Chem.*, 1991, **30**, 1046. 515

91IC1776 H. R. Allcock, J. S. Rutt and M. Parvez; *Inorg. Chem.*, 1991, **30**, 1776. 558

91IC2228 N. E. Miller; *Inorg. Chem.*, 1991, **30**, 2228. 492

91IC2433 M. R. M. D. Charandabi, D. A. Feakes, M. L. Ettel and K. W. Morse; *Inorg. Chem.*, 1991, **30**, 2433. 514

91IZV2593 V. A. Dorokhov *et al.*; *Izv. Akad. Nauk SSSR Ser. Khim.*, 1991, 2593. 504

91JA7654 W. Adam, R. Curci, M. E. González Nuñez and R. Mello; *J. Am. Chem. Soc.*, 1991, **113**, 7654. 230

91JA7790 W. Ando, H. Yoshida, K. Kurishima and M. Sugiyama; *J. Am. Chem. Soc.*, 1991, **113**, 7790. 383

91JA8899 Y. Ito, M. Suginome, T. Matsura and M. Murakami; *J. Am. Chem. Soc.*, 1991, **113**, 8899. 384, 671

91JA9379 H. Jun, V. G. Young, Jr. and R. J. Angelici; *J. Am. Chem. Soc.*, 1991, **113**, 9379. 681, 690

91JAN1083 W.-J. Kim, K.-Y. Ko, M. H. Jung, M. Kim, K.-I. Lee and J. H. Kim; *J. Antibiot.*, 1991, **44**, 1083. 428

91JAP03060419 T. Kagawa, T. Morooka, T. Uotani and K. Tsuzuki; *Jpn. Pat.* 03 060 419 (1991) (*Chem. Abstr.*, 1991, **115**, 52 903). 529

91JAP03193759 T. Kagawa, T. Uotani and K. Tsuzuki; *Jpn. Pat.* 03 193 759 (1991) (*Chem. Abstr.*, 1991, **116**, 58 982). 533

91JAP03218336 N. Hirowatari, M. Tsuda and M. Tanaka; *Jpn. Pat.* 03 218 336 (*Chem. Abstr.*, 1992, **116**, 58 978). 533

91JCR(S)160 R. H. Khan and R. C. Rastogi; *J. Chem. Res.* (*S*), 1991, 160. 627

91JCS(D)1017 J. Evans, P. M. Stroud and M. Webster; *J. Chem. Soc., Dalton Trans.*, 1991, 1017. 344

91JCS(P1)37 G. Schneider, L. Hackler, J. Szanyi and P. Sohar; *J. Chem. Soc. Perkin Trans. 1*, 1991, 37. 489

91JCS(P1)157 A. J. Moore and M. R. Bryce; *J. Chem. Soc., Perkin Trans. 1*, 1991, 157. 120, 123

91JCS(P1)787 M. J. Kelly and S. M. Roberts; *J. Chem. Soc., Perkin Trans. 1*, 1991, 787. 543

91JCS(P1)897 M. Ashwell, W. Clegg and R. F. W. Jackson; *J. Chem. Soc., Perkin Trans. 1*, 1991, 897. 126

91JCS(P1)2495 L. A. Silks, III, J. Peng, J. D. Odom and R. B. Dunlap; *J. Chem. Soc., Perkin Trans. 1*, 1991, 2495. 595

91JCS(P1)3091 C. T. Hewkin, R. F. W. Jackson and W. Clegg; *J. Chem. Soc., Perkin Trans. 1*, 1991, 3091. 98, 99, 101

91JCS(P2)501 M. S. F. Lie Ken Jie, C. Yan-Kit, S. H. Chau and B. F. Y. Yan; *J. Chem. Soc., Perkin Trans. 2*, 1991, 501. 481, 482

91JFC(52)229 G. Pawelke; *J. Fluorine Chem.*, 1991, **52**, 229. 15

91JFC(53)181 L. Bragante and D. D. DesMarteau; *J. Fluorine Chem.*, 1991, **53**, 181. 604

91JFC(55)313 G. L. Gard, N. N. Hamel, J. Mohtasham, A. Waterfeld and R. Mews; *J. Fluorine Chem.*, 1991, **55**, 313. 251

91JHC481 J. Szabo, L. Fodor, G. Bernath, P. Sohar, G. S. Talpas and E. Bani-Akoto; *J. Heterocycl. Chem.*, 1991, **28**, 481. 111

91JHC1013 M. S. South; *J. Heterocycl. Chem.*, 1991, **28**, 1013. 82

91JMC98 W. D. Kingsbury *et al.*; *J. Med. Chem.*, 1991, **34**, 98. 487

91JMC569 M. L. Edwards, D. M. Stemerick, A. J. Bitonti, J. A. Dumont, P. P. McCann, P. Bey and A. Sjoerdsma; *J. Med. Chem.*, 1991, **34**, 569. 487

91JMC2049 J. R. Cashman and Y. Liu; *J. Med. Chem.*, 1991, **34**, 2049. 478

91JMC3187 M. J. Kukla, H. J. Breslin, C. J. Diamond, P. P. Grous, C. Y. Ho, M. Miranda, J. D. Rodgers, R. G. Sherrill, E. De Clercq, R. Pauwels, K. Andries, L. J. Moens, M. A. C. Janssen and P. A. J. Janssen; *J. Med. Chem.*, 1991, **34**, 3187. 598

91JMOC389 J. F. Knifton and R. G. Duranleau; *J. Mol. Catal.*, 1991, **67**, 389. 467

91JOC103 J. Castaner, J. Riera, J. Carilla, A. Robert, E. Molins and C. Miravitlles; *J. Org. Chem.*, 1991, **56**, 103. 30

91JOC751 D. J. Mathre, T. K. Jones, L. C. Xavier, T. J. Blacklock, R. A. Reamer, J. J. Mohan, E. T. T. Jones, K. Hoogsteen, M. W. Baum and E. J. J. Grabowski; *J. Org. Chem.*, 1991, **56**, 751. 446

91JOC891 D. A. Laufer and E. Al-Farhan; *J. Org. Chem.*, 1991, **56**, 891. 416, 417

91JOC1801 F. Castan, A. Baceiredo, D. Bigg and G. Bertrand; *J. Org. Chem.*, 1991, **56**, 1801. 668

91JOC1816 E. C. Taylor and R. Doetzer; *J. Org. Chem.*, 1991, **56**, 1816. 121

91JOC2143 A. R. Katritzky, Z. Yang and J. N. Lam; *J. Org. Chem.*, 1991, **56**, 2143. 170

91JOC2849 S. Iwasa, M. Yamamoto, S. Kohmoto and K. Yamada; *J. Org. Chem.*, 1991, **56**, 2849. 550

91JOC5684 M. Jørgensen, K. A. Lerstrup and K. Bechgaard; *J. Org. Chem.*, 1991, **56**, 5684. 561

91JOC5919 E. Juaristi and M. A. Aguilar; *J. Org. Chem.*, 1991, **56**, 5919. 120

91JOC6733 L. A. Silks, III, J. Peng, J. D. Odom and R. B. Dunlap; *J. Org. Chem.*, 1991, **56**, 6733. 595

91JOM(403)293 S. S. Al-Juaid, Y. Derouiche, P. B. Hitchcock, P. D. Lickiss and A. G. Brook; *J. Organomet. Chem.*, 1991, **403**, 293. 515

91JOM(405)149 F. I. Aigbirhio, S. S. Al-Juaid, C. Eaborn, A. Habtemariam, P. B. Hitchcock and J. D. Smith; *J. Organomet. Chem.*, 1991, **405**, 149. 176, 394

91JOM(407)1 D. Naumann, W. Strauss and W. Tyrra; *J. Organomet. Chem.*, 1991, **407**, 1. 245

91JOM(411)19 D. Grdenić, B. Korpar-Čolig, D. Matkovic-Čalogović, M. Sikirica and Z. Popović; *J. Organomet. Chem.*, 1991, **411**, 19. 207

91JOM(417)C47 S. Pasenok, D. Naumann and W. Tyrra; *J. Organomet. Chem.*, 1991, **417**, C47. 243

91JOM(417)305 T. Mandai, H. Murayama, T. Nakata, H. Yamaoki, M. Ogawa, M. Kawada and J. Tsuji; *J. Organomet. Chem.*, 1991, **417**, 305. 461

91JOM(419)C1 J. P. Picard, S. Grelier, J. Dunoguès, J.-M. Aizpurua and C. Palomo; *J. Organomet. Chem.*, 1991, **419**, C1. 162, 163

91JOM(419)9 R. J. P. Corriu, G. F. Lanneau and V. D. Mehta; *J. Organomet. Chem.*, 1991, **419**, 9. 576, 577

91JOM(419)357 R. G. Sutherland and C. Zhang; *J. Organomet. Chem.*, 1991, **419**, 357. 30

91JOM(420)C12 D. Miguel, V. Riera, J. A. Perez-Martinez, V. Riera and S. Garcia-Granda; *J. Organomet. Chem.*, 1991, **420**, C12. 328

91JOM(421)175 M. Westerhausen, B. Rademacher and W. Poll; *J. Organomet. Chem.*, 1991, **421**, 175. 190, 192, 195, 393

91JPR139 S. S. Zlotskij, L. S. Rol'nik, I. R. Kabibulin, D. L. Rachmankulov and H.-J. Timpe; *J. Prakt. Chem.*, 1991, **333**, 139. 555

91JPR187 M. Hans, A. Gulube and H. Dehne; *J. Prakt. Chem.*, 1991, **333**, 187. 617, 635

91KGS416 A. S. Fisyuk and B. V. Unkovskii; *Khim. Geterotsikl. Soedin.*, 1991, 416 (*Chem. Abstr.*, 1991, **115**, 136 029). 566

91LA405 W. Hanefeld and H. J. von Gösseln; *Liebigs Ann. Chem.*, 1991, 405. 540

91M1047 P. Perjési and P. Sohár; *Monatsh. Chem.*, 1991, **122**, 1047. 562

91MIP54636 C. Huszar, F. Sperber, A. Nemeth, P. Sarkozi, E. Somfai and L. Pali; *Hung. Teljes HU* 54 636 (1991) (*Chem. Abstr.*, 1991, **115**, 114 138). 645

91MIP 601-01 C. G. Krespan, A. C. Sievert, F. J. Weigert (Du Pont); *W. Pat.* 9 105 753 (1991) (*Chem. Abstr.* 1991, **115**, 70904).

91MIP9110639 F. J. Lund; *W. Pat.* 9 110 639 (1991) (*Chem. Abstr.*, 1991, **115**, 182 648) 477

91MI 601-01 M. Kotora and M. Hajek; *React. Kinet. Catal. Lett.*, 1991, **44**, 415. 7

91MI 601-02 T. Haga, Y. Tsujii, K. Hayashi, F. Kimura. N. Sakashita and K. Fujikawa; *ACS Symp. Ser.*, 1991, **443**. 9

B-91MI 601-03 J. T. Welch and S. Eswarakrishnan; "Fluorine in Bioorganic Chemistry," Wiley-Interscience, New York, 1991. 9

91MI 604-01 K. T. Kang, J. S. U, I. N. Yoon and C. H. Park; *J. Korean Chem. Soc.*, 1991, **35**, 292 (*Chem. Abstr.*, 1991, **115**, 159047). 128

91MI 606-01 D. W. H. Rankin; in "Front. Organosilicon Chemistry, Proceedings of the International Symposium on Organosilicon Chemistry," 9th conf., Cambridge, 1990, p. 253 (*Chem. Abstr.*, 1991, **115**, 114 606). 172

91MI 606-02 N. Wiberg; in "Front. Organosilicon Chemistry, Proceedings of the International Symposium on Organosilicon Chemistry," 9th conf., Cambridge, 1990, p. 263 (*Chem. Abstr.*, 1991, **115**, 114 607). 172

91MI 607-01 I. B. Sladkov and A. V. Bogacheva; *Zh. Prikl. Khim.* (*St. Petersburg*), 1991, **64**, 2435 (*Chem. Abstr.*, 1992, **117**, 14730). 223, 226, 227, 228

B-91MI 607-02 "Kirk-Othmer Encyclopedia of Chemical Technology," 3rd edn., Wiley, New York, 1990, **1**, 1001. 215

91MI 613-01 M. Aggarwal, R. A. Geanangel and M. A. Ghuman; *Main Gp. Met. Chem.*, 1991, **14**, 263. 390

B-91MI 615-01 W. Schneider and W. Diller; in "Ullmann's Encyclopedia of Industrial Chemistry," ed. B. Elvers, S. Hawkins and G. Schulz; Verlag Chemie, Weinheim, 1991, vol. A19, p. 411. 460, 462

91MI 615-02 J. Saha, R. M. Ruprecht and A. Rosowsky; *Nucleosides Nucleotides*, 1991, **10**, 1465. 483, 489

B-91MI 615-03 T. W. Greene and P. G. M. Wuts; "Protective Groups in Organic Synthesis," 2nd edn., Wiley, New York, 1991, p. 315. 487

91MI 620-01 K. Pye and S. Smith; *Eur. Polym. J.*, 1991, **27**, 155. 604

B-91MI 621-01 Y. Yamamoto and S. Kojima; in "The Chemistry of Amidines and Imidates," ed. S. Patai and Z. Rappoport, Wiley, Chichester, 1991, vol. 2, p. 485.

91MM2302 C. Li and J. Kohn; *Macromolecules*, 1991, **24**, 2302. 624

91NJC393 G. R. Gillette, A. Igau, A. Baceiredo and G. Bertrand; *Nouv. J. Chim.*, 1991, **15**, 393. 125, 668, 669

91OM366 L. Balas, B. Jousseaume, H. A. Shin, J.-B. Verlhal and F. Wallian; *Organometallics*, 1991, **10**, 366. 449

91OM384 B. Alvarez, D. Miguel, V. Riera, J. A. Miguel and S. Garcia-Granda; *Organometallics*, 1991, **10**, 384. 331

91OM551 D. Seyferth and H. Lang; *Organometallics*, 1991, **10**, 551. 190, 381, 392

91OM1647 P. Jutzi, A. Becker, H. G. Stammler and B. Neumann; *Organometallics*, 1991, **10**, 1647. 187, 387

91OM1683 D. Miguel, V. Riera, J. A. Miguel, M. Gomez and X. Solans; *Organometallics*, 1991, **10**, 1683. 331

91OM1704 C. J. Schaverien and J. B. van Mechelen; *Organometallics*, 1991, **10**, 1704. 190, 193, 195

91OM2101 C. Koellemann and F. Sladky; *Organometallics*, 1991, **10**, 2101. 357, 358

91OM2384 M. R. Churchill, C. H. Lake, W. G. Feighery and J. B. Keister; *Organometallics*, 1991, **10**, 2384. 343

91OM2485 Y. Chi, S.-H. Chuang, L.-K. Liu and Y.-S. Wen; *Organometallics*, 1991, **10**, 2485. 348

91OM2781 F. L. Joslin, M. P. Johnson, J. T. Mague and D. M. Roundhill; *Organometallics*, 1991, **10**, 2781. 516

91OM2884 H. H. Karsch, K. Zellner and G. Müller; *Organometallics*, 1991, **10**, 2884. 374

91OM3005 B. Alvarez, S. Garcia-Granda, Y. Jeannin, D. Miguel, J. A. Miguel and V. Riera; *Organometallics*, 1991, **10**, 3005. 331

91OM3205 M. P. Arthur, H. P. Goodwin, A. Baceiredo, K. B. Dillon and G. Bertrand; *Organometallics*, 1991, **10**, 3205. 670, 675

91OM3466 A. D. Fanta, D. J. DeYoung, J. Belzner and R. West; *Organometallics*, 1991, **10**, 3466. 125

91OM3838 P. Jutzi, A. Becker, C. Leue, H. G. Stammler and B. Neumann; *Organometallics*, 1991, **10**, 3838. 387

91OPP721 J. Zmitek, B. Jenko, D. Milivojevic and J. Anzel; *Org. Prep. Proced. Int.*, 1991, **23**, 721. 632

91PJS422 M. A. Butt, R. Perveen, M. A. Ahmad, S. Parveen and G. R. Khan; *Pak. J. Sci. Ind. Res.*, 1991, **34**, 422. 558

91POL823 M. Rahman, M. Akbar Ali, P. Bhattacharjee and M. Nazimuddin; *Polyhedron*, 1991, **10**, 823. 563

91POL1153 M. Lazraq, C. Couret, J. Escudié, J. Satgé and M. Soufiaoui; *Polyhedron*, 1991, **10**, 1153. 186, 188

91PS(55)185 P. Majewski; *Phosphorus Sulfur*, 1991, **55**, 185. 242

91PS(56)117 L. H. Merwin, W. A. Dollase, G. Hägele and H. Blum; *Phosphorus Sulfur*, 1991, **56**, 117. 158

91PS(57)211 A. A. Fahmy, T. S. Hafez, A. F. El-Farargy and M. M. Hamad; *Phosphorus Sulfur*, 1991, **57**, 211. 117

91PS(60)287 K. Diemert, T. Hahn, W. Kuchen; *Phosphorus Sulfur*, 1991, **60**, 287. 489

91PS(61)361 A. S. Ionkin, N. V. Nikolaeva, V. M. Nekhoroshkov, Yu. Ya. Efremov and B. A. Arbuzov; *Phosphorus Sulfur*, 1991, **61**, 361. 160

91PS(62)35 H. Gross, I. Keitel and B. Costisella; *Phosphorus Sulfur*, 1991, **62**, 35. 155

91PS(63)51 S. Cossu, O. De Lucchi, E. Piga and G. Valle; *Phosphorus Sulfur*, 1991, **63**, 51. 556

91PS(63)95 A. Köckritz, G. Röhr and M. Schnell; *Phosphorus Sulfur*, 1991, **63**, 95. 156

91RCR162 E. N. Dianova and E. Ya. Zabotina; *Russ. Chem. Rev.* (*Engl. Transl.*), 1991, **60**, 162. 678, 691

91RCR391 O. I. Kolodyazhnyi; *Russ. Chem. Rev.* (*Engl. Transl.*), 1991, **60**, 391. 694, 700, 701, 702

91RCR1318 A. O. Gukasyan, L. Kh. Calstyanand and A. A. Avetisyan; *Russ. Chem. Rev.*, 1991, **60**, 1318. 25

91S26 A. J. Moore and M. R. Bryce; *Synthesis*, 1991, 26. 698

91S86 R. M. Giuliano, T. T. Duong, T. W. Deisenroth, W. G. McMahon and W. J. Boyko; *Synthesis*, 1991, 86. 632

91S147 J. Garin, E. Meléndez, F. L. Merchán, T. Tejero, S. Uriel and J. Ayestarán; *Synthesis*, 1991, 147. 563

91S220 S. P. Maybhate, P. P. Rajamohanan and S. Rajappa; *Synthesis*, 1991, 220. 627

91S265 P. Renaut, D. Thomas and F. D. Bellamy; *Synthesis*, 1991, 265. 558

91S347 P. A. M. van der Klein, A. E. J. de Nooy, G. A. van der Marel and J. H. van Boom; *Synthesis*, 1991, 347. 317

91S405 A. B. Khare and C. E. McKenna; *Synthesis*, 1991, 405. 666

91S465 R. Jaeger; *Synthesis*, 1991, 465. 633

91S637 H. Ahlbrecht, C. Schmitt and D. Kornetzky; *Synthesis*, 1991, 637. 561, 563

91S753 M. Haake and B. Schümmelfeder; *Synthesis*, 1991, 753. 625, 632

91S1099 E. Fuchs, B. Breit, U. Bergsträßer, J. Hoffmann, H. Heydt and M. Regitz; *Synthesis*, 1991, 1099. 512, 582

91SA(A)849 B. Wrackmeyer and H. Zhou; *Spectrochim. Acta, Part A*, 1991, **47**, 849. 209, 401, 405

91SA(A)1619 C. O. Della Védova; *Spectrochim. Acta, Part A*, 1991, **47**, 1619. 432

91SC285 R. Cortez, I. A. Rivero, R. Somanathan, G. Aguirre and F. Ramirez; *Synth. Commun.*, 1991, **21**, 285. 416

91SC2189 R. F. Cunico and C.-P. Zhang; *Synth. Commun.*, 1991, **21**, 2189. 24

91SL160 D. Waldmüller, M. Braun and A. Steigel; *Synlett*, 1991, 160. 72

91SL381 D. B. Grotjahn and K. H. Dötz; *Synlett*, 1991, 381. 716, 719, 722

91SRI401 A. K. Saxena, C. S. Bisaria, S. Saini and L. M. Pande; *Synth. React. Inorg. Metal-Org. Chem.*, 1991, **21**, 401. 270

91SUL143 J. Nielsen and A. Senning; *Sulfur Letters*, 1991, **12**, 143. 535

91SUL247 A. Senning; *Sulfur Letters*, 1991, **12**, 247. 301

91T121 A. L. J. Beckwith, V. W. Bowry and C. H. Schiesser; *Tetrahedron*, 1991, **47**, 121. 543, 550, 551

91T549 N. Muller; *Tetrahedron*, 1991, **47**, 549. 20

91T615 J.-i. Yoshida, S.-i. Matsunaga, T. Murata and S. Isoe; *Tetrahedron*, 1991, **47**, 615. 129

91T3171 B. De Ancos, F. Farina, M. C. Maestro, M. R. Martín and M. M. Vicioso; *Tetrahedron*, 1991, **47**, 3171. 124

91T3207 J.-P. Bégué and D. Bonnet-Delphon; *Tetrahedron*, 1991, **47**, 3207. 9

91T5149 A. Marra, F. Gauffeny and P. Sinaÿ; *Tetrahedron*, 1991, **47**, 5149. 549

91T6381 C. J. France, I. M. McFarlane, C. G. Newton and P. Pitchen; *Tetrahedron*, 1991, **47**, 6381. 543, 550

91T6915 A. Sood, C. K. Sood, I. H. Hall and B. F. Spielvogel; *Tetrahedron*, 1991, **47**, 6915. 492, 515

91T10053 S. J. Allcock, T. L. Gilchrist, S. J. Shuttleworth and F. D. King; *Tetrahedron*, 1991, **47**, 10053.

91TL91 H. Urata and T. Fuchikami; *Tetrahedron Lett.*, 1991, **32**, 91. 22

91TL259 G. Pattenden and S. J. Reynolds; *Tetrahedron Lett.*, 1991, **32**, 259. 447

91TL375 K. Uneyama and T. Kitagawa; *Tetrahedron Lett.*, 1991, **32**, 375. 21

91TL597 A. I. Meyers, W. R. Leonard, Jr. and J. L. Romine; *Tetrahedron Lett.*, 1991, **32**, 597. 107

91TL1019 Z.-Y. Yang and D. J. Burton; *Tetrahedron Lett.*, 1991, **32**, 1019. 58

91TL2703 D. H. R. Barton, J. C. Jaszberenyi and C. Tachdjian; *Tetrahedron Lett.*, 1991, **32**, 2703. 541, 550

91TL2775 M. Sanchez, V. Romanenko, M.-R. Mazieres, A. Gudima and L. Markowskii; *Tetrahedron Lett.*, 1991, **32**, 2775. 660

91TL2821 J. De Brabander, K. Vanhessche and M. Vandewalle; *Tetrahedron Lett.*, 1991, **32**, 2821. 128

91TL2971 B. F. Bonini, S. Masiero, G. Mazzanti and P. Zani; *Tetrahedron Lett.*, 1991, **32**, 2971. 127

91TL4189 M. Mikolajczyk, M. Mikana, P. Graczyk, M. W. Wieczorek and G. Bujacz; *Tetrahedron Lett.*, 1991, **32**, 4189. 124

91TL4251 A. K. Ghosh, T. T. Duong and S. P. McKee; *Tetrahedron Lett.*, 1991, **32**, 4251. 417

91TL4329 T. W. K. Yung and M. P. Sammes; *Tetrahedron Lett.*, 1991, **32**, 4329. 309

91TL4371 T. Uyehara, N. Chiba, I. Susuki and Y. Yamamoto; *Tetrahedron Lett.*, 1991, **32**, 4371. 426

91TL4791 S. G. Davies and A. A. Mortlock; *Tetrahedron Lett.*, 1991, **32**, 4791. 503

91TL4823 J. Come, L. Burr and R. Chem; *Tetrahedron Lett.*, 1991, **32**, 4823. 598

91TL5963 F. Tellier and R. Sauvêtre; *Tetrahedron Lett.*, 1991, **32**, 5963. 19

91TL5965 G. Etemad-Moghadam, M. Rifqui, P. Layrolle, J. Berlan and M. Koenig; *Tetrahedron Lett.*, 1991, **32**, 5965. 264

91TL5987 Y. Tominaga, K. Ogata, S. Kohra, M. Hojo and A. Hosomi; *Tetrahedron Lett.*, 1991, **32**, 5987. 634

91TL6329 G. Solladié and V. Beil; *Tetrahedron Lett.*, 1991, **32**, 6329. 317

91TL6425 G. M. Blackburn and S. P. Langston; *Tetrahedron Lett.*, 1991, **32**, 6425. 264

91TL6801 B. F. Bonini, S. Masiero, G. Mazzanti and P. Zani; *Tetrahedron Lett.*, 1991, **32**, 6801. 127

91TL6867 T. Mizuno, I. Nishiguchi, T. Okushi and T. Hirashima; *Tetrahedron Lett.*, 1991, 6867. 494

91TL7329 P. Graczyk and M. Mikolajczyk; *Tetrahedron Lett.*, 1991, **32**, 7329. 121

91TL7425 K. Uenyama and M. Kanai; *Tetrahedron Lett.*, 1991, **32**, 7425. 21
91TL7453 W. Birberg and H. Lönn; *Tetrahedron Lett.*, 1991, **32**, 7453. 546
91TL7525 B. R. Langlois, E. Laurent and N. Roidot; *Tetrahedron Lett.*, 1991, **32**, 7525. 11
91TL7557 K. Harano, H. Kiyonaga and T. Hisano; *Tetrahedron Lett.*, 1991, 7557. 473
91ZAAC(600)145 C. O. Della Védova and A. Haas; *Z. Anorg. Allg. Chem.*, 1991, **600**, 145. 432
91ZAAC(595)225 W. Uhl and J. E. O. Schnepf; *Z. Anorg. Allg. Chem.*, 1991, **595**, 225. 195, 197
91ZAAC(605)131 P. K. Verma; *Z. Anorg. Allg. Chem.*, 1991, **605**, 131. 608
91ZN(B)339 H. Eckert, B. Forster and C. Seidel; *Z. Naturforsch., Teil B*, 1991, **46**, 339. 487
91ZN(B)978 J. Grobe, Le Van Duc and T. Grosspietsch; *Z. Naturforsch., Teil B*, 1991, **46**, 978. 242
91ZN(B)1539 B. Neumüller; *Z. Naturforsch., Teil B*, 1991, **46**, 1539. 195
91ZOB79 M. I. Povolotskii, V. V. Negrebetskii, A. V. Ruban, V. D. Romanenko, A. N. Chernega, P. V. Muzyka and L. N. Markovskii; *Zh. Obshch. Khim.*, 1991, **61**, 79 (*Chem. Abstr.*, 1991, **115**, 71745). 693
91ZOB401 L. N. Markovskii, V. D. Romanenko, A. V. Ruban, T. V. Sarina, M. I. Povolotskii and L. F. Lur'e; *Zh. Obshch. Khim.*, 1991, **61**, 401 (*Chem. Abstr.*, 1991, **115**, 92418). 655, 685
91ZOB643 M. Y. Dmitrichenko, V. G. Rozinov, V. I. Donskikh, G. V. Ratovskii, V. G. Efremov, G. V. Dolgushin, V. Y. Vitkovskii and S. V. Shilin; *Zh. Obshch. Khim.*, 1991, **61**, 643 (*Chem. Abstr.*, 1991, **115**, 136226). 620
91ZOB776 B. I. Buzykin, M. P. Sokolov and M. P. Sokolov; *Zh. Obshch. Khim.*, 1991, **61**, 776 (*Chem. Abstr.*, 1991, **115**, 114650). 660
91ZOB1202 V. D. Sheludyakov, A. B. Lebeveda, A. D. Kirilin, G. N. Turkel'taub, A. V. Kisin, I. S. Nikishina, G. G. Baukova and A. V. Lebedev; *Zh. Obshch. Khim.*, 1991, **61**, 1202 (*Chem. Abstr.*, 1992, **116**, 21096). 626
91ZOR533 A. B. Koldobskii and V. V. Lunin; *Zh. Org. Khim.*, 1991, **27**, 533 (*Chem. Abstr.*, 1991, **115**, 183141). 611

92ACR90 F. Mathey, *Acc. Chem. Res.*, 1992, **25**, 90. 692
92AF1453 E. Palaska, S. Sarac, C. Safak, H. Erdogan, K. Erol and R. S. Alpan; *Arzneim.-Forsch.*, 1992, **42**, 1453 (*Chem. Abstr.*, 1993, **118**, 147519). 561
92AG81 D. Miguel, J. A. Perez-Martinez, V. Riera and S. Garcia-Granda; *Angew. Chem.*, 1992, **104**, 81. 327
92AG(E)584 A. Maercker; *Angew. Chem., Int. Ed. Engl.*, 1992, **31**, 584. 205
92AG(E)758 T. Wettling, B. Geissler, R. Schneider, S. Barth, P. Binger and M. Regitz; *Angew. Chem., Int. Ed. Engl.*, 1992, **31**, 758. 153
92AG(E)866 K. Banert and S. Groth; *Angew. Chem., Int. Ed. Engl.*, 1992, **31**, 866. 436, 533, 534
92AG(E)879 M. Birkel, J. Schulz, U. Bergsträsser and M. Regitz; *Angew. Chem., Int. Ed. Engl.*, 1992, **31**, 879. 153
92AG(E)1055 B. Breit, U. Bergsträsser, G. Maas and M. Regitz; *Angew. Chem., Int. Ed. Engl.*, 1992, **31**, 1055. 154
92AG(E)1064 H. J. Bestmann, R. Dostalek and B. Bauroth; *Angew. Chem., Int. Ed. Engl.*, 1992, **31**, 1064. 520
92AG(E)1238 P. Willershausen, C. Kybart, N. Stamatis, W. Massa, M. Bühl, P. von R. Schleyer and A. Berndt; *Angew. Chem., Int. Ed. Engl.*, 1992, **31**, 1238. 201, 715
92AG(E)1364 W. Uhl, W. Hiller, M. Layh and W. Schwarz; *Angew. Chem. Int. Ed. Engl.*, 1992, **31**, 1364. 395
92AG(E)1384 P. Willershausen, G. Schmidt-Lukasch, C. Kybart, J. Allwohn, W. Massa, M. L. McKee, P. von R. Schleyer and A. Berndt; *Angew. Chem., Int. Ed. Engl.*, 1992, **31**, 1384. 188
92AP(325)173 W. Hanefeld and H. J. van Gösseln; *Arch. Pharm.* (*Weinheim, Ger.*), 1992, **325**, 173. 566
92AP(325)267 M. K. Das, P. K. Maiti, S. Roy, M. Mittakanti, K. W. Morse and I. H. Hall; *Arch. Pharm.* (*Weinheim, Ger.*), 1992, **325**, 267. 492
92BAU370 Yu. V. Seifman; *Bull. Acad. Sci. USSR Ser. Sci. Chim.*, 1992, **41**, 370. 26
92BMC407 S. Halazy, A. Eggenspiller, A. Ehrhard and C. Danzin; *Bioorg. Med. Chem. Lett.*, 1992, **2**, 407. 58, 58
92BMC1607 E. J. Iwanowicz, J. Lin, D. G. M. Roberts, I. M. Michel and S. M. Seiler; *Bioorg. Med. Chem. Lett.*, 1992, **2**, 1607. 317
92BSF367 J.-M. Sotiropoulos, A. Baceiredo and G. Bertrand; *Bull. Soc. Chim. Fr.*, 1992, **129**, 367. 668, 669, 671
92C78 A. Haas and C. Limberg; *Chimia*, 1992, **46**, 78. 237, 256
92C130 T. Berclaz, G. Bernardinelli, G. Michel and N. Rajalakshmi; *Chimia*, 1992, **46**, 130. 335
92CB535 D. Viets, A. Waterfeld, R. Mews, I. Weiss and H. Oberhammer; *Chem. Ber.*, 1992, **125**, 535. 253
92CB567 J. Grobe; D. Le Van, U. Althoff and G. Lange; *Chem. Ber.*, 1992, **125**, 567. 700
92CB571 A. Haas and H.-W. Praas; *Chem. Ber.*, 1992, **125**, 571. 80, 237
92CB1053 N. W. Mitzel, A. Schier, H. Beruda and H. Schmidbaur; *Chem. Ber.*, 1992, **125**, 1053. 702
92CB1333 H. H. Karsch, G. Baumgartner, S. Gamper, J. Lachmann and G. Müller; *Chem. Ber.*, 1992, **125**, 1333. 167, 371
92CB1547 W. Uhl, M. Layh, G. Becker, K. W. Klinkhammer and T. Hildenbrand; *Chem. Ber.*, 1992, **125**, 1547. 195
92CB1807 W. Maringgele, H. Knop, D. Bromm, A. Meller, S. Dielkus, R. Herbst-Irmer and G. M. Sheldrick; *Chem. Ber.*, 1992, **125**, 1807. 184
92CB2487 R. Kupfer, S. Meier and E.-U. Würthwein; *Chem. Ber.*, 1992, **125**, 2487. 730
92CB2493 R. Reiser, C. Süling and G. Schröder; *Chem. Ber.*, 1992, **125**, 2493. 469
92CC53 J. M. Paratian, S. Sibille and S. Périchon; *J. Chem. Soc., Chem. Commun.*, 1992, 53. 22
92CC807 D.-B. Su, J.-X. Duan and Q.-Y. Chen; *J. Chem. Soc., Chem. Commun.*, 1992, 807. 219, 227
92CC870 K. Ando, C. Hatano, N. Akadegawa, A. Shigihara and H. Takayama; *J. Chem. Soc., Chem. Commun.*, 1992, 870. 135

92CC1017 D. Viets, W. Heilemann, A. Waterfeld, R. Mews, S. Besser, R. Herbst-Irmer, G. M. Sheldrick and W. D. Stohrer; *J. Chem. Soc., Chem. Commun.*, 1992, 1017. 252

92CC1274 F. Castan, A. Baceiredo, F. Dahan, N. Auner and G. Bertrand; *J. Chem. Soc., Chem. Commun*, 1992, 1274. 659

92CC1410 Y. Misaki, H. Nishikawa, K. Nomura, T. Yamabe, H. Yamochi, G. Saito, T. Sato and M. Shiro; *J. Chem. Soc., Chem. Commun.*, 1992, 1410. 555

92CC1539 T. Chivers, M. Parvez, M. Peach and R. Vollmerhaus; *J. Chem. Soc., Chem. Commun.*, 1992, 1539. 93

92CCC1291 F. Adamek, M. Hajek and Z. Janousek; *Collect. Czech. Chem. Commun.*, 1992, **57**, 1291. 7

92CCC1699 A. Krutosíková, M. Dandárová and V. Konecny; *Collect. Czech. Chem. Commun.*, 1992, **57**, 1699. 560

92CCC1947 J. Burkhard and Z. Arnold; *Collect. Czech Chem. Commun.*, 1992, **57**, 1947. 635

92CJC2335 M. Quimpère, L. Ruest and P. Deslongchamps; *Can. J. Chem.*, 1992, **70**, 2335. 134

92CL827 M. Kuroboshi and T. Hiyama; *Chem. Lett.*, 1992, 827. 10

92CL867 A. Sekiguchi, T. Nakanishi, C. Kabuto and H. Sakurai; *Chem. Lett.*, 1992, 867. 193

92CL1389 K. Shimada, S. Oikawa and Y. Takikawa; *Chem. Lett.*, 1992, 1389. 498

92CM979 N. H. Dryden, J. G. Shapter, L. L. Coatsworth, P. R. North and R. J. Puddephatt; *Chem. Mater.*, 1992, **4**, 979. 247

92CPB2279 S. Harusawa, S. Tomii, C. Takehisa, H. Ohishi, R. Yoneda and T. Kurihara; *Chem. Pharm. Bull.*, 1992, **40**, 2279. 544

92CRV209 V. I. Bregadze; *Chem. Rev.*, 1992, **92**, 209. 169

92CRV225 B. Stibr; *Chem. Rev.*, 1992, **92**, 225. 406

92CRV251 R. N. Grimes; *Chem. Rev.*, 1992, **92**, 251. 406

92EUP496413 V. A. Petrov, D. D. DesMarteau and W. Navarrini (Ausimont SpA); *Eur. Pat.* 496413 (1992) (*Chem. Abstr.*, 1992, **117**, 191831). 279

92EUP473105 H. Aoyama, S. Kono, T. Yasuhara, S. Ueda and S. Koyama (Daikin Ind.); *Eur. Pat. Appl.* 473105 (1992) (*Chem. Abstr.*, 1992, **116**, 213988). 4

92EUP486948 D. Kempf, L. M. Codalovi, D. W. Norbeck, J. J. Plattner, H. L. Sham, S. J. Wittenberger and Chen Zhao (Abbott Laboratories); *Eur. Pat.* 486 948 (1992) (*Chem. Abstr.*, 1993, **118**, 192283). 417

92EUP499157 W. Navarrini and S. Fontana (Ausimont SpA); *Eur. Pat.* 499 157 (1992) (*Chem. Abstr.*, 1992, **117**, 236286). 250

92EUP499158 W. Navarrini and L. Bragante (Ausimont SpA); *Eur. Pat.* 499 158 (1992) (*Chem. Abstr.*, 1992, **117**, 234003). 250

92EUP515258 G. Drivon (Elf Atochem.); *Eur. Pat. Appl.* 515258 (1992) (*Chem. Abstr.*, 1993, **118**, 168678). 32

92GEP4023107 E. Bartmann, V. Reiffenrath, R. Hittich, B. Reiger and B. Scheuble (Merck Patent GmbH); *Ger. Pat.* 4 023 107, (*Chem. Abstr.*, 1992, **116**, 193775). 421, 422

92GEP4025227 A. Winterfeldt, G. Bartels and R. Knieps (Riedel de Haen); *Ger. Pat.* 4025227 (1992) (*Chem. Abstr.*, 1992, **116**, 173550). 32

92GEP4215648 W. Klauck, W. Maier and H. Berthauer, *Ger. Pat.* 4 215 648 (1992) (*Chem. Abstr.*, 1994, **120**, 325 310). 488

92GEP4219418 K. Haeberle, E. Wistuba, H. J. Fricke, O. Aydin, N. Kokel and U. Licht, *Ger. Pat.* 4 219 418 (1992) (*Chem. Abstr.*, 1994, **120**, 325 318). 488

92HAC115 R. M. Kamalov, G. S. Stepanov and L. F. Chertanova, A. A. Gazikasheva, M. A. Pudovik and A. N. Pudovik; *Heteroatom. Chem.*, 1992, **3**, 115. 121

92HCA907 E. Walvogel; *Helv. Chim. Acta*, 1992, **75**, 907. 546, 552

92ICA527 G. Sarwar, N. R. Patel, Y. Y. Zheng, E. O. John, R. L. Kirchmeier and J. M. Shreeve; *Inorg. Chim. Acta*, 1992, **198–200**, 527. 262

92IC329 E. O. John, R. L. Kirchmeier and J. M. Shreeve; *Inorg. Chem.*, 1992, **31**, 329. 260, 261

92IC488 Y. Y. Zheng, N. R. Patel, R. L. Kirchmeier and J. M. Shreeve; *Inorg. Chem.*, 1992, **31**, 488. 259, 643, 651, 653

92IC3493 V. D. Romanenko, A. O. Gudima, A. N. Chernega and G. Bertrand; *Inorg. Chem.*, 1992, **31**, 3493. 680, 681, 689

92IC4911 C. H. Sutton, M. W. Baize, W. J. Mills and L. J. Todd; *Inorg. Chem.*, 1992, **31**, 4911. 515

92IC4917 Y. Chen, N. R. Patel, R. L. Kirchmeier and J. M. Shreeve; *Inorg. Chem.*, 1992, **31**, 4917. 260

92IZV425 G. D. Kolomnikova, D. Y. Prikhodchenko, P. V. Petrovskii, L. F. Kasukhin and Y. G. Gololobov; *Izv. Akad. Nauk SSSR Ser. Khim.*, 1992, 425 (*Chem. Abstr.*, 1993, **118**, 124 641). 635

92JA10573 P. L. H. M. Cobben, R. J. M. Egberink, J. G. Bomer, P. Bergveld, W. Verboom and D. N. Reinhoudt; *J. Am. Chem. Soc.*, 1992, **114**, 10 573. 562

92JA1906 E. J. Corey and J. O. Link; *J. Am. Chem. Soc.*, 1992, **114**, 1906. 42

92JA4111 J.-M. Mattalia, B. Vacher, A. Samat and M. Chanon; *J. Am. Chem. Soc.*, 1992, **114**, 4111. 550

92JA6059 R. Reau, G. Veneziani and G. Bertrand; *J. Am. Chem. Soc.*, 1992, **114**, 6059. 665, 670, 675

92JA7302 M. Hogenbirk, G. Schat, O. S. Akkerman and F. Bickelhaupt; *J. Am. Chem. Soc.*, 1992, **114**, 7302. 399

92JA7909 J. Boivin, J. Camara and S. Z. Zard; *J. Am. Chem. Soc.*, 1992, **114**, 7909. 551

92JA8008 A. B. Smith, T. A. Rano, N. Chida, G. A. Sulikowski and J. L. Wood; *J. Am. Chem. Soc.*, 1992, **114**, 8008. 488

92JA9386 R. A. Moss and P. Balcerzak; *J. Am. Chem. Soc.*, 1992, **114**, 9386.

92JAP04054156 H. Sasaki, K. Komya and S. Fukuoka, *Jpn. Pat.* 04 054 156 (1992) (*Chem. Abstr.*, 1992, **116**, 237 798). 465

92JAP04108765 T. Okawa; *Jpn. Pat.* 04 108 765 (1992) (*Chem. Abstr.*, 1992, **117**, 150 547). 465

92JAP04110383 T. Teraoka *et al.* (Daikin Industries); *Jpn. Pat.* 04 110 383 (1992) (*Chem. Abstr.*, 1992, **117**, 133849). 261

92JAP04149153 J. Odera; *Jpn. Pat.* 04 149 153 (1992) (*Chem. Abstr.*, 1992, **117**, 233 700). 126

92JAP04149143 S. Kumai and O. Yokokoji (Asahi Glass); *Jpn. Pat.* 04149143 (1992) (*Chem. Abstr.*, 1992, **117**, 191510). 29

92JAP04164039 T. Kawasaki, G. Yanase, R. Takei, Y. Sato, K. Onishi and T. Tanuma (Asahi Glass); *Jpn. Pat.* 04164039 (1992) (*Chem. Abstr.*, 1992, **117**, 211995). 4

92JAP04356446 H. Watanabe, K. Murayama, K. Ida, H. Wada and Y. Kasori; *Jpn. Pat.* 04 356 446 (1992) (*Chem. Abstr.*, 1993, **119**, 8359). 465

92JAP05339226 T. Taniguchi, M. Kataoka, H. Fujino, K. Kuroda and K. Nitsuta; *Jap. Pat.* 05 339 226 (1992) (*Chem. Abstr.*, 1994, **120**, 324413). 488

92JAP05339542 M. Sugishima, K. Kasei and Y. Nakayama; *Jap. Pat.* 05 339 542 (1992) (*Chem. Abstr.*, 1994, **120**, 325910). 488

92JAP06033008 S. Yamashita; *Jpn. Pat.* 06 033 008 (1992) (*Chem. Abstr.*, 1994, **120**, 325941). 488

92JAP(K)04164085 M. Higashiura, H. Nishino and Y. Kodera (Toyobo Co. Ltd); *Jpn. Kokai* 04164085 (92164085) (1992) (*Chem. Abstr.*, 1992, **117**, 191857). 300

92JCS(D)1015 F. I. Aigbirhio, N. H. Buttrus, C. Eaborn, S. H. Gupta, P. B. Hitchcock, J. D. Smith and A. C. Sullivan; *J. Chem. Soc., Dalton Trans.*, 1992, 1015. 174, 176, 381

92JCS(D)2307 A. Galindo, E. Guitiérez-Puebla, A. Monge, M. A. Muñoz, A. Pastor, C. Ruiz and E. Carmona; *J. Chem. Soc., Dalton Trans.*, 1992, 2307. 327

92JCS(P1)47 F. Saczewski and M. Gdaniec; *J. Chem. Soc., Perkin Trans. 1*, 1992, 47. 547

92JCS(P1)3179 L. Crombie and S. R. M. Jarrett; *J. Chem. Soc., Perkin Trans. 1*, 1992, 3179. 641

92JCS(P2)835 J. Vepsäläinen, H. Nupponen, E. Pohjala, M. Ahlgren and P. Vainiotalo; *J. Chem. Soc., Perkin Trans. 2*, 1992, 835. 263

92JC241 K. R. Kim, J.-H. Kim, C.-H. Oh and T. J. Mabry; *J. Chromatogr.*, 1992, **605**, 241. 480

92JFC(57)293 Y. Y. Zheng, E. O. John, R. L. Kirchmeier and J. M. Shreeve; *J. Fluorine Chem.*, 1992, **57**, 293. 259, 614, 624

92JFC(58)143 W. Navarrini, V. Tortelli and A. Zedda; *J. Fluorine Chem.*, 1992, **58**, 143. 250

92JFC(58)290 W. Navarrini and S. Fontana; *J. Fluorine Chem.*, 1992, **58**, 290. 250

92JFC(59)91 S. Munavalli, D. I. Rossman, D. K. Rohrbaugh, C. P. Ferguson and L. J. Szefraniec; *J. Fluorine Chem.*, 1992, **59**, 91. 528

92JFC(59)417 K. Morimoto, K. Makino and G. Sakata; *J. Fluorine Chem.*, 1992, **59**, 417. 229

92JHC383 K. Tabakovic and I. Tabakovic; *J. Heterocycl. Chem.*, 1992, **29**, 383. 562

92JHC1551 G. B. Okide; *J. Heterocycl. Chem.*, 1992, **29**, 1551. 621

92JMC3641 M. J. L. Lee, J. A. Elberling and H. T. Nagasawa; *J. Med. Chem.*, 1992, **35**, 3641. 447

92JMC3731 P. E. Finke, S. K. Shah, B. M. Asher *et al.*; *J. Med. Chem.*, 1992, **35**, 3731. 485

92JMOC51 M. Kotora and M. Hajek; *J. Mol. Catal.*, 1992, **77**, 51. 7

92JOC178 H. B. Wood, H.-P. Buser and B. Ganem; *J. Org. Chem.*, 1992, **57**, 178. 118, 666

92JOC1053 Y. Sanemitsu, S. Kawamura and Y. Tanabe; *J. Org. Chem.*, 1992, **57**, 1053. 478

92JOC1696 M. R. Bryce, M. A. Coffin and W. Clegg; *J. Org. Chem.*, 1992, **57**, 1696. 85, 729

92JOC2497 M. S. Bernatowicz, Y. Wu and G. R. Matsueda; *J. Org. Chem.*, 1992, **57**, 2497. 641

92JOC2523 S. F. Martin, D. Daniel, R. J. Cherney and S. Liras; *J. Org. Chem.*, 1992, **57**, 2523. 472

92JOC2755 R. Wilder and S. Mobashery; *J. Org. Chem.*, 1992, **57**, 2755. 486

92JOC3026 K. Baum, N. V. Nguyen, R. Gilardi, J. L. Flippen-Anderson and C. George; *J. Org. Chem.*, 1992, **57**, 3026. 150

92JOC4352 R. A. Volkmann, P. R. Kelbaugh, D. M. Nason and V. J. Jasys; *J. Org. Chem.*, 1992, **57**, 4352. 553

92JOC4507 A. Bulpin and S. Masson; *J. Org. Chem.*, 1992, **57**, 4507. 121, 316

92JOC4749 J. L. Adcock, H. Luo and S. S. Zuberi; *J. Org. Chem.*, 1992, **57**, 4749. 230

92JOC5136 J. A. King, Jr. and G. L. Bryant, Jr.; *J. Org. Chem.*, 1992, **57**, 5136. 443

92JOC6803 M. D. Bachi, E. Bosch, D. Denenmark and D. Girsh; *J. Org. Chem.*, 1992, **57**, 6803. 543, 550, 557

92JOC7359 M. Komatsu, Y. Kajihara, M. Kobayashi, S. Itoh and Y. Ohshiro; *J. Org. Chem.*, 1992, **57**, 7359. 489

92JOM(427)9 A. G. Avent, S. G. Bott, J. A. Ladd, P. D. Lickiss and A. Pidcock; *J. Organomet. Chem.*, 1992, **427**, 9. 381

92JOM(427)213 H. Poleschner, R. Radeglia and J. Fuchs; *J. Organomet. Chem.*, 1992, **427**, 213. 590, 592, 593

92JOM(429)1 H. Bock, M. Kremer and H. Schmidbaur; *J. Organomet. Chem.*, 1992, **429**, 1. 379

92JOM(433)C11 R. Bartsch, P. B. Hitchcock and J. F. Nixon; *J. Organomet. Chem.*, 1992, **433**, C11. 154

92JOM(433)63 R. Eujen and U. Thurmann; *J. Organomet. Chem.*, 1992, **433**, 63. 245

92JOM(434)159 R. Eujen, N. Jahn and U. Thurmann; *J. Organomet. Chem.*, 1992, **434**, 159. 245

92JOM(437)41 S. S. Al-Juaid, C. Eaborn, A. Habtemariam, P. B. Hitchcock and J. D. Smith, *J. Organomet. Chem.*, 1992, **437**, 41. 392, 394

92JOM(437)323 J. Borgdorff, E. J. Ditzel, N. W. Duffy, B. H. Robinson and J. Simpson; *J. Organomet. Chem.*, 1992, **437**, 323. 402

92JOM(438)57 R. Eujen and A. Patorra; *J. Organomet. Chem.*, 1992, **438**, 57. 245

92JST(268)389 H. G. Ang, W. L. Kwik, Y. W. Lee, S. Liedle and H. Oberhammer; *J. Mol. Struct.*, 1992, **268**, 389. 243

92KGS1117 O. J. Neiland, B. J. Adamsone and I. K. Raiskuma; *Khim. Geterotsikl. Soedin.*, 1992, 1117 (*Chem. Abstr.*, 1993, **119**, 95460). 555

92MC52 O. A. Luk'yanov, T. B. Salamonov, Y. T. Struchkov, Yu. N. Burtsev and V. S. Kuz'min; *Mendeleev Chem. J.* (*Engl. Transl.*), 1992, 52. 284

B-92MI 601-01 G. K. S. Prakash; in "Synthetic Fluorine Chemistry," ed. G. A. Olah, R. D. Chambers and G. K. S. Prakash, Wiley-Interscience, New York, 1992, chap. 10. 13, 22

92MI 601-02 J.-P. Bosmans; *Janssen Chim. Acta.*, 1992, **10(1)**, 22. 14

B-92MI 601-03 G. K. S. Prakash; in "Synthetic Fluorine Chemistry," ed. G. A. Olah, R. D. Chambers and G. K. S. Prakash, Wiley-Interscience, New York, 1992, p. 240. 14

92MI 604-01 J. M. Fang and M. Y. Chen; *J. Chin. Chem. Soc.*, 1992, **39**, 431. 134

B-92MI 606-01 A. Sekiguchi and H. Sakurai; in "The Chemistry of Inorganic Ring Systems," ed. R. Steudel, Elsevier, Amsterdam, 1992, vol. 14, p. 101. 172

92MI 606-02 J. R. Baran, Jr.; Dissertation, University of Texas, Austin, 1992 (*Chem. Abstr.*, 1994, **120**, 270535). 204

B-92MI 607-01 M. Hudlický; "Chemistry of Organic Fluorine Compounds," 2nd edn., Ellis Horwood, London, 1992. 212, 229

B-92MI 607-02 G. A. Olah, R. D. Chambers and G. K. Surya Prakash (eds.); "Synthetic Fluorine Chemistry," Wiley-Interscience, New York, 1992. 212

92MI 607-03 W. Fu, X. Li and X. Jiang; *Chin J. Chem.*, 1992, **10**, 189 (*Chem. Abstr.*, 1992, **117**, 233392). 221, 222, 226

B-92MI 607-04 M. Hudlický; "Chemistry of Organic Fluorine Compounds," 2nd edn., Ellis Horwood, London, 1992, p. 100. 225

92MI 608-01 M. Salvi-Narkhede, H. B. Wang, J. I. Adcock and W. A. Van Hook; *J. Chem. Thermodyn.*, 1992, **24**, 1065 (*Chem. Abstr.*, 1993, **118**, 176299). 251

92MI 614-01 C. Davies and M. Kilner; *J. Catalysis*, 1992, **136**, 403. 448

B-92MI 615-01 "Ullmann's Encyclopedia of Industrial Chemistry," eds. B. Elvers, S. Hawkins and G. Schulz, Verlag Chemie, Weinheim, 1992, vol. A21 p. 210. 471

92MI 615-02 P. Metha, T. Konakahara and B. Gold; *J. Labelled Compd. Radiopharm.*, 1992, **31**, 925.

92MI 616-01 M. C. Miller, III, S. D. Wyrick, I. H. Hall, A. Sood and B. F. Spielvogel; *J. Labelled Compd. Radiopharm.*, 1992, **31**, 595. 514

92MI 617-01 X. Zhu and S. Shan; *Huaxue Shiji*, 1992, **14**, 113 (*Chem. Abstr.*, 1992, **117**, 111115). 538

92MI 617-02 G. C. Chiang; *Res. Discl.*, 1992, **342**, 786 (*Chem. Abstr.*, 1992, **118**, 6611). 538

92MI 617-03 D. Koscik, P. Kutschy, M. Dzurilla and M. Mazanek; *Chem. Pap.*, 1992, **46**, 405 (*Chem. Abstr.*, 1993, **119**, 72559). 566

92MI 617-04 F. D. Tropper, F. O. Andersson, S. Cao and R. Roy; *J. Carbohydr. Chem.*, 1992, **11**, 741. 546

92MI 617-05 P. Smid, W. P. A. Jörning, A. M. G. van Dauren, G. J. P. H. Boons, G. A. van der Marel and J. H. van Boom; *J. Carbohydr. Chem.*, 1992, **11**, 849. 549

92MI 621-01 V. I. Kharchenko, L. N. Alekseiko, V. V. Pen'kovskii and S. A. Pavlenko; *Zh. Fiz. Khim*, 1992, **66**, 487 (*Chem. Abstr.*, 1992, **117**, 151062). 671, 672

92MI 621-02 T. Iwakawa, H. Tamura, M. Masuko, A. Murabayashi and Y. Hayase; *J. Pestic. Sci.* (*Int. Ed.*), 1992, **17**, 131 (*Chem. Abstr.*, 1992, **117**, 250745). 652

92MI 622-01 V. D. Romanenko, L. N. Markovskii and A. V. Ruban; "Chemistry of Low-Coordinated Pentavalent Phosphorus Compounds," Naukova Dumka, Kiev, 1992. 692, 693

92MM3829 T. Takata, T. Ariga and T. Endo; *Macromolecules*, 1992, **25**, 3829. 298, 300

92OM26 J. T. Mague and C. L. Lloyd; *Organometallics*, 1992, **11**, 26. 151

92OM145 M. Ates, H. J. Breunig, K. Ebert, S. Gülec, R. Kaller and M. Dräger; *Organometallics*, 1992, **11**, 145. 169

92OM171 F. Ozawa, T. Son, S. Ebina, K. Osakada and A. Yamamoto; *Organometallics*, 1992, **11**, 171. 470

92OM2748 G. Anselme, H. Ranaivonjatovo, J. Escudié, C. Couret and J. Satgé; *Organometallics*, 1992, **11**, 2748. 724

92OM2938 C. L. Smith, L. M. James and K. L. Sibley; *Organometallics*, 1992, **11**, 2938. 174, 175, 176, 289, 380, 391, 393, 395

92OM2947 D. C. Gordon, R. U. Kirss and D. W. Brown; *Organometallics*, 1992, **11**, 2947. 237

92OM3464 D. Seyferth and J. L. Robison; *Organometallics*, 1992, **11**, 3464. 190

92OPP200 K. Harano, H. Nakagawa, H. Kiyonaga and T. Hisano; *Org. Prep. Proced. Int.*, 1992, **24**, 200. 472

92PHA251 F. Oertel, D. Vogel and P. H. Richter; *Pharmazie*, 1992, **47**, 251. 556

92POL893 H. C. Raubenheimer, R. Otte, L. Linford, W. E. Van Zyl, A. Lombard and G. J. Kruger; *Polyhedron*, 1992, **11**, 893. 540

92POL2713 E. Cuyas, D. Miguel, J. A. Perez-Martinez, V. Riera and S. Garcia-Granda; *Polyhedron*, 1992, **11**, 2713. 329, 330

92PS(66)257 A. S. Ionkin, N. V. Nikolaeva, R. Z. Musin, Y. Y. Efremov and B. A. Arbuzov; *Phosphorus Sulfur*, 1992, **66**, 257. 691

92PS(69)75 C. Yuan and C. Li; *Phosphorus Sulfur*, 1992, **69**, 75. 118

92PS(70)183 J. Vepsäläinen, E. Pohjala, H. Nupponen, P. Vainiotalo and M. Ahlgren; *Phosphorus Sulfur*, 1992, **70**, 183. 263

92PS(73)81 H. Wilhelm, E. Herrmann, G. Ohms and P. G. Jones; *Phosphorus Sulfur*, 1992, **73**, 81. 581

92RCR75 L. E. Deev, T. I. Nazarenko, K. I. Pashkevich and V. G. Ponomarev; *Russ. Chem. Rev.* (*Engl. Transl.*), 1992, **61**, 75. 219

92RCR517 A. Yu Sizov, A. K. Kolomiets and A. V. Fokin; *Russian Chem. Rev.* (*Engl. Trans.*), 1992, **61**, 517. 236

92S112 F. Beaulieu and V. Snieckus; *Synthesis*, 1992, 112. 556

92S297 P. Molina, A. Lorenzo and E. Aller; *Synthesis*, 1992, **3**, 297. 504

92S380 M. Mittakanti, D. A. Feakes and K. W. Morse; *Synthesis*, 1992, 380. 491

92S787 H.-J. Bestmann, A. Bomhard, R. Dostalek, R. Pichl, R. Riemer and R. Zimmermann; *Synthesis*, 1992, **8**, 787. 702

92S919 L. D. S. Yadav and S. Sharma; *Synthesis*, 1992, 919. 562

92S1216 F. Hintze and D. Hoppe; *Synthesis*, 1992, 1216. 447

92SA(A)1179 C. O. Della Védova; *Spectrochim. Acta.*, *Part A*, 1992, **48**, 1179. 453

92SC1191 S. Nagarajan, T.-L. Ho and G. E. DuBois; *Synth. Commun.*, 1992, **22**, 1191. 606, 645

92SC1359 D. Villemin, A. Ben Alloum and F. Thibault-Starzyk; *Synth. Commun.*, 1992, **22**, 1359. 121, 325

92SC2381 A. Couture, E. Deniau and P. Grandclaudon; *Synth. Commun.*, 1992, **22**, 2381. 122

92SL843 J.-i. Yoshida, Y. Morita, M. Itoh, Y. Ishichi and S. Isoe; *Synlett*, 1992, 843. 131

92SL975 G. G. Cox, J. J. Kulagowski, C. J. Moody and E. R. H. B. Sie; *Synlett*, 1992, 975. 119

92SUL275 A. Senning; *Sulfur Letters*, 1992, **14**, 275. 530

92T189 D. J. Burton and Z.-Y. Yang; *Tetrahedron*, 1992, **48**, 189. 9, 14, 22, 235, 266

92T2501 R. Bellato, A. Gatti and P. Venturello; *Tetrahedron*, 1992, **48**, 2501. 126

92T6555 M. A. McClinton and D. A. McClinton; *Tetrahedron*, 1992, **48**, 6555. 9, 14, 228

92T6585 M. A. McClinton and D. A. McClinton; *Tetrahedron*, 1992, **48**, 6585. 11

92T6883 W. R. Bowman, D. S. Brown, C. A. Burns, B. A. Marples and N. A. Zaidi; *Tetrahedron*, 1992, **48**, 6883. 544, 557

92T7505 G. L'abbé and S. Leurs; *Tetrahedron*, 1992, **48**, 7505. 573

92T8031 R. A. Batey, J. D. Harling and W. B. Motherwell; *Tetrahedron*, 1992, **48**, 8031. 557

92T8271 A. El-Hamid Ismail, A. Hamed, I. Zeid and J. C. Jochims; *Tetrahedron*, 1992, **48**, 8271. 618, 621

92T8697 M. Mikolajczyk and P. Balczewski; *Tetrahedron*, 1992, **48**, 8697. 121

92T9433 S. Harusawa, H. Osaki, H. Fujii, R. Yoneda and T. Kurihara; *Tetrahedron Lett.*, 1992, **33**, 9433. 542

92TA1515 M. Mikolajczyk and W. H. Midura; *Tetrahedron Asymmetry*, 1992, **3**, 1515. 123, 124

92TL213 K. Nohair, I. Lachaise, J.-P. Paugam and J.-Y. Nédélec; *Tetrahedron Lett.*, 1992, **33**, 213. 7

92TL261 A. A. Freer, D. D. McNicol and P. R. Mallinson; *Tetrahedron Lett.*, 1992, **33**, 261. 528

92TL385 R. Ramage and G. Raphy; *Tetrahedron Lett.*, 1992, **33**, 385. 425

92TL683 P. Vinczer, S. Struhar, L. Novak and C. Szantay; *Tetrahedron Lett.*, 1992, **33**, 683. 695

92TL1025 S. Knapp, A. B. J. Naughton and T. G. M. Dhar; *Tetrahedron Lett.*, 1992, **33**, 1025. 629

92TL1839 S. F. Martin, D. W. Dean and A. S. Wagman; *Tetrahedron Lett.*, 1992, **33**, 1839. 58, 542

92TL2339 F. Szönyi and A. Cambon; *Tetrahedron Lett.*, 1992, **33**, 2339. 608

92TL2543 S. Harusawa, H. Osaki, S. Takemura, R. Yoneda and T. Kurihara; *Tetrahedron Lett.*, 1992, **33**, 2543. 542

92TL2629 C.-N. Hsiao and T. Kolasa; *Tetrahedron Lett.*, 1992, **33**, 2629. 446

92TL2781 A. K. Ghosh, T. T. Duong, S. P. McKee and W. J. Thompson; *Tetrahedron Lett.*, 1992, **33**, 2781. 417

92TL2981 V. D. Romanenko, M. Sanchez, T. V. Sarina, M. R. Mazieres and R. Wolf; *Tetrahedron Lett.*, 1992, **33**, 2981. 690

92TL3113 J. C. Rohloff, J. Robinson, III and J. O. Gardner; *Tetrahedron Lett.*, 1992, **33**, 3113. 611

92TL3435 E. J. Corey, J. O. Link and Y. Shao; *Tetrahedron Lett.*, 1992, **33**, 3435. 25, 26, 31

92TL3483 O. Tamura, M. Hashimoto, Y. Kobayashi, T. Katoh, K. Nakatani, M. Kamada, I. Hayakawa, T. Akiba and S. Terashima; *Tetrahedron Lett.*, 1992, **33**, 3483. 447

92TL3903 C. Palomo, J.-M. Aizpurua, M. Legido, J. P. Picard, J. Dunoguès and T. Constantieux; *Tetrahedron Lett.*, 1992, **33**, 3903. 163

92TL4173 M. Kuroboshi, K. Suzuki and T. Hiyama; *Tetrahedron Lett.*, 1992, **33**, 4173. 229, 230

92TL4963 K. C. Oh, H. Lee and K. Kim; *Tetrahedron Lett.*, 1992, **33**, 4963. 651

92TL5261 J. M. Contelles, P. Ruiz, B. Sánchez and M. L. Jimeno; *Tetrahedron Lett.*, 1992, **33**, 5261. 550, 557

92TL5375 W. S. Shin, K. Lee and D. Y. Oh; *Tetrahedron Lett.*, 1992, **33**, 5375. 123

92TL5933 M. A. Poss, E. Iwanowicz, J. A. Reid, J. Lin and Z. Gu; *Tetrahedron Lett.*, 1992, **33**, 5933. 641

92TL6155 F. Berrée, E. Marchand and G. Morel; *Tetrahedron Lett.*, 1992, **33**, 6155. 617

92ZAAC(608)69 D. Naumann, H. Butler, J. Fischer, J. Hanke, J. Mogias and B. Wilkes; *Z. Anorg. Allg. Chem.*, 1992, **608**, 69. 237

92ZAAC(609)150 C. O. Della Védova; *Z. Anorg. Allg. Chem.*, 1992, **609**, 150. 442

92ZAAC(613)67 W. Uhl, E. Schnepf and J. Wagner; *Z. Anorg. Allg. Chem.*, 1992, **613**, 67. 195

92ZAAC(617)53 D. Lentz and R. Marschall; *Z. Anorg. Allg. Chem.*, 1992, **617**, 53. 373

92ZAAC(617)136 R. Minkwitz, W. Meckstroth and H. Preut; *Z. Anorg. Allg. Chem.*, 1992, **617**, 136. 610, 727

92ZN(B)321 J. Grobe, D. Le Van, S. Martin and J. Szameitat; *Z. Naturforsch., Teil B*, 1992, **47**, 321. 265

92ZN(B)351 G. Pawelke, R. Dammel and W. Poll; *Z. Naturforsch., Teil B*, 1992, **47**, 351. 604

92ZN(B)369 W. Gombler and G. Bollmann; *Z. Naturforsch., Teil B*, 1992, **47**, 369. 252, 274, 275

92ZN(B)805 N. Auner and E. Penzenstadler; *Z. Naturforsch., Teil B*, 1992, **47**, 805. 174, 192

92ZOB306 M. Y. Dmitrichenko, V. G. Rozinov, V. I. Donskikh, G. V. Ratovskii, V. G. Eremov and G. V. Dolgushin; *Zh. Obshch. Khim.*, 1992, **62**, 306 (*Chem. Abstr.*, 1993, **118**, 169178). 620

92ZOB324 Y. I. Matveev and V. I. Gorbatenko; *Zh. Obshch. Khim.*, 1992, **62**, 324 (*Chem. Abstr.*, 1993, **118**, 39026). 620

92ZOB474 V. D. Romanenko, L. S. Kachkovskaya, A. V. Ruban and M. I. Povolotskii; *Zh. Obshch. Khim.*, 1992, **62**, 474 (*Chem. Abstr.*, 1992, **117**, 203948). 692

92ZOB948 A. P. Marchenko, G. N. Koidan, G. O. Baram, V. A. Oleinik and A. M. Pinchuk; *Zh. Obshch. Khim.*, 1992, **62**, 948 (*Chem. Abstr.*, 1993, **118**, 147650). 242, 682

92ZOR607 M. V. Vovk; *Zh. Org. Khim.*, 1992, **28**, 607 (*Chem. Abstr.*, 1992, **118**, 168748). 539

93ACS125 M. Strasser and I. Ugi; *Acta Chem. Scand.*, 1993, **47**, 125. 488

93AG429 A. Ansorge, D. J. Brauer, H. Bürger, T. Hagen and G. Pawelke; *Angew. Chem.*, 1993, **105**, 429. 243

93AG1489 H. J. Reich and R. R. Dykstra; *Angew. Chem.*, 1993, **105**, 1489. 327, 336

93AG(E)103 P. B. Hitchcock, J. A. Johnson and J. F. Nixon; *Angew. Chem., Int. Ed. Engl.*, 1993, **32**, 103. 155

93AG(E)554 K. A. Hughes, P. G. Dopico, M. Sabat and M. G. Finn; *Angew. Chem., Int. Ed. Engl.*, 1993, **32**, 554. 706

93AG(E)767 G. A. Olah; *Angew. Chem., Int. Ed. Engl.*, 1993, **32**, 767. 729

93AG(E)905 A. Russo and D. D. DesMarteau; *Angew. Chem., Int. Ed. Engl.*, 1993, **32**, 905. 250

93AG(E)985 A. Berndt; *Angew. Chem., Int. Ed. Engl.*, 1993, **32**, 985. 187, 188, 201, 712, 714, 715, 716

93AG(E)1424 E. Lindner, C. Haase, H. A. Mayer, M. Kemmler, R. Fawzi and M. Steimann; *Angew. Chem., Int. Ed. Engl.*, 1993, **32**, 1424. 154

93AG(E)1469 H. J. Reich and R. R. Dykstra; *Angew. Chem., Int. Ed. Engl.*, 1993, **32**, 1469. 135

93AOC(35)211 J. A. Morrison; *Adv. Organomet. Chem.*, 1993, **35**, 211. 244, 717

93BCJ148 N. Fukada, M. Takano, K. Nakai, S. Kuboike and Y. Takeda; *Bull. Chem. Soc. Jpn.*, 1993, **66**, 148. 636

93BSF599 Y. Sasson, F. Kitson and O. W. Webster; *Bull. Soc. Chem. Fr.*, 1993, **130**, 599. 221, 222, 226

93CAR(242)281 I. I. Cubero, M. T. P. López-Espinosa, A. C. Richardson and K. H. Aamlid; *Carbohydr. Res.*, 1993, **242**, 281. 550

93CAR(246)119 T. Ekberg and G. Magnusson; *Carbohydr. Res.*, 1993, **246**, 119. 550

93CB537 M. Frasch and W. Sundermeyer; *Chem. Ber.*, 1993, **126**, 537. 128, 132

93CB1355 M. Ostrowski, J. Jeske, P. G. Jones and W.-W. duMont; *Chem. Ber.*, 1993, **126**, 1355. 349

93CB2227 H. Bock, J. Meuret and H. Schödel; *Chem. Ber.*, 1993, **22**, 2227. 174

93CB2247 G. Ossig, A. Meller, S. Freitag, R. Herbst-Irmer and G. M. Sheldrick; *Chem. Ber.*, 1993, **126**, 2247. 196

93CB2637 W. Uhl, H. H. Karsch, U. Schütz and A. Vester; *Chem. Ber.*, 1993, **126**, 2637. 197

93CC311 R. Bartsch, P. B. Hitchcock and J. F. Nixon; *J. Chem. Soc., Chem. Commun.*, 1993, 311. 154

93CC673 U. Heim, H. Pritzkow, H. Schönberg and H. Grützmacher; *J. Chem. Soc., Chem. Commun.*, 1993, 673. 691

93CC918 Q.-Y. Chen and J.-X. Duan; *J. Chem. Soc., Chem. Commun.*, 1993, 918. 232, 234

93CC1354 G. Alcaraz, R. Reed, A. Baceiredo and G. Bertrand; *J. Chem. Soc., Chem. Commun.*, 1993, 1354. 691, 704

93CC1585 M. Driess and H. Pritzkow; *J. Chem. Soc., Chem. Commun.*, 1993, 1585. 691

93CL13 S. Nakano, A. Miyashita and H. Nohira; *Chem. Lett.*, 1993, 13. 115

93CL267 A. Sekiguchi, M. Ichinohe, T. Nakanishi and H. Sakurai; *Chem. Lett.*, 1993, 267. 193

93CL1401 T. Wakamiya, K. Saruta, S. Kusumoto, K. Nakajima, K. Yoshizawa-Kumagaye, S. Imajoh-Ohmi and S. Kanegasaki; *Chem. Lett.*, 1993, 1401. 487

93CM1321 M. Danek, S. Patnaik, K. F. Jensen, D. C. Gordon, D. W. Brown and R. U. Kirss; *Chem. Mater.*, 1993, **5**, 1321. 237

93EUP528498 T. Koyama, M. Tonosaki, N. Yamada and K. Mori; *Eur. Pat.* 528 498 (1993) (*Chem. Abstr.*, 1993, **119**, 138 751). 465

93EUP534545 F. Rivetti and U. Romano; *Eur. Pat.* 534 545 (1993) (*Chem. Abstr.*, 1993, **119**, 138 752). 465

93EUP543234 F. J. Mais and H. J. Buysch; *Eur. Pat.* 543 234 (1993) (*Chem. Abstr.*, 1993, **119**, 141 608). 467

93EUP547699 K. Chapman and M. Maccoss; *Eur. Pat.* 547 699 (1993) (*Chem. Abstr.*, 1994, **120**, 77 638). 487

93EUP558996 E. Wolters, H. Landscheidt, A. Klausener and L. Puppe; *Eur. Pat.* 558 996 (1993) (*Chem. Abstr.*, 1993, **119**, 225 595). 465

93EUP559001 E. Wolters, H. Landscheidt, A. Klausener and L. Puppe; *Eur. Pat.* 559 001 (1993) (*Chem. Abstr.*, 1994, **120**, 8213). 465

93EUP559212 K. Nishihira, S. Tanaka, T. Kawashita, Y. Nishida, T. Matsuzaki and K. Abe; *Eur. Pat.* 559 212 (1993) (*Chem. Abstr.*, 1993, **119**, 203 009). 465

93EUP560730 G. Kottirsch and R. Metternich; *Eur. Pat.* 560 730 (1993) (*Chem. Abstr.*, 1994, **120**, 135 147). 487

93EUP561363 S. Kuze, R. Okumura and Y. Suwabe; *Eur. Pat.* 561 363 (1993) (*Chem. Abstr.*, 1994, **120**, 135 446). 471

93GEP4134688 H. Landscheidt, E. Wolfers, A. Klausener, H. U. Blank and U. Birkenstock; *Ger. Pat.* 4 134 688 (1993) (*Chem. Abstr.*, 1993, **119**, 116 821). 465

93GEP4140446 U. Hess and M. Schulze; *Ger. Pat.* 4 140 446 (1993) (*Chem. Abstr.*, 1993, **119**, 180 384). 477

93HCA995 M. Hürzeler, B. Bernet and A. Vasella; *Helv. Chim. Acta*, 1993, **76**, 995. 549

93HCA2407 R. Neidlein, P. Greulich and W. Kramer; *Helv. Chim. Acta*, 1993, **76**, 2407. 118

93H(35)77 K. Hartke and K. Timpe; *Heterocycles*, 1993, **35**, 77. 546, 554, 555, 567

93H(36)473 G. Fantin, M. Fogagnolo, A. Medici and P. Pedrini; *Heterocycles*, 1993, **36**, 473. 637

93H(36)1521 W. Zieliński and M. Mazik; *Heterocycles*, 1993, **36**, 1521. 621

93IC2308 L. H. Gade, C. Becker and J. W. Lauher; *Inorg. Chem.*, 1993, **32**, 2308. 174

93IC4802 N. R. Patel, R. L. Kirchmeier and J. M. Shreeve; *Inorg. Chem.*, 1993, **32**, 4802. 259

93IC5126 P. J. Bonasia, D. E. Gindelberger and J. Arnold; *Inorg. Chem.*, 1993, **32**, 5126. 349, 355, 357, 358

93IC5569 A. Galindo, E. Guitiérez-Puebla, A. Monge, A. Pastor, A. Pizzano, C. Ruiz, L. Sanchez and E. Carmona; *Inorg. Chem.*, 1993, **32**, 5569. 327, 330

93IC5883 M. W. Kim, D. S. Uh, S. Kim and Y. Do; *Inorg. Chem.*, 1993, **32**, 5883. 340

93IZV2513 V. V. Avdonin, E. Kirpicheva, Yu. I. Rubtsov, M. A. Fadeev, G. V. Ovechko and L. T. Eremenko; *Izv. Akad. Nauk SSSR Ser. Khim.*, 1993, 2513. 250

93JA545 V. Christou, S. P. Wuller and J. Arnold; *J. Am. Chem. Soc.*, 1993, **115**, 545. 357, 358

93JA2156 T. Umemoto and S. Ishihara; *J. Am. Chem. Soc.*, 1993, **115**, 2156. 13, 237

93JA6777 P. J. Bonasia, V. Christou and J. Arnold; *J. Am. Chem. Soc.*, 1993, **115**, 6777. 355, 357, 358

93JA8444 J. Levillain, S. Masson, A. Hudson and A. Alberti; *J. Am. Chem. Soc.*, 1993, **115**, 8444. 734

93JA8898 V. Dupont, A. Lecoq, J. P. Mangeot, A. Aubry, G. Boussard and M. Marraud; *J. Am. Chem. Soc.*, 1993, **115**, 8898. 488

93JAP05058990 T. Kagawa, S. Kuranashi and K. Tsuzuki; *Jpn. Pat.*, 05 058 990 (1993) (*Chem. Abstr.*, 1993, **119**, 95 126). 533

93JAP05255197 H. Watanabe and K. Murayama; *Jpn. Pat.* 05 255 197 (1993) (*Chem. Abstr.*, 1994, **120**, 163 453). 465

93JAP05255201 T. Arai, H. Kaneko and K. Aoki; *Jpn. Pat.* 05 255 201 (1993) (*Chem. Abstr.*, 1994, **120**, 76 895). 465

93JAP05310644 T. Arai, H. Kaneko and M. Yamada; *Jpn. Pat.* 05 310 644 (1993) (*Chem. Abstr.*, 1994, **120**, 191 140). 465

93JAP(K)05062690 A. Saito and K. Kuryama (Yuasa Koohoreeshon Kk); *Jpn. Kokai* 05062690 (1993) (*Chem. Abstr.*, 1993, **119**, 12 073). 317

93JCS(D)663 H. G. Ang, W. L. Kwik, Y. W. Lee and A. L. Rheingold; *J. Chem. Soc., Dalton Trans.*, 1993, 663. 243

93JCS(D)1031 E. A. V. Ebsworth, N. Robertson and L. J. Yellowlees; *J. Chem. Soc., Dalton Trans.* 1993, 1031. 457

93JCS(D)2547 R. Boese, A. Haas and C. Limberg; *J. Chem. Soc., Dalton Trans.*, 1993, 2547. 256

93JCS(D)2653 B. S. Jolly, M. F. Lappert, L. M. Engelhardt, A. H. White and C. L. Raston; *J. Chem. Soc., Dalton Trans.*, 1993, 2653. 196

93JCS(D)2891 S. P. Mallela, J. Myrczek, I. Bernal and R. A. Geanangel; *J. Chem. Soc., Dalton Trans.*, 1993, 2891. 396

93JCS(D)3259 A. G. Avent, C. Eaborn, P. B. Hitchcock, G. A. Lawless, P. D. Lickiss, M. Mallien, J. D. Smith, A. D. Webb and B. Wrackmeyer; *J. Chem. Soc., Dalton Trans.*, 1993, 3259. 390

93JCS(P1)505 V. A. Petrov and D. D. DesMarteau; *J. Chem. Soc., Perkin Trans. 1*, 1993, 505. 614

93JCS(P2)59 C. Eaborn, P. D. Lickiss, S. T. Najim and W. A. Stańczyk; *J. Chem. Soc., Perkin Trans. 2*, 1993, 59. 381

93JCS(P2)391 C. Eaborn, P. D. Lickiss and S. T. Najim; *J. Chem. Soc., Perkin Trans. 2*, 1993, 391. 381

93JFC(60)85 S. Munavalli, D. I. Rossman, D. K. Rohrbaugh, C. P. Ferguson and H. D. Banks; *J. Fluorine Chem.*, 1993, **60**, 85. 528

93JFC(60)175 S. Zhu and A. Li; *J. Fluorine Chem.*, 1993, **60**, 175 (*Chem. Abstr.*, 1993, **119**, 48911). 699

93JFC(60)233 W. Dmowski and J. Porwisiak; *J. Fluorine Chem.*, 1993, **60**, 233. 10

93JFC(61)1 F. J. Weigert; *J. Fluorine Chem.*, 1993, **61**, 1. 12

93JFC(61)133 M. Van Der Puy; *J. Fluorine Chem.*, 1993, **61**, 133. 7

93JMC1090 M. D. Mullican, M. W. Wilson, D. T. Connor, C. R. Kostlan, D. J. Schrier and R. D. Dyer; *J. Med. Chem.*, 1993, **36**, 1090. 558, 564

93JMC1802 D. H. Boschelli, D. T. Connor, D. A. Bornemeier, R. D. Dyer, J. A. Kennedy, P. J. Kuipers, G. C. Okonkwo, D. J. Schrier and C. D. Wright; *J. Med. Chem.*, 1993, **36**, 1802. 558, 564

93JOC1748 D. Fabbri, G. Delogu and O. De Lucchi; *J. Org. Chem.*, 1993, **58**, 1748. 556

93JOC2894 J. Marco-Contelles, P. Ruiz-Fernández and B. Sánches; *J. Org. Chem.*, 1993, **58**, 2894. 473, 544, 550

93JOC3703 J. J.-W. Duan and A. B. Smith, III; *J. Org. Chem.*, 1993, **58**, 3703. 469

93JOC3798 A. G. Chaudhary, J. M. Rimoldi and D. G. I. Kingston; *J. Org. Chem.*, 1993, **58**, 3798. 550

93JOC4105 K. K. Laali, M. Regitz, M. Birkel, P. J. Stang and C. M. Crittell; *J. Org. Chem.*, 1993, **58**, 4105. 153

93JOC4159 C. E. Burgos-Lepley, S. A. Mizsak, R. A. Nugent and R. A. Johnson; *J. Org. Chem.*, 1993, **58**, 4159. 118

93JOC4754 V. A. Petrov and D. D. DesMarteau; *J. Org. Chem.*, 1993, **58**, 4754. 256

93JOC6930 S. J. Goede, B. L. M. van Baar and F. Bickelhaupt; *J. Org. Chem.*, 1993, **58**, 6930. 610

93JOM(443)C39 W.-P. Leung, H.-K. Lee, Z.-Y. Zhou and T. C. W. Mak; *J. Organomet. Chem.*, 1993. **443**, C39. 195

93JOM(446)25 A. Ansorge, D. J. Brauer, H. Bürger, B. Krumm and G. Pawelke; *J. Organomet. Chem.*, 1993, **446**, 25. 243

93JOM(448)219 H. G. Ang, Y. M. Cai and W. L. Kwik; *J. Organomet. Chem.*, 1993, **448**, 219. 242

93JOM(451)45 C. Eaborn and M. N. Romanelli; *J. Organomet. Chem.*, 1993, **451**, 45. 381

93JOM(451)243 P. Giannoccaro, N. Ravasio and M. Aresta; *J. Organomet. Chem.*, 1993, **451**, 243. 428

93JOM(453)299 T. G. Appleton, J. R. Hall, C. H. L. Kennard, M. T. Mathieson, D. W. Neale, G. Smith and T. C. W. Mak; *J. Organomet. Chem.*, 1993, **453**, 299. 247

93JOM(453)307 T. G. Appleton, J. R. Hall, M. T. Mathieson and D. W. Neale; *J. Organomet. Chem.*, 1993, **453**, 307. 247

93JOM(456)19 H. Bürger, T. Hagen and G. Pawelke; *J. Organomet. Chem.*, 1993, **456**, 19. 243

93JOM(459)55 U. Schubert, P. Wülfert and S. Mock; *J. Organomet. Chem.*, 1993, **459**, 55. 720, 723

93JOM(462)31 H. Bock, J. Meuret and K. Ruppert; *J. Organomet. Chem.*, 1993, **462**, 31. 174

93JOM(462)45 S. S. Al-Juaid, C. Eaborn, A. Habtemariam, P. B. Hitchcock, J. D. Smith, K. Tavakkoli and A. D. Webb; *J. Organomet. Chem.*, 1993, **462**, 45. 173, 394

93JPO319 V. A. Mikhailov, D. S. Yufit, E. Y. Balabanov, Y. T. Struchkov, A. A. Popov, V. A. Savelova, I. V. Kolesnikova, T. D. Petrova and V. E. Platonov; *J. Phys. Org. Chem.*, 1993, **6**, 319. 619

B-93MI 607-01 Aldrich-Chemie; "Katalog, Handbuch, Feinchemikalien," Aldrich-Chemie, Steinheim, Germany, 1993. 213

B-93MI 607-02 K. Weissermel and H.-J. Arpe; "Industrial Organic Chemistry," VCH, Weinheim, 1993. 215, 220, 224

B-93MI 607-03 "Kirk-Othmer Encyclopedia of Chemical Technology," 4th edn., Wiley, New York, 1993, **5**, 57. 214

B-93MI 607-04 "Kirk-Othmer Encyclopedia of Chemical Technology," 4th edn., Wiley, New York, 1993, **5**, 1026. 216

93MI 608-01 S. Devotta, S. Gopichand and V. R. Pendyala; *Int. J. Refrig.*, 1993, **16**, 84 (*Chem. Abstr.*, 1993, **118**, 236 676). 251

B-93MI 614-01 M. L. Richardson (ed.); "The Dictionary of Substances and their Effects," Royal Society of Chemistry, Cambridge, 1993, vol. 3, p. 615. 411, 442

B-93MI 614-02 "Phosgene Toxicity Monograph," Institution of Chemical Engeineers, 1993, Rugby, England. 411

93MI 615-01 A. Loffet and H. X. Zhang; *Int. J. Peptide Protein Res.*, 1993, **42**, 346. 480

93MI 615-02 J. H. Boal, R. P. Iyer and W. Egan; *Nucleosides Nucleotides*, 1993, **12**, 1075. 489

93MI 615-03 M. Jasko, A. Shipitsin, E. Shirokova, A. Krayevsky, B. Polsky, P. Baron, C. MacLow, M. Ostrander and B. O'Hara; *Nucleosides Nucleotides*, 1993, **12**, 879. 489

B-93MI 616-01 P. R. Raithby; "Transition Metal Cluster Carbonyls: Structure, Synthesis, Reactivity," Ellis Horwood, Chichester, 1993. 522

B-93MI 621-01 J. V. Greenhill and P. Lue; in "Progress in Medicinal Chemistry," ed. G. P. Ellis and D. K. Luscombe, Elsevier Science, Amsterdam, 1993, vol. 30, p. 203. 641

93MI 601-01 L. E. Manzer and V. N. M. Rao; *Adv. Catalysis*, 1993, **39**, 329. 8

93MIP9305026 G. B. Dreyer, J. G. Gleason, T. D. Meek and S. K. Thompson; *W. Pat.* 9 305 026 (1993) (*Chem. Abstr.*, 1994, **120**, 107 748). 487

93OM1394 D. Miguel, J. A. Perez-Martinez, V. Riera and S. Garcia-Granda; *Organometallics*, 1993, **12**, 1394. 328

93OM2401 M. Rahim, C. White, A. L. Rheingold and K. J. Ahmed; *Organometallics*, 1993, **12**, 2401. 517

93OM2757 H. H. Karsch, G. Grauvogl, M. Kawecki and P. Bissinger; *Organometallics*, 1993, **12**, 2757. 371

93OM2806 Th. R. Hoye, K. Chen and J. R. Vyvyan; *Organometallics*, 1993, **12**, 2806. 719

93OM2888 D. Miguel, J. A. Perez-Martinez, V. Riera and S. Garcia-Granda; *Organometallics*, 1993, **12**, 2888. 329, 330

93OM2994 C. Baldoli, L. Lattuada, E. Licandro, S. Maiorana and A. Papagni; *Organometallics*, 1993, **12**, 2994. 723

93OM3405 A. J. Arduengo, III, H. V. R. Dias, J. C. Calabrese and F. Davidson; *Organometallics*, 1993, **12**, 3405. 722, 733

93OM4265 H. Jun and R. J. Angelici; *Organometallics*, 1993, **12**, 4265. 690

93OM4267 E. D. Jemmis, G. Subramanian and B. V. Prasad; *Organometallics*, 1993, **12**, 4267. 322

93OM4930 H. Bürger and P. Moritz; *Organometallics*, 1993, **12**, 4930. 243

93OPP83 A. R. Katritzky, M. K. Bernard, Q. H. Long, L. Xie, N. Malhotra and M. Beltzer; *Org. Prep. Proced. Int.*, 1993, **25**, 83. 558

93POL13 M. Di Vaira, D. Rovai and P. Stoppioni; *Polyhedron*, 1993, **12**, 13. 552

93PS(74)279 M. R. Bryce, A. J. Moore, M. A. Coffin, G. J. Marshallsay, G. Cooke, P. J. Skabara, A. S. Batsanov, J. A. K. Howard and W. Clegg; *Phosphorus Sulfur*, 1993, **74**, 279. 698

93PS(76)21 G. Grützmacher, U. Heim, H. Schönberg and H. Pritzkow; *Phosphorus Sulfur*, 1993, **76**, 21. 691, 733

93PS(76)265 J. Grobe, D. Le Van, M. Hegemann, B. Lüth and J. Nientiedt; *Phosphorus Sulfur*, 1993, **76**, 265. 684

93PS(83)65 K. Diemert, T. Hahn, W. Kuchen and P. Tommes; *Phosphorus Sulfur*, 1993, **83**, 65. 489

93RHA232 I. I. Maletina, A. A. Mironova, V. V. Orda and L. M. Yagupolskii; *Rev. Heteroatom. Chem.*, 1993, **8**, 232. 14

93S59 B. I. Buzykin and Z. A. Bredikhina; *Synthesis*, 1993, **1**, 59. 660

93S103 T. Konakahara, T. Ozaki, K. Sato and B. Gold; *Synthesis*, 1993, 103. 417, 426

93S483 S. Gronowitz, A.-B. Hörnfeldt and M. Temciuc; *Synthesis*, 1993, 483. 566

93S1109 N. Greeves and J. S. Torode; *Synthesis*, 1993, 1109. 461

93SC657 M. S. Bernatowicz and G. R. Matsueda; *Synth. Commun.*, 1993, **23**, 657. 641

93SC2839 E. Falb, A. Nudelman and A. Hassner; *Synth. Commun.*, 1993, **23**, 2839. 486

93T3035 W. H. Watson, E. E. Eduok, R. P. Kashyap and M. Krawiec; *Tetrahedron*, 1993, **49**, 3035. 317

93T3227 V. I. Gorbatenko; *Tetrahedron*, 1993, **49**, 3227. 451, 606

93T8487 K. Matsumoto, Y. Aoki, K. Oshima and K. Utimoto; *Tetrahedron*, 1993, **49**, 8487. 60, 62

93T10725 M. Pozo and V. Gotor; *Tetrahedron*, 1993, **49**, 10 725. 461

93TH61301 S. M. Whittaker, PhD. Thesis, University of Salford, 1993. 380, 381

93TL149 G. M. Blackburn and M.-J. Guo; *Tetrahedron Lett.*, 1993, **34**, 149. 242

93TL239 J. A. Marx and K. Turnbull; *Tetrahedron Lett.*, 1993, **34**, 239. 453

93TL395 R. M. Burk and M. B. Roof; *Tetrahedron. Lett.*, 1993, 395. 467

93TL1311 K. Uneyama, Y. Tokunaga and K. Maeda; *Tetrahedron Lett.*, 1993, **34**, 1311. 256

93TL1859 R. Rathore, Z. Lin and J. K. Kochi; *Tetrahedron Lett.*, 1993, **34**, 1859. 369

93TL2733 D. H. R. Barton, S. I. Parekh and C.-L. Tse; *Tetrahedron Lett.*, 1993, **34**, 2733. 550

93TL2973 S. B. Nielsen and A. Senning; *Tetrahedron Lett.*, 1993, **34**, 2973. 232

93TL3389 M. S. Bernatowicz, Y. Wu and G. R. Matsueda; *Tetrahedron Lett.*, 1993, **34**, 3389. 641

93TL3745 K. Nishiyama, M. Oba, M. Oshimi, T. Sugawara and R. Ueno; *Tetrahedron Lett.*, 1993, **34**, 3745. 558

93TL4551 J. Vepsäläinen, H. Nupponen and E. Pohjala; *Tetrahedron Lett.*, 1993, **34**, 4551. 263

93TL6447 K. Ramadas, N. Srinivasan and N. Janarthanan; *Tetrahedron Lett.*, 1993, **34**, 6447. 580

93ZAAC(619)675 M. Westerhausen, B. Rademacher and W. Schwarz; *Z. Anorg. Allg. Chem.*, 1993, **619**, 675. 195

93ZAAC(619)1494 G. Fritz, S. Lauble, R. Befurt, K. Peters, E. M. Peters and H. G. v. Schnering; *Z. Anorg. Allg. Chem.*, 1993, **619**, 1494. 173, 383

93ZAAC(619)1693 M. Ostrowski, I. Wagner, W.-W. duMont, P. G. Jones and J. Jeske; *Z. Anorg. Allg. Chem.*, 1993, **619**, 1693. 355, 356

93ZN(B)19 R. Minkwitz, W. Meckstroth and H. Preut; *Z. Naturforsch., Teil B*, 1993, **48**, 19. 613

93ZN(B)58 J. Grobe, D. Le Van and G. Lange; *Z. Naturforsch., Teil. B*, 1993, **48**, 58. 242, 683

93ZN(B)935 H. Bürger, T. Hagen and G. Pawelke; *Z. Naturforsch., Teil B*, 1993, **48**, 935. 243

93ZN(B)1000 P. Link and W. Voelter; *Z. Naturforsch., Teil B*, 1993, **48**, 1000. 487

93ZOB633 G. A. Kutyrev, A. V. Mironov, R. G. Islanov and R. A. Cherkasov; *Zh. Obshch. Khim.*, 1993, **63**, 633 (*Chem. Abstr.*, 1993, **119**, 203 490). 583

93ZOB1240 G. O. Baram, G. N. Koidan, A. P. Marchenko and A. M. Pinchuk; *Zh. Obshch. Khim.*, 1993, **63**, 1240. 242

94AG(E)323 D. Naumann, R. Möckel and W. Tyrra; *Angew. Chem., Int. Ed. Engl.*, 1994, **33**, 323. 49

94AG(E)578 A. H. Cowley, F. P. Gabbaï, C. J. Carrano, L. M. Mokry, M. R. Bond and G. Bertrand; *Angew. Chem., Int. Ed. Engl.*, 1994, **33**, 578. 704

94AG(E)672 H. G. Raubenheimer, S. Cronje, P. H. van Rooyen, P. J. Olivier and J. G. Toerien; *Angew. Chem., Int. Ed. Engl.*, 1994, **33**, 672. 723

94AG(E)1268 S. S. Al-Juaid, C. Eaborn, P. B. Hitchcock, K. Izod, M. Mallien and J. D. Smith; *Angew. Chem., Int. Ed. Engl.*, 1994, **33**, 1268. 393

94CB533 S. Reimann-Andersen, H. Pritzkow and W. Sundermeyer; *Chem. Ber.*, 1994, **127**, 533. 306

94CB597 R. Boese, A. Haas, C. Krüger, G. Möller and A. Waterfeld; *Chem. Ber.*, 1994, **127**, 597. 729

94CCR427 J. Escudié, C. Couret, H. Ranaivonjatovo and J. Satge; *Coord. Chem. Rev.*, 1994, **130**, 427. 709

94CL495 K. Otsuka, T. Yagi and I. Yamanaka; *Chem. Lett.*, 1994, 495. 469

94EUP581240 N. Manada, M. Murakami, K. Abe, Y. Yamamoto and T. Kurafuji; *Eur. Pat.* 581 240 (1994) (*Chem. Abstr.*, 1994, **120**, 191 141). 465

94EUP583789 T. Yano and H. Miura; *Eur. Pat.* 583 789 (1994) (*Chem. Abstr.*, 1994, **120**, 195 093). 467

94JAP06041019 S. Myamoto, J. Watanuki and Y. Sakane; *Jpn. Pat.* 06 041 019 (1994) (*Chem. Abstr.*, 1994, **120**, 269 667). 461

94JOC2393 H. K. Nair, R. D. Guneratne, A. S. Modak and D. J. Burton; *J. Org. Chem.*, 1994, **59**, 2393. 58, 264

94JOC4691 L. Guziec and F. S. Guziec, Jr.; *J. Org. Chem.*, 1994, **59**, 4691. 598

94JOM(465)153 R. Eujen, N. Jahn and U. Thurmann; *J. Organomet. Chem.*, 1994, **465**, 153. 65

94JOM(466)35 C. Eaborn, K. L. Jones and P. D. Lickiss; *J. Organomet. Chem.*, 1994, **466**, 35. 381, 392

94JOM(467)231 E. M. Lopez, D. Miguel, J. A. Perez-Martinez and V. Riera; *J. Organomet. Chem.*, 1994, **467**, 231. 329

94JOM(469)129 S. S. Al-Juaid, C. Eaborn, P. B. Hitchcock, A. J. Jagger and J. D. Smith; *J. Organomet. Chem.*, 1994, **469**, 129. 393

94JOM(469)135 M. Westerhausen, B. Rademacher, W. Schwarz, J. Weidlein and S. Henkel, *J. Organomet. Chem.*, 1994, **469**, 135. 393

94JOM(480)199 S. S. Al-Juaid, C. Eaborn, P. B. Hitchcock, K. Kundu, C. A. McGeary and J. D. Smith; *J. Organomet. Chem.*, 1994, **480**, 199. 393

B-94MI 601-01 W.-Y. Huang; "Macromolecular Symposia, Fluorinated Polymers," Hüthig and Wepf, Basel, 1994, p. 67. 21

94OM69 C. J. Schaverien; *Organometallics*, 1994, **13**, 69. 195

94OM434 W. Ando, T. Ohtaki and Y. Kabe; *Organometallics*, 1994, **13**, 434. 187

94OM436 C. Leue, R. Réau, B. Neumann, H.-G. Stammler, P. Jutzi and G. Bertrand; *Organometallics*, 1994, **13**, 436. 187

94OM753 C. Eaborn, P. B. Hitchcock, K. Izod, A. J. Jagger and J. D. Smith; *Organometallics*, 1994, **13**, 753. 393

94OM1336 D. Miguel, J. A. Perez-Martinez, V. Riera and S. Garcia-Grande; *Organometallics*, 1994, **13**, 1336. 329, 330

94S145 A. A. Kolomeitsev, K. Yu. Chabanenko, G.-V. Röschenthaler and Yu. L. Yagupolskii; *Synthesis*, 1994, 145. 234

94S983 H. Ahlbrecht and C. Schmitt; *Synthesis*, 1994, 983. 561

94SC305 S.-K. Kang, J.-H. Jeon, K.-S. Nam, C.-H. Park and H.-W. Lee; *Synth. Commun.*, 1994, **24**, 305. 468

94SL251 M. Kuroboshi and T. Hiyama; *Synlett*, 1994, 251. 251

94T3195 G. G. Cox, D. J. Miller, C. J. Moody, E.-R. H. B. Sie and J. J. Kulagowski; *Tetrahedron*, 1994, **50**, 3195. 628, 636

94TA937 T. Hayashi and V. A. Soloshonok; *Tetrahedron Asymmetry*, 1994, **5**, 937. 9

94TL977 D. S. Dodd and A. P. Kozikowski; *Tetrahedron Lett.*, 1994, **35**, 977. 487

94TL2365 A. Guirado, A. Zapata and J. Gálvez; *Tetrahedron Lett.*, 1994, **35**, 2365. 624

94TL3227 J. Matulic-Adamic and N. Usman; *Tetrahedron Lett.*, 1994, **35**, 3227. 58

94TL3537 D. Villemin, F. Sauvaget and M. Hajek; *Tetrahedron Lett.*, 1994, **35**, 3537. 58

95UP 613-01 S. S. Al-Juaid, C. Eaborn, P. B. Hitchcock, P. D. Lickiss, J. D. Smith, K. Tavakolli and A. Webb; *J. Chem. Soc., Dalton Trans.*, 1995, in press. 394

Subject Index

Every effort has been made to index as comprehensively as possible, and to standardize the terms used in the index in line with the IUPAC Recommendations. In view of the diverse nature of the terminology employed by the different authors, the reader is advised to search for related entries under the appropriate headings.

The index entries are presented in letter-by-letter alphabetical sequence. Compounds are normally indexed under the parent compound name, with the substituent component separated by a comma of inversion. An entry with a prefix/locant is filed after the same entry without any attachments, and in alphanumerical sequence. For example, 'dienes', '1,3-dienes', and '1-bromo-1,4-dienes' will be filed as:

Dienes
1,3-Dienes
1,4-Dienes, 1-bromo-

The Index is arranged in set-out style, with a maximum of three levels of heading. Location references refer to page number; major coverage of a subject is indicated by bold, elided page numbers: for example,

Halides, alkynyl, synthesis of, **1234–55**
  properties of, 345

See cross-references direct the user to the preferred term; for example,

Mercaptans *See* Thiols

See also cross-references provide the user with guideposts to terms of related interest, from the broader term to the narrower term, and appear at the end of the main heading to which they refer, e.g.

Thiones
  See also
    Thioketone